Coulson and Richardson's

# CHEMICAL ENGINEERING

## VOLUME 2
## FIFTH EDITION

*Particle Technology and Separation Processes*

**Related Butterworth-Heinemann Titles in the Chemical Engineering Series by**
J. M. COULSON & J. F. RICHARDSON

Chemical Engineering, Volume 1, Sixth edition
Fluid Flow, Heat Transfer and Mass Transfer
(with J. R. Backhurst and J. H. Harker)

Chemical Engineering, Volume 3, Third edition
Chemical and Biochemical Reaction Engineering, and Control
(edited by J. F. Richardson and D. G. Peacock)

Chemical Engineering, Volume 6, Third edition
Chemical Engineering Design
(R. K. Sinnott)

Chemical Engineering, Solutions to Problems in Volume 1
(J. R. Backhurst, J. H. Harker and J. F. Richardson)

Chemical Engineering, Solutions to Problems in Volume 2
(J. R. Backhurst, J. H. Harker and J. F. Richardson)

Coulson and Richardson's

# CHEMICAL ENGINEERING

**VOLUME 2**
**FIFTH EDITION**

## *Particle Technology and Separation Processes*

J. F. RICHARDSON
*University of Wales Swansea*

and

J. H. HARKER
*University of Newcastle upon Tyne*

*with*

J. R. BACKHURST
*University of Newcastle upon Tyne*

AMSTERDAM BOSTON HEIDELBERG LONDON NEW YORK OXFORD
PARIS SAN DIEGO SAN FRANCISCO SINGAPORE SYDNEY TOKYO

Butterworth-Heinemann
An imprint of Elsevier
Linacre House, Jordan Hill, Oxford OX2 8DP
200 Wheeler Road, Burlington, MA 01803

First published 1955
Reprinted (with revisions) 1956, 1959, 1960
Reprinted 1962
Second edition 1968
Reprinted 1976
Third edition (SI units) 1978
Reprinted (with revisions) 1980, 1983, 1985, 1987, 1989
Fourth edition 1991
Reprinted (with revisions) 1993, 1996, 1997, 1998, 1998, 2001
Fifth edition 2002
Reprinted 2003



**British Library Cataloguing in Publication Data**
A catalogue record for this book is available from the British Library

**Library of Congress Cataloguing in Publication Data**
A catalogue record for this book is available from the Library of Congress

ISBN 0 7506 4445 1



Typeset by Laserwords Private Limited, Chennai, India
Printed and bound in Great Britain by MPG Books Ltd, Bodmin, Cornwall

# Contents

# *Preface to the Fifth Edition*

It is now 47 years since Volume 2 was first published in 1955, and during the intervening time the profession of chemical engineering has grown to maturity in the UK, and worldwide; the Institution of Chemical Engineers, for instance, has moved on from its 33rd to its 80th year of existence. No longer are the heavy chemical and petroleum-based industries the main fields of industrial applications of the discipline, but chemical engineering has now penetrated into areas, such as pharmaceuticals, health care, foodstuffs, and biotechnology, where the general level of sophistication of the products is much greater, and the scale of production often much smaller, though the unit value of the products is generally much higher. This change has led to a move away from large-scale continuous plants to smaller-scale batch processing, often in multipurpose plants. Furthermore, there is an increased emphasis on product purity, and the need for more refined separation technology, especially in the pharmaceutical industry where it is often necessary to carry out the difficult separation of stereo-isomers, one of which may have the desired therapeutic properties while the other is extremely malignant. Many of these large molecules are fragile and are liable to be broken down by the harsh solvents commonly used in the manufacture of bulk chemicals. The general principles involved in processing these more specialised materials are essentially the same as in bulk chemical manufacture, but special care must often be taken to ensure that processing conditions are mild.

One big change over the years in the chemical and processing industries is the emphasis on designing products with properties that are specified, often in precise detail, by the customer. Chemical composition is often of relatively little importance provided that the product has the desired attributes. Hence *product design*, a multidisciplinary activity, has become a necessary precursor to *process design*.

Although undergraduate courses now generally take into account these new requirements, the basic principles of chemical engineering remain largely unchanged and this is particularly the case with the two main topics of Volume 2, *Particle Mechanics* and *Separation Processes*. In preparing this new edition, the authors have faced a typical engineering situation where a compromise has to be reached on size. The knowledge-base has increased to such an extent that many of the individual chapters appear to merit expansion into separate books. At the same time, as far as students and those from other disciplines are concerned, there is still a need for a an integrated concise treatment in which there is a consistency of approach across the board and, most importantly, a degree of uniformity in the use of symbols. It has to be remembered that the learning capacity of students is certainly no greater than it was in the past, and a book of manageable proportions is still needed.

The advice that academic staffs worldwide have given in relation to revising the book has been that the layout should be retained substantially unchanged—*better the devil we know, with all his faults!* With this in mind the basic structure has been maintained. However, the old Chapter 8 on *Gas Cleaning*, which probably did not merit a chapter on its own, has been incorporated into Chapter 1, where it sits comfortably beside other topics involving the separation of solid particles from fluids. This has left Chapter 8 free to accommodate *Membrane Separations* (formerly Chapter 20) which then follows on logically from *Filtration* in Chapter 7. The new Chapter 20 then provides an opportunity to look to the future, and to introduce the topics of *Product Design* and the *Use of Intensified Fields* (particularly centrifugal in place of gravitational) and *miniaturisation*, with all the advantages of reduced hold-up, leading to a reduction in the amount of out-of-specification material produced during the changeover between products in the case multipurpose plants, and in improved safety where the materials have potentially hazardous properties.

Other significant changes are the replacement of the existing chapter on *Crystallisation* by an entirely new chapter written with expert guidance from Professor J. W. Mullin, the author of the standard textbook on that topic. The other chapters have all been updated and additional Examples and Solutions incorporated in the text. Several additional Problems have been added at the end, and solutions are available in the Solutions Manual, and now on the Butterworth-Heinemann website.

We are, as usual, indebted to both reviewers and readers for their suggestions and for pointing out errors in earlier editions. These have all been taken into account. Please keep it up in future! We aim to be error-free but are not always as successful as we would like to be! Unfortunately, the new edition is somewhat longer than the previous one, almost inevitably so with the great expansion in the amount of information available. Whenever in the past we have cut out material which we have regarded as being out-of-date, there is inevitably somebody who writes to say that he now has to keep both the old and the new editions because he finds that something which he had always found particularly useful in the past no longer appears in the revised edition. It seems that you cannot win, but we keep trying!

J. F. RICHARDSON
J. H. HARKER

# Preface to the Fourth Edition

Details of the current restructuring of this Chemical Engineering Series, coinciding with the publication of the Fourth Edition of Volumes 1 and 2 and to be followed by new editions of the other volumes, have been set out in the Preface to the Fourth Edition of Volume 1. The revision involves the inclusion in Volume 1 of material on non-Newtonian flow (previously in Volume 3) and the transference from Volume 2 to Volume 1 of *Pneumatic and Hydraulic Conveying* and *Liquid Mixing*. In addition, Volume 6, written by Mr. R. K. Sinnott, which first appeared in 1983, nearly thirty years after the first volumes, acquires some of the design-orientated material from Volume 2, particularly that related to the hydraulics of packed and plate columns.

The new sub-title of Volume 2, *Particle Technology and Separation Processes*, reflects both the emphasis of the new edition and the current importance of these two topics in Chemical Engineering. *Particle Technology* covers the basic properties of systems of particles and their preparation by comminution (Chapters 1 and 2). Subsequent chapters deal with the interaction between fluids and particles, under conditions ranging from those applicable to single isolated particles, to systems of particles freely suspended in fluids, as in sedimentation and fluidisation; and to packed beds and columns where particles are held in a fixed configuration relative to one another. The behaviour of particles in both gravitational and centrifugal fields is also covered. It will be noted that *Centrifugal Separations* are now brought together again in a single chapter, as in the original scheme of the first two editions, because the dispersal of the material between other chapters in the Third Edition was considered to be not entirely satisfactory.

Fluid–solids Separation Processes are discussed in the earlier chapters under the headings of Sedimentation, Filtration, Gas Cleaning and Centrifugal Separations. The remaining separations involve applications of mass-transfer processes, in the presence of solid particles in Leaching (solid–liquid extraction), Drying and Crystallisation. In Distillation, Gas Absorption and Liquid–Liquid Extraction, interactions occur between two fluid streams with mass transfer taking place across a phase boundary. Usually these operations are carried out as continuous countercurrent flow processes, either stagewise (as in a plate-column) or with differential contacting (as in a packed-column). There is a case therefore for a generalised treatment of countercurrent contacting processes with each of the individual operations, such as Distillation, treated as particular cases. Although this approach has considerable merit, both conceptually and in terms of economy of space, it has not been adopted here, because the authors' experience of teaching suggests that the student more readily grasps the principles involved, by considering each topic in turn, provided of course that the teacher makes a serious attempt to emphasise the common features.

The new edition concludes with four chapters which are newcomers to Volume 2, each written by a specialist author from the Chemical Engineering Department at Swansea—

Adsorption and Ion Exchange (Chapters 17 and 18)
(topics previously covered in Volume 3)
*by J. H. Bowen*

Chromatographic Separations (Chapter 19)
*by J. R. Conder*

and

Membrane Separations (Chapter 20)
*by W. R. Bowen.*

These techniques are of particular interest in that they provide a means of separating molecular species which are difficult to separate by other techniques and which may be present in very low concentrations. Such species include large molecules, sub-micrometre size particles, stereo-isomers and the products from bioreactors (Volume 3). The separations can be *highly specific* and may depend on molecular size and shape, and the configuration of the constituent chemical groups of the molecules.

Again I would express our deep sense of loss on the death of our colleague, Professor John Coulson, in January 1990. His two former colleagues at Newcastle, Dr. John Backhurst and the Reverend Dr. John Harker, have played a substantial part in the preparation of this new edition both by updating the sections originally attributable to him, and by obtaining new illustrations and descriptions of industrial equipment.

Finally, may I again thank our readers who, in the past, have made such helpful suggestions and have drawn to our attention errors, many of which would never have been spotted by the authors. Would they please continue their good work!

Swansea
July 1990

J. F. Richardson

## Note to Fourth Edition — Revised Impression 1993

In this reprint corrections and minor revisions have been incorporated. The principal changes are as follows:

(1) Addition of an account of the construction and operation of the Szego Grinding Mill (Chapter 2).

(2) Inclusion of the Yoshioka method for the design of thickeners (Chapter 5).

(3) Incorporation of Geldart's classification of powders in relation to fluidisation characteristics (Chapter 6).

(4) The substitution of a more logical approach to filtration of slurries yielding compressible cakes and redefinition of the specific resistance (Chapter 7).

(5) Revision of the nomenclature for the underflow streams of washing thickeners to bring it into line with that used for other stagewise processes, including distillation and absorption (Chapter 10).

(6) A small addition to the selection of dryers and the inclusion of Examples (Chapter 16).

JFR

# *Preface to the 1983 Reprint of the Third Edition*

In this volume, there is an account of the basic theory underlying the various Unit Operations, and typical items of equipment are described. The equipment items are the essential components of a complete chemical plant, and the way in which such a plant is designed is the subject of Volume 6 of the series which has just appeared. The new volume includes material on flowsheeting, heat and material balances, piping, mechanical construction and costing. It completes the Series and forms an introduction to the very broad subject of Chemical Engineering Design.

# Preface to Third Edition

In producing a third edition, we have taken the opportunity, not only of updating the material but also of expressing the values of all the physical properties and characteristics of the systems in the SI System of units, as has already been done in Volumes 1 and 3. The SI system, which is described in detail in Volume 1, is widely adopted in Europe and is now gaining support elsewhere in the world. However, because some readers will still be more familiar with the British system, based on the foot, pound and second, the old units have been retained as alternatives wherever this can be done without causing confusion.

The material has, to some extent, been re-arranged and the first chapter now relates to the characteristics of particles and their behaviour in bulk, the blending of solids, and classification according to size or composition of material. The following chapters describe the behaviour of particles moving in a fluid and the effects of both gravitational and centrifugal forces and of the interactions between neighbouring particles. The old chapter on centrifuges has now been eliminated and the material dispersed into the appropriate parts of other chapters. Important applications which are considered include flow in granular beds and packed columns, fluidisation, transport of suspended particles, filtration and gas cleaning. An example of the updating which has been carried out is the addition of a short section on fluidised bed combustion, potentially the most important commercial application of the technique of fluidisation. In addition, we have included an entirely new section on flocculation, which has been prepared for us by Dr. D. J. A. Williams of University College, Swansea, to whom we are much indebted.

Mass transfer operations play a dominant role in chemical processing and this is reflected in the continued attention given to the operations of solid–liquid extraction, distillation, gas absorption and liquid–liquid extraction. The last of these subjects, together with material on liquid–liquid mixing, is now dealt within a single chapter on liquid–liquid systems, the remainder of the material which appeared in the former chapter on mixing having been included earlier under the heading of solids blending. The volume concludes with chapters on evaporation, crystallisation and drying.

Volumes 1, 2 and 3 form an integrated series with the fundamentals of fluid flow, heat transfer and mass transfer in the first volume, the physical operations of chemical engineering in this, the second volume, and in the third volume, the basis of chemical and biochemical reactor design, some of the physical operations which are now gaining in importance and the underlying theory of both process control and computation. The solutions to the problems listed in Volumes 1 and 2 are now available as Volumes 4 and 5 respectively. Furthermore, an additional volume in the series is in course of preparation and will provide an introduction to chemical engineering design and indicate how the principles enunciated in the earlier volumes can be translated into chemical plant.

We welcome the collaboration of J. R. Backhurst and J. H. Harker as co-authors in the preparation of this edition, following their assistance in the editing of the latest edition of Volume 1 and their authorship of Volumes 4 and 5. We also look forward to the appearance of R. K. Sinnott's volume on chemical engineering design.

# Preface to Second Edition

This text deals with the physical operations used in the chemical and allied industries. These operations are conveniently designated "unit operations" to indicate that each single operation, such as filtration, is used in a wide range of industries, and frequently under varying conditions of temperature and pressure.

Since the publication of the first edition in 1955 there has been a substantial increase in the relevant technical literature but the majority of developments have originated in research work in government and university laboratories rather than in industrial companies. As a result, correlations based on laboratory data have not always been adequately confirmed on the industrial scale. However, the section on absorption towers contains data obtained on industrial equipment and most of the expressions used in the chapters on distillation and evaporation are based on results from industrial practice.

In carrying out this revision we have made substantial alteration to Chapters 1, 5, 6, 7, 12, 13 and 15* and have taken the opportunity of presenting the volume paged separately from Volume 1. The revision has been possible only as the result of the kind co-operation and help of Professor J. D. Thornton (Chapter 12), Mr. J. Porter (Chapter 13), Mr. K. E. Peet (Chapter 10) and Dr. B. Waldie (Chapter 1), all of the University at Newcastle, and Dr. N. Dombrowski of the University of Leeds (Chapter 15). We want in particular to express our appreciation of the considerable amount of work carried out by Mr. D. G. Peacock of the School of Pharmacy, University of London. He has not only checked through the entire revision but has made numerous additions to many chapters and has overhauled the index.

We should like to thank the companies who have kindly provided illustrations of their equipment and also the many readers of the previous edition who have made useful comments and helpful suggestions.

Chemical engineering is no longer confined to purely physical processes and the unit operations, and a number of important new topics, including reactor design, automatic control of plants, biochemical engineering, and the use of computers for both process design and control of chemical plant will be covered in a forthcoming Volume 3 which is in course of preparation.

Chemical engineering has grown in complexity and stature since the first edition of the text, and we hope that the new edition will prove of value to the new generation of university students as well as forming a helpful reference book for those working in industry.

In presenting this new edition we wish to express our gratitude to Pergamon Press who have taken considerable trouble in coping with the technical details.

J. M. COULSON
J. F. RICHARDSON

* N.B. Chapter numbers are altered in the current (third) edition.

# *Preface to First Edition*

In presenting Volume 2 of *Chemical Engineering*, it has been our intention to cover what we believe to be the more important unit operations used in the chemical and process industries. These unit operations, which are mainly physical in nature, have been classified, as far as possible, according to the underlying mechanism of the transfer operation. In only a few cases is it possible to give design procedures when a chemical reaction takes place in addition to a physical process. This difficulty arises from the fact that, when we try to design such units as absorption towers in which there is a chemical reaction, we are not yet in a position to offer a thoroughly rigorous method of solution. We have not given an account of the transportation of materials in such equipment as belt conveyors or bucket elevators, which we feel lie more distinctly in the field of mechanical engineering.

In presenting a good deal of information in this book, we have been much indebted to facilities made available to us by Professor Newitt, in whose department we have been working for many years. The reader will find a number of gaps, and a number of principles which are as yet not thoroughly developed. Chemical engineering is a field in which there is still much research to be done, and, if this work will in any way stimulate activities in this direction, we shall feel very much rewarded. It is hoped that the form of presentation will be found useful in indicating the kind of information which has been made available by research workers up to the present day. Chemical engineering is in its infancy, and we must not suppose that the approach presented here must necessarily be looked upon as correct in the years to come. One of the advantages of this subject is that its boundaries are not sharply defined.

Finally, we should like to thank the following friends for valuable comments and suggestions: Mr. G. H. Anderson, Mr. R. W. Corben, Mr. W. J. De Coursey, Dr. M. Guter, Dr. L. L. Katan, Dr. R. Lessing, Dr. D. J. Rasbash, Dr. H. Sawistowski, Dr. W. Smith, Mr. D. Train, Mr. M. E. O'K. Trowbridge, Mr. F. E. Warner and Dr. W. N. Zaki.

# Acknowledgements

The authors and publishers acknowledge with thanks the assistance given by the following companies and individuals in providing illustrations and data for this volume and giving their permission for reproduction. Everyone was most helpful and some firms went to considerable trouble to provide exactly what was required. We are extremely grateful to them all.

Robinson Milling Systems Ltd for Fig. 1.16.
Baker Perkins Ltd for Figs. 1.23, 2.8, 2.34, 13.39, 13.40.
Buss (UK) Ltd, Cheadle Hulme, Cheshire for Figs. 1.24, 14.24.
Dorr-Oliver Co Ltd, Croydon, Surrey for Figs. 1.26, 1.29, 7.15, 7.22, 10.8, 10.9.
Denver Process Equipment Ltd, Leatherhead, Surrey for Figs. 1.27, 1.30, 1.32, 1.47, 1.48.
Wilfley Mining Machinery Co Ltd for Fig. 1.33.
NEI International Combustion Ltd, Derby for Figs. 1.34, 1.35, 1.36, 1.37, 2.10, 2.20–2.24, 2.27, 2.30, 14.18, 16.23.
Lockers Engineers Ltd, Warrington for Figs. 1.40, 1.41, 1.42.
Master Magnets Ltd, Birmingham for Figs. 1.43, 1.44, 1.45.
AAF Ltd, Cramlington, Northumberland for Figs. 1.54, 1.68, 1.70, 1.72.
Vaba Process Plant Ltd, Rotherham, Yorks, successors to Edgar Allen Co Ltd for Figs. 2.4, 2.7, 2.13, 2.14, 2.29.
Hadfields Ltd for Fig. 2.5.
Hosokawa Micron Ltd, Runcorn, Cheshire for Fig. 2.12.
Babcock & Wilcox Ltd for Fig. 2.19.
Premier Colloid Mills for Figs. 2.33.
Amandus–Kahl, Hamburg, Germany for Fig. 2.35.
McGraw-Hill Book Co for Fig. 3.3.
Norton Chemical Process Products (Europe) Ltd, Stoke-on-Trent, Staffs for Table 4.3.
Glitsch UK Ltd, Kirkby Stephen, Cheshire for Figs. 11.51, 11.52, 11.53.
Sulzer (UK) Ltd, Farnborough, Hants for Figs. 1.67, 4.13, 4.14, 4.15.
Johnson-Progress Ltd, Stoke-on-Trent, Staffs for Figs. 7.6, 7.7, 7.9, 7.10.
Filtration Systems Ltd, Mirfield, W. Yorks, for Fig. 7.11.
Stockdale Filtration Systems Ltd, Macclesfield, Cheshire for Fig. 7.12, and Tables 7.1, 7.2, 7.3.
Stella-Meta Filters Ltd, Whitchurch, Hants for Figs. 7.13, 7.14.
Delfilt Ltd, Bath, Avon for Fig. 7.16.
Charlestown Engineering Ltd, St Austell, Cornwall for Figs. 7.24, 7.25.
Institution of Mechanical Engineers for Fig. 1.59.
Sturtevant Engineering Co Ltd for Fig. 1.60.
Thomas Broadbent & Sons Ltd, Huddersfield, West Yorkshire for Figs. 9.13, 9.14.

Amicon Ltd for Figs. 8.12.
P. C. I. for Fig. 8.9.
A/S De Danske Sukkerfabrikker, Denmark for Fig. 8.10.
Dr Huabin Yin for Fig. 8.1.
Dr Nrdal Hilal for Fig. 8.2.
Alfa-Laval Sharples Ltd, Camberley, Surrey, for Figs. 9.8, 9.10, 9.11, 9.12, 9.16, 9.17, 13.37, 14.20.
Sulzer Escher Wyss Ltd, Zurich, Switzerland for Fig. 9.15.
APV Mitchell Ltd for Fig. 12.17.
Davy Powergas Ltd for Fig. 13.26.
Swenson Evaporator Co for Figs. 14.17, 14.18, 14.19, 14.20.
APV Baker Ltd, Crawley, Sussex for Figs. 14.21, 14.22, 14.23, 14.24, 14.25, 14.26.
The Editor and Publishers of *Chemical and Process Engineering* for Figs. 14.28, 14.30.
APV Pasilac Ltd, Carlisle, Cumbria for Figs. 16.10, 16.11.
Buflovak Equipment Division of Blaw-Knox Co Ltd for Figs. 16.17, 16.18, and Table 16.3.
Dr N. Dombrowski for Figs. 16.19, 16.22.
Ventilex for Fig. 16.29.

## INDEX

The authors are indebted to our colleague, Dr. D. G. Peacock, formerly of the School of Pharmacy, University of London, for the skill and patience which has shown in the preparation of the Index.

J.F.R.
J.H.H.

# INTRODUCTION

The understanding of the design and construction of chemical plant is frequently regarded as the essence of chemical engineering and it is this area which is covered in Volume 6 of this series. Starting from the original conception of the process by the chemist, it is necessary to appreciate the chemical, physical and many of the engineering features in order to develop the laboratory process to an industrial scale. This volume is concerned mainly with the physical nature of the processes that take place in industrial units, and, in particular, with determining the factors that influence the rate of transfer of material. The basic principles underlying these operations, namely fluid dynamics, and heat and mass transfer, are discussed in Volume 1, and it is the application of these principles that forms the main part of Volume 2.

Throughout what are conveniently regarded as the process industries, there are many physical operations that are common to a number of the individual industries, and may be regarded as *unit operations*. Some of these operations involve particulate solids and many of them are aimed at achieving a separation of the components of a mixture. Thus, the separation of solids from a suspension by filtration, the separation of liquids by distillation, and the removal of water by evaporation and drying are typical of such operations. The problem of designing a distillation unit for the fermentation industry, the petroleum industry or the organic chemical industry is, in principle, the same, and it is mainly in the details of construction that the differences will occur. The concentration of solutions by evaporation is again a typical operation that is basically similar in the handling of sugar, or salt, or fruit juices, though there will be differences in the most suitable arrangement. This form of classification has been used here, but the operations involved have been grouped according to the mechanism of the transfer operation, so that the operations involving solids in fluids are considered together and then the diffusion processes of distillation, absorption and liquid-liquid extraction are taken in successive chapters. In examining many of these unit operations, it is found that the rate of heat transfer or the nature of the fluid flow is the governing feature. The transportation of a solid or a fluid stream between processing units is another instance of the importance of understanding fluid dynamics.

One of the difficult problems of design is that of maintaining conditions of similarity between laboratory units and the larger-scale industrial plants. Thus, if a mixture is to be maintained at a certain temperature during the course of an exothermic reaction, then on the laboratory scale there is rarely any real difficulty in maintaining isothermal conditions. On the other hand, in a large reactor the ratio of the external surface to the volume — which is inversely proportional to the linear dimension of the unit — is in most cases of a different order, and the problem of removing the heat of reaction becomes a major item in design. Some of the general problems associated with *scaling-up* are considered as they arise in many of the chapters. Again, the introduction and removal of the reactants may present difficult problems on the large scale, especially if they contain corrosive liquids or abrasive solids. The general tendency with many industrial units is to provide a continuous process, frequently involving a series of stages. Thus, exothermic reactions may be carried out in a series of reactors with interstage cooling.

The planning of a process plant will involve determining the most economic method, and later the most economic arrangement of the individual operations used in the process. This amounts to designing a process so as to provide the best combination of capital and operating costs. In this volume the question of costs has not been considered in any detail, but the aim has been to indicate the conditions under which various types of units will operate in the most economical manner. Without a thorough knowledge of the physical principles involved in the various operations, it is not possible to select the most suitable one for a given process. This aspect of the design can be considered by taking one or two simple illustrations of separation processes. The particles in a solid-solid system may be separated, first according to size, and secondly according to the material. Generally, sieving is the most satisfactory method of classifying relatively coarse materials according to size, but the method is impracticable for very fine particles and a form of settling process is generally used. In the first of these processes, the size of the particle is used directly as the basis for the separation, and the second depends on the variation with size of the behaviour of particles in a fluid. A mixed material can also be separated into its components by means of settling methods, because the shape and density of particles also affect their behaviour in a fluid. Other methods of separation depend on differences in surface properties (froth flotation), magnetic properties (magnetic separation), and on differences in solubility in a solvent (leaching). For the separation of miscible liquids, three commonly used methods are:

1. Distillation — depending on difference in volatility.
2. Liquid–liquid extraction — depending on difference in solubility in a liquid solvent.
3. Freezing — depending on difference in melting point.

The problem of selecting the most appropriate operation will be further complicated by such factors as the concentration of liquid solution at which crystals start to form. Thus, in the separation of a mixture of ortho-, meta-, and para-mononitrotoluenes, the decision must be made as to whether it is better to carry out the separation by distillation followed by crystallisation, or in the reverse order. The same kind of consideration will arise when concentrating a solution of a solid; then it must be decided whether to stop the evaporation process when a certain concentration of solid has been reached and then to proceed with filtration followed by drying, or whether to continue to concentration by evaporation to such an extent that the filtration stage can be omitted before moving on to drying.

In many operations, for instance in a distillation column, it is necessary to understand the fluid dynamics of the unit, as well as the heat and mass transfer relationships. These factors are frequently interdependent in a complex manner, and it is essential to consider the individual contributions of each of the mechanisms. Again, in a chemical reaction the final rate of the process may be governed either by a heat transfer process or by the chemical kinetics, and it is essential to decide which is the controlling factor; this problem is discussed in Volume 3, which deals with both chemical and biochemical reactions and their control.

Two factors of overriding importance have not so far been mentioned. Firstly, the plant must be operated in such a way that it does not present an unacceptable hazard to the workforce or to the surrounding population. *Safety considerations* must be in the forefront in the selection of the most appropriate process route and design, and must also be reflected in all the aspects of plant operation and maintenance. An inherently safe plant is to be preferred to one with inherent hazards, but designed to minimise the risk of the hazard being released. Safety considerations must be taken into account at an early stage of design; they are not an add-on at the end. Similarly control systems, the integrity of which play a major part in safe operation of plant, must be designed into the plant, not built on after the design is complete.

The second consideration relates to the *environment*. The engineer has the responsibility for conserving natural resources, including raw materials and energy sources, and at the same time ensuring that effluents (solids, liquids and gases) do not give rise to unacceptable environmental effects. As with safety, effluent control must feature as a major factor in the design of every plant.

The topics discussed in this volume form an important part of any chemical engineering project. They must not, however, be considered in isolation because, for example, a difficult separation problem may often be better solved by adjustment of conditions in the preceding reactor, rather than by the use of highly sophisticated separation techniques.

CHAPTER 1

# Particulate Solids

## 1.1. INTRODUCTION

In Volume 1, the behaviour of fluids, both liquids and gases is considered, with particular reference to their flow properties and their heat and mass transfer characteristics. Once the composition, temperature and pressure of a fluid have been specified, then its relevant physical properties, such as density, viscosity, thermal conductivity and molecular diffusivity, are defined. In the early chapters of this volume consideration is given to the properties and behaviour of systems containing solid particles. Such systems are generally more complicated, not only because of the complex geometrical arrangements which are possible, but also because of the basic problem of defining completely the physical state of the material.

The three most important characteristics of an individual particle are its composition, its size and its shape. Composition determines such properties as density and conductivity, provided that the particle is completely uniform. In many cases, however, the particle is porous or it may consist of a continuous matrix in which small particles of a second material are distributed. Particle size is important in that this affects properties such as the surface per unit volume and the rate at which a particle will settle in a fluid. A particle shape may be regular, such as spherical or cubic, or it may be irregular as, for example, with a piece of broken glass. Regular shapes are capable of precise definition by mathematical equations. Irregular shapes are not and the properties of irregular particles are usually expressed in terms of some particular characteristics of a regular shaped particle.

Large quantities of particles are handled on the industrial scale, and it is frequently necessary to define the system as a whole. Thus, in place of particle size, it is necessary to know the distribution of particle sizes in the mixture and to be able to define a mean size which in some way represents the behaviour of the particulate mass as a whole. Important operations relating to systems of particles include storage in hoppers, flow through orifices and pipes, and metering of flows. It is frequently necessary to reduce the size of particles, or alternatively to form them into aggregates or sinters. Sometimes it may be necessary to mix two or more solids, and there may be a requirement to separate a mixture into its components or according to the sizes of the particles.

In some cases the interaction between the particles and the surrounding fluid is of little significance, although at other times this can have a dominating effect on the behaviour of the system. Thus, in filtration or the flow of fluids through beds of granular particles, the characterisation of the porous mass as a whole is the principal feature, and the resistance to flow is dominated by the size and shape of the free space between the particles. In such situations, the particles are in physical contact with adjoining particles and there is little relative movement between the particles. In processes such as the sedimentation of particles in a liquid, however, each particle is completely surrounded by fluid and is free to move relative to other particles. Only very simple cases are capable of a precise theoretical analysis and Stokes' law, which gives the drag on an isolated spherical particle due to its motion relative to the surrounding fluid at very low velocities, is the most important theoretical relation in this area of study. Indeed very many empirical laws are based on the concept of defining correction factors to be applied to Stokes' law.

## 1.2. PARTICLE CHARACTERISATION

### 1.2.1. Single particles

The simplest shape of a particle is the sphere in that, because of its symmetry, any question of orientation does not have to be considered, since the particle looks exactly the same from whatever direction it is viewed and behaves in the same manner in a fluid, irrespective of its orientation. No other particle has this characteristic. Frequently, the size of a particle of irregular shape is defined in terms of the size of an equivalent sphere although the particle is represented by a sphere of different size according to the property selected. Some of the important sizes of equivalent spheres are:

(a) The sphere of the same volume as the particle.
(b) The sphere of the same surface area as the particle.
(c) The sphere of the same surface area per unit volume as the particle.
(d) The sphere of the same area as the particle when projected on to a plane perpendicular to its direction of motion.
(e) The sphere of the same projected area as the particle, as viewed from above, when lying in its position of maximum stability such as on a microscope slide for example.
(f) The sphere which will just pass through the same size of square aperture as the particle, such as on a screen for example.
(g) The sphere with the same settling velocity as the particle in a specified fluid.

Several definitions depend on the measurement of a particle in a particular orientation. Thus Feret's statistical diameter is the mean distance apart of two parallel lines which are tangential to the particle in an arbitrarily fixed direction, irrespective of the orientation of each particle coming up for inspection. This is shown in Figure 1.1.

A measure of particle shape which is frequently used is the sphericity, $\psi$, defined as:

$$\psi = \frac{\text{surface area of sphere of same volume as particle}}{\text{surface area of particle}} \tag{1.1}$$

Another method of indicating shape is to use the factor by which the cube of the size of the particle must be multiplied to give the volume. In this case the particle size is usually defined by method (e).

Other properties of the particle which may be of importance are whether it is crystalline or amorphous, whether it is porous, and the properties of its surface, including roughness and presence of adsorbed films.

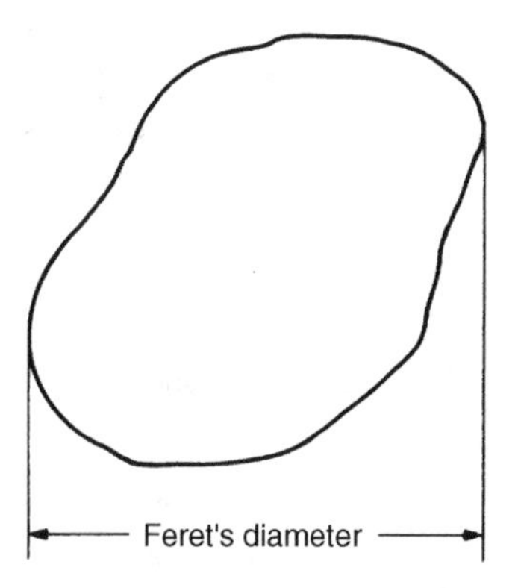


Figure 1.1. Feret's diameter

Hardness may also be important if the particle is subjected to heavy loading.

### 1.2.2. Measurement of particle size

Measurement of particle size and of particle size distribution is a highly specialised topic, and considerable skill is needed in the making of accurate measurements and in their interpretation. For details of the experimental techniques, reference should be made to a specialised text, and that of ALLEN[1] is highly recommended.

No attempt is made to give a detailed account or critical assessment of the various methods of measuring particle size, which may be seen from Figure 1.2 to cover a range of $10^7$ in linear dimension, or $10^{21}$ in volume! Only a brief account is given of some of the principal methods of measurement and, for further details, it is necessary to refer to one of the specialist texts on particle size measurement, the outstanding example of which is the two-volume monograph by ALLEN[1], with HERDAN[2] providing additional information. It may be noted that both the size range in the sample and the particle shape may be as important, or even more so, than a single characteristic linear dimension which at best can represent only one single property of an individual particle or of an assembly of particles. The ability to make accurate and reliable measurements of particle size is acquired only after many years of practical experimental experience. For a comprehensive review of methods and experimental details it is recommended that the work of Allen be consulted and also Wardle's work on Instrumentation and Control discussed in Volume 3.

Before a size analysis can be carried out, it is necessary to collect a representative sample of the solids, and then to reduce this to the quantity which is required for the chosen method of analysis. Again, the work of Allen gives information on how this is best carried out. Samples will generally need to be taken from the bulk of the powder, whether this is in a static heap, in the form of an airborne dust, in a flowing or falling stream, or on a conveyor belt, and in each case the precautions which need to be taken to obtain a representative sample are different.

A wide range of measuring techniques is available both for single particles and for systems of particles. In practice, each method is applicable to a finite range of sizes and gives a particular equivalent size, dependent on the nature of the method. The principles

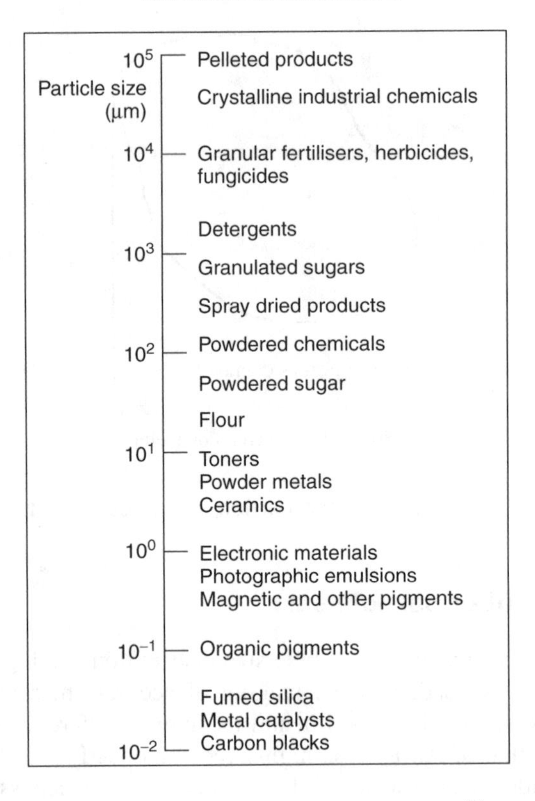


Figure 1.2. Sizes of typical powder products[1]

of some of the chief methods are now considered together with an indication of the size range to which they are applicable.

## *Sieving* (>50 μm)

Sieve analysis may be carried out using a nest of sieves, each lower sieve being of smaller aperture size. Generally, sieve series are arranged so that the ratio of aperture sizes on consecutive sieves is 2, $2^{1/2}$ or $2^{1/4}$ according to the closeness of sizing that is required. The sieves may either be mounted on a vibrator, which should be designed to give a degree of vertical movement in addition to the horizontal vibration, or may be hand shaken. Whether or not a particle passes through an aperture depends not only upon its size, but also on the probability that it will be presented at the required orientation at the surface of the screen. The sizing is based purely on the linear dimensions of the particle and the lower limit of size which can be used is determined by two principal factors. The first is that the proportion of free space on the screen surface becomes very small as the size of the aperture is reduced. The second is that attractive forces between particles become larger at small particle sizes, and consequently particles tend to stick together and block the screen. Sieves are available in a number of standard series. There are several standard series of screen and the sizes of the openings are determined by the thickness of wire used. In the U.K., British Standard (B.S.)[3] screens are made in sizes from 300-mesh upwards, although these are too fragile for some work. The Institute of Mining and Metallurgy (I.M.M.)[4] screens are more robust, with the thickness of the wire approximately equal to the size of the apertures. The Tyler series, which is standard in the United States, is intermediate between the two British series. Details of the three series of screens[3] are given in Table 1.1, together with the American Society for Testing Materials (ASTM) series[5].

Table 1.1. Standard sieve sizes

| British fine mesh (B.S.S. 410)[3] | | | I.M.M.[4] | | | U.S. Tyler[5] | | | U.S. A.S.T.M.[5] | | |
|---|---|---|---|---|---|---|---|---|---|---|---|
| Sieve no. | Nominal aperture in. | Nominal aperture μm | Sieve no. | Nominal aperture in. | Nominal aperture μm | Sieve no. | Nominal aperture in. | Nominal aperture μm | Sieve no. | Nominal aperture in. | Nominal aperture μm |
| | | | | | | 325 | 0.0017 | 43 | 325 | 0.0017 | 44 |
| | | | | | | 270 | 0.0021 | 53 | 270 | 0.0021 | 53 |
| 300 | 0.0021 | 53 | | | | 250 | 0.0024 | 61 | 230 | 0.0024 | 61 |
| 240 | 0.0026 | 66 | 200 | 0.0025 | 63 | 200 | 0.0029 | 74 | 200 | 0.0029 | 74 |
| 200 | 0.0030 | 76 | | | | | | | 170 | 0.0034 | 88 |
| 170 | 0.0035 | 89 | 150 | 0.0033 | 84 | 170 | 0.0035 | 89 | | | |
| 150 | 0.0041 | 104 | | | | 150 | 0.0041 | 104 | 140 | 0.0041 | 104 |
| 120 | 0.0049 | 124 | 120 | 0.0042 | 107 | 115 | 0.0049 | 125 | 120 | 0.0049 | 125 |
| 100 | 0.0060 | 152 | 100 | 0.0050 | 127 | 100 | 0.0058 | 147 | 100 | 0.0059 | 150 |
| | | | 90 | 0.0055 | 139 | 80 | 0.0069 | 175 | 80 | 0.0070 | 177 |
| 85 | 0.0070 | 178 | 80 | 0.0062 | 157 | 65 | 0.0082 | 208 | 70 | 0.0083 | 210 |
| | | | 70 | 0.0071 | 180 | | | | 60 | 0.0098 | 250 |
| 72 | 0.0083 | 211 | 60 | 0.0083 | 211 | 60 | 0.0097 | 246 | 50 | 0.0117 | 297 |
| 60 | 0.0099 | 251 | | | | | | | 45 | 0.0138 | 350 |
| 52 | 0.0116 | 295 | 50 | 0.0100 | 254 | 48 | 0.0116 | 295 | 40 | 0.0165 | 420 |
| | | | 40 | 0.0125 | 347 | 42 | 0.0133 | 351 | 35 | 0.0197 | 500 |
| 44 | 0.0139 | 353 | | | | 35 | 0.0164 | 417 | 30 | 0.0232 | 590 |
| 36 | 0.0166 | 422 | 30 | 0.0166 | 422 | 32 | 0.0195 | 495 | | | |
| 30 | 0.0197 | 500 | | | | 28 | 0.0232 | 589 | | | |
| 25 | 0.0236 | 600 | | | | | | | | | |
| 22 | 0.0275 | 699 | 20 | 0.0250 | 635 | 24 | 0.0276 | 701 | 25 | 0.0280 | 710 |
| 18 | 0.0336 | 853 | 16 | 0.0312 | 792 | 20 | 0.0328 | 833 | 20 | 0.0331 | 840 |
| 16 | 0.0395 | 1003 | | | | 16 | 0.0390 | 991 | 18 | 0.0394 | 1000 |
| 14 | 0.0474 | 1204 | 12 | 0.0416 | 1056 | 14 | 0.0460 | 1168 | 16 | 0.0469 | 1190 |
| 12 | 0.0553 | 1405 | 10 | 0.0500 | 1270 | 12 | 0.0550 | 1397 | | | |
| 10 | 0.0660 | 1676 | 8 | 0.0620 | 1574 | 10 | 0.0650 | 1651 | 14 | 0.0555 | 1410 |
| 8 | 0.0810 | 2057 | | | | 9 | 0.0780 | 1981 | 12 | 0.0661 | 1680 |
| 7 | 0.0949 | 2411 | | | | 8 | 0.0930 | 2362 | 10 | 0.0787 | 2000 |
| 6 | 0.1107 | 2812 | 5 | 0.1000 | 2540 | 7 | 0.1100 | 2794 | 8 | 0.0937 | 2380 |
| 5 | 0.1320 | 3353 | | | | 6 | 0.1310 | 3327 | | | |
| | | | | | | 5 | 0.1560 | 3962 | 7 | 0.1110 | 2839 |
| | | | | | | 4 | 0.1850 | 4699 | | | |
| | | | | | | | | | 6 | 0.1320 | 3360 |
| | | | | | | | | | 5 | 0.1570 | 4000 |
| | | | | | | | | | 4 | 0.1870 | 4760 |

The efficiency of screening is defined as the ratio of the mass of material which passes the screen to that which is capable of passing. This will differ according to the size of the material. It may be assumed that the rate of passage of particles of a given size through the screen is proportional to the number or mass of particles of that size on the screen at any instant. Thus, if $w$ is the mass of particles of a particular size on the screen at a time $t$, then:

$$\frac{dw}{dt} = -kw \tag{1.2}$$

where $k$ is a constant for a given size and shape of particle and for a given screen.

Thus, the mass of particles $(w_1 - w_2)$ passing the screen in time $t$ is given by:

$$\ln \frac{w_2}{w_1} = -kt$$

or:

$$w_2 = w_1 \, e^{-kt} \tag{1.3}$$

If the screen contains a large proportion of material just a little larger than the maximum size of particle which will pass, its capacity is considerably reduced. Screening is generally continued either for a predetermined time or until the rate of screening falls off to a certain fixed value.

Screening may be carried out with either wet or dry material. In wet screening, material is washed evenly over the screen and clogging is prevented. In addition, small particles are washed off the surface of large ones. This has the obvious disadvantage, however, that it may be necessary to dry the material afterwards. With dry screening, the material is sometimes brushed lightly over the screen so as to form a thin even sheet. It is important that any agitation is not so vigorous that size reduction occurs, because screens are usually quite fragile and easily damaged by rough treatment. In general, the larger and the more abrasive the solids the more robust is the screen required.

## *Microscopic analysis* (1–100 μm)

Microscopic examination permits measurement of the projected area of the particle and also enables an assessment to be made of its two-dimensional shape. In general, the third dimension cannot be determined except when using special stereomicroscopes. The apparent size of particle is compared with that of circles engraved on a graticule in the eyepiece as shown in Figure 1.3. Automatic methods of scanning have been developed. By using the electron microscope[7], the lower limit of size can be reduced to about 0.001 μm.

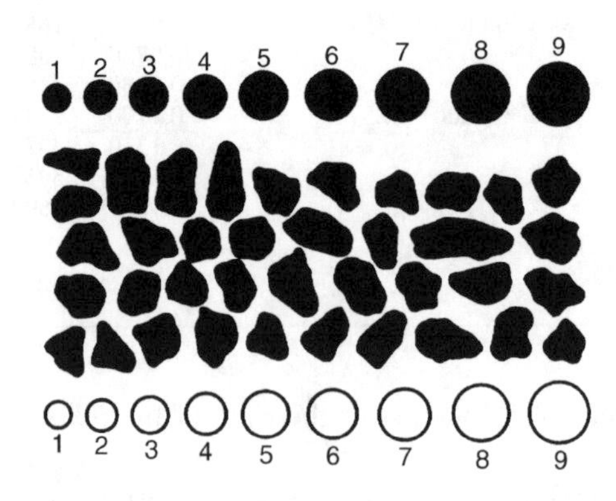


Figure 1.3. Particle profiles and comparison circles

### *Sedimentation and elutriation methods* (>1 μm)

These methods depend on the fact that the terminal falling velocity of a particle in a fluid increases with size. Sedimentation methods are of two main types. In the first, the pipette method, samples are abstracted from the settling suspension at a fixed horizontal level at intervals of time. Each sample contains a representative sample of the suspension, with the exception of particles larger than a critical size, all of which will have settled below the level of the sampling point. The most commonly used equipment, the Andreason pipette, is described by ALLEN[1]. In the second method, which involves the use of the sedimentation balance, particles settle on an immersed balance pan which is continuously weighed. The largest particles are deposited preferentially and consequently the rate of increase of weight falls off progressively as particles settle out.

Sedimentation analyses must be carried out at concentrations which are sufficiently low for interactive effects between particles to be negligible so that their terminal falling velocities can be taken as equal to those of isolated particles. Careful temperature control (preferably to ±0.1 deg K) is necessary to suppress convection currents. The lower limit of particle size is set by the increasing importance of Brownian motion for progressively smaller particles. It is possible however, to replace gravitational forces by centrifugal forces and this reduces the lower size limit to about 0.05 μm.

The elutriation method is really a reverse sedimentation process in which the particles are dispersed in an upward flowing stream of fluid. All particles with terminal falling velocities less than the upward velocity of the fluid will be carried away. A complete size analysis can be obtained by using successively higher fluid velocities. Figure 1.4 shows the standard elutriator (BS 893)[6] for particles with settling velocities between 7 and 70 mm/s.

### *Permeability methods* (>1 μm)

These methods depend on the fact that at low flowrates the flow through a packed bed is directly proportional to the pressure difference, the proportionality constant being proportional to the square of the specific surface (surface : volume ratio) of the powder. From this method it is possible to obtain the diameter of the sphere with the same specific surface as the powder. The reliability of the method is dependent upon the care with which the sample of powder is packed. Further details are given in Chapter 4.

### *Electronic particle counters*

A suspension of particles in an electrolyte is drawn through a small orifice on either side of which is positioned an electrode. A constant electrical current supply is connected to the electrodes and the electrolyte within the orifice constitutes the main resistive component of the circuit. As particles enter the orifice they displace an equivalent volume of electrolyte, thereby producing a change in the electrical resistance of the circuit, the magnitude of which is related to the displaced volume. The consequent voltage pulse across the electrodes is fed to a multi-channel analyser. The distribution of pulses arising from the passage of many thousands of particles is then processed to provide a particle (volume) size distribution.

The main disadvantage of the method is that the suspension medium must be so highly conducting that its ionic strength may be such that surface active additives may be

correlated with itself at different time intervals using a digital correlator and associated computer software. The relationship of the (so-called) auto-correlation function to the time intervals is processed to provide estimates of an average particle size and variance (polydispersity index). Analysis of the signals at different scattering angles enables more detailed information to be obtained about the size distribution of this fine, and usually problematical, end of the size spectrum.

The technique allows fine particles to be examined in a liquid environment so that estimates can be made of their effective hydrodynamic sizes. This is not possible using other techniques.

Provided that fluid motion is uniform in the illuminated region of the suspension, then similar information may also be extracted by analysis of laser light scattering from particles undergoing electrophoretic motion, that is migratory motion in an electric field, superimposed on that motion.

Instrumentation and data processing techniques for systems employing dynamic light scattering for the examination of fine particle motion are currently under development.

### 1.2.3. Particle size distribution

Most particulate systems of practical interest consist of particles of a wide range of sizes and it is necessary to be able to give a quantitative indication of the mean size and of the spread of sizes. The results of a size analysis can most conveniently be represented by means of a *cumulative mass fraction curve*, in which the proportion of particles ($x$) smaller than a certain size ($d$) is plotted against that size ($d$). In most practical determinations of particle size, the size analysis will be obtained as a series of steps, each step representing the proportion of particles lying within a certain small range of size. From these results a cumulative size distribution can be built up and this can then be approximated by a smooth curve provided that the size intervals are sufficiently small. A typical curve for size distribution on a cumulative basis is shown in Figure 1.5. This curve rises from zero to unity over the range from the smallest to the largest particle size present.

The distribution of particle sizes can be seen more readily by plotting a *size frequency curve*, such as that shown in Figure 1.6, in which the slope ($dx/dd$) of the cumulative

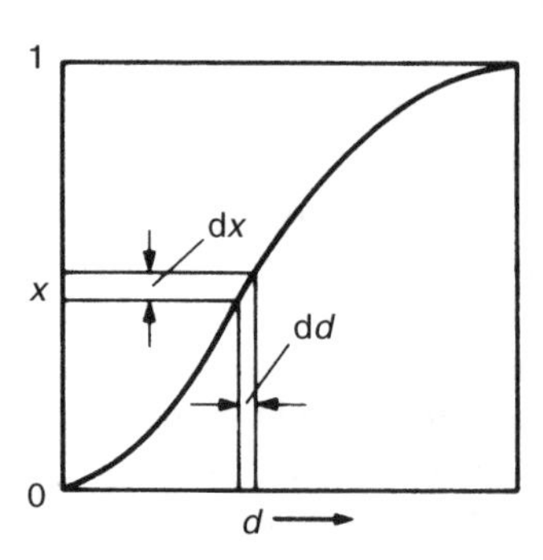


Figure 1.5. Size distribution curve — cumulative basis

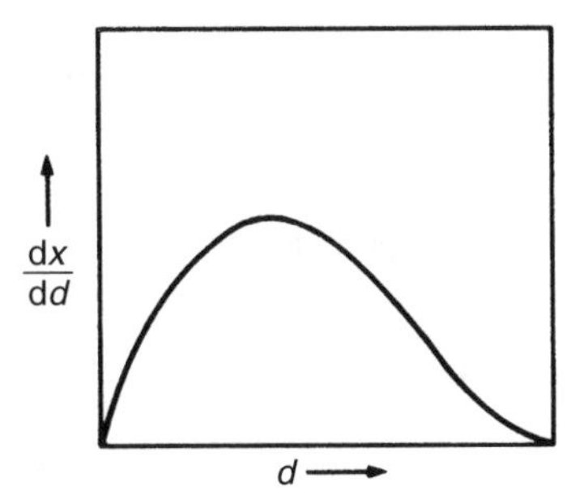


Figure 1.6. Size distribution curve — frequency basis

curve (Figure 1.5) is plotted against particle size ($d$). The most frequently occurring size is then shown by the maximum of the curve. For naturally occurring materials the curve will generally have a single peak. For mixtures of particles, there may be as many peaks as components in the mixture. Again, if the particles are formed by crushing larger particles, the curve may have two peaks, one characteristic of the material and the other characteristic of the equipment.

### 1.2.4. Mean particle size

The expression of the particle size of a powder in terms of a single linear dimension is often required. For coarse particles, BOND[9,10] has somewhat arbitrarily chosen the size of the opening through which 80 per cent of the material will pass. This size $d_{80}$ is a useful rough comparative measure for the size of material which has been through a crusher.

A mean size will describe only one particular characteristic of the powder and it is important to decide what that characteristic is before the mean is calculated. Thus, it may be desirable to define the size of particle such that its mass or its surface or its length is the mean value for all the particles in the system. In the following discussion it is assumed that each of the particles has the same shape.

Considering unit mass of particles consisting of $n_1$ particles of characteristic dimension $d_1$, constituting a mass fraction $x_1$, $n_2$ particles of size $d_2$, and so on, then:

$$x_1 = n_1 k_1 d_1^3 \rho_s \tag{1.4}$$

and:

$$\Sigma x_1 = 1 = \rho_s k_1 \Sigma(n_1 d_1^3) \tag{1.5}$$

Thus:

$$n_1 = \frac{1}{\rho_s k_1} \frac{x_1}{d_1^3} \tag{1.6}$$

If the size distribution can be represented by a continuous function, then:

$$\mathrm{d}x = \rho_s k_1 d^3 \, \mathrm{d}n$$

or:

$$\frac{\mathrm{d}x}{\mathrm{d}n} = \rho_s k_1 d^3 \tag{1.7}$$

and:
$$\int_0^1 \mathrm{d}x = 1 = \rho_s k_1 \int d^3\,\mathrm{d}n \tag{1.8}$$

where $\rho_s$ is the density of the particles, and
$k_1$ is a constant whose value depends on the shape of the particle.

### Mean sizes based on volume

The mean abscissa in Figure 1.5 is defined as the *volume mean diameter* $d_v$, or as the *mass mean diameter*, where:

$$d_v = \frac{\int_0^1 d\,\mathrm{d}x}{\int_0^1 \mathrm{d}x} = \int_0^1 d\,\mathrm{d}x. \tag{1.9}$$

Expressing this relation in finite difference form, then:

$$d_v = \frac{\Sigma(d_1 x_1)}{\Sigma x_1} = \Sigma(x_1 d_1) \tag{1.10}$$

which, in terms of particle numbers, rather than mass fractions gives:

$$d_v = \frac{\rho_s k_1 \Sigma(n_1 d_1^4)}{\rho_s k_1 \Sigma(n_1 d_1^3)} = \frac{\Sigma(n_1 d_1^4)}{\Sigma(n_1 d_1^3)} \tag{1.11}$$

Another mean size based on volume is the *mean volume diameter* $d'_v$. If all the particles are of diameter $d'_v$, then the total volume of particles is the same as in the mixture.

Thus:
$$k_1 d'^3_v \Sigma n_1 = \Sigma(k_1 n_1 d_1^3)$$

or:
$$d'_v = \sqrt[3]{\left(\frac{\Sigma(n_1 d_1^3)}{\Sigma n_1}\right)} \tag{1.12}$$

Substituting from equation 1.6 gives:

$$d'_v = \sqrt[3]{\left(\frac{\Sigma x_1}{\Sigma(x_1/d_1^3)}\right)} = \sqrt[3]{\left(\frac{1}{\Sigma(x_1/d_1^3)}\right)} \tag{1.13}$$

### Mean sizes based on surface

In Figure 1.5, if, instead of fraction of total mass, the surface in each fraction is plotted against size, then a similar curve is obtained although the mean abscissa $d_s$ is then the *surface mean diameter*.

Thus:
$$d_s = \frac{\Sigma[(n_1 d_1) S_1]}{\Sigma(n_1 S_1)} = \frac{\Sigma(n_1 k_2 d_1^3)}{\Sigma(n_1 k_2 d_1^2)} = \frac{\Sigma(n_1 d_1^3)}{\Sigma(n_1 d_1^2)} \tag{1.14}$$

where $S_1 = k_2 d_1^2$, and $k_2$ is a constant whose value depends on particle shape. $d_s$ is also known as the *Sauter mean diameter* and is the diameter of the particle with the same specific surface as the powder.

Substituting for $n_1$ from equation 1.6 gives:

$$d_s = \frac{\Sigma x_1}{\Sigma\left(\dfrac{x_1}{d_1}\right)} = \frac{1}{\Sigma\left(\dfrac{x_1}{d_1}\right)} \tag{1.15}$$

The *mean surface diameter* is defined as the size of particle $d'_s$ which is such that if all the particles are of this size, the total surface will be the same as in the mixture.

Thus: $$k_2 d'^2_s \Sigma n_1 = \Sigma(k_2 n_1 d_1^2)$$

or: $$d'_s = \sqrt{\left(\frac{\Sigma(n_1 d_1^2)}{\Sigma n_1}\right)} \tag{1.16}$$

Substituting for $n_1$ gives:

$$d'_s = \sqrt{\left(\frac{\Sigma(x_1/d_1)}{\Sigma(x_1/d_1^3)}\right)} \tag{1.17}$$

### Mean dimensions based on length

A *length mean diameter* may be defined as:

$$d_l = \frac{\Sigma[(n_1 d_1) d_1]}{\Sigma(n_1 d_1)} = \frac{\Sigma(n_1 d_1^2)}{\Sigma(n_1 d_1)} = \frac{\Sigma\left(\dfrac{x_1}{d_1}\right)}{\Sigma\left(\dfrac{x_1}{d_1^2}\right)} \tag{1.18}$$

A *mean length diameter* or arithmetic mean diameter may also be defined by:

$$d'_l \Sigma n_1 = \Sigma(n_1 d_1)$$

$$d'_l = \frac{\Sigma(n_1 d_1)}{\Sigma n_1} = \frac{\Sigma\left(\dfrac{x_1}{d_1^2}\right)}{\Sigma\left(\dfrac{x_1}{d_1^3}\right)} \tag{1.19}$$

## Example 1.1

The size analysis of a powdered material on a mass basis is represented by a straight line from 0 per cent mass at 1 μm particle size to 100 per cent mass at 101 μm particle size as shown in Figure 1.7. Calculate the surface mean diameter of the particles constituting the system.

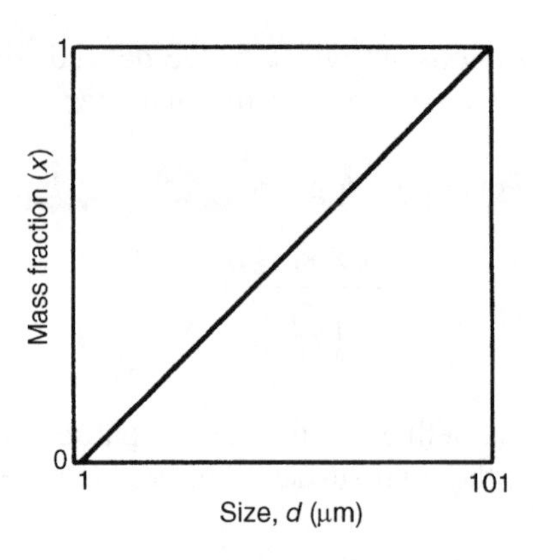


Figure 1.7. Size analysis of powder

## Solution

From equation 1.15, the surface mean diameter is given by:

$$d_s = \frac{1}{\Sigma(x_1/d_1)}$$

Since the size analysis is represented by the continuous curve:

$$d = 100x + 1$$

then:

$$d_s = \frac{1}{\displaystyle\int_0^1 \frac{dx}{d}}$$

$$= \frac{1}{\displaystyle\int_0^1 \frac{dx}{100x + 1}}$$

$$= (100/\ln 101)$$

$$= \underline{\underline{21.7\ \mu\text{m}}}$$

## Example 1.2

The equations giving the number distribution curve for a powdered material are $dn/dd = d$ for the size range 0–10 μm and $dn/dd = 100{,}000/d^4$ for the size range 10–100 μm where $d$ is in μm. Sketch the number, surface and mass distribution curves and calculate the surface mean diameter for the powder. Explain briefly how the data required for the construction of these curves may be obtained experimentally.

## Solution

Note: The equations for the number distributions are valid only for $d$ expressed in μm.

For the range, $d = 0 - 10$ μm, $dn/dd = d$

On integration:

$$n = 0.5d^2 + c_1 \qquad \text{(i)}$$

where $c_1$ is the constant of integration.

For the range, $d = 10 - 100$ μm, $dn/dd = 10^5 d^{-4}$

On integration: $$n = c_2 - (0.33 \times 10^5 d^{-3}) \qquad \text{(ii)}$$

where $c_2$ is the constant of integration.

When $d = 0$, $n = 0$, and from (i): $c_1 = 0$

When $d = 10$ μm, in (i): $n = (0.5 \times 10^2) = 50$

In (ii): $50 = c_2 - (0.33 \times 10^5 \times 10^{-3})$, and $c_2 = 83.0$.

Thus for $d = 0 - 10$ μm: $$n = 0.5d^2$$

and for $d = 10 - 100$ μm: $$n = 83.0 - (0.33 \times 10^5 d^{-3})$$

Using these equations, the following values of $n$ are obtained:

| $d$(μm) | $n$ | $d$(μm) | $n$ |
|---|---|---|---|
| 0 | 0 | 10 | 50.0 |
| 2.5 | 3.1 | 25 | 80.9 |
| 5.0 | 12.5 | 50 | 82.7 |
| 7.5 | 28.1 | 75 | 82.9 |
| 10.0 | 50.0 | 100 | 83.0 |

and these data are plotted in Figure 1.8.

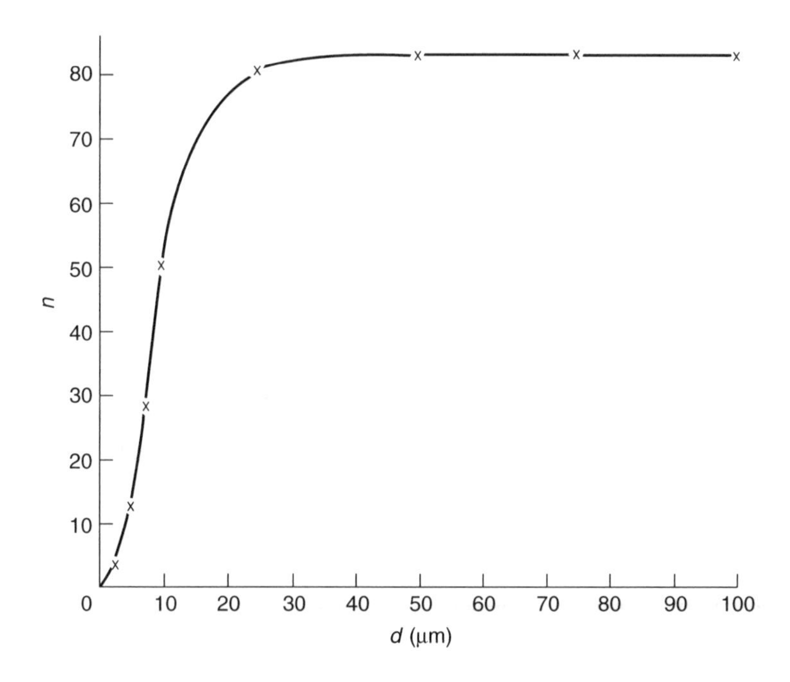


Figure 1.8. Plot of data for Example 1.2

## 1.3. PARTICULATE SOLIDS IN BULK

### 1.3.1. General characteristics

The properties of solids in bulk are a function of the properties of the individual particles including their shapes and sizes and size distribution, and of the way in which the particles interact with one another. By the very nature of a particulate material, it is always interspersed with a fluid, generally air, and the interaction between the fluid and the particles may have a considerable effect on the behaviour of the bulk material. Particulate solids present considerably greater problems than fluids in storage, in removal at a controlled rate from storage, and when introduced into vessels or reactors where they become involved in a process. Although there has recently been a considerable amount of work carried out on the properties and behaviour of solids in bulk, there is still a considerable lack of understanding of all the factors determining their behaviour.

One of the most important characteristics of any particulate mass is its voidage, the fraction of the total volume which is made up of the free space between the particles and is filled with fluid. Clearly, a low voidage corresponds to a high density of packing of the particles. The way in which the particles pack depends not only on their physical properties, including shape and size distribution, but also on the way in which the particulate mass has been introduced to its particular location. In general, isometric particles, which have approximately the same linear dimension in each of the three principal directions, will pack more densely than long thin particles or plates. The more rapidly material is poured on to a surface or into a vessel, the more densely will it pack. If it is then subjected to vibration, further consolidation may occur. The packing density or voidage is important in that it determines the bulk density of the material, and hence the volume taken up by a given mass: It affects the tendency for agglomeration of the particles, and it critically influences the resistance which the material offers to the percolation of fluid through it — as, for example, in filtration as discussed in Chapter 7.

### 1.3.2. Agglomeration

Because it is necessary in processing plant to transfer material from storage to process, it is important to know how the particulate material will flow. If a significant amount of the material is in the form of particles smaller than 10 μm or if the particles deviate substantially from isometric form, it may be inferred that the flow characteristics will be poor. If the particles tend to agglomerate, poor flow properties may again be expected. Agglomeration arises from interaction between particles, as a result of which they adhere to one another to form clusters. The main mechanisms giving rise to agglomeration are:

(1) *Mechanical interlocking*. This can occur particularly if the particles are long and thin in shape, in which case large masses may become completely interlocked.
(2) *Surface attraction*. Surface forces, including van der Waals' forces, may give rise to substantial bonds between particles, particularly where particles are very fine (<10 μm), with the result that their surface per unit volume is high. In general, freshly formed surface, such as that resulting from particle fracture, gives rise to high surface forces.

(3) *Plastic welding.* When irregular particles are in contact, the forces between the particles will be borne on extremely small surfaces and the very high pressures developed may give rise to plastic welding.
(4) *Electrostatic attraction.* Particles may become charged as they are fed into equipment and significant electrostatic charges may be built up, particularly on fine solids.
(5) *Effect of moisture.* Moisture may have two effects. Firstly, it will tend to collect near the points of contact between particles and give rise to surface tension effects. Secondly, it may dissolve a little of the solid, which then acts as a bonding agent on subsequent evaporation.
(6) *Temperature fluctuations* give rise to changes in particle structure and to greater cohesiveness.

Because interparticle forces in very fine powders make them very difficult to handle, the effective particle size is frequently increased by agglomeration. This topic is discussed in Section 2.4 on particle size enlargement in Chapter 2.

### 1.3.3. Resistance to shear and tensile forces

A particulate mass may offer a significant resistance to both shear and tensile forces, and this is specially marked when there is a significant amount of agglomeration. Even in non-agglomerating powders there is some resistance to relative movement between the particles and it is always necessary for the bed to dilate, that is for the particles to move apart, to some extent before internal movement can take place. The greater the density of packing, the higher will be this resistance to shear and tension.

The resistance of a particulate mass to shear may be measured in a shear cell such as that described by JENIKE *et al.*[11,12] The powder is contained in a shallow cylindrical cell (with a vertical axis) which is split horizontally. The lower half of the cell is fixed and the upper half is subjected to a shear force which is applied slowly and continuously measured. The shearing is carried out for a range of normal loads, and the relationship between shear force and normal force is measured to give the shear strength at different degrees of compaction.

A method of measuring tensile strength has been developed by ASHTON *et al.*[13] who also used a cylindrical cell split diametrically. One half of the cell is fixed and the other, which is movable, is connected to a spring, the other end of which is driven at a slow constant speed. A slowly increasing force is thus exerted on the powder compact and the point at which failure occurs determines the tensile strength; this has been shown to depend on the degree of compaction of the powder.

The magnitude of the shear and tensile strength of the powder has a considerable effect on the way in which the powder will flow, and particularly on the way in which it will discharge from a storage hopper through an outlet nozzle.

### 1.3.4. Angles of repose and of friction

A rapid method of assessing the behaviour of a particulate mass is to measure its *angle of repose*. If solid is poured from a nozzle on to a plane surface, it will form an approximately conical heap and the angle between the sloping side of the cone and the horizontal is the angle of repose. When this is determined in this manner it is sometimes referred to as the *dynamic angle of repose* or the *poured angle*. In practice, the heap will not be exactly conical and there will be irregularities in the sloping surface. In addition, there will be a tendency for large particles to roll down from the top and collect at the base, thus giving a greater angle at the top and a smaller angle at the bottom.

The angle of repose may also be measured using a plane sheet to which is stuck a layer of particles from the powder. Loose powder is then poured on to the sheet which is then tilted until the powder slides. The angle of slide is known as the *static angle of repose* or the *drained angle*.

Angles of repose vary from about 20° with free-flowing solids, to about 60° with solids with poor flow characteristics. In extreme cases of highly agglomerated solids, angles of repose up to nearly 90° can be obtained. Generally, material which contains no particles smaller than 100 μm has a low angle of repose. Powders with low angles of repose tend to pack rapidly to give a high packing density almost immediately. If the angle of repose is large, a loose structure is formed initially and the material subsequently consolidates if subjected to vibration.

An angle which is similar to the static angle of repose is the *angle of slide* which is measured in the same manner as the drained angle except that the surface is smooth and is not coated with a layer of particles.

A measure of the frictional forces within the particulate mass is the *angle of friction*. This can be measured in a number of ways, two of which are described. In the first, the powder is contained in a two-dimensional bed, as shown in Figure 1.12 with transparent walls and is allowed to flow out through a slot in the centre of the base. It is found that a triangular wedge of material in the centre flows out leaving stationary material at the outside. The angle between the cleavage separating stationary and flowing material and the horizontal is the angle of friction. A simple alternative method of measuring the angle of friction, as described by ZENZ[14], employs a vertical tube, open at the top, with a loosely fitting piston in the base as shown in Figure 1.13. With small quantities of solid in the tube, the piston will freely move upwards, but when a certain critical amount is exceeded no force, however large, will force the solids upwards in the tube. With the largest movable core of solids in the tube, the ratio of its length to diameter is the tangent of the angle of friction.

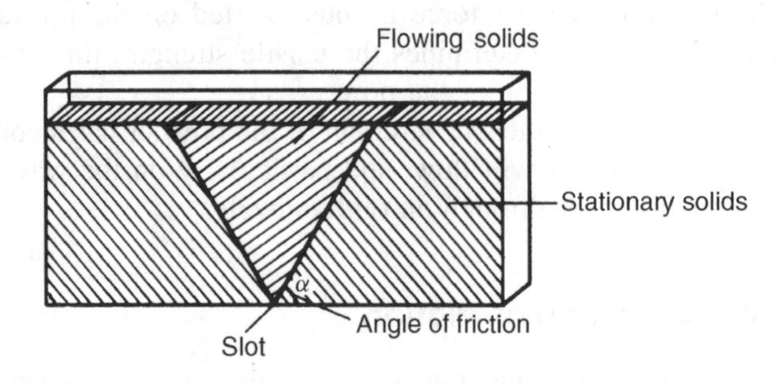


Figure 1.12. Angle of friction — flow through slot

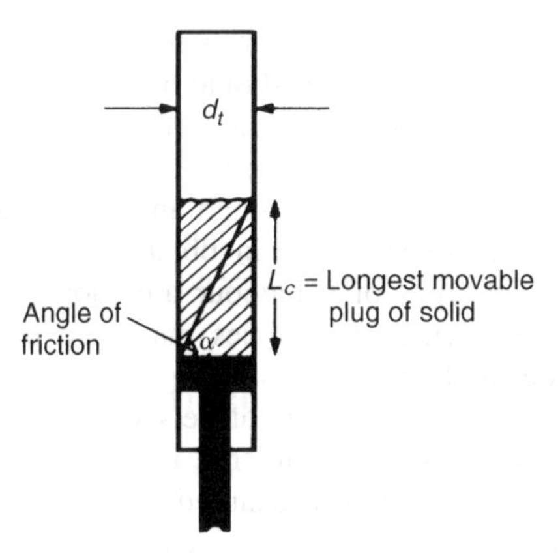


Figure 1.13. Angle of friction — tube test

The angle of friction is important in its effect on the design of bins and hoppers. If the pressure at the base of a column of solids is measured as a function of depth, it is found to increase approximately linearly with height up to a certain critical point beyond which it remains constant. A typical curve is shown in Figure 1.14. The point of discontinuity on the curve is given by:

$$L_c/d_t = \tan\alpha \tag{1.31}$$

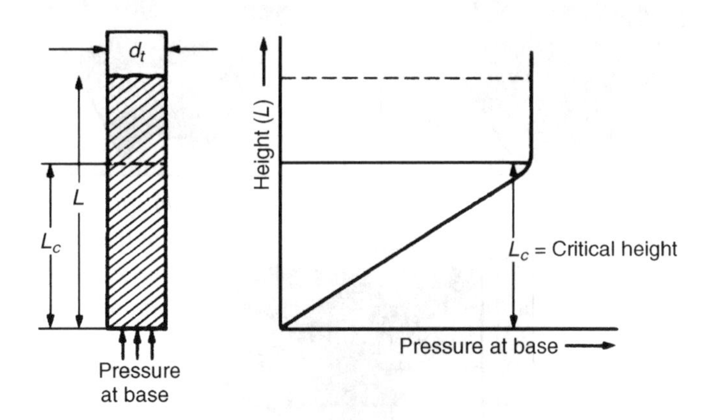


Figure 1.14. Angle of friction — pressure at base of column

For heights greater than $L_c$ the mass of additional solids is supported by frictional forces at the walls of the hopper. It may thus be seen that hoppers must be designed to resist the considerable pressures due to the solids acting on the walls.

### 1.3.5. Flow of solids in hoppers

Solids may be stored in heaps or in sacks although subsequent handling problems may be serious with large-scale operations. Frequently, solids are stored in hoppers which are

Alternatively, the level of the solids in the hopper may be continuously monitored using transducers covered by flexible diaphragms flush with the walls of the hopper. The diaphragm responds to the presence of the solids and thus indicates whether there are solids present at a particular level.

The problems associated with the measurement and control of the flowrate of solids are much more complicated than those in the corresponding situation with liquids. The flow characteristics will depend, not only on particle size, size range and shape, but also on how densely the particles are packed. In addition, surface and electrical properties and moisture content all exert a strong influence on flow behaviour, and the combined effect of these factors is almost impossible to predict in advance. It is therefore desirable to carry out a preliminary qualitative assessment before making a selection of the, most appropriate technique for controlling and measuring flowrate for any particular application.

A detailed description of the various methods of measuring solids flowrates has been given by LIPTAK[17] and Table 1.2, taken from this work, gives a summary of the principal types of solids flowmeters in common use.

Table 1.2. Different types of solids flowmeter[17]

| Type of Meter | Flowrate (kg/s) | Accuracy (per cent FSD[a] over 10:1 range) | Type of Material |
|---|---|---|---|
| Gravimetric belt | <25 (or <0.3 $m^3/s$) | $\pm 0.5R$ | Dependent on mechanism used to feed belt. Vertical gate feeder suitable for non-fluidised materials with particle size <3 mm. |
| Belt with nucleonic sensor | <25 (or <0.3 $m^3/s$) | $\pm 0.5$ to $\pm 1$ | Preferred when material is difficult to handle, e.g. corrosive, hot, dusty, or abrasive. Accuracy greatly improved when particle size, bulk density, and moisture content are constant, when belt load is 70–100 per cent of maximum. |
| Vertical gravimetric | Limited capacity | $\pm 0.5$ for 5:1 turndown. $\pm 1.0$ for 20:1 turndown. | Dry and free-flowing powders with particle size <2.5 mm. |
| Loss-in-weight | Depends on size of hopper | $\pm 1.0R$ | Liquids, slurries, and solids. |
| Dual hopper | 0.13–40 | $\pm 0.5R$ | Free-flowing bulk solids. |
| Impulse | 0.4–400 | $\pm 1$ to $\pm 2$ | Free-flowing powders. Granules/pellets <13 mm in size. |
| Volumetric | <0.3 $m^3/s$ | $\pm 2$ to $\pm 4$ | Solids of uniform size. |

[a]FSD = full-scale deflection.

Methods include:

a) Fitting an orifice plate at the discharge point from the hopper. The flow of solids through orifices is discussed briefly in Section 1.3.6.
b) Using a belt-type feeder in which the mass of material on the belt is continuously measured, by load cells for example or by a nuclear densitometer which measures

the degree of absorption of gamma rays transmitted vertically through the solids on the belt which is travelling at a controlled speed.

c) Applying an impulse method in which a solids stream impacts vertically on a sensing plate orientated at an angle to the vertical. The horizontal component of the resulting force is measured by as load cell attached to the plate.

The rate of feed of solids may be controlled using screw feeders, rotating tables or vibrating feeders, such as magnetically vibrated troughs. Volumetric rates may be controlled by regulating the speeds of rotation of star feeders or rotary vaned valves.

### 1.3.8. Conveying of solids

The variety of requirements in connection with the conveying of solids has led to the development of a wide range of equipment. This includes:

(a) *Gravity chutes*—down which the solids fall under the action of gravity.

(b) *Air slides*—where the particles, which are maintained partially suspended in a channel by the upward flow of air through a porous distributor, flow at a small angle to the horizontal.

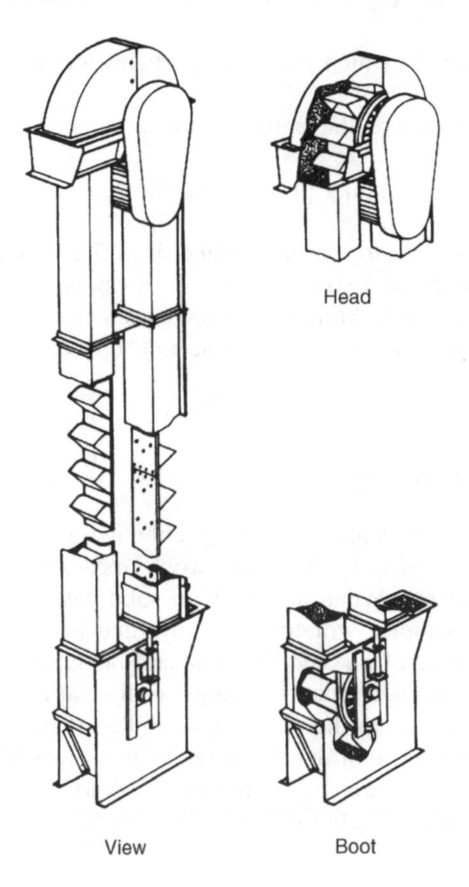


Figure 1.16. Bucket elevator

(c) *Belt conveyors* — where the solids are conveyed horizontally, or at small angles to the horizontal, on a continuous moving belt.
(d) *Screw conveyors* — in which the solids are moved along a pipe or channel by a rotating helical impeller, as in a screw lift elevator.
(e) *Bucket elevators* — in which the particles are carried upwards in buckets attached to a continuously moving vertical belt, as illustrated in Figure 1.16.
(f) *Vibrating conveyors* — in which the particles are subjected to an asymmetric vibration and travel in a series of steps over a table. During the forward stroke of the table the particles are carried forward in contact with it, but the acceleration in the reverse stroke is so high that the table slips under the particles. With fine powders, vibration of sufficient intensity results in a fluid-like behaviour.
(g) *Pneumatic/hydraulic conveying installations* — in which the particles are transported in a stream of air/water. Their operation is described in Volume 1, Chapter 5.

## 1.4. BLENDING OF SOLID PARTICLES

In the mixing of solid particles, the following three mechanisms may be involved:

(a) Convective mixing, in which groups of particles are moved from one position to another,
(b) Diffusion mixing, where the particles are distributed over a freshly developed interface, and
(c) Shear mixing, where slipping planes are formed.

These mechanisms operate to varying extents in different kinds of mixers and with different kinds of particles. A trough mixer with a ribbon spiral involves almost pure convective mixing, and a simple barrel-mixer involves mainly a form of diffusion mixing.

The mixing of pastes is discussed in the section on Non-Newtonian Technology in Volume 1, Chapter 7.

### 1.4.1. The degree of mixing

It is difficult to quantify the degree of mixing, although any index should be related to the properties of the required mix, should be easy to measure, and should be suitable for a variety of different mixers. When dealing with solid particles, the statistical variation in composition among samples withdrawn at any time from a mix is commonly used as a measure of the degree of mixing. The standard deviation $s$ (the square root of the mean of the squares of the individual deviations) or the variance $s^2$ is generally used. A particulate material cannot attain the perfect mixing that is possible with two fluids, and the best that can be obtained is a degree of randomness in which two similar particles may well be side by side. No amount of mixing will lead to the formation of a uniform mosaic as shown in Figure 1.17, but only to a condition, such as shown in Figure 1.18,

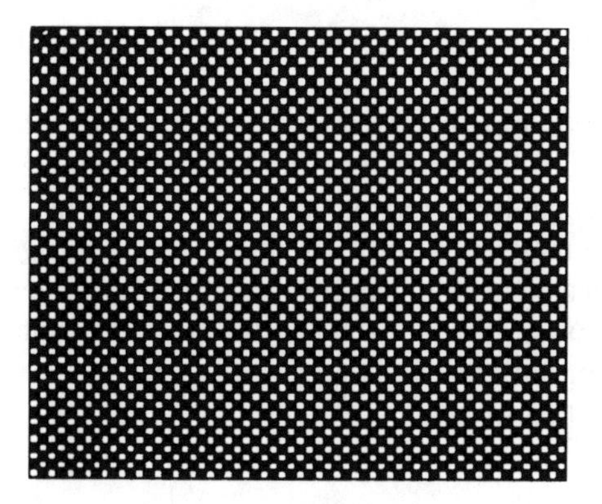

Figure 1.17. Uniform mosaic

Figure 1.18. Overall but not point uniformity in mix

where there is an overall uniformity but not point uniformity. For a completely random mix of uniform particles distinguishable, say, only by colour, LACEY[18,19] has shown that:

$$s_r^2 = \frac{p(1-p)}{n} \tag{1.33}$$

where $s_r^2$ is the variance for the mixture, $p$ is the overall proportion of particles of one colour, and $n$ is the number of particles in each sample.

This equation illustrates the importance of the size of the sample in relation to the size of the particles. In an incompletely randomised material, $s^2$ will be greater, and in a completely unmixed system, indicated by the suffix 0, it may be shown that:

$$s_0^2 = p(1-p) \tag{1.34}$$

which is independent of the number of particles in the sample. Only a definite number of samples can in practice be taken from a mixture, and hence $s$ will itself be subject to random errors. This analysis has been extended to systems containing particles of different sizes by BUSLIK[20].

When a material is partly mixed, then the degree of mixing may be represented by some term $b$, and several methods have been suggested for expressing $b$ in terms of measurable quantities. If $s$ is obtained from examination of a large number of samples then, as suggested by Lacey, $b$ may be defined as being equal to $s_r/s$, or $(s_0 - s)/(s_0 - s_r)$, as suggested by KRAMERS[21] where, as before, $s_0$ is the value of $s$ for the unmixed material. This form of expression is useful in that $b = 0$ for an unmixed material and 1 for a completely randomised material where $s = s_r$. If $s^2$ is used instead of $s$, then this expression may be modified to give:

$$b = \frac{(s_0^2 - s^2)}{(s_0^2 - s_r^2)} \tag{1.35}$$

or:

$$1 - b = \frac{(s^2 - s_r^2)}{(s_0^2 - s_r^2)} \tag{1.36}$$

For diffusive mixing, $b$ will be independent of sample size provided the sample is small. For convective mixing, Kramers has suggested that for groups of particles randomly distributed, each group behaves as a unit containing $n_g$ particles. As mixing proceeds $n_g$ becomes smaller. The number of groups will then be $n_p/n_g$, where $n_p$ is the number of particles in each sample. Applying equation 1.33:

$$s^2 = \frac{p(1-p)}{n_p/n_g} = n_g s_r^2$$

which gives:

$$1 - b = \frac{(n_g s_r^2 - s_r^2)}{(n_p s_r^2 - s_r^2)} = \frac{(n_g - 1)}{(n_p - 1)} \tag{1.37}$$

Thus, with convective mixing, $1 - b$ depends on the size of the sample.

## Example 1.3

The performance of a solids mixer was assessed by calculating the variance occurring in the mass fraction of a component amongst a selection of samples withdrawn from the mixture. The quality was tested at intervals of 30 s and the data obtained are:

| sample variance (−) | 0.025 | 0.006 | 0.015 | 0.018 | 0.019 |
|---|---|---|---|---|---|
| mixing time (s) | 30 | 60 | 90 | 120 | 150 |

If the component analysed represents 20 per cent of the mixture by mass and each of the samples removed contains approximately 100 particles, comment on the quality of the mixture produced and present the data in graphical form showing the variation of the mixing index with time.

## Solution

For a completely unmixed system:

$$s_0^2 = p(1 - p) = 0.20(1 - 0.20) = 0.16 \qquad \text{(equation 1.34)}$$

For a completely random mixture:

$$s_r^2 = p(1 - p)/n = 0.20(1 - 0.20)/100 = 0.0016 \qquad \text{(equation 1.33)}$$

The degree of mixing $b$ is given by equation 1.35 as: $b = (s_0^2 - s^2)/(s_0^2 - s_r^2)$ In this case, $b = (0.16 - s^2)/(0.16 - 0.0016) = 1.01 - 6.313s^2$ The calculated data are therefore:

| $t$(s) | 30 | 60 | 90 | 120 | 150 |
|---|---|---|---|---|---|
| $s^2$ | 0.025 | 0.006 | 0.015 | 0.018 | 0.019 |
| $b$ | 0.852 | 0.972 | 0.915 | 0.896 | 0.890 |

These data are plotted in Figure 1.19 from which it is clear that the degree of mixing is a maximum at $t = 60$s.

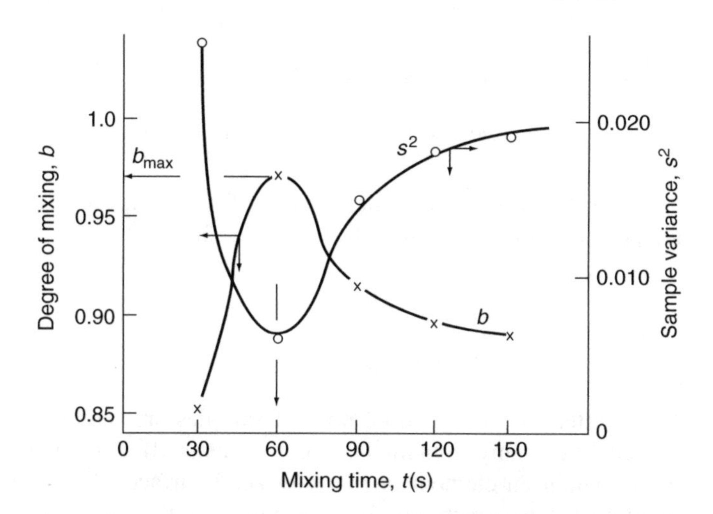


Figure 1.19. Example 1.3. Degree of mixing as a function of mixing time

### 1.4.2. The rate of mixing

Expressions for the rate of mixing may be developed for any one of the possible mechanisms. Since mixing involves obtaining an equilibrium condition of uniform randomness, the relation between $b$ and time might be expected to take the general form:

$$b = 1 - e^{-ct} \qquad (1.38)$$

where $c$ is a constant depending on the nature of the particles and the physical action of the mixer.

Considering a cylindrical vessel (Figure 1.20) in which one substance **A** is poured into the bottom and the second **B** on top, then the boundary surface between the two materials is a minimum. The process of mixing consists of making some of **A** enter the space occupied by **B**, and some of **B** enter the lower section originally filled by **A**. This may be considered as the diffusion of **A** across the initial boundary into **B**, and of **B** into **A**. This process will continue until there is a maximum degree of dispersion, and a maximum interfacial area between the two materials. This type of process is somewhat akin to that of diffusion, and tentative use may be made of the relationship given by Fick's law discussed in Volume 1, Chapter 10. This law may be applied as follows.

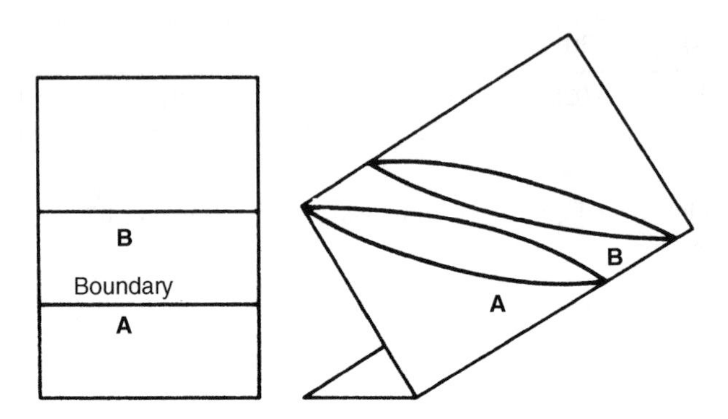


Figure 1.20. Initial arrangement in mixing test

If $a$ is the area of the interface per unit volume of the mix, and $a_m$ the maximum interfacial surface per unit volume that can be obtained, then:

$$\frac{da}{dt} = c(a_m - a) \tag{1.39}$$

and:

$$a = a_m(1 - e^{-ct}) \tag{1.40}$$

It is assumed that, after any time $t$, a number of samples are removed from the mix, and examined to see how many contain both components. If a sample contains both components, it will contain an element of the interfacial surface. If $Y$ is the fraction of the samples containing the two materials in approximately the proportion in the whole mix, then:

$$Y = 1 - e^{-ct} \tag{1.41}$$

COULSON and MAITRA[22] examined the mixing of a number of pairs of materials in a simple drum mixer, and expressed their results as a plot of $\ln(100/X)$ against $t$, where $X$ is the percentage of the samples that is unmixed and $Y = 1 - (X/100)$. It was shown that the constant $c$ depends on:

(a) the total volume of the material,
(b) the inclination of the drum,

(c) the speed of rotation of the drum,
(d) the particle size of each component,
(e) the density of each component, and
(f) the relative volume of each component.

Whilst the precise values of $c$ are only applicable to the particular mixer under examination, they do bring out the effect of these variables. Thus, Figure 1.21 shows the effect of speed of rotation, where the best results are obtained when the mixture is just not

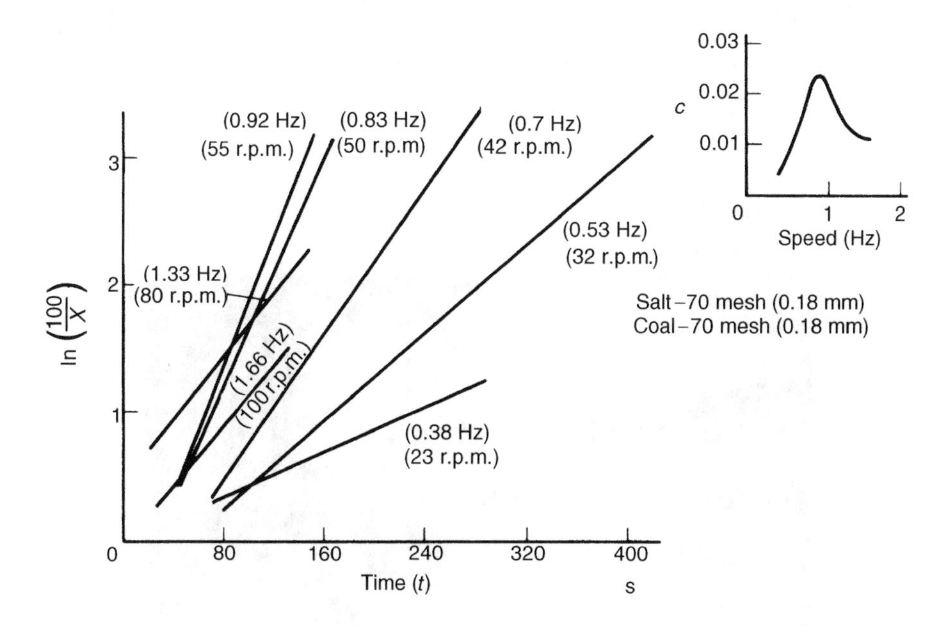


Figure 1.21. Effect of speed on rate of mixing

taken round by centrifugal action. If fine particles are put in at the bottom and coarse on the top, then no mixing occurs on rotation and the coarse material remains on top. If the coarse particles are put at the bottom and the fine on top, then on rotation mixing occurs to an appreciable extent, although on further rotation $Y$ falls, and the coarse particles settle out on the top. This is shown in Figure 1.22 which shows a maximum degree of mixing. Again, with particles of the same size but differing density, the denser migrate to the bottom and the lighter to the top. Thus, if the lighter particles are put in at the bottom and the heavier at the top, rotation of the drum will give improved mixing to a given value, after which the heavier particles will settle to the bottom. This kind of analysis of simple mixers has been made by BLUMBERG and MARITZ[23] and by KRAMERS[21], although the results of their experiments are little more than qualitative.

BROTHMAN *et al.*[24] have given an alternative theory for the process of mixing of solid particles, which has been described as a shuffling process. It is proposed that, as mixing takes place, there will be an increasing chance that a sample of a given size intercepts more than one part of the interface, and hence the number of mixed samples will increase

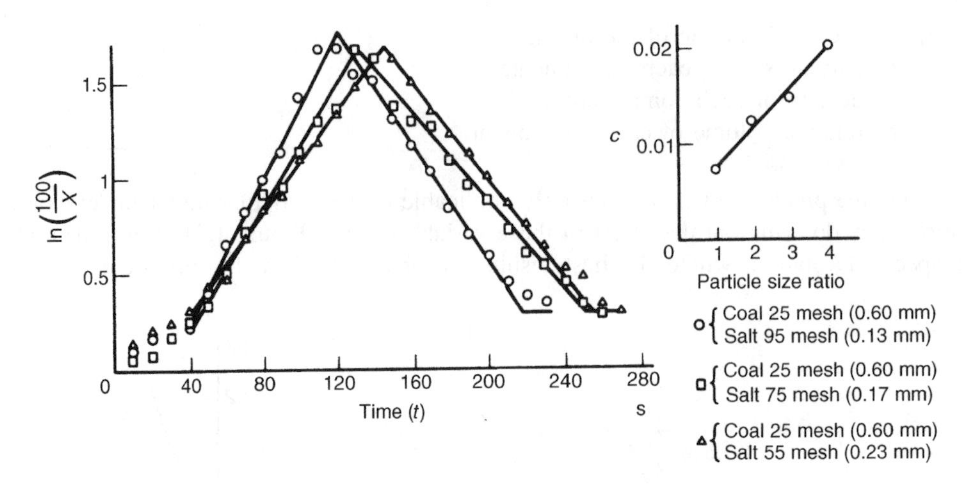


Figure 1.22. Mixing and subsequent separation of solid particles

Figure 1.23. The Oblicone blender

less rapidly than the surface. This is further described by HERDAN[2]. The application of some of these ideas to continuous mixing has been attempted by DANCKWERTS[25,26].

Two different types of industrial mixers are illustrated in Figures 1.23 and 1.24. The Oblicone blender, which is available in sizes from 0.3 to 3 m, is used where ease of cleaning and 100 per cent discharge is essential, such as in the pharmaceutical, food or metal powder industries.

The Buss kneader shown in Figure 1.24 is a continuous mixing or kneading machine in which the characteristic feature is that a reciprocating motion is superimposed on the rotation of the kneading screw, resulting in the interaction of specially profiled screw flights with rows of kneading teeth in the casing. The motion of the screw causes the kneading teeth to pass between the flights of the screw at each stroke backwards and forwards. In this manner, a positive exchange of material occurs both in axial and radial directions, resulting in a homogeneous distribution of all components in a short casing length and with a short residence time.

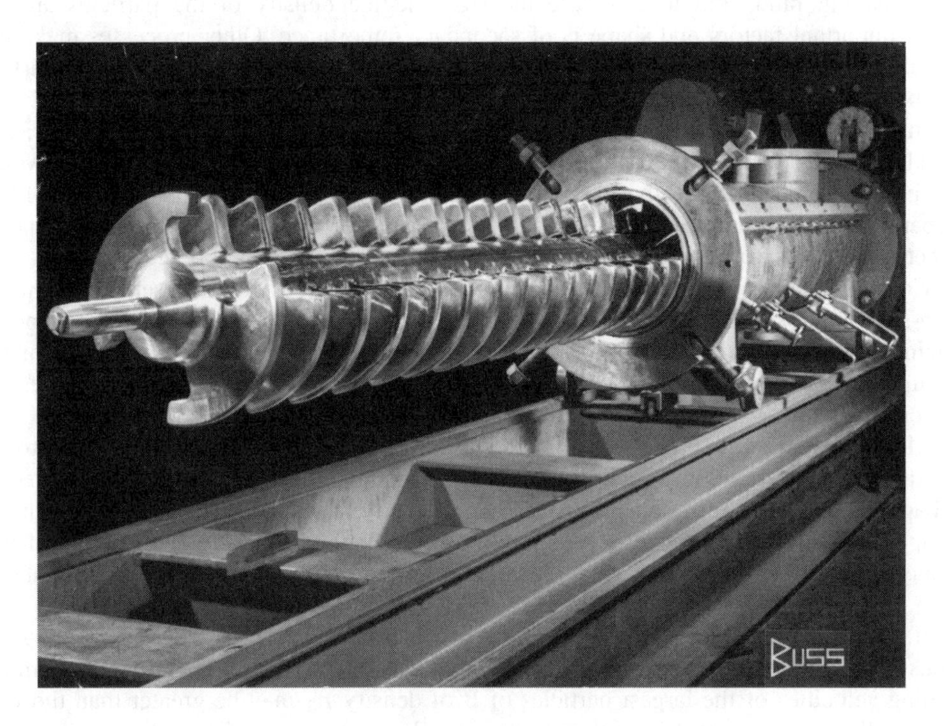


Figure 1.24. The Buss kneader

## 1.5. CLASSIFICATION OF SOLID PARTICLES

### 1.5.1. Introduction

The problem of separating solid particles according to their physical properties is of great importance with large-scale operations in the mining industry, where it is necessary to separate the valuable constituents in a mineral from the adhering gangue, as it is called, which is usually of a lower density. In this case, it is first necessary to crush the material so that each individual particle contains only one constituent. There is a similar problem in coal washing plants in which dirt is separated from the clean coal. The processing industries are more usually concerned with separating a single material, such as the product from a size reduction plant, into a number of size fractions, or in obtaining a uniform material for incorporation in a system in which a chemical reaction takes place. As similar problems are involved in separating a mixture into its constituents and into size fractions, the two processes are considered together.

Separation depends on the selection of a process in which the behaviour of the material is influenced to a very marked degree by some physical property. Thus, if a material is to be separated into various size fractions, a sieving method may be used because this process depends primarily on the size of the particles, though other physical properties such as the shape of the particles and their tendency to agglomerate may also be involved. Other methods of separation depend on the differences in the behaviour of the particles in a moving fluid, and in this case the size and the density of the particles are the most important factors and shape is of secondary importance. Other processes make use of differences in electrical or magnetic properties of the materials or in their surface properties.

In general, large particles are separated into size fractions by means of screens, and small particles, which would clog the fine apertures of the screen or for which it would be impracticable to make the openings sufficiently fine, are separated in a fluid. Fluid separation is commonly used for separating a mixture of two materials though magnetic, electrostatic and froth flotation methods are also used where appropriate.

Considerable development has taken place into techniques for size separation in the sub-sieve range. As discussed by WORK[27], the emphasis has been on techniques lending themselves to automatic working. Many of the methods of separation and much of the equipment has been developed for use in the mining and metallurgical industries, as described by TAGGART[28].

Most processes which depend on differences in the behaviour of particles in a stream of fluid separate materials according to their terminal falling velocities, as reported in Chapter 3, which in turn depend primarily on density and size and to a lesser extent on shape. Thus, in many cases it is possible to use the method to separate a mixture of two materials into its constituents, or to separate a mixture of particles of the same material into a number of size fractions.

If, for example, it is desired to separate particles of a relatively dense material **A** of density $\rho_A$ from particles of a less dense material **B** and the size range is large, the terminal falling velocities of the largest particles of **B** of density $\rho_B$ may be greater than those of the smallest particles of **A**, and therefore a complete separation will not be possible. The maximum range of sizes that can be separated is calculated from the ratio of the sizes of the particles of the two materials which have the same terminal falling velocities. It is shown in Chapter 3 that this condition is given by equation 3.32 as:

$$\frac{d_B}{d_A} = \left(\frac{\rho_A - \rho}{\rho_B - \rho}\right)^j \tag{1.42}$$

where $j = 0.5$ for fine particles, where Stokes' law applies, and
$j = 1$ for coarse particles where Newton's law applies.

It is seen that this size range becomes wider with increase in the density of the separating fluid and, when the fluid has the same density as the less dense material, complete separation is possible whatever the relative sizes. Although water is the most commonly used fluid, an effective specific gravity greater than unity is obtained if *hindered settling* takes place. Frequently, the fluid density is increased artificially by forming a suspension of small particles of a very dense solid, such as galena, ferro-silicon, magnetite, sand or clay, in the water. These suspensions have effective densities of up to about 3500 kg/m$^3$. Alternatively, zinc chloride or calcium chloride solutions may be used. On the small scale, liquids with a range of densities are produced by mixing benzene and carbon tetrachloride.

If the particles are allowed to settle in the fluid for only a very short time, they will not attain their terminal falling velocities and a better degree of separation can be obtained. A particle of material **A** will have an initial acceleration $g[1 - (\rho/\rho_A)]$, because there is no fluid friction when the relative velocity is zero. Thus the initial velocity is a function of density only and it is unaffected by size and shape. A very small particle of the denser material will therefore always commence settling at a greater rate than a large particle of the less dense material. Theoretically, therefore, it should be possible to separate materials completely, irrespective of the size range, provided that the periods of settling are sufficiently short. In practice, the required periods will often be so short that it is impossible to make use of this principle alone, although a better degree of separation may be obtained if the particles are not allowed to become fully accelerated. As the time of settling increases, the larger particles of the less dense material catch up and overtake the small heavy particles.

Size separation equipment in which particles move in a fluid stream is now considered, noting that most of the plant utilises the difference in the terminal falling velocities of the particles: In the hydraulic jig, however, the particles are allowed to settle for only very brief periods at a time, and this equipment may therefore be used when the size range of the material is large.

## Example 1.4

A mixture of quartz and galena of a size range from 0.015 mm to 0.065 mm is to be separated into two pure fractions using a hindered settling process. What is the minimum apparent density of the fluid that will give this separation? The density of galena is 7500 kg/m$^3$ and the density of quartz is 2650 kg/m$^3$.

## Solution

Assuming the galena and quartz particles are of similar shapes, then from equation 1.42, the required density of fluid when Stokes' law applies is given by:

$$\frac{0.065}{0.015} = \left(\frac{7500 - \rho}{2650 - \rho}\right)^{0.5}$$

and:

$$\rho = 2377 \text{ kg/m}^3$$

The required density of fluid when Newton's law applies is given by:

$$\frac{0.065}{0.015} = \left(\frac{7500 - \rho}{2650 - \rho}\right)^{1.0}$$

and hence: $\rho = 1196 \text{ kg/m}^3$

Thus, the required density of the fluid is between 1196 and 2377 kg/m$^3$.

## 1.5.2. Gravity settling

### *The settling tank*

Material is introduced in suspension into a tank containing a relatively large volume of water moving at a low velocity, as shown in Figure 1.25. The particles soon enter the slowly moving water and, because the small particles settle at a lower rate, they are carried further forward before they reach the bottom of the tank. The very fine particles are carried away in the liquid overflow. Receptacles at various distances from the inlet collect different grades of particles according to their terminal falling velocities, with the particles of high terminal falling velocity collecting near the inlet. The positions at which the particles are collected may be calculated on the assumption that they rapidly reach their terminal falling velocities, and attain the same horizontal velocity as the fluid.

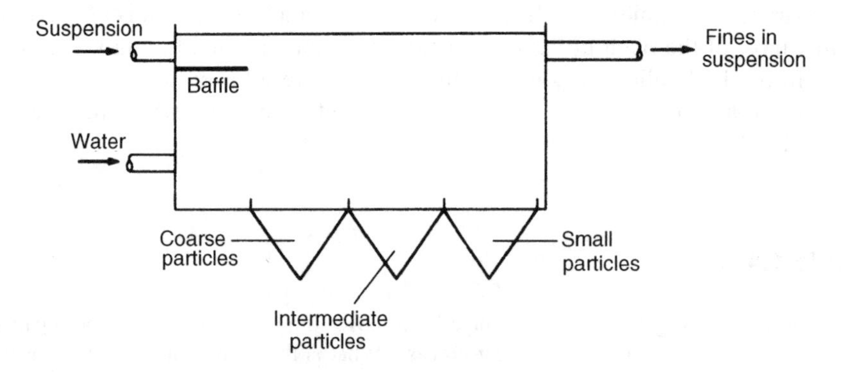


Figure 1.25. Gravity settling tank

If the material is introduced in solid form down a chute, the position at which the particles are deposited will be determined by the rate at which they lose the horizontal component of their velocities. The larger particles will therefore be carried further forward and the smaller particles will be deposited near the inlet. The position at which the particles are deposited may be calculated using the relations given in Chapter 3.

### *The elutriator*

Material may be separated by means of an elutriator, which consists of a vertical tube up which fluid is passed at a controlled velocity. The particles are introduced, often through a side tube, and the smaller particles are carried over in the fluid stream while the large particles settle against the upward current. Further size fractions may be collected if the overflow from the first tube is passed vertically upwards through a second tube of greater cross-section, and any number of such tubes can be arranged in series.

### *The Spitzkasten*

This plant consists of a series of vessels of conical shape arranged in series. A suspension of the material is fed into the top of the first vessel and the larger particles settle, while the smaller ones are carried over in the liquid overflow and enter the top of a second conical vessel of greater cross-sectional area. The bases of the vessels are fitted with wide diameter outlets, and a stream of water can be introduced near the outlet so that the particles have to settle against a slowly rising stream of liquid. The size of the smallest particle which is collected in each of the vessels is influenced by the upward velocity at the bottom outlet of each of the vessels. The size of each successive vessel is increased, partly because the amount of liquid to be handled includes all the water used for classifying in the previous vessels, and partly because it is desired to reduce, in stages, the surface velocity of the fluid flowing from one vessel to the next. The Spitzkasten thus combines the principles used in the settling tank and in the elutriator.

The size of the material collected in each of the units is determined by the rate of feeding of suspension, the upward velocity of the liquid in the vessel and the diameter of the vessel. The equipment can also be used for separating a mixture of materials into its constituents, provided that the size range is not large. The individual units can be made of wood or sheet metal.

The Hydrosizer, illustrated in Figure 1.26, works on the same principle although it has a number of compartments trapezoidal in section. It is suitable for use with materials finer

Figure 1.26. Hydrosizer

than about 4-mesh (*ca*. 5 mm) and it works at high concentrations in order to obtain the advantages of hindered settling. Altogether eight sharply classified fractions are produced, ranging from coarse nearest to the feed inlet to fine from the eighth compartment, at feed rates in the range 2–15 kg/s.

### *The double cone classifier*

This classifier, shown in Figure 1.27, consists of a conical vessel, with a second hollow cone of greater angle arranged apex downwards inside it, so that there is an annular space of approximately constant cross-section between the two cones. The bottom portion of the inner cone is cut away and its position relative to the outer cone can be regulated by a screw adjustment. Water is passed in an upward direction, as in the Spitzkasten, and overflows into a launder arranged round the whole of the periphery of the outer cone. The material to be separated is fed in suspension to the centre of the inner cone, and the liquid level is maintained slightly higher than the overflow level, so that there is a continuous flow of liquid downwards in the centre cone. The particles are therefore brought into the annular space where they are subjected to a classifying action; the smaller particles are carried away in the overflow, and the larger particles settle against the liquid stream and are taken off at the bottom.

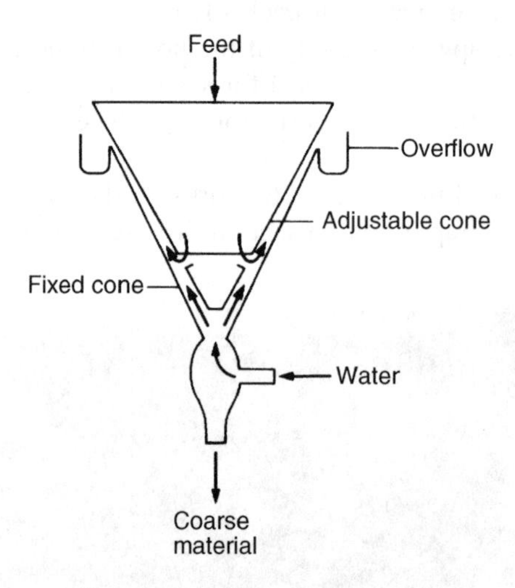


Figure 1.27. Double cone classifier

### *The mechanical classifier*

Several forms of classifier exist in which the material of lower settling velocity is carried away in a liquid overflow, and the material of higher settling velocity is deposited on the bottom of the equipment and is dragged upwards against the flow of the liquid, by some mechanical means. During the course of the raking action, the solids are turned over so that any small particles trapped under larger ones are brought to the top again.

The mechanical classifier is extensively used where it is necessary to separate a large amount of fines from the oversize material. Arrangement of a number of mechanical classifiers in series makes it possible for a material to be separated into several size fractions.

In the rake classifier, which consists of a shallow rectangular tank inclined to the horizontal, the feed is introduced in the form of a suspension near the middle of the tank and water for classifying is added at the upper end, as shown in Figure 1.28. The liquid, together with the material of low terminal falling velocity, flows down the tank under a baffle and is then discharged over an overflow weir. The heavy material settles to the bottom and is then dragged upwards by means of a rake. It is thus separated from the liquid and is discharged at the upper end of the tank. Figure 1.29 shows the Dorr–Oliver rake classifier. Figure 1.30 shows the Denver classifier which is similar in action although a trough is used which is semicircular in cross-section, and the material which settles to the bottom is continuously moved to the upper end by means of a rotating helical scraper.

### *The bowl classifier*

The bowl classifier, which is used for fine materials, consists of a shallow bowl with a concave bottom as shown in Figure 1.31. The suspension is fed into the centre of the

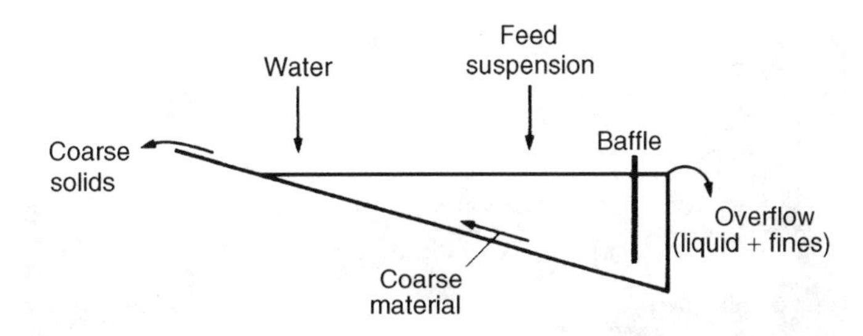


Figure 1.28. The principle of operation of the mechanical classifier

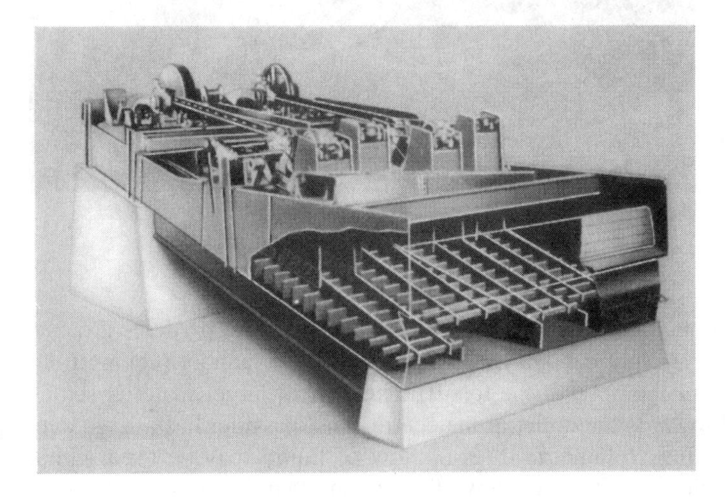

Figure 1.29. A rake classifier

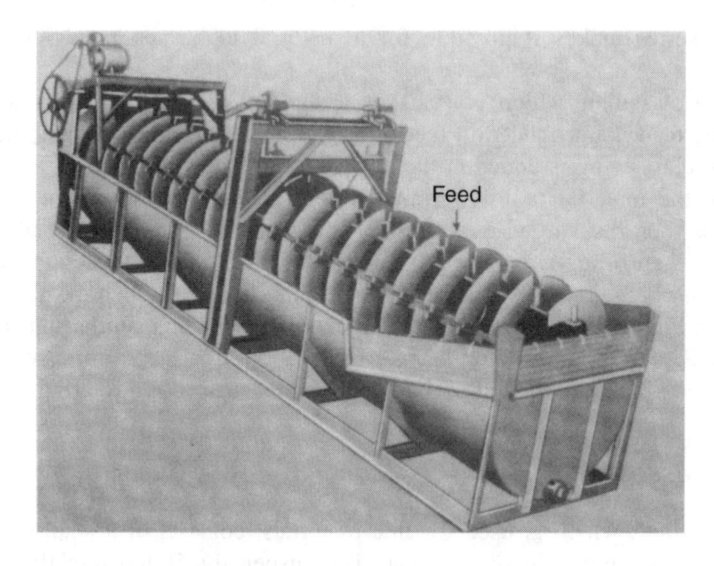


Figure 1.30. The Denver classifier

Figure 1.31. Bowl classifier

bowl near the liquid surface, and the liquid and the fine particles are carried in a radial direction and pass into the overflow which is an open launder running round the whole of the periphery of the bowl at the top. The heavier or larger material settles to the bottom and is raked towards the outlet at the centre. The classifier has a large overflow area, and consequently high volumetric rates of flow of liquid may be used without producing a high linear velocity at the overflow. The action is similar to that of a thickener which is effecting incomplete clarification.

### *The hydraulic jig*

The hydraulic jig operates by allowing material to settle for brief periods so that the particles do not attain their terminal falling velocities, and is therefore suitable for separating materials of wide size range into their constituents. The material to be separated is fed dry, or more usually in suspension, over a screen and is subjected to a pulsating action by liquid which is set in oscillation by means of a reciprocating plunger. The particles on the screen constitute a suspension of high concentration and therefore the advantages of hindered settling are obtained. The jig usually consists of a rectangular-section tank with a tapered bottom, divided into two portions by a vertical baffle. In one section, the plunger operates in a vertical direction; the other incorporates the screen over which the separation is carried out. In addition, a stream of liquid is fed to the jig during the upward stroke.

The particles on the screen are brought into suspension during the downward stroke of the plunger (Figure 1.32*a*). As the water passes upwards the bed opens up, starting at the top, and thus tends to rise *en masse*. During the upward stroke the input of water is adjusted so that there is virtually no flow through the bed (Figure 1.32*b*). During this period differential settling takes place and the denser material tends to collect near the screen and the lighter material above it. After a short time the material becomes divided into three strata, the bottom layer consisting of the large particles of the heavy material, the next of large particles of the lighter material together with small particles of the heavy material, and the top stratum of small particles of the light material. Large particles wedge at an earlier stage than small ones and therefore the small particles of the denser material

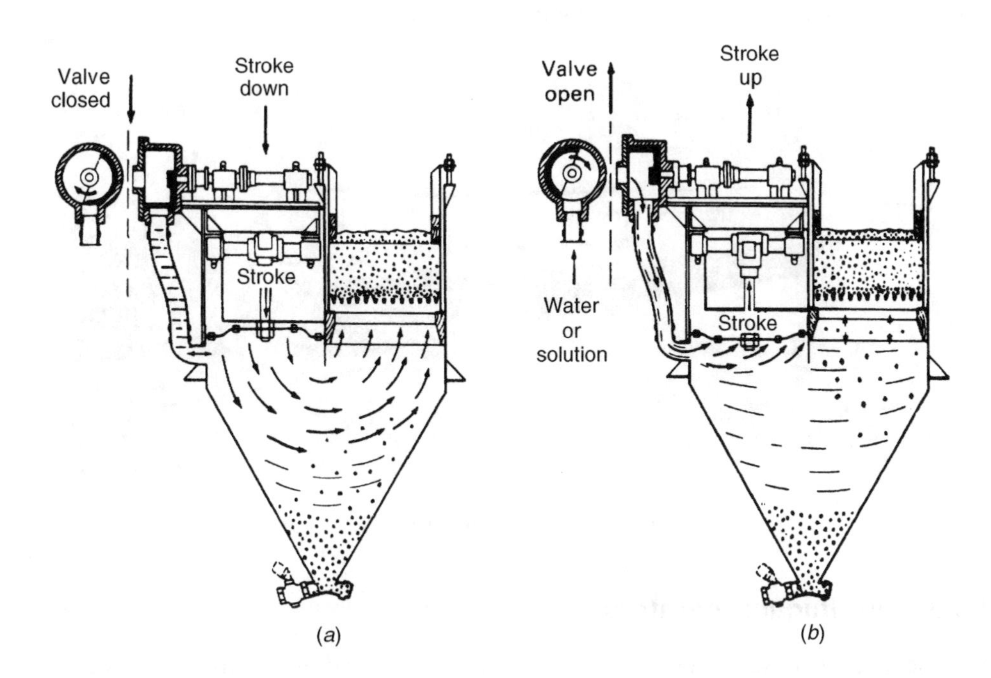


Figure 1.32. The Denver hydraulic jig (*a*) Downward stroke (*b*) Upward stroke

are able to fall through the spaces between the larger particles of the light material. Many of these small particles then fall through the supporting gauze.

Four separate fractions are obtained from the jig and the successful operation of the plant depends on their rapid removal. The small particles of the heavy material which have fallen through the gauze are taken off at the bottom of the tank. The larger particles of each of the materials are retained on the gauze in two layers, the denser material at the bottom and the less dense material on top. These two fractions are removed through gates at the side of the jig. The remaining material, consisting of small particles of the lighter material, is carried away in the liquid overflow.

### *Riffled tables*

The riffled table, of which the Wilfley table shown in Figure 1.33 is a typical example, consists of a flat table which is inclined at an angle of about 3° to the horizontal. Running approximately parallel to the top edge is a series of slats, or riffles as they are termed, about 6 mm in height. The material to be separated is fed on to one of the top corners and a reciprocating motion, consisting of a slow movement in the forward direction and a very rapid return stroke, causes it to move across the table. The particles also tend to move downwards under the combined action of gravity and of a stream of water which is introduced along the top edge, but are opposed by the riffles behind which the smaller particles or the denser material tend to be held. Thus, the large particles and the less dense material are carried downwards, and the remainder is carried parallel to the riffles. In many cases, each riffle is tapered, with the result that a number of fractions can be collected. Riffled tables can be used for separating materials down to about 50 μm in size, provided that the difference in the densities is large.

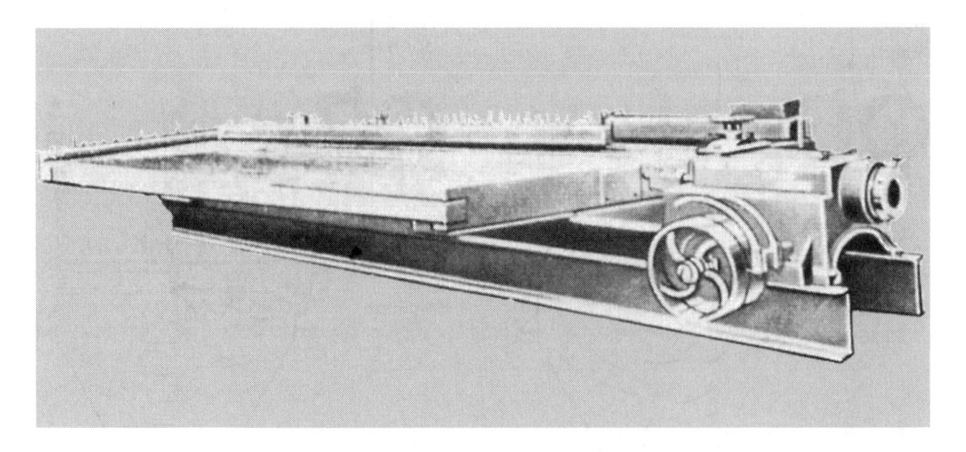

Figure 1.33. Wilfley riffled table

## 1.5.3. Centrifugal separators

The use of cyclone separators for the removal of suspended dust particles from gases is discussed in Section 1.6. By suitable choice of operating conditions, it is also possible to

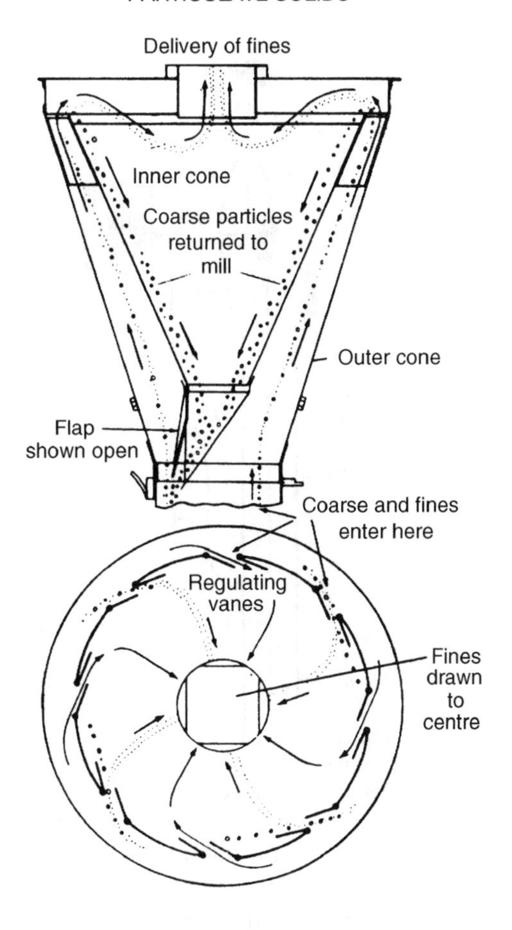


Figure 1.34. NEI International Combustion vacuum air separator

use centrifugal methods for the classification of solid particles according to their terminal falling velocities. Figure 1.34 shows a typical air separation unit which is similar in construction to the double cone-classifier. The solids are fed to the bottom of the annular space between the cones and are carried upwards in the air stream and enter the inner cone through a series of ports fitted with adjustable vanes. As may be seen in the diagram, the suspension enters approximately tangentially and is therefore subjected to the action of centrifugal force. The coarse solids are thrown outwards against the walls and fall to the bottom under the action of gravity, while the small particles are removed by means of an exhaust fan. This type of separator is widely used for separating the oversize material from the product from a ball mill, and is suitable for materials as fine as 50 μm.

A mechanical air separator is shown in Figures 1.35 and 1.36. The material is introduced at the top through the hollow shaft and falls on to the rotating disc which throws it outwards. Very large particles fall into the inner cone and the remainder are lifted by the air current produced by the rotating vanes above the disc. Because a rotary motion has been imparted to the air stream, the coarser particles are thrown to the walls of the inner cone and, together with the very large particles, are withdrawn from the bottom. The fine

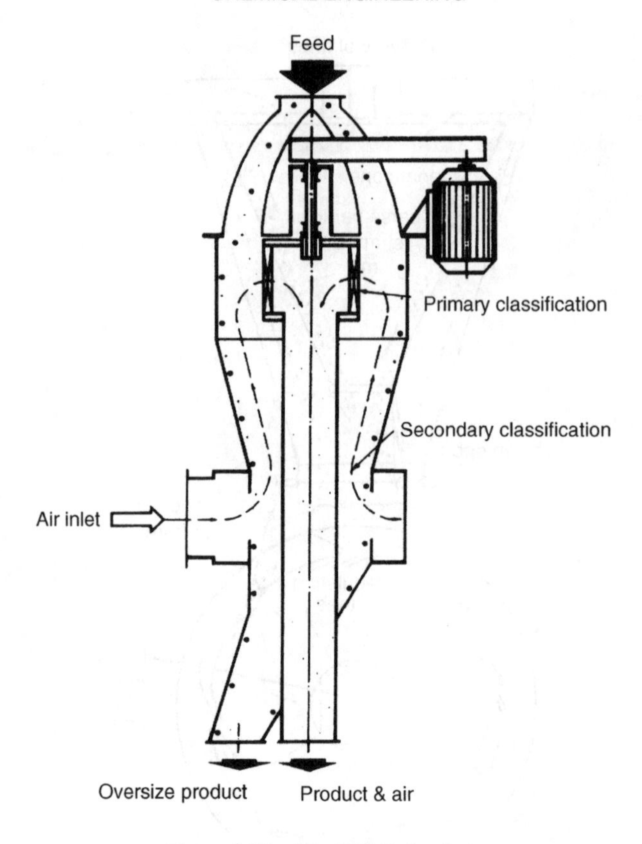


Figure 1.35. The NEI Delta sizer

particles remain in suspension and are carried down the space between the two cones and are collected from the bottom of the outer cone. The air returns to the inner cone through a series of deflector vanes, and blows any fines from the surfaces of the larger particles as they fall. The size at which the "cut" is made is controlled by the number of fixed and rotating blades, and the speed of rotation. The Delta sizer is designed to separate dry powders of mixed particle size within the range 5 to 200 μm into distinct fractions at a high extraction efficiency and low power consumption. The equipment shown in Figure 1.36 has a feed rate from 2 to 7 kg/s depending upon the desired cut point and the bulk density of the material.

### 1.5.4. The hydrocyclone or liquid cyclone

In the hydrocyclone, or hydraulic cyclone, which is discussed extensively in the literature[29–35], separation is effected in the centrifugal field generated as a result of introducing the feed at a high tangential velocity into the separator. The hydrocyclone may be used for:

(a) separating particles (suspended in a liquid of lower density) by size or density, or more generally, by terminal falling velocity;

Figure 1.36. NEI Delta sizer ultrafine classifier

(b) the removal of suspended solids from a liquid;
(c) separating immiscible liquids of different densities;
(d) dewatering of suspensions to give a more concentrated product;
(e) breaking down liquid–liquid or gas–liquid dispersions; and
(f) the removal of dissolved gases from liquids.

In this section, the general design of the hydrocyclone and its application in the grading of solid particles, or their separation from a liquid, is considered and then the special features required in hydrocyclones required for the separation of immiscible liquids will be addressed. The use of cyclones for separating suspended particles from gases is discussed in Section 1.6.2.

### *General principles and applications in solids classification*

A variety of geometrical designs exists, the most common being of the form shown in Figure 1.37, with an upper cylindrical portion of between 20 and 300 mm in diameter and a lower conical portion, although larger units are occasionally employed. The fluid is introduced near the top of the cylindrical section, and overflow is removed through a centrally located offtake pipe at the top, usually terminating approximately at the level corresponding to the junction of the cylindrical and tapered portions of the shell. Other configurations include an entirely cylindrical shell, a conical shell with no cylindrical

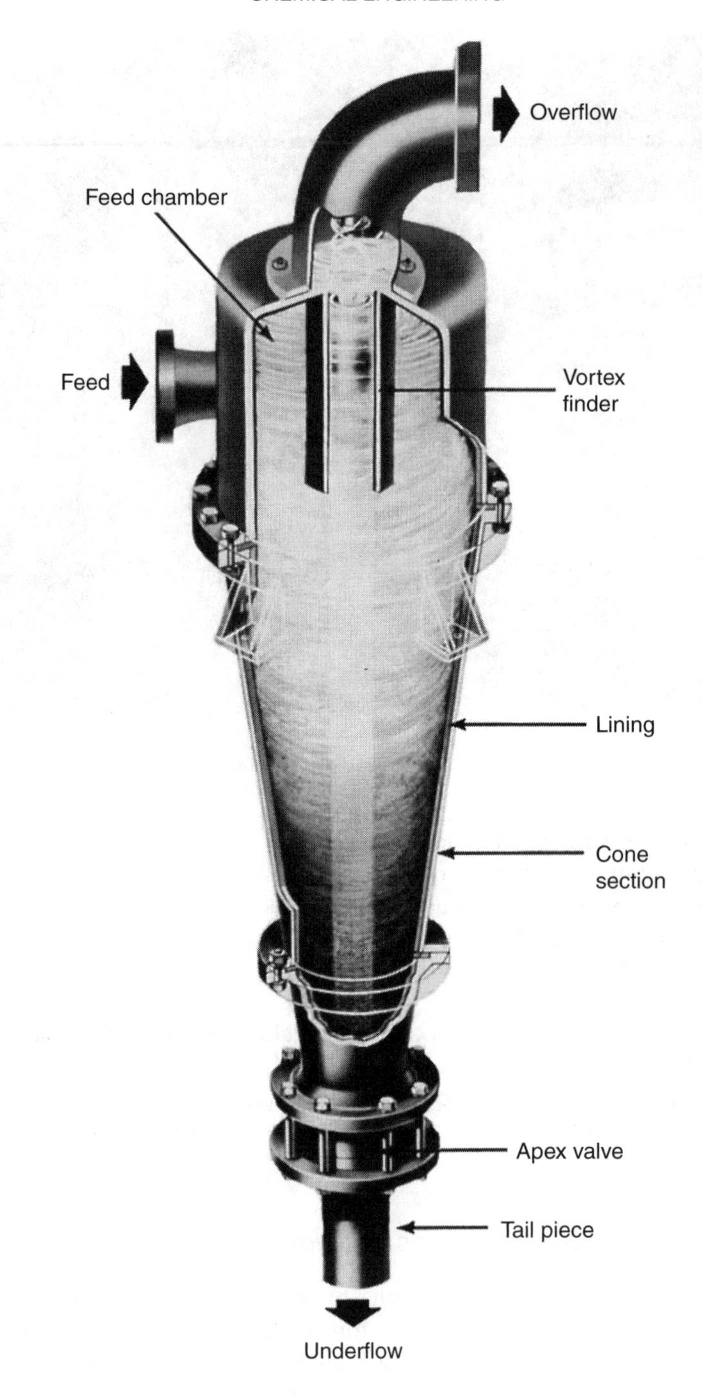


Figure 1.37. Liquid cyclone or hydrocyclone

portion and curved, as opposed to straight, sides to the tapered section. Generally, a long tapered section is preferred.

As flow patterns are influenced only slightly by gravitational forces, hydrocyclones may be operated with their axes inclined at any angle, including the horizontal, although the removal of the underflow is facilitated, with the axis vertical.

Much of the separating power of the hydrocyclone is associated with the interaction of the primary vortex which follows the walls, and the secondary vortex, revolving about a low pressure core, which moves concentrically in a countercurrent direction as shown in Figure 1.37. The separating force is greatest in this secondary vortex which causes medium sized particles to be rejected outwards to join the primary vortex flow in which they are then carried back towards the apex. It is this secondary vortex which exerts the predominant influence in determining the largest size or heaviest particle which will remain in the overflow stream. The tangential fluid velocity is a maximum at a radius roughly equal to that of the overflow discharge pipe or "vortex finder".

The flow patterns in the hydrocyclone are complex, and much development work has been necessary to determine the most effective geometry, as theoretical considerations alone will not allow the accurate prediction of the size cut which will be obtained. A mathematical model has been proposed by RHODES *et al.*[36], and predictions of streamlines from their work are shown in Figure 1.38. SALCUDEAN and GARTSHORE[37] have also carried out numerical simulations of the three-dimensional flow in a hydrocyclone and have used the results to predict cut sizes. Good agreement has been obtained with experimental measurements.

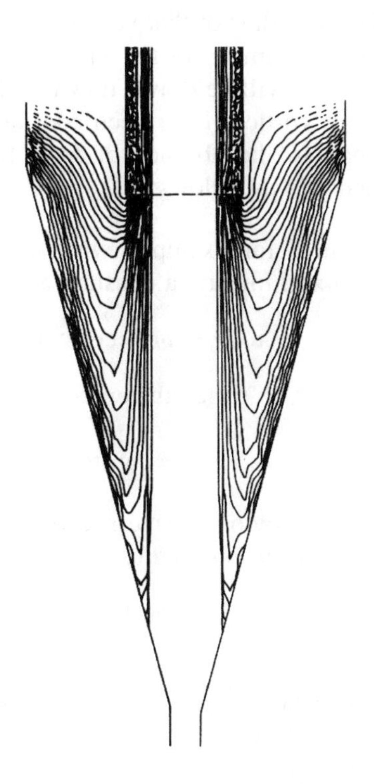

Figure 1.38. Predicted streamlines in a hydrocyclone[36]

Near the top of the hydrocyclone there will be some short-circuiting of the flow between the inlet and the overflow, although the effects are reduced as a result of the formation of circulating eddies, often referred to as the mantle, which tend to act as a barrier. Within the secondary vortex the pressure is low and there is a depression in the liquid surface in the region of the axis. Frequently a gas core is formed, and any gas dispersed in the form of fine bubbles, or coming out of solution, tends to migrate to this core. In pressurised systems, the gas core may be very much reduced in size, and sometimes completely eliminated.

The effectiveness of a hydrocyclone as a separator depends primarily on the liquid flow pattern and, in particular, on the values of the three principal components of velocity (tangential, $u_t$, axial $u_a$, and radial $u_r$) throughout the body of the separator. The efficiency with which fractionation takes place is highest at very low concentrations, and therefore there is frequently a compromise between selectivity and throughput in selecting the optimum concentration. Turbulent flow conditions should be avoided wherever possible as they give rise to undesirable mixing patterns which reduce the separating capacity.

Most studies of hydrocyclone performance for particle classification have been carried out at particle concentrations of about 1 per cent by volume. The simplest theory for the classification of particles is based on the concept that particles will tend to orbit at the radius at which the centrifugal force is exactly balanced by the fluid friction force on the particles. Thus, the orbits will be of increasing radius as the particle size increases. Unfortunately, there is scant information on how the radial velocity component varies with location. In general, a particle will be conveyed in the secondary vortex to the overflow, if its orbital radius is less than the radius of that vortex. Alternatively, if the orbital radius would have been greater than the diameter of the shell at a particular height, the particle will be deposited on the walls and will be drawn downwards to the bottom outlet.

The general characteristics of vortices have been considered in Volume 1, Chapter 2, where it has been shown that, in the absence of fluid friction, the relation between tangential fluid velocity $u_t$ and $r$ is as follows:

(a) in a *forced vortex* (formed, for example, when a rotating member, such as an impeller in a pump or mixer, imparts a constant angular velocity $\omega$ to the fluid):

$$u_t/r = \omega = \text{constant} \tag{1.43}$$

(b) in a *free* (or *natural*) vortex, in which the energy per unit mass of fluid is constant throughout:

$$u_t r = \text{constant} \tag{1.44}$$

Laboratory measurements on hydrocylones have shown that, in the primary vortex, the relation between tangential velocity $u_t$ and radius $r$, is approximately of the form:

$$u_t r^n = \text{constant} \tag{1.45}$$

where $n = 0.5 - 0.9$.

Thus, as might have been be expected, the primary vortex in the hydrocyclone is more akin to a free ($n = 1$) than to a forced ($n = -1$) vortex.

Typical profiles of the components $u_t$, $u_r$ and $u_a$ of the liquid velocity in a hydrocyclone, as given by KELSALL[38] and SVAROVSKY[34], are shown in Figure 1.39. These profiles are

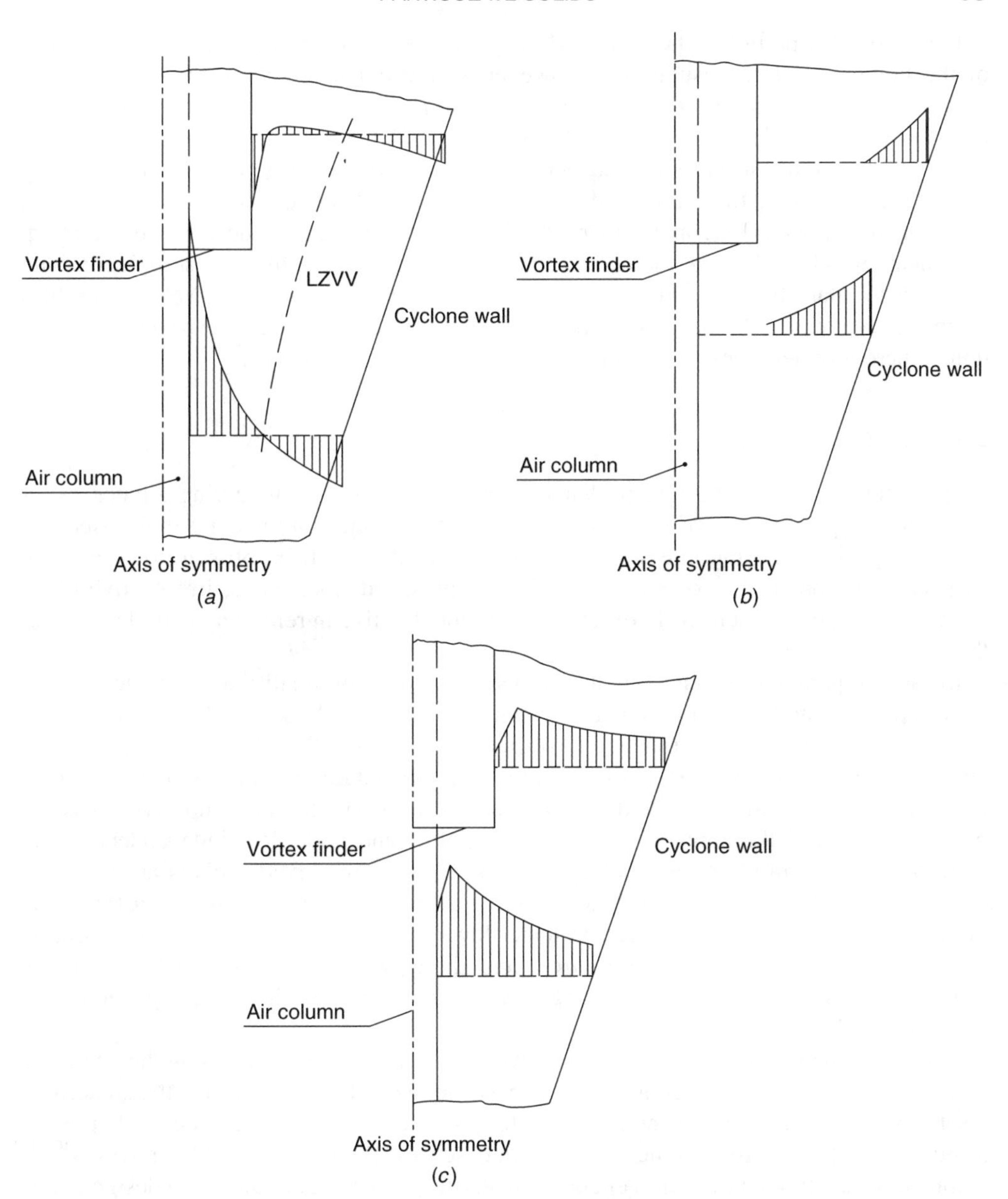


Figure 1.39. Typical velocity distributions in a hydrocyclone[38]. (*a*) axial (*b*) radial (*c*) tangential (broken line LZVV is the locus of zero axial velocity)

quite critically dependent on the geometry of the separator. Entry velocities, corresponding to $u_t$ near the walls at that level, will be up to about 5 m/s.

Qualitatively similar velocity profiles are set up within gas cyclones, discussed in Section 1.6.2, which are extensively used for the removal of suspended solids from gases. In this case, the velocities are generally considerably higher.

For suspended particles, the axial and tangential velocities will differ little from those of the liquid. As already suggested, however, in the radial direction, the particles will tend to rotate at the radius at which the centrifugal force is balanced by the drag force of the fluid on the particle. This means that the radius of the particle orbit in a liquid will increase with both particle size and density, or more particularly with terminal falling velocity. As a result, the larger and denser particles will move selectively towards the walls of the hydrocyclone and will be discharged at the apex of the cone. For a sharp fractionation to be achieved, the residence times of the particles must be sufficiently long for them to attain their equilibrium orbits within the hydrocyclone[39], Velocity profiles do in practice tend to be flatter than those predicted by equation 1.44, especially in large units where turbulence may develop.

### *Design Considerations*

Hydrocyclones are 20–500 mm in diameter, with the smaller units giving a much better separation. Typical values of length to diameter ratios range from about 5 to 20. Because of the very high shearing stresses which are set up, flocs will be broken down and the suspension in the secondary vortex will be completely deflocculated, irrespective of its condition on entry. Generally, hydrocylones are not effective in removing particles smaller than about 2–3 μm.

Because separating power is greatest in hydrocyclones of small diameter, the cut size being approximately proportional to the diameter of the cylindrical shell raised to the power of 1.5, it is common practice to operate banks of small hydrocyclones in parallel inside a large containing vessel. Furthermore, this procedure also makes scale-up easier to carry out. With units connected only in parallel (or as single units), however, it is not possible to control the compositions of the overflow and underflow independently, and therefore some form of series-parallel operation is often employed. This is an important consideration when the hydrocyclone is used for thickening a suspension. In this case, there is the requirement to produce an underflow of the appropriate solids concentration and for the overflow to be particle-free, and these two conditions cannot usually be satisfied in a single unit. In thickening, the particle concentrations are high and little classification by size occurs.

In general, the performance of the hydrocyclone is improved by increasing the operating pressure, and the principal control variable is the size of the orifice on the underflow discharge. Several theoretical and practical studies have been made in an attempt to present a sound basis for design, and these have been described BRADLEY[29] and SAVROVSKY[30,35] among others, although they are generally not entirely satisfactory, and some design charts and formulae have been given by ZANKER[40]. In practice, tests with the actual materials to be used are desirable for the evaluation for the various parameters. Since systems are usually scaled-up by increasing the number of units in parallel, it is seldom necessary to carry out tests on large units.

The optimum design of a hydrocyclone for a given function depends upon reconciling a number of conflicting factors and reference should be made to specialist publications. Because it is simple in construction and has no moving parts, maintenance costs are low. The chief problem arises from the abrasive effect of the solids; materials of construction, such as polyurethane, show less wear than metals and ceramics.

### *Effect of non-Newtonian fluid properties*

There have been very few studies of the effects of non-Newtonian properties on flow patterns in hydrocyclones, although DYAKOWSKI *et al.*[41] have carried out numerical simulations for power-law fluids, and these have been validated by experimental measurements in which velocity profiles were obtained by laser–doppler anemometry.

### *Liquid–liquid separations*

For the separation of two liquids, the hydrocyclone is normally operated at a pressure high enough to suppress the formation of the gas core by appropriate throttling of the outlets on the underflow and overflow streams. Hydrocyclones are now used extensively for separating mixtures of oil (as the less dense component) from water. A wide variety of designs has been developed to cope with different proportions of the light and heavy components. COLMAN[42] has described a plant for cleaning up large quantities of sea or river water following a oil spill, the proportion of the light component may be very low (2–3 per cent). A separator with an enlarged cylindrical top section may be fitted with two tangential inlets, and the conical section may be very long with an included angle as small as 1–2°. The gradual taper leads to the formation of a secondary vortex of very small diameter to accommodate the relatively small proportion of the lighter (oil) phase. By using this configuration, it has been possible to recover a high proportion (up to 97 per cent) of the oil in the feed. In this case, the correct setting of the orifice on the underflow discharge is of critical importance.

### *Liquid–gas separations*

Hydrocyclones are used for removing entrained gas bubbles from liquids, and the extracted gas collects in the gas-core of the secondary vortex before leaving through the vortex finder. NEBRENSKY[43] points out that because of the low pressure in the region close to the axis, they will also remove dissolved gases[43].

### *Applications*

Hydrocyclones are now being used for a very wide range of applications and are displacing other types of separation equipment in many areas. They are compact and have low maintenance costs, having no moving parts, and have substantially lower liquid holdups than gravity-driven separators. Since any aggregrates tend to be broken down in the high shear fields, they usually give cleaner separations. However, they are relatively inflexible in that a given unit will operate satisfactorily over only a narrow range of flowrates and particle concentrations. This is not usually a serious drawback as hydrocyclones are usually operated with banks of small units in parallel, and their number can be adjusted to suit the current flowrate.

## 1.5.5. Sieves or screens

Sieves or screens are used on a large scale for the separation of particles according to their sizes, and on a small scale for the production of closely graded materials in carrying out size analyses. The method is applicable for particles of a size as small as about 50 μm, but not for very fine materials because of the difficulty of producing accurately woven fine gauze of sufficient strength, and the fact that the screens become clogged. Woven wire cloth is generally used for fine sizes and perforated plates for the larger meshes. Some large industrial screens are formed either from a series of parallel rods or from H-shaped links bolted together, though square or circular openings are more usual. Screens may be operated on both a wet or a dry basis. With coarse solids the screen surface may be continuously washed by means of a flowing stream of water which tends to keep the particles apart, to remove the finer particles from the surface of larger particles and to keep the screen free of adhering materials. Fine screens are normally operated wet, with the solids fed continuously as a suspension. Concentrated suspensions, particularly when flocculated, have high effective viscosities and frequently exhibit shear-thinning non-Newtonian characteristics. By maintaining a high cross-flow velocity over the surface of the screen, or by rapid vibration, the apparent viscosity of the suspension may be reduced and the screening rate substantially increased.

The only large screen that is hand operated is the *grizzly*. This has a plane screening surface composed of longitudinal bars up to 3 m long, fixed in a rectangular framework. In some cases, a reciprocating motion is imparted to alternate bars so as to reduce the risk of clogging. The grizzly is usually inclined at an angle to the horizontal and the greater the angle then the greater is the throughput although the screening efficiency is reduced. If the grizzly is used for wet screening, a very much smaller angle is employed, In some screens, the longitudinal bars are replaced by a perforated plate.

Mechanically operated screens are vibrated by means of an electromagnetic device as shown in Figure 1.40, or mechanically as shown in Figure 1.41. In the former case the screen itself is vibrated, and in the latter, the whole assembly. Because very rapid accelerations and retardations are produced, the power consumption and the wear on the bearings are high. These screens are sometimes mounted in a multi-deck fashion with

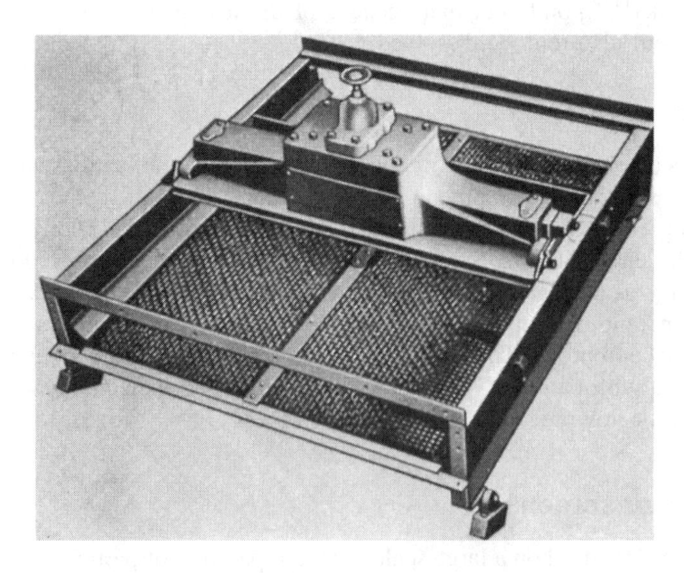

Figure 1.40. Hummer electromagnetic screen

Figure 1.41. Tyrock mechanical screen

the coarsest screen on top, either horizontally or inclined at angles up to 45°. With the horizontal machine, the vibratory motion fulfils the additional function of moving the particles across the screen.

The screen area which is required for a given operation cannot be predicted without testing the material under similar conditions on a small plant. In particular, the tendency of the material to clog the screening surface can only be determined experimentally.

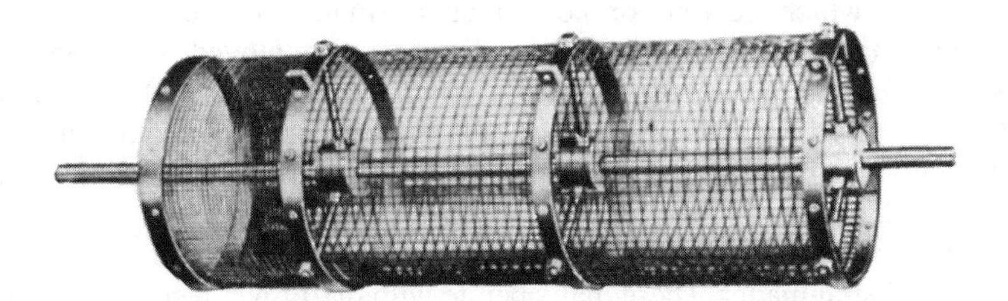

Figure 1.42. Trommel

A very large mechanically operated screen is the *trommel*, shown in Figure 1.42, which consists of a slowly rotating perforated cylinder with its axis at a slight angle to the horizontal. The material to be screened is fed in at the top and gradually moves down the screen and passes over apertures of gradually increasing size, with the result that all the material has to pass over the finest screen. There is therefore a tendency for blockage of the apertures by the large material and for oversize particles to be forced through. Further, the relatively fragile fine screen is subjected to the abrasive action of the large particles.

These difficulties are obviated to some extent by arranging the screens in the form of concentric cylinders, with the coarsest in the centre. The disadvantage of all screens of this type is that only a small fraction of the screening area is in use at any one time. The speed of rotation of the trommel should not be so high that the material is carried completely round in contact with the screening surface. The lowest speed at which this occurs is known as the critical speed and is analogous to the critical speed of the ball mill, discussed in Chapter 2. Speeds of between one-third and a half of the critical speed are usually recommended. In a modified form of the trommel, the screen surfaces are in the form of truncated cones. Such screens are mounted with their axes horizontal and the material always flows away from the apex of the cone.

Screening of suspensions of very fine particles can present difficulties because of the tendency of the apertures to clog and thus impede the passage of further particles. Furthermore, if flocculated, the suspensions can exhibit very high apparent viscosities, and flowrates can fall to unacceptably low levels. Many such suspensions exhibit highly shear-thinning non-Newtonian characteristics with their apparent viscosities showing a marked decrease as their velocity of flow is increased (See Volume 1, Chapter 3). With such materials the rate of screening may be substantially increased by imparting a high velocity to the suspension in a direction parallel to the screen surface. Such motion has been successfully imparted by imposing a rotational or a reciprocating motion to the screen.

### 1.5.6. Magnetic separators

In the magnetic separator, material is passed through the field of an electromagnet which causes the retention or retardation of the magnetic constituent. It is important that the material should be supplied as a thin sheet in order that all the particles are subjected to a field of the same intensity and so that the free movement of individual particles is not impeded. The two main types of equipment are:

(a) Eliminators, which are used for the removal of small quantities of magnetic material from the charge to a plant. These are frequently employed, for example, for the removal of stray pieces of scrap iron from the feed to crushing equipment. A common type of eliminator is a magnetic pulley incorporated in a belt conveyor so that the non-magnetic material is discharged in the normal manner and the magnetic material adheres to the belt and falls off from the underside.

(b) Concentrators, which are used for the separation of magnetic ores from the accompanying mineral matter. These may operate with dry or wet feeds and an example of the latter is the Mastermag wet drum separator, the principle of operation of which is shown in Figure 1.43. An industrial machine is shown in operation in Figure 1.44. A slurry containing the magnetic component is fed between the rotating magnet drum cover and the casing. The stationary magnet system has several radial poles which attract the magnetic material to the drum face, and the rotating cover carries the magnetic material from one pole to another, at the same time gyrating the magnetic particles, allowing the non-magnetics to fall back into the slurry mainstream. The clean magnetic product is discharged clear of the slurry tailings. Operations can be co- or counter-current and the recovery of magnetic material can be as high as 99.5 per cent.

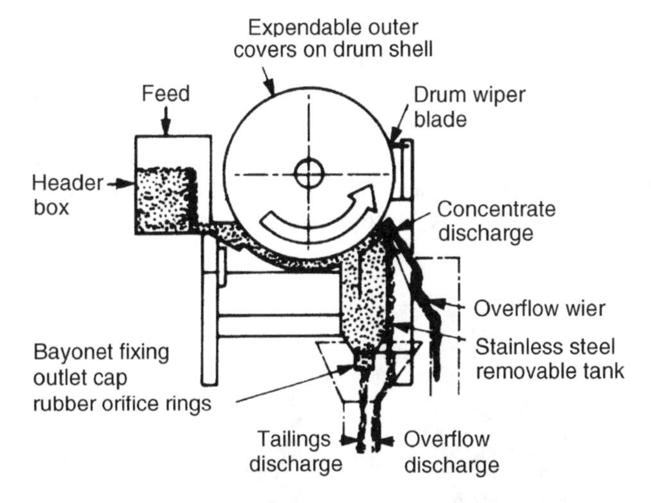


Figure 1.43. Principle of operation of a wet drum separator

Figure 1.44. A wet drum separator in action

An example of a concentrator operating on a dry feed is a rotating disc separator. The material is fed continuously in a thin layer beneath a rotating magnetic disc which picks up the magnetic material in the zone of high magnetic intensity. The captured particles are carried by the disc to the discharge chutes where they are released. The non-magnetic material is then passed to a second magnetic separation zone where secondary separation occurs in the same way, leaving a clean non-magnetic product to emerge from the discharge end of the machine. A Mastermagnet disc separator is shown in Figure 1.45.

Figure 1.45. Rotating disc separator

### High gradient magnetic separation

The removal of small quantities of finely dispersed ferromagnetic materials from fine minerals, such as china clay, may be effectively carried out in a high gradient magnetic field. The suspension of mineral is passed through a matrix of ferromagnetic wires which is magnetised by the application of an external magnetic field. The removal of the weakly magnetic particles containing iron may considerably improve the "brightness" of the mineral, and thereby enhance its value as a coating or filler material for paper, or for use in the manufacture of high quality porcelain. In cases where the magnetic susceptibility of the contaminating component is too low, adsorption may first be carried out on to the surface of a material with the necessary magnetic properties. The magnetic field is generated in the gap between the poles of an electromagnet into which a loose matrix of fine stainless steel wire, usually of voidage of about 0.95, is inserted.

The attractive force on a particle is proportional to its magnetic susceptibility and to the product of the field strength and its gradient, and the fine wire matrix is used to minimise the distance between adjacent magnetised surfaces. The attractive forces which bind the particles must be sufficiently strong to ensure that the particles are not removed by the hydrodynamic drag exerted by the flowing suspension. As the deposit of separated particles builds up, the capture rate progressively diminishes and, at the appropriate stage, the particles are released by reducing the magnetic field strength to zero and flushing out with water. Commercial machines usually have two reciprocating canisters, in one of which particles are being collected from a stream of suspension, and in the other released into a waste stream. The dead time during which the canisters are being exchanged may be as short as 10 s.

Magnetic fields of very high intensity may be obtained by the use of superconducting magnets which operate most effectively at the temperature of liquid helium, and conservation of both gas and "cold" is therefore of paramount importance. The reciprocating canister system employed in the china clay industry is described by SVAROVSKY[30] and involves the use a single superconducting magnet and two canisters. At any time one is in the magnetic field while the other is withdrawn for cleaning. The whole system needs delicate magnetic balancing so that the two canisters can be moved without the use of very large forces and, for this to be the case, the amount of iron in the magnetic field must be maintained at a constant value throughout the transfer process. The superconducting magnet then remains at high field strength, thereby reducing the demand for liquid helium.

### *Magnetic micro-organisms*

Micro-organisms can play an important role in the removal of certain heavy metal ions from effluent solutions. In the case of uranyl ions which are paramagnetic, the cells which have adsorbed the ions may be concentrated using a high gradient magnetic separation process. If the ions themselves are not magnetic, it may be possible to precipitate a magnetic deposit on the surfaces of the cells. Some micro-organisms incorporate a magnetic component in their cellular structure and are capable of taking up non-magnetic pollutants and are then themselves recoverable in a magnetic field. Such organisms are referred to a being *magnetotactic*.

## 1.5.7. Electrostatic separators

Electrostatic separators, in which differences in the electrical properties of the materials are exploited in order to effect a separation, are sometimes used with small quantities of fine material. The solids are fed from a hopper on to a rotating drum, which is either charged or earthed, and an electrode bearing the opposite charge is situated at a small

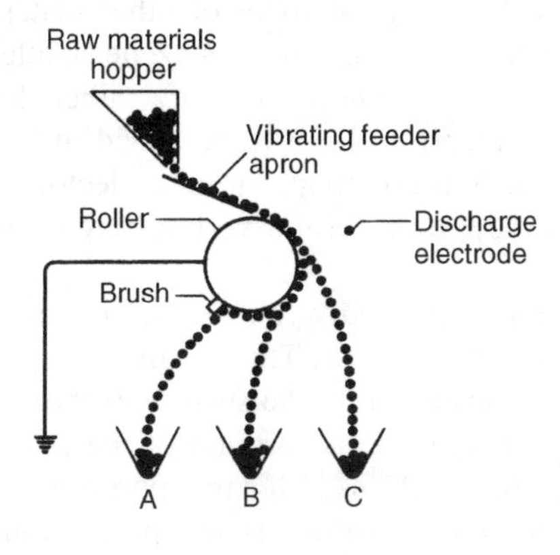


Figure 1.46. Electrostatic separator

distance from the drum, as shown in Figure 1.46. The point at which the material leaves the drum is determined by the charge it acquires, and by suitable arrangement of the collecting bins (A, B, C) a sharp classification can be obtained.

## 1.5.8. Flotation

### *Froth flotation*

The separation of a mixture using flotation methods depends on differences in the surface properties of the materials involved. If the mixture is suspended in an aerated liquid, the gas bubbles will tend to adhere preferentially to one of the constituents–the one which is more difficult to wet by the liquid–and its effective density may be reduced to such an extent that it will rise to the surface. If a suitable frothing agent is added to the liquid, the particles are held in the surface by means of the froth until they can be discharged over a weir. Froth flotation is widely used in the metallurgical industries where, in general, the ore is difficult to wet and the residual earth is readily wetted. Both the theory and practical application of froth flotation are discussed by CLARKE and WILSON[44].

The process depends on the existence, or development, of a selective affinity of one of the constituents for the envelopes of the gas bubbles. In general, this affinity must be induced, and the reagents which increase the angle of contact between the liquid and one of the materials are known as *promoters* and *collectors*. Promoters are selectively adsorbed on the surface of one material and form a monomolecular layer. The use of excess material destroys the effect. Concentrations of the order of 0.05 kg/Mg of solids are usually required. A commonly used promoter is sodium ethyl xanthate:

$$\begin{matrix} NaS - \\ \quad \\ S = \end{matrix} C - O - C_2H_5$$

In solution, this material ionises giving positively charged sodium ions and negatively charged xanthate ions. The polar end of the xanthate ion is adsorbed and the new surface of the particles is therefore made up of the non-polar part of the radical, so that the contact angle with water is increased. A large number of other materials are used in particular cases, such as other xanthates and diazoaminobenzene. Collectors are materials which form surface films on the particles. These films are rather thicker than those produced by promoters, and collectors therefore have to be added in higher concentrations–about 0.5 kg/Mg of solids. Pine oil is a commonly used collector, and petroleum compounds are frequently used though they often form a very greasy froth which is then difficult to disperse.

In many cases, it is necessary to modify the surface of the other constituent so that it does not adsorb the collector or promoter. This is effected by means of materials known as *modifiers* which are either adsorbed on the surface of the particles or react chemically at the surface, and thereby prevent the adsorption of the collector or promoter. Mineral acids, alkalis, and salts are frequently used for this purpose.

An essential requirement of the process is the production of a froth of a stability sufficient to retain the particles of the one constituent in the surface so that they can be discharged over the overflow weir. On the other hand, the froth should not be so stable that it is difficult to break down at a later stage. The frothing agent should reduce the interfacial tension between the liquid and gas, and the quantities required are less when the frothing agent is adsorbed at the interface. Liquid soaps, soluble oils, cresol and alcohols, in the range between amyl and octyl alcohols, are frequently used as frothing agents. In many cases, the stability of the foam is increased by adding a stabiliser–usually a mineral oil–which increases the viscosity of the liquid film forming the bubble, and therefore reduces the rate at which liquid can drain from the envelope. Pine oil is sometimes used as a frothing agent. This produces a stable froth and, in addition, acts as a collector.

The reagents which are used in froth flotation are usually specific in their action, and the choice of suitable reagents is usually made as a result of tests on small-scale equipment. Mixtures of more than two components may be separated in stages using different reagents at each stage. Sometimes the behaviour of a system is considerably influenced by change in the pH of the solution, and it is then possible to remove materials at successive stages by progressive alterations in the pH. In general, froth flotation processes may be used for particles of 5–250 μm. The tendency of various materials to respond to froth flotation methods is given by DOUGHTY[45]. HANUMANTH and WILLIAMS[46,47] have published details of the design and operation of laboratory flotation cells and the effect of froth height on the flotation of china clay.

It is advantageous to generate bubbles of micron-size when the particles to be floated are very small. The generation of such bubbles is almost impossible in conventional equipment which relies on mechanical means of breaking down the gas. If air, or another gas, is dissolved under pressure in the suspension before it is introduced into the cell, numerous microbubbles are formed when the pressure is reduced and these then attach themselves to the hydrophobic particles. Similar effects can be obtained by operating the cells under vacuum, or producing gas bubbles electrolytically. Dissolved and electro filtration are discussed later.

### *The Denver DR flotation machine*

The Denver DR flotation machine, which is an example of a typical froth flotation unit used in the mining industry, is illustrated in Figure 1.47. The pulp is introduced through a feed box and is distributed over the entire width of the first cell. Circulation of the pulp through each cell is such that, as the pulp comes into contact with the impeller, it is subjected to intense agitation and aeration. Low pressure air for this purpose is introduced down the standpipe surrounding the shaft and is thoroughly disseminated throughout the pulp in the form of minute bubbles when it leaves the impeller/diffuser zone, thus assuring maximum contact with the solids, as shown in Figure 1.47. Each unit is suspended in an essentially open trough and generates a "ring doughnut" circulation pattern, with the liquid being discharged radially from the impeller, through the diffuser, across the base of the tank, and then rising vertically as it returns to the eye of the impeller through the recirculation well. This design gives strong vertical flows in the base zone of the tank in order to suspend coarse solids and, by recirculation through the well, isolates the upper zone which remains relatively quiescent.

Froth baffles are placed between each unit mechanism to prevent migration of froth as the liquid flows along the tank. The liquor level is controlled at the end of each bank

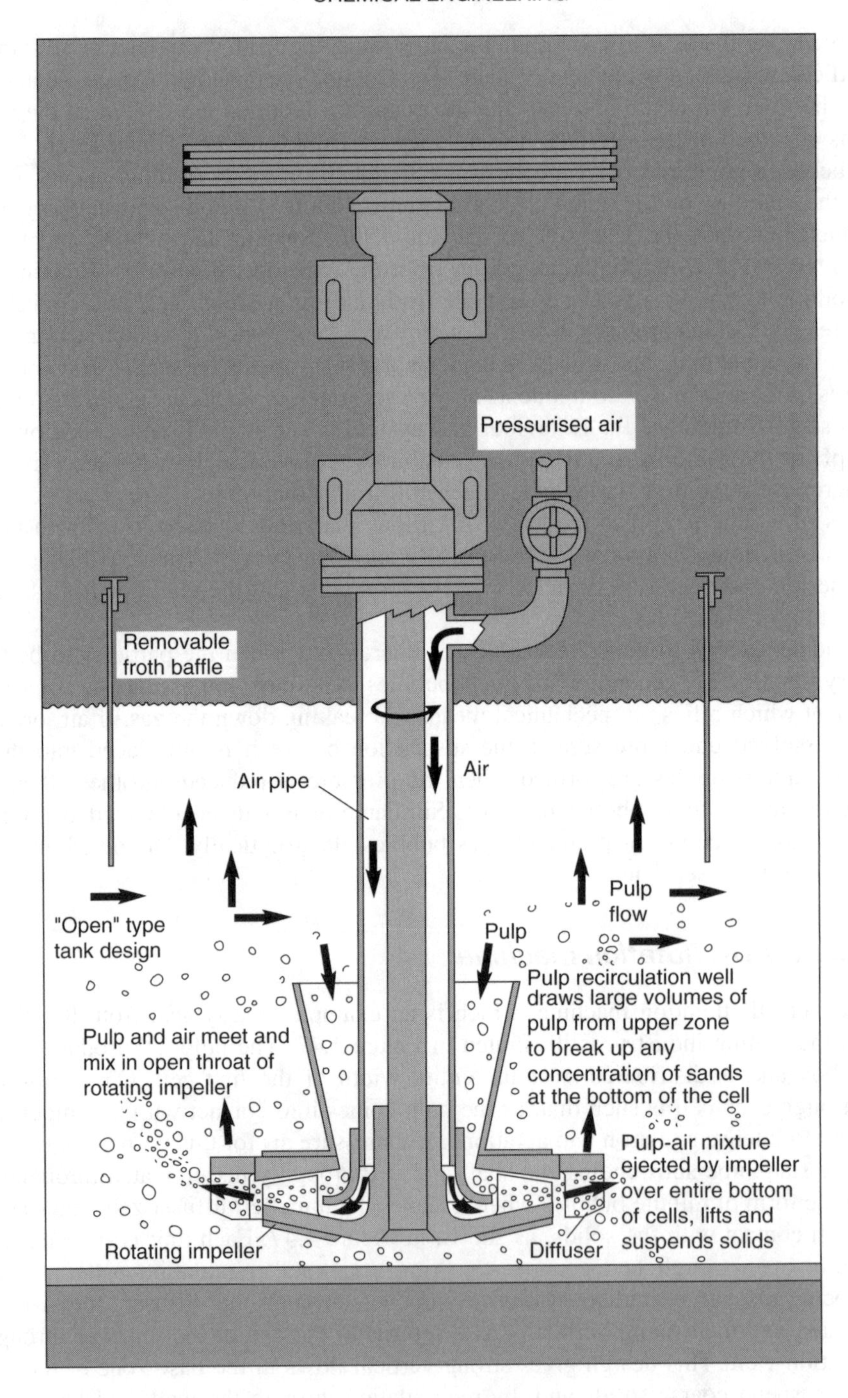


Figure 1.47. Typical cross-section through a Denver DR flotation machine

section by a combination of weir overflows and dart valves which can be automated. These units operate with a fully flooded impeller, and a low pressure air supply is required to deliver air into the eye of the impeller where it is mixed with the recirculating liquor at the tip of the air bell. Butterfly valves are used to adjust and control the quantity of air delivered into each unit.

Each cell is provided with an individually controlled air valve. Air pressure is between 108 and 124 $kN/m^2$ (7 and 23 $kN/m^2$ gauge) depending on the depth and size of the machine and the pulp density. Typical energy requirements for this machine range from 3.1 $kW/m^3$ of cell volume for a 2.8 $m^3$ unit to 1.2 $kW/m^3$ for a 42 $m^3$ unit.

In the froth flotation cell used for coal washing, illustrated in Figure 1.48, the suspension contains about 10 per cent of solids, together with the necessary reagents. The liquid flows along the cell bank and passes over a weir, and directly enters the unit via a feed pipe and feed hood. Liquor is discharged radially from the impeller, through the diffuser, and

Figure 1.48. A Denver flotation cell for coal washing

is directed along the cell base and recirculated through ports in the feed hood. The zone of maximum turbulence is confined to the base of the tank; a quiescent zone exists in the upper part of the cell. These units induce sufficient air to ensure effective flotation without the need for an external air blower.

### Dissolved Air Flotation

The efficiency with which air bubbles attach themselves to fine particles is generally considerably improved if the bubbles themselves are very small. In dissolved air flotation, described in detail by GOCHIN[(48)] and by ZABEL[(49)], the suspension is saturated with air under pressure before being introduced through a nozzle into the flotation chamber. Because shear rates are much lower than they are in dispersed-air flotation, flocculated suspensions may be processed without deflocculation occurring. If the flocs are very weak, however, they may be broken down in the nozzle, and it may then be desirable to avoid passing the suspension through the high-pressure system by recycling part of the effluent from the flotation cell. However, in this case higher pressures will be needed in order to maintain the same ratio of released air to suspension.

The suspension must be saturated, or nearly saturated, with air at high pressures in a contacting device, such as a packed column, before it is introduced into the flotation cell. At a pressure of 500 kN/m$^2$, the solubility of air in water at 291 K is about 0.1 m$^3$ air/m$^3$ water and, on reducing the pressure to 100 kN/m$^2$, from Henry's law approximately 0.08 m$^3$ air/m$^3$ water will be released as a dispersion of fine bubbles in the liquid.

Table 1.3 gives data on the solubility of air in water as a function of temperature and pressure.

Table 1.3. Solubility of air in water and air released (litres free air/m$^3$ water)

| Temperature (K) | Pressure 500 kN/m$^2$ (5 bar) | Pressure 100 kN/m$^2$ (1 bar) | Air released on reducing pressure |
|---|---|---|---|
| 273 | 0.145 | 0.029 | 0.116 |
| 283 | 0.115 | 0.023 | 0.092 |
| 293 | 0.095 | 0.019 | 0.076 |
| 303 | 0.080 | 0.016 | 0.064 |

### Electroflotation

KUHN *et al.*[(50)] described the electroflotation process which represents an interesting recent development for the treatment of dilute suspensions. Gas bubbles are generated electrolytically within the suspension, and attach themselves to the suspended particles which then rise to the surface. Because the bubbles are very small, they have a high surface to volume ratio and are therefore very effective in suspensions of fine particles.

The method has been developed particularly for the treatment of dilute industrial effluents, including suspensions and colloids, especially those containing small quantities of organic materials. It allows the dilute suspension to be separated into a concentrated slurry and a clear liquid and, at the same time, permits oxidation of unwanted organic materials at the positive electrode.

Typically, the electrodes are of lead dioxide on a titanium substrate in the form of horizontal perforated plates, usually 5–40 mm apart, depending on the conductivity of the liquid. A potential difference of 5–10 V may be applied to give current densities of the order of 100 $A/m^2$. Frequently, the conductivity of the suspension itself is adequate, though it may be necessary to add ionic materials, such as sodium chloride or sulphuric acid. Electrode fouling can usually be prevented by periodically reversing the polarity of the electrodes. Occasionally, consumable iron or aluminium anodes may be used because the ions released into the suspension may then assist flocculation of the suspended solids.

An electroflotation plant usually consists of a steel or concrete tank with a sloping bottom as shown in Figure 1.49; liquid depth may typically be about 1 m. Since the flotation process is much faster than sedimentation with fine particles, flotation can be achieved with much shorter retention times — usually about 1 hour (3.6 ks) — and the land area required may be only about one-eighth of that for a sedimentation tank.

The floated sludge, which frequently contains 95 per cent of the solids, forms a blanket on the liquid surface and may be continuously removed by means of slowly moving brushes or scrapers mounted across the top of the tank.

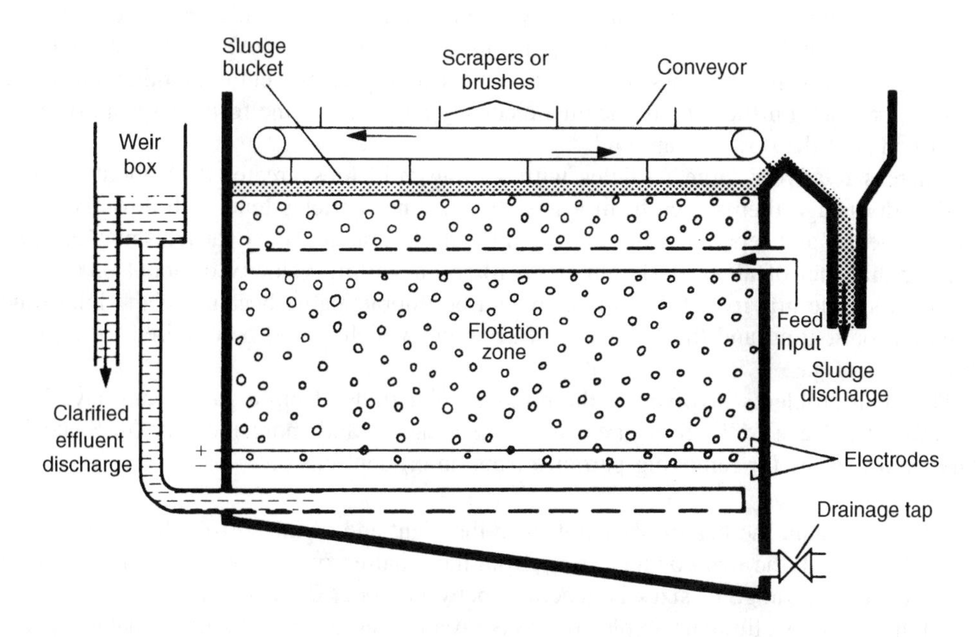


Figure 1.49. Schematic diagram of an electroflotation plant

## 1.6. SEPARATION OF SUSPENDED SOLID PARTICLES FROM FLUIDS

### 1.6.1. Introduction

Solids are removed from fluids in order to purify the fluid although, in some cases, and particularly with liquids, it is the solid material that is the product. Where solids are to be removed from liquids, a variety of operations are available including:

(a) Sedimentation, in which the solids are allowed to settle by gravity through the liquid from which they are removed, usually as a pumpable sludge. Such an operation is discussed in Chapter 5.
(b) Filtration, in which the solids are collected on a medium, such as a porous material or a layer of fine particles, through which the liquid is pumped. The theory, practice and available equipment for filtration are discussed in Chapter 7.
(c) Centrifugal separation in which the solids are forced on to the walls of a vessel which is rotated to provide the centrifugal force. Such an operation forms the basis of the discussion in Chapter 8.

Since the removal of solids from liquids in these ways is discussed elsewhere, this section is concerned only with the removal of suspended particles and liquid mists from gases.

The need to remove suspended dust and mist from a gas arises not only in the treatment of effluent gas from a plant before it is discharged into the atmosphere, but also in processes where solids or liquids are carried over in the vapour or gas stream. For example, in an evaporator it is frequently necessary to eliminate droplets which become entrained in the vapour, and in a plant involving a fluidised solid the removal of fine particles is necessary, first to prevent loss of material, and secondly to prevent contamination of the gaseous product. Further, in all pneumatic conveying plants, some form of separator must be provided at the downstream end.

Whereas relatively large particles with settling velocities greater than about 0.3 m/s readily disengage themselves from a gas stream, fine particles tend to follow the same path as the gas and separation is therefore difficult. In practice, dust particles may have an average diameter of about 0.01 mm (10 μm) and a settling velocity of about 0.003 m/s, so that a simple gravity settling vessel would be impracticable because of the long time required for settling and the large size of separator which would be required for a given throughput of gas.

The main problems involved in the removal of particles from a gas stream have been reviewed by ASHMAN[51], STAIRMAND and NONHOBEL[52] and more recently by SWIFT[53]. The main reasons for removing particles from an effluent gas are:

(a) To maintain the health of operators in the plant and of the surrounding population. In general, the main danger arises from the inhaling of dust particles, and the most dangerous range of sizes is generally between about 0.5 and 3 μm.
(b) In order to eliminate explosion risks. A number of carbonaceous materials and finely powdered metals give rise to explosive mixtures with air, and flame can be propagated over large distances.
(c) In order to reduce the loss of valuable materials.
(d) Because the gas itself may be required for use in a further process such as, for example, blast furnace gas used for firing stoves.

The size range of commercial aerosols and the methods available for determination of particle size and for removing the particles from the gas are shown in Figure 1.50, which is taken from the work of Ashman. It may noted that the ranges over which the various items of equipment operate overlap to some extent, and the choice of equipment depends

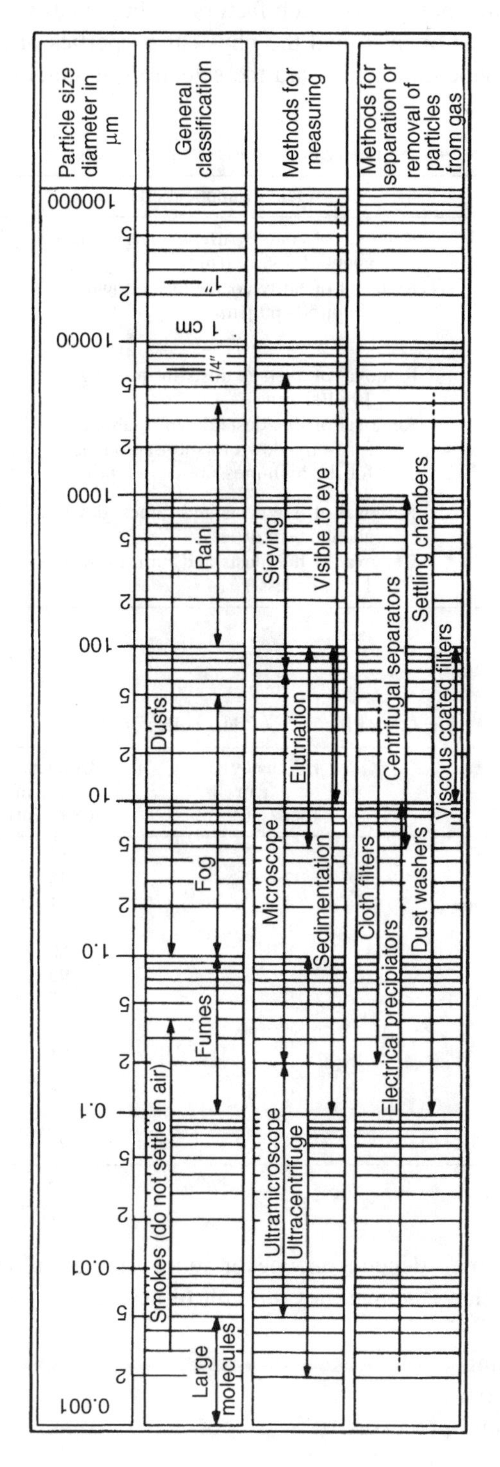


Figure 1.50. Characteristics of aerosols and separators[51]

not only on the particle size but also on such factors as the quantity of gas to be handled, the concentration of the dust or mist, and the physical properties of the particles.

Tables 1.4 and 1.5 provide data which aid the selection of suitable equipment.

Table 1.4. Summary of dust arrester performance[52]

| Type of equipment | Field of application | Pressure loss |
|---|---|---|
| Settling chambers | Removal of coarse particles, larger than about 100–150 μm | Below 50 $N/m^2$ |
| Scroll collectors, Shutter collectors, Low-pressure-drop cyclones | Removal of fairly coarse dusts down to about 50–60 μm | Below 250 $N/m^2$ |
| High-efficiency cyclones | Removal of average dusts in the range 10–100 μm | 250–1000 $N/m^2$ |
| Wet washers (including spray towers, venturi scrubbers, etc.) | Removal of fine dusts down to about 5 μm (or down to sub-micron sizes for the high-pressure-drop type) | 250–600 $N/m^2$ or more |
| Bag filters | Removal of fine dusts and fumes, down to about 1 μm or less | 100–1000 $N/m^2$ |
| Electrostatic precipitators | Removal of fine dusts and fumes down to 1 μm or less | 50–250 $N/m^2$ |

Table 1.5. Efficiency of dust collectors*[52]

| Dust collector | Efficiency at 5 μm (per cent) | Efficiency at 2 μm (per cent) | Efficiency at 1 μm (per cent) |
|---|---|---|---|
| Medium-efficiency cyclone | 27 | 14 | 8 |
| High-efficiency cyclone | 73 | 46 | 27 |
| Low-pressure-drop cellular cyclone | 42 | 21 | 13 |
| Tubular cyclone | 89 | 77 | 40 |
| Irrigated cyclone | 87 | 60 | 42 |
| Electrostatic precipitator | 99 | 95 | 86 |
| Irrigated electrostatic precipitator | 98 | 97 | 92 |
| Fabric filter | 99.8 | 99.5 | 99 |
| Spray tower | 94 | 87 | 55 |
| Wet impingement scrubber | 97 | 95 | 80 |
| Self-induced spray deduster | 93 | 75 | 40 |
| Disintegrator | 98 | 95 | 91 |
| Venturi scrubber | 99.8 | 99 | 97 |

*For dust of density 2700 $kg/m^3$.

Separation equipment may depend on one or more of the following principles, the relative importance of which is sometimes difficult to assess:

(a) Gravitational settling.
(b) Centrifugal separation.
(c) Inertia or momentum processes.
(d) Filtration.

(e) Electrostatic precipitation.
(f) Washing with a liquid.
(g) Agglomeration of solid particles and coalescence of liquid droplets.

## Example 1.5

The collection efficiency of a cyclone is 45 per cent over the size range 0–5 μm, 80 per cent over the size range 5–10 μm, and 96 per cent for particles exceeding 10 μm. Calculate the efficiency of collection for a dust with a mass distribution of 50 per cent 0–5 μm, 30 per cent 5–10 μm and 20 per cent above 10 μm.

## Solution

*For the collector:*

| Size (μm) | 0–5 | 5–10 | >10 |
|---|---|---|---|
| Efficiency (per cent) | 45 | 80 | 96 |

*For the dust:*

| | | | |
|---|---|---|---|
| Mass (per cent) | 50 | 30 | 20 |

On the basis of 100 kg dust:

| | | | |
|---|---|---|---|
| Mass at inlet (kg) | 50 | 30 | 20 |
| Mass retained (kg) | 22.5 | 24.0 | 19.2—a total of 65.7 kg |

Thus: Overall efficiency = 100(65.7/100) = 65.7 per cent

## Example 1.6

The size distribution by mass of the dust carried in a gas, together with the efficiency of collection over each size range, is as follows:

| Size range (μm) | 0–5 | 5–10 | 10–20 | 20–40 | 40–80 | 80–160 |
|---|---|---|---|---|---|---|
| Mass (per cent) | 10 | 15 | 35 | 20 | 10 | 10 |
| Efficiency (per cent) | 20 | 40 | 80 | 90 | 95 | 100 |

Calculate the overall efficiency of the collector, and the percentage by mass of the emitted dust that is smaller than 20 μm in diameter. If the dust burden is 18 $g/m^3$ at entry and the gas flow 0.3 $m^3/s$, calculate the mass flow of dust emitted.

## Solution

Taking 1 $m^3$ of gas as the basis of calculation, the following table of data may be completed noting that the inlet dust concentration is 18 $g/m^3$. Thus:

| Size range (μm) | 0–5 | 5–10 | 10–20 | 20–40 | 40–80 | 80–160 | Total |
|---|---|---|---|---|---|---|---|
| Mass in size range (g) | 1.8 | 2.7 | 6.3 | 3.6 | 1.8 | 1.8 | 18.0 |
| Mass retained (g) | 0.36 | 1.08 | 5.04 | 3.24 | 1.71 | 1.80 | 13.23 |
| Mass emitted (g) | 1.44 | 1.62 | 1.26 | 0.36 | 0.09 | 0 | 4.77 |

Hence: overall efficiency $= 100(13.23/18.0) = \underline{\underline{73.5 \text{ per cent}}}$

dust $< 20\ \mu\text{m}$ emitted $= 100[(1.44 + 1.62 + 1.26)/4.77] = \underline{\underline{90.1 \text{ per cent}}}$

The inlet gas flow $= 0.3\ \text{m}^3/\text{s}$ and hence:

mass emitted $= (0.3 \times 4.77) = 1.43$ g/s or: $\underline{\underline{1.43 \times 10^{-3} \text{ kg/s}}}$ (0.12 tonne/day)

## 1.6.2. Gas cleaning equipment

The classification given in Section 1.6.1 is now used to describe commercially available equipment.

### *a) Gravity separators*

If the particles are large, they will settle out of the gas stream if the cross-sectional area is increased. The velocity will then fall so that the eddy currents which are maintaining the particles in suspension are suppressed. In most cases, however, it is necessary to introduce baffles or screens as shown in Figure 1.51, or to force the gas over a series of trays as shown in Figure 1.52 which depicts a separator used for removing dust from sulphur dioxide produced by the combustion of pyrites. This equipment is suitable when the concentration of particles is high, because it is easily cleaned by opening the doors at the side. Gravity separators are seldom used as they are very bulky and will not remove particles smaller than 50–100 μm.

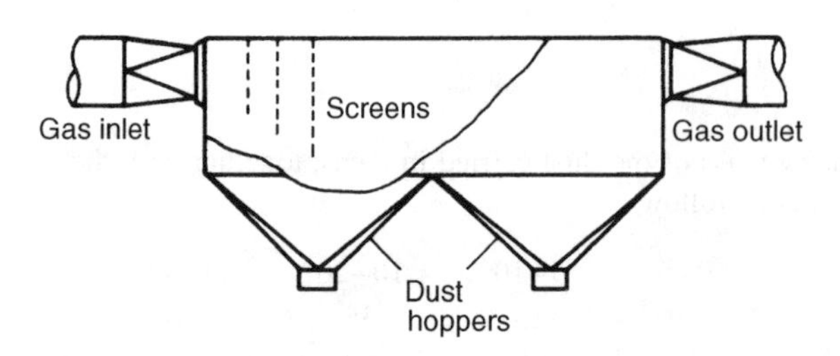


Figure 1.51. Settling chamber

Gravity separation is little used since the equipment must be very large in order to reduce the gas velocity to a reasonably low value which allows the finer particles to settle. For example, a settling chamber designed to remove particles with a diameter of 20 μm and density 2000 kg/m$^3$ from a gas stream flowing at 10 m$^3$/s would have a volume of about 3000 m$^3$. Clearly, this very large volume imposes a severe limitation and this type of equipment is normally restricted to small plants as a pre-separator which reduces the load on a more efficient secondary collector.

### *b) Centrifugal separators*

The rate of settling of suspended particles in a gas stream may be greatly increased if centrifugal rather than gravitational forces are employed. In the cyclone separator shown

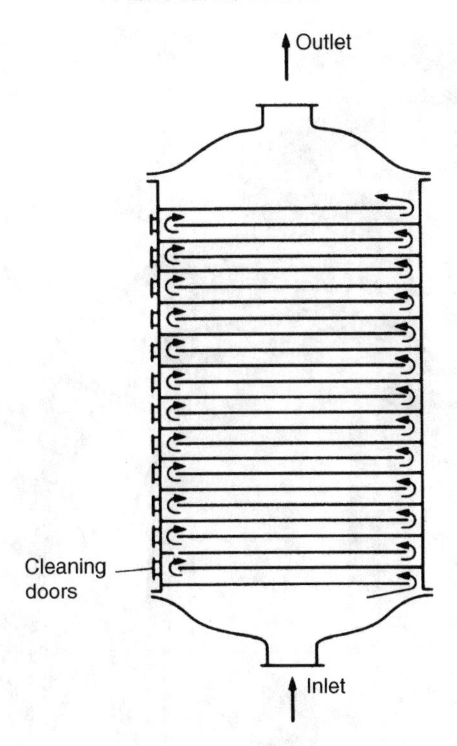


Figure 1.52. Tray separator

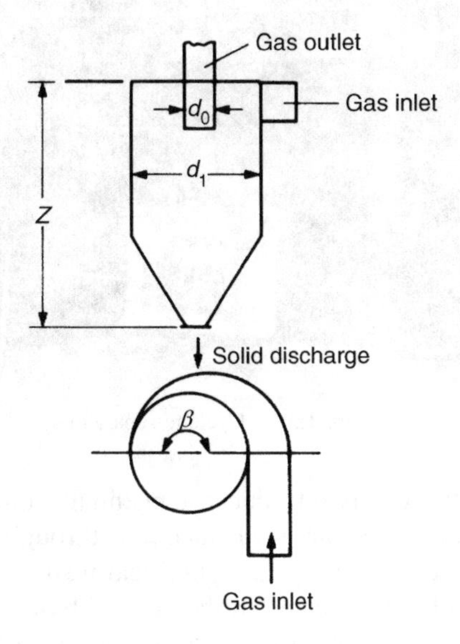


Figure 1.53. Cyclone separator

Figure 1.54. Cyclone separator

in Figures 1.53 and 1.54 the gas is introduced tangentially into a cylindrical vessel at a velocity of about 30 m/s and the clean gas is taken off through a central outlet at the top. The solids are thrown outwards against the cylindrical wall of the vessel, and then move away from the gas inlet and are collected in the conical base of the plant. This separator is very effective unless the gas contains a large proportion of particles less than about 10 μm in diameter and is equally effective when used with either dust- or mist-laden gases. It is now the most commonly used general purpose separator. The use of cyclones for liquid–solid systems is considered in Section 1.5.4.

Because the rotating motion of the gas in the cyclone separator arises from its tangential entry and no additional energy is imparted within the separator body, a free vortex is established. The energy per unit mass of gas is then independent of its radius of rotation and the velocity distribution in the gas may be calculated approximately by methods discussed in Volume 1, Chapter 2.

There have been several studies in which the flow patterns within the body of the cyclone separator have been modelled using a Computational Fluid Dynamics (CFD) technique. A recent example is that of SLACK *et al.*[54] in which the computed three-dimensional flow fields have been plotted and compared with the results of experimental studies in which a backscatter laser Doppler anemometry system was used to measure flowfields. Agreement between the computed and experimental results was very good. When using very fine grid meshes, the existence of time-dependent vortices was identified. These had the potentiality of adversely affecting the separation efficiency, as well as leading to increased erosion at the walls.

The effects of the dimensions of the separator on its efficiency have been examined experimentally by TER LINDEN[55] and STAIRMAND[56,57] who studied the flow pattern in a glass cyclone separator with the aid of smoke injected into the gas stream. The gas was found to move spirally downwards, gradually approaching the central portion of the separator, and then to rise and leave through the central outlet at the top. The magnitude and direction of the gas velocity have been explored by means of small pitot tubes, and the pressure distribution has been measured. The tangential component of the velocity of the gas appears to predominate throughout the whole depth, except within a highly turbulent central core of diameter about 0.4 times that of the gas outlet pipe. The radial component of the velocity acts inwards, and the axial component is away from the gas inlet near the walls of the separator, but is in the opposite direction in the central core. Pressure measurements indicate a relatively high pressure throughout, except for a region of reduced pressure corresponding to the central core. Any particle is therefore subjected to two opposing forces in the radial direction, the centrifugal force which tends to throw it to the walls and the drag of the fluid which tends to carry the particle away through the gas outlet. Both of these forces are a function of the radius of rotation and of the size of the particles, with the result that particles of different sizes tend to rotate at different radii. As the outward force on the particles increases with the tangential velocity and the inward force increases with the radial component, the separator should be designed so as to make the tangential velocity as high as possible and the radial velocity low. This is generally effected by introducing the gas stream with a high tangential velocity, with as little shock as possible, and by making the height of the separator large.

The radius at which a particle will rotate within the body of a cyclone corresponds to the position where the net radial force on the particle is zero. The two forces acting are the centrifugal force outwards and the frictional drag of the gas acting inwards.

Considering a spherical particle of diameter $d$ rotating at a radius $r$, then the centrifugal force is:

$$\frac{mu_t^2}{r} = \frac{\pi d^3 \rho_s u_t^2}{6r} \tag{1.46}$$

where: $m$ is the mass of the particle, and

$u_t$ is the tangential component of the velocity of the gas.

It is assumed here that there is no slip between the gas and the particle in the tangential direction.

If the radial velocity is low, the inward radial force due to friction will, from Chapter 3, be equal to $3\pi\mu d u_r$, where $\mu$ is the viscosity of the gas, and $u_r$ is the radial component of the velocity of the gas.

The radius $r$ at which the particle will rotate at equilibrium, is then given by:

$$\frac{\pi d^3 \rho_s u_t^2}{6r} = 3\pi\mu d u_r$$

or:

$$\frac{u_t^2}{r} = \frac{18\mu}{d^2 \rho_s} u_r \tag{1.47}$$

The free-falling velocity of the particle $u_0$, when the density of the particle is large compared with that of the gas, is given from equation 3.23 as:

$$u_0 = \frac{d^2 g \rho_s}{18\mu} \tag{1.48}$$

Substituting in equation 1.47:

$$\frac{u_t^2}{r} = \frac{u_r}{u_0} g$$

or:

$$u_0 = \frac{u_r}{u_t^2} r g \tag{1.49}$$

Thus the higher the terminal falling velocity of the particle, the greater is the radius at which it will rotate and the easier it is to separate. If it is assumed that a particle will be separated provided it tends to rotate outside the central core of diameter $0.4d_0$, the terminal falling velocity of the smallest particle which will be retained is found by substituting $r = 0.2d_0$ in equation 1.49 to give:

$$u_0 = \frac{u_r}{u_t^2} 0.2 d_0 g \tag{1.50}$$

In order to calculate $u_0$, it is necessary to evaluate $u_r$ and $u_t$ for the region outside the central core. The radial velocity $u_r$ is found to be approximately constant at a given radius and to be given by the volumetric rate of flow of gas divided by the cylindrical area for flow at the radius $r$. Thus, if $G$ is the mass rate of flow of gas through the separator and $\rho$ is its density, the linear velocity in a radial direction at a distance $r$ from the centre is given by:

$$u_r = \frac{G}{2\pi r Z \rho} \tag{1.51}$$

where $Z$ is the depth of the separator.

For a free vortex, it is shown in Volume 1, Chapter 2, that the product of the tangential velocity and the radius of rotation is a constant. Because of fluid friction effects, this relation does not hold exactly in a cyclone separator where it is found experimentally that the tangential velocity is more nearly inversely proportional to the square root of radius, rather than to the radius. This relation appears to hold at all depths in the body of the separator. If $u_t$ is the tangential component of the velocity at a radius $r$, and $u_{t0}$ is the corresponding value at the circumference of the separator, then:

$$u_t = u_{t0}\sqrt{(d_t/2r)} \tag{1.52}$$

It is found that $u_{t0}$ is approximately equal to the velocity with which the gas stream enters the cyclone separator. If these values for $u_r$ and $u_t$ are now substituted into equation 1.50, the terminal falling velocity of the smallest particle which the separator will retain is given by:

$$\begin{aligned} u_0 &= \left(\frac{G}{2\pi \times 0.2d_0\rho Z}\right)\left(\frac{2 \times 0.2d_0}{d_t}\right)\left(\frac{1}{u_{t0}^2}\right)0.2d_0g \\ &= \frac{0.2Gd_0g}{\pi\rho Zd_tu_{t0}^2} \end{aligned} \tag{1.53}$$

If the cross-sectional area of the inlet is $A_i$, $G = A_i\rho u_{t0}$ and:

$$u_0 = \frac{0.2A_i^2d_0\rho g}{\pi Zd_tG} \tag{1.54}$$

A small inlet and outlet therefore result in the separation of smaller particles but, as the pressure drop over the separator varies with the square of the inlet velocity and the square of the outlet velocity[56], the practical limit is set by the permissible drop in pressure. The depth and diameter of the body should be as large as possible because the former determines the radial component of the gas velocity and the latter controls the tangential component at any radius. In general, the larger the particles, the larger should be the diameter of the separator because the greater is the radius at which they rotate. The larger the diameter, the greater too is the inlet velocity which can be used without causing turbulence within the separator. The factor which ultimately settles the maximum size is, of course, the cost. Because the separating power is directly related to the throughput of gas, the cyclone separator is not very flexible though its efficiency can be improved at low throughputs by restricting the area of the inlet with a damper and thereby increasing the velocity. Generally, however, it is better to use a number of cyclones in parallel and to keep the load on each approximately the same whatever the total throughput.

Because the vertical component of the velocity in the cyclone is downwards everywhere outside the central core, the particles will rotate at a constant distance from the centre and move continuously downwards until they settle on the conical base of the plant. Continuous removal of the solids is desirable so that the particles do not get entrained again in the gas stream due to the relatively low pressures in the central core. Entrainment is reduced to a minimum if the separator has a deep conical base of small angle.

Though the sizes of particles which will be retained or lost by the separator can be calculated, it is found in practice that some smaller particles are retained and some larger particles are lost. The small particles which are retained have in most cases collided with other particles and adhered to form agglomerates which behave as large particles. Relatively large particles are lost because of eddy motion within the cyclone separator, and because they tend to bounce off the walls of the cylinder back into the central core of fluid. If the gas contains a fair proportion of large particles, it is desirable to remove these in a preliminary separator before the gas is fed into the cyclone.

The efficiency of the cyclone separator is greater for large than for small particles, and it increases with the throughput until the point is reached where excessive turbulence is created. Figure 1.55 shows the efficiency of collection plotted against particle size for an experimental separator for which the theoretical "cut" occurs at about 10 μm. It may be noted that an appreciable quantity of fine material is collected, largely as a result of agglomeration, and that some of the coarse material is lost with the result that a sharp cut is not obtained.

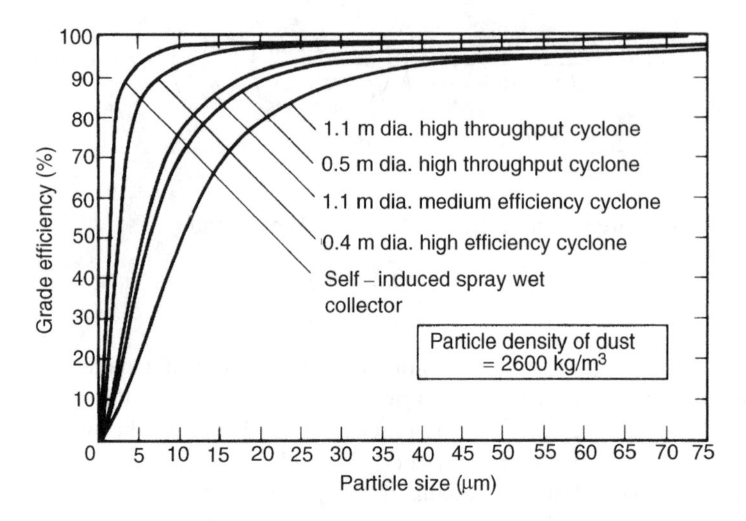


Figure 1.55. Typical grade efficiency curves for cyclones and a self-induced spray wet collector(53)

The effect of the arrangement and size of the gas inlet and outlet has been investigated and it has been found that the inlet angle $\beta$ in Figure 1.53 should be of the order of 180°. Further, the depth of the inlet pipe should be small, and a square section is generally preferable to a circular one because a greater area is then obtained for a given depth. The outlet pipe should extend downwards well below the inlet in order to prevent short-circuiting.

Various modifications may be made to improve the operation of the cyclone separator in special cases. If there is a large proportion of fine material present, a bag filter may be attached to the clean gas outlet. Alternatively, the smaller particles may be removed by means of a spray of water which is injected into the separator. In some cases, the removal of the solid material is facilitated by running a stream of water down the walls and this also reduces the risk of the particles becoming re-entrained in the gas stream. The main difficulty lies in wetting the particles with the liquid.

Because the separation of the solid particles which have been thrown out to the walls is dependent on the flow of gas parallel to the axis rather than to the effect of gravity, the cyclone can be mounted in any desired direction. In many cases horizontal cyclone separators are used, and occasionally the separator is fixed at the junction of two mutually perpendicular pipes and the axis is then a quadrant of a circle. The cyclone separator is usually mounted vertically, except where there is a shortage of headroom, because removal of the solids is more readily achieved especially if large particles are present.

A double cyclone separator is sometimes used when the range of the size of particles in the gas stream is large. This consists of two cyclone separators, one inside the other. The gas stream is introduced tangentially into the outer separator and the larger particles are deposited. The partially cleaned gas then passes into the inner separator through tangential openings and the finer particles are deposited there because the separating force in the inner separator is greater than that in the outer cylinder. In Figure 1.56, a multicyclone is illustrated in which the gas is subjected to further action in a series of tubular units, the number of which can be varied with the throughput. This separator is therefore rather more flexible than the simple cyclone.

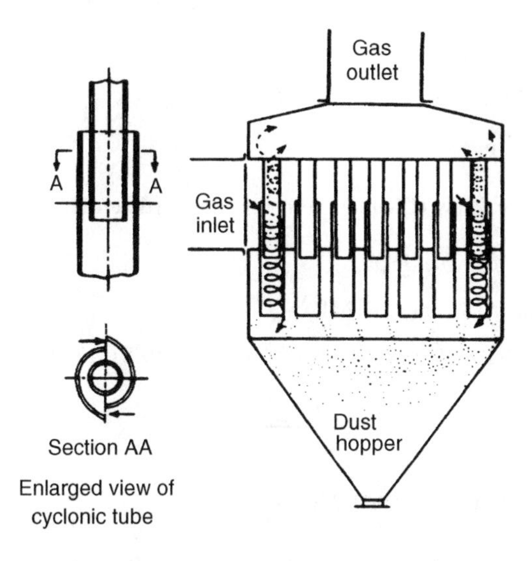


Figure 1.56. Multi-tube cyclone separator

SZEKELY and CARR[58] have studied the heat transfer between the walls of a cyclone and a gas–solid suspension, and have shown that the mechanism of heat transfer is quite different from that occurring in a fluidised bed. There is a high rate of heat transfer directly from the wall to the particles, but the transfer direct to the gas is actually reduced. Overall, the heat transfer rate at the walls is slightly greater than that obtained in the absence of the particles.

## Example 1.7

A cyclone separator, 0.3 m in diameter and 1.2 m long, has a circular inlet 75 mm in diameter and an outlet of the same size. If the gas enters at a velocity of 1.5 m/s, at what particle size will the theoretical cut occur?

The viscosity of air is 0.018 mN s/m$^2$, the density of air is 1.3 kg/m$^3$ and the density of the particles is 2700 kg/m$^3$.

## Solution

Using the data provided:

cross-sectional area at the gas inlet, $A_i = (\pi/4)(0.075)^2 = 4.42 \times 10^{-3}$ m$^2$

gas outlet diameter, $d_0 = 0.075$ m

gas density, $\rho = 1.30$ kg/m$^3$

height of separator, $Z = 1.2$ m, separator diameter, $d_t = 0.3$ m.

Thus: mass flow of gas, $G = (1.5 \times 4.42 \times 10^{-3} \times 1.30) = 8.62 \times 10^{-3}$ kg/s

The terminal velocity of the smallest particle retained by the separator,

$$u_0 = 0.2A_i^2 d_0 \rho g/(\pi Z d_t G) \quad \text{(equation 1.54)}$$

or: $$u_0 = [0.2 \times (4.42 \times 10^{-3})^2 \times 0.075 \times 1.3 \times 9.81]/[\pi \times 1.2 \times 0.3 \times 8.62 \times 10^{-3}]$$

$$= 3.83 \times 10^{-4} \text{ m/s}$$

Use is now made of Stokes' law (Chapter 3) to find the particle diameter, as follows:

$$u_0 = d^2 g(\rho_s - \rho)/18\mu \quad \text{(equation 3.24)}$$

or: $$d = [u_0 \times 18\mu/g(\rho_s - \rho)]^{0.5}$$

$$= [(3.83 \times 10^{-4} \times 18 \times 0.018 \times 10^{-3})/(9.81(2700 - 1.30))]^{0.5}$$

$$= 2.17 \times 10^{-6} \text{ m or } \underline{\underline{2.17\ \mu\text{m}}}$$

### c) Inertia or momentum separators

Momentum separators rely on the fact that the momentum of the particles is far greater than that of the gas, so that the particles do not follow the same path as the gas if the direction of motion is suddenly changed. Equipment developed for the separation of particles from mine gases is shown in Figure 1.57, from which it may be seen that the direction of the gas is changed suddenly at the end of each baffle. Again, the separator shown in Figure 1.58 consists of a number of vessels—up to about thirty connected in series—in each of which the gas impinges on a central baffle. The dust drops to the bottom and the velocity of the gas must be arranged so that it is sufficient for effective separation to be made without the danger of re-entraining the particles at the bottom of each vessel.

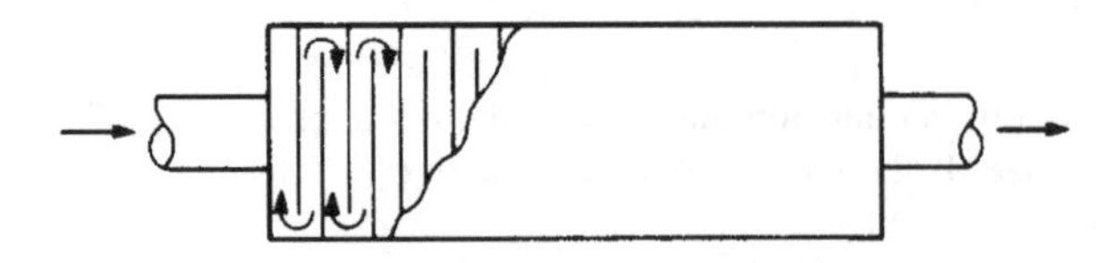

Figure 1.57. Baffled separator

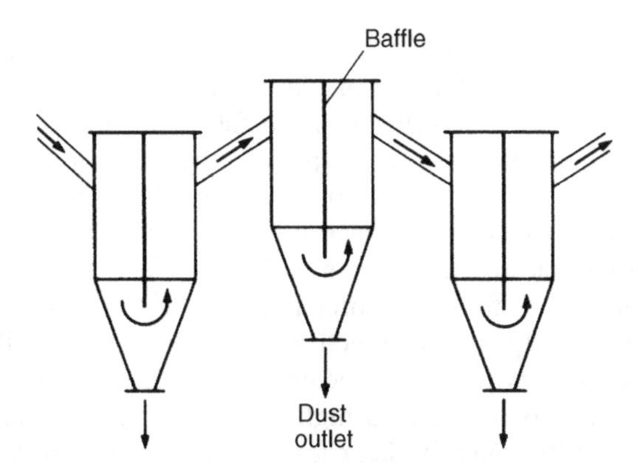


Figure 1.58. Battery of momentum separators

As an alternative to the use of rigid baffles, the separator may be packed with a loose fibrous material. In this case, the separation will be attributable partly to gravitational settling on to the packing, partly to inertia effects and partly to filtration. If the packing is moistened with a viscous liquid, the efficiency is improved because the film of liquid acts as an effective filter and prevents the particles being picked up again in the gas stream. The viscous filter as shown in Figure 1.59 consists of a series of corrugated plates, mounted in a frame and covered with a non-drying oil. These units are then arranged in banks to give the required area. They are readily cleaned and offer a low resistance to flow. Packs of slag wool offer a higher resistance though they are very effective.

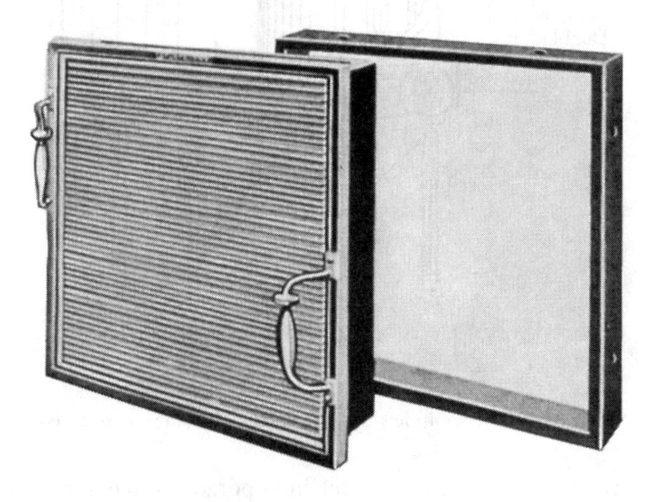

Figure 1.59. Viscous filter

### d) Fabric filters

Fabric filters include all types of bag filters in which the filter medium is in the form of a woven or felted textile fabric which may be arranged as a tube or supported on a suitable framework. As reported by DORMAN[59], type of filter is capable of removing particles of a size down to 1 μm or less by the use of glass fibre paper or pads. In a normal fabric filter, particles smaller than the apertures in the fabric are trapped by impingement on the fine "hairs" which span the apertures. Typically, the main strands of the material may have a diameter of 500 μm, spaced 100–200 μm apart. The individual textile fibres with a diameter of 5–10 μm crisscross the aperture and form effective impingement targets capable of removing particles of sizes down to 1 μm.

In the course of operation, filtration efficiency will be low until a loose "floc" builds up on the fabric surface and it is this which provides the effective filter for the removal of fine particles. The cloth will require cleaning from time to time to avoid excessive build-up of solids which gives rise to a high pressure drop. The velocity at which the gases pass through the filter must be kept low, typically 0.005 to 0.03 m/s, in order to avoid compaction of the floc and consequently high pressure drops, or to avoid local breakdown of the filter bed which would allow large particles to pass the filter.

There are three main types of bag filter. The simplest, which is shown in Figure 1.60, consists of a number of elements assembled together in a "bag-house". This is the cheapest type of unit and operates with a velocity of about 0.01 m/s across the bag surface.

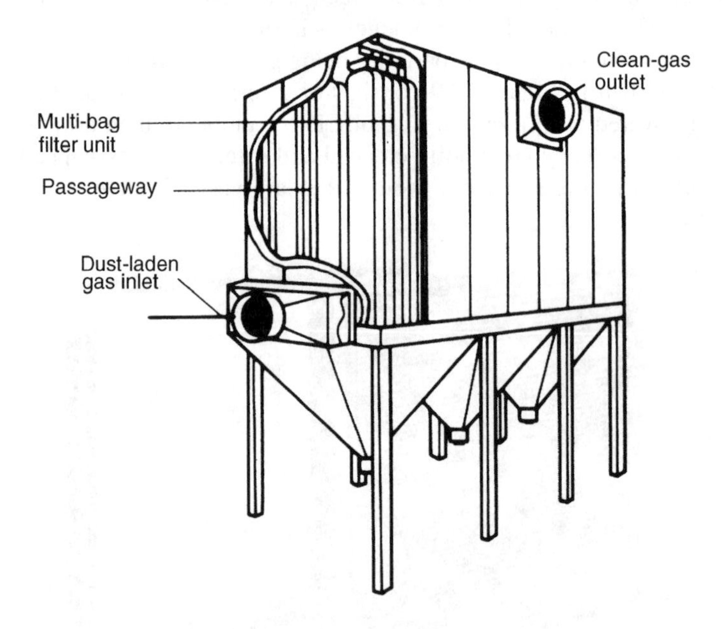


Figure 1.60. General view of "bag-house" (low face-velocity type)

A more sophisticated and robust version incorporates some form of automatic bag-shaking mechanism which may be operated by mechanical, vibratory or air-pulsed methods. A heavier fabric allows higher face velocities, up to 0.02 m/s, to be used and permits operation under more difficult conditions than the simpler bag-house type can handle.

The third type of bag filter is the reverse-jet filter, described by HERSEY[60] and illustrated in Figures 1.61 and 1.62. With face velocities of about 0.05 m/s and with the capability

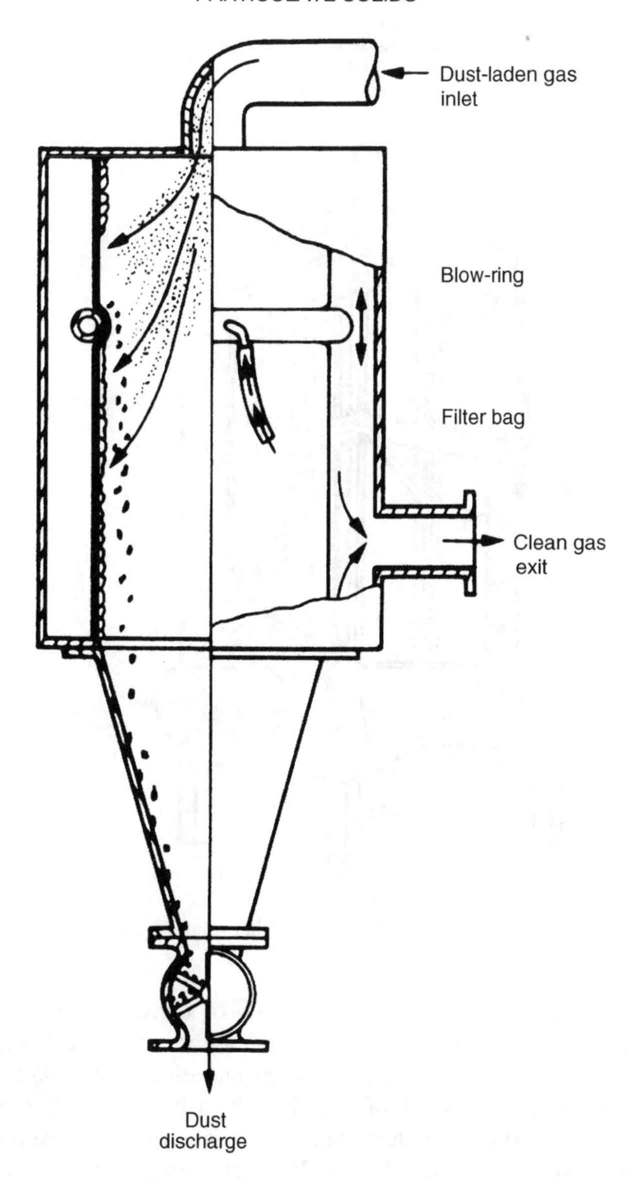


Figure 1.61. Reverse-jet filter[52]

of dealing with high dust concentrations at high efficiencies, this type of filter can deal with difficult mixtures in an economic and compact unit. Use of the blow ring enables the cake to be dislodged in a cleaning cycle which takes only a few seconds.

### e) Electrostatic precipitators

Electrostatic precipitators, such as that shown in Figure 1.63, are capable of collecting very fine particles, $<2\ \mu m$, at high efficiencies. Their capital and operating costs are high,

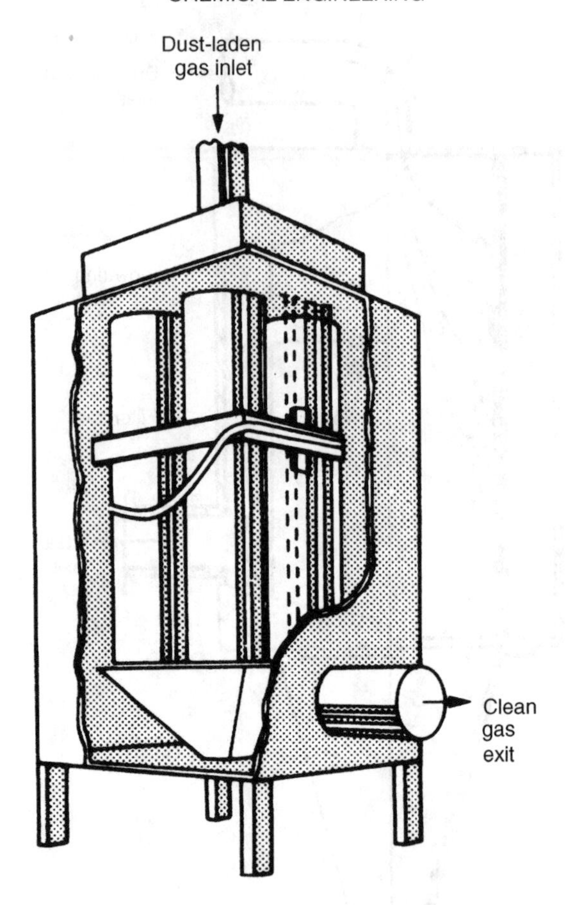


Figure 1.62. General view of reverse-jet filter (high face-velocity type)[52]

however, and electrostatic precipitation should only be considered in place of alternative processes, such as filtration, where the gases are hot or corrosive. Electrostatic precipitators are used extensively in the metallurgical, cement and electrical power industries. Their main application is in the removal of the fine fly ash formed in the combustion of pulverised coal in power-station boilers. The basic principle of operation is simple. The gas is ionised in passing between a high-voltage electrode and an earthed at grounded electrode. The dust particles become charged and are attracted to the earthed electrode. The precipitated dust is removed from the electrodes mechanically, usually by vibration, or by washing. Wires are normally used for the high-voltage electrode shown in Figure 1.64 and plates or tubes for the earthed electrode shown in Figure 1.65. A typical design is shown in Figure 1.63. A full description of the construction, design and application of electrostatic precipitators is given by SCHNEIDER *et al.*[61] and by ROSE and WOOD[62].

If the gas is passed between two electrodes charged to a potential difference of 10–60 kV, it is subjected to the action of a corona discharge. Ions which are emitted by the smaller electrode — on which the charge density is greater — attach themselves to the particles which are then carried to the larger electrode under the action of the

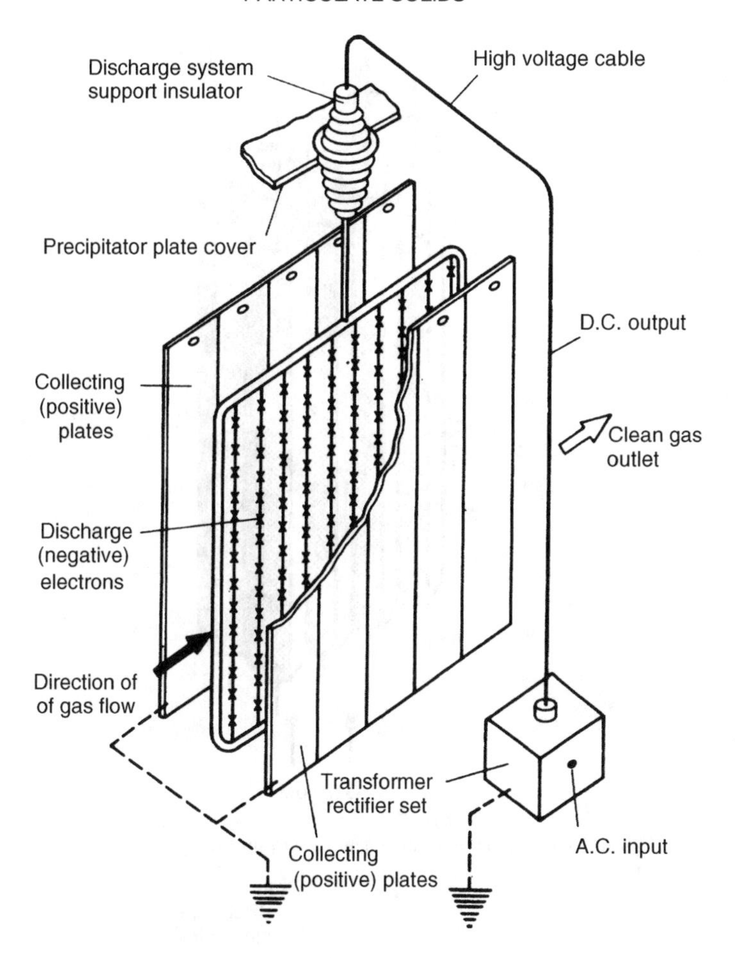


Figure 1.63. Electrostatic precipitator

electric field. The smaller electrode is known as the discharge electrode and the larger one, which is usually earthed, as the receiving electrode. Most industrial gases are sufficiently conducting to be readily ionised, the most important conducting gases being carbon dioxide, carbon monoxide, sulphur dioxide, and water vapour, though if the conductivity is low, water vapour can be added.

The gas velocity over the electrodes is usually between about 0.6 and 3 m/s, with an average contact time of about 2 s. The maximum velocity is determined by the maximum distance through which any particle must move in order to reach the receiving electrode, and by the attractive force acting on the particle. This force is given by the product of the charge on the particle and the strength of the electrical field, although calculation of the path of a particle is difficult because it gradually becomes charged as it enters the field and the force therefore increases during the period of charging. This rate of charging cannot be estimated with any degree of certainty. The particle moves towards the collecting electrode under the action of the accelerating force due to the electrical field and the retarding force of fluid friction, and the maximum rate of passage of gas is

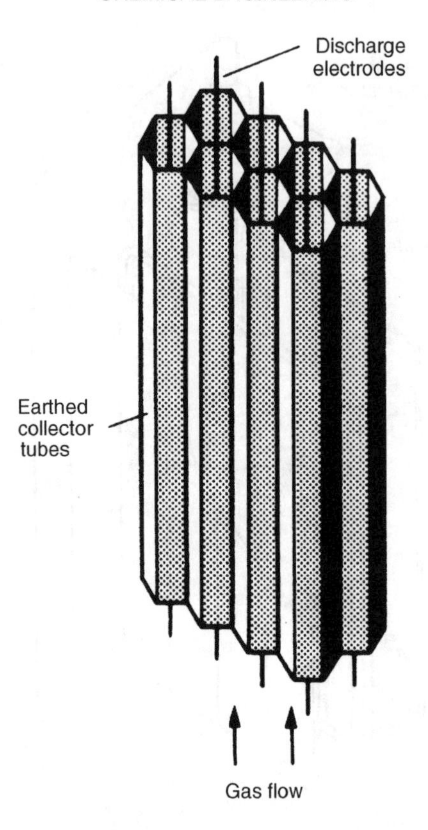


Figure 1.64. General arrangement of wire-in-tube precipitator

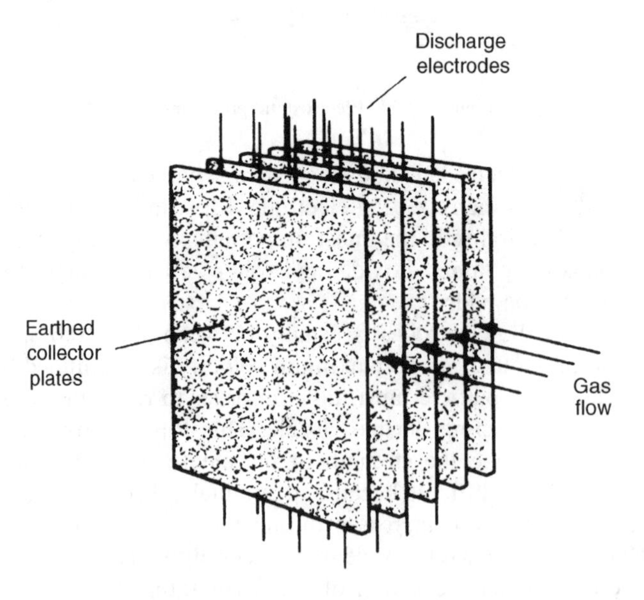


Figure 1.65. General arrangement of wire-and-plate precipitator

that which will just allow the most unfavourably located particle to reach the collecting electrode before the gas leaves the precipitator. Collection efficiencies of nearly 100 per cent can be obtained at low gas velocities although the economic limit is usually about 99 per cent.

Electrostatic precipitators are made in a very wide range of sizes and will handle gas flows up to about 50 $m^3/s$. Although they operate more satisfactorily at low temperatures, they can be used up to about 800 K. Pressure drops over the separator are low.

## 1.6.3. Liquid washing

If the gas contains an appreciable proportion of fine particles, liquid washing provides an effective method of cleaning which gives a gas of high purity. In the spray column illustrated in Figure 1.66, the gas flows upwards through a set of primary sprays to the main part of the column where it flows countercurrently to a water spray which is redistributed at intervals. In some cases packed columns are used for gas washing, though it is generally better to arrange the packing on a series of trays to facilitate cleaning.

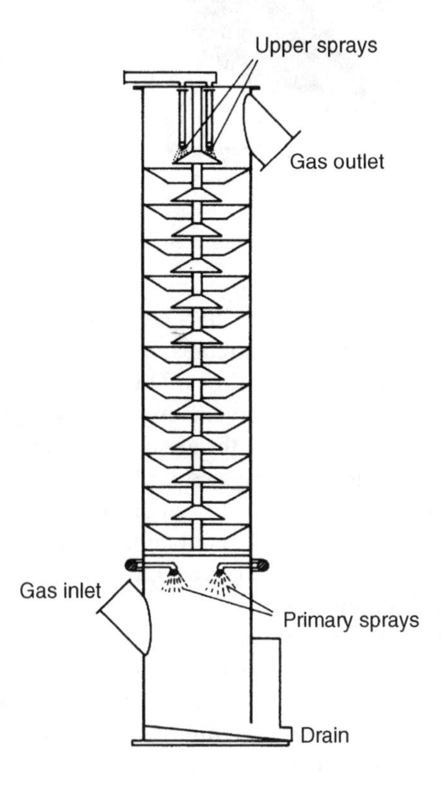


Figure 1.66. Spray washer

A co-current washer is shown in Figure 1.67 where the gas and the water flow down through a packed section before entering a disengagement space. The gas leaves through a mist eliminator and the bulk of the water is recirculated.

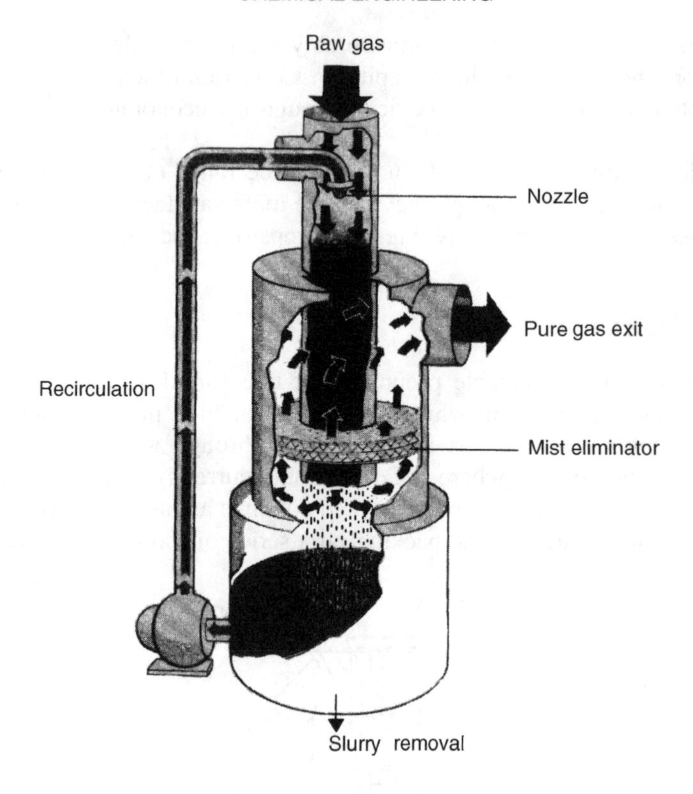


Figure 1.67. The Sulzer co-current scrubber

Figure 1.68 shows a wet gas scrubber designed initially for the removal of small quantities of hydrogen sulphide from gas streams, though used for cleaning gases, vapours and mists from process and ventilation air flows. This works on the absorption principle of dissolving impurities, such as acids and solvents in cleaning liquors like water or chemical solutions. The contaminated air is drawn through a packing zone filled with saddles or any other suitable packings which are irrigated with the liquor. Due to the design of the saddles and the way they pack, the impurities achieve good contact with the absorbing liquor and hence a high cleaning efficiency is obtained. A demister following the packing zone removes any entrained liquid droplets, leaving the exhaust air containing less than 2 per cent of the original contaminant.

Figure 1.69 shows a venturi scrubber in which water is injected at the throat and the separation is then carried out in a cyclone separator.

The AAF KINPACTOR scrubber, illustrated in Figure 1.70, operates in the same way and is capable of handling gas flowrates as high as 30 $m^3$/s.

In general, venturi scrubbers work by injecting water into a venturi throat through which contaminated air is passing at 60–100 m/s. The air breaks the water into small droplets which are then accelerated to the air velocity. Both are then decelerated relatively gently in a tapered section of ducting. The difference between the water and dust densities gives different rates of deceleration, as does the larger size of water droplets, all resulting in collisions between dust particles and droplets. Thus the dust particles are encapsulated

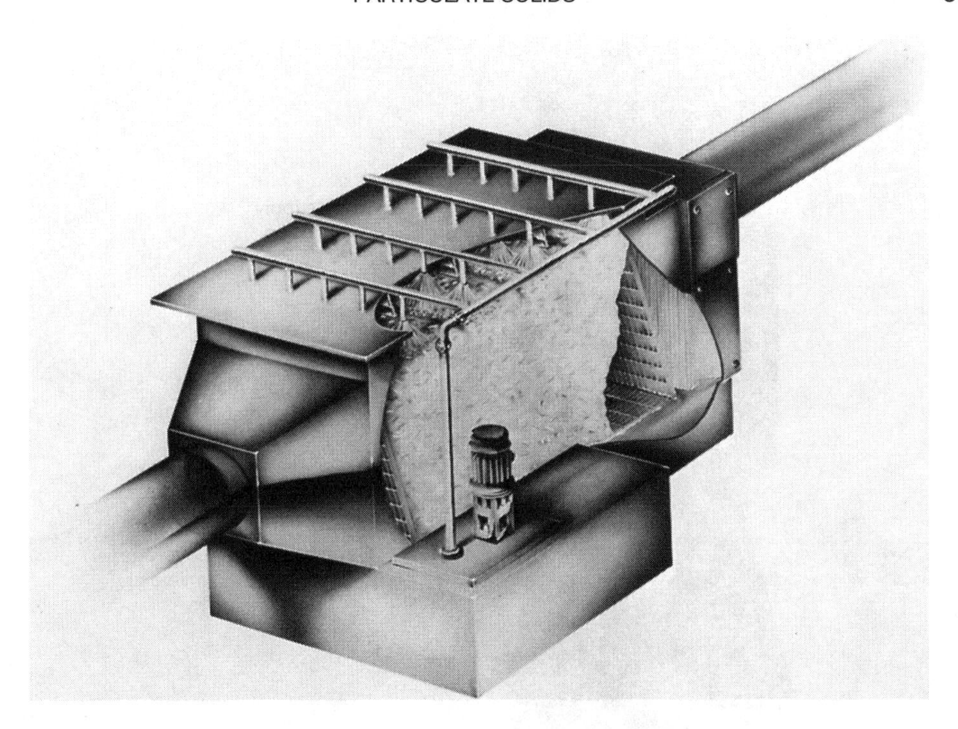

Figure 1.68. The AAF HS gas scrubber

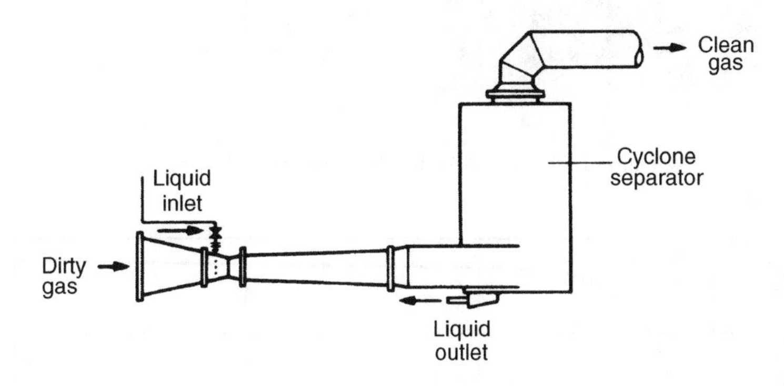


Figure 1.69. Venturi washer with cyclone separator

in the droplets which are of a size easily collected by a conventional spray eliminator system. The end result is a very good grade efficiency curve shown in Figure 1.71, but at the price of a very high pressure drop, from 300 mm of water upwards. Water/gas ratios of up to 1 $m^3$ water/1000 $m^3$ gas may be used, although commercial venturi scrubbers tend to use considerable flows of water.

The main difficulty in operating any gas washer is the wetting of the particles because it is often necessary for the liquid to displace an adsorbed layer of gas. In some cases the gas is sprayed with a mixture of water and oil to facilitate wetting and it may then

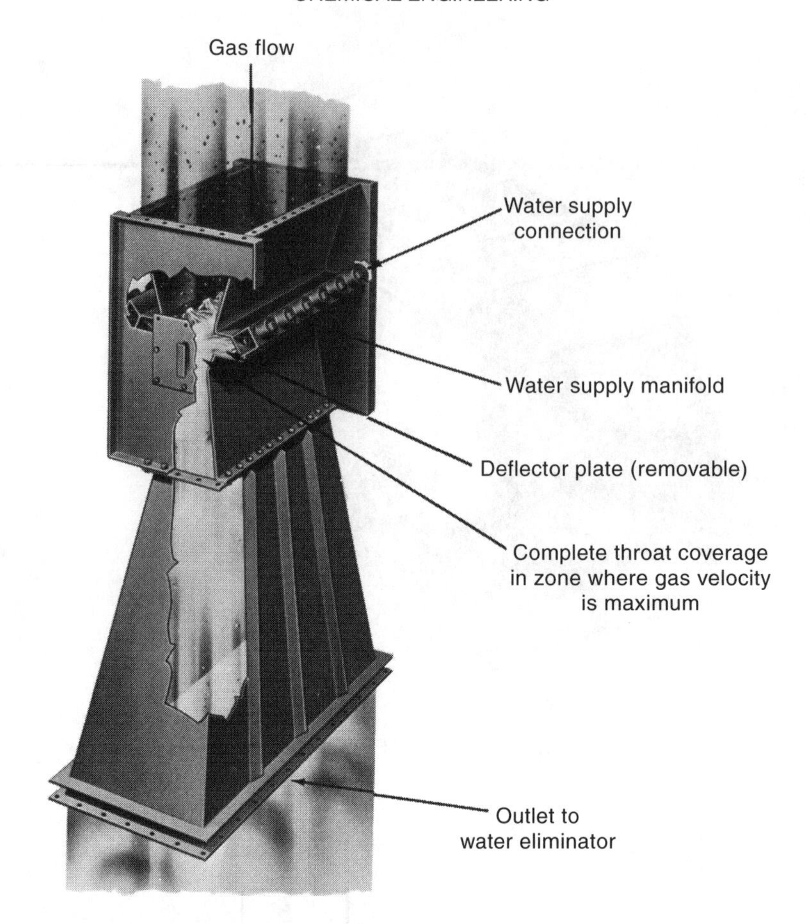


Figure 1.70. The AAF KINPACTOR venturi scrubber

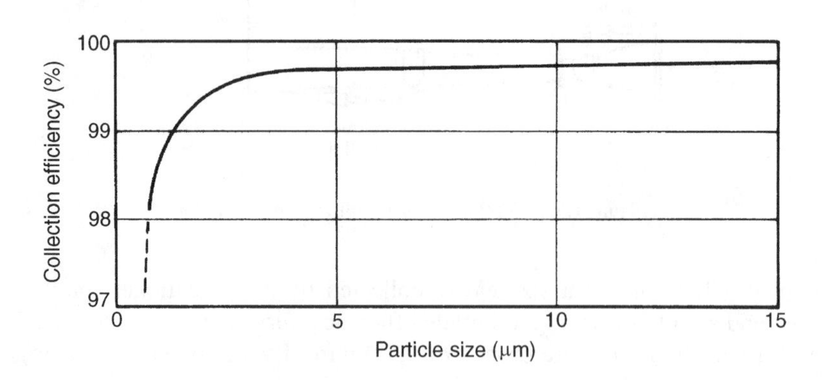


Figure 1.71. Typical grade efficiency curve for a medium velocity venturi scrubber[53]

be bubbled through foam. This method of separation cannot be used, of course, where a dry gas is required.

### *g) Agglomeration and coalescence*

Separation of particles or droplets is often carried out by first effecting an increase in the effective size of the individual particles by causing them to agglomerate or coalesce, and then separating the enlarged particles. A number of methods are available. Thus, if the dust- or mist-laden gas is brought into contact with a supersaturated vapour, condensation occurs on the particles which act as nuclei. Again if the gas is brought into an ultrasonic field, the vibrational energy of the particles is increased so that they collide and agglomerate. Ultrasonic agglomeration[4,63] has been found very satisfactory in causing fine dust particles to form agglomerates, about 10 μm in diameter, which can then be removed in a cyclone separator.

### *h) Odour removal*

In addition to the removal of solids from gases, the problem of small quantities of compounds causing odour nuisance must frequently be handled. Activated carbon filters are used for this purpose and, in continuous operation, gaseous impurities from the air-stream are adsorbed on the activated carbon granules. The carbon will frequently adsorb contaminants until they reach a level of 20–30 per cent of the mass of the carbon.

When saturation level is reached, the carbon is regenerated by the injection of live steam for a period of 30 to 60 minutes. Both the desorbed substances and steam are condensed in the lower part of the unit.

In case of solvent recovery some units feature a water separator permitting recycling of the recovered solvent. Adsorption processes are discussed in Chapter 17.

## 1.7. FURTHER READING

ALLEN, T.: *Particle Size Measurement, Volumes 1 and 2*. 5th edn. (Chapman and Hall, London, 1997)

BEDDOW, J. K.: *Particle Characterization in Technology*, Vol 1: *Application and Micro-analysis*, Vol 2: *Morphological Analysis*. (CRC Press, Boca Raton, FL, 1984)

BOHREN, C. F. and HUFFMAN, D. R.: *Absorption and Scattering of Light by Small Particles* (Wiley, New York, 1983)

BRADLEY, D.: *The Hydrocyclone*. (Pergamon Press, Oxford, 1965)

CLARKE, A. N. and WILSON, D. J.: *Foam Flotation, Theory and Applications*. (Marcel Dekker, New York, 1983)

DALLAVALLE, J. M.: *Micromerities*. 2nd edn. (Pitman, London 1948)

DORMAN, R. G.: *Dust Control and Air Cleaning*. (Pergamon Press, Oxford, 1974)

DULLIEN, F. A. L.: *Introduction to Industrial Gas Cleaning*. (Academic Press, San Diego, 1989).

*Dust and Fume Control, A User Guide*. 2nd edn. (Institution of Chemical Engineers, 1992)

FINCH, J. A. and DOBBY, G. S.: *Column Flotation*. (Pergamon Press, Oxford, 1990)

HERDAN, G.: *Small Particle Statistics*. 2nd edn. (Butterworths, London, 1960)

LEWIS PUBLICATIONS INC. *Wet Scrubbers: a Practical Handbook*. (1986)

NONHEBEL, G. (ed.): *Gas Purification Processes for Air Pollution Control*. 2nd edn (Newnes–Butterworths, London, 1972)

RHODES, M.: *Introduction to Particle Technology*. (Wiley, 1998)

RUMPF, H.: *Particle Technology*. (Chapman and Hall, London, 1990)

ŠTERBAČEK, Z. and TAUSK, P.: *Mixing in the Chemical Industry*. (translated from Czech by K. MAYER) (Pergamon Press, Oxford, 1965)

SVAROVSKY, L. (ed.).: *Solid-Liquid Separation*. 3rd edn (Butterworths, Oxford, 1990); *Solid-Liquid Separation*. 4th edn. (Butterworth-Heinemann, Oxford, 2000)

SVAROVSKY, L.: *Hydrocyclones*. (Holt, Rinehart and Winston, 1984)
TAGGART, A. F.: *Handbook of Mineral Dressing*. (Wiley, New York, 1945)
WHITE, H. J.: *Industrial Electrostatic Precipitation*. (Addison-Wesley Publ. Co. Inc. and Pergamon Press, London, 1963)

## 1.8. REFERENCES

1. ALLEN, T.: *Particle Size Measurement, Volume 1, Powder Sampling and Particle Size Measurement; Volume 2, Surface Area and Pore Size Determination*, 5th edn. (Chapman and Hall, London, 1997)
2. HERDAN, G.: *Small Particle Statistics*, 2nd edn. (Butterworths, 1960)
3. BS410: 1962: British Standard 410 (British Standards Institution, London). Specification for test sieves.
4. DALLAVALLE, J. M.: *Micromerities*, 2nd edn. (Pitman, 1948)
5. ROSE, J. W. and COOPER, J. R. (Eds.): *Technical Data on Fuel*, 7th edn. (British National Committee, World Energy Conference, 1977).
6. BS893: 1940: British Standard 893 (British Standards Institution, London). The method of testing dust extraction plant and the emission of solids from chimneys of electric power stations.
7. GREENE, R. S. B., MURPHY, P. J., POSNER, A. M. and QUIRK, J. P.: *Clays and Clay Minerals* **22** (1974) 187. Preparation techniques for electron microscopic examination of colloidal particles.
8. BOHREN, C. F. and HUFFMAN, D. R.: *Absorption and Scattering of Light by Small Particles* (Wiley, 1983).
9. BOND, F. C.: *Brit. Chem. Engr.* **8** (1963) 631. Some recent advances in grinding theory and practice.
10. BOND, F. C.: *Chem. Eng. Albany* **71** No. 6 (16 Mar. 1964) 134. Costs of process equipment.
11. JENIKE, A. W., ELSEY, P. J., and WOOLLEY, R. H.: *Proc. Amer. Soc. Test. Mat.* **60** (1960) 1168. Flow properties of bulk solids.
12. JENIKE, A. W.: *Trans. I. Chem. E.* **40** (1962) 264. Gravity flow of solids.
13. ASHTON, M. D., FARLEY, R., and VALENTIN, F. H. H.: *J. Sci. Instrum.* **41** (1964) 763. An improved apparatus for measuring the tensile strength of powders.
14. ZENZ, F. A. and OTHMER, D. F.: *Fluidization and Fluid Particle Systems* (Reinhold, 1960).
15. WEIGAND, J.: *Applied Rheology* **9** (1999) 204. Viscosity measurements on powders with a new viscometer.
16. BROWN, R. L.: *Nature* **191**. No. 4787 (1961) 458. Minimum energy theorem for flow of dry granules through apertures.
17. LIPTAK, B. G. *et al.*: in LIPTAK, B. G. (ed.): *Instrument Engineer's Handbook - Process Instrumentation and Analysis*. 3rd edn. Chapter 2, Flow Measurement (Butterworth-Heinemann, Boston, 1995).
18. LACEY, P. M. C.: *Trans. Inst. Chem. Eng.* **21** (1943) 53. The mixing of solid particles.
19. LACEY, P. M. C.: *J. Appl. Chem.* **4** (1954) 257. Developments in the theory of particle mixing.
20. BUSLIK, D.: *Bull. Am. Soc. Testing Mat.* **165** (1950) 66. Mixing and sampling with special reference to multisized granular particles.
21. KRAMERS, H.: Private Communication.
22. COULSON, J. M. and MAITRA, N. K.: *Ind. Chemist* **26** (1950) 55. The mixing of solid particles.
23. BLUMBERG, R. and MARITZ, J. S.: *Chem. Eng. Sci.* **2** (1953) 240. Mixing of solid particles.
24. BROTHMAN, A., WOLLAN, G. N., and FELDMAN, S. M.: *Chem. Met. Eng.* **52** (iv) (1945) 102. New analysis provides formula to solve mixing problems.
25. DANCKWERTS, P. V.: *Appl. Scient. Res.* **3A** (1952) 279. The definition and measurement of some characteristics of mixtures.
26. DANCKWERTS, P. V.: *Research* (London) **6** (1953) 355. Theory of mixtures and mixing.
27. WORK, L. T.: *Ind. Eng. Chem.* **57**, No. 11 (Nov. 1965) 97. Annual review–size reduction (21 refs.).
28. TAGGART, A. F.: *Handbook of Mineral Dressing* (Wiley, 1945).
29. BRADLEY, D.: *The Hydrocyclone* (Pergamon Press, 1965)
30. SVAROVSKY, L. (ed.): *Solid–Liquid Separation*, 3rd edn. (Butterworths, 1990)
31. TRAWINSKI, H. F.: In *Hydrocyclones Conference* Cambridge, U.K. (1980), BHRA Fluid Engineering 179. The application of hydrocyclones as versatile separators in chemical and mineral industries.
32. TRAWINSKI, H. F.: *Eng. Min. Jl* (Sept, 1976) 115. Theory, applications and practical operation of hydrocyclones.
33. TRAWINSKI, H. F.: *Filt and Sepn.* **22** (1985) 22. Practical hydrocyclone operation.
34. SVAROVSKY, L.: *Hydrocyclones* (Holt, Rinehart and Winston, 1984)
35. MOZLEY, R.: *Filt and Sepn.* **20** (1983) 474. Selection and operation of high performance hydrocyclones.
36. RHODES, N., PERICLEOUS, K. A., and DRAKE, S. N.: *Solid Liquid Flow* **1** (1989) 35. The prediction of hydrocyclone performance with a mathematical model.
37. SALCUDEAN, P. He. and GARTSHORE, I. S.: *Ch.E.R.D.* **77 A5** (1999) 429. A numerical simulation of hydrocyclones.

38. KELSALL, D. F.: *Trans. I. Chem. E.* **30** (1957) 87. A study of the motion of solid particles in a hydraulic cyclone.
39. RIETEMA, K.: *Chem. Eng. Sci.* **15** (1961) 298. Performance and design of hydraulic cyclones.
40. ZANKER, A.: In *Separation Techniques* Vol 2 *Gas Liquid Solid Systems* (McGraw-Hill, 1980), p. 178. Hydrocyclones: dimensions and performance.
41. DYAKOWSKI, T., HORNUNG, G. and WILLIAMS, R. A.: *Ch.E.R.D.* **72 A4** (1994) 513. Simulation of non-Newtonian flow in a hydrocyclone.
42. COLMAN, D. A., THEW, M. T. and CORNEY, D. R.: *Int. Conf. on Hydrocyclones* (1980) BHRA (Cranfield) Paper 11, 143–166. Hydrocyclones for oil-water separation.
43. NEBRENSKY, J. R., MORGAN, G. E. and OSWALD, B. J.: *Intl Conf. on Hydrocyclones*, Paper 12:167–178 BHRA, Cranfield.
44. CLARKE, A. N. and WILSON, D. J.: *Foam Flotation, Theory and Applications* (Marcel Dekker, 1983).
45. DOUGHTY, F. I. C.: *Mining Magazine* **81** (1949) 268. Floatability of minerals.
46. HANUMANTH, G. S. and WILLIAMS, D. J. A.: *Minerals Engineering* **1** (1988) 177. Design and operation characteristics of an improved laboratory flotation cell.
47. HANUMANTH, G. S. and WILLIAMS, D. J. A.: *Powder Tech.* **60** (1990) 133. An experimental study of the effects of froth height on flotation of china clay.
48. GOCHIN, R. J.: in SVAROVSKY, L.: *Solid-Liquid Separation*, 3rd edn. (Butterworths, 1990), 591–613 Flotation.
49. ZABEL, T.: *J. Am. Water Works Assocn.* **77** (1985) 42–76. The advantages of dissolved air flotation for water treatment.
50. KUHN, A. T.: in BOCKRIS, J. O'M. (ed.): *Electrochemistry of Cleaner Environments* (Plenum Press, 1972). The electrochemical treatment of aqueous effluent streams.
51. ASHMAN, R.: *Proc. Inst. Mech. Eng.* **1B** (1952) 157. Control and recovery of dust and fume in industry.
52. STAIRMAND, C. J.: in NONHEBEL, G. (ed.): *Gas Purification Processes for Air Pollution Control* 2nd edn. (Newnes-Butterworths, London, 1972)
53. SWIFT, P.: *Chem. Engr., London* No 403 (May 1984) 22. A user's guide to dust control.
54. SLACK, M. D., PRASAD, R. O., BAKKER, A. and BOYSAN, F.: *Trans. I. Chem. E.* **78** (Part A) (2000) 1098. Advances in cyclone modelling using unstructured grids.
55. TER LINDEN, A. J.: *Proc. Inst. Mech. Eng.* **160** (1949) 233. Cyclone dust collectors.
56. STAIRMAND, C. J.: *Engineering* **168** (1949) 409. Pressure drop in cyclone separators.
57. STAIRMAND, C. J.: *J. Inst. Fuel* **29** (1956) 58. The design and performance of modern gas cleaning equipment.
58. SZEKELY, J. and CARR, R.: *Chem. Eng. Sci.* **21** (1966) 1119. Heat transfer in a cyclone.
59. DORMAN, R. G. and MAGGS, F. A. P.: *Chem. Engr. London*. No 314 (Oct 1976) 671. Filtration of fine particles and vapours from gases.
60. HERSEY, H. J., Jr.: *Ind. Chem. Mfr.* **31** (1955) 138. Reverse-jet filters.
61. SCHNEIDER, G. G., HORZELLA, T. I., COOPER, J. and STRIEGL, P. J.: *Chem. Eng. Albany* **82** (May 26, 1975) 94. Selecting and specifying electrostatic precipitators.
62. ROSE, H. E. and WOOD, A. J.: *An Introduction to Electrostatic Precipitation in Theory and Practice*, 2nd edn. (Constable, 1966)
63. WHITE, S. T.: *Heat Vent.* **45** (Sept. 1948) 59. Inaudible sound: a new tool for air cleaning.

## 1.9. NOMENCLATURE

| | | Units in SI System | Dimensions in **M, L, T** |
|---|---|---|---|
| $A_i$ | Cross-sectional area of gas inlet to cyclone separator | $m^2$ | $\mathbf{L}^2$ |
| $a$ | Interfacial area per unit volume between constituents | $m^2/m^3$ | $\mathbf{L}^{-1}$ |
| $a_m$ | Maximum interfacial area per unit volume | $m^2/m^3$ | $\mathbf{L}^{-1}$ |
| $b$ | Degree of mixing | — | — |
| $C_f$ | Volumetric concentration of solids in feed | — | — |
| $C_o$ | Volumetric concentration of solids in overflow | — | — |
| $C_u$ | Volumetric concentration of solids in underflow | — | — |
| $c$ | Coefficient of $t$ (Equation 1.39) | $s^{-1}$ | $\mathbf{T}^{-1}$ |
| $d$ | Particle size, diameter of sphere or characteristic dimension of particle | m | **L** |
| $d_a$ | Particle size corresponding to $G(d) = E$ | m | **L** |
| $d_A, d_B$ | Equivalent spherical diameters of particles **A, B** | m | **L** |
| $d_{eff}$ | Effective diameter of orifice | m | **L** |

| | | Units in SI System | Dimensions in **M, L, T** |
|---|---|---|---|
| $d_{max}$ | Maximum size of particle | m | **L** |
| $d_l$ | Length mean diameter | m | **L** |
| $d_s$ | Surface mean diameter (Sauter mean) | m | **L** |
| $d_t$ | Tube diameter *or* diameter of cyclone separator | m | **L** |
| $d_v$ | Volume mean diameter | m | **L** |
| $d_0$ | Diameter of outlet of cyclone separator | m | **L** |
| $d_{80}$ | Size of opening through which 80 per cent of material passes | m | **L** |
| $d'_l$ | Mean length diameter | m | **L** |
| $d'_s$ | Mean surface diameter | m | **L** |
| $E$ | Separation efficiency | — | — |
| $F(d)$ | Size distribution of particles in feed | — | — |
| $F_o(d)$ | Size distribution of particles in overflow | — | — |
| $F_u(d)$ | Size distribution of particles in underflow | — | — |
| $G$ | Mass rate of flow of solids or of fluid | kg/s | $\mathbf{MT^{-1}}$ |
| $G(d)$ | Grade efficiency | — | — |
| $g$ | acceleration due to gravity | m/s$^2$ | $\mathbf{LT^{-2}}$ |
| $j$ | Exponent in equation 1.42 | — | — |
| $k$ | Constant in equation 1.2 | s$^{-1}$ | $\mathbf{T^{-1}}$ |
| $k_1$ | Constant depending on particle shape (volume/$d^3$) | — | — |
| $L$ | Length of solids plug or length of bowl | m | **L** |
| $L_c$ | Critical length of solids plug | m | **L** |
| $M$ | Mass feedrate of solids | kg/s | $\mathbf{ML^{-1}}$ |
| $m$ | Mass of particle | kg | **M** |
| $n$ | Number of particles per unit mass | kg$^{-1}$ | $\mathbf{M^{-1}}$ |
| $n_g$ | Number of particles in a group | — | — |
| $n_p$ | Number of particles in a sample | — | — |
| $p$ | Overall proportion of particles of one type | — | — |
| $Q_f$ | Volumetric feedrate of solids | m$^3$/s | $\mathbf{L^3T^{-1}}$ |
| $Q_o$ | Volumetric feedrate of solids in overflow | m$^3$/s | $\mathbf{L^3T^{-1}}$ |
| $Q_{fu}$ | Volumetric feedrate of solids in underflow | m$^3$/s | $\mathbf{L^3T^{-1}}$ |
| $R$ | Bowl radius | m | **L** |
| $R_f$ | Minimum value of $G(d)$ | — | — |
| $r$ | Radius at which particle rotates or radius in centrifuge bowl or hydrocyclone | m | **L** |
| $s$ | Standard deviation | — | — |
| $s_r$ | Standard deviation of random mixture | — | — |
| $s_o$ | Standard deviation of unmixed material | — | — |
| $t$ | Time | s | **T** |
| $t_R$ | Residence time | s | **T** |
| $u_r$ | Radial component of velocity of gas in separator | m/s | $\mathbf{LT^{-1}}$ |
| $u_t$ | Tangential component of gas in separator | m/s | $\mathbf{LT^{-1}}$ |
| $u_{t0}$ | Tangential component of gas velocity at circumference | m/s | $\mathbf{LT^{-1}}$ |
| $u_0$ | Free falling velocity of particle | m/s | $\mathbf{LT^{-1}}$ |
| $w$ | Mass of particles on screen at time $t$ | kg | **M** |
| $X$ | Percentage of samples in which particles are not mixed | — | — |
| $x$ | Proportion of particles smaller than size $d$ | — | — |
| $Y$ | Fraction of samples in which particles are mixed | — | — |
| $Z$ | Total height of separator | m | **L** |
| $\alpha$ | Angle of friction | — | — |
| $\beta$ | Angle between cone wall and horizontal *or* angle of entry into cyclone separator | — | — |
| $\mu$ | Viscosity of fluid | Ns/m$^2$ | $\mathbf{ML^{-1}T^{-1}}$ |
| $\rho$ | Density of fluid | kg/m$^3$ | $\mathbf{ML^{-3}}$ |
| $\rho_A, \rho_B$ | Density of material **A, B** | kg/m$^3$ | $\mathbf{ML^{-3}}$ |
| $\rho_s$ | Density of solid or particle | kg/m$^3$ | $\mathbf{ML^{-3}}$ |
| $\psi$ | Sphericity of particle | — | — |
| $\omega$ | Angular velocity | s$^{-1}$ | $\mathbf{T^{-1}}$ |

CHAPTER 2

# *Particle Size Reduction and Enlargement*

## 2.1. INTRODUCTION

Materials are rarely found in the size range required, and it is often necessary either to decrease or to increase the particle size. When, for example, the starting material is too coarse, and possibly in the form of large rocks, and the final product needs to be a fine powder, the particle size will have to be progressively reduced in stages. The most appropriate type of machine at each stage depends, not only on the size of the feed and of the product, but also on such properties as compressive strength, brittleness and stickiness. For example, the first stage in the process may require the use of a large jaw crusher and the final stage a sand grinder, two machines of very different characters.

At the other end of the spectrum, many very fine powders are frequently to difficult handle, and may also give rise to hazardous dust clouds when they are transported. It may therefore be necessary to increase the particle size. Examples of size enlargement processes include granulation for the preparation of fertilisers, and compaction using compressive forces to form the tablets required for the administration of pharmaceuticals.

In this Chapter, the two processes of size reduction and size enlargement are considered in Sections 2.2 and 2.4, respectively.

## 2.2. SIZE REDUCTION OF SOLIDS

### 2.2.1. Introduction

In the materials processing industry, size reduction or *comminution* is usually carried out in order to increase the surface area because, in most reactions involving solid particles, the rate of reactions is directly proportional to the area of contact with a second phase. Thus the rate of combustion of solid particles is proportional to the area presented to the gas, though a number of secondary factors may also be involved. For example, the free flow of gas may be impeded because of the higher resistance to flow of a bed of small particles. In leaching, not only is the rate of extraction increased by virtue of the increased area of contact between the solvent and the solid, but the distance the solvent has to penetrate into the particles in order to gain access to the more remote pockets of solute is also reduced. This factor is also important in the drying of porous solids, where reduction in size causes both an increase in area and a reduction in the distance the moisture must travel within the particles in order to reach the surface. In this case, the capillary forces acting on the moisture are also affected.

There are a number of other reasons for carrying out size reduction. It may, for example, be necessary to break a material into very small particles in order to separate two constituents, especially where one is dispersed in small isolated pockets. In addition, the properties of a material may be considerably influenced by the particle size and, for example, the chemical reactivity of fine particles is greater than that of coarse particles, and the colour and covering power of a pigment is considerably affected by the size of the particles. In addition, far more intimate mixing of solids can be achieved if the particle size is small.

## 2.2.2. Mechanism of size reduction

Whilst the mechanism of the process of size reduction is extremely complex, in recent years a number of attempts have been made at a more detailed analysis of the problem. If a single lump of material is subjected to a sudden impact, it will generally break so as to yield a few relatively large particles and a number of fine particles, with relatively few particles of intermediate size. If the energy in the blow is increased, the larger particles will be of a rather smaller size and more numerous and, whereas the number of fine particles will be appreciably increased, their size will not be much altered. It therefore appears that the size of the fine particles is closely connected with the internal structure of the material, and the size of the larger particles is more closely connected with the process by which the size reduction is effected.

This effect is well illustrated by a series of experiments on the grinding of coal in a small mill, carried out by HEYWOOD[1]. The results are shown in Figure 2.1, in which the distribution of particle size in the product is shown as a function of the number of

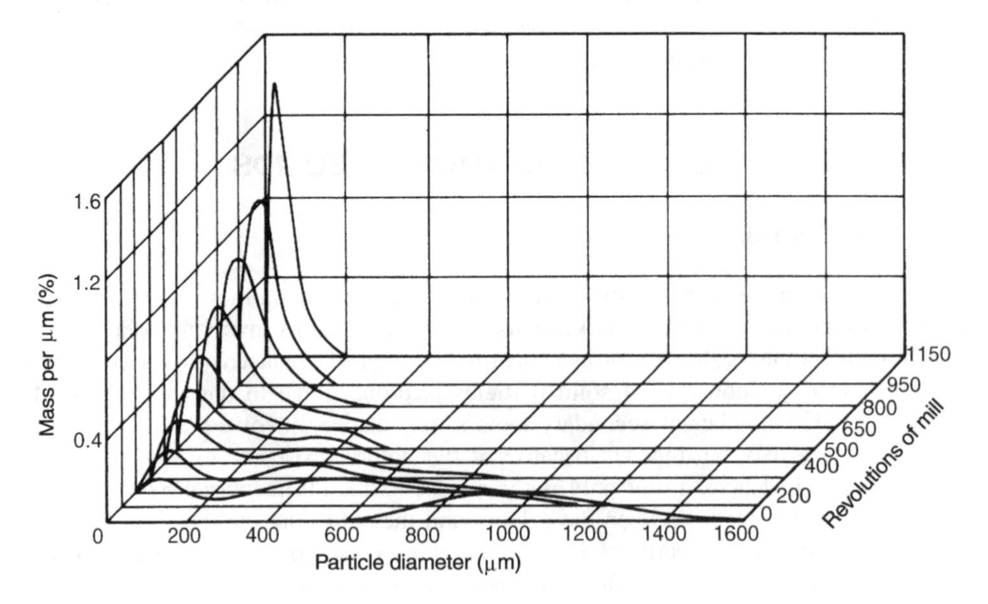


Figure 2.1. Effect of progressive grinding on size distribution

revolutions of the mill. The initial size distribution shows a single mode corresponding to a relatively coarse size, but as the degree of crushing is gradually increased this mode progressively decreases in magnitude and a second mode develops at a particular size. This process continues until the first mode has completely disappeared. Here the second mode is characteristic of the material and is known as the *persistent mode*, and the first is known as the *transitory mode*. There appears to be a *grind limit* for a particular material and machine. After some time there seems to be little change in particle size if grinding is continued, though the particles may show some irreversible plastic deformation which results in a change in shape rather than in size.

The energy required to effect size reduction is related to the internal structure of the material and the process consists of two parts, first opening up any small fissures which are already present, and secondly forming new surface. A material such as coal contains a number of small cracks and tends first to break along these, and therefore the large pieces are broken up more readily than the small ones. Since a very much greater increase in surface results from crushing a given quantity of fine as opposed to coarse material, fine grinding requires very much more power. Very fine grinding can be impeded by the tendency of some relatively soft materials, including gypsum and some limestones, to form aggregates. These are groups of relatively weakly adhering particles held together by cohesive and van der Waals forces. Materials, such as quartz and clinker, form agglomerates in which the forces causing adhesion may be chemical in nature, and the bonds are then very much stronger.

In considering energy utilisation, size reduction is a very inefficient process and only between 0.1 and 2.0 per cent of the energy supplied to the machine appears as increased surface energy in the solids. The efficiency of the process is very much influenced by the manner in which the load is applied and its magnitude. In addition the nature of the force exerted is also very important depending, for example, on whether it is predominantly a compressive, an impact or a shearing force. If the applied force is insufficient for the elastic limit to be exceeded, and the material is compressed, energy is stored in the particle. When the load is removed, the particle expands again to its original condition without doing useful work. The energy appears as heat and no size reduction is effected. A somewhat greater force will cause the particle to fracture, however, and in order to obtain the most effective utilisation of energy the force should be only slightly in excess of the crushing strength of the material. The surface of the particles will generally be of a very irregular nature so that the force is initially taken on the high spots, with the result that very high stresses and temperatures may be set up locally in the material. As soon as a small amount of breakdown of material takes place, the point of application of the force alters. BEMROSE and BRIDGEWATER[2] and HESS and SCHÖNERT[3] have studied the breakage of single particles. All large lumps of material contain cracks and size reduction occurs as a result of crack propagation that occurs above a critical parameter, $F$, where:

$$F = \frac{\tau^2 a}{Y} \tag{2.1}$$

where: $a$ = crack length,
$\tau$ = stress, and
$Y$ = Young's modulus.

HESS[3] suggests that at lower values of $F$, elastic deformation occurs without fracture and the energy input is completely ineffective in achieving size reduction. Fundamental studies of the application of fracture mechanics to particle size reduction have been carried out, by SCHÖNERT[4]. In essence, an energy balance is applied to the process of crack extension within a particle by equating the loss of energy from the strain field within the particle to the increase in surface energy when the crack propagates. Because of plastic deformation near the tip of the crack, however, the energy requirement is at least ten times greater and, in addition kinetic energy is associated with the sudden acceleration of material as the crack passes through it. Orders of magnitude of the surface fracture energy per unit volume are:

| | |
|---|---|
| glass | $1-10$ J/m$^2$ |
| plastics | $10-10^3$ J/m$^2$ |
| metals | $10^3-10^5$ J/m$^2$ |

All of these values are several orders of magnitude higher than the thermodynamic surface energy which is about $10^{-1}$ J/m$^2$. Where a crack is initially present in a material, the stresses near the tip of the crack are considerably greater than those in the bulk of the material. Calculation of the actual value is well nigh impossible as the crack surfaces are usually steeply curved and rough. The presence of a crack modifies the stress field in its immediate location, with the increase in energy being approximately proportional to $\sqrt{(a/l)}$ where $a$ is the crack length and $l$ is the distance from the crack tip. Changes in crack length are accompanied by modifications in stress distribution in the surrounding material and hence in its energy content.

During the course of the size reduction processes, much energy is expended in causing plastic deformation and this energy may be regarded as a waste as it does not result in fracture. Only part of it is retained in the system as a result of elastic recovery. It is not possible, however, to achieve the stress levels necessary for fracture to occur without first passing through the condition of plastic deformation and, in this sense, this must be regarded as a necessary state which must be achieved before fracture can possibly occur.

The nature of the flaws in the particles changes with their size. If, as is customary, fine particles are produced by crushing large particles, the weakest flaws will be progressively eliminated as the size is reduced, and thus small particles tend to be stronger and to require more energy for fracture to occur. In addition, as the capacity of the particle for storing energy is proportional to its volume ($\propto d^3$) and the energy requirement for propagating geometrically similar cracks is proportional to the surface area ($\propto d^2$), the energy available per unit crack area increases linearly with particle size ($d$). Thus, breakage will occur at lower levels of stress in large particles. This is illustrated in Figure 2.2 which shows the results of experimental measurements of the compressive strengths for shearing two types of glass. It may be noted from Figure 2.2 that, for quartz glass, the compressive strength of 2 $\mu$m particles is about three times that of 100 $\mu$m particles.

The exact method by which fracture occurs is not known, although it is suggested by PIRET[5] that the compressive force produces small flaws in the material. If the energy concentration exceeds a certain critical value, these flaws will grow rapidly and will generally branch, and the particles will break up. The probability of fracture of a particle

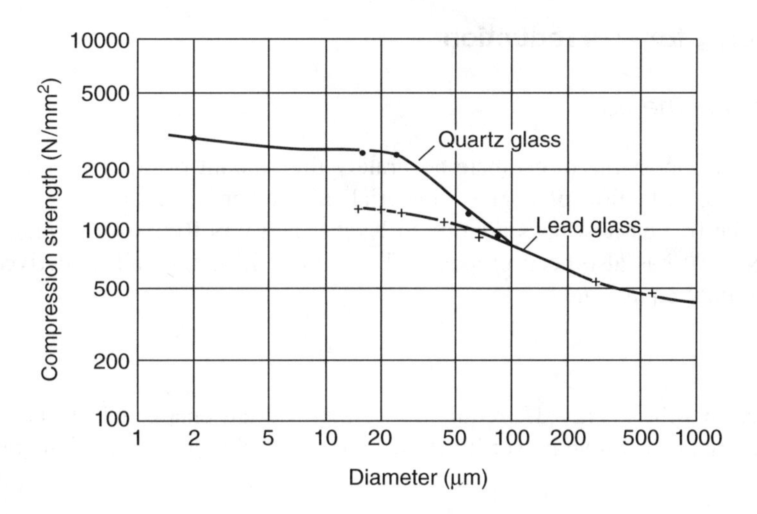


Figure 2.2. Compressive strength of glass spheres as a function of their diameter

in an assembly of particles increases with the number of contact points, up to a number of around ten, although the probability then decreases for further increase in number. The rate of application of the force is important because there is generally a time lag between attainment of maximum load and fracture. Thus, a rather smaller force will cause fracture provided it is maintained for a sufficient time. This is a phenomenon similar to the ignition lag which is obtained with a combustible gas–oxidant mixture. Here the interval between introducing the ignition source and the occurrence of ignition is a function of the temperature of the source, and when it is near the minimum ignition temperature delays of several seconds may be encountered. The greater the rate at which the load is applied, the less effectively is the energy utilised and the higher is the proportion of fine material which is produced. If the particle shows any viscoelastic behaviour, a high rate of application of the force is needed for fracture to occur. The efficiency of utilisation of energy as supplied by a falling mass has been compared with that of energy applied slowly by means of hydraulic pressure. Up to three or four times more surface can be produced per unit of energy if it is applied by the latter method. PIRET(5) suggests that there is a close similarity between the crushing operation and a chemical reaction. In both cases a critical energy level must be exceeded before the process will start, and in both cases time is an important variable.

The method of application of the force to the particles may affect the breakage pattern. PRASHER(6) suggests that four basic patterns may be identified, though it is sometimes difficult to identify the dominant mode in any given machine. The four basic patterns are:

(a) *Impact* — particle concussion by a single rigid force.
(b) *Compression* — particle disintegration by two rigid forces.
(c) *Shear* — produced by a fluid or by particle–particle interaction.
(d) *Attrition* — arising from particles scraping against one another or against a rigid surface.

### 2.2.3. Energy for size reduction

#### *Energy requirements*

Although it is impossible to estimate accurately the amount of energy required in order to effect a size reduction of a given material, a number of empirical laws have been proposed. The two earliest laws are due to KICK[7] and VON RITTINGER[8], and a third law due to BOND[9,10] has also been proposed. These three laws may all be derived from the basic differential equation:

$$\frac{dE}{dL} = -CL^p \tag{2.2}$$

which states that the energy $dE$ required to effect a small change $dL$ in the size of unit mass of material is a simple power function of the size. If $p = -2$, then integration gives:

$$E = C\left(\frac{1}{L_2} - \frac{1}{L_1}\right)$$

Writing $C = K_R f_c$, where $f_c$ is the crushing strength of the material, then *Rittinger's law*, first postulated in 1867, is obtained as:

$$E = K_R f_c \left(\frac{1}{L_2} - \frac{1}{L_1}\right) \tag{2.3}$$

Since the surface of unit mass of material is proportional to $1/L$, the interpretation of this law is that the energy required for size reduction is directly proportional to the increase in surface.

If $p = -1$, then:

$$E = C \ln \frac{L_1}{L_2}$$

and, writing $C = K_K f_c$:

$$E = K_K f_c \ln \frac{L_1}{L_2} \tag{2.4}$$

which is known as *Kick's law*. This supposes that the energy required is directly related to the reduction ratio $L_1/L_2$ which means that the energy required to crush a given amount of material from a 50 mm to a 25 mm size is the same as that required to reduce the size from 12 mm to 6 mm. In equations 2.3 and 2.4, $K_R$ and $K_K$ are known respectively as Rittinger's constant and Kick's constant. It may be noted that neither of these constants is dimensionless.

Neither of these two laws permits an accurate calculation of the energy requirements. Rittinger's law is applicable mainly to that part of the process where new surface is being created and holds most accurately for fine grinding where the increase in surface per unit mass of material is large. Kick's law, more closely relates to the energy required to effect elastic deformation before fracture occurs, and is more accurate than Rittinger's law for coarse crushing where the amount of surface produced is considerably less.

Bond has suggested a law intermediate between Rittinger's and Kick's laws, by putting $p = -3/2$ in equation 2.1. Thus:

$$E = 2C\left(\frac{1}{L_2^{1/2}} - \frac{1}{L_1^{1/2}}\right)$$

$$= 2C\sqrt{\left(\frac{1}{L_2}\right)}\left(1 - \frac{1}{q^{1/2}}\right) \tag{2.5}$$

where: $$q = \frac{L_1}{L_2}$$

the reduction ratio. Writing $C = 5E_i$, then:

$$E = E_i\sqrt{\left(\frac{100}{L_2}\right)}\left(1 - \frac{1}{q^{1/2}}\right) \tag{2.6}$$

Bond terms $E_i$ the *work index*, and expresses it as the amount of energy required to reduce unit mass of material from an infinite particle size to a size $L_2$ of 100 μm, that is $q = \infty$. The size of material is taken as the size of the square hole through which 80 per cent of the material will pass. Expressions for the work index are given in the original papers[8,9] for various types of materials and various forms of size reduction equipment.

Austin and Klimpel[11] have reviewed these three laws and their applicability, and Cutting[12] has described laboratory work to assess grindability using rod mill tests.

## Example 2.1

A material is crushed in a Blake jaw crusher such that the average size of particle is reduced from 50 mm to 10 mm with the consumption of energy of 13.0 kW/(kg/s). What would be the consumption of energy needed to crush the same material of average size 75 mm to an average size of 25 mm:

a) assuming Rittinger's law applies?
b) assuming Kick's law applies?

Which of these results would be regarded as being more reliable and why?

## Solution

a) *Rittinger's law.*

This is given by: $E = K_R f_c[(1/L_2) - (1/L_1)]$ (equation 2.3)

Thus: $13.0\ K_R f_c[(1/10) - (1/50)]$

and: $K_R f_c = (13.0 \times 50/4) = 162.5$ kW/(kg mm)

Thus the energy required to crush 75 mm material to 25 mm is:

$$E = 162.5[(1/25) - (1/75)] = \underline{\underline{4.33 \text{ kJ/kg}}}$$

b) *Kick's law.*

This is given by: $E = K_K f_c \ln(L_1/L_2)$ (equation 2.4)

Thus: $13.0 = K_K f_c \ln(50/10)$

and: $K_K f_c = (13.0/1.609) = 8/08$ kW/(kg/s)

Thus the energy required to crush 75 mm material to 25 mm is given by:

$$E = 8.08 \ln(75/25) = \underline{\underline{8.88 \text{ kJ/kg}}}$$

The size range involved by be considered as that for coarse crushing and, because Kick's law more closely relates the energy required to effect elastic deformation before fracture occurs, this would be taken as given the more reliable result.

## *Energy utilisation*

One of the first important investigations into the distribution of the energy fed into a crusher was carried out by OWENS[13] who concluded that energy was utilised as follows:

(a) In producing elastic deformation of the particles before fracture occurs.
(b) In producing inelastic deformation which results in size reduction.
(c) In causing elastic distortion of the equipment.
(d) In friction between particles, and between particles and the machine.
(e) In noise, heat and vibration in the plant, and
(f) In friction losses in the plant itself.

Owens estimated that only about 10 per cent of the total power is usefully employed.

In an investigation by the U.S. BUREAU OF MINES[14], in which a drop weight type of crusher was used, it was found that the increase in surface was directly proportional to the input of energy and that the rate of application of the load was an important factor.

This conclusion was substantiated in a more recent investigation of the power consumption in a size reduction process which is reported in three papers by KWONG *et al.*[15], ADAMS *et al.*[16] and JOHNSON *et al.*[17]. A sample of material was crushed by placing it in a cavity in a steel mortar, placing a steel plunger over the sample and dropping a steel ball of known weight on the plunger over the sample from a measured height. Any bouncing of the ball was prevented by three soft aluminium cushion wires under the mortar, and these wires were calibrated so that the energy absorbed by the system could be determined from their deformation. Losses in the plunger and ball were assumed to be proportional to the energy absorbed by the wires, and the energy actually used for size reduction was then obtained as the difference between the energy of the ball on striking the plunger and the energy absorbed. Surfaces were measured by a water or air permeability method or by gas adsorption. The latter method gave a value approximately double that obtained from the former indicating that, in these experiments, the internal surface was approximately the same as the external surface. The experimental results showed that, provided the new surface did not exceed about 40 $m^2/kg$, the new surface produced was directly proportional to the energy input. For a given energy input the new surface produced was independent of:

(a) The velocity of impact,
(b) The mass and arrangement of the sample,
(c) The initial particle size, and
(d) The moisture content of the sample.

Between 30 and 50 per cent of the energy of the ball on impact was absorbed by the material, although no indication was obtained of how this was utilised. An extension of the range of the experiments, in which up to 120 $m^2$ of new surface was produced per kilogram of material, showed that the linear relationship between energy and new surface no longer held rigidly. In further tests in which the crushing was effected slowly, using a hydraulic press, it was found, however, that the linear relationship still held for the larger increases in surface.

In order to determine the efficiency of the surface production process, tests were carried out with sodium chloride and it was found that 90 J was required to produce 1 $m^2$ of new surface. As the theoretical value of the surface energy of sodium chloride is only 0.08 $J/m^2$, the efficiency of the process is about 0.1 per cent. Zeleny and Piret[18] have reported calorimetric studies on the crushing of glass and quartz. It was found that a fairly constant energy was required of 77 $J/m^2$ of new surface created, compared with a surface-energy value of less than 5 $J/m^2$. In some cases over 50 per cent of the energy supplied was used to produce plastic deformation of the steel crusher surfaces.

The apparent efficiency of the size reduction operation depends on the type of equipment used. Thus, for instance, a ball mill is rather less efficient than a drop weight type of crusher because of the ineffective collisions that take place in the ball mill.

Further work[5] on the crushing of quartz showed that more surface was created per unit of energy with single particles than with a collection of particles. This appears to be attributable to the fact that the crushing strength of apparently identical particles may vary by a factor as large as 20, and it is necessary to provide a sufficient energy concentration to crush the strongest particle. Some recent developments, including research and mathematical modelling, are described by Prasher[6].

### 2.2.4. Methods of operating crushers

There are two distinct methods of feeding material to a crusher. The first, known as *free crushing*, involves feeding the material at a comparatively low rate so that the product can readily escape. Its residence time in the machine is therefore short and the production of appreciable quantities of undersize material is avoided. The second method is known as *choke feeding*. In this case, the machine is kept full of material and discharge of the product is impeded so that the material remains in the crusher for a longer period. This results in a higher degree of crushing, although the capacity of the machine is reduced and energy consumption is high because of the cushioning action produced by the accumulated product. This method is therefore used only when a comparatively small amount of materials is to be crushed and when it is desired to complete the whole of the size reduction in one operation.

If the plant is operated, as in *choke feeding*, so that the material is passed only once through the equipment, the process is known as *open circuit grinding*. If, on the other hand, the product contains material which is insufficiently crushed, it may be necessary to separate the product and return the oversize material for a second crushing. This system which is generally to be preferred, is known as *closed circuit grinding*. A flow-sheet for a typical closed circuit grinding process, in which a coarse crusher, an intermediate crusher and a fine grinder are used, is shown in Figure 2.3. In many plants, the product is continuously removed, either by allowing the material to fall on to a screen or by subjecting it to the action of a stream of fluid, such that the small particles are carried away and the oversize material falls back to be crushed again.

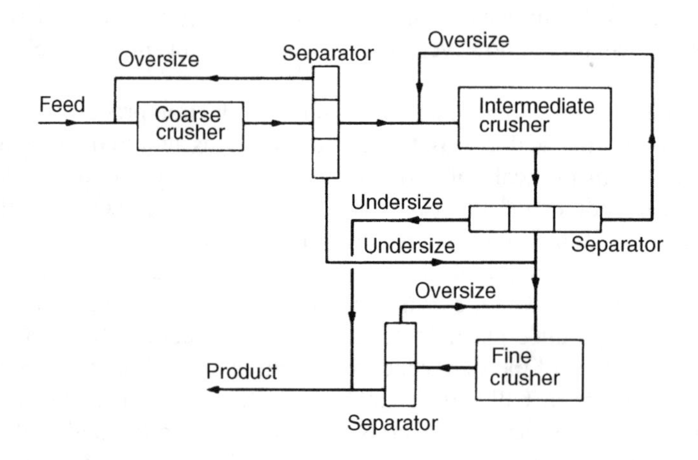


Figure 2.3. Flow diagram for closed circuit grinding system

The desirability of using a number of size reduction units when the particle size is to be considerably reduced arises from the fact that it is not generally economical to effect a large reduction ratio in a single machine. The equipment used is usually divided into classes as given in Table 2.1, according to the size of the feed and the product.

A greater size reduction ratio can be obtained in fine crushers than in coarse crushers.

Table 2.1. Classification of size reduction equipment

| | Feed size | Product size |
|---|---|---|
| Coarse crushers | 1500–40 mm | 50–5 mm |
| Intermediate crushers | 50–5 mm | 5–0.1 mm |
| Fine crushers | 5–2 mm | 0.1 mm |
| Colloid mills | 0.2 mm | down to 0.01 μm |

The equipment may also be classified, to some extent, according to the nature of the force which is applied though, as a number of forces are generally involved, it is a less convenient basis.

Grinding may be carried out either wet or dry, although wet grinding is generally applicable only with low speed mills. The advantages of wet grinding are:

(a) The power consumption is reduced by about 20–30 per cent.
(b) The capacity of the plant is increased.
(c) The removal of the product is facilitated and the amount of fines is reduced.
(d) Dust formation is eliminated.
(e) The solids are more easily handled.

Against this, the wear on the grinding medium is generally about 20 per cent greater, and it may be necessary to dry the product.

The separators in Figure 2.3 may be either a cyclone type, as typified by the Bradley microsizer or a mechanical air separator. Cyclone separators, the theory of operation and application of which are fully discussed in Chapter 1, may be used. Alternatively, a *whizzer* type of air separator such as the NEI air separator shown in Figures 1.29 and 1.30 is often included as an integral part of the mill, as shown in the examples of the NEI pendulum mill in Figure 2.21. Oversize particles drop down the inner case and are returned directly to the mill, whilst the fine material is removed as a separate product stream.

### 2.2.5. Nature of the material to be crushed

The choice of a machine for a given crushing operation is influenced by the nature of the product required and the quantity and size of material to be handled. The more important properties of the feed apart from its size are as follows:

*Hardness*. The hardness of the material affects the power consumption and the wear on the machine. With hard and abrasive materials it is necessary to use a low-speed machine and to protect the bearings from the abrasive dusts that are produced. Pressure lubrication is recommended. Materials are arranged in order of increasing hardness in the *Mohr* scale in which the first four items rank as soft and the remainder as hard. The Mohr Scale of Hardness is:

1. Talc
2. Rock salt or gypsum
3. Calcite
4. Fluorspar
5. Apatite
6. Felspar
7. Quartz
8. Topaz
9. Carborundum
10. Diamond.

*Structure*. Normal granular materials such as coal, ores and rocks can be effectively crushed employing the normal forces of compression, impact, and so on. With fibrous materials a tearing action is required.

*Moisture content*. It is found that materials do not flow well if they contain between about 5 and 50 per cent of moisture. Under these conditions the material tends to cake together in the form of balls. In general, grinding can be carried out satisfactorily outside these limits.

*Crushing strength*. The power required for crushing is almost directly proportional to the crushing strength of the material.

*Friability*. The friability of the material is its tendency to fracture during normal handling. In general, a crystalline material will break along well-defined planes and the power required for crushing will increase as the particle size is reduced.

*Stickiness*. A sticky material will tend to clog the grinding equipment and it should therefore be ground in a plant that can be cleaned easily.

*Soapiness*. In general, this is a measure of the coefficient of friction of the surface of the material. If the coefficient of friction is low, the crushing may be more difficult.

*Explosive materials* must be ground wet or in the presence of an inert atmosphere.

*Materials yielding dusts that are harmful to the health* must be ground under conditions where the dust is not allowed to escape.

WORK[19] has presented a guide to equipment selection based on size and abrasiveness of material.

## 2.3. TYPES OF CRUSHING EQUIPMENT

The most important coarse, intermediate and fine crushers may be classified as in Table 2.2.

Table 2.2. Crushing equipment

| Coarse crushers | Intermediate crushers | Fine crushers |
|---|---|---|
| Stag jaw crusher | Crushing rolls | Buhrstone mill |
| Dodge jaw crusher | Disc crusher | Roller mill |
| Gyratory crusher | Edge runner mill | NEI pendulum mill |
| Other coarse crushers | Hammer mill | Griffin mill |
| | Single roll crusher | Ring roller mill (Lopulco) |
| | Pin mill | Ball mill |
| | Symons disc crusher | Tube mill |
| | | Hardinge mill |
| | | Babcock mill |

The features of these crushers are now considered in detail.

### 2.3.1. Coarse crushers

The Stag jaw crusher shown in Figure 2.4, has a fixed jaw and a moving jaw pivoted at the top with the crushing faces formed of manganese steel. Since the maximum movement of the jaw is at the bottom, there is little tendency for the machine to clog, though some uncrushed material may fall through and have to be returned to the crusher. The maximum pressure is exerted on the large material which is introduced at the top. The machine is usually protected so that it is not damaged if lumps of metal inadvertently enter, by making one of the toggle plates in the driving mechanism relatively weak so that, if any large stresses are set up, this is the first part to fail. Easy renewal of the damaged part is then possible.

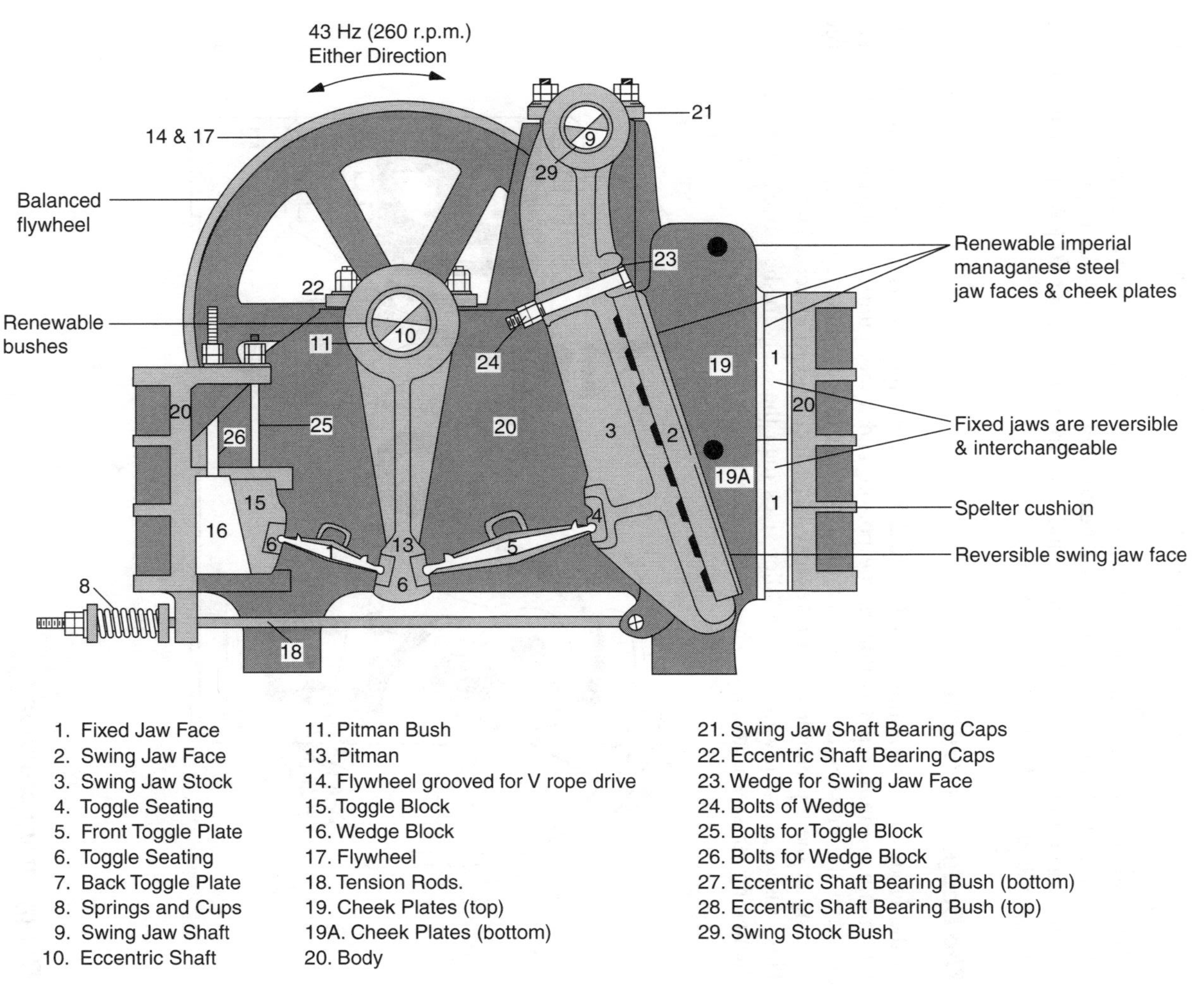
43 Hz (260 r.p.m.)
Either Direction
14 & 17
21
29
9
Balanced flywheel
22
Renewable bushes
11
10
24
23
Renewable imperial managanese steel jaw faces & cheek plates
19
1
20
Fixed jaws are reversible & interchangeable
25
26
20
3
2
19A
15
16
6
7
13
5
4
6
1
Spelter cushion
Reversible swing jaw face
8
18
1. Fixed Jaw Face
2. Swing Jaw Face
3. Swing Jaw Stock
4. Toggle Seating
5. Front Toggle Plate
6. Toggle Seating
7. Back Toggle Plate
8. Springs and Cups
9. Swing Jaw Shaft
10. Eccentric Shaft
11. Pitman Bush
13. Pitman
14. Flywheel grooved for V rope drive
15. Toggle Block
16. Wedge Block
17. Flywheel
18. Tension Rods.
19. Cheek Plates (top)
19A. Cheek Plates (bottom)
20. Body
21. Swing Jaw Shaft Bearing Caps
22. Eccentric Shaft Bearing Caps
23. Wedge for Swing Jaw Face
24. Bolts of Wedge
25. Bolts for Toggle Block
26. Bolts for Wedge Block
27. Eccentric Shaft Bearing Bush (bottom)
28. Eccentric Shaft Bearing Bush (top)
29. Swing Stock Bush

Figure 2.4. Typical cross-section of Stag jaw crusher

Stag crushers are made with jaw widths varying from about 150 mm to 1.0 m and the running speed is about 4 Hz (240 rpm) with the smaller machines running at the higher speeds. The speed of operation should not be so high that a large quantity of fines is produced as a result of material being repeatedly crushed because it cannot escape sufficiently quickly. The angle of nip, the angle between the jaws, is usually about 30°.

Because the crushing action is intermittent, the loading on the machine is uneven and the crusher therefore incorporates a heavy flywheel. The power requirements of the crusher depend upon size and capacity and vary from 7 to about 70 kW, the latter figure corresponding to a feed rate of 10 kg/s.

### The Dodge jaw crusher

In the Dodge crusher, shown in Figure 2.5, the moving jaw is pivoted at the bottom. The minimum movement is thus at the bottom and a more uniform product is obtained, although the crusher is less widely used because of its tendency to choke. The large opening at the top enables it to take very large feed and to effect a large size reduction. This crusher is usually made in smaller sizes than the Stag crusher, because of the high fluctuating stresses that are produced in the members of the machine.

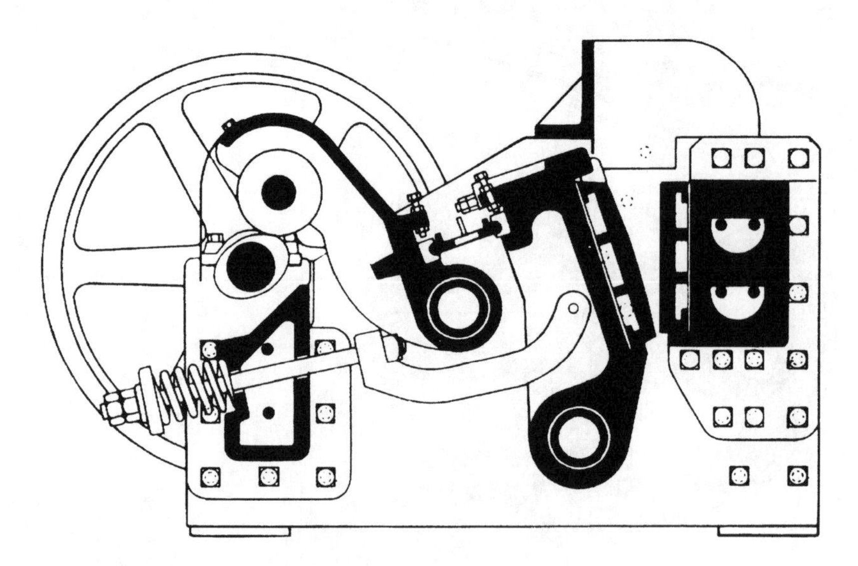

Figure 2.5. Dodge crusher

### The gyratory crusher

The gyratory crusher shown in Figure 2.6 employs a crushing head, in the form of a truncated cone, mounted on a shaft, the upper end of which is held in a flexible bearing, whilst the lower end is driven eccentrically so as to describe a circle. The crushing action takes place round the whole of the cone and, since the maximum movement is at the

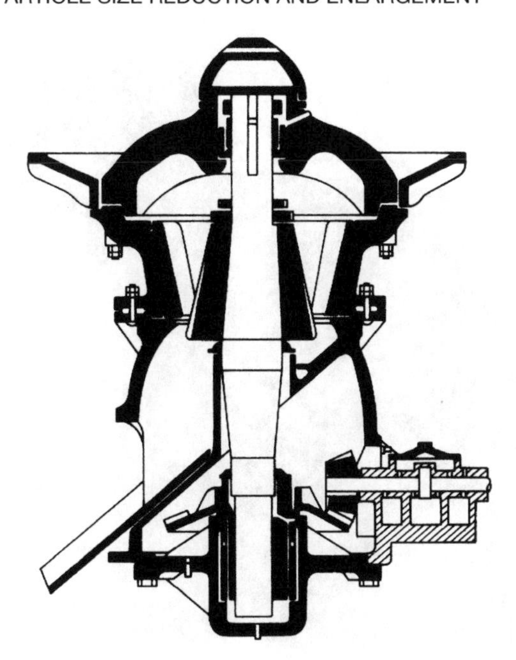

Figure 2.6. Gyratory crusher

bottom, the characteristics of the machine are similar to those of the Stag crusher. As the crusher is continuous in action, the fluctuations in the stresses are smaller than in jaw crushers and the power consumption is lower. This unit has a large capacity per unit area of grinding surface, particularly if it is used to produce a small size reduction. It does not, however, take such a large size of feed as a jaw crusher, although it gives a rather finer and more uniform product. Because the capital cost is high, the crusher is suitable only where large quantities of material are to be handled.

The jaw crushers and the gyratory crusher all employ a predominantly compressive force.

### *Other coarse crushers*

Friable materials, such as coal, may be broken up without the application of large forces, and therefore less robust plant may be used. A common form of coal breaker consists of a large hollow cylinder with perforated walls. The axis is at a small angle to the horizontal and the feed is introduced at the top. The cylinder is rotated and the coal is lifted by means of arms attached to the inner surface and then falls against the cylindrical surface. The coal breaks by impact and passes through the perforations as soon as the size has been sufficiently reduced. This type of equipment is less expensive and has a higher throughput than the jaw or gyratory crusher. Another coarse rotary breaker, the rotary coal breaker, is similar in action to the hammer mill described later, and is shown in Figure 2.7. The crushing action depends upon the transference of kinetic energy from hammers to the material and these pulverisers are essentially high speed machines with a speed of rotation of about 10 Hz (600 rpm) giving hammer tip velocities of about 40 m/s.

Figure 2.7. Rotary coal breaker

### 2.3.2. Intermediate crushers

#### *The edge runner mill*

In the edge runner mill shown in Figure 2.8 a heavy cast iron or granite wheel, or *muller* as it is called, is mounted on a horizontal shaft which is rotated in a horizontal plane in

Figure 2.8. Edge runner mill

a heavy pan. Alternatively, the muller remains stationary and the pan is rotated, and in some cases the mill incorporates two mullers. Material is fed to the centre of the pan and is worked outwards by the action of the muller, whilst a scraper continuously removes material that has adhered to the sides of the pan, and returns it to the crushing zone. In many models the outer rim of the bottom of the pan is perforated, so that the product may be removed continuously as soon as its size has been sufficiently reduced. The mill may be operated wet or dry and it is used extensively for the grinding of paints, clays and sticky materials.

### *The hammer mill*

The hammer mill is an impact mill employing a high speed rotating disc, to which are fixed a number of hammer bars which are swung outwards by centrifugal force. An industrial model is illustrated in Figure 2.9 and a laboratory model in Figure 2.10. Material is fed in, either at the top or at the centre, and it is thrown out centrifugally and crushed by being beaten between the hammer bars, or against breaker plates fixed around the periphery of the cylindrical casing. The material is beaten until it is small enough to fall through the screen which forms the lower portion of the casing. Since the hammer bars are hinged, the presence of any hard material does not cause damage to the equipment. The bars are readily replaced when they are worn out. The machine is suitable for the crushing of both brittle and fibrous materials, and, in the latter case, it is usual to employ a screen

Figure 2.9. Swing claw hammer mill

Figure 2.10. The Raymond laboratory hammer mill

with cutting edges. The hammer mill is suitable for hard materials although, since a large amount of fines is produced, it is advisable to employ positive pressure lubrication to the bearings in order to prevent the entry of dust. The size of the product is regulated by the size of the screen and the speed of rotation.

A number of similar machines are available, and in some the hammer bars are rigidly fixed in position. Since a large current of air is produced, the dust must be separated in a cyclone separator or a bag filter.

### *The pin-type mill*

The Alpine pin disc mill shown in Figure 2.11 is a form of pin mill and consists of two vertical steel plates with horizontal projections on their near faces. One disc may be stationary whilst the other disc is rotated at high speed; sometimes, the two discs may be rotated in opposite directions. The material is gravity fed in through a hopper or air conveyed to the centre of the discs, and is thrown outwards by centrifugal action and broken against of the projections before it is discharged to the outer body of the mill and

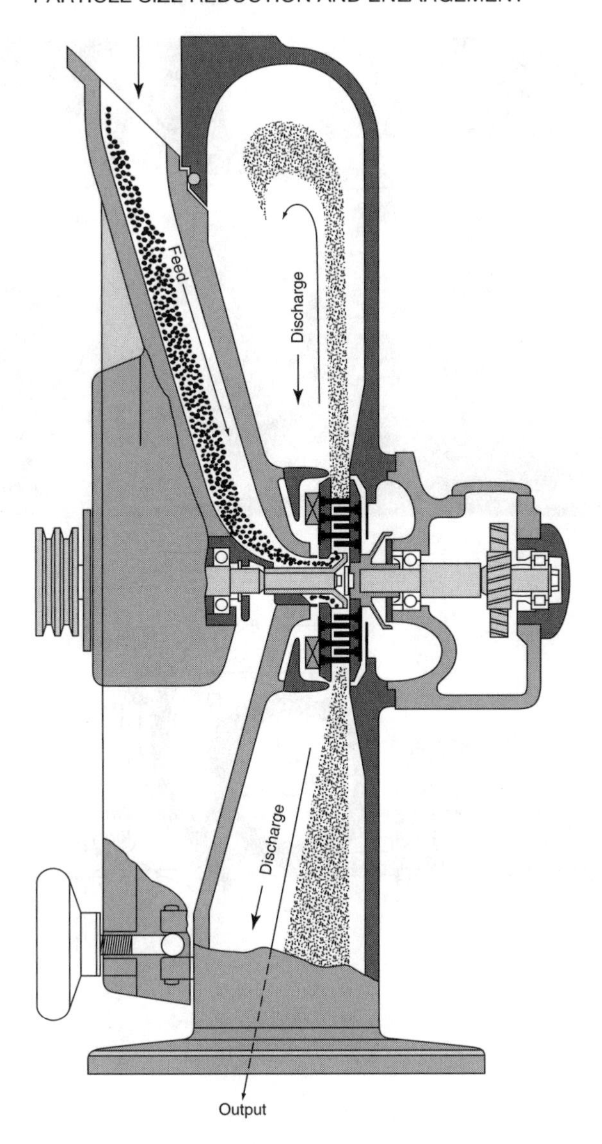


Figure 2.11. Alpine pin mill with both discs and sets of pins rotating

falls under gravity from the bottom of the casing. Alternatively, the pins may be replaced by swing beaters or plate beaters, depending on the setup and application. The mill gives a fairly uniform fine product with little dust and is extensively used with chemicals, fertilisers and other materials that are non-abrasive, brittle or crystalline. Control of the size of the product is effected by means of the speed and the spacing of the projections and a product size of 20 μm is readily attainable.

The Alpine universal mill with turbine beater and grinding track shown in Figure 2.12 is suitable for both brittle and tough materials. The high airflow from the turbine keeps the temperature rise to a minimum.

Figure 2.12. Alpine Universal mill

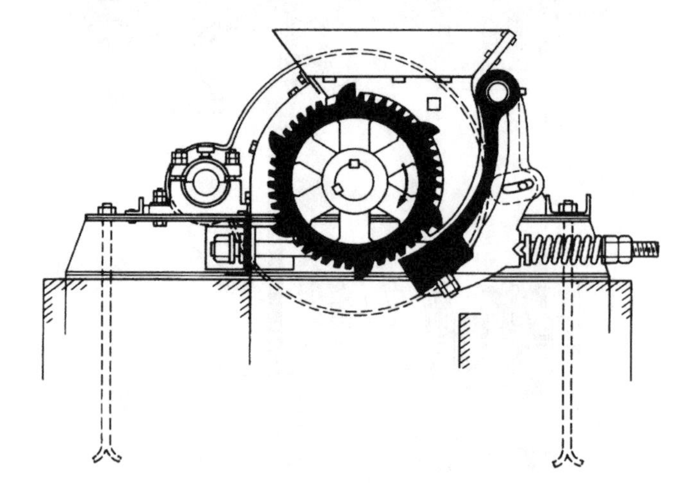

Figure 2.13. Single roll crusher

### *The single roll crusher*

The single roll crusher shown in Figure 2.13 consists of a toothed crushing roll which rotates close to a breaker plate. Figure 2.14 is an illustration of an industrial model. The material is crushed by compression and shearing between the two surfaces. It is used extensively for crushing coal.

### *Crushing rolls*

Two rolls, one in adjustable bearings, rotate in opposite directions as shown in Figure 2.15 and the clearance between them can be adjusted according to the size of feed and the

Figure 2.14. The Stag single roll crusher

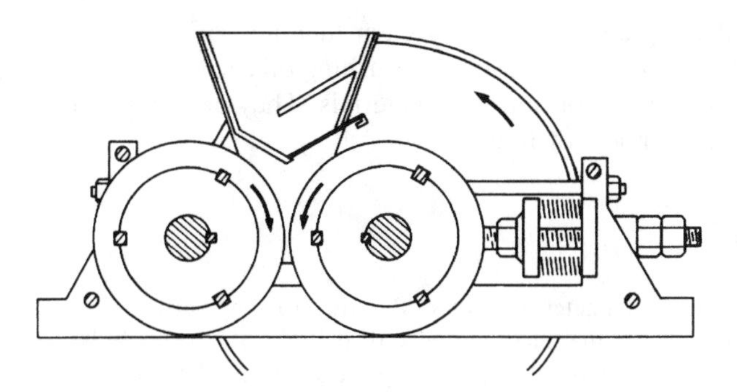

Figure 2.15. Crushing rolls

required size of product. The machine is protected, by spring loading, against damage from very hard material. Both rolls may be driven, or one directly and the other by friction with the solids. The crushing rolls, which may vary from a few centimetres up to about 1.2 m in diameter, are suitable for effecting a small size reduction ratio, 4 : 1 in a single operation, and it is therefore common to employ a number of pairs of rolls in series, one above the other. Roll shells with either smooth or ridged surfaces are held in place by keys to the main shaft. The capacity is usually between one-tenth and one-third of that calculated on the assumption that a continuous ribbon of the material forms between the rolls.

An idealised system where a spherical or cylindrical particle of radius $r_2$ is being fed to crushing rolls of radius $r_1$ is shown in Figure 2.16. $2\alpha$ is the angle of nip, the angle between the two common tangents to the particle and each of the rolls, and $2b$ is the distance between the rolls. It may be seen from the geometry of the system that the angle

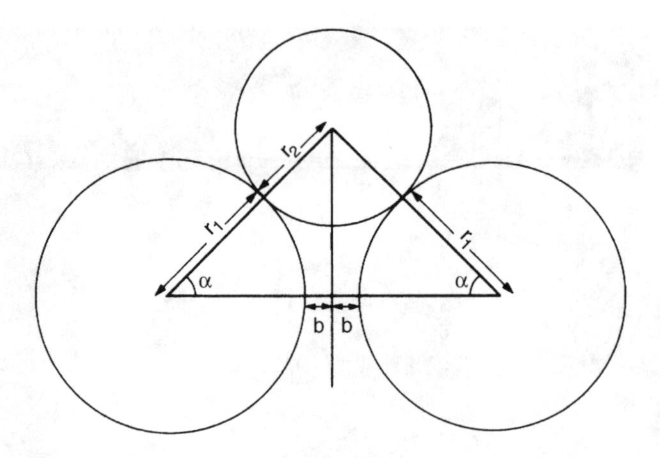


Figure 2.16. Particle fed to crushing rolls

of nip is given by:

$$\cos\alpha = \frac{(r_1 + b)}{(r_2 + r_1)} \tag{2.7}$$

For steel rolls, the angle of nip is not greater than about 32°.

Crushing rolls are extensively used for crushing oil seeds and in the gunpowder industry and they are also suitable for abrasive materials. They are simple in construction and do not give a large percentage of fines.

## Example 2.2

If crushing rolls, 1 m in diameter, are set so that the crushing surfaces are 12.5 mm apart and the angle of nip is 31°, what is the maximum size of particle which should be fed to the rolls?

If the actual capacity of the machine is 12 per cent of the theoretical, calculate the throughput in kg/s when running at 2.0 Hz if the working face of the rolls is 0.4 m long and the bulk density of the feed is 2500kg/m$^3$.

## Solution

The particle size may be obtained from:

$$\cos\alpha = (r_1 + b)/(r_1 + r_2) \qquad \text{(equation 2.7)}$$

In this case: $2\alpha = 31°$ and $\cos\alpha = 0.964$, $b = (12.5/2) = 6.25$ mm or 0.00625 m and:

$$r_1 = (1.0/2) = 0.5 \text{ m}$$

Thus:

$$0.964 = (0.5 + 0.00625)/(0.5 + r_2)$$

and:

$$r_2 = 0.025 \text{ m or } \underline{\underline{25 \text{ mm}}}$$

The cross sectional area for flow $= (0.0125 \times 0.4) = 0.005$ m$^2$

and the volumetric flowrate = $(2.0 \times 0.005) = 0.010\ m^3/s$.

Thus, the actual throughput = $(0.010 \times 12)/100 = 0.0012\ m^3/s$

or: $(0.0012 \times 2500) = \underline{\underline{3.0\ kg/s}}$

### *The Symons disc crusher*

The disc crusher shown in Figure 2.17 employs two saucer-shaped discs mounted on horizontal shafts, one of which is rotated and the other is mounted in an eccentric bearing so that the two crushing faces continuously approach and recede. The material is fed into the centre between the two discs, and the product is discharged by centrifugal action as soon as it is fine enough to escape through the opening between the faces.

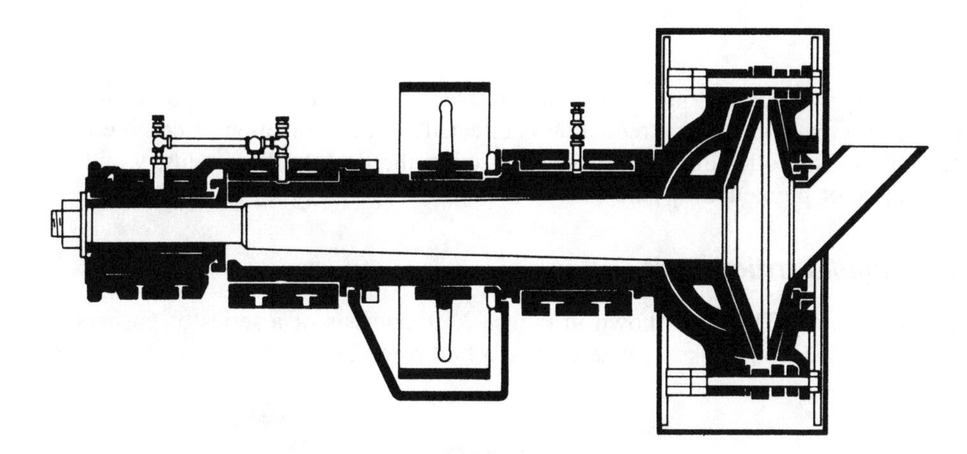

Figure 2.17. Symons disc crusher

## 2.3.3. Fine crushers

### *The buhrstone mill*

The buhrstone mill shown in Figure 2.18 is one of the oldest forms of fine crushing equipment, though it has been very largely superseded now by roller mills. Grinding takes place between two heavy horizontal wheels, one of which is stationary and the other is driven. The surface of the stones is carefully dressed so that the material is continuously worked outwards from the centre of the circumference of the wheels. Size reduction takes place by a shearing action between the edges of the grooves on the two grinding stones. This equipment has been used for the grinding of grain, pigments for paints, pharmaceuticals, cosmetics and printer's ink, although it is now used only where the quantity of material is very small.

### *Roller mills*

The roller mill consists of a pair of rollers that rotate at different speeds, a 3 : 1 ratio for example, in opposite directions. As in crushing rolls, one of the rollers is held in a

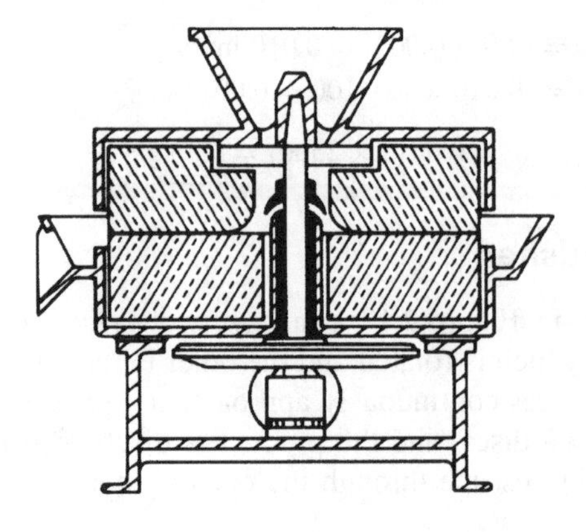

Figure 2.18. Buhrstone mill

fixed bearing whereas the other has an adjustable spring-loaded bearing. Since the rollers rotate at different speeds, size reduction is effected by a combination of compressive and shear forces. The roller mill is extensively used in the flour milling industry and for the manufacture of pigments for paints.

### Centrifugal attrition mills

*The Babcock mill.* This mill shown in Figure 2.19, consists of a series of pushers which cause heavy cast iron balls to rotate against a bull ring like a ball race, with the pressure

Figure 2.19. Babcock mill

of the balls on the bull ring being produced by a loading applied from above. Material fed into the mill falls on the bull ring, and the product is continuously removed in an upward stream of air which carries the ground material between the blades of the classifier, which is shown towards the top of the photograph. The oversize material falls back and is reground. The air stream is produced by means of an external blower which may involve a considerable additional power consumption. The fineness of the product is controlled by the rate of feeding and the air velocity. This machine is used mainly for preparation of pulverised coal and sometimes for cement.

*The Lopulco mill or ring-roll pulveriser.* These mills are manufactured in large numbers for the production of industrial minerals such as limestone and gypsum. A slightly concave circular bull ring rotates at high speed and the feed is thrown outwards by centrifugal action under the crushing rollers, which are shaped like truncated cones, as shown in Figure 2.20. The rollers are spring-loaded and the strength of the springs determines the grinding force available. There is a clearance between the rollers and the bull ring so that there is no wear on the grinding heads if the plant is operated with a light load, and quiet operation is obtained. The product is continuously removed by means of a stream of air produced by an external fan and is carried into a separator fitted above the grinding mechanism. In this separator, the cross-sectional area for flow is gradually increased and, as the air velocity fails, the oversize material falls back and is ground again. The product is separated in a cyclone separator from the air which is then recycled as shown in Figure 2.21. Chemicals, dyes, cements, and phosphate rocks are also ground in the Lopulco mill. As the risk of sparking is reduced by the maintenance of a clearance between the grinding media, the mill may be used for grinding explosive materials.

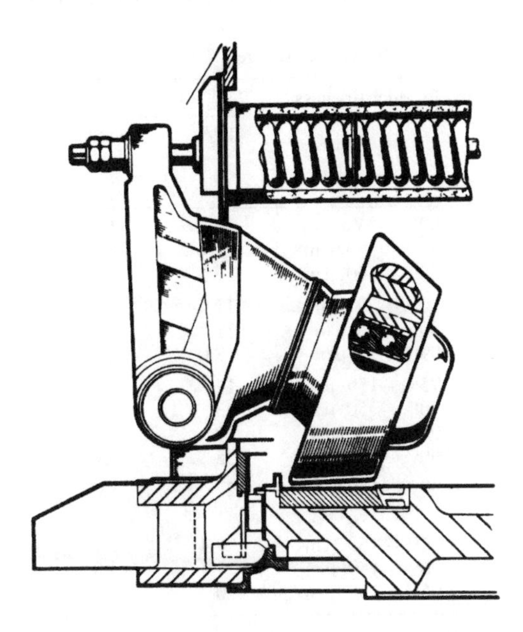

Figure 2.20. Crushing roller in a Lopulco mill

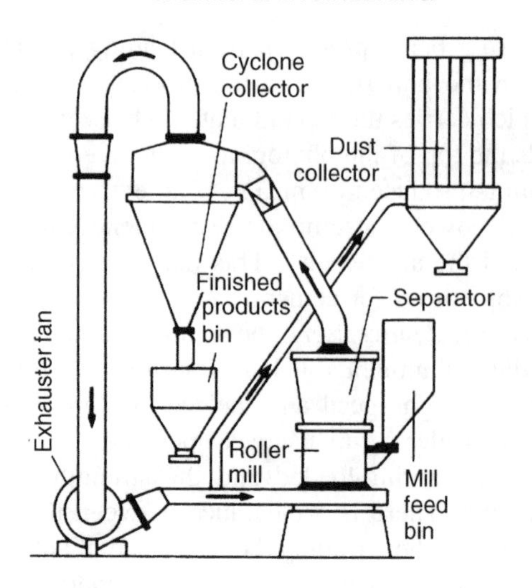


Figure 2.21. Typical air classification for a Lopulco or pendulum mill

Some typical figures for power consumption with the Lopulco mill are given in Table 2.3.

*The NEI pendulum mill.* The NEI pendulum mill shown in Figure 2.22, is slightly less economical in operation than the Lopulco mill, although it gives a rather finer and more uniform product. A central shaft driven by a bevel gear carries a yoke at the top and terminates in a foot-step bearing at the bottom. On the yoke are pivoted a number of heavy arms, shown in Figures 2.23 and 2.24, carrying the rollers which are thrown outwards by centrifugal action and bear on a circular bull ring. Both the rollers and the bull ring

Table 2.3. Typical power requirements for grinding

| Material | Mill size | Product fineness | Output | | Power consumption | |
|---|---|---|---|---|---|---|
| | | | kg/s | tonne/h | MJ/tonne | HP h/tonne |
| | | | | | (mill and fan installed) | |
| Gypsum | LM12 | 99%—75 μm | 4.5 | 1.3 | 134 | 50 |
| | LM14 | 99%—150 μm | 11.2 | 3.2 | 78 | 29 |
| | LM16 | 80%—150 μm | 27.5 | 7.8 | 48 | 18 |
| Limestone | LM12 | 70%—75 μm | 11.0 | 3.1 | 55 | 20.5 |
| | LM14 | 80%—75 μm | 11.5 | 3.2 | 75 | 28 |
| | LM16 | 99%—75 μm | 8.0 | 2.3 | 166 | 62 |
| Phosphate (Morocco) | LM12 | 75%—150 μm | 12 | 3.4 | 50 | 18.5 |
| | LM14 | 90%—150 μm | 13 | 3.7 | 67 | 25 |
| | LM16 | 90%—150 μm | 19 | 5.4 | 62 | 23 |
| | LM16/3 | 90%—150 μm | 35 | 9.9 | 54 | 20 |
| Phosphate (Nauru) | LM16/3 | 97%—150 μm | 27 | 7.6 | 70 | 26 |
| Coal | LM12 | 96%—150 μm | 9.4 | 2.7 | 56 | 21 |
| | LM14 | 96%—150 μm | 13.4 | 3.8 | 56 | 21 |
| | LM16 | 96%—150 μm | 20 | 5.6 | 56 | 21 |

100-mesh B.S.S. ≡ 150 μm 200-mesh B.S.S. ≡ 75 μm

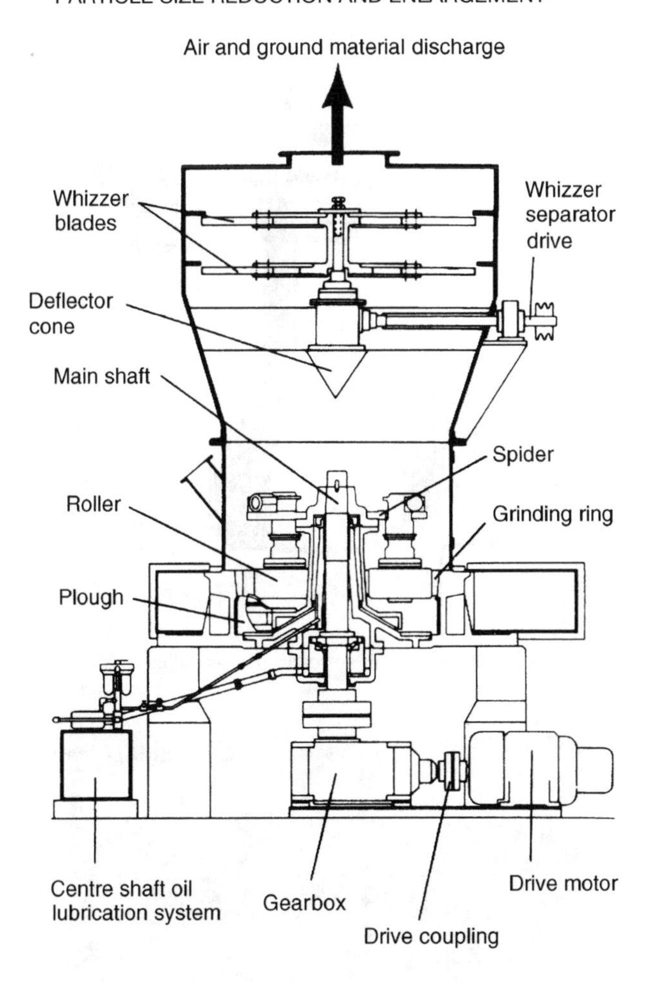


Figure 2.22. Sectional arrangement of an NEI pendulum 5-roller mill

are readily replaceable. The material, which is introduced by means of an automatic feed device, is forced on to the bull ring by means of a plough which rotates on the central shaft. The ground material is removed by means of an air current, as in the case of the Lopulco mill, and the oversize material falls back and is again brought on to the bull ring by the plough.

As the mill operates at high speeds, it is not suitable for use with abrasive materials, nor will it handle materials that soften during milling. Although the power consumption and maintenance costs are low, this machine does not compare favourably with the Lopulco mill under most conditions. It has the added disadvantage that wear will take place if the machine is run without any feed, because no clearance is maintained between the grinding heads and the bull ring. Originally used for the preparation of pulverised coal, the pendulum mill is now used extensively in the fine grinding of softer materials such as sulphur, bentonite and ball clay as well as coal, barytes, limestone and phosphate rock. In many industries, the sizing of the raw materials must be carried out within fine limits.

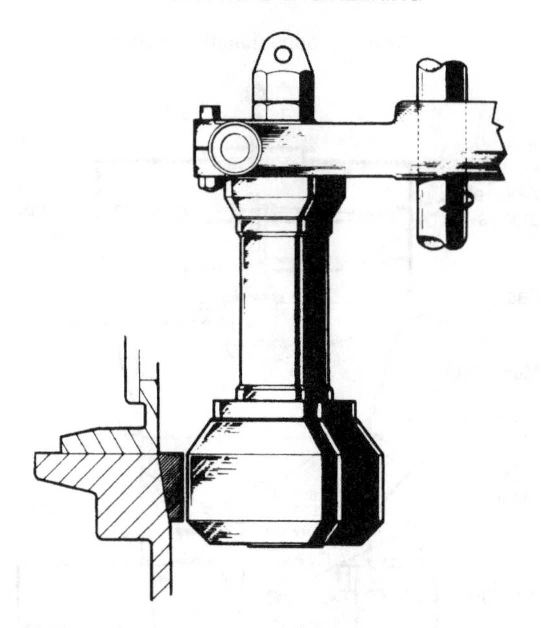

Figure 2.23. Crushing roller in a pendulum mill

Figure 2.24. Crushing heads of a pendulum mill

For example, the pottery industry might require a product whose size lies between 55 and 65 μm, and the pendulum mill is capable of achieving this. A comparison between the Lopulco and pendulum mills has shown that, whereas the Lopulco mill would give a product 98 per cent of which was below 50 μm in size, the pendulum mill would give 100 per cent below this size; in the latter case, however, the power consumption is considerably higher.

The crushing force in a pendulum mill may be obtained as follows.

Considering one arm rotating under uniform conditions, the sum of the moments of all the forces, acting on the roller and arm, about the point of support O shown in Figure 2.25 will be zero.

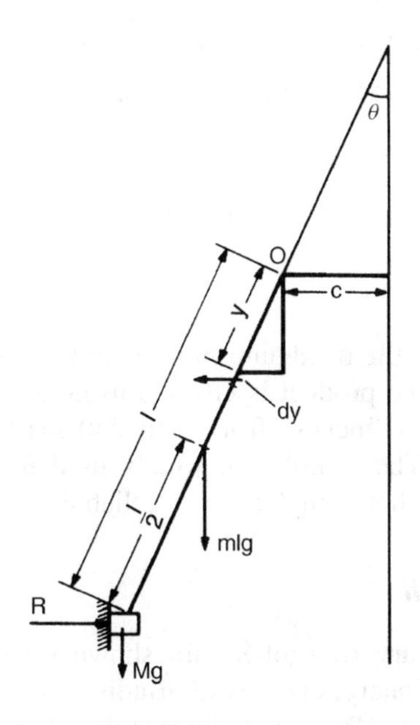


Figure 2.25. Forces in the pendulum mill

Using the following notation:

$M$ = the mass of the grinding head,
$m$ = the mass per unit length of the arm,
$\omega$ = the angular speed of rotation,
$\theta$ = the angle between the arm and the vertical,
$l$ = the length of the arm,
$c$ = the radius of the yoke, and
$R$ = the normal reaction of the bull ring on the grinding head.

For a length $dy$ of the rod at a distance $y$ from O:

Mass of the element $= m\,dy$

Centrifugal force acting on element $= m\,dy(c + y\sin\theta)\omega^2$

Moment of the force about O $= -m\,\mathrm{d}y(c + y\sin\theta)\omega^2 y\cos\theta$

Total moment of the whole arm $= -m\omega^2\cos\theta\left(\frac{1}{2}cl^2 + \frac{1}{3}l^3\sin\theta\right)$

Moment of the centrifugal force acting on the grinding head

$$= -M\omega^2 l\cos\theta(c + l\sin\theta)$$

Moment of the weight of the arm $= \frac{1}{2}ml^2 g\sin\theta$

Moment of the weight of the grinding head $= Mgl\sin\theta$

Moment of the normal reaction of the bull ring on the grinding head $= Rl\cos\theta$

Total moment about O $= \frac{1}{2}ml^2 g\sin\theta + Mgl\sin\theta + Rl\cos\theta$

$$-ml^2\omega^2\cos\theta\left(\frac{1}{2}c + \frac{1}{3}l\sin\theta\right) - M\omega^2 l\cos\theta(c + l\sin\theta) = 0$$

Thus: $R = M'\omega^2\left(\frac{1}{2}c + \frac{1}{3}l\sin\theta\right) + M\omega^2(c + l\sin\theta) - \frac{1}{2}M'g\tan\theta - Mg\tan\theta$

where $M' = ml =$ the mass of arm.

Thus: $$R = M'\left(\frac{1}{2}\omega^2 c + \frac{1}{3}l\omega^2\sin\theta - \frac{1}{2}g\tan\theta\right) + M(\omega^2 c + \omega^2 l\sin\theta - g\tan\theta) \quad (2.8)$$

### *The Griffin mill*

The Griffin mill is similar to the pendulum mill, other than it employs only one grinding head and the separation of the product is effected using a screen mesh fitted around the grinding chamber. A product fineness from 8 to 240 mesh at an output from 0.15 to 1.5 kg/s may be obtained. These mills are widely used for dry fine grinding in many industries and are noted for their simplicity and reliability.

### *The Szego grinding mill*

The Szego Mill is a planetary ring-roller mill shown schematically in Figure 2.26. It consists principally of a stationary, cylindrical grinding surface inside which a number of helically grooved rollers rotate. These radially mobile rollers are suspended from flanges connected to the central drive shaft; they are pushed outward by centrifugal force and roll on the grinding surface.

The material is fed by gravity, or pumped into a top feed cylinder and is discharged at the bottom of the mill. The particles, upon entering the grinding section, are repeatedly crushed between the rollers and the stationary grinding surface. The crushing force is created mainly by the radial acceleration of the rollers. Shearing action is induced by the high velocity gradients generated in the mill, and hence the primary forces acting on the particles are the crushing and shearing force caused by rotational motion of the rollers. Attrition is also important, particularly in dry grinding, and impaction also occurs at higher rotational speeds.

An important feature is the ability of the roller grooves to aid the transport of material through the mill, thus providing a means to control residence time and mill capacity. This transporting action is particularly important with materials which would not readily flow by gravity. Pastes and sticky materials fall into this category.

The mill has several design variables which may be utilised to meet specific product requirements. The important variables are the number of rollers, their mass, diameter and

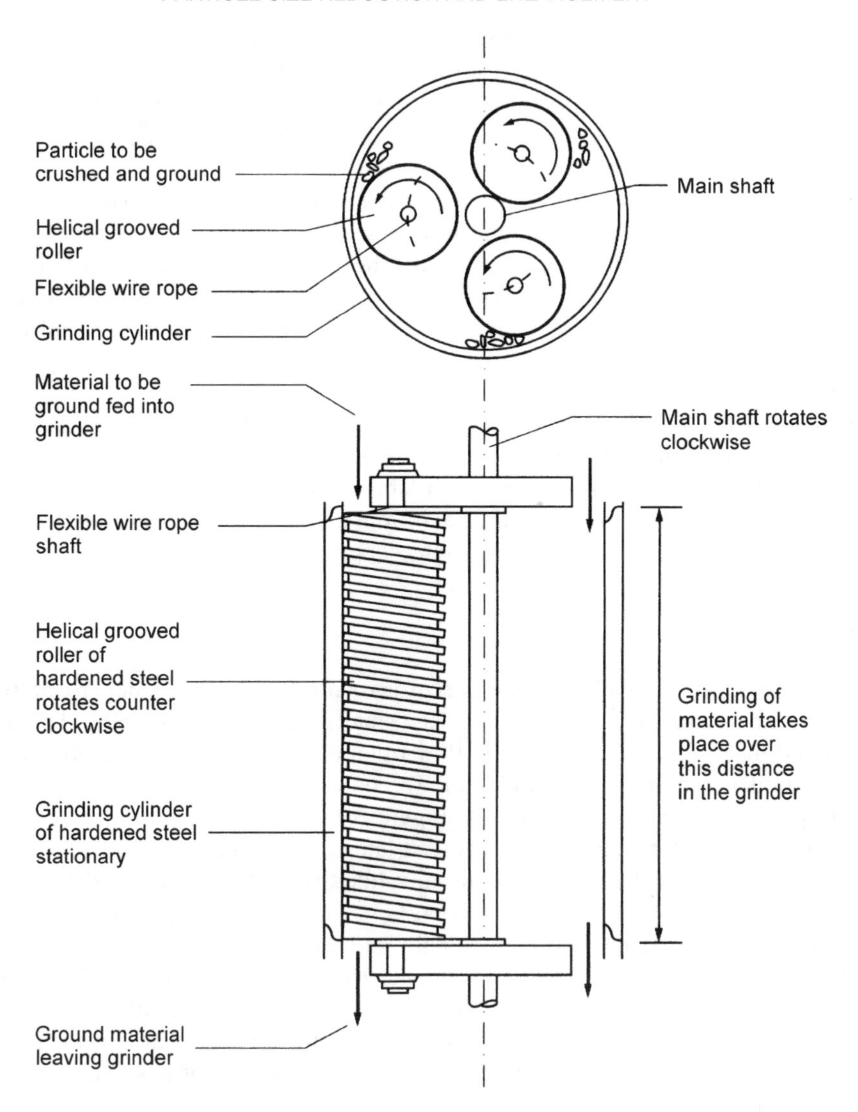


Figure 2.26. The Szego mill

length, and the shape, size and number of starts of the helical grooves on the rollers. As the number of rollers is increased, the product becomes finer. Heavier rollers and higher rotational speeds generate greater crushing forces and give higher reduction ratios. The ridge:groove size ratio may also be changed to increase or decrease the effective pressure acting on the particles. The common groove shapes are rectangular and tapered and the latter will decrease the chance of particles getting stuck in the grooves.

The Szego mill has been used for dry as well as wet grinding of coal, in both water and oil, for the preparation of coal-slurry fuels. To grind coal in water to the standard boiler size (80 per cent $< 74$ μm) takes about 7 MJ/Mg (20 kWh/tonne). For 'micronized' grinding, to a 15 μm median size, 2–3 passes through the mill are required and the specific energy requirement increases to 20 MJ/Mg (50 kWh/tonne). The mill has been used for simultaneous grinding of coal in oil and water where the carbonaceous matter agglomerates with oil as the bridging liquid, while the inorganic mineral matter stays suspended in the aqueous phase. Separating out the agglomerates allows good coal beneficiation.

The mill has also been used to grind industrial minerals and technical ceramics including limestone, lead zirconates, metal powders, fibrous materials, such as paper, wood chips and peat, and chemicals and agricultural products, such as grains and oilseeds.

Since its grinding volume is some 30–40 times smaller than that of a ball mill for the same task, the Szego mill is compact for its capacity. It is characterised by relatively low specific energy consumption, typically 20–30 per cent lower than in a ball mill, and flexibility of operation. It can give a large reduction ratio, typically 10–20, and grinds material down to a 15-45 $\mu$m size range.

The mill has been developed, partly at the University of Toronto, Canada[20,21] and commercialised by General Comminution Inc. in Toronto. Capacities range from 20 kg to 10 tonnes of dry material per hour. A small laboratory pilot mill has an inside diameter of 160 mm and fits on a bench. The large, 640 mm diameter unit has external dimensions of about 2 m $\times$ 2 m $\times$ 1 m in terms of height, length and width.

### *The ball mill*

In its simplest form, the ball mill consists of a rotating hollow cylinder, partially filled with balls, with its axis either horizontal or at a small angle to the horizontal. The material to be ground may be fed in through a hollow trunnion at one end and the product leaves through a similar trunnion at the other end. The outlet is normally covered with a coarse screen to prevent the escape of the balls. Figure 2.27 shows a section of an example of the Hardinge ball mill which is also discussed later in this chapter.

The inner surface of the cylinder is usually lined with an abrasion-resistant material such as manganese steel, stoneware or rubber. Less wear takes place in rubber-lined mills, and the coefficient of friction between the balls and the cylinder is greater than with steel or stoneware linings. The balls are therefore carried further in contact with the cylinder and thus drop on to the feed from a greater height. In some cases, lifter bars are fitted to the inside of the cylinder. Another type of ball mill is used to an increasing extent, where the mill is vibrated instead of being rotated, and the rate of passage of material is controlled by the slope of the mill.

The ball mill is used for the grinding of a wide range of materials, including coal, pigments, and felspar for pottery, and it copes with feed up to about 50 mm in size. The efficiency of grinding increases with the hold-up in the mill, until the voids between the balls are filled. Further increase in the quantity then lowers the efficiency.

The balls are usually made of flint or steel and occupy between 30 and 50 per cent of the volume of the mill. The diameter of ball used will vary between 12 mm and 125 mm and the optimum diameter is approximately proportional to the square root of the size of the feed, with the proportionality constant being a function of the nature of the material.

During grinding, the balls wear and are constantly replaced by new ones so that the mill contains balls of various ages, and hence of various sizes. This is advantageous since the large balls deal effectively with the feed and the small ones are responsible for giving a fine product. The maximum rate of wear of steel balls, using very abrasive materials,

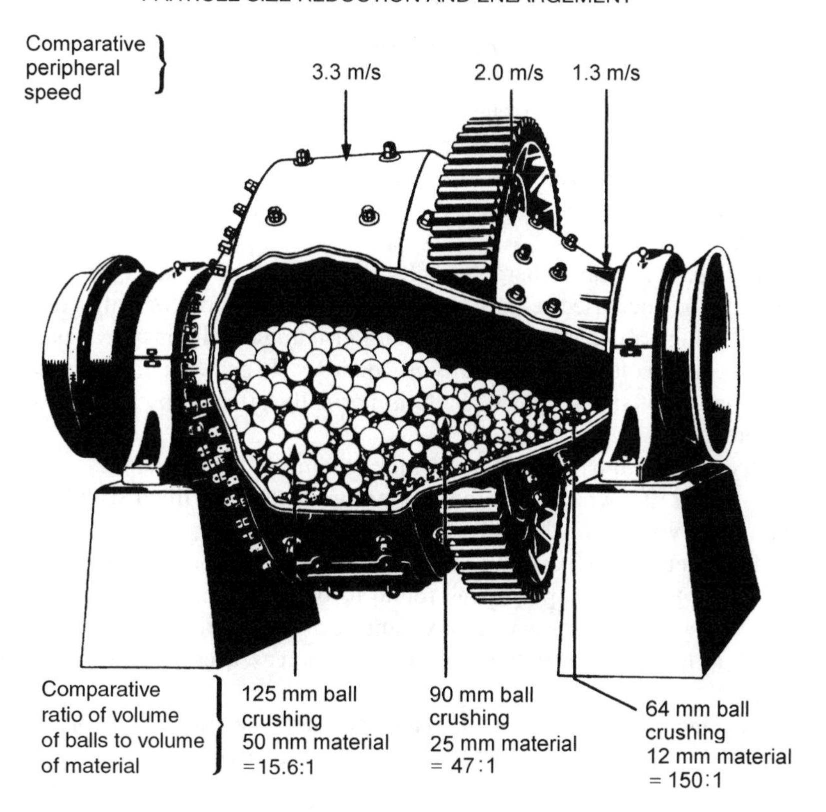


Figure 2.27. Cut-away view of the Hardinge conical ball mill showing how energy is proportioned to the work required

is about 0.3 kg/Mg of material for dry grinding, and 1–1.5 kg/Mg for wet grinding. The normal charge of balls is about 5 $Mg/m^3$. In small mills where very fine grinding is required, pebbles are often used in place of balls.

In the compound mill, the cylinder is divided into a number of compartments by vertical perforated plates. The material flows axially along the mill and can pass from one compartment to the next only when its size has been reduced to less than that of the perforations in the plate. Each compartment is supplied with balls of a different size. The large balls are at the entry end and thus operate on the feed material, whilst the small balls come into contact with the material immediately before it is discharged. This results in economical operation and the formation of a uniform product. It also gives an improved residence time distribution for the material, since a single stage ball mill approximates closely to a completely mixed system.

In wet grinding the power consumption is generally about 30 per cent lower than that for dry grinding and, additionally, the continuous removal of product as it is formed is facilitated. The rheological properties of the slurry are important and the performance tends to improve as the apparent viscosity increases, reaching an optimum at about 0.2 Pa.s. At very high volumetric concentrations (*ca.* 50 volume per cent), the fluid may exhibit shear-thickening behaviour or have a yield stress, and the behaviour may then be adversely affected.

### *Factors influencing the size of the product*

(a) *The rate of feed.* With high rates of feed, less size reduction is effected since the material is in the mill for a shorter time.
(b) *The properties of the feed material.* The larger the feed the larger is the product under given operating conditions. A smaller size reduction is obtained with a hard material.
(c) *Weight of balls.* A heavy charge of balls produces a fine product. The weight of the charge can be increased, either by increasing the number of balls, or by using a material of higher density. Since optimum grinding conditions are usually obtained when the bulk volume of the balls is equal to 50 per cent of the volume of the mill, variation of the weight of balls is normally effected by the use of materials of different densities.
The effect on grinding performance of the loading of balls in a mill has been studied by KANO *et al.*[22], who varied the proportion of the mill filled with balls from 0.2 to 0.8 of the volume of the mill. The grinding rates were found to be a maximum at loadings between 0.3 and 0.4. The effect of relative rotational speed, the ratio of actual speed to critical speed, was found to be complex. At loadings between 0.4 and 0.8, the grinding rate was a maximum at a relative rotation speed of about 0.8, although at lower loadings the grinding rate increased up to relative speeds of 1.1 to 1.6.
(d) *The diameter of the balls.* Small balls facilitate the production of fine material although they do not deal so effectively with the larger particles in the feed. The limiting size reduction obtained with a given size of balls is known as the free grinding limit. For most economical operation, the smallest possible balls should be used.
(e) *The slope of the mill.* An increase in the slope of the mill increases the capacity of the plant because the retention time is reduced, although a coarser product is obtained.
(f) *Discharge freedom.* Increasing the freedom of discharge of the product has the same effect as increasing the slope. In some mills, the product is discharged through openings in the lining.
(g) *The speed of rotation of the mill.* At low speeds of rotation, the balls simply roll over one another and little crushing action is obtained. At slightly higher speeds, the balls are projected short distances across the mill, and at still higher speeds they are thrown greater distances and considerable wear of the lining of the mill takes place. At very high speeds, the balls are carried right round in contact with the sides of the mill and little relative movement or grinding takes place again. The minimum speed at which the balls are carried round in this manner is called the critical speed of the mill and, under these conditions, there will be no resultant force acting on the ball when it is situated in contact with the lining of the mill in the uppermost position, that is the centrifugal force will be exactly equal to the weight of the ball. If the mill is rotating at the critical angular velocity $\omega_c$, then:

$$r\omega_c^2 = g$$

or:

$$\omega_c = \sqrt{\frac{g}{r}} \tag{2.9}$$

The corresponding critical rotational speed, $N_c$ in revolutions per unit time, is given by:

$$N_c = \frac{\omega_c}{2\pi} = \frac{1}{2\pi}\sqrt{\frac{g}{r}} \tag{2.10}$$

In this equation, $r$ is the radius of the mill less that of the particle. It is found that the optimum speed is between one-half and three-quarters of the critical speed. Figure 2.28 illustrates conditions in a ball mill operating at the correct rate.

(h) *The level of material in the mill.* Power consumption is reduced by maintaining a low level of material in the mill, and this can be controlled most satisfactorily by fitting a suitable discharge opening for the product. If the level of material is raised, the cushioning action is increased and power is wasted by the production of an excessive quantity of undersize material.

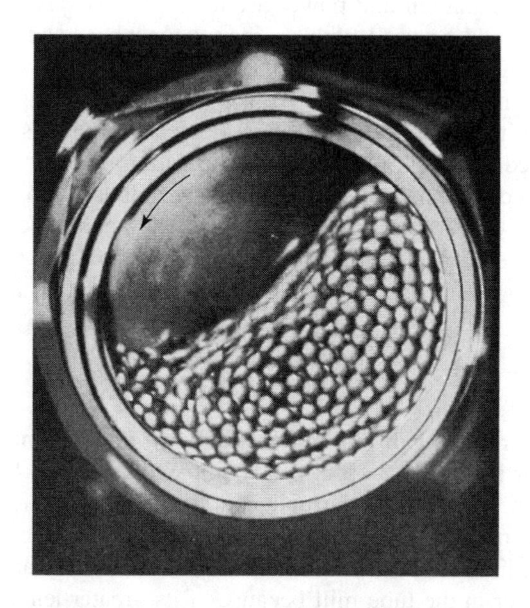

Figure 2.28. A ball mill operating at the correct speed

## Example 2.3

A ball mill, 1.2 m in diameter, is run at 0.80 Hz and it is found that the mill is not working properly. Should any modification in the conditions of operation be suggested?

## Solution

The angular velocity is given by:

$$\omega_c = \sqrt{(g/r)} \qquad \text{(equation 2.10)}$$

In this equation, $r =$ (radius of the mill $-$ radius of the particle). For small particles, $r = 0.6$ m and hence:

$$\omega_c = \sqrt{(9.81/0.6)} = 4.04 \text{ rad/s}$$

The actual speed $= (2\pi \times 0.80) = 5.02$ rad/s and hence it may be concluded that the speed of rotation is too high and that the balls are being carried round in contact with the sides of the mill with little relative movement or grinding taking place.

The optimum speed of rotation lies in the range $(0.5-0.75)\omega_c$, say $0.6\omega_c$ or:

$$(0.6 \times 4.04) = 2.42 \text{ rad/s}$$

This is equivalent to: $(2.42/2\pi) = 0.39$ Hz, or, in simple terms:

the speed of rotation should be halved.

### *Advantages of the ball mill*

(i) The mill may be used wet or dry although wet grinding facilitates the removal of the product.
(ii) The costs of installation and power are low.
(iii) The ball mill may be used with an inert atmosphere and therefore can be used for the grinding of explosive materials.
(iv) The grinding medium is cheap.
(v) The mill is suitable for materials of all degrees of hardness.
(vi) It may be used for batch or continuous operation.
(vii) It may be used for open or closed circuit grinding. With open circuit grinding, a wide range of particle sizes is obtained in the product. With closed circuit grinding, the use of an external separator can be obviated by continuous removal of the product by means of a current of air or through a screen, as shown in Figure 2.29.

### ***The tube mill***

The tube mill is similar to the ball mill in construction and operation, although the ratio of length to the diameter is usually 3 or 4 : 1, as compared with 1 or 1.5 : 1 for the ball mill. The mill is filled with pebbles, rather smaller in size than the balls used in the ball mill, and the inside of the mill is so shaped that a layer of pebbles becomes trapped in it to form a self-renewing lining. The characteristics of the two mills are similar although the material remains longer in the tube mill because of its greater length, and a finer product is therefore obtained.

### ***The rod mill***

In the rod mill, high carbon steel rods about 50 mm diameter and extending the whole length of the mill are used in place of balls. This mill gives a very uniform fine product and power consumption is low, although it is not suitable for very tough materials and the feed should not exceed about 25 mm in size. It is particularly useful with sticky materials which would hold the balls together in aggregates, because the greater weight of the rods causes them to pull apart again. Worn rods must be removed from time to time and replaced by new ones, which are rather cheaper than balls.

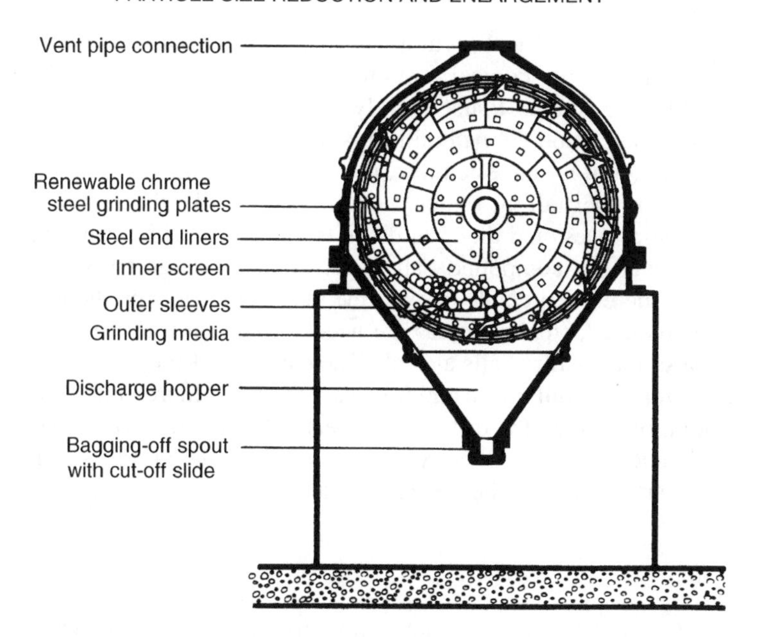


Figure 2.29. End view of ball mill showing screens

### *The Hardinge mill*

The Hardinge mill, shown in Figure 2.30, is a ball mill in which the balls segregate themselves according to size. The main portion of the mill is cylindrical as in the ordinary ball mill, although the outlet end is conical and tapers towards the discharge point. The

Figure 2.30. The Hardinge mill

large balls collect in the cylindrical portion while the smaller balls, in order of decreasing size, locate themselves in the conical portion as shown in Figure 2.27. The material is therefore crushed by the action of successively smaller balls, and the mill is thus similar in characteristics to the compound ball mill. It is not known exactly how balls of different sizes segregate although it is suggested that, if the balls are initially mixed, the large ones will attain a slightly higher falling velocity and therefore strike the sloping surface of the mill before the smaller ones, and then run down towards the cylindrical section. The mill has an advantage over the compound ball mill in that the large balls are raised to the greatest height and therefore are able to exert the maximum force on the feed. As the size of the material is reduced, smaller forces are needed to cause fracture and it is therefore unnecessary to raise the smaller balls as high. The capacity of the Hardinge mill is higher than that of a ball mill of similar size and it gives a finer and more uniform product with a lower consumption of power. It is difficult to select an optimum speed, however, because of the variation in shell diameter. It is extensively used for the grinding of materials such as cement, fuels, carborundum, silica, talc, slate and barytes.

### *The sand mill*

The sand mill, or stirred ball mill, achieves fine grinding by continuously agitating the bed of grinding medium and charge by means of rotating bars which function as paddles. Because of its high density, zircon sand is frequently used as the grinding medium. A fluid medium, liquid or gas, is continuously passed through the bed to remove the fines, as shown in Figure 2.31. A very fine product can be obtained at a relatively low energy input, and the mill is used for fine grades of ceramic oxides and china clay, and in the preparation of coal slurries.

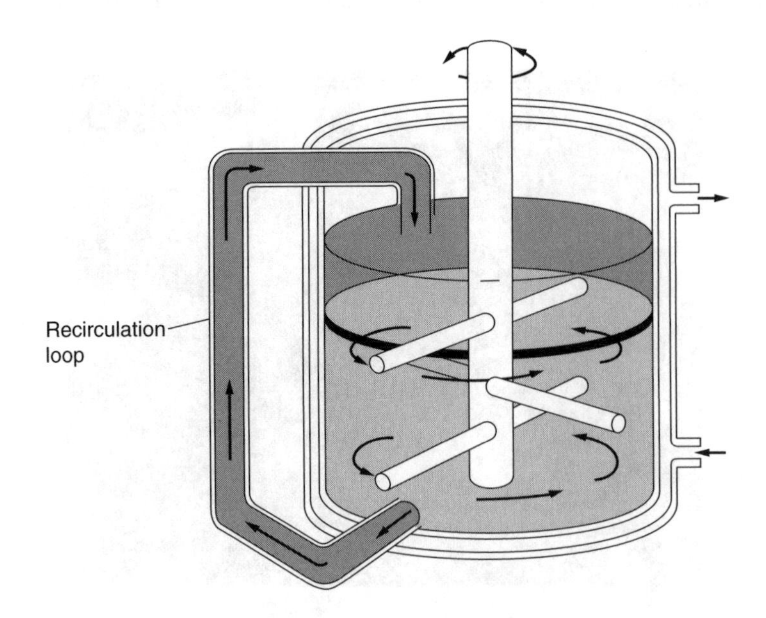


Figure 2.31. Typical sand mill

In cases where it is important that the product should not be contaminated with fine fragments of the grinding medium, *autogenous grinding* is used where coarse particles of the material are to be ground form the grinding medium.

### The planetary mill

A serious limitation of the ball or tube mill is that it operates effectively only below its critical speed, as given by equation 2.10. In the planetary mill, described by BRADLEY *et al.*[23], this constraint is obviated by rotating the mill simultaneously about its own axis and about an axis of gyration, as shown in Figure 2.32. In practice, several cylinders are incorporated in the machine, all rotating about the same axis of gyration.

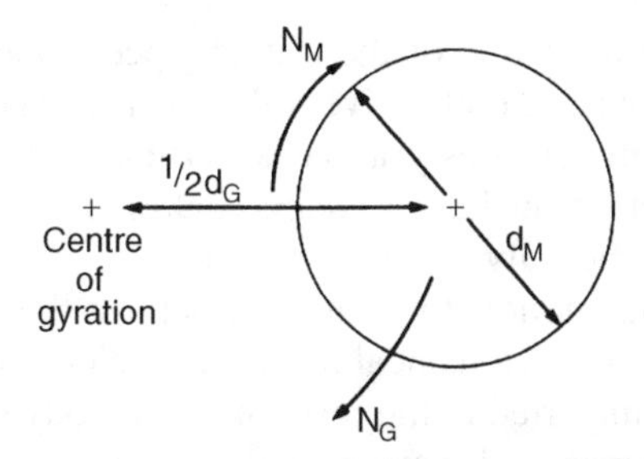


Figure 2.32. Geometry of the planetary mill

A particle of mass $M$ in contact with the cylindrical wall is subject to the following two centrifugal forces acting simultaneously:

(a) A centrifugal force attributable to rotation about the axis of gyration (radius $d_G/2) = (Md_G/2)(2\pi N_G)^2$, where $N_G$ = revolutions per unit time about the axis of gyration.
(b) A centrifugal force attributable to rotation of the cylinder (radius $d_M/2) = (Md_M/2)[2\pi(N_M + N_G)]^2$, where $N_M$ = revolutions per unit time about the axis of the cylinder.
In unit time, the total number of revolutions of the cylinder is $N_M + N_G$.

In addition, the gravitational force acts on the particle although, under normal operating conditions, this is small compared with the centrifugal forces.

At the critical condition:

$$\left(M\frac{d_G}{2}\right)(2\pi N_G)^2 = \left(M\frac{d_M}{2}\right)[2\pi(N_M + N_G)]^2 \tag{2.11}$$

If $N_M/N_G = s$, say, which is determined by the gear ratio of the mill, then substituting into equation 2.11 and simplifying:

$$(s_c + 1)^2 = \frac{d_G}{d_M} \tag{2.12}$$

and:

$$s_c = \pm\sqrt{(d_G/d_M)} - 1 \tag{2.13}$$

where $s_c$ is the value of $s$ which gives rise to the critical condition for a given value of $d_G/d_M$.
The positive sign applies when $s_c > -1$ and the negative sign when $s_c < -1$.

It may be noted that it is necessary to take account of *Coriolis* forces in calculating the trajectory of a particle once it ceases to be in contact with the wall.

It is possible to achieve accelerations up to about $15g$ in practical operation. For further details of the operation of the planetary mill, reference may be made to BRADLEY *et al.*[23] and KITSCHEN *et al.*[24] The planetary mill is used for the preparation of stabilised slurries of coal in both oil and water, and it can also handle paste-like materials.

### The vibration mill

Another way of increasing the value of the critical speed, and so improving the performance of a mill, is to increase the effective value of the acceleration due to gravity $g$. The rotation of the mill simultaneously about a vertical and a horizontal axis has been used to simulate the effect of an increased gravitational acceleration, although clearly such techniques are applicable only to small machines.

By imparting a vibrating motion to a mill, either by the rotation of out-of-balance weights or by the use of electro-mechanical devices, accelerations many times the gravitational acceleration may be imparted to the machine. The body of the machine is generally supported on powerful springs and caused to vibrate in a vertical direction. Vibration frequencies of 6–60 Hz are common. In some machines the grinding takes place in two stages, the material falling from an upper to a lower chamber when its size has been reduced below a certain value.

The vibration mill has a very much higher capacity than a conventional mill of the same size, and consequently either smaller equipment may be used or a much greater throughput obtained. Vibration mills are well suited to incorporation in continuous grinding systems.

### Colloid mills

Colloidal suspensions, emulsions and solid dispersions are produced by means of colloid mills or dispersion mills. Droplets or particles of sizes less than 1 μm may be formed, and solids suspensions consisting of discrete solid particles are obtainable with feed material of approximately 100-mesh or 50 μm in size.

As shown in Figure 2.33, the mill consists of a flat rotor and stator manufactured in a chemically inert synthetic abrasive material, and the mill can be set to operate at clearances from virtually zero to 1.25 mm, although in practice the maximum clearance used is about 0.3 mm. When duty demands, steel working surfaces may be fitted, and in such cases the minimum setting between rotor and stator must be 0.50–0.75 mm, otherwise 'pick up' between the steel surfaces occurs.

The gap setting between rotor and stator is not necessarily in direct proportion to the droplet size or particle size of the end product. The thin film of material continually passing between the working surfaces is subjected to a high degree of shear, and consequently the energy absorbed within this film is frequently sufficient to reduce the dispersed phase to a particle size far smaller than the gap setting used. The rotor speed varies with the physical size of the mill and the clearance necessary to achieve the desired result, although

Figure 2.33. Rotor and stator of a colloid mill

peripheral speeds of the rotor of 18–36 m/s are usual. The required operating conditions and size of mill can only be found by experiment.

Some of the energy imparted to the film of material appears in the form of heat, and a jacketed shroud is frequently fitted round the periphery of the working surfaces so that some of the heat may be removed by coolant. This jacket may also be used for circulation of a heating medium to maintain a desired temperature of the material being processed.

In all colloid mills, the power consumption is very high, and the material should therefore be ground as finely as possible before it is fed to the mill.

## *Fluid energy mills*

Another form of mill which does not give quite such a fine product is the jet pulveriser, in which the solid is pulverised in jets of high pressure superheated steam or compressed air, supplied at pressures up to 3.5 MN/m$^2$ (35 bar). The pulverising takes place in a shallow cylindrical chamber with a number of jets arranged tangentially at equal intervals around the circumference. The solid is thrown to the outside walls of the chamber, and the fine particles are formed by the shearing action resulting from the differential velocities within the fluid streams. The jet pulveriser will give a product with a particle size of 1–10 $\mu$m.

The microniser, probably the best known of this type of pulveriser, effects comminution by bombarding the particles of material against each other. Pre-ground material, of about 500 $\mu$m in size, is fed into a shallow circular grinding chamber which may be horizontal or vertical, the periphery of which is fitted with a number of jets, equally spaced, and arranged tangentially to a common circle.

Gaseous fluid, often compressed air at approximately 800 kN/m$^2$ (8 bar) or superheated steam at pressures of 800–1600 kN/m$^2$ (8–16 bar) and temperatures ranging from 480 to 810 K, issues through these jets, thereby promoting high-speed reduction in the size of the contents of the grinding chamber, with turbulence and bombardment of the particles against each other. An intense centrifugal classifying action within the grinding chamber causes the coarser particles to concentrate towards the periphery of the chamber whilst the finer particles leave the chamber, with the fluid, through the central opening.

The majority of applications for fluid energy mills are for producing powders in the sub-sieve range, of the order of 20 μm and less, and it may be noted that the power consumption per kilogram is proportionately higher than for conventional milling systems which grind to a top size of about 44 μm.

A section through a typical microniser is shown in Figure 2.34.

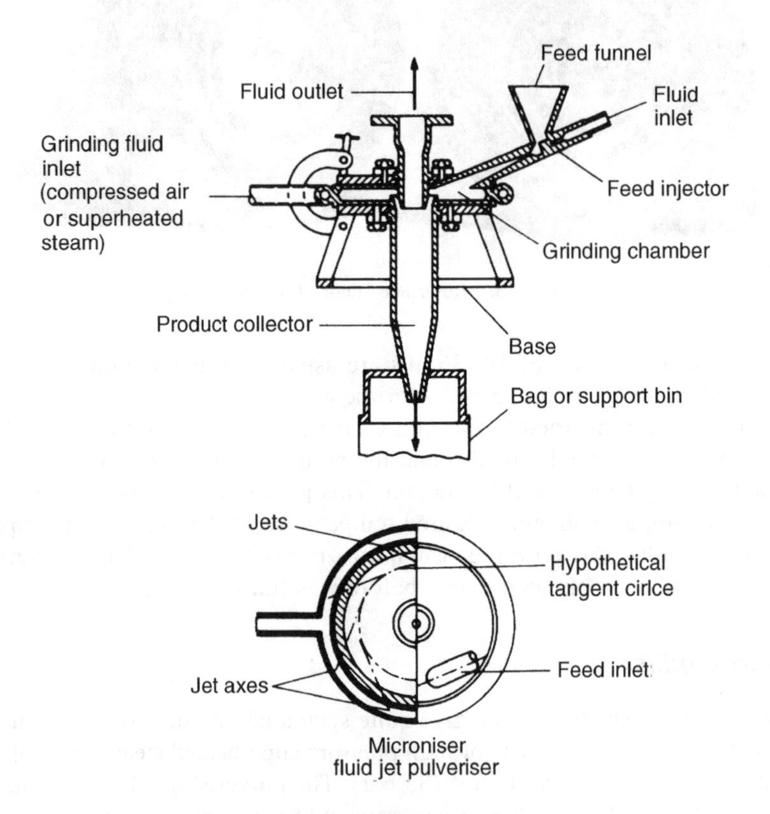


Figure 2.34. Section through a microniser

Another pulveriser in this group is the Wheeler fluid energy mill which is in the form of a vertical loop. The pre-ground feed material is injected towards the bottom of the loop in which are situated the nozzles for admitting the compressed air or superheated steam. Size reduction occurs as a result of bombardment of the particles against each other, and classification is effected by arranging for the fluid to leave the circulating gas stream through vanes which are situated just downstream of the top of the loop and on the inner face of the loop. Oversize particles continue their downward path with the circulating fluid and re-enter the reduction chamber for further grinding.

### 2.3.4. Specialised applications

#### *Specialised techniques*

Several techniques have been developed for specialised applications, as discussed by PRASHER[6]. These include:

(i) *Electrohydraulic crushing* in which an underwater discharge is generated by the release of energy from a high-voltage capacitor. The spark length depends on the nature of the material to be crushed, though it is commonly 15–80 mm.

(ii) *Ultrasonic grinding* in which the material to be ground is fed between a drive roll and a curved plate, both of which are ultrasonically activated. Experimental work has been carried out to produce coal particles smaller than 10 $\mu$m.

(iii) *Cryogenic grinding* Size reduction is difficult to achieve by conventional means with many materials, such as plastics, rubber, waxes and textile-based products, as they tend to distort rather than to fracture when subjected to compressive forces. However, it is frequently possible to effect a change in the structure of the material by subjecting it to very low temperatures. In *cryogenic grinding*, the material is cooled with liquid nitrogen at a temperature of about −196°C (77K) to render it brittle before it enters the grinder. According to BUTCHER[25], cooling causes the crystal lattice to shrink and to give rise to microscopic cracks which act as nuclei and then grow, thereby reducing the energy input required to cause fracturing to occur. As about 99 per cent of the energy for size reduction finishes up as heat, economy in the use of liquid nitrogen is a critical consideration. It is therefore desirable, if practicable, to effect a preliminary reduction of particle size in order to reduce the time taken for the material to be cooled by the liquid nitrogen. Precooling of the feed material by using a conventional refrigeration plant is also advantageous. Both rotary cutting mills and pin mills show improved energy utilisation with non-brittle materials as the temperature is reduced. The method is of new importance with the expanding market for frozen foods.

(iv) *Explosive shattering* in which energy is transmitted to particles as shock waves set up on suddenly releasing steam from an explosion chamber containing the solid to be compressed. Equipment based on this technique is still at the development stage.

## 2.4. SIZE ENLARGEMENT OF PARTICLES

### 2.4.1. Agglomeration and granulation

There are many situations where fine particles are difficult to handle, mainly due to the fact that particles in bulk do not flow readily because of their tendency to adhere together as conglomerates as a result of the action of surface forces. In such cases, the finer the particles, the greater is their specific surface, and the gravitational forces acting on the particles may not be great enough to keep them apart during flow. The flowability of particulate systems can sometimes be improved by the use of "glidants", which are very fine powders which are capable of reducing interparticle friction by forming surface layers on the particles, thereby combating the effects of friction arising from surface roughness; they can also reduce the effects of electrostatic charges. However, the optimisation of particle size is by far the most important method of improving flow properties. As discussed in Chapter 6, fine particles are often difficult to fluidise in gases because channelling, rather than even dispersion of the particles, tends to occur.

Fine particles may be difficult to discharge from hoppers as particles may cling to the walls and also form bridges at the point of discharge. Although such problems may be minimised, either by vibration or by mechanical stirring, it is very difficult to overcome them entirely, and the only satisfactory solution may be to increase the particle size by forming them into aggregates. In addition, very fine particles often give rise to serious environmental and health problems, particularly as they may form dust clouds when loaded into vehicles, and in windy conditions may become dispersed over long distances. Although the particles involved may, in themselves, be inert, serious respiratory problems may result if these are inhaled. In such situations, particle size may be a critical factor since very fine particles may be exhaled, and very large particles may have a negligible effect on health. In this respect, it may be noted that the particular health hazard imposed by asbestos is largely associated with the size range and shape of the particles and their tendency to collect in the lungs.

The size of particles may be increased from molecular dimensions by growing them by crystallisation from both solutions and melts as discussed in Chapter 15. Here, dissolving and recrystallising may provide a mechanism for controlling both particle size and shape. It may be noted, as also discussed in Chapter 15, that fine particles may also be condensed out from both vapours and gases.

A desired particle size may also be achieved by building up from fine particles, and one example of this is the production of fertiliser granules by agglomeration, or by a repeated coating process. Another example is the formation of pellets or pills for medicinal purposes by the compression of a particulate mass, often with the inclusion of a binding agent that will impart the necessary strength to the pellet. VON SMOLUCHOWSKI[26] has characterised suspensions by a 'half-time', *t*, defined as the time taken to halve the number of original particles in a mono-disperse system. WALTON[27] arbitrarily defined an agglomerating system as one in which more than 10 per cent of the original number of particles has agglomerated in less than 1000 s.

### 2.4.2. Growth mechanisms

There are essentially two types of processes that can cause agglomeration of particles when they are suspended in a fluid:

(a) *perikinetic processes* which are attributable to Brownian movement can therefore occur even in a static fluid. Double-layer repulsive forces and van der Waals attractive forces may both operate independently in disperse systems. Repulsion forces decrease exponentially with distance across the ionic double-layer, although attraction forces decrease, at larger distances from the particle surface, and are inversely proportional to the distance. Consequently, as TADROS[28] points out, attraction normally predominates both at very small and very large distances, and repulsion over intermediate distances. Fine particles may also be held together by electrostatic forces.

(b) *orthokinetic processes* occur where the perikinetic process is supplemented by the action of eddy currents which may be set up in stirred vessels or in flowing systems. In these circumstances, the effects of the perikinetic mechanism are generally negligible.

According to Söhnel and Mullin[29], the change in agglomerate size as a function of time may be represented by equations of the form:

For *perikinetic processes*:

$$d_t^3 = A_1 + B_1 t \tag{2.14}$$

and for *orthokinetic processes*:

$$\log \frac{d_t}{d_o} = A_2 + B_2 t \tag{2.15}$$

where $d_t$ is the agglomerate size at time $t$. Thus, the shapes of the plots of $d_t$ against $t$ give an indication of the relative importance of the two processes. Equations 2.14 and 2.15 will apply only in the initial stages of the enlargement process, since otherwise they would indicate an indefinite increase of size $d_t$ with time $t$. The dimensions of $A_1$ are $\mathbf{L}^3$, of $B_1$ are $\mathbf{L}^3\mathbf{T}^{-1}$ and of $B_2$ are $\mathbf{T}^{-1}$. $A_2$ is dimensionless. The limiting size is that at which the rates of aggregation and of breakdown of aggregates are in balance.

The stability of the aggregates may be increased by the effects of mechanical interlocking that may occur, especially between particles in the form of long fibres.

It is often desirable to add a liquid binder to fill the pore spaces between particles in order to increase the strength of the aggregate. The amount of binder is a function of the voidage of the particulate mass, a parameter that is strongly influenced by the size distribution and shape of the particles. Wide size distributions generally lead to close packing requiring smaller amounts of binder and, as a result, the formation of strong aggregates.

As discussed in Chapter 15, the size distribution of particles in an agglomeration process is essentially determined by a *population balance* that depends on the kinetics of the various processes taking place simultaneously, some of which result in particle growth and some in particle degradation. In a batch process, an equilibrium condition will eventually be established with the net rates of formation and destruction of particles of each size reaching an equilibrium condition. In a continuous process, there is the additional complication that the residence time distribution of particles of each size has an important influence.

In general, starting with a mixture of particles of uniform size, the following stages may be identified:

(a) *Nucleation* in which fresh particles are formed, generally by attrition.
(b) *Layering* or *coating* as material is deposited on the surfaces of the nuclei, thus increasing both the size and total mass of the particles.
(c) *Coalescence* of particles which results in an increase in particle size but not in the total mass of particles.
(d) *Attrition*. This results in degradation and the formation of small particles, thus generating nuclei that re-enter the cycle again.

It is difficult to build these four stages into a mathematical model because the kinetics of each processes is not generally known, and therefore more empirical methods have to be adopted.

### 2.4.3. Size enlargement processes

Processes commonly used for size enlargement are listed in Table 2.4, taken from Perry[30]. For comprehensive overall reviews, reference may be made to PERRY and to the work of BROWNING[31].

(a) *Spray drying* (as discussed in Chapter 16).
In this case, particle size is largely determined by the size of the droplet of liquid or suspension, which may be controlled by a suitably designed spray nozzle. The aggregates of dried material are held together as a result of the deposition of small amounts of solute on the surface of the particles. For a given nozzle, the drop sizes will be a function of both flowrate and liquid properties, particularly viscosity, and to a lesser extent of outlet temperature. In general, viscous liquids tend to form large drops yielding large aggregates.

(b) *Prilling* in which relatively coarse droplets are introduced into the top of a tall, narrow tower and allowed to fall against an upward flow of air. This results in somewhat larger particles than those formed in spray dryers.

(c) *Fluidised beds* (as discussed in Chapter 6). In this case, an atomised liquid or suspension is sprayed on to a bed of hot fluidised particles and layers of solid build up to give enlarged particles the size of which is largely dependent on their residence time, that is the time over which successive layers of solids are deposited.
*Spouted beds* (as discussed in Chapter 6). These are used, particularly with large particles. In this case, the rapid circulation within the bed gives rise to a high level of inter-particle impacts. These processes are discussed by MORTENSEN and HOVMAND[32] and by MATHUR and EPSTEIN[33].

(d) *Drum and pan agglomerators*.
In *drum agglomerators*, particles are 'tumbled' in an open cylinder with roughened walls and subjected to a combination of gravitational and centrifugal forces. In order to aid agglomeration, liquid may be sprayed on to the surface of the bed or introduced through distribution pipes under the bed. Mean retention times in the equipment are in the range 60 to 120 s. A similar action is achieved in a *paddle mixer* where centrifugal forces dominate. In the *pan agglomerator*, a classifying action may be achieved which results in the fines having a preferentially longer retention time. Larger, denser and stronger agglomerates are produced as compared with those from the drum agglomerator.

(e) *Pug mills and extruders*.
*Pug mills* impart a complex kneading action that is a combination of ribbing and shearing and mixing. Densification and extrusion are both achieved in a single operation. The feed, which generally has only a small water content, is subjected to a high energy input which leads to a considerable rise in temperature. The action is similar to that occurring in an *extruder*. High degrees of compaction are achieved, leading to the production of pellets with low porosity with the result that less binder is required.

(f) *Elevated temperatures*.
With many materials, agglomeration may be achieved by heating as a result of which softening occurs in the surface layers. For the formation of porous metal sheets and discs, high temperatures are required.

Table 2.4. Size-enlargement methods and applications

| Method | Product size (mm) | Granule density | Scale of operation | Additional comments | Typical application |
|---|---|---|---|---|---|
| Tumbling granulators<br>Drums<br>Discs | 0.5 to 20 | Moderate | 0.5–800 tonne/h | Very spherical granules | Fertilisers, iron ore, ferrous ore, agricultural chemicals |
| Mixer granulators | | | | | |
| Continuous high shear (e.g. Shugi mixer) | 0.1 to 2 | Low to high | Up to 50 tonne/h | Handles very cohesive materials well, both batch and continuous | Chemicals, detergents, clays, carbon black |
| Batch high shear (e.g. paddle mixer) | 0.1 to 2 | High | Up to 500 kg batch | | Pharmaceuticals, ceramics |
| Fluidised granulators<br>Fluidised beds<br>Spouted beds<br>Wurster coaters | 0.1 to 2 | Low (agglomerated)<br>Moderate (layered) | 100–900 kg batch<br>50 tonne/h continuous | Flexible, relatively easy to scale, difficult for cohesive powders, good for coating applications | Continuous: fertilisers, inorganic salts, detergents<br>Batch: pharmaceuticals, agricultural chemicals, nuclear wastes |
| Centrifugal granulators | 0.3 to 3 | Moderate to high | Up to 200 kg batch | Powder layering and coating applications | Pharmaceuticals, agricultural chemicals |
| Spray methods | | | | | |
| Spray drying | 0.05 to 0.5 | Low | | Morphology of spray dried powders can vary widely | Instant foods, dyes, detergents, ceramics |
| Prilling | 0.7 to 2 | Moderate | | | Urea, ammonium nitrate |
| Pressure compaction | | High to very high | | Very narrow size distributions, very sensitive to powder flow and mechanical properties | Pharmaceuticals, catalysts, inorganic chemicals, organic chemicals plastic performs, metal parts, ceramics, clay minerals, animal feeds |
| Extrusion | >0.5 | | Up to 5 tonne/h | | |
| Roll press | >1 | | Up to 50 tonne/h | | |
| Tablet press | 10 | | Up to 1 tonne/h | | |
| Molding press | | | | | |
| Pellet mill | | | | | |
| Thermal processes | | | | | |
| Sintering | 2 to 50 | High to very high | Up to 100 tonne/h | Strongest bonding | Ferrous & non-ferrous ores, cement clinker minerals, ceramics |
| Liquid systems | | | | | |
| Immiscible wetting in mixers | <0.3 | Low | Up to 10 tonne/h | Wet processing based on flocculation properties of particulate feed | Coal fines, soot and oil removal from water |
| Sol–gel processes | | | | | Metal dicarbide, silica hydrogels |
| Pellet flocculation | | | | | Waste sludges and slurries |

Reprinted from *Granulation and Coating Technologies for High-Value-Added Industries*, Ennis and Litster (1996) with permission of E & G Associates. 

(g) *Pressure compaction.*

If a material is subjected to very high compaction forces, it may be formed into sheets, briquettes or tablets. In the tableting machines used for producing pills of pharmaceuticals, the powder is compressed into dies, either with or without the addition of a binder.

Powder compaction may also be achieved in roll processes, including briquetting, in which compression takes place between two rollers rotating at the same speed — that is without producing any shearing action. In pellet mills, a moist feed is forced through die holes where the resistance force is attributable to the friction between the powder and the walls of the dies.

A commercial pelleting process, used for powdery, lumpy and pasty products, is illustrated in Figure 2.35.

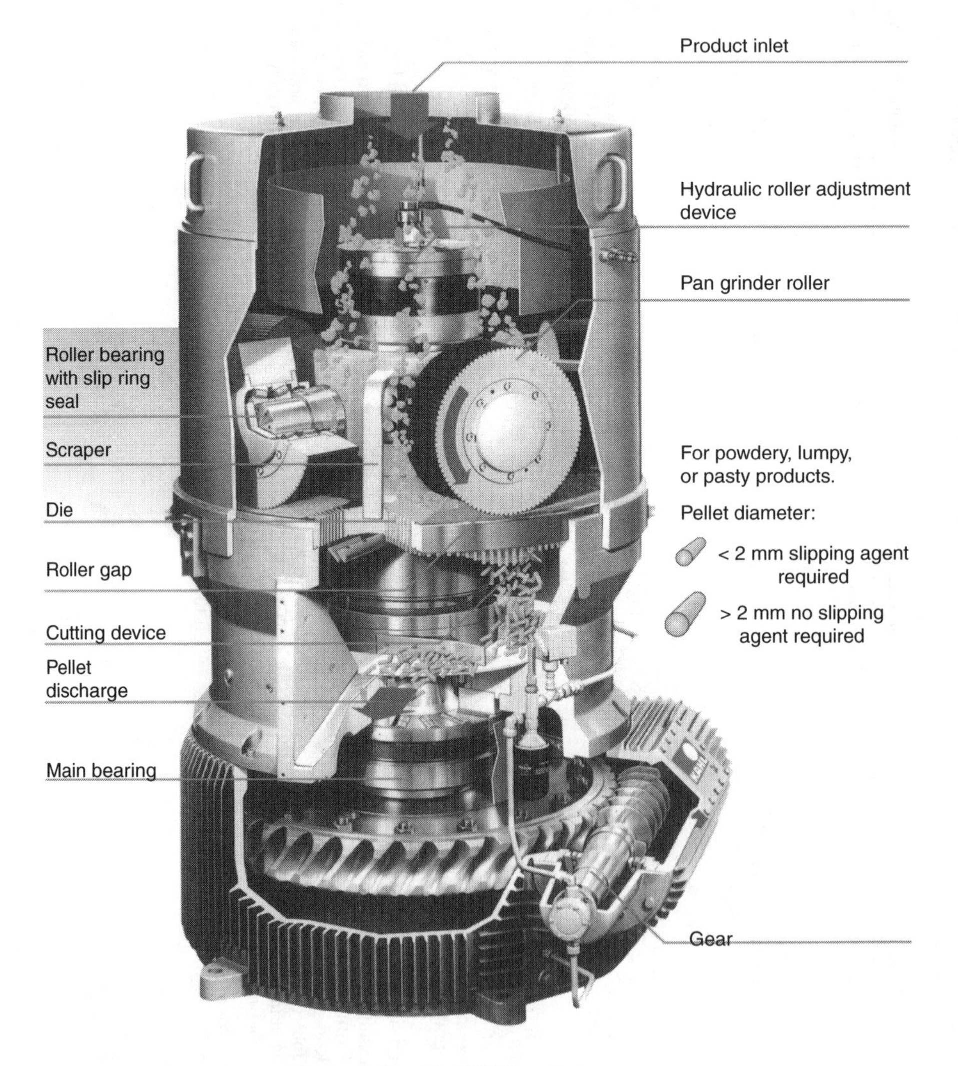


Figure 2.35. The KAHL pelleting press

## 2.5. FURTHER READING

BOND, F. C.: *Brit. Chem. Eng.* **6** (1961) 378–85, 543–8. Crushing and grinding calculations.
CAPES, C. E., GERMAIN, R. J. and COLEMAN, R. D.: *Ind. Eng. Chem. Proc. Des. Dev.* **16** (1977) 517. Bonding requirements for agglomeration by tumbling.
CAPES, C. E.: *Particle Size Enlargement (Handbook of Power Technology*, Volume 1: eds. Williams, J. C. and Allen, T.) (Elsevier Scientific Publishing Company, 1980).
KOSSEN, N. W. F. and HEERTJES, A. M.: *Chem. Eng. Sci.* **20** (1965) 593. The determination of contact angle for systems with a powder.
KRUIS, F. E., MAISEL, S. A. and FISSAN, H.: *A.I.Ch.E.Jl.* **46** (2000) 1735. Direct simulation Monte Carlo method for particle coagulation and aggregation.
LAWN, B. R. and WILSHAW, T. R.: *Fracture of Brittle Solids* (Cambridge University Press, Cambridge, 1975)
LOWRISON, G. C.: *Crushing and Grinding* (Butterworths, London, 1974)
MARSHALL, V. C. (ed): *Comminution* (Institution of Chemical Engineers, London, 1974)
MATHUR, K. B. and EPSTEIN, N.: *Spouted Beds* (Academic Press, New York, 1974)
NEWITT, D. M. and CONWAY-JONES, J. M.: *Trans. I. Chem. E.* **36** (1958) 422. A contribution to the theory and practice of granulation.
PERRY, R. H., GREEN, D. W. and MALONEY, J. O.: *Perry's Chemical Engineers' Handbook.* 7th edn. (McGraw-Hill, 1997). 20–56. Principles of size enlargements
PRASHNER, C. L.: *Crushing and Grinding Process Handbook* (Wiley, New York, 1987)
SHERRINGTON P. J.: *The Chemical Engineer* No. 220 (1968) CE201. The granulation of sand as an aid to understanding fertilizer granulation.
TRAIN, D. and LEWIS, C. J.: *Trans. I. Chem. E.* **40** (1962) 235. Agglomeration of solids by compaction.
TRAIN, D.: *Trans. I. Chem. E.* **35** (1957) 258. Transmission of forces through a powder mass during the process of pelleting.

## 2.6. REFERENCES

1. HEYWOOD, H.: *J. Imp. Coll. Chem. Eng. Soc.* **6** (1950–2) 26. Some notes on grinding research.
2. BEMROSE, C. R. and BRIDGWATER, J.: *Powder Technology* **49** (1987) 97. A review of attrition and attrition test methods.
3. HESS, W. and SCHÖNERT, K.: Proc. 1981 Powtech Conf. on Particle Technology, Birmingham 1981. EFCE Event No. 241 pp D2/I/1 - D2/I/9. Plastic Transition in Small Particles.
4. SCHÖNERT, K.: *Trans. Soc. Mining Engineers AIME* **252** (1972) 21–26. Role of fracture physics in understanding comminution phenomena.
5. PIRET, E. L.: *Chem. Eng. Prog.* **49** (1953) 56. Fundamental aspects of crushing.
6. PRASHER, C. L.: *Crushing and Grinding Process Handbook* (Wiley, 1987)
7. KICK, F.: *Das Gesetz der proportionalen Widerstande und seine Anwendungen*. (Leipzig, 1885)
8. VON RITTINGER, P. R.: *Lehrbuch der Aufbereitungskunde in ihrer neuesten Entwicklung und Ausbildung systematisch dargestellt.* (Ernst und Korn, 1867)
9. BOND, F. C.: *Min. Engng. N.Y.* **4** (1952) 484. Third theory of communition.
10. BOND, F. C.: *Chem. Eng., Albany* **59** (October 1952) 169. New grinding theory aids equipment selection.
11. AUSTIN, L. G. and KLIMPEL, R. R.: *Ind. Eng. Chem.* **56** No. 11 (November 1964) 18–29. The theory of grinding operations (53 refs.).
12. CUTTING, G. W.: *The Chemical Engineer* No. 325 (October 1977) 702–704. Grindability assessments using laboratory rod mill tests.
13. OWENS, J. S.: *Trans. Inst. Min. Met.* **42** (1933) 407. Notes on power used in crushing ore, with special reference to rolls and their behaviour.
14. GROSS, J.: *U.S. Bur Mines Bull.* **402** (1938). Crushing and grinding.
15. KWONG, J. N. S., ADAMS, J. T., JOHNSON, J. F. and PIRET, E. L.: *Chem. Eng. Prog.* **45** (1949) 508. Energy–new surface relationship in crushing. I. Application of permeability methods to an investigation of the crushing of some brittle solids.
16. ADAMS, J. T., JOHNSON, J. F. and PIRET, E. L.: *Chem. Eng. Prog.* **45** (1949) 655. Energy — new surface relationship in the crushing of solids. II. Application of permeability methods to an investigation of the crushing of halite.
17. JOHNSON, J. F., AXELSON, J. and PIRET, E. L.: *Chem. Eng. Prog.* **45** (1949) 708. Energy–new surface relationship in crushing. III. Application of gas adsorption methods to an investigation of the crushing of quartz.

18. ZELENY, R. A. and PIRET, E. L.: *Ind. Eng. Chem. Proc. Des. & Development* **1**, No. 1 (January, 1962) 37–41. Dissipation of energy in single particle crushing.
19. WORK, L. T.: *Ind. Eng. Chem.* **55** No. 2 (February, 1963) 56–58. Trends in particle size technology.
20. GANDOLFI, E. A. J., PAPACHRISTODOULOU, G. and TRASS, O.: *Powder Technology* **40** (1984) 269–282. Preparation of coal slurry fuels with the Szego mill.
21. KOKA, V. R. and TRASS, O.: *Powder Technology* **51** (1987) 201–204. Determination of breakage parameters and modelling of coal breakage in the Szego mill.
22. KANO, J., MIO, H. and SAITO, F.: *A.I.Ch.E.Jl.* **46** (2000) 1694. Correlation of grinding rate of gibbsite with impact energy balls.
23. BRADLEY, A. A., HINDE, A. L., LLOYD, P. J. D. and SCHYMURA, K.: *Proc. Europ. Symp. Particle Technol.* (Amsterdam, 1980) 153. Development in centrifugal milling.
24. KITSCHEN, L. P., LLOYD, P. J. D. and HARTMANN, R.: *Proc. 14th Int. Mineral Process. Congr.* (Toronto, 1982) I-9. 1. The centrifugal mill: experience with a new grinding system and its application.
25. BUTCHER, C.: *The Chemical Engineer* No 713 (23 November, 2000) Cryogenic grinding: an independent voice.
26. VON SMOLUCHOWSKI, M.: *Z. Physik. Chem.* **92** (1917) 129. Versuch einer mathematischen Theorie der Koagulationskinetik kolloider Losungen.
27. WALTON, J. S.: *The Formation and Properties of Precipitates* (Interscience, New York, 1967).
28. TADROS, T. F.: *Chem. Ind. (London)* **7** (1985) 210–218. Rheology of concentrated suspensions.
29. SÖHNEL, O. and MULLIN, J. W.: *A.I.Ch.E. Symposium Series* No. 284 **87** (1991) 182–190. Agglomeration of batch precipitated suspensions.
30. PERRY, R. H., GREEN, D. W. and MALONEY, J. O. (eds): *Perry's Chemical Engineers' Handbook.* 7th edn. (McGraw-Hill Book Company, New York, 1997).
31. BROWNING, J. E.: *Chem. Eng., Albany* **74** No. 25 (4th. December, 1967) 147. Agglomeration: Growing larger in applications and technology.
32. MORTENSEN S. and HOVMAND S.: in *Fluidization Technology*, Volume II, KEAIRNS D. L. (ed.) page 519. Particle formation and agglomeration in a spray granulator. (Hemisphere, Washington, 1976).
33. MATHUR, K. B. and EPSTEIN, N.: *Spouted Beds* (Academic Press, New York, 1974).

## 2.7. NOMENCLATURE

| | | Units in SI System | Dimensions in M, L, T |
|---|---|---|---|
| $A_1$ | Parameter in equation 2.14 | $m^3$ | $\mathbf{L}^3$ |
| $A_2$ | Parameter in equation 2.15 | — | — |
| $B_1$ | Coefficient in equation 2.14 | $m^3/s$ | $\mathbf{L}^3\mathbf{T}^{-1}$ |
| $B_2$ | Parameter in equation 2.15 | $s^{-1}$ | $\mathbf{T}^{-1}$ |
| $a$ | Crack length | m | **L** |
| $b$ | Half distance between crushing rolls | m | **L** |
| $C$ | A coefficient | — | $\mathbf{L}^{1-p}\mathbf{T}^{-2}$ |
| $c$ | Radius of yoke of pendulum mill | m | **L** |
| $d_0$ | Initial size of agglomerate | m | **L** |
| $d_c$ | Twice radius of gyration of planetary mill | m | **L** |
| $d_M$ | Diameter of cylindrical mill unit | m | **L** |
| $d_t$ | Size of agglomerate at time $t$ | m | **L** |
| $E$ | Energy per unit mass | J/kg | $\mathbf{L}^2\mathbf{T}^{-2}$ |
| $E_i$ | Work index | J/kg | $\mathbf{L}^2\mathbf{T}^{-2}$ |
| $F$ | Parameter in equation 2.1 | N/m | $\mathbf{M}\mathbf{T}^{-2}$ |
| $f_c$ | Crushing strength of material | $N/m^2$ | $\mathbf{M}\mathbf{L}^{-1}\mathbf{T}^{-2}$ |
| $g$ | Acceleration due to gravity | $m/s^2$ | $\mathbf{L}\mathbf{T}^{-2}$ |
| $K_K$ | Kick's constant | $m^3/kg$ | $\mathbf{M}^{-1}\mathbf{L}^3$ |
| $K_R$ | Rittinger's constant | $m^4/kg$ | $\mathbf{M}^{-1}\mathbf{L}^4$ |
| $L$ | Characteristic linear dimension | m | **L** |
| $l$ | Length of arm of pendulum mill | m | **L** |
| $M$ | Mass of crushing head in pendulum mill, or of particle | kg | **M** |
| $M'$ | Mass of arm of pendulum mill | kg | **M** |
| $m$ | Mass per unit length of arm of pendulum mill | kg/m | $\mathbf{M}\mathbf{L}^{-1}$ |

| | | Units in SI System | Dimensions in **M**, **L**, **T** |
|---|---|---|---|
| $N_c$ | Critical speed of rotation of ball mill (rev/time) | $s^{-1}$ | $\mathbf{T}^{-1}$ |
| $N_G$ | Speed of rotation of planetary mill about axis of gyration (rev/time) | $s^{-1}$ | $\mathbf{T}^{-1}$ |
| $N_M$ | Speed of rotation of cylindrical mill unit about own axis (rev/time) | $s^{-1}$ | $\mathbf{T}^{-1}$ |
| $p$ | A constant used as an index in equation 2.2 | — | — |
| $q$ | Size reduction factor $L_1/L_2$ | — | — |
| $R$ | Normal reaction | N | $\mathbf{MLT}^{-2}$ |
| $r$ | Radius of ball mill minus radius of particle | m | $\mathbf{L}$ |
| $r_1$ | Radius of crushing rolls | m | $\mathbf{L}$ |
| $r_2$ | Radius of particle in feed | m | $\mathbf{L}$ |
| $s$ | Gear ratio in planetary mill $(N_M/N_G)$ | — | — |
| $s_c$ | Value of $s$ at critical speed for given value of $d_G/d_M$ | — | — |
| $t$ | Time | s | $\mathbf{T}$ |
| $Y$ | Young's modulus | $N/m^2$ | $\mathbf{ML}^{-1}\mathbf{T}^{-2}$ |
| $y$ | Distance along arm from point of support | m | $\mathbf{L}$ |
| $\alpha$ | Half angle of nip | — | — |
| $\theta$ | Angle between axis and vertical | — | — |
| $\rho_s$ | Density of solid material | $kg/m^3$ | $\mathbf{ML}^{-3}$ |
| $\omega$ | Angular velocity | $s^{-1}$ | $\mathbf{T}^{-1}$ |
| $\omega_c$ | Critical speed of rotation of ball mill | $s^{-1}$ | $\mathbf{T}^{-1}$ |
| $\tau$ | Stress | N | $\mathbf{ML}^{-1}\mathbf{T}^{-2}$ |

CHAPTER 3

# *Motion of Particles in a Fluid*

## 3.1. INTRODUCTION

Processes for the separation of particles of various sizes and shapes often depend on the variation in the behaviour of the particles when they are subjected to the action of a moving fluid. Further, many of the methods for the determination of the sizes of particles in the sub-sieve ranges involve relative motion between the particles and a fluid.

The flow problems considered in Volume 1 are unidirectional, with the fluid flowing along a pipe or channel, and the effect of an obstruction is discussed only in so far as it causes an alteration in the forward velocity of the fluid. In this chapter, the force exerted on a body as a result of the flow of fluid past it is considered and, as the fluid is generally diverted all round it, the resulting three-dimensional flow is more complex. The flow of fluid relative to an infinitely long cylinder, a spherical particle and a non-spherical particle is considered, followed by a discussion of the motion of particles in both gravitational and centrifugal fields.

## 3.2. FLOW PAST A CYLINDER AND A SPHERE

The flow of fluid past an infinitely long cylinder, in a direction perpendicular to its axis, is considered in the first instance because this involves only two-directional flow, with no flow parallel to the axis. For a non-viscous fluid flowing past a cylinder, as shown in Figure 3.1, the velocity and direction of flow varies round the circumference. Thus at A and D the fluid is brought to rest and at B and C the velocity is at a maximum. Since the fluid is non-viscous, there is no drag, and an infinite velocity gradient exists at the surface of the cylinder. If the fluid is incompressible and the cylinder is small, the sum of the kinetic energy and the pressure energy is constant at all points on the surface. The kinetic energy is a maximum at B and C and zero at A and D, so that the pressure falls from A to B and from A to C and rises again from B to D and from C to D; the pressure at A and D being the same. No net force is therefore exerted by the fluid on the cylinder. It is found that, although the predicted pressure variation for a non-viscous fluid agrees well with the results obtained with a viscous fluid over the front face, very considerable differences occur at the rear face.

It is shown in Volume 1, Chapter 11 that, when a viscous fluid flows over a surface, the fluid is retarded in the boundary layer which is formed near the surface and that the boundary layer increases in thickness with increase in distance from the leading edge. If the pressure is falling in the direction of flow, the retardation of the fluid is less and the

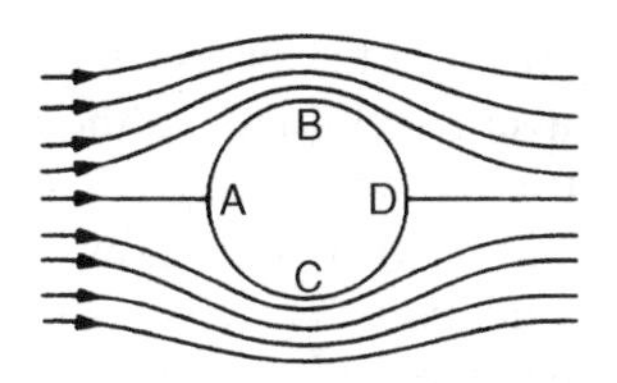


Figure 3.1. Flow round a cylinder

boundary layer is thinner in consequence. If the pressure is rising, however, there will be a greater retardation and the thickness of the boundary layer increases more rapidly. The force acting on the fluid at some point in the boundary layer may then be sufficient to bring it to rest or to cause flow in the reverse direction with the result that an eddy current is set up. A region of reverse flow then exists near the surface where the boundary layer has separated as shown in Figure 3.2. The velocity rises from zero at the surface to a maximum negative value and falls again to zero. It then increases in the positive direction until it reaches the main stream velocity at the edge of the boundary layer, as shown in Figure 3.2. At PQ the velocity in the $X$-direction is zero and the direction of flow in the eddies must be in the $Y$-direction.

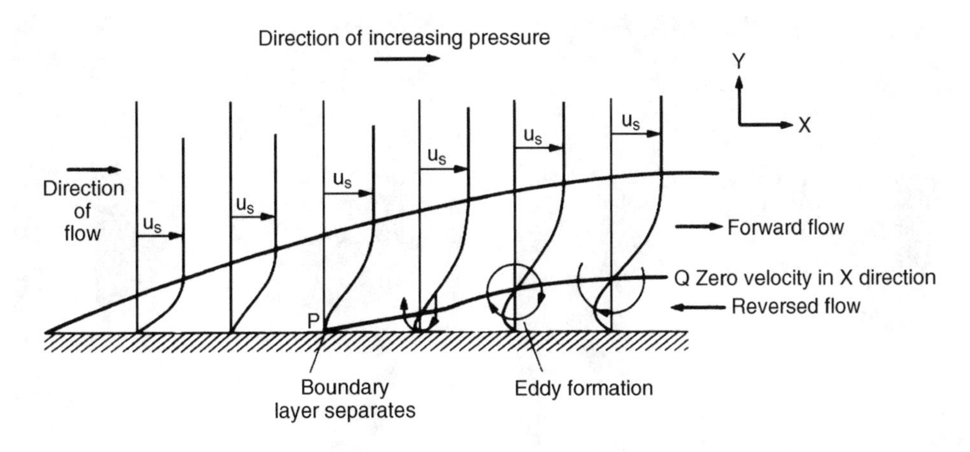


Figure 3.2. Flow of fluid over a surface against a pressure gradient

For the flow of a viscous fluid past the cylinder, the pressure decreases from A to B and from A to C so that the boundary layer is thin and the flow is similar to that obtained with a non-viscous fluid. From B to D and from C to D the pressure is rising and therefore the boundary layer rapidly thickens with the result that it tends to separate from the surface. If separation occurs, eddies are formed in the wake of the cylinder and energy is thereby dissipated and an additional force, known as form drag, is set up. In this way, on the forward surface of the cylinder, the pressure distribution is similar to that obtained with the ideal fluid of zero viscosity, although on the rear surface, the boundary layer is thickening rapidly and pressure variations are very different in the two cases.

All bodies immersed in a fluid are subject to a buoyancy force. In a flowing fluid, there is an additional force which is made up of two components: the skin friction (or viscous drag) and the form drag (due to the pressure distribution). At low rates of flow no separation of the boundary layer takes place, although as the velocity is increased, separation occurs and the skin friction forms a gradually decreasing proportion of the total drag. If the velocity of the fluid is very high, however, or if turbulence is artificially induced, the flow within the boundary layer will change from streamline to turbulent before separation takes place. Since the rate of transfer of momentum through a fluid in turbulent motion is much greater than that in a fluid flowing under streamline conditions, separation is less likely to occur, because the fast-moving fluid outside the boundary layer is able to keep the fluid within the boundary layer moving in the forward direction. If separation does occur, this takes place nearer to D in Figure 3.1, the resulting eddies are smaller and the total drag will be reduced.

Turbulence may arise either from an increased fluid velocity or from artificial roughening of the forward face of the immersed body. Prandtl roughened the forward face of a sphere by fixing a hoop to it, with the result that the drag was considerably reduced. Further experiments have been carried out in which sand particles have been stuck to the front face, as shown in Figure 3.3. The tendency for separation, and hence the magnitude of the form drag, are also dependent on the shape of the body.

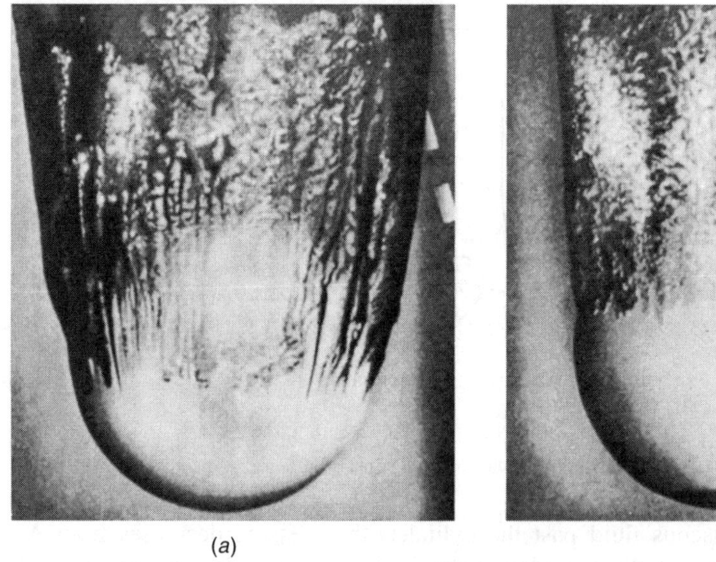


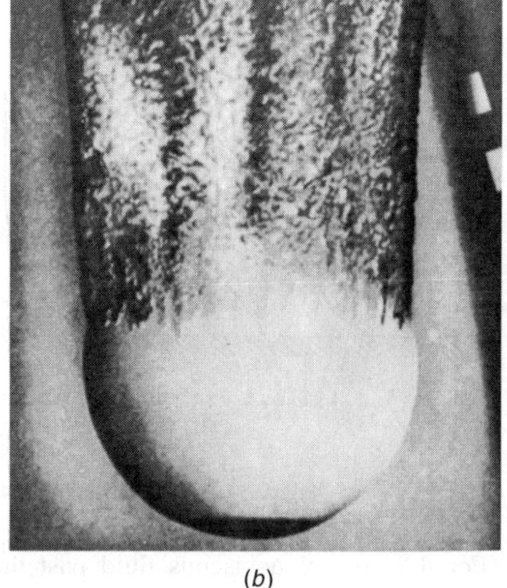


Figure 3.3. Effect of roughening front face of a sphere (*a*) 216 mm diameter ball entering water at 298 K (*b*) As above, except for 100 mm diameter patch of sand on nose

Conditions of flow relative to a spherical particle are similar to those relative to a cylinder, except that the flow pattern is three-directional. The flow is characterised by the Reynolds number $Re'(= ud\rho/\mu)$ in which $\rho$ is the density of the fluid, $\mu$ is the viscosity of the fluid, $d$ is the diameter of the sphere, and $u$ is the velocity of the fluid relative to the particle.

For the case of *creeping flow*, that is flow at very low velocities relative to the sphere, the drag force $F$ on the particle was obtained in 1851 by STOKES[1] who solved the hydrodynamic equations of motion, the Navier–Stokes equations, to give:

$$F = 3\pi\mu d u \tag{3.1}$$

Equation 3.1, which is known as Stokes' law is applicable only at very low values of the particle Reynolds number and deviations become progressively greater as $Re'$ increases. Skin friction constitutes two-thirds of the total drag on the particle as given by equation 3.1. Thus, the total force $F$ is made up of two components:

$$\left.\begin{array}{ll}\text{(i) skin friction:} & 2\pi\mu d u \\ \text{(ii) form drag:} & \pi\mu d u\end{array}\right\} \text{ total } 3\pi\mu d u$$

As $Re'$ increases, skin friction becomes proportionately less and, at values greater than about 20, *flow separation* occurs with the formation of vortices in the wake of the sphere. At high Reynolds numbers, the size of the vortices progressively increases until, at values of between 100 and 200, instabilities in the flow give rise to *vortex shedding*. The effect of these changes in the nature of the flow on the force exerted on the particle is now considered.

## 3.3. THE DRAG FORCE ON A SPHERICAL PARTICLE

### 3.3.1. Drag coefficients

The most satisfactory way of representing the relation between drag force and velocity involves the use of two dimensionless groups, similar to those used for correlating information on the pressure drop for flow of fluids in pipes.

The first group is the particle Reynolds number $Re'(= ud\rho/\mu)$.

The second is the group $R'/\rho u^2$, in which $R'$ is the force per unit projected area of particle in a plane perpendicular to the direction of motion. For a sphere, the projected area is that of a circle of the same diameter as the sphere.

Thus:
$$R' = \frac{F}{(\pi d^2/4)} \tag{3.2}$$

and
$$\frac{R'}{\rho u^2} = \frac{4F}{\pi d^2 \rho u^2} \tag{3.3}$$

$R'/\rho u^2$ is a form of *drag coefficient*, often denoted by the symbol $C'_D$. Frequently, a drag coefficient $C_D$ is defined as the ratio of $R'$ to $\frac{1}{2}\rho u^2$.

Thus:
$$C_D = 2C'_D = \frac{2R'}{\rho u^2} \tag{3.4}$$

It is seen that $C'_D$ is analogous to the friction factor $\phi(= R/\rho u^2)$ for pipe flow, and $C_D$ is analogous to the Fanning friction factor $f$.

When the force $F$ is given by Stokes' law (equation 3.1), then:

$$\frac{R'}{\rho u^2} = 12\frac{\mu}{ud\rho} = 12Re'^{-1} \tag{3.5}$$

Equations 3.1 and 3.5 are applicable only at very low values of the Reynolds number $Re'$. GOLDSTEIN[2] has shown that, for values of $Re'$ up to about 2, the relation between $R'/\rho u^2$ and $Re'$ is given by an infinite series of which equation 3.5 is just the first term.

Thus:
$$\frac{R'}{\rho u^2} = \frac{12}{Re'}\left\{1 + \frac{3}{16}Re' - \frac{19}{1280}Re'^2 + \frac{71}{20{,}480}Re'^3 - \frac{30{,}179}{34{,}406{,}400}Re'^4 + \frac{122{,}519}{560{,}742{,}400}Re'^5 - \cdots\right\} \tag{3.6}$$

OSEEN[3] employs just the first two terms of equation 3.6 to give:

$$\frac{R'}{\rho u^2} = 12Re'^{-1}\left(1 + \frac{3}{16}Re'\right) \tag{3.7}$$

The correction factors for Stokes' law from both equation 3.6 and equation 3.7 are given in Table 3.1. It is seen that the correction becomes progressively greater as $Re'$ increases.

Table 3.1. Correction factors for Stokes' law

| $Re'$ | Goldstein eqn. 3.6 | Oseen eqn. 3.7 | Schiller & Naumann eqn. 3.9 | Wadell eqn. 3.12 | Khan & Richardson eqn. 3.13 |
|---|---|---|---|---|---|
| 0.01 | 1.002 | 1.002 | 1.007 | 0.983 | 1.038 |
| 0.03 | 1.006 | 1.006 | 1.013 | 1.00 | 1.009 |
| 0.1 | 1.019 | 1.019 | 1.03 | 1.042 | 1.006 |
| 0.2 | 1.037 | 1.037 | 1.05 | 1.067 | 1.021 |
| 0.3 | 1.055 | 1.056 | 1.07 | 1.115 | 1.038 |
| 0.6 | 1.108 | 1.113 | 1.11 | 1.346 | 1.085 |
| 1 | 1.18 | 1.19 | 1.15 | 1.675 | 1.137 |
| 2 | 1.40 | 1.38 | 1.24 | 1.917 | 1.240 |

Several workers have used numerical methods for solving the equations of motion for flow at higher Reynolds numbers relative to spherical and cylindrical particles. These include, JENSON[4], and LE CLAIR, HAMIELEC and PRUPPACHER[5].

The relation between $R'/\rho u^2$ and $Re'$ is conveniently given in graphical form by means of a logarithmic plot as shown in Figure 3.4. The graph may be divided into four regions as shown. The four regions are now considered in turn.

### *Region (a) ($10^{-4} < Re' < 0.2$)*

In this region, the relationship between $\frac{R'}{\rho u^2}$ and $Re'$ is a straight line of slope $-1$ represented by equation 3.5:

$$\frac{R'}{\rho u^2} = 12Re'^{-1} \qquad \text{(equation 3.5)}$$

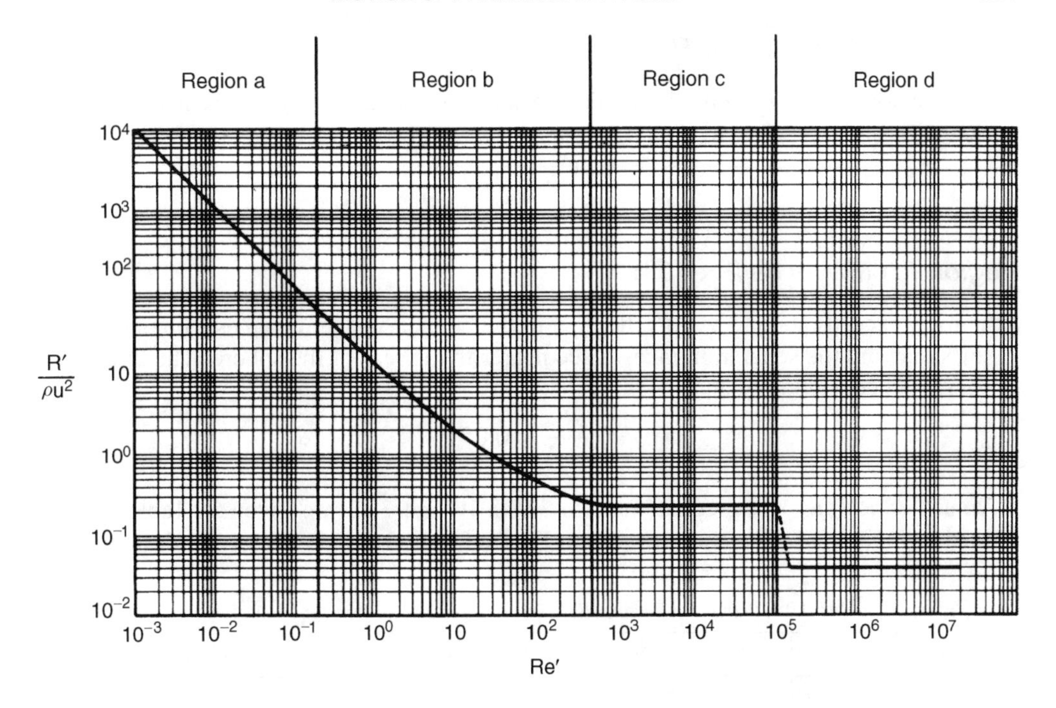


Figure 3.4. $R'/\rho u^2$ versus $Re'$ for spherical particles

The limit of $10^{-4}$ is imposed because reliable experimental measurements have not been made at lower values of $Re'$, although the equation could be applicable down to very low values of $Re'$, provided that the dimensions of the particle are large compared with the mean free path of the fluid molecules so that the fluid behaves as a continuum.

The upper limit of $Re' = 0.2$ corresponds to the condition where the error arising from the application of Stokes' law is about 4 per cent. This limit should be reduced if a greater accuracy is required, and it may be raised if a lower level of accuracy is acceptable.

## *Region* (*b*) (*0.2 < Re' < 500–1000*)

In this region, the slope of the curve changes progressively from $-1$ to 0 as $Re'$ increases. Several workers have suggested approximate equations for flow in this intermediate region. Dallavelle[6] proposed that $R'/\rho u^2$ may be regarded as being composed of two component parts, one due to Stokes' law and the other, a constant, due to additional non-viscous effects.

Thus:
$$\frac{R'}{\rho u^2} = 12Re'^{-1} + 0.22 \tag{3.8}$$

Schiller and Naumann[7] gave the following simple equation which gives a reasonable approximation for values of $Re'$ up to about 1000:

$$\frac{R'}{\rho u^2} = 12Re'^{-1}(1 + 0.15Re'^{0.687}) \tag{3.9}$$

### Region (c) (500–1000 < $Re'$ < ca $2 \times 10^5$)

In this region, *Newton's law* is applicable and the value of $R'/\rho u^2$ is approximately constant giving:

$$\frac{R'}{\rho u^2} = 0.22 \tag{3.10}$$

### Region (d) ($Re'$ > ca $2 \times 10^5$)

When $Re'$ exceeds about $2 \times 10^5$, the flow in the boundary layer changes from streamline to turbulent and the separation takes place nearer to the rear of the sphere. The drag force is decreased considerably and:

$$\frac{R'}{\rho u^2} = 0.05 \tag{3.11}$$

Values of $R'/\rho u^2$ using equations 3.5, 3.9, 3.10 and 3.11 are given in Table 3.2 and plotted in Figure 3.4. The curve shown in Figure 3.4 is really continuous and its division

Table 3.2. $R'/\rho u^2$, $(R'/\rho u^2)Re'^2$ and $(R'/\rho u^2)Re'^{-1}$ as a function of $Re'$

| $Re'$ | $R'/\rho u^2$ | $(R'/\rho u^2)Re'^2$ | $(R'/\rho u^2)Re'^{-1}$ |
|---|---|---|---|
| $10^{-3}$ | 12,000 | | |
| $2 \times 10^{-3}$ | 6000 | | |
| $5 \times 10^{-3}$ | 2400 | | |
| $10^{-2}$ | 1200 | $1.20 \times 10^{-1}$ | $1.20 \times 10^{5}$ |
| $2 \times 10^{-2}$ | 600 | $2.40 \times 10^{-1}$ | $3.00 \times 10^{4}$ |
| $5 \times 10^{-2}$ | 240 | $6.00 \times 10^{-1}$ | $4.80 \times 10^{3}$ |
| $10^{-1}$ | 124 | 1.24 | $1.24 \times 10^{3}$ |
| $2 \times 10^{-1}$ | 63 | 2.52 | $3.15 \times 10^{2}$ |
| $5 \times 10^{-1}$ | 26.3 | 6.4 | $5.26 \times 10$ |
| $10^{0}$ | 13.8 | $1.38 \times 10$ | $1.38 \times 10$ |
| $2 \times 10^{0}$ | 7.45 | $2.98 \times 10$ | 3.73 |
| $5 \times 10^{0}$ | 3.49 | $8.73 \times 10$ | $7.00 \times 10^{-1}$ |
| 10 | 2.08 | $2.08 \times 10^{2}$ | $2.08 \times 10^{-1}$ |
| $2 \times 10$ | 1.30 | $5.20 \times 10^{2}$ | $6.50 \times 10^{-2}$ |
| $5 \times 10$ | 0.768 | $1.92 \times 10^{3}$ | $1.54 \times 10^{-2}$ |
| $10^{2}$ | 0.547 | $5.47 \times 10^{3}$ | $5.47 \times 10^{-3}$ |
| $2 \times 10^{2}$ | 0.404 | $1.62 \times 10^{4}$ | $2.02 \times 10^{-3}$ |
| $5 \times 10^{2}$ | 0.283 | $7.08 \times 10^{4}$ | $5.70 \times 10^{-4}$ |
| $10^{3}$ | 0.221 | $2.21 \times 10^{5}$ | $2.21 \times 10^{-4}$ |
| $2 \times 10^{3}$ | 0.22 | $8.8 \times 10^{5}$ | $1.1 \times 10^{-4}$ |
| $5 \times 10^{3}$ | 0.22 | $5.5 \times 10^{6}$ | $4.4 \times 10^{-5}$ |
| $10^{4}$ | 0.22 | $2.2 \times 10^{7}$ | $2.2 \times 10^{-5}$ |
| $2 \times 10^{4}$ | 0.22 | | |
| $5 \times 10^{4}$ | 0.22 | | |
| $10^{5}$ | 0.22 | | |
| $2 \times 10^{5}$ | 0.05 | | |
| $5 \times 10^{5}$ | 0.05 | | |
| $10^{6}$ | 0.05 | | |
| $2 \times 10^{6}$ | 0.05 | | |
| $5 \times 10^{6}$ | 0.05 | | |
| $10^{7}$ | 0.05 | | |

into four regions is merely a convenient means by which a series of simple equations can be assigned to limited ranges of values of $Re'$.

A comprehensive review of the various equations proposed to relate drag coefficient to particle Reynolds number has been carried out by CLIFT, GRACE and WEBER[8]. One of the earliest equations applicable over a wide range of values of $Re'$ is that due to WADELL[9] which may be written as:

$$\frac{R'}{\rho u^2} = \left(0.445 + \frac{3.39}{\sqrt{Re'}}\right)^2 \tag{3.12}$$

Subsequently, KHAN and RICHARDSON[10] have examined the experimental data and suggest that a very good correlation between $R'/\rho u^2$ and $Re'$, for values of $Re'$ up to $10^5$, is given by:

$$\frac{R'}{\rho u^2} = [1.84Re'^{-0.31} + 0.293Re'^{0.06}]^{3.45} \tag{3.13}$$

In Table 3.3, values of $R'/\rho u^2$, calculated from equations 3.12 and 3.13, together with values from the Schiller and Naumann equation 3.9, are given as a function of $Re'$ over the range $10^{-2} < Re' < 10^5$. Values are plotted in Figure 3.5 from which it will be noted that equation 3.13 gives a shallow minimum at $Re'$ of about $10^4$, with values rising to 0.21 at $Re' = 10^5$. This agrees with the limited experimental data which are available in this range.

Table 3.3. Values of drag coefficient $R'/\rho u^2$ as a function of $Re'$

| $Re'$ | Schiller & Naumann eqn. 3.9 | Wadell eqn. 3.12 | Khan & Richardson eqn. 3.13 |
|---|---|---|---|
| 0.01 | 1208 | 1179 | 1246 |
| 0.1 | 124 | 125 | 121 |
| 1 | 13.8 | 14.7 | 13.7 |
| 10 | 2.07 | 2.3 | 2.09 |
| 100 | 0.55 | 0.62 | 0.52 |
| 500 | 0.281 | 0.356 | 0.283 |
| 1000 | 0.219 | 0.305 | 0.234 |
| 3000 | 0.151 | 0.257 | 0.200 |
| 10,000 | 0.10 | 0.229 | 0.187 |
| 30,000 | — | 0.216 | 0.191 |
| 100,000 | — | 0.208 | 0.210 |

For values of $Re' < 2$, correction factors for Stokes' law have been calculated from equations 3.9, 3.12 and these 3.13 and are these included in Table 3.1.

### 3.3.2. Total force on a particle

The force on a spherical particle may be expressed using equations 3.5, 3.9, 3.10 and 3.11 for each of the regions $a$, $b$, $c$ and $d$ as follows.

In region ($a$):

$$R' = 12\rho u^2\left(\frac{\mu}{ud\rho}\right) = \frac{12u\mu}{d} \tag{3.14}$$

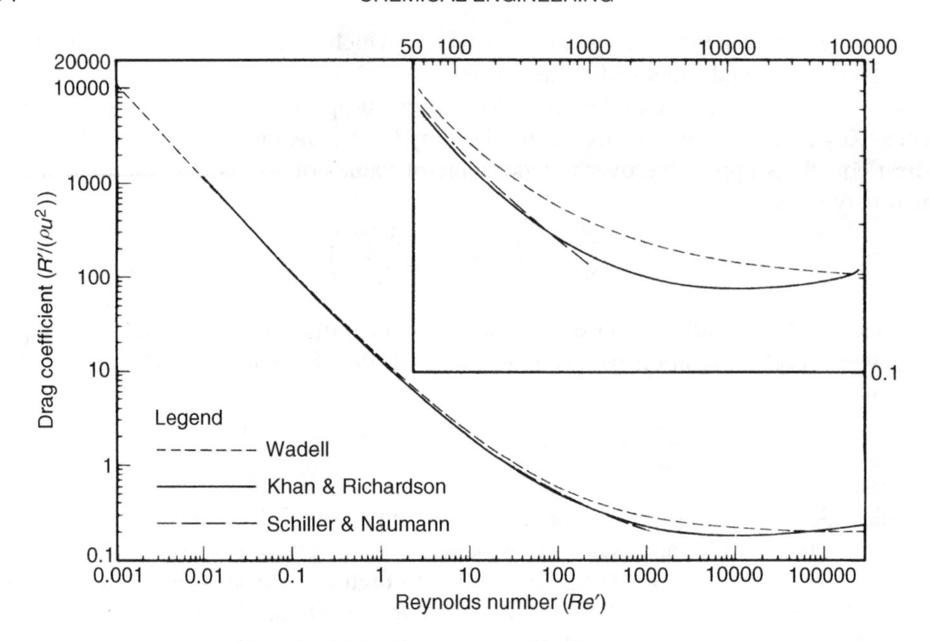


Figure 3.5. $R'/\rho u^2$ versus $Re'$ for spherical particles from equations 3.9 (SCHILLER and NAUMANN[7]), 3.12 (WADELL[9]) and 3.13 (KHAN and RICHARDSON[10]). The enlarged section covers the range of $Re'$ from 50 to $10^5$

The projected area of the particle is $\pi d^2/4$. Thus the total force on the particle is given by:

$$F = \frac{12u\mu}{d}\tfrac{1}{4}\pi d^2 = 3\pi\mu d u \tag{3.15}$$

This is the expression originally obtained by STOKES[1] already given as equation 3.1.

In region ($b$), from equation 3.9:

$$R' = \frac{12u\mu}{d}(1 + 0.15Re'^{0.687}) \tag{3.16}$$

and therefore:

$$F = 3\pi\mu d u(1 + 0.15Re'^{0.687}) \tag{3.17}$$

In region ($c$):

$$R' = 0.22\rho u^2 \tag{3.18}$$

and:

$$F = 0.22\rho u^2 \tfrac{1}{4}\pi d^2 = 0.055\pi d^2 \rho u^2 \tag{3.19}$$

This relation is often known as Newton's law.

In region ($d$):

$$R' = 0.05\rho u^2 \tag{3.20}$$

$$F = 0.0125\pi d^2 \rho u^2 \tag{3.21}$$

Alternatively using equation 3.13, which is applicable over the first three regions ($a$), ($b$) and ($c$) gives:

$$F = \frac{\pi}{4} d^2 \rho u^2 (1.84Re'^{-0.31} + 0.293Re'^{0.06})^{3.45} \tag{3.22}$$

### 3.3.3. Terminal falling velocities

If a spherical particle is allowed to settle in a fluid under gravity, its velocity will increase until the accelerating force is exactly balanced by the resistance force. Although this state is approached exponentially, the effective acceleration period is generally of short duration for very small particles. If this terminal falling velocity is such that the corresponding value of $Re'$ is less than 0.2, the drag force on the particle is given by equation 3.15. If the corresponding value of $Re'$ lies between 0.2 and 500, the drag force is given approximately by Schiller and Naumann in equation 3.17. It may be noted, however, that if the particle has started from rest, the drag force is given by equation 3.15 until $Re'$ exceeds 0.2. Again if the terminal falling velocity corresponds to a value of $Re'$ greater than about 500, the drag on the particle is given by equation 3.19. Under terminal falling conditions, velocities are rarely high enough for $Re'$ to approach $10^5$, with the small particles generally used in industry.

The accelerating force due to gravity is given by:

$$= (\tfrac{1}{6}\pi d^3)(\rho_s - \rho)g \tag{3.23}$$

where $\rho_s$ is the density of the solid.

The terminal falling velocity $u_0$ corresponding to region $(a)$ is given by:

$$(\tfrac{1}{6}\pi d^3)(\rho_s - \rho)g = 3\pi\mu d u_0$$

and:

$$u_0 = \frac{d^2 g}{18\mu}(\rho_s - \rho) \tag{3.24}$$

The terminal falling velocity corresponding to region $(c)$ is given by:

$$(\tfrac{1}{6}\pi d^3)(\rho_s - \rho)g = 0.055\pi d^2 \rho u_0^2$$

or:

$$u_0^2 = 3dg\frac{(\rho_s - \rho)}{\rho} \tag{3.25}$$

In the expressions given for the drag force and the terminal falling velocity, the following assumptions have been made:

(a) That the settling is not affected by the presence of other particles in the fluid. This condition is known as "free settling". When the interference of other particles is appreciable, the process is known as "hindered settling".
(b) That the walls of the containing vessel do not exert an appreciable retarding effect.
(c) That the fluid can be considered as a continuous medium, that is the particle is large compared with the mean free path of the molecules of the fluid, otherwise the particles may occasionally "slip" between the molecules and thus attain a velocity higher than that calculated.

These factors are considered further in Sections 3.3.4 and 3.3.5 and in Chapter 5.

From equations 3.24 and 3.25, it is seen that terminal falling velocity of a particle in a given fluid becomes greater as both particle size and density are increased. If for a particle of material **A** of diameter $d_A$ and density $\rho_A$, Stokes' law is applicable, then the terminal falling velocity $u_{0A}$ is given by equation 3.24 as:

$$u_{0A} = \frac{d_A^2 g}{18\mu}(\rho_A - \rho) \tag{3.26}$$

Similarly, for a particle of material **B**:

$$u_{0B} = \frac{d_B^2 g}{18\mu}(\rho_B - \rho) \tag{3.27}$$

The condition for the two terminal velocities to be equal is then:

$$\frac{d_B}{d_A} = \left(\frac{\rho_A - \rho}{\rho_B - \rho}\right)^{1/2} \tag{3.28}$$

If Newton's law is applicable, equation 3.25 holds and:

$$u_{0A}^2 = \frac{3d_A g(\rho_A - \rho)}{\rho} \tag{3.29}$$

and

$$u_{0B}^2 = \frac{3d_B g(\rho_B - \rho)}{\rho} \tag{3.30}$$

For equal settling velocities:

$$\frac{d_B}{d_A} = \left(\frac{\rho_A - \rho}{\rho_B - \rho}\right) \tag{3.31}$$

In general, the relationship for equal settling velocities is:

$$\frac{d_B}{d_A} = \left(\frac{\rho_A - \rho}{\rho_B - \rho}\right)^S \tag{3.32}$$

where $S = \frac{1}{2}$ for the Stokes' law region, $S = 1$ for Newton's law and, as an approximation, $\frac{1}{2} < S < 1$ for the intermediate region.

This method of calculating the terminal falling velocity is satisfactory provided that it is known which equation should be used for the calculation of drag force or drag coefficient. It has already been seen that the equations give the drag coefficient in terms of the particle Reynolds number $Re'_0$ $(= u_0 d\rho/\mu)$ which is itself a function of the terminal falling velocity $u_0$ which is to be determined. The problem is analogous to that discussed in Volume 1, where the calculation of the velocity of flow in a pipe in terms of a known pressure difference presents difficulties, because the unknown velocity appears in both the friction factor and the Reynolds number.

The problem is most effectively solved by the generation of a new dimensionless group which is independent of the particle velocity. The resistance force per unit projected area of the particle under terminal falling conditions $R'_0$ is given by:

$$R'_0 \tfrac{1}{4}\pi d^2 = \tfrac{1}{6}\pi d^3(\rho_s - \rho)g$$

or:

$$R'_0 = \tfrac{2}{3}d(\rho_s - \rho)g \tag{3.33}$$

Thus:
$$\frac{R'_0}{\rho u_0^2} = \frac{2dg}{3\rho u_0^2}(\rho_s - \rho) \tag{3.34}$$

The dimensionless group $(R'_0/\rho u_0^2)Re_0'^2$ does not involve $u_0$ since:

$$\frac{R'_0}{\rho u_0^2}\frac{u_0^2 d^2 \rho^2}{\mu^2} = \frac{2dg(\rho_s - \rho)}{3\rho u_0^2}\frac{u_0^2 d^2 \rho^2}{\mu^2}$$
$$= \frac{2d^3(\rho_s - \rho)\rho g}{3\mu^2} \tag{3.35}$$

The group $\dfrac{d^3\rho(\rho_s - \rho)g}{\mu^2}$ is known as the Galileo number $Ga$ or sometimes the Archimedes number $Ar$.

Thus:
$$\frac{R'_0}{\rho u_0^2}Re_0'^2 = \tfrac{2}{3}Ga \tag{3.36}$$

Using equations 3.5, 3.9 and 3.10 to express $R'/\rho u^2$ in terms of $Re'$ over the appropriate range of $Re'$, then:

$$Ga = 18Re'_0 \qquad (Ga < 3.6) \tag{3.37}$$
$$Ga = 18Re'_0 + 2.7Re_0'^{1.687} \qquad (3.6 < Ga < ca.\ 10^5) \tag{3.38}$$
$$Ga = \tfrac{1}{3}Re_0'^2 \qquad (Ga > ca.\ 10^5) \tag{3.39}$$

$(R'_0/\rho u_0^2)Re_0'^2$ can be evaluated if the properties of the fluid and the particle are known.

In Table 3.4, values of log $Re'$ are given as a function of $\log\{(R'/\rho u^2)Re'^2\}$ and the data taken from tables given by HEYWOOD(11), are represented in graphical form in Figure 3.6. In order to determine the terminal falling velocity of a particle, $(R'_0/\rho u_0^2)Re_0'^2$ is evaluated and the corresponding value of $Re'_0$, and hence of the terminal velocity, is found either from Table 3.4 or from Figure 3.6.

Table 3.4. Values of log $Re'$ as a function of $\log\{(R'/\rho u^2)Re'^2\}$ for spherical particles

| $\log\{(R'/\rho u^2)Re'^2\}$ | 0.0 | 0.1 | 0.2 | 0.3 | 0.4 | 0.5 | 0.6 | 0.7 | 0.8 | 0.9 |
|---|---|---|---|---|---|---|---|---|---|---|
| $\bar{2}$ | | | | | | | | $\bar{3}.620$ | $\bar{3}.720$ | $\bar{3}.819$ |
| $\bar{1}$ | $\bar{3}.919$ | $\bar{2}.018$ | $\bar{2}.117$ | $\bar{2}.216$ | $\bar{2}.315$ | $\bar{2}.414$ | $\bar{2}.513$ | $\bar{2}.612$ | $\bar{2}.711$ | $\bar{2}.810$ |
| 0 | $\bar{2}.908$ | $\bar{1}.007$ | $\bar{1}.105$ | $\bar{1}.203$ | $\bar{1}.301$ | $\bar{1}.398$ | $\bar{1}.495$ | $\bar{1}.591$ | $\bar{1}.686$ | $\bar{1}.781$ |
| 1 | $\bar{1}.874$ | $\bar{1}.967$ | 0.008 | 0.148 | 0.236 | 0.324 | 0.410 | 0.495 | 0.577 | 0.659 |
| 2 | 0.738 | 0.817 | 0.895 | 0.972 | 1.048 | 1.124 | 1.199 | 1.273 | 1.346 | 1.419 |
| 3 | 1.491 | 1.562 | 1.632 | 1.702 | 1.771 | 1.839 | 1.907 | 1.974 | 2.040 | 2.106 |
| 4 | 2.171 | 2.236 | 2.300 | 2.363 | 2.425 | 2.487 | 2.548 | 2.608 | 2.667 | 2.725 |
| 5 | 2.783 | 2.841 | 2.899 | 2.956 | 3.013 | 3.070 | 3.127 | 3.183 | 3.239 | 3.295 |

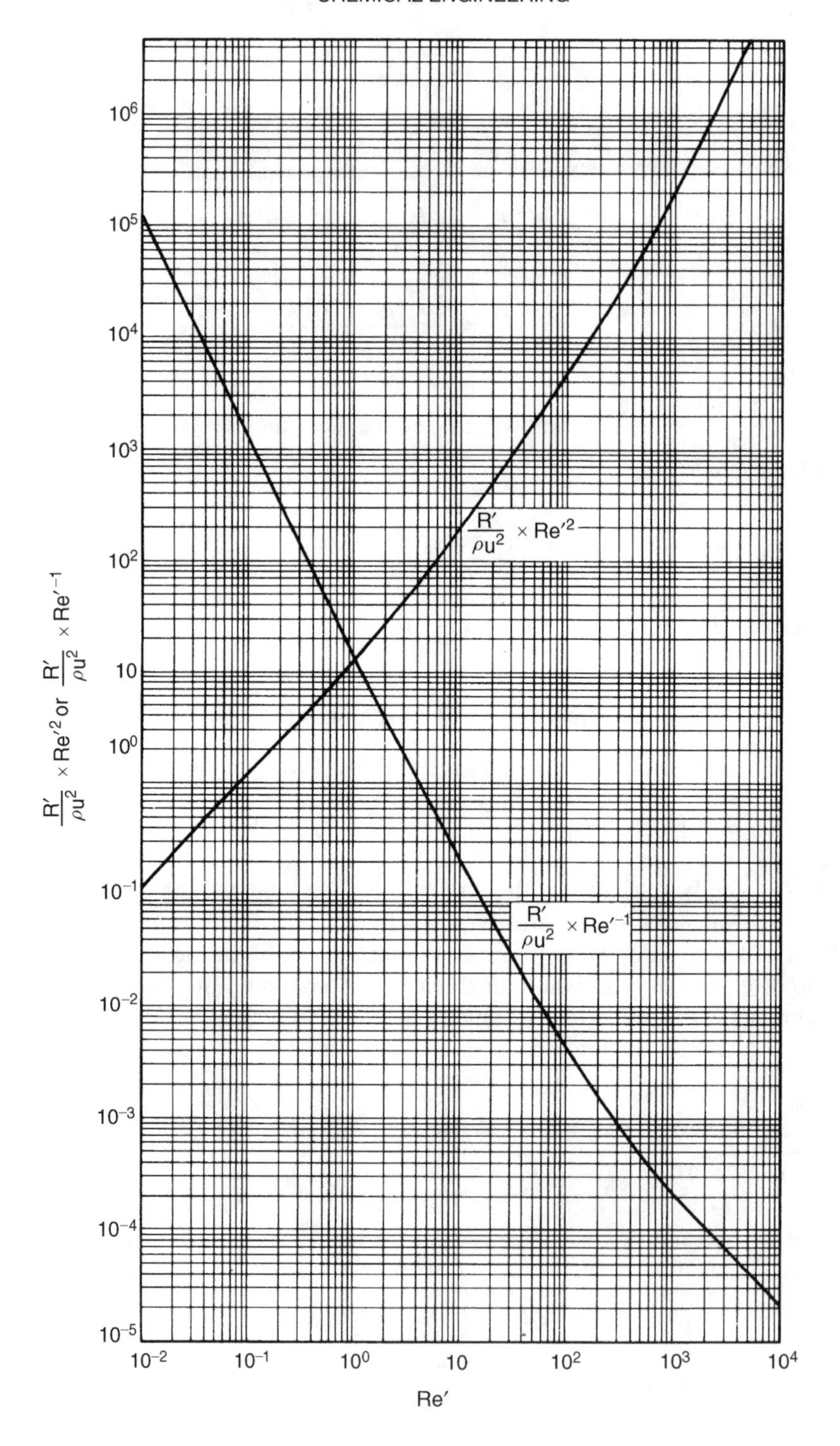


Figure 3.6. $(R'/\rho u^2)Re'^2$ and $(R'/\rho u^2)Re'^{-1}$ versus $Re'$ for spherical particles

## Example 3.1

What is the terminal velocity of a spherical steel particle, 0.40 mm in diameter, settling in an oil of density 820 kg/m$^3$ and viscosity 10 mN s/m$^2$? The density of steel is 7870 kg/m$^3$.

## Solution

For a sphere:

$$\frac{R'_0}{\rho u_0^2} Re_0'^2 = \frac{2d^3(\rho_s - \rho)\rho g}{3\mu^2} \qquad \text{(equation 3.35)}$$

$$= \frac{2 \times 0.0004^3 \times 820(7870 - 820)9.81}{3(10 \times 10^{-3})^2}$$

$$= 24.2$$

$$\log_{10} 24.2 = 1.384$$

From Table 3.4: $\log_{10} Re'_0 = 0.222$

Thus: $Re'_0 = 1.667$

and: $$u_0 = \frac{1.667 \times 10 \times 10^{-3}}{820 \times 0.0004}$$

$$= 0.051 \text{ m/s or } \underline{\underline{51 \text{ mm/s}}}$$

## Example 3.2

A finely ground mixture of galena and limestone in the proportion of 1 to 4 by mass is subjected to elutriation by an upward-flowing stream of water flowing at a velocity of 5 mm/s. Assuming that the size distribution for each material is the same, and is as shown in the following table, estimate the percentage of galena in the material carried away and in the material left behind. The viscosity of water is 1 mN s/m$^2$ and Stokes' equation (3.1) may be used.

| Diameter (μm) | 20 | 30 | 40 | 50 | 60 | 70 | 80 | 100 |
|---|---|---|---|---|---|---|---|---|
| Undersize (per cent by mass) | 15 | 28 | 48 | 54 | 64 | 72 | 78 | 88 |

The densities of galena and limestone are 7500 and 2700 kg/m$^3$, respectively.

## Solution

The first step is to determine the size of a particle which has a settling velocity equal to that of the upward flow of fluid, that is 5 mm/s.

Taking the largest particle, $d = (100 \times 10^{-6}) = 0.0001$ m

and: $$Re' = (5 \times 10^{-3} \times 0.0001 \times 1000)/(1 \times 10^{-3}) = 0.5$$

Thus, for the bulk of particles, the flow will be within region ($a$) in Figure 3.4 and the settling velocity is given by Stokes' equation:

$$u_0 = (d^2 g/18\mu)(\rho_s - \rho) \qquad \text{(equation 3.24)}$$

For a particle of galena settling in water at 5 mm/s:

$$(5 \times 10^{-3}) = ((d^2 \times 9.81)/(18 \times 10^{-3}))(7500 - 1000) = 3.54 \times 10^6 d^2$$

and: $$d = 3.76 \times 10^{-5} \text{ m or } 37.6 \ \mu\text{m}$$

For a particle of limestone settling at 5 nmm/s:

$$(5 \times 10^{-3}) = ((d^2 \times 9.81)/(18 \times 10^{-3}))(2700 - 1000) = 9.27 \times 10^5 d^2$$

and: $$d = 7.35 \times 10^{-5} \text{ m or } 73.5 \ \mu\text{m}$$

Thus particles of galena of less than 37.6 μm and particles of limestone of less than 73.5 μm will be removed in the water stream.

Interpolation of the data given shows that 43 per cent of the galena and 74 per cent of the limestone will be removed in this way.

In 100 kg feed, there is 20 kg galena and 80 kg limestone.

Therefore galena removed = (20 × 0.43) = 8.6 kg, leaving 11.4 kg, and limestone removed = (80 × 0.74) = 59.2 kg, leaving 20.8 kg.

Hence in the *material removed*:

$$\text{concentration of galena} = (8.6 \times 100)/(8.6 + 59.2) = \underline{\underline{12.7 \text{ per cent}}} \text{ by mass}$$

and in the *material remaining*:

$$\text{concentration of galena} = (11.4 \times 100)/(11.4 + 20.8) = \underline{\underline{35.4 \text{ per cent}}} \text{ by mass}$$

As an alternative, the data used for the generation of equation 3.13 for the relation between drag coefficient and particle Reynolds number may be expressed as an explicit relation between $Re'_0$ (the value of $Re'$ at the terminal falling condition of the particle) and the Galileo number $Ga$. The equation takes the form[(10)]:

$$Re'_0 = (2.33Ga^{0.018} - 1.53Ga^{-0.016})^{13.3} \tag{3.40}$$

The Galileo number is readily calculated from the properties of the particle and the fluid, and the corresponding value of $Re'_0$, from which $u_0$ can be found, is evaluated from equation 3.40.

A similar difficulty is encountered in calculating the size of a sphere having a given terminal falling velocity, since $Re'_0$ and $R'_0/\rho u^2$ are both functions of the diameter $d$ of the particle. This calculation is similarly facilitated by the use of another combination, $(R'_0/\rho u_0^2)Re_0'^{-1}$, which is independent of diameter. This is given by:

$$\frac{R'_0}{\rho u_0^2} Re_0'^{-1} = \frac{2\mu g}{3\rho^2 u_0^3}(\rho_s - \rho) \tag{3.41}$$

Log $Re'$ is given as a function of $\log[(R'/\rho u^2)Re'^{-1}]$ in Table 3.5 and the functions are plotted in Figure 3.6. The diameter of a sphere of known terminal falling velocity may be calculated by evaluating $(R'_0/\rho u_0^2)Re_0'^{-1}$, and then finding the corresponding value of $Re'_0$, from which the diameter may be calculated.

As an alternative to this procedure, the data used for the generation of equation 3.13 may be expressed to give $Re'_0$ as an explicit function of $\{(R'/\rho u_0^2)Re_0'^{-1}\}$, which from

Table 3.5. Values of log $Re'$ as a function of $\log\{(R'/\rho u^2)Re'^{-1}\}$ for spherical particles

| $\log\{(R'/\rho u^2)Re'^{-1}\}$ | 0.0 | 0.1 | 0.2 | 0.3 | 0.4 | 0.5 | 0.6 | 0.7 | 0.8 | 0.9 |
|---|---|---|---|---|---|---|---|---|---|---|
| $\bar{5}$ | | | | | | | | | | 3.401 |
| $\bar{4}$ | 3.316 | 3.231 | 3.148 | 3.065 | 2.984 | 2.903 | 2.824 | 2.745 | 2.668 | 2.591 |
| $\bar{3}$ | 2.517 | 2.443 | 2.372 | 2.300 | 2.231 | 2.162 | 2.095 | 2.027 | 1.961 | 1.894 |
| $\bar{2}$ | 1.829 | 1.763 | 1.699 | 1.634 | 1.571 | 1.508 | 1.496 | 1.383 | 1.322 | 1.260 |
| $\bar{1}$ | 1.200 | 1.140 | 1.081 | 1.022 | 0.963 | 0.904 | 0.846 | 0.788 | 0.730 | 0.672 |
| 0 | 0.616 | 0.560 | 0.505 | 0.449 | 0.394 | 0.339 | 0.286 | 0.232 | 0.178 | 0.125 |
| 1 | 0.072 | 0.019 | $\bar{1}.969$ | $\bar{1}.919$ | $\bar{1}.865$ | $\bar{1}.811$ | $\bar{1}.760$ | $\bar{1}.708$ | $\bar{1}.656$ | $\bar{1}.605$ |
| 2 | $\bar{1}.554$ | $\bar{1}.503$ | $\bar{1}.452$ | $\bar{1}.401$ | $\bar{1}.350$ | $\bar{1}.299$ | $\bar{1}.249$ | $\bar{1}.198$ | $\bar{1}.148$ | $\bar{1}.097$ |
| 3 | $\bar{1}.047$ | $\bar{2}.996$ | $\bar{2}.946$ | $\bar{2}.895$ | $\bar{2}.845$ | $\bar{2}.794$ | $\bar{2}.744$ | $\bar{2}.694$ | $\bar{2}.644$ | $\bar{2}.594$ |
| 4 | $\bar{2}.544$ | $\bar{2}.493$ | $\bar{2}.443$ | $\bar{2}.393$ | $\bar{2}.343$ | $\bar{2}.292$ | | | | |

equation 3.40 is equal to $2/3[(\mu g/\rho^2 u_0^3)(\rho_s - \rho)]$. Then writing $K_D = (\mu g/\rho^2 u_0^3)(\rho_s - \rho)]$, $Re'_0$ may be obtained from:

$$Re'_0 = (1.47\ K_D^{-0.14} + 0.11\ K_D^{0.4})^{3.56} \tag{3.42}$$

$d$ may then be evaluated since it is the only unknown quantity involved in the Reynolds number.

### 3.3.4. Rising velocities of light particles

Although there appears to be no problem in using the standard relations between drag coefficient and particle Reynolds number for the calculation of terminal falling velocities of particles denser than the liquid, Karamanev, Chavarie and Mayer[12] have shown experimentally that, for light particles rising in a denser liquid, an *overestimate* of the terminal rising velocity may result. This can occur in the Newton's law region and may be associated with an increase in the drag coefficient $C_D{}'$ from the customary value of 0.22 for a spherical particle up to a value as high as 0.48. Vortex shedding behind the rising particle may cause it to take a longer spiral path thus reducing its vertical component of velocity. A similar effect is not observed with a falling dense particle because its inertia is too high for vortex-shedding to have a significant effect. Further experimental work by Dewsbury, Karamav and Margaritis[13] with shear-thinning power-law solutions of CMC (carboxymethylcellulose) has shown similar effects.

### 3.3.5. Effect of boundaries

The discussion so far relates to the motion of a single spherical particle in an effectively infinite expanse of fluid. If other particles are present in the neighbourhood of the sphere, the sedimentation velocity will be decreased, and the effect will become progressively more marked as the concentration is increased. There are three contributory factors. First, as the particles settle, they will displace an equal volume of fluid, and this gives rise to an upward flow of liquid. Secondly, the buoyancy force is influenced because the suspension has a higher density than the fluid. Finally, the flow pattern of the liquid relative to the particle is changed and velocity gradients are affected. The settling of concentrated suspensions is discussed in detail in Chapter 5.

The boundaries of the vessel containing the fluid in which the particle is settling will also affect its settling velocity. If the ratio of diameter of the particle ($d$) to that of the tube ($d_t$) is significant, the motion of the particle is retarded. Two effects arise. First, as the particle moves downwards it displaces an equal volume of liquid which must rise through the annular region between the particle and the wall. Secondly, the velocity profile in the fluid is affected by the presence of the tube boundary. There have been several studies[14–18] of the influence of the walls, most of them in connection with the use of the "falling sphere" method of determining viscosity, in which the viscosity is calculated from the settling velocity of the sphere. The resulting correction factors have been tabulated by CLIFT, GRACE and WEBER[8]. The effect is difficult to quantify accurately because the particle will not normally follow a precisely uniform vertical path through the fluid. It is therefore useful also to take into account work on the sedimentation of suspensions of uniform spherical particles at various concentrations, and to extrapolate the results to zero concentration to obtain the free falling velocity for different values of the ratio $d/d_t$. The correction factor for the influence of the walls of the tube on the settling velocity of a particle situated at the axis of the tube was calculated by LADENBURG[14] who has given the equation:

$$\frac{u_{0t}}{u_0} = \left(1 + 2.4\frac{d}{d_t}\right)^{-1} \quad (d/d_t < 0.1) \tag{3.43}$$

where $u_{0t}$ is the settling velocity in the tube, and
$u_0$ is the free falling velocity in an infinite expanse of fluid.

Equation 3.43 was obtained for the Stokes' law regime. It overestimates the wall effect, however, at higher particle Reynolds number ($Re' > 0.2$).

Similar effects are obtained with non-cylindrical vessels although, in the absence of adequate data, it is best to use the correlations for cylinders, basing the vessel size on its hydraulic mean diameter which is four times the ratio of the cross-sectional area to the wetted perimeter.

The particles also suffer a retardation as they approach the bottom of the containing vessel because the lower boundary then influences the flow pattern of the fluid relative to the particle. This problem has been studied by LADENBURG[14], TANNER[19] and SUTTERBY[20]. Ladenburg gives the following equation:

$$\frac{u_{0t}}{u_0} = \left(1 + 1.65\frac{d}{L'}\right)^{-1} \tag{3.44}$$

where $L'$ is the distance between the centre of the particle and the lower boundary, for the Stokes' law regime.

### 3.3.6. Behaviour of very fine particles

Very fine particles, particularly in the sub-micron range ($d < 1\ \mu\text{m}$), are very readily affected by natural convection currents in the fluid, and great care must be taken in making measurements to ensure that temperature gradients are eliminated.

The behaviour is also affected by Brownian motion. The molecules of the fluid bombard each particle in a random manner. If the particle is small, the net resultant force acting at any instant may be large enough to cause a change in its direction of motion. This effect has been studied by DAVIES[21], who has developed an expression for the combined effects of gravitation and Brownian motion on particles suspended in a fluid.

In the preceding treatment, it has been assumed that the fluid constitutes a continuum and that the size of the particles is small compared with the mean free path $\lambda$ of the molecules. Particles of diameter $d < 0.1$ μm in gases at atmospheric pressure (and for larger particles in gases at low pressures) can "slip" between the molecules and therefore attain higher than predicted settling velocities. According to CUNNINGHAM[22] the slip factor is given by:

$$1 + \beta\frac{\lambda}{d} \tag{3.45}$$

DAVIES[23] gives the following expression for $\beta$:

$$\beta = 1.764 + 0.562\,\mathrm{e}^{-0.785(d/\lambda)} \tag{3.46}$$

### 3.3.7. Effect of turbulence in the fluid

If a particle is moving in a fluid which is in laminar flow, the drag coefficient is approximately equal to that in a still fluid, provided that the local relative velocity at the particular location of the particle is used in the calculation of the drag force. When the velocity gradient is sufficiently large to give a significant variation of velocity across the diameter of the particle, however, the estimated force may be somewhat in error.

When the fluid is in turbulent flow, or where turbulence is generated by some external agent such as an agitator, the drag coefficient may be substantially increased. BRUCATO *et al.*[24] have shown that the increase in drag coefficient may be expressed in terms of the Kolmogoroff scale of the eddies ($\lambda_E$) given by:

$$\lambda_E = [(\mu/\rho)^3/\varepsilon]^{1/4} \tag{3.47}$$

where $\varepsilon$ is the mechanical power generated per unit mass of fluid by an agitator, for example.

The increase in the drag coefficient $C_D$ over that in the absence of turbulence $C_{D0}$ is given by:

$$\psi = (C_D - C_{D0})/C_{D0} = 8.76 \times 10^{-4}(d/\lambda_E)^3 \tag{3.48}$$

Values of $\psi$ of up to about 30, have been reported.

### 3.3.8. Effect of motion of the fluid

If the fluid is moving relative to some surface other than that of the particle, there will be a superimposed velocity distribution and the drag on the particle may be altered. Thus, if the particle is situated at the axis of a vertical tube up which fluid is flowing in streamline motion, the velocity near the particle will be twice the mean velocity because of the parabolic velocity profile in the fluid. The drag force is then determined by the difference in the velocities of the fluid and the particle at the axis.

The effect of turbulence in the fluid stream has been studied by RICHARDSON and MEIKLE[25] who suspended a particle on a thread at the centre of a vertical pipe up which water was passed under conditions of turbulent flow. The upper end of the thread was attached to a lever fixed on a coil free to rotate in the field of an electromagnet. By passing a current through the coil it was possible to bring the level back to a null position. After calibration, the current required could be related to the force acting on the sphere.

The results were expressed as the friction factor ($R'/\rho u^2$), which was found to have a constant value of 0.40 for particle Reynolds numbers ($Re'$) over the range from 3000 to 9000, and for tube Reynolds numbers ($Re$) from 12,000 to 26,000. Thus the value of $R'/\rho u^2$ has been approximately doubled as a result of turbulence in the fluid.

By surrounding the particle with a fixed array of similar particles on a hexagonal spacing the effect of neighbouring particles was measured. The results are discussed in Chapter 5.

ROWE and HENWOOD[26] made similar studies by supporting a spherical particle 12.7 mm diameter, in water, at the end of a 100 mm length of fine nichrome wire. The force exerted by the water when flowing in a 150 mm square duct was calculated from the measured deflection of the wire. The experiments were carried out at low Reynolds numbers with respect to the duct ($<1200$), corresponding to between 32 and 96 relative to the particle. The experimental values of the drag force were about 10 per cent higher than those calculated from the Schiller and Naumann equation. The work was then extended to cover the measurement of the force on a particle surrounded by an assemblage of particles, as described in Chapter 5.

If $Re'$ is of the order of $10^5$, the drag on the sphere may be reduced if the fluid stream is turbulent. The flow in the boundary layer changes from streamline to turbulent and the size of the eddies in the wake of the particle is reduced. The higher the turbulence of the fluid, the lower is the value of $Re'$ at which the transition from region ($c$) to region ($d$) occurs. The value of $Re'$ at which $R'/\rho u^2$ is 0.15 is known as the *turbulence number* and is taken as an indication of the degree of turbulence in the fluid.

## 3.4. NON-SPHERICAL PARTICLES

### 3.4.1. Effect of particle shape and orientation on drag

There are two difficulties which soon become apparent when attempting to assess the very large amount of experimental data which are available on drag coefficients and terminal falling velocities for non-spherical particles. The first is that an infinite number of non-spherical shapes exists, and the second is that each of these shapes is associated with an infinite number of orientations which the particle is free to take up in the fluid, and the orientation may oscillate during the course of settling.

In a recent comprehensive study, CHHABRA, AGARWAL, and SINHA[27] have found that the most satisfactory characteristic linear dimension to use is the diameter of the sphere of equal volume and that the most relevant characteristic shape is the sphericity, (surface area of particle / surface area of sphere of equal volume). The limitation of this whole approach is that mean errors are often as high as about 16 per cent, and maximum errors may be of the order of 100 per cent. The extent of the errors may be reduced however, by using separate shape factors in the Stokes' and Newton's law regions. Another problem is that, when settling, a non-spherical particle will not travel vertically in a fixed orientation unless it has a plane of symmetry which is horizontal. In general, the resistance force to movement in the gravitational field will not act vertically and the particle will tend to spiral, to rotate and to wobble.

A spherical particle is unique in that it presents the same area to the oncoming fluid whatever its orientation. For non-spherical particles, the orientation must be specified before the drag force can be calculated. The experimental data for the drag can be correlated in the same way as for the sphere, by plotting the dimensionless group $R'/\rho u^2$ against the Reynolds number, $Re' = ud'\rho/\mu$, using logarithmic coordinates, and a separate curve is obtained for each shape of particle and for each orientation. In these groups, $R'$ is taken, as before, as the resistance force per unit area of particle, projected on to a plane perpendicular to the direction of flow. $d'$ is defined as the diameter of the circle having the same area as the projected area of the particle and is therefore a function of the orientation, as well as the shape, of the particle.

The curve for $R'/\rho u^2$ against $Re'$ may be divided into four regions, $(a)$, $(b)$, $(c)$ and $(d)$, as before. In region $(a)$ the flow is entirely streamline and, although no theoretical expressions have been developed for the drag on the particle, the practical data suggest that a law of the form:

$$\frac{R'}{\rho u^2} = K\,Re'^{-1} \tag{3.49}$$

is applicable. The constant $K$ varies somewhat according to the shape and orientation of the particle although it always has a value of about 12. In this region, a particle falling freely in the fluid under the action of gravity will normally move with its longest surface nearly parallel to the direction of motion.

At higher values of $Re'$, the linear relation between $R'/\rho u^2$ and $Re'^{-1}$ no longer holds and the slope of the curve gradually changes until $R'/\rho u^2$ becomes independent of $Re'$ in region $(c)$. Region $(b)$ represents transition conditions and commences at a lower value of $Re'$, and a correspondingly higher value of $R'/\rho u^2$, than in the case of the sphere. A freely falling particle will tend to change its orientation as the value of $Re'$ changes and some instability may be apparent. In region $(c)$ the particle tends to fall so that it is presenting the maximum possible surface to the oncoming fluid. Typical values of $R'/\rho u^2$ for non-spherical particles in region $(c)$ are given in Table 3.6.

Table 3.6. Drag coefficients for non-spherical particles

| Configuration | Length/breadth | $R'/\rho u^2$ |
|---|---|---|
| Thin rectangular plates with their planes perpendicular to the direction of motion | 1–5 | 0.6 |
| | 20 | 0.75 |
| | $\infty$ | 0.95 |
| Cylinders with axes parallel to the direction of motion | 1 | 0.45 |
| Cylinders with axes perpendicular to the direction of motion | 1 | 0.3 |
| | 5 | 0.35 |
| | 20 | 0.45 |
| | $\infty$ | 0.6 |

It may be noted that all these values of $R'/\rho u^2$ are higher than the value of 0.22 for a sphere. CLIFT, GRACE and WEBER[8] have critically reviewed the information available on non-spherical particles.

### 3.4.2. Terminal falling velocities

HEYWOOD[11] has developed an approximate method for calculating the terminal falling velocity of a non-spherical particle, or for calculating its size from its terminal falling velocity. The method is an adaptation of his method for spheres.

A mean projected diameter of the particle $d_p$ is defined as the diameter of a circle having the same area as the particle when viewed from above and lying in its most stable position. Heywood selected this particular dimension because it is easily measured by microscopic examination.

If $d_p$ is the mean projected diameter, the mean projected area is $\pi d_p^2/4$ and the volume is $k'd_p^3$, where $k'$ is a constant whose value depends on the shape of the particle. For a spherical particle, $k'$ is equal to $\pi/6$. For rounded isometric particles, that is particles in which the dimension in three mutually perpendicular directions is approximately the same, $k'$ is about 0.5, and for angular particles $k'$ is about 0.4. For most minerals $k'$ lies between 0.2 and 0.5.

The method of calculating the terminal falling velocity consists in evaluating $(R_0'/\rho u^2) Re_0'^2$, using $d_p$ as the characteristic linear dimension of the particle and $\pi d_p^2/4$ as the projected area in a plane perpendicular to the direction of motion. The corresponding value of $Re_0'$ is then found from Table 3.4 or from Figure 3.6, which both refer to spherical particles, and a correction is then applied to the value of $\log Re_0'$ to account for the deviation from spherical shape. Values of this correction factor, which is a function both of $k'$ and of $(R'/\rho u^2)Re'^2$, are given in Table 3.7. A similar procedure is adopted for calculating the size of a particle of given terminal velocity, using Tables 3.5 and 3.8.

Table 3.7. Corrections to $\log Re'$ as a function of $\log\{(R'/\rho u^2)Re'^2\}$ for non-spherical particles

| $\log\{(R'/\rho u^2)Re'^2\}$ | $k' = 0.4$ | $k' = 0.3$ | $k' = 0.2$ | $k' = 0.1$ |
|---|---|---|---|---|
| $\bar{2}$ | −0.022 | −0.002 | +0.032 | +0.131 |
| $\bar{1}$ | −0.023 | −0.003 | +0.030 | +0.131 |
| 0 | −0.025 | −0.005 | +0.026 | +0.129 |
| 1 | −0.027 | −0.010 | +0.021 | +0.122 |
| 2 | −0.031 | −0.016 | +0.012 | +0.111 |
| 2.5 | −0.033 | −0.020 | 0.000 | +0.080 |
| 3 | −0.038 | −0.032 | −0.022 | +0.025 |
| 3.5 | −0.051 | −0.052 | −0.056 | −0.040 |
| 4 | −0.068 | −0.074 | −0.089 | −0.098 |
| 4.5 | −0.083 | −0.093 | −0.114 | −0.146 |
| 5 | −0.097 | −0.110 | −0.135 | −0.186 |
| 5.5 | −0.109 | −0.125 | −0.154 | −0.224 |
| 6 | −0.120 | −0.134 | −0.172 | −0.255 |

Table 3.8. Corrections to $\log Re'$ as a function of $\{\log(R'/\rho u^2)Re'^{-1}\}$ for non-spherical particles

| $\log\{(R'/\rho u^2)Re'^{-1}\}$ | $k' = 0.4$ | $k' = 0.3$ | $k' = 0.2$ | $k' = 0.1$ |
|---|---|---|---|---|
| $\bar{4}$ | +0.185 | +0.217 | +0.289 | |
| $\bar{4}.5$ | +0.149 | +0.175 | +0.231 | |
| $\bar{3}$ | +0.114 | +0.133 | +0.173 | +0.282 |
| $\bar{3}.5$ | +0.082 | +0.095 | +0.119 | +0.170 |
| $\bar{2}$ | +0.056 | +0.061 | +0.072 | +0.062 |
| $\bar{2}.5$ | +0.038 | +0.034 | +0.033 | −0.018 |
| $\bar{1}$ | +0.028 | +0.018 | +0.007 | −0.053 |
| $\bar{1}.5$ | +0.024 | +0.013 | −0.003 | −0.061 |
| 0 | +0.022 | +0.011 | −0.007 | −0.062 |
| 1 | +0.019 | +0.009 | −0.008 | −0.063 |
| 2 | +0.017 | +0.007 | −0.010 | −0.064 |
| 3 | +0.015 | +0.005 | −0.012 | −0.065 |
| 4 | +0.013 | +0.003 | −0.013 | −0.066 |
| 5 | +0.012 | +0.002 | −0.014 | −0.066 |

The method is only approximate because it is assumed that $k'$ completely defines the shape of the particle, whereas there are many different shapes of particle for which the value of $k'$ is the same. Further, it assumes that the diameter $d_p$ is the same as the mean projected diameter $d'$. This is very nearly so in regions $(b)$ and $(c)$, although in region $(a)$ the particle tends to settle so that the longest face is parallel to the direction of motion and some error may therefore be introduced in the calculation, as indicated by HEISS and COULL[28].

For a non-spherical particle:

$$\text{total drag force, } F = R'_0 \tfrac{1}{4}\pi d_p^2 = (\rho_s - \rho) g k' d_p^3 \tag{3.50}$$

Thus:

$$\frac{R'_0}{\rho u_0^2} = \frac{4k' d_p g}{\pi \rho u_0^2}(\rho_s - \rho) \tag{3.51}$$

$$\frac{R'_0}{\rho u_0^2} Re_0'^2 = \frac{4k' \rho d_p^3 g}{\mu^2 \pi}(\rho_s - \rho) \tag{3.52}$$

and:

$$\frac{R'_0}{\rho u_0^2} Re_0'^{-1} = \frac{4k' \mu g}{\pi \rho^2 u_0^3}(\rho_s - \rho) \tag{3.53}$$

Provided $k'$ is known, the appropriate dimensionless group may be evaluated and the terminal falling velocity, or diameter, calculated.

## Example 3.3

What will be the terminal velocities of mica plates, 1 mm thick and ranging in area from 6 to 600 mm$^2$, settling in an oil of density 820 kg/m$^3$ and viscosity 10 mN s/m$^2$? The density of mica is 3000 kg/m$^3$.

## Solution

| | smallest particles | largest particles |
|---|---|---|
| $A'$ | $6 \times 10^{-6}$ m$^2$ | $6 \times 10^{-4}$ m$^2$ |
| $d_p$ | $\sqrt{(4 \times 6 \times 10^{-6}/\pi)} = 2.76 \times 10^{-3}$ m | $\sqrt{(4 \times 6 \times 10^{-4}/\pi)} = 2.76 \times 10^{-2}$ m |
| $d_p^3$ | $2.103 \times 10^{-8}$ m$^3$ | $2.103 \times 10^{-5}$ m$^3$ |
| volume | $6 \times 10^{-9}$ m$^3$ | $6 \times 10^{-7}$ m$^3$ |
| $k'$ | 0.285 | 0.0285 |

$$\left(\frac{R_0'}{\rho u^2}\right) Re_0'^2 = \frac{4k'}{\mu^2 \pi}(\rho_s - \rho)\rho d_p^3 g \quad \text{(equation 3.52)}$$

$= (4 \times 0.285/\pi \times 0.01^2)(3000 - 820)(820 \times 2.103 \times 10^{-8} \times 9.81)$

$= 1340$ for the smallest particles and, similarly, 134,000 for the largest particles.

Thus:

| | smallest particles | largest particles |
|---|---|---|
| $\log\left(\frac{R_0'}{\rho u_0^2} Re_0'^2\right)$ | 3.127 | 5.127 |
| $\log Re_0'$ | 1.581 | 2.857 (from Table 3.4) |
| Correction from Table 3.6 | −0.038 | −0.300 (estimated) |
| Corrected $\log Re_0'$ | 1.543 | 2.557 |
| $Re_0'$ | 34.9 | 361 |
| $u_0$ | 0.154 m/s | 0.159 m/s |

Thus it is seen that all the mica particles settle at approximately the same velocity.

## 3.5. MOTION OF BUBBLES AND DROPS

The drag force acting on a gas bubble or a liquid droplet will not, in general, be the same as that acting on a rigid particle of the same shape and size because circulating currents are set up inside the bubble. The velocity gradient at the surface is thereby reduced and the drag force is therefore less than for the rigid particle. HADAMARD[29] showed that, if the effects of surface energy are neglected, the terminal falling velocity of a drop, as calculated from Stokes' law, must be multiplied by a factor $Q$, to account for the internal circulation, where:

$$Q = \frac{3\mu + 3\mu_l}{2\mu + 3\mu_l} \quad (3.54)$$

In this equation, $\mu$ is the viscosity of the continuous fluid and $\mu_l$ is the viscosity of the fluid forming the drop or bubble. This expression applies only in the range for which Stokes' law is valid.

If $\mu_l/\mu$ is large, $Q$ approaches unity. If $\mu_l/\mu$ is small, $Q$ approaches a value of 1.5. Thus the effect of circulation is small when a liquid drop falls in a gas although is large when a gas bubble rises in a liquid. If the fluid within the drop is very viscous, the amount of energy which has to be transferred in order to induce circulation is large and circulation effects are therefore small.

Hadamard's work was later substantiated by BOND[30] and by BOND and NEWTON[31] who showed that equation 3.51 is valid provided that surface tension forces do not play a large role. With very small droplets, the surface tension forces tend to nullify the tendency for circulation, and the droplet falls at a velocity close to that of a solid sphere.

In addition, drops and bubbles are subject to deformation because of the differences in the pressures acting on various parts of the surface. Thus, when a drop is settling in a still fluid, both the hydrostatic and the impact pressures will be greater on the forward face than on the rear face and will tend to flatten the drop, whereas the viscous drag will tend to elongate it. Deformation of the drop is opposed by the surface tension forces so that very small drops retain their spherical shape, whereas large drops may be considerably deformed and the resistance to their motion thereby increased. For drops above a certain size, the deformation is so great that the drag force increases at the same rate as the volume, and the terminal falling velocity therefore becomes independent of size.

GARNER and SKELLAND[32,33] have shown the importance of circulation within a drop in determining the coefficient of mass transfer between the drop and the surrounding medium. The critical Reynolds number at which circulation commences has been shown[32] to increase at a rate proportional to the logarithm of viscosity of the liquid constituting the drop and to increase with interfacial tension. The circulation rate may be influenced by mass transfer because of the effect of concentration of diffusing material on both the interfacial tension and on the viscosity of the surface layers. As a result of circulation the falling velocity may be up to 50 per cent greater than for a rigid sphere, whereas oscillation of the drop between oblate and prolate forms will reduce the velocity of fall[33]. Terminal falling velocities of droplets have also been calculated by HAMIELEC and JOHNSON[34] from approximate velocity profiles at the interface and the values so obtained compare well with experimental values for droplet Reynolds numbers up to 80.

## 3.6. DRAG FORCES AND SETTLING VELOCITIES FOR PARTICLES IN NON-NEWTONIAN FLUIDS

Only a very limited amount of data is available on the motion of particles in non-Newtonian fluids and the following discussion is restricted to their behaviour in shear-thinning *power-law* fluids and in fluids exhibiting a *yield-stress*, both of which are discussed in Volume 1, Chapter 3.

### 3.6.1. Power-law fluids

Because most shear-thinning fluids, particularly polymer solutions and flocculated suspensions, have high apparent viscosities, even relatively coarse particles may have velocities in the *creeping-flow* of Stokes' law regime. CHHABRA[35,36] has proposed that both theoretical and experimental results for the drag force $F$ on an isolated spherical particle of diameter $d$ moving at a velocity $u$ may be expressed as a modified form of Stokes' law:

$$F = 3\pi \mu_c d u Y \tag{3.55}$$

where the apparent viscosity $\mu_c$ is evaluated at a characteristic shear rate $u/d$, and $Y$ is a correction factor which is a function of the rheological properties of the fluid. The best available theoretical estimates values of $Y$ for power-law fluids are given in Table 3.9.

Table 3.9. Values of $Y$ for power-law fluids[35]

| $n$ | 1 | 0.9 | 0.8 | 0.7 | 0.6 | 0.5 | 0.4 | 0.3 | 0.2 | 0.1 |
|---|---|---|---|---|---|---|---|---|---|---|
| $Y$ | 1 | 1.14 | 1.24 | 1.32 | 1.38 | 1.42 | 1.44 | 1.46 | 1.41 | 1.35 |

Several expressions of varying forms and complexity have been proposed[35,36] for the prediction of the drag on a sphere moving through a power-law fluid. These are based on a combination of numerical solutions of the equations of motion and extensive experimental results. In the absence of wall effects, dimensional analysis yields the following functional relationship between the variables for the interaction between a single isolated particle and a fluid:

$$2C'_D = C_D = f(Re'_n, n) \tag{3.56}$$

where $C_D$ and $C'_D$ are drag coefficients defined by equation 3.4, $n$ is the power-law index and $Re'_n$ is the particle Reynolds number given by:

$$Re'_n = (u^{2-n} d^n \rho)/k \tag{3.57}$$

where $k$ is the consistency coefficient in the power-law relation. Combining equations 3.55 and 3.56:

$$C_D = (24 Re'^{-1}_n)Y \tag{3.58}$$

$$C'_D = (12 Re'^{-1}_n)Y \tag{3.59}$$

From Table 3.9 it is seen that, depending on the value of $n$, the drag on a sphere in a power-law fluid may be up to 46 per cent higher than that in a Newtonian fluid at the same particle Reynolds number. Practical measurements lie in the range $1 < Y < 1.8$, with considerable divergences between the results of the various workers.

In view of the general uncertainty concerning the value of $Y$, it may be noted that the unmodified Stokes' law expression gives a acceptable first approximation.

The terminal settling velocity $u_0$ of a particle in the gravitational field is then given by equating the buoyant weight of the particle to the drag force to give:

$$u_0 = \left\{ \frac{g d^{n+1} (\rho_s - \rho)}{18kY} \right\}^{1/n} \tag{3.60}$$

where $(\rho_s - \rho)$ is the density difference between the particle and the fluid.

From equation 3.59, it is readily seen that in a shear-thinning fluid ($n < 1$) the terminal velocity is more strongly dependent on $d$, $g$ and $\rho_s - \rho$ than in a Newtonian fluid and a small change in any of these variables produces a larger change in $u_0$.

Outside the creeping flow regime, experimental results for drag on spheres in power-law fluids have been presented by TRIPATHI *et al.*[37] and GRAHAM[38] for values of $Re'_n$

up to 100, and these are reasonably well correlated by the following expressions with an average error of about 10 per cent:

$$C_D = [(35.2Re_n'^{-1.03})2^n] + n[1 - (20.9Re_n'^{-1.11})2^n] \quad (3.61a)$$

$$(0.2 < (2^{-n}Re_n') < 24)$$

$$C_D = [(37Re_n'^{-1.1})2^n] + [0.36n + 0.25] \quad (3.61b)$$

$$(24 < (2^{-n}Re_n') < 100)$$

$$C_D' = [(17.6Re_n'^{-1.03})2^n] + n[0.5 - (10.5Re_n'^{-1.11})2^n] \quad (3.62a)$$

$$[0.2 < (2^{-n}Re_n') < 24]$$

$$C_D' = [(18.5Re_n'^{-1.1})2^n] + [0.18n + 0.125] \quad (3.62b)$$

$$[24 < (2^{-n}Re_n') < 100]$$

It may be noted that these two equations do not reduce exactly to the relation for a Newtonian fluid ($n = 1$).

Extensive comparisons of predictions and experimental results for drag on spheres suggest that the influence of non-Newtonian characteristics progressively diminishes as the value of the Reynolds number increases, with inertial effects then becoming dominant, and the standard curve for Newtonian fluids may be used with little error. Experimentally determined values of the drag coefficient for power-law fluids ($1 < Re_n' < 1000$; $0.4 < n < 1$) are within 30 per cent of those given by the standard drag curve[37,38].

While equations 3.62*a* and 3.62*b* are convenient for estimating the value of the drag coefficient, they need to be re-arranged in order to enable the settling velocity $u_0$ of a sphere of given diameter and density to be calculated, since both $C_D(C_D')$ and $Re_n'$ are functions of the unknown settling velocity. By analogy with the procedure used for Newtonian fluids (equation 3.36), the dimensionless Galileo number $Ga_n$ which is independent of $u_0$ may be defined by:

$$\tfrac{2}{3}Ga_n = C_D' Re'^{[2/(2-n)]} = gd^{[(n+2)/(2-n)]}(s_r - 1)(\rho/k)^{[2/(2-n)]} \quad (3.63)$$

where $s_r$ is the ratio of the densities of the particle and of the fluid ($\rho_s/\rho$).

Equation 3.56 may be written as:

$$Re_n' = f(Ga_n, n) \quad (3.64)$$

Experimental results comprising about 1000 data points from a large number of sources cover the following range of variables:

$1 < d < 20$ (mm); $1190 < \rho_s < 16,600$ (kg/m$^3$); $990 < \rho < 1190$ (kg/m$^3$);

$0.4 < n < 1$; and $1 < Re_n' < 10^4$.

These are satisfactorily correlated by:

$$Re'_n = r_1(\tfrac{2}{3}Ga_n)^{r_2} \quad (3.65)$$

where $r_1 = 0.1\{\exp[(0.5/n) - 0.73n]$ and $r_2 = (0.954/n) - 0.16$

Thus, in equation 3.65 only $Re'_n$ includes the terminal falling velocity which may then be calculated for a spherical particle in a power-law fluid.

### 3.6.2. Fluids with a yield stress

Much less is known about the settling of particles in fluids exhibiting a yield stress. BARNES[39] suggests that this is partly due to the fact that considerable confusion exists in the literature as to whether or not the fluids used in the experiments do have a true yield stress[39]. Irrespective of this uncertainty, which usually arises from the inappropriateness of the rheological techniques used for their characterisation, many industrially important materials, notably particulate suspensions, have rheological properties closely approximating to viscoelastic behaviour.

By virtue of its yield stress, an unsheared viscoelastic material is capable of supporting the immersed weight of a particle for an indefinite period of time, provided that the immersed weight of the particle does not exceed the maximum upward force which can be exerted by virtue of the yield stress of the fluid. The conditions for the static equilibrium of a sphere are now discussed.

#### *Static equilibrium*

Many investigators[35] have reported experimental results on the necessary conditions for the static equilibrium of a sphere. The results of all such studies may be represented by a factor $Z$ which is proportional to the ratio of the forces due to the yield stress $\tau_Y$ and those due to gravity.

Thus:
$$Z = \frac{\tau_Y}{dg(\rho_s - \rho)} \tag{3.66}$$

The critical value of $Z$ which indicates the point at which the particle starts to settle from rest appears to lie in the range $0.04 < Z < 0.2$.

#### *Drag force*

Under conditions where a spherical particle is not completely supported by the forces attributable to the yield stress, it will settle at a velocity such that the total force exerted by the fluid on the particle balances its weight.

For a fluid whose rheological properties may be represented by the Herschel-Bulkley model discussed in Volume 1, Chapter 3, the shear stress $\tau$ is a function of the shear rate $\dot{\gamma}$ or:

$$\tau = \tau_Y + k'_{\text{HB}}\dot{\gamma}^{n_{\text{HB}}} \tag{3.67}$$

From dimensional considerations, the drag coefficient is a function of the Reynolds number for the flow relative to the particle, the exponent, $n_{\text{HB}}$, and the so-called Bingham number $Bi$ which is proportional to the ratio of the yield stress to the viscous stress attributable to the settling of the sphere. Thus:

$$C'_D = R'/pu^2 \text{f}(Re' m_{\text{HB}}, n_{\text{HB}}, Bi) \tag{3.68}$$

where:
$$Bi = \tau_Y/k_{\text{HB}}/m_{\text{HB}}(u/d) \tag{3.69}$$

Using the scant data in the literature and their own experimental results, ATAPATTU *et al.*[40] suggest the following expression for the drag on a sphere moving through a Herschel-Bulkley fluid in the creeping flow regime:

$$C'_D = 12Re'_{\text{HB}}(1 + Bi)^{-1} \tag{3.70}$$

It may be noted that an iterative solution to equation 3.70 is required for the calculation of the unknown settling velocity $u_0$, since this term appears in all three dimensionless groups, $Re'_{\text{HB}}$, $Bi$, and $C'_D$, for a given combination of properties of sphere and fluid.

The effect of particle shape on the forces acting when the particle is moving in a shear-thinning fluid has been investigated by TRIPATHI *et al.*[37], and by VENUMADHAV AND CHHABRA[41]. In addition,, some information is available on the effects of viscoelasticity of the fluid[35].

It is seen therefore that, in the absence of any entirely satisfactory theoretical approach or reliable experimental data, it is necessary to adopt a highly pragmatic approach to the estimation of the drag force on a particle in a non-Newtonian fluid.

## 3.7. ACCELERATING MOTION OF A PARTICLE IN THE GRAVITATIONAL FIELD

### 3.7.1. General equations of motion

The motion of a particle through a fluid may be traced because the value of the drag factor $R'/\rho u^2$ for a given value of the Reynolds number is fixed. The behaviour of a particle undergoing acceleration or retardation has been the subject of a very large number of investigations, which have been critically reviewed by TOROBIN and GAUVIN[42]. The results of different workers are not consistent, although it is shown that the drag factor is often related, not only to the Reynolds number, but also to the number of particle diameters traversed by the particle since the initiation of the motion.

A relatively simple approach to the problem, which gives results closely in accord with practical measurements, is to consider the mass of fluid which is effectively given the same acceleration as the particle, as discussed by MIRONER[43]. This is only an approximation because elements of fluid at different distances from the particle will not all be subject to the same acceleration. For a spherical particle, this added or *hydrodynamic* mass is equal to the mass of fluid whose volume is equal to one half of that of the sphere. This can give rise to a very significant effect in the case of movement through a liquid, and can result in accelerations substantially less than those predicted when the added mass is neglected. For movement through gases, the contribution of the added mass term is generally negligible. Added mass is most important for the motion of gas bubbles in a liquid because in that case the surrounding liquid has a much greater density than the bubble. The total mass of particle and associated fluid is sometimes referred to as the *virtual mass* ($m'$).

Thus: Virtual mass = Mass of particle + Added mass.

For a sphere:

$$m' = \frac{\pi}{6}d^3\rho_s + \frac{\pi}{12}d^3\rho$$

or:
$$m' = \frac{\pi}{6}d^3\rho_s\left(1+\frac{\rho}{2\rho_s}\right) = m\left(1+\frac{\rho}{2\rho_s}\right) \tag{3.71}$$

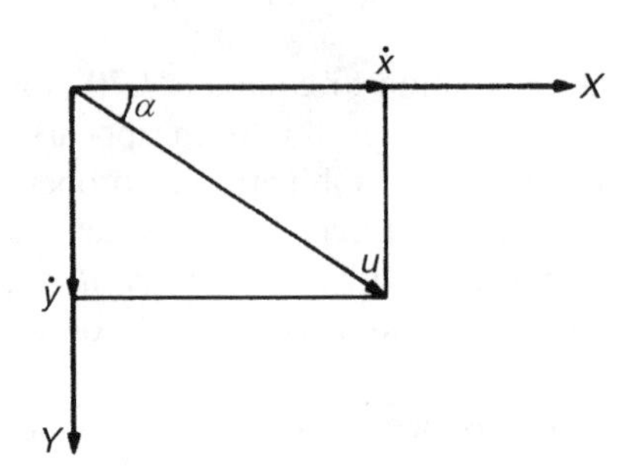


Figure 3.7. Two-dimensional motion of a particle

Considering the motion of a particle of mass $m$ in the earth's gravitational field, at some time $t$ the particle will be moving at an angle $\alpha$ to the horizontal with a velocity $u$ as shown in Figure 3.7. The velocity $u$ may then be resolved into two components, $\dot{x}$ and $\dot{y}$, in the horizontal and vertical directions. $\dot{x}$ and $\ddot{x}$ will be taken to denote the first and second derivatives of the displacement $x$ in the $X$-direction with respect to time, and $\dot{y}$ and $\ddot{y}$ the corresponding derivatives of $y$.

Thus:
$$\cos\alpha = \frac{\dot{x}}{u} \tag{3.72}$$
$$\sin\alpha = \frac{\dot{y}}{u} \tag{3.73}$$
and:
$$u = \sqrt{(\dot{x}^2 + \dot{y}^2)} \tag{3.74}$$

There are two forces acting on the body:

(a) In the vertical direction, the apparent weight of the particle,
$$W' = mg\left(1-\frac{\rho}{\rho_s}\right) \tag{3.75}$$

(b) The drag force which is equal to $R'A'$ and acts in such a direction as to oppose the motion of the particle. Its direction therefore changes as $\alpha$ changes. Here $A'$ is the projected area of the particle on a plane at right angles to the direction of motion and its value varies with the orientation of the particle in the fluid. The drag force can be expressed by:
$$F = \frac{R'}{\rho u^2}\rho u^2 A' \tag{3.76}$$

This has a component in the $X$-direction of:

$$\frac{R'}{\rho u^2}\rho u^2 A' \cos\alpha = \frac{R'}{\rho u^2} A' \rho \dot{x} \surd(\dot{x}^2 + \dot{y}^2)$$

and in the $Y$-direction of:

$$\frac{R'}{\rho u^2}\rho u^2 A' \sin\alpha = \frac{R'}{\rho u^2} A' \rho \dot{y} \surd(\dot{x}^2 + \dot{y}^2)$$

The equations of motion in the $X$- and $Y$-directions are therefore:

$$m\ddot{x} = -\frac{R'}{\rho u^2}\rho A' \dot{x} \surd(\dot{x}^2 + \dot{y}^2) \tag{3.77}$$

and:

$$m\ddot{y} = -\frac{R'}{\rho u^2}\rho A' \dot{y} \surd(\dot{x}^2 + \dot{y}^2) + mg\left(1 - \frac{\rho}{\rho_s}\right) \tag{3.78}$$

If allowance is now made for the *added mass*, $m' = m[1 + (\rho/2\rho_s)]$ must be substituted for $m$. Equations 3.77 and 3.78 refer to conditions where the particle is moving in the positive sense in the $X$-direction and in the positive (downward) sense in the $Y$-direction. If the particle is moving in the negative $X$-direction, the form of solution is unchanged, except that all increments of $x$ will be negative. If, however, the particle is initially moving upwards, the sign of only the frictional term in equation 3.78 is changed and the form of solution will, in general, be different from that for downward movement. Care must therefore be exercised in the application of the equation, particularly if a change of sense may occur during the motion of the particle.

It may be noted that the equations of motion for the two directions ($X$ and $Y$) are coupled, with $\dot{x}$ and $\dot{y}$ appearing in each of the equations. General solutions are therefore not possible, except as will be seen later for motion in the Stokes' law region.

Putting $m = \frac{\pi}{6}d^3\rho_s$ and $A' = \frac{\pi}{4}d^2$ for a spherical particle, then:

$$\ddot{x} = -\frac{R'}{\rho u^2}\frac{3}{2d}\frac{\rho}{\rho_s}\dot{x}\surd(\dot{x}^2 + \dot{y}^2) \tag{3.79}$$

If allowance is made for the added mass and $m'$ is substituted for $m$ in equation 3.78, then:

$$\ddot{x} = -\frac{R'}{\rho u^2}\frac{3}{d}\frac{\rho}{(2\rho_s + \rho)}\dot{x}\surd(\dot{x}^2 + \dot{y}^2) \tag{3.80}$$

Similarly:

$$\ddot{y} = \mp\frac{R'}{\rho u^2}\frac{3}{2d}\frac{\rho}{\rho_s}\dot{y}\surd(\dot{x}^2 + \dot{y}^2) + g\left(1 - \frac{\rho}{\rho_s}\right) \tag{3.81}$$

and allowing for the added mass:

$$\ddot{y} = \mp\frac{R'}{\rho u^2}\frac{3}{d}\frac{\rho}{(2\rho_s + \rho)}\dot{y}\surd(\dot{x}^2 + \dot{y}^2) + \frac{2g(\rho_s - \rho)}{(2\rho_s + \rho)} \tag{3.82}$$

where the minus sign in equations 3.81 and 3.82 is applicable for downward motion and the positive sign for upward motion (with downwards taken as the positive sense).

### 3.7.2. Motion of a sphere in the Stokes' law region

Under these conditions, from equation 3.5:

$$\frac{R'}{\rho u^2} = 12Re'^{-1} = \frac{12\mu}{d\rho\sqrt{(\dot{x}^2 + \dot{y}^2)}} \tag{3.83}$$

Substituting in equations 3.79, 3.80, 3.81 and 3.82 gives:

$$\ddot{x} = -\frac{18\mu}{d^2\rho_s}\dot{x} = -a\dot{x} \tag{3.84}$$

or:

$$\ddot{x} = -\frac{36\mu}{d^2(2\rho_s + \rho)}\dot{x} = -a'\dot{x} \tag{3.85}$$

if allowance is made for the *added mass*.

$$\ddot{y} = -\frac{18\mu}{d^2\rho_s}\dot{y} + g\left(1 - \frac{\rho}{\rho_s}\right) = -a\dot{y} + b \tag{3.86}$$

or:

$$\ddot{y} = -\frac{36\mu}{d^2(2\rho_s + \rho)}\dot{y} + \frac{2g(\rho_s - \rho)}{(2\rho_s + \rho)} = -a'y' + b' \tag{3.87}$$

if allowance is made for the *added mass*.

In this particular case, the equations of motion for the $X$ and $Y$ directions are mutually independent and therefore can be integrated separately. Furthermore, because the frictional term is now a linear function of velocity, the sign will automatically adjust to take account of whether motion is downwards or upwards.

The equations are now integrated, ignoring the effects of *added mass* which can be accounted for by replacing $a$ by $a'$ and $b$ by $b'$. For the $Y$-direction, integrating equation 3.86 with respect to $t$:

$$\dot{y} = -ay + bt + \text{constant}.$$

The axes are chosen so that the particle is at the origin at time $t = 0$. If the initial component of the velocity of the particle in the $Y$-direction is $v$, then, when $t = 0$, $y = 0$ and $\dot{y} = v$, and the constant $= v$,

or:

$$\dot{y} + ay = bt + v$$

Thus:

$$e^{at}\dot{y} + e^{at}ay = (bt + v)e^{at}$$

Thus:

$$e^{at}y = (bt + v)\frac{e^{at}}{a} - \int b\frac{e^{at}}{a}\,dt$$

$$= (bt + v)\frac{e^{at}}{a} - \frac{b}{a^2}e^{at} + \text{constant}$$

When $t = 0$, $y = 0$, and the constant $= \dfrac{b}{a^2} - \dfrac{v}{a}$.

Thus:
$$y = \frac{b}{a}t + \frac{v}{a} - \frac{b}{a^2} + \left(\frac{b}{a^2} - \frac{v}{a}\right)e^{-at} \tag{3.88}$$

where:
$$a = 18\frac{\mu}{d^2\rho_s} \tag{3.89}$$

and:
$$b = \left(1 - \frac{\rho}{\rho_s}\right)g. \tag{3.90}$$

It may be noted that $b/a = u_0$, the terminal falling velocity of the particle. This equation enables the displacement of the particle in the $Y$-direction to be calculated at any time $t$.

For the $X$-direction, equation 3.84 is of the same form as equation 3.86 with $b = 0$. Substituting $b = 0$ and writing $w$ as the initial velocity in the $X$-direction, equation 3.88 becomes:
$$x = \frac{w}{a}(1 - e^{-at}) \tag{3.91}$$

Thus the displacement in the $X$-direction may also be calculated for any time $t$.

By eliminating $t$ between equations 3.89, 3.90 and 3.91, a relation between the displacements in the $X$- and $Y$-directions is obtained. Equations of this form are useful for calculating the trajectories of particles in size-separation equipment.

From equation 3.91:
$$e^{-at} = 1 - \frac{ax}{w}$$

and:
$$t = -\frac{1}{a}\ln\left(1 - \frac{ax}{w}\right) \tag{3.92}$$

Substituting in equation 3.88 gives:
$$y = \frac{b}{a}\left\{-\frac{1}{a}\ln\left(1 - \frac{ax}{w}\right)\right\} + \frac{v}{a} - \frac{b}{a^2} + \left(\frac{b}{a^2} - \frac{v}{a}\right)\left(1 - \frac{ax}{w}\right)$$
$$= -\frac{b}{a^2}\ln\left(1 - \frac{ax}{w}\right) - \frac{bx}{aw} + \frac{vx}{w} \tag{3.93}$$

The values of $a$ and $b$ can now be substituted and the final relation is:
$$y = -\frac{g\rho_s(\rho_s - \rho)d^4}{324\mu^2}\left\{\ln\left(1 - \frac{18\mu x}{w\rho_s d^2}\right) + \frac{18\mu x}{w\rho_s d^2}\left(1 - \frac{18v\mu}{d^2(\rho_s - \rho)g}\right)\right\} \tag{3.94}$$

If allowance is made for *added mass*, $a'$ and $b'$ are substituted for $a$ and $b$, respectively.

Then:
$$y = -\frac{g(2\rho_s + \rho)(\rho_s - \rho)d^4}{648\mu^2}\left\{\ln\left(1 - \frac{36\mu x}{w(2\rho_s + \rho)d^2}\right)\right.$$
$$\left. + \frac{36\mu x}{w(2\rho_s + \rho)d^2}\left(1 - \frac{18v\mu}{d^2(\rho_s - \rho)g}\right)\right\} \tag{3.95}$$

### 3.7.3. Vertical motion (general case)

For the *Stokes' law regime* equations 3.88, 3.89 and 3.90 are applicable.

For the *Newton's law regime*, $R'/\rho u^2$ is a constant and equal to 0.22 for a spherical particle. Therefore, substituting in equation 3.81 and putting $\dot{x} = 0$ for vertical motion, and using the *negative sign for downward motion* (and neglecting the effect of *added mass*):

$$\ddot{y} = -\frac{1}{3d}\frac{\rho}{\rho_s}\dot{y}^2 + g\left(1 - \frac{\rho}{\rho_s}\right) \tag{3.96}$$

$$\ddot{y} = -c\dot{y}^2 + b \tag{3.97}$$

Thus: $$\frac{\mathrm{d}\dot{y}}{b - c\dot{y}^2} = \mathrm{d}t$$

and: $$\frac{\mathrm{d}\dot{y}}{f^2 - \dot{y}^2} = c\,\mathrm{d}t \tag{3.98}$$

where: $$c = \frac{1}{3d}\frac{\rho}{\rho_s} \tag{3.99}$$

and: $$f = \sqrt{\left(\frac{b}{c}\right)} = \sqrt{\left(\frac{3d(\rho_s - \rho)g}{\rho}\right)} \tag{3.100}$$

Integrating equation 3.98 gives:

$$\frac{1}{2f}\ln\left(\frac{f + \dot{y}}{f - \dot{y}}\right) = ct + \text{constant}$$

When $t = 0$, $\dot{y} = v$, say, and therefore the constant $= \dfrac{1}{2f}\ln\left(\dfrac{f + v}{f - v}\right)$.

Thus: $$\frac{1}{2f}\ln\left(\frac{f + \dot{y}}{f - \dot{y}}\right)\left(\frac{f - v}{f + v}\right) = ct$$

Thus: $$\left(\frac{f + \dot{y}}{f - \dot{y}}\right)\left(\frac{f - v}{f + v}\right) = \mathrm{e}^{2fct}$$

and: $$(f - \dot{y}) = \frac{2f}{1 + \left(\dfrac{f + v}{f - v}\right)\mathrm{e}^{2fct}}$$

Thus: $$y = ft - 2f\int\frac{\mathrm{d}t}{1 + \left(\dfrac{f + v}{f - v}\right)\mathrm{e}^{2fct}}$$

$$= ft - 2f\int\frac{\mathrm{d}t}{1 + j\,\mathrm{e}^{pt}} \quad \text{(say)}$$

Putting: $$s = 1 + j\,\mathrm{e}^{pt}$$

then: $$\mathrm{d}s = pj\,\mathrm{e}^{pt}\,\mathrm{d}t = p(s - 1)\,\mathrm{d}t$$

Thus:

$$y = ft - 2f \int \frac{\mathrm{d}s}{ps(s-1)}$$

$$= ft - 2f\left\{\frac{1}{p}\ln\frac{s-1}{s} + \text{constant}\right\}$$

$$= ft - \frac{1}{c}\ln\frac{1}{1+\left(\dfrac{f-v}{f+v}\right)\mathrm{e}^{-2fct}} + \text{constant}$$

When $t = 0$, $y = 0$ and:

$$\text{constant} = \frac{1}{c}\ln\frac{1}{1+\left(\dfrac{f-v}{f+v}\right)} = \frac{1}{c}\ln\frac{f+v}{2f}$$

Thus:

$$y = ft + \frac{1}{c}\ln\frac{f+v}{2f}\left\{1+\frac{f-v}{f+v}\mathrm{e}^{-2fct}\right\}$$

Thus for downward motion:

$$y = ft + \frac{1}{c}\ln\frac{1}{2f}\{f+v+(f-v)\mathrm{e}^{-2fct}\} \tag{3.101}$$

If the *added mass* is taken into account, $f$ remains unchanged, but $c$ must be replaced by $c'$ in equations 3.97 and 3.101, where:

$$c' = \frac{2\rho}{3d(2\rho_s+\rho)}. \tag{3.102}$$

For *vertical upwards motion* in the *Newton's law* regime; the positive sign in equation 3.81 applies and thus, by analogy with equation 3.98:

$$\frac{\mathrm{d}\dot{y}}{f^2+\dot{y}^2} = c\,\mathrm{d}t \tag{3.103}$$

Integrating:

$$\frac{1}{f}\tan^{-1}\frac{\dot{y}}{f} = ct + \text{constant}$$

When $t = 0$, $\dot{y} = v$, say, and the constant $= (1/f)\tan^{-1}(v/f)$, $v$ is a negative quantity. Thus:

$$\frac{1}{f}\tan^{-1}\frac{\dot{y}}{f} = ct + \frac{1}{f}\tan^{-1}\frac{v}{f}$$

$$\frac{\dot{y}}{f} = \tan\left(fct + \tan^{-1}\frac{v}{f}\right)$$

and:

$$y = \frac{f}{-fc}\ln\cos\left(fct + \tan^{-1}\frac{v}{f}\right) + \text{constant}$$

When $t = 0$, $y = 0$, and:

$$\text{constant} = \frac{1}{c} \ln \cos \tan^{-1} \frac{v}{f},$$

then:

$$y = -\frac{1}{c} \ln \frac{\cos\left(fct + \tan^{-1} \dfrac{v}{f}\right)}{\cos \tan^{-1} \dfrac{v}{f}}$$

$$= -\frac{1}{c} \ln \frac{\cos fct \cos\left[\tan^{-1} \dfrac{v}{f}\right] - \sin fct \sin\left[\tan^{-1} \dfrac{v}{f}\right]}{\cos\left[\tan^{-1} \dfrac{v}{f}\right]}$$

or:

$$y = -\frac{1}{c} \ln\left(\cos fct - \frac{v}{f} \sin fct\right) \tag{3.104}$$

The relation between $y$ and $t$ may also be obtained graphically, though the process is more tedious than that of using the analytical solution appropriate to the particular case in question. When $Re'$ lies between 0.2 and 500 there is no analytical solution to the problem and a numerical or graphical method must be used.

When the spherical particle is moving downwards, that is when its velocity is positive:

$$\ddot{y} = -\frac{3}{2d} \frac{\rho}{\rho_s} \frac{R'}{\rho u^2} \dot{y}^2 + \left(1 - \frac{\rho}{\rho_s}\right) g \qquad \text{(from equation 3.81)}$$

$$\frac{\mu}{\rho d} \frac{dRe'}{dt} = -\frac{3}{2d} \frac{\rho}{\rho_s} \frac{R'}{\rho u^2} Re'^2 \frac{\mu^2}{d^2 \rho^2} + \left(1 - \frac{\rho}{\rho_s}\right) g$$

and:

$$t = \int_{Re'_1}^{Re'_2} \frac{dRe'}{\dfrac{d\rho(\rho_s - \rho)g}{\mu\rho_s} - \dfrac{3\mu}{2d^2\rho_s} \dfrac{R'}{\rho u^2} Re'^2} \tag{3.105}$$

If the particle is moving upwards, the corresponding expression for $t$ is:

$$t = \int_{Re'_1}^{Re'_2} \frac{dRe'}{\dfrac{d\rho(\rho_s - \rho)g}{\mu\rho_s} + \dfrac{3\mu}{2d^2\rho_s} \dfrac{R'}{\rho u^2} Re'^2} \tag{3.106}$$

Equations 3.105 and 3.106 do not allow for *added mass*. If this is taken into account then:

$$t = \int_{Re'_1}^{Re'_2} \frac{dRe'}{\dfrac{2d(\rho_s - \rho)g}{\mu(2\rho_s + \rho)} \pm \dfrac{3\mu}{d^2(2\rho_s + \rho)} \dfrac{R'}{\rho u^2} Re'^2} \tag{3.107}$$

where the positive sign applies to upward motion and the negative sign to downward motion.

From these equations, $Re'$ may be obtained as a function of $t$. The velocity $\dot{y}$ may then be calculated. By means of a second graphical integration, the displacement $y$ may be found at any time $t$.

In using the various relations which have been obtained, it must be noted that the law of motion of the particle will change as the relative velocity between the particle and the fluid changes. If, for example, a particle is initially moving upwards with a velocity $v$, so that the corresponding value of $Re'$ is greater than about 500, the relation between $y$ and $t$ will be given by equation 3.104. The velocity of the particle will progressively decrease and, when $Re'$ is less than 500, the motion is obtained by application of equation 3.106. The upward velocity will then fall still further until $Re'$ falls below 0.2. While the particle is moving under these conditions, its velocity will fall to zero and will then gradually increase in the downward direction. The same equation (3.88) may be applied for the whole of the time the Reynolds group is less than 0.2, irrespective of sense. Then for higher downward velocities, the particle motion is given by equations 3.105 and 3.101.

Unidimensional motion in the vertical direction, under the action of gravity, occurs frequently in elutriation and other size separation equipment, as described in Chapter 1.

## Example 3.4

A material of density 2500 kg/m$^3$ is fed to a size separation plant where the separating fluid is water rising with a velocity of 1.2 m/s. The upward vertical component of the velocity of the particles is 6 m/s. How far will an approximately spherical particle, 6 mm diameter, rise relative to the walls of the plant before it comes to rest relative to the fluid?

## Solution

Initial velocity of particle relative to fluid, $v = (6.0 - 1.2) = 4.8$ m/s.

Thus:
$$Re' = (6 \times 10^{-3} \times 4.8 \times 1000)/(1 \times 10^{-3}) = 28{,}800$$

When the particle has been retarded to a velocity such that $Re' = 500$, the minimum value for which equation 3.104 is applicable:

$$\dot{y} = (4.8 \times 500)/28{,}800 = 0.083 \text{ m/s}$$

In this solution, the effect of *added mass* is not taken into account. Allowance may be made by adjustment of the values of the constants in the equations as indicated in Section 3.7.3.

When $Re'$ is greater than 500, the relation between the displacement of the particle $y$ and the time $t$ is:

$$y = -\frac{1}{c}\ln\left(\cos fct - \frac{v}{f}\sin fct\right) \quad \text{(equation 3.104)}$$

where:
$$c = \frac{1}{3d}\frac{\rho}{\rho_s} = (0.33/6 \times 10^{-3})(1000/2500) = 22.0 \quad \text{(equation 3.99)}$$

$$f = \sqrt{\left(\frac{3d(\rho_s - \rho)g}{\rho}\right)} = \sqrt{[(6 \times 10^{-3} \times 1500 \times 9.81)/(0.33 \times 1000)]} \quad \text{(equation 3.100)}$$

$$= 0.517$$

and: $$v = -4.8 \text{ m/s}$$

Thus:
$$y = -\frac{1}{22.0}\ln\left(\cos 0.517 \times 22t + \frac{4.8}{0.517}\sin 0.517 \times 22t\right)$$
$$= -0.0455\ln(\cos 11.37t + 9.28\sin 11.37t)$$
$$\dot{y} = -\frac{0.0455(-11.37\sin 11.37t + 9.28 \times 11.37\cos 11.37t)}{\cos 11.37t + 9.28\sin 11.37t}$$
$$= -\frac{0.517(9.28\cos 11.37t - \sin 11.37t)}{\cos 11.37t + 9.28\sin 11.37t}$$

The time taken for the velocity of the particle relative to the fluid to fall from 4.8 m/s to 0.083 m/s is given by:

$$-0.083 = -\frac{0.517(9.28\cos 11.37t - \sin 11.37t)}{\cos 11.37t + 9.28\sin 11.37t}$$

or: $$\cos 11.37t + 9.28\sin 11.37t = -6.23\sin 11.37t + 57.8\cos 11.37t$$

or: $$56.8\cos 11.37 = 15.51\sin 11.37t$$

$\therefore$ $$\sin 11.37t = 3.66\cos 11.37t$$

squaring: $$1 - \cos^2 11.37t = 13.4\cos^2 11.37t$$

$\therefore$ $$\cos 11.37t = 0.264 \qquad \text{(i)}$$

and: $$\sin 11.37t = \sqrt{(1 - 0.264^2)} = 0.965$$

The distance moved by the particle relative to the fluid during this period is therefore given by:

$$y = -0.0455\ln(0.264 + 9.28 \times 0.965)$$
$$= -0.101 \text{ m}$$

If equation 3.104 is applied for a relative velocity down to zero, the time taken for the particle to come to rest is given by:

$$9.28\cos 11.37t = \sin 11.37t$$

squaring: $$1 - \cos^2 11.37t = 86.1\cos^2 11.37t$$

and: $$\cos 11.37t = 0.107$$

and: $$\sin 11.37t = \sqrt{(1 - 0.107^2)} = 0.994$$

The corresponding distance the particle moves relative to the fluid is then given by:

$$y = -0.0455\ln(0.107 + 9.28 \times 0.994)$$
$$= -0.102 \text{ m}$$

that is the particle moves only a very small distance with a velocity of less than 0.083 m/s.

If form drag were neglected for all velocities less than 0.083 m/s, the distance moved by the particle would be given by:

$$y = \frac{b}{a}t + \frac{v}{a} - \frac{b}{a^2} + \left(\frac{b}{a^2} - \frac{v}{a}\right)e^{-at} \qquad \text{(equation 3.88)}$$

and: $$\dot{y} = \frac{b}{a} - \left(\frac{b}{a} - v\right)e^{-at}$$

where: $$a = 18\frac{\mu}{d^2\rho_s} = \left(\frac{18 \times 0.001}{0.006^2 \times 2500}\right) = 0.20 \qquad \text{(equation 3.89)}$$

$$b = \left(1 - \frac{\rho}{\rho_s}\right)g = [1 - (1000/2500)]9.81 = 5.89 \qquad \text{(equation 3.90)}$$

Thus: $$b/a = 29.43$$

and: $$v = -0.083 \text{ m/s}$$

Thus: $$y = 29.43t - \left(\frac{0.083}{0.20} + \frac{29.43}{0.20}\right)(1 - e^{-0.20t})$$

$$= 29.43t - \frac{29.51}{0.20}(1 - e^{-0.20t})$$

and: $$\dot{y} = 29.43 - 29.51\,e^{-0.20t}$$

When the particle comes to rest in the fluid, $\dot{y} = 0$ and:

$$e^{-0.20t} = 29.43/29.51$$

and: $$t = 0.0141 \text{ s}$$

The corresponding distance moved by the particle is given by:

$$y = 29.43 \times 0.0141 - (29.51/0.20)(1 - e^{-0.20 \times 0.0141})$$

$$= 0.41442 - 0.41550 = -0.00108 \text{ m}$$

Thus whether the resistance force is calculated by equation 3.15 or equation 3.19, the particle moves a negligible distance with a velocity relative to the fluid of less than 0.083 m/s. Further, the time is also negligible, and thus the fluid also has moved through only a very small distance.

It may therefore be taken that the particle moves through 0.102 m before it comes to rest in the fluid. The time taken for the particle to move this distance, on the assumption that the drag force corresponds to that given by equation 3.19, is given by:

$$\cos 11.37t = 0.264 \qquad \text{(from equation (i) above)}$$

$\therefore$ $$11.37t = 1.304$$

and: $$t = 0.115 \text{ s}$$

The distance travelled by the fluid in this time = $(1.2 \times 0.115) = 0.138$ m

Thus the total distance moved by the particle relative to the walls of the plant

$$= 0.102 + 0.138 = 0.240 \text{ m or } \underline{\underline{240 \text{ mm}}}$$

## Example 3.5

Salt of density 2350 kg/m$^3$ is charged to the top of a reactor containing a 3 m depth of aqueous liquid of density 1100 kg/m$^3$ and viscosity 2 mN s/m$^2$, and the crystals must dissolve completely before reaching the bottom. If the rate of dissolution of the crystals is given by:

$$-\mathrm{d}d/\mathrm{d}t = (3 \times 10^{-6}) + (2 \times 10^{-4}u)$$

where $d$ is the size of the crystal (m) at time $t$ (s) and $u$ its velocity in the fluid (m/s), calculate the maximum size of crystal which should be charged. The inertia of the particles may be neglected and the resistance force may be taken as that given by Stokes' law ($3\pi\mu du$), where $d$ is taken as the equivalent spherical diameter of the particle.

## Solution

Assuming that the salt always travels at its terminal velocity, then for the Stokes law region this is given by equation 3.24. $u_0 = (d^2/g/18\mu)(\rho_8 - \rho)$ or, in this case, $u_0 = (d^2 \times 9.81)/(18 \times 2 \times 10^{-3})(2350 - 1100) = 3.406 \times 10^5 d^2$ m/s

The rate of dissolution: $-\mathrm{d}d/\mathrm{d}t = (3 \times 10^{-6}) + (2 \times 10^{-4}u)$ m/s

and substituting: $\mathrm{d}d/\mathrm{d}t = (-3 \times 10^{-6}) - (2 \times 10^{-4} \times 3.406 \times 10^5 d^2)$
$= -3 \times 10^{-6} - 68.1d^2$

The velocity at any point $h$ from the top of the reactor is $u = \mathrm{d}h/\mathrm{d}t$ and:

$$\frac{\mathrm{d}h}{\mathrm{d}d} = \frac{\mathrm{d}h}{\mathrm{d}t}\frac{\mathrm{d}t}{\mathrm{d}d} = 3.406 \times 10^5 d^2/(-3 \times 10^{-6} - 68.1d^2)$$

Thus:
$$\int_0^3 \mathrm{d}h = -\int_d^0 \frac{(3.406 \times 10^5 d^2 \mathrm{d}d)}{(3 \times 10^{-6} + 68.1d^2)}$$

or:
$$3 = 3.406 \times 10^5 \left(\int_0^d \frac{\mathrm{d}d}{C_2} - \frac{C_1}{C_2^2}\int_0^d \frac{\mathrm{d}d}{(C_1/C_2) + d^2}\right)$$

where $C_1 = 3 \times 10^{-6}$ and $C_2 = 68.1$.

Thus:
$$3 = (3.406 \times 10^5)\left\{\left[\frac{d}{C_2}\right]_0^d - \left[\frac{C_1}{C_2^2}\frac{1}{(C_1/C_2)^{0.5}}\tan^{-1}\left(\frac{d}{(C_1/C_2)^{0.5}}\right)\right]_0^d\right\}$$

and:
$$3 = (3.406 \times 10^5/C_2)[d - (C_1/C_2)^{0.5}\tan^{-1} d(C_1/C_2)^{-0.5}]$$

Substituting for $C_1$ and $C_2$: $d = (6 \times 10^{-4}) + (2.1 \times 10^{-4})\tan^{-1}(4.76 \times 10^3 d)$ and, solving by trial and error: $d = 8.8 \times 10^{-4}$ m or 0.88 mm

The integration may also be carried out numerically with the following results:

| $d$ | $d^2$ | $\left(\frac{3.406 \times 10^5 d^2}{3 \times 10^{-6} + 68.1d^2}\right)$ | Interval of $d$ | Mean value of function in interval | Integral over interval | Total integral |
|---|---|---|---|---|---|---|
| 0 | 0 | 0 | | | | |
| $1 \times 10^{-4}$ | $1 \times 10^{-8}$ | $9.25 \times 10^2$ | $1 \times 10^{-4}$ | $4.63 \times 10^2$ | 0.0463 | 0.0463 |
| $2 \times 10^{-4}$ | $4 \times 10^{-8}$ | $2.38 \times 10^3$ | $1 \times 10^{-4}$ | $1.65 \times 10^3$ | 0.1653 | 0.2116 |
| $3 \times 10^{-4}$ | $9 \times 10^{-8}$ | $3.358 \times 10^3$ | $1 \times 10^{-4}$ | $2.86 \times 10^3$ | 0.2869 | 0.4985 |
| $4 \times 10^{-4}$ | $1.6 \times 10^{-7}$ | $3.922 \times 10^3$ | $1 \times 10^{-4}$ | $3.64 \times 10^3$ | 0.364 | 0.8625 |
| $5 \times 10^{-4}$ | $2.5 \times 10^{-7}$ | $4.25 \times 10^3$ | $1 \times 10^{-4}$ | $4.09 \times 10^3$ | 0.409 | 1.2715 |
| $6 \times 10^{-4}$ | $3.6 \times 10^{-7}$ | $4.46 \times 10^3$ | $1 \times 10^{-4}$ | $4.35 \times 10^3$ | 0.435 | 1.706 |
| $7 \times 10^{-4}$ | $4.9 \times 10^{-7}$ | $4.589 \times 10^3$ | $1 \times 10^{-4}$ | $4.52 \times 10^3$ | 0.452 | 2.158 |
| $8 \times 10^{-4}$ | $6.4 \times 10^{-7}$ | $4.679 \times 10^3$ | $1 \times 10^{-4}$ | $4.634 \times 10^3$ | 0.463 | 2.621 |
| $9 \times 10^{-4}$ | $8.1 \times 10^{-7}$ | $4.74 \times 10^3$ | $1 \times 10^{-4}$ | $4.709 \times 10^3$ | 0.471 | 3.09 |

From which $d$ = 0.9 mm.

The acceleration of the particle to its terminal velocity has been neglected, and in practice the time taken to reach the bottom of the reactor will be slightly larger, allowing a somewhat larger crystal to dissolve completely.

## 3.8. MOTION OF PARTICLES IN A CENTRIFUGAL FIELD

In most practical cases where a particle is moving in a fluid under the action of a centrifugal field, gravitational effects are comparatively small and may be neglected. The equation of motion for the particles is similar to that for motion in the gravitational field, except that the gravitational acceleration $g$ must be replaced by the centrifugal acceleration $r\omega^2$, where $r$ is the radius of rotation and $\omega$ is the angular velocity. It may be noted, however, that in this case the acceleration is a function of the position $r$ of the particle.

For a spherical particle in a fluid, the equation of motion for the Stokes' law region is:

$$\frac{\pi}{6}d^3(\rho_s - \rho)r\omega^2 - 3\pi\mu d\frac{\mathrm{d}r}{\mathrm{d}t} = \frac{\pi}{6}d^3\rho_s\frac{\mathrm{d}^2r}{\mathrm{d}t^2} \tag{3.108}$$

As the particle moves outwards, the accelerating force increases and therefore it never acquires an equilibrium velocity in the fluid.

If the inertial terms on the right-hand side of equation 3.108 are neglected, then:

$$\frac{\mathrm{d}r}{\mathrm{d}t} = \frac{d^2(\rho_s - \rho)r\omega^2}{18\mu} \tag{3.109}$$

$$= \frac{d^2(\rho_s - \rho)g}{18\mu}\frac{r\omega^2}{g}$$

$$= u_0\left(\frac{r\omega^2}{g}\right) \tag{3.110}$$

Thus, the instantaneous velocity $(\mathrm{d}r/\mathrm{d}t)$ is equal to the terminal velocity $u_0$ in the gravitational field, increased by a factor of $r\omega^2/g$.

Returning to the exact form (equation 3.108), this may be re-arranged to give:

$$\frac{\mathrm{d}^2r}{\mathrm{d}t^2} + \frac{18\mu}{d^2\rho_s}\frac{\mathrm{d}r}{\mathrm{d}t} - \frac{\rho_s - \rho}{\rho_s}\omega^2 r = 0 \tag{3.111}$$

or:

$$\frac{\mathrm{d}^2r}{\mathrm{d}t^2} + a\frac{\mathrm{d}r}{\mathrm{d}t} - qr = 0 \tag{3.112}$$

The solution of equation 3.112 takes the form:

$$r = B_1\,\mathrm{e}^{-[a/2+\sqrt{(a^2/4+q)}]t} + B_2\,\mathrm{e}^{-[a/2-\sqrt{(a^2/4+q)}]t} \tag{3.113}$$

$$= \mathrm{e}^{-at/2}\{B_1\,\mathrm{e}^{-kt} + B_2\,\mathrm{e}^{kt}\} \tag{3.114}$$

where

$$a = \frac{18\mu}{d^2\rho_s}, \quad q = \left(1 - \frac{\rho}{\rho_s}\right)\omega^2 \text{ and } k = \sqrt{[(a^2/4) + q]} \tag{3.115}$$

The effects of added mass, which have not been taken into account in these equations, require the replacement of $a$ by $a'$ and $q$ by $q'$, where:

$$a' = \frac{36\mu}{d^2(2\rho_s + \rho)} \quad \text{and} \quad q' = \frac{2(\rho_s - \rho)}{(2\rho_s + \rho)}\omega^2$$

Equation 3.114 requires the specification of two boundary conditions in order that the constants $B_1$ and $B_2$ may be evaluated.

If the particle starts ($t = 0$) at a radius $r_1$ with zero velocity $(\mathrm{d}r/\mathrm{d}t) = 0$, then from equation 3.114:

$$\begin{aligned}\frac{\mathrm{d}r}{\mathrm{d}t} &= \mathrm{e}^{-at/2}[-kB_1\,\mathrm{e}^{-kt} + kB_2\mathrm{e}^{kt}] - \frac{a}{2}\mathrm{e}^{-at/2}[B_1\,\mathrm{e}^{-kt} + B_2\,\mathrm{e}^{kt}] \\ &= \mathrm{e}^{-at/2}\left\{\left(k - \frac{a}{2}\right)B_2\,\mathrm{e}^{kt} - \left(k + \frac{a}{2}\right)B_1\,\mathrm{e}^{-kt}\right\} \qquad (3.116)\end{aligned}$$

Substituting the boundary conditions into equations 3.114 and 3.116:

$$r_1 = B_1 + B_2$$

$$0 = B_2\left(k - \frac{a}{2}\right) - B_1\left(k + \frac{a}{2}\right)$$

Hence:

$$B_1 = \frac{k - a/2}{2k}r_1 \quad \text{and} \quad B_2 = \frac{k + a/2}{2k}r_1 \qquad (3.117)$$

Thus:

$$r = \mathrm{e}^{-at/2}\left\{\frac{k - a/2}{2k}r_1\,\mathrm{e}^{-kt} + \frac{k + a/2}{2k}r_1\,\mathrm{e}^{kt}\right\}$$

and:

$$\frac{r}{r_1} = \mathrm{e}^{-at/2}\left\{\cosh kt + \frac{a}{2k}\sinh kt\right\} \qquad (3.118)$$

Hence $r/r_1$ may be directly calculated at any value of $t$, although a numerical solution is required to determine $t$ for any particular value of $r/r_1$.

If the effects of particle acceleration may be neglected, equation 3.112 simplifies to:

$$a\frac{\mathrm{d}r}{\mathrm{d}t} - qr = 0$$

Direct integration gives:

$$\ln\frac{r}{r_1} = \frac{q}{a}t = \frac{d^2(\rho_s - \rho)\omega^2}{18\mu}t \qquad (3.119)$$

Thus the time taken for a particle to move to a radius $r$ from an initial radius $r_1$ is given by:

$$t = \frac{18\mu}{d^2\omega^2(\rho_s - \rho)}\ln\frac{r}{r_1} \qquad (3.120)$$

For a suspension fed to a centrifuge, the time taken for a particle initially situated in the liquid surface ($r_1 = r_0$) to reach the wall of the bowl ($r = R$) is given by:

$$t = \frac{18\mu}{d^2\omega^2(\rho_s - \rho)}\ln\frac{R}{r_0} \qquad (3.121)$$

If $h$ is the thickness of the liquid layer at the walls then:

$$h = R - r_0$$

Then:
$$\ln \frac{R}{r_0} = \ln \frac{R}{R-h} = -\ln\left(1 - \frac{h}{R}\right)$$

$$= \frac{h}{R} + \frac{1}{2}\left(\frac{h}{R}\right)^2 + \cdots$$

If $h$ is small compared with $R$, then:

$$\ln \frac{R}{r_0} \approx \frac{h}{R}$$

Equation 3.120 then becomes:

$$t = \frac{18\mu h}{d^2\omega^2(\rho_s - \rho)R} \tag{3.122}$$

For the Newton's law region, the equation of motion is:

$$\frac{\pi}{6}d^3(\rho_s - \rho)r\omega^2 - 0.22\frac{\pi}{4}d^2\rho\left(\frac{dr}{dt}\right)^2 = \frac{\pi}{6}d^3\rho_s\frac{d^2r}{dt^2}$$

This equation can only be solved numerically. If the acceleration term may be neglected, then:

$$\left(\frac{dr}{dt}\right)^2 = 3d\omega^2\left(\frac{\rho_s - \rho}{\rho}\right)r$$

Thus:
$$r^{-1/2}\frac{dr}{dt} = \left\{3d\omega^2\left(\frac{\rho_s - \rho}{\rho}\right)\right\}^{1/2} \tag{3.123}$$

Integration gives:

$$2(r^{1/2} - r_1^{1/2}) = \left\{3d\omega^2\frac{\rho_s - \rho}{\rho}\right\}^{1/2} t$$

or:
$$t = \left[\frac{\rho}{3d\omega^2(\rho_s - \rho)}\right]^{1/2} 2(r^{1/2} - r_1^{1/2}) \tag{3.124}$$

## 3.9. FURTHER READING

CHHABRA, R.P.: *Bubbles, Drops, and Particles*. (CRC Press, 1992)
CLIFT, R., GRACE, J. R. and WEBER, M. E.: *Bubbles, Drops and Particles* (Academic Press, 1978).
CURLE, N. and DAVIES, H. J.: *Modern Fluid Dynamics*. Volume 1. *Incompressible Flow* (Van Nostrand, 1968).
DALLAVALLE, J. M.: *Micromeritics*, 2nd edn. (Pitman, 1948).
HERDAN, G.: *Small Particle Statistics*, 2nd edn. (Butterworths, 1960).
MIRONER, A.: *Engineering Fluid Mechanics* (McGraw Hill, 1979).
ORR, C.: *Particulate Technology* (Macmillan, 1966).

## 3.10. REFERENCES


1. STOKES, G. G.: *Trans. Cam. Phil. Soc.* **9** (1851) 8. On the effect of the internal friction of fluids on the motion of pendulums.
2. GOLDSTEIN, S.: *Proc. Roy. Soc.* **A 123** (1929) 225. The steady flow of viscous fluid past a fixed spherical obstacle at small Reynolds numbers.
3. OSEEN, C. W.: *Ark. Mat. Astr. Fys.* **9**, No 16 (1913) 1–15. Über den Gültigkeitsbereich der Stokesschen Widerstandsformel.
4. JENSON, V. G.: *Proc. Roy. Soc.* **249A** (1959) 346. Viscous flow around a sphere at low Reynolds numbers (<40).
5. LE CLAIR, B. P., HAMIELEC, A. E. and PRUPPACHER, H. R.: *J. Atmos. Sci.* **27** (1970) 308. A numerical study of the drag on a sphere at low and intermediate Reynolds numbers.
6. DALLAVALLE, J. M.: *Micromeritics*, 2nd edn. (Pitman, 1948)
7. SCHILLER, L. and NAUMANN, A.: *Z. Ver. deut. Ing.* **77** (1933) 318. Über die grundlegenden Berechnungen der Schwerkraftaufbereitung.
8. CLIFT, R., GRACE, J. R. and WEBER, M. E.: *Bubbles, Drops and Particles* (Academic Press, 1978).
9. WADELL, H.: *J. Franklin. Inst.* **217** (1934) 459. The coefficient of resistance as a function of Reynolds' number for solids of various shapes.
10. KHAN, A. R. and RICHARDSON, J. F.: *Chem. Eng. Comm.* **62** (1987) 135. The resistance to motion of a solid sphere in a fluid.
11. HEYWOOD, H.: *J. Imp. Coll. Chem. Eng. Soc.* **4** (1948) 17. Calculation of particle terminal velocities.
12. KARAMANEV, D. G., CHAVARIE, C. and MAYER, R. C.: *AIChEJl* **42** (1996) 1789. Free rise of a light solid sphere in liquid.
13. DEWSBURY, K. H., KARAMANEV, D. G. and MARGARITIS, A. *AIChEJl* **46** (2000) 46. Dynamic behavior of freely using buoyant solid sphere in non-Newtonian liquids.
14. LADENBURG, R.: *Ann. Phys.* **23** (1907) 447. Über den Einfluss von Wänden auf die Bewegung einer Kugel in einer reibenden Flüssigkeit.
15. FRANCIS, A. W.: *Physics* **4** (1933) 403. Wall effects in falling ball method for viscosity.
16. HABERMAN, W. L. and SAYRE, R. M.: *David Taylor Model Basin Report No* 1143 (Oct 1958) Motion of rigid and fluid spheres in stationary and moving liquids inside cylindrical tubes.
17. FIDLERIS, V. and WHITMORE, R. L.: *Brit. J. App. Phys.* **12** (1961) 490. Experimental determination of the wall effect for spheres falling axially in cylindrical vessels.
18. KHAN, A. R. and RICHARDSON, J. F.: *Chem. Eng. Comm.* **78** (1989) 111. Fluid-particle interactions and flow characteristics of fluidized beds and settling suspensions of spherical particles.
19. TANNER, R. I.: *J. Fluid. Mech.* **17** (1963) 161. End effects in falling ball viscometry.
20. SUTTERBY, J. L.: *Trans. Soc. Rheol.* **17** (1973) 559. Falling sphere viscometry. I. Wall and inertial corrections to Stokes' law in long tubes.
21. DAVIES, C. N.: *Proc. Roy. Soc.* **A200** (1949) 100. The sedimentation and diffusion of small particles.
22. CUNNINGHAM, E.: *Proc. Roy. Soc.* **A83** (1910) 357. Velocity of steady fall of spherical particles.
23. DAVIES, C. N.: *Proc. Phys. Soc.* **57** (1945) 259. Definitive equations for the fluid resistance of spheres.
24. BRUCATO, A., GRISAFI, F. and MONTANTE, G.: *Chem. Eng. Sci* **53** (1998) 3295. Particle drag coefficients in turbulent fluids.
25. RICHARDSON, J. F. and MEIKLE, R. A.: *Trans. Inst. Chem. Eng.* **39** (1961) 357. Sedimentation and fluidisation. Part IV. Drag force on individual particles in an assemblage.
26. ROWE, P. N. and HENWOOD, G. N.: *Trans. Inst. Chem. Eng.* **39** (1961) 43. Drag forces in a hydraulic model of a fluidised bed. Part 1.
27. CHHABRA, R. P., AGARWAL, L. and SINHA, N. K. *Powder Technology* **101** (1999) 288. Drag on non-spherical particles: an evaluation of available methods.
28. HEISS, J. F. and COULL, J.: *Chem. Eng. Prog.* **48** (1952) 133. The effect of orientation and shape on the settling velocity of non-isometric particles in a viscous medium.
29. HADAMARD, J.: *Comptes rendus* **152** (1911) 1735. Mouvement permanent lent d'une sphère liquide et visqueuse dans un liquide visqueux.
30. BOND, W. N.: *Phil. Mag*, 7th. ser. **4** (1927) 889. Bubbles and drops and Stokes' law.
31. BOND, W. N. and NEWTON, D. A.: *Phil Mag.* 7th ser. **5** (1928) 794. Bubbles, drops and Stokes' law (Paper 2).
32. GARNER, F. H. and SKELLAND, A. H. P.: *Trans. Inst. Chem. Eng.* **29** (1951) 315. Liquid–liquid mixing as affected by the internal circulation within drops.
33. GARNER, F. H. and SKELLAND, A. H. P.: *Chem. Eng. Sci.* **4** (1955) 149. Some factors affecting drop behaviour in liquid–liquid systems.
34. HAMIELEC, A. E. and JOHNSON, A. I.: *Can. J. Chem. Eng.* **40** (1962) 41. Viscous flow around fluid spheres at intermediate Reynolds numbers.

35. CHHABRA, R. P.: *Bubbles, Drops and Particles in non-Newtonian Fluids*, CRC Press, Boca Raton, Fl (1993).
36. CHHABRA, R. P. and RICHARDSON, J. F.: *Non-Newtonian Flow in the Process Industries*, Butterworth–Heinemann, Oxford (1999).
37. TRIPATHI, A., CHHABRA, R. P. and SUNDARARAJAN, T.: *Ind. Eng. Chem. Res.* **34** (1994) 403. Power law fluid flow over spherical particles.
38. GRAHAM, D. I. and JONES, T. E. R.: *J. Non-Newt. Fluid Mech.* **54** (1994) 465. Settling and transport of spherical particles in power-law fluids at finite Reynolds number.
39. BARNES, H. A.: *J. Non-Newt. Fluid. Mech.* **81** (1999) 133. The yield stress - a review.
40. ATAPATTU, D. D., CHHABRA, R. P. and UHLHERR, P. H. T.: *J. Non-Newt. Fluid Mech.* **59** (1995) 245. Creeping sphere motion in Herschel - Bulkley fluids; Flow field and drag.
41. VENUMADHAV, G. and CHHABRA, R.P.: *Powder Technol.* **78** (1994) 77. Settling velocities of single nonspherical particles in non-Newtonian fluids.
42. TOROBIN, L. B. and GAUVIN, W. H.: *Can. J. Chem. Eng.* **38** (1959) 129, 167, 224. Fundamental aspects of solids-gas flow. Part I: Introductory concepts and idealized sphere-motion in viscous regime. Part II: The sphere wake in steady laminar fluids. Part III: Accelerated motion of a particle in a fluid.
43. MIRONER, A.: *Engineering Fluid Mechanics* (McGraw Hill, 1979).

## 3.11. NOMENCLATURE

| | | Units in SI System | Dimensions in **M**, **L**, **T** |
|---|---|---|---|
| $A'$ | Projected area of particle in plane perpendicular to direction of motion | m$^2$ | $\mathbf{L}^2$ |
| $a$ | $18\mu/d^2\rho_s$ | s$^{-1}$ | $\mathbf{T}^{-1}$ |
| $a'$ | $36\mu/d^2(2\rho_s+\rho)$ | s$^{-1}$ | $\mathbf{T}^{-1}$ |
| $B_1, B_2$ | Coefficients in equation 3.111 | m | **L** |
| $b$ | $[1-(\rho/\rho_s)]g$ | m/s$^2$ | $\mathbf{LT}^{-2}$ |
| $b'$ | $2g(\rho_s-\rho)/(2\rho_s+\rho)$ | m/s$^2$ | $\mathbf{LT}^{-2}$ |
| $C_D$ | Drag coefficient $2R'/\rho u^2$ | — | — |
| $C'_D$ | Drag coefficient $R'/\rho u^2$ | — | — |
| $C_{D_0}$ | Drag coefficient in the absence of turbulence | — | — |
| $c$ | $\rho/3d\rho_s$ | m$^{-1}$ | $\mathbf{L}^{-1}$ |
| $d$ | Diameter of sphere or characteristic dimension of particle | m | **L** |
| $d_p$ | Mean projected diameter of particle | m | **L** |
| $d_t$ | Diameter of tube or vessel | m | **L** |
| $d'$ | Linear dimension of particle | m | **L** |
| $F$ | Total force on particle | N | $\mathbf{MLT}^{-2}$ |
| $f$ | $\sqrt{(b/c)}$ | m/s | $\mathbf{LT}^{-1}$ |
| $g$ | Acceleration due to gravity | m/s$^2$ | $\mathbf{LT}^{-2}$ |
| $h$ | Thickness of liquid layer | m | **L** |
| $j$ | $(f+v)/(f-v)$ | — | — |
| $K$ | Constant for given shape and orientation of particle | — | — |
| $K_D$ | $\frac{2}{3}(R'/\rho u^2)(Re_0'^{-1})$-see equation 3.42 | — | — |
| $k$ | Consistency coefficient for power-law fluid | Ns$^n$/m$^2$ | $\mathbf{ML}^{-1}\mathbf{T}^{n-2}$ |
| $k_{HB}$ | Consistency coefficient for Herschel-Bulkley fluid (Equation 3.67) | Ns$^n$/m$^2$ | $\mathbf{ML}^{-1}\mathbf{T}^{n-2}$ |
| $k'$ | Constant for calculating volume of particle | — | — |
| $L'$ | Distance of particle from bottom of container | m | **L** |
| $m$ | Mass of particle | kg | **M** |
| $m'$ | Virtual mass (mass + added mass) | kg | **M** |
| $n$ | Power-law index for non-Newtonian fluid | — | — |
| $n_{HB}$ | Power-law index for Herschel-Bulkley fluid (Equation 3.67) | — | — |
| $q$ | $[1-(\rho/\rho_s)]\omega^2$ (equations 3.111 and 3.107) | s$^{-2}$ | $\mathbf{T}^{-2}$ |
| $q'$ | $2(\rho_s-\rho)\omega^2/(2\rho_s+\rho)$ | s$^{-2}$ | $\mathbf{T}^{-2}$ |
| $p$ | $2fc$ | s$^{-1}$ | $\mathbf{T}^{-1}$ |
| $Q$ | Correction factor for velocity of bubble | — | — |

| | | Units in SI System | Dimensions in **M**, **L**, **T** |
|---|---|---|---|
| $R$ | Radius of basket, *or* | m | **L** |
| | Shear stress at wall of pipe | N/m$^2$ | $\mathbf{ML^{-1}\,T^{-2}}$ |
| $R'$ | Resistance per unit projected area of particle | N/m$^2$ | $\mathbf{ML^{-1}\,T^{-2}}$ |
| $R'_0$ | Resistance per unit projected area of particle at free falling condition | N/m$^2$ | $\mathbf{ML^{-1}\,T^{-2}}$ |
| $r$ | Radius of rotation | m | **L** |
| $r_0$ | Radius of inner surface of liquid | m | **L** |
| $r_1$ | Coefficient in equation 3.65 | — | — |
| $r_2$ | Exponent in equation 3.65 | — | — |
| $S$ | Index in equation 3.32 | — | — |
| $s$ | $1 + j\,e^{pt}$ | — | — |
| $s_r$ | Density ratio $(\rho_s/\rho)$ | — | — |
| $t$ | Time | s | **T** |
| $u$ | Velocity of fluid relative to particle | m/s | $\mathbf{LT^{-1}}$ |
| $u_0$ | Terminal falling velocity of particle | m/s | $\mathbf{LT^{-1}}$ |
| $u_{0t}$ | Terminal falling velocity of particle in vessel | m/s | $\mathbf{LT^{-1}}$ |
| $v$ | Initial component of velocity of particle in $Y$-direction | m/s | $\mathbf{LT^{-1}}$ |
| $W'$ | Apparent (buoyant) weight of particle | N | $\mathbf{MLT^{-2}}$ |
| $w$ | Initial component of velocity of particle in $X$-direction | m/s | $\mathbf{LT^{-1}}$ |
| $x$ | Displacement of particle in $X$-direction at time $t$ | m | **L** |
| $\dot{x}$ | Velocity of particle in $X$-direction at time $t$ | m/s | $\mathbf{LT^{-1}}$ |
| $\ddot{x}$ | Acceleration of particle in $X$-direction at time $t$ | m/s$^2$ | $\mathbf{LT^{-2}}$ |
| $Y$ | Correction factor in Stokes' law for power-law fluid | — | — |
| $y$ | Displacement of particle in $Y$-direction at time $t$ | m | **L** |
| $\dot{y}$ | Velocity of particle in $Y$ direction at time $t$ | m/s | $\mathbf{LT^{-1}}$ |
| $\ddot{y}$ | Acceleration of particle in $Y$-direction at time $t$ | m/s$^2$ | $\mathbf{LT^{-2}}$ |
| $Z$ | Ratio of forces due to yield stress and to gravity (equation 3.66) | — | — |
| $\alpha$ | Angle between direction of motion of particle and horizontal | — | — |
| $\beta$ | Coefficient in equations 3.45 and 3.46 | — | — |
| $\dot{\gamma}$ | Shear rate | s$^{-1}$ | $\mathbf{T^{-1}}$ |
| $\lambda$ | Mean free path | m | **L** |
| $\lambda_E$ | Kolmogoroff scale of turbulence (equation 3.46) | m | **L** |
| $\mu$ | Viscosity of fluid | N s/m$^2$ | $\mathbf{ML^{-1}\,T^{-1}}$ |
| $\mu_1$ | Viscosity of fluid in drop or bubble | N s/m$^2$ | $\mathbf{ML^{-1}\,T^{-1}}$ |
| $\mu_c$ | Viscosity at shear rate $(u/d)$ | Ns/m$^2$ | $\mathbf{ML^{-1}T^{-1}}$ |
| $\phi$ | Pipe friction factor $R/\rho u^2$ | — | — |
| $\tau_Y$ | Yield stress | N/m$^2$ | $\mathbf{ML^{-1}T^{-2}}$ |
| $\rho$ | Density of fluid | kg/m$^3$ | $\mathbf{ML^{-3}}$ |
| $\rho_s$ | Density of solid | kg/m$^3$ | $\mathbf{ML^{-3}}$ |
| $\omega$ | Angular velocity | rad/s | $\mathbf{T^{-1}}$ |
| $\psi$ | $(C_D - C_{D_0})/C_{D_0}$ | — | — |
| $Bi$ | Bingham number (equation 3.69) | — | — |
| $Ga$ | Galileo number $d^3(\rho_s - \rho)\rho g/\mu^2$ | — | — |
| $Ga_n$ | Galileo number for power-law fluid (equation 3.63) | — | — |
| $Re'$ | Reynolds number $ud\rho/\mu$ or $ud'\rho/\mu$ | — | — |
| $Re'_0$ | Reynolds number $u_0 d\rho/\mu$ | — | — |
| $Re'_{HB}$ | Reynolds number for spherical particle in a Herschel-Bulkley fluid | — | — |
| $Re'_n$ | Reynolds number for spherical particle in power-law fluid | — | — |

Suffixes

| | |
|---|---|
| **A**, **B** | Particle **A**, **B** |

CHAPTER 4

# Flow of Fluids through Granular Beds and Packed Columns

## 4.1. INTRODUCTION

The flow of fluids through beds composed of stationary granular particles is a frequent occurrence in the chemical industry and therefore expressions are needed to predict pressure drop across beds due to the resistance caused by the presence of the particles. For example, in fixed bed catalytic reactors, such as $SO_2$–$SO_3$ converters, and drying columns containing silica gel or molecular sieves, gases are passed through a bed of particles. In the case of gas absorption into a liquid, the gas flows upwards against a falling liquid stream, the fluids being contained in a vertical column packed with shaped particles. In the filtration of a suspension, liquid flows at a relatively low velocity through the spaces between the particles which have been retained by the filter medium and, as a result of the continuous deposition of solids, the resistance to flow increases progressively throughout the operation. Furthermore, deep bed filtration is used on a very large scale in water treatment, for example, where the quantity of solids to be removed is small. In all these instances it is necessary to estimate the size of the equipment required, and design expressions are required for the drop in pressure for a fluid flowing through a packing, either alone or as a two-phase system. The corresponding expressions for fluidised beds are discussed in Chapter 6. The drop in pressure for flow through a bed of small particles provides a convenient method for obtaining a measure of the external surface area of a powder, for example cement or pigment.

The flow of either a single phase through a bed of particles or the more complex flow of two fluid phases is approached by using the concepts developed in Volume 1 for the flow of an incompressible fluid through regular pipes or ducts. It is found, however, that the problem is not in practice capable of complete analytical solution and the use of experimental data obtained for a variety of different systems is essential. Later in the chapter some aspects of the design of industrial packed columns involving countercurrent flow of liquids and gases are described.

## 4.2. FLOW OF A SINGLE FLUID THROUGH A GRANULAR BED

### 4.2.1. Darcy's law and permeability

The first experimental work on the subject was carried out by DARCY[1] in 1830 in Dijon when he examined the rate of flow of water from the local fountains through beds of sand of various thicknesses. It was shown that the average velocity, as measured over the whole area of the bed, was directly proportional to the driving pressure and inversely proportional to the thickness of the bed. This relation, often termed Darcy's law, has subsequently been confirmed by a number of workers and can be written as follows:

$$u_c = K\frac{(-\Delta P)}{l} \tag{4.1}$$

where $-\Delta P$ is the pressure drop across the bed,
$l$ is the thickness of the bed,
$u_c$ is the average velocity of flow of the fluid, defined as $(1/A)(\mathrm{d}V/\mathrm{dt})$,
$A$ is the total cross sectional area of the bed,
$V$ is the volume of fluid flowing in time t, and
$K$ is a constant depending on the physical properties of the bed and fluid.

The linear relation between the rate of flow and the pressure difference leads one to suppose that the flow was streamline, as discussed in Volume 1, Chapter 3. This would be expected because the Reynolds number for the flow through the pore spaces in a granular material is low, since both the velocity of the fluid and the width of the channels are normally small. The resistance to flow then arises mainly from viscous drag. Equation 4.1 can then be expressed as:

$$u_c = \frac{K(-\Delta P)}{l} = B\frac{(-\Delta P)}{\mu l} \tag{4.2}$$

where $\mu$ is the viscosity of the fluid and $B$ is termed the permeability coefficient for the bed, and depends only on the properties of the bed.

The value of the permeability coefficient is frequently used to give an indication of the ease with which a fluid will flow through a bed of particles or a filter medium. Some values of $B$ for various packings, taken from EISENKLAM[2], are shown in Table 4.1, and it can be seen that $B$ can vary over a wide range of values. It should be noted that these values of $B$ apply only to the laminar flow regime.

### 4.2.2. Specific surface and voidage

The general structure of a bed of particles can often be characterised by the specific surface area of the bed $S_B$ and the fractional voidage of the bed $e$.

$S_B$ is the surface area presented to the fluid per unit volume of bed when the particles are packed in a bed. Its units are (length)$^{-1}$.

$e$ is the fraction of the volume of the bed not occupied by solid material and is termed the fractional voidage, voidage, or porosity. It is dimensionless. Thus the fractional volume of the bed occupied by solid material is $(1 - e)$.

$S$ is the specific surface area of the particles and is the surface area of a particle divided by its volume. Its units are again (length)$^{-1}$. For a sphere, for example:

$$S = \frac{\pi d^2}{\pi(d^3/6)} = \frac{6}{d} \tag{4.3}$$

Table 4.1. Properties of beds of some regular-shaped materials[2]

| Solid constituents | | Porous mass | | |
|---|---|---|---|---|
| No. | Description | Specific surface area $S$(m$^2$/m$^3$) | Fractional voidage, $e$ (−) | Permeability coefficient $B$ (m$^2$) |
| | Spheres | | | |
| 1 | 0.794 mm diam. ($\frac{1}{32}$ in.) | 7600 | 0.393 | $6.2 \times 10^{-10}$ |
| 2 | 1.588 mm diam. ($\frac{1}{16}$ in.) | 3759 | 0.405 | $2.8 \times 10^{-9}$ |
| 3 | 3.175 mm diam. ($\frac{1}{8}$ in.) | 1895 | 0.393 | $9.4 \times 10^{-9}$ |
| 4 | 6.35 mm diam. ($\frac{1}{4}$ in.) | 948 | 0.405 | $4.9 \times 10^{-8}$ |
| 5 | 7.94 mm diam. ($\frac{5}{16}$ in.) | 756 | 0.416 | $9.4 \times 10^{-8}$ |
| | Cubes | | | |
| 6 | 3.175 mm ($\frac{1}{8}$ in.) | 1860 | 0.190 | $4.6 \times 10^{-10}$ |
| 7 | 3.175 mm ($\frac{1}{8}$ in.) | 1860 | 0.425 | $1.5 \times 10^{-8}$ |
| 8 | 6.35 mm ($\frac{1}{4}$ in.) | 1078 | 0.318 | $1.4 \times 10^{-8}$ |
| 9 | 6.35 mm ($\frac{1}{4}$ in.) | 1078 | 0.455 | $6.9 \times 10^{-8}$ |
| | Hexagonal prisms | | | |
| 10 | 4.76 mm × 4.76 mm thick ($\frac{3}{16}$ in. × $\frac{3}{16}$ in.) | 1262 | 0.355 | $1.3 \times 10^{-8}$ |
| 11 | 4.76 mm × 4.76 mm thick ($\frac{3}{16}$ in. × $\frac{3}{16}$ in.) | 1262 | 0.472 | $5.9 \times 10^{-8}$ |
| | Triangular pyramids | | | |
| 12 | 6.35 mm length × 2.87 mm ht. ($\frac{1}{4}$ in. × 0.113 in.) | 2410 | 0.361 | $6.0 \times 10^{-9}$ |
| 13 | 6.35 mm length × 2.87 mm ht. ($\frac{1}{4}$ in. × 0.113 in.) | 2410 | 0.518 | $1.9 \times 10^{-8}$ |
| | Cylinders | | | |
| 14 | 3.175 mm × 3.175 mm diam. ($\frac{1}{8}$ in. × $\frac{1}{8}$ in.) | 1840 | 0.401 | $1.1 \times 10^{-8}$ |
| 15 | 3.175 mm × 6.35 mm diam. ($\frac{1}{8}$ in. × $\frac{1}{4}$ in.) | 1585 | 0.397 | $1.2 \times 10^{-8}$ |
| 16 | 6.35 mm × 6.35 mm diam. ($\frac{1}{4}$ in. × $\frac{1}{4}$ in.) | 945 | 0.410 | $4.6 \times 10^{-8}$ |
| | Plates | | | |
| 17 | 6.35 mm × 6.35 mm × 0.794 mm ($\frac{1}{4}$in. × $\frac{1}{4}$in. × $\frac{1}{32}$ in.) | 3033 | 0.410 | $5.0 \times 10^{-9}$ |
| 18 | 6.35 mm × 6.35 mm × 1.59 mm ($\frac{1}{4}$in. × $\frac{1}{4}$in. × $\frac{1}{16}$ in.) | 1984 | 0.409 | $1.1 \times 10^{-8}$ |
| | Discs | | | |
| 19 | 3.175 mm diam. × 1.59 mm ($\frac{1}{8}$ in. × $\frac{1}{16}$ in.) | 2540 | 0.398 | $6.3 \times 10^{-9}$ |
| | Porcelain Berl saddles | | | |
| 20 | 6 mm (0.236 in.) | 2450 | 0.685 | $9.8 \times 10^{-8}$ |
| 21 | 6 mm (0.236 in.) | 2450 | 0.750 | $1.73 \times 10^{-7}$ |
| 22 | 6 mm (0.236 in.) | 2450 | 0.790 | $2.94 \times 10^{-7}$ |
| 23 | 6 mm (0.236 in.) | 2450 | 0.832 | $3.94 \times 10^{-7}$ |
| 24 | Lessing rings (6 mm) | 5950 | 0.870 | $1.71 \times 10^{-7}$ |
| 25 | Lessing rings (6 mm) | 5950 | 0.889 | $2.79 \times 10^{-7}$ |

It can be seen that $S$ and $S_B$ are not equal due to the voidage which is present when the particles are packed into a bed. If point contact occurs between particles so that only a very small fraction of surface area is lost by overlapping, then:

$$S_B = S(1 - e) \tag{4.4}$$

Some values of $S$ and $e$ for different beds of particles are listed in Table 4.1. Values of $e$ much higher than those shown in Table 4.1, sometimes up to about 0.95, are possible in beds of fibres[3] and some ring packings. For a given shape of particle, $S$ increases as the particle size is reduced, as shown in Table 4.1.

As $e$ is increased, flow through the bed becomes easier and so the permeability coefficient $B$ increases; a relation between $B$, $e$, and $S$ is developed in a later section of this chapter. If the particles are randomly packed, then $e$ should be approximately constant throughout the bed and the resistance to flow the same in all directions. Often near containing walls, $e$ is higher, and corrections for this should be made if the particle size is a significant fraction of the size of the containing vessel. This correction is discussed in more detail later.

### 4.2.3. General expressions for flow through beds in terms of Carman–Kozeny equations

#### *Streamline flow — Carman–Kozeny equation*

Many attempts have been made to obtain general expressions for pressure drop and mean velocity for flow through packings in terms of voidage and specific surface, as these quantities are often known or can be measured. Alternatively, measurements of the pressure drop, velocity, and voidage provide a convenient way of measuring the surface area of some particulate materials, as described later.

The analogy between streamline flow through a tube and streamline flow through the pores in a bed of particles is a useful starting point for deriving a general expression.

From Volume 1, Chapter 3, the equation for streamline flow through a circular tube is:

$$u = \frac{d_t^2}{32\mu}\frac{(-\Delta P)}{l_t} \tag{4.5}$$

where: $\mu$ is the viscosity of the fluid,
$u$ is the mean velocity of the fluid,
$d_t$ is the diameter of the tube, and
$l_t$ is the length of the tube.

If the free space in the bed is assumed to consist of a series of tortuous channels, equation 4.5 may be rewritten for flow through a bed as:

$$u_1 = \frac{d_m'^2}{K'\mu}\frac{(-\Delta P)}{l'} \tag{4.6}$$

where: $d_m'$ is some equivalent diameter of the pore channels,
$K'$ is a dimensionless constant whose value depends on the structure of the bed,
$l'$ is the length of channel, and
$u_1$ is the average velocity through the pore channels.

It should be noted that $u_1$ and $l'$ in equation 4.6 now represent conditions in the pores and are not the same as $u_c$ and $l$ in equations 4.1 and 4.2. However, it is a reasonable assumption that $l'$ is directly proportional to $l$. Dupuit[4] related $u_c$ and $u_1$ by the following argument.

In a cube of side $X$, the volume of free space is $eX^3$ so that the mean cross-sectional area for flow is the free volume divided by the height, or $eX^2$. The volume flowrate through this cube is $u_c X^2$, so that the average linear velocity through the pores, $u_1$, is given by:

$$u_1 = \frac{u_c X^2}{eX^2} = \frac{u_c}{e} \tag{4.7}$$

Although equation 4.7 is reasonably true for random packings, it does not apply to all regular packings. Thus with a bed of spheres arranged in cubic packing, $e = 0.476$, but the fractional free area varies continuously, from 0.215 in a plane across the diameters to 1.0 between successive layers.

For equation 4.6 to be generally useful, an expression is needed for $d'_m$, the equivalent diameter of the pore space. Kozeny[5,6] proposed that $d'_m$ may be taken as:

$$d'_m = \frac{e}{S_B} = \frac{e}{S(1-e)} \tag{4.8}$$

where:

$$\frac{e}{S_B} = \frac{\text{volume of voids filled with fluid}}{\text{wetted surface area of the bed}}$$

$$= \frac{\text{cross-sectional area normal to flow}}{\text{wetted perimeter}}$$

The hydraulic mean diameter for such a flow passage has been shown in Volume 1, Chapter 3 to be:

$$4\left(\frac{\text{cross-sectional area}}{\text{wetted perimeter}}\right)$$

It is then seen that:

$$\frac{e}{S_B} = \frac{1}{4}\,(\text{hydraulic mean diameter})$$

Then taking $u_1 = u_c/e$ and $l' \propto l$, equation 4.6 becomes:

$$u_c = \frac{1}{K''}\frac{e^3}{S_B^2}\frac{1}{\mu}\frac{(-\Delta P)}{l}$$

$$= \frac{1}{K''}\frac{e^3}{S^2(1-e)^2}\frac{1}{\mu}\frac{(-\Delta P)}{l} \tag{4.9}$$

$K''$ is generally known as Kozeny's constant and a commonly accepted value for $K''$ is 5. As will be shown later, however, $K''$ is dependent on porosity, particle shape, and other factors. Comparison with equation 4.2 shows that $B$ the permeability coefficient is given by:

$$B = \frac{1}{K''}\frac{e^3}{S^2(1-e)^2} \tag{4.10}$$

Inserting a value of 5 for $K''$ in equation 4.9:

$$u_c = \frac{1}{5}\frac{e^3}{(1-e)^2}\frac{-\Delta P}{S^2 \mu l} \tag{4.11}$$

For spheres: $S = 6/d$ and: (equation 4.3)

$$u_c = \frac{1}{180}\frac{e^3}{(1-e)^2}\frac{-\Delta P d^2}{\mu l} \tag{4.12}$$

$$= 0.0055\frac{e^3}{(1-e)^2}\frac{-\Delta P d^2}{\mu l} \tag{4.12a}$$

For non-spherical particles, the Sauter mean diameter $d_s$ should be used in place of $d$. This is given in Chapter 1, equation 1.15.

## *Streamline and turbulent flow*

Equation 4.9 applies to streamline flow conditions, though CARMAN[7] and others have extended the analogy with pipe flow to cover both streamline and turbulent flow conditions through packed beds. In this treatment a modified friction factor $R_1/\rho u_1^2$ is plotted against a modified Reynolds number $Re_1$. This is analogous to plotting $R/\rho u^2$ against $Re$ for flow through a pipe as in Volume 1, Chapter 3.

The modified Reynolds number $Re_1$ is obtained by taking the same velocity and characteristic linear dimension $d'_m$ as were used in deriving equation 4.9. Thus:

$$Re_1 = \frac{u_c}{e}\frac{e}{S(1-e)}\frac{\rho}{\mu}$$

$$= \frac{u_c \rho}{S(1-e)\mu} \tag{4.13}$$

The friction factor, which is plotted against the modified Reynolds number, is $R_1/\rho u_1^2$, where $R_1$ is the component of the drag force per unit area of particle surface in the direction of motion. $R_1$ can be related to the properties of the bed and pressure gradient as follows. Considering the forces acting on the fluid in a bed of unit cross-sectional area and thickness $l$, the volume of particles in the bed is $l(1-e)$ and therefore the total surface is $Sl(1-e)$. Thus the resistance force is $R_1 Sl(1-e)$. This force on the fluid must be equal to that produced by a pressure difference of $\Delta P$ across the bed. Then, since the free cross-section of fluid is equal to $e$:

$$(-\Delta P)e = R_1 Sl(1-e)$$

and

$$R_1 = \frac{e}{S(1-e)}\frac{(-\Delta P)}{l} \tag{4.14}$$

Thus

$$\frac{R_1}{\rho u_1^2} = \frac{e^3}{S(1-e)}\frac{(-\Delta P)}{l}\frac{1}{\rho u_c^2} \tag{4.15}$$

Carman found that when $R_1/\rho u_1^2$ was plotted against $Re_1$ using logarithmic coordinates, his data for the flow through randomly packed beds of solid particles could be correlated approximately by a single curve (curve A, Figure 4.1), whose general equation is:

$$\frac{R_1}{\rho u_1^2} = 5Re_1^{-1} + 0.4Re_1^{-0.1} \tag{4.16}$$

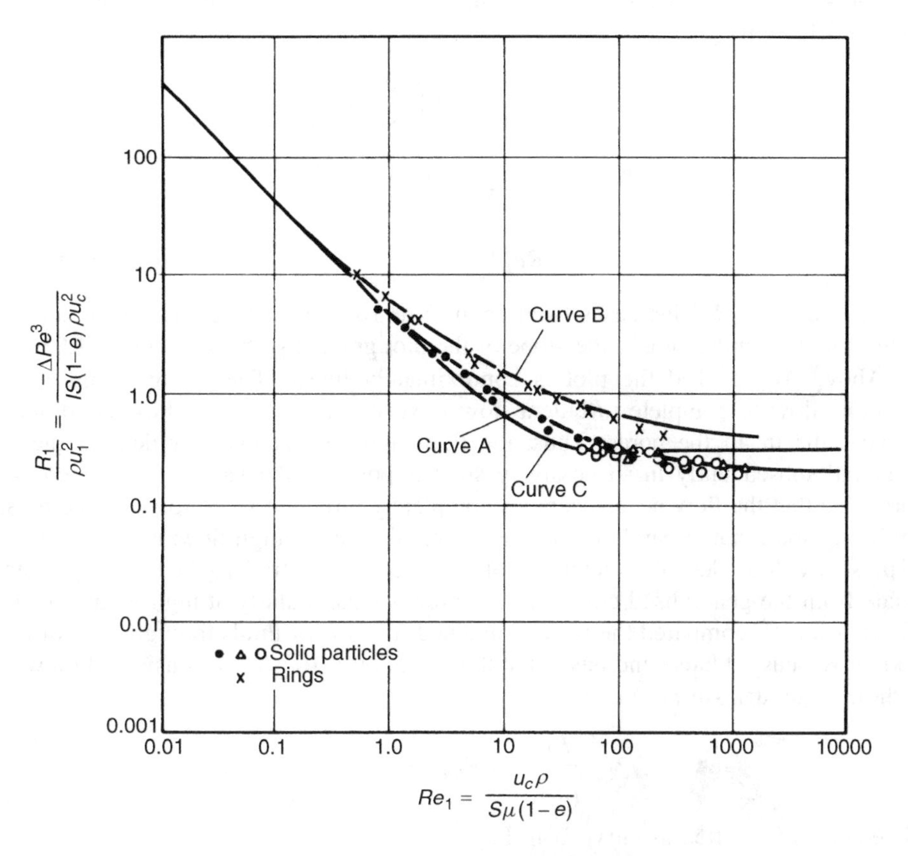


Figure 4.1. Carman's graph of $R_1/\rho u_1^2$ against $Re_1$

The form of equation 4.16 is similar to that of equation 4.17 proposed by FORCHHEIMER[8] who suggested that the resistance to flow should be considered in two parts: that due to the viscous drag at the surface of the particles, and that due to loss in turbulent eddies and at the sudden changes in the cross-section of the channels. Thus:

$$(-\Delta P) = \alpha u_c + \alpha' u_c^{n'} \tag{4.17}$$

The first term in this equation will predominate at low rates of flow where the losses are mainly attributable to skin friction, and the second term will become significant at high flowrates and in very thin beds where the enlargement and contraction losses become very important. At very high flowrates the effects of viscous forces are negligible.

From equation 4.16 it can be seen that for values of $Re_1$ less than about 2, the second term is small and, approximately:

$$\frac{R_1}{\rho u_1^2} = 5Re_1^{-1} \tag{4.18}$$

Equation 4.18 can be obtained from equation 4.11 by substituting for $-\Delta P/l$ from equation 4.15. This gives:

$$u_c = \frac{1}{5}\left(\frac{1}{1-e}\right)\left(\frac{\rho u_c^2}{S\mu}\right)\left(\frac{R_1}{\rho u_1^2}\right)$$

Thus:

$$\frac{R_1}{\rho u_1^2} = 5\left(\frac{S(1-e)\mu}{u_c \rho}\right)$$

$$= 5Re_1^{-1} \quad \text{(from equation 4.13)}$$

As the value of $Re_1$ increases from about 2 to 100, the second term in equation 4.16 becomes more significant and the slope of the plot gradually changes from $-1.0$ to about $-\frac{1}{4}$. Above $Re_1$ of 100 the plot is approximately linear. The change from complete streamline flow to complete turbulent flow is very gradual because flow conditions are not the same in all the pores. Thus, the flow starts to become turbulent in the larger pores, and subsequently in successively smaller pores as the value of $Re_1$ increases. It is probable that the flow never becomes completely turbulent since some of the passages may be so small that streamline conditions prevail even at high flowrates.

Rings, which as described later are often used in industrial packed columns, tend to deviate from the generalised curve A on Figure 4.1 particularly at high values of $Re_1$.

Sawistowski[9] compared the results obtained for flow of fluids through beds of hollow packings (discussed later) and has noted that equation 4.16 gives a consistently low result for these materials. He proposed:

$$\frac{R_1}{\rho u_1^2} = 5Re_1^{-1} + Re_1^{-0.1} \tag{4.19}$$

This equation is plotted as curve B in Figure 4.1.

For flow through ring packings which as described later are often used in industrial packed columns, Ergun[10] obtained a good semi-empirical correlation for pressure drop as follows:

$$\frac{-\Delta P}{l} = 150\frac{(1-e)^2}{e^3}\frac{\mu u_c}{d^2} + 1.75\frac{(1-e)}{e^3}\frac{\rho u_c^2}{d} \tag{4.20}$$

Writing $d = 6/S$ (from equation 4.3):

$$\frac{-\Delta P}{Sl\rho u_c^2}\frac{e^3}{1-e} = 4.17\frac{\mu S(1-e)}{\rho u_c} + 0.29$$

or:

$$\frac{R_1}{\rho u_1^2} = 4.17Re_1^{-1} + 0.29 \tag{4.21}$$

This equation is plotted as curve C in Figure 4.1. The form of equation 4.21 is somewhat similar to that of equations 4.16 and 4.17, in that the first term represents viscous losses which are most significant at low velocities and the second term represents kinetic energy losses which become more significant at high velocities. The equation is thus applicable over a wide range of velocities and was found by Ergun to correlate experimental data well for values of $Re_1/(1-e)$ from 1 to over 2000.

The form of the above equations suggests that the only properties of the bed on which the pressure gradient depends are its specific surface $S$ (or particle size $d$) and its voidage $e$. However, the structure of the bed depends additionally on the particle size distribution, the particle shape and the way in which the bed has been formed; in addition both the walls of the container and the nature of the bed support can considerably affect the way the particles pack. It would be expected, therefore, that experimentally determined values of pressure gradient would show a considerable scatter relative to the values predicted by the equations. The importance of some of these factors is discussed in the next section.

Furthermore, the rheology of the fluid is important in determining how it flows through a packed bed. Only Newtonian fluid behaviour has been considered hitherto. For non-Newtonian fluids, the effect of continual changes in the shape and cross-section of the flow passages may be considerable and no simple relation may exist between pressure gradient and flowrate. This problem has been the subject of extensive studies by several workers including KEMBLOWSKI *et al.*[11].

In some applications, there may be simultaneous flow of two immiscible liquids, or of a liquid and a gas. In general, one of the liquids (or the liquid in the case of liquid–gas systems) will preferentially wet the particles and flow as a continuous film over the surface of the particles, while the other phase flows through the remaining free space. The problem is complex and the exact nature of the flow depends on the physical properties of the two phases, including their surface tensions. An analysis has been made by several workers including BOTSET[12] and GLASER and LITT[13].

## Dependence of $K''$ on bed structure

*Tortuosity.* Although it was implied in the derivation of equation 4.9 that a single value of the Kozeny constant $K''$ applied to all packed beds, in practice this assumption does not hold.

CARMAN[7] has shown that:

$$K'' = \left(\frac{l'}{l}\right)^2 \times K_0 \tag{4.22}$$

where $(l'/l)$ is the tortuosity and is a measure of the fluid path length through the bed compared with the actual depth of the bed,

$K_0$ is a factor which depends on the shape of the cross-section of a channel through which fluid is passing.

For streamline fluid flow through a circular pipe where Poiseuille's equation applies (given in Volume 1, Chapter 3), $K_0$ is equal to 2.0, and for streamline flow through a rectangle where the ratio of the lengths of the sides is 10 : 1, $K_0 = 2.65$. CARMAN[14] has listed values of $K_0$ for other cross-sections. From equation 4.22 it can be seen that if, say,

$K_0$ were constant, then $K''$ would increase with increase in tortuosity. The reason for $K''$ being near to 5.0 for many different beds is probably that changes in tortuosity from one bed to another have been compensated by changes in $K_0$ in the opposite direction.

*Wall effect.* In a packed bed, the particles will not pack as closely in the region near the wall as in the centre of the bed, so that the actual resistance to flow in a bed of small diameter is less than it would be in an infinite container for the same flowrate per unit area of bed cross-section. A correction factor $f_w$ for this effect has been determined experimentally by Coulson[15]. This takes the form:

$$f_w = \left(1 + \frac{1}{2}\frac{S_c}{S}\right)^2 \tag{4.23}$$

where $S_c$ is the surface of the container per unit volume of bed.

Equation 4.9 then becomes:

$$u_c = \frac{1}{K''}\frac{e^3}{S^2(1-e)^2}\frac{1}{\mu}\frac{(-\Delta P)}{l}f_w \tag{4.24}$$

The values of $K''$ shown on Figure 4.2 apply to equation 4.24.

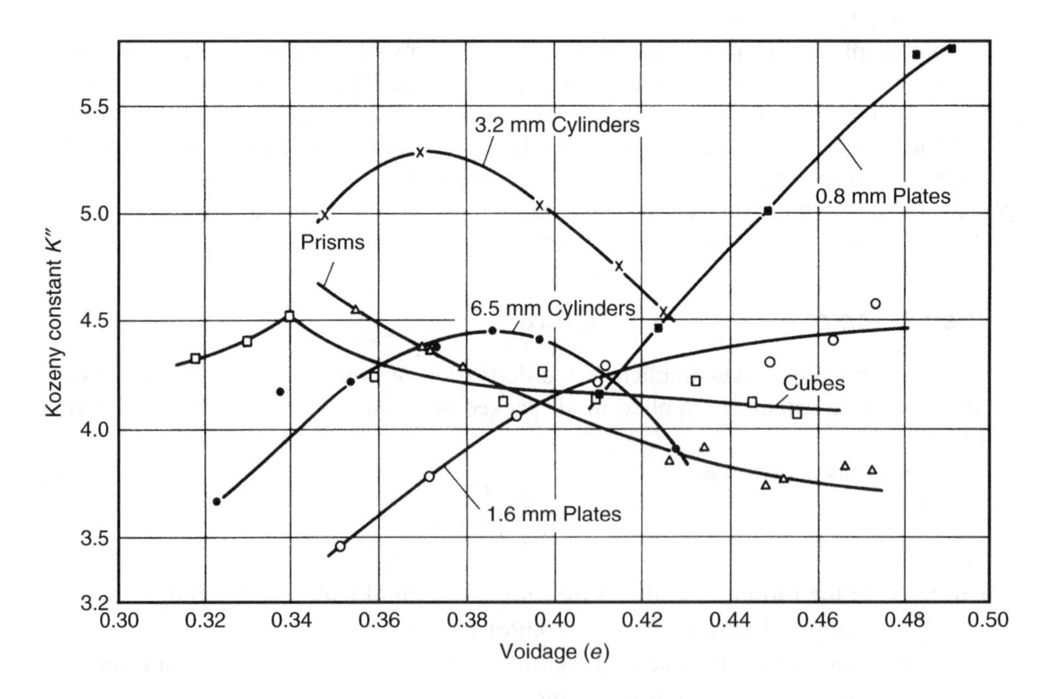


Figure 4.2. Variation of Kozeny's constant $K''$ with voidage for various shapes

*Non-spherical particles.* Coulson[15] and Wyllie and Gregory[16] have each determined values of $K''$ for particles of many different sizes and shapes, including prisms, cubes, and plates. Some of these values for $K''$ are shown in Figure 4.2 where it is seen that they lie between 3 and 6 with the extreme values only occurring with thin plates. This variation of $K''$ with plates probably arises, not only from the fact that area contact is obtained between the particles, but also because the plates tend to give greater tortuosities. For normal granular materials KIHN[17] and PIRIE[18] have found that $K''$ is reasonably constant and does not vary so widely as the $K''$ values for extreme shapes in Figure 4.2.

*Spherical particles.* Equation 4.24 has been tested with spherical particles over a wide range of sizes and $K''$ has been found to be about $4.8 \pm 0.3$[15,19].

For beds composed of spheres of mixed sizes the porosity of the packing can change very rapidly if the smaller spheres are able to fill the pores between the larger ones. Thus COULSON[15] found that, with a mixture of spheres of size ratio 2 : 1, a bed behaves much in accordance with equation 4.19 but, if the size ratio is 5 : 1 and the smaller particles form less than 30 per cent by volume of the larger ones, then $K''$ falls very rapidly, emphasising that only for uniform sized particles can bed behaviour be predicted with confidence.

*Beds with high voidage.* Spheres and particles which are approximately isometric do not pack to give beds with voidages in excess of about 0.6. With fibres and some ring packings, however, values of $e$ near unity can be obtained and for these high values $K''$ rises rapidly. Some values are given in Table 4.2.

Table 4.2. Experimental values of $K''$ for beds of high porosity

| Voidage $e$ | Experimental value of $K''$ BRINKMAN[3] | DAVIES[21] | Silk fibres LORD[20] |
|---|---|---|---|
| 0.5 | 5.5 | | |
| 0.6 | 4.3 | | |
| 0.8 | 5.4 | 6.7 | 5.35 |
| 0.9 | 8.8 | 9.7 | 6.8 |
| 0.95 | 15.2 | 15.3 | 9.2 |
| 0.98 | 32.8 | 27.6 | 15.3 |

Deviations from the Carman–Kozeny equation (4.9) become more pronounced in these beds of fibres as the voidage increases, because the nature of the flow changes from one of channel flow to one in which the fibres behave as a series of obstacles in an otherwise unobstructed passage. The flow pattern is also different in expanded fluidised beds and the Carman–Kozeny equation does not apply there either. As fine spherical particles move far apart in a fluidised bed, Stokes' law can be applied, whereas the Carman–Kozeny equation leads to no such limiting resistance. This problem is further discussed by CARMAN[14].

*Effect of bed support.* The structure of the bed, and hence $K''$, is markedly influenced by the nature of the support. For example, the initial condition in a filtration may affect the whole of a filter cake. Figure 4.3 shows the difference in orientation of two beds of cubical particles. The importance of the packing support should not be overlooked in considering the drop in pressure through the column since the support may itself form an important resistance, and by orientating the particles as indicated may also affect the total pressure drop.

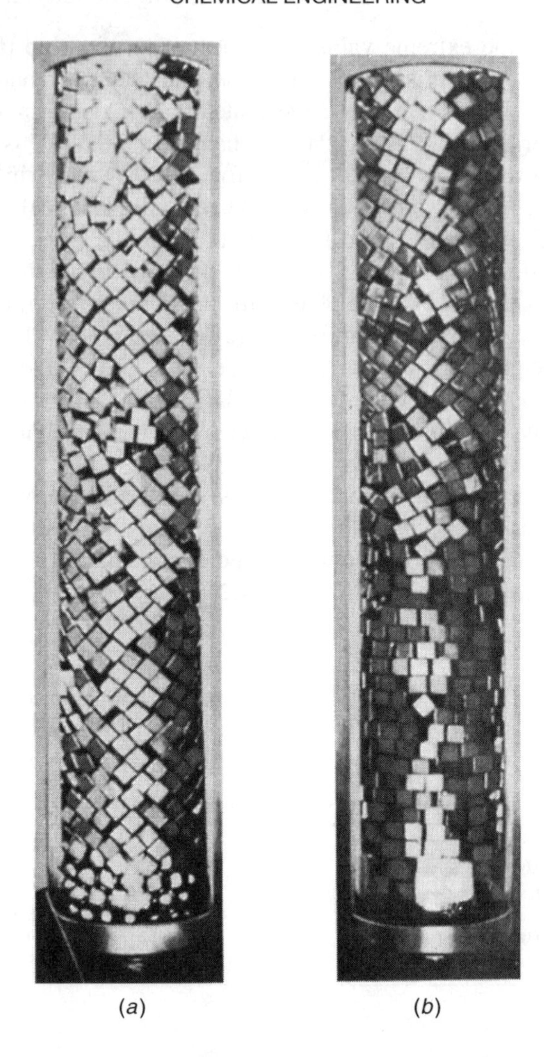


Figure 4.3. Packing of cubes, stacked on (*a*) Plane surface, and (*b*) On bed of spheres

## *The application of Carman–Kozeny equations*

Equations 4.9 and 4.16, which involve $e/S_B$ as a measure of the effective pore diameter, are developed from a relatively sound theoretical basis and are recommended for beds of small particles when they are nearly spherical in shape. The correction factor for wall effects, given by equation 4.23, should be included where appropriate. With larger particles which will frequently be far from spherical in shape, the correlations are not so reliable. As shown in Figure 4.1, deviations can occur for rings at higher values of $Re_1$. Efforts to correct for non-sphericity, though frequently useful, are not universally effective, and in such cases it will often be more rewarding to use correlations, such as equation 4.19, which are based on experimental data for large packings.

### *Use of Carman–Kozeny equation for measurement of particle surface*

The Carman–Kozeny equation relates the drop in pressure through a bed to the specific surface of the material and can therefore be used as a means of calculating $S$ from measurements of the drop in pressure. This method is strictly only suitable for beds of uniformly packed particles and it is not a suitable method for measuring the size distribution of particles in the subsieve range. A convenient form of apparatus developed by LEA and NURSE[(22)] is shown diagrammatically in Figure 4.4. In this apparatus, air or another suitable gas flows through the bed contained in a cell (25 mm diameter, 87 mm deep), and the pressure drop is obtained from $h_1$ and the gas flowrate from $h_2$.

If the diameters of the particles are below about 5 μm, then *slip* will occur and this must be allowed for, as discussed by CARMAN and MALHERBE[(23)].

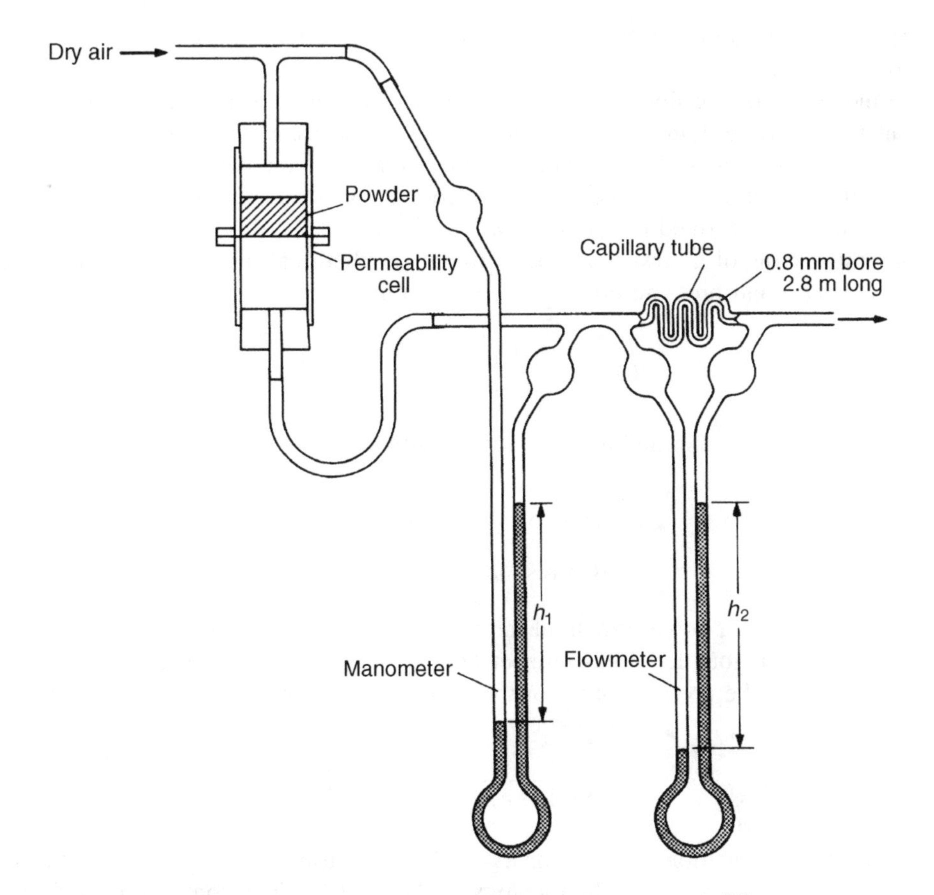


Figure 4.4. The permeability apparatus of LEA and NURSE[(22)]

The method has been successfully developed for measurement of the surface area of cement and for such materials as pigments, fine metal powders, pulverised coal, and fine fibres.

### 4.2.4. Non-Newtonian fluids

There is only a very limited amount of published work on the flow of non-Newtonian fluids through packed beds, and there are serious discrepancies between the results and conclusions of different workers. The range of voidages studied is very narrow, in most cases falling in the range $0.35 < e < 0.41$. For a detailed account of the current situation, reference should be made to work of CHHABRA *et al.*[24] and of KEMBLOWSKI *et al.*[25].

Most published work relates to the flow of shear-thinning fluids whose rheological behaviour follows the two-parameter *power-law* model (discussed in Volume 1, Chapter 3), in which the shear stress $\tau$ and shear rate $\dot{\gamma}$ are related by:

$$\tau = k\dot{\gamma}^n \tag{4.25}$$

where $k$ is known as the consistency coefficient and $n$ ($< 1$ for a shear-thinning fluid) is the power-law index.

The modelling of the flow of a non-Newtonian fluid through a packed bed follows a similar, though more complex, procedure to that adopted earlier in this chapter for the flow of a Newtonian fluid. It first involves a consideration of the flow through a cylindrical tube and then adapting this to the flow in the complex geometry existing in a packed bed. The procedure is described in detail elsewhere[24,25].

For laminar flow of a power-law fluid through a cylindrical tube, the relation between mean velocity $u$ and pressure drop $-\Delta P$ is given by:

$$u = \left(\frac{-\Delta P}{4kl}\right)^{1/n} \frac{n}{6n+2} d_t^{(n+1)/n} \tag{4.26}$$

and the so-called Metzner and Reed Reynolds number by:

$$Re_{MR} = 8\left(\frac{n}{6n+2}\right)^n \frac{\rho u^{2-n} d_t^n}{k} \tag{4.27}$$

(Corresponding to Volume 1, equations 3.136 and 3.140)

For laminar flow of a power-law fluid through a packed bed, KEMBLOWSKI *et al.*[25] have developed an analogous Reynolds number $(Re_1)_n$, which they have used as the basis for the calculation of the pressure drop for the flow of power-law fluids:

$$(Re_1)_n = \frac{\rho u_c^{2-n}}{kS^n(1-e)^n}\left(\frac{4n}{3n+1}\right)^n \left(\frac{b\sqrt{2}}{e^2}\right)^{1-n} \tag{4.28}$$

The last term in equation 4.28 is not a simple geometric characterisation of the flow passages, as it also depends on the rheology of the fluid ($n$). The constant $b$ is a function of the shape of the particles constituting the bed, having a value of about 15 for particles of spherical, or near-spherical, shapes; there are insufficient reliable data available to permit values of $b$ to be quoted for other shapes. Substitution of $n = 1$ and of $\mu$ for $k$ in equation 4.28 reduces it to equation 4.13, obtained earlier for Newtonian fluids.

Using this definition of the Reynolds number in place of $Re_1$ the value of the friction group $(R_1/\rho u_1^2)$ may be calculated from equation 4.18, developed previously for Newtonian fluids, and hence the superficial velocity $u_c$ for a power-law fluid may be calculated as a function of the pressure difference for values of the Reynolds number less than 2 to give:

$$u_c = \left(\frac{-\Delta P}{5kl}\right)^{1/n} \frac{1}{S^{(n+1)/n}} \frac{e^3}{(1-e)^2} \left(\frac{4n}{3n+1}\right)^{1/n} \left(\frac{b\sqrt{2}}{e^2}\right)^{(1-n)/n} \tag{4.29}$$

For Newtonian fluids ($n = 1$), equation 4.29 reduces to equation 4.9.

For polymer solutions, equation 4.29 applies only to flow through unconsolidated media since, otherwise, the pore dimensions may be of the same order of magnitude as those of the polymer molecules and additional complications, such as pore blocking and adsorption, may arise.

If the fluid has significant elastic properties, the flow may be appreciably affected because of the rapid changes in the magnitude and direction of flow as the fluid traverses the complex flow path between the particles in the granular bed, as discussed by CHHABRA[24].

### 4.2.5. Molecular flow

In the relations given earlier, it is assumed that the fluid can be regarded as a continuum and that there is no slip between the wall of the capillary and the fluid layers in contact with it. However, when conditions are such that the mean free path of the molecules of a gas is a significant fraction of the capillary diameter, the flowrate at a given value of the pressure gradient becomes greater than the predicted value. If the mean free path exceeds the capillary diameter, the flowrate becomes independent of the viscosity and the process is one of diffusion. Whereas these considerations apply only at very low pressures in normal tubes, in fine-pored materials the pore diameter and the mean free path may be of the same order of magnitude even at atmospheric pressure.

## 4.3. DISPERSION

Dispersion is the general term which is used to describe the various types of self-induced mixing processes which can occur during the flow of a fluid through a pipe or vessel. The effects of dispersion are particularly important in packed beds, though they are also present under the simple flow conditions which exist in a straight tube or pipe. Dispersion can arise from the effects of molecular diffusion or as the result of the flow pattern existing within the fluid. An important consequence of dispersion is that the flow in a packed bed reactor deviates from plug flow, with an important effect on the characteristics of the reactor.

It is of interest to consider first what is happening in pipe flow. Random molecular movement gives rise to a mixing process which can be described by Fick's law (given in Volume 1, Chapter 10). If concentration differences exist, the rate of transfer of a component is proportional to the product of the molecular diffusivity and the concentration gradient. If the fluid is in laminar flow, a parabolic velocity profile is set up over the cross-section and the fluid at the centre moves with twice the mean velocity in the pipe. This can give rise to dispersion since elements of fluid will take different times to traverse the length of the pipe, according to their radial positions. When the fluid leaves the pipe, elements that have been within the pipe for very different periods of time will be mixed together. Thus, if the concentration of a tracer material in the fluid is suddenly changed, the effect will first be seen in the outlet stream after an interval required for the fluid at the axis to traverse the length of the pipe. Then, as time increases, the effect will be evident in the fluid issuing at progressively greater distances from the centre. Because the fluid velocity approaches zero at the pipe wall, the fluid near the wall will reflect the change over only a very long period.

If the fluid in the pipe is in turbulent flow, the effects of molecular diffusion will be supplemented by the action of the turbulent eddies, and a much higher rate of transfer of material will occur within the fluid. Because the turbulent eddies also give rise to momentum transfer, the velocity profile is much flatter and the dispersion due to the effects of the different velocities of the fluid elements will be correspondingly less.

In a packed bed, the effects of dispersion will generally be greater than in a straight tube. The fluid is flowing successively through constrictions in the flow channels and then through broader passages or cells. Radial mixing readily takes place in the cells because the fluid enters them with an excess of kinetic energy, much of which is converted into rotational motion within the cells. Furthermore, the velocity profile is continuously changing within the fluid as it proceeds through the bed. Wall effects can be important in a packed bed because the bed voidage will be higher near the wall and flow will occur preferentially in that region.

At low rates of flow the effects of molecular diffusion predominate and cell mixing contributes relatively little to the dispersion. At high rates, on the other hand, a realistic model is presented by considering the bed to consist of a series of mixing cells, the dimension of each of which is of the same order as the size of the particles forming the bed. Whatever the mechanism, however, the rate of dispersion can be conveniently described by means of a dispersion coefficient. The process is generally anisotropic, except at very low flowrates; that is the dispersion rate is different in the longitudinal and radial directions, and therefore separate dispersion coefficients $D_L$ and $D_R$ are generally used to represent the behaviour in the two directions. The process is normally linear, with the rate of dispersion proportional to the product of the corresponding dispersion coefficient and concentration gradient. The principal factors governing dispersion in packed beds are discussed in a critical review by GUNN[26].

The differential equation for dispersion in a cylindrical bed of voidage $e$ may be obtained by taking a material balance over an annular element of height $\delta l$, inner radius $r$, and outer radius $r + \delta r$ (as shown in Figure 4.5). On the basis of a dispersion model it is seen that if $C$ is concentration of a reference material as a function of axial position $l$, radial position $r$, time $t$, and $D_L$ and $D_R$ are the axial and radial dispersion coefficients, then:

Rate of entry of reference material due to flow in axial direction:

$$= u_c(2\pi r \delta r)C$$

Corresponding efflux rate:

$$= u_c(2\pi r \delta r)\left(C + \frac{\partial C}{\partial l}\delta l\right)$$

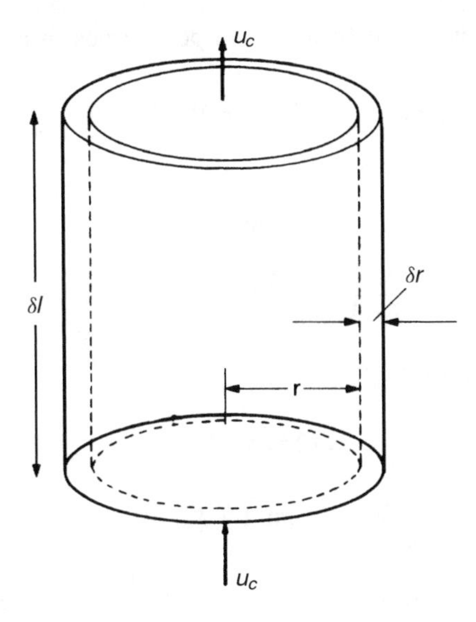


Figure 4.5. Dispersion in packed beds

Net accumulation rate in element due to flow in axial direction:

$$= -u_c(2\pi r \delta r)\frac{\partial C}{\partial l}\delta l \tag{4.30}$$

Rate of diffusion in axial direction across inlet boundary:

$$= -(2\pi r\,\delta r\,e)D_L\frac{\partial C}{\partial l}$$

Corresponding rate at outlet boundary:

$$= -(2\pi r\,\delta r\,e)D_L\left(\frac{\partial C}{\partial l} + \frac{\partial^2 C}{\partial l^2}\delta l\right)$$

Net accumulation rate due to diffusion from boundaries in axial direction:

$$= (2\pi r\,\delta r\,e)D_L\frac{\partial^2 C}{\partial l^2}\delta l \tag{4.31}$$

Diffusion in radial direction at radius $r$:

$$= (2\pi r\,\delta l\,e)D_R\frac{\partial C}{\partial r}$$

Corresponding rate at radius $r + \delta r$:

$$= [2\pi(r + \delta r)\,\delta l\,e]D_R\left[\frac{\partial C}{\partial r} + \frac{\partial^2 C}{\partial r^2}\delta r\right]$$

Net accumulation rate due to diffusion from boundaries in radial direction:

$$= -[2\pi r\delta l\, e]D_R\frac{\partial C}{\partial r} + [2\pi(r+\delta r)\delta l\, e]D_R\left(\frac{\partial C}{\partial r} + \frac{\partial^2 C}{\partial r^2}\delta r\right)$$

$$= 2\pi\,\delta l\, e\, D_R\left[\frac{\partial C}{\partial r}\delta r + r\delta r\frac{\partial^2 C}{\partial r^2} + (\delta r)^2\frac{\partial^2 C}{\partial r^2}\right]$$

$$= 2\pi\,\delta l\, e\, D_R\left[\delta r\frac{\partial}{\partial r}\left(r\frac{\partial C}{\partial r}\right)\right] \text{ (ignoring the last term)} \tag{4.32}$$

Now the total accumulation rate:

$$= (2\pi r\delta r\delta l)e\frac{\partial C}{\partial t} \tag{4.33}$$

Thus, from equations 4.33, 4.30, 4.31 and 4.32:

$$(2\pi r\delta r\delta l)e\frac{\partial C}{\partial t} = -u_c(2\pi r\delta r)\frac{\partial C}{\partial l}\delta l + (2\pi r\delta r e)D_L\frac{\partial^2 C}{\partial l^2}\delta l + 2\pi\,\delta l e D_R\left[\delta r\frac{\partial}{\partial r}\left(r\frac{\partial C}{\partial r}\right)\right]$$

On dividing through by $(2\pi r\delta r\delta l)e$:

$$\frac{\partial C}{\partial t} + \frac{1}{e}u_c\frac{\partial C}{\partial l} = D_L\frac{\partial^2 C}{\partial l^2} + \frac{1}{r}D_R\frac{\partial}{\partial r}\left(r\frac{\partial C}{\partial r}\right) \tag{4.34}$$

Longitudinal dispersion coefficients can be readily obtained by injecting a pulse of tracer into the bed in such a way that radial concentration gradients are eliminated, and measuring the change in shape of the pulse as it passes through the bed. Since $\partial C/\partial r$ is then zero, equation 4.34 becomes:

$$\frac{\partial C}{\partial t} + \frac{u_c}{e}\frac{\partial C}{\partial l} = D_L\frac{\partial^2 C}{\partial l^2} \tag{4.35}$$

Values of $D_L$ can be calculated from the change in shape of a pulse of tracer as it passes between two locations in the bed, and a typical procedure is described by EDWARDS and RICHARDSON[27]. GUNN and PRYCE[28], on the other hand, imparted a sinusoidal variation to the concentration of tracer in the gas introduced into the bed. The results obtained by a number of workers are shown in Figure 4.6 as a Peclet number $Pe(= u_c d/eD_L)$ plotted against the particle Reynolds number $(Re'_c = u_c d\rho/\mu)$.

For gases, at low Reynolds numbers (<1), the Peclet number increases linearly with Reynolds number, giving:

$$\frac{u_c d}{eD_L} = K\frac{u_c d\rho}{\mu} = K\,Sc^{-1}\frac{u_c d}{D} \tag{4.36}$$

or:

$$\frac{D_L}{D} = \text{constant, } \gamma \text{ which has a value of approximately 0.7} \tag{4.37}$$

since $Sc$, the Schmidt number, is approximately constant for gases and the voidage of a randomly packed bed is usually about 0.4. This is consistent with the hypothesis that, at low Reynolds numbers, molecular diffusion predominates. The factor 0.7 is a tortuosity

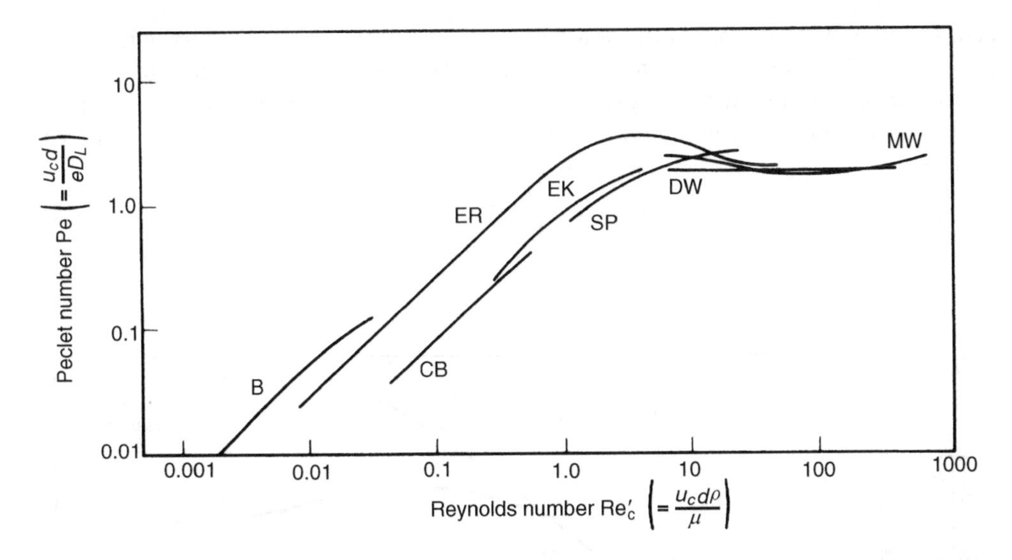


Figure 4.6. Longitudinal dispersion in gases in packed beds. ER — EDWARDS and RICHARDSON[27]; B — BLACKWELL *et al.*[29]; CB — CARBERRY and BRETTON[32]; DW — DE MARIA and WHITE[33]; MW — MCHENRY and WILHELM[34]; SP — SINCLAIR and POTTER[35]; EK — EVANS and KENNEY[36], $N_2 + He$ in $N_2 + H_2$

factor which allows for the fact that the molecules must negotiate a tortuous path because of the presence of the particles.

At Reynolds numbers greater than about 10 the Peclet number becomes approximately constant, giving:

$$D_L \approx \frac{1}{2}\frac{u_c}{e}d \tag{4.38}$$

This equation is predicted by the mixing cell model, and turbulence theories put forward by ARIS and AMUNDSON[30] and by PRAUSNITZ[31].

In the intermediate range of Reynolds numbers, the effects of molecular diffusivity and of macroscopic mixing are approximately additive, and the dispersion coefficient is given by an equation of the form:

$$D_L = \gamma D + \frac{1}{2}\frac{u_c d}{e} \tag{4.39}$$

However, the two mechanisms interact and molecular diffusion can reduce the effects of convective dispersion. This can be explained by the fact that with streamline flow in a tube molecular diffusion will tend to smooth out the concentration profile arising from the velocity distribution over the cross-section. Similarly radial dispersion can give rise to lower values of longitudinal dispersion than predicted by equation 4.39. As a result the curves of Peclet versus Reynolds number tend to pass through a maximum as shown in Figure 4.6.

A comparison of the effects of axial and radial mixing is seen in Figure 4.7, which shows results obtained by GUNN and PRYCE[28] for dispersion of argon into air. The values of $D_L$ were obtained as indicated earlier, and $D_R$ was determined by injecting a steady stream of tracer at the axis and measuring the radial concentration gradient across the bed. It is seen that molecular diffusion dominates at low Reynolds numbers, with both the axial and radial dispersion coefficients $D_L$ and $D_R$ equal to approximately 0.7 times the molecular diffusivity. At high Reynolds numbers, however, the ratio of the longitudinal dispersion coefficient to the radial dispersion coefficient approaches a value of about 5. That is:

$$\frac{D_L}{D_R} \approx 5 \tag{4.40}$$

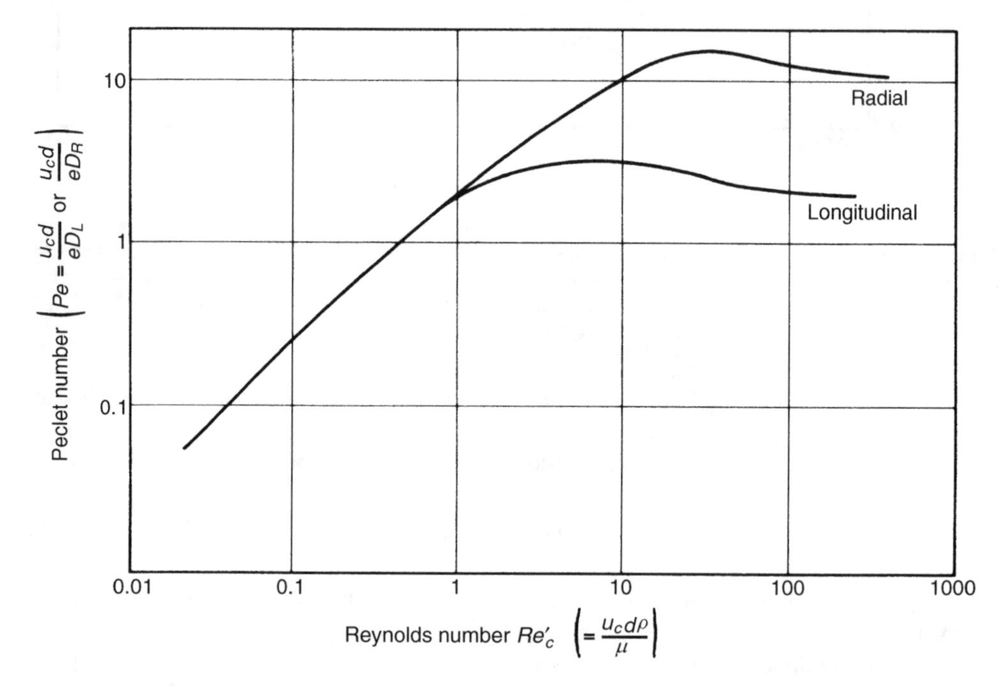


Figure 4.7. Longitudinal and radial mixing coefficients for argon in air[28]

The experimental results for dispersion coefficients in gases show that they can be satisfactorily represented as Peclet number expressed as a function of particle Reynolds number, and that similar correlations are obtained, irrespective of the gases used. However, it might be expected that the Schmidt number would be an important variable, but it is not possible to test this hypothesis with gases as the values of Schmidt number are all approximately the same and equal to about unity.

With liquids, however, the Schmidt number is variable and it is generally about three orders of magnitude greater than for a gas. Results for longitudinal dispersion available in the literature, and plotted in Figure 4.8, show that over the range of Reynolds numbers studied ($10^{-2} < Re'_c < 10^3$) the Peclet number shows little variation and is of the order of unity. Comparison of these results with the corresponding ones for gases (shown in Figure 4.6) shows that the effect of molecular diffusion in liquids is insignificant at

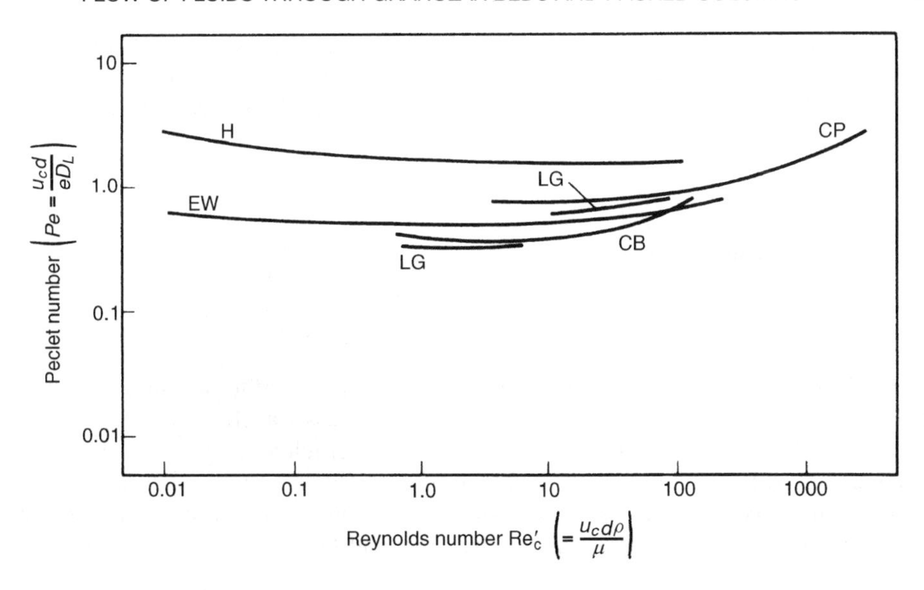


Figure 4.8. Longitudinal dispersion in liquids in packed beds. CP — CAIRNS and PRAUSNITZ[37], CB — CARBERRY and BRETTON[32]; EW — EBACH and WHITE[38]; H — HIBY[39]; LG — LILES and GEANKOPLIS[40]

Reynolds numbers up to unity. This difference can be attributed to the very different magnitudes of the Schmidt numbers.

## 4.4. HEAT TRANSFER IN PACKED BEDS

For heat and mass transfer through a stationary or streamline fluid to a single spherical particle, it has been shown in Volume 1, Chapter 9, that the heat and mass transfer coefficients reach limiting low values given by:

$$Nu' = Sh' = 2 \tag{4.41}$$

where $Nu'(= hd/k)$ and $Sh'(= h_D d/D)$ are the Nusselt and Sherwood numbers with respect to the fluid, respectively.

KRAMERS[41] has shown that, for conditions of forced convection, the heat transfer coefficient can be represented by:

$$Nu' = 2.0 + 1.3Pr^{0.15} + 0.66Pr^{0.31}Re_c'^{0.5} \tag{4.42}$$

where $Re'_c$ is the particle Reynolds number $u_c d\rho/\mu$ based on the superficial velocity $u_c$ of the fluid, and $Pr$ is the Prandtl number $C_p\mu/k$.

This expression has been obtained on the basis of experimental results obtained with fluids of Prandtl numbers ranging from 0.7 to 380.

For natural convection, RANZ and MARSHALL[42] have given:

$$Nu' = 2.0 + 0.6Pr^{1/3}Gr'^{1/4} \tag{4.43}$$

where $Gr'$ is the Grashof number discussed in Volume 1, Chapter 9.

Results for packed beds are much more difficult to obtain because the driving force cannot be measured very readily, GUPTA and THODOS[43] suggest that the $j$-factor for heat transfer, $j_h$ (Volume 1, Chapter 9), forms the most satisfactory basis of correlation for experimental results and have proposed that:

$$ej_h = 2.06 Re_c'^{-0.575} \quad (4.44)$$

where: $e$ is the voidage of the bed,
$j_h = St' Pr^{2/3}$, and
$St'$ = Stanton number $h/C_p \rho u_c$.

The $j$-factors for heat and mass transfer, $j_h$ and $j_d$, are found to be equal, and therefore equation 4.44 can also be used for the calculation of mass transfer rates.

Reproducible correlations for the heat transfer coefficient between a fluid flowing through a packed bed and the cylindrical wall of the container are very difficult to obtain. The main difficulty is that a wide range of packing conditions can occur in the vicinity of the walls. However, the results quoted by ZENZ and OTHMER[44] suggest that:

$$Nu \propto Re_c'^{0.7-0.9} \quad (4.45)$$

It may be noted that in this expression the Nusselt number with respect to the tube wall $Nu$ is related to the Reynolds number with respect to the particle $Re_c'$.

## 4.5. PACKED COLUMNS

Since packed columns consist of shaped particles contained within a column, their behaviour will in many ways be similar to that of packed beds which have already been considered. There are, however, several important differences which make the direct application of the equations for pressure gradient difficult. First, the size of the packing elements in the column will generally be very much larger and the Reynolds number will therefore be such that the flow is turbulent. Secondly, the packing elements will normally be hollow, and therefore have a large amount of internal surface which will offer a higher flow resistance than their external surface. The shapes too are specially designed to produce good mass transfer characteristics with relatively small pressure gradients. Although some of the general principles already discussed can be used to predict pressure gradient as a function of flowrate, it is necessary to rely heavily on the literature issued by the manufacturers of the packings.

In general, packed towers are used for bringing two phases in contact with one another and there will be strong interaction between the fluids. Normally one of the fluids will preferentially wet the packing and will flow as a film over its surface; the second fluid then passes through the remaining volume of the column. With gas (or vapour)–liquid systems, the liquid will normally be the wetting fluid and the gas or vapour will rise through the column making close contact with the down-flowing liquid and having little direct contact with the packing elements. An example of the liquid–gas system is an absorption process where a soluble gas is scrubbed from a mixture of gases by means of a liquid, as shown in Figure 4.9. In a packed column used for distillation, the more volatile component of, say, a binary mixture is progressively transferred to the vapour phase and the less volatile condenses out in the liquid. Packed columns have also been used extensively for liquid–liquid extraction processes where a solute is transferred from one solvent to another, as discussed in Chapter 12. Some principles involved in the design and operation of packed columns will be illustrated by considering columns for gas absorption. In this chapter an outline of the construction of the column and the flow characteristics will be dealt with, whereas the magnitude of the mass transfer coefficients is discussed later in Chapters 11, 12, and 13. The full design process is discussed in Volume 6.

In order to obtain a good rate of transfer per unit volume of the tower, a packing is selected which will promote a high interfacial area between the two phases and a high degree of turbulence in the fluids. Usually increased area and turbulence are achieved at the expense of increased capital cost and/or pressure drop, and a balance must be made between these factors when arriving at an economic design.

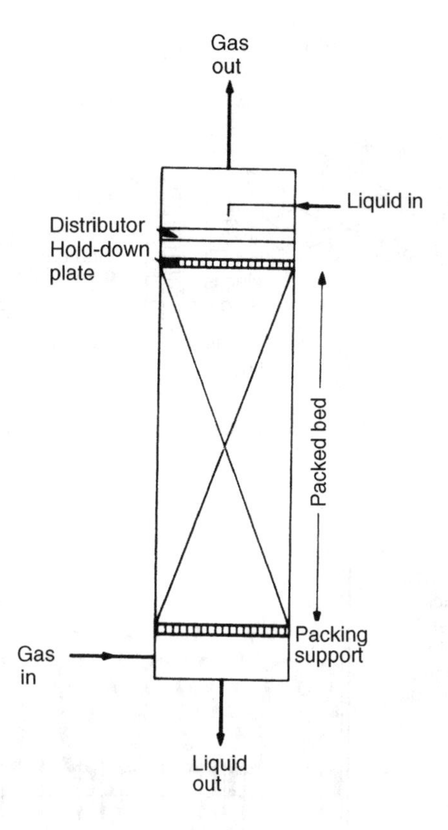


Figure 4.9. Packed absorption column

### 4.5.1. General description

The construction of packed towers is relatively straightforward. The shell of the column may be constructed from metal, ceramics, glass, or plastics material, or from metal with a corrosion-resistant lining. The column should be mounted truly vertically to help uniform liquid distribution. Detailed information on the mechanical design and mounting of industrial scale column shells is given by BROWNELL and YOUNG[45], MOLYNEUX[46] and in BS 5500[47], as well as in Volume 6.

The bed of packing rests on a support plate which should be designed to have at least 75 per cent free area for the passage of the gas so as to offer as low a resistance as possible. The simplest support is a grid with relatively widely spaced bars on which a few layers of large Raschig or partition rings are stacked. One such arrangement is shown in Figure 4.10. The gas injection plate described by LEVA[48] shown in Figure 4.11 is designed to provide separate passageways for gas and liquid so that they need not vie for passage through the same opening. This is achieved by providing the gas inlets to the bed at a point above the level at which liquid leaves the bed.

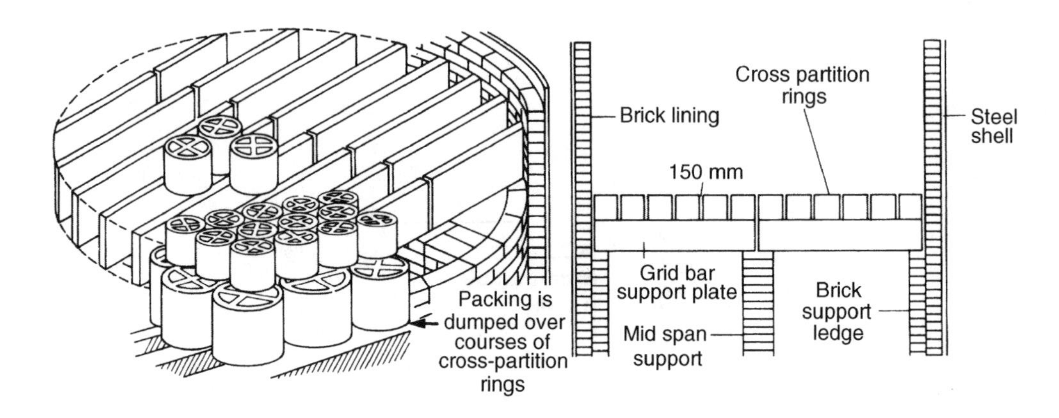


Figure 4.10. Grid bar supports for packed towers

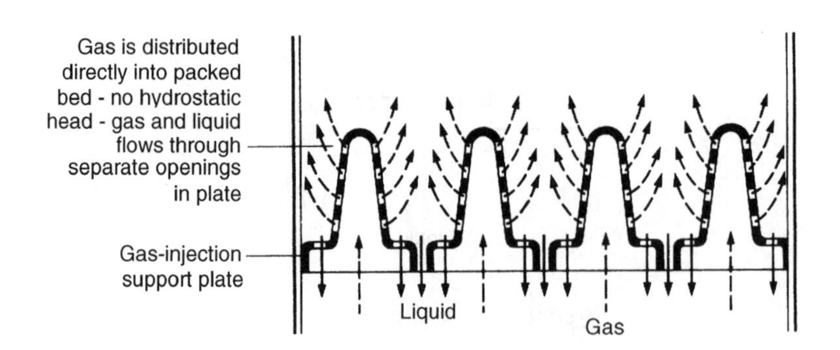


Figure 4.11. The gas injection plate[48]

At the top of the packed bed a liquid distributor of suitable design provides for the uniform irrigation of the packing which is necessary for satisfactory operation. Four examples of different distributors are shown in Figure 4.12[49], and may be described as follows:

(a) A simple orifice type which gives very fine distribution though it must be correctly sized for a particular duty and should not be used where there is any risk of the holes plugging
(b) The notched chimney type of distributor, which has a good range of flexibility for the medium and upper flowrates, and is not prone to blockage
(c) The notched trough distributor which is specially suitable for the larger sizes of tower, and, because of its large free area, it is also suitable for the higher gas rates
(d) The perforated ring type of distributor for use with absorption columns where high gas rates and relatively small liquid rates are encountered. This type is especially suitable where pressure loss must be minimised. For the larger size of tower, where installation through manholes is necessary, it may be made up in flanged sections.

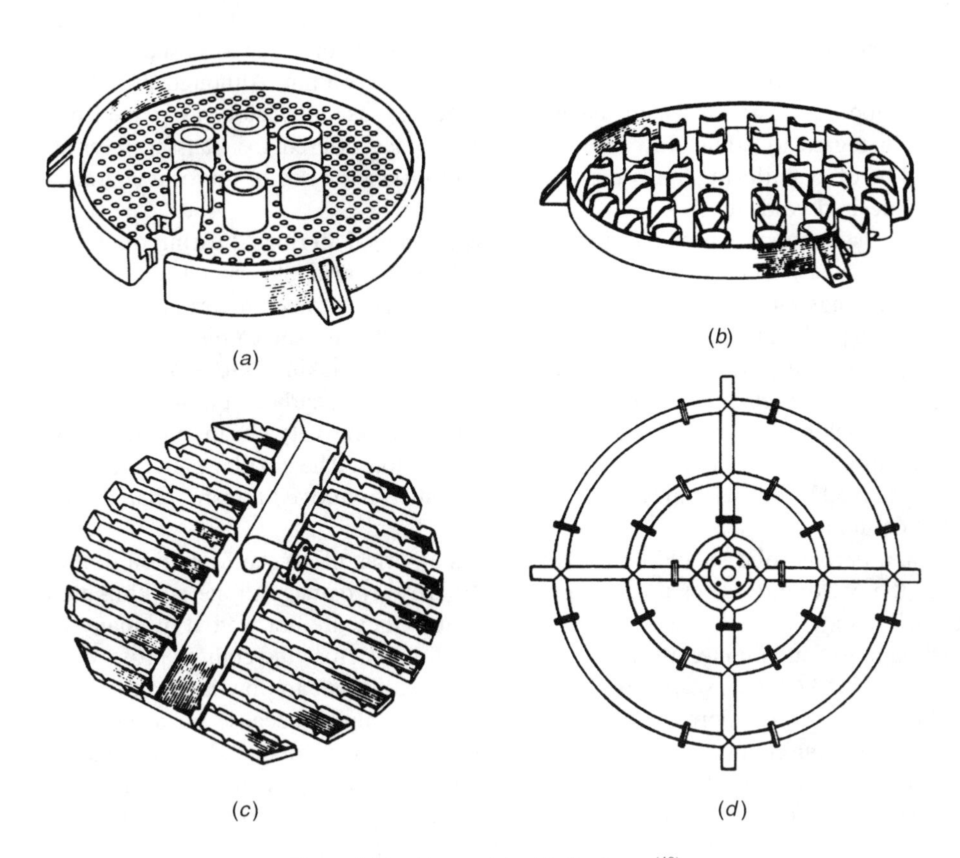


Figure 4.12. Types of liquid distributor[49]

Uniform liquid flow is essential if the best use is to be made of the packing and, if the tower is high, re-distributing plates are necessary. These plates are needed at intervals of about $2\frac{1}{2}$–3 column diameters for Raschig rings and about 5–10 column diameters for Pall rings, but are usually not more than 6 m apart[50]. A "hold-down" plate is often placed at the top of a packed column to minimise movement and breakage of the packing caused by surges in flowrates. The gas inlet should also be designed for uniform flow over the cross-section and the gas exit should be separate from the liquid inlet. Further details on internal fittings are given by LEVA[48].

Columns for both absorption and distillation vary in diameter from about 25 mm for small laboratory purposes to over 4.5 m for large industrial operations; these industrial columns may be 30 m or more in height. Columns may operate at pressures ranging from high vacuum to high pressure, the optimum pressure depending on both the chemical and the physical properties of the system.

## 4.5.2. Packings

Packings can be divided into four main classes — broken solids, shaped packings, grids, and structured packings. Broken solids are the cheapest form and are used in sizes from about 10 mm to 100 mm according to the size of the column. Although they frequently form a good corrosion-resistant material they are not as satisfactory as shaped packings either in regard to liquid flow or to effective surface offered for transfer. The packing should be of as uniform size as possible so as to produce a bed of uniform characteristics with a desired voidage.

The most commonly used packings are Raschig rings, Pall rings, Lessing rings, and Berl saddles. Newer packings include Nutter rings, Intalox and Intalox metal saddles, Hy-Pak, and Mini rings and, because of their high performance characteristics and low pressure drop, these packings now account for a large share of the market. Commonly used packing elements are illustrated in Figure 4.13. Most of these packings are available in a wide range of materials such as ceramics, metals, glass, plastics, carbon, and sometimes rubber. Ceramic packings are resistant to corrosion and comparatively cheap, but are heavy and may require a stronger packing support and foundations. The smaller metal rings are also available made from wire mesh, and these give much-improved transfer characteristics in small columns.

A non-porous solid should be used if there is any risk of crystal formation in the pores when the packing dries, as this can give rise to serious damage to the packing elements. However, some plastics are not very good because they are not wetted by many liquids. Channelling, that is non-uniform distribution of liquid across the column cross-section, is much less marked with shaped packings, and their resistance to flow is much less. Shaped packings also give a more effective surface per unit volume because surface contacts are reduced to a minimum and the film flow is much improved compared with broken solids. On the other hand, the shaped packings are more expensive, particularly when small sizes are used. The voidage obtainable with these packings varies from about 0.45 to 0.95. Ring packings are either dumped into a tower, dropped in small quantities, or may be individually stacked if 75 mm or larger in size. To obtain high and uniform voidage and to prevent breakage, it is often found better to dump the packings into a tower full of liquid. Stacked packings, as shown in Figure 4.10, have the advantage that the flow channels are vertical and there is much less tendency for the liquid to flow to the walls than with

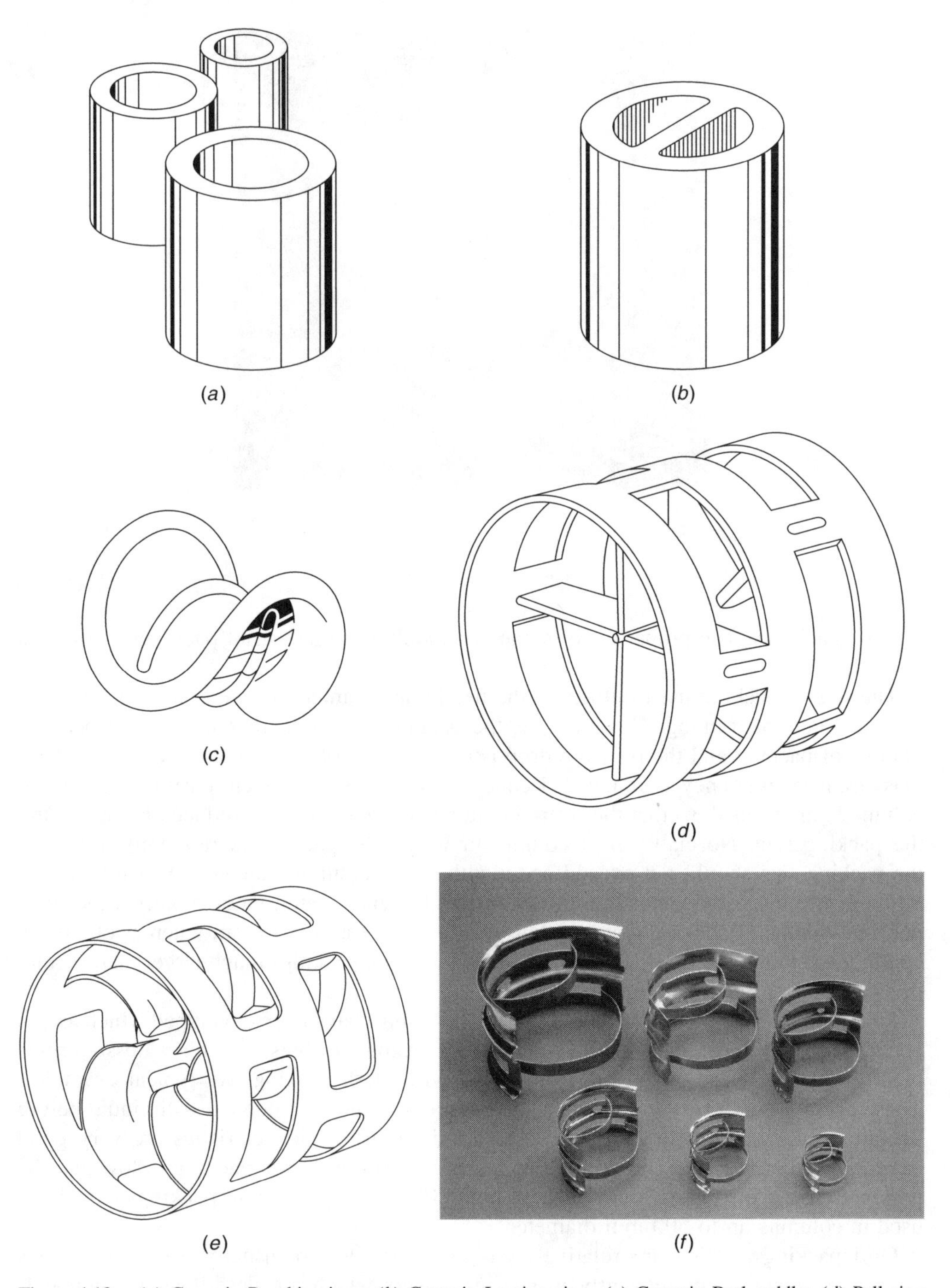


Figure 4.13. (*a*) Ceramic Raschig rings; (*b*) Ceramic Lessing ring; (*c*) Ceramic Berl saddle; (*d*) Pall ring (plastic); (*e*) Pall ring (metal); (*f*) Metal Nutter rings; (*g*) Plastic Nutter ring

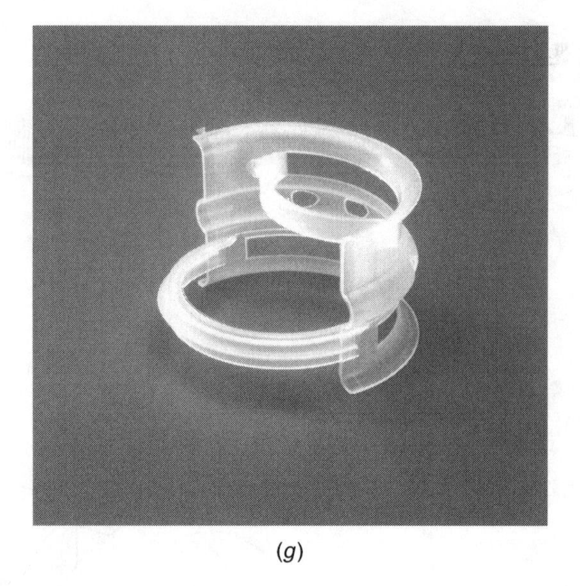

(*g*)

Figure 4.13. *continued*

random packings. The properties of some commonly used industrial packings are shown in Table 4.3.

The size of packing used influences the height and diameter of a column, the pressure drop and cost of packing. Generally, as the packing size is increased, the cost per unit volume of packing and the pressure drop per unit height of packing are reduced, and the mass transfer efficiency is reduced. Reduced mass transfer efficiency results in a taller column being needed, so that the overall column cost is not always reduced by increasing the packing size. Normally, in a column in which the packing is randomly arranged, the packing size should not exceed one-eighth of the column diameter. Above this size, liquid distribution, and hence the mass transfer efficiency, deteriorates rapidly. Since cost per unit volume of packing does not fall much for sizes above 50 mm whereas efficiency continues to fall, there is seldom any advantage in using packings much larger than 50 mm in a randomly packed column.

For laboratory purposes a number of special packings have been developed which are, in general, too expensive for large diameter towers. Dixon packings, which are Lessing rings made from wire mesh, and KnitMesh, a fine wire mesh packing, are typical examples. These packings give very high interfacial areas and, if they are flooded with liquid before operation, all of the surface is active so that the transfer characteristics are very good even at low liquid rates. The volume of liquid held up in such a packing is low and the pressure drop is also low. Some of these high efficiency woven wire packings have been used in columns up to 500 mm diameter.

Grid packings, which are relatively easy to fabricate, are usually used in columns of square section, and frequently in cooling towers which are described in Volume 1, Chapter 13. They may be made from wood, plastics, carbon, or ceramic materials, and, because of the relatively large spaces between the individual grids, they give low pressure drops. Further advantages lie in their ease of assembly, their ability to accept fluids with

Table 4.3. Design data for various packings

| | Size | | Wall thickness | | Number | | Bed density | | Contact surface $S_B$ | | Free space % | Packing factor $F$ | |
|---|---|---|---|---|---|---|---|---|---|---|---|---|---|
| | (in.) | (mm) | (in.) | (mm) | (/ft$^3$) | (/m$^3$) | (lb/ft$^3$) | (kg/m$^3$) | (ft$^2$/ft$^3$) | (m$^2$/m$^3$) | (100 $e$) | (ft$^2$/ft$^3$) | (m$^2$/m$^3$) |
| Ceramic Raschig Rings | 0.25 | 6 | 0.03 | 0.8 | 85,600 | 3,020,000 | 60 | 960 | 242 | 794 | 62 | 1600 | 5250 |
| | 0.38 | 9 | 0.05 | 1.3 | 24,700 | 872,000 | 61 | 970 | 157 | 575 | 67 | 1000 | 3280 |
| | 0.50 | 12 | 0.07 | 1.8 | 10,700 | 377,000 | 55 | 880 | 112 | 368 | 64 | 640 | 2100 |
| | 0.75 | 19 | 0.09 | 2.3 | 3090 | 109,000 | 50 | 800 | 73 | 240 | 72 | 255 | 840 |
| | 1.0 | 25 | 0.14 | 3.6 | 1350 | 47,600 | 42 | 670 | 58 | 190 | 71 | 160 | 525 |
| | 1.25 | 31 | | | 670 | 23,600 | 46 | 730 | | | 71 | 125 | 410 |
| | 1.5 | 38 | | | 387 | 13,600 | 43 | 680 | | | 73 | 95 | 310 |
| | 2.0 | 50 | 0.25 | 6.4 | 164 | 5790 | 41 | 650 | 29 | 95 | 74 | 65 | 210 |
| | 3.0 | 76 | | | 50 | 1765 | 35 | 560 | | | 78 | 36 | 120 |
| Metal Raschig Rings | 0.25 | 6 | 0.03 | 0.8 | 88,000 | 3,100,000 | 133 | 2130 | | | 72 | 700 | 2300 |
| | 0.38 | 9 | 0.03 | 0.8 | 27,000 | 953,000 | 94 | 1500 | | | 81 | 390 | 1280 |
| | 0.50 | 12 | 0.03 | 0.8 | 11,400 | 402,000 | 75 | 1200 | 127 | 417 | 85 | 300 | 980 |
| | 0.75 | 19 | 0.03 | 0.8 | 3340 | 117,000 | 52 | 830 | 84 | 276 | 89 | 185 | 605 |
| (Bed densities | 0.75 | 19 | 0.06 | 1.6 | 3140 | 110,000 | 94 | 1500 | | | 80 | 230 | 750 |
| are for mild | 1.0 | 25 | 0.03 | 0.8 | 1430 | 50,000 | 39 | 620 | 63 | 207 | 92 | 115 | 375 |
| steel; multiply | 1.0 | 25 | 0.06 | 1.6 | 1310 | 46,200 | 71 | 1130 | | | 86 | 137 | 450 |
| by 1.105, 1.12, 1.37, | 1.25 | 31 | 0.06 | 1.6 | 725 | 25,600 | 62 | 990 | | | 87 | 110 | 360 |
| 1.115 for stainless | 1.5 | 38 | 0.06 | 1.6 | 400 | 14,100 | 49 | 780 | | | 90 | 83 | 270 |
| steel, copper, aluminium, | 2.0 | 50 | 0.06 | 1.6 | 168 | 5930 | 37 | 590 | 31 | 102 | 92 | 57 | 190 |
| and monel respectively) | 3.0 | 76 | 0.06 | 1.6 | 51 | 1800 | 25 | 400 | 22 | 72 | 95 | 32 | 105 |
| Carbon Raschig Rings | 0.25 | 6 | 0.06 | 1.6 | 85,000 | 3,000,000 | 46 | 730 | 212 | 696 | 55 | 1600 | 5250 |
| | 0.50 | 12 | 0.06 | 1.6 | 10,600 | 374,000 | 27 | 430 | 114 | 374 | 74 | 410 | 1350 |
| | 0.75 | 19 | 0.12 | 3.2 | 3140 | 110,000 | 34 | 540 | 75 | 246 | 67 | 280 | 920 |
| | 1.0 | 25 | 0.12 | 3.2 | 1325 | 46,000 | 27 | 430 | 57 | 187 | 74 | 160 | 525 |
| | 1.25 | 31 | | | 678 | 23,000 | 31 | 490 | | | 69 | 125 | 410 |
| | 1.5 | 38 | | | 392 | 13,800 | 34 | 540 | | | 67 | 130 | 425 |
| | 2.0 | 50 | 0.25 | 6.4 | 166 | 5860 | 27 | 430 | 29 | 95 | 74 | 65 | 210 |
| | 3.0 | 76 | 0.31 | 8.0 | 49 | 1730 | 23 | 370 | 19 | 62 | 78 | 36 | 120 |
| Metal Pall Rings | 0.62 | 15 | 0.02 | 0.5 | 5950 | 210,000 | 37 | 590 | 104 | 341 | 93 | 70 | 230 |
| (Bed densities are | 1.0 | 25 | 0.025 | 0.6 | 1400 | 49,000 | 30 | 480 | 64 | 210 | 94 | 48 | 160 |
| for mild steel) | 1.25 | 31 | 0.03 | 0.8 | 375 | 13,000 | 24 | 380 | 39 | 128 | 95 | 28 | 92 |
| | 2.0 | 50 | 0.035 | 0.9 | 170 | 6000 | 22 | 350 | 31 | 102 | 96 | 20 | 66 |
| | 3.5 | 76 | 0.05 | 1.2 | 33 | 1160 | 17 | 270 | 20 | 65 | 97 | 16 | 52 |

(*continued overleaf*)

Table 4.3. (*continued*)

| | Size | | Wall thickness | | Number | | Bed density | | Contact surface $S_B$ | | Free space % | Packing factor $F$ | |
|---|---|---|---|---|---|---|---|---|---|---|---|---|---|
| | (in.) | (mm) | (in.) | (mm) | (/ft$^3$) | (/m$^3$) | (lb/ft$^3$) | (kg/m$^3$) | (ft$^2$/ft$^3$) | (m$^2$/m$^3$) | (100 $e$) | (ft$^2$/ft$^3$) | (m$^2$/m$^3$) |
| Plastic Pall Rings | 0.62 | 16 | 0.03 | 0.8 | 6050 | 213,000 | 7.0 | 112 | 104 | 341 | 87 | 97 | 320 |
| (Bed densities are | 1.0 | 25 | 0.04 | 1.0 | 1440 | 50,800 | 5.5 | 88 | 63 | 207 | 90 | 52 | 170 |
| for polypropylene) | 1.5 | 38 | 0.04 | 1.0 | 390 | 14,000 | 4.75 | 76 | 39 | 128 | 91 | 40 | 130 |
| | 2.0 | 50 | 0.06 | 1.5 | 180 | 6350 | 4.25 | 68 | 31 | 102 | 92 | 25 | 82 |
| | 3.5 | 88 | 0.06 | 1.5 | 33 | 1160 | 4.0 | 64 | 26 | 85 | 92 | 16 | 52 |
| Ceramic Intalox Saddles | 0.25 | 6 | | | 117,500 | 4,150,000 | 54 | 860 | | | 65 | 725 | 2400 |
| | 0.38 | 9 | | | 49,800 | 1,750,000 | 50 | 800 | | | 67 | 330 | 1080 |
| | 0.50 | 12 | | | 18,300 | 646,000 | 46 | 730 | | | 71 | 200 | 660 |
| | 0.75 | 19 | | | 5640 | 199,000 | 44 | 700 | | | 73 | 145 | 475 |
| | 1.0 | 25 | | | 2150 | 76,000 | 42 | 670 | | | 73 | 92 | 300 |
| | 1.5 | 38 | | | 675 | 23,800 | 39 | 620 | 59 | 194 | 76 | 52 | 170 |
| | 2.0 | 50 | | | 250 | 8820 | 38 | 600 | | | 76 | 40 | 130 |
| | 3.0 | 76 | | | 52 | 1830 | 36 | 570 | | | 79 | 22 | 72 |
| Plastic Super Intalox | No. 1 | | | | 1620 | 57,200 | 6.0 | 96 | 63 | 207 | 90 | 33 | 108 |
| | No. 2 | | | | 190 | 6710 | 3.75 | 60 | 33 | 108 | 93 | 21 | 210 |
| | No. 3 | | | | 42 | 1480 | 3.25 | 52 | 27 | 88 | 94 | 16 | 52 |
| Intalox Metal | | 25 | | 4770 | | 168,000 | | | | | 96.7 | 41 | 135 |
| | | 40 | | 1420 | | 50,100 | | | | | 97.3 | 25 | 82 |
| | | 50 | | 416 | | 14,600 | | | | | 97.8 | 16 | 52 |
| | | 70 | | 131 | | 4600 | | | | | 98.1 | 13 | 43 |
| Hy-Pak | No. 1 | | | | 850 | 30,000 | 19 | 304 | | | 96 | 43 | 140 |
| (Bed densities are | No. 2 | | | | 107 | 3770 | 14 | 224 | | | 97 | 18 | 59 |
| for mild steel) | No. 3 | | | | 31 | 1090 | 13 | 208 | | | 97 | 15 | 49 |
| Plastic Cascade Mini Rings | No. 1 | | | | | | | | | | | 25 | 82 |
| | No. 2 | | | | | | | | | | | 15 | 49 |
| | No. 3 | | | | | | | | | | | 12 | 39 |
| Metal Cascade Mini Rings | No. 0 | | | | | | | | | | | 55 | 180 |
| | No. 1 | | | | | | | | | | | 34 | 110 |
| | No. 2 | | | | | | | | | | | 22 | 72 |
| | No. 3 | | | | | | | | | | | 14 | 46 |
| | No. 4 | | | | | | | | | | | 10 | 33 |
| Ceramic Cascade Mini Rings | No. 2 | | | | | | | | | | | 38 | 125 |
| | No. 3 | | | | | | | | | | | 24 | 79 |
| | No. 5 | | | | | | | | | | | 18 | 59 |

The packing factor $F$ replaces the term $S_B/e^3$. Use of the given value of $F$ in Figure 4.18 permits more predictable performance of designs incorporating packed beds since the values quoted are derived from operating characteristics of the packings rather than from their physical dimensions.

suspended solids, and their ease of wetting even at very low liquid rates. The main problem is that of obtaining good liquid distribution since, at high liquid rates, the liquid tends to cascade from one grid to the next without being broken up into fine droplets which are desirable for a high interfacial surface. An example of a cooling tower packing, 'Coolflo 3'[51] is shown in Figure 4.14. This is similar to the structured packings described later, and consists of vacuum formed PVC sheets clamped together within a metal and plastics frame to form a module which can be 0.6 m or 1.2 m in depth. Structured packings may be broadly classified into either the knitted or the non-knitted type, and both types may be assembled in a segmented way or in a spiral form. In the latter, corrugated strips or ribbons coil about a centre axis to form a flat cake of the requisite tower diameter which is usually less than 1 m. These elements are then stacked one upon the other to provide the necessary bed depth. In the rigid type of structured packing, these corrugated sheets of metal or plastic are assembled to form intersecting open channels. The sheets may, in addition, be perforated and they provide uniform liquid flow over both sides while vapour flows upwards and provides intimate contact with the liquid. One such type of packing, Mellapak[52] is shown in Figure 4.15, and others such as Gempak[53] are also available. Low pressure drops of typically 50 $N/m^2$ per theoretical stage are possible with HETP's, discussed in Chapters 10 and 11, ranging from 0.2 to 0.6 m, voidages in excess of 95 per cent, and high specific surface areas. The resulting higher capacity and efficiency with structured packings is, however, achieved at higher initial capital cost than with the other packings discussed in this section[54].

Figure 4.14. Visco Coolflo 3 extended surface, cooling tower packing

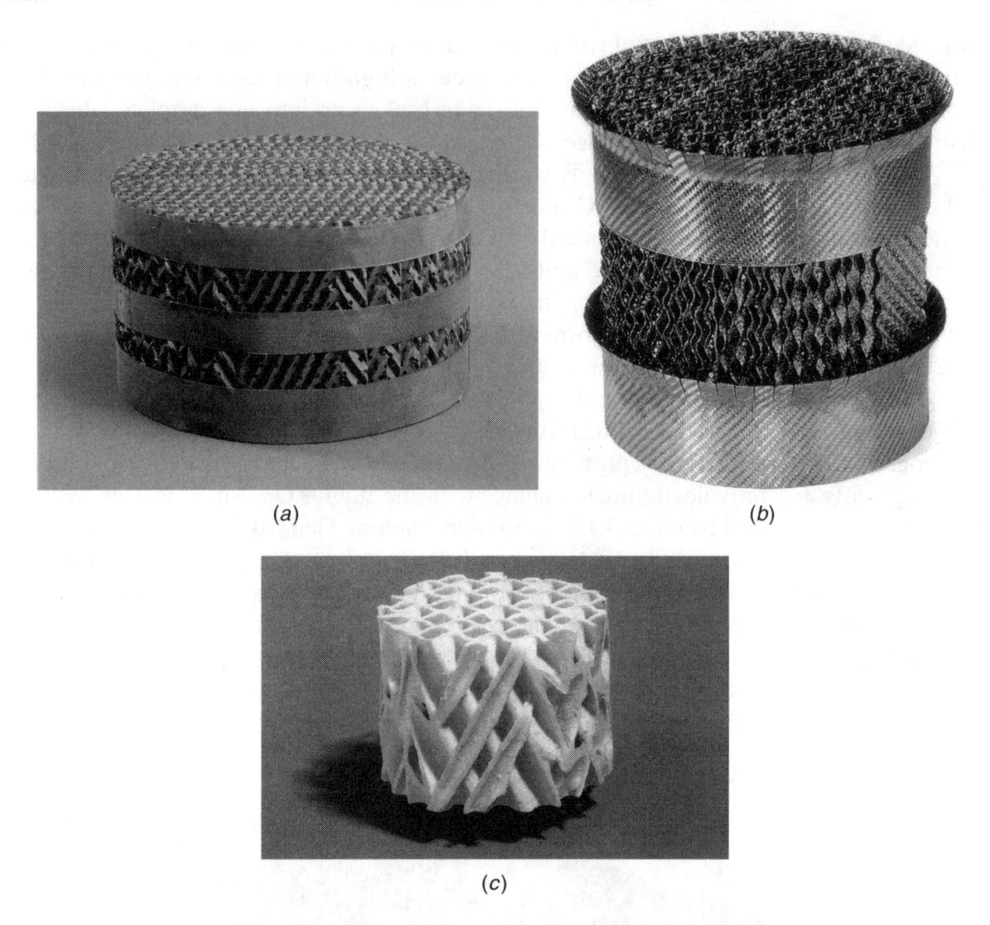


Figure 4.15. Structured packings (*a*) metal gauze (*b*) carbon (*c*) corrosion-resistant plastic

## 4.5.3. Fluid flow in packed columns

### *Pressure drop*

It is important to be able to predict the drop in pressure for the flow of the two fluid streams through a packed column. Earlier in this chapter the drop in pressure arising from the flow of a single phase through granular beds is considered and the same general form of approach is usefully adopted for the flow of two fluids through packed columns. It was noted that the expressions for flow through ring-type packings are less reliable than those for flow through beds of solid particles. For the typical absorption column there is no very accurate expression, but there are several correlations that are useful for design purposes. In the majority of cases the gas flow is turbulent and the general form of the relation between the drop in pressure $-\Delta P$ and the volumetric gas flowrate per unit area of column $u_G$ is shown on curve A of Figure 4.16. $-\Delta P$ is then proportional to $u_G^{1.8}$ approximately, in agreement with the curve A of Figure 4.1 at high Reynolds numbers. If, in addition to the gas flow, liquid flows down the tower, the passage of the gas is not

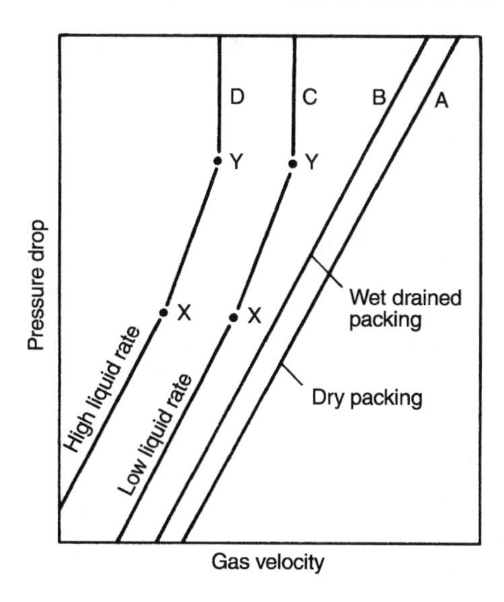


Figure 4.16. Pressure drops in wet packings (logarithmic axes)

significantly affected at low liquid rates and the pressure drop line is similar to line A, although for a given value of $u_G$ the value of $-\Delta P$ is somewhat increased. When the gas rate reaches a certain value, the pressure drop then rises very much more quickly and is proportional to $u_G^{2.5}$, as shown by the section XY on curve C. Over this section the liquid flow is interfering with the gas flow and the hold-up of liquid is progressively increasing. The free space in the packings is therefore being continuously taken up by the liquid, and thus the resistance to flow rises quickly. At gas flows beyond Y, $-\Delta P$ rises very steeply and the liquid is held up in the column. The point X is known as the loading point, and point Y as the flooding point for the given liquid flow. If the flowrate of liquid is increased, a similar plot D is obtained in which the loading point is achieved at a lower gas rate though at a similar value of $-\Delta P$. Whilst it is advantageous to have a reasonable hold-up in the column as this promotes interphase contact, it is not practicable to operate under flooding conditions, and columns are best operated over the section XY. Since this is a section with a relatively short range in gas flow, the safe practice is to design for operation at the loading point X. It is of interest to note that, if a column is flooded and then allowed to drain, the value of $-\Delta P$ for a given gas flow is increased over that for an entirely dry packing as shown by curve B. Rose and Young[55] correlated their experimental pressure drop data for Raschig rings by the following equation:

$$-\Delta P_w = -\Delta P_d \left(1 + \frac{3.30}{d_n}\right) \tag{4.46}$$

where: $-\Delta P_w$ is the pressure drop across the wet drained column.
$-\Delta P_d$ is the pressure drop across the dry column, and
$d_n$ is the nominal size of the Raschig rings in mm.

This effect will thus be most significant for small packings.

There are several ways of calculating the pressure drop across a packed column when gas and liquid are flowing simultaneously and the column is operating below the loading point.

One approach is to calculate the pressure drop for gas flow only and then multiply this pressure drop by a factor which accounts for the effect of the liquid flow. Equation 4.19 may be used for predicting the pressure drop for the gas only, and then the pressure drop with gas and liquid flowing is obtained by using the correction factors for the liquid flow rate given by SHERWOOD and PIGFORD[56].

Another approach is that of MORRIS and JACKSON[57] who arranged experimental data for a wide range of ring and grid packings in a graphical form convenient for the calculation of the number of velocity heads $N$ lost per unit height of packing. $N$ is substituted in the equation:

$$-\Delta P = \tfrac{1}{2} N \rho_G u_G^2 l \tag{4.47}$$

where: $-\Delta P$ = pressure drop,
$\rho_G$ = gas density,
$u_G$ = gas velocity, based on the empty column cross-sectional area, and
$l$ = height of packing.

Equation 4.47 is in consistent units. For example, with $\rho_G$ in kg/m$^3$, $u_G$ in m/s, $l$ in m, and $N$ in m$^{-1}$, $-\Delta P$ is then in N/m$^2$.

Empirical correlations of experimental data for pressure drop have also been presented by LEVA[58], and by ECKERT *et al.*[59] for Pall rings. Where the data are available, the most accurate method of obtaining the pressure drop for flow through a bed of packing is from the manufacturer's own literature. This is usually presented as a logarithmic plot of gas rate against pressure drop, with a parameter of liquid flowrate on the graphs, as shown in Figure 4.16, although it should be stressed that all of these methods apply only to conditions at or below the loading point X on Figure 4.16. If applied to conditions above the loading point the calculated pressure drop would be too low. It is therefore necessary first to check whether the column is operating at or below the loading point, and methods of predicting loading points are now considered.

### Loading and flooding points

Although the loading and flooding points have been shown on Figure 4.16, there is no completely generalised expression for calculating the onset of loading, although one of the following semi-empirical correlations will often be adequate. MORRIS and JACKSON[57] gave their results in the form of plots of $\psi(u_G/u_L)$ at the loading rate for various wetting rates $L_W$ (m$^3$/s m). $u_G$ and $u_L$ are average gas and liquid velocities based on the empty column and $\psi = (\sqrt{(\rho_G/\rho_A)})$ is a gas density correction factor, where $\rho_A$ is the density of air at 293 K.

A useful graphical correlation for flooding rates was first presented by SHERWOOD *et al.*[60] and later developed by LOBO *et al.*[61] for random-dumped packings, as shown in Figure 4.17 in which:

$$\left(\frac{u_G^2 S_B}{g e^3}\right)\left(\frac{\rho_G}{\rho_L}\right)\left(\frac{\mu_L}{\mu_w}\right)^{0.2} \text{ is plotted against } \frac{L'}{G'}\sqrt{\left(\frac{\rho_G}{\rho_L}\right)}$$

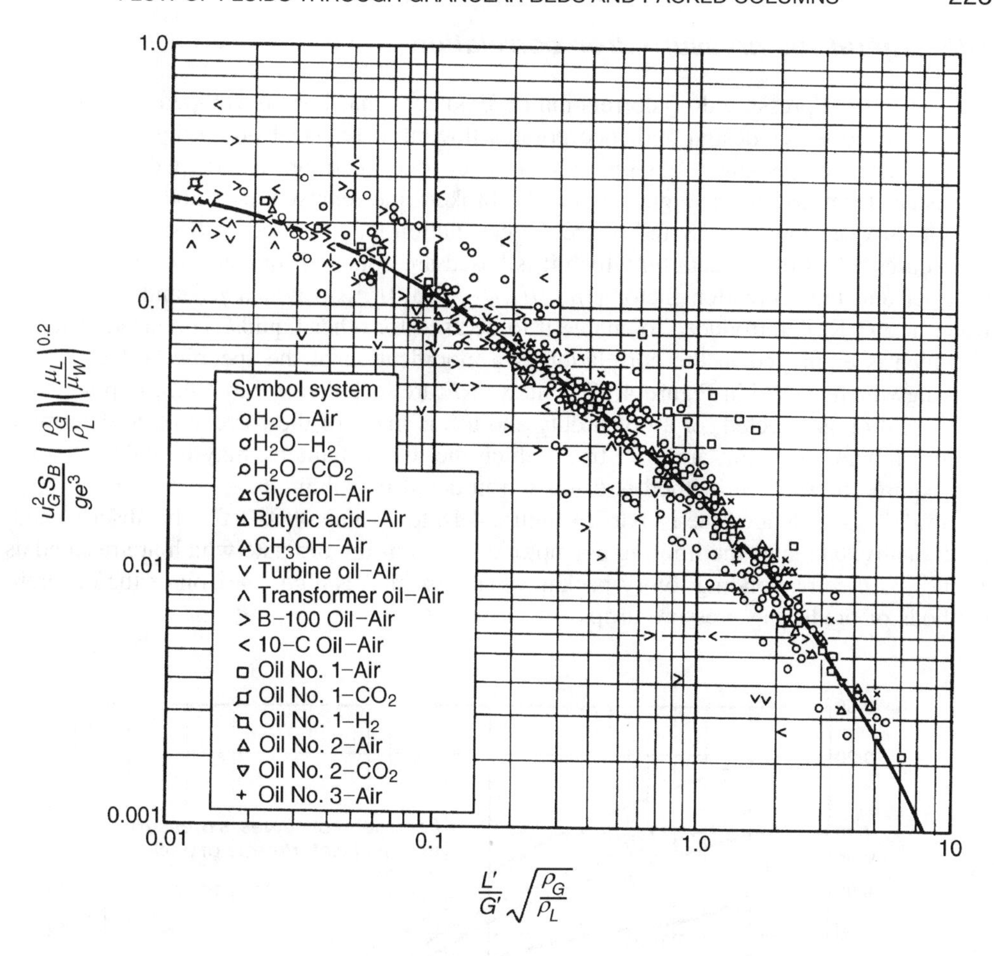


Figure 4.17. Generalised correlation for flooding rates in packed towers[61]

where: $u_G$ is the velocity of the gas, calculated over the whole cross-section of the bed,
$S_B$ is the surface area of the packing per unit volume of bed,
$g$ is the acceleration due to gravity,
$L'$ is the mass rate of flow per unit area of the liquid,
$G'$ is the mass rate of flow per unit area of the gas, and
$\mu_w$ is the viscosity of water at 293 K approximately 1 mN s/m$^2$, and
suffix $G$ refers to the gas and suffix $L$ to the liquid.

The area inside the curve represents possible conditions of operation. In these expressions, the ratios $\rho_G/\rho_L$ and $\mu_L/\mu_w$ have been introduced so that the relationship can be applied for a wide range of liquids and gases. It may be noted that, if the effective value of $g$ is increased by using a rotating bed, then higher flowrates can be achieved before the onset of flooding.

### The generalised pressure drop correlation

The generalised pressure drop correlation by ECKERT[(62)] has been developed as a practical aid to packed tower design and incorporates flowrates, physical properties of the fluid, a wide range of packings and pressure drop on one chart presented in dimensionally consistent form, as shown Figure 4.18. The broken line representing flooding lies above the top curve in Figure 4.18 and hence the correlation may be used with safety in design procedures. Most of the data on which it is based are obtained for cases where the liquid is water and the correction factor $[(\mu_L/\mu_w)/(\rho_w/\rho_L)]^{0.1}$, in which $\mu_w$ and $\rho_w$ refer to water at 293 K, is introduced to enable it to be used for other liquids. The packing factor $F$ which is employed in the correlation is a modification of the specific surface of the packing which is used in Figure 4.17. Values of $F$ are included in Table 4.3. In practice, a pressure drop is selected for a given duty and use is made of the correlation to determine the gas flowrate per unit area $G'$, from which the tower diameter may be calculated for the required flows. The method is discussed in detail in Volume 6.

LEVA[(63)] has extended the isobars in Figure 4.18 to a value of 0.01 for the abscissa and, additionally, tabulates the limiting asymptotic values of the ordinate which are reached as $L'/G' \to 0$, corresponding to dry packages. He also makes a comparison of the different methods of predicting pressure drops.

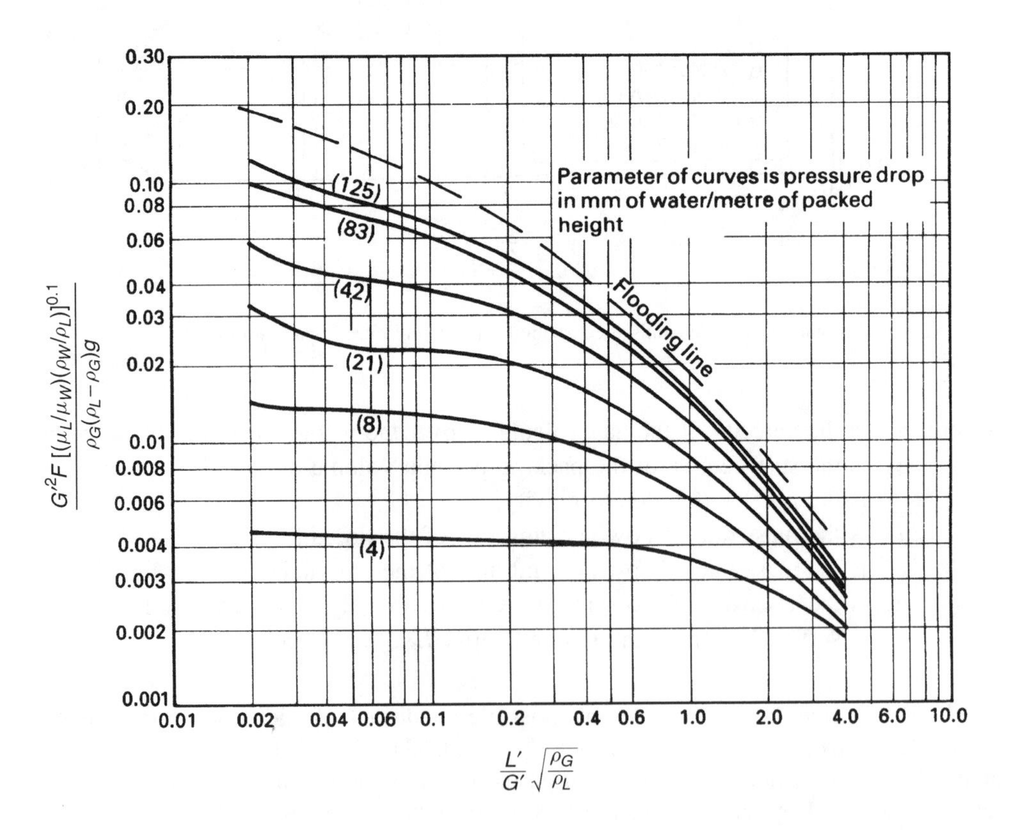


Figure 4.18. Generalised pressure drop correlation (adapted from a figure by the Norton Co. with permission)

### *Liquid distribution*

Provision of a packing with a high surface area per unit volume may not result in good contacting of gas and liquid unless the liquid is distributed uniformly over the surface of the packing. The need for liquid distribution and redistribution and correct packing size has been noted previously. The effective wetted area decreases as the liquid rate is decreased and, for a given packing, there is a minimum liquid rate for effective use of the surface area of the packing. A useful measure of the effectiveness of wetting of the available area is the wetting rate $L_w$ defined as:

$$\frac{\text{Volumetric liquid rate per unit cross-sectional area of column}}{\text{Packing surface area per unit volume of column}}$$

or:

$$L_W = \frac{L}{A\rho_L S_B} = \frac{u_L}{S_B} \tag{4.48}$$

Thus the wetting rate is analogous to the volumetric liquid rate per unit length of circumference in a wetted-wall column in which the liquid flows down the surface of a cylinder. If the liquid rate were too low, a continuous liquid film would not be formed around the circumference of the cylinder and some of the area would be ineffective.

Similar effects occur in a packed column, although the flow patterns and arrangement of the surfaces are then obviously much more complex. MORRIS and JACKSON[57] have recommended minimum wetting rates of $2 \times 10^{-5}$ $m^3/s$ m for rings 25–75 mm in diameter and grids of pitch less than 50 mm, and $3.3 \times 10^{-5}$ $m^3/s$ m for larger packings.

The distribution of liquid over packings has been studied experimentally by many workers and, for instance, TOUR and LERMAN[64,65] showed that for a single point feed the distribution is given by:

$$Q_x = c\exp(-a^2x^2) \tag{4.49}$$

where $Q_x$ is the fraction of the liquid collected at a distance $x$ from the centre and $c$ and $a$ are constants depending on the packing arrangement. NORMAN[66], MANNING and CANNON[67] and others have shown that this maldistribution is one cause of falling transfer coefficients with tall towers.

A nomograph which relates liquid rate, tower diameter and packing size is given in Figure 4.19[49]. The wetting rate $L_W$ may be obtained as an absolute value from the inner right-hand axis or as *wetting fraction* from the outer scale. A value of wetting fraction exceeding unity on that scale indicates that the packing is satisfactorily wet. It should be noted that many organic liquids have favourable wetting properties and wetting may be effective at much lower rates, though materials such as plastics and polished stainless steel are difficult to wet. Figure 4.19 does, however, represent the best available data on the subject of wetting. In the example shown in Figure 4.19, the arrowed line corresponds to the case of a liquid flow of 0.018 $m^3/s$ in a column of 1.6 m diameter and a packing size of 25 mm, which gives an approximate wetting rate of $5 \times 10^{-5}$ $m^3/m$ s, corresponding to a total wetting of more than 1 on the outside right-hand scale; this is satisfactory.

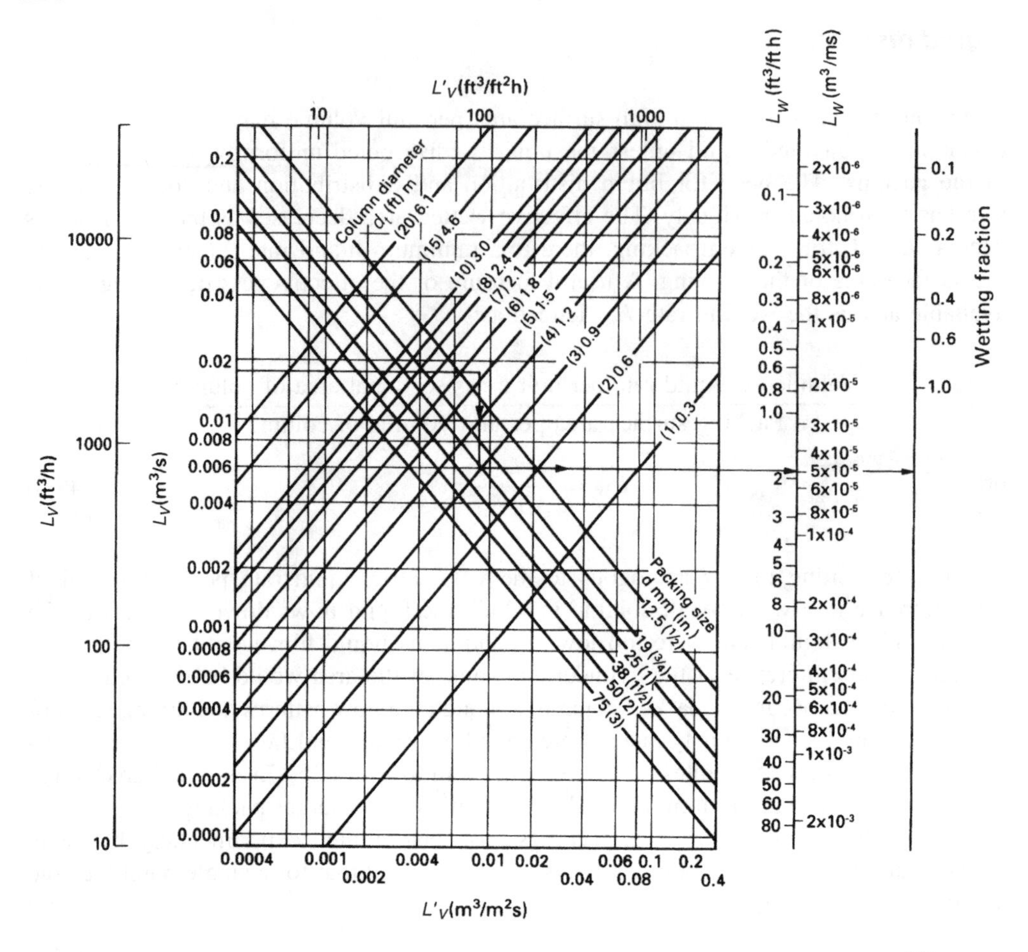


Figure 4.19. Nomograph for the estimation of the degree of wetting in a packed column[49]

## Hold-up

In many industrial applications of packed columns, it is desirable to know the volumetric hold-up of the liquid phase in the column. This information might be needed, for example, if the liquid were involved in a chemical reaction or if a control system for the column were being designed. For gas–liquid systems the hold-up of liquid $H_w$ for conditions below the loading point has been found[48] to vary approximately as the 0.6 power of the liquid rate, and for rings and saddles this is given approximately by:

$$H_w = 0.143 \left( \frac{L'}{d} \right)^{0.6} \tag{4.50}$$

where: $L'$ is the liquid flowrate (kg/m$^2$s),

$d$ is the equivalent diameter of the packing (mm), and

$H_w$ is the hold-up (m$^3$ of liquid/m$^3$ of column).

Thus with 25 mm Raschig rings, $L'$ of 1.0 kg/m$^2$ s and $d = 20$ mm, $H_w$ has a value of 0.021 m$^3$/m$^3$.

For further information on the design of packed columns, reference should be made to Volume 6. In addition, manufacturers' data are often available on hold-up for specific packings and these should be consulted whenever possible for design purposes.

### Economic design of packed columns

In designing industrial scale packed columns a balance must be made between the capital cost of the column and ancillary equipment on the one side, and the running costs on the other. Generally, reducing the diameter of the column will reduce the capital cost though increase the cost of pumping the gas through the column due to the increased pressure drop.

For columns operating at atmospheric and sub-atmospheric pressure and where the mass transfer rate is controlled by transfer through the gas film, MORRIS and JACKSON[57] calculated ranges of economic gas velocities for various packings under specified conditions. In selecting a gas velocity, and hence the column cross-sectional area, it is necessary to check that the liquid rate is above the minimum wetting rate, as discussed in the previous section. Selection of the appropriate packing will help in achieving the minimum wetting rate. The loading condition should also be calculated to ensure that the column would not be operating above this condition.

For columns operating at high pressures, the capital cost of the column shell becomes much more significant and it may be more economic to operate at gas velocities above the loading condition. MORRIS and JACKSON[57] suggest a gas velocity about 75–80 per cent of the flooding velocity for normal systems, and less than 40 per cent of the flooding rate if foaming is likely to occur. The height of the column would have to be taken into consideration in making an economically optimum design. The height is usually determined by the mass transfer duty of the columns and mass transfer rates per unit height of packing are discussed in Chapter 11.

### Vacuum columns

SAWISTOWSKI[9] has shown that the curve in Figure 4.17 may be converted to a straight line by plotting:

$$\ln\left\{\ln\left[\frac{u_G^2 S_B}{g e^3}\left(\frac{\rho_G}{\rho_L}\right)\left(\frac{\mu_L}{\mu_w}\right)^{0.2}\right]\right\} \quad \text{against} \quad \ln\left\{\frac{L'}{G'}\sqrt{\frac{\rho_G}{\rho_L}}\right\}$$

The equation of the curve is then found to be:

$$\ln\left\{\frac{u_G^2 S_B}{g e^3}\left(\frac{\rho_G}{\rho_L}\right)\left(\frac{\mu_L}{\mu_w}\right)^{0.2}\right\} = -4\left(\frac{L'}{G'}\right)^{1/4}\left(\frac{\rho_G}{\rho_L}\right)^{1/8} \tag{4.51}$$

When a column is operating under reduced pressure and the pressure drop is of the same order of magnitude as the absolute pressure, it is not immediately obvious whether the onset of flooding will be determined by conditions at the top or the bottom of the column. If $G_F$ is the gas flowrate under flooding conditions in the column, then:

$$G_F = u_G \rho_G A \tag{4.52}$$

Substituting in equation 4.51 gives:

$$\ln\left\{\frac{G_F^2 S_B}{A^2 g e^3}\frac{1}{\rho_G \rho_L}\left(\frac{\mu_L}{\mu_w}\right)^{0.2}\right\} = -4\left(\frac{L'}{G'}\right)^{1/4}\left(\frac{\rho_G}{\rho_L}\right)^{1/8} \tag{4.53}$$

For a column operating at a given reflux ratio, $L'/G'$ is constant and the only variables over the length of the column are, now, the minimum flooding rate $G_F$ and the gas density $\rho_G$. In order to find the condition for a minimum or maximum value of $G_F$, $\mathrm{d}(G_F^2)/\mathrm{d}\rho_G$ is obtained from equation 4.53 and equated to zero. Thus:

$$\left(\frac{G_F^2}{\rho_G}\right)^{-1}\left\{G_F^2(-\rho_G)^{-2} + \rho_G^{-1}\frac{\mathrm{d}G_F^2}{\mathrm{d}\rho_G}\right\} = -4\left(\frac{L'}{G'}\right)^{1/4}\rho_L^{-1/8}\left(\tfrac{1}{8}\rho_G^{-7/8}\right) \tag{4.54}$$

This gives $(L'/G')\sqrt{(\rho_G/\rho_L)} = 16$, when $\mathrm{d}G_F^2/\mathrm{d}\rho_G = 0$. As the second differential coefficient is negative at this point, $G_F^2$ is a maximum.

A value of $(L'/G')\sqrt{(\rho_G/\rho_L)} = 16$ is well in excess of the normal operating range of the column (especially of a distillation column operating at reduced pressure), as seen in Figure 4.17. Thus, over the whole operating range of a column, the value of $G'$ which just gives rise to flooding increases with gas density and hence with the absolute pressure. The tendency for a column to flood will always be greater therefore at the low pressure end, that is at the top.

Calculation of the pressure drop and flooding rate is particularly important for vacuum columns, in which the pressure may increase severalfold from the top to the bottom of the column. When a heat-sensitive liquid is distilled, the maximum temperature, and hence the pressure, at the bottom of the column is limited and hence the vapour rate must not exceed a certain value. In a vacuum column, the throughput is very low because of the high specific volume of the vapour, and the liquid reflux rate is generally so low that the liquid flow has little effect on the pressure drop. The pressure drop can be calculated by applying equation 4.15 over a differential height and integrating. Thus:

$$-\frac{\mathrm{d}P}{\mathrm{d}l} = \left(\frac{R_1}{\rho u_1^2}\right) S\rho_G u^2\left(\frac{1-e}{e^3}\right) \tag{4.55}$$

Writing $G' = \rho_G u$ and $P/\rho_G = P_0/\rho_0$ for isothermal operation, where $\rho_0$ is the vapour density at some arbitrary pressure $P_0$:

$$-\frac{\mathrm{d}P}{\mathrm{d}l} = \left(\frac{R_1}{\rho u_1^2}\right) S\frac{(1-e)}{e^3}\frac{G'^2}{P}\frac{P_0}{\rho_0} \tag{4.56}$$

The Reynolds number, and hence $R_1/\rho u_1^2$, will remain approximately constant over the column.

Integrating:

$$P_1^2 - P_2^2 = 2\left(\frac{R_1}{\rho u_1^2}\right)\frac{(1-e)}{e^3}\frac{SG'^2 P_0}{\rho_0}l \tag{4.57}$$

It may be noted that, when the pressure at the top of the column is small compared with that at the bottom, the pressure drop is directly proportional to the vapour rate.

## Example 4.1

Two-heat sensitive organic liquids, of average molecular weight of 155 kg/kmol, are to be separated by vacuum distillation in a 100 mm diameter column packed with 6 mm stoneware Raschig rings. The number of theoretical plates required is 16 and it has been found that the HETP is 150 mm. If the product rate is 0.005 kg/s at a reflux ratio of 8, calculate the pressure in the condenser so that the temperature in the still does not exceed 395 K, which is equivalent to a pressure of 8 kN/m$^2$. It may be assumed that $a = 800$ m$^2$/m$^3$, $\mu = 0.02$ mN s/m$^2$, $e = 0.72$ and that temperature changes and the correction for liquid flow may be neglected.

## Solution

The modified Reynolds number $Re_1$ is defined by:

$$Re_1 = \frac{u_c \rho}{S(1-e)\mu} = \frac{G'}{S(1-e)\mu} \quad \text{(equation 4.13)}$$

The Ergun equation may be rewritten as:

$$\frac{R_1}{\rho u_1^2} = \frac{4.17}{Re_1} + 0.29 \quad \text{(equation 4.21)}$$

Hence:

$$\frac{R}{\rho u_1^2} = \frac{4.17S(1-e)\mu}{G'} + 0.29$$

Equation 4.15, written in differential form, becomes:

$$\frac{R_1}{\rho u_1^2} = \frac{e^3}{S(1-e)}\left(-\frac{dP}{dl}\right)\frac{1}{\rho u_c^2}$$

$$= \frac{e^3}{S(1-e)}\left(-\frac{dP}{dl}\right)\frac{\rho}{G'^2}$$

$$-\rho\frac{dP}{dl} = \frac{R_1}{\rho u_1^2}\frac{S(1-e)}{e^3}G'^2$$

Thus:

$$-\int_{P_C}^{P_S} \rho\, dP = \frac{R_1}{\rho u_1^2}\frac{S(1-e)}{e^3}G'^2\int_0^l dl$$

(where suffix $C$ refers to the condenser and $S$ to the still)

$$= \frac{R_1}{\rho u_1^2}\frac{S(1-e)}{e^3}G'^2 l$$

In this example: $a = 800$ m$^2$/m$^3$ $= S(1-e)$

Product rate = 0.5 g/s and, if the reflux ratio = 8, then:

Vapour rate = 4.5 g/s

and: $$G' = 4.5 \times 10^{-3}/(\pi/4)(0.1)^2 = 0.573 \text{ kg/m}^2\text{s},$$

$$\mu = 0.02 \times 10^{-3}\text{Ns/m}^2 \text{ and } e = 0.72$$

Hence: $$Re_1 = 0.573/800 \times 0.28 \times 0.02 \times 10^{-3} = 128$$

$$\frac{R_1}{\rho u_1^2} = (4.17/128) + 0.29 = 0.32$$

Since: $$l = (16 \times 0.15) = 2.4 \text{ m}$$

$$-\int_{P_C}^{P_S} \rho \mathrm{d}P = (0.32 \times 800 \times 0.28) \times (0.573)^2 \times 2.4/(0.72)^3$$

giving: $$-\int_{P_C}^{P_S} \rho \mathrm{d}P = 151.3 \text{ N/m}^2$$

$$\rho = \rho_S \times P/P_S$$

The vapour density in the still is given by:

$$\rho_S = \left(\frac{155}{22.4}\right)\left(\frac{273}{395}\right)\left(\frac{P_S}{101.3 \times 10^3}\right) = 4.73 \times 10^{-5} P_S \text{ kg/m}^3$$

$\therefore$ $$\rho = 4.73 \times 10^{-5} P \text{ kg/m}^3$$

$\therefore$ $$-\int_{P_C}^{P_S} \rho \mathrm{d}P = -\int_{P_C}^{P_S} 4.73 \times 10^{-5} P \mathrm{d}P = 4.73 \times 10^{-5}(P_S^2 - P_C^2)$$

Thus: $$151.3 = (4.73 \times 10^{-5})(P_S^2 - P_C^2)$$

Since $$P_S = 8000 \text{ N/m}^2$$

then: $$151.3 = (4.73 \times 10^{-5})(8 \times 10^3)^2 - (4.73 \times 10^{-5})P_C^2$$

and: $$P_C = 7790 \text{ N/m}^2 \text{ or } \underline{\underline{7.79 \text{ kN/m}^2}}$$

## 4.6. FURTHER READING

BACKHURST, J. R. and HARKER, J. H.: *Process Plant Design* (Heinemann Educational Books, 1973).
CHHABRA, R.P. and RICHARDSON, J.F: *Non-Newtonian Flow in the Process Industries* (Butterworth-Heinemann, 1999)
DULLIEN, F. A. L.: *Porous Media: Fluid Transport and Pore Structure* (Academic Press, New York, 1979).
LEVA, M.: *Tower Packings and Packed Tower Design*, 2nd edn. (U.S. Stoneware Co., 1953).
MORRIS, G. A. and JACKSON, J.: *Absorption Towers* (Butterworths, 1953).
NORMAN, W. S.: *Absorption, Distillation and Cooling Towers* (Longmans, 1961).
STRIGLE, R. F.: *Random Packings and Packed Towers. Design and Applications* (Gulf Publishing Company, 1987).

## 4.7. REFERENCES

1. DARCY, H. P. G.: *Les Fontaines publiques de la ville de Dijon. Exposition et application à suivre et des formules à employer dans les questions de distribution d'eau* (Victor Dalamont, 1856).

2. EISENKLAM, P.: Chapter 9 "Porous Masses" in CREMER, H. W. and DAVIES, T.: *Chemical Engineering Practice*, Vol. 2 (Butterworths, 1956).
3. BRINKMAN, H. C.: *Appl. Scient. Res.* **1A** (1948) 81–86. On the permeability of media consisting of closely packed porous particles.
4. DUPUIT, A. J. E. J.: *Etudes théoriques et pratiques sur le mouvement des eaux* (1863).
5. KOZENY, J.: *Sitzb. Akad. Wiss., Wien, Math.-naturw. Kl.* **136** (Abt, IIa) (1927) 271–306. Über kapillare Leitung des Wassers im Boden (Aufstieg, Versicherung, und Anwendung auf die Bewässerung).
6. KOZENY, J.: *Z. Pfl.-Ernähr. Düng. Bodenk*, **28A** (1933) 54–56. Über Bodendurchlässigkeit.
7. CARMAN, P. C.: *Trans. Inst. Chem. Eng.* **15** (1937) 150–66. Fluid flow through granular beds.
8. FORCHHEIMER, P.: *Hydraulik* (Teubner, 1930).
9. SAWISTOWSKI, H.: *Chem. Eng. Sci.* **6** (1957) 138. Flooding velocities in packed columns operating at reduced pressures.
10. ERGUN, S.: *Chem. Eng. Prog.* **48** (1952) 89–94. Fluid flow through packed columns.
11. KEMBLOWSKI, Z., DZIUBINSKI, M. and MERTL, J.: *Adv. in Transport Proc.* **5** (1987) 117. Flow of non-Newtonian fluids through granular media.
12. BOTSET, H. G.: *Trans. Am. Inst. Mining Met. Engrs.* **136** (1940) 41. Flow of gas–liquid mixtures through consolidated sand.
13. GLASER, M. B. and LITT, M.: *A.I.Ch.E.Jl.* **9** (1963) 103. A physical model for mixed phase flow through beds of porous particles.
14. CARMAN, P. C.: *Flow of Gases Through Porous Media* (Butterworths, 1956).
15. COULSON, J. M.: *Trans. Inst. Chem. Eng.* **27** (1949) 237–57. The flow of fluids through granular beds; effect of particle shape and voids in streamline flow.
16. WYLLIE, M. R. J. and GREGORY, K. R.: *Ind. Eng. Chem.* **47** (1955) 1379–88. Fluid flow through unconsolidated porous aggregates — effect of porosity and particle shape on Kozeny–Carman constants.
17. KIHN, E.: University of London, Ph.D. Thesis (1939). Streamline flow of fluids through beds of granular materials.
18. PIRIE, J. M.: in discussion of COULSON[15].
19. MUSKAT, M. and BOTSET, H. G.: *Physics* **1** (1931) 27–47. Flow of gas through porous material.
20. LORD, E.: *J. Text. Inst.* **46** (1951) T191. Air flow through plugs of textile fibres. Part I. General flow relations.
21. DAVIES, C. N.: in discussion of HUTCHISON, H. P., NIXON, I. S., and DENBIGH, K. G.: *Discns. Faraday Soc.* **3** (1948) 86–129. The thermosis of liquids through porous materials.
22. LEA, F. M. and NURSE, R. W.: *Trans. Inst. Chem. Eng.* **25** (1947). Supplement: Symposium on Particle Size Analysis, pp. 47–63. Permeability methods of fineness measurement.
23. CARMAN, P. C. and MALHERBE, P. LE R.: *J. Soc. Chem. Ind. Trans.* **69** (1950) 134T–143T. Routine measurement of surface of paint pigments and other fine powders, I.
24. CHHABRA, R. P., COMITI, J. C. and MACHĂC, I: *Chem. Eng. Sci.* **50** (2001) 1. Flow of non-Newtonian fluids in fixed and fluidised beds.
25. KEMBLOWSKI, Z., DZIUBINSKI, M. and SEK, J.: *in Advances in Transport Processes* **5**, eds MASHELKAR, R. A., MUJUMDAR, A. S. and KAMAL, M. R., (Wiley Eastern Limited, New Delhi.) (1987) 117. Flow of non-Newtonian fluids through granular media.
26. GUNN, D. J.: *The Chemical Engineer* No 219 (1968) CE153. Mixing in packed and fluidised beds.
27. EDWARDS, M. F. and RICHARDSON, J. F.: *Chem. Eng. Sci.* **22** (1968) 109. Gas dispersion in packed beds.
28. GUNN, D. J. and PRYCE, C.: *Trans. Inst. Chem. Eng.* **47** (1969) T341. Dispersion in packed beds.
29. BLACKWELL, R. J., RAYNE, J. R. and TERRY, M. W.: *J. Petrol. Technol.* **11** (1959) 1–8. Factors influencing the efficiency of miscible displacement.
30. ARIS, R. and AMUNDSON, N. R.: *A.I.Ch.E.Jl.* **3** (1957) 280. Some remarks on longitudinal mixing or diffusion in fixed beds.
31. PRAUSNITZ, J. M.: *A.I.Ch.E.Jl.* **4** (1958) 14M. Longitudinal dispersion in a packed bed.
32. CARBERRY, J. J. and BRETTON, R. H.: *A.I.Ch.E.Jl.* **4** (1958) 367. Axial dispersion of mass in flow through fixed beds.
33. DE MARIA, F. and WHITE, R. R.: *A.I.Ch.E.Jl.* **6** (1960) 473. Transient response study of gas flowing through irrigated packing.
34. MCHENRY, K. W. and WILHELM, R. H.: *A.I.Ch.E.Jl.* **3** (1957) 83. Axial mixing of binary gas mixtures flowing in a random bed of spheres.
35. SINCLAIR, R. J. and POTTER, O. E.: *Trans. Inst. Chem. Eng.* **43** (1965) T3. The dispersion of gas in flow through a bed of packed solids.
36. EVANS, E. V. and KENNEY, C. N.: *Trans. Inst. Chem. Eng.* **44** (1966) T189. Gaseous dispersion in packed beds at low Reynolds numbers.
37. CAIRNS, E. J. and PRAUSNITZ, J. M.: *Chem. Eng. Sci.* **12** (1960) 20. Longitudinal mixing in packed beds.
38. EBACH, E. E. and WHITE, R. R.: *A.I.Ch.E.Jl.* **4** (1958) 161. Mixing of fluids flowing through beds of packed solids.

39. HIBY, J. W.: *Proceedings of the Symposium on Interaction between Fluids and Particles*. I. Chem. E., London. (1962) 312. Longitudinal and transverse mixing during single-phase flow through granular beds.
40. LILES, A. W. and GEANKOPLIS, C. J.: *A.I.Ch.E.Jl.* **6** (1960) 591. Axial diffusion of liquids in packed beds and end effects.
41. KRAMERS, H.: *Physica* **12** (1946) 61. Heat transfer from spheres to flowing media.
42. RANZ, W. E. and MARSHALL, W. R.: *Chem. Eng. Prog.* **48** (1952) 141, 173. Evaporation from drops.
43. GUPTA, A. S. and THODOS, G.: *A.I.Ch.E.Jl.* **9** (1963) 751. Direct analogy between mass and heat transfer to beds of spheres.
44. ZENZ, F. A. and OTHMER, D. F.: *Fluidization and Fluid-particle Systems* (Reinhold, 1960).
45. BROWNELL, L. E. and YOUNG, E. H.: *Process Equipment Design, Vessel Design* (Chapman & Hall, 1959).
46. MOLYNEUX, F.: *Chemical Plant Design*, Vol. 1 (Butterworths, 1963).
47. BS 5500: 1978: *Fusion Welded Pressure Vessels* (British Standards Institution, London).
48. LEVA, M.: *Tower Packings and Packed Tower Design* (U.S. Stoneware Co., 1953).
49. Norton Chemical Process Products Div., Box 350, Akron, Ohio; Hydronyl Ltd., King St., Fenton, Stokeon-Trent, U.K.
50. ECKERT, J. S.: *Chem. Eng. Prog.* **57** No. 9 (1961) 54. Design techniques for designing packed towers.
51. Coolflo 3 is a product of Visco Ltd., Croydon Surrey.
52. Mellapak is a registered trademark of Sulzer (UK) Ltd., Farnborough, Hants.
53. Gempak is a registered trademark of Glitsch (UK) Ltd., Kirkby Stephen, Cumbria.
54. CHENG, G. K., *Chem. Eng. Albany*, **91** No 5 (March 5 1984) 40. Packed column internals.
55. ROSE, H. E. and YOUNG, P. H.: *Proc. Inst. Mech. Eng.* **1B** (1952) 114. Hydraulic characteristics of packed towers operating under countercurrent flow conditions.
56. SHERWOOD, T. K. and PIGFORD, R. L.: *Absorption and Extraction* (McGraw-Hill, 1952).
57. MORRIS, G. A. and JACKSON, J.: *Absorption Towers* (Butterworths, 1953).
58. LEVA, M.: *Chem. Eng. Prog.* Symp. Ser. No. 10, **50** (1954) 51–59. Flow through irrigated dumped packing Pressure drop, loading, flooding.
59. ECKERT, J. S., FOOTE, E. H., and HUNTINGTON, R. L.: *Chem. Eng. Prog.* **54,** No. 1 (Jan. 1958) 70–5. Pall rings — new type of tower packing.
60. SHERWOOD, T. K., SHIPLEY, G. H., and HOLLOWAY, F. A. L.: *Ind. Eng. Chem.* **30** (1938) 765–9. Flooding velocities in packed columns.
61. LOBO, W. E., FRIEND, L., HASHMALL, F., and ZENZ, F.: *Trans. Am. Inst. Chem. Eng.* **41** (1945) 693–710. Limiting capacity of dumped tower packings.
62. ECKERT, J. S.: *Chem. Eng. Prog.* **59** No 5 (1963) 76. Tower packings — comparative performance.
63. LEVA, M.: *Chem. Eng. Prog.* **88** No. 1 (1992) 65. Reconsider packed-tower pressure-drop correlations.
64. TOUR, R. S. and LERMAN, F.: *Trans. Am. Inst. Chem. Eng.* **35** (1939) 709–18. An improved device to demonstrate the laws of frequency distribution. With special reference to liquid flow in packed towers.
65. TOUR, R. S. and LERMAN, F.: *Trans. Am. Inst. Chem. Eng.* **35** (1939) 719–42. The unconfined distribution of liquid in tower packing.
66. NORMAN, W. S.: *Trans. Inst. Chem. Eng.* **29** (1951) 226–39. The performance of grid-packed towers.
67. MANNING, R. E. and CANNON, M. R.: *Ind Eng. Chem.* **49** (1957) 347–9. Distillation improvement by control of phase channelling in packed columns.

## 4.8. NOMENCLATURE

| | | Units in SI System | Dimensions in **M**, **L**, **T**, $\boldsymbol{\theta}$ |
|---|---|---|---|
| $A$ | Total cross-sectional area of bed or column | $m^2$ | $\mathbf{L}^2$ |
| $a$ | Coefficient in equation 4.49 | $m^{-1}$ | $\mathbf{L}^{-1}$ |
| $B$ | Permeability coefficient (equation 4.2) | $m^2$ | $\mathbf{L}^2$ |
| $b$ | Constant in equation 4.28 (15 for spherical particles) | — | — |
| $C$ | Concentration | $kg/m^3$ | $\mathbf{ML}^{-3}$ |
| $C_p$ | Specific heat at constant pressure | J/kg K | $\mathbf{L}^2\mathbf{T}^{-2}\,\boldsymbol{\theta}^{-1}$ |
| $c$ | Coefficient in equation 4.49 | — | — |
| $D$ | Molecular diffusivity | $m^2/s$ | $\mathbf{L}^2\mathbf{T}^{-1}$ |
| $D_L$ | Axial dispersion coefficient | $m^2/s$ | $\mathbf{L}^2\mathbf{T}^{-1}$ |
| $D_R$ | Radial dispersion coefficient | $m^2/s$ | $\mathbf{L}^2\mathbf{T}^{-1}$ |
| $d$ | Diameter of particle | m | $\mathbf{L}$ |
| $d_t$ | Diameter of tube or column | m | $\mathbf{L}$ |

| | | Units in SI System | Dimensions in **M**, **L**, **T**, $\boldsymbol{\theta}$ |
|---|---|---|---|
| $d'_m$ | Equivalent diameter of pore space $= e/S_B$ as used by Kozeny | m | **L** |
| $d_n$ | Nominal packing size (e.g. diameter for a Raschig ring) | m | **L** |
| $d_t$ | Tube diameter | — | **L** |
| $e$ | Fractional voidage of bed of particles or packing | — | — |
| $F$ | Packing factor | $m^2/m^3$ | $\mathbf{L}^{-1}$ |
| $f_w$ | Wall correction factor (equation 4.23) | — | — |
| $G$ | Gas mass flowrate | kg/s | $\mathbf{MT}^{-1}$ |
| $G_F$ | Gas mass flowrate under flooding conditions | kg/s | $\mathbf{MT}^{-1}$ |
| $G'$ | Gas mass velocity | kg/s $m^2$ | $\mathbf{ML}^{-2}\mathbf{T}^{-1}$ |
| $g$ | Acceleration due to gravity | $m/s^2$ | $\mathbf{LT}^{-2}$ |
| $H_w$ | Liquid hold-up in bed, volume of liquid per unit volume of bed | — | — |
| $h$ | Heat transfer coefficient | W/$m^2$ K | $\mathbf{MT}^{-3}\boldsymbol{\theta}^{-1}$ |
| $h_D$ | Mass transfer coefficient | m/s | $\mathbf{LT}^{-1}$ |
| $j_d$ | $j$-factor for mass transfer | — | — |
| $j_h$ | $j$-factor for heat transfer | — | — |
| $K$ | Constant in flow equations 4.1 and 4.2 | $m^3$s/kg | $\mathbf{M}^{-1}\mathbf{L}^3\mathbf{T}$ |
| $K'$ | Dimensionless constant in equation 4.6 | — | — |
| $K''$ | Kozeny constant in equation 4.9 | — | — |
| $K_0$ | Shape factor in equation 4.22 | — | — |
| $k$ | Thermal conductivity | W/m K | $\mathbf{MLT}^{-3}\boldsymbol{\theta}^{-1}$ |
| $k$ | Consistency coefficient for power-law fluid | $Ns^n/m^2$ | $\mathbf{ML}^{-1}\mathbf{T}^{n-2}$ |
| $L$ | Liquid mass flowrate | kg/s | $\mathbf{MT}^{-1}$ |
| $L'$ | Liquid mass velocity | kg/s $m^2$ | $\mathbf{ML}^{-2}\mathbf{T}^{-1}$ |
| $L_v$ | Volumetric liquid rate | $m^3/s$ | $\mathbf{L}^3\mathbf{T}^{-1}$ |
| $L'_v$ | Volumetric liquid rate per unit area | m/s | $\mathbf{LT}^{-1}$ |
| $L_w$ | Wetting rate ($u_L/S_B$) | $m^2/s$ | $\mathbf{L}^2\mathbf{T}^{-1}$ |
| $l$ | Length of bed or height of column packing | m | **L** |
| $l'$ | Length of flow passage through bed | m | **L** |
| $l_t$ | Length of circular tube | m | **L** |
| $N$ | Number of velocity heads lost through unit height of bed (equation 4.47) | $m^{-1}$ | $\mathbf{L}^{-1}$ |
| $n$ | Flow behaviour index for power-law-fluid | — | — |
| $n'$ | Exponent in equation 4.17 | — | — |
| $P$ | Pressure | $N/m^2$ | $\mathbf{ML}^{-1}\mathbf{T}^{-2}$ |
| $-\Delta P$ | Pressure drop across bed or column | $N/m^2$ | $\mathbf{ML}^{-1}\mathbf{T}^{-2}$ |
| $-\Delta P_d$ | Pressure drop across bed of dry packing | $N/m^2$ | $\mathbf{ML}^{-1}\mathbf{T}^{-2}$ |
| $-\Delta P_w$ | Pressure drop across bed of wet packing | $N/m^2$ | $\mathbf{ML}^{-1}\mathbf{T}^{-2}$ |
| $Q_x$ | Fraction of liquid collected at distance $x$ from centre line of packing in equation 4.49 | — | — |
| $R$ | Drag force per unit area of tube wall | $N/m^2$ | $\mathbf{ML}^{-1}\mathbf{T}^{-2}$ |
| $R_1$ | Drag force per unit surface area of particles | $N/m^2$ | $\mathbf{ML}^{-1}\mathbf{T}^{-2}$ |
| $r$ | Radius | m | **L** |
| $S$ | Surface area per unit volume of particle or packing | $m^2/m^3$ | $\mathbf{L}^{-1}$ |
| $S_B$ | Surface area per unit volume of bed (specific surface) | $m^2/m^3$ | $\mathbf{L}^{-1}$ |
| $S_c$ | Surface area of container per unit volume of bed | $m^2/m^3$ | $\mathbf{L}^{-1}$ |
| $t$ | Time | s | **T** |
| $u_c$ | Average fluid velocity based on cross-sectional area $A$ of empty column | m/s | $\mathbf{LT}^{-1}$ |
| $u_t$ | Mean velocity of fluid in tube | m/s | $\mathbf{LT}^{-1}$ |
| $u_G$ | Volumetric flowrate of gas per unit area of cross-section | m/s | $\mathbf{LT}^{-1}$ |
| $u_L$ | Volumetric flowrate of liquid per unit area of cross-section | m/s | $\mathbf{LT}^{-1}$ |
| $u_1$ | Mean velocity in pore channel | m/s | $\mathbf{LT}^{-1}$ |
| $V$ | Volume of fluid flowing in time $t$ | $m^3$ | $\mathbf{L}^3$ |
| $X$ | Side of cube | m | **L** |

| | | Units in SI System | Dimensions in **M**, **L**, **T**, $\theta$ |
|---|---|---|---|
| $x$ | Distance from centre | m | **L** |
| $\alpha$ | Coefficient in equation 4.17 | (N/m$^2$)/(m/s) | $\mathbf{ML^{-2}T^{-1}}$ |
| $\alpha'$ | Coefficient in equation 4.17 | (N/m$^2$)/(m/s)$^{n'}$ | $\mathbf{ML^{-(n'+1)}T^{(n'-2)}}$ |
| $\gamma$ | Coefficient of $D$ in equations 4.37, 4.39 | — | — |
| $\dot{\gamma}$ | Shear rate | s$^{-1}$ | $\mathbf{T^{-1}}$ |
| $\mu$ | Fluid viscosity | Ns/m$^2$ | $\mathbf{ML^{-1}T^{-1}}$ |
| $\rho$ | Density of fluid | kg/m$^3$ | $\mathbf{ML^{-3}}$ |
| $\tau$ | Shear stress | N/m$^2$ | $\mathbf{ML^{-1}T^{-2}}$ |
| $\psi$ | Density correction factor $\sqrt{(\rho_G/\rho_A)}$ | — | — |
| $Gr'$ | Grashof number (particle) (Volume 1, Chapter 9) | — | — |
| $Nu$ | Nusselt number for tube wall $(hd_t/k)$ | — | — |
| $Nu'$ | Nusselt number (particle) $(hd/k)$ | — | — |
| $Pe$ | Peclet number $(u_c d/eD_L)$ or $(u_c d/eD_R)$ | — | — |
| $Pr$ | Prandtl number $(C_p\mu/k)$ | — | — |
| $Re$ | Reynolds number for flow through tube $(ud_t\rho/\mu)$ | — | — |
| $Re_1$ | Modified Reynolds number based on pore size as used by Carman (equation 4.13) | — | — |
| $Re'_c$ | Modified Reynolds number based on particle size $(u_c d\rho/\mu)$ | — | — |
| $Re_{MR}$ | Metzner and Reed Reynolds number (equation 4.27) | — | — |
| $(Re_1)_n$ | Reynolds number for power-law fluid in a granular bed (equation 4.28) | — | — |
| $Sc$ | Schmidt number $(\mu/\rho D)$ | — | — |
| $Sh'$ | Sherwood number (particle) $(h_D d/D)$ | — | — |
| $St'$ | Stanton number (particle) $(h/C_p\rho u_c)$ | — | — |
| Suffix | $A$ refers to air at 293 K | — | — |
| | $G$ refers to gas | | |
| | $L$ refers to liquid | | |
| | $w$ refers to water at 293 K | | |
| | 0 refers to standard conditions | | |

CHAPTER 5

# Sedimentation

## 5.1. INTRODUCTION

In Chapter 3 consideration is given to the forces acting on an isolated particle moving relative to a fluid and it is seen that the frictional drag may be expressed in terms of a friction factor which is, in turn, a function of the particle Reynolds number. If the particle is settling in the gravitational field, it rapidly reaches its terminal falling velocity when the frictional force has become equal to the net gravitational force. In a centrifugal field the particle may reach a very much higher velocity because the centrifugal force may be many thousands of times greater than the gravitational force.

In practice, the concentrations of suspensions used in industry will usually be high enough for there to be significant interaction between the particles, and the frictional force exerted at a given velocity of the particles relative to the fluid may be greatly increased as a result of modifications of the flow pattern, so that *hindered settling* takes place. As a corollary, the sedimentation rate of a particle in a concentrated suspension may be considerably less than its terminal falling velocity under *free settling* conditions when the effects of mutual interference are negligible. In this chapter, the behaviour of concentrated suspensions in a gravitational field is discussed and the equipment used industrially for concentrating or *thickening* such suspensions will be described. Sedimentation in a centrifugal field is considered in Chapter 9.

It is important to note that suspensions of fine particles tend to behave rather differently from coarse suspensions in that a high degree of flocculation may occur as a result of the very high specific surface of the particles. For this reason, fine and coarse suspensions are considered separately, and the factors giving rise to flocculation are discussed in Section 5.2.2.

Although the sedimentation velocity of particles tends to decrease steadily as the concentration of the suspension is increased, it has been shown by KAYE and BOARDMAN(1) that particles in very dilute suspensions may settle at velocities up to 1.5 times the normal terminal falling velocities, due to the formation of clusters of particles which settle in well-defined streams. This effect is important when particle size is determined by a method involving the measurement of the settling velocity of particles in dilute concentration, though is not significant with concentrated suspensions.

## 5.2. SEDIMENTATION OF FINE PARTICLES

### 5.2.1. Experimental studies

The sedimentation of metallurgical slimes has been studied by COE and CLEVENGER(2), who concluded that a concentrated suspension may settle in one of two different ways.

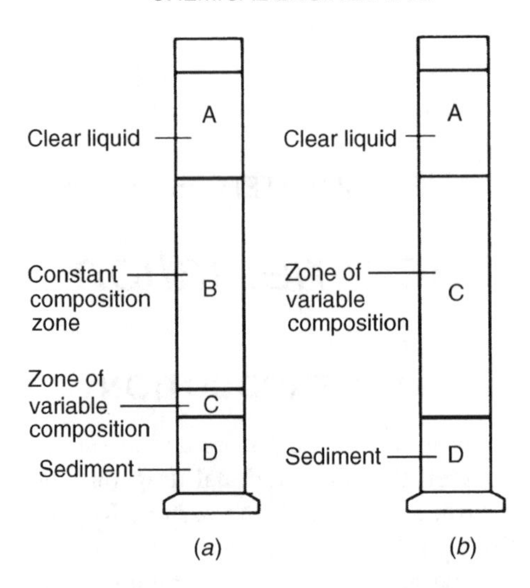


Figure 5.1. Sedimentation of concentrated suspensions (*a*) Type 1 settling (*b*) Type 2 settling

In the first, after an initial brief acceleration period, the interface between the clear liquid and the suspension moves downwards at a constant rate and a layer of sediment builds up at the bottom of the container. When this interface approaches the layer of sediment, its rate of fall decreases until the "critical settling point" is reached when a direct interface is formed between the sediment and the clear liquid. Further sedimentation then results solely from a consolidation of the sediment, with liquid being forced upwards around the solids which are then forming a loose bed with the particles in contact with one another. Since the flow area is gradually being reduced, the rate progressively diminishes. In Figure 5.1*a*, a stage in the sedimentation process is illustrated. A is clear liquid, B is suspension of the original concentration, C is a layer through which the concentration gradually increases, and D is sediment. The sedimentation rate remains constant until the upper interface corresponds with the top of zone C and it then falls until the critical settling point is reached when both zones B and C will have disappeared. A second and rather less common mode of sedimentation as shown in Figure 5.1*b*, is obtained when the range of particle size is very great. The sedimentation rate progressively decreases throughout the whole operation because there is no zone of constant composition, and zone C extends from the top interface to the layer of sediment.

The main reasons for the modification of the settling rate of particles in a concentrated suspension are as follows:

(a) If a significant size range of particles is present, the large particles are settling relative to a suspension of smaller ones so that the effective density and viscosity of the fluid are increased.
(b) The upward velocity of the fluid displaced during settling is appreciable in a concentrated suspension and the apparent settling velocity is less than the actual velocity relative to the fluid.
(c) The velocity gradients in the fluid close to the particles are increased as a result of the change in the area and shape of the flow spaces.

(d) The smaller particles tend to be dragged downwards by the motion of the large particles and are therefore accelerated.
(e) Because the particles are closer together in a concentrated suspension, flocculation is more marked in an ionised solvent and the effective size of the small particles is increased.

If the range of particle size is not more than about 6 : 1, a concentrated suspension settles with a sharp interface and all the particles fall at the same velocity. This is in contrast with the behaviour of a dilute suspension, for which the rates of settling of the particles can be calculated by the methods given in Chapter 3, and where the settling velocity is greater for the large particles. The two types of settling are often referred to as *sludge line settling* and *selective settling* respectively. The overall result is that in a concentrated suspension the large particles are retarded and the small ones accelerated.

Several attempts have been made to predict the apparent settling velocity of a concentrated suspension. In 1926 ROBINSON[3] suggested a modification of Stokes' law and used the density ($\rho_c$) and viscosity ($\mu_c$) of the suspension in place of the properties of the fluid to give:

$$u_c = \frac{K'' d^2(\rho_s - \rho_c)g}{\mu_c} \tag{5.1}$$

where $K''$ is a constant.

The effective buoyancy force is readily calculated since:

$$(\rho_s - \rho_c) = \rho_s - \{\rho_s(1 - e) + \rho e\} = e(\rho_s - \rho) \tag{5.2}$$

where $e$ is the voidage of the suspension.

Robinson determined the viscosity of the suspension $\mu_c$ experimentally, although it may be obtained approximately from the following formula of EINSTEIN[4]:

$$\mu_c = \mu(1 + k''C) \tag{5.3}$$

where: $k''$ is a constant for a given shape of particle (2.5 for spheres),
$C$ is the volumetric concentration of particles, and
$\mu$ is the viscosity of the fluid.

This equation holds for values of $C$ up to 0.02. For more concentrated suspensions, VAND[5] gives the equation:

$$\mu_c = \mu\, \mathrm{e}^{k''C/(1-a'C)} \tag{5.4}$$

in which $a'$ is a second constant, equal to $(39/64) = 0.609$ for spheres.

STEINOUR[6], who studied the sedimentation of small uniform particles, adopted a similar approach, using the viscosity of the fluid, the density of the suspension and a function of the voidage of the suspension to take account of the character of the flow spaces, and obtained the following expression for the velocity of the particle relative to the fluid $u_p$:

$$u_p = \frac{d^2(\rho_s - \rho_c)g}{18\mu} \mathrm{f}(e) \tag{5.5}$$

Since the fraction of the area available for flow of the displaced fluid is $e$, its upward velocity is $u_c(1-e)/e$ so that:

$$u_p = u_c + u_c\frac{1-e}{e} = \frac{u_c}{e} \tag{5.6}$$

From his experiments on the sedimentation of tapioca in oil, Steinour found:

$$\text{f}(e) = 10^{-1.82(1-e)} \tag{5.7}$$

Substituting in equation 5.5, from equations 5.2, 5.6 and 5.7:

$$u_c = \frac{e^2 d^2(\rho_s - \rho)g}{18\mu} 10^{-1.82(1-e)} \tag{5.8}$$

HAWKSLEY[7] also used a similar method and gave:

$$u_p = \frac{u_c}{e} = \frac{d^2(\rho_s - \rho_c)g}{18\mu_c} \tag{5.9}$$

In each of these cases, it is correctly assumed that the upthrust acting on the particles is determined by the density of the suspension rather than that of the fluid. The use of an effective viscosity, however, is valid only for a large particle settling in a fine suspension. For the sedimentation of uniform particles the increased drag is attributable to a steepening of the velocity gradients rather than to a change in viscosity.

The rate of sedimentation of a suspension of fine particles is difficult to predict because of the large number of factors involved. Thus, for instance, the presence of an ionised solute in the liquid and the nature of the surface of the particles will affect the degree of flocculation and hence the mean size and density of the flocs. The flocculation of a suspension is usually completed quite rapidly so that it is not possible to detect an increase in the sedimentation rate in the early stages after the formation of the suspension. Most fine suspensions flocculate readily in tap water and it is generally necessary to add a deflocculating agent to maintain the particles individually dispersed. The factors involved in flocculation are discussed later in this chapter. A further factor influencing the sedimentation rate is the degree of agitation of the suspension. Gentle stirring may produce accelerated settling if the suspension behaves as a non-Newtonian fluid in which the apparent viscosity is a function of the rate of shear. The change in apparent viscosity can probably be attributed to the re-orientation of the particles. The effect of stirring is, however, most marked on the consolidation of the final sediment, in which "bridge formation" by the particles can be prevented by gentle stirring. During these final stages of consolidation of the sediment, liquid is being squeezed out through a bed of particles which are gradually becoming more tightly packed.

A number of empirical equations have been obtained for the rate of sedimentation of suspensions, as a result of tests carried out in vertical tubes. For a given solid and liquid, the main factors which affect the process are the height of the suspension, the diameter of the containing vessel, and the volumetric concentration. An attempt at co-ordinating the results obtained under a variety of conditions has been made by WALLIS[8].

### Height of suspension

The height of suspension does not generally affect either the rate of sedimentation or the consistency of the sediment ultimately obtained. If, however, the position of the sludge line is plotted as a function of time for two different initial heights of slurry, curves of the form shown in Figure 5.2 are obtained in which the ratio $OA' : OA''$ is everywhere constant. Thus, if the curve is obtained for any one initial height, the curves can be drawn for any other height.

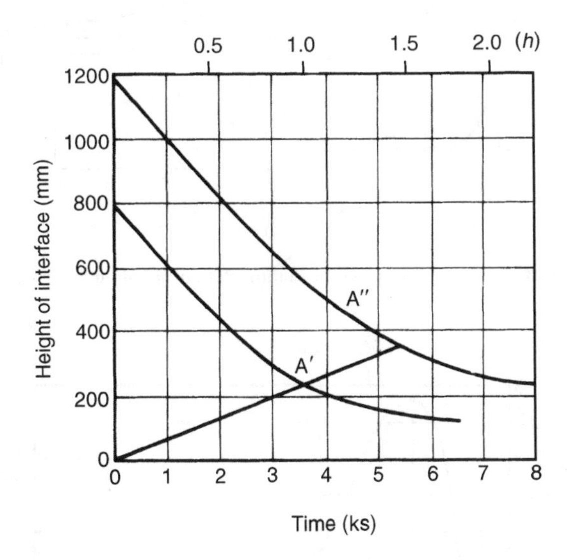


Figure 5.2. Effect of height on sedimentation of a 3 per cent (by volume) suspension of calcium carbonate

### Diameter of vessel

If the ratio of the diameter of the vessel to the diameter of the particle is greater than about 100, the walls of the container appear to have no effect on the rate of sedimentation. For smaller values, the sedimentation rate may be reduced because of the retarding influence of the walls.

### Concentration of suspension

As already indicated, the higher the concentration, the lower is the rate of fall of the sludge line because the greater is the upward velocity of the displaced fluid and the steeper are the velocity gradients in the fluid. Typical curves for the sedimentation of a suspension of precipitated calcium carbonate in water are shown in Figure 5.3, and in Figure 5.4 the mass rate of sedimentation ($kg/m^2s$) is plotted against the concentration. This curve has a maximum value, corresponding to a volumetric concentration of about 2 per cent. EGOLF and MCCABE[9], WORK and KOHLER[10], and others have given empirical expressions for the rate of sedimentation at the various stages, although these are generally applicable over a narrow range of conditions and involve constants which need to be determined experimentally for each suspension.

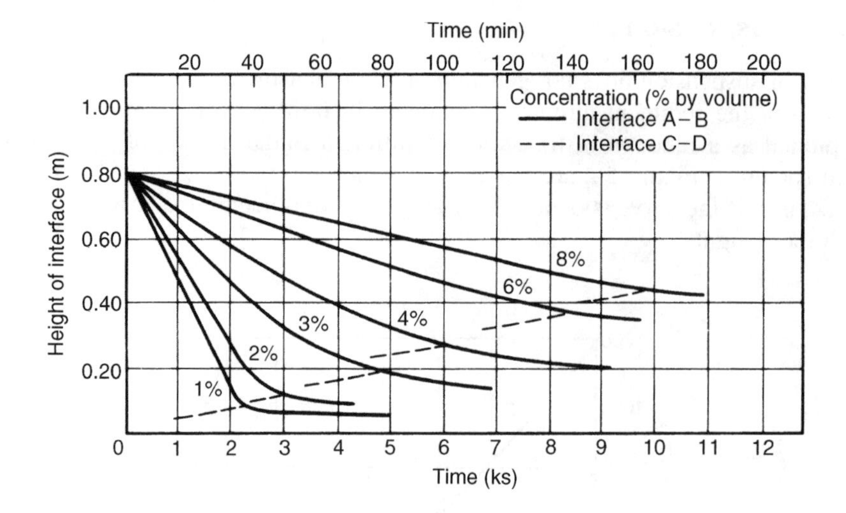


Figure 5.3. Effect of concentration on the sedimentation of calcium carbonate suspensions

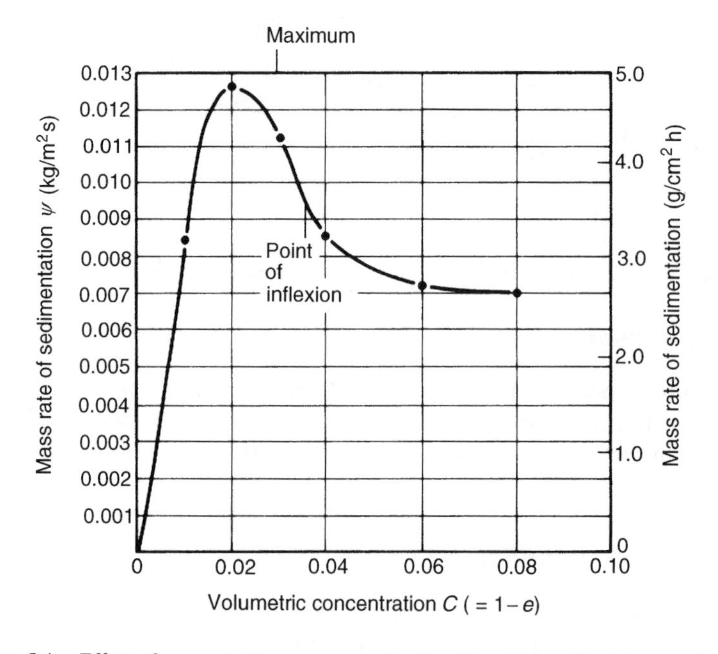


Figure 5.4. Effect of concentration on mass rate of sedimentation of calcium carbonate

The final consolidation of the sediment is the slowest part of the process because the displaced fluid has to flow through the small spaces between the particles. As consolidation occurs, the rate falls off because the resistance to the flow of liquid progressively increases. The porosity of the sediment is smallest at the bottom because the compressive force due to the weight of particles is greatest and because the lower portion was formed at an earlier stage in the sedimentation process. The rate of sedimentation during this period is given approximately by:

$$-\frac{dH}{dt} = b(H - H_\infty) \tag{5.10}$$

where: $H$ is the height of the sludge line at time $t$,
$H_\infty$ is the final height of the sediment, and
$b$ is a constant for a given suspension.

The time taken for the sludge line to fall from a height $H_c$, corresponding to the critical settling point, to a height $H$ is given by:

$$-bt = \ln(H - H_\infty) - \ln(H_c - H_\infty) \tag{5.11}$$

Thus, if $\ln(H - H_\infty)$ is plotted against $t$, a straight line of slope $-b$ is obtained.

The values of $H_\infty$ are determined largely by the surface film of liquid adhering to the particles.

### Shape of vessel

Provided that the walls of the vessel are vertical and that the cross-sectional area does not vary with depth, the shape of the vessel has little effect on the sedimentation rate. However, if parts of the walls of the vessel face downwards, as in an inclined tube, or if part of the cross-section is obstructed for a portion of the height, the effect on the sedimentation process may be considerable.

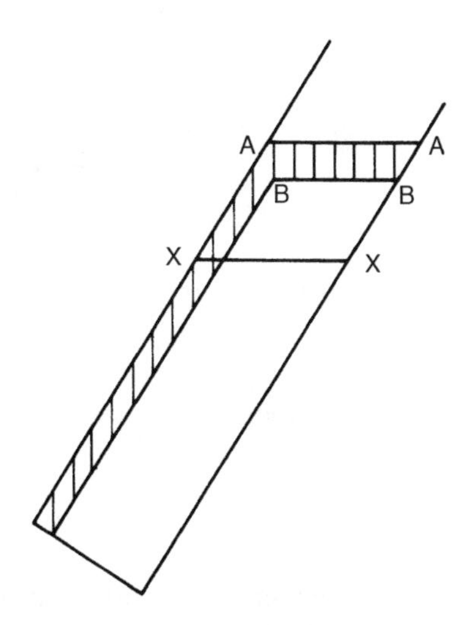


Figure 5.5. Sedimentation in an inclined tube

PEARCE[11] studied the effect of a downward-facing surface by considering an inclined tube as shown in Figure 5.5. Starting with a suspension reaching a level AA, if the sludge line falls to a new level BB, then material will tend to settle out from the whole of the shaded area. This configuration is not stable and the system tends to adjust itself so that the sludge line takes up a new level XX, the volume corresponding to the area AAXX being equal to that corresponding to the shaded area. By applying this principle, it is seen that it is possible to obtain an accelerated rate of settling in an inclined tank by inserting a series of inclined plates. The phenomenon has been studied further by several workers including SCHAFLINGER[12].

The effect of a non-uniform cross-section was considered by ROBINS[13], who studied the effect of reducing the area in part of the vessel by immersing a solid body, as shown in Figure 5.6. If the cross-sectional area, sedimentation velocity, and fractional volumetric concentration are $C$, $u_c$, and $A$ below the obstruction, and $C'$, $u_c'$, and $A'$ at the horizontal level of the obstruction, and $\psi$ and $\psi'$ are the corresponding rates of deposition of solids per unit area, then:

$$A\psi = ACu_c \tag{5.12}$$

and:

$$A'\psi' = A'C'u_c' \tag{5.13}$$

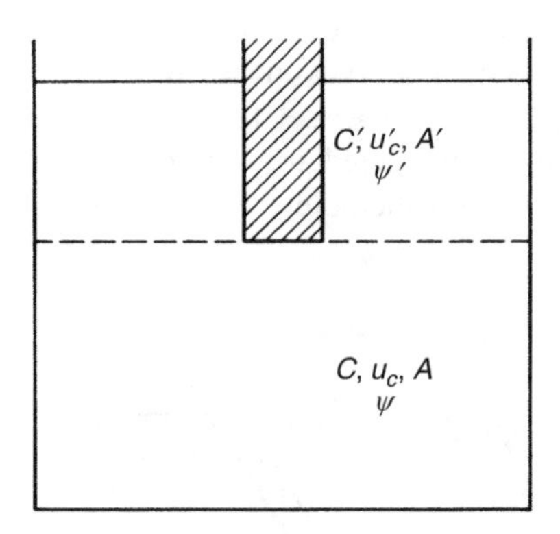


Figure 5.6. Sedimentation in partially obstructed vessel

For continuity at the bottom of the obstruction:

$$\psi' = \frac{A}{A'}\psi \tag{5.14}$$

A plot of $\psi$ versus $C$ will have the same general form as Figure 5.4 and a typical curve is given in Figure 5.7. If the concentration $C$ is appreciably greater than the value $C_m$ at which $\psi$ is a maximum, $C'$ will be less than $C$ and the system will be stable. On the other hand, if $C$ is less than $C_m$, $C'$ will be greater than $C$ and mixing will take place because of the greater density of the upper portion of the suspension. The range of values of $C$ for which equation 5.14 is valid in practice, and for which mixing currents are absent, may be very small.

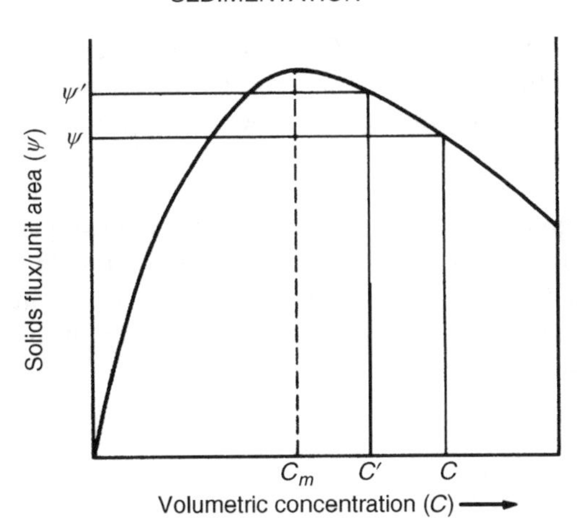


Figure 5.7. Solids flux per unit area as a function of volumetric concentration

## 5.2.2. Flocculation

### Introduction

The behaviour of suspensions of fine particles is very considerably influenced by whether the particles flocculate. The overall effect of flocculation is to create large conglomerations of elementary particles with occluded liquid. The flocs, which easily become distorted, are effectively enlarged particles of a density intermediate between that of the constituent particles and the liquid.

The tendency of the particulate phase of colloidal dispersions to aggregate is an important physical property which finds practical application in solid–liquid separation processes, such as sedimentation and filtration. The aggregation of colloids is known as coagulation, or flocculation. Particles dispersed in liquid media collide due to their relative motion; and the stability (that is stability against aggregation) of the dispersion is determined by the interaction between particles during these collisions. Attractive and repulsive forces can be operative between the particles; these forces may react in different ways depending on environmental conditions, such as salt concentration and pH. The commonly occurring forces between colloidal particles are van der Waals forces, electrostatic forces and forces due to adsorbed macromolecules. In the absence of macromolecules, aggregation is largely due to van der Waals attractive forces, whereas stability is due to repulsive interaction between similarly charged electrical double-layers.

### The electrical double-layer

Most particles acquire a surface electric charge when in contact with a polar medium. Ions of opposite charge (counter-ions) in the medium are attracted towards the surface and ions of like charge (co-ions) are repelled, and this process, together with the mixing tendency due to thermal motion, results in the creation of an electrical double-layer which comprises the charged surface and a neutralising excess of counter-ions over co-ions distributed in a diffuse manner in the polar medium. The quantitative theory of the electrical double-layer, which deals with the distribution of ions and the magnitude of electric potentials, is beyond the scope of this text although an understanding of it is essential in an analysis of colloid stability[14,15].

For present purposes, the electrical double-layer is represented in terms of Stern's model (Figure 5.8) wherein the double-layer is divided into two parts separated by a plane (Stern plane) located at a distance of about one hydrated-ion radius from the surface. The potential changes from $\psi_0$ (surface) to $\psi_\delta$ (Stern potential) in the Stern layer and decays to zero in the diffuse double-layer; quantitative treatment of the diffuse double-layer follows the Gouy–Chapman theory[16,17].

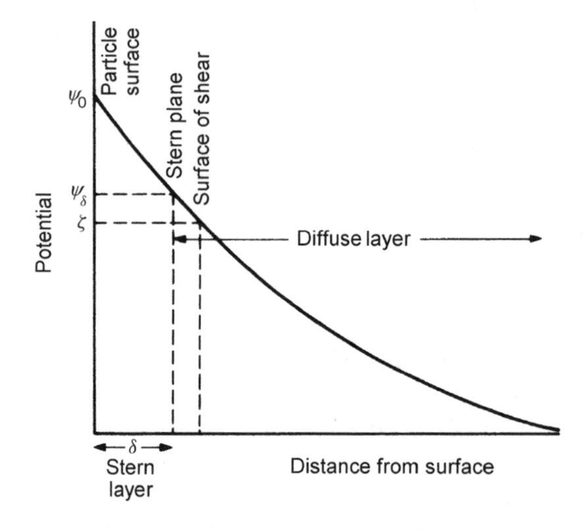


Figure 5.8. Stern's model

$\psi_\delta$ can be estimated from electrokinetic measurements, such as electrophoresis, streaming potential. In such measurements, surface and liquid move tangentially with respect to each other. For example, in electrophoresis the liquid is stationary and the particles move under the influence of an applied electric field. A thin layer of liquid, a few molecules thick, moves together with the particle so that the actual hydrodynamic boundary between the moving unit and the stationary liquid is a *slipping plane* inside the solution. The potential at the slipping plane is termed the *zeta potential*, $\zeta$, as shown in Figure 5.8.

LYKLEMA[18] considers that the slipping plane may be identified with the Stern plane so that $\psi_\delta \simeq \zeta$. Thus, since the surface potential $\psi_0$ is inaccessible, zeta potentials find practical application in the calculation of $V_R$ from equation 5.16. In practice, electrokinetic measurements must be carried out with considerable care if reliable estimates of $\zeta$ are to be made[19].

## Interactions between particles

The interplay of forces between particles in lyophobic sols may be interpreted in terms of the theory of DERJAGUIN and LANDAU[20] and VERWEY and OVERBEEK[14]. Their theory (the DLVO theory) considers that the potential energy of interaction between a pair of particles consists of two components:

(a) a repulsive component $V_R$ arising from the overlap of the electrical double-layers, and
(b) a component $V_A$ due to van der Waals attraction arising from electromagnetic effects.

These are considered to be additives so that the total potential energy of interaction $V_T$ is given by:

$$V_T = V_R + V_A \tag{5.15}$$

In general, as pointed out by Gregory[21], the calculation of $V_R$ is complex but a useful approximation for identical spheres of radius $a$ is given by[14]:

$$V_R = \frac{64\pi a n_i K T \gamma^2 e^{-\kappa H_s}}{\kappa^2} \tag{5.16}$$

where:

$$\gamma = \frac{\exp(Ze_c\psi_\delta/2KT) - 1}{\exp(Ze_c\psi_\delta/2KT) + 1} \tag{5.17}$$

and:

$$\kappa = \left(\frac{2e_c^2 n_i Z^2}{\varepsilon K T}\right)^{1/2} \tag{5.18}$$

For identical spheres with $H_s \leq 10$–$20$ nm (100–200 Å) and when $H_s \ll a$, the energy of attraction $V_A$ is given by the approximate expression[27]:

$$V_A = -\frac{\mathscr{A} a}{12 H_s} \tag{5.19}$$

where $\mathscr{A}$ is the Hamaker[22] constant whose value depends on the nature of the material of the particles. The presence of liquid between particles reduces $V_A$, and an effective Hamaker constant is calculated from:

$$\mathscr{A} = (\mathscr{A}_2^{1/2} - \mathscr{A}_1^{1/2})^2 \tag{5.20}$$

where subscripts 1 and 2 refer to dispersion medium and particles respectively. Equation 5.19 is based on the assumption of complete additivity of intermolecular interactions; this assumption is avoided in the theoretical treatment of Lifshitz[23] which is based on macroscopic properties of materials[24]. Tables of $\mathscr{A}$ are available in the literature[25]; values are generally found to lie in the range $0.1 \times 10^{-20}$ to $10 \times 10^{-20}$ J.

The general form of $V_T$ versus distance of separation between particle surfaces $H_s$ is shown schematically in Figure 5.9. At very small distances of separation, repulsion due to overlapping electron clouds (Born repulsion)[14] predominates, and consequently a deep minimum (primary minimum) occurs in the potential energy curve. For smooth surfaces this limits the distance of closest approach ($H_{s\min}$) to $\sim 0.4$ nm (4 Å). Aggregation of particles occurring in this primary minimum, for example, aggregation of lyophobic sols in the presence of NaCl, is termed *coagulation*[24].

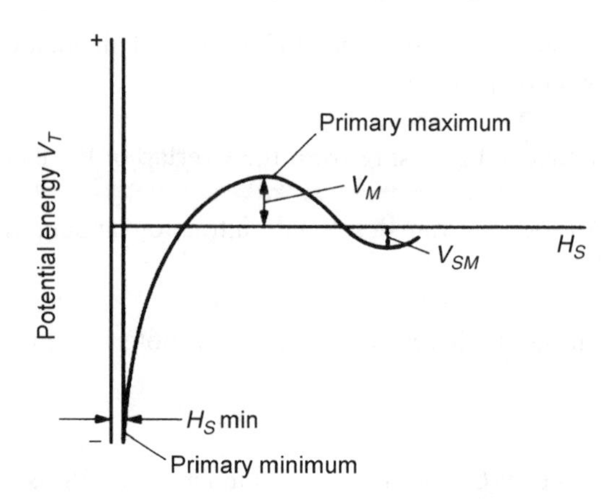


Figure 5.9. Potential energy as a function of separation

At high surface potentials, low ionic strengths and intermediate distance the electrical repulsion term is dominant and so a maximum (primary maximum) occurs in the potential energy curve. At larger distances of separation $V_R$ decays more rapidly than $V_A$ and a secondary minimum appears. If the potential energy maximum is large compared with the thermal energy $KT$ of the particles ($\sim 4.2 \times 10^{-20}$ J) the system should be stable, otherwise the particles would coagulate. The height of this energy barrier to coagulation depends upon the magnitude of $\psi_\delta$ (and $\zeta$) and upon the range of the repulsive forces (that is, upon $1/\kappa$). If the depth of the secondary minimum is large compared with $KT$, it should produce a loose, easily reversible form of aggregation which is termed flocculation. This term also describes aggregation of particles in the presence of polymers[(24)], as discussed later in this chapter.

It is of interest to note that both $V_A$ and $V_R$ increase as particle radius $a$ becomes larger and thus $V_M$ in Figure 5.9 would be expected to increase with the sol becoming more stable; also if $a$ increases then $V_{SM}$ increases and may become large enough to produce "secondary minimum" flocculation.

## Coagulation concentrations

Coagulation concentrations are the electrolyte concentrations required just to coagulate a sol. Clearly $V_M$ in Figure 5.9 must be reduced, preferably to zero, to allow coagulation to occur. This can be achieved by increasing the ionic strength of the solution, thus increasing $\kappa$ and thereby reducing $V_R$ in equation 5.16. The addition of salts with multivalent ions (such as $Al^{3+}$, $Ca^{2+}$, $Fe^{3+}$) is most effective because of the effect of charge number $Z$ and $\kappa$ (equation 5.18). Taking as a criterion that $V_T = 0$ and $dV_T/dH_s = 0$ for the same value of $H_s$, it may be shown[(14)] that the coagulation concentration $c'_c$ is given by:

$$c'_c = \frac{9.75 B^2 \varepsilon^3 K^5 T^5 \gamma^4}{e_c^2 \mathbf{N} \mathscr{A}^2 Z^6} \tag{5.21}$$

where $B = 3.917 \times 10^{39}$ coulomb$^{-2}$. At high values of surface potentials, $\gamma \approx 1$ and equation 5.21 predicts that the coagulation concentration should be inversely proportional to the sixth power of the valency $Z$. Thus coagulation concentrations of those electrolytes whose counter-ions have charge numbers 1, 2, 3 should be in the ratio 100 : 1.6 : 0.13. It may be noted that, if an ion is specifically adsorbed on the particles, $\psi_\delta$ can be drastically reduced and coagulation effected without any great increase in ionic strength. For example, minute traces of certain hydrolysed metal ions can cause coagulation of negatively charged particles[26]. In such cases charge reversal often occurs and the particles can be restabilised if excess coagulant is added.

### *Kinetics of coagulation*

The rate of coagulation of particles in a liquid depends on the frequency of collisions between particles due to their relative motion. When this motion is due to Brownian movement coagulation is termed *perikinetic*; when the relative motion is caused by velocity gradients coagulation is termed *orthokinetic*.

Modern analyses of perikinesis and orthokinesis take account of hydrodynamic forces as well as interparticle forces. In particular, the frequency of binary collisions between spherical particles has received considerable attention[27–30].

The frequency of binary encounters during perikinesis is determined by considering the process as that of diffusion of spheres (radius $a_2$ and number concentration $n_2$) towards a central reference sphere of radius $a_1$, whence the frequency of collision $I$ is given by[27]:

$$I = \frac{4\pi D_{12}^{(\infty)} n_2(a_1 + a_2)}{1 + \dfrac{a_2}{a_1} \displaystyle\int_{1+(a_2/a_1)}^{\infty} (D_{12}^{(\infty)}/D_{12}) \exp(V_T/KT) \frac{ds}{s^2}} \tag{5.22}$$

where $s = a_r/a_1$ and the coordinate $a_r$ has its origin at the centre of sphere 1.

Details of $D_{12}$, the relative diffusivity between unequal particles, are given by SPIELMAN[27] who illustrates the dependence of $D_{12}$ on the relative separation $a_r$ between particle centres. At infinite separation, where hydrodynamic effects vanish:

$$D_{12} = D_{12}^{(\infty)} = D_1 + D_2 \tag{5.23}$$

where $D_1$ and $D_2$ are absolute diffusion coefficients given by the Stokes–Einstein equation.

$$\left.\begin{aligned} D_1 &= KT/(6\pi\mu a_1) \\ D_2 &= KT/(6\pi\mu a_2) \end{aligned}\right\} \tag{5.24}$$

When long-range particle interactions and hydrodynamics effects are ignored, equation 5.22 becomes equivalent to the solution of VON SMOLUCHOWSKI[31] who obtained the collision frequency $I_s$ as:

$$I_s = 4\pi D_{12}^{(\infty)} n_2(a_1 + a_2) \tag{5.25}$$

and who assumed an attractive potential only given by:

$$V_A = -\infty \quad s \leq +\frac{a_2}{a_1}$$

$$V_A = 0 \qquad s > 1 + \frac{a_2}{a_1} \tag{5.26}$$

Thus:

$$I = \alpha_p I_s \tag{5.27}$$

where the ratio $\alpha_p$ is the reciprocal of the denominator in equation 5.22; values of $\alpha_p$ are tabulated by SPIELMAN[27].

Assuming an attractive potential only, given by equation 5.26, Smoluchowski showed that the frequency of collisions per unit volume between particles of radii $a_1$ and $a_2$ in the presence of a laminar shear gradient $\dot{\gamma}$ is given by:

$$J_s = \tfrac{4}{3} n_1 n_2 (a_1 + a_2)^3 \dot{\gamma} \tag{5.28}$$

Analyses of the orthokinetic encounters between equi-sized spheres[30] have shown that, as with perikinetic encounters, equation 5.28 can be modified to include a ratio $\alpha_0$ to give the collision frequency $J$ as:

$$J = \alpha_0 J_s \tag{5.29}$$

where $\alpha_0$, which is a function of $\dot{\gamma}$, corrects the Smoluchowski relation for hydrodynamic interactions and interparticle forces. ZEICHNER and SCHOWALTER[30] present $\alpha_0^{-1}$ graphically as a function of a dimensionless parameter $N_F (= 6\pi \mu a^3 \dot{\gamma} / \mathscr{A})$ for the condition $V_R = 0$, whence it is possible to show that, for values of $N_F > 10$, $J$ is proportional to $\dot{\gamma}$ raised to the 0.77 power instead of the first power as given by equation 5.28.

Perikinetic coagulation is normally too slow for economic practical use in such processes as wastewater treatment, and orthokinetic coagulation is often used to produce rapid growth of aggregate of floc size. In such situations floc–floc collisions occur under non-uniform turbulent flow conditions. A rigorous analysis of the kinetics of coagulation under these conditions is not available at present. A widely used method of evaluating a mean shear gradient in such practical situations is given by CAMP and STEIN[32], who propose that:

$$\dot{\gamma} = [\mathbf{P}/\mu]^{1/2} \tag{5.30}$$

where $\mathbf{P}$ = power input/unit volume of fluid.

### *Effect of polymers on stability*

The stability of colloidal dispersions is strongly influenced by the presence of adsorbed polymers. Sols can be stabilised or destabilised depending on a number of factors including the relative amounts of polymer and sol, the mechanism of adsorption of polymer and the method of mixing polymer and dispersion[33]. Adsorption of polymer on to colloidal particles may increase their stability by decreasing $V_A$[34,35], increasing $V_R$[36] or by introducing a *steric* component of repulsion $V_S$[37,38].

Flocculation is readily produced by linear homopolymers of high molecular weight. Although they may be non-ionic, they are commonly polyelectrolytes; polyacrylamides and their derivatives are widely used in practical situations[39]. Flocculation by certain high molecular weight polymers can be interpreted in terms of a *bridging* mechanism; polymer molecules may be long and flexible enough to adsorb on to several particles. The precise nature of the attachment between polymer and particle surface depends on the nature of the surfaces of particle and polymer, and on the chemical properties of the solution. Various types of interaction between polymer segments and particle surfaces may be envisaged. In the case of polyelectrolytes, the strongest of these interactions would be ionic association between a charged site on the surface and an oppositely charged polymer segment, such as polyacrylic acid and positively charged silver iodide particles[40–43].

Polymers may show an optimum flocculation concentration which depends on molecular weight and concentration of solids in suspension. Overdosing with flocculant may lead to restabilisation[44], as a consequence of particle surfaces becoming saturated with polymer. Optimum flocculant concentrations may be determined by a range of techniques including sedimentation rate, sedimentation volume, filtration rate and clarity of supernatant liquid.

### *Effect of flocculation on sedimentation*

In a flocculated, or coagulated, suspension the aggregates of fine particles or flocs are the basic structural units and in a low shear rate process, such as gravity sedimentation, their settling rates and sediment volumes depend largely on volumetric concentration of floc and on interparticle forces. The type of settling behaviour exhibited by flocculated suspensions depends largely on the initial solids concentration and chemical environment. Two kinds of batch settling curve are frequently seen. At low initial solids concentration the flocs may be regarded as discrete units consisting of particles and immobilised fluid. The flocs settle initially at a constant settling rate though as they accumulate on the bottom of the vessel they deform under the weight of the overlying flocs. The curves shown earlier in Figure 5.3 for calcium carbonate suspensions relate to this type of sedimentation. When the solids concentration is very high the maximum settling rate is not immediately reached and thus may increase with increasing initial height of suspension[45]. Such behaviour appears to be characteristic of structural flocculation associated with a continuous network of flocs extending to the walls of the vessel. In particular, the first type of behaviour, giving rise to a constant settling velocity of the flocs, has been interpreted quantitatively by assuming that the flocs consist of aggregates of particles and occluded liquid. The flocs are considerably larger than the fundamental particles and of density intermediate between that of the water and the particles themselves. MICHAELS and BOLGER[45] found good agreement between their experimental results and predicted sedimentation rates. The latter were calculated from the free settling velocity of an individual floc, corrected for the volumetric concentration of the flocs using equation 5.71 (Section 5.3.2) which has been developed for the sedimentation of systems of fully-dispersed mono-size particles.

## 5.2.3. The Kynch theory of sedimentation

The behaviour of concentrated suspensions during sedimentation has been analysed by KYNCH[46], largely using considerations of continuity. The basic assumptions which are made are as follows:

(a) Particle concentration is uniform across any horizontal layer,
(b) Wall effects can be ignored,
(c) There is no differential settling of particles as a result of differences in shape, size, or composition,

(d) The velocity of fall of particles depends only on the local concentration of particles,
(e) The initial concentration is either uniform or increases towards the bottom of the suspension, and
(f) The sedimentation velocity tends to zero as the concentration approaches a limiting value corresponding to that of the sediment layer deposited at the bottom of the vessel.

If at some horizontal level where the volumetric concentration of particles is $C$ and the sedimentation velocity is $u_c$, the volumetric rate of sedimentation per unit area, or flux, is given by:

$$\psi = Cu_c \tag{5.31}$$

Then a material balance taken between a height $H$ above the bottom, at which the concentration is $C$ and the mass flux is $\psi$, and a height $H + \mathrm{d}H$, where the concentration is $C + (\partial C/\partial H)\mathrm{d}H$ and the mass flux is $\psi + (\partial\psi/\partial H)\mathrm{d}H$, gives:

$$\left\{\left(\psi + \frac{\partial\psi}{\partial H}\mathrm{d}H\right) - \psi\right\}\mathrm{d}t = \frac{\partial}{\partial t}(C\mathrm{d}H)\,\mathrm{d}t$$

That is:

$$\frac{\partial\psi}{\partial H} = \frac{\partial C}{\partial t} \tag{5.32}$$

Hence:

$$\frac{\partial\psi}{\partial H} = \frac{\partial\psi}{\partial C}\cdot\frac{\partial C}{\partial H} = \frac{\mathrm{d}\psi}{\mathrm{d}C}\cdot\frac{\partial C}{\partial H} \quad \text{(since } \psi \text{ depends only on } C\text{)} \tag{5.33}$$

Thus:

$$\frac{\partial C}{\partial t} - \frac{\mathrm{d}\psi}{\mathrm{d}C}\cdot\frac{\partial C}{\partial H} = 0 \tag{5.34}$$

In general, the concentration of particles will be a function of position and time and thus:

$$C = \mathrm{f}(H, t)$$

and:

$$\mathrm{d}C = \frac{\partial C}{\partial H}\mathrm{d}H + \frac{\partial C}{\partial t}\mathrm{d}t$$

Conditions of constant concentration are therefore defined by:

$$\frac{\partial C}{\partial H}\mathrm{d}H + \frac{\partial C}{\partial t}\mathrm{d}t = 0$$

Thus:

$$\frac{\partial C}{\partial H} = -\frac{\partial C}{\partial t}\bigg/\frac{\mathrm{d}H}{\mathrm{d}t} \tag{5.35}$$

Substituting in equation 5.34 gives the following relation for constant concentration:

$$\frac{\partial C}{\partial t} - \frac{\mathrm{d}\psi}{\mathrm{d}C}\left(-\frac{\partial C}{\partial t}\bigg/\frac{\mathrm{d}H}{\mathrm{d}t}\right) = 0$$

or:

$$-\frac{\mathrm{d}\psi}{\mathrm{d}C} = \frac{\mathrm{d}H}{\mathrm{d}t} = u_w \tag{5.36}$$

Since equation 5.36 refers to a constant concentration, $\mathrm{d}\psi/\mathrm{d}C$ is constant and $u_w(= \mathrm{d}H/\mathrm{d}t)$ is therefore also constant for any given concentration and is the velocity of propagation of a zone of constant concentration $C$. Thus lines of constant slope, on a plot of $H$ versus $t$, will refer to zones of constant composition each of which will be propagated at a constant rate, dependent only on the concentration. Then since $u_w = -(\mathrm{d}\psi/\mathrm{d}C)$ (equation 5.36) when $\mathrm{d}\psi/\mathrm{d}C$ is negative (as it is at volumetric concentrations greater than 0.02 in Figure 5.4), $u_w$ is positive and the wave will propagate upwards. At lower concentrations $\mathrm{d}\psi/\mathrm{d}C$ is positive and the wave will propagate downwards. Thus, waves originating at the base of the sedimentation column will propagate upwards to the suspension interface if $\mathrm{d}\psi/\mathrm{d}C$ is negative, but will be prevented from propagating if $\mathrm{d}\psi/\mathrm{d}C$ is positive because of the presence of the base. Although Kynch's arguments may be applied to any suspension in which the initial concentration increases continuously from top to bottom, consideration will be confined to suspensions initially of uniform concentration.

In an initially uniform suspension of concentration $C_0$, the interface between the suspension and the supernatant liquid will fall at a constant rate until a zone of composition, greater than $C_0$, has propagated from the bottom to the free surface. The sedimentation rate will then fall off progressively as zones of successively greater concentrations reach the surface, until eventually sedimentation will cease when the $C_{max}$ zone reaches the surface. This assumes that the propagation velocity decreases progressively with increase of concentration.

However, if zones of higher concentration propagate at velocities greater than those of lower concentrations, they will automatically overtake them, giving rise to a sudden discontinuity in concentration. In particular, if the propagation velocity of the suspension of maximum possible concentration $C_{max}$ exceeds that of all of the intermediate concentrations between $C_0$ and $C_{max}$, sedimentation will take place at a constant rate, corresponding to the initial uniform concentration $C_0$, and will then cease abruptly as the concentration at the interface changes from $C_0$ to $C_{max}$.

Since the propagation velocity $u_w$ is equal to $-(\mathrm{d}\psi/\mathrm{d}C)$, the sedimentation behaviour will be affected by the shape of the plot of $\psi$ versus $C$. If this is consistently concave to the time-axis, $\mathrm{d}\psi/\mathrm{d}C$ will become increasingly negative as $C$ increases, $u_w$ will increase monotonically, and consequently there will be a discontinuity because the rate of propagation of a zone of concentration $C_{max}$ exceeds that for all lower concentrations; this is the condition referred to in the previous paragraph. On the other hand, if there is a point of inflexion in the curve (as at $C = 0.033$ in Figure 5.4), the propagation rate will increase progressively up to the condition given by this point of inflexion (concentration $C_i$) and will then decrease as the concentration is further increased. There will again be a discontinuity, although this time when the wave corresponding to concentration $C_i$ reaches the interface. The sedimentation rate will then fall off gradually as zones of successively higher concentration reach the interface, and sedimentation will finally cease when the concentration at the interface reaches $C_{max}$.

It is possible to apply this analysis to obtain the relationship between flux of solids and concentration over the range where $-(\mathrm{d}\psi/\mathrm{d}C)$ is decreasing with increase of concentration, using the results of a single sedimentation test. By taking a suspension of initial concentration $C_0(\geq C_i)$, it is possible to obtain the $\psi - C$ curve over the concentration range $C_0$ to $C_{max}$ from a single experiment. Figure 5.10 shows a typical sedimentation curve for such a suspension. The $H$-axis represents the initial condition ($t = 0$) and lines such as KP, OB representing constant concentrations have slopes of $u_w(= \mathrm{d}H/\mathrm{d}t)$. Lines from all points between A and O corresponding to the top and bottom of the suspension,

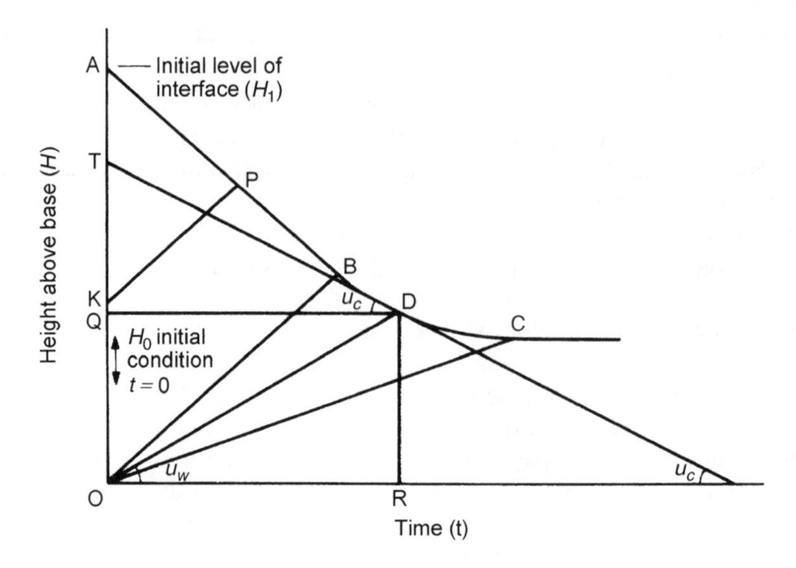


Figure 5.10. Construction for Kynch theory

respectively, will be parallel because the concentration is constant and their location and slope are determined by the initial concentration of the suspension. As solids become deposited at the bottom the concentration there will rapidly rise to the maximum possible value $C_{max}$ (ignoring the effects of possible sediment consolidation) and the line OC represents the line of constant concentration $C_{max}$. Other lines, such as OD of greater slope, all originate at the base of the suspension and correspond to intermediate concentrations.

Considering a line such as KP which refers to the propagation of a wave corresponding to the initial uniform composition from an initial position K in the suspension, this line terminates on the curve ABDC at P which is the position of the top interface of the suspension at time $t$. The location of P is determined by the fact that KP represents the upward propagation of a zone of constant composition at a velocity $u_w$ through which particles are falling at a sedimentation velocity $u_c$. Thus the total volume of particles passing per unit area through the plane in time $t$ is given by:

$$V = C_0(u_c + u_w)t \tag{5.37}$$

Since $P$ corresponds to the surface of the suspension, $V$ must be equal to the total volume of particles which was originally above the level indicated by K.

Thus: $$C_0(u_c + u_w)t = C_0(H_t - H_0)$$

or: $$(u_c + u_w)t = H_t - H_0 \tag{5.38}$$

Because the concentration of particles is initially uniform and the sedimentation rate is a function solely of the particle concentration, the line APB will be straight, having a slope $(-\mathrm{d}H/\mathrm{d}t)$ equal to $u_c$.

After point B, the sedimentation curve has a decreasing negative slope, reflecting the increasing concentration of solids at the interface. Line OD represents the locus of points of some concentration $C$, where $C_0 < C < C_{max}$. It corresponds to the propagation of a wave at a velocity $u_w$, from the bottom of the suspension. Thus when the wave reaches the interface, point D, all the particles in the suspension must have passed through the plane of the wave. Thus considering unit area:

$$C(u_c + u_w)t = C_0 H_t \tag{5.39}$$

In Figure 5.10:

$$H_t = \text{OA}$$

By drawing a tangent to the curve ABDC at D, the point T is located.

Then:

$$u_c t = \text{QT (since } -u_c \text{ is the slope of the curve at D)}$$

$$u_w t = \text{RD} = \text{OQ (since } u_w \text{ is the slope of line OD)}$$

and:

$$(u_c + u_w)t = \text{OT}$$

Thus the concentration $C$ corresponding to the line OD is given by:

$$C = C_0 \frac{\text{OA}}{\text{OT}} \tag{5.40}$$

and the corresponding solids flux is given by:

$$\psi = C u_c = C_0 \frac{\text{OA}}{\text{OT}} u_c \tag{5.41}$$

Thus, by drawing the tangent at a series of points on the curve BDC and measuring the corresponding slope $-u_c$ and intercept OT, it is possible to establish the solids flux $\psi$ for any concentration $C(C_i < C < C_{max})$.

It is shown in Section 5.3.3 that, for coarse particles, the point of inflexion does not occur at a concentration which would be obtained in practice in a suspension, and therefore the particles will settle throughout at a constant rate until an interface forms between the clear liquid and the sediment when sedimentation will abruptly cease. With a highly flocculated suspension the point of inflexion may occur at a very low volumetric concentration. In these circumstances, there will be a wide range of concentrations for which the constant rate sedimentation is followed by a period of falling rate.

### 5.2.4. The thickener

The thickener is the industrial unit in which the concentration of a suspension is increased by sedimentation, with the formation of a clear liquid. In most cases, the concentration of the suspension is high and hindered settling takes place. Thickeners may operate as batch or continuous units, and consist of tanks from which the clear liquid is taken off at the top and the thickened liquor at the bottom.

In order to obtain the largest possible throughput from a thickener of given size, the rate of sedimentation should be as high as possible. In many cases, the rate may be artificially increased by the addition of small quantities of an electrolyte, which causes precipitation of colloidal particles and the formation of flocs. The suspension is also frequently heated because this lowers the viscosity of the liquid, and encourages the larger particles in the suspension to grow in size at the expense of the more soluble small particles. Further, the thickener frequently incorporates a slow stirrer, which causes a reduction in the apparent viscosity of the suspension and also aids in the consolidation of the sediment.

The batch thickener usually consists of a cylindrical tank with a slightly conical bottom. After sedimentation has proceeded for an adequate time, the thickened liquor is withdrawn from the bottom and the clear liquid is taken off through an adjustable offtake pipe from the upper part of the tank. The conditions prevailing in the batch thickener are similar to those in the ordinary laboratory sedimentation tube, and during the initial stages there will generally be a zone in which the concentration of the suspension is the same as that in the feed.

The continuous thickener consists of a cylindrical tank with a flat bottom. The suspension is fed in at the centre, at a depth of from 0.3 to 1 m below the surface of the liquid, with as little disturbance as possible. The thickened liquor is continuously removed through an outlet at the bottom, and any solids which are deposited on the floor of the tank may be directed towards the outlet by means of a slowly rotating rake mechanism incorporating scrapers. The rakes are often hinged so that the arms fold up automatically if the torque exceeds a certain value; this prevents it from being damaged if it is overloaded. The raking action can increase the degree of thickening achieved in a thickener of given size. The clarified liquid is continuously removed from an overflow which runs round the whole of the upper edge of the tank. The solids are therefore moving continuously downwards, and then inwards towards the thickened liquor outlet; the liquid is moving upwards and radially outwards as shown in Figure 5.11. In general, there will be no region of constant composition in the continuous thickener.

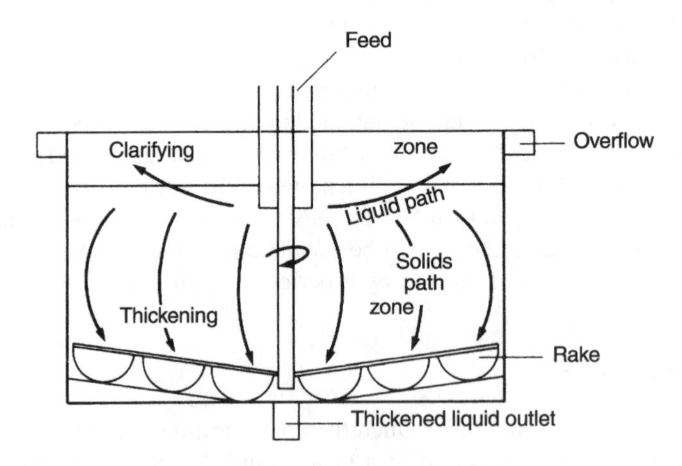


Figure 5.11. Flow in continuous thickener

Thickeners may vary from a few metres to several hundred metres in diameter. Small ones are made of wood or metal and the rakes rotate at about 0.02 Hz (1 rpm). Very large thickeners generally consist of large concrete tanks, and the stirrers and rakes are driven by means of traction motors which drive on a rail running round the whole circumference; the speed of rotation may be as low as 0.002 Hz (0.1 rpm).

The thickener has a twofold function. First, it must produce a clarified liquid, and therefore the upward velocity of the liquid must, at all times, be less than the settling velocity of the particles. Thus, for a given throughput, the clarifying capacity is determined by the diameter of the tank. Secondly, the thickener is required to produce a given degree of thickening of the suspension. This is controlled by the time of residence of the particles in the tank, and hence by the depth below the feed inlet.

There are therefore two distinct requirements in the design — first, the provision of an adequate diameter to obtain satisfactory clarification and, secondly, sufficient depth to achieve the required degree of thickening of the underflow. Frequently, the high diameter : height ratios which are employed result in only the first requirement being adequately met. The Dorr thickener is an example of a relatively shallow equipment employing a rake mechanism. In order to save ground space, a number of trays may be mounted above one another and a common drive shaft employed. A four-tray thickener is illustrated in Figure 5.12; this type of equipment generally gives a better performance for clarification than for thickening.

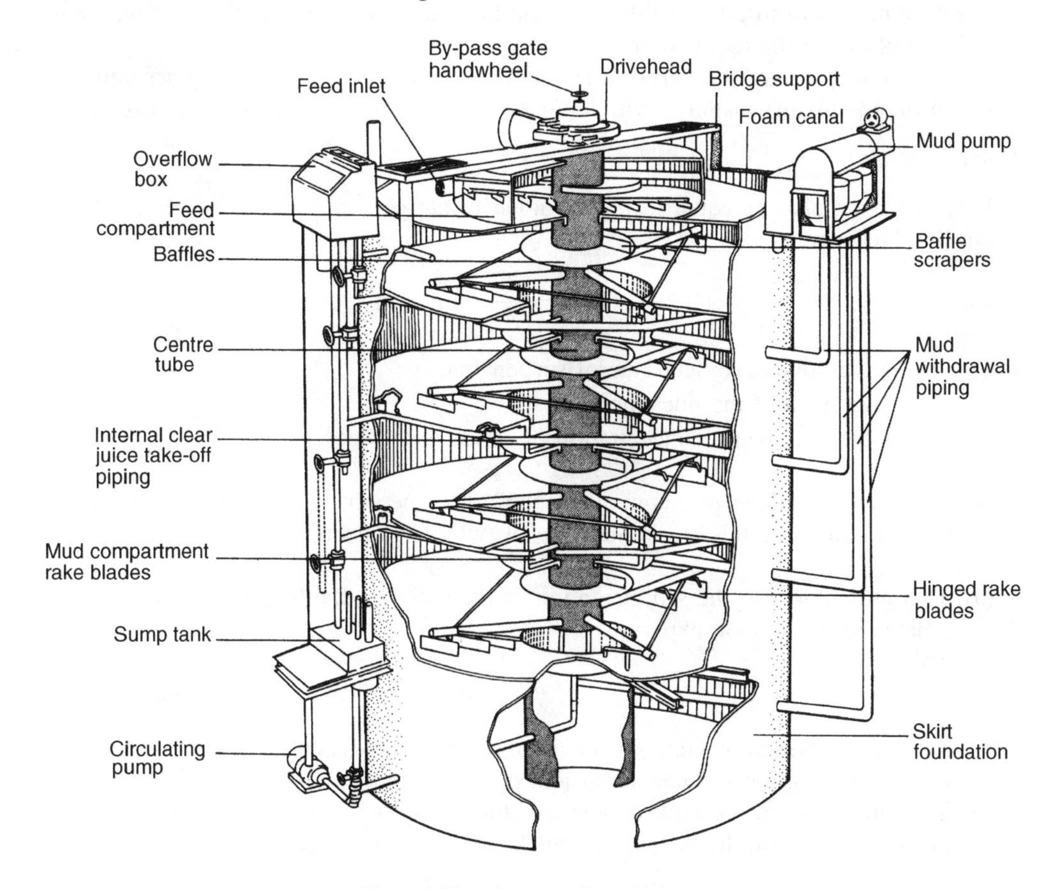


Figure 5.12. Four-tray Dorr thickener

The satisfactory operation of the thickener as a clarifier depends upon the existence of a zone of negligible solids content towards the top. In this zone conditions approach those under which free settling takes place, and the rate of sedimentation of any particles which have been carried to this height is therefore sufficient for them to settle against the upward current of liquid. If this upper zone is too shallow, some of the smaller particles may escape in the liquid overflow. The volumetric rate of flow of liquid upwards through the clarification zone is equal to the difference between the rate of feed of liquid in the slurry and the rate of removal in the underflow. Thus the required concentration of solids in the underflow, as well as the throughput, determines the conditions in the clarification zone.

### Thickening zone

In a continuous thickener, the area required for thickening must be such that the total solids flux (volumetric flowrate per unit area) at any level does not exceed the rate at which the solids can be transmitted downwards. If this condition is not met, solids will build up and steady-state operation will not be possible. If no solids escape in the overflow, this flux must be constant at all depths below the feed point. In the design of a thickener, it is therefore necessary to establish the concentration at which the total flux is a *minimum* in order to calculate the required area.

The total flux $\psi_T$ may be expressed as the product of the volumetric rate per unit area at which thickened suspension is withdrawn ($u_u$) and its volumetric concentration $C_u$.

Thus:
$$\psi_T = u_u C_u \tag{5.42}$$

This flux must also be equal to the volumetric rate per unit area at which solids are fed to the thickener.

Thus:
$$\psi_T = \frac{Q_0}{A} C_0 \tag{5.43}$$

where: $Q_0$ is the volumetric feed rate of suspension,
$A$ is the area of the thickener, and
$C_0$ is the volumetric concentration of solids in the feed.

At *any horizontal plane* in a continuous thickener operating under steady-state conditions, the total flux of solids $\psi_T$ is made up of two components:

(a) That attributable to the sedimentation of the solids in the liquid — as measured in a batch sedimentation experiment.
This is given by:
$$\psi = u_c C \qquad \text{(equation 5.31)}$$
where $u_c$ is the sedimentation velocity of solids at a concentration $C$. $\psi$ corresponds to the flux in a batch thickener at that concentration.

(b) That arising from the bulk downward flow of the suspension which is drawn off as underflow from the base of the thickener which is given by:
$$\psi_u = u_u C \tag{5.44}$$
Thus the total flux:
$$\psi_T = \psi + \psi_u = \psi + u_u C \tag{5.45}$$

Figure 5.13 shows a typical plot of sedimentation flux $\psi$ against volumetric concentration $C$; this relationship needs to be based on experimental measurements of $u_c$ as a function of $C$. The curve must always pass through a maximum, and usually exhibits a point of inflexion at high concentrations. At a given withdrawal rate per unit area ($u_u$) the bulk flux ($\psi_u$) is given by a straight line, of slope $u_u$, passing through the origin as given by equation 5.44. The total solids flux $\psi_T$, obtained as the summation of the two curves, passes through a maximum, followed by a minimum ($\psi_{TL}$) at a higher concentration ($C_L$). For all concentrations exceeding $C_M$ (shown in Figure 5.13), $\psi_{TL}$ is the parameter which determines the capacity of the thickener when operating at the fixed withdrawal rate $u_u$.

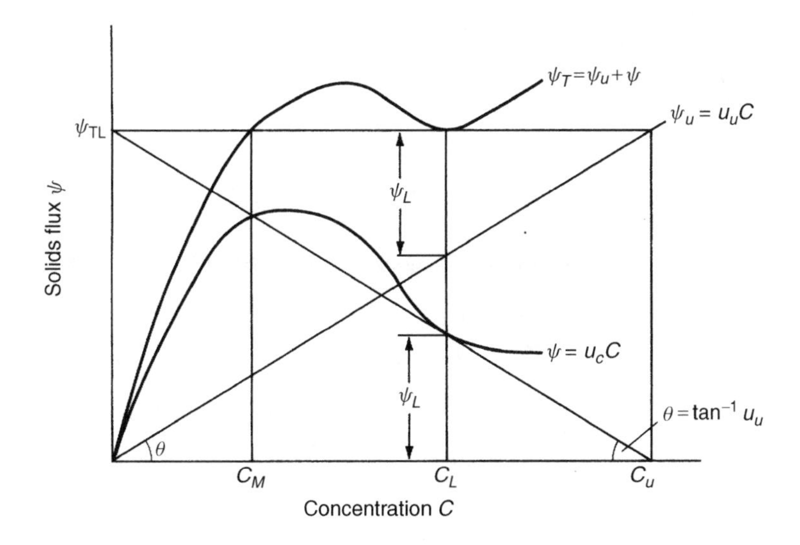


Figure 5.13. Solids fluxes as functions of concentration and Yoshioka construction[47]

It may be noted that, because no further sedimentation occurs below the level of the exit of the thickener, there will at that position be a discontinuity, and the solids concentration will undergo a step change from its value in the suspension to that in the underflow ($C_u$).

It is now necessary to determine the limiting total flux $\psi_{TL}$ for a specified concentration $C_u$ of overflow. The required area of the thickener is then obtained by substituting this value into equation 5.43 to give:

$$A = \frac{Q_0 C_0}{\psi_{TL}} \tag{5.46}$$

A simple construction, proposed by YOSHIOKA *et al.*[47] has been described by HASSETT[48] amongst others. By differentiation of equation 5.45 at a fixed value of $u_u$:

$$\frac{\partial \psi_T}{\partial C} = \frac{\partial \psi}{\partial C} + u_u \tag{5.47}$$

The minimum value of $\psi_T(=\psi_{TL})$ occurs when $\dfrac{\partial \psi_T}{\partial C} = 0$;

that is when:

$$\frac{\partial \psi}{\partial C} = -u_u \tag{5.48}$$

If a tangent is drawn from the point on the abscissa corresponding to the required underflow concentration $C_u$, it will meet the $\psi$ curve at a concentration value $C_L$ at which $\psi_T$ has the minimum value $\psi_{TL}$ and will intersect the ordinate axis at a value equal to $\psi_{TL}$. The construction is dependent on the fact that the slopes of the tangent and of the $\psi_u$ line are equal and opposite ($\mp u_u$). Thus, in order to determine both $C_L$ and $\psi_{TL}$, it is not necessary to plot the total-flux curve ($\psi_T$), but only to draw the tangent to the batch sedimentation ($\psi$) curve. The value of $\psi_{TL}$ determined in this way is then inserted in equation 5.46 to obtain the required area $A$.

From the geometry of Figure 5.13:

$$\frac{\psi_{TL}}{C_u} = \frac{\psi_L}{C_u - C_L} \tag{5.49}$$

where $\psi_L$ is the value of $\psi$ at the concentration $C_L$.

Since $\psi_L = u_{cL} C_L$ (from equation 5.31), again from equation 5.46:

$$A = \frac{Q_0 C_0}{\psi_{TL}}$$

$$= Q_0 C_0 \left[ \frac{\dfrac{1}{C_L} - \dfrac{1}{C_u}}{u_{cL}} \right] \tag{5.50}$$

where $u_{cL}$ is the value of $u_c$ at the concentration $C_L$. Thus, the minimum necessary area of the thickener may be obtained from the maximum value of

$$\left[ \frac{\dfrac{1}{C} - \dfrac{1}{C_u}}{u_c} \right] \text{ which is designated } \left[ \frac{\dfrac{1}{C} - \dfrac{1}{C_u}}{u_c} \right]_{\max}.$$

Concentrations may also be expressed as mass per unit volume (using $c$ in place of $C$) to give:

$$A = Q_0 c_0 \left[ \frac{[(1/c) - (1/c_u)]}{u_c} \right]_{\max} \tag{5.51}$$

## *Overflow*

The liquid flowrate in the overflow ($Q'$) is the difference between that in the feed and in the underflow.

Thus:
$$Q' = Q_0(1 - C_0) - (Q_0 - Q')(1 - C_u)$$

or:
$$\frac{Q'}{Q_0} = 1 - \frac{C_0}{C_u} \tag{5.52}$$

At any depth below the feed point, the upward liquid velocity must not exceed the settling velocity of the particles. ($u_c$) Where the concentration is $C$, the required area is therefore given by:

$$A = Q_0 \frac{1}{u_c}\left[1 - \frac{C}{C_u}\right] \tag{5.53}$$

It is therefore necessary to calculate the *maximum* value of $A$ for all the values of $C$ which may be encountered.

Equation 5.53 can usefully be rearranged in terms of the mass ratio of liquid to solid in the feed ($Y$) and the corresponding value ($U$) in the underflow to give:

$$Y = \frac{1 - C}{C}\frac{\rho}{\rho_s} \quad \text{and} \quad U = \frac{1 - C_u}{C_u}\frac{\rho}{\rho_s}$$

Then:
$$C = \frac{1}{1 + Y(\rho_s/\rho)} \qquad C_u = \frac{1}{1 + U(\rho_s/\rho)}$$

and:
$$A = \frac{Q_0}{u_c}\left\{1 - \frac{1 + U(\rho_s/\rho)}{1 + Y(\rho_s/\rho)}\right\}$$

$$= \frac{Q_0(Y - U)C\rho_s}{u_c\rho} \tag{5.54}$$

The values of $A$ should be calculated for the whole range of concentrations present in the thickener, and the design should then be based on the maximum value so obtained.

The above procedure for the determination of the required cross-sectional area of a continuous thickener is illustrated in Example 5.2

Great care should be used, however, in applying the results of batch sedimentation tests carried out in the laboratory to the design of large continuous industrial thickeners as the conditions in the two cases are different. In a batch experiment, the suspension is initially well-mixed and the motion of both fluid and particles takes place in the vertical direction only. In a continuous thickener, the feed is normally introduced near the centre at some depth (usually between 0.3 and 1 metre) below the free surface, and there is a significant radial velocity component in the upper regions of the suspension. The liquid flows predominantly outwards and upwards towards the overflow; this is generally located round the whole of the periphery of the tank. In addition, the precise design of the feed device will exert some influence on the flow pattern in its immediate vicinity.

### *Underflow*

In many operations the prime requirement of the thickener is to produce a thickened product of as high a concentration as possible. This necessitates the provision of sufficient depth to allow time for the required degree of consolidation to take place; the critical dimension is the vertical distance between the feed point and the outlet at the base of the tank. In most cases, the time of compression of the sediment will be large compared with the time taken for the critical settling conditions to be reached.

The time required to concentrate the sediment after it has reached the critical condition can be determined approximately by allowing a sample of the slurry at its critical composition to settle in a vertical glass tube, and measuring the time taken for the interface between the sediment and the clear liquid to fall to such a level that the concentration is that required in the underflow from the thickener. The use of data so obtained assumes that the average concentration in the sediment in the laboratory test is the same as that which would be obtained in the thickener after the same time. This is not quite so because, in the thickener, the various parts of the sediment have been under compression for different times. Further, it assumes that the time taken for the sediment to increase in concentration by a given amount is independent of its depth.

The top surface of the suspension should always be sufficiently far below the level of the overflow weir to provide a clear zone deep enough to allow any entrained particles to settle out. As they will be present only in very low concentrations, they will settle approximately at their terminal falling velocities. Provided that this requirement is met, the depth of the thickener does not have any appreciable effect on its clarifying capacity.

However, it is the depth of the thickener below the clarifying zone that determines the residence time of the particles and the degree of thickening which is achieved at any give throughput. COMINGS[49] has carried out an experimental study of the effect of underflow rate on the depth of the thickening, (compression) zone and has concluded that it is the retention time of the particles within the thickening zone, rather than its depth *per se*, which determines the underflow concentration.

An approximate estimate of the mean residence time of particles in the thickening zone may be made from the results of a batch settling experiment on a suspension which is of the *critical concentration* (the concentration at which the settling rate starts to fall off with time). Following a comprehensive experimental programme of work, ROBERTS[50] has found that the rate of sedimentation $\mathrm{d}H/\mathrm{d}t$ decreases linearly with the difference between its height at any time $t$ and its ultimate height $H_\infty$ which would be achieved after an infinite time of settling:

Thus:
$$-\frac{\mathrm{d}H}{\mathrm{d}t} = k(H - H_\infty) \tag{5.55}$$

where $k$ is a constant.

On integration:
$$\ln\frac{H - H_\infty}{H_c - H_\infty} = -kt \tag{5.56}$$

or:
$$\frac{H - H_\infty}{H_c - H_\infty} = \mathrm{e}^{-kt} \tag{5.57}$$

where: $t$ is time from the start of the experiment,

$H$ is the height of the interface at time $t$, and

$H_c$ is the initial height corresponding to the critical height at which the constant settling rate gives way to a diminishing rate.

$H_\infty$, which can be seen from equation 5.57 to be approached exponentially, cannot be estimated with any precision, and a trial and error method must be used to determine at what particular value the plot of the left hand side of equation 5.56 against $t$ yields a straight line. The slope of this line is equal to $-k$, which is constant for any particular suspension.

If the required fractional volumetric concentration in the underflow is $C_u$, and $C_c$ is the value of the critical concentration, a simple material balance gives the value of $H_u$ the height of the interface when the concentration is $C_u$.

Thus:
$$H_u = H_c \frac{C_c}{C_u} \tag{5.58}$$

The corresponding value of the residence time $t_R$ to reach this condition is obtained by substituting from equation 5.58 into equation 5.56, giving:

$$\begin{aligned} t_R &= \frac{1}{k} \ln \frac{H_c - H_\infty}{H_u - H_\infty} \\ &= \frac{1}{k} \ln \frac{H_c - H_\infty}{H_c(C_c/C_u) - H_\infty} \end{aligned} \tag{5.59}$$

Thus, equation 5.59 may be used to calculate the time required for the concentration to be increased to such a value that the height of the suspension is reduced from $H_0$ to a desired value $H_u$ at which height the concentration corresponds with the value $C_u$ required in the underflow from the thickener.

In a batch sedimentation experiment, the sediment builds up gradually and the solids which are deposited in the early stages are those which are subjected to the compressive forces for the longest period of time. In the continuous thickener, on the other hand, all of the particles are retained for the same length of time with fresh particles continuously being deposited at the top of the sediment and others being removed at the same rate in the underflow, with the inventory thus remaining constant. Residence time distributions are therefore not the same in batch and continuous systems. Therefore, the value of $t_R$ calculated from equation 5.59 will be subject to some inaccuracy because of the mismatch between the models for batch and continuous operation.

An approximate value for the depth of the thickening zone is then found by adding the volume of the liquid in the sediment to the corresponding volume of solid, and dividing by the area which has already been calculated, in order to determine the clarifying capacity of the thickener. The required depth of the thickening region is thus:

$$\left\{ \frac{W t_R}{A \rho_s} + W \frac{t_R}{A \rho} X \right\} = \frac{W t_R}{A \rho_s} \left( 1 + \frac{\rho_s}{\rho} X \right) \tag{5.60}$$

where: $t_R$ is the required time of retention of the solids, as determined experimentally,
$W$ is the mass rate of feed of solids to the thickener,
$X$ is the average value of the mass ratio of liquid to solids in the thickening portion, and
$\rho$ and $\rho_s$ are the densities of the liquid and solid respectively.

This method of design is only approximate and therefore, in a large tank, about 1 metre should be added to the calculated depth as a safety margin and to allow for the depth required for the suspension to reach the critical concentration. In addition, the bottom of the tanks may be slightly pitched to assist the flow of material towards the thickened liquor outlet.

The use of slowly rotating rakes is beneficial in large thickeners to direct the underflow to the central outlet at the bottom of the tank. At the same time the slow motion of the rakes tends to give a gentle agitation to the sediment which facilitates its consolidation and water removal. The height requirement of the rakes must be added to that needed for the thickening zone to achieve the desired underflow concentration.

Additional height (up to 1 m) should also be allowed for the depth of submergence of the feed pipe and to accommodate fluctuations in the feed rate to the thickener (*ca.* 0.5 m).

The limiting operating conditions for continuous thickeners has been studied by a number of workers including TILLER and CHEN[51], and the height of the compression zone has been the subject of a paper by FONT[52].

A paper by FITCH[53] gives an analysis of existing theories and identifies the domains in which the various models which have been proposed give reasonable approximations to observed behaviour.

The importance of using deep thickeners for producing thickened suspensions of high concentrations has been emphasised by DELL and KELEGHAN[54] who used a tall tank in the form of an inverted cone. They found that consolidation was greatly facilitated by the use of stirring which created channels through the sediment for the escape of water in an upwards direction, and eliminated frictional support of the solids by the walls of the vessel. The conical shape is clearly uneconomic for large equipment, however, because of the costs of both fabrication and supports, and because of the large area at the top.

CHANDLER[55] has reported on the use of deep cylindrical tanks of height to diameter ratios of 1.5 and up to 19 m in height, fitted with steep conical bases of half angles of about 30°. Using this method for the separation of "red mud" from caustic liquors in the aluminium industry, it was found possible to dispense with the conical section because the slurry tended to form its own cone at the bottom of the tank as a result of the build up of stagnant solids. Rakes were not used because of the very severe loading which would have been imposed at the high concentrations achieved in the sediments. With flocculated slurries, the water liberated as the result of compaction was able to pass upwards through channels in the sediment—an essential feature of the operation because the resistance to flow through the sediment itself would have been much too large to permit such a high degree of thickening. The system was found to be highly effective, and much more economic than using large diameter tanks where it has been found that the majority of the solids tend to move downwards through the central zone, leaving the greater part of the cross-section ineffective.

## Example 5.1

A slurry containing 5 kg of water/kg of solids is to be thickened to a sludge containing 1.5 kg of water/kg of solids in a continuous operation. Laboratory tests using five different concentrations of the slurry yielded the following data:

| Concentration (kg water/kg solid) | 5.0 | 4.2 | 3.7 | 3.1 | 2.5 |
|---|---|---|---|---|---|
| Rate of sedimentation (mm/s) | 0.20 | 0.12 | 0.094 | 0.070 | 0.050 |

Calculate the minimum area of a thickener required to effect the separation of a flow of 1.33 kg/s of solids.

## Solution

*Basis: 1 kg solids*

$$\text{Mass rate of feed of solids} = 1.33\ \text{kg/s}$$

1.5 kg water is carried away in the underflow, with the balance in the overflow. Thus, $U = 1.5$ kg water/kg solids

| concentration $(Y)$ (kg water/kg solids) | water to overflow $(Y - U)$ (kg water/kg solids) | sedimentation rate $u_c$ (m/s) | $\dfrac{(Y-U)}{u_c}$ (s/m) |
|---|---|---|---|
| 5.0 | 3.5 | $2.00 \times 10^{-4}$ | $1.75 \times 10^4$ |
| 4.2 | 2.7 | $1.20 \times 10^{-4}$ | $2.25 \times 10^4$ |
| 3.7 | 2.2 | $0.94 \times 10^{-4}$ | $2.34 \times 10^4$ |
| 3.1 | 1.6 | $0.70 \times 10^{-4}$ | $2.29 \times 10^4$ |
| 2.5 | 1.0 | $0.50 \times 10^{-4}$ | $2.00 \times 10^4$ |

Maximum value of $\dfrac{(Y-U)}{u_c} = 2.34 \times 10^4$ s/m.

From equation 5.54:

$$A = \left(\frac{Y-U}{u_c}\right)\left(\frac{QC\rho_s}{\rho}\right)$$

$QC\rho_s = 1.33$ kg/s, and taking $\rho$ as 1000 kg/m$^3$, then:

$$A = 2.34 \times 10^4 \times \left(\frac{1.33}{1000}\right)$$

$$= \underline{\underline{31.2\ \text{m}^2}}$$

## Example 5.2

A batch test on the sedimentation of a slurry containing 200 kg solids/m$^3$ gave the results shown in Figure 5.14 for the position of the interface between slurry and clear liquid as a function of time.

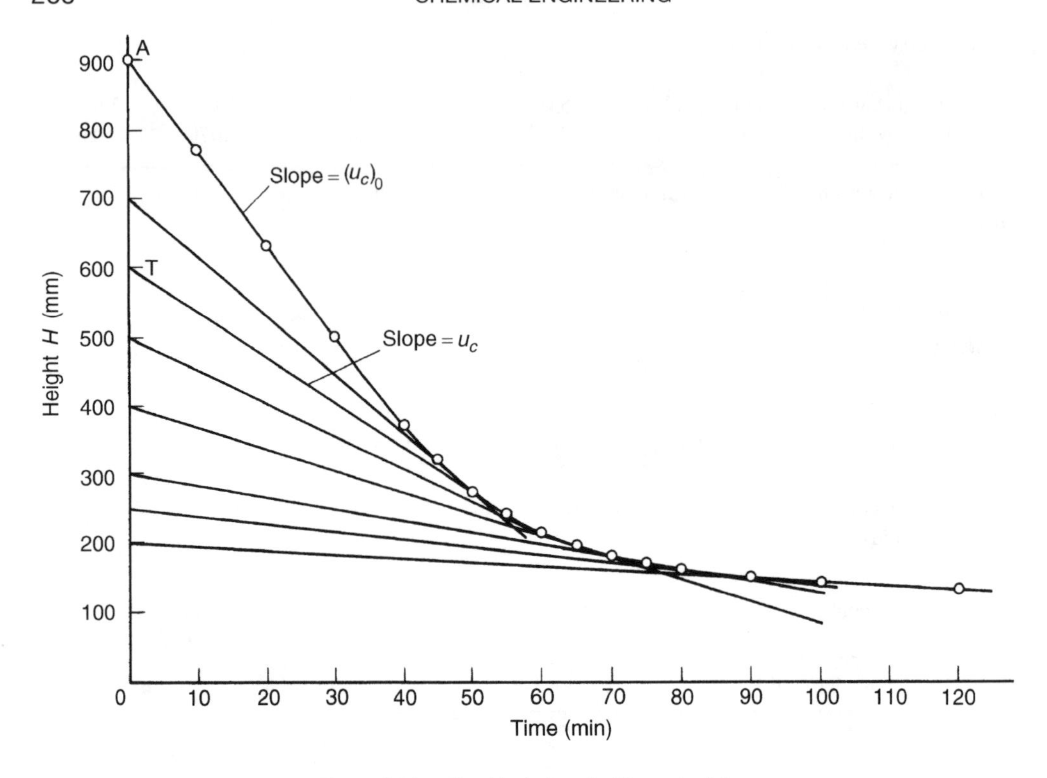


Figure 5.14. Graphical data for Example 5.2

Using the Kynch theory, tabulate the sedimentation velocity and solids flux due to sedimentation as a function of concentration. What area of tank will be required to give an underflow concentration of 1200 kg/m$^3$ for a feed rate of 2 m$^3$/min of slurry?

## Solution

On the diagram the height $H$ above the base is plotted against time. The initial height is given by OA (90 cm) and the initial constant slope of the curve gives the sedimentation velocity $(u_c)_0$ for a concentration of 200 kg/m$^3$ ($c_0$).

For some other height, such as OT (600 mm), the slope of the tangent gives the sedimentation velocity $u_c$ for a concentration of:

$$c = c_0 \frac{\text{OA}}{\text{OT}} = 200 \times \frac{\text{OA}}{\text{OT}} \text{ kg/m}^3 \qquad \text{(from equation 5.40)}$$

Thus, for each height, the corresponding concentration may be calculated and the slope of the tangent measured to give the sedimentation velocity. The solids flux in kg/m$^2$s is then:

$$c(\text{kg/m}^3) \times u_c(\text{mm/min}) \times \frac{1}{1000 \times 60}$$

These quantities are tabulated as follows:

| $H$ (mm) | $c$ (kg/m$^3$) | $u_c$ (mm/min) | Sedimentation flux (kg/m$^2$s) | $1000\left(\frac{1}{c}-\frac{1}{c_u}\right)=x$ (m$^3$/kg × 10$^3$) | $\frac{u_c}{(1/c)-(1/c_u)}$ (kg/m$^2$s) | $\frac{(1/c)-(1/c_u)}{u_c}$ (m$^2$/s kg) |
|---|---|---|---|---|---|---|
| 900 | 200 | 13.4 | 0.0447 | 4.167 | 0.0536 | 18.7 |
| 800 | 225 | 10.76 | 0.0403 | 3.611 | 0.0497 | 20.1 |
| 700 | 257 | 8.6 | 0.0368 | 3.058 | 0.0468 | 21.4 |
| 600 | 300 | 6.6 | 0.0330 | 2.500 | 0.0440 | 22.7 |
| 500 | 360 | 4.9 | 0.0294 | 1.944 | 0.0420 | 23.8 |
| 400 | 450 | 3.2 | 0.0240 | 1.389 | 0.0383 | 26.1 |
| 300 | 600 | 1.8 | 0.0180 | 0.833 | 0.0360 | 27.8 |
| 260 | 692 | 1.21 | 0.0140 | 0.612 | 0.0330 | 30.3 |
| 250 | 720 | 1.11 | 0.0133 | 0.556 | 0.0333 | 30.0 |
| 220 | 818 | 0.80 | 0.0109 | 0.389 | 0.0342 | 29.2 |
| 200 | 900 | 0.60 | 0.0090 | 0.278 | 0.0358 | 27.9 |
| 180 | 1000 | 0.40 | 0.0067 | 0.167 | 0.0398 | 25.1 |

Now $c_u = 1200$ kg/m$^3$ and the maximum value of $\dfrac{(1/c)-(1/c_u)}{u_c} \approx 30.3$ m$^2$/kg.

From equation 5.51, the area $A$ required is given by:

$$A = Q_0 c_0 \left[\frac{(1/c - (1/c_u))}{u_c}\right]_{max}$$

$$= \frac{2}{60} \times 200 \times 30.3$$

$$= \underline{\underline{202 \text{ m}^2}}$$

## 5.3. SEDIMENTATION OF COARSE PARTICLES

### 5.3.1. Introduction

Coarse particles have a much lower specific surface, and consequently surface forces and electrical interactions between particles are of very much less significance than in the fine particle systems considered in the previous sections. Flocculation will be absent and, generally, the particles will not influence the rheology of the liquid. The dividing point between coarse and fine particles is somewhat arbitrary although is of the order of 0.1 mm (100 μm).

There has been considerable discussion in the literature as to whether the buoyancy force acting on a particle in a suspension is determined by the density of the liquid or by that of the suspension. For the sedimentation of a coarse particle in a suspension of much finer particles, the suspension behaves effectively as a continuum. As the large particle settles, it displaces an equal volume of suspension in which there will be comparatively little relative movement between the fine particles and the liquid. The effective buoyancy may, in these circumstances, be attributed to a fluid of the same density as the suspension of fines. On the other hand, the pressure distribution around a small particle settling in a suspension of much coarser particles, will hardly be affected by the presence of the large particles, which will therefore contribute little to the buoyancy force. If the surrounding particles are all of a comparable size, an intermediate situation will exist.

In a sedimenting suspension, however, the gravitational force on any individual particle is balanced by a combination of buoyancy and fluid friction forces, both of which are influenced by the flow field and pressure distribution in the vicinity of the particle. It is a somewhat academic exercise to apportion the total force between its two constituent parts, both of which are influenced by the presence of neighbouring particles, the concentration of which will determine the magnitude of their effect.

### 5.3.2. Suspensions of uniform particles

Several experimental and analytical studies have been made of the sedimentation of uniform particles. The following approach is a slight modification of that originally used by RICHARDSON and ZAKI[56].

For a single spherical particle settling at its terminal velocity $u_0$, a force balance gives:

$$R_0' \frac{\pi}{4} d^2 = \frac{\pi}{6} d^3 (\rho_s - \rho) g \tag{5.61}$$

or:

$$R_0' = \frac{2}{3} d (\rho_s - \rho) g \tag{5.62}$$

(see equation 3.33)

where: $R_0'$ is the drag force per unit projected area of particle,
$d$ is the particle diameter,
$\rho_s$ is the density of the particle,
$\rho$ is the density of the liquid, and
$g$ is the acceleration due to gravity

$R'$ has been shown in Chapter 3 to be determined by the diameter of the particle, the viscosity and density of the liquid, and the velocity of the particle relative to the liquid. Thus, at the terminal falling condition:

$$R_0' = \mathrm{f}(\rho, \mu, u_0, d)$$

In a tube of diameter $d_t$, the wall effects may be significant and:

$$R_0' = \mathrm{f}_1(\rho, \mu, u_0, d, d_t) \tag{5.63}$$

In a concentrated suspension, the drag force on a particle will be a function of its velocity $u_p$ relative to the liquid and will be influenced by the concentration of particles; that is, it will be a function of the voidage $e$ of the suspension.

Thus:

$$R_0' = \mathrm{f}_2(\rho, \mu, u_p, d, d_t, e) \tag{5.64}$$

The argument holds equally well if the buoyancy force is attributable to the density of the suspension as opposed to that of the liquid. The force balance equation (5.61) then becomes:

$$R'_{0s}\frac{\pi}{4}d^2 = \frac{\pi}{6}d^3(\rho_s - \rho_c)g \qquad \text{(from equation 5.2)}$$

$$= \frac{\pi}{6}d^3(\rho_s - \rho)eg \qquad (5.65)$$

where $\rho_c$ is the density of the suspension.

Comparing equations 5.65 and 5.61:

$$R'_{0s} = eR'_0 \qquad (5.66)$$

Thus, no additional variable would be introduced into equation 5.64.

Rearranging equations 5.63 and 5.64 to give $u_0$ and $u_p$ explicitly:

$$u_0 = \mathrm{f}_3(R'_0, \rho, \mu, d, d_t) \qquad (5.67)$$

$$u_p = \mathrm{f}_4(R'_0, \rho, \mu, d, d_t, e) \qquad (5.68)$$

Dividing equation 5.68 by equation 5.67 gives:

$$\frac{u_p}{u_0} = \mathrm{f}_5(R'_0, \rho, \mu, d, d_t, e) \qquad (5.69)$$

since all the variables in function $\mathrm{f}_3$ also appear in function $\mathrm{f}_4$.

Since $u_p/u_0$ is dimensionless function, $f_5$ must also be dimensionless and equation 5.64 becomes:

$$\frac{u_p}{u_0} = \mathrm{f}_5\left(\frac{R'_0 d^2\rho}{\mu^2}, \frac{d}{d_t}, e\right)$$

because $R'_0 d^2\rho/\mu^2$ is the only possible dimensionless derivative of $R'_0$.

The sedimentation velocity $u_c$ of the particles relative to the walls of the containing vessel will be less than the velocity $u_p$ of the particles relative to the fluid.

From equation 5.6: $$u_c = eu_p$$

Thus: $$\frac{u_c}{u_0} = \mathrm{f}_6\left(\frac{R'_0 d^2\rho}{\mu^2}, \frac{d}{d_t}, e\right)$$

From equation 5.62:

$$\frac{R'_0 d^2\rho}{\mu^2} = \frac{2}{3}\frac{d^3\rho(\rho_s - \rho)g}{\mu^2} \qquad \text{(see equation 3.36)}$$

$$= \frac{2}{3}Ga$$

where: $$Ga = \frac{d^3\rho(\rho_s - \rho)g}{\mu^2} \text{ is the Galileo number}$$

Thus: $$\frac{u_c}{u_0} = f_7\left(Ga, \frac{d}{d_t}, e\right) \tag{5.70}$$

However: $$\frac{R_0' d^2 \rho}{\mu^2} = \frac{R_0'}{\rho u_0^2}\left(\frac{u_0 d\rho}{\mu}\right)^2$$

Since there is a unique relation between $R_0'/\rho u_0^2$ and $u_0 d\rho/\mu$ for spherical particles:

$$Ga = f_8\left(\frac{u_0 d\rho}{\mu}\right) = f_8(Re_0') \tag{5.71}$$

and: $$\frac{u_c}{u_0} = f_9\left(Re_0', \frac{d}{d_t}, e\right) \tag{5.72}$$

Equation 5.72 is just an alternative way of expressing equation 5.70.

Two special cases of equations 5.70 and 5.72 are now considered.

(a) *At low values of* $Re_0'(< ca.\,0.2)$. This corresponds to the Stokes' law region in which drag force, and hence $R_0'$, is independent of the density of the liquid. Equation 5.69 then becomes:

$$\frac{u_p}{u_0} = f_{10}(R_0', \mu, d, d_t, e) \tag{5.73}$$

Now $R_0'$, $\mu$ and $d$ cannot be formed into a dimensionless group and therefore, in this region, $u_p/u_0$ must be independent of $R_0'$ and $\mu$.

(b) Similarly, *at high values of* $Re_0'(> ca.\,500)$, the Newton's law region, drag force is independent of viscosity, and for similar reasons to (a) above, $u_p/u_0$ is independent of $R_0'$ and $\rho$.

Thus ($Re_0' < 0.2$ *and* $Re_0' > 500$) both equations 5.70 and 5.72 become:

$$\frac{u_c}{u_0} = f_{11}\left(\frac{d}{d_t}, e\right) \tag{5.74}$$

The form of the functions ($f_7$, $f_9$ and $f_{11}$) has been obtained from experimental results for both the sedimentation and the fluidisation (Chapter 6) of suspensions of uniform spheres. For a given suspension, a plot of log $u_c$ versus log $e$ does, in most cases, give a good straight line, as shown in Figure 5.15.

Thus: $$\log u_c = n \log e + \log u_i \tag{5.75}$$

where $n$ is the slope and log $u_i$ is the intercept at $e = 1$, that is at infinite dilution.

Hence: $$\frac{u_c}{u_i} = e^n \tag{5.76}$$

Now: $$\frac{u_i}{u_0} = f\left(\frac{d}{d_t}\right) \text{ only} \tag{5.77}$$

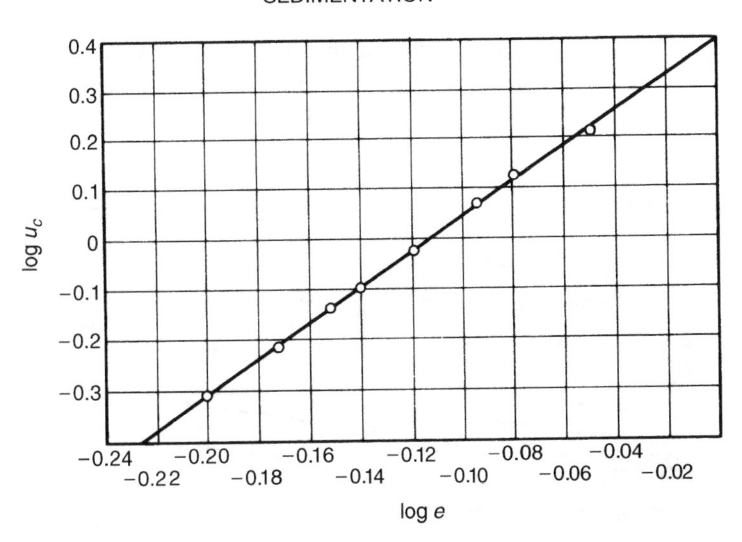


Figure 5.15. Sedimentation velocity as a function of voidage

since $u_0$ and $u_i$ are the free falling velocities in an infinite fluid and in a tube of diameter $d_t$, respectively.

Substituting from equations 5.76 and 5.77 in equation 5.70 gives:

$$e^n = \text{f}_{12}\left(Ga, \frac{d}{d_t}, e\right)$$

Since $n$ is found experimentally to be independent of $e$ (see Figure 5.15):

$$n = \text{f}_{13}\left(Ga, \frac{d}{d_t}\right) \tag{5.78}$$

Equations 5.72 and 5.74 similarly become, respectively:

$$n = \text{f}_{14}\left(Re'_0, \frac{d}{d_t}\right) \tag{5.79}$$

and:

$$n = \text{f}_{15}\left(\frac{d}{d_t}\right) \tag{5.80}$$

Equation 5.80 is applicable for:

$$\begin{cases} Re'_0 < 0.2 \\ Ga < 3.6 \end{cases} \quad \text{and} \quad \begin{cases} Re'_0 > 500 \\ Ga > 8.3 \times 10^4 \end{cases}$$

Experimental results generally confirm the validity of equation 5.80 over these ranges, with $n \approx 4.8$ at low Reynolds numbers and $\approx 2.4$ at high values. Equation 5.78 is to be preferred to equation 5.79 as the Galileo number can be calculated directly from the properties of the particles and of the fluid, whereas equation 5.79 necessitates the calculation of the terminal falling velocity $u_0$.

Values of the exponent $n$ were determined by RICHARDSON and ZAKI[56] by plotting $n$ against $d/d_t$ with $Re'_0$ as parameter, as shown in Figure 5.16. Equations for evaluation of $n$ are given in Table 5.1, over the stated ranges of $Ga$ and $Re'_0$.

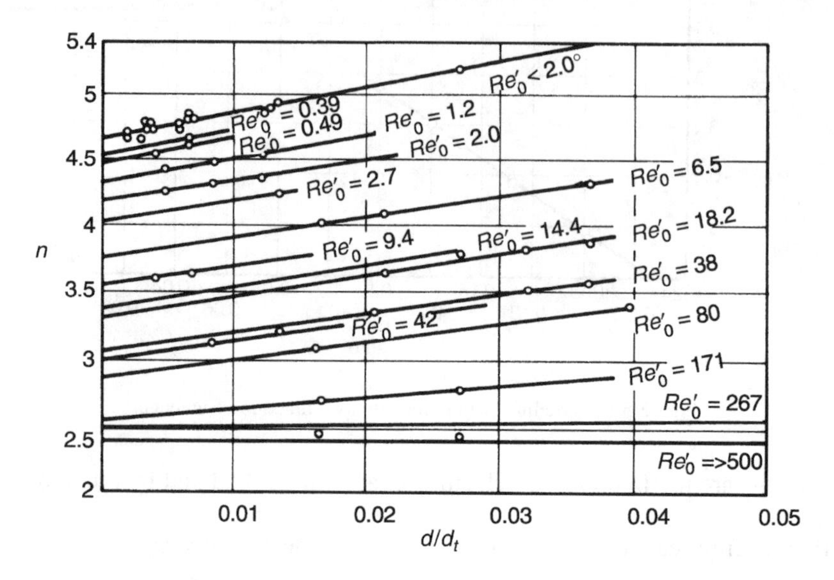


Figure 5.16. Dependence of exponent $n$ on the ratio of particle-to-container diameter[56]

As an alternative to the use of Table 5.1, it is convenient to be able to use a single equation for the calculation of $n$ over the whole range of values of $Ga$ or $Re'_0$ of interest. Thus, GARSIDE and AL-DIBOUNI[57] have proposed:

$$\frac{5.1 - n}{n - 2.7} = 0.1 Re_0'^{0.9} \tag{5.81}$$

Table 5.1. $n$ as a function of $Ga$ or $Re'_0$ and $d/d_t$

| Range of $Ga$ | Range of $Re'_0$ | $n$ as function of $Ga, d/d_t$ | $n$ as function of $Re'_0, d/d_t$ |
|---|---|---|---|
| 0–3.6 | 0–0.2 | $4.6 + 20\,d/d_t$ | $4.6 + 20\,d/d_t$ |
| 3.6–21 | 0.2–1 | $(4.8 + 20\,d/d_t)\,Ga^{-0.03}$ | $(4.4 + 18\,d/d_t)\,Re_0'^{-0.03}$ |
| 21–2.4 $\times 10^4$ | 1–200 | $(5.5 + 23\,d/d_t)\,Ga^{-0.075}$ | $(4.4 + 18\,d/d_t)\,Re_0'^{-0.1}$ |
| 2.4 $\times 10^4$–8.3 $\times 10^4$ | 200–500 | $5.5Ga^{-0.075}$ | $4.4Re_0'^{-0.1}$ |
| >8.3 $\times 10^4$ | >500 | 2.4 | 2.4 |

They suggest that particle interaction effects are underestimated by equation 5.71, particularly at high voidages ($e > 0.9$) and for the turbulent region, and that for a given value of voidage ($e$) the equation predicts too high a value for the sedimentation (or fluidisation) velocity ($u_c$). They prefer the following equation which, they claim, overall gives a higher degree of accuracy:

$$\frac{u_c/u_0 - e^{5.14}}{e^{2.68} - u_c/u_0} = 0.06Re'_0 \tag{5.82}$$

Equation 5.82 is less easy to use than equation 5.71 and it is doubtful whether it gives any significantly better accuracy, there being evidence[58] that with viscous oils equation 5.76 is more satisfactory.

ROWE[59] has subsequently given:

$$\frac{4.7 - n}{n - 2.35} = 0.175Re_0'^{0.75} \tag{5.83}$$

More recently KHAN and RICHARDSON[60] have examined the published experimental results for both sedimentation and fluidisation of uniform spherical particles and recommend the following equation from which $n$ may be calculated in terms of both the Galileo number $Ga$ and the particle to vessel diameter ratio $d/d_t$:

$$\frac{4.8 - n}{n - 2.4} = 0.043Ga^{0.57}\left[1 - 1.24\left(\frac{d}{d_t}\right)^{0.27}\right] \tag{5.84}$$

A comparison of calculated and experimental values of $n$ is given in Chapter 6, Figure 6.6.

As already indicated, it is preferable to employ an equation in incorporating $Ga$ which can be directly calculated from the properties of the solid and of the liquid.

Values of $n$ from equation 5.84 are plotted against Galileo number for $d/d_t = 0$ and $d/d_t = 0.1$ in Figure 5.17. It may be noted that the value of $n$ is critically dependent on $d/d_t$ in the intermediate range of Galileo number, but is relatively insensitive at the two extremities of the curves where $n$ attains a constant value of about 4.8 at low values of $Ga$ and about 2.4 at high values. It is seen that $n$ becomes independent of $Ga$ in these regions, as predicted by dimensional analysis and by equation 5.80.

In Chapter 3, equation 3.40 was proposed for the calculation of the free falling velocity of a particle in an infinite medium[61]. This equation which was shown to apply over the whole range of values of $Ga$ of interest takes the form:

$$Re'_0 = [2.33Ga^{0.018} - 1.53Ga^{-0.016}]^{13.3} \tag{5.85}$$

(see equation 3.40)

The correction factor for the effect of the tube wall is given by substituting $u_i$ for $u_{0t}$ in equation 3.43 to give:

$$\frac{u_i}{u_0} = \left[1 + 1.24\frac{d}{d_t}\right]^{-1} \tag{5.86}$$

Thus, combining equations 5.84, 5.85 and 5.86:

$$u_c = \frac{\mu}{\rho d}[2.33Ga^{0.018} - 1.53Ga^{-0.016}]^{13.3}\left[1 + 1.24\frac{d}{d_t}\right]^{-1} e^n \tag{5.87}$$

where $n$ is given by equation 5.84.

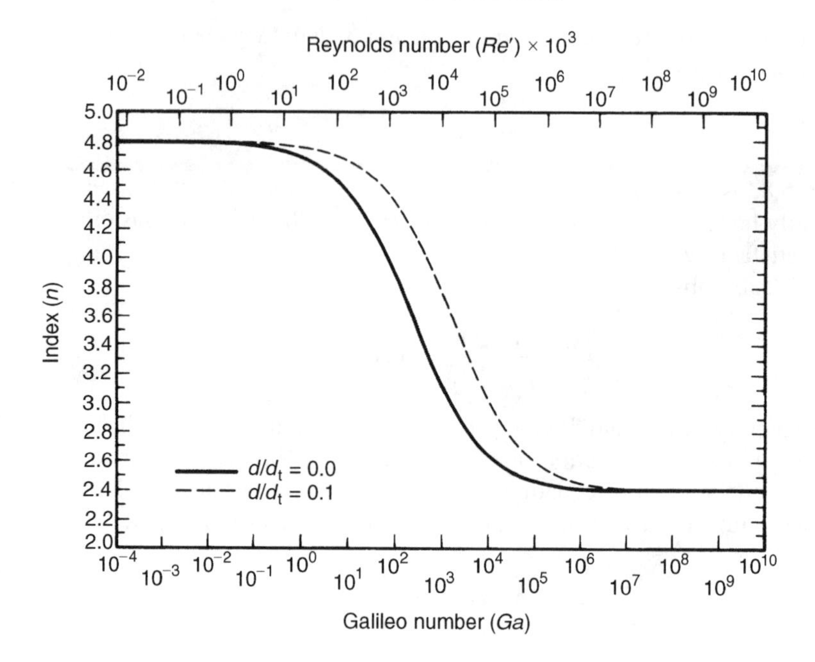


Figure 5.17. Sedimentation index $n$ as function of Galileo number $Ga$ (from equation 5.84)

Equation 5.87 permits the calculation of the sedimentation velocity of uniform spherical particles in a liquid at any voidage $e$ in a vessel of diameter $d_t$.

For a more detailed account of the various approaches to the problem of sedimentation of concentrated suspensions, reference may be made to the work KHAN and RICHARDSON(62).

## 5.3.3. Solids flux in batch sedimentation

In a sedimenting suspension, the sedimentation velocity $u_c$ is a function of fractional volumetric concentration $C$, and the volumetric rate of sedimentation per unit area or flux $\psi$ is equal to the product $u_cC$.

Thus:

$$\psi = u_cC = u_c(1 - e) \tag{5.88}$$

Then, if the relation between settling velocity and concentration can be expressed in terms of a terminal falling velocity ($u_0$) for the particles, substituting for $u_0$ using equation 5.71 gives:

$$\psi = u_0e^n(1 - e) \tag{5.89}$$

From the form of the function, it is seen that $\psi$ should have a maximum at some value of $e$ lying between 0 and 1.

Differentiating;

$$\frac{d\psi}{de} = u_0\{ne^{n-1}(1 - e) + e^n(-1)\} = u_0\{ne^{n-1} - (n + 1)e^n\} \tag{5.90}$$

When $d\psi/de$ is zero, then:

$$ne^{n-1}(1-e) - e^n = 0$$

or:

$$n(1-e) = e$$

and:

$$e = \frac{n}{1+n} \tag{5.91}$$

If $n$ ranges from 2.4 to 4.8 as for suspensions of uniform spheres, the maximum flux should occur at a voidage between 0.71 and 0.83 (volumetric concentration 0.29 to 0.17). Furthermore, there will be a point of inflexion if $d^2\psi/de^2$ is zero for real values of $e$. Differentiating equation 5.90:

$$\frac{d^2\psi}{de^2} = u_0\{n(n-1)e^{n-2} - (n+1)ne^{n-1}\} \tag{5.92}$$

When $d^2\psi/de^2 = 0$:

$$n - 1 - (n+1)e = 0$$

or:

$$e = \frac{n-1}{n+1} \tag{5.93}$$

If $n$ ranges from 2.4 to 4.8, there should be a point of inflexion in the curve occuring at values of $e$ between 0.41 and 0.65, corresponding to very high concentrations ($C = 0.59$ to 0.35). It may be noted that the point of inflexion occurs at a value of voidage $e$ below that at which the mass rate of sedimentation $\psi$ is a maximum. For coarse particles ($n \approx 2.4$), the point of inflexion is not of practical interest, since the concentration at which it occurs ($C \approx 0.59$) usually corresponds to a packed bed rather than a suspension.

The form of variation of flux ($\psi$) with voidage ($e$) and volumetric concentration ($C$) is shown in Figure 5.18 for a value of $n = 4.8$. This corresponds to the sedimentation of uniform spheres for which the free-falling velocity is given by Stokes' law. It may be compared with Figure 5.4 obtained for a flocculated suspension of calcium carbonate.

## Example 5.3

For the sedimentation of a suspension of uniform particles in a liquid, the relation between observed sedimentation velocity $u_c$ and fractional volumetric concentration $C$ is given by:

$$\frac{u_c}{u_0} = (1 - C)^{4.8}$$

where $u_0$ is the free falling velocity of an individual particle.

Calculate the concentration at which the rate of deposition of particles per unit area will be a maximum, and determine this maximum flux for 0.1 mm spheres of glass (density 2600 kg/m$^3$) settling in water (density 1000 kg/m$^3$, viscosity 1 mNs/m$^2$).

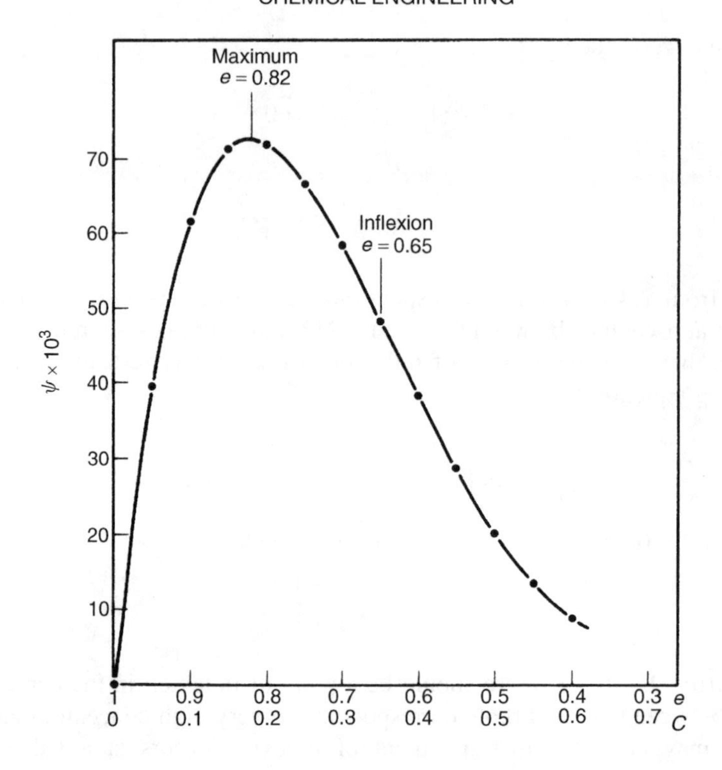


Figure 5.18. Flux ($\psi$)-concentration ($C$) curve for a suspension for which $n = 4.8$

It may be assumed that the resistance force $F$ on an isolated sphere is given by Stokes' law:

$$F = 3\pi\mu du$$

where: $\mu$ = fluid viscosity,
$d$ = particle diameter,
$u$ = velocity of particle relative to fluid.

## Solution

The mass rate of sedimentation per unit area $= \psi = u_c C$

$$= u_0(1 - C)^{4.8} C.$$

The maximum flux occurs when $\dfrac{d\psi}{dC} = 0.$

that is, when: $$-[4.8(1 - C)^{3.8} C] - (1 - C)^{4.8} = 0$$

or: $$-4.8C + (1 - C) = 0$$

Then: $$C = \frac{1}{5.8} = 0.172$$

The maximum flux $\psi_{max} = u_0(1 - 0.172)^{4.8} \times 0.172$

$$= 0.0695\ u_0$$

The terminal falling velocity $u_0$ is given by a force balance, as explained in Chapter 3:

$$\text{gravitational force} = \text{resistance force}$$

$$\frac{\pi}{6}d^3(\rho_s - \rho)g = 3\pi\mu d u_0$$

$$u_0 = \frac{d^2 g}{18\mu}(\rho_s - \rho) \qquad \text{(equation 3.24)}$$

$$= \frac{(10^{-4})^2 9.81}{(18 \times 10^{-3})}(2600 - 1000)$$

$$= 0.00872 \text{ m/s } 8.72 \text{ mm/s}$$

The maximum flux $\psi_{max} = (0.0695 \times 0.00872)$

$$= \underline{\underline{6.06 \times 10^{-4} \text{ m}^3/\text{m}^2\text{s}}}$$

### 5.3.4. Comparison of sedimentation with flow through fixed beds

RICHARDSON and MEIKLE[63] have shown that, at high concentrations, the results of sedimentation and fluidisation experiments can be represented in a manner similar to that used by CARMAN[64] for fixed beds, discussed in Chapter 4. Using the interstitial velocity and a linear dimension given by the reciprocal of the surface of particles per unit volume of fluid, the Reynolds number is defined as:

$$Re_1 = \frac{(u_c/e)e/[(1-e)S]\rho}{\mu} = \frac{u_c\rho}{S\mu(1-e)} \qquad (5.94)$$

The friction group $\phi''$ is defined in terms of the resistance force per unit particle surface ($R_1$) and the interstitial velocity.

Thus:
$$\phi'' = \frac{R_1}{\rho(u_c/e)^2} \qquad (5.95)$$

A force balance on an element of system containing unit volume of particles gives:

$$R_1 S = (\rho_s - \rho_c)g = e(\rho_s - \rho)g \qquad (5.96)$$

$\therefore$
$$\phi'' = \frac{e^3(\rho_s - \rho)g}{S\rho u_c^2} \qquad (5.97)$$

In Figure 5.19, $\phi''$ is plotted against $Re_1$ using results obtained in experiments on sedimentation and fluidisation. On a logarithmic scale a linear relation is obtained, for values of $Re_1$ less than 1, with the following equation:

$$\phi'' = 3.36 Re_1^{-1} \qquad (5.98)$$

Results for flow through a fixed bed are also shown. These may be represented by:

$$\phi'' = 5 Re_1^{-1} \qquad \text{(equation 4.18)}$$

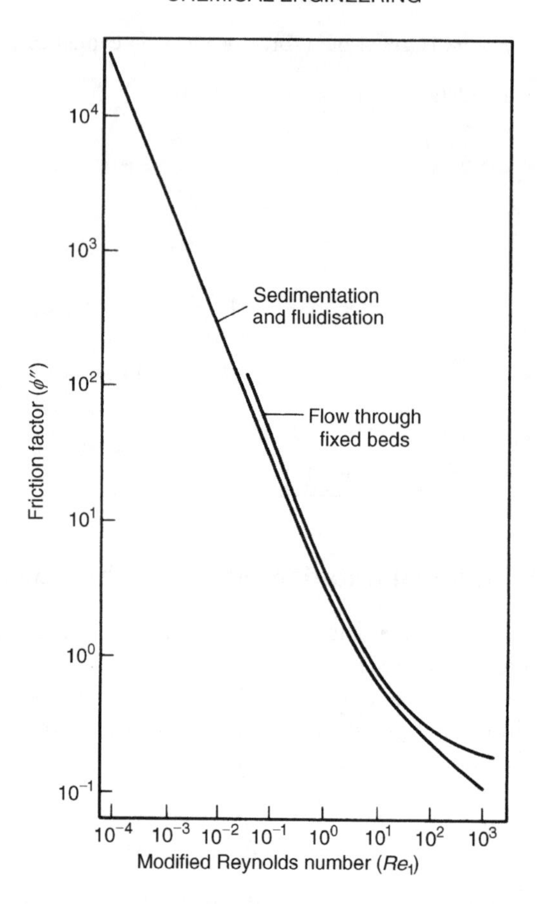


Figure 5.19. Correlation of data on sedimentation and fluidisation and comparison with results for flow through fixed granular beds

Because equation 4.18 is applicable to low Reynolds numbers at which the flow is streamline, it appears that the flow of fluid at high concentrations of particles in a sedimenting or fluidised system is also streamline. The resistance to flow in the latter case appears to be about 30 per cent lower, presumably because the particles are free to move relative to one another.

Equation 5.98 can be arranged to give:

$$u_c = \frac{e^3}{3.36(1-e)} \frac{(\rho_s - \rho)g}{S^2\mu} \tag{5.99}$$

In the Stokes' law region, from equation 3.24:

$$u_0 = \frac{d^2 g}{18\mu}(\rho_s - \rho) = \frac{2g}{S^2\mu}(\rho_s - \rho) \tag{5.100}$$

Dividing:

$$\frac{u_c}{u_0} = \frac{e^3}{6.7(1-e)} \tag{5.101}$$

From equations 5.76 and 5.84 and neglecting the tube wall effect:

$$\frac{u_c}{u_0} = e^{4.8} (\text{for } Re_1 < 0.2) \tag{5.102}$$

The functions $e^3/[6.7(1 - e)]$ and $e^{4.8}$ are both plotted as a function of $e$ in Figure 5.20, from which it is seen that they correspond closely for voidages less than 0.75.

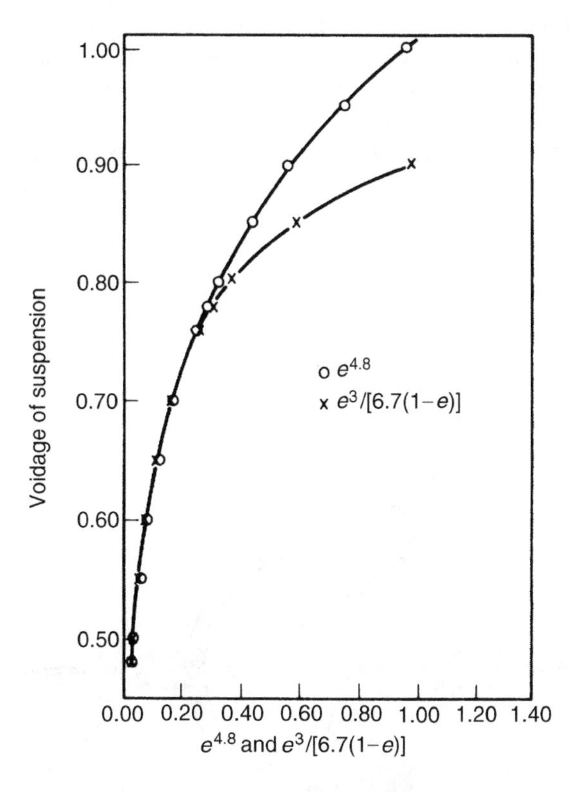


Figure 5.20. Comparison of two functions of voidage used as correction factors for sedimentation velocities

The application of the relations obtained for monodisperse systems to fine suspensions containing particles of a wide range of sizes was examined by RICHARDSON and SHABI[65] who studied the behaviour of aqueous suspensions of zirconia of particle size ranging from less than 1 to 20 μm. A portion of the solid forming the suspension was irradiated and the change in concentration with time at various depths below the surface was followed by means of a Geiger-Müller counter. It was found that selective settling took place initially and that the effect of concentration on the falling rate of a particle was the same as in a monodisperse system, provided that the total concentration of particles of all sizes present was used in the calculation of the correction factor. Agglomeration was found to occur rapidly at high concentrations, and the conclusions were therefore applicable only to the initial stages of settling.

### 5.3.5. Model experiments

The effect of surrounding a particle with an array of similar particles has been studied by RICHARDSON and MEIKLE[66] and by ROWE and HENWOOD[67]. In both investigations, the effect of the proximity and arrangement of fixed neighbouring particles on the drag was examined.

Meikle surrounded the test particle by a uniplanar hexagonal arrangement of identical spheres which were held in position by means of rods passing through glands in the tube walls, as shown in Figure 5.21. This arrangement permitted the positioning of the spheres to be altered without interfering with the flow pattern in the hexagonal space. For each arrangement of the spheres, the drag force was found to be proportional to the square of the liquid velocity. The effective voidage $e$ of the system was calculated by considering a hexagonal cell located symmetrically about the test sphere, of depth equal to the diameter of the particle, as shown in Figure 5.22. The drag force $F_c$ on a particle of volume $v$ settling in a suspension of voidage $e$ is given by:

$$F_c = v(\rho_s - \rho_c)g = ev(\rho_s - \rho)g \tag{5.103}$$

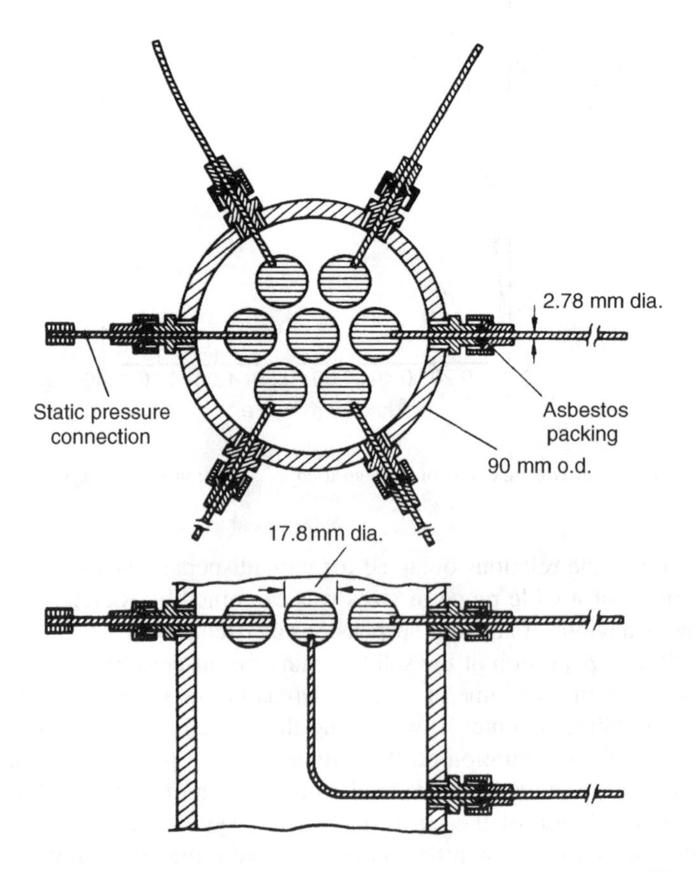


Figure 5.21. Arrangement of hexagonal spacing of particles in tube[66]

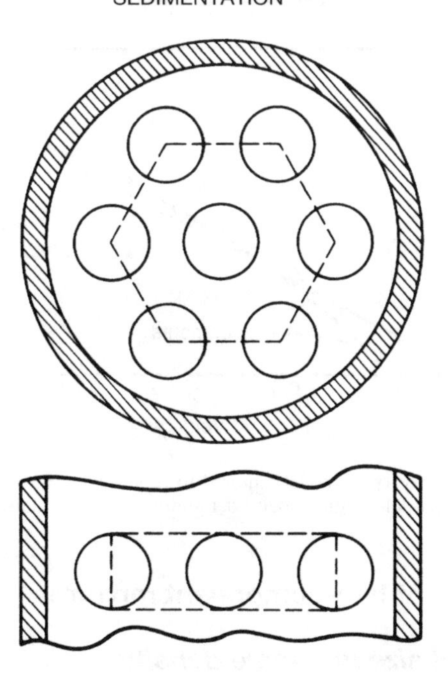

Figure 5.22. Hexagonal cell of particles[66]

Thus, for any voidage the drag force on a sedimenting particle can be calculated, and the corresponding velocity required to produce this force on a particle at the same voidage in the model is obtained from the experimental results. All the experiments were carried out at particle Reynolds number greater than 500, and under these conditions the observed sedimentation velocity is given by equations 5.76 and 5.84 as:

$$\frac{u_c}{u_0} = e^{2.4} \tag{5.104}$$

Writing $u_c = eu_p$, where $u_p$ is the velocity of the particle relative to the fluid:

$$\frac{u_p}{u_0} = e^{1.4} \tag{5.105}$$

In Figure 5.23 the velocity of the fluid relative to the particle, as calculated for the model experiments and from equation 5.100, is plotted against voidage. It may be seen that reasonable agreement is obtained at voidages between 0.45 and 0.90, indicating that the model does fairly closely represent the conditions in a suspension.

Rowe and Henwood[67] made assemblies of particles from cast blocks of polythene and found that the force on a single particle was increased by an order of magnitude as a result of surrounding it with a close assembly. They also found that adjacent spheres tend to repel one another, that surfaces facing downstream tend to expel particles, whereas those facing upstream attract particles. This was offered as an explanation for the stable upper surface and rather diffuse lower surface of a bubble in a fluidised bed.

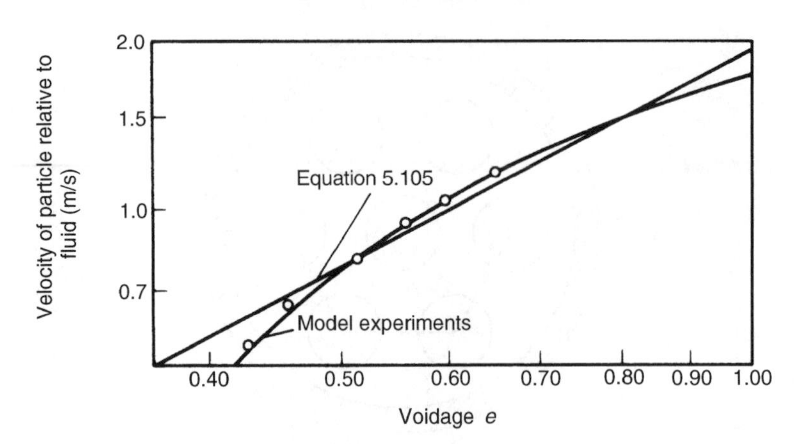


Figure 5.23. Comparison of velocity in model required to produce a drag force equal to buoyant weight of particle, with velocity determined experimentally

## 5.3.6. Sedimentation of two-component mixtures

### *Particles of different size but same density*

Several workers[68,69,70] have studied the sedimentation of suspensions formed of particles of two different sizes, but of the same densities. The large particles have higher settling velocities than the small ones and therefore four zones form, in order, from the top downwards:

(a) Clear liquid.
(b) Suspension of the fine particles.
(c) Suspension of mixed sizes.
(d) Sediment layer.

For relatively coarse particles, the rates of fall of the interface between (a) and (b) and between (b) and (c) can be calculated approximately if the relation between sedimentation velocity and voidage, or concentration, is given by equation 5.76. Richardson and Shabi[65] have shown that, in a suspension of particles of mixed sizes, it is the total concentration which controls the sedimentation rate of each species.

A suspension of a mixture of large particles of terminal falling velocity $u_{0L}$ and of small particles of terminal falling velocity $u_{0S}$ may be considered, in which the fractional volumetric concentrations are $C_L$ and $C_S$, respectively. If the value of $n$ in equation 5.76 is the same for each particle. For each of the spheres settling on its own:

$$\frac{u_{cL}}{u_{0L}} = e^n \tag{5.106}$$

and:

$$\frac{u_{cS}}{u_{0S}} = e^n \tag{5.107}$$

Then, considering the velocities of the particles relative to the fluid, $u_{cL}/e$ and $u_{cS}/e$, respectively:

$$\frac{u_{cL}}{e} = u_{0L}e^{n-1} \tag{5.108}$$

and:

$$\frac{u_{cS}}{e} = u_{0S}e^{n-1} \tag{5.109}$$

When the particles of two sizes are settling together, the upflow of displaced fluid is caused by the combined effects of the sedimentation of the large and small particles. If this upward velocity is $u_F$, the sedimentation rates $u_{ML}$ and $u_{MS}$ will be obtained by deducting $u_F$ from the velocities relative to the fluid.

Thus:

$$u_{ML} = u_{0L}e^{n-1} - u_F \tag{5.110}$$

and:

$$u_{MS} = u_{0S}e^{n-1} - u_F \tag{5.111}$$

Then, since the volumetric flow of displaced fluid upwards must be equal to the total volumetric flowrate of particles downwards, then:

$$u_F e = (u_{0L}e^{n-1} - u_F)C_L + (u_{0S}e^{n-1} - u_F)C_S$$

$$\therefore u_F = e^{n-1}(u_{0L}C_L + u_{0S}C_S) \tag{5.112}$$

$$(\text{since } e + C_L + C_S = 1)$$

Then substituting from equation 5.112 into equations 5.110 and 5.111:

$$u_{ML} = e^{n-1}[u_{0L}(1 - C_L) - u_{0S}C_S] \tag{5.113}$$

and:

$$u_{MS} = e^{n-1}[u_{0S}(1 - C_S) - u_{0L}C_L] \tag{5.114}$$

Equation 5.113 gives the rate of fall of the interface between zones (b) and (c); that is the apparent rate of settling of the zone of mixed particles.

The velocity of fall $u_f$ of the interface between zones (a) and (b) is the sedimentation rate of the suspension composed only of fine particles, and will therefore depend on the free-falling velocity $u_{0S}$ and the concentration $C_f$ of this zone.

Now:

$$C_f = \frac{\text{Volumetric rate at which solids are entering zone}}{\text{Total volumetric growth rate of zone}}$$

$$= \frac{(u_{ML} - u_{MS})C_S}{u_{ML} - u_f} \tag{5.115}$$

Thus:

$$u_f = u_{0S}(1 - C_f)^n \tag{5.116}$$

$u_f$ and $C_f$ are determined by solving equations 5.115 and 5.116 simultaneously. Further study has been carried out of the behaviour of particles of mixed sizes by several groups of workers including SELIM, KOTHARI and TURIAN[71].

### Particles of equal terminal falling velocities

By studying suspensions containing two different solid components, it is possible to obtain a fuller understanding of the process of sedimentation of a complex mixture. RICHARDSON and MEIKLE[63] investigated the sedimentation characteristics of suspensions of glass ballotini and polystyrene particles in a 22 per cent by mass ethanol–water mixture. The free-falling velocity, and the effect of concentration on sedimentation rate, were identical for each of the two solids alone in the liquid.

The properties of the components are given in Table 5.2.

Table 5.2. Properties of solids and liquids in the sedimentation of two-component mixtures

| | Glass ballotini | Polystyrene | 22% ethanol in water |
|---|---|---|---|
| Density ($kg/m^3$) | $\rho_B = 1921$ | $\rho_P = 1045$ | $\rho = 969$ |
| Particle size (μm) | 71.1 | 387 | — |
| Viscosity (mN $s/m^2$) | — | — | 1.741 |
| Free-falling velocity ($u_0$) (mm/s) | 3.24 | 3.24 | — |

The sedimentation of mixtures containing equal volumes of the two solids was then studied, and it was found that segregation of the two components tended to take place, the degree of segregation increasing with concentration. This arises because the sedimentation velocity of an individual particle in the suspension is different from its free-falling velocity, first because the buoyancy force is greater, and, secondly, because the flow pattern is different. For a monodisperse suspension of either constituent of the mixture, the effect of flow pattern as determined by concentration is the same, but the buoyant weights of the two species of particles are altered in different proportions. The settling velocity of a particle of polystyrene or ballotini in the mixture ($u_{PM}$ or $u_{BM}$) can be written in terms of its free-falling velocity ($u_{P0}$ or $u_{B0}$) as:

$$u_{PM} = u_{P0}\frac{\rho_P - \rho_c}{\rho_P - \rho}\,\mathrm{f}(e) \tag{5.117}$$

and:

$$u_{BM} = u_{B0}\frac{\rho_B - \rho_c}{\rho_B - \rho}\,\mathrm{f}(e) \tag{5.118}$$

In these equations f($e$) represents the effects of concentration, other than those associated with an alteration of buoyancy arising from the fact that the suspension has a higher density than the liquid. In a uniform suspension, the density of the suspension is given by:

$$\rho_c = \rho e + \frac{1-e}{2}(\rho_B + \rho_P) \tag{5.119}$$

Substitution of the numerical values of the densities in equations 5.117, 5.118 and 5.119 gives:

$$u_{PM} = u_{P0}(13.35e - 12.33)\,\mathrm{f}(e) \tag{5.120}$$

and:

$$u_{BM} = u_{B0}(0.481 + 0.520)\,\mathrm{f}(e) \tag{5.121}$$

Noting that $u_{P0}$ and $u_{B0}$ are equal, it is seen that the rate of fall of the polystyrene becomes progressively less than that of the ballotini as the concentration is increased. When $e = 0.924$, the polystyrene particles should remain suspended in the mixed suspension, and settling should occur only when the ballotini have separated out from that region. At lower values of $e$, the polystyrene particles should rise in the mixture. At a voidage of 0.854, the polystyrene should rise at the rate at which the ballotini are settling, to that there will then be no net displacement of liquid. When the ballotini are moving at the higher rate, the net displacement of liquid will be upwards and the zone of polystyrene suspension will become more dilute. When the polystyrene is moving at the higher rate, the converse is true.

Because there is no net displacement of liquid at a voidage of 0.854, corresponding to a total volumetric concentration of particles of 14.6 per cent, the concentration of each zone of suspension, after separation, should be equal. This has been confirmed within the limits of experimental accuracy. It has also been confirmed experimentally that upward movement of the polystyrene particles does not take place at concentrations less than 8 per cent ($e > 0.92$).

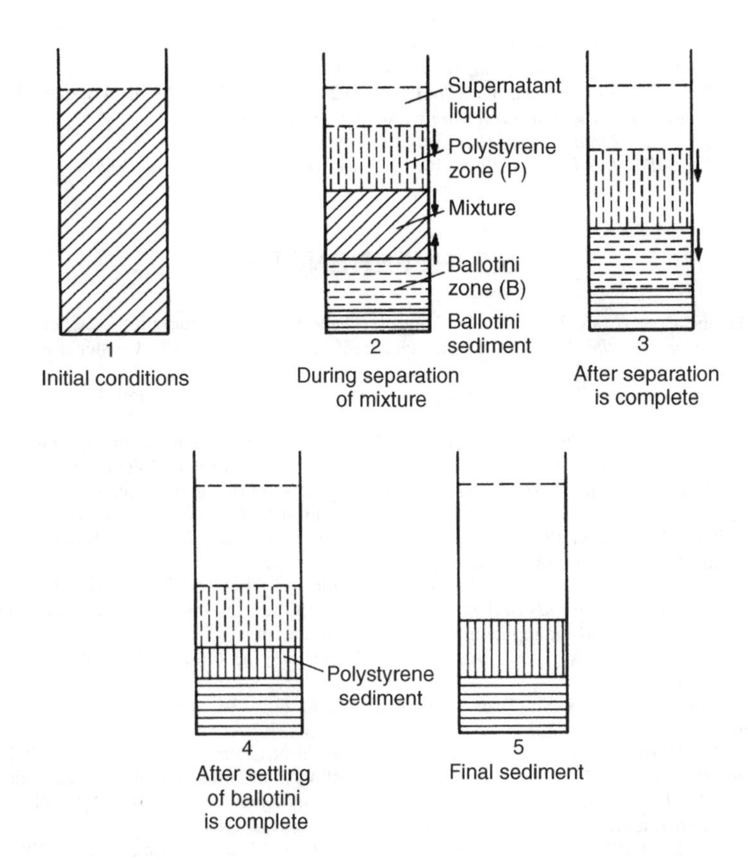


Figure 5.24. Settling behaviour of a mixture consisting of equal volumes of polystyrene and ballotini at a volumetric concentrations exceeding 15 per cent

On the above basis, the differences between the velocities of the two types of particles should increase with concentration and, consequently, the concentration of the two separated zones should become increasingly different. This is found to be so at volumetric concentrations up to 20 per cent. At higher concentrations, the effect is not observed, probably because the normal corrections for settling velocities cannot be applied when particles are moving in opposite directions in suspensions of high concentration.

The behaviour of a suspension of high concentration ($e < 0.85$) is shown in Figure 5.24, where it is seen that the glass ballotini settle out completely before any deposition of polystyrene particles occurs.

These experiments clearly show that the tendency for segregation to occur in a two-component mixture becomes progressively greater as the concentration is increased. This behaviour is in distinct contrast to that observed in the sedimentation of a suspension of multi-sized particles of a given material, when segregation becomes less as the concentration is increased.

## 5.4. FURTHER READING

CLIFT, R., GRACE, J. R. and WEBER, M. E.: *Bubbles, Drops and Particles* (Academic Press, 1978).
DALLAVALLE, J. M.: *Micromeritics*, 2nd edn. (Pitman, 1948).
ORR, C.: *Particulate Technology* (Macmillan, New York, 1966).
OTTEWILL, R. H.: Particulate dispersions. In *Colloid Science* **2**, 173–219 (The Chemical Society, London, 1973).
SMITH, A. L. (ed.): *Particle Growth in Suspensions* (Academic Press, 1973).
SVAROVSKY, L. (ed.): *Solid–Liquid Separation*, 3rd edn, (Butterworth, 1990).
SVAROVSKY, L. (ed.): *Solid–Liquid Separation*, 4th edn. (Butterworth-Heinemann, 2000).

## 5.5. REFERENCES

1. KAYE, B. H. and BOARDMAN, R. P.: Third Congress of the European Federation of Chemical Engineering (1962). *Symposium on the Interaction between Fluids and Particles* 17. Cluster formation in dilute suspensions.
2. COE, H. S. and CLEVENGER, G. H.: *Trans. Am. Inst. Min. Met. Eng.* **55** (1916) 356. Methods for determining the capacities of slime-settling tanks.
3. ROBINSON, C. S.: *Ind. Eng. Chem.* **18** (1926) 869. Some factors influencing sedimentation.
4. EINSTEIN, A.: *Ann. Phys.* **19** (1906) 289–306. Eine neue Bestimmung der Molekuldimensionen.
5. VAND, V.: *J. Phys. Coll. Chem.* **52** (1948) 277. Viscosity of solutions and suspensions.
6. STEINOUR, H. H.: *Ind. Eng. Chem.* **36** (1944) 618, 840, and 901. Rate of sedimentation.
7. HAWKSLEY, P. G. W.: *Inst. of Phys. Symposium* (1950) 114. The effect of concentration on the settling of suspensions and flow through porous media.
8. WALLIS, G. B.: Third Congress of the European Federation of Chemical Engineering (1962). *Symposium on the Interaction between Fluids and Particles*, 9. A simplified one-dimensional representation of two-component vertical flow and its application to batch sedimentation.
9. EGOLF, C. B. and MCCABE, W. L.: *Trans. Am. Inst. Chem. Eng.* **33** (1937) 620. Rate of sedimentation of flocculated particles.
10. WORK, L. T. and KOHLER, A. S.: *Trans. Am. Inst. Chem. Eng.* **36** (1940) 701. The sedimentation of suspensions.
11. PEARCE, K. W.: Third Congress of the European Federation of Chemical Engineering (1962). *Symposium on the Interaction between Fluids and Particles*, 30. Settling in the presence of downward-facing surfaces.
12. SCHAFLINGER, U.: *Intl. Jl. Multiphase Flow* **11** (1985) 783. Influence of non-uniform particle size on settling beneath downward-facing walls.
13. ROBINS, W. H. M.: Third Congress of the European Federation of Chemical Engineering (1962). *Symposium on the Interaction between Fluids and Particles*, 26. The effect of immersed bodies on the sedimentation of suspensions.

14. Verwey, E. J. W. and Overbeek, J. Th. G.: *Theory of the Stability of Lyophobic Colloids* (Elsevier, Amsterdam, 1948).
15. Sparnaay, M. J.: *The Electrical Double Layer* (Pergamon Press, Oxford, 1972).
16. Shaw, D. J.: *Introduction to Colloid and Surface Chemistry* (Butterworths, London, 1970).
17. Smith, A. L.: Electrical phenomena associated with the solid–liquid interface, in *Dispersion of Powders in Liquids*, ed. G. D. Parfitt (Applied Science Publishers, London, 1973).
18. Lyklema, J.: *J. Coll. and Interface Sci.* **58** (1977) 242. Water at interfaces: A colloid chemical approach.
19. Williams, D. J. A. and Williams, K. P.: *J. Coll. and Interface Sci.* **65** (1978) 79. Electrophoresis and zeta potential of kaolinite.
20. Derjaguin, B. V. and Landau, L.: *Acta Physicochim.* **14** (1941) 663. Theory of the stability of strongly charged lyophobic solutions and of the adhesion of strongly charged particles in solution of electrolytes.
21. Gregory, J.: Interfacial phenomena, in *The Scientific Basis of Filtration*, ed. K. J. Ives (Noordhoff, Leyden, 1975).
22. Hamaker, H. C.: *Physica* **4** (1937) 1058. The London-van der Waals attraction between spherical particles.
23. Lifshitz, E. M.: *Soviet Physics. J.E.T.P.* **2** (1956) 73. Theory of molecular attractive forces between solids.
24. Ottewill, R. H.: Particulate dispersions. In *Colloid Science* **2** (The Chemical Society, London, 1973) 173.
25. Visser, J.: *Adv. Colloid Interface Sci.* **3** (1972) 331. On Hamaker constants: a comparison between Hamaker constants and Lifshitz–van der Waals constants.
26. Matijevic, E., Janauer, G. E. and Kerker, M.: *J. Coll. Sci.* **19** (1964) 333. Reversal of charge of hydrophobic colloids by hydrolyzed metal ions, I. Aluminium nitrate.
27. Spielman, L. A.: *J. Coll. and Interface Sci.* **33** (1970) 562. Viscous interactions in Brownian coagulation.
28. Honig, E. P., Roebersen, G. J., and Wiersema, P. H.: *J. Coll. and Interface Sci.* **36** (1971) 97. Effect of hydrodynamic interaction on the coagulation rate of hydrophobic colloids.
29. van de Ven, T. G. M. and Mason, S. G.: *J. Coll. and Interface Sci.* **57** (1976) 505. The microrheology of colloidal dispersions. IV. Pairs of interacting spheres in shear flow.
30. Zeichner, G. R. and Schowalter, W. R.: *A.I.Ch.E.Jl* **23** (1977) 243. Use of trajectory analysis to study stability of colloidal dispersions in flow fields.
31. von Smoluchowski, M.: *Z. Physik. Chem.* **92** (1917) 129. Versuch einer mathematischen Theorie der Koagulationskinetik Kolloider Lösungen.
32. Camp, T. R. and Stein, P. C.: *J. Boston Soc. Civ. Eng.* **30** (1943) 219. Velocity gradients and internal work in fluid motion.
33. La Mer, V. K. and Healy, T. W.: *Rev. Pure Appl. Chem.* (Australia) **13** (1963) 112. Adsorption–flocculation reactions of macromolecules at the solid–liquid interface.
34. Vold, M. J.: *J. Coll. Sci.* **16** (1961) 1. The effect of adsorption on the van der Waals interaction of spherical particles.
35. Osmond, D. W. J., Vincent, B., and Waite, F. A.: *J. Coll. and Interface Sci.* **42** (1973) 262. The van der Waals attraction between colloid particles having adsorbed layers, I. A re-appraisal of the "Vold" effect.
36. Vincent, B.: *Adv. Coll. Interface Sci.* **4** (1974) 193. The effect of adsorbed polymers on dispersion stability.
37. Napper, D. H.: *J. Coll. and Interface Sci.* **58** (1977) 390. Steric stabilisation.
38. Dobbie, J. W., Evans, R. E., Gibson, D. V., Smitham, J. B. and Napper, D. H.: *J. Coll. and Interface Sci.* **45** (1973) 557. Enhanced steric stabilisation.
39. O'Gorman, J. V. and Kitchener, J. A.: *Int. J. Min. Proc.* **1** (1974) 33. The flocculation and dewatering of Kimberlite clay slimes.
40. Williams, D. J. A. and Ottewill, R. H.: *Kolloid-Z. u. Z. Polymere* **243** (1971) 141. The stability of silver iodide solutions in the presence of polyacrylic acids of various molecular weights.
41. Gregory, J.: *Trans. Faraday Soc.* **65** (1969) 2260. Flocculation of polystyrene particles with cationic polyelectrolytes.
42. Gregory, J.: *J. Coll. and Interface Sci.* **55** (1976) 35. The effect of cationic polymers on the colloid stability of latex particles.
43. Griot, O. and Kitchener, J. A.: *Trans. Faraday Soc.* **61** (1965) 1026. Role of surface silanol groups in the flocculation of silica by polycrylamide.
44. la Mer, V. K.: *Disc. Faraday Soc.* **42** (1966) 248. Filtration of colloidal dispersions flocculated by anionic and cationic polyelectrolytes.
45. Michaels, A. S. and Bolger, J. C.: *Ind. Eng. Chem. Fundamentals* **1** (1962) 24. Settling rates and sediment volumes of flocculated kaolinite suspensions.
46. Kynch, G. J.: *Trans. Faraday Soc.* **48** (1952) 166. A theory of sedimentation.
47. Yoshioka, N., Hotta, Y., Tanaka, S., Naito, S. and Tsugami, S.: *Kagaku Kogaku (J. Soc. Chem. Eng., Japan* **2**) **21** (1957) 66. Continuous thickening of homogeneous flocculated slurries.
48. Hassett, N. J.: *Industrial Chemist* **37** (1961) 25. Theories of the operation of continuous thickeners.
49. Comings, E.W.: *Ind.Eng.Chem.* **32** (1940) 1663. Thickening calcium carbonate slurries.
50. Roberts, E.J.: *Trans. Amer. Inst. Min.,Met. Engrs.* **184** (1949) 61. (*Trans. Min. Engrs* **1** (1949) 61) Thickening–Art or Science?

51. TILLER, F. M. and CHEN, W.: *Chem. Eng. Sci.* **43** (1988) 1695. Limiting operating conditions for continuous thickeners.
52. FONT, R.: *A.I.Ch.E.Jl.* **36** (1990) 3. Calculation of the compression zone height in continuous thickeners.
53. FITCH, B.: *A.I.Ch.E.Jl.* **39** (1993) 27. Thickening theories — an analysis.
54. DELL, C. C. and KELEGHAN, W. T. H.: *Powder Technology* **7** (1973) 189. The dewatering of polyclay suspensions.
55. CHANDLER, J. L.: *World Filtration Congress* III. (Downingtown, Pennsylvania). *Proceedings* **1** (1982) 372. Dewatering by deep thickeners without rakes.
56. RICHARDSON, J. F. and ZAKI, W. N.: *Trans. Inst. Chem. Eng.* **32** (1954) 35. Sedimentation and fluidisation: Part I.
57. GARSIDE, J. and AL-DIBOUNI, M. R.: *Ind. Eng. Chem. Proc. Des. Dev.* **16** (1977) 206 Velocity–voidage relationships for fluidization and sedimentation in solid–liquid systems.
58. KHAN, A. R.: University of Wales, Ph.D. thesis (1978). Heat transfer from immersed surfaces to liquid-fluidised beds.
59. ROWE, P. N.: *Chem. Eng. Sci.* **42** (1987) 2795. A convenient empirical equation for estimation of the Richardson-Zaki exponent.
60. KHAN, A. R. and RICHARDSON, J. F.: *Chem. Eng. Comm.* **78** (1989) 111. Fluid-particle interactions and flow characteristics of fluidized beds and settling suspensions of spherical particles.
61. KHAN, A. R. and RICHARDSON, J. F.: *Chem. Eng. Comm.* **62** (1987) 135. The resistance to motion of a solid sphere in a fluid.
62. KHAN, A. R. and RICHARDSON, J. F.: *Chem. Eng. Sci.* **45** (1990) 255. Pressure gradient and friction factor for sedimentation and fluidisation of uniform spheres in liquids.
63. RICHARDSON, J. F. and MEIKLE, R. A.: *Trans. Inst. Chem. Eng.* **39** (1961) 348. Sedimentation and fluidisation. Part III. The sedimentation of uniform fine particles and of two-component mixtures of solids.
64. CARMAN, P C.: *Trans. Inst. Chem. Eng.* **15** (1937) 150. Fluid flow through granular beds.
65. RICHARDSON, J. F. and SHABI, F. A.: *Trans. Inst. Chem. Eng.* **38** (1960) 33. The determination of concentration distribution in a sedimenting suspension using radioactive solids.
66. RICHARDSON, J. F. and MEIKLE, R. A.: *Trans. Inst. Chem. Eng.* **39** (1961) 357. Sedimentation and fluidisation. Part IV. Drag force on individual particles in an assemblage.
67. ROWE, P. N. and HENWOOD, G. N.: *Trans. Inst. Chem. Eng.* **39** (1961) 43. Drag forces in a hydraulic model of a fluidised bed. Part I.
68. SMITH, T. N.: *Trans. Inst. Chem. Eng.* **45** (1967) T311. The differential sedimentation of particles of various species.
69. LOCKETT, M. J. and AL-HABBOOBY, H. M.: *Trans. Inst. Chem. Eng.* **51** (1973) 281. Differential settling by size of two particle species in a liquid.
70. RICHARDSON, J. F. and MIRZA, S.: *Chem. Eng. Sci.* **34** (1979) 447. Sedimentation of suspensions of particles of two or more sizes.
71. SELIM, M. S., KOTHARI, A. C. and TURIAN, R. M.: *A.I.Ch.E.Jl.* **29** (1983) 1029. Sedimentation of multisized particles in concentrated suspensions.


## 5.6. NOMENCLATURE

| | | Units in SI System | Dimensions in **M, N, L, T, $\theta$, A** |
|---|---|---|---|
| $A$ | Cross-sectional area of vessel or tube | $m^2$ | $\mathbf{L}^2$ |
| $A'$ | Cross-sectional area at level of obstruction | $m^2$ | $\mathbf{L}^2$ |
| $\mathscr{A}$ | Hamaker constant | J | $\mathbf{ML}^2\mathbf{T}^{-2}$ |
| $a$ | Particle radius | m | $\mathbf{L}$ |
| $a_r$ | Separation distance between particle centres | m | $\mathbf{L}$ |
| $a$ | Constant in Vand's equation (5.4) | — | — |
| $B$ | Constant in equation 5.21 ($= 3.917 \times 10^{39}$ $C^{-2}$) | $C^{-2}$ | $\mathbf{T}^{-2}\mathbf{A}^{-2}$ |
| $b$ | Constant in equation 5.10 | $s^{-1}$ | $\mathbf{T}^{-1}$ |
| $C$ | Fractional volumetric concentration $(1 - e)$ | — | — |
| $C_0$ | Initial uniform concentration | — | — |
| $C_c$ | Critical concentration | — | — |
| $C_f$ | Concentration of fines in upper zone | — | — |
| $C_i$ | Concentration corresponding to point of inflexion on $\psi - C$ curve | — | — |

| | | Units in SI System | Dimensions in **M, N, L, T, θ, A** |
|---|---|---|---|
| $C_L$ | Limit value of C | — | — |
| $C_M$ | Concentration on Figure 5.14 | — | — |
| $C_m$ | Value of $C$ at which $\psi$ is a maximum | — | — |
| $C_{max}$ | Concentration corresponding to sediment layer | — | — |
| $C_u$ | Concentration of underflow in continuous thickener | — | — |
| $C'$ | Value of $C$ at constriction | — | — |
| $c$ | Concentration of solids (mass/volume) | kg/m$^3$ | $\mathbf{ML^{-3}}$ |
| $c_u$ | Concentration of solids in underflow (mass/volume) | kg/m$^3$ | $\mathbf{ML^{-3}}$ |
| $c_c'$ | Critical coagulation concentration | kmol/m$^3$ | $\mathbf{NL^{-3}}$ |
| $D_1, D_2$ | Absolute diffusivities | m$^2$/s | $\mathbf{L^2T^{-1}}$ |
| $D_{12}$ | Relative diffusivity | m$^2$/s | $\mathbf{L^2T^{-1}}$ |
| $d$ | Diameter of sphere or equivalent spherical diameter | m | **L** |
| $d_t$ | Diameter of tube or vessel | m | **L** |
| $e$ | Voidage of suspension ($= 1 - C$) | — | — |
| $e_c$ | Elementary charge ($= 1.60 \times 10^{-19}$ C) | C | **TA** |
| $F_c$ | Drag force on particle of volume $v$ | N | $\mathbf{MLT^{-2}}$ |
| $g$ | Acceleration due to gravity | m/s$^2$ | $\mathbf{LT^{-2}}$ |
| $H$ | Height in suspension above base | m | **L** |
| $H_0$ | Value of $H$ at $t = 0$ | m | **L** |
| $H_c$ | Height corresponding to critical settling point | m | **L** |
| $H_s$ | Separation between particle surfaces | m | **L** |
| $H_t$ | Initial total depth of suspension | m | **L** |
| $H_u$ | Height of interface when underflow concentration is reached | m | **L** |
| $H_\infty$ | Final height of sediment | m | **L** |
| $I$ | Collision frequency calculated from equation 5.22 | s$^{-1}$ | $\mathbf{T^{-1}}$ |
| $I_s$ | Collision frequency calculated from equation 5.25 | s$^{-1}$ | $\mathbf{T^{-1}}$ |
| $J$ | Collision frequency calculated from equation 5.29 | m$^{-3}$s$^{-1}$ | $\mathbf{L^{-3}T^{-1}}$ |
| $J_s$ | Collision frequency calculated from equation 5.28 | m$^{-3}$s$^{-1}$ | $\mathbf{L^{-3}T^{-1}}$ |
| $K$ | Boltzmann's constant ($= 1.38 \times 10^{-23}$ J/K) | J/K | $\mathbf{ML^2T^{-2}\theta^{-1}}$ |
| $K''$ | Constant in Robinson's equation (5.1) | — | — |
| $k$ | Constant in equation 5.55 | s$^{-1}$ | $\mathbf{T^{-1}}$ |
| $k''$ | Constant in Einstein's equation (5.3) | — | — |
| **N** | Avogadro number ($6.023 \times 10^{26}$ molecules per kmol *or* $6.023 \times 10^{23}$ molecules per mol) | kmol$^{-1}$ | $\mathbf{N^{-1}}$ |
| $N_F$ | Dimensionless parameter ($6\pi\mu a^3\dot{\gamma}/\mathscr{A}$) | — | — |
| $n$ | Index in equations 5.70 and 5.71 | — | — |
| $n_i$ | Number of ions per unit volume | m$^{-3}$ | $\mathbf{L^{-3}}$ |
| $n_1, n_2$ | Number of particles per unit volume | m$^{-3}$ | $\mathbf{L^{-3}}$ |
| **P** | Power input per unit volume of fluid | W/m$^3$ | $\mathbf{ML^{-1}T^{-3}}$ |
| $Q_0$ | Volumetric feed rate of suspension to continuous thickener | m$^3$/s | $\mathbf{L^3T^{-1}}$ |
| $Q'$ | Volumetric flowrate of overflow from continuous thickener | m$^3$/s | $\mathbf{L^3T^{-1}}$ |
| $R_0'$ | Drag force per unit projected area of isolated spherical particle | N/m$^2$ | $\mathbf{ML^{-1}T^{-2}}$ |
| $R_c'$ | Drag force per unit projected area of spherical particle in suspension | N/m$^2$ | $\mathbf{ML^{-1}T^{-2}}$ |
| $R_1$ | Drag force per unit surface of particle | N/m$^2$ | $\mathbf{ML^{-1}T^{-2}}$ |
| $S$ | Specific surface | m$^2$/m$^3$ | $\mathbf{L^{-1}}$ |
| $s$ | Dimensionless separation distance ($= a_r/a_1$) | — | — |
| $T$ | Absolute temperature | K | $\boldsymbol{\theta}$ |
| $t$ | Time | s | **T** |
| $t_R$ | Residence time | s | **T** |
| $U$ | Mass ratio of liquid to solid in underflow from continuous thickener | — | — |
| $u$ | Velocity | m/s | $\mathbf{LT^{-1}}$ |
| $u_0$ | Free-falling velocity of particle | m/s | $\mathbf{LT^{-1}}$ |
| $u_c$ | Sedimentation velocity of particle in suspension | m/s | $\mathbf{LT^{-1}}$ |
| $u_{cL}$ | Limit value of $u_c$ | m/s | $\mathbf{LT^{-1}}$ |

| | | Units in SI System | Dimensions in M, N, L, T, $\theta$, A |
|---|---|---|---|
| $u'_c$ | Sedimentation velocity at level of obstruction | m/s | $\mathbf{LT^{-1}}$ |
| $u_F$ | Velocity of displaced fluid | m/s | $\mathbf{LT^{-1}}$ |
| $u_f$ | Velocity of sedimentation of small particles in upper layer | m/s | $\mathbf{LT^{-1}}$ |
| $u_i$ | Sedimentation velocity at infinite dilution | m/s | $\mathbf{LT^{-1}}$ |
| $u_p$ | Velocity of particle relative to fluid | m/s | $\mathbf{LT^{-1}}$ |
| $u_u$ | Sedimentation velocity at concentration $C_u$ of underflow | m/s | $\mathbf{LT^{-1}}$ |
| $u_w$ | Velocity of propagation of concentration wave | m/s | $\mathbf{LT^{-1}}$ |
| $V$ | Volume of particles passing plane per unit area | m/s | $\mathbf{LT^{-1}}$ |
| $V_A$ | van der Waals attraction energy | J | $\mathbf{ML^2T^{-2}}$ |
| $V_R$ | Electrical repulsion energy | J | $\mathbf{ML^2T^{-2}}$ |
| $V_T$ | Total potential energy of interaction | J | $\mathbf{ML^2T^{-2}}$ |
| $v$ | Volume of particle | m$^3$ | $\mathbf{L^3}$ |
| $X$ | Average value of mass ratio of liquid to solids | — | — |
| $Y$ | Mass ratio of liquid to solids in feed to continuous thickener | — | — |
| $Z$ | Valence of ion | — | — |
| $\alpha_0$ | Ratio defined by equation 5.29 | — | — |
| $\alpha_p$ | Ratio defined by equation 5.27 | — | — |
| $\gamma$ | Constant defined by equation 5.17 | — | — |
| $\dot{\gamma}$ | Shear rate | s$^{-1}$ | $\mathbf{T^{-1}}$ |
| $\delta$ | Stern layer thickness | m | $\mathbf{L}$ |
| $\varepsilon$ | Permittivity | s$^4$A$^2$/kg m$^3$ | $\mathbf{M^{-1}L^{-3}T^4A^2}$ |
| $\kappa$ | Debye-Hückel parameter (equation 5.18) | m$^{-1}$ | $\mathbf{L^{-1}}$ |
| $\zeta$ | Electrokinetic or zeta potential | V | $\mathbf{ML^2T^{-3}A^{-1}}$ |
| $\mu$ | Viscosity of fluid | Ns/m$^2$ | $\mathbf{ML^{-1}T^{-1}}$ |
| $\mu_c$ | Viscosity of suspension | Ns/m$^2$ | $\mathbf{ML^{-1}T^{-1}}$ |
| $\rho$ | Density of fluid | kg/m$^3$ | $\mathbf{ML^{-3}}$ |
| $\rho_c$ | Density of suspension | kg/m$^3$ | $\mathbf{ML^{-3}}$ |
| $\rho_s$ | Density of solid | kg/m$^3$ | $\mathbf{ML^{-3}}$ |
| $\Phi''$ | Friction factor defined by equation 5.90 | — | — |
| $\psi$ | Volumetric rate of *sedimentation* per unit area | m/s | $\mathbf{LT^{-1}}$ |
| $\psi_L$ | Limit value of $\psi$ | m/s | $\mathbf{LT^{-1}}$ |
| $\psi_T$ | *Total* volumetric rate of solids movement per unit area | m/s | $\mathbf{LT^{-1}}$ |
| $\psi_{TL}$ | Limit value of $\psi_T$ | m/s | $\mathbf{LT^{-1}}$ |
| $\psi_u$ | Solids volumetric flowrate per unit area due to underflow | m/s | $\mathbf{LT^{-1}}$ |
| $\psi'$ | Mass rate of sedimentation per unit area at level of obstruction | kg/m$^2$s | $\mathbf{ML^{-2}T^{-1}}$ |
| $\psi_s$ | Stern plane potential | V | $\mathbf{ML^2T^{-3}A^{-1}}$ |
| $\psi_0$ | Surface potential | V | $\mathbf{ML^2T^{-3}A^{-1}}$ |
| $Ga$ | Galileo number $(d^3g(\rho_s - \rho)\rho/\mu^2)$ | — | — |
| $Re'$ | Reynolds number $(ud\rho/\mu)$ | — | — |
| $Re'_0$ | Reynolds number for particle under terminal falling conditions $(u_0 d\rho/\mu)$ | — | — |
| $Re_1$ | Bed Reynolds number $(u_c\rho/S\mu(1-e))$ | — | — |

Suffix

| | |
|---|---|
| $B$ | Glass ballotini particles |
| $L$ | Large particles |
| $M$ | Mixture |
| $P$ | Polystyrene particles |
| $S$ | Small particles |
| 0 | Value under free-falling conditions |

CHAPTER 6

# Fluidisation

## 6.1. CHARACTERISTICS OF FLUIDISED SYSTEMS

### 6.1.1. General behaviour of gas solids and liquid solids systems

When a fluid is passed downwards through a bed of solids, no relative movement between the particles takes place, unless the initial orientation of the particles is unstable, and where the flow is streamline, the pressure drop across the bed is directly proportional to the rate of flow, although at higher rates the pressure drop rises more rapidly. The pressure drop under these conditions may be obtained using the equations in Chapter 4.

When a fluid is passed upwards through a bed, the pressure drop is the same as that for downward flow at relatively low rates. When, however, the frictional drag on the particles becomes equal to their apparent weight, that is the actual weight less the buoyancy force, the particles become rearranged thus offering less resistance to the flow of fluid and the bed starts to expand with a corresponding increase in voidage. This process continues with increase in velocity, with the total frictional force remaining equal to the weight of the particles, until the bed has assumed its loosest stable form of packing. If the velocity is then increased further, the individual particles separate from one another and become freely supported in the fluid. At this stage, the bed is described as *fluidised*. Further increase in the velocity causes the particles to separate still further from one another, although the pressure difference remains approximately equal to the weight per unit area of the bed. In practice, the transition from the fixed to the fluidised bed condition is not uniform mainly due to irregularities in the packing and, over a range of velocities, fixed and fluidised bed regions may co-exist. In addition, with gases, surface-related forces give rise to the formation of conglomerates of particles through which there is a minimal flow and, as a result, much of the gas may pass through the bed in channels. This is an unstable condition since the channels that offer a relatively low resistance to flow, tend to open up as the gas flowrate is increased and regions of the bed may remain in an unfluidised state even though the overall superficial velocity may be much higher than the minimum fluidising velocity.

Up to this point, the system behaves in a similar way with both liquids and gases, although at high fluid velocities, there is usually a fairly sharp distinction between the behaviour of the two systems. With a liquid, the bed continues to expand as the velocity is increased and it maintains its uniform character, with the degree of agitation of the particles increasing progressively. This type of fluidisation is known as *particulate fluidisation*. With a gas, however, uniform fluidisation is frequently obtained only at low velocities. At higher velocities two separate *phases* may form — a continuous phase, often referred to as the *dense* or *emulsion* phase, and a discontinuous phase known as the *lean* or *bubble* phase. The fluidisation is then said to be *aggregative*. At much higher velocities, the bubbles tend to break down — a feature that leads to a much more chaotic structure. When gas bubbles pass through a relatively high-density fluidised bed the system closely resembles a boiling liquid, with the lean phase corresponding to the vapour and the dense or continuous phase corresponding to the liquid. The bed is then often referred to as a *boiling bed*, as opposed to the *quiescent bed* usually formed at low flowrates. As the gas flowrate is increased, the velocity relative to the particles in the dense phase does not change appreciably, and streamline flow may persist even at very high overall rates of flow because a high proportion of the total flow is then in the form of bubbles. At high flowrates in deep beds, coalescence of the bubbles takes place, and in narrow vessels, slugs of gas occupying the whole cross-section may be produced. These slugs of gas alternate with slugs of fluidised solids that are carried upwards and subsequently collapse, releasing the solids which fall back.

In an early attempt to differentiate between the conditions leading to particulate or aggregative fluidisation, WILHELM and KWAUK[1] suggested using the value of the Froude number ($u_{mf}^2/gd$) as a criterion, where:

$u_{mf}$ is the minimum velocity of flow, calculated over the whole cross-section of the bed, at which fluidisation takes place,
$d$ is the diameter of the particles, and
$g$ is the acceleration due to gravity.

At values of a Froude group of less than unity, particulate fluidisation normally occurs and, at higher values, aggregative fluidisation takes place. Much lower values of the Froude number are encountered with liquids because the minimum velocity required to produce fluidisation is less. A theoretical justification for using the Froude group as a means of distinguishing between particulate and aggregative fluidisation has been provided by JACKSON[2] and MURRAY[3].

Although the possibility of forming fluidised beds had been known for many years, the subject remained of academic interest until the adoption of fluidised catalysts by the petroleum industry for the cracking of heavy hydrocarbons and for the synthesis of fuels from natural gas or from carbon monoxide and hydrogen. In many ways, the fluidised bed behaves as a single fluid of a density equal to that of the mixture of solids and fluid. Such a bed will flow, it is capable of transmitting hydrostatic forces, and solid objects with densities less than that of the bed will float at the surface. Intimate mixing occurs within the bed and heat transfer rates are very high with the result that uniform temperatures are quickly attained throughout the system. The easy control of temperature is the feature that has led to the use of fluidised solids for highly exothermic processes, where uniformity of temperature is important.

In order to understand the properties of a fluidised system, it is necessary to study the flow patterns of both the solids and the fluid. The mode of formation and behaviour of fluid bubbles is of particular importance because these usually account for the flow of a high proportion of the fluid in a gas–solids system.

In any study of the properties of a fluidised system, it is necessary to select conditions which are reproducible and the lack of agreement between the results of many workers, particularly those relating to heat transfer, is largely attributable to the existence of widely different uncontrolled conditions within the bed. The fluidisation should be of good quality, that is to say, that the bed should be free from irregularities and channelling. Many solids, particularly those of appreciably non-isometric shape and those that have a tendency to form agglomerates will never fluidise readily in a gas. Furthermore, the fluid must be evenly distributed at the bottom of the bed and it is usually necessary to provide a distributor across which the pressure drop is equal to at least that across the bed. This condition is much more readily achieved in a small laboratory apparatus than in large-scale industrial equipment.

As already indicated, when a liquid is the fluidising agent, substantially uniform conditions pervade in the bed, although with a gas, bubble formation tends to occur except at very low fluidising velocities. In an attempt to improve the reproducibility of conditions within a bed, much of the earlier research work with gas fluidised systems was carried out at gas velocities sufficiently low for bubble formation to be absent. In recent years, however, it has been recognised that bubbles normally tend to form in such systems, that they exert an important influence on the flow pattern of both gas and solids, and that the behaviour of individual bubbles can often be predicted with reasonable accuracy.

## 6.1.2. Effect of fluid velocity on pressure gradient and pressure drop

When a fluid flows slowly upwards through a bed of very fine particles the flow is streamline and a linear relation exists between pressure gradient and flowrate as discussed in Chapter 4, Section 4.2.3. If the pressure gradient $(-\Delta P/l)$ is plotted against the superficial velocity $(u_c)$ using logarithmic co-ordinates a straight line of unit slope is obtained, as shown in Figure 6.1. As the superficial velocity approaches the *minimum* fluidising velocity $(u_{mf})$, the bed starts to expand and when the particles are no longer in physical contact with one another the bed *is fluidised.* The pressure *gradient* then becomes lower because of the increased voidage and, consequently, the weight of particles per unit height of bed is smaller. This fall continues until the velocity is high enough for transport of the material to take place, and the pressure gradient then starts to increase again because the frictional drag of the fluid at the walls of the tube starts to become significant. When the bed is composed of large particles, the flow will be laminar only at very low velocities and the slope $s$ of the lower part of the curve will be greater $(1 < s < 2)$ and may not

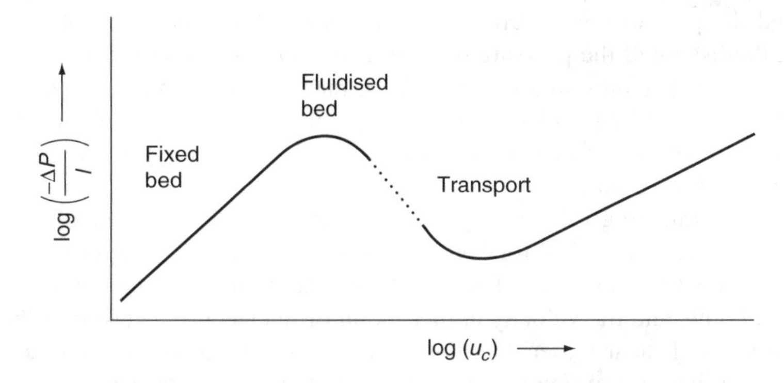


Figure 6.1. Pressure gradient within a bed as a function of fluid velocity

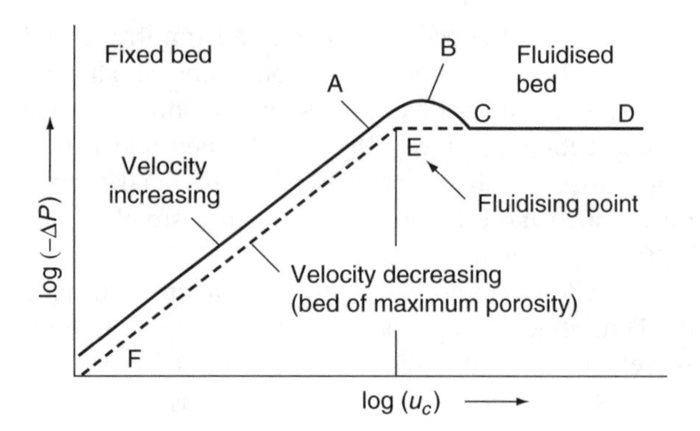


Figure 6.2. Pressure drop over fixed and fluidised beds

be constant, particularly if there is a progressive change in flow regime as the velocity increases.

If the pressure across the whole bed instead of the pressure gradient is plotted against velocity, also using logarithmic coordinates as shown in Figure 6.2, a linear relation is again obtained up to the point where expansion of the bed starts to take place (A), although the slope of the curve then gradually diminishes as the bed expands and its porosity increases. As the velocity is further increased, the pressure drop passes through a maximum value (B) and then falls slightly and attains an approximately constant value that is independent of the fluid velocity (CD). If the fluid velocity is reduced again, the bed contracts until it reaches the condition where the particles are just resting on one another (E). The porosity then has the maximum stable value which can occur for a fixed bed of the particles. If the velocity is further decreased, the structure of the bed then remains unaffected provided that the bed is not subjected to vibration. The pressure drop (EF) across this reformed fixed bed at any fluid velocity is then less than that before fluidisation. If the velocity is now increased again, it might be expected that the curve (FE) would be retraced and that the slope would suddenly change from 1 to 0 at the fluidising point. This condition is difficult to reproduce, however, because the bed tends to become consolidated again unless it is completely free from vibration. In the absence of channelling, it is the shape and size of the particles that determine both the maximum porosity and the pressure drop across a given height of fluidised bed of a given depth. In an ideal fluidised bed the pressure drop corresponding to ECD is equal to the buoyant weight of particles per unit area. In practice, it may deviate appreciably from this value as a result of channelling and the effect of particle-wall friction. Point B lies above CD because the frictional forces between the particles have to be overcome before bed rearrangement can take place.

The minimum fluidising velocity, $u_{mf}$, may be determined experimentally by measuring the pressure drop across the bed for both increasing and decreasing velocities and plotting the results as shown in Figure 6.2. The two 'best' straight lines are then drawn through the experimental points and the velocity at their point of intersection is taken as the minimum fluidising velocity. Linear rather than logarithmic plots are generally used, although it is necessary to use logarithmic plots if the plot of pressure gradient against velocity in the fixed bed is not linear.

The theoretical value of the minimum fluidising velocity may be calculated from the equations given in Chapter 4 for the relation between pressure drop and velocity in a fixed packed bed, with the pressure drop through the bed put equal to the apparent weight of particles per unit area, and the porosity set at the maximum value that can be attained in the fixed bed.

In a fluidised bed, the total frictional force on the particles must equal the effective weight of the bed. Thus, in a bed of unit cross-sectional area, depth $l$, and porosity $e$, the additional pressure drop across the bed attributable to the layout weight of the particles is given by:

$$-\Delta P = (1 - e)(\rho_s - \rho)lg \tag{6.1}$$

where: $g$ is the acceleration due to gravity and

$\rho_s$ and $\rho$ are the densities of the particles and the fluid respectively.

This relation applies from the initial expansion of the bed until transport of solids takes place. There may be some discrepancy between the calculated and measured minimum velocities for fluidisation. This may be attributable to channelling, as a result of which the drag force acting on the bed is reduced, to the action of electrostatic forces in case of gaseous fluidisation — particularly important in the case of sands — to agglomeration which is often considerable with small particles, or to friction between the fluid and the walls of the containing vessel. This last factor is of greatest importance with beds of small diameters. Leva *et al.*[4] introduced a term, $(G_F - G_E)/G_F$, which is a fluidisation efficiency, in which $G_F$ is the minimum flowrate required to produce fluidisation and $G_E$ is the rate required to produce the initial expansion of the bed.

If flow conditions within the bed are streamline, the relation between fluid velocity $u_c$, pressure drop $(-\Delta P)$ and voidage $e$ is given, for a fixed bed of spherical particles of diameter $d$, by the Carman-Kozeny equation (4.12*a*) which takes the form:

$$u_c = 0.0055\left(\frac{e^3}{(1-e)^2}\right)\left(\frac{-\Delta P d^2}{\mu l}\right) \tag{6.2}$$

For a fluidised bed, the buoyant weight of the particles is counterbalanced by the frictional drag. Substituting for $-\Delta P$ from equation 6.1 into equation 6.2 gives:

$$u_c = 0.0055\left(\frac{e^3}{1-e}\right)\left(\frac{d^2(\rho_s - \rho)g}{\mu}\right) \tag{6.3}$$

There is evidence in the work reported in Chapter 5 on sedimentation[5] to suggest that where the particles are free to adjust their orientations with respect to one another and to the fluid, as in sedimentation and fluidisation, the equations for pressure drop in fixed beds overestimate the values where the particles can 'choose' their orientation. A value of 3.36 rather than 5 for the Carman-Kozeny constant is in closer accord with experimental data. The coefficient in equation 6.3 then takes on the higher value of 0.0089. The experimental evidence is limited to a few measurements however and equation 6.3, with its possible inaccuracies, is used here.

### 6.1.3. Minimum fluidising velocity

As the upward velocity of flow of fluid through a packed bed of uniform spheres is increased, the point of *incipient fluidisation* is reached when the particles are just supported in the fluid. The corresponding value of the *minimum fluidising velocity* $(u_{mf})$ is then obtained by substituting $e_{mf}$ into equation 6.3 to give:

$$u_{mf} = 0.0055 \left( \frac{e_{mf}^3}{1 - e_{mf}} \right) \frac{d^2(\rho_s - \rho)g}{\mu} \tag{6.4}$$

Since equation 6.4 is based on the Carman–Kozeny equation, it applies only to conditions of laminar flow, and hence to low values of the Reynolds number for flow in the bed. In practice, this restricts its application to fine particles.

The value of $e_{mf}$ will be a function of the shape, size distribution and surface properties of the particles. Substituting a typical value of 0.4 for $e_{mf}$ in equation 6.4 gives:

$$(u_{mf})_{e_{mf}=0.4} = 0.00059 \left( \frac{d^2(\rho_s - \rho)g}{\mu} \right) \tag{6.5}$$

When the flow regime at the point of incipient fluidisation is outside the range over which the Carman-Kozeny equation is applicable, it is necessary to use one of the more general equations for the pressure gradient in the bed, such as the Ergun equation given in equation 4.20 as:

$$\frac{-\Delta P}{l} = 150 \left( \frac{(1-e)^2}{e^3} \right) \left( \frac{\mu u_c}{d^2} \right) + 1.75 \left( \frac{(1-e)}{e^3} \right) \left( \frac{\rho u_c^2}{d} \right) \tag{6.6}$$

where $d$ is the diameter of the sphere with the same volume:surface area ratio as the particles.

Substituting $e = e_{mf}$ at the incipient fluidisation point and for $-\Delta P$ from equation 6.1, equation 6.6 is then applicable at the minimum fluidisation velocity $u_{mf}$, and gives:

$$(1 - e_{mf})(\rho_s - \rho)g = 150 \left( \frac{(1 - e_{mf})^2}{e_{mf}^3} \right) \left( \frac{\mu u_{mf}}{d^2} \right) + 1.75 \left( \frac{(1 - e_{mf})}{e_{mf}^3} \right) \left( \frac{\rho u_{mf}^2}{d} \right) \tag{6.7}$$

Multiplying both sides by $\dfrac{\rho d^3}{\mu^2(1 - e_{mf})}$ gives:

$$\frac{\rho(\rho_s - \rho)gd^3}{\mu^2} = 150 \left( \frac{1 - e_{mf}}{e_{mf}^3} \right) \left( \frac{u_{mf} d\rho}{\mu} \right) + \left( \frac{1.75}{e_{mf}^3} \right) \left( \frac{u_{mf} d\rho}{\mu} \right)^2 \tag{6.8}$$

In equation 6.8:

$$\frac{d^3 \rho(\rho_s - \rho)g}{\mu^2} = Ga \tag{6.9}$$

where $Ga$ is the 'Galileo number'.

and:

$$\frac{u_{mf} d\rho}{\mu} = Re'_{mf}. \tag{6.10}$$

where $Re_{mf}$ is the Reynolds number at the minimum fluidising velocity and equation 6.8 then becomes:

$$Ga = 150\left(\frac{1-e_{mf}}{e_{mf}^3}\right)Re'_{mf} + \left(\frac{1.75}{e_{mf}^3}\right)Re'^2_{mf} \tag{6.11}$$

For a typical value of $e_{mf} = 0.4$:

$$Ga = 1406Re'_{mf} + 27.3Re'^2_{mf} \tag{6.12}$$

Thus:

$$Re'^2_{mf} + 51.4Re'_{mf} - 0.0366Ga = 0 \tag{6.13}$$

and:

$$(Re'_{mf})_{e_{mf}=0.4} = 25.7\{\sqrt{(1 + 5.53 \times 10^{-5}Ga)} - 1\} \tag{6.14}$$

and, similarly for $e_{mf} = 0.45$:

$$(Re'_{mf})_{e_{mf}=0.45} = 23.6\{\sqrt{(1 + 9.39 \times 10^{-5}\,Ga)} - 1\} \tag{6.14a}$$

By definition:

$$u_{mf} = \frac{\mu}{d\rho}Re'_{mf} \tag{6.15}$$

It is probable that the Ergun equation, like the Carman-Kozeny equation, also overpredicts pressure drop for fluidised systems, although no experimental evidence is available on the basis of which the values of the coefficients may be amended.

WEN and YU[(6)] have examined the relationship between voidage at the minimum fluidising velocity, $e_{mf}$, and particle shape, $\phi_s$, which is defined as the ratio of the diameter of the sphere of the same specific as the particle $d$, as used in the Ergun equation to the diameter of the sphere with the same volume as the particle $d_p$.

Thus:

$$\phi_s = d/d_p \tag{6.16}$$

where:

$$d = 6V_p/A_p \text{ and } d_p = (6V_p/\pi)^{1/3}.$$

In practice the particle size $d$ can be determined only by measuring both the volumes $V_p$ and the areas $A_p$ of the particles. Since this operation involves a somewhat tedious experimental technique, it is more convenient to measure the particle volume only and then work in terms of $d_p$ and the shape factor.

The minimum fluidising velocity is a function of both $e_{mf}$ and $\phi_s$, neither of which is easily measured or estimated, and Wen and Yu have shown that these two quantities are, in practice, inter-related. These authors have published experimental data of $e_{mf}$ and $\phi_s$ for a wide range of well-characterised particles, and it has been shown that the relation between these two quantities is essentially independent of particle size over a wide range. It has also been established that the following two expressions give reasonably good correlations between $e_{mf}$ and $\phi_s$, as shown in Figure 6.3:

$$\left(\frac{1-e_{mf}}{e_{mf}^3}\right)\frac{1}{\phi_s^2} = 11 \tag{6.17}$$

$$\left(\frac{1}{e_{mf}{}^3}\frac{1}{\phi_s}\right) = 14 \tag{6.18}$$

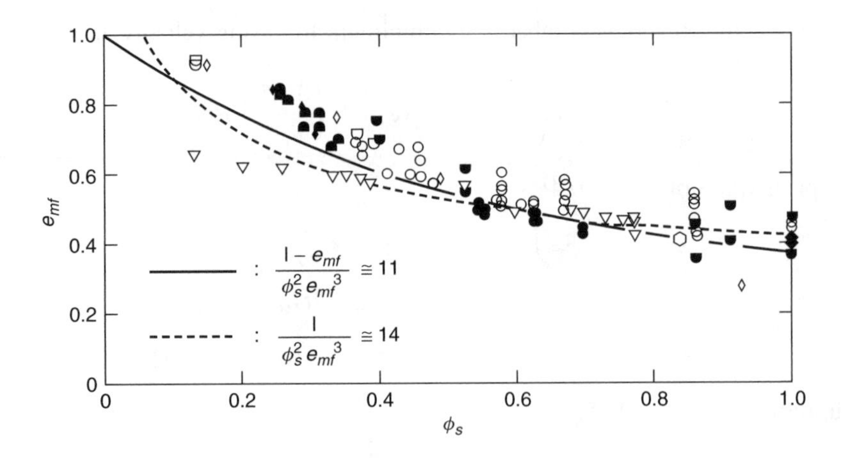


Figure 6.3. Relation between $e_{mf}$ and $\phi_s$

NIVEN[7] discusses the significance of the two dimensionless groups in equations 6.17 and 6.18, and also suggests that $d$ and $u_{mf}$ in equations 6.8, 6.9 and 6.10 are more appropriately replaced by a mean linear dimension of the pores and the mean pore velocity at the point of incipient fluidisation.

Using equation 6.16 to substitute for $\dfrac{\phi_s}{d_p}$ for $d$ in equation 6.6 gives:

$$(1 - e_{mf})(\rho_s - \rho)g = 150\left(\frac{(1 - e_{mf})^2}{e_{mf}^3}\right)\left(\frac{\mu u_{mf}}{\phi_s^2 d_p{}^2}\right) + 1.75\left(\frac{1 - e_{mf}}{e_{mf}^3}\right)\frac{\rho u_{mf}{}^2}{\phi_s d_p}$$

Thus:

$$\frac{(\rho_s - \rho)\rho g d_p^3}{\mu^2} = 150\left(\frac{1 - e_{mf}}{e_{mf}^3}\right)\frac{1}{\phi_s^2}\left(\frac{\rho d_p u_{mf}}{\mu}\right) + 1.75\left(\frac{1}{e_{mf}^3 \phi_s}\right)\left(\frac{\rho^2 d_p^2 u_{mf}^2}{\mu^2}\right)$$

Substituting from equations 6.17 and 6.18:

$$Ga_p = (150 \times 11)Re'_{mfp} + (1.75 \times 14)Re'^2_{mfp}$$

where $Ga_p$ and $Re_{mfp}$ are the Galileo number and the particle Reynolds number at the point of incipient fluidisation, in both cases with the linear dimension of the particles expressed as $d_p$.

Thus: $$Re'^2_{mfp} + 67.3Re'_{mfp} - 0.0408Ga_p = 0$$

giving: $$Re'_{mfp} = 33.65[\sqrt{(1 + 6.18 \times 10^{-5}Ga_p)} - 1] \qquad (6.19)$$

where: $$u_{mf} = \left(\frac{\mu}{d_p \rho}\right) Re'_{mfp} \qquad (6.20)$$

### Example 6.1

A bed consists of uniform spherical particles of diameter 3 mm and density 4200 kg/m$^3$. What will be the minimum fluidising velocity in a liquid of viscosity 3 mNs/m$^2$ and density 1100 kg/m$^3$?

## Solution

By definition:

$$\text{Galileo number, } Ga = d^3\rho(\rho_s - \rho)g/\mu^2$$
$$= ((3 \times 10^{-3})^3 \times 1100 \times (4200 - 1100) \times 9.81)/(3 \times 10^{-3})^2$$
$$= 1.003 \times 10^5$$

Assuming a value of 0.4 for $e_{mf}$, equation 6.14 gives:

$$Re'_{mf} = 25.7\{\surd(1 + (5.53 \times 10^{-5})(1.003 \times 10^5)) - 1\} = 40$$

and: $$u_{mf} = (40 \times 3 \times 10^{-3})/(3 \times 10^{-3} \times 1100) = 0.0364 \text{ m/s or } \underline{\underline{36.4 \text{ mm/s}}}$$

## Example 6.2

Oil, of density 900 kg/m$^3$ and viscosity 3 mNs/m$^2$, is passed vertically upwards through a bed of catalyst consisting of approximately spherical particles of diameter 0.1 mm and density 2600 kg/m$^3$. At approximately what mass rate of flow per unit area of bed will (a) fluidisation, and (b) transport of particles occur?

## Solution

(a) Equations 4.9 and 6.1 may be used to determine the fluidising velocity, $u_{mf}$.

$$u = (1/K'')(e^3/(S^2(1 - e)^2)(1/\mu)(-\Delta P/l) \qquad \text{(equation 4.9)}$$

$$-\Delta P = (1 - e)(\rho_s - \rho)lg \qquad \text{(equation 6.1)}$$

where $S$ = surface area/volume, which, for a sphere, $= \pi d^2/(\pi d^3/6) = 6/d$.

Substituting $K'' = 5$, $S = 6/d$ and $-\Delta P/l$ from equation 6.1 into equation 4.9 gives:

$$u_{mf} = 0.0055(e^3/(1 - e))(d^2(\rho_s - \rho)g)/\mu$$

Hence : $$G'_{mf} = \rho u = (0.0055e^3/(1 - e))(d^2(\rho_s - \rho)g)/\mu$$

In this problem, $\rho_s = 2600$ kg/m$^3$, $\rho = 900$ kg/m$^3$, $\mu = 3.0 \times 10^{-3}$ Ns/m$^2$ and $d = 0.1$ mm $= 1.0 \times 10^{-4}$ m.

As no value of the voidage is available, $e$ will be estimated by considering eight closely packed spheres of diameter $d$ in a cube of side $2d$. Thus:

$$\text{volume of spheres} = 8(\pi/6)d^3$$

$$\text{volume of the enclosure} = (2d)^3 = 8d^3$$

and hence: $$\text{voidage, } e = [8d^3 - 8(\pi/6)d^3]/8d^3 = 0.478, \text{ say, } 0.48.$$

Thus : $$G'_{mf} = 0.0055(0.48)^3(10^{-4})^2((900 \times 1700) \times 9.81)/((1 - 0.48) \times 3 \times 10^{-3}$$
$$= \underline{\underline{0.059 \text{ kg/m}^2\text{s}}}$$

(b) Transport of the particles will occur when the fluid velocity is equal to the terminal falling velocity of the particle.

Using Stokes' law : $u_0 = d^2 g(\rho_s - \rho)/18\mu$ (equation 3.24)

$$= ((10^{-4})^2 \times 9.81 \times 1700)/(18 \times 3 \times 10^{-3})$$

$$= 0.0031 \text{ m/s}$$

The Reynolds number $= ((10^{-4} \times 0.0031 \times 900)/(3 \times 10^{-3}) = 0.093$ and hence Stokes' law applies.

The required mass flow $= (0.0031 \times 900) = 2.78$ kg/m$^2$s

An alternative approach is to make use of Figure 3.6 and equation 3.35,

$$(R/\rho u^2)Re^2 = 2d^3\rho g(\rho_s - \rho)/3\mu^2$$

$$= (2 \times (10^{-4})^3 \times (900 \times 9.81) \times 1700)/(3(3 \times 10^{-3})^2) = 1.11$$

From Figure 3.6, $Re = 0.09$

Hence: $u_0 = Re(\mu/\rho d) = (0.09 \times 3 \times 10^{-3})/(900 \times 10^{-4}) = 0.003$ m/s

and: $G' = (0.003 \times 900) = \underline{\underline{2.7 \text{ kg/m}^2\text{s}}}$

### 6.1.4. Minimum fluidising velocity in terms of terminal failing velocity

The minimum fluidising velocity, $u_{mf}$, may be expressed in terms of the free-falling velocity $u_0$ of the particles in the fluid. The Ergun equation (equation 6.11) relates the Galileo number $Ga$ to the Reynolds number $Re'_{mf}$ in terms of the voidage $e_{mf}$ at the incipient fluidisation point.

In Chapter 3, relations are given that permit the calculation of $Re'_0(u_0 d\rho/\mu)$, the particle Reynolds number for a sphere at its terminal falling velocity $u_0$, also as a function of Galileo number. Thus, it is possible to express $Re'_{mf}$ in terms of $Re'_0$ and $u_{mf}$ in terms of $u_0$.

For a spherical particle the Reynolds number $Re'_0$ is expressed in terms of the Galileo number $Ga$ by equation 3.40 which covers the whole range of values of $Re'$ of interest. This takes the form:

$$Re'_0 = (2.33Ga^{0.018} - 1.53Ga^{-0.016})^{13.3} \tag{6.21}$$

Equation 6.21 applies when the particle motion is not significantly affected by the walls of the container, that is when $d/d_t$ tends to zero.

Thus, for any value of $Ga$, $Re'_0$ may be calculated from equation 6.21 and $Re'_{mf}$ from equation 6.11 for a given value of $e_{mf}$. The ratio $Re'_0/Re'_{mf}(= u_0/u_{mf})$ may then be plotted against $Ga$ with $e_{mf}$ as the parameter. Such a plot is given in Figure 6.4 which includes some experimental data. Some scatter is evident, largely attributable to the fact that the diameter of the vessel ($d_t$) was not always large compared with that of the particle. Nevertheless, it is seen that the experimental results straddle the curves covering a range

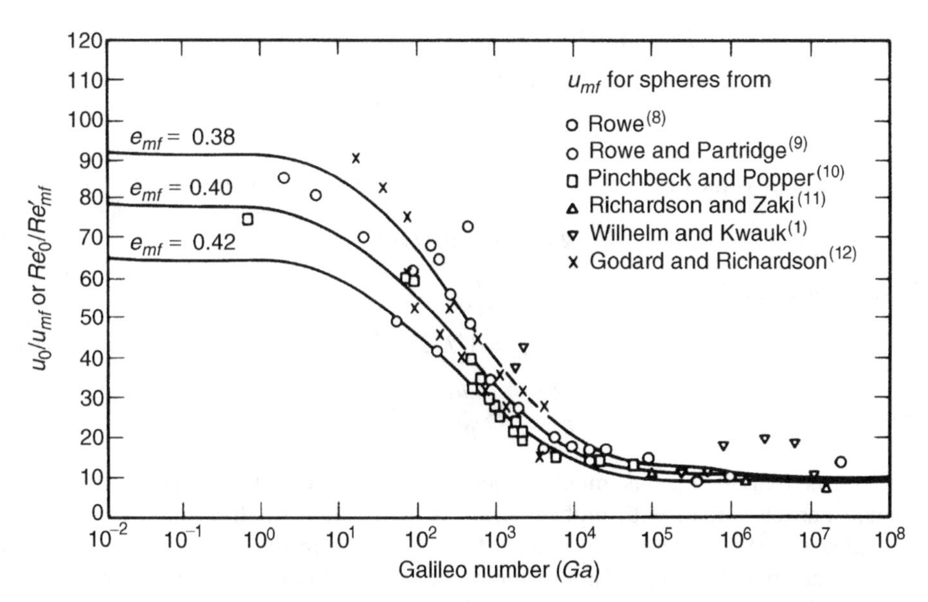


Figure 6.4. Ratio of terminal falling velocity to minimum fluidising velocity, as a function of Galileo number

of values of $e_{mf}$ from about 0.38 to 0.42. The agreement between the experimental and calculated values is quite good, especially in view of the uncertainty of the actual values of $e_{mf}$ in the experimental work, and the fact that the Ergun equation does not necessarily give an accurate prediction of pressure drop in a fixed bed, especially near the incipient fluidisation points.

It is seen in Chapter 3 that it is also possible to express $Re'_0$ in terms of $Ga$ by means of three simple equations, each covering a limited range of values of $Ga$ (equations 3.37, 3.38 and 3.39) as follows:

$$Ga = 18Re'_0 \qquad (Ga < 3.6) \tag{6.22}$$

$$Ga = 18Re'_0 + 2.7Re_0'^{1.687} \qquad (3.6 < \text{Ga} < 10^5) \tag{6.23}$$

$$Ga = \tfrac{1}{3}Re_0'^2 \qquad (Ga > \text{ca.}10^5) \tag{6.24}$$

It is convenient to use equations 6.22 and 6.24 as these enable very simple relations for $Re'_0/Re'_{mf}$ to be obtained at both low and high values of $Ga$.

Taking a typical value of $e_{mf}$ of 0.4, the relation between $Re'_{mf}$ and $Ga$ is given by equation 6.13.

For low values of $Re'_{mf}(<0.003)$ and of $Ga(<3.6)$, the first term may be neglected and:

$$Re'_{mf} = 0.000712Ga \tag{6.25}$$

Equation 6.22 gives:
$$Re'_0 = 0.0556Ga \tag{6.26}$$

Combining equations 6.25 and 6.26:

$$\frac{Re'_0}{Re'_{mf}} = \frac{u_0}{u_{mf}} = 78 \tag{6.27}$$

Again, for high values of $Re'_{mf}(>\sim 200)$ and $Ga(>10^5)$, equation 6.13 gives:

$$Re'_{mf} = 0.191 Ga^{1/2} \tag{6.28}$$

Equation 6.24 gives:

$$Re'_0 = 1.732 Ga^{1/2} \tag{6.29}$$

Thus :

$$\frac{Re'_0}{Re'_{mf}} = \frac{u_0}{u_{mf}} = 9.1 \tag{6.30}$$

This shows that $u_0/u_{mf}$ is much larger for low values of $Ga$, generally obtained with small particles, than with high values. For particulate fluidisation with liquids, the theoretical range of fluidising velocities is from a minimum of $u_{mf}$ to a maximum of $u_0$. It is thus seen that there is a far greater range of velocities possible in the streamline flow region. In practice, it is possible to achieve flow velocities greatly in excess of $u_0$ for gases, because a high proportion of the gas can pass through the bed as bubbles and effectively by-pass the particles.

## 6.2. LIQUID-SOLIDS SYSTEMS

### 6.2.1. Bed expansion

Liquid-fluidised systems are generally characterised by the regular expansion of the bed that takes place as the velocity increases from the minimum fluidisation velocity to the terminal falling velocity of the particles. The general relation between velocity and volumetric concentration or voidage is found to be similar to that between sedimentation velocity and concentration for particles in a suspension. The two systems are hydrodynamically similar in that in the fluidised bed the particles undergo no net movement and are maintained in suspension by the upward flow of liquid, whereas in the sedimenting suspension the particles move downwards and the only flow of liquid is the upward flow of that liquid which is displaced by the settling particles. RICHARDSON and ZAKI[11] showed that, for sedimentation or fluidisation of uniform particles:

$$\frac{u_c}{u_i} = e^n = (1 - C)^n \tag{6.31}$$

where: $u_c$ is the observed sedimentation velocity or the empty tube fluidisation velocity,

$u_i$ is the corresponding velocity at infinite dilution,

$e$ is the voidage of the system,

$C$ is the volumetric fractional concentration of solids, and

$n$ is an index.

The existence of a relationship of the form of equation 6.31 had been established six years earlier by WILHELM and KWAUK[1] who fluidised particles of glass, sand and lead shot with water. On plotting particle Reynolds number against bed voidage using logarithmic scales, good straight lines were obtained over the range of conditions for which the bed was fluidised.

A similar equation had previously been given by LEWIS and BOWERMAN[13].

Equation 6.31 is similar to equation 5.71 for a sedimenting suspension. Values of the index $n$ range from 2.4 to 4.8 and are the same for sedimentation and for fluidisation at a given value of the Galileo number $Ga$. These may be calculated from equation 6.32, which is identical to equation 5.84 in Chapter 5:

$$\frac{(4.8 - n)}{(n - 2.4)} = 0.043Ga^{0.57}\left[1 - 1.24\left(\frac{d}{d_t}\right)^{0.27}\right] \tag{6.32}$$

RICHARDSON and ZAKI[11] found that $u_i$ corresponded closely to $u_0$, the free settling velocity of a particle in an infinite medium, for work on sedimentation as discussed in Chapter 5, although $u_i$ was somewhat less than $u_0$ in fluidisation. The following equation for fluidisation was presented:

$$\log_{10} u_0 = \log_{10} u_i + \frac{d}{d_t} \tag{6.33}$$

The difference is likely to be attributed to the fact that $d/d_t$ was very small in the sedimentation experiments. More recently, KHAN and RICHARDSON[14] have proposed the following relation to account for the effect of the walls of the vessel in fluidisation:

$$\frac{u_i}{u_0} = 1 - 1.15\left(\frac{d}{d_t}\right)^{0.6} \tag{6.34}$$

If logarithmic co-ordinates are used to plot the voidage $e$ of the bed against the superficial velocity $u_c$ (Figure 6.5), the resulting curve can be represented approximately by two straight lines joined by a short transitional curve. At low velocities the voidage remains constant corresponding to that of the fixed bed, and for the fluidised state there is a linear relation between $\log u_c$ and $\log e$. The curve shown refers to the fluidisation of steel spheres in water. It should be noted that whereas, in the absence of channelling, the pressure drop across a bed of a given expansion is directly proportional to its depth, the fluidising velocity is independent of depth.

An alternative way of calculating the index $n$ in equation 6.31 for the expansion of particulately fluidised systems is now considered. Neglecting effects due to the container wall then:

$$\frac{u_c}{u_0} = \frac{Re'_c}{Re'_0} = e^n \tag{6.35}$$

where $Re'_c$ is the Reynolds number $u_c d\rho/\mu$.

Taking logarithms:
$$n = \frac{\log(u_c/u_0)}{\log e} = \frac{-\log(Re'_0/Re'_c)}{\log e} \tag{6.36}$$

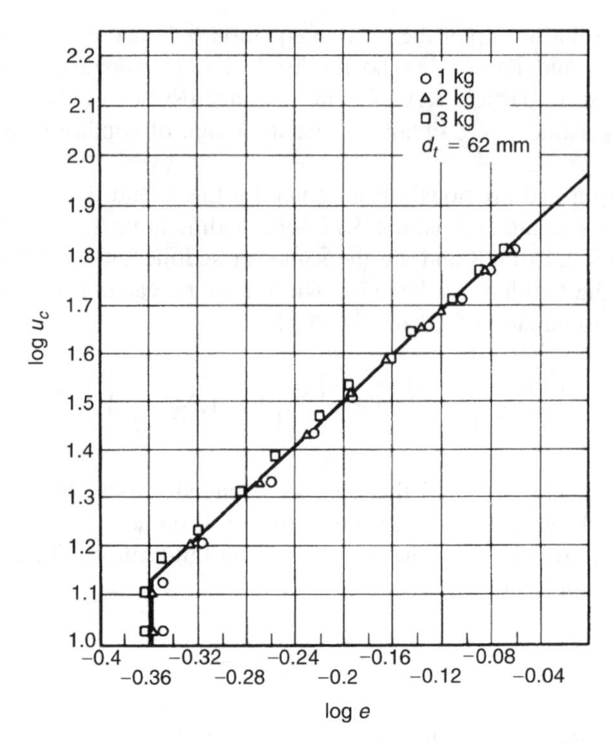


Figure 6.5. Relation between fluid velocity ($u_c$) and voidage ($e$) for the fluidisation of 6.4 mm steel spheres in water

On the assumption that equation 6.31 may be applied at the point of incipient fluidisation:

$$n = \frac{\log(u_{mf}/u_0)}{\log e_{mf}} = \frac{-\log(Re'_0/Re'_{mf})}{\log e_{mf}} \tag{6.37}$$

For a typical value of $e_{mf}$ of 0.4, $Re'_{mf}$ is given by equation 6.14. Furthermore, $Re'_0$ is given by equation 6.21. Substitution into equation 6.37 then gives:

$$n = 2.51 \log \left\{ \frac{(1.83Ga^{0.018} - 1.2Ga^{-0.016})^{13.3}}{\sqrt{(1 + 5.53 \times 10^{-5}Ga)} - 1} \right\} \tag{6.38}$$

Equation 6.38 which applies to low values of $d/d_t$ is plotted in Figure 6.6, together with experimental points from the literature, annotated according to the $d/d_t$ range which is applicable[14]. The scatter, and the low experimental values of $n$, are attributable partly to the wider range of $d/d_t$ values covered and also inaccuracies in the experimental measurements which are obtained from the results of a number of workers. For $e_{mf} = 0.43$, the calculated values of $n$ are virtually unchanged over the range $10 < Ga < 10^5$.

An alternative method of calculating the value of $Re'_{mf}$ (and hence $u_{mf}$) is to substitute for $Re'_0$ from equation 6.21 into equation 6.35, and to put the voidage $e$ equal to its value $e_{mf}$ at the minimum fluidising velocity.

In this way:

$$Re'_{mf} = (2.33Ga^{0.018} - 1.53Ga^{0.016})^{13.3} e^n_{mf} \tag{6.39}$$

where $n$ is given by equation 6.32.

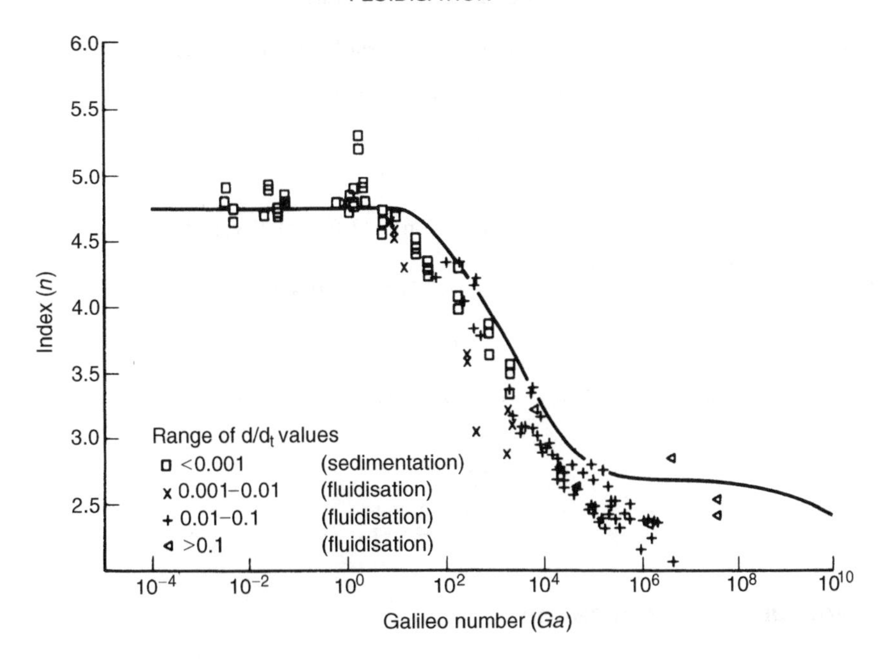


Figure 6.6. Comparison of values of the index $n$ calculated from equation 6.37 with experimental data

The same procedure may be adopted for calculating the minimum fluidising for a shear-thinning non-Newtonian fluid which exhibits *power-law* behaviour, although it is necessary to use the modified Reynolds number $(Re_1)_n$ given in Chapter 4, equation 4.28.

For inelastic fluids exhibiting power-law behaviour, the bed expansion which occurs as the velocity is increased above the minimum fluidising velocity follows a similar pattern to that obtained with a Newtonian liquid, with the exponent in equation 6.31 differing by no more than about 10 per cent. There is some evidence, however, that with viscoelastic polymer solutions the exponent may be considerably higher. Reference may be made to work by SRINIVAS and CHHABRA[15] for further details.

## Example 6.3

Glass particles of 4 mm diameter are fluidised by water at a velocity of 0.25 m/s. What will be the voidage of the bed?

The density of glass = 2500 kg/m$^3$, the density of water = 1000 kg/m$^3$, and the viscosity of water = 1mNs/m$^2$.

## Solution

Galileo number for particles in water, $Ga = \dfrac{d^3\rho(\rho_s - \rho)g}{\mu^2}$ (equation 6.9)

$$= \frac{(4 \times 10^{-3})^3 \times 1000 \times 1500 \times 9.81}{(1 \times 10^{-3})^2} = 9.42 \times 10^5$$

Reynolds number $Re'_0$ at terminal falling velocity is given by equation 6.21:

$$Re'_0 = (2.33Ga^{0.018} - 1.53Ga^{-0.016})^{13.3}$$

Thus:
$$u_0 = 1800\left(\frac{1 \times 10^{-3}}{4 \times 10^{-3} \times 1000}\right) = 0.45 \text{ m/s}$$

The value of $n$ in equation 6.31 is given by equation 6.32 for small values of $d/d_t$ as:

$$\frac{(4.8 - n)}{(n - 2.4)} = 0.043Ga^{0.57} = 109.5$$

$\therefore$
$$n = 2.42$$

The voidage $e$ at a velocity of 0.25 m/s is then given by equation 6.31 as:

$$\frac{0.25}{0.45} = e^{2.42}$$

and:
$$e = \underline{\underline{0.784}}$$

## 6.2.2. Non-uniform fluidisation

Regular and even expansion of the bed does not always occur when particles are fluidised by a liquid. This is particularly so for solids of high densities, and non-uniformities are most marked with deep beds of small particles. In such cases, there are significant deviations from the relation between bed voidage and velocity predicted by equation 6.31.

STEWART (referred to in STEWART and DAVIDSON[16]) has shown that well-defined bubbles of liquid and slugs are formed when tungsten beads (density 19,300 kg/m$^3$, and particle sizes 776 and 930 μm) are fluidised with water. SIMPSON and RODGER[17], HARRISON *et al.*[18], LAWTHER and BERGLIN[19] and RICHARDSON and SMITH[20] have observed that lead shot fluidised by water gives rise to non-uniform fluidised beds. ANDERSON and JACKSON[21] have shown that this system would be expected to be transitional in behaviour. HASSETT[22] and LAWSON and HASSETT[23] have also noted instabilities and non-uniformities in liquid-solids systems, particularly in beds of narrow diameter. Similar observations have also been made by CAIRNS and PRAUSNITZ[24], by KRAMERS *et al.*[25] and by REUTER[26], who have published photographs of bubbles in liquid-solids systems. GIBILARO *et al.*[27] have made experimental measurements of one dimensional waves in liquid–solids fluidised beds. BAILEY[28] has studied the fluidisation of lead shot with water and has reported the occurrence of non-uniformities, though not of well-defined bubbles. He has shown that the logarithmic plots of voidage against velocity are no longer linear and that the deviations from the line given by equation 6.31 increase with:

(a) increase in bed weight per unit area,
(b) decrease in particle size.

The deviation passes through a maximum as the velocity is increased, as shown in Figure 6.7.

The importance of particle density in determining the nature of fluidised systems is well established, and increase in density generally results in a less uniform fluidised system.

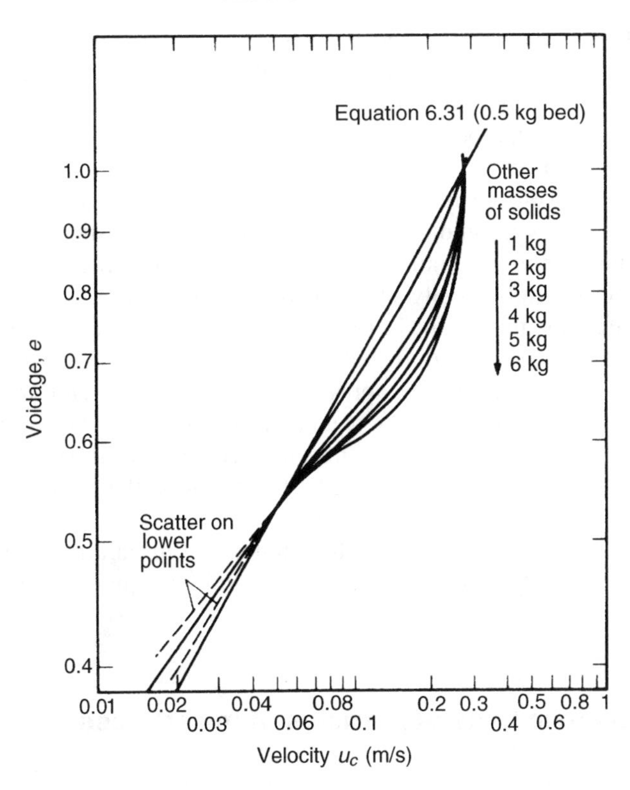


Figure 6.7. Bed expansion for the fluidisation of 0.5–0.6 mm lead shot in water in a 100 mm tube

It is, however, surprising that a reduction in particle size should also cause increased deviations from the ideal system. It may be noted from Figure 6.7 that, over a wide range of liquid velocities, the mean voidage of the bed is less than that predicted by equation 6.31. This may be explained in terms of part of the fluid taking a low resistance path through the bed, remaining there for less than the average residence time, and not therefore contributing fully to the expansion of the bed. The effect of partial channelling will certainly be more marked with fine solids than with coarse, since the ratio of the resistance of the bed to that of the channel will be much greater, and a comparatively small channel will accommodate the flow of a proportionately larger amount of fluid.

A simple model may be built up to represent what is happening under these circumstances. The bed may be considered to be divided into two portions, one with uniformly dispersed particles and the other consisting of fluid channels. It is assumed that the voidage of the region of uniformly dispersed particles is determined according to equation 6.31 by the flowrate through that part of the bed. If, then, a fraction $f$ of the fluid introduced to the bottom of the bed flows through the channels at a velocity $u_f$, it can be readily shown that the relation between the mean voidage of the bed $e$ and the mean superficial velocity of the liquid $u_c$ is given by:

$$e = f\frac{u_c}{u_f} + \left(\frac{u_c}{u_i}(1-f)\right)^{1/n}\left(1 - f\frac{u_c}{u_f}\right)^{1-(1/n)} \tag{6.40}$$

Equation 6.40 gives the relation between all possible corresponding values of $u_f$ and $f$. For a typical experiment on the fluidisation of 5 kg of lead shot ($d = 0.55$ mm) in a 100 mm diameter tube with water flowing at a mean superficial velocity of 0.158 m/s, the measured voidage of the bed was 0.676. This would give a value of $f = 0.53$ for a channel velocity $u_f = 1.58$ m/s, or 0.68 for a channel velocity $u_f = 0.80$ m/s.

Local variations in voidage in a liquid-solids fluidised bed have also been observed by VOLPICELLI *et al.*[29] who fluidised steel, aluminium and plastic beads with water and glycerol in a column only 3.55 mm thick, using particles ranging from 2.86 to 3.18 mm diameter which thus gave effectively a bed one particle thick. This system facilitated observation of flow patterns within the bed. It was found that the velocity–voidage relationship was of the same form as equation 6.31, but that it was necessary to use the actual measured falling velocity of the particle in the apparatus to represent $u_i$. Non-uniformities within the bed were not apparent at voidages near unity or near $u_{mf}$, but rose to a maximum value intermediately; this is generally in line with Bailey's work (Figure 6.7). The local variations of voidage were found to be highly dependent on the arrangement of the liquid distributor.

More recent work by FOSCOLO *et al.*[30] has shown that instabilities can also arise in the fluidisation of particles where densities are only slightly greater than that of the fluidising liquid.

### 6.2.3. Segregation in beds of particles of mixed sizes

When a bed consists of particles with a significant size range, stratification occurs, with the largest particles forming a bed of low voidage near the bottom and the smallest particles forming a bed of high voidage near the top. If the particles are in the form of sharp-cut size fractions, segregation will be virtually complete with what is, in effect, a number of fluidised beds of different voidages, one above the other. If the size range is small, there will be a continuous variation in both composition and concentration of particles throughout the depth of the bed.

It has been shown experimentally[11] that a mixture of equal masses of 1.0 mm and 0.5 mm particles when fluidised by water will segregate almost completely over the whole range of velocities for which particles of each size will, on its own, form a fluidised bed. WEN and YU[7] have shown that this behaviour is confined to mixtures in which the ratio of the minimum fluidising velocities of the components exceeds about 2. The tendency for classification has been examined experimentally and theoretically by several workers, including JOTTRAND[31], PRUDEN and EPSTEIN[32], KENNEDY and BRETTON[33], AL-DIBOUNI and GARSIDE[34], JUMA and RICHARDSON[35], GIBILARO *et al.*[36] and MORITOMI *et al.*[37].

In a mixture of large and small particles fluidised by a liquid, one of three situations may exist:

(a) Complete (or virtually complete) segregation, with a high voidage bed of small particles above a bed of lower voidage containing the large particles.
(b) Beds of small and of large particles as described in (a), but separated by a transition region in which the proportion of small particles and the voidage both increase from bottom to top.

(c) A bed in which there are no fully segregated regions, but with the transition region described in (b) extending over the whole extent of the bed. If the range of particle sizes in the bed is small, this transition region may be of nearly constant composition, with little segregation occurring.

At any level in the transition region, there will be a balance between the mixing effects attributable to (a) axial dispersion and to (b) the segregating effect which will depend on the difference between the interstitial velocity of the liquid and that interstitial velocity which would be required to produce a bed of the same voidage for particles of that size on their own. On this basis a model may be set up to give the vertical concentration profile of each component in terms of the axial mixing coefficients for the large and the small particles.

Experimental measurements[35] of concentration profiles within the bed have been made using a pressure transducer attached to a probe whose vertical position in the bed could be varied. The voidage $e$ of the bed at a given height may then be calculated from the local value of the pressure gradient using equation 6.1, from which:

$$-\frac{\mathrm{d}P}{\mathrm{d}l} = (1 - e)(\rho_s - \rho)g \tag{6.41}$$

It has been established that the tendency for segregation increases, not only with the ratio of the sizes of particles (largest:smallest), but also as the liquid velocity is raised. Thus, for a given mixture, there is more segregation in a highly expanded bed than in a bed of low voidage.

### *Binary mixtures - particles differing in both size and density*

The behaviour of a fluidised bed consisting of particles differing in both size and density, can be extremely complex. This situation may be illustrated by considering the simplest case — the fluidisation of a binary mixture of spherical particles. If the heavy particles are also the larger, they will always constitute the denser bottom layer. On the other hand, if the mixture consists of small high density particles H, and larger particles of lower density L, the relative densities of the two layers is a function of the fluidising velocity. In either case, for both species of solids to be fluidised simultaneously, the superficial velocity $u_c$ of the liquid must lie between the minimum fluidising velocity $u_{mf}$ and the terminal falling velocity $u_0$ for each of the solids of $u_{mf} < u_c < u_0$. In general, segregation tends to occur, resulting in the formation of two fluidised beds of different densities, possibly separated by a transition zone, with the bed of higher density forming the bottom layer. In all cases, the interface between the two beds may be diffuse as a result of the effect of dispersion.

The densities of the two beds are then given by:

$$\rho_{bH} = (1 - e_H)\rho_{sH} + e_H\rho = \rho_{sH} - e_H(\rho_{sH} - \rho) \tag{6.42a}$$

$$\rho_{bL} = (1 - e_L)\rho_{sL} + e_L\rho = \rho_{sL} - e_L(\rho_{sL} - \rho) \tag{6.42b}$$

where the suffix $L$ refers to the light particles and the suffix $H$ to the heavy particles.

Applying equation 6.31 to each fluidised bed gives:

$$\frac{u_c}{u_{0H}} = e_H^{n_H} \text{ and } \frac{u_c}{u_{0L}} = e_L^{n_L}$$

Noting that the superficial velocity $u_c$ is the same in each case, and assuming that $n_H \approx n_L \approx n$, then:

$$\frac{e_L}{e_H} = \left(\frac{u_{0H}}{u_{0L}}\right)^{1/n} \tag{6.43}$$

As the fluidising velocity is progressively increased, the voidages of both beds increases, although not generally at the same rate Two cases are considered:

(a) $u_{0H} > u_{0L}$

From equation 6.43, $e_H < e_L$ and therefore, from equations 6.42*a* and 6.42b, $\rho_{bh} > \rho_{bl}$ at all fluidising velocities $u_c$, and the heavy particles will always form the bottom layer.

(b) $u_{0H} < u_{0L}$

If, with increase in velocity, the density of the upper zone decreases more rapidly than that of the bottom zone, the two beds will maintain the same relative orientation. If the reverse situation applies, there may be a velocity $u_{\text{INV}}$ where the densities of the two layers become equal, with virtually complete mixing of the two species taking place. Any further increase in velocity above $u_{\text{INV}}$ then causes the beds to invert, as shown diagrammatically in Figure 6.8(*a*).

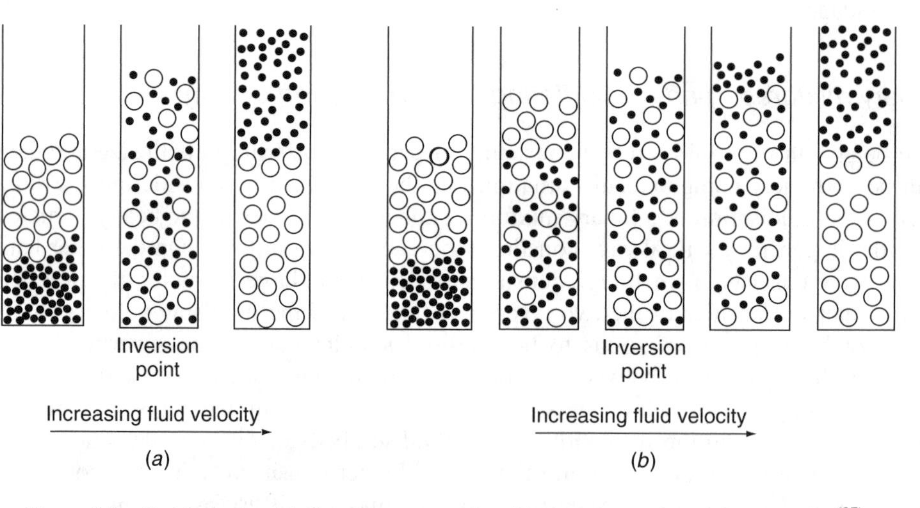


Figure 6.8. Bed inversion (*a*) Complete segregation (*b*) Complete and partial segregation[37]

The relative rates at which the bed densities change as the fluidising velocity is increased may be obtained by differentiating equations 6.42*a* and 6.42*b* with respect to $u_c$, and dividing to give:

$$r = -\frac{d\rho_{bH}}{du_c} \Big/ -\frac{d\rho_{bL}}{du_c} = \frac{d\rho_{bH}}{d\rho_{bL}} = \frac{(\rho_{sH} - \rho)}{(\rho_{sL} - \rho)} \left(\frac{u_{0L}}{u_{0H}}\right)^{1/n} = \frac{(\rho_{sH} - \rho)}{(\rho_{sL} - \rho)} \frac{e_H}{e_L} \tag{6.44}$$

As $e_H > e_L$ and $\rho_{sH} > \rho_{sL}$, then from equation 6.44, $r$, which is independent of fluidising velocity, must be greater than unity. It is thus the bed of heavy particles which expands more rapidly as the velocity is increased, and which must therefore be forming the bottom layer at low velocities if inversion is possible. That is, it is the small heavy particles which move from the lower to the upper layer, and vice versa, as the velocity is increased beyond the inversion velocity $u_{INV}$. RICHARDSON and AFIATIN[38] have analysed the range of conditions over which segregation of spherical particles can occur, and have shown these diagrammatically in Figure 6.9 for the Stokes' law region (*a*) and for the Newton's law region (*b*).

It has been observed by several workers, including by MORITOMI *et al.*[37] and EPSTEIN and PRUDEN[39], that a sharp transition between two mono-component layers does not always occur and that, on each side of the transition point, there may be a condition where the lower zone consists of a mixture of both species of particles, the proportion of heavy particles becoming progressively smaller as the velocity is increased. This situation, depicted in Figure 6.8*b*, can arise when, at a given fluidising velocity, there is a stable two-component bed which has a higher density than a bed composed of either of the two species on its own. Figure 6.10, taken from the work of EPSTEIN and PRUDEN[39], shows how the bed densities for the two mono-component layers change as the liquid velocity is increased, with point C then defining the inversion point when complete segregation can take place. Between points A and D (corresponding to velocities $u_{cA}$ and $u_{cB}$), however, a two-component bed (represented by curve ABD) may be formed which has a density greater than that of either mono-component bed over this velocity range. In moving along this curve from A to D, the proportion of light particles in the lower layer decreases progressively from unity to zero, as shown on the top scale of the diagram. This proportion

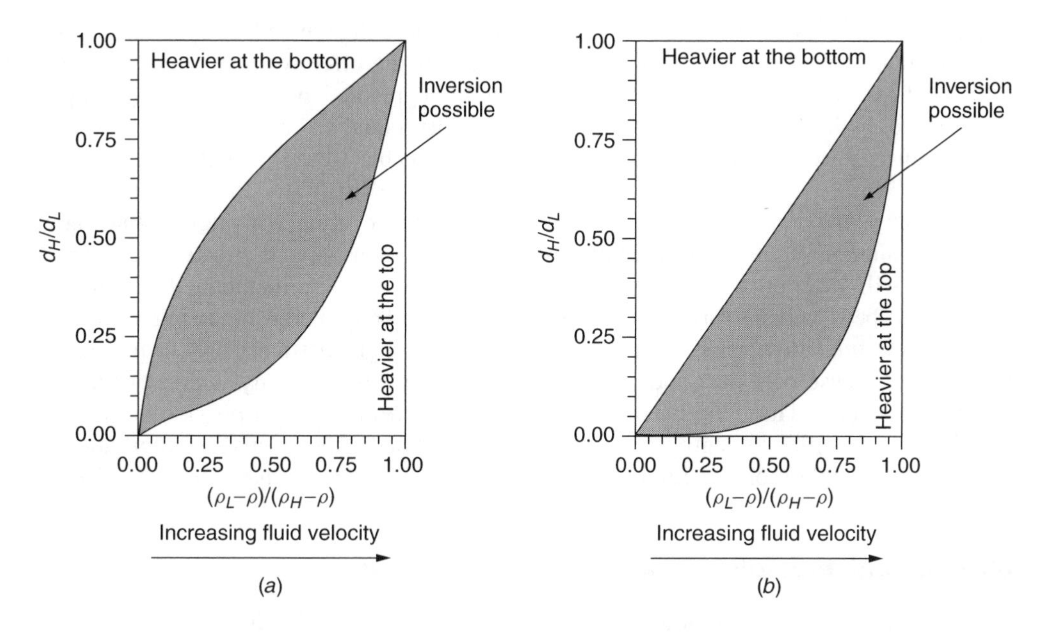


Figure 6.9. The possibility of inversion (*a*) Stokes' law region (*b*) Newton's law region[38]

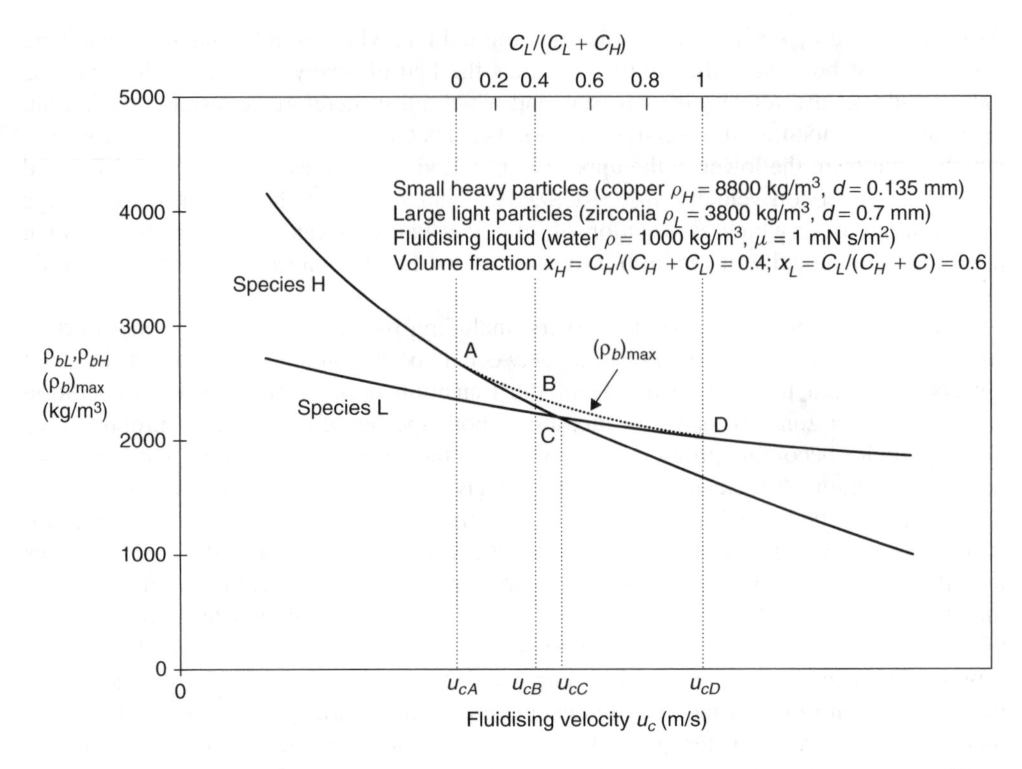


Figure 6.10. Bed densities as a function of fluidising velocity, showing the mixed particle region[39]

is equal to that in the total mix of solids at point B, where the whole bed is the of uniform composition, and the velocity $u_{cB}$ therefore represents the effective inversion velocity.

If the flow of fluidising liquid to a completely segregated bed is suddenly stopped, the particles will all then start to settle at a velocity equal to that at which they have been fluidised, because equation 6.31 is equally applicable to sedimentation and fluidisation.

Thus, since the voidages of the two beds will both be greater at higher fluidisation velocities, the subsequent sedimentation velocity will then also be greater. Particles in both beds will settle at the same velocity and segregation will be maintained. Eventually, two packed beds will be formed, one above the other. Thus, if the fluidising velocity is less than the transition velocity, a packed bed of large light particles will form above a bed of small dense particles, and conversely, if the fluidising velocity is greater than the inversion velocity. Thus, fluidisation followed by sedimentation can provide a means of forming two completely segregated mono-component beds, the relative configuration of which depends solely on the liquid velocity at which the particles have been fluidised.

## 6.2.4. Liquid and solids mixing

Kramers *et al.*[25] have studied longitudinal dispersion in the liquid in a fluidised bed composed of glass spheres of 0.5 mm and 1 mm diameter. A step change was introduced by feeding a normal solution of potassium chloride into the system. The concentration at the top of the bed was measured as a function of time by means of a small conductivity cell. On the assumption that the flow pattern could be regarded as longitudinal diffusion superimposed on piston flow, an eddy longitudinal diffusivity was calculated. This was found to range from $10^{-4}$ to $10^{-3}$ $m^2/s$, increasing with both voidage and particle size.

The movement of individual particles in a liquid–solid fluidised bed has been measured by Handley *et al.*[40] Carlos[41,42], and Latif[43]. In all cases, the method involved fluidising transparent particles in a liquid of the same refractive index so that the whole system became transparent. The movement of coloured tracer particles, whose other physical properties were identical to those of the bed particles, could then be followed photographically.

Handley fluidised soda glass particles using methyl benzoate, and obtained data on the flow pattern of the solids and the distribution of vertical velocity components of the particles. It was found that a bulk circulation of solids was superimposed on their random movement. Particles normally tended to move upwards in the centre of the bed and downwards at the walls, following a circulation pattern which was less marked in regions remote from the distributor.

Carlos and Latif both fluidised glass particles in dimethyl phthalate. Data on the movement of the tracer particle, in the form of spatial co-ordinates as a function of time, were used as direct input to a computer programmed to calculate vertical, radial, tangential and radial velocities of the particle as a function of location. When plotted as a histogram, the total velocity distribution was found to be of the same form as that predicted by the kinetic theory for the molecules in a gas. A typical result is shown in Figure 6.11[41]. Effective diffusion or mixing coefficients for the particles were then calculated from the product of the mean velocity and mean free path of the particles, using the simple kinetic theory.

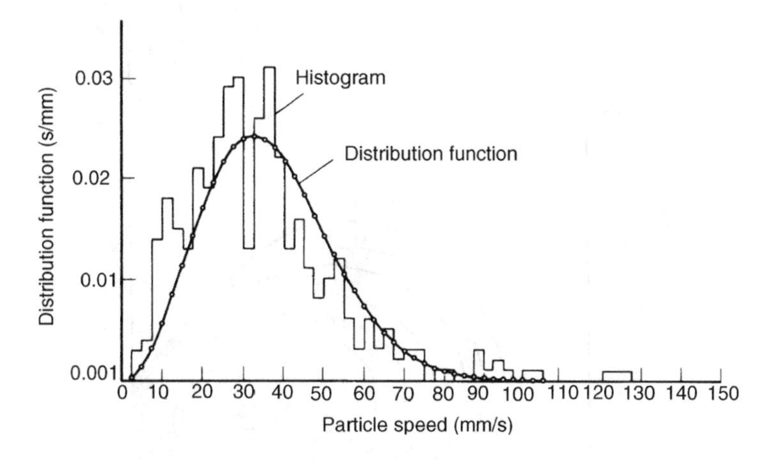


Figure 6.11. Distribution of particle speeds in fluidised bed[41]

Solids mixing was also studied by Carlos[42] in the same apparatus, starting with a bed composed of transparent particles and a layer of tracer particles at the base of the bed. The concentration of particles in a control zone was then determined at various intervals of time after the commencement of fluidisation. The mixing process was described by a diffusion-type equation. This was then used to calculate the mixing coefficient. A comparison of the values of mixing coefficient obtained by the two methods then enabled the *persistence of velocity* factor to be calculated. A typical value of the mixing coefficient was $1.5 \times 10^{-3}$ $m^2/s$ for 9 mm glass ballotini fluidised at a velocity of twice the minimum fluidising velocity.

LATIF[43] represented the circulation currents of the particles in a fluidised bed, by plotting stream functions for the particles on the assumption that the particles could be regarded as behaving as a continuum. A typical result for the fluidisation of 6-mm glass particles by dimethyl phthalate is shown in Figure 6.12; in this case the velocity has been adjusted to give a bed voidage of 0·65. Because the bed is symmetrical about its axis, the pattern over only a radial slice is shown. It may be noted that the circulation patterns are concentrated mainly in the lower portion of the bed, with particles moving upwards in the centre and downwards at the walls. As the bed voidage is decreased, the circulation patterns tend to occupy progressively smaller portions of the bed, but there is a tendency for a small reverse circulation pattern to develop in the upper regions of the bed.

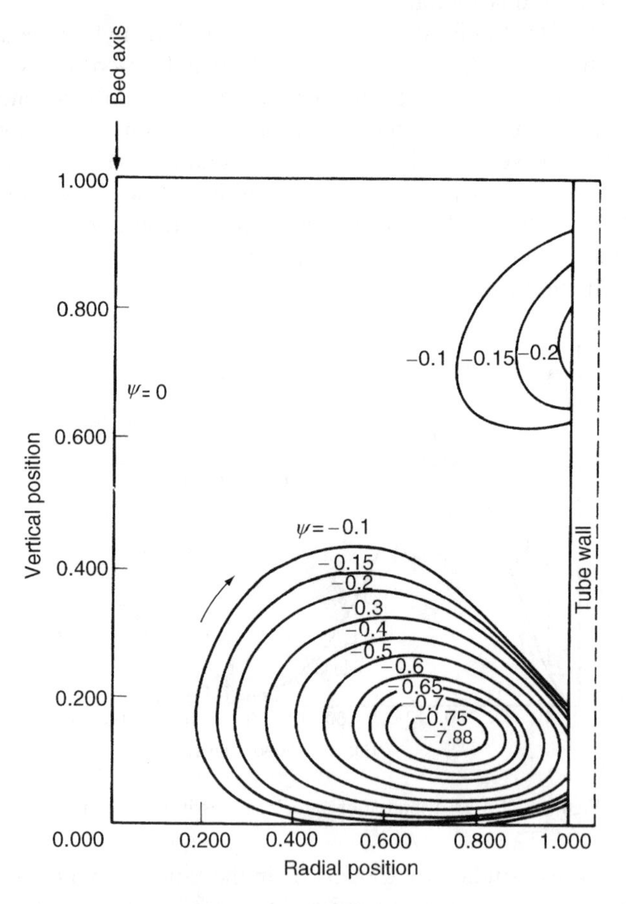


Figure 6.12. Particle stream functions, $\psi(e = 0.65)$ (Radial position is expressed as fraction of radial distance from centre line, and axial position as fraction of bed height measured from the bottom)[43]

Later work on axial dispersion of particles has been carried out by DORGELO *et al.* [44] who used an random-walk approach.

## 6.3. GAS–SOLIDS SYSTEMS

### 6.3.1. General behaviour

In general, the behaviour of gas-fluidised systems is considerably more complex than that of liquid-fluidised systems which exhibit a gradual transition from fixed bed to fluidised bed followed by particle transport, without a series of transition regions, and with bed expansion and pressure drop conforming reasonably closely to values calculated for ideal systems.

Part of the complication with gas–solid systems arises from the fact that the purely hydrodynamic forces acting on the particles are relatively small compared with frictional forces between particles, electrostatic forces and surface forces which play a much more dominant role when the particles are very fine. As the gas velocity in a fluidised bed is increased, the system tends to go through various stages:

(a) *Fixed bed* in which the particles remain in contact with one another and the structure of the bed remains stable until the velocity is increased to the point where the pressure drop is equal to the weight per unit area of the particles.
(b) *Particulate* and regular predictable expansion over a limited range of gas velocities.
(c) A *bubbling* region characterised by a high proportion of the gas passing through the bed as bubbles which cause rapid mixing in the dense particulate phase.
(d) A *turbulent* chaotic region in which the gas bubbles tend to coalesce and lose their identity.
(e) A region where the dominant pattern is one of *vertically upward transport of particles*, essentially gas–solids transport or pneumatic conveying. This condition, sometimes referred to as *fast fluidisation*, lies outside the range of true fluidisation.

### 6.3.2. Particulate fluidisation

Although fine particles generally form fluidised beds more readily than coarse particles, surface-related forces tend to predominate with very fine particles. It is very difficult to fluidise some very fine particles as they tend to form large stable conglommerates that are almost entirely by-passed by the gas. In some extreme cases, particularly with small diameter beds, the whole of the particulate mass may be lifted as a solid 'piston'. The uniformity of the fluidised bed is often critically influenced by the characteristics of the gas distributor or bed support. Fine mesh distributors are generally to be preferred to a series of nozzles at the base of the bed, although the former are generally more difficult to install in larger beds because they are less robust.

Good distribution of gas over the whole cross-section of the bed may often be difficult to achieve, although this is enhanced by ensuring that the pressure drop across the distributor is large compared with that across the bed of particles. In general, the quality of gas distribution improves with increased flowrate because the pressure drop across the bed when it is fluidised is, theoretically, independent of the flowrate. The pressure drop across the distributor will increase, however, approximately in proportion to the square of the flowrate, and therefore the fraction of the total pressure drop that occurs across the distributor increases rapidly as the flowrate increases.

Apart from the non-uniformities which characterise many gas–solid fluidised beds, it is in the low fluidising-velocity region that the behaviour of the gas–solid and liquid–solid beds are most similar. At low gas rates the bed may exhibit a regular expansion as the flowrate increases, with the relation between fluidising velocity and voidage following the form of equation 6.31, although, in general, the values of the exponent $n$ are higher than those for liquid-solids systems partly because particles have a tendency to form small agglomerates thereby increasing the effective particle size. The range of velocities over which particulate expansion occurs is, however, quite narrow in most cases.

### 6.3.3. Bubbling fluidisation

The region of particulate fluidisation usually comes to an abrupt end as the gas velocity is increased, with the formation of gas bubbles. These bubbles are usually responsible for the flow of almost all of the gas in excess of that flowing at the minimum fluidising velocity. If bed expansion has occurred before bubbling commences, the excess gas will be transferred to the bubbles whilst the continuous phase reverts to its voidage at the minimum fluidising velocity and, in this way, it contracts. Thus, the expanded bed appears to be in a meta-stable condition which is analogous to that of a supersaturated solution reverting to its saturated concentration when fed with small seed crystals, with the excess solute being deposited on to the seed crystals which then increase in size as a result, as discussed in Chapter 15.

The upper limit of gas velocity for particulate expansion is termed the *minimum bubbling* velocity, $u_{mb}$. Determining this can present difficulties as its value may depend on the nature of the distributor, on the presence of even tiny obstructions in the bed, and even on the immediate pre-history of the bed. The ratio $u_{mb}/u_{mf}$, which gives a measure of the degree of expansion which may be effected, usually has a high value for fine light particles and a low value for large dense particles.

For cracker catalyst ($d = 55$ μm, density $= 950$ kg/m$^3$) fluidised by air, values of $u_{mb}/u_{mf}$ of up to 2.8 have been found by Davies and Richardson[45]. During the course of this work it was found that there is a minimum size of bubble which is stable. Small bubbles injected into a non-bubbling bed tend to become assimilated in the dense phase, whilst, on the other hand, larger bubbles tend to grow at the expense of the gas flow in the dense phase. If a bubble larger than the critical size is injected into an expanded bed, the bed will initially expand by an amount equal to the volume of the injected bubble. When, however, the bubble breaks the surface, the bed will fall back below the level existing before injection and will therefore have acquired a reduced voidage.

Thus, the bubbling region, which is an important feature of beds operating at gas velocities in excess of the minimum fluidising velocity, is usually characterised by two phases — a continuous emulsion phase with a voidage approximately equal to that of a bed at its minimum fluidising velocity, and a discontinous or bubble phase that accounts for most of the excess flow of gas. This is sometimes referred to as the *two-phase theory of fluidisation*.

The bubbles exert a very strong influence on the flow pattern in the bed and provide the mechanisim for the high degree of mixing of solids which occurs. The properties and behaviour of the bubbles are describe later in this Section.

When the gas flowrate is increased to a level at which the bubbles become very large and unstable, the bubbles tend to lose their identity and the flow pattern changes to a chaotic form without well-defined regions of high and low concentrations of particles. This is commonly described as the *turbulent* region which has, until fairly recently, been the subject of relatively few studies.

### Categorisation of Solids

The ease with which a powder can be fluidised by a gas is highly dependent on the properties of the particles. Whilst it is not possible to forecast just how a given powder will fluidise without carrying out tests on a sample, it is possible to indicate some trends. In general, fine low density particles fluidise more evenly than large dense ones, provided that they are not so small that the London–van der Waals attractive forces are great enough for the particles to adhere together strongly. For very fine particles, these attractive forces can be three or more orders of magnitude greater than their weight. Generally, the more nearly spherical the particles then the better they will fluidise. In this respect, long needle-shaped particles are the most difficult to fluidise. Particles of mixed sizes will usually fluidise more evenly than those of a uniform size. Furthermore, the presence of a small proportion of fines will frequently aid the fluidisation of coarse particles by coating them with a 'lubricating' layer.

In classifying particles into four groups, GELDART[46] has used the following criteria:

(a) Whether or not, as the gas flowrate is increased, the fluidised bed will expand significantly before bubbling takes place. This property may be quantified by the ratio $u_{mb}/u_{mf}$, where $u_{mb}$ is the minimum velocity at which bubbling occurs. This assessment can only be qualitative as the value of $u_{mb}$ is very critically dependent on the conditions under which it is measured.

(b) Whether the rising velocity of the majority of the bubbles, is greater or less than the interstitial gas velocity. The significance of this factor is discussed in Section 6.3.5.

(c) Whether the adhesive forces between particles are so great that the bed tends to channel rather than to fluidise. Channelling depends on a number of factors, including the degree to which the bed has consolidated and the condition of the surface of the particles at the time. With powders that channel badly, it is sometimes possible to initiate fluidisation by mechanical stirring, as discussed in Section 6.3.4.

The classes into which powders are grouped are given in Table 6.1, which is taken from the work of GELDART[46], and in Figure 6.13. In they are located approximately on a particle density–particle size chart.

### The Effect of Pressure

The effect of pressure on the behaviour of the bed is important because many industrial processes, including fluidised bed combustion which is discussed in Section 6.8.4., are

Table 6.1. Categorisation of Powders in Relation to Fluidisation Characteristics[46]

| | Typical particle size (μm) | Fluidisation/Powder Characteristics | Examples of Materials |
|---|---|---|---|
| Group A | 30–100 | Particulate expansion of bed will take place over significant velocity range. Small particle size and low density ($\rho$, <1400 kg/m$^3$). | Cracker catalyst |
| Group B | 100–800 | Bubbling occurs at velocity $>u_{mf}$. Most bubbles have velocities greater than interstitial gas velocity. No evidence of maximum bubble size. | Sand. |
| Group C | 20 | Fine cohesive powders, difficult to fluidise and readily form channels. | Flour<br>Fine silica |
| Group D | 1000 | All but largest bubbles rise at velocities less than interstitial gas velocity. Can be made to form spouted beds. Particles large and dense. | Wheat<br>Metal shot |

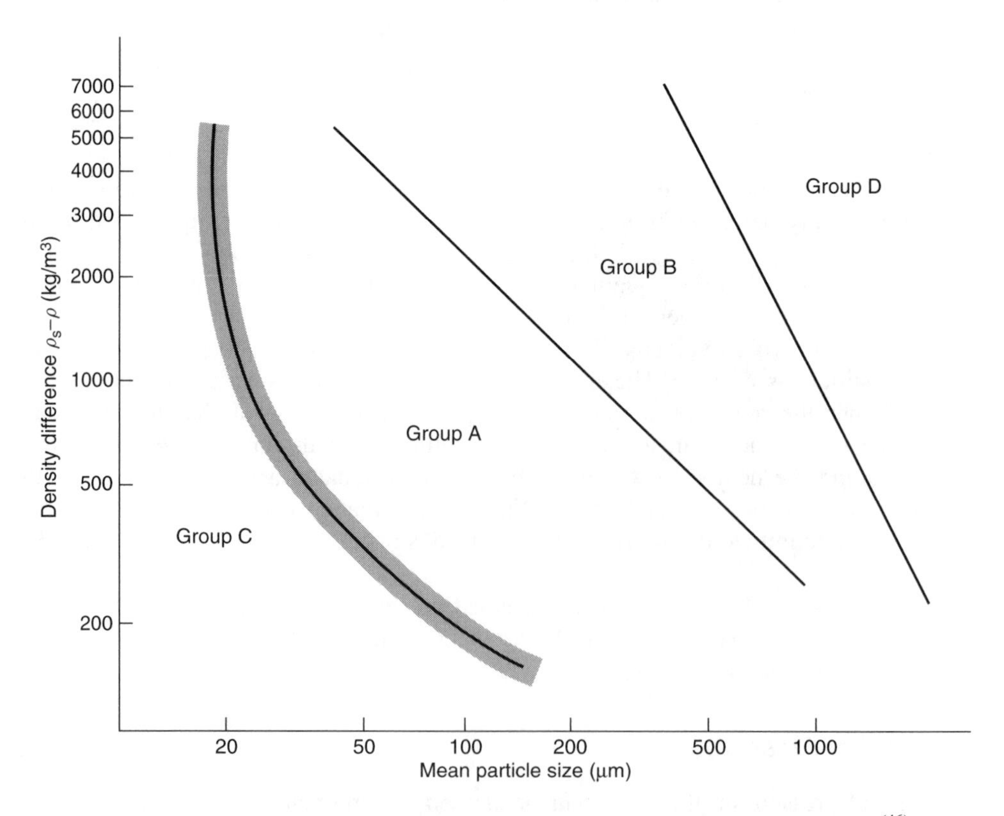


Figure 6.13. Powder classification diagram for fluidisation by air at ambient conditions[46]

carried out at elevated pressures. Several workers have reported measurements of bed expansion as a function of gas rate for elevated pressures when very much higher values of the ratio $u_{mb}/u_{mf}$ may be obtained[17,47,48,49].

Because minimum fluidising velocity is not very sensitive to the pressure in the bed, much greater mass flowrates of gas may be obtained by increasing the operating pressure.

The influence of pressure, over the range 100–1600 kN/m$^2$, on the fluidisation of three grades of sand in the particle size range 0.3 to 1 mm has been studied by OLOWSON and ALMSTEDT[50] and it was showed that the minimum fluidising velocity became less as the pressure was increased. The effect, most marked with the coarse solids, was in agreement with that predicted by standard relations such as equation 6.14. For fine particles, the minimum fluidising velocity is independent of gas density (equation 6.5 with $\rho_s >> \rho$), and hence of pressure.

## Tapered Beds

Where there is a wide range of particle sizes in the powder, fluidisation will be more even in a bed that is tapered so as to provide the minimum cross-section at the bottom. If the pressure gradient is low and the gas does not therefore expand significantly, the velocity will decrease in the direction of flow. Coarse particles which will then tend to become fluidised near the bottom of the bed assist in the dispersion of the fluidising gas. At the same time, the carry-over of fines from the top will be reduced because of the lower velocity at the exit.

When deep beds of solids are fluidised by a gas, the use of a tapered bed can counter-balance the effects of gas expansion. For example, the pressure drop over a 5 m deep bed of solids of density 4000 kg/m$^3$ is about $10^5$ N/m$^2$. Thus, with atmospheric pressure at the outlet, the volumetric flowrate will double from the bottom to the top of an isothermal cylindrical bed. If the area at the outlet is twice that at the base, the velocity will be maintained approximately constant throughout.

## The Effect of Magnetic and Electrical Fields

Magnetic particles may form much more stable beds when subjected to a magnetic field. SAXENA and SHRIVASTAVA[51] have examined the complex behaviour of spherical steel particles of a range of sizes when subjected to fields of different strengths, considering in particular the bed pressure drop, the quality of fluidisation and the structure of the surface of the bed.

Dielectric particles show a reduced tendency for bubbling and a larger range of velocities over which particulate expansion occurs when an alternating electrical field is applied.

## The Effect of Baffles

It is possible substantially to eliminate the fluctuations which are characteristic of beds of coarse solids by incorporating baffles into the bed. The nature and arrangement of the baffles is critical, and it is generally desirable to avoid downward-facing horizontal surfaces because these can give rise to regimes of defluidisation by blocking the upward flow of gas. For this reason, horizontal tubes immersed in a fluidised bed tend to exhibit low values of heat transfer coefficients because of the partial defluidisation that occurs.

### 6.3.4. The effect of stirring

Stirring can be effective in improving the quality of fluidisation. In particular, if the agitator blades lift the particles as they rotate, fluidisation can be effected at somewhat lower gas velocities. In addition, the fluidisation of very fine particles that tend to aggregate can be substantially improved by a slow stirrer fitted with blades that provide only a small clearance at the gas distributor. GODARD and RICHARDSON[52] found that it was possible to fluidise fine silica particles ($d = 0.05$ μm) only if the stirrer was situated less than 10 mm from the support. In the absence of stirring, the fluidising gas passed through channels in the bed and the solids were completely unfluidised. Once fluidisation had been established, the stirrer could be stopped and uniform fluidisation would then be maintained indefinitely. With such fine solids, a high degree of bed expansion could be achieved with $u_{mb}/u_{mf}$ ratios up to 18. Over this range, equation 6.31 is followed with a value of $n$ ranging from 6.7 to 7.5 as compared with a value of 3.7 calculated from equation 6.32 for a liquid at the same value of the Galileo number.

The rotation of a paddle in a fluidised bed provides a means of measuring an effective viscosity of the bed in terms of the torque required to rotate the paddle at a controlled speed[53].

### 6.3.5. Properties of bubbles in the bed

The formation of bubbles at orifices in a fluidised bed, including measurement of their size, the conditions under which they will coalesce with one another, and their rate of rise in the bed has been investigated. DAVIDSON *et al.*[54] injected air from an orifice into a fluidised bed composed of particles of sand (0.3–0.5 mm) and glass ballotini (0.15 mm) fluidised by air at a velocity just above the minimum required for fluidisation. By varying the depth of the injection point from the free surface, it was shown that the injected bubble rises through the bed with a constant velocity, which is dependent only on the volume of the bubble. In addition, this velocity of rise corresponds with that of a spherical cap bubble in an inviscid liquid of zero surface tension, as determined from the equation of DAVIES and TAYLOR[55]:

$$u_b = 0.792 V_B^{1/6} g^{1/2} \tag{6.45}$$

The velocity of rise is independent of the velocity of the fluidising air and of the properties of the particles making up the bed. Equation 6.45 is applicable provided that the density of the gas in the bubbles may be neglected in comparison with the density of the solids. In other cases, the expression must be multiplied by a factor of $(1 - \rho/\rho_c)^{1/2}$.

HARRISON *et al.*[18] applied these results to the problem of explaining why gas and liquid fluidised systems behave differently. Photographs of bubbles in beds of lead shot fluidised with air and with water have shown that an injected bubble is stable in the former case though it tends to collapse in the latter. The water–lead system tends to give rise to inhomogeneities, and it is therefore interesting to note that bubbles as such are apparently not stable. As the bubble rises in a bed, internal circulation currents are set up because of the shear stresses existing at the boundary of the bubble. These circulation velocities are of the same order of magnitude as the rising velocity of the bubble. If the circulation velocity is appreciably greater than the falling velocity of the particles, the bubble will tend to draw in particles at the wake and will therefore be destroyed. On the other hand, if the rising velocity is lower, particles will not be drawn in at the wake and the bubble will be stable.

As a first approximation, a bubble is assumed to be stable if its rising velocity is less than the free-falling velocity of the particles, and therefore, for any system, the limiting size of stable bubble may be calculated using equation 6.45. If this is of the same order of size as the particle diameter, the bubble will not readily be detected. On the other hand, if it is more than about ten times the particle diameter, it will be visible and the system will be seen to contain stable bubbles. On the basis of this argument, large bubbles are generally stable in gases, whereas in liquids the largest size of stable bubble is comparable with the diameter of the particles. It should be possible to achieve fluidisation free of bubbles with very light particles by using a gas of high density. LEUNG[48] succeeded in reaching this condition by fluidising hollow phenolic microballoons at pressures of about 4500 $kN/m^2$. It was found possible to form stable bubbles with glycerine–water mixtures and lead shot of 0.77 mm particle size. This transitional region has also been studied by SIMPSON and RODGER[17].

HARRISON and LEUNG[56] have shown that the frequency of formation of bubbles at an orifice (size range 1.2–25 mm) is independent of the bed depth, the flowrate of gas and the properties of the particles constituting the continuous phase, although the frequency of formation depends on the injection rate of gas, tending to a frequency of 18–21 $s^{-1}$ at high flowrates.

In a further paper[57], it has been shown that a wake extends for about 1.1 bubble diameters behind each rising bubble. If a second bubble follows in this wake, its velocity is increased by an amount equal to the velocity of the leading bubble, and in this way coalescence takes place.

The differences between liquid and gas fluidised systems have also been studied theoretically by JACKSON[2] who showed that small discontinuities tend to grow in a fluidised bed, although the rate of growth is greater in a gas–solids system.

ROWE and WACE[58] and WACE and BURNETT[59] have examined the influence of gas bubbles on the flow of gas in their vicinity. By constructing a thin bed 300 mm wide, 375 mm deep, and only 25 mm across, it was possible to take photographs through the Perspex wall, showing the behaviour of a thin filament of nitrogen dioxide gas injected into the bed. In a bed consisting of 0.20 mm ballotini, it was found that the filament tended to be drawn towards a rising bubble, as shown in Figure 6.14, and through it if sufficiently close. This establishes that there is a flow of gas from the continuous phase into the bubble and out through the roof of the bubble as shown in the figure, and that the gas tends to flow in definite streamlines. As a result, the gas is accelerated towards the bubble and is given a horizontal velocity component, and the gas velocity in the continuous phase close to the bubble is reduced.

In a bubbling bed, there is a tendency for the bubbles to form 'chains' and for successive bubbles to follow a similar track, thus creating a relatively low-resistance path for the gas flow. The bubbles are usually considerably larger at the top than at the bottom of the bed,

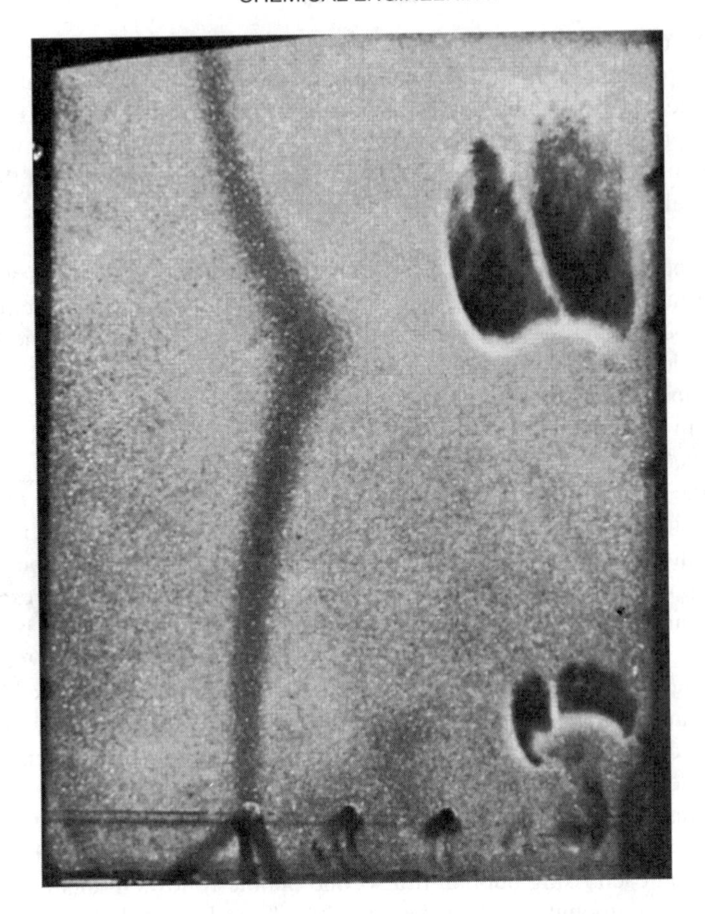

Figure 6.14. Photograph of tracer and bubble[58]

mainly as a result of coalescence, with larger bubbles catching up and absorbing smaller ones, particularly in the bubble-chains. In addition, the gas expands as the hydrostatic pressure falls, although this is usually a much smaller effect except in very deep beds.

The pressure distribution round a stationary bubble has been measured by inserting a gauze sphere 50 mm in diameter in the bed and measuring the pressure throughout the bed using the pressure inside the sphere as a datum, as shown in Figure 6.15. It has been found that near the bottom of the bubble the pressure was less than that remote from it at the same horizontal level, and that the situation was reversed, though to a smaller degree, towards the top of the bubble. Although the pressure distribution is somewhat modified in a moving bubble, the model serves qualitatively to explain the observed flow patterns of the tracer gas.

When the rate of rise of the bubble exceeds the velocity of the gas in the continuous phase, the gas leaving the top of the bubble is recycled and it re-enters the base. As a result, the gas in the bubble comes into contact with only those solid particles which immediately surround the bubble. DAVIDSON[60] has analysed this problem and shown that if the inertia of the gas is neglected, the diameter $d_c$ of the cloud of recycling gas

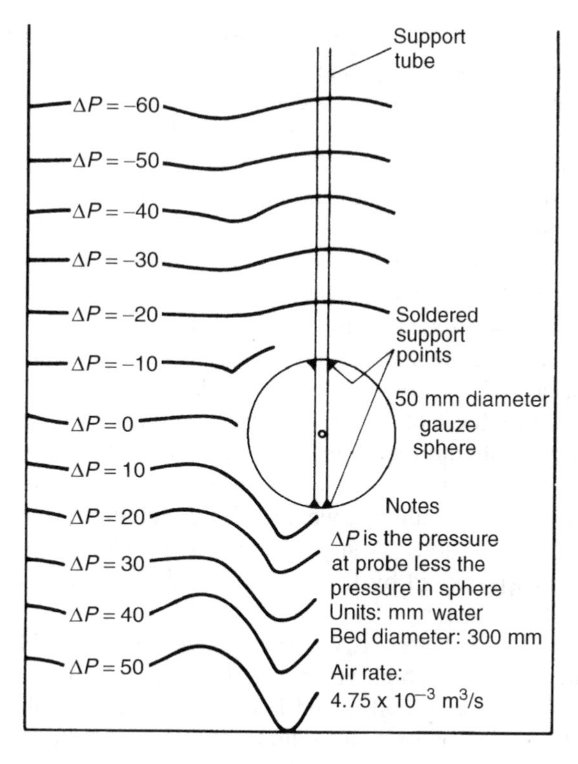


Figure 6.15. Isobars round a gauze sphere in a bed of mixed sand[58]

surrounding a bubble of diameter $d_b$ is given by:

$$\frac{d_c}{d_b} = \left(\frac{\alpha + 2}{\alpha - 1}\right)^{1/3} \tag{6.46}$$

where $\alpha$ is the ratio of the linear velocity of the gas in the bubble to that in the emulsion phase, that is:

$$\alpha = e\left(\frac{u_b}{u_c}\right) \tag{6.47}$$

The corresponding expression for a thin, essentially two-dimensional, bed is:

$$\frac{d_c}{d_b} = \left(\frac{\alpha + 2}{\alpha - 1}\right)^{1/2} \tag{6.48}$$

ROWE, PARTRIDGE, and LYALL[61] have studied the behaviour of the cloud surrounding a bubble by using nitrogen dioxide gas as a tracer, and found that Davidson's theory consistently overestimated the size of the cloud. A more sophisticated theory, developed by MURRAY[3], predicted the observed size of the cloud much more closely over forward face, although neither theory satisfactorily represented conditions in the wake of bubble. The gas cloud was found to break off at intervals, and the gas so detached became dispersed in the emulsion phase. In addition, it was shown by X-ray photographs that the bubbles would from time to time become split by fingers of particles falling through the roof of the bubble.

In most practical gas–solid systems, the particles are fine and the rising velocities of the individual bubbles are considerably greater than the velocity of the gas in the continuous phase. Thus, $\alpha$ in equation 6.47 is considerably greater than unity. Furthermore, the bubbles will tend to flow in well-defined paths at velocities considerably in excess of the rising velocity of individual bubbles. The whole pattern is, in practice, very complex because of the large number of bubbles present simultaneously, and also because of the size range of the bubbles.

The work of ROWE and HENWOOD[62] on the drag force exerted by a fluid on a particle, discussed in Chapter 5, showed why a bubble tended to be stable during its rise. Surfaces containing particles that face downstream, corresponding to the bottom of a bubble, tend to expel particles and are therefore diffuse. Surfaces facing upstream tend to attract particles and thus the top surface of a bubble will be sharp. Particles in a close-packed array were found to be subjected to a force 68.5 times greater than that on an isolated particle for the same relative velocity. It should therefore be possible to evaluate the minimum fluidising velocity as the fluid velocity at which the drag force acting on a single isolated particle would be equal to 1/68.5 of its buoyant weight. Values calculated on this basis agree well with experimental determinations. A similar method of calculating minimum fluidising velocities has also been proposed by DAVIES[45] who gives a mean value of 71.3 instead of 68.5 for the drag force factor.

From Figure 6.4 it is seen that for $e_{mf} = 0.40$, $u_0/u_{mf}$ is about 78 at low values of the Galileo number and about 9 for high values. In the first case, the drag on the particle is directly proportional to velocity and in the latter case proportional to the square of the velocity. Thus the force on a particle in a fluidised bed of voidage 0.4 is about 80 times that on an isolated particle for the same velocity.

### 6.3.6. Turbulent fluidisation

The upper limit of velocity for a bubbling fluidised bed is reached when the bubbles become so large that they are unstable and break down and merge with the continuous phase to give a highly turbulent chaotic type of flow. Pressure fluctuations become less violent and the bed is no longer influenced by the effect of large bubbles bursting at the surface, thereby causing a sudden reduction of the inventory of gas within the bed. The transition point occurs at a flowrate at which the standard deviation of the pressure fluctuations reaches a maximum. This occurs where gas bubbles occupy roughly half of the total volume of the bed, a value which corresponds to very much more than half of the overall volumetric flow because, for beds composed of fine particles, the velocity of rise of the bubbles is much greater than the interstitial velocity of the gas in the continuous phase. The breakdown of the bubbles tends to occur when they reach a maximum sustainable size which is often reached as a result of coalescence. This may frequently occur when a following bubble is drawn into the wake of a bubble that is already close to its maximum stable size.

The turbulent region has been the subject of comparatively few studies until recent years. A comprehensive critical review of the present state of knowledge has been presented by BI, ELLIS, ABBA and GRACE[63] who emphasise the apparent inconsistencies between the experimental results and correlations of different workers. Bi *et al.* define the regime of turbulent fluidisation as 'that in which there is no clear continuous phase, but, instead, either, via intermittency or by interspersing voids and dense regions, a competition between dense and dilute phases takes place in which neither gains the ascendancy'.

The onset of *turbulent* fluidisation appears to be almost independent of bed height, or height at the minimum fluidisation velocity, if this condition is sufficiently well defined. It is, however, strongly influenced by the bed diameter which clearly imposes a maximum on the size of the bubble which can form. The critical fluidising velocity tends to become smaller as the column diameter and gas density, and hence pressure, increase. Particle size distribution appears to assert a strong influence on the transition velocity. With particles of wide size distributions, pressure fluctuations in the bed are smaller and the transition velocity tends to be lower.

As already noted, the onset of the turbulent region seems to occur at the superficial gas velocity at which the standard deviation of the fluctuations of local pressure in the bed is a maximum. It may also be identified by an abrupt change in the degree of bed expansion as the fluidising velocity is increased. Both of these methods give results that are in reasonable accord with visual observations. Measurements based on differential pressures in the bed, rather than on point values, tend to give higher values for the transition velocity. Thus, BI and GRACE[64] give the following correlations for the Reynolds number at the commencement of turbulent fluidisation, $Re_c$, in terms of the Galileo (or Archimedes) number $Ga$ of the gas–solids system. The Reynolds number is based on a 'mean' particle diameter and the superficial velocity of the fluidising gas. The correlations (based, respectively, on differential pressures and on point values) take the form:

$$Re_c = 1.24Ga^{0.447} \tag{6.49}$$

and:

$$Re_c = 0.56Ga^{0.461} \tag{6.50}$$

Equations 6.49 and 6.50 should be used with caution in view of the not inconsiderable scatter of the experimental data.

The upper boundary of the turbulent region, based on velocity, is set by the condition at which particle entrainment increases so rapidly as to give rise to what is essentially a gas–solids transport system, sometimes referred to as *fast-fluidisation*. There is no sharp transition because entrainment is always occurring once the gas velocity exceeds the terminal falling velocity of the smallest particles though, in the bubbling region in particular, a high proportion of the particles may still be retained in the bed at high superficial velocities because these do not necessarily 'see' the high proportion of the gas which rises in the bubbles.

Because of the complex flow pattern, it is not generally possible to determine visually the velocity at which the system ceases, in effect, to be a fluidised bed. From measurements of pressure fluctuations, the following approximate relation may be used, although with considerable care:

$$Re_c \approx 2Ga^{0.45} \tag{6.51}$$

It may be seen by comparison with equations 6.49 and 6.50 that the turbulent fluidised bed region covers an approximately 2–4 fold range of fluidising velocities.

### 6.3.7. Gas and solids mixing

*Flow Pattern of Solids*

The movement of the solid particles in a gas-fluidised bed has been studied by a number of workers using a variety of techniques, although very few satisfactory quantitative studies have been made. In general, it has been found that a high degree of mixing is usually present and that, for most practical purposes, the solids in a bed may be regarded as completely mixed. One of the earliest quantitative investigations was that of BART[65] who added some particles impregnated with salt to a bed 32 mm diameter composed of spheres of cracker catalyst. The tracer particles were fed continuously to the middle of the bed and a sample was removed at the top. MASSIMILLA and BRACALE[66] fluidised glass beads in a bed of diameter 100 mm and showed that, in the absence of bubbles, solids mixing could be represented by a diffusional type of equation. Most other workers report almost complete mixing of solids, and fail to account for any deviations by assuming a diffusional model.

MAY[67] studied the flow patterns of solids in beds up to 1.5 m in diameter by introducing radioactive particles near the top of the bed, and monitoring the radioactivity at different levels by means of a series of scintillation counters. With small beds, it was found that the solids mixing could be represented as a diffusional type of process, although in large-diameter beds there appeared to be an overall circulation superimposed. Values of longitudinal diffusivity ranged from 0.0045 $m^2/s$ in a bed of 0.3 m diameter to about 0.015 $m^2/s$ in a bed 1.5 m in diameter. May also investigated the residence time distribution of gas in a fluidised bed by causing a step-change in the concentration of helium in the fluidising air and following the change in concentration in the outlet air using a mass spectrometer.

SUTHERLAND[68] used nickel particles as tracers in a bed of copper shot. The nickel particles were virtually identical to the copper shot in all the relevant physical properties although, being magnetic, they could be readily separated again from the bed. About 1 per cent of nickel particles was added to the top of a fluidised bed of copper spheres (140 mm diameter and up to 2 m deep). Mixing was found to be rapid although the results were difficult to interpret because the flow patterns were complex. It was established, however, that vertical mixing was less in a bed that was tapered towards the base. In this case, the formation of bubbles was reduced, because the gas velocity was maintained at a nearly constant value by increasing the flow area in the direction in which the pressure was falling. Gas bubbles appear to play a very large part in promoting circulation of the solids, although the type of flow obtained depends very much on the geometry of the system.

ROWE and PARTRIDGE[69] made a photographic study of the movement of solids produced by the rise of a single bubble. As the bubble forms and rises it gathers a wake of particles, and then draws up a spout of solids behind it as shown in Figure 6.16. The wake grows by the addition of particles and then becomes so large that it sheds a fragment of itself, usually in the form of a complete ring. This process is repeated at intervals as the bubble rises, as shown in Figure 6.17. A single bubble causes roughly half its own volume of particles to be raised in the spout through a distance of half its diameter, and one-third of its volume in the wake through a bubble diameter. The overall effect is to move a quantity of particles equal to a bubble volume through one and a half diameters. In a fluidised bed

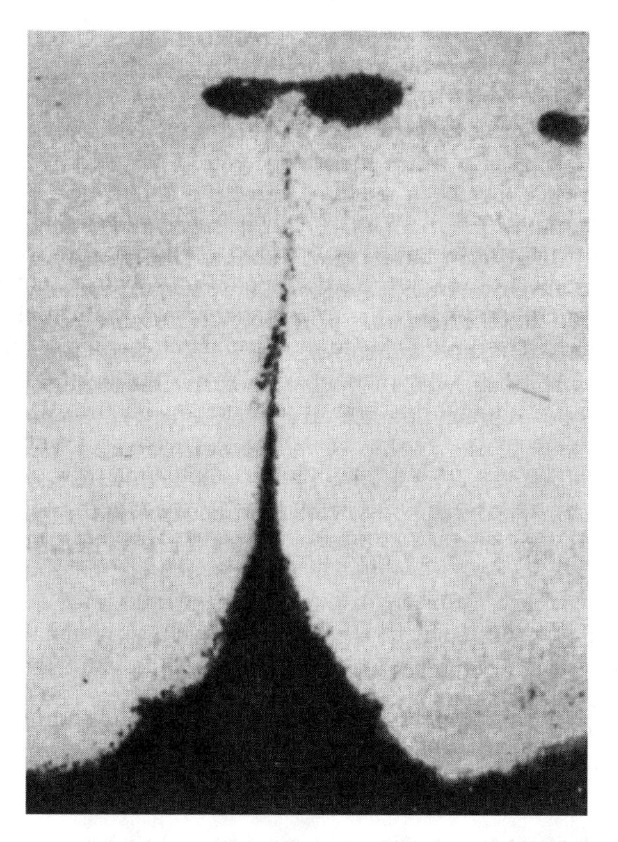

Figure 6.16. Photograph of the solids displacement caused by a single bubble[69]

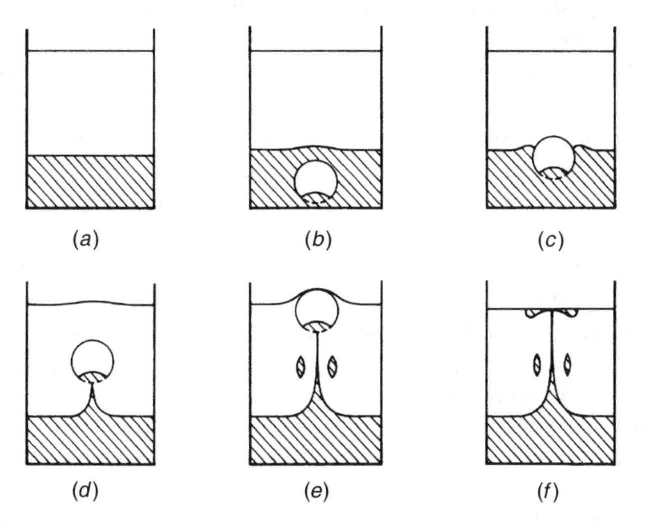


Figure 6.17. Stages of particle movement caused by a bubble[69]

there will normally be many bubbles present simultaneously and their interaction results in a very complex process which gives rise to a high degree of mixing.

### *Flow Pattern of Gas*

The existence and the movement of gas bubbles in a fluidised bed has an appreciable influence on the flow pattern within the bed. Several studies have been made of gas–flow patterns, although, in most cases, these have suffered from the disadvantage of having been carried out in small equipment where the results tend to be specific to the equipment used.

Some measurements have been made in industrial equipment and, for example, ASKINS *et al.*[70] measured the gas composition in a catalytic regenerator, and concluded that much of the gas passed through the bed in the form of bubbles. DANCKWERTS *et al.*[71] used a tracer injection technique, also in a catalytic regenerator, to study the residence time distribution of gas, and concluded that the flow was approximately piston-type and that the amount of back-mixing was small. GILLILAND and co-workers[72,73,74] carried out a series of laboratory-scale investigations, although the results were not conclusive. In general, information on the flow patterns within the bed is limited since these are very much dependent on the scale and geometry of the equipment and on the quality of fluidisation obtained. The gas flow pattern may be regarded, however, as a combination of piston-type flow, of back-mixing, and of by-passing. Back-mixing appears to be associated primarily with the movement of the solids and to be relatively unimportant at high gas rates. This has been confirmed by LANNEAU[75].

## 6.3.8. Transfer between continuous and bubble phases

There is no well-defined boundary between the bubble and continuous phase, although the bubble exhibits many properties in common with a gas bubble in a liquid. If the gas in the bubble did not interchange with the gas in the continuous phase, there would be the possibility of a large proportion of the gas in the bed effectively bypassing the continuous phase and of not coming into contact with the solids. This would obviously have a serious effect where a chemical reaction involving the solids was being carried out. The rate of interchange between the bubble and continuous phases has been studied by SZEKELY[76] and by DAVIES[45].

In a preliminary investigation, SZEKELY[76] injected bubbles of air containing a known concentration of carbon tetrachloride vapour into a 86 mm diameter bed of silica-alumina microspheres fluidised by air. By assuming that the vapour at the outside of the bubble was instantaneously adsorbed on the particles and by operating with very low concentrations of adsorbed material, it was possible to determine the transfer rate between the two phases from the vapour concentrations in the exit gas. By studying beds of differing depths, it was established that most of the mass transfer occurs during the formation of the bubble rather than during its rise. Mass transfer coefficients were calculated on the assumption that the bubbles were spherical during formation and then rose as spherical cap bubbles. Values of mass transfer coefficients of the order of 20 or 30 mm/s were obtained. The variation of the average value of the coefficient with depth is shown in Figure 6.18. At

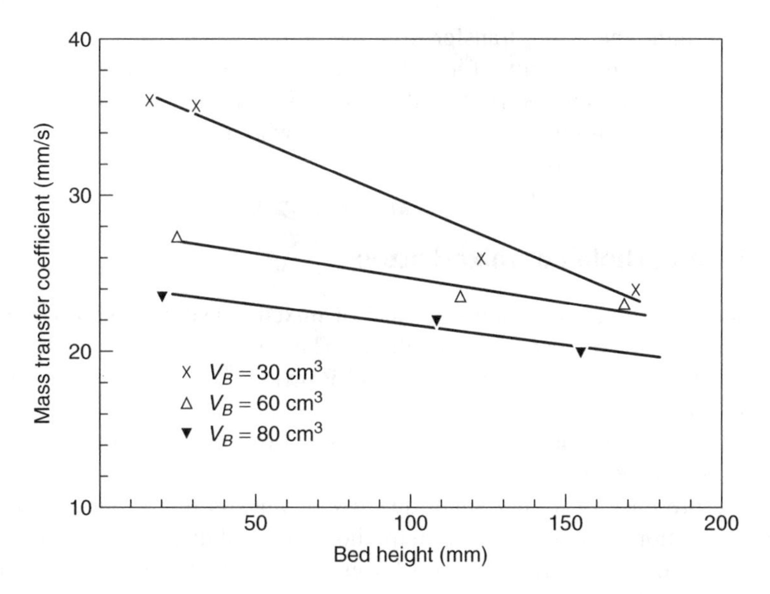


Figure 6.18. Mass transfer coefficient between the bubble and continuous phases as function of the bed height for various bubble volumes[76]

low bed depths the coefficient is low because the gas is able to bypass the bed during the formation of the bubble.

DAVIES[45] has found that, with a range of fine materials, a bubble injected into a fluidised bed tends to grow as a result of a net transfer of gas to it from the continuous phase. The effect becomes progressively less marked as the minimum fluidising velocity of the system increases. It is found that the growth in bubble volume from $V_{B1}$ to $V_{B2}$ in a height of bed, $\Delta z$, for a system with a minimum fluidising velocity $u_{mf}$, is given by:

$$\ln\left(\frac{V_{B1}}{V_{B2}}\right) = \frac{1}{K}\Delta z \tag{6.52}$$

where $K$ represents the distance the bubble must travel for its volume to increase by a factor of e (exponential). Typically, $K$ has a value of 900 mm for cracker catalyst of mean particle size 55 μm fluidised at a velocity of $2.3u_{mf}$.

As a result of gas flow into a bubble, the mean residence time of the gas in such systems is reduced because the bubble rises more rapidly than the gas in the continuous phase. Thus, the injection of a single bubble into the bed will initially cause the bed to expand by an amount equal to the volume of the bubble. When this bubble breaks the surface of the bed, however, the bed volume decreases to a value less than its initial value, the gas content of the bed being reduced. If the value of $u_c$ is only slightly in excess of $u_{mf}$, the gas in a small injected bubble may become dispersed throughout the continuous phase so that no bubble appears at the surface of the bed.

Equation 6.52 enables the net increase in volume of a bubble to be calculated. The bubble grows because more gas enters through the base than leaves through the cap. By injecting bubbles of carbon dioxide, which is not adsorbed by the particles, and analysing the concentration in the continuous phase at various heights in the bed, it has been possible to determine the actual transfer from the bubble when rising in a bed of cracker catalyst fluidised at an air velocity of $u_{mf}$. For bubbles in the size range of 80 – 250 cm$^3$, the transfer velocity through the roof of the bubble is constant at about 20 mm/s which compares very closely with the results of SZEKELY[76] who used a quite different technique and made different assumptions.

### 6.3.9. Beds of particles of mixed sizes

The behaviour of particles of mixed sizes and of mixed densities is highly complex and it has been the subject of many investigations. The works of CHEN and KEAIRNS[77] of HOFFMAN, JANSSEN and PRINS[78] and of WU and BAEYENS[79] indicate the current state of knowledge in this area.

The most uniform fluidisation might be expected to occur when all the particles are of approximately the same size so that there is no great difference in their terminal falling velocities. The presence of a very small quantity of fines is often found to improve the quality of fluidisation of gas–solids systems however, although, if fines are present in the bed, bubble formation may occur at a lower velocity. If the sizes differ appreciably, elutriation occurs and the smaller particles are continuously removed from the system. If the particles forming the bed are initially of the same size, fines will often be produced as a result of mechanical attrition or as a result of breakage due to high thermal stresses. If the particles themselves take part in a chemical reaction, their sizes may alter as a result of the elimination of part of the material, such as during carbonisation or combustion for example. Any fines which are produced should be recovered using a cyclone separator. Final traces of fine material may then be eliminated with an electrostatic precipitator as discussed in Chapter 1.

LEVA[80] measured the rate of elutriation from a bed composed of particles of two different sizes fluidised in air and found that, if the height of the containing vessel above the top of the bed was small, the rate of elutriation was high. If the height was greater than a certain value, the rate was not affected. This is due to the fact that the small particles were expelled from the bed with a velocity higher than the equilibrium value in the unobstructed area above the bed, because the linear velocity of the fluid in the bed is much higher than that in the empty tube. The tests showed that the concentration of the fine particles in the bed varied with the time of elutriation according to a law of the form:

$$C = C_0 e^{-Mt} \tag{6.53}$$

where: $C$ is the concentration of particles at time $t$,
$C_0$ is the initial concentration of particles, and
$M$ is a constant.

Further work on elutriation has been carried out by PEMBERTON and DAVIDSON[81].

A bed, containing particles of different sizes, behaves in a similar manner to a mixture of liquids of different volatilities. Thus the finer particles, when associated with the fluidising medium, correspond to the lower boiling liquid and are more readily elutriated and the rate of their removal from the system and the degree of separation are affected by the height of the reflux column. A law analogous to Henry's Law for the solubility of gases in liquids is obeyed, with the concentration of solids of given size in the bed bearing, at equilibrium, a constant relation to the concentration of solids in the gas which is passing upwards through the bed. Thus, if clean gas is passed upwards through a bed containing fine particles, these particles are continuously removed. On the other hand, if a dust-laden gas is passed through the bed, particles will be deposited until the equilibrium condition is reached. Fluidised beds are therefore sometimes used for removing suspended dusts and mist droplets from gases, having the advantage that the resistance to flow is in many cases less than in equipment employing a fixed filter medium. In this way, clay particles and small glass beads may be used for the removal of sulphuric acid mists.

In fluidisation with a liquid, a bed of particles of mixed sizes will become sharply stratified with the small particles on top and the large ones at the bottom. The pressure drop is that which would be expected for each of the layers in series. If the size range is small however, no appreciable segregation will occur.

### 6.3.10. The centrifugal fluidised bed

One serious limitation of the fluidised bed is that the gas throughput must be kept below the level at which a high level of entrainment of particles takes place. In addition, at high gas rates there can be significant by-passing because of the rapid rate of rise of large bubbles even though there is interchange of gas with the continuous phase. One way of increasing the gas throughput and of reducing the tendency for the formation of large bubbles is to operate at high pressures. Another method is to operate in a centrifugal field[82,83].

In the centrifugal fluidised bed, the solids are rotated in a basket and the gravitational field is replaced by a centrifugal field that is usually sufficiently strong for gravitational effects to be neglected. Gas is fed in at the periphery and travels inwards through the bed. At a radius in the bed, the centrifugal acceleration is $r\omega^2$ for an angular speed of rotation of $\omega$. Thus $r\omega^2/g = E$ is a measure of the enhancement factor for the accelerating force on the particles. From equation 6.13, it may be seen that, since the effective value of the Galileo number is increased by the factor $E$, the minimum fluidising velocity is also increased. If the bed is shallow, it behaves essentially as a fluidised bed, although it has a very much higher minimum fluidising velocity and a wider range of operating flowrates, and less tendency for bubbling. From equation 6.1 it is seen that the pressure drop over the bed is increased by the factor $E$.

In a deep bed, the situation is more complex and the value of $E$ may be considerably greater at the bed inlet than at its inner surface. In addition, the gas velocity increases towards the centre because of the progressively reducing area of flow. These factors combine so that the value of the minimum fluidising velocity increases with radius in the bed. Thus, as flowrate is increased, fluidisation will first occur at the inner surface of the bed and it will then take place progressively further outwards, until eventually the whole bed will become fluidised. In effect, there are therefore two minimum fluidising velocities of interest, that at which fluidisation first occurs at the inner surface, and that at which the whole bed becomes fluidised.

The equipment required for a centrifugal fluidised bed is much more complex than that used for a conventional fluidised bed, and therefore it has found use in highly specialised situations only. One of these is in zero-gravity situations, such as on a spacecraft, where carbon dioxide needs to be absorbed from the atmosphere.

### 6.3.11. The spouted bed

The spouted bed represents the ultimate in channelling in a fluidised bed. Gas enters at the conical base, as shown in Figure 6.19, with much of it flowing rapidly upwards through a centre core, and the remainder percolating through the surrounding solids which may not be in a fluidised state. The solids therefore form a wall for the central spout of gas, gradually become entrained, and then disengage from the gas stream in the space above the bed where the gas velocity falls as the flow area increases from the spout area to the vessel cross-sectional area. As a result of the erosion of the walls of the spout, the surrounding solids gradually settle to give a solids circulation pattern consisting of rapid upward movement in the spout and slow downward movement in the surrounding bed.

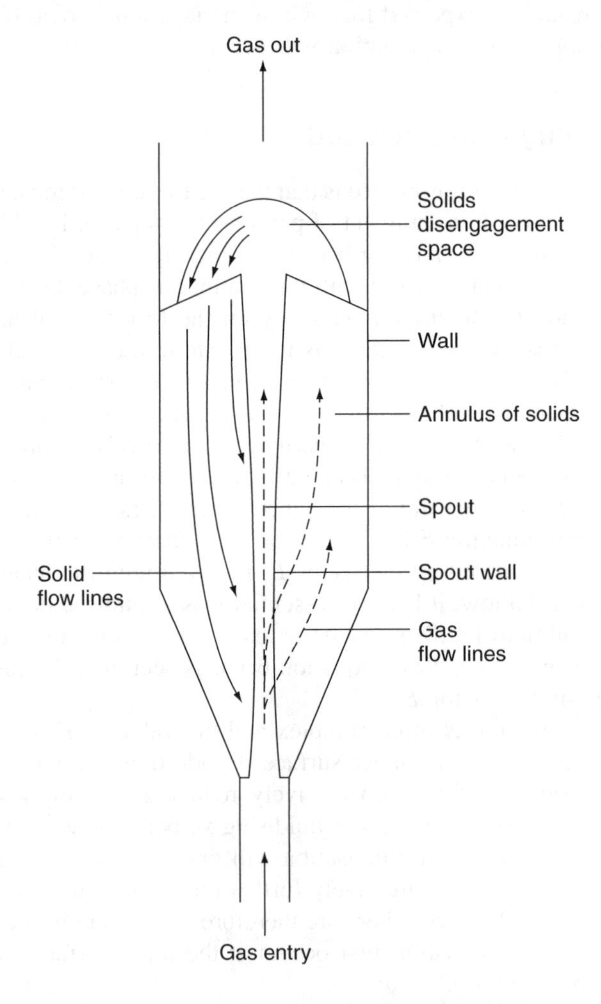


Figure 6.19. Schematic layout of a spouted bed

The pressure drop over a spouted bed is normally lower than that for a fluidised bed, because part of the weight of the solids is supported by the frictional force between the solids and the wall of the vessel. Good contacting is achieved between the solids and the gas, and high rates of heat and mass transfer occur because the relative velocity between the gas and the particles at the wall of the spout is considerably higher than in a fluidised bed. Further, spouted beds can be operated with sticky materials in which the adhesion between particles is too strong for the possibility of fluidisation taking place. In this way, spouted beds are frequently used for the drying of relatively coarse or sticky solids.

A critical minimum volumetric flow rate of gas per unit area is needed to maintain the spouted condition, and its value tends to increase as the bed becomes deeper. At elevated pressures, lower gas velocities are required although, because of the increased gas density, mass flowrates are higher. In general, a higher gas flowrate is required for the start-up of the spouted bed compared with that needed for steady state operation.

The effect of operating parameters cannot be predicted reliably, although a substantial amount of work has been published on the operation of spouted beds[84,85,86].

## 6.4. GAS–LIQUID–SOLIDS FLUIDISED BEDS

If a gas is passed through a liquid–solids fluidised bed, it is possible to disperse the gas in the form of small bubbles and thereby obtain good contacting between the gas, the liquid and the solid. Such systems are often referred to as *three-phase fluidised beds*. An important application is in a biological fluidised bed reactor in which oxygen transfer to the biomass takes place, first by its dissolution from air which is bubbled through the bed, and then its subsequent transfer from the solution to the biomass particles. Good overall mass transfer characteristics are obtained in this way. Three-phase fluidised beds are also used for carrying out gas–liquid reactions which are catalysed by solids, such as hydrogenations.

The hydrodynamics of three-phase fluidised systems is complex and reference should be made to specialised publications for a detailed treatment[87–91]. Only a brief summary of their characteristics is given here.

If a gas is introduced at the bottom of a bed of solids fluidised by a liquid, the expansion of the bed may either decrease or increase, depending on the nature of the solids and particularly their inertia. It is generally found that the minimum fluidising velocity of the liquid is usually reduced by the presence of the gas stream. Measurements are difficult to make accurately, however, because of the fluctuating flow pattern which develops.

At *low gas rates* it is found that with relatively large particles of high inertia such as, for example, 6 mm glass beads fluidised by water, any large gas bubbles are split up by the solids to give a dispersion of fine gas bubbles ($\sim$ 2 mm in diameter) which therefore present a high interfacial area for mass transfer. Even smaller bubbles are produced if the surface tension of the liquid is reduced. On the other hand, small particles appear to be unable to overcome the surface tension forces and do not penetrate into the large bubbles. As discussed in Section 6.3.3, large gas bubbles in a fluidised bed have extensive wakes. The same effect is observed in three-phase systems and, in this case, the bubbles draw liquid rapidly upwards in their wake, thus reducing the liquid flow on the remainder of the bed and causing the bed to contract.

At very *high gas rates*, even large particles do not break down the large gas bubbles and the amount of liquid dragged up in their wake can be so great that some regions of the bed may become completely defluidised. This effect occurs when the ratio of the volumetric flowrate of liquid to that of gas is less than about 0.4.

## 6.5. HEAT TRANSFER TO A BOUNDARY SURFACE

### 6.5.1. Mechanisms involved

The good heat transfer properties of fluidised systems have led to their adoption in circumstances where close control of temperature is required. The presence of the particles in a fluidised system results in an increase of up to one-hundredfold in the heat transfer coefficient, as compared with the value obtained with a gas alone at the same velocity. In a liquid-fluidised system the increase is not so marked.

Many investigations of heat transfer between a gas-fluidised system and a heat transfer surface have been carried out, although the agreement between the correlations proposed by different workers is very poor, with differences of 1 or even 2 orders of magnitude occurring at times. The reasons for these large discrepancies appear to be associated with the critical dependence of heat transfer coefficients on the geometry of the system, on the quality of fluidisation, and consequently on the flow patterns obtained. Much of the work was carried out before any real understanding existed of the nature of the flow patterns within the bed. There is however, almost universal agreement that the one property which has virtually no influence on the process is the thermal conductivity of the solids.

Three main mechanisms have been suggested for the improvement in the heat transfer coefficients brought about by the presence of the solids. First, the particles, with a heat capacity per unit volume many times greater than that of the gas, act as heat-transferring agents. As a result of their rapid movement within the bed, the particles pass from the bulk of the bed to the layers of gas in close contact with the heat transfer surface, exchanging heat at this point and then returning to the body of the bed. Because of their short residence time and their high heat capacity, they change little in temperature, and this fact, coupled with the extremely short physical contact time of the particle with the surface, accounts for the unimportance of their thermal conductivity. The second mechanism is the erosion of the laminar sub-layer at the heat transfer surface by the particles, and the consequent reduction in its effective thickness. The third mechanism, suggested by Mickley and Fairbanks[92], is that "packets" of particles move to the heat transfer surface, where an unsteady state heat transfer process takes place.

### 6.5.2. Liquid-solids systems

The heat transfer characteristics of liquid–solid fluidised systems, in which the heat capacity per unit volume of the solids is of the same order as that of the fluid are of considerable interest. The first investigation into such a system was carried out by Lemlich and Caldas[93], although most of their results were obtained in the transitional region between streamline and turbulent flow and are therefore difficult to assess. Mitson[94] and Smith[20] measured heat transfer coefficients for systems in which a number of different solids were fluidised by water in a 50 mm diameter brass tube, fitted with an annular heating jacket.

The apparatus was fitted with thermocouples to measure the wall and liquid temperatures at various points along the heating section. Heat transfer coefficients at the tube wall were calculated and plotted against volumetric concentration, as shown in Figure 6.20 for gravel. The coefficients were found to increase with concentration and to pass through a maximum at a volumetric concentration of 25–30 per cent. Since in a liquid–solid fluidised system there is a unique relation between concentration and velocity, as the concentration is increased the velocity necessarily falls, and the heat transfer coefficient for liquid alone at the corresponding velocity shows a continuous decrease as the concentration is increased. The difference in the two values, that is the increase in coefficient attributable to the presence of the particles, $(h - h_l)$, is plotted against concentration in Figure 6.21 for ballotini. These curves also pass through a maximum. Experimental results

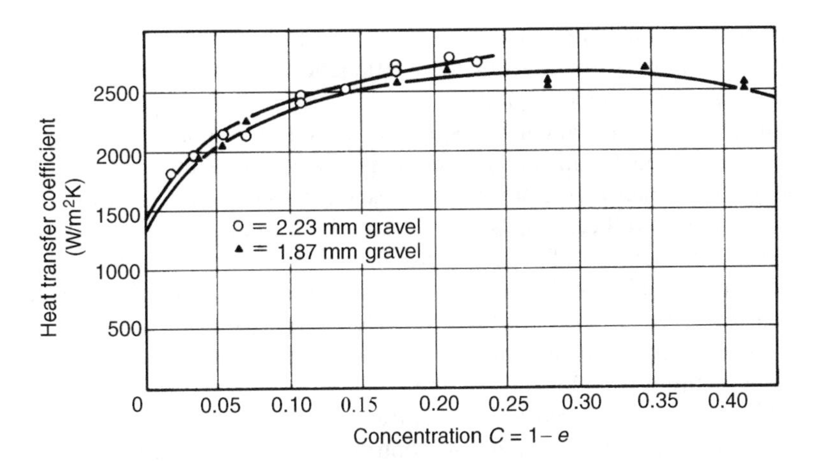


Figure 6.20. Heat transfer coefficients for gravel particles fluidised in water[20]

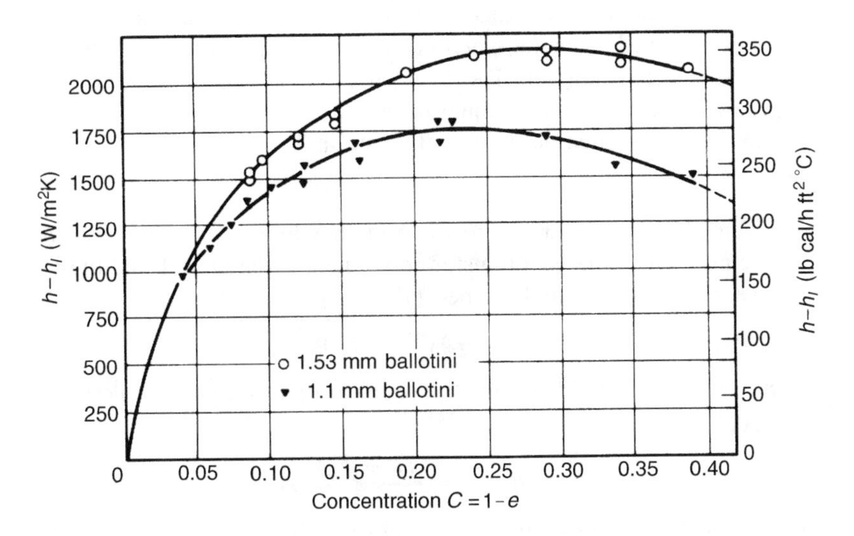


Figure 6.21. The increase in heat transfer coefficient caused by glass ballotini fluidised in water[20]

in the region of turbulent flow may be conveniently correlated in terms of the specific heat $C_s$ of the solid (kJ/kg K) by the equation:

$$(h - h_l) = 24.4(1 + 1.71C_s^{2.12})(1 - e)^m \left(\frac{u_c}{e}\right)^{1.15} \tag{6.54}$$

Here the film coefficients are in kW/m$^2$ K and the fluidising velocity $u_c$ is in m/s. As equation 6.54 is not in dimensionally consistent units (because some of the relevant properties were not varied), the coefficient, 24.4, is valid only for the units stated. The value of the index $m$ is given by:

$$m = 0.079 \left(\frac{u_0 d\rho}{\mu}\right)^{0.36} \tag{6.55}$$

where $u_0$ is the terminal falling velocity of the particle.

The maximum value of the ratio of the coefficient for the fluidised system to that for liquid alone at the same velocity is about 3.

In a modified system in which a suspension of solids is conveyed through the heat transfer section, the heat transfer coefficient is greater than that obtained with liquid alone, though lower than that obtained at the same concentration in a fluidised system. Similar conclusions have been reached by JEPSON, POLL, and SMITH[95] who measured the heat transfer to a suspension of solids in gas.

KANG, FAN and KIM[96] measured coefficients for heat transfer from a cone-shaped heater to beds of glass particles fluidised by water. They also found that the heat transfer coefficient passed through a maximum as the liquid velocity was increased. The heat transfer rate was strongly influenced by the axial dispersion coefficient for the particles, indicating the importance of convective heat transfer by the particles. The region adjacent to the surface of the heater was found to contribute the greater part of the resistance to heat transfer.

Average values of heat-transfer coefficients to liquid–solids systems[97–99] have been measured using small electrically heated surfaces immersed in the bed. The temperature of the element is obtained from its electrical resistance, provided that the temperature coefficient of resistance is known. The heat supplied is obtained from the measured applied voltage and resistance and is equal to $V^2/R$,

where: $V$ is the voltage applied across the element, and

$R$ is its resistance.

The energy given up by the element to the bed in which it is immersed, $Q$, may be expressed as the product of the heat transfer coefficient $h$, the area $A$ and the temperature difference between the element and the bed $(T_E - T_B)$.

Thus:

$$Q = hA(T_E - T_B) \tag{6.56}$$

At equilibrium, the energy supplied must be equal to that given up to the bed and:

$$\frac{V^2}{R} = hA(T_E - T_B) \tag{6.57}$$

Thus the heat-transfer coefficient may be obtained from the slope of the straight line relating $V^2$ and $T_E$. This method has been successfully used for measuring the heat transfer coefficient from a small heated surface, 25 mm square, to a wide range of fluidised systems. Measurements have been made of the coefficient for heat transfer between the 25 mm square surface and fluidised beds formed in a tube of 100 mm diameter. Uniform particles of sizes between 3 mm and 9 mm were fluidised by liquids consisting of mixtures of kerosene and lubricating oil whose viscosities ranged from 1.55 to 940 mN s/m$^2$. The heat transfer coefficient increases as the voidage, and hence the velocity, is increased and passes through a maximum. This effect has been noted by most of the workers in the area. The voidage $e_{max}$ at which the maximum heat transfer coefficient occurs becomes progressively greater as the viscosity of the liquid is increased.

The experimental results obtained for a wide range of systems[(96–99)] are correlated by equation 6.58, in terms of the Nusselt number ($Nu = hd/k$) for the particle expressed as a function of the Reynolds number ($Re'_c = u_c d\rho/\mu$) for the particle, the Prandtl number $Pr$ for the liquid, and the voidage of the bed. This takes the form:

$$Nu' = (0.0325Re'_c + 1.19Re_c'^{0.43})Pr^{0.37}(1-e)^{0.725} \tag{6.58}$$

A plot showing agreement between the equation and the experimental results is given in Figure 6.22.

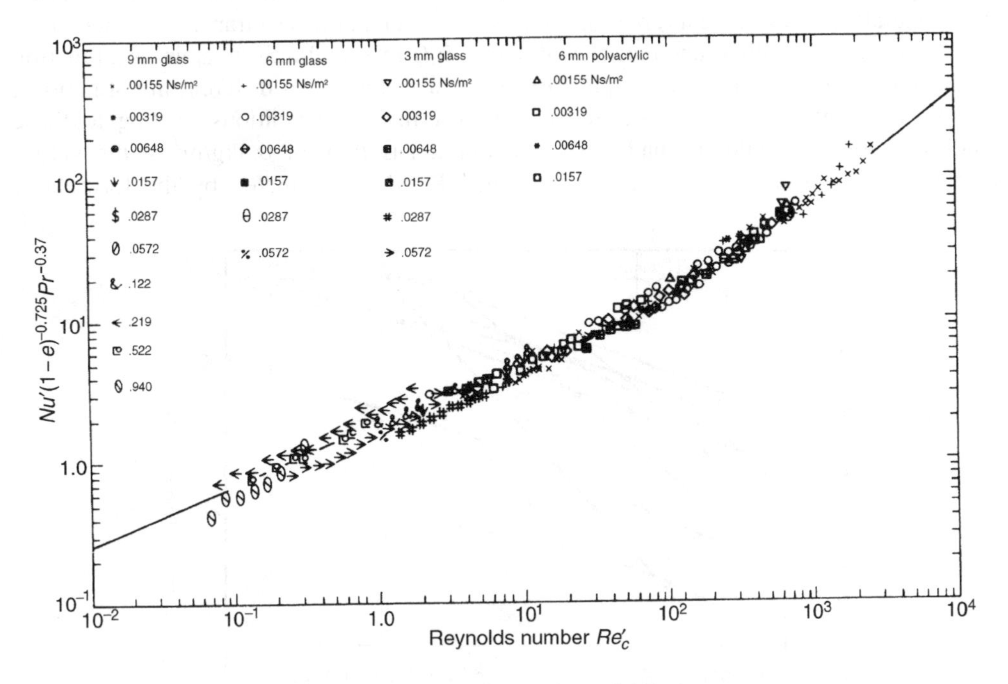


Figure 6.22. Data for heat transfer to fluidised systems

Equation 6.58 covers the following range of variables:

$$10^{-1} < Re'_c < 10^3$$

$$22 < Pr < 14,000$$

$$0.4 < e < 0.9$$

This equation predicts that the heat transfer coefficient should pass through a maximum as the velocity of the liquid increases. It may be noted that for liquid–solids fluidised beds, $(1 - e)$ falls as $Re$ increases, and a maximum value of the heat transfer coefficient is usually obtained at a voidage $e$ of about 0.6 to 0.8.

Although the parameters in equation 6.58 were varied over a wide range, the heat capacities per unit volume and the thermal conductivities of the liquids were almost constant, as they are for most organic liquids, and the dimensions of the surface and of the tube were not varied. Nevertheless, for the purposes of comparison with other results, it is useful to work in terms of dimensionless groups.

The maximum values of the heat transfer coefficients, and of the corresponding Nusselt numbers, may be predicted satisfactorily from equation 6.58 by differentiating with respect to voidage and putting the derivative equal to zero.

## 6.5.3. Gas-solids systems

With gas-solids systems, the heat-transfer coefficient to a surface is very much dependent on the geometrical arrangement and the quality of fluidisation; furthermore, in many cases the temperature measurements are suspect. LEVA[(100)] plotted the heat transfer coefficient for a bed, composed of silica sand particles of diameter 0.15 mm fluidised in air, as a function of gas rate, using the correlations put forward as a result of ten different investigations, as shown in Figure 6.23. For details of the experimental conditions relating to these studies reference should be made to Leva. For a gas flow of 0.3kg/m$^2$ s, the values of the coefficient ranged from about 75 W/m$^2$ K when calculated by the formula of

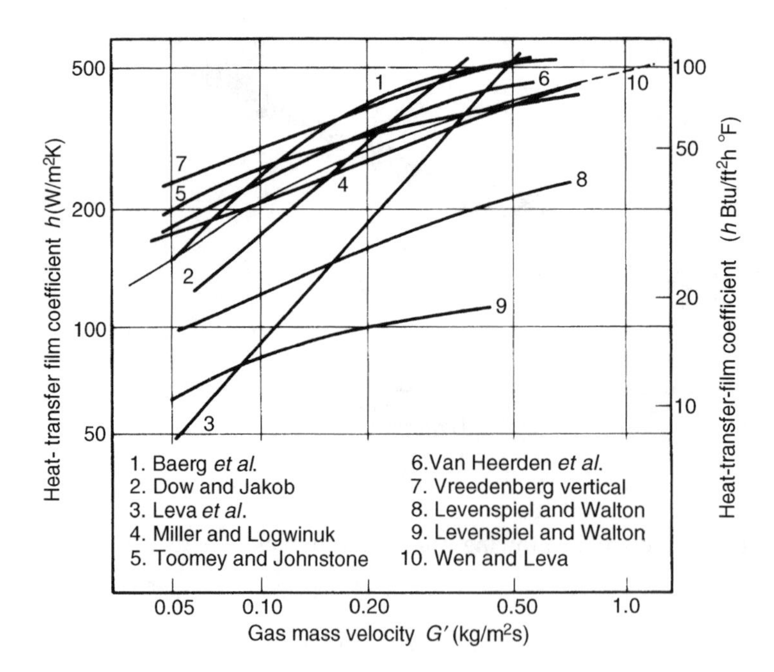


Figure 6.23. Comparison of heat transfer correlations. Silica sand (0.15 mm) fluidised in air[(100)]

LEVENSPIEL and WALTON[101] to about 340 W/m$^2$ K when VREEDENBERG'S[102] expression was used. The equation of Dow and JAKOB[103], gives a value of about 200 W/m$^2$ K for this case. The equation takes the form:

$$\frac{hd_t}{k} = 0.55\left(\frac{d_t}{l}\right)^{0.65}\left(\frac{d_t}{d}\right)^{0.17}\left\{\frac{(1-e)\rho_s C_s}{e\rho C_p}\right\}^{0.25}\left(\frac{u_c d_t \rho}{\mu}\right)^{0.80} \tag{6.59}$$

where: $h$ is the heat transfer coefficient,
$k$ is the thermal conductivity of the gas,
$d$ is the diameter of the particle,
$d_t$ is the diameter of the tube,
$l$ is the depth of the bed,
$e$ is the voidage of the bed,
$\rho_s$ is the density of the solid,
$\rho$ is the density of the gas,
$C_s$ is the specific heat of the solid,
$C_p$ is the specific heat of the gas at constant pressure,
$\mu$ is the viscosity of the gas, and
$u_c$ is the superficial velocity based on the empty tube.

The Nusselt number with respect to the tube $Nu(= hd_t/k)$ is expressed as a function of four dimensionless groups: the ratio of tube diameter to length, the ratio of tube to particle diameter, the ratio of the heat capacity per unit volume of the solid to that of the fluid, and the tube Reynolds number, $Re_c = (u_c d_t \rho/\mu)$. However, equation 6.59 and other equations quoted in the literature should be used with extreme caution, as the value of the heat transfer coefficient will be highly dependent on the flow patterns of gas and solid and the precise geometry of the system.

### *Mechanism of Heat Transfer*

The mechanism of heat transfer to a surface has been studied in a fixed and a fluidised bed by BOTTERILL and co-workers[104–107]. An apparatus was constructed in which particle replacement at a heat transfer surface was obtained by means of a rotating stirrer with blades close to the surface. A steady-state system was employed using an annular bed, with heat supplied at the inner surface and removed at the outer wall by means of a jacket. The average residence time of the particle at the surface in a fixed bed was calculated, assuming the stirrer to be perfectly efficient, and in a fluidised bed it was shown that the mixing effects of the gas and the stirrer were additive. The heat transfer process was considered to be one of unsteady-state thermal conduction during the period of residence of the particle and its surrounding layer of fluid at the surface. Virtually all the heat transferred between the particle and the surface passes through the intervening fluid as there is only point contact between the particle and the surface. It is shown that the heat transfer rate falls off in an exponential manner with time and that, with gases, the heat transfer is confined to the region surrounding the point of contact. With liquids, however, the thermal conductivities and diffusivities of the two phases are comparable, and appreciable heat flow occurs over a more extended region. Thus, the heat transfer process is capable of being broken down into a series of unsteady state stages, at the completion of each of which the particle with its attendant fluid layer becomes mixed again with the bulk of the bed. This process is very similar to that assumed in the penetration theory of mass transfer proposed by HIGBIE[108] and by DANCKWERTS[109], and described in Chapter 10 of Volume 1. The theory shows how the heat-carrying effect of the particles, and their disruption of the laminar sub-layer at the surface, must be considered as component parts of a single mechanism. Reasonable agreement is obtained by Botterill *et al.* between the practical measurements of heat transfer coefficient and the calculated values.

MICKLEY, FAIRBANKS, and HAWTHORN[110] made measurements of heat transfer between an air–fluidised bed 104 mm diameter and a concentric heater 6.4 mm diameter and 600 mm long, divided into six 100 mm lengths each independently controllable. The mean value of the heat transfer coefficient was found to decrease with increase in height, and this was probably attributable to an increased tendency for slugging. The instantaneous value of the heat transfer coefficient was found by replacing a small segment of one of the heaters, of arc 120° and depth 7.9 mm, by a piece of platinum foil of low heat capacity with a thermocouple behind it. The coefficient was found to fluctuate from less than 50 to about 600W/m$^2$ K. A typical curve is shown in Figure 6.24 from which it is seen that the coefficient rises rapidly to a peak, and then falls off slowly before there is a sudden drop in value as a bubble passes the foil. It is suggested that heat transfer takes place by unsteady-state conduction to a packet of particles which are then displaced by a gas slug. Coefficients were calculated on the assumption that heat flowed for a mean exposure time into a mixture whose physical and thermal properties corresponded with those of an element of the continuous phase, and that the heat flow into the gas slug could be neglected. The fraction of the time during which the surface was in contact with a gas slug ranged from zero to 40 per cent with an easily fluidised solid, but appeared to continue increasing with gas rate to much higher values for a solid which gave uneven fluidisation. It is easy on this basis to see why very variable values of heat transfer coefficients have been obtained by different workers. For good heat transfer, rapid replacement of the solids is required, without an appreciable coverage of the surface by gas alone.

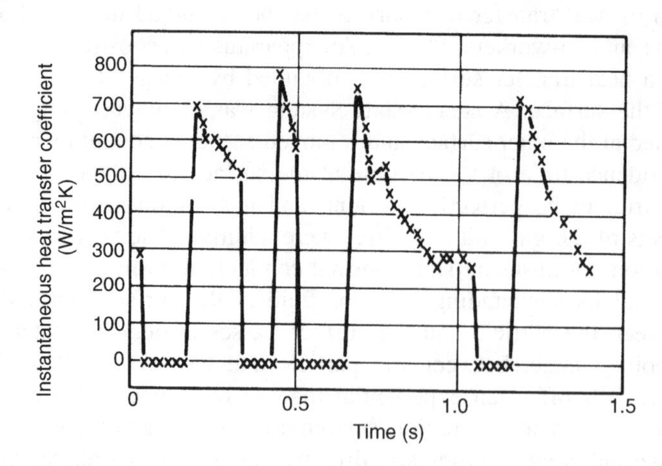


Figure 6.24. Instantaneous heat transfer coefficients for glass beads fluidised in air[110]

Bock and Molerus[111] also concluded that the heat transfer coefficient decreases with increase in contact time between elements of bed and the heat transfer surface. In order to observe the effects of long contact times, tests were also carried out with non-fluidised solids. Vertical single tubes and a vertical tube bundle were used. It was established that it was necessary to allow for the existence of a "gas-gap" between the fluidised bed and the surface to account for the observed values of transfer coefficients. The importance of having precise information on the hydrodynamics of the bed before a reasonable prediction can be made of the heat transfer coefficient was emphasised.

Heat transfer coefficients for a number of gas–solids systems were measured by Richardson and Shakiri[112] using an electrically heated element 25 mm square over a range of pressures from sub-atmospheric (0.03 $MN/m^2$) to elevated pressures (up to 1.5 $MN/m^2$). The measuring technique was essentially similar to that employed earlier for liquid–solids systems, as described in Section 6.5.2.

In all cases there was a similar pattern as the fluidising velocity was increased, as typified by the results shown in Figure 6.25 for the fluidisation of 78 μm glass spheres by air at atmospheric pressure. Region AB corresponds to a fixed bed. As the velocity is increased beyond the minimum fluidising velocity (B), the frequency of formation of small bubbles increases rapidly and the rate of replacement of solids in the vicinity of the surface rises. This is reflected in a rapid increase in the heat transfer coefficient (BC). A point is then reached where the rate of bubble formation is so high that an appreciable proportion of the heat transfer surface is blanketed by gas. The coefficient thus rises to a maximum (C) then falls as the gas rate is further increased (CD). This fall occurs over only a comparatively small velocity range, following which the coefficient passes through a minimum (D), and then starts to rise again (DE) because of the effect of the increased rate of circulation of the solids in the bed. Values of the heat transfer coefficient show considerable fluctuations at all velocities, as a result of the somewhat random behaviour

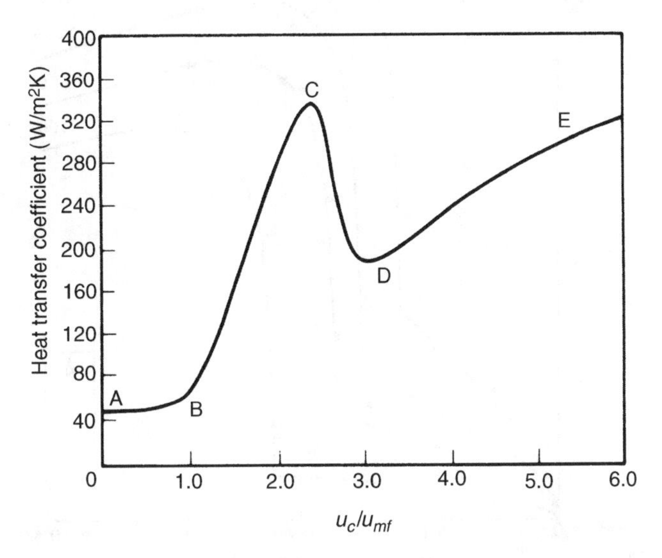


Figure 6.25. Heat transfer coefficient from a surface to 78 μm glass spheres fluidised by air. AB = fixed bed; B = minimum fluidising velocity; C = maximum coefficient; CD = falling coefficient; D = minimum coefficient; DE = final region of increasing coefficient

of the gas bubbles which are responsible for the continual replacement of the solids in the vicinity of the heat transfer surface.

The nature of the curves depicted in Figures 6.24 and 6.25 shows just how difficult it is to compare heat transfer coefficients obtained under even marginally different operating conditions. Any correlation should therefore be treated with extreme caution. Some attempts have been made to correlate values of the maximum heat transfer coefficient (Point C in Figure 6.25), although it is difficult then to relate these to the mean coefficient experienced under practical circumstances.

### *Effect of Pressure*

The operation of fluidised beds at elevated pressures not only permits greater mass flowrates through a bed of a given diameter and gives rise to more uniform fluidisation, but also results in improved heat transfer coefficients. XAVIER *et al.*[113] fluidised a variety of solid particles in the size range 0.06–0.7 mm in a bed of 100 mm diameter. The heat transfer surface was provided by a small vertical flat element arranged near the top of the axis of the bed, as shown in the illustration on the right of Figure 6.26. Experimental results for the fluidisation of glass beads 0.475 mm diameter by nitrogen at pressures from 100 to 2000 $kN/m^2$ are shown. It is seen that the plots of heat transfer coefficient as a function of fluidising velocity all retain the general shape of those in Figure 6.25, and show an improvement in heat transfer at all velocities as the pressure is increased. The discrepancy between the lines and the points arises from the fact that the former were calculated from the model proposed by Xavier *et al.*

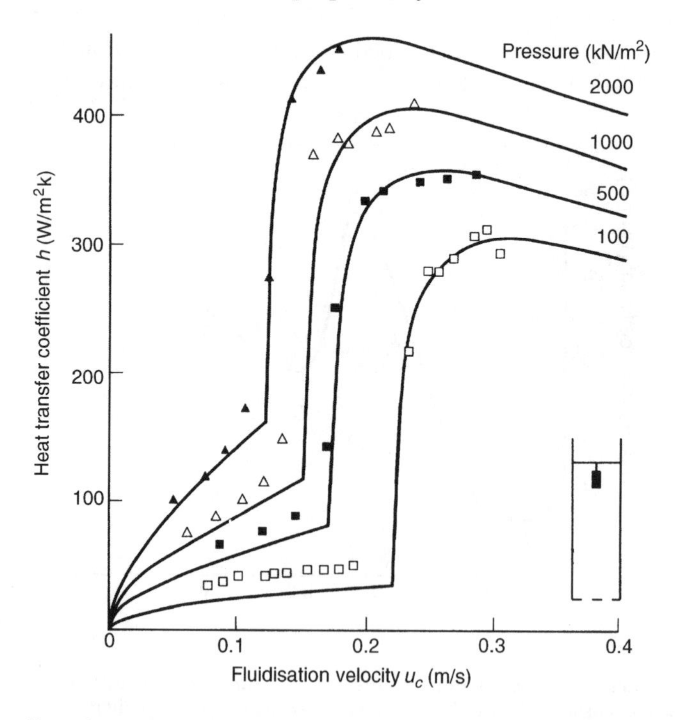


Figure 6.26. The effect of pressure on heat transfer between a 40 mm vertical flat surface and a fluidised bed

# 6.6. MASS AND HEAT TRANSFER BETWEEN FLUID AND PARTICLES

## 6.6.1. Introduction

The calculation of coefficients for the transfer of heat or mass between the particles and the fluid stream requires a knowledge of the heat or mass flow, the interfacial area, and the driving force expressed either as a temperature or a concentration difference. Many early investigations are unsatisfactory in that one or more of these variables was inaccurately determined. This applies particularly to the driving force, which was frequently based on completely erroneous assumptions about the nature of the flow in the bed.

One difficulty in making measurements of transfer coefficients is that equilibrium is rapidly attained between particles and fluidising medium. This has in some cases been obviated by the use of very shallow beds. In addition, in measurements of mass transfer, the methods of analysis have been inaccurate, and the particles used have frequently been of such a nature that it has not been possible to obtain fluidisation of good quality.

## 6.6.2. Mass transfer between fluid and particles

BAKHTIAR[114] adsorbed toluene and iso-octane vapours from a vapour-laden air stream on to the surface of synthetic alumina microspheres and followed the change of concentration of the outlet gas with time, using a sonic gas analyser. It was found that equilibrium was attained between outlet gas and solids in all cases, and therefore transfer coefficients could not be calculated. The progress of the adsorption process was still followed, however.

SZEKELY[115] modified the system so that equilibrium was not achieved at the outlet. Thin beds and low concentrations of vapour were used, so that the slope of the adsorption isotherm was greater. Particles of charcoal of different pore structures, and of silica gel, were fluidised by means of air or hydrogen containing a known concentration of carbon tetrachloride or water vapour. A small glass apparatus was used so that it could be readily dismantled, and the adsorption process was followed by weighing the bed at intervals. The inlet concentration was known and the outlet concentration was determined as a function of time from a material balance, using the information obtained from the periodic weighings. The driving force was then obtained at the inlet and the outlet of the bed, on the assumption that the solids were completely mixed and that the partial pressure of vapour at their surface was given by the adsorption isotherm.

At any height $z$ above the bottom of the bed, the mass transfer rate per unit time, on the assumption of *piston flow* of gas, is given by:

$$dN_A = h_D \Delta C a' dz \tag{6.60}$$

where $a'$ is the transfer area per unit height of bed.

Integrating over the whole depth of the bed gives:

$$N_A = h_D a' \int_0^Z \Delta C dz \tag{6.61}$$

The integration may be carried out only if the variation of driving force throughout the depth of the bed may be estimated. It was not possible to make measurements of the concentration profiles within the bed, although as the value of $\Delta C$ did not vary greatly from the inlet to the outlet, no serious error was introduced by using the logarithmic mean value $\Delta C_{lm}$.

Thus:
$$N_A \approx h_D a' Z \Delta C_{lm} \tag{6.62}$$

Values of mass transfer coefficients were calculated using equation 6.62, and it was found that the coefficient progressively became less as each experiment proceeded and as the solids became saturated. This effect was attributed to the gradual build up of the resistance to transfer in the solids. In all cases the transfer coefficient was plotted against the relative saturation of the bed, and the values were extrapolated back to zero relative saturation, corresponding to the commencement of the test. These maximum extrapolated values were then correlated by plotting the corresponding value of the Sherwood number ($Sh' = h_D d/D$) against the particle Reynolds number ($Re'_c = u_c d\rho/\mu$) to give two lines as shown in Figure 6.27, which could be represented by the following equations:

$$(0.1 < Re'_c < 15) \qquad \frac{h_D d}{D} = Sh' = 0.37 Re_c'^{1.2} \tag{6.63}$$

$$(15 < Re'_c < 250) \qquad \frac{h_D d}{D} = Sh' = 2.01 Re_c'^{0.5} \tag{6.64}$$

These correlations are applicable to all the systems employed, provided that the initial maximum values of the transfer coefficients are used. This suggests that the extrapolation gives the true gas-film coefficient. This is borne out by the fact that the coefficient remained unchanged for a considerable period when the pores were large, though it fell off extremely rapidly with solids with a fine pore structure. It was not possible, to relate the behaviour of the system quantitatively to the pore size distribution however.

The values of Sherwood number fall below the theoretical minimum value of 2 for mass transfer to a spherical particle and this indicates that the assumption of piston flow of gases is not valid at low values of the Reynolds number. In order to obtain realistic values in this region, information on the axial dispersion coefficient is required.

A study of mass transfer between a liquid and a particle forming part of an assemblage of particles was made by MULLIN and TRELEAVEN[116], who subjected a sphere of benzoic acid to the action of a stream of water. For a fixed sphere, or a sphere free to circulate in the liquid, the mass transfer coefficient was given, for $50 < Re'_c < 700$, by:

$$Sh' = 0.94 Re_c'^{1/2} Sc^{1/3} \tag{6.65}$$

The presence of adjacent spheres caused an increase in the coefficient because the turbulence was thereby increased. The effect became progressively greater as the concentration increased, although the results were not influenced by whether or not the surrounding particles were free to move. This suggests that the transfer coefficient was the same in a fixed or a fluidised bed.

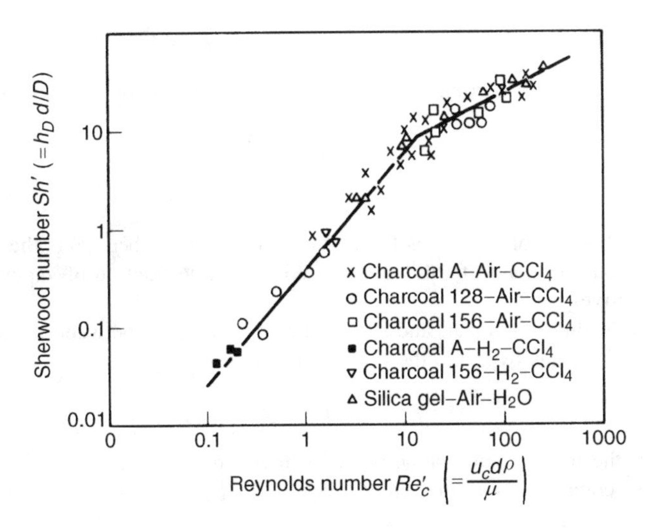


Figure 6.27. Sherwood number as a function of Reynolds number for adsorption experiments[115]

The results of earlier work by CHU, KALIL, and WETTEROTH[117] suggested that transfer coefficients were similar in fixed and fluidised beds. Apparent differences at low Reynolds numbers were probably due to the fact that there could be appreciable back-mixing of fluid in the fluidised bed.

## Example 6.4

In a fluidised bed, *iso*-octane vapour is adsorbed from an air stream on to the surface of alumina microspheres. The mole fraction of *iso*-octane in the inlet gas is $1.442 \times 10^{-2}$ and the mole fraction in the outlet gas is found to vary with time as follows:

| Time from start (s) | Mole fraction in outlet gas ($\times 10^2$) |
|---|---|
| 250 | 0.223 |
| 500 | 0.601 |
| 750 | 0.857 |
| 1000 | 1.062 |
| 1250 | 1.207 |
| 1500 | 1.287 |
| 1750 | 1.338 |
| 2000 | 1.373 |

Show that the results may be interpreted on the assumptions that the solids are completely mixed, that the gas leaves in equilibrium with the solids, and that the adsorption isotherm is linear over the range considered. If the gas flowrate is $0.679 \times 10^{-6}$ kmol/s and the mass of solids in the bed is 4.66 g, calculate the slope of the adsorption isotherm. What evidence do the results provide concerning the flow pattern of the gas?

## Solution

A mass balance over a bed of particles at any time $t$ after the start of the experiment, gives;

$$G_m(y_0 - y) = \frac{d(WF)}{dt}$$

where:

$G_m$ is the molar flowrate of gas, $W$ is the mass of solids in the bed, $F$ is the number of moles of vapour adsorbed on unit mass of solid, and $y_0$, $y$ is the mole fraction of vapour in the inlet and outlet stream respectively.

If the adsorption isotherm is linear, and if equilibrium is reached between the outlet gas and the solids and if none of the gas bypasses the bed, then $F$ is given by:

$$F = f + by$$

where $f$ and $b$ are the intercept and slope of the isotherm respectively.

Combining these equations and integrating gives:

$$\ln(1 - y/y_0) = -(G_m/Wb)t$$

If the assumptions outlined previously are valid, a plot of $\ln(1 - y/y_0)$ against $t$ should yield a straight line of slope $-G_m/Wb$. As $y_0 = 0.01442$, the following table may be produced:

| Time (s) | $y$ | $y/y_0$ | $1 - (y/y_0)$ | $\ln(1 - (y/y_0))$ |
|---|---|---|---|---|
| 250 | 0.00223 | 0.155 | 0.845 | −0.168 |
| 500 | 0.00601 | 0.417 | 0.583 | −0.539 |
| 750 | 0.00857 | 0.594 | 0.406 | −0.902 |
| 1000 | 0.0106 | 0.736 | 0.263 | −1.33 |
| 1250 | 0.0121 | 0.837 | 0.163 | −1.81 |
| 1500 | 0.0129 | 0.893 | 0.107 | −2.23 |
| 1750 | 0.0134 | 0.928 | 0.072 | −2.63 |
| 2000 | 0.0137 | 0.952 | 0.048 | −3.04 |

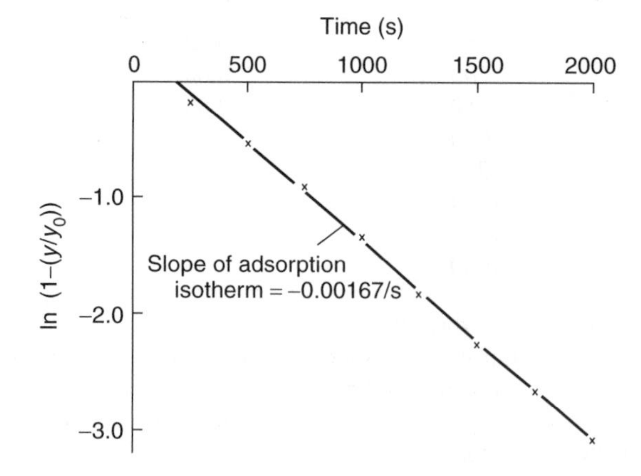


Figure 6.28. Adsorption isotherm for Example 6.4

These data are plotted in Figure 6.28 and a straight line is obtained, with a slope of $-0.00167/\text{s}$ If $G_m = 0.679 \times 10^{-6}$ kmol/s and $W = 4.66$ g, then:

$$-0.00167 = (-0.679 \times 10^{-6})/4.66b$$

from which: $b = 87.3 \times 10^{-6}$ kmol/g or $\underline{\underline{0.0873}}$ kmol/kg

## 6.6.3. Heat transfer between fluid and particles

In measuring heat transfer coefficients, many workers failed to measure any temperature difference between gas and solid in a fluidised bed. Frequently, an incorrect area for transfer was assumed, since it was not appreciated that thermal equilibrium existed everywhere in a fluidised bed, except within a thin layer immediately above the gas distributor. KETTENRING, MANDERFIELD, and SMITH[118] and HEERTJES and MCKIBBINS[119] measured heat transfer coefficients for the evaporation of water from particles of alumina or silica gel fluidised by heated air. In the former investigation, it is probable that considerable errors arose from the conduction of heat along the leads of the thermocouples used for measuring the gas temperature. Heertjes and McKibbins found that any temperature gradient was confined to the bottom part of the bed. A suction thermocouple was used for measuring gas temperatures, although this probably caused some disturbance to the flow pattern in the bed. FRANTZ[120] has reviewed many of the investigations in this field.

AYERS[121] used a steady-state system in which spherical particles were fluidised in a rectangular bed by means of hot air. A continuous flow of solids was maintained across the bed, and the particles on leaving the system were cooled and then returned to the bed. Temperature gradients within the bed were measured using a fine thermocouple assembly, with a junction formed by welding together 40-gauge wires of copper and constantan. The thermo-junction leads were held in an approximately isothermal plane to minimise the effect of heat conduction. After steady-state conditions had been reached, it was found that the temperature gradient was confined to a shallow zone, not more than 2.5 mm deep at the bottom of the bed. Elsewhere, the temperature was uniform and equilibrium existed between the gas and the solids. A typical temperature profile is shown in Figure 6.29.

At any height $z$ above the bottom of the bed, the heat transfer rate between the particles and the fluid, on the assumption of complete mixing of the solids and *piston flow* of the gas, is given by:

$$\mathrm{d}Q = h \Delta T a' \mathrm{d}z \tag{6.66}$$

Integrating gives:

$$Q = ha' \int_0^Z \Delta T \,\mathrm{d}z \tag{6.67}$$

In equation 6.67, $Q$ may be obtained from the change in temperature of the gas stream, and $a'$, the area for transfer per unit height of bed, from measurements of the surface of solids in the bed. In this case, the integration may be carried out graphically, because the relation between $\Delta T$ and $z$ was obtained from the readings of the thermocouple. The temperature recorded by the thermocouple was assumed to be the gas temperature, and,

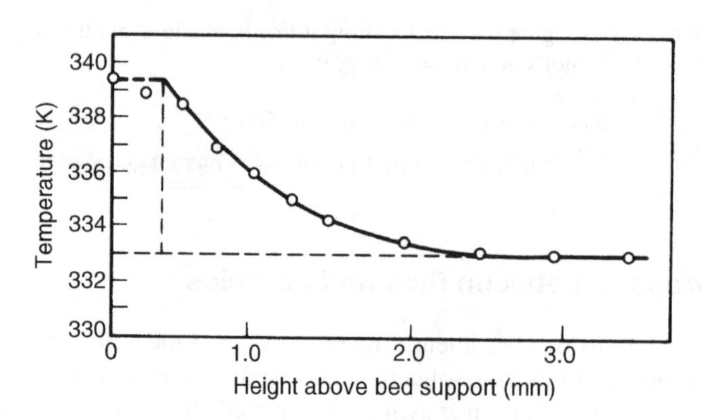


Figure 6.29. Vertical temperature gradient in fluidised bed[121]

if the solids were completely mixed, their temperature would be the same as that of the gas in the upper portion of the bed.

The results for the heat transfer coefficient were satisfactorily correlated by equation 6.68 as shown in Figure 6.30. Since the resistance to heat transfer in the solid could be neglected compared with that in the gas, the coefficients which were calculated were gas-film coefficients, correlated by:

$$Nu' = \frac{hd}{k} = 0.054\left(\frac{u_c\, d\rho}{e\mu}\right)^{1.28} = 0.054\left(\frac{Re'_c}{e}\right)^{1.28} \tag{6.68}$$

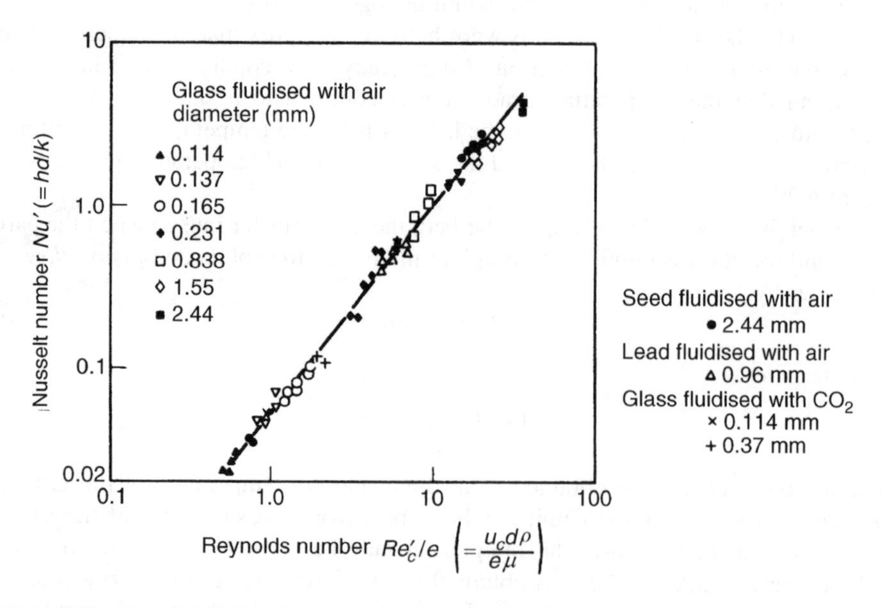


Figure 6.30. Correlation of experimental results for heat transfer to particles in a fluidised bed[121]

Taking an average value of 0.57 for the voidage of the bed, this equation may be rewritten as:

$$Nu' = 0.11 Re_c'^{1.28} \tag{6.69}$$

This equation was found to be applicable for values of $Re_e'$ from 0.25 to 18.

As in the case of mass transfer, the assumption of piston flow is not valid, certainly not at low values of the Reynolds number (< 0) at which the Nusselt number is less than the theoretical minimum value of 2. This question is discussed further at the end of this section.

## Example 6.5

Cold particles of glass ballotini are fluidised with heated air in a bed in which a constant flow of particles is maintained in a horizontal direction. When steady conditions have been reached, the temperatures recorded by a bare thermocouple immersed in the bed are as follows:

| Distance above bed support (mm) | Temperature (K) |
|---|---|
| 0 | 339.5 |
| 0.64 | 337.7 |
| 1.27 | 335.0 |
| 1.91 | 333.6 |
| 2.54 | 333.3 |
| 3.81 | 333.2 |

Calculate the coefficient for heat transfer between the gas and the particles, and the corresponding values of the particle Reynolds and Nusselt numbers. Comment on the results and on any assumptions made.

The gas flowrate is 0.2 kg/m$^2$ s, the specific heat of air is 0.88 kJ/kg K, the viscosity of air is 0.015 mNs/m$^2$, the particle diameter is 0.25 mm and the thermal conductivity of air is 0.03 W/m K.

## Solution

For the system described in this problem, the rate of heat transfer between the particles and the fluid is given by:

$$dQ = ha'\Delta T\, dz \quad \text{(equation 6.66)}$$

and on integration:

$$Q = ha' \int_0^Z \Delta T\, dz \quad \text{(equation 6.67)}$$

where $Q$ is the heat transferred, $h$ is the heat transfer coefficient, $a'$ is the area for transfer/unit height of bed, and $\Delta T$ is the temperature difference at height $z$.

From the data given, $\Delta T$ may be plotted against $z$ as shown in Figure 6.31 where the area under the curve gives the value of the integral as 8.82 mm K.

Thus: Heat transferred $= 0.2 \times 0.88(339.5 - 332.2)$

$= 1.11$ kW/m$^2$ of bed cross-section.

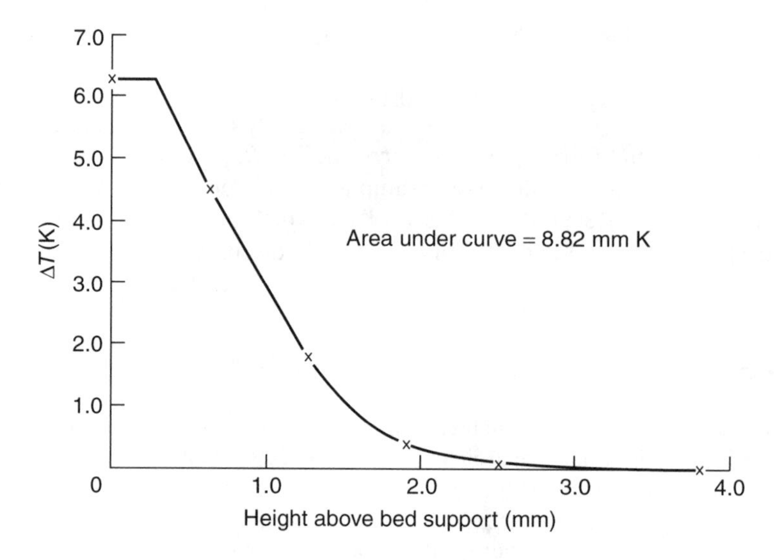


Figure 6.31. Temperature rise as a function of bed height for Example 6.5

If the bed voidage = 0.57, and a bed 1 $m^2$ × 1 m high, is considered with a volume = 1 $m^3$,

then: Volume of particles $= (1 - 0.57) \times 1 = 0.43$ $m^3$.

Volume of 1 particle $= (\pi/6)(0.25 \times 10^{-3})^3 = 8.18 \times 10^{-12}$ $m^3$.

Thus: number of particles $= 0.43/(8.18 \times 10^{-12}) = 5.26 \times 10^{10}$ per $m^3$.

Area of particles $a'$ $= 5.26 \times 10^{10} \times (\pi/4)(0.25 \times 10^{-3})^2 = 1.032 \times 10^4$ $m^2/m^3$.

Substituting in equation 6.67 gives:

$$1100 = h \times (1.03 \times 10^4 \times 8.82 \times 10^{-3})$$

and :

$$h = \underline{\underline{12.2 \text{ W/m}^2 \text{ K}}}$$

From equation 6.69: $Nu = 0.11 Re^{1.28}$

$$Re = G'd/\mu = (0.2 \times 0.25 \times 10^{-3})/(0.015 \times 10^{-3}) = \underline{\underline{3.33}}$$

Thus:

$$Nu = 0.11 \times (3.33)^{1.28} = \underline{\underline{0.513}}$$

and:

$$h = (0.513 \times 0.03)/(0.25 \times 10^{-3}) = \underline{\underline{61.6 \text{ W/m}^2 \text{ K}}}$$

## Example 6.6

Ballotini particles, 0.25 mm in diameter, are fluidised by hot air flowing at the rate of 0.2 kg/$m^2$ s to give a bed of voidage 0.5 and a cross-flow of particles is maintained to remove the heat. Under steady state conditions, a small bare thermocouple immersed in the bed gives the following temperature data:

| Distance above bed support (mm) | Temperature (K) | (°C) |
|---|---|---|
| 0 | 339.5 | 66.3 |
| 0.625 | 337.7 | 64.5 |
| 1.25 | 335.0 | 61.8 |
| 1.875 | 333.6 | 60.4 |
| 2.5 | 333.3 | 60.1 |
| 3.75 | 333.2 | 60.0 |

Assuming plug flow of the gas and complete mixing of the solids, calculate the coefficient for heat transfer between the particles and the gas. This specific heat capacity of air is 0.85 kJ/kg K.

A fluidised bed of total volume 0.1 $m^3$ containing the same particles is maintained at an approximately uniform temperature of 423 K (150 °C) by external heating, and a dilute aqueous solution at 373 K (100 °C) is fed to the bed at the rate of 0.1 kg/s so that the water is completely evaporated at atmospheric pressure. If the heat transfer coefficient is the same as that previously determined, what volumetric fraction of the bed is effectively carrying out the evaporation? The latent heat of vaporisation of water is 2.6 MJ/kg.

## Solution

Considering unit area of the bed with voidage $e$ and a mass flowrate $G'$, then a heat balance over an increment of height d$z$ gives:

$$G'C_p \mathrm{d}T = hS(1-e)\,\mathrm{d}z\,(T_p - T)$$

where: $C_p$ = specific heat, (J/kg K)
$T_p$ = Particle temperature, (K)
$S(1-e)$ = surface area/volume of bed ($m^2/m^3$)
$h$ = heat transfer coefficient (W/$m^2$ K)

Thus:

$$\int_0^Z \frac{\mathrm{d}T}{T_p - T} = \frac{hS(1-e)z}{G'C_p}$$

or:

$$G'C_p \Delta T = hS(1-e)\int_0^Z (T_p - T)\,\mathrm{d}z$$

$\int (T_p - T)\mathrm{d}z$ may be found from a plot of the experimental data shown in Figure 6.32 as 6.31 K mm

$$S(1-e) = \frac{6}{d}(1-e) = 6(1-0.5)/(0.25 \times 10^{-3}) = 1.2 \times 10^4/\mathrm{m}$$

$$G' = 0.2\,\mathrm{kg/m^2 s},\ C_p = 850\,\mathrm{J/kg\ K}$$

Hence:

$$(0.2 \times 850 \times 6.3) = h \times (1.2 \times 10^4 \times 6.31 \times 10^{-3})$$

and:

$$h = \underline{\underline{14.1\,\mathrm{W/m^2\ K}}}$$

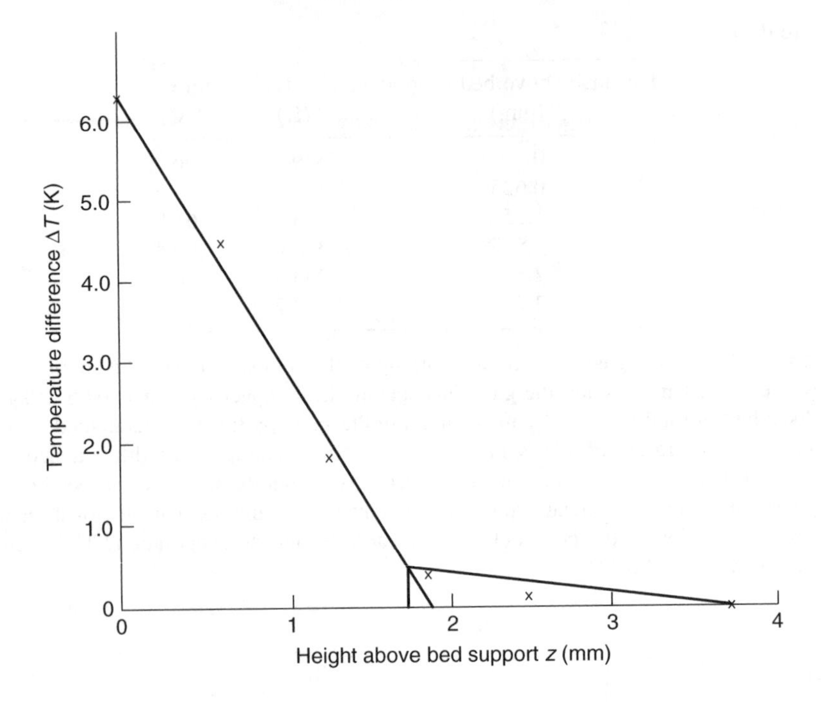


Figure 6.32. $\Delta T$ as a function of bed height $z$ in Example 6.6

If the evaporation rate is 0.1 kg/s at a temperature difference, $\Delta T = 50$ deg K, the heat flow $= (0.1 \times 2.6 \times 10^6) = 2.6 \times 10^5$ W

If the effective area of the bed is $A$ then:

$$(14.1 \times A \times 50) = 2.6 \times 10^5 \text{ and } A = 369 \text{ m}^2$$

The surface area of the bed $= (1.2 \times 10^4 \times 0.1) = 1200$ m$^2$

Hence, the fraction of bed which is used $= (369/1200) = 0.31$ or 31 per cent

## 6.6.4. Analysis of results for heat and mass transfer to particles

A comparison of equations 6.63 and 6.69 shows that similar forms of equations describe the processes of heat and mass transfer. The values of the coefficients are however different in the two cases, largely to the fact that the average value for the Prandtl number, $Pr$, in the heat transfer work was lower than the value of the Schmidt number, $Sc$, in the mass transfer tests.

It is convenient to express results for tests on heat transfer and mass transfer to particles in the form of $j$-factors. If the concentration of the diffusing component is small, then the $j$-factor for mass transfer may be defined by:

$$j_d' = \frac{h_D}{u_c} Sc^{0.67} \tag{6.70}$$

where: $h_D$ is the mass transfer coefficient,
$u_c$ is the fluidising velocity,
$Sc$ is the Schmidt number $(\mu/\rho D)$,
$\mu$ is the fluid viscosity,
$\rho$ is the fluid density, and
$D$ is the diffusivity of the transferred component in the fluid.

The corresponding relation for heat transfer is:

$$j'_h = \frac{h}{C_p \rho u_c} Pr^{0.67} \tag{6.71}$$

where: $h$ is the heat transfer coefficient,
$C_p$ is the specific heat of the fluid at constant pressure,
$Pr$ is the Prandtl number $(C_p\mu/k)$, and
$k$ is the thermal conductivity of the fluid.

The significance of $j$-factors is discussed in detail in Volume 1, Chapters 9 and 10.
Rearranging equations 6.64, 6.65, and 6.70 in the form of 6.70, and 6.71 and substituting mean values of 2.0 and 0.7 respectively for $Sc$ and $Pr$, gives:

$(0.1 < Re'_c < 15)$

$$j'_d = \frac{Sh'}{ScRe'_c} Sc^{0.67} = 0.37 Re'^{0.2}_c Sc^{-0.33} = 0.29 Re'^{0.2}_c \tag{6.72}$$

$(15 < Re'_c < 250)$

$$j'_d = \frac{Sh'}{ScRe'_c} Sc^{0.67} = 2.01 Re'^{-0.5}_c Sc^{-0.33} = 1.59 Re'^{-0.5}_c \tag{6.73}$$

$(0.25 < Re'_c < 18)$

$$j'_h = \frac{Nu'}{PrRe'_c} Pr^{0.67} = 0.11 Re'^{0.28}_c Pr^{-0.33} = 0.13 Re'^{0.28}_c \tag{6.74}$$

These relations are plotted in Figure 6.33 as lines A, B and C respectively.

RESNICK and WHITE[122] fluidised naphthalene crystals of five different size ranges (between 1000 and 250 μm) in air, hydrogen, and carbon dioxide at a temperature of 298 K. The gas was passed through a sintered disc, which served as the bed support, at rates of between 0.01 and 1.5 kg/m$^2$ s. Because of the nature of the surface and of the shape of the particles, uneven fluidisation would have been obtained. The rate of vaporisation was determined by a gravimetric analysis of the outlet gas, and mass transfer coefficients were calculated. These were expressed as $j$-factors and plotted against Reynolds number $Re'_c (= u_c d\rho/\mu)$ in Figure 6.34. It may be seen that, whilst separate curves were obtained for each size fraction of particles, each curve was of the same general shape, showing a maximum in the fluidisation region, roughly at the transition between bubbling and slugging conditions.

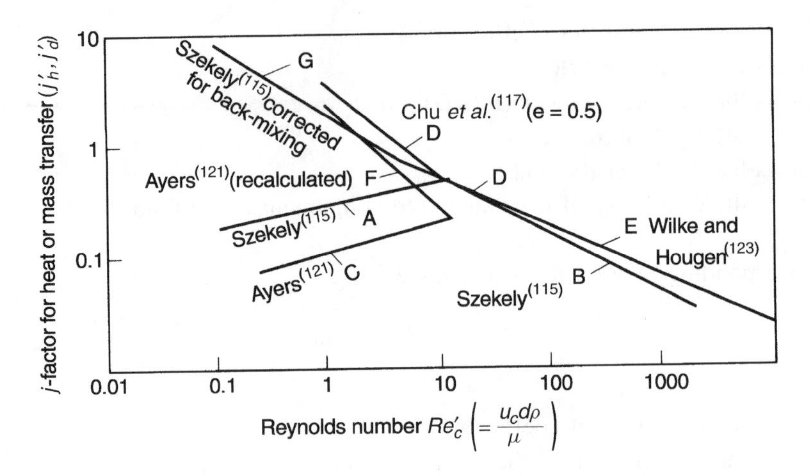


Figure 6.33. Heat and mass transfer results expressed as $j$-factors

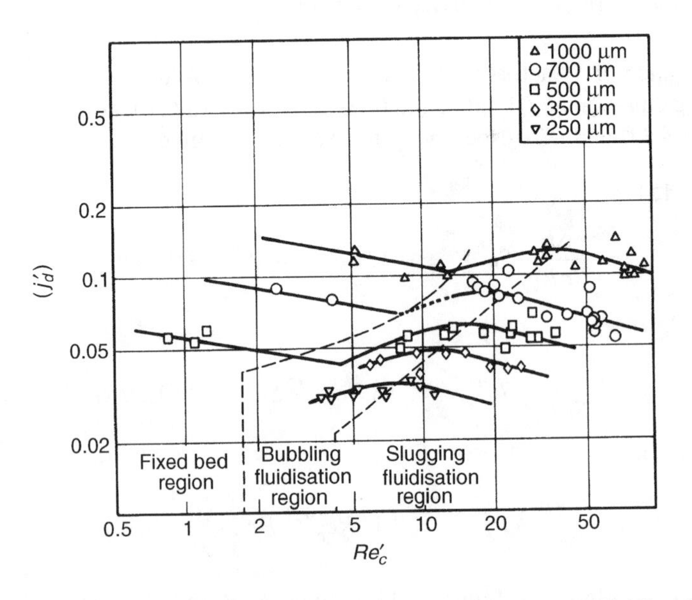


Figure 6.34. $j'_d$, for the transfer of naphthalene vapour to air in fixed and fluidised beds[122]

Chu, Kalil, and Wetteroth[117] obtained an improved quality of fluidisation by coating spherical particles with naphthalene, although it is probable that some attrition occurred. Tests were carried out with particles ranging in size from 0.75 to 12.5 mm and voidages from 0.25 to 0.97. Fixed beds were also used. Again, it was found that particle size was an important parameter in the relation between $j$-factor and Reynolds number. When plotted as shown in Figure 6.35 against a modified Reynolds number $Re_1^*[= (u_c d\rho/(1 - e)\mu)]$, however, a single correlation was obtained. In addition, it was possible to represent with a single curve the results of a number of workers, obtained in fixed and fluidised beds with both liquids and gases as the fluidising media. A range of 0.6–1400 for the Schmidt

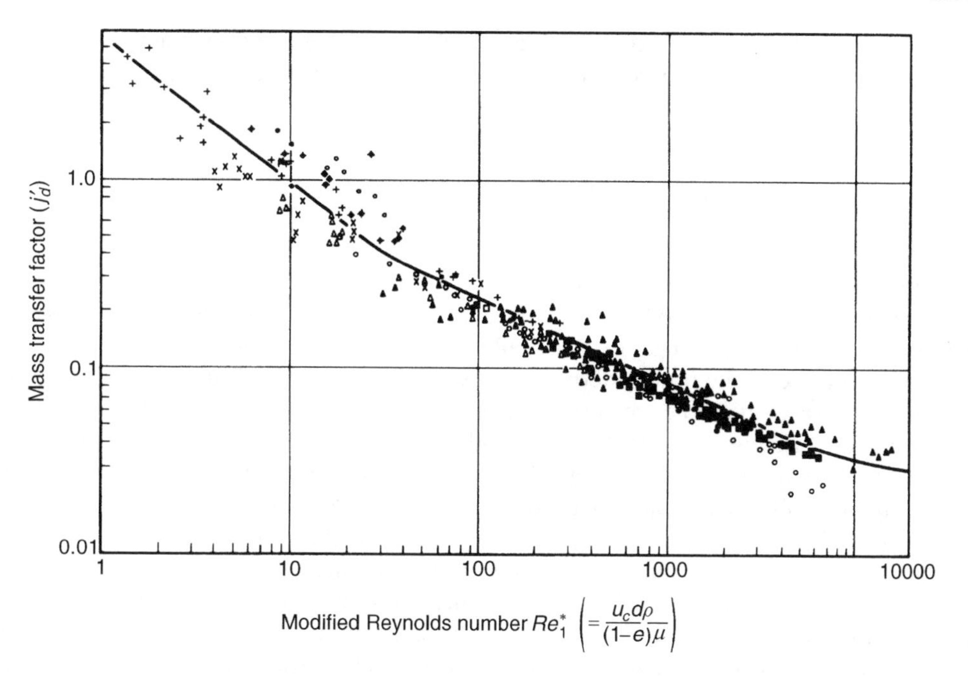


| *Symbol* | *System* | *Schmidt No.* | *Type of particle* | *State of bed* | *Ref.* |
|---|---|---|---|---|---|
| ⊗ | Naphthalene-air | 2.57 | Spheres, cylinders | Fixed, fluidised | 117 |
| ■ | Water-air | 0.60 | Spheres, cylinders | Fixed | 123 |
| ○ | 2-naphthol-water | 1400 | Modified spheres | Fixed, fluidised | 124 |
| ● | Isobutyl alcohol-water | 866 | Spheres | Fixed | 125 |
| ⌀ | Methyl ethyl ketone-water | 776 | Spheres | Fixed | 125 |
| + | Salicylic acid-benzene | 368 | Modified spheres | Fixed | 126 |
| × | Succinic acid-*n*-butyl alcohol | 690 | Modified spheres | Fixed | 126 |
| Δ | Succinic acid-acetone | 164 | Modified spheres | Fixed | 126 |

Figure 6.35. $j$-factor, $j'_d$, for fixed and fluidised beds[117]

number was covered. It may be noted that the results for fluidised systems are confined to values obtained at relatively high values of the Reynolds number. The curve may be represented approximately by the equations:

$$(1 < Re_1^* < 30) \qquad j'_d = 5.7 Re_1^{*-0.78} \tag{6.75}$$

$$(30 < Re_1^* < 5000) \qquad j'_d = 1.77 Re_1^{*-0.44} \tag{6.76}$$

These two relations are also shown as curve D in Figure 6.33 for a voidage of 0.5.

A number of other workers have measured mass transfer rates. McCune and Wilhelm[124] studied transfer between naphthol particles and water in fixed and fluidised beds. Hsu and Molstad[127] absorbed carbon tetrachloride vapour on activated carbon particles in very shallow beds which were sometimes less than one particle diameter deep. WILKE and HOUGEN[123] dried Celite particles (size range approximately 3–19 mm) in a fixed bed by means of a stream of air, and found that their results were represented by:

$$(50 < Re'_c < 250) \qquad j'_d = 1.82 Re'^{-0.51}_c \tag{6.77}$$

$$(Re'_c > 350) \qquad j'_d = 0.99 Re'^{-0.41}_c \tag{6.78}$$

These relations are plotted as curve E in Figure 6.33.

It may be seen that the general trend of the results of different workers is similar but that the agreement is not good. In most cases a direct comparison of results is not possible because the experimental data are not available in the required form.

The importance of the flow pattern on the experimental data is clearly apparent, and the reasons for discrepancies between the results of different workers are largely attributable to the rather different characters of the fluidised systems. It is of particular interest to note that, at high values of the Reynolds number when the effects of back-mixing are unimportant, similar results are obtained in fixed and fluidised beds. This conclusion was also reached by MULLIN and TRELEAVEN[116] in their tests with models.

There is apparently an inherent anomaly in the heat and mass transfer results in that, at low Reynolds numbers, the Nusselt and Sherwood numbers (Figures. 6.30 and 6.27) are very low, and substantially below the theoretical minimum value of 2 for transfer by thermal conduction or molecular diffusion to a spherical particle when the driving force is spread over an infinite distance (Volume 1, Chapter 9). The most probable explanation is that at low Reynolds numbers there is appreciable back-mixing of gas associated with the circulation of the solids. If this is represented as a diffusional type of process with a longitudinal diffusivity of $D_L$, the basic equation for the heat transfer process is:

$$D_L C_p \rho \frac{d^2T}{dz^2} - u_c C_p \rho \frac{dT}{dz} - ha(T - T_s) = 0 \tag{6.79}$$

Equation 6.79 may be obtained in a similar manner to equation 4.34, but with the addition of the last term which represents the transfer of sensible heat from the gas to the solids. The time derivative is zero because a steady-state process is considered.

On integration, equation 6.79 gives a relation between $h$ and $D_L$, and $h$ may only be evaluated if $D_L$ is known. If it is assumed that at low Reynolds numbers the value of the Nusselt or Sherwood numbers approaches the theoretical minimum value of 2, it is possible to estimate the values of $D_L$ at low Reynolds numbers, and then to extrapolate these values over the whole range of Reynolds numbers used. This provides a means of recalculating all the results using equation 6.79. It is then found that the results for low Reynolds numbers are substantially modified and the anomaly is eliminated, whereas the effect at high Reynolds numbers is small. Recalculated values[115] of Nusselt number for heat transfer tests[121] are shown in Figure 6.36. This confirms the view already expressed, and borne out by the work of LANNEAU[75], that back-mixing is of importance only at low flowrates.

Recalculated values of $j'_h$ and $j'_d$ obtained from the results of AYERS[121] and SZEKELY[115] are shown in Figure 6.32 as curves F and G. It will be seen that the curves B, D, E, F, and G follow the same trend.

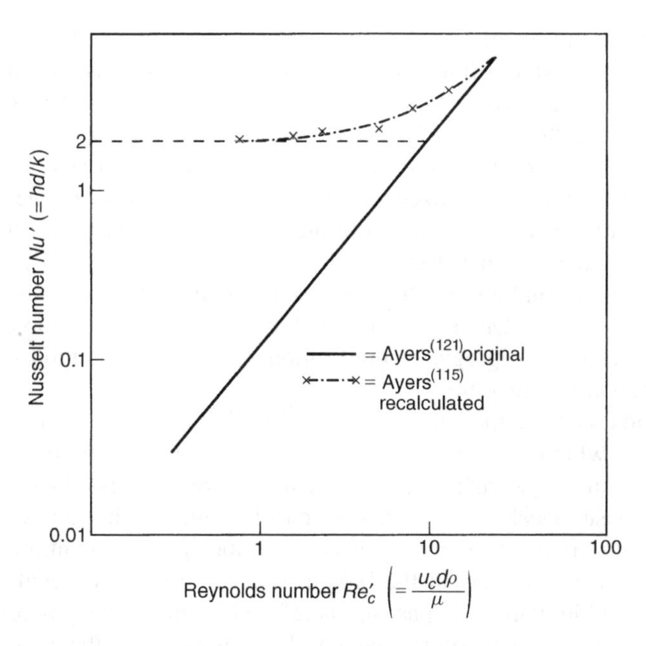


Figure 6.36. Recalculated values of Nusselt number, taking into account the effects of back-mixing[115]

CORNISH[128] considered the minimum possible value of the Nusselt number in a multiple particle system. By regarding an individual particle as a source and the remote fluid as a sink, it was shown that values of Nusselt number less than 2 may then be obtained. In a fluidised system, however, the inter-particle fluid is usually regarded as the sink and under these circumstances the theoretical lower limit of 2 for the Nusselt number applies. ZABRODSKY[129] has also discussed the fallacy of Cornish's argument.

## 6.7. SUMMARY OF THE PROPERTIES OF FLUIDISED BEDS

Data on, and the understanding of, fluidised systems are increasing at a very high rate, with as many as a hundred papers appearing in any given year.

Fluidised beds may be divided into two classes. In the first, there is a uniform dispersion of the particles within the fluid and the bed expands in a regular manner as the fluid velocity is increased. This behaviour, termed *particulate fluidisation*, is exhibited by most liquid–solids systems, the only important exceptions being those composed of fine particles of high density. This behaviour is also exhibited by certain gas–solids systems over a very small range of velocities just in excess of the minimum fluidising velocity — particularly where the particles are approximately spherical and have very low free-falling velocities. In particulate fluidisation the rate of movement of the particles is comparatively low, and the fluid is predominantly in piston-type flow with some back-mixing, particularly at low flowrates. Overall turbulence normally exists in the system.

In the other form of fluidisation, *aggregative fluidisation*, two phases are present in the bed — a continuous or emulsion phase, and a discontinuous or bubble phase. This is the pattern normally encountered with gas-solids systems. Bubbles tend to form at gas rates above the minimum fluidising rate and grow as they rise through the bed. The bubbles grow because the hydrostatic pressure is falling, as a result of coalescence with other bubbles, and by flow of gas from the continuous to the bubble phase. The rate of rise of the bubble is approximately proportional to the one-sixth power of its volume. If the rising velocity of the bubble exceeds the free-falling velocity of the particles, it will tend to draw in particles at its wake and to destroy itself. There is therefore a maximum stable bubble size in a given system. If this exceeds about 10 particle diameters, the bubble will be obvious and aggregative fluidisation will exist; this is the usual condition with a gas–solids system. Otherwise, the bubble will not be observable and particulate fluidisation will occur. In aggregative fluidisation, the flow of the fluid in the continuous phase is predominantly streamline.

In a gas–solids system, the gas distributes itself between the bubble phase and the continuous phase which generally has a voidage a little greater than at the point of incipient fluidisation. If the rising velocity of the bubbles is less than that of the gas in the continuous phase, it behaves as a rising void through which the gas will tend to flow preferentially. If the rising velocity exceeds the velocity in the continuous phase — and this is the usual case — the gas in the bubble is continuously recycled through a cloud surrounding the bubble. Partial by-passing therefore occurs and the gas comes into contact with only a limited quantity of solids. The gas cloud surrounding the bubble detaches itself from time to time, however.

The bubbles appear to be responsible for a large amount of mixing of the solids. A rising bubble draws up a spout of particles behind it and carries a wake of particles equal to about one-third of the volume of the bubble and wake together. This wake detaches itself at intervals. The pattern in a bed containing a large number of bubbles is, of course, very much more complex.

One of the most important properties of the fluidised bed is its good heat transfer characteristics. For a liquid–solids system, the presence of the particles may increase the coefficient by a factor of 2 or 3. In a gas-solids system, the factor may be about two orders of magnitude, with the coefficient being raised by the presence of the particles, from a value for the gas to one normally associated with a liquid. The improved heat transfer is associated with the movement of the particles between the main body of the bed and the heat transfer surface. The particles act as heat-transferring elements and bring material at the bulk temperature in close proximity to the heat transfer surface. A rapid circulation therefore gives a high heat transfer coefficient. In a gas-solids system, the amount of bubbling within the bed should be sufficient to give adequate mixing, and at the same time should not be sufficient to cause an appreciable blanketing of the heat transfer surface by gas.

## 6.8. APPLICATIONS OF THE FLUIDISED SOLIDS TECHNIQUE

### 6.8.1. General

The use of the fluidised solids technique was developed mainly by the petroleum and chemical industries, for processes where the very high heat transfer coefficients and the high degree of uniformity of temperature within the bed enabled the development of processes which would otherwise be impracticable. Fluidised solids are now used quite extensively in many industries where it is desirable to bring about intimate contact between small solid particles and a gas stream. In many cases it is possible to produce the same degree of contact between the two phases with a very much lower pressure drop over the system. Drying of finely divided solids is carried out in a fluidised system, and some carbonisation and gasification processes are in operation. Fluidised beds are employed in gas purification work, in the removal of suspended dusts and mists from gases, in lime burning and in the manufacture of phthalic anhydride.

### 6.8.2. Fluidised Bed Catalytic Cracking

The existence of a large surplus of high boiling material after the distillation of crude oil led to the introduction of a cracking process to convert these materials into compounds of lower molecular weight and lower boiling point — in particular into petroleum spirit. The cracking was initially carried out using a fixed catalyst, although local variations in temperature in the bed led to a relatively inefficient process, and the deposition of carbon on the surface of the catalyst particles necessitated taking the catalyst bed out of service periodically so that the carbon could be burned off. Many of these difficulties are obviated by the use of a fluidised catalyst, since it is possible to remove catalyst continuously from the reaction vessel and to supply regenerated catalyst to the plant. The high heat transfer coefficients for fluidised systems account for the very uniform temperatures within the reactors and make it possible to control conditions very closely. The fluidised system has one serious drawback however, in that some longitudinal mixing occurs which gives rise to a number of side reactions.

A diagram of the plant used for the catalytic cracking process is given in Figure 6.37[130]. Hot oil vapour, containing the required amount of regenerated catalyst, is introduced into the reactor which is at a uniform temperature of about 775 K. As the velocity of the vapour falls in the reactor, because of the greater cross-sectional area for flow, a fluidised bed is formed and the solid particles are maintained in suspension. The vapours escape from the top, of the unit and the flowrate is such that the vapour remains in the reactor for about 20 s. It is necessary to provide a cyclone separator in the gas outlet to remove entrained catalyst particles and droplets of heavy oil. The vapour from the cracked material is then passed to the fractionating unit, whilst the catalyst particles and the heavy residue are returned to the bed. Some of the catalyst is continuously removed from the bottom of the reactor and, together with any fresh catalyst required, is conveyed in a stream of hot air into the regenerator where the carbon deposit and any adhering film of heavy oil are burned off at about 875 K. In the regenerator the particles are again suspended as a fluidised bed. The hot gases leave the regenerator through a cyclone separator, from which the solids return to the bed, and then flow to waste through an electrostatic precipitator which removes any very fine particles still in suspension. The temperature in the regenerator remains constant to within about 3 deg K, even where the dimensions of the fluidised bed are as high as 6 m deep and 15 m in diameter. Catalyst is continuously returned from the regenerator to the reactor by introducing it into the supply of hot vapour. The complete time cycle for the catalyst material is about 600 s. By this process a product consisting of between about 50 per cent and 75 per cent of petroleum spirit of high octane number is obtained. The

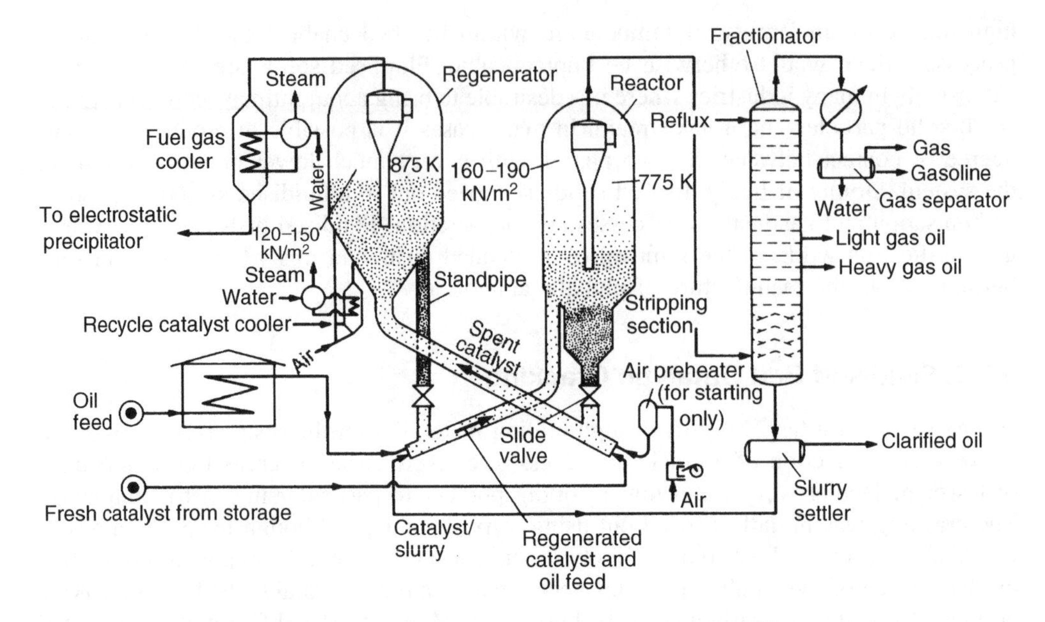


Figure 6.37. Fluidised-bed catalytic cracking plant[130]

quality of the product may be controlled by the proportion of catalyst used and the exact temperature in the reactor.

More recent tests have shown that much of the cracking takes place in the transfer line in which the regenerated catalyst is conveyed into the reactor in the stream of oil vapour. The chemical reaction involved is very fast, and the performance of the reactor is not sensitive to the hydrodynamic conditions.

From the diagram of the catalytic cracking plant in Figure 6.37 it may be noted that there is a complete absence of moving parts in the reactor and the regenerator. The relative positions of components are such that the catalyst is returned to the reactor under the action of gravity.

## 6.8.3. Applications in the Chemical and Process Industries

Fluidised catalysts are also used in the synthesis of high-grade fuels from mixtures of carbon monoxide and hydrogen, obtained either by coal carbonisation or by partial oxidation of methane. An important application in the chemical industry is the oxidation of naphthalene to phthalic anhydride, as discussed by RILEY[131]. The kinetics of this reaction are much slower than those of catalytic cracking, and considerable difficulties have been experienced in correctly designing the system.

Purely physical operations are also frequently carried out in fluidised beds. Thus, fluidised bed dryers, discussed in Chapter 16, are successfully used, frequently for heat-sensitive materials which must not be subjected to elevated temperatures for prolonged periods.

The design of a fluidised bed for the carrying out of an exothermic reaction involving a long reaction time has been considered by ROWE[132] and, as an example, the reaction between gaseous hydrogen fluoride and solid uranium dioxide to give solid uranium tetrafluoride and water vapour is considered. This is a complex reaction which, as an approximation, may be considered as first order with respect to hydrogen fluoride. The problem here is to obtain the required time of contact between the two phases by the most economical method. The amount of gas in the bubbles must be sufficient to give an adequate heat transfer coefficient, although the gas in the bubbles does, however, have a shorter contact time with the solids because of its greater velocity of rise. In order to increase the contact time of the gas, the bed may be deep, although this results in a large pressure drop. Bubble size may be reduced by the incorporation of baffles and, as described by VOLK *et al.*[133], this is frequently an effective manner both of increasing the contact time and of permitting a more reliable scale-up from small scale experiments. The most effective control is obtained by careful selection of the particle size of the solids. If the particle size is increased, there will be a higher gas flow in the emulsion phase and less in the bubble phase. Thus, the ratio of bubble to emulsion phase gas velocity will be reduced, and the size of the gas cloud will increase. If the particle size is increased too much, however, there will be insufficient bubble phase to give good mixing. In practice, an overall gas flowrate of about twice that required for incipient fluidisation is frequently suitable.

## 6.8.4. Fluidised Bed Combustion

An important application of fluidisation which has attracted considerable interest in recent years is fluidised bed combustion. The combustible material is held in a fluidised bed of inert material and the air for combustion is the fluidising gas. The system has been developed for steam raising on a very large scale for electricity generation and for incineration of domestic refuse.

The particular features of fluidised combustion of coal which have given rise to the current interest are first its suitability for use with very low-grade coals, including those with very high ash contents and sulphur concentrations, and secondly the very low concentrations of sulphur dioxide which are attained in the stack gases. This situation arises from the very much lower bed temperatures ($\sim$1200 K) than those existing in conventional grate-type furnaces, and the possibility of reacting the sulphur in the coal with limestone or dolomite to enable its discharge as part of the ash.

Much of the basic research, development studies and design features of large-scale fluidised bed combustors is discussed in the Proceedings of the Symposium on Fluidised Bed Combustion, organised by the Institute of Fuel as long ago as 1975[134], and subsequently in the literature[135–139]. Pilot scale furnaces with ratings up to 0.5 MW have been operated and large-scale furnaces have outputs of up to 30 MW.

The bed material normally consists initially of an inert material, such as sand or ash, of particle size between 500 and 1500 $\mu$m. This gradually becomes replaced by ash from the coal and additives used for sulphur removal. Ash is continuously removed from the bottom of the bed and, in addition, there is a considerable carry-over by elutriation and this flyash must be collected in cyclone separators. Bed depths are usually kept below about 0.6 m in order to limit power requirements.

Coal has a lower density than the bed material and therefore tends to float, although in a vigorously bubbling bed, the coal can become well mixed with the remainder of the material and the degree of mixing determines the number of feed points which are required. In general, the combustible material does not exceed about 5 per cent of the total solids content of the bed. The maximum size of coal which has been successfully used is 25 mm, which is about two orders of magnitude greater than the particle size in pulverised fuel. The ability to use coal directly from the colliery eliminates the need for pulverising equipment, with its high capital and operating costs.

The coal, when it first enters the bed, gives up its volatiles and the combustion process thus involves both vapour and char. Mixing of the volatiles with air in the bed is not usually very good, with the result that there is considerable flame burning above the surface of the bed. Because the air rates are chosen to give vigorously bubbling beds, much of the oxygen for combustion must pass from the bubble phase to the dense phase before it can react with the char. Then in the dense phase there will be a significant diffusional resistance to transfer of oxygen to the surface of the particles. The combustion process is, as a result, virtually entirely diffusion-controlled at temperatures above about 1120 K. The reaction is one of oxidation to carbon monoxide near the surface of the particles and subsequent reaction to carbon dioxide. Despite the diffusional limitations, up to 90 per cent utilisation of the inlet oxygen may be achieved. The residence time of the gas in the bed is of the order of one second, whereas the coal particles may remain in the bed for many minutes. There is generally a significant amount of carry-over char in the flyash which is then usually recycled to a burner operating at a rather higher temperature than the main bed. An additional advantage of using a large size of feed coal is that the proportion carried over is correspondingly small.

Whilst fluidised bed furnaces can be operated in the range 1075–1225 K, most operate close to 1175 K. Some of the tubes are immersed in the bed and others are above the free surface. Heat transfer to the immersed tubes is good. Tube areas are usually 6–10 $m^2/m^3$ of furnace, and transfer coefficients usually range from 300 to 500 $W/m^2$ K. The radiation component of heat transfer is highly important and heat releases in large furnaces are about $10^6 W/m^3$ of furnace.

One of the major advantages of fluidised bed combustion of coal is that it is possible to absorb the sulphur dioxide formed. Generally limestone or dolomite is added and thus breaks up in the bed to yield calcium oxide or magnesium and calcium oxide, which then react with the sulphur dioxide as follows:

$$CaO + SO_2 + \tfrac{1}{2}O_2 \rightarrow CaSO_4$$

It is possible to regenerate the solid in a separate reactor using a reducing gas consisting of hydrogen and carbon monoxide. There is some evidence that the reactivity of the limestone or dolomite is improved by the addition of chloride, although its use is not generally favoured because of corrosion problems. Fluidised combustion also gives less pollution because less oxides of nitrogen are formed at the lower temperatures existing in the beds.

Corrosion and erosion of the tubes immersed in the bed are at a low level, although there is evidence that the addition of limestone or dolomite causes some sulphide penetration. The chief operating problem is corrosion by chlorine.

Experimental and pilot scale work has been carried out on pressurised operation and plants have been operated up to 600 $kN/m^2$, and in at least one case up to 1 $MN/m^2$ pressure. High pressure operation permits the use of smaller beds. The fluidising velocity required to produce a given condition in the bed is largely independent of pressure, and thus the mass rate of feed of oxygen to the bed is approximately linearly related to the pressure. It is practicable to use deeper beds for pressure operation. Because of the low temperature of operation of fluidised beds, the ash is friable and relatively non-erosive, so that the combustion products can be passed directly through a gas turbine. This combination of a gas turbine is an essential feature of the economic operation of pressurised combustors. In general, it is better to use dolomite in place of limestone as an absorbent for sulphur dioxide, because the higher pressures of carbon dioxide lead to inhibition of the breakdown of calcium carbonate to oxide.

It appears likely that fluidised bed combustion of coal may, in the near future, be one of the most important applications of fluidised systems and it may well be that many new coal-fired generating stations will incorporate fluidised bed combustors.

Fluidised bed combustors are now common-place for large-scale operations and are extensively used in large electricity generating stations. More recently, smaller scale units have been developed for use by individual industrial concerns and in operating, as combined heat and power units, can give overall thermal efficiencies of up to 80 per cent.

Pressurised fluidised combustion units operating at 1–2 MN/m$^2$ (10–20 bar) are now gaining favour because they give improved combustion and the hot gases can be expanded through a gas turbine, which may be used for generating up to 20 per cent of the total electrical power. Thus, there are marked improvements in efficiency. Because of the reduced specific volume of the high pressure air, it is necessary to use tall narrow combustion units in order to obtain the required gas velocity through the bed, with the minimum fluidising velocity being approximately proportional to the square root of the density (or pressure) of the fluidising air. The flowsheet of a typical pressurised fluidised bed combustor is given in Figure 6.38[138], where it is seen that the exhaust gas from

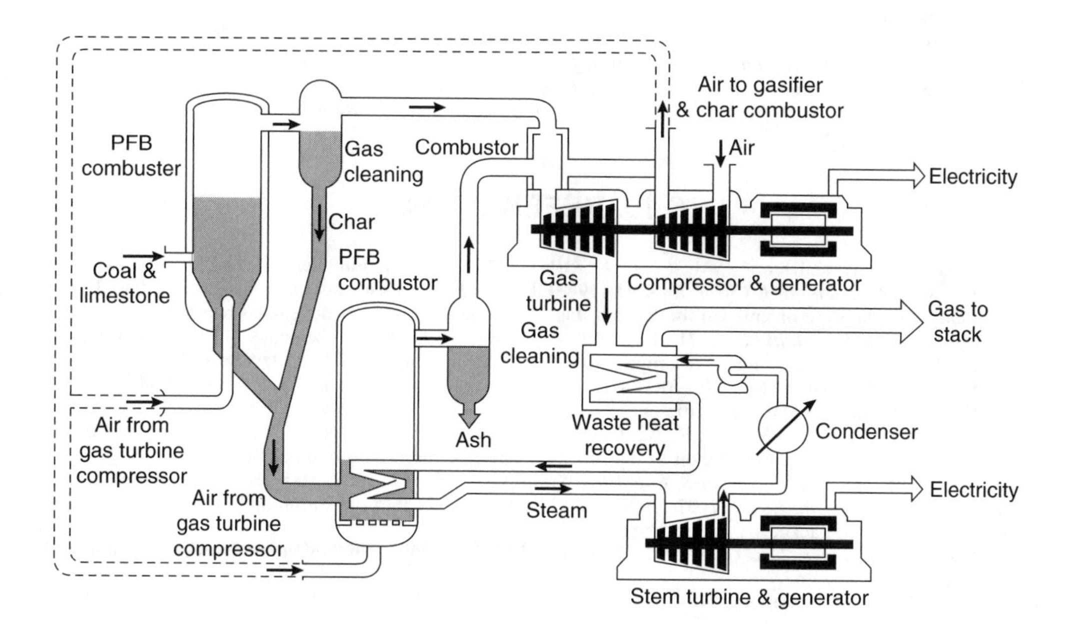


Figure 6.38. Pressurised fluidised bed combustor with waste heat recovery[138]

the gas turbine is used to preheat the boiler feed water. Sometimes gas velocities are sufficient to entrain part of the fuel from the bed which is then recycled from the outlet of the cyclone to the combustion unit.

Fuel may constitute only about 1 per cent of the total volume of the solids in the bed and the particle size needs to be reduced to about 10 mm, compared with 100 μm in pulverised coal burners.

## 6.9. FURTHER READING


BOTTERILL, J. S. M.: *Fluid Bed Heat Transfer* (Academic Press, 1975).
DALLAVALLE, J. M.: *Micromeritics*, 2nd edn. (Pitman, 1943).
DAVIDSON, J. F. and HARRISON, D. (eds.): *Fluidized Particles* (Cambridge, 1963).
DAVIDSON, J. F. and HARRISON, D. (eds.): *Fluidization* (Academic Press, 1971).
DAVIDSON, J. F., CLIFT, R. and HARRISON, D. (eds.): *Fluidization*, 2nd edn. (Academic Press, 1985).
FAN, L-S.: *Gas–Liquid–Solid Fluidization Engineering* (Butterworths, 1989).
HALOW, J. S. (ed.): *Fluidization Theories and Applications*. A.I.Ch.E. Symposium Series 161 (1977).
KEAIRNS, D. L. *et al*(eds.): *Fluidized Bed Fundamentals and Applications*. A.I.Ch.E. Symposium Series **69** No 128 (1973).
KEAIRNS, D. L. (ed.): *Fluidization and Fluid–Particle Systems*. A.I.Ch.E. Symposium Series **70** No 141 (1974).
KEAIRNS, D. L. (ed.): *Fluidization Technology*. Proc. Int. Fluidisation Conf., Pacific Grove, California, 1975.
KUNII, D. and LEVENSPIEL, O.: *Fluidization Engineering*, 2nd edn. (Butterworth-Heinemann, 1991).
KWAUK, M.: In *Advances in Chemical Engineering* 17 by WEI, J., ANDERSON, J. L., BISCHOFF, K. B. and SEIRFELD, J. H. (eds.), Academic Press (1992), 207. Particulate fluidization: an overview.
KWAUK, M.: *Fluidization. Idealized and Bubbleless, with Applications* (Science Press and Ellis Horwood, 1992).
LEVA, M.: *Fluidization* (McGraw-Hill, 1959).
MOLERUS, O.: *Principles of Flow in Disperse Systems* (Chapman and Hall, Berlin, 1993).
MOLERUS, O and WIRTH, K.E.: *Heat Transfer in Fluidized Beds*. (Chapman and Hall, 1997).
OSTERGAARD, K.: *Studies of Gas–Liquid Fluidisation*. Thesis (Technical University of Denmark, 1968).
SAXENA, S. C.: *Advances in Heat Transfer* **19** (1989) 97–190. Heat transfer between immersed surfaces and gas fluidized beds.
VALK, M. (ed.): *Atmospheric Fluidized Bed Coal Combustion* (Elsevier, Amsterdam, 1994).
ZABRODSKY, S. S.: *Hydrodynamics and Heat Transfer in Fluidized Beds* (The M.I.T. Press, 1966).
ZANDI, I. (ed.): *Advances in Solid–Liquid Flow in Pipes and its Application* (Pergamon Press, Oxford, 1971).
ZENZ, F. A. and OTHMER, D. F.: *Fluidization and Fluid–Particle Systems* (Reinhold, 1960).


## 6.10. REFERENCES


1. WILHELM, R. H. and KWAUK, M.: *Chem. Eng. Prog.* **44** (1948) 201. Fluidization of solid particles.
2. JACKSON, R.: *Trans. Inst. Chem. Eng.* **41** (1963) 13, 22. The mechanisms of fluidised beds. Part 1. The stability of the state of uniform fluidisation. Part 2. The motion of fully developed bubbles.
3. MURRAY, J. D.: *J. Fluid Mech.* **21** (1965) 465; **22** (1965) 57. On the mathematics of fluidization. Part 1. Fundamental equations and wave propagation. Part 2. Steady motion of fully developed bubbles.
4. LEVA, M., GRUMMER, M., WEINTRAUB, M., and POLLCHIK, M.: *Chem. Eng. Prog.* **44** (1948) 619. Fluidization of non-vesicular particles.
5. RICHARDSON, J. F. and MEIKLE, R. A.: *Trans. Inst. Chem. Eng.* **39** (1961) 348. Sedimentation and fluidisation. Part III. The sedimentation of uniform fine particles and of two-component mixtures of solids.
6. WEN, C. Y. and YU, Y. H.: *Chem. Eng. Prog. Symp. Series* No. 62 (1960) 100. Mechanics of fluidization.
7. NIVEN, R. K.: *Chem. Eng. Sci.* **57** (2002) 527. Physical insight into the Ergun and Wen & Yu equations for fluid flow in packed on fludised beds.
8. ROWE, P. N.: *Trans. Inst. Chem. Eng.* **39** (1961) 175. Drag forces in a hydraulic model of a fluidised bed — Part II.
9. ROWE, P. N. and PARTRIDGE, B. A.: *Trans. Inst. Chem. Eng.* **43** (1965) T157. An X-ray study of bubbles in fluidised beds.

10. PINCHBECK, P. H. and POPPER, F.: *Chem. Eng. Sci.* **6** (1956) 57. Critical and terminal velocities in fluidization.
11. RICHARDSON, J. F. and ZAKI, W. N.: *Trans. Inst. Chem. Eng.* **32** (1954) 35. Sedimentation and fluidisation. Part 1.
12. GODARD, K. E. and RICHARDSON, J. F.: *Chem. Eng. Sci.* **24** (1969) 363. Correlation of data for minimum fluidising velocity and bed expansion in particulately fluidised beds.
13. LEWIS, E. W. and BOWERMAN, E. W.: *Chem. Eng. Prog.* **48** (1952) 603. Fluidization of solid particles in liquids.
14. KHAN, A. R. and RICHARDSON, J. F.: *Chem. Eng. Commun.* **78** (1989) 111. Fluid–particle interactions and flow characteristics of fluidized beds and settling suspensions of spherical particles.
15. SRINIVAS, B. K. and CHHABRA, R.P.: *Chem. Eng. & Processing* **29** (1991) 121–131. An experimental study of non-Newtonian fluid flow in fluidised beds: minimum fluidisation velocity and bed expansion.
16. STEWART, P. S. B. and DAVIDSON, J. F.: *Powder Tech.* **1** (1967) 61. Slug flow in fluidised beds.
17. SIMPSON, H. C. and RODGER, B. W.: *Chem. Eng. Sci.* **16** (1961) 153. The fluidization of light solids by gases under pressure and heavy solids by water.
18. HARRISON, D., DAVIDSON, J. F. and DE KOCK, J. W.: *Trans. Inst. Chem. Eng.* **39** (1961) 202. On the nature of aggregative and particulate fluidisation.
19. LAWTHER, K. P. and BERGLIN, C. L. W.: United Kingdom Atomic Energy Authority Report, A.E.R.E., CE/R 2360 (1957). Fluidisation of lead shot with water.
20. RICHARDSON, J. F. and SMITH, J. W.: *Trans. Inst. Chem. Eng.* **40** (1962) 13. Heat transfer to liquid fluidised systems and to suspensions of coarse particles in vertical transport.
21. ANDERSON, T. B. and JACKSON, R.: *Chem. Eng. Sci.* **19** (1964) 509. The nature of aggregative and particulate fluidization.
22. HASSETT, N. J.: *Brit. Chem. Eng.* **6** (1961) 777. The mechanism of fluidization.
23. LAWSON, A. and HASSETT, N. J.: *Proc. Intl. Symp. on Fluidization*, Netherlands Univ. Press, Eindhoven, (1967) 113. Discontinuities and flow patterns in liquid–fluidized beds.
24. CAIRNS, E. J. and PRAUSNITZ, J. M.: *A.I.Ch.E. Jl.* **6** (1960) 400. Longitudinal mixing in fluidization.
25. KRAMERS, H., WESTERMANN, M. D., DE GROOT, J. H. and DUPONT, F. A. A.: Third Congress of the European Federation of Chemical Engineering (1962). *The Interaction between Fluids and Particles* 114. The longitudinal dispersion of liquid in a fluidised bed.
26. REUTER, H.: *Chem. Eng. Prog. Symp. Series* No. 62 (1966) 92. On the nature of bubbles in gas and liquid fluidized beds.
27. GIBILARO, L. G., DI FELICE, R., HOSSAIN, I. and FOSCOLO, P. U.: *Chem. Eng. Sci.* **44** (1989) 101. The experimental determination of one-dimensional wave velocities in liquid fluidized beds.
28. BAILEY, C.: *Private Communication.*
29. VOLPICELLI, G., MASSIMILLA, L. and ZENZ, F. A.: *Chem. Eng. Prog. Symp. Series* No. 62, (1966) 63. Non-homogeneities in solid–liquid fluidization.
30. FOSCOLO, P. U., DI FELICE, R. and GIBILARO, L. G.: *A.I.Ch.E.Jl.* **35** (1989). 1921. The pressure field in an unsteady state fluidized bed.
31. JOTTRAND, R.: *Chem. Eng. Sci.* **3** (1954) 12. Etude de quelques aspects de la fluidisation dans les liquides.
32. PRUDEN, B. B. and EPSTEIN, N.: *Chem. Eng. Sci.* **19** (1964) 696. Stratification by size in particulate fluidization and in hindered settling.
33. KENNEDY, S. C. and BRETTON, R. H.: *A.I.Ch.E. Jl* **12** (1966) 24. Axial dispersion of spheres fluidized with liquids.
34. AL-DIBOUNI, M. R. and GARSIDE, J.: *Trans. Inst. Chem. Eng.* **57** (1979) 94. Particle mixing and classification in liquid fluidised beds.
35. JUMA, A. K. A. and RICHARDSON, J. F.: *Chem. Eng. Sci.* **34** (1979) 137. Particle segregation in liquid-solid fluidised beds.
36. GIBILARO, L. G., HOSSAIN, I. and WALDRAM, S. P.: *Chem. Eng. Sci.* **40** (1985) 2333. On the Kennedy and Bretton model for mixing and segregation in liquid fluidized beds.
37. MORITOMI, H., YAMAGISHI, T. and CHIBA, T.: *Chem. Eng. Sci.* **41** (1986) 297. Prediction of complete mixing of liquid-fluidized binary solid particles.
38. RICHARDSON, J. F. and AFIATIN, E.: *T & S 9th International Conference on Solid Particles* (2–5 September, 1997, Crakow, Poland) 486. Fluidisation and sedimentation of mixtures of particles.
39. EPSTEIN, N. and PRUDEN, B.B.: *Chem. Eng. Sci.* **54** (1999) 401. Liquid fluidisation of binary particle mixtures - III. Stratification by size and related topics.
40. HANDLEY, D., DORAISAMY, A., BUTCHER, K. L., and FRANKLIN, N. L.: *Trans. Inst. Chem. Eng.* **44** (1966) T260. A study of the fluid and particle mechanics in liquid-fluidised beds.
41. CARLOS, C. R. and RICHARDSON, J. F.: *Chem. Eng. Sci.* **22** (1967) 705. Particle speed distribution in a fluidised system.

42. Carlos, C. R.: University of Wales, Ph.D. thesis (1967). Solids mixing in fluidised beds.
43. Latif, B. A. J. and Richardson, J. F.: *Chem. Eng. Sci.* **27** (1972) 1933. Circulation patterns and velocity distributions for particles in a liquid fluidised bed.
44. Dorgelo, E. A. H., van der Meer, A. P. and Wesselingh, J. A.: *Chem. Eng. Sci.* **40** (1985) 2105. Measurement of the axial dispersion of particles in a liquid fluidized bed applying a random walk method.
45. Davies, L. and Richardson, J. F.: *Trans. Inst. Chem. Eng.* **44** (1966) T293. Gas interchange between bubbles and the continuous phase in a fluidised bed.
46. Geldart, D.: *Powder Technology* **7** (1973) 285. Types of fluidization.
47. Godard, K. E. and Richardson, J. F.: *I. Chem. E. Symposium Series* No. 30 (1968) 126. The behaviour of bubble-free fluidised beds.
48. Leung, L. S.: University of Cambridge, Ph.D. thesis (1961). Bubbles in fluidised beds.
49. Jacob, K. V. and Weiner, A. W.: *A.I.Ch.E. Jl.* **33** (1987) 1698. High pressure particulate expansion and minimum bubbling of fine carbon powders.
50. Olowson, P. A. and Almstedt, A. E.: *Chem. Eng. Sci.* **46** (1991) 637. Influence of pressure on the minimum fluidization velocity.
51. Saxena, S. C. and Shrivastava, S.: *Chem. Eng. Sci.* **45** (1990) 1125. The influence of an external magnetic field on an air-fluidized bed of ferromagnetic particles.
52. Godard, K. and Richardson, J. F.: *Chem. Eng. Sci.* **24** (1969) 194. The use of slow speed stirring to initiate particulate fluidisation.
53. Matheson, G. L., Herbst, W. A., and Holt, P. H.: *Ind. Eng. Chem.* **41** (1949) 1099–1104. Characteristics of fluid–solid systems.
54. Davidson, J. F., Paul, R. C., Smith, M. J. S. and Duxbury, H. A.: *Trans. Inst. Chem. Eng.* **37** (1959) 323. The rise of bubbles in a fluidised bed.
55. Davies, R. M. and Taylor, G. I.: *Proc. Roy. Soc.* **A200** (1950) 375. The mechanics of large bubbles rising through extended liquids and liquids in tubes.
56. Harrison, D. and Leung, L. S.: *Trans. Inst. Chem. Eng.* **39** (1961) 409. Bubble formation at an orifice in a fluidised bed.
57. Harrison, D. and Leung, L. S.: Third Congress of the European Federation of Chemical Engineering (1962). *The Interaction between Fluids and Particles* 127. The coalescence of bubbles in fluidised beds.
58. Rowe, P. N. and Wace, P. F.: *Nature* **188** (1960) 737. Gas-flow patterns in fluidised beds.
59. Wace, P. F. and Burnett, S. T.: *Trans. Inst. Chem. Eng.* **39** (1961) 168. Flow patterns in gas-fluidised beds.
60. Davidson, J. F.: *Trans. Inst. Chem. Eng.* **39** (1961) 230. In discussion of Symposium on Fluidisation.
61. Rowe, P. N., Partridge, B. A., and Lyall, E.: *Chem. Eng. Sci.* **19** (1964) 973; **20** (1965) 1151. Cloud formation around bubbles in gas fluidized beds.
62. Rowe, P. N. and Henwood, G. N.: *Trans. Inst. Chem. Eng.* **39** (1961) 43. Drag forces in a hydraulic model of a fluidised bed. Part 1.
63. Bi, H. T., Ellis, N., Abba, I. A. and Grace, J. R.: *Chem. Eng. Sci.* **55** (2000) 4789. A state-of-the-art review of gas-solid turbulent fluidization.
64. Bi, H. T. and Grace J. R.: *Chem. Eng. Jl.* **57** (1995) 261. Effects of measurement method on velocities used to demarcate the transition to turbulent fluidization.
65. Bart, R.: *Massachusetts Institute of Technology, Sc.D. thesis (1950)*. Mixing of fluidized solids in small diameter columns.
66. Massimilla, L. and Bracale, S.: *La Ricerca Scientifica* **27** (1957) 1509. Il mesolamento della fase solida nei sistemi: Solido–gas fluidizzati, liberi e frenati.
67. May, W. G.: *Chem. Eng. Prog.* **55** (1959) 49. Fluidized bed reactor studies.
68. Sutherland, K. S.: *Trans. Inst. Chem. Eng.* **39** (1961) 188. Solids mixing studies in gas fluidised beds. Part 1. A preliminary comparison of tapered and non-tapered beds.
69. Rowe, P. N. and Partridge, B. A.: Third Congress of the European Federation of Chemical Engineering (1962). *The Interaction between Fluids and Particles* 135. Particle movement caused by bubbles in a fluidised bed.
70. Askins, J. W., Hinds, G. P., and Kunreuther, F.: *Chem. Eng. Prog.* **47** (1951) 401. Fluid catalyst–gas mixing in commercial equipment.
71. Danckwerts, P. V., Jenkins, J. W., and Place, G.: *Chem. Eng. Sci.* **3** (1954) 26. The distribution of residence times in an industrial fluidised reactor.
72. Gilliland, E. R. and Mason, E. A.: *Ind. Eng. Chem.* **41** (1949) 1191. Gas and solid mixing in fluidized beds.
73. Gilliland, E. R. and Mason, E. A.: *Ind. Eng. Chem.* **44** (1952) 218. Gas mixing in beds of fluidized solids.

74. GILLILAND, E. R., MASON, E. A., and OLIVER, R. C.: *Ind. Eng. Chem.* **45** (1953) 1177. Gas-flow patterns in beds of fluidized solids.
75. LANNEAU, K. P.: *Trans. Inst. Chem. Eng.* **38** (1960) 125. Gas–solids contacting in fluidized beds.
76. SZEKELY, J.: Third Congress of the European Federation of Chemical Engineering (1962). *The Interaction between Fluids and Particles* 197. Mass transfer between the dense phase and lean phase in a gas–solid fluidised system.
77. CHEN, J. L.-P., and KEAIRNS, D. L.: *Can. J. Chem. Eng.* **53** (1975) 395. Particle segregation in a fluidized bed.
78. HOFFMAN, A. C., JANSSEN, L. P. B. M. and PRINS, J. *Chem. Eng. Sci.* **48** (1993) 1583. Particle segregation in fluidized binary mixtures.
79. WU, S. Y. and BAEYENS, J.: *Powder Technol.* **98** (1998) 139. Segregation by size difference in gas fluidized beds.
80. LEVA, M.: *Chem. Eng. Prog.* **47** (1951) 39. Elutriation of fines from fluidized systems.
81. PEMBERTON, S. T. and DAVIDSON, J. F.: *Chem. Eng. Sci.* **41** (1986) 243, 253. Elutriation from fluidized beds. I Particle ejection from the dense phase into the freeboard. II Disengagement of particles from gas in the freeboard.
82. CHEN, Y-M: *A.I.Ch.E.Jl.* **33** (1987) 722. Fundamentals of a centrifugal fluidized bed.
83. FAN, L. T., CHANG, C. C. and YU, Y. S.: *A.I.Ch.E.Jl.* **31** (1985) 999. Incipient fluidization condition for a centrifugal fluidized bed.
84. BRIDGWATER, J.: In *Fluidization* by DAVIDSON, J. F., CLIFT, R. and HARRISON, D. (eds.), 2nd edn, (Academic Press, 1985), 201. Spouted beds.
85. MATHUR, K. B. and EPSTEIN, N.: *Spouted Beds* (Academic Press, 1974).
86. MATHUR, K. B. and EPSTEIN, N.: In *Advances in Chemical Engineering* **9** by DREW, T. B., COKELET, G. R., HOOPES, J. W. and VERMEULEN, T. (eds.), Academic Press (1974), 111. Dynamics of spouting beds.
87. DARTON, R. C.: In *Fluidization* by DAVIDSON, J. F., CLIFT, R. and HARRISON, D. (eds.), 2nd edn, (Academic Press, 1985), 495. The physical behaviour of three-phase fluidized beds.
88. EPSTEIN, N.: *Canad. J. Chem. Eng.* **59** (1981) 649. Three phase fluidization: some knowledge gaps.
89. LEE, J. C. and BUCKLEY, P. S.: In *Biological Fluidised Bed Treatment of Water and Wastewater* by COOPER, P. F. and ATKINSON, B., (eds.), Ellis Horwood, Chichester (1981), 62. Fluid mechanics and aeration characteristics of fluidised beds.
90. FAN, L-S.: *Gas–Liquid–Solid Fluidization Engineering* (Butterworth, 1989).
91. MUROYAMA, K. and FAN, L-S.: *A.I.Ch.E.Jl.* **31** (1985). 1. Fundamentals of gas–liquid–solid fluidization.
92. MICKLEY, H. S. and FAIRBANKS, D. F.: *A.I.Ch.E.Jl.* **1** (1955) 374. Mechanism of heat transfer to fluidized beds.
93. LEMLICH, R. and CALDAS, I.: *A.I.Ch.E.Jl.* **4** (1958) 376. Heat transfer to a liquid fluidized bed.
94. RICHARDSON, J. F. and MITSON, A. E.: *Trans. Inst. Chem. Eng.* **36** (1958) 270. Sedimentation and fluidisation. Part II. Heat transfer from a tube wall to a liquid-fluidised system.
95. JEPSON, G., POLL, A., and SMITH, W.: *Trans. Inst. Chem. Eng.* **41** (1963) 207. Heat transfer from gas to wall in a gas–solids transport line.
96. KANG, Y., FAN, L. T. and KIM, S. D.: *A.I.Ch.E.Jl.* **37** (1991) 1101. Immersed heater-to-bed heat transfer in liquid–solid fluidized beds.
97. ROMANI, M. N. and RICHARDSON, J. F.: *Letters in Heat and Mass Transfer* **1** (1974) 55. Heat transfer from immersed surfaces to liquid-fluidized beds.
98. RICHARDSON, J. F., ROMANI, M. N., SHAKIRI, K. J.: *Chem. Eng. Sci.* **31** (1976) 619. Heat transfer from immersed surfaces in liquid fluidised beds.
99. KHAN, A. R., JUMA, A. K. A. and RICHARDSON, J. F.: *Chem. Eng. Sci.* **38** (1983) 2053. Heat transfer from a plane surface to liquids and to liquid-fluidised beds.
100. LEVA, M.: *Fluidization* (McGraw-Hill, 1959).
101. LEVENSPIEL, O. and WALTON, J. S.: *Proc. Heat Transf. Fluid Mech. Inst. Berkeley, California* (1949) 139–46. Heat transfer coefficients in beds of moving solids.
102. VREEDENBERG, H. A.: *J. Appl. Chem.* **2** (1952) S26. Heat transfer between fluidised beds and vertically inserted tubes.
103. DOW, W. M. and JAKOB, M.: *Chem. Eng. Prog.* **47** (1951) 637. Heat transfer between a vertical tube and a fluidized air–solid mixture.
104. BOTTERILL, J. S. M., REDISH, K. A., ROSS, D. K., and WILLIAMS, J. R.: Third Congress of the European Federation of Chemical Engineering (1962). *The Interaction between Fluids and Particles* 183. The mechanism of heat transfer to fluidised beds.
105. BOTTERILL, J. S. M. and WILLIAMS, J. R.: *Trans. Inst. Chem. Eng.* **41** (1963) 217. The mechanism of heat transfer to gas-fluidised beds.

106. BOTTERILL, J. S. M., BRUNDRETT, G. W., CAIN, G. L., and ELLIOTT, D. E.: *Chem. Eng. Prog.* Symp. Ser. No. 62, **62** (1966) 1. Heat transfer to gas-fluidized beds.
107. WILLIAMS, J. R.: University of Birmingham, Ph.D. thesis (1962). The mechanism of heat transfer to fluidised beds.
108. HIGBIE, R.: *Trans. Am. Inst. Chem. Eng.* **31** (1935) 365. The rate of absorption of a pure gas into a still liquid during periods of exposure.
109. DANCKWERTS, P. V.: *Ind. Eng. Chem.* **43** (1951) 1460. Significance of liquid-film coefficients in gas absorption.
110. MICKLEY, H. S., FAIRBANKS, D. F., and HAWTHORN, R. D.: *Chem. Eng. Prog.* Symp. Ser. No. 32, **57** (1961) 51. The relation between the transfer coefficient and thermal fluctuations in fluidized-bed heat transfer.
111. BOCK, H.-J. and MOLERUS, O.: In *Fluidization* by GRACE, J. R. and MATSEN, J. M. (eds.), Plenum Press, New York (1980), 217. Influence of hydrodynamics on heat transfer in fluidized beds.
112. RICHARDSON, J. F. and SHAKIRI, K. J.: *Chem. Eng. Sci.* **34** (1979) 1019. Heat transfer between a gas–solid fluidised bed and a small immersed surface.
113. XAVIER, A. M., KING, D. F., DAVIDSON, J. F. and HARRISON, D.: In *Fluidization* by GRACE, J. R. and MATSEN, J. M. (eds.), Plenum Press, New York, (1980), 209. Surface–bed heat transfer in a fluidised bed at high pressure.
114. RICHARDSON, J. F. and BAKHTIAR, A. G.: *Trans. Inst. Chem. Eng.* **36** (1958) 283. Mass transfer between fluidised particles and gas.
115. RICHARDSON, J. F. and SZEKELY, J.: *Trans. Inst. Chem. Eng.* **39** (1961) 212. Mass transfer in a fluidised bed.
116. MULLIN, J. W. and TRELEAVEN, C. R.: Third Congress of the European Federation of Chemical Engineering (1962). *The Interaction between Fluids and Particles* 203. Solids–liquid mass transfer in multi-particulate systems.
117. CHU, J. C., KALIL, J., and WETTEROTH, W. A.: *Chem. Eng. Prog.* **49** (1953) 141. Mass transfer in a fluidized bed.
118. KETTENRING, K. N., MANDERFIELD, E. L., and SMITH, J. M.: *Chem. Eng. Prog.* **46** (1950) 139. Heat and mass transfer in fluidized systems.
119. HEERTJES, P. M. and MCKIBBINS, S. W.: *Chem. Eng. Sci.* **5** (1956) 161. The partial coefficient of heat transfer in a drying fluidized bed.
120. FRANTZ, J. F.: *Chem. Eng. Prog.* **57** (1961) 35. Fluid-to-particle heat transfer in fluidized beds.
121. RICHARDSON, J. F. and AYERS, P.: *Trans. Inst. Chem. Eng.* **37** (1959) 314. Heat transfer between particles and a gas in a fluidised bed.
122. RESNICK, W. and WHITE, R. R.: *Chem. Eng. Prog.* **45** (1949) 377. Mass transfer in systems of gas and fluidized solids.
123. WILKE, C. R. and HOUGEN, O. A.: *Trans. Am. Inst. Chem. Eng.* **41** (1945) 445. Mass transfer in the flow of gases through granular solids extended to low modified Reynolds Numbers.
124. MCCUNE, L. K. and WILHELM, R. H.: *Ind. Eng. Chem.* **41** (1949) 1124. Mass and momentum transfer in solid–liquid system. Fixed and fluidized beds.
125. HOBSON, M. and THODOS, G.: *Chem. Eng. Prog.* **45** (1949) 517. Mass transfer in flow of liquids through granular solids.
126. GAFFNEY, B. J. and DREW, T. B.: *Ind. Eng. Chem.* **42** (1950) 1120. Mass transfer from packing to organic solvents in single phase flow through a column.
127. HSU, C. T. and MOLSTAD, M. C.: *Ind. Eng. Chem.* **47** (1955) 1550. Rate of mass transfer from gas stream to porous solid in fluidized beds.
128. CORNISH, A. R. H.: *Trans. Inst. Chem. Eng.* **43** (1965) T332. Note on minimum possible rate of heat transfer from a sphere when other spheres are adjacent to it.
129. ZABRODSKY, S. S.: *J. Heat and Mass Transfer* **10** (1967) 1793. On solid-to-fluid heat transfer in fluidized systems.
130. WINDEBANK, C. S.: *J. Imp. Coll. Chem. Eng. Soc.* **4** (1948) 31. The fluid catalyst technique in modern petroleum refining.
131. RILEY, H. L.: *Trans. Inst. Chem. Eng.* **37** (1959) 305. Design of fluidised reactors for naphthalene oxidation: a review of patent literature.
132. ROWE, P. N.: *Soc. Chem. Ind., Fluidisation* (1964) 15. A theoretical study of a batch reaction in a gas-fluidised bed.
133. VOLK, W., JOHNSON, C. A., and STOTLER, H. H.: *Chem. Eng. Prog.* Symp. Ser. No. 38, **58** (1962) 38. Effect of reactor internals on quality of fluidization.
134. INSTITUTE OF FUEL: Symposium Series No. 1: *Fluidised Combustion* **1** (1975).
135. TURNBULL, E. and DAVIDSON, J. F.: *A.I.Ch.E.Jl.* **30** (1984) 881. Fluidized combustion of char and volatiles from coal.

136. WU, R. L., GRACE, J. R., LIM, C. J. and BRERETON, M. H.: *A.I.Ch.E.Jl.* **35** (1989) 1685. Surface-to-surface heat transfer in a circulating-fluidized-bed combustor.
137. AGARWAL, P. K. and LA NAUZE, R. D.: *Chem. Eng. Res. Des.* **67** (1989) 457. Transfer processes local to the coal particle. A review of drying, devolatilization and mass transfer in fluidized bed combustion.
138. REDMAN, J.: *The Chemical Engineer* No. 467 (Dec. 1989) 32. Fluidised bed combustion, $SO_x$ and $NO_x$.
139. VALK, M. (ed.): *Atmospheric fluidized bed coal combustion* (Elsivier, Amsterdam, 1994).

# 6.11. NOMENCLATURE

| | | Units in SI System | Dimensions in **M, N, L, T, $\theta$, A** |
|---|---|---|---|
| $A$ | Area of heat transfer surface | $m^2$ | $\mathbf{L}^2$ |
| $A_p$ | Area of particle | $m^2$ | $\mathbf{L}^2$ |
| $a$ | Area for transfer per unit volume of bed | $m^2/m^3$ | $\mathbf{L}^{-1}$ |
| $a'$ | Area for transfer per unit height of bed | $m^2/m$ | $\mathbf{L}$ |
| $C$ | Fractional volumetric concentration of solids | — | — |
| $C_0$ | Value of $C$ at $t = 0$ | — | — |
| $\Delta C$ | Driving force expressed as a molar concentration difference | kmol/m$^3$ | $\mathbf{NL}^{-3}$ |
| $\Delta C_{lm}$ | Logarithmic mean value of $C$ | kmol/m$^3$ | $\mathbf{NL}^{-3}$ |
| $C_p$ | Specific heat of gas at constant pressure | J/kg K | $\mathbf{L}^2\mathbf{T}^{-2}\boldsymbol{\theta}^{-1}$ |
| $C_s$ | Specific heat of solid particle | J/kg K | $\mathbf{L}^2\mathbf{T}^{-2}\boldsymbol{\theta}^{-1}$ |
| $D$ | Gas phase diffusivity | $m^2/s$ | $\mathbf{L}^2\mathbf{T}^{-1}$ |
| $D_L$ | Longitudinal diffusivity | $m^2/s$ | $\mathbf{L}^2\mathbf{T}^{-}1$ |
| $d$ | Particle diameter or diameter of sphere with same surface¡w¿as particle | m | $\mathbf{L}$ |
| $d_b$ | Bubble diameter | m | $\mathbf{L}$ |
| $d_p$ | Diameter of sphere of same volume as particle | m | $\mathbf{L}$ |
| $d_c$ | Cloud diameter | m | $\mathbf{L}$ |
| $d_t$ | Tube diameter | m | $\mathbf{L}$ |
| $E$ | Factor $r\omega^2/g$ for centrifugal fluidised bed | — | — |
| $e$ | Voidage | — | — |
| $e_{mf}$ | Voidage corresponding to minimum fluidising velocity | — | — |
| $f$ | Fraction of fluid passing through channels in bed | — | — |
| $G$ | Mass flowrate of fluid | kg/s | $\mathbf{MT}^{-1}$ |
| $G_E$ | Mass flowrate of fluid to cause initial expansion of bed | kg/s | $\mathbf{MT}^{-1}$ |
| $G_F$ | Mass flowrate of fluid to initiate fluidisation | kg/s | $\mathbf{MT}^{-1}$ |
| $G'$ | Mass flowrate of fluid per unit area | kg/m$^2$s | $\mathbf{ML}^{-2}\mathbf{T}^{-1}$ |
| $g$ | Acceleration due to gravity | 9.81m/s$^2$ | $\mathbf{LT}^{-2}$ |
| $h$ | Heat transfer coefficient | W/m$^2$K | $\mathbf{MT}^{-3}\boldsymbol{\theta}^{-1}$ |
| $h_D$ | Mass transfer coefficient | m/s | $\mathbf{LT}^{-1}$ |
| $h_L$ | Heat transfer coefficient for liquid alone at same rate as in bed | W/m$^2$K | $\mathbf{MT}^{-3}\boldsymbol{\theta}^{-1}$ |
| $j'_d$ | $j$-factor for mass transfer to particles | — | — |
| $j'_h$ | $j$-factor for heat transfer to particles | — | — |
| $K$ | Distance travelled by bubble to increase its volume by a factor e | m | $\mathbf{L}$ |
| $k$ | Thermal conductivity of fluid | W/m K | $\mathbf{MLT}^{-3}\boldsymbol{\theta}^{-1}$ |
| $l$ | Depth of fluidised bed | m | $\mathbf{L}$ |
| $M$ | Constant in equation 6.53 | 1/s | $\mathbf{T}^{-1}$ |
| $m$ | Index of $(1 - e)$ in equation 6.55 | — | — |

| | | Units in SI System | Dimensions in $\mathbf{M, N, L, T, \theta, A}$ |
|---|---|---|---|
| $N_A$ | Molar rate of transfer of diffusing component | kmol/s | $\mathbf{NT^{-1}}$ |
| $n$ | Index of $e$ in equation 6.31 | — | — |
| $P$ | Pressure | N/m$^2$ | $\mathbf{ML^{-1}T^{-2}}$ |
| $-\Delta P$ | Pressure drop across bed due to the presence of solids | N/m$^2$ | $\mathbf{ML^{-1}T^{-2}}$ |
| $Q$ | Rate of transfer of heat | W | $\mathbf{ML^2T^{-3}}$ |
| $R$ | Electrical resistance | W/A$^2$ | $\mathbf{ML^2T^{-3}A^{-2}}$ |
| $r$ | Ratio of rates of change of bed density with velocity for two species (equation 6.44) | | |
| $S$ | Specific surface of particles | m$^2$/m$^3$ | $\mathbf{L^{-1}}$ |
| $T$ | Temperature of gas | K | $\boldsymbol{\theta}$ |
| $T_B$ | Bed temperature | K | $\boldsymbol{\theta}$ |
| $T_E$ | Element temperature | K | $\boldsymbol{\theta}$ |
| $T_s$ | Temperature of solid | K | $\boldsymbol{\theta}$ |
| $\Delta T$ | Temperature driving force | deg K | $\boldsymbol{\theta}$ |
| $t$ | Time | s | $\mathbf{T}$ |
| $u_b$ | Velocity of rise of bubble | m/s | $\mathbf{LT^{-1}}$ |
| $u_c$ | Superficial velocity of fluid (open tube) | m/s | $\mathbf{LT^{-1}}$ |
| $u_i$ | Value of $u_c$ at infinite dilution | m/s | $\mathbf{LT^{-1}}$ |
| $u_{mb}$ | Minimum value of $u_c$ at which bubbling occurs | m/s | $\mathbf{LT^{-1}}$ |
| $u_{mf}$ | Minimum value of $u_c$ at which fluidisation occurs | m/s | $\mathbf{LT^{-1}}$ |
| $u_0$ | Free-falling velocity of particle in infinite fluid | m/s | $\mathbf{LT^{-1}}$ |
| $V$ | Voltage applied to element | W/A | $\mathbf{ML^2T^{-3}A^{-1}}$ |
| $V_B$ | Volume of bubble | m$^3$ | $\mathbf{L^3}$ |
| $V_p$ | Volume of particle | m$^3$ | $\mathbf{L^3}$ |
| $W$ | Mass of solids in bed | kg | $\mathbf{M}$ |
| $x$ | Volume fraction of spheres | — | — |
| $y$ | Mole fraction of vapour in gas stream | — | — |
| $y_0$ | Mole fraction of vapour in inlet gas | — | — |
| $y^*$ | Mole fraction of vapour in equilibrium with solids | — | — |
| $Z$ | Total height of bed | m | $\mathbf{L}$ |
| $z$ | Height above bottom of bed | m | $\mathbf{L}$ |
| $\alpha$ | Ratio of gas velocities in bubble and emulsion phases | — | — |
| $\mu$ | Viscosity of fluid | Ns/m$^2$ | $\mathbf{ML^{-1}T^{-1}}$ |
| $\rho$ | Density of fluid | kg/m$^3$ | $\mathbf{ML^{-3}}$ |
| $\rho_b$ | Bed density | kg/m$^3$ | $\mathbf{ML^{-3}}$ |
| $\rho_c$ | Density of suspension | kg/m$^3$ | $\mathbf{ML^{-3}}$ |
| $\rho_s$ | Density of particle | kg/m$^3$ | $\mathbf{ML^{-3}}$ |
| $\psi$ | Particle stream function | — | — |
| $\phi_s$ | Ratio of diameter of the same specific surface as particles to that of same volume (equation 6.16) | — | — |
| $\omega$ | Angular speed of rotation | s$^{-1}$ | $\mathbf{T^{-1}}$ |
| $Ga$ | Galileo number $[d^3\rho(\rho_s-\rho)g/\mu^2]$ (equation 6.9) | — | — |
| $Ga_p$ | Galileo number at minimum fluidising velocity | — | — |
| $Nu$ | Nusselt number $(hd_t/k)$ | — | — |
| $Nu'$ | Nusselt number $(hd/k)$ | — | — |
| $Pr$ | Prandtl number $(C_p\mu/k)$ | — | — |

| | | Units in SI System | Dimensions in **M, N, L, T, $\theta$, A** |
|---|---|---|---|
| $Re_c$ | Tube Reynolds number $(u_c d_t \rho/\mu)$ | — | — |
| $Re'_c$ | Particle Reynolds number $(u_c d\rho/\mu)$ | — | — |
| $Re'_{mf}$ | Particle Reynolds number at minimum fluidising velocity $(u_{mf}\rho d/\mu)$ | — | — |
| $Re'_{mfh}$ | Particle Reynolds number $Re'_h$ at minimum fluidising velocity | — | — |
| $Re_1$ | Bed Reynolds number $(u_c\rho/S\mu(1-e))$ | — | — |
| $Re'_p$ | Particle Reynolds number with $d_p$ as characteristic diameter | — | — |
| $Re_{1n}$ | $Re_1$ for power-law fluid | — | — |
| $Re'_0$ | Particle Reynolds number $(u_0 d\rho/\mu)$ | — | — |
| $Re^*_1$ | Particle Reynolds number $(u_c d\rho/(1-e)\mu)$ | — | — |
| $Sc$ | Schmidt number $(\mu/\rho D)$ | — | — |
| $Sh'$ | Sherwood number $(h_D d/D)$ | — | — |

CHAPTER 7

# Liquid Filtration

## 7.1. INTRODUCTION

The separation of solids from a suspension in a liquid by means of a porous medium or screen which retains the solids and allows the liquid to pass is termed filtration.

In general, the pores of the medium are larger than the particles which are to be removed, and the filter works efficiently only after an initial deposit has been trapped in the medium. In the laboratory, filtration is often carried out using a form of Buchner funnel, and the liquid is sucked through the thin layer of particles using a source of vacuum. In even simpler cases the suspension is poured into a conical funnel fitted with a filter paper. In the industrial equivalent, difficulties are encountered in the mechanical handling of much larger quantities of suspension and solids. A thicker layer of solids has to form and, in order to achieve a high rate of passage of liquid through the solids, higher pressures are needed, and a far greater area has to be provided. A typical filtration operation is illustrated in Figure 7.1, which shows the filter medium, in this case a cloth, its support and the layer of solids, or filter cake, which has already formed.

Volumes of the suspensions to be handled vary from the extremely large quantities involved in water purification and ore handling in the mining industry to relatively small quantities, as in the fine chemical industry where the variety of solids is considerable. In most industrial applications it is the solids that are required and their physical size and properties are of paramount importance. Thus, the main factors to be considered when selecting equipment and operating conditions are:

(a) The properties of the fluid, particularly its viscosity, density and corrosive properties.
(b) The nature of the solid — its particle size and shape, size distribution, and packing characteristics.
(c) The concentration of solids in suspension.
(d) The quantity of material to be handled, and its value.
(e) Whether the valuable product is the solid, the fluid, or both.
(f) Whether it is necessary to wash the filtered solids.
(g) Whether very slight contamination caused by contact of the suspension or filtrate with the various components of the equipment is detrimental to the product.
(h) Whether the feed liquor may be heated.
(i) Whether any form of pretreatment might be helpful.

Filtration is essentially a mechanical operation and is less demanding in energy than evaporation or drying where the high latent heat of the liquid, which is usually water, has to be provided. In the typical operation shown in Figure 7.1, the cake gradually builds up

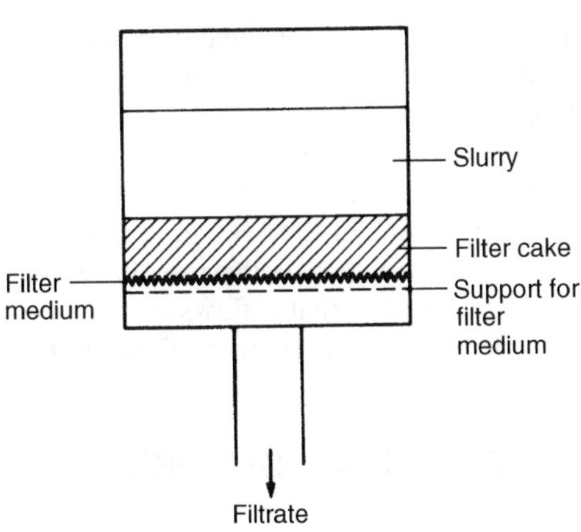


Figure 7.1. Principle of filtration

on the medium and the resistance to flow progressively increases. During the initial period of flow, particles are deposited in the surface layers of the cloth to form the true filtering medium. This initial deposit may be formed from a special initial flow of precoat material which is discussed later. The most important factors on which the rate of filtration then depends will be:

(a) The drop in pressure from the feed to the far side of the filter medium.
(b) The area of the filtering surface.
(c) The viscosity of the filtrate.
(d) The resistance of the filter cake.
(e) The resistance of the filter medium and initial layers of cake.

Two basic types of filtration processes may be identified, although there are cases where the two types appear to merge. In the first, frequently referred to as *cake filtration*, the particles from the suspension, which usually has a high proportion of solids, are deposited on the surface of a porous septum which should ideally offer only a small resistance to flow. As the solids build up on the septum, the initial layers form the effective filter medium, preventing the particles from embedding themselves in the filter cloth, and ensuring that a particle-free filtrate is obtained.

In the second type of filtration, *depth* or *deep-bed filtration*, the particles penetrate into the pores of the filter medium, where impacts between the particles and the surface of the medium are largely responsible for their removal and retention. This configuration is commonly used for the removal of fine particles from very dilute suspensions, where the recovery of the particles is not of primary importance. Typical examples here include air and water filtration. The filter bed gradually becomes clogged with particles, and its resistance to flow eventually reaches an unacceptably high level. For continued operation, it is therefore necessary to remove the accumulated solids, and it is important that this can be readily achieved. For this reason, the filter commonly consists of a bed of particulate solids, such as sand, which can be cleaned by back-flushing, often accompanied by fluidisation. In this chapter, the emphasis is on cake filtration although deep-bed filtration, which has been discussed in detail by IVES[1,2] is considered in the section on bed filters.

There are two principal modes under which deep bed filtration may be carried out. In the first, *dead-end filtration* which is illustrated in Figure 7.1, the slurry is filtered in such a way that it is fed perpendicularly to the filter medium and there is little flow parallel to the surface of the medium. In the second, termed *cross-flow filtration* which is discussed in Section 7.3.5. and which is used particularly for very dilute suspensions, the slurry is continuously recirculated so that it flows essentially across the surface of the filter medium at a rate considerably in excess of the flowrate through the filter cake.

## 7.2. FILTRATION THEORY

### 7.2.1. Introduction

Equations are given in Chapter 4 for the calculation of the rate of flow of a fluid through a bed of granular material, and these are now applied to the flow of filtrate through a filter cake. Some differences in general behaviour may be expected, however, because the cases so far considered relate to uniform fixed beds, whereas in filtration the bed is steadily growing in thickness. Thus, if the filtration pressure is constant, the rate of flow progressively diminishes whereas, if the flowrate is to be maintained constant, the pressure must be gradually increased.

The mechanical details of the equipment, particularly of the flow channel and the support for the medium, influence the way the cake is built up and the ease with which it may be removed. A uniform structure is very desirable for good washing and cakes formed from particles of very mixed sizes and shapes present special problems. Although filter cakes are complex in their structure and cannot truly be regarded as composed of rigid non-deformable particles, the method of relating the flow parameters developed in Chapter 4 is useful in describing the flow within the filter cake. The general theory of filtration and its importance in design has been considered by SUTTLE[3]. It may be noted that there are two quite different methods of operating a batch filter. If the pressure is kept constant then the rate of flow progressively diminishes, whereas if the flowrate is kept constant then the pressure must be gradually increased. Because the particles forming the cake are small and the flow through the bed is slow, streamline conditions are almost invariably obtained, and, at any instant, the flowrate of the filtrate may be represented by the following form of equation 4.9:

$$u_c = \frac{1}{A}\frac{\mathrm{d}V}{\mathrm{d}t} = \frac{1}{5}\frac{e^3}{(1-e)^2}\frac{-\Delta P}{S^2\mu l} \tag{7.1}$$

where $V$ is the volume of filtrate which has passed in time $t$, $A$ is the total cross-sectional area of the filter cake, $u_c$ is the superficial velocity of the filtrate, $l$ is the cake thickness, $S$ is the specific surface of the particles, $e$ is the voidage, $\mu$ is the viscosity of the filtrate, and $\Delta P$ is the applied pressure difference.

In deriving this equation it is assumed that the cake is uniform and that the voidage is constant throughout. In the deposition of a filter cake this is unlikely to be the case and the voidage, $e$ will depend on the nature of the support, including its geometry and surface structure, and on the rate of deposition. The initial stages in the formation of the cake are therefore of special importance for the following reasons:

(a) For any filtration pressure, the rate of flow is greatest at the beginning of the process since the resistance is then a minimum.
(b) High initial rates of filtration may result in plugging of the pores of the filter cloth and cause a very high resistance to flow.
(c) The orientation of the particle in the initial layers may appreciably influence the structure of the whole filter cake.

Filter cakes may be divided into two classes — incompressible cakes and compressible cakes. In the case of an incompressible cake, the resistance to flow of a given volume of cake is not appreciably affected either by the pressure difference across the cake or by the rate of deposition of material. On the other hand, with a compressible cake, increase of the pressure difference or of the rate of flow causes the formation of a denser cake with a higher resistance. For incompressible cakes $e$ in equation 7.1 may be taken as constant and the quantity $e^3/[5(1-e)^2S^2]$ is then a property of the particles forming the cake and should be constant for a given material.

Thus:
$$\frac{1}{A}\frac{dV}{dt} = \frac{-\Delta P}{\mathbf{r}\mu l} \tag{7.2}$$

where:
$$\mathbf{r} = \frac{5(1-e)^2S^2}{e^3} \tag{7.3}$$

It may be noted that, when there is a hydrostatic pressure component such as with a horizontal filter surface, this should be included in the calculation of $-\Delta P$.

Equation 7.2 is the basic filtration equation and $\mathbf{r}$ is termed the specific resistance which is seen to depend on $e$ and $S$. For incompressible cakes, $\mathbf{r}$ is taken as constant, although it depends on rate of deposition, the nature of the particles, and on the forces between the particles. $\mathbf{r}$ has the dimensions of $\mathbf{L}^{-2}$ and the units m$^{-2}$ in the SI system.

### 7.2.2. Relation between thickness of cake and volume of filtrate

In equation 7.2, the variables $l$ and $V$ are connected, and the relation between them may be obtained by making a material balance between the solids in both the slurry and the cake as follows.

Mass of solids in filter cake $= (1-e)Al\rho_s$, where $\rho_s$ is the density of the solids

Mass of liquid retained in the filter cake $= eAl\rho$, where $\rho$ is the density of the filtrate.

If $J$ is the mass fraction of solids in the original suspension then:

$$(1-e)lA\rho_s = \frac{(V+eAl)\rho J}{1-J}$$

or:
$$(1-J)(1-e)Al\rho_s = JV\rho + AeJl\rho$$

so that:
$$l = \frac{JV\rho}{A\{(1-J)(1-e)\rho_s - Je\rho\}} \tag{7.4}$$

and:
$$V = \frac{\{\rho_s(1-e)(1-J) - e\rho J\}Al}{\rho J} \tag{7.5}$$

If $v$ is the volume of cake deposited by unit volume of filtrate then:

$$v = \frac{lA}{V} \quad \text{or} \quad l = \frac{vV}{A} \tag{7.6}$$

and from equation 7.5:

$$v = \frac{J\rho}{(1-J)(1-e)\rho_s - Je\rho} \tag{7.7}$$

Substituting for $l$ in equation 7.2:

$$\frac{1}{A}\frac{dV}{dt} = \frac{(-\Delta P)}{\mathbf{r}\mu}\frac{A}{vV}$$

or:
$$\frac{dV}{dt} = \frac{A^2(-\Delta P)}{\mathbf{r}\mu vV} \tag{7.8}$$

Equation 7.8 may be regarded as the basic relation between $-\Delta P$, $V$, and $t$. Two important types of operation are: (i) where the pressure difference is maintained constant and (ii) where the rate of filtration is maintained constant.

*For a filtration at constant rate*

$$\frac{dV}{dt} = \frac{V}{t} = \text{constant}$$

so that:
$$\frac{V}{t} = \frac{A^2(-\Delta P)}{\mathbf{r}\mu Vv} \tag{7.9}$$

or:
$$\frac{t}{V} = \frac{\mathbf{r}\mu v}{A^2(-\Delta P)}V \tag{7.10}$$

and $-\Delta P$ is directly proportional to $V$.

*For a filtration at constant pressure difference*

$$\frac{V^2}{2} = \frac{A^2(-\Delta P)t}{\mathbf{r}\mu v} \tag{7.11}$$

or:
$$\frac{t}{V} = \frac{\mathbf{r}\mu v}{2A^2(-\Delta P)}V \tag{7.12}$$

Thus for a constant pressure filtration, there is a linear relation between $V^2$ and $t$ or between $t/V$ and $V$.

Filtration at constant pressure is more frequently adopted in practice, although the pressure difference is normally gradually built up to its ultimate value.

If this takes a time $t_1$ during which a volume $V_1$ of filtrate passes, then integration of equation 7.12 gives:

$$\frac{1}{2}(V^2 - V_1^2) = \frac{A^2(-\Delta P)}{\mathbf{r}\mu v}(t - t_1) \tag{7.13}$$

or:
$$\frac{t - t_1}{V - V_1} = \frac{\mathbf{r}\mu v}{2A^2(-\Delta P)}(V - V_1) + \frac{\mathbf{r}\mu v V_1}{A^2(-\Delta P)} \tag{7.14}$$

Thus, there where is a linear relation between $V^2$ and $t$ and between $(t - t_1)/(V - V_1)$ and $(V - V_1)$, where $(t - t_1)$ represents the time of the constant pressure filtration and $(V - V_1)$ the corresponding volume of filtrate obtained.

RUTH *et al.*[(4–7)] have made measurements on the flow in a filter cake and have concluded that the resistance is somewhat greater than that indicated by equation 7.1. It was assumed that part of the pore space is rendered ineffective for the flow of filtrate because of the adsorption of ions on the surface of the particles. This is not borne out by GRACE[(8)] or by HOFFING and LOCKHART[(9)] who determined the relation between flowrate and pressure difference, both by means of permeability tests on a fixed bed and by filtration tests using suspensions of quartz and diatomaceous earth.

Typical values of the specific resistance **r** of filter cakes, taken from the work of CARMAN[(10)], are given in Table 7.1. In the absence of details of the physical properties of the particles and of the conditions under which they had been formed, these values are approximate although they do provide an indication of the orders of magnitude.

Table 7.1. Typical Values of Specific Resistance, **r**[(10)]

| Material | Upstream filtration pressure ($kN/m^2$) | **r** $m^{-2}$ |
|---|---|---|
| High-grade kieselguhr | — | $2 \times 10^{12}$ |
| Ordinary kieselguhr | 270 | $1.6 \times 10^{14}$ |
| | 780 | $2.0 \times 10^{14}$ |
| Carboraffin charcoal | 110 | $4 \times 10^{13}$ |
| | 170 | $8 \times 10^{13}$ |
| Calcium carbonate (precipitated) | 270 | $3.5 \times 10^{14}$ |
| | 780 | $4.0 \times 10^{14}$ |
| Ferric oxide (pigment) | 270 | $2.5 \times 10^{15}$ |
| | 780 | $4.2 \times 10^{15}$ |
| Mica clay | 270 | $7.5 \times 10^{14}$ |
| | 780 | $13 \times 10^{14}$ |
| Colloidal clay | 270 | $8 \times 10^{15}$ |
| | 780 | $10 \times 10^{15}$ |
| Gelatinous magnesium hydroxide | 270 | $5 \times 10^{15}$ |
| | 780 | $11 \times 10^{15}$ |
| Gelatinous aluminium hydroxide | 270 | $3.5 \times 10^{16}$ |
| | 780 | $6.0 \times 10^{16}$ |
| Gelatinous ferric hydroxide | 270 | $3.0 \times 10^{16}$ |
| | 780 | $9.0 \times 10^{16}$ |
| Thixotropic mud | 650 | $2.3 \times 10^{17}$ |

### 7.2.3. Flow of liquid through the cloth

Experimental work on the flow of the liquid under streamline conditions[(10)] has shown that the flowrate is directly proportional to the pressure difference. It is the resistance of the cloth plus initial layers of deposited particles that is important since the latter, not only form the true medium, but also tend to block the pores of the cloth thus increasing its resistance. Cloths may have to be discarded because of high resistance well before they are mechanically worn. No true analysis of the buildup of resistance is possible because the resistance will depend on the way in which the pressure is developed and small variations in support geometry can have an important influence. It is therefore usual to combine the resistance of the cloth with that of the first few layers of particles and suppose that this corresponds to a thickness $L$ of cake as deposited at a later stage. The resistance to flow through the cake and cloth combined is now considered.

## 7.2.4. Flow of filtrate through the cloth and cake combined

If the filter cloth and the initial layers of cake are together equivalent to a thickness $L$ of cake as deposited at a later stage in the process, and if $-\Delta P$ is the pressure drop across the cake and cloth combined, then:

$$\frac{1}{A}\frac{dV}{dt} = \frac{(-\Delta P)}{\mathbf{r}\mu(l + L)} \tag{7.15}$$

which may be compared with equation 7.2.

Thus:

$$\frac{dV}{dt} = \frac{A(-\Delta P)}{\mathbf{r}\mu\left(\dfrac{Vv}{A} + L\right)} = \frac{A^2(-\Delta P)}{\mathbf{r}\mu v\left(V + \dfrac{LA}{v}\right)} \tag{7.16}$$

This equation may be integrated between the limits $t = 0$, $V = 0$ and $t = t_1$, $V = V_1$ for constant rate filtration, and $t = t_1$, $V = V_1$ and $t = t$, $V = V$ for a subsequent constant pressure filtration.

For the period of *constant rate filtration*:

$$\frac{V_1}{t_1} = \frac{A^2(-\Delta P)}{\mathbf{r}\mu v\left(V_1 + \dfrac{LA}{v}\right)}$$

or:

$$\frac{t_1}{V_1} = \frac{\mathbf{r}\mu v}{A^2(-\Delta P)}V_1 + \frac{\mathbf{r}\mu L}{A(-\Delta P)}$$

or:

$$V_1^2 + \frac{LA}{v}V_1 = \frac{A^2(-\Delta P)}{\mathbf{r}\mu v}t_1 \tag{7.17}$$

For a subsequent *constant pressure filtration*:

$$\frac{1}{2}(V^2 - V_1^2) + \frac{LA}{v}(V - V_1) = \frac{A^2(-\Delta P)}{\mathbf{r}\mu v}(t - t_1) \tag{7.18}$$

or:

$$(V - V_1 + 2V_1)(V - V_1) + \frac{2LA}{v}(V - V_1) = \frac{2A^2(-\Delta P)}{\mathbf{r}\mu v}(t - t_1)$$

or:

$$\frac{t - t_1}{V - V_1} = \frac{\mathbf{r}\mu v}{2A^2(-\Delta P)}(V - V_1) + \frac{\mathbf{r}\mu v V_1}{A^2(-\Delta P)} + \frac{\mathbf{r}\mu L}{A(-\Delta P)} \tag{7.19}$$

Thus there is a linear relation between $(t - t_1)/(V - V_1)$ and $V - V_1$, as shown in Figure 7.2, and the slope is proportional to the specific resistance, as in the case of the flow of the filtrate through the filter cake alone given by equation 7.14, although the line does not now go through the origin.

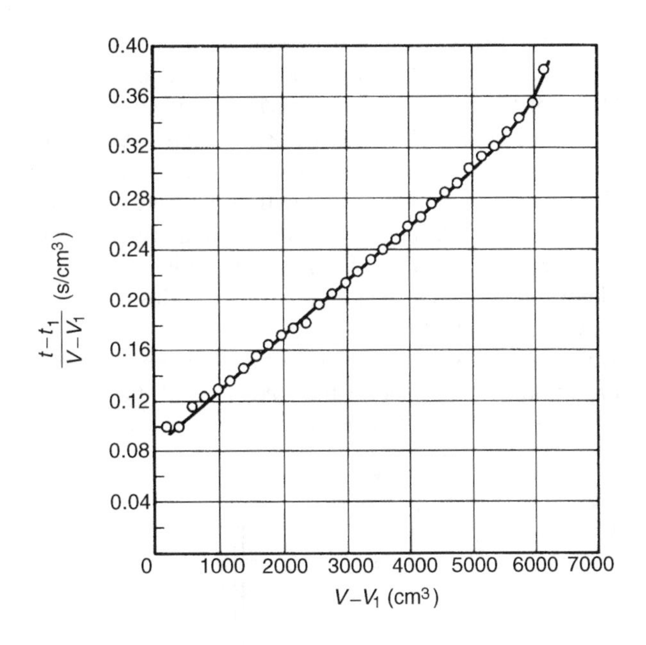


Figure 7.2. A typical filtration curve

The intercept on the $(t - t_1)/(V - V_1)$ axis should enable $L$, the equivalent thickness of the cloth, to be calculated although reproducible results are not obtained because this resistance is critically dependent on the exact manner in which the operation is commenced. The time at which measurement of $V$ and $t$ is commenced does not affect the slope of the curve, only the intercept. It may be noted that a linear relation between $t$ and $V^2$ is no longer obtained when the cloth resistance is appreciable.

## 7.2.5. Compressible filter cakes

Nearly all filter cakes are compressible to at least some extent although in many cases the degree of compressibility is so small that the cake may, for practical purposes, be regarded as incompressible. The evidence for compressibility is that the specific resistance is a function of the pressure difference across the cake. Compressibility may be a reversible or an irreversible process. Most filter cakes are inelastic and the greater resistance offered to flow at high pressure differences is caused by the more compact packing of the particles forming the filter cake. Thus the specific resistance of the cake corresponds to that for the highest pressure difference to which the cake is subjected, even though this maximum pressure difference may be maintained for only a short time. It is therefore important that the filtration pressure should not be allowed to exceed the normal operating pressure at any stage. In elastic filter cakes the elasticity is attributable to compression of the particles themselves. This is less usual, although some forms of carbon can give rise to elastic cakes.

As the filtrate flows through the filter cake, it exerts a drag force on the particles and this force is transmitted through successive layers of particles right up to the filter cloth. The magnitude of this force increases progressively from the surface of the filter cake to the filter cloth since at any point it is equal to the summation of the forces on all the particles up to that point. If the cake is compressible, then its voidage will decrease progressively in the direction of flow of the filtrate, giving rise to a corresponding increase in the local value of the specific resistance, $\mathbf{r}_z$, of the filter cake. The structure of the cake is, however, complex and may change during the course of the filtration process. If the feed suspension is flocculated, the flocs may become deformed within the cake, and this may give rise to a change in the effective value of the specific surface, $S$. In addition, the particles themselves may show a degree of compressibility. Whenever possible, experimental measurements should be made to determine how the specific resistance varies over the range of conditions which will be employed in practice.

It is usually possible to express the voidage $e_z$ at a depth $z$ as a function of the difference between the pressure at the free surface of the cake $P_1$ and the pressure $P_z$ at that depth, that is $e_z$ as a function of $(P_1 - P_z)$. The nomenclature is as defined in Figure 7.3.

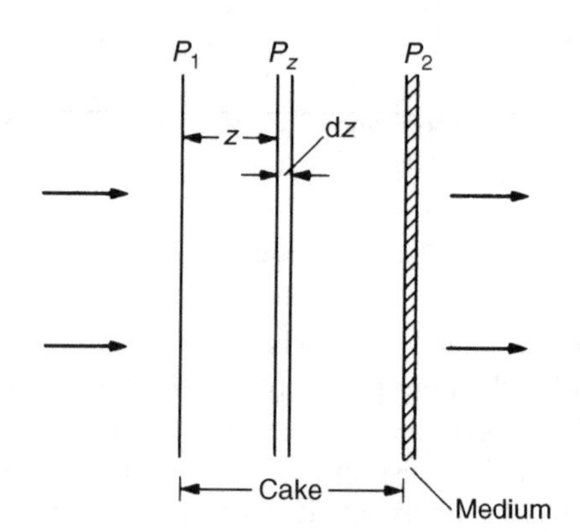


Figure 7.3. Flow through a compressible filter cake

For a compressible cake, equation 7.1 may be written as:

$$\frac{1}{A}\frac{\mathrm{d}V}{\mathrm{d}t} = \frac{e_z^3}{5(1-e_z)^2 S^2}\frac{1}{\mu}\left(-\frac{\mathrm{d}P_z}{\mathrm{d}z}\right) \tag{7.20}$$

where $e_z$ is now a function of depth $z$ from the surface of the cake.

In a compressible cake, the volume $v$ of cake deposited per unit area as a result of the flow of unit volume of filtrate will not be constant, but will vary during the filtration cycle. If the particles themselves are not compressible, however, the volume of *particles* $(v')$ will be *almost* independent of the conditions under which the cake is formed assuming a dilute feed suspension. Any small variations in $v'$ arise because the volume of filtrate retained in the cake is a function of its voidage, although the effect will be very small, except possibly for the filtration of very highly concentrated suspensions. The increase in cake thickness, $\mathrm{d}z$ resulting from the flow of a volume of filtrate $\mathrm{d}V$ is given by:

$$\mathrm{d}z = \mathrm{d}V \frac{v'}{(1 - e_z)A} \tag{7.21}$$

By comparison with equation 7.6, it may seen that:

$$\frac{v'}{v} = 1 - e_z \tag{7.22}$$

Substituting from equation 7.21 into equation 7.20 gives:

$$\frac{1}{A}\frac{\mathrm{d}V}{\mathrm{d}t} = \frac{e_z^3}{5(1 - e_z)^2 S^2}\frac{(1 - e_z)A}{v'}\frac{1}{\mu}\left(-\frac{\mathrm{d}P_z}{\mathrm{d}V}\right)$$

Thus:

$$\frac{\mathrm{d}V}{\mathrm{d}t} = \frac{e_z^3}{5(1 - e_z)S^2}\frac{A^2}{\mu v'}\left(-\frac{\mathrm{d}P_z}{\mathrm{d}V}\right) \tag{7.23}$$

$$= \frac{A^2}{\mu v' \mathbf{r}_z}\left(-\frac{\mathrm{d}P_z}{\mathrm{d}V}\right) \tag{7.24}$$

where:

$$\mathbf{r}_z = \frac{5(1 - e_z)S^2}{e_z^3} \tag{7.25}$$

Comparing equations 7.8 and 7.24 shows that for an incompressible cake:

$$v'\mathbf{r}_z = v\mathbf{r}$$

or:

$$\mathbf{r}_z = \mathbf{r}\frac{v}{v'}$$

At any instant in a constant pressure filtration, integration of equation 7.24 through the whole depth of the cake gives:

$$\int_0^V \frac{\mathrm{d}V}{\mathrm{d}t}\mathrm{d}V = \frac{A^2}{\mu v'}\int_{P_1}^{P_2} \frac{(-\mathrm{d}P_z)}{\mathbf{r}_z} \tag{7.26}$$

At any time $t$, $\mathrm{d}V/\mathrm{d}t$ is approximately constant throughout the cake, unless the rate of change of holdup of liquid within the cake is comparable with the filtration rate $\mathrm{d}V/\mathrm{d}t$, such as is the case with very highly compressible cakes and concentrated slurries, and therefore:

$$V\frac{\mathrm{d}V}{\mathrm{d}t} = \frac{A^2}{\mu v' V}\int_{P_1}^{P_2} \frac{(-\mathrm{d}P_z)}{\mathbf{r}_z} \tag{7.27}$$

$\mathbf{r}_z$ has been shown to be a function of the pressure difference $(P_1 - P_z)$ although it is independent of the absolute value of the pressure. Experimental studies frequently show that the relation between $\mathbf{r}_z$ and $(P_1 - P_z)$ is of the form:

$$\mathbf{r}_z = \mathbf{r}'(P_1 - P_z)^{n'} \tag{7.28}$$

where $\mathbf{r}'$ is independent of $P_z$ and $0 < n' < 1$.

Thus:
$$\int_{P_1}^{P_2} \frac{(-\mathrm{d}P_z)}{\mathbf{r}_z} = \frac{1}{\mathbf{r}'}\int_{P_2}^{P_1} \frac{\mathrm{d}P}{(P_1 - P_z)^{n'}}$$
$$= \frac{1}{\mathbf{r}'}\frac{(P_1 - P_2)^{1-n'}}{1-n'}$$
$$= \frac{1}{\mathbf{r}'}\frac{(-\Delta P)^{1-n'}}{1-n'} \tag{7.29}$$

Thus:
$$\frac{\mathrm{d}V}{\mathrm{d}t} = \frac{A^2}{V\mu v'\mathbf{r}'}\frac{(-\Delta P)}{(1-n')(-\Delta P)^{n'}}$$
$$= \frac{A^2(-\Delta P)}{V\mu v'\mathbf{r}''(-\Delta P)^{n'}} \tag{7.30}$$

where $\mathbf{r}'' = (1 - n')\mathbf{r}'$

and:
$$\frac{\mathrm{d}V}{\mathrm{d}t} = \frac{A^2(-\Delta P)}{V\mu v'\bar{\mathbf{r}}} \tag{7.31}$$

where $\bar{\mathbf{r}}$ is the mean resistance defined by:

$$\bar{\mathbf{r}} = \bar{\mathbf{r}}''(-\Delta P)^{n'} \tag{7.32}$$

HEERTJES[11] has studied the effect of pressure on the porosity of a filter cake and suggested that, as the pressure is increased above atmospheric, the porosity decreases in proportion to some power of the excess pressure.

GRACE[8] has related the anticipated resistance to the physical properties of the feed slurry. VALLEROY and MALONEY[12] have examined the resistance of an incompressible bed of spherical particles when measured in a permeability cell, a vacuum filter, and a centrifuge, and emphasised the need for caution in applying laboratory data to units of different geometry.

TILLER and HUANG[13] give further details of the problem of developing a usable design relationship for filter equipment. Studies by TILLER and SHIRATO[14], TILLER and YEH[15] and RUSHTON and HAMEED[16] show the difficulty in presenting practical conditions in a way which can be used analytically. It is very important to note that tests on slurries must be made with equipment that is geometrically similar to that proposed. This means that specific resistance is very difficult to define in practice, since it is determined by the nature of the filtering unit and the way in which the cake is initially formed and then built up.

## 7.3. FILTRATION PRACTICE

### 7.3.1. The filter medium

The function of the filter medium is generally to act as a support for the filter cake, and the initial layers of cake provide the true filter. The filter medium should be mechanically strong, resistant to the corrosive action of the fluid, and offer as little resistance as possible to the flow of filtrate. Woven materials are commonly used, though granular materials and porous solids are useful for filtration of corrosive liquids in batch units. An important feature in the selection of a woven material is the ease of cake removal, since this is a key factor in the operation of modern automatic units. EHLERS(17) has discussed the selection of woven synthetic materials and WROTNOWSKI(18) that of non-woven materials. Further details of some more recent materials are given in the literature(19) and a useful summary is presented in Volume 6.

### 7.3.2. Blocking filtration

In the previous discussion it is assumed that there is a well-defined boundary between the filter cake and the filter cloth. The initial stages in the build-up of the filter cake are important, however, because these may have a large effect on the flow resistance and may seriously affect the useful life of the cloth.

The blocking of the pores of the filter medium by particles is a complex phenomenon, partly because of the complicated nature of the surface structure of the usual types of filter media, and partly because the lines of movement of the particles are not well defined. At the start of filtration, the manner in which the cake forms will lie between two extremes — the penetration of the pores by particles and the shielding of the entry to the pores by the particles forming bridges. HEERTJES(11) considered a number of idealised cases in which suspensions of specified pore size distributions were filtered on a cloth with a regular pore distribution. First, it was assumed that an individual particle was capable on its own of blocking a single pore, then, as filtration proceeded, successive pores would be blocked, so that the apparent value of the specific resistance of the filter cake would depend on the amount of solids deposited.

The pore and particle size distributions might, however, be such that more than one particle could enter a particular pore. In this case, the resistance of the pore increases in stages as successive particles are trapped until the pore is completely blocked. In practice, however, it is much more likely that many of the pores will never become completely blocked and a cake of relatively low resistance will form over the entry to the partially blocked pore.

One of the most important variables affecting the tendency for blocking is the concentration of particles. The greater the concentration, the smaller will be the average distance between the particles, and the smaller will be the tendency for the particle to be drawn in to the streamlines directed towards the open pores. Instead, the particles in the concentrated suspension tend to distribute themselves fairly evenly over the filter surface and form bridges. As a result, suspensions of high concentration generally give rise to cakes of lower resistance than those formed from dilute suspensions.

### 7.3.3. Effect of particle sedimentation on filtration

There are two important effects due to particle sedimentation which may affect the rate of filtration. First, if the sediment particles are all settling at approximately the same rate, as is frequently the case in a concentrated suspension in which the particle size distribution is not very wide, a more rapid build-up of particles will occur on an upward-facing surface and a correspondingly reduced rate of build-up will take place if the filter surface is facing downwards. Thus, there will be a tendency for accelerated filtration with downward-facing filter surfaces and reduced filtration rates for upward-facing surfaces. On the other hand, if the suspension is relatively dilute, so that the large particles are settling at a higher rate than the small ones, there will be a preferential deposition of large particles on an upward-facing surface during the initial stages of filtration, giving rise to a low resistance cake. Conversely, for a downward-facing surface, fine particles will initially be deposited preferentially and the cake resistance will be correspondingly increased. It is thus seen that there can be complex interactions where sedimentation is occurring at an appreciable rate, and that the orientation of the filter surface is an important factor.

### 7.3.4. Delayed cake filtration

In the filtration of a slurry, the resistance of the filter cake progressively increases and consequently, in a constant pressure operation, the rate of filtration falls. If the build-up of solids can be reduced, the effective cake thickness will be less and the rate of flow of filtrate will be increased.

In practice, it is sometimes possible to incorporate moving blades in the filter equipment so that the thickness of the cake is limited to the clearance between the filter medium and the blades. Filtrate then flows through the cake at an approximately constant rate and the solids are retained in suspension. Thus the solids concentration in the feed vessel increases until the particles are in permanent physical contact with one another. At this stage the boundary between the slurry and the cake becomes ill-defined, and a significant resistance to the flow of liquid develops within the slurry itself with a consequent reduction in the flowrate of filtrate.

By the use of this technique, a much higher rate of filtration can be achieved than is possible in a filter operated in a conventional manner. In addition, the resulting cake usually has a lower porosity because the blades effectively break down the bridges or arches which give rise to a structure in the filter cake, and the final cake is significantly drier as a result.

If the scrapers are in the form of rotating blades, the outcome differs according to whether they are moving at low or at high speed. At low speeds, the cake thickness is reduced to the clearance depth each time the scraper blade passes, although cake then builds up again until the next passage of the scraper. If the blade is operated at high speed, there is little time for solids to build up between successive passages of the blade and the cake reaches an approximately constant thickness. Since particles tend to be swept across the surface of the cake by the moving slurry, they will be trapped in the cake only if the drag force which the filtrate exerts on them is great enough. As the thickness of the cake increases the pressure gradient becomes less and there is a smaller force retaining particles in the cake surface. Thus the thickness of the cake tends to reach an equilibrium value, which can be considerably less than the clearance between the medium and the blades.

Experimental results for the effect of stirrer speed on the rate of filtration of a 10 per cent by mass suspension of clay are shown in Figure 7.4 taken from the work of TILLER and CHENG[20], in which the filtrate volume collected per unit cross-section of filter is plotted against time, for several stirrer speeds.

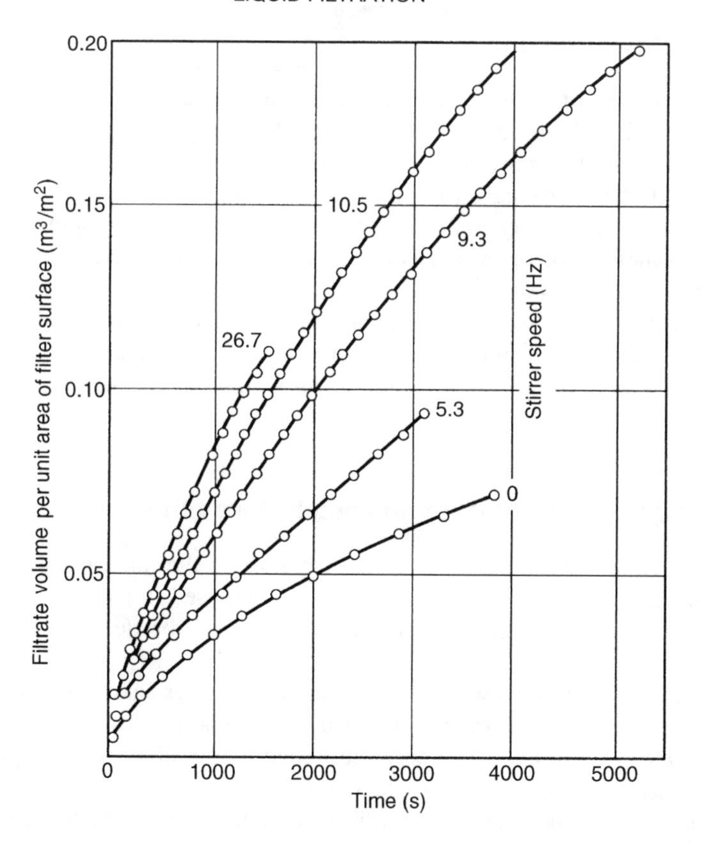


Figure 7.4. Volume as function of time for delayed cake and constant pressure filtration as a function of stirrer speed[20]

The concentration of solids in the slurry in the feed vessel to the filter at any time can be calculated by noting that the volumetric rate of feed of slurry must be equal to the rate at which filtrate leaves the vessel. For a rate of flow of filtrate of $dV/dt$ out of the filter, the rate of flow of slurry into the vessel must also be $dV/dt$ and the corresponding influx of solids is $(1 - e_0)\,dV/dt$, where $(1 - e_0)$ is the volume fraction of solids in the feed slurry. At any time $t$, the volume of solids in the vessel is $\mathbf{V}(1 - e_V)$, where $\mathbf{V}$ is its volume and $(1 - e_V)$ is the volume fraction of solids at that time. Thus a material balance on the solids gives:

$$(1 - e_0)\frac{dV}{dt} = \frac{d}{dt}[\mathbf{V}(1 - e_V)] \tag{7.33}$$

$$\frac{d(1 - e_V)}{dt} = \frac{1}{\mathbf{V}}(1 - e_0)\frac{dV}{dt} \tag{7.34}$$

For a constant filtration rate $dV/dt$, the fractional solids hold-up $(1 - e_V)$ increases linearly with time, until it reaches a limiting value when the resistance to flow of liquid within the slurry becomes significant. The filtration rate then drops rapidly to a near zero value.

### 7.3.5. Cross-flow filtration

An alternative method of reducing the resistance to filtration is to recirculate the slurry and thereby maintain a high velocity of flow parallel to the surface of the filter medium. Typical recirculation rates may be 10–20 times the filtration rate. By this means the cake is prevented from forming during the early stages of filtration. This can be particularly beneficial when the slurry is flocculated and exhibits shear-thinning non-Newtonian properties. This method of operation is discussed by MACKLEY and SHERMAN[21] and by HOLDICH, CUMMING and ISMAIL[22].

In cases where a dilute solution containing small quantities of solids which tend to blind the filter cloth is to be filtered, cross-flow filtration is extensively used. This is the normal mode of operation for ultrafiltration using membranes, a topic which is discussed in Chapter 8.

### 7.3.6. Preliminary treatment of slurries before filtration

If a slurry is dilute and the solid particles settle readily in the fluid, it may be desirable to effect a preliminary concentration in a thickener as discussed in Chapter 5. The thickened suspension is then fed from the thickener to the filter and the quantity of material to be handled is thereby reduced.

Theoretical treatment has shown that the nature of the filter cake has a very pronounced effect on the rate of flow of filtrate and that it is, in general, desirable that the particles forming the filter cake should have as large a size as possible. More rapid filtration is therefore obtained if a suitable agent is added to the slurry to cause coagulation. If the solid material is formed in a chemical reaction by precipitation, the particle size can generally be controlled to a certain extent by the actual conditions of formation. For example, the particle size of the resultant precipitate may be controlled by varying the temperature and concentration, and sometimes the pH, of the reacting solutions. As indicated by GRACE[8], a flocculated suspension gives rise to a more porous cake although the compressibility is greater. In many cases, crystal shape may be altered by adding traces of material which is selectively adsorbed on particular faces as noted in Chapter 15.

Filter aids are extensively used where the filter cake is relatively impermeable to the flow of filtrate. These are materials which pack to form beds of very high voidages and therefore they are capable of increasing the porosity of the filter cake if added to the slurry before filtration. Apart from economic considerations, there is an optimum quantity of filter aid which should be added in any given case. Whereas the presence of the filter aid reduces the specific resistance of the filter cake, it also results in the formation of a thicker cake. The actual quantity used will therefore depend on the nature of the material. The use of filter aids is normally restricted to operations in which the filtrate is valuable and the cake is a waste product. In some circumstances, however, the filter aid must be readily separable from the rest of the filter cake by physical or chemical means. Filter cakes incorporating filter aid are usually very compressible and care should therefore be taken to ensure that the good effect of the filter aid is not destroyed by employing too high a filtration pressure. Kieselguhr, which is a commonly used filter aid, has a voidage of about 0.85. Addition of relatively small quantities increases the voidage of most filter cakes, and the resulting porosity normally lies between that of the filter aid and that of the filter solids. Sometimes the filter medium is "precoated" with filter aid, and a thin layer of the filter aid is removed with the cake at the end of each cycle.

In some cases the filtration time can be reduced by diluting the suspension in order to reduce the viscosity of the filtrate. This does, of course, increase the bulk to be filtered and is applicable only when the value of the filtrate is not affected by dilution. Raising the temperature may be advantageous in that the viscosity of the filtrate is reduced.

### 7.3.7. Washing of the filter cake

When the wash liquid is miscible with the filtrate and has similar physical properties, the rate of washing at the same pressure difference will be about the same as the final rate of filtration. If the viscosity of the wash liquid is less, a somewhat greater rate will be obtained. Channelling sometimes occurs, however, with the result that much of the cake is incompletely washed and the fluid passes preferentially through the channels, which are gradually enlarged by its continued passage. This does not occur during filtration because channels are self-sealing by virtue of deposition of solids from the slurry. Channelling is most marked with compressible filter cakes and can be minimised by using a smaller pressure difference for washing than for filtration.

Washing may be regarded as taking place in two stages. First, filtrate is displaced from the filter cake by wash liquid during the period of *displacement washing* and in this way up to 90 per cent of the filtrate may be removed. During the second stage, *diffusional washing*, solvent diffuses into the wash liquid from the less accessible voids and the following relation applies:

$$\left(\frac{\text{volume of wash liquid passed}}{\text{cake thickness}}\right) = (\text{constant}) \times \log\left(\frac{\text{initial concentration of solute}}{\text{concentration at particular time}}\right) \tag{7.35}$$

Although an immiscible liquid is seldom used for washing, air is often used to effect partial drying of the filter cake. The rate of flow of air must normally be determined experimentally.

## 7.4. FILTRATION EQUIPMENT

### 7.4.1. Filter selection

The most suitable filter for any given operation is the one which will fulfil the requirements at minimum overall cost. Since the cost of the equipment is closely related to the filtering area, it is normally desirable to obtain a high overall rate of filtration. This involves the use of relatively high pressures although the maximum pressures are often limited by mechanical design considerations. Although a higher throughput from a given filtering surface is obtained from a continuous filter than from a batch operated filter, it may sometimes be necessary to use a batch filter, particularly if the filter cake has a high resistance, since most continuous filters operate under reduced pressure and the maximum filtration pressure is therefore limited. Other features which are desirable in a filter include ease of discharge of the filter cake in a convenient physical form, and a method of observing the quality of the filtrate obtained from each section of the plant. These factors are important in considering the types of equipment available. The most common types are filter presses, leaf filters, and continuous rotary filters. In addition, there are filters for special purposes, such as bag filters, and the disc type of filter which is used for the removal of small quantities of solids from a fluid.

The most important factors in filter selection are the specific resistance of the filter cake, the quantity to be filtered, and the solids concentration. For free-filtering materials, a rotary vacuum filter is generally the most satisfactory since it has a very high capacity for its size and does not require any significant manual attention. If the cake has to be washed, the rotary drum is to be preferred to the rotary leaf. If a high degree of washing is required, however, it is usually desirable to repulp the filter cake and to filter a second time.

For large-scale filtration, there are three principal cases where a rotary vacuum filter will not be used. Firstly, if the specific resistance is high, a positive pressure filter will be required, and a filter press may well be suitable, particularly if the solid content is not so high that frequent dismantling of the press is necessary. Secondly, when efficient washing is required, a leaf filter is effective, because very thin cakes can be prepared and the risk of channelling during washing is reduced to a minimum. Finally, where only very small quantities of solids are present in the liquid, an edge filter may be employed.

Whilst it may be possible to predict qualitatively the effect of the physical properties of the fluid and the solid on the filtration characteristics of a suspension, it is necessary in all cases to carry out a test on a sample before the large-scale plant can be designed. A simple vacuum filter with a filter area of 0.0065 $m^2$ is used to obtain laboratory data, as illustrated in Figure 7.5. The information on filtration rates and specific resistance obtained in this way can be directly applied to industrial filters provided due account is taken of the compressibility of the filter cake. It cannot be stressed too

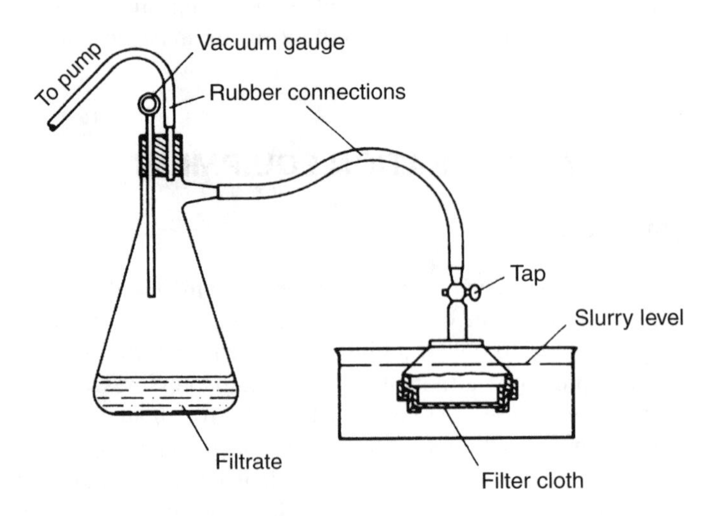


Figure 7.5. Laboratory test filter

strongly that data from any laboratory test cell must not be used without practical experience in the design of industrial units where the geometry of the flow channel is very different. The laying down of the cake influences the structure to a very a marked extent.

A "compressibility–permeability" test cell has been developed by RUTH[7] and GRACE[8] for testing the behaviour of slurries under various conditions of filtration. A useful guide to the selection of a filter type based on slurry characteristics is given in Volume 6.

### 7.4.2. Bed filters

Bed filters provide an example of the application of the principles of *deep bed filtration* in which the particles penetrate into the interstices of the filter bed where they are trapped following impingement on the surfaces of the material of the bed.

For the purification of water supplies and for waste water treatment where the solid content is about 10 g/m$^3$ or less, as noted by CLEASBY[23] granular bed filters have largely replaced the former very slow sand filters. The beds are formed from granular material of grain size 0.6–1.2 mm in beds 0.6–1.8 m deep. The very fine particles of solids are removed by mechanical action although the particles finally adhere as a result of surface electric forces or adsorption, as IVES[24] points out. This operation has been analysed by IWASAKI[25] who proposes the following equation:

$$-\frac{\partial C}{\partial l} = \lambda C \qquad (7.36)$$

On integration:

$$C/C_0 = e^{-\lambda l} \qquad (7.37)$$

where: $C$ is the volume concentration of solids in suspension in the filter,
$C_0$ is the value of $C$ at the surface of the filter,
$l$ is the depth of the filter, and
$\lambda$ is the filter coefficient.

If $u_c$ is the superficial flowrate of the slurry, then the rate of flow of solids through the filter at depth $l$ is $u_c C$ per unit area. Thus the rate of accumulation of solids in a distance $dl = -u_c(\partial C/\partial l)\, dl$. If $\sigma$ is the volume of solids deposited per unit volume of filter at a depth $l$, the rate of accumulation may also be expressed as $(\partial\sigma/\partial t)\, dl$.

Thus:

$$-\frac{\partial C}{\partial l} = u_c \frac{\partial \sigma}{\partial t} \qquad (7.38)$$

The problem is discussed further by IVES[24] and by SPIELMAN and FRIEDLANDER[26]. The backwashing of these beds has presented problems and several techniques have been adopted. These include a backflow of air followed by water, the flowrate of which may be high enough to give rise to fluidisation, with the maximum hydrodynamic shear occurring at a voidage of about 0.7.

### 7.4.3. Bag filters

Bag filters have now been almost entirely superseded for liquid filtration by other types of filter, although one of the few remaining types is the Taylor bag filter which has been widely used in the sugar industry. A number of long thin bags are attached to a horizontal feed tray and the liquid flows under the action of gravity so that the rate of filtration per unit area is very low. It is possible, however, to arrange a large filtering area in the plant of up to about 700 $m^2$. The filter is usually arranged in two sections so that each may be inspected separately without interrupting the operation.

Bag filters are still extensively used for the removal of dust particles from gases and can be operated either as pressure filters or as suction filters. Their use is discussed in Chapter 1.

### 7.4.4. The filter press

The filter press is one of two main types, the *plate and frame press* and the *recessed plate* or *chamber press*.

#### *The plate and frame press*

This type of filter consists of plates and frames arranged alternately and supported on a pair of rails as shown in Figure 7.6. The plates have a ribbed surface and the edges stand slightly proud and are carefully machined. The hollow frame is separated from the plate by the filter cloth, and the press is closed either by means of a hand screw or hydraulically, using the minimum pressure in order to reduce wear on the cloths. A chamber is therefore formed between each pair of successive plates as shown in Figure 7.7. The slurry is introduced through a port in each frame and the filtrate passes through the cloth on each side so that two cakes are formed simultaneously in each chamber, and these join when the frame is full. The frames are usually square and may be 100 mm–2.5 m across and 10 mm–75 mm thick.

The slurry may be fed to the press through the continuous channel formed by the holes in the corners of the plates and frames, in which case it is necessary to cut corresponding holes in the cloths which themselves act as gaskets. Cutting of the cloth can be avoided by feeding through a channel at the side although rubber bushes must then be fitted so that a leak-tight joint is formed.

The filtrate runs down the ribbed surface of the plates and is then discharged through a cock into an open launder so that the filtrate from each plate may be inspected and any plate can be isolated if it is not giving a clear filtrate. In some cases the filtrate is removed through a closed channel although it is not then possible to observe the discharge from each plate separately.

In many filter presses, provision is made for steam heating so that the viscosity of the filtrate is reduced and a higher rate of filtration obtained. Materials, such as waxes, that solidify at normal temperatures may also be filtered in steam-heated presses. Steam heating also facilitates the production of a dry cake.

*Optimum time cycle*. The optimum thickness of cake to be formed in a filter press depends on the resistance offered by the filter cake and on the time taken to dismantle

Figure 7.6. A large filter press with 2 m by 1.5 m plates

and refit the press. Although the production of a thin filter cake results in a high average rate of filtration, it is necessary to dismantle the press more often and a greater time is therefore spent on this operation. For a filtration carried out entirely at constant pressure, a rearrangement equation 7.19 gives:

$$\frac{t}{V} = \frac{\mathbf{r}\mu v}{2A^2(-\Delta P)}V + \frac{\mathbf{r}\mu L}{A(-\Delta P)} \tag{7.39}$$

$$= B_1 V + B_2 \tag{7.40}$$

where $B_1$ and $B_2$ are constants.

Thus the time of filtration $t$ is given by:

$$t = B_1 V^2 + B_2 V \tag{7.41}$$

The time of dismantling and assembling the press, say $t'$, is substantially independent of the thickness of cake produced. The total time of a cycle in which a volume $V$ of filtrate is collected is then $(t + t')$ and the overall rate of filtration is given by:

$$W = \frac{V}{B_1 V^2 + B_2 V + t'}$$

$W$ is a maximum when $\mathrm{d}W/\mathrm{d}V = 0$.

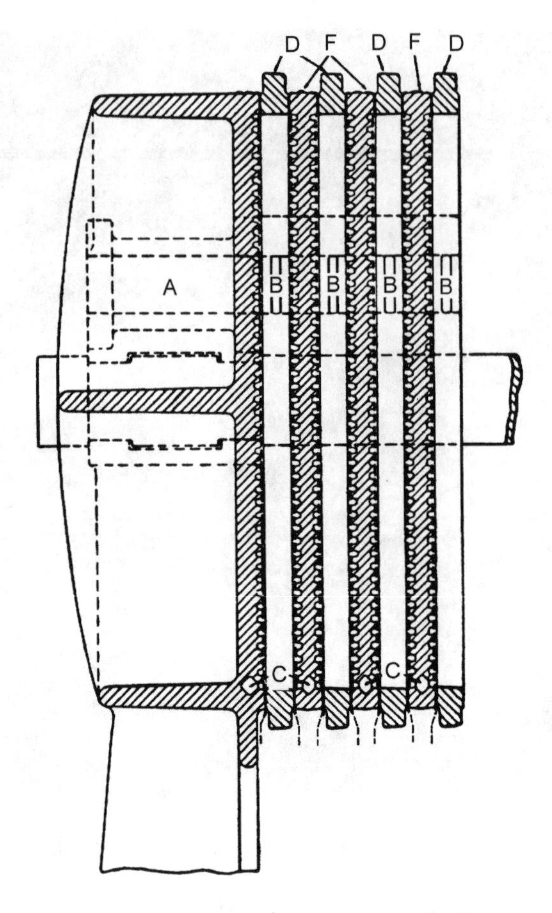


Figure 7.7. Plate and frame press. A–inlet passage. B–feed ports. C–filtrate outlet. D–frames. F–plates

Differentiating $W$ with respect to $V$ and equating to zero:

$$B_1V^2 + B_2V + t' - V(2B_1V + B_2) = 0$$

or:
$$t' = B_1V^2 \tag{7.42}$$

or:
$$V = \sqrt{\left(\frac{t'}{B_1}\right)} \tag{7.43}$$

If the resistance of the filter medium is neglected, $t = B_1V^2$ and the time during which filtration is carried out is exactly equal to the time the press is out of service. In practice, in order to obtain the maximum overall rate of filtration, the filtration time must always be somewhat greater in order to allow for the resistance of the cloth, represented by the term $B_2V$. In general, the lower the specific resistance of the cake, the greater will be the economic thickness of the frame.

The application of these equations is illustrated later in Example 7.5 which is based on the work of HARKER[27].

It is shown in Example 7.5, which appears later in the chapter, that, provided the cloth resistance is very low, adopting a filtration time equal to the downtime will give the maximum throughput. Where the cloth resistance is appreciable, then the term $B_2(t'/B_1)^{0.5}$ becomes significant and a longer filtration time is desirable. It may be seen in Figure 7.8, which is based on data from Example 7.5, that neither of these values represents the minimum cost condition however, except for the unique situation where $t' = $ (cost of shutdown)/(cost during filtering), and a decision has to be made as to whether cost or throughput is the overiding consideration. In practice, operating schedules are probably the dominating feature, although significant savings may be made by operating at the minimum cost condition.

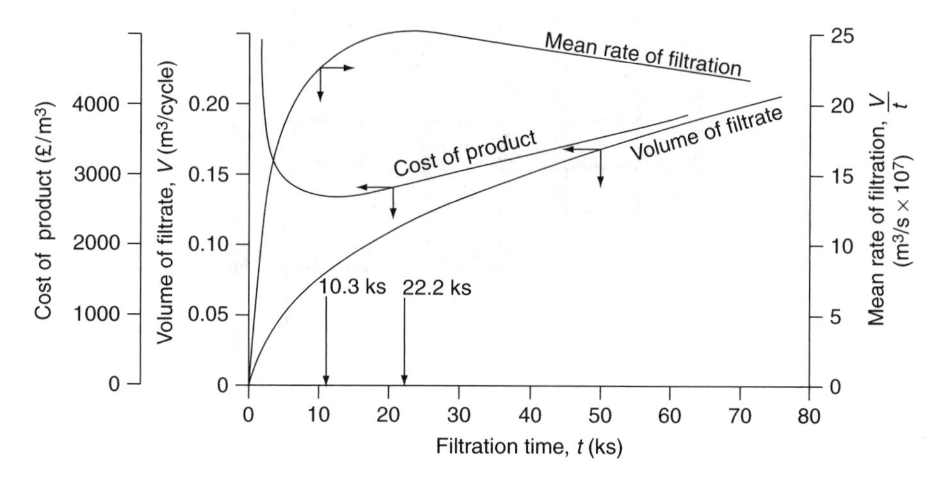


Figure 7.8. Optimisation of plate and frame press (data from Example 7.5)[27]

*Washing*

Two methods of washing may be employed, "simple" washing and "through" or "thorough" washing. With simple washing, the wash liquid is fed in through the same channel as the slurry although, as its velocity near the point of entry is high, erosion of the cake takes place. The channels which are thus formed gradually enlarge and uneven washing is usually obtained. Simple washing may be used only when the frame is not completely full.

In *thorough* washing, the wash liquid is introduced through a separate channel behind the filter cloth on alternate plates, known as washing plates shown in Figure 7.9, and flows through the whole thickness of the cake, first in the opposite direction and then in the same direction as the filtrate. The area during washing is one-half of that during filtration and, in addition, the wash liquid has to flow through twice the thickness, so that the rate of washing should therefore be about one-quarter of the final rate of filtration. The wash liquid is usually discharged through the same channel as the filtrate though sometimes a separate outlet is provided. Even with thorough washing some channelling occurs and several inlets are often provided so that the liquid is well distributed. If the cake is appreciably compressible, the minimum pressure should be used during washing,

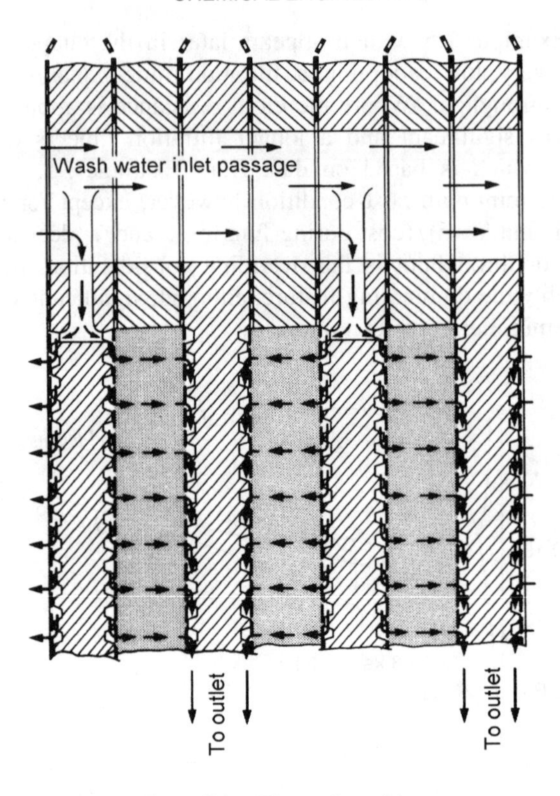


Figure 7.9. Thorough washing

and in no case should the final filtration pressure be exceeded. After washing, the cake may be made easier to handle by removing excess liquid with compressed air.

For ease in identification, small buttons are embossed on the sides of the plates and frames, one on the non-washing plates, two on the frames and three on the washing plates as shown in Figure 7.10.

## Example 7.1

A slurry is filtered in a plate and frame press containing 12 frames, each 0.3 m square and 25 mm thick. During the first 180 s the pressure difference for filtration is slowly raised to the final value of 400 $kN/m^2$ and, during this period, the rate of filtration is maintained constant. After the initial period, filtration is carried out at constant pressure and the cakes are completely formed in a further 900 s. The cakes are then washed with a pressure difference of 275 $kN/m^2$ for 600 s using *thorough washing* (See the plate and frame press in Section 7.4.4). What is the volume of filtrate collected per cycle and how much wash water is used?

A sample of the slurry had previously been tested with a leaf filter of 0.05 $m^2$ filtering surface using a vacuum giving a pressure difference of 71.3 $kN/m^2$. The volume of filtrate collected in the first 300 s, was 250 $cm^3$ and, after a further 300 s, an additional 150 $cm^3$ was collected. It may be assumed that the cake is incompressible and that the cloth resistance is the same in the leaf as in the filter press.

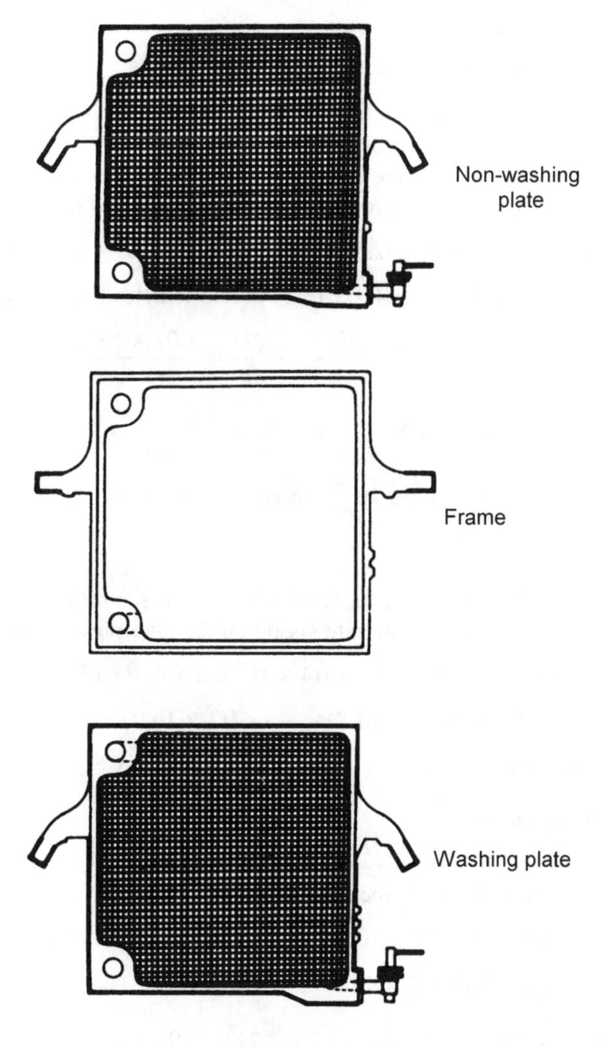


Figure 7.10. Plates and frames

## Solution

In the leaf filter, filtration is at constant pressure from the start.

Thus: $$V^2 + 2\frac{AL}{v}V = 2\frac{(-\Delta P)A^2}{\mathbf{r}\mu v}t \qquad \text{(from equation 7.18)}$$

In the filter press, a volume $V_1$ of filtrate is obtained under constant rate conditions in time $t_1$, and filtration is then carried out at constant pressure.

Thus: $$V_1^2 + \frac{AL}{v}V_1 = \frac{(-\Delta P)A^2}{\mathbf{r}\mu v}t_1 \qquad \text{(from equation 7.17)}$$

and:

$$(V^2 - V_1^2) + 2\frac{AL}{v}(V - V_1) = 2\frac{(-\Delta P)A^2}{\mathbf{r}\mu v}(t - t_1) \quad \text{(from equation 7.18)}$$

*For the leaf filter*

When $t = 300$ s, $V = 250\text{cm}^3 = 2.5 \times 10^{-4}$ m$^3$ and when $t = 600$ s, $V = 400$ cm$^3 = 4 \times 10^{-4}$ m$^3$, $A = 0.05$ m$^2$ and $-\Delta P = 71.3$ kN/m$^2$ or $7.13 \times 10^4$ N/m$^2$.

Thus: $$(2.5 \times 10^{-4})^2 + 2(0.05L/v)2.5 \times 10^{-4} = 2(7.13 \times 10^4 \times 0.05^2/\mathbf{r}\mu v)300$$

and: $$(4 \times 10^{-4})^2 + 2(0.05L/v)4 \times 10^{-4} = 2(7.13 \times 10^4 \times 0.05^2/\mathbf{r}\mu v)600$$

That is: $$6.25 \times 10^{-8} + 2.5 \times 10^{-5}\frac{L}{v} = \frac{1.07 \times 10^5}{\mathbf{r}\mu v}$$

and: $$16 \times 10^{-8} + 4 \times 10^{-5}\frac{L}{v} = \frac{2.14 \times 10^5}{\mathbf{r}\mu v}$$

Hence: $$L/v = 3.5 \times 10^{-3} \text{ and } \mathbf{r}\mu v = 7.13 \times 10^{11}$$

*For the filter press*

$A = (12 \times 2 \times 0.3^2) = 2.16$ m$^2$, $-\Delta P = 400$ kN/m$^2 = 4 \times 10^5$ N/m$^2$, $t = 180$ s. The volume of filtrate $V_1$ collected during the constant rate period on the filter press is given by:

$$V_1^2 + (2.16 \times 3.5 \times 10^{-3} V_1) = [(4 \times 10^5 \times 2.16^2)/(7.13 \times 10^{11})]180$$

$$V_1^2 + (7.56 \times 10^{-3} V_1) - (4.711 \times 10^{-4}) = 0$$

or: $$V_1 = -(3.78 \times 10^{-3}) + \sqrt{(1.429 \times 10^{-5} + 4.711 \times 10^{-4})} = 1.825 \times 10^{-2} \text{ m}^3$$

For the constant pressure period:

$$(t - t_1) = 900 \text{ s}$$

The total volume of filtrate collected is therefore given by:

$$(V^2 - 3.33 \times 10^{-4}) + (1.512 \times 10^{-2})(V - 1.825 \times 10^{-2}) = 5.235 \times 10^{-6} \times 900$$

or: $$V^2 + (1.512 \times 10^{-2} V) - (4.712 \times 10^{-3}) = 0$$

Thus: $$V = -0.756 \times 10^{-2} + \sqrt{(0.572 \times 10^{-4} + 4.712 \times 10^{-3})}$$

$$= 6.15 \times 10^{-2} \text{ or } \underline{\underline{0.062 \text{ m}^3}}$$

The final rate of filtration is given by:

$$\frac{-\Delta P A^2}{\mathbf{r}\mu v(V + AL/v)} = \frac{4 \times 10^5 \times 2.16^2}{7.13 \times 10^{11}(6.15 \times 10^{-2} + 2.16 \times 3.5 \times 10^{-3})} = 3.79 \times 10^{-5}\text{m}^3/\text{s}$$

(from equation 7.16)

If the viscosity of the filtrate is the same as that of the wash-water, then:

$$\text{Rate of washing at 400 kN/m}^2 = \tfrac{1}{4} \times 3.79 \times 10^{-5} = 9.5 \times 10^{-6} \text{ m}^3/\text{s}$$

$$\text{Rate of washing at 275 kN/m}^2 = 9.5 \times 10^{-6} \times (275/400) = 6.5 \times 10^{-6} \text{ m}^3/\text{s}$$

Thus the amount of wash-water passing in 600 s $= (600 \times 6.5 \times 10^{-6})$

$$= 3.9 \times 10^{-3}\text{m}^3 \text{ or } \underline{\underline{0.004 \text{ m}^3}}$$

### *The recessed plate filter press*

The recessed type of press is similar to the plate and frame type except that the use of frames is obviated by recessing the ribbed surface of the plates so that the individual filter chambers are formed between successive plates. In this type of press therefore the thickness of the cake cannot be varied and it is equal to twice the depth of the recess on individual plates.

Figure 7.11. A recessed chamber plate, 2 m square

The feed channel shown in Figure 7.11 usually differs from that employed on the plate and frame press. All the chambers are connected by means of a comparatively large hole in the centre of each of the plates and the cloths are secured in position by means of screwed unions. Slurries containing relatively large solid particles may readily be handled in this type of press without fear of blocking the feed channels. As described by CHERRY[28], developments in filter presses have been towards the fabrication of larger units, made possible by mechanisation and the use of newer lighter materials of construction. The plates of wood used in earlier times were limited in size because of limitations of pressures and large cast-iron plates presented difficulty in handling. Large plates are now frequently made of rubber mouldings or of polypropylene although distortion may be a problem, particularly if the temperature is high.

The second area of advance is in mechanisation which enables the opening and closing to be done automatically by a ram driven hydraulically or by an electric motor. Plate transportation is effected by fitting triggers to two endless chains operating the plates, and labour costs have consequently been reduced very considerably. Improved designs have given better drainage which has led to improved washing. Much shorter time cycles are now obtained and the cakes are thinner, more uniform, and drier. These advantages have been rather more readily obtained with recessed plates where the cloth is subjected to less wear.

### Advantages of the filter press

(a) Because of its basic simplicity the filter press is versatile and may be used for a wide range of materials under varying operating conditions of cake thickness and pressure.
(b) Maintenance cost is low.
(c) It provides a large filtering area on a small floor space and few additional associated units are needed.
(d) Most joints are external and leakage is easily detected.
(e) High pressure operation is usually possible.
(f) It is equally suitable whether the cake or the liquid is the main product.

### Disadvantages of the filter press

(a) It is intermittent in operation and continual dismantling is apt to cause high wear on the cloths.
(b) Despite the improvements mentioned previously, it is fairly heavy on labour.

## Example 7.2

A slurry containing 100 kg of whiting, of density 3000 kg/m$^3$, per m$^3$ of water, and, is filtered in a plate and frame press, which takes 900 s to dismantle, clean, and re-assemble. If the cake is incompressible and has a voidage of 0.4, what is the optimum thickness of cake for a filtration pressure $(-\Delta P)$ of 1000 kN/m$^2$? The density of the whiting is 3000 kg/m$^3$. If the cake is washed at 500 kN/m$^2$ and the total volume of wash water employed is 25 per cent of that of the filtrate, how is the optimum thickness of the cake affected? The resistance of the filter medium may be neglected and the viscosity of water is 1 mNs/m$^2$. In an experiment, a pressure difference of 165 kN/m$^2$ produced a flow of water of 0.02 cm$^3$/s through a centimetre cube of filter cake.

## Solution

The basic filtration equation may be written as:

$$\frac{1}{A}\frac{dV}{dt} = \frac{(-\Delta P)}{\mathbf{r}\mu l} \qquad \text{(equation 7.2)}$$

where $\mathbf{r}$ is defined as the specific resistance of the cake.

The slurry contains 100 kg whiting/m$^3$ of water.

Volume of 100 kg whiting = (100/3000) = 0.0333 m$^3$.

Volume of cake = 0.0333/(1 − 0.4) = 0.0556 m$^3$.

Volume of liquid in cake = $(0.0556 \times 0.4) = 0.0222\ \text{m}^3$.

Volume of filtrate = $(1 - 0.0222) = 0.978\ \text{m}^3$.

Thus: volume of cake/volume of filtrate $v = 0.0569$

In the experiment:

$$A = 10^{-4}\ \text{m}^2, \quad (-\Delta P) = 1.65 \times 10^5\ \text{N/m}^2, \quad l = 0.01\ \text{m},$$

$$\frac{dV}{dt} = 2 \times 10^{-8}\ \text{m}^3/\text{s}, \quad \mu = 10^{-3}\ \text{Ns/m}^2$$

Inserting these values in equation 7.2 gives:

$$\left(\frac{1}{10^{-4}}\right)(2 \times 10^{-8}) = \frac{1}{\mathbf{r}}\frac{(1.65 \times 10^5)}{(10^{-3})(10^{-2})}, \quad \text{and } \mathbf{r} = 8.25 \times 10^{13}\ \text{m}^{-2}$$

From equation 7.2:

$$V^2 = \frac{2A^2(-\Delta P)t}{\mathbf{r}\mu v} \qquad \text{(equation 7.11)}$$

But: $L$ = half frame thickness = $Vv/A$ (equation 7.6)

Thus:

$$L^2 = \frac{2A(-\Delta P)vt}{\mathbf{r}\mu}$$

$$= \frac{2 \times (1 \times 10^6) \times 0.0569}{(8.25 \times 10^{13})(1 \times 10^{-3})} t_f$$

$$= 1.380 \times 10^{-6}\, t_f \text{ (where } t_f \text{ in the filtration time)}$$

$$L = 1.161 \times 10^{-3} t_f^{1/2}$$

It is shown in Section 7.4.4 that if the resistance of the filter medium is neglected, the optimum cake thickness occurs when the filtration time is equal to the downtime,

Thus: $t = 900\ \text{s}, \quad t^{1/2} = 30$

$\therefore$ $L_{opt} = 34.8 \times 10^{-3}\ \text{m} = 34.8$ or $35\ \text{mm}$

and: optimum frame thickness = <u>70 mm</u>

For the washing process, if the filtration pressure is halved, the rate of washing is halved. The wash water has twice the thickness to penetrate and half the area for flow that is available to the filtrate, so that, considering these factors, the washing rate is one-eighth of the final filtration rate.

$$\text{The final filtration rate } \frac{dV}{dt} = \frac{A^2(-\Delta P)}{\mathbf{r}\mu v V}$$

$$= \frac{1 \times 10^6 A^2}{(8.25 \times 10^{13}) \times 10^{-3} \times 0.0569V} = \frac{(2.13 \times 10^{-4} A^2)}{V}$$

and: The washing rate = (final rate of filtration/8) = $2.66 \times 10^{-5} A^2/V$

The volume of wash water $= V/4$.

Hence: washing time $t_w = (V/4)/(2.66 \times 10^{-5} A^2/V)$

That is:

$$t_w = 940V^2/A^2$$

$$V^2 = L^2A^2/v^2$$

Therefore:

$$t_w = \left(\frac{L^2A^2}{(0.0569)^2}\right)\left(\frac{940}{A^2}\right) = 2.90 \times 10^5 L^2$$

The filtration time $t_f$ was shown earlier to be: $t_f = L^2/1.380 \times 10^{-6} = 7.25 \times 10^5 L^2$

Thus:

$$\text{total cycle time} = L^2(2.90 \times 10^5 + 7.25 \times 10^5) + 900$$

$$= 1.015 \times 10^6 L^2 + 900$$

The rate of cake production is then:

$$= \frac{L}{1.025 \times 10^6 L^2 + 900} = R$$

For $\mathrm{d}R/\mathrm{d}L = 0$, then: $1.025 \times 10^6 L^2 + 900 - 2.050 \times 10^6 L^2 = 0$

$$L^2 = \frac{900}{1.025 \times 10^6} \quad \text{and} \quad L = 29.6 \times 10^{-3}\ \text{m} = 29.6\ \text{mm}$$

Thus: Frame thickness $= 59.2 \approx \underline{\underline{60\ \text{mm}}}$

## 7.4.5. Pressure leaf filters

Pressure leaf filters are designed for final discharge of solids in either a dry or wet state, under totally enclosed conditions, with fully automatic operation.

Each type of pressure leaf filter features a pressure vessel in which are located one or more filter elements or leaves of circular or rectangular construction. The filter media may be in the form of a synthetic fibre or other fabrics, or metallic mesh. Supports and intermediate drainage members are in coarse mesh with all components held together by edge binding. Leaf outlets are connected individually to an outlet manifold which passes through the wall of the pressure vessel.

The material to be filtered is fed into the vessel under pressure, and separation takes place with the solids being deposited on the leaf surface, and the liquid passing through the drainage system and out of the filter. Cycle times are determined by pressure, cake capacity or batch quantity. Where particularly fine solids must be removed, a layer of precoat material may be deposited on the leaves prior to filtration, using diatomaceous earth, Perlite, or other suitable precoat materials.

Cake washing, for recovery of mother liquor or for removal of solubles, may be carried out before discharge of the solids as a slurry or a dry cake.

Pressure leaf filters are supplied in a wide range of size and materials of construction. One typical design is the "Verti-jet" unit with a vertical tank and vertical leaf filter, as shown in Figure 7.12, with rectangular leaves mounted individually but connected to a common outlet manifold. For sluice cleaning either a stationary or oscillating jet system

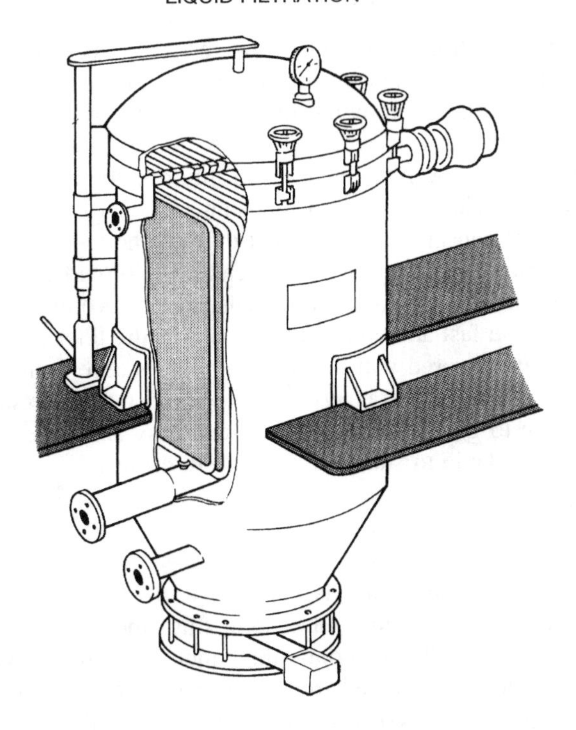

Figure 7.12. "Verti-jet" pressure leaf filter

using high efficiency spray nozzles is fitted so as to give complete cake removal. For recovery of dry solids, vibration of the leaves allows automatic discharge of the solids through a bottom discharge port provided with a quick opening door.

In the "Auto jet" design, circular leaves are mounted on a horizontal shaft which serves as the filtrate outlet manifold. The leaves are rotated during the cleaning cycle although, in addition, extra low speed continuous rotation during operation ensures uniform cake build-up in difficult applications. The leaves are of metallic or plastics construction covered with fabric or wire cloth for direct or precoat operation, and rotation of the leaves during cleaning promotes fast efficient sluice discharge with minimum power consumption. As an alternative, the leaves may be rotated over knife blades which remove the cake in a dry state. Units of this type are used for handling foodstuffs and also for the processing of minerals and effluents.

For the handling of edible oils, molten sulphur, effluents and foodstuffs, a Filtra-Matic unit is used in which either the bundle is retracted from the shell as a unit, or the filter tank is retracted leaving the filter leaves and filter cover in position. Such units are available in cylindrical, conical or trough shell configurations, and cleaning may be either wet or dry, manual or automatic. In the latter case, for dry discharge, vibration systems are used and for wet removal spray jets mounted in an overhead manifold sweep the entire leaf surface in an oscillating motion. In this design, the heavy duty leaves covered with cloth or screen are all interchangeable and, whether round or rectangular, are all the same size to give uniform precoat, cake build-up and filtration. In horizontal tray pressure filters, used in batch processes and intermittent flows, the trays are mounted horizontally with connections to a vertical filtrate manifold at the rear, and such units are ideally suited where cake washing and positive cake drying are required. In many cases, the accumulated cake may be sluiced off without removing trays from the filter and a special recovery leaf is provided where heel filtration is required in which a thin layer of cake is left semi-permanently in contact with the filter medium to improve the clarity of the filtrate. This system is used in various clarification processes and is ideal for handling high flows of liquids with a low solids concentration. In most designs the tubes are mounted vertically from a tube sheet at the top of the tank and cleaning is provided with a self-contained internal "air-pump" backwash, thus avoiding the use of large volumes of sluicing liquid or separate pumps to provide fast and complete removal of the filter cake. The heavy gauge perforated tube cores are covered with a seamless cloth sleeve sealed at either end by a clamping device. As an alternative, heavy gauge wire is wound around the centre core, with controlled spacing to give reliable filtration and easy cake release. Tubular element units of this type are available in standard sizes up to 40 $m^2$.

### Cartridge filters

One particular design of pressure filter is the *filter cartridge*, typified by the Metafilter which employs a filter bed deposited on a base of rings mounted on a fluted rod, and is extensively used for clarifying liquids containing small quantities of very fine suspended solids. The rings are accurately pressed from sheet metal of very uniform thickness and are made in a large number of corrosion-resistant metals, though stainless steels are usually employed. The standard rings are 22 mm in external diameter, 16 mm in internal diameter and 0.8 mm thick, and are scalloped on one side, as shown in Figure 7.13, so that the edges of the discs are separated by a distance of 0.025–0.25 mm according to

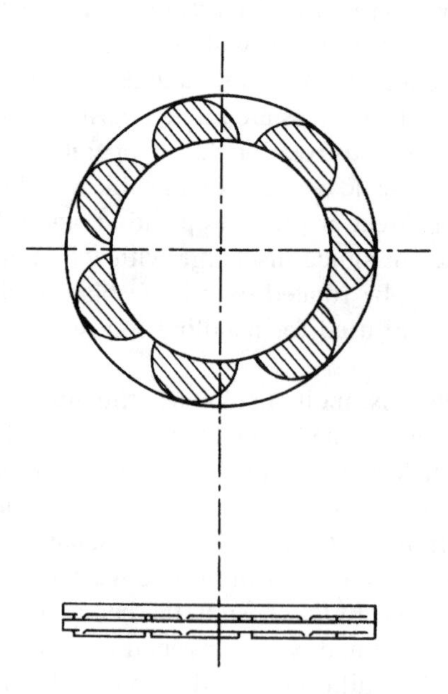

Figure 7.13. Rings for metafilter (Stella Meta)

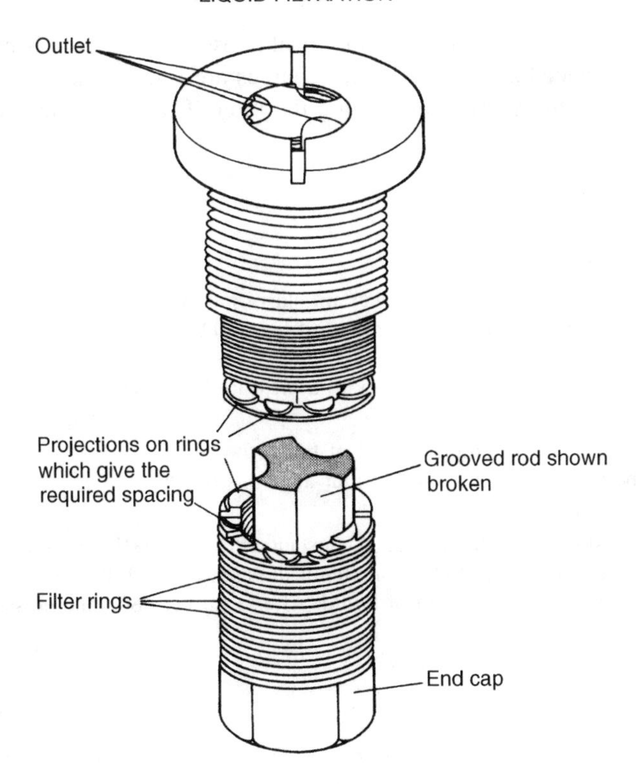


Figure 7.14. Metafilter pack (Stella Meta)

requirements. The pack is formed by mounting the rings, all the same way up, on the drainage rod and tightening them together by a nut at one end against a boss at the other as shown in Figure 7.14. The packs are mounted in the body of the filter which operates under either positive or reduced pressure.

The bed is formed by feeding a dilute suspension of material, to the filter usually a form of kieselguhr, which is strained by the packs to form a bed about 3 mm thick. Kieselguhr is available in a number of grades and forms a bed of loose structure which is capable of trapping particles much smaller than the channels. During filtration, the solids build up mainly on the surface and do not generally penetrate more than 0.5 mm into the bed. The filtrate passes between the discs and leaves through the fluted drainage rod, and operation is continued until the resistance becomes too high. The filter is then cleaned by back-flushing, which causes the filter cake to crack and peel away. In some cases the cleaning may be incomplete as a result of channelling. If for any reason the spaces between the rings become blocked, the rings may be quickly removed and washed.

The Metafilter is widely used for filtering domestic water, beer, organic solvents and oils. The filtration characteristics of clay-like materials can often be improved by the continuous introduction of a small quantity of filter aid to the slurry as it enters the filter. On the other hand, when the suspended solid is relatively coarse, the Metafilter will operate successfully as a strainer, without the use of a filter bed.

The Metafilter is very robust and is economical in use because there is no filter cloth and the bed is easily replaced and hence labour charges are low. Mono pumps or diaphragm pumps are most commonly used for feeding the filter. These are discussed in Volume 1, Chapter 8.

## Example 7.3

The relation between flow and head for a certain slurry pump may be represented approximately by a straight line, the maximum flow at zero head being 0.0015 $m^2/s$ and the maximum head at zero flow 760 m of liquid.

Using this pump to feed a particular slurry to a pressure leaf filter:

(a) How long will it take to produce 1 $m^3$ of filtrate?
(b) What will be the pressure across the filter after this time?

A sample of the slurry was filtered at a constant rate of 0.00015 $m^3/s$ through a leaf filter covered with a similar filter cloth but of one-tenth the area of the full-scale unit, and after 625 s the pressure across the filter was 360 m of liquid. After a further 480 s the pressure was 600 m of liquid.

## Solution

For constant rate filtration through the filter leaf:

$$V^2 + \frac{LA}{v}V = \frac{A^2(-\Delta P)t}{\mathbf{r}\mu v} \qquad \text{(equation 7.17)}$$

At a constant rate of 0.00015 $m^3/s$, then, when the time = 625 s:

$$V = 0.094 \text{ m}^3,\ (-\Delta P) = 3530 \text{ kN/m}^2$$

and, at $t = 1105$ s: $$V = 0.166 \text{ m}^3 \text{ and } (-\Delta P) = 5890 \text{ kN/m}^2$$

Substituting these values into equation 7.17 gives:

$$(0.094)^2 + LA/v \times 0.094 = (A^2/\mathbf{r}\mu v) \times 3530 \times 625$$

or: $$0.0088 + 0.094LA/v = 2.21 \times 10^6 A^2/\mathbf{r}\mu v \qquad \text{(i)}$$

and: $$(0.166)^2 + LA/v \times 0.166 = (A^2/\mathbf{r}\mu v) \times 5890 \times 1105$$

or: $$0.0276 + 0.166LA/v = 6.51 \times 10^6 A^2/\mathbf{r}\mu v \qquad \text{(ii)}$$

Equations (i) and (ii) may be solved simultaneously to give:

$$LA/v = 0.0154 \text{ and } A^2/\mathbf{r}\mu v = 4.64 \times 10^{-9}$$

As the full-size plant is 10 times that of the leaf filter, then:

$$LA/v = 0.154 \text{ and } A^2/\mathbf{r}\mu v = 4.64 \times 10^{-7}$$

If the pump develops 760 m (7460 kN/m$^2$) at zero flow and has zero head at $Q = 0.0015$ m$^3$/s, its performance may be expressed as:

$$-\Delta P = 7460 - (7460/0.0015)Q$$

or:

$$-\Delta P = 7460 - 4.97 \times 10^6 Q \text{ (kN/m}^2\text{)}$$

$$\frac{\mathrm{d}V}{\mathrm{d}t} = \frac{A^2(-\Delta P)}{\mathbf{r}\mu v(V + LA/v)} \qquad \text{(equation 7.16)}$$

Substituting for $(-\Delta P)$ and the filtration constants gives:

$$\frac{\mathrm{d}V}{\mathrm{d}t} = \frac{A^2}{\mathbf{r}\mu v}\frac{(7460 - 4.97 \times 10^6 \mathrm{d}V/\mathrm{d}t)}{(V + 0.154)}$$

Since $Q = \mathrm{d}V/\mathrm{d}t$, then:

$$\frac{\mathrm{d}V}{\mathrm{d}t} = \frac{4.67 \times 10^{-7}[7460 - 4.97 \times 10^6(\mathrm{d}V/\mathrm{d}t)]}{(V + 0.154)}$$

$$(V + 0.154)\mathrm{d}V = 3.46 \times 10^{-3} - 2.31\mathrm{d}V/\mathrm{d}t$$

The time to collect 1m$^3$ is then given by:

$$\int_0^1 (V + 0.154 + 2.31)\mathrm{d}V = \int_0^t (3.46 \times 10^{-3})\mathrm{d}t$$

and:

$$\underline{\underline{t = 857 \text{ s}}}$$

The pressure at this time is found by substituting in equation 7.17 with $V = 1$ m$^3$ and $t = 857$ s to give:

$$1^2 + 0.154 \times 1 = 4.64 \times 10^{-7} \times 857(-\Delta P)$$

and:

$$-\Delta P = \underline{\underline{2902 \text{ kN/m}^2}}$$

### 7.4.6. Vacuum filters

If in a horizontal filter both sides are open to atmosphere then, as slurry is introduced above the filter medium, filtration will occur at an adequate rate as long as the hydrostatic head of slurry above the filter medium is sufficient. The unit then operates as a gravity filter. If, however, the top of the vessel is enclosed and the slurry is introduced under pressure, then the unit operates as a pressure filter and the driving force is the amount by which the applied pressure exceeds atmospheric pressure below the filter medium. Greatly enhanced filtration rates may then be achieved. Similarly, if the pressure beneath the filter is reduced below atmospheric, the unit operates as a vacuum filter and the driving force may again be appreciably greater than that available with a gravity filter. The maximum theoretical driving force is then 101 kN/m$^2$ less the vapour pressure of the filtrate, and much higher flowrates may be obtained than with gravity filtration. With, for example, a head of liquor of 0.3 m in a gravity filter, the maximum driving force is about 3 kN/m$^2$, whereas in theory the driving force of a vacuum filter could be some thirty times greater.

Vacuum filtration has many advantages, not the least being the fact that the feed slurry, often containing abrasive solids in a corrosive liquid, can be delivered by gravity flow or by low pressure pumps which need only to overcome the resistance in the feed pipework. In addition, vacuum filtration equipment does not have to withstand high pressures and can therefore be manufactured in a wider range of materials. When, because of corrosive conditions or the need to prevent product contamination, expensive materials of construction are required, the low pressures involved allow considerable economy in manufacturing costs. A further advantage is that, after the bulk of the mother liquor has been filtered from the solids, the cake is accessible to mechanical dewatering, flood washing and sampling, and cake removal can be carried out quickly as it is easily automated. Because of the ease of solids removal, a vacuum filter may be operated with very short cycle times of around 60 s, and the resulting thin filter cakes give very high flowrates per unit area, typically six times as great as those obtained with a pressure filter operating at the same differential pressure. Where delicate solids are handled, very low differential pressures may be used and, since the feed and the filtrate are not handled under positive pressure, there is no possibility of liquor leakage from the system as might be the case with high pressure differentials. This is an important advantage especially where corrosive or hazardous materials are involved.

### *Batch type vacuum filters*

Batch vacuum filters were developed from gravity filters and, in essence, Buchner funnels as used in laboratories and Nutsche type filters as used in industry are similar to gravity filters, except that they feature a vacuum pump or some other vacuum generating equipment to reduce the pressure under the filter medium, thereby increasing the driving force across the filter medium.

Simple filters operating in this manner are often referred to as pan filters, and they incorporate certain features which are now being incorporated into continuously operating machines. These filters enable the initial cake formation to be made, either under gravity or at very low differential pressures, so that uniform distribution of the solids over the filter medium is obtained. Preferential settling occurs so that the heavier, and usually coarser, solids contact the filter medium first, thereby protecting it from the slower settling fine solids. Following this, the filtrate drainage section may be subjected to higher vacuum to give greater differential pressures, higher flowrates per unit area and greater total flows. A typical cycle for a batch type vacuum filter might be:

(a) Feeding of slurry,
(b) Cake formation and filtration, initially under gravity if required, followed by vacuum operation,
(c) Initial drying for maximum removal of mother liquor,
(d) Cake washing and drying,
(e) Manual discharge of solids,
(f) Cloth washing.

The advantages of such units include simplicity in both design and operation, low cost per unit area and flexibility with regard to filtration, washing and drying periods. In addition, very sharp separation of the mother liquor and various wash liquors can be obtained due to the wash liquors being completely contained within the pan. The filter cakes remain in position between batches without the necessity for power to be used for their retention. In addition, the flow through the filter medium can be stopped or be restricted to a very low rate in order to obtain leaching from the solids in the filter bed.

Against this, it may be necessary with a batch vacuum filter to provide holding vessels for feed, and possibly filtrate and wash. The discharge of solids is intermittent and this may lead to complications in feeding the solids to a continuous dryer or to some other continuous process, and cake hoppers and a feeding arrangement may have to be provided.

Since the liquor flowrate decreases as filtration proceeds, facilities to cope with this must be incorporated in the feed and filtrate handling sections of the system and the discharge of solids must be a manual operation if the solids cannot be sluiced off.

Equipment has been developed to overcome the disadvantages of manual discharge of solids from the simple batch pan type vacuum filter, and such equipment incorporates scrapers or rakes for the removal of the solids. Other pan type vacuum filters are arranged so that the pan is tipped over, when the solids are discharged, either under their own weight or sometimes with the assistance of blow-back air admitted to the underside of the filter cloth when the pan is inverted.

Further developments eliminate, or reduce to acceptable proportions, the cyclic variations noted previously and one simple development which achieves this to a limited extent involves a number of individual tipping pans arranged in parallel. For example, with three pans discharging into a common slurry tank, the operation is arranged so that whilst one pan is being charged with feed material and, whilst solids–mother liquor separation is being carried out in it, the second pan is at the washing stage and third pan is at the drying and discharging stage. This arrangement obviously reduces cyclic variations though it necessitates diversion of the feed and wash liquor from one pan to another. The operation can, however, be carried out automatically by the use of a process timer or similar control unit.

### *Classification of batch and continuous vacuum filters*

The disadvantages of cyclic operation, attributable to the need to remove the cake from batch filters, have been largely overcome by the development of a wide range of continuous vacuum filters as summarised in Table 7.2 which includes data on the selection for a given duty. This is further detailed in Table 7.3 and is based on the work of G. H. Duffield of Stockdale Engineering and E. Davies of EDACS Ltd.

### *Horizontal continuous filters*

One way of limiting still further cyclic variations is to use vacuum filters of the horizontal continuous type, and many designs have been developed which in the main consist of pans arranged in line or in a circle so that each in turn goes through the same cycle of operations, namely: feeding, dewatering, washing, drying, automatic solids discharge and cloth washing. The pans move past the stationary feed and wash areas where the feed and wash liquors are continuously introduced to the filter. Alternatively, a plain filter surface may be arranged as a linear belt or as a circular table, in place of pans.

All filters of this type have the distinct advantage of continuous operation and, as a result, the feed and wash liquors may be fed to the equipment at steady rates. Dense, quick settling solids can be handled, and the cake thickness — and the washing and drying times — can be varied independently between quite wide limits. The cake can be flooded with wash liquor, which can be taken away separately and re-used to obtain countercurrent

Table 7.2. Classification of vacuum filters

| Type of vacuum filter | Note | Usual max. area ($m^2$) | Characteristics of suitable slurry (See notes below) | | | | | Performance index (See notes below) | | |
|---|---|---|---|---|---|---|---|---|---|---|
| | | | A | B | C | D | E | Cake dryness | Cake washing | Filtrate clarity |
| (A) *Cake filters* | | | | | | | | | | |
| Single tipping pan | 1 | 3 | X | X | | | | 4–7 | 8–9 | 8 |
| Multi tipping pan | 2 | 12 | X | X | | | | 4–7 | 8–9 | 8 |
| Horizontal linear tipping pan | 3 | 25 | X | X | | | | 4–7 | 8–9 | 7 |
| Horizontal linear belt | 4 | 20 | X | X | | | | 5–8 | 7–8 | 6 |
| Rotary tipping pan | 3 | 200 | X | X | | | | 4–7 | 8–9 | 7 |
| Horizontal rotary table | 4 | 15 | X | X | | | | 4–7 | 7–8 | 7 |
| Rotary drum—string discharge | 5 | 80 | | X | X | | | 5–8 | 6 | 8 |
| Rotary drum—knife discharge | 5 | 80 | | X | X | | | 5–8 | 6 | 8 |
| Rotary drum—roller discharge | 5 | 80 | | | X | X | | 5–6 | 5 | 8 |
| Rotary drum—belt discharge | 5 | 80 | | | X | X | X | 5–8 | 6 | 7 |
| Top feed drum | 6 | 10 | X | X | | | | 1–2 | 1 | 6 |
| Hopper/dewaterer | 6 | 10 | X | | | | | 1–2 | 1 | 6 |
| Internal drum | 7 | 12 | | X | X | | | 3–4 | 1 | 6 |
| Single comp. drum | 8 | 15 | | | X | | | 2–3 | 4 | 5 |
| Disc | 9 | 300 | | X | X | | | 2–3 | 1 | 6 |
| (B) *Media filters* | | | | | | | | | | |
| Rotary drum precoat | 10 | 80 | | | | X | X | — | 6 | 9 |

NOTES

*Filters*

1. For small batch production. Has very wide application, is very adaptable and can be automated.
2. Usually 2 to 4 pans, for medium size batch production. Very wide application, very adaptable, can be automated.
3. For free-draining materials where very good washing is required with sharp separation between mother liquor and wash liquors.
4. For free-draining materials where very good washing is required.
5. Wide range of types and size available. Generally suitable for most slurries in categories B and C. Can usually be fitted with various mechanical devices to improve the washing and drying.
6. Restricted to very free-draining materials not requiring washing.
7. Restricted to very free-draining materials not requiring washing, but where the solids can be retained by vacuum alone.
8. Allows use of high drum speed and is capable of very high flow rates.
9. Large throughputs for small floor space.
10. Suitable for almost any clarification and for handling materials which blind normal filter media.

*Slurries*

A. High solids concentration, normally greater than 20 per cent, having solids which are free-draining and fast settling, giving difficulty in mechanical agitation and giving high filtration rates.
B. Rapid cake formation with reasonably fast settling solids which can be kept in suspension by mechanical agitation.
C. Lower solids concentration with solids giving slow cake formation and thin filter cakes which can be difficult to discharge.
D. Low solids concentration with solids giving slow cake formation and a filter cake having very poor mechanical strength.
E. Very low solids concentration (i.e. clarification duty), or containing solids which blind normal filter media. Filtrate usually required.
X. Slurries handled.

*Performance index*

9 = the highest possible performance.
1 = very poor or negligible performance.

Table 7.3. Operating data for some vacuum filter applications

| Application | Type of vacuum filter frequently used (see note 2) | Solids content of feed (per cent by mass) | Solids handling rate kg dry solids/s $m^2$ filter surface (see note 3) | Moisture content of cake (per cent by mass) | Air flow | |
|---|---|---|---|---|---|---|
| | | | | | Filter surface (see note 4) ($m^3/s\ m^2$) | Vacuum ($kN/m^2$ pressure, below atmospheric) |
| *Chemicals* | | | | | | |
| Aluminium hydrate | Top feed drum | 25–50 | 0.12–0.2 | 10–17 | 0.025 | 84 |
| Barium nitrate | ” | 80 | 0.34 | 5 | 0.13 | 67 |
| Barium sulphate | Drum | 40 | 0.013 | 30 | 0.005 | 33 |
| Sodium bicarbonate | ” | 50 | 0.5 | 12 | 0.15 | 61 |
| Calcium carbonate | ” | 50 | 0.03 | 22 | 0.010 | 33 |
| Calcium (pptd) | ” | 30 | 0.04 | 40 | 0.010 | 26 |
| Calcium sulphate | Tipping pan | 35 | 0.16 | 30 | 0.025 | 41 |
| Caustic lime mud | Drum | 30 | 0.20 | 50 | 0.03 | 51 |
| Sodium hypochlorite | Belt disch. drum | 12 | 0.04 | 30 | 0.015 | 33 |
| Titanium dioxide | Drum | 30 | 0.03 | 40 | 0.010 | 33 |
| Zinc stearate | ” | 5 | 0.007 | 65 | 0.015 | 33 |
| *Minerals* | | | | | | |
| Frothed coal (coarse) | Top feed drum | 30 | 0.20 | 18 | 0.020 | 61 |
| Frothed coal (fine) | Drum or Disc | 35 | 0.10 | 22 | 0.015 | 51 |
| Frothed coal tailings | Drum | 40 | 0.05 | 30 | 0.010 | 26 |
| Copper concentrates | ” | 50 | 0.07 | 10 | 0.010 | 30 |
| Lead concentrates | ” | 70 | 0.27 | 12 | 0.015 | 26 |
| Zinc concentrates | ” | 70 | 0.20 | 10 | 0.010 | 33 |
| Flue dust (blast furnace) | ” | 40 | 0.04 | 20 | 0.015 | 33 |
| Fluorspar | ” | 50 | 0.27 | 12 | 0.025 | 51 |

Table 7.3. *(continued)*

| Application | Type of vacuum filter frequently used (see note 2) | Solids content of feed (per cent by mass) | Solids handling rate kg dry solids/s $m^2$ filter surface (see note 3) | Moisture content of cake (per cent by mass) | Air flow | |
|---|---|---|---|---|---|---|
| | | | | | Filter surface (see note 4) ($m^3/s\ m^2$) | Vacuum ($kN/m^2$ pressure, below atmospheric) |
| *Paper* | | | | | | |
| Kraft pulp | Deep sub. drum or Disc | 1.5 | 0.05 | 90 | Barometric leg | |
| Effluents | Drum | 3 | 0.005 | 80 | 0.015 | 41 |
| Bleach washer | ” | 1.0 | 0.20 (filtrate) | 90 | 0.025 | 67 |
| *Foodstuffs* | | | | | | |
| Starch | Drum | 50 | 0.05 | 40 | 0.015 | 16 |
| Gluten | Belt discharge drum | 15 | 0.007 | 48 | 0.010 | 33 |
| Salt | Top feed drum | 30 | 0.5 | 6 | 0.20 | 91 |
| Sugar cane mud | Drum | 10 | 0.03 | 30 | 0.005 | 33 |
| Glucose (spent carbon) | ” | 15 | 0.02 | 50 | 0.0025 | 41 |
| Glucose (44%) | Drum—Precoat | 2 | 0.14 (filtrate) | 50 | 0.0025 | 33 |
| *Effluents* | | | | | | |
| Primary sewage | Drum | 8 | 0.01 | 65 | 0.008 | 33 |
| Primary sewage, digested | ” | 6 | 0.005 | 65 | 0.008 | 33 |
| Neutralised $H_2SO_4$ pickle | ” | 8 | 0.013 | 50 | 0.010 | 41 |
| Plating shop effluent | Drum—Precoat | 2 | 0.10 (filtrate) | 80 | 0.013 | 33 |

NOTES

1. The information given should only be used as a general guide, for slight differences in the nature, size range and concentration of solids, and in the nature and temperature of liquor in which they are suspended can significantly affect the performance of any filter.
2. It should not be assumed the type of filter stated is the only suitable unit for each application. Other types may be suitable, and the ultimate selection will normally be a compromise based on consideration of many factors regarding the process and the design features of the filter.
3. The handling rate (in $kg/m^2s$) generally refers to dry solids except where specifically referred to as filtrate.
4. The air volumes stated are measured at the operating vacuum (i.e. they refer to attenuated air).

washing of the cake if desired. Disadvantages of this equipment are the high cost per unit area, mechanical complexity, and the fact that some effectively use only 50 per cent of the total filtration area. With the horizontal linear pan and the rotary tipping pan filters, a truly continuous discharge of solids is not obtained since these units operate as short-time-cycle batch-discharge machines.

The ***horizontal pan filter*** incorporates a number of open top pans which move in a horizontal plane through the feed, filtration, washing and drying. At the discharge point the pans pass over a roller and then return to the feed end of the machine in the inverted position. The design of this type of filter varies greatly with different manufacturers, though one frequently used type has the walls of the pans built up on an open type of horizontal linear conveyor. The conveyor, within the walls of the pans, serves as the support for the filter media and for the liquor drainage system, and elsewhere the conveyor serves as the mechanical linkage between adjacent pans. The conveyor passes over vacuum boxes in the feed, filtration, washing and drying zones.

The ***horizontal rotary tipping pan filter*** has the pans located in a circle moving in a horizontal plane and, in effect, the pans are supported between the spokes of a wheel turning in a vertical shaft. Each pan is connected to a single, centrally-mounted rotary valve which controls the vacuum applied to the feed, filtration, washing and drying zones and from which the mother liquor and the various wash liquors can be diverted to individual vessels. At the discharge position, the pans are inverted through 180° and, in some designs, cake removal is assisted by the application of compressed air to the underside of the filter medium via the rotary valve. One example of this type of filter is the Prayon continuous filter, developed particularly for the vacuum filtration of the highly corrosive liquids formed in the production of phosphoric acid and phosphates. It is made in five sizes providing filtering surfaces from about 2–40 $m^2$ and it consists of a number of horizontal cells held in a rotating frame, and each cell passes in turn under a slurry feed pipe and a series of wash liquor inlets. Connections are arranged so as to provide three countercurrent washes. The timing of the cycle is regulated to give the required times for filtration, washing and cake drying. The discharge of the cake is by gravity as the cell is automatically turned upside-down, and can be facilitated if necessary by washing or by the application of back-pressure. The cloth is cleaned by spraying with wash liquid and is allowed to drain before it reaches the feed point again. The filter has a number of advantages including:

(a) A large filtration area, up to 85 per cent of which is effective at any instant.
(b) Low operating and maintenance costs and simple cloth replacement.
(c) Clean discharge of solids.
(d) Fabrication in a wide range of corrosion-resistant materials.
(e) Countercurrent washing.
(f) Operation at a pressure down to 15 $kN/m^2$.

The ***horizontal rotary table filter*** incorporates, as its name suggests, a horizontal rotary circular table divided into segments or drainage compartments, each of which is "ported" to a rotary valve underneath the table. The top surface of the table is fitted with a filter medium and walls are provided around the periphery of the table and around the hub. The feed slurry is distributed across the table through a feed box and vacuum is applied to the segments in the feed position to extract the mother liquor. On rotation of the table, the filter cake then passes through the washing and drying zones. At the discharge position the vacuum on the segments can be broken to atmospheric pressure and the filter cake removed by a horizontal rotary scroll which lifts the solids over the periphery wall of the table. The scroll cannot be in contact with the filter medium and therefore a "heel" of solids remains on the table after passing the discharge point. This heel can either be disturbed by blow-back air and re-slurried with the incoming feed, or, in certain applications, the heel can be retained intact on the filter medium in order to obtain better retention of solids from the incoming feed than would be achieved with a clean filter cloth.

### *Horizontal belt or band filters*

In the horizontal belt or band filter, shown in Figure 7.15, an endless belt arranged in the horizontal plane and running over pulleys at about 0.05 m/s at the feed and discharge ends (similar to the horizontal band conveyor). The principle is applied in the Landskröna band filter, as described by PARRISH and OGILVIE[29]. In the filtration, washing and drying zones, where the filter medium passes over vacuum boxes, it may be supported on a separate endless belt which also serves as the drainage member and as the valve. This belt can be provided with side walls to contain the feed slurry and wash liquors, or a flat belt can be used in conjunction with rigid static walls, against which the belt slides. Rubber, or similar, wiper blades which drag against the cake surface can be used to isolate the filtration and washing zones from each other. In some designs the belts move continuously, in others the belts are moved along in stages.

Figure 7.15. Rigid belt filter

The applications of horizontal belt filters are discussed by BLENDULF and BOND[30]. A typical filter for dewatering concentrates is shown in Figure 7.16, and it may be noted that this type of equipment is the most expensive per unit area.

Figure 7.16. Delkor horizontal belt filter dewatering copper flotation concentrate (filter area 21 $m^2$)

An interesting development is the ADPEC filter described by BOSLEY[31] which has an intermittently moving belt. The slurry is fed to one end and the vacuum applied to the underside boxes and filtering occurs with the belt stationary. As the discharge roll at the other end moves inwards it trips the system so that the vacuum is relieved and the belt moved forward, thus avoiding the problem of pulling the belt continuously over the sections under vacuum.

Band filters have several advantages over rotary vacuum filters which are described in the following section. These include:

(a) Some gravitational drainage of liquid occurs because the cake and the belt are vertically separated at all stages of the filtration operation. In addition, a vacuum is not needed to keep the cake in place and this simply supplements the effect of gravity on the flow of filtrate.
(b) Because the feed is from the top, it is not necessary to agitate the suspension in the feed trough.

(c) Large particles in the slurry tend to reach the surface of the filter cloth first and to form a layer which therefore protects the cloth from the blinding effect of the fines, although this does lead to the formation of a non-uniform cake.
(d) Washing is more effective because of the improved structure of the filter cake.
(e) The fitting of an impervious cover over the cake towards its exit will reduce the loss of vacuum due to air flow through cracks in the cake.

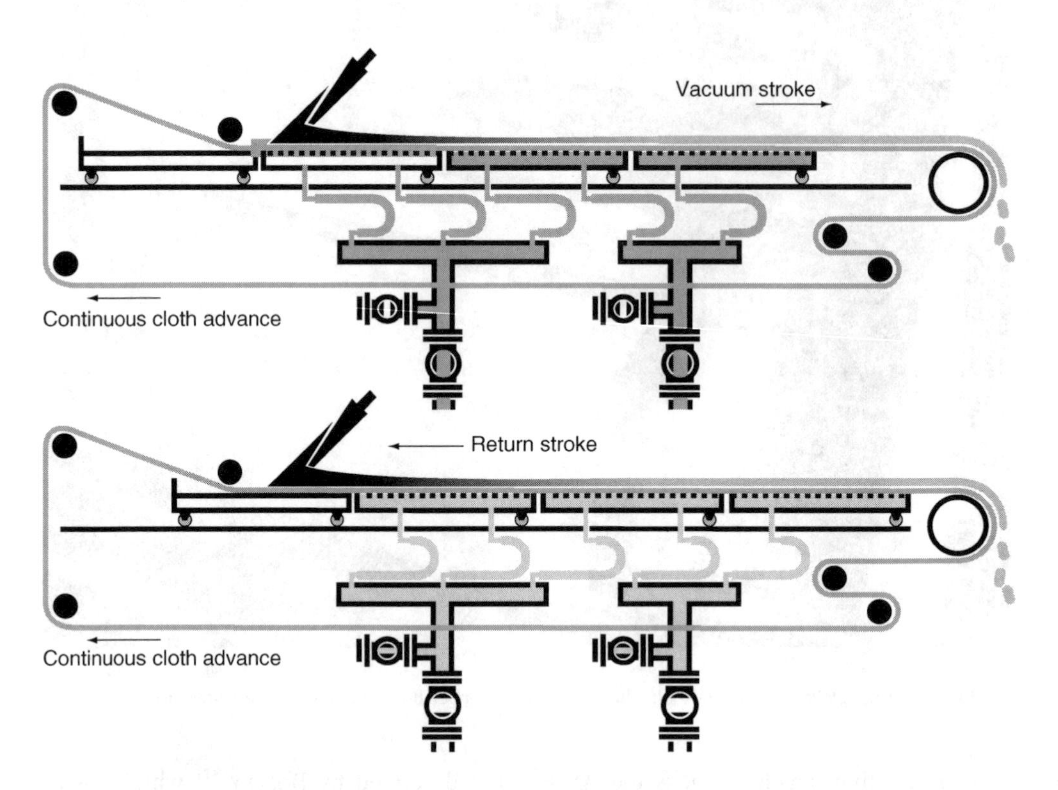


Figure 7.17. Principle of the Pannevis vacuum belt filter

Another development by Pannevis in Holland, shown in Figure 7.17, incorporates vacuum trays which support the filtercloth as it moves at a constant speed which may be varied as necessary. The vacuum tray is evacuated, filtration takes place and the tray is pulled forward by the cloth in such away that cloth and tray move at the same velocity thus offering negligible sealing problems, minimum wear, low driving power requirements and the possibility of high vacuum operation. At the end of the vacuum stroke, the vacuum is released, the tray is vented and pulled back quickly by pneumatic action to its starting position. Although an intermittent vacuum is required, the Pannevis design is in effect a continuous unit since the suspension feed, wash liquor feed and cake discharge are all fully continuous. A vacuum tray is typically 1.4 m long and 150 mm deep and the width and length of the belt are 0.25–3.0 m and 2.8–25 m, respectively.

### *Rotary drum filters*

Because of its versatility and simplicity, one of the most widely used vacuum filters is the *rotary drum filter* and a filter of this type was patented in England in 1872 by William and James Hart. The basic design varies with different manufacturers, although essentially all drum type vacuum filters may be divided into two categories:

(a) Those where vacuum is created within compartments formed on the periphery of the drum, and
(b) Those where vacuum is applied to the whole of the interior of the drum.

The most frequently used continuous drum type filters fall into the first category. These give maximum versatility, low cost per unit area, and also allow a wide variation of the respective time periods devoted to filtration, washing and drying.

Essentially, a multi-compartment drum type vacuum filter consists of a drum rotating about a horizontal axis, arranged so that the drum is partially submerged in the trough into which the material to be filtered is fed. The periphery of the drum is divided into compartments, each of which is provided with a number of drain lines. These pass through the inside of the drum and terminate as a ring of ports covered by a rotary valve, through which vacuum is applied. The surface of the drum is covered with a filter fabric, and the drum is arranged to rotate at low speed, usually in the range 0.0016–0.004 Hz (0.1–0.25 rpm) or up to 0.05 Hz (3 rpm) for very free filtering materials.

As the drum rotates, each compartment undergoes the same cycle of operations and the duration of each of these is determined by the drum speed, the submergence of the drum and the arrangement of the valve. The normal cycle of operations consists of filtration, drying and discharge. It is also possible, however, to introduce other operations into the basic cycle, including:

(a) Separation of initial dirty filtrate — which may be an advantage if a relatively open filter fabric is used.
(b) Washing of the filter cake.
(c) Mechanical dewatering of the filter cake.
(d) Cloth cleaning.

Figure 7.18*a* shows a typical layout of a rotary drum installation and Figure 7.18*b* shows the sequence of cake formation, washing and dewatering. A large rotary drum vacuum filter is shown in Figure 7.19.

In order to achieve consistent performance of a continuous filter, it is necessary to maintain the filter medium in a clean condition. With a drum type vacuum filter this requires the complete and continuous removal of the filter cake from the drum surface, and the operating conditions are often influenced by the need to form a fully dischargeable cake. Again, in order to achieve high capacity and good cake washing and/or drying, it is very often desirable to operate with very thin cakes. Therefore, the cake discharge system of most drum type vacuum filters must be arranged so as to ensure the complete and continuous removal of extremely thin filter cakes. The most effective way of achieving this is determined to a large extent by the physical nature of the solids being handled.

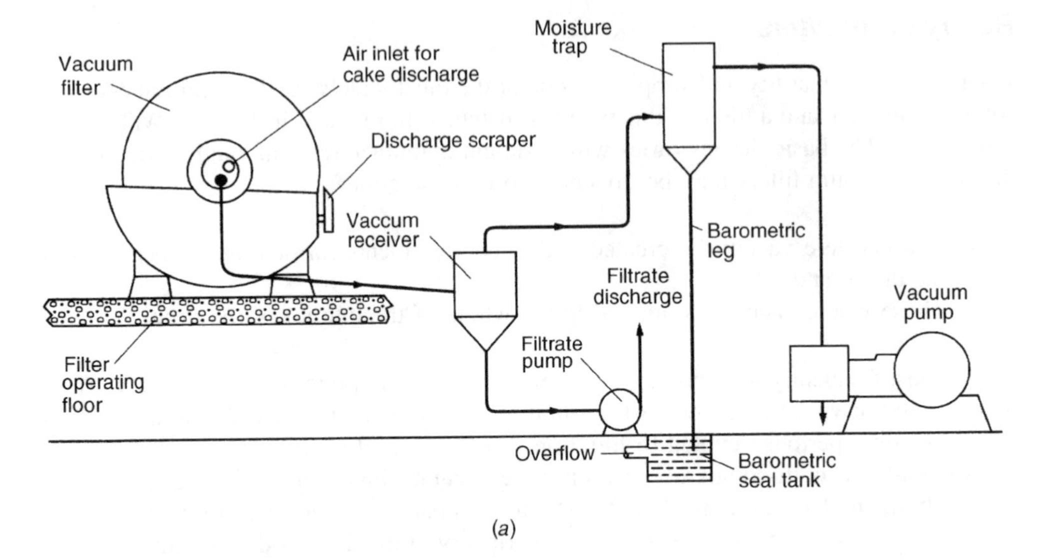


(*a*)

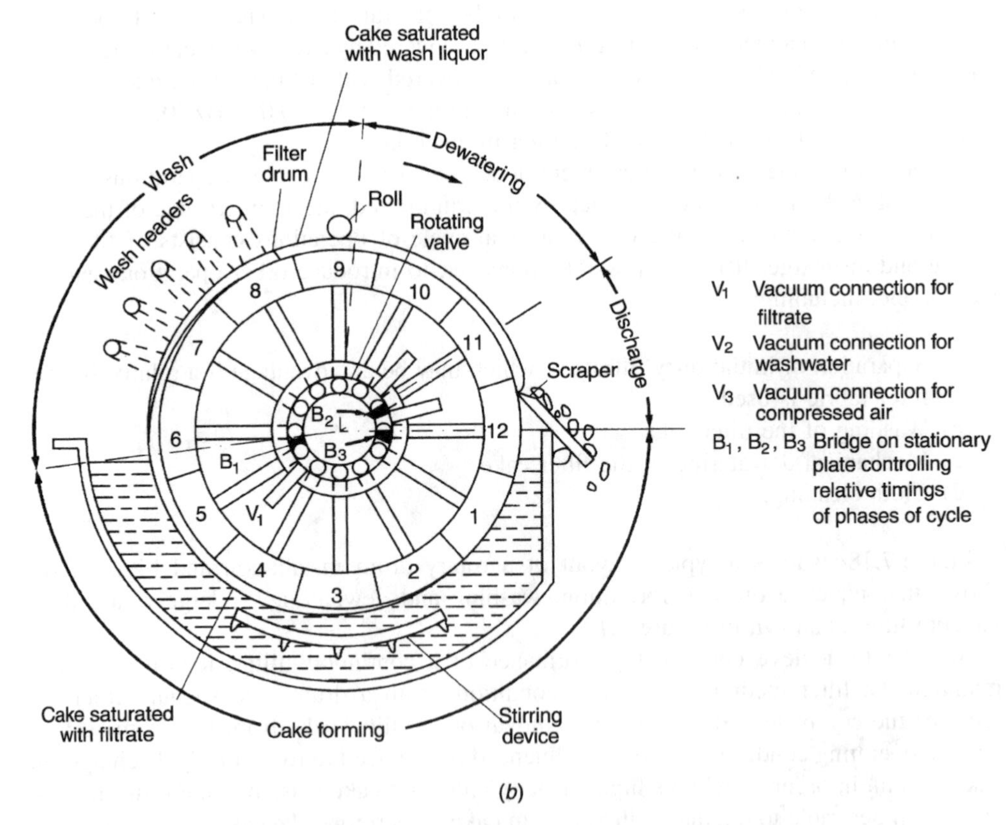


(*b*)

Figure 7.18. Typical layout of rotary drum filter installation

Figure 7.19. Rotary vacuum drum filter used in a zinc leaching operation

The various discharge systems that are suitable for drum type filters include the string discharge technique which is effective for an extremely wide range of materials. Essentially this involves forming the filter cake on an open type of conveyor which is in contact with the filter medium in the filtration, washing and drying zones. Consequently, the solids which are trapped by the filter medium form a cake on top of the "open" conveyor. From the discharge point on the drum the conveyor transports the filter cake to a discharge roll, at which point the cake is dislodged. The conveyor then passes through an aligning mechanism and over a return roll which guides it back into contact with the filter drum at the commencement of the cycle and just above the level of feed liquor in the filter trough.

In the string discharge system, the conveyor consists of a number of endless strings which are spaced at a pitch of approximately 12 mm over the width of the filter drum. The string spacing may, however, be in the range 6–25 mm depending on the mechanical properties of the solids.

The advantages of the string discharge system are:

(a) Thin and sticky filter cakes down to about 1.5 mm of materials such as clay may be effectively discharged.
(b) The filter cloth is almost free of mechanical wear and tear so that thin and delicate cloths may be used, and these can be selected almost solely for their filtration properties. Such cloths are usually less prone to plugging than the stronger and thicker cloths required for other discharge systems.
(c) The cloth can be attached to the drum in a simple manner so that fitting and subsequent replacement can be carried out quickly. Normally, the cloth is loosely

wrapped around the drum and it is secured to the drum at the edges and, once across the drum, at the overlap, by a simple caulking system. The use of wire winding, clamping bars, and the necessity of securing the cloth at every panel which is essential with other discharge systems, is avoided.

(d) The use of compressed air, to loosen cake from the drum surface, is avoided and consequently there is no possibility of blowing back into the filter cake moisture which has previously been removed under vacuum. This is a possibility with knife discharge filters operating with blow-back.

(e) If required, the path of the discharge strings can be altered so that the filter cake is conveyed by the strings to a convenient point for feeding a continuous dryer-extruder, or other processing equipment.

A typical string discharge mechanism is shown in Figure 7.20.

Figure 7.20. String discharge mechanism on a filter handling silica gel

The knife discharge system incorporates a knife which is arranged so that the surface of the drum runs on or near to the knife edge. The cake is dislodged from the cloth either by its own weight, with thick and heavy cakes, or by applying compressed air to the underside of the filter cloth. The blow-back air can either be admitted at low pressure for a long period, or at high pressure and instantaneously by means of a mechanical "blow-off timer". With suitable solids it is possible to operate with the knife spaced from the drum so that not all of the cake is removed. The heel of solids is then retained on the drum and this acts as the filter media. This discharge system is particularly suitable for friable cakes that do not have the mechanical properties to bridge over the strings of a string discharge system.

With the roller cake discharge system, the cake is transferred from the drum to a discharge roll from which it is removed by a knife. This is a relatively simple method of removing thin and sticky filter cakes without having a knife rubbing against and wearing the filter cloth. Certain applications require facilities for washing the cloth either continuously or intermittently without dilution of the trough contents, and such a feature is provided by the belt discharge system where the filter cloth also acts as cake conveyor. In such a system, the cake is completely supported between the drum and the discharge roll so that thin cakes and cakes of low mechanical strength can be handled and higher drum speeds, and hence higher filtration rates, can be achieved.

The performance of drum type vacuum filters for given feed conditions can be controlled by three main variables–drum speed, vacuum (if necessary with differential vacuum applied to the filtration, washing and drying zones) and the percentage of drum surface submerged in the feed slurry. Most drum filters have facilities which allow for easy manual adjustment of these variables, although automatic adjustment of any, or all, of them can be actuated by changes in the quality and/or quantity of the feed or cake. For maximum throughput, a drum filter should be operated at the highest submergence and at the highest possible drum speed. The limiting conditions affecting submergence are:

(a) Any increase in submergence limits the proportion of the drum area available for washing and/or drying.
(b) Drum submergence above approximately 40 per cent entails the use of glands where the drum shaft and valve hub pass through the trough.
(c) High submergence may complicate the geometry of the discharge system.

Under all conditions, it is essential to operate with combinations of drum speed, submergence and vacuum, which, for the feed conditions that apply, will ensure that a fully dischargeable cake is formed. If this is not done then progressive deterioration in the effectiveness of the filter medium will occur and this will adversely affect the performance of the machine. Typical filtration cycles for drum filters are given in Table 7.4.

Drum type vacuum filters can be fitted, if required, with simple hoods to limit the escape of toxic or obnoxious vapours. These may be arranged for complete sealing, and for operation under a nitrogen or similar blanket, although this complicates access to the drum and necessitates a design in which the vacuum system and the cake receiving system are arranged to prevent gas loss. The cake moves through the washing and drying zones in the form of a continuous sheet and, because the cake and filter medium are adequately supported on the drum shell, it is possible to fit the filter with various devices that will improve the quality of the cake regarding both washing and drying, prior to discharge.

Simple rolls, extending over the full width of the filter, can be so arranged that any irregularities or cracks in the cake are eliminated, and subsequent washing and drying is therefore applied to a uniform surface. Otherwise wash liquors and air tend to short circuit, or "channel", the deposited solids. The cake compression system may also incorporate a

Table 7.4. Filtration cycles for drum-type rotary vacuum filter (times in seconds)

| Speed | | 25 per cent submergence | | | | | 37.5 per cent submergence | | | | | 50 per cent sub. | | 66.7 per cent sub. | | 75 per cent sub. | |
|---|---|---|---|---|---|---|---|---|---|---|---|---|---|---|---|---|---|
| | | A | B | C | | | A | B | C | | | A | B | A | B | A | B |
| (rev/min) | (Hz) | Filter | Dewater | Initial dewater | Wash | Final dewater | Filter | Dewater | Initial dewater | Wash | Final dewater | Filter | Dewater | Filter | Dewater | Filter | Dewater |
| 3 | 0.05 | 5 | 10 | 2 | 5 | 3 | 7 | 8 | 2 | 4 | 2 | 8 | 5 | 12 | 2 | 13 | 1 |
| $2\frac{1}{2}$ | 0.04 | 6 | 12 | 2 | 7 | 3 | 9 | 10 | 2 | 5 | 3 | 11 | 6.5 | 16 | 2.2 | 17 | 1.8 |
| 2 | 0.033 | 7 | 15 | 3 | 8 | 4 | 12 | 14 | 2 | 9 | 3 | 13 | 7.5 | 19 | 3 | 21 | 2 |
| $1\frac{1}{2}$ | 0.025 | 10 | 23 | 5 | 12 | 6 | 16 | 18 | 3 | 11 | 4 | 16 | 10 | 24 | 4 | 28 | 3 |
| 1 | 0.017 | 15 | 30 | 7 | 16 | 7 | 22 | 24 | 3 | 16 | 5 | 27 | 15 | 38 | 5 | 42 | 4 |
| $\frac{2}{3}$ | 0.011 | 22 | 45 | 10 | 24 | 11 | 33 | 37 | 5 | 24 | 7 | 40 | 22 | 55 | 8 | 63 | 6 |
| $\frac{1}{2}$ | 0.008 | 30 | 60 | 13 | 32 | 15 | 44 | 48 | 6 | 32 | 10 | 54 | 30 | 76 | 10 | 84 | 8 |
| $\frac{1}{3}$ | 0.0055 | 45 | 91 | 20 | 48 | 23 | 66 | 72 | 9 | 48 | 15 | 80 | 44 | 110 | 16 | 126 | 12 |
| $\frac{1}{4}$ | 0.004 | 60 | 120 | 26 | 64 | 30 | 88 | 96 | 12 | 64 | 20 | 110 | 60 | 150 | 21 | 170 | 17 |
| $\frac{1}{5}$ | 0.0033 | 75 | 150 | 32 | 80 | 38 | 110 | 120 | 15 | 80 | 25 | 135 | 75 | 180 | 27 | 210 | 21 |
| $\frac{1}{6}$ | 0.0028 | 90 | 180 | 39 | 96 | 45 | 132 | 144 | 18 | 96 | 30 | 160 | 90 | 220 | 32 | 280 | 28 |
| $\frac{1}{7}$ | 0.0024 | 105 | 210 | 45 | 112 | 54 | 154 | 168 | 21 | 112 | 35 | 190 | 105 | 260 | 37 | 310 | 31 |
| $\frac{1}{8}$ | 0.0200 | 120 | 240 | 52 | 128 | 60 | 176 | 192 | 24 | 128 | 40 | 215 | 120 | 300 | 42 | 340 | 34 |
| $\frac{1}{10}$ | 0.0017 | 135 | 300 | 65 | 160 | 75 | 220 | 240 | 30 | 160 | 50 | 240 | 135 | 360 | 54 | 420 | 42 |

(Columns A + B or A + C should be used but *not* all three)

compression or wash blanket which limits still further any tendency for the air or wash liquor to "channel". Wash blankets also avoid disturbance of the cake which might occur when high pressure sprays are directed on to the cake, and they allow the wash liquor to be applied much nearer to the point where the cake emerges from the slurry. Compression rolls, with or without blankets, can be arranged with pneumatic or hydraulically loaded cylinders or they can be weighted or fitted with springs to increase further the pressure applied to the cake as shown in Figure 7.21. In some applications, the rolls can be arranged to give a controlled cake thickness, and so give some rearrangement of the solids in the filter cake with the advantages of the elimination of cracks, compaction of the cake, and the liberation of moisture from thixotropic materials, with the liberated moisture finally being drawn into the vacuum system prior to discharge.

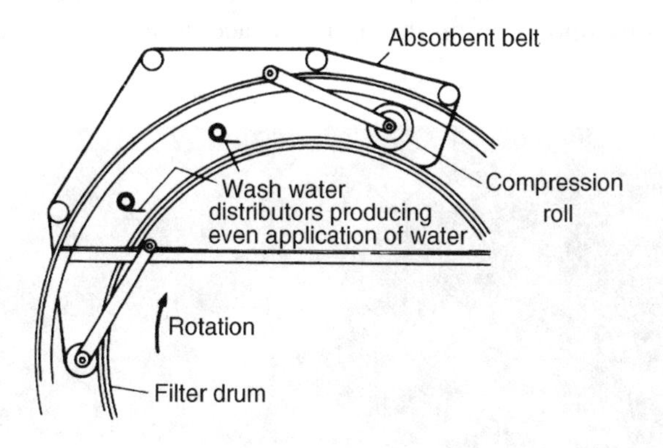


Figure 7.21. Cake compression

Other devices which find varying degrees of success in the treatment of thixotropic materials include rotary beaters and pulsating valves to interrupt the air flow in the drying zone. Both of these result in movement of the cake, giving a reduction in moisture content.

A drum type vacuum filter can easily be arranged so that the trough and hood, if fitted is thermally insulated. In addition, the filter can be equipped with heat exchange equipment, such as jackets for heating fluid or refrigerant, and electric resistance wires for heating. These features can be of particular significance in applications such as the dewaxing of lubricating oil which is normally carried out at temperatures below 263 K (−10 °C), and for salt drying which may be effected at temperatures approaching 673 K (400 °C) on a filter/dryer.

In most applications, bottom feed drum filters, that is those where the drum is partially suspended in the feed slurry, must be fitted with agitators to keep the solids in suspension. Normally, pendulum type rakes, which are pivoted on or above the drum axis, are used so as to avoid the use of glands, although rotary paddles fitted at low level in the filter trough also find application. Solids suspension may also be improved by giving careful attention to the flow pattern within the trough such as, for example, by feeding slurry into the trough at low level and at a rate in excess of that at which the material can be filtered. With this technique a constant overflow is provided so that upward movement within the trough assists in conveying solids toward the filter surface. Air sparge pipes may also be used to prevent solids settlement, although this results in higher operating costs than with a mechanical agitation system.

Top-feed drum filters are used for the treatment of very fast settling solids that cannot easily be kept in suspension. A drum filter may be preferable, rather than a horizontal pan or similar filter, because of its low cost, simplicity and reliability, and because of the relative ease with which it can be fitted with appropriate accessory features. A top feed drum filter incorporates multi-compartments, drainlines and a rotary valve identical to those in a conventional drum filter. The feed slurry is introduced at, or just before, top dead centre, and the cake is discharged from 90° to 180° from the feed point. The feed slurry can be distributed across the drum by sprays or a weir box, or it may be contained within a 3-sided head-box which is sealed against the moving drum at each end and across the back on the ascending filter surface. For materials which do not cake together sufficiently to be self-supporting on the drum surface, a hopper-type top feed filter may

Figure 7.22. Dorr–Oliver press-belt drum filter

be used which is similar to a conventional top feed filter except that the walls extend above the drum surface around all sides of each compartment and these support the filter cake. Because larger particles settle more rapidly these are deposited near the filter cloth and smaller particles form the outer portions of the filter cake. A cake of relatively high porosity is thereby obtained and high filtration rates are achieved.

Modern plant utilises drums with surfaces of 60–100 $m^2$ as compared with the 20 $m^2$ of the older cast iron drums. Construction materials such as stainless steel, titanium, epoxy resins and plastics all give much improved corrosion resistance for many slurries and hence longer life. The replacement of the knife system by some form of belt has given better cake discharge and permitted the use of thinner filtering media, such as synthetic fibres. The belt provides some support for the cake and materially assists the effect of compressed air for lifting off the cake. Drying can be improved by totally covering the filter with a hood. Improvements have also been made in techniques for reducing cake cracking.

Recent developments in rotary filters include equipment marketed by Dorr–Oliver, illustrated in Figure 7.22, which combines vacuum filtration with pressure filtration in which filter cake moisture is reduced by 20–200 per cent. A combined filtering and drying plant has been developed in which a continuous belt, rather like a bed-spring in construction, passes round the underside of the drum filter and the filter cake is deposited on the belt to which it adheres. The belt leaves the filter towards the top and is then carried through a cabinet dryer. It then returns to the underside of the drum filter after the cake has been removed by agitation. The metal belt, which assists the drying of the material by virtue of its good heat-conducting properties, is formed into loops which are carried through the dryer on a slat conveyor. With finely divided materials the total loss of solid from the belt is as little as 1 or 2 per cent.

## Example 7.4

A slurry containing 40 per cent by mass solid is to be filtered on a rotary drum filter 2 m diameter and 2 m long which normally operates with 40 per cent of its surface immersed in the slurry and under a pressure of 17 kN/$m^2$. A laboratory test on a sample of the slurry using a leaf filter of area 200 $cm^2$ and covered with a similar cloth to that on the drum produced 300 $cm^3$ of filtrate in the first 60 s and 140 $cm^3$ in the next 60 s, when the leaf was under an absolute pressure of 17 kN/$m^2$. The bulk density of the dry cake was 1500 kg/$m^3$ and the density of the filtrate was 1000 kg/$m^3$. The minimum thickness of cake which could be readily removed from the cloth was 5 mm.

At what speed should the drum rotate for maximum throughput and what is this throughout in terms of the mass of the slurry fed to the unit per unit time?

## Solution

For the leaf filter:

$$A = 0.02 \text{ m}^2,$$

$$(-\Delta P) = (101.3 - 17) = 84.3 \text{ kN/m}^2 \text{ or } 84,300 \text{ N/m}^2$$

When $t = 60$ s: $V = 0.0003 \text{ m}^3$

When $t = 120$ s: $V = 0.00044 \text{ m}^3$

These values are substituted into the constant pressure filtration equation:

$$V^2 + \frac{2LAV}{v} = \frac{2(-\Delta P)A^2 t}{\mathbf{r}\mu v} \quad \text{(equation 7.18)}$$

which enables the filtration constants to be determined as:

$$L/v = 2.19 \times 10^{-3} \text{ and } \mathbf{r}\mu v = 3.48 \times 10^{10}$$

For the rotary filter, equation 7.18 applies as the whole operation is at constant pressure The maximum throughput will be attained when the cake thickness is a minimum, that is 5 mm or 0.005 m.

Area of filtering surface $= (2\pi \times 2) = 4\pi$ m$^2$.

Bulk volume of cake deposited $= (4\pi \times 0.005) = 0.063$ m$^3$/revolution

If the rate of filtrate production $= w$ kg/s, then the volume flow is $0.001\ w$ m$^3$/s

For a 40 per cent slurry: $S/(S + w) = 0.4$, and the mass of solids $= 0.66w$.

Thus:
volume of solids deposited $= (0.66w/1500) = 4.4 \times 10^{-4} w$ m$^3$/s

If one revolution takes $t$ s, then: $4.4 \times 10^{-4} wt = 0.063$
and the mass of filtrate produced per revolution = 143 kg.

Rate of production of filtrate $= 0.001w$ m$^3$/s $= V/t$

Thus: $$V^2 = 1 \times 10^{-6} w^2 t^2 = 1 \times 10^{-6} (143)^2 = 0.02 \text{ m}^6$$

and: $$V = 0.141 \text{ m}^3$$

Substituting $V = 0.141$ m$^3$ and the constants into equation 7.18, gives:

$$(0.141)^2 + 2 \times 2.19 \times 10^{-3} \times 0.141 = 2 \times 84,300 \times (4\pi)^2 t/(3.48 \times 10^{10})$$

from which $t = 26.95$ s, which is equal to time of submergence/revolution.

Thus: $$\text{time for 1 revolution} = (26.9/0.4) = 67.3 \text{ s}$$

and: $$\text{speed} = (1/67.3) = \underline{\underline{0.015 \text{ Hz}}}$$

$$w = (143/67.3) = 2.11 \text{ kg/s}$$

$$S = (0.66 \times 2.11) \text{ kg/s}$$

and: $$\text{mass of slurry} = (1.66 \times 2.11) = \underline{\underline{3.5 \text{ kg/s}}}$$

## Example 7.5

A plate and frame press with a filtration area of 2.2 m$^2$ is operated with a pressure drop of 413 kN/m$^2$ and with a downtime of 21.6 ks (6 h). In a test with a small leaf filter 0.05 m$^2$ in area, operating with a pressure difference of 70 kN/m$^2$, 0.00025 m$^3$ of filtrate was obtained in 300 s and a total of 0.00040 m$^3$ in 600 s. Estimate the optimum filtration time for maximum throughput.

If the operating cost during filtration is £10/ks and the cost of a shutdown is £100, what is the optimum filtration time for minimum cost?

## Solution

Substituting $V = 0.00025$ m$^3$ at $t = 300$ s in equation 7.39 gives:

$$(300/0.00025) = 0.00025\mathbf{r}\mu v/(2 \times 0.05^2 \times 70 \times 10^3) + \mathbf{r}\mu L/(0.05 \times 70 \times 10^3)$$

or:
$$1.2 \times 10^6 = 7.14 \times 10^{-6}\mathbf{r}\mu v + 2.86 \times 10^{-4}\mathbf{r}\mu L \quad \text{(i)}$$

Substituting $V = 0.00040$ m$^3$ at $t = 400$ s in equation 7.39 gives:

$$1.0 \times 10^6 = 11.42 \times 10^{-6}\mathbf{r}\mu v + 2.86 \times 10^{-4}\mathbf{r}\mu L \quad \text{(ii)}$$

Solving equations (i) and (ii) simultaneously gives:

$$\mathbf{r}\mu v = 7 \times 10^{12} \text{ Ns/m}^4 \text{ and } \mathbf{r}\mu L = 4.6 \times 10^9 \text{ N/sm}^3$$

Thus, for the plate and frame filter:

$$B_1 = \frac{\mathbf{r}\mu v}{2A^2(-\Delta P)} = 7 \times 10^{12}/(2 \times 2.2^2 \times 413 \times 10^3) = 1.75 \times 10^6 \text{ s/m}^6$$

and:
$$B_2 = \frac{\mathbf{r}\mu L}{A(-\Delta P)} = 4.6 \times 10^9/(2.2 \times 413 \times 10^3) = 5.06 \times 10^3 \text{ s/m}^3$$

Substituting for $V$ from equation 7.43 into equation 7.41, the filtration time for maximum throughput is:

$$\begin{aligned} t &= t' + B_2(t'/B_1)^{0.5} \\ &= 21.6 \times 10^3 + 5.06 \times 10^3[(21.6 \times 10^3)/(1.75 \times 10^6)]^{0.5} \\ &= 2.216 \times 10^4 \text{ s or } \underline{\underline{22.2 \text{ ks}}} \text{ (6.2 h)} \end{aligned}$$

From equation 7.43:
$$V = [(21.6 \times 10^3)/(1.75 \times 10^6)]^{0.5} = 0.111 \text{ m}^3$$

and the mean rate of filtration is:

$$V/(t + t') = 0.111/(2.216 \times 10^4 + 21.6 \times 10^3) = \underline{\underline{2.54 \times 10^{-6} \text{ m}^3/\text{s}}}$$

The total cost is:

$$C = 0.01t + 100 \text{ £/cycle}$$

or:
$$C = (0.01t + 100)/V \text{ £/m}^3$$

Substituting for $t$ from equation 7.41 gives:

$$C = (0.01B_1V^2 + 0.01B_2V + 100)/V$$

Differentiating and putting $dC/dV = 0$:

$$V = (100/0.01B_1)^{0.5} \text{ m}^3$$

and from equation 7.41, the optimum filtration time for minimum cost is:

$$t = (100/0.01) + B_2(100/0.01B_1^{0.5}) \text{ s}$$

Substituting for $B_1$ and $B_2$:

$$t = 10^4 + 5.06 \times 10^3(10^4/1.75 \times 10^6)^{0.5}$$

$$= 1.03 \times 10^4 \text{ s or } \underline{\underline{10.3 \text{ ks}}} \text{ (2.86 h)}$$

## Example 7.6

A slurry, containing 0.2 kg of solid per kilogram of water, is fed to a rotary drum filter 0.6 m long and 0.6 m diameter. The drum rotates at one revolution in 360 s and 20 per cent of the filtering surface is in contact with the slurry at any instant. If filtrate is produced at the rate of 0.125 kg/s and the cake has a voidage of 0.5, what thickness of cake is produced when filtering with a pressure difference of 65 kN/m$^2$? The density of the solids is 3000 kg/m$^3$.

The rotary filter breaks down and the operation has to be carried out temporarily in a plate and frame press with frames 0.3 m square. The press takes 120 s to dismantle and 120 s to reassemble and, in addition, 120 s is required to remove the cake from each frame. If filtration is to be carried out at the same overall rate as before, with an operating pressure difference of 175 kN/m$^2$, what is the minimum number of frames that needs to be used and what is the thickness of each? It may be assumed that the cakes are incompressible and that the resistance of the filter medium may be neglected.

## Solution

### *Drum filter*

$$\text{Area of filtering surface} = (0.6 \times 0.6\pi) = 0.36\pi \text{ m}^2$$

$$\text{Rate of filtration} = 0.125 \text{ kg/s}$$

$$= (0.125/1000) = 1.25 \times 10^{-4} \text{ m}^3\text{/s of filtrate}$$

1 kg or $10^{-3}$ m$^3$ water is associated with 0.2 kg of solids $= 0.2/(3 \times 10^3) = 6.67 \times 10^{-5}$ m$^3$ of solids in the slurry.

Since the cake porosity is 0.5, $6.67 \times 10^{-5}$ m$^3$ of water is held in the filter cake and $(10^{-3} - 6.67 \times 10^{-5}) = 9.33 \times 10^{-4}$ m$^3$ appears as filtrate, per kg of total water in the slurry.

Volume of cake deposited by unit volume of filtrate, $v = (6.67 \times 10^{-5} \times 2)/(9.33 \times 10^{-4}) = 0.143$.

Volumetric rate of deposition of solids $= (1.25 \times 10^{-4} \times 0.143) = 1.79 \times 10^{-5}$ m$^3$/s. One revolution takes 360 s. Therefore the given piece of filtering surface is immersed for $(360 \times 0.2) = 72$ s

The bulk volume of cake deposited per revolution $= (1.79 \times 10^{-5} \times 360) = 6.44 \times 10^{-3}$ m$^3$. Thickness of cake produced $= (6.44 \times 10^{-3})/(0.36\pi) = 5.7 \times 10^{-3}$ m or $\underline{\underline{5.7 \text{ mm}}}$

## Properties of filter cake

$$\frac{dV}{dt} = \frac{(-\Delta P)A}{\mathbf{r}\mu l} = \frac{(-\Delta P)A^2}{\mathbf{r}\mu V v} \qquad \text{(from equations 7.2 and 7.8)}$$

At constant pressure:

$$V^2 = \frac{2}{\mathbf{r}\mu v}(-\Delta P)A^2 t = K(-\Delta P)A^2 t \text{ (say)} \qquad \text{(from equation 7.11)}$$

Expressing pressures, areas, times and volumes in N/m$^2$, m$^2$, s and m$^3$ respectively, then for one revolution of the drum:

$$(1.25 \times 10^{-4} \times 360)^2 = K(6.5 \times 10^4)(0.36\pi)^2 \times 72$$

since each element of area is immersed for one-fifth of a cycle,

and: $$K = 3.38 \times 10^{-10}$$

## Filter press

Using a filter press with $n$ frames of thickness $b$ m, the total time, for one complete cycle of the press $= (t_f + 120n + 240)$ s, where $t_f$ is the time during which filtration is occurring.

$$\text{Overall rate of filtration} = \frac{V_f}{t_f + 120n + 240} = 1.25 \times 10^{-4}\ \text{m}^3/\text{s}$$

where $V_f$ is the total volume of filtrate per cycle.

The volume of frames/volume of cake deposited by unit volume of filtrate, $v$, is given by:

$$V_f = 0.3^2 nb/0.143 = 0.629nb$$

But: $$V_f^2 = (3.38 \times 10^{-10}) \times (1.75 \times 10^5)(2n \times 0.3 \times 0.3)^2 t_f \quad \text{(from equation 7.11)}$$

$$= 0.629\ nb^2$$

and: $$t_f = 2.064 \times 10^5\ b^2$$

Thus: $$1.25 \times 10^{-4} = \frac{0.629nb}{2.064 \times 10^5 b^2 + 120n + 240}$$

That is: $$25.8b^2 + 0.015n + 0.030 = 0.629nb$$

or: $$n = \frac{0.030 + 25.8b^2}{0.629b - 0.015}$$

$n$ is a minimum when $dn/db = 0$, that is when:

$$(0.629b - 0.015) \times 51.6b - (0.030 + 25.8b^2) \times 0.629 = 0$$

$$b^2 - 0.0458b - 0.001162 = 0$$

Thus: $b = 0.0229 \pm \sqrt{(0.000525 + 0.001162)}$ and taking the positive root:

$d = 0.0640$ m or $\underline{\underline{64\ \text{mm}}}$

Hence:

$$n = \frac{(0.030 + 25.8 \times 0.0640^2)}{(0.629 \times 0.0640 - 0.015)}$$

$$= 5.4$$

Thus a minimum of 6 frames must be used.

The sizes of frames which will give exactly the required rate of filtration when six are used are given by:

$$0.030 + 25.8b^2 = 3.774b - 0.090$$

or:

$$b^2 - 0.146b + 0.00465 = 0$$

and:

$$b = 0.073 \pm \sqrt{(0.005329 - 0.00465)}$$

$$= 0.047 \text{ or } 0.099 \text{ m}$$

Thus, 6 frames of thickness either 47 mm or 99 mm will give exactly the required filtration rate; intermediate sizes give higher rates.

Thus any frame thickness between 47 and 99 mm will be satisfactory. In practice, 50 mm (2 in) frames would probably be used.

## *Rotary disc filters*

In essence, *the disc filter* operates in a manner similar to a bottom-feed drum filter. The principal differences to be noted are the compartments which are formed on both faces of vertical discs. These comprise eight or more segments, each of which is connected to the horizontal shaft of the machine. The filtrate and air are drawn through the filter medium, into the drainage system of the segments, and finally through passages in the rotary shaft to a valve located at one or both ends.

The discs are arranged on the shaft about 0.3 m apart as shown in Figure 7.23, resulting in economy of space, and consequently a far greater filtration area can be accommodated in a given floor space than is possible with a drum type filter. A disc filter area of approximately 80 m$^2$ can be achieved with eight discs, each approximately 2.5 m in diameter, in the same space as a 2.5 m diameter × 2.5 m drum filter which would have a filtration area of 20 m$^2$. Because of the large areas and consequently the large liquor and air flows, two rotary valves, one at each end of the main shaft, are frequently fitted. Cake discharge is usually achieved using a knife or wire, and discharge may be assisted by blow-back air. The cake falls through vertical openings in the filter trough.

In general, disc filters have the advantages that, not only is the cost per unit filtration area low and large filtration areas can be accommodated in a small floor space, but the trough can be divided into two sections so that the different slurries can be handled at the same time in the same machine. The disadvantages of these units is that heavy filter cloths are necessary, with a great tendency to "blind", and cake washing is virtually impossible due to the ease with which deposited solids can be disturbed from the vertical faces. In contrast to a drum type filter, in which all parts of the filter medium take the same path through the slurry, each part of a disc segment takes a different path, depending on its radial distance from the hub. Consequently, if homogeneous conditions cannot be maintained in the filter trough, then very uneven cakes may be formed, a situation which

Figure 7.23. Rotary disc filter

favours preferential air flow and results in a widely variable moisture content across each segment. Disc filters also have the limitation that, because every segment must be fully submerged in slurry prior to the application of vacuum, very little variation in submergence level can be accommodated.

## Precoat filters

All the filters described previously have been in the category of cake filters for they all rely on the solids present in the feed slurry having properties which are suitable for them to act as the actual filter medium and to form cakes which are relatively easily discharged. Applications arise where extremely thin filter cakes, perhaps only a fraction of a millimetre thick, have to be discharged, where extremely good filtrate clarity is essential, or where a particularly blinding substance is present, and filtration through a permanent filter medium alone, such as a filter fabric, may not be satisfactory. A much more efficient performance is achieved by utilising a bed of easily filtered material which is *precoated* on to what is otherwise a standard drum type filter.

After the precoat is established the solids to be removed from the filter feed are trapped on the surface of the precoated bed. This thin layer of slime is removed by a knife which is caused to advance slowly towards the drum. The knife also removes a thin layer of the precoated bed so that a new surface of the filter medium is exposed. This procedure allows steady filtration rates to be achieved.

A *precoated rotary drum vacuum filter* is the only filtration equipment from which extremely thin filter cakes can be positively and continuously removed in a semi-dry state.

The usual precoat materials, for most production processes, are conventional filter aids, such as diatomaceous earth and expanded Perlite, although frothed coal, calcium sulphate and other solids, which form very permeable filter cakes, may be used where such materials are compatible with the materials being processed. Wood flour and fly-ash have also been used in some applications, particularly in effluent treatment plants, where some contamination of the filtered liquor is of no great consequence. The establishment of the precoat bed should preferably be carried out using a precoat slurry of low solids concentration, at high drum speed and with low submergence, so that the bed is built up in the form of a thin layer with each revolution, thereby ensuring a uniform and compacted bed which will not easily be disturbed from the drum. For long periods of operation, the maximum bed thickness should be used. This is often determined by the mechanical design of the filter and by the cake formation properties of the precoat material. Normally the bed thickness is 75–100 mm and, with most materials this can be established on the drum in a period of about one hour. The type and grade of precoat material used, in addition to process considerations, can influence the rate at which the bed must be removed and the cost of precoat vacuum filter operation. Generally, for maximum economy, a precoat filter should operate at the highest possible submergence, consistent with the required degree of cake washing and drying, when required, and with the lowest possible knife advance rate. Some precoat filters are capable of operation at up to 70 per cent drum submergence though this, particularly with the facility to accommodate 100 mm thick beds, necessitates special design features to give adequate clearance between the drum and the trough while at the same time giving acceptable angles for the discharge knife. Knife advance rates are 0.013–0.13 mm per revolution, and drum speeds are frequently 0.002–0.03 Hz (0.1–2 rev/min). By careful control of all the variables, and by correct sizing of the equipment, it is possible to obtain precoat beds which last 60–240 hours.

To obtain maximum economy in operation, a precoat type vacuum filter should have:

(a) Easy variation of the knife advance rate.
(b) Very accurate control of the knife advance rate.
(c) A very rigid knife assembly (due to the need to advance at extremely low rates over a face width up to approximately 6 m).
(d) Rapid advance and retraction of the knife.

Mechanical and hydraulic systems for control of the knife are now used. The latter allows easy selection of the rate and direction of knife movement and the system is particularly suited to remote control or automatic operation.

## Example 7.7

A sludge is filtered in a plate and frame press fitted with 25 mm frames. For the first 600 s the slurry pump runs at maximum capacity. During this period the pressure rises to 415 kN/m$^2$ and 25 per cent of the total filtrate is obtained. The filtration takes a further 3600 s to complete at constant pressure and 900 s is required for emptying and resetting the press.

It is found that if the cloths are precoated with filter aid to a depth of 1.6 mm, the cloth resistance is reduced to 25 per cent of its former value. What will be the increase in the overall throughput of the press if the precoat can be applied in 180 s?

## Solution

### Case 1

$$\frac{dV}{dt} = \frac{A^2(-\Delta P)}{v\mathbf{r}\mu\left(V + \dfrac{AL}{v}\right)} = \frac{a}{V + b} \quad \text{(equation 7.16)}$$

For constant rate filtration:

$$\frac{V_0}{t_0} = \frac{a}{V_0 + b}$$

or:

$$V_0^2 + bV_0 = at_0$$

For constant pressure filtration:

$$\frac{1}{2}(V^2 - V_0^2) + b(V - V_0) = a(t - t_0)$$

$$t_0 = 600 \text{ s},\ t - t_0 = 3600 \text{ s},\ V_0 = V/4$$

$$\frac{V^2}{16} + b\frac{V}{4} = 600a$$

and:

$$\frac{1}{2}(V^2 - V^2/16) + b(V - V/4) = 3600a$$

Thus:

$$3600a = \frac{15}{32}V^2 + \frac{3}{4}bV = \frac{3}{8}V^2 + \frac{3}{2}bV$$

and:

$$b = \frac{V}{8}$$

Thus:

$$a = \frac{1}{600}\left(\frac{V^2}{16} + \frac{V^2}{32}\right) = \frac{3}{19,200}V^2$$

Total cycle time $= (900 + 4200) = 5100$ s
Filtration rate $= V/5100 = 0.000196V$

### Case 2

$$\frac{V_1}{t_1} = \frac{a}{V_1 + \dfrac{b}{4}} = \frac{V_0}{t_0} = \frac{a}{V_0 + b}$$

$$\frac{1}{2}\left(\frac{49}{64}V^2 - V_1^2\right) + \frac{b}{4}\left(\frac{7}{8}V - V_1\right) = a(t - t_1)$$

Thus:

$$\frac{V}{2400} = \frac{\dfrac{3}{19,200}V^2}{V_1 + \dfrac{V}{32}}$$

and:

$$t_1 = \frac{t_0}{V_0}V_1 = \frac{600}{V/4} \times \frac{11}{32}V = \frac{3300}{4} \text{ s} = 825 \text{ s}$$

Substituting gives:

$$\frac{1}{2}\left(\frac{49}{64}V^2 - \frac{121}{1024}V_1^2\right) + \frac{1}{4}\frac{V}{8}\left(\frac{7}{8}V - \frac{11}{32}V\right) = \frac{3}{19,200}V^2(t - t_1)$$

or:

$$\frac{49}{128} - \frac{121}{2048} + \frac{17}{1024} = \frac{3}{19,200}(t - t_1)$$

$$t - t_1 = \left(\frac{19,200}{3}\right)\left(\frac{784 - 121 + 34}{2048}\right)$$

$$= 2178 \text{ s}$$

$$\text{Cycle time} = (180 + 900 + 825 + 2178) = 4083 \text{ s}$$

$$\text{Filtration rate} = \left(\frac{7}{8} \times \frac{V}{4083}\right) = 0.000214V$$

$$\text{Increase} = \frac{(0.000214 - 0.000196)V}{0.000196V} \times 100 = \underline{\underline{9.1 \text{ per cent}}}$$

### 7.4.7. The tube press

One of the major problems in coal preparation is the dewatering of fine coal to a moisture content sufficiently low enough both to meet market requirements and to ease the problems of handlability encountered at some collieries and there is an interest in different types of dewatering equipment in addition to the conventional rotary vacuum filter. As a result, a tube press built by the former English China Clays Limited, now Imerys, was installed at a colliery where it was used to treat raw slurry which was blended into the final product. As described by GWILLIAM[(32)] and BROWN[(33)]. It was found that the product from the tube press was very low in moisture content. In the china clay industry, for which the tube press was first developed, the formation of a very dry cake reduces the load on the drying plant where the cost of removing a given quantity of water is an order of magnitude greater than in filtration.

In essence, the tube press is an automatic membrane filter press operating at pressures up to 10 $MN/m^2$ (100 bar) and consists of two concentric cylindrical tubes, the inner filter candle and the outer hydraulic casing. Cylindrical tubes are used as these have the ability to withstand high pressure without a reinforcing structure and are available commercially. Between the filter candle and the hydraulic casing is a flexible membrane which is fastened to both ends of the hydraulic casing. Hydraulic pressure is exerted on to the flexible membrane, applying filtration pressure to the slurry which is contained between the flexible membrane and filter candle. The membrane, being flexible, allows for the slurry feed volume to be varied to suit each individual application and this permits near optimum filter cake thicknesses to be achieved. The filter candle is perforated and around its outer circumference is fitted a drain mesh, felt backing cloth and top filter cloth forming the filter media. The backing cloth protects the filter cloth against the hydraulic high pressure whilst the mesh allows the filtrate to run into the centre of the filter candle and be drained away. Under high pressure conditions, the resistance of the filter cloth is low compared with the resistance of the filter cake. This allows a tightly woven filter cloth to be used and any imperfections in this cloth are compensated for by the felt backing cloth. In this way, clear filtrate is obtained from the tube press.

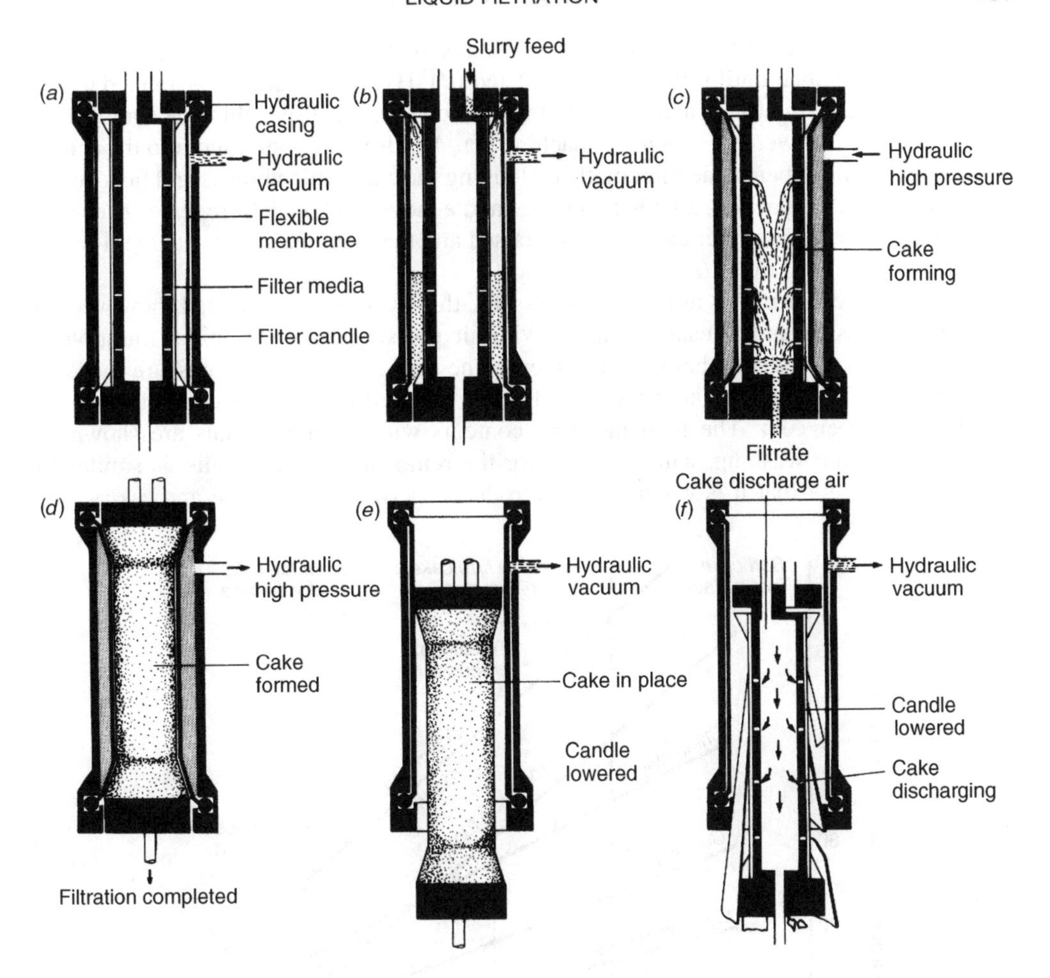


Figure 7.24. Sequence of operation of the tube press

The sequence of operation of the tube press, illustrated in Figure 7.24, is as follows. The empty tube filter press is closed and hydraulic vacuum is applied to the flexible membrane, dilating it against the outer hydraulic casing (*a*). Slurry is fed into the machine, partially filling the volume contained within the flexible membrane, the filter media and the end pieces (*b*). The exact volume of slurry fed into the tube press depends upon the characteristics of each individual slurry and corresponds to the amount required to give optimum cake thickness. The machine will automatically discharge a cake of 16–19 mm maximum thickness. The minimum thickness is dependent on the cake discharge characteristics although is normally 4–5 mm. Low pressure hydraulic fluid is fed into the tube press between the outer hydraulic casing and the flexible membrane. The membrane contracts, reducing the volume between the membrane and filter cloth and expelling the entrained air. The low pressure hydraulic flow is stopped and high pressure hydraulic fluid is applied. Filtration commences and the high pressure hydraulic fluid continues contracting the membrane, reducing the suspension volume and expelling the filtrate through the filter medium into the centre of the filter candle and away through the filtrate drain line ($c$). Filter cake is formed around the cloth and high pressure is maintained until filtration is completed ($d$). Hydraulic vacuum is applied to the flexible membrane, dilating it against the outer hydraulic casing. When this is accomplished the filter candle is lowered to open the machine ($e$). A pulse of air admitted into the centre of the filter candle expands the filter cloth, dislodging the cake which breaks and falls out of the machine ($f$). This action, together with the candle movement, may be repeated to ensure complete discharge. The filter candle is then raised and the machine closes leaving it ready to commence the next cycle ($a$).

The unique design of the tube press allows for this cycle to be amended, however, to include air pressing and/or cake washing. With air pressing, once the initial filtration is complete, air is introduced between the membrane and the cake. The pressure cycle is then repeated. Typically an air press will further reduce the moisture content of china clay by 2.5–8 per cent. The final moisture contents with other materials are shown in Figure 7.25. Water washing, which is used for the removal of soluble salts, is similar to air pressing, except that it is water that is introduced between membrane and cake.

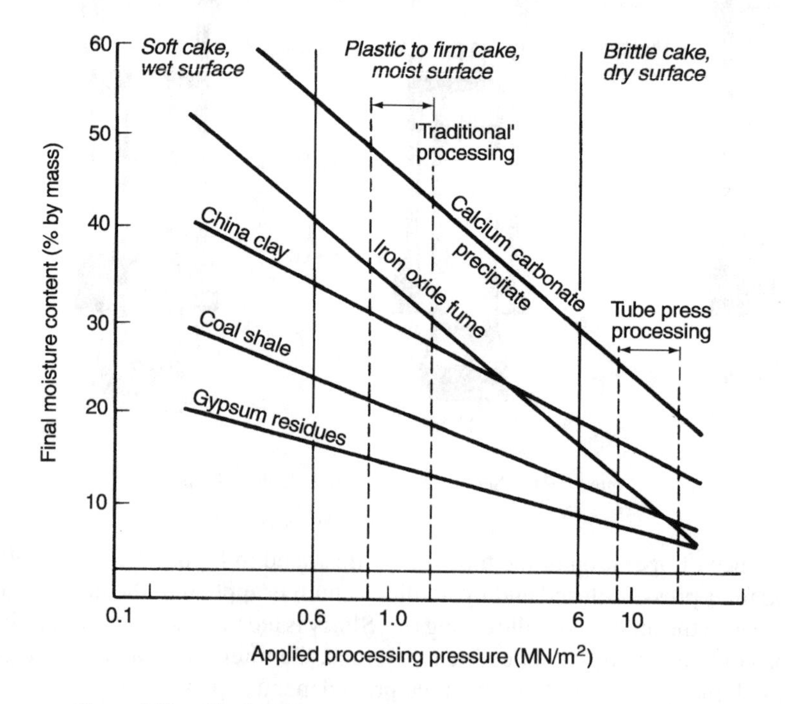


Figure 7.25. Final moisture content as a function of applied pressure

## 7.5. FURTHER READING

IVES, K. J. (ed.): *The Scientific Basis of Filtration* (Noordhoff, Leyden, 1975).
MATTESON, M. S. and ORR, C.: *Filtration Principles and Practice*, 2nd edn. (Marcel Dekker, New York, 1987).
PURCHAS, D. B.: *Industrial Filtration of Liquids*, 2nd edn. (Leonard Hill, London, 1971).
SUTTLE, H. K. (ed.): *Process Engineering Technique Evaluation–Filtration* (Morgan–Grampian (Publishers) Ltd., London, 1969).
SVAROVSKY, L. (ed:): *Solid–Liquid Separation*, 4th edn. (Butterworth-Heinemann, Oxford, 2000).

## 7.6. REFERENCES


1. IVES, K. J. (ed.): *The Scientific Basis of Filtration* (Noordhoff, Leyden, 1975).
2. IVES, K. J.: *Trans. Inst. Chem. E.* **48** (1970) T94. Advances in deep-bed filtration.
3. SUTTLE, H. K.: *The Chemical Engineer* No. 314 (Oct. 1976) 675. Development of industrial filtration.
4. RUTH, B. F., MONTILLON, G. H. and MONTONNA, R. E.: *Ind. Eng. Chem.* **25** (1933) 76 and 153. Studies in filtration. I. Critical analysis of filtration theory. II. Fundamental axiom of constant-pressure filtration.
5. RUTH, B. F.: *Ind. Eng. Chem.* **27** (1935) 708 and 806. Studies in filtration. III. Derivation of general filtration equations. IV. Nature of fluid flow through filter septa and its importance in the filtration equation.
6. RUTH, B. F. and KEMPE, L.: *Trans. Am. Inst. Chem. Eng.* **33** (1937) 34. An extension of the testing methods and equations of batch filtration practice to the field of continuous filtration.
7. RUTH, B. F.: *Ind. Eng. Chem.* **38** (1946) 564. Correlating filtration theory with practice.
8. GRACE, H. P.: *Chem. Eng. Prog.* **49** (1953) 303, 367, and 427. Resistance and compressibility of filter cakes.
9. HOFFING, E. H. and LOCKHART, F. J.: *Chem. Eng. Prog.* **47** (1951) 3. Resistance to filtration.
10. CARMAN, P. C.: *Trans. Inst. Chem. Eng.* **16** (1938) 168. Fundamental principles of industrial filtration.
11. HEERTJES, P. M.: *Chem. Eng. Sci.* **6** (1957) 190 and 269. Studies in filtration.
12. VALLEROY, V. V. and MALONEY, J. O.: *A.I.Ch.E.Jl.* **6** (1960) 382. Comparison of the specific resistances of cakes formed in filters and centrifuges.
13. TILLER, F. M. and HUANG, C. J.: *Ind. Eng. Chem.* **53** (1961) 529. Filtration equipment. Theory.
14. TILLER, F. M. and SHIRATO, M.: *A.I.Ch.E.Jl.* **10** (1964) 61. The role of porosity in filtration: Part VI. New definition of filtration resistance.
15. TILLER, F. M. and YEH, C. S.: *A.I.Ch.E.Jl.* **31** (1985) 1241. The role of porosity in filtration: Part X. Deposition of compressible cakes on external radial surfaces.
16. RUSHTON, A. and HAMEED, M. S.: *Filtn. and Sepn.* **6** (1969) 136. The effect of concentration in rotary vacuum filtration.
17. EHLERS, S.: *Ind. Eng. Chem.* **53** (1961) 552. The selection of filter fabrics re-examined.
18. WROTNOWSKI, A. C.: *Chem. Eng. Prog.* **58** No. 12 (Dec. 1962) 61. Nonwoven filter media.
19. PURCHAS, D. B.: *Industrial Filtration of Liquids*, 2nd edn. (Leonard Hill, London, 1971).
20. TILLER, F. M. and CHENG, K. S.: *Filt. and Sepn.* **14** (1977) 13. Delayed cake filtration.
21. MACKLEY, M. R. and SHERMAN, N. E.: *Chem. Eng. Sci.* **47** (1997) 3067. Cross-flow cake filtration mechanics and kinetics.
22. HOLDICH, R. G., CUMMING, I. W. and ISMAIL, B.: *Chem. Eng. Res. Des.* **73** (1995) 20. The variation of cross-flow filtration rate with wall shear stress and the effect of deposit thickness.
23. CLEASBY, J. L.: *The Chemical Engineer* No. 314 (Oct. 1976) 663. Filtration with granular beds.
24. IVES, K. J.: *Proc. Int. Water Supply Assn. Eighth Congress*, Vienna. Vol. 1. (Internatl. Water Supply Assn., London, 1969). Special Subject No. 7. Theory of Filtration.
25. IWASAKI, T.: *J. Am. Water Works Assn.* **29** (1937) 1591. Some notes on sand filtration.
26. SPIELMAN, L. A. and FRIEDLANDER, S. K.: *J. Colloid and Interface. Sci.* **46** (1974) 22. Role of the electrical double layer in particle deposition by convective diffusion.
27. HARKER, J. H.: *Processing* **2** (1979) 35. Getting the best out of batch filtration.
28. CHERRY, G. B.: *Filtn. and Sepn.* **11** (1974) 181. New developments in filter plates and filter presses.
29. PARRISH, P. and OGILVIE, H.: *Calcium Superphosphates and Compound Fertilisers. Their Chemistry and Manufacture* (Hutchison, 1939).
30. BLENDULF, K. A. G. and BOND, A. P.: *I. Chem. E. Symposium Series* No. 59, (1980) 1:5/1. The development and application of horizontal belt filters in the mineral, processing and chemical industries.
31. BOSLEY, R.: *Filtn. and Sepn.* **11** (1974) 138. Vacuum filtration equipment innovations.
32. GWILLIAM, R. D.: *Filt. and Sepn.* **8** (1971) 173. The E.C.C. tube filter press.
33. BROWN, A.: *Filt. and Sepn.* **16** (1979) 468. The tube press.


## 7.7. NOMENCLATURE

| | | Units in SI System | Dimensions in **M, L, T** |
|---|---|---|---|
| $A$ | Cross-sectional area of bed or filtration area | $m^2$ | $\mathbf{L}^2$ |
| $B_1$ | Coefficient | $s/m^6$ | $\mathbf{L}^{-6}\mathbf{T}$ |
| $B_2$ | Coefficient | $s/m^3$ | $\mathbf{L}^{-3}\mathbf{T}$ |
| $b$ | Frame thickness | m | $\mathbf{L}$ |
| $C$ | Volume concentration of solids in the filter | — | — |

| | | Units in SI System | Dimensions in **M**, **L**, **T** |
|---|---|---|---|
| $C'$ | Total cost of filtration per unit volume | £/m$^3$ | $\mathbf{L}^{-3}$ |
| $C_0$ | Value of $C$ at filter surface | — | — |
| $e$ | Voidage of bed or filter cake | — | — |
| $e_0$ | Liquid fraction in feed slurry | — | — |
| $e_V$ | Liquid fraction in slurry in vessel | — | — |
| $e_z$ | Voidage at distance $z$ from surface | — | — |
| $J$ | Mass fraction of solids in slurry | — | — |
| $L$ | Thickness of filter cake with same resistance as cloth | m | $\mathbf{L}$ |
| $l$ | Thickness of filter cake or bed | m | $\mathbf{L}$ |
| $n'$ | Compressibility index | — | — |
| $P_1$ | Pressure at downstream face of cake | N/m$^2$ | $\mathbf{ML}^{-1}\mathbf{T}^{-2}$ |
| $P_2$ | Pressure at upstream face of cake | N/m$^2$ | $\mathbf{ML}^{-1}\mathbf{T}^{-2}$ |
| $P_z$ | Pressure at distance $z$ from surface | N/m$^2$ | $\mathbf{ML}^{-1}\mathbf{T}^{-2}$ |
| $-\Delta P$ | Total drop in pressure | N/m$^2$ | $\mathbf{ML}^{-1}\mathbf{T}^{-2}$ |
| $-\Delta P'$ | Drop in pressure across cake | N/m$^2$ | $\mathbf{ML}^{-1}\mathbf{T}^{-2}$ |
| $-\Delta P''$ | Pressure drop across cloth | N/m$^2$ | $\mathbf{ML}^{-1}\mathbf{T}^{-2}$ |
| $R$ | Rate of cake production | m/s | $\mathbf{LT}^{-1}$ |
| $\mathbf{r}$ | Specific resistance of filter cake | m$^{-2}$ | $\mathbf{L}^{-2}$ |
| $\bar{\mathbf{r}}$ | Mean value of $\mathbf{r}_z$ | m$^{-2}$ | $\mathbf{L}^{-2}$ |
| $\mathbf{r}_z$ | Specific resistance of compressible cake at distance $z$ from surface (equation 7.25) | m$^{-2}$ | $\mathbf{L}^{-2}$ |
| $\bar{\mathbf{r}}'$, $\bar{\mathbf{r}}''$ | Functions of $\mathbf{r}_z$ independent of $\Delta P$ | m$^{-2}$/(N/m$^2$)$^{n'}$ | $\mathbf{M}^{-n'}\mathbf{L}^{n'-2}\mathbf{T}^{2n'}$ |
| $S$ | Specific surface | m$^{-1}$ | $\mathbf{L}^{-1}$ |
| $t$ | Time | s | $\mathbf{T}$ |
| $t'$ | Time of dismantling filter press | s | $\mathbf{T}$ |
| $t_1$ | Time at beginning of operation | s | $\mathbf{T}$ |
| $u_c$ | Mean velocity of flow calculated over the whole area | m/s | $\mathbf{LT}^{-1}$ |
| $V$ | Volume of liquid flowing in time $t$ | m$^3$ | $\mathbf{L}^3$ |
| $V_1$ | Volume of liquid passing in time $t_1$ | m$^3$ | $\mathbf{L}^3$ |
| $\mathbf{V}$ | Volume of vessel | m$^3$ | $\mathbf{L}^3$ |
| $v$ | Volume of cake deposited by unit volume of filtrate | — | — |
| $v'$ | Volume of solids deposited by unit volume of filtrate | — | — |
| $w$ | Mass rate of production of filtrate | kg/s | $\mathbf{MT}^{-1}$ |
| $W$ | Overall volumetric rate of filtration | m$^3$/s | $\mathbf{L}^3\mathbf{T}^{-1}$ |
| $z$ | Distance from surface of filter cake | m | $\mathbf{L}$ |
| $\lambda$ | Filter coefficient | m$^{-1}$ | $\mathbf{L}^{-1}$ |
| $\mu$ | Viscosity of fluid | Ns/m$^2$ | $\mathbf{ML}^{-1}\mathbf{T}^{-1}$ |
| $\rho$ | Density of fluid | kg/m$^3$ | $\mathbf{ML}^{-3}$ |
| $\rho_s$ | Density of solids | kg/m$^3$ | $\mathbf{ML}^{-3}$ |
| $\sigma$ | Volume of solids deposited per unit volume of filter | — | — |

CHAPTER 8

# Membrane Separation Processes

## 8.1. INTRODUCTION

Whilst effective product separation is crucial to economic operation in the process industries, certain types of materials are inherently difficult and expensive to separate. Important examples include:

(a) Finely dispersed solids, especially those which are compressible, and which have a density close to that of the liquid phase, have high viscosity, or are gelatinous.
(b) Low molecular weight, non-volatile organics or pharmaceuticals and dissolved salts.
(c) Biological materials which are very sensitive to their physical and chemical environment.

The processing of these categories of materials has become increasingly important in recent years, especially with the growth of the newer biotechnological industries and with the increasingly sophisticated nature of processing in the food industries. When difficulties arise in the processing of materials of biological origin, it is worth asking, how does nature solve the problem? The solution which nature has developed is likely to be both highly effective and energy efficient, though it may be slow in process terms. Nature separates biologically active materials by means of membranes. As STRATHMANN[1] has pointed out, a membrane may be defined as "an interphase separating two phases and selectively controlling the transport of materials between those phases". A membrane is an interphase rather than an interface because it occupies a finite, though normally small, element of space. Human beings are all surrounded by a membrane, the skin, and membranes control the separation of materials at all levels of life, down to the outer layers of bacteria and subcellular components.

As discussed by LONSDALE[2], since the 1960s a new technology using synthetic membranes for process separations has been rapidly developed by materials scientists, physical chemists and chemical engineers. Such membrane separations have been widely applied to a range of conventionally difficult separations. They potentially offer the advantages of ambient temperature operation, relatively low capital and running costs, and modular construction. In this chapter, the nature and scope of membrane separation processes are outlined, and then those processes most frequently used industrially are described more fully.

## 8.2. CLASSIFICATION OF MEMBRANE PROCESSES

Industrial membrane processes may be classified according to the size range of materials which they are to separate and the driving force used in separation. There is always a degree of arbitrariness about such classifications, and the distinctions which are typically drawn are shown in Table 8.1. This chapter is primarily concerned with the pressure driven processes, microfiltration (MF), ultrafiltration (UF), nanofiltration (NF) and reverse osmosis (RO). These are already well-established large-scale industrial processes. For example, reverse osmosis is used world-wide for the desalination of brackish water, with more than 1,000 units in operation. Plants capable of producing up to $10^5$ $m^3$/day of drinking water are in operation. As a further example, it is now standard practice to include an ultrafiltration unit in paint plants in the car industry. The resulting recovery of paint from wash waters can produce savings of 10–30 per cent in paint usage, and allows recycling of the wash waters. The use of reverse osmosis and ultrafiltration in the dairy industry has led to substantial changes in production techniques and the development of new types of cheeses and related products. Nanofiltration is a process, with characteristics between those of ultrafiltration and reverse osmosis, which is finding increasing application in pharmaceutical processing and water treatment. Electrodialysis is a purely electrically driven separation process used extensively for the desalination or concentration of brackish water. There are about 300 such plants in operation. Economics presently favour reverse osmosis, however, rather than electrodialysis for such separations. The major use of dialysis is in hemodialysis of patients with renal failure, where it is most appropriate to use such a gentle technique. Hemodialysis poses many interesting problems of a chemical engineering nature, although dialysis is a relatively slow process not really suited to large-scale industrial separations.

Table 8.1. Classification of membrane separation processes for liquid systems

| Name of process | Driving force | Separation size range | Examples of materials separated |
|---|---|---|---|
| Microfiltration | Pressure gradient | 10–0.1 μm | Small particles, large colloids, microbial cells |
| Ultrafiltration | Pressure gradient | <0.1 μm–5 nm | Emulsions, colloids, macromolecules, proteins |
| Nanofiltration | Pressure gradient | ~1 nm | Dissolved salts, organics |
| Reverse osmosis (hyperfiltration) | Pressure gradient | <1 nm | Dissolved salts, small organics |
| Electrodialysis | Electric field gradient | <5 nm | Dissolved salts |
| Dialysis | Concentration gradient | <5 nm | Treatment of renal failure |

## 8.3. THE NATURE OF SYNTHETIC MEMBRANES

Membranes used for the pressure-driven separation processes, microfiltration, ultrafiltration and reverse osmosis, as well as those used for dialysis, are most commonly made of polymeric materials[1]. Initially most such membranes were cellulosic in nature. These are now being replaced by polyamide, polysulphone, polycarbonate and a number of other advanced polymers. These synthetic polymers have improved chemical stability and better resistance to microbial degradation. Membranes have most commonly been produced by a form of phase inversion known as immersion precipitation. This process has four main steps: (a) the polymer is dissolved in a solvent to 10–30 per cent by mass, (b) the resulting solution is cast on a suitable support as a film of thickness, approximately 100 μm, (c) the film is quenched by immersion in a non-solvent bath, typically water or an aqueous solution, (d) the resulting membrane is annealed by heating. The third step gives a polymer-rich phase forming the membrane, and a polymer-depleted phase forming the pores. The ultimate membrane structure results as a combination of phase separation and mass transfer, variation of the production conditions giving membranes with different separation characteristics. Most microfiltration membranes have a symmetric pore structure, and they can have a porosity as high as 80 per cent. Ultrafiltration and reverse osmosis membranes have an asymmetric structure comprising a 1–2 μm thick top layer of finest pore size supported by a ∼100 μm thick more openly porous matrix, as shown in Figure 8.1. Such an asymmetric structure is essential if reasonable membrane permeation rates are to be obtained. Another important type of polymeric membrane is the thin-film composite membrane. This consists of an extremely thin layer, typically ∼1 μm, of finest pore structure deposited on a more openly porous matrix. The thin layer is formed by phase inversion or interfacial polymerisation on to an existing microporous structure. Polymeric membranes are most commonly produced in the form of flat sheets, but they are also widely produced as tubes of diameter 10–25 mm and in the form of hollow fibres of diameter 0.1–2.0 mm.

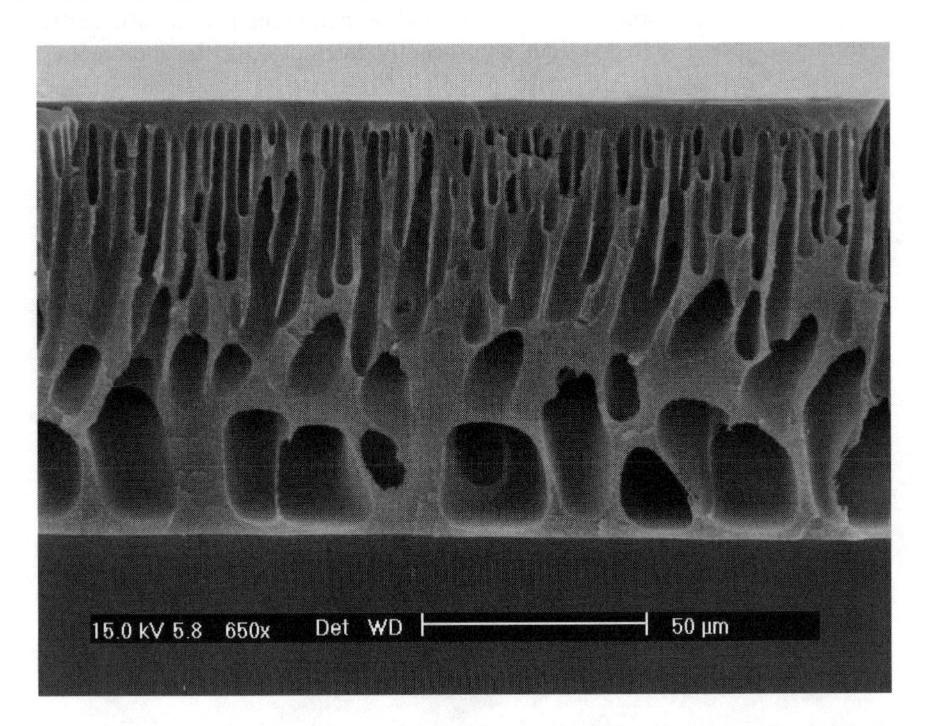


Figure 8.1. Electron micrograph of a section of an asymmetric ultrafiltration membrane showing finely porous "skin" layer on more openly porous supporting matrix (courtesy of Dr Huabing Yin)

A significant recent advance has been the development of microfiltration and ultrafiltration membranes composed of inorganic oxide materials. These are presently produced by two main techniques: (a) deposition of colloidal metal oxide on to a supporting material such as carbon, and (b) as purely ceramic materials by high temperature sintering of spray-dried oxide microspheres. Other innovative production techniques lead to the formation of membranes with very regular pore structures. Zirconia, alumina and titania are the materials most commonly used. The main advantages of inorganic membranes compared with the polymeric types are their higher temperature stability, allowing steam sterilisation in biotechnological and food applications, increased resistance to fouling, and narrower pore size distribution.

The physical characterisation of a membrane structure is important if the correct membrane is to be selected for a given application. The pore structure of microfiltration membranes is relatively easy to characterise, atomic force microscopy and electron microscopy being the most convenient methods and allowing the three-dimensional structure of the membrane to be determined. The limit of resolution of a simple electron microscope is about 10 nm, and that of an atomic force microscope is $<1$ nm, as shown in Figure 8.2. Additional characterisation techniques, such as the bubble point, mercury intrusion or permeability methods, use measurements of the permeability of membranes to fluids. Both the maximum pore size and the pore size distribution may be determined. A parameter often quoted in manufacturer's literature is the nominal molecular weight cut-off (MWCO) of a membrane. This is based on studies of how solute molecules are rejected by membranes. A solute will pass through a membrane if it is sufficiently small to pass through a pore, if it does not significantly interact with the membrane and if

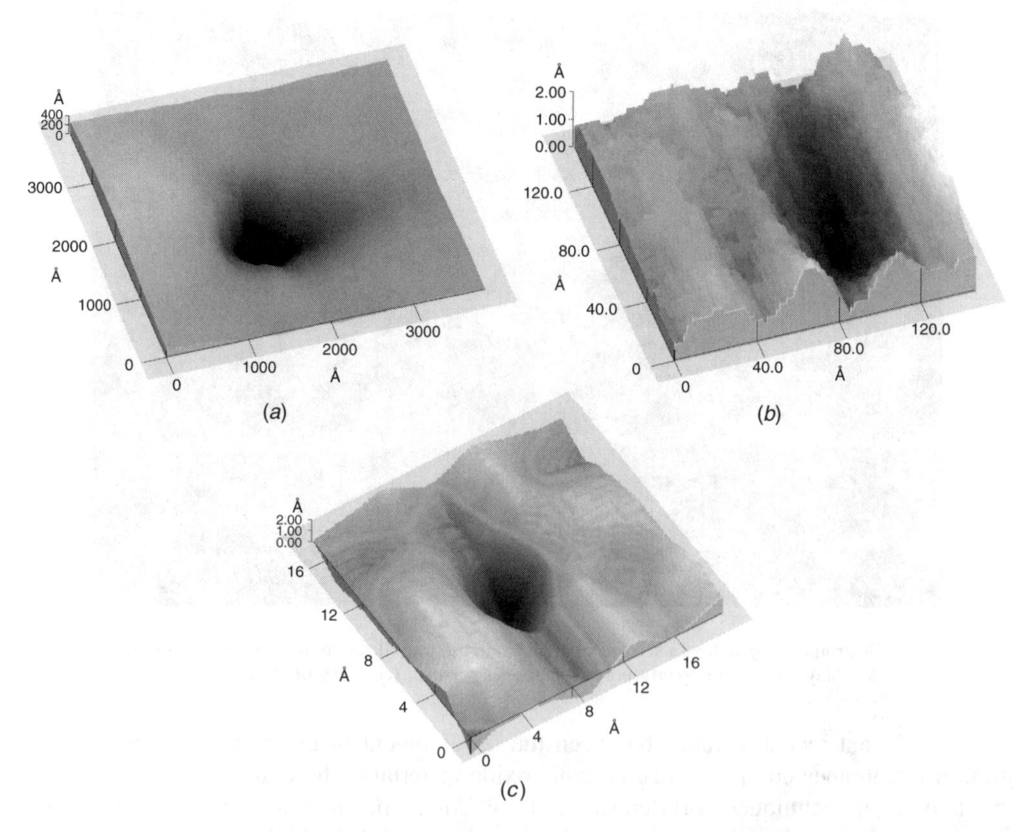


Figure 8.2. AFM images of single pores in (a) microfiltration, (b) ultrafiltration and (c) nanofiltration membranes (courtesy Dr Nichal Nidal)

it does not interact with other, larger solutes. It is possible to define a solute rejection coefficient $R$ by:

$$R = 1 - (C_p/C_f) \tag{8.1}$$

where $C_f$ is the concentration of solute in the feed stream and $C_p$ is the concentration of solute in the permeate. For a given ultrafiltration membrane with a distribution of pore sizes there is a relationship between $R$ and the solute molecular weight, as shown in Figure 8.3. The nominal molecular weight cut-off is normally defined as the molecular weight of a solute for which $R = 0.95$. Values of MWCO typically lie in the range 2000–100,000 kg/kmol with values of the order of 10,000 being most common. High resolution electron microscopy does not allow the resolution of an extensive pore structure in the separating layer of reverse-osmosis membranes. As discussed later, it is generally considered that reverse osmosis membranes do not contain pores and that they operate mainly by a "solution-diffusion" mechanism.

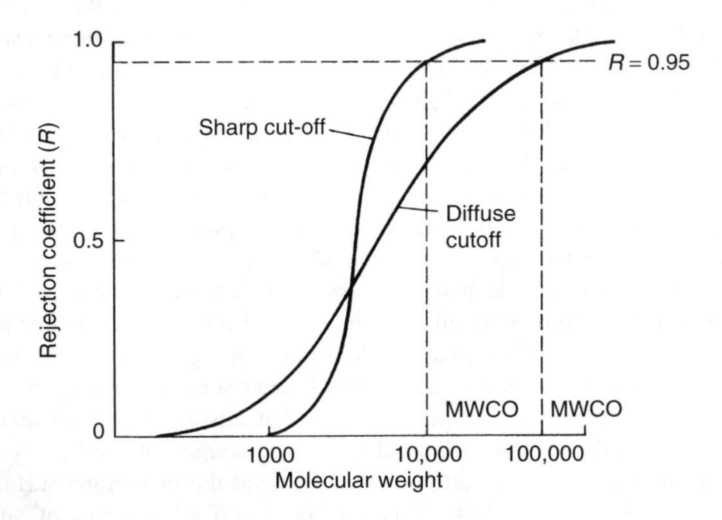


Figure 8.3. Dependence of rejection coefficient on molecular weight for ultrafiltration membranes

Ion-exchange membranes, which are used for electrodialysis, usually consist of highly swollen charged gels prepared either by dispersing a conventional ion-exchange material in a polymer matrix, or from a homogenous polymer in which electrically charged groups such as sulphonic, carboxylic or quarternised amine groups have been introduced as discussed by LACEY[3]. The first type is referred to as a heterogeneous membrane, while the second type is termed a homogeneous membrane. A membrane with fixed positive charges is referred to as an anion exchange membrane since it may bind and hence selectively transport anions from the surrounding solution. Similarly, a membrane containing fixed negative charges is termed a cation exchange membrane. Ion-exchange membranes exclude, that is, do not bind and do not allow the transport of, ions which bear charges of the same sign as the membrane.

## 8.4. GENERAL MEMBRANE EQUATION

It is not possible at present to provide an equation, or set of equations, that allows the prediction from first principles of the membrane permeation rate and solute rejection for a given real separation. Research aimed at providing such a prediction for model systems is under way, although the physical properties of real systems, both the membrane and the solute, are complex. An analogous situation exists for conventional filtration processes. The *general membrane equation* is an attempt to state the factors which may be important in determining the membrane permeation rate for pressure driven processes. This takes the form:

$$J = \frac{|\Delta P| - |\Delta \Pi|}{(R_m + R_c)\mu} \tag{8.2}$$

where $J$ is the membrane flux*, expressed as volumetric rate per unit area, $|\Delta P|$ is the pressure difference applied across the membrane, the transmembrane pressure, $\Delta\Pi$ is the difference in osmotic pressure across the membrane, $R_m$ is the resistance of the membrane, and $R_c$ is the resistance of layers deposited on the membrane, the filter cake and gel foulants. If the membrane is only exposed to pure solvent, say water, then equation 8.2 reduces to $J = |\Delta P|/R_m\mu$. For microfiltration and ultrafiltration membranes where solvent flow is most often essentially laminar through an arrangement of tortuous channels, this is analogous to the Carman–Kozeny equation discussed in Chapter 4. Knowledge of such water fluxes is useful for characterising new membranes and also for assessing the effectiveness of membrane cleaning procedures. In the processing of solutes, equation 8.2 shows that the transmembrane pressure must exceed the osmotic pressure for flow to occur. It is generally assumed that the osmotic pressure of most retained solutes is likely to be negligible in the cases of microfiltration. The resistance $R_c$ is due to the formation of a filter cake, the formation of a gel when the concentration of macromolecules at the membrane surface exceeds their solubility giving rise to a precipitation, or due to materials in the process feed that adsorb on the membrane surface producing an additional barrier to solvent flow. The separation of a solute by a membrane gives rise to an increased concentration of that solute at the membrane surface, an effect known as concentration polarisation. This may be described in terms of an increase in $\Delta\Pi$. It is within the framework of this equation that the factors influencing membrane permeation rate will be discussed in the following sections.

## 8.5. CROSS-FLOW MICROFILTRATION

The solids–liquid separation of slurries containing particles below 10 μm is difficult by conventional filtration techniques. A conventional approach would be to use a slurry thickener in which the formation of a filter cake is restricted and the product is discharged continuously as a concentrated slurry. Such filters use filter cloths as the filtration medium

* Membrane flux is denoted by $J$, the usual symbol in the literature on membranes. It corresponds with $u_c$, as used in Chapters 4 and 7 for flow in packed beds and filtration.

and are limited to concentrating particles above 5 μm in size. *Dead end* or *frontal* membrane microfiltration, in which the particle containing fluid is pumped directly through a polymeric membrane, is used for the industrial clarification and sterilisation of liquids. Such a process allows the removal of particles down to 0.1 μm or less, but is only suitable for feeds containing very low concentrations of particles as otherwise the membrane becomes too rapidly clogged.

The concept of *cross-flow* microfiltration, described by BERTERA, STEVEN and METCALFE[4], is shown in Figure 8.4 which represents a cross-section through a rectangular or tubular membrane module. The particle-containing fluid to be filtered is pumped at a velocity in the range 1–8 m/s parallel to the face of the membrane and with a pressure difference of 0.1–0.5 MN/m$^2$ (MPa) across the membrane. The liquid permeates through the membrane and the feed emerges in a more concentrated form at the exit of the module.

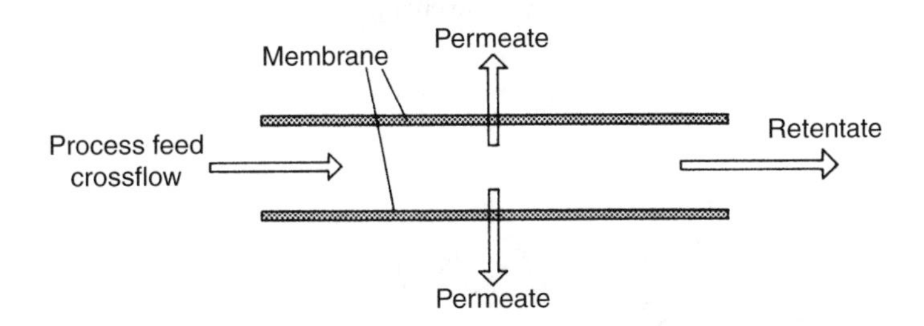


Figure 8.4. The concept of cross-flow filtration[4]

All of the membrane processes listed in Table 8.1 are operated with such a cross-flow of the process feed. The advantages of cross-flow filtration over conventional filtration are:

(a) A higher overall liquid removal rate is achieved by prevention of the formation of an extensive filter cake.
(b) The process feed remains in the form of a mobile slurry suitable for further processing.
(c) The solids content of the product slurry may be varied over a wide range.
(d) It may be possible to fractionate particles of different sizes.

A flow diagram of a simple cross-flow system[4] is shown in Figure 8.5. This is the system likely to be used for batch processing or development rigs and is, in essence, a basic pump recirculation loop. The process feed is concentrated by pumping it from the tank and across the membrane in the module at an appropriate velocity. The partially concentrated *retentate* is recycled into the tank for further processing while the *permeate* is stored or discarded as required. In cross-flow filtration applications, product washing is frequently necessary and is achieved by a process known as *diafiltration* in which wash water is added to the tank at a rate equal to the permeation rate.

In practice, the membrane permeation rate falls with time due to membrane fouling; that is blocking of the membrane surface and pores by the particulate materials, as shown in Figure 8.6. The rate of fouling depends on the nature of the materials being processed, the nature of the membrane, the cross-flow velocity and the applied pressure. For example, increasing the cross-flow velocity results in a decreased rate of fouling. Backflushing

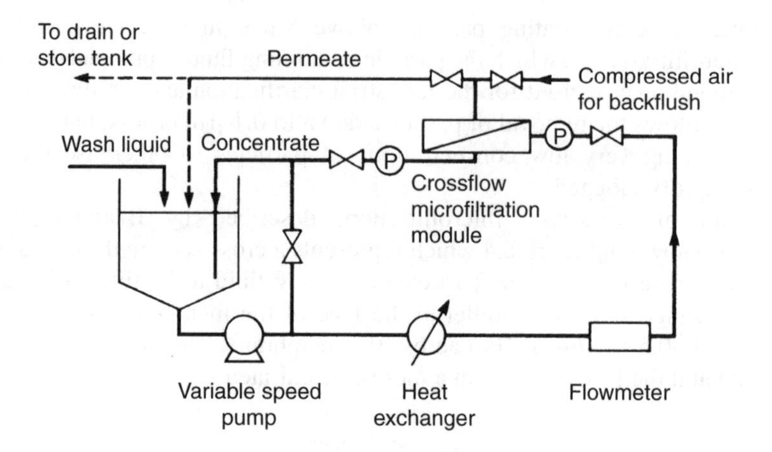


Figure 8.5. Flow diagram for a simple cross-flow system[4]

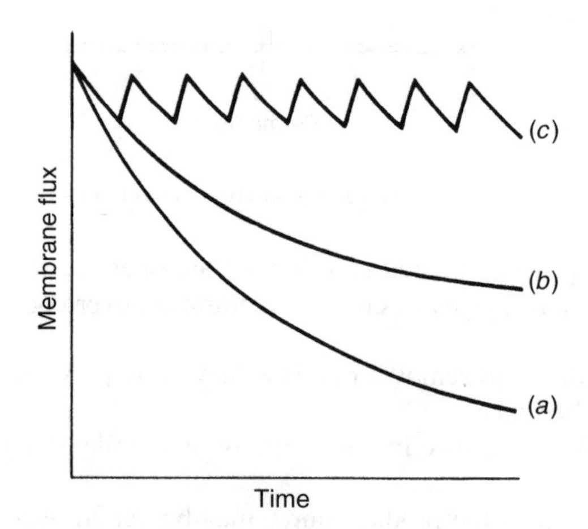


Figure 8.6. The time-dependence of membrane permeation rate during cross-flow filtration: (*a*) Low cross-flow velocity, (*b*) Increased cross-flow velocity, (*c*) Backflushing at the bottom of each "saw-tooth"

the membrane using permeate is often used to control fouling as shown in Figure 8.6*c*. Further means of controlling membrane fouling are discussed in Section 8.9.

Ideally, cross-flow microfiltration would be the pressure-driven removal of the process liquid through a porous medium without the deposition of particulate material. The flux decrease occurring during cross-flow microfiltration shows that this is not the case. If the decrease is due to particle deposition resulting from incomplete removal by the cross-flow liquid, then a description analogous to that of generalised cake filtration theory, discussed in Chapter 7, should apply. Equation 8.2 may then be written as:

$$J = \frac{|\Delta P|}{(R_m + R_c)\mu} \tag{8.3}$$

where $R_c$ now represents the resistance of the cake, which if all filtered particles remain in the cake, may be written as:

$$R_c = \frac{\mathbf{r}VC_b}{A_m} = \frac{\mathbf{r}V_s}{A_m} \tag{8.4}$$

where $\mathbf{r}$ is the specific resistance of the deposit, $V$ the total volume filtered, $V_s$ the volume of *particles* deposited, $C_b$ the bulk concentration of particles in the feed (particle volume/feed volume) and $A_m$ the membrane area. The specific resistance may theoretically be related to the particle properties for spherical particles by the Carman relationship, discussed in Chapter 4, as:

$$\mathbf{r} = 180\left(\frac{1-e}{e^3}\right)\left(\frac{1}{d_s^2}\right) \tag{8.5}$$

where $e$ is the void volume of the cake and $d_s$ the mean particle diameter.

Combining equations 8.3 and 8.4 gives:

$$J = \frac{1}{A_m}\frac{\mathrm{d}V}{\mathrm{d}t} = \frac{|\Delta P|}{(R_m + \mathbf{r}VC_b/A_m)\mu} \tag{8.6}$$

Solution of equation 8.6 for $V$ at constant pressure gives:

$$\frac{t}{V} = \frac{R_m\mu}{|\Delta P|A_m} + \frac{C_b\mathbf{r}\mu V}{2|\Delta P|A_m^2} \tag{8.7}$$

yielding a straight line on plotting $t/V$ against $V$.

SCHNEIDER and KLEIN[5] have pointed out that the early stages of cross-flow microfiltration often follow such a pattern although the growth of the cake is limited by the cross-flow of the process liquid. There are a number of ways of accounting for the control of cake growth. A useful method is to rewrite the resistance model to allow for the dynamics of polarisation in the film layer as discussed by FANE[6]. Equation 8.3 is then written as:

$$J = \frac{1}{A_m}\frac{\mathrm{d}V}{\mathrm{d}t} = \frac{|\Delta P|}{(R_m + R_{sd} - R_{sr})\mu} \tag{8.8}$$

where $R_{sd}$ is the resistance that would be caused by deposition of all filtered particles and $R_{sr}$ is the resistance removed by cross-flow. Assuming the removal of solute by cross-flow to be constant and equal to the convective particle transport at steady state ($=J_{ss}C_b$), then:

$$\frac{1}{A_m}\frac{\mathrm{d}V}{\mathrm{d}t} = \frac{|\Delta P|}{(R_m + (V/A_m - J_{ss}t)\mathbf{r}C_b)\mu} \tag{8.9}$$

where $J_{ss}$ can be obtained experimentally or from the film-model given in equation 8.15.

In a number of cases, a steady rate of filtration is never achieved and it is then possible to describe the time dependence of filtration by introducing an efficiency factor $\beta$ representing the fraction of filtered particles remaining in the filter cake rather than being swept along by the bulk flow. Equation 8.4 then becomes:

$$R_c = \frac{\beta\mathbf{r}VC_b}{A_m} \tag{8.10}$$

where $0 < \beta < 1$. This is analogous to a *scour model* describing shear erosion at a surface. The layers deposited on the membrane during cross-flow microfiltration are sometimes thought to constitute dynamically formed membranes with their own rejection and permeation characteristics.

In the following section, film and gel-polarisation models are developed for ultrafiltration. These models are also widely applied to cross-flow microfiltration, although even these cannot be simply applied, and there is at present no generally accepted mathematical description of the process.

## 8.6. ULTRAFILTRATION

Ultrafiltration is one of the most widely used of the pressure-driven membrane separation processes. The solutes retained or rejected by ultrafiltration membranes are those with molecular weights of $10^3$ or greater, depending mostly on the MWCO of the membrane chosen. The process liquid, dissolved salts and low molecular weight organic molecules (500–1000 kg/kmol) generally pass through the membrane. The pressure difference applied across the membrane is usually in the range 0.1–0.7 MN/m$^2$ and membrane permeation rates are typically 0.01–0.2 m$^3$/m$^2$ h. In industry, ultrafiltration is always operated in the cross-flow mode.

The separation of process liquid and solute that takes place at the membrane during ultrafiltration gives rise to an increase in solute concentration close to the membrane surface, as shown in Figure 8.7. This is termed concentration polarisation and takes place within the boundary film generated by the applied cross-flow. With a greater concentration at the membrane, there will be a tendency for solute to diffuse back into the bulk feed according to Fick's Law, discussed in Volume 1, Chapter 10. At steady state, the rate of back-diffusion will be equal to the rate of removal of solute at the membrane, minus the rate of solute leakage through the membrane:

$$J(C - C_p) = -D\frac{dC}{dy} \tag{8.11}$$

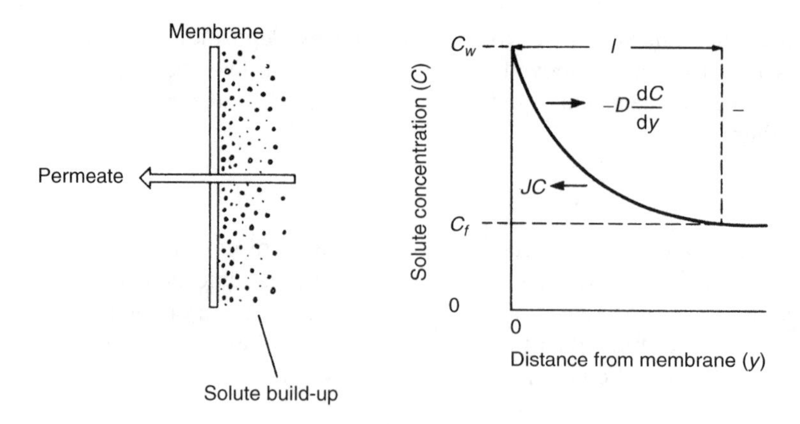


Figure 8.7. Concentration polarisation at a membrane surface

Here solute concentrations $C$ and $C_p$ in the permeate are expressed as mass fractions, $D$ is the diffusion coefficient of the solute and $y$ is the distance from the membrane. Rearranging and integrating from $C = C_f$ when $y = l$ the thickness of the film, to $C = C_w$, the concentration of solute at the membrane wall, when $y = 0$, gives:

$$-\int_{C_w}^{C_f} \frac{dC}{C - C_p} = \frac{J}{D}\int_0^l dy \tag{8.12}$$

or:

$$\frac{C_w - C_p}{C_f - C_p} = \exp\left(\frac{Jl}{D}\right) \tag{8.13}$$

If it is further assumed that the membrane completely rejects the solute, that is, $R = 1$ and $C_p = 0$, then:

$$\frac{C_w}{C_f} = \exp\left(\frac{Jl}{D}\right) \tag{8.14}$$

where the ratio $C_w/C_f$ is known as the polarisation modulus. It may be noted that it has been assumed that $l$ is independent of $J$ and that $D$ is constant over the whole range of $C$ at the interface. The film thickness is usually incorporated in an overall mass transfer coefficient $h_D$, where $h_D = D/l$, giving:

$$J = h_D \ln\left(\frac{C_w}{C_f}\right) \tag{8.15}$$

The mass transfer coefficient is usually obtained from correlations for flow in non-porous ducts. One case is that of laminar flow in channels of circular cross-section where the parabolic velocity profile is assumed to be developed at the channel entrance. Here the solution of LÉVÊQUE[7], discussed by BLATT *et al.*[8], is most widely used. This takes the form:

$$Sh = 1.62\left(Re\,Sc\,\frac{d_m}{L}\right)^{1/3} \tag{8.16}$$

where $Sh$ is the Sherwood number $(h_D d_m/D)$, $d_m$ is the hydraulic diameter, $L$ is the channel length, $Re$ is the Reynolds number $(u d_m \rho/\mu)$, $Sc$ the *Schmidt* number $(\mu/\rho D)$, with $u$ being the cross-flow velocity, $\rho$ the fluid density and $\mu$ the fluid viscosity. This gives:

$$h_D = 1.62\left(\frac{uD^2}{d_m L}\right)^{1/3} \tag{8.17}$$

or for tubular systems:

$$h_D = 0.81\left(\frac{\dot{\gamma}}{L}D^2\right)^{1/3} \tag{8.18}$$

where $\dot{\gamma}$, the shear rate at the membrane surface equals $8u/d_m$, as shown in Volume 1, Chapter 3.

For the case of turbulent flow the DITTUS–BOELTER[9] correlation given in Volume 1, Chapters 9 and 10, is used:

$$Sh = 0.023 Re^{0.8} Sc^{0.33} \tag{8.19}$$

which for tubular systems gives:

$$h_D = 0.023 \frac{u^{0.8} D^{0.67}}{d_m^{0.2}} \left(\frac{\rho}{\mu}\right)^{0.47} \tag{8.20}$$

and for thin rectangular flow channels, with channel height $b$:

$$h_D = 0.02 \frac{u^{0.8} D^{0.67}}{b^{0.2}} \left(\frac{\rho}{\mu}\right)^{0.47} \tag{8.21}$$

For both laminar and turbulent flow it is clear that the mass transfer coefficient and hence the membrane permeation rate may be increased, where these equations are valid, by increasing the cross-flow velocity or decreasing the channel height. The effects are greatest for turbulent flow. For laminar flow the mass transfer coefficient is decreased if the channel length is increased. This is due to the boundary layer increasing along the membrane module. The mass transfer coefficient is, therefore, averaged along the membrane length.

This boundary-layer theory applies to mass-transfer controlled systems where the membrane permeation rate is independent of pressure, for there is no pressure term in the model. In such cases it has been proposed that, as the concentration at the membrane increases, the solute eventually precipitates on the membrane surface. This layer of precipitated solute is known as the *gel-layer*, and the theory has thus become known as the *gel-polarisation* model proposed by MICHAELS[(10)]. Under such conditions $C_w$ in equation 8.15 becomes replaced by a constant $C_G$ the concentration of solute in the gel-layer, and:

$$J = h_D \ln\left(\frac{C_G}{C_f}\right) \tag{8.22}$$

If an increase in pressure occurs under these conditions, this produces a temporary increase in flux which brings more solute to the gel-layer and increases its thickness, subsequently reducing the flux to the initial level.

The agreement between theoretical and experimental ultrafiltration rates for macromolecular solutions can be said to be within 15–30 per cent, as discussed by PORTER[(11)]. Process patterns diagnostic of gel-polarisation type behaviour are shown in Figure 8.8. The dependence of the membrane permeation rate on the applied pressure is shown in Figure 8.8*a*. There is an initial pressure-dependent region followed by a pressure-independent region. The convergence of plots of the membrane permeation rate against ln $C_f$, as shown in Figure 8.8*b*, is a test of equation 8.15. Finally, the slope of plots of the membrane permeation rate against the average cross-flow velocity confirms the usefulness of the correlations for laminar and turbulent flow. The gel-polarisation model also suggests, however, that in the pressure-independent region the gel concentration should be independent of membrane permeability and membrane type. As pointed out by LE and HOWELL[(12)], neither of these is observed in practice. This shows the need for a more detailed understanding of the nature of membrane-solute interactions. Further, for colloidal suspensions, experimental membrane permeation rates are often one to two orders of magnitude higher than those indicated by the Lévêque and Dittus–Boelter correlations[(8)]. This has been termed, *the flux paradox for colloidal suspensions* by

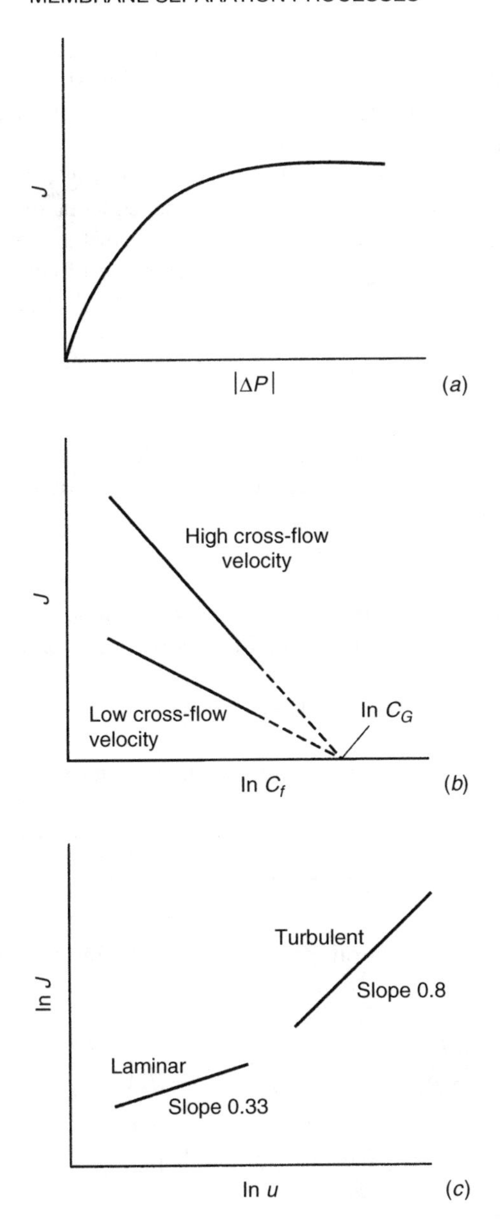


Figure 8.8. Dependence of membrane flux $J$ on ($a$) Applied pressure difference $|\Delta P|$, ($b$) Feed solute concentration $C_f$, ($c$) Cross-flow velocity ($u$) for ultrafiltration

Green and Belfort[13] and by Porter[14]. The *paradox* is most convincingly explained in terms of the *tubular-pinch effect* described by Sergre and Silberberg[15]. There is clear visual evidence that particles flowing through a tube migrate away from the tube wall and axis, reaching equilibrium at some eccentric radial position. The difficulty has been to produce a quantitative model incorporating lift forces to describe the effect, as pointed out by BELFORT[16]. This is an area of both considerable mathematical complexity and controversy, though new models such as that proposed by ALTENA and BELFORT[17] appear to allow the prediction of membrane permeation rates for both macromolecular solutions and colloidal suspensions.

The explanation of the pressure-independent region during the ultrafiltration of macromolecules requires the arbitrary introduction of the concept of a gel-layer in the film model. A more complete description of the dependence of the membrane permeation rate on the applied pressure may be given by considering the effect of the osmotic pressure of the macromolecules as described by WIJMANS *et al.*[18]. Equation 8.2 may then be written as:

$$J = \frac{(|\Delta P| - |\Delta\Pi|)}{R_m \mu} \tag{8.23}$$

where $|\Delta\Pi|$ is difference in osmotic pressure across the membrane. The osmotic pressure of concentrated solutions is best represented in terms of a polynomial as:

$$\Pi = a_1 C + a_2 C^2 + a_3 C^3 \tag{8.24}$$

where $a_1 a_2$ and $a_3$ are coefficients and $C$ is the solute concentration expressed as mass fractions. In the present case the difference in osmotic pressure across the membrane can be approximated as:

$$|\Delta\Pi| = \Pi = a C_w^n \tag{8.25}$$

where $C_w$ is the concentration at the membrane surface and $n > 1$. Then, from equations 20.15 and 20.23:

$$J = \frac{(|\Delta P| - a C_f^n \exp(nJ/h_D))}{R_m \mu} \tag{8.26}$$

Taking derivatives of this equation provides valuable insights into the ultrafiltration process. This gives:

$$\begin{aligned}
\frac{\partial J}{\partial |\Delta P|} &= \left( R_m \mu + a C_f^n \frac{n}{h_D} \exp\left(\frac{nJ}{h_D}\right)\right)^{-1} \\
&= \left( R_m \mu + \frac{n}{h_D} |\Delta\Pi| \right)^{-1} \\
&= \left( R_m \mu + \frac{n}{h_D} (|\Delta P| - J R_m \mu) \right)^{-1}
\end{aligned} \tag{8.27}$$

which gives the asymptotes:

$$\frac{\partial J}{\partial |\Delta P|} \to (R_m \mu)^{-1} \quad \text{for} \quad |\Delta P| \to 0 \quad \text{or} \quad |\Delta\Pi| \to 0$$

and:

$$\frac{\partial J}{\partial |\Delta P|} \to 0 \quad \text{for} \quad |\Delta P| \to \infty \quad \text{or} \quad |\Delta P| \gg J R_m \mu$$

Thus, the basic features of the flux-pressure profiles (Figure 8.8$a$) are accounted for without further assumptions:

(a) at low $|\Delta P|$ the slope is similar to that for pure solvent flow,
(b) as $|\Delta P|$ increases, the slope declines and approaches zero at high pressure.

The relationship between flux and solute concentration can be examined by rearranging equation 8.26, taking logarithms and differentiating to give:

$$\frac{\partial J}{\partial \ln C_f} = -\left(\frac{1}{h_D} + \frac{1}{n\left(\dfrac{|\Delta P|}{R_m \mu} - J\right)}\right)^{-1}$$

$$= -h_D\left(1 + \frac{R_m \mu h_D}{|\Delta \Pi| n}\right)^{-1} \tag{8.28}$$

which shows that when polarisation is significant, that is $|\Delta P| \gg J R_m \mu$ or $|\Delta \Pi| n / R_m \mu k \gg 1$:

$$\frac{\partial J}{\partial (\ln C_f)} \rightarrow -h_D$$

This is the same prediction for the limiting slope of a plot of $J$ against $\ln C_f$ as for the gel-polarisation model. The value of the slope of such plots at all other conditions is less in magnitude than $h_D$.

Finally, from equation 8.26, when $J \rightarrow 0$:

$$|\Delta P| = a C_f^n = \Pi \qquad (C_f \rightarrow C_{f,\text{lim}}) \tag{8.29}$$

that is, the limiting concentration is that giving an osmotic pressure equal to the applied pressure. This also implies that $C_{f,\text{lim}} = \text{f}|\Delta P|$, an important difference from the gel-polarisation model which predicts that $C_{f,\text{lim}} = C_g \neq \text{f}|\Delta P|$.

Osmotic pressure models can be developed from a very fundamental basis. For example, it is becoming possible to predict the rate of ultrafiltration of proteins starting from a knowledge of the sequence and three-dimensional structure of the molecule[(19)].

## Example 8.1

Obtain expressions for the optimum concentration for minimum process time in the diafiltration of a solution of protein content $S$ in an initial volume $V_0$.

(a) If the gel-polarisation model applies.
(b) If the osmotic pressure model applies.

It may be assumed that the extent of diafiltration is given by:

$$V_d = \frac{\text{Volume of liquid permeated}}{\text{Initial feed volume}} = \frac{V_p}{V_0}$$

## Solution

(a) *Assuming the gel–polarisation model applies*

The membrane permeation rate, $J = h_D \ln(C_G/C_f)$ (equation 8.22)

where $C_G$ and $C_f$ are the gel and the bulk concentrations respectively.

In this case: $C_f = S/V_0$

and the volume $V_d$ liquid permeated, $V_p = V_d S/C_f$

The process time per unit area, $t = V_p/J$

$$= V_d S/(C_f h_D \ln(C_G/C_f))$$

Assuming $C_f$ and $h_D$ are constant, then:

$$dt/dC_f = -V_d S/[h_D C_f^2 \ln(C_G/C_f)] + V_d S/\{h_D C_f^2[\ln(C_G/C_f)]^2\}$$

If, at the optimum concentration $C_f^*$ and $dt/dC_f = 0$, then:

$$1 = \ln(C_G/C_f^*)$$

and: $$\underline{\underline{C_f^* = C_G/e}}$$

(b) *Assuming the osmotic pressure model applies*

$$J = h_D \ln(C_G/C_f) \quad \text{(equation 8.22)}$$

Substituting for $|\Delta\Pi|$ at $C = C_w$, then:

$$J = [|\Delta P| - a_i C_f^n \exp(nJ/h_D)]/R_m\mu \quad \text{(equation 8.26)}$$

If $|\Delta\Pi|$ is very much greater than $JR_m\mu$, then:

$$J = (h_D/n)\ln(|\Delta P|/(aC_f^n))$$

As before: $V_d = V_p/V_0$ and $C_f = S/V_0$.

Thus: $$t = V_p/J$$

$$= (V_d S/C_f)/[(h_D/n)\ln(|\Delta P|/aC_f^n)]$$

and: $$dt/dC_f = (V_d nS/h_D)[-\ln(|\Delta P|/aC_f^n)/C_f^2 + n/C_f^2]/\ln(|\Delta P|/aC_f^n)$$

The process time, *t*, is a minimum when $dt/dC_f = 0$, that is when:

$$\underline{\underline{C_f^n = |\Delta P|/ae^n}}$$

## 8.7. REVERSE OSMOSIS

A classical demonstration of osmosis is to stretch a parchment membrane over the mouth of a tube, fill the tube with a sugar solution, and then hold it in a beaker of water. The level of solution in the tube rises gradually until it reaches a steady level. The static head developed would be equivalent to the osmotic pressure of the solution if the parchment were a perfect semipermeable membrane, such a membrane having the property of allowing the solvent to pass through but preventing the solute from passing through.

The pure solvent has a higher chemical potential than the solvent in the solution and so diffuses through until the difference is cancelled out by the pressure head. If an additional pressure is applied to the liquid column on the solution side of the membrane then it is possible to force water back through the membrane. This pressure-driven transport of water from a solution through a membrane is known as *reverse osmosis*. It may be noted that it is not quite the reverse of osmosis because, for all real membranes, there is always a certain transport of the solute along its chemical potential gradient, and this is not reversed. The phenomenon of reverse osmosis has been extensively developed as an industrial process for the concentration of low molecular weight solutes and especially for the desalination, or more generally demineralisation, of water.

Many models have been developed to explain the semi-permeability of reverse osmosis membranes and to rationalise the observed behaviour of separation equipment. These have included the postulation of preferential adsorption of the solute at the solution–membrane interface, hydrogen bonding of water in the membrane structure, and the exclusion of ions by the membrane due to dielectric effects. They are all useful in explaining aspects of membrane behaviour, although the most common approach has been to make use of the theories of the thermodynamics of irreversible processes proposed by SPIEGLER and KEDEM[20]. This gives a phenomenological description of the relative motion of solution components within the membrane, and does not allow for a microscopic explanation of the flow and rejection properties of the membrane. In the case of reverse osmosis however, the thermodynamic approach is combined with a macroscopic *solution–diffusion* description of membrane transport as discussed by SOLTANIEH and GILL[21]. This implies that the membrane is non-porous and that solvent and solutes can only be transported across the membrane by first dissolving in, and subsequently diffusing through, the membrane.

For any change to occur a chemical potential gradient must exist. For a membrane system, such as the one under consideration, HAASE[22] and BELFORT[23] have derived the following simplified equation for constant temperature:

$$d\mu_i = v_i \, dP + \left(\frac{\partial \mu_i}{\partial C_i}\right)_T dC_i + z_i \, \mathbf{F} \, d\phi \tag{8.30}$$

where $\mu_i$ is the chemical potential of component $i$, $v_i$ is the partial molar volume of component $i$, $z_i$ is the valence of component $i$, $\phi$ is the electrical potential and $\mathbf{F}$ is Faraday's constant. This equation may be applied to any membrane process. For ultrafiltration, only the pressure forces are usually considered. For electrodialysis, the electrical and concentration forces are more important, whereas, in the present case of reverse osmosis both pressure and concentration forces need to be considered. Integrating across the thickness of the membrane for a two-component system with subscript 1 used to designate the solvent (water) and subscript 2 used to designate the solute, for the solvent:

$$\begin{aligned}\Delta\mu_1 &= \int \left(\frac{\partial \mu_1}{\partial C_1}\right)_{P,T} dC_1 + \int v_1 \, dP \\ &= \int \left(\frac{\partial \mu_1}{\partial C_2}\right)_{P,T} dC_2 + \int v_1 \, dP \end{aligned} \tag{8.31}$$

When $\Delta\mu_1$ becomes small, only the osmotic pressure difference $\Delta\Pi$ remains. Thus, for constant $v_1$:

$$\Delta\mu_1 = v_1(|\Delta P| - |\Delta\Pi|) \tag{8.32}$$

For the solute:

$$\Delta\mu_2 = \int \left(\frac{\partial\mu_2}{\partial C_2}\right)_{P,T} \mathrm{d}C_2 + \int v_2\,\mathrm{d}P \tag{8.33}$$

and for dilute solutions, $\mu_2 = \mu_2^0 + \mathbf{R}T \ln C_2$, and for constant $v_2$:

$$\Delta\mu_2 = \mathbf{R}T\Delta \ln \mathbf{C}_2 + v_2|\Delta P| \tag{8.34}$$

where the second term on the right-hand side is negligible compared with the first term. In the present case:

$$\Delta\mu_2 = \mathbf{R}T\Delta \ln C_2 \approx \left(\frac{\mathbf{R}T}{C_2}\right)\Delta C_2 \text{ (for low values of } C_2) \tag{8.35}$$

Incorporating the model of diffusion across the membrane, and writing Fick's law in the generalised form[24], (using $\mu = \mu^0 + \mathbf{R}T \ln C$)*:

$$J = -\frac{DC}{\mathbf{R}T}\frac{\mathrm{d}\mu}{\mathrm{d}y} \tag{8.36}$$

where $y$ is distance in the direction of transfer.

It is found for the solvent that:

$$J_1 = K_1(|\Delta P| - |\Delta\Pi|) \tag{8.37}$$

where the permeability coefficient is described in terms of a diffusion coefficient, water concentration, partial molar volume of water, absolute temperature and effective membrane thickness. For the solute it is found that:

$$J_2 = K_2|\Delta C_2| \tag{8.38}$$

where $K_2$ is described in terms of a diffusion coefficient, distribution coefficient and effective membrane thickness. It is clear from these equations that solvent (water) flow only occurs if $|\Delta P| > |\Delta\Pi|$, though solute flow is independent of $|\Delta P|$. Thus, increasing the operating pressure increases the effective separation. This explains why reverse osmosis plants operate at relatively high pressure. For example, the osmotic pressure of brackish water containing 1.5–12 kg/m$^3$ salts is 0.1–0.7 MN/m$^2$ (MPa) and the osmotic pressure of sea water containing 30–50 kg/m$^3$ salts is 2.3–3.7 MN/m$^2$ (MPa). In practice, desalination plants operate at 3–8 MN/m$^2$ (MPa).

The rejection of dissolved ions at reverse osmosis membranes depends on valence. Typically, a membrane which rejects 93 per cent of $Na^+$ or $Cl^-$ will reject 98 per cent of $Ca^{2+}$ or $SO_4{}^{2-}$ when rejections are measured on solutions of a single salt. With mixtures of salts in solution, the rejection of a single ion is influenced by its

* Equation 8.36 represents the general form of Fick's law. The form used in previous chapters where driving force is expressed as a concentration gradient is a simplification of this equation.

relative proportion in the mixture. Thus, for 0.1 kg/m$^3$ $Cl^-$ in the presence of 1 kg/m$^3$ $SO_4{}^{2-}$ there would be only 50–70 per cent rejection compared with 93 per cent for solutions of a single salt. The rejection of organic molecules depends on molecular weight. Those with molecular weights less than 100 kg/kmol may not be rejected, those with molecular weights of about 150 kg/kmol may have about the same rejection as NaCl, and those with molecular weights greater than some 300 kg/kmol are effectively entirely rejected.

The thermodynamic approach does not make explicit the effects of concentration at the membrane. A good deal of the analysis of concentration polarisation given for ultrafiltration also applies to reverse osmosis. The control of the boundary layer is just as important. The main effects of concentration polarisation in this case are, however, a reduced value of solvent permeation rate as a result of an increased osmotic pressure at the membrane surface given in equation 8.37, and a decrease in solute rejection given in equation 8.38. In many applications it is usual to pretreat feeds in order to remove colloidal material before reverse osmosis. The components which must then be retained by reverse osmosis have higher diffusion coefficients than those encountered in ultrafiltration. Hence, the polarisation modulus given in equation 8.14 is lower, and the concentration of solutes at the membrane seldom results in the formation of a gel. For the case of turbulent flow the DITTUS–BOELTER correlation[9] may be used, as was the case for ultrafiltration giving a polarisation modulus of:

$$\frac{C_w}{C_f} = \exp\left(\frac{Jd^{0.2}}{0.023u^{0.8}D_L^{0.67}}\frac{\mu^{0.47}}{\rho^{0.47}}\right) \tag{8.39}$$

where solvent permeation rate, cross-flow velocity and solute diffusion coefficient have the greatest importance. The results are more complex for the case of laminar flow as $|\Delta\Pi|$ increases down the channel as the boundary layer is developed and water is removed. As a consequence, results do not appear in a closed form and finite-difference methods have been applied[10].

## 8.8. MEMBRANE MODULES AND PLANT CONFIGURATION

Membrane equipment for industrial scale operation of microfiltration, ultrafiltration and reverse osmosis is supplied in the form of modules. The area of membrane contained in these basic modules is in the range 1–20 m$^2$. The modules may be connected together in series or in parallel to form a plant of the required performance. The four most common types of membrane modules are tubular, flat sheet, spiral wound and hollow fibre, as shown in Figures 8.9–8.12.

(a) *Tubular modules* are widely used where it is advantageous to have a turbulent flow regime, for example, in the concentration of high solids content feeds. The membrane is cast on the inside of a porous support tube which is often housed in a perforated stainless steel pipe as shown in Figure 8.10. Individual modules contain a cluster of tubes in series held within a stainless steel permeate shroud. The tubes are generally 10–25 mm in diameter and 1–6 m in length. The feed

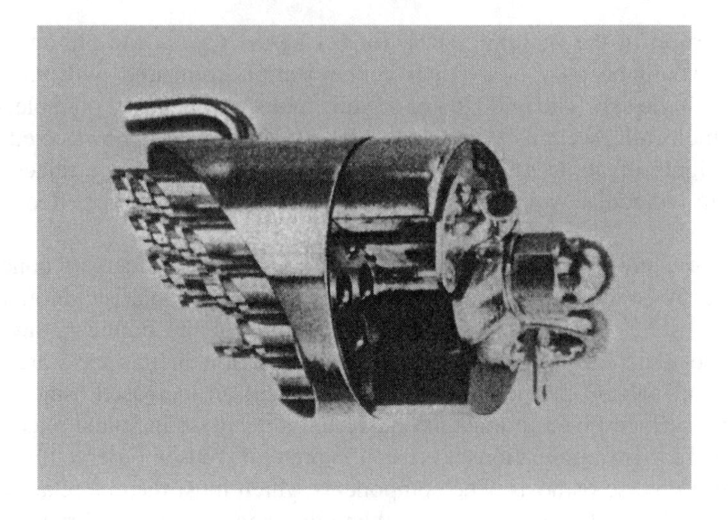

Figure 8.9. Tubular module

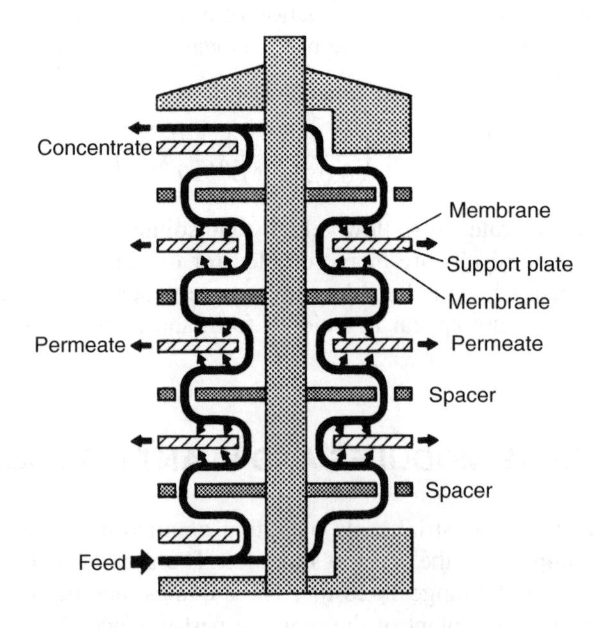


Figure 8.10. Schematic diagram of flat-sheet module

is pumped through the tubes at Reynolds numbers greater than 10,000. Tubular modules are easily cleaned and a good deal of operating data exist for them. Their main disadvantages are the relatively low membrane surface area contained in a module of given overall dimensions and their high volumetric hold-up.

(b) *Flat-sheet modules* are similar in some ways to conventional filter presses. An example is shown in Figure 8.10. This consists of a series of annular membrane discs of outer diameter 0.3 m placed on either side of polysulphone support plates

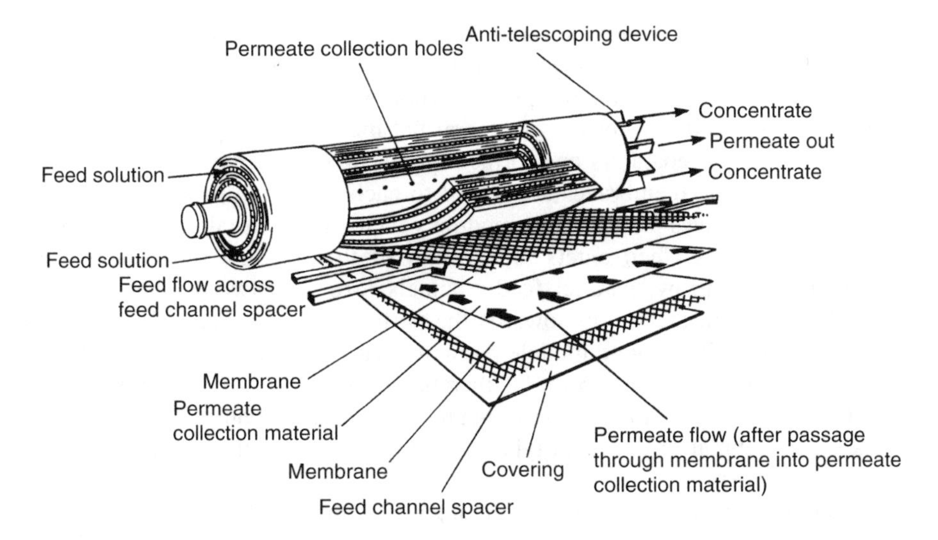


Figure 8.11. Schematic diagram of spiral-wound module

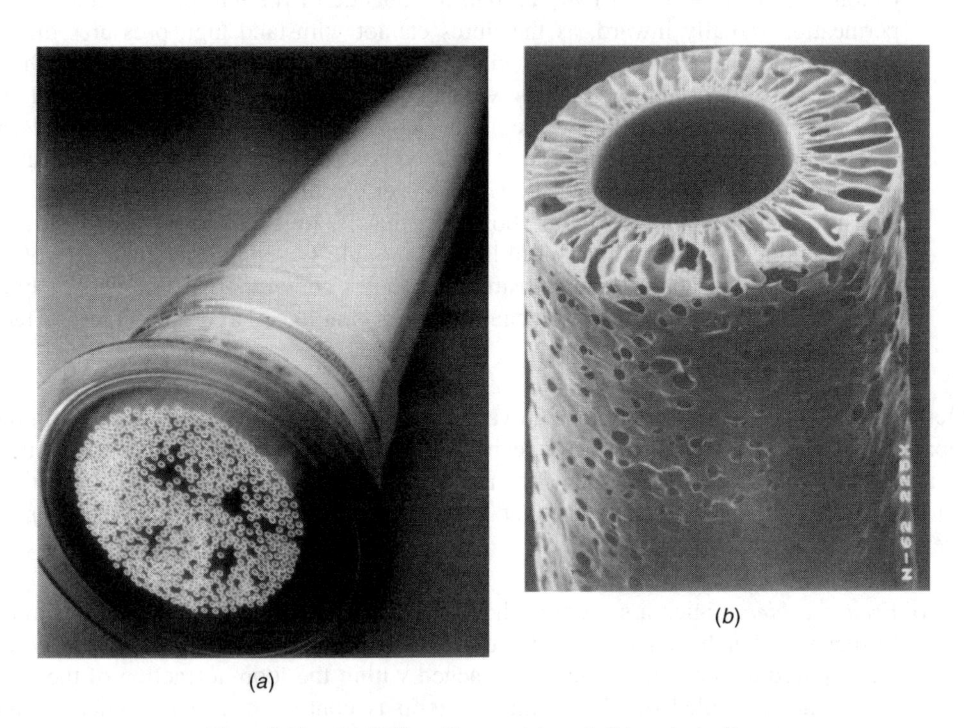


Figure 8.12. (*a*) Hollow-fibre module and, (*b*), a single fibre

which also provide channels through which permeate can be withdrawn. The sandwiches of membrane and support plate are separated from one another by spacer plates which have central and peripheral holes, through which the feed liquor is directed over the surface of the membranes, The flow is laminar. A single module contains 19 $m^2$ of membrane area. Permeate is collected from each membrane pair so that damaged membranes can be easily identified, though replacement of membranes requires dismantling of the whole stack.

(c) *Spiral-wound modules* consist of several flat membranes separated by turbulence-promoting mesh separators and formed into a *Swiss roll*, as shown in Figure 8.11. The edges of the membranes are sealed to each other and to a central perforated tube. This produces a cylindrical module which can be installed within a pressure tube. The process feed enters at one end of the pressure tube and encounters a number of narrow, parallel feed channels formed between adjacent sheets of membrane. Permeate spirals towards the perforated central tube for collection. A standard size spiral-wound module has a diameter of some 0.1 m, a length of about 0.9 m and contains about 5 $m^2$ of membrane area. Up to six such modules may be installed in series in a single pressure tube. These modules make better use of space than tubular or flat-sheet types, although they are rather prone to fouling and difficult to clean.

(d) *Hollow-fibre modules*, shown in Figure 8.12, consist of bundles of fine fibres, 0.1–2.0 mm in diameter, sealed in a tube. For reverse-osmosis desalination applications, the feed flow is usually around the outside of the unsupported fibres with permeation radially inward, as the fibres cannot withstand high pressures differences in the opposite direction. This gives very compact units capable of high pressure operation, although the flow channels are less than 0.1 mm wide and are therefore readily fouled and difficult to clean. The flow is usually reversed for biotechnological applications so that the feed passes down the centre of the fibres giving better controlled laminar flow and easier cleaning. This limits the operating pressure to less than 0.2 $MN/m^2$ however, that is, to microfiltration and ultrafiltration applications. A single ultrafiltration module typically contains up to 3000 fibres and be 1 m long. Reverse osmosis modules contain larger numbers of finer fibres. This is a very effective means of incorporating a large membrane surface area in a small volume.

Membrane modules can be configured in various ways to produce a plant of the required separation capability. A simple batch recirculation system has already been described in Section 8.5. Such an arrangement is most suitable for small-scale batch operation, but larger scale plants will operate either as *feed and bleed* or *continuous single-pass* operations, as shown in Figure 8.13.

(a) *Feed and bleed*. Such a system is shown in Figure 8.13(*a*). The start-up is similar to that in a batch system in that the retentate is initially totally recycled. When the required solute concentration is reached within the loop, a fraction of the loop is continuously bled off. Feed into the loop is controlled at a rate equal to the permeate plus concentrate flowrates. The main advantage is that the final concentration is then continuously available as feed is pumped into the loop. The main disadvantage is that the loop is operating continuously at a concentration equivalent to the final concentration in the batch system and the flux is therefore lower than the average flux in the batch mode, with a correspondingly higher membrane area requirement.

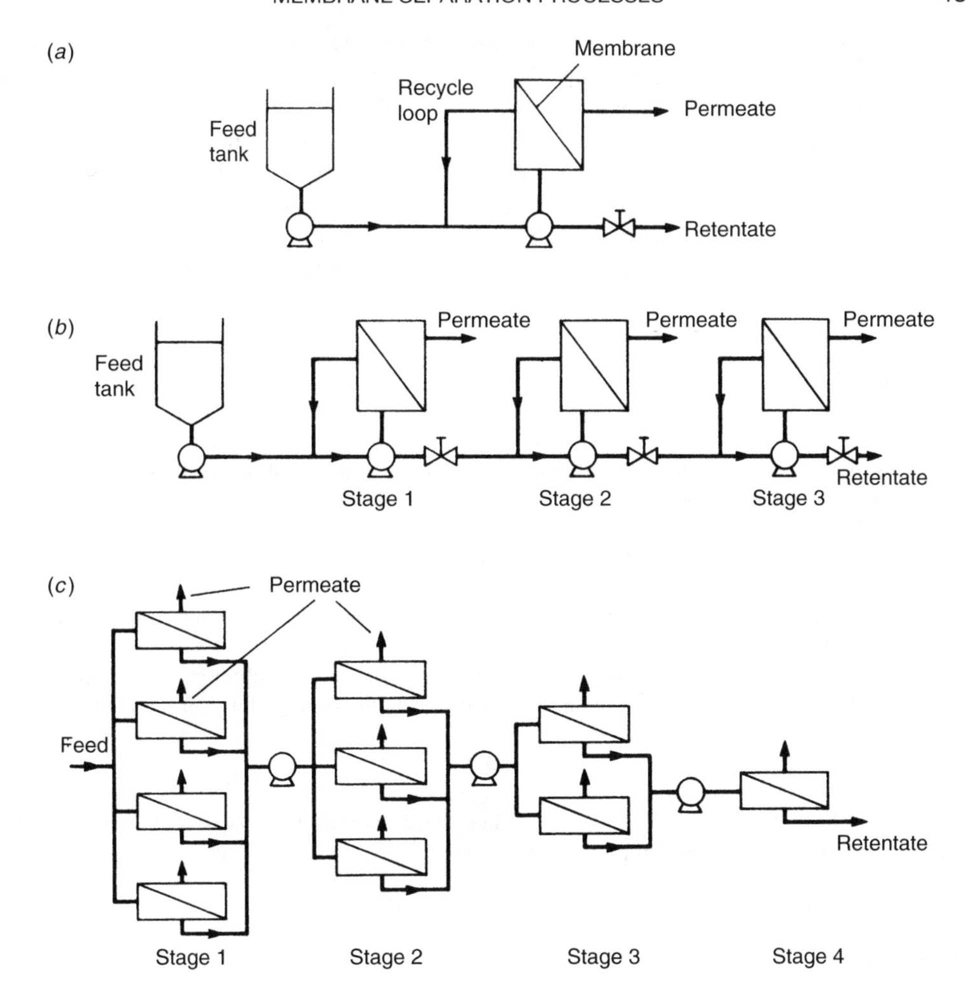


Figure 8.13. Schematic flow-diagrams of (*a*) Single-stage "feed and bleed", (*b*) Multiple-stage "feed and bleed", (*c*) Continuous single-pass membrane plants

Large-scale plants usually use multiple stages operated in series to overcome the low-flux disadvantage of the feed and bleed operation and yet to maintain its continuous nature as shown in Figure 8.13(*b*). Only the final stage is operating at the highest concentration and lowest flux, while the other stages are operating at lower concentrations with higher flux. Thus, the total membrane area is less than that required for a single-stage operation. Usually a minimum of three stages is required. The residence time, volume hold-up and tankage required are much less than for the same duty in batch operation. Feed and bleed systems also require less frequent sterilisation than batch processes in biotechnological applications, allowing longer effective operating times.

(b) *Continuous single pass*. In such a system the concentration of the feed stream increases gradually along the length of several stages of membrane modules arranged in series as shown in Figure 8.13($c$). The feed only reaches its final concentration at the last stage. There is no recycle and the system has a low residence time. Such systems must however, either be applied on a very large scale or have only a low overall concentration factor, due to the need to maintain high cross-flow velocities to control concentration polarisation. The smallest possible single-pass system will have a single module in the final stage, with a typical feed flowrate of 0.1 $m^3$/min. Such systems are used in large-scale reverse osmosis desalination plants but are unlikely to be used in biotechnological applications.

## Example 8.2

As part of a downstream processing sequence, 10 $m^3$ of a process fluid containing 20 kg $m^{-3}$ of an enzyme is to be concentrated to 200 kg/$m^3$ by means of ultrafiltration. Tests have shown that the enzyme is completely retained by a 10,000 MWCO surface-modified polysulphone membrane with a filtration flux given by:

$$J = 0.04\ln(250/C_f)$$

where $J$ is the flux in m/h and $C_b$ is the enzyme concentration in kg $m^{-3}$. Four hours is available for carrying out the process.

(a) Calculate the area of membrane needed to carry out the concentration as a simple batch process, (b) Use the following approximation for estimating the average flux during a simple batch process:

$$J_{av} = J_f + 0.27(J_i - J_f)$$

where $J_{av}$ is the average flux, $J_i$ is the initial flux and $J_f$ is the final flux. Is this approximation suitable for design purposes in the present case?

## Solution

(a) From the gel–polarisation model:

$$J = \frac{1}{A}\frac{\mathrm{d}V}{\mathrm{d}t} = h_D \ln\left(\frac{C_G}{C_f}\right)$$

Also:

$$C_f = C_0\left(\frac{V_0}{V}\right)$$

where $C_0$ and $V_0$ are the initial concentration and volume, respectively and $C_f$ and $V$ are the values at subsequent times.

Combining these equations gives:

$$\frac{\mathrm{d}V}{\mathrm{d}t} = A\left(h_D \ln\left(\frac{C_G}{C_0}\right) - h_D \ln\left(\frac{V_0}{V}\right)\right)$$

$$\int_{V_0}^{V_t} \frac{\mathrm{d}V}{\left(J_0 - h_D \ln\left(\frac{V_0}{V}\right)\right)} = A\int_0^t \mathrm{d}t$$

| $V$ | $\left(J_0 - h_D \ln(V_0/V)\right)^{-1}$ |
|---|---|
| 10 | 9.90 |
| 5 | 13.64 |
| 3 | 18.92 |
| 2 | 27.30 |
| 1 | 112.40 |

The data are plotted in Figure 8.14.

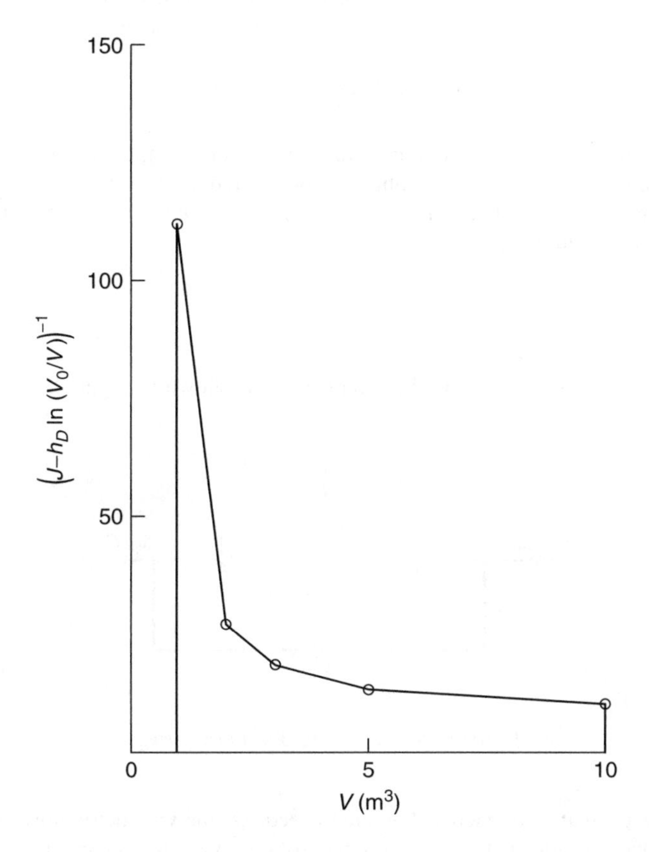


Figure 8.14. Graphical integration for Example 8.2

The area under the curve = 184.4

Operation for four hours gives $\underline{\underline{46.1 \text{ m}^2 \text{ membrane area}}}$

(b)

$$J_0 = 0.04 \ln(250/20) = 0.101 \text{ m/h}$$

$$J_f = 0.04 \ln(250/200) = 0.008 \text{ m/h}$$

$$J_{av} = 0.008 + 0.27(0.101 - 0.008) = 0.033 \text{ m/h}$$

For the removal of 9 $m^3$ filtrate in 4 hours:

Area $= (9/4)/0.033 = 68.2\ m^2$ membrane

The approximation is not suitable for design purposes.

## Example 8.3

An ultrafiltration plant is required to treat 50 $m^3$/day of a protein-containing waste stream. The waste contains 0.5 kg/$m^3$ of protein which has to be concentrated to 20 kg/$m^3$ so as to allow recycling to the main process stream. The tubular membranes to be used are available as 30 $m^2$ modules. Pilot plant studies show that the flux $J$ through these membranes is given by:

$$J = 0.02 \ln\left(\frac{30}{C_f}\right) \qquad \text{m/h.}$$

where $C_f$ is the concentration of protein in kg/$m^3$. Due to fouling, the flux never exceeds 0.04 m/h.

Estimate the minimum number of membrane modules required for the operation of this process (a) as a single *feed and bleed* stage, and (b) as two *feed and bleed* stages in series. Operation for 20 h/day may be assumed.

## Solution

(a) with a single *feed and bleed* stage, the arrangement is shown in Figure 8.15:

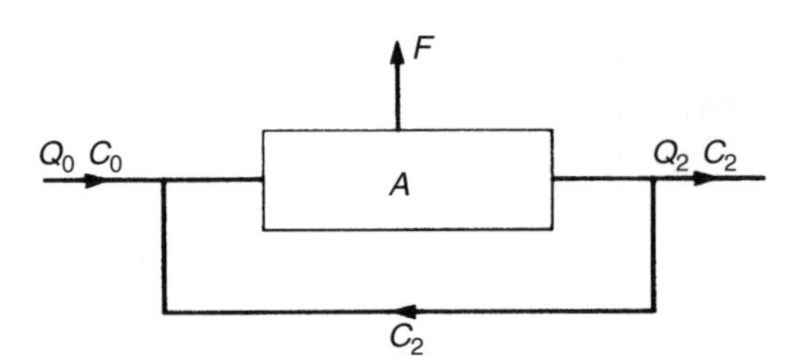


Figure 8.15. Single 'feed and bleed' stage

It is assumed that $Q_0$ is the volumetric flowrate of feed, $Q_2$ the volumetric flowrate of concentrate, $C_0$ the solute concentration in the feed, $C_2$ the solute concentration in the concentrate, $F$ the volumetric flowrate of membrane permeate, and $A$ the required membrane area. It is also assumed that there is no loss of solute through the membrane.

The concentration ($C_l$) at which the flux becomes fouling-limited is:

$$0.04 = 0.02 \ln\left(\frac{30}{C_l}\right)$$

or: $$C_l \approx 4\ \text{kg/m}^3$$

That is, below this concentration the membrane flux is 0.04 m/h.

This does not pose a constraint for the single stage as the concentration of solute $C_2$ will be that of the final concentrate, 20 kg/m$^3$.

Conservation of solute gives:

$$Q_0C_0 = Q_2C_2 \tag{i}$$

A fluid balance gives:

$$Q_0 = F + Q_2 \tag{ii}$$

Combining these equations and substituting known values:

$$2.438 = A\,0.02\ln\left(\frac{30}{20}\right)$$

and:

$$\underline{\underline{A = 301\ \text{m}^2}}$$

Thus, $\underline{\underline{\text{10 modules}}}$ will almost meet the specification for the single-stage process.

(b) with two *feed and bleed* stages in series, the arrangement is shown in Figure 8.16:

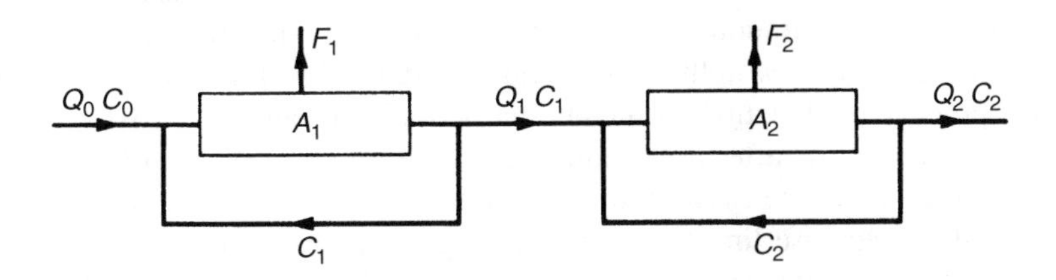


Figure 8.16. Two 'feed and bleed' stages in series

In addition to the symbols previously defined, $Q_1$ will be taken as the volumetric flowrate of retenate at the intermediate point, $C_1$ the concentration of solute in the retentate at this point, $F_1$ and $F_2$ the volumetric flowrates of membrane permeate in the first and second stages respectively, and $A_1$ and $A_2$ the required membrane areas in these respective stages.

Conservation of solute gives:

$$Q_0C_0 = Q_1C_1 = Q_2C_2 \tag{iii}$$

A fluid balance on stage 1 gives:

$$Q_0 = Q_1 + F_1 \tag{iv}$$

A fluid balance on stage 2 gives:

$$Q_1 = Q_2 + F_2 \tag{v}$$

Substituting given values in equations (iv) and (v) gives:

$$2.5 = \frac{1.25}{C_1} + 0.02\,A_1\ln\left(\frac{30}{C_1}\right) \tag{vi}$$

or:

$$\frac{1.25}{C_1} = 0.0625 + 0.00811A_1 \tag{vii}$$

The procedure is to use trial and error to estimate the value of $C_1$ that gives the optimum values of $A_1$ and $A_2$. Thus:

If $C_1 = 5$ kg/m$^3$, then, $A_1 = 63$ m$^2$ and $A_2 = 23$ m$^2$.
That is, an arrangement of 3 modules $-1$ module is required.

If $C_1 = 4$ kg/m$^3$, then $A_1 = 55$ m$^2$ and $A_2 = 31$ m$^2$.
That is, an arrangement of 2 modules $-1$ module is almost sufficient.

If $C_1 = 4.5$ kg/m$^3$, then $A_1 = 59$ m$^2$ and $A_2 = 27$ m$^2$.
That is, an arrangement of 2 modules $-1$ module which meets the requirement.

This arrangement requires the minimum number of modules.

## 8.9. MEMBRANE FOULING

A limitation to the more widespread use of membrane separation processes is membrane fouling, as would be expected in the industrial application of very finely porous materials. Fouling results in a continuous decline in membrane permeation rate, an increased rejection of low molecular weight solutes and eventually blocking of flow channels. On start-up of a process, a reduction in membrane permeation rate to 30–10 per cent of the pure water permeation rate after a few minutes of operation is common for ultrafiltration. Such a rapid decrease may be even more extreme for microfiltration. This is often followed by a more gradual decrease throughout processing. Fouling is partly due to blocking or reduction in effective diameter of membrane pores, and partly due to the formation of a slowly thickening layer on the membrane surface.

The extent of membrane fouling depends on the nature of the membrane used and on the properties of the process feed. The first means of control is therefore careful choice of membrane type. Secondly, a module design which provides suitable hydrodynamic conditions for the particular application should be chosen. Process feed pretreatment is also important. The type of pretreatment used in reverse osmosis for desalination applications is outlined in Section 8.11. In biotechnological applications pretreatment might include prefiltration, pasteurisation to destroy bacteria, or adjustment of pH or ionic strength to prevent protein precipitation. When membrane fouling has occurred, backflushing of the membrane may substantially restore the permeation rate. This is seldom totally effective however, so that chemical cleaning is eventually required. This involves interruption of the separation process, and consequently time losses due to the extensive nature of cleaning required. Thus, a typical cleaning procedure might involve: flushing with filtered water at 35–50 °C to displace residual retentate; recirculation or back-flushing with a cleaning agent, possibly at elevated temperature; rinsing with water to remove the cleaning agent; sterilisation by recirculation of a solution of 50–100 ppm of chlorine for 10–30 minutes (600–1800s) at (293–303 K) (20–30 °C); and flushing with water to remove sterilising solution. More recent approaches to the control of membrane fouling include the use of more sophisticated hydrodynamic control effected by pulsated feed flows or non-planar membrane surfaces, and the application of further perturbations at the membrane surface, such as continuous or pulsated electric fields.

## 8.10. ELECTRODIALYSIS

As discussed by PLETCHER[24], electrodialysis is an electrically driven membrane separation process. The main use of electrodialysis is in the production of drinking water by the desalination of sea-water or brackish water. Another large-scale application is in the production of sodium chloride for table salt, the principal method in Japan, with production exceeding $10^6$ tonne per annum.

In electrodialysis, cation exchange membranes are alternated with anion exchange membranes in a parallel array to form solution compartments[3] of thickness ~1 mm, as shown in Figure 8.17. A single membrane stack will typically contain 100–400 membranes each of area 0.5–2.0 $m^2$. The process feed is pumped through the solution compartments. Cations are transported towards the cathode and anions towards the anode when an electrical potential is applied to the electrodes. However, to a first approximation the cations can be transported across the cation exchange membranes but not across the anion exchange membranes, and conversely for the anions. The net result is ion depletion and ion concentration in alternate compartments throughout the stack. The power requirements of a stack are typically 100 A at 150 V.

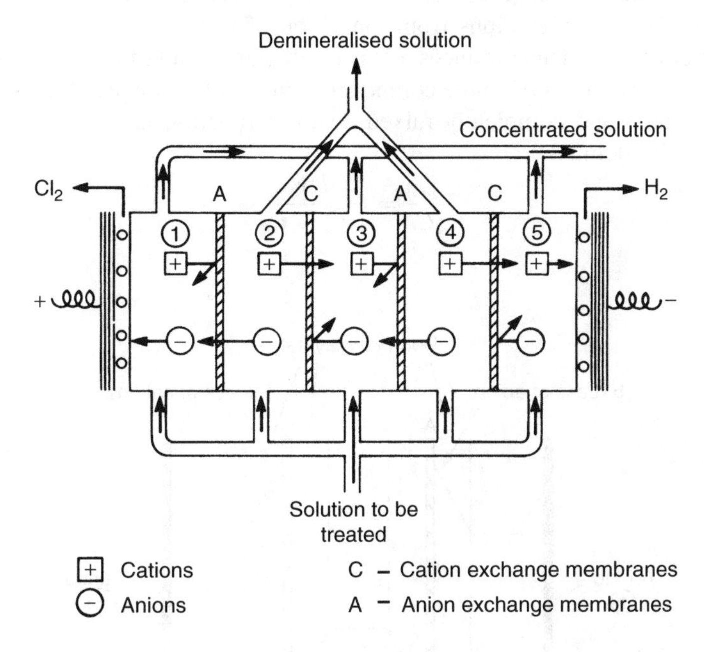


Figure 8.17. Schematic flow-diagram for an electrodialysis stack

The membranes in electrodialysis stacks are kept apart by spacers which define the flow channels for the process feed. There are two basic types[3], (*a*) tortuous path, causing the solution to flow in long narrow channels making several 180° bends between entrance and exit, and typically operating with a channel length-to-width ratio of 100:1 with a cross-flow velocity of 0.3–1.0 m/s (*b*) sheet flow, with a straight path from entrance to exit ports and a cross-flow velocity of 0.05–0.15 m/s. In both cases the spacer screens are also designed as turbulence promoters. These help reduce the concentration polarisation which occurs at the membranes, though with a rather different effect from that in the pressure driven membrane processes.

In order to understand the occurrence of concentration polarisation in electrodialysis, it is necessary to investigate how ion transport occurs. The electric current is carried through the system of membranes and solution channels by the anions and cations. The fraction of the current carried by a given ion is termed its transference number, designated $\mathbf{t}^+$ for cations and $\mathbf{t}^-$ for anions. If the transference number of anions through the solution ($\mathbf{t}_s^-$) is 0.5 and the transference number of anions through the anion exchange membrane ($\mathbf{t}_m^-$) is 1.0; then only half as many ions will be transferred electrically through the solution to the static side of the boundary film on the side of the membrane that anions enter, as will be transferred through the membrane. The solution at the membrane interface will be depleted of anions, and for similar reasons the boundary film on the other side of the membrane will accumulate anions as shown in Figure 8.18. The same effects will occur for cations at the cation exchange membranes. In practice, it is usually found that $\mathbf{t}_m \geq 0.90$. With any given thickness of boundary film a current density (and hence ion flux) can be reached at which the concentrations of electrolytes at the membrane interfaces on the depleting sides will approach zero. At such a current density, known as the *limiting current density*, $H^+$ and $OH^-$ ions from ionisation of water will begin to be transferred through the membranes. This produces a loss of efficiency and often causes fouling of the membranes by precipitation of solute components due to pH changes. With the idealisation of completely static and completely mixed zones, a function known as the *polarisation parameter* may be defined as:

$$\frac{i_{\text{lim}}}{zC_i} = \frac{D\mathbf{F}}{l(\mathbf{t}_m - \mathbf{t}_s)} \tag{8.40}$$

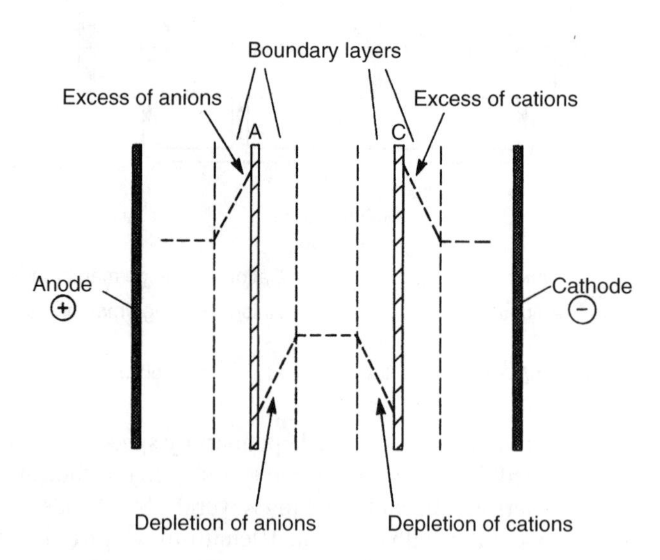


Figure 8.18. Schematic representation of concentration polarisation with demineralisation of the central compartment

where $i_{lim}$ is the limiting current density, $D$ the diffusion coefficient, $\mathbf{F}$ is Faraday's constant, $l$ the equivalent film thickness, $C_i$ is ion concentration (kmol/m$^3$) and $\mathbf{t}_m$ and $\mathbf{t}_s$ the transference numbers in the membrane and solution respectively.

The value of $i_{lim}$ is determined by the discontinuity in the dependence of cell current on applied cell voltage which occurs when the interfacial concentration approaches zero. The polarisation parameter is convenient in the design and scale-up of electrodialysis equipment. It can be easily measured in small-scale stacks at a given value of bulk concentration and then used to predict limiting current densities in larger stacks at other concentrations. Most stacks use operating values of the polarisation parameter that are 50–70 per cent of the limiting values.

## 8.11. REVERSE OSMOSIS WATER TREATMENT PLANT

The largest scale applications of membrane separation processes are those which form the key step in the desalination, or more generally demineralisation, of brackish water in the production of drinking water. In this section an outline is given of such a plant capable of producing 70,000 m$^3$/day of drinking water for a large city in the Middle East, as described by FINLAY and FERGUSON[(25)]. The water to be processed is obtained from a deep well with a total dissolved solids (TDS) content of 1.4 kg/m$^3$; that is, it is of moderate salinity and hardness. The plant specification required that the product water should have a maximum TDS of 0.5 kg/m$^3$. A flow diagram of the overall process is shown in Figure 8.19. Part of the flow bypasses the main pretreatment and demineralisation stages. This is possible because demineralisation by reverse osmosis produces a permeate with a TDS of about 0.22 kg/m$^3$, significantly better than the product specification. The minimum plant output is 59,000 m$^3$/day, requiring reverse osmosis demineralisation of 51,000 m$^3$/day of pretreated water, of which 6,000 m$^3$/day is rejected, leaving 45,000 m$^3$/day to be blended with 14,000 m$^3$/day of slipstream. Wash-water and other losses are less than 1,300 m$^3$/day. Hence, a total flow of 66,500 m$^3$/day is required from the cooling towers, with an overall water loss of less than 11 per cent. The main process steps are as follows:

### Pretreatment

(a) Evaporative cooling to reduce the feed water temperature from 50–55° C to 30–35° C which is more compatible with satisfactory operation of the reverse osmosis unit and more suitable for final use.
(b) Precipitation softening by addition of slaked lime ($Ca(OH)_2$) and sodium aluminate or ferric chloride. The net result is part-removal of calcium, silica and especially colloids. The clarifiers used ensure completion of these processes within the tank.
(c) Acidification to optimise removal of residual coagulant,
(d) Prechlorination to ensure a disinfected supply to the reverse osmosis plant.
(e) Rapid gravity filtration to reduce further the content of particulate material. The first stage is upflow through a 1.5 m deep gravel bed, and the second stage downflow

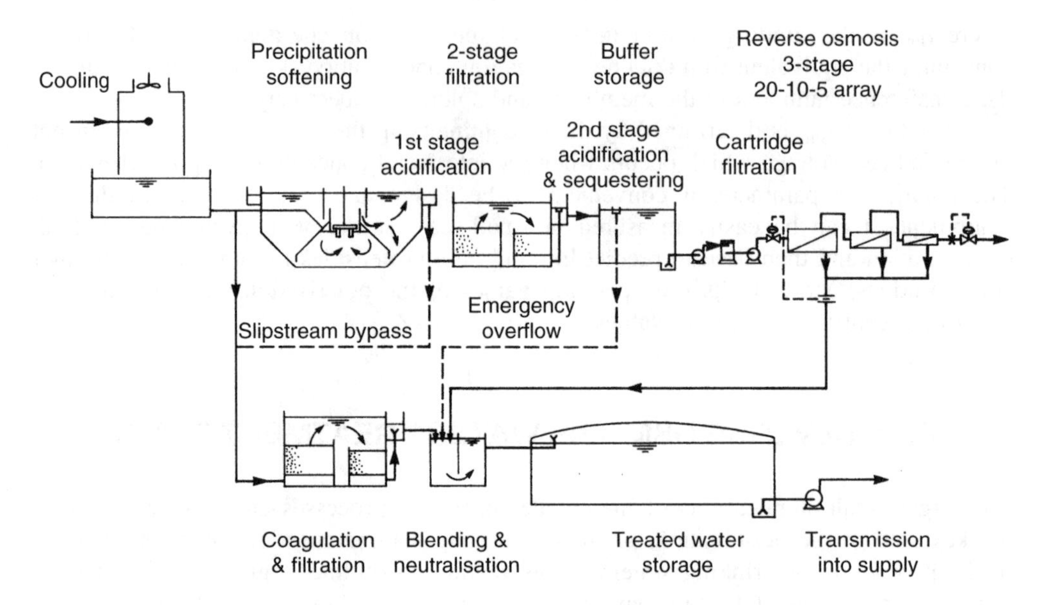


Figure 8.19. Outline flow-diagram of a large-scale reverse-osmosis plant for the demineralisation of brackish waters[26]

through a 1.1 m deep sand bed of effective particle size 0.9 mm. Identical units are used to filter the cooled and coagulated water in the slipstream.

(f) Acidification to reduce the pH to 5.0 for optimum life of the reverse osmosis membrane.

(g) Sequestering, addition of sodium hexametaphosphate to retard the precipitation of calcium sulphate which otherwise will exceed its solubility limit in the reject stream.

(h) Cartridge filtration with elements rated at 25 μm to protect the high pressure pumps and reverse osmosis membranes in the event of a break-through of particulate material.

### *Demineralisation by reverse osmosis*

Reverse osmosis was chosen for the demineralisation step as it gave an economic solution in terms of both capital and running costs, allowed a high water recovery rate, was modular in construction and so could be easily extended, could cope with reasonable variations in feed salinity and had a proven track record in relatively large-scale installations. The pre-booster pumps, cartridge filters and high pressure pumps are arranged in seven parallel streams, one of which is on standby. A total of thirteen reverse osmosis stacks is installed, any twelve of which will meet the required throughput. Each stack contains 210 reverse osmosis modules accommodated in 35 pressure vessels arranged in a series–parallel array of 20–10–5 to achieve the desired water recovery. These modules are of the spirally-wound type. The inlet to the reverse osmosis unit is instrumented for flowrate, pH, residual chlorine, turbidity, temperature and conductivity. The permeate from the unit is blended with the slipstream flow, with pH adjustment if necessary, to maintain a final water TDS $< 0.5$ kg/m$^3$. The reject is discharged to evaporation ponds.

In temperate climate zones it may be more appropriate to install a nanofiltration process rather than reverse osmosis. Nanofiltration allows the production of drinking water from polluted rivers. As for reverse osmosis, pretreatment is important to control fouling of the membranes. One of the largest such plants produces 140,000 m$^3$/day of water for the North Paris region[26].

## 8.12. PERVAPORATION

Two industrially important categories of separation problems are the separation of liquid mixtures which form an azeotrope and/or where there are only small differences in boiling characteristics. Pervaporation is a membrane process which shows promise for both of such separations as described by RAUTENBACH and ALBRECHT[27]. The process differs from other membrane processes in that there is a phase change from liquid to vapour in the permeate. The feed mixture is a liquid. The driving force in the membrane is achieved by lowering the activity of the permeating components at the permeate side. Components in the mixture permeate through the membrane and evaporate as a result of the partial pressure on the permeate side being held lower than the saturation vapour pressures. The driving force is controlled by applying a vacuum on the permeate side, as shown in Figure 8.20.

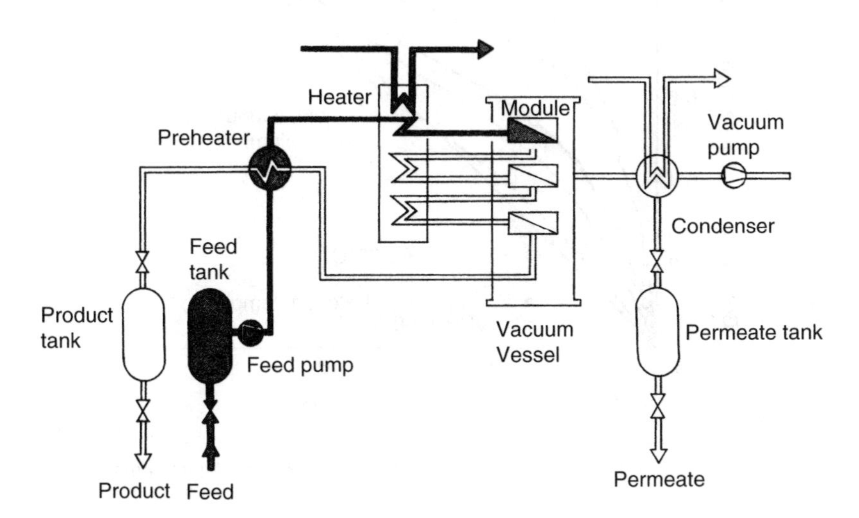


Figure 8.20. Schematic diagram of a pervaporation unit

Dense membranes are used for pervaporation, as for reverse osmosis, and the process can be described by a *solution–diffusion* model. That is, in an ideal case there is equilibrium at the membrane interfaces and diffusional transport of components through the bulk of the membrane. The activity of a component on the feed side of the membrane is proportional to the composition of that component in the feed solution.

The composition at the permeate-phase interface depends on the partial pressure and saturation vapour pressure of the component. Solvent composition within the membrane may vary considerably between the feed and permeate sides interface in pervaporation. By lowering the pressure at the permeate side, very low concentrations can be achieved while the solvent concentration on the feed-side can be up to 90 per cent by mass. Thus, in contrast to reverse osmosis, where such differences are not observed in practice, the modelling of material transport in pervaporation must take into account the concentration dependence of the diffusion coefficients.

Polyethylene (PE) is a standard material for separating organic mixtures, although selectivity and fluxes are too low for commercial application. A further development is the use of composite membranes in which a substantially insoluble support material is combined with an additive in which only one of the components of the mixture is highly soluble. An example is a membrane for separating benzene from cyclohexane consisting of a cellulose acetate support matrix and incorporating polyphosphonates to improve the preferential permeability of benzene (CA-PPN). Such mixtures are usually separated by extractive distillation as the equilibrium curve is very shallow and shows an azeotropic point. As shown in Figure 8.21 the separation characteristics are much more favourable in the case of pervaporation[28].

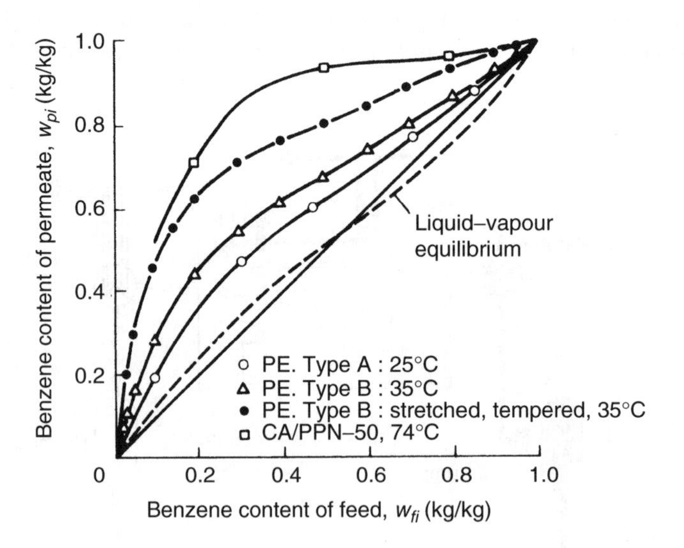


Figure 8.21. Comparison of selectivity of pervaporation membranes and liquid–vapour equilibrium for benzene/cyclohexane mixtures

At present, there is one main commercial application of pervaporation, the production of high purity alcohol by a hybrid process which also incorporates distillation. Such separations use cellulose-acetate-based composite-membranes, with an active layer of polyvinyl alcohol, for example. Membrane fluxes are in the range 0.45–2.2 kg/m$^2$ h. Pervaporation may have special potential if used following a conventional separation processes, such as distillation, when high product purities are required.

## 8.13. LIQUID MEMBRANES

As discussed by FRANKEMFELD and LI[28] and DEL CERRO and BOEY[29], liquid membrane extraction[28,29] involves the transport of solutes across thin layers of liquid interposed between two otherwise miscible liquid phases. There are two types of liquid membranes, emulsion liquid membranes (ELM) and supported liquid membranes (SLM). They are conceptually similar, but substantially different in their engineering.

ELM are multiple emulsions of the water/oil/water or oil/water/oil types shown in Figure 8.22. The membrane phase is that which is interposed between the continuous external phase and the encapsulated internal phase of the emulsion. Preparation of an ELM first involves emulsification of the inner phase in an immiscible solvent which may contain a surfactant and, depending on the application, other additives. The resulting emulsion is then dispersed in the continuous phase, either in mixing vessels or in column extractors. Globule diameters are in the range 0.1–2.0 mm, while the internal emulsion droplets are 1–10 μm in diameter. The interfacial areas for mass transfer can be as high as 3000 $m^2/m^3$. Mass transport can be from the continuous phase to the inner encapsulated droplets, or the reverse. The position of equilibrium can be enhanced by additives which cause a chemical or enzymatic reaction in the receiving phase. After extraction of the required product, the emulsion is broken into its component parts, usually by thermal or electrical treatment. In SLM, the liquid separating layer is held within a solid microporous inert support by capillary forces. The support is typically a microporous membrane of the type used for the more conventional pressure-driven membrane process. Large areas for mass transfer, up to about 1000 $m^2/m^3$, can be achieved using spiral-wound or hollow fibre modules.

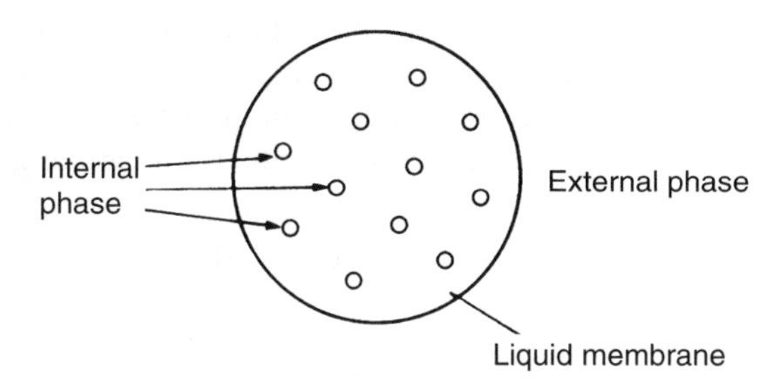


Figure 8.22. Schematic diagram of emulsion liquid-membrane

It is valuable to compare liquid membrane extraction with conventional solvent extraction in which the required product first partitions from an aqueous feed into an immiscible organic solvent. After separation of the phases, the product is extracted from the enriched solvent phase by contact with a stripping solution. Extraction and stripping occur simultaneously on either side of the membrane in liquid membrane extraction. As the liquid membrane is thin and large interfacial areas can be created, only short contact times are required to achieve a separation. Advantages include reductions both of solvent inventory and of equipment capacity.

The development of liquid-membrane extraction has been mainly in the fields of hydrometallurgy and waste-water treatment. There are also potential advantages for their use in biotechnology, such as extraction from fermentation broths, and biomedical engineering, such as blood oxygenation.

## 8.14. GAS SEPARATIONS

The main emphasis in this chapter is on the use of membranes for separations in liquid systems. As discussed by KOROS and CHERN[30] and KESTING and FRITZSCHE[31], gas mixtures may also be separated by membranes and both porous and non-porous membranes may be used. In the former case, Knudsen flow can result in separation, though the effect is relatively small. Much better separation is achieved with non-porous polymer membranes where the transport mechanism is based on sorption and diffusion. As for reverse osmosis and pervaporation, the transport equations for gas permeation through dense polymer membranes are based on Fick's Law, material transport being a function of the partial pressure difference across the membrane.

A number of polymers are suitable for gas permeation. The elimination of pores is crucial to the successful operation of membranes. Composite membranes have proved to be most suitable, for example silicone-coated polysulphone. The relative permeability of a number of gases in such a membrane is shown in Figure 8.23. Systems with high packing density, such as hollow fibre or spiral wound modules, are used for gas permeation. Using very fine fibres, membrane-area packing-densities of up to 50,000 $m^2/m^3$ are achieved. Modules with high, up to 14.8 $MN/m^2$ or low, 0.8–0.9 $MN/m^2$ feed-side operating pressures are available. The most important application of membrane gas separation is the generation of $N_2$ from air. The production of oxygen from air is also significant. Other substantial applications are the recovery of hydrogen from refinery off-gases (CO, $N_2$, $C_1$, $C_2$) and vapours ($C_{3+}$, $CO_2$), and the removal of carbon dioxide from natural gas.

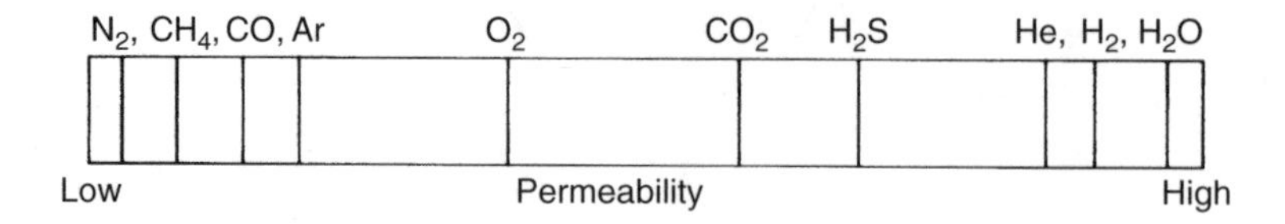


Figure 8.23. Permeability of polysulphone–silicone membranes

## 8.15. FURTHER READING

BELFORT, G. (ed.): *Synthetic Membrane Processes* (Academic Press, Orlando, 1984).
CHERYAN, M.: *Ultrafiltration Handbook* (Technomic Publishing Company, Pennsylvania, 1998).

RAUTENBACH, R. and ALBRECHT, R.: *Membrane Processes* (Wiley, 1989).
SCHWEITZER, P. A., (ed.): *Handbook of Separation Techniques for Chemical Engineers*, 2nd edn. (McGraw Hill, New York, 1988).
ZEMAN, L. J. and ZYDNEY, A. L., *Microfiltration and Ultrafiltration. Principles and Applications* (Marcel Dekker, New York, 1996).

## 8.16. REFERENCES

1. STRATHMANN, H.: *J. Memb. Sci.* **9** (1981) 121. Membrane separation processes (review).
2. LONSDALE, H. K.: *J. Memb. Sci.* **10** (1982) 81. The growth of membrane technology (review).
3. LACEY, R. E.: In *Handbook of Separation Techniques for Chemical Engineers*, SCHWEITZER, P. A. (ed.) 2nd edn. (McGraw-Hill, New York, 1988). Dialysis and electrodialysis.
4. BERTERA, R. STEVEN, H. and METCALFE, M.: *The Chemical Engineer*, No. 401 (June, 1984). 10. Development studies of crossflow microfiltration.
5. SCHNEIDER, K. and KLEIN, W.: *Desalination*, **41** (1983) 271. The concentration of suspensions by means of cross flow microfiltration.
6. FANE, A. G.: In *Progress in Filtration and Separation* Vol. 4, WAKEMAN, R. J. (ed.) (Elsevier, Amsterdam, 1986). Ultrafiltration: Factors influencing flux and rejection.
7. LÉVÊQUE, M. D.: *Ann. Mines*, **13**, April 1928, 201. Les lois de la transmission de chaleur pour convection.
8. BLATT, W. F., DRAVID, A., MICHAELS, A. S. and NELSON, L.: In *Membrane Science and Technology*, FLINN, J. E. (ed.) (Plenum, New York, 1970). Solute polarisation and cake formation in membrane ultrafiltration: Causes, consequences and control techniques.
9. DITTUS, F. W. and BOELTER, L. M. K.: *Univ. Calif. Berkeley, Publ. Eng.* **2** (1930) 443. Heat transfer in automobile radiators of the tubular type.
10. MICHAELS, A. S.: *Chem. Eng. Prog.*, **64** (1968) 31. New separation technique for the CPI.
11. PORTER, M. C.: In *Handbook of Separation Techniques for Chemical Engineers*, SCHWEITZER, P. A. (ed.) 2nd edn. (McGraw-Hill, New York, 1988). Membrane Filtration.
12. LE, M. S. and HOWELL, J. A.: Ultrafiltration, in *Comprehensive Biotechnology*, **2**, MOO-YOUNG, M, (ed.) (Pergamon, Oxford, 1985).
13. GREEN, G and BELFORT, G.: *Desalination* **35** (1980) 129. Fouling of ultrafiltration membranes: lateral migration and particle trajectory model.
14. PORTER, M. C.: *Ind. Eng. Chem. Prod. Res. Develop*, **11** (1972) 234. Concentration polarisation with membrane ultrafiltration.
15. SERGRE, G. and SILBERBERG, A.: *Nature* **189** (1961) 209. Radial particle displacements in Poiseuille flow of suspensions.
16. BELFORT, G.: In *Advanced Biochemical Engineering*, BUNGAY, H. R. and BELFORT, G. (eds.) (John Wiley, New York, 1987). Membrane Separation Technology: An Overview.
17. ALTENA, F. W. and BELFORT, G.: *Chem. Eng. Sci.* **39** (1984) 343. Lateral migration of spherical particles in porous flow channels: application to membrane filtration.
18. WIJMANS, J. G., NAKAO, S. and SMOLDERS, C. A.: *Journal of Membrane Science* **20** (1984) 115. Flux limitation in ultrafiltration; osmotic pressure model and gel layer model.
19. BOWEN, W. R. and WILLIAMS, P. M.: *Chem. Eng. Sci.* **56** (2001) 3083. Prediction of the rate of cross-flow ultrafiltration of colloids with concentration–dependent diffusion coefficient and viscosity–theory and experiment.
20. SPIEGLER, K. S. and KEDEM, O.: *Desalination* **1** (1966) 311. Thermodynamics of hyperfiltration (reverse osmosis): criteria for efficient membranes.
21. SOLTANIEH, M. and GILL, W. N.: *Chem. Eng. Commun.* **12** (1981) 279. Review of reverse osmosis membranes and transport models.
22. HAASE, R.: *Thermodynamics of Irreversible Processes* (Addison Wesley, Massachusetts, 1969).
23. BELFORT, G.: In *Synthetic Membrane Processes*, BELFORT, G. (ed.) (Academic Press, Orlando, 1984). Desalting experience by hyperfiltration (reverse osmosis) in the United States.
24. PLETCHER, D.: *Industrial Electrochemistry* (Chapman and Hall, London, 1982).
25. FINLAY, W. S. and FERGUSON, P. V.: Design and operation of a turnkey reverse osmosis water treatment plant, presented at the *International Congress on Desalination and Water Re-use* (Tokyo, 1977).
26. VENTRESQUE, C., GISCLON, V., BABLON, G. and CHAGNEAU, G.: *Desalination* **131** (2001). An outstanding feat of modern technology: the Mery-sur-Oise nanofiltration treatment plant.
27. RAUTENBACH, R. and ALBRECHT, R.: *Membrane Processes* (Wiley, Chichester, 1989).
28. FRANKENFELD, J. W. and LI, N. N.: In *Handbook of Separation Process Technology*, ROUSSEAU, R. W. (ed.) (Wiley, New York, 1987) pp. 840–861. Recent advances in membrane technology.

29. DEL CERRO, C. and BOEY, D.: *Chemistry and Industry* (November 1988) 681–687. Liquid membrane extraction.
30. KOROS, W. J. and CHERN, R. T.: In *Handbook of Separation Process Technology*, ROUSSEAU, R. W. (ed.) (Wiley, New York, 1987) pp. 862–953. Separation of gaseous mixtures using polymer membranes.
31. KESTING, R. E. and FRITZSCHE, A. K. *Polymeric Gas Separation Membranes*, (Wiley, New York, 1993).

## 8.17. NOMENCLATURE

| | | Units in SI System | Dimensions in **M**, **N**, **L**, **T**, $\boldsymbol{\theta}$, **A** |
|---|---|---|---|
| $A_m$ | Membrane area | m$^2$ | $\mathbf{L}^2$ |
| $a \ldots a_n$ | Osmotic coefficients | N/m$^2$ | $\mathbf{ML}^{-1}\mathbf{T}^{-2}$ |
| $b$ | Channel height | m | **L** |
| $C$ | Solute concentration (mass fraction) | — | — |
| $C_b$ | Bulk concentration of particles | m$^3$/m$^3$ | — |
| $C_i$ | Concentration of ions | kmol/m$^3$ | $\mathbf{NL}^{-3}$ |
| $C_G$ | Solute concentration in gel layer | — | — |
| $C_f$ | Solute concentration in bulk of feed | — | — |
| $C_p$ | Solute concentration in permeate | — | — |
| $C_w$ | Solute concentration at membrane | — | — |
| $D_L$ | Diffusion coefficient for liquid | m$^2$/s | $\mathbf{L}^2\mathbf{T}^{-1}$ |
| $d_e$ | Channel hydraulic diameter | m | **L** |
| $e$ | Void fraction (porosity) | — | — |
| **F** | Faraday's constant | $9.649 \times 10^7$ C/kmol | $\mathbf{N}^{-1}\mathbf{TA}$ |
| $h_D$ | Mass transfer coefficient | m/s | $\mathbf{LT}^{-1}$ |
| $i_{\lim}$ | Limiting current density | A/m$^2$ | $\mathbf{L}^{-2}\mathbf{A}$ |
| $J$ | Membrane permeation rate (flux) | m/s | $\mathbf{LT}^{-1}$ |
| $J_{ss}$ | Steady state membrane permeation rate | m/s | $\mathbf{LT}^{-1}$ |
| $l$ | Thickness of boundary film | m | **L** |
| $\Delta P$ | Hydraulic pressure difference | N/m$^2$ | $\mathbf{ML}^{-1}\mathbf{T}^{-2}$ |
| **R** | Universal gas constant | 8314 J/kmolK | $\mathbf{MN}^{-1}\mathbf{L}^2\mathbf{T}^{-2}\boldsymbol{\theta}^{-1}$ |
| $R$ | Rejection coefficient | — | — |
| $R_c$ | Resistance of deposited layers | m$^{-1}$ | $\mathbf{L}^{-1}$ |
| $R_f$ | Resistance of film layer | m$^{-1}$ | $\mathbf{L}^{-1}$ |
| $R_m$ | Resistance of membrane | m$^{-1}$ | $\mathbf{L}^{-1}$ |
| **r** | Specific resistance of deposit | m$^{-2}$ | $\mathbf{L}^{-2}$ |
| $T$ | Absolute temperature | K | $\boldsymbol{\theta}$ |
| $t$ | Time | s | **T** |
| $\mathbf{t}_m$ | Transference number in membrane | — | — |
| $\mathbf{t}_s$ | Transference number in solution | — | — |
| $u$ | Cross-flow velocity | m/s | $\mathbf{LT}^{-1}$ |
| $v$ | Partial molar volume | m$^3$/kmol | $\mathbf{N}^{-1}\mathbf{L}^3$ |
| $V$ | Volume of filtrate | m$^3$ | $\mathbf{L}^3$ |
| $V_s$ | Skeletal volume of particles deposited | m$^3$ | $\mathbf{L}^3$ |
| $y$ | Distance from membrane | m | **L** |
| $z$ | Valence of ion | — | — |
| $\beta$ | Efficiency factor | — | — |
| $\phi$ | Electrical potential | V | $\mathbf{ML}^2\mathbf{T}^{-3}\mathbf{A}^{-1}$ |
| $\dot{\gamma}$ | Shear rate at membrane surface | s$^{-1}$ | $\mathbf{T}^{-1}$ |
| $\mu$ | Viscosity of permeate | Ns/m$^2$ | $\mathbf{ML}^{-1}\mathbf{T}^{-1}$ |
| $\mu_i$ | Chemical potential | J/kmol | $\mathbf{MN}^{-1}\mathbf{L}^2\mathbf{T}^{-2}$ |
| $\Pi$ | Osmotic pressure | N/m$^2$ | $\mathbf{ML}^{-1}\mathbf{T}^{-2}$ |
| $\rho$ | Fluid density | kg/m$^3$ | $\mathbf{ML}^{-3}$ |

CHAPTER 9

# *Centrifugal Separations*

## 9.1. INTRODUCTION

There is now a wide range of situations where centrifugal force is used in place of the gravitational force in order to effect separations. The resulting accelerations may be several thousand times that attributable to gravity. Some of the benefits include far greater rates of separation; the possibility of achieving separations which are either not practically feasible, or actually impossible, in the gravitational field; and a substantial reduction of the size of the equipment. Recent developments in the use of centrifugal fields in *process intensification* are discussed in Chapter 20.

Centrifugal fields can be generated in two distinctly different ways:

(a) By introducing a fluid with a high tangential velocity into a cylindrical or conical vessel, as in the *hydrocyclone* and in the *cyclone separator* (Chapter 1). In this case, the flow pattern in the body of the separator approximates to a *free vortex* in which the tangential velocity varies *inversely* with the radius (see Volume 1, Chapter 2). Generally, the larger and heavier particles will collect and be removed near the walls of the separator, and the smaller and lighter particles will be taken off through an outlet near the axis of the vessel.

(b) By the use of the *centrifuge*. In this case the fluid is introduced into some form of rotating bowl and is rapidly accelerated. Because the frictional drag within the fluid ensures that there is very little *rotational slip* or relative motion between fluid layers within the bowl, all the fluid tends to rotate at a constant angular velocity $\omega$ and a *forced vortex* is established. Under these conditions, the tangential velocity will be *directly* proportional to the radius at which the fluid is rotating.

In this chapter, attention is focused on the operation of the centrifuge. Some of the areas where it is extensively used are as follows:

(a) *For separating particles on the basis of their size or density.* This is effectively using a centrifugal field to achieve a higher rate of sedimentation than could be achieved under gravity.

(b) *For separating immiscible liquids of different densities*, which may be in the form of dispersions or even emulsions in the feed stream. This is the equivalent of a gravitational decantation process.

(c) *For filtration of a suspension.* In this case centrifugal force replaces the force of gravity or the force attributable to an applied pressure difference across the filter.

(d) *For the drying of solids and, in particular, crystals.* Liquid may be adhering to the surface of solid particles and may be trapped between groups of particles. Drainage may be slow in the gravitational field, especially if the liquid has a high viscosity. Furthermore, liquid is held in place by surface tension forces which must be exceeded before liquid can be freed. This is particularly important with fine particles. Thus, processes which are not possible in the gravitational field can be carried out in the centrifuge.

(e) *For breaking down of emulsions and colloidal suspensions.* A colloid or emulsion may be quite stable in the gravitational field where the dispersive forces, such as those due to Brownian motion, are large compared with the gravitational forces acting on the fine particles or droplets. In a centrifugal field which may be several thousand times more powerful, however, the dispersive forces are no longer sufficient to maintain the particles in suspension and separation is effected.

(f) *For the separation of gases.* In the nuclear industry isotopes are separated in the gas centrifuge in which the accelerating forces are sufficiently great to overcome the dispersive effects of molecular motion. Because of the very small difference in density between isotopes and between compounds of different isotopes, fields of very high intensity are needed.

(g) *For mass transfer processes.* Because far greater efficiencies and higher throughputs can be obtained before flooding occurs, centrifugal packed bed contactors are finding favour and are replacing ordinary packed columns in situations where compactness is important, or where it is desirable to reduce the holdup of materials undergoing processing because of their hazardous properties. An important application is the use of inert gases in the desorption of oxygen from sea water in order to reduce its corrosiveness; in North Sea oil rigs the sea water is used as a coolant in heat exchangers. In addition, centrifugal contactors for liquid–liquid extraction processes now have important applications discussed in Chapter 13. These are additional areas where centrifugal fields, and the employment of centrifuges, is gaining in importance.

## 9.2. SHAPE OF THE FREE SURFACE OF THE LIQUID

For an element of liquid in a centrifuge bowl which is rotating at an angular velocity of $\omega$, the centrifugal acceleration is $r\omega^2$, compared with the gravitational acceleration of $g$. The ratio $r\omega^2/g$ is one measure of the separating effect obtained in a centrifuge relative to that arising from the gravitational field; values of $r\omega^2/g$ may be very high (up to $10^4$) in some industrial centrifuges and more than an order of magnitude greater in the ultra-centrifuge, discussed in Section 9.8. In practice, the axis of rotation may be vertical, horizontal, or intermediate, and the orientation is usually determined by the means adopted for introducing feed and, removing product streams, from the centrifuge.

Figure 9.1 shows an element of the free surface of the liquid in a bowl which is rotating at a radius $r_0$ about a vertical axis at a very low speed; the centrifugal and gravitational fields will then be of the same order of magnitude. The centrifugal force per unit mass is $r_0\omega^2$ and the corresponding gravitational force is $g$. These two forces are perpendicular

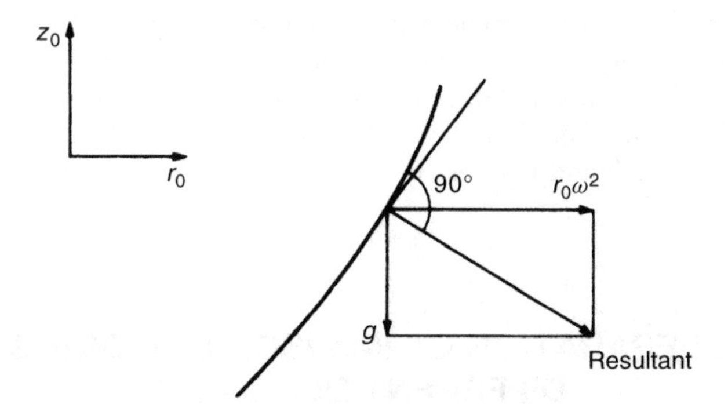


Figure 9.1. Element of surface of liquid

to one another and may be combined as shown to give the resultant force which must, at equilibrium, be at right angles to the free surface. Thus, the slope at this point is given by:

$$\frac{dz_0}{dr_0} = \frac{\text{radial component of force}}{\text{axial component of force}} = \frac{r_0\omega^2}{g} \tag{9.1}$$

where $z_0$ is the axial coordinate of the free surface of the liquid.

Equation 9.1 may be integrated to give:

$$z_0 = \frac{\omega^2}{2g}r_0^2 + \text{constant}$$

If $z_a$ is the value of $z_0$ which corresponds to the position where the free surface is at the axis of rotation ($r_0 = 0$), then:

$$z_0 - z_a = \frac{\omega^2}{2g}r_0^2 \tag{9.2}$$

Equations 9.1 and 9.2 correspond with equations 2.80 and 2.79 in Volume 1, Chapter 2. Taking the base of the bowl as the origin for the measurement of $z_0$, positive values of $z_a$ correspond to conditions where the whole of the bottom of the bowl is covered by liquid. Negative values of $z_a$ imply that the paraboloid of revolution describing the free surface would cut the axis of rotation below the bottom, and therefore the central portion of the bowl will be dry.

Normally $r_0\omega^2 \gg g$, the surface is nearly vertical, and $z_0$ has a very large negative value. Thus, in practice, the free surface of the liquid will be effectively concentric with the walls of the bowl. It is seen therefore that the operation of a high speed centrifuge is independent of the orientation of the axis of rotation.

## 9.3. CENTRIFUGAL PRESSURE

A force balance on a sector of fluid in the rotating bowl, carried out as in Volume 1, Chapter 2, gives the pressure gradient at a radius $r$:

$$\frac{\partial P}{\partial r} = \rho\omega^2 r \tag{9.3}$$

Unlike the vertical pressure gradient in a column of liquid which is constant at all heights, the centrifugal pressure gradient is a function of radius of rotation $r$, and increases towards the wall of the basket. Integration of equation 9.3 at a given height gives the pressure $P$ exerted by the liquid on the walls of the bowl of radius $R$ when the radius of the inner surface of the liquid is $r_0$ as:

$$P = \frac{1}{2}\rho\omega^2(R^2 - r_0^2) \tag{9.4}$$

## 9.4. SEPARATION OF IMMISCIBLE LIQUIDS OF DIFFERENT DENSITIES

The problem of the continuous separation of a mixture of two liquids of different densities is most readily understood by first considering the operation of a gravity settler, as shown in Figure 9.2. For equilibrium, the hydrostatic pressure exerted by a height $z$ of the denser liquid must equal that due to a height $z_2$ of the heavier liquid and a height $z_1$ of the lighter liquid in the separator.

Thus:
$$z\rho_2 g = z_2\rho_2 g + z_1\rho_1 g$$

or:
$$z = z_2 + z_1\frac{\rho_1}{\rho_2} \tag{9.5}$$

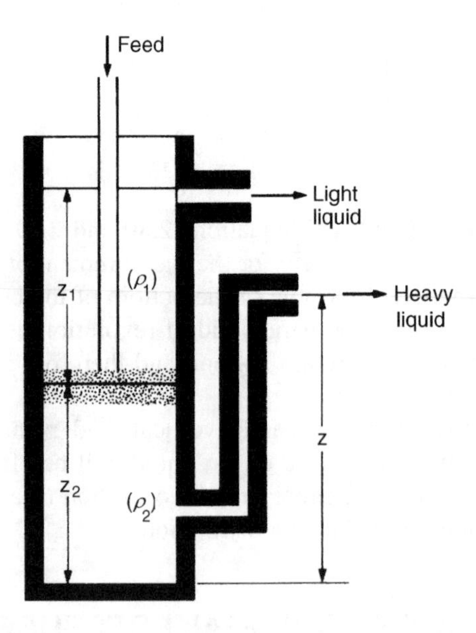


Figure 9.2. Gravity separation of two immiscible liquids

For the centrifuge it is necessary to position the overflow on the same principle, as shown in Figure 9.3. In this case the radius $r_i$ of the weir for the less dense liquid will correspond approximately to the radius of the inner surface of the liquid in the bowl. That

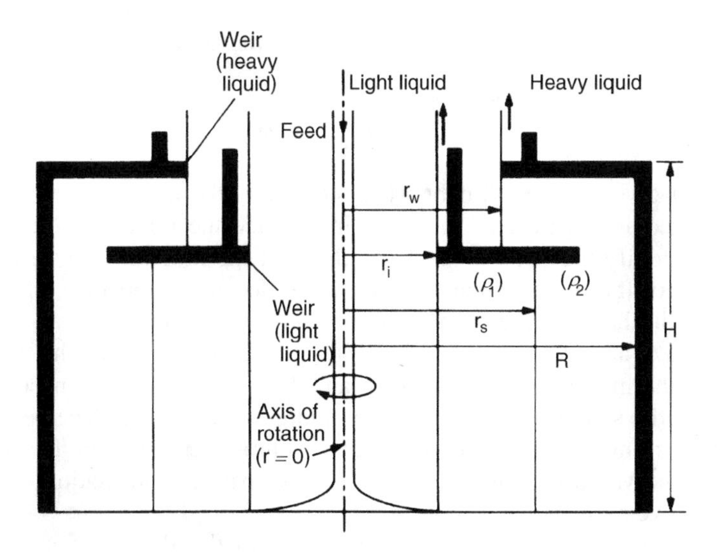


Figure 9.3. Separation of two immiscible liquids in a centrifuge

of the outer weir $r_w$ must be such that the pressure developed at the wall of the bowl of radius $R$ by the heavy liquid alone as it flows over the weir is equal to that due to the two liquids within the bowl. Thus, applying equation 9.4 and denoting the densities of the light and heavy liquids by $\rho_1$ and $\rho_2$ respectively and the radius of the interface between the two liquids in the bowl as $r_s$:

$$\tfrac{1}{2}\rho_2\omega^2(R^2 - r_w^2) = \tfrac{1}{2}\rho_2\omega^2(R^2 - r_s^2) + \tfrac{1}{2}\rho_1\omega^2(r_s^2 - r_i^2)$$

or:

$$\frac{r_s^2 - r_i^2}{r_s^2 - r_w^2} = \frac{\rho_2}{\rho_1} \tag{9.6}$$

If $Q_1$ and $Q_2$ are the volumetric rates of feed of the light and heavy liquids respectively, on the assumption that there is no slip between the liquids in the bowl and that the same, then residence time is required for the two phases, then:

$$\frac{Q_1}{Q_2} = \frac{r_s^2 - r_i^2}{R^2 - r_s^2} \tag{9.7}$$

Equation 9.7 enables the value of $r_s$ to be calculated for a given operating condition.

The retention time $t_R$ necessary to give adequate separation of the liquids will depend on their densities and interfacial tension, and on the form of the dispersion, and can only be determined experimentally for that mixture. The retention time is given by:

$$t_R = \frac{V'}{Q_1 + Q_2} = \frac{V'}{Q} \tag{9.8}$$

where $Q$ is the total feed rate of liquid, $Q_1$ and $Q_2$ refer to the light and heavy liquids respectively, and $V'$ is the volumetric holdup of liquid in the bowl.

Approximately:

$$V' \approx \pi(R^2 - r_i^2)H \tag{9.9}$$

where $H$ is the axial length (or clarifying length) of the bowl.

Thus:
$$t_R = \frac{Q}{\pi(R^2 - r_i^2)H} \tag{9.10}$$

Equation 9.10 gives the relation between $Q$ and $r_i$ for a given retention time, and determines the required setting of the weir. In practice the relative values of $r_i$ and $r_w$ are adjusted to modify the relative residence times of the individual phases to give extra separating time for the more difficult phase. In the decanter centrifuge, $r_i$ is adjusted to influence either discharged cake dryness or the conveying efficiency.

In equations 9.6 and 9.7, $\rho_2/\rho_1$ and $Q_1/Q_2$ are determined by the properties and composition of the mixture to be separated, and $R$ is fixed for a given bowl; $r_i$ and $r_w$ are governed by the settings of the weirs. The radius $r_s$ of the interface between the two liquids is then the only unknown and this may be eliminated between the two equations. Substitution of the value $r_i$ from equation 9.10 then permits calculation of the required radius $r_w$ of the weir for the heavy liquid.

This treatment gives only a simplified approach to the design of a system for the separation of two liquids. It will need modification to take account of the geometric configuration of the centrifuge — a topic which is discussed in Section 9.8.

## 9.5. SEDIMENTATION IN A CENTRIFUGAL FIELD

Centrifuges are extensively used for separating fine solids from suspension in a liquid. As a result of the far greater separating power compared with that available using gravity, fine solids and even colloids may be separated. Furthermore, it is possible to break down emulsions and to separate dispersions of fine liquid droplets, though in this case the suspended phase is in the form of liquid droplets which will coalesce following separation. Centrifuges may be used for batch operation when dealing with small quantities of suspension although, on the large scale, arrangements must sometimes be incorporated for the continuous removal of the separated constituents. Some of the methods of achieving this are discussed along with the various types of equipment in Section 9.8. When centrifuges are used for *polishing*, the removal of the very small quantities of finely divided solids needs be carried out only infrequently, and manual techniques are then often used.

Because centrifuges are normally used for separating fine particles and droplets, it is necessary to consider only the Stokes' law region in calculating the drag between the particle and the liquid.

It is seen in Chapter 3 that, as a particle moves outwards towards the walls of the bowl of a centrifuge, the accelerating force progressively increases and therefore the particle never reaches an equilibrium velocity as is the case in the gravitational field. Neglecting the inertia of the particle, then:

$$\frac{dr}{dt} = \frac{d^2(\rho_s - \rho)r\omega^2}{18\mu} \tag{9.11}$$

(equation 3.109)

$$= u_0 \frac{r\omega^2}{g} \tag{9.12}$$

(equation 3.110)

At the walls of the bowl of radius $r$, $\mathrm{d}r/\mathrm{d}t$ is given by:

$$\left(\frac{\mathrm{d}r}{\mathrm{d}t}\right)_{r=R} = \frac{d^2(\rho_s - \rho)R\omega^2}{18\mu} \tag{9.13}$$

The time taken to settle through a liquid layer of thickness $h$ at the walls of the bowl is given by integration of equation 9.11 between the limits $r = r_0$ (the radius of the inner surface of the liquid), and $r = R$ to give equation 3.121. This equation may be simplified where $R - r_0\,(= h)$ is small compared with $R$, as in equation 3.122.

Then:
$$t_R = \frac{18\mu h}{d^2(\rho_s - \rho)R\omega^2} \tag{9.14}$$

(equation 3.100)

It may be seen that equation 9.14 is the time taken to settle through the distance $h$ at a velocity given by equation 9.13. $t_R$ is then the minimum retention time required for all particles of size greater than $d$ to be deposited on the walls of the bowl. Thus, the maximum throughput $Q$ at which all particles larger than $d$ will be retained is given by substitution for $t_R$ from equation 9.8 to give:

$$Q = \frac{d^2(\rho_s - \rho)R\omega^2 V'}{18\mu h} \tag{9.15}$$

or:
$$Q = \frac{d^2(\rho_s - \rho)g}{18\mu}\frac{R\omega^2 V'}{hg} \tag{9.16}$$

From equation 3.24:
$$\frac{d^2(\rho_s - \rho)g}{18\mu} = u_0$$

where $u_0$ is the terminal falling velocity of the particle in the gravitational field and hence:

$$Q = u_0 \frac{R\omega^2 V'}{hg} \tag{9.17}$$

Writing the capacity term as:

$$\Sigma = \frac{R\omega^2 V'}{hg}$$

$$= \frac{\pi R(R^2 - r_0^2)H\omega^2}{hg}$$

$$= \pi R(R + r_0)H\frac{\omega^2}{g} \tag{9.18}$$

Then:
$$Q = u_0\Sigma \tag{9.19}$$

Equation 9.18 implies that the greater the depth $h$ of liquid in the bowl, that is the lower the value of $r_0$, the smaller will be the value of $\Sigma$, but this is seldom borne out in practice in decanter centrifuges. This is probably due to high turbulence experienced and the effects of the scrolling mechanism. $\Sigma$ is independent of the properties of the fluid and the particles and depends only on the dimensions of the centrifuge, the location of the overflow weir and the speed of rotation. It is equal to the cross-sectional area of a gravity settling tank with the same clarifying capacity as the centrifuge. Thus, $\Sigma$ is a measure of the capacity of the centrifuge and gives a quantification of its performance for clarification. This treatment is attributable to the work of AMBLER[1].

For cases where the thickness $h$ of the liquid layer at the walls is comparable in order of magnitude with the radius $R$ of the bowl, it is necessary to use equation 3.121 in place of equation 9.14 for the required residence time in the centrifuge or:

$$t_R = \frac{18\mu}{d^2(\rho_s - \rho)\omega^2} \ln \frac{R}{r_0} \tag{9.20}$$

(from equation 3.121)

Then:

$$Q = \frac{d^2(\rho_s - \rho)\omega^2 V'}{18\mu \ln(R/r_0)} \tag{9.21}$$

$$= \frac{d^2(\rho_s - \rho)g}{18\mu} \frac{\omega^2 V'}{g \ln(R/r_0)} \tag{9.22}$$

$$= u_0 \Sigma \tag{9.23}$$

In this case:

$$\Sigma = \frac{\omega^2 V'}{g \ln(R/r_0)}$$

$$= \frac{\pi(R^2 - r_i^2)H}{\ln(R/r_0)} \frac{\omega^2}{g} \tag{9.24}$$

A similar analysis can be carried out with various geometrical arrangements of the bowl of the centrifuge. Thus, for example, for a disc machine (described later) the value of $\Sigma$ is very much greater than for a cylindrical bowl of the same size. Values of $\Sigma$ for different arrangements are quoted by HAYTER[2] and by TROWBRIDGE[3].

This treatment leads to the calculation of the condition where all particles larger than a certain size are retained in the centrifuge. Other definitions are sometimes used, such as for example, the size of the particle which will just move half the radial distance from the surface of the liquid to the wall, or the condition when just half of the particles of the specified size will be removed from the suspension.

## Example 9.1

In a test on a centrifuge all particles of a mineral of density 2800 kg/m$^3$ and of size 5 $\mu$m, equivalent spherical diameter, were separated from suspension in water fed at a volumetric throughput rate of 0.25 m$^3$/s. Calculate the value of the capacity factor $\Sigma$.

What will be the corresponding size cut for a suspension of coal particles in oil fed at the rate of 0.04 $m^3/s$? The density of coal is 1300 $kg/m^3$ and the density of the oil is 850 $kg/m^3$ and its viscosity is 0.01 $Ns/m^2$.

It may be assumed that Stokes' law is applicable.

## Solution

The terminal falling velocity of particles of diameter 5 μm in water, of density $\rho = 1000$ $kg/m^3$ and, of viscosity $\mu = 10^{-3}$ $Ns/m^2$, is given by:

$$u_0 = \frac{d^2(\rho_s - \rho)g}{18\mu} = \frac{25 \times 10^{-12} \times (2800 - 1000) \times 9.81}{18 \times 10^{-3}} \quad \text{(equation 3.24)}$$

$$= 2.45 \times 10^{-5} \text{ m/s}$$

From the definition of $\Sigma$:

$$Q = u_0\Sigma \quad \text{(equation 9.19)}$$

and:

$$\Sigma = \frac{0.25}{(2.45 \times 10^{-5})} = 1.02 \times 10^4 \text{ m}^2.$$

For the coal-in-oil mixture:

$$u_0 = \frac{Q}{\Sigma} = \frac{0.04}{(1.02 \times 10^4)} = 3.92 \times 10^{-6} \text{ m/s}.$$

From equation 3.24:

$$d^2 = \frac{18\mu u_0}{(\rho_s - \rho)g}$$

$$= \frac{18 \times 10^{-2} \times 3.92 \times 10^{-6}}{(1300 - 850) \times 9.81}$$

and:

$$d = 4.0 \times 10^{-6} \text{ m or } \underline{\underline{4 \ \mu\text{m}}}.$$

## Example 9.2

A centrifuge is fitted with a conical disc stack with an included angle of $2\theta$, and there are $n$ flow passages between the discs. A suspension enters at radius $r_1$ and leaves at radius $r_2$. Obtain an expression for the separating power $\Sigma$ of the centrifuge. It may be assumed that the resistance force acting on the particles is given by Stokes' law.

## Solution

For two discs AA′ and BB′, as shown in Figure 9.4, the particle which is most unfavourably placed for collection will enter at point A at radius $r_1$, and be deposited on the upper plate at point B′ at

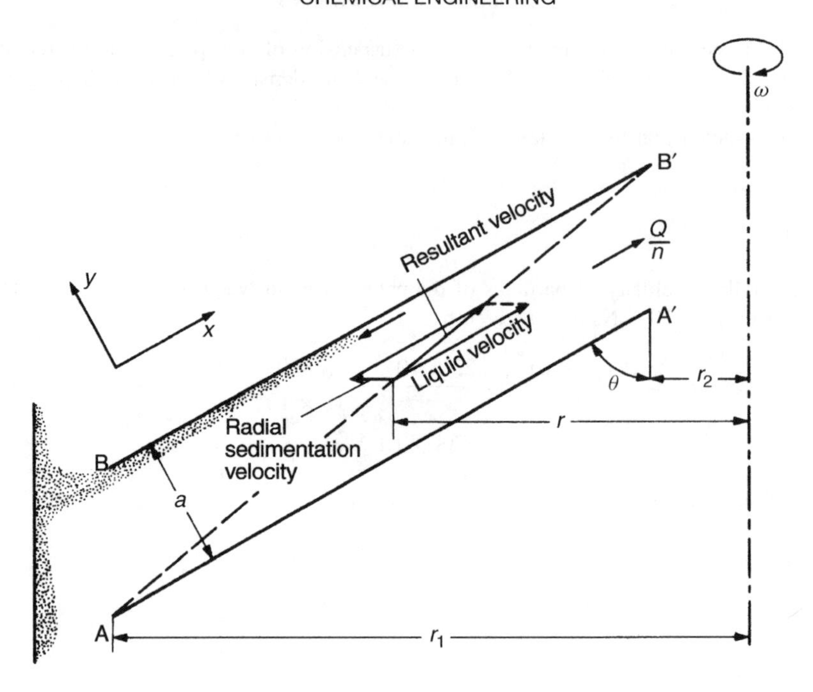


Figure 9.4. Path of limit particle through separation channel

radius $r_2$. It is assumed that the suspension is evenly divided between the discs. The particles will move in not quite in a straight line because both velocity components are a function of $r$.

At radius $r$, the velocity of the liquid in the flow channel is:

$$\frac{Q}{2\pi r a n} = \frac{dx}{dt} \tag{i}$$

where $x$ is the distance parallel to the discs and $a$ is the spacing.

At radius $r$, the centrifugal sedimentation velocity of a particle, whose diameter is $d$ is given by:

$$\frac{dr}{dt} = \frac{d^2 r\omega^2(\rho_s - \rho)}{18\mu} \qquad \text{(equation 9.11)}$$

$$= u_0\frac{r\omega^2}{g} \qquad \text{(equation 9.12) (ii)}$$

From the geometry of the system:

$$-\frac{dr}{dx} = \sin\theta$$

Thus from equation (i):

$$-\frac{dr}{dt} = \frac{Q}{2\pi r a n}\sin\theta \tag{iii}$$

and:

$$\frac{dy}{dr} = \cos\theta$$

Thus from equation (ii):

$$\frac{dy}{dt} = u_0 \frac{r\omega^2}{g} \cos\theta \qquad \text{(iv)}$$

Dividing equation (iv) by equation (iii) gives:

$$-\frac{dy}{dr} = \left(\frac{u_0 r\omega^2 \cos\theta}{g}\right) \cdot \left(\frac{2\pi ran}{Q\sin\theta}\right)$$

$$= \frac{2\pi nau_0\omega^2 \cot\theta}{Qg} r^2$$

The particle must move through distance $a$ in the $y$ direction as its radial position changes from $r_1$ to $r_2$.

Thus:

$$-\int_0^a dy = \frac{2\pi nau_0\omega^2 \cot\theta}{Qg} \int_{r_1}^{r_2} r^2 \, dr$$

$$a = \frac{2\pi nau_0\omega^2 \cot\theta}{3Qg}(r_1^3 - r_2^3)$$

$$Q = u_0 \frac{2\pi n\omega^2 \cot\theta}{3g}(r_1^3 - r_2^3)$$

$$= u_0 \Sigma \qquad \text{(from the definition of } \Sigma\text{, equation 9.19)}$$

or:

$$\Sigma = \frac{2\pi\omega^2 n \cot\theta (r_1^3 - r_2^3)}{3g}$$

## 9.6. FILTRATION IN A CENTRIFUGE

When filtration is carried out in a centrifuge, it is necessary to use a perforated bowl to permit removal of the filtrate. The driving force is the centrifugal pressure due to the liquid and suspended solids, and this will not be affected by the presence of solid particles deposited on the walls. The resulting force must overcome the friction caused by the flow of liquid through the filter cake, the cloth, and the supporting gauze and perforations. The resistance of the filter cake will increase as solids are deposited although the other resistances will remain approximately constant throughout the process. Considering filtration in a bowl of radius $R$ and supposing that the suspension is introduced at such a rate that the inner radius of the liquid surface remains constant as shown in Figure 9.5, then at some time $t$ after the commencement of filtration, a filter cake of thickness $l$ will have been built up and the radius of the interface between the cake and the suspension will be $r'$.

If $dP'$ is the pressure difference across a small thickness $dl$ of cake, the velocity of flow of the filtrate is given by equation 7.2, and:

$$\frac{1}{A}\frac{dV}{dt} = u_c = \frac{1}{\mathbf{r}\mu}\left(\frac{-dP'}{dl}\right) \qquad (9.25)$$

where $\mathbf{r}$ is the specific resistance of the filter cake and $\mu$ is the viscosity of the filtrate.

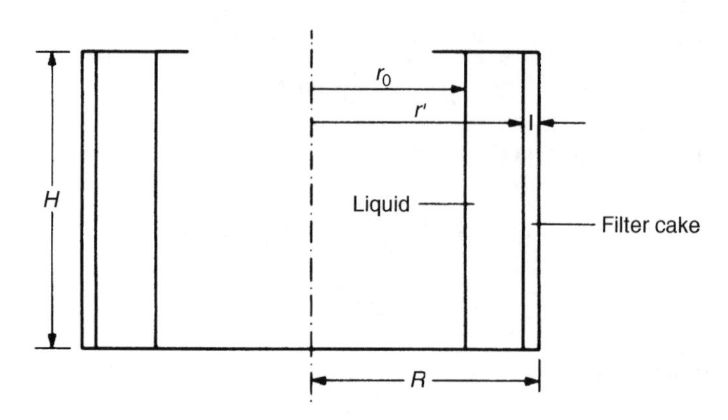


Figure 9.5. Filtration in a centrifuge

If the centrifugal force is large compared with the gravitational force, the filtrate will flow in an approximately radial direction, and will be evenly distributed over the axial length of the bowl. The area available for flow will increase towards the walls of the bowl. If $\mathrm{d}V$ is the volume of filtrate flowing through the filter cake in time $\mathrm{d}t$, then:

$$u_c = \frac{1}{2\pi r' H}\frac{\mathrm{d}V}{\mathrm{d}t}$$

Thus:

$$\frac{1}{\mathbf{r}\mu}\left(\frac{-\mathrm{d}P'}{\mathrm{d}l}\right) = \frac{1}{2\pi r' H}\frac{\mathrm{d}V}{\mathrm{d}t} \tag{9.26}$$

$$-\mathrm{d}P' = \frac{\mathbf{r}\mu\,\mathrm{d}l}{2\pi r' H}\frac{\mathrm{d}V}{\mathrm{d}t}$$

and thus the total pressure drop through the cake at time $t$ is given by:

$$\begin{aligned} -\Delta P' &= \frac{\mathbf{r}\mu}{2\pi H}\frac{\mathrm{d}V}{\mathrm{d}t}\int_0^l \frac{\mathrm{d}l}{r'} \\ &= \frac{\mathbf{r}\mu}{2\pi H}\frac{\mathrm{d}V}{\mathrm{d}t}\int_{r'}^R \frac{\mathrm{d}r'}{r'} \\ &= \frac{\mathbf{r}\mu}{2\pi H}\frac{\mathrm{d}V}{\mathrm{d}t}\ln\frac{R}{r'} \end{aligned} \tag{9.27}$$

If the resistance of the cloth is negligible, $-\Delta P'$ is equal to the centrifugal pressure. More generally, if the cloth, considered together with the supporting wall of the basket, is equivalent in resistance to a cake of thickness $L$, situated at the walls of the basket, the pressure drop $-\Delta P''$ across the cloth is given by:

$$\frac{-\Delta P''}{\mathbf{r}\mu L} = \frac{1}{2\pi HR}\frac{\mathrm{d}V}{\mathrm{d}t}$$

$$-\Delta P'' = \frac{\mathbf{r}\mu}{2\pi H}\frac{\mathrm{d}V}{\mathrm{d}t}\frac{L}{R} \tag{9.28}$$

Thus the total pressure drop across the filter cake and the cloth $(-\Delta P)$, say, is given by:

$$(-\Delta P) = (-\Delta P') + (-\Delta P'')$$

Thus:
$$-\Delta P = \frac{\mathbf{r}\mu}{2\pi H}\frac{\mathrm{d}V}{\mathrm{d}t}\left(\ln\frac{R}{r'} + \frac{L}{R}\right) \tag{9.29}$$

Before this equation can be integrated it is necessary to establish the relation between $r'$ and $V$. If $v$ is the bulk volume of incompressible cake deposited by the passage of unit volume of filtrate, then:

$$v\,\mathrm{d}V = -2\pi r' H\,\mathrm{d}r'$$

$$\frac{\mathrm{d}V}{\mathrm{d}t} = -\frac{2\pi H r'}{v}\frac{\mathrm{d}r'}{\mathrm{d}t} \tag{9.30}$$

and substituting for $\mathrm{d}V/\mathrm{d}t$ in the previous equation gives:

$$-\Delta P = -\frac{\mathbf{r}\mu}{2\pi H}\frac{2\pi H r'}{v}\frac{\mathrm{d}r'}{\mathrm{d}t}\left(\ln\frac{R}{r'} + \frac{L}{R}\right) \tag{9.31}$$

Thus:
$$\frac{v(-\Delta P)}{\mathbf{r}\mu}\,\mathrm{d}t = \left(\ln\frac{r'}{R} - \frac{L}{R}\right) r'\,\mathrm{d}r'$$

This may be integrated between the limits $r' = R$ and $r' = r'$ as $t$ goes from 0 to $t$. $-\Delta P$ is constant because the inner radius $r_0$ of the liquid is maintained constant:

$$\frac{(-\Delta P)vt}{\mathbf{r}\mu} = \int_R^{r'}\left\{\left(\ln\frac{r'}{R} - \frac{L}{R}\right)r'\right\}\mathrm{d}r'$$

$$= \frac{1}{4}(R^2 - r'^2) + \frac{L}{2R}(R^2 - r'^2) + \tfrac{1}{2}r'^2\ln\frac{r'}{R} \tag{9.32}$$

and:
$$(R^2 - r'^2)\left(1 + 2\frac{L}{R}\right) + 2r'^2\ln\frac{r'}{R} = \frac{4(-\Delta P)vt}{\mathbf{r}\mu}$$

$$= \frac{2vt\rho\omega^2}{\mathbf{r}\mu}(R^2 - r_0^2) \tag{9.33}$$

since:
$$-\Delta P = \tfrac{1}{2}\rho\omega^2(R^2 - r_0^2) \qquad \text{(from equation 9.4)}$$

From this equation, the time $t$ taken to build up the cake to a given thickness $r'$ may be calculated. The corresponding volume of cake is given by:

$$Vv = \pi(R^2 - r'^2)H \tag{9.34}$$

and the volume of filtrate is:
$$V = \frac{\pi}{v}(R^2 - r'^2)H \tag{9.35}$$

Haruni and Storrow[4] have carried out an extensive investigation into the flow of liquid through a cake formed in a centrifuge and have concluded that, although the results of tests on a filtration plant and a centrifuge are often difficult to compare because of the effects of the compressibility of the cake, it is frequently possible to predict the flowrate in a centrifuge to within 20 per cent. They have also shown that, when the thickness varies with height in the basket, the flowrate can be calculated on the assumption that the cake has a uniform thickness equal to the mean value; this gives a slightly high value in most cases.

## Example 9.3

When an aqueous slurry is filtered in a plate and frame press, fitted with two 50 mm thick frames each 150 mm square, operating with a pressure difference of 350 kN/m$^2$, the frames are filled in 3600 s (1 h). How long will it take to produce the same volume of filtrate as is obtained from a single cycle when using a centrifuge with a perforated basket, 300 mm diameter and 200 mm deep? The radius of the inner surface of the slurry is maintained constant at 75 mm and the speed of rotation is 65 Hz (3900 rpm).

It may be assumed that the filter cake is incompressible, that the resistance of the cloth is equivalent to 3 mm of cake in both cases, and that the liquid in the slurry has the same density as water.

## Solution

*In the filter press*

Noting that $V = 0$ when $t = 0$, then:

$$V^2 + 2\frac{AL}{v}V = \frac{2(-\Delta P)A^2 t}{\mathbf{r}\mu v} \quad \text{(from equation 7.18)}$$

and:

$$V = \frac{lA}{v} \quad \text{(from equation 7.6)}$$

Thus:

$$\frac{l^2A^2}{v^2} + \left(\frac{2AL}{v}\right)\left(\frac{lA}{v}\right) = \frac{2(-\Delta P)A^2 t}{\mathbf{r}\mu v}$$

or:

$$l^2 + 2Ll = \frac{2(-\Delta P)vt}{\mathbf{r}\mu}$$

*For one cycle*

$$l = 25 \text{ mm} = 0.025 \text{ m}; \quad L = 3 \text{ mm} = 0.003 \text{ m}$$

$$-\Delta P = 350 \text{ kN/m}^2 = 3.5 \times 10^5 \text{ N/m}^2$$

$$t = 3600 \text{ s}$$

$\therefore$

$$0.025^2 + (2 \times 0.003 \times 0.025) = 2 \times 3.5 \times 10^5 \times 3600 \times \left(\frac{v}{\mathbf{r}\mu}\right)$$

and:

$$\left(\frac{\mathbf{r}\mu}{v}\right) = 3.25 \times 10^{12}$$

*In the centrifuge*

$$(R^2 - r'^2)\left(1 + 2\frac{L}{R}\right) + 2r'^2 \ln\frac{r'}{R} = \frac{2vt\rho\omega^2}{r\mu}(R^2 - r_0^2) \quad \text{(equation 9.33)}$$

$R = 0.15$ m, $H = 0.20$ m, and the volume of cake $= 2 \times 0.050 \times 0.15^2 = 0.00225$ m$^3$

Thus: $$\pi(R^2 - r'^2) \times 0.20 = 0.00225$$

and: $$(R^2 - r^2) = 0.00358$$

Thus: $$r'^2 = (0.15^2 - 0.00358) = 0.0189 \text{ m}^2$$

and: $$r' = 0.138 \text{ m}$$

$$r_0 = 75 \text{ mm} = 0.075 \text{ m}$$

and: $$\omega = 65 \times 2\pi = 408.4 \text{ rad/s}$$

The time taken to produce the same volume of filtrate or cake as in one cycle of the filter press is therefore given by:

$$(0.15^2 - 0.138^2)(1 + 2 \times 0.003/0.15) + 2(0.0189)\ln(0.138/0.15)$$

$$= \frac{2 \times t \times 1000 \times 408.4^2}{3.25 \times 10^{12}}(0.15^2 - 0.075^2)$$

or: $$0.00359 - 0.00315 = 1.732 \times 10^{-6}t$$

from which: $$t = \frac{4.4 \times 10^{-4}}{1.732 \times 10^{-6}}$$

$$= \underline{\underline{254 \text{ s or } 4.25 \text{ min.}}}$$

## 9.7. MECHANICAL DESIGN

Features of the mechanical design of centrifuges are discussed in Volume 6, where, in particular, the following items are considered:

(a) The mechanical strength of the bowl, which will be determined by the dimensions of the bowl, the material of construction and the speed of operation.
(b) The implications of the critical speed of rotation, which can cause large deflections of the shaft and vibration.
(c) The slow gyratory motion, known as *precession*, which can occur when the bowl or basket is tilted.

## 9.8. CENTRIFUGAL EQUIPMENT

### 9.8.1. Classification of centrifuges

Centrifuges may be grouped into two distinct categories–those that utilise the principle of filtration and those that utilise the principle of sedimentation, both enhanced by the use of a centrifugal field. These two classes may be further subdivided according to the method of discharge, particularly the solids discharge. This may be batch, continuous or a combination of both. Further subdivisions may be made according to source of manufacture or mechanical features, such as the method of bearing suspension, axis orientation or containment. In size, centrifuges vary from the small batch laboratory tube spinner to the large continuous machines with several tons of rotating mass, developing a few thousands of $g$ relative centrifugal force. RECORDS[5] has suggested the following classification.

### *Filtration centrifuges*

(a) Batch discharge, vertical axis, perforate basket.
(b) Knife discharge, pendulum suspension, vertical axis, perforate basket.
(c) Peeler: horizontal axis, knife discharge at speed.
(d) Pusher.
(e) Scroll discharge.

### *Sedimentation centrifuges*

(a) Bottle spinner.
(b) Tubular bowl.
(c) Decanter — scroll discharge.
(d) Imperforate bowl — skimmer pipe discharge, sometimes also a knife.
(e) Disc machine.
  (i) Batch.
  (ii) Nozzle discharge.
  (iii) Opening bowl (solids ejecting).
  (iv) Valve discharge.

### *Liquid–liquid separation centrifuges*

(a) Tubular bowl.
(b) Three phase scroll discharge decanter.
(c) Disc machine.

Clearly with such a wide range of equipment it is not realistic to describe examples of machinery in each category. Discussion will therefore be limited to the more important types of machine and the conditions under which they may be used.

## 9.8.2. Simple bowl centrifuges

Most batch centrifuges are mounted with their axes vertical and, because of the possibility of uneven loading of the machine, the bowl is normally supported in bearings and the centrifuge itself is supported by resilient mountings. In this way the inevitable out-of-balance forces on the supporting structure are reduced to, typically, 5 per cent or less.

In the case of the underdriven batch machine, where the drive and bearings are underneath, as shown in Figure 9.6, access to the bowl is direct, while product is discharged from the top in manual machines and from the bottom in automatically controlled machines. In the case of the overdriven batch centrifuge, where the bowl is suspended from above, a valve is often incorporated in the bottom of the bowl for easy discharge of the solids. With a purpose designed drive, the unit can handle higher

Figure 9.6. Underdriven batch centrifuge

throughputs of freely filtering feed slurry by cycling more quickly than the underdriven machine.

Batch centrifuges with imperforate bowls are used either for producing an accelerated separation of solid particles in a liquid, or for separating mixtures of two liquids. In the former case, the solids are deposited on the wall of the bowl and the liquid is removed through an overflow or skimming tube. The suspension is continuously fed in until a suitable depth of solids has been built up on the wall; this deposit is then removed either by hand or by a mechanical scraper. With the bowl mounted about a horizontal axis, solids are more readily discharged because they can be allowed to fall directly into a chute.

In the centrifuge shown in Figure 9.7, the liquid is taken off through a skimming tube and the solids, which may be washed before discharge if desired, are removed by a cutter. This machine is often mounted vertically and the cutter knife sometimes extends over the full depth of the bowl.

Perforated bowls are used when relatively large particles are to be separated from a liquid, as for example in the separation of crystals from a mother liquor. The mother liquor passes through the bed of particles and then through the perforations in the bowl. When the centrifuge is used for filtration, a coarse gauze is laid over the inner surface of the bowl and the filter cloth rests on the gauze. Space is thus provided behind the cloth for the filtrate to flow to the perforations.

When a mixture of liquids is to be separated, the denser liquid collects at the imperforate bowl wall and the less dense liquid forms an inner layer. Overflow weirs are arranged

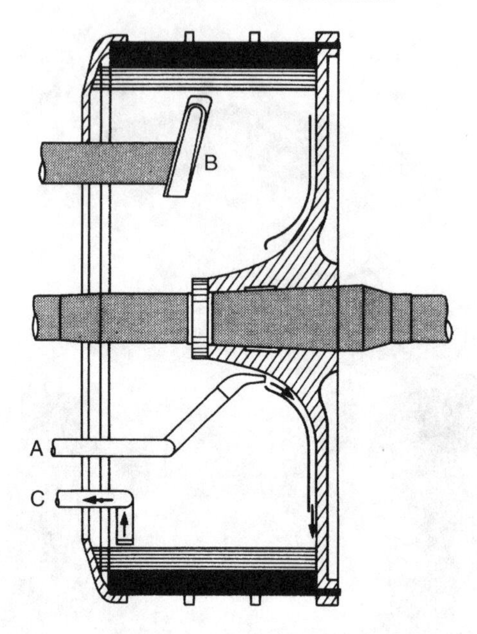


Figure 9.7. Horizontally mounted bowl with automatic discharge of solids. A–Feed, B–Cutter, C–Skimming tube

so that the two constituents are continuously removed. The design of the weirs has been considered in a previous section.

## 9.8.3. Disc centrifuges — general

For a given rate of feed to the centrifuge, the degree of separation obtained will depend on the thickness of the liquid layer formed at the wall of the bowl and on the total depth of the bowl, since both these factors control the time the mixture remains in the machine. A high degree of separation could therefore be obtained with a long bowl of small diameter, although the speed required would then need to be very high. By comparison the introduction of conical discs into the bowl, as illustrated in Figure 9.8, enables the liquid stream to be split into a large number of very thin layers and permits a bowl of greater diameter. The separation of a mixture of water and dirt from a relatively low density oil takes place as shown in Figure 9.9 with the dirt and water collecting close to the undersides of the discs and moving radially outwards, and with the oil moving inwards along the top sides.

A disc bowl, although still a high speed machine, can thus be run at lower speeds relative to the excessive speeds required for a long bowl of small diameter. Also its size is very much smaller relative to a bowl without discs, as may be seen from Figure 9.10. The separation of two liquids in a disc-type bowl is illustrated in the left-hand side of Figure 9.8. Liquid enters through the distributor AB, passes through C, and is distributed between the discs E through the holes D. The denser liquid is taken off through F and I and the less dense liquid through G.

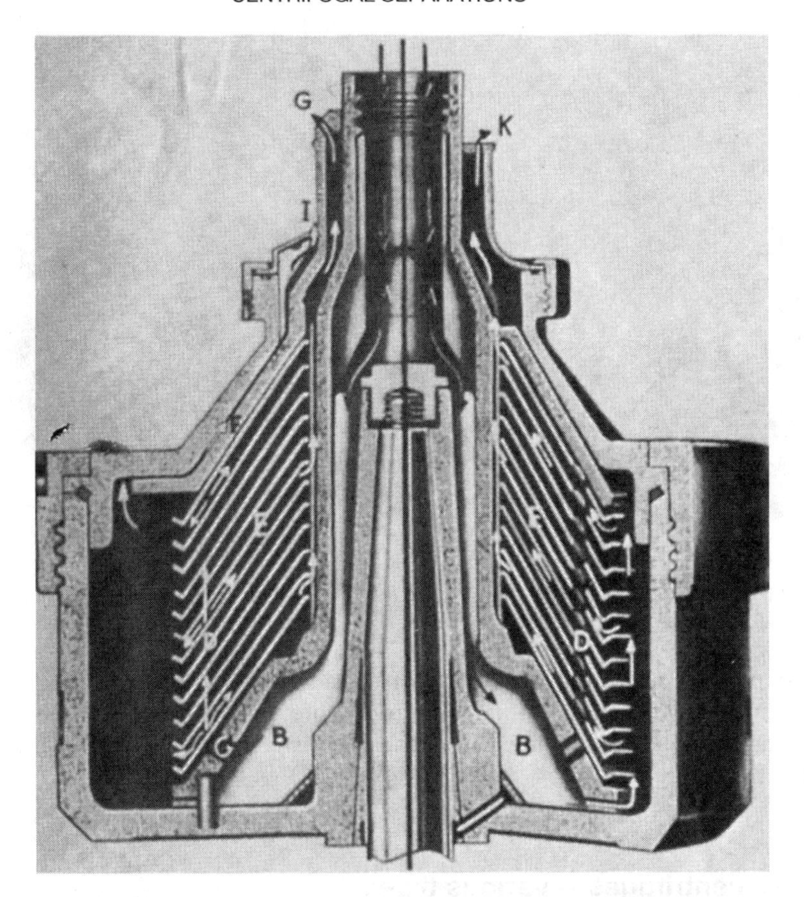


Figure 9.8. Bowl with conical discs (left-hand side for separating liquids, right-hand side for separating solid from liquid)

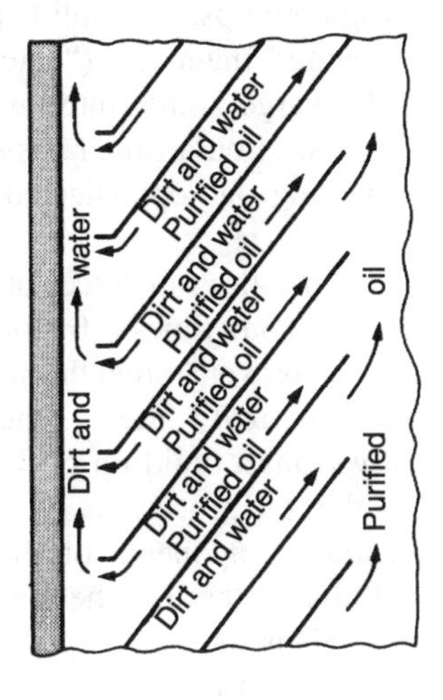


Figure 9.9. Separation of water and dirt from oil in disc bowl

Figure 9.10. Two bowls of equal capacity; with discs (left) and without discs (right)

A disc-type bowl is often used for the separation of fine solids from a liquid and its construction is shown in the right-hand side of Figure 9.8. Here there is only one liquid outlet K, and the solids are retained in the space between the ends of the discs and the wall of the bowl.

### 9.8.4. Disc centrifuges — various types

Disc centrifuges vary considerably according to the type of discharge. The liquid(s) can discharge freely or through centrifugal pumps. The solids may be allowed to accumulate within the bowl and then be discharged manually ("batch bowl"). A large number of disc machines ("opening bowl") discharge solids intermittently when ports are opened automatically at the periphery of the bowl on a timed basis or are actuated from a signal from a liquid clarity meter. The opening on sophisticated machines may be adjusted to discharge only the compacted solids.

Another type of disc centrifuge has nozzles or orifices at the periphery where thickened solids are discharged continuously, and sometimes a fraction of this discharge is recycled to ensure that the nozzles are able to prevent breakthrough of the clarified supernatant liquid. Sometimes these nozzles are internally fitted to the bowl. Others are opened and closed electrically or hydrostatically from a build up of the sludge within the bowl.

Centrifuges of these types are used in the processing of yeast, starch, meat, fish products and fruit juices. They form essential components of the process of rendering in the extraction of oils and fats from cellular materials. The raw material, consisting of bones, animal fat, fish offal, or vegetable seeds is first disintegrated and then, after a preliminary gravitational separation, the final separation of water, oil, and suspended solids is carried out in a number of valve nozzle centrifuges.

## 9.8.5. Decanting centrifuges

The widely used decanting type continuous centrifuge, shown in Figure 9.11, is mounted on a horizontal axis. The mixture of solids and liquids is fed to the machine through a stationary pipe which passes through one of the support bearings, shown in Figure 9.12. The feed pipe discharges the mixture near the centre of the machine. The heavier solids settle on to the wall of the imperforate or solid bowl under the influence of the centrifugal

Figure 9.11. Solid-bowl decanter centrifuge

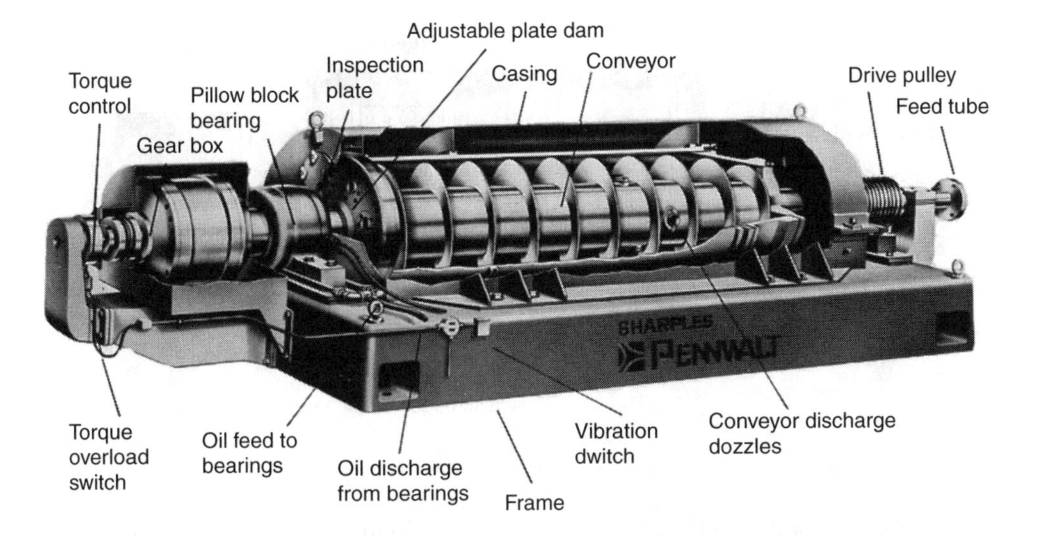


Figure 9.12. Solid-bowl decanter centrifuge–principles of operation

force. Inside the bowl is a close fitting helical scroll which rotates at a slightly different speed from the solid bowl. Typically the speed differential is in the range 0.5 to 100 rpm (0.01–2 Hz).

At one end of the bowl a conical section is attached giving a smaller diameter. The liquid runs round the helical scroll and is discharged over weir plates fitted at the parallel end of the bowl. The solids are moved by the conveying action of the helical scroll up the gentle slope of the conical section, out of the liquid and finally out of the machine. Decanters of this type are known as solid bowl decanters.

A variant of the decanting centrifuge is the screen bowl decanter, shown in Figures 9.13 and 9.14. In this unit a further perforated section is attached to the smaller diameter end of the conical section. This is known as the screen and allows further drying and/or washing of the solids to take place.

Figure 9.13. Screen bowl decanter centrifuge (cover removed)

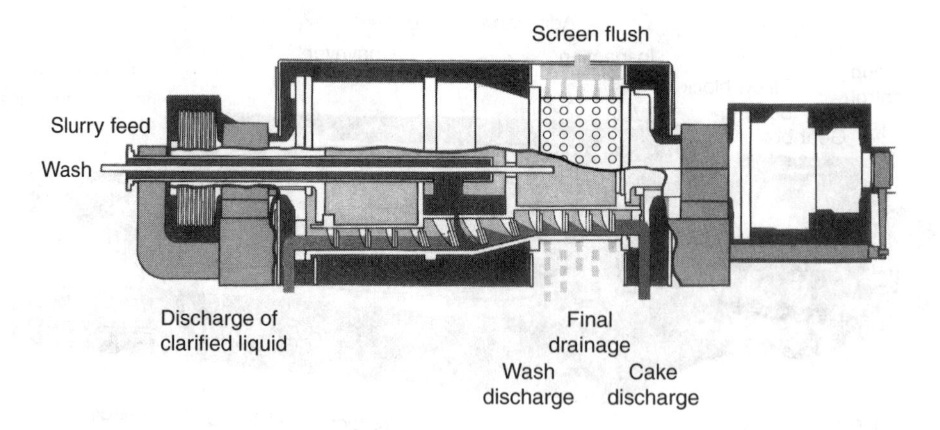


Figure 9.14. Screen bowl decanter centrifuge-principles of operation

Machines may be tailored to meet specific process requirement by altering the diameter of the liquid discharge, the differential speed of the helical scroll, the position at which the feed enters the machine and the rotational speed of the bowl. Decanting centrifuges are available in a wide range of diameters and lengths. Diameters are 0.2–1.5 m, and lengths are typically 1.5–5 times the diameter. The longer machines are used when clear liquids are required. Shorter bowl designs are used to produce the driest solids. Continuous throughputs in excess of 250 tonnes/h (70 kg/s) of feed can be handled by a single machine.

### 9.8.6. Pusher-type centrifuges

This type of centrifuge is used for the separation of suspensions and is fitted either with a perforated or imperforate bowl. The feed is introduced through a conical funnel and the cake is formed in the space between the flange and the end of the bowl. The solids are intermittently moved along the surface of the bowl by means of a reciprocating pusher. The pusher comes forward and returns immediately and then waits until a further layer of solids has been built up before advancing again. In this machine the thickness of filter cake cannot exceed the distance between the surface of the bowl and the flange of the funnel. The liquid either passes through the holes in the bowl or, in the case of an imperforate bowl, is taken away through an overflow. The solids are washed by means of a spray, as shown in Figure 9.15.

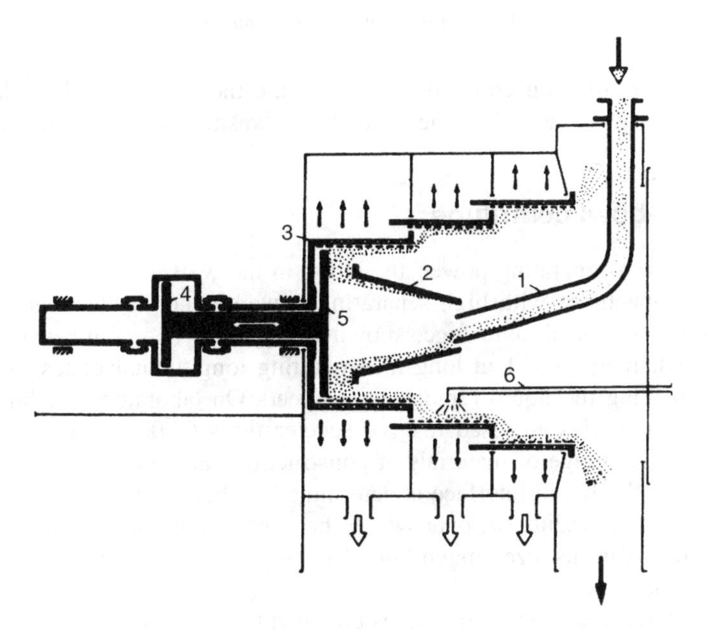


Figure 9.15. Pusher-type centrifuge. 1. Inlet; 2. Inlet funnel; 3. Bowl; 4. Piston; 5. Pusher disc; 6. Washing spray

A form of pusher type centrifuge which is particularly suitable for filtering slurries of low concentrations is shown in Figure 9.16. A perforated pusher cone gently accelerates the feed and secures a large amount of preliminary drainage near the apex of the cone. The

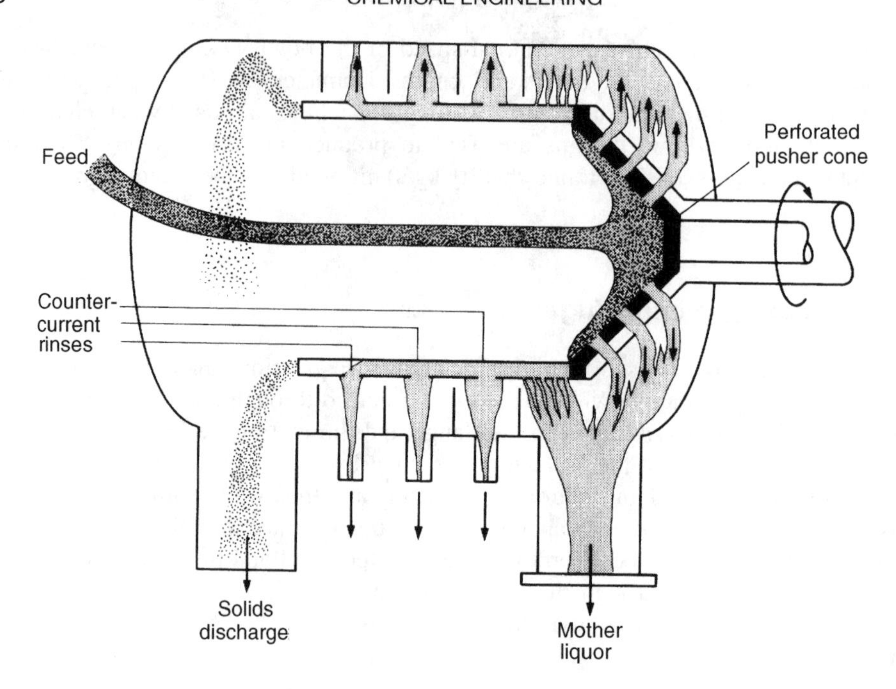


Figure 9.16. Pusher-type centrifuge

solids from the partially concentrated suspension are then evenly laid on the cylindrical surface and, in this way, the risk of the solids being washed out of the bowl is minimised.

## 9.8.7. Tubular-bowl centrifuge

Because, for a given separating power, the stress in the wall is a minimum for machines of small radius, machines with high separating powers generally use very tall bowls of small diameters. A typical centrifuge, shown in Figure 9.17, would consist of a bowl about 100 mm diameter and 1 m long incorporating longitudinal plates to act as accelerator blades to bring the liquid rapidly up to speed. On laboratory machines speeds up to 50,000 rpm (1000 Hz) are used to give accelerations 60,000 times the gravitational acceleration. A wide range of materials of construction can be used.

The position of the liquid interface is determined by balancing centrifugal forces as in Figure 9.3. The lip, of radius $r_w$, over which the denser liquid leaves the bowl is part of a removable ring. Various sizes maybe fitted to provide for the separation of liquids of various relative densities.

Often the material fed to these machines contains traces of denser solids in addition to the two liquid phases. These solids are deposited on the inner wall of the bowl, and the machine is dismantled periodically to remove them. A common application is the removal of water and suspended solids from lubricating oil.

The super-centrifuge is used for clarification of oils and fruit juices and for the removal of oversize and undersize particles from pigmented liquids. The liquid is continuously discharged, but the solids are retained in the bowl and must be removed periodically.

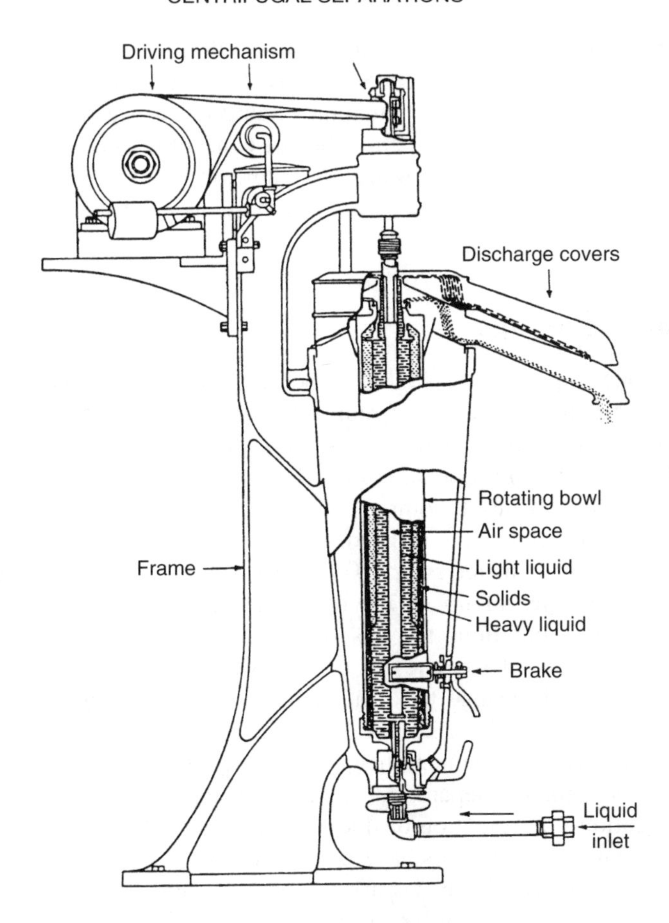


Figure 9.17. Sectional view of the super-centrifuge

### 9.8.8. The ultra-centrifuge

For separation of colloidal particles and for breaking down emulsions, the ultra-centrifuge is used. This operates at speeds up to 30 rpm (1600 Hz) and produces a force of 100,000 times the force of gravity for a continuous liquid flow machine, and as high as 500,000 times for gas phase separation, although these machines are very small. The bowl is usually driven by means of a small air turbine. The ultra-centrifuge is often run either at low pressures or in an atmosphere of hydrogen in order to reduce frictional losses, and a fivefold increase in the maximum speed can be attained by this means.

### 9.8.9. Multistage centrifuges

Among many specialist types of centrifuge is the multistage machine which consists of series of bowls mounted concentrically on a vertical axis. The feed suspension is introduced into the innermost bowl and the overflow then passes successively to each larger bowl in turn. As the separating force is directly proportional to the radius of rotation, the largest particles are separated out at the first stage, and progressively finer particles are recovered at each subsequent stage. The finest particles are collected in the outermost vessel from which the remaining liquid or suspension is discharged. The multistage system may also incorporate a series of concentric vertical baffles in a single bowl, with the suspension flowing upwards and downwards through successive annular channels. Both the design and operation are complex, and the machines are somewhat inflexible.

### 9.8.10. The gas centrifuge

A specialised, though nevertheless highly important, function for which the centrifuge has been developed is the separation of radioactive isotopes as described by FISHMAN(6). The concentration of uranium-235 from less than 1 per cent up to about 5 per cent is achieved by subjecting uranium hexafluoride ($UF_6$) to an intense centrifugal field. The small differences in density of the components of the mixture necessitate the use of very high accelerations in order to obtain the desired separation. Mechanical considerations dictate the use of small diameter rotors (0.1–0.2 m) rotating at speeds up to 2000 Hz ($10^5$ rpm), giving velocities of up to 700 m/s and accelerations of up to $10^6$ $g$ at the periphery. Under these conditions the gas flow can vary from free molecular flow in the low pressure region at the centre, to a high Mach number flow at the periphery. Furthermore, pressure will change by a factor of 10 over a distance of about 2 mm as compared with a distance of about 20 km in the earth's gravitational field.

For use in a radioactive environment the gas centrifuge must be completely maintenance free. It has been used for the separation of xenon isotopes and consideration has been given to its application for separation of fluorohydrocarbons. Worldwide, in the region of a quarter of a million gas centrifuges have been manufactured. As an order of magnitude figure, an investment of £1000 is necessary to obtain 0.3 g/s (10 g/h) of product.

Further information is given in the papers by WHITLEY(7,8).

## 9.9. FURTHER READING

AMBLER, C. M.: In *Encyclopedia of Chemical Processing and Design*, Vol. 7. MCKETTA, J. J. ed. (Marcel Dekker, New York, 1978).

HSU, H-W.: *Ind. Eng. Chem. Fundamentals* **25** (1986) 588. Separations by liquid centrifugation.

LAVANCHY, A. C. and KEITH, F. W.: In *Encyclopedia of Chemical Technology*, KIRK, R. E. and OHMER, D. F. (eds.) Vol. 4, (Wiley-Interscience, 1979) 710 Centrifugal separation.

MULLIN, J. W.: In *Chemical Engineering Practice*, CREMER, H. W. and DAVIES, T. (eds.) Vol. 6, (Butterworths, 1958) p 528. Centrifuging.

ZEITSCH, K. *Centrifugal Filtration* (Butterworth & Co, 1981).

## 9.10. REFERENCES

1. AMBLER, C. A.: *Chem. Eng. Prog.* **48** (1952) 150. The evaluation of centrifuge performance.
2. HAYTER, A. J.: *J. Soc. Cosmet. Chem.* (1962) 152. Progress in centrifugal separations.
3. TROWBRIDGE, M. E. O'K.: *The Chemical Engineer* No. 162 (Aug. 1962) A73. Problems in the scaling-up of centrifugal separation equipment.
4. HARUNI, M. M. and STORROW, J. A.: *Ind. Eng. Chem.* **44** (1952) 2751: *Chem. Eng. Sci.* **1** (1952) 154: **2** (1953) 97, 108, 164 and 203. Hydroextraction.

5. RECORDS, F. A.: Alfa Laval Sharples Ltd, Private communication (March 1990).
6. FISHMAN, A. M.: *A. I. Chem. E. Symposium Series* No. 169, 73 (1977). Developments in Uranium enrichment, 43. The centar gas centrifuge enrichment project: Economics and engineering considerations.
7. WHITLEY, S.: *A Summary of the Development of the Gas Centrifuge* British Nuclear Fuels plc, Enrichment Division, Capenhurst, (June 1988).
8. WHITLEY, S.: *Reviews of Modern Physics* **56** (1984) 41, 67. Review of the gas centrifuges until 1962. Part I: Principles of separation physics. Part II: Principles of high-speed rotation.

## 9.11. NOMENCLATURE

| | | Units in SI System | Dimensions in **M, L, T** |
|---|---|---|---|
| $A$ | Cross-sectional area of filter | m$^2$ | $\mathbf{L}^2$ |
| $a$ | Distance between discs in stack | m | **L** |
| $d$ | Diameter (or equivalent diameter of particle) | m | **L** |
| $g$ | Acceleration due to gravity | m/s$^2$ | $\mathbf{LT}^{-2}$ |
| $H$ | Length (axial) of centrifuge bowl | m | **L** |
| $h$ | Depth of liquid at wall of bowl | m | **L** |
| $L$ | Thickness of filter cake with same resistance as cloth | m | **L** |
| $l$ | Thickness of filter cake | m | **L** |
| $n$ | Number of passages between discs in bowl | — | — |
| $P$ | Pressure | N/m$^2$ | $\mathbf{ML}^{-1}\mathbf{T}^{-2}$ |
| $P'$ | Pressure in cake | N/M$^2$ | $\mathbf{ML}^{-1}\mathbf{T}^{-2}$ |
| $P''$ | Pressure in cloth | N/m$^{-2}$ | $\mathbf{ML}^{-1}\mathbf{T}^{-2}$ |
| $-\Delta P$ | Pressure drop (total) | N/M$^2$ | $\mathbf{ML}^{-1}\mathbf{T}^{-2}$ |
| $-\Delta P'$ | Pressure drop over filter cake | N/M$^2$ | $\mathbf{ML}^{-1}\mathbf{T}^{-2}$ |
| $-\Delta P''$ | Pressure drop over filter cloth | N/M$^2$ | $\mathbf{ML}^{-1}\mathbf{T}^{-2}$ |
| $Q$ | Volumetric feed rate to centrifuge | m$^3$/s | $\mathbf{L}^3\mathbf{T}^{-1}$ |
| $R$ | Radius of bowl of centrifuge | m | **L** |
| $\mathbf{r}$ | Specific resistance of filter cake | m$^{-2}$ | $\mathbf{L}^{-2}$ |
| $r$ | Radius of rotation | m | **L** |
| $r_0$ | Radius of inner surface of liquid in bowl | m | **L** |
| $r_1$ | Radius at inlet to disc bowl centrifuge | m | **L** |
| $r_2$ | Radius at outlet of disc bowl centrifuge | m | **L** |
| $r_i$ | Radius of weir for lighter liquid | m | **L** |
| $r_s$ | Radius of interface between liquids in bowl | m | **L** |
| $r_w$ | Radius of overflow weir for heavier liquid | m | **L** |
| $r'$ | Radius at interface between liquid and filter cake | m | **L** |
| t | Time | s | **T** |
| $t_R$ | Retention time in centrifuge | s | **T** |
| $u_0$ | Terminal falling velocity of particle | m/s | $\mathbf{LT}^{-1}$ |
| $u_c$ | Superficial filtration velocity $\left[\frac{1}{A}\frac{dV}{dt}\right]$ | m/s | $\mathbf{LT}^{-1}$ |
| $V$ | Volume of filtrate passing in time $t$ | m$^3$ | $\mathbf{L}^3$ |
| $V'$ | Volumetric capacity of centrifuge bowl | m$^3$ | $\mathbf{L}^3$ |
| $v$ | Volume of cake deposited by passage of unit volume of filtrate | — | — |
| $x$ | Distance parallel to discs in disc bowl centrifuge | m | **L** |
| $y$ | Distance perpendicular to discs in disc bowl centrifuge | m | **L** |
| $z$ | Vertical height | m | **L** |
| $z_0$ | Vertical height of free surface of liquid (at radius $r_0$) | m | **L** |
| $z_a$ | Value of $z_0$ at vertical axis of rotation ($r_0 = 0$) | m | **L** |
| $\mu$ | Viscosity of liquid | Ns/m$^2$ | $\mathbf{ML}^{-1}\mathbf{T}^{-1}$ |
| $\rho$ | Density of liquid | kg/m$^3$ | $\mathbf{ML}^{-3}$ |
| $\rho_s$ | Density of particles | kg/m$^3$ | $\mathbf{ML}^{-3}$ |
| $\theta$ | Half included angle between discs | — | — |
| $\omega$ | Angular velocity | rad/s | $\mathbf{s}^{-1}$ |
| $\Sigma$ | Capacity term for centrifuge | m$^2$ | $\mathbf{L}^2$ |

CHAPTER 10

# Leaching

## 10.1. INTRODUCTION

### 10.1.1. General principles

Leaching is concerned with the extraction of a soluble constituent from a solid by means of a solvent. The process may be used either for the production of a concentrated solution of a valuable solid material, or in order to remove an insoluble solid, such as a pigment, from a soluble material with which it is contaminated. The method used for the extraction is determined by the proportion of soluble constituent present, its distribution throughout the solid, the nature of the solid and the particle size.

If the solute is uniformly dispersed in the solid, the material near the surface will be dissolved first, leaving a porous structure in the solid residue. The solvent will then have to penetrate this outer layer before it can reach further solute, and the process will become progressively more difficult and the extraction rate will fall. If the solute forms a very high proportion of the solid, the porous structure may break down almost immediately to give a fine deposit of insoluble residue, and access of solvent to the solute will not be impeded. Generally, the process can be considered in three parts: first the change of phase of the solute as it dissolves in the solvent, secondly its diffusion through the solvent in the pores of the solid to the outside of the particle, and thirdly the transfer of the solute from the solution in contact with the particles to the main bulk of the solution. Any one of these three processes may be responsible for limiting the extraction rate, though the first process usually occurs so rapidly that it has a negligible effect on the overall rate.

In some cases the soluble material is distributed in small isolated pockets in a material which is impermeable to the solvent such as gold dispersed in rock, for example. In such cases the material is crushed so that all the soluble material is exposed to the solvent. If the solid has a cellular structure, the extraction rate will generally be comparatively low because the cell walls provide an additional resistance. In the extraction of sugar from beet, the cell walls perform the important function of impeding the extraction of undesirable constituents of relatively high molecular weight, and the beet should therefore be prepared in long strips so that a relatively small proportion of the cells is ruptured. In the extraction of oil from seeds, the solute is itself liquid.

### 10.1.2. Factors influencing the rate of extraction

The selection of the equipment for an extraction process is influenced by the factors which are responsible for limiting the extraction rate. Thus, if the diffusion of the solute through the porous structure of the residual solids is the controlling factor, the material should be of small size so that the distance the solute has to travel is small. On the other hand, if diffusion of the solute from the surface of the particles to the bulk of the solution is the controlling factor, a high degree of agitation of the fluid is required.

There are four important factors to be considered:

*Particle size.* Particle size influences the extraction rate in a number of ways. The smaller the size, the greater is the interfacial area between the solid and liquid, and therefore the higher is the rate of transfer of material and the smaller is the distance the solute must diffuse within the solid as already indicated. On the other hand, the surface may not be so effectively used with a very fine material if circulation of the liquid is impeded, and separation of the particles from the liquid and drainage of the solid residue are made more difficult. It is generally desirable that the range of particle size should be small so that each particle requires approximately the same time for extraction and, in particular, the production of a large amount of fine material should be avoided as this may wedge in the interstices of the larger particles and impede the flow of the solvent.

*Solvent.* The liquid chosen should be a good selective solvent and its viscosity should be sufficiently low for it to circulate freely. Generally, a relatively pure solvent will be used initially, although as the extraction proceeds the concentration of solute will increase and the rate of extraction will progressively decrease, first because the concentration gradient will be reduced, and secondly because the solution will generally become more viscous.

*Temperature.* In most cases, the solubility of the material which is being extracted will increase with temperature to give a higher rate of extraction. Further, the diffusion coefficient will be expected to increase with rise in temperature and this will also improve the rate of extraction. In some cases, the upper limit of temperature is determined by secondary considerations, such as, for example, the necessity to avoid enzyme action during the extraction of sugar.

*Agitation of the fluid.* Agitation of the solvent is important because this increases the eddy diffusion and therefore the transfer of material from the surface of the particles to the bulk of the solution, as discussed in the following section. Further, agitation of suspensions of fine particles prevents sedimentation and more effective use is made of the interfacial surface.

## 10.2. MASS TRANSFER IN LEACHING OPERATIONS

Mass transfer rates within the porous residue are difficult to assess because it is impossible to define the shape of the channels through which transfer must take place. It is possible, however, to obtain an approximate indication of the rate of transfer from the particles to the bulk of the liquid. Using the concept of a thin film as providing the resistance to transfer, the equation for mass transfer may be written as:

$$\frac{\mathrm{d}M}{\mathrm{d}t} = \frac{k'A(c_s - c)}{b} \tag{10.1}$$

where: $A$ is the area of the solid–liquid interface,

$b$ is the effective thickness of the liquid film surrounding the particles,

$c$ is the concentration of the solute in the bulk of the solution at time $t$,

$c_s$ is the concentration of the saturated solution in contact with the particles,
$M$ is the mass of solute transferred in time $t$, and
$k'$ is the diffusion coefficient. (This is approximately equal to the liquid phase diffusivity $D_L$, discussed in Volume 1, Chapter 10, and is usually assumed constant.)

For a batch process in which $V$, the total volume of solution, is assumed to remain constant, then:

$$\mathrm{d}M = V\,\mathrm{d}c$$

and:

$$\frac{\mathrm{d}c}{\mathrm{d}t} = \frac{k'A(c_s - c)}{bV}$$

The time $t$ taken for the concentration of the solution to rise from its initial value $c_0$ to a value $c$ is found by integration, on the assumption that both $b$ and $A$ remain constant. Rearranging:

$$\int_{c_0}^{c} \frac{\mathrm{d}c}{c_s - c} = \int \frac{k'A}{Vb}\,\mathrm{d}t$$

and:

$$\ln \frac{c_s - c_0}{c_s - c} = \frac{k'A}{Vb}t \tag{10.2}$$

If pure solvent is used initially, $c_0 = 0$, and:

$$1 - \frac{c}{c_s} = \mathrm{e}^{-(k'A/bV)t}$$

or:

$$c = c_s(1 - \mathrm{e}^{-(k'A/bV)t}) \tag{10.3}$$

which shows that the solution approaches a saturated condition exponentially.

In most cases the interfacial area will tend to increase during the extraction and, when the soluble material forms a very high proportion of the total solid, complete disintegration of the particles may occur. Although this results in an increase in the interfacial area, the rate of extraction will probably be reduced because the free flow of the solvent will be impeded and the effective value of $b$ will be increased.

Work on the rate of dissolution of regular shaped solids in liquids has been carried out by LINTON and SHERWOOD[1], to which reference is made in Volume 1. Benzoic acid, cinnamic acid, and $\beta$-naphthol were used as solutes, and water as the solvent. For streamline flow, the results were satisfactorily correlated on the assumption that transfer took place as a result of molecular diffusion alone. For turbulent flow through small tubes cast from each of the materials, the rate of mass transfer could be predicted from the pressure drop by using the '$j$-factor' for mass transfer. In experiments with benzoic acid, unduly high rates of transfer were obtained because the area of the solids was increased as a result of pitting.

The effect of agitation, as produced by a rotary stirrer, for example, on mass transfer rates has been investigated by HIXSON and BAUM[2] who measured the rate of dissolution of pure salts in water. The degree of agitation is expressed by means of a dimensionless group $(Nd^2\rho/\mu)$ in which:

$N$ is the number of revolutions of the stirrer per unit time,
$d$ is the diameter of the vessel,
$\rho$ is the density of the liquid, and
$\mu$ is its viscosity.

This group is referred to in Volume 1, Chapter 7 in the discussion of the power requirements for agitators.

For values of $(Nd^2\rho/\mu)$ less than 67,000, the results are correlated by:

$$\frac{K_L d}{D_L} = 2.7 \times 10^{-5} \left(\frac{Nd^2\rho}{\mu}\right)^{1.4} \left(\frac{\mu}{\rho D_L}\right)^{0.5} \tag{10.4}$$

and for higher values of $(Nd^2\rho/\mu)$ by:

$$\frac{K_L d}{D_L} = 0.16 \left(\frac{Nd^2\rho}{\mu}\right)^{0.62} \left(\frac{\mu}{\rho D_L}\right)^{0.5} \tag{10.5}$$

where $K_L$ is the mass transfer coefficient, equal to $k'/b$ in equation 10.1.

Further experimental work has been carried out on the rates of melting of a solid in a liquid, using a single component system, and Hixson and Baum express their results for the heat transfer coefficient as:

$$\frac{hd}{k} = 0.207 \left(\frac{Nd^2\rho}{\mu}\right)^{0.63} \left(\frac{C_p\mu}{k}\right)^{0.5} \tag{10.6}$$

for values of $(Nd^2\rho/\mu)$ greater than 67,000.

In equation 10.6:

$h$ is the heat transfer coefficient,
$k$ is the thermal conductivity of the liquid, and
$C_p$ is the specific heat of the liquid.

It may be seen from equations 10.5 and 10.6 that at high degrees of agitation the ratio of the heat and mass transfer coefficients is almost independent of the speed of the agitator and:

$$\frac{K_L}{h} = 0.77 \left(\frac{D_L}{\rho C_p k}\right)^{0.5} \tag{10.7}$$

Piret *et al.*[3] attempted to reproduce the conditions in a porous solid using banks of capillary tubes, beds of glass beads and porous spheres, and measured the rate of transfer of a salt as solute through water to the outside of the system. It was shown that the rate of mass transfer is that predicted for an unsteady transfer process and that the shape of the pores could be satisfactorily taken into account.

In a theoretical study, Chorny and Krasuk[4] analysed the diffusion process in extraction from simple regular solids, assuming constant diffusivity.

## Example 10.1

In a pilot scale test using a vessel 1 $m^3$ in volume, a solute was leached from an inert solid and the water was 75 per cent saturated in 100 s. If, in a full-scale unit, 500 kg of the inert solid containing, as before, 28 per cent by mass of the water-soluble component, is agitated with 100 $m^3$ of water, how long will it take for all the solute to dissolve, assuming conditions are equivalent to those in the pilot scale vessel? Water is saturated with the solute at a concentration of 2.5 $kg/m^3$.

## Solution

For the *pilot-scale* vessel:

$$c = (2.5 \times 75/100) = 1.875\ \text{kg/m}^3$$

$$c_s = 2.5\ \text{kg/m}^3, \quad V = 1.0\ \text{m}^3 \text{ and } t = 10\ \text{s}$$

Thus, in equation 10.3:

$$1.875 = 2.5(1 - e^{-(k'A/1.0b)100})$$

and:

$$k'A/b = 0.139\ \text{m}^3\text{/s}$$

For the *full-scale* vessel:

$$c = (500 \times 28/100)/100 = 1.40\ \text{kg/m}^3$$

$$c_s = 2.5\ \text{kg/m}^3, \qquad V = 100\ \text{m}^3$$

Thus:

$$1.40 = 2.5(1 - e^{-0.139t/100})$$

and:

$$t = \underline{\underline{591\ \text{s}}}\ (9.9\ \text{min})$$

# 10.3. EQUIPMENT FOR LEACHING

## 10.3.1. Processes involved

Three distinct processes are usually involved in leaching operations:

(a) Dissolving the soluble constituent.
(b) Separating the solution, so formed, from the insoluble solid residue.
(c) Washing the solid residue in order to free it of unwanted soluble matter or to obtain as much of the soluble material as possible as the product.

Leaching has in the past been carried out mainly as a batch process although many continuous plants have also been developed. The type of equipment employed depends on the nature of the solid — whether it is granular or cellular and whether it is coarse or fine. The normal distinction between coarse and fine solids is that the former have sufficiently large settling velocities for them to be readily separable from the liquid, whereas the latter can be maintained in suspension with the aid of only a small amount of agitation.

Generally, the solvent is allowed to percolate through beds of coarse materials, whereas fine solids offer too high a resistance.

As already pointed out, the rate of extraction will, in general, be a function of the relative velocity between the liquid and the solid. In some plants the solid is stationary and the liquid flows through the bed of particles, whilst in some continuous plants the solid and liquid move countercurrently.

### 10.3.2. Extraction from cellular materials

With seeds such as soya beans, containing only about 15 per cent of oil, solvent extraction is often used because mechanical methods are not very efficient. Light petroleum fractions are generally used as solvents. Trichlorethylene has been used where fire risks are serious, and acetone or ether where the material is very wet. A batch plant for the extraction of oil from seeds is illustrated in Figure 10.1. This consists of a vertical cylindrical vessel divided into two sections by a slanting partition. The upper section is filled with the charge of seeds which is sprayed with fresh solvent via a distributor. The solvent percolates through the bed of solids and drains into the lower compartment where, together with any water extracted from the seeds, it is continuously boiled off by means of a steam coil. The vapours are passed to an external condenser, and the mixed liquid is passed to a separating box from which the solvent is continuously fed back to the plant and the water is run to waste. By this means a concentrated solution of the oil is produced by the continued application of pure solvent to the seeds.

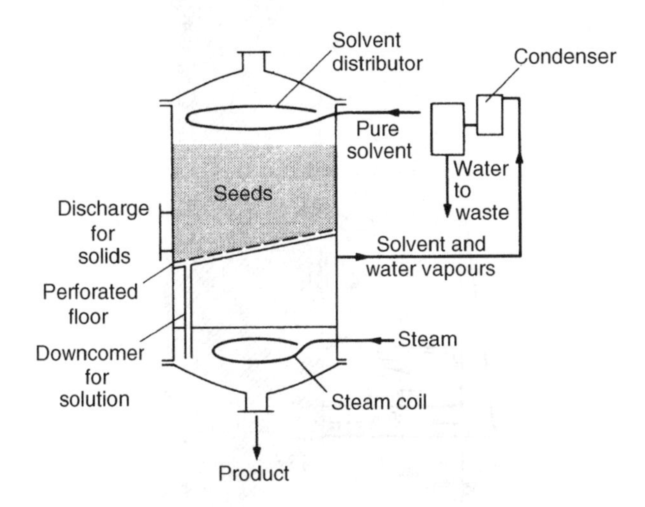


Figure 10.1. Batch plant for extraction of oil from seeds

The Bollmann continuous moving bed extractor, as shown in Figures 10.2 and 10.3, which is described by Goss[5], consists of a series of perforated baskets, arranged as in a bucket elevator and contained in a vapour-tight vessel, is widely used with seeds which do not disintegrate on extraction. Solid is fed into the top basket on the downward side and is discharged from the top basket on the upward side, as shown in Figure 10.3. The

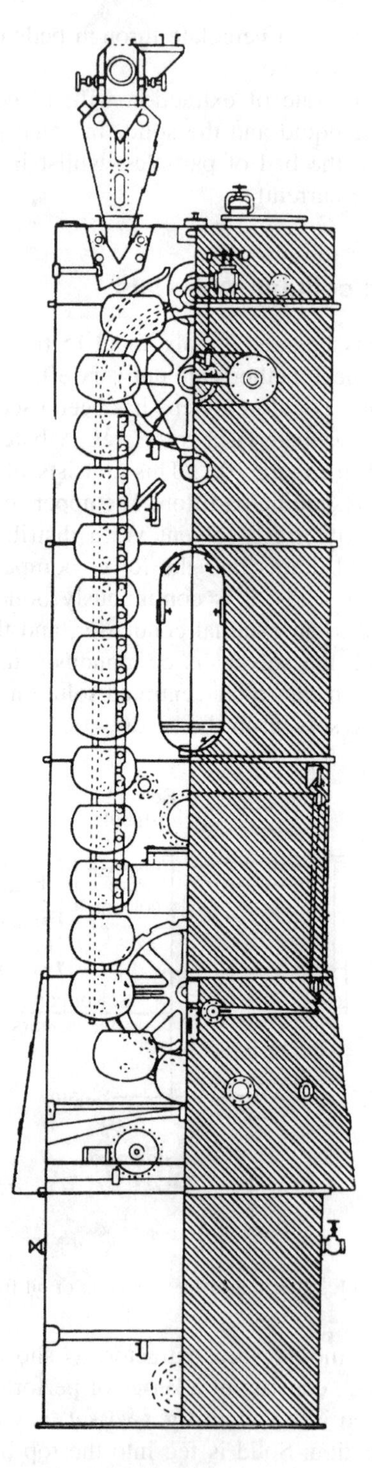

Figure 10.2. Bollmann extractor

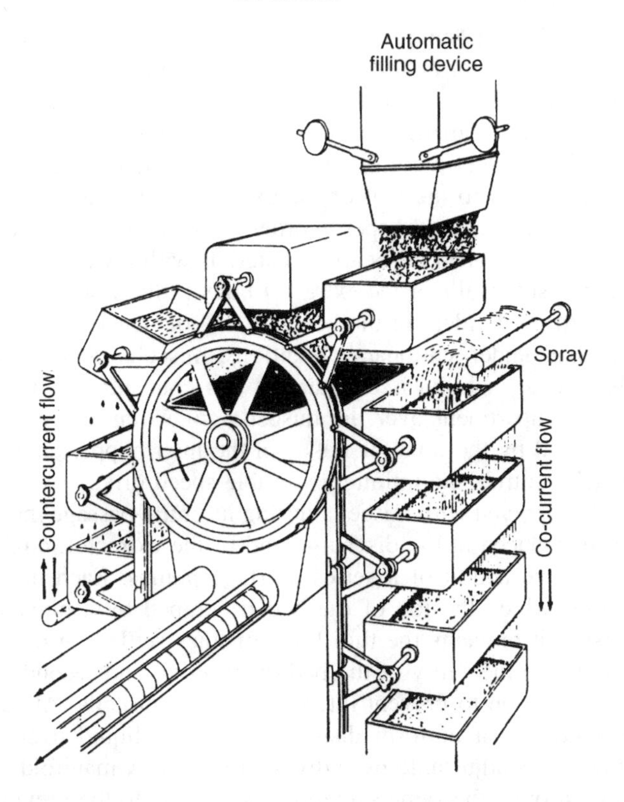


Figure 10.3. Bollmann extractor — filling and emptying of baskets

solvent is sprayed on to the solid which is about to be discarded, and passes downwards through the baskets so that countercurrent flow is achieved. The solvent is finally allowed to flow down through the remaining baskets in co-current flow. A typical extractor moves at about 0.3 mHz (1 revolution per hour), each basket containing some 350 kg of seeds. Generally, about equal masses of seeds and solvent are used and the final solution, known as miscella, contains about 25 per cent of oil by mass.

The Bonotto extractor[5] consists of a tall cylindrical vessel with a series of slowly rotating horizontal trays. The solid is fed continuously on to the top tray near its outside edge and a stationary scraper, attached to the shell of the plant, causes it to move towards the centre of the plate. It then falls through an opening on to the plate beneath, and another scraper moves the solids outwards on this plate which has a similar opening near its periphery. By this means the solid is moved across each plate, in opposite directions on alternate plates, until it reaches the bottom of the tower from which it is removed by means of a screw conveyor. The extracting liquid is introduced at the bottom and flows upwards so that continuous countercurrent flow is obtained, though a certain amount of mixing of solvent and solution takes place when the density of the solution rises as the concentration increases.

A more recently developed continuous extractor is the horizontal perforated belt extractor, probably the simplest percolation extractor from a mechanical view point. Here the basic principle is the extraction of an intermediate bed depth on a continuous belt without partitions. The extractor is fitted with a slow moving "perforated belt" running on sprockets at each end of the extractor. A series of specially designed screens form a flat surface attached to the chains, made of wedge-bar type grids when non-powdery products are processed, or stainless steel mesh cloths for fine particle products. The flakes are fed into the hopper and flow on to the belt of the extractor, and the level is controlled by a damper at the outlet of the feeding hopper in order to maintain a constant flake bed height. This height can be adjusted when oil-bearing materials with lower percolation rates are to be processed. The two side walls of the extractor body provide support for the bed on the moving belt and, with no dividers in the belt, the bed of material becomes a continuous mass. This means that, under stationary conditions, miscella concentration in each and every point of the bed of material is constant as this concentration is not related to the position of a given compartment over the miscella collecting hopper. The belt speed is automatically controlled by the level of flakes in the inlet hopper, and is measured by a nuclear sensor which controls the infinitely variable speed drive.

The raw material is sprayed with miscella during its entire passage through the extractor and fresh solvent, introduced at the discharge end of the extractor, circulates against the flow of flakes, under the action of a series of stage pumps. Each miscella wash has a draining section after which the top of the bed is scraped by a hinged rake which has two functions. Firstly, it prevents the thin layer of fines settled on the upper part of the bed from reducing the permeability of the bed of material, and secondly, it forms a flake pile at each draining section to prevent intermingling of miscella. At each wash section, a spray distributor ensures a uniform distribution of the liquid over the width of the bed and liquid flowrate is adjustable by individual valves. A manifold is fitted to permit miscella circulation through the same stage, or progression to the previous one. Before it is discharged, liquid in the material bed drains into the final collecting hoppers. Discharge of material into the outlet hopper is regulated by a rotary scraper which ensures an even feed of the extracted meal to the drainage section. The belt is effectively cleaned twice, first by fresh solvent just after material discharge and then at the other end of the return span, by means of miscella.

## 10.3.3. Leaching of coarse solids

A simple batch plant used for coarse solids consists of a cylindrical vessel in which the solids rest on a perforated support. The solvent is sprayed over the solids and, after extraction is complete, the residue is allowed to drain. If the solid contains a high proportion of solute such that it disintegrates, it is treated with solvent in a tank and the solution is decanted.

In a simple countercurrent system, the solid is contained in a number of tanks and the solvent flows through each in turn. The first vessel contains solid which is almost completely extracted and the last contains fresh solid. After some time, the first tank is disconnected and a fresh charge is introduced at the far end of the battery. The solvent may flow by gravity or be fed by positive pressure, and is generally heated before it enters each tank. The system is unsatisfactory in that it involves frequent interruption while the tanks are recharged, and countercurrent flow is not obtained within the units themselves.

A continuous unit in which countercurrent flow is obtained is the tray classifier, such as the Dorr classifier described in Chapter 1 (see Figures 1.24 and 10.4). Solid is introduced near the bottom of a sloping tank and is gradually moved up by means of a rake. The solvent enters at the top and flows in the opposite direction to the solid, and passes under a baffle before finally being discharged over a weir (Figure 10.4). The classifier operates satisfactorily provided the solid does not disintegrate, and the solids are given ample time to drain before they are discharged. A number of these units may be connected in series to give countercurrent flow.

A plant has been successfully developed in Australia for the extraction of potassium sulphate from alums containing about 25 per cent soluble constituents. After roasting, the material which is then soft and porous with a size range from 12 mm to very fine particles, with 95 per cent greater than 100-mesh (0.15 mm), is leached at 373 K with a solution that is saturated at 303 K, the flow being as shown in Figure 10.5. The make-up water, which is used for washing the extracted solid, is required to replace that removed in the residue of spent solid, in association with the crystals, and by evaporation in the leaching tank and the crystalliser.

The leaching plant, shown in Figure 10.6, consists of an open tank, 3 m in diameter, into the outer portion of which the solid is continuously introduced from an annular hopper. Inside the tank a 1.8 m diameter vertical pipe rotates very slowly at the rate of

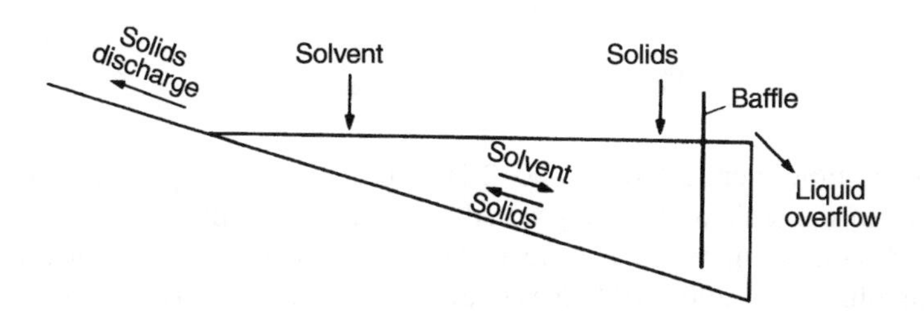


Figure 10.4. Flow of solids and liquids in Dorr classifier

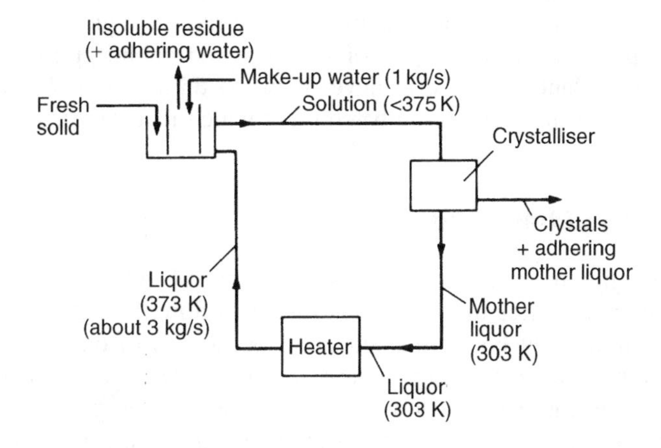


Figure 10.5. Flow diagram for continuous leaching plant

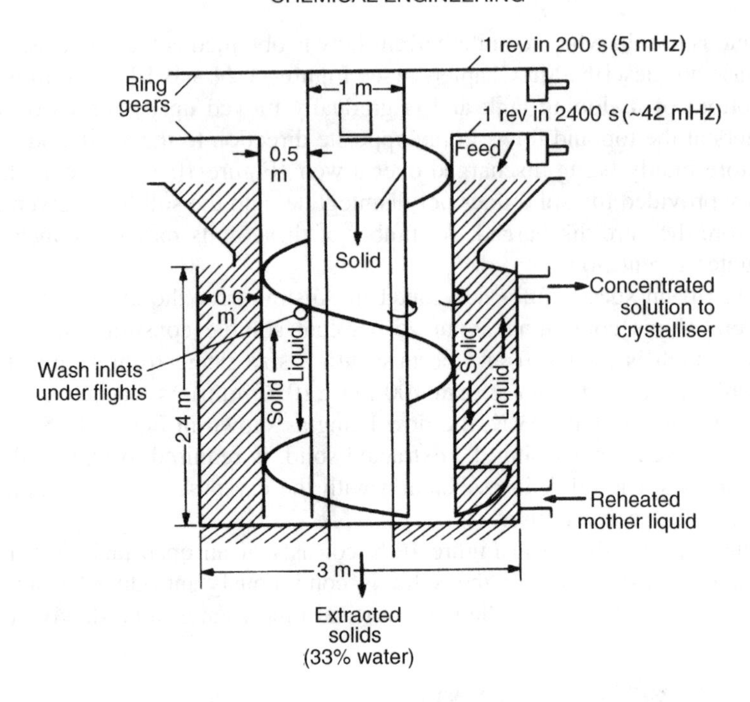


Figure 10.6. Continuous leaching tank

about one revolution every 2400 s (0.0042 Hz). It carries three ploughs stretching to the circumference of the tank, and these gradually take the solid through holes into the inside of the pipe. A hollow shaft, about 1 m in diameter, rotates in the centre of the tank at about one revolution in 200 s (0.005 Hz) and carries a screw conveyor which lifts the solid and finally discharges it through an opening, so that it falls down the shaft and is deflected into a waste pipe passing through the bottom of the tank. Leaching takes place in the outer portion of the tank where the reheated mother liquor rises through the descending solid. The make-up water is introduced under the flutes of the screw elevator, flows down over the solid and then joins the reheated mother liquor. Thus countercurrent extraction takes place in the outer part of the tank and countercurrent washing in the central portion. The plant described achieves between 85 and 90 per cent extraction, as compared with only 50 per cent in the batch plant which it replaced.

### 10.3.4. Leaching of fine solids

Whereas coarse solids may be leached by causing the solvent to pass through a bed of the material, fine solids offer too high a resistance to flow. Particles of less than about 200-mesh (0.075 mm) may be maintained in suspension with only a small amount of agitation, and as the total surface area is large, an adequate extraction can be effected in a reasonable time. Because of the low settling velocity of the particles and their large surface, the subsequent separation and washing operations are more difficult for fine materials than with coarse solids.

Agitation may be achieved either by the use of a mechanical stirrer or by means of compressed air. If a paddle stirrer is used, precautions must be taken to prevent the whole of the liquid being swirled, with very little relative motion occurring between solids and liquid. The stirrer is often placed inside a central tube, as shown in Figure 10.7, and the shape of the blades arranged so that the liquid is lifted upwards through the tube. The liquid then discharges at the top and flows downwards outside the tube, thus giving continuous circulation. Other types of stirrers are discussed in Volume 1, Chapter 7, in the context of liquid–liquid mixing.

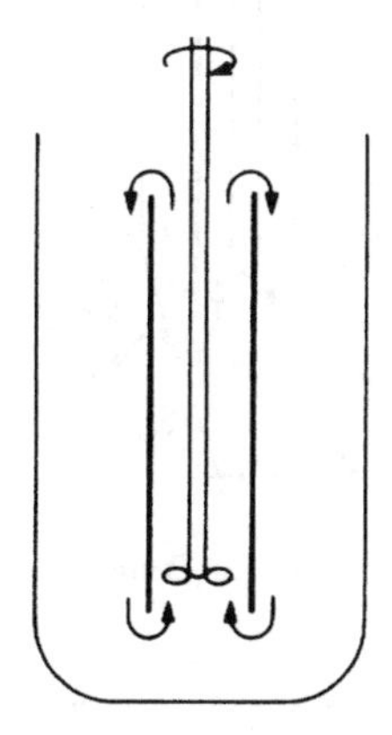

Figure 10.7. Simple stirred tank

An example of an agitated vessel in which compressed air is used is the Pachuca tank, shown in Figure 10.8. This is a cylindrical tank with a conical bottom, fitted with a central pipe connected to an air supply. Continuous circulation is obtained with the central pipe acting as an air lift. Additional air jets are provided in the conical portion of the base and are used for dislodging any material which settles out.

The Dorr agitator which is illustrated in Figure 10.9, also uses compressed air for stirring, and consists of a cylindrical flat-bottomed tank fitted with a central air lift inside a hollow shaft which slowly rotates. To the bottom of the shaft are fitted rakes which drag the solid material to the centre as it settles, so that it is picked up by the air lift. At the upper end of the shaft the air lift discharges into a perforated launder which distributes the suspension evenly over the surface of the liquid in the vessel. When the shaft is not rotating the rakes automatically fold up so as to prevent the plant from seizing up if it is shut down full of slurry. This type of agitator can be used for batch or continuous operation. In the latter case the entry and delivery points are situated at opposite sides of the tank. The discharge pipe often takes the form of a flexible connection which can be arranged to take off the product from any desired depth. Many of these agitators are heated by steam coils. If the soluble material dissolves very rapidly, extraction can be carried out in a thickener, such as the Dorr thickener described in Chapter 5. Thickeners are also extensively used for separating the discharge from an agitator, and are frequently connected in series to give countercurrent washing of the residue.

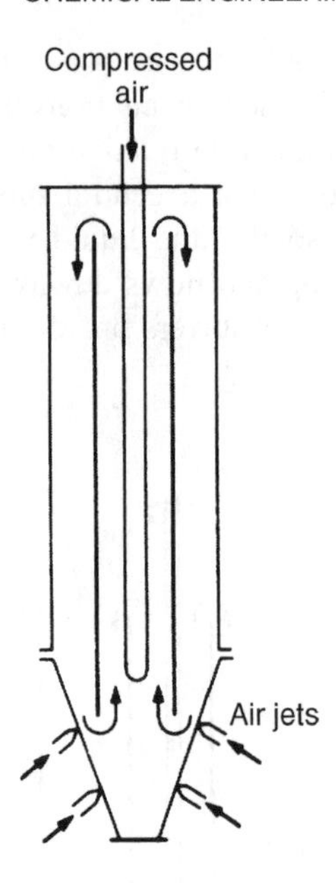


Figure 10.8. Pachuca tank

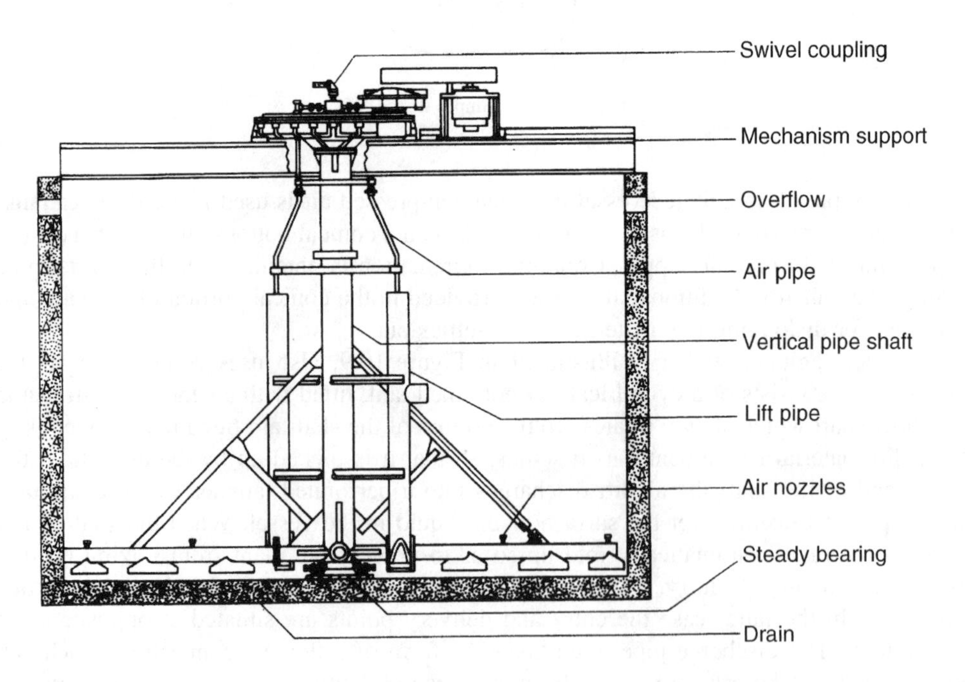


Figure 10.9. Dorr agitator

### 10.3.5. Batch leaching in stirred tanks

The batch dissolution of solids in liquids is very often carried out in tanks agitated by co-axial impellers including turbines, paddles and propellers, a system which may also be used for the leaching of fine solids. In this case, the controlling rate in the mass transfer process is the rate of transfer of material into or from the interior of the solid particles, rather than the rate of transfer to or from the surface of the particles, and therefore the main function of the agitator is to supply unexhausted solvent to the particles which remain in the tank long enough for the diffusive process to be completed. This is achieved most efficiently if the agitator is used to circulate solids across the bottom of the tank, or barely to suspend them above the bottom of the tank. After the operation is completed, the leached solids must be separated from the extract and this may be achieved by settling followed by decantation, or externally by filters, centrifuges or thickeners. The difficulties involved in separating the solids and the extract is one of the main disadvantages of batch operation coupled with the fact that batch stirred tanks provide only one equilibrium stage. The design of agitators in order to produce a batch suspension of closely sized particles has been discussed by BOHNET and NIESMARK[6] and the general approach is to select the type and geometry of impeller and tank, to specify the rotational speed required for acceptable performance, and then to determine the shaft power required to drive the impeller at that speed.

Ores of gold, uranium and other metals are often batch-leached in *Pachua tanks* which are described in Section 10.3.4.

## 10.4. COUNTERCURRENT WASHING OF SOLIDS

Where the residual solid after separation is still mixed with an appreciable amount of solution, it is generally desirable to pass it through a battery of washers, arranged to give countercurrent flow of the solids and the solvent as shown in Figure 10.10. If the solids are relatively coarse a number of classifiers may be used and, with the more usual case of fine solids, thickeners are generally employed. In each unit a liquid, referred to as the overflow, and a mixture of insoluble residue and solution, referred to as the underflow, are brought into contact so that intimate mixing is achieved and the solution leaving in the overflow has the same composition as that associated with the solids in the underflow. Each unit then represents an ideal stage. In some cases perfect mixing may not be achieved, and allowance must be made for the reduced efficiency of the stage.

By means of a series of material balances, the compositions of all the streams in the system shown in Figure 10.10 may be calculated on the assumption that the whole of the solute has been dissolved and that equilibrium has been reached in each of the thickeners. In many cases it may only be necessary to determine the compositions of the streams entering and leaving, and the simplified methods given later may then be used.

In order to define such a system completely, the following six quantities must be specified. The first four relate to the quantities and compositions of the materials used. Quantity (e) specifies the manner in which each unit operates, and quantity (f) involves either the number of units or a specification of the duty required of the plant.

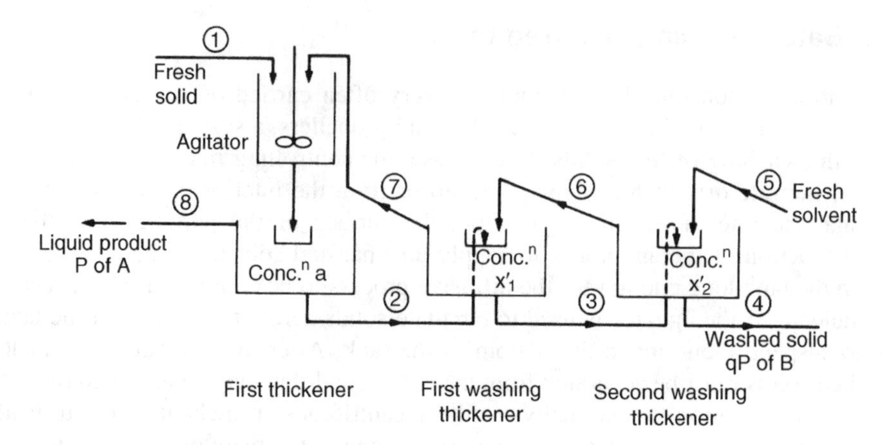


Figure 10.10. Agitator and washing system

(a) The composition of the solvent fed to the system — in particular, the concentration of soluble material already present.
(b) The quantity of solvent used; alternatively the concentration of the solution to be produced may be specified, and the corresponding amount of solvent calculated from a material balance.
(c) The composition of the solid to be leached.
(d) The amount of solid fed to the system; alternatively, the amount of soluble or insoluble material required as product may be specified, and the necessary amount of solid feed then calculated from a material balance.
(e) The amount of liquid discharged with the solid in the underflow from each of the thickeners.
(f) The number and arrangement of the units; the purity of the product from the plant can then be calculated. Alternatively, the required purity of the washed solid may be stated, and the number of units can then be calculated.

In the following example, a solid consisting of a soluble constituent **A** and an insoluble constituent **B** is considered. Leaching is carried out with a pure solvent **S** and a solution is produced containing a mass $a$ of **A**, per unit mass of **S** and the total mass of **A** in solution is $P$. It will be assumed that the quantity of solvent removed in the underflow from each of the thickeners is the same, and that this is independent of the concentration of the solution in that thickener. It will be assumed that unit mass of the insoluble material **B** removes a mass $s$ of solvent **S** in association with it. Perfect mixing in each thickener will be assumed and any adsorption of solute on the surface of the insoluble solid will be neglected. In a given thickener, therefore, the ratio of solute to solvent will be the same in the underflow as in the overflow.

The compositions of the various streams may be calculated in terms of three unknowns: $x_1'$ and $x_2'$, the ratios of solute to solvent in the first and second washing thickeners respectively, and $qP$, the amount of insoluble solid **B** in the underflow streams. An overall material balance is made and then a balance is made on the agitator and its thickener combined, the first washing thickener and the second washing thickener. The procedure is as follows.

In the overall balance, streams 1 and 8 in Figure 10.10 can be recorded immediately. Stream 4 is then obtained since the whole of **B** appears in the underflow, the ratio **S/B** is $s$ and the ratio **A/S** is $x_2'$. Stream 5 is obtained by difference. As pure solvent is fed to the system, a relation is obtained by equating the solute content of this stream to zero. This enables $q$ to be eliminated in terms of $x_2'$. For the agitator and separating thickener, streams 1 and 8 have already been determined, and stream 2 is obtained in the same way as stream 4 on the previous balance. Stream 7 is then obtained by difference. It is then possible to proceed in this manner from unit to unit because two streams are always common to consecutive units. Further equations can then be obtained since the ratios **A/S** for the overflows from the two washing thickeners are equal to $x_1'$ and $x_2'$ respectively. Solution of these simultaneous equations for $x_1'$ and $x_2'$ gives each of the streams in terms of known quantities. It will be apparent that this procedure would be laborious if a large number of units were involved, and therefore alternative methods are used to obtain the compositions of the end streams. The method described would also involve a number of trial and error solutions if it were necessary to calculate the number of units required to give a certain degree of washing. The application of the method is illustrated in the following example.

## Example 10.2

Caustic soda is manufactured by the lime-soda process. A solution of sodium carbonate in water, containing 0.25 kg/s $Na_2CO_3$, is treated with the theoretical requirement of lime, and after the reaction is complete the $CaCO_3$ sludge, containing 1 part of $CaCO_3$ per 9 parts of water, by mass, is fed continuously to three thickeners in series and washed countercurrently, as shown in Figure 10.11. Calculate the necessary rate of feed of neutral water to the thickeners so that the calcium carbonate, on drying, contains only 1 per cent of sodium hydroxide. The solid discharged from each thickener contains 1 part by mass of calcium carbonate to 3 of water. The concentrated wash liquid is mixed with the contents of the agitator before being fed to the first thickener.

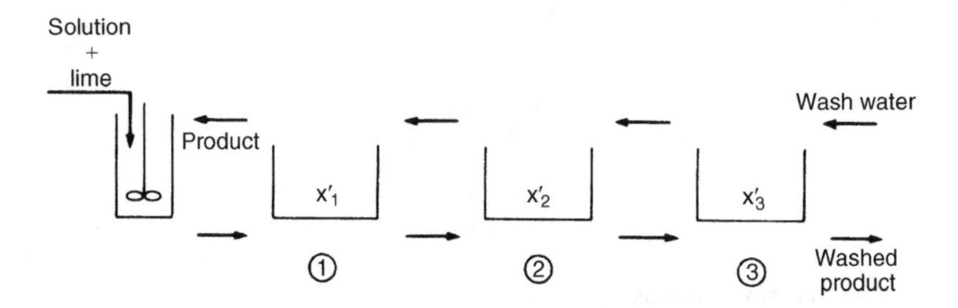


Figure 10.11. Flow diagram for countercurrent washing of calcium carbonate

## Solution

$$\begin{array}{llll} Na_2CO_3 + Ca(OH)_2 & = & 2NaOH & + CaCO_3 \\ 106\text{ kg} & & 80\text{ kg} & 100\text{ kg} \end{array}$$

If $x_1'$, $x_2'$, $x_3'$ are the solute: solvent ratios in thickeners 1, 2, and 3, respectively, the quantities of $CaCO_3$. NaOH, and water in each of the streams can be calculated for every 100 kg of calcium carbonate.

| | $CaCO_3$ | NaOH | Water |
|---|---|---|---|
| *Overall balance* | | | |
| Feed from reactor | 100 | 80 | 900 |
| Feed as washwater | — | — | $W_f$ (say) |
| Product-underflow | 100 | $300x'_3$ | 300 |
| Product-overflow | — | $80-300x'_3$ | $600 + W_f$ |
| *Thickener* 1 | | | |
| Feed from reactor | 100 | 80 | 900 |
| Feed-overflow | — | $300(x'_1 - x'_3)$ | $W_f$ |
| Product-underflow | 100 | $300x'_1$ | 300 |
| Product-overflow | — | $80-300x'_3$ | $600 + W_f$ |
| *Thickener* 2 | | | |
| Feed-underflow | 100 | $300x'_1$ | 300 |
| Feed-overflow | — | $300(x'_2 - x'_3)$ | $W_f$ |
| Product-underflow | 100 | $300x'_2$ | 300 |
| Product-overflow | — | $300(x'_1 - x'_3)$ | $W_f$ |
| *Thickener* 3 | | | |
| Feed-underflow | 100 | $300x'_2$ | 300 |
| Feed-water | — | — | $W_f$ |
| Product-underflow | 100 | $300x'_3$ | 300 |
| Product-overflow | — | $300(x'_2 - x'_3)$ | $W_f$ |

Since the final underflow must contain only 1 per cent of NaOH, then:

$$\frac{300x'_3}{100} = 0.01$$

If the equilibrium is achieved in each of the thickeners, the ratio of NaOH to water will be the same in the underflow and the overflow and:

$$\frac{300(x'_2 - x'_3)}{W_f} = x'_3$$

$$\frac{300(x'_1 - x'_3)}{W_f} = x'_2$$

$$\frac{80 - 300x'_3}{600 + W_f} = x'_1$$

Solution of these four simultaneous equations gives:

$$x'_3 = 0.0033, \quad x'_2 = 0.0142, \quad x'_1 = 0.05, \quad \text{and} \quad W_f = 980.$$

Thus the amount of water required for washing 100 kg $CaCO_3$ is 980 kg.
The solution fed to reactor contains 0.25 kg/s $Na_2CO_3$. This is equivalent to 0.236 kg/s $CaCO_3$, and hence the actual water required:

$$= (980 \times 0.236)/100$$

$$= \underline{\underline{0.23 \text{ kg/s}}}$$

## 10.5. CALCULATION OF THE NUMBER OF STAGES

### 10.5.1. Batch processes

The solid residue obtained from a batch leaching process may be washed by mixing it with liquid, allowing the mixture to settle, and then decanting the solution. This process can then be repeated until the solid is adequately washed. Suppose that, in each decantation operation, the ratio $R'$ of the amount of solvent decanted to that remaining in association with insoluble solid is a constant and independent of the concentration of the solution, then, after the first washing, the fraction of the soluble material remaining behind with the solid in the vat is $1/(R' + 1)$. After the second washing, a fraction $1/(R' + 1)$ of this remains behind, or $1/(R' + 1)^2$ of the solute originally present is retained. Similarly after $m$ washing operations, the fraction of the solute retained by the insoluble residue is $1/(R' + 1)^m$.

### 10.5.2. Countercurrent washing

If, in a battery of thickeners arranged in series for countercurrent washing, the amount of solvent removed with the insoluble solid in the underflow is constant, and independent of the concentration of the solution in the thickener, then the amount of solvent leaving each thickener in the underflow will then be the same, and therefore the amount of solvent in the overflow will also be the same. Hence the ratio of the solvent discharged in the overflow to that in the underflow is constant. This will be taken as $R$, where:

$$R = \frac{\text{Amount of solvent discharged in the overflow}}{\text{Amount of solvent discharged in the underflow}} \tag{10.8}$$

If perfect mixing occurs in each of the thickeners and solute is not preferentially adsorbed on the surface of the solid, the concentration of the solution in the overflow will be the same as that in the underflow. If it is assumed that all the solute has been brought into solution in the agitators, then:

$$R = \frac{\text{Amount of solute discharged in the overflow}}{\text{Amount of solute discharged in the underflow}} \tag{10.9}$$

and:

$$R = \frac{\text{Amount of solution discharged in the overflow}}{\text{Amount of solution discharged in the underflow}} \tag{10.10}$$

It may be noted that these relations apply only to the washing thickeners and not, in general, to the primary thickener in which the product from the agitators is first separated.

A system is now considered consisting of $n$ washing thickeners arranged for countercurrent washing of a solid from a leaching plant, in which the whole of the soluble material is dissolved. The suspension is separated in a thickener and the underflow from this thickener is fed to the washing system as shown in Figure 10.12.

The argument as follows is based on unit mass of insoluble solid.

$L_1, \ldots, L_h, \ldots, L_n$ are the amounts of solute in the overflows from washing thickeners 1 to $n$, respectively.

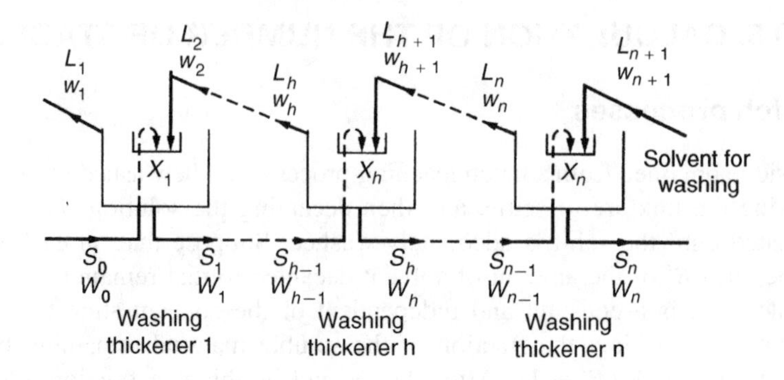


Figure 10.12. Series of thickeners arranged for countercurrent washing

$w_1, \ldots, w_h, \ldots, w_n$ are the corresponding quantities of solution.
An amount $w_{n+1}$ of wash liquid, fed to the $n$th thickener, contains an amount $L_{n+1}$ of solute.
$S_1, \ldots, S_h, \ldots, S_n$ and $W_1, \ldots, W_h, \ldots, W_n$ are the amounts of solute and solution in the underflows from the thickeners.
$S_0$ and $W_0$ are the amounts of solute and solution with the solids which are fed to the system for washing. The solvent associated with these solids is taken as the same as that in the underflows from the washing thickeners.
Taking a solute balance on thickener $h$, then:

$$S_{h-1} - S_h = L_h - L_{h+1} \tag{10.11}$$

Taking a balance on solution, then:

$$W_{h-1} - W_h = w_h - w_{h+1} \tag{10.12}$$

Also:
$$R = \frac{L_h}{S_h} = \frac{w_h}{W_h}$$

or:
$$L_h = RS_h \tag{10.13}$$

and:
$$w_h = RW_h \tag{10.14}$$

Taking a balance on solute for each of the thickeners in turn:
*Thickener n:*

$$S_{n-1}S_n = L_n - L_{n+1} = RS_n - L_{n+1} = RS_n - L_{n+1}$$

*Thickener n − 1*:

$$S_{n-2} - S_{n-1} = L_{n-1} - L_n = RS_{n-1} - RS_n = R^2S_n - RL_{n+1}$$

. . . . . . . . . . . . . . . . . . . . . . . . . . . . . . . . . . . . . . . . . . . . . . . . . . . . . . . . . . . .

*Thickener* 2:

$$S_1 - S_2 = L_2 - L_3 = RS_2 - RS_3 = R^{n-1}S_n - R^{n-2}L_{n+1}$$

*Thickener* 1:

$$S_0 - S_1 = L_1 - L_2 = RS_1 - RS_2 = R^n S_n - R^{n-1} L_{n+1}$$

Adding over the whole system:

$$S_0 - S_n = (R + R^2 + \cdots + R^n)S_n - (1 + R + \cdots + R^{n-1})L_{n+1}$$

or: $$S_0 = \frac{R^{n+1} - 1}{R - 1} S_n - \frac{R^n - 1}{R - 1} L_{n+1}$$

and: $$(R - 1)S_0 = (R^{n+1} - 1)S_n - (R^n - 1)L_{n+1} \qquad (10.15)$$

Thus the amount of solute associated with the washed solid may be calculated in terms of the composition of the solid and of the wash liquid fed to the system. In many cases, the amount of solute associated with the washed solid residue must not exceed a certain value. It is then possible to calculate directly the minimum number of thickeners necessary in order to achieve this.

If the liquid fed to the washing system is pure solvent, $L_{n+1}$ will be equal to zero and:

$$\frac{S_n}{S_0} = \frac{R - 1}{R^{n+1} - 1} \qquad (10.16)$$

In this equation, $(S_n/S_0)$ represents the fraction of the solute fed to the washing system which remains associated with the washed solids. If in a given case it is required that this fraction should not exceed a value $f$, the minimum number of washing thickeners required is given by:

$$f = \frac{R - 1}{R^{n+1} - 1}$$

or: $$R^{n+1} = 1 + (R - 1)\frac{1}{f}$$

$$(n + 1)\log R = \log\left(1 + (R - 1)\frac{1}{f}\right)$$

and: $$n = \frac{\log\left(1 + (R - 1)\frac{1}{f}\right)}{\log R} - 1 \qquad (10.17)$$

In general, $n$ will rarely be a whole number and the number of stages to be specified will be taken as the next higher number.

It is sometimes more convenient to work in terms of the total amount of *solution* entering and leaving each thickener and in this case:

$$W_{h-1} - W_h = w_h - w_{h+1} \qquad \text{(equation 10.12)}$$

and: $$w_h = RW_h \qquad \text{(equation 10.14)}$$

Using the same method as before, then:

$$(R - 1)W_0 = (R^{n+1} - 1)W_n - (R^n - 1)w_{n+1} \qquad (10.18)$$

### 10.5.3. Washing with variable underflow

In the systems considered so far, the quantity of solvent, or of solution, removed in association with the insoluble solids has been assumed to be constant and independent of the concentration of solution in the thickener. A similar countercurrent system is now considered in which the amount of solvent or solution in the underflow is a function of the concentration of the solution. This treatment which is equally applicable to the washing thickeners alone, or to the whole system involving agitator and thickeners, is attributable to RUTH[7].

The same notation will be employed as that used previously and shown in Figure 10.12, that is:

$L$ and $w$ denote solute and solution, respectively, in the liquid overflows and
$S$ and $W$ denote solute and solution in the underflows.

The concentration of the solution in each thickener, defined as the ratio of solute to solution, is denoted by the symbol $X$. Considering the overflow from thickener $h$, then:

$$X_h = \frac{L_h}{w_h} \tag{10.19}$$

and for the underflow:

$$X_h = \frac{S_h}{W_h} \tag{10.20}$$

It is seen in Section 10.4 that in order to define the system, it is necessary to specify the following quantities or other quantities from which they can be calculated by a material balance:

(a) The composition of the liquid used for washing, $X_{n+1}$.
(b) The quantity of wash liquid employed, $w_{n+1}$. Thus $L_{n+1}$ can be calculated from the relation $L_{n+1} = w_{n+1}X_{n+1}$ (from equation 10.19).
(c) The composition of the solid to be washed, $S_0$ and $W_0$.
(d) The quantity of insoluble solid to be washed; this is taken as unity.
(e) The quantity of solution removed by the solid in the underflow from the thickeners; this will vary according to the concentration of the solution in the thickener, and it is therefore necessary to know the relation between $W_h$ and $X_h$. This must be determined experimentally under conditions similar to those under which the plant will operate. The data for $W_h$ should then be plotted against $X_{h+1}$. On the same graph it is convenient to plot values calculated for $S_h$ ($= W_hX_h$, from equation 10.20).
(f) The required purity of the washed solid $S_n$. This automatically defines $X_n$ and $W_n$ whose values can be read off from the graph referred to under (e). Alternatively, the number of thickeners in the washing system may be given, and a calculation made of the purity of the product required. This problem is slightly more complicated and it will therefore be dealt with separately.

The solution of the problem depends on the application of material balances with respect to solute and to solution, first over the system as a whole and then over the first $h$ thickeners.

### Balance on the system as a whole

*Solute*

$$L_{n+1} + S_0 = L_1 + S_n$$

Thus the solute in the liquid overflow from the system as a whole is given by:

$$L_1 = L_{n+1} + S_0 - S_n \qquad (10.21)$$

*Solution*

$$w_{n+1} + W_0 = w_1 + W_n$$

Thus the solution discharged in the liquid overflow is given by:

$$w_1 = w_{n+1} + W_0 - W_n \qquad (10.22)$$

The concentration of the solution discharged from the system is obtained by substituting from equations 10.21 and 10.22 in 10.19:

$$X_1 = \frac{L_{n+1} + S_0 - S_n}{w_{n+1} + W_0 - W_n} \qquad (10.23)$$

### Balance on the first h thickeners

*Solute*

$$L_{h+1} + S_0 = L_1 + S_h$$

Thus the amount of solute in the liquid fed to thickener $h$,

$$\begin{aligned} L_{h+1} &= L_1 + S_h - S_0 \\ &= L_{n+1} - S_n + S_h \text{ (from equation 10.21)} \end{aligned} \qquad (10.24)$$

*Solution*

$$w_{h+1} + W_0 = w_1 + W_h$$

Thus the amount of solution fed to thickener $h$,

$$\begin{aligned} w_{h+1} &= w_1 + W_h - W_0 \\ &= w_{n+1} - W_n + W_h \text{ (from equation 10.22)} \end{aligned} \qquad (10.25)$$

Thus the concentration of the solution fed to thickener $h$ is given by substituting from equations 10.24 and 10.25 in equation 10.19 to give:

$$X_{h+1} = \frac{L_{n+1} - S_n + S_h}{w_{n+1} - W_n + W_h} \qquad (10.26)$$

In equation 10.23, all the quantities except $X_1$ are known, and therefore $X_1$ can be calculated. It may be noted that if, instead of the quantity of wash liquid fed to the system, the concentration of the solution leaving the system had been given, equation 10.23 could have been used to calculate $w_{n+1}$. When $X_1$ has been evaluated, the solution of the problem depends on the application of equation 10.26 in successive stages. The only unknown quantities in equation 10.26 are $X_{h+1}$, $S_h$ and $W_h$.

Applying equation 10.26 to the first stage ($h = 1$), then:

$$X_2 = \frac{L_{n+1} - S_n + S_1}{w_{n+1} - W_n + W_1} \tag{10.27}$$

Since $X_1$ is now known, the values of $S_1$ and $W_1$ can be obtained from a graph in which $S_h$ and $W_h$ are plotted against $X_h$. After substituting these values in the equation, $X_2$ can be calculated. The next step is to apply equation 10.26 for $h = 2$. $X_2$ is now known so that $S_2$ and $W_2$ can be obtained from the graph, and the value of $X_3$ can then be calculated. It is thus possible to apply equation 10.26 in this way in successive stages until the value obtained for $S_h$ is brought down to the specified value of $S_n$. The number of washing thickeners required to reduce the solute associated with the washed solid to a specified figure is thus readily calculated. In general, of course, it will not be possible to choose the number of thickeners so that $S_h$ is exactly equal to $S_n$.

It may be seen that the purity of the washed solid must be known before equations 10.23 and 10.26 can be applied. If in a given problem it is necessary to calculate the degree of washing obtained by the use of a certain number of washing thickeners, an initial assumption of the values of $S_n$ and $W_n$ must be made before the problem can be attempted. As a first step, an average value for $R$ may be taken and $S_n$ calculated from equation 10.16. The method, as already given, should then be applied for the number of thickeners specified in the problem, and the calculated and assumed values of $S_n$ compared. If the calculated value is higher than the assumed value, the latter is too low. The calculated values of $S_n$ can then be plotted against the corresponding assumed values. The correct solution is then denoted by the point at which the two values agree.

## Example 10.3

A plant produces 8640 tonnes per day (100 kg/s) of titanium dioxide pigment which must be 99.9 per cent pure when dried. The pigment is produced by precipitation and the material, as prepared, is contaminated with 1 kg of salt solution, containing 0.55 kg of salt/kg of pigment. The material is washed countercurrently with water in a number of thickeners arranged in series. How many thickeners will be required if water is added at the rate of 17,400 tonnes per day (200 kg/s) and the solid discharged from each thickener removes 0.5 kg of solvent/kg of pigment?

What will be the required number of thickeners if the amount of solution removed in association with the pigment varies with the concentration of the solution in the thickener, as follows?

| Concentration of solution (kg solute/kg solution) | Amount of solution removed (kg solution/kg pigment) |
|---|---|
| 0 | 0.30 |
| 0.1 | 0.32 |
| 0.2 | 0.34 |
| 0.3 | 0.36 |
| 0.4 | 0.38 |
| 0.5 | 0.40 |

The concentrated wash liquor is mixed with the material fed to the first thickener.

## Solution

*Part* 1

Overall balance (units: kg/s)

| | $TiO_2$ | Salt | Water |
|---|---|---|---|
| Feed from reactor | 100 | 55 | 45 |
| Wash liquor added | — | — | 200 |
| Washed solid | 100 | 0.1 | 50 |
| Liquid product | — | 54.9 | 195 |

Solvent in underflow from final washing thickener = 50 kg/s.
The solvent in the overflow will be the same as that supplied for washing (200 kg/s).

$$\left(\frac{\text{Solvent discharged in overflow}}{\text{Solvent discharged in underflow}}\right) = 4 \text{ for the washing thickeners.}$$

Liquid product from plant contains 54.9 kg of salt in 195 kg of solvent.
This ratio will be the same in the underflow from the first thickener.
Thus the material fed to the washing thickeners consists of 100 kg $TiO_2$, 50 kg solvent and $50 \times (54.9/195) = 14$ kg salt.
The required number of thickeners for washing is given by equation 10.16, as:

$$\frac{(4-1)}{(4^{n+1}-1)} = \frac{0.1}{14}$$

Thus: $$4^{n+1} = 421, \quad \text{and } n+1 = 4.35$$

or: $$\underline{\underline{4 < n+1 < 5}}$$

*Part* 2

From an inspection of the data, it is seen that $W_h = 0.30 + 0.2X_h$.

Thus: $$S_h = W_h X_h = 0.30X_h + 0.2X_h^2 = 5W_h^2 - 1.5W_h$$

Considering the passage of unit quantity of $TiO_2$ through the plant, then:

$$L_{n+1} = 0, \quad w_{n+1} = 2, \quad X_{n+1} = 0$$

since 200 kg/s of pure solvent is used.

$$S_n = 0.001 \text{ and therefore } W_n = 0.3007$$

$$S_0 = 0.55 \text{ and } W_0 = 1.00$$

Thus the concentration in the first thickener is given by equation 10.23 as:

$$X_1 = \frac{L_{n+1} + S_0 - S_n}{w_{n+1} + W_0 - W_n} = \frac{(0 + 0.55 - 0.001)}{(2 + 1 - 0.3007)} = \frac{0.549}{2.6993} = 0.203$$

From equation 10.26:

$$X_{h+1} = \frac{L_{n+1} - S_n + S_h}{w_{n+1} - W_n + W_h} = \frac{(0 - 0.001 + S_h)}{(2 - 0.3007 + W_h)} = \frac{-0.001 + S_h}{1.7 + W_h}$$

Since: $X_1 = 0.203, \quad W_1 = (0.30 + 0.2 \times 0.203) = 0.3406$

and: $S_1 = (0.3406 \times 0.203) = 0.0691$

Thus: $$X_2 = \frac{(0.0691 - 0.001)}{(1.7 + 0.3406)} = \frac{0.0681}{2.0406} = 0.0334$$

Since: $X_2 = 0.0334, \quad W_2 = 0.30 + 0.2 \times 0.0334 = 0.30668$

and: $S_2 = (0.3067 \times 0.0334) = 0.01025$

Thus: $$X_3 = \frac{(0.01025 - 0.001)}{(1.7 + 0.3067)} = \frac{0.00925}{2.067} = 0.00447$$

Since: $X_3 = 0.00447, \quad W_3 = 0.30089 \text{ and } S_3 = 0.0013$

By the same method, $X_4 = 0.000150$

and: $W_4 = 0.30003$ and $S_4 = 0.000045$

Thus $S_4$ is less than $S_n$ and therefore 4 thickeners are required.

## 10.6. NUMBER OF STAGES FOR COUNTERCURRENT WASHING BY GRAPHICAL METHODS

### 10.6.1. Introduction

It is sometimes convenient to use graphical constructions for the solution of countercurrent leaching or washing problems. This may be done by a method similar to the McCabe–Thiele method for distillation which is discussed in Chapter 11, with the overflow and underflow streams corresponding to the vapour and liquid respectively. The basis of this method is now given, although a generally more convenient method involves the use of triangular diagrams which will be discussed in some detail.

For the countercurrent washing system shown in Figure 10.13, the ratio of solute to solvent in the overflow at any stage $y_h''$ may be related to the ratio of solute to insoluble solid in the underflow $S_h$ by means of a simple material balance. Using this notation:

$$y_h'' = \frac{L_h}{w_h - L_h} = \frac{L_h}{Z_h} \tag{10.28}$$

where $Z_h$ is the amount of solvent in the overflow per unit mass of insoluble solid in the underflow. If the solvent in the underflow is constant throughout the system, $Z_h$ will not

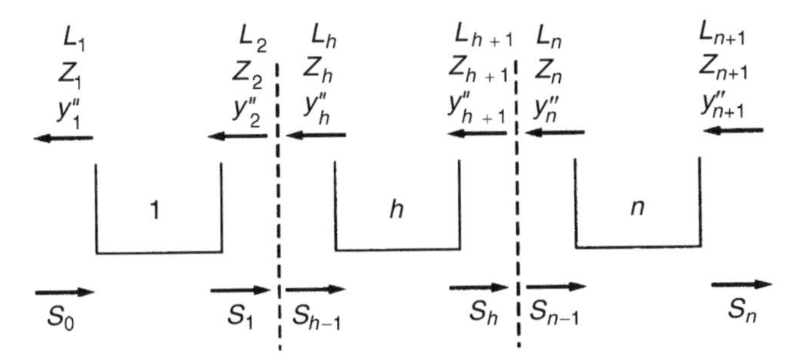


Figure 10.13. Countercurrent washing system

be a function of concentration. Dropping the suffix of $Z$ therefore, and taking a balance on solute over thickeners $h$ to $n$ inclusive:

$$Z(y''_h - y''_{n+1}) = S_{h-1} - S_n$$

or:
$$y''_h = \frac{S_{h-1}}{Z} - \frac{S_n}{Z} + y''_{n+1} \tag{10.29}$$

Thus a linear relation exists between $y''_h$ and $S_{h-1}$.

If the ratio of solvent in the overflow to solvent in the underflow from any thickener is equal to $R$, then:

$$\frac{L_h}{S_h} = \frac{w_h - L_h}{W_h - S_h} = \frac{Zy''_h}{S_h} = R$$

or:
$$y''_h = \frac{R}{Z}S_h \tag{10.30}$$

This is the equation of a straight line of slope $R/Z$ which passes through the origin. Equations 10.29 and 10.30 may be represented on a $y''$–$S$ diagram, as shown in Figure 10.14. If pure solvent is used for washing, $y''_{n+1} = 0$, and the intercept on the $S$-axis is $S_n$. As the two lines represent the relation between $y''_h$ and both $S_{h-1}$ and $S_h$, the change in composition of the underflow and overflow streams can be determined by a series of stepwise constructions, with the number of steps required to change the composition of the overflow from $y''_{n+1}$ to $y''_1$ being the number of stages required.

For a variable underflow the relation between $y''_h$ and $S_h$ must be determined experimentally as the two curves are no longer straight lines, although the procedure is similar once these have been drawn. Further, it is assumed that each thickener represents an ideal stage and that the ratio of solute to solvent is the same in the overflow and the underflow. If each stage is only 80 per cent efficient, for example, equation 10.30 is no longer applicable, but the same method can be used except that each of the vertical steps will extend only 80 per cent of the way to the curve of $y''_h$ versus $S_h$.

Further use of graphical methods is discussed by SCHEIBEL[8], though here attention is confined to the use of right-angled triangular diagrams. Equilateral triangles will also be used in liquid–liquid extraction, as illustrated in Chapter 13.

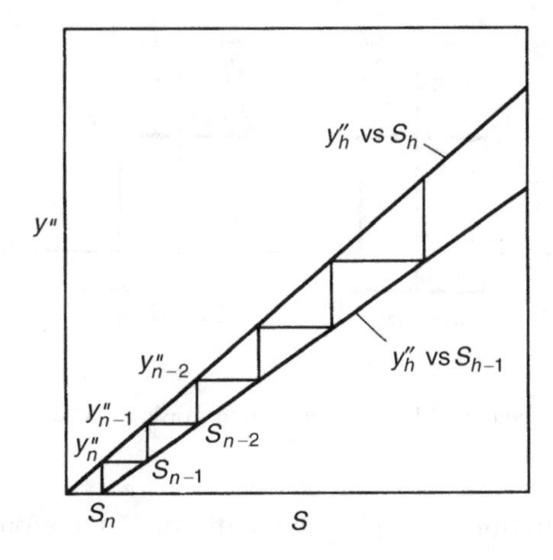


Figure 10.14. Graphical construction for determining the number of thickeners

## 10.6.2. The use of right-angled triangular diagrams

If a total mass $F$ of material is fed to a thickener and separated into a mass $w'$ of overflow and a mass $W'$ of underflow, and the whole of the insoluble solid appears in the underflow, a material balance then gives:

$$F = W' + w' \tag{10.31}$$

If $z$, $y$, and $x$ are the fractional compositions of $F$, $w'$, and $W'$, respectively, with respect to any one component in the mixture. the solute **A**, the insoluble solid **B**, or the solvent **S**, then, taking a material balance on any component:

$$F_z = W'x + w'y$$

or from equation 10.31:

$$(W' + w')z = W'x + w'y \tag{10.32}$$

and:

$$z = \frac{(W'x + w'y)}{(W' + w')} \tag{10.33}$$

Thus for the solute and solvent respectively:

$$z_A = \frac{(W'x_A + w'y_A)}{(W' + w')}$$

$$z_S = \frac{(W'x_S + w'y_S)}{(W' + w')}$$

The advantages of using a right-angled triangular diagram to represent the composition of the three-component mixture triangular diagram are discussed by ELGIN[9]. The proportion of solute **A** in the mixture is plotted as the abscissa, and the proportion of solvent as the ordinate; the proportion of insoluble solid is then obtained by difference. If point a $(z_A, z_S)$ represents the composition of the material fed to the thickener, point b $(x_A, x_S)$ the composition of the underflow and point c $(y_A, y_S)$ the composition of the overflow as shown in Figure 10.15, then the slope of the line ab is given by:

$$\frac{z_S - x_S}{z_A - x_A} = \frac{\dfrac{W'x_S + w'y_S}{W' + w'} - x_S}{\dfrac{W'x_A + w'y_A}{W' + w'} - x_A} = \frac{w'y_S - w'x_S}{w'y_A - w'x_A} = \frac{(y_S - x_S)}{(y_A - x_A)} \tag{10.34}$$

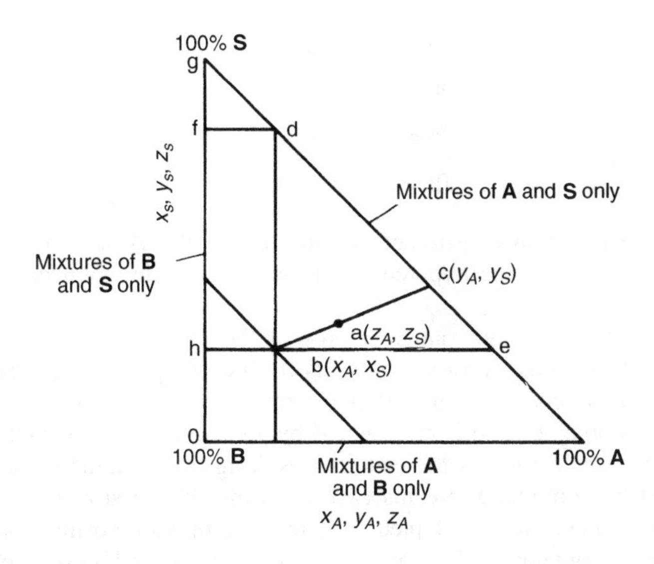


Figure 10.15. Representation of a three-component system on a right-angled triangular diagram

The slope of the line bc is also $(y_S - x_S)/(y_A - x_A)$, however, so that, a, b, and c lie on the same straight line. Thus, if two streams are mixed, the composition of the mixture will be given by some point on the line joining the points representing the compositions of the constituent streams. Similarly, if one stream is subtracted from another, the composition of the resulting stream will lie at some point on the corresponding straight line produced. The location of the point will depend on the relative quantities in the two streams.

From equation 10.32:

$$(W' + w')z = W'x + w'y$$

or:

$$w'(z - y) = W'(x - z)$$

and:

$$\frac{w'}{W'} = \frac{x - z}{z - y} = \frac{x_A - z_A}{z_A - y_A} = \frac{x_S - z_S}{z_S - y_S} \tag{10.35}$$

Thus the point representing the mixture divides the line bc so that ba/ac $= w'/W'$; that is a is nearer to the point corresponding to the larger stream.

The proportion of insoluble solid in the underflow, for example, is given by the relation:

$$x_A + x_S + x_B = 1$$

or:

$$x_S = -x_A + (1 - x_B) \tag{10.36}$$

Lines representing constant values of $x_B$ are therefore straight lines of slope $-1$; that is they are parallel to the hypotenuse of the right-angled triangle: the intercept on either axis is $(1 - x_B)$. Thus the hypotenuse represents mixtures containing no insoluble solid, and therefore the compositions of all possible overflows are represented by the hypotenuse. Further, it can be seen from the geometry of the diagram that:

$$\text{fd} = \text{fg} = x_A$$

and:

$$\text{Oh} = x_S$$

thus:

$$\text{fh} = \text{bd} = \text{be}$$

and:

$$\text{fh} = 1 - x_A - x_S = x_B$$

The proportion of the third constituent, the insoluble solid **B**, is therefore given by the distance of the point from the hypotenuse, measured in a direction parallel to either of the main axes.

Points which lie within the triangle represent the compositions of real mixtures of the three components. Each vertex represents a pure component and each of the sides represents a two-component mixture. If two streams are mixed, the composition of the resultant stream is obtained and is represented by an addition point which must lie within the diagram. The composition of the material resulting from mixing a number of streams can be obtained by combining streams two at a time. If one stream is subtracted from another, a similar procedure is adopted to determine the composition of the remaining stream. The point so obtained is known as a *difference point* and must lie on the extension of the line joining the two given points, on the side nearer the one representing the mixed stream.

If an attempt is made to remove from a stream more of a given component than is actually present, the composition of the resulting stream will be imaginary and will be represented by some point outside the triangle. The concept of an imaginary difference point is useful in its application to countercurrent flow processes.

### *Effect of saturation*

When the solute is initially present as a solid, the amount that can be dissolved in a given amount of solvent is limited by the solubility of the material. A saturated solution of the solute will be represented by some point, such as A, on the hypotenuse of the triangular diagram (Figure 10.16). The line OA represents the compositions of all possible mixtures of saturated solution with insoluble solid, since $x_S/x_A$ is constant at all points on this line. The part of the triangle above OA therefore represents unsaturated solutions mixed with the insoluble solid **B**. If a mixture, represented by some point N, is separated into solid and liquid, it will yield a solid, of composition represented by O, that is pure component **B**, and an unsaturated solution (N′). Again the lower part of the diagram represents mixtures

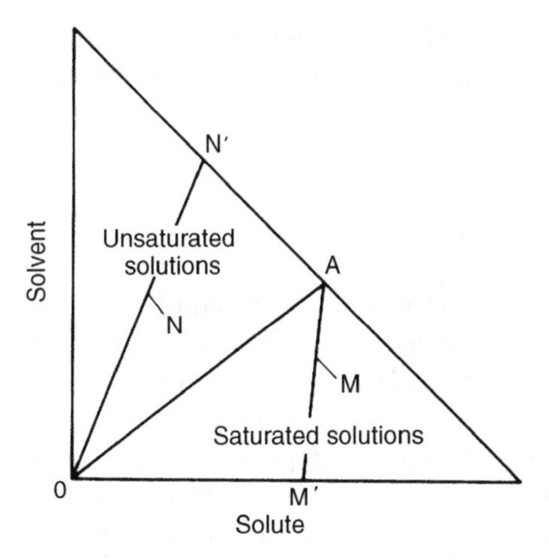


Figure 10.16. Effect of saturation — solid state

of insoluble solid, undissolved solute, and saturated solution. Thus, if a mixture M is separated into a liquid and a solid fraction, it will yield a liquid, indicated by point A (that is, a saturated solution), and a solid, consisting of **B** together with undissolved **A** (M′).

If the solute is initially in a liquid form and the solvent is completely miscible with it, the whole of the triangle will represent unsaturated conditions. If the solvent and solute are not completely miscible, the area can be divided into three distinct regions, as shown in Figure 10.17. In region 1, the solvent is present as an unsaturated solution in the solute. In region 2, the liquid consists of two phases — a saturated solution of **A** in **S** and a saturated solution of **S** in **A**, in various proportions. In region 3, the liquid consists of an unsaturated solution of solute in solvent.

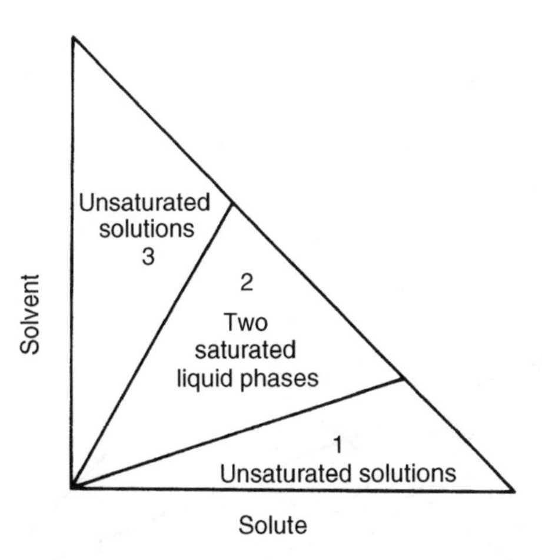


Figure 10.17. Effect of saturation — liquid solute

In any leaching process, in which solvent is used to wash the adhering film of solution from the surface of the insoluble solid, the solution formed as a result of mixing in any of the units will consist of unsaturated solution. In an extracting plant, sufficient solvent will normally be added in order to dissolve the solute completely, so that, in practice, the solutions considered will rarely reach saturation.

### Representation of underflow

Considering the multistage industrial unit, in any equilibrium stage, the quantity of solution in the underflow may be a function of the concentration of the solution in the thickener, and the concentration of the overflow solution will be the same as that in the underflow. If the curved line EF (Figure 10.18) represents the experimentally determined composition of the underflow for various concentrations, any point f on this line represents the composition of a mixture of pure **B** with a solution of composition g, and Of/fg is the ratio of solution to solids in the underflow. If the amount of solution removed in the underflow is not affected by its concentration, the fractional composition of the underflow with respect to the insoluble material **B** ($x_B$) is a constant, and is represented by a straight line, through E, parallel to the hypotenuse, such as EF′. Point E represents the composition of the underflow when the solution is infinitely weak, that is when it contains pure solvent. If K is the mass of solution removed in the underflow per unit mass of solids, the ordinate of E is given by:

$$x_S = \frac{K}{K + 1}$$

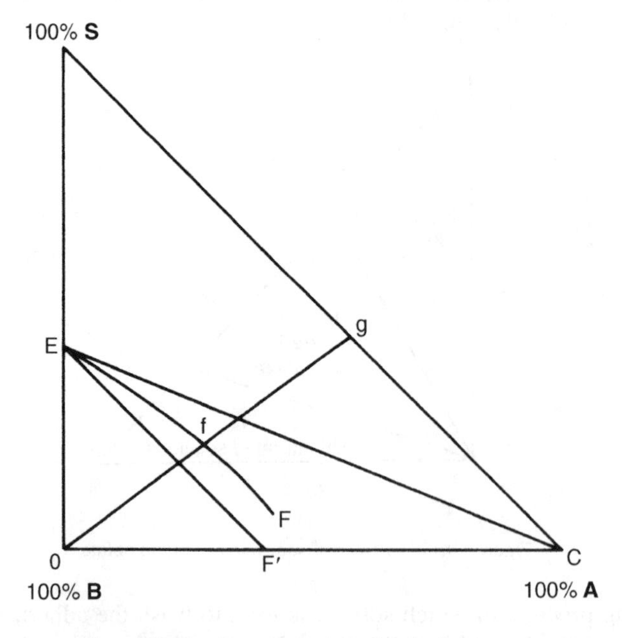


Figure 10.18. Representation of underflow stream

The equation of the line EF′ is therefore:

$$x_S = -x_A + \frac{K}{K+1}$$

If the ratio of solvent to insoluble solid in the underflow is constant (and equal to $s$), the line EF will be a straight line passing through the vertex C of the triangle, which corresponds to pure solute **A**. E is then given by the coordinates:

$$x_A = 0, \quad x_S = \frac{S}{S+1}$$

### 10.6.3. Countercurrent systems

Considering a countercurrent system consisting of $n$ thickeners, as shown in Figure 10.19, the net flow to the right must necessarily be constant throughout the system, if no material enters or leaves at intermediate points.

Thus: $$\text{Net flow to the right} = F' = W'_{h-1} - w'_h, \text{ etc.} \tag{10.37}$$

The point representing the stream $F'$ will be a difference point, since it will represent the composition of the stream which must be added to $w'_h$ to give $W'_{h-1}$. In general, in a countercurrent flow system of this sort, the net flow of all the constituents will not be in the same direction. Thus one or more of the fractional compositions in this difference stream will be negative, and the difference point will lie outside the triangle. A balance on the whole system, as shown in Figure 10.19, gives:

$$W'_0 - w'_1 = W'_n - w'_{n+1}$$

and:

$$W'_0 x_0 - w'_1 y_1 = W'_n x_n - w'_{n+1} y_{n+1}$$

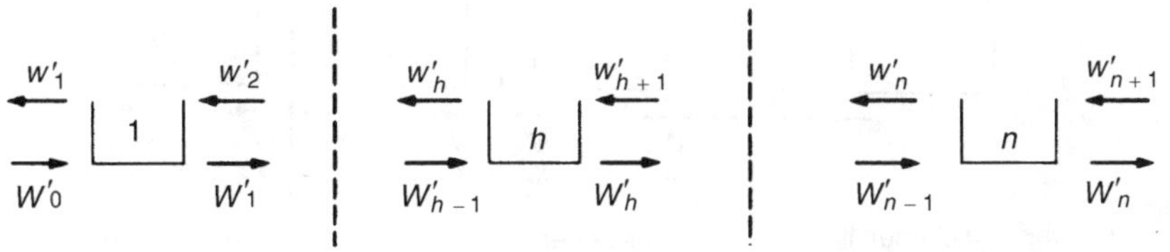

Figure 10.19. Countercurrent extraction system

for any of three components.

The total net flow of material to the right at some intermediate point

$$= W'_{h-1} - w'_h$$

and the net flow of one of the constituents

$$= W'_{h-1} x_{h-1} - w'_h y_h$$

The fractional composition of the stream flowing to the right with respect to one of the components is given by:

$$x_d = \frac{W'_{h-1}x_{h-1} - w'_h y_h}{W'_{h-1} - w'_h} \tag{10.38}$$

If the direction of flow of this component is towards the right $x_d$ is positive, and if its direction of flow is to the left $x_d$ is negative. For a countercurrent washing system, as shown in the diagram, the net flow of solvent at any stage will be to the left so that $x_{dS}$ is negative. The solute and insoluble residue will flow to the right, making $x_{dA}$ and $x_{dB}$ positive.

Considering a system as shown in Figure 10.20, in which a dry solid is extracted with a pure solvent, the compositions of the solid and solvent, and their flowrates, are specified. Assuming it is desired to wash the residual solid so that it has not less than a certain degree of purity, the number of thickeners required to achieve this must be calculated. The compositions of the solid to be extracted, the washed solid and the solvent ($x_0$, $x_n$ and $y_{n+1}$ respectively) are therefore given. The composition of the concentrated solution leaving the system $y_1$ can then be calculated from a material balance over the whole plant.

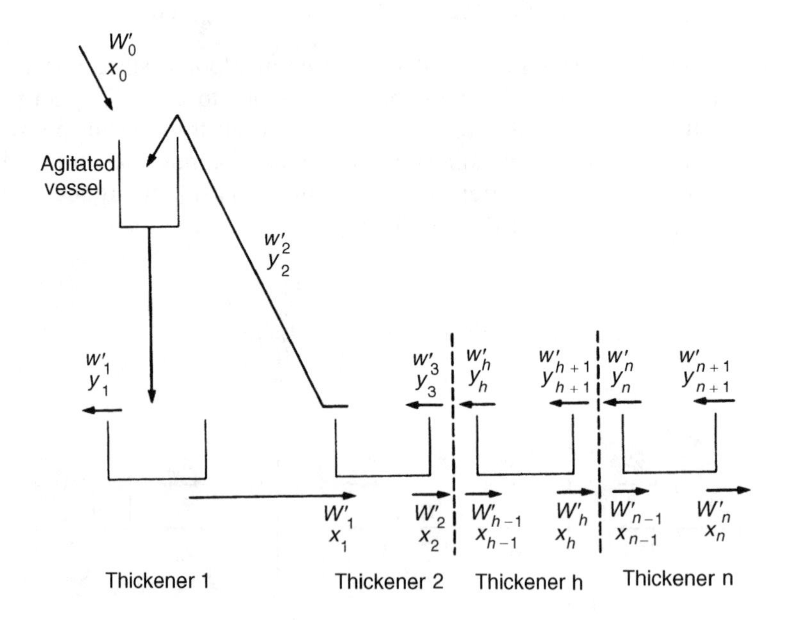


Figure 10.20. Countercurrent washing system and agitator

The difference point, which represents the composition of the net stream of material flowing to the right at all stages, must lie on the straight line through points $x_1$ and $y_1$ produced, as shown in Figure 10.21. Point $x_1$ has co-ordinates ($x_{A0}$, $x_{S0}$); $x_{A0}$ is the composition of the dry solid fed to the system; and $x_{S0}$ is zero because this solid has no solvent associated with it. Point $y_1$ represents the composition of the concentrated solution discharged from the plant, and will lie on the hypotenuse of the triangle. The difference point will lie on the line through points $x_n$ and $y_{n+1}$. Now $x_n$, the composition of the

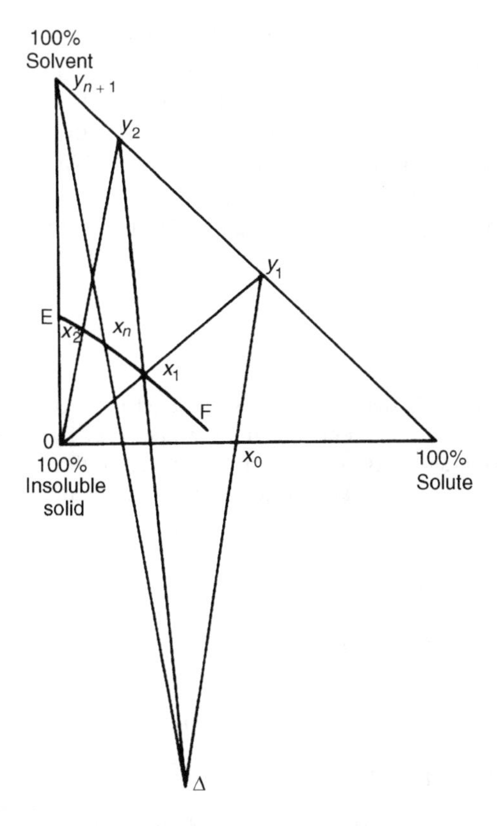


Figure 10.21. Graphical method of solution with a triangular diagram

final product in the underflow, will lie on the line EF which represents the compositions of all possible underflows; this line may be constructed from the experimental data on the amount of solution discharged in the underflow for various concentrations in the thickeners. Because pure solvent is used for washing, $y_{n+1}$ will lie at the top vertex of the triangle. The difference point is then obtained as the point of intersection of these two straight lines. It is denoted by $\Delta$.

As the difference point represents the difference between the underflow and the overflow at any point in the system, it is now possible to calculate the compositions of all the streams, by considering each thickener in turn.

For the agitator and thickener 1, the underflow, of composition $x_1$, will contain insoluble solid mixed with solution of the same concentration as that in the overflow $y_1$, on the assumption that equilibrium conditions are reached in the thickener. All such mixtures of solution and insoluble solid are represented by compositions on the line $0y_1$. As this stream is an underflow, its composition must also be given by a point on the line EF. Thus $x_1$ is given by the point of intersection of EF and $0y_1$. The composition $y_2$ of the overflow stream from thickener 2 must lie on the hypotenuse of the triangle and also on the line through points $\Delta$ and $x_1$. The composition $y_2$ is therefore determined. In this manner it is possible to find the compositions of all the streams in the system. The procedure is repeated until the amount of solute in the underflow has been reduced to a value not greater than $x_n$, and the number of thickeners is then readily counted. For the simple example illustrated, only two thickeners would be required.

This method of calculation may be applied to any system, provided that streams of material do not enter or leave at some intermediate point. If the washing system alone were considered, such as that shown in Figure 10.12, the insoluble solid would be introduced, not as fresh solid free of solvent, but as the underflow from the thickener in which the mixture from the agitator is separated. Thus $x_0$ would lie on the line EF instead of on the A-axis of the diagram.

## Example 10.4

Seeds, containing 20 per cent by mass of oil, are extracted in a countercurrent plant, and 90 per cent of the oil is recovered in a solution containing 50 per cent by mass of oil. If the seeds are extracted with fresh solvent and 1 kg of solution is removed in the underflow in association with every 2 kg of insoluble matter, how many ideal stages are required?

## Solution

This example will be solved using the graphical method.

Since the seeds contain 20 per cent of oil, then:

$$x_{A0} = 0.2 \quad \text{and} \quad x_{B0} = 0.8$$

The final solution contains 50 per cent of oil.

Thus:
$$y_{A1} = 0.5 \quad \text{and} \quad y_{S1} = 0.5$$

The solvent which is used for extraction is pure and hence;

$$y_{Sn+1} = 1$$

1 kg of insoluble solid in the washed product is associated with 0.5 kg of solution and 0.025 kg oil.

Thus:
$$x_{An} = 0.0167, \quad x_{Bn} = 0.6667 \quad \text{and} \quad x_{Sn} = 0.3166$$

The mass fraction of insoluble material in the underflow is constant and equal to 0.667. The composition of the underflow is therefore represented, on the diagram Figure 10.22, by a straight line parallel to the hypotenuse of the triangle with an intercept of 0.333 on the two main axes.

The difference point is now found by drawing in the two lines connecting $x_0$ and $y_1$ and $x_n$ and $y_{n+1}$.

The graphical construction described in the text is then used and it is seen from Figure 10.22 that $x_n$ lies in between $x_4$ and $x_5$.

Thus 5 thickeners are adequate and for the required degree of extraction.

## Example 10.5

Halibut oil is extracted from granulated halibut livers in a countercurrent multi-batch arrangement using ether as the solvent. The solids charge contains 0.35 kg oil/kg exhausted livers and it is

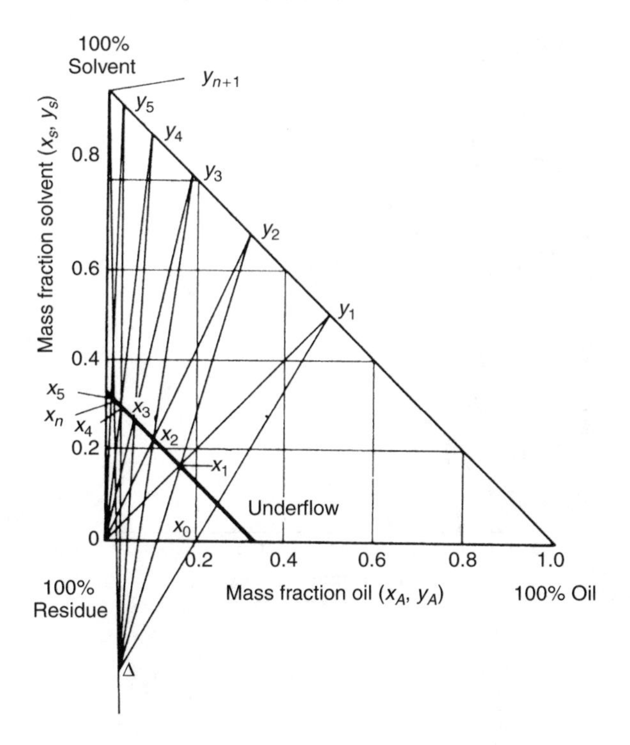


Figure 10.22. Graphical solution to Example 10.3

desired to obtain a 90 per cent oil recovery. How many theoretical stages are required if 50 kg ether is used/100 kg untreated solids? The entrainment data are:

| | | | | | | | | |
|---|---|---|---|---|---|---|---|---|
| Concentration of overflow (kg oil/kg solution) | 0 | 0.1 | 0.2 | 0.3 | 0.4 | 0.5 | 0.6 | 0.67 |
| Entrainment (kg solution/kg extracted livers) | 0.28 | 0.34 | 0.40 | 0.47 | 0.55 | 0.66 | 0.80 | 0.96 |

## Solution

The entrainment data may be expressed in terms of mass fractions as follows:

| Overflow concentration (kg oil/kg soln.) | Entrainment (kg soln./ kg livers) | Ratio (kg/kg extracted livers) | | | Mass fraction | |
|---|---|---|---|---|---|---|
| | | Oil | Ether | Underflow | $x_A$ | $x_s$ |
| 0 | 0.28 | 0 | 0.280 | 1.280 | 0 | 0.219 |
| 0.1 | 0.34 | 0.034 | 0.306 | 1.340 | 0.025 | 0.228 |
| 0.2 | 0.40 | 0.080 | 0.320 | 1.400 | 0.057 | 0.228 |
| 0.3 | 0.47 | 0.141 | 0.329 | 1.470 | 0.096 | 0.223 |
| 0.4 | 0.55 | 0.220 | 0.330 | 1.550 | 0.142 | 0.212 |
| 0.5 | 0.66 | 0.330 | 0.330 | 1.660 | 0.199 | 0.198 |
| 0.6 | 0.80 | 0.480 | 0.320 | 1.880 | 0.255 | 0.170 |
| 0.67 | 0.96 | 0.643 | 0.317 | 1.960 | 0.328 | 0.162 |

and these are plotted in Figure 10.23.

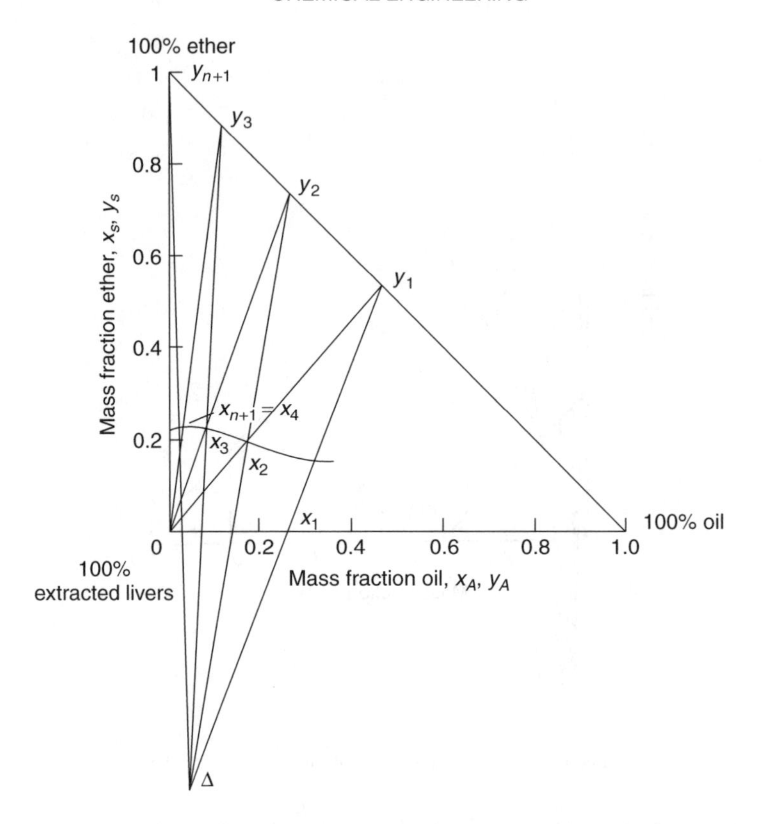


Figure 10.23. Graphical construction for Example 10.5

On the *basis of 100 kg untreated solids:*

In the *underflow feed:*

0.35 kg oil is associated with each kg of exhausted livers.

Thus: mass of livers fed $= 100/(1 + 0.35) = 74$ kg containing $(100 - 74) = 26$ kg oil

and hence: $x_A = 0.26, \quad x_s = 0$

This point is marked in as $x_1$.

In the *overflow feed*, pure ether is used and $y_s = 1.0$, $x_s = 0$, which is marked in as the point $y_{n+1}$.

Since the recovery of oil is 90 per cent, the overall mass balance becomes:

| | Exhausted livers | Oil | Ether |
|---|---|---|---|
| Underflow feed | 74 | 26 | — |
| Overflow feed | — | — | 50 |
| Underflow product | 74 | 2.6 | $e$ (say) |
| Overflow product | — | 23.4 | $(50 - e)$ |

In the *underflow product*:

the ratio (oil/exhausted livers) $= (2.6/74) = 0.035$ kg/kg

which, from the entrainment data, is equivalent to $x_A = 0.025$, $x_s = 0.228$ which is marked in as $x_{n+1}$.

The ratio (ether/exhausted livers) = 0.306 kg/kg or $e = (0.306 \times 74) = 22.6$ kg

In the *overflow product:*

$$\text{the mass of ether} = (50 - 22.6) = 27.4 \text{ kg}$$

and: $y_A = 23.4/(23.4 + 27.4) = 0.46$, from which $y_s = 0.54$ which is marked in as $y_1$.

Following the construction described in the text, it is found that point $x_4$ coincides exactly with $x_{n+1}$, as shown in Figure 10.23, and hence 3 ideal stages are required.

## 10.6.4. Non-ideal stages

If each stage in the extraction or washing system is not perfectly efficient, the ratio of solute to solvent in the overflow will be less than that in the underflow, and rather more units will be required than the number calculated by the method given. If the efficiency is independent of the concentration, allowance is simply made by dividing the theoretical number of ideal units by the efficiency. On the other hand, if the efficiency varies appreciably, account must be taken of this at each stage in the graphical construction. In Figure 10.24, $\bar{x}_{h-1}$ represents the composition of the underflow fed to thickener $h$, and this composition would be changed to $\bar{x}_h$, say, in an ideal stage. The proportion of solute in the underflow is then reduced by an amount represented by AB. If the efficiency is less than unity, the change in the proportion of solute will be represented by BC, where BC/AB is equal to the efficiency at the concentration considered, and the actual composition of the underflow is given by $x_h$. By this method it is possible to make allowance for a variable efficiency at each stage.

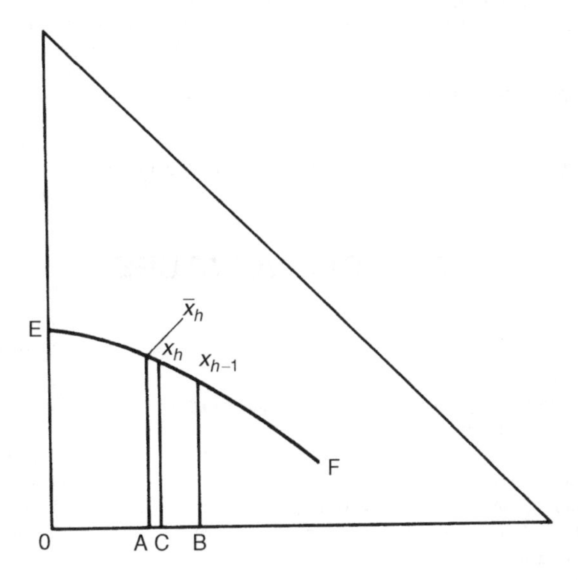


Figure 10.24. Effect of stage efficiency

## 10.7. FURTHER READING

BACKHURST, J. R., HARKER, J. H., and PORTER, J. E.: *Problems in Heat and Mass Transfer* (Edward Arnold, London, 1974).
BENNETT, C. O. and MYERS, J. E.: *Momentum, Heat and Mass Transfer*. 3rd edn (McGraw-Hill, New York, 1982).
CHEN, NING HSING: *Chem. Eng., Albany* **71**, No. 24 (23 Nov. 1964) 125–8. Calculating theoretical stages in counter-current leaching.
HENLEY, E. J. and STAFFIN, H. K.: *Stagewise Process Design* (John Wiley, New York, 1963).
HINES, A. L. and MADDOX, R. N.: *Mass Transfer Fundamentals and Applications* (Prentice-Hall, Englewood Cliffs, 1985).
KARNOFSKY, G.: *Chem. Eng., Albany* **57** (Aug. 1950) 109. The Rotocel extractor.
KING, C. J.: *Separation Processes*, 2nd edn. (McGraw-Hill, New York, 1980).
MCCABE, W. L., SMITH, J. C. and HARRIOTT, P.: *Unit Operations of Chemical Engineering*, 4th edn. (McGraw-Hill, New York, 1984).
MOLYNEUX, F.: *Ind. Chemist* **37**, No. 440 (Oct. 1961) 485–92. Prediction of "A" factor and efficiency in leaching calculations.
PAYNE, K. R.: *Ind. Chemist* **39**, No. 10 (Oct. 1963) 532–5. Isolation of alkaloids by batch solvent extraction.
PERRY, R. H., GREEN, D. W. and MALONEY, J. O. (eds.): *Perry's Chemical Engineers' Handbook*, 7th edn (McGraw-Hill Book Company, New York, 1997).
SAWISTOWSKI, H. and SMITH, W.: *Mass Transfer Process Calculations* (Interscience, London, 1963).
TREYBAL, R. E.: *Mass Transfer Operations*, 3rd edn. (McGraw-Hill, New York, 1980).

## 10.8. REFERENCES

1. LINTON, W. H. and SHERWOOD, T. K.: *Chem. Eng. Prog.* **46** (1950) 258. Mass transfer from solid shapes to water in streamline and turbulent flow.
2. HIXSON, A. W. and BAUM, S. J.: *Ind. Eng. Chem.* **33** (1941) 478, 1433. Agitation: mass transfer coefficients in liquid–solid agitated systems. Agitation: heat and mass transfer coefficients in liquid–solid systems.
3. PIRET, E. L., EBEL, R. A., KIANG, C. T. and ARMSTRONG, W. P.: *Chem. Eng. Prog.* **47** (1951) 405 and 628. Diffusion rates in extraction of porous solids–1. Single phase extractions; 2. Two-phase extractions.
4. CHORNY, R. C. and KRASUK, J. H.: *Ind. Eng. Chem. Process Design and Development* **5**, No. 2 (Apr. 1966) 206–8. Extraction for different geometries. Constant diffusivity.
5. GOSS, W. H.: *J. Am. Oil Chem. Soc.* **23** (1946) 348. Solvent extraction of oilseeds.
6. BOHNET, M. and NIESMARK, G.: *German Chem. Eng.* **3** (1980) 57. Distribution of solids in stirred suspensions.
7. RUTH, B. F.: *Chem. Eng. Prog.* **44** (1948) 71. Semigraphical methods of solving leaching and extraction problems.
8. SCHEIBEL, E. G.: *Chem. Eng. Prog.* **49** (1953) 354. Calculation of leaching operations.
9. ELGIN, J. C.: *Trans. Am. Inst. Chem. Eng.* **32** (1936) 451. Graphical calculation of leaching operations.

## 10.9. NOMENCLATURE

| | | Units in SI System | Dimensions in **M, L, T, θ** |
|---|---|---|---|
| $A$ | Area of solid–liquid interface | $m^2$ | $\mathbf{L}^2$ |
| $a$ | Mass of solute per unit mass of solvent in final overflow | kg/kg | — |
| $b$ | Thickness of liquid film | m | $\mathbf{L}$ |
| $C_p$ | Specific heat of solution | J/kg K | $\mathbf{L}^2\mathbf{T}^{-2}\boldsymbol{\theta}^{-1}$ |
| $c$ | Concentration of solute in solvent | $kg/m^3$ | $\mathbf{ML}^{-3}$ |
| $c_o$ | Initial concentration of solute in solvent | $kg/m^3$ | $\mathbf{ML}^{-3}$ |
| $c_g$ | Concentration of solute in solvent in contact with solid | $kg/m^3$ | $\mathbf{ML}^{-3}$ |
| $D_L$ | Liquid phase diffusivity | $m^2/s$ | $\mathbf{L}^2\mathbf{T}^{-1}$ |
| $d$ | Diameter of vessel | m | $\mathbf{L}$ |
| $F$ | Total mass of material fed to thickener | kg | $\mathbf{M}$ |

| | | Units in SI System | Dimensions in **M, L, T, $\boldsymbol{\theta}$** |
|---|---|---|---|
| $F'$ | Difference between underflow and overflow | kg | **M** |
| $f$ | Fraction of solute remaining with solids after washing | — | — |
| $h$ | Heat transfer coefficient | W/m$^2$ K | $\mathbf{MT^{-3}\theta^{-1}}$ |
| $K_L$ | Mass transfer coefficient | m/s | $\mathbf{LT^{-1}}$ |
| $K$ | Solution per unit mass of insoluble solid in underflow | kg/kg | — |
| $k$ | Thermal conductivity | W/m K | $\mathbf{MLT^{-3}\theta^{-1}}$ |
| $k'$ | A diffusion constant | m$^2$/s | $\mathbf{L^2T^{-1}}$ |
| $L$ | Mass of solute in overflow per unit mass of insoluble | kg/kg | — |
| $M$ | Mass of solute transferred in time $t$ | kg | **M** |
| $m$ | Number of batch washing thickeners | — | — |
| $N$ | Number of revolutions of stirrer in unit time | Hz | $\mathbf{T^{-1}}$ |
| $n$ | Number of countercurrent washing thickeners | — | — |
| $P$ | Mass of solute in overflow | kg | **M** |
| $q$ | Insoluble in underflow per unit mass of solute in overflow | kg/kg | — |
| $R$ | Solvent ratio overflow: underflow | kg/kg | — |
| $R'$ | Ratio of solvent decanted to solvent retained | kg/kg | — |
| $S$ | Solute in underflow per unit mass insoluble | kg/kg | — |
| $s$ | Solvent in underflow per unit mass insoluble | kg/kg | — |
| $t$ | Time | s | **T** |
| $V$ | Volume of solvent used for extraction | m$^3$ | $\mathbf{L^3}$ |
| $W$ | Mass of solution in underflow per unit mass of insoluble | kg/kg | — |
| $W'$ | Total mass of underflow | kg | **M** |
| $w$ | Mass of solution in overflow per unit mass of insoluble | kg/kg | — |
| $w'$ | Total mass of overflow | kg | **M** |
| $X$ | Mass of solute per unit mass of solution | kg/kg | — |
| $x$ | Fractional composition of underflow | — | — |
| $\bar{x}$ | Value of $x$ in ideal stage | — | — |
| $x_d$ | Fractional composition of difference stream | — | — |
| $x'$ | Mass of solute per unit mass of solvent | kg/kg | — |
| $y$ | Fractional composition of overflow | — | — |
| $y''$ | Ratio of solute to solvent in overflow | kg/kg | — |
| $Z$ | Solvent in overflow per unit mass insoluble | kg/kg | — |
| $z$ | Fractional composition of feed | — | — |
| $\mu$ | Viscosity of solution or liquid | Ns/m$^2$ | $\mathbf{ML^{-1}T^{-1}}$ |
| $\rho$ | Density of solution or liquid | kg/m$^3$ | $\mathbf{ML^{-3}}$ |

Suffixes

$A$, $B$, $S$ refer to solute, insoluble solid, solvent respectively

$1, \ldots, h, \ldots, n$ refer to liquid overflow or underflow from units $1, \ldots, h, \ldots, n$

0 refers to the liquid underflow feed to unit 1.

CHAPTER 11

# Distillation

## 11.1. INTRODUCTION

The separation of liquid mixtures into their various components is one of the major operations in the process industries, and distillation, the most widely used method of achieving this end, is the key operation in any oil refinery. In processing, the demand for purer products, coupled with the need for greater efficiency, has promoted continued research into the techniques of distillation. In engineering terms, distillation columns have to be designed with a larger range in capacity than any other types of processing equipment, with single columns 0.3–10 m in diameter and 3–75 m in height. Designers are required to achieve the desired product quality at minimum cost and also to provide constant purity of product even though there may be variations in feed composition. A distillation unit should be considered together with its associated control system, and it is often operated in association with several other separate units.

The vertical cylindrical column provides, in a compact form and with the minimum of ground requirements, a large number of separate stages of vaporisation and condensation. In this chapter the basic problems of design are considered and it may be seen that not only the physical and chemical properties, but also the fluid dynamics inside the unit, determine the number of stages required and the overall layout of the unit.

The separation of benzene from a mixture with toluene, for example, requires only a simple single unit as shown in Figure 11.1, and virtually pure products may be obtained. A more complex arrangement is shown in Figure 11.2 where the columns for the purification of crude styrene formed by the dehydrogenation of ethyl benzene are shown. It may be seen that, in this case, several columns are required and that it is necessary to recycle some of the streams to the reactor.

In this chapter consideration is given to the theory of the process, methods of distillation and calculation of the number of stages required for both binary and multicomponent systems, and discussion on design methods is included for plate and packed columns incorporating a variety of column internals.

## 11.2. VAPOUR–LIQUID EQUILIBRIUM

The composition of the vapour in equilibrium with a liquid of given composition is determined experimentally using an equilibrium still. The results are conveniently shown on a temperature–composition diagram as shown in Figure 11.3. In the normal case shown in Figure 11.3*a*, the curve ABC shows the composition of the liquid which boils at any

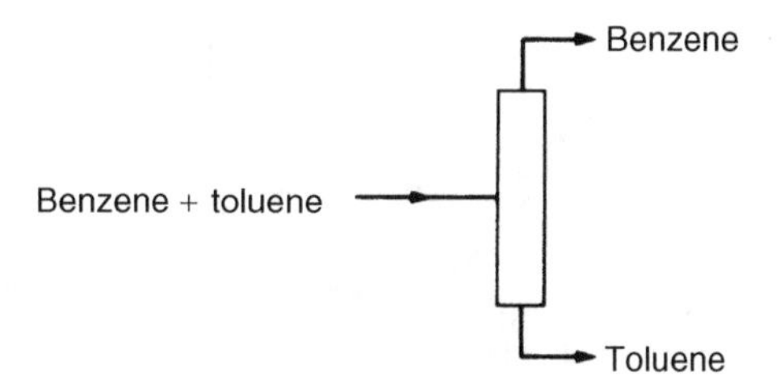


Figure 11.1. Separation of a binary mixture

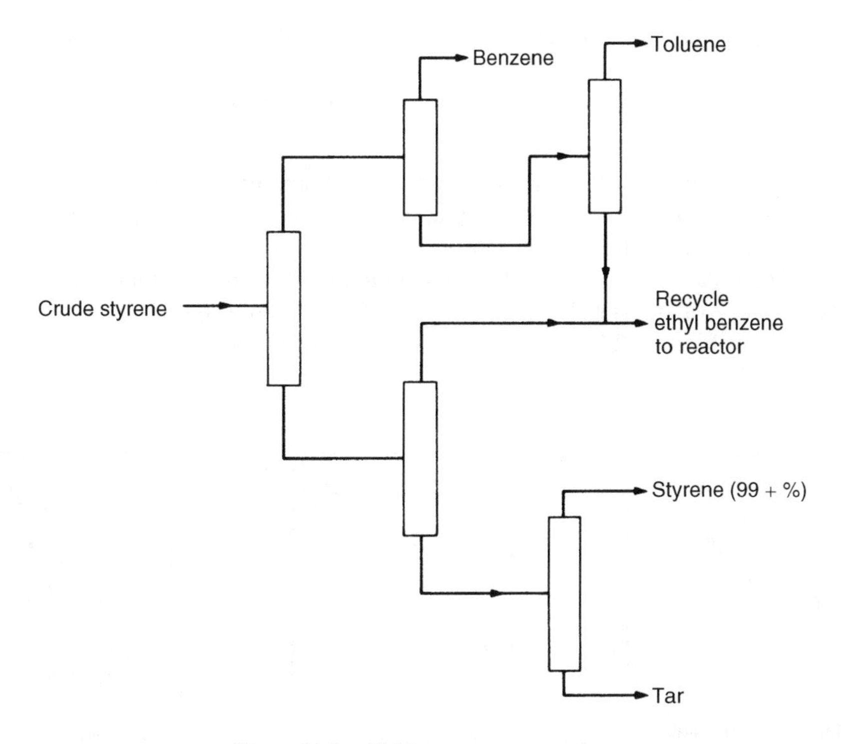


Figure 11.2. Multicomponent separation

given temperature, and the curve ADC the corresponding composition of the vapour at that temperature. Thus, a liquid of composition $x_1$ will boil at temperature $T_1$, and the vapour in equilibrium is indicated by point D of composition $y_1$. It is seen that for any liquid composition $x$ the vapour formed will be richer in the more volatile component, where $x$ is the mole fraction of the more volatile component in the liquid, and $y$ in the vapour. Examples of mixtures giving this type of curve are benzene–toluene, $n$-heptane–toluene, and carbon disulphide–carbon tetrachloride.

In Figures 11.3*b* and *c*, there is a critical composition $x_g$ where the vapour has the same composition as the liquid, so that no change occurs on boiling. Such critical mixtures are called azeotropes. Special methods which are necessary to effect separation of these are discussed in Section 11.8. For compositions other than $x_g$, the vapour formed has a

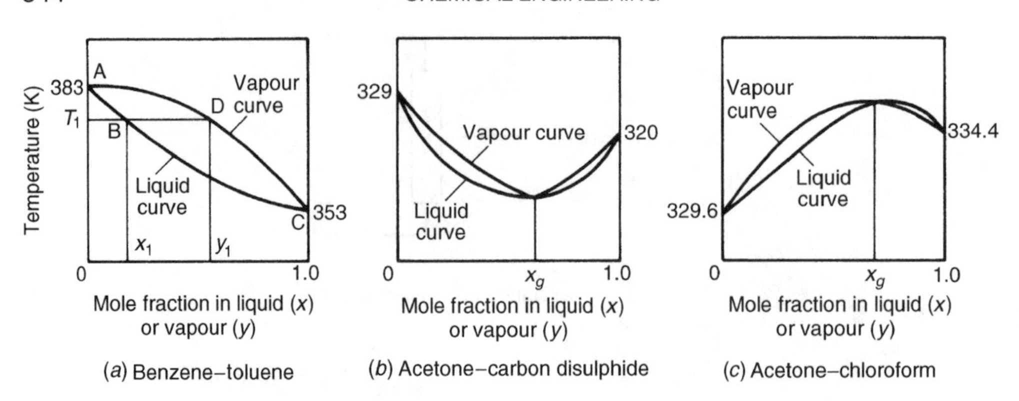


Figure 11.3. Temperature composition diagrams

different composition from that of the liquid. It is important to note that these diagrams are for constant pressure conditions, and that the composition of the vapour in equilibrium with a given liquid will change with pressure.

For distillation purposes it is more convenient to plot $y$ against $x$ at a constant pressure, since the majority of industrial distillations take place at substantially constant pressure. This is shown in Figure 11.4 where it should be noted that the temperature varies along each of the curves.

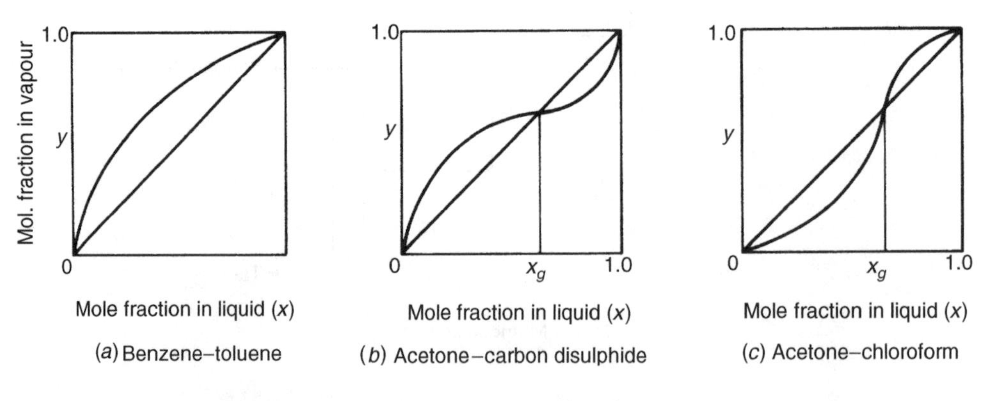


Figure 11.4. Vapour composition as a function of liquid composition at constant pressure

### 11.2.1. Partial vaporisation and partial condensation

If a mixture of benzene and toluene is heated in a vessel, closed in such a way that the pressure remains atmospheric and no material can escape and the mole fraction of the more volatile component in the liquid, that is benzene, is plotted as abscissa, and the temperature at which the mixture boils as ordinate, then the boiling curve is obtained as shown by ABCJ in Figure 11.5. The corresponding dew point curve ADEJ shows the temperature at which a vapour of composition $y$ starts to condense.

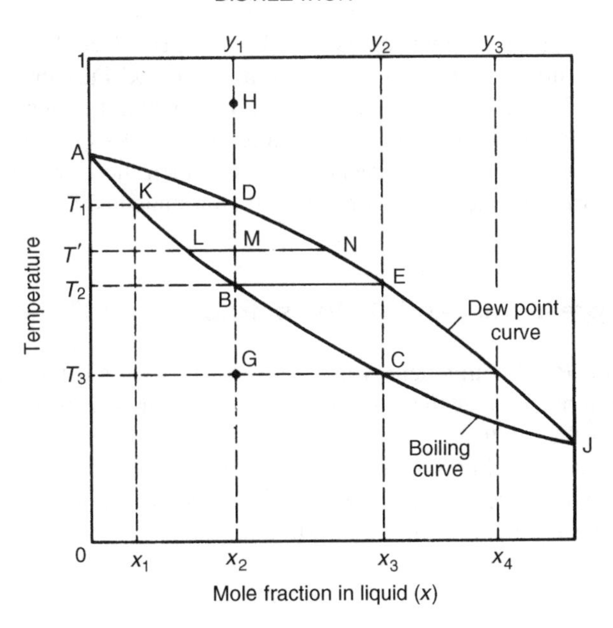


Figure 11.5. Effect of partial vaporisation and condensation at the boiling point

If a mixture of composition $x_2$ is at a temperature $T_3$ below its boiling point, $T_2$, as shown by point G on the diagram, then on heating at constant pressure the following changes will occur:

(a) When the temperature reaches $T_2$, the liquid will boil, as shown by point B, and some vapour of composition $y_2$, shown by point E, is formed.

(b) On further heating the composition of the liquid will change because of the loss of the more volatile component to the vapour and the boiling point will therefore rise to some temperature $T'$. At this temperature the liquid will have a composition represented by point L, and the vapour a composition represented by point N. Since no material is lost from the system, there will be a change in the proportion of liquid to vapour, where the ratio is:

$$\frac{\text{Liquid}}{\text{Vapour}} = \frac{\text{MN}}{\text{ML}}$$

(c) On further heating to a temperature $T_1$, all of the liquid is vaporised to give vapour D of the same composition $y_1$ as the original liquid.

It may be seen that partial vaporisation of the liquid gives a vapour richer in the more volatile component than the liquid. If the vapour initially formed, as for instance at point E, is at once removed by condensation, then a liquid of composition $x_3$ is obtained, represented by point C. The step BEC may be regarded as representing an ideal stage, since the liquid passes from composition $x_2$ to a liquid of composition $x_3$, which represents a greater enrichment in the more volatile component than can be obtained by any other single stage of vaporisation.

Starting with superheated vapour represented by point H, on cooling to D condensation commences, and the first drop of liquid has a composition K. Further cooling to $T'$ gives liquid L and vapour N. Thus, partial condensation brings about enrichment of the vapour in the more volatile component in the same manner as partial vaporisation. The industrial distillation column is, in essence, a series of units in which these two processes of partial vaporisation and partial condensation are effected simultaneously.

### 11.2.2. Partial pressures, and Dalton's, Raoult's and Henry's laws

The partial pressure $P_A$ of component **A** in a mixture of vapours is the pressure that would be exerted by component **A** at the same temperature, if present in the same volumetric concentration as in the mixture.

By Dalton's law of partial pressures, $P = \Sigma P_A$, that is the total pressure is equal to the summation of the partial pressures. Since in an ideal gas or vapour the partial pressure is proportional to the mole fraction of the constituent, then:

$$P_A = y_A P \tag{11.1}$$

For an *ideal mixture*, the partial pressure is related to the concentration in the liquid phase by Raoult's law which may be written as:

$$P_A = P_A^{\circ} x_A \tag{11.2}$$

where $P_A^{\circ}$ is the vapour pressure of pure **A** at the same temperature. This relation is usually found to be true only for high values of $x_A$, or correspondingly low values of $x_B$, although mixtures of organic isomers and some hydrocarbons follow the law closely.

For low values of $x_A$, a linear relation between $P_A$ and $x_A$ again exists, although the proportionality factor is Henry's constant $\mathscr{H}'$, and not the vapour pressure $P_A^0$ of the pure material.

For a liquid solute **A** in a solvent liquid **B**, Henry's law takes the form:

$$P_A = \mathscr{H}' x_A \tag{11.3}$$

If the mixture follows Raoult's law, then the vapour pressure of a mixture may be obtained graphically from a knowledge of the vapour pressure of the two components. Thus, in Figure 11.6. OA represents the partial pressure $P_A$ of **A** in a mixture, and CB the partial pressure of **B**, with the total pressure being shown by the line BA. In a mixture of composition D, the partial pressure $P_A$ is given by DE, $P_B$ by DF, and the total pressure $P$ by DG, from the geometry of Figure 11.6.

Figure 11.7 shows the partial pressure of one component **A** plotted against the mole fraction for a mixture that is not ideal. It is found that over the range OC the mixture follows Henry's law, and over BA it follows Raoult's law. Although most mixtures show wide divergences from ideality, one of the laws is usually followed at very high and very low concentrations.

If the mixture follows Raoult's law, then the values of $y_A$ for various values of $x_A$ may be calculated from a knowledge of the vapour pressures of the two components at

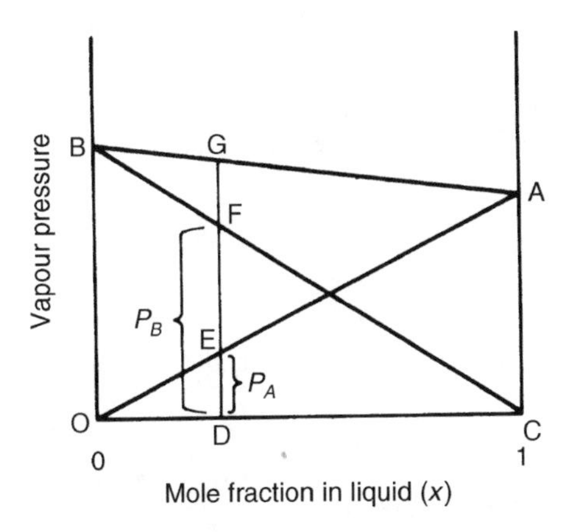


Figure 11.6. Partial pressures of ideal mixtures

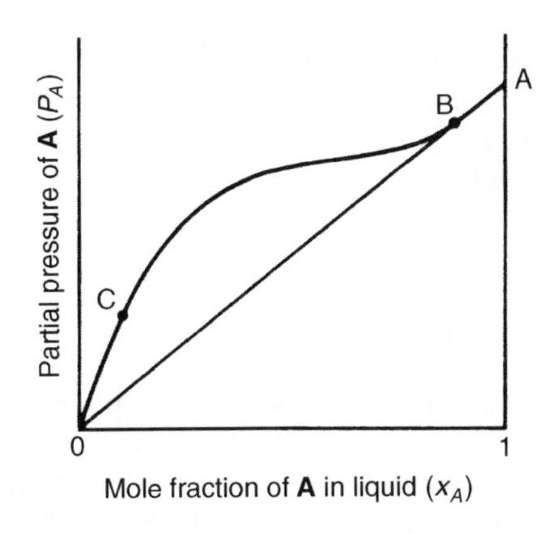


Figure 11.7. Partial pressures of non-ideal mixtures

various temperatures.

Thus: $$P_A = P_A^\circ x_A$$

and: $$P_A = P y_A$$

so that: $$y_A = \frac{P_A^\circ x_A}{P}, \quad \text{and} \quad y_B = \frac{P_B^\circ x_B}{P} \tag{11.4}$$

But: $$y_A + y_B = 1$$

$$\frac{P_A^\circ x_A}{P} + \frac{P_B^\circ(1 - x_A)}{P} = 1$$

giving: $$x_A = \frac{P - P_B^\circ}{P_A^\circ - P_B^\circ} \tag{11.5}$$

## Example 11.1

The vapour pressures of $n$-heptane and toluene at 373 K are 106 and 73.7 kN/m$^2$ respectively. What are the mole fractions of $n$-heptane in the vapour and in the liquid phase at 373 K if the total pressure is 101.3 kN/m$^2$?

## Solution

$$\text{At 373 K, } P_A^\circ = 106 \text{ kN/m}^2 \text{ and } P_B^\circ = 73.7 \text{ kN/m}^2$$

Thus, in equation 11.5:

$$x_A = (P - P_B^\circ)/(P_A^\circ P_B^\circ) = \frac{(101.3 - 73.7)}{(106 - 73.7)} = \underline{\underline{0.856}}$$

and, in equation 11.4:

$$y_A = P_B^\circ x_B/P = \frac{(106 \times 0.856)}{101.3} = \underline{\underline{0.896}}$$

Equilibrium data usually have to be determined by tedious laboratory methods. Proposals have been made which enable the complete diagram to be deduced with reasonable accuracy from a relatively small number of experimental values. Some of these methods are discussed by ROBINSON and GILLIAND[1] and by THORNTON and GARNER[2].

One of the most widely used correlations of saturated vapour pressure is that proposed by ANTOINE[3]. This takes the form:

$$\ln P^\circ = k_1 - k_2/(T + k_3) \tag{11.6}$$

where the constants, $k_1$, $k_2$ and $k_3$ must be determined experimentally[4,5,6] although many values of these constants are available in the literature[6,7,8,9,10]. Equation 11.6 is valid only over limited ranges of both temperature and pressure, although the correlation interval may be extended by using the equation proposed by RIEDEL[11]. This takes the form:

$$\ln P^\circ = k_4 - k_5/T + k_6 \ln T + k_7 T^6 \tag{11.7}$$

If only two values of the vapour pressure at temperatures $T_1$ and $T_2$ are known, then the Clapeyron equation may be used:

$$\ln P^\circ = k_8 - k_9/T \tag{11.8}$$

where:

$$k_8 = \ln P_1^\circ + k_9/T_1 \tag{11.9}$$

and:

$$k_9 = \ln(P_2/P_1)/[(1/T_1) - (1/T_2)] \tag{11.10}$$

Equation 11.8 may be used for the evaluation of vapour pressure over a small range of temperature, although large errors may be introduced over large temperature intervals. If the critical values of temperature and pressure are available along with one other vapour pressure point such as, for example, the normal boiling point, then a reduced form of the Riedel equation may be used; this takes the form:

$$\ln P_r^\circ = k_9 - k_{10}/T_r + k_{11} \ln T_r + k_{12} T_r^6 \tag{11.11}$$

where: $P_r$ = reduced vapour pressure = $(P^\circ/P_c)$, $T_r$ = reduced temperature = $(T/T_c)$, $k_9 = -35c_1$, $k_{10} = -36c_1$, $k_{11} = 42c_1 + c_2$, $k_{12} = -c_1$ and $c_1 = 0.0838(3.758 - c_2)$. $c_2$ is determined by inserting the other known vapour pressure point into equation 11.11 and solving for $c_2$. This gives:

$$c_2 = [(0.315c_5 - \ln P^\circ_{r1})/(0.0838c_5 - \ln T_{r1})] \tag{11.12}$$

where: $c_5 = -35 + 36T_{r1} + 42\ln T_{r1} - T^6_{r1}$.

## Example 11.2

The following data have been reported for acetone by AMBROSE *et al.*[12]: $P_c = 4700$ kN/m$^2$, $T_c = 508.1$ K, $P^\circ_1 = 100.666$ kN/m$^2$ when $T_1 = 329.026$ K. What is $P^\circ$ when $T = 350.874$ K?

## Solution

$T_{r1} = (329.026/508.1) = 0.64756$, $P_{r1} = (100.666/4700.0) = 0.021418$ and hence, in equation 11.12:

$$c_5 = -35 + (36/0.064756) + 42\ln 0.64756 - (0.64756)^6 = 2.2687$$

and: $c_2 = [((0.315 \times 2.2687) - \ln 0.021418)/((0.0838 \times 2.2687) - \ln 0.64756)] = 7.2970$

$$c_1 = 0.0838(3.758 - 7.2970) = -0.29657$$

$$k_9 = -35(-0.29657) = 10.380$$

$$k_{10} = -36(-0/29657) = 10.677$$

$$k_{11} = 42(-0.29657) + 7.2970 = -5.1589$$

$$k_{12} = 1(10.29657) = 0.29657$$

Substituting these values into equation 11.11 together with a value of $T_r = (350.874/508.1) = 0.69056$, then:

$$\ln P^\circ_r = 10.380 - (10.677/0.69056) - 5.1589\ln 0.69056 + 0.29657(0.69056)^6 = -3.1391$$

From which: $P^\circ_r = 0.043322$

and: $P^\circ = (0.043322 \times 4700.0) = \underline{\underline{203.61 \text{ kN/m}^2}}$

This may be compared with an experimental value of 201.571 kN/m$^2$.

## Example 11.3

The constants in the Antoine equation, Equation 11.6, are:

| | | | |
|---|---|---|---|
| For benzene: | $k_1 = 6.90565$ | $k_2 = 1211.033$ | $k_3 = 220.79$ |
| For toluene: | $k_1 = 6.95334$ | $k_2 = 1343.943$ | $k_3 = 219.377$ |

where $P^\circ$ is in mm Hg, $T$ is in °C and $\log_{10}$ is used instead of $\log_e$.

Determine the vapour phase composition of a mixture in equilibrium with a liquid mixture of 0.5 mole fraction benzene and 0.5 mole fraction of toluene at 338 K. Will the liquid vaporise at a pressure of 101.3 kN/m$^2$?

## Solution

The saturation vapour pressure of benzene at 338 K = 65°C is given by:

$$\log_{10} P_B^\circ = 6.90565 - [1211.033/(65 + 220.70)] = 2.668157$$

from which: $P_B^\circ = 465.75$ mm Hg or 62.10 kN/m$^2$

Similarly for toluene at 338 K = 65°C:

$$\log_{10} P_T^\circ = 6.95334 - [1343.943/(65 + 219.377)] = 2.22742$$

and: $P_T^\circ = 168.82$ mm Hg or 22.5 kN/m$^2$

The partial pressures in the mixture are:

$$P_B = (0.50 \times 62.10) = 31.05 \text{ kN/m}^2$$

and: $P_T = (0.50 \times 22.51) = 11.255$ kN/m$^2$ – a total pressure of 42.305 kN/m$^2$

Using equation 11.1, the composition of the vapour phase is:

$$y_B = (31.05/42.305) = \underline{\underline{0.734}}$$

and:

$$y_T = (11.255/42.305) = \underline{\underline{0.266}}$$

Since the total pressure is only 42.305 kN/m$^2$, then with a total pressure of 101.3 kN/m$^2$, the liquid will not vaporise unless the pressure is decreased.

## Example 11.4

What is the boiling point of a equimolar mixture of benzene and toluene at 101.3 kN/m$^2$?

## Solution

The saturation vapour pressures are calculated as a function of temperature using the Antoine equation, equation 11.6, and the constants given in Example 11.3, and then, from Raoult's Law, Equation 11.1, the actual vapour pressures are given by:

$$P_B = x_B P_B^\circ \text{ and } P_T = x_T P_T^\circ$$

It then remains, by a process of trial and error, to determine at which temperature: $(P_B + P_T) = 101.3$ kN/m$^2$. The data, with pressures in kN/m$^2$, are:

| $T$(K) | $P_B^\circ$ | $P_T^\circ$ | $P_B$ | $P_T$ | $(P_B + P_T)$ |
|---|---|---|---|---|---|
| 373 | 180.006 | 74.152 | 90.003 | 37.076 | 127.079 |
| 353 | 100.988 | 38.815 | 50.494 | 77.631 | 128.125 |
| 363 | 136.087 | 54.213 | 68.044 | 27.106 | 95.150 |
| 365 | 144.125 | 57.810 | 72.062 | 28.905 | 100.967 |
| 365.1 | 144.534 | 57.996 | 72.267 | 28.998 | 101.265 |

101.265 kN/m$^2$ is essentially 101.3 kN/m$^2$ and hence, at this pressure, the boiling or the bubble point of the equimolar mixture is 365.1 K which lies between the boiling points of pure benzene, 353.3 K, and pure toluene, 383.8 K.

## Example 11.5

What is the dew point of a equimolar mixture of benzene and toluene at 101.3 kN/m$^2$?

## Solution

From Raoult's Law, equations 11.1 and 11.2:

$$P_B = x_B P_B^{\circ} = y_B P$$

and:

$$P_T = x_T P_T^{\circ} = y_T P$$

Since the total pressure is 101.3 kN/m$^2$, $P_B = P_T = 50.65$ kN/m$^2$ and hence:

$$x_B = P_B/P_B^{\circ} = 50.65/P_B^{\circ} \text{ and } x_T = 50.65/P_T^{\circ}$$

It now remains to estimate the saturation vapour pressures as a function of temperature, using the data of Example 11.3, and then determine, by a process of trial and error, when $(x_B + x_T) = 1.0$. The data, with pressures in kN/m$^2$ are:

| $T$(K) | $P_B^{\circ}$ | $x_B$ | $P_T^0$ | $x_T$ | $(x_B + x_T)$ |
|---|---|---|---|---|---|
| 373.2 | 180.006 | 0.2813 | 74.152 | 0.6831 | 0.9644 |
| 371.2 | 170.451 | 0.2872 | 69.760 | 0.7261 | 1.0233 |
| 371.7 | 172.803 | 0.2931 | 70.838 | 0.7150 | 1.0081 |
| 371.9 | 173.751 | 0.2915 | 71.273 | 0.7107 | 1.0021 |
| 372.0 | 174.225 | 0.2907 | 71.491 | 0.7085 | 0.9992 |

As 0.9992 is near enough to 1.000, the dew point may be taken as 372.0 K.

### 11.2.3. Relative volatility

The relationship between the composition of the vapour $y_A$ and of the liquid $x_A$ in equilibrium may also be expressed in a way, which is particularly useful in distillation calculations. If the ratio of the partial pressure to the mole fraction in the liquid is defined as the volatility, then:

$$\text{Volatility of } \mathbf{A} = \frac{P_A}{x_A} \text{ and volatility of } \mathbf{B} = \frac{P_B}{x_B}$$

The ratio of these two volatilities is known as the relative volatility $\alpha$ given by:

$$\alpha = \frac{P_A x_B}{x_A P_B}$$

Substituting $Py_A$ for $P_A$, and $Py_B$ for $P_B$:

$$\alpha = \frac{y_A x_B}{y_B x_A} \tag{11.13}$$

or:

$$\frac{y_A}{y_B} = \alpha \frac{x_A}{x_B} \tag{11.14}$$

This gives a relation between the ratio of **A** and **B** in the vapour to that in the liquid.

Since with a binary mixture $y_B = 1 - y_A$, and $x_B = 1 - x_A$ then:

$$\alpha = \left(\frac{y_A}{1 - y_A}\right)\left(\frac{1 - x_A}{x_A}\right)$$

or:

$$y_A = \frac{\alpha x_A}{1 + (\alpha - 1)x_A} \tag{11.15}$$

and:

$$x_A = \frac{y_A}{\alpha - (\alpha - 1)y_A} \tag{11.16}$$

This relation enables the composition of the vapour to be calculated for any desired value of $x$, if $\alpha$ is known. For separation to be achieved, $\alpha$ must not equal 1 and, considering the more volatile component, as $\alpha$ increases above unity, $y$ increases and the separation becomes much easier. Equation 11.14 is useful in the calculation of plate enrichment and finds wide application in multicomponent distillation.

From the definition of the volatility of a component, it is seen that for an ideal system the volatility is numerically equal to the vapour pressure of the pure component. Thus the relative volatility $\alpha$ may be expressed as:

$$\alpha = \frac{P_A^\circ}{P_B^\circ} \tag{11.17}$$

This also follows by applying equation 11.1 from which $P_A/P_B = y_A/y_B$, so that:

$$\alpha = \frac{P_A x_B}{P_B x_A} = \frac{P_A^\circ x_A x_B}{P_B^\circ x_B x_A} = \frac{P_A^\circ}{P_B^\circ}$$

Whilst $\alpha$ does vary somewhat with temperature, it remains remarkably steady for many systems, and a few values to illustrate this point are given in Table 11.1.

Table 11.1. Relative volatility of mixtures of benzene and toluene

| Temperature (K) | 353 | 363 | 373 | 383 |
|---|---|---|---|---|
| $\alpha$ (–) | 2.62 | 2.44 | 2.40 | 2.39 |

It may be seen that $\alpha$ increases as the temperature falls, so that it is sometimes worthwhile reducing the boiling point by operating at reduced pressure. When Equation 11.16 is used to construct the equilibrium curve, an average value of $\alpha$ must be taken over the whole column. As FRANK[13] points out, this is valid if the relative volatilities at the top and bottom of the column differ by less than 15 per cent. If they differ by more than this amount, the equilibrium curve must be constructed incrementally by calculating the relative volatility at several points along the column.

Another frequently used relationship for vapour–liquid equilibrium is the simple equation:

$$y_A = Kx_A \tag{11.18}$$

For many systems $K$ is constant over an appreciable temperature range and Equation 11.11 may be used to determine the vapour composition at any stage. The method is particularly suited to multicomponent systems, discussed further in Section 11.7.1.

### 11.2.4. Non-ideal systems

Equation 11.4 relates $x_A$, $y_A$, $P_A^\circ$ and $P$. For a *non-ideal* system the term $\gamma$, the activity coefficient, is introduced to give:

$$y_A = \frac{\gamma_1 P_A^\circ x_A}{P} \quad \text{and} \quad y_B = \frac{\gamma_2 P_B^\circ x_B}{P} \tag{11.19}$$

or in Equation 11.18:

$$y_A = K\gamma_1 x_A \quad \text{and} \quad y_B = K\gamma_2 x_B \tag{11.20}$$

The liquid phase activity coefficients $\gamma_1$ and $\gamma_2$ depend upon temperature, pressure and concentration. Typical values taken from Perry's Chemical Engineers' Handbook[14] are shown in Figure 11.8 for the systems $n$-propanol–water and acetone–chloroform. In the former, the activity coefficients are considered positive, that is greater than unity, whilst in the latter, they are fractional so that the logarithms of the values are negative. In both cases, $\gamma$ approaches unity as the liquid concentration approaches unity and the highest values of $\gamma$ occur as the concentration approaches zero.

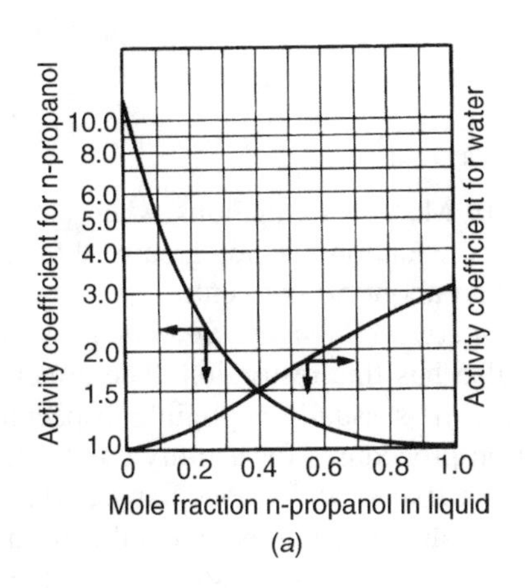


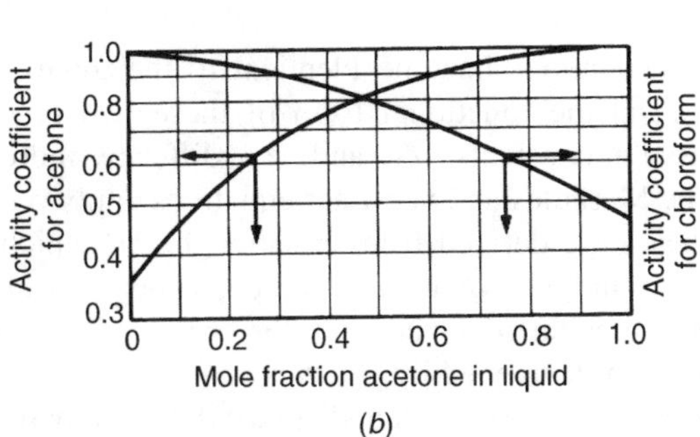


Figure 11.8. Activity coefficient data

The fundamental thermodynamic equation relating activity coefficients and composition is the *Gibbs–Duhem* relation which may be expressed as:

$$x_1\left(\frac{\partial \ln \gamma_1}{\partial x_1}\right)_{T,P} - x_2\left(\frac{\partial \ln \gamma_2}{\partial x_2}\right)_{T,P} = 0 \qquad (11.21)$$

This equation relates the slopes of the curves in Figure 11.8 and provides a means of testing experimental data. It is more convenient, however, to utilise integrated forms of these relations. A large number of different solutions to the basic Gibbs–Duhem equation are available, each of which gives a different functional relationship between log $\gamma$ and $x$. Most binary systems may be characterised, however, by either the three- or four-suffix equations of Margules, or by the two-suffix van Laar equations, given as follows in the manner of WOHL[15,16]. The three-suffix Margules binary equations are:

$$\log \gamma_1 = x_2^2[\mathscr{A}_{12} + 2x_1(\mathscr{A}_{21} - \mathscr{A}_{12})] \qquad (11.22)$$

$$\log \gamma_2 = x_1^2[\mathscr{A}_{21} + 2x_2(\mathscr{A}_{12} - \mathscr{A}_{21})] \qquad (11.23)$$

Constants $\mathscr{A}_{12}$ and $\mathscr{A}_{21}$ are the limiting values of $\log \gamma$ as the composition of the component considered approaches zero. For example, in Equation 11.22, $\mathscr{A}_{12} = \log \gamma_1$ when $x_1 = 0$.

The four-suffix Margules binary equations are:

$$\log \gamma_1 = x_2^2[\mathscr{A}_{12} + 2x_1(\mathscr{A}_{21} - \mathscr{A}_{12} - \mathscr{A}_D) + 3\mathscr{A}_D x_1^2] \qquad (11.24)$$

$$\log \gamma_2 = x_1^2[\mathscr{A}_{21} + 2x_2(\mathscr{A}_{12} - \mathscr{A}_{21} - \mathscr{A}_D) + 3\mathscr{A}_D x_2^2] \qquad (11.25)$$

$\mathscr{A}_{12}$ and $\mathscr{A}_{21}$ have the same significance as before and $\mathscr{A}_D$ is a third constant. Equations 11.24 and 11.25 are more complex than equations 11.22 and 11.23 though, because they contain an additional constant $\mathscr{A}_D$, they are more flexible. When $\mathscr{A}_D$ becomes zero in equations 11.24 and 11.25, they become identical to the three-suffix equations.

The two-suffix van Laar binary equations are:

$$\log \gamma_1 = \frac{\mathscr{A}_{12}}{[1 + (\mathscr{A}_{12}x_1/\mathscr{A}_{21}x_2)]^2} \qquad (11.26)$$

$$\log \gamma_2 = \frac{\mathscr{A}_{21}}{[1 + (\mathscr{A}_{21}x_2/\mathscr{A}_{12}x_1)]^2} \qquad (11.27)$$

These equations become identical to the three-suffix Margules equations when $\mathscr{A}_{12} = \mathscr{A}_{21}$, and the functional form of these two types of equations is not greatly different unless the constants $\mathscr{A}_{12}$ and $\mathscr{A}_{21}$ differ by more than about 50 per cent.

The Margules and van Laar equations apply only at *constant temperature and pressure*, as they were derived from equation 11.21, which also has this restriction. The effect of pressure upon $\gamma$ values and the constants $\mathscr{A}_{12}$ and $\mathscr{A}_{21}$ is usually negligible, especially at pressures far removed from the critical. Correlation procedures for activity coefficients have been developed by BALZHISER *et al.*[17], FRENDENSLUND *et al.*[18], PRAUNSITZ *et al.*[19], REID *et al.*[20], VAN NESS and ABBOTT[21] and WALAS[22] and actual experimental data may be obtained from the PPDS system of the NATIONAL ENGINEERING LABORATORY, UK[23]. When the liquid and vapour compositions are the same, that is $x_A = y_A$, point $x_g$ in

Figures 11.3 and 11.4, the system is said to form an azeotrope, a condition which is discussed in Section 11.8.

## 11.3. METHODS OF DISTILLATION — TWO COMPONENT MIXTURES

From curve *a* of Figure 11.4 it is seen that, for a binary mixture with a normal $y - x$ curve, the vapour is always richer in the more volatile component than the liquid from which it is formed. There are three main methods used in distillation practice which all rely on this basic fact. These are:

(a) Differential distillation.
(b) Flash or equilibrium distillation, and
(c) Rectification.

Of these, rectification is much the most important, and it differs from the other two methods in that part of the vapour is condensed and returned as liquid to the still, whereas, in the other methods, all the vapour is either removed as such, or is condensed as product.

### 11.3.1. Differential distillation

The simplest example of batch distillation is a single stage, differential distillation, starting with a still pot, initially full, heated at a constant rate. In this process the vapour formed on boiling the liquid is removed at once from the system. Since this vapour is richer in the more volatile component than the liquid, it follows that the liquid remaining becomes steadily weaker in this component, with the result that the composition of the product progressively alters. Thus, whilst the vapour formed over a short period is in equilibrium with the liquid, the total vapour formed is not in equilibrium with the residual liquid. At the end of the process the liquid which has not been vaporised is removed as the bottom product. The analysis of this process was first proposed by RAYLEIGH[24].

If $S$ is the number of moles of material in the still, $x$ is the mole fraction of component **A** and an amount $\mathrm{d}S$, containing a mole fraction $y$ of **A**, is vaporised, then a material balance on component **A** gives:

$$y\,\mathrm{d}S = \mathrm{d}(Sx)$$

$$= S\,\mathrm{d}x + x\,\mathrm{d}S$$

$$\int_{S_0}^{S} \frac{\mathrm{d}S}{S} = \int_{x_0}^{x} \left(\frac{\mathrm{d}x}{y - x}\right)$$

and:

$$\ln\frac{S}{S_0} = \int_{x_0}^{x} \left(\frac{\mathrm{d}x}{y - x}\right) \tag{11.28}$$

The integral on the right-hand side of this equation may be solved graphically if the equilibrium relationship between $y$ and $x$ is available. In some cases a direct integration is possible. Thus, if over the range concerned the equilibrium relationship is a straight line of the form $y = mx + c$, then:

$$\ln \frac{S}{S_0} = \left(\frac{1}{m-1}\right) \ln \left[\frac{(m-1)x + c}{(m-1)x_0 + c}\right]$$

or:

$$\frac{S}{S_0} = \left(\frac{y - x}{y_0 - x_0}\right)^{1/(m-1)}$$

and:

$$\left(\frac{y - x}{y_0 - x_0}\right) = \left(\frac{S}{S_0}\right)^{m-1} \tag{11.29}$$

From this equation the amount of liquid to be distilled in order to obtain a liquid of given concentration in the still may be calculated, and from this the average composition of the distillate may be found by a mass balance.

Alternatively, if the relative volatility is assumed constant over the range concerned, then $y = \alpha x/(1 + (\alpha - 1)x)$, equation 11.15 may be substituted in equation 11.28. This leads to the solution:

$$\ln \frac{S}{S_0} = \left(\frac{1}{\alpha - 1}\right) \ln \left[\frac{x(1 - x_0)}{x_0(1 - x)}\right] + \ln \left[\frac{1 - x_0}{1 - x}\right] \tag{11.30}$$

As this process consists of only a single stage, a complete separation is impossible unless the relative volatility is infinite. Application is restricted to conditions where a preliminary separation is to be followed by a more rigorous distillation, where high purities are not required, or where the mixture is very easily separated.

### 11.3.2. Flash or equilibrium distillation

Flash or equilibrium distillation, frequently carried out as a continuous process, consists of vaporising a definite fraction of the liquid feed in such a way that the vapour evolved is in equilibrium with the residual liquid. The feed is usually pumped through a fired heater and enters the still through a valve where the pressure is reduced. The still is essentially a separator in which the liquid and vapour produced by the reduction in pressure have sufficient time to reach equilibrium. The vapour is removed from the top of the separator and is then usually condensed, while the liquid leaves from the bottom.

In a typical pipe still where, for example, a crude oil might enter at 440 K and at about 900 kN/m$^2$, and leave at 520 K and 400 kN/m$^2$, some 15 per cent may be vaporised in the process. The vapour and liquid streams may contain many components in such an application, although the process may be analysed simply for a binary mixture of **A** and **B** as follows:

If $F$ = moles per unit time of feed of mole fraction $x_f$ of **A**,
$V$ = moles per unit time of vapour formed with $y$ the mole fraction of **A**, and
$S$ = moles per unit time of liquid with $x$ the mole fraction of **A**,

then an overall mass balance gives:

$$F = V + S$$

and for the more volatile component:

$$Fx_f = Vy + Sx$$

Thus:

$$\frac{V}{F} = \left(\frac{x_f - x}{y - x}\right)$$

or:

$$y = \frac{F}{V}x_f - x\left(\frac{F}{V} - 1\right) \tag{11.31}$$

Equation 11.31 represents a straight line of slope:

$$-\left(\frac{F - V}{V}\right) = \frac{-S}{V}$$

passing through the point $(x_f, x_f)$. The values of $x$ and $y$ required must satisfy, not only the equation, but also the appropriate equilibrium data. Thus these values may be determined graphically using an $x - y$ diagram as shown in Figure 11.9.

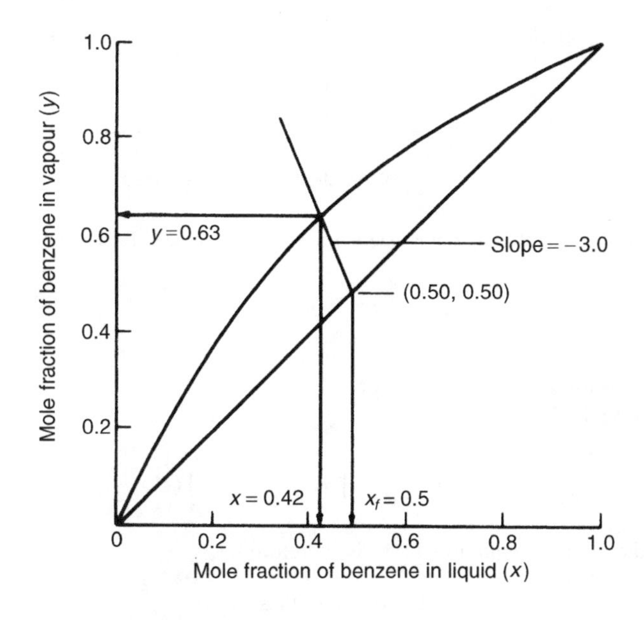


Figure 11.9. Equilibrium data for benzene–toluene for Example 11.6

In practice, the quantity vaporised is not fixed directly but it depends upon the enthalpy of the hot incoming feed and the enthalpies of the vapour and liquid leaving the separator. For a given feed condition, the fraction vaporised may be increased by lowering the pressure in the separator.

## Example 11.6

An equimolar mixture of benzene and toluene is subjected to flash distillation at 100 kN/m$^2$ in the separator. Using the equilibrium data given in Figure 11.9, determine the composition of the liquid and vapour leaving the separator when the feed is 25 per cent vaporised. For this condition, the boiling point diagram in Figure 11.10 may be used to determine the temperature of the exit liquid stream.

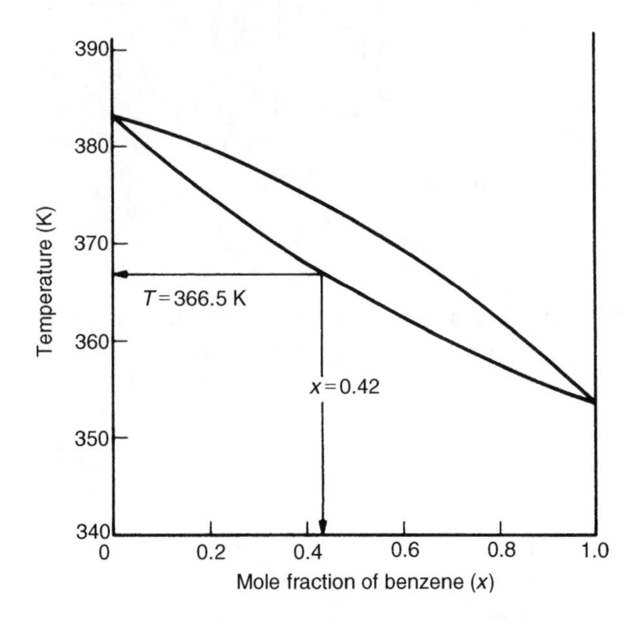


Figure 11.10. Boiling point diagram for benzene–toluene for Example 11.6

## Solution

The fractional vaporisation $= V/F = f$ (say)
The slope of equation 11.31 is:

$$-\left(\frac{F-V}{V}\right) = -\left(\frac{1-f}{f}\right)$$

When $f = 0.25$, the slope of equation 11.31 is therefore:

$$-(1 - 0.25)/0.25 = -3.0$$

and the construction is made as shown in Figure 11.9 to give $x = \underline{\underline{0.42}}$ and $y = \underline{\underline{0.63}}$.

From the boiling point diagram, in Figure 11.10 the liquid temperature when $x = 0.42$ is seen to be $\underline{\underline{366.5 \text{ K}}}$.

### 11.3.3. Rectification

In the two processes considered, the vapour leaving the still at any time is in equilibrium with the liquid remaining, and normally there will be only a small increase in concentration of the more volatile component. The essential merit of rectification is that it enables a vapour to be obtained that is substantially richer in the more volatile component than is the liquid left in the still. This is achieved by an arrangement known as a fractionating column which enables successive vaporisation and condensation to be accomplished in one unit. Detailed consideration of this process is given in Section 11.4.

### 11.3.4. Batch distillation

In batch distillation, which is considered in detail in Section 11.6, the more volatile component is evaporated from the still which therefore becomes progressively richer in the less volatile constituent. Distillation is continued, either until the residue of the still contains a material with an acceptably low content of the volatile material, or until the distillate is no longer sufficiently pure in respect of the volatile content.

## 11.4. THE FRACTIONATING COLUMN

### 11.4.1. The fractionating process

The operation of a typical fractionating column may be followed by reference to Figure 11.11. The column consists of a cylindrical structure divided into sections by

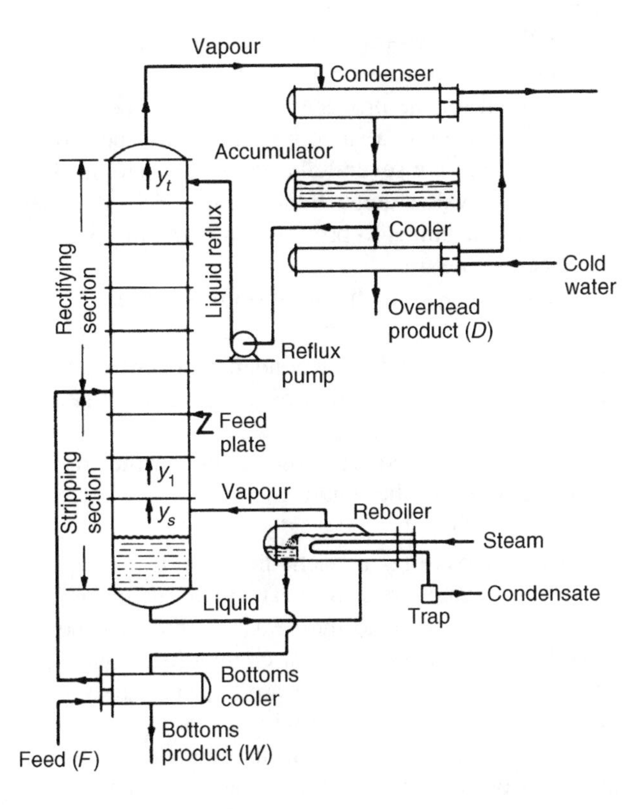


Figure 11.11. Continuous fractionating column with rectifying and stripping sections

a series of perforated trays which permit the upward flow of vapour. The liquid reflux flows across each tray, over a weir and down a downcomer to the tray below. The vapour rising from the top tray passes to a condenser and then through an accumulator or reflux drum and a reflux divider, where part is withdrawn as the overhead product D, and the remainder is returned to the top tray as reflux R.

The liquid in the base of the column is frequently heated, either by condensing steam or by a hot oil stream, and the vapour rises through the perforations to the bottom tray. A more commonly used arrangement with an external reboiler is shown in Figure 11.11 where the liquid from the still passes into the reboiler where it flows over the tubes and weir and leaves as the bottom product by way of a bottoms cooler, which preheats the incoming feed. The vapour generated in the reboiler is returned to the bottom of the column with a composition $y_s$, and enters the bottom tray where it is partially condensed and then revaporised to give vapour of composition $y_1$. This operation of partial condensation of the rising vapour and partial vaporisation of the reflux liquid is repeated on each tray. Vapour of composition $y_t$ from the top tray is condensed to give the top product D and the reflux R, both of the same composition $y_t$. The feed stream is introduced on some intermediate tray where the liquid has approximately the same composition as the feed. The part of the column above the feed point is known as the rectifying section and the lower portion is known as the stripping section. The vapour rising from an ideal tray will be in equilibrium with the liquid leaving, although in practice a smaller degree of enrichment will occur.

In analysing the operation on each tray it is important to note that the vapour rising to it, and the reflux flowing down to it, are not in equilibrium, and adequate rates of mass and heat transfer are essential for the proper functioning of the tray.

The tray as described is known as a sieve tray and it has perforations of up to about 12 mm diameter, although there are several alternative arrangements for promoting mass transfer on the tray, such as valve units, bubble caps and other devices described in Section 11.10.1. In all cases the aim is to promote good mixing of vapour and liquid with a low drop in pressure across the tray.

On each tray the system tends to reach equilibrium because:

(a) Some of the less volatile component condenses from the rising vapour into the liquid thus increasing the concentration of the more volatile component (MVC) in the vapour.
(b) Some of the MVC is vaporised from the liquid on the tray thus decreasing the concentration of the MVC in the liquid.

The number of molecules passing in each direction from vapour to liquid and in reverse is approximately the same since the heat given out by one mole of the vapour on condensing is approximately equal to the heat required to vaporise one mole of the liquid. The problem is thus one of equimolecular counterdiffusion, described in Volume 1, Chapter 10. If the molar heats of vaporisation are approximately constant, the flows of liquid and vapour in each part of the column will not vary from tray to tray. This is the concept of constant molar overflow which is discussed under the heat balance heading in Section 11.4.2. Conditions of varying molar overflow, arising from unequal molar latent heats of the components, are discussed in Section 11.5.

In the arrangement discussed, the feed is introduced continuously to the column and two product streams are obtained, one at the top much richer than the feed in the MVC and the second from the base of the column weaker in the MVC. For the separation of small quantities of mixtures, a batch still may be used. Here the column rises directly from a large drum which acts as the still and reboiler and holds the charge of feed. The trays in the column form a rectifying column and distillation is continued until it is no longer possible to obtain the desired product quality from the column. The concentration of the MVC steadily falls in the liquid remaining in the still so that enrichment to the desired level of the MVC is not possible. This problem is discussed in more detail in Section 11.6.

A complete unit will normally consist of a feed tank, a feed heater, a column with boiler, a condenser, an arrangement for returning part of the condensed liquid as reflux, and coolers to cool the two products before passing them to storage. The reflux liquor may be allowed to flow back by gravity to the top plate of the column or, as in larger units, it is run back to a drum from which it is pumped to the top of the column. The control of the reflux on very small units is conveniently effected by hand-operated valves, and with the larger units by adjusting the delivery from a pump. In many cases the reflux is divided by means of an electromagnetically operated device which diverts the top product either to the product line or to the reflux line for controlled time intervals.

### 11.4.2. Number of plates required in a distillation column

In order to develop a method for the design of distillation units to give the desired fractionation, it is necessary, in the first instance, to develop an analytical approach which enables the necessary number of trays to be calculated. First the heat and material flows over the trays, the condenser, and the reboiler must be established. Thermodynamic data are required to establish how much mass transfer is needed to establish equilibrium between the streams leaving each tray. The required diameter of the column will be dictated by the necessity to accommodate the desired flowrates, to operate within the available drop in pressure, while at the same time effecting the desired degree of mixing of the streams on each tray.

Four streams are involved in the transfer of heat and material across a plate, as shown in Figure 11.12 in which plate $n$ receives liquid $L_{n+1}$ from plate $n+1$ above, and vapour $V_{n-1}$ from plate $n-1$ below. Plate $n$ supplies liquid $L_n$ to plate $n-1$, and vapour $V_n$ to plate $n+1$.

The action of the plate is to bring about mixing so that the vapour $V_n$, of composition $y_n$, approaches equilibrium with the liquid $L_n$, of composition $x_n$. The streams $L_{n+1}$ and $V_{n-1}$ cannot be in equilibrium and, during the interchange process on the plate, some of the more volatile component is vaporised from the liquid $L_{n+1}$, decreasing its concentration to $x_n$, and some of the less volatile component is condensed from $V_{n-1}$, increasing the vapour concentration to $y_n$. The heat required to vaporise the more volatile component from the liquid is supplied by partial condensation of the vapour $V_{n-1}$. Thus the resulting effect is that the more volatile component is passed from the liquid running down the column to the vapour rising up, whilst the less volatile component is transferred in the opposite direction.

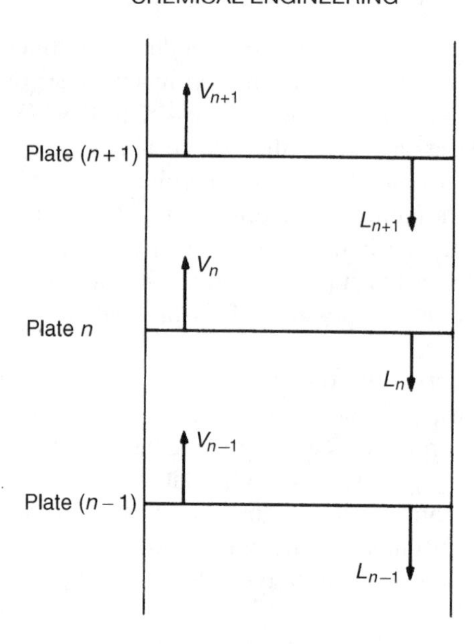


Figure 11.12. Material balance over a plate

### Heat balance over a plate

A heat balance across plate $n$ may be written as:

$$L_{n+1}H_{n+1}^L + V_{n-1}H_{n-1}^V = V_n H_n^V + L_n H_n^L + \text{losses} + \text{heat of mixing} \tag{11.32}$$

where: $H_n^L$ is the enthalpy per mole of the liquid on plate $n$, and
$H_n^V$ is the enthalpy per mole of the vapour rising from plate $n$.

This equation is difficult to handle for the majority of mixtures, and some simplifying assumptions are usually made. Thus, with good lagging, the heat losses will be small and may be neglected, and for an ideal system the heat of mixing is zero. For such mixtures, the molar heat of vaporisation may be taken as constant and independent of the composition. Thus, one mole of vapour $V_{n-1}$ on condensing releases sufficient heat to liberate one mole of vapour $V_n$. It follows that $V_n = V_{n-1}$, so that the molar vapour flow is constant up the column unless material enters or is withdrawn from the section. The temperature change from one plate to the next will be small, and $H_n^L$ may be taken as equal to $H_{n+1}^L$. Applying these simplifications to equation 11.32, it is seen that $L_n = L_{n+1}$, so that the moles of liquid reflux are also constant in this section of the column. Thus $V_n$ and $L_n$ are constant over the rectifying section, and $V_m$ and $L_m$ are constant over the stripping section.

For these conditions there are two basic methods for determining the number of plates required. The first is due to SOREL[25] and later modified by LEWIS[26], and the second is due to McCABE and THIELE[27]. The Lewis method is used here for binary systems, and also in Section 11.7.4 for calculations involving multicomponent mixtures. This method is also the basis of modern computerised methods. The McCabe–Thiele method is particularly important since it introduces the idea of the operating line which is an important common concept in multistage operations. The best assessment of these methods and their various applications is given by UNDERWOOD[28].

When the molar heat of vaporisation varies appreciably and the heat of mixing is no longer negligible, these methods have to be modified, and alternative techniques are discussed in Section 11.5.

### *Calculation of number of plates using the Lewis–Sorel method*

If a unit is operating as shown in Figure 11.13, so that a binary feed F is distilled to give a top product D and a bottom product W, with $x_f$, $x_d$, and $x_w$ as the corresponding mole fractions of the more volatile component, and the vapour $V_t$ rising from the top plate is condensed, and part is run back as liquid at its boiling point to the column as reflux, the remainder being withdrawn as product, then a material balance above plate $n$, indicated by the loop I in Figure 11.13 gives:

$$V_n = L_{n+1} + D \tag{11.33}$$

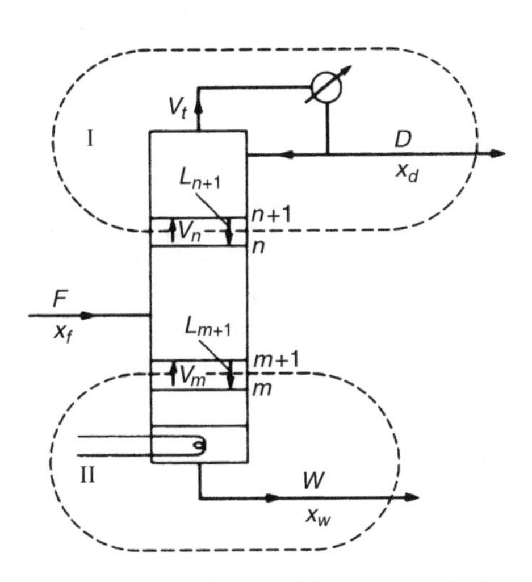


Figure 11.13. Material balances at top and bottom of column

Expressing this balance for the more volatile component gives:

$$y_n V_n = L_{n+1}x_{n+1} + Dx_d$$

Thus:

$$y_n = \frac{L_{n+1}}{V_n}x_{n+1} + \frac{D}{V_n}x_d \tag{11.34}$$

This equation relates the composition of the vapour rising to the plate to the composition of the liquid on any plate above the feed plate. Since the molar liquid overflow is constant, $L_n$ may be replaced by $L_{n+1}$ and:

$$y_n = \frac{L_n}{V_n}x_{n+1} + \frac{D}{V_n}x_d \tag{11.35}$$

Similarly, taking a material balance for the total streams and for the more volatile component from the bottom to above plate $m$, as indicated by the loop II in Figure 11.13, and noting that $L_m = L_{m+1}$ gives:

$$L_m = V_m + W \tag{11.36}$$

and:
$$y_m V_m = L_m x_{m+1} - W x_w$$

Thus:
$$y_m = \frac{L_m}{V_m} x_{m+1} - \frac{W}{V_m} x_w \tag{11.37}$$

This equation, which is similar to equation 11.35, gives the corresponding relation between the compositions of the vapour rising to a plate and the liquid on the plate, for the section below the feed plate. These two equations are the equations of the operating lines.

In order to calculate the change in composition from one plate to the next, the equilibrium data are used to find the composition of the vapour above the liquid, and the enrichment line to calculate the composition of the liquid on the next plate. This method may then be repeated up the column, using equation 11.37 for sections below the feed point, and equation 11.35 for sections above the feed point.

## Example 11.7

A mixture of benzene and toluene containing 40 mole per cent benzene is to be separated to give a product containing 90 mole per cent benzene at the top, and a bottom product containing not more than 10 mole per cent benzene. The feed enters the column at its boiling point, and the vapour leaving the column which is condensed but not cooled, provides reflux and product. It is proposed to operate the unit with a reflux ratio of 3 kmol/kmol product. It is required to find the number of theoretical plates needed and the position of entry for the feed. The equilibrium diagram at 100 kN/m$^2$ is shown in Figure 11.14.

## Solution

For 100 kmol of feed, an overall mass balance gives:

$$100 = D + W$$

A balance on the MVC, benzene, gives:

$$(100 \times 0.4) = 0.9\ D + 0.1\ W$$

Thus:
$$40 = 0.9(100 - W) + 0.1\ W$$

and:
$$W = 62.5 \quad \text{and} \quad D = 37.5 \text{ kmol}$$

Using the notation of Figure 11.13 then:

$$L_n = 3D = 112.5$$

and:
$$V_n = L_n + D = 150$$

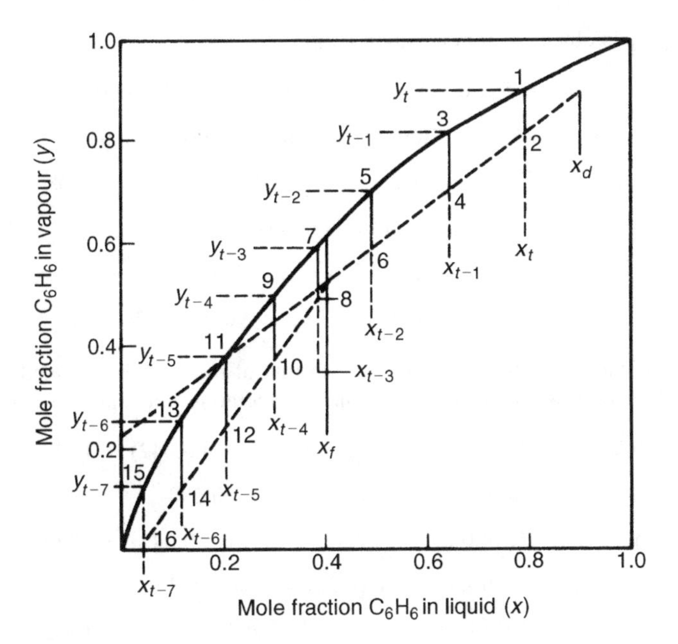


Figure 11.14. Calculation of the number of plates by the Lewis–Sorel method for Example 11.7

Thus, the top operating line from equation 11.35 is:

$$y_n = \left(\frac{112.5}{150}\right) x_{n+1} + \frac{(37.5 \times 0.9)}{150}$$

or: $$y_n = 0.75x_{n+1} + 0.225 \qquad \text{(i)}$$

Since the feed is all liquid at its boiling point, this will all run down as increased reflux to the plate below.

Thus: $$L_m = L_n + F$$

$$= (112.5 + 100) = 212.5$$

Also: $$V_m = L_m - W$$

$$= 212.5 - 62.5 = 150 = V_n$$

Thus: $$y_m = \left(\frac{212.5}{150}\right) x_{m+1} - \left(\frac{62.5}{150}\right) \times 0.1 \qquad \text{(equation 11.37)}$$

or: $$y_m = 1.415x_{m+1} - 0.042 \qquad \text{(ii)}$$

With the two equations (i) and (ii) and the equilibrium curve, the composition on the various plates may be calculated by working either from the still up to the condenser, or in the reverse direction. Since all the vapour from the column is condensed, the composition of the vapour $y_t$ from the top plate must equal that of the product $x_d$, and that of the liquid returned as reflux $x_r$. The composition $x_t$ of the liquid on the top plate is found from the equilibrium curve and, since it is in equilibrium with vapour of composition, $y_t = 0.90$, $x_t = 0.79$.

The value of $y_{t-1}$ is obtained from equation (i) as:

$$y_{t-1} = (0.75 \times 0.79) + 0.225 = (0.593 + 0.225) = 0.818$$

$x_{t-1}$ is obtained from the equilibrium curve as 0.644

$$y_{t-2} = (0.75 \times 0.644) + 0.225 = (0.483 + 0.225) = 0.708$$

$x_{t-2}$ from equilibrium curve = 0.492

$$y_{t-3} = (0.75 \times 0.492) + 0.225 = (0.369 + 0.225) = 0.594$$

$x_{t-3}$ from the equilibrium curve = 0.382

This last value of composition is sufficiently near to that of the feed for the feed to be introduced on plate $(t - 3)$. For the lower part of the column, the operating line equation (ii) will be used.

Thus:

$$y_{t-4} = (1.415 \times 0.382) - 0.042 = (0.540 - 0.042) = 0.498$$

$x_{t-4}$ from the equilibrium curve = 0.298

$$y_{t-5} = (1.415 \times 0.298) - 0.042 = (0.421 - 0.042) = 0.379$$

$x_{t-5}$ from the equilibrium curve = 0.208

$$y_{t-6} = (1.415 \times 0.208) - 0.042 = (0.294 - 0.042) = 0.252$$

$x_{t-6}$ from the equilibrium curve = 0.120

$$y_{t-7} = (1.415 \times 0.120) - 0.042 = (0.169 - 0.042) = 0.127$$

$x_{t-7}$ from the equilibrium curve = 0.048

This liquid $x_{t-7}$ is slightly weaker than the minimum required and it may be withdrawn as the bottom product. Thus, $x_{t-7}$ will correspond to the reboiler, and there will be seven plates in the column.

## *The method of McCabe and Thiele*

The simplifying assumptions of constant molar heat of vaporisation, no heat losses, and no heat of mixing, lead to a constant molar vapour flow and a constant molar reflux flow in any section of the column, that is $V_n = V_{n+1}$, $L_n = L_{n+1}$, and so on. Using these simplifications, the two enrichment equations are obtained:

$$y_n = \frac{L_n}{V_n} x_{n+1} + \frac{D}{V_n} x_d \qquad \text{(equation 11.35)}$$

and:

$$y_m = \frac{L_m}{V_m} x_{m+1} - \frac{W}{V_m} x_w \qquad \text{(equation 11.37)}$$

These equations are used in the Lewis–Sorel method to calculate the relation between the composition of the liquid on a plate and the composition of the vapour rising to that plate. McCabe and Thiele[27] pointed out that, since these equations represent straight lines connecting $y_n$ with $x_{n+1}$ and $y_m$ with $x_{m+1}$, they can be drawn on the same diagram as the equilibrium curve to give a simple graphical solution for the number of stages required. Thus, the line of equation 11.35 will pass through the points 2, 4 and 6 shown in Figure 11.14, and similarly the line of equation 11.37 will pass through points 8, 10, 12 and 14.

If $x_{n+1} = x_d$ in equation 11.35, then:

$$y_n = \frac{L_n}{V_n} x_d + \frac{D}{V_n} x_d = x_d \tag{11.38}$$

and this equation represents a line passing through the point $y_n = x_{n+1} = x_d$. If $x_{n+1}$ is put equal to zero, then $y_n = Dx_d/V_n$, giving a second easily determined point. The top operating line is therefore drawn through two points of coordinates $(x_d, x_d)$ and $(0, (Dx_d/V_n))$.

For the bottom operating line, equation 11.30, if $x_{m+1} = x_w$, then:

$$y_m = \frac{L_m}{V_m} x_w - \frac{W}{V_m} x_w \tag{11.39}$$

Since $V_m = L_m - W$, it follows that $y_m = x_w$. Thus the bottom operating line passes through the point C, that is $(x_w, x_w)$, and has a slope $L_m/V_m$. When the two operating lines have been drawn in, the number of stages required may be found by drawing steps between the operating line and the equilibrium curve starting from point A.

This method is one of the most important concepts in chemical engineering and is an invaluable tool for the solution of distillation problems. The assumption of constant molar overflow is not limiting since in very few systems do the molar heats of vaporisation differ by more than 10 per cent. The method does have limitations, however, and should not be employed when the relative volatility is less than 1.3 or greater than 5, when the reflux ratio is less than 1.1 times the minimum, or when more than twenty-five theoretical trays are required(13). In these circumstances, the Ponchon–Savarit method described in Section 11.5 should be used.

## Example 11.8. The McCabe-Thiele Method

Example 11.7 is now worked using this method. Thus, with a feed composition, $x_f = 0.4$, the top composition, $x_d$ is to have a value of 0.9 and the bottom composition, $x_w$ is to be 0.10. The reflux ratio, $L_n/D = 3$.

## Solution

a) From a material balance for a feed of 100 kmol:

$$V_n = V_m = 150;\ L_n = 112.5;\ L_m = 212.5;\ D = 37.5 \text{ and } W = 62.5 \text{ kmol}$$

b) The equilibrium curve and the diagonal line are drawn in as shown in Figure 11.15.

c) The equation of the top operating line is:

$$y_n = 0.75x_{n+1} + 0.225 \tag{i}$$

Thus, the line AB is drawn through the two points A (0.9, 0.9) and B (0, 0.225).

d) The equation of the bottom operating line is:

$$y_m = 1.415x_{m+1} - 0.042 \quad \text{(ii)}$$

This equation is represented by the line CD drawn through C (0.1, 0.1) at a slope of 1.415.

e) Starting at point A, the horizontal line is drawn to cut the equilibrium line at point 1. The vertical line is dropped through 1 to the operating line at point 2 and this procedure is repeated to obtain points 3–6.

f) A horizontal line is drawn through point 6 to cut the equilibrium line at point 7 and a vertical line is drawn through point 7 to the lower enrichment line at point 8. This procedure is repeated in order to obtain points 9–16.

g) The number of stages are then counted, that is points 2, 4, 6, 8, 10, 12, and 14 which gives the number of plates required as 7.

## *Enrichment in still and condenser*

Point 16 in Figure 11.15 represents the concentration of the liquor in the still. The concentration of the vapour is represented by point 15, so that the enrichment represented by the increment 16–15 is achieved in the boiler or still body. Again, the concentration on the top plate is given by point 2, but the vapour from this plate has a concentration given by point 1, and the condenser by completely condensing the vapour gives a product of equal concentration, represented by point A. The still and condenser together, therefore, provide enrichment (16 − 15) + (1 − A), which is equivalent to one ideal stage. Thus, the actual number of theoretical plates required is one less than the number of stages shown on the diagram. From a liquid in the still, point 16 to the product, point A, there are eight steps, although the column need only contain seven theoretical plates.

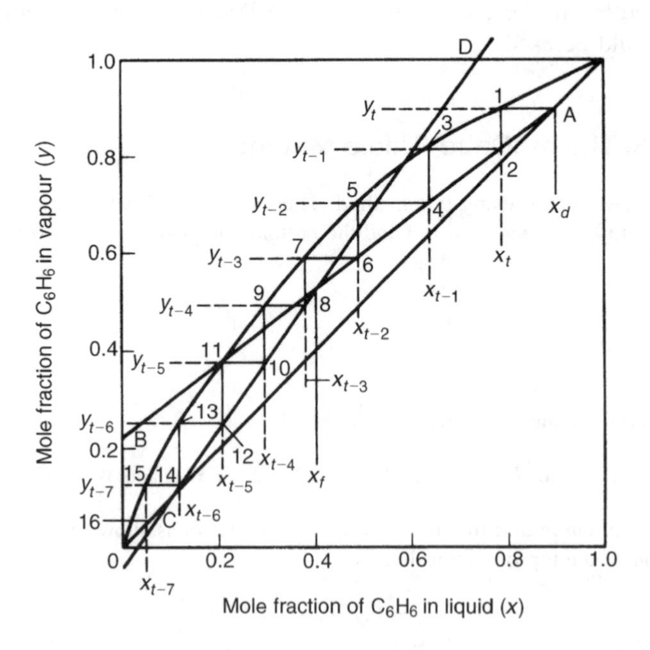


Figure 11.15. Determination of number of plates by the McCabe–Thiele method (Example 11.8)

### *The intersection of the operating lines*

It is seen from the example shown in Figure 11.15 in which the feed enters as liquid at its boiling point that the two operating lines intersect at a point having an $X$-coordinate of $x_f$. The locus of the point of intersection of the operating lines is of considerable importance since, as will be seen, it is dependent on the temperature and physical condition of the feed.

If the two operating lines intersect at a point with coordinates $(x_q, y_q)$, then from equations 11.35 and 11.37:

$$V_n y_q = L_n x_q + D x_d \tag{11.40}$$

and:
$$V_m y_q = L_m x_q - W x_w \tag{11.41}$$

or:
$$y_q(V_m - V_n) = (L_m - L_n)x_q - (D x_d + W x_w) \tag{11.42}$$

A material balance over the feed plate gives:

$$F + L_n + V_m = L_m + V_n$$

or:
$$V_m - V_n = L_m - L_n - F \tag{11.43}$$

To obtain a relation between $L_n$ and $L_m$, it is necessary to make an enthalpy balance over the feed plate, and to consider what happens when the feed enters the column. If the feed is all in the form of liquid at its boiling point, the reflux $L_m$ overflowing to the plate below will be $L_n + F$. If however the feed is a liquid at a temperature $T_f$, that is less than the boiling point, some vapour rising from the plate below will condense to provide sufficient heat to bring the feed liquor to the boiling point.

If $H_f$ is the enthalpy per mole of feed, and $H_{fs}$ is the enthalpy of one mole of feed at its boiling point, then the heat to be supplied to bring feed to the boiling point is $F(H_{fs} - H_f)$, and the number of moles of vapour to be condensed to provide this heat is $F(H_{fs} - H_f)/\lambda$, where $\lambda$ is the molar latent heat of the vapour.

The reflux liquor is then:

$$\begin{aligned} L_m &= L_n + F + \frac{F(H_{fs} - H_f)}{\lambda} \\ &= L_n + F\left(\frac{\lambda + H_{fs} - H_f}{\lambda}\right) \\ &= L_n + qF \end{aligned} \tag{11.44}$$

where:
$$q = \frac{\text{heat to vaporise 1 mole of feed}}{\text{molar latent heat of the feed}}$$

Thus, from equation 11.43:

$$V_m - V_n = qF - F \tag{11.45}$$

A material balance of the more volatile component over the whole column gives:

$$F x_f = D x_d + W x_w$$

Thus, from equation 11.42:

$$F(q-1)y_q = qFx_q - Fx_f$$

or:

$$y_q = \left(\frac{q}{q-1}\right)x_q - \left(\frac{x_f}{q-1}\right) \tag{11.46}$$

This equation is commonly known as the equation of the $q$-line. If $x_q = x_f$, then $y_q = x_f$. Thus, the point of intersection of the two operating lines lies on the straight line of slope $q/(q-1)$ passing through the point $(x_f, x_f)$. When $y_q = 0$, $x_q = x_f/q$. The line may thus be drawn through two easily determined points. From the definition of $q$, it follows that the slope of the $q$-line is governed by the nature of the feed as follows.

| | | |
|---|---|---|
| (a) Cold feed as liquor | $q > 1$ | $q$ line / |
| (b) Feed at boiling point | $q = 1$ | $q$ line \| |
| (c) Feed partly vapour | $0 < q < 1$ | $q$ line \ |
| (d) Feed saturated vapour | $q = 0$ | $q$ line — |
| (e) Feed superheated vapour | $q < 0$ | $q$ line / |

These various conditions are indicated in Figure 11.16.

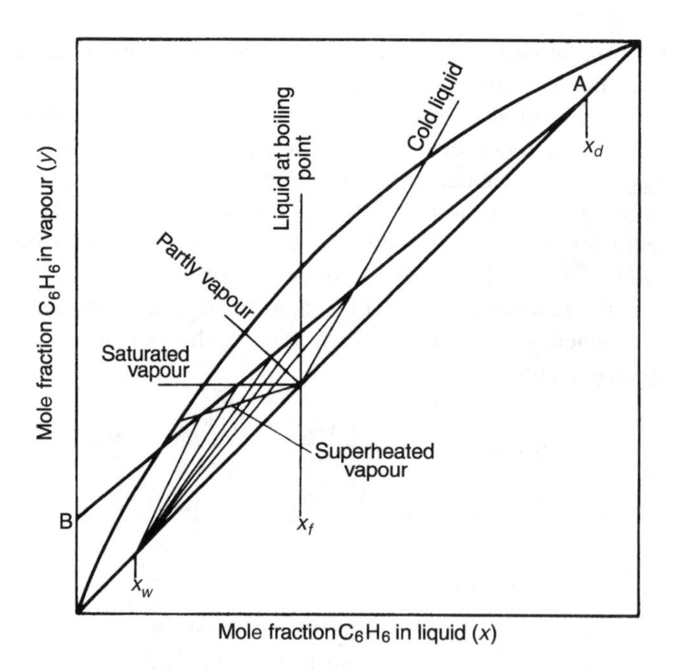


Figure 11.16. Effect of the condition of the feed on the intersection of the operating lines for a fixed reflux ratio

Altering the slope of the $q$-line will alter the liquid concentration at which the two operating lines cut each other for a given reflux ratio. This will mean a slight alteration in the number of plates required for the given separation. Whilst the change in the number of plates is usually rather small, if the feed is cold, there will be an increase in reflux flow below the feed plate, and hence an increased heat consumption from the boiler per mole of distillate.

### 11.4.3. The importance of the reflux ratio

#### *Influence on the number of plates required*

The ratio $L_n/D$, that is the ratio of the top overflow to the quantity of product, is denoted by $R$, and this enables the equation of the operating line to be expressed in another way, which is often more convenient. Thus, introducing $R$ in equation 11.35 gives:

$$y_n = \left(\frac{L_n}{L_n + D}\right) x_{n+1} + \left(\frac{D}{L_n + D}\right) x_d \tag{11.47}$$

$$= \left(\frac{R}{R+1}\right) x_{n+1} + \left(\frac{x_d}{R+1}\right) \tag{11.48}$$

Any change in the reflux ratio $R$ will therefore modify the slope of the operating line and, as may be seen from Figure 11.15, this will alter the number of plates required for a given separation. If $R$ is known, the top line is most easily drawn by joining point A $(x_d, x_d)$ to B $(0, x_d/(R+1))$ as shown in Figure 11.17. This method avoids the calculation of the actual flow rates $L_n$ and $V_n$, when the number of plates only is to be estimated.

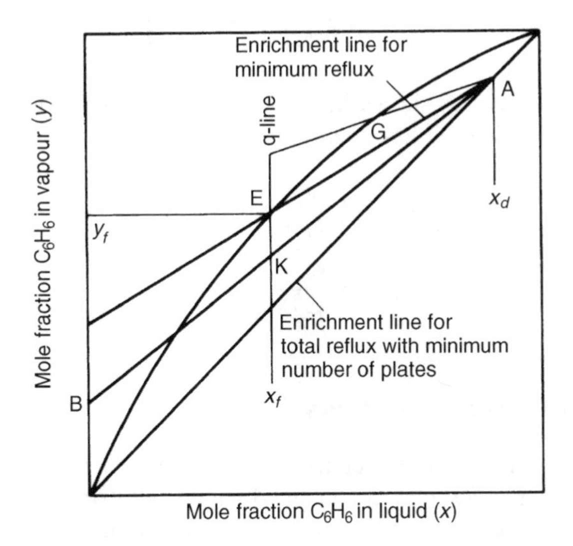


Figure 11.17. Influence of reflux ratio on the number of plates required for a given separation

If no product is withdrawn from the still, that is $D = 0$, then the column is said to operate under conditions of total reflux and, as seen from equation 11.47, the top operating line has its maximum slope of unity, and coincides with the line $x = y$. If the reflux ratio is reduced, the slope of the operating line is reduced and more stages are required to pass from $x_f$ to $x_d$, as shown by the line AK in Figure 11.17. Further reduction in $R$ will eventually bring the operating line to AE, where an infinite number of stages is needed to pass from $x_d$ to $x_f$. This arises from the fact that under these conditions the steps become very close together at liquid compositions near to $x_f$, and no enrichment occurs from the feed plate to the plate above. These conditions are known as *minimum reflux*, and the reflux ratio is denoted by $R_m$. Any small increase in $R$ beyond $R_m$ will give a workable system, although a large number of plates will be required. It is important to note that any line such as AG, which is equivalent to a smaller value of $R$ than $R_m$, represents an impossible condition, since it is impossible to pass beyond point G towards $x_f$. Two important deductions may be made. Firstly that the minimum number of plates is required for a given separation at conditions of total reflux, and secondly that there is a minimum reflux ratio below which it is impossible to obtain the desired enrichment, however many plates are used.

### Calculation of the minimum reflux ratio

Figure 11.17 represents conditions where the $q$-line is vertical, and the point E lies on the equilibrium curve and has co-ordinates $(x_f, y_f)$. The slope of the line AE is then given by:

$$\left(\frac{R_m}{R_m + 1}\right) = \left(\frac{x_d - y_f}{x_d - x_f}\right)$$

or:

$$R_m = \left(\frac{x_d - y_f}{y_f - x_f}\right) \tag{11.49}$$

If the $q$-line is horizontal as shown in Figure 11.18, the enrichment line for minimum reflux is given by AC, where C has coordinates $(x_c, y_c)$. Thus:

$$\left(\frac{R_m}{R_m + 1}\right) = \left(\frac{x_d - y_c}{x_d - x_c}\right)$$

or, since $y_c = x_f$:

$$R_m = \left(\frac{x_d - y_c}{y_c - x_c}\right) = \left(\frac{x_d - x_f}{x_f - x_c}\right) \tag{11.50}$$

### Underwood and Fenske equations

For ideal mixtures, or where over the concentration range concerned the relative volatility may be taken as constant, $R_m$ may be obtained analytically from the physical properties of the system as discussed by UNDERWOOD[28]. Thus, if $x_{nA}$ and $x_{nB}$ are the mole fractions of two components **A** and **B** in the liquid on any plate $n$, then a material balance over the top portion of the column above plate $n$ gives:

$$V_n y_{nA} = L_n x_{(n+1)A} + D x_{dA} \tag{11.51}$$

and:

$$V_n y_{nB} = L_n x_{(n+1)B} + D x_{dB} \tag{11.52}$$

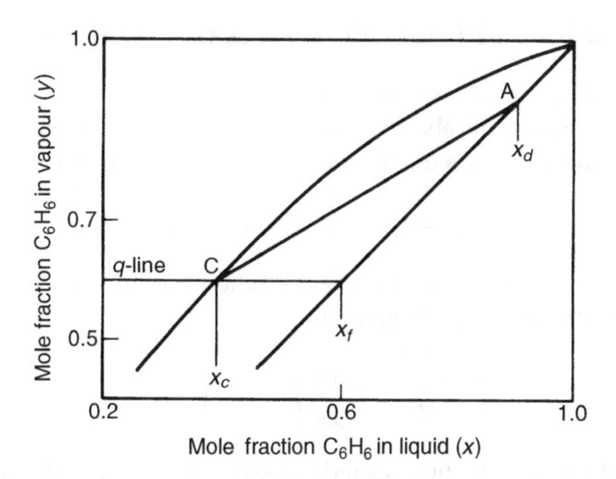


Figure 11.18. Minimum reflux ratio with feed as saturated vapour

Under conditions of minimum reflux, a column has to have an infinite number of plates, or alternatively the composition on plate $n$ is equal to that on plate $n+1$. Dividing equation 11.51 by equation 11.52 and using the relations $x_{(n+1)A} = x_{nA}$ and $x_{(n+1)B} = x_{nB}$, then:

$$\frac{\alpha x_{nA}}{x_{nB}} = \frac{y_{nA}}{y_{nB}} = \frac{L_n x_{nA} + D x_{dA}}{L_n x_{nB} + D x_{dB}}$$

Thus:
$$R_m = \left(\frac{L_n}{D}\right)_{\min} = \frac{1}{\alpha - 1}\left[\frac{x_{dA}}{x_{nA}} - \alpha\left(\frac{x_{dB}}{x_{nB}}\right)\right] \tag{11.53}$$

In this analysis, $\alpha$ is taken as the volatility of **A** relative to **B**. There is, in general, therefore a different value of $R_m$ for each plate. In order to produce any separation of the feed, the minimum relevant value of $R_m$ is that for the feed plate, so that the minimum reflux ratio for the desired separation is given by:

$$R_m = \frac{1}{(\alpha - 1)}\left[\frac{x_{dA}}{x_{fA}} - \alpha\frac{x_{dB}}{x_{fB}}\right] \tag{11.54}$$

For a binary system, this becomes:

$$R_m = \frac{1}{(\alpha - 1)}\left[\frac{x_{dA}}{x_{fA}} - \alpha\frac{(1 - x_{dA})}{(1 - x_{fA})}\right] \tag{11.55}$$

This relation may be obtained by putting $y = \alpha x/[1 + (\alpha - 1)x]$ from equation 11.15, in equation 11.49 to give:

$$R_m = \frac{x_d - \left(\dfrac{\alpha x_f}{1 + (\alpha - 1)x_f}\right)}{\left(\dfrac{\alpha x_f}{1 + (\alpha - 1)x_f}\right) - x_f} = \frac{1}{(\alpha - 1)}\left[\frac{x_d}{x_f} - \frac{\alpha(1 - x_d)}{(1 - x_f)}\right] \tag{11.56}$$

### *The number of plates at total reflux. Fenske's method*

For conditions in which the relative volatility is constant, FENSKE[29] derived an equation for calculating the required number of plates for a desired separation. Since no product is withdrawn from the still, the equations of the two operating lines become:

$$y_n = x_{n+1} \quad \text{and} \quad y_m = x_{m+1} \tag{11.57}$$

If for two components **A** and **B**, the concentrations in the still are $x_{sA}$ and $x_{sB}$, then the composition on the first plate is given by:

$$\left(\frac{x_A}{x_B}\right)_1 = \left(\frac{y_A}{y_B}\right)_s = \alpha_s\left(\frac{x_A}{x_B}\right)_s$$

where the subscript outside the bracket indicates the plate, and $s$ the still.

For plate 2: $$\left(\frac{x_A}{x_B}\right)_2 = \left(\frac{y_A}{y_B}\right)_1 = \alpha_1\left(\frac{x_A}{x_B}\right)_1 = \alpha_1\alpha_s\left(\frac{x_A}{x_B}\right)_s$$

and for plate $n$:

$$\left(\frac{x_A}{x_B}\right)_n = \left(\frac{y_A}{y_B}\right)_{n-1} = \alpha_1\alpha_2\alpha_3\ldots\alpha_{n-1}\alpha_s\left(\frac{x_A}{x_B}\right)_s$$

If an average value of $\alpha$ is used, then:

$$\left(\frac{x_A}{x_B}\right)_n = \alpha_{av}^n\left(\frac{x_A}{x_B}\right)_s$$

In most cases total condensation occurs in the condenser, so that:

$$\left(\frac{x_A}{x_B}\right)_d = \left(\frac{y_A}{y_B}\right)_n = \alpha_n\left(\frac{x_A}{x_B}\right)_n = \alpha_{av}^{n+1}\left(\frac{x_A}{x_B}\right)_s$$

$$n+1 = \frac{\log\left[\left(\frac{x_A}{x_B}\right)_d\left(\frac{x_B}{x_A}\right)_s\right]}{\log\alpha_{av}} \tag{11.58}$$

and $n$ is the required number of theoretical plates in the column.

It is important to note that, in this derivation, only the relative volatilities of two components have been used. The same relation may be applied to two components of a multicomponent mixture, as is seen in Section 11.7.6.

## Example 11.9

For the separation of a mixture of benzene and toluene, considered in Example 11.7, $x_d = 0.9$, $x_w = 0.1$, and $x_f = 0.4$. If the mean volatility of benzene relative to toluene is 2.4, what is the number of plates required at total reflux?

## Solution

The number of plates at total reflux is given by:

$$n + 1 = \frac{\log\left[\left(\frac{0.9}{0.1}\right)\left(\frac{0.9}{0.1}\right)\right]}{\log 2.4} = 5.0 \qquad \text{(equation 11.58)}$$

Thus the number of theoretical plates in the column is $\underline{\underline{4}}$, a value which is independent of the feed composition.

If the feed is liquid at its boiling point, then the minimum reflux ratio $R_m$ is given by:

$$R_m = \frac{1}{\alpha - 1}\left[\frac{x_d}{x_f} - \alpha\frac{(1 - x_d)}{(1 - x_f)}\right] \qquad \text{(equation 11.56)}$$

$$= \frac{1}{2.4 - 1}\left[\frac{0.9}{0.4} - \frac{(2.4 \times 0.1)}{0.6}\right]$$

$$= \underline{\underline{1.32}}$$

Using the graphical construction shown in Figure 11.18, with $y_f = 0.61$, the value of $R_m$ is:

$$R_m = \frac{x_d - y_f}{y_f - x_f} = \frac{(0.9 - 0.61)}{(0.61 - 0.4)} = \underline{\underline{1.38}}$$

### *Selection of economic reflux ratio*

The cost of a distillation unit includes the capital cost of the column, determined largely by the number and diameter of the plates, and the operating costs, determined by the steam and cooling water requirements. The depreciation charges may be taken as a percentage of the capital cost, and the two together taken as the overall charges. The steam required will be proportional to $V_m$, which may be taken as $V_n$ where the feed is liquid at its boiling point. From a material balance over the top portion of the column, $V_n = D(R + 1)$, and hence the steam required per mole of product is proportional to $(R + 1)$. This will be a minimum when $R$ equals $R_m$, and will steadily rise as $R$ is increased. The relationship between the number of plates $n$ and the reflux ratio $R$, as derived by GILLILAND[30], is discussed in Section 11.7.7.

The reduction in the required number of plates as $R$ is increased beyond $R_m$ will tend to reduce the cost of the column. For a column separating a benzene–toluene mixture, for example, where $x_f = 0.79$, $x_d = 0.99$ and $x_w = 0.01$, the numbers of theoretical plates as given by the McCabe–Thiele method for various values of $R$ are given as follows. The minimum reflux ratio for this case is 0.81.

| Reflux ratio $R$ | 0.81 | 0.9 | 1.0 | 1.1 | 1.2 |
|---|---|---|---|---|---|
| Number of plates | $\infty$ | 25 | 22 | 19 | 18 |

Thus, an increase in $R$, at values near $R_m$, gives a marked reduction in the number of plates, although at higher values of $R$, further increases have little effect on the number of plates. Increasing the reflux ratio from $R_m$ therefore affects the capital and operating costs of a column as follows:

(a) The operating costs rise and are approximately proportional to $(R + 1)$.
(b) The capital cost initially falls since the number of plates falls off rapidly at this stage.
(c) The capital cost rises at high values of $R$, since there is then only a very small reduction in the number of plates, although the diameter, and hence the area, continually increases because the vapour load becomes greater. The associated condenser and reboiler will also be larger and hence more expensive.

The total charges may be obtained by adding the fixed and operating charges as shown in Figure 11.19, where curve A shows the steam costs and B the fixed costs. The final total is shown by curve C which has a minimum value corresponding to the economic reflux ratio. There is no simple relation between $R_m$ and the optimum value, although practical values are generally 1.1–1.5 times the minimum, with much higher values being employed, particularly in the case of vacuum distillation. It may be noted that, for a fixed degree of enrichment from the feed to the top product, the number of trays required increases rapidly as the difficulty of separation increases, that is as the relative volatility approaches unity. A demand for a higher purity of product necessitates a very considerable increase in the number of trays, particularly when $\alpha$ is near unity. In these circumstances only a limited improvement in product purity may be obtained by increasing the reflux ratio. The designer must be careful to consider the increase in cost of plant resulting from specification of a higher degree of purity of production and at the same time assess the highest degree of purity that may be obtained with the proposed plant.

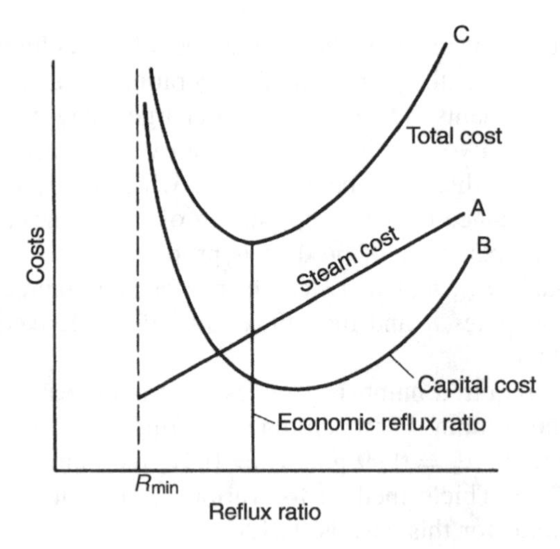


Figure 11.19. Influence of reflux ratio on capital and operating costs of a still

In general, the greater the reflux ratio, the lower is the number of plates or transfer units required although the requirements of steam in the reboiler and cooling water in the condenser are both increased and a column of larger diameter is required in order to achieve acceptable vapour velocities. An optimum value of the reflux ratio may be obtained by using the following argument which is based on the work of COLBURN[31].

The annual capital cost of a distillation column, $c_c$ per mole of distillate, including depreciation, interest and overheads, may be written as:

$$c_c = c_a An/(Et_a D) \tag{11.59}$$

where $c_a$ is the annual cost of the column per unit area of plate, $A$ is the cross-sectional area of the column, $n$ is the number of theoretical plates, $E$ is the plate efficiency, $t_a$ is the annual period of operation and $D$ is the molar flowrate of distillate. The cross-sectional area of the column is given by:

$$A = V/u' \tag{11.60}$$

where $V$ is the molar flow of vapour and $u'$ is the allowable molar vapour velocity per unit area. Since $V = D(R+1)$, where $R$ is the reflux ratio, then the cost of the column is:

$$c_c = c_a n(R+1)/(Et_a u') \tag{11.61}$$

The annual cost of the reboiler and the condenser, $c_h$ per mole of distillate may be written as:

$$c_h = c_b A_h/(t_a D) \tag{11.62}$$

where $c_b$ is the annual cost of the heat exchange equipment per unit area including depreciation and interest and $A_h$ is the area for heat transfer. $A_h = V/N''$ where $N''$ is the vapour handling capacity of the boiler and condenser in terms of molar flow per unit area. Thus $A_h = D(R+1)/N''$ and the cost of the reboiler and the condenser is:

$$c_h = c_b(R+1)/(t_a N'') \tag{11.63}$$

As far as operating costs are concerned, the important annual variable costs are that of the steam in the reboiler and that of the cooling water in the condenser. These may be written as:

$$c_w = c_d V/D = c_3(R+1) \tag{11.64}$$

where $c_d$ is the annual cost of the steam and the cooling water. The total annual cost, $c$ per mole of distillate, is the cost of the steam and the cooling water plus the costs of the column, reboiler and condenser, or:

$$c = (R+1)[(c_a n/Et_a u') + (c_b/t_a N'') + c_d] \tag{11.65}$$

As the number of plates, $n$, is a function of $R$, equation 11.65 may be differentiated with respect to $R$ to give:

$$\mathrm{d}c/\mathrm{d}R = c_a n/(Et_a V') + [(c_a/(Et_a u'))(R+1)\,\mathrm{d}n/\mathrm{d}R] + c_b/(t_a N'') + c_d \tag{11.66}$$

Equating to zero for minimum cost, the optimum value of the reflux ratio is:

$$R_{\text{opt}} + 1 = (n_{\text{opt}} + F)/(-\mathrm{d}n/\mathrm{d}R) \tag{11.67}$$

where $n_{\text{opt}}$ is the optimum number of theoretical plates corresponding to $R_{\text{opt}}$ and the cost factor, $F$, is:

$$F = [c_d + c_b/(t_a N'')][(Et_a u')/c_a] \tag{11.68}$$

Because there is no simple equation relating $n$ and $dn/dR$, it is not possible to obtain an expression for $R_{opt}$ although a method of solution is given in the Example 11.18 which is based on the work of HARKER[32].

In practice, values of 110–150 per cent of the minimum reflux ratio are used although higher values are sometimes employed particularly in vacuum distillation. Where a high purity product is required, only limited improvements can be obtained by increasing the reflux ratio and since there is a very large increase in the number of trays required, an arrangement by which the minimum acceptable purity is achieved in the product is usually adopted.

### 11.4.4. Location of feed point in a continuous still

From Figure 11.20 it may be seen that, when stepping off plates down the top operating line AB, the bottom operating line CE cannot be used until the value of $x_n$ on any plate is less than $x_e$. Again it is essential to pass to the lower line CE by the time $x_n = x_b$. The best conditions are those where the minimum number of plates is used. From the geometry of the figure, the largest steps in the enriching section occur down to the point of intersection of the operating lines at $x = x_q$. Below this value of $x$, the steps are larger on the lower operating line. Thus, although the column will operate for a feed composition between $x_e$ and $x_b$, the minimum number of plates will be required if $x_f = x_q$. For a binary mixture at its boiling point, this is equivalent to making $x_f$ equal to the composition of the liquid on the feed plate.

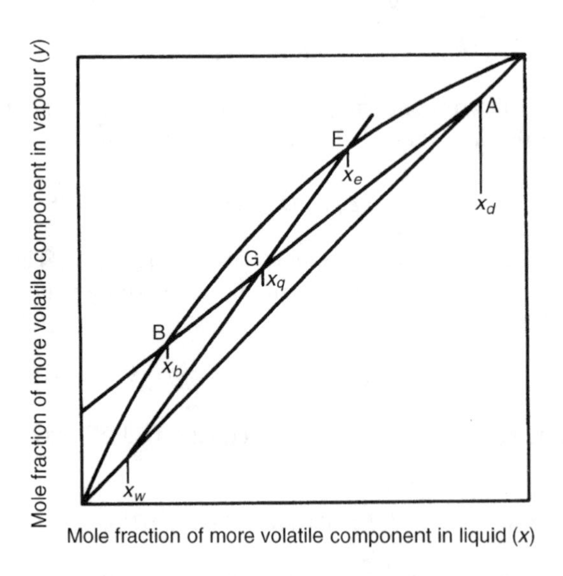


Figure 11.20. Location of feed point

### 11.4.5. Multiple feeds and sidestreams

In general, a sidestream is defined as any product stream other than the overhead product and the residue such as the streams S′, S″, and S‴ in Figure 11.21. In a similar way,

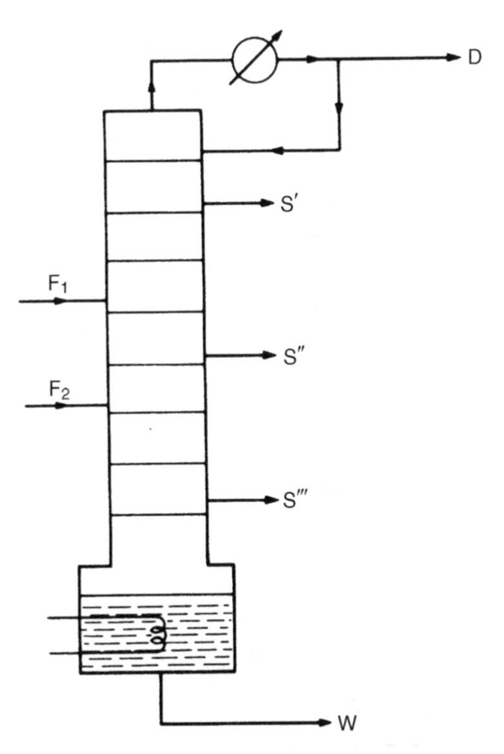


Figure 11.21. Column with multiple feeds and sidestreams

$F_1$ and $F_2$ are separate feed streams to the column. Sidestreams are most often removed with multicomponent systems, although they may be used with binary mixtures. A binary system is now considered, with one sidestream, as shown in Figure 11.22. $S'$ represents the rate of removal of the sidestream and $x_{s'}$ its composition.

Assuming constant molar overflow, then for the part of the column above the sidestream the operating line is given by:

$$y_n = \frac{L_n}{V_n}x_{n+1} + \frac{Dx_d}{V_n} \quad \text{(equation 11.35)}$$

as before. Balances for the part of the tower above a plate between the feed plate and the sidestream give:

$$V_s = L_s + S' + D \tag{11.69}$$

and:

$$V_s y_n = L_s x_{n+1} + S'x_{s'} + Dx_d \tag{11.70}$$

Thus:

$$y_n = \frac{L_s}{V_s}x_{n+1} + \frac{S'x_{s'} + Dx_d}{V_s} \tag{11.71}$$

Since the sidestream is normally removed as a liquid, $L_s = (L_n - S')$ and $V_s = V_n$.

The line represented by equation 11.35 has a slope $L_n/V_n$ and passes through the point $(x_d, x_d)$. Equation 11.71 represents a line of slope $L_s/V_s$, which passes through the point

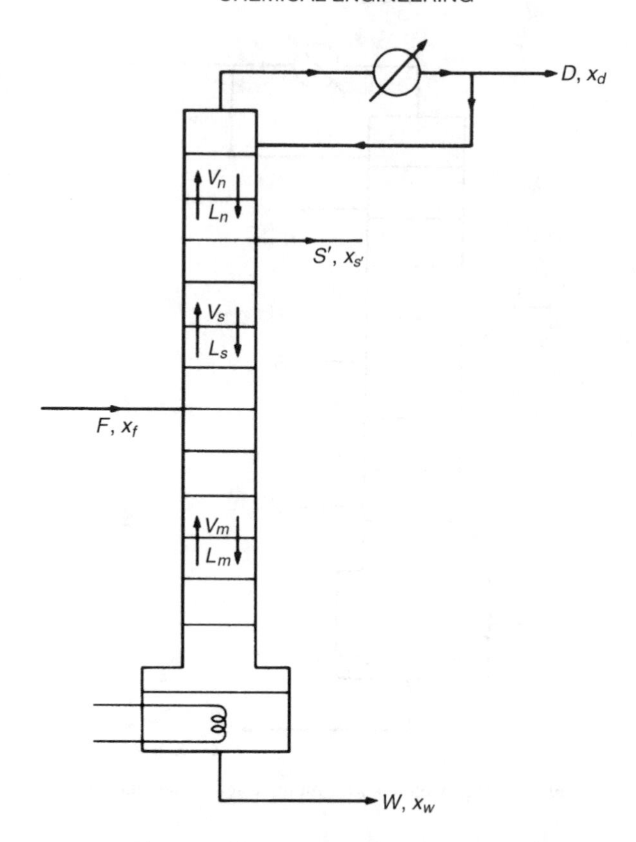


Figure 11.22. Column with a sidestream

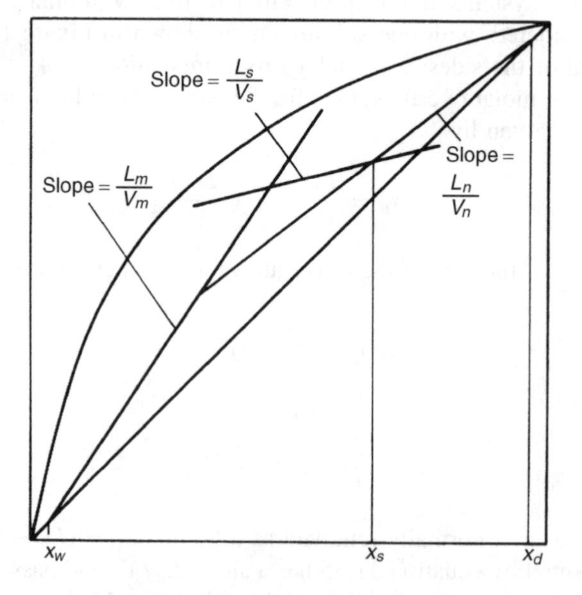


Figure 11.23. Effect of a sidestream

$y = x = (S'x_{s'} + Dx_d)/(S' + D)$, which is the mean molar composition of the overhead product and sidestream. Since $x_{s'} < x_d$, and $L_s < L_n$, this additional operating line cuts the $y = x$ line at a lower value than the upper operating line though it has a smaller slope, as shown in Figure 11.23. The two lines intersect at $x = x_{s'}$. Plates are stepped off as before between the appropriate operating line and the equilibrium curve. It may be seen that the removal of a sidestream increases the number of plates required, due to the decrease in liquid rate below the sidestream.

The effect of any additional sidestream or feed is to introduce an additional operating line for each stream. In all other respects the method of calculation is identical with that used for the straight separation of a binary mixture.

The Ponchon–Savarit method, using an enthalpy–composition diagram, may also be used to handle sidestreams and multiple feeds, though only for binary systems. This is dealt with in Section 11.5.

## 11.5. CONDITIONS FOR VARYING OVERFLOW IN NON-IDEAL BINARY SYSTEMS

### 11.5.1. The heat balance

In previous sections, the case of constant molar latent heat has been considered with no heat of mixing, and hence a constant molar rate of reflux in the column. These simplifying assumptions are extremely useful in that they enable a simple geometrical method to be used for finding the change in concentration on the plates and, whilst they are rarely entirely true in industrial conditions, they often provide a convenient start for design purposes. For a non-ideal system, where the molar latent heat is no longer constant and where there is a substantial heat of mixing, the calculations become much more tedious. For binary mixtures of this kind a graphical model has been developed by RUHEMANN[33], PONCHON[34], and SAVARIT[35], based on the use of an enthalpy–composition chart. A typical enthalpy–composition or $H - x$ chart is shown in Figure 11.24, where the upper curve V is the dew-point curve, and the lower curve L the boiling-point curve. The use of this diagram is based on the geometrical properties, as illustrated in Figure 11.25. A quantity of mixture in any physical state is known as a "phase" and is denoted by mass, composition and enthalpy. The phase is shown upon the diagram by a point which shows enthalpy and composition, though it does not show the mass. If $m$ is the mass, $x$ the composition and $H$ the enthalpy per unit mass, then the addition of two phases A and B to give phase C is governed by:

$$m_A + m_B = m_C \tag{11.72}$$

$$m_A x_A + m_B x_B = m_C x_C \tag{11.73}$$

and:

$$m_A H_A + m_B H_B = m_C H_C \tag{11.74}$$

Similarly, if an amount $Q$ of heat is added to a mass $m_A$ of a phase, the increase in enthalpy from $H_A$ to $H_C$ will be given by:

$$H_A + \frac{Q}{m_A} = H_C \tag{11.75}$$

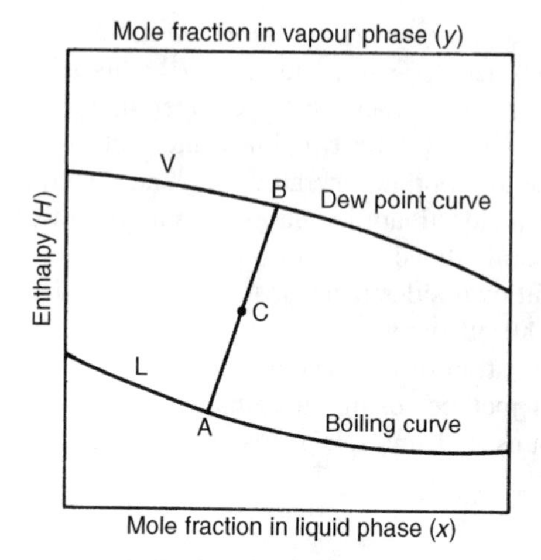


Figure 11.24. Enthalpy–composition diagram, showing the enthalpies of liquid and vapour

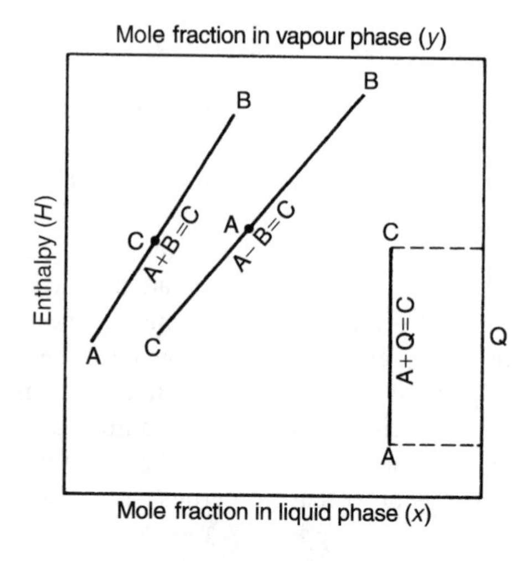


Figure 11.25. Combination and separation of a mixture on an enthalpy–composition diagram

Thus, the addition of two phases A and B is shown on the diagram by point C on the straight line joining the two phases, whilst the difference (A − B) is found by a point C on the extension of the line AB. If, as shown in Figure 11.24, a phase represented by C in the region between the dew-point and boiling-point curves is considered, then this phase will divide into two phases A and B at the ends of a tie line through the point C, so that:

$$\frac{m_A}{m_B} = \frac{\text{CB}}{\text{CA}} \tag{11.76}$$

The $H - x$ chart, therefore, enables the effect of adding two phases, with or without the addition of heat, to be determined geometrically. The diagram may be drawn for unit mass or for one mole of material, although as a constant molar reflux does not now apply, it is more convenient to use unit mass as the basis. Thus, working with unit mass of product, the mass of the individual streams as proportions of the product are calculated.

Figure 11.26 represents a continuous distillation unit operating with a feed F of composition $x_f$, and giving a top product D of composition $x_d$ and a bottom product W of composition $x_w$. In this analysis, the quantities in the streams V of rising vapour and L of reflux are given in mass units, such as kg/s, and the composition of the streams as mass fractions, $x$ referring to the liquid and $y$ to the vapour streams as usual.

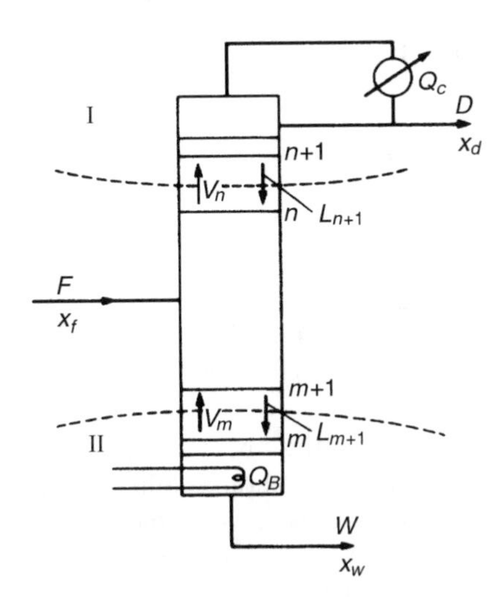


Figure 11.26. Continuous distillation column

The plates are numbered from the bottom upwards, subscript $n$ indicating the rectifying and $m$ the stripping section.

$H^V$ and $H^L$ represent the enthalpy per unit mass of a vapour and liquid stream respectively.

$Q_C$ is the heat removed in the condenser. In this case no cooling of product is considered.

$Q_B$ is the heat added in the boiler.

The following relationships are then obtained by taking material and heat balances:

$$V_n = L_{n+1} + D$$

or:
$$V_n - L_{n+1} = D \tag{11.77}$$

$$V_n y_n = L_{n+1} x_{n+1} + D x_d$$

or:
$$V_n y_n - L_{n+1} x_{n+1} = D x_d \tag{11.78}$$

$$V_n H_n^V = L_{n+1} H_{n+1}^L + D H_d^L + Q_c$$

or:
$$V_n H_n^V - L_{n+1} H_{n+1}^L = D H_d^L + Q_c \tag{11.79}$$

Putting $H'_d = H^L_d + Q_C/D$, then equation 11.79 may be written as:

$$V_n H^V_n = L_{n+1} H^L_{n+1} + D H'_d$$

or:
$$V_n H^V_n - L_{n+1} H^L_{n+1} = D H'_d \qquad (11.80)$$

From equations 11.77 and 11.78:

$$\frac{L_{n+1}}{D} = \frac{x_d - y_n}{y_n - x_{n+1}} \qquad (11.81)$$

and from equations 11.77 and 11.80:

$$\frac{L_{n+1}}{D} = \frac{H'_d - H^V_n}{H^V_n - H^L_{n+1}} \qquad (11.82)$$

or:
$$\frac{H'_d - H^V_n}{H^V_n - H^L_{n+1}} = \frac{x_d - y_n}{y_n - x_{n+1}} \qquad (11.83)$$

and:
$$y_n = \left[\frac{H'_d - H^V_n}{H'_d - H^L_{n+1}}\right] x_{n+1} + \left[\frac{H^V_n - H^L_{n+1}}{H'_d - H^L_{n+1}}\right] x_d \qquad (11.84)$$

Equation 11.84 represents any operating line relating the composition of the vapour $y_n$ rising from a plate to the composition of the liquid reflux entering the plate, or alternatively it represents the relation between the composition of the vapour and liquid streams between any two plates. From equation 11.83, it may be seen that all such operating lines pass through a common pole N of coordinates $x_d$ and $H'_d$.

Alternatively, noting that the right-hand side of equations 11.77, 11.78 and 11.79 are independent of conditions below the feed plate, a stream N may be defined with mass equal to the difference between the vapour and liquid streams between two plates, of composition $x_d$ and of enthalpy $H'_d$. The three quantities $V_n$, $L_{n+1}$, and N are then on a straight line passing through N, as shown in Figure 11.27.

Below the feed plate a similar series of equations for material and heat balances may be written as:

$$V_m + W = L_{m+1}$$

or:
$$-V_m + L_{m+1} = W \qquad (11.85)$$

$$V_m y_m + W x_w = L_{m+1} x_{m+1}$$

or:
$$-V_m y_m + L_{m+1} x_{m+1} = W x_w \qquad (11.86)$$

$$V_m H^V_m + W H^L_w = L_{m+1} H^L_{m+1} + Q_B$$

or:
$$-V_m H^V_m + L_{m+1} H^L_{m+1} = W H^L_w - Q_B \qquad (11.87)$$

Putting:
$$H'_w = H^L_w - \frac{Q_B}{W} \qquad (11.88)$$

then:
$$-V H^V_m + L_{m+1} H^L_{m+1} = W H'_w \qquad (11.89)$$

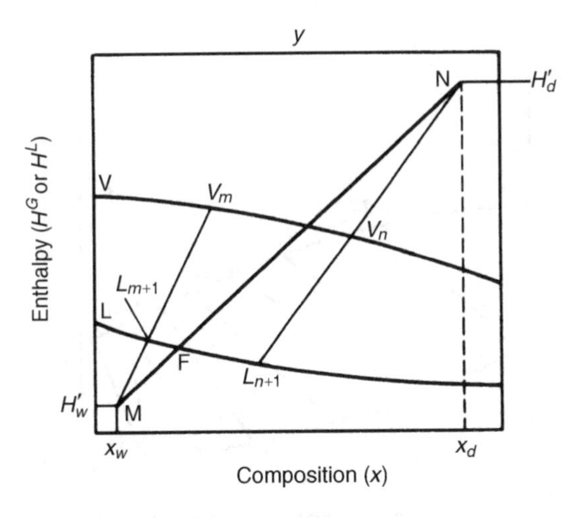


Figure 11.27. Enthalpy–composition diagram

Then:

$$\frac{L_{m+1}}{W} = \frac{-x_w + y_m}{y_m - x_{m+1}} \tag{11.90}$$

and:

$$\frac{L_{m+1}}{W} = \frac{-H'_w + H^V_m}{H^V_m - H^L_{m+1}} \tag{11.91}$$

Thus:

$$\frac{-H'_w + H^V_m}{H^V_m - H^L_{m+1}} = \frac{-x_w + y_m}{y_m - x_{m+1}} \tag{11.92}$$

Equation 11.92 represents any operating line below the feed plate, and it shows that all such lines pass through a common pole M of coordinates $x_w$ and $H'_w$. As with the rectifying section, a stream M may be defined by mass $L_{m+1} - V_m$, composition $x_w$ and enthalpy $H'_w$. Thus:

$$F = M + N \tag{11.93}$$

and:

$$Fx_f = Mx_w + Nx_d \tag{11.94}$$

It therefore follows that phases F, M, and N are on a straight line on the $H - x$ chart, as shown in Figures 11.27 and 11.28.

## 11.5.2. Determination of the number of plates on the *H* − *x* diagram

The determination of the number of plates necessary for a desired separation is shown in Figure 11.28. The position of the feed $(F, x_f)$ is shown at F on the boiling line and the pole N is located as $(x_d, H'_d)$, where:

$$H'_d = H^L_d + \frac{Q_C}{D} \tag{11.95}$$

Pole M is located as on the extension of NF cutting the ordinate at $x_w$ in M.

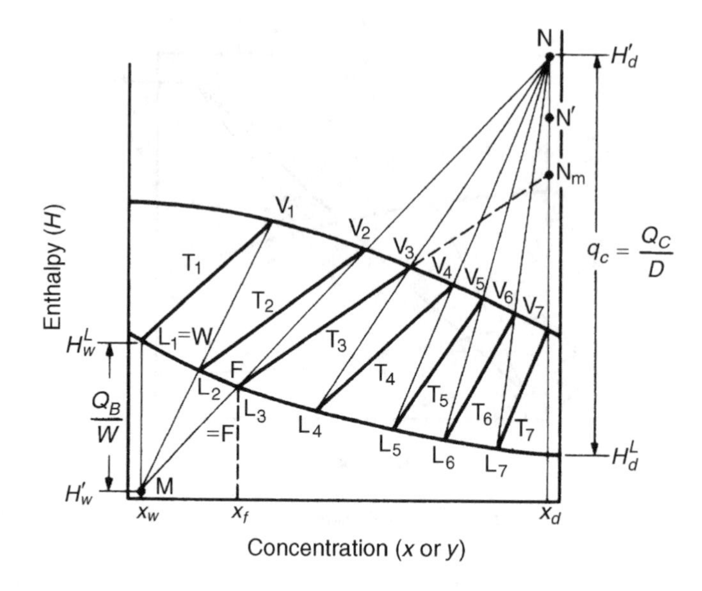


Figure 11.28. Determination of the number of plates using the enthalpy–composition diagram

The condition of the vapour leaving the top plate is shown at $V_7$ on the dew-point curve with abscissa $x_d$. The condition of the liquid on the top plate is then found by drawing the tie line $T_7$ from $V_7$ to $L_7$ on the boiling curve. The condition $V_6$ of the vapour on the second plate is found, from equation 11.77, by drawing $L_7N$ to cut the dew-point curve on $V_6$. $L_6$ is then found on the tie line $T_6$. The conditions of vapour and liquid $V_5$, $V_4$, $V_3$ and $L_5$, $L_4$ are found in the same way. Tie line $T_3$ gives $L_3$, which has the same composition as the feed. $V_2$ is then found using the line $MFV_2$, as this represents the vapour on the top plate of the stripping section. $L_2$, $L_1$ and $V_1$ are then found by a similar construction. $L_1$ has the required composition of the bottoms, $x_w$.

Alternatively, calculations may start with the feed condition and proceed up and down the column.

### 11.5.3. Minimum reflux ratio

The pole N has coordinates $[x_d, H_d^L + Q_C/D]$. $Q_C/D$ is the heat removed in the condenser per unit mass of product, as liquid at its boiling point and is represented as shown in Figure 11.28. The number of plates in the rectifying section is determined, for a given feed $x_f$ and product $x_d$, by the height of this pole N. As N is lowered to say N′ the heat $q_c$ falls, although the number of plates required increases. When N lies at $N_m$ on the isothermal through F, $q_c$ is a minimum although the number of plates required becomes infinite. Since the tie lines have different slopes, it follows that there is a minimum reflux for each plate, and the tie line cutting the vertical axis at the highest value of H will give the minimum practical reflux. This will frequently correspond to the tie line through F.

From equations 11.83 and 11.95 and writing $Q_C/D = q_c$, then:

$$\frac{H_d^L + q_c - H_n^V}{H_n^V - H_{n+1}^L} = \frac{x_d - y_n}{y_n - x_{n+1}} \tag{11.96}$$

or:
$$q_c = (H_n^V - H_{n+1}^L)\left(\frac{x_d - y_n}{y_n - x_{n+1}}\right) + H_n^V - H_d^L \tag{11.97}$$

and:
$$(q_c)_{\min} = (H_f^V - H_{f+1}^L)\left(\frac{x_d - y_f}{y_f - x_{f+1}}\right) + H_f^V - H_d^L \tag{11.98}$$

The advantage of the $H - x$ chart lies in the fact that the heat quantities required for the distillation are clearly indicated. Thus, the higher the reflux ratio the more heat must be removed per mole of product, and point N rises. This immediately shows that both $q_c$ and $Q_B$ are increased. The use of this method is illustrated by considering the separation of ammonia from an ammonia–water mixture, as occurs in the ammonia absorption unit for refrigeration.

## Example 11.10

It is required to separate 1 kg/s (3.6 tonnes/h) of a solution of ammonia in water, containing 30 per cent by mass of ammonia, to give a top product of 99.5 per cent purity and a weak solution containing 10 per cent by mass of ammonia.

Calculate the heat required in the boiler and the heat to be rejected in the condenser, assuming a reflux 8 per cent in excess of the minimum and a column pressure of 1000 kN/m$^2$. The plates may be assumed to have an ideal efficiency of 60 per cent.

## Solution

Taking a material balance for the whole throughput and for the ammonia gives:

$$D + W = 1.0$$

$$0.995D + 0.1W = (1.0 \times 0.3)$$

Thus:
$$D = 0.22 \text{ kg/s}$$

and:
$$W = 0.78 \text{ kg/s}$$

The enthalpy–composition chart for this system is shown in Figure 11.29. It is assumed that the feed F and the bottom product W are both liquids at their boiling points.

### *Location of the poles* N *and* M

$N_m$ for minimum reflux is found by drawing a tie-line through F, representing the feed, to cut the line $x = 0.995$ at $N_m$.

$$\text{The minimum reflux ratio, } R_m = \frac{\text{length } N_mA}{\text{length AL}}$$

$$= \frac{(1952 - 1547)}{(1547 - 295)} = 0.323$$

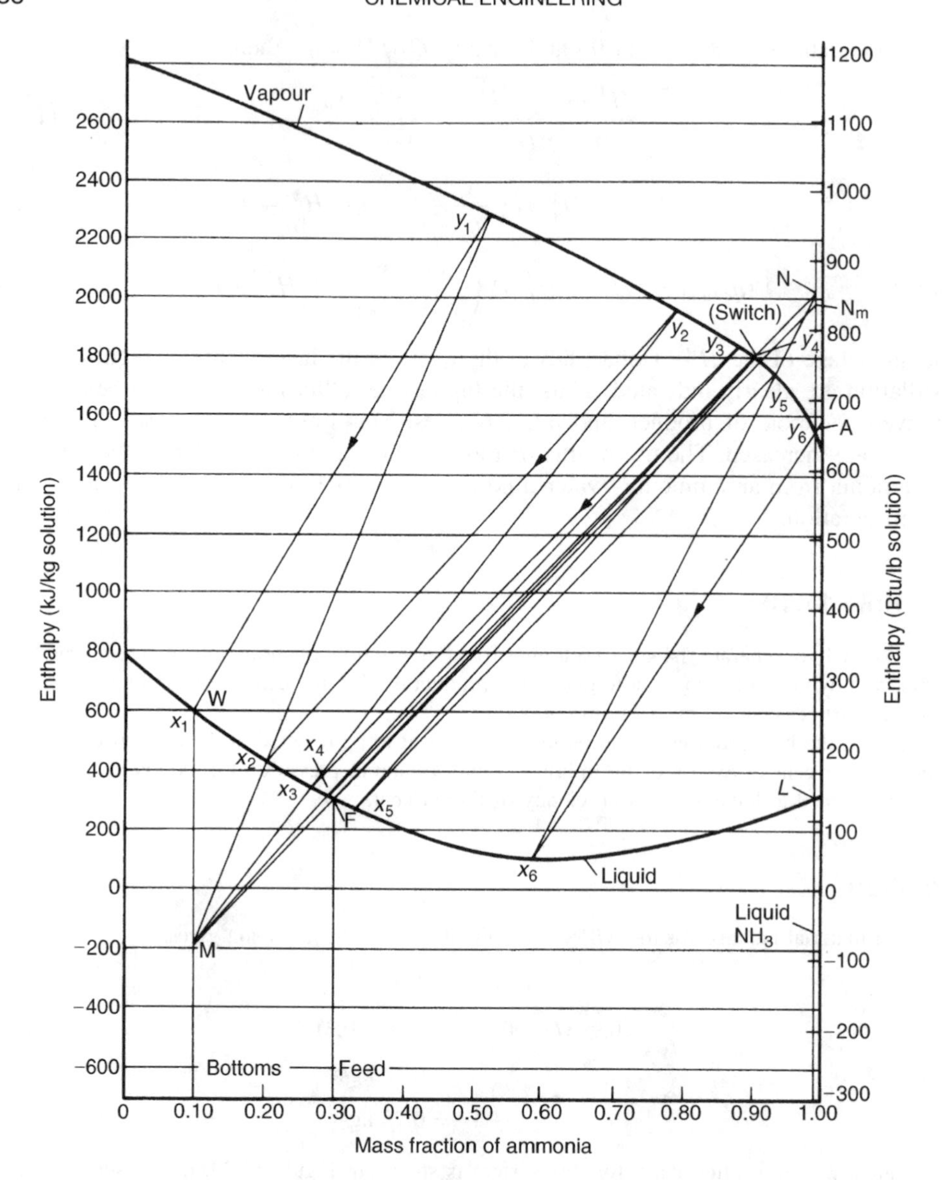


Figure 11.29. Enthalpy–composition diagram for ammonia–water at 1.0 $MN/m^2$ pressure (Example 11.10)

Since the actual reflux is 8 per cent above the minimum, then:

$$NA = 1.08\ N_m A$$

$$= (1.08 \times 405) = 437$$

Point N therefore has an ordinate of $(437 + 1547) = 1984$ and an abscissa of 0.995.
Point M is found by drawing NF to cut the line $x = 0.10$, through W, at M.
The number of theoretical plates is found, as on the diagram, to be 5+.

The number of plates to be provided = (5/0.6) = 8.33, say 9.
The feed is introduced just below the third ideal plate from the top, or just below the fifth actual plate.
The heat input at the boiler per unit mass of bottom product is:

$$\frac{Q_B}{W} = 582 - (-209) = 791$$

$$\text{Heat input to boiler} = (791 \times 0.78) = \underline{\underline{617 \text{ kW}}}$$

$$\begin{aligned}\text{Condenser duty} &= \text{length NL} \times D \\ &= (1984 - 296) \times 0.22 \\ &= \underline{\underline{372 \text{ kW}}}\end{aligned}$$

## 11.5.4. Multiple feeds and sidestreams

The enthalpy–composition approach may also be used for multiple feeds and sidestreams for binary systems. For the condition of constant molar overflow, each additional sidestream or feed adds a further operating line and pole point to the system.

Taking the same system as used in Figure 11.22, with one sidestream only, the procedure is as shown in Figure 11.30.

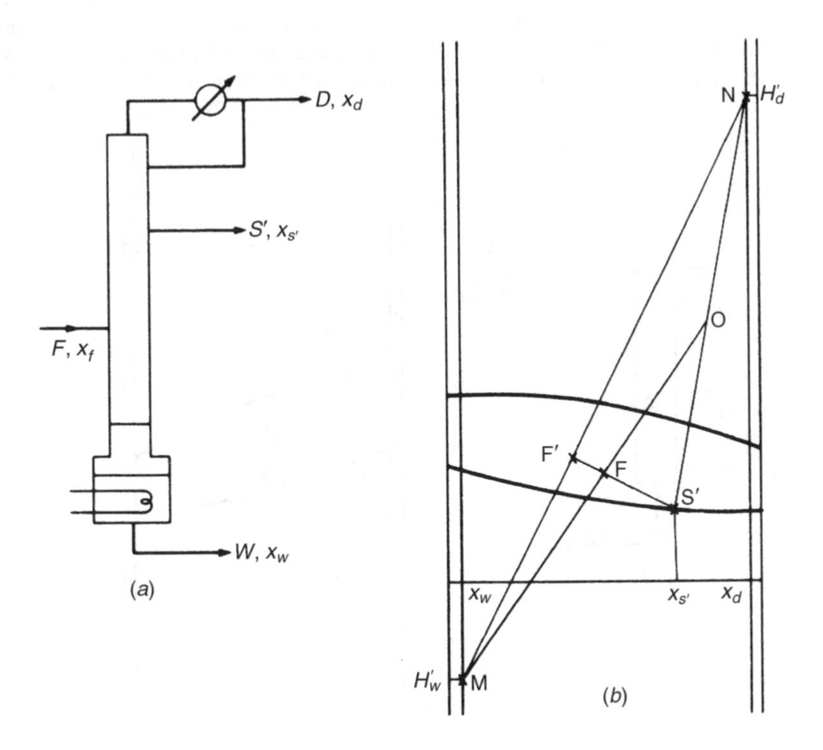


Figure 11.30. Enthalpy–composition diagram for a system with one sidestream

The upper pole point N is located as before. The effect of removing a sidestream $S'$ from the system is to produce an effective feed $F'$, where $F' = F - S'$ and where $F'S'/F'F = F/S'$. Thus, once $S'$ and $F$ have been located in the diagram, the position of $F'$ may also be determined. The position of the lower pole point M, which must lie on the intersection of $x = x_w$ and the straight line drawn through NF′, may then be found. N relates to the section of the column above the sidestream and M to that part below the feed plate. A third pole point must be defined to handle that part of the column between the feed and the sidestream.

The pole point for the intermediate section must be on the limiting operating line for the upper part of the column, that is NS′. This must also lie on the limiting operating line for the lower part of the column, that is MF or its extension. Thus the intersection of NS′ and MF extended gives the position of the intermediate pole point O.

The number of stages required is determined in the same manner as before, using the upper pole point N for that part of the column between the sidestream and the top, the intermediate pole point O between the feed and the sidestream, and the lower pole point M between the feed and the bottom.

For the case of multiple feeds, the procedure is similar and may be followed by reference to Figure 11.31.

## Example 11.11

A mixture containing equal parts by mass of carbon tetrachloride and toluene is to be fractionated to give an overhead product containing 95 mass per cent carbon tetrachloride, a bottom product of 5 mass per cent carbon tetrachloride, and a sidestream containing 80 mass per cent carbon tetrachloride. Both the feed and sidestream may be regarded as liquids at their boiling points.

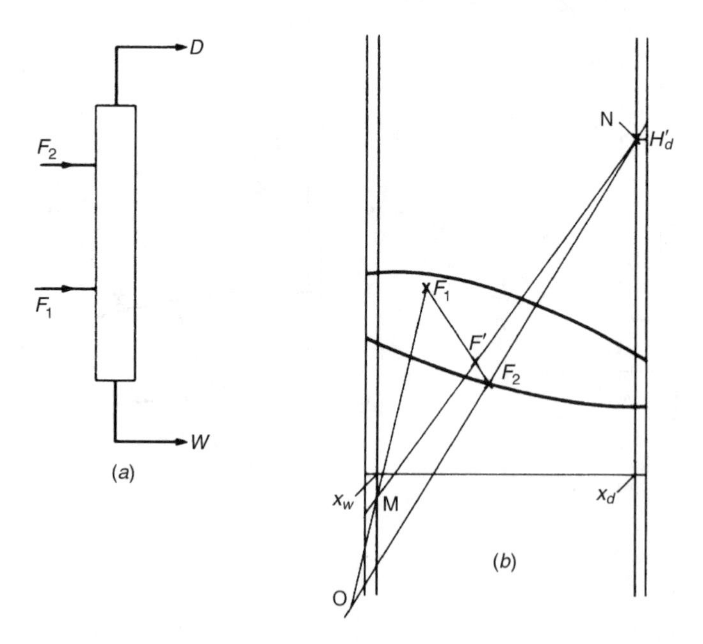


Figure 11.31. Enthalpy–composition diagram for a system with two feeds

The rate of withdrawal of the sidestream is 10 per cent of the column feed rate and the external reflux ratio is 2.5. Using the enthalpy composition method, determine the number of theoretical stages required, and the amounts of bottom product and distillate as percentages of the feed rate.

It may be assumed that the enthalpies of liquid and vapour are linear functions of composition. Enthalpy and equilibrium data are provided.

## Solution

*Basis: 100 kg feed.*

An overall material balance gives:

$$F = D + W + S'$$

or:
$$100 = D + W + 10$$

$$Fx_f = Dx_d + Wx_w + S'x_{s'}$$

and:
$$50 = 0.95D + 0.05W + 8$$

Thus:
$$\underline{\underline{D = 41.7 \text{ per cent}; \quad W = 48.3 \text{ per cent}}}$$

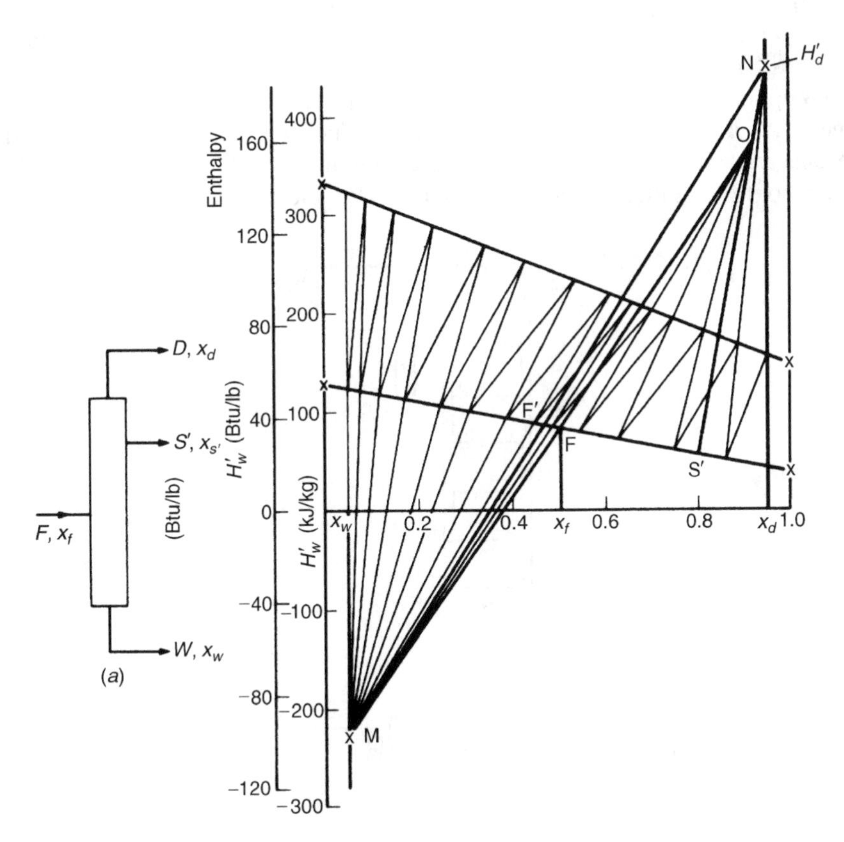


Figure 11.32. Enthalpy–composition diagram for carbon tetrachloride–toluene separation with one sidestream — Example 11.11

From the enthalpy data and the reflux ratio, the upper pole point M may be located as shown in Figure 11.32. Points F and S′ are located on the liquid line, and the position of the effective feed, such that F′S′/F′F = 10. NF′ is joined and extended to cut $x = x_w$ at M, the lower pole point.

MF is Joined and extended to cut NS′ at O, the immediate pole point. The number of stages required is then obtained from the figure and

13 theoretical stages are required.

# 11.6. BATCH DISTILLATION

## 11.6.1. The process

In the previous sections conditions have been considered in which there has been a continuous feed to the still and a continuous withdrawal of products from the top and bottom. In many instances processes are carried out in batches, and it is more convenient to distil each batch separately. In these cases the whole of a batch is run into the boiler of the still and, on heating, the vapour is passed into a fractionation column, as shown in Figure 11.33. As with continuous distillation, the composition of the top product depends on the still composition, the number of plates in the column and on the reflux ratio used. When the still is operating, since the top product will be relatively rich in the more volatile component, the liquid remaining in the still will become steadily weaker in this component. As a result, the purity of the top product will steadily fall. Thus, the still may be charged with $S_1$ mols of a mixture containing a mole fraction $x_{s1}$ of the more volatile component. Initially, with a reflux ratio $R_1$, the top product has a composition

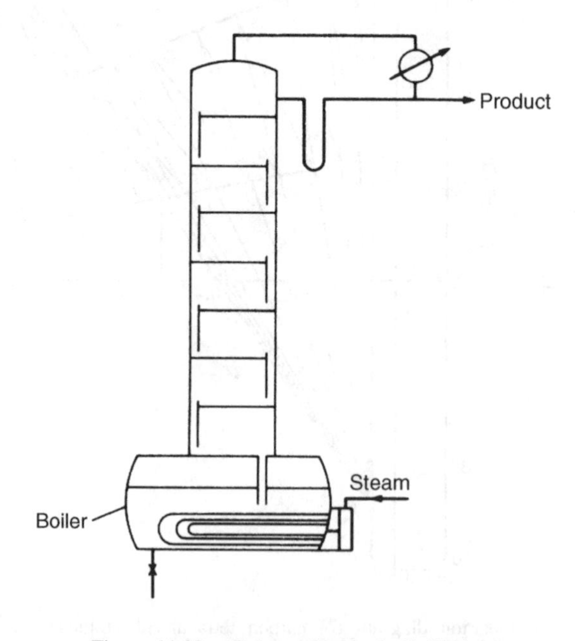


Figure 11.33. Column for batch distillation

$x_{d1}$. If after a certain interval of time the composition of the top product starts to fall, then, if the reflux ratio is increased to a new value $R_2$, it will be possible to obtain the same composition at the top as before, although the composition in the still is weakened to $x_{s2}$. This method of operating a batch still requires a continuous increase in the reflux ratio to maintain a constant quality of the top product.

An alternative method of operation is to work with a constant reflux ratio and allow the composition of the top product to fall. For example, if a product of composition 0.9 with respect to the more volatile component is required, the composition initially obtained may be 0.95, and distillation is allowed to continue until the composition has fallen to some value below 0.9, say 0.82. The total product obtained will then have the required composition, provided the amounts of a given purity are correctly chosen.

One of the added merits of batch distillation is that more than one product may be obtained. Thus, a binary mixture of alcohol and water may be distilled to obtain initially a high quality alcohol. As the composition in the still weakens with respect to alcohol, a second product may be removed from the top with a reduced concentration of alcohol. In this way it is possible to obtain not only two different quality products, but also to reduce the alcohol in the still to a minimum value. This method of operation is particularly useful for handling small quantities of multi-component organic mixtures, since it is possible to obtain the different components at reasonable degrees of purity, in turn. To obtain the maximum recovery of a valuable component, the charge remaining in the still after the first distillation may be added to the next batch.

### 11.6.2. Operation at constant product composition

The case of a column with four ideal plates used to separate a mixture of ethyl alcohol and water may be considered. Initially there are $S_1$ moles of liquor of mole fraction $x_{s1}$ with respect to the more volatile component, alcohol, in the still. The top product is to contain a mole fraction $x_d$, and this necessitates a reflux ratio $R_1$. If the distillation is to be continued until there are $S_2$ moles in the still, of mole fraction $x_{s2}$, then, for the same number of plates the reflux ratio will have been increased to $R_2$. If the amount of product obtained is $D_b$ moles, then a material balance gives:

$$S_1 x_{s1} - S_2 x_{s2} = D_b x_d \tag{11.96}$$

and:

$$S_1 - S_2 = D_b \tag{11.97}$$

Thus:

$$S_1 x_{s1} - (S_1 - D_b) x_{s2} = D_b x_d$$

and:

$$S_1 x_{s1} - S_1 x_{s2} = D_b x_d - D_b x_{s2}$$

and:

$$D_b = S_1 \left[ \frac{x_{s1} - x_{s2}}{x_d - x_{s2}} \right] = \left( \frac{a}{b} \right) S_1 \tag{11.98}$$

where $a$ and $b$ are as shown in Figure 11.34. If $\phi$ is the intercept on the $Y$-axis for any operating line, equation 11.48, then:

$$\frac{x_d}{R+1} = \phi, \quad \text{or} \quad R = \frac{x_d}{\phi} - 1 \tag{11.99}$$

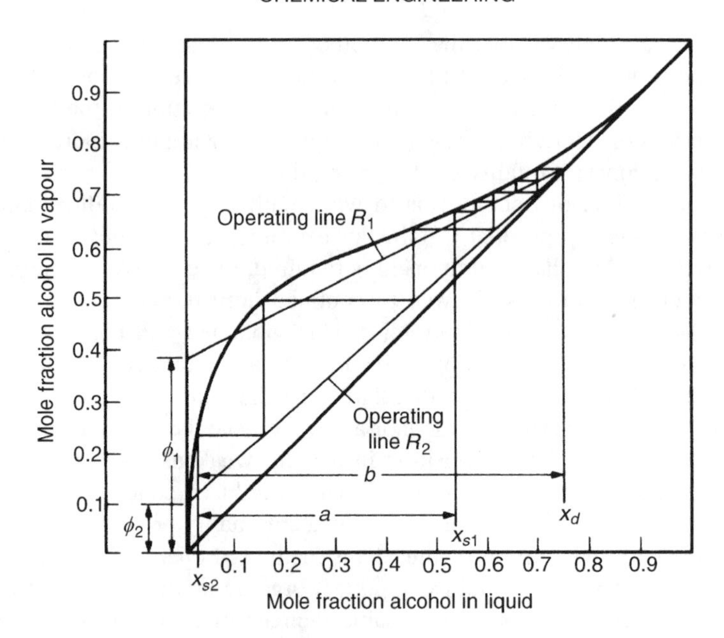


Figure 11.34. Graphical representation of batch distillation

These equations enable the final reflux ratio to be determined for any desired end concentration in the still, and they also give the total quantity of distillate obtained. What is important, in comparing the operation at constant reflux ratio with that at constant product composition, is the difference in the total amount of steam used in the distillation, for a given quantity of product, $D_b$.

If the reflux ratio $R$ is assumed to be adjusted continuously to keep the top product at constant quality, then at any moment the reflux ratio is given by $R = \mathrm{d}L_b/\mathrm{d}D_b$. During the course of the distillation, the total reflux liquor flowing down the column is given by:

$$\int_0^{L_b} \mathrm{d}L_b = \int_{R=R_1}^{R=R_2} R\,\mathrm{d}D_b \tag{11.100}$$

To provide the reflux $\mathrm{d}L_b$ the removal of a quantity of heat equal to $\lambda\,\mathrm{d}L_b$ in the condenser is required, where $\lambda$ is the latent heat per mole. Thus, the heat to be supplied in the boiler to provide this reflux during the total distillation $Q_R$ is given by:

$$Q_R = \lambda \int_0^{L_b} \mathrm{d}L_b = \lambda \int_{R=R_1}^{R=R_2} R\,\mathrm{d}D_b \tag{11.101}$$

This equation may be integrated graphically if the relation between $R$ and $D_b$ is known. For any desired value of $R$, $x_s$ may be obtained by drawing the operating line, and marking off the steps corresponding to the given number of stages. The amount of product $D_b$ is then obtained from equation 11.88 and, if the corresponding values of $R$ and $D_b$ are plotted, graphical integration will give the value of $\int R\,\mathrm{d}D_b$.

The minimum reflux ratio $R_m$ may be found for any given still concentration $x_s$ from equation 11.56.

### 11.6.3. Operation at constant reflux ratio

If the same column is operated at a constant reflux ratio $R$, the concentration of the more volatile component in the top product will continuously fall. Over a small interval of time $\mathrm{d}t$, the top-product composition with respect to the more volatile component will change from $x_d$ to $x_d + \mathrm{d}x_d$, where $\mathrm{d}x_d$ is negative for the more volatile component. If in this time the amount of product obtained is $\mathrm{d}D_b$, then a material balance on the more volatile component gives:

More volatile component removed in product $= \mathrm{d}D_b \left[ x_d + \dfrac{\mathrm{d}x_d}{2} \right]$

which, neglecting second-order terms, gives: $= x_d\, \mathrm{d}D_b$ (11.102)

and: $$x_d\, \mathrm{d}D_b = -\mathrm{d}(Sx_s)$$

But $\mathrm{d}D_b = -\mathrm{d}S$,

and hence: $$-x_d\, \mathrm{d}S = -S\, \mathrm{d}x_s - x_s\, \mathrm{d}S$$

and: $$S\, \mathrm{d}x_s = \mathrm{d}S(x_d - x_s)$$

Thus: $$\int_{S_1}^{S_2} \frac{\mathrm{d}S}{S} = \int_{x_{s1}}^{x_{s2}} \frac{\mathrm{d}x_s}{x_d - x_s}$$

$$\ln \frac{S_1}{S_2} = \int_{x_{s2}}^{x_{s1}} \frac{\mathrm{d}x_s}{x_d - x_s} \tag{11.103}$$

The right-hand side of this equation may be integrated by plotting $1/(x_d - x_s)$ against $x_s$. This enables the ratio of the initial to final quantity in the still to be found for any desired change in $x_s$, and hence the amount of distillate $D_b$. The heat to be supplied to provide the reflux is $Q_R = \lambda R D_b$ and hence the reboil heat required per mole of product may be compared with that from the first method.

### Example 11.12

A mixture of ethyl alcohol and water with 0.55 mole fraction of alcohol is distilled to give a top product of 0.75 mole fraction of alcohol. The column has four ideal plates and the distillation is stopped when the reflux ratio has to be increased beyond 4.0.

What is the amount of distillate obtained, and the heat required per kmol of product?

### Solution

For various values of $R$ the corresponding values of the intercept $\phi$ and the concentration in the still $x_s$ are calculated. Values of $x_s$ are found as shown in Figure 11.35 for the two values of $R$ of 0.85 and 4. The amount of product is then found from equation 11.98. Thus, for $R = 4$:

$$D_b = 100 \left[ \frac{0.55 - 0.05}{0.75 - 0.05} \right] = 100 \left( \frac{0.5}{0.7} \right) = \underline{\underline{71.4 \text{ kmol}}}$$

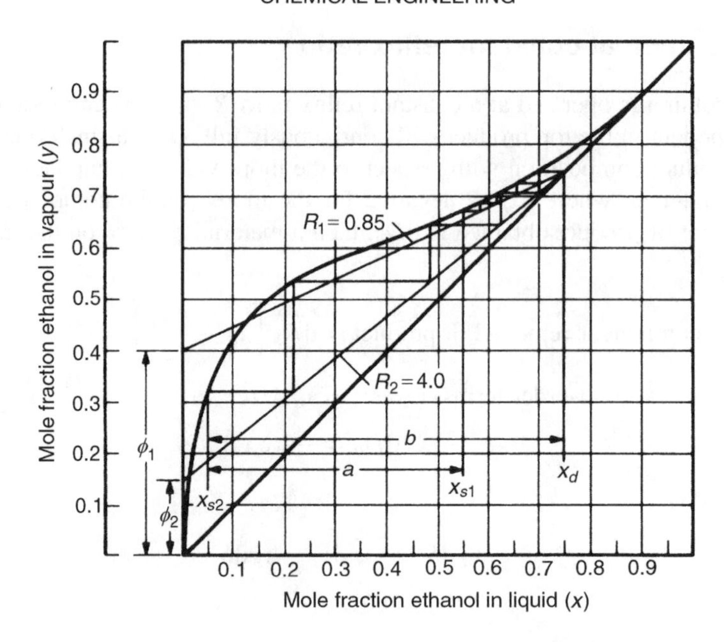


Figure 11.35. Batch distillation–constant product composition

Values of $D_b$ found in this way are:

| $R$ | $\phi$ | $x_s$ | $D_b$ |
|---|---|---|---|
| 0.85 | 0.405 | 0.55 | 0 |
| 1.0 | 0.375 | 0.50 | 20.0 |
| 1.5 | 0.3 | 0.37 | 47.4 |
| 2.0 | 0.25 | 0.20 | 63.8 |
| 3.0 | 0.187 | 0.075 | 70.5 |
| 4.0 | 0.15 | 0.05 | 71.4 |

The relation between $D_b$ and $R$ is shown in Figure 11.36 and the $\int_{R=0.85}^{R=4.0} R\,dD_b$ is given by area OABC as 96 kmol.

Assuming an average latent heat for the alcohol–water mixtures of 4000 kJ/kmol, the heat to be supplied to provide the reflux, $Q_R$ is $(96 \times 4000)/1000$ or approximately 380 MJ.

The heat to be supplied to provide the reflux per kmol of product is then $(380/71.4) = 5.32$ MJ and the total heat is $(5.32 + 4.0) = 9.32$ MJ/kmol product.

## Example 11.13

If the same batch as in Example 11.12 is distilled with a constant reflux ratio of $R = 2.1$, what will be the heat required and the average composition of the distillate if the distillation is stopped when the composition in the still has fallen to 0.105 mole fraction of ethanol?

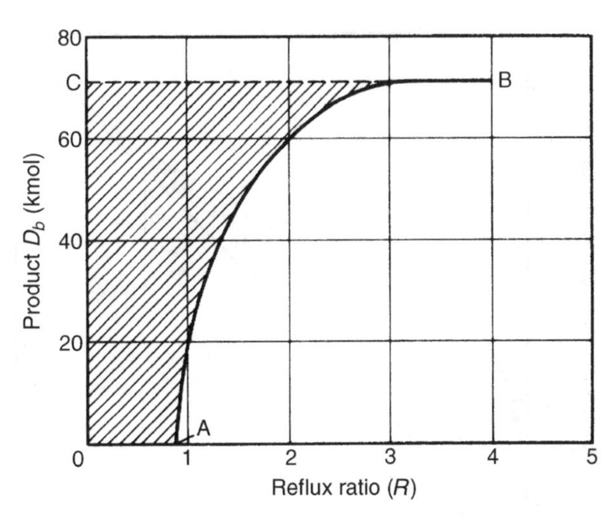


Figure 11.36. Graphical integration for Example 11.12

## Solution

The initial composition of the top product will be 0.78, as shown in Figure 11.37, and the final composition will be 0.74. Values of $x_d$, $x_s$, $x_d - x_s$ and of $1/(x_d - x_s)$ for various values of $x_s$ and a constant reflux ratio are:

| $x_s$ | $x_d$ | $x_d - x_s$ | $1/(x_d - x_s)$ |
|---|---|---|---|
| 0.550 | 0.780 | 0.230 | 4.35 |
| 0.500 | 0.775 | 0.275 | 3.65 |
| 0.425 | 0.770 | 0.345 | 2.90 |
| 0.310 | 0.760 | 0.450 | 2.22 |
| 0.225 | 0.750 | 0.525 | 1.91 |
| 0.105 | 0.740 | 0.635 | 1.58 |

Values of $x_s$ and $1/(x_d - x_s)$ are plotted in Figure 11.38 from which $\int_{0.105}^{0.55} (\mathrm{d}x_s/(x_d - x_s)) = 1.1$.

From equation 11.103: $\ln(S_1/S_2) = 1.1$ and $(S_1/S_2) = 3.0$.

Product obtained, $D_b = S_1 - S_2 = (100 - 100/3) = 66.7$ kmol.

Amount of ethanol in product $= x_1 S_1 - x_2 S_2$

$$= (0.55 \times 100) - (0.105 \times 33.3) = 51.5 \text{ kmol}$$

Thus: average composition of product $= (51.5/66.7) =$ 0.77 mole fraction ethanol.

The heat required to provide the reflux $= (4000 \times 2.1 \times 66.7) = 560{,}380$ kJ.

Heat required to provide reflux per kmol of product $= (560{,}380/66.7) =$ 8400 kJ.

Thus in Example 11.12 the total heat required per kmol of product is $(5320 + 4000) = 9320$ kJ and at constant reflux ratio (Example 11.13) it is $(8400 + 4000) = 12{,}400$ kJ, although the average quality of product is 0.77 for the second case and only 0.75 for the first.

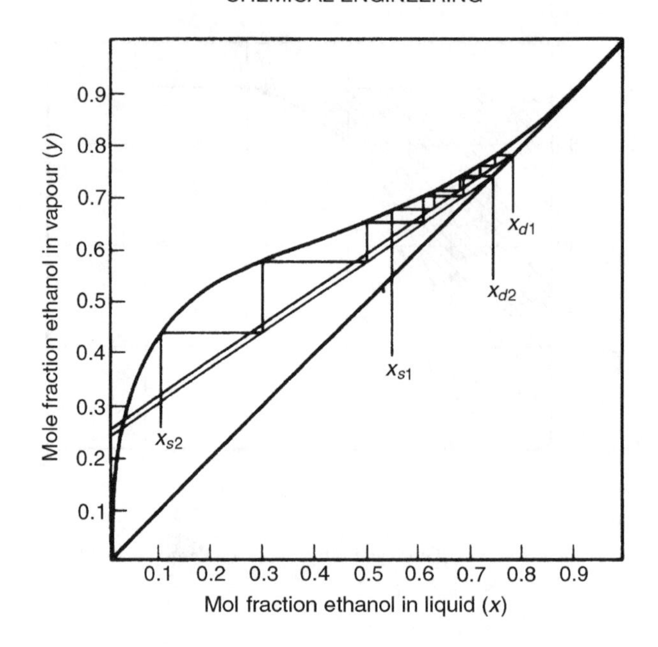


Figure 11.37. Batch distillation–constant reflux ratio (Example 11.13)

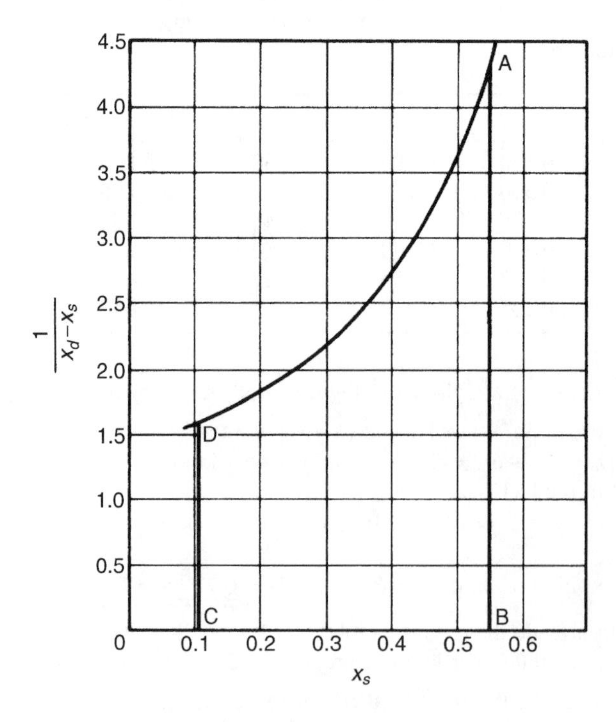


Figure 11.38. Graphical integration for Example 11.13

### 11.6.4. Batch or continuous distillation

A discussion on the relative merits of batch and continuous distillation is given by ELLIS[36], who shows that when a large number of plates is used and the reflux ratio approaches the minimum value, then continuous distillation has the lowest reflux requirement and hence operating costs. If a smaller number of plates is used and high purity product is not required, then batch distillation is probably more attractive.

## 11.7. MULTICOMPONENT MIXTURES

### 11.7.1. Equilibrium data

For a binary mixture under constant pressure conditions the vapour–liquid equilibrium curve for either component is unique so that, if the concentration of either component is known in the liquid phase, the compositions of the liquid and of the vapour are fixed. It is on the basis of this single equilibrium curve that the McCabe–Thiele method was developed for the rapid determination of the number of theoretical plates required for a given separation. With a ternary system the conditions of equilibrium are more complex, for at constant pressure the mole fraction of two of the components in the liquid phase must be given before the composition of the vapour in equilibrium can be determined, even for an ideal system. Thus, the mole fraction $y_A$ in the vapour depends not only on $x_A$ in the liquid, but also on the relative proportions of the other two components.

Determining the equilibrium relationships for a multicomponent mixture experimentally requires a considerable quantity of data, and one of two methods of simplification is usually adopted. For many systems, particularly those consisting of chemically similar substances, the relative volatilities of the components remain constant over a wide range of temperature and composition. This is illustrated in Table 11.2 for mixtures of phenol, ortho and meta-cresols, and xylenols, where the volatilities are shown relative to ortho-cresol.

Table 11.2. Volatilities relative to *o*-cresol

| | Temperature (K) | | |
|---|---|---|---|
| | 353 | 393 | 453 |
| phenol | 1.25 | 1.25 | 1.25 |
| *o*-cresol | 1 | 1 | 1 |
| *m*-cresol | 0.57 | 0.62 | 0.70 |
| xylenols | 0.30 | 0.38 | 0.42 |

An alternative method, particularly useful for the separation of multicomponent mixtures of hydrocarbons, is to use the simple relation $y_A = Kx_A$. $K$ values have been measured for a wide range of hydrocarbons at various pressures, and some values are shown in Figure 11.39.

Some progress has been made in presenting methods for calculating ternary data from known data for the binary mixtures, though as yet no entirely satisfactory method is available.

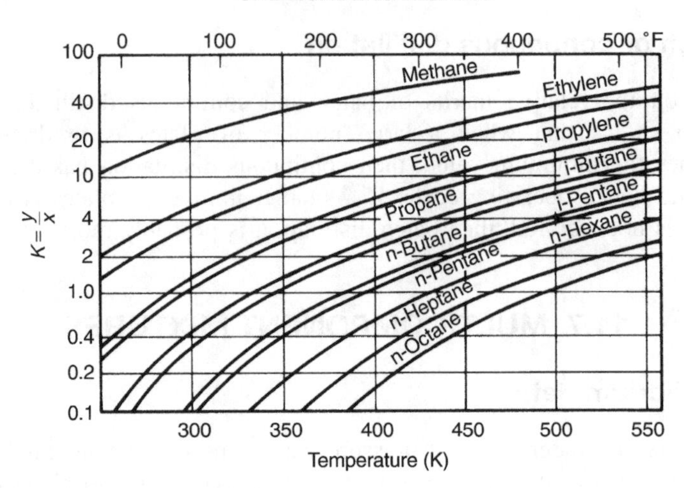


Figure 11.39. Vapour–liquid equilibrium data for hydrocarbons

## 11.7.2. Feed and product compositions

With a binary system, if the feed composition $x_f$ and the top product composition $x_d$ are known for one component, then the composition of the bottoms $x_w$ may have any desired value, and a material balance will determine the amounts of the top and bottom products $D$ and $W$. This freedom of selecting the compositions does not apply for mixtures with three or more components. GILLILAND and REED[37] have determined the number of degrees of freedom for the continuous distillation of a multicomponent mixture. For the common case in which the feed composition, nature of the feed, and operating pressure are given, there remain only four variables that may be selected. If the reflux ratio $R$ is fixed and the number of plates above and below the feed plate are chosen to give the best use of the plates, then only two variables remain. The complete composition of neither the top nor bottom product can then be fixed at will. This means that some degree of trial and error is unavoidable in calculating the number of plates required for any desired separation. Thus, if a trial composition is taken, and it is found that for the given bottom composition the desired top composition is not obtained with the selected reflux ratio, then an adjustment must be made in the bottom composition. An exact fit in a calculation of this kind is not essential since the equilibrium data and the plate efficiency are known with only limited accuracy. This problem is frequently simplified if a sharp cut is to be made between the components, so that all of the more volatile components appear in the top and all of the less volatile in the bottom product.

## 11.7.3. Light and heavy key components

In the fractionation of multicomponent mixtures, the essential requirement is often the separation of two components. Such components are called the key components and by concentrating attention on these it is possible to simplify the handling of complex mixtures. If a four-component mixture **A–B–C–D**, in which **A** is the most volatile and **D** the least volatile, is to be separated as shown in Table 11.3, then **B** is the lightest component appearing in the bottoms and is termed the light key component. **C** is the

Table 11.3. Separation of multicomponent mixture

| Feed | Top product | Bottoms |
|---|---|---|
| **A** | **A** | |
| **B** | **B** | **B** |
| **C** | **C** | **C** |
| **D** | | **D** |

heaviest component appearing in the distillate and is called the heavy key component. The main purpose of the fractionation is the separation of **B** from **C**.

## 11.7.4. The calculation of the number of plates required for a given separation

One of the most successful methods for calculating the number of plates necessary for a given separation is due to LEWIS and MATHESON[38]. This is based on the Lewis–Sorel method, described previously for binary mixtures. If the composition of the liquid on any plate is known, then the composition of the vapour in equilibrium is calculated from a knowledge of the vapour pressures or relative volatilities of the individual components. The composition of the liquid on the plate above is then found by using an operating equation, as for binary mixtures, although in this case there will be a separate equation for each component.

If a mixture of components **A**, **B**, **C**, **D**, and so on has mole fractions $x_A$, $x_B$, $x_C$, $x_D$, and so on in the liquid and $y_A$, $y_B$, $y_C$, $y_D$, and so on in the vapour, then:

$$y_A + y_B + y_C + y_D + \cdots = 1 \tag{11.104}$$

and:

$$\frac{y_A}{y_B} + \frac{y_B}{y_B} + \frac{y_C}{y_B} + \frac{y_D}{y_B} + \cdots = \frac{1}{y_B}$$

$$\alpha_{AB}\frac{x_A}{x_B} + \alpha_{BB}\frac{x_B}{x_B} + \alpha_{CB}\frac{x_C}{x_B} + \alpha_{DB}\frac{x_D}{x_B} + \cdots = \frac{1}{y_B}$$

$$\Sigma(\alpha_{AB}x_A) = \frac{x_B}{y_B} \tag{11.105}$$

$$y_B = \frac{x_B}{\Sigma(\alpha_{AB}x_A)} \tag{11.106}$$

and, similarly:

$$y_A = \frac{x_A\alpha_{AB}}{\Sigma(\alpha_{AB}x_A)}; \quad y_C = \frac{x_C\alpha_{CB}}{\Sigma(\alpha_{AB}x_A)}; \quad y_D = \frac{x_D\alpha_{DB}}{\Sigma(\alpha_{AB}x_A)} \tag{11.107}$$

Thus, the composition of the vapour is conveniently found from that of the liquid by use of the relative volatilities of the components. Examples 11.14–11.17 which follow illustrate typical calculations using multicomponent systems. Such solutions are now computerised, as discussed further in Volume 6.

### Example 11.14

A mixture of ortho, meta, and para-mononitrotoluenes containing 60, 4, and 36 mole per cent respectively of the three isomers is to be continuously distilled to give a top product of 98 mole per cent ortho, and the bottom is to contain 12.5 mole per cent ortho. The mixture is to be distilled at a temperature of 410 K requiring a pressure in the boiler of about 6.0 kN/m$^2$. If a reflux ratio of 5 is used, how many ideal plates will be required and what will be the approximate compositions of the product streams? The volatility of ortho relative to the para isomer may be taken as 1.70 and of the meta as 1.16 over the temperature range of 380–415 K.

## Solution

As a first estimate, it is supposed that the distillate contains 0.6 mole per cent meta and 1.4 mole per cent para. A material balance then gives the composition of the bottoms.

For 100 kmol of feed with $D$ and $W$ kmol of product and bottoms, respectively and $x_{do}$ and $x_{wo}$ the mole fraction of the ortho in the distillate and bottoms, then an overall material balance gives:

$$100 = D + W$$

An ortho balance gives:

$$60 = Dx_{do} + Wx_{wo}$$

and:

$$60 = (100 - W)0.98 + 0.125\ W$$

from which:

$$D = 55.56 \text{ kmol} \quad \text{and} \quad W = 44.44. \text{ kmol}$$

The compositions and amounts of the streams are then be obtained as:

| Component | Feed | | Distillate | | Bottoms | |
|---|---|---|---|---|---|---|
| | (kmol) | (mole per cent) | (kmol) | (mole per cent) | (kmol) | (mole per cent) |
| Ortho $o$ | 60 | 60 | 54.44 | 98.0 | 5.56 | 12.5 |
| Meta $m$ | 4 | 4 | 0.33 | 0.6 | 3.67 | 8.3 |
| Para $p$ | 36 | 36 | 0.79 | 1.4 | 35.21 | 79.2 |
| | 100 | 100 | 55.56 | 100 | 44.44 | 100 |

*Equations of operating lines.*

The liquid and vapour streams in the column are obtained as follows:
Above the feed-point:

$$\text{Liquid downflow, } L_n = 5D = 277.8$$

$$\text{Vapour up, } V_n = 6D = 333.4$$

Below the feed-point, assuming the feed is liquid at its boiling point then:

$$\text{Liquid downflow, } L_m = L_n + F = (277.8 + 100) = 377.8$$

$$\text{Vapour up, } V_m = L_m - W = (377.8 - 44.44) = 333.4$$

The equations for the operating lines may then be written as:

below the feed plate:

$$y_m = \frac{L_m}{V_m}x_{m+1} - \frac{W}{V_m}x_w \qquad \text{(equation 11.37)}$$

ortho:

$$y_{mo} = \left(\frac{377.8}{333.4}\right)x_{m+1} - \left(\frac{44.44}{333.4}\right)x_w$$

$$
\left.\begin{aligned}
&= 1.133x_{m+1} - 0.0166 \\
\text{meta:}\quad y_{mm} &= 1.133x_{m+1} - 0.011 \\
\text{para:}\quad y_{mp} &= 1.133x_{m+1} - 0.105
\end{aligned}\right\} \quad \text{(i)}
$$

and above the feed plate:

$$y_n = \frac{L_n}{V_n}x_{n+1} + \frac{D}{V_n}x_d \quad \text{(equation 11.35)}$$

ortho:

$$y_{no} = \left(\frac{277.8}{333.4}\right)x_{n+1} + \left(\frac{55.56}{333.4}\right)0.98$$

$$
\left.\begin{aligned}
&= 0.833x_{n+1} + 0.163 \\
\text{meta:}\quad y_{nm} &= 0.833_{x+1} + 0.001 \\
\text{para:}\quad y_{np} &= 0.833x_{n+1} + 0.002
\end{aligned}\right\} \quad \text{(ii)}
$$

*Composition of liquid on first plate.*

The temperature of distillation is fixed by safety considerations at 410 K and, from a knowledge of the vapour pressures of the three components, the pressure in the still is found to be about 6 kN/m$^2$. The composition of the vapour in the still is found from the relation $y_{so} = \alpha_o x_{so}/\Sigma\alpha x_s$.

The liquid composition on the first plate is then found from equation (i) and for ortho:

$$0.191 = (1.133x_1 - 0.0166)$$

and:

$$x_1 = 0.183$$

The values of the compositions as found in this way are shown in the following table. The liquid on plate 7 has a composition with the ratio of the concentrations of ortho and para about that in the feed, and the feed will therefore be introduced on this plate. Above this plate the same method is used but the operating equations are equation (ii). The vapour from the sixteenth plate has the required concentration of the ortho isomer, and the values the meta and para are sufficiently near to take this as showing that sixteen ideal plates will be required.

Using the relation $v_{mo} = \alpha_o x_{mo}/\Sigma\alpha x_m$:

Plate compositions below the feed plate

| Component | $x_s$ | $\alpha x_s$ | $y_s$ | $x_1$ | $\alpha x_1$ | $y_1$ | $x_2$ |
|---|---|---|---|---|---|---|---|
| $o$ | 0.125 | 0.211 | 0.191 | 0.183 | 0.308 | 0.270 | 0.253 |
| $m$ | 0.083 | 0.096 | 0.088 | 0.088 | 0.102 | 0.090 | 0.089 |
| $p$ | 0.792 | 0.792 | 0.721 | 0.729 | 0.729 | 0.640 | 0.658 |
| | 1 | 1.099 | 1 | 1 | 1.139 | 1 | 1 |

| Component | $\alpha x_2$ | $y_2$ | $x_3$ | $\alpha x_3$ | $y_3$ | $x_4$ | $\alpha x_4$ |
|---|---|---|---|---|---|---|---|
| $o$ | 0.430 | 0.357 | 0.330 | 0.561 | 0.450 | 0.411 | 0.698 |
| $m$ | 0.103 | 0.086 | 0.086 | 0.100 | 0.080 | 0.080 | 0.093 |
| $p$ | 0.658 | 0.557 | 0.584 | 0.584 | 0.470 | 0.509 | 0.509 |
| | 1.191 | 1 | 1 | 1.245 | 1 | 1 | 1.300 |

| Component | $y_4$ | $x_5$ | $\alpha x_5$ | $y_5$ | $x_6$ | $\alpha x_6$ | $y_6$ |
|---|---|---|---|---|---|---|---|
| $o$ | 0.537 | 0.488 | 0.830 | 0.613 | 0.556 | 0.944 | 0.674 |
| $m$ | 0.071 | 0.072 | 0.083 | 0.061 | 0.063 | 0.073 | 0.052 |
| $p$ | 0.392 | 0.440 | 0.440 | 0.326 | 0.381 | 0.381 | 0.274 |
| | 1 | 1 | 1.353 | 1 | 1 | 1.398 | 1 |

| | $x_{-}$ |
|---|---|
| $o$ | 0.609 |
| $m$ | 0.055 |
| $p$ | 0.336 |
| | 1 |

Plate compositions above the feed plate

| Component | $x_7$ | $\alpha x_7$ | $y_7$ | $x_8$ | $\alpha x_8$ | $y_8$ | $x_9$ |
|---|---|---|---|---|---|---|---|
| $o$ | 0.609 | 1.035 | 0.721 | 0.669 | 1.136 | 0.770 | 0.728 |
| $m$ | 0.055 | 0.064 | 0.044 | 0.051 | 0.059 | 0.040 | 0.047 |
| $p$ | 0.336 | 0.336 | 0.235 | 0.280 | 0.280 | 0.190 | 0.225 |
| | 1 | 1.435 | 1 | 1 | 1.475 | 1 | 1 |

| | $\alpha x_9$ | $y_9$ | $x_{10}$ | $\alpha x_{10}$ | $y_{10}$ | $x_{11}$ | $\alpha x_{11}$ |
|---|---|---|---|---|---|---|---|
| $o$ | 1.238 | 0.816 | 0.782 | 1.330 | 0.856 | 0.832 | 1.415 |
| $m$ | 0.054 | 0.035 | 0.041 | 0.047 | 0.030 | 0.035 | 0.040 |
| $p$ | 0.225 | 0.149 | 0.177 | 0.177 | 0.144 | 0.133 | 0.133 |
| | 1.517 | 1 | 1 | 1.554 | 1 | 1 | 1.588 |

| | $y_{11}$ | $x_{12}$ | $\alpha x_{12}$ | $y_{12}$ | $x_{13}$ | $\alpha x_{13}$ | $y_{13}$ |
|---|---|---|---|---|---|---|---|
| $o$ | 0.891 | 0.874 | 1.485 | 0.920 | 0.907 | 1.542 | 0.940 |
| $m$ | 0.025 | 0.029 | 0.033 | 0.020 | 0.023 | 0.027 | 0.017 |
| $p$ | 0.084 | 0.097 | 0.097 | 0.060 | 0.070 | 0.070 | 0.043 |
| | 1 | 1 | 1.615 | 1 | 1 | 1.639 | 1 |

| | $x_{14}$ | $\alpha x_{14}$ | $y_{14}$ | $x_{15}$ | $\alpha x_{15}$ | $y_{15}$ | $x_{16}$ |
|---|---|---|---|---|---|---|---|
| $o$ | 0.932 | 1.585 | 0.957 | 0.953 | 1.620 | 0.970 | 0.968 |
| $m$ | 0.019 | 0.022 | 0.013 | 0.014 | 0.016 | 0.010 | 0.010 |
| $p$ | 0.049 | 0.049 | 0.030 | 0.033 | 0.033 | 0.020 | 0.022 |
| | 1 | 1.656 | 1 | 1 | 1.669 | 1 | 1 |

| | $\alpha x_{16}$ | $y_{16}$ |
|---|---|---|
| $o$ | 1.632 | 0.980 |
| $m$ | 0.012 | 0.007 |
| $p$ | 0.022 | 0.013 |
| | 1.666 | 1 |

### 11.7.5. Minimum reflux ratio

In the distillation of binary mixtures, the minimum reflux ratio is given by the operating line which joins the product-composition to the point where the $q$-line cuts the equilibrium curve. Thus, if Figure 11.40 represents the McCabe–Thiele diagram for a binary mixture of the two key components of a multicomponent mixture, then DF and WF give the minimum reflux ratios for the rectifying and stripping sections respectively. Moving along these operating lines, the change in composition on adjacent plates becomes less and less until it becomes negligible at the feed plate. At F the compositions are said to be "pinched".

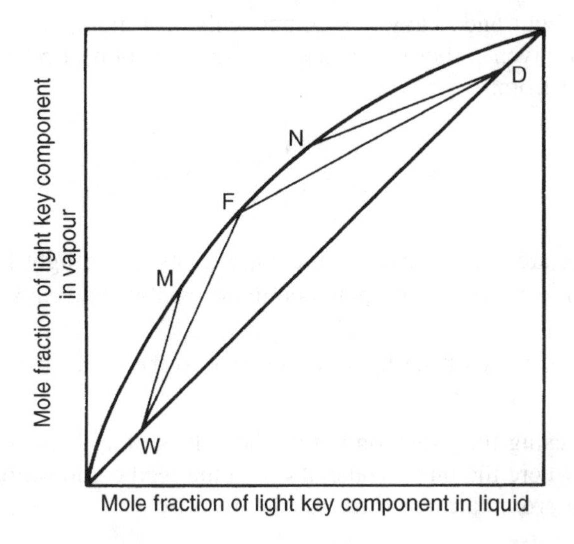


Figure 11.40. McCabe–Thiele diagram for two key components

In a multicomponent mixture, the pinch does not necessarily occur at the position corresponding to the feed plate. In general, there are relatively few plates above the feed plate in which the concentrations of components heavier than the heavy key are reduced to negligible proportions, and then a true pinched condition occurs with only the heavy key and more volatile components present. Similarly, there is a region in which the concentrations of materials lighter than the light key are reduced to very low values, and thus there is a second pinch below the feed plate. In multicomponent distillation there are thus two pinched-in-regions. In locating these pinched-in regions it may be noted that:

(a) If there are no components lighter than the light key, then all of the components appear in the bottoms and the pinch in the stripping section will be near the feed plate.
(b) If there are no components heavier than the heavy key, then all of the components will appear in the top and the upper pinch is also at the feed plate.

If both of these conditions are true, then the two pinches coincide at the feed plate, as for a binary system. For the general case, a number of proposals have been made for locating the pinched regions and hence the minimum reflux ratio $R_m$. Of these, the methods of Colburn and Underwood are considered. It may be noted that between the feed plate and the enriching pinch, the concentrations of components heavier than the heavy key fall off rapidly, so that the upper pinch may be regarded as containing the heaviest key and lighter components. Similarly the lower pinch has the lightest key and heavier components.

### Colburn's method for minimum reflux

If **A** and **B** are the light and heavy key components of a multicomponent mixture, then applying the method given earlier for binary mixtures, equation 11.53, the minimum reflux ratio $R_m$ is obtained from:

$$R_m = \frac{1}{(\alpha_{AB} - 1)}\left[\frac{x_{dA}}{x_{nA}} - \alpha_{AB}\frac{x_{dB}}{x_{nB}}\right] \tag{11.108}$$

where: $x_{dA}$ and $x_{nA}$ are the top and pinch compositions of the light key component,
$x_{dB}$ and $x_{nB}$ are the top and pinch compositions of the heavy key component, and
$\alpha_{AB}$ is the volatility of the light key relative to the heavy key component.

The difficulty in using this equation is that the values of $x_{nA}$ and $x_{nB}$ are known only in the special case where the pinch coincides with the feed composition. Colburn[39] has suggested that an approximate value for $x_{nA}$ is given by:

$$x_{nA} \approx \frac{r_f}{(1 + r_f)\left(1 + \Sigma\alpha x_{fh}\right)} \tag{11.109}$$

and:

$$x_{nB} \approx \frac{x_{nA}}{r_f} \tag{11.110}$$

where: $r_f$ is the estimated ratio of the key components on the feed plate. For an all liquid feed at its boiling point, $r_f$ equals the ratio of the key components in the feed. Otherwise $r_f$ is the ratio of the key components in the liquid part of the feed.
$x_{fh}$ is the mole fraction of each component in the liquid portion of feed heavier than the heavy key, and
$\alpha$ is the volatility of the component relative to the heavy key.

Using this approximate value for $R_m$, equation 11.109 may be rearranged to give the concentrations of all the light components in the upper pinch as:

$$x_n = \frac{x_d}{(\alpha - 1)R_m + \alpha(x_{dB}/x_{nB})} \tag{11.111}$$

The concentration of the heavy key in the upper pinch is then obtained by difference, after obtaining the values for all the light components. The second term in the denominator is usually negligible, as the concentration of the heavy key in the top product is small.

A similar condition occurs in the stripping section, and the concentration of all components heavier than the light key is given by:

$$x_m = \frac{\alpha_{AB} x_w}{(\alpha_{AB} - \alpha)(L_m/W) + \alpha(x_{wA}/x_{mA})} \tag{11.112}$$

where: $x_m$ and $x_w$ are the compositions of a given heavy component at the pinch and in the bottoms,
$x_{mA}$ and $x_{wA}$ are the compositions of the light key component at the pinch and in the bottoms,
$L_m/W$ is the molar ratio of the liquid in the stripping section to the bottom product,
$\alpha_{AB}$ is the volatility of the light key relative to the heavy key, and
$\alpha$ is the volatility of the component relative to the heavy key.

Again, the second term in the denominator may usually be neglected.

The essence of Colburn's method is that an empirical relation between the compositions at the two pinches for the condition of minimum reflux is provided. This enables the assumed value of $R_m$ to be checked. This relation may be written as:

$$\frac{r_m}{r_n} = \psi = \frac{1}{(1 - \Sigma b_m \alpha x_m)(1 - \Sigma b_n x_n)}. \tag{11.113}$$

where: $r_m$ is the ratio of the light key to the heavy key in the stripping pinch,
$r_n$ is the ratio of the light key to the heavy key in the upper pinch,
$\Sigma b_m \alpha x_m$ is the summation of $b_m \alpha x_m$ for all components heavier than the heavy key in the lower pinch,
$\Sigma b_n x_n$ is the summation of $b_n x_n$ for all components lighter than the light key in the upper pinch, and
$b_m, b_n$ are the factors shown in Figure 11.41.

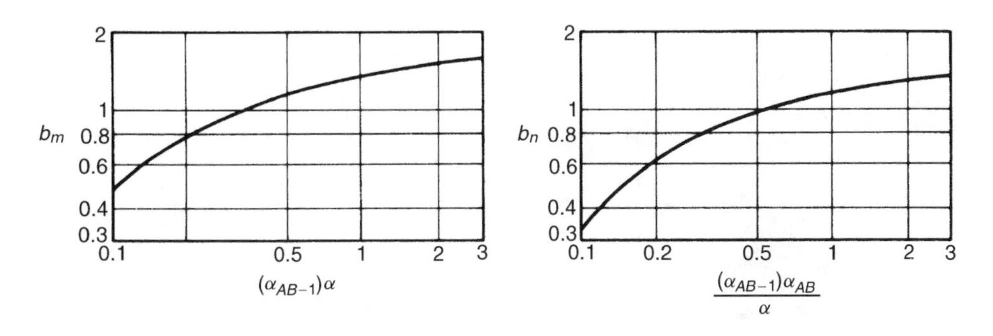


Figure 11.41. Factors in Colburn's solution

## Example 11.15

A mixture of $n$-$C_4$ to $n$-$C_7$ hydrocarbons is to be distilled to give top and bottom products as follows. The distillation is effected at 800 kN/m$^2$ and the feed is at 372 K. The equilibrium values, $K$, are shown in Figure 11.39. It is required to find the minimum reflux ratio. No cooling occurs in the condenser.

| Component | Feed | Distillate | | Bottoms | |
|---|---|---|---|---|---|
| | $F$ | $D$ | $x_d$ | $W$ | $x_w$ |
| $n$-$C_4$ (light key) | 40 | 39 | 0.975 | 1 | 0.017 |
| $C_5$ (heavy key) | 23 | 1 | 0.025 | 22 | 0.367 |
| $C_6$ | 17 | | | 17 | 0.283 |
| $C_7$ | 20 | | | 20 | 0.333 |
| Totals: | $F = 100$ | $D = 40$ | 1.000 | $W = 60$ | 1.000 |

## Solution

1. *Estimation of top temperature $T_d$.*

By a dew-point calculation, $\Sigma x_d = \Sigma(x_d/K)$.

| Component | $x_d$ | $T_d = 344$ K | | $T_d = 343$ K | |
|---|---|---|---|---|---|
| | | $K$ | $x_d/K$ | $K$ | $x_d/K$ |
| $n$-$C_4$ | 0.975 | 1.05 | 0.929 | 1.04 | 0.938 |
| $C_5$ | 0.025 | 0.41 | 0.061 | 0.405 | 0.062 |
| | 1 | | 0.990 | | 1.000 |

Hence the top temperature $T_d = 343$ K.

2. *Estimation of still temperature $T_s$.*

$\Sigma x_w = \Sigma K x_w$.

| Component | $x_w$ | $T_s = 419$ K | | $T_s = 416$ K | |
|---|---|---|---|---|---|
| | | $K$ | $Kx_w$ | $K$ | $Kx_w$ |
| $n$-$C_4$ | 0.017 | 3.05 | 0.052 | 2.93 | 0.050 |
| $C_5$ | 0.367 | 1.6 | 0.586 | 1.54 | 0.565 |
| $C_6$ | 0.283 | 0.87 | 0.246 | 0.82 | 0.232 |
| $C_7$ | 0.333 | 0.49 | 0.163 | 0.46 | 0.153 |
| | 1 | | 1.047 | | 1.000 |

Hence the still temperature $T_s = 416$ K.

3. *Calculation of feed condition.*

To determine the nature of the feed, its boiling point $T_B$ must be found, that is where $\Sigma K x_f = 1$.

| Component | $x_f$ | $T_B = 377$ K | | $T_B = 376$ K | |
|---|---|---|---|---|---|
| | | $K$ | $Kx_f$ | $K$ | $Kx_f$ |
| $n$-$C_4$ | 0.40 | 1.80 | 0.720 | 1.78 | 0.712 |
| $C_5$ | 0.23 | 0.81 | 0.186 | 0.79 | 0.182 |
| $C_6$ | 0.17 | 0.39 | 0.066 | 0.38 | 0.065 |
| $C_7$ | 0.20 | 0.19 | 0.038 | 0.185 | 0.037 |
| | 1 | | 1.010 | | 0.996 |

Hence $T_B$ is approximately 376 K, and since the feed is at 372 K it may be assumed to be all liquid at its boiling point.

4. *Calculation of pinch temperatures.*

The temperatures of the pinches are taken in the first place at one third and two thirds of the difference between the still and top temperatures.

Thus the upper pinch temperature, $T_n = 343 + 0.33(416 - 343) = 367$ K

and the lower pinch temperature, $T_m = 343 + 0.67(416 - 343) = 391$ K

5. *Calculation of approximate minimum reflux ratio.*

The calculations may be laid out as:

| Component | $T_n = 367$ K | $T_m = 391$ K | $x'_{fh}$ | $\alpha x_{fh}$ |
|---|---|---|---|---|
| | $\alpha$ | $\alpha$ | | |
| $n$-$C_4$ | 2.38 | 2.00 | | |
| $C_5$ | 1.00 | 1.00 | | |
| $C_6$ | 0.455 | 0.464 | 0.17 | 0.077 |
| $C_7$ | 0.220 | 0.254 | 0.20 | 0.044 |
| | | | | 0.121 |

Thus:

$$r_f = \frac{x_{f4}}{x_{f5}} = \left(\frac{0.40}{0.23}\right) = 1.740$$

From equation 11.109:

$$x_{n4} = \frac{1.740}{(1 + 1.740)(1 + 0.121)} = 0.565$$

and:

$$x_{n5} = \left(\frac{0.565}{1.740}\right) = 0.325$$

From equation 11.108:

$$R_m = \left(\frac{1}{(2.38 - 1)}\right)\left(\frac{0.975}{0.563}\right) - 2.38\left(\frac{0.025}{0.325}\right)$$

$$= 1.12$$

6. *The streams in the column*

$$L_n = DR_m = (40 \times 1.12) = 44.8 \text{ kmol}$$

$$V_n = L_n + D = (44.8 + 40) = 84.8 \text{ kmol}$$

$$L_m = L_n + F = (44.8 + 100) = 144.8 \text{ kmol}$$

$$V_m = L_m - W = (144.8 - 60) = 84.4 \text{ kmol}$$

$$L_m/W = (144.8/60) = 2.41$$

7. *Check on minimum reflux ratio*

$$x_n = \frac{x_d}{(\alpha - 1)R_m} \text{ for components lighter than heavy key}$$

For $n$-$C_4$:

$$x_n = \frac{0.975}{(2.38 - 1)1.12} = 0.630$$

and for $n$-$C_5$:

$$x_n = (1 - 0.630) = 0.370$$

Temperature check for upper pinch. $\Sigma K x_n = 1.0$

$$\Sigma K x_n = (1.62 \times 0.630) + (0.68 \times 0.370) = 1.273$$

Thus an upper pinch temperature of 367 K is incorrect. A value of $T_n = 355$ K will be tried.

| Component | $K_{355}$ | $\alpha$ | $x_n$ | $Kx_n$ |
|---|---|---|---|---|
| $n$-$C_4$ | 1.35 | 2.55 | 0.562 | 0.759 |
| $C_5$ | 0.53 | 1 | 0.438 | 0.233 |
| | | | 1.000 | 0.992 |

This is sufficiently near, and the upper pinch temperature will be taken as 355 K.

Thus: $r_n = (0.562/0.438) = 1.282$

Since there is no component lighter than light key $\Sigma b_n x_n = 0$. In the lower pinch, (from equation 11.112):

$$x_m = \frac{\alpha_{AB} x_w}{(\alpha_{AB} - \alpha) L_m/W}$$

| Component | $K_{391}$ | $\alpha$ | $\alpha_{AB} - \alpha$ | $x_m$ | $Kx_m$ |
|---|---|---|---|---|---|
| $n$-$C_4^*$ | 2.2 | 2 | | 0.384 | 0.845 |
| $C_5$ | 1.1 | 1 | 1.000 | 0.305 | 0.335 |
| $C_6$ | 0.51 | 0.464 | 1.536 | 0.153 | 0.078 |
| $C_7$ | 0.28 | 0.254 | 1.746 | 0.158 | 0.044 |
| | | | | 1.000 | 1.302 |

$^*x_m$ by difference.

As the temperatures do not check $T_m$ is not 391 K and a value of 372 K will be tried.

| Component | $K_{372}$ | $\alpha$ | $\alpha_{AB} - \alpha$ | $x_m$ | $Kx_m$ | $b_m$ | $b_m\alpha_m x_m$ |
|---|---|---|---|---|---|---|---|
| $n$-$C_4^*$ | 1.70 | 2.30 | — | 0.428 | 0.729 | | |
| $C_5$ | 0.74 | 1 | 1.30 | 0.270 | 0.200 | | |
| $C_6$ | 0.35 | 0.47 | 1.83 | 0.148 | 0.052 | 1.22 | 0.085 |
| $C_7$ | 0.17 | 0.23 | 2.07 | 0.154 | 0.026 | 0.91 | 0.032 |
| | | | | 1.000 | 1.007 | | 0.117 |

$^*x_m$ by difference.

This is sufficiently near, and $T_m = 372$ K.

Thus:

$$r_m = (0.428/0.270) = 1.586$$

$$\alpha r_m/r_n = (1.586/1.282) = 1.235$$

$$\psi = 1/(1 - 0.117) = (1/0.883)$$

$$= 1.132$$

Hence $R_m$ is not quite equal to 1.12.

8. *Second approximation to reflux ratio*

A value of $R_m = 1.08$ will be tried.

$$L_n = (40 \times 1.08) = 43.2$$

$$V_n = (40 + 43.2) = 83.2 = V_m$$

$$L_m = 143.2$$

Taking $T_n = 355$ K as before, then:

| Component | $K_{355}$ | $\alpha$ | $x_n$ | $Kx_n$ | |
|---|---|---|---|---|---|
| $n$-$C_4$ | 1.35 | 2.55 | 0.582 | 0.785 | |
| $C_5$ | 0.53 | 1 | 0.418 | 0.221 | |
| | | | 1.000 | 1.006 | Checks |

Thus:

$$r_n = (0.582/0.418) = 1.393$$

Taking $T_m = 372$ K as before, then:

| Component | $K_{372}$ | $\alpha$ | $\alpha_{AB} - \alpha$ | $x_m$ | $Kx_m$ | $b_m$ | $b_m\alpha_m x_m$ |
|---|---|---|---|---|---|---|---|
| $n$-$C_4^*$ | 1.70 | 2.30 | — | 0.424 | 0.721 | | |
| $C_5$ | 0.74 | 1 | 1.30 | 0.272 | 0.201 | | |
| $C_6$ | 0.35 | 0.47 | 1.83 | 0.149 | 0.052 | 1.22 | 0.085 |
| $C_7$ | 0.17 | 0.23 | 2.07 | 0.155 | 0.026 | 0.91 | 0.034 |
| | | | | 1.000 | 1.000 | | 0.119 |

$^*x_m$ by difference.

Thus: $r_m = (0.424/0.272) = 1.558 \qquad \alpha r_m/r_n = (1.558/1.393) = 1.12$

and:

$$\psi = 1/(1 - 0.119) = 1.13$$

Thus: $$R_m = \underline{\underline{1.08 \text{ is near enough.}}}$$

Since there are no components lighter than the light key, the lower pinch is expected to be near the feed plate as it is, although the general method of taking the pinch temperature as one-third and two-thirds up the column was used above.

The small change in $R$ from 1.12 to 1.08 gives a change in $r_n$ though very little change in $r_m$. It is seen that the first estimation for $R_m$ of 1.12 based on equation 11.108 for locating the upper pinch composition is nearly correct but that it gives the wrong pinch composition.

### *Minimum reflux ratio, using Underwood's method*

For conditions where the relative volatilities remain constant, UNDERWOOD[40] developed the following two equations from which $R_m$ may be calculated:

$$\frac{\alpha_A x_{fA}}{\alpha_A - \theta} + \frac{\alpha_B x_{fB}}{\alpha_B - \theta} + \frac{\alpha_C x_{fC}}{\alpha_C - \theta} + \cdots = 1 - q \tag{11.114}$$

and:

$$\frac{\alpha_A x_{dA}}{\alpha_A - \theta} + \frac{\alpha_B x_{dB}}{\alpha_B - \theta} + \frac{\alpha_C x_{dC}}{\alpha_C - \theta} + \cdots = R_m + 1 \tag{11.115}$$

where: $x_{fA}$, $x_{fB}, x_{fC}, x_{dA}, x_{dB}, x_{dC}$, etc., are the mole fractions of components **A**, **B**, **C**, etc., in the feed and distillate, **A** being the light and **B** the heavy key,

$q$ is the ratio of the heat required to vaporise 1 mole of the feed to the molar latent heat of the feed, as in equation 11.44,

$\alpha_A$, $\alpha_B, \alpha_C$, etc., are the volatilities with respect to the least volatile component, and

$\theta$ is the root of equation 11.114, which lies between the values of $\alpha_A$ and $\alpha_B$.

If one component in the system has a relative volatility falling between those of the light and heavy keys, it is necessary to solve for two values of $\theta$.

## Example 11.16

A mixture of hexane, heptane, and octane is to be separated to give the following products. What will be the value of the minimum reflux ratio, if the feed is liquid at its boiling point?

| Component | Feed | | Product | | Bottoms | | Relative volatility |
|---|---|---|---|---|---|---|---|
| | $F$ (kmol) | $x_f$ | $D$ (kmol) | $x_d$ | $W$ (kmol) | $x_w$ | |
| Hexane | 40 | 0.40 | 40 | 0.534 | 0 | 0 | 2.70 |
| Heptane | 35 | 0.35 | 34 | 0.453 | 1 | 0.04 | 2.22 |
| Octane | 25 | 0.25 | 1 | 0.013 | 24 | 0.96 | 1.0 |

## Solution

From equation 11.114, the light key (**A**) is heptane and the heavy key (**B**) is octane. With $q = 1$ then:

$$\left(\frac{2.70 \times 0.40}{2.70 - \theta}\right) + \left(\frac{2.22 \times 0.35}{2.22 - \theta}\right) + \left(\frac{1 \times 0.25}{1 - \theta}\right) = 0$$

The required value of $\theta$ must satisfy the relation $\alpha_B < \theta < \alpha_A$, that is $1.0 < \theta < 2.22$. Assuming $\theta = 1.15$, then:

$$\Sigma \frac{\alpha x_f}{\alpha - \theta} = -0.243$$

Assuming $\theta = 1.17$, then:

$$\Sigma \frac{\alpha x_f}{\alpha - \theta} = -0.024$$

This is near enough and from equation 11.115:

$$\left(\frac{2.70 \times 0.534}{2.70 - 1.17}\right) + \left(\frac{2.22 \times 0.453}{2.22 - 1.17}\right) + \left(\frac{1.00 \times 0.013}{1.00 - 1.17}\right) = 1.827$$

Thus:

$$R_m = \underline{\underline{0.827}}$$

## 11.7.6. Number of plates at total reflux

The number of plates required for a desired separation under conditions of total reflux can be found by applying Fenske's equation, equation 11.59, to the two key components.

Thus:

$$n + 1 = \frac{\log\left[\left(\frac{x_A}{x_B}\right)_d \left(\frac{x_B}{x_A}\right)_s\right]}{\log(\alpha_{AB})_{\text{av}}} \tag{11.116}$$

## Example 11.17

For the separation of hexane, heptane, and octane as in Example 11.16, determine the number of theoretical plates required.

## Solution

$$n + 1 = \frac{-\log\left[\frac{0.0453}{0.013} \times \frac{0.960}{0.040}\right]}{-\log 2.22} = \frac{(\log 835)}{(\log 2.22)}$$

$$= 8.5.$$

Thus, the minimum number of plates = $\underline{\underline{7.5, \text{ say } 8}}$.

## 11.7.7. Relation between reflux ratio and number of plates

GILLILAND[30] has given an empirical relation between the reflux ratio $R$ and the number of plates $n$, in which only the minimum reflux ratio $R_m$ and the number of plates at total reflux $n_m$ are required. This is shown in Figure 11.42, where the group $[(n+1)-(n_m+1)]/(n+2)$ is plotted against $(R-R_m)/(R+1)$.

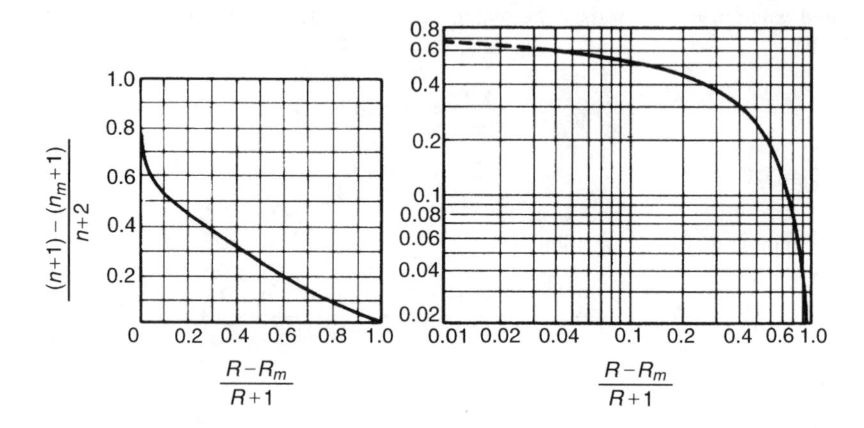


Figure 11.42. Relation between reflux ratio and number of plates

## Example 11.18

Using the data of Example 11.17, investigate the change in $n$ with $R$ using Figure 11.42. It may be assumed that $R_m = 0.83$ and $(n_m + 1) = 8.5$

## Solution

| $R$ | $\dfrac{R-R_m}{R+1}$ | $\dfrac{(n+1)-(n_m+1)}{n+2}$ | $n$ |
|---|---|---|---|
| 1 | $\dfrac{0.17}{2} = 0.085$ | 0.55 | 19 |
| 2 | $\dfrac{1.17}{3} = 0.390$ | 0.32 | 11.8 |
| 5 | $\dfrac{4.17}{6} = 0.695$ | 0.15 | 9.1 |
| 10 | $\dfrac{9.17}{11} = 0.833$ | 0.08 | 8.2 |

This Example shows that the number of plates falls off rapidly at first, though more slowly later, and a value of $R \approx 2$ is probably the most economic.

For calculations of the type illustrated by Example 11.18 a convenient nomograph has recently been produced by ZANKER[41] which relates $R$, $R_m$, $n$, and $n_m$ so that any variable may be quickly found if the other three are known.

## Example 11.19

In the separation of a mixture of 100 kmol of hexane, heptane and octane, the flows and concentrations are:

| | hexane | heptane | octane |
|---|---|---|---|
| Relative volatility | 2.70 | 2.22 | 1.0 |
| Feed: | | | |
| kmol | 40 | 35 | 25 |
| mole fraction | 0.40 | 0.35 | 0.25 |
| Overheads: | | | |
| kmol | 40 | 34 | 1 |
| mole fraction | 0.534 | 0.453 | 0.013 |
| Bottom product: | | | |
| kmol | 0 | 1 | 24 |
| mole fraction | 0 | 0.04 | 0.96 |

Assuming operation for 8000 h/year, a plate efficiency of 0.95, allowable vapour velocities of $u' = 2 \times 10^{-3}$ and $N'' = 1.35 \times 10^{-3}$ kmol/m$^2$ s respectively and the following incremental costs:

$c_a =$ £400/m$^2$ year plate.

$c_b =$ £25/m$^2$ year, both of which allow 50 per cent/year for depreciation, interest and maintenance.

and $c_d =$ £0.05/kmol based of an overall coefficient of 0.5 kW/m$^2$ deg K and a temperature difference of 15 deg K in both the condenser and the reboiler.

Estimate the optimum reflux ratio.

## Solution

The minimum reflux ratio, $R_m$ is calculated using Underwood's method (Example 11.16) as 0.83 and, using Fenske's method, Example 11.17, the number of plates at total reflux is $n_m = 8$. The following data have been taken from Figure 11.42, attributable to GILLILAND[30]:

| $(R - R_m)/(R + 1)$ | $[(n + 1) - (n_m + 1)]/(n + 2)$ |
|---|---|
| 0 | 0.75 |
| 0.02 | 0.62 |
| 0.04 | 0.60 |
| 0.06 | 0.57 |
| 0.08 | 0.55 |
| 0.1 | 0.52 |
| 0.2 | 0.45 |
| 0.4 | 0.30 |
| 0.6 | 0.18 |
| 0.8 | 0.09 |
| 1.0 | 0 |

Substituting $R_m = 0.83$ and $n_m = 8$, the following data are obtained, together with values of $-\mathrm{d}n/\mathrm{d}R$ obtained from a plot of $R$ against $n$:

| $R$ | $n$ | $-dn/dR$ |
|---|---|---|
| 0.92 | 28.6 | 110.0 |
| 1.08 | 22.8 | 34.9 |
| 1.25 | 16.9 | 9.8 |
| 1.75 | 13.5 | 3.8 |
| 2.5 | 11.7 | 1.7 |
| 3.5 | 10.5 | 0.6 |
| 5.0 | 9.8 | 0.4 |
| 7.0 | 9.2 | 0.2 |
| 9.0 | 8.95 | 0.05 |

On the basis of the cost data, $F = 7.72$ and hence from equation 11.67:

$$R_{opt} + 1 = (n_{opt} + 7.72)/(-dn/dR)$$

Values of $R$ are now selected and substituted in this equation with the corresponding values of $n$ and $-dn/dR$ taken from the previous table until a balance is attained. This occurs when $\underline{\underline{R = 1.25}}$ or approximately 150 per cent of the minimum reflux condition.

### 11.7.8. Multiple Column Systems

In considering multicomponent systems, a feed of say **A**, **B** and **C** has been fed to a single column and the top product, mainly pure **A**, is obtained with bottoms of mainly **B** and **C**. If each component is required pure, then a two-column system is required in which the bottom stream is fed to the second column to separate **B** and **C**. With more components in the feed, more columns will be required and their arrangement becomes more complex. This is a typical problem in the petrochemical industry and a paper by ELICECHE and SARGENT[42] offers particularly helpful advice.

## 11.8. AZEOTROPIC AND EXTRACTIVE DISTILLATION

In the systems considered previously, the vapour becomes steadily richer in the more volatile component on successive plates. There are two types of mixture where this steady increase in the concentration of the more volatile component, either does not take place, or it takes place so slowly that an uneconomic number of plates is required.

If, for example, a mixture of ethanol and water is distilled, the concentration of the alcohol steadily increases until it reaches 96 per cent by mass, when the composition of the vapour equals that of the liquid, and no further enrichment occurs. This mixture is called an azeotrope, and it cannot be separated by straightforward distillation. Such a condition is shown in the $y - x$ curves of Fig. 11.4 where it is seen that the equilibrium curve crosses the diagonal, indicating the existence of an azeotrope. A large number of azeotropic mixtures have been found, some of which are of great industrial importance, such as water-nitric acid, water-hydrochloric acid, and water-alcohols. The problem of non-ideality is discussed in Section 11.2.4 where the determination of the equilibrium data is considered. When the activity coefficient is greater than unity, giving a positive deviation from Raoult's law, the molecules of the components in the system repel each other and exert a higher partial pressure than if their behaviour were ideal. This leads to the formation of a "minimum boiling" azeotrope shown in Figure 11.43*a*. For values of the activity coefficient less than unity, negative deviation from Raoult's law results in a lower partial pressure and the formation of a "maximum boiling" azeotrope, as shown in Figure 11.43*b*.

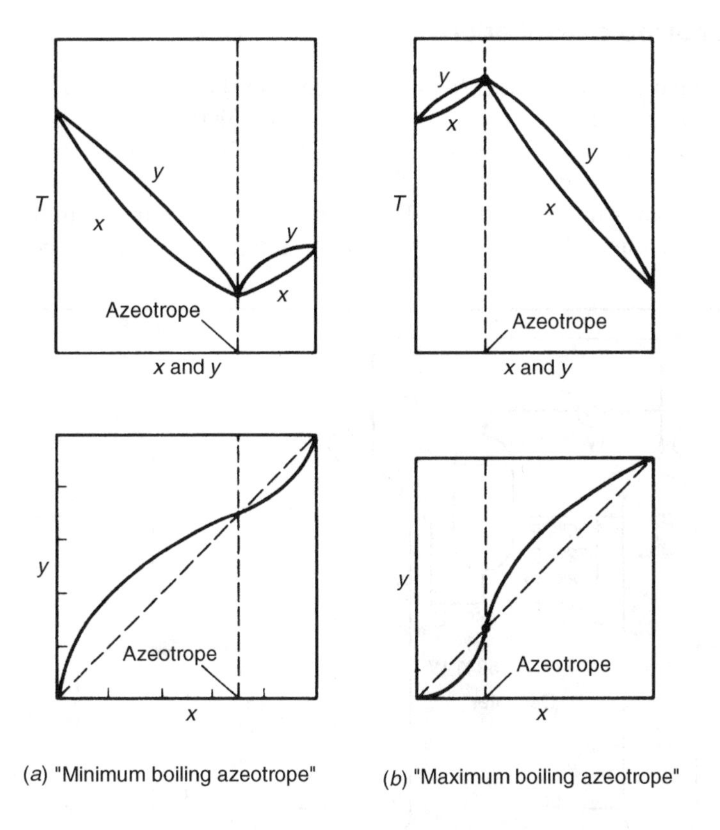


Figure 11.43. Types of azeotropic behaviour

The second type of problem occurs where the relative volatility of a binary mixture is very low, in which case continuous distillation of the mixture to give nearly pure products will require high reflux ratios with correspondingly high heat requirements. In addition, it will necessitate a tower of large cross-section containing many trays. An example of the second type of problem is the separation of *n*-heptane from methyl cyclohexane in which the relative volatility is only 1.08 and a large number of plates is required to achieve separation.

The principle of azeotropic and of extraction distillation lies in the addition of a new substance to the mixture so as to increase the relative volatility of the two key components, and thus make separation relatively easy. Benedict and Rubin[43] have defined these two processes as follows. In azeotropic distillation the substance added forms an azeotrope with one or more of the components in the mixture, and as a result is present on most of the plates of the column in appreciable concentrations. With extractive distillation the substance added is relatively non-volatile compared with the components to be separated, and it is therefore fed continuously near the top of the column. This extractive agent runs down the column as reflux and is present in appreciable concentrations on all the plates. The third component added to the binary mixture is sometimes known as the *entrainer* or the *solvent*.

### 11.8.1. Azeotropic distillation

YOUNG[44], found in 1902, that if benzene is added to the ethanol–water azeotrope, then a ternary azeotrope is formed with a boiling point of 338.0 K, that is less than that of the binary azeotrope, 351.3 K. The industrial production of ethanol from the azeotrope, using this principle, has been described by GUINOT and CLARK[45] and the general arrangement of the plant is as shown in Figure 11.44. This requires the use of three atmospheric pressure fractionating columns, and a continuous two-phase liquid separator or decanter.

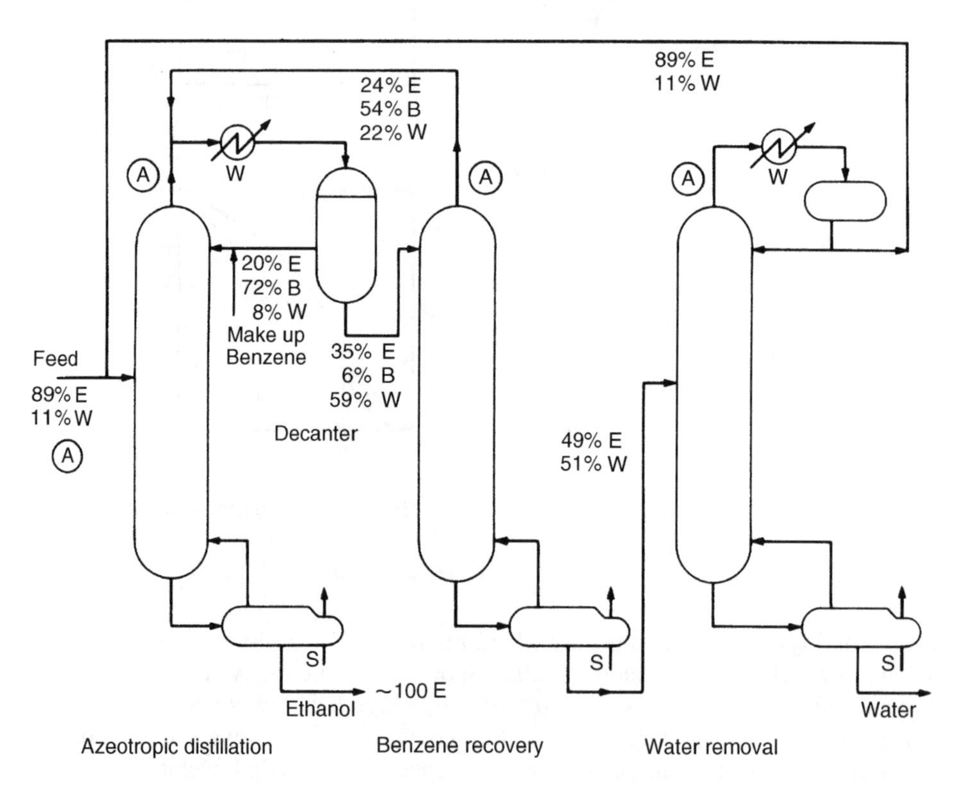


Figure 11.44. Azeotropic distillation for the separation of ethanol from water using benzene as entrainer. Compositions are given in mole per cent. E = Ethanol, B = Benzene, W = Water, S = Steam

The azeotrope in the ethanol–water binary system has a composition of 89 mole per cent of ethanol[14]. Starting with a mixture containing a lower proportion of ethanol, it is not possible to obtain a product richer in ethanol than this by normal binary distillation. Near azeotropic conditions exist at points marked Ⓐ in Figure 11.44. The addition of the relatively non-polar benzene entrainer serves to volatilise water, a highly polar molecule, to a greater extent than ethanol, a moderately polar molecule, and a virtually pure ethanol product may be obtained. Equilibrium conditions for this system have been discussed by NORMAN[46] who shows how the number of plates required may be determined.

The first tower in Figure 11.44 gives the ternary azeotrope as an overhead vapour, and nearly pure ethanol as bottom product. The ternary azeotrope is condensed and splits into two liquid phases in the decanter. The benzene-rich phase from the decanter serves as reflux, while the water–ethanol-rich phase passes to two towers, one for benzene recovery and the other for water removal. The azeotropic overheads from these successive towers are returned to appropriate points in the primary tower.

Figure 11.45 shows a composition profile for the azeotropic distillation column in the process shown in Figure 11.44. This is taken from a solution presented by ROBINSON and GILLILAND[1].

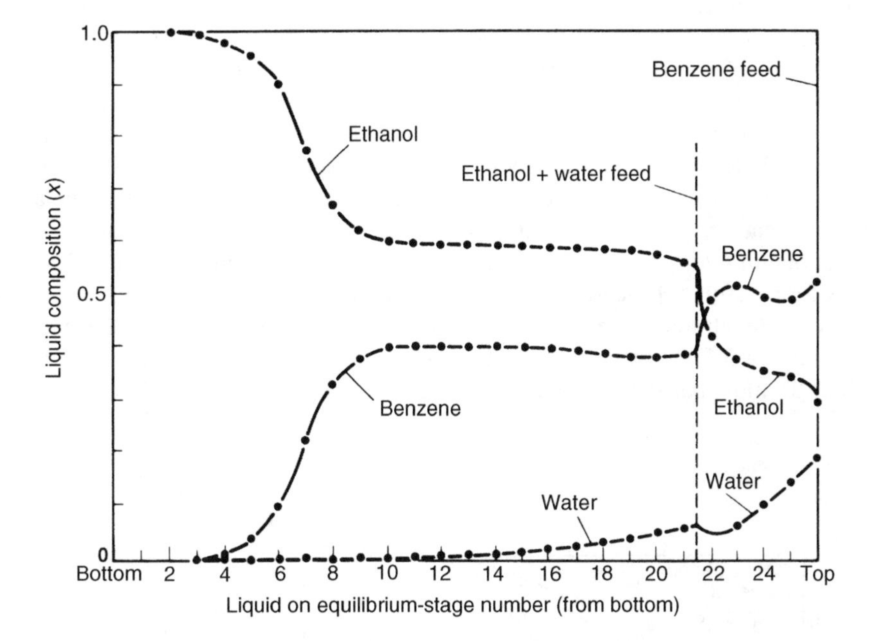


Figure 11.45. Composition profile for the azeotropic distillation of ethanol and water, with benzene as entrainer

## 11.8.2. Extractive distillation

Extractive distillation is a method of rectification similar in purpose to azeotropic distillation. To a binary mixture which is difficult or impossible to separate by ordinary means, a third component, termed a *solvent*, is added which alters the relative volatility of the original constituents, thus permitting the separation. The added solvent is, however, of low volatility and is itself not appreciably vaporised in the fractionator.

For a non-ideal binary mixture the partial pressure may be expressed as:

$$P_A = \gamma_A P_A^\circ x_A \tag{11.117}$$

$$P_B = \gamma_B P_B^\circ x_B \tag{11.118}$$

where $\gamma_A$ and $\gamma_B$ are the activity coefficients for the two components. The relative volatility $\alpha$ may thus be written as:

$$\alpha = \frac{P_A}{P_B}\frac{x_B}{x_A}$$

$$= \frac{\gamma_A}{\gamma_B}\frac{P_A^\circ}{P_B^\circ} \tag{11.119}$$

The solvent added to the mixture in extractive distillation differentially affects the activities of the two components, and hence the relative volatility, $\alpha$.

Such a process depends upon the difference in departure from ideality between the solvent and the components of the binary mixture to be separated. In the following example, both toluene and *iso*-octane separately form non-ideal liquid solutions with phenol, although the extent of the non-ideality with *iso*-octane is greater than that with toluene. When all three substances are present, therefore, the toluene and *iso*-octane themselves behave as a non-ideal mixture, and their relative volatility becomes high.

An example of extractive distillation given by TREYBAL[47] is the separation of toluene, boiling point 384K, from paraffin hydrocarbons of approximately the same molecular weight. This is either very difficult or impossible, owing to low relative volatility or azeotrope formation, yet such a separation is necessary in the recovery of toluene from certain petroleum hydrocarbon mixtures. Using *iso*-octane of boiling point 372.5 K, as an example of a paraffin hydrocarbon, Figure 11.46*a* shows that *iso*-octane in this mixture is the more volatile, although the separation is obviously difficult. In the presence of phenol, boiling point 454.6 K, however, the relative volatility of *iso*-octane increases, so that, with as much as 83 mole per cent phenol in the liquid, the separation from toluene is relatively

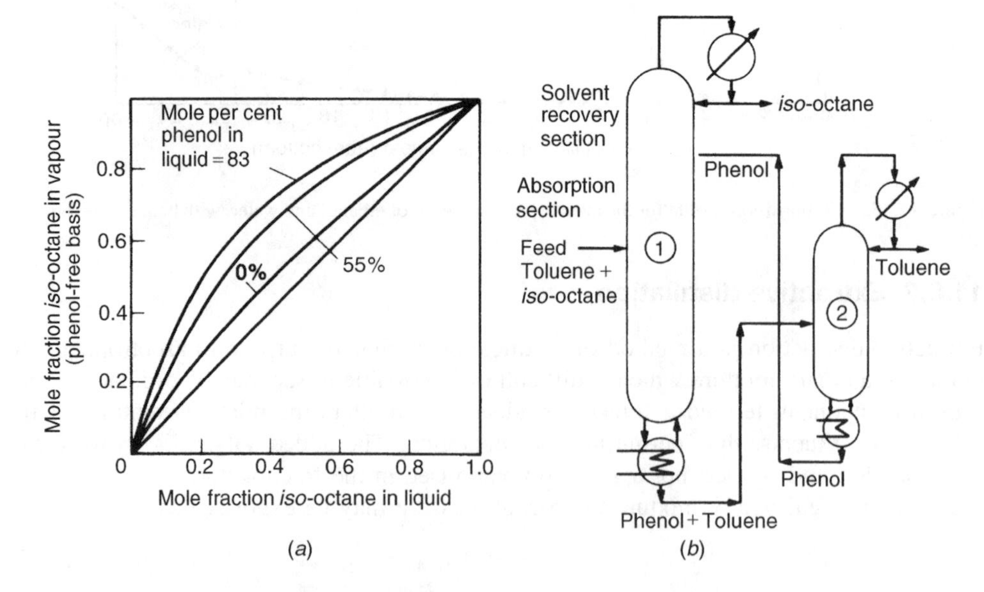


Figure 11.46. Extractive distillation of toluene-**iso**-octane with phenol

easy. A flowsheet of a process for accomplishing this is shown in Figure 11.46*b*, where the binary mixture is introduced more or less centrally into the extractive distillation tower (1), and phenol as the solvent is introduced near the top so as to be present in high concentration upon most of the trays in the tower. Under these conditions *iso*-octane is readily distilled as an overhead product, while toluene and phenol are removed as a residue. Although phenol is relatively high-boiling, its vapour pressure is nevertheless sufficient for some to appear in the overhead product. The solvent-recovery section of the tower, which may be relatively short, serves to separate the phenol from the *iso*-octane. The residue from the tower must be rectified in the auxiliary tower (2) to separate toluene from the phenol which is recycled, but this is a relatively easy separation. In practice, the paraffin hydrocarbon is a mixture rather than pure *iso*-octane, although the principle of the operation remains the same.

The solvent to be used is selected on the basis of selectivity, volatility, ease of separation from the top and bottom products, and the cost. The selectivity is most easily assessed by determining the effect on the relative volatility of the two key components of addition of the solvent. The more volatile the solvent, the greater the percentage of solvent in the vapour, and the poorer the separation for a given heat consumption in the boiler. It is important to note that the solvent must not form an azeotrope with any of the components. Some of the problems of selecting the solvent are discussed by Scheibel[48] who points out that use may be made of the fact that, when two compounds show deviations from Raoult's law, then one of these compounds shows the same type of deviation with any member of the homologous series of the other component. Thus the azeotropic mixture acetone (b.p. 329.6 K)–methanol (b.p. 337.9 K) has 20 mole per cent acetone and boils at 328.9 K., that is less than the boiling point of either component. Thus any member of the series ethanol (b.p. 357.5 K), propanol (b.p. 370.4 K), water (b.p. 373.2 K), butanol (b.p. 391.0 K) may be used as an extractive agent, or in the series of ketones, methyl *n*-propyl ketone (b.p. 375 K) and methyl *iso*-butyl ketone (b.p. 389.2 K). The advantage of using a solvent from the alcohol series is that the more volatile acetone will be taken overhead, though water would have the advantage of cheapness. Pratt[49] has given details of a method of calculation for extractive distillation, using the system acetonitrile–trichloroethylene–water as an example.

Extractive distillation is usually more desirable than azeotropic distillation since no large quantities of solvent have to be vaporised. In addition, a greater choice of added component is possible since the process is not dependent upon the accident of azeotrope formation. It cannot, however, be conveniently carried out in batch operations.

Azeotropic and extractive-distillation equipment may be designed using the general methods for multicomponent distillation, and detailed discussion is available elsewhere[1,42] and presented by Hoffman[50] and Smith[51].

## 11.9. STEAM DISTILLATION

Where a material to be distilled has a high boiling point, and particularly where decomposition might occur if direct distillation is employed, the process of steam distillation may be used. Steam is passed directly into the liquid in the still and the solubility of the steam in the liquid must be very low. Steam distillation is perhaps the most common example of differential distillation.

Two cases may be considered. The steam may be superheated and so provide sufficient heat to vaporise the material concerned, without itself condensing. Alternatively, some of the steam may condense, producing a liquid water phase. In either case, assuming the ideal gas laws to apply, the composition of the vapour produced may be obtained from:

$$\left(\frac{m_A}{M_A}\right) \Big/ \left(\frac{m_B}{M_B}\right) = \frac{P_A}{P_B} = \frac{y_A}{y_B} = \frac{P_A}{P - P_A} \tag{11.120}$$

where the subscript $A$ refers to the component being recovered, and $B$ to steam, and:

$m$ = mass,
$M$ = molecular weight,
$P_A, P_B$ = partial pressure of **A**, **B**, and
$P$ = total pressure.

If there is no liquid phase present, then from the phase rule there will be two degrees of freedom, both the total pressure and the operating temperature can be fixed independently, and $P_B = P - P_A$, which must not exceed the vapour pressure of pure water, if no liquid phase is to appear.

When a liquid water phase is present, there will be only one degree of freedom, and selecting the temperature or pressure fixes the system, with the water and the other component each exerting a partial pressure equal to its vapour pressure at the boiling point of the mixture. In this case, the distillation temperature will always be less than that of boiling water at the total pressure in question. Consequently, a high boiling organic material may be steam-distilled at temperatures below 373 K at atmospheric pressure. By using reduced operating pressures, the distillation temperature may be reduced still further, with a consequent economy of steam.

A convenient method of calculating the temperature and composition of the vapour, for the case where the liquid water phase is present, is by using the diagram shown in Figure 11.47 which is due to HAUSBRAND[52], where the parameter $(P - P_B)$ is plotted for total pressures of 101.3, 40 and 9.3 kN/m$^2$, and the vapour pressures of a number of other materials are plotted directly against temperature. The intersection of the two appropriate curves gives the temperature of distillation, and the molar ratio of water to organic material is given by $(P - P_A)/P_A$. Thus, if nitrobenzene is distilled at atmospheric pressure with live saturated steam, the boiling point will be about 372 K and the mass-ratio of water to nitrobenzene in the vapour will be:

$$\frac{(101.3 - 2.7)}{2.7}\left(\frac{18}{123}\right) = 5.34$$

Where there is no liquid water phase present, the steam consumption will be high unless the steam is very highly superheated. With a water phase present, the boiling point of the mixture will be low, and consequently $P_A$ will have a low value. Thus, on a molar basis, the steam consumption will again be high, although due to the relatively low molecular weight of steam, the consumption may not be excessive. Steam economy may be effected by using indirect heating of the still, having no liquid water phase present, or by operating under reduced pressure.

In an operation of this kind, illustrated in Figure 11.48, it is essential that the separation of the material being distilled from the water should be a relatively simple operation.

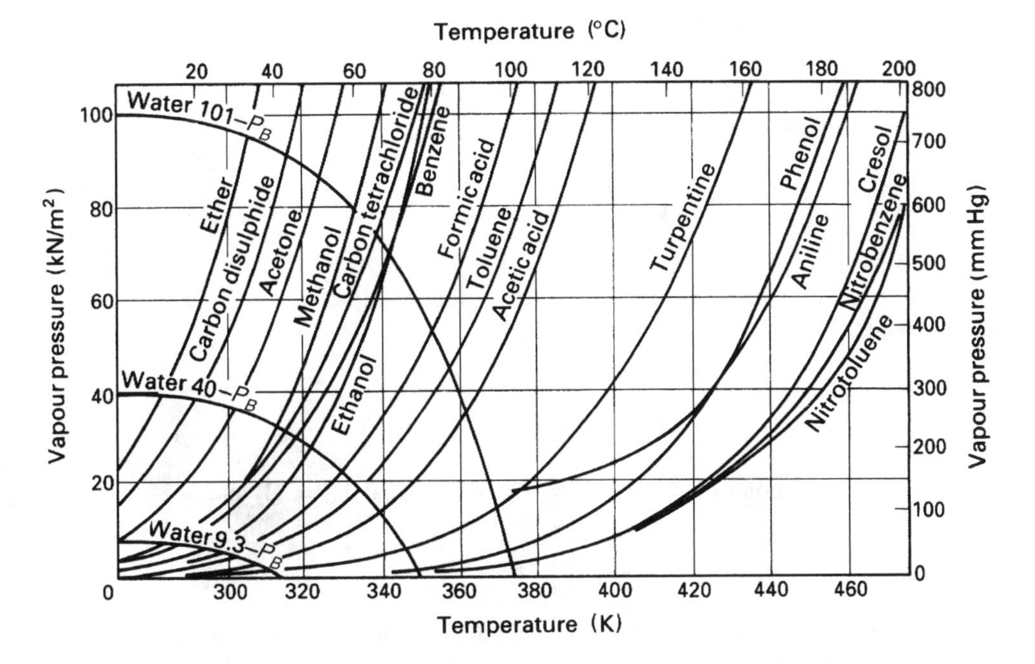


Figure 11.47. Vapour pressure curves for steam distillation calculations

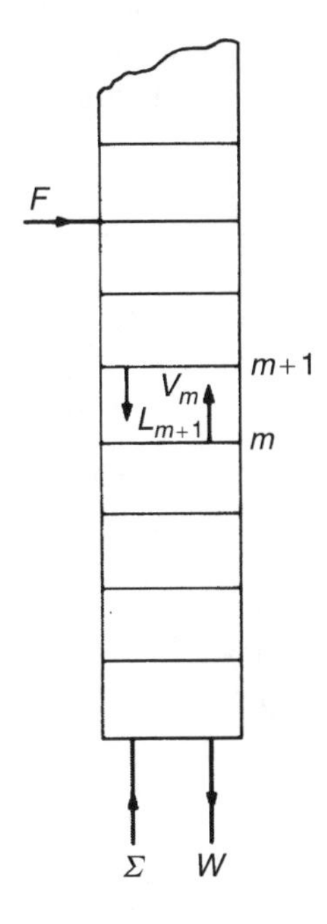


Figure 11.48. Steam distillation

In determining the number of stages required to effect a steam distillation, the steam flow must be included in the operating-line equation for the lower part of the column. Using indirect heating and assuming constant molar overflow, the lower operating line for the organic material is given by:

$$y_m = \frac{L_m}{V_m} x_{m+1} - \frac{W}{V_m} x_w \qquad \text{(equation 11.37)}$$

This has a slope $L_m/V_m$ and cuts the $y = x$ line at $x = x_w$.

If $\Sigma$ kmol of live steam is used as shown in Figure 11.48, then considering that part of the column below the plate $(m + 1)$:

$$V_m = L_{m+1} + \Sigma - W \qquad (11.121)$$

and:

$$V_m y_m = L_{m+1} x_{m+1} - W x_w \qquad (11.122)$$

Assuming constant molar overflow, $L_m = L_{m+1} = W$ and $V_m = \Sigma$, and:

$$y_m = \frac{L_m}{V_m} x_{m+1} - \frac{W}{V_m} x_w \qquad (11.123)$$

This equation also has a slope of $L_m/V_m$ although it cuts the $y = x$ line at $x = W x_w/(W - \Sigma)$ as shown in Figure 11.49. When $x = x_w$, $y = 0$, corresponding to the composition of the vapour, steam, rising to the bottom plate. For a given external reflux ratio and feed condition, $L_m/V_m$ will be the same whether direct or indirect steam is used and the lower operating line must cut the $y = x$ line at the same value in each case, though, of course, $x_m$ for the indirect steam will be higher than $x_w$ for direct steam.

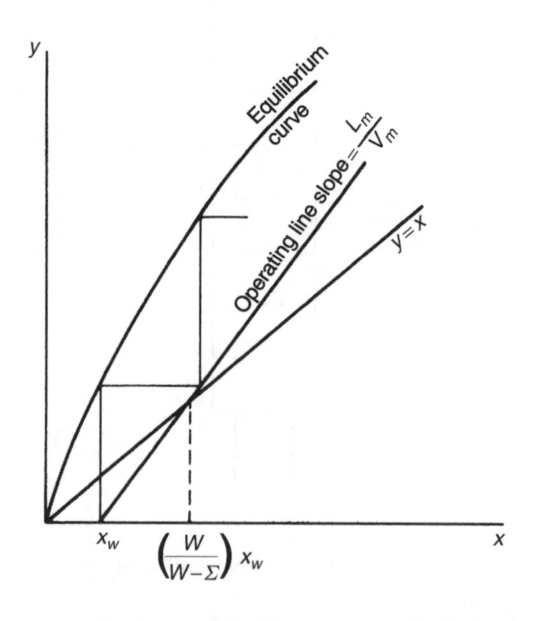


Figure 11.49. Operating lines for steam distillation

When stepping off the theoretical stages, the bottom step should start at $y = 0, x = x_w$. In this way, the use of direct steam, although eliminating the still, dilutes the bottom material, and so increases the number of stages required in the lower part of the column.

## 11.10. PLATE COLUMNS

Distillation may be carried out in plate columns in which each plate constitutes a single stage, or in packed columns where mass transfer is between a vapour and liquid in continuous countercurrent flow. Plate columns are now considered, and packed columns are discussed in Section 11.11.

The number of theoretical stages required to effect a required separation, and the corresponding rates for the liquid and vapour phases, may be determined by the procedures described previously. In order to translate these quantities into an actual design the following factors should be considered:

(a) The type of plate or tray.
(b) The vapour velocity, which is the major factor in determining the diameter of the column.
(c) The plate spacing, which is the major factor fixing the height of the column when the number of stages is known.

Design methods are discussed later, although it is first necessary to consider the range of trays available and some of their important features.

### 11.10.1. Types of trays

The main requirement of a tray is that it should provide intimate mixing between the liquid and vapour streams, that it should be suitable for handling the desired rates of vapour and liquid without excessive entrainment or flooding, that it should be stable in operation, and that it should be reasonably easy to erect and maintain. In many cases, particularly with vacuum distillation, it is essential that the drop in pressure over the tray should be a minimum.

The arrangements for the liquid flow over the tray depend largely on the ratio of liquid to vapour flow. Three layouts are shown in Figure 11.50, of which the cross-flow arrangement is much the most frequently used. Considering these in turn:

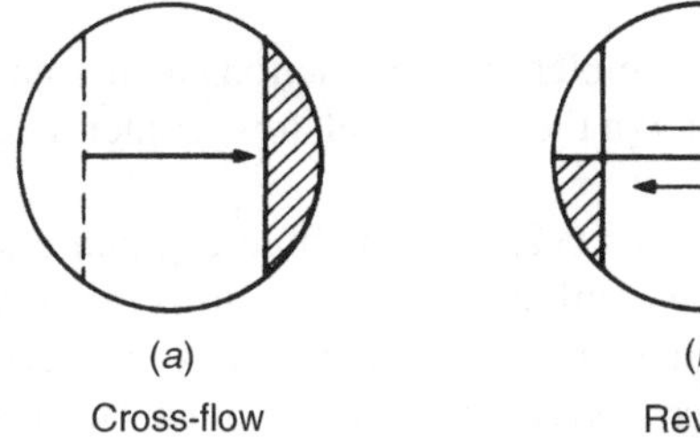


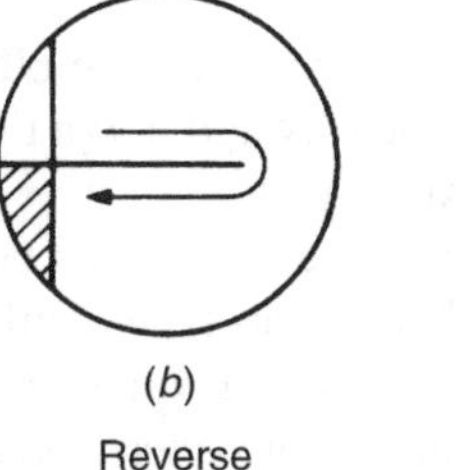


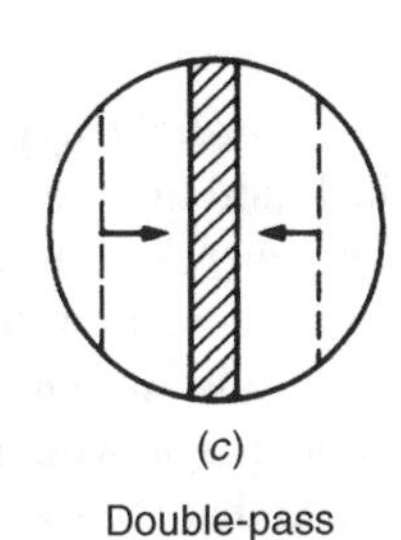


Figure 11.50. Arrangements for liquid flow over a tray

(a) *Cross-flow*. Normal, with a good length of liquid path giving a good opportunity for mass transfer.
(b) *Reverse*. Downcomers are much reduced in area, and there is a very long liquid path. This design is suitable for low liquid–vapour ratios.
(c) *Double-pass*. As the liquid flow splits into two directions, this system will handle high liquid–vapour ratios.

The liquid reflux flows across each tray and enters the downcomer by way of a weir, the height of which largely determines the amount of liquid on the tray. The downcomer extends beneath the liquid surface on the tray below, thus forming a vapour seal. The vapour flows upwards through risers into caps, or through simple perforations in the tray.

*The bubble-cap tray*. This is the most widely used tray because of its range of operation, although it is being superseded by newer types, such as the valve tray discussed later. The general construction is shown in Figure 11.51. The individual caps are mounted on risers and have rectangular or triangular slots cut around their sides. The caps are held in position by some form of spider, and the areas of the riser and the annular space around the riser should be about equal. With small trays, the reflux passes to the tray below over two or three circular weirs, and with the larger trays through segmental downcomers.

Figure 11.51. A bubble tray

*Sieve or perforated trays*. These are much simpler in construction, with small holes in the tray. The liquid flows across the tray and down the segmental downcomer. Figure 11.52 indicates the general form of tray layout.

*Valve trays*. These may be regarded as a cross between a bubble-cap and a sieve tray. The construction is similar to that of cap types, although there are no risers and no slots. It may be noted that with most types of valve tray the opening may be varied by the vapour flow, so that the trays can operate over a wide range of flowrates. Because of their flexibility and price, valve trays are tending to replace bubble-cap trays. Figure 11.53 shows a typical tray.

Figure 11.52. A perforated plate tray

Figure 11.53. A valve tray

These three types of trays have a common feature in that they all have separate downcomers for the passage of liquid from each tray to the one below. There is another class of tray which has no separate downcomers and yet it still employs a tray type of construction giving a hydrodynamic performance between that of a packed and a plate column. Two examples of this type of device are the Kittel plate and a Turbogrid tray[53]. Design data for these trays are sparse in the literature and the manufacturer's recommendations should be sought.

### 11.10.2. Factors determining column performance

The performance of a column may be judged in relation to two separate but related criteria. First, if the vapour and liquid leaving a tray are in equilibrium this constitutes a theoretical tray and provides a standard of performance. Secondly, the relative performance of, say, two columns of the same diameter must be considered in relation to their capacity for liquid and vapour flow. The main features are:

(a) Liquid and vapour velocities.
(b) Physical properties of the liquid and vapour.
(c) Extent of entrainment of liquid by rising vapour streams.
(d) The hydraulics of the flow of liquid and vapour across and through the tray.

It has been found by CAREY[54], CAREY *et al.*[55] and SOUDERS and BROWN[56] that the vapour velocity is a prime factor in determining the diameter of a column. KIRSCHBAUM[57] using an equimolar mixture of ethanol and water on a 400 mm diameter plate containing 15 bubble caps, obtained the following results:

(a) For all plate spacings the efficiency $E_{Mv}$, which is defined in equation 11.124 as the ratio of the actual change in liquid composition on a plate to that which would be obtained if the liquid left in equilibrium with the vapour, decreases as the velocity is increased due mainly to the reduction in contact time between the phases.
(b) The decrease in efficiency is much less with high plate spacings.
(c) The capacity is limited by the ability of the downcomers to carry the reflux, rather than that of the caps to handle the vapour.

The effects of liquid viscosity on tray efficiency have been studied by DRICKAMER and BRADFORD[58] and O'CONNELL[59] and these are discussed in Section 11.10.5. Surface tension influences operation with sieve trays, in relation both to foaming and to the stability of bubbles.

### 11.10.3. Operating ranges for trays

For a given tray layout there are certain limits for the flows of vapour and liquid within which stable operation is obtained. The range is shown in Figure 11.54, which relates to a bubble-cap plate. The region of satisfactory operation is bounded by areas where

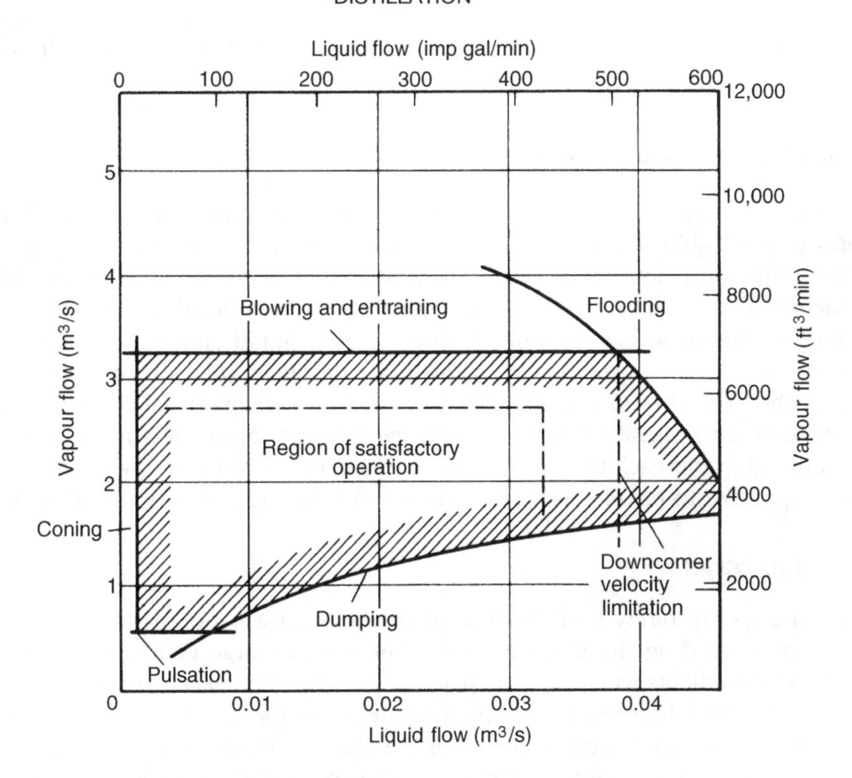


Figure 11.54. Capacity graph for a typical bubble-cap tray

undesirable phenomena occur. Coning occurs at low liquid rates, where the vapour forces the liquid back from the slots and passes out as a continuous stream, with a consequent loss in efficiency. Low vapour rates result in pulsating vapour flow or dumping. With low liquid rates, vapour passes through the slots intermittently, though with higher liquid rates some slots dump liquid rather than passing vapour. Both pulsating vapour flow and dumping, which may be referred to jointly as weeping, result in poor efficiency. At very high vapour rates, the vapour bubbles carry liquid as spray or droplets to the plate above, giving excessive entrainment. With high liquid rates, a point is reached where the drop in pressure across the plate equals the liquid head in the downcomer. Beyond this point, the liquid builds up and floods the tray.

The extent of entrainment of the liquid by the vapour rising over a plate has been studied by many workers. The entrainment has been found to vary with the vapour velocity in the slot or perforation, and the spacing used. STRANG[60], using an air–water system, found that entrainment was small until a critical vapour velocity was reached, above which it increased rapidly. Similar results from PEAVY and BAKER[61] and COLBURN[62] have shown the effect on tray efficiency, which is not seriously affected until the entrainment exceeds 0.1 kmol of liquid per kmol of vapour. The entrainment on sieve trays is discussed in Section 11.10.4.

The design of the tray fittings and the downcomers influences the column performance. It is convenient to consider this separately for bubble cap trays and sieve trays. In design, the important factors are the diameter of the tower, the tray spacing and the detailed design of the tray.

## 11.10.4. General design methods

In designing a column for a given separation, the number of stages required and the flowrates of the liquid and vapour streams must first be determined using the general methods outlined previously. In the mechanical design of the column, tower diameter, tray spacing, and the detailed layout of each tray is considered. Initially, a diameter is established, based on the criterion of absence from liquid entrainment in the vapour stream, and then the weirs and the downcomers are designed to handle the required liquid flow. It is then possible to consider the tray geometry in more detail, and, finally, to examine the general operating conditions for the tray and to establish its optimum range of operation. This approach to design is covered in detail in Volume 6 where the different methods applicable to bubble caps, sieve trays and valve trays are discussed in detail.

### *Bubble cap trays*

Bubble-cap trays are rarely used for new installations on account of their high cost and their high pressure drop. In addition, difficulties arise in large columns because of the large hydraulic gradients which are set up across the trays. Bubble cap trays are capable of dealing with very low liquid rates and are therefore useful for operation at low reflux ratios. There are still many bubble-cap columns in use and the design considerations presented in Volume 6 are given to enable, in particular, existing equipment to be assessed for new applications and duties.

### *Sieve trays*

Sieve trays offer several advantages over bubble-cap trays, and their simpler and cheaper construction has led to their increasing use. The general form of the flow on a sieve tray is typical of a cross-flow system with perforations in the tray taking the place of the more complex bubble caps. The hydraulic flow conditions for such a tray are discussed in Volume 6 in the same manner as for the bubble-cap tray, by considering entrainment, flooding, pressure loss, and so on. The key differences in operation between these two types of tray should be noted. With the sieve tray the vapour passes vertically through the holes into the liquid on the tray, whereas with the bubble cap the vapour issues in an approximately horizontal direction from the slots. With the sieve plate the vapour velocity through the perforations must be greater than a certain minimum value in order to prevent the weeping of the liquid stream down through the holes. At the other extreme, a very high vapour velocity leads to excessive entrainment and loss of tray efficiency. The capacity graph for a sieve tray is similar to that shown in Figure 11.54 for bubble cap trays.

### *Valve trays*

The valve tray, which may be regarded as intermediate between the bubble cap and the sieve tray, offers advantages over both. The important feature of the tray is that liftable caps act as variable orifices which adjust themselves to changes in vapour flow. The valves are either metal discs of up to about 38 mm diameter, or metal strips which are raised above the openings in the tray deck as vapour passes through the trays. The caps are restrained by legs or spiders which limit the vertical movement and some types are capable of forming a total liquid seal when the vapour flow is insufficient to lift the cap.

Advantages claimed for valve trays include:

(a) Operation at the same capacity and efficiency as sieve trays.
(b) A low pressure drop which is fairly constant over a large portion of the operating range.
(c) A high turndown ratio, that is it can be operated at a small fraction of design capacity.
(d) A relatively simple construction which leads to a cost of only 20 per cent higher than that of a comparable sieve tray.

Valve trays, because of their proprietary nature, are usually designed by manufacturers, although it is possible to obtain an estimate of design and performance from published literature[63] and from the methods summarised in Volume 6.

### 11.10.5. Plate efficiency

The number of ideal stages required for a desired separation may be calculated by one of the methods discussed previously, although in practice more trays are required than ideal stages. The ratio $n/n_p$ of the number of ideal stages $n$ to the number of actual trays $n_p$ represents the overall efficiency $E$ of the column, which may be 30–100 per cent[4]. The main reason for loss in efficiency is that the kinetics for the rate of approach to equilibrium, and the flow pattern on the plate, may not permit an equilibrium between the vapour and liquid to be attained. Some empirical equations have been developed from which values of efficiency may be calculated, and this approach is of considerable value in giving a general picture of the problem. The proportion of liquid and vapour, and the physical properties of the mixtures on the trays, will vary up the column, and conditions on individual trays must be examined, as suggested by MURPHREE[64]. For a single ideal tray, the vapour leaving is in equilibrium with the liquid leaving, and the ratio of the actual change in composition achieved to that which would occur if equilibrium between $y_n$ and $x_n$ were attained is known as the Murphree plate efficiency $E_M$. Using the notation shown in Figure 11.55, the plate efficiency expressed in vapour terms is given by:

$$E_{Mv} = \frac{y_n - y_{n-1}}{y_e - y_{n-1}} \tag{11.124}$$

where $y_e$ is the composition of the vapour that would be in equilibrium with the liquid of composition $x_n$ actually leaving the plate. This equation gives the efficiency in vapour terms, although if the concentrations in the liquid streams are used then the plate efficiency $E_{Ml}$ is given by:

$$E_{Ml} = \frac{x_{n+1} - x_n}{x_{n+1} - x_e} \tag{11.125}$$

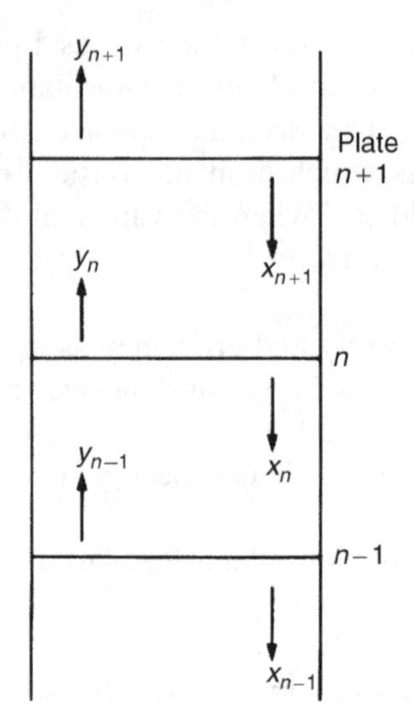


Figure 11.55. Compositions of liquid and vapour streams from plates

where $x_e$ is the composition of the liquid that would be in equilibrium with the composition $y_n$ of the vapour actually leaving the plate.

The ratio $E_{Mv}$ is shown graphically in Figure 11.56 where for any operating line AB the enrichment that would be achieved by an ideal plate is BC, and that achieved with an actual plate is BD. The ratio BD/BC then represents the plate efficiency. The efficiency may vary from point to point on a tray. Local values of the Murphree efficiency are designated $E_{mv}$ and $E_{ml}$.

## Empirical expressions for plate efficiency

The efficiency of the individual plates is expected to depend on the physical properties of the mixture, the geometrical arrangements of the trays, and the flowrates of the two

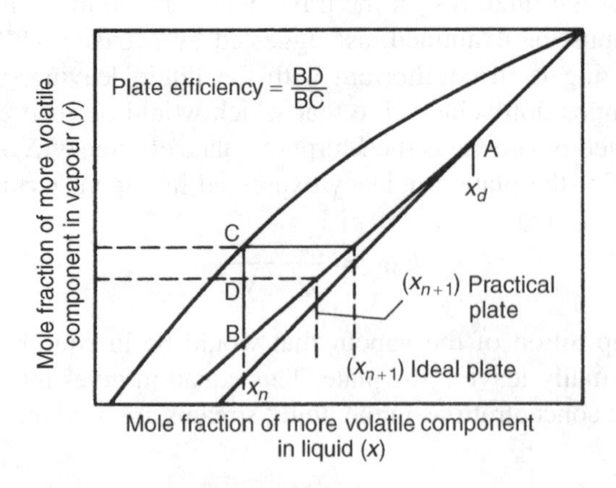


Figure 11.56. Graphical representation of plate efficiency $E_{Mr}$

phases. A simple empirical relationship for the overall efficiency, $E$, of columns handling petroleum hydrocarbons is given by DRICKAMER and BRADFORD[58] who relate efficiency of the column to the average viscosity of the feed by:

$$E = 0.17 - 0.616 \log_{10} \Sigma[x_f(\mu_L/\mu_w)] \quad (11.126)$$

where: $x_f$ is the mole fraction of the component in the feed,
$\mu_L$ is the viscosity at the mean tower temperature, and
$\mu_w$ is the viscosity of water at 293 K (approximately 1 mNs/m$^2$).

Further work, mainly with larger towers 3 m in diameter, suggested that higher efficiencies were obtained with larger diameters because of the longer liquid path. Thus, compared with a 0.9 m diameter tray, one of 3 m diameter might give up to 25 per cent greater efficiency.

## Example 11.20

Using equation 11.126, determine the plate efficiency for the following data on the separation of a stream of $C_3$ to $C_6$ hydrocarbons.

| Component | Mole fraction in feed $x_f$ | $\mu_L$ (mNs / m$^2$) | $x_f(\mu_L/\mu_w)$ |
|---|---|---|---|
| $C_3$ | 0.2 | 0.048 | 0.0096 |
| $C_4$ | 0.3 | 0.112 | 0.0336 |
| $C_5$ | 0.2 | 0.145 | 0.0290 |
| $C_6$ | 0.3 | 0.188 | 0.0564 |
| | | | 0.1286 |

## Solution

From equation 11.126:

$$E = (0.17 - 0.616 \log 0.1286)$$

$$= \underline{\underline{0.72}}.$$

O'CONNELL[59] found that a rather better relation may obtained by plotting the overall efficiency as a function of the product of the viscosity and the relative volatility of the key components. This relation has also been presented by LOCKHART and LEGGETT[65], as shown in Figure 11.57. Thus, in Example 11.19, taking $C_3$ and $C_5$ as key components, the relative volatility $\alpha$ is about 1.76 and the mean viscosity about 0.4 mN s/m$^2$, giving a product of 0.246. From Figure 11.57, $E$ is then found as 70 per cent.

CHU *et al.*[66] has given a more complex correlation for overall efficiency $E$ by including the relative flowrates $L$ and $V$ of the phases and the effective submergence of the liquid $h_L$. This takes the form:

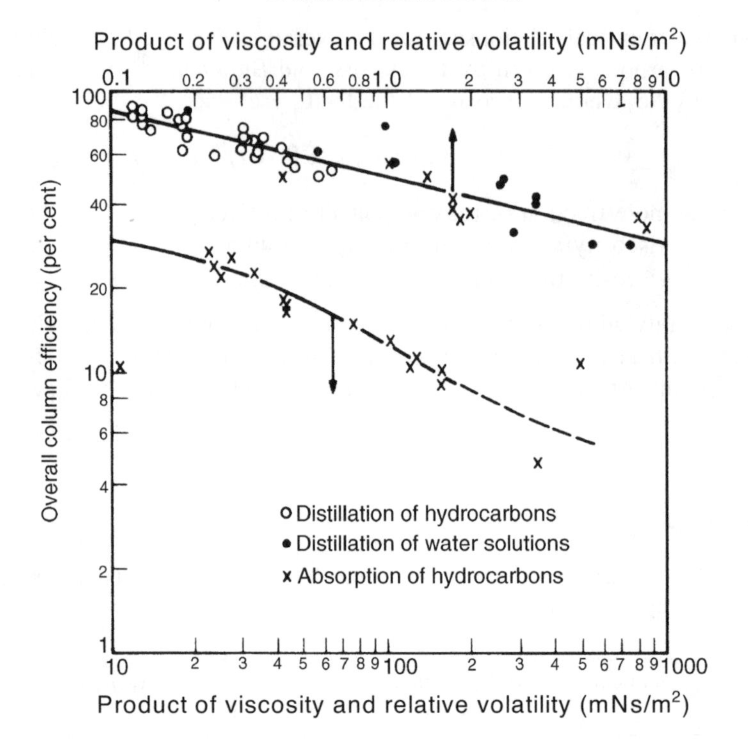


Figure 11.57. Overall column efficiency $E$ as function of viscosity–relative volatility product $(\mu_l x_f)$[65]

$$\log_{10} E = 1.67 + 0.30 \log_{10}(L/V) - 0.25 \log_{10}(\mu_L \alpha) + 0.30 h_L \tag{11.127}$$

where: $L, V$ are the liquid and vapour flowrates (kmol/s),
$\mu_L$ is the viscosity of the liquid feed (mNs/m$^2$),
$\alpha$ is the relative volatility of the key components, and
$h_L$ is the effective submergence (m), taken as the distance from the top of the lot to the weir lip plus half the slot height.

## *Expressions for plate efficiency related to mass transfer*

By assuming that the vapour issuing from slots is in the form of spherical bubbles, Chu[66], Geddes[67], and Bakowski[68] derived methods for expressing the efficiency $E$ in terms of transfer coefficients, $k_g$, $k_l$ and tray parameters such as the slot dimensions. These methods have proved very difficult to use because of the unreliability of data for calculating transfer coefficients, and the greater problem of calculating the interfacial areas. Probably the most successful analysis for determining efficiency in terms of mass transfer functions is that obtained by the American Institute of Chemical Engineers Research Project[69], and an outline of this work is given here. Whilst this is a complex analysis containing parameters which are known only approximately, the method does outline some of the important factors involved and shows that some parameters are of little importance.

### *Plate efficiency in terms of transfer units*

The process of mass transfer across a phase boundary is discussed in Volume 1, Chapter 10. A resistance to mass transfer exists within the fluid on each side of the interface, and the overall transfer rate of a component in a mixture depends on the sum of these resistances and the total driving force.

The concept of a transfer unit for a countercurrent mass transfer process, introduced in Volume 1, is developed further for distillation in packed columns in Section 11.11. The number of transfer units is defined as the integrated value of the ratio of the change in composition to the driving force. Thus, considering the vapour phase, the number of overall gas transfer, units $\mathbf{N}_{OG}$ is given by:

$$\mathbf{N}_{OG} = \int \frac{\mathrm{d}y}{y_e - y} \tag{11.128}$$

For the liquid phase, the corresponding number of overall liquid transfer units $\mathbf{N}_{OL}$ is given by:

$$\mathbf{N}_{OL} = \int \frac{\mathrm{d}x}{x - x_e} \tag{11.129}$$

Equations 11.128 and 11.129 are derived in the same way as equations 11.157 and 11.158.

Noting equation 11.164, the relation between $\mathbf{N}_{OG}$ and $\mathbf{N}_{OL}$ is given by:

$$\frac{\mathbf{N}_{OL}}{\mathbf{N}_{OG}} = \frac{mG'}{L'} = \varepsilon \tag{11.130}$$

where: $m$ is the slope of the vapour–liquid equilibrium line ($y_e$ versus $x$), and
$G'$ and $L'$ are the molar rates of flow of vapour and liquid, respectively, per unit cross-section of column.

The equation for transfer units may be applied to the mass transfer over a tray, and thus relate the local Murphree efficiency $E_{mv}$ to the overall transfer units $\mathbf{N}_{OG}$. With the notation in Figure 11.55, the vapour $y_{n-1}$ rises from plate $n-1$, crosses the liquid on plate $n$ and leaves with composition $y_n$. The liquid flowing from plate $n+1$ through the downcomer crosses tray $n$ and leaves with composition $x_n$. It is supposed for this argument that there is no change in the composition of the liquid in a vertical plane through the liquid. Applying the mass transfer equation for the flow of vapour on a vertical path and over a small element of plate area gives:

$$\mathrm{N}_{OG} = \int \frac{\mathrm{d}y}{y_e - y} = -\ln\left(\frac{y_e - y_n}{y_e - y_{n-1}}\right) \tag{11.131}$$

or:

$$\exp(-\mathbf{N}_{OG}) = \left(\frac{y_e - y_n}{y_e - y_{n-1}}\right)$$

and:

$$1 - \exp(-\mathbf{N}_{OG}) = \left(\frac{y_n - y_{n-1}}{y_e - y_{n-1}}\right) = E_{mv} \tag{11.132}$$

This analysis refers to a small area for vertical flow, and $E_{mv}$ is therefore the *point or local* Murphree efficiency. The relation between this point efficiency and the tray efficiency depends on the nature of the liquid mixing on the tray. If there is complete mixing of the liquid, $x = x_n$ for the liquid, and $y_e$ and $y$ will also be constant over a horizontal plane. The tray efficiency $E_{Mv} = E_{mv}$. With no mixing of the liquid, the liquid may be considered to be in plug flow. If $y_e = mx + b$ and $E_{mv}$ is taken as constant over tray, it may be shown[69] that:

$$E_{mv} = \frac{1}{\varepsilon}[\exp(\varepsilon E_{Mv}) - 1] \tag{11.133}$$

where:

$$\varepsilon = \frac{mG'}{L'}$$

Intermediate cases where partial mixing of liquid occurs are dealt with in the A.I.Ch.E. MANUAL[69].

### Plate efficiency in terms of liquid concentrations

With the same concept for tray layout as in Figure 11.55, relations for $E_{ml}$ and $E_{Ml}$ may be derived. Assuming that the vapour concentration does not change in a horizontal plane, a similar analysis to that above gives:

$$E_{ml} = 1 - \exp(-\mathbf{N}_{OL}) \tag{11.134}$$

The efficiencies $E_{mv}$ and $E_{ml}$ may be related by using the relation between $\mathbf{N}_{OG}$ and $\mathbf{N}_{OL}$ given in equation 11.130 to give:

$$\ln(1 - E_{ml}) = \varepsilon \ln(1 - E_{mv}) \tag{11.135}$$

### Effect of entrainment on efficiency

For conditions where the entrainment may be assumed constant across a tray, COLBURN[62] has suggested that the following expression gives, for entrainment $e'$ (moles/unit time. unit area), a correction to $E_{Mv}$, so that the new value of efficiency $E_a$ is given by:

$$E_a = \frac{E_{Mv}}{1 + (e'E_{Mv})/L'} \tag{11.136}$$

### Experimental work from A.I.Ch.E. programme

Having noted the way in which the tray efficiencies may be related to the values of $\mathbf{N}_{OG}$ and $\mathbf{N}_{OL}$, experimentally determined results are now required for expressing the mass transfer in terms of degree of mixing, entrainment, geometrical arrangements on the trays and the operating conditions including mass flowrates. These are provided from experimental work, which gives expressions for the number of film transfer units $\mathbf{N}_G$ and $\mathbf{N}_L$, outlined in Section 11.11.3.

### *Gas phase transfer*

The value of $\mathbf{N}_G$ is expressed in terms of weir height $h_w$, gas flow expressed as $\bar{F}$, liquid flow $L_p$ and the Schmidt number $Sc_v$ for the vapour phase. The two key relations are:

$$\mathbf{N}_G = [0.776 + 0.0046h_w - 0.24\bar{F} + 105L_p]Sc_v^{-0.5} \quad (11.137)$$

and:

$$\mathbf{N}_G = -\ln(1 - E_{mv}) \quad (11.138)$$

Equation 11.138 gives the point efficiency for cases where all the resistance occurs in the gas phase. In these equations:

$h_w$ is the exit weir height (mm),
$\bar{F} = u\sqrt{\rho_v}$, where $u$ is the vapour rate (m/s) based on the bubbling area, and $\rho_v$ is vapour density (kg/m$^3$),
$L_p$ is the liquid flow (m$^3$/s per m liquid flow path),
$\mu_v$ is the vapour viscosity (N s/m$^2$),
$D_v$ is the vapour diffusivity (m$^2$/s), and
$Sc_v$, is the Schmidt number $\mu_v/\rho_v D_v$.

### *Liquid phase transfer*

The value of $\mathbf{N}_L$ is expressed in terms of the $\bar{F}$-factor for vapour flow, the time of contact $t_L(s)$, and the liquid diffusivity $D_L(m^2/s)$. Experimental work gives:

$$\mathbf{N}_L = [4.13 \times 10^8 D_L]^{0.5}[0.21\bar{F} + 0.15]t_L \quad (11.139)$$

The residence time $t_L$ in seconds is expressed by:

$$t_L = Z_c Z_L/L_p \quad (11.140)$$

where $Z_c$ is the hold-up of liquid on the tray in m$^3$/m$^2$ of effective cross-section, given by:

$$Z_c = 0.043 + 1.91 \times 10^{-4}h_w - 0.013\bar{F} + 2.5L_p \quad (11.141)$$

and $Z_L$ is the distance between the weirs in metres.

### *Relationships for $\mathbf{N}_G$ and $\mathbf{N}_L$*

From a knowledge of $\mathbf{N}_G$ and $\mathbf{N}_L$, the value of $\mathbf{N}_{OG}$ is obtained from equation 11.142 which is derived in the same way as equation 11.163.

Thus:

$$\frac{1}{\mathbf{N}_{OG}} = \frac{1}{\mathbf{N}_G} + \frac{mG'}{L'}\frac{1}{\mathbf{N}_L} \quad (11.142)$$

The point efficiency $E_{mv}$ is then obtained from:

$$E_{mv} = 1 - \exp(-\mathbf{N}_{OG}) \quad \text{(equation 11.132)}$$

Whilst these expressions are difficult to use and involve some inconsistent assumptions about the liquid and vapour flow, they do bring out some useful features in relation to the tray efficiency. Thus $\mathbf{N}_G$ varies linearly with $h_w$, $\bar{F}$, and $L_p$, although the important relation between $\mathbf{N}_G$ and $E_{mv}$ is complex. The A.I.Ch.E. MANUAL[69] gives guideline figures.

## 11.11. PACKED COLUMNS FOR DISTILLATION

In bubble cap and perforated plate columns, a large interfacial area between the rising vapour and the reflux is obtained by causing the vapour to bubble through the liquid. An alternative arrangement, which also provides the necessary large interfacial area for diffusion, is the packed column, in which the cylindrical shell of the column is filled with some form of packing. A common arrangement for distillation is as indicated in Figure 11.58, where the packing may consist of rings, saddles, or other shaped particles, all of which are designed to provide a high interfacial area for transfer. These are referred to in Chapter 4. In packed columns the vapour flows steadily up and the reflux steadily down the column, giving a true countercurrent system in contrast to the conditions in bubble cap columns, where the process of enrichment is stagewise.

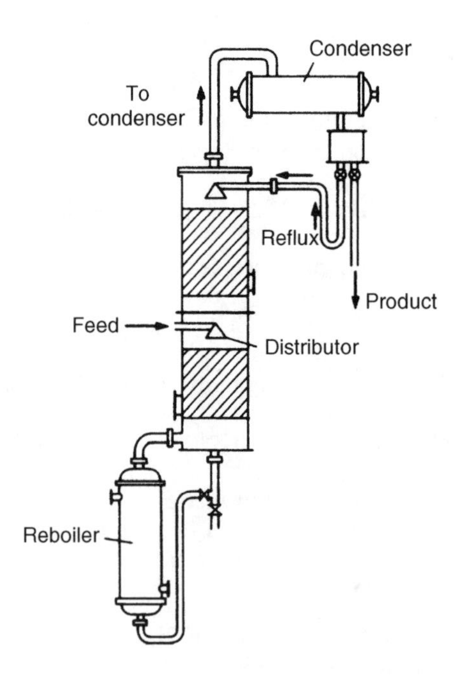


Figure 11.58. A packed column for distillation

### 11.11.1. Packings

The selection of a suitable packing material is based on the same arguments as for absorption towers considered in Chapter 12, although for industrial units the most usual packings are rings, and the material of construction is determined by the corrosive nature of the fluids, or otherwise. It is important to note that in a distillation system operating at high reflux ratios the mass of reflux is approximately equal to the mass of vapour, although at low reflux ratios the flow of the liquid is only a small fraction of that of the vapour. Since the vapour has a much lower density than the liquid, the process is really one in which a small quantity of liquid passes through the vapour, and the establishment of good distribution of the liquid is more difficult than in absorption towers, where the two streams are more nearly balanced.

In Chapter 4 the characteristics of packings and their influence on column hydraulics are considered, and in Chapter 12 the mass transfer aspects are covered. The packing for a particular application may be selected using this information, although it may be noted that, in the case of vacuum distillation for example, pressure-drop considerations may be of overriding importance, and there may be problems associated with the wetting of the packing because of the low liquid loadings. For distillation in packed towers, it is normal practice to increase the calculated height of packing by 40 per cent to allow for liquid mal-distribution and wetting problems.

### 11.11.2. Calculation of enrichment in packed columns

With plate columns, the vapour leaving an ideal plate is richer in the more volatile component than the vapour entering the plate, by one equilibrium step. PETERS[70] suggests that this same enrichment of the vapour will occur in a certain height of packing, which is termed the *height equivalent of a theoretical plate* (HETP). As all sections of the packing are physically the same, it is assumed that one equilibrium stage is represented by a given height of packing. Thus the required height of packing for any desired separation is given by HETP $\times$ (No. of ideal stages required).

This is a simple method of representation which has been widely used as a method of design. Despite this fact, there have been few developments in the theory. MURCH[71] gives the following relationships for the HETP from an analysis of the results of a number of workers. Columns 50–750 mm diameter and packed over heights of 0.9–3.0 m with rings, saddles, and other packings have been considered. Most of the results were for conditions of total reflux, with a vapour rate of 0.18–2.5 kg/m$^2$s which corresponded to 25–80 per cent of flooding. The relationship is:

$$\text{HETP} = C_1 G'^{C_2} d_c^{C_3} Z^{1/3} \left(\frac{\alpha \mu_L}{\rho_L}\right) \tag{11.143}$$

where the values of $C_1, C_2, C_3$ varied with packings as given in Table 11.4.

It may also be noted that the mixtures considered were mainly hydrocarbons with values of relative volatilities only slightly in excess of 3. In equation 11.143:

$G'$ = mass velocity of vapour (kg/m$^2$ s of tower area),
$d_c$ = column diameter (m),
$Z$ = packed height (m),
$\alpha$ = relative volatility,
$\mu_L$ = liquid viscosity (N s/m$^2$), and
$\rho_L$ = liquid density (kg/m$^3$).

Table 11.4. Constants for use in equation 11.143

| Type of packing | Size (mm) | $C_1$ ($\times 10^{-5}$) | $C_2$ | $C_3$ |
|---|---|---|---|---|
| Rings | 6 | | | 1.24 |
| | 9 | 0.77 | −0.37 | 1.24 |
| | 12.5 | 7.43 | −0.24 | 1.24 |
| | 25 | 1.26 | −0.10 | 1.24 |
| | 50 | 1.80 | 0 | 1.24 |
| Saddles | 12.5 | 0.75 | −0.45 | 1.11 |
| | 25 | 0.80 | −0.14 | 1.11 |
| Raschig rings of protruded metal | 6 | 0.28 | 0.25 | 0.30 |
| | 9 | 0.29 | 0.50 | 0.30 |
| | 19 | 0.45 | 0.30 | 0.30 |
| | 25 | 0.92 | 0.12 | 0.30 |

ELLIS[36] presented the following dimensionally consistent equation for the HETP ($Z_t$) of packed columns using 25 and 50 mm Raschig rings:

$$Z_t = 18d_r + 12m\left[\frac{G'}{L'} - 1\right] \tag{11.144}$$

where: $d_r$ is the diameter of the rings.
$m$ is the average slope of equilibrium curve,
$G''$ is the vapour flowrate, and
$L''$ is the liquor flowrate.

In practice, the HETP concept is used to convert empirically the number of theoretical stages to packing height. As most data in the literature have been derived from small-scale operations, these do not provide a good guide to the values which will be obtained on full-scale plant. The values given in Table 11.5 may, however, be used as a guide.

Table 11.5. Values of HETP[14] for full-scale plant

| Type of packing, application | HETP (m) |
|---|---|
| 25 mm diam. packing | 0.46 |
| 38 mm diam. packing | 0.66 |
| 50 mm diam. packing | 0.9 |
| Absorption duty | 1.5–1.8 |
| Small diameter columns (<0.6 m diam.) | column diameter |
| Vacuum columns | values as above + 0.1 m |

For a particular type of packing, the ratio, HETP/pressure drop, is fairly constant for all sizes so that there is no advantage in attempting to improve the HETP by using a smaller packing, since the disadvantages of the higher pressure drop will offset the savings made by reduction of packed height.

Further data on HETP for packings smaller than 38 mm is presented in Figure 11.59, where it may be seen that some of the newer packings, such as Pall rings and Mini rings, give a relatively constant value of HETP over a wide range of vapour rates.

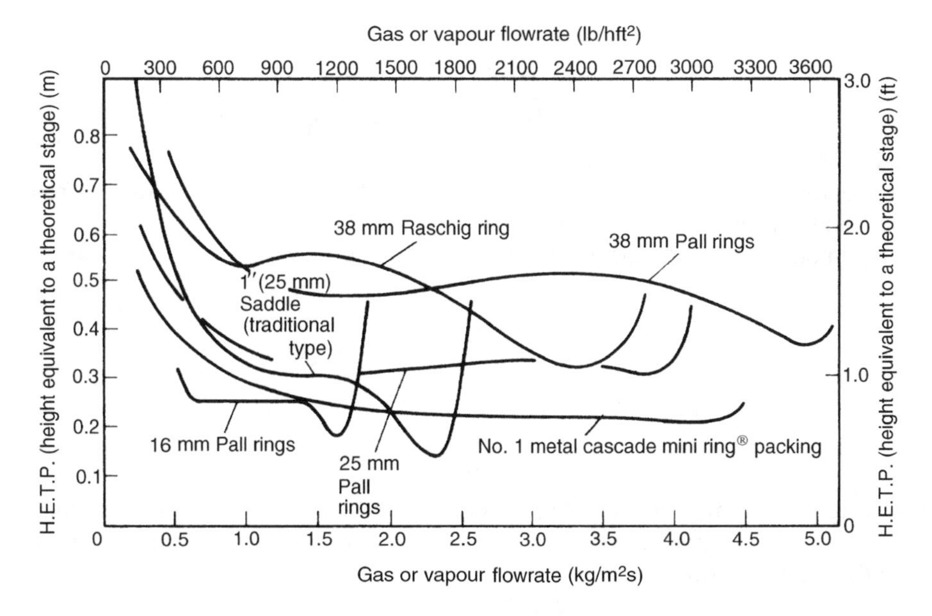


Figure 11.59. Height equivalent to a theoretical stage for common packings

## 11.11.3. The method of transfer units

The proposals of Peters[70] are really the application of the stagewise mechanism for the plate column to the packed tower, where the process is one of continuous countercurrent mass transfer. The degree of separation is represented by the rate of change of composition of a component with height of packing, that is $\mathrm{d}y/\mathrm{d}Z$. This rate of change of composition is dependent upon the equipment, the operating conditions, and on the diffusional potential across the two films. Over the vapour film this driving force is measured by $y_i - y$ or $(\Delta y)_f$ where $y_i$ is the mole fraction of the diffusing component at the interface and $y$ the value for the vapour.

The performance of the column may therefore be represented by $J$, the change in composition with height for unit driving force where:

$$J = \frac{\mathrm{d}y/\mathrm{d}Z}{(\Delta y)_f} \tag{11.145}$$

A relation between $\mathrm{d}y/\mathrm{d}Z$ and $(\Delta y)_f$ may be obtained on the basis of the two-film theory of mass transfer. For the vapour film, Fick's law, Volume 1, Chapter 10, gives:

$$N_A = -D_v \frac{\mathrm{d}C}{\mathrm{d}z} \tag{11.146}$$

where: $N_A$ is the molar rate of transfer per unit area of interface of component **A**,
$D_v$ is the vapour diffusion coefficient,
$C_A$ is the concentration of **A** in moles per unit volume, and
$z$ is the distance in direction of diffusion.

For an ideal gas, this gives:

$$N_A = -\frac{D_v}{\mathbf{R}T}\frac{\mathrm{d}P_A}{\mathrm{d}z} = -\frac{D_v P}{\mathbf{R}T}\frac{\mathrm{d}y}{\mathrm{d}z} \tag{11.147}$$

where $y$ is the mole fraction of **A**.

The negative sign occurs because $z$ is taken in the direction of diffusion from the interface, and $y$ decreases in this direction.

For equimolecular counterdiffusion, integration gives:

$$N_A = -\frac{D_v P}{\mathbf{R}Tz}(\Delta y)_f \tag{11.148}$$

or:

$$N_A = k'_g(\Delta y)_f \tag{11.149}$$

where:

$$k'_g = -\frac{D_v P}{\mathbf{R}Tz} \tag{11.150}$$

Considering the column shown in Figure 11.60, in which the concentration of the more volatile component increases from $y_b$ to $y_t$, the rate of transfer over a small height of the column $\mathrm{d}Z$ may be written as:

$$k'_g aA(y_i - y)\,\mathrm{d}Z \tag{11.151}$$

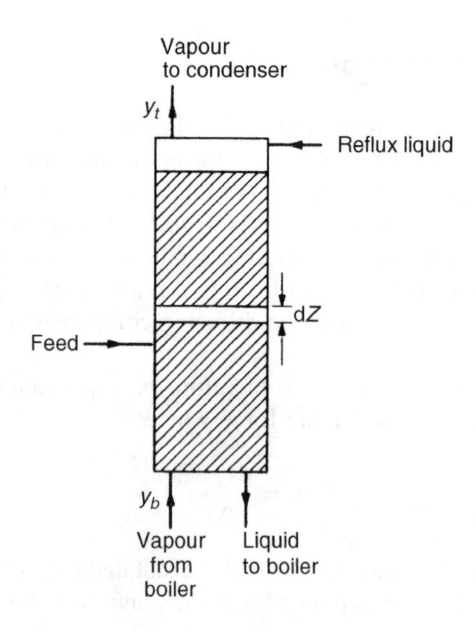


Figure 11.60. Arrangement for a packed column

where: $a$ is the active interfacial area for transfer per unit volume of the column and,
$A$ is the cross-sectional area of the column.

The moles of component **A** diffusing = (total moles of vapour × change in mole fraction)
$= G'A\,\mathrm{d}y$

where: $G'$ is the vapour rate in the column in moles/unit time and unit cross-section,
and $k'_g$ is known as the gas-film transfer coefficient, and is measured as moles/unit time unit area-unit mole fraction difference, and hence:

$$G'A\,\mathrm{d}y = k'_g a A(y_i - y)\,\mathrm{d}Z \tag{11.152}$$

or:

$$J = \frac{\mathrm{d}y/\mathrm{d}Z}{y_i - y} = \frac{k'_g a}{G'} \quad \text{(from equation 11.145)}$$

and:

$$\int_{y_b}^{y_t} \frac{\mathrm{d}y}{y_i - y} = \frac{k'_g a}{G'} Z \tag{11.153}$$

where $k'_g a$ is taken as constant over the column.

The group on the left-hand side of this equation represents the integrated ratio of the change in composition to the driving force tending to bring this change about. This group has been defined by CHILTON and COLBURN[72] as the number of transfer units $\mathbf{N}_G$. The quantity $G'/(k'_g a)$, which is the reciprocal of the efficiency $J$ and has the dimensions of length, is defined as the height of a transfer unit, $\mathbf{H}_G$. Equation 11.153 may be written as:

$$Z = \mathbf{H}_G \mathbf{N}_G \tag{11.154}$$

The concentrations $y_i$, $y$ refer to conditions on either side of the gas film, and hence $\mathbf{N}_G$ is the number of gas-film transfer units, and $\mathbf{H}_G$ the height of a gas-film transfer unit.

For packed columns $(k'_g a)/G'$ represents a useful value for the efficiency, and the performance of a packed column is commonly represented by the simple term $\mathbf{H}_G$, where a low value of $\mathbf{H}_G$ corresponds to an efficient column. If $\mathbf{H}_G$ is known, the necessary height of a column is found from equation 11.154, since $\mathbf{N}_G$ is determined from the change in concentration required and the shape of the equilibrium curve. Application of this technique is discussed in Chapter 12 and reference may also be made to Volume 1, Chapter 10.

The same number of moles pass through the liquid film and a similar series of equations may be obtained in terms of concentrations across the liquid film, that is:

$$G'\,\mathrm{d}y = L'\,\mathrm{d}x \tag{11.155}$$

where: $L'$ is the molar flowrate of liquid/unit area, and
$x$ is the mole fraction of the more volatile component in the liquid.

If $k'_l$ is the mass transfer coefficient for the liquid phase in moles/unit time, unit area, unit mole fraction driving force, then:

$$AL'\,\mathrm{d}x = k'_l a A(x - x_i)\,\mathrm{d}Z$$

Thus:

$$\int_{x_b}^{x_t} \frac{\mathrm{d}x}{x - x_i} = \frac{k'_l a}{L'} Z \tag{11.156}$$

or:

$$\mathbf{N}_L = \frac{1}{\mathbf{H}_L} Z$$

where: $\mathbf{N}_L$ is the number of liquid film transfer units and
$\mathbf{H}_L$ is the height of a liquid film transfer unit, which for distillation applications is presented in Table 11.6, taken from the work of GILLILAND and SHERWOOD[73], as a function of type and size of packing.

Table 11.6. Values of $H_L$ for distillation[73]

| Packing size (mm) | | 12 | 18 | 25 | 40 | 50 |
|---|---|---|---|---|---|---|
| (in.) | | 0.5 | 0.75 | 1.0 | 1.5 | 2.0 |
| Raschig type | m | 0.073 | 0.092 | 0.104 | 0.143 | 0.177 |
| | ft | 0.24 | 0.30 | 0.34 | 0.47 | 0.58 |
| Intalox | m | 0.061 | 0.079 | 0.089 | 0.122 | — |
| | ft | 0.20 | 0.26 | 0.29 | 0.40 | — |
| Pall rings | m | — | — | — | 0.122 | 0.150 |
| | ft | — | — | — | 0.40 | 0.49 |

### *Overall transfer coefficients and transfer units*

The driving force over the gas film is taken as $(y_i - y)$ and over the liquid film as $(x - x_i)$. If $y_e$ is the concentration in the gas phase in equilibrium with concentration $x$ in the liquid phase, then $(y_e - y)$ is taken as the overall driving force expressed in terms of $y$. Similarly $(x - x_e)$ is taken as the overall driving force in terms of $x$, where $x_e$ is the concentration in the liquid in equilibrium with a concentration $y$ in the vapour.

The overall driving forces $(\Delta y)_o$ and $(\Delta x)_o$ may then be written as:

$$(\Delta y)_o = y_e - y = (y_e - y_i) + (y_i - y)$$

$$(\Delta x)_o = x - x_e = (x - x_i) + (x_i - x_e)$$

By analogy with the derivation for film coefficients, a series of overall transfer coefficients and overall transfer units based on these overall driving forces may be defined.

Thus, the number of overall gas transfer units is given by:

$$\mathbf{N}_{OG} = \int \frac{dy}{y_e - y} \tag{11.157}$$

and the number of overall liquid transfer units is given by:

$$\mathbf{N}_{OL} = \int \frac{dx}{x - x_e} \tag{11.158}$$

The heights of the overall transfer units are:

$$\mathbf{H}_{OG} = \frac{G'}{K'_g a} \tag{11.159}$$

and:

$$\mathbf{H}_{OL} = \frac{L'}{K'_l a} \tag{11.160}$$

where $K'_g a$ and $K'_l a$ are overall transfer coefficients, based on gas or liquid concentrations in moles/unit time-unit volume-unit mole fraction driving force.

### Relation between overall and film transfer units

From these definitions, the following equation may be written:

$$\frac{\mathrm{d}Z}{\mathbf{H}_{OG}} = \frac{\mathrm{d}y}{y_e - y} \qquad \frac{\mathrm{d}Z}{\mathbf{H}_G} = \frac{\mathrm{d}y}{y_i - y}$$

$$\frac{\mathrm{d}Z}{\mathbf{H}_{OL}} = \frac{\mathrm{d}x}{x - x_e} \qquad \frac{\mathrm{d}Z}{\mathbf{H}_L} = \frac{\mathrm{d}x}{x - x_i}$$

Thus:

$$\frac{\mathbf{H}_{OG}}{y_e - y} = \frac{\mathbf{H}_G}{y_i - y}$$

and:

$$\mathbf{H}_{OG} = \mathbf{H}_G \left[\frac{y_e - y_i + y_i - y}{y_i - y}\right] = \mathbf{H}_G \left[1 + \frac{y_e - y_i}{y_i - y}\right] \tag{11.161}$$

If the equilibrium curve is linear over the range $y = y_e$ to $y = y_i$, then assuming equilibrium at the interface:

$$y_e - y_i = m(x - x_i)$$

and:

$$\mathbf{H}_{OG} = \mathbf{H}_G + m\left(\frac{x - x_i}{y_i - y}\right)\mathbf{H}_G$$

But:

$$\frac{x - x_i}{y_i - y} = \frac{\mathrm{d}x}{\mathrm{d}y}\frac{\mathbf{H}_L}{\mathbf{H}_G} = \frac{G'}{L'}\frac{\mathbf{H}_L}{\mathbf{H}_G}$$

since $G'\,\mathrm{d}y = L'\,\mathrm{d}x$.

Thus:

$$\mathbf{H}_{OG} = \mathbf{H}_G + \frac{mG'}{L'}\mathbf{H}_L \tag{11.162}$$

Similarly:

$$\mathbf{H}_{OL} = \mathbf{H}_L + \frac{L'}{mG'}\mathbf{H}_G \tag{11.163}$$

Dividing equation 11.162 by equation 11.163 gives:

$$\frac{\mathbf{H}_{OG}}{\mathbf{H}_{OL}} = \frac{mG'}{L'} = \frac{\mathbf{N}_{OL}}{\mathbf{N}_{OG}} \tag{11.164}$$

This form of relationship may be written in terms of transfer coefficients, as discussed in Chapter 12, to give:

$$\frac{1}{K'_g a} = \frac{1}{k'_g a} + \frac{m}{k'_l a} \tag{11.165}$$

### Relation of HTU to HETP

In a theoretical plate, the mole fraction of the more volatile component in the vapour will increase from $y$ to $y_e$, so that the total mass transfer is $G'(y_e - y)$. If the equilibrium curve may be considered linear over the height of column equivalent to a theoretical plate $Z_t$, the logarithmic mean driving force may be used. Thus, referring to Figure 11.61:

$$G'(y_e - y) = K_g'aZ_t(y_e - y)\frac{(mG'/L') - 1}{\ln(mG'/L')}$$

Thus:

$$Z_t = \frac{G'}{K_g'a}\frac{\ln(mG'/L')}{(mG'/L') - 1} = \mathbf{H}_{OG}\frac{\ln(mG'/L')}{(mG'/L') - 1}$$

$$= \mathbf{H}_{OG}\frac{\ln\{1 - [1 - (mG'/L')]\}}{(mG'/L') - 1}$$

$$= \mathbf{H}_{OG}\left\{1 + \tfrac{1}{2}[1 - (mG'/L')] + \tfrac{1}{3}[1 - (mG'/L')]^2 + \cdots\right\} \quad (11.166)$$

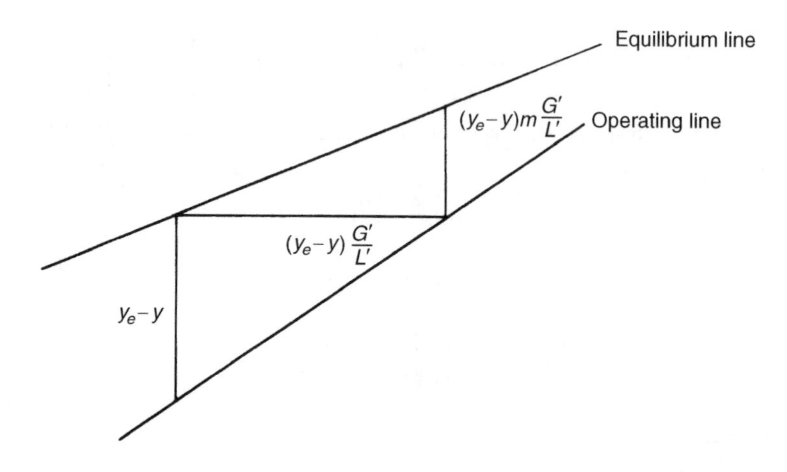


Figure 11.61. Height equivalent of a theoretical plate

If the operating and equilibrium lines are parallel, $mG'/L' = 1$, and:

$$Z_t = \mathbf{H}_{OG} \quad (11.167)$$

Thus, the ratio of the height equivalent to a theoretical plate to the height of the transfer unit $(Z_t/\mathbf{H}_{OG})$ may be greater or less than unity, according to whether the slope of the operating line is greater or less than that of the equilibrium curve.

### *Experimental determination of transfer units*

There have been a number of reports presented by Furnas and Taylor(74), Duncan *et al.*(75) and Sawistowski and Smith(76) on the influence of flow parameters and physical properties on the value of the height of a transfer unit. Most of the work has been carried out in small laboratory columns and great care must be exercised if these data are applied to large diameter units. Some general indication of the values of $\mathbf{H}_{OG}$ are given in Table 11.7, which gives values obtained by Furnas and Taylor(74) for experiments with

Table 11.7. Values of the height of the transfer unit $\mathbf{H}_{OG}$[74]

| Packing | Depth of packing (m) | Liquid rate (kg/m$^2$s) | $\mathbf{H}_{OG}$ (m) |
|---|---|---|---|
| 50 mm Raschig rings | 3.0 | 1.06 | 0.670 |
| 25 mm Raschig rings | 3.0 | 1.02 | 0.366 |
| 25 mm Berl saddles | 2.75 | 0.195 | 0.427 |
| 25 mm Berl saddles | 2.75 | 1.25 | 0.335 |
| 12.5 mm Berl saddles | 3.0 | 0.25 | 0.457 |
| 12.5 mm Berl saddles | 3.0 | 1.196 | 0.274 |
| 9.5 mm Raschig rings | 2.44 | 0.416 | 0.396 |
| 9.5 mm Raschig rings | 2.44 | 0.780 | 0.305 |

ethanol–water mixtures at atmospheric pressure in a column 305 mm diameter operating at total reflux.

## *Values of HTU in terms of flowrates and physical properties*

### *Wetted-wall columns.*

The proposals made for calculating transfer coefficients from physical data of the system and the liquid and vapour rates are all related to conditions existing in a simpler unit in the form of a wetted-wall column. In the wetted-wall column, discussed in Chapter 12, vapour rising from the boiler passes up the column which is lagged to prevent heat loss. The liquid flows down the walls, and it thus provides the simplest form of equipment giving countercurrent flow. The mass transfer in the unit may be expressed by means of the $j$-factor of Chilton and Colburn which is discussed in Volume 1, Chapter 10. Thus:

$$j_d = \frac{k_g'}{G'}\left(\frac{\mu}{\rho D}\right)_v^{2/3} = 0.023\left(\frac{d_c u \rho}{\mu}\right)_v^{-0.17} \tag{11.168}$$

where the flowrate and physical properties refer to the vapour. This type of unit has been studied by GILLILAND and SHERWOOD[73], CHARI and STORROW[77], SUROWIEC and FURNAS[78] and others.

For a wetted-wall column:

$$\frac{\text{area of interface}}{\text{volume of column}} = (4/d_c) = a$$

Thus:

$$\mathbf{H}_G = \frac{G'}{k_g' a} = 10.9 d_c Re_v^{0.17} Sc_v^{2/3} \tag{11.169}$$

where the linear characteristic length is taken as the diameter of the column $d_c$. $Re_v$ and $Sc_v$ are the Reynolds and Schmidt numbers with respect to the vapour. Surowiec and Furnas were able to express their results, obtained with alcohol and water, in this form. For transfer through the liquid film, an expression was derived based on the analysis of heat transfer from a tube to a liquid flowing under viscous conditions down the inside of the tube.

The equation was presented as:

$$\mathbf{H}_L = B'Z\left(\frac{M}{M_m}\right) Re_l^{8/9} Sc_l^{10/9}\left(\frac{D_L^2}{gZ^3}\right)^{2/9} \tag{11.170}$$

where: $B'$ is a constant,
$M_m$ is the mean molecular weight of the liquid,
$M$ is the point value of the molecular weight,
$Z$ is the height of the tube, and
$Re_l$ and $Sc_l$ are the Reynolds and Schmidt numbers with respect to the liquid.

It has been suggested, however, that for mass transfer, the transfer in the liquid phase is from a vapour–liquid interface where the liquid velocity is a maximum to the wall where it is zero. With a liquid flowing inside the tube the heat transfer is from a layer of zero velocity at the wall to the fluid all the way to the centre of the tube where it is moving with a maximum velocity. HATTA[(79)] based his analysis on the more closely related process of diffusion of a gas into a liquid, and obtained the expression:

$$\mathbf{H}_L = B''Z\frac{M}{M_m} Re_l^{2/3} Sc_l^{5/6}\left(\frac{D_L^2}{gZ^3}\right)^{1/6} \tag{11.171}$$

It may be seen that, despite the difference in the arguments, the two equations are really of a similar nature.

*Packed columns.*

The application of the ideas for wetted-wall columns to the more complex case of packed columns requires the assumptions: (a) that the mechanism is unchanged and (b) that the expressions are valid over the much wider ranges of flowrates used in packed columns. This has been attempted by SAWISTOWSKI and SMITH[(76)] and PRATT[(80)]. Pratt started from the basic equation:

$$j_d = \frac{k_g'}{G'}\left(\frac{\mu}{\rho D}\right)_v^{2/3} = \text{const}Re^{-0.2} \tag{11.172}$$

and suggested, from the examination of the available data, that the importance of the degree of wetting may be taken into account by writing this as:

$$\frac{k_g}{G'}e\left(\frac{\mu}{\rho D}\right)_v^{2/3} = p\omega\left(\frac{d_e G'}{\mu e}\right)^{-0.25} \tag{11.173}$$

where: $G'$ is the mass velocity (mass rate per unit area),
$d_e$ is the hydraulic mean diameter for the packing,
$p$ is a constant,
$\omega$ is the fraction of the packing wetted, and
$e$ is the fractional voidage of the packing.

PRATT[(80)] gives several plots of $p\omega$ against $L_p$, as shown in Figure 11.62, where $L_p$ is the liquid rate based on the periphery of the packing, in $m^3$/s m. The periphery is taken as $a^{-1}(m^3/m^2)$, although this is only correct for geometrical systems such as stacked rings.

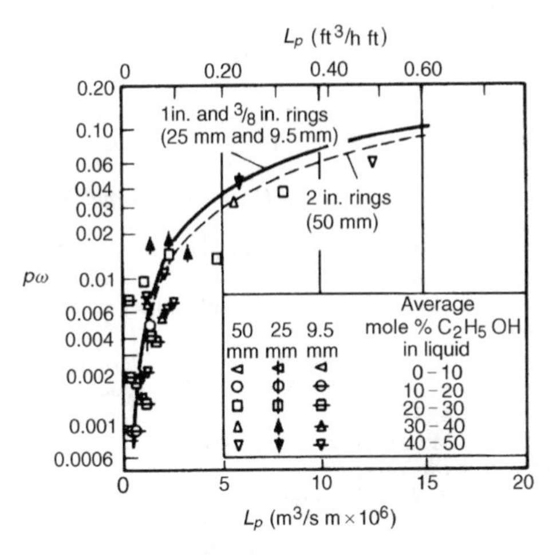


Figure 11.62. Effect of liquid rate on the degree of wetting of packing[80]

The problems of the wetting of packings is discussed in Chapter 4 where other methods are presented.

## 11.12. FURTHER READING

BACKHURST, J. R. and HARKER, J. H.: *Process Plant Design* (Heinemann Educational Books, London, 1973).

BILLET, R.: *Distillation Engineering* (Heyden and Sons Ltd., 1979).

BUCKLEY, P. S., LUYBEN, W. L. and SHUNTA, J. P.: *Design of Distillation Column Control Systems* (Edward Arnold, New York, 1985).

HOFFMAN, E. J.: *Azeotropic and Extractive Distillation* (Interscience Publishers, Inc., New York, 1964).

HOLLAND, C. D.: *Fundamentals of Multicomponent Distillation* (McGraw-Hill Book Co., New York, 1981).

KING, C. J.: *Separation Processes*, 2nd edn. (McGraw-Hill Book Co., New York, 1981).

LOCKETT, M. J.: *Distillation Fundamentals* (Cambridge University Press, 1986).

LOCKETT, M. J.: *Distillation Tray Fundamentals* (Cambridge University Press, 1986).

MCCABE, W. L., SMITH, J. C. and HARRIOTT, P.: *Unit Operations of Chemical Engineering*, 4th ed. (McGraw-Hill Book Co., New York, 1985).

SAWISTOWSKI, H. and SMITH, W.: *Mass Transfer Process Calculations* (Wiley, Chichester, 1963).

SHERWOOD, T. K., PIGFORD, R. L., and WILKE, C. R.: *Mass Transfer* (McGraw-Hill Book Co., New York, 1974).

SMITH, B. D.: *Design of Equilibrium Stage Processes* (McGraw-Hill Book Co., New York, 1963).

TREYBAL, R. E.: *Mass Transfer Operations*, 3rd edn. (McGraw-Hill Book Co., New York, 1980).

WANKAT, P. C.: *Equilibrium Staged Separations: Separations for Chemical Engineers* (Elsevier, New York, 1988).

## 11.13. REFERENCES

1. ROBINSON, C. S. and GILLILAND, E. R.: *Elements of Fractional Distillation*, 4th edn. (McGraw-Hill, New York, 1950).
2. THORNTON, J. D. and GARNER, F. H.: *J. Appl. Chem. Suppl.* **1** (1951) 61. Vapour–liquid equilibria in hydrocarbon-non-hydrocarbon systems. 1: The system benzene–cyclohexane–furfuraldehyde.
3. ANTOINE, C.: *Comptes Rendus* **107** (1888) 681, 836. Tensions des vapeurs: nouvelle relation entre les tensions et les températures.
4. BOUBLIK, T., FRIED, V. and HALA, E.: *The Vapour Pressure of Pure Substances* (Elsevier, New York, 1973).
5. JORDAN, T. E.: *Vapour Pressure of Organic Compounds* (Interscience, New York, 1954).

6. STULL, D. R.: *Ind. Eng. Chem.* **39** (1947) 517–540. Vapour pressures of pure substances–organic compounds. *Ind. Eng. Chem.* **39** (1947) 540–550. Vapour pressures of pure substances–inorganic compounds.
7. OHE, S.: *Computer Aided Data Book of Vapour Pressure* (Data Book Publishing Co., Tokyo, 1976).
8. REID, R. C., PRAUSNITZ, J. M. and SHERWOOD, T. K.: *The Properties of Liquids and Gases*, 3rd edn. (McGraw-Hill, New York, 1977).
9. ZWOLINSKI, B. J.: *Selected Values of Properties of Hydrocarbons and Related Compounds* (API Research Project 44, College Station, Texas).
10. ZWOLINSKI, B. J.: *Selected Values of Properties of Chemical Compounds* (Thermodynamics Research Center, Texas A & M University, College Station, Texas).
11. RIEDEL, L.: *Chem. Ing. Tech.* **26** (1954) 259–264. Liquid density in the saturated state. Extension of the theory of corresponding states.
12. AMBROSE, D., SPRAKE, C. H. S. and TOWNSEND, R.: *J. Chem. Thermodyn.* **6** (1974) 693–700. Thermodynamic properties of organic oxygen compounds XXXIII. The vapour pressure of acetone.
13. FRANK, O.: *Chem. Eng. Albany* **84** (14 Mar. 1977) 111, Distillation design.
14. PERRY, R. H., GREEN, D. W. and MALONEY, J. O. (eds.): *Perry's Chemical Engineers' Handbook*, 7th edn. (McGraw-Hill Book Company, New York, 1997).
15. WOHL, K.: *Trans. Amer. Inst. Chem. Eng.* **42** (1946) 215. Thermodynamic evaluation of binary and ternary liquid systems.
16. WOHL, K.: *Chem. Eng. Prog.* **49** (1953) 218. Thermodynamic evaluation of binary and ternary liquid systems.
17. BALZHISER, R. E., SAMUELS, M. R. and ELIASSEN, J. D.: *Chemical Engineering Thermodynamics: The Study of Energy, Entropy, and Equilibrium* (Prentice-Hall, Englewood Cliffs, NJ, 1972).
18. FREDENSLUND, A., BMEHLING, J. and RASMUSSEN, P.: *Vapor-Liquid Equilibria Using UNIFAC: A Group-Contribution Method* (Elsevier, Amsterdam, 1977).
19. PRAUSNTIZ, J. M., ANDERSON, T. F., GRENS, E. A., ECKERT, C. A., HSIEH, R. and O'CONNELL, J. P.: *Computer Calculations for Multicomponent Vapor–Liquid and Liquid–Liquid Equilibria* (Prentice-Hall, Englewood Cliffs, NJ, 1980).
20. REID, R. C., PRAUSNITZ, J. M. and POLING, B. E.: *The Properties of Gases and Liquids*, 4th edn. (McGraw-Hill, New York, 1987).
21. VAN NESS, H. C. and ABBOTT, M. M.: *Classical Thermodynamics of Non-Electrolyte Solutions*. (McGraw-Hill, New York, 1982).
22. WALAS, S. M.: *Phase Equilibria in Chemical Engineering* (Butterworth, Boston, 1985).
23. The National Engineering Laboratory, UK.
24. RAYLEIGH, LORD: *Phil. Mag.* **4**, (vi) No. 23 (1902) 521. On the distillation of binary mixtures.
25. SOREL, E.: *Distillation et Rectification Industrielle* (G. Carré et C. Naud, 1899).
26. LEWIS, W. K.: *Ind. Eng. Chem.* **1** (1909) 522. The theory of fractional distillation.
27. MCCABE, W. L. and THIELE, E. W.: *Ind. Eng. Chem.* **17** (1925) 605. Graphical design of fractionating columns.
28. UNDERWOOD, A. J. V.: *Trans. Inst. Chem. Eng.* **10** (1932) 112. The theory and practice of testing stills.
29. FENSKE, M. R.: *Ind. Eng. Chem.* **24** (1932) 482. Fractionation of straight-run Pennsylvania gasoline.
30. GILLILAND, E. R.: *Ind. Eng. Chem.* **32** (1940) 1220. Multicomponent rectification. Estimation of the number of theoretical plates as a function of the reflux ratio.
31. COLBURN, A. P.: *Division of Chemical Engineering Lecture Notes,* (University of Delaware, Newark, U.S.A., 1943).
32. HARKER, J. H.: *Processing* (April 1979) 39. Economic balance in distillation.
33. RUHEMANN, M.: *Trans. Inst. Chem. Eng.* **25** (1947) 143. The ammonia absorption machine. *Ibid.* 152. A study of the generator and rectifier of an ammonia absorption machine.
34. PONCHON, M.: *Technique Moderne* **13** (1921) 20 and 55. Etude graphique de la distillation fractionnée industrielle.
35. SAVARIT, P.: *Arts et Métiers* **75** (1922) 65. Eléments de distillation.
36. ELLIS, S. R. M.: *Birmingham University Chemical Engineer* **5**, No. 1 (1953) 21. H.E.T.P. values in ring packed columns.
37. GILLILAND, E. R. and REED, C. E.: *Ind. Eng. Chem.* **34** (1942) 551. Degrees of freedom in multicomponent absorption and rectification columns.
38. LEWIS, W. K. and MATHESON, G. L.: *Ind. Eng. Chem.* **24** (1932) 494. Studies in distillation. Design of rectifying columns for natural and refining gasoline.
39. COLBURN, A. P.: *Trans. Am. Inst. Chem. Eng.* **37** (1941) 805. The calculation of minimum reflux ratio in the distillation of multicomponent mixtures.
40. UNDERWOOD, A. J. V.: *J. Inst. Petroleum* **32** (1946) 614. Fractional distillation of multi-component mixtures — calculation of minimum reflux ratio.
41. ZANKER, A.: *Hydrocarbon Processing* **56**, No. 5 (1977) 263. Nomograph replaces Gilliland plot.

42. ELICECHE, A. M. and SARGENT, R. W. H.: In *Cost Savings in Distillation*. I. Chem. E Symposium Series No. 61 (1981) 1. Synthesis and design of distillation systems.
43. BENEDICT, M. and RUBIN, L. C.: *Trans. Am. Inst. Chem. Eng.* **41** (1945) 353. Extractive and azeotropic distillation.
44. YOUNG, S.: *Fractional Distillation* (Macmillan, London, 1902).
45. GUINOT, H. and CLARK, F. W.: *Trans. Inst. Chem. Eng.* **16** (1938) 189. Azeotropic distillation in industry.
46. NORMAN, W. S.: *Trans. Inst. Chem. Eng.* **23** (1945) 66. The dehydration of ethanol by azeotropic distillation. *Ibid.* 89. Design calculations for azeotropic dehydration columns.
47. TREYBAL, R. E.: *Mass Transfer Operations*, 2nd edn. (McGraw-Hill Book Co., New York, 1968).
48. SCHEIBEL, E. G.: *Chem. Eng. Prog.* **44** (1948) 927. Principles of extractive distillation.
49. PRATT, H. R. C.: *Trans. Inst. Chem. Eng.* **25** (1947) 43. Continuous purification and azeotropic dehydration of acetonitrile produced by the catalytic acetic acid–ammonia reaction.
50. HOFFMAN, E. J.: *Azeotropic and Extractive Distillation* (Interscience Publishers Inc., New York, 1964).
51. SMITH, B. D.: *Design of Equilibrium Stage Processes* (McGraw-Hill, New York, 1963).
52. HAUSBRAND, E.: *Principles and Practice of Industrial Distillation*, 6th edn., translated by TRIPP, E. H. (Wiley, 1926).
53. Engineering Staff, Shell Development Company, Emeryville, California: *Chem. Eng. Prog.* **50** (1954) 57. Turbogrid distillation trays.
54. CAREY, J. S.: *Chem. Met. Eng.* **46** (1939) 314. Plate-type distillation columns.
55. CAREY, J. S., GRISWOLD, J., LEWIS, W. K. and MCADAMS, W. H.: *Trans. Am. Inst. Chem. Eng.* **30** (1934) 504. Plate efficiencies in rectification of binary mixtures.
56. SOUDERS, M. and BROWN, G. G.: *Ind. Eng. Chem.* **26** (1934) 98. Design of fractionating columns.
57. KIRSCHBAUM, E.: *Distillation and Rectification* (Chemical Publishing Co., 1948).
58. DRICKAMER, H. G. and BRADFORD, J. R.: *Trans. Am. Inst. Chem. Eng.* **39** (1943) 319. Overall plate efficiency of commercial hydrocarbon fractionating columns as a function of viscosity.
59. O'CONNELL, H. E.: *Trans. Am. Inst. Chem. Eng.* **42** (1946) 741. Plate efficiency of fractionating columns and absorbers.
60. STRANG, L. C.: *Trans. Inst. Chem. Eng.* **12** (1934) 169. Entrainment in a bubble-cap fractionating column.
61. PEAVY, C. C. and BAKER, E. M.: *Ind. Eng. Chem.* **29** (1937) 1056. Efficiency and capacity of a bubble-plate fractionating column.
62. COLBURN, A. P.: *Ind. Eng. Chem.* **28** (1936) 526. Effect of entertainment on plate efficiency in distillation.
63. *Ballast Tray Manual.* Bulletin No. 4900 (revised) (Fritz Glitsch and Sons Inc., Dallas, Texas, 1970).
64. MURPHREE, E. V.: *Ind. Eng. Chem.* **17** (1925) 747. Rectifying column calculations — with particular reference to N component mixtures.
65. LOCKHART, F. J. and LEGGETT, C. W.: *Advances in Petroleum Chemistry and Refining*, by KOBE J. A. and MCKEYTTA, J. J. Vol. 1 (Interscience, 1958). Chapter 6, New fractionating-tray designs.
66. CHU, J. C., DONOVAN, J. R., BOSEWELL, B. C. and FURMEISTER, L. C.: *J. Appl. Chem.* **1** (1951) 529. Plate efficiency correlation in distilling columns and gas absorbers.
67. GEDDES, R. L.: *Trans. Am. Inst. Chem. Eng.* **42** (1946) 79. Local efficiencies of bubble plate fractionators.
68. BAKOWSKI, S.: *Chem. Eng. Sci.* **1** (1951/2) 266. A new method for predicting the plate efficiency of bubble-cap columns.
69. *Bubble Tray Design Manual* (American Institute of Chemical Engineers, New York, 1958).
70. PETERS, W. A.: *Ind. Eng. Chem.* **14** (1922) 476. The efficiency and capacity of fractionating columns.
71. MURCH, D. P.: *Ind. Eng. Chem.* **45** (1953) 2616. Height of equivalent theoretical plate in packed fractionation columns.
72. CHILTON, T. H. and COLBURN, A. P.: *Ind. Eng. Chem.* **27** (1935) 255, 904. Distillation and absorption in packed columns.
73. GILLILAND, E. R. and SHERWOOD, T. K.: *Ind. Eng. Chem.* **26** (1934) 516. Diffusion of vapours into air streams.
74. FURNAS, C. C. and TAYLOR, M. L.: *Trans. Am. Inst. Chem. Eng.* **36** (1940) 135. Distillation in packed columns.
75. DUNCAN, D. W., KOFFOLT, J. H. and WITHROW, J. R.: *Trans. Am. Inst. Chem. Eng.* **38** (1942) 259. The effect of operating variables on the performance of a packed column still.
76. SAWISTOWSKI, H. and SMITH, W.: *Ind. Eng. Chem.* **51** (1959) 915. Performance of packed distillation columns.
77. CHARI, K. S. and STORROW, J. A.: *J. Appl. Chem.* **1** (1951) 45. Film resistances in rectification.
78. SUROWIEC, A. J. and FURNAS, C. C.: *Trans. Am. Inst. Chem. Eng.* **38** (1942) 53. Distillation in a wetted-wall tower.
79. HATTA, S.: *J. Soc. Chem. Ind. Japan* **37** (1934) 275. On the theory of absorption of gases by liquids flowing as a thin layer.
80. PRATT, H. R. C.: *Trans. Inst. Chem. Eng.* **29** (1951) 195. The performance of packed absorption and distillation columns with particular reference to wetting.

## 11.14. NOMENCLATURE

| | | Units in SI System | Dimensions in M, N, L, T, $\theta$ |
|---|---|---|---|
| **A** | Component **A** | — | — |
| $A$ | Cross-sectional area of column | $m^2$ | $\mathbf{L}^2$ |
| $\mathscr{A}_{12}, \mathscr{A}_{21}$ | Margules and Van Laar constants | — | — |
| $\mathscr{A}_D$ | Margules third constant | — | — |
| $a$ | Interfacial surface per unit volume of column | $m^2/m^3$ | $\mathbf{L}^{-1}$ |
| **B** | Component **B** | — | — |
| $B'$ | Constant in equation 11.170 | — | — |
| $B''$ | Constant in equation 11.171 | — | — |
| $b_m, b_n$ | Factors in equation 11.113 | — | — |
| $C$ | Concentration in mols/unit volume | $kmol/m^3$ | $\mathbf{NL}^{-3}$ |
| $c$ | Total annual cost | £/year | |
| $c_a$ | Annual cost of column per unit area of plate | £/$m^2$year | |
| $c_b$ | Annual cost of heat exchanger equipment per unit area including depreciation and interest | £/$m^2$year | |
| $c_c$ | Annual cost of column per kmol of distillate | £/kmol year | |
| $c_d$ | Annual cost of steam and cooling water | £/year | |
| $c_h$ | Annual cost of reboiler and condenser | £/year | |
| $c_w$ | Annual variable cost | £/year | |
| $c_1$ | Function in equation 11.11 | K | $\theta$ |
| $c_2$ | Function in equation 11.11 | — | — |
| $c_5$ | Function in equation 11.12 | — | — |
| $D$ | Moles (or mass) of product per unit time | kmol/s or (kg/s) | $\mathbf{NT}^{-1}$ ($\mathbf{MT}^{-1}$) |
| $D_b$ | Moles of product in batch distillation | kmol | **N** |
| $D_L$ | Diffusivity in the liquid phase | $m^2/s$ | $\mathbf{L}^2\ \mathbf{T}^{-1}$ |
| $D_v$ | Diffusivity in the vapour phase | $m^2/s$ | $\mathbf{L}^2\ \mathbf{T}^{-1}$ |
| $d$ | Bubble diameter | m | **L** |
| $d_c$ | Column diameter | m | **L** |
| $d_r$ | Diameter of ring | m | **L** |
| $E$ | Average overall plate efficiency, $n/n_p$ | — | — |
| $E_a$ | Plate efficiency allowing for entrainment $e$ | — | — |
| $E_m$ | Local Murphree plate efficiency | — | — |
| $E_M$ | Average Murphree plate efficiency | — | — |
| $e$ | Fractional voidage of packing | — | — |
| $e'$ | Entrainment (mols per unit time and unit cross-section) | $kmol/m^2s$ | $\mathbf{NL}^{-2}\ \mathbf{T}^{-1}$ |
| $F$ | Cost factor in equation 11.67 | £/year | |
| $F$ | Molar or mass feed per unit time | kmol/s, kg/s | $\mathbf{NT}^{-1}$, $\mathbf{MT}^{-1}$ |
| $F_{lv}$ | Parameter $L'/G'\sqrt{(\rho_v/\rho_L)}$ | — | — |
| $\bar{F}$ | Parameter $u\sqrt{\rho_v}$ used in equation 11.137 | $kg^{1/2}/m^{1/2}s$ | $\mathbf{M}^{1/2}\mathbf{L}^{-1/2}\mathbf{T}^{-1}$ |
| $G'$ | Molar flowrate of vapour per unit time and unit cross-section | $kmol/m^2s$ | $\mathbf{NL}^{-2}\mathbf{T}^{-1}$ |
| $g$ | Acceleration due to gravity | $m/s^2$ | $\mathbf{LT}^{-2}$ |
| $H$ | Enthalpy per mole or unit mass | J/kmol, J/kg | $\mathbf{MN}^{-1}\mathbf{L}^2\mathbf{T}^{-2}$, $\mathbf{L}^2\mathbf{T}^{-2}$ |
| $H^L$ | Enthalpy per mole or unit mass of liquid | J/kmol, J/kg | $\mathbf{MN}^{-1}\mathbf{L}^2\mathbf{T}^{-2}$, $\mathbf{L}^2\mathbf{T}^{-2}$ |
| $H^V$ | Enthalpy per mole or unit mass of vapour | J/kmol, J/kg | $\mathbf{MN}^{-1}\mathbf{L}^2\mathbf{T}^{-2}$, $\mathbf{L}^2\mathbf{T}^{-2}$ |
| $H'_d$ | $H^L_d + (Q_c/D)$ | J/kmol, J/kg | $\mathbf{MN}^{-1}\mathbf{L}^2\mathbf{T}^{-2}$, $\mathbf{L}^2\mathbf{T}^{-2}$ |
| $H'_w$ | $H^L_w - (Q_B/W)$ | J/kmol, J/kg | $\mathbf{MN}^{-1}\mathbf{L}^2\mathbf{T}^{-2}$, $\mathbf{L}^2\mathbf{T}^{-2}$ |
| $\mathbf{H}$ | Height of transfer unit | m | **L** |
| $\mathbf{H}_G$ | Height of transfer unit — gas film | m | **L** |
| $\mathbf{H}_{OG}$ | Height of transfer unit — overall (gas concentrations) | m | **L** |
| $\mathbf{H}_L$ | Height of transfer unit — liquid film | m | **L** |
| $\mathbf{H}_{OL}$ | Height of transfer unit — overall (liquid concentration) | m | **L** |

| | | Units in SI System | Dimensions in $\mathbf{M, N, L, T}, \theta$ |
|---|---|---|---|
| $H'$ | Henry's constant $(P_A/x_A)$ | N/m$^2$ | $\mathbf{ML^{-1}T^{-2}}$ |
| $J$ | Rate of change in composition with height for unit driving force | m$^{-1}$ | $\mathbf{L^{-1}}$ |
| $j_d$ | $j$-factor for mass transfer | — | — |
| $K$ | Equilibrium constant $(y/x)$ or coefficient | — | — |
| $K_g$ | Overall mass transfer coefficient | m/s | $\mathbf{LT^{-1}}$ |
| $K'_g$ | Overall mass transfer coefficient (mol/unit time-unit area-unit mole fraction driving force) | kmol/m$^2$s | $\mathbf{NL^{-2}T^{-1}}$ |
| $K_l$ | Overall mass transfer coefficient | m/s | $\mathbf{LT^{-1}}$ |
| $K'_l$ | Overall mass transfer coefficient | kmol/m$^2$s | $\mathbf{NL^{-2}T^{-1}}$ |
| $k_g, k'_g$ | Film coefficients corresponding to $K_g$, $K'_g$ | m/s, kmol/m$^2$s | $\mathbf{LT^{-1}}$, $\mathbf{NL^{-2}T^{-1}}$ |
| $k_l, k'_l$ | Film coefficients corresponding to $K_l$, $K'_l$ | m/s, kmol/m$^2$s | $\mathbf{LT^{-1}}$, $\mathbf{NL^{-2}T^{-1}}$ |
| $k_1$ | Constant in equation 11.6 | — | — |
| $k_2$ | Constant in equation 11.6 | K | $\theta$ |
| $k_3$ | Constant in equation 11.6 | K | $\theta$ |
| $k_4$ | Constant in equation 11.7 | — | — |
| $k_5$ | Constant in equation 11.7 | K | $\theta$ |
| $k_6$ | Constant in equation 11.7 | — | — |
| $k_7$ | Constant in equation 11.7 | K$^{-6}$ | $\theta^{-6}$ |
| $k_8$ | Constant in equation 11.8 | — | — |
| $k_9$ | Constant in equation 11.8 | K | $\theta$ |
| $k_{10}$ | Constant in equation 11.11 | — | — |
| $k_{11}$ | Constant in equation 11.11 | — | — |
| $k_{12}$ | Constant in equation 11.11 | K$^{-6}$ | $\theta^{-6}$ |
| $L$ | Liquid flowrate in mass or moles/unit time | kg/s, kmol/s | $\mathbf{MT^{-1}}$, $\mathbf{NT^{-1}}$ |
| $L'$ | Liquid flowrate, in moles/unit time-unit area | kmol/m$^2$s | $\mathbf{NL^{-2}T^{-1}}$ |
| $L_b$ | Mole of liquid in batch distillation | kmol | $\mathbf{N}$ |
| $L_p$ | Liquid rate per unit periphery | m$^3$/ms | $\mathbf{L^2T^{-1}}$ |
| $l_m$ | Metal thickness | m | $\mathbf{L}$ |
| $M$ | Molecular weight | kg/kmol | $\mathbf{MN^{-1}}$ |
| | *or* difference stream below feed plate | kmol/s | $\mathbf{NT^{-1}}$ |
| $M_m$ | Mean molecular weight | kg/kmol | $\mathbf{MN^{-1}}$ |
| $m$ | Mass of material | kg | $\mathbf{M}$ |
| | *or* gradient of equilibrium curve | — | — |
| $N$ | Molar rate of transfer per unit area of interface | kmol/m$^2$s | $\mathbf{NL^{-2}T^{-1}}$ |
| | *or* difference stream above feed plate | kmol/s | $\mathbf{NT^{-1}}$ |
| $N''$ | Vapour handling capacity of boiler and condenser | kmol/m$^2$s | $\mathbf{NL^{-2}T^{-1}}$ |
| $\mathbf{N}$ | Number of transfer units | — | — |
| $\mathbf{N}_G, \mathbf{N}_{OG}$ | Number of gas film and overall gas transfer units | — | — |
| $\mathbf{N}_L, \mathbf{N}_{OL}$ | Number of liquid film and overall liquid transfer units | — | — |
| $n$ | Number of (theoretical) plates | — | — |
| $n_m$ | Number of plates at total reflux | — | — |
| $n_p$ | Number of actual plates | — | — |
| $P$ | Total pressure | N/m$^2$ | $\mathbf{ML^{-1}T^{-2}}$ |
| $P_A, P_B$ | Partial pressure of **A**, **B** | N/m$^2$ | $\mathbf{ML^{-1}T^{-2}}$ |
| $P_A^\circ, P_B^\circ$ | Vapour pressure of **A**, **B** | N/m$^2$ | $\mathbf{ML^{-1}T^{-2}}$ |
| $-\Delta P_{dry}$ | Pressure drop over dry tray expressed as head | m | $\mathbf{L}$ |
| $-\Delta P_T$ | Total pressure drop expressed as head | m | $\mathbf{L}$ |
| $p$ | Constant in equation 11.159 | — | — |
| $P_c$ | Critical pressure | N/m$^2$ | $\mathbf{ML^{-1}T^{-2}}$ |
| $P_r$ | Reduced pressure | — | — |
| $Q_B$ | Heat per unit time supplied to boiler for bottom product $W$ | W | $\mathbf{ML^2T^{-3}}$ |
| $Q_C$ | Heat per unit time removed in condenser for product $D$ | W | $\mathbf{ML^2T^{-3}}$ |

| | | Units in SI System | Dimensions in $\mathbf{M}$, $\mathbf{N}$, $\mathbf{L}$, $\mathbf{T}$, $\theta$ |
|---|---|---|---|
| $Q_R$ | Heat to still in batch distillation to provide reflux | J | $\mathbf{ML^2T^{-2}}$ |
| $Q'$ | Liquid rate (volumetric) | m$^3$/s | $\mathbf{L^3T^{-1}}$ |
| $q$ | Heat to vaporise one mole of feed divided by molar latent heat | — | — |
| $q_c$ | $Q_c/D$ | W/kmol | $\mathbf{MN^{-1}L^2T^{-3}}$ |
| $R$ | Reflux ratio | — | — |
| $R_m$ | Minimum reflux ratio | — | — |
| $R'$ | Resistance force per unit area, as used in Chapter 4 | N/m$^2$ | $\mathbf{ML^{-1}T^{-2}}$ |
| $\mathbf{R}$ | Universal gas constant | J/kmolK | $\mathbf{MN^{-1}L^2T^{-2}}\ \boldsymbol{\theta}^{-1}$ |
| $r$ | Radius of an orifice | m | $\mathbf{L}$ |
| $r_m$, $r_n$, $r_f$ | Ratio of concentrations of key components on plates $m, n$ and in the feed | — | — |
| $S$ | Molar quantity of material in the still | kmol | $\mathbf{N}$ |
| $S'$, $S''$, $S'''$ | Molar quantity or mass of material per unit time in sidestreams 1, 2, 3 | kmol/s | $\mathbf{NT^{-1}}$ |
| $T$ | Absolute temperature | K | $\boldsymbol{\theta}$ |
| $T_B$ | Boiling point | K | $\boldsymbol{\theta}$ |
| $t$ | Time | s | $\mathbf{T}$ |
| $T_c$ | Critical temperature | K | $\boldsymbol{\theta}$ |
| $T_r$ | Reduced temperature | — | — |
| $t_a$ | Annual period of operation | h/year | — |
| $u$ | Vapour velocity | m/s | $\mathbf{LT^{-1}}$ |
| $u'$ | Allowable vapour velocity | m/s | $\mathbf{LT^{-1}}$ |
| $V$ | Vapour flow in mols or mass per unit time | kmol/s, kg/s | $\mathbf{NT^{-1}}$, $\mathbf{MT^{-1}}$ |
| $v$ | Volumetric vapour flowrate | m$^3$/s | $\mathbf{L^3T^{-1}}$ |
| $W$ | Bottom product in mols or mass per unit time | kmol/s kg/s | $\mathbf{NT^{-1}}$, $\mathbf{MT^{-1}}$ |
| $x$ | Mole or mass fraction of a component in the liquid phase | — | — |
| $y$ | Mole fraction of a component in the gas phase | — | — |
| $Z$ | Height of packed column | m | $\mathbf{L}$ |
| $Z_p$ | Plate spacing | m | $\mathbf{L}$ |
| $Z_t$ | Height of equivalent theoretical plate | m | $\mathbf{L}$ |
| $z$ | Distance | m | $\mathbf{L}$ |
| $\alpha$ | Relative volatility or volatility relative to heavy key | — | — |
| $\alpha_{AB}$ | Volatility of **A** relative to **B** | — | — |
| $\gamma$ | Activity coefficient | — | — |
| $\varepsilon$ | $m(G'/L')$ | — | — |
| $\lambda$ | Latent heat per mole | J/kmol | $\mathbf{MN^{-1}L^2T^{-2}}$ |
| $\mu_L$ | Viscosity of liquid | Ns/m$^2$ | $\mathbf{ML^{-1}T^{-1}}$ |
| $\mu_v$ | Viscosity of vapour | Ns/m$^2$ | $\mathbf{ML^{-1}T^{-1}}$ |
| $\rho_L$ | Density of liquid | kg/m$^3$ | $\mathbf{ML^{-3}}$ |
| $\rho_v$ | Density of vapour | kg/m$^3$ | $\mathbf{ML^{-3}}$ |
| $\rho_m$ | Density of metal | kg/m$^3$ | $\mathbf{ML^{-3}}$ |
| $\Sigma$ | Mols per unit time of steam for steam distillation | kmol/s | $\mathbf{NT^{-1}}$ |
| $\sigma$ | Surface tension | J/m$^2$ (or N/m) | $\mathbf{MT^{-2}}$ |
| $\theta$ | Root of equation 11.115 | — | — |
| $\phi$ | Intercept of operating line on $Y$-axis $[x_d/(R+1)]$ | — | — |
| $\psi$ | $r_m/r_n$ at minimum reflux or fractional entrainment | — | — |
| $\omega$ | Fraction of packing wetted | — | — |
| $Re$ | Reynolds number | — | — |
| $Sc$ | Schmidt number | — | — |

| | | Units in SI System | Dimensions in $\mathbf{M}, \mathbf{N}, \mathbf{L}, \mathbf{T}, \theta$ |
|---|---|---|---|
| Suffixes | | | |
| 1, 2 | Inlet, outlet | | |
| $A, B, C, D$ | Materials **A**, **B**, **C**, **D** | | |
| $b$ | Bottom | | |
| $c$ | Intersection of equilibrium and operating lines at minimum reflux | | |
| $d$ | Top product | | |
| $e$ | Equilibrium | | |
| $f$ | Feed | | |
| $fs$ | Feed at its boiling point | | |
| $g$ | Azeotrope | | |
| $h$ | Component heavier than heavy key | | |
| $i$ | Interface | | |
| $L, l$ | Liquid | | |
| $m, n$ | Plates $m, n$ below and above feed plate respectively | | |
| $q$ | Intersection of operating lines | | |
| $S', S'', S'''$ | Sidestreams, 1, 2, 3 | | |
| $s$ | Still | | |
| $t$ | Top | | |
| $v$ | Vapour | | |
| $w$ | Bottom product | | |
| av | Average | | |

CHAPTER 12

# Absorption of Gases

## 12.1. INTRODUCTION

The removal of one or more selected components from a mixture of gases by absorption into a suitable liquid is the second major operation of chemical engineering that is based on interphase mass transfer controlled largely by rates of diffusion. Thus, acetone can be recovered from an acetone–air mixture by passing the gas stream into water in which the acetone dissolves while the air passes out. Similarly, ammonia may be removed from an ammonia–air mixture by absorption in water. In each of these examples the process of absorption of the gas in the liquid may be treated as a physical process, the chemical reaction having no appreciable effect. When oxides of nitrogen are absorbed in water to give nitric acid, however, or when carbon dioxide is absorbed in a solution of sodium hydroxide, a chemical reaction occurs, the nature of which influences the actual rate of absorption. Absorption processes are therefore conveniently divided into two groups, those in which the process is solely physical and those where a chemical reaction is occurring. In considering the design of equipment to achieve gas absorption, the main requirement is that the gas should be brought into intimate contact with the liquid, and the effectiveness of the equipment will largely be determined by the success with which it promotes contact between the two phases. The general form of equipment is similar to that described for distillation in Chapter 11, and packed and plate towers are generally used for large installations. The method of operation, as will be seen later, is not the same. In absorption, the feed is a gas introduced at the bottom of the column, and the solvent is fed to the top, as a liquid; the absorbed gas and solvent leave at the bottom, and the unabsorbed components leave as gas from the top. The essential difference between distillation and absorption is that in the former the vapour has to be produced in each stage by partial vaporisation of the liquid which is therefore at its boiling point, whereas in absorption the liquid is well below its boiling point. In distillation there is a diffusion of molecules in both directions, so that for an ideal system equimolecular counterdiffusion takes place, though in absorption gas molecules are diffusing into the liquid, with negligible transfer in the reverse direction, as discussed in Volume 1, Chapter 10. In general, the ratio of the liquid to the gas flowrate is considerably greater in absorption than in distillation with the result that layout of the trays is different in the two cases. Furthermore, with the higher liquid rates in absorption, packed columns are much more commonly used.

## 12.2. CONDITIONS OF EQUILIBRIUM BETWEEN LIQUID AND GAS

When two phases are brought into contact they eventually reach equilibrium. Thus, water in contact with air evaporates until the air is saturated with water vapour, and the air is absorbed by the water until it becomes saturated with the individual gases. In any mixture of gases, the degree to which each gas is absorbed is determined by its partial pressure. At a given temperature and concentration, each dissolved gas exerts a definite partial pressure. Three types of gases may be considered from this aspect — a very soluble one, such as ammonia, a moderately soluble one, such as sulphur dioxide, and a slightly soluble one, such as oxygen. The values in Table 12.1 show the concentrations in kilograms per 1000 kg of water that are required to develop a partial pressure of 1.3, 6.7, 13.3, 26.7, and 66.7 kN/m$^2$ at 303 K. It may be seen that a slightly soluble gas requires a much higher partial pressure of the gas in contact with the liquid to give a solution of a given concentration. Conversely, with a very soluble gas a given concentration in the liquid phase is obtained with a lower partial pressure in the vapour phase. At 293 K a solution of 4 kg of sulphur dioxide per 1000 kg of water exerts a partial pressure of 2.7 kN/m$^2$. If a gas is in contact with this solution with a partial pressure $SO_2$ greater than 2.7 kN/m$^2$, sulphur dioxide will be absorbed. The most concentrated solution that can be obtained is that in which the partial pressure of the solute gas is equal to its partial pressure in the gas phase. These equilibrium conditions fix the limits of operation of an absorption unit. Thus, in an ammonia–air mixture containing 13.1 per cent of ammonia, the partial pressure of the ammonia is 13.3 kN/m$^2$ and the maximum concentration of the ammonia in the water at 303 K is 93 kg per 1000 kg of water.

Table 12.1. Partial pressures and concentrations of aqueous solutions of gases at 303 K

| Partial pressure of solute in gas phase (kN/m$^2$) | Concentration of solute in water kg/1000 kg water | | |
|---|---|---|---|
| | Ammonia | Sulphur dioxide | Oxygen |
| 1.3 | 11 | 1.9 | — |
| 6.7 | 50 | 6.8 | — |
| 13.3 | 93 | 12 | 0.008 |
| 26.7 | 160 | 24.4 | 0.013 |
| 66.7 | 315 | 56 | 0.033 |

Whilst the solubility of a gas is not substantially affected by the total pressure in the system for pressures up to about 500 kN/m$^2$, it is important to note that the solubility falls with a rise of temperature. Thus, for a concentration of 25 per cent by mass of ammonia in water, the equilibrium partial pressure of the ammonia is 30.3 kN/m$^2$ at 293 K and 46.9 kN/m$^2$ at 303 K.

In many instances the absorption is accompanied by the evolution of heat, and it is therefore necessary to fit coolers to the equipment to keep the temperature sufficiently low for an adequate degree of absorption to be obtained.

For dilute concentrations of most gases, and over a wide range for some gases, the equilibrium relationship is given by Henry's law. This law, as used in Chapter 11, can be written as:

$$P_A = \mathscr{H} C_A \tag{12.1}$$

where: $P_A$ is the partial pressure of the component **A** in the gas phase,
$C_A$ is the concentration of the component in the liquid, and
$\mathscr{H}$ is Henry's constant.

## 12.3. THE MECHANISM OF ABSORPTION

### 12.3.1. The two-film theory

The most useful concept of the process of absorption is given by the two-film theory due to WHITMAN[1], and this is explained fully in Volume 1, Chapter 10. According to this theory, material is transferred in the bulk of the phases by convection currents, and concentration differences are regarded as negligible except in the vicinity of the interface between the phases. On either side of this interface it is supposed that the currents die out and that there exists a thin film of fluid through which the transfer is effected solely by molecular diffusion. This film will be slightly thicker than the laminar sub-layer, because it offers a resistance equivalent to that of the whole boundary layer. According to Fick's law (Volume 1, equation 10.1) the rate of transfer by diffusion is proportional to the concentration gradient and to the area of interface over which the diffusion is occurring. Fick's law is limited to cases where the concentration of the absorbed component is low. At high concentrations, bulk flow occurs and the mass transfer rate, which is increased by a factor $C_T/C_B$, is governed by Stefan's law, equation 12.2. Under these circumstances, the concentration gradient is no longer constant throughout the film and the lines AB and DE are curved. This question has been discussed in Chapter 10 of Volume 1, but some of the important features will be given here.

The direction of transfer of material across the interface is not dependent solely on the concentration difference, but also on the equilibrium relationship. Thus, for a mixture of ammonia or hydrogen chloride and air which is in equilibrium with an aqueous solution, the concentration in the water is many times greater than that in the air. There is, therefore, a very large concentration gradient across the interface, although this is not the controlling factor in the mass transfer, as it is generally assumed that there is no resistance at the interface itself, where equilibrium conditions will exist. The controlling factor will be the rate of diffusion through the two films where all the resistance is considered to lie. The change in concentration of a component through the gas and liquid phases is illustrated in Figure 12.1. $P_{AG}$ represents the partial pressure in the bulk of the gas phase and $P_{Ai}$ the partial pressure at the interface. $C_{AL}$ is the concentration in the bulk of the liquid phase and $C_{Ai}$ the concentration at the interface. Thus, according to this theory, the concentrations at the interface are in equilibrium, and the resistance to transfer is centred in the thin films on either side. This type of problem is encountered in heat transfer across a tube, where the main resistance to transfer is shown to lie in the thin films on either side of the wall; here the transfer is by conduction.

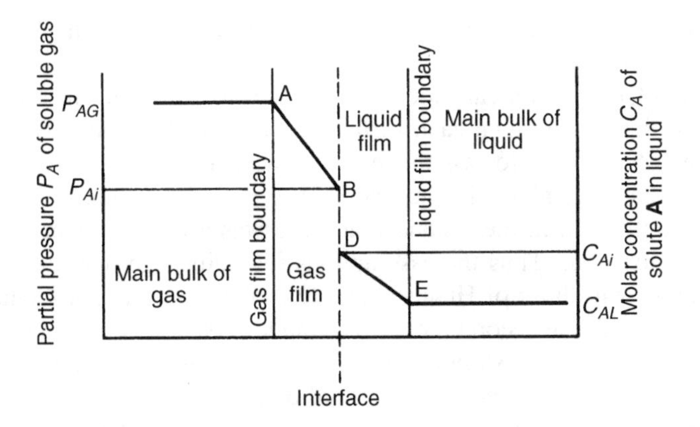


Figure 12.1. Concentration profile for absorbed component **A**

## 12.3.2. Application of mass transfer theories

The preceding analysis of the process of absorption is based on the two-film theory of WHITMAN[1]. It is supposed that the two films have negligible capacity, but offer all the resistance to mass transfer. Any turbulence disappears at the interface or free surface, and the flow is thus considered to be laminar and parallel to the surface.

An alternative theory described in detail in Volume 1, Chapter 10, has been put forward by HIGBIE[2], and later extended by DANCKWERTS[3] and DANCKWERTS and KENNEDY[4] in which the liquid surface is considered to be composed of a large number of small elements each of which is exposed to the gas phase for an interval of time, after which they are replaced by fresh elements arising from the bulk of the liquid.

All three of these proposals give the mass transfer rate $N'_A$ directly proportional to the concentration difference $(C_{Ai} - C_{AL})$ so that they do not directly enable a decision to be made between the theories. However, in the Higbie–Danckwerts theory $N'_A \propto \sqrt{D_L}$ whereas $N'_A \propto D_L$ in the two-film theory. DANCKWERTS[3] applied this theory to the problem of absorption coupled with chemical reaction but, although in this case the three proposals give somewhat different results, it has not been possible to distinguish between them.

The application of the penetration theory to the interpretation of experimental results obtained in wetted-wall columns has been studied by LYNN, STRAATEMEIER, and KRAMERS[5]. They absorbed pure sulphur dioxide in water and various aqueous solutions of salts and found that, in the presence of a trace of Teepol which suppressed ripple formation, the rate of absorption was closely predicted by the theory. In very short columns, however, the rate was overestimated because of the formation of a region in which the surface was stagnant over the bottom one centimetre length of column. The studies were extended to columns containing spheres and again the penetration theory was found to hold, there being very little mixing of the surface layers with the bulk of the fluid as it flowed from one layer of spheres to the next.

Absorption experiments in columns packed with spheres, 37.8 mm diameter, were also carried out by DAVIDSON *et al.*[6] who absorbed pure carbon dioxide into water. When a small amount of surface active agent was present in the water no appreciable mixing was found between the layers of spheres. With pure water, however, the liquid was almost completely mixed in this region.

DAVIDSON[7] built up theoretical models of the surfaces existing in a packed bed, and assumed that the liquid ran down each surface in laminar flow and was then fully mixed before it commenced to run down the next surface. The angles of inclination of the surfaces were taken as random. In the first theory it was assumed that all the surfaces were of equal length, and in the second that there was a random distribution of surface lengths up to a maximum. Thus the assumptions regarding age distribution of the liquid surfaces were similar to those of HIGBIE[2] and DANCKWERTS[3]. Experimental results were in good agreement with the second theory. All random packings of a given size appeared to be equivalent to a series of sloping surfaces, and therefore the most effective packing would be that which gave the largest interfacial area.

In an attempt to test the surface renewal theory of gas absorption, DANCKWERTS and KENNEDY[8] measured the transient rate of absorption of carbon dioxide into various solutions by means of a rotating drum which carried a film of liquid through the gas. Results so obtained were compared with those for absorption in a packed column and it was shown that exposure times of at least one second were required to give a strict comparison; this was longer than could be obtained with the rotating drum. ROBERTS and DANCKWERTS[9] therefore used a wetted-wall column to extend the times of contact up to 1.3 s. The column was carefully designed to eliminate entry and exit effects and the formation of ripples. The experimental results and conclusions are reported by DANCKWERTS, KENNEDY, and ROBERTS[10] who showed that they could be used, on the basis of the penetration theory model, to predict the performance of a packed column to within about 10 per cent.

There have been many recent studies of the mechanism of mass transfer in a gas absorption system. Many of these have been directed towards investigating whether there is a significant resistance to mass transfer at the interface itself. In order to obtain results which can readily be interpreted, it is essential to operate with a system of simple geometry. For that reason a laminar jet has been used by a number of workers.

CULLEN and DAVIDSON[11] studied the absorption of carbon dioxide into a laminar jet of water. When the water issued with a uniform velocity over the cross-section, the measured rate of absorption corresponded closely with the theoretical value. When the velocity profile in the water was parabolic, the measured rate was lower than the calculated value; this was attributed to a hydrodynamic entry effect.

The possible existence of an interface resistance in mass transfer has been examined by RAIMONDI and TOOR[12] who absorbed carbon dioxide into a laminar jet of water with a flat velocity profile, using contact times down to 1 ms. They found that the rate of absorption was not more than 4 per cent less than that predicted on the assumption of instantaneous saturation of the surface layers of liquid. Thus, the effects of interfacial resistance could not have been significant. When the jet was formed at the outlet of a long capillary tube so that a parabolic velocity profile was established, absorption rates were lower than predicted because of the reduced surface velocity. The presence of surface-active agents appeared to cause an interfacial resistance, although this effect is probably attributable to a modification of the hydrodynamic pattern.

STERNLING and SCRIVEN[13] have examined interfacial phenomena in gas absorption and have explained the interfacial turbulence which has been noted by a number of workers in terms of the Marangoni effect which gives rise to movement at the interface due to local variations in interfacial tension. Some systems have been shown to give rise to stable interfaces when the solute is transferred in one direction, although instabilities develop during transfer in the reverse direction.

GOODRIDGE and ROBB[14] used a laminar jet to study the rate of absorption of carbon dioxide into sodium carbonate solutions containing a number of additives including glycerol, sucrose, glucose, and arsenites. For the short times of exposure used, absorption rates into sodium carbonate solution or aqueous glycerol corresponded to those predicted on the basis of pure physical absorption. In the presence of the additives, however, the process was accelerated as the result of chemical reaction.

Absorption of gases and vapour by drops has been studied by GARNER and KENDRICK[15] and GARNER and LANE[16] who developed a vertical wind tunnel in which drops could be suspended for considerable periods of time in the rising gas stream. During the formation of each drop the rate of mass transfer was very high because of the high initial turbulence. After the initial turbulence had subsided, the mass transfer rate approached the rate for molecular diffusion provided that the circulation had stopped completely. In a drop with stable natural circulation the rate was found to approach 2.5 times the rate for molecular diffusion.

### 12.3.3. Diffusion through a stagnant gas

The process of absorption may be regarded as the diffusion of a soluble gas **A** into a liquid. The molecules of **A** have to diffuse through a stagnant gas film and then through a stagnant liquid film before entering the main bulk of liquid. The absorption of a gas consisting of a soluble component **A** and an insoluble component **B** is a problem of mass transfer through a stationary gas to which Stefan's law (Volume 1, Chapter 10) applies:

$$N'_A = -D_V \frac{C_T}{C_B} \frac{dC_A}{dz} \tag{12.2}$$

where $N'_A$ is the overall rate of mass transfer (moles/unit area and unit time),
$D_V$ is the gas-phase diffusivity,
$z$ is distance in the direction of mass transfer, and
$C_A$, $C_B$, and $C_T$ are the molar concentrations of **A**, **B**, and total gas, respectively.

Integrating over the whole thickness $z_G$ of the film, and representing concentrations at each side of the interface by suffixes 1 and 2:

$$N'_A = \frac{D_V C_T}{z_G} \ln \frac{C_{B2}}{C_{B1}} \tag{12.3}$$

Since $C_T = P/\mathbf{R}T$, where **R** is the gas constant, $T$ the absolute temperature, and $P$ the total pressure. For an ideal gas, then:

$$N'_A = \frac{D_V P}{\mathbf{R}T z_G} \ln \frac{P_{B2}}{P_{B1}} \tag{12.4}$$

Writing $P_{Bm}$ as the log mean of the partial pressures $P_{B1}$ and $P_{B2}$, then:

$$P_{Bm} = \frac{P_{B2} - P_{B1}}{\ln(P_{B2}/P_{B1})} \tag{12.5}$$

$$N'_A = \frac{D_V P}{\mathbf{R}T z_G} \frac{P_{B2} - P_{B1}}{P_{Bm}}$$

$$= \frac{D_V P}{\mathbf{R}T z_G} \left[\frac{P_{A1} - P_{A2}}{P_{Bm}}\right] \tag{12.6}$$

Hence the rate of absorption of **A** per unit time per unit area is given by:

$$N'_A = k'_G P \left[\frac{P_{A1} - P_{A2}}{P_{Bm}}\right] \tag{12.7}$$

or:

$$N'_A = k_G(P_{A1} - P_{A2}) \tag{12.8}$$

where:

$$k'_G = \frac{D_V}{\mathbf{R}T z_G}, \quad \text{and} \quad k_G = \frac{D_V P}{\mathbf{R}T z_G P_{Bm}} = \frac{k'_G P}{P_{Bm}} \tag{12.9}$$

In the great majority of industrial processes the film thickness is not known, so that the rate equation of immediate use is equation 12.8 using $k_G$. $k_G$ is known as the gas-film transfer coefficient for absorption and is a direct measure of the rate of absorption per unit area of interface with a driving force of unit partial pressure difference.

## 12.3.4. Diffusion in the liquid phase

The rate of diffusion in liquids is much slower than in gases, and mixtures of liquids may take a long time to reach equilibrium unless agitated. This is partly due to the much closer spacing of the molecules, as a result of which the molecular attractions are more important.

Whilst there is at present no theoretical basis for the rate of diffusion in liquids comparable with the kinetic theory for gases, the basic equation is taken as similar to that for gases, or for dilute concentrations:

$$N'_A = -D_L \frac{\mathrm{d}C_A}{\mathrm{d}z} \tag{12.10}$$

On integration:

$$N'_A = -D_L \left[\frac{C_{A2} - C_{A1}}{z_L}\right] \tag{12.11}$$

where: $C_A$, $C_B$ are the molar concentrations of **A** and **B**,
$z_L$ is the thickness of liquid film through which diffusion occurs, and
$D_L$ is the diffusivity in the liquid phase.

Since the film thickness is rarely known, equation 12.11 is usually rewritten as:

$$N'_A = k_L(C_{A1} - C_{A2}) \tag{12.12}$$

which is similar to equation 12.8 for gases.

In equation 12.12, $k_L$ is the liquid-film transfer coefficient, which is usually expressed in kmol/s m$^2$(kmol/m$^3$) = m/s. For dilute concentrations:

$$k_L = \frac{D_L}{z_L}$$

### 12.3.5. Rate of absorption

In a steady-state process of absorption, the rate of transfer of material through the gas film will be the same as that through the liquid film, and the general equation for mass transfer of a component **A** may be written as:

$$N'_A = k_G(P_{AG} - P_{Ai}) = k_L(C_{Ai} - C_{AL}) \tag{12.13}$$

where $P_{AG}$ is the partial pressure in the bulk of the gas, $C_{AL}$ is the concentration in the bulk of the liquid, and $P_{Ai}$ and $C_{Ai}$ are the values of concentration at the interface where equilibrium conditions are assumed to exist. Therefore:

$$\frac{k_G}{k_L} = \frac{C_{Ai} - C_{AL}}{P_{AG} - P_{Ai}} \tag{12.14}$$

These conditions may be illustrated graphically as in Figure 12.2, where ABF is the equilibrium curve for the soluble component **A**.

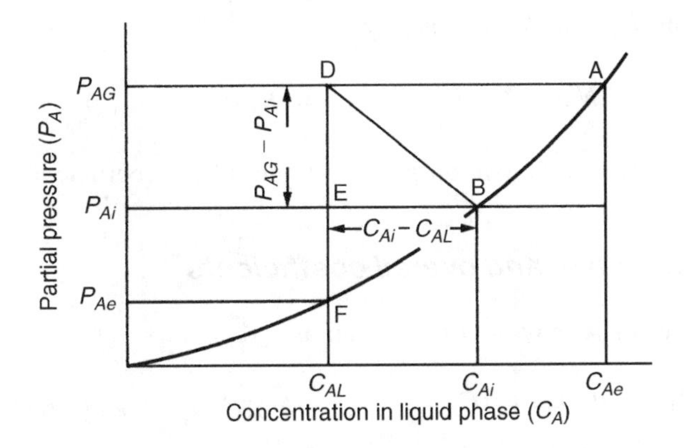


Figure 12.2. Driving forces in the gas and liquid phases

Point D ($C_{AL}$, $P_{AG}$) represents conditions in the bulk of the gas and liquid.
$P_{AG}$) is the partial pressure of **A** in the main bulk of the gas stream, and
$C_{AL}$) is the average concentration of **A** in the main bulk of the liquid stream.

Point A ($C_{Ae}$, $P_{AG}$) represents a concentration of $C_{Ae}$ in the liquid in equilibrium with $P_{AG}$ in the gas.

Point B $(C_{Ai}, P_{Ai})$ represents the concentration of $C_{Ai}$ in the liquid in equilibrium with $P_{Ai}$ in the gas, and gives conditions at the interface.

Point F $(C_{AL}, P_{Ae})$ represents a partial pressure $P_{Ae}$ in the gas phase in equilibrium with $C_{AL}$ in the liquid.

Then, the driving force causing transfer in the gas phase is:

$$(P_{AG} - P_{Ai}) \equiv \text{DE}$$

and the driving force causing transfer in the liquid phase is:

$$(C_{Ai} - C_{AL}) \equiv \text{BE}$$

Then:

$$\frac{P_{AG} - P_{Ai}}{C_{Ai} - C_{AL}} = \frac{k_L}{k_G}$$

and the concentrations at the interface (point B) are found by drawing a line through D of slope $-k_L/k_G$ to cut the equilibrium curve in B.

### *Overall coefficients*

In order to obtain a direct measurement of the values of $k_L$ and $k_G$ the measurement of the concentration at the interface would be necessary. These values can only be obtained in very special circumstances, and it has been found of considerable value to use two overall coefficients $K_G$ and $K_L$ defined by:

$$N'_A = K_G(P_{AG} - P_{Ae}) = K_L(C_{Ae} - C_{AL}) \tag{12.15}$$

$K_G$ and $K_L$ are known as the overall gas and liquid phase coefficients, respectively.

### *Relation between film and overall coefficients*

The rate of transfer of **A** may now be written as:

$$N'_A = k_G[P_{AG} - P_{Ai}] = k_L[C_{Ai} - C_{AL}] = K_G[P_{AG} - P_{Ae}] = K_L[C_{Ae} - C_{AL}]$$

Thus:

$$\frac{1}{K_G} = \frac{1}{k_G}\left[\frac{P_{AG} - P_{Ae}}{P_{AG} - P_{Ai}}\right] \tag{12.16}$$

$$= \frac{1}{k_G}\left[\frac{P_{AG} - P_{Ai}}{P_{AG} - P_{Ai}}\right] + \frac{1}{k_G}\left[\frac{P_{Ai} - P_{Ae}}{P_{AG} - P_{Ai}}\right]$$

From the previous discussion:

$$\frac{1}{k_G} = \frac{1}{k_L}\left[\frac{P_{AG} - P_{Ai}}{C_{Ai} - C_{AL}}\right]$$

Thus:
$$\frac{1}{K_G} = \frac{1}{k_G} + \frac{1}{k_L}\left[\frac{P_{AG} - P_{Ai}}{C_{Ai} - C_{AL}}\right]\left[\frac{P_{Ai} - P_{Ae}}{P_{AG} - P_{Ai}}\right]$$
$$= \frac{1}{k_G} + \frac{1}{k_L}\left[\frac{P_{Ai} - P_{Ae}}{C_{Ai} - C_{AL}}\right]$$

$(P_{Ai} - P_{Ae})/(C_{Ai} - C_{AL})$ is the average slope of the equilibrium curve and, when the solution obeys Henry's law, $\mathscr{H} = \mathrm{d}P_A/\mathrm{d}C_A \approx (P_{Ai} - P_{Ae})/(C_{Ai} - C_{AL})$.

Therefore:
$$\frac{1}{K_G} = \frac{1}{k_G} + \frac{\mathscr{H}}{k_L} \tag{12.17}$$

Similarly:
$$\frac{1}{K_L} = \frac{1}{k_L} + \frac{1}{\mathscr{H}k_G} \tag{12.18}$$

and:
$$\frac{1}{K_G} = \frac{\mathscr{H}}{K_L} \tag{12.19}$$

A more detailed discussion of the relationship between film and overall coefficients is given in Volume 1, Chapter 10.

The validity of using equations 12.17 and 12.18 in order to obtain an overall transfer coefficient has been examined in detail by King[17]. He has pointed out that the equilibrium constant $\mathscr{H}$ must be constant, there must be no significant interfacial resistance, and there must be no interdependence of the values of the two film-coefficients.

### *Rates of absorption in terms of mole fractions*

The mass transfer equations can be written as:

$$N'_A = k''_G(y_A - y_{Ai}) = K''_G(y_A - y_{Ae}) \tag{12.20}$$

and:
$$N'_A = k''_L(x_{Ai} - x_A) = K''_L(x_{Ae} - x_A) \tag{12.21}$$

where $x_A$, $y_A$ are the mole fractions of the soluble component **A** in the liquid and gas phases, respectively.

$k''_G$, $k''_L$, $K''_G$, $K''_L$ are transfer coefficients defined in terms of mole fractions by equations 12.20 and 12.21.

If $m$ is the slope of the equilibrium curve [approximately $(y_{Ai} - y_{Ae})/(x_{Ai} - x_A)$], it can then be shown that:

$$\frac{1}{K''_G} = \frac{1}{k''_G} + \frac{m}{k''_L} \tag{12.22}$$

which is similar to equation 11.151 used for distillation.

### *Factors influencing the transfer coefficient*

The influence of the solubility of the gas on the shape of the equilibrium curve, and the effect on the film and overall coefficients, may be seen by considering three cases in turn — very soluble, almost insoluble, and moderately soluble gases.

(a) *Very soluble gas*. Here the equilibrium curve lies close to the concentration-axis and the points E and F are very close to one another as shown in Figure 12.2. The driving force over the gas film (DE) is then approximately equal to the overall driving force (DF), so that $k_G$ is approximately equal to $K_G$.

(b) *Almost insoluble gas*. Here the equilibrium curve rises very steeply so that the driving force $(C_{Ai} - C_{AL})$ (EB) in the liquid film becomes approximately equal to the overall driving force $(C_{Ae} - C_{AL})$ (AD). In this case $k_L$ will be approximately equal to $K_L$.

(c) *Moderately soluble gas*. Here both films offer an appreciable resistance, and the point B at the interface must be located by drawing a line through D of slope $-(k_L/k_G) = -(P_{AG} - P_{Ai})/(C_{Ai} - C_{AL})$.

In most experimental work, the concentration at the interface cannot be measured directly, and only the overall coefficients are therefore found. To obtain values for the film coefficients, the relations between $k_G$, $k_L$ and $K_G$ are utilised as discussed previously.

## 12.4. DETERMINATION OF TRANSFER COEFFICIENTS

In the design of an absorption tower, the most important single factor is the value of the transfer coefficient or the height of the transfer unit. Whilst the total flowrates of the gas and liquid streams are fixed by the process, it is necessary to determine the most suitable flow per unit area through the column. The gas flow is limited by the fact that the flooding rate must not be exceeded and there will be a serious drop in performance if the liquid rate is very low. It is convenient to examine the effects of flowrates of the gas and liquid on the transfer coefficients, and also to investigate the influence of variables such as temperature, pressure, and diffusivity.

In the laboratory, wetted-wall columns have been used by a number of workers and they have proved valuable in determining the importance of the various factors, and have served as a basis from which correlations have been developed for packed towers.

### 12.4.1. Wetted-wall columns

In many early studies, the rate of vaporisation of liquids into an air stream was measured in a wetted-wall column, similar to that shown in Figure 12.3. Logarithmic plots of $d/z_G$ and $Re = du\rho/\mu$ gave a series of approximately straight lines and $d/z_G$ was proportional to $Re^{0.83}$

where: $d$ is the diameter of tube,
$z_G$ is the thickness of gas film,
$u$ is the gas velocity,
$\rho$ is the gas density,
$\mu$ is the gas viscosity, and
$B$ is a constant.

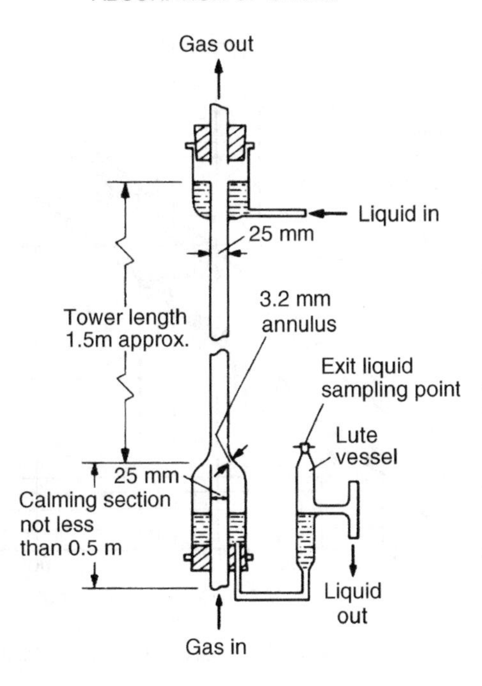


Figure 12.3. Diagram of a typical laboratory wetted-wall column

The unknown film thickness $z_G$ may be eliminated as follows:

$$k_G = \frac{D_V P}{\mathbf{R}T z_G P_{Bm}} \qquad \text{(equation 12.9)}$$

Thus:

$$\frac{k_G \mathbf{R}T P_{Bm}}{D_V P} = \frac{1}{z_G} = \frac{B}{d} Re^{0.83}$$

or:

$$\frac{h_D d P_{Bm}}{D_V P} = B Re^{0.83} \qquad (12.23)$$

where $h_D = k_G \mathbf{R}T$ is the mass transfer coefficient with the driving force expressed as a molar concentration difference.

GILLILAND and SHERWOOD's data[18], expressed by equation 12.23, are shown in Figure 12.4 for a number of systems. To allow for the variation in the physical properties, the Schmidt Group $Sc$ is introduced, and the general equation for mass transfer in a wetted-wall column is then given by:

$$\frac{h_D d}{D_V} \frac{P_{Bm}}{P} = B' Re^{0.83} Sc^{0.44} \qquad (12.24)$$

Values of $B'$ 0.021–0.027 have been reported and a mean value of 0.023 may be taken, which means that equation 12.24 very similar to the general heat transfer equation for forced convection in tubes (Volume 1, Chapter 9). The data shown in Figure 12.4 are

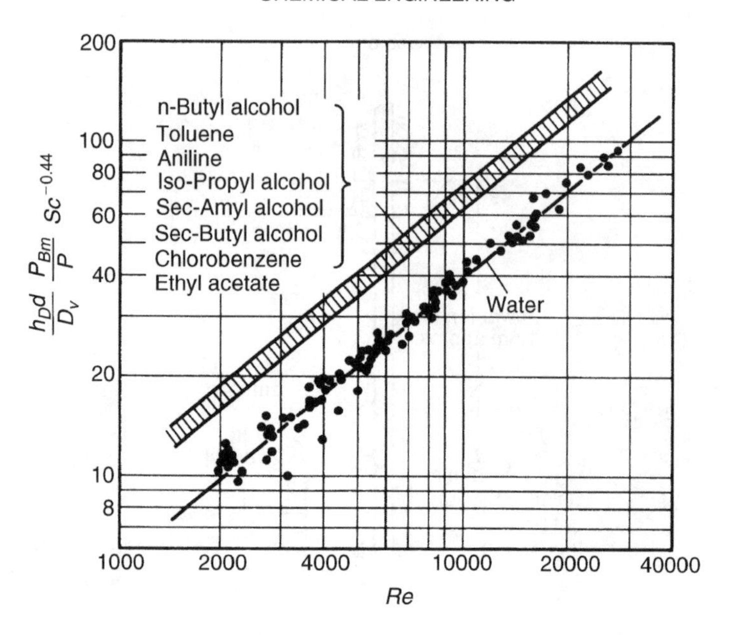


Figure 12.4. Vaporisation of liquids in a wetted-wall column

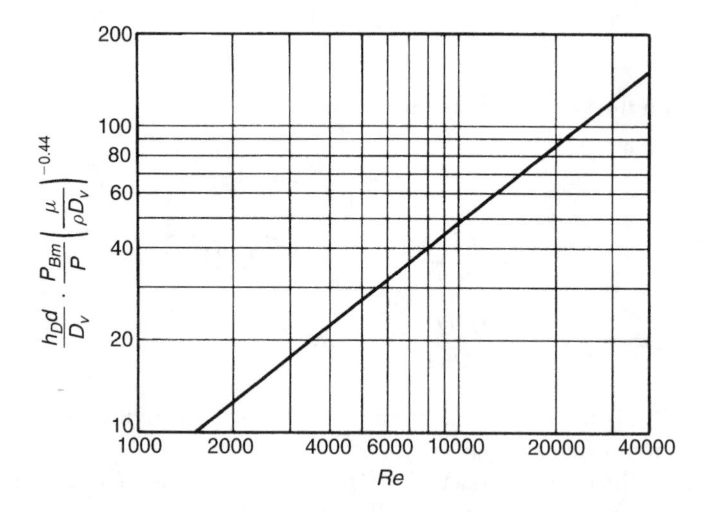


Figure 12.5. Correlation of data on the vaporisation of liquids in wetted-wall columns

replotted as $(h_D d/D_V)(P_{Bm}/P)Sc^{-0.44}$ in Figure 12.5 and, in this way, they may be correlated by means of a single line.

In comparing the results of various workers, it is important to ensure that the inlet arrangements for the air are similar. Modifications of the inlet give rise to various values for the index on the Reynolds number, as found by HOLLINGS and SILVER[(19)]. A good calming length is necessary before the inlet to the measuring section, if the results are to be reproducible.

Equation 12.24 is frequently rearranged as:

$$\frac{h_D d}{D_V}\frac{P_{Bm}}{P}\frac{\mu}{du\rho}\left[\frac{\mu}{\rho D_V}\right]^{0.56} = B' Re^{-0.17}\left(\frac{\mu}{\rho D_V}\right)$$

or:

$$\frac{h_D}{u}\frac{P_{Bm}}{P}\left[\frac{\mu}{\rho D_V}\right]^{0.56} = B' Re^{-0.17} = j_d \qquad (12.25)$$

where $j_d$ is the $j$-factor for mass transfer as introduced by CHILTON and COLBURN[20] and discussed in Volume 1, Chapter 10. The main feature of this type of work is that $h_D \propto G'^{0.8}$, $D_V^{0.56}$ and $P/P_{Bm}$. This form of relation is the basis for correlating data on packed towers.

## Example 12.1

The overall liquid transfer coefficient, $K_L a$, for the absorption of $SO_2$ in water in a column is 0.003 kmol/s m$^3$ (kmol/m$^3$). By assuming an expression for the absorption of $NH_3$ in water at the same liquor rate and varying gas rates, derive an expression for the overall liquid film coefficient $K_L a$ for absorption of $NH_3$ in water in this equipment at the same water rate though with varying gas rates. The diffusivities of $SO_2$ and $NH_3$ in air at 273 K are 0.103 and 0.170 cm$^2$/s. $SO_2$ dissolves in water, and Henry's constant is equal to 50 (kN/m$^2$)/(kmol/m$^3$). All data are expressed for the same temperature.

## Solution

From equation 12.18:

$$\frac{1}{K_L a} = \frac{1}{k_L a} + \frac{1}{\mathscr{H} k_G a} = \frac{1}{0.003} = 333.3$$

For the absorption of a moderately soluble gas it is reasonable to assume that the liquid and gas phase resistances are of the same order of magnitude, assuming them to be equal.

$$\frac{1}{k_L a} = \frac{1}{\mathscr{H} k_G a} = \left(\frac{333}{2}\right) = 166.7$$

or:

$$k_L a = \mathscr{H} k_G a = 0.006 \text{ kmol/s m}^3(\text{kmol/m}^3)$$

Thus, for $SO_2$: $k_G a = 0.006/\mathscr{H} = 0.006/50 = 0.00012$ kmol/s m$^3$(kN/m$^2$)

From equation 12.25: $k_G a$ is proportional to (diffusivity)$^{0.56}$.

Hence for $NH_3$ : $k_G a = 0.00012(0.17/0.103)^{0.56} = 0.00016$ kmol/s m$^3$(kN/m$^2$)

For a very soluble gas such as $NH_3$, $k_G a \simeq K_G a$.

For $NH_3$ the liquid-film resistance will be small, and:

$$k_G a = K_G a = \underline{\underline{0.00016 \text{ kmol/s m}^3(\text{kN/m}^2)}}$$

In early work on wetted-wall columns, MORRIS and JACKSON[21] represented the experimental data for the mass transfer coefficient for the gas film $h_D$ in a form similar to equation 12.25, though with slightly different indices, to give:

$$\frac{h_D}{u} = 0.04 \left[\frac{ud\rho}{\mu}\right]^{-0.25} \left[\frac{\mu}{\rho D_V}\right]^{-0.5} \left[\frac{P}{P_{Bm}}\right] \tag{12.26}$$

The velocity $u$ of the gas is strictly the velocity relative to the surface of the falling liquid film, though little error is introduced if it is taken as the superficial velocity in the column.

## Compounding of film coefficients

Assuming $k_G$ is approximately proportional to $G'^{0.8}$, equation 12.17 may be rearranged to give:

$$\frac{1}{K_G} = \frac{1}{k_G} + \frac{\mathscr{H}}{k_L} = \frac{1}{\psi u^{0.8}} + \frac{\mathscr{H}}{k_L} \tag{12.27}$$

If $k_L$ is assumed to be independent of the gas velocity, then a plot of $1/K_G$ against $1/u^{0.8}$ will give a straight line with a positive intercept on the vertical axis representing the liquid film resistance $\mathscr{H}/k_L$, as shown for ammonia and for sulphur dioxide in Figure 12.6. It may be seen that in each case a straight line is obtained. The lines for ammonia pass almost through the origin showing that the liquid film resistance is very small, although the line for sulphur dioxide gives a large intercept on the vertical axis, indicating a high value of the liquid film resistance.

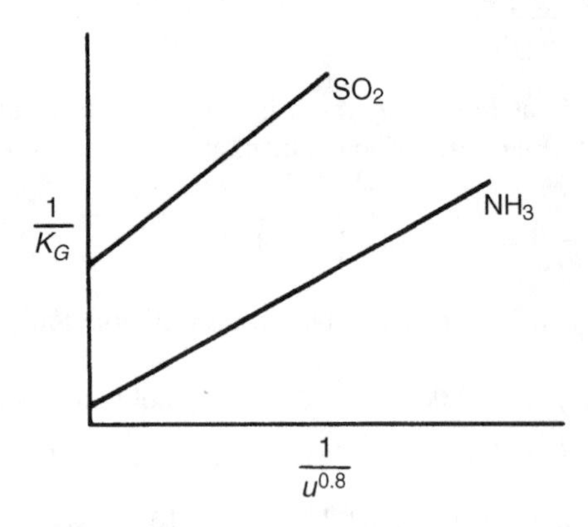


Figure 12.6. Plot of $1/K_G$ versus $1/u^{0.8}$ for ammonia and for sulphur dioxide

For a constant value of $Re$, the film thickness $z_G$ should be independent of temperature, since $\mu/\rho D_V$ is almost independent of temperature. $k_G$ will then vary as $\sqrt{T}$, because

$D_V \propto T^{3/2}$ and $k_G \propto D_V/T$. This is somewhat difficult to test accurately since the diffusivity in the liquid phase also depends on temperature. Thus, the data for sulphur dioxide, shown in Figure 12.6, qualitatively support the theory for different temperatures, although the increase in value of $k_L$ masks the influence of temperature on $k_G$.

## Example 12.2

A wetted-wall column is used for absorbing sulphur dioxide from air by means of a caustic soda solution. At an air flow of 2 kg/m$^2$s, corresponding to a Reynolds number of 5160, the friction factor $R/\rho u^2$ is 0.0200.

Calculate the mass transfer coefficient in kg $SO_2$/s m$^2$(kN/m$^2$) under these conditions if the tower is at atmospheric pressure. At the temperature of absorption the following values may be used:

The diffusion coefficient for $SO_2 = 0.116 \times 10^{-4}$ m$^2$/s, the viscosity of gas = 0.018 mNs/m$^2$, and the density of gas stream = 1.154 kg/m$^3$.

## Solution

For wetted-wall columns, the data are correlated by:

$$\left(\frac{h_d}{u}\right)\left(\frac{P_{Bm}}{P}\right)\left(\frac{\mu}{\rho D}\right)^{0.56} = B' Re^{-0.17} = j_d \qquad \text{(equation 12.25)}$$

From Volume 1, Chapter 10: $j_d \simeq R/\rho u^2$

In this problem: $G' = 2.0$ kg/m$^2$s, $Re = 5160$ and $R/\rho u^2 = 0.020$

$$D = 0.116 \times 10^{-4}\ \text{m}^2/\text{s}, \quad \mu = 1.8 \times 10^{-5}\ \text{Ns/m}^2, \text{ and } \rho = 1.154\ \text{kg/m}^3$$

Substituting these values gives:

$$\left(\frac{\mu}{\rho D}\right)^{0.56} = \left(\frac{1.8 \times 10^{-5}}{1.154 \times 0.116 \times 10^{-4}}\right)^{0.56} = 1.18$$

Thus:
$$\left(\frac{h_d}{u}\right)\left(\frac{P_{Bm}}{P}\right) = (0.020/1.18) = 0.0169$$

$$G' = \rho u = 2.0\ \text{kg/m}^2\text{s}$$

and:
$$u = (2.0/1.154) = 1.73\ \text{m/s}$$

Thus:
$$h_d(P_{Bm}/P) = (0.0169 \times 1.73) = 0.0293$$

$d$ may be obtained from $d = Re\mu/\rho u = 0.046$ m (46 mm), which is the same order of size of wetted-wall column as that which was originally used in the research work.

$$k_G = \left(\frac{h_d}{\mathbf{R}T}\right)\left(\frac{P_{Bm}}{P}\right)$$

$\mathbf{R} = 8314$ m$^3$(N/m$^2$)/K kmol and $T$ will be taken as 298 K, and hence:

$$k_G = [0.0293/(8314 \times 298)] = 1.18 \times 10^{-8} \text{ kmol/m}^2\text{s(N/m}^2)$$

$$= \underline{\underline{7.56 \times 10^{-4} \text{ kg SO}_2/\text{m}^2\text{s(kN/m}^2)}}$$

## 12.4.2. Coefficients in packed towers

The majority of published data on transfer coefficients in packed towers are for rather small laboratory units, and there is still some uncertainty in extending the data for use in industrial units. One of the great difficulties in correlating the performance of packed towers is the problem of assessing the effective wetted area for interphase transfer. It is convenient to consider separately conditions where the gas-film is controlling, and then where the liquid film is controlling. The general method of expressing results is based on that used for wetted-wall columns.

### *Gas-film controlled processes*

The absorption of ammonia in water has been extensively studied by a number of workers. KOWALKE *et al.*[22] used a tower of 0.4 m internal diameter with a packing 1.2 m deep, and expressed their results as:

$$K_G a = \alpha G'^{0.8} \tag{12.28}$$

where $K_G$ is expressed in kmol/s m$^2$ (kN/m$^2$) and $a$ is the interfacial surface per unit volume of tower (m$^2$/m$^3$). Thus $K_G a$ is a transfer coefficient based on unit volume of tower. $G'$ is in kg/s m$^2$, and varies with the nature of the packing and the liquid rate. It was noted that $\alpha$ increased with $L'$ for values up to 1.1 kg/s m$^2$, after which further increase gave no significant increase in $K_G a$. It was thought that the initial increase in the coefficient was occasioned by a more effective wetting of the packing. On increasing the liquid rate so that the column approached flooding conditions, it was found that $K_G a$ decreased. Other measurements by BORDEN and SQUIRES[23] and NORMAN[24] confirm the applicability of equation 12.28.

FELLINGER[25] used a 450 mm diameter column with downcomers and risers in an attempt to avoid the problem of determining any entrance or exit effects. Some of the results for $\mathbf{H}_{OG}$ are shown in Table 12.2, taken from Perry's Chemical Engineers' Handbook[26]. Further discussion on the use of transfer units is included in Section 12.8.8 and in Chapter 11.

Table 12.2. Height of the transfer unit $\mathbf{H}_{OG}$ in metres

| Raschig rings size (mm) | $G'$ (kg/m$^2$s) | $\mathbf{H}_{OG}$ ($L' = 0.65$ kg/m$^2$s) | $\mathbf{H}_{OG}$ ($L' = 1.95$ kg/m$^2$s) |
|---|---|---|---|
| 9.5 | 0.26 | 0.37 | 0.23 |
| | 0.78 | 0.60 | 0.32 |
| 25 | 0.26 | 0.40 | 0.22 |
| | 0.78 | 0.64 | 0.34 |
| 50 | 0.26 | 0.60 | 0.34 |
| | 0.78 | 1.04 | 0.58 |

MOLSTAD *et al.*[27] also measured the absorption of ammonia in water using a tower of 384 mm side packed with wood grids, or with rings or saddles, and obtained $K_G a$ by direct experiment. The value of $k_G a$ was then calculated from the following relation based on equation 12.17:

$$\frac{1}{K_G a} = \frac{1}{k_G a} + \frac{\mathscr{H}}{k_L a} \tag{12.29}$$

The simplest method of representing data for gas-film coefficients is to relate the Sherwood number $[(h_D d/D_V)(P_{Bm}/P)]$ to the Reynolds number $(Re)$ and the Schmidt number $(\mu/\rho D_V)$. The indices used vary between investigators though VAN KREVELEN and HOFTIJZER[28] have given the following expression, which is claimed to be valid over a wide range of Reynolds numbers:

$$\frac{h_D d}{D_V}\frac{P_{Bm}}{P} = 0.2 Re^{0.8}\left(\frac{\mu}{\rho D_V}\right)^{0.33} \tag{12.30}$$

Later work suggests that 0.11 is a more realistic value for the coefficient.

SEMMELBAUER[29] has recommended the following correlation for $100 < (Re)_G < 10{,}000$ and $0.01 \text{ m} < d_p < 0.05$ m:

$$(Sh)_G = \beta (Re)_G^{0.59} (Sc)_G^{0.33} \tag{12.31}$$

where: $\beta = 0.69$ for Raschig rings and 0.86 for Berl saddles,
$(Sh)_G = h_D d_p/D_G$,
$(Re)_G = G' d_p/\mu_G$,
$(Sc)_G = \mu_G/\rho_G D_G$, and
$d_p$ = packing size.

### *Processes controlled by liquid-film resistance*

The absorption of carbon dioxide, oxygen, and hydrogen in water are three examples in which most, if not all, of the resistance to transfer lies in the liquid phase. SHERWOOD and HOLLOWAY[30] measured values of $k_L a$ for these systems using a tower of 500 mm diameter packed with 37 mm rings. The results were expressed in the form:

$$\frac{k_L a}{D_L} = \beta\left[\frac{L'}{\mu_L}\right]^{0.75}\left[\frac{\mu_L}{\rho_L D_L}\right]^{0.50} \tag{12.32}$$

It may be noted that this equation has no term for characteristic length on the right-hand side and therefore it is not a dimensionally consistent equation. If values of $k_L a$ are plotted against value $L'$ on logarithmic scales as shown in Figure 12.7, a slope of about 0.75 is obtained for values of $L'$ 0.5–20 kg/s m$^2$. Beyond this value of $L'$, it was found that $k_L a$ tended to fall because the loading point for the column was reached. These values of $k_L a$ were found to be affected by the gas rate. Subsequently, COOPER *et al.*[31] established that, at the high liquid rates and low gas rates used in practice, the transfer rates were much lower than given by equation 12.32. This was believed to be due to maldistribution at gas velocities as low as 0.03 m/s. The results of COOPER *et al.*[31] and SHERWOOD and

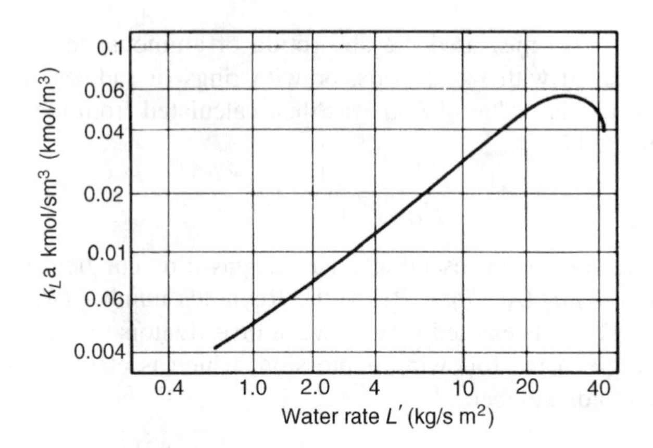


Figure 12.7. Variation of liquid-film coefficient with liquid flow for the absorption of oxygen in water

HOLLOWAY[30] are compared in Figure 12.8, where the height of the transfer unit $\mathbf{H}_{OL}$ is plotted against the liquid rate for various gas velocities.

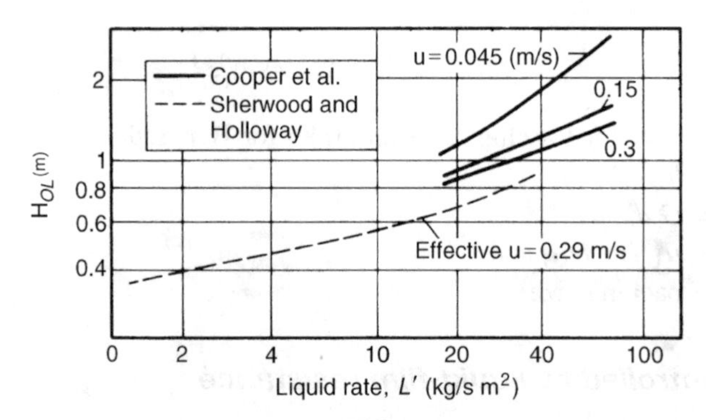


Figure 12.8. Effect of liquid rate on height of transfer unit $\mathbf{H}_{OL}$. Comparison of the results of Sherwood and Holloway[30], and Cooper *et al.*[31]

In an equation similar to equation 12.31, SEMMELBAUER[29] produced the following correlation for the liquid film mass transfer coefficient $k_L$ for $3 < Re_L < 3000$ and $0.01 \text{ m} < d_p < 0.05$ m:

$$(Sh)_L = \beta'(Re)_L^{0.59}(Sc)_L^{0.5}(d_p^3 g\rho_L^2/\mu_L^2)^{0.17} \qquad (12.33)$$

where: $\beta' = 0.32$ and 0.25 for Raschig rings and Berl saddles, respectively.
$(Sh)_L = k_L d_p/D_L$,
$(Re)_L = L'd_p/\mu_L$, and
$(Sc)_L = \mu_L/\rho_L D_L$.

NONHEBEL[32] emphasises that values of the individual film mass transfer coefficients obtained from this equation must be used with caution when designing large-scale towers and appropriately large safety factors should be incorporated.

### 12.4.3. Coefficients in spray towers

It is difficult to compare the performance of various spray towers since the type of spray distributor used influences the results. Data from HIXSON and SCOTT[33] and others show that $K_G a$ varies as $G'^{0.8}$, and is also affected by the liquid rate. More reliable data with spray columns might be expected if the liquid were introduced in the form of individual drops through a single jet into a tube full of gas. Unfortunately the drops tend to alter in size and shape and it is not possible to get the true interfacial area very accurately. This has been investigated by WHITMAN *et al.*[34], who found that $k_G$ for the absorption of ammonia in water was about 0.035 kmol/s m$^2$ (N/m$^2$), compared with 0.00025 for the absorption of carbon dioxide in water.

Some values obtained by PIGFORD and PYLE[35] for the height of a transfer unit $\mathbf{H}_L$ for the stripping of oxygen from water are shown in Figure 12.9. For short heights, the efficiency of the spray chamber approximates closely to that of a packed tower although, for heights greater than 1.2 m, the efficiency of the spray tower drops off rather rapidly. Whilst it might be possible to obtain a very large active interface by producing small drops, in practice it is impossible to prevent these coalescing, and hence the effective interfacial surface falls off with height, and spray towers are not used extensively.

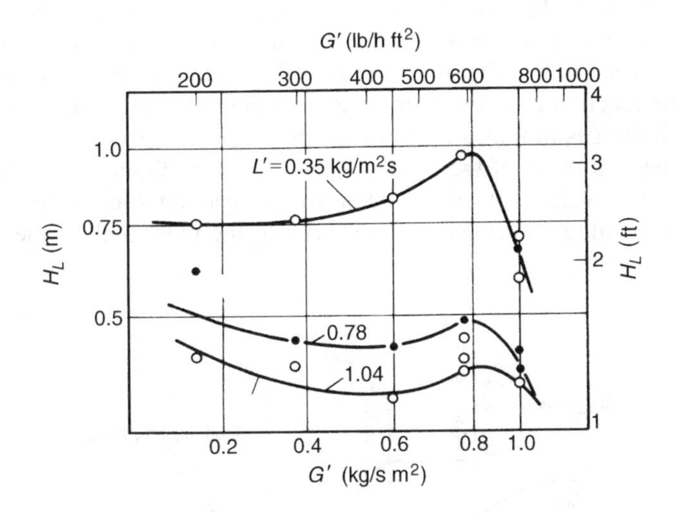


Figure 12.9. Height of the transfer unit $\mathbf{H}_L$ for stripping of oxygen from water in a spray tower

## 12.5. ABSORPTION ASSOCIATED WITH CHEMICAL REACTION

In the instances so far considered, the process of absorption of the gas in the liquid has been entirely a physical one. There are, however, a number of cases in which the gas, on absorption, reacts chemically with a component of the liquid phase[36]. The topic of mass transfer accompanied by chemical reaction is treated in detail in Volume 1, Chapter 10.

In the absorption of carbon dioxide by caustic soda, the carbon dioxide reacts directly with the caustic soda and the process of mass transfer is thus made much more complicated. Again, when carbon dioxide is absorbed in an ethanolamine solution, there is direct chemical reaction between the amine and the gas. In such processes the conditions in the gas phase are similar to those already discussed, though in the liquid phase there is a liquid film followed by a reaction zone. The process of diffusion and chemical reaction may still be represented by an extension of the film theory by a method due to HATTA[37]. In the case considered, the chemical reaction is irreversible and of the type in which a solute gas **A** is absorbed from a mixture by a substance **B** in the liquid phase, which combines with **A** according to the equation $\mathbf{A} + \mathbf{B} \rightarrow \mathbf{AB}$. As the gas approaches the liquid interface, it dissolves and reacts at once with **B**. The new product **AB**, thus formed, diffuses towards the main body of the liquid. The concentration of **B** at the interface falls; this results in diffusion of **B** from the bulk of the liquid phase to the interface. Since the chemical reaction is rapid, **B** is removed very quickly, so that it is necessary for the gas **A** to diffuse through part of the liquid film before meeting **B**. There is thus a zone of reaction between **A** and **B** which moves away from the gas–liquid interface, taking up some position towards the bulk of the liquid. The final position of this reaction zone will be such that the rate of diffusion of **A** from the gas–liquid interface is equal to the rate of diffusion of **B** from the main body of the liquid. When this condition has been reached, the concentrations of **A**, **B**, and **AB** may be indicated as shown in Figure 12.10, where the concentrations are shown as ordinates and the positions of a plane relative to the interface as abscissae. In this Figure, the plane of the interface between gas and liquid is shown by U, the reaction zone by R, and the outer boundary of liquid film by S. Then **A** diffuses through the gas film as a result of the driving force $(P_{AG} - P_{Ai})$ and diffuses to the reaction zone as a result of the driving force $C_{Ai}$ in the liquid phase. The component **B** diffuses from the main body of the liquid to the reaction zone under a driving force $q$, and the non-volatile product **AB** diffuses back to the main bulk of the liquid under a driving force $(m - n)$.

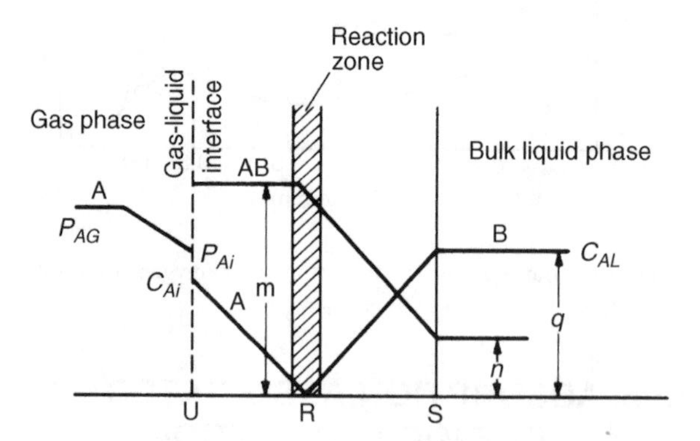


Figure 12.10. Concentration profile for absorption with chemical reaction

The difference between a physical absorption, and one in which a chemical reaction occurs, can also be shown by Figures 12.11*a* and 12.11*b*, taken from a paper by VAN KREVELEN and HOFTIJZER[28]. Figure 12.11*a* shows the normal concentration profile for

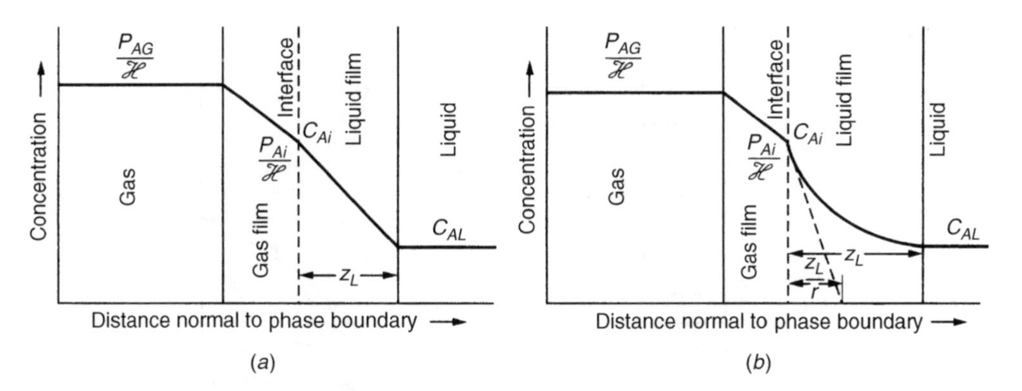


Figure 12.11. Concentration profiles for absorption ($a$) without chemical reaction, ($b$) with chemical reaction. The scales for concentration in the two phases are not the same and are chosen so that $P_{Ai}/\mathscr{H}$ in the gas phase and $C_{Ai}$ for the liquid phase are at the same position in the diagrams

physical absorption whilst Figure 12.11$b$ shows the profile modified by the chemical reaction. For transfer in the gas phase:

$$N'_A = k_G(P_{AG} - P_{Ai}) \tag{12.34}$$

and in the liquid phase:

$$N'_A = k_L(C_{Ai} - C_{AL}) \tag{12.35}$$

The effect of the chemical reaction is to accelerate the removal of **A** from the interface, and supposing that it is now $r$ times as great then:

$$N''_A = rk_L(C_{Ai} - C_{AL}) \tag{12.36}$$

In Figure 12.11$a$, the concentration profile through the liquid film of thickness $z_L$ is represented by a straight line such that $k_L = D_L/z_L$. In $b$, component **A** is removed by chemical reaction, so that the concentration profile is curved. The dotted line gives the concentration profile if, for the same rate of absorption, **A** were removed only by diffusion. The effective diffusion path is $1/r$ times the total film thickness $z_L$.

Thus:

$$N''_A = \frac{rD_L}{z_L}(C_{Ai} - C_{AL}) = rk_L(C_{Ai} - C_{AL}) \tag{12.37}$$

van Krevelen and Hoftyzer[28] showed that the factor $r$ may be related to $C_{Ai}$, $D_L$, $k_L$, to the concentration of **B** in the bulk liquid $C_{BL}$, and to the second-order reaction rate constant $\boldsymbol{k}_2$ for the absorption of $CO_2$ in alkaline solutions. Their relationship is shown in Figure 12.12, in which $r$, that is $N''_A/k_LC_{Ai}$, is plotted against $(\boldsymbol{k}_2D_LC_{BL})^{1/2}/k_L$ for various values of $C_{BL}/iC_{Ai}$, where $i$ is the number of kmol of **B** combining with 1 kmol of **A**.

Figure 12.2 illustrates three conditions:

(a) If $\boldsymbol{k}_2$ is very small, $r \simeq 1$, and conditions are those of physical absorption.

(b) If $\boldsymbol{k}_2$ is very large, $r \simeq C_{BL}/iC_{Ai}$, and the rate of the process is determined by the transport of **B** towards the phase boundary.

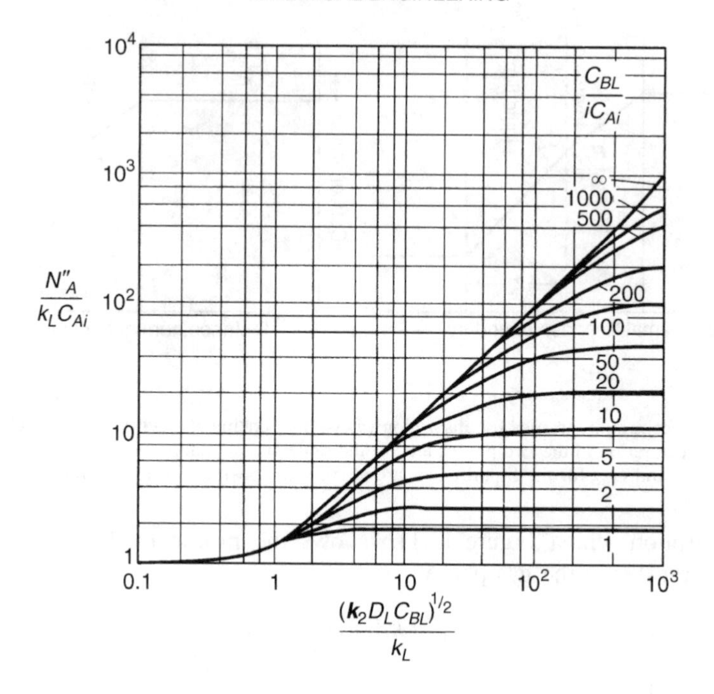


Figure 12.12. $N''_A/k_L C_{Ai}$ versus $(\boldsymbol{k}_2 D_L C_{BL})^{1/2}/k_L$ for various values of $C_{BL}/iC_{Ai}$

(c) At moderate values of $\boldsymbol{k}_2$, $r \simeq (jD_L C_{BL})^{1/2}/k_L$, and the rate of the process is determined by the rate of the chemical reaction.

Thus, from equation 12.37:

$$N''_A = k_L(C_{Ai} - C_{AL})\frac{(\boldsymbol{k}_2 D_L C_{BL})^{1/2}}{k_L} = (C_{Ai} - C_{AL})(\boldsymbol{k}_2 D_L C_{BL})^{1/2} \tag{12.38}$$

and the controlling parameter is now $\boldsymbol{k}_2$.

The results of this work have been confirmed by NIJSING, HENDRIKSZ, and KRAMERS[38].

As an illustration of combined absorption and chemical reaction, the results of TEPE and DODGE[39] on the absorption of carbon dioxide by sodium hydroxide solution may be considered. A 150 mm diameter tower filled to a depth of 915 mm with 12.5 mm carbon Raschig rings was used. Some of the results are indicated in Figure 12.13. $K_G a$ increases rapidly with increasing sodium hydroxide concentration up to a value of about 2 kmol/m$^3$. Changes in the gas rate were found to have negligible effect on $K_G a$, indicating that the major resistance to absorption was in the liquid phase. The influence of the liquid rate was rather low, and was proportional to $L'^{0.28}$. It may be assumed that, in this case, the final rate of the process is controlled by the resistance to diffusion in the liquid, by the rate of the chemical reaction, or by both together.

CRYDER and MALONEY[40] presented data on the absorption of carbon dioxide in diethanolamine solution, using a 200 mm tower filled with 20 mm rings, and some of their data are shown in Figure 12.14. The coefficient $K_G a$ is found to be independent of

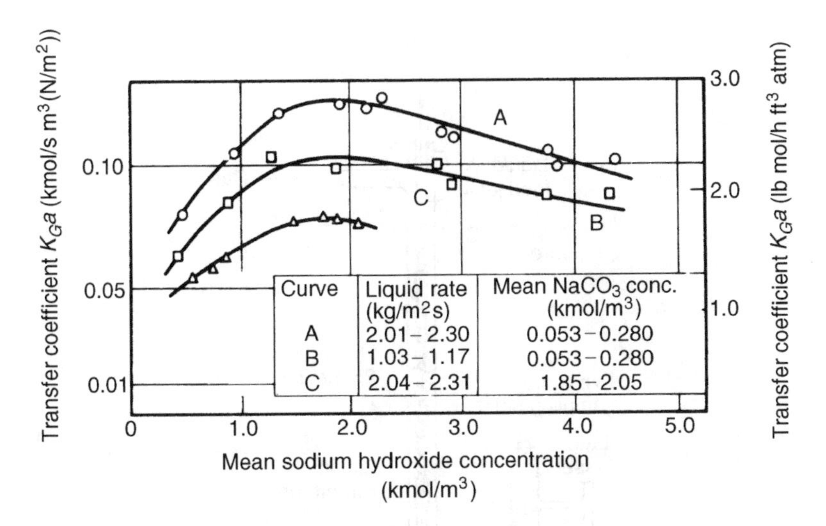


Figure 12.13. Absorption of carbon dioxide in sodium hydroxide solution $G' = 0.24–0.25$ kg/m$^2$s, temperature = 298 K

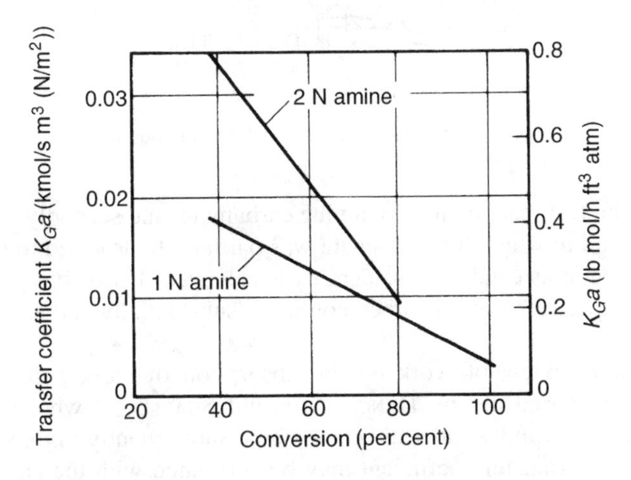


Figure 12.14. Absorption of carbon dioxide in diethanolamine solutions. Liquid rate = 1.85 kg/m$^2$s

the gas rate but to increase with the liquid rate, as expected in a process controlled by the resistance in the liquid phase.

It is difficult to deduce the size of tower required for an absorption combined with a chemical reaction, and a laboratory scale experiment should be carried out in all cases. STEPHENS and MORRIS[41] have used a small disc-type tower illustrated in Figure 12.15 for preliminary experiments of this kind. It was found that a simple wetted-wall column was unsatisfactory where chemical reactions took place. In this unit a series of discs, supported by means of a wire, was arranged one on top of the other as shown.

The absorption of carbon dioxide into aqueous amine solutions has been investigated by DANCKWERTS and MCNEIL[42] using a stirred cell. It was found that the reaction proceeded

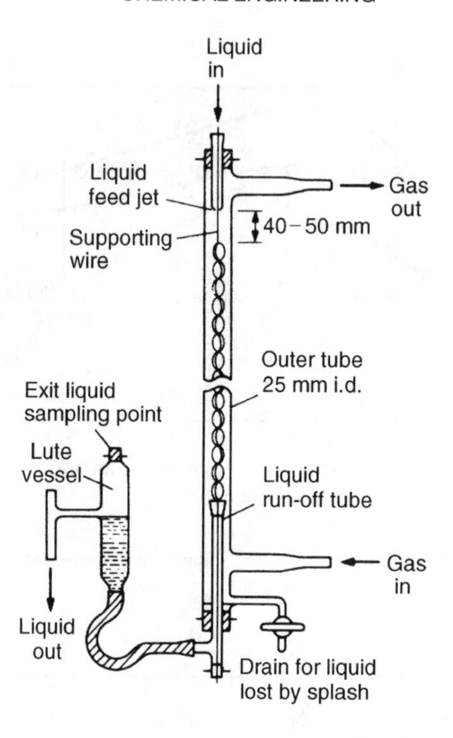


Figure 12.15. Small disc-tower for absorption tests

in two stages: first a fast reaction to give amine carbamate, and secondly a slow reaction in the bulk of the liquid in which the carbamate was partially hydrolysed to bicarbonate. The use of sodium arsenite as catalyst considerably accelerated this second reaction, showing that the overall capacity of an absorber could be substantially increased by a suitable catalyst.

A comprehensive review of work on the absorption of carbon dioxide by alkaline solutions has been carried out by DANCKWERTS and SHARMA(43) who applied results of research to the design of industrial scale equipment. Subsequently, SAHAY and SHARMA(44) showed that the mass transfer coefficient may be correlated with the gas and liquid rates and the gas and liquid compositions by:

$$K_G a = \text{const. } L'^{a_1} G'^{a_2} \exp(a_3 F' + a_4 y) \tag{12.39}$$

where: $a_1, a_2, a_3, a_4$ are experimentally determined constants,
$F'$ = fractional conversion of the liquid, and
$y$ = mole fraction of $CO_2$ in the gas.

ECKERT(45), by using the same reaction, determined the mass transfer performance of packings in terms of $K_G a$ as:

$$K_G a = \frac{N}{V(\Delta P_A)_{\text{lm}}} \tag{12.40}$$

where: N = number of moles of $CO_2$ absorbed,

$V$ = packed volume, and

$(\Delta P_A)_{lm}$ = log mean driving force.

Data obtained from this work are limited by the conditions under which they were obtained. It is both difficult and dangerous to extrapolate over the entire range of conditions encountered on a full-scale plant.

## 12.6. ABSORPTION ACCOMPANIED BY THE LIBERATION OF HEAT

In some absorption processes, especially where a chemical reaction occurs, there is a liberation of heat. This generally gives rise to an increase in the temperature of the liquid, with the result that the position of the equilibrium curve is adversely affected.

In the case of plate columns, a heat balance may be performed over each plate and the resulting temperature determined. For adiabatic operation, where no heat is removed from the system, the temperature of the streams leaving the absorber will be higher than those entering, due to the heat of solution. This rise in temperature lowers the solubility of the solute gas so that a large value of $L_m/G_m$ and a larger number of trays will be required than for isothermal operation.

For packed columns, the temperature rise will affect the equilibrium curve, and differential equations for heat and mass transfer, together with heat and mass balances, must be integrated numerically. An example of this procedure is given in Volume 1, Chapter 13, for the case of water cooling. For gas absorption under non-isothermal conditions, reference may be made to specialist texts[46,47] for a detailed description of the methods available. As an approximation, it is sometimes assumed that all the heat evolved is taken up by the liquid, and that temperature rise of the gas may be neglected. This method gives an overestimate of the rise in temperature of the liquid and results in the design of a tower which is taller than necessary. Figure 12.16 shows the effect of the temperature rise on the equilibrium curve for an adiabatic absorption process of ammonia in water. If the amount of heat liberated is very large, it may be necessary to cool the liquid. This is most conveniently done in a plate column, either with heat exchangers connected between consecutive plates, or with cooling coils on the plate, as shown in Figure 12.17.

The overall heat transfer coefficient between the gas–liquid dispersion on the tray and the cooling medium in the tubes is dependent upon the gas velocity, as pointed out by Poll and Smith[48], but is usually in the range 500–2000 W/m$^2$ K.

With packed towers it is considerably more difficult to arrange for cooling, and it is usually necessary to remove the liquid stream at intervals down the column and to cool externally. Coggan and Bourne[49] have presented a computer programme to enable the economic decision to be made between an adiabatic absorption tower, or a smaller isothermal column with interstage cooling.

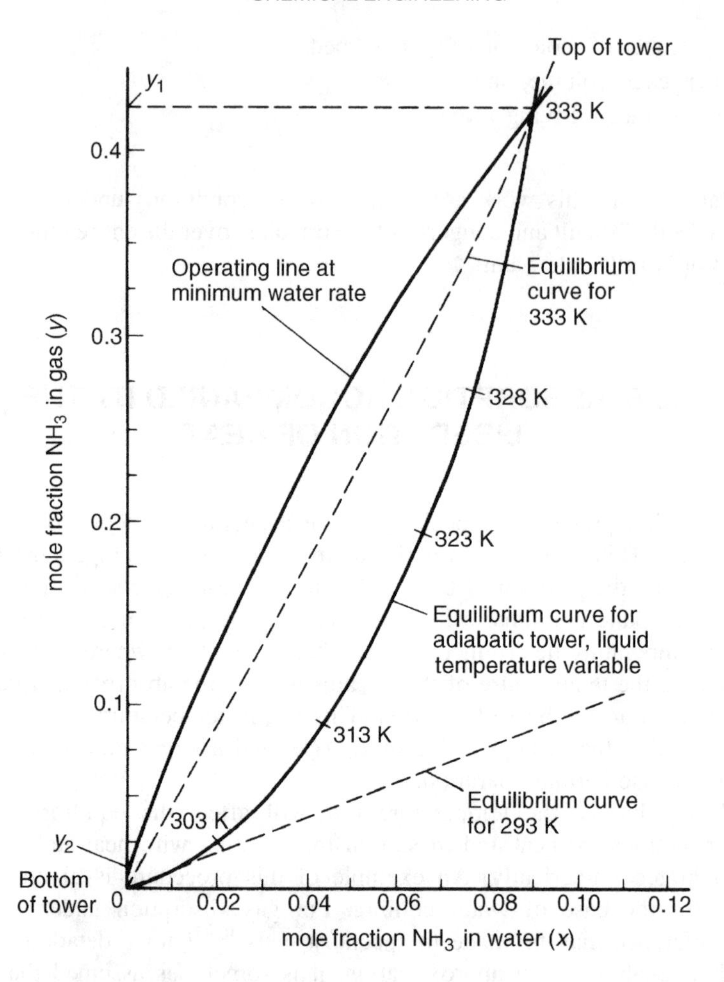


Figure 12.16. Equilibrium curve modified to allow for the heat of solution of the solute[46]

## 12.7. PACKED TOWERS FOR GAS ABSORPTION

From the analysis given already of the diffusional nature of absorption, one of the outstanding requirements is to provide as large an interfacial area of contact as possible between the phases. For this purpose, columns similar to those used for distillation are suitable. However, whereas distillation columns are usually tall and thin absorption columns are more likely to be short and fat. In addition, equipment may be used in which gas is passed into a liquid which is agitated by a stirrer. A few special forms of units have also been used, although it is the packed column which is most frequently used for gas absorption applications.

Figure 12.17. Glitsch "truss type" bubble-tray in stainless steel for a 1.9 m absorption column

## 12.7.1. Construction

The essential features of a packed column, as discussed in Chapter 4, are the shell, the arrangements for the gas and liquid inlets and outlets and the packing with its necessary supporting and redistributing systems. Reference may be made to Chapter 4 and to Volume 6 for details of these aspects, whilst this section is largely concerned with the determination of the height of packing for a particular duty. In installations where the gas is fed from a previous stage of a process where it is under pressure, there is no need to use a blower for the transfer of the gas through the column. When this is not the case, a simple blower is commonly used, and such blowers have been described in Volume 1, Chapter 8. The pressure drop across the column may be calculated by the methods presented in Chapter 4 of this volume and the blower sized accordingly. A pressure drop exceeding 30 mm of water per metre of packing is said to improve gas distribution though process conditions may not permit a figure as high as this. The packed height should not normally exceed 6 m in any section of the tower and for some packings a much lower height must be used.

In the design of an absorption tower it is necessary to take into account the characteristics of the packing elements and the flow behaviour discussed in Chapter 4, together with the considerations given in the following sections concerning the performance of columns under operating conditions.

## 12.7.2. Mass transfer coefficients and specific area in packed towers

Traditional methods of assessing the capacity of tower packings, which involve the use of the specific surface area $S$ and the voidage $e$, developed from the fact that these properties could be readily defined and measured for a packed bed of granular material such as granite, limestone, and coke which were some of the earliest forms of tower packings. The values of $S$ and $e$ enabled a reasonable prediction of hydraulic performance to be made. With the introduction of Raschig rings and other specially shaped packings, it was necessary to introduce a basis for comparing their relative efficiencies. Although the commonly published values of specific surface area $S$ provide a reasonable basis of comparison, papers such as that by SHULMAN *et al.*[50] showed that the total area offered by Raschig rings was not used, and varied considerably with hydraulic loading.

Further evidence of the importance of the wetted fraction of the total area came with the introduction of the Pall type ring. A Pall ring having the same surface area as a Raschig ring is up to 60 per cent more efficient, though many still argue the relative merits of packings purely on the basis of surface area.

The selection of a tower packing is based on its hydraulic capacity, which determines the required cross-sectional area of the tower, and the efficiency, $K_Ga$ typically, which governs the packing height. Here $a$ is the area of surface per unit volume of column and is therefore equal to $S(1 - e)$. Table 12.3[51] shows the capacity of the commonly available tower packings relative to 25 mm Raschig rings, for which a considerable amount of information is published in the literature. The table lists the packings in order of relative efficiency, $K_Ga$, evaluated at the same approach to the hydraulic capacity limit determined by flooding in each case.

## 12.7.3. Capacity of packed towers

The drop in pressure for the flow of gas and liquid over packings is discussed in Chapter 4. It is important to note that, during operation, the tower does not reach flooding conditions. In addition, every effort should be made to have as high a liquid rate as possible, in order to attain satisfactory wetting of the packing.

With low liquid rates, the whole of the surface of the packing is not completely wetted. This may be seen very readily by allowing a coloured liquid to flow over packing contained in a glass tube. From the flow patterns, it is obvious how little of the surface is wetted until the rate is quite high. This difficulty of wetting can sometimes be overcome by having considerable recirculation of the liquid over the tower, although in other cases, such as vacuum distillation, poor wetting will have to be accepted because of the low volume of liquid available. In selecting a packing, it is desirable to choose the form which will give as near complete wetting as possible. The minimum liquid rate below which the packing will no longer perform satisfactorily is known as the minimum wetting rate, discussed in Chapter 4.

The following treatment is a particular application of the more general approach adopted in Volume 1, Chapter 10.

Figure 12.18 illustrates the conditions that occur during the steady operation of a countercurrent gas–liquid absorption tower. It is convenient to express the concentration of the streams in terms of moles of solute gas per mole of inert gas in the gas phase, and as moles of solute gas per mole of solute free liquid in the liquid phase. The actual area of interface between the two phases is not known, and the term $a$ is introduced as the interfacial area per unit volume of the column. On this basis the general equation, 12.13,

Table 12.3. Capacity of commonly available packings relative to 25 mm Raschig rings[51]

| Relative $K_G a$ | Raschig rings | Traditional saddles | Pall rings | Ceramic Pall rings | Ceramic cascade mini ring® (3) | Super Intalox® saddles | Hypak® (1) | Tellerettes® (2) | Cascade mini-ring® (3) |
|---|---|---|---|---|---|---|---|---|---|
| Materials available for this relative $K_G a$ | Ceramic | Ceramic Plastic (P) | Metal (M) | Ceramic | Ceramic | Ceramic Plastic | Metal | Plastic | Metal (M) Plastic (P) |
| 0.6–0.7 | 75 mm | | | | | | | | |
| 0.7–0.8 | 50 mm | | | | | | | | |
| 0.8–0.9 | 37 mm | | | | | | | | |
| 0.9–1.0 | 25 mm | | | | | | | | |
| 1.0–1.1 | 12 mm | 75 mm | 87 mm | | No. 5 | No. 3 | | Size L | |
| 1.1–1.2 | | 50 mm | 50 mm | | | | No. 3 | | |
| 1.2–1.3 | | 37 mm | 50 mm | | No. 3 | No. 2 | | | |
| 1.3–1.4 | | 25 mm | 37 mm | 50 mm | | | No. 2 | | No. 4 (M) |
| 1.4–1.5 | | | | 37 mm | | | | | No. 3 (P) |
| 1.5–1.6 | | | 25 mm | 25 mm | No. 2 | No. 1 | | | No. 3 (M) |
| 1.6–1.7 | | | 25 mm | | | | No. 1 | Size S | No. 2 (P) |
| 1.7–1.8 | | | 16 mm | | | | | | |
| 1.8–1.9 | | | | | | | | | No. 2 (M) |
| 1.9–2.0 | | | | | | | | | No. 1 (P) |
| 2.0–2.1 | | | | | | | | | |
| 2.1–2.2 | | | | | | | | | No. 1 (M) |

Gas capacity before hydraulic limit (flooding) relative to 25 mm Raschig rings (also approx. the reciprocal of tower cross-sectional area relative to 25 mm Raschig rings for the same pressure drop throughout loading range). All relative capacity figures are valid for the same liquid to gas mass rate ratio:

(1) Trade Mark of Norton Company, U.S.A. (Hydronyl U.K.).
(2) Trade Mark of Ceilcote Company.
(3) Trade Mark of Mass Transfer Ltd. (& Inc.).

*Note:*
Relative $K_G a$ valid for all systems controlled by mass transfer coefficient ($K_G$) and wetted area ($a$) per unit volume of column. Some variation should be expected when liquid *reaction* rate is controlling (not liquid *diffusion* rate). In these cases liquid hold-up becomes more important. In general a packing having high liquid hold-up which is clearly greater than that in the falling film has poor capacity.

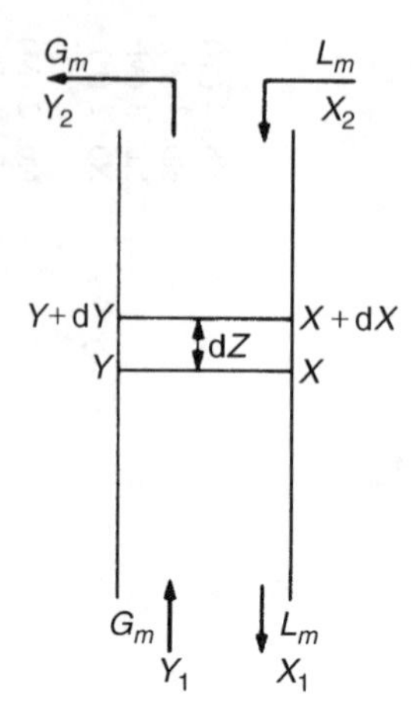


Figure 12.18. Countercurrent absorption tower

for mass transfer can be written as:

$$N'_A A\,dZa = k_G a(P_{AG} - P_{Ai})A\,dZ$$
$$= k_L a(C_{Ai} - C_{AL})A\,dZ \quad (12.41)$$

where: $N'_A$ = kmol of solute absorbed per unit time and unit interfacial area,
$a$ = surface area of interface per unit volume of column,
$A$ = cross-sectional area of column, and
$Z$ = height of packed section.

$$\text{The interfacial area for transfer} = a\,dV = aA\,dZ \quad (12.42)$$

## 12.7.4. Height of column based on conditions in the gas film

If $G_m$ = moles of inert gas/(unit time) (unit cross-section of tower),
$L_m$ = moles of solute-free liquor/(unit time) (unit cross-section of tower),
$Y$ = moles of solute gas **A**/mole of inert gas **B** in gas phase, and
$X$ = moles of solute **A**/mole of inert solvent in liquid phase.

and at any plane at which the molar ratios of the diffusing material in the gas and liquid phases are $Y$ and $X$, then over a small height $dZ$, the moles of gas leaving the gas phase will equal the moles taken up by the liquid.

Thus:
$$AG_m\,dY = AL_m\,dX \quad (12.43)$$

But:
$$G_m A\,dY = N'_A(a\,dV) = k_G a(P_{Ai} - P_{AG})A\,dZ \quad (12.44)$$

It may be noted that, in a gas absorption process, gas and liquid concentrations will decrease in the upwards direction and both d$X$ and d$Y$ will be negative.

Since:
$$P_{AG} = \frac{Y}{1+Y}P$$

$$G_m\,\mathrm{d}Y = k_G a P\left[\frac{Y_i}{1+Y_i} - \frac{Y}{1+Y}\right]\mathrm{d}Z$$

$$= k_G a P\left[\frac{Y_i - Y}{(1+Y)(1+Y_i)}\right]\mathrm{d}Z$$

Hence the height of column $Z$ required to achieve a change in $Y$ from $Y_1$ at the bottom to $Y_2$ at the top of the column is given by:

$$\int_0^Z \mathrm{d}Z = Z = \frac{G_m}{k_G a P}\int_{Y_1}^{Y_2}\frac{(1+Y)(1+Y_i)\,\mathrm{d}Y}{Y_i - Y} \tag{12.45}$$

which for dilute mixtures may be written as:

$$Z = \frac{G_m}{k_G a P}\int_{Y_1}^{Y_2}\frac{\mathrm{d}Y}{Y_i - Y} \tag{12.46}$$

In this analysis it has been assumed that $k_G$ is a constant throughout the column, and provided the concentration changes are not too large this will be reasonably true.

### 12.7.5. Height of column based on conditions in liquid film

A similar analysis may be made in terms of the liquid film. Thus from equations 12.41 and 12.42:

$$AL_m\,\mathrm{d}X = k_L a(C_{Ai} - C_{AL})A\,\mathrm{d}Z \tag{12.47}$$

where the concentrations $C$ are in terms of moles of solute per unit volume of liquor. If $C_T$ = (moles of solute + solvent) (volume of liquid), then:

$$\frac{C_A}{C_T - C_A} = \frac{\text{moles of solute}}{\text{moles of solvent}} = X$$

whence:
$$C_A = \frac{X}{1+X}C_T \tag{12.48}$$

The transfer equation (12.47) may now be written as:

$$L_m\,\mathrm{d}X = k_L a C_T\left[\frac{X}{1+X} - \frac{X_i}{1+X_i}\right]\mathrm{d}Z$$

$$= k_L a C_T\left[\frac{X - X_i}{(1+X_i)(1+X)}\right]\mathrm{d}Z$$

Thus:
$$\int_0^Z \mathrm{d}Z = Z = \frac{L_m}{k_L a C_T}\int_{X_1}^{X_2}\frac{(1+X_i)(1+X)\,\mathrm{d}X}{X - X_i} \tag{12.49}$$

and for dilute concentrations this gives:

$$Z = \frac{L_m}{k_L a C_T} \int_{X_1}^{X_2} \frac{dX}{X - X_i} \tag{12.50}$$

where $C_T$ and $k_L$ have been taken as constant over the column.

### 12.7.6. Height based on overall coefficients

If the driving force based on the gas concentration is written as $(Y - Y_e)$ and the overall gas transfer coefficient as $K_G$, then the height of the tower for dilute concentrations becomes:

$$Z = \frac{G_m}{K_G a P} \int_{Y_1}^{Y_2} \frac{dY}{Y_e - Y} \tag{12.51}$$

or in terms of the liquor concentration as:

$$Z = \frac{L_m}{K_L a C_T} \int_{X_1}^{X_2} \frac{dX}{X - X_e} \tag{12.52}$$

#### *Equations for dilute concentrations*

As the mole fraction is approximately equal to the molar ratio at dilute concentrations then considering the gas film:

$$Z = \frac{G_m}{K_G a P} \int_{Y_1}^{Y_2} \frac{dY}{Y_e - Y} = \frac{G_m}{K_G a P} \int_{y_1}^{y_2} \frac{dy}{y_e - y} \tag{12.53}$$

and considering the liquid film:

$$Z = \frac{L_m}{K_L a C_T} \int_{X_1}^{X_2} \frac{dX}{X - X_e} = \frac{L_m}{K_L a C_T} \int_{x_1}^{x_2} \frac{dx}{x - x_e} \tag{12.54}$$

### 12.7.7. The operating line and graphical integration for the height of a column

Taking a material balance on the solute from the bottom of the column to any plane where the mole ratios are $Y$ and $X$ gives for unit area of cross-section:

$$G_m(Y_1 - Y) = L_m(X_1 - X) \tag{12.55}$$

or:

$$Y_1 - Y = \frac{L_m}{G_m}(X_1 - X) \tag{12.56}$$

This is the equation of a straight line of slope $L_m/G_m$, which passes through the point $(X_1, Y_1)$. It may be seen by making a material balance over the whole column that the same line passes through the point $(X_2, Y_2)$. This line, known as the operating line,

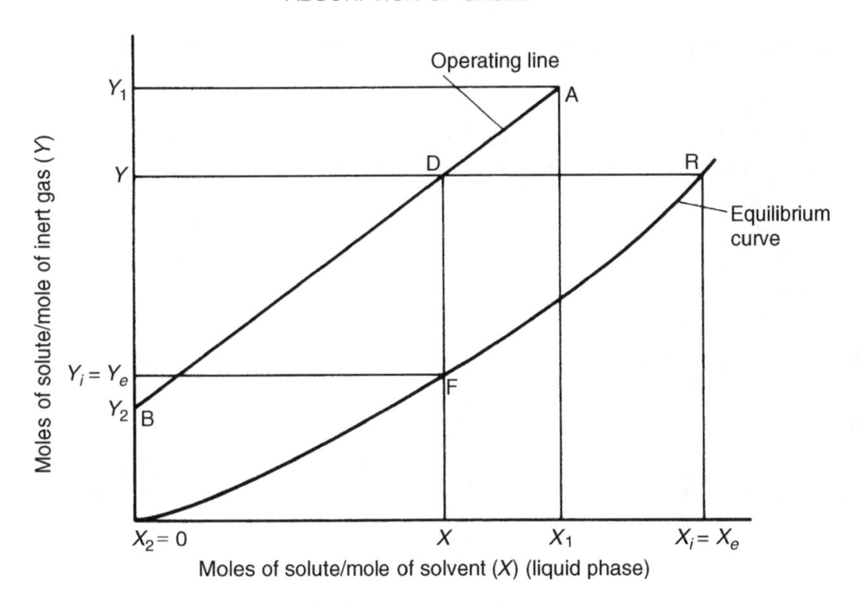


Figure 12.19. Driving force in gas and liquid-film controlled processes. The Figure shows the operating line BDA and the equilibrium curve FR

represents the conditions at any point in the column. It is similar to the operating line used in Chapter 11. Figure 12.19 illustrates typical conditions for the case of moist air and sulphuric acid or caustic soda solution, where the main resistance lies in the gas phase.

The equilibrium curve is represented by the line FR, and the operating line is given by AB, A corresponding to the concentrations at the bottom of the column and B to those at the top of the column. D represents the condition of the bulk of the liquid and gas at any point in the column, and has coordinates $X$ and $Y$. Then, if the gas film is controlling the process, $Y_i$ equals $Y_e$, and is given by a point F on the equilibrium curve, with coordinates $X$ and $Y_i$. The driving force causing transfer is then given by the distance DF. It is therefore possible to evaluate the expression:

$$\int_{Y_1}^{Y_2} \frac{dY}{Y_i - Y}$$

by selecting values of $Y$, reading off from the Figure the corresponding values of $Y_i$, and thus calculating $1/(Y_i - Y)$. It may be noted that, for gas absorption, $Y > Y_i$ and $Y_i - Y$ and $dY$ in the integral are both negative.

If the liquid film controls the process, $X_i$ equals $X_e$ and the driving force $X_i - X$ is given in Figure 12.19 by the line DR. The evaluation of the integral:

$$\int_{X_1}^{X_2} \frac{dX}{X - X_i}$$

may be effected in the same way as for the gas film.

### Special case when equilibrium curve is a straight line

If over the range of concentrations considered the equilibrium curve is a straight line, it is permissible to use a mean value of the driving force over the column. For dilute concentrations, over a small height $\mathrm{d}Z$ of column, the absorption is given by:

$$N'_A Aa\,\mathrm{d}Z = G_m A\,\mathrm{d}y = K_G a A P(y_e - y)\,\mathrm{d}Z \tag{12.57}$$

If:

$$y_e = mx + c \tag{12.58}$$

then:

$$y_{e2} = mx_2 + c$$

and:

$$y_{e1} = mx_1 + c$$

so that:

$$m = \frac{y_{e1} - y_{e2}}{x_1 - x_2} \tag{12.59}$$

Further, taking a material balance over the lower portion of the columns gives:

$$L_m(x_1 - x) = G_m(y_1 - y)$$

and:

$$x = x_1 - \frac{G_m}{L_m}(y_1 - y) \tag{12.60}$$

From equation 12.57:

$$\int_0^Z \frac{K_G a P}{G_m}\,\mathrm{d}Z = \int_{y_1}^{y_2} \frac{\mathrm{d}y}{y_e - y} \tag{12.61}$$

$$= \int_{y_1}^{y_2} \frac{\mathrm{d}y}{m[x_1 + (G_m/L_m)(y - y_1)] + c - y}$$

(from equations 12.58 and 12.60)

$$= \frac{1}{1 - (mG_m/L_m)} \ln \frac{mx_1 + c - y_1}{y_2 - m[x_1 + (G_m/L_m)(y_2 - y_1)] - c}$$

$$= \frac{1}{1 - \dfrac{y_{e1} - y_{e2}}{x_1 - x_2} \cdot \dfrac{x_1 - x_2}{y_1 - y_2}} \ln \frac{y_{e1} - y_1}{y_{e1} - y_{e2} - \left(\dfrac{y_{e1} - y_{e2}}{x_1 - x_2}\dfrac{x_1 - x_2}{y_1 - y_2} y_1 - y_2\right)}$$

(from equations 12.58, 12.59, and 12.60)

$$= \frac{y_1 - y_2}{(y - y_e)_1 - (y - y_e)_2} \ln \frac{(y - y_e)_1}{(y - y_e)_2}$$

$$= \frac{y_1 - y_2}{(y - y_e)_{\mathrm{lm}}}$$

where $(y - y_e)_{\mathrm{lm}}$ is the logarithmic mean value of $y - y_e$.

Substituting in equation 12.61:

$$\frac{K_G a P}{G_m} Z = \frac{y_1 - y_2}{(y - y_e)_{\mathrm{lm}}}$$

Thus:

$$aAZN'_A = G_m(y_1 - y_2)A = K_G a A P(y - y_e)_{\mathrm{lm}} Z \tag{12.62}$$

and in terms of mole ratios:

$$aAZN'_A = G_m(Y_1 - Y_2)A = K_G aAP(Y - Y_e)_{\text{lm}}Z \tag{12.63}$$

Thus, the logarithmic mean of the driving forces at the top and the bottom of the column may be used.

For concentrated solutions:

$$aAZN'_A = G_m(Y_1 - Y_2)A = K_G aA\phi P(Y - Y_e)_{\text{lm}}Z \tag{12.64}$$

It is necessary to introduce the factor $\phi$ since $Y$ is not directly proportional to $P$. The value of $\phi$ may be found from the relation:

$$\phi Y = \frac{Y}{1 + Y} \tag{12.65}$$

from which $\phi = 1/(1 + Y)$. Although the value of $\phi$ will change slightly over the column, a mean value will generally be acceptable.

It is of interest to note from Figure 12.20, that, as long as the ratio $k_L/k_G$ remains constant (that is, if the slope of DE is constant), then the ratio of DQ, the driving force through the gas phase, divided by DF, the driving force assuming all the resistance to be in the gas phase, will be a constant. Thus, the use of the driving force DF is satisfactory even if the resistance does not lie wholly in the gas phase. The coefficient $k_G$ on this basis is not an accurate value for the gas-film coefficient, although is proportional to it. It follows that, if the equilibrium curve is straight, either the gas-film or the liquid-film coefficient may be used. This simplification is of considerable value.

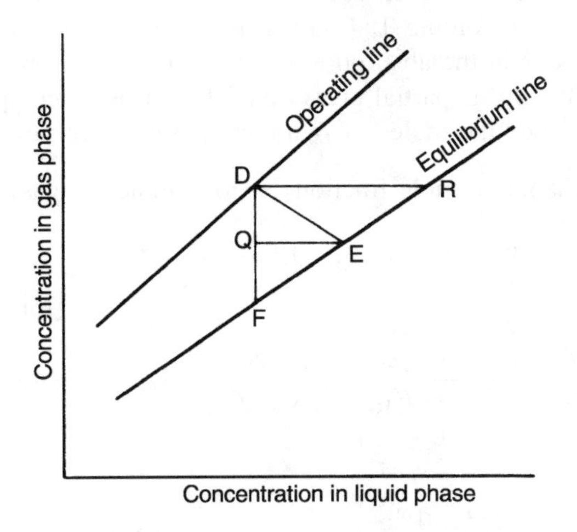


Figure 12.20. Driving force when equilibrium curve is a straight line

### 12.7.8. Capacity of tower in terms of partial pressures for high concentrations

A material balance taken between the bottom of the column and some plane where the partial pressure in the gas phase is $P_{AG}$ and the concentration in the liquid is $X$ gives:

$$L_m(X_1 - X) = G_m\left[\frac{P_{AG1}}{P - P_{AG1}} - \frac{P_{AG}}{P - P_{AG}}\right] \tag{12.66}$$

Over a small height of the column $dZ$, therefore:

$$-L_m\,dX = \frac{-G_m P}{(P - P_{AG})^2}\,dP_{AG} = k_G a(P_{AG} - P_{Ai})\,dZ \tag{12.67}$$

$$= \frac{k_G P_{Bm}}{P} a P \frac{(P_{AG} - P_{Ai})}{P_{Bm}}\,dZ$$

$$= k'_G a P \frac{(P_{AG} - P_{Ai})}{P_{Bm}}\,dZ$$

Thus:

$$\int_0^Z dZ = G_m \int_{P_{AG_1}}^{P_{AG_2}} \frac{-P_{Bm}\,dP_{AG}}{k'_G a(P - P_{AG})^2(P_{AG} - P_{Ai})} \tag{12.68}$$

The advantage of using $k'_G$ instead of $k_G$ is that $k'_G$ is independent of concentration, although this equation is almost unmanageable in practice. If a substantial amount of the gas is absorbed from a concentrated mixture, $k'_G$ will still change as a result of a reduced gas velocity, although it is independent of concentration.

### 12.7.9. The transfer unit

The group $\int(dy/y_e - y)$, which is used in Chapter 11, has been defined by CHILTON and COLBURN[52] as the number of overall gas transfer units $\mathbf{N}_{OG}$. The concept of the transfer unit is also introduced in Volume 1, Chapter 10. The application of this group to the countercurrent conditions in the absorption tower is now considered.

Over a small height $dZ$, the partial pressure of the diffusing component **A** will change by an amount $dP_{AG}$. Then the moles of **A** transferred are given by:

(change in mole fraction) × (total moles of gas)

Therefore:

$$K_G a(P_{AG} - P_{Ae})\,dZ = \frac{-dP_{AG}}{P} G'_m \tag{12.69}$$

(for dilute concentrations)

Thus:

$$\int_{P_{AG_2}}^{P_{AG_1}} \frac{dP_{AG}}{P_{Ae} - P_{AG}} = \int_0^Z \frac{K_G a P}{G'_m}\,dZ \tag{12.70}$$

or in terms of mole fractions:

$$\mathbf{N}_{OG} = \int_{y_1}^{y_2} \frac{dy}{y_e - y} = \int_0^Z K_G a \frac{P}{G'_m}\,dZ = K_G a \frac{P}{G'_m} Z \tag{12.71}$$

The number of overall gas transfer units $\mathbf{N}_{OG}$ is an integrated value of the change in composition per unit driving force, and therefore represents the difficulty of the separation.

In many cases in gas absorption, $(y - y_e)$ is very small at the top of the column, and consequently $1/(y - y_e)$ is very much greater at the top than at the bottom of the column. Thus, equation 12.77 may lead to the use of an integral which is difficult to evaluate graphically because of the very steep slope of the curve.

Now:
$$\mathbf{N}_{OG} = \int_{y_1}^{y_2} \frac{\mathrm{d}y}{y_e - y} = \int_{y_1}^{y_2} \frac{y\,\mathrm{d}(\ln y)}{y_e - y} \tag{12.72}$$

In these circumstances, the new form of the integral is much more readily evaluated, as pointed out by RACKETT[53].

As in Chapter 11 (equation 11.140), equation 12.77 may be written as:

$$\mathbf{N}_{OG} = \frac{\text{Height of column}}{\text{Height of transfer unit}} = \frac{Z}{\mathbf{H}_{OG}} \tag{12.73}$$

The height of the overall gas transfer unit is then $\mathbf{H}_{OG} = \dfrac{G'_m}{PK_Ga}$ (12.74)

If the driving force is taken over the gas-film only, the height of a gas-film transfer unit $\mathbf{H}_G = G'_m/Pk_Ga$ is obtained. Similarly for the liquid film, the height of the overall liquid-phase transfer unit $\mathbf{H}_{OL}$ is given by:

$$\mathbf{H}_{OL} = \frac{L'_m}{K_L a C_T} \tag{12.75}$$

The height of the liquid-film transfer unit is given by:

$$\mathbf{H}_L = \frac{L'_m}{k_L a C_T} \tag{12.76}$$

where $C_T$ is the mean molar density of the liquid.

In this analysis, it is assumed that the total number of moles of gas and liquid remain the same. This is true in absorption only when a small change in concentration takes place. With distillation, the total number of moles of gas and liquid does remain more nearly constant so that no difficulty then arises. In Chapter 11, the following relationships between individual and overall heights of transfer units are obtained and methods of obtaining the values of $\mathbf{H}_G$ and $\mathbf{H}_L$ are discussed:

$$\mathbf{H}_{OG} = \mathbf{H}_G + \frac{mG'_m}{L'_m}\mathbf{H}_L \qquad \text{(equation 11.148)}$$

$$\mathbf{H}_{OL} = \mathbf{H}_L + \frac{L'_m}{mG'_m}\mathbf{H}_G \qquad \text{(equation 11.149)}$$

For absorption duties, SEMMELBAUER[29] presented the following equations to evaluate $\mathbf{H}_G$ and $\mathbf{H}_L$ for Raschig rings and Berl saddles:

$$\mathbf{H}_G = \beta\left[\frac{G'^{0.41}\mu_G^{0.26}\mu_L^{0.46}\sigma^{0.5}}{L'^{0.46}\rho_G^{0.67}\rho_L^{0.5}D_G^{0.67}d_p^{0.05}}\right] \tag{12.77}$$

$$\mathbf{H}_L = \beta \left[ \frac{\mu_L^{0.88} \sigma^{0.5}}{L'^{0.05} \rho_L^{1.33} D_L^{0.5} d_p^{0.55}} \right] \tag{12.78}$$

where $\beta = 30$ for Raschig rings and $\beta = 21$ for Berl saddles respectively, and $L'$ and $G'$ are *mass* flowrates per unit area.

The limits of validity and the units for the terms in equations 12.77 and 12.78 are given in Table 12.4.

Table 12.4. Range of application of equations 12.77 and 12.78

| | | | | | |
|---|---|---|---|---|---|
| $L'$ | 0.1–10 | kg/m²s | $\mu_L$ | 0.2–2 | mN s/m² |
| $G'$ | 0.1–1.0 | kg/m²s | $\mu_G$ | 0.005–0.03 | mN s/m² |
| $d_p$ | 0.006–0.06 | m | $\sigma$ | (20–200) $\times 10^{-3}$ | J/m² |
| $\rho_L$ | 600–1400 | kg/m³ | $T$ | 273–373 | K |
| $\rho_G$ | 0.4–4 | kg/m³ | $d/d_p$ | 2.5–25 | — |
| $D_L$ | ($3$–$30 \times 10^{-10}$) | m²/s | $h_p/d_p$ | 10–100 | — |
| $D_G$ | ($3$–$90 \times 10^{-6}$) | m²/s | | | |

For a range of packings, MORRIS and JACKSON[21] have presented values of the heights of the individual film transfer-units as shown in Table 12.5. For Pall rings and Intalox saddles, the nomographs in Figures 12.21 and 12.22[54] may be used though Figure 12.22 must not be used to estimate $\mathbf{H}_L$ for distillation applications. Table 11.6 gives the value as a function of size and type of packing. It is, however, satisfactory for absorption and stripping duties.

## Concentrated solutions

With concentrated solutions, allowance must be made for the change in the total number of moles flowing, because the molar flow will decrease up the column if the amount of absorption is large.

COLBURN[55] has shown that, under these conditions, the number of transfer units is given by:

$$\mathbf{N}_{OG} = \int_{y_1}^{y_2} \frac{dy}{y_e - y} \frac{(1-y)_{\text{lm}}}{1-y} \tag{12.79}$$

where $(1-y)_{\text{lm}}$ is the logarithmic mean of $(1-y)$ and $(1-y_i)$.

## Example 12.3

An acetone–air mixture containing 0.015 mole fraction of acetone has the mole fraction reduced to 1 per cent of this value by countercurrent absorption with water in a packed tower. The gas flowrate $G'$ is 1 kg/m²s of air and the water enters at 1.6 kg/m²s. For this system, Henry's law holds and $y_e = 1.75x$, where $y_e$ is the mole fraction of acetone in the vapour in equilibrium with a mole fraction $x$ in the liquid. How many overall transfer units are required?

Table 12.5. Height of a transfer unit for various packings[54]

| Material | Size (mm) | | | Height of a transfer unit (m) | |
|---|---|---|---|---|---|
| Grids | Pitch | Height | Thickness | $\mathbf{H}_G$ | $\mathbf{H}_L$ |
| Plain grids | | | | | |
| Metal | 25 | 25 | 1.6 | 1 | 0.5 |
| | 25 | 50 | 1.6 | 1.2 | 0.6 |
| Wood | 25 | 25 | 6.4 | 0.9 | 0.5 |
| | 25 | 50 | 6.4 | 1.2 | 0.6 |
| Serrated grids | | | | | |
| Wood | 100 | 100 | 13 | 6.8 | 0.7 |
| | 50 | 50 | 9.5 | 1.8 | 0.6 |
| | 38 | 38 | 4.8 | 1.6 | 0.6 |
| Solid material | nominal size | | | | |
| Coke | 75 | | | 0.7 | 0.9 |
| | 38 | | | 0.25 | 0.8 |
| | 25 | | | 0.2 | 0.7 |
| Quartz | 50 | | | 0.5 | 0.8 |
| | 25 | | | 0.16 | 0.8 |
| | Diameter | Height | Thickness | | |
| Stacked Raschig rings | | | | | |
| Stoneware | 100 | 100 | 9.5 | 1.8 | 0.7 |
| | 75 | 75 | 9.5 | 1.1 | 0.6 |
| | 75 | 75 | 6.4 | 1.4 | 0.6 |
| | 50 | 50 | 6.4 | 0.7 | 0.6 |
| | 50 | 50 | 4.8 | 0.8 | 0.6 |
| Random Raschig rings | | | | | |
| Metal | 50 | 50 | 1.6 | 0.5 | 0.6 |
| | 25 | 25 | 1.6 | 0.2 | 0.5 |
| | 13 | 13 | 0.8 | 0.1 | 0.5 |
| Stoneware | 75 | 75 | 9.5 | 0.8 | 0.7 |
| | 50 | 50 | 6.4 | 0.5 | 0.6 |
| | 50 | 50 | 4.8 | 0.5 | 0.6 |
| | 38 | 38 | 4.8 | 0.3 | 0.6 |
| | 25 | 25 | 2.5 | 0.2 | 0.5 |
| | 19 | 19 | 2.5 | 0.15 | 0.5 |
| | 13 | 13 | 1.6 | 0.1 | 0.5 |
| Carbon | 50 | 50 | 6.4 | 0.5 | 0.6 |
| | 25 | 25 | 4.8 | 0.2 | 0.5 |
| | 13 | 13 | 3.2 | 0.1 | 0.5 |

## Solution

As the system is dilute, mole fractions are approximately equal to mole ratios.

At the bottom of the tower: $y_1 = 0.015$, $G' = 1.0$ kg/m$^2$s, $x_1$ is unknown

At the top of the tower: $y_2 = 0.00015$, $x_2 = 0$ and $L' = 1.6$ kg/m$^2$s

Thus: $L_m = (1.6/18) = 0.0889$ kmol/m$^2$s

and: $G_m = (1.0/29) = 0.0345$ kmol/m$^2$s

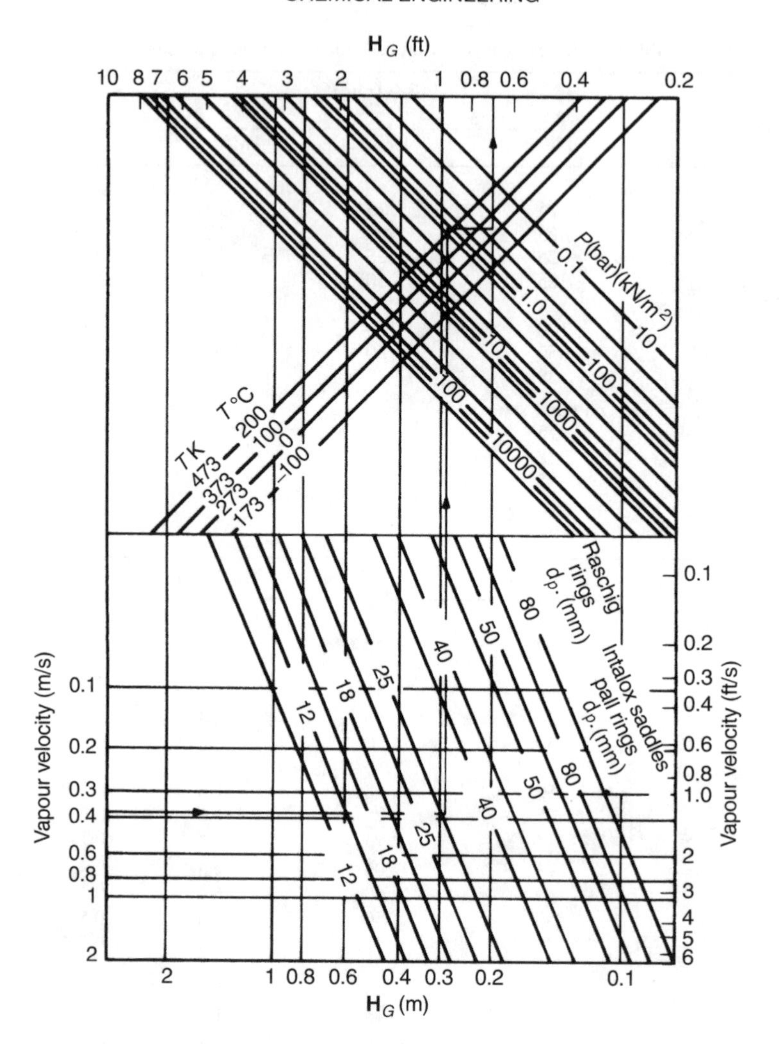


Figure 12.21. Nomograph for the estimation of the height of a gas-phase transfer unit[54]

An overall mass balance gives:

$$G_m(y_1 - y_2) = L_m(x_1 - x_2)$$

or: $0.0345(0.015 - 0.00015) = 0.0889(x_1 - 0)$ and $x_1 = 0.00576$

Thus: $y_{e_1} = (1.75 \times 0.00576) = 0.0101$

The number of overall transfer units is defined by:

$$\mathbf{N}_{OG} = \int_{y_2}^{y_1} \frac{\mathrm{d}y}{y - y_e} = \frac{y_1 - y_2}{(y - y_e)_{\mathrm{lm}}} \quad \text{(from equations 12.60 and 12.70)}$$

Top driving force, $(y_2 - y_{e2}) = 0.00015$ since $x_2 = 0$.

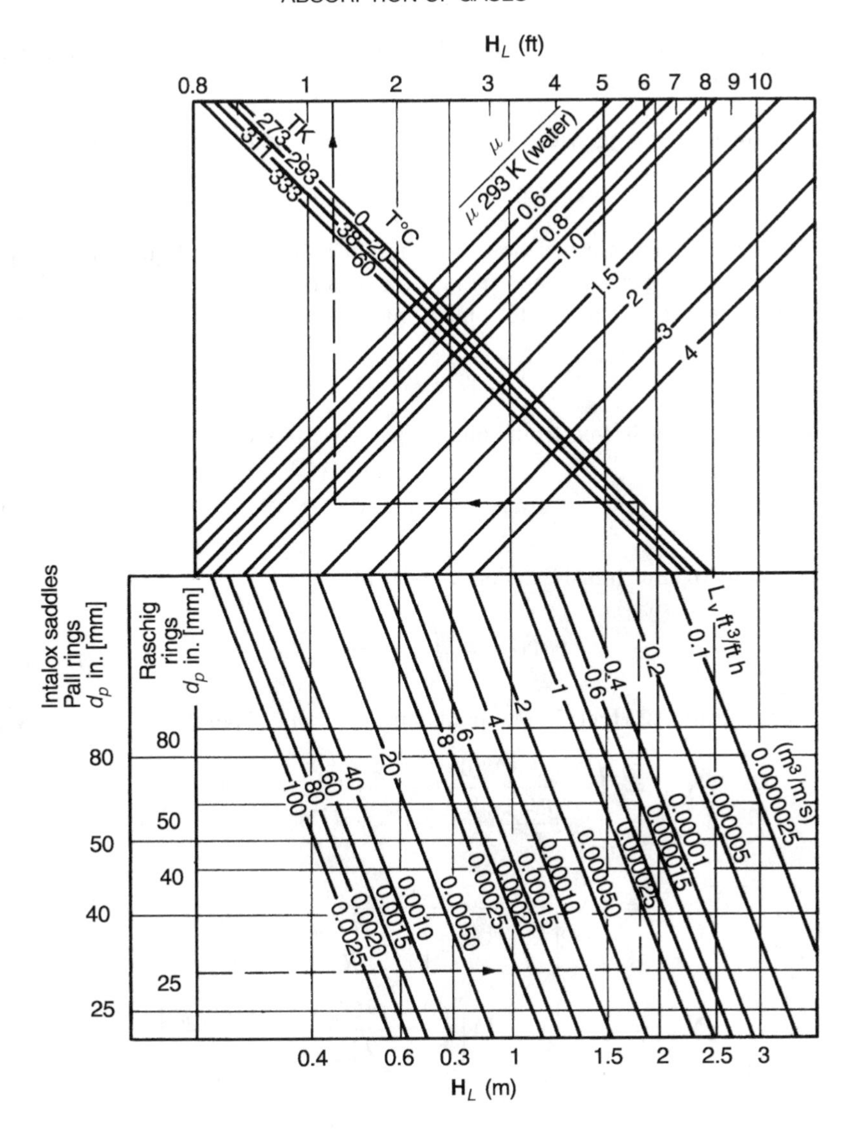


Figure 12.22. Nomograph for the estimation of the height of a liquid-phase transfer unit[54]

Bottom driving force, $= (y_1 - y_{e1}) = (0.015 - 0.0101) = 0.0049$.

Thus: $(y - y_e)_{lm} = (0.0049 - 0.00015)/\ln(0.0049/0.00015) = 0.00136$

and: $N_{OG} = (0.015 - 0.00015)/0.00136 = \underline{\underline{10.92}}$

Also: $N_{OL} = N_{OG} \times mG_m/L_m$

$= (10.92 \times 1.75 \times 0.0345)/0.0889 = \underline{\underline{7.42}}$

## 12.7.10. The importance of liquid and gas flowrates and the slope of the equilibrium curve

For a packed tower operating with dilute concentrations, since $x \simeq X_1$ and $y \simeq Y_1$, then:

$$G'_m(y_1 - y_2) = L'_m(x_1 - x_2) \tag{12.80}$$

where, as before, $x$ and $y$ are the mole fractions of solute in the liquid and gas phases, and $G'_m$ and $L'_m$ are the gas and liquid molar flowrates per unit area on a solute free basis.

A material balance between the top and some plane where the mole fractions are $x$, $y$ gives:

$$G'_m(y - y_2) = L'_m(x - x_2) \tag{12.81}$$

If the entering solvent is free from solute, then $x_2 = 0$ and:

$$x = \frac{G'_m}{L'_m}(y - y_2) \tag{12.82}$$

But the number of overall transfer units is given by:

$$\mathbf{N}_{OG} = \int_{y_1}^{y_2} \frac{dy}{y_e - y}$$

For dilute concentrations, Henry's law holds and $y_e = mx$. Therefore:

Thus:
$$\mathbf{N}_{OG} = \int_{y_1}^{y_2} \frac{dy}{\dfrac{mG'_m}{L'_m}(y - y_2) - y}$$

$$= \int_{y_1}^{y_2} \frac{dy}{\left[\dfrac{mG'_m}{L'_m} - 1\right] y - \dfrac{mG_m}{L_m} y_2}$$

and:
$$\mathbf{N}_{OG} = \frac{1}{1 - \dfrac{mG'_m}{L'_m}} \ln\left[\left(1 - \frac{mG'_m}{L'_m}\right)\frac{y_1}{y_2} + \frac{mG'_m}{L'_m}\right] \tag{12.83}$$

Colburn[55] has shown that this equation may usefully be plotted as shown in Figure 12.23 which is taken from his paper. In this plot the number of transfer units $\mathbf{N}_{OG}$ is shown for values of $y_1/y_2$ using $mG'_m/L'_m$ as a parameter and it may be seen that the greater $mG_m/L_m$, the greater is the value of $\mathbf{N}_{OG}$ for a given ratio of $y_1/y_2$. From equation 12.82:

$$\frac{L'_m}{G'_m} = \frac{y_1 - y_2}{x_1} = \frac{y_1 - y_2}{y_{e1}/m}$$

Thus:
$$\frac{mG'_m}{L'_m} = \frac{y_{e1}}{y_1 - y_2}$$

where $y_{e1}$ is the value of $y$ in equilibrium with $x_1$.

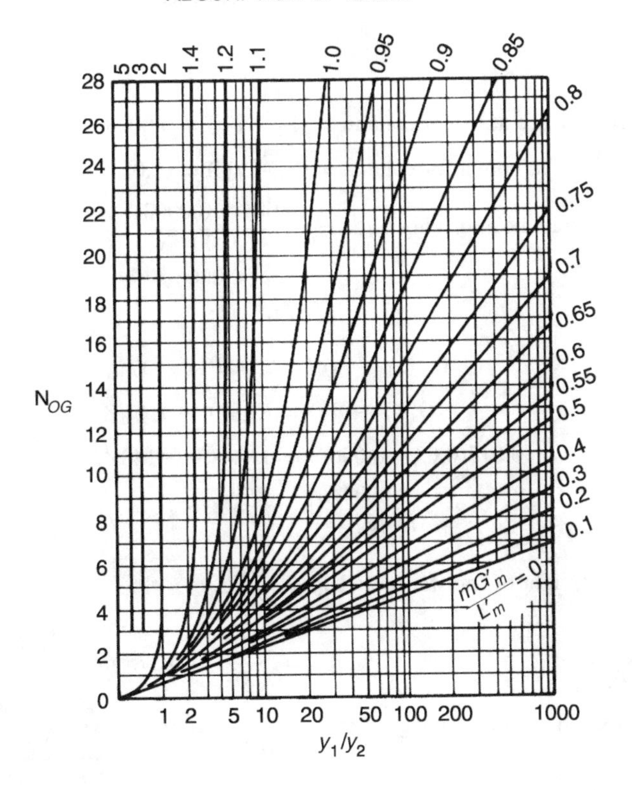


Figure 12.23. Number of transfer units $\mathbf{N}_{OG}$ as a function of $y_1$, $y_2$, with $mG'_m/L'_m$ as parameter

On this basis, the lower the value of $mG'_m/L'_m$, the lower will be $y_{e1}$, and hence the weaker the exit liquid. Colburn has suggested that the economic range for $mG'_m/L'_m$ is 0.7–0.8. If the value of $\mathbf{H}_{OG}$ is known, the quickest way of obtaining a good indication of the required height of the column is by using Figure 12.23.

## Example 12.4

Gas, from a petroleum distillation column, has a concentration of $H_2S$ reduced from 0.03 (kmol $H_2S$/kmol of inert hydrocarbon gas) to 1 per cent of this value by scrubbing with a triethanolamine–water solvent in a countercurrent tower, operating at 300 K and atmospheric pressure. The equilibrium relation for the solution may be taken as $Y_e = 2X$.

The solvent enters the tower free of $H_2S$ and leaves containing 0.013 kmol of $H_2S$/kmol of solvent. If the flow of inert gas is 0.015 kmol/s m$^2$ of tower cross-section, calculate:

(a) the height of the absorber necessary, and
(b) the number of transfer units $\mathbf{N}_{OG}$ required.

The overall coefficient for absorption $K''_G a$ may be taken as 0.04 kmol/s m$^3$ (unit mole fraction driving force).

## Solution

$$\text{Driving force at top of column} = (Y_2 - Y_{2e}) = 0.0003$$

$$\text{Driving force at bottom of column} = (Y_1 - Y_{1e}) = (0.03 - 0.026) = 0.004$$

$$\text{Logarithmic mean driving force} = \frac{(0.004 - 0.0003)}{\ln\left(\dfrac{0.004}{0.0003}\right)} = 0.00143$$

From equation 12.62: $$G'_m(Y_1 - Y_2)S = K_G a P(Y - Y_e)_{\text{lm}} S Z$$

That is: $$G'_m(Y_1 - Y_2) = K''_G a(Y - Y_e)_{\text{lm}} Z$$

Thus: $$0.015(0.03 - 0.0003) = 0.04 \times 0.00143Z$$

and: $$Z = \frac{0.000446}{0.0000572} = 7.79 = 7.8 \text{ m (say)}$$

$$\text{Height of transfer unit } \mathbf{H}_{OG} = \frac{G'_m}{K''_G a} = \frac{0.015}{0.04} = 0.375 \text{ m}$$

$$\text{Number of transfer units } \mathbf{N}_{OG} = \frac{7.79}{0.375} = 20.7 = \underline{\underline{21 \text{ (say)}}}$$

## Example 12.5

Ammonia is to be removed from a 10 per cent ammonia–air mixture by countercurrent scrubbing with water in a packed tower at 293 K so that 99 per cent of the ammonia is removed when working at a total pressure of 101.3 $kN/m^2$.

If the gas rate is 0.95 $kg/m^2s$ of tower cross-section and the liquid rate is 0.65 $kg/m^2s$, find the necessary height of the tower if the absorption coefficient $K_G a = 0.001$ $kmol/m^3s(kN/m^2)$ partial pressure difference. The equilibrium data are:

| kmol $NH_3$/kmol water: | 0.021 | 0.031 | 0.042 | 0.053 | 0.079 | 0.106 | 0.159 |
|---|---|---|---|---|---|---|---|
| Partial pressure $NH_3$: | | | | | | | |
| (mm Hg) | 12.0 | 18.2 | 24.9 | 31.7 | 50.0 | 69.6 | 114.0 |
| ($kN/m^2$) | 1.6 | 2.4 | 3.3 | 4.2 | 6.7 | 9.3 | 15.2 |

## Solution

If the compositions of the gas are given as per cent by volume, at the bottom of the tower are $y_1 = 0.10$ and $Y_1 = 0.10/(1 - 0.10) = 0.111$.

At the top of the tower: $y_2 = 0.001 \simeq Y_2$.

Mass flowrate of gas $= 0.95$ $kg/m^2s$.

Mass per cent of air $= [0.9 \times 29/(0.1 \times 17 + 0.9 \times 29)] \times 100 = 93.8$.

Thus: mass flowrate of air $= (0.938 \times 0.95) = 0.891$ kg/m$^2$s

and: $G'_m = (0.891/29) = 0.0307$ kmol/m$^2$s

$L'_m = 0.65/18 = 0.036$ kmol/m$^2$s

A mass balance between a plane in the tower where the compositions are $X$ and $Y$ and the top of the tower gives:

$$G'_m(Y - Y_2) = L'_m(X - X_2)$$

But: $X_2 = 0$

Thus: $0.0307(Y - 0.001) = 0.036X$, or $Y = 1.173X + 0.001$

This is the equation of the operating line in terms of mole ratios.

The given equilibrium data may be converted to the same basis since:

$$P_G = yP = \frac{YP}{1+Y}$$

and:
$$Y = \frac{P_G}{P - P_G}$$

Using these equations, the following data are obtained:

| | | | | | | | |
|---|---|---|---|---|---|---|---|
| kmol $NH_3$/kmol $H_2O$ | 0.021 | 0.031 | 0.042 | 0.053 | 0.079 | 0.106 | 0.159 |
| Partial pressure $P_G$ (mm) | 12 | 18.2 | 24.9 | 31.7 | 50.0 | 69.6 | 114.0 |
| $P - P_G = 760 - P_G$ (mm) | 748 | 741.8 | 735.1 | 728.3 | 710 | 690.4 | 646 |
| $Y = P_G/(P - P_G)$ | 0.016 | 0.0245 | 0.0339 | 0.0435 | 0.0704 | 0.101 | 0.176 |

These data are plotted in Figure 12.24.
From a mass balance over the column, the height $Z$ is given by:

$$Z = \frac{G'_m}{k_G a P}\int_{Y_2}^{Y_1} \frac{(1+Y)(1+Y_i)}{(Y - Y_i)} dY \quad \text{(equation 12.45)}$$

Figure 12.25 may be used to evaluate the integral as follows:

| $Y$ | $Y_i$ | $(1+Y)(1+Y_i)$ | $\frac{(1+Y)(1+Y_i)}{(Y-Y_i)}$ |
|---|---|---|---|
| 0.111 | 0.089 | 1.21 | 55.0 |
| 0.10 | 0.078 | 1.185 | 53.8 |
| 0.08 | 0.059 | 1.14 | 54.3 |
| 0.06 | 0.042 | 1.11 | 61.4 |
| 0.04 | 0.027 | 1.067 | 82.0 |
| 0.02 | 0.013 | 1.035 | 148 |
| 0.01 | 0.006 | 1.016 | 254 |
| 0.005 | 0.0026 | 1.010 | 421 |
| 0.001 | 0 | 1.0 | 1000 |

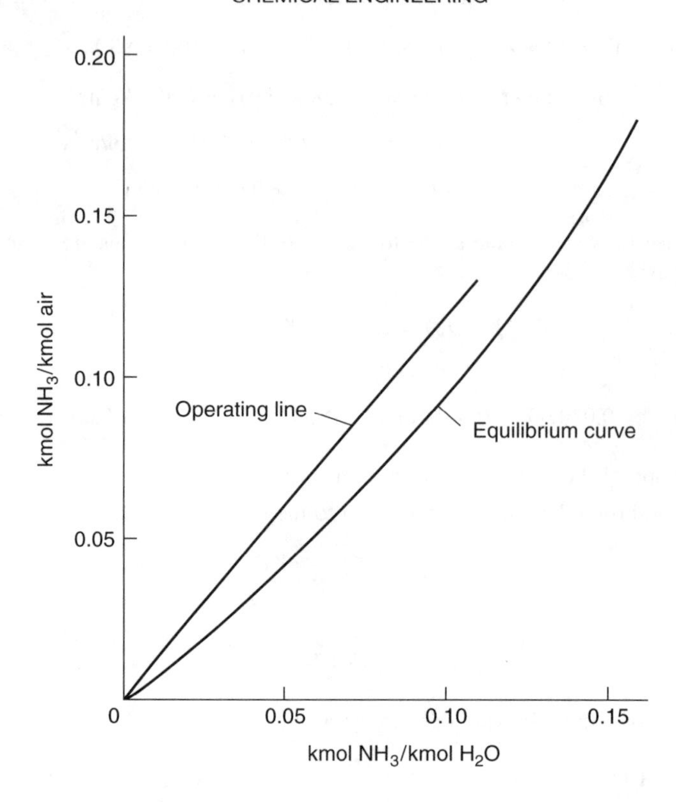


Figure 12.24. Operating and equilibrium lines for Example 12.5

The area under the curve in Figure 12.25 is 12.6. For a very soluble gas $k_{Ga} \simeq k_{Ga}$ so that:

$$Z = \frac{0.0307}{(0.001 \times 101.3)} \times 12.6 = \underline{\underline{3.82 \text{ m}}}$$

If the equilibrium line is assumed to be straight, then:

$$G'_m(Y_2 - Y_1) = K_G a Z \Delta P_{lm}$$

Top driving force $= \Delta Y_2 = 0.022$. Bottom driving force $= \Delta Y_1 = 0.001$.

Thus: $$\Delta Y_{\text{lm}} = 0.0068, \quad \Delta P_{\text{lm}} = 0.688 \text{ kN/m}^2$$

and: $$Z = \frac{(0.0307 \times 0.11)}{(0.001 \times 0.688)} = \underline{\underline{4.91 \text{ m}}}.$$

## 12.8. PLATE TOWERS FOR GAS ABSORPTION

Bubble-cap columns or sieve trays, of similar construction to those described in Chapter 11 on distillation, are sometimes used for gas absorption, particularly when the load is more than can be handled in a packed tower of about 1 m diameter and when there is any

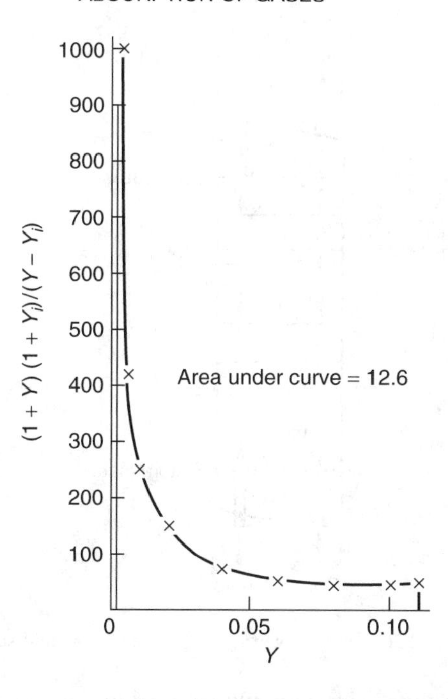


Figure 12.25. Determination of column height for Example 12.5

likelihood of deposition of solids which would quickly choke a packing. Plate towers are particularly useful when the liquid rate is sufficient to flood a packed tower. Since the ratio of liquid rate to gas rate is greater than with distillation, the slot area will be rather less and the downcomers rather larger. On the whole, plate efficiencies have been found to be less than with the distillation equipment, and to range from 20 to 80 per cent.

The plate column is a common type of equipment for large installations, although when the diameter of the column is less than 2 m, packed columns are more often used. For the handling of very corrosive fluids, packed columns are frequently preferred for larger units. The essential arrangement of such a unit is shown in Figure 12.26, where:

$L'_m$ is the molar rate of flow per unit area of solute free liquid,
$G'_m$ is the molar rate of flow per unit area of inert gas,
$n$ refers to the plate numbered from the bottom upwards (and suffix $n$ refers to material leaving plate $n$),
$x$ is the mole fraction of the absorbed component in the liquid,
$y$ is the mole fraction of the absorbed component in the gas, and
$s$ is the total number of plates in the column.

It may be assumed that dilute solutions are used so that mole fractions and mole ratios are approximately equal. Each plate is taken as an "ideal" unit, so that the gas leaving of composition $y_n$ is in equilibrium with the liquid of composition $x_n$ leaving the plate.

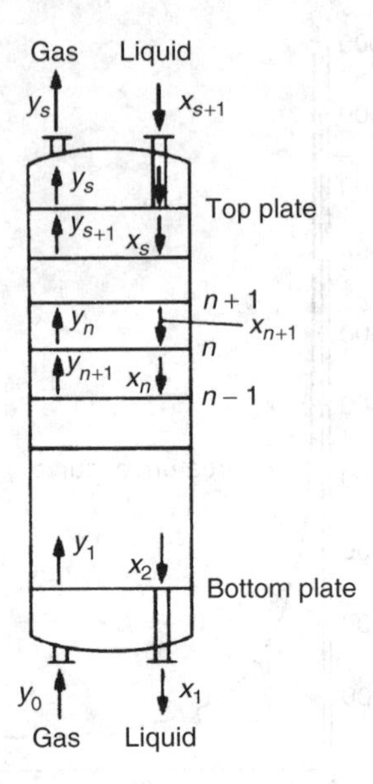


Figure 12.26. Plate tower — nomenclature for fluid streams

A material balance for the absorbed component from the bottom to a plane above plate $n$ gives:

$$G'_m y_n + L'_m x_1 = G'_m y_0 + L'_m x_{n+1} \tag{12.84}$$

or:

$$y_n = \frac{L'_m}{G'_m} x_{n+1} + y_0 - \frac{L'_m}{G'_m} x_1 \tag{12.85}$$

This is the equation of a straight line of slope $L'_m/G'_m$, relating the composition of the gas entering a plate to the liquid leaving the plate, and is known as the *operating line*. As shown in Figure 12.27, such a line passes through two points $B(x_{s+1}, y_s)$ and $A(x_1, y_0)$, representing the terminal concentrations in the column. The equilibrium curve is shown in this figure as PQR.

Point A represents conditions at the bottom of the tower. The gas rising from the bottom plate is in equilibrium with a liquid of concentration $x_1$ and is shown as point 3 on the operating line. Then point 4 indicates the concentration of the liquid on the second plate from the bottom. In this way steps may be drawn to point B, giving the gas $y_s$ rising from the top plate and the liquid $x_{s+1}$ entering the top of the absorber.

## 12.8.1. Number of plates by use of absorption factor

If the equilibrium curve can be represented by the relation $y_e = mx$, then the number of plates required for a given degree of absorption can conveniently be found by a method due to KREMSER[56] and SOUDERS and BROWN[57]. The same treatment is applicable for concentrated solutions provided concentrations are expressed as mole ratios, and if the equilibrium curve can be represented approximately by $Y_e = mX$.

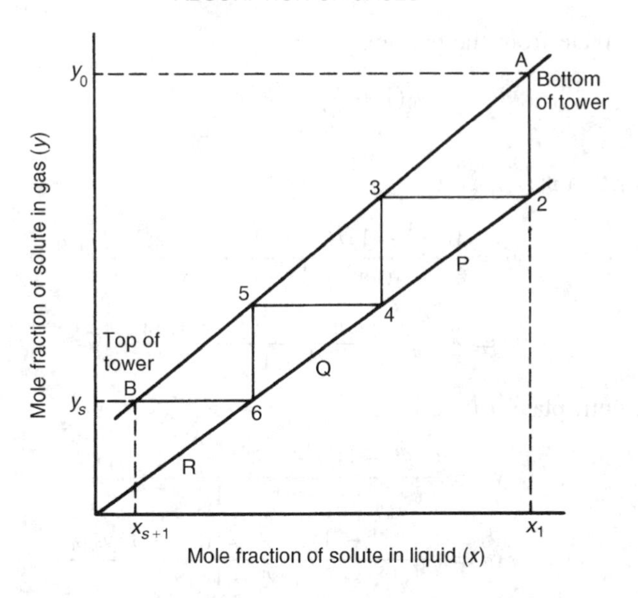


Figure 12.27. Diagrammatic representation of changes in a plate column

A material balance over plate $n$ gives:

$$L'_m(x_n - x_{n+1}) = G'_m(y_{n-1} - y_n) \tag{12.86}$$

For an ideal plate, $y_n = mx_n$;

and:

$$\frac{L'_m}{mG'_m}(y_n - y_{n+1}) = y_{n-1} - y_n \tag{12.87}$$

This group $L'_m/mG'_m$, which will be taken as constant, is called the *absorption factor* $\mathscr{A}$.

Thus:

$$y_n = \frac{y_{n-1} + \mathscr{A}y_{n+1}}{1 + \mathscr{A}} \tag{12.88}$$

Applying this relation to the bottom plate and taking $y_0$ as the mole fraction of absorbed component in the gas entering the column, then:

$$y_1 = \frac{y_0 + \mathscr{A}y_2}{1 + \mathscr{A}}$$

And for the second plate from the bottom:

$$y_2 = \frac{y_1 + \mathscr{A}y_3}{1 + \mathscr{A}}$$

$$= \frac{\mathscr{A}(1 + \mathscr{A})y_3 + \mathscr{A}y_2 + y_0}{(1 + \mathscr{A})^2}$$

Simplifying:

$$y_2 = \frac{y_0(1 + \mathscr{A}) + \mathscr{A}^2 y_3}{\mathscr{A}^2 + \mathscr{A} + 1}$$

And for the third plate from the bottom:

$$y_3 = \frac{y_0(1 + \mathscr{A} + \mathscr{A}^2) + \mathscr{A}^3 y_4}{\mathscr{A}^3 + \mathscr{A}^2 + \mathscr{A} + 1}$$

which may be written as:

$$y_3 = \frac{[(\mathscr{A}^3 - 1)/(\mathscr{A} - 1)]y_0 + \mathscr{A}^3 y_4}{(\mathscr{A}^4 - 1)/(\mathscr{A} - 1)}$$

$$= \frac{(\mathscr{A}^3 - 1)y_0 + \mathscr{A}^3(\mathscr{A} - 1)y_4}{\mathscr{A}^4 - 1}$$

Proceeding thus until plate $n$ is reached:

$$y_n = \frac{(\mathscr{A}^n - 1)y_0 + \mathscr{A}^n(\mathscr{A} - 1)y_{n+1}}{\mathscr{A}^{n+1} - 1}$$

$$y_0 = \frac{(\mathscr{A}^{n+1} - 1)y_n - \mathscr{A}^n(\mathscr{A} - 1)y_{n+1}}{\mathscr{A}^n - 1}$$

Thus:

$$y_0 - y_n = \frac{(\mathscr{A}^{n+1} - \mathscr{A}^n)y_n - \mathscr{A}^n(\mathscr{A} - 1)y_{n+1}}{\mathscr{A}^n - 1}$$

and:

$$y_0 - y_{n+1} = \frac{(\mathscr{A}^{n+1} - 1)y_n - (\mathscr{A}^{n+1} - 1)y_{n+1}}{\mathscr{A}^n - 1}$$

Dividing:

$$\frac{y_0 - y_n}{y_0 - y_{n+1}} = \frac{(\mathscr{A}^{n+1} - \mathscr{A}^n)y_n - \mathscr{A}^n(\mathscr{A} - 1)y_{n+1}}{(\mathscr{A}^{n+1} - 1)y_n - (\mathscr{A}^{n+1} - 1)y_{n+1}}$$

$$= \frac{\mathscr{A}^n(\mathscr{A} - 1)(y_n - y_{n+1})}{(\mathscr{A}^{n+1} - 1)(y_n - y_{n+1})}$$

or:

$$\frac{y_0 - y_n}{y_0 - y_{n+1}} = \frac{\mathscr{A}^{n+1} - \mathscr{A}}{\mathscr{A}^{n+1} - 1}$$

Applying this relation over the whole column and putting $n = s$ gives:

$(y_0 - y_s)$ = actual change in composition of gas, and

$(y_0 - y_{s+1})$ = maximum possible change in composition of gas, that is if the gas leaving the absorber is in equilibrium with the entering liquid (or $y_s = mx_{s+1}$).

Then:

$$\frac{y_0 - y_s}{y_0 - mx_{s+1}} = \frac{(L'_m/mG'_m)^{s+1} - (L'_m/mG'_m)}{(L'_m/mG'_m)^{s+1} - 1} \quad (12.89)$$

This equation is conveniently represented, as suggested by SOUDERS and BROWN[57], by Figure 12.28, and it is easy to use such a diagram to determine the number of plates required.

A high degree of absorption can be obtained, either by using a large number of plates, or by using a high absorption factor $L'_m/mG'_m$. Since $m$ is fixed by the system, this means that $L'_m/G'_m$ must be large if a high degree of absorption is to be obtained, although this

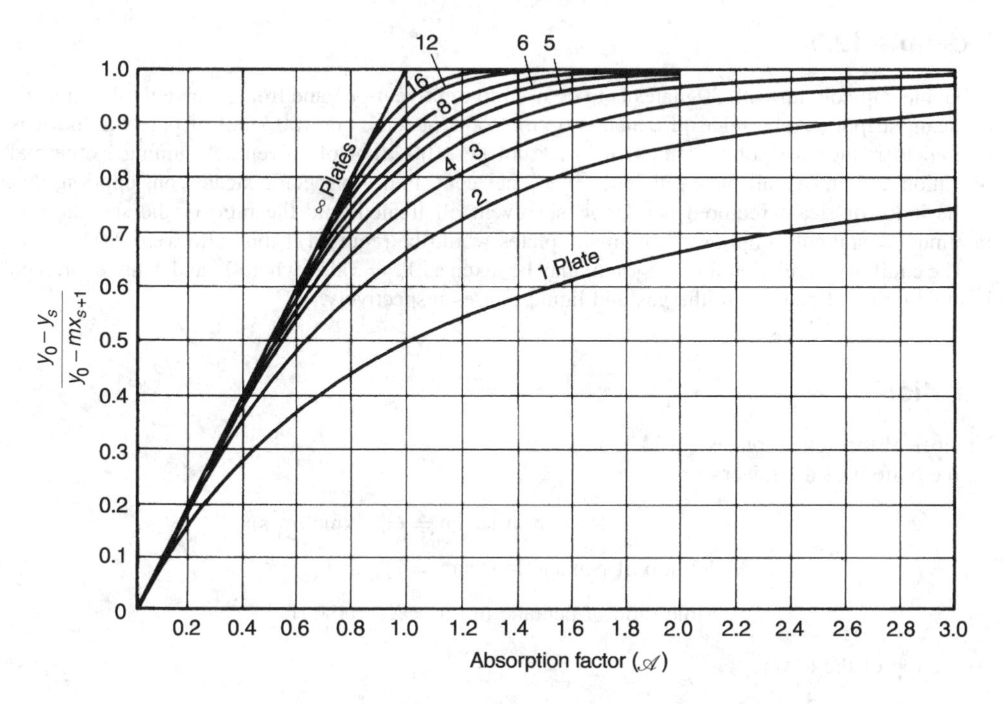


Figure 12.28. Graphical representation of the effect of the absorption factor and the number of plates on the degree of absorption

will result in a low value of $x$ for the liquid leaving at the bottom. This problem is to some extent met by recirculating the liquid over the tower, although the advantages of a countercurrent flow system are then lost. A value of $mG'_m/L'_m$ of about 0.7–0.8 is probably the most economic, that is $L'_m/mG'_m$ 1.3.

It is important to note that, if $L'_m/mG'_m$ is less than 1, then a very large number of plates are required to achieve a high recovery, and even an infinite number will not give complete recovery. $L'_m/mG'_m$ is the ratio of the slope of the operating line $L_m/G_m$ to the slope of the equilibrium curve $m$, so that if $L'_m/G'_m < m$, or $L'_m/mG'_m < 1$, then the operating line will never cut the equilibrium curve and the gas leaving the top of the column will not therefore reach equilibrium with the entering liquid.

## 12.8.2. Tray types for absorption

It has already been noted that trays which are suitable for distillation may be used for absorption duties though in general lower efficiencies will be obtained. In Chapter 11, the design of trays for common contacting devices is considered and the methods presented in that chapter are generally applicable. The most commonly used tray types are shown in Figure 11.50*a* with the crossflow tray being the most popular.

At high liquid flowrates, the liquid gradient on the tray can become excessive and lead to poor vapour distribution across the plate. This problem may be overcome by the shortening of the liquid flow-path as in the case of the double-pass and cascade trays. The whole design process is discussed in Volume 6.

## Example 12.6

A bubble-cap column with 30 plates is to be used to remove n-pentane from a solvent oil by means of steam stripping. The inlet oil contains 6 kmol of n-pentane per 100 kmol of pure oil and it is desired to reduce the solute content to 0.1 kmol per 100 kmol of solvent. Assuming isothermal operation and an overall plate efficiency of 30 per cent, find the specific steam consumption, that is the kmol of steam required per kmol of solvent oil treated, and the ratio of the specific and minimum steam consumptions. How many plates would be required if this ratio were 2.0?

The equilibrium relation for the system may be taken as $Y_e = 3.0X$, where $Y_e$ and $X$ are expressed in mole ratios of pentane in the gas and liquid phases respectively.

## Solution

Number of theoretical plates $= (30 \times 0.3) = 9$.
At the bottom of the tower:

$$\text{Flowrate of steam} = G'_m \text{ (kmol/m}^2\text{s)}$$

$$\text{Mole ratio of pentane in steam} = Y_1, \text{ and}$$

$$\text{Mole ratio of pentane in oil} = X_1 = 0.001$$

At the top of the tower:

$$\text{exit steam composition} = Y_2, \text{ inlet oil composition} = X_2 = 0.06,$$

$$\text{flowrate of oil} = L'_m \text{ (kmol/m}^2\text{s)}$$

The minimum steam consumption occurs when the exit steam stream is in equilibrium with the inlet oil, that is when:

$$Y_{e2} = (0.06 \times 3) = 0.18$$

$$L'_{\text{min}}(X_2 - X_1) = G'_{\text{min}}(Y_2 - Y_1)$$

If $Y_1 = 0$, that is the inlet steam is pentane-free, then:

$$L'_{\text{min}}(0.06 - 0.001) = (G'_{\text{min}} \times 0.18)$$

and:

$$(G'/L')_{\text{min}} = (0.06 - 0.001)/0.18 = 0.328$$

The operating line may be fixed by trial and error as it passes through the point (0.001, 0), and 9 theoretical plates are required for the separation. Thus it is a matter of selecting the operating line which, with 9 steps, will give $X_2 = 0.001$ when $X_1 = 0.06$. This is tedious but possible, and the problem may be better solved analytically since the equilibrium line is straight.

Use may be made of the absorption factor method where

$$\frac{Y_1 - Y_2}{Y_1 - mX_2} = \frac{\mathscr{A}^{N+1} - \mathscr{A}}{\mathscr{A}^{N+1} - 1} \qquad \text{(equation 12.89)}$$

where $\mathscr{A}$ is the absorption factor $= L'_m/mG'_m$ and $N$ is the number of theoretical plates.

The corresponding expression for a stripping operation is:

$$\frac{X_2 - X_1}{X_2 - Y_1/m} = \frac{(1/\mathscr{A})^{N+1} - (1/\mathscr{A})}{(1/\mathscr{A})^{N+1} - 1}$$

In this problem, $N = 9$, $X_2 = 0.06$, $X_1 = 0.001$, and $Y_1 = 0$

Thus: $$\frac{(0.06 - 0.001)}{0.06} = 0.983 = \frac{(1/\mathscr{A})^{10} - (1/\mathscr{A})}{(1/\mathscr{A})^{10} - 1} \text{ from which } (1/\mathscr{A}) = 1.37$$

Thus: $$\frac{mG'_m}{L'_m} = 1.37, \frac{G'_m}{L'_m} = \frac{1.37}{3} = 0.457$$

and: $$\frac{\text{actual } G'_m/L'_m}{\text{minimum } G'_m/L'_m} = \frac{0.457}{0.328} = \underline{\underline{1.39}}$$

If (actual $G'_m/L'_m$)/(min $G'_m/L'_m$) = 2, actual $G'_m/L'_m = 0.656$.

Thus: $$1/\mathscr{A} = mG'_m/L'_m = 1.968$$

and: $$0.983 = \frac{(1.968)^{N+1} - 1.968}{(1.968)^{N+1} - 1} \text{ from which } N = 4.9$$

The actual number of plates = (4.9/0.3) = 16.3 (say 17).

## 12.9. OTHER EQUIPMENT FOR GAS ABSORPTION

### 12.9.1. The use of vessels with agitators

A gas may be dissolved in a liquid by dispersing it through holes in a pipe immersed in the liquid which is stirred with some form of agitator, as shown in Figure 12.29. Although this type of equipment will give only one theoretical stage per unit, but it often provides a useful method of saturating a liquid with a gas. COOPER *et al.*[58] have studied the absorption of oxygen from air in an aqueous solution of sodium sulphite using simple vessels of 0.15 to 2.44 m diameter fitted with four simple baffles. Air was just below the agitator which was a vaned-disc or flat-paddle. It was found that the absorption coefficient $K_G a$ varied almost directly with $\mathbf{P}_V$, the power input per unit volume. For constant values of $\mathbf{P}_V$, the following relation was obtained:

$$K_G a \propto u_s^{0.67} \tag{12.90}$$

where $u_s$ is the superficial gas velocity based on the volume of gas at inlet and the cross-section of tank. A general correlation was obtained by plotting $K_G a/u_s^{0.67}$ against the power input per unit volume $\mathbf{P}_V$, as shown in Figure 12.30 taken from this investigation. AYERST and HERBERT[59] have given some data on the use of this type of unit for the absorption of carbon dioxide into ammoniacal solutions.

The interfacial area, $a$, was the subject of an investigation by WESTERTERP *et al.*[60] though the correlations proposed are complex. Maximum values of $a$ are about 1000 $m^2/m^3$. Further work on the interfacial area in agitated vessels has been reviewed and summarised by SRIDAR and POTTER[61] who found that the correlation of CALDERBANK[62] was applicable for most situations. Calderbank proposed that, for pure

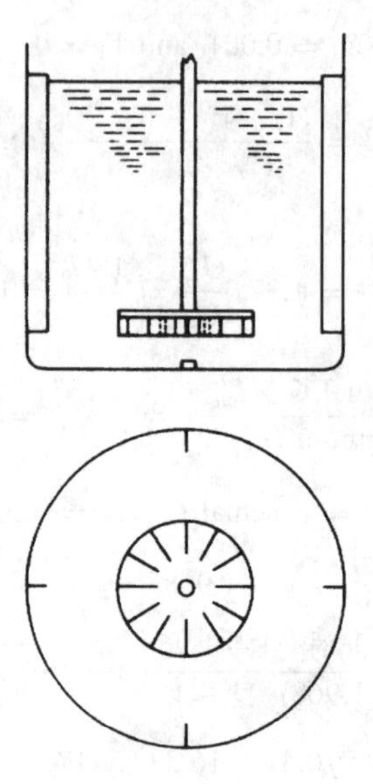

Figure 12.29. Vessel fitted with vaned-disc agitator

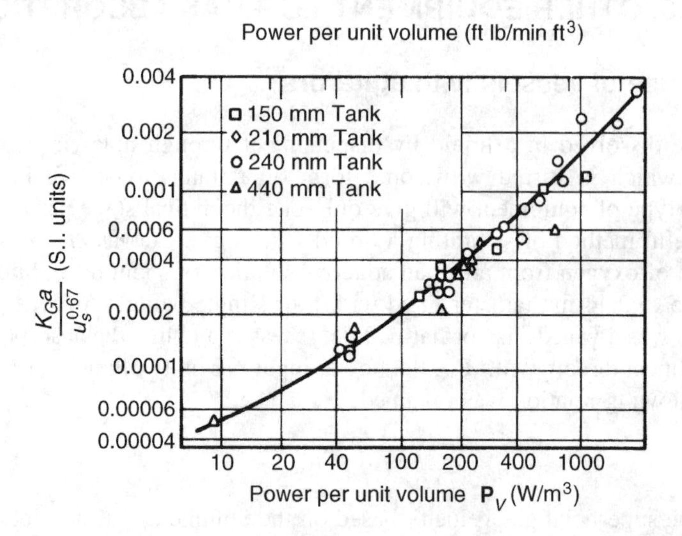


Figure 12.30. General correlation of data for a vessel (height = diameter) with vaned-disc agitator

liquids, the specific interfacial area, that is the surface area per unit volume of aerated suspension, $a(\text{m}^2/\text{m}^3)$ is given by:

$$a = 24{,}200\ (\mathbf{P}_V)^{0.4}\left(\frac{\rho_L^{0.2}}{\sigma^{0.6}}\right)\left(\frac{u_s}{u_0}\right)^{0.5} \tag{12.91}$$

where surface aeration is negligible, that is when:

$$\left(\frac{N'd_t^2\rho_L}{\mu_L}\right)^{0.7}\left(\frac{N'd_i}{u_s}\right)^{0.3} < 25,000 \tag{12.92}$$

When the surface aeration is significant, then the interfacial area is:

$$\frac{a'}{a} = 10^{-4}\left\{\left[\left(\frac{N'd_i^2\rho_L}{\mu_L}\right)^{0.7}\left(\frac{N'd_t}{u_s}\right)^{0.2}\right] - 25,000\right\} \tag{12.93}$$

In these equations, $a'$ is the specific interfacial area for a significant degree of surface aeration ($m^2/m^3$), $\mathbf{P}_V$ is the agitator power per unit volume of vessel ($W/m^3$), $\rho_L$ is the liquid density, $\sigma$ is the surface tension (N/m), $u_s$ is the superficial gas velocity (m/s), $u_0$ is the terminal bubble-rise velocity (m/s), $N'$ is the impeller speed (Hz), $d_i$ is the impeller diameter (m), $d_t$ is the tank diameter (m), $\mu_L$ is the liquid viscosity ($Ns/m^2$) and $d_0$ is the Sauter mean bubble diameter defined in Chapter 1, Section 1.2.4.

The effects of gas hold-up and bubble diameter have also been studied by Sridhar and Potter and, again, the correlations obtained by Calderbank are recommended.

The liquid-phase mass transfer coefficient, $k_L$, in agitated vessels has been measured and data correlated by several workers. SIDEMAN *et al.*[63] and VALENTIN[64] have presented reviews of the early work and more recent work has been published by YAGI and YOSHIDA[65], ZLOKARNIK[66], VAN'T RIET[67] and HOKER, LANGER and UDO[68]. For small bubbles (< 2.5 mm diameter) produced in well-agitated vessels, CALDERBANK[62] suggests the following correlation for bubbles in agitated electrolytes:

$$k_L = 0.31\left(\frac{\Delta\rho\mu_L g}{\rho_L^2}\right)^{1/3}(Sc)^{-2/3} \tag{12.94}$$

where: $\Delta\rho$ = density difference between gas and liquid,
$\rho_L, \mu_L$ = density and viscosity of the liquid, and
$Sc$ = Schmidt number for transport in the liquid.

JOSHI and SHARMA[69] and FUKADA *et al.*[70] have investigated the performance of vessels with multiple impellers on horizontal shafts.

Several investigations have been carried out into the power requirements for agitation of aerated liquids including those of YUNG *et al.*[71] and LUONG and VOLESKY[72] and it is generally concluded that the power required is less for an aerated system than for a non-aerated system.

Although, as described by BJERLE *et al.*[73], liquid jet-type absorbers are also used, one relatively recent application of mass transfer in agitated tanks with chemical reaction is the absorption of pollutants from flue gases and, in particular, the scrubbing of sulphur dioxide by a slurry containing fine limestone particles. In this case, the concentration of sulphur dioxide is usually very low and the mechanism of the absorption is complicated due to the presence of solids in the liquid phase where the rate of solid dissolution may significantly affect the absorption rate.

Studies on the dissolution of solids in the liquid phase include that of HIXSON and BAUM[74] whose correlation of data in terms of Reynolds, Sherwood and Schmidt numbers, discussed in detail in Section 10.2 in connection with mass transfer during leaching, is one of the most frequently used methods for calculating the mass transfer coefficient for the solid dissolution.

Further work on the absorption of sulphur dioxide by UCHIDA *et al.*[75] has shown that the absorption rate changes with the surface area of the limestone particles which in turn varies with the size and the number of particles, and that the rate of dissolution plays a very important role on the absorption. It was further found the absorption rate does not vary significantly with temperature and that the reactions involved may be considered as being instantaneous.

## 12.9.2. The centrifugal absorber

In an attempt to obtain the benefits of repeated spray formations, a centrifugal type absorber has been developed from the ideas of Piazza for a still head. The principle of the unit is shown in Figure 12.31. A set of stationary concentric rings intermeshes with a second set of rings attached to a rotating plate. Liquid fed to the centre of the plate

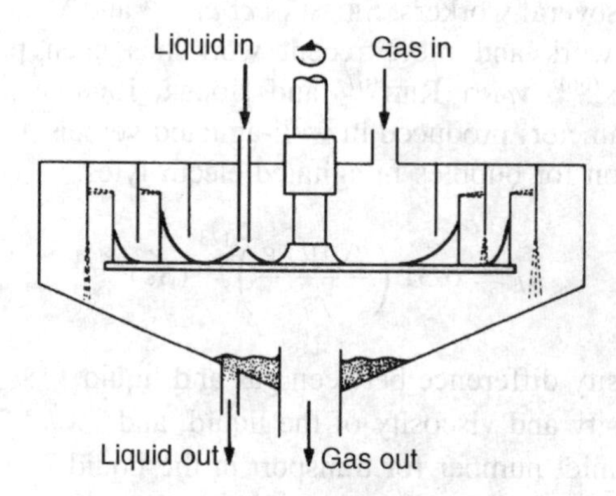


Figure 12.31. The centrifugal absorber

is carried up the first ring, splashes over to the baffle and falls into the through between the rings. It then runs up the second ring and in a similar way passes from ring to ring through the unit. The gas stream can be introduced at the top to give cocurrent flow, or at the bottom if countercurrent flow is desired. Some of the features of this unit are discussed by AHMED[76] who found that the depth of the ring was not very important and that most of the transfer took place as the gas mixed with the liquid spray leaving the top of the rings. CHAMBERS and WALL[77] have given some particulars of the performance of the 510 mm diameter unit shown in Figure 12.32, for the absorption of carbon dioxide from air containing 10–15 per cent of carbon dioxide, using mono-ethanolamine solution. Some values of absorption rates are given in Table 12.7.

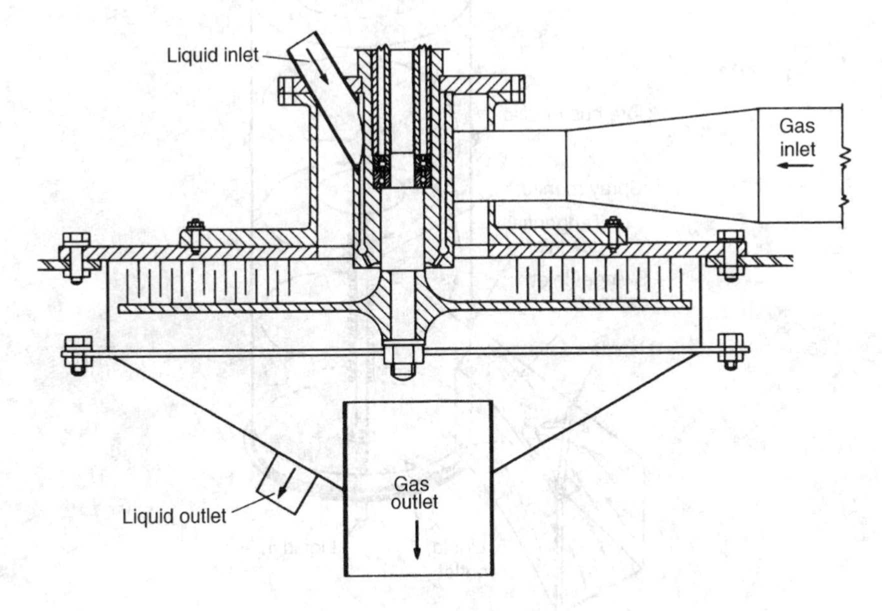


Figure 12.32. Details of a 510 mm diameter centrifugal absorber

Table 12.7. Results for absorption in a 510 mm diameter absorber

| Gas flow ($m^3/s$) | Liquid flow ($m^3/s$) | per cent $CO_2$ in gas in | per cent $CO_2$ in gas out | Absorption rate (kg/s) |
|---|---|---|---|---|
| 0.016 | $1.07 \times 10^{-4}$ | 16.3 | 2.3 | 0.0044 |
| 0.024 | $1.07 \times 10^{-4}$ | 15.8 | 4.5 | 0.0055 |
| 0.031 | $1.07 \times 10^{-4}$ | 14.3 | 6.6 | 0.0051 |
| 0.039 | $1.07 \times 10^{-4}$ | 16.3 | 8.7 | 0.0065 |

## 12.9.3. Spray towers

In the spray tower, the gas enters at the bottom and the liquid is introduced through a series of sprays at the top. The performance of these units is generally rather poor, because the droplets tend to coalesce after they have fallen through a few metres, and the interfacial surface is thereby seriously reduced. Although there is considerable turbulence in the gas phase, there is little circulation of the liquid within the drops, and the resistance of the equivalent liquid film tends to be high. Spray towers are therefore useful only where the

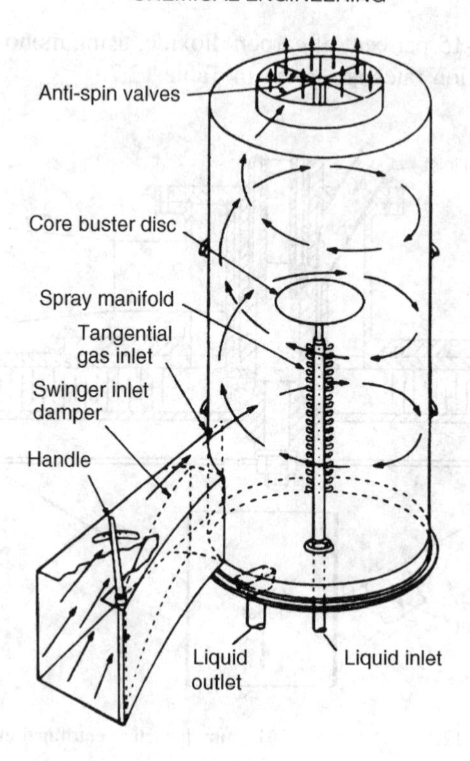


Figure 12.33. Centrifugal spray tower(78)

main resistance to mass transfer lies within the gas phase, and have consequently been used with moderate success for the absorption of ammonia in water. They are also used as air humidifiers, in which case the whole of the resistance lies within the gas phase.

### *Centrifugal spray tower*

Figure 12.33, taken from the work of Kleinschmidt and Anthony(78), illustrates a spray tower in which the gas stream enters tangentially, so that the liquid drops are subjected to centrifugal force before they are taken out of the gas stream at the top.

## 12.10. FURTHER READING

Hobler, T.: *Mass Transfer and Absorbers* (Pergamon Press, Oxford, 1966).
McCabe, W. L., Smith, J. C. and Harriott, P.: *Unit Operations of Chemical Engineering*, 4th edn. (McGraw-Hill, New York, 1984).
Norman, W. S.: *Absorption, Distillation and Cooling Towers* (Longmans, London, 1961).
Sherwood, T. K. and Pigford, R. L.: *Absorption and Extraction* (McGraw-Hill Book Co., New York, 1952).
Sherwood, T. K., Pigford, R. L. and Wilke, C. R.: *Mass Transfer* (McGraw-Hill Book Co., New York, 1975).
Smith, B. D.: *Design of Equilibrium Stage Processes* (McGraw-Hill Book Co., New York, 1963).
Treybal, R. E.: *Mass Transfer Operations*, 3rd edn. (McGraw-Hill Book Co., New York, 1980).

WANKAT, P. C.: *Equilibrium Staged Separations: Separations for Chemical Engineers* (Elsevier, New York, 1988).
ZARZYCKI, R. and CHACUK, A.: *Absorption. Fundamentals and Applications* (Pergamon Press, Oxford, 1993).
ZENZ, F. A.: *Design of Gas Absorption Towers*. In SCHWEITZER, P. A.: *Handbook of Separation Techniques for Chemical Engineers* 2nd edn. (McGraw Hill, New York, 1988).

## 12.11. REFERENCES

1. WHITMAN, W. G.: *Chem. Met. Eng.* **29** (1923) 147. The two-film theory of absorption.
2. HIGBIE, R.: *Trans. Am. Inst. Chem. Eng.* **31** (1935) 365. The rate of absorption of pure gas into a still liquid during short periods of exposure.
3. DANCKWERTS, P. V.: *Ind. Eng. Chem.* **43** (1951) 1460. Significance of liquid-film coefficients in gas absorption.
4. DANCKWERTS, P. V. and KENNEDY, A. M.: *Trans. Inst. Chem. Eng.* **32** (1954) S49. Kinetics of liquid-film processes in gas absorption.
5. LYNN, S., STRAATEMEIER, J. R., and KRAMERS, H.: *Chem. Eng. Sci.* **4** (1955) 49, 58, 63. Absorption studies in the light of the penetration theory. I. Long wetted-wall columns. II. Absorption by short wetted-wall columns. III. Absorption by wetted-spheres, singly and in columns.
6. DAVIDSON, J. F., CULLEN, E. J., HANSON, D., and ROBERTS, D.: *Trans. Inst. Chem. Eng.* **37** (1959) 122. The hold-up and liquid film coefficient of packed towers. Part I. Behaviour of a string of spheres.
7. DAVIDSON, J. F.: *Trans. Inst. Chem. Eng.* **37** (1959) 131. The hold-up and liquid film coefficient of packed towers. Part II: Statistical models of the random packing.
8. DANCKWERTS, P. V. and KENNEDY, A. M.: *Chem. Eng. Sci.* **8** (1958) 201. The kinetics of absorption of carbon dioxide into neutral and alkaline solutions.
9. ROBERTS, D. and DANCKWERTS, P. V.: *Chem. Eng. Sci.* **17** (1962) 961. Kinetics of $CO_2$ absorption in alkaline solutions. I. Transient absorption rates and catalysis by arsenite.
10. DANCKWERTS, P. V., KENNEDY, A. M. and ROBERTS, D.: *Chem. Eng. Sci.* **18** (1963) 63. Kinetics of $CO_2$ absorption in alkaline solutions. II. Absorption in a packed column and tests of surface renewal models.
11. CULLEN, E. J. and DAVIDSON, J. F.: *Trans. Faraday Soc.* **53** (1957) 113. Absorption of gases in liquid jets.
12. RAIMONDI, P. and TOOR, H. L.: *A.I.Ch.E.Jl.* **5** (1959) 86. Interfacial resistance in gas absorption.
13. STERNLING, C. V. and SCRIVEN, L. E.: *A.I.Ch.E.Jl.* **5** (1959) 514. Interfacial turbulence: Hydrodynamic instability and the Marangoni effect.
14. GOODRIDGE, F. and ROBB, I. D.: *Ind. Eng. Chem. Fundamentals* **4** (1965) 49. Mechanism of interfacial resistance.
15. GARNER, F. H. and KENDRICK, P.: *Trans. Inst. Chem. Eng.* **37** (1959) 155. Mass transfer to drops of liquid suspended in a gas stream. Part I — A wind tunnel for the study of individual liquid drops.
16. GARNER, F. H. and LANE, J. J.: *Trans. Inst. Chem. Eng.* **37** (1959) 162. Mass transfer to drops of liquid suspended in a gas stream. Part II: Experimental work and results.
17. KING, C. J.: *A.I.Ch.E.Jl.* **10** (1964) 671. The additivity of individual phase resistances in mass transfer operations.
18. GILLIAND, E. R. and SHERWOOD, T. K.: *Ind. Eng. Chem.* **26** (1934) 516. Diffusion of vapours into air streams.
19. HOLLINGS, H. and SILVER, L.: *Trans. Inst. Chem. Eng.* **12** (1934) 49. The washing of gas.
20. CHILTON, T. H. and COLBURN, A. P.: *Ind. Eng. Chem.* **26** (1934) 1183. Mass transfer (absorption) coefficients — prediction from data on heat transfer and fluid friction.
21. MORRIS, G. A. and JACKSON, J.: *Absorption Towers* (Butterworths, London, 1953).
22. KOWALKE, O. L., HOUGEN, O. A., and WATSON, K. M.: *Bull. Univ. Wisconsin Eng. Sta. Ser.* No. 68 (1925). Transfer coefficients of ammonia in absorption towers.
23. BORDEN, H. M. and SQUIRES, W.: Massachusetts Institute of Technology, S. M. thesis (1937). Absorption of ammonia in a ring-packed tower. (cited in Reference 30).
24. NORMAN, W. S.: *Trans. Inst. Chem. Eng.* **29** (1951) 226. The performance of grid-packed towers.
25. FELLINGER, L. L.: Massachusetts Institute of Technology. D.Sc. thesis (1941). Absorption of ammonia by water and acids in various standard packings.
26. PERRY, R. H., GREEN, D. W., and MALONEY, J. O. (eds.): *Perry's Chemical Engineers' Handbook*. 7th edn. (McGraw-Hill Book Company, New York, 1997).
27. MOLSTAD, M. C., MCKINNEY, J. F. and ABBEY, R. G.: *Trans. Am. Inst. Chem. Eng.* **39** (1943) 605. Performance of drip-point grid tower packings, III. Gas-film mass transfer coefficients: additional liquid-film mass transfer coefficients.

28. van Krevelen, D. W. and Hoftijzer, P. J.: *Rec. Trav. Chim* **67** (1948) 563. Kinetics of gas–liquid reactions. Part I. General theory.
29. Semmelbauer, R.: *Chem. Eng. Sci.* **22** (1967) 1237. Die Berechnung der Schütthöhe bei Absorptionsvorgängen in Füllkörperkolonnen. (Calculation of the height of packing in packed towers.)
30. Sherwood, T. K. and Holloway, F. A. L.: *Trans. Am. Inst. Chem. Eng.* **36** (1940). Performance of packed towers. 21 — Experimental studies of absorption and desorption, 39, 181 — liquid film data for several packings.
31. Cooper, C. M., Christl, R. J., and Peery, L. C.: *Trans. Am. Inst. Chem. Eng.* **37** (1941) 979. Packed tower performance at high liquor rates — The effect of gas and liquor rates upon performance in a tower packed with two-inch rings.
32. Nonhebel, G.: *Gas Purification Processes for Air Pollution Control*, 2nd edn. (Newnes–Butterworth, London, 1972).
33. Hixson, A. W. and Scott, C. E.: *Ind. Eng. Chem.* **27** (1935) 307. Absorption of gases in spray towers.
34. Whitman, W. G., Long, L., and Wang, H. Y.: *Ind. Eng. Chem.* **18** (1926) 363. Absorption of gases by a liquid drop.
35. Pigford, R. L. and Pyle, C.: *Ind. Eng. Chem.* **43** (1951) 1649. Performance characteristics of spray-type absorption equipment.
36. Norman, W. S.: *Absorption, Distillation and Cooling Towers* (Longmans, London, 1961).
37. Hatta, S.: *Tech. Repts. Tohoku Imp. Univ.* **10** (1932) 119. On the absorption velocity of gases by liquids. II. Theoretical considerations of gas absorption due to chemical reaction.
38. Nijsing, R. A. T. O., Hendriksz, R. H. and Kramers, H.: *Chem. Eng. Sci.* **10** (1959) 88. Absorption of $CO_2$ in jets and falling films of electrolyte solutions, with and without chemical reaction.
39. Tepe, J. B. and Dodge, B. F.: *Trans. Am. Inst. Chem. Eng.* **39** (1943) 255. Absorption of carbon dioxide by sodium hydroxide solutions in a packed column.
40. Cryder, D. S. and Maloney, J. O.: *Trans. Am. Inst. Chem. Eng.* **37** (1941) 827. The rate of absorption of carbon dioxide in diethanolamine solutions.
41. Stephens, E. J. and Morris, G. A.: *Chem. Eng. Prog.* **47** (1951) 232. Determination of liquid-film absorption coefficients. A new type of column and its application to problems of absorption in presence of chemical reaction.
42. Danckwerts, P. V. and McNeil, K. M.: *Trans. Inst. Chem. Eng.* **45** (1967) 32. The absorption of carbon dioxide into aqueous amine solutions and the effects of catalysis.
43. Danckwerts, P. V. and Sharma, M. M.: *Chem. Engr. London* No. **202** (Oct. 1966) CE244. The absorption of carbon dioxide into solutions of alkalis and amines (with some notes on hydrogen sulphide and carbonyl sulphide).
44. Sahay, B. N. and Sharma, M. M.: *Chem.Eng. Sci.* **28** (1973) 41. Effective interfacial areas and liquid and gas side mass transfer coefficients in a packed column.
45. Eckert, J. S.: *Chem. Engg.* **82** (14 April 1975) 70. How tower packings behave.
46. Sherwood, T. K., Pigford, R. L., and Wilke, C. R.: *Mass Transfer* (McGraw-Hill Book Company, New York, 1980).
47. Treybal, R. E.: *Mass Transfer Operations*, 3rd edn. (McGraw-Hill Book Co., New York, 1980).
48. Poll, A. and Smith, W.: *Chem. Engg.* **71** (26 Oct. 1964) 111. Froth contact heat exchanger.
49. Coggan, C. G. and Bourne, J. R.: *Trans. I. Chem. E.* **47** (1969) T96, T160. The design of gas absorbers with heat effects.
50. Shulman, H. L., Ullrich, C. F., Proulx, A. Z. and Zimmerman, J. O.: *A.I.Ch.E.Jl.* **1** (1955) 2, 253. Interfacial areas — gas and liquid phase mass transfer rates.
51. Eastham, I. E.: Private communication (1977).
52. Chilton, T. H. and Colburn, A. P.: *Ind. Eng. Chem.* **27** (1935) 255. Distillation and absorption in packed columns.
53. Rackett, H. G.: *Chem. Eng. Albany* **71** (21 Dec. 1964) 108. Modified graphical integration for determining transfer units.
54. Norton Chemical Process Products Div., Box 350, Akron, Ohio; Hydronyl Ltd., King St., Fenton, Stoke-on-Trent, U.K.
55. Colburn, A. P.: *Trans. Am. Inst. Chem. Eng.* **35** (1939) 211. The simplified calculation of diffusional processes. General consideration of two-film resistances.
56. Kremser, A.: *Nat. Petroleum News* **22** (21 May 1930) 43. Theoretical analysis of absorption processes.
57. Souders, M. and Brown, G. G.: *Ind. Eng. Chem.* **24** (1932) 519. Fundamental design of high pressure equipment involving paraffin hydrocarbons. IV. Fundamental design of absorbing and stripping columns for complex vapours.
58. Cooper, C. M., Fernstrom, G. A., and Millers, S. A.: *Ind. Eng. Chem.* **36** (1944) 504. Performance of agitated gas–liquid contactors.
59. Ayerst, R. R. and Herbert, L. S.: *Trans. Inst. Chem. Eng.* **32** (1954) S68. A study of the absorption of carbon dioxide in ammonia solutions in agitated vessels.

60. WESTERTERP, K. R., VAN DIERENDONCK, L. L. and DE KRAA, J. R.: *Chem. Eng. Sci.* **18** (1963) 157. Interfacial areas in agitated gas–liquid contactors.
61. SRIDHAR, T. and POTTER, O.E: *Chem. Eng. Sci.* **35** (1980) 683. Interfacial areas in gas–liquid stirred vessels.
62. CALDERBANK, P. H.: *Chem. Engnr.* No. 212 (Oct. 1967) CE 209. Gas absorption from bubbles.
63. SIDEMAN, S. O., HORTACSU, O., and FULTON, J. W.: *Ind. Eng. Chem.* **58** (July 1966) 32. Mass transfer in gas–liquid contacting systems.
63. SIDEMAN, S. O., HORTACSU, O. and FULTON, J. W.: *Ind. Eng. Chem.* **58** (July 1966) 32. Mass transfer in gas–liquid contacting systems.
64. VALENTIN, F. H. H.: *Brit. Chem. Eng.* **12** (1967) 1213. Mass transfer in agitated tanks.
65. YAGI, H. and YOSHIDA, F.: *Ind. Eng. Chem. Proc. Des. Dev.* **14** (1975) 488. Gas absorption by Newtonian and non-Newtonian fluids in sparged agitation vessels.
66. ZLOKARNIK, M.: *Adv. Biochem. Eng.* **8** (1978) 133. Sorption characteristics for gas–liquid contacting in mixing vessels.
67. VAN'T RIET, K.: *Ind. Eng. Chem. Proc. Des. Dev.* **18** (1979) 357. Review of measuring methods and results in nonviscous gas–liquid mass transfer in stirred tanks.
68. HÖCKER, H., LANGER, G. and UDO, W.: *Germ. Chem. Eng.* **4** (1981) 51. Mass transfer in aerated Newtonian and non-Newtonian liquids.
69. JOSHI, J. B. and SHARMA, M. M.: *Can. J. Chem. Eng.* **54** (1976) 460. Mass transfer characteristics of horizontal agitated contactors.
70. FUKUDA, H., IDOGAWA, K., IKEDA, K. and ENDOH, K.: *J. Chem. Eng. Japan* **13** (1980) 298. Volumetric gas-phase mass transfer coefficients in baffled horizontal stirred tanks.
71. YUNG, C. H., WONG, C. W. and CHANG, C. L.: *Can. J. Chem. Eng.* **59** (1979) 672. Gas holdup and aerated power consumption in mechanically stirred tanks.
72. LUONG, H. T. and VOLESKY, B.: *AIChE Jl.* **25** (1970) 893. Mechanical power requirements of gas–liquid agitated systems.
73. BJERLE, I., BENGTSSON, S. and FÄRNKVIST, K.: *Chem. Eng. Sci.* **27** (1972) 1853. Absorption of $SO_2$ in $CaCO_3$-slurry in a laminar jet absorber.
74. HIXSON, A. W. and BAUM, S. J.: *Ind. Eng. Chem.* **33** (1941) 478. Agitation: heat and mass transfer coefficients in liquid-solid systems.
75. UCHIDA, S., MORIGUCHI, H., MAEJIMA, H., KOIDE, K. and KAGEYAMA, S.: *Can. J. Chem. Eng.* **56** (1978) 690. Absorption of sulphur dioxide into limestone slurry in a stirred tank reactor.
76. AHMED, N.: University of London, Ph.D. thesis (1949). Design of gas scrubber based upon thin films and sprays.
77. CHAMBERS, H. H. and WALL, R. C.: *Trans. Inst. Chem. Eng.* **32** (1954) S96. Some factors affecting the design of centrifugal gas absorbers.
78. KLEINSCHMIDT, R. V. and ANTHONY, A. W.: *Trans. Am. Soc. Mech. Eng.* **63** (1941) 349. Recent development of Pease–Anthony gas scrubber.

## 12.12. NOMENCLATURE

| | | Units in SI System | Dimensions in M, L, T $\theta$ |
|---|---|---|---|
| $A$ | Cross-sectional area of column | m$^2$ | $\mathbf{L}^2$ |
| $\mathscr{A}$ | Absorption factor | — | — |
| $a$ | Surface area of interface per unit volume of column | m$^2$/m$^3$ | $\mathbf{L}^{-1}$ |
| $a_1, a_2 \ldots$ | Constants in equation 12.39 | — | — |
| $a'$ | Specific surface area (equation 12.94) | m$^{-1}$ | $\mathbf{L}^{-1}$ |
| $B$ | A constant in equation 12.23 | — | — |
| $B'$ | A constant in equation 12.24 | — | — |
| $C$ | Molar concentration | kmol/m$^3$ | $\mathbf{NL}^{-3}$ |
| $C_A$, $C_B$ | Molar concentrations of **A**, **B** | kmol/m$^3$ | $\mathbf{NL}^{-3}$ |
| $C_{AL}$, $C_{BL}$ | Molar concentrations of **A**, **B** in bulk of liquid phase | kmol/m$^3$ | $\mathbf{NL}^{-3}$ |
| $C_{Ae}$ | Molar concentration of **A** in liquid phase in equilibrium with partial pressure $P_{AG}$ in gas phase | kmol/m$^3$ | $\mathbf{NL}^{-3}$ |

| | | Units in SI System | Dimensions in M, L, T θ |
|---|---|---|---|
| $C_{Ai}$ | Molar concentration of **A** at interface | kmol/m$^3$ | $\mathbf{NL}^{-3}$ |
| $C_{AL}$ | Molar concentration of **A** in bulk of liquid | kmol/m$^3$ | $\mathbf{NL}^{-3}$ |
| $C_T$ | Total molar concentration | kmol/m$^3$ | $\mathbf{NL}^{-3}$ |
| $c$ | Constant term in equation of equilibrium line | — | — |
| $c_G$ | Gas mixture constant $(\rho_r/\mu_r)^{0.25}/(D_{Vr})^{0.5}$ in cgs units | $[(\text{cm}^2/\text{s})^{-3/4}]$ | $\mathbf{L}^{-3/2}\mathbf{T}^{3/4}$ |
| $D_L$ | Liquid phase diffusivity | m$^2$/s | $\mathbf{L}^2\mathbf{T}^{-1}$ |
| $D_V$ | Vapour phase diffusivity | m$^2$/s | $\mathbf{L}^2\mathbf{T}^{-1}$ |
| $d$ | Column diameter | m | **L** |
| $d_i$ | Impeller diameter | m | L |
| $d_0$ | Sauter mean diameter | m | L |
| $d_p$ | Packing size | m | L |
| $d_t$ | Tank diameter | m | L |
| $e$ | Voidage | — | — |
| $F'$ | Fractional conversion (equation 12.39) | — | — |
| $f$ | Fraction of surface renewed per unit time | s$^{-1}$ | $\mathbf{T}^{-1}$ |
| $G'_m$ | Molar rate of flow of inert gas per unit cross-section | kmol/m$^2$s | $\mathbf{NL}^{-2}\mathbf{T}^{-1}$ |
| $G'$ | Gas flowrate (mass) per unit cross-section | kg/m$^2$s | $\mathbf{ML}^{-2}\mathbf{T}^{-1}$ |
| **H** | Height of transfer unit | m | **L** |
| $h$ | Heat transfer coefficient | W/m$^2$K | $\mathbf{MT}^{-3}\boldsymbol{\theta}^{-1}$ |
| $h_D$ | Mass transfer coefficient $(D_V/z_G)$ | m/s | $\mathbf{LT}^{-1}$ |
| $h_p$ | Height of packing | m | **L** |
| $\mathscr{H}$ | Henry's constant | (N/m$^2$)/(kmol/m$^3$) | $\mathbf{MN}^{-1}\mathbf{L}^2\mathbf{T}^{-2}$ |
| $i$ | Number of mole of **B** reacting with 1 mole of **A** | — | — |
| $j_d$ | $j$-factor for mass transfer | — | — |
| $K_G$ | Overall gas-phase transfer coefficient | s/m | $\mathbf{L}^{-1}\mathbf{T}$ |
| $K_L$ | Overall liquid-phase transfer coefficient | m/s | $\mathbf{LT}^{-1}$ |
| $K''_G$ | Overall gas-phase transfer coefficient in terms of mole fractions | kmol/m$^2$s | $\mathbf{NL}^{-2}\mathbf{T}^{-1}$ |
| $K''_L$ | Overall liquid-phase transfer coefficient in terms of mole fractions | kmol/m$^2$s | $\mathbf{NL}^{-2}\mathbf{T}^{-1}$ |
| $k$ | Thermal conductivity | W/m K | $\mathbf{MLT}^{-3}\boldsymbol{\theta}^{-1}$ |
| $k_G$ | Gas-film transfer coefficient $(D_V P/\mathbf{R}T z_G P_{Bm})$ | s/m | $\mathbf{L}^{-1}\mathbf{T}$ |
| $k'_G$ | Gas-film transfer coefficient $(D_V/\mathbf{R}T z_G)$ | s/m | $\mathbf{L}^{-1}\mathbf{T}$ |
| $k''_G$ | Gas-film transfer coefficient in terms of mole fractions | kmol/m$^2$s | $\mathbf{NL}^{-2}\mathbf{T}^{-1}$ |
| $k_L$ | Liquid-film transfer coefficient | m/s | $\mathbf{LT}^{-1}$ |
| $k''_L$ | Liquid-film transfer coefficient in terms of mole fractions | kmol/m$^2$s | $\mathbf{NL}^{-2}\mathbf{T}^{-1}$ |
| $\boldsymbol{k}_2$ | Reaction rate constant for second-order reaction | m$^3$/kmols | $\mathbf{N}^{-1}\mathbf{L}^3\mathbf{T}^{-1}$ |
| $L'_m$ | Molar rate of flow of solute-free liquor per unit cross-section | kmol/s m$^2$ | $\mathbf{NL}^{-2}\mathbf{T}^{-1}$ |
| $L_v$ | Volumetric liquid rate | m$^3$/s | $\mathbf{L}^3\mathbf{T}^{-1}$ |
| $L'$ | Liquid flowrate (mass) per unit cross-section | kg/s m$^2$ | $\mathbf{ML}^{-2}\mathbf{T}^{-1}$ |
| $m$ | Slope of equilibrium line | — | — |
| $N_A$, $N_B$ | Molar rate of diffusion of **A**, **B** per unit area | kmol/s m$^2$ | $\mathbf{NL}^{-2}\mathbf{T}^{-1}$ |

| | | Units in SI System | Dimensions in **M**, **L**, **T** $\theta$ |
|---|---|---|---|
| $N'_A$, $N'_B$ | Molar rate of absorption of **A**, **B** per unit area | kmol/s m$^2$ | $\mathbf{NL^{-2}T^{-1}}$ |
| $N''_A$ | Molar rate of absorption of **A** per unit area with chemical reaction | kmol/s m$^2$ | $\mathbf{NL^{-2}T^{-1}}$ |
| $\mathbf{N}$ | Number of transfer units | — | — |
| $\mathbf{N'}$ | Impeller speed | s$^{-1}$, (Hz) | $\mathbf{T^{-1}}$ |
| $n$ | Number of plates from bottom | — | — |
| $P$ | Total pressure | N/m$^2$ | $\mathbf{ML^{-1}T^{-2}}$ |
| $P_A$, $P_B$ | Partial pressures of **A** and **B** | N/m$^2$ | $\mathbf{ML^{-1}T^{-2}}$ |
| $P_{Bm}$ | Logarithmic mean value of $P_B$ | N/m$^2$ | $\mathbf{ML^{-1}T^{-2}}$ |
| $P_{Ae}$ | Partial pressure of **A** in equilibrium with concentration $C_{AL}$ in liquid phase | N/m$^2$ | $\mathbf{ML^{-1}T^{-2}}$ |
| $P_{AG}$ | Partial pressure of **A** in bulk of gas phase | N/m$^2$ | $\mathbf{ML^{-1}T^{-2}}$ |
| $P_{Ai}$ | Partial pressure of **A** at interface | N/m$^2$ | $\mathbf{ML^{-1}T^{-2}}$ |
| $\Delta P_{Alm}$ | Log mean driving force for **A** | N/m$^2$ | $\mathbf{ML^{-1}T^{-2}}$ |
| $\mathbf{P}_V$ | Power input per unit volume | W/m$^3$ | $\mathbf{ML^{-1}T^{-3}}$ |
| $\mathbf{R}$ | Universal gas constant | J/kmol K | $\mathbf{NM^{-1}L^2T^{-2}}\theta^{-1}$ |
| $r$ | Ratio of effective film thickness for absorption without and with chemical reaction | — | — |
| $S$ | Specific surface of packing | m$^{-1}$ | $\mathbf{L^{-1}}$ |
| $s$ | Total number of plates in column | — | — |
| $T$ | Absolute temperature | K | $\theta$ |
| $t$ | Time | s | $\mathbf{T}$ |
| $u$ | Gas velocity | m/s | $\mathbf{LT^{-1}}$ |
| $u_0$ | Terminal rise velocity | m/s | $\mathbf{LT^{-1}}$ |
| $u_s$ | Superficial gas velocity (based on inlet conditions) | m/s | $\mathbf{LT^{-1}}$ |
| $V$ | Volume of packed section of column | m$^3$ | $\mathbf{L^3}$ |
| $X$ | Moles of solute gas **A** per mole of solvent in liquid phase | — | — |
| $x$ | Mole fraction of **A** in liquid phase | — | — |
| $Y$ | Molar ratio of solute gas **A** to inert gas **B** in gas phase | — | — |
| $y$ | Mole fraction of **A** in gas phase | — | — |
| $Z$ | Height of packed column | m | $\mathbf{L}$ |
| $z$ | Distance of direction of mass transfer | m | $\mathbf{L}$ |
| $z_G$ | Thickness of gas film | m | $\mathbf{L}$ |
| $z_L$ | Thickness of liquid film | m | $\mathbf{L}$ |
| $\alpha$ | A coefficient in equation 12.28 | s$^{1.8}$/kg$^{0.8}$m$^{0.4}$ | $\mathbf{M^{-0.8}L^{-0.4}T^{1.8}}$ |
| $\beta$ | A coefficient | 1/m$^{1.25}$ | $\mathbf{L^{-5/4}}$ |
| $\beta'$ | A coefficient (equation 12.33) | — | — |
| $\mu$ | Viscosity of gas | Ns/m$^2$ | $\mathbf{ML^{-1}T^{-1}}$ |
| $\mu_L$ | Viscosity of liquid | Ns/m$^2$ | $\mathbf{ML^{-1}T^{-1}}$ |
| $\rho$ | Density of gas | kg/m$^3$ | $\mathbf{ML^{-3}}$ |
| $\rho_L$ | Density of liquid | kg/m$^3$ | $\mathbf{ML^{-3}}$ |
| $\sigma$ | Surface tension | J/m$^2$ | $\mathbf{MT^{-2}}$ |
| $\phi$ | Correction factor for concentrated solutions | — | — |
| $Ga$ | Galileo number | — | — |
| $Pr$ | Prandtl number | — | — |
| $Re$ | Reynolds number | — | — |
| $Sc$ | Schmidt number | — | — |
| $Sh$ | Sherwood number | — | — |

| | | Units in SI System | Dimensions in **M**, **L**, **T** $\theta$ |
|---|---|---|---|

Suffixes

| | |
|---|---|
| 1 | denotes conditions at bottom of packed column, or at plane 1 |
| 2 | denotes conditions at top of packed column, or at plane 2 |
| $A$ | denotes soluble gas |
| $B$ | denotes insoluble gas |
| $e$ | denotes equilibrium value |
| $f$ | denotes film value |
| $i$ | denotes value at interface |
| $G$ | denotes gas phase |
| $L$ | denotes liquid phase |
| lm | denotes logarithmic mean value |
| $n$ | denotes values on plate $n$ |
| $r$ | denotes reference state (293 K, 101.3 kN/m$^2$) |

$LG$, $OG$, $L$, $OL$ refer to gas film, overall gas, liquid film, and overall liquid transfer units

CHAPTER 13

# Liquid–Liquid Extraction

## 13.1. INTRODUCTION

The separation of the components of a liquid mixture by treatment with a solvent in which one or more of the desired components is preferentially soluble is known as liquid–liquid extraction — an operation which is used, for example, in the processing of coal tar liquids and in the production of fuels in the nuclear industry, and which has been applied extensively to the separation of hydrocarbons in the petroleum industry. In this operation, it is essential that the liquid-mixture feed and solvent are at least partially if not completely immiscible and, in essence, three stages are involved:

(a) Bringing the feed mixture and the solvent into intimate contact,
(b) Separation of the resulting two phases, and
(c) Removal and recovery of the solvent from each phase.

It is possible to combine stages (a) and (b) into a single piece of equipment such as a column which is then operated continuously. Such an operation is known as differential contacting. Liquid–liquid extraction is also carried out in stagewise equipment, the prime example being a mixer–settler unit in which the main features are the mixing of the two liquid phases by agitation, followed by settling in a separate vessel by gravity. This mixing of two liquids by agitation is of considerable importance and the topic is discussed in some detail in Volume 1, Chapter 7.

Extraction is in many ways complementary to distillation and is preferable in the following cases:

(a) Where distillation would require excessive amounts of heat, such as, for example, when the relative volatility is near unity.
(b) When the formation of azeotropes limits the degree of separation obtainable in distillation.
(c) When heating must be avoided.
(d) When the components to be separated are quite different in nature.

Important applications of liquid–liquid extraction include the separation of aromatics from kerosene-based fuel oils to improve their burning qualities and the separation of aromatics from paraffin and naphthenic compounds to improve the temperature-viscosity characteristics of lubricating oils. It may also be used to obtain, for example, relatively pure compounds such as benzene, toluene, and xylene from catalytically produced reformates in the oil industry, in the production of anhydrous acetic acid, in the extraction of phenol from coal tar liquors, and in the metallurgical and biotechnology industries.

In all extraction processes, the important feature is the selective nature of the solvent, in that the separation of compounds is based on differences in solubilities, rather than differences in volatilities as in distillation. In recent years, it has become possible to use computerised techniques to aid in the choice of a solvent with the required selectivity and to "design" appropriate molecular structures.

A recent and extremely important development lies in the application of the technique of liquid extraction to *metallurgical processes*. The successful development of methods for the purification of uranium fuel and for the recovery of spent fuel elements in the nuclear power industry by extraction methods, mainly based on packed, including pulsed, columns as discussed in Section 13.5 has led to their application to other metallurgical processes. Of these, the recovery of copper from acid leach liquors and subsequent electro-winning from these liquors is the most extensive, although further applications to nickel and other metals are being developed. In many of these processes, some form of chemical complex is formed between the solute and the solvent so that the kinetics of the process become important. The extraction operation may be either a physical operation, as discussed previously, or a chemical operation. Chemical operations have been classified by HANSON[1] as follows:

(a) Those involving cation exchange such as, for example, the extraction of metals by carboxylic acids;
(b) Those involving anion exchange, such as the extraction of anions involving a metal with amines, and
(c) Those involving the formation of an additive compound, for example, extraction with neutral organo-phosphorus compounds. An important operation of this type is the purification of uranium from the nitrate with tri-*n*-butyl phosphate.

This process of metal purification is of particular interest in that it involves the application of principles of both chemistry and chemical engineering and necessitates the cost evaluation of alternatives.

A whole new technology with respect to extraction, developed within the last decade, has been the use of *supercritical* or *near supercritical fluids* as solvent.

In *biotechnology*, many of the usual organic solvents will degrade a sensitive product, such as a protein; this has led to the use of "mild" aqueous-based extractants, such as water–polyethyleneglycol–phosphate mixtures, which will partition and concentrate the product in one of the two aqueous layers which are formed.

The use of supercritical fluids and aqueous-based extractants is discussed in Section 13.8

## 13.2. EXTRACTION PROCESSES

The three steps outlined in Section 13.1, necessary in all liquid–liquid extraction operations, may be carried out either as a batch or as a continuous process.

In the single-stage batch process illustrated in Figure 13.1, the solvent and solution are mixed together and then allowed to separate into the two phases — the *extract* E containing the required solute in the added solvent and the *raffinate* R, the weaker solution with some associated solvent. With this simple arrangement mixing and separation occur in the same vessel.

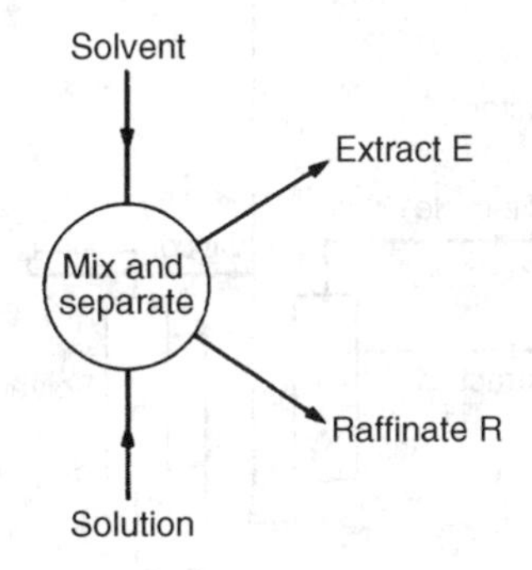


Figure 13.1. Single-stage batch extraction

A continuous two-stage operation is shown in Figure 13.2, where the mixers and separators are shown as separate vessels. There are three main forms of equipment. First there is the mixer-settler as shown in Figure 13.1, secondly, there is the column type of design with trays or packing as in distillation and, thirdly, there are a variety of units incorporating rotating devices such as the Scheibel and the Podbielniak extractors. In all cases, the extraction units are followed by distillation or a similar operation in order to recover the solvent and the solute. Some indication of the form of these alternative arrangements may be seen by considering two of the processes referred to in Section 13.1.

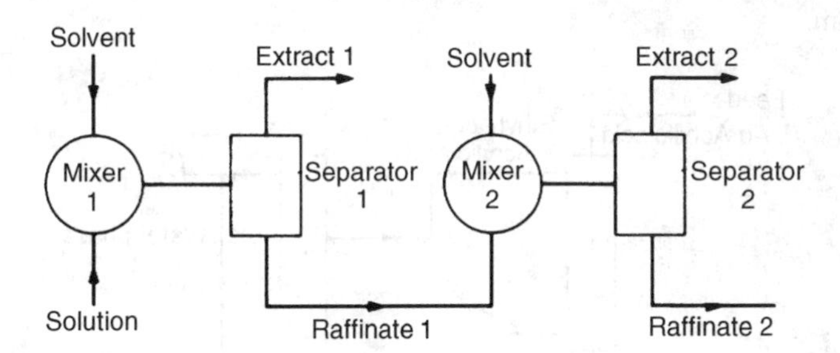


Figure 13.2. Multiple-contact system with fresh solvent

One system for separating benzene, toluene, and xylene groups from light feed-stocks is shown in Figure 13.3, where n-methylpyrolidone (NMP) with the addition of some glycol is used as the solvent. The feed is passed to a multistage extractor arranged as a tower from which an aromatics-free raffinate is obtained at the top. The extract stream containing the solvent, aromatics, and low boiling non-aromatics is distilled to provide the extractor recycle stream as a top product, and a mixture of aromatics and solvent at the bottom. This stream passes to a stripper from which the glycol and the aromatics

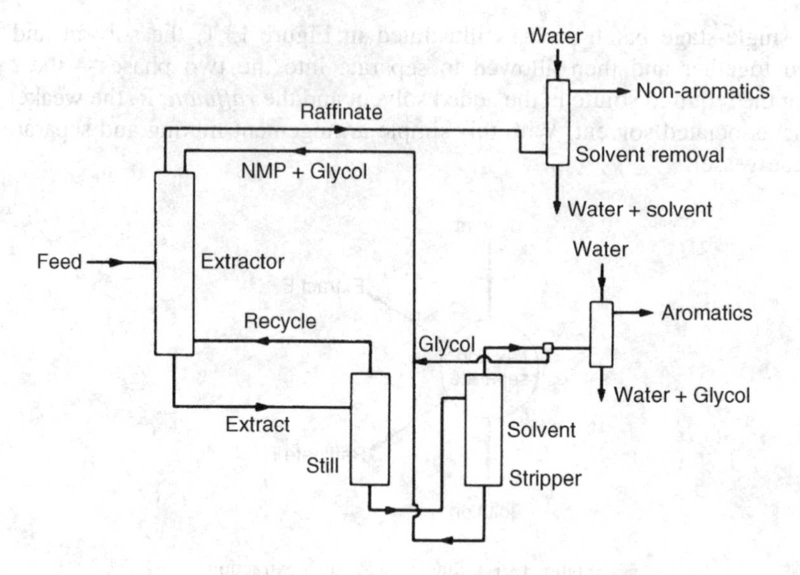


Figure 13.3. Process for benzene, toluene and xylene recovery

are recovered. This is a complex system illustrating the need for careful recycling and recovery of solvent.

The concentration of acrylic acid by extraction with ethyl acetate[2] is a rather different illustration of this technique. As shown in Figure 13.4, the dilute acrylic acid solution of concentration about 20 per cent is fed to the top of the extraction column 1, the ethyl acetate solvent being fed in at the base. The acetate containing the dissolved acrylic acid and water leaves from the top and is fed to the distillation column 2, where the acetate is removed as an azeotrope with water and the dry acrylic acid is recovered as product from the bottom.

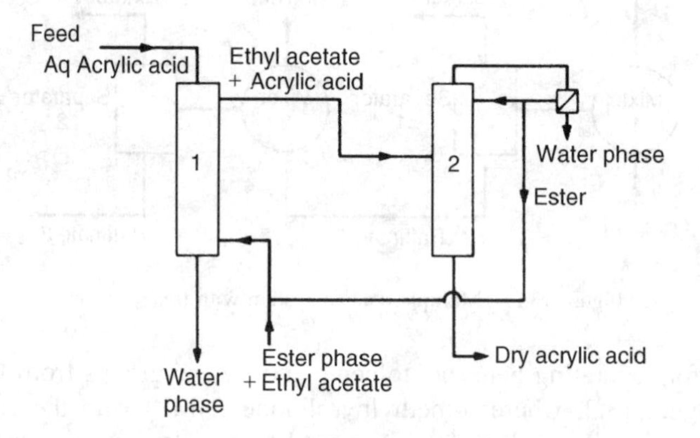


Figure 13.4. Concentration of acrylic acid by extraction with ethyl acetate[2]

It may be seen from these illustrations that successful extraction processes should not be judged simply by the performance of the extraction unit alone, but by assessment of the recovery achieved by the whole plant. This aspect of the process may be complex if chemical reactions are involved. The sections of the plant for mixing and for separation must be considered together when assessing capital cost. The cost of the organic solvents used in the metallurgical processes may also be high.

The mechanism of transfer of solute from one phase to the second is one of molecular and eddy diffusion and the concepts of phase equilibrium, interfacial area, and surface renewal are all similar in principle to those met in distillation and absorption, even though, in liquid–liquid extraction, dispersion is effected by mechanical means including pumping and agitation, except in standard packed columns.

In formulating design criteria for extraction equipment, it is necessary to take into account the equilibrium conditions for the distribution of solute between the phases as this determines the maximum degree of separation possible in a single stage. The resistance to diffusion and, in the case of chemical effects, the kinetics are also important in that these determine the residence time required to bring about near equilibrium in a stage-wise unit, or the height of a transfer unit in a differential contactor. The transfer rate is given by the accepted equation:

$$\text{Rate per unit interfacial area} = k\Delta C \tag{13.1}$$

where $k$ is a mass transfer coefficient and $\Delta C$ a concentration driving force. A high value of $k$ can be obtained only if turbulent or eddy conditions prevail and, although these may be readily achieved in the continuous phase by some form of agitation, it is very difficult to generate eddies in the drops which constitute the dispersed phase.

## 13.3. EQUILIBRIUM DATA

The equilibrium condition for the distribution of one solute between two liquid phases is conveniently considered in terms of the distribution law. Thus, at equilibrium, the ratio of the concentrations of the solute in the two phases is given by $C_E/C_R = K'$, where $K'$ is the distribution constant. This relation will apply accurately only if both solvents are immiscible, and if there is no association or dissociation of the solute. If the solute forms molecules of different molecular weights, then the distribution law holds for each molecular species. Where the concentrations are small, the distribution law usually holds provided no chemical reaction occurs.

The addition of a new solvent to a binary mixture of a solute in a solvent may lead to the formation of several types of mixture:

(a) A homogeneous solution may be formed and the selected solvent is then unsuitable.
(b) The solvent may be completely immiscible with the initial solvent.
(c) The solvent may be partially miscible with the original solvent resulting in the formation of one pair of partially miscible liquids.
(d) The new solvent may lead to the formation of two or three partially miscible liquids.

Of these possibilities, types (b), (c), and (d) all give rise to systems that may be used, although those of types (b) and (c) are the most promising. With conditions of type (b), the equilibrium relation is conveniently shown by a plot of the concentration of solute in one phase against the concentration in the second phase. Conditions given by (c) and (d) are usually represented by triangular diagrams. Equilateral triangles are used, although it is also possible to use right-angled isosceles triangles, which are discussed in Chapter 10.

The system, acetone (**A**)–Water (**B**)–methyl isobutyl ketone (**C**), as shown in Figure 13.5, is of type (c). Here the solute **A** is completely miscible with the two solvents **B** and **C**, although the two solvents are only partially miscible with each other. A mixture indicated by point H consists of the three components **A**, **B** and **C** in the ratio of the perpendiculars HL, HJ, HK. The distance BN represents the solubility of solvent **C** in **B**, and MC that of **B** in **C**. The area under the curved line NPFQM, the binodal solubility curve, represents a two-phase region which will split up into two layers in equilibrium with each other. These layers have compositions represented by points P and Q, and PQ is known as a "tie line". Such lines, two of which are shown in the diagram, connect the compositions of two phases in equilibrium with each other, and these compositions must be found by practical measurement. There is one point on the binodal curve at F which represents a single phase that does not split into two phases. F is known as a *plait* point, and this must also be found by experimental measurement. The plait point is fixed if either the temperature or the pressure is fixed. Within the area under the curve, the temperature and composition of one phase will fix the composition of the other. Applying the phase rule to the three-components system at constant temperature and pressure, the number of degrees of freedom is equal to 3 minus the number of phases. In the area where there is only one liquid phase, there are two degrees of freedom and two compositions must be stated. In a system where there are two liquid phases, there is only one degree of freedom.

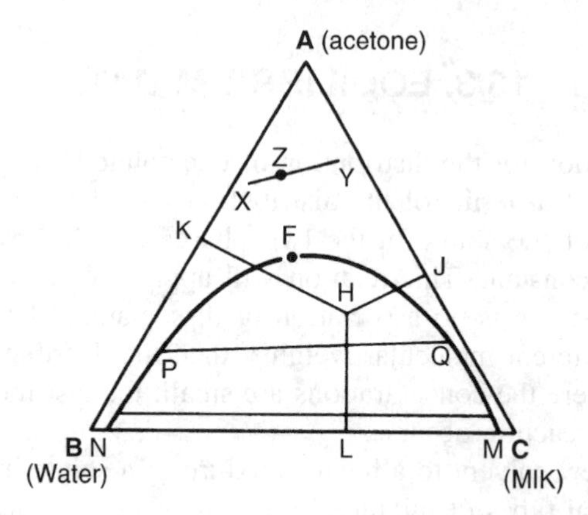


Figure 13.5. Equilibrium relationship for acetone distributed between water and methyl isobutyl ketone

One of the most useful features of this method of representation is that, if a solution of composition X is mixed with one of composition Y, then the resulting mixture will have a composition shown by Z on a line XY, such that:

$$XZ/ZY = (\text{amount of Y})/(\text{amount of X}).$$

Similarly, if an extract Y is removed, from a mixture Z the remaining liquor will have composition X.

In Figure 13.6 two separate two-phase regions are formed, whilst in Figure 13.7 the two-phase regions merge on varying the temperature. Aniline (**A**), water (**B**), and phenol (**C**) represent a system of the latter type. Under the conditions shown in Figures 13.6 and 13.7, **A** and **C** are miscible in all proportions, although **B** and **A**, and **B** and **C** are only partially miscible.

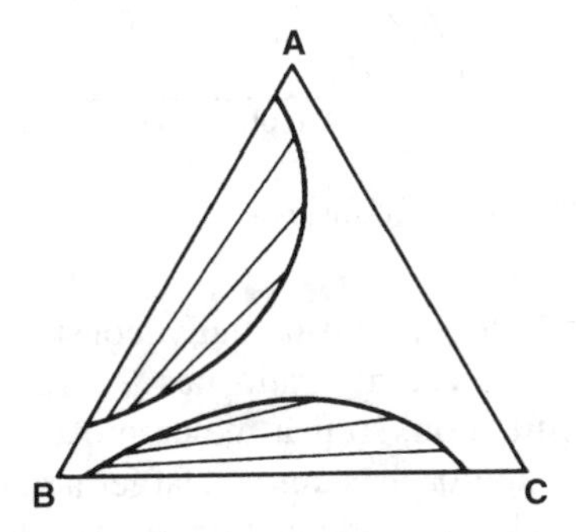


Figure 13.6. Equilibrium relationship for the aniline-water-phenol system

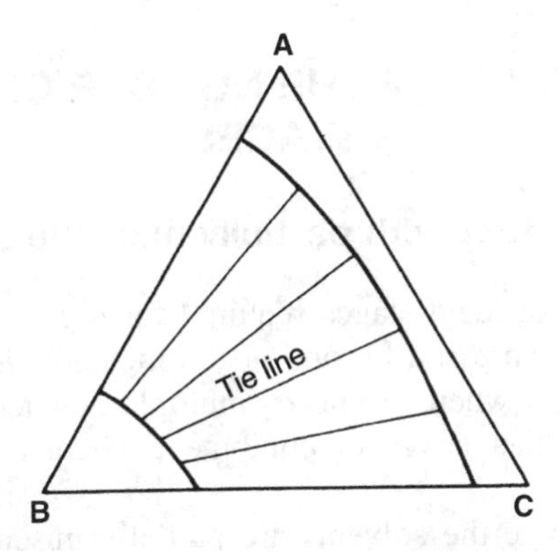


Figure 13.7. Equilibrium relationship for the aniline-water-phenol system at a higher temperature

Whilst these diagrams are of considerable use in presenting equilibrium data, Figure 13.8 is in many ways more useful for determining the selectivity of a solvent, and the number of stages that are likely to be required. In Figure 13.8 the percentage of solute in one phase is plotted against the percentage in the second phase in equilibrium with it. This is equivalent to plotting the compositions at either end of a tie line. The important factor in assessing the value of a solvent is the ratio of the concentrations of the desired component in the two phases, rather than the actual concentrations. A selectivity ratio may be defined in terms of either mass or mole fractions as:

$$\beta = \left[\frac{x_A}{x_B}\right]_E \Big/ \left[\frac{x_A}{x_B}\right]_R \tag{13.2}$$

where $x_A$ and $x_B$ are the mass or mole fractions of **A** and **B** in the two phases E and R.

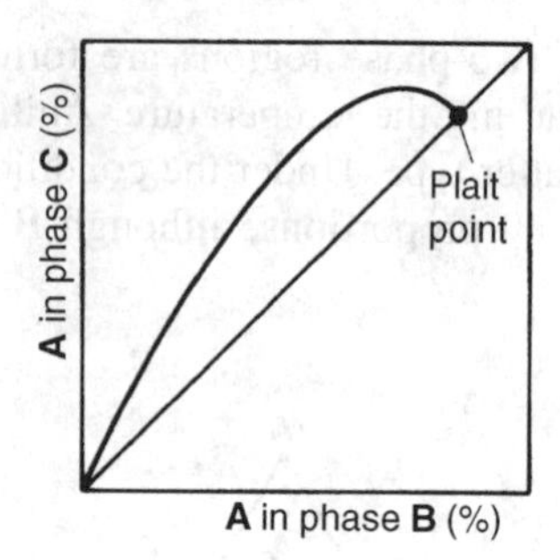


Figure 13.8. Equilibrium distribution of solute **A** in phases **B** and **C**

For a few systems $\beta$ tends to be substantially constant, although it more usually varies with concentration. The selectivity ratio has the same significance in extraction as relative volatility has in distillation, so that the ease of separation is directly related to the numerical value of $\beta$. As $\beta$ approaches unity, a larger number of stages is necessary for a given degree of separation and the capital and operating costs increase correspondingly. When $\beta = 1$ any separation is impossible.

## 13.4. CALCULATION OF THE NUMBER OF THEORETICAL STAGES

### 13.4.1. Co-current contact with partially miscible solvents

In calculating the number of ideal stages required for a given degree of separation, the conditions of equilibrium expressed by one of the methods discussed in Section 13.3 is used. The number of stages where single or multiple contact equipment is involved is considered first, and then the design of equipment where the concentration change is continuous is discussed.

For the general case where the solvents are partially miscible, the feed solution F is brought into contact with the selective solvent **S**, to give raffinate $R_1$ and an extract $E_1$. The addition of streams F and **S** is shown on the triangular diagram in Figure 13.9, by the point M, where FM/MS $= S/F$. This mixture M breaks down to give extract $E_1$ and raffinate $R_1$, at opposite ends of a tie line through M.

If a second stage is used, then the raffinate $R_1$ is treated with a further quantity of solvent **S**, and extract $E_2$ and raffinate $R_2$ are obtained as shown in the same figure.

The complete process consists in carrying out the extraction, and recovering the solvent from the raffinate and extract obtained. Thus, for a single-stage system as shown in Figure 13.10, the raffinate R is passed into the distillation column where it is separated to give purified raffinate R′ and solvent $S_R$. The extract E is passed to another distillation unit to give extract E′ and a solvent stream $S_E$. These recovered solvents $S_R$ and $S_E$ are pumped back to the extraction process as shown. This cycle may be represented on a diagram, as shown in Figure 13.11, by showing the removal of $S_R$ from R to give composition R′, and the removal of $S_E$ from E to give composition E′. It has been assumed in this case that perfect separation is obtained in the stills, so that pure solvent

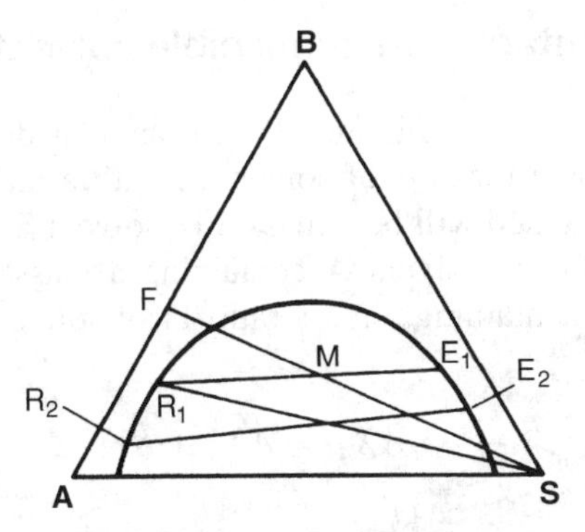


Figure 13.9. Multiple contact with fresh solvent used at each stage

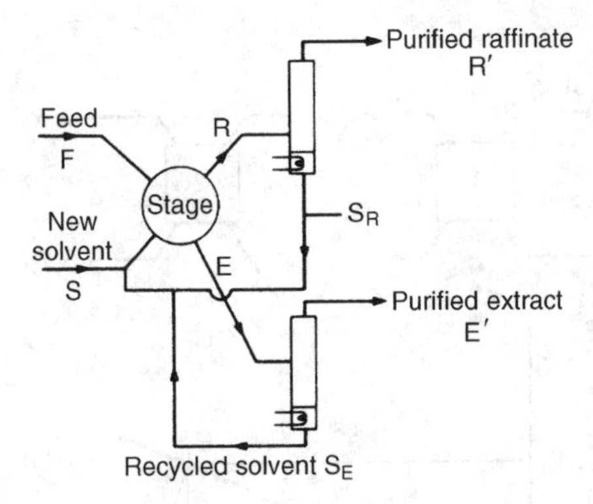


Figure 13.10. Single-stage process with solvent recovery

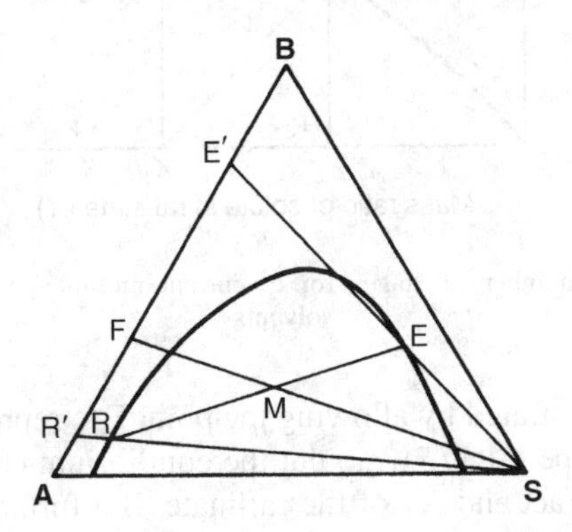


Figure 13.11. Representation of process shown in Figure 13.10

is obtained in the streams $S_R$ and $S_E$, although the same form of diagram can be used where imperfect separation is obtained. It may be noted that, when ES is a tangent to the binodal curve, then the maximum concentration of solute **B** in the extract E′ is obtained. It also follows that E′ then represents the maximum possible concentration of **B** in the feed. Sufficient solvent **S** must be used to bring the mixture M within the two-phase area.

### 13.4.2. Co-current contact with immiscible solvents

In this case, which is illustrated in Figure 13.12, triangular diagrams are not required. If the initial solution contains a mass $A$ of solvent **A** with a mass ratio $X_f$ of solute, then the selective solvent to be added will be a mass $S$ of solvent **S**. On mixing and separating, a raffinate is obtained with the solvent **A** containing a mass ratio $X_1$ of solute, and an extract with the solvent **S** containing a mass ratio $Y_1$ of solute. A material balance on the solute gives:

$$AX_f = AX_1 + SY_1$$

or:

$$\frac{Y_1}{X_1 - X_f} = -\frac{A}{S} \tag{13.3}$$

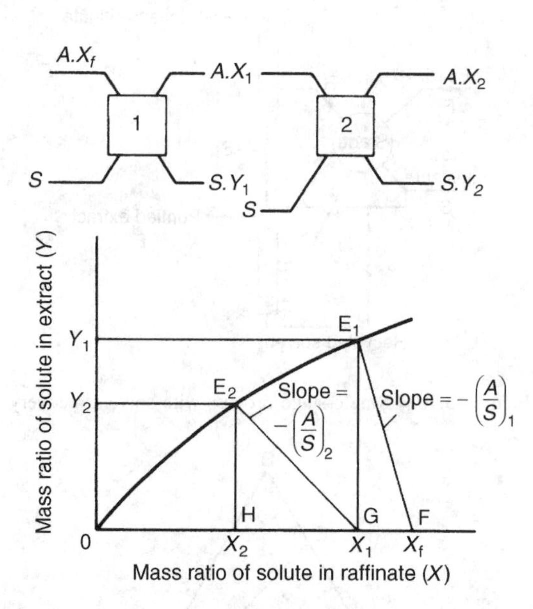


Figure 13.12. Calculation of number of stages for co-current multiple-contact process, using immiscible solvents

This process may be illustrated by allowing the point F to represent the feed solution and drawing a line $FE_1$, of slope $-(A/S)_1$, to cut the equilibrium curve at $E_1$. This then gives composition $Y_1$ of the extract and $X_1$ of the raffinate. If a further stage is then carried out by the addition of solvent **S** to the stream $AX_1$, then point $E_2$ is found on the equilibrium curve by drawing $GE_2$ of slope $-(A/S)_2$. Point $E_2$ then gives the compositions $X_2$ and $Y_2$ of the final extract and raffinate. This system may be used for any number of stages, with any assumed variation in the proportion of solvent **S** to raffinate from stage to stage.

*If the distribution law is followed*, then the equilibrium curve becomes a straight line given by $Y = mX$. The material balance on the solute may then be rewritten as:

$$AX_f = AX_1 + SY_1 = AX_1 + SmX_1 = (A + Sm)X_1$$

or:
$$X_1 = \left[\frac{A}{A + Sm}\right] X_f. \tag{13.4}$$

If a further mass $S$ of **S** is added to raffinate $AX_1$ to give an extract of composition $Y_2$ and a raffinate $X_2$ in a second stage, then:

$$AX_1 = AX_2 + SmX_2 = X_2(A + Sm)$$

and:
$$X_2 = \left[\frac{A}{A + Sm}\right] X_1 = \left[\frac{A}{A + Sm}\right]^2 X_f \tag{13.5}$$

For $n$ stages:

$$X_n = \left[\frac{A}{A + Sm}\right]^n X_f \tag{13.6}$$

and the number of stages is given by:

$$n = \frac{\log X_n/X_f}{\log\left[\dfrac{A}{A + Sm}\right]} \tag{13.7}$$

### 13.4.3. Countercurrent contact with immiscible solvents

If a series of mixing and separating vessels is arranged so that the flow is countercurrent, then the conditions of flow may be represented as shown in Figure 13.13, where each circle corresponds to a mixer and a separator. The initial solution F of the solute **B** in solvent **A** is fed to the first unit and leaves as raffinate $R_1$. This stream passes through the units and leaves from the $n$th unit as stream $R_n$. The fresh solvent **S** enters the $n$th unit and passes in the reverse direction through the units, leaving as extract $E_1$.

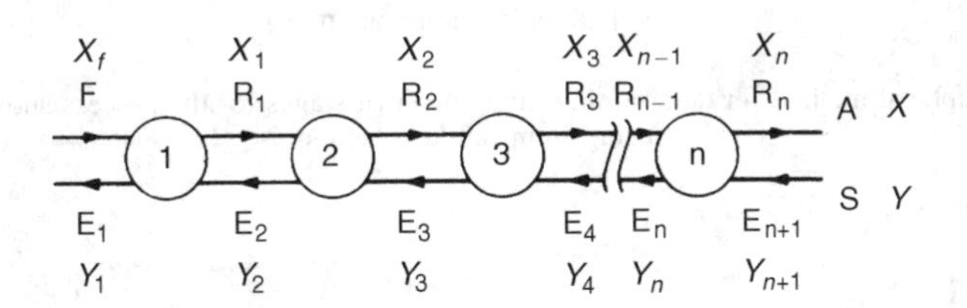


Figure 13.13. Arrangement for multiple-contact extraction in countercurrent flow

The following definitions may be made:

$X$ = the ratio of solute to solvent in the raffinate streams, and

$Y$ = the ratio of the solute to solvent in the extract streams.

If the two solvents are immiscible, the solvent in the raffinate streams remains as $A$, and the added solvent in the extract streams as $S$. The material balances for the solute may then be written as

(a) For the 1st stage: $AX_f + SY_2 = AX_1 + SY_1$

(b) For the $n$th stage: $AX_{n-1} + SY_{n+1} = AX_n + SY_n$

(c) For the whole unit: $AX_f + SY_{n+1} = AX_n + SY_1$

or:
$$Y_{n+1} = \frac{A}{S}(X_n - X_f) + Y_1 \tag{13.8}$$

This is the equation of a straight line of slope $A/S$, known as the *operating line*, which passes through the points $(X_f, Y_1)$ and $(X_n, Y_{n+1})$. In Figure 13.14, the equilibrium relation, $Y_n$ against $X_n$, and the operating line are drawn in, and the number of stages required to pass from $X_f$ to $X_n$ is found by drawing in steps between the operating line and the equilibrium curve. In this example, four stages are required, and $(X_n, Y_{n+1})$ corresponds to $(X_4, Y_5)$. It may be noted that the operating line connects the compositions of the raffinate stream leaving and the fresh solvent stream entering a unit, $X_n$ and $Y_{n+1}$, respectively.

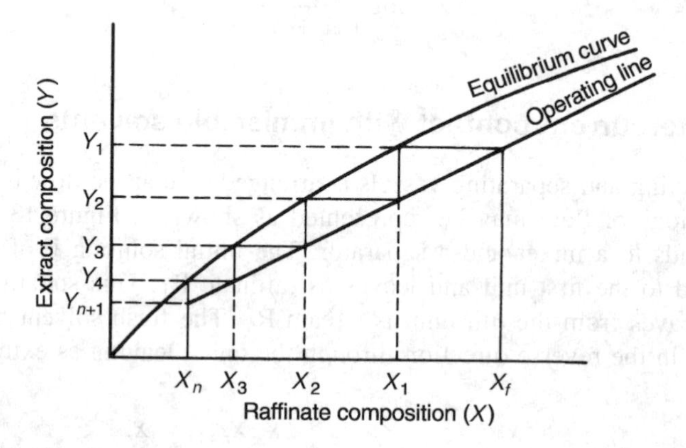


Figure 13.14. Graphical method for determining the number of stages for the process shown in Figure 13.13, using immiscible solvents

## Example 13.1

160 cm$^3$/s of a solvent **S** is used to treat 400cm$^3$/s of a 10 per cent by mass solution of **A** in **B**, in a three-stage countercurrent multiple-contact liquid–liquid extraction plant. What is the composition of the final raffinate?

Using the same total amount of solvent, evenly distributed between the three stages, what would be the composition of the final raffinate if the equipment were used in a simple multiple-contact arrangement?

Equilibrium data:

| | | | |
|---|---|---|---|
| kg **A**/kg **B**: | 0.05 | 0.10 | 0.15 |
| kg **A**/kg **S**: | 0.069 | 0.159 | 0.258 |
| Densities (kg/m$^3$): | $\rho_A = 1200$, | $\rho_B = 1000$, | $\rho_S = 800$ |

## Solution

(a) *Countercurrent operation*

Considering the solvent **S**, $160\text{cm}^3/\text{s} = 1.6 \times 10^{-4}\text{m}^3/\text{s}$

and: mass flowrate $= (1.6 \times 10^{-4} \times 800) = 0.128$ kg/s

Considering the solution, $400\text{cm}^3/\text{s} = 4 \times 10^{-4}\ \text{m}^3/\text{s}$
containing, say, $a$ m$^3$/s **A** and $(5 \times 10^{-4} - a)$ m$^3$ /s **B**.

Thus: mass flowrate of **A** $= 1200a$ kg/s

and: mass flowrate of **B** $= (4 \times 10^{-4} - a)1000 = (0.4 - 1000a)$ kg/s

a total of: $(0.4 + 200a)$ kg/s

The concentration of the solution is:

$$0.10 = 1200a/(0.4 + 200a)$$

Thus: $$a = 3.39 \times 10^{-5}\text{m}^3/\text{s}$$

mass flowrate of **A** $= 0.041$ kg/s, mass flowrate of **B** $= 0.366$ kg/s

and: ratio of **A**/**B** in the feed, $X_f = (0.041/0.366) = 0.112$ kg/kg

The equilibrium data are plotted in Figure 13.15 and the value of $X_f = 0.112$ kg/kg is marked in. The slope of the equilibrium line is:

$$(\text{mass flowrate of } \mathbf{B})/(\text{mass flowrate of } \mathbf{S}) = (0.366/0.128) = 2.86$$

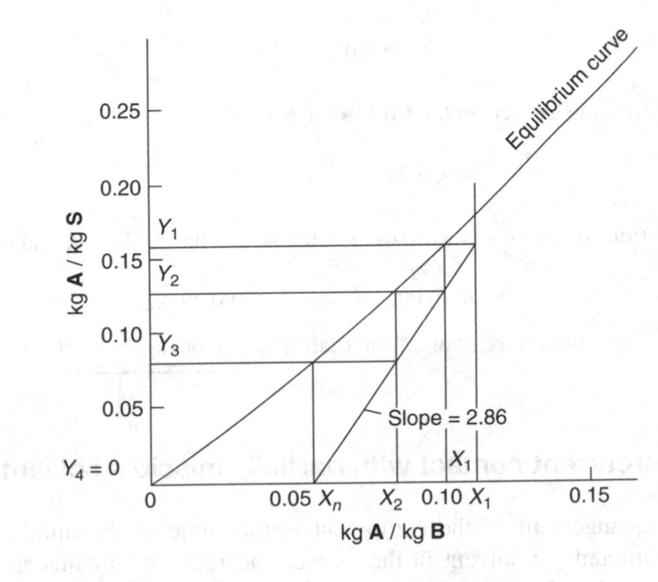


Figure 13.15. Construction for Example 13.1

Since pure solvent is added, $Y_{n+1} = Y_4 = 0$ and a line of slope 2.86 is drawn in such that stepping off from $X_f = 0.112$ kg/kg to $Y_4 = 0$ gives exactly three stages.

When $Y_4 = 0$, $X_n = X_3 = 0.057$ kg/kg,

Thus: the composition of final raffinate is $\underline{\underline{0.057 \text{ kg } \mathbf{A}/\text{kg } \mathbf{B}}}$

(b) *Multiple contact*

In this case, $(0.128/3) = 0.0427$ kg/s of pure solvent **S** is fed to each stage.

*Stage 1*

$$X_f = (0.041/0.366) = 0.112\text{kg/kg}$$

and from the equilibrium curve, the extract contains 0.18 **A**/kg **S** and $(0.18 \times 0.0427) = 0.0077$ kg/s **A**.

Thus: raffinate from stage 1 contains $(0.041 - 0.0077) = 0.0333$ kg/s **A** and 0.366 kg/s **B**

and:
$$X_1 = (0.0333/0.366) = 0.091\text{kg/kg}$$

*Stage 2*

$$X_1 = 0.091\text{kg/kg}$$

and from Figure 13.15 the extract contains 0.14 kg **A**/kg **S**

or:
$$(0.14 \times 0.0427) = 0.0060 \text{ kg/s } \mathbf{A}$$

Thus: the raffinate from stage 2 contains $(0.0333 - 0.0060) = 0.0273$ kg/s **A** and 0.366 kg/s **B**

Thus:
$$X_2 = (0.0273/0.366) = 0.075 \text{ kg/kg}$$

*Stage 3*

$$X_2 = 0.075 \text{ kg/kg}$$

and from Figure 13.15, the extract contains 0.114 kg **A**/kg **S**

or:
$$(0.114 \times 0.0427) = 0.0049 \text{ kg/s } \mathbf{A}.$$

Thus: the raffinate from stage 3 contains $(0.0273 - 0.0049) = 0.0224$ kg/s **A** and 0.366 kg/s **B**

and:
$$X_3 = (0.0224/0.366) = 0.061 \text{ kg/kg}$$

Thus: the composition of final raffinate $= \underline{\underline{0.061\text{kg } \mathbf{A}/\text{kg } \mathbf{B}}}$

## 13.4.4. Countercurrent contact with partially miscible solvents

In this case the arrangement of the equipment is the same as for immiscible solvents although, as the amounts of solvent in the extract and raffinate streams are varying, the material balance is taken for the total streams entering and leaving each stage.

With the notation as shown in Figure 13.13, if the feed F, the final extract $E_1$, the fresh solvent **S** = stream $E_{n+1}$ and, the final raffinate $R_n$ are fixed, then making material balances:

(a) *Over the first unit*

$$F + E_2 = R_1 + E_1$$

and: $$F - E_1 = R_1 - E_2 = P, \text{ say—the difference stream} \tag{13.9}$$

(b) *Over stages* 1 *to n*

$$F + E_{n+1} = R_n + E_1 = M, \text{ say} \tag{13.10}$$

and: $$F - E_1 = R_n - E_{n+1} = P. \tag{13.11}$$

(c) *Over the unit n*

$$R_{n-1} + E_{n+1} = E_n + R_n$$

and: $$R_{n-1} - E_n = R_n - E_{n+1} = P \tag{13.12}$$

Thus the difference in quantity between the raffinate leaving a stage $R_n$, and the extract entering from next stage $E_{n+1}$, is constant. Similarly, it can be shown that the difference between the amounts of each component in the raffinate and the extract streams is constant. This means that, with the notation of a triangular diagram, lines joining any two points representing $R_n$ and $E_{n+1}$ pass through a common pole. The number of stages required to so from an initial concentration F to a final raffinate concentration $R_n$ may then be found using a triangular diagram, shown in Figure 13.16.

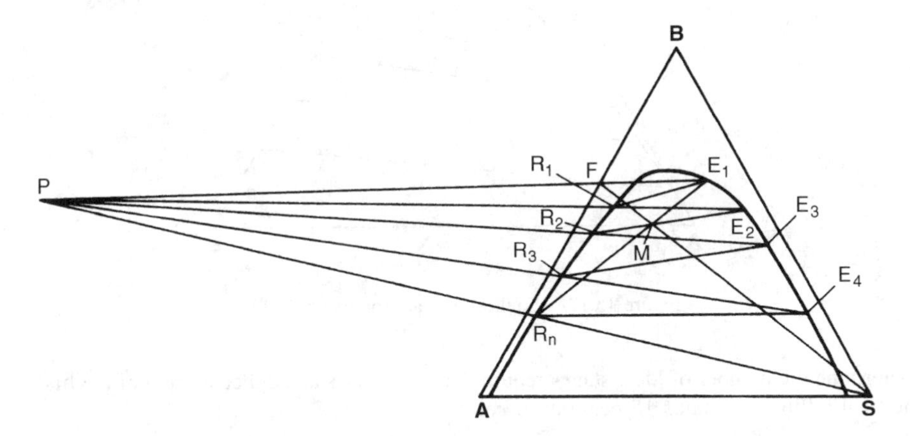


Figure 13.16. Graphical method for determining the number of stages for the process shown in Figure 13.13, using partially miscible solvents

If the points F and S representing the compositions of the feed and fresh solvent **S** are joined, then the composition of a mixture of F and **S** is shown by point M where:

$$\frac{\text{MS}}{\text{MF}} = \frac{\text{mass of F}}{\text{mass of } \mathbf{S}}$$

A line is drawn from $R_n$ through M to give $E_1$ on the binodal curve and $E_1F$ and $SR_n$ to meet at the pole P. It may be noted that P represents an imaginary mixture, as described for the leaching problems discussed in Chapter 10.

In an ideal stage, the extract $E_1$ leaves in equilibrium with the raffinate $R_1$, so that the point $R_1$ is at the end of the tie line through $E_1$. To determine the extract $E_2$, $PR_1$ is drawn to cut the binodal curve at $E_2$. The points $R_2$, $E_3$, $R_3$, $E_4$, and so on, may be found in the same way. If the final tie line, say $ER_4$, does not pass through $R_n$, then the amount of solvent added is incorrect for the desired change in composition. In general, this does not invalidate the method, since it gives the required number of ideal stages with sufficient accuracy.

## Example 13.2

A 50 per cent solution of solute **C** in solvent **A** is extracted with a second solvent **B** in a countercurrent multiple contact extraction unit. The mass of **B** is 25 per cent of that of the feed solution, and the equilibrium data are as given in Figure 13.17.

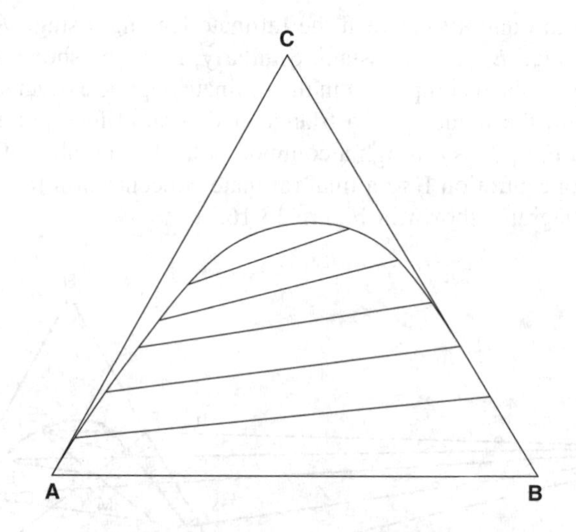


Figure 13.17. Equlibrium data for Example 13.2.

Determine the number of ideal stages required and the mass and concentration of the first extract if the final raffinate contains 15 per cent of solute **C.**

## Solution

The equilibrium data are replotted in Figure 13.18 and F, representing the feed, is drawn in on AC at $C = 0.50$, $A = 0.50$. FB is joined and M located such that FM/MB $= 0.25$. $R_n$ is located on the equilibrium curve such that $C = 0.15$. In fact $B = 0.01$ and $A = 0.84$. $E_1$ is located by projecting $R_nM$ on to the curve and the pole P by projecting $E_1F$ and $BR_n$. $R_1$ is found by projecting from

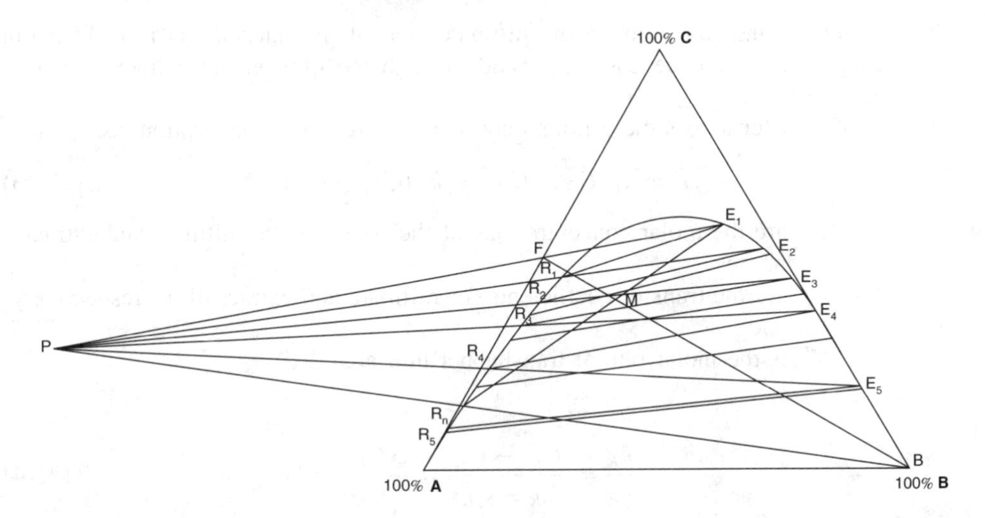


Figure 13.18. Graphical construction for Example 13.2.

$E_1$ along a tie-line and $E_2$ as the projection of $PR_1$. The working is continued in this way and it is found that $R_5$ is below $R_n$ and hence 5 ideal stages are required.

From Figure 13.18 the concentration of extract $E_1$ is 9 per cent **A,** 58 per cent **C,** and 33 per cent **B.**

## 13.4.5. Continuous extraction in columns

As Sherwood and Pigford[3] point out, the use of spray towers, packed towers or mechanical columns enables continuous countercurrent extraction to be obtained in a similar manner to that in gas absorption or distillation. Applying the two-film theory of mass transfer, explained in detail in Volume 1, Chapter 10, the concentration gradients for transfer to a desired solute from a raffinate to an extract phase are as shown in Figure 13.19, which is similar to Figure 12.1 for gas absorption.

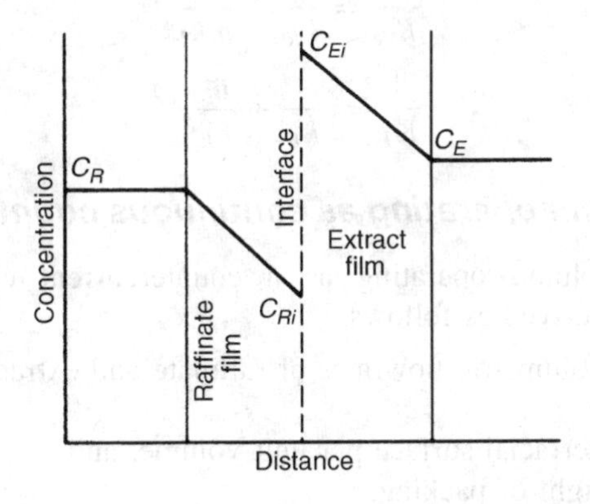


Figure 13.19. Concentration profile near an interface

The transfer through the film on the raffinate side of the interface is brought about by a concentration difference $C_R - C_{Ri}$, and through the film on the extract side by a concentration difference $C_{Ei} - C_E$.

The rate of transfer across these films may be expressed, as in absorption, as:

$$N' = k_R(C_R - C_{Ri}) = k_E(C_{Ei} - C_E) \tag{13.13}$$

where: $C_R$, $C_E$ are the molar concentrations of the solute in the raffinate and extract phases.
$k_R$, $k_E$ are the transfer coefficients for raffinate and extract films, respectively, and
$N'$ is the molar rate of transfer per unit area.

Then:

$$\frac{k_R}{k_E} = \frac{C_{Ei} - C_E}{C_R - C_{Ri}} = \frac{\Delta C_E}{\Delta C_R} \tag{13.14}$$

If the equilibrium curve may be taken as a straight line of slope $m$, then assuming equilibrium at the interface gives:

$$C_{Ei} = mC_{Ri} \tag{13.15}$$

$$C_E = mC_R^* \tag{13.16}$$

and:

$$C_E^* = mC_R \tag{13.17}$$

where $C_E^*$ is the concentration in phase E in equilibrium with $C_R$ in phase R, and $C_R^*$ is the concentration in phase R in equilibrium with $C_E$ in phase E.

The relations for mass transfer may also be written in terms of overall transfer coefficients $K_R$ and $K_E$ defined by:

$$N' = K_R(C_R - C_R^*) = K_E(C_E^* - C_E) \tag{13.18}$$

and by a similar reasoning to that used for absorption, as discussed in Chapter 12 (equation 12.22):

$$\frac{1}{K_R} = \frac{1}{k_R} + \frac{1}{mk_E} \tag{13.19}$$

and:

$$\frac{1}{K_E} = \frac{1}{k_E} + \frac{m}{k_R} \tag{13.20}$$

### Capacity of a column operating as continuous countercurrent unit

The capacity of a column operating as a countercurrent extractor, as shown in Figure 13.20, may be derived as follows.

If $L'_R$, $L'_E$ are the volumetric flowrates of raffinate and extract phases per unit area,
$a$ is the interfacial surface per unit volume, and
$Z$ is the height of packing,

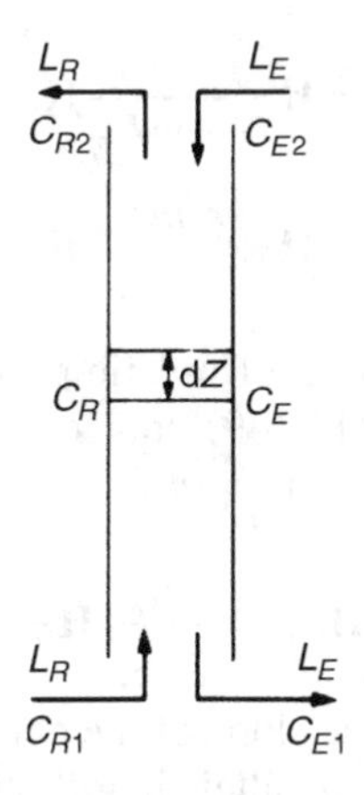


Figure 13.20. Countercurrent flow in a packed column

then, over a small height $\mathrm{d}Z$, a material balance gives:

$$L'_R\,\mathrm{d}C_R = L'_E\,\mathrm{d}C_E \tag{13.21}$$

From equation 13.13:

$$L'_R\,\mathrm{d}C_R = k_R(C_R - C_{Ri})a\,\mathrm{d}Z$$

and:

$$\int_{C_{R2}}^{C_{R1}} \frac{\mathrm{d}C_R}{C_R - C_{Ri}} = \frac{k_R a}{L'_R} Z \tag{13.22}$$

The integral on the left-hand side of this equation is known as the number of raffinate-film transfer units, $\mathbf{N}_R$, and the height of the raffinate-film transfer unit is:

$$\mathbf{H}_R = \frac{L'_R}{k_R a} \tag{13.23}$$

In a similar manner, and by analogy with absorption (equations 12.80, 12.81, 12.82):

$$\mathbf{H}_E = \frac{L'_E}{k_E a} = \text{height of extract-film transfer unit} \tag{13.24}$$

$$\mathbf{H}_{OR} = \frac{L'_R}{K_R a} = \text{height of overall transfer unit based on concentration in raffinate phase} \tag{13.25}$$

$$\mathbf{H}_{OE} = \frac{L'_E}{K_E a} = \text{height of overall transfer unit based on concentration in extract phase} \tag{13.26}$$

Since:

$$\frac{1}{K_R} = \frac{1}{k_R} + \frac{1}{mk_E} \qquad \text{(equation 13.18)}$$

$$\frac{L'_R}{K_R} = \frac{L'_R}{k_R} + \left(\frac{L'_R}{mk_E} \times \frac{L'_E}{L'_E}\right)$$

Thus:
$$\mathbf{H}_{OR} = \mathbf{H}_R + \frac{L'_R}{mL'_E}\mathbf{H}_E \tag{13.27}$$

and:
$$\mathbf{H}_{OE} = \mathbf{H}_E + \frac{mL'_E}{L'_R}\mathbf{H}_R \tag{13.28}$$

These equations are the same form of relation as already obtained for distillation (Chapter 11, equations 11.148 and 11.149) and for absorption (Chapter 12), although it is only with dilute solutions that the group $mL'_E/L'_R$ is constant. If equations 13.27 and 13.28 are combined, then:

$$\mathbf{H}_{OR} = \frac{L'_R}{mL'_E}\mathbf{H}_{OE} \tag{13.29}$$

The group $L'_R/mL'_E$ is the ratio of the slope of the operating line to that of the equilibrium curve so that, when these two are parallel, it follows that $\mathbf{H}_{OR}$ and $\mathbf{H}_{OE}$ are numerically equal.

In deriving these relationships, it is assumed that $L'_R$ and $L'_E$ are constant throughout the tower. This is not the case if a large part of the solute is transferred from a concentrated solution to the other phase.

It is also assumed that the transfer coefficients are independent of concentration. For dilute solutions and where the equilibrium relation is a straight line, a simple expression may be obtained for determining the required height of a column, by the same method as given in Chapter 12.

Thus $L'_R\,\mathrm{d}C_R = K_R(C_R - C_R^*)a\,\mathrm{d}Z$ may be integrated over the height $Z$ and expressed as:

$$L'_R(C_{R1} - C_{R2}) = K_R(\Delta C_R)_{\mathrm{lm}}aZ \tag{13.30}$$

where $(\Delta C_R)_{\mathrm{lm}}$ is the logarithmic mean of $(C_R - C_R^*)_1$ and $(C_R - C_R^*)_2$. This simple relation has been used by workers in the determination of $K_R$ or $K_E$ in small laboratory columns, although care should be taken when applying these results to other conditions.

Equations 13.27 and 13.28 have been used as a basis of correlating mass transfer measurements in continuous countercurrent contactors. For example, LEIBSON and BECKMANN[4] plotted $\mathbf{H}_{OE}$ against $mL'_E/L'_R$, as shown in Figure 13.21, for a variety of column packings and obtained good straight lines. Caution must, be exercised, however, in drawing conclusions from such plots. Although equation 13.28 suggests that the intercepts and slopes of the lines are numerically equal to $\mathbf{H}_E$ and $\mathbf{H}_R$ respectively, this is true only provided that both these quantities are independent of the flow ratio $L'_E/L'_R$. This is not always the case and, with packed towers, the height of the continuous phase film transfer unit does in fact depend upon the flow ratio, as discussed by GAYLER and PRATT[5]. Under these conditions neither equation 13.27 nor 13.28 can be used to apportion the individual resistances to mass transfer between the two phases, and the film coefficients have to be determined by direct measurement.

## Example 13.3

In the extraction of acetic acid from an aqueous solution with benzene in a packed column of height 1.4 m and of cross-sectional area 0.0045 m$^2$, the concentrations measured at the inlet and outlet of the column are as shown in Figure 13.22.

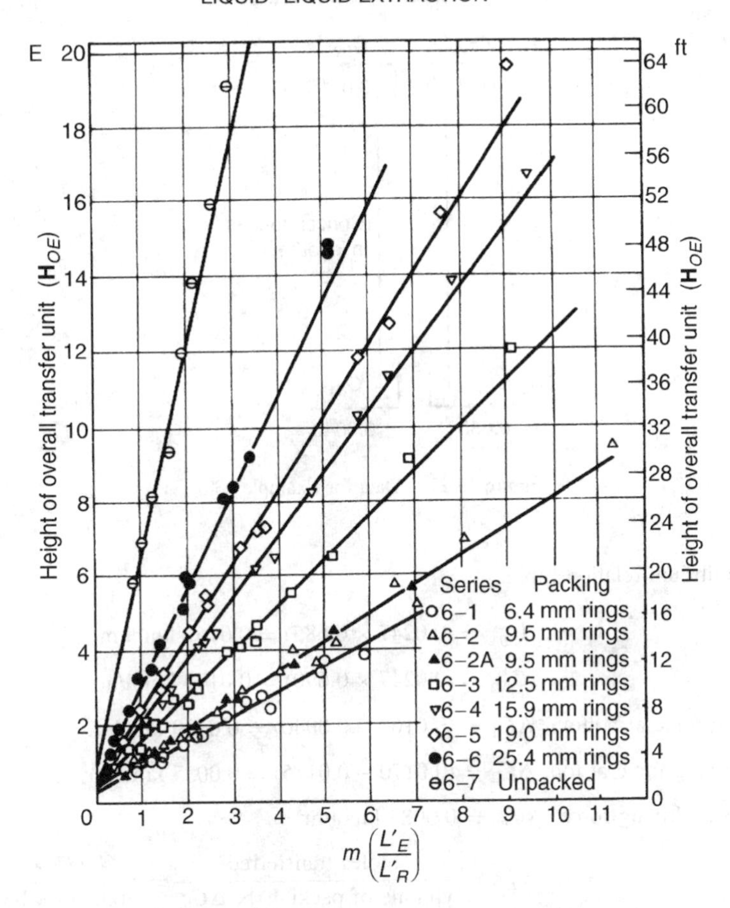


Figure 13.21. Height of the transfer unit $\mathbf{H}_{OE}$ as a function of $mL_E/L_R$ for the transfer of diethylamine from water to dispersed toluene using various packings[4]

Determine the overall transfer coefficient and the height of the transfer unit.

| | |
|---|---|
| Acid concentration in inlet water phase, $C_{W2}$ | $= 0.690\ \text{kmol/m}^3$ |
| Acid concentration in outlet water phase, $C_{W1}$ | $= 0.685\ \text{kmol/m}^3$ |
| Flowrate of benzene phase | $= 5.7\ \text{cm}^3\text{s or } 1.27 \times 10^{-3}\ \text{m}^3/\text{m}^2\text{s}$ |
| Inlet benzene phase concentration, $C_{B1}$ | $= 0.0040\ \text{kmol/m}^3$ |
| Outlet benzene phase concentration, $C_{B2}$ | $= 0.0115\ \text{kmol/m}^3$ |
| The equilibrium relationship for this system is: $\dfrac{C_B^*}{C_W^*}$ | $= 0.0247$ |

## Solution

The acid transferred to the benzene phase is:

$$5.7 \times 10^{-6}(0.0115 - 0.0040) = 4.275 \times 10^{-8}\ \text{kmol/s}$$

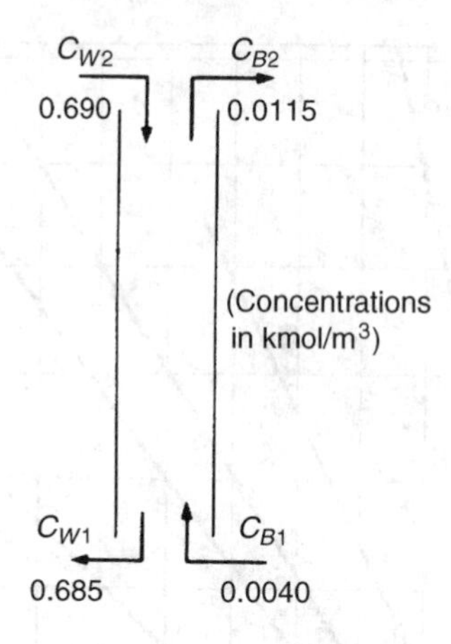


Figure 13.22. Data for Example 13.3

From the equilibrium relationship:

$$\mathbf{C}^*_{B1} = (0.0247 \times 0.685) = 0.0169 \text{ kmol/m}^3$$

and: $$C^*_{B2} = (0.0247 \times 0.690) = 0.0170 \text{ kmol/m}^3$$

Thus: Driving force at bottom, $\Delta C_1 = (0.0169 - 0.0040) = 0.0129 \text{ kmol/m}^3$

and: Driving force at top, $\Delta C_2 = (0.0170 - 0.0115) = 0.0055 \text{ kmol/m}^3$

Thus: Log mean driving force, $\Delta C_{\text{lm}} = 0.0087 \text{ kmol/m}^3$

Thus: $$K_B a = \frac{\text{moles transferred}}{\text{volume of packing } \times \Delta C_{\text{lm}}} = \frac{(4.275 \times 10^{-8})}{(0.0063 \times 0.0087)}$$

$$= 7.8 \times 10^{-4} \text{ kmol/s m}^3 (\text{kmol/m}^3)$$

and: $$\mathbf{H}_{OB} = (1.27 \times 10^{-3})/(7.8 \times 10^{-4}) = \underline{\underline{1.63 \text{ m}}}$$

## 13.5. CLASSIFICATION OF EXTRACTION EQUIPMENT

In most industrial applications, multistage countercurrent contacting is required. The hydrodynamic driving force necessary to induce countercurrent flow and subsequent phase separation may be derived from the differential effects of either gravity or centrifugal force on the two phases of different densities. Essentially there are two types of design by which effective multistage operation may be obtained:

(a) *stage-wise contactors*, in which the equipment includes a series of physical stages in which the phases are mixed and separated, and
(b) *differential contactors*, in which the phases are continuously brought into contact with complete phase separation only at the exits from the unit.

The three factors, the inducement of countercurrent flow, stage-wise or differential contacting and the means of effecting phase separation are the basis of a classification of contactors proposed by HANSON[1] which is summarised in Table 13.1.

Table 13.1. Classification of contactors

| Countercurrent flow produced by | Phase interdispersion by | Differential contactors | Stagewise contactors |
|---|---|---|---|
| Gravity | Gravity | GROUP A<br>Spray column<br>Packed column | GROUP B<br>Perforated plate column |
| | Pulsation | GROUP C<br>Pulsed packed column<br>Pulsating plate column | GROUP D<br>Pulsed sieve plate column<br>Controlled cycling column |
| | Mechanical agitation | GROUP E<br>Rotating disc contactor<br>Oldshue-Rushton column<br>Zeihl column<br>Graesser contractor | GROUP F<br>Scheibel column<br>Mixer-settlers |
| Centrifugal force | Centrifugal force | GROUP G<br>Podbielniak<br>Quadronic<br>De Laval | GROUP H<br>Westfalia<br>Robatel |

Typical regions for application of contactors of different types are given in Table 13.2. The choice of a contactor for a particular application requires the consideration of several factors including chemical stability, the value of the products and the rate of phase separation. Occasionally, the extraction system may be chemically unstable and the contact time must then be kept to a minimum by using equipment such as a centrifugal contactor.

Table 13.2. Typical regions of application of contactor groups listed in Table 13.1[1]

| | System criterion | Modest throughput | High throughput |
|---|---|---|---|
| Small number of stages required | Chemically stable<br>Easy phase separation<br>Low value | A, B or mixer–settler | E or F |
| | Chemically stable<br>Appreciable value | C or D | E or F (not mixer–settler) |
| | Chemically unstable<br>Slow phase separation | G or H | G or H |
| Large number of stages required | Chemically stable<br>Easy phase separation<br>Low value | B, C, D or mixer–settler | E or F |
| | Chemically stable<br>Appreciable value | C or D | E or F (not mixer–settler) |
| | Chemically unstable<br>Slow phase separation | G or H | G or H |

The more important types of stage-wise and differential contactors are discussed in Sections 13.6 and 13.7 respectively.

## 13.6. STAGE-WISE EQUIPMENT FOR EXTRACTION

### 13.6.1. The mixer–settler

In the mixer-settler, the solution and solvent are mixed by some form of agitator in the *mixer*, and then transferred to the *settler* where the two phases separate to give an extract and a raffinate. The mixer unit, which is usually a circular or square vessel with a stirrer, may be designed on the principles given in Volume 1, Chapter 7. In the settler the separation is often gravity-controlled, and the liquid densities and the form of the dispersion are important parameters. It is necessary to establish the principles which determine the size of these units and to have an understanding of the criteria governing their internal construction. Whilst the mixer and settler are first considered as separate items, it is important to appreciate that they are essential component parts of a single processing unit.

*The mixer.* As a result of the agitation achieved in a mixer, the two phases are brought to, or near to, equilibrium so that one theoretical stage is frequently obtained in a single mixer where a physical extraction process is taking place. Where a chemical reaction occurs, the kinetics must be established so that the residence time and the hold-up may be calculated. The hold-up is the key parameter in determining size, and scale-up is acceptably reliable, although a reasonably accurate estimate of the power required is important with large units. For a circular vessel, baffles are required to give the optimum degree of agitation and the propeller, which should be about one-third of the diameter of the vessel, should be mounted just below the interface and operate with a tip speed of 3–15 m/s, depending on the nature of the propeller or turbine. A shroud around the propeller helps to give good initial mixing of the streams, and it also provides some pumping action and hence improves circulation.

As discussed in Volume 1, Chapter 7, the two key parameters determining power are the Reynolds number for the agitator and the power number. The Reynolds number should exceed $10^4$ for optimum agitation. This gives a power number of about 6 for a fully baffled tank. It is important to note that the power number and the tip speed cannot both be kept constant in scale-up.

*The settler.* In this unit, gravitational settling frequently occurs and, in addition, coalescence of droplets must take place. Baffles are fitted at the inlet in order to aid distribution. The rates of sedimentation and coalescence increase with drop size, and therefore excessive agitation resulting in the formation of very small drops should be avoided. The height of the dispersion band $Z_B$ is influenced by the throughput since a minimum residence time is required for coalescence to occur. This height $Z_B$ is related to the dispersed and continuous phase superficial velocities, $u_d$ and $u_c$ by:

$$Z_B = \text{constant } (u_d + u_c) \tag{13.31}$$

Pilot tests may be necessary to achieve satisfactory design, although the sizing of the settler is a difficult problem when the throughputs are large.

#### *Combined mixer–settler units*

Recent work has emphasised the need to consider the combined mixer–settler operation, particularly in metal extraction systems where the throughput may be very large. Thus

WARWICK and SCUFFHAM[6] give details of a design, shown in Figure 13.23, in which the two operations are effected in the one combined unit. The impeller has swept-back vanes with double shrouds, and the two phases meet in the draught tube. A baffle on top of the agitator reduces air intake and a baffle on the inlet to the settler is important in controlling the flow pattern. This arrangement gives a good performance and is mechanically neat. Raising the impeller above the draught tube increases internal recirculation which in turn improves the stage efficiency, as shown in Figure 13.24. The effect of agitation on the thickness of the dispersion band is shown in Figure 13.25. The depth of the dispersion

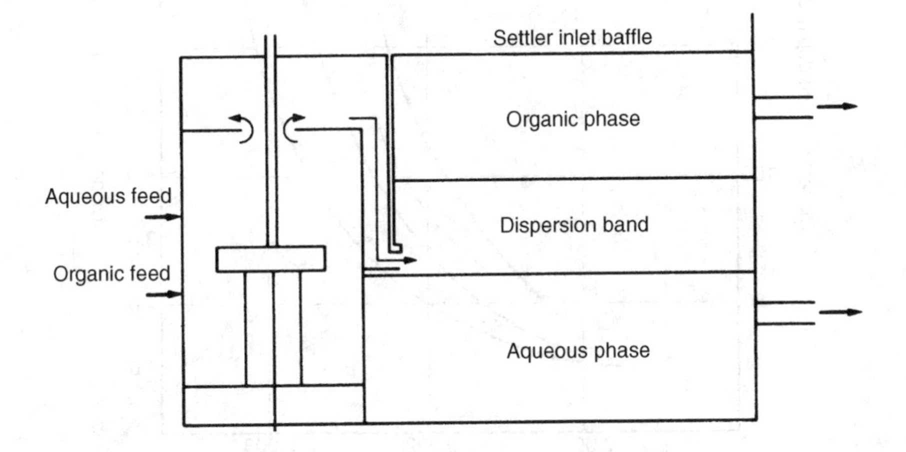


Figure 13.23. Mixer–settler

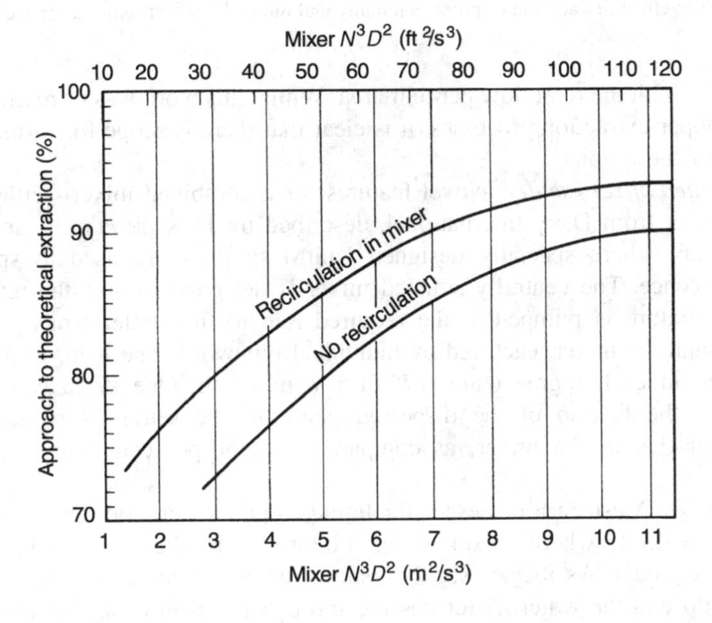


Figure 13.24. The effect of variation of mixer internal recirculation on extraction efficiency[6]

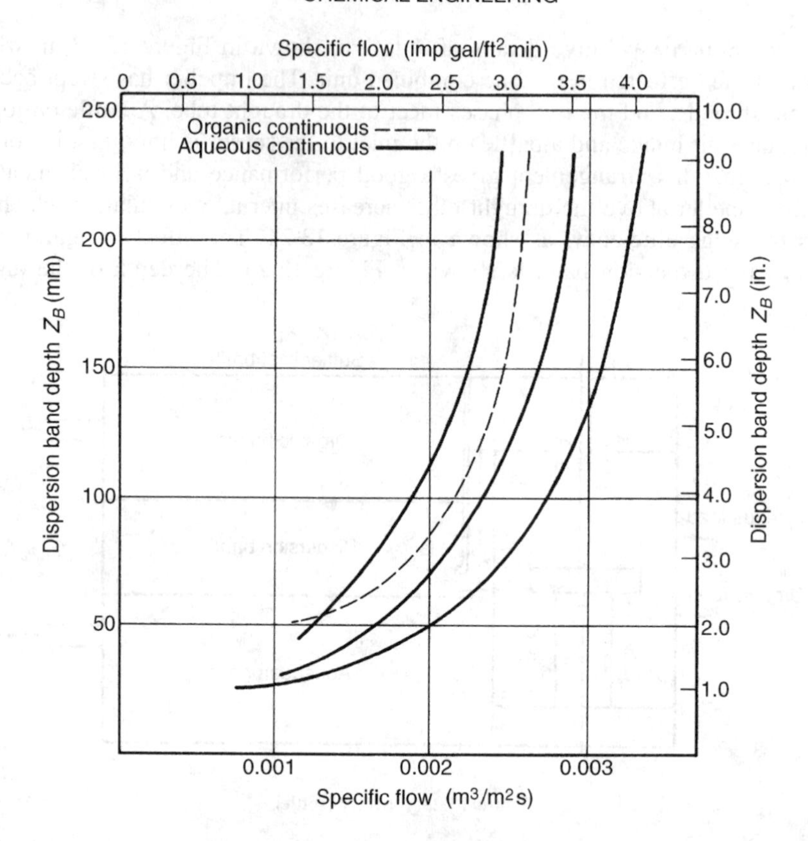


Figure 13.25. The effect of variation of phase continuity and mixer $N^3 D^2$ on settler dispersion band depth[6]

band $Z_B$ varies with the total flow per unit area. Whilst this work was primarily aimed at a design for copper-extraction processes, it is clear that there is scope for further important applications of these units.

*The segmented mixer–settler*. Novel features for a combined mixer–settler are incorporated in a unit from Davy International, described by Jackson *et al.*[7] and illustrated in Figure 13.26, where specially designed KnitMesh pads are used to speed up the rate of coalescence. The centrally situated mixer is designed to give the required hold-up, and the mixture is pumped at the required rate to the settler which is formed in segments around the mixer, each fed by individual pipework. The KnitMesh pads which are positioned in each segment are 0.75–1.5 m in depth. One key advantage of this design is that the holdup of the dispersed phase in the settler is reduced to about 20 per cent of that in the mixer, as compared with 50 per cent with simple gravity settlers.

The use of KnitMesh in a coalescer for liquid–liquid separation applications is illustrated in Figure 13.27 where an oil–water mixture enters the unit and passes through the coalescer element. As it does so, the water droplets coalesce and separation occurs between the oil and the water. After passing through the KnitMesh, the two phases are readily removed from the top and bottom of the unit.

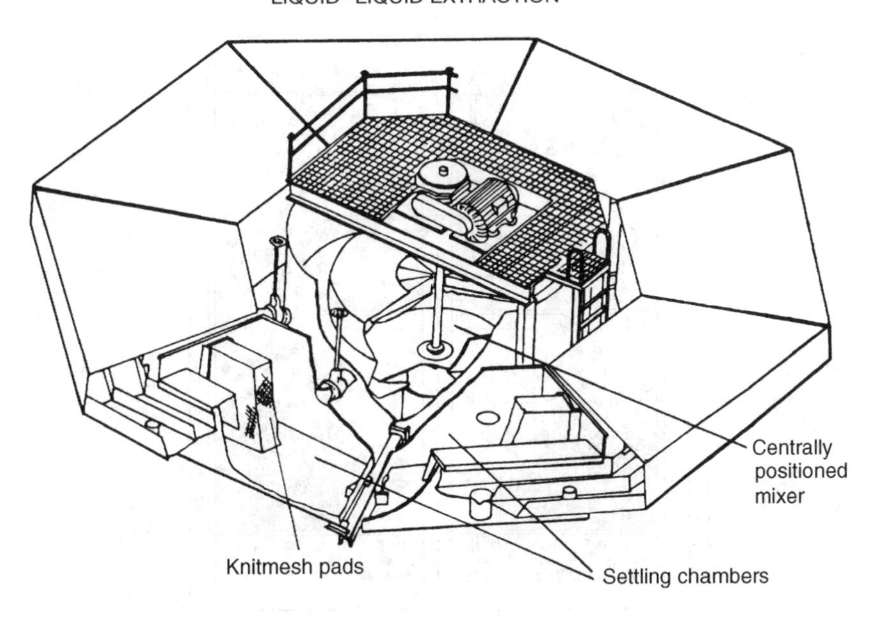


Figure 13.26. Segmented mixer–settler[7]

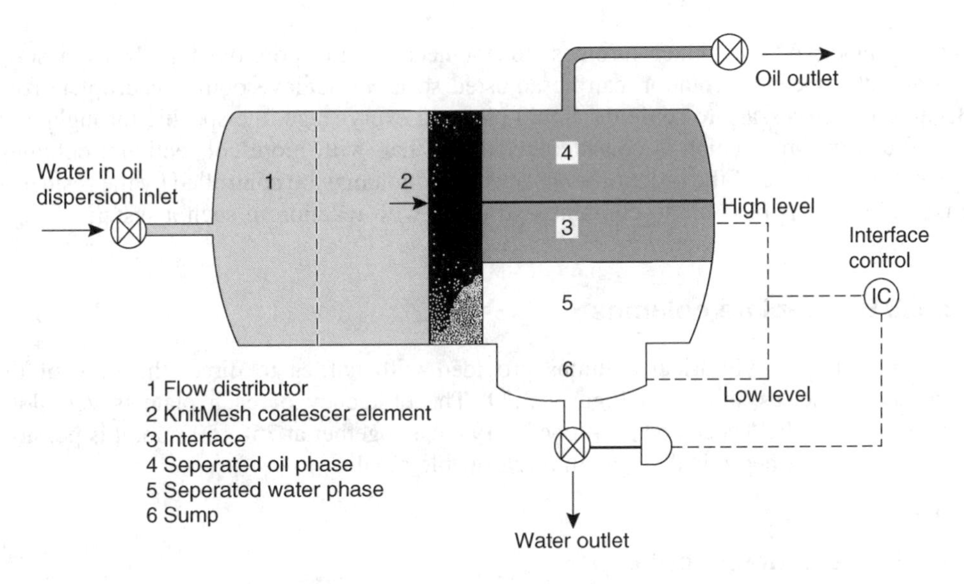


Figure 13.27. Flow in a KnitMesh separator

Kühni have recently developed a mixer-settler column which, as its name suggests is a series of mixer–settlers in the form of a column. The unit consists of a number of stages installed one on top of another, each hydraulically separated, and each with a mixing and settling zone as shown in Figure 13.28. With this design, it is possible to eliminate some of the main disadvantages of conventional mixer–settlers whilst maintaining stagewise

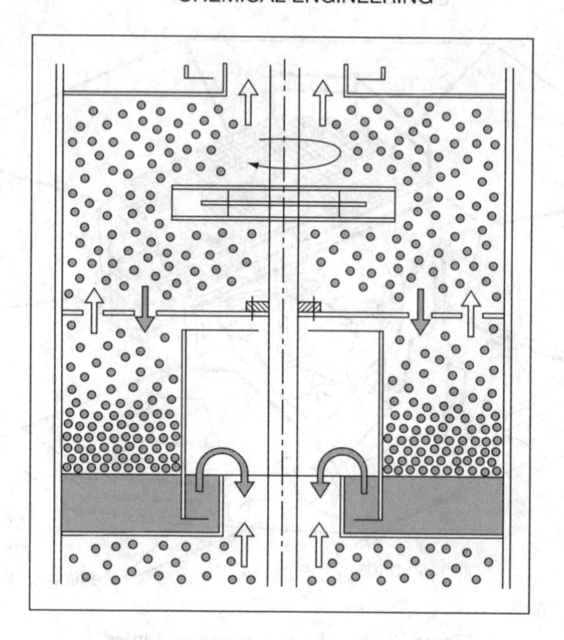

Figure 13.28. Kühni mixer–settler column.

phase contact. As the mixer turbines do not need to transport the liquids from stage to stage, the speed of rotation can be adjusted so as to achieve optimum droplet sizes. Because it is necessary to settle the liquid phase in every stage, the specific throughput is only 0.003 $m^3/m^2s$ which is considerably lower than with more conventional columns. Residence times are of the order of 900 s and extraction may be controlled by the residence time and pH. It is possible to combine extraction with reaction in such a system.

## 13.6.2. Baffle-plate columns

These are simple cylindrical columns provided with baffles to direct the flow of the dispersed phase, as shown in Figure 13.29. The efficiency of each plate is very low, though since the baffles can be positioned very close together at 75–150 mm, it is possible to obtain several theoretical stages in a reasonable height.

## 13.6.3. The Scheibel column

One of the problems with perforated plate and indeed packed columns is that redispersion of the liquids after each stage is very poor. To overcome this, Scheibel and Karr[8] introduced a unit, shown in Figure 13.30, in which a series of agitators is mounted on a central rotating shaft. Between the agitators is fitted a wire mesh section which successfully breaks up any emulsions. Some results for a column 292 mm diameter, with 100 mm diameter agitators and with packing sections 230 and 340 mm, are shown in Figure 13.31. It is found that one theoretical stage is obtained in a height of 0.45–0.75 m. This is a

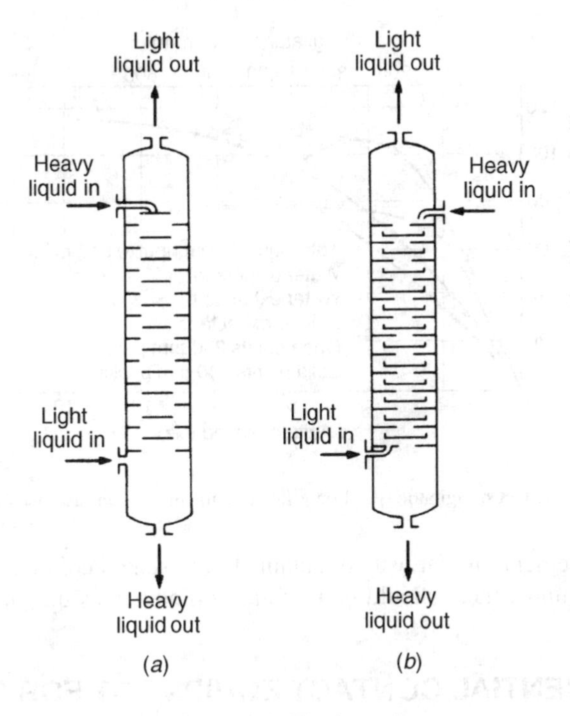


Figure 13.29. Baffle-plate column

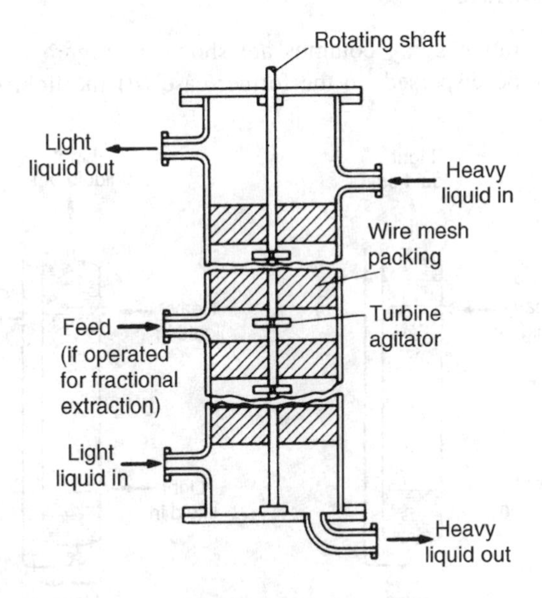


Figure 13.30. Scheibel column[8]

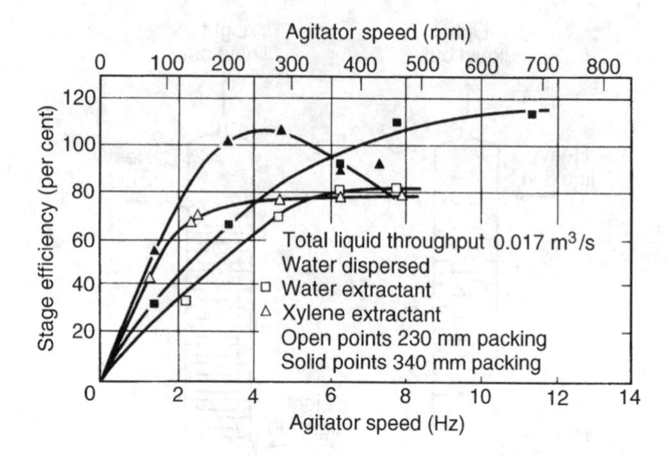


Figure 13.31. Effect of agitator speed on efficiency for the system acetone–xylene–ester[8]

significant improvement on that usually obtained in a packed column. Although there are few data on large units, there should be no fall in efficiency as the diameter is increased.

# 13.7. DIFFERENTIAL CONTACT EQUIPMENT FOR EXTRACTION

## 13.7.1. Spray columns

Two methods of operating spray columns are shown in Figure 13.32. Either the light or heavy phase may be dispersed. In the former case (*a*) the light phase enters from a

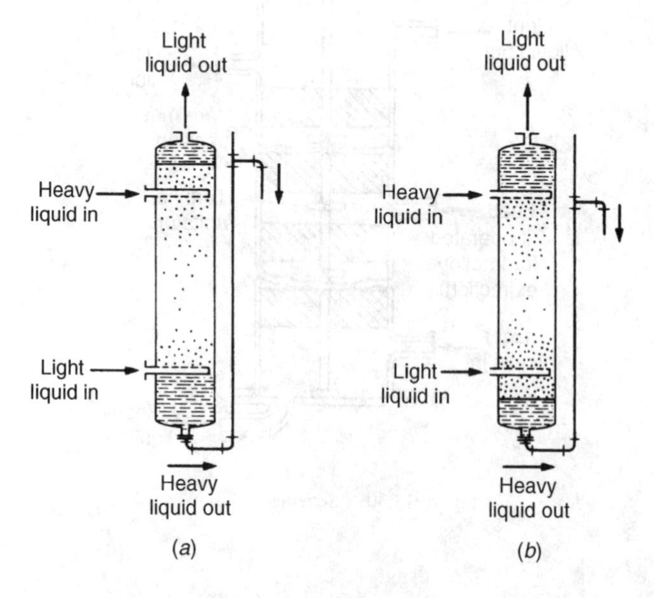


Figure 13.32. Spray towers

distributor at the bottom of the column and the droplets rise through the heavier phase, finally coalescing to form a liquid–liquid interface at the top of the tower. Alternatively, the heavier phase may be dispersed, in which case the interface is held at the bottom of the tower as shown in (*b*). Although spray towers are simple in construction, they are inefficient because considerable recirculation of the continuous phase takes place. As a result, true countercurrent flow is not maintained and up to 6 m may be required for the height of one theoretical stage. There is very little turbulence in the continuous phase and lack of interface renewal, and appreciable axial mixing results in poor performance.

Because the droplets of dispersed phase rise or fall through the continuous phase under the influence of gravity, it will be apparent that there is a limit to the amount of dispersed phase that can pass through the tower for any given flowrate of continuous phase. Thus referring to Figure 13.32(*a*), any additional light phase fed to the bottom of the tower, in excess of that which can pass upwards under the influence of gravity, will be rejected from the bottom of the unit and the tower is then said to be flooded, as discussed by BLANDING and ELGIN[9]. It is therefore important to be able to predict the conditions under which flooding will occur, so that the diameter of tower may be calculated for any required throughput. Although no complete analysis of this problem has, as yet, been achieved, it may be treated approximately in the following way.

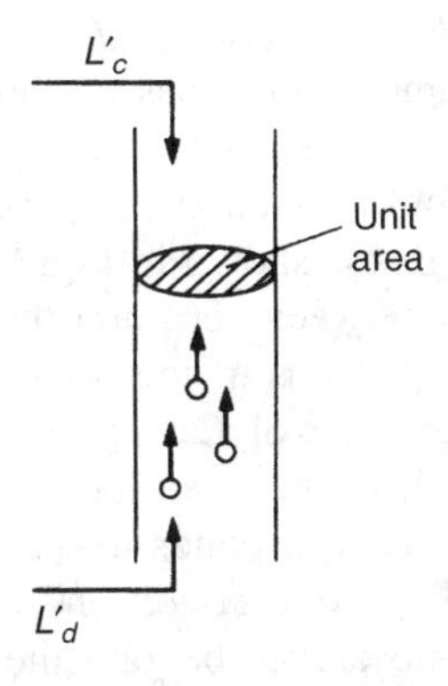


Figure 13.33. Section of a spray tower

*Dispersed phase hold-up.* Figure 13.33 represents a section of a spray tower of unit cross-sectional area. The light phase is assumed to be dispersed, and the volumetric flowrates per unit area of the two phases are $L'_d$ and $L'_c$ respectively. The superficial velocities $u_d, u_c$ of the phases are therefore also equal to $L'_d$ and $L'_c$. Under steady-state conditions the amount of dispersed phase held up in the tower in the form of droplets is conveniently expressed in terms of the fractional hold-up $j$, that is the fractional volume of the two-phase dispersion occupied by the dispersed phase. This may also be thought of as the fraction of the cross-sectional area of the tower occupied by the dispersed phase. The velocity of the dispersed phase relative to the tower is therefore $L'_d/j$. Similarly, the relative velocity of the continuous phase is equal to $L'_c/(1 - j)$. If the overall flow is regarded as strictly countercurrent, the sum of these two velocities will be equal to the velocity of the dispersed phase relative to the continuous phase, $u_r$ or:

$$\frac{L'_d}{j} + \frac{L'_c}{(1-j)} = u_r \tag{13.32}$$

In the case of spray towers it has been shown by THORNTON[10] that $u_r$ is well represented by $\bar{u}_0(1-j)$ where $\bar{u}_0$ is the velocity of a single droplet relative to the continuous phase, and is termed the droplet characteristic velocity. The term $(1-j)$ is a correction to $\bar{u}_0$ which takes into account the way in which the characteristic velocity is modified when there is a finite population of droplets present, as opposed to a single droplet. It must be seen therefore that for very dilute dispersions, that is as $j \to 0$, $\bar{u}_0(1-j) \to \bar{u}_0$. On the other hand, as the fractional hold-up increases, the relative velocity of the dispersed phase decreases due to interactions between the droplets. Substituting for $u_r$, equation 13.32 may be written as:

$$\frac{u_d}{j} + \frac{u_c}{1-j} = \bar{u}_0(1-j) \tag{13.33}$$

This equation relates the hold-up to the flowrates of the phases and column diameter through the characteristic velocity, $\bar{u}_0$. It therefore gives a method of calculating the hold-up for a given set of flowrates if $\bar{u}_0$ is known. Conversely, equation 13.33 may be used to calculate $\bar{u}_0$ from experimental hold-up measurements made at different flowrates. Thus, if hold-up data are plotted with $L'_d + (j/(1-j))L'_c$ as the ordinate against $j(1-j)$ as the abscissa, a linear plot is obtained which passes through the origin and which has a slope equal to $\bar{u}_0$[10].

*Flooding-point condition.* A plot of equation 13.33 in the form of $u_d$ and $u_c$ against $j$ for a typical value of $\bar{u}_0 = 0.042$ m/s is shown in Figure 13.34. Although equation 13.33 is cubic, only the root which lies between zero and the flooding-point values of $u_d$ and $u_c$ is realisable in practice. These portions of the hold-up curves are shown by full lines in Figure 13.34. If the flowrate of one of the phases is kept constant, an increase in the flowrate of the other phase will result in an increased value of the hold-up until the flooding-point is reached. The latter corresponds to the maxima in the two curves shown in Figure 13.34, so that the flooding-point condition is given by $\mathrm{d}u_d/\mathrm{d}j = \mathrm{d}u_c/\mathrm{d}j = 0$. Since those portions of the hold-up curves beyond the flooding-point are unrealistic in practice, the value of $j$ corresponding to the maximum flowrates also represents the limiting hold-up at flooding, although this condition is not obtainable mathematically from equation 13.33.

Carrying out the differentiation described previously gives:

$$u_{df} = 2\bar{u}_0 j_f^2(1-j_f) \quad \text{when } \mathrm{d}u_c/\mathrm{d}j = 0 \tag{13.34}$$

and:

$$u_{cf} = \bar{u}_0(1-j_f)^2(1-2j_f) \text{ when } \mathrm{d}u_d/\mathrm{d}j = 0 \tag{13.35}$$

The value of $j_f$, the limiting hold-up at the flooding-point, may be obtained by eliminating $\bar{u}_0$ between equations 13.34 and 13.35 and solving for $j_f$.

Thus:

$$j_f = \frac{(r^2+8r)^{0.5} - 3r}{4(1-r)} \tag{13.36}$$

where:

$$r = u_{df}/u_{cf}. \tag{13.37}$$

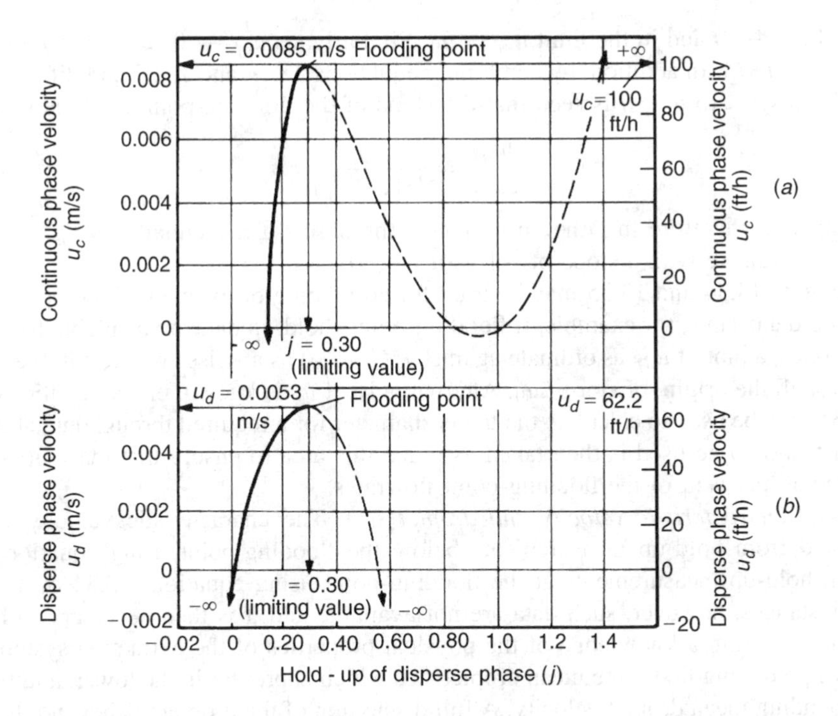


Figure 13.34. Spray tower ($a$) Continuous phase velocity ($u_c$) as function of hold-up of disperse phase ($j$)($u_c = 0.0085$ m/s, $\bar{u}_0 = 0.042$ m/s) ($b$) Disperse phase velocity ($u_d$) as a function of hold-up of disperse phase ($j$) ($u_d = 0.0053$ m/s, $\bar{u}_0 = 0.0042$ m/s)

The derivation of equations 13.34 and 13.35 has been carried out assuming that $\bar{u}_0$ is constant and independent of the flowrates, up to and including the flooding-point. This in turn assumes that the droplet size is constant and that no coalescence occurs as the hold-up increases. Whilst this assumption is essentially valid in properly designed spray towers, this is certainly not the case with packed towers. Equations 13.34 and 13.35 cannot therefore be used to predict the flooding-point in packed towers and a more empirical procedure must be adopted.

A typical form of flooding-point curve for spray towers is shown in Figure 13.35 where values of $u_{df}$ are plotted against $u_{cf}$. The limiting values of each flowrate as the other approaches zero may be determined readily from equations 13.34 and 13.35. Thus, when

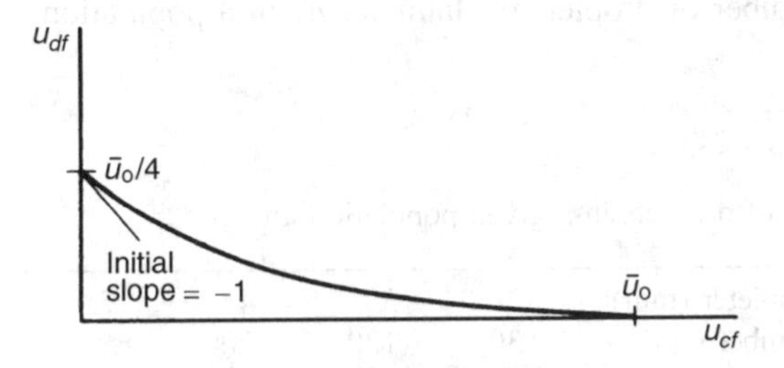


Figure 13.35. Flooding-point curve for spray tower

$u_{df} \to 0$, $j_f \to 0$ and in the limit $u_{cf} = \bar{u}_0$. Similarly, as $u_{cf} \to 0$, $j_f \to 0.50$ and in the limit $u_{df} = \bar{u}_0/4$. In addition, by combining equations 13.34 and 13.35 and differentiating $u_{df}$ with respect to $u_{cf}$, it is seen that the slope of the flooding-point curve is given by:

$$\frac{du_{df}}{du_{cf}} = -\frac{j_f}{(1 - j_f)} \tag{13.38}$$

It is apparent therefore that the curve meets the abscissa tangentially as $u_{df} \to 0$, and meets the ordinate with a slope of $-1$ as $u_{cf} \to 0$.

Equations 13.34 and 13.35 may be used for correlating or extending incomplete flooding-point data. Thus, for example, if flooding-point–hold-up data are available for a range of flowrates, a plot of $u_{df}$ as ordinate against $j_f^2(1 - j_f)$ as abscissa will result in a straight line through the origin, of slope $2\bar{u}_0$. Alternatively, if $\bar{u}_0$ is known, either equations 13.34 or 13.35 may be used to calculate the tower diameter for a required throughput. The actual area of tower to be used is then taken as twice this area to ensure that the unit operates at below 50 per cent of the flooding-point flowrates.

*Droplet characteristic velocity and droplet size.* The characteristic velocity may be calculated from hold-up measurements below the flooding-point using equation 13.33, or from hold-up measurements at the flooding-point using equations 13.34 or 13.35. In many instances, however, such data are not available and it is then necessary to be able to predict $\bar{u}_0$ from a knowledge of the physical properties of the extraction system. This involves predicting firstly the mean droplet size which is present in the tower, and then the corresponding mean droplet velocity. A full discussion of the problem is beyond the scope of this chapter although for low nozzle velocities, the mean droplet size may be established by means of the correlation proposed by HAYWORTH and TREYBAL(11). The corresponding droplet velocity may then be calculated by the methods of HU and KINTNER(12) or KLEE and TREYBAL(13).

*Interfacial area.* The interfacial area per unit volume of tower, or specific area, is given by:

$$a = \frac{\text{Total area}}{\text{Volume}} = \frac{\pi d_s^2}{\pi \dfrac{d_s^3}{6}\dfrac{1}{j}} = \frac{6j}{d_s} \tag{13.39}$$

where $j$ is the fractional hold-up and $d_s$ is the mean droplet size, defined in Chapter 1, equation 1.14, as:

$$d_s = \frac{\Sigma n_1 d_1^3}{\Sigma n_1 d_1^2} \tag{13.40}$$

where $n_1$ is the number of droplets of diameter $d_1$ in a population.

## Example 13.4

The number and size of droplets in a given population are:

| diameter (mm) | 2 | 3 | 4 | 5 | 6 |
|---|---|---|---|---|---|
| number (–) | 30 | 120 | 200 | 80 | 20 |

Estimate the surface mean droplet size.

## Solution

The mean droplet size is given by equations 13.40 and 1.14 as:

$$d_s = \Sigma n_1 d_1^3 / \Sigma n_1 d_1^2$$

where $n_1$ is the number of droplets of diameter $d_1$ in a population.
Thus:

Table 13.3. Calculation of mean droplet Diameter

| $n_1$ | $d_1$ mm | $d_1^2$ | $d_1^3$ | $n_1 d_1^3$ | $n_1 d_1^2$ |
|---|---|---|---|---|---|
| 30 | 2 | 4 | 8 | 240 | 120 |
| 120 | 3 | 9 | 27 | 3240 | 1080 |
| 200 | 4 | 16 | 64 | 12800 | 3200 |
| 80 | 5 | 25 | 125 | 100000 | 2000 |
| 200 | 6 | 36 | 216 | 4320 | 720 |
| Total | | | | 30600 | 7120 |

Thus: $\Sigma n_1 d_1^3 = 30600$ and $\Sigma n_1 d_1^2 = 7120$

and: $d_s = (30600/7120) = \underline{\underline{4.30 \text{ mm}}}$

*Mass transfer*. It is not yet possible to predict the mass transfer coefficient with a high degree of accuracy because the mechanisms of solute transfer are but imperfectly understood as discussed LIGHT and CONWAY[14], COULSON and SKINNER[15] and GARNER and HALE[16]. In addition, the flow in spray towers is not strictly countercurrent due to recirculation of the continuous phase, and consequently the effective overall driving force for mass transfer is not the same as that for true countercurrent flow.

As a first approximation, the dispersed-phase film coefficients may be calculated using the HANDLOS and BARON[17] model for circulating liquid spheres. The continuous phase film coefficients may be estimated from the correlation of RUBY and ELGIN[18], and the overall coefficients then calculated using equations 13.20 or 13.21. Such procedures are, however, only approximate at best, and further work is required before generalised correlations can be developed which take into account the effect of recirculation and the different mass transfer rates at the dispersed phase entry nozzles, during droplet rise and during coalescence at the top of the tower. Typical data[18–21] and standard texts[22–24] for HTU values covering a range of extraction systems are available in the literature. In this respect, scale-up from pilot data raises problems. Reference may also be made to the work of HANSON[1] for further discussion on mass transfer in liquid–liquid systems.

Hitherto no mention has been made of interfacial effects accompanying the mass transfer process. Under certain conditions the presence of an undistributed solute gives rise to the Marangoni effect, discussed by DAVIES and RIDEAL[25] and GROOTHUIS and ZUIDERWEG[26], which results in interfacial turbulence and droplet coalescence. These effects are generally more obvious when solute transfer takes place from an organic solvent droplet into an aqueous continuous phase. In the reverse direction of transfer, interfacial disturbances of this nature are frequently absent. The existence of interfacial phenomena of this type presents yet one more obstacle to a quantitative interpretation of the mass transfer process. In addition, when coalescence is marked, the droplet size is no longer constant up the tower and the derivations of equations 13.34 and 13.35, which are based upon constant $\bar{u}_0$ values, are no longer valid. From a practical point of view, these equations may still be used for design purposes, since the effect of droplet coalescence is to enhance the flooding-point beyond the value predicted by equations 13.34 and 13.35, so that the tower operates in practice below 50 per cent of the flooding-point.

### 13.7.2. Packed columns

Although packed columns are similar to those used for distillation and absorption, it must be noted that the flowrates of the phases are very different and two liquid phases are always present. The packing increases the interfacial area, and considerably increases mass transfer rates compared with those obtained with spray columns because of the continuous coalescence and break-up of the drops, though the HTU values are still high. Packed columns are unsuitable for use with dirty liquids, suspensions, or high viscosity liquids. They have proved to be satisfactory in the petroleum industry, though at present they cannot be scaled-up to cope with the very high flows encountered in metallurgical processes. They are economical in the use of ground space.

*Minimum packing size*. For reproducible results the minimum packing size should be such that the mean void height is not less than the mean droplet diameter. As reported by GAYLER *et al.*[27], this critical packing size is given by:

$$d_{\text{crit}} = 2.42\left(\frac{\sigma}{\Delta\rho g}\right)^{0.5} \tag{13.41}$$

*Dispersed phase hold-up*. Three regimes of flow may be distinguished with packings greater than the critical size:

(a) Region of linear hold-up. At low dispersed phase flowrates the droplets move freely within the interstices of the packing and the hold-up increases linearly with dispersed phase flowrate.
(b) Region of rapidly increasing hold-up. Above a hold-up of approximately 10 per cent, corresponding to the lower transition point, the hold-up increases more rapidly with increasing dispersed phase flow. This is due to the onset of hindered movement of the droplets within the packing voids.
(c) Region of constant hold-up. At higher dispersed phase flowrates, the upper transition point is encountered above which droplet coalescence occurs. In this region the hold-up remains constant as the flowrate is increased until the flooding-point is reached. It is apparent therefore that a plot of $u_d$ against $j$ is of similar form to that shown in Figure 13.34, except that the point corresponding to the flooding point for spray towers coincides with the upper transition point for packed towers.

Below the upper transition point, hold-up data may be correlated by an equation analogous to that used for spray towers, except as GAYLER *et al.*[27] and GAYLER and PRATT[28] point out, a correction must be introduced to take the packing voidage $e$ into account. Thus:

$$\frac{u_d}{j} + \frac{u_c}{(1-j)} = e\bar{u}_0(1-j) \tag{13.42}$$

*Droplet size and interfacial area.* In the absence of interfacial effects accompanying mass transfer, the droplets break down by impact with elements of packing and finally reach an equilibrium size which is independent of the packing size. Conversely, small droplets gradually coalesce until the equilibrium size is attained. PRATT and his coworkers[5,29] showed that the mean droplet size attained in the tower is well represented by:

$$d_0 = 0.92\left(\frac{\sigma}{\Delta\rho g}\right)^{0.50}\left(\frac{\bar{u}_0 e j}{u_d}\right) \tag{13.43}$$

Correcting equation 13.39 for the packing voidage, the specific area in the tower is given by:

$$a = \frac{6ej}{d_0} \tag{13.44}$$

*Droplet characteristic velocity.* GAYLER, ROBERTS and PRATT[27] proposed a graphical correlation for $\bar{u}_0$ which takes into account the fact that the droplets are periodically halted by collisions with elements of packing and then accelerate to some fraction of their terminal velocity, $u_0$, before being deflected or stopped again by a further collision. Figure 13.36 enables $u_0$ to be determined from a knowledge of the physical properties of the system. This value is then used in conjunction with Figure 13.37 to evaluate $\bar{u}_0$ for any given column diameter $d_c$ and nominal packing size $d_p$.

Once $\bar{u}_0$ has been determined, the hold-up for any particular set of flowrates may be calculated from equation 13.42, the mean droplet size from equation 13.43, and the specific area from equation 13.44.

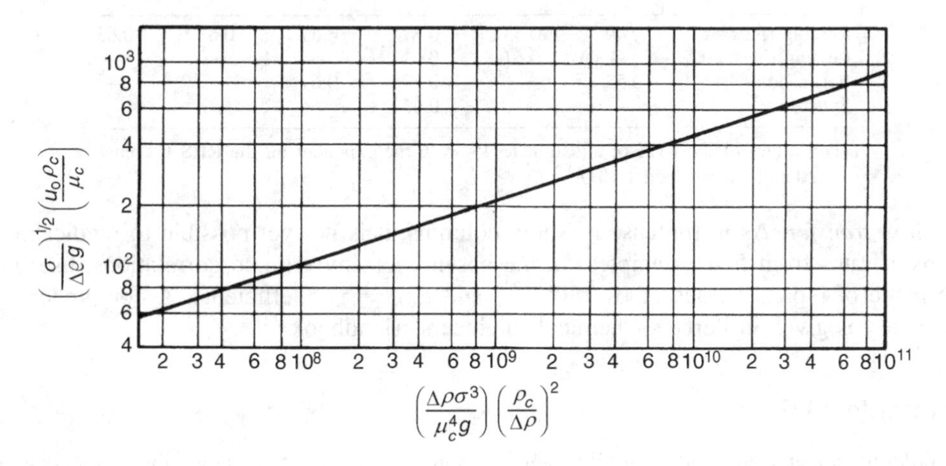


Figure 13.36. Chart for calculation of droplet terminal velocity, $u_0$[27]

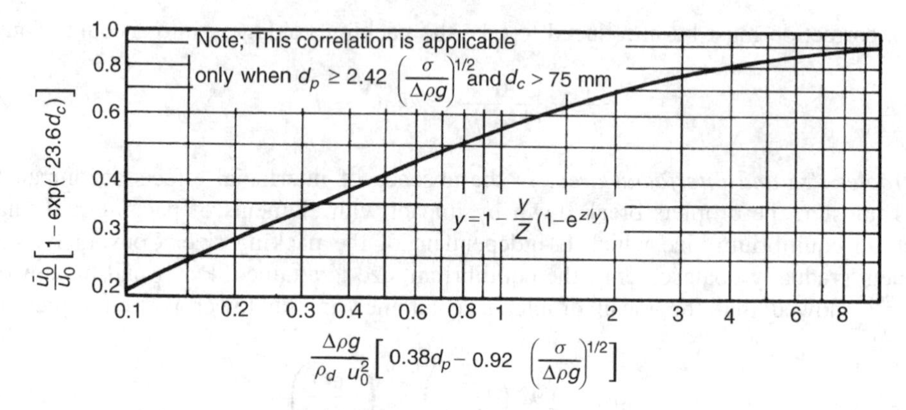


Figure 13.37. Chart for calculation of velocity ($\bar{u}_0$) of a single droplet relative to continuous phase ($d_c$ and $d_p$ are in m)[27]

*Flooding-point.* Because the flooding-point is no longer synonymous with that for spray towers, equations 13.34 and 13.35 predict only the upper transition point. Dell and Pratt[30] adopted a semi-empirical approach for the flooding-point by consideration of the forces acting on the separate dispersed and continuous phase channels which form when coalescence sets in just below the flooding-point. The following expression correlates data to within ±20 per cent:

$$1 + 0.835\left(\frac{\rho_d}{\rho_c}\right)^{1/4}\left(\frac{u_d}{u_c}\right)^{1/2} = J\left[\frac{u_c^2 a_p}{g e^3}\left(\frac{\rho_c}{\Delta\rho}\right)\left(\frac{\sigma}{\sigma_T}\right)^{1/4}\right]^{-1/4} \tag{13.45}$$

Values of the constant $J$ are given in Table 13.4.

Table 13.4. Values of constant $J$ for use in equation 13.45[30]

| Type of packing | Nominal packing size, $d_p$ (mm) | | | | | |
|---|---|---|---|---|---|---|
| | 6 | 9 | 12.5 | 19 | 25 | 57.5 |
| Raschig rings* | 0.49 | 0.53 | 0.57 | 0.61 | 0.63 | 0.63 |
| Lessing rings | — | 0.61 | 0.61 | — | — | — |
| Berl saddles | 0.52 | — | 0.67 | 0.67 | — | — |
| Spheres | — | — | 0.61 | — | — | — |

*For values of the LHS of equation 13.45 <1.6 the exponent on the RHS becomes −0.20 and the constant 1.10 $J^{0.8}$.

*Mass transfer.* As in the case of spray columns, it is not yet possible to predict mass transfer rates from first principles. In the absence of any reliable correlations, use may be made of typical values of overall[20,31] and film[32,33] coefficients. A comprehensive summary is given in Perry's Chemical Engineers' Handbook[22].

## Example 13.5

In order to extract acetic acid from dilute aqueous solution with isopropyl ether, the two immiscible phases are passed countercurrently through a packed column 3 m in length and 75 mm in diameter.

It is found that, if 0.5 $kg/m^2s$ of the pure ether is used to extract 0.25 $kg/m^2s$ of 4.0 per cent acid by mass, then the ether phase leaves the column with a concentration of 1.0 per cent acid by mass.

Calculate:

(a) the number of overall transfer units based on the raffinate phase, and
(b) the overall extraction coefficient based on the raffinate phase.

The equilibrium relationship is given by:

$$\text{(kg acid/kg isopropyl ether)} = 0.3 \text{ (kg acid/kg water)}.$$

## Solution

$$\text{Cross-sectional area of packing} = (\pi/4)0.075^2 = 0.0044 \text{ m}^2$$

and: $$\text{volume of packing} = (0.0044 \times 3) = 0.0133 \text{ m}^3.$$

The concentration of acid in the extract = 1.0 per cent or 0.01 kg/kg.

Thus: $$\text{mass of acid transferred to the ether} = 0.05(0.01 - 0) = 0.005 \text{ kg/m}^2\text{s}$$

or: $$(0.005 \times 0.0044) = 0.000022 \text{ kg/s}.$$

$$\text{Acid in the aqueous feed} = (0.25 \times 0.04) = 0.01 \text{ kg/m}^3 \text{ s}.$$

Thus: $$\text{acid in raffinate} = (0.01 - 0.005) = 0.005 \text{ kg/m}^2 \text{ s}$$

and: $$\text{concentration of acid in the raffinate} = (0.005/0.25) = 0.02 \text{ kg/kg or 2.0 per cent.}$$

*At the top of the column:*

$$C_{R_2} = 0.040 \text{ kg/kg and } C^*_{R_2} = (0.040 \times 0.3) = 0.012\text{kg/kg}$$

Thus: $$\Delta C_2 = (0.012 - 0.010) = 0.002 \text{ kg/kg}$$

*At the bottom of the column:*

$$C_{R_1} = 0.20 \text{ kg/kg and } C^*_{E_1} = (0.020 \times 0.3) = 0.006 \text{ kg/kg}$$

Thus: $$\Delta C_1 = (0.006 - 0) = 0.006 \text{ kg/kg}$$

and the logarithmic mean driving force is:

$$(\Delta C_R)_{\text{lm}} = (0.006 - 0.002)/\ln(0.006/0.002) = 0.0036 \text{ kg/kg}$$

$$K_R a = \text{mass transferred}/[\text{volume of packing} \times (\Delta C_R)_{\text{lm}}]$$

$$= 0.000022/(0.0133 \times 0.0036) = \underline{\underline{0.461 \text{ kg/m}^3\text{s(kg/kg)}}}$$

From equation 13.25:

the height of an overall transfer unit, $$\mathbf{H}_{OR} = L'_R/K_R a$$

$$= (0.25/0.461) = 0.54 \text{ m}$$

and: $$\text{the number of overall transfer units} = (3/0.54) = \underline{\underline{5.53}}$$

### 13.7.3. Rotary annular columns and rotary disc columns

With these columns as described by THORNTON and PRATT[34] and VERMIJS and KRAMERS[35], mechanical energy is provided to form the dispersed phase. The equipment is particularly suitable for installations where a moderate number of stages is required, and where the throughput is considerable. A well dispersed system is obtained with this arrangement. Figure 13.38 shows a rotary annular column.

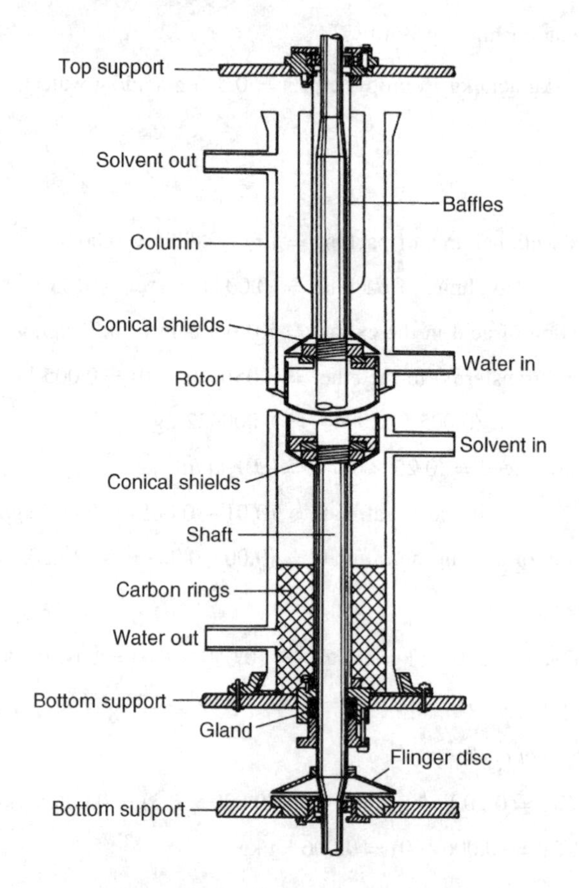


Figure 13.38. Rotary annular column

Flooding-point data may be correlated by equations 13.34 and 13.35 using the droplet characteristic velocity concept as discussed by THORNTON and PRATT[34], since coalescence is absent.

### 13.7.4. Pulsed columns

In order to prevent coalescence of the dispersed drops, VAN DIJCK[36] and others have devised methods of providing the whole of the continuous phase with a pulsed motion. This may be done, either by some mechanical device, or by the introduction of compressed air.

The pulsation markedly improves performance of packed columns and the HTU is about half that of an unpulsed column. There are advantages in using gauze-type packings since the pulsation operation often breaks ceramic rings. Perforated plates, as used in distillation, may also be used for pulsed extraction. Pulsed packed columns have been used in the nuclear industry though they are limited in size since the pulsation system is difficult to arrange and the pulsation itself demands strengthening of the column.

Flooding-point data have been correlated by equations analogous to 13.34 and 13.35. This procedure is permissible since, as THORNTON[37] and LOGSDAIL and THORNTON[38] report, pulsed columns can be operated up to the flooding-point with no droplet coalescence.

### 13.7.5. Centrifugal extractors

If separation is difficult in a mixer–settler unit, a centrifugal extractor may be used in which the mixing and the separation stages are contained in the same unit which operates as a differential contactor.

In the *Podbielniak* contactor, the first of the rotating machines to be developed, the heavy phase is driven outwards by centrifugal force and the light phase is displaced inwards. Referring to Figure 13.39 which illustrates a unit produced by Baker Perkins, the heavy phases enters at D, passes to J and is driven out at B. The light phase enters at A and is displaced inwards towards the shaft and leaves at C. The two liquids intermix in zone E where they are flowing countercurrently through the perforated concentric elements and are separated in the spaces between. In zones F and G the perforated elements are surfaces on which the small droplets of entrained liquid can coalesce, the large drops then being driven out by centrifugal force.

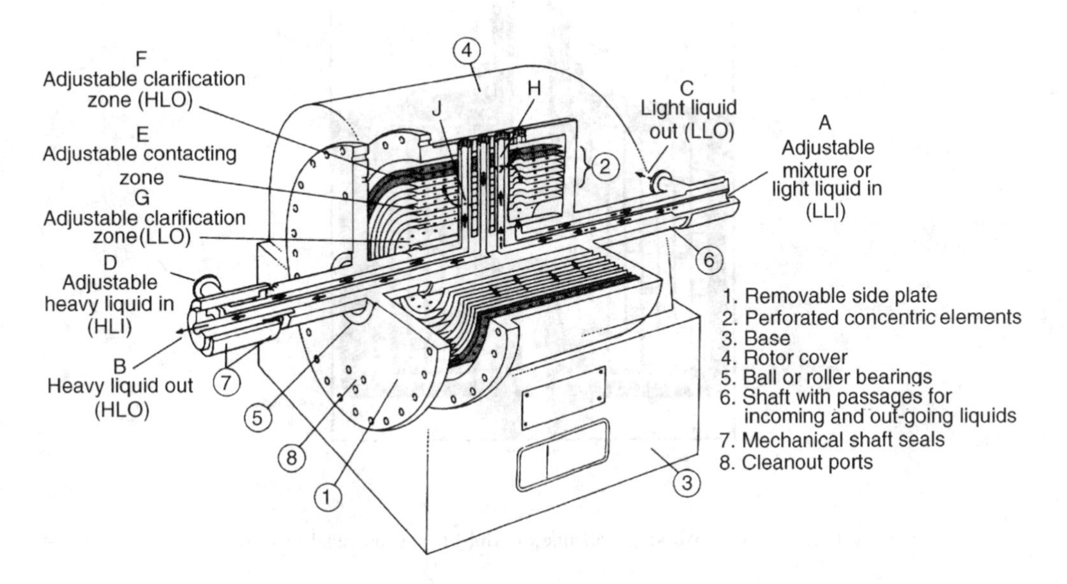


Figure 13.39. Podbielniak contactor

The contactor finds extensive use where high performance phase separation and countercurrent extraction or washing in the one unit are required. Particularly important applications are the removal of acid sludges from hydrocarbons, shown in Figure 13.40, hydrogen peroxide extraction, sulphonate soap and antibiotics extraction, the extraction of rare earths such as uranium and vanadium from leach liquors, and the washing of refined edible oils.

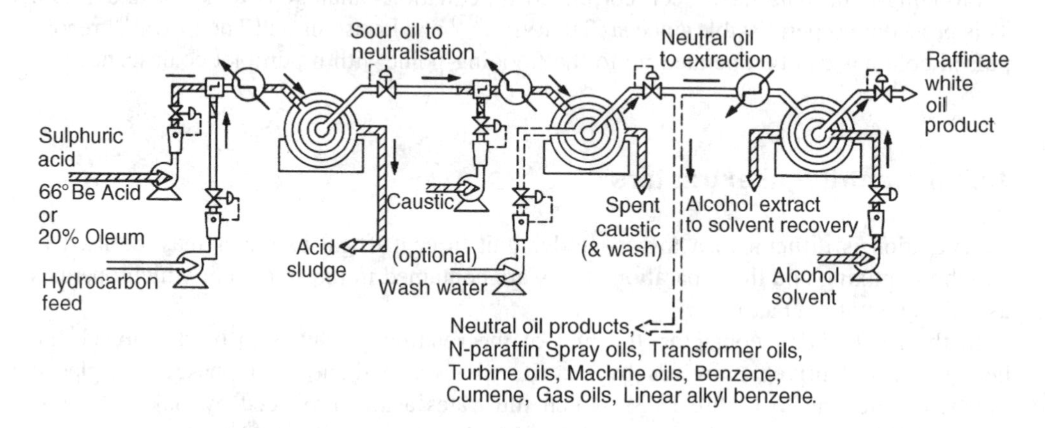


Figure 13.40. Application of the Podbielniak contactor to the acid treatment of hydrocarbon feeds

The *Alfa-Laval* contactor shown in Figure 13.41, has a vertical spindle and the rotor is fitted with concentric cylindrical inserts with helical wings forming a series of spiral

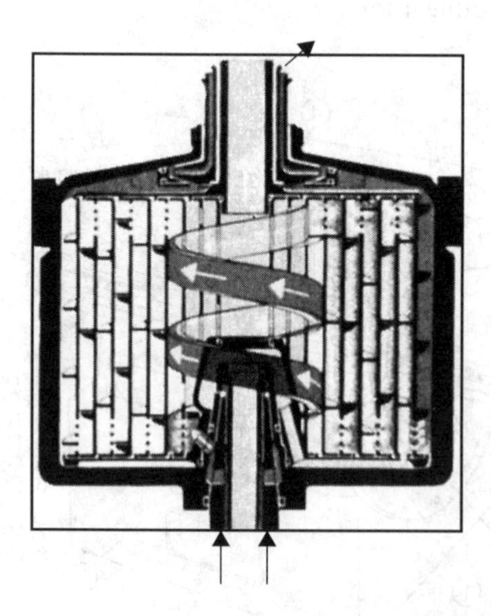

Figure 13.41. Working principle of Alfa-Laval centrifugal extractor

passages. The two phases are fed into the bottom, the light phase being led to the periphery from which it flows inwards along the spiral, with the heavy phase flowing countercurrently. High shear forces are thus generated giving high extraction rates.

These units give many ideal stages, run continuously and take up a minimum space. For these reasons they have been adopted in many drug extractions, though they are unsuitable for medium or large throughputs.

To some extent, the *Scheibel column* may be considered as a centrifugal device, although as coalescence takes place between the layers of mesh packing, in that sense it is a stage-wise contactor, and it has therefore been described in Section 13.6.3.

## 13.8. USE OF SPECIALISED FLUIDS

With the widening use of the liquid–liquid extraction for the separation of complex mixtures into their components, it has been necessary to develop fluids with highly selective characteristics. The metallurgical, nuclear, biotechnolgy and food industries are now major users of the technique, and many of the recent developments have originated in those fields. Some of the characteristics and properties of two classes of fluids of increasing importance–supercritical fluids and aqueous two-phase systems are described in this section.

Supercritical fluids can be highly selective and their solvent power can be controlled by adjustment of the operating pressure. With fluids such as carbon dioxide, there is no residual contamination of the product as the solvent evaporates completely at the end of the operation.

The application of two aqueous phase systems is largely confined to the biotechnology industry where the highly complex fragile molecules would be degraded by the action of most of the solvents in common use in the chemical industry. Because biochemical and micrbiological products are normally produced in the aqueous phase, the solvent is very benign because interfacial tensions are very low.

### 13.8.1. Supercritical fluids

A supercritical fluid is a substance that is above its critical point, that is above the highest temperature and pressure at which its vapour and liquid can co-exist at equilibrium. Although some materials decompose at a temperature below what would be their critical temperature, the supercritical regions of many common gases and liquids are easily attained. The two most popular and inexpensive fluids in this respect are carbon dioxide and water, which are non-toxic and non-flammable and therefore, as pointed out by BRENECKE[39], essentially environmentally benign solvents that can be used even for food processing without undue regulatory restrictions. As presented by REID *et al.*[40], the critical point of carbon dioxide, 304.3 K and 7.3 MN/m$^2$, is readily accessible although that of water at 647.2 K and 22.09 MN/m$^2$ is somewhat more challenging. Data for other supercritical fluids are:

| | $T_c$ (K) | $P_c$ (MN/m$^2$) |
|---|---|---|
| ethane | 305.4 | 4.88 |
| ethylene | 282.9 | 5.04 |
| propane | 369.9 | 4.25 |
| ammonia | 405.7 | 11.28 |

Since the densities of these fluids change dramatically with very small changes in temperature or applied pressure, any density-dependent property such as the solubility of a heavy organic solute for example, may be manipulated, as FRIEDRICH *et al.*[41] have pointed out, over wide ranges. This feature can be utilised in simple separation schemes in which a compound is extracted at high pressure where its solubility is high and then a reduction of the pressure causes the solute to come out of solution with the supercritical fluid being recycled by repressurisation.

The dependence of solubility on pressure can be exploited in the separation of complex mixtures. If the pressure is initially high enough for all the components to be dissolved and is then reduced in stages, precipitation of successive components may be achieved in each stage. Other advantages of supercritical fluids include the fact that their viscosities are less than those of typical liquids and, as the diffusivities of the solutes are closer to those of gases rather than liquids, mass transfer resistances are considerably less than those in normal liquids. Thus, in general, supercritical fluids combine the advantages of having the diffusivities of gases and the solvent power of liquids.

One of the main commercial applications of supercritical fluids is in food processing where supercritical carbon dioxide, which is used on a large scale in the de-caffeination of tea and coffee, is particularly favoured as it leaves no residue in the product. In addition, MCHUGH and KRUKONIS[42] have described the extraction of hops, spices and flavours using supercritical carbon dioxide, which may also be used for the extraction of a whole range of natural products, including pharmaceutical compounds, health supplements and fragrances. JENNINGS *et al.*[43] have described the extraction of taxol, an anti-cancer agent, from a slow-growing variety of the yew tree; FAVATI *et al.*[44] have found that gamma-linolenic acid, a health aid, can be extracted from evening primrose oil seeds; and BORCH-JENSEN *et al.*[45] have shown that the valuable eicosapentaenoic and docosahexaenoic acids in fish oils may be fractionated with supercritical carbon dioxide. MOYLER[46] lists over 80 varieties of seeds, roots, leaves, flowers and fruits which may be extracted with either liquid or supercritical carbon dioxide, and many supercritical carbon dioxide extracts are currently available including celery, ginger, paprika, rosemary, sage and vanilla.

Supercritical fluids also find application in the areas of pollution prevention and remediation, and supercritical carbon dioxide is used as a replacement solvent for many hazardous solvents in both extraction and separation processes and also as a reaction medium and in materials processing. Although carbon dioxide is considered as a 'greenhouse gas', there is actually no net increase in the amount of the gas if it is removed from the environment, used as the solvent instead of a hazardous substance, and returned to the environment. In this way, most of the uses of supercritical carbon dioxide may be considered as environmentally friendly. Because the solubilities of oils and greases in carbon dioxide are high, it is particularly suited to the cleaning of machinery[47] and, as discussed in the literature[48], it is used as a solvent in textile dyeing operations where it is used to treat any dye-laden wastewater. With the addition of appropriate chelating agents, as described by SAITO *et al.*[49], metals can be effectively extracted from solutions and soils with supercritical carbon dioxide which is also used, as discussed by YAZDI and BEEKMAN[50] to recover uranium from the aqueous solutions produced in the reprocessing of spent nuclear fuels. The main application of supercritical water has been in the oxidation of hazardous organic materials since water is readily miscible with both oxygen and organics and very high degrees of destruction can be achieved with very short residence times. This technology also finds use in the destruction of chemical weapons and stockpiled explosives, as well as in the clean-up of industrial and municipal wastes.

In the use of supercritical carbon dioxide for bioseparations, JOHNSTON *et al.*[51] have reported that proteins can be solublised in reverse micelles formed in carbon dioxide. In general, bioseparations can offset the high cost of attaining the necessary pressures since the products are of high value, they are present in the broth in low concentrations and conventional solvents often lead to recovery problems.

Although perhaps beyond the area of solvent extraction, it may be noted that various processes have been developed in which the solute is dissolved in the supercritical fluid and then the solution is expanded through a nozzle. This gives very high degree of supersaturation and results in the growth and nucleation of very fine particles, as discussed in Chapter 15. Products formed in this way include drug-polymers and materials which promote the gradual release of flavours and fragrances, as discussed by HUTCHENSON and FORSTER[52]. The selective precipitation of solutes is used for a wide variety of products including foods, proteins and explosives, as described by CHANG and RANDOLPH[53]. Spray drying of a solution in a supercritical liquid has been used for the production of microspheres and microporous fibres[54]. Again, beyond the field of solvent extraction, it may be noted that supercritical fluids have found increasing use in providing benign solvents in which to carry out chemical reactions. A review of this work has been provided by SAVAGE *et al.*[55], and the processes described include the polymerisation of highly fluorinated acrylic polymers, the production of formic acid, brominations of alkylaromatics and phase-transfer catalysis. A general review of the exploitation of supercritical fluids in reaction chemistry has been provided by CLIFFORD and BARTLE[56].

### 13.8.2. Aqueous two-phase systems

The possibility of a three component system consisting of water and two organic components separating into two phases has been known for some time, though it is only since biochemical processes, including fermentation, have been used for the production of expensive highly complex molecules that it has assumed such a great industrial importance. Two-phase aqueous systems are now extensively used for the separation of product from the associated cell debris. Essentially, the addition of a polymer, such a polyethylene glycol or dextran, usually with a phosphate, to a fermentation product causes phase separation to occur, and the way in which the phases separate is determined by the structure and molecular weight of the polymer, and its concentration. A good separation depends on operating under conditions where the tie-line on the phase equilibrium diagram is long. One of the inevitable problems with a system in which both interfacial tension and difference in density of the phases are low lies in the difficulty of obtaining a sharp separation of the liquid layers.

One of the most useful sources of information about these systems is the work of ZASLAVSKY[57]. The partition of proteins and other compounds in aqueous two-phase systems is influenced by a large number of parameters including the types of polymer composing the two-phase system, the mean molecular weight of the polymers, the molecular weight distribution of the polymers, the length of a tie-line, the types of ions composing or added to the system, ionic strength, pH and temperature. Most of these factors do not act independently and, in the main, conditions for a desired partition have to be determined experimentally, as discussed by KULA *et al.*[58], although a thermodynamic interpretation has been presented by BROOKS *et al.*[59]. Where an efficient separation of different substances in an aqueous two-phase system cannot be achieved in a single step, a multi-stage procedure, known as countercurrent distribution, is used. The equipment consists of a series of compartments containing defined volumes of top and bottom phases, in which a substrate partitions itself accordingly. Phase-mixing is followed by phase-settling, and moving the top phases to the adjacent bottom phases. MATTIASSON and KAUL[60] have proposed that a centrifugal separation step be introduced so as to reduce the phase settling time, and a continuous cross-current extraction unit has been developed by HUSTEDT *et al.*[61]. On comparing the large-scale recovery of enzymes by partitioning in aqueous phase systems with other separation processes such as solid–liquid separation by centrifugation, filtration using a drum filter and cross-flow filtration, it is found that partitioning requires larger amounts of chemicals when operating without a recycle, although, as KRONER *et al.*[62] have reported that labour and investment costs are considerably lower. One disadvantage of using two-phase aqueous extraction for many separations is the need for highly purified dextran as one of the phase-forming polymers which is very expensive. KRONER *et al.*[63] have investigated the use of crude dextran as a cheaper alternative and found only a small change in the partition coefficients for a number of enzymes, all of which were recovered with high yields. Crude dextran also gave a wider variation in system parameters and operating costs were some times reduced as shorter residence times were required. Polyethylene glycol–salt systems are preferred for large-scale enzyme extractions even though most proteins partition strongly into the salt-rich bottom phase. Protein isolation and purification are now carried out on a large-scale and HUDSTEDT and PAPAMICHAEL[64] have developed a two-stage aqueous two-phase extraction system for the isolation of $\beta$-D-Galactosidase from Escherichia Coli. WOODROW and QUIRK[65] have investigated the partitioning behaviour of acylamidase using a polyethylene glycol–dextran system. Other extractive bioconversions and fermentations with aqueous-phase systems include the production of toxin factor, production of butanol and acetone, cyclic fermentation production of alcohol, the production of glucose-6-phosphate from glucose, the production of L-methionine from racemic mixtures, the production of glucose from starch, deacetylation of penicillin-G, saccharification of cellulose and the production of ethanol and fermentable sugars from cellulose.

## 13.9. FURTHER READING

BACKHURST, J. R., HARKER, J. H. and PORTER, J. E.: *Problems in Heat and Mass Transfer* (Edward Arnold, London, 1974).
BLUMBERG, R.: *Liquid–Liquid Extraction* (Harcourt Brace Jovanovich, London, 1988).
FAHIEN, R.: *Transport Operations* (McGraw-Hill, New York, 1982).

FRANCIS, A. W.: *Handbook for Components in Solvent Extraction* (Gordon & Breach, New York, 1972).
GODFREY, J. C. and SLATER, M. J.: *Liquid–liquid Extraction Equipment* (Wiley, New York, 1994).
HANSON, C. (ed.): *Recent Advances in Liquid–Liquid Extraction* (Pergamon Press, Oxford, 1971).
HINES, A. L. and MADDOX, R. N.: *Mass Transfer Fundamentals and Applications* (Prentice-Hall, Englewood Cliffs, 1985).
HOLLAND, F. A. and CHAPMAN, F. S.: *Liquid Mixing and Processing in Stirred Tanks* (Reinhold, New York, 1966).
JAMRACK, W. D.: *Base Metal Extraction by Chemical Engineering Techniques* (Pergamon Press, Oxford, 1963).
KING, C. J.: *Separation Processes* 2nd. edn. (McGraw-Hill, New York, 1980).
LO, T. C., BAIRD, M. I. and HANSON, C. (eds): *Handbook of Solvent Extraction* (John Wiley, 1983).
MARCUS, Y. (ed): *Solvent Extraction Reviews* (Marcel Dekker, New York, 1971).
MCCABE, W. L., SMITH, J. C. and HARRIOTT, P.: *Unit Operations in Chemical Engineering*. 4th edn. (McGraw-Hill, New York, (1984)).
MCHUGH, M. A. and KRUKONIS, V. J.: *Supercritical Fluid Extraction — Principles and Practice* (Butterworths, Boston, 1986).
PAUL, P. M. F. and WISE, W. S.: *The Principles of Gas Extraction* (Mills and Boon, London, 1971).
PAULAITIS, M. E., PENNINGER, S. M. L., GRAY, R. D. and DAVIDSON, P. (eds): *Chemical Engineering at Supercritical Fluid Conditions* (Ann Arbor Science, Ann Arbon, 1983).
PRATT, H. R. C.: *Countercurrent Separation Processes* (Elsevier, Amsterdam, 1967).
SCHWEITZER, M. (ed): *Handbook of Separation Techniques for Chemical Engineers* (McGraw-Hill, New York, 1979).
SHERWOOD, T. K., PIGFORD, R. L. and WILKE, C. R.: *Mass Transfer* (McGraw-Hill, New York, 1975).
SQUIRES, T. G. and PAULAITIS, M. E. (eds): *Supercritical Fluids — Chemical and Engineering Principles and Applications*. ACS Symposium Series 329 (American Chemical Society, Washington, 1987).
TREYBAL, R. E.: *Mass Transfer Operations*. 3rd edn. (McGraw-Hill, New York, 1980).
WALTER, H., BROOKS, D. E. and FISHER, D.: *Partitioning in Aqueous Two-Phase Systems*. (Academic Press, New York, 1985).
WANKAT, P. C.: *Equilibrium Staged Separations: Separations for Chemical Engineers* (Elsevier, New York, 1988).
ZASLAVSKY, B. Y.: *Aqueous Two-Phase Partitioning* (Marcel Dekker, 1994). Solvent Extraction, Proc. of the Int. Solvent Extraction Conference. The Hague, 1971. (Soc. Chem. Ind., London, 1971).

## 13.10. REFERENCES

1. HANSON, C.: *Het Ingenieursblad* **41**, 15–16 (Aug. 1972) 408–17. The technology of solvent extraction in metallurgical processes.
2. British Patent No. 995472: Distillers Company Limited (29 April 1964). Acrylic acid recovery.
3. SHERWOOD, T. K. and PIGFORD, R. L.: *Absorption and Extraction*, 2nd edn. (McGraw-Hill, New York, 1952).
4. LEIBSON, I and BECKMANN, R. B.: *Chem. Eng. Prog.* **49** (1953) 405. The effect of packing size and column diameter on mass transfer in liquid–liquid extraction.
5. GAYLER, R. and PRATT, H. R. C.: *Trans. Inst. Chem. Eng.* **31** (1953) 69. Liquid–liquid extraction. Part V — Further studies of droplet behaviour in packed columns.
6. WARWICK, G. C. I. and SCUFFHAM, J. B.: *Het Ingenieursblad* **41,** 15–16 (Aug. 1972) 442–449. The design for mixer–settlers for metallurgical duties.
7. JACKSON, I. D., NEWRICK, G. M. and WARWICK, G. C. I.: *I. Chem. E.* Symposium Series **42**, 15.1–15.8. A recent development in the design of hydrometallurgical mixer–settlers.
8. SCHEIBEL, E. G. and KARR, A. E.: *Ind. Eng. Chem.* **42** (1950) 1048. Semicommercial multistage extraction columns.
9. BLANDING, F. H. and ELGIN, J. C.: *Trans. Am. Inst. Chem. Eng.* **38** (1942) 305. Limiting flow in liquid–liquid extraction columns.
10. THORNTON, J. D.: *Chem. Eng. Sci.* **5** (1956) 201. Spray liquid–liquid extraction columns.
11. HAYWORTH, C. B. and TREYBAL, R. E.: *Ind. Eng. Chem.* **42** (1950) 1174. Drop formation in two-liquid-phase systems.
12. HU, S. and KINTNER, R. C.: *A.I.Ch.E.Jl.* **1** (1955) 42. The fall of single liquid drops through water.
13. KLEE, A. J. and TREYBAL, R. E.: *A.I.Ch.E.Jl.* **2** (1956) 44. Rate of rise or fall of liquid drops.
14. LICHT, W. and CONWAY, J. B.: *Ind. Eng. Chem.* **42** (1950) 1151. Mechanism of solute transfer in spray towers.

15. COULSON, J. M. and SKINNER, S. J.: *Chem. Eng. Sci.* **1** (1952) 197. The mechanism of liquid–liquid extraction across stationary and moving interfaces. Part 1. Mass transfer into single dispersed drops.
16. GARNER, F. H. and HALE, A. A.: *Chem. Eng. Sci.* **2** (1953) 157. The effect of surface agents in liquid extraction processes.
17. HANDLOS, A. E. and BARON, T.: *A.I.Ch.E.Jl.* **3** (1957) 127. Mass and heat transfer from drops in liquid–liquid extraction.
18. RUBY, C. L. and ELGIN, J. C.: Mass transfer — Transport properties. *Chem. Eng. Prog.* Symp. Series No. 16, **51** (1955) 17. Mass transfer between liquid drops and a continuous liquid phase in a countercurrent fluidized system. Liquid–liquid extraction in a spray tower.
19. APPEL, F. J. and ELGIN, J. C.: *Ind. Eng. Chem.* **29** (1973) 451. Countercurrent extraction of benzoic acid between toluene and water.
20. SHERWOOD, T. K., EVANS, J. E. and LONGCOR, J. V. A.: *Trans. Am. Inst. Chem. Eng.* **35** (1939) 597. Extraction in spray and packed columns.
21. ROW, S. B., KOFFOLT, J. H., and WITHROW, J. R.: *Trans. Am. Inst. Chem. Eng.* **37** (1941) 559. Characteristics and performance of a nine-inch liquid–liquid extraction column.
22. PERRY, R. H., GREEN, D. W. and MALONEY, J. O. (eds.): *Perry's Chemical Engineers' Handbook.* 7th edn. (McGraw-Hill Book Company, New York, 1997).
23. TREYBAL, R. E.: *Liquid Extraction.* 2nd edn. (McGraw-Hill, New York, 1963).
24. SHERWOOD, T. K., PIGFORD, R. L. WILKE, C. R.: *Mass Transfer* (McGraw-Hill, New York, 1975).
25. DAVIES, J. T. and RIDEAL, E. K.: *Interfacial Phenomena* (Academic Press, 1961).
26. GROOTHUIS, H. and ZUIDERWEG, F. J.: *Chem. Eng. Sci.* **12** (1960) 288. Influence of mass transfer on coalescence of drops.
27. GAYLER, R., ROBERTS, N. W. and PRATT, H. R. C.: *Trans. Inst. Chem. Eng.* **31** (1953) 57. Liquid–liquid extraction. Part IV. A further study of hold-up in packed columns.
28. GAYLER, R. and PRATT, H. R. C.: *Trans. Inst. Chem. Eng.* **29** (1951) 110. Symposium on liquid–liquid extraction. Part II. Hold-up and pressure drop in packed columns.
29. LEWIS, J. B., JONES, I. and PRATT, H. R. C.: *Trans. Inst. Chem. Eng.* **29** (1951) 126. Symposium on liquid–liquid extraction. Part III. A study of droplet behaviour in packed columns.
30. DELL, F. R. and PRATT, H. R. C.: *Trans. Inst. Chem. Eng.* **29** (1951) 89 Symposium on liquid–liquid extraction. Part I. Flooding rates for packed columns.
31. PRATT, H. R. C. and GLOVER, S. T.: *Trans. Inst. Chem. Eng.* **24** (1946) 54. Liquid–liquid extraction: Removal of acetone and acetaldehyde from vinyl acetate with water in a packed column.
32. COLBURN, A. P. and WELSH, D. G.: *Trans. Am. Inst. Chem. Eng.* **38** (1942) 179. Experimental study of individual transfer resistances in countercurrent liquid–liquid extraction.
33. LADDHA, G. S. and SMITH, J. M.: *Chem. Eng. Prog.* **46** (1950) 195. Mass transfer resistances in liquid–liquid extraction.
34. THORNTON, J. D. and PRATT, H. R. C.: *Trans. Inst. Chem. Eng.* **31** (1953) 289. Liquid–liquid extraction. Part VII. Flooding rates and mass transfer data for rotary annular columns.
35. VERMIJS, H. J. A. and KRAMERS, H.: *Chem. Eng. Sci.* **3** (1954) 55. Liquid–liquid extraction in a "rotating disc contactor".
36. VAN DIJCK, W. J. D.: *U.S. Patent* 2,011,186 (1935). Intimately contacting fluids (immiscible liquids).
37. THORNTON, J. D.: *Trans. Inst. Chem. Eng.* **35** (1957) 316. Liquid–liquid extraction. Part XIII. The effect of pulse wave-form and plate geometry on the performance and throughput of a pulsed column.
38. LOGSDAIL, D. H. and THORNTON, J. D.: *Trans. Inst. Chem. Eng.* **35** (1957) 331. Liquid–liquid extraction. Part XIV. The effect of column diameter upon the performance and throughput of pulsed plate columns.
39. BRENECKE, J. F.: *Chem. & Industry* (4 November 1996) 831. New applications of supercritical fluids.
40. REID, R. C., PRAUSNITZ, J. M. and POLING, B. E.: *The Properties of Liquids and Gases*. 4th. edn. (McGraw-Hill, New York, 1987)
41. FRIEDRICH, J. P., LIST, G. R. and HEAKIN, A. J.: *J. Am. Oil. Chem. Soc.* **59** (1982) 288–292. Petroleum-free extraction of oil from soybeans with supercritical $CO_2$.
42. MCHUGH, M. A. and KRUKONIS, V. J.: *Supercritical Fluid Extraction - Principles and Practice*. 2nd. edn. (Butterworth-Heinemann, Oxford, 1994)
43. JENNINGS, D. W., CHANG, F., BAZOOK, V.: *J. Chem. Eng. Data*. **37** (1992) 337–338. Vapor-liquid equilibria for carbon dioxide plus 1-pentanol.
44. FAVATI, F., KING, J. W. and MAZZANTI, M.: *J. Am. Oil. Chem. Soc.* **68** (1991) 422–427. Supercritical carbon dioxide extraction of evening primrose oil.
45. BORCH-JENSEN, C., STABY, A. and MOLLERUP, J. M.: *Ind. Eng. Chem. Res.* **33** (1994) 1574–1579. Phase equilibria of urea-fractioned fish oil fatty acid ethyl esters and supercritical carbon dioxide.
46. MOYLER, D. A. in KING, M. B. and BOTT, T. R. (eds): *Extraction of Natural Products using near-critical Solvents*. (Chapman & Hall, Glasgow, 1993)
47. *Electronic Materials Technology News* **9** (March 1995) No. 3.
48. STEINER, R.: *Chemical Engineering* **100** (3) (March 1993) 114–119. Carbon dioxide's expanding role.

49. SAITO, N., IKUSHIMA, Y. and GOTO, T.: *Bull. Chem. Soc. Japan* **63** (1990) 1532–1534. Liquid-solid extraction of acetylacetone chelates with supercritical carbon dioxide.
50. YAZDI, A. V. and BEEKMAN, E. J.: *Mater. Res.* **10** (1995) 530–537. Design of highly $CO_2$-soluble chelating agents for carbon dioxide extraction of heavy metals.
51. JOHNSTON, K. P., HARRISON, K. L., CLARKE, M. J.: *Science* **271** (1996) 624–626. Water in carbon dioxide microemulsions: An environment for hydrophiles including proteins.
52. HUTCHENSON, K. W. and FOSTER, N. R. in HUTCHENSON, K. W. and FOSTER, N. R.: *Innovations in Supercritical Fluids*. ACS Symposium Series 608, (American Chemical Society, Washington, 1995)
53. CHANG, C. M. J., RANDOLPH, A. D. and CROFT, N. E.: *Biotech. Prog.* **7** (1991) 275. Separation of beta-caretone mixtures precipitated from liquid solvents with high pressure $CO_2$.
54. DIXON, D. J., JOHNSTON, K. P. and BODMEIER, R. A.: *AIChE Jl.* **39** (1993) 127. Polymeric materials formed by precipitation with a compressed fluid antisolvent.
55. SAVAGE, P. E., GOPLAN, S., MIZAN, T. I., MARTINO, C. J. and BROCK, E. E.: *AIChE Jl.* **41** (1995) 1723–1778. Reactions at supercritical conditions - applications and fundamentals.
56. CLIFFORD, T. and BARTLE, K.: *Chem. Ind.* (17 June, 1996) 449–452. Chemical reactions in supercritical fluids.
57. ZASLAVSKY, B. Y.: *Aqueous Two-phase Partitioning*. (Marcel Dekker, London, 1994)
58. KULA, M-R. in WINGARD, L. B., KATCHALSKI-KATZIRE, E. and GOLDSTEIN, L. (eds.): *Applied Biochemistry and Bioengineering. Vol 2: Enzyme Technology*. (Academic Press, New York, 1979) 71–95.
59. BROOKS, D. E., SHARP, K. A. and FISHER, D. in WALTER, H., BROOKS, D. E. and FISHER, D. (eds.): *Partitioning in Aqueous Two Phase Systems, Theory, Methods, Uses and Applications to Biotechnology*. (Academic Press, New York, 1985) 11–84.
60. MATTIASSON, B. and KAUL, R.: *Jl. Am. Chem. Soc.* **314** (1986) 79–92. Use of aqueous 2-phase systems for recovery and purification in biotechnology.
61. HUSTEDT, H., KRONER, K. H. and PAPMICHAEL, N.: *Proc. Biochem.* (October, 1988) 129–137. Continuous cross-current aqueous 2-phase extraction of enzymes from biomass - automated recovery in production scale.
62. KRONER, K. H., HUSTEDT, H. and KULA, M-R.: *Proc. Biochem.* No. 10 (1984) 170–179. Extractive enzyme recovery - economic considerations.
63. KRONER, K. H., HUSTEDT, H. and KULA, M-R.: *Biotech. Bioeng.* **24** (1982) 1015–1045. Evaluation of crude dextrain as phase-forming polymer for the extraction of enzymes in aqueous 2-phase systems in large scale.
64. HUDSTEDT, H. and PAPAMICHAEL, N.: *Abstr. Pap. Am. Chem. Soc.* **195** (June, 1988) 198. IEC Part 1. Design of economical aqueous 2-phase extraction processes.
65. WOODROW, J. R. and QUIRK, A. V.: *Enzyme. Microb. Tech.* **8** (1986) 183–187. Anomalous partitioning of aryl acylamidase in aqueous 2-phase systems.

## 13.11. NOMENCLATURE

| | | Units in SI System | Dimensions in **M, N, L, T** |
|---|---|---|---|
| $A$ | Mass of solvent **A** | kg | **M** |
| $a$ | Interfacial area per unit volume of tower | $m^2/m^3$ | $\mathbf{L}^{-1}$ |
| $a_p$ | Superficial area of packing per unit volume of tower | $m^2/m^3$ | $\mathbf{L}^{-1}$ |
| $B$ | Mass of solvent **B** | kg | **M** |
| $b$ | index | — | — |
| $C$ | Concentration of solute | $kmol/m^3$ | $\mathbf{NL}^{-3}$ |
| $C^*$ | Value of concentration in equilibrium with second phase | $kmol/m^3$ | $\mathbf{NL}^{-3}$ |
| $D_T$ | Diameter of tank | m | **L** |
| $d$ | Particle size | m | **L** |
| $d_c$ | Diameter of column | m | **L** |
| $d_p$ | Nominal size of packing | m | **L** |
| $d_s$ | Surface mean (Sauter mean) diameter of drop | m | **L** |
| $d_{crit}$ | Critical packing size | m | **L** |
| $E$ | Mass of extract | kg | **M** |
| $e$ | Voidage of packing | — | — |
| $F$ | Mass of feed | kg | **M** |
| $g$ | Acceleration due to gravity | $m/s^2$ | $\mathbf{LT}^{-2}$ |

| | | Units in SI System | Dimensions in **M**, **N**, **L**, **T** |
|---|---|---|---|
| **H** | Height of the transfer unit | m | **L** |
| $J$ | Coefficient in equation 13.45 for flooding data | — | — |
| $j$ | Fractional hold-up of dispersed phase | — | — |
| $K$ | Overall mass transfer coefficient | m/s | $\mathbf{LT^{-1}}$ |
| $K'$ | Constant | — | — |
| $k$ | Mass transfer coefficient | m/s | $\mathbf{LT^{-1}}$ |
| $L'$ | Volumetric rate of flow per unit area | $m^3/m^2s$ | $\mathbf{LT^{-1}}$ |
| $M$ | Total flow of material through the system | kg | **M** |
| $m$ | Slope of the equilibrium line | — | — |
| $N$ | Revolutions per unit time | $s^{-1}$ | $\mathbf{T^{-1}}$ |
| $N'$ | Molar rate of transfer per unit area | $kmol/m^2s$ | $\mathbf{NL^{-2}\ T^{-1}}$ |
| **N** | Number of transfer units | — | — |
| $n$ | Number of stages | — | — |
| $n_1, n_2, \dots$ | Numbers of particles or droplets | — | — |
| $P$ | Difference stream | kg | **M** |
| $R$ | Mass of raffinate | kg | **M** |
| $r$ | Ratio $u_{df}/u_{cf}$ | — | — |
| $S$ | Added solvent in extract stream | kg | **M** |
| $u$ | Volumetric flowrate per unit area | $m^3/m^2s$ | $\mathbf{LT^{-1}}$ |
| $u_r$ | Velocity of dispersed phase relative to continuous phase | m/s | $\mathbf{LT^{-1}}$ |
| $u_0$ | Terminal falling velocity of droplet | m/s | $\mathbf{LT^{-1}}$ |
| $\bar{u}_0$ | Velocity of single droplet relative to continuous phase | m/s | $\mathbf{LT^{-1}}$ |
| $X_f$ | Mass ratio of solute in feed | kg/kg (kmol/kmol) | — |
| $X_1,\ X_2$ | Mass ratio of solute in raffinate | kg/kg (kmol/kmol) | — |
| $x_A,\ x_B$ | Mass (or mole) fraction of **A**, **B** | kg/kg (kmol/kmol) | — |
| $Y_1,\ Y_2$ | Mass ratio of solute in extract | kg/kg (kmol/kmol) | — |
| $Z$ | Height of packing | m | **L** |
| $Z_B$ | Height of dispersion band | m | **L** |
| $\beta$ | Selectivity ratio (equation 13.2) | — | — |
| $\rho$ | Density | $kg/m^3$ | $\mathbf{ML^{-3}}$ |
| $\Delta\rho$ | Density difference between phases | $kg/m^3$ | $\mathbf{ML^{-3}}$ |
| $\sigma$ | Interfacial tension | $J/m^2$ or N/m | $\mathbf{MT^{-2}}$ |
| $\sigma_T$ | Interfacial tension, water-air at 288 K (0.073 N/m) | $J/m^2$ or N/m | $\mathbf{MT^{-2}}$ |

Suffixes

| | |
|---|---|
| $c$, $d$ | continuous and disperse phases |
| $E$, $R$ | extract and raffinate phases |
| $O$ | overall (transfer units) |
| $f$ | limiting value at flooding point |
| $i$ | value of interface |
| lm | logarithmic mean value |
| 1, 2 | values at bottom, top of column |

CHAPTER 14

# Evaporation

## 14.1. INTRODUCTION

Evaporation, a widely used method for the concentration of aqueous solutions, involves the removal of water from a solution by boiling the liquor in a suitable vessel, an evaporator, and withdrawing the vapour. If the solution contains dissolved solids, the resulting strong liquor may become saturated so that crystals are deposited. Liquors which are to be evaporated may be classified as follows:

(a) Those which can be heated to high temperatures without decomposition, and those that can be heated only to a temperature of about 330 K.
(b) Those which yield solids on concentration, in which case crystal size and shape may be important, and those which do not.
(c) Those which, at a given pressure, boil at about the same temperature as water, and those which have a much higher boiling point.

Evaporation is achieved by adding heat to the solution to vaporise the solvent. The heat is supplied principally to provide the latent heat of vaporisation, and, by adopting methods for recovery of heat from the vapour, it has been possible to achieve great economy in heat utilisation. Whilst the normal heating medium is generally low pressure exhaust steam from turbines, special heat transfer fluids or flue gases are also used.

The design of an evaporation unit requires the practical application of data on heat transfer to boiling liquids, together with a realisation of what happens to the liquid during concentration. In addition to the three main features outlined above, liquors which have an inverse solubility curve and which are therefore likely to deposit scale on the heating surface merit special attention.

## 14.2. HEAT TRANSFER IN EVAPORATORS

### 14.2.1. Heat transfer coefficients

The rate equation for heat transfer takes the form:

$$Q = UA\Delta T \tag{14.1}$$

where: $Q$ is the heat transferred per unit time,
$U$ is the overall coefficient of heat transfer,
$A$ is the heat transfer surface, and
$\Delta T$ is the temperature difference between the two streams.

In applying this equation to evaporators, there may be some difficulty in deciding the correct value for the temperature difference because of what is known as the *boiling point rise* (BPR). If water is boiled in an evaporator under a given pressure, then the temperature of the liquor may be determined from steam tables and the temperature difference is readily calculated. At the same pressure, a solution has a boiling point greater than that of water, and the difference between its boiling point and that of water is the BPR. For example, at atmospheric pressure (101.3 kN/m$^2$), a 25 per cent solution of sodium chloride boils at 381 K and shows a BPR of 8 deg K. If steam at 389 K were used to concentrate the salt solution, the overall temperature difference would not be $(389 - 373) = 16$ deg K, but $(389 - 381) = 8$ deg K. Such solutions usually require more heat to vaporise unit mass of water, so that the reduction in capacity of a unit may be considerable. The value of the BPR cannot be calculated from physical data of the liquor, though Dühring's rule is often used to find the change in BPR with pressure. If the boiling point of the solution is plotted against that of water at the same pressure, then a straight line is obtained, as shown for sodium chloride in Figure 14.1. Thus, if the pressure is fixed, the boiling point of water is found from steam tables, and the boiling point of the solution from Figure 14.1. The boiling point rise is much greater with strong electrolytes, such as salt and caustic soda.

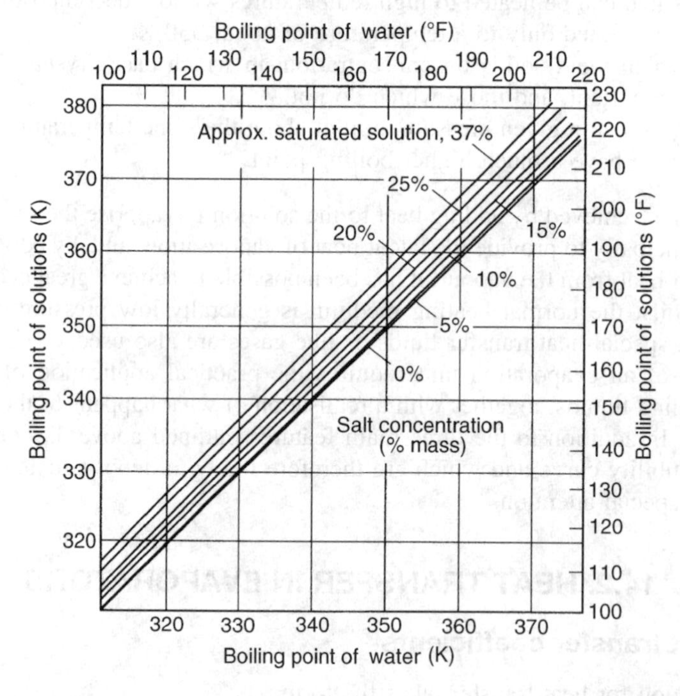


Figure 14.1. Boiling point of solutions of sodium chloride as a function of the boiling point of water. Dühring lines

Overall heat transfer coefficients for any form of evaporator depend on the value of the film coefficients on the heating side and for the liquor, together with allowances for scale deposits and the tube wall. For condensing steam, which is a common heating medium, film coefficients are approximately 6 kW/m$^2$ K. There is no entirely satisfactory general method for calculating transfer coefficients for the boiling film. Design equations of sufficient accuracy are available in the literature, however, although this information should be used with caution.

## 14.2.2. Boiling at a submerged surface

The heat transfer processes occurring in evaporation equipment may be classified under two general headings. The first of these is concerned with boiling at a submerged surface. A typical example of this is the horizontal tube evaporator considered in Section 14.7, where the basic heat transfer process is assumed to be nucleate boiling with convection induced predominantly by the growing and departing vapour bubbles. The second category includes two-phase forced-convection boiling processes occurring in closed conduits. In this case convection is induced by the flow which results from natural or forced circulation effects.

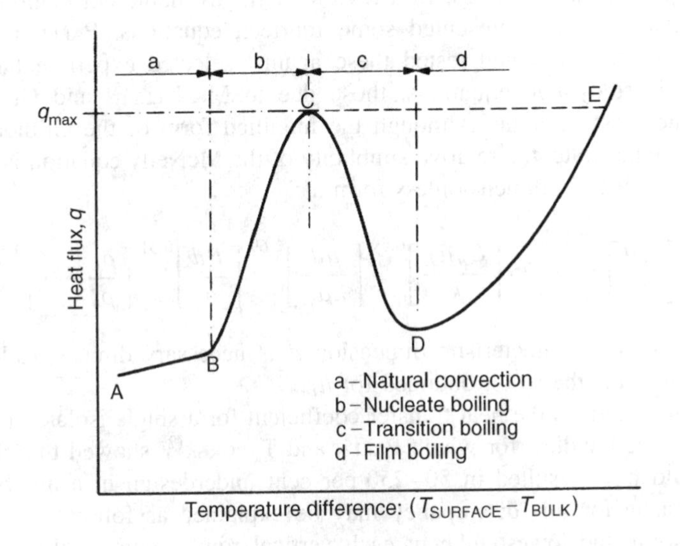


Figure 14.2. Typical characteristic for boiling at a submerged surface

As detailed in Volume 1, Chapter 9 and in Volume 6, the heat flux–temperature difference characteristic observed when heat is transferred from a surface to a liquid at its boiling point, is as shown in Figure 14.2. In the range AB, although the liquid in the vicinity of the surface will be slightly superheated, there is no vapour formed and heat transfer is by natural convection with evaporation from the free surface. Boiling commences at B with bubble columns initiated at preferred sites of nucleation centres on the surface. Over the nucleate boiling region, BC, the bubble sites become more numerous with increasing flux until, at C, the surface is completely covered. In the majority of commercial evaporation processes the heating medium is a fluid and therefore the controlling parameter is the overall temperature difference. If an attempt is made to increase the heat flux beyond that at C, by increasing the temperature difference, the nucleate boiling mechanism will partially collapse and portions of the surface will be exposed to vapour blanketing. In the region of transition boiling CD the average heat transfer coefficient, and frequently the heat flux, will decrease with increasing temperature difference, due to the increasing proportion of the surface exposed to vapour. This self-compensating behaviour is not exhibited if heat flux rather than temperature difference is the controlling parameter. In this case an attempt to increase the heat flux beyond point C will cause the nucleate boiling regime to collapse completely, exposing the whole surface to a vapour film. The inferior heat transfer characteristics of the vapour mean that the surface temperature must rise to E in order to dissipate the heat. In many instances this temperature exceeds the melting point of the surface and results can be disastrous. For obvious reasons the point C is generally known as *burnout*, although the terms *departure from nucleate boiling (DNB point)* and *maximum heat flux* are in common usage. In the design of evaporators, a method of predicting the heat transfer coefficient in nucleate boiling $h_b$, and the maximum heat flux which might be expected before $h_b$ begins to decrease, is of extreme importance. The complexity of the nucleate boiling process has been the subject of many studies. In a review of the available correlations for nucleate boiling, WESTWATER[1] has presented some fourteen equations. PALEN and TABOREK[2] reduced this list to seven and tested these against selected experimental data[3,4]. As a result of this study two equations, those due to MCNELLY[5] and GILMOUR[6], were selected as the most accurate. Although the modified form of the Gilmour equation is somewhat more accurate, the relative simplicity of the McNelly equation is attractive and this equation is given in dimensionless form as:

$$\left[\frac{h_b d}{k}\right] = 0.225\left[\frac{C_p \mu_L}{k}\right]^{0.69}\left[\frac{qd}{\lambda \mu_L}\right]^{0.69}\left[\frac{Pd}{\sigma}\right]^{0.31}\left[\frac{\rho_L}{\rho_v} - 1\right]^{0.31} \quad (14.2)$$

The inclusion of the characteristic dimension $d$ is necessary dimensionally, though its value does not affect the result obtained for $h_b$.

This equation predicts the heat transfer coefficient for a single isolated tube and is not applicable to tube bundles, for which PALEN and TABOREK[2] showed that the use of this equation would have resulted in 50–250 per cent underdesign in a number of specific cases. The reason for this discrepancy may be explained as follows. In the case of a tube bundle, only the lowest tube in each vertical row is completely irrigated by the liquid with higher tubes being exposed to liquid–vapour mixtures. This partial vapour blanketing results in a lower average heat transfer coefficient for tube bundles than the value given by equation 14.2. In order to calculate these average values of $h$ for a tube bundle, equations of the form $h = C_s h_b$ have been suggested[2] where the surface factor $C_s$ is less than 1 and is, as might be expected, a function of the number of tubes in a vertical row, the pitch of the tubes, and the basic value of $h_b$. The factor $C_s$ can only be determined by statistical analysis of experimental data and further work is necessary before it can be predicted from a physical model for the process.

The single tube values for $h_b$ have been correlated by equation 14.2, which applies to the true nucleate boiling regime and takes no account of the factors which eventually lead to the maximum heat flux being approached. As discussed in Volume 1, Chapter 9, equations for *maximum flux*, often a limiting factor in evaporation processes, have been tested by PALEN and TABOREK[2], though the simplified equation of ZUBER[7] is recommended. This takes the form:

$$q_{\max} = \frac{\pi}{24}\lambda\rho_v\left[\frac{\sigma g(\rho_L - \rho_v)}{\rho_v^2}\right]^{1/4}\left[\frac{\rho_L + \rho_v}{\rho_L}\right]^{1/2} \tag{14.3}$$

where: $q_{\max}$ is the maximum heat flux,
$\lambda$ is the latent heat of vaporisation,
$\rho_L$ is the density of liquid,
$\rho_v$ is the density of vapour,
$\sigma$ is the interfacial tension, and
$g$ is the acceleration due to gravity.

### 14.2.3. Forced convection boiling

The performance of evaporators operating with forced convection depends very much on what happens when a liquid is vaporised during flow through a vertical tube. If the liquid enters the tube below its boiling point, then the first section operates as a normal heater and the heat transfer rates are determined by the well-established equations for single phase flow. When the liquid temperature reaches the boiling point corresponding to the local pressure, boiling commences. At this stage the vapour bubbles are dispersed in the continuous liquid phase although progressive vaporisation of the liquid gives rise to a number of characteristic flow patterns which are shown in Figure 14.3. Over the initial boiling section convective heat transfer occurs with vapour bubbles dispersed in the liquid. Higher up, the tube bubbles become more numerous and elongated, and bubble coalescence occurs and eventually the bubbles form slugs which later collapse to give an annular flow regime in which vapour forms the central core with a thin film of liquid carried up the wall. In the final stage, dispersed flow with liquid entrainment in the vapour core occurs. In general, the conditions existing in the tube are those of annular flow. With further evaporation, the rising liquid film becomes progressively thinner and this thinning, together with the increasing vapour core velocity, eventually causes breakdown of the liquid film, leading to dry wall conditions.

For boiling in a tube, there is therefore a contribution from nucleate boiling arising from bubble formation, together with forced convection boiling due to the high velocity liquid–vapour mixture. Such a system is inherently complex since certain parameters influence these two basic processes in different ways.

Dengler and Addoms[8] measured heat transfer to water boiling in a 6 m tube and found that the heat flux increased steadily up the tube as the percentage of vapour increased, as shown in Figure 14.4. Where convection was predominant, the data were correlated using the ratio of the observed two-phase heat transfer coefficient ($h_{tp}$) to that which would be obtained had the same total mass flow been all liquid ($h_L$) as the ordinate. As discussed in Volume 6, Chapter 12, this ratio was plotted against the reciprocal of $X_{tt}$, the parameter for two-phase turbulent flow developed by Lockhart and Martinelli[9]. The liquid coefficient $h_L$ is given by:

$$h_L = 0.023\left[\frac{k}{d_t}\right]\left[\frac{4W}{\pi d_t \mu_L}\right]^{0.8}\left[\frac{C_p\mu_L}{k}\right]^{0.4} \tag{14.4}$$

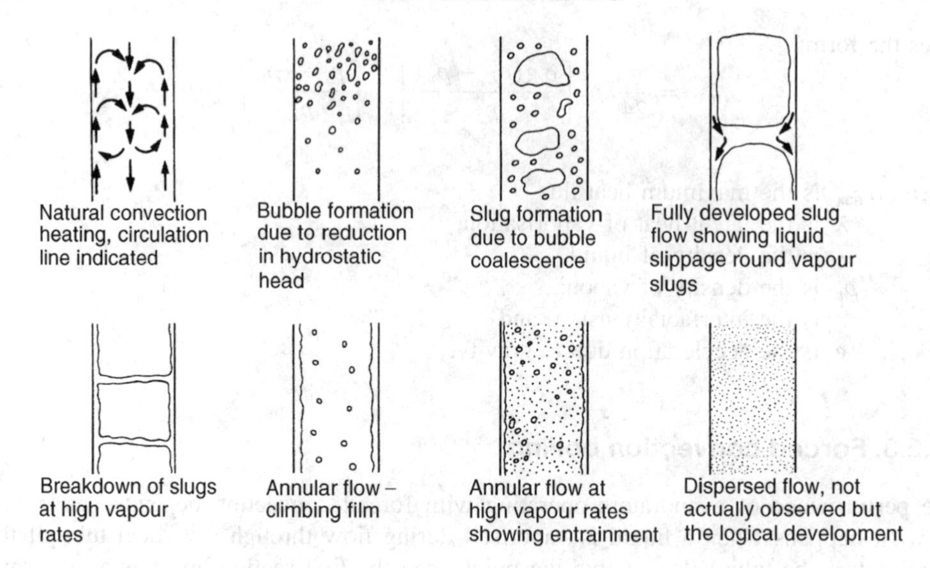


Figure 14.3. The nature of two-phase flow in an evaporator tube

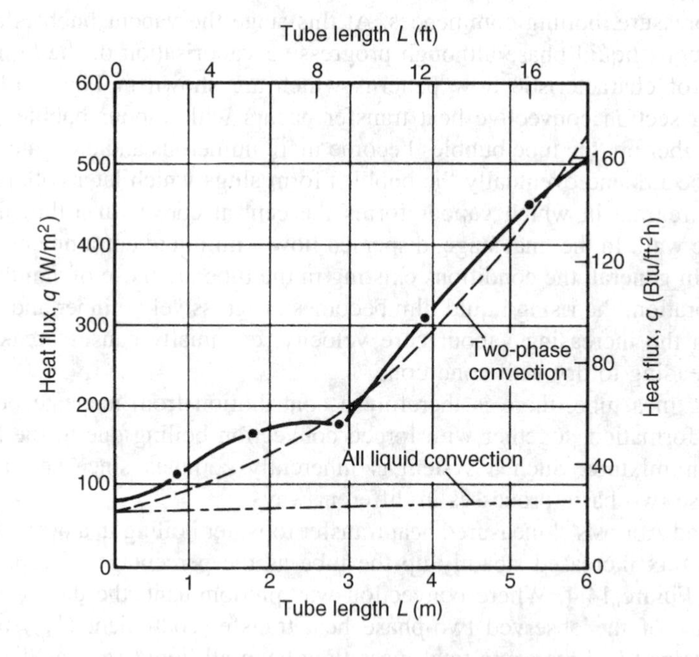


Figure 14.4. Variation of the heat flux to water in an evaporator tube[8]

where $W$ is the total mass rate of flow. The parameter $1/X_{tt}$ is given by:

$$\frac{1}{X_{tt}} = \left[\frac{y}{1-y}\right]^{0.9} \left[\frac{\rho_L}{\rho_v}\right]^{0.5} \left[\frac{\mu_v}{\mu_L}\right]^{0.1} \tag{14.5}$$

$1/X_{tt}$ is strongly dependent on the mass fraction of vapour $y$. The density and viscosity terms give a quantitative correction for the effect of pressure in the absence of nucleate boiling.

Eighty-five per cent of the purely convective data for two-phase flow were correlated to within 20 per cent by the expression:

$$\frac{h_{tp}}{h_L} = 3.5\left[\frac{1}{X_{tt}}\right]^{0.5} \quad \text{where } 0.25 < \frac{1}{X_{tt}} < 70 \tag{14.6}$$

Similar results for a range of organic liquids are reported by GUERRIERI and TALTY[10], though, in this work, $h_L$ is based on the point mass flowrate of the unvaporised part of the stream, that is, $W$ is replaced by $W(1-y)$ in equation 14.4.

One unusual characteristic of equation 14.2 is the dependence of $h_b$ on the heat flux $q$. The calculation of $h_b$ presents no difficulty in situations where the controlling parameter is the heat flux, as is the case with electrical heating. If a value of $q$ is selected, this together with a knowledge of operating conditions and the physical properties of the boiling liquid permits the direct calculation of $h_b$. The surface temperature of the heater may now be calculated from $q$ and $h_b$ and the process is described completely. Considering the evaluation of a process involving heat transfer from steam condensing at temperature $T_c$ to a liquid boiling at temperature $T_b$, assuming that the condensing coefficient is constant and specified as $h_c$, and also that the thermal resistance of the intervening wall is negligible, an initial estimate of the wall temperature $T_w$ may be made. The heat flux $q$ for the condensing film may now be calculated since $q = h_c(T_c - T_w)$, and the value of $h_b$ may then be determined from equation 14.2 using this value for the heat flux. A heat balance across the wall tests the accuracy of the estimated value of $T_w$ since $h_c(T_c - T_w)$ must equal $h_b(T_w - T_b)$, assuming the intervening wall to be plane. If the error in this heat balance is unacceptable, further values of $T_w$ must be assumed until the heat balance falls within specified limits of accuracy.

A more refined design procedure would include the estimation of the steam-side coefficient $h_c$ by one of the methods discussed in Volume 1, Chapter 9. Whilst such iterative procedures are laborious when carried out by hand, they are ideally handled by computers which enable a rapid evaluation to any degree of accuracy to be easily achieved.

### 14.2.4. Vacuum operation

With a number of heat sensitive liquids it is necessary to work at low temperatures, and this is effected by boiling under a vacuum, as indeed is the case in the last unit of a multi-effect system. Operation under a vacuum increases the temperature difference between the steam and boiling liquid as shown in Table 14.1 and therefore tends to increase the heat flux. At the same time, the reduced boiling point usually results in a more viscous material and a lower film heat transfer coefficient.

For a standard evaporator using steam at 135 kN/m$^2$ and 380 K with a total heat content of 2685 kJ/kg, evaporating a liquor such as water, the capacity under vacuum is $(101.3/13.5) = 7.5$ times great than that at atmospheric pressure. The advantage in capacity for the same unit is therefore considerable, though there is no real change in the consumption of steam in the unit. In practice, the advantages are not as great as this since

Table 14.1. Advantages of vacuum operation

| | Atmospheric pressure (101.3 $kN/m^2$) | Vacuum Operation (13.5 $kN/m^2$) |
|---|---|---|
| Boiling point | 373 K | 325 K |
| Temperature drop to liquor | 7 deg K | 55 deg K |
| Heat lost in condensate | 419 kJ/kg | 216 kJ/kg |
| Heat used | 2266 kJ/kg | 2469 kJ/kg |

operation at a lower boiling point reduces the value of the heat transfer coefficient and additional energy is required to achieve and maintain the vacuum.

## 14.3. SINGLE-EFFECT EVAPORATORS

Single-effect evaporators are used when the throughput is low, when a cheap supply of steam is available, when expensive materials of construction must be used as is the case with corrosive feedstocks and when the vapour is so contaminated so that it cannot be reused. Single effect units may be operated in batch, semi-batch or continuous batch modes or continuously. In strict terms, batch units require that filling, evaporating and emptying are consecutive steps. Such a method of operation is rarely used since it requires that the vessel is large enough to hold the entire charge of feed and that the heating element is low enough to ensure that it is not uncovered when the volume is reduced to that of the product. Semi-batch is the more usual mode of operation in which feed is added continuously in order to maintain a constant level until the entire charge reaches the required product density. Batch-operated evaporators often have a continuous feed and, over at least part of the cycle, a continuous discharge. Often a feed drawn from a storage tank is returned until the entire contents of the tank reach the desired concentration. The final evaporation is then achieved by batch operation. In essence, continuous evaporators have a continuous feed and discharge and concentrations of both feed and discharge remain constant.

The heat requirements of single-effect continuous evaporators may be obtained from mass and energy balances. If enthalpy data or heat capacity and heat of solution data are not available, heat requirements may be taken as the sum of the heat needed to raise the feed from feed to product temperature and the heat required to evaporate the water. The latent heat of water is taken at the vapour head pressure instead of the product temperature in order to compensate, at least to some extent, for the heat of solution. If sufficient vapour pressure data are available for the liquor, methods are available for calculating the true latent heat from the slope of the Dühring line and detailed by OTHMER[11]. The heat requirements in batch operation are generally similar to those in continuous evaporation. Whilst the temperature and sometimes the pressure of the vapour will change during the course of the cycle which results in changes in enthalpy, since the enthalpy of water vapour changes only slightly with temperature, the differences between continuous and batch heat requirements are almost negligible for all practical purposes. The variation of the fluid properties, such as viscosity and boiling point rise, have a much greater effect on heat transfer, although these can only be estimated by a step-wise calculation. In estimating the boiling temperature, the effect of temperature on the heat transfer characteristics of the type of unit involved must be taken into account. At low temperatures some evaporator types show a marked drop in the heat transfer coefficient which is often more than enough to offset any gain in available temperature difference. The temperature and cost of the cooling water fed to the condenser are also of importance in this respect.

## Example 14.1

A single-effect evaporator is used to concentrate 7 kg/s of a solution from 10 to 50 per cent solids. Steam is available at 205 $kN/m^2$ and evaporation takes place at 13.5 $kN/m^2$. If the overall coefficient of heat transfer is 3 $kW/m^2$ deg K, estimate the heating surface required and the amount of steam used if the feed to the evaporator is at 294 K and the condensate leaves the heating space at 352.7 K. The specific heats of 10 and 50 per cent solutions are 3.76 and 3.14 kJ/kg deg K respectively.

## Solution

Assuming that the steam is dry and saturated at 205 $kN/m^2$, then from the Steam Tables in the Appendix, the steam temperature = 394 K at which the total enthalpy = 2530 kJ/kg.

At 13.5 $kN/m^2$, water boils at 325 K and, in the absence of data on the boiling point elevation, this will be taken as the temperature of evaporation, assuming an aqueous solution. The total enthalpy of steam at 325 K is 2594 kJ/kg.

Thus the feed, containing 10 per cent solids, has to be heated from 294 to 325 K at which temperature the evaporation takes place.

$$\text{In the feed, mass of dry solids} = (7 \times 10)/100 = 0.7 \text{ kg/s}$$

and, for x kg/s of water in the product:

$$(0.7 \times 100)/(0.7 + x) = 50$$

from which: $$x = 0.7 \text{ kg/s}$$

Thus: $$\text{water to be evaporated} = (7.0 - 0.7) - 0.7 = 5.6 \text{ kg/s}$$

Summarising:

| Stream | Solids (kg/s) | Liquid (kg/s) | Total (kg/s) |
|---|---|---|---|
| Feed | 0.7 | 6.3 | 7.0 |
| Product | 0.7 | 0.7 | 1.4 |
| Evaporation | | 5.6 | 5.6 |

Using a datum of 273 K:

$$\text{Heat entering with the feed} = (7.0 \times 3.76)(294 - 273) = 552.7 \text{ kW}$$

$$\text{Heat leaving with the product} = (1.4 \times 3.14)(325 - 273) = 228.6 \text{ kW}$$

$$\text{Heat leaving with the evaporated water} = (5.6 \times 2594) = 14,526 \text{ kW}$$

Thus:

$$\text{Heat transferred from the steam} = (14526 + 228.6) - 552.7 = 14,202 \text{ kW}$$

The enthalpy of the condensed steam leaving at 352.7 K = 4.18(352.7 − 273) = 333.2 kJ/kg

The heat transferred from 1 kg steam = (2530 − 333.2) = 2196.8 kJ/kg

and hence:

$$\text{Steam required} = (14,202/2196.8) = \underline{\underline{6.47 \text{ kg/s}}}$$

As the preheating of the solution and the sub-cooling of the condensate represent but a small proportion of the heat load, the temperature driving force may be taken as the difference between the temperatures of the condensing steam and the evaporating water, or:

$$\Delta T = (394 - 325) = 69 \text{ deg K}$$

Thus:

$$\text{Heat transfer area, } A = Q/U\Delta T \quad \text{(equation 14.1)}$$

$$= 14,202/(3 \times 69) = \underline{\underline{68.6 \text{ m}^2}}$$

## 14.4. MULTIPLE-EFFECT EVAPORATORS

The single effect evaporator uses rather more than 1 kg of steam to evaporate 1 kg of water. Three methods have been introduced which enable the performance to be improved, either by direct reduction in the steam consumption, or by improved energy efficiency of the whole unit. These are:

(a) Multiple effect operation
(b) Recompression of the vapour rising from the evaporator
(c) Evaporation at low temperatures using a heat pump cycle.

The first of these is considered in this section and (b) and (c) are considered in Section 14.5.

### 14.4.1. General principles

If an evaporator, fed with steam at 399 K with a total heat of 2714 kJ/kg, is evaporating water at 373 K, then each kilogram of water vapour produced will have a total heat content of 2675 kJ. If this heat is allowed to go to waste, by condensing it in a tubular condenser or by direct contact in a jet condenser for example, such a system makes very poor use of steam. The vapour produced is, however, suitable for passing to the calandria of a similar unit, provided the boiling temperature in the second unit is reduced so that an adequate temperature difference is maintained. This, as discussed in Section 14.2.4, can be effected by applying a vacuum to the second effect in order to reduce the boiling point of the liquor. This is the principle reached in the multiple effect systems which were introduced by Rillieux in about 1830.

For three evaporators arranged as shown in Figure 14.5, in which the temperatures and pressures are $T_1$, $T_2$, $T_3$, and $P_1$, $P_2$, $P_3$, respectively, in each unit, if the liquor has no

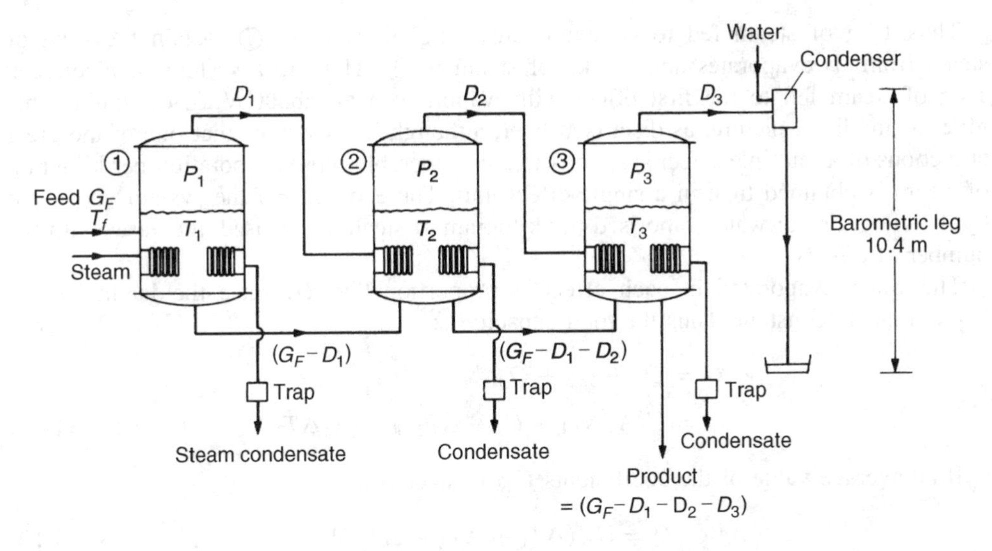


Figure 14.5. Forward-feed arrangement for a triple-effect evaporator

boiling point rise, then the heat transmitted per unit time across each effect is:

Effect 1 $\quad Q_1 = U_1 A_1 \Delta T_1$, where $\Delta T_1 = (T_0 - T_1)$,

Effect 2 $\quad Q_2 = U_2 A_2 \Delta T_2$, where $\Delta T_2 = (T_1 - T_2)$,

Effect 3 $\quad Q_3 = U_3 A_3 \Delta T_3$, where $\Delta T_3 = (T_2 - T_3)$.

Neglecting the heat required to heat the feed from $T_f$ to $T_1$, the heat $Q_1$ transferred across where $A_1$ appears as latent heat in the vapour $D_1$ and is used as steam in the second effect, and:

$$Q_1 = Q_2 = Q_3$$

So that:
$$U_1 A_1 \Delta T_1 = U_2 A_2 \Delta T_2 = U_3 A_3 \Delta T_3 \tag{14.7}$$

If, as is commonly the case, the individual effects are identical, $A_1 = A_2 = A_3$, and:

$$U_1 \Delta T_1 = U_2 \Delta T_2 = U_3 \Delta T_3 \tag{14.8}$$

On this analysis, the difference in temperature across each effect is inversely proportional to the heat transfer coefficient. This represents a simplification, however, since:

(a) the heat required to heat the feed from $T_f$ to $T_1$ has been neglected, and
(b) the liquor passing from stages ① to ② carries heat into the second effect, and this is responsible for some evaporation. This is also the case in the third effect.

The latent heat required to evaporate 1 kg of water in ①, is approximately equal to the heat obtained in condensing 1 kg of steam at $T_0$.

Thus 1 kg of steam fed to ① evaporates 1 kg of water in ①. Again the 1 kg of steam from ① evaporates about 1 kg of steam in ②. Thus, in a system of $N$ effects, 1 kg of steam fed to the first effect will evaporate in all about $N$ kg of liquid. This gives a simplified picture, as discussed later, although it does show that one of the great attractions of a multiple-effect system is that considerably more evaporation per kilogram of steam is obtained than in a single-effect unit. The economy of the system, measured by the kilograms of water vaporised per kilogram of steam condensed, increases with the number of effects.

The water evaporated in each effect is proportional to $Q$, since the latent heat is approximately constant. Thus the total capacity is:

$$\begin{aligned} Q &= Q_1 + Q_2 + Q_3 \\ &= U_1 A_1 \Delta T_1 + U_1 A_2 \Delta T_2 + U_3 A_3 \Delta T_3 \end{aligned} \tag{14.9}$$

If an average value of the coefficients $U_{av}$ is taken, then:

$$Q = U_{av}(\Delta T_1 + \Delta T_2 + \Delta T_3)A \tag{14.10}$$

assuming the area of each effect is the same. A single-effect evaporator operating with a temperature difference $\Sigma\Delta T$, with this average coefficient $U_{av}$, would, however, have the same capacity $Q = U_{av} A \Sigma\Delta T$. Thus, it is seen that the capacity of a multiple-effect system is the same as that of a single effect, operating with the same total temperature difference and having an area $A$ equal to that of one of the multiple-effect units. The value of the multiple-effect system is that better use is made of steam although, in order to achieve this, a much higher capital outlay is required for the increased number of units and accessories.

### 14.4.2. The calculation of multiple-effect systems

In the equations considered in Section 14.4.1, various simplifying assumptions have been made which are now considered further in the calculation of a multiple-effect system. In particular, the temperature distribution in such a system and the heat transfer area required in each effect are determined. The method illustrated in Example 14.2 is essentially based on that of HAUSBRAND[12].

### Example 14.2A (Forward-feed)

4 kg/s (14.4 tonne/hour) of a liquor containing 10 per cent solids is fed at 294 K to the first effect of a triple-effect unit. Liquor with 50 per cent solids is to be withdrawn from the third effect, which is at a pressure of 13 $kN/m^2$ ($\sim$0.13 bar). The liquor may be assumed to have a specific heat of 4.18 kJ/kg K and to have no boiling point rise. Saturated dry steam at 205 $kN/m^2$ is fed to the heating element of the first effect, and the condensate is removed at the steam temperature in each effect as shown in Figure 14.5.

If the three units are to have equal areas, estimate the area, the temperature differences and the steam consumption. Heat transfer coefficients of 3.1, 2.0 and 1.1 $kW/m^2$ K for the first, second, and third effects respectively, may be assumed.

## Solution 1

A precise theoretical solution is neither necessary nor possible, since during the operation of the evaporator, variations of the liquor levels, for example, will alter the heat transfer coefficients and hence the temperature distribution. It is necessary to assume values of heat transfer coefficients, although, as noted previously, these will only be approximate and will be based on practical experience with similar liquors in similar types of evaporators.

Temperature of dry saturated steam at 205 kN/m$^2$ = 394 K.

At a pressure of 13 kN/m$^2$ (0.13 bar), the boiling point of water is 325 K, so that the total temperature difference $\Sigma\Delta T = (394 - 325) = 69$ deg K.

*First Approximation.*

Assuming that: $U_1\Delta T_1 = U_2\Delta T_2 = U_3\Delta T_3$ (equation 14.8)

then substituting the values of $U_1$, $U_2$ and $U_3$ and $\Sigma\Delta T = 69$ deg K gives:

$$\Delta T_1 = 13 \text{ deg K}, \quad \Delta T_2 = 20 \text{ deg K}, \quad \Delta T_3 = 36 \text{ deg K}$$

Since the feed is cold, it will be necessary to have a greater value of $\Delta T_1$ than given by this analysis. It will be assumed that $\Delta T_1 = 18$ deg K, $\Delta T_2 = 17$ deg K, $\Delta T_3 = 34$ deg K.

If the latent heats are given by $\lambda_0$, $\lambda_1$, $\lambda_2$ and $\lambda_3$, then from the Steam Tables in the Appendix:

For steam to 1: $T_0 = 394$ K and $\lambda_0 = 2200$ kJ/kg
For steam to 2: $T_1 = 376$ K and $\lambda_1 = 2249$ kJ/kg
For steam to 3: $T_2 = 359$ K and $\lambda_2 = 2293$ kJ/kg
$T_3 = 325$ K and $\lambda_3 = 2377$ kJ/kg

Assuming that the condensate leaves at the steam temperature, then heat balances across each effect may be made as follows:

Effect 1:

$$D_0\lambda_0 = G_F C_p(T_1 - T_f) + D_1\lambda_1, \text{ or } 2200\ D_0 = 4 \times 4.18(376 - 294) + 2249\ D_1$$

Effect 2:

$$D_1\lambda_1 + (G_F - D_1)C_p(T_1 - T_2) = D_2\lambda_2, \text{ or } 2249\ D_1 + (4 - D_1)4.18(376 - 359) = 2293\ D_2$$

Effect 3:

$$D_2\lambda_2 + (G_F - D_1 - D_2)C_p(T_2 - T_3) = D_3\lambda_3,$$
$$\text{or } 2293\ D_2 + (4 - D_1 - D_2)4.18(359 - 325) = 2377\ D_3$$

where $G_F$ is the mass flowrate of liquor fed to the system, and $C_p$ is the specific heat capacity of the liquid, which is assumed to be constant.

A material balance over the evaporator is:

| | Solids (kg/s) | Liquor (kg/s) | Total (kg/s) |
|---|---|---|---|
| Feed | 0.4 | 3.6 | 4.0 |
| Product | 0.4 | 0.4 | 0.8 |
| Evaporation | | 3.2 | 3.2 |

Making use of the previous equations and the fact that $(D_1 + D_2 + D_3) = 3.2$ kg/s, the evaporation in each unit is, $D_1 \approx 0.991$, $D_2 \approx 1.065$, $D_3 \approx 1.144$, $D_0 \approx 1.635$ kg/s. The area of the surface of each calandria necessary to transmit the necessary heat under the given temperature difference may then be obtained as:

$$A_1 = \frac{D_0\lambda_0}{U_1\Delta T_1} = \frac{(1.635 \times 2200)}{(3.1 \times 18)} = 64.5 \text{ m}^2$$

$$A_2 = \frac{D_1\lambda_1}{U_2\Delta T_2} = \frac{(0.991 \times 2249)}{(2.0 \times 17)} = 65.6 \text{ m}^2$$

$$A_3 = \frac{D_2\lambda_2}{U_3\Delta T_3} = \frac{(1.085 \times 2293)}{(1.1 \times 34)} = 65.3 \text{ m}^2$$

These three calculated areas are approximately equal, so that the temperature differences assumed may be taken as nearly correct. In practice, $\Delta T_1$ would have to be a little larger since $A_1$ is the smallest area. It may be noted that, on the basis of these calculations, the economy is given by $e = (3.2/1.635) = \underline{\underline{2.0}}$. Thus, a triple effect unit working under these conditions gives a reduction in steam utilisation compared with a single effect, though not as large an economy as might be expected.

A simplified method of solving problems of multiple effect evaporation, suggested by STORROW[13], is particularly useful for systems with a large number of effects because it obviates the necessity for solving many simultaneous equations. Essentially the method depends on obtaining only an approximate value for those heat quantities which are a small proportion of the whole. Example 14.2A is now solved by this method.

## Solution 2

From Figure 14.5 it may be seen that for a feed $G_F$ to the first effect, vapour $D_1$ and liquor $(G_F - D_1)$ are fed forward to the second effect. In the first effect, steam is condensed partly in order to raise the feed to its boiling point and partly to effect evaporation. In the second effect, further vapour is produced mainly as a result of condensation of the vapour from the first effect and to a smaller extent by flash vaporisation of the concentrated liquor which is fed forward. As the amount of vapour produced by the latter means is generally only comparatively small, this may be estimated only approximately. Similarly, the vapour produced by flash evaporation in the third effect will be a small proportion of the total and only an approximate evaluation is required.

### *Vapour production by flash vaporisation — approximate evaluation*

If the heat transferred in each effect is the same, then:

$$U_1\Delta T_1 = U_2\Delta T_2 = U_3\Delta T_3 \qquad \text{(equation 14.8)}$$

or:

$$3.1\Delta T_1 = 2.0\Delta T_2 = 1.1\Delta T_3$$

Steam temperature = 394 K. Temperature in condenser = 325 K.

Thus: $\Sigma\Delta T = (394 - 325) = 69$ deg K

Solving: $\Delta T_1 = 13$ deg K $\quad \Delta T_2 = 20$ deg K $\quad \Delta T_3 = 36$ deg K

These values of $\Delta T$ will be valid provided the feed is approximately at its boiling point.

Weighting the temperature differences to allow for the fact that the feed enters at ambient temperature gives:

$$\Delta T_1 = 18 \text{ deg K} \quad \Delta T_2 = 18 \text{ deg K} \quad \Delta T_3 = 33 \text{ deg K}$$

and the temperatures in each effect are:

$$T_1 = 376 \text{ K} \quad T_2 = 358 \text{ K} \quad \text{and} \quad T_3 = 325 \text{ K}$$

The total evaporation $(D_1 + D_2 + D_3)$ is obtained from a material balance:

| | Solids (kg/s) | Liquor (kg/s) | Total (kg/s) |
|---|---|---|---|
| Feed | 0.4 | 3.6 | 4.0 |
| Product | 0.4 | 0.4 | 0.8 |
| Evaporation | | 3.2 | 3.2 |

Assuming, as an approximation, equal evaporation in each effect, or $D_1 = D_2 = D_3 = 1.07$ kg/s, then the latent heat of flash vaporisation in the second effect is given by:

$$4.18(4.0 - 1.07)(376 - 358) = 220.5 \text{ kW}$$

and latent heat of flash vaporisation in the third effect is:

$$4.18(4.0 - 2 \times 1.07)(358 - 325) = 256.6 \text{ kW}$$

## *Final calculation of temperature differences*

Subsequent calculations are considerably simplified if it is assumed that the latent heat of vaporisation is the same at all temperatures in the multiple-effect system, since under these conditions the condensation of 1 kg of steam gives rise to the formation of 1 kg of vapour.

Thus: At 394 K, the latent heat = 2200 kJ/kg

At 325 K, the latent heat = 2377 kJ/kg

Mean value, $\lambda$ = 2289 kJ/kg

The amounts of heat transferred in each effect $(Q_1, Q_2, Q_3)$ and in the condenser $(Q_c)$ are related by:

$$Q_1 - G_F C_p(T_1 - T_f) = Q_2 = (Q_3 - 220.5) = (Q_c - 220.5 - 256.6)$$

or: $$Q_1 - 4.0 \times 4.18(394 - \Delta T_1 - 294) = Q_2 = (Q_3 - 220.5) = (Q_c - 477.1) \text{ kW}$$

$$\text{Total evaporation} = (Q_2 + Q_3 + Q_c)/2289 = 3.2 \text{ kg/s}$$

Thus: $$Q_2 + (Q_2 + 220.5) + (Q_2 + 477.1) = 7325 \text{ kW}$$

or: $$Q_2 = 2209 \text{ kW}$$

$$Q_3 = 2430 \text{ kW}$$

and: $$Q_1 = 2209 + 4.0 \times 4.18(394 - \Delta T_1 - 294)$$
$$= (3881 - 16.72\Delta T_1) \text{ kW}$$

Applying the heat transfer equations, then:

$$3881 - 16.72\Delta T_1 = 3.1A\Delta T_1, \quad \text{or} \quad A\Delta T_1 = (1252 - 5.4\Delta T_1) \text{ m}^2\text{K}$$
$$2209 = 2.0A\Delta T_2, \quad \text{or} \quad A\Delta T_2 = 1105 \text{ m}^2\text{K}$$
$$2430 = 1.1A\Delta T_3, \quad \text{or} \quad A\Delta T_3 = 2209 \text{ m}^2\text{K}$$

Further: $$\Delta T_1 + \Delta T_2 + \Delta T_3 = 69 \text{ deg K}$$

Values of $\Delta T_1$, $\Delta T_2$, $\Delta T_3$ are now chosen by trial and error to give equal values of $A$ in each effect, as follows:

| $\Delta T_1$ (deg K) | $A_1$ ($m^2$) | $\Delta T_2$ (deg K) | $A_2$ ($m^2$) | $\Delta T_3$ (deg K) | $A_3$ ($m^2$) |
|---|---|---|---|---|---|
| 18 | 64.2 | 18 | 61.4 | 33 | 66.9 |
| 19 | 60.5 | 17 | 65.0 | 33 | 66.9 |
| 18 | 64.2 | 17.5 | 63.1 | 33.5 | 65.9 |
| 18 | 64.2 | 17 | 65.0 | 34 | 64.9 |

The areas, as calculated in the last line, are approximately equal, so that the assumed temperature differences are acceptable and:

$$\text{Steam consumption} = (Q_1/2289) = (3580/2289) = 1.56 \text{ kg/s}$$
$$\text{Economy} = (3.2/1.56) \approx \underline{\underline{2.0}} \text{ kg/kg}$$

The calculation of areas in multiple-effect systems is relatively straightforward for one or two configurations, although it becomes tedious in the extreme where a wide range of operating conditions is to be investigated. Fortunately the calculations involved lend themselves admirably to processing by computer, and in this respect reference should be made to work such as that by Stewart and Beveridge[14].

### 14.4.3. Comparison of forward and backward feeds

In the unit considered in Example 14.2A, the weak liquor is fed to effect ① and flows on to ② and then to ③. The steam is also fed to ①, and the process is known as forward-feed since the feed is to the same unit as the steam and travels down the unit in the same direction as the steam or vapour. It is possible, however, to introduce the weak liquor to effect ③ and cause it to travel from ③ to ② to ①, whilst the steam and vapour still travel in the direction of ① to ② to ③. This system, shown in Figure 14.6, is known as backward-feed. A further arrangement for the feed is known as parallel-feed, which is shown in Figure 14.7. In this case, the liquor is fed to each of the three effects in parallel although the steam is fed only to the first effect. This arrangement is commonly used in the concentration of salt solutions, where the deposition of crystals makes it difficult to use the standard forward-feed arrangement. The effect of backward-feed on the temperature distribution, the areas of surface required, and the economy of the unit is of importance, and Example 14.2A is now considered for this flow arrangement.

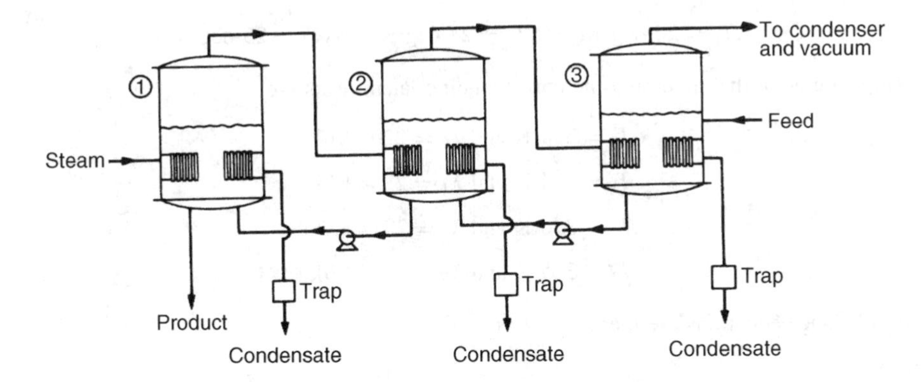


Figure 14.6. Backward-feed arrangement for a triple-effect evaporator

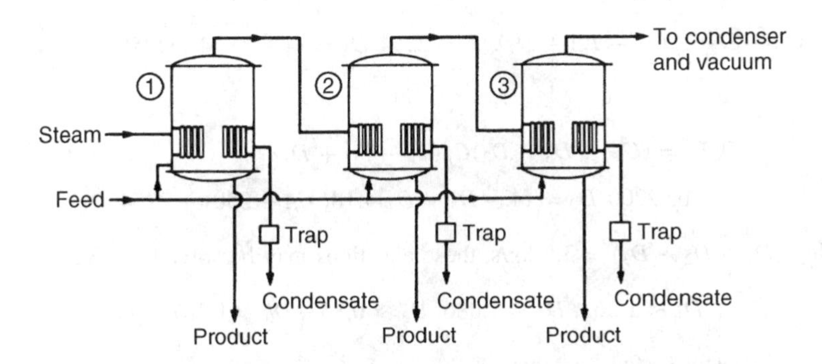


Figure 14.7. Parallel-feed arrangement for a triple-effect evaporator

## Example 14.2B (Backward-Feed)

Since the dilute liquor is now at the lowest temperature and the concentrated liquor at the highest, the heat transfer coefficients will not be the same as in the case of forward-feed. In effect ①, the liquor is now much more concentrated than in the former case, and hence $U_1$ will not be as large as before. Again, on the same argument, $U_3$ will be larger than before. Although it is unlikely to be exactly the same, $U_2$ will be taken as being unaltered by the arrangement. Taking values of $U_1 = 2.5$, $U_2 = 2.0$ and $U_3 = 1.6$ kW/m$^2$ K, the temperature distribution may be determined in the same manner as for forward feed, by taking heat balances across each unit.

## Solution 1

In this case, it is more difficult to make a reasonable first estimate of the temperature differences because the liquid temperature is increasing as it passes from effect to effect ($3 \rightarrow 2 \rightarrow 1$) and sensible heat must be added at each stage. It may therefore be necessary to make several trial and error solutions before achieving the conditions for equal areas. In addition, the values of $U_1$, $U_2$ and $U_3$ may be different from those in forward-feed, depending as they do on concentration as well as on temperature.

Taking: $\Delta T_1 = 20$ deg K, $\Delta T_2 = 24$ deg K, $\Delta T_3 = 25$ deg K

The temperatures in the effect and the corresponding latent heats are:

$$T_0 = 394 \text{ K and } \lambda_0 = 2200 \text{ kJ/kg}$$
$$T_1 = 374 \text{ K and } \lambda_1 = 2254 \text{ kJ/kg}$$
$$T_2 = 350 \text{ K and } \lambda_2 = 2314 \text{ kJ/kg}$$
$$T_3 = 325 \text{ K and } \lambda_3 = 2377 \text{ kJ/kg}$$

The heat balance equations are then:

Effect 3:

$$D_2\lambda_2 = G_F C_p(T_3 - T_f) + D_3\lambda_3, \text{ or } 2314\ D_2 = 4 \times 4.18(325 - 294) + 2377\ D_3$$

Effect 2:

$$D_1\lambda_1 = (G_F - D_3)C_p(T_2 - T_3) + D_2\lambda_2, \text{ or } 2254\ D_1 = (4 - D_3)4.18(350 - 325) + 2314\ D_2$$

Effect 1:

$$D_0\lambda_0 = (G_F - D_3 - D_2)C_p(T_1 - T_2) + D_1\lambda_1,$$
$$\text{or } 2200\ D_0 = (4 - D_3 - D_2)4.18(374 - 350) + 2254\ D_1$$

Again taking $(D_1 + D_2 + D_3) = 3.2$ kg/s, these equations may be solved to give:

$$D_1 \approx 1.261,\ D_2 \approx 1.086,\ D_3 \approx 0.853,\ D_0 \approx 1.387 \text{ kg/s}$$

The areas of transfer surface are then:

$$A_1 = \frac{D_0\lambda_0}{U_1\Delta T_1} = \frac{(1.387 \times 2200)}{(2.5 \times 20)} = 61.0 \text{ m}^2$$

$$A_2 = \frac{D_1\lambda_1}{U_2\Delta T_2} = \frac{(1.261 \times 2254)}{(2.00 \times 24)} = 59.2 \text{ m}^2$$

$$A_3 = \frac{D_2\lambda_2}{U_3\Delta T_3} = \frac{(1.086 \times 2314)}{(1.6 \times 25)} = 62.8 \text{ m}^2$$

These three areas are approximately equal, so that the temperature differences suggested are sufficiently acceptable for design purposes. The economy for this system is $(3.2/1.387) = \underline{\underline{2.3}}$ kg/kg.

## Solution 2

Using Storrow's method, as in Example 14.2A, the temperatures in the effects will be taken as:

$$T_1 = 374 \text{ K}, \quad T_2 = 350 \text{ K}, \quad T_3 = 325 \text{ K}$$

With backward-feed, as shown in Figure 14.6, the liquid has to be raised to its boiling point as it enters each effect.

The heat required to raise the feed to the second effect to its boiling point is:

$$= 4.18(4.0 - 1.07)(350 - 325)$$
$$= 306.2 \text{ kW}$$

The heat required to raise the feed to the first effect to its boiling point is:

$$= 4.18(4.0 - 2 \times 1.07)(374 - 350)$$
$$= 186.6 \text{ kW}$$

Assuming a constant value of 2289 kJ/kg for the latent heat in all the stages, the relation between the heat transferred in each effect and in the condenser is:

$$Q_1 - 186.6 = Q_2 = (Q_3 + 306.2) = (Q_c + 306.2 + 4 \times 4.18(325 - 294))$$
$$= Q_c + 824.5$$

$$\text{Total evaporation} = (Q_2 + Q_3 + Q_c)/2289 = 3.2 \text{ kg/s}$$

and: $$Q_2 + (Q_2 - 306.2) + (Q_2 - 824.5) = 7325 \text{ kW}$$

Thus: $$Q_2 = 2819 = A\Delta T_2 \times 2.0 \text{ kW}$$

$$Q_3 = 2512 = A\Delta T_3 \times 1.6 \text{ kW}$$

and: $$Q_1 = 3006 = A\Delta T_1 \times 2.5 \text{ kW}$$

or: $$A\Delta T_1 = 1202 \text{ m}^2 \text{ K}$$

$$A\Delta T_2 = 1410 \text{ m}^2 \text{ K}$$

$$A\Delta T_3 = 1570 \text{ m}^2 \text{ K}$$

and: $$\Delta T_1 + \Delta T_2 + \Delta T_3 = 69 \text{ deg K}$$

Thus:

| $\Delta T_1$ (deg K) | $A_1$ (m$^2$) | $\Delta T_2$ (deg K) | $A_2$ (m$^2$) | $\Delta T_3$ (deg K) | $A_3$ (m$^2$) |
|---|---|---|---|---|---|
| 20 | 60.1 | 24 | 58.9 | 25 | 62.8 |

The areas are approximately equal and the assumed values of $\Delta T$ are therefore acceptable.

$$\text{Economy} = \frac{3.2}{(3006/2289)} = \underline{\underline{2.4}} \text{ kg/kg}$$

On the basis of heat transfer area and thermal considerations, a comparison of the two methods of feed is:

| | Forward | Backward |
|---|---|---|
| Total steam used $D_0$ (kg) | 1.635 | 1.387 |
| Economy (kg/kg) | 2.0 | 2.3 |
| Condenser load $D_3$ (kg) | 1.44 | 0.853 |
| Heat transfer surface per effect $A$ (m$^2$) | 65.1 | 61.0 |

For the conditions of Example 14.2, the backward feed system shows a reduction in steam consumption, an improved economy, a reduction in condenser load, and a small reduction in heat transfer area.

## Effect of feed system on economy

In the case of forward feed systems, all the liquor has to be heated from $T_f$ to $T_1$ by steam although, in the case of backward feed, the heating of the feed in the last effect is done with steam that has already evaporated $(N - 1)$ times its own mass of water, assuming ideal conditions. The feed temperature must therefore be regarded as a major feature in this class of problem. WEBRE[15] has examined the effect of feed temperature on the economy and the evaporation in each effect, for the case of a liquor fed at the rate of 12.5 kg/s to a triple-effect evaporator in which a concentrated product was obtained at a flowrate of 8.75 kg/s. Neglecting boiling-point rise and working with a fixed vacuum on the third effect, the curves shown in Figures 14.8 and 14.9 for the three methods of forward, backward and parallel feed were prepared.

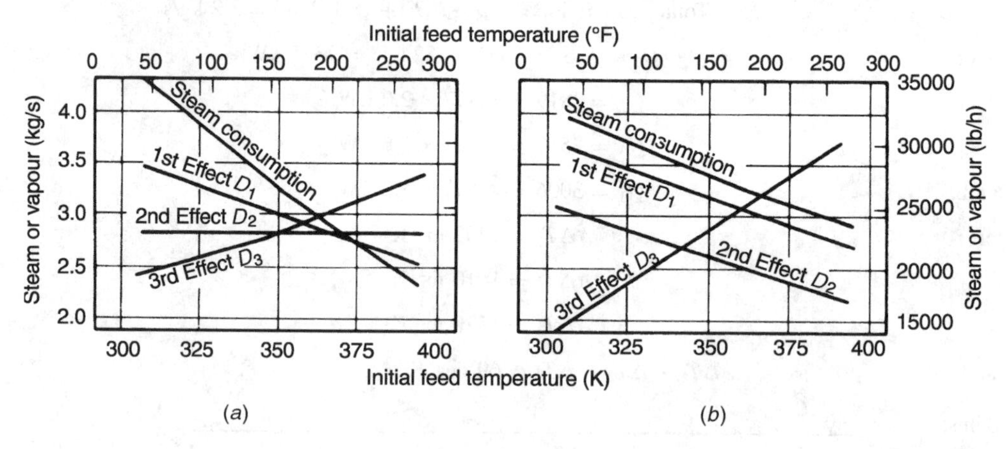


Figure 14.8. Effect of feed temperature on the operation of a triple effect evaporator (*a*) Forward feed (*b*) Backward feed

Figure 14.8*a* illustrates the drop in steam consumption as the feed temperature is increased with forward feed. It may be seen that, for these conditions, $D_1$ falls, $D_2$ remains constant and $D_3$ rises with increase in the feed temperature $T_f$. With backward feed shown in Figure 14.8*b*, the fall in steam consumption is not so marked and it may be seen that, whereas $D_1$ and $D_2$ fall, the load on the condenser $D_3$ increases. The results are conveniently interpreted in Figure 14.9, which shows that the economy increases with $T_f$ for a forward-feed system to a marked extent, whilst the corresponding increase with the backward-feed system is relatively small. At low values of $T_f$, the backward feed gives the higher economy. At some intermediate value, the two systems give the same value of economy, whilst for high values of $T_f$ the forward-feed system is more economical in steam.

These results, whilst showing the influence of $T_f$ on the economy, should not be interpreted too rigidly, since the values for the coefficients for the two systems and

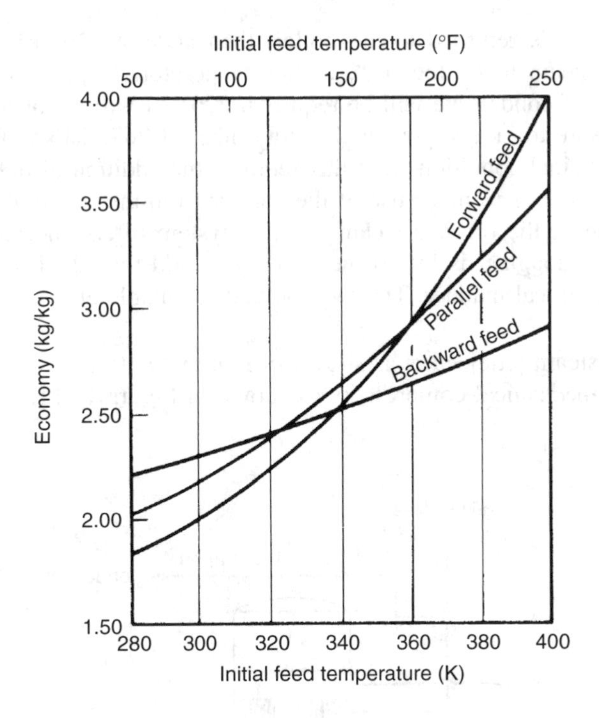


Figure 14.9. Economy of triple-effect evaporators

the influence of boiling-point rise may make a substantial difference to these curves. In general, however, it will be found that with cold feeds the backward-feed system is more economical. Despite this fact, the forward-feed system is the most common, largely because it is the simplest to operate, whilst backward feed requires the use of pumps between each effect.

The main criticism of the forward-feed system is that the most concentrated liquor is in the last effect, where the temperature is lowest. The viscosity is therefore high and low values of $U$ are obtained. In order to compensate for this, a large temperature difference is required, and this limits the number of effects. It is sometimes found, as in the sugar industry, that it is preferable to run a multiple-effect system up to a certain concentration, and to run a separate effect for the final stage where the crystals are formed.

## 14.5. IMPROVED EFFICIENCY IN EVAPORATION

### 14.5.1. Vapour compression evaporators

Considering an evaporator fed with saturated steam at 387 K, equivalent to 165 kN/m$^2$, concentrating a liquor boiling at 373 K at atmospheric pressure, if the condensate leaves at 377 K, then:

1 kg of steam at 387 K has a total heat of 2698 kJ.
1 kg of condensate at 377 K has a total heat of 437 kJ and
the heat given up is 2261 kJ/kg steam.

If this condensate is returned to the boiler, then at least 2261 kJ/kg must be added to yield 1 kg of steam to be fed back to the evaporator. In practice, of course, more heat per kilogram of condensate will be required. 2261 kJ will vaporise 1 kg of liquid at atmospheric pressure to give vapour with a total heat of 2675 kJ/kg. To regenerate 1 kg of steam in the original condition from this requires the addition of only 23 kJ. The idea of vapour compression is to make use of the vapour from the evaporator, and to upgrade it to the condition of the original steam. Such a system offers enormous advantages in thermal economy, though it is by no means easy to add the 23 kJ to each kilogram of vapour in an economical manner. The two methods available are:

(a) the use of steam-jet ejectors as shown in Figure 14.10, and:
(b) the use of mechanical compressors as shown in Figure 14.11.

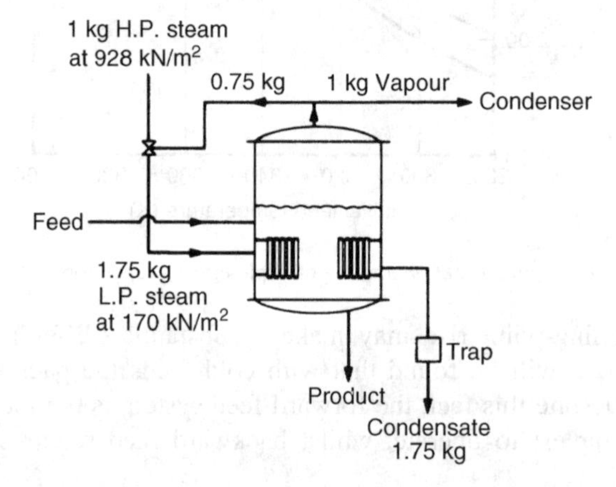


Figure 14.10. Vapour compression evaporator with high pressure steam-jet compression

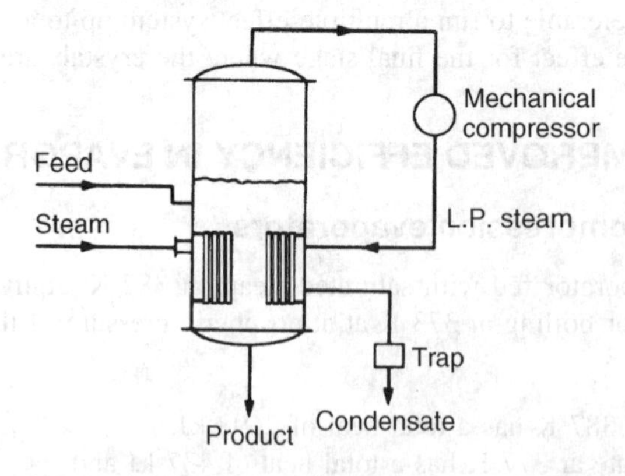


Figure 14.11. Vapour compression evaporator with a mechanical compressor

In selecting a compressor for this type of operation, the main difficulty is the very large volume of vapour to be handled. Rotary compressors of the Rootes type, described in Volume 1, Chapter 8, are suitable for small and medium size units, though these have not often been applied to large installations. Mechanical compressors have been used extensively in evaporation systems for the purification of sea water.

The use of an ejector, fed with high-pressure steam, is illustrated in Figure 14.10. High-pressure steam is injected through a nozzle and the low-pressure vapours are drawn in through a second inlet at right angles, the issuing jet of steam passing out to the calandria, as shown. These units are relatively simple in construction and can be made of corrosion-resistant material. They have no moving parts and for this reason will have a long life. They have the great advantage over mechanical compressors in that they can handle large volumes of vapour and can therefore be arranged to operate at very low pressures. The disadvantage of the steam-jet ejector is that it works at maximum efficiency at only one specific condition. Some indication of the performance of these units is shown in Figure 14.12, where the pressure of the mixture, for different amounts of vapour compressed per kilogram of live steam, is shown for a series of different pressures. With an ejector of these characteristics using steam at 965 kN/m$^2$, 0.75 kg vapour/kg steam can be compressed to give 1.75 kg of vapour at 170 kN/m$^2$. An evaporator unit, as shown in Figure 14.11, will therefore give 1.75 kg of vapour/kg high pressure steam. Of the 1.75 kg of vapour, 0.75 kg is taken to the compressor and the remaining 1 kg to the condenser. Ideally, this single-effect unit gives an economy of 1.75, or approximately the economy of a double-effect unit.

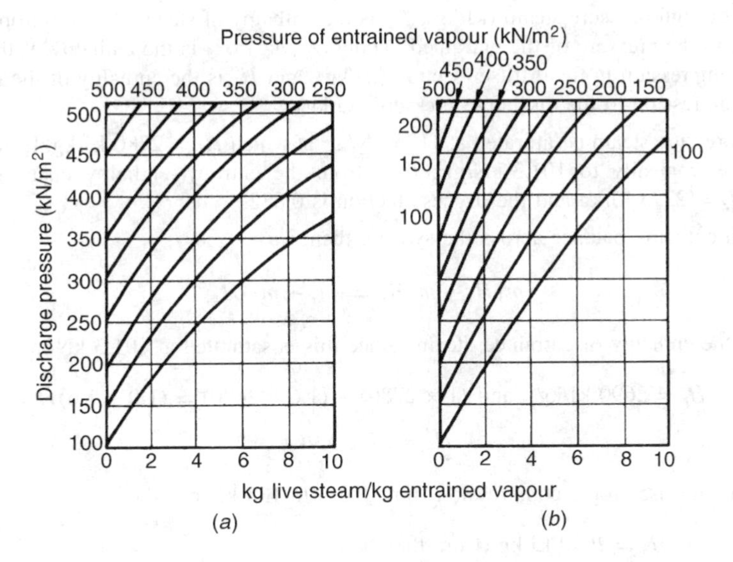


Figure 14.12. Performance of a steam jet ejector, (*a*) 790 kN/m$^2$ operating pressure, (*b*) 1135 kN/m$^2$ operating pressure

Vapour compression may be applied to the vapour from the first effect of a multiple-effect system, thus giving increased utilisation of the steam. Such a device is not suitable for use with liquors with a high boiling-point rise, for in these cases the vapour, although initially superheated, has to be compressed to such a great degree, in order to give the desired temperature difference across the calandria, that the efficiency is reduced. The application of these compressors depends on the steam load of the plant. If there is plenty of low-pressure steam available, then the use of vapour compression can rarely be advocated. If, however, high-pressure steam is available, then it may be used to advantage in a vapour compression unit. It will, in fact, be far superior to the practice of passing high-pressure steam through a reducing valve to feed an evaporator.

## Example 14.3

Saturated steam leaving an evaporator at atmospheric pressure is compressed by means of saturated steam at 1135 kN/m$^2$ in a steam jet to a pressure of 135 kN/m$^2$. If 1 kg of the high-pressure steam compresses 1.6 kg of the vapour produced at atmospheric pressure, comment on the efficiency of the compressor.

## Solution

The efficiency of an ejector $\eta'$ is given by:

$$\eta' = (m_1 + m_2)(H_4 - H_3)/[m_1(H_1 - H_2)]$$

where $m_1$ is the mass of high-pressure steam (kg), $m_2$ is the mass of entrained steam (kg), $H_1$ is the enthalpy of high-pressure steam (kJ/kg), $H_2$ is the enthalpy of steam after isentropic expansion in the nozzle to the pressure of the entrained vapours (kJ/kg), $H_3$ is the enthalpy of the mixture at the start of compression in the diffuser section (kJ/kg), and $H_4$ is the enthalpy of the mixture after isentropic compression to the discharge pressure (kJ/kg).

The high-pressure steam is saturated at 1135 kN/m$^2$ at which $H_1 = 2780$ kJ/kg. If this is allowed to expand isentropically to 101.3 kN/m$^2$, then from the entropy–enthalpy chart, given in the Appendix, $H_2 = 2375$ kJ/kg and the dryness faction is 0.882.

Making an enthalpy balance across the system, then:

$$m_1 H_1 + m_2 H_e = (m_1 + m_2)H_4$$

where $H_e$ is the enthalpy of entrained steam. Since this is saturated at 101.3 kN/m$^2$, then:

$$H_e = 2690 \text{ kJ/kg} \quad \text{and } (1 \times 2780) + (1.6 \times 2690) = (1.0 + 1.6)H_4$$

from which:

$$H_4 = 2725 \text{ kJ/kg}$$

Again assuming isentropic compression from 101.3 to 135 kN/m$^2$, then:

$$H_3 = 2640 \text{ kJ/kg (from the chart)}$$

and:

$$\eta' = (1.0 + 1.6)(2725 - 2640)/[1.0(2780 - 2375)] = \underline{\underline{0.55}}$$

This value is low, since in good design overall efficiencies approach 0.75–0.80. Obviously the higher the efficiency the greater the entrainment ratio or the higher the saving in live steam. The low efficiency is borne out by examination of Figure 14.12$b$, which applies for an operating pressure of 1135 kN/m$^2$.

Since the pressure of entrained vapour = 101.3 kN/m$^2$ and the discharge pressure = 135 kN/m$^2$, the required flow of live steam = 0.5 kg/kg entrained vapour.

In this case the ratio is (1.0/1.6) = 0.63 kg/kg.

## Example 14.4

Distilled water is produced from sea water by evaporation in a single-effect evaporator working on the vapour compression system. The vapour produced is compressed by a mechanical compressor at 50 per cent efficiency and then returned to the calandria of the evaporator. Additional steam, dry and saturated at 650 kN/m$^2$, is bled into the steam space through a throttling valve. The distilled water is withdrawn as condensate from the steam space. 50 per cent of the sea water is evaporated in the plant. The energy supplied in addition to that necessary to compress the vapour may be assumed to appear as superheat in the vapour.

Using the following data, calculate the quantity of additional steam required in kg/s.

Production of distillate = 0.125 kg/s, pressure in vapour space = 101.3 kN/m$^2$, temperature difference from steam to liquor = 8 deg K, boiling point rise of sea water = 1.1 deg K, specific heat capacity of sea water = 4.18 kJ/kg deg K. The sea water enters the evaporator at 344 K from an external heater.

## Solution

The pressure in the vapour space is 101.3 kN/m$^2$ at which pressure, water boils at 373 K. The sea water is therefore boiling at (373 + 1.1) = 374.1 K and the temperature in the steam space is (374.1 + 8) = 382.1 K. At this temperature, steam is saturated at 120 kN/m$^2$ and has sensible and total enthalpies of 439 and 2683 kJ/kg respectively.

Making a *mass balance*, there are two inlet streams — the additional steam, say $G_x$ kg/s, and the sea water feed, say $G_y$ kg/s. The two outlet streams are the distilled water product, 0.125 kg/s, and the concentrated sea water, 0.5 $G_y$ kg/s.

Thus: $$(G_x + G_y) = (0.125 + 0.5\ G_y) \text{ or } (G_x + 0.5\ G_y) = 0.125 \quad \text{(i)}$$

Making an *energy balance*, energy is supplied by the compressor and in the steam and inlet sea water and is removed by the sea water and the product. At 650 kN/m$^2$, the total enthalpy of the steam = 2761 kJ/kg. Thus the energy in this stream = $2761G_x$ kW. The sea water enters at 344 K.

Thus: $$\text{enthalpy of feed} = [G_y \times 4.18(344 - 273)] = 296.8G_y \text{ kW}$$

The sea water leaves the plant at 374.1 K and hence:

the enthalpy of the concentrated sea water = $(0.5G_y \times 4.18)(374.1 - 273) = 211.3G_y$ kW

The product has an enthalpy of 439 kJ/kg or $(439 \times 0.125) = 54.9$ kW

Making a balance:

$$(E + 2761G_x + 296.8G_y) = (211.3G_y + 54.9)$$

and: $$(E + 2761G_x + 85.5G_y) = 54.9 \quad \text{(ii)}$$

where $E$ is the power supplied to the compressor.

Substituting from equation (i) into equation (ii) gives:

$$(E + 2761G_x) + 85.5(0.25 - 2G_x) = 54.9$$

and: $$(E + 2590G_x) = 33.5 \quad \text{(iii)}$$

For a single-stage isentropic compression, the work done in compressing a volume $V_1$ of gas at pressure $P_1$ to a volume $V_2$ at pressure $P_2$ is given by equation 8.32 in Volume 1 as:

$$[P_1V_1/(\gamma - 1)][(P_2/P_1)^{\gamma-1/\gamma} - 1]$$

In the compressor, $0.5G_y$ kg/s vapour is compressed from $P_1 = 101.3$ kN/m$^2$, the pressure in the vapour space, to $P_2 = 120$ kN/m$^2$, the pressure in the calandria.

At 101.3 kN/m$^2$ and 374.1 K, the density of steam = (18/22.4)(273/374.1) = 0.586 kg/m$^3$ and hence the volumetric flowrate at pressure $P_1$ is $(0.5\ G_y/0.586) = 0.853\ G_y$ m$^3$/s

Taking $\gamma = 1.3$ for steam, then:

$$(E' \times 0.5G_y) = [(101.3 \times 0.853G_y)/(1.3 - 1)][(120/101.3)^{0..3/1..3} - 1]$$

$$0.5E'G_y = 288.0G_y(1.185^{0.231} - 1) = 11.5G_y$$

and: $$E' = 23.0 \text{ kW/(kg/s)}$$

As the compressor is 50 per cent efficient, then:

$$E = (E/0.5) = 46.0 \text{ kW/(kg/s)}$$

$$= (46.0 \times 0.5G_y) = 23.0G_y \text{ kW}$$

Substituting in equation (ii) gives:

$$(E + 2761G_x) + 85.5(E/23.0) = 54.9$$

Thus: $$2761G_x = (54.9 + 4.72E)$$

From equation (iii): $$E = (33.5 - 2590G_x)$$

and in equation (iv): $$2761G_x = 54.9 + 4.72(33.5 - 2590G_x)$$

from which: $$\underline{\underline{G_x = 0.014 \text{ kg/s}}}$$

## Example 14.5

An evaporator operating on the thermo-recompression principle employs a steam ejector to maintain atmospheric pressure over the boiling liquid. The ejector uses 0.14 kg/s of steam at 650 kN/m$^2$ and superheated by 100 deg K and produces a pressure in the steam chest of 205 kN/m$^2$. A condenser removes surplus vapour from the atmospheric pressure line.

What is the capacity and economy of the system and how could the economy be improved?

### *Data*

Properties of the ejector:

nozzle efficiency = 0.95, efficiency of momentum transfer = 0.80, efficiency of compression = 0.90.

The feed enters the evaporator at 295 K and concentrated liquor is withdrawn at the rate of 0.025 kg/s. This concentrated liquor exhibits a boiling-point rise of 10 deg K. The plant is sufficiently well lagged so that heat losses to the surroundings are negligible.

## Solution

It is assumed that $P_1$ is the pressure of live steam = 650 kN/m$^2$ and $P_2$ is the pressure of entrained steam = 101.3 kN/m$^2$.

The enthalpy of the live steam at 650 kN/m$^2$ and (435 + 100) = 535 K, $H_1$ = 2970 kJ/kg.

Therefore $H_2$, the enthalpy after isentropic expansion from 650 to 101.3 kN/m$^2$, using an enthalpy–entropy chart, is $H_2$ = 2605 kJ/kg and the dryness fraction, $x_2$ = 0.97. The enthalpy of the steam after actual expansion to 101.3 kN/m$^2$ is given by $H_2'$, where:

$$(H_1 - H_2') = 0.95(2970 - 2605) = 347 \text{ kJ/kg}$$

and:

$$H_2' = (2970 - 347) = 2623 \text{ kJ/kg}$$

At $P_2$ = 101.3 kN/m$^2$, $\lambda$ = 2258 kJ/kg

and the dryness after expansion but before entrainment $x_2'$ is given by:

$$(x_2' - x_2)\lambda = (1 - e_1)(H_1 - H_2)$$

or:

$$(x_2' - 0.97)2258 = (1 - 0.95)(2970 - 2605) \text{ and } x_2' = 0.978.$$

If $x_2''$ is the dryness after expansion *and* entrainment, then:

$$(x_2'' - x_2')\lambda = (1 - e_3)(H_1 - H_2')$$

or:

$$(x_2'' - 0.978)2258 = (1 - 0.80)(2970 - 2623) \text{ and } x_2'' = 1.00$$

Assuming that the steam at the discharge pressure $P_3$ = 205 kN/m$^2$ is also saturated, that is $x_3$ = 1.00, then from the steam chart in the Appendix, $H_3$ the enthalpy of the mixture at the start of compression in the diffuser section at 101.3 kN/m$^2$ is $H_3$ = 2675 kJ/kg. Again assuming the entrained steam is also saturated, the enthalpy of the mixture after isentropic compression in the diffuser from 101.3 to 205 kN/m$^2$, $H_4$ = 2810 kJ/kg.

The entrainment ratio is given by:

$$(m_2/m_1) = \{[(H_1 - H_2)/(H_4 - H_3)]\eta_1\eta_2\eta_3 - 1\}$$

where $\eta_1$, $\eta_2$ and $\eta_3$ are the efficiency of the nozzle, momentum transfer and compression, respectively.

Thus:

$$(m_2/m_1) = \{[(2970 - 2605)/(2810 - 2675)]0.95 \times 0.80 \times 0.90 - 1\}$$

$$= 0.85 \text{ kg vapour entrained/kg live steam}$$

It was assumed that $x_3$ = 1.0. This may be checked as follows:

$$x_3 = [x_2 + x_4(m_2/m_1)]/(1 + m_2/m_1)$$

$$= (1.0 + 1.0 \times 0.85)/(1 + 0.85) = 1.0$$

Thus with a flow of 0.14 kg/s live steam, the vapour entrained at 101.3 kN/m$^2$ is (0.14 × 0.85) = 0.12 kg/s, giving 0.26 kg/s steam saturated at 205 kN/m$^2$ to the calandria.

Allowing for a 10 deg K boiling-point rise, the temperature of boiling liquor in the unit is $T_1'$ = 383 K and taking the specific heat capacity as 4.18 kJ/kg K, then:

$$D_0\lambda_0 = G_F C_p(T_1' - T_f) + D_1\lambda_1$$

or:

$$0.26 \times 2200 = (G_F \times 4.18)(393 - 295) + (D_1 \times 2258)$$

$$572 = (368G_F + 2258D_1)$$

But: $(G_F - D_1) = 0.025$ kg/s and $D_1 = 0.214$ kg/s

Thus: the economy of system $= (0.214/0.14) = \underline{\underline{1.53}}$

The capacity, in terms of the throughput of solution, is:

$$G_F = (0.214 + 0.025) = \underline{\underline{0.239 \text{ kg/s}}}$$

Apart from increasing the efficiency of the ejector, the economy of the system might be improved by operating with a higher live-steam pressure, increasing the pressure in the vapour space, and by using the vapour not returned to the ejector to preheat the feed solution.

## 14.5.2. The heat pump cycle

The evaporation of citrus juices at temperatures up to 328 K, or of pharmaceutical products at even lower temperatures, has led to the development of an evaporator incorporating a heat-pump cycle using a separate working fluid. The use of the heat pump cycle, with ammonia as the working fluid is shown in Figure 14.13. In this arrangement, ammonia

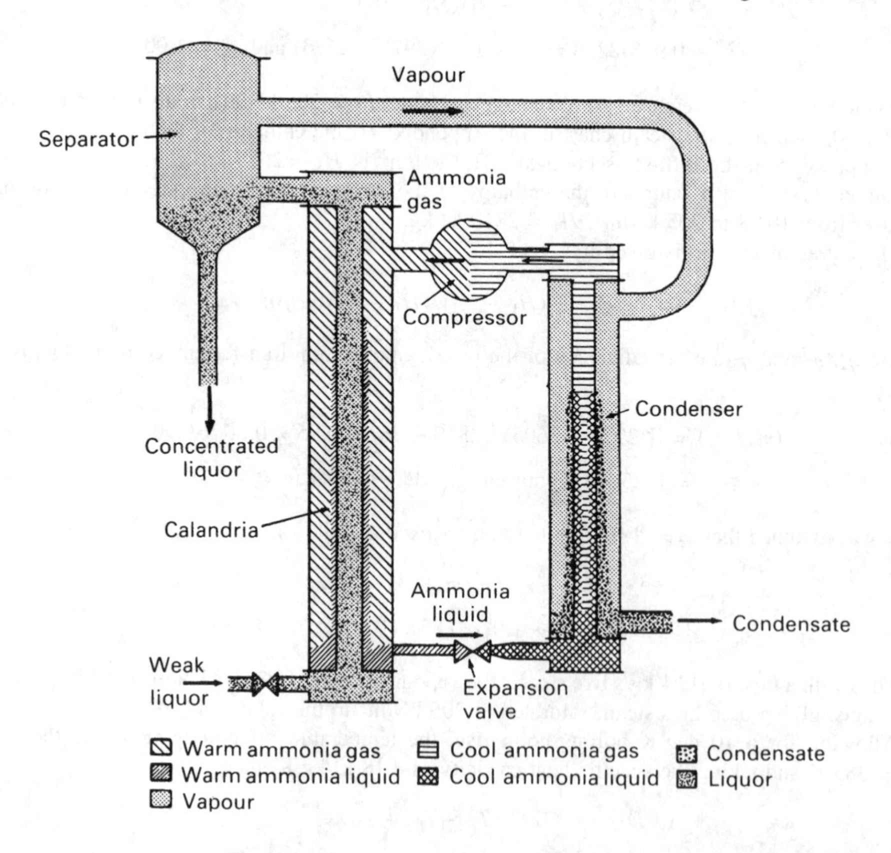


Figure 14.13. Heat pump cycle using ammonia

gas vaporises the feed liquor at 288–313 K. The ammonia is condensed and the liquid ammonia is then passed through an expansion valve, where it is cooled to a much lower temperature. The cooled liquid ammonia then enters the condenser where it condenses the vapour leaving the separator. The ammonia is vaporised and leaves as low pressure gas, to be compressed in a mechanical compressor and then passed to the evaporator for a second cycle. The excess heat introduced by the compressor must be removed from the ammonia by means of a cooler.

The main advantage of this form of unit is the very great reduction in the volume of gas handled by the compressor. Thus, 1 kg of water vapour at, say, 311 K, with a volume of 22 $m^3$ and latent heat about 2560 kJ/kg, passes this heat to ammonia at a temperature of say 305 K. About 2.1 kg of ammonia will be vaporised to give a vapour with a volume of only about 0.22 $m^3$ at the high pressure used in the ammonia cycle.

SCHWARZ[16] gives a comparison of the various units used for low temperature evaporation. The three types in general use are the single-effect single-pass, the single-effect with recirculation, and the multiple-effect with recirculation. Each of these types may involve vapour compression or the addition of a second heat transfer medium. Schwarz suggests that multiple-effect units are the most economical, in terms of capital and operating costs. It is important to note that the single-effect, single-pass system offers the minimum hold-up, and hence a very short transit time. With film-type units, there seems little to be gained by recirculation, since over 70 per cent vaporisation can be achieved in one pass. The figures in Table 14.2 show the comparison between a double-effect unit with vapour compression on the first effect, and a unit with an ammonia refrigeration cycle, both units giving 1.25 kg/s (4.5 tonne/h) of evaporation.

Table 14.2. Comparison of refrigeration and vapour compression systems

| System | Steam at 963 $kN/m^2$ (kg/s) | Water at 300 K ($m^3/s$) | Power (kW) |
|---|---|---|---|
| Refrigeration cycle | 0.062 | 0.019 | 320* |
| Vapour compression | 0.95 | 0.076 | 20 |
| Ratio of steam system to refrigeration | 15.1 | 4 | 0.06 |

*Includes 300 kW compressor.

The utilities required for the refrigeration system other than power are therefore very much less than for recompression with steam, although the capital cost and the cost of power will be much higher.

REAVELL[17] has given a comparison of costs for the concentration of a feed of a heat-sensitive protein liquor at 1.70 kg/s from 10 per cent to 50 per cent solids, on the basis of a 288 ks (160 hour) week. These data are shown in Table 14.3. It may be noted that, when using the double-effect evaporation with vapour compression, a lower temperature can be used in the first effect than when a triple-effect unit is used. In determining these figures no account has been taken of depreciation, although if this is 15 per cent of the capital costs it does not make a significant difference to the comparison.

The use of a heat pump cycle is the subject of Problem 14.22 at the end of this Volume, and a detailed discussion of the topic is given in the Solutions Manual.

Table 14.3. Comparison of various systems for the concentration of a protein liquid

| Type | Approx. installed cost (£) | Cost of steam (£/year) | Net saving compared with single effect (£/year) |
|---|---|---|---|
| Single effect | 50,000 | 403,000 | — |
| Double effect | 70,000 | 214,000 | 189,000 |
| Double effect with vapour compression | 90,000 | 137,000 | 266,000 |
| Triple effect | 100,000 | 143,000 | 260,000 |

## Example 14.6

For the concentration of fruit juice by evaporation it is proposed to use a falling-film evaporator and to incorporate a heat pump cycle with ammonia as the medium. The ammonia in vapour form will enter the evaporator at 312 K and the water will be evaporated from the juices at 287 K. The ammonia in the vapour–liquid mixture will enter the condenser at 278 K and the vapour will then pass to the compressor. It is estimated that the work for compressing the ammonia will be 150 kJ/kg of ammonia and that 2.28 kg of ammonia will be cycled/kg water evaporated. The following proposals are available for driving the compressor:

(a) to use a diesel engine drive taking 0.4 kg of fuel/MJ; the calorific value being 42 MJ/kg and the cost £0.02/kg;
(b) to pass steam, costing £0.01/10 kg through a turbine which operates at 70 per cent isentropic efficiency, between 700 and 101.3 kN/m$^2$.

Explain by means of a diagram how this plant will work, and include all necessary major items of equipment required. Which method should be adopted for driving the compressor?

A simplified flow diagram of the plant is given in Figure 14.14.

## Solution

*Considering the ammonia cycle*

Ammonia gas will leave the condenser, probably saturated at low pressure, and enter the compressor which it leaves at high pressure and 312 K. In the calandria heat will be transferred to the liquor and the ammonia gas will be cooled to saturation, condense, and indeed may possibly leave the unit at 278 K as slightly sub-cooled liquid though still at high pressure. This liquid will then be allowed to expand adiabatically in the throttling valve to the lower pressure during which some vaporisation will occur and the vapour—liquid mixture will enter the condenser with a dryness fraction of, say, 0.1–0.2. In the condenser heat will be transferred from the condensing vapours, and the liquid ammonia will leave the condenser, probably just saturated, though still at the low pressure. The cycle will then be repeated.

*Considering the liquor stream*

Weak liquor will enter the plant and pass to the calandria where it will be drawn up as a thin film by the partial vacuum caused by ultimate condensation of vapour in the condenser. Vaporisation will take place due to heat transfer from condensing ammonia in the calandria, and the vapour and concentrated liquor will then pass to a separator from which the concentrated liquor will be

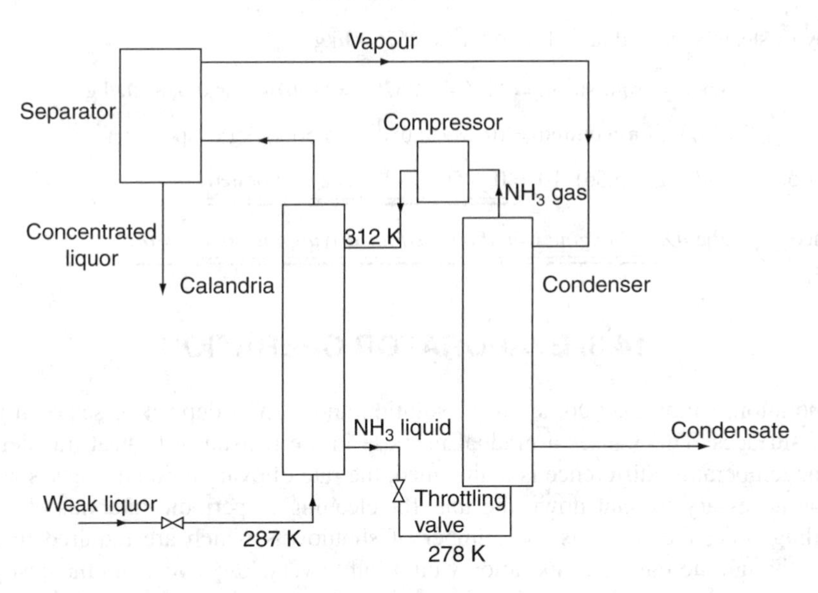


Figure 14.14. Flow diagram for Example 14.6

drawn off as product. The vapours will pass to the condenser where they will be condensed by heat transfer to the evaporating ammonia and leave the plant as condensate. A final point is that any excess heat introduced by the compressor must be removed from the ammonia by means of a cooler.

Fuller details of the cycle and salient features of operation are given in Section 14.5.2.

*Choice of compressor drive* (basis 1 kg water evaporated)

(a) *Diesel engine*

For 1 kg evaporation, ammonia circulated = 2.28 kg and the work done in compressing the ammonia

$$= (150 \times 2.28)$$

$$= 342 \text{ kJ or } 0.342 \text{ MJ/kg evaporation}$$

For an output of 1 MJ, the engine consumes 0.4 kg fuel.

Thus: $\text{fuel consumption} = (0.4 \times 0.342) = 0.137 \text{ kg/kg water evaporated}$

and: $\text{cost} = (0.02 \times 0.137) = \underline{\underline{0.00274 \text{ £/kg water evaporated}}}$

(b) *Turbine*

The work required is 0.342 MJ/kg evaporation.

Therefore with an efficiency of 70 per cent:

$$\text{energy required from steam} = (0.342 \times 100/70) = 0.489 \text{ MJ/kg.}$$

Enthalpy of steam saturated at 700 kN/m$^2$ = 2764 kJ/kg.

Enthalpy of steam saturated at 101.3 kN/m$^2$ = 2676 kJ/kg.

Thus: energy from steam = (2764 – 2676) = 88 kJ/kg or 0.088 MJ/kg

and: steam required = (0.489/0.088) = 5.56 kg/kg evaporation

at a cost of: $(0.01 \times 5.56)/10 = \underline{\underline{0.0056\ \text{£/kg water evaporated}}}$

and hence: the *Diesel engine would be used for driving the compressor.*

## 14.6. EVAPORATOR OPERATION

In evaporation, solids may come out of solution and form a deposit or scale on the heat transfer surfaces. This causes a gradual increase in the resistance to heat transfer and, if the same temperature difference is maintained, the rate of evaporation decreases with time and it is necessary to shut down the unit for cleaning at periodic intervals. The longer the boiling time, the lower is the number of shutdowns which are required in a given period although the rate of evaporation would fall to very low levels and the cost per unit mass of material handled would become very high. A far better approach is to make a balance which gives a minimum number of shutdowns whilst maintaining an acceptable throughput.

It has long been established[18] that, with scale formation, the overall coefficient of heat transfer may be expressed as a function of the boiling time by an equation of the form:

$$1/U^2 = at_b + b \tag{14.11}$$

where $t_b$ is the boiling time. If $Q_b$ is the total heat transferred in this time, then:

$$\frac{dQ_b}{dt_b} = UA\Delta T$$

and substituting for $U$ from equation 14.11 gives:

$$\frac{dQ_b}{dt_b} = \frac{A\Delta T}{(at_b + b)^{0.5}} \tag{14.12}$$

Integrating between 0 and $Q_b$ and 0 and $t_b$ gives:

$$Q_b = (2A\Delta T/a)[(at_b + b)^{0.5} - b^{0.5}] \tag{14.13}$$

There are two conditions for which an optimum value of the boiling time may be sought–the time whereby the heat transferred and hence the solvent evaporated is a maximum and secondly, the time for which the cost per unit mass of solvent evaporated is a minimum. These are now considered in turn.

### *Maximum heat transfer*

If the time taken to empty, clean and refill the unit is $t_c$, then the total time for one cycle is $t = (t_b + t_c)$ and the number of cycles in a period $t_P$ is $t_P/(t_b + t_c)$. The total heat transferred during this period is the product of the heat transferred per cycle and the number of cycles in the period or:

$$Q_P = (2A\Delta T/a)[(at_b + b)^{0.5} - b^{0.5}][t_P/(t_b + t_c)] \qquad (14.14)$$

The optimum value of the boiling time which gives the maximum heat transferred per cycle is obtained by differentiating equation 14.14 and equating to zero which gives:

$$t_{bopt} = t_c + (2/a)(abt_c)^{0.5} \qquad (14.15)$$

## *Minimum cost*

Taking $C_c$ as the cost of a shutdown and the variable cost during operation including a labour component as $C_b$, then the total cost during period $t_P$ is:

$$C_T = (C_c + t_b C_b)t_P/(t_b + t_c)$$

and substituting from equation 14.14:

$$C_T = [aQ_P(C_c + t_b C_b)]/2A\Delta T[at_b + b)^{0.5} - b^{0.5}] \qquad (14.16)$$

The optimum value of the boiling time to give minimum cost is obtained by differentiating equation 14.16 and equating to zero to give:

$$t_{bopt} = (C_c/C_b) + 2(abC_c C_b)^{0.5}/(aC_b) \qquad (14.17)$$

In using this equation, it must be ensured that the required evaporation is achieved. If this is greater than that given by equation 14.17, then it is not possible to work at minimum cost conditions. The use of these equations is illustrated in the following example which is based on the work of HARKER[19].

## Example 14.7

In an evaporator handling an aqueous salt solution, the overall coefficient $U$ (kW/m$^2$ deg K) is given by a form of equation 14.14 as:

$$1/U^2 = 7 \times 10^{-5} t_b + 0.2,$$

the heat transfer area is 40 m$^2$, the temperature driving force is 40 deg K and the latent heat of vaporisation of water is 2300 kJ/kg. If the down-time for cleaning is 15 ks (4.17 h), the cost of a shutdown is £600 and the operating cost during boiling is £18/ks (£64.6/h), estimate the optimum boiling times to give a) maximum throughput and b) minimum cost.

## Solution

(a) *Maximum throughput*

The boiling time to give maximum heat transfer and hence maximum throughput is given by equation 14.15:

$$t_{bopt} = (15 \times 10^3) + (2/(7 \times 10^{-5}))(7 \times 10^{-5} \times 0.2 \times 15 \times 10^3)^{0.5}$$

$$= 2.81 \times 10^4 \text{ s or } \underline{\underline{28.1 \text{ ks}}} \text{ (7.8 h)}$$

The heat transferred during boiling is given by equation 14.13:

$$Q_b = (2 \times 40 \times 40)(7 \times 10^{-5})[((7 \times 10^{-5} \times 2.81 \times 10^4) + 0.2)^{0.5} - 0.2^{0.5}] = 4.67 \times 10^7 \text{ kJ}$$

and the water vaporated $= (4.67 \times 10^7)/2300 = 2.03 \times 10^4$ kg

Rate of evaporation during boiling $= (2.03 \times 10^4)/(2.81 \times 10^4) = 0.723$ kg/s

Mean rate of evaporation during the cycle $= 2.03/[(2.8 \times 10^4) + (15 \times 10^3)] = \underline{\underline{0.471\text{kg/s}}}.$

Cost of the operation $= ((2.81 \times 10^4 \times 18)/1000) + 600 = 1105.8$ £/cycle

or: $(1105.8/(2.03 \times 10^4) = 0.055$ £/kg.

(b) *Minimum cost*

The boiling time to give minimum cost is given by equation 14.17:

$$t_{bopt} = (600/0.018) + [2(7 \times 10^{-5} \times 0.2 \times 600 \times 0.018)^{0.5}]/(7 \times 10^{-5} \times 0.018)$$

$$= 5.28 \times 10^4 \text{ s or } \underline{\underline{52.8 \text{ ks}}}(14.7 \text{ h})$$

The heat transferred during one boiling period is given by equation 14.13:

$$Q_b = [(2 \times 40 \times 40)/(7 \times 10^{-5})][7 \times 10^{-5} \times 5.28 \times 10^4 + 0.2)^{0.5} - 0.2^{0.5}] = 6.97 \times 10^7 \text{ kJ}$$

and the water evaporated $= (6.97 \times 10^7)/2300 = 3.03 \times 10^4$ kg

Rate of evaporation during boiling $= (3.03 \times 10^4)/(5.28 \times 10^4) = 0.574$ kg/s

Mean rate of evaporation during the cycle $= (3.03 \times 10^4)/[(5.28 \times 10^4) + (15 \times 10^3)] = \underline{\underline{0.45 \text{ kg/s}}}$

In this case, cost of one cycle $= (5.28 \times 10^4 \times 0.018) + 600 =$ £1550.4

or: $1550.4/(3.03 \times 10^4) = \underline{\underline{0.0512 \text{ £/kg}}}$

Thus, the maximum throughput is 0.471 kg/s and the throughput to give minimum cost, 0.0512 £/kg, is 0.45 kg/s. If the desired throughput is between 0.45 and 0.471 kg/s, then this can be achieved although minimum cost operation is not possible. If a throughput of less than 0.45 kg/s is required, say 0.35 kg/s, then a total cycle time of $(3.03 \times 10^4)/0.35 = 8.65 \times 10^4$ s or 86.5 ks is required. This could be achieved by boiling at 0.423 kg/s for 71.5 ks followed by a shutdown of 15 ks, which gives a cost of 0.0624 £/kg. This is not the optimum boiling time for minimum cost and an alternative approach might be to boil for 52.8 ks at the optimum value, 0.45 kg/s, and, with a shutdown of 15 ks, a total cost of 0.0654 £/kg is estimated which is again higher than the minimum value. It would be, in fact, more cost effective to operate with the optimum boiling time of 52.8 ks and the down-time of 15 ks and to *close the plant down* for the remaining 18.7 ks of the 86.5 ks cycle. In this way, the minimum cost of 0.0512 £/kg would be achieved. In practice, the plant would probably not be closed down each cycle but rather for the equivalent period say once per month or indeed once a year. In all such considerations, it should be noted that, when a plant is shut down, there is no return on the capital costs and overheads which still have to be paid and this may affect the economics.

Whilst calculated optimum cycle times may not exactly correspond to convenient operating schedules, this is not important as slight variations in the boiling times will not affect the economics greatly.

# 14.7. EQUIPMENT FOR EVAPORATION

## 14.7.1. Evaporator selection

The rapid development of the process industries and of new products has provided many liquids with a wide range of physical and chemical properties all of which require concentration by evaporation. The type of equipment used depends largely on the method of applying heat to the liquor and the method of agitation. Heating may be either direct or indirect. Direct heating is represented by solar evaporation and by submerged combustion of a fuel. In indirect heating, the heat, generally provided by the condensation of steam, passes through the heating surface of the evaporator.

Some of the problems arising during evaporation include:

(a) High product viscosity.
(b) Heat sensitivity.
(c) Scale formation and deposition.

Equipment has been developed in an attempt to overcome one or more of these problems. In view of the large number of types of evaporator which are available, the selection of equipment for a particular application can only be made after a detailed analysis of all relevant factors has been made. These will, of course, include the properties of the liquid to be evaporated, capital and running costs, capacity, holdup, and residence time characteristics. Evaporator selection considered in detail in Volume 6, has been discussed by Moore and Hesler[20] and Parker[21]. Parker has attempted to test the suitability of each basic design for dealing with the problems encountered in practice, and the basic information is presented in the form shown in Figure 14.15. The factors considered include the ability to handle liquids in three viscosity ranges, to deal with foaming, scaling or fouling, crystal production, solids in suspension, and heat sensitive materials. A comparison of residence time and holding volume relative to the wiped film unit is also given. It is of interest to note that the agitated or wiped film evaporator is the only one which is shown to be applicable over the whole range of conditions covered.

## 14.7.2. Evaporators with direct heating

The use of solar heat for the production of Glauber's salt has been described by Holland[22,23]. Brine is pumped in hot weather to reservoirs of 100,000 $m^2$ in area to a depth of 3–5 m, and salt is deposited. Later in the year, the mother liquor is drained off and the salt is stacked mechanically, and conveyed to special evaporators in which hot gases enter at 1150–1250 K through a suitable refractory duct and leave at about 330 K. The salt crystals melt in their water of crystallisation and are then dried in the stream of hot gas. Bloch *et al.*[24], who examined the mechanism of evaporation of salt brines by direct solar energy, found that the rate of evaporation increased with the depth of brine. The addition of dyes, such as 2-naphthol green, enables the solar energy to be absorbed in a much shallower depth of brine, and this technique has been used to obtain a significant increase in the rate of production in the Dead Sea area.

| Operational category | Evaporator type | Feed condition[a] | | | | | | | Suitable for heat-sensitive products | Retention time[b] (s) | Holding volume[c] ($m^3$) |
|---|---|---|---|---|---|---|---|---|---|---|---|
| | | Very viscous (above 2000 mN $s/m^2$) | Med. viscosity (100–1000 mN $s/m^2$) | Low viscosity to water (max. 100 mN $s/m^2$) | Foaming | Scaling or fouling | Crystal producing | Solids in suspension | | | |
| Recirculating | Calandria[d] (short vertical tube) | | ■ | ■ | ■ | ■ | ■ | ■ | No | 168 | 3.03 |
| | Forced circulation | | ■ | ■ | ■ | ■ | ■ | ▨ | Yes | 41.6 | 12.8 |
| | Falling film | | ▨ | ■ | | | | | No[e] | Not available | Not available |
| | Natural circulation (thermo-siphon) | | ▨ | ■ | ■ | | | | No[e] | 16 | 10.1 |
| Single pass | Agitated film (vertical or horizontal) | ■ | ■ | ■ | ■ | ■ | ■ | ■ | Yes | 1.0 | 1.0 |
| | Tubular (long tube) Falling film | | ▨ | ■ | | | | | Yes | Not available | Not available |
| | Tubular (long tube) Rising film | | ▨ | ■ | ■ | | | ▨ | | | |
| Single pass special type | Rising-Falling concentrator | ▨ | ■ | ■ | ■ | | | | Yes | 0.45 | 0.79 |
| | Plate (can be recirculating) | ▨ | ■ | ■ | | | | | Yes | Not available | Not available |

■ = Applicable to conditions noted    ▨ = Applicable over lower portion of range noted

a. Viscosities are at operating temperatures
b. Based on agitated film evaporator = 1.0
c. Based on agitated film evaporator = 1.0, proportioned to equal surface
d. Special disengagement arrangement required for foamy liquids
e. May be used in special cases

Figure 14.15. Evaporator selection (after PARKER[21])

The submerged combustion of a gas, such as natural gas, has been used for the concentration of very corrosive liquors, including spent pickle liquors, weak phosphoric and sulphuric acids. A suitable burner for direct immersion in the liquor, as developed by SWINDIN[25], is shown in Figure 14.16. The depth of immersion of the burner is determined by the time of heat absorption and, for example, a 50 mm burner may be immersed by 250 mm and a 175 mm burner by about 450 mm. The efficiency of heat absorption is measured by the difference between the temperature of the liquid and that of the gases leaving the surface, values of 2–5 deg K being obtained in practice. The great attraction of this technique, apart from the ability to handle corrosive liquors, is the very great heat release obtained and the almost instantaneous transmission of the heat to the liquid, typically 70 $MW/m^3$.

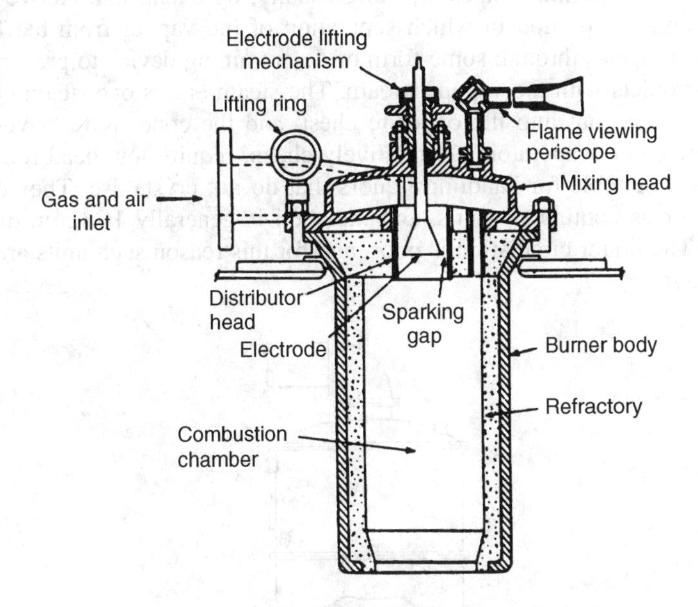


Figure 14.16. Burner for submerged combustion[25]

### 14.7.3. Natural circulation evaporators

Whilst each of the previous types of evaporator is of considerable importance in a given industry, it is the steam-heated evaporator that is the most widely used unit in the process industries and this is now considered in detail. In Chapter 9 of Volume 1, it is shown that the movement of the liquid over the heating surface has a marked influence on the rate of heat transfer, and it is thus convenient to classify evaporators according to the method of agitation or the nature of the circulation of the liquor over the heating surface. On this basis evaporators may be divided into three main types:

(a) Natural circulation units.
(b) Forced circulation units.
(c) Film-type units.

The developments that have taken place have, in the main, originated from the sugar and salt industries where the cost of evaporation represents a major factor in the process economics. In recent years, particular attention has been given to obtaining the most efficient use of the heating medium, and the main techniques that have been developed are the use of the multiple-effect unit, and of various forms of vapour compression units. With natural-circulation evaporators, circulation of the liquor is achieved by convection currents arising from the heating surface. This group of evaporators may be subdivided according to whether the tubes are horizontal with the steam inside, or vertical with the steam outside.

Rillieux is usually credited with first using *horizontal tubes*, and a unit of this type is shown in Figure 14.17. The horizontal tubes extend between two tube plates to which they are fastened either by packing plates or, more usually, by expansion. Above the heating section is a cylindrical portion in which separation of the vapour from the liquid takes place. The vapour leaves through some form of de-entraining device to prevent the carry-over of liquid droplets with the vapour stream. The steam enters one steam chest, passes through the tubes and out into the opposite chest, and the condensate leaves through a steam trap. Horizontal evaporators are relatively cheap, require low head room, are easy to install, and are suitable for handling liquors that do not crystallise. They can be used either as batch or as continuous units, and the shell is generally 1–3.5 m diameter and 2.5–4 m high. The liquor circulation is poor, and for this reason such units are unsuitable for viscous liquors.

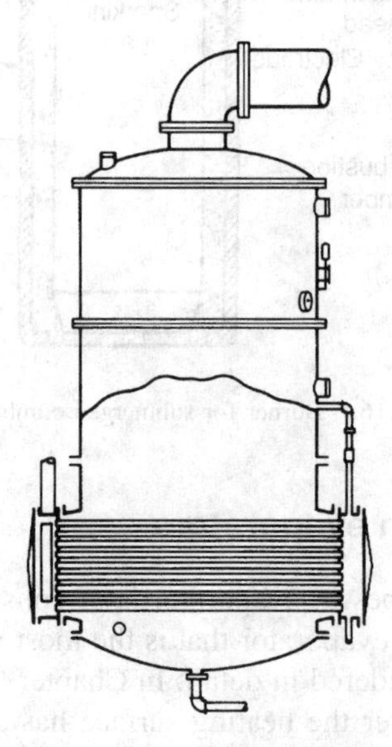

Figure 14.17. Natural circulation evaporator with horizontal tubes

The use of *vertical tubes* is associated with Robert, and this type is sometimes known as the Robert or Standard Evaporator. A typical form of vertical evaporator is illustrated in Figure 14.18, in which a vertical cylindrical body is used, with the tubes held between two horizontal tube plates which extend right across the body. The lower portion of the evaporator is frequently spoken of as the calandria section shown in Figure 14.19. Tubes

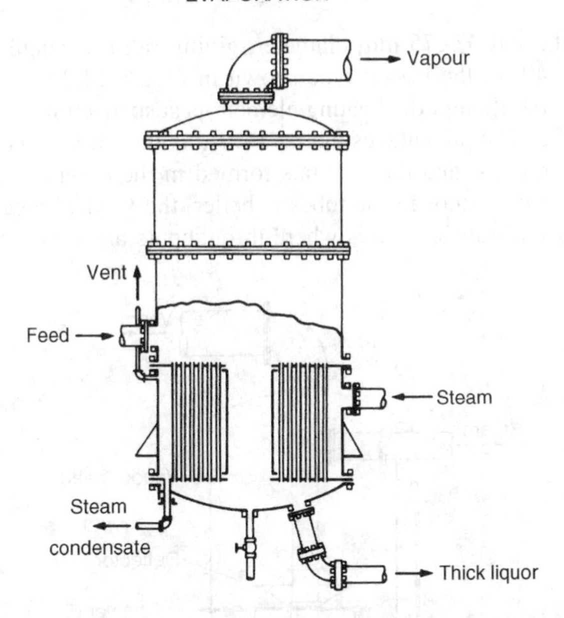


Figure 14.18. Evaporator with vertical tubes and a large central downcomer

Figure 14.19. Calandria for an evaporator

are 1–2 m in length and 37–75 mm diameter, giving ratio of length to inside diameter of the tubes of 20–40. In the basket type shown in Figure 14.20 vertical tubes are used with the steam outside, though the heating element is suspended in the body so as to give an annular downtake. The advantages claimed for this design are that the heating unit is easily removed for repairs, and that crystals formed in the downcomer do not break up. As the circulation of the liquor in the tubes is better, the vertical tube evaporator is used widely in the sugar and salt industries where throughputs are very large.

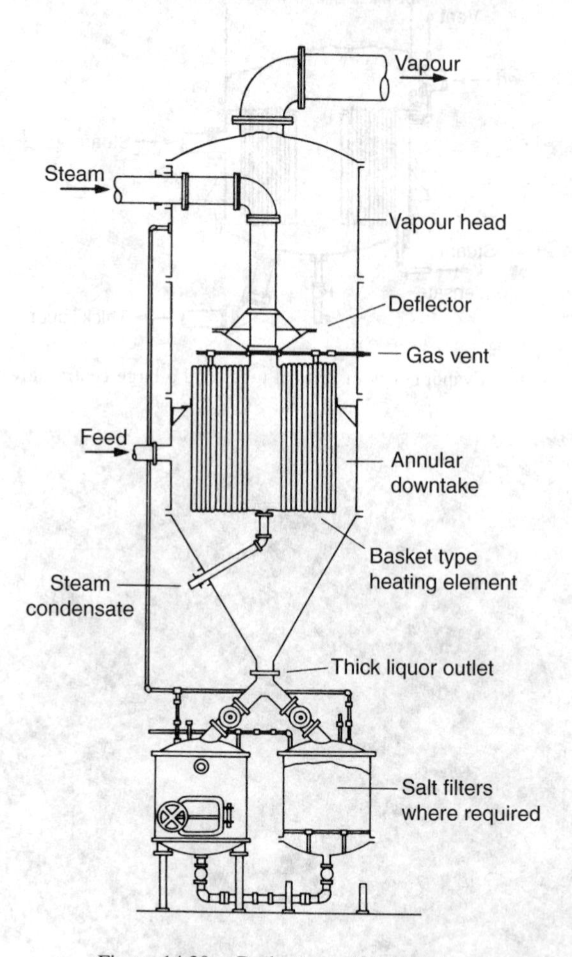


Figure 14.20. Basket type of evaporator

### 14.7.4. Forced circulation evaporators

Increasing the velocity of flow of the liquor through tubes results in a significant increase in the liquid-film transfer coefficient. This is achieved in the forced circulation units where a propeller or other impeller is mounted in the central downcomer, or a circulating pump is mounted outside the evaporator body. In the concentration of strong brines, for example, an internal impeller, often a turbine impeller, is fitted in the downtake, and this form of construction is particularly useful where crystallisation takes place. Forced circulation enables higher degrees of concentration to be achieved, since the heat transfer rate can be maintained in spite of the increased viscosity of the liquid. Because pumping costs increase roughly as the cube of the velocity, the added cost of operation of this type of unit may make it uneconomic, although many forced circulation evaporators are running with a liquor flow through the tubes of 2–5 m/s which is a marked increase on the value for natural circulation. Where stainless steel or expensive alloys such as Monel are to be used, forced circulation is favoured because the units can be made smaller and cheaper than those relying on natural circulation. In the type illustrated in Figure 14.21, there is an external circulating pump, usually of the centrifugal type when crystals are present, though otherwise vane types may be used. The liquor is either introduced at the bottom and pumped straight through the calandria, or it is introduced in the separating section. In most units, boiling does not take place in the tubes, because the hydrostatic head of liquid raises the boiling point above that in the separating space. Thus the liquor enters the bottom of the tubes and is heated as it rises and at the same time the pressure falls. In the separator the pressure is sufficiently low for boiling to occur. Forced circulation evaporators work well on materials such as meat extracts, salt, caustic soda, alum and other crystallising materials and also with glues, alcohols, and foam-forming materials.

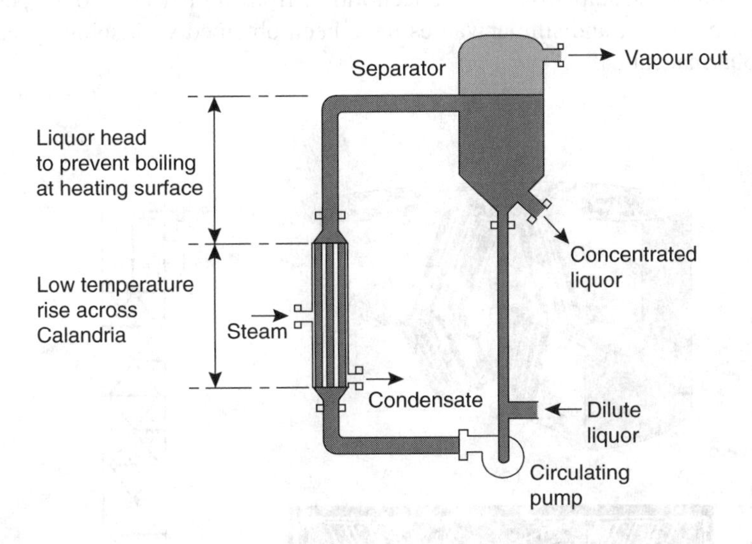


Figure 14.21. Forced circulation evaporator with an external pump

For certain applications multi-pass arrangements are used. When a plate heat exchanger is used instead of the tubular unit, boiling on the heating surfaces is avoided by increasing the static head using a line restriction between the plate pack and the separator. Compared with tubular units, lower circulation rates and reduced liquid retention times are important advantages. Plate-type units are discussed further in Section 14.7.7.

For the handling of corrosive fluids, forced-circulation evaporators have been constructed in a variety of inert materials, and particularly in graphite where the unique combination of chemical inertness coupled with excellent thermal conductivity gives the material important advantages. Graphite differs from most constructional materials in its high anisotropy which results in directionally preferred thermal conductivity, and in the difference between its relatively good compressive strength and its poor tensile or torsional strength. Although it is easily machinable, it is not ductile or malleable and cannot be cast or welded. The use of cements in assembly is undesirable because they are usually less chemically or thermally stable. There are also problems of differential expansion. In order to exploit the advantages of this material and to avoid the foregoing problems, special constructional techniques are necessary. The Polybloc system, which is described by HILLIARD[26–28], is based on the use of robust blocks assembled exclusively under compression. Heat transfer occurs between fluids passing through holes drilled in the blocks and positioned so as to exploit preferred anisotropic crystal orientation for the highest thermal conductivity in the direction of heat flow. Inert gaskets eliminate the need for cements and enable units of varying size to be assembled simply by stacking the required number of blocks as shown in Figure 14.22. A similar form of construction has been adopted by the Powell Duffryn Company. In commercial installations high values of overall transfer coefficients have been achieved and, for example, a value of 1.1 $kW/m^2K$ has been obtained for concentrating thick fruit juice containing syrup, and also for concentrating 40 per cent sulphuric acid to 60 per cent. A value of 0.8 $kW/m^2K$ has been obtained for sulphuric acid concentration from 60 per cent to 74 per cent at a pressure of 1.5 $kN/m^2$, and similar values have been obtained with spinning-bath liquors and some pharmaceuticals.

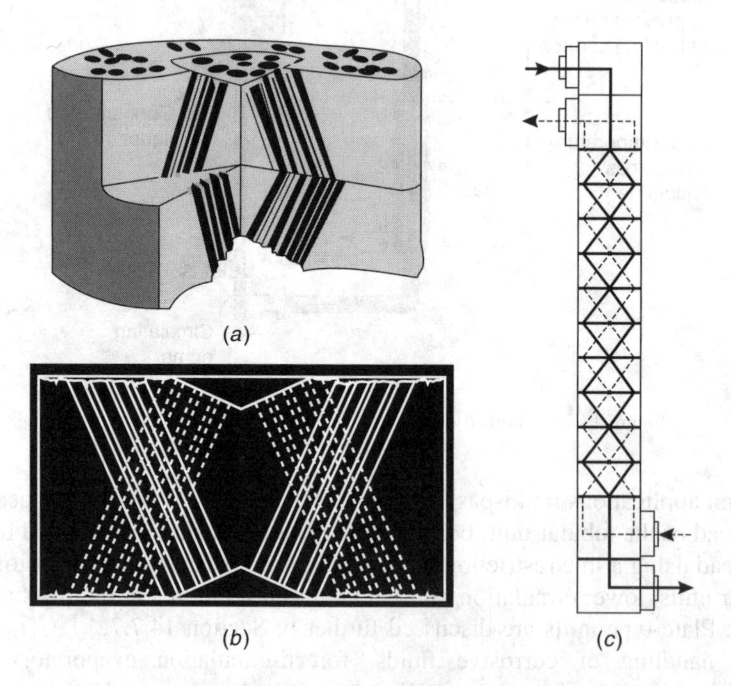


Figure 14.22. The Polybloc system (after HILLIARD[27]). (*a*) Cutaway section of x-flow block as used for two corrosive fluids. (*b*) Section through x-flow block. (*c*) Stacked Polybloc exchanger

### 14.7.5. Film-type units

In all the units so far discussed, the liquor remains for some considerable time in the evaporator, up to several hours with batch operation, and this may be undesirable as many liquors decompose if kept at temperatures at or near their boiling points for any length of time. The temperature can be reduced by operating under a vacuum, as discussed previously, though there are many liquors which are very heat-sensitive, such as orange juice, blood plasma, liver extracts and vitamins. If a unit is designed so that the residence time is only a few seconds, then these dangers are very much reduced. This is the principle of the Kestner long tube evaporator, introduced in 1909, which is fitted with tubes of 38 to 50 mm diameter, mounted in a simple vertical steam chest. The liquor enters at the bottom, and a mixture of vapour and entrained liquor leaves at the top and enters a separator, usually of the tangential type. The vapour passes out from the top and the liquid from the bottom of the separator. In the early models the thick liquid was recirculated through the unit, although the once-through system is now normally used.

An alternative name for the long-tube evaporator is the *climbing film evaporator*. The progressive evaporation of a liquid, whilst it passes through a tube, gives rise to a number of flow regimes discussed in Section 14.2.3. In the long-tube evaporator the annular flow or climbing-film regime is utilised throughout almost all the tube length, the climbing film being maintained by drag induced by the vapour core which moves at a high velocity relative to the liquid film. With many viscous materials, however, heat transfer rates in this unit are low because there is little turbulence in the film, and the thickness of the film is too great to permit much evaporation from the film as a result of conduction through it. In evaporators of this type it is essential that the feed should enter the tubes as near as possible to its boiling point. If the feed is subcooled, the initial sections will act merely as a feed heater thus reducing the overall performance of the unit. Pressure drop over the tube length will be attributable to the hydrostatic heads of the single-phase and two-phase regions, friction losses in these regions, and losses due to the acceleration of the vapour phase. The first published analysis of the operation of this type of unit was given by BADGER and his associates[29–31] who fitted a small thermocouple inside the experimental tube, 32 mm outside diameter and 5.65 m long, so that the couple could be moved up and down the centre of the tube. In this way, it was found that the temperature rose slightly from the bottom of the tube to the point where boiling commenced, after which the change in temperature was relatively small. Applying this technique, it was possible to determine the heat transfer coefficients in the non-boiling and boiling sections of the tube.

A *falling-film evaporator* with the liquid film moving downwards, operates in a similar manner, as shown in Figure 14.23. The falling-film evaporator is the simplest and most commonly used type of film-evaporator in which the liquid flows under gravitational force as a thin film on the inside of heated vertical tubes and the resulting vapour normally flows co-currently with the liquid in the centre of the tubes. A complete evaporator stage consists of the evaporator, a separator to separate the vapours from the residual liquid, and a condenser. Where high evaporation ratios are required, part of the concentrated liquid is recycled back to the evaporator inlet in order to ensure that the tubes are sufficiently wetted. An essential part of every falling-film evaporator is the liquid distribution system since the liquid feed must not only be evenly distributed to all the tubes, but also form a

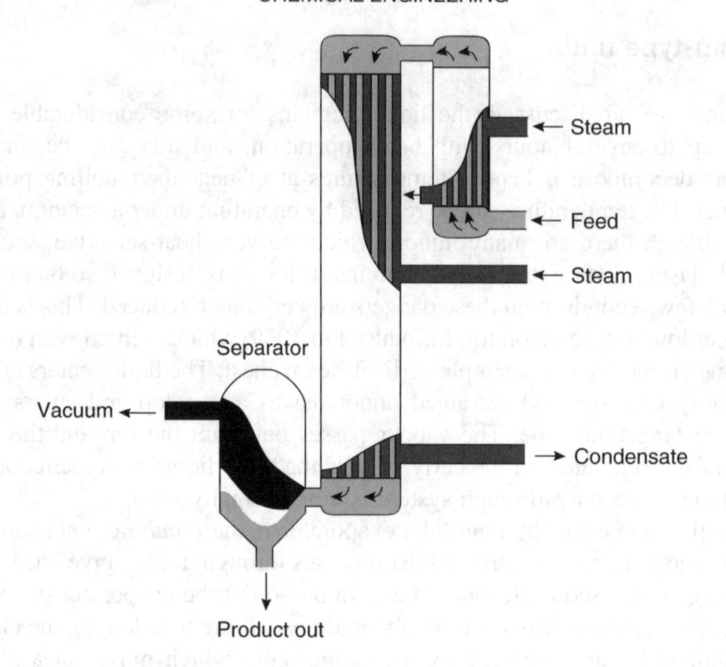


Figure 14.23. Single-effect falling-film evaporator

continuous film of the inner circumference of the tubes. Kuhni has developed a two-stage unit in which, after an initial pre-distribution, the liquid is directed tangentially onto the tubes through slits in specially designed tube inserts. The advantages of falling film evaporators include:

(a) high heat transfer coefficients, 2000–5000 $W/m^2K$ for water and 500–1000 $W/m^2K$ for organics,
(b) short residence times on the heated surface, 5–10 s without recirculation,
(c) low pressure drops, 0.2–0.5 $kN/m^2$,
(d) suitablity for vacuum operation,
(e) high evaporation ratios, c. 70 per cent without and 95 per cent with recirculation,
(f) wide operating range, up to 400 per cent of the minimum throughput,
(g) low susceptibility to fouling,
(h) minimum cost operation.

## 14.7.6. Thin-layer or wiped-film evaporators

This type of unit, known also as a thin-film evaporator is shown in Figure 14.24. It consists of a vertical tube, the lower portion of which is surrounded by a jacket which contains the heating medium. The upper part of the tube is not jacketed and this acts as a separator. A rotor, driven by an external motor, has blades which extend nearly to the bottom of the tube, mounted so that there is a clearance of only about 1.3 mm between their tips and the inner surface of the tube. The liquor to be concentrated is picked up as it enters by the

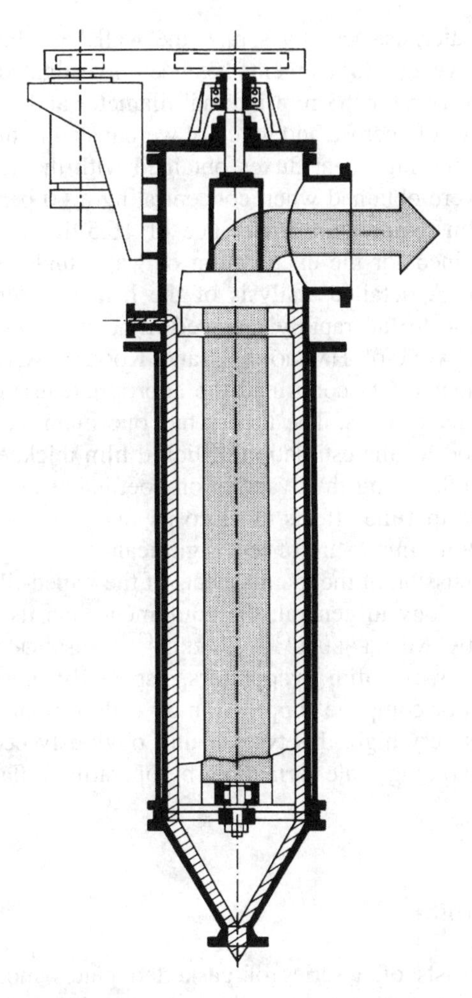

Figure 14.24. Thin-film evaporator

rotating blades and thrown against the tube wall. This action provides a thin film of liquid and sufficient agitation to give good heat transfer, even with very viscous liquids. The film flows down by gravity, becoming concentrated as it falls. The concentrated liquor is taken off at the bottom by a pump, and the vapour leaves the top of the unit where it is passed to a condenser. Development of this basic design has been devoted mainly to the modification of the blade system. An early alternative was the use of a hinged blade. In this type of unit the blade is forced on to the wall under centrifugal action, the thickness of the film being governed by a balance between this force and the hydrodynamic forces produced in the liquid film on which the blade rides. The first experimental comparison of the fixed and hinged blade wiped-film evaporators was that of BRESSLER[32]. For each type of blade there appeared to be an optimum wiper speed beyond which an increase had no further effect on heat transfer. This optimum was reached at a lower speed with the hinged blade. Other agitator designs in which the blades, usually made from rubber, graphite or synthetic materials, actually scrape the wall have been studied. The use of nylon brushes as the active agitator elements has been investigated by McManus[33] using a small steam heated evaporator, 63 mm internal diameter and 762 mm long. Water and various aqueous solutions of sucrose and glycerol were tested in the evaporator. A notable feature of the unit was the high heat fluxes obtained with the viscous solutions. Values as high as 70 $kW/m^2$ were obtained when concentrating a 60 per cent sucrose feedstock to 73 per cent, at a film temperature difference of 16.5 deg K with a wiper speed of 8.3 Hz. The fluxes obtained for the evaporation of water under similar conditions were nearly 4.5 times higher. A detailed analysis of the heat transfer mechanism, based on unsteady-state conduction to the rapidly renewed film, was presented. Similar analyses are to be found in the work of Harriott[34] and Kool[35]. Close agreement between the theory and experimental data confirmed the appropriateness of the model chosen to represent the heat transfer process. The theory has one main disadvantage, however, in that a satisfactory method for the estimation of liquid film thickness is not available. The most important factor influencing the evaporation coefficient is the thermal conductivity of the film material, and that the effects of viscosity and wiper speed which is inversely proportional to the heating time $t$, are of less significance.

A comprehensive discussion of the main aspects of the wiped-film evaporator technique covering thin-film technology in general, the equipment, and its economics and process applications is given by Mutzenburg[36], Parker[37], Fischer[38], and Ryley[39]. An additional advantage of wiped-film evaporators, especially those producing a scraped surface, is the reduction or complete suppression of scale formation though, in processes where the throughput is very high, this type of unit obviously becomes uneconomic and the traditional way of avoiding scale formation, by operating a flash evaporation process, is more suitable.

## 14.7.7. Plate-type units

A plate evaporator consists of a series of gasketted plates mounted within a support frame. Film-type plate evaporators can be climbing-film, falling-film or a combination of these. Figure 14.25 shows the flow and plate arrangement of an APV falling-film plate evaporator. Each unit comprises a product plate and a steam plate, and this arrangement is repeated to provide the required heat transfer area. Product flow down each side of the plate may be in series where this is advantageous in terms of wetting rates.

Both the vapour evaporated from the boiling film and the concentrated product are discharged from the evaporator to a vapour–liquid separator from which the product is pumped, the vapour passing to the next effect of the evaporator, or the condenser. Compared with tubular evaporators, plate evaporators can offer important advantages in terms of headroom, floorspace, accessibility and flexibility.

APV, whose "Paraflow" plate heat exchanger is illustrated in Volume 1, Chapter 9, supply climbing and falling-film plate evaporators with evaporative capacities up to 10 kg/s. Such units offer the advantages of short contact and residence times and low liquor hold-up, and hence are widely used for the concentration of heat-sensitive materials.

For applications where viscosities or product concentrations are high, APV have developed the "Paravap" evaporator in which corrugated-plate heat exchanger plates are

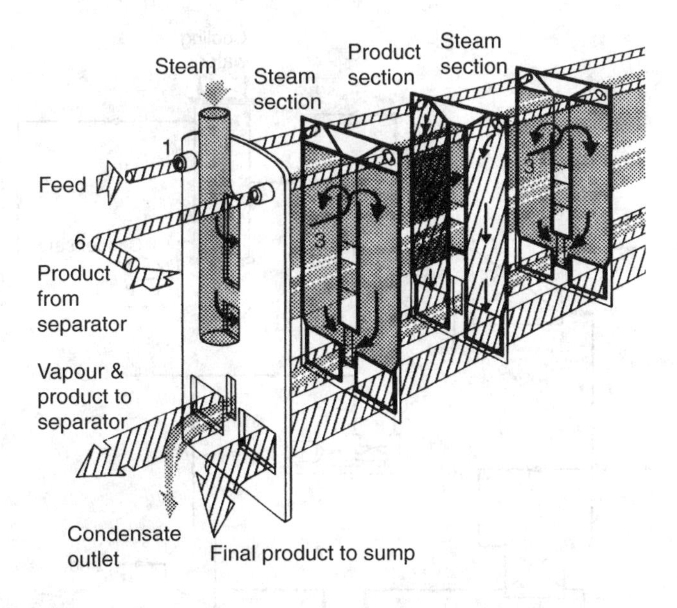


Figure 14.25. Flow and plate arrangement for two-stage operation

used in a climbing-film arrangement, thereby increasing turbulence in the liquid film compared with standard plate evaporators. A typical arrangement of an APV "Paravap" plant is shown in Figure 14.26. Feed liquor from balance tank 1 is pumped 2 through the feed preheater 3 to the evaporator 4 where it boils. Concentrated product and evaporated vapour are discharged to the separator 5 from which product is pumped 6. Vapour passes to the condenser 7 from which condensate is pumped 8, vacuum being maintained by a liquid ring pump 9. Single-pass operation is used for low concentration ratios between feed and product, whilst higher ratios require the recirculation of some of the product. This can be to the balance tank or feed pipework, although in some cases it is necessary to use in-line devices to achieve satisfactory mixing.

For some products it has been found advantageous to pressurise the plate heat exchanger, with an orifice or valve preventing boiling until the liquor enters the separator, in what is known as the APV "Paraflash" system. This is a special case of the forced circulation evaporator described earlier.

APV "Paravap" and "Paraflash" evaporators are used for products with viscosities up to 5 $Ns/m^2$ and concentrations in excess of 99 per cent by mass. Evaporation rates are up to 4 kg/s.

It may be noted that the gasket is a key component in plate heat exchangers, and this may limit the maximum temperature which can be used and may indeed prevent the use of this type of equipment with some corrosive fluids.

## 14.7.8. Flash evaporators

In the flash evaporator, boiling in the actual tubes is suppressed and the superheated liquor is flashed into a separator operating at reduced pressure. Whilst the high heat transfer rates

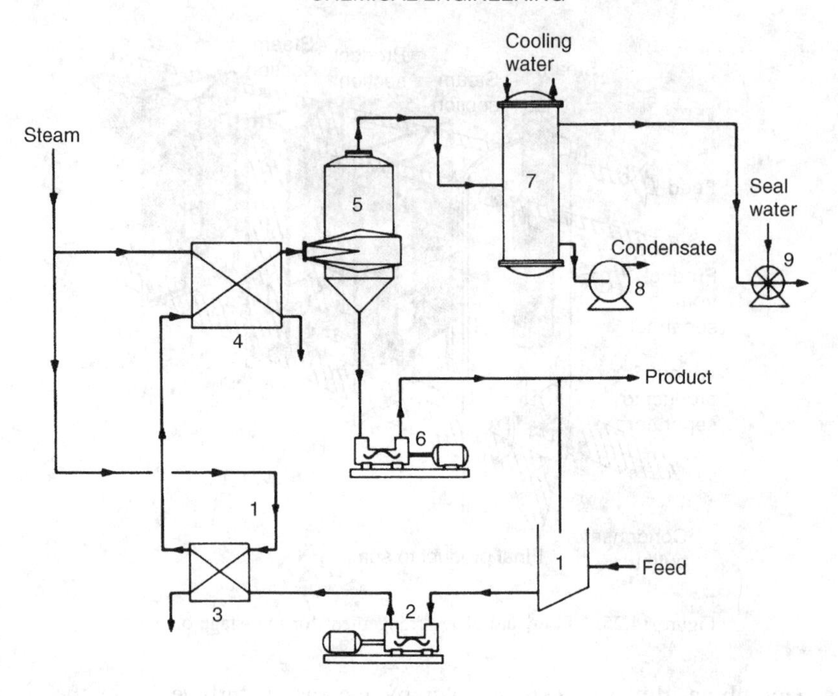


Figure 14.26. Flowsheet for a typical "Paravap" evaporator installation

associated with boiling in tubes cannot be utilised, the thermodynamic and economical advantages of the system when operated in a multistage configuration outweigh this consideration. These advantages, stated independently by FRANKEL[40] and SILVER[41], have been important in the past decade in the intensive effort to devise economic processes for the desalination of sea water. This topic is discussed further by BAKER[42], who considers multistage flash evaporation with heat input supplied by a conventional steam boiler, by a gas-turbine cycle, or by vapour recompression. The combined power–water plant is also considered. Attempts to reduce scale formation in flash evaporators to even lower levels have resulted in a number of novel developments. In one unit described by WOODWARD[43], sea water is heated by a countercurrent spray of hot immiscible oil. In this respect the process is similar to liquid–liquid extraction, the extracted quantity being heat in this case. The sea water is heated under pressure and subsequently flashed into a low pressure chamber. A similar direct contact system is discussed by WILKE *et al.*[44]. Yet another arrangement which avoids the intervening metallic wall of the conventional heat exchanger is described by OTHMER *et al.*[45]. In this process direct mass transfer between brine and pure water is utilised in the desalination operation.

The formation of solids in evaporators is not always undesirable and, indeed, this is precisely what is required in the *evaporator-crystalliser* discussed in Chapter 15. The evaporator–crystalliser is a unit in which crystallisation takes place largely as a result of the removal of solvent by evaporation. Cooling of the liquor may, in some cases, produce further crystallisation thus establishing conditions similar to those in vacuum crystallisation. The true evaporator–crystalliser is distinguished, however, by its use of an external heat source. Crystallisation by evaporation is practised on salt solutions having a small change of solubility with temperature, such as sodium chloride and ammonium sulphate, which cannot be dealt with economically by other means, as well as those with inverted solubility curves. It is also widely used in the production of many other crystalline materials, as outlined by BAMFORTH[46]. The problem of design for crystallising equipment is extremely complicated and consequently design data are extremely meagre and unreliable. This topic is discussed further in Chapter 15.

The *development of unwanted foams* is a problem that evaporation has in common with a number of processes, and a considerable amount of effort has been devoted to the study of defoaming techniques using chemical, thermal, or mechanical methods. Chemical techniques involve the addition of substances, called antifoams, to foam-producing solutions to eliminate completely, or at least to reduce drastically, the resultant foam. Antifoams are, in general, slightly soluble in foaming solutions and can cause a decrease in surface tension. Their ability to produce an expanded surface film is, however, one explanation of their foam-inhibiting characteristic, as discussed by BECKERMAN[47]. Foams may be caused to collapse by raising or lowering the temperature. Many foams collapse at high temperature due to a decrease in surface tension, solvent evaporation, or chemical degradation of the foam-producing agents; at low temperatures freezing or a reduction in surface elasticity may be responsible. Other methods which are neither chemical nor thermal may be classified as mechanical. Tensile, shear, or compressive forces may be used to destroy foams, and such methods are discussed in some detail by GOLDBERG and RUBIN[48]. The ultimate choice of defoaming procedure depends on the process under consideration and the convenience with which a technique may be applied.

### 14.7.9. Ancillary equipment

One important component of any evaporator installation is the *equipment for condensing the vapour* leaving the last effect of a multiple-effect unit, achieved either by direct contact with a jet of water, or in a normal tubular exchanger. If $M$ is the mass of cooling water used per unit mass of vapour in a jet condenser, and $H$ is the enthalpy per unit mass of vapour, then a heat balance gives:

$$\underset{\text{(Heat in)}}{H + MC_pT_i} = \underset{\text{(Heat out)}}{C_pT_e + MC_pT_e} \tag{14.18}$$

where $T_i$ and $T_e$ are the inlet and outlet temperatures of the water, above a standard datum temperature, and where the condensate is assumed to leave at the same temperature as the cooling water. From equation 14.18:

$$M = \frac{H - C_pT_e}{C_p(T_e - T_i)} \tag{14.19}$$

If, for example, $T_e = 316$ K, $T_i = 302$ K, and the pressure $= 87.8$ kN/m$^2$, then:

$$M = \frac{2596 - 4.18(316 - 273)}{4.18(316 - 302)} = 41.5 \text{ kg/kg}$$

The water is then conveniently discharged at atmospheric pressure, without the aid of a pump, by allowing it to flow down a vertical pipe, known as a barometric leg, of sufficient length for the pressure at the bottom to be slightly in excess of atmospheric pressure. For a jet condenser with a barometric leg, a chart for determining the water requirement has been prepared by ARROWSMITH(49), and this is shown in Figure 14.27.

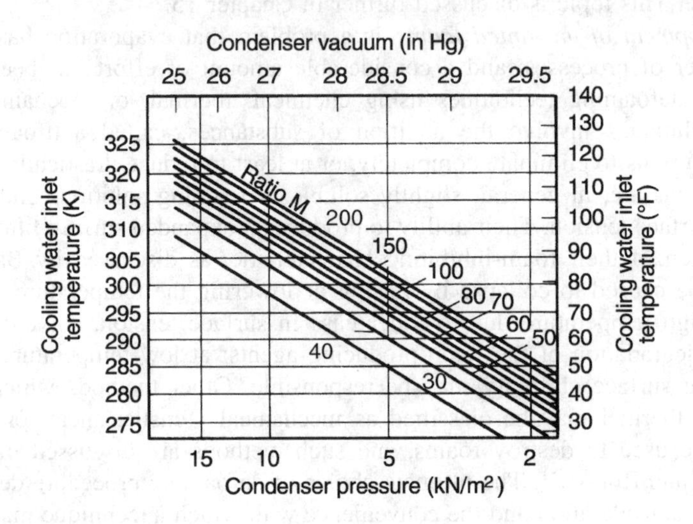


Figure 14.27. Ratio ($M$) of cooling water to vapour required under various conditions(49)

*Jet condensers* may be either of the countercurrent or parallel flow type. In the countercurrent unit, the water leaves at the bottom through a barometric leg, and any entrained gases leave at the top. This provides what is known as a *dry vacuum system*, since the pump has to handle only the non-condensable gases. The cooling water will generally be heated to within 3–6 deg K of the vapour temperature. With the parallel flow system, the temperature difference will be rather greater and, therefore, more cooling water will be required. In this case, the water and gas will be withdrawn from the condenser and passed through a wet vacuum system. As there is no barometric leg, the unit can be mounted at floor level, although the pump displacement is about one and a half times that for the dry vacuum system.

Air is introduced into a jet condenser from the cooling water, as a result of the evolution of non-condensable gases in the evaporator, and as a result of leakages. The volume of air to be removed is frequently about 15 per cent of that of the cooling water. The most convenient way of obtaining a vacuum is usually by means of a steam jet ejector. Part of the momentum of a high velocity steam jet is transferred to the gas entering the ejector, and the mixture is then compressed in the diverging portion of the ejector by conversion of kinetic energy into pressure energy. Good performance by the ejector is obtained largely by correct proportioning of the steam nozzle and diffuser, and poor ejectors will use much more high pressure steam than a well designed unit. The amount of steam required increases with the compression ratio. Thus, a single-stage ejector will remove air from a system at a pressure of 17 kN/m$^2$ where a compression ratio of 6 is required. To remove air from a system at 3.4 kN/m$^2$ would involve a compression ratio of 30, and a single-stage unit would be uneconomic in steam consumption. A two-stage ejector is shown in Figure 14.28. The first stage withdraws air from the high vacuum vessel and compresses it to say 20 kN/m$^2$, and the second stage compresses the discharge from the first ejector to atmospheric pressure. A further improvement is obtained if a condenser is inserted after the first stage, as this will reduce the amount of vapour to be handled in the final stage. An indication of the number of stages required for various conditions is shown in Figure 14.29[49]. The higher the steam pressure the smaller is the consumption, and pressures of 790–1135 kN/m$^2$ are commonly used in multistage units. Typical performance curves are shown in Figure 14.30[49], where the air duty for a given steam consumption and given steam pressure is shown as a function of the number of stages and the operating pressure.

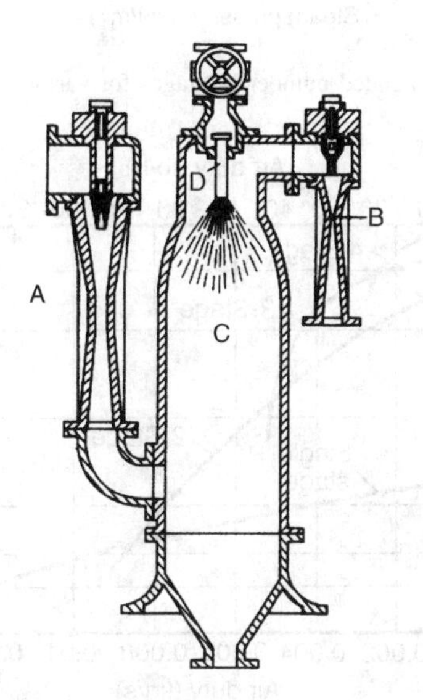


Figure 14.28. Two-stage ejector with condenser: *A*—First stage. *B*—Second stage. *C*—Condenser. *D*—Water spray

In operating an evaporator, it is important to minimise entrainment of the liquid in the vapour passing over to the condenser. Entrainment is reduced by having a considerable headroom, of say 1.8 m, above the boiling liquid, though the addition of some form of de-entrainer is usually essential. Figure 14.31 shows three *methods of reducing entrainment*. The simplest is to take the vapour from an upturned pipe as in *a*, and this has been found to give quite good results in small units. The deflector type *b* is a common form of de-entrainer and the tangential separator *c* is the type usually fitted to climbing-film units. This problem is particularly important in the concentration of radioactive waste liquors and has been discussed by McCullough[50], who cites the case of a batch evaporator of the forced-circulation type in which the vapours are passed to a 3.6 m diameter separator and then through four bubble cap trays to give complete elimination of entrained liquor. A good entrainment separator will reduce the amount of liquid carried over to 10–20 kg/10$^6$ kg of vapour.

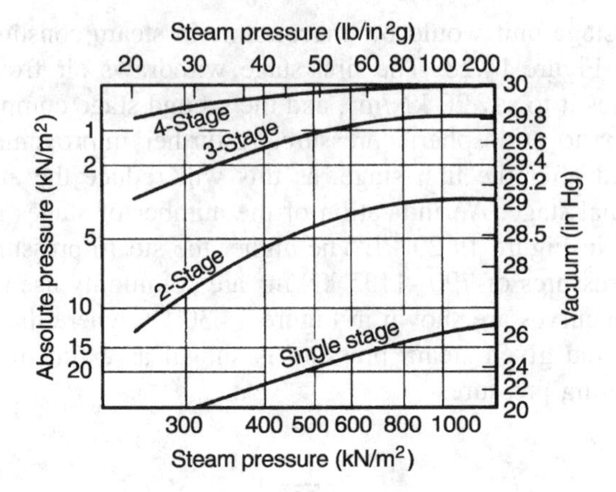


Figure 14.29. Recommended number of stages for various operating conditions[49]

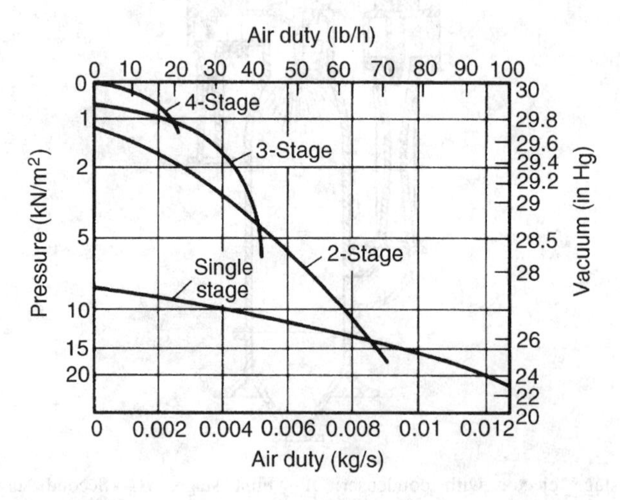


Figure 14.30. Air duty of steam jet ejectors, for a given steam consumption and steam pressure[49]

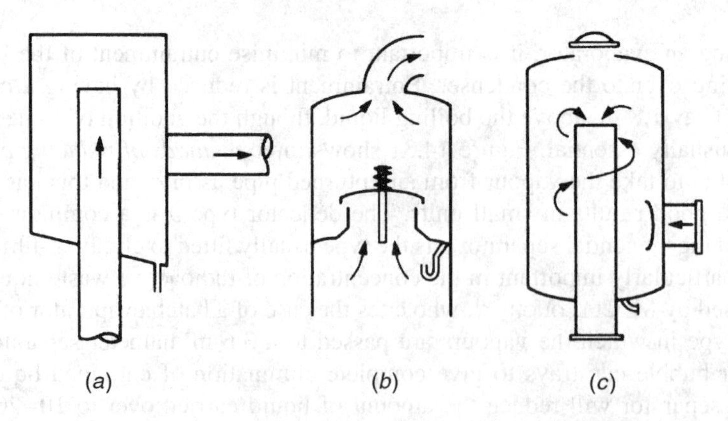


Figure 14.31. Entrainment separators: (*a*) Upturned pipe, (*b*) Deflector type, (*c*) Tangential type

## 14.8. FURTHER READING

AZBEL, D. S.: *Fundamentals of Heat Transfer in Process Engineering* (Noyes, New York, 1984).
AZBEL, D. S.: *Heat Transfer Applications in Process Engineering* (Noyes, New York, 1984).
BACKHURST, J. R. and HARKER, J. H.: *Process Plant Design* (Heinemann, London, 1973).
BADGER, W. L.: *Heat Transfer and Evaporation* (Chemical Catalog Co., 1926).
BILLET, R.: *Evaporation Technology: Principles, Applications, Economics*. (VCH, London, 1989).
HAUSBRAND, E.: *Evaporating, Condensing and Cooling Apparatus*. Translated from the second revised German edition by A. C. WRIGHT, Fifth English edition revised by B. HEASTIE (E. Benn, London, 1933).
HOLLAND, F. A., MOORES, R. M., WATSON, F. A., and WILKINSON, J. K.: *Heat Transfer* (Heinemann, London, 1970).
KERN, D. Q.: *Process Heat Transfer* (McGraw-Hill, New York, 1950).
KING, C. J.: *Separation Processes* 2nd edn. (McGraw-Hill, New York, 1980).
KREITH, F.: *Principles of Heat Transfer*, 2nd edn. (International Textbook Co., London, 1965).
MCADAMS, W. H.: *Heat Transmission*, 3rd edn. (McGraw-Hill, New York, 1954).
MCCABE, W. L., SMITH, J. C. and HARRIOTT, P.: *Unit Operations of Chemical Engineering*, 4th edn. (McGraw-Hill, New York, 1984).
MCKENNA, B. M. (ed): *Engineering and Food*, Vol. 2. *Process Applications* (Elsevier Applied Science, New York, 1984).
MINTON, P. E.: *Handbook of Evaporation Technology*. (Noyes, New York, 1987).
PERRY, R. H. and GREEN, D. W. (eds): *Perry's Chemical Engineers' Handbook*. 6th edn. (McGraw-Hill Book Company, New York, 1984).
PETERS, M. S. and TIMMERHAUS, K. D.: *Plant Design and Economics for Chemical Engineers*, 3rd edn. (McGraw-Hill, New York, 1984).
SMITH, R. A.: *Vaporisors, Selection, Design and Operation*. (Longman, London, 1987).
TYNER, M.: *Process Engineering Calculations* (Ronald Press, New York, 1960).

## 14.9. REFERENCES

1. WESTWATER, J. W.: *Petro/Chem. Engr.* **33**, No. 9 (Aug. 1961) 186–9 and **33**, No. 10 (Sept. 1961) 219–26. Nucleate pool boiling.
2. PALEN, J. W. and TABOREK, J. J.: *Chem. Eng. Prog.* **58**, No. 7 (July 1962) 37–46. Refinery kettle reboilers: proposed method for design and optimization.
3. CRYDER, D. S. and GILLILAND, E. R.: *Ind. Eng. Chem.* **24** (1932) 1382–7. Heat transmission from metal surfaces to boiling liquids. I. Effect of physical properties of boiling liquid on liquid film coefficient.
4. CICHELLI, M. T. and BONILLA, C. F.: *Trans. Am. Inst. Chem. Eng.* **41** (1945) 755–87. Heat transfer to liquids boiling under pressure.
5. MCNELLY, M. J.: *J. Imp. Coll. Chem. Eng. Soc.* **7** (1953) 18. A correlation of the rates of heat transfer to nucleate boiling liquids.
6. GILMOUR, C. H.: *Chem. Eng. Prog.* Symp. Ser. No. 29, **55** (1959) 67–78. Performance of vaporizers: Heat transfer analysis of plant data.
7. ZUBER, N.: *Trans. Am. Soc. Mech. Eng.* **80** (1958) 711–20. On the stability of boiling heat transfer.
8. DENGLER, C. E. and ADDOMS, J. N.: *Chem. Eng. Prog.* Symp. Ser. No. 18, **52** (1956) 95–103. Heat transfer mechanism for vaporization of water in a vertical tube.
9. LOCKHART, R. W. and MARTINELLI, R. C.: *Chem. Eng. Prog.* **45** (1949) 39–48. Proposed correlation of data for isothermal two-phase, two-component flow in pipes.
10. GUERRIERI, S. A. and TALTY, R. D.: *Chem. Eng. Prog.* Symp. Ser. No. 18, **52** (1956) 69–77. A study of heat transfer to organic liquids in single-tube, natural-circulation, vertical-tube boilers.
11. OTHMER, D.F.: *Ind. Eng. Chem.* **32** (1940) 841. Correlating vapor pressure and latent heat date. A new plot.
12. HAUSBRAND, E.: *Evaporating, Condensing and Cooling Apparatus.* Translated from the second revised German edition by A. C. Wright. Fifth English edition revised by B. HEASTIE (E. Benn, London. 1933).
13. STORROW, J. A.: *Ind. Chemist* **27** (1951) 298. Design calculations for multiple-effect evaporators — Part 3.
14. STEWART, G. and BEVERIDGE, G. S. G.: *Computers and Chemical Engineering* **1** (1977). 3. Steady-state cascade simulation in multiple effect evaporation.

15. WEBRE, A. L.: *Chem. Met. Eng.* **27** (1922) 1073. Evaporation — A study of the various operating cycles in triple effect units.
16. SCHWARZ, H. W.: *Food Technol.* **5** (1951) 476. Comparison of low temperature (e.g. 15–24 citrus juice) evaporators.
17. REAVELL, B. N.: *Ind. Chemist* **29** (1953) 475. Developments in evaporation with special reference to heat sensitive liquors.
18. MCCABE, W.L. and ROBINSON, C.S.: *Ind. Eng. Chem.* **16** (1924) 478. Evaporator scale formation.
19. HARKER, J.H.: *Processing* **12** (December 1978) 31–32. Finding the economic balance in evaporator operation.
20. MOORE, J. G. and HESLER, W. E.: *Chem. Eng. Prog.* **59**, No. 2 (Feb. 1963) 87–92. Equipment for the food industry — 2: Evaporation of heat sensitive materials.
21. PARKER, N. H.: *Chem. Eng., Albany* **70**, No. 15 (22 July 1963) 135–40. How to specify evaporators.
22. HOLLAND, A. A.: *Chem. Eng., Albany* **55** (xii) (1948) 121. More Saskatchewan salt cake.
23. HOLLAND, A. A.: *Chem. Eng., Albany* **58** (i) (1951) 106. New type evaporator.
24. BLOCH, M. R., FARKAS, L., and SPIEGLER, K. S.: *Ind. Eng. Chem.* **43** (1951) 1544. Solar evaporation of salt brines.
25. SWINDIN, N.: *Trans. Inst. Chem. Eng.* **27** (1949) 209. Recent developments in submerged combustion.
26. HILLIARD, A.: *Brit. Chem. Eng.* **4** (1959) 138–43. Considerations on the design of graphite heat exchangers.
27. HILLIARD, A.: *Brit. Chem. Eng.* **8** (1963) 234–7. The X-flow Polybloc system of construction for graphite.
28. HILLIARD, A.: *Ind. Chemist* **39** (1963) 525–31. Effect of anisotropy on design considerations for graphite.
29. BROOKS, C. H. and BADGER, W. L.: *Trans. Am. Inst. Chem. Eng.* **33** (1937) 392. Heat transfer coefficients in the boiling section of a long-tube, natural circulation evaporator.
30. STROEBE, G. W., BAKER, E. M., and BADGER, W. L.: *Trans. Am. Inst. Chem. Eng.* **35** (1939) 17. Boiling film heat transfer coefficients in a long-tube vertical evaporator.
31. CESSNA, O. C., LIENTZ, J. R., and BADGER, W. L.: *Trans. Am. Inst. Chem. Eng.* **36** (1940) 759. Heat transfer in a long-tube vertical evaporator.
32. BRESSLER, R.: *Z. Ver. deut. Ing.* **100**, No. 15 (1958) 630–8. Versuche über die Verdampfung von dünnen Flüssigkeitsfilmen.
33. MCMANUS, T.: University of Durham, Ph.D. Thesis (1963). The influence of agitation on the boiling of liquids in tubes.
34. HARRIOTT, P.: *Chem. Eng. Prog.* Symp. Ser. No. 29, **55** (1959) 137–9. Heat transfer in scraped surface exchangers.
35. KOOL, J.: *Trans. Inst. Chem. Eng.* **36** (1958) 253–8. Heat transfer in scraped vessels and pipes handling viscous materials.
36. MUTZENBURG, A. B.: *Chem. Eng., Albany* **72**, No. 19 (13 Sept. 1965) 175–8. Agitated thin film evaporators. Part I. Thin film technology.
37. PARKER, N.: *Chem. Eng., Albany* **72**, No. 19 (13 Sept. 1965) 179–85. Agitated thin film evaporators. Part 2. Equipment and economics.
38. FISCHER, R.: *Chem. Eng., Albany* **72**, No. 19 (13 Sept. 1965) 186–90. Agitated thin film evaporators. Part 3. Process applications.
39. RYLEY, J. T.: *Ind. Chemist* **38** (1962) 311–19. Controlled film processing.
40. FRANKEL, A.: *Proc. Inst. Mech. Eng.* **174**, No. 7 (1960) 312–24. Flash evaporators for the distillation of sea-water.
41. SILVER, R. S.: *Proc. Inst. Mech. Eng.* **179**, Pt. 1. No. 5 (1964–5) 135–54. Nominated Lecture: Fresh water from the sea.
42. BAKER, R. A.: *Chem. Eng. Prog.* **59**, No. 6 (June 1963) 80–3. The flash evaporator.
43. WOODWARD, T.: *Chem. Eng. Prog.* **57**, No. 1 (Jan. 1961) 52–7. Heat transfer in a spray column.
44. WILKE, C. R., CHENG, C. T., LEDESMA, V. L., and PORTER, J. W.: *Chem. Eng. Prog.* **59**, No. 12 (Dec. 1963) 69–75. Direct contact heat transfer for sea water evaporation.
45. OTHMER, D. F., BENENATI, R. F., and GOULANDRIS, G. C.: *Chem. Eng. Prog.* **57**, No. 1 (Jan. 1961) 47–51. Vapour reheat flash evaporation without metallic surfaces.
46. BAMFORTH, A. W.: *Industrial Crystallization* (Leonard Hill, London, 1965).
47. BECKERMAN, J. J.: *Foams, Theory and Industrial Applications* (Reinhold, New York, 1953).
48. GOLDBERG, M. and RUBIN, E.: *Ind. Eng. Chem. Process Design and Development* **6** (1967) 195–200. Mechanical foam breaking.
49. ARROWSMITH, G.: *Trans. Inst. Chem. Eng.* **27** (1949) 101. Production of vacuum for industrial chemical processes.
50. MCCULLOUGH, G. E.: *Ind. Eng. Chem.* **43** (1951) 1505. Concentration of radioactive liquid waste by evaporation.

## 14.10. NOMENCLATURE

| | | Units in SI System | Dimensions in **M, L, T, $\theta$** |
|---|---|---|---|
| $A$ | Heat transfer surface | $m^2$ | $\mathbf{L}^2$ |
| $a$ | Constant in Equation 14.11 | $m^4K^2/W^2$ | $\mathbf{M}^{-2}\mathbf{T}^5\theta^2$ |
| $b$ | Constant in Equation 14.11 | $m^4K^2/W^2$ | $\mathbf{M}^{-2}\mathbf{T}^6\theta^2$ |
| $C_b$ | Variable cost during operation | £/s | $\mathbf{T}^{-1}$ |
| $C_c$ | Total cost of a shutdown | £ | — |
| $C_p$ | Specific heat of liquid at constant pressure | J/kg K | $\mathbf{L}^2\mathbf{T}^{-2}\theta^{-1}$ |
| $C_s$ | Surface factor | — | — |
| $C_T$ | Total cost during period $t_P$ | £ | — |
| $D$ | Liquid evaporated or steam condensed per unit time | kg/s | $\mathbf{MT}^{-1}$ |
| $d$ | A characteristic dimension | m | $\mathbf{L}$ |
| $d_t$ | Tube diameter | m | $\mathbf{L}$ |
| $E$ | Power to compressor | W | $\mathbf{ML}^2\mathbf{T}^{-3}$ |
| $E'$ | Net work done on unit mass | J/kg | $\mathbf{L}^2\mathbf{T}^{-2}$ |
| $G_F$ | Mass rate of feed | kg/s | $\mathbf{MT}^{-1}$ |
| $G_x$ | mass flow of extra steam dryness fraction | kg/s | $\mathbf{MT}^{-1}$ |
| $G_y$ | mass flow of sea water | kg/s | $\mathbf{MT}^{-1}$ |
| $g$ | Acceleration due to gravity | $m/s^2$ | $\mathbf{LT}^{-2}$ |
| $H$ | Enthalpy per unit mass of vapour | J/kg | $\mathbf{L}^2\mathbf{T}^{-2}$ |
| $h$ | Average value of $h_b$ for a tube bundle | $W/m^2$ K | $\mathbf{MT}^{-3}\theta^{-1}$ |
| $h_b$ | Film heat transfer coefficient for boiling liquid | $W/m^2$ K | $\mathbf{MT}^{-3}\theta^{-1}$ |
| $h_c$ | Film heat transfer coefficient for condensing steam | $W/m^2$ K | $\mathbf{MT}^{-3}\theta^{-1}$ |
| $h_L$ | Liquid-film heat transfer coefficient | $W/m^2$ K | $\mathbf{MT}^{-3}\theta^{-1}$ |
| $h_{tp}$ | Heat transfer coefficient for two phase mixture | $W/m^2$ K | $\mathbf{MT}^{-3}\theta^{-1}$ |
| $k$ | Thermal conductivity of liquid | W/m K | $\mathbf{MLT}^{-3}\theta^{-1}$ |
| $m$ | Mass | kg | $\mathbf{M}$ |
| $M$ | Mass of cooling water per unit mass of vapour | kg/kg | — |
| $N$ | Number of effects | — | — |
| $P$ | Pressure | $N/m^2$ | $\mathbf{ML}^{-1}\mathbf{T}^{-2}$ |
| $Q$ | Heat transferred per unit time | W | $\mathbf{ML}^2\mathbf{T}^{-3}$ |
| $Q_b$ | Total heat transferred during boiling time | J | $ML^2T^{-2}$ |
| $q$ | Heat flux per unit area | $W/m^2$ | $\mathbf{MT}^{-3}$ |
| $T$ | Temperature | K | $\theta$ |
| $T_b$ | Boiling temperature of liquid | K | $\theta$ |
| $T_c$ | Condensing temperature of steam | K | $\theta$ |
| $T_f$ | Feed temperature | K | $\theta$ |
| $T_w$ | Heater wall temperature | K | $\theta$ |
| $\Delta T$ | Temperature difference | K | $\theta$ |
| $t$ | Time | s | $\mathbf{T}$ |
| $t_b$ | Boiling time | s | $\mathbf{T}$ |
| $t_c$ | Time for emptying, cleaning and refilling unit | s | $\mathbf{T}$ |
| $t_P$ | Total production time | s | $\mathbf{T}$ |
| $U$ | Overall heat transfer coefficient | $W/m^2$ K | $\mathbf{MT}^{-3}\theta^{-1}$ |
| $V$ | Volume | $m^3$ | $\mathbf{L}^3$ |
| $G_F$ | Feed rate | kg/s | $\mathbf{MT}^{-1}$ |
| $X_{tt}$ | Lockhart and Martinelli's parameter(equations 14.5 and 14.6) | — | — |
| $y$ | Mass fraction of vapour | — | — |
| $Z$ | Hydrostatic head | m | $\mathbf{L}$ |
| $\gamma$ | Ratio of specific heat at constant pressure to specific heat at constant volume | — | — |
| $\lambda$ | Latent heat of vaporisation per unit mass | J/kg | $\mathbf{L}^2\mathbf{T}^{-2}$ |
| $\eta$ | Economy | — | — |

| | | Units in SI System | Dimensions in **M, L, T, $\theta$** |
|---|---|---|---|
| $\eta'$ | Efficiency of ejector | — | — |
| $\mu_L$ | Viscosity of liquid | $Ns/m^2$ | $\mathbf{ML^{-1}T^{-1}}$ |
| $\mu_v$ | Viscosity of vapour | $Ns/m^2$ | $\mathbf{ML^{-1}T^{-1}}$ |
| $\rho_L$ | Density of liquid | $kg/m^3$ | $\mathbf{ML^{-3}}$ |
| $\rho_v$ | Density of vapour | $kg/m^3$ | $\mathbf{ML^{-3}}$ |
| $\sigma$ | Interfacial tension | $J/m^2$ | $\mathbf{MT^{-2}}$ |

Suffixes

| | |
|---|---|
| 0 | refers to the steam side of the first effect |
| 1, 2, 3 | refer to the first, second and third effects |
| av | refers to an average value |
| $c$ | refers to the condenser |
| $i$ and $e$ | refer to the inlet and exit cooling water |

CHAPTER 15

# Crystallisation

## 15.1. INTRODUCTION

Crystallisation, one of the oldest of unit operations, is used to produce vast quantities of materials, including sodium chloride, sodium and aluminium sulphates and sucrose which all have production rates in excess of $10^8$ tonne/year on a world basis. Many organic liquids are purified by crystallisation rather than by distillation since, as shown by MULLIN[1] in Table 15.1, enthalpies of crystallisation are generally much lower than enthalpies of vaporisation and crystallisation may be carried out closer to ambient temperature thereby reducing energy requirements. Against this, crystallisation is rarely the last stage in a process and solvent separation, washing and drying stages are usually required. Crystallisation is also a key operation in the freeze-concentration of fruit juices, the desalination of sea water, the recovery of valuable materials such as metal salts from electroplating processes, the production of materials for the electronic industries and in biotechnological operations such as the processing of proteins.

Table 15.1. Energy requirements for crystallisation and distillation[1]

| Substance | Melting point (K) | Enthalpy of crystallisation (kJ/kg) | Boiling point (K) | Enthalpy of vaporisation (kJ/kg) |
|---|---|---|---|---|
| *o*-cresol | 304 | 115 | 464 | 410 |
| *m*-cresol | 285 | 117 | 476 | 423 |
| *p*-cresol | 306 | 110 | 475 | 435 |
| *o*-xylene | 246 | 128 | 414 | 347 |
| *m*-xylene | 225 | 109 | 412 | 343 |
| *p*-xylene | 286 | 161 | 411 | 340 |
| *o*-nitrotoluene | 268.9 | 120 | 495 | 344 |
| *m*-nitrotoluene | 288.6 | 109 | 506 | 364 |
| *p*-nitrotoluene | 325 | 113 | 511 | 366 |
| water | 273 | 334 | 373 | 2260 |

Although crystals can be grown from the liquid phase — either a solution or a melt — and also from the vapour phase, a degree of supersaturation, which depends on the characteristics of the system, is essential in all cases for crystal formation or growth to take place. Some solutes are readily deposited from a cooled solution whereas others crystallise only after removal of solvent. The addition of a substance to a system in order to alter equilibrium conditions is often used in precipitation processes where supersaturation is sometimes achieved by chemical reaction between two or more substances and one of the reaction products is precipitated.

## 15.2. CRYSTALLISATION FUNDAMENTALS

In evaluating a crystallisation operation, data on phase equilibria are important as this indicates the composition of product which might be anticipated and the degree of supersaturation gives some idea of the driving force available. The rates of nuclei formation and crystal growth are equally important as these determine the residence time in, and the capacity of a crystalliser. These parameters also enable estimates to be made of crystal sizes, essential for the specification of liquor flows through beds of crystals and also the mode and degree of agitation required. It is these considerations that form the major part of this Section.

### 15.2.1. Phase equilibria

#### *One-component systems*

Temperature and pressure are the two variables that affect phase equilibria in a one-component system. The phase diagram in Figure 15.1 shows the equilibria between the solid, liquid, and vapour states of water where all three phases are in equilibrium at the *triple point*, 0.06 N/m$^2$ and 273.3 K. The *sublimation curve* indicates the vapour pressure of ice, the *vaporisation curve* the vapour pressure of liquid water, and the *fusion curve* the effect of pressure on the melting point of ice. The fusion curve for ice is unusual in that, in most one component systems, increased pressure increases the melting point, whilst the opposite occurs here.

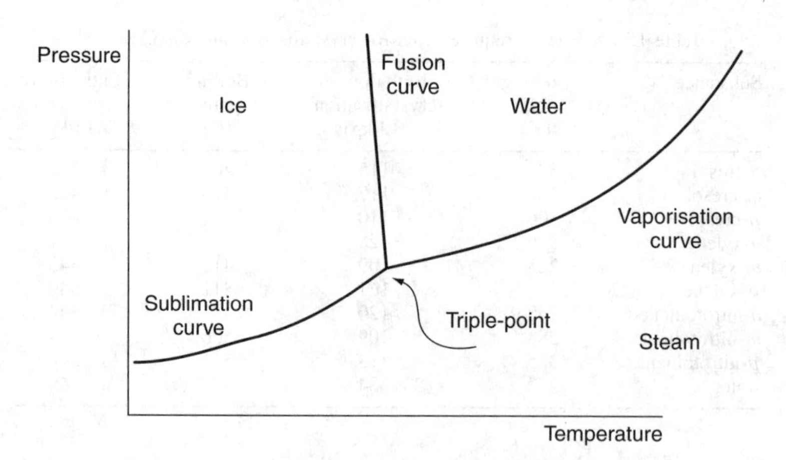


Figure 15.1. Phase diagram for water

A single substance may crystallise in more than one of seven crystal systems, all of which differ in their lattice arrangement, and exhibit not only different basic shapes but also different physical properties. A substance capable of forming more than one different crystal is said to exhibit *polymorphism*, and the different forms are called *polymorphs*. Calcium carbonate, for example, has three polymorphs — calcite (hexagonal), aragonite (tetragonal), and vaterite (trigonal). Although each polymorph is composed of the same single substance, it constitutes a separate phase. Since only one polymorph is thermodynamically stable at a specified temperature and pressure, all the other polymorphs are potentially capable of being transformed into the stable polymorph. Some polymorphic transformations are rapid and reversible and polymorphs may be *enantiotropic* (interconvertible) or *monotropic* (incapable of transformation). Graphite and carbon, for example, are monotropic at ambient temperature and pressure, whereas ammonium nitrate has five enantiotropic polymorphs over the temperature range 255–398 K. Figure 15.2*a*, taken from MULLIN[2], shows the phase reactions exhibited by two enantiotropic forms, $\alpha$ and $\beta$, of the same substance. The point of intersection of the two vapour pressure curves is the transition point at which the two forms can co-exist in equilibrium at the specified temperature and pressure. The triple point at which vapour, liquid and $\beta$ solid can co-exist may be considered as the melting point of the $\beta$ form. On slow heating, solid $\alpha$ changes into solid $\beta$ and finally melts with the reverse process taking place on slow cooling. Rapid heating or cooling can, however, result in

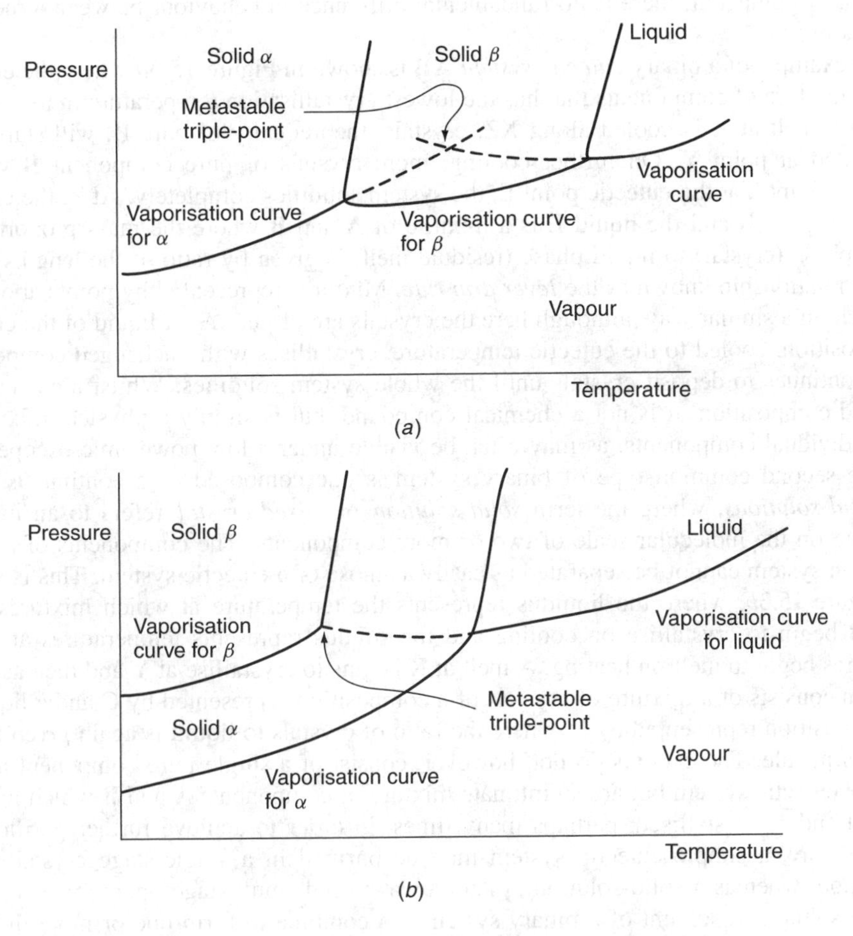


Figure 15.2. Phase diagram for polymorphic substances[2]

different behaviour where the vapour pressure of the $\alpha$ form follows a continuation of the vaporisation curve, and changes in the liquid are represented by the liquid vaporisation curve. The two curves intersect at a *metastable triple point* where the liquid, vapour, and a solid can coexist in metastable equilibrium. Figure 15.2*b* shows the pressure–temperature curves for a monotropic substance for which the vapour pressure curves of the $\alpha$ and $\beta$ forms do not intersect, and hence there is no transition point. In this case, solid $\beta$ is the metastable form, and the metastable triple point is as shown.

### Two-component systems

Temperature, pressure, and concentration can affect phase equilibria in a two-component or binary system, although the effect of pressure is usually negligible and data can be shown on a two-dimensional temperature–concentration plot. Three basic types of binary system — eutectics, solid solutions, and systems with compound formation — are considered and, although the terminology used is specific to melt systems, the types of behaviour described may also be exhibited by aqueous solutions of salts, since, as MULLIN[3] points out, there is no fundamental difference in behaviour between a melt and a solution.

An example of a binary *eutectic system* **AB** is shown in Figure 15.3*a* where the eutectic is the mixture of components that has the lowest crystallisation temperature in the system. When a melt at X is cooled along XZ, crystals, theoretically of pure **B**, will start to be deposited at point Y. On further cooling, more crystals of pure component **B** will be deposited until, at the eutectic point E, the system solidifies completely. At Z, the crystals C are of pure **B** and the liquid L is a mixture of **A** and **B** where the mass proportion of solid phase (crystal) to liquid phase (residual melt) is given by ratio of the lengths LZ to CZ; a relationship known as the *lever arm rule*. Mixtures represented by points above AE perform in a similar way, although here the crystals are of pure **A**. A liquid of the eutectic composition, cooled to the eutectic temperature, crystallises with unchanged composition and continues to deposit crystals until the whole system solidifies. Whilst a eutectic has a fixed composition, it is not a chemical compound, but is simply a physical mixture of the individual components, as may often be visible under a low-power microscope.

The second common type of binary system is one composed of a continuous series of *solid solutions*, where the term *solid solution* or *mixed crystal* refers to an intimate mixture on the molecular scale of two or more components. The components of a solid-solution system cannot be separated as easily as those of a eutectic system. This is shown in Figure 15.3*b*, where the liquidus represents the temperature at which mixtures of **A** and **B** begin to crystallise on cooling and the solidus represents temperatures at which mixtures begin to melt on heating. A melt at X begins to crystallise at Y and then at Z, the system consists of a mixture of crystals of a composition represented by C and a liquid of a composition represented by L, where the ratio of crystals to liquid is again given by the lever arm rule. The crystals do not, however, consist of a single pure component as in a simple eutectic system but are an intimate mixture of components **A** and **B** which must be heated and re-crystallised, perhaps many times, in order to achieve further purification. In this way, a simple eutectic system may be purified in a single-stage crystallisation operation, whereas a solid-solution system always needs multistage operation.

The solute and solvent of a binary system can combine to form one or more different *compounds* such as, for example, hydrates in aqueous solutions. If the compound can

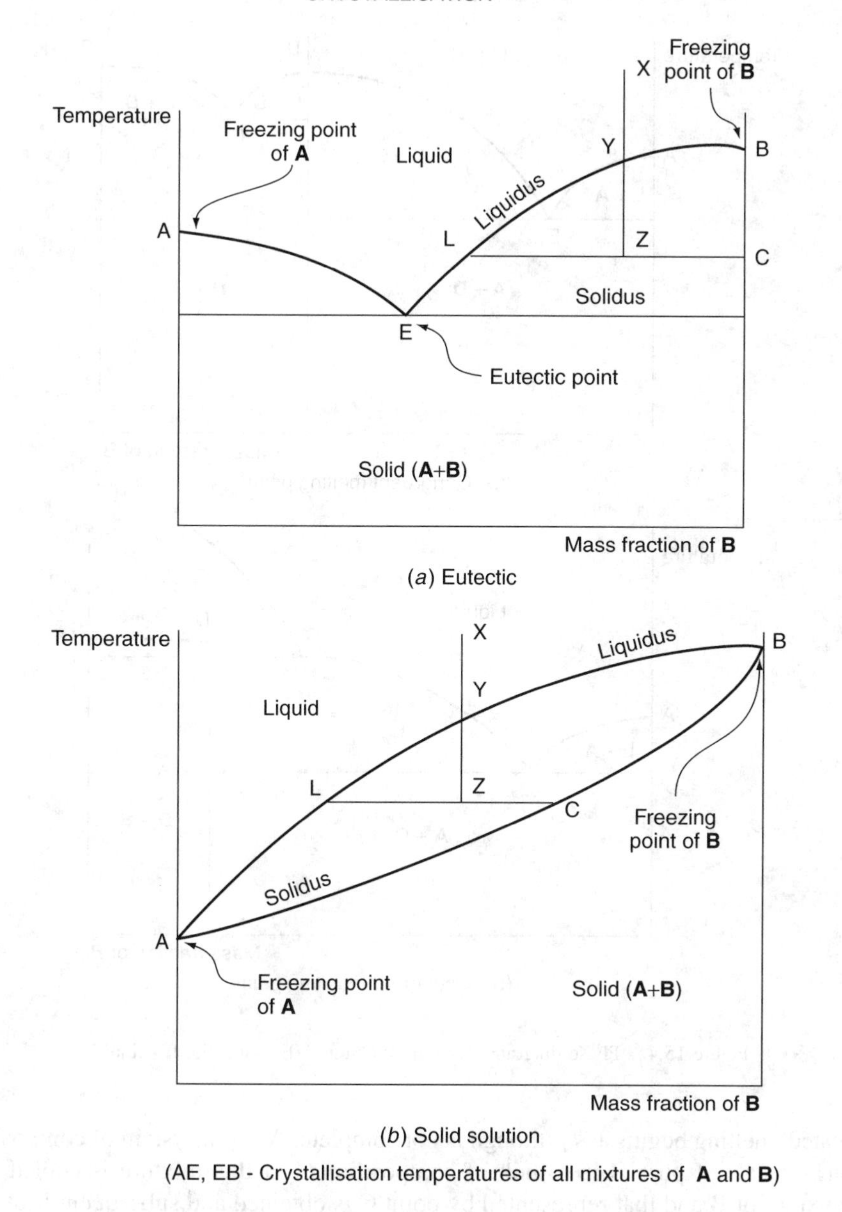


(AE, EB - Crystallisation temperatures of all mixtures of **A** and **B**)

Figure 15.3. Phase diagrams for binary systems

co-exist in stable equilibrium with a liquid phase of the same composition, then it has a *congruent* melting point, that is where melting occurs without change in composition. If this is not the case, then the melting point is *incongruent*. In Figure 15.4*a*, the heating–cooling cycle follows the vertical line through point D since melting and crystallisation occur without any change of composition. In Figure 15.4*b*, however, compound **D** decomposes at a temperature $T_1$ which is below its theoretical melting point $T_2$. Thus, if

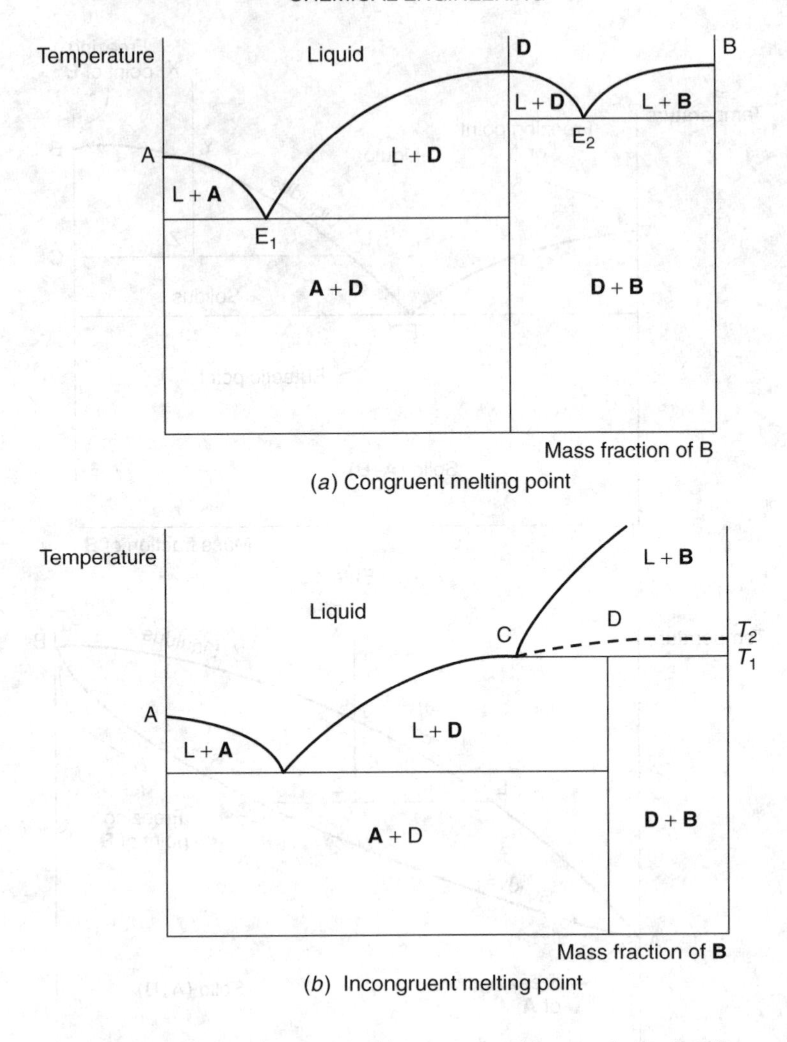


Figure 15.4. Phase diagrams for binary systems (E - eutectic, L - liquid)

**D** is heated, melting begins at $T_1$, though is not complete. At $T_1$, a system of composition D contains crystals of pure **B** in a melt of composition C. If this mixture is cooled, then a solid mixture of **B** and that represented by point C is obtained and subsequent heating and cooling cycles result in further decomposition of the compound represented by D.

There is current interest in the use of inorganic-salt hydrates as heat-storage materials, particularly for storage of solar heat in domestic and industrial space heating, where, ideally, the hydrate should have a congruent melting point so that sequences of crystallisation–melting–crystallisation can be repeated indefinitely. Incongruently melting hydrate systems tend to stratify on repeated temperature cycling with a consequent loss of efficiency as melting gives a liquid phase that contains crystals of a lower hydrate or of the anhydrous salt, which settle to the bottom of the container and fail to re-dissolve on subsequent heating. Calcium chloride hexahydrate, whilst not a true congruently melting hydrate, appears to be one of the most promising materials[4,5] as are sodium sulphate deca-hydrate, sodium acetate tri-hydrate, and sodium thiosulphate penta-hydrate which all do have incongruent melting points[6,7].

### Three-component systems

Phase equilibria in three-component systems are affected by temperature, pressure, and the concentrations of any two of the three components. Since the effect of pressure is usually negligible, phase equilibria may be plotted on an isothermal triangular diagram and, as an example, the temperature–concentration space model for *o*-, *m*-, and *p*-nitrophenol is shown in Figure 15.5*a*[3]. The three components are **O**, **M**, and **P**, respectively and points O′, M′, and P′ represent the melting points of the pure components *o*- (318 K), *m*- (370 K) and *p*-nitrophenol (387 K). The vertical faces of the prism represent temperature–concentration diagrams for the three binary eutectic systems **O-M**, **O-P**, and **M-P**, which are all similar to that shown in Figure 15.4. The binary eutectics are represented by points A (304.7 K; 72.5 per cent **O**, 27.5 per cent **M**), B (306.7 K; 75.5 per cent **O**, 24.5 per cent **P**), and C (334.7 K; 54.8 per cent **M**, 45.2 per cent **P**) and AD within the prism represents the effect of adding **P** to the **O-M** binary eutectic at A. Similarly, curves BD and CD denote the lowering of freezing points of the binary eutectics represented by points B and C, respectively, upon adding the third component. Point D is a ternary eutectic point (294.7 K; 57.7 per cent **O**, 23.2 per cent **M**, 19.1 per cent **P**) at which the liquid freezes to form a solid mixture of the three components. The section above the freezing point surfaces formed by the liquidus curves represents the homogeneous liquid phase, the section below these surfaces down to a temperature, D denotes solid and liquid phases in equilibrium and the section below this temperature represents a completely solidified system.

Figure 15.5*b* is the projection of AD, BD, and CD in Figure 15.5*a* on to the triangular base of the prism. Again **O**, **M** and **P** are the pure components, points A, B, and C represent the three binary eutectic points and D is the ternary eutectic point. The diagram is divided by AD, BD, and CD into three regions which denote the three liquidus surfaces in the space model and the temperature falls from the apexes and sides of the triangle toward the eutectic point D. Several isotherms showing points on the liquidus surfaces are shown. When, for example, a molten mixture with a composition X is cooled, solidification starts when the temperature is reduced to 353 K and since X lies in the region ADCM, pure *m*-nitrophenol is deposited. The composition of the remaining melt changes along line MX′ and at X′, equivalent to 323 K, *p*-nitrophenol also starts to crystallise. On further cooling, both *m* and *p*-nitrophenol are deposited and the composition of the liquid phase changes in the direction X′D. When the melt composition and temperature reach point D, *o*-nitrophenol also crystallises out and the system solidifies without further change in composition.

Many different types of phase behaviour are encountered in ternary systems that consist of water and two solid solutes. For example, the system $KNO_3$–$NaNO_3$–$H_2O$ which does not form hydrates or combine chemically at 323 K is shown in Figure 15.6, which is taken from MULLIN[3]. Point A represents the solubility of $KNO_3$ in water at 323 K (46.2 kg/100 kg solution), C the solubility of $NaNO_3$ (53.2 kg/100 kg solution), AB is the composition of saturated ternary solutions in equilibrium with solid $KNO_3$ and BC

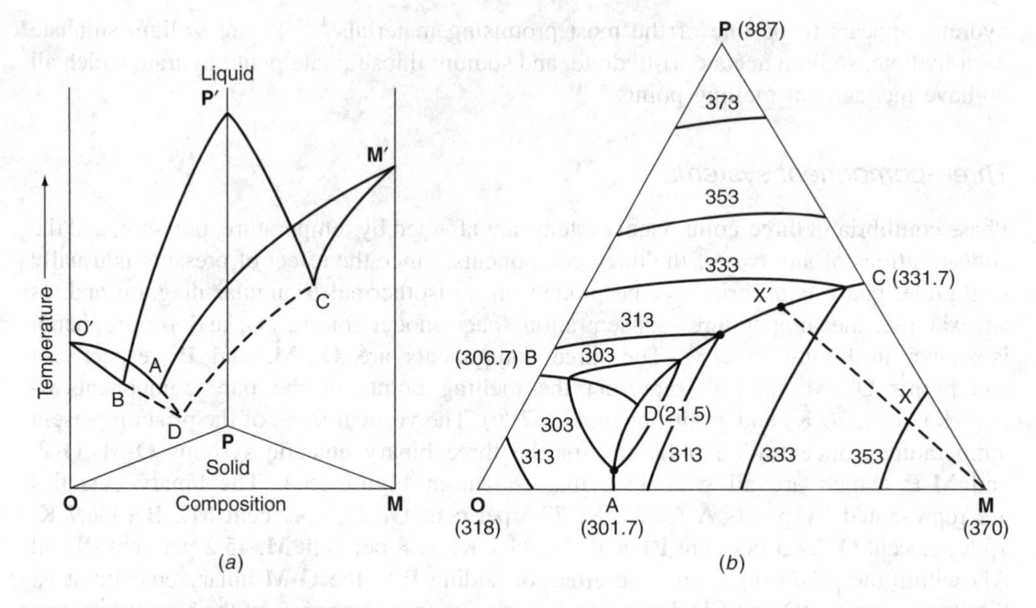


Figure 15.5. Eutectic formation in the ternary system *o*-, *m*- and *p*-nitrophenol[3] a) Temperature–concentration space model; b) Projection on a triangular diagram. (Numerical values represent temperatures in K)

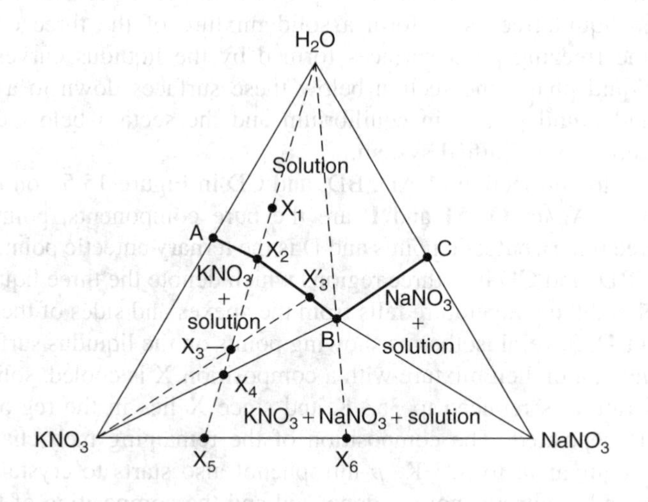


Figure 15.6. Phase diagram for the ternary system $KNO_3$–$NaNO_3$–$H_2O$ at 323 K[3]

those in equilibrium with solid $NaNO_3$. The area above the line ABC is the region of unsaturated homogeneous solutions. At point B, the solution is saturated with both $KNO_3$ and $NaNO_3$. If, for example, water is evaporated isothermally from an unsaturated solution at $X_1$, the solution concentration increases along $X_1X_2$ and pure $KNO_3$ is deposited when the concentration reaches $X_2$. If more water is evaporated to give a system of composition $X_3$, the solution composition is represented by $X_3'$ on the saturation curve AB, and by point B when composition $X_4$ is reached. Further removal of water causes deposition of $NaNO_3$. After this, all solutions in contact with solid have a constant composition B, which is referred to as the *eutonic point* or *drying-up point* of the system. After complete evaporation of water, the composition of the solid residue is indicated by $X_5$. Similarly, if an unsaturated solution, represented by a point to the right of B is evaporated isothermally, only $NaNO_3$ is deposited until the solution composition reaches B. $KNO_3$ is then also deposited and the solution composition remains constant until evaporation is complete. If water is removed isothermally from a solution of composition B, the composition of deposited solid is given by $X_6$ and it remains unchanged throughout the evaporation process.

### Multi-component systems

The more components in a system, the more complex are the phase equilibria and it is more difficult to represent phases graphically. Descriptions of multi-component solid–liquid diagrams and their uses have been given by MULLIN[3], FINDLAY and CAMPBELL[8], RICCI[9], NULL[10] and NYVLT[11] and techniques for predicting multi-component solid–liquid phase equilibria have been presented by HORMEYER *et al.*[12], KUSIK *et al.*[13], and SANDER *et al.*[14].

### Phase transformations

Metastable crystalline phases frequently crystallise to a more stable phase in accordance with Ostwald's rule of stages, and the more common types of phase transformation that occur in crystallising and precipitating systems include those between polymorphs and solvates. Transformations can occur in the solid state, particularly at temperatures near the melting point of the crystalline solid, and because of the intervention of a solvent. A stable phase has a lower solubility than a metastable phase, as indicated by the solubility curves in Figures 15.7*a* and 15.7*b* for enantiotropic and monotropic systems respectively and,

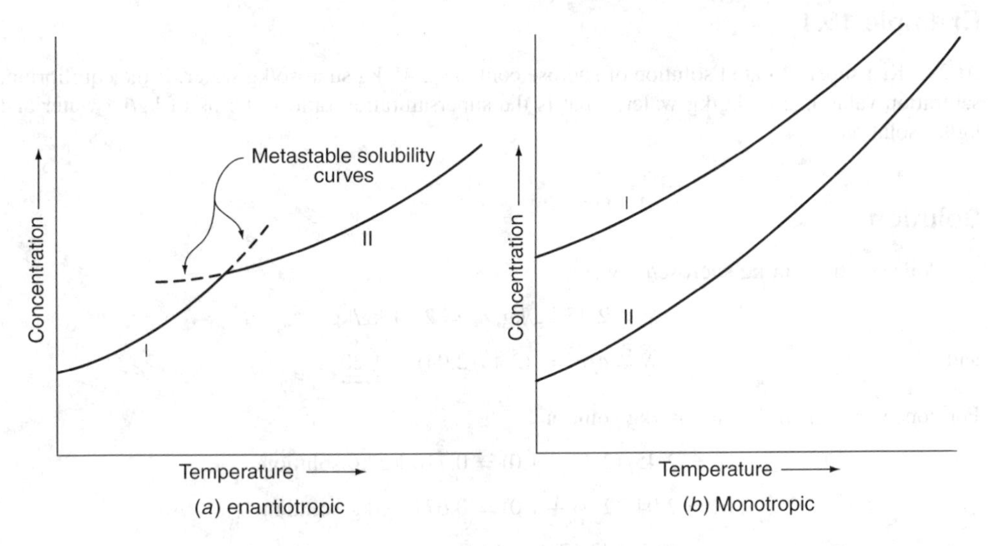


Figure 15.7. Solubility curves for substances with two polymorphs I and II[2]

whilst transformation cannot occur between the metastable (I) and stable (II) phases in the monotropic system in the temperature range shown, it is possible above the transition temperature in an enantiotropic system. Polymorphic transformation adds complexity to a phase diagram, as illustrated by NANCOLLAS *et al.*[15] and NANCOLLAS and REDDY[16] who have studied dissolution–recrystallisation transformations in hydrate systems, and CARDEW *et al.*[17] and CARDEW and DAVEY[18] who have presented theoretical analyses of both solid state and solvent-mediated transformations in an attempt to predict their kinetics.

## 15.2.2. Solubility and saturation

### *Supersaturation*

A solution that is in thermodynamic equilibrium with the solid phase of its solute at a given temperature is a saturated solution, and a solution containing more dissolved solute than that given by the equilibrium saturation value is said to be supersaturated. The degree of supersaturation may be expressed by:

$$\Delta c = c - c^* \tag{15.1}$$

where $c$ and $c^*$ are the solution concentration and the equilibrium saturation value respectively. The supersaturation ratio, $S$, and the relative supersaturation, $\varphi$ are then:

$$S = c/c^* \tag{15.2}$$

and:

$$\varphi = \Delta c/c^* = S - 1 \tag{15.3}$$

Solution concentrations may be expressed as mass of anhydrate/mass of solvent or as mass of hydrate/mass of free solvent, and the choice affects the values of $S$ and $\varphi$ as shown in the following example which is based on the data of MULLIN[3].

### Example 15.1

At 293 K, a supersaturated solution of sucrose contains 2.45 kg sucrose/kg water. If the equilibrium saturation value is 2.04 kg/kg water, what is the supersaturation ratio in terms of kg/kg water and kg/kg solution?

### Solution

For concentrations in kg sucrose/kg water:

$$c = 2.45 \text{ kg/kg}, \; c^* = 2.04 \text{ kg/kg}$$

and:

$$S = c/c^* = (2.45/2.04) = \underline{\underline{1.20}}$$

For concentrations in kg sucrose/kg solution:

$$c = 2.45/(2.45 + 1.0) = 0.710 \text{ kg/kg solution,}$$

$$c^* = 2.04/(2.04 + 1.0) = 0.671 \text{ kg/kg solution}$$

and:

$$S = (0.710/0.671) = \underline{\underline{1.06}}$$

Whilst the fundamental driving force for crystallisation, the true thermodynamic supersaturation, is the difference in chemical potential, in practice supersaturation is generally expressed in terms of solution concentrations as given in equations 15.1–15.3. MULLIN and SÖHNEL[19] has presented a method of determining the relationship between concentration-based and activity-based supersaturation by using concentration-dependent activity-coefficients.

In considering the state of supersaturation, OSTWALD[20] introduced the terms *labile* and *metastable* supersaturation to describe conditions under which spontaneous (primary) nucleation would or would not occur, and MIERS and ISAAC[21] have represented the metastable zone by means of a solubility–supersolubility diagram, as shown in Figure 15.8.

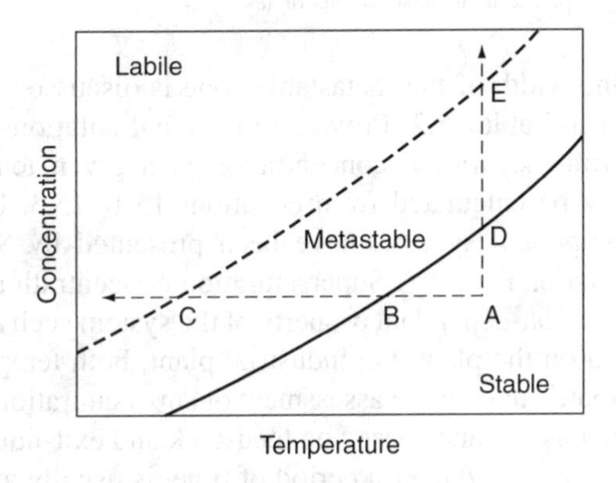


Figure 15.8. Solubility supersolubility diagram

Whilst the (continuous) solubility curve can be determined accurately, the position of the (broken) supersolubility curve is less certain as it is influenced by factors such as the rate at which the supersaturation is generated, the degree of agitation and the presence of crystals or impurities. In the stable unsaturated zone, crystallisation is impossible. In the metastable supersaturated zone, spontaneous nucleation is improbable although a crystal would grow, and in the unstable or labile saturated zone, spontaneous nucleation is probable but not inevitable. If a solution at A is cooled without loss of solvent along ABC, spontaneous nucleation cannot occur until C is reached. Although the tendency to nucleate increases once the labile zone is reached, some solutions become too viscous to permit nucleation and set to a glass. Supersaturation can also be achieved by removing solvent and ADE represents such an operation carried out at constant temperature. Because the solution near the evaporating surface is more highly supersaturated than the bulk solution, penetration into the labile zone rarely occurs and crystals at the surface fall into the solution and induce nucleation, often before bulk conditions at E have been reached. Industrial crystallisers often combine cooling and evaporation. The width of the metastable zone is often expressed as a temperature difference, $\Delta T$ which is related to the corresponding concentration difference, $\Delta c$ by the point slope of the solubility curve, $\mathrm{d}c^*/\mathrm{d}T$ or:

$$\Delta c \simeq \frac{\mathrm{d}c^*}{\mathrm{d}T} \Delta T \tag{15.4}$$

Table 15.2. Maximum allowable supercooling $\Delta T_{max}$ for aqueous salt Solutions at 298 K[3]

| Substance | deg K | Substance | deg K | Substance | deg K |
|---|---|---|---|---|---|
| $NH_4Cl$ | 0.7 | $Na_2CO_3.10H_2O$ | 0.6 | $Na_2S_2O_3.5H_2O$ | 1.0 |
| $NH_4NO_3$ | 0.6 | $Na_2CrO_4.10H_2O$ | 1.6 | K alum | 4.0 |
| $(NH_4)_2SO_4$ | 1.8 | NaCl | 4.0 | KBr | 1.1 |
| $NH_4H_2PO_4$ | 2.5 | $Na_2B_4O_7.10H_2O$ | 4.0 | KCl | 1.1 |
| $CuSO_4.5H_2O$ | 1.4 | NaI | 1.0 | KI | 0.6 |
| $FeSO_4.7H_2O$ | 0.5 | $NaHPO_4.12H_2O$ | 0.4 | $KH_2PO_4$ | 9.0 |
| $MgSO_4.7H_2O$ | 1.0 | $NaNO_3$ | 0.9 | $KNO_3$ | 0.4 |
| $NiSO_4.7H_2O$ | 4.0 | $NaNO_2$ | 0.9 | $KNO_2$ | 0.8 |
| $NaBr.2H_2O$ | 0.9 | $Na_2SO_4.10H_2O$ | 0.3 | $K_2SO_4$ | 6.0 |

Data measured in the presence of crystals with slow cooling and moderate agitation. The working value for a normal crystalliser may be 50 per cent of these values or less.

The measurement of the width of the metastable zone is discussed in Section 15.2.4, and typical data are shown in Table 15.2. Provided the actual solution concentration and the corresponding equilibrium saturation concentration at a given temperature are known, the supersaturation may be calculated from equations 15.1–15.3. Data on the solubility for two- and three-component systems have been presented by SEIDELL and LINKE[22], STEPHEN *et al.*[23] and BROUL *et al.*[24]. Supersaturation concentrations may be determined by measuring a concentration-dependent property of the system such as density or refractive index, preferably in situ on the plant. On industrial plant, both temperature and feedstock concentration can fluctuate, making the assessment of supersaturation difficult. Under these conditions, the use of a mass balance based on feedstock and exit-liquor concentrations and crystal production rates, averaged over a period of time, is usually an adequate approach.

### *Prediction of solubilities*

Techniques are available for estimating binary and multi-component solubility behaviour. One example is the van't Hoff relationship which, as stated by MOYERS and ROUSSEAU[25], takes the following form for an ideal solution:

$$\ln x = \frac{H_f}{\mathbf{R}T}\left(\frac{T}{T_M} - 1\right) \tag{15.5}$$

where $x$ is the mole fraction of solute in solution, $H_f$ is the heat of fusion and $T_M$ is the melting point of the pure component. One interesting consequence of this equation is that solubility depends only on the properties of the solute occurring in the equation. Another equation frequently used for ideal systems incorporates cryoscopic constants, values of which have been obtained empirically for a wide variety of materials in the course of the American Petroleum Research Project 44[26]. This takes the form:

$$\ln(1/x) = z_1(T_M - T)[1 + z_2(T_M - T)\ldots..] \tag{15.6}$$

where: $$z_1 = \frac{H_f}{\mathbf{R}T_M^2} \text{ and } z_2 = \frac{1}{T_M} - \frac{C_p}{2H_f}/2H_f.$$

MOYERS and ROUSSEAU[25] have used equations 15.5 and 15.6. to calculate the freezing point data for *o*- and *p*-xylene shown in Table 15.3.

Table 15.3. Calculated freezing point curves for *o*- and *p*-xylene[25]

| Data | | $p$-xylene | | $o$-xylene |
|---|---|---|---|---|
| $T_M$ (K) | | 286.41 | | 247.97 |
| $\Delta H_f$ (kJ/kmol) | | 17120 | | 13605 |
| **A** (mole fraction/deg K) | | 0.02599 | | 0.02659 |
| **B** (mole fraction/deg K) | | 0.0028 | | 0.0030 |

| Temperature (K) | Mole fraction in solution $p$-xylene Eqn. 15.5 | Eqn. 15.6 | $o$-xylene Eqn. 15.5 | Eqn. 15.6 |
|---|---|---|---|---|
| 286.41 | 1.00 | 1.00 | | |
| 280 | 0.848 | 0.844 | | |
| 270 | 0.646 | 0.640 | | |
| 260 | 0.482 | 0.478 | | |
| 249.97 | | | 1.00 | 1.00 |
| 240 | 0.249 | 0.256 | 0.803 | 0.805 |
| 235 | | | 0.695 | 0.699 |
| 230 | 0.172 | 0.183 | | |

## *Crystal size and solubility*

If *very small* solute particles are dispersed in a solution, the solute concentration may exceed the normal equilibrium saturation value. The relationship between particle size and solubility first applied to solid–liquid systems by OSTWALD[20] may be expressed as:

$$\ln \frac{c_r}{c^*} = \frac{2\,M\sigma}{n_i \mathbf{R} T \rho_s r} \tag{15.7}$$

where $c_r$ is the solubility of particles of radius $r$, $\rho_s$ the density of the solid, $M$ the relative molecular mass of the solute in solution, $\sigma$ the interfacial tension of the crystallisation surface in contact with its solution and $n_i$ the moles of ions formed from one mole of electrolyte. For a non-electrolyte, $n_i = 1$ and for most inorganic salts in water, the solubility increase is really only significant for particles of less than 1 μm. The use of this equation is illustrated in the following example which is again based on data from MULLIN[3].

## Example 15.2

Compare the increase in solubility above the normal equilibrium values of 1, 0.1 and 0.01 μm particles of barium sulphate and sucrose at 298 K. The relevant properties of these materials are:

| | barium sulphate | sucrose |
|---|---|---|
| relative molecular mass (kg/kmol) | 233 | 342 |
| number of ions (–) | 2 | 1 |
| solid density (kg/m$^3$) | 4500 | 1590 |
| interfacial tension (J/m$^2$) | 0.13 | 0.01 |

## Solution

Taking the gas constant, **R** as 8314 J/kmol K, then in equation 15.7:

For barium sulphate:

$$\ln(c_r/c^*) = (2 \times 233 \times 0.13)/(2 \times 8314 \times 298 \times 4500r) = 2.72 \times 10^{-9}/r$$

For sucrose:

$$\ln(c_r/c^*) = (2 \times 342 \times 0.01)/(1 \times 8314 \times 298 \times 1590\,r) = 1.736 \times 10^{-9}/r$$

Substituting $0.5 \times 10^{-7}$, $0.5 \times 10^{-8}$ and $0.5 \times 10^{-9}$ m for $r$ gives the following data:

| | particle size $d$ (μm) | $r$ (μm) | $c_r/c^*$ | increase (per cent) |
|---|---|---|---|---|
| barium sulphate | 1 | 0.5 | 1.005 | 0.5 |
| | 0.1 | 0.05 | 1.06 | 6 |
| | 0.01 | 0.005 | 1.72 | 72 |
| sucrose | 1 | 0.5 | 1.004 | 0.4 |
| | 0.1 | 0.05 | 1.035 | 3.5 |
| | 0.01 | 0.005 | 1.415 | 41.5 |

### *Effect of impurities*

Industrial solutions invariably contain dissolved impurities that can increase or decrease the solubility of the prime solute considerably, and it is important that the solubility data used to design crystallisation processes relate to the actual system used. Impurities can also have profound effects on other characteristics, such as nucleation and growth.

## 15.2.3. Crystal nucleation

Nucleation, the creation of crystalline bodies within a supersaturated fluid, is a complex event, since nuclei may be generated by many different mechanisms. Most nucleation classification schemes distinguish between *primary nucleation* - in the absence of crystals and *secondary nucleation* - in the presence of crystals. STRICKLAND-CONSTABLE[27] and KASHCHIEV[28] have reviewed nucleation, and GARSIDE and DAVEY[29] have considered secondary nucleation in particular.

### *Primary nucleation*

Classical theories of primary nucleation are based on sequences of bimolecular collisions and interactions in a supersaturated fluid that result in the build-up of lattice-structured bodies which may or may not achieve thermodynamic stability. Such primary nucleation is known as *homogeneous*, although the terms *spontaneous* and *classical* have also been used. As discussed by UBBELHODE[30] and GARTEN and HEAD[31], ordered solute-clustering can occur in supersaturated solutions prior to the onset of homogeneous nucleation, and BERGLUND *et al.*[32] has detected the presence of quasi-solid-phase species even in unsaturated solutions. MULLIN and LECI[33] discussed the development of concentration gradients in supersaturated solutions of citric acid under the influence of gravity, and LARSON and GARSIDE[34] estimated the size of the clusters at 4–10 nm. Primary nucleation may also be initiated by suspended particles of foreign substances, and this mechanism is generally referred to as *heterogeneous* nucleation. In industrial crystallisation, most primary nucleation is almost certainly heterogeneous, rather than homogeneous, in that it is induced by foreign solid particles invariably present in working solutions. Although the mechanism of heterogeneous nucleation is not fully understood, it probably begins with adsorption of the crystallising species on the surface of solid particles, thus creating apparently crystalline bodies, larger than the critical nucleus size, which then grow into macro-crystals.

*Homogeneous nucleation.* A consideration of the energy involved in solid-phase formation and in creation of the surface of an arbitrary spherical crystal of radius $r$ in a supersaturated fluid gives:

$$\Delta G = 4\pi r^2\sigma + (4\pi/3)r^3\Delta G_v \tag{15.8}$$

where $\Delta G$ is the overall excess free energy associated with the formation of the crystalline body, $\sigma$ is the interfacial tension between the crystal and its surrounding supersaturated fluid, and $\Delta G_v$ is the free energy change per unit volume associated with the phase change. The term $4\pi r^2\sigma$, which represents the surface contribution, is positive and is proportional to $r^2$ and the term $(4\pi/3)r^3\Delta G_v$ which represents the volume contribution, is negative and is proportional to $r^3$. Any crystal smaller than the critical nucleus size $r_c$ is unstable and tends to dissolve whilst any crystal larger than $r_c$ is stable and tends to grow. Combining equations 15.7 and 15.8, and expressing the rate of nucleation $J$ in the form of an Arrhenius reaction rate equation, gives the nucleation rate as:

$$J = F\exp\left[-\frac{16\pi\sigma^3 v^2}{3\,\mathbf{k}^3T^3(\ln S)^2}\right] \tag{15.9}$$

where $F$ is a pre-exponential factor, $v$ is molar volume, $\mathbf{k}$ is the Boltzmann constant and $S$ is the supersaturation ratio. Since equation 15.9 predicts an explosive increase in the nucleation rate beyond some so-called critical value of $S$, it not only demonstrates the powerful effect of supersaturation on homogeneous nucleation, but also indicates the possibility of nucleation at any level of supersaturation.

*Heterogeneous nucleation.* The presence of foreign particles or heteronuclei enhances the nucleation rate of a given solution, and equations similar to those for homogeneous nucleation have been proposed to express this enhancement. The result is simply a displacement of the nucleation rate against supersaturation curve, as shown in Figure 15.9, indicating that nucleation occurs more readily at a lower degree of supersaturation. For primary nucleation in industrial crystallisation, classical relationships similar to those based on equation 15.9 have little use, and all that can be justified is a simple empirical relationship such as:

$$J = K_N(\Delta c)^n \tag{15.10}$$

which relates the primary nucleation rate $J$ to the supersaturation $\Delta c$ from equation 15.1. The primary nucleation rate constant $K_N$, and the order of the nucleation process $n$, which is usually greater than 2, depend on the physical properties and hydrodynamics of the system.

## Secondary nucleation

Secondary nucleation can, by definition, take place only if crystals of the species under consideration are already present. Since this is usually the case in industrial crystallisers, secondary nucleation has a profound influence on virtually all industrial crystallisation processes.

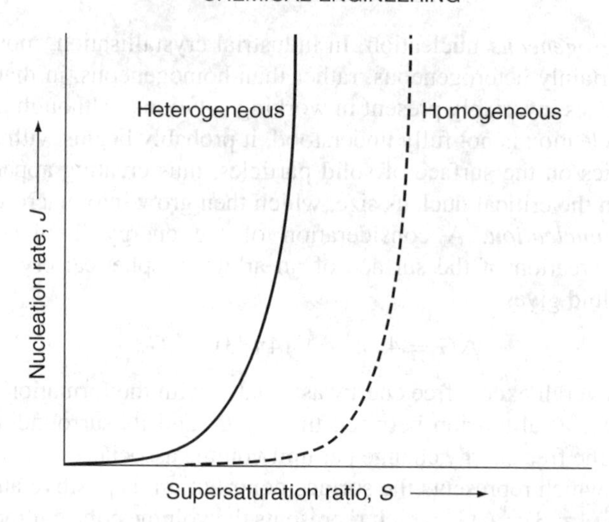


Figure 15.9. Effect of supersaturation on the rates of homogeneous and heterogeneous nucleation.

Apart from deliberate or accidental introduction of tiny seed crystals to the system, and productive interactions between existing crystals and quasi-crystalline embryos or clusters in solution, the most influential mode of new crystal generation in an industrial crystalliser is contact secondary nucleation between the existing crystals themselves, between crystals and the walls or other internal parts of the crystalliser, or between crystals and the mechanical agitator. Secondary nucleation rates (in $m^{-3}s^{-1}$) are most commonly correlated by empirical relationships such as:

$$B = K_b \rho_m^j N^l \Delta c^b \tag{15.11}$$

where $B$ is the rate of secondary nucleation or birthrate, $K_b$ *is* the birthrate constant, $\rho_m$ is the slurry concentration or magma density and $N$ is a term that gives some measure of the intensity of agitation in the system such as the rotational speed of an impeller. The exponents $j$, $l$, and $b$ vary according to the operating conditions.

## *Nucleation measurements*

One of the earliest attempts to derive nucleation kinetics for solution crystallisation was proposed by NYVLT[35] and NYVLT *et al.*[36] whose method is based on the measurement of metastable zone widths shown in Figure 15.8, using a simple apparatus, shown in Figure 15.10, consisting of a 50 ml flask fitted with a thermometer and a magnetic stirrer, located in an external cooling bath. Nucleation is detected visually and both primary and secondary nucleation can be studied in this way. Typical results[3] shown in Figure 15.11 demonstrate that seeding has a considerable influence on the nucleation process, and the difference between the slopes of the two lines indicates that primary and secondary nucleation occur by different mechanisms. Solution turbulence also affects nucleation and, in general, agitation reduces the metastable zone width. For example, the metastable zone width for gently agitated potassium sulphate solutions is about 12 deg K whilst vigorous agitation reduces this to around 8 deg K. The presence of crystals also induces secondary

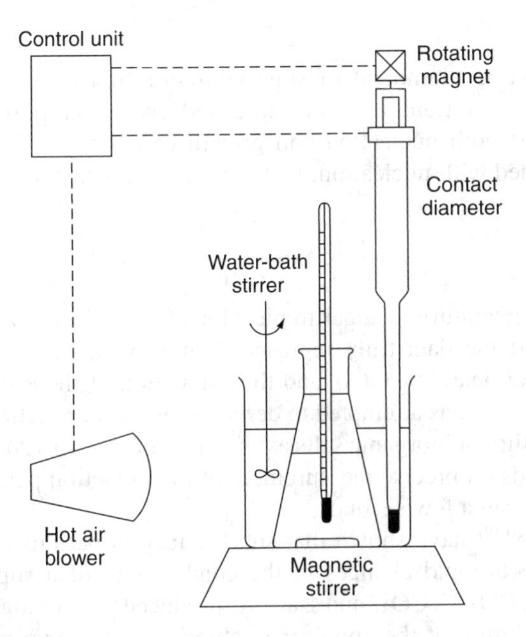


Figure 15.10. Simple apparatus for measuring metastable zone widths[36]

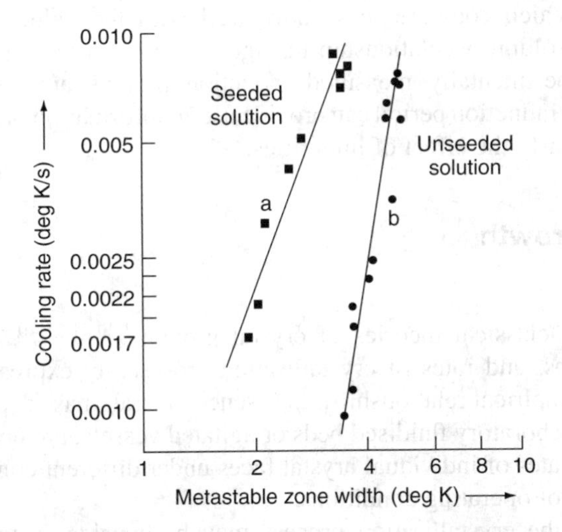


Figure 15.11. Metastable zone width of aqueous ammonium[3]

nucleation at a supercooling of around 4 deg K. The relation between supercooling $\Delta T$ and supersaturation $\Delta c$ is given by equation 15.4. Useful information on secondary nucleation kinetics for crystalliser operation and design can be determined only from model experiments that employ techniques such as those developed for MSMPR (mixed-suspension mixed-product removal) crystallisers. As discussed by, NYVLT *et al.*[36] and RANDOLPH and LARSON[37], in a real crystalliser, both nucleation and growth proceed together and interact with other system parameters in a complex manner.

### *Induction periods*

A delay occurs between attainment of supersaturation and detection of the first newly created crystals in a solution, and this so-called *induction period*, $t_i$ is a complex quantity that involves both nucleation and growth components. If it is assumed that $t_i$ is essentially concerned with nucleation, that is $t_i \propto 1/J$, then MULLIN[3] has shown, from equation 15.9, that:

$$\frac{1}{t_i} \propto \exp \frac{\sigma^3}{T^3(\log S)^2} \tag{15.12}$$

Thus, for a given temperature, a logarithmic plot of $t_i$ against $(\log S)^{-2}$ should yield a straight line which, if the data truly represent homogeneous nucleation, will allow the calculation of the interfacial tension $\sigma$ and the evaluation of the effect of temperature on $\sigma$. NIELSEN and SÖHNEL[38] has attempted to derive a general correlation between interfacial tension and the solubility of inorganic salts as shown in Figure 15.12*a*, although the success of this method depends on precise measurement of the induction period $t_i$, which presents problems if $t_i$ is less than a few seconds.

SÖHNEL and MULLIN[39] have shown that short induction periods can be determined by a technique that detects rapid changes in the conductivity of a supersaturated solution. Typical results for $CaCO_3$, $SrCO_3$, and $BaCO_3$, produced by mixing an aqueous solution of $Na_2CO_3$ with a solution of the appropriate chloride, are shown in Figure 15.12*b*. The slopes of the linear, high-supersaturation regions are used to calculate the interfacial tensions (0.08–0.12 J/m$^2$), which compare reasonably well with the values predicted from the interfacial tension–solubility relationship in Figure 15.12*a*. Although interfacial tensions evaluated from experimentally measured induction periods are somewhat unreliable, measurements of the induction period can provide useful information on other crystallisation phenomena, particularly the effect of impurities.

## 15.2.4. Crystal growth

### *Fundamentals*

As with nucleation, classical theories of crystal growth[3,20,21,35,40–42] have not led to working relationships, and rates of crystallisation are usually expressed in terms of the supersaturation by empirical relationships. In essence, overall mass deposition rates, which can be measured in laboratory fluidised beds or agitated vessels, are needed for crystalliser design, and growth rates of individual crystal faces under different conditions are required for the specification of operating conditions.

In simple terms, the crystallisation process may be considered to take place in two stages—a diffusional step in which solute is transported from the bulk fluid through the solution boundary layer adjacent to the crystal surface, and a deposition step in which adsorbed solute ions or molecules at the crystal surface are deposited and integrated into the crystal lattice. These two stages which are shown in Figure 15.13, may be described by:

$$\mathrm{d}m/\mathrm{d}t = k_d A(c - c_i) = k_r A(c_i - c^*)^i \tag{15.13}$$

where $m$ is the mass deposited in time $t$, $A$ is the crystal surface area, $c$, $c_i$ and $c^*$ are the solute concentrations in the bulk solution, at the interface and at equilibrium saturation and $k_d$ and $k_r$ are the diffusion and deposition or reaction mass transfer coefficients.

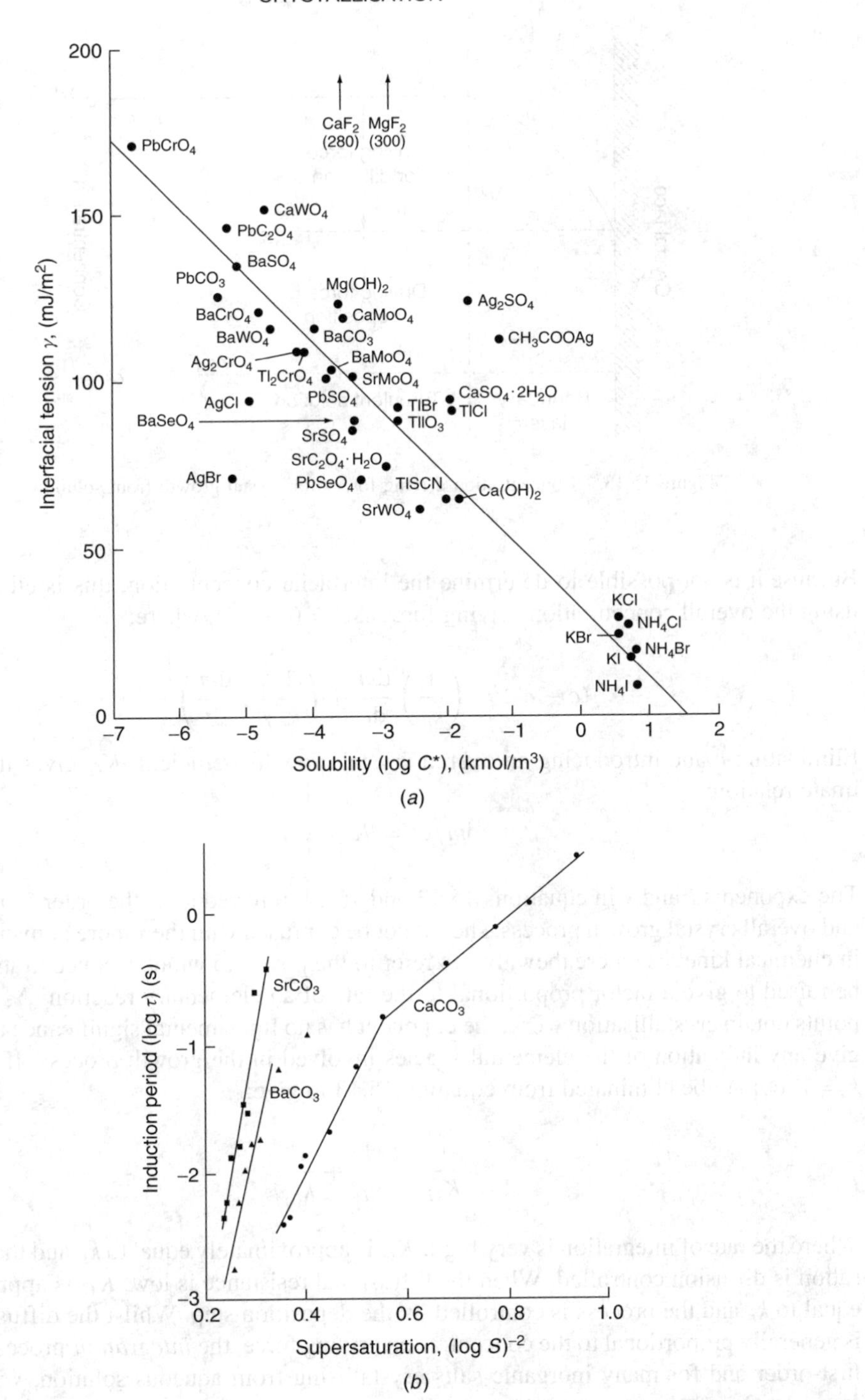


Figure 15.12. (*a*) Interfacial tension as a function of solubility[38] (*b*) Induction period as a function of initial supersaturation[39]

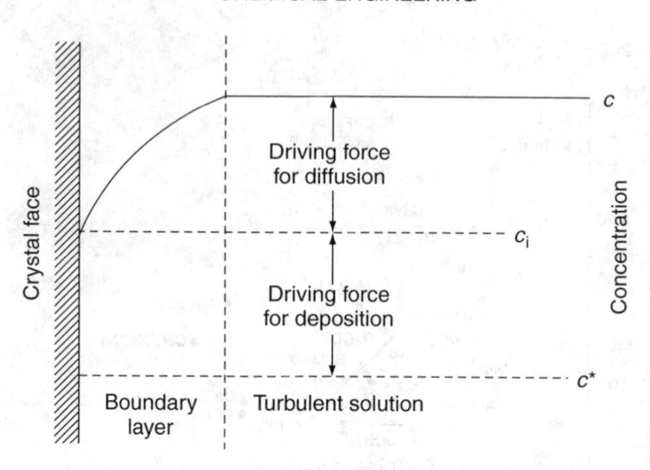


Figure 15.13. Concentration driving forces for crystal growth from solution

Because it is not possible to determine the interfacial concentration, this is eliminated by using the overall concentration driving force $\Delta c = (c - c^*)$, where:

$$(c - c^*) = \left(\frac{1}{k_d}\right)\frac{\mathrm{d}m}{\mathrm{d}t} + \left(\frac{1}{k_r}\right)\left(\frac{\mathrm{d}m}{\mathrm{d}t}\right)^{\frac{1}{i}}$$

Eliminating $c$ and introducing an overall crystal growth coefficient, $K_G$ gives the approximate relation:

$$\mathrm{d}m/\mathrm{d}t = K_G(\Delta c)^s \tag{15.14}$$

The exponents $i$ and $s$ in equations 15.13 and 15.14, referred to as the order of integration and overall crystal growth process, should not be confused with their more conventional use in chemical kinetics where they always refer to the power to which a concentration should be raised to give a factor proportional to the rate of an elementary reaction. As MULLIN[3] points out, in crystallisation work, the exponent has no fundamental significance and cannot give any indication of the elemental species involved in the growth process. If $i = 1$ and $s = 1$, $c_i$ may be eliminated from equation 15.13 to give:

$$\frac{1}{K_G} = \frac{1}{k_D} + \frac{1}{k_r} \tag{15.15}$$

Where the rate of integration is very high, $K_G$ is approximately equal to $k_d$ and the crystallisation is diffusion controlled. When the diffusional resistance is low, $K_G$ is approximately equal to $k_r$ and the process is controlled by the deposition step. Whilst the diffusional step is generally proportional to the concentration driving force, the *integration* process is rarely first-order and for many inorganic salts crystallising from aqueous solution, $s$ lies in the range 1–2.

Comprehensive reviews of theories of crystal growth have been presented by GARSIDE[43], NIELSEN[44], PAMPLIN[45] and KALDIS and SCHEEL[46].

### Measurement of growth rate

Methods used for the measurement of crystal growth rates are either a) direct measurement of the linear growth rate of a chosen crystal face or b) indirect estimation of an overall linear growth rate from mass deposition rates measured on individual crystals or on groups of freely suspended crystals[3,35,41,47,48].

*Face growth rates.* Different crystal faces grow at different rates and faces with a high value of $s$ grow faster than faces with low values. Changes in growth environment such as temperature, supersaturation pH, and impurities can have a profound effect on growth, and differences in individual face growth rates give rise to habit changes in crystals. For the measurement of individual crystal-face growth-rates, a fixed crystal in a glass cell is observed with a travelling microscope under precisely controlled conditions of solution temperature, supersaturation and liquid velocity[3]. The solution velocity past the fixed crystal is often an important growth-determining parameter, sometimes responsible for the so-called size-dependent growth effect often observed in agitated and fluidised-bed crystallisers. Large crystals have higher settling velocities than small crystals and, if their growth is diffusion-controlled, they tend to grow faster. Salts that exhibit solution velocity dependent growth rates include the alums, nickel ammonium sulphate, and potassium sulphate, although salts such as ammonium sulphate and ammonium or potassium dihydrogen phosphate are not affected by solution velocity.

*Overall growth rates.* In the laboratory, growth rate data for crystalliser design can be measured in fluidised beds or in agitated vessels, and crystal growth rates measured by growing large numbers of carefully sized seeds in fluidised suspension under strictly controlled conditions. A warm undersaturated solution of known concentration is circulated in the crystalliser and then supersaturated by cooling to the working temperature. About 5 g of closely sized seed crystals with a narrow size distribution and a mean size of around 500 μm is introduced into the crystalliser, and the upward solution velocity is adjusted so that the crystals are maintained in a reasonably uniform fluidised state in the growth zone. The crystals are allowed to grow at a constant temperature until their total mass is some 10 g, when they are removed, washed, dried, and weighed. The final solution concentration is measured, and the mean of the initial and final supersaturations is taken as the average for the run, an assumption which does not involve any significant error because the solution concentration is usually not allowed to change by more than about 1 per cent during a run. The overall crystal growth rate is then calculated in terms of mass deposited per unit area per unit time at a specified supersaturation.

### Expression of growth rate

Because the rate of growth depends, in a complex way, on temperature, supersaturation, size, habit, system turbulence and so on, there is no simple was of expressing the rate of crystal growth, although, under carefully defined conditions, growth may be expressed as an overall mass deposition rate, $R_G$ (kg/m$^2$ s), an overall linear growth rate, $G_d (= \mathrm{d}d/\mathrm{d}t)$ (m/s) or as a mean linear velocity, $u' (= \mathrm{d}r/\mathrm{d}t)$ (m/s). Here $d$ is some characteristic size of the crystal such as the equivalent aperture size, and $r$ is the radius corresponding to the equivalent sphere where $r = 0.5d$. The relationships between these quantities are:

$$R_G = K_G \Delta c^s = \left(\frac{1}{A}\right)\frac{dm}{dt} = \frac{3\alpha\rho \frac{dd}{dt}}{\beta} = \left(\frac{6\alpha\rho}{\beta}\right)\frac{dr}{dt} = \frac{6\alpha\rho u'}{\beta} \tag{15.16}$$

where $\rho$ is the density of the crystal and the volume and surface shape factors, $\alpha$ and $\beta$, are related to particle mass $m$ and surface area $A$, respectively, by:

$$m = \alpha\rho d^3 \tag{15.17}$$

and:

$$A = \beta d^2 \tag{15.18}$$

Values of $6\alpha/\beta$ are 1 for spheres and cubes and 0.816 for octahedra and typical values of the mean linear growth velocity, $u'(= 0.5\, G_d)$ for crystals 0.5–1 mm growing in the presence of other crystals are given in Table 15.4 which is taken from MULLIN[3].

Table 15.4. Mean over-all crystal growth rates expressed as a linear velocity[3]

| Substance | deg K | Supersaturation ratio $S$ | $u'$ (m/s) |
|---|---|---|---|
| $NH_4NO_3$ | 313 | 1.05 | $8.5 \times 10^{-7}$ |
| $(NH_4)_2SO_4$ | 303 | 1.05 | $2.5 \times 10^{-7}$ |
| | 333 | 1.05 | $4.0 \times 10^{-7}$ |
| $MgSO_4.7H_2O$ | 293 | 1.02 | $4.5 \times 10^{-8}$ |
| | 303 | 1.01 | $8.0 \times 10^{-8}$ |
| | 303 | 1.02 | $1.5 \times 10^{-7}$ |
| KCl | 313 | 1.01 | $6.0 \times 10^{-7}$ |
| $KNO_3$ | 293 | 1.05 | $4.5 \times 10^{-8}$ |
| | 313 | 1.05 | $1.5 \times 10^{-7}$ |
| $K_2SO_4$ | 293 | 1.09 | $2.8 \times 10^{-8}$ |
| | 293 | 1.18 | $1.7 \times 10^{-7}$ |
| | 303 | 1.07 | $4.2 \times 10^{-8}$ |
| | 323 | 1.06 | $7.0 \times 10^{-8}$ |
| | 323 | 1.12 | $3.2 \times 10^{-7}$ |
| NaCl | 323 | 1.002 | $2.5 \times 10^{-8}$ |
| | 323 | 1.003 | $6.5 \times 10^{-8}$ |
| | 343 | 1.002 | $9.0 \times 10^{-8}$ |
| | 343 | 1.003 | $1.5 \times 10^{-7}$ |

### Dependence of growth rate on crystal size

Experimental evidence indicates that crystal growth kinetics often depend on crystal size, possibly because the size depends on the surface deposition kinetics and different crystals of the same size can also have different growth rates because of differences in surface structure or perfection. In addition, as discussed by WHITE *et al.*[49], JONES and MULLIN[50], JANSE and DE JONG[51] and GARSIDE and JANČIĆ[52], small crystals of many substances grow much more slowly than larger crystals, and some do not grow at all. The behaviour of very small crystals has considerable influence on the performance of continuously operated crystallisers because new crystals with a size of 1–10 μm are constantly generated by secondary nucleation and these then grow to populate the full crystal size distribution.

### Growth - nucleation interactions

Crystal nucleation and growth in a crystalliser cannot be considered in isolation because they interact with one another and with other system parameters in a complex manner. For a complete description of the crystal size distribution of the product in a continuously operated crystalliser, both the nucleation and the growth processes must be quantified, and the laws of conservation of mass, energy, and crystal population must be applied. The importance of population balance, in which all particles are accounted for, was first stressed in the pioneering work of RANDOLPH and LARSON(37).

### Crystal habit modification

Differences in the face growth-rates of crystals give rise to changes in their habit or shape. Although the growth kinetics of individual crystal faces usually depend to various extents on supersaturation so that crystal habit can sometimes be controlled by adjusting operating conditions, the most common cause of habit modification is the presence of impurities. Although a soluble impurity will often remain in the liquid phase so that pure crystals are formed, in many cases, both the rate of nucleation and the crystal growth rate are affected. More usually, the effect is one of retardation, thought to be due to the adsorption of the impurity on the surface of the nucleus or crystal. Materials with large molecules such as tannin, dextrin or sodium hexametaphosphate, added in small quantities to boiler feed water, prevent the nucleation and growth of calcium carbonate crystals and hence reduce scaling. In a similar way, the addition of 0.1 per cent of HCl and 0.1 per cent $PbCl_2$ prevent the growth of sodium chloride crystals. In some cases the adsorption occurs preferentially on one particular face of the crystal, thus modifying the crystal shape. One example is that sodium chloride crystallised from solutions containing traces of urea forms octohedral instead of the usual cubic crystals. In a similar way, dyes are preferentially adsorbed on inorganic crystals(53), although the effect is not always predictable. GARRETT(54) has described a number of uses of additives as habit modifiers, and industrial applications of habit modification are reported in several reviews(3,55,56) in which the factors that must be considered in selecting a suitable habit modifier are discussed. In the main, solid impurities act as condensation nuclei and cause dislocations in the crystal structure.

### Inclusions in crystals

Inclusions are small pockets of solid, liquid, or gaseous impurities trapped in crystals that usually occur randomly although a regular pattern may be sometimes observed. As described by MULLIN(3), a simple technique for observing inclusions is to immerse the crystal in an inert liquid of similar refractive index or, alternatively, in its own saturated solution when, if the inclusion is a liquid, concentration streamlines will be seen as the two fluids meet and, if it is a vapour, a bubble will be released. Industrial crystals may contain significant amounts of included mother liquor that can significantly affect product purity and stored crystals may cake because of liquid seepage from inclusions in broken crystals. In order to minimise inclusions, the crystallising system should be free of dirt and other solid debris, vigorous agitation or boiling should be avoided, and ultrasonic irradiation may be used to suppress adherence of bubbles to a growing crystal face. As fast crystal growth is probably the most common cause of inclusion formation, high supersaturation levels should be avoided. DEICHA[57], POWERS[58], WILCOX and KUO[59] and SASKA and MYERSON[60] have published detailed accounts of crystal inclusion.

### 15.2.5. Crystal yield

The yield of crystals produced by a given degree of cooling may be estimated from the concentration of the initial solution and the solubility at the final temperature, allowing for any evaporation, by making solvent and solute balances. For the solvent, usually water, the initial solvent present is equal to the sum of the final solvent in the mother liquor, the water of crystallisation within the crystals and any water evaporated, or:

$$w_1 = w_2 + y\frac{R-1}{R} + w_1 E \tag{15.19}$$

where $w_1$ and $w_2$ are the initial and final masses of solvent in the liquor, $y$ is the yield of crystals, $R$ is the ratio (molecular mass of hydrate/molecular mass of anhydrous salt) and $E$ is the ratio (mass of solvent evaporated/mass of solvent in the initial solution). For the solute:

$$w_1 c_1 = w_2 c_2 + y/R \tag{15.20}$$

where $c_1$ and $c_2$ are the initial and final concentrations of the solution expressed as (mass of anhydrous salt/mass of solvent). Substituting for $w_2$ from equation 15.19:

$$w_1 c_1 = c_2\left[w_1(1-E) - y\frac{R-1}{R}\right] + \frac{y}{R} \tag{15.21}$$

from which the yield for aqueous solutions is given by:

$$y = Rw_1\frac{c_1 - c_2(1-E)}{1 - c_2(R-1)} \tag{15.22}$$

The actual yield may differ slightly from that given by this equation since, for example, when crystals are washed with fresh solvent on the filter, losses may occur through dissolution. On the other hand, if mother liquor is retained by the crystals, an extra quantity of crystalline material will be deposited on drying. Since published solubility data usually refer to pure solvents and solutes that are rarely encountered industrially, solubilities should always be checked against the actual working liquors.

Before equation 15.22 can be applied to vacuum or adiabatic cooling crystallisation, the quantity $E$ must be estimated, where, from a heat balance:

$$E = \frac{qR(c_1 - c_2) + C_p(T_1 - T_2)(1 + c_1)[1 - c_2(R-1)]}{\lambda[1 - c_2(R-1)] - qRc_2} \tag{15.23}$$

In this equation, $\lambda$ is the latent heat of evaporation of the solvent (J/kg), $q$ is the heat of crystallisation of the product (J/kg), $T_1$ is the initial temperature of the solution (K), $T_2$ is the final temperature of the solution (K) and $C_p$ is the specific heat capacity of the solution (J/kg K).

## Example 15.3

What is the theoretical yield of crystals which may be obtained by cooling a solution containing 1000 kg of sodium sulphate (molecular mass = 142 kg/kmol) in 5000 kg water to 283 K? The solubility of sodium sulphate at 283 K is 9 kg anhydrous salt/100 kg water and the deposited crystals will consist of the deca-hydrate (molecular mass = 322 kg/kmol). It may be assumed that 2 per cent of the water will be lost by evaporation during cooling.

## Solution

The ratio, $R = (322/142) = 2.27$

The initial concentration, $c_1 = (1000/5000) = 0.2$ kg $Na_2SO_4$/kg water

The solubility, $c_2 = (9/100) = 0.09$ kg $Na_2SO_4$/kg water

The initial mass of water, $w_1 = 5000$ kg and the water lost by evaporation,
$E = (2/100) = 0.02$ kg/kg

Thus, in equation 15.22:

$$\text{yield, } y = (5000 \times 2.27)[0.2 - 0.09(1 - 0.02)]/[1 - 0.09(2.27 - 1)]$$

$$= \underline{\underline{1432 \text{ kg } Na_2SO_4.10H_2O}}$$

## Example 15.4

What is the yield of sodium acetate crystals ($CH_3COONa.3H_2O$) obtainable from a vacuum crystalliser operating at 1.33 kN/m$^2$ when it is supplied with 0.56 kg/s of a 40 per cent aqueous solution of the salt at 353 K? The boiling point elevation of the solution is 11.5 deg K.

Data:

Heat of crystallisation, $q = 144$ kJ/kg trihydrate
Heat capacity of the solution, $C_p = 3.5$ kJ/kg deg K
Latent heat of water at 1.33 kN/m$^2$, $\lambda = 2.46$ MJ/kg
Boiling point of water at 1.33 kN/m$^2$ = 290.7 K
Solubility of sodium acetate at 290.7 K, $c_2 = 0.539$ kg/kg water.

## Solution

Equilibrium liquor temperature = $(290.7 + 11.5) = 302.2$ K.
Initial concentration, $c_1 = 40/(100 - 40) = 0.667$ kg/kg water
Final concentration, $c_2 = 0.539$ kg/kg water
Ratio of molecular masses, $R = (136/82) = 1.66$

Thus, in equation 15.23:

$$E = \{144 \times 1.66(0.667 - 0.539) + 3.5(353 - 302.2)(1 + 0.667)[1 - 0.539(1.66 - 1)]\}/$$
$$\{2460[1 - 0.539(1.66 - 1)] - (144 \times 1.66 \times 0.539)\}$$
$$= 0.153 \text{ kg/kg water originally present.}$$

The yield is then given by equation 15.22 as:

$$y = (0.56(100 - 40)/100)1.66[0.667 - 0.539(1 - 0.153)]/[1 - 0.539(1.66 - 1)]$$
$$= \underline{\underline{0.183 \text{ kg/s}}}$$

## 15.2.6. Caking of crystals

Crystalline materials frequently cake or cement together on storage and crystal size, shape, moisture content, and storage conditions can all contribute to the caking tendency. In general, caking is caused by a dampening of the crystal surfaces in storage because of inefficient drying or an increase in atmospheric humidity above some critical value that depends on both substance and temperature. The presence of a hygroscopic trace impurity in the crystals, can also greatly influence their tendency to absorb atmospheric moisture. Moisture may also be released from inclusions if crystals fracture under storage conditions and, if crystal surface moisture later evaporates, adjacent crystals become firmly joined together with a cement of re-crystallised solute. Caking may be minimised by efficient drying, packaging in airtight containers, and avoiding compaction on storage. In addition, crystals may be coated with an inert dust that acts as a moisture barrier. Small crystals are more prone to cake than large crystals because of the greater number of contact points per unit mass, although actual size is less important than size distribution and shape and the narrower the size distribution and the more granular the shape, the lower is the tendency of crystals to cake. Crystal size distribution can be controlled by adjusting operating conditions in a crystalliser and crystal shape may be influenced by the use of habit modifiers. A comprehensive account of the inhibition of caking by trace additives has been given by PHOENIX(61).

## 15.2.7. Washing of crystals

The product from a crystalliser must be subjected to efficient solid–liquid separation in order to remove mother liquor and, whilst centrifugal filtration can reduce the liquor content of granular crystals to 5–10 per cent, small irregular crystals may retain more than 50 per cent. After filtration, the product is usually washed to reduce the amount of liquor retained still further and, where the crystals are very soluble in the liquor, another liquid in which the substance is relatively insoluble is used for the washing, although this two-solvent method means that a solvent recovery unit is required. When simple washing is inadequate, two stages may be required for the removal of mother liquor with the crystals removed from the filter, re-dispersed in wash liquor and filtered again. This may cause a loss of yield although this is much less than the loss after a complete re-crystallisation.

If, for simplicity, it is assumed that the soluble impurity is in solution and that solution concentrations are constant throughout the dispersion vessel, then wash liquor requirements for decantation washing may be estimated as follows.

If, in *batch operation*, $c_{io}$ and $c_{in}$ denote the impurity concentrations in the crystalline material (kg impurity/kg product) initially and after washing stage $n$ respectively, and $F$

is the fraction of liquid removed at each decantation, then a mass balance gives:

$$c_{in} = c_{io}(1 - F)^n \tag{15.24}$$

or:

$$\ln(c_{in}/c_{io}) = n \ln(1 - F) \tag{15.25}$$

For *continuous operation*, where fresh wash-liquid enters the vessel continuously and liquor is withdrawn through a filter screen, then a mass balance gives:

$$V_L \, dc = -c_i \, dV_W \tag{15.26}$$

or:

$$\ln(c_{in}/c_{io}) = -V_W/V_L \tag{15.27}$$

where $c_{io}$ and $c_{in}$ are the initial and final concentrations and $V_L$ and $V_W$ are the volumes of liquor in the vessel and of the wash-water respectively. Combining equations 15.25 and 15.27:

$$n \ln(1 - F) = -V_W/V_L \tag{15.28}$$

or:

$$\frac{1}{nF}\frac{V_w}{V_L} = -\ln\frac{1 - F}{F} \tag{15.29}$$

As MULLIN[3] points out, this equation can be used for comparing batch and continuous processing since $V_W$ and $nFV_L$ represent the wash liquor requirements for both cases.

## 15.3. CRYSTALLISATION FROM SOLUTIONS

Solution crystallisers are usually classified according to the method by which supersaturation is achieved, that is by cooling, evaporation, vacuum, reaction and salting out. The term *controlled* denotes supersaturation control whilst *classifying* refers to classification of product size.

### 15.3.1. Cooling crystallisers

#### *Non-agitated vessels*

The simplest type of cooling crystalliser is an unstirred tank in which a hot feedstock solution is charged to an open vessel and allowed to cool, often for several days, by natural convection. Metallic rods may be suspended in the solution so that large crystals can grow on them thereby reducing the amount of product that sinks to the bottom of the unit. The product is usually removed manually. Because cooling is slow, large interlocked crystals are usually produced. These retain mother liquor and thus the dried crystals are generally impure. Because of the uncontrolled nature of the process, product crystals range from a fine dust to large agglomerates. Labour costs are high, but the method is economical for small batches since capital, operating, and maintenance costs are low, although productivity is low and space requirements are high.

#### *Agitated vessels*

Installation of an agitator in an open-tank crystalliser gives smaller and more uniform crystals and reduces batch times. Because less liquor is retained by the crystals after filtration and more efficient washing is possible, the final product has a higher purity. Water jackets are usually preferred to coils for cooling because the latter often become encrusted with crystals and the inner surfaces of the crystalliser should be smooth and flat to minimise encrustation. Operating costs of agitated coolers are higher than for simple tanks and, although the productivity is higher, product handling costs are still high. Tank crystallisers vary from shallow pans to large cylindrical tanks.

The typical agitated cooling crystalliser, shown in Figure 15.14*a*, has an upper conical section which reduces the upward velocity of liquor and prevents the crystalline product from being swept out with the spent liquor. An agitator, located in the lower region of a draught tube circulates the crystal slurry through the growth zone of the crystalliser; cooling surfaces may be provided if required. External circulation, as shown in Figure 15.14*b*, allows good mixing inside the unit and promotes high rates of heat transfer between liquor and coolant, and an internal agitator may be installed in the crystallisation tank if required. Because the liquor velocity in the tubes is high, low temperature differences are usually adequate, and encrustation on heat transfer surfaces is reduced considerably. Batch or continuous operation may be employed.

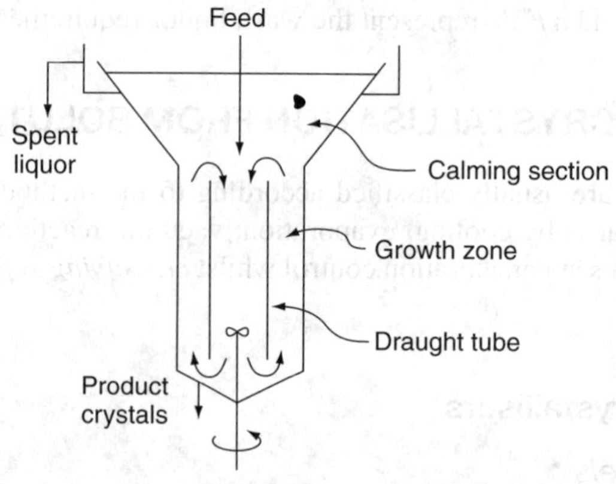


(*a*) Internal circulation through a draught tube

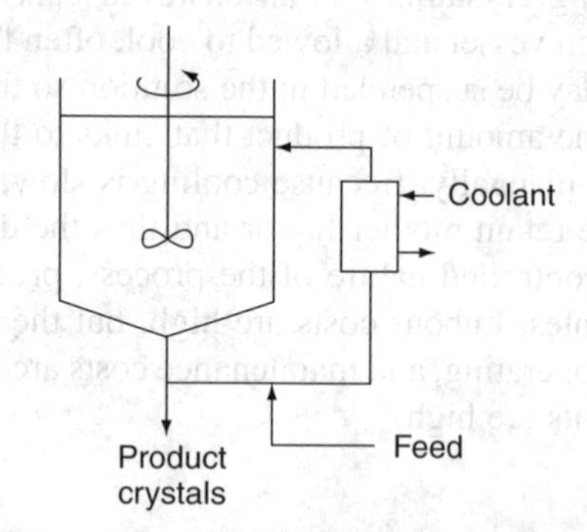


(*b*) External circulation through a heat exchanger

Figure 15.14. Cooling crystallisers

### Scraped-surface crystallisers

The Swenson-Walker scraped-surface unit, which is used for processing inorganic salts that have a high temperature solubility coefficient with water, is a shallow semi-cylindrical trough, about 600 mm wide and 3–12 m long, fitted with a water-cooled jacket. A helical scraper rotating at 0.8–1.6 Hz, keeps the cooling surfaces clean and enhances growth of crystals by moving them through the solution which flows down the sloping trough. Several units may be connected in series and the capacity is determined by the heat transfer rate which should exceed 60 kW[1] for economic operation, with heat transfer coefficients in the range 50–150 W/m$^2$ deg K. High coefficients and hence high production rates are obtained with double-pipe, scraped-surface units such as Votator and Armstrong crystallisers in which spring-loaded internal agitators scrape the heat transfer surfaces. With turbulent flow in the tube, coefficients of 50–700 W/m$^2$ deg K are achieved. Such units range from 75 to 600 mm in diameter and 0.3 to 3 m long. They are used mainly for processing fats, waxes and other organic melts, as outlined in Section 15.4, although the processing of inorganic solutions such as sodium sulphate from viscose spin-bath liquors, has been reported by ARMSTRONG[62].

## Example 15.5

A solution containing 23 per cent by mass of sodium phosphate is cooled from 313 to 298 K in a Swenson-Walker crystalliser to form crystals of $Na_3PO_4.12H_2O$. The solubility of $Na_3PO_4$ at 298 K is 15.5 kg/100 kg water, and the required product rate of crystals is 0.063 kg/s. The mean heat capacity of the solution is 3.2 kJ/kg deg K and the heat of crystallisation is 146.5 kJ/kg. If cooling water enters and leaves at 288 and 293 K, respectively, and the overall coefficient of heat transfer is 140 W/m$^2$ deg K, what length of crystalliser is required?

## Solution

The molecular mass of hydrate/molecular mass of anhydrate, $R = (380/164) = 2.32$

It will be assumed that the evaporation is negligible and that $E = 0$.

The initial concentration, $c_1 = 0.23$ kg/kg solution or $0.23/(1 - 0.23) = 0.30$ kg/kg water

The final concentration, $c_2 = 15.5$ kg/kg water or 0.155 kg/kg water

In 1 kg of the initial feed solution, there is 0.23 kg salt and 0.77 kg water and hence $w_1 = 0.77$ kg

The yield is given by equation 15.22:

$$y = 2.32 \times 0.77[0.30 - 0.155(1 - 0)]/[1 - 0.155(2.32 - 1)]$$
$$= 0.33 \text{ kg}$$

In order to produce 0.063 kg/s of crystals, the required feed is:

$$= (1 \times 0.063/0.33) = 0.193 \text{ kg/s}$$

The heat required to cool the solution $= 0.193 \times 3.2(313 - 298) = 9.3$ kW

Heat of crystallisation = $(0.063 \times 146.5) = 9.2$ kW; a total of $(9.3 + 9.2) = 18.5$ kW

Assuming countercurrent flow, $\Delta T_1 = (313 - 293) = 20$ deg K

$\Delta T_2 = (298 - 288) = 10$ deg K

and the logarithmic mean, $\Delta T_m = (20 - 10)/\ln(20/10) = 14.4$ deg K

The heat transfer area required, $A' = Q/U\Delta T_m = 18.5/(0.14 \times 14.4) = 9.2\ \text{m}^2$

Assuming that the area available is, typically, 1 $\text{m}^2$/m length, the length of exchanger required = 9.2 m. In practice 3 lengths, each of 3 m length would be specified.

### *Direct-contact cooling*

The occurrence of crystal encrustation in conventional heat exchangers can be avoided by using direct-contact cooling (DCC) in which supersaturation is achieved by allowing the process liquor to come into contact with a cold heat-transfer medium. Other potential advantages of DCC include better heat transfer and lower cooling loads, although disadvantages include product contamination from the coolant and the cost of extra processing required to recover the coolant for further use. Since a solid, liquid, or gaseous coolant can be used with transfer of sensible or latent heat, the coolant may or may not boil during the operation, and it can be either miscible or immiscible with the process liquor, several types of DCC crystallisation are possible:

(a) immiscible, boiling, solid or liquid coolant where heat is removed mainly by transfer of latent heat of sublimation or vaporisation;
(b) immiscible, non-boiling, solid, liquid, or gaseous coolant with mainly sensible heat transfer;
(c) miscible, boiling, liquid coolant with mainly latent heat transfer; and
(d) miscible, non-boiling, liquid coolant with mainly sensible heat transfer.

Crystallisation processes employing DCC have been used successfully in the de-waxing of lubricating oils(63), the desalination of water(64), and the production of inorganic salts from aqueous solution(65).

## 15.3.2. Evaporating crystallisers

If the solubility of a solute in a solvent is not appreciably decreased by lowering the temperature, the appropriate degree of solution supersaturation can be achieved by evaporating some of the solvent and the oldest and simplest technique, the use of solar energy, is still employed commercially throughout the world(66). Common salt is produced widely from brine in steam-heated evaporators, multiple-effect evaporator-crystallisers are used in sugar refining and many types of forced-circulation evaporating crystallisers are in large-scale use(3,40,67). Evaporating crystallisers are usually operated under reduced pressure to aid solvent removal, minimise heat consumption, or decrease the operating temperature of the solution, and these are described as *reduced-pressure evaporating crystallisers*.

### 15.3.3. Vacuum (adiabatic cooling) crystallisers

A vacuum crystalliser operates on a slightly different principle from the reduced-pressure unit since supersaturation is achieved by simultaneous evaporation and adiabatic cooling of the feedstock. A hot, saturated solution is fed into an insulated vessel maintained under reduced pressure. If the feed liquor temperature is higher than the boiling point of the solution under the low pressure existing in the vessel, the liquor cools adiabatically to this temperature and the sensible heat and any heat of crystallisation liberated by the solution evaporate solvent and concentrate the solution.

### 15.3.4. Continuous crystallisers

The majority of continuously operated crystallisers are of three basic types: forced-circulation, fluidised-bed and draft-tube agitated units.

#### *Forced-circulation crystallisers*

A *Swenson forced-circulation crystalliser* operating at reduced pressure is shown in Figure 15.15. A high recirculation rate through the external heat exchanger is used to provide good heat transfer with minimal encrustation. The crystal magma is circulated from the lower conical section of the evaporator body, through the vertical tubular heat exchanger, and reintroduced tangentially into the evaporator below the liquor level to create a swirling action and prevent flashing. Feed-stock enters on the pump inlet side of the circulation system and product crystal magma is removed below the conical section.

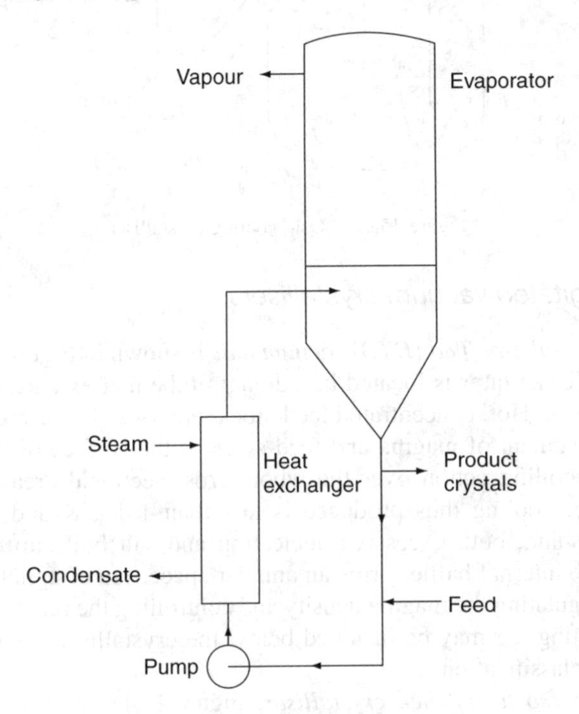


Figure 15.15. Forced-circulation Swenson crystalliser

### Fluidised-bed crystallisers

In an *Oslo fluidised-bed crystalliser*, a bed of crystals is suspended in the vessel by the upward flow of supersaturated liquor in the annular region surrounding a central downcomer, as shown in Figure 15.16. Although originally designed as classifying crystallisers, fluidised-bed Oslo units are frequently operated in a mixed-suspension mode to improve productivity, although this reduces product crystal size[68]. With the classifying mode of operation, hot, concentrated feed solution is fed into the vessel at a point directly above the inlet to the circulation pipe. Saturated solution from the upper regions of the crystalliser, together with the small amount of feedstock, is circulated through the tubes of the heat exchanger and cooled by forced circulation of water or brine. In this way, the solution becomes supersaturated although care must be taken to avoid spontaneous nucleation. Product crystal magma is removed from the lower regions of the vessel.

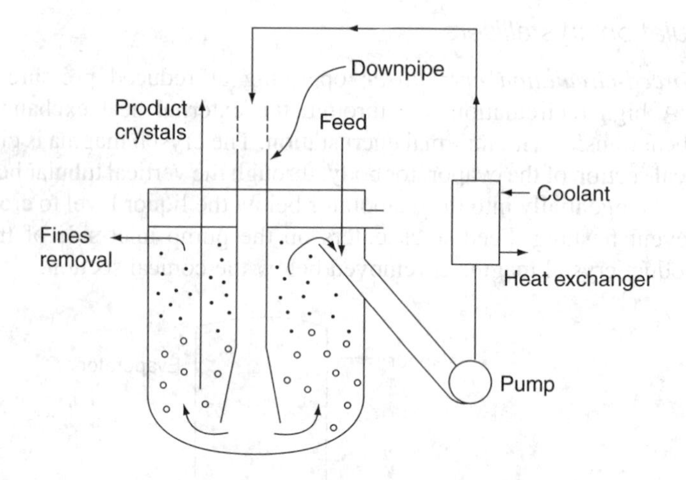


Figure 15.16. Oslo cooling crystalliser

### Draught-tube agitated vacuum crystallisers

A *Swenson draught-tube-baffled (DTB) vacuum unit* is shown in Figure 15.17. A relatively slow-speed propellor agitator is located in a draught tube that extends to a small distance below the liquor level. Hot, concentrated feed-stock, enters at the base of the draught tube, and the steady movement of magma and feed-stock to the surface of the liquor produces a gentle, uniform boiling action over the whole cross-sectional area of the crystalliser. The degree of supercooling thus produced is less than 1 deg K and, in the absence of violent vapour flashing, both excessive nucleation and salt build-up on the inner walls are minimised. The internal baffle forms an annular space free of agitation and provides a settling zone for regulating the magma density and controlling the removal of excess nuclei. An integral elutriating leg may be installed below the crystallisation zone to effect some degree of product classification.

The *Standard-Messo turbulence crystalliser*, Figure 15.18, is a draught-tube vacuum unit in which two liquor flow circuits are created by concentric pipes: an outer ejector

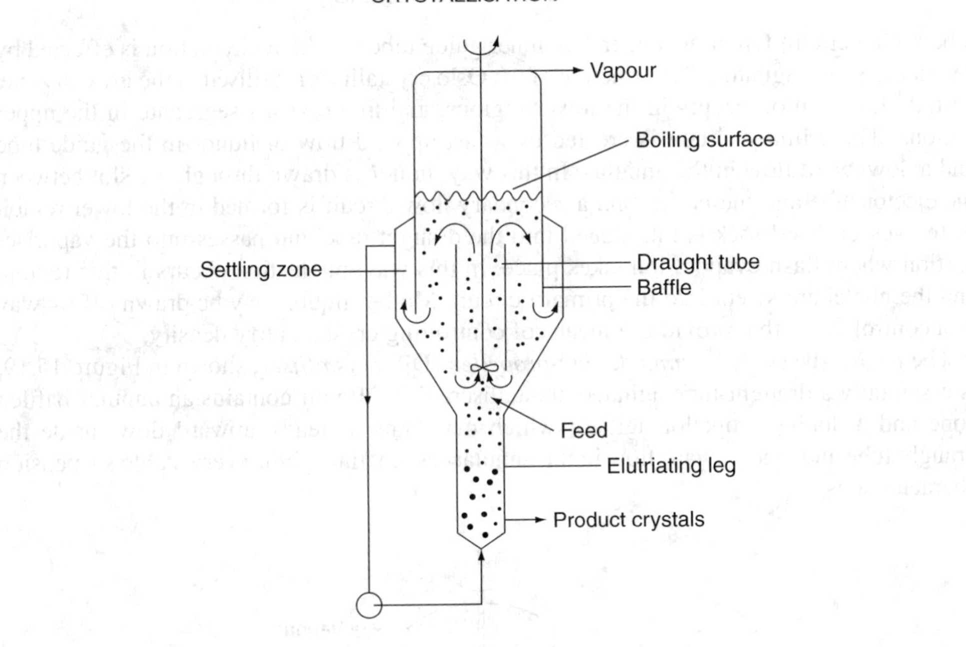


Figure 15.17. Swenson draught-tube-baffled (DTB) crystalliser

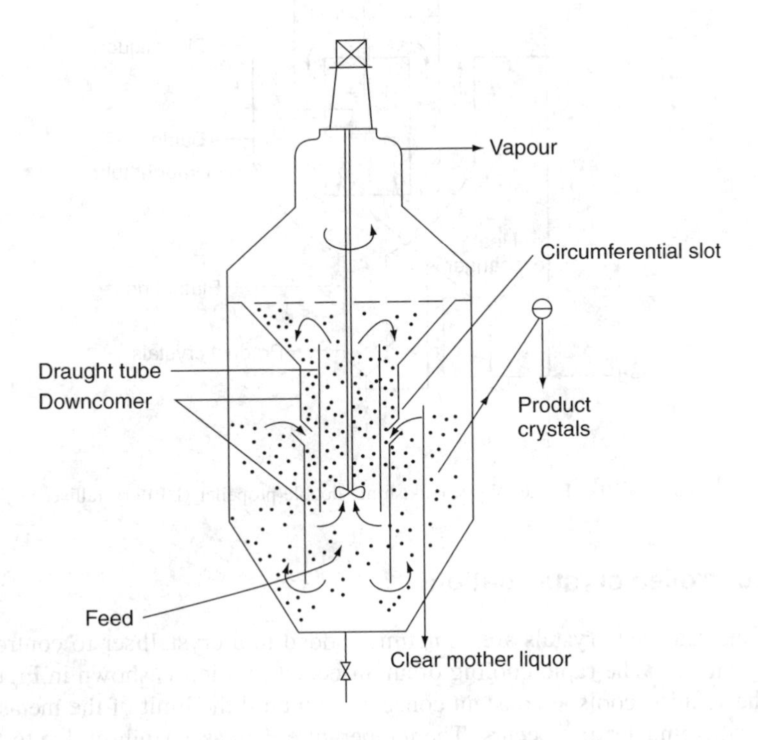


Figure 15.18. Standard-Messo turbulence crystalliser

tube with a circumferential slot, and an inner guide tube in which circulation is effected by a variable-speed agitator. The principle of the Oslo crystalliser is utilised in the growth zone, partial classification occurs in the lower regions, and fine crystals segregate in the upper regions. The primary circuit is created by a fast upward flow of liquor in the guide tube and a downward flow in the annulus. In this way, liquor is drawn through the slot between the ejector tube and the baffle, and a secondary flow circuit is formed in the lower region of the vessel. Feedstock is introduced into the draught tube and passes into the vaporiser section where flash evaporation takes place. In this way, nucleation occurs in this region, and the nuclei are swept into the primary circuit. Mother liquor may be drawn off by way of a control valve that provides a means of controlling crystal slurry density.

The *Escher-Wyss Tsukishima double-propeller* (DP) *crystalliser*, shown in Figure 15.19, is essentially a draught-tube agitated crystalliser. The DP unit contains an annular baffled zone and a double-propellor agitator which maintains a steady upward flow inside the draught tube and a downward flow in the annular region, thus giving very stable suspension characteristics.

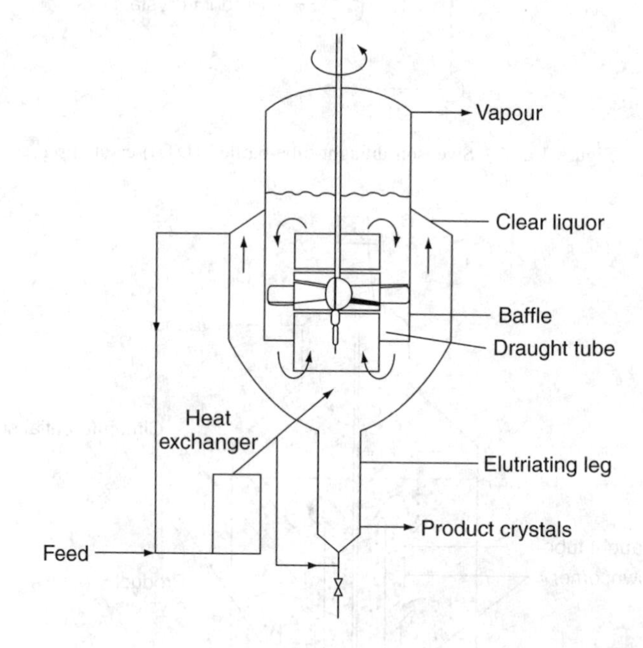


Figure 15.19. Escher-Wyss Tsukishima double-propeller (DP) crystalliser

## 15.3.5. Controlled crystallisation

Carefully selected seed crystals are sometimes added to a crystalliser to control the final product crystal size. The rapid cooling of an unseeded solution is shown in Figure 15.20*a* in which the solution cools at constant concentration until the limit of the metastable zone is reached, where nucleation occurs. The temperature increases slightly due to the release of latent heat of crystallisation, but on cooling more nucleation occurs. The temperature and concentration subsequently fall and, in such a process, nucleation and growth cannot be controlled. The slow cooling of a seeded solution, in which temperature and solution composition are controlled within the metastable zone throughout the cooling cycle, is shown in Figure 15.20*b*. Crystal growth occurs at a controlled rate depositing only on the added seeds and spontaneous nucleation is avoided because the system is never allowed to become labile. Many large-scale crystallisers are operated on this batch operating method that is known as *controlled crystallisation*.

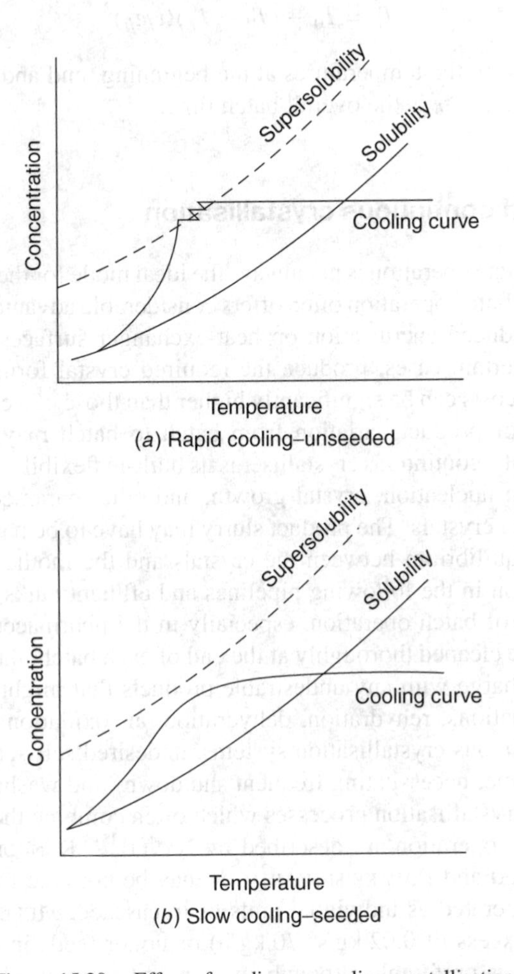


Figure 15.20. Effect of seeding on cooling crystallisation

If crystallisation occurs only on the added seeds, the mass $m_s$ of seeds of size $d_s$ that can be added to a crystalliser depends on the required crystal yield $y$ and the product crystal size $d_p$, as follows:

$$m_s = y\left(\frac{d_s^3}{d_p^3 - d_s^3}\right) \tag{15.30}$$

The product crystal size from a batch crystalliser can also be controlled by adjusting the rates of cooling or evaporation. Natural cooling, for example, produces a supersaturation peak in the early stages of the process when rapid, uncontrolled heavy nucleation inevitably occurs, although nucleation can be controlled within acceptable limits by following a cooling path that maintains a constant low level of supersaturation. As MULLIN and NYVLT[69] has pointed out, the calculation of optimum cooling curves for different operating conditions is complex, although the following simplified relationship is usually adequate for general application:

$$T_t = T_0 - (T_0 - T_f)(t/t_b)^3 \tag{15.31}$$

where $T_0$, $T_f$, and $T_t$ are the temperatures at the beginning, end and any time $t$ during the process, respectively, and $t_b$ is the overall batch time.

### 15.3.6. Batch and continuous crystallisation

Continuous, steady-state operation is not always the ideal mode for the operation of crystallisation processes, and batch operation often offers considerable advantages such as simplicity of equipment and reduced encrustation on heat-exchanger surfaces. Whilst only a batch crystalliser can, in certain cases, produce the required crystal form, size distribution, or purity, the operating costs can be significantly higher than those of a comparable continuous unit, and problems of product variation from batch to batch may be encountered. The particular attraction of a continuous crystalliser is its built-in flexibility for control of temperature, supersaturation nucleation, crystal growth, and other parameters that influence the size distribution of the crystals. The product slurry may have to be passed to a holding tank, however, to allow equilibrium between the crystals and the mother liquor to be reached if unwanted deposition in the following pipelines and effluent tanks is to be avoided. One important advantage of batch operation, especially in the pharmaceutical industry, is that the crystalliser can be cleaned thoroughly at the end of each batch, thus preventing contamination of the next charge with any undesirable products that might have been formed as a result of transformations, rehydration, dehydration, air oxidation and so on during the batch cycle. In continuous crystallisation systems, undesired self-seeding may occur after a certain operating time, necessitating frequent shutdowns and washouts.

Semi-continuous crystallisation processes which often combine the best features of both batch and continuous operation are described by NYVLT[35], RANDOLPH[37], ROBINSON and ROBERTS[70] and ABBEG and BALAKRISHNAM[71]. It may be possible to use a series of tanks which can then be operated as individual units or in cascade. MULLIN[3] suggests that for production rates in excess of 0.02 kg/s (70 kg/h) or liquor feeds in excess of 0.005 m$^3$/s, continuous operation is preferable although sugar may be produced batch-wise at around 0.25 kg/s (900 kg/h) per crystalliser.

### 15.3.7. Crystalliser selection

The temperature–solubility relationship for solute and solvent is of prime importance in the selection of a crystalliser and, for solutions that yield appreciable amounts of crystals on cooling, either a simple cooling or a vacuum cooling unit is appropriate. An evaporating crystalliser would be used for solutions that change little in composition on cooling and salting-out would be used in certain cases. The shape, size and size distribution of the product is also an important factor and for large uniform crystals, a controlled suspension unit fitted with suitable traps for fines, permitting the discharge of a partially classified product, would be suitable. This simplifies washing and drying operations and screening of the final product may not be necessary. Simple cooling-crystallisers are relatively inexpensive, though the initial cost of a mechanical unit is fairly high although no costly vacuum or condensing equipment is required. Heavy crystal slurries can be handled in cooling units without liquor circulation, though cooling surfaces can become coated with crystals thus reducing the heat transfer efficiency. Vacuum crystallisers with no cooling surfaces do not have this disadvantage but they cannot be used when the liquor has a high boiling point elevation. In terms of space, both vacuum and evaporating units usually require a considerable height.

Once a particular class of unit has been decided upon, the choice of a specific unit depends on initial and operating costs, the space available, the type and size of the product, the characteristics of the feed liquor, the need for corrosion resistance and so on. Particular attention must be paid to liquor mixing zones since the circulation loop includes many regions where flow streams of different temperature and composition mix. These are all points at which temporary high supersaturations may occur causing heavy nucleation and hence encrustation, poor performance and operating instabilities. As TOUSSAINT and DONDERS[72] stresses, it is essential that the compositions and enthalpies of mixer streams are always such that, at equilibrium, only one phase exists under the local conditions of temperature and pressure.

### 15.3.8. Crystalliser modelling and design

#### *Population balance*

Growth and nucleation interact in a crystalliser in which both contribute to the final crystal size distribution (CSD) of the product. The importance of the population balance[37] is widely acknowledged. This is most easily appreciated by reference to the simple, idealised case of a mixed-suspension, mixed-product removal (MSMPR) crystalliser operated continuously in the steady state, where no crystals are present in the feed stream, all crystals are of the same shape, no crystals break down by attrition, and crystal growth rate is independent of crystal size. The crystal size distribution for steady state operation in terms of crystal size $d$ and population density $n'$ (number of crystals per unit size per unit volume of the system), derived directly from the population balance over the system[37] is:

$$n' = n^\circ \exp(-d/G_d t_r) \tag{15.32}$$

where $n^\circ$ is the population density of nuclei and $t_r$ is the residence time. Rates of nucleation $B$ and growth $G_d(= \mathrm{d}d/\mathrm{d}t)$ are conventionally written in terms of supersaturation as:

$$B = k_1 \Delta c^b \tag{15.33}$$

and:

$$G_d = k_2 \Delta c^s \tag{15.34}$$

These empirical expressions may be combined to give:

$$B = k_3 G^i \quad (15.35)$$

where: $$i = b/s \text{ and } k_3 = k_1/k_2^i \quad (15.36)$$

where $b$ and $s$ are the kinetic orders of nucleation and growth, respectively, and $i$ is the relative kinetic order. The relationship between nucleation and growth may be expressed as:

$$B = n^\circ G_d \quad (15.37)$$

or: $$n^\circ = k_4 G_d{}^{i-1} \quad (15.38)$$

In this way, experimental measurement of crystal size distribution, recorded on a number basis, in a steady-state MSMPR crystalliser can be used to quantify nucleation and growth rates. A plot of log $n$ against $d$ should give a straight line of slope $-(G_d t_r)^{-1}$ with an intercept at $d = 0$ equal to $n^\circ$ and, if the residence time $t_r$ is known, the crystal growth rate $G_d$ can be calculated. Similarly, a plot of log $n^\circ$ against log $G_d$ should give a straight line of slope $(i - 1)$ and, if the order of the growth process $s$ is known, the order of nucleation $b$ may be calculated. Such plots are shown in Figure 15.21.

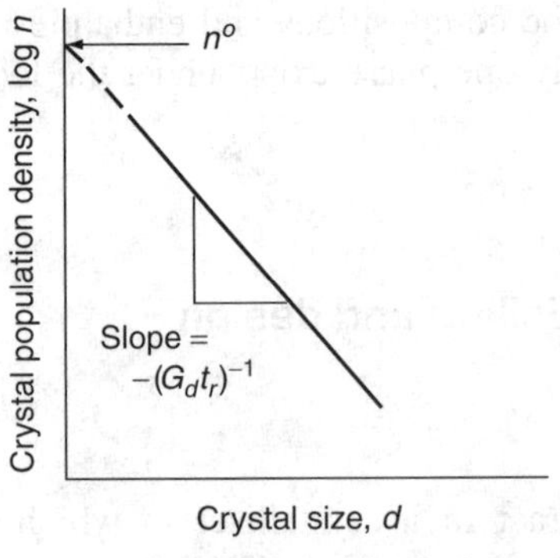


(a) Crystal size distribution

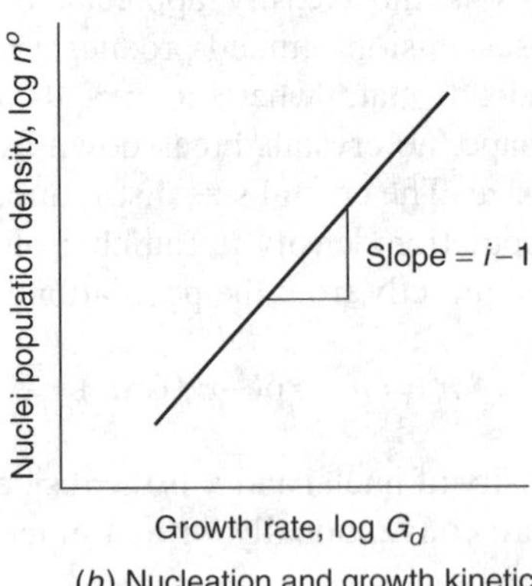


(b) Nucleation and growth kinetics

Figure 15.21. Population plots for a continuous mixed-suspension mixed-product removal (MSMPR) crystalliser

The mass of crystals per unit volume of the system, the so-called magma density, $\rho_m$ is given by:

$$\rho_m = 6\alpha\rho n^\circ(G_d t_r)^4 \tag{15.39}$$

where $\alpha$ is the volume shape factor defined by $\alpha = \text{volume}/d^3$ and $\rho$ is the crystal density.

The peak of the mass distribution, the dominant size $d_D$ of the CSD, is given by MULLIN[1] as:

$$d_D = 3\ G_d t_r \tag{15.40}$$

and this can be related to the crystallisation kinetics by:

$$d_D \propto t_r^{(i-1)/(i+3)} \tag{15.41}$$

This interesting relationship[37] enables the effect of changes in residence time to be evaluated. For example, if $i = 3$, a typical value for many inorganic salt systems, a doubling of the residence time would increase the dominant product crystal size by only 26 per cent. This could be achieved either by doubling the volume of the crystalliser or by halving the volumetric feed rate, and hence the production rate. Thus, residence time adjustment is usually not a very effective means of controlling product crystal size.

CSD modelling based on population balance considerations may be applied to crystalliser configurations other than MSMPR[37] and this has become a distinct, self-contained branch of reaction engineering[56,59,60,73].

## Example 15.6

An MSMPR crystalliser operates with a steady nucleation rate of $n^\circ = 10^{13}/\text{m}^4$, a growth rate, $G_d = 10^{-8}$ m/s and a mixed-product removal rate, based on clear liquor of 0.00017 m$^3$/s. The volume of the vessel, again based on clear liquor, is 4 m$^3$, the crystal density is 2660 kg/m$^3$ and the volumetric shape factor is 0.7. Determine:

(a) the solids content in the crystalliser
(b) the crystal production rate
(c) the percent of nuclei removed in the discharge by the time they have grown to 100 μm.
(d) the liquor flowrate which passes through a trap which removes 90 per cent of the original nuclei by the time they have grown to 100 μm

## Solution

Draw-down time = (4/0.00017) = 23530 s

(a) From a mass balance, the total mass of solids is:

$$c_s = 6\alpha\rho n^\circ(G_d t_r)^4 \qquad \text{(equation 15.39)}$$

$$= (6 \times 0.6 \times 2660 \times 10^{13})(10^{-8} \times 23530)^4$$

$$= \underline{\underline{343 \text{ kg/m}^3}}$$

(b) The production rate = (343 × 0.00017) = $\underline{\underline{0.058 \text{ kg/s (200 kg/h)}}}$

(c) The crystal population decreases exponentially with size or:

$$n/n^\circ = \exp(-L/G_d t_r) \quad \text{(equation 15.32)}$$
$$= \exp[(-100 \times 10^{-6})/(10^{-8} \times 23530)]$$
$$= 0.66 \text{ or } 66 \text{ per cent}$$

Thus: $(100 - 66) =$ 34 per cent have been discharged by the time they reach 100 μm.

(d) If $(100 - 90) = 10$ per cent of the nuclei remain and grow to >100 μm, then in equation 15.32:

$$(1/0.10) = \exp[(-100 \times 10^{-6})/(10^{-8} t_r)]$$

and: $$t_r = 4343 \text{ s}$$

Thus: $$4343 = 4/(0.00017 + Q_F)$$

and: $$Q_F = 0.00075 \text{ m}^3/\text{s} \ (2.68 \text{ m}^3/\text{h})$$

## *Design procedures*

MULLIN[3] has given details of a procedure for the design of classifying crystallisers in which the calculation steps are as follows.

(a) The maximum allowable supersaturation is obtained and hence the working saturation, noting that this is usually about 30 per cent of the maximum.
(b) The solution circulation rate is obtained from a materials balance.
(c) The maximum linear growth-rate is obtained based on the supersaturation in the lowest layer which contains the product crystals and assuming that $(\beta/\alpha) = 6$.
(d) The crystal growth time is calculated from the growth rate for different relative desupersaturations (100 per cent desuperation corresponding to the reduction of the degree of supersaturation to zero).
(e) The mass of crystals in suspension and the suspension volume are calculated assuming a value for the voidage which is often about 0.85.
(f) The solution up-flow velocity is calculated for very small crystals (< 0.1 mm) using Stokes' Law although strictly this procedure should not be used for particles other than spheres or for Re > 0.3. In a real situation, laboratory measurements of the velocity are usually required.
(g) The crystalliser area and diameter are first calculated and then the height which is taken as (volume of suspension/cross-sectional area).
(h) A separation intensity (S.I.), defined by GRIFFITHS[74] as the mass of equivalent 1 mm crystals produced in 1 m$^3$ of crystalliser volume in 1 s, is calculated. Typical values are 0.015 kg/m$^3$ s at 300 K and up to 0.05 at higher temperatures and, for crystals >1 mm, the intensity is given by:

$$\text{S.I.} = d_p P'/V \quad (15.42)$$

where $d_p$ is the product crystal size, $P'$ (kg/s) is the crystal production rate and $V$ (m$^3$) is the suspension volume.

MULLIN[3] has used this procedure for the design of a unit for the crystallisation of potassium sulphate at 293 K. The data are given in Table 15.5 from which it will be noted that the cross-sectional area depends linearly on the relative degree of de-supersaturation and the production rate depends linearly on the area but is independent of the height. If the production rate is fixed, then the crystalliser height may be adjusted by altering the sizes of the seed or product crystals. MULLIN and NYVLT[75] have proposed a similar procedure for mixed particle-size in a crystalliser fitted with a classifier at the product outlet which controls the minimum size of product crystals.

Table 15.5. Design of a continuous classifying crystalliser[3]

Basic Data:

Substance: potassium sulphate at 293 K
Product: 0.278 kg/s of 1 mm crystals
Growth constant: $k_d = 0.75\Delta c^{-2}$ kg/m$^2$s
Nucleation constant: $k_n = 2 \times 10^8 \Delta c^{-7.3}$ kg/s
Crystal size: Smallest in fluidised bed = 0.3 mm, (free settling velocity = 40 mm/s)
Smallest in system = 0.1 mm
Crystal density = 2660 kg/m$^3$, Solution density = 1082 kg/m$^3$
Solution viscosity = 0.0012 Ns/m$^2$, Solubility, $c^* = 0.1117$ kg/kg water

| Desupersaturation | 1.0 | 0.9 | 0.5 | 0.1 |
|---|---|---|---|---|
| Maximum growth rate (μm/s) | 5.6 | 5.6 | 5.6 | 5.6 |
| Up-flow velocity (m/s) | 0.04 | 0.04 | 0.04 | 0.04 |
| Circulation rate (m$^3$/s) | 0.029 | 0.032 | 0.058 | 0.286 |
| Crystal residence time (ks) | 1469 | 907 | 51.8 | 12.6 |
| Mass of crystals (Mg) | 145 | 90 | 5.1 | 1.25 |
| Volume of crystal suspension (m$^3$) | 364 | 225 | 12.8 | 3.15 |
| Cross-sectional area of crystalliser (m$^2$) | 0.72 | 0.80 | 1.45 | 7.2 |
| Crystalliser diameter (m) | 0.96 | 1.01 | 1.36 | 3.02 |
| Crystalliser height (m) | 505 | 281 | 8.8 | 0.44 |
| Height/diameter | 525 | 280 | 6.5 | 0.15 |
| Separation intensity | 3.0 | 4.5 | 78 | 320 |
| Economically possible | no | no | yes | no |

## *Scale-up problems*

Crystalliser design is usually based on data measured on laboratory or pilot-scale units or, in difficult cases, both. One of the main problems in scaling up is characterisation of the particle–fluid hydrodynamics and the assessment of its effects on the kinetics of nucleation and crystal growth. In fluidised-bed crystallisers, for example, the crystal suspension velocity must be evaluated — a parameter which is related to crystal size, size distribution, and shape, as well as bed voidage and other system properties — such as density differences between particles and liquid and viscosity of the solution. Possible ways of estimating suspension velocity are discussed in the literature[3,41,43,76,77], although, as MULLIN[3] points out, determination of suspension velocities on actual crystal samples is often advisable. In agitated vessels, the 'just-suspended' agitator speed $N_{JS}$, that is the minimum rotational speed necessary to keep all crystals in suspension, must be determined since, not only do all the crystals have to be kept in suspension, but the development of 'dead spaces' in the vessel must also be avoided as these are unproductive zones and regions of high supersaturation in which vessel surfaces can become encrusted. Fluid and crystal properties, together with vessel and agitator geometries, are important in establishing $N_{JS}$ values[3,43]. As discussed in Volume 1, Section 7.3, agitated vessel crystallisers are often scaled up successfully on the crude basis of either *constant power input per unit volume* or *constant agitator tip speed*, although BENNETT *et al.*[78] have suggested that, in draught-tube agitated vessels, the quantity (tip speed)$^2$/(vessel volume/volumetric circulation rate) should be kept constant.

## 15.4. CRYSTALLISATION FROM MELTS

### 15.4.1. Basic techniques

A *melt* is a liquid or a liquid mixture at a temperature near its freezing point and *melt crystallisation* is the process of separating the components of a liquid mixture by cooling until crystallised solid is deposited from the liquid phase. Where the crystallisation process is used to separate, or partially separate, the components, the composition of the crystallised solid will differ from that of the liquid mixture from which it is deposited. The ease or difficulty of separating one component from a multi-component mixture by crystallisation may be represented by a phase diagram as shown in Figures 15.4 and 15.5, both of which depict binary systems — the former depicts a eutectic, and the latter a continuous series of solid solutions. These two systems behave quite differently on freezing since a eutectic system can deposit a pure component, whereas a solid solution can only deposit a mixture of components.

Two basic techniques of melt crystallisation are:

(a) gradual deposition of a crystalline layer on a chilled surface in a static or laminar-flowing melt, and
(b) fast generation of discrete crystals in the body of an agitated vessel.

Gradual deposition (a) occurs in the Proabd refiner[79] which essentially utilises a batch cooling process in which a static liquid feedstock is progressively crystallised on to extensive cooling surfaces, such as fin-tube heat-exchangers, supplied with a cold heat-transfer fluid located inside a crystallisation tank. As crystallisation proceeds, the liquid becomes increasingly impure and crystallisation may be continued until virtually the entire charge has solidified. When the crystallised mass is then slowly melted by circulating a hot fluid through the heat exchanger, the impure fraction melts first and drains out of the tank. As melting proceeds, the melt run-off becomes progressively richer in the desired component, and fractions may be taken off during the melting stage. A typical flow diagram, based on a scheme for the purification of naphthalene, is shown in Figure 15.22 where the circulating fluid is usually cold water that is heated during the melting stage by steam injection. Another example of gradual deposition occurs in the rotary drum crystalliser which consists of a horizontal cylinder, partially immersed in the melt, or otherwise supplied with feedstock. The coolant enters and leaves the inside of the hollow drum through trunnions and, as the drum rotates, a crystalline layer forms on the cold surface and is removed with a scraper knife. Two feed and discharge arrangements are shown in Figure 15.23. Rotary drum behaviour and design have been discussed by GELPERIN and

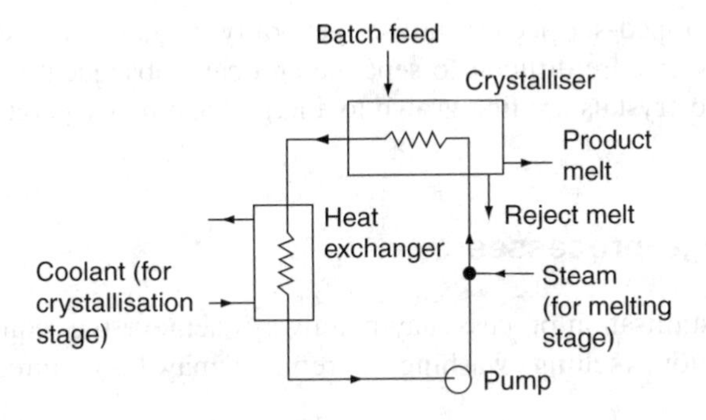


Figure 15.22. Batch cooling crystallisation of melts: flow diagram for the Proabd refiner

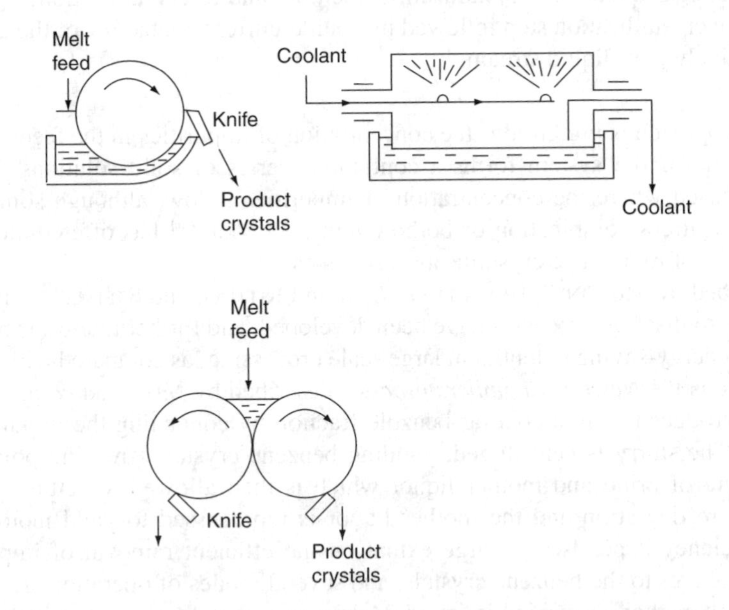


Figure 15.23. Feed and discharge arrangements for drum crystallisers

NOSOV[80] and PONOMARENKO *et al.*[81]. WINTERMANTEL[82] has shown that the structure and impurity levels of growing crystal layers are determined primarily by mass-transfer effects at the layer front.

Fast-melt crystallisation (b) takes place in the scraped-surface heat exchanger, which consists of a cylindrical tube surrounded by a heat-exchange jacket. The tube is surrounded by close-clearance scraper blades and rotates at relatively low speed. Two basic types are available: the large (>200 mm in diameter, >3 m long) slow-speed (< 0.15 Hz) unit, and the small (<150 mm in diameter, <1.5 m long) high-speed (> 8 Hz) machine. Both types can handle viscous magmas, operating at temperatures as low as 190 K, and are widely used in the manufacture of margarine (crystallisation of triglycerides), de-waxing of lubricating oils (crystallisation of higher n-alkanes), and large-scale processing of many organic substances, such as naphthalene, *p*-xylene, chlorobenzenes and so on. The magma emerging from a scraped-surface crystalliser generally contains very small crystals, often less than 10 μm, which are difficult to separate and can subsequently cause reprocessing problems unless the crystals are first grown to a larger size in a separate holdup tank.

### 15.4.2. Multistage-processes

A single-stage crystallisation process may not always achieve the required product purity and further separation, melting, washing, or refining may be required. Two approaches are used:

a) a repeating sequence of crystallisation, melting, and re-crystallisation;
b) a single crystallisation step followed by countercurrent contacting of the crystals with a relatively pure liquid stream.

The first approach is preferred if the concentration of impurities in the feedstock is high, and is essential if the system forms a continuous series of solid solutions. The second approach is used where the concentration of impurities is low, although some industrial operations require a combination of both systems. ATWOOD[83] has offered an analysis of different types of multistage crystallisation schemes.

As described by MULLIN[3], BENNETT *et al.*[78] and RITTNER and STEINER[84], many industrial melt-crystallisation processes have been developed, and further interest is being stimulated by the energy-saving potential in large scale processing, as compared with distillation. One example is the *Newton-Chambers process*, described by MOLINARI *et al.*[79], in which benzene is produced from a coal-tar benzole fraction by contacting the impure feedstock with brine. The slurry is centrifuged, yielding benzene crystals (freezing point 278.6 K) and a mixture of brine and mother liquor which is then allowed to settle. The brine is returned for refrigeration and the mother liquor is reprocessed to yield motor fuel. The process efficiency depends to a large extent on the efficient removal of impure mother liquor that adheres to the benzene crystals, and several modes of operation are possible. In the thaw–melt method shown in Figure 15.24, benzene crystals are washed in the centrifuge with brine at a temperature above 279 K. Some of the benzene crystals partially melt, which helps to wash away the adhering mother liquor. The thawed liquor may then be recycled. Multistage operation can be employed, in which the first crop of crystals is removed as product and the second, from the liquor, is melted for recycle. The *Sulzer MWB process*, described by FISCHER *et al.*[85], involves crystallisation on a cold surface and, since it may be operated effectively as a multistage separation device, it can be used to purify solid solutions. In an effective multistage countercurrent scheme, only one crystalliser, a vertical multi-tube heat exchanger, is required and the crystals do not have to be transported since they remain deposited on the internal heat-exchange surfaces in the vessel until they are melted for further processing. The intermediate storage tanks and crystalliser are linked by a control system consisting of a programme timer, actuating valves, pumps, and cooling loop, as shown in Figure 15.25. This process has been used on a large scale for the purification of organics, such as chloronitrobenzenes, nitrotoluenes, cresols, and xylenols, and in the separation of fatty acids.

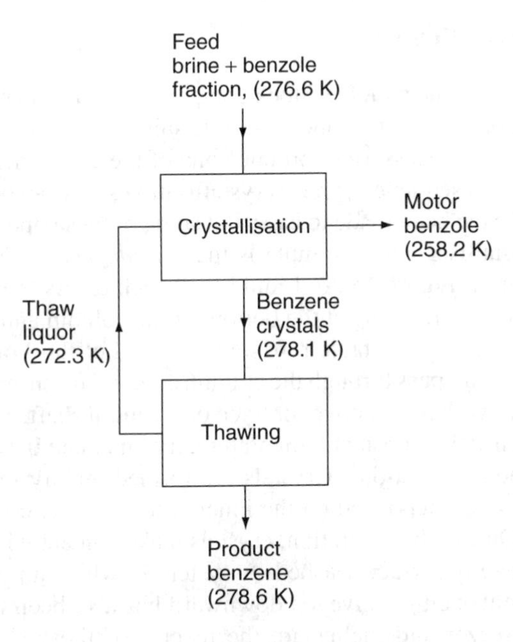


Figure 15.24. Newton — Chambers process for the purification of benzene

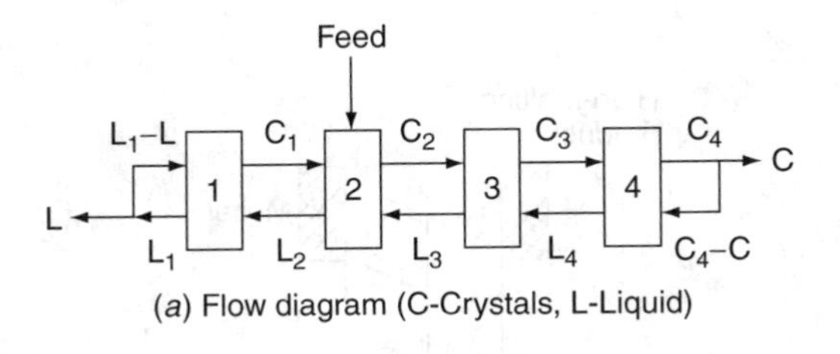


(*a*) Flow diagram (C-Crystals, L-Liquid)

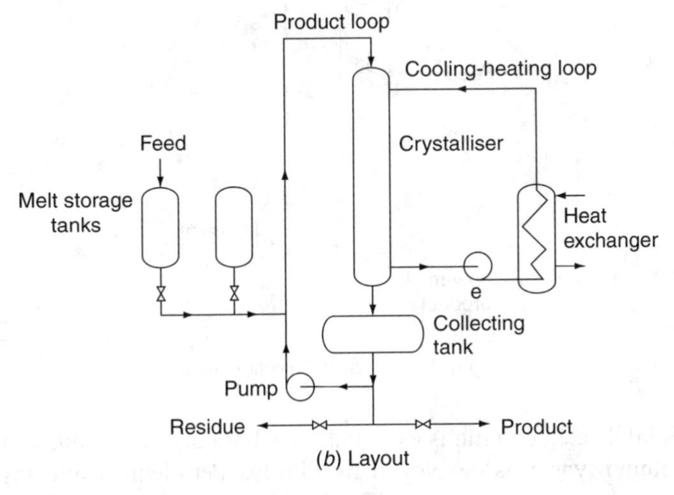


(*b*) Layout

Figure 15.25. Sulzer MWB process

### 15.4.3. Column crystallisers

Because the components of the melt feedstock components can form both eutectic and solid-solution systems with one another, sequences of washing, partial or complete melting, and re-crystallisation are often necessary to produce one of the components in near-pure form. Because the operation of a sequence of melt crystallisation steps can be time-consuming and costly, many attempts have been made to carry out some of these operations in a single unit, such as a column crystalliser. One example is the *Schildknecht column* developed in the 1950's which is shown in Figure 15.26. Liquid feedstock enters the column continuously at an intermediate point and freezing at the bottom of the column and melting at the top are achieved using, respectively, a suitable refrigerant and a hot fluid or an electrical heating element. Crystals and liquid pass through the column countercurrently, and the solid phase is transported downward by a helical conveyor fixed on a central shaft. The purification zone is usually operated at a virtually constant temperature, intermediate between the temperatures of the freezing and melting sections. Crystals are formed mainly in the freezing section, although some may also be deposited on the inner surface of the column and removed by the helical conveyor. During this operation, crystals make contact with the counter-flowing liquid melt and are thereby surface-washed. A system in which an upward flowing liquid is in contact with crystals being conveyed downward has also been used and, in this case, the locations of the freezer and melter are the reverse of those shown in Figure 15.26. GATES and POWERS[86] and HENRY and POWERS[87] have discussed the modelling of column crystallisers.

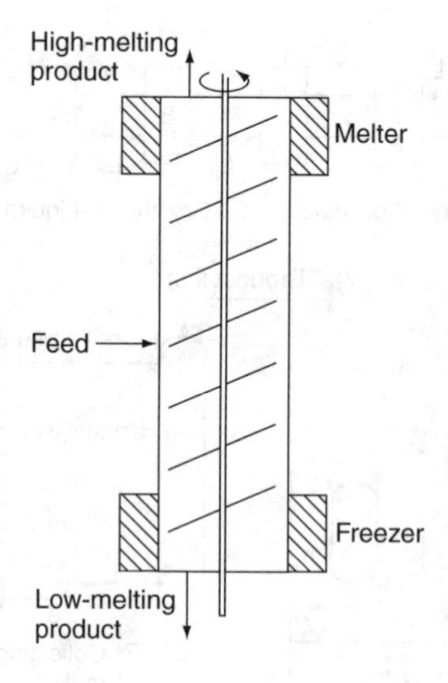


Figure 15.26. Schildknecht column

Whilst the Schildknecht column is essentially a laboratory-scale unit, a melt-crystalliser of the wash-column type was developed by Phillips Petroleum Company in the 1960s for large-scale production of *p*-xylene. The key features of this *Phillips pulsed-column crystalliser*, as described by MCKAY *et al.*[88], are shown in Figure 15.27. A cold slurry

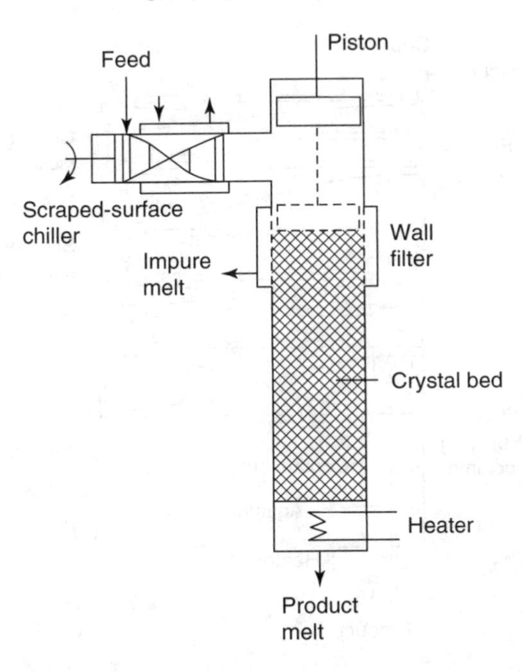


Figure 15.27. Phillips pulsed-column crystalliser[88]

feed, produced in a scraped-surface chiller, enters at the top of the column and crystals are pulsed downward in the vertical bed by a piston, whilst impure mother liquor leaves through a filter. The upward-flowing wash liquor is generated at the bottom by a heater that melts pure crystals before they are removed from the column.

Whilst the *Brodie purifier*[89] developed in the late 1960s incorporates several features of the column crystallisers described previously, it also has the potential to deal effectively with solid-solution systems. As shown in Figure 15.28, it is essentially a centre-fed column that can convey crystals from one end to the other. As the crystals move through the unit, their temperature is gradually increased along the flow path and thus they are subjected to partial melting which encourages the release of low-melting impurities. The interconnected scraped-surface heat exchangers are of progressively smaller diameter so as to maintain reasonably constant axial flow velocities and to prevent back-mixing. The vertical purifying column acts as a countercurrent washer in which descending, nearly pure, crystals meet an upward stream of pure melt. The Brodie purifier has been used in the large-scale production of high-purity 1,4-dichlorobenzene and naphthalene.

The *Tsukishima Kikai (TSK) countercurrent cooling crystallisation process* described by Takegami[90] is, in effect, a development of Brodie technology. A typical system, consisting of three conventional cooling crystallisers connected in series is shown in Figure 15.29. Feed enters the first-stage vessel and partially crystallises, and the slurry is then concentrated in a hydrocyclone before passing into a Brodie purifying column. After passage through a settling zone in the crystalliser, clear liquid overflows to the next stage. Slurry pumping and overflow of clear liquid in each stage result in countercurrent flow of liquid and solid. This process has been applied in the large-scale production of $p$-xylene.

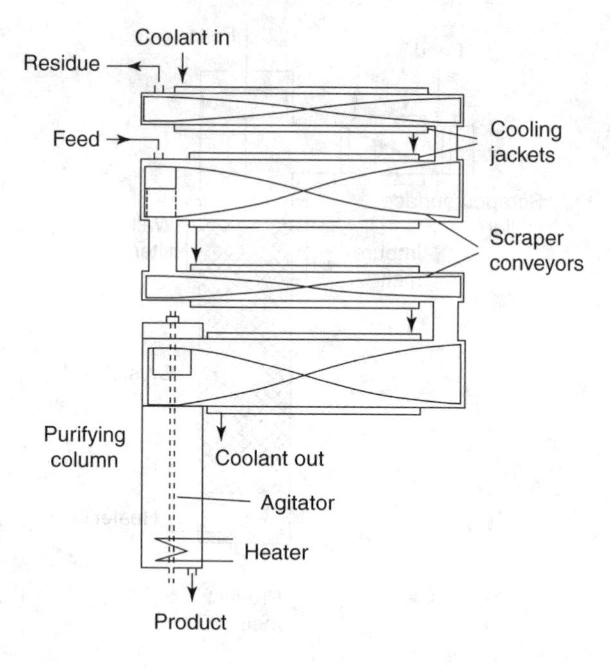


Figure 15.28. Brodie purifier[89]

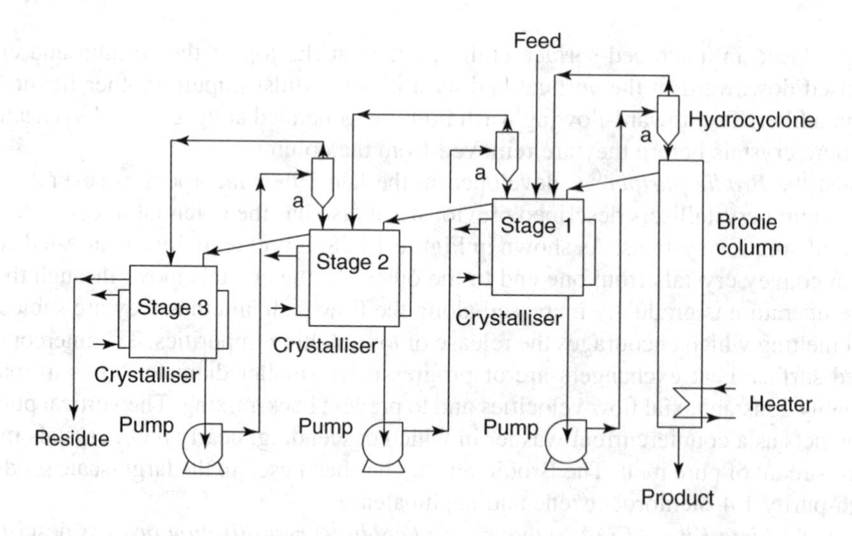


Figure 15.29. Tsukishima Kikai (TSK) countercurrent cooling crystallisation process[90]

Units have been developed by Sulzer Chemtech which consist of vertical tubes where the product flows as a film down the inside surface of the tubes, and the liquid used for cooling and heating is distributed so as to wet the external surface of the tubes. During the initial freezing stage, the heat transfer medium chills the tubes, partial melting is then induced by raising the temperature of the heat transfer medium and higher temperatures are then applied for the final melting stage. A distribution system equalises the flow through the tubes and optimum performance is achieved by accurate control of the heating and cooling profiles. Sulzer also produce a unit for static melt crystallisation which employs cooled plates immersed in a stagnant melt. After a crystal layer has formed, sweating is induced, as with the falling-film device, and the sweated fraction is removed. The remaining crystal layer is then melted and passed to storage. A higher degree of purity may be obtained by using the intermediate product as feedstock and repeating the procedure and, in a similar way, the residue drained from the first phase of operation may be further depleted by additional melt-freezing processes to give an enhanced yield. A relatively new development is the use of a heat pump in which one crystalliser operates in the crystallisation mode as an evaporator, and a further identical unit operates in the sweating or melting mode as a condenser. In this way, energy costs are reduced due to the use of the enthalpy of condensation for crystal melting. Auxiliary exchangers are required only for the elimination of excess energy and for the start-up operation.

A further development, discussed by MORITOKI[(91)], is *high-pressure crystallisation*, which is considered in Section 15.9.

### 15.4.4. Prilling and granulation

Prilling, a melt-spray crystallisation process in which solid spherical granules are formed, is used particularly in the manufacture of fertilisers such as ammonium nitrate and urea. SHEARON and DUNWOODY[(92)] describe the prilling of ammonium nitrate, in which a very concentrated solution containing about 5 per cent of water is sprayed at 415 K into the top of a 30 m high, 6 m diameter tower, and the droplets fall countercurrently through an upwardly flowing air stream that enters the base of the tower at 293 K. The solidified droplets, which leave the tower at 353 K and contain about 4 per cent water, must be dried to an acceptable moisture content at a temperature below 353 K in order to prevent any polymorphic transitions. NUNNELLY and CARTNEY[(93)] describe a melt granulation technique for urea in which molten urea is sprayed at 420 K on to cascading granules in a rotary drum. Seed granules of less than 0.5 mm diameter can be built up to the product size of 2–3 mm. Heat released by the solidifying melt is removed by the evaporation of a fine mist of water sprayed into the air as it passes through the granulation drum.

An important application of granulation is in improving the 'flowability' of very fine (submicron) particles which stick together because of the large surface forces acting in materials with very high surface/volume ratios.

## 15.5. CRYSTALLISATION FROM VAPOURS

### 15.5.1. Introduction

The term *sublimation* strictly refers to the phase change: solid $\rightarrow$ vapour, with no intervention of a liquid phase. In industrial applications, however, the term usually includes the reverse process of condensation or *desublimation*: solid $\rightarrow$ vapour $\rightarrow$ solid. In practice, it is sometimes desirable to vaporise a substance from the liquid state and hence the complete series of phase changes is then: solid → liquid → vapour → solid, and, on the condensation side of the process, with the supersaturated vapour condensing directly to the crystalline solid state without the creation of a liquid phase.

Common organic compounds that can be purified led by sublimation include[94]:

> 2-aminophenol, anthracene, anthranilic acid, anthraquinone, benzanthrone, benzoic acid, 1,4-benzoquinone, camphor, cyanuric chloride, *iso*-phthalic acid, naphthalene, 2-napththol, phthalic anhydride, phthalimide, pyrogallol, salicylic acid, terephthalic acid and thymol

and the following elemental and inorganic substances for which the process is suitable include:

> aluminum chloride, arsenic, arsenic(III) oxide, calcium, chromium(III) chloride, hafnium tetrachloride, iodine, iron(III) chloride, magnesium, molybdenum trioxide, sulphur, titanium tetrachloride, uranium hexafluoride and zirconium tetrachloride.

In addition, the sublimation of ice in freeze-drying, discussed in Chapter 16, has become an important operation particularly in the biological and food industries. The various industrial applications of sublimation techniques are discussed by several authors[3,40,95,96,97], and the principles underlying vaporisation and condensation[98] and the techniques for growing crystals from the vapour phase[99] are also presented in the literature.

### 15.5.2. Fundamentals

*Phase equilibria*

A sublimation process is controlled primarily by the conditions under which phase equilibria occur in a single-component system, and the phase diagram of a simple one-component system is shown in Figure 15.30 where the *sublimation curve* is dependent on the vapour pressure of the solid, the *vaporisation curve* on the vapour pressure of the liquid, and the *fusion curve* on the effect of pressure on the melting point. The slopes of these three curves can be expressed quantitatively by the Clapeyron equation:

$$(\mathrm{d}P/\mathrm{d}T)_{\text{sub}} = \lambda_s/T(v_g - v_s) \tag{15.43}$$

$$(\mathrm{d}P/\mathrm{d}T)_{\text{vap}} = \lambda_v/T(v_g - v_l) \tag{15.44}$$

$$(\mathrm{d}P/\mathrm{d}T)_{\text{fus}} = \lambda_f/T(v_l - v_s) \tag{15.45}$$

where $P$ is the vapour pressure, and $v_s$, $v_l$ and $v_g$ are the molar volumes of the solid, liquid, and gas phases, respectively. The molar latent heats (enthalpies) of sublimation, vaporisation, and fusion ($\lambda_s$, $\lambda_v$ and $\lambda_f$ respectively) are related at a given temperature by:

$$\lambda_s = \lambda_v + \lambda_f \tag{15.46}$$

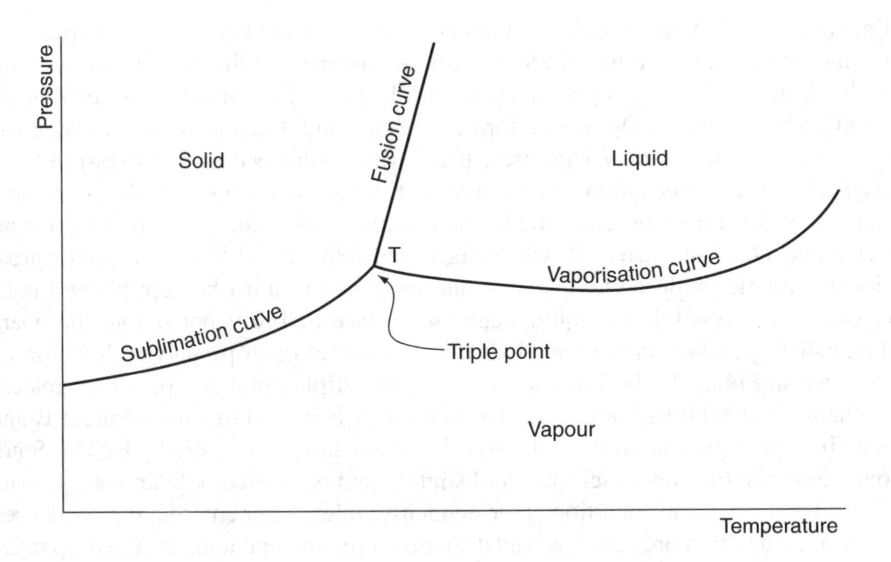


Figure 15.30. Phase diagram for a single-component system

Although there are few data available on sublimation-desublimation, a considerable amount of information can be calculated using the Clausius-Clapeyron equation provided that information on vapour pressure is available at two or more temperatures. In this way:

$$\ln \frac{P_1}{P_2} = \frac{\lambda'_v}{\mathbf{R}} \left( \frac{1}{T_2} - \frac{1}{T_1} \right) \tag{15.47}$$

where $\lambda'_v$ is latent heat of vaporisation per mole.

## Example 15.7

The vapour pressures of naphthalene at 463 and 433 K are 0.780 and 0.220 kN/m$^2$ respectively. If $\lambda'_v$ does not vary greatly over the temperature range considered, what is the vapour pressure at 393 K?

## Solution

In equation 15.47: $\ln(780/220) = [\lambda'_v(463 - 433)]/[8314 \times 463 \times 433]$

and: $\lambda'_v = 70340$ kJ/kmol

Thus: $\ln(220/P) = [70340(433 - 393)]/[8.314 \times 433 \times 393]$

and: $P = \underline{\underline{30 \text{ kN/m}^2}}$

The position of the *triple point* T, which represents the temperature and pressure at which the solid, liquid, and gas phases co-exist in equilibrium, is of the utmost importance in sublimation-desublimation processes. If it occurs at a pressure above atmospheric, the solid cannot melt under normal atmospheric conditions, and true sublimation (solid → vapour) is easily achieved. For example, since the triple point for carbon dioxide is at 216 K and 500 kN/m$^2$, liquid $CO_2$ is not formed when solid $CO_2$ is heated at atmospheric pressure and the solid simply vaporises. If the triple point occurs at a pressure less than atmospheric, certain precautions are necessary if the phase changes solid → vapour and vapour → solid are to be controlled. For example, since the triple point for water is 273.21 K and 0.6 kN/m$^2$, ice melts when it is heated above 273.2 K at atmospheric pressure. For ice to sublime, both the temperature and the pressure must be kept below the triple-point values. If the solid → liquid stage takes place before vaporisation, the operation is often called *pseudo-sublimation*. Both true sublimation and pseudo-sublimation cycles are depicted in Figure 15.31. For a substance with a triple point at a pressure greater than atmospheric, true sublimation occurs. The solid at A is heated to a temperature B and the increase in vapour pressure is given by AB. The condensation is given by BCDE. Since the vapour passing to the condenser may cool slightly and be diluted with an inert gas such as air, C can be taken as the condition at the condenser inlet. After entering the condenser, the vapour is mixed with more inert gas, and the partial pressure and temperature drop to D. The vapour then cools at essentially constant pressure to E which is the condenser temperature. When the triple point of the substance occurs at a pressure lower than atmospheric, heating may result in the temperature and vapour pressure of the solid exceeding the values at the triple-point, and the solid will then melt in the vaporiser along AB′. In the condensation stage, the partial pressure in the vapour stream entering the condenser must be reduced below the pressure at the triple-point to prevent initial condensation to a liquid by diluting the vapour with an inert gas, although the frictional pressure drop in the vapour line is often sufficient to effect the required drop in partial pressure. C′ then represents the conditions at the entry into the condenser and the condensation path is C′DE.

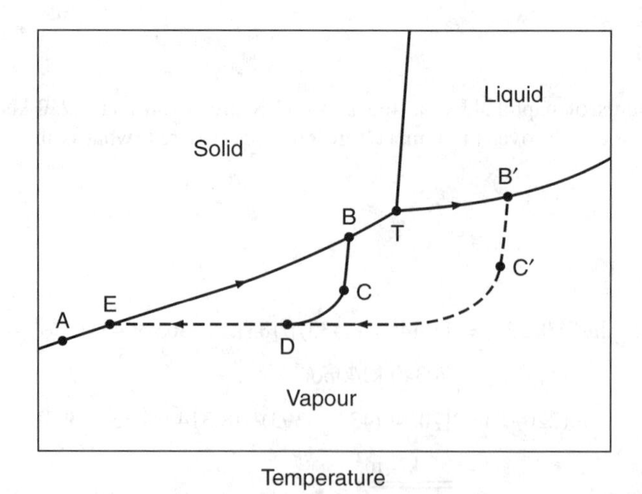


Figure 15.31. Phase diagram showing true sublimation (ABCDE) and pseudosublimation (AB′C′DE) cycles.[94]

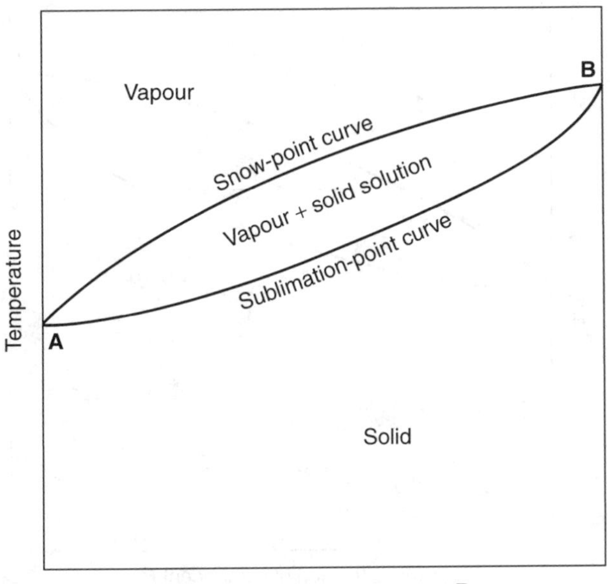


Figure 15.32. Phase diagram for a two-component solid-solution system at a pressure below the triple points of the two components **A** and **B**

*Fractional sublimation*. If two or more sublimable substances form true solid solutions, their separation by fractional sublimation is theoretically possible. The phase diagram for a binary solid-solution system at a pressure below the triple-point pressures of the two components is shown in Figure 15.32, where points A and B represent the equilibrium sublimation temperatures of pure components **A** and **B**, respectively, at a given pressure. The lower curve represents the sublimation temperatures of mixtures of **A** and **B**, while the upper curve represents the solid-phase condensation temperatures, generally called *snow points*. Figure 15.33 shows that if a solid solution at S is heated to some temperature X, the resulting vapour phase at Y and residual solid solution at Z contain different proportions of the original components, quantified by the *lever arm rule*. The sublimate and the residual solid may then each be subjected again to this procedure and, therefore, the possibility of fractionation exists, although the practical difficulties may be considerable. GILLOT and GOLDBERGER[100], VITOVEC *et al.*[101] and EGGERS *et al.*[102] have described experimental studies of fractional sublimation, and nucleation and growth rates, of organic condensates. MATSOUKA *et al.*[103] has also applied the procedure to the fractionation of mixed vapours.

## *Vaporisation and condensation*

*Vaporisation*. The maximum theoretical vaporisation rate $v'$ (kg/m$^2$ s) from the surface of a pure liquid or solid is limited by its vapour pressure and is given by the Hertz-Knudsen equation[104], which can be derived from the kinetic theory of gases:

$$v' = P_s(M'/2\pi \mathbf{R} T_s)^{0.5} \tag{15.48}$$

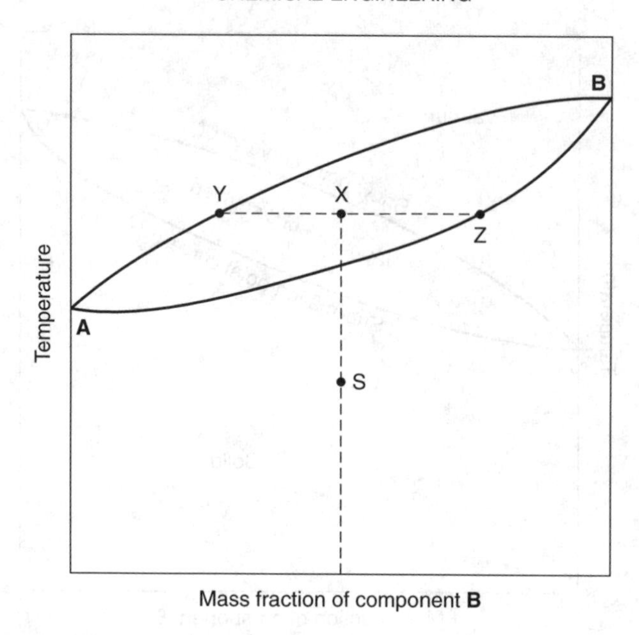


Figure 15.33. Fractional sublimation of two components **A** and **B** demonstrated on a solid-solution phase diagram

where $P_s$ is the vapour pressure at the surface temperature $T_s$, $M'$ is the molecular weight and $\mathbf{R}$ is the gas constant. In practice, the actual vaporisation rate may be lower than predicted by equation 15.48, and a correction factor $e$, generally referred to as an evaporation coefficient, is included to give:

$$v' = eP_s(M'/2\pi\mathbf{R}T_s)^{0.5} \tag{15.49}$$

A laboratory technique used to measure values of $e$ for sublimable solid materials is described by PLEWES and KLASSEN[105].

Sublimation rates of pure solids into turbulent air streams have been successfully correlated by the Gilliland–Sherwood equation[102]:

$$d'/x' = 0.023Re^{0.38}Sc^{0.44} \tag{15.50}$$

where $d'$ *is* a characteristic dimension of the vaporisation chamber, $x'$ is the effective film thickness for mass transfer at the vapour–solid interface, and $Re$ and $Sc$ are the dimensionless Reynolds and Schmidt numbers, respectively.

*Condensation* is generally a transient operation in which, as discussed by UEDA and TAKASHIMA[106], simultaneous heat and mass transfer are further complicated by the effects of spontaneous condensation in the bulk gaseous phase. After the creation of supersaturation in the vapour phase, nucleation normally occurs which may be homogeneous in special circumstances, but more usually heterogeneous. This process is followed by both crystal growth and agglomeration which lead to the formation of the final crystal product. As a rate process, the condensation of solids from vapours is less well understood than vaporisation[98]. STRICKLAND-CONSTABLE[107] has described a simple laboratory technique for measuring kinetics in subliming systems which has been used to compare the rates of solid evaporation and crystal growth of benzophenone under comparable conditions. The two most common ways of creating the supersaturation necessary for crystal nucleation and subsequent growth are: (a) cooling by a metal surface to give either a glassy or multi-crystalline deposit that requires mechanical removal and (b) dilution with an inert gas to produce a loose crystalline mass that is easy to handle. BILIK and KRUPICZKA[108] have described the measurement and correlation of heat transfer rates during condensation of phthalic anhydride in a pilot plant connected to an industrial desublimation unit, CIBOROWSKI and WRENSKI[109] have reported on the condensation of several sublimable materials in a fluidised bed, and KNUTH and WEINSPACH[110] have summarised an extensive study on heat- and mass-transfer processes in a fluidised-bed desublimation unit.

### 15.5.3. Sublimation processes

#### *Simple and vacuum sublimation*

*Simple sublimation* is a batch-wise process in which the solid material is vaporised and then diffuses towards a condenser under the action of a driving force attributable to difference in partial pressures at the vaporising and condensing surfaces. The vapour path between the vaporiser and the condenser should be as short as possible in order to reduce mass-transfer resistance. Simple sublimation has been used for centuries, often in very crude equipment, for the commercial production of ammonium chloride, iodine, and flowers of sulphur.

*Vacuum sublimation* is a development of simple sublimation, which is particularly useful if the pressure at the triple-point is lower than atmospheric, where the transfer of vapour from the vaporiser to the condenser is enhanced by the increased driving force attributable to the lower pressure in the condenser. Iodine, pyrogallol, and many metals have been purified by vacuum sublimation processes in which the exit gases from the condenser are usually passed through a cyclone or scrubber to protect the vacuum equipment and to minimise product loss.

#### *Entrainer sublimation*

In entrainer sublimation, an entrainer gas is blown into the vaporisation chamber of a sublimer in order to increase the vapour flowrate to the condensing equipment, thereby increasing the yield. Air is the most commonly used entrainer, though superheated steam can be employed for substances such as anthracene that are relatively insoluble in water. If steam is used, the vapour may be cooled and condensed by direct contact with a spray of cold water. Although the recovery of the sublimate is efficient, the product is wet. The use of an entrainer gas in a sublimation process also provides the heat needed for sublimation and an efficient means of temperature control. If necessary, it may also provide dilution for the fractional condensation at the desublimation stage. Entrainer sublimation, whether by gas flow over a static bed of solid particles or through a fluidised bed, is ideally suited to continuous operation.

A general-purpose, continuous entrainer–sublimation plant is shown in Figure 15.34. The impure feedstock is pulverised in a mill and, if necessary, a suitable entrainer gas,

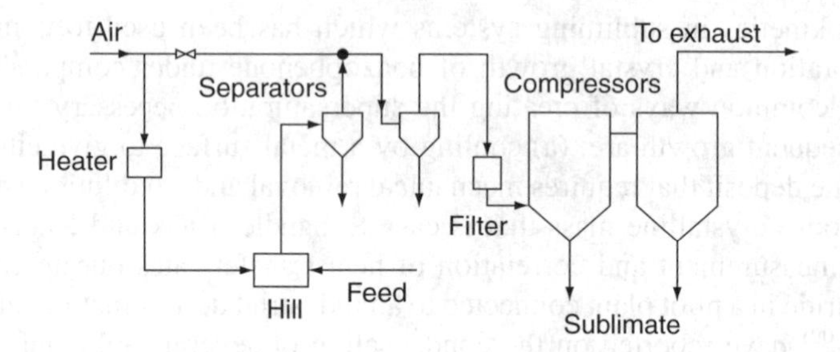


Figure 15.34. General-purpose continuous sublimation unit[3]

such as hot air, is used to blow the fine particles, which volatilise readily, into a series of separators, such as cyclones, where nonvolatile solid impurities are removed. A filter may also be located in the vapour line to remove final traces of inert, solid impurities. The vapour then passes to a series of condensers from which the sublimate is subsequently discharged. The exhaust gases may be recycled, or passed to the atmosphere through a cyclone or wet scrubber to recover any entrained product.

Although the application of fluidisation techniques to sublimation–desublimation processes was first proposed by MATZ[111], the technique has not yet been widely adopted for large-scale commercial use, despite its obvious advantage of improving both heat and mass transfer rates. CEDRO[112] has, however, reported on a fluidised-bed de-sublimation unit operating in the United States for the production of aluminum chloride at the rate of 3 kg/s (11 tonne/h).

The *product yield* from an entrainer–sublimation process may be estimated as follows. The mass flowrate $G'$ of the inert gas and the mass sublimation rate $S'$ are related by:

$$\frac{G'}{S'} = \frac{\rho_g P_g}{\rho_s P_s'} \tag{15.51}$$

where $P_g$ and $P_s'$ are the partial pressures of the inert gas and vaporised material, respectively, in the vapour stream, and $\rho_g$ and $\rho_s$ are their respective vapour densities. The total pressure $P_t$ of the system is the sum of the partial pressures of the components or:

$$P_t = P_g + P_s' \tag{15.52}$$

From equation 15.51:

$$S' = G'\left(\frac{\rho_s}{\rho_g}\right)\left(\frac{P_s'}{P_t - P_s'}\right) \tag{15.53}$$

or, in terms of the molecular weights of the inert gas, $M_g$ and of the material being sublimed, $M_s$:

$$S' = G'\left(\frac{M_s}{M_g}\right)\left(\frac{P_s'}{P_t - P_s'}\right) \tag{15.54}$$

The theoretical maximum yield from an entrainer sublimation process is the difference between the calculated sublimation rates corresponding to the conditions in the vaporisation and condensation stages.

## Example 15.8

Salicylic acid ($M_s = 138$ kg/kmol) is to be purified by entrainer sublimation with air ($M_g = 29$ kg/kmol) at 423 K. The vapour is fed at 101.5 kN/m$^2$ to a series of condensers, the internal temperature and pressure of the last condenser being 313 K and 100 kN/m$^2$ respectively. The air flowrate is 0.56 kg/s and the pressure drop between the vaporiser and the last condenser is 1.5 kN/m$^2$. The vapour pressures of salicylic acid at 423 and 313 K are 1.44 and 0.0023 kN/m$^2$ respectively. What are the mass sublimation rates in the vaporiser and condenser?

## Solution

Under saturated conditions:

*Vaporisation stage*: $P_t = 101.5$ kN/m$^2$, $P_s' = 1.44$ kN/m$^2$

Thus, in equation 15.54:

$$S_v' = 0.56(138/29)(1.44/(101.5 - 1.44)) = 0.038 \text{ kg/s (38 g/s)}$$

*Condensation stage*: $P_t = 100$ kN/m$^2$, $P_s' = 0.0023$ kN/m$^2$

Thus, in equation 15.54:

$$S' = 0.56(138/29)(0.0023/(100 - 0.0023)) = \underline{\underline{0.000061 \text{ kg/s (0.061 g/s)}}}$$

In this example, the loss from the condenser exit gases is only 0.061 g/s whilst the theoretical maximum yield is 38 g/s. This maximum yield is obtained, however, only if the air is saturated with salicylic acid vapour at 423 K, and saturation is approached only if the air and salicylic acid are in contact for a sufficiently long period of time at the required temperature. A fluidised-bed vaporiser may achieve these optimum conditions though, if air is simply blown over bins or trays containing the solid, saturation will not be achieved and the actual rate of sublimation will be lower than that calculated. In some cases, the degree of saturation achieved may be as low as 10 per cent of the calculated value. The actual loss of product in the exit gases from the condenser is then greater than the calculated value. There are other losses which can be minimised by using an efficient scrubber.

### *Comparison of entrainer-sublimation and crystallisation*

An analysis by Kudela and Sampson[97] of the processes for commercial purification of naphthalene suggests that sublimation is potentially more economical than conventional melt crystallisation. In the sublimation method, the feedstock is completely vaporised in a nitrogen stream and then partially condensed. Heat is removed by vaporising water at the top of the condenser and, in order to prevent deposition of sublimate on the vessel walls, the inner wall of the condenser is sufficiently permeable to allow it to pass some of the entrainer gas. Impurities remain in the vapour stream and are subsequently condensed in a cooler located after the compressor used to circulate the entrainer. The stream carrying impurities and wastewater from the separator is washed with benzene in an extractor. Sublimation gives a higher yield of naphthalene at a lower cost, and with a smaller space requirement, than crystallisation[94]. Although steam and electricity consumption is higher for sublimation, this is offset by a much lower cooling-water requirement.

### *Fractional sublimation*

As discussed in Section 15.5.2, the separation of two or more sublimable substances by fractional sublimation is theoretically possible if the substances form true solid solutions. GILLOT and GOLDBERGER[100] have reported the development of a laboratory-scale process known as thin-film fractional sublimation which has been applied successfully to the separation of volatile solid mixtures such as hafnium and zirconium tetrachlorides, 1,4-dibromobenzene and 1-bromo-4-chlorobenzene, and anthracene and carbazole. A stream of inert, non-volatile solids fed to the top of a vertical fractionation column falls countercurrently to the rising supersaturated vapour which is mixed with an entrainer gas. The temperature of the incoming solids is maintained well below the snow-point temperature of the vapour, and thus the solids become coated with a thin film (10 μm) of sublimate which acts as a reflux for the enriching section of the column above the feed entry point.

## 15.5.4. Sublimation equipment

Very few standard forms of sublimation or de-sublimation equipment are in common use and most industrial units, particularly on the condensation side of the process, have been developed on an ad hoc basis for a specific substance and purpose. The most useful source of information on sublimation equipment is the patent literature, although as HOLDEN and BRYANT[95] and KEMP *et al.*[96] point out, it is not clear whether a process has been, or even can be, put into practice.

### *Vaporisers*

A variety of types of vaporisation units has been used or proposed for large-scale operation[95], the design depending on the manner in which the solid feedstock is to be vaporised. These include:

(a) a bed of dry solids without entrainer gas,
(b) dry solids suspended in a dense non-volatile, liquid;
(c) solids suspended in a boiling (entrainer) liquid where the entrainer vapour is formed in situ;
(d) entrainer gas flowing through a fixed bed of solid particles;
(e) entrainer gas bubbling through molten feedstock such that vaporisation takes place above the triple-point pressure;
(f) entrainer gas flowing through a dense phase of solid particles in a fluidised bed;
(g) entrainer gas flowing through a dilute phase of solid particles, such as in a transfer-line vaporiser where the solid and gas phases are in co-current flow, or in a raining solids unit where the solids and entrainer may be in countercurrent flow.

### *Condensers*

Sublimate condensers are usually large, air-cooled chambers which tend to have very low heat-transfer coefficients (5–10 W/m$^2$ deg K) because sublimate deposits on the condenser walls act as an insulator, and vapour velocities in the chambers are generally very low.

Quenching the vapour with cold air in the chamber may increase the rate of heat removal although excessive nucleation is likely and the product crystals will be very small. Condenser walls may be kept free of solid by using internal scrapers, brushes, and other devices, and all vapour lines in sublimation units should be of large diameter, be adequately insulated, and if necessary, be provided with supplementary heating to minimise blockage due to the buildup of sublimate. One of the main hazards of air-entrainment sublimation is the risk of explosion since many solids that are considered safe in their normal state can form explosive mixtures with air. All electrical equipment should therefore be flame-proof, and all parts of the plant should be efficiently earthed to avoid build-up of static electricity.

The method of calculating the density of deposited layers of sublimate and of other variables and the optimisation of sublimate condenser design, has been discussed by WINTERMANTEL *et al.*[113]. It is generally assumed that the growth rate of sublimate layers is governed mainly by heat and mass transfer. The model which is based on conditions in the diffusion boundary layer takes account of factors such as growth rate, mass transfer, and concentrations in the gas. The model shows a reasonably good fit to experimental data.

In a variant of the large-chamber de-sublimation condenser, the crystallisation chamber may be fitted with gas-permeable walls as described by VITOVEC *et al.*[101]. The vapour and the entrainer gas are cooled by evaporation of water dispersed in the pores of the walls, and an inert gas passes through the porous walls into the cooling space and protects the internal walls from solid deposits. Crystallisation takes place in the bulk vapour–gas mixture as a result of direct contact with the dispersed water. This arrangement has been used, for example, for the partial separation of a mixture of phthalic anhydride and naphthalene by using nitrogen as the entrainer. Although fluidised-bed condensers have been considered for large-scale application, most of the published reports are concerned with laboratory-scale investigations[110].

## 15.6. FRACTIONAL CRYSTALLISATION

A single crystallisation operation performed on a solution or a melt may fail to produce a pure crystalline product for a variety of reasons including:

(a) the impurity may have solubility characteristics similar to those of the desired pure component, and both substances consequently co-crystallise,
(b) the impurity may be present in such large amounts that the crystals inevitably become contaminated.
(c) a pure substance cannot be produced in a single crystallisation stage if the impurity and the required substance form a solid solution.

Re-crystallisation from a solution or a melt is, therefore, widely employed to increase crystal purity.

### Example 15.9

Explain how fractional crystallisation may be applied to a mixture of sodium chloride and sodium nitrate given the following data. At 293 K, the solubility of sodium chloride is 36 kg/100 kg water and of sodium nitrate 88 kg/100 kg water. Whilst at this temperature, a saturated solution comprising both salts will contain 25 kg sodium chloride and 59 kg of sodium nitrate per 100 kg of water. At 373 K, these values, again per 100 kg of water, are 40 and 176 and 17 and 160, respectively.

## Solution

The data enable a plot of kg NaCl/100 kg of water to be drawn against kg $NaNO_3/100$ kg of water as shown in Figure 15.35. On the diagram, points C and E represent solutions saturated with respect to both NaCl and $NaNO_3$ at 293 K and 373 K respectively. Fractional crystallisation may then be applied to this system as follows:

(a) A solution saturated with both NaCl and $NaNO_3$ is made up at 373 K. This is represented by point E, and, on the basis of 100 kg water, this contains 17 kg NaCl and 160 kg $NaNO_3$.
(b) The solution is separated from any residual sold and then cooled to 293 K. In so doing, the composition of the solution moves along EG.
(c) Point G lies on CB which represents solutions saturated with $NaNO_3$ but not with NaCl. Thus the solution still contains 17 kg NaCl and in addition is saturated with 68 kg $NaNO_3$. That is $(168 - 68) = 92$ kg of pure $NaNO_3$ crystals have come out of solution and this may be drained and washed.

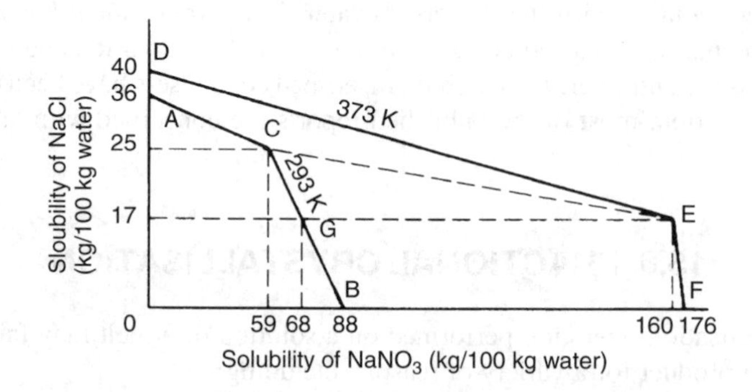


Figure 15.35. Effect of sodium chloride on the solubility of sodium nitrate

In this way, relatively pure $NaNO_3$, depending on the choice of conditions and particle size, has been separated from a mixture of $NaNO_3$ and NaCl.

The amount of $NaNO_3$ recovered from the saturated solution at 373 K is:

$$(92 \times 100)/160 = \underline{\underline{57.5\%}}$$

An alternative approach is to note that points C and B represent 59 and 88 kg $NaNO_3/100$ kg water and assuming CB to be a straight line, then by similar triangles:

$$\text{concentration of } NaNO_3 = 59 + [(88 - 59)(25 - 17)]/24$$

$$= 68.3 \text{ kg/100 kg water}$$

and:

$$\text{yield of } NaNO_3 = (160 - 68.3) = \underline{\underline{91.7 \text{ kg/100 kg water}}}.$$

whilst all the sodium chloride remains in solution.

If the cycle is then repeated, during the evaporation stage the sodium chloride is precipitated (and removed!) whilst the concentration of the nitrate re-attains 160 kg/100 kg water. On cooling again, the amount of sodium nitrate which crystallises out is 91.7 kg/100 kg water,

or: $(91.7 \times 100)/160 = \underline{\underline{57.3 \text{ per cent of the nitrate in solution}}}$, as before.

The same percentage of the chloride will be precipitated on re-evaporation.

### 15.6.1. Recrystallisation from solutions

Most of the impurities from a crystalline mass can often be removed by dissolving the crystals in a small amount of fresh hot solvent and cooling the solution to produce a fresh crop of purer crystals. The solubility of the impurities in the solvent must, however, be greater than that of the main product. Re-crystallisation may have to be repeated many times before crystals of the desired purity are obtained. A simple recrystallisation scheme is:

$$\begin{array}{ccccccc} S & & S & & & & \\ \downarrow & & \downarrow & & & & \\ \mathbf{AB} & \rightarrow & X_1 & \rightarrow & X_2 & \rightarrow & X_3 \\ & \downarrow & & \downarrow & & \downarrow & \\ & L_1 & & L_2 & & L_3 & \end{array}$$

An impure crystalline mass **AB**, where **A** is the less soluble, desired component, is dissolved in the minimum amount of hot solvent **S** and then cooled. The first crop of crystals $X_1$ will contain less impurity **B** than the original mixture, and **B** is concentrated in the liquor $L_1$. To achieve a higher degree of crystal purity, the procedure can be repeated. In each stage of such a sequence, losses of the desired component **A** can be considerable, and the final amount of 'pure' crystals may easily be a small fraction of the starting mixture **AB**. Many schemes have been designed to increase both the yield and the separation efficiency of fractional re-crystallisation. The choice of solvent depends on the characteristics of the required substance **A** and the impurity **B**. Ideally, **B** should be very soluble in the solvent at the lowest temperature employed and **A** should have a high temperature coefficient of solubility, so that high yields of **A** can be obtained from operation within a small temperature range.

### 15.6.2. Recrystallisation from melts

Schemes for recrystallisation from melts are similar to those for solutions, although a solvent is not normally added. Usually, simple sequences of heating (melting) and cooling (partial crystallisation) are followed by separation of the purified crystals from the residual melt. Selected melt fractions may be mixed at intervals according to the type of scheme employed, and fresh feed-stock may be added at different stages if necessary. As BAILEY[114] reports, several such schemes have been proposed for purification of fats and waxes.

As described in Section 15.2.1, eutectic systems can be purified in theory by single-stage crystallisation, whereas solid solutions always require multistage operations. Countercurrent fractional crystallisation processes in column crystallisers are described in Section 15.4.3.

### 15.6.3. Recrystallisation schemes

A number of fractional crystallisation schemes have been devised by MULLIN[3] and GORDON *et al.*[115], and the use of such schemes has been discussed by JOY and PAYNE[116] and SALUTSKY and SITES[117].

## 15.7. FREEZE CRYSTALLISATION

Crystallisation by freezing, or *freeze crystallisation*, is a process in which heat is removed from a solution to form crystals of the solvent rather than of the solute. This is followed by separation of crystals from the concentrated solution, washing the crystals with near-pure solvent, and finally melting the crystals to produce virtually pure solvent. The product of freeze crystallisation can be either the melted crystals, as in water desalination, or the concentrated solution, as in the concentration of fruit juice or coffee extracts. Freeze crystallisation is applicable in principle to a variety of solvents and solutions although, because it is most commonly applied to aqueous systems, the following comments refer exclusively to the freezing of water.

One of the more obvious advantages of freezing over evaporation for removal of water from solutions is the potential for saving heat energy resulting from the fact that the enthalpy of crystallisation of ice, 334 kJ/kg, is only one-seventh of the enthalpy of vaporisation of water, 2260 kJ/kg, although it has to be acknowledged that the cost of producing 'cold' is many times more than the cost of producing 'heat'. Process energy consumption may be reduced below that predicted by the phase-change enthalpy, however, by utilising energy recycle methods, such as multiple-effect or vapour compression, as commonly employed in evaporation as discussed in Chapter 14. In freeze-crystallisation plants operating by direct heat exchange, vapour compression has been used to recover refrigeration energy by using the crystals to condense the refrigerant evaporated in the crystalliser. Another advantage of freeze crystallisation, important in many food applications, is that the volatile flavour components normally lost during conventional evaporation can be retained in the freeze-concentrated product. Despite earlier enthusiasm, large-scale applications in desalination, effluent treatment, dilute liquor concentration and solvent recovery and so on have not been developed as yet.

All freeze separation processes depend on the formation of pure solvent crystals from solution, as described for eutectic systems in Section 15.2.1. which allows single-stage operation. Solid-solution systems, requiring multistage-operation, are not usually economic. Several types of freeze crystallisation processes may be designated according to the kind of refrigeration system used as follows:.

(a) In *indirect-contact freezing*, the liquid feedstock is crystallised in a scraped-surface heat exchanger as described in Section 15.4, fitted with internal rotating scraper blades and an external heat-transfer jacket through which a liquid refrigerant is passed. The resulting ice-brine slurry passes to a wash column where the ice crystals are separated and washed before melting. VAN PELT and VAN NISTELROOY[118] have described one of the commercial systems which are based on this type of freezing process.

(b) *Direct-contact freezing* processes utilise inert, immiscible refrigerants and are suitable for desalination. A typical scheme taken from BARDUHN[64] is shown in Figure 15.36. Sea-water, at a temperature close to its freezing point, is fed continuously into the crystallisation vessel where it comes in direct contact with a liquid refrigerant such as *n*-butane which vaporises and causes ice crystals to form due to the exchange of latent heat. The ice-brine slurry is fed to a wash column where it is washed countercurrently with fresh water. The emerging brine-free ice is melted by the enthalpy of the condensation of the vapour released from the compressed refrigerant. A major part of the energy input is that required for the compressors.

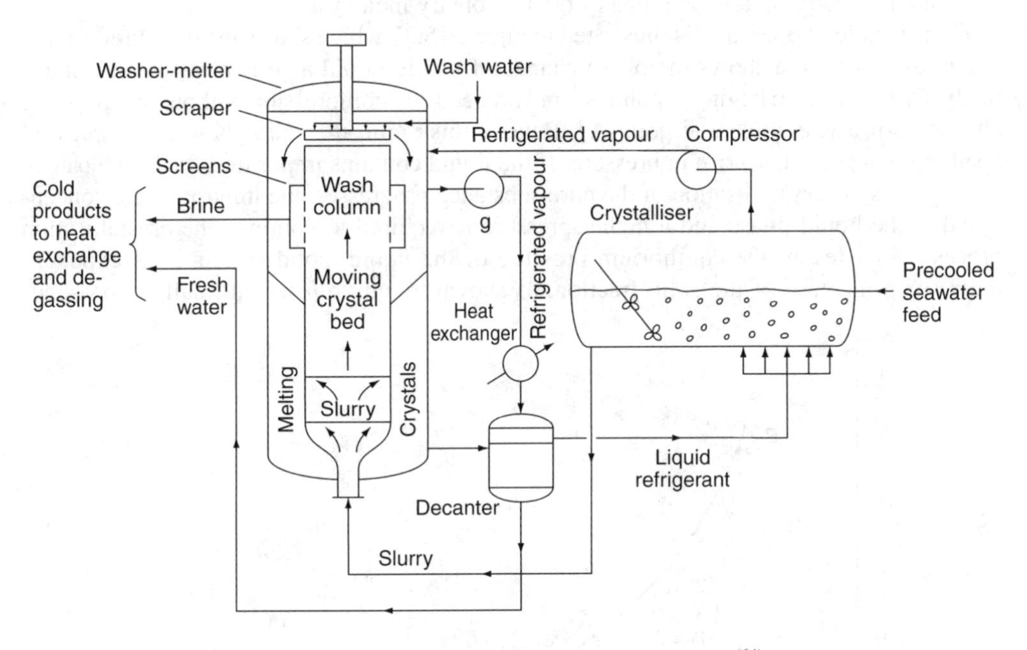


Figure 15.36. Desalination of seawater by freezing[64]

(c) *Vacuum freezing processes* do not require a conventional heat exchanger, and the problems of scale formation on heat-transfer surfaces are avoided. Cooling is effected by flash evaporating some of the solvent as the liquid feedstock enters a crystallisation vessel maintained at reduced pressure. Although vacuum freezing is potentially attractive for aqueous systems it has not, as yet, achieved widespread commercial success.

THIJSSEN and SPICER[119] has given a general review of freeze concentration as an industrial separation process and BUSHNELL and EAGEN[63] have discussed the status of freeze desalination. The potential of freeze crystallisation in the recycling and re-use of wastewater has been reviewed by HEIST[120], and the kinetics of ice crystallisation in aqueous sugar solutions and fruit juice are considered by OMRAN and KING[121].

## 15.8. HIGH PRESSURE CRYSTALLISATION

As noted previously, high pressure crystallisation in which an impure liquid feedstock is subjected to pressures of up to 300 MN/m$^2$ in a relatively small chamber, 0.001 m$^3$ in volume, under adiabatic conditions is a relatively recent development. As the pressure and temperature of the charge increase, fractional crystallisation takes place and the impurities are concentrated in the liquid phase which is then discharged from the pressure chamber. At the end of the cycle, further purification is possible since residual impurities in the compressed crystalline plug may then be 'sweated out' when the pressure is released. MORITOKI[(91)] has claimed that a single-cycle operation lasting less than 300 s is capable of substantially purifying a wide range of organic binary melt systems.

The principle of operation is illustrated in Figure 15.37 which shows the pressure-volume relationship. Curve a shows the phase change of a pure liquid as it is pressurised isothermally. Crystallisation begins at point $A_1$ and proceeds by compression without any pressure change until it is complete at point $A_2$. Beyond this point, the solid phase is compressed resulting in a very sharp rise in pressure. If the liquid contains impurities, these nucleate at point $B_1$. As the crystallisation of the pure substance progresses, the impurities are concentrated in the liquid phase and a higher pressure is required to continue the crystallisation process. As a result, the equilibrium pressure of the liquid–solid system rises exponentially with increase of the solid fraction, as shown by curve *b* which finally approaches

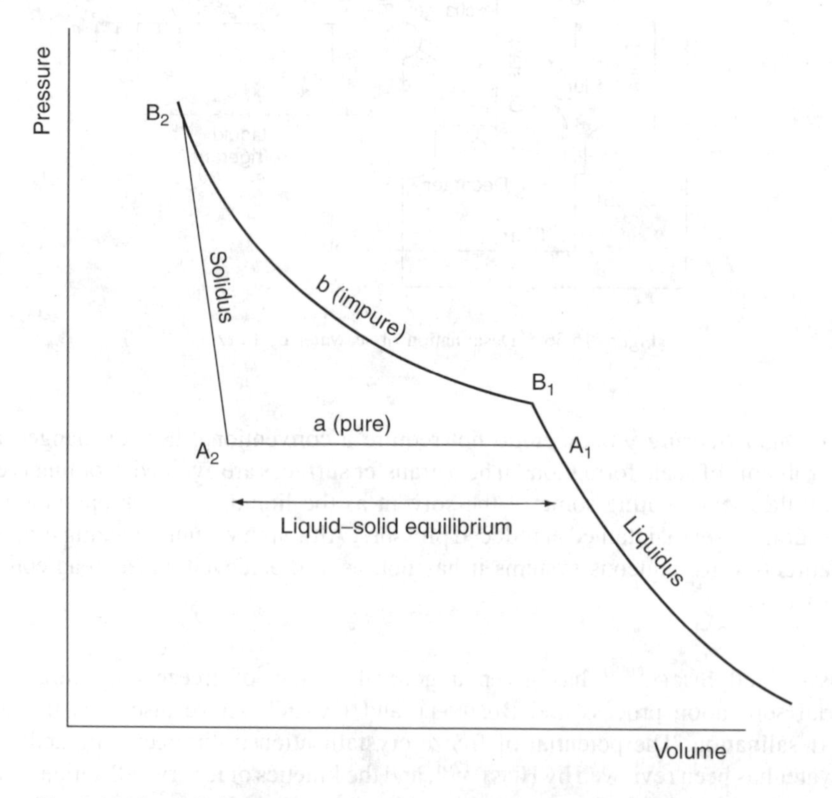


Figure 15.37. Relationship between pressure and volume for isothermal conditions

the solidus curve. A liquid–solid equilibrium line in terms of pressure and temperature is shown in Figure 15.38*a*. The liquid–solid equilibrium line moves from line a to line b with increase in impurities and line c represents the liquid–solid equilibrium for eutectic composition. On the industrial scale, a liquid is adiabatically compressed first from point A to point B, accompanied by heat generation and then to point C at which nucleation occurs accompanied by a temperature rise due to the release of the latent heat. Again, it is during this step that the impurities are concentrated into the mother liquor. At this stage, the liquid is separated from the solid phase and removed from the vessel at point C which is at a slightly lower pressure than the eutectic line. When the greater part of the liquid has been removed, its pressure decreases at first gradually and then rapidly to atmospheric pressure whilst the crystals are maintained at the initial separation pressure. In this way, the crystals are compacted and their surfaces are purified by slight melting, or by the so-called 'sweating' phenomenon. After the separation at point D, the crystals are highly purified and the line representing the equilibrium state gradually approaches line a. The basic pattern of operation as a function of time is shown in Figure 15.38*b*.

Pilot scale investigations by Kobe Steel have shown, for example, that the impurity level in a feed of mesitylene is reduced from 0.52 to 0.002 per cent in a single operating cycle at 15 $MN/m^2$ and a concentration of greater than 99 per cent *p*-xylene is obtained from a mixture of *p*-xylene and mesitylene containing 80 per cent *p*-xylene. It has also been shown that whilst crystals of cumenealdehyde are very difficult to obtain by cooling, nucleation and crystal growth occur at pressures of 50–70 $MN/m^2$ and, where the crystals obtained are then used as seed material, cumenealdehyde is then easily crystallised and purified at pressures below 20 $MN/m^2$. In this work, even though the capacity of the pilot unit was only 0.0015 $m^3$, some 360 tonne/year of raw material could be processed in 120 s cycles over a period of 8000 h.

Kobe Steel claim that, in terms of running costs, not only is the energy consumption low, being 10–50 per cent compared with conventional processes, but high pressure operation is ideally suited to the separation of isomers which are difficult to purify by other processes. These substances may have close boiling points or may be easily decomposed by temperature elevation. In this respect, recent work on supercritical fluids, as described by POLOAKOFF *et al.*(122), is of great importance.

As discussed in Chapter 14, supercritical fluids are gases that are compressed until their densities are close to those of liquids. They are extremely non-ideal gases in which interactions between molecules of a supercritical fluid and a potential solute can provide a 'solvation energy' for many solids to dissolve. The higher the pressure and hence the density of the supercritical fluid, the greater are the solvent–solute interactions and hence the higher the solubility of the solid. In other words the solvent power of a supercritical fluid is 'tunable' and this is the key factor in the use of supercritical fluids for a wide range of processes. As discussed in Chapter 14, supercritical fluids in common use include ethene, ethane and propane together with supercritical carbon dioxide and these have all been widely discussed in the literature(123–127). Although supercritical fluids have considerable advantages in the field of process intensification and can also replace environmentally undesirable solvents and indeed most organics, their most important property as far as crystallisation is concerned is that they can be tuned to dissolve the desired product, or indeed any impurities, which are then separated by crystallisation at high pressure. It is in this way that crystallisation is moving from the simple production of

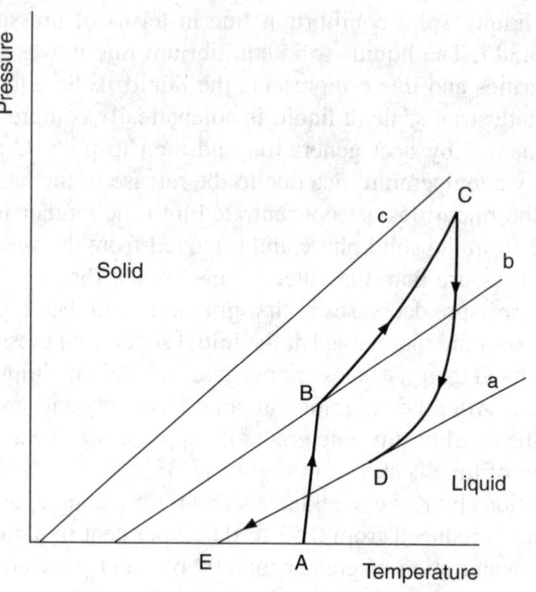


(*a*) Adiabatic application of pressure

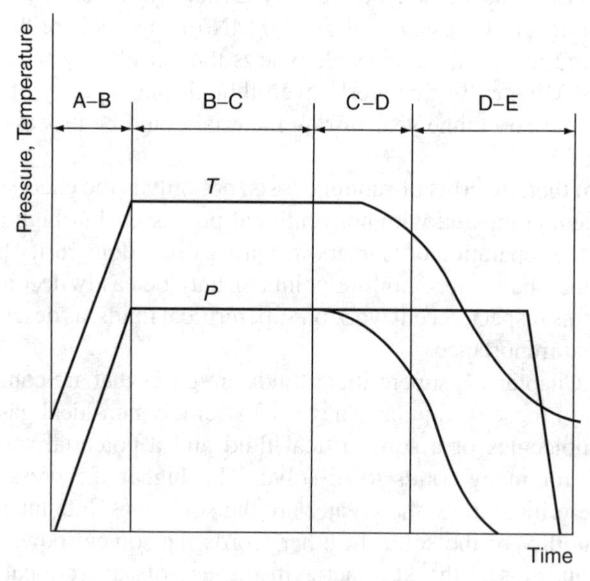


A–B–C: Pressurising step
C: Liquid-phase discharge
C–D–E: Solid phase compaction/sweating step

(*b*) Pressure and temperature variation during a cycle

Figure 15.38. High pressure crystallisation

hydrates from salt solutions towards a fully fledged separation technique which has and will have many advantages in comparison with more traditional operations in the years to come.

## 15.9. FURTHER READING


BAILEY, A. E.: *Solidification of Fats and Waxes* (Interscience, New York, 1950).
BAMFORTH, A. W.: *Industrial Crystallisation* (Leonard Hill, London, 1965).
BARTOLOMAI, A. (ed.): *Food Factories: Processes, Equipment and Costs* (VCH, New York, 1987).
BUCKLEY, H. E.: *Crystal Growth* (Wiley, New York, 1951).
BRUIN, S. (ed.): *Preconcentration and Drying of Food Materials* (Elsevier, New York, 1988).
FAKTOR, M. M. and GARRETT, D. E.: *Growth of Crystals from the Vapour* (Chapman and Hall, London, 1974).
FINDLAY, A. and CAMPBELL, A. N.: *The Phase Rule and its Applications* 9th. edition. (Longman, London, 1951).
JANCIC, S. J. and GROOTSCHOLTEN, A. M.: *Industrial Crystallisation* (Reidel, Dordrecht, 1984)
KALDIS, E. and SCHEEL, H. J.: *Crystal Growth and Materials* (North-Holland, Amsterdam, 1977).
LARSON, R.: *Constitutive Equations for Polymer Melts and Solutions* (Butterworth, London, 1988).
LAWSON, W. D. and NIELSEN, S.: *Preparation of Single Crystals* (Butterworths, London, 1958).
LUI, Y. A., MCGEE, H. A. and EPPERLY, W. R. (eds.): *Recent Developments in Chemical Process and Plant Design* (Wiley, New York, 1987).
MATZ, G.: *Kristallisation* 2nd. edition (Springer Verlag, Berlin, 1969).
MERSTMANN, A. (ed.): *Crystallization Technology Handbook* (Marcel Dekker, New York, 1995).
MOYERS, C. G. and ROUSSEAU, R. W. in ROUSSEAU, R. W.: *Handbook of Separation Process Technology* (John Wiley, New York)
MULLIN, J. W.: *Crystallization* 3rd. edition (Butterworth-Heinemann, Oxford, 1993, 1997).
MULLIN, J. W.: *Crystallization* 4th edition (Butterworth-Heinemann, Oxford, 2001).
MULLIN, J. W.: 'Crystallisation' in Kirk-Othmer: *Encyclopedia of Chemical Technology*, Volume 7, 3rd. edition (John Wiley & Sons, New York, 1979).
MULLIN, J. W. 'Crystallisation and Precipitation' in *Ullmann's Encyclopedia of Industrial Chemistry*, Volume B2 (VCH Verlagsgesellschaft mbH, Weinheim, 1988)
MYERSON, A. S.: *Handbook of Industrial Crystallization*, 2nd edition (Butterworth-Heinemann, Oxford, 2000)
NULL, H. R.: *Phase Equilibrium in Process Design* (Wiley-Interscience, New York, 1970).
NYVLT, J.: *Industrial Crystallisation from Solutions* (Butterworths, London, 1971)
NYVLT, J.: *Solid–Liquid Phase Equilibria* (Elsevier-North Holland, New York, 1977).
NYVLT, J.: *Industrial Crystallisation* (Verlag Chemie, Weinheim, NY. 1978).
PAMPLIN, B. R. (ed.): *Crystal Growth*, 2nd. edition (Pergamon Press, Oxford, 1980).
RANDOLPH, D. and LARSON, M. A.: *Theory of Particulate Processes* (Academic Press, New York, 1971).
RUTNER, E., GOLDFINGER, P. and HIRTH, J. P. (eds.): *Condensation and Evaporation of Solids* (Gordon and Breach, New York, 1964).
SCHWEITZER, P. A. (ed.): *Handbook of Separation Techniques for Chemical Engineers* 2nd. edition (McGraw-Hill, New York, 1988).
STRICKLAND-CONSTABLE, R. F.: *Kinetics and Mechanism of Crystallisation* (Academic Press, London, 1968).
THIJSSEN, H. A. in SPICER, A. (ed.): *Advances in Preconcentration and Dehydration of Foods* (Wiley-Interscience, New York, 1974).
UBBELOHDE, A. R.: *Melting and Crystal Structure* (OUP, Oxford, 1965).
WALTON, A. G.: *The Formation and Properties of Precipitates* (Interscience, New York, 1967).
WALAS, S. M.: *Chemical Process Equipment:Selection and Design* (Butterworth, London, 1989).
WISNIAK, J.: *Phase Diagrams.* Physical Sciences Data 10 (Elsevier-North Holland, New York, 1981).
ZETTLEMOYER, A. C. (ed.): *Nucleation* (Marcel Dekker, New York, 1969).
*AIChE Symp. Ser.* (a) 65 (1969) no. 95, Crystallization from solutions and melts; (b) 67 (1971) no. 110, Factors affecting size distribution; (c) 68 (1972) no. 121, Crystallization from solutions: Nucleation phenomena in growing crystal systems; (d) 72 (1976) no. 153, Analysis and design of crystallisation processes; (e) 76 (1980) no. 193, Design, control and analysis of crystallisation processes; (f) 78 (1982) no. 215, Nucleation, growth and impurity effects in crystallisation process engineering; (g) 80 (1984) no. 240, Advances in crystallisation from solutions.

## 15.10. REFERENCES


1. MULLIN, J. W.: 'Crystallization' in Kirk-Othmer: *Encyclopedia of Chemical Technology*, Volume 7, 3rd. edition (John Wiley & Sons, New York, 1979)
2. MULLIN, J. W. 'Crystallization and Precipitation' in *Ullmann's Encyclopedia of Industrial Chemistry*, Volume B2 (VCH Verlagsgesellschaft mbH, Weinheim, 1988)
3. MULLIN, J. W.: *Crystallization* 4th. edn. (Butterworth-Heinemann, Oxford, 2001)
4. FEILCHENFELD, H. and SARIG, S.: *Ind. Eng. Chem. Process. Prod. Res. Dev.* **24** (1985) 130–133. Calcium chloride hexahydrate: A phase-changing material for energy storage.
5. KIMURAH, H.: *J. Cryst. Growth* **73** (1985) 53–62. Impurity effect on growth rates of $CaCl_2.6H_2O$ crystals.
6. GRONVOLD, F. and MEISINGSET, K. K.: *J. Chem. Thermodyn.* **14** (1982) 1083–1098. Thermodynamic properties and phase transitions of salt hydrates between 270 and 400K: $NH_4Al(SO_4)_2.12H_2O$, $KAl(SO_4)_2.12H_2O$, $Al_2(SO_4)_3.17H_2O$, $ZnSO_4.7H_2O$, $NaSO_4.10H_2O$ and $Na_2S_2O_3.5H_2O$.
7. KIMURA, H. and KAI, J.: *Solar Energy* **35** (1985) 527–534. Phase change stability of sodium acetate trihydrate and its mixtures.
8. FINDLAY, A. and CAMPBELL, A. N.: *The Phase Rule and its Applications*, 9th. edition (Longman, London, 1951).
9. RICCI, J. E.: *The Phase Rule and Heterogeneous Equilibrium* (van Nostrand, New York, 1951).
10. NULL, H. R.: *Phase Equilibrium in Process Design* (Wiley-Interscience, New York, 1970).
11. NYVLT, J.: *Solid-Liquid Phase Equilibrium* (Academia, Prague, Elsevier, Amsterdam, 1977).
12. HÖRMEYER, H.: *Germ. Chem. Eng.* **6** (1983) 277–281. Calculation of crystallisation equilibria.
13. KUSIK, C. L., MEISSNER, H. P. and FIELD, E. L.: *A.I.Ch.E. Jl.* **25** (1979) 759–762. Estimation of phase diagrams and solubilities for aqueous multi-ion systems.
14. SANDER, B., RASMUSSEN, P. and FREDENSLUND, A.: *Chem. Eng. Sci.* **41** (1986) 1197–1202. Calculation of solid-liquid equilibria in aqueous solutions of nitrate salts using an extended uniquac equation.
15. NANCOLLAS, G. H., REDDY, M. M. and TSAI, F.: *J. Cryst. Growth* **20** (1973) 125–134. Crystal growth kinetics of minerals encountered in water treatment processes.
16. NANCOLLAS, G. H. and REDDY, M. M.: *Soc. Pet. Eng. J.* (April 1974) 117–123. The kinetics of crystallisation of scale-forming materials.
17. CARDEW, P. T., DAVEY, R. J. and RUDDICK, A. J.: *J. Chem. Soc. Faraday Trans. 2*, **80** (1984) 659–668. Kinetics of polymorphic solid-state transformations.
18. CARDEW, P. T. and DAVEY, R. J.: *Proc. R. Soc. London Ser. A* **398** (1985) 415–428. The kinetics of solvent-mediated phase transformations.
19. MULLIN, J. W. and SÖHNEL, O.: *Chem. Eng. Sci.* **32** (1977) 683–686. Expressions of supersaturation in crystallisation studies.
20. OSTWALD, Wilhelm: *Z. Phys. Chem. (Leipzig)* **34** (1900) 493–503. Über die vermeintliche Isomeric des roten und gelben Quecksilberoxyds und die Oberflaschen spannung fester Körper.
21. MIERS, H. A. and ISAAC, F.: *J. Chem. Soc.* **89** (1906) 413–454. Refractive indices of crystallizing solutions.
22. SEIDELL, A. in LINKE, W. F. (ed.): *Solubilities of Inorganic and Metal Organic Compounds*, 4th. edition vol. 1 (van Nostrand, New York, 1958) vol. 2 (*Am. Chem. Soc.*, Washington, D.C., 1965).
23. STEPHEN, H and STEPHEN, T.: *Solubilities of Inorganic and Organic Compounds* (Pergamon Press, London, 1963).
24. BROUL, M., NYVLT, J. and SÖHNEL, O.: *Solubilities in Binary Aqueous Solutions* (Academia, Prague, 1981).
25. MOYERS, C. G. and ROUSSEAU, R. W. in ROUSSEAU, R. W.: *Handbook of Separation Process Technology* (John Wiley, New York)
26. American Petroleum Institute: *Selected Values of Properties of Hydrocarbons* Research Project No. 44 (Amer. Pet. Institute)
27. STRICKLAND-CONSTABLE, R. F.: *Kinetics and Mechanism of Crystallization*, (Academic Press, London, 1968)
28. KASHCHIEV, D.: *Nucleation—basic theory with applications*. (Butterworth-Heinemann, Oxford, 2000)
29. GARSIDE, J. and DAVEY, R. J.: *Chem. Eng. Commun.* **4** (1980) 393–424. Secondary contact nucleation: kinetics, growth and scale-up.
30. UBBLEOHDE, A. R.: *Melting and Crystal Structure* (Oxford University Press, Oxford, 1965).
31. GARTEN, V. A. and HEAD, R. B.: *Phil. Mag.* **8** (1963) 1793–1803 and **14** (1966) 1243–1253. Homogeneous nucleation and the phenomenon of crystalloluminescence.
32. BERGLUND, K. A. in JANČIĆ, S. J. and DE JONG, E. J. (eds.): *Industrial Crystallisation 84* (Elsevier, Amsterdam, 1984)
33. MULLIN, J. W. and LECI, C. L.: *Phil. Mag.* **19** (1969) 1075–1077. Evidence of molecular cluster formation in supersaturated solutions of citric acid.
34. LARSON, M. A. and GARSIDE, J.: *Chem. Eng. Sci.* **41** (1986) 1285–1289. Solute clustering in supersaturated solutions.

35. NYVLT, J.: *Industrial Crystallization from Solutions* (Butterworth, London, 1971).
36. NYVLT, J., SÖHNEL, O., MATUCHOVA, M. and BROUL, M.: *The Kinetics of Industrial Crystallization* (Academia, Prague, 1985).
37. RANDOLPH, A. D and LARSON, M. A.: *Theory of Particulate Processes* (Academic Press, New York, 1971).
38. NIELSEN, A. E. and SÖHNEL, O.: *J. Cryst. Growth* **11** (1971) 233–242. Interfacial tensions, electrolyte crystal-aqueous solution, from nucleation data.
39. SÖHNEL, O. and MULLIN, J. W.: *J. Cryst. Growth* **44** (1978) 377–382. A method for the determination of precipitation induction periods.
40. MATZ, G.: *Kristallisation* 2nd. edition (Springer Verlag, Berlin, 1969).
41. JANČIČ, S. J. and GROOTSCHOLTEN, P. A. M.: *Industrial Crystallization* (Reidel, Dordecht, 1984).
42. LAUDICE, R. A.: The growth of single crystals (Prentice-Hall, Englewood Cliffs, N.J., 1970).
43. GARSIDE, J.: *Chem. Eng. Sci.* **40** (1985) 3–26. Industrial crystallisation from solution.
44. NIELSEN, A. E.: *J. Cryst. Growth* **67** (1984) 289–310. Electrolyte crystal growth mechanisms.
45. PAMPLIN, B. R. (ed): *Crystal Growth* 2nd. edition (Pergamon Press, Oxford, 1980).
46. KALDIS, E. and SCHEEL, H. J.: *Crystal Growth and Materials* (North-Holland, Amsterdam, 1977).
47. TENGLER, T. and MERSMANN, A.: *Germ. Chem. Eng.* **7** (1984) 248–259. Influence of temperature, saturation and flow velocity on crystal growth from solutions.
48. WÖHLK, W.: *Meszanordnungen zur Bestimmung von Kristallwachstumsgeschwindigkeiten*, (Fortschr. Ber. VDI Z., Reihe **3** 1982 no. 71).
49. WHITE, E. T., BENDIG L. L. and LARSON, M. A.: *A.I.Ch.E. Symp. Ser.* **72** No. 153 (1976) 41–47. Analysis and design of crystallization systems.
50. JONES, A. G. and MULLIN, J. W.: *Chem. Eng. Sci.* **29** (1974) 105–118. Programmed cooling crystallisation of potassium sulphate solutions.
51. JANSE, A. H. and DE JONG, E. J. in MULLIN, J. W. (ed): *Industrial Crystallization 75* (Plenum Publishing, London, 1976).
52. GARSIDE, J. and JANČIČ, S. J.: *AIChE J.* **25** (1979) 948–958. Measurement and scale-up of secondary nucleation kinetics for the potash alum-water system.
53. WHETSTONE, J.: *Trans. Faraday. Soc.* **51** (1955) 973–80 and 1142–53. The crystal habit modification of inorganic salts with dyes (in two parts).
54. GARRETT, D. E.: *Brit. Chem. Eng.* **4** (1959) 673–677. Industrial crystallisation: influence of chemical environment.
55. BOTSARIS, G. D. in JANČIČ, S. J. and DE JONG, E. J. (eds.): *Industrial Crystallization 81* (North-Holland, Amsterdam, 1982).
56. DAVEY, R. J. in JANČIČ, S. J. and DE JONG, E. J. (eds.): *Industrial Crystallization 78* (North-Holland, Amsterdam, 1979).
57. DEICHA, G.: *Lacunes des cristeaux et leurs inclusions fluides* (Masson, Paris, 1955).
58. POWERS, H. E. C.: *Sugar Technol. Rev.* **1** (1970) 85–190. Sucrose crystals: inclusions and structure.
59. WILCOX, W. R. and KUO, V. H. S.: *J. Cryst. Growth* **19** (1973) 221–229. Nucleation of monosodium urate crystals.
60. SASKA, M. and MYERSON, A. S.: *J. Crys. Growth* **67** (1984) 380–382. A crystal-growth model with concentration dependent diffusion.
61. PHOENIX, L.: *Brit. Chem. Eng.* **11** (1966) 34–38. How trace additives inhibit the caking of inorganic salts.
62. ARMSTRONG, A. J.: *Chem. Proc. Eng.* **51**(11)(1970) 59. Scraped surface crystallisers.
63. BUSHNELL, J. D. and EAGEN, J. F.: *Oil Gas Jl.* **73** (1975) 42, 80–84. Dewax process produces benefits.
64. BARDUHN, A. J.: *Chem. Eng. Prog.* **71** (1975) 11, 80–87. The status of freeze-desalination.
65. MULLIN, J. W. and WILLIAMS, J. R.: *Chem. Eng. Res. Des.* **62** (1984) 296–302. A comparison between indirect contact cooling methods for the crystallisation of potassium sulphate.
66. FINKELSTEIN, E.: *J. Heat Recovery Syst.* **3** (1983) 431–437. On solar ponds: critique, physical fundamentals and engineering aspects.
67. BAMFORTH, A. W.: *Industrial Crystallisation* (Leonard Hill, London, 1965).
68. SAEMAN, W. C.: *A.I.Ch.E.Jl.* **2** (1956) 107–112. Crystal size distribution in mixed systems.
69. MULLIN, J. W. and NYVLT, J.: *Chem. Eng. Sci.* **26** (1971) 369–377. Programmed cooling of batch crystallizers.
70. ROBINSON, J. and ROBERTS, J. E.: *Can. J. Chem. Eng.* **35** (1957) 105–112. A mathematical study of crystal growth in a cascade of agitators.
71. ABEGG, C. F. and BALAKRISHNAM, N. S.: *A.I.Ch.E. Symp. Ser.* **72** No. 153 (1976) 88–94. The tanks-in-series concept as a model for imperfectly mixed crystallizers.
72. TOUSSAINT, A. G. and DONDERS, A. J. M.: *Chem. Eng. Sci.* **29** (1974) 237–245. The mixing criterion in crystallisation by cooling.
73. WHITE, E. T. and RANDOLPH, A. D.: *AIChE Jl* **33** (1984) 686–689. Graphical solution of the material balance constraint for MSMPR crystallizers.

74. GRIFFITHS, H.: *Trans. Instn. Chem. Engrs.* **25** (1947) 14–18. Crystallisation.
75. MULLIN, J. W. and NYVLT, J.: *Trans. Instn. Chem. Engrs.* **48** (1970) 7–14. Design of classifying crystalisers.
76. MERSMANN, A. and KIND, M.: *Intl Chem. Eng.* **29** (1989) 616–626. Modeling of chemical process equipment: the design of crystallizers.
77. DE JONG, E. J.: *Intl Chem. Eng.* **24** (1984) 419–431. Development of crystallizers.
78. BENNETT, R. C., FIEDELMAN, H. and RANDOLPH, A. D.: *Chem. Eng. Prog.* **69** (July, 1973) 86–93. Crystallizer influenced nucleation.
79. MOLINARI, J. G. D. in ZEIF, M. and WILCOX, W. R. (eds.): *Fractional Solidification* (Dekker, New York, 1967) 393–400.
80. GEL'PERIN, N. I. and NOSOV, G. A.: *Sov. Chem. Ind.* (Eng. Transl.) **9** (1977) 713–717. Fractional crystallisation of melts in drum crystallizers.
81. PONOMARENKO, K., POTEBYNA, G. F. and BEI, V. I.: *Theor. Found. Chem. Eng.* (Engl. Transl.) **13** (1980) 724–729. Crystallization of melts on a thin moving wall.
82. WINTERMANTEL, K.: *Chem. Ing. Tech.* **58** (1986) 6, 498–499. Effective separation power in freezing out layers of crystals from melts and solutions - a unified presentation.
83. ATWOOD, G. R. in Reference 42a, 112–121. The progressive mode of multistage crystallisation.
84. RITTNER, S. and STEINER, R.: *Chem. Ing. Tech.* **57** (1985) 2, 91–102. Melt crystallisation of organic substances and its large scale application.
85. FISCHER, O., JANČIĆ, S. J. and SAXER, K. in JANČIĆ, S. J. and DE JONG, E. J. (eds.): *Industrial Crystallization 84* (Elsevier, Amsterdam, 1984).
86. GATES, W. C. and POWERS, J. E.: *AIChE Jl* **16** (1970) 648–657. Determination of the mechanics causing and limiting separation by column crystallization.
87. HENRY, J. D. and POWERS, J. E.: *AIChE Jl* **16** (1970) 1055–1063. Experimental and theoretical investigation of continuous flow column crystallisation.
88. MCKAY, D. L. in ZEIT, F. and WILCOX, W. R. (eds.): *Fractional Solidification* (Marcel Dekker, New York, 1967) 427–439.
89. BRODIE, J. A.: *Mech. Chem. Trans. Inst. Engrs. Australia* **7** (1971) 1, 37–44. A continuous multistage melt purification process.
90. TAKEGAMI, K., NAKAMARU, N. and MORITA, M. in JANČIĆ, S. J. and DE JONG, E. J. (eds.): *Industrial Crystallization 84* (Elsevier, Amsterdam, 1984). 143–146.
91. MORITOKI, M.: *Int. Chem. Eng.* **20** (1980) 394–401. Crystallization and sweating of p-cresol by application of high pressure.
92. SHEARON, W. H. and DUNWOODY, W. B.: *Ind. Eng. Chem.* **45** (1953) 496–504. Ammonium nitrate.
93. NUNELLY, L. M. and CARTNEY, F. T.: *Ind. Eng. Chem. Prod. Res. Dev.* **21** (1982) 617–620. Granulation of urea by the falling-curtain process.
94. MULLIN, J. W.: Sublimation in *Ullmann's Encyclopedia of Industrial Chemistry* (VCH Verlagsgesellschaft mbH, D-6940 Weinheim, 1988).
95. HOLDEN, C. A. and BRYANT, H. S.: *Sep. Sci.* **4** (1969) (1) 1–13. Purification by sublimation.
96. KEMP, S. D. in CREMER, H. W. and DAVIES, T. (eds.): *Chemical Engineering Practice*, **6**, (Butterworth, London, 1958) 567–600. Sublimation and vacuum freeze drying.
97. KUDELA, L. and SAMPSON, M. J.: *Chem. Eng. (N. Y.)* **93** (1986) 12, 93–98. Understanding sublimation technology.
98. RUTNER, E., GOLDFINGER, P. and HIRTH, J. P. (eds.): *Condensation and Evaporation of Solids* (Gordon and Breach, New York, 1964).
99. FAKTOR, M. M. and GARRETT, I.: *Growth of Crystals from the Vapour* (Chapman and Hall, London, 1974).
100. GILLOT, J. and GOLDBERGER, W. M.: *Chem. Eng. Prog. Symp. Ser.* **65** (1969) (91) 36–42. Separation by thin-film fractional sublimation.
101. VITOVEC, J., SMOLIK, J. and KUGLER, K.: *Collect. Czech. Chem. Commun.* **43** (1978) 396–400. Separation of sublimable compounds by partial crystallisation from their vapour mixture.
102. EGGERS, H. H., OLLMANN, D., HEINZ, D., DROBOT, D. W. and NIKOLAJEW, A. W.: *Z. Phys. Chem. (Leipzig)* **267** (1986) 353–363. Sublimation und Desublimation im System $AlCl_3$–$FeCl_3$.
103. MATSOUKA, M. in JANČIĆ, S. J. and DE JONG, E. J. (eds.): *Industrial Crystallization 84* (Elsevier, Amsterdam 1984) 357–360. Rates of nucleation and growth of organic condensates of mixed vapours.
104. SHERWOOD, T. K. and JOHANNES, C.: *AIChE Jl.* **8** (1962) 590–593. The maximum rate of sublimation of solids.
105. PLEWES, A. C. and KLASSEN, J.: *Can. J. Technol.* **29** (1951) 322. Mass transfer rates from the solid to the gas phase.
106. UEDA, H. and TAKASHIMA, Y.: *J. Chem. Eng. Japan* **10** (1977) 6–12. Desublimation of two-component systems.
107. STRICKLAND-CONSTABLE, R. F.: *Kinetics and Mechanism of Crystallization* (Academic Press, London, 1968).

108. BILIK, R. J. and KRUPICZKA, R.: *Chem. Eng. Jl.* **26** (1983) 169–180. Heat transfer in the desublimation of phthalic anhydride.
109. CIBOROWSKI, J. and WRENSKI, S.: *Chem. Eng. Sci.* **17** (1962) 481–489. Condensation of subliminable materials in a fluidized bed.
110. KNUTH, M. and WEINSPACH, P. M.: *Chem. Ing. Tech.* **48** (1976) 893. Experimentelle Untersuchung des Wärme und Stoffübergangs an die Partikeln einer Wirbelschicht bei der Desublimation.
111. MATZ, G.: *Chem. Ing. Tech.* **30** (1958) 319–329. Fleitzbett-Sublimation.
112. CEDRO, V.: letter to *Chem. Eng. (N. Y.)* **93** (1986) 21, 5. Fluid beds and sublimation.
113. WINTERMANTEL, K., HOLZKNECH, H. and THOMA, P.: *Chem.Eng. Technol.* **10** (1987) 205–210. Density of sublimed layers.
114. BAILEY, A. E.: *Solidification of Fats and Waxes* (Interscience, New York, 1950).
115. GORDON, L., SALUTSKY, M. L. and WILLARD, H. H.: *Precipitation from Homogeneous Solution* (Wiley-Interscience, New York, 1959).
116. JOY, E. F. and PAYNE, J. H.: *Ind. Eng. Chem.* **47** (1955) 2157–2161. Fractional precipitation or crystallisation systems.
117. SALUTSKY, M. L. and SITES, J. G.: *Ind. Eng. Chem.* **47** (1955) 2162–2166. Ra-Ba separation process.
118. VAN PELT, W. H. S. M. and VAN NISTELROOY, M. G. J.: Food Eng. (November, 1975) 77–79. Procedure for the concentration of beer.
119. THIJSSEN, H. A. C. in SPICER, A. (ed.): *Advances in Preconcentration and Dehydration of Foods* (Wiley-Interscience, New York, 1974).
120. HEIST, J. A.: *AIChE Symp. Ser.* **77** (1984) (209) 259–272. Freeze crystallisation: waste water recycling and reuse.
121. OMRAN, A. M.: and KING, C. J.: *AIChE Jl.* **20** (1974) 799–801. Kinetics of ice crystallisation in sugar solutions and fruit juices.
122. POLOAKOFF, M., MEEHAN, N. J. and ROSS, S. K.: *Chem. & Industry* **10** (1999) 750–752 A supercritical success story.
123. JESSOP, P. G. and LEITNER, W. (eds): *Chemical Synthesis using supercritical fluids* (Wiley-VCH, Weinheim 1999)
124. MCHUGH, M. A. and KRUKONIS, V. J.: *Supercritical Fluid Extraction: Principles and Practice* (2nd edition) (Butterworth-Heinemann, Boston, 1994)
125. DARR, J. A. and POLIAKOFF, M.: *Chem. Rev.* **99** (1999) 495. New directions in inorganic and metal-organic coordination chemistry in supercritical fluids.
126. JESSOP, P. G., IKARIYA, T. and NOYORI, R.: *Chem. Rev.* **99** (1999) 475. Homogeneous catalysis in supercritical fluids.
127. BAIKER, A.: *Chem. Rev.* **99** (2) (1999) 453–473. Supercritical fluids in heterogeneous catalysis.

## 15.11. NOMENCLATURE

| | | Units in SI System | Dimensions in **M, N, L, T,** $\boldsymbol{\theta}$ |
|---|---|---|---|
| $A$ | crystal surface area | m$^2$ | $\mathbf{L}^2$ |
| $A'$ | heat transfer area | m$^2$ | $\mathbf{L}^2$ |
| $a_o$ | amount of component **A** in original solid | kg | **M** |
| $B$ | secondary nucleation rate | 1/m$^3$ s | $\mathbf{L}^{-3}\mathbf{T}^{-1}$ |
| $b$ | kinetic order of nucleation or exponent in equation 15.11 | — | — |
| $b_o$ | amount of component **B** in original solid | kg | **M** |
| $C$ | concentration of solute | kmol/m$^3$ | $\mathbf{NL}^{-3}$ |
| $C^*$ | saturated concentration of solute | kmol/m$^3$ | $\mathbf{NL}^{-3}$ |
| $C_p$ | specific heat capacity | J/kg K | $\mathbf{L}^2\mathbf{T}^{-2}\boldsymbol{\theta}^{-1}$ |
| $c$ | solution concentration | kg/kg | — |
| $c^*$ | equilibrium concentration | kg/kg | — |
| $c_i$ | concentration at interface | kg/kg | — |
| $c_{io}$ | initial concentration of impurity | kg/kg | — |
| $c_{in}$ | impurity concentration after stage $n$ | kg/kg | — |
| $c_r$ | solubility of particles of radius $r$ | kg/kg | — |

| | | Units in SI System | Dimensions in M, N, L, T, $\theta$ |
|---|---|---|---|
| $c_s$ | slurry concentration or magma density | kg/m$^3$ | $\mathbf{MT^{-3}}$ |
| $D$ | molecular diffusivity | m$^2$/s | $\mathbf{L^2T^{-1}}$ |
| $d$ | characteristic size of crystal | m | $\mathbf{L}$ |
| $d'$ | characteristic dimension of vaporisation chamber | m | $\mathbf{L}$ |
| $d_D$ | dominant size of crystal size distribution | m | $\mathbf{L}$ |
| $d_p$ | product crystal size | m | $\mathbf{L}$ |
| $d_s$ | size of seed crystals | m | $\mathbf{L}$ |
| $E$ | mass of solvent evaporated/mass of solvent in initial solution | kg/kg | — |
| $e$ | evaporation coefficient (equation 15.49) | — | — |
| $F$ | fraction of liquid removed after each decantation | — | — |
| $F'$ | pre-exponential factor in equation 15.9 | 1/m$^3$ s | $\mathbf{L^{-3}T^{-1}}$ |
| $\Delta G$ | excess free energy | J | $\mathbf{ML^2T^{-2}}$ |
| $G'$ | flowrate of inert gas | kg/s | $\mathbf{MT^{-1}}$ |
| $G_d$ | overall growth rate | m/s | $\mathbf{LT^{-1}}$ |
| $\Delta G_v$ | free energy change per unit volume | J/m$^3$ | $\mathbf{ML^{-1}T^{-2}}$ |
| $H_f$ | heat of fusion | J/kg | $\mathbf{L^2T^{-2}}$ |
| $i$ | order of integration or relative kinetic order | — | — |
| $J$ | rate of nucleation | 1/m$^3$ s | $\mathbf{L^{-3}T^{-1}}$ |
| $j$ | exponent in equation 15.11 | — | — |
| $K_b$ | birthrate constant | — | — |
| $K_G$ | overall crystal growth coefficient | kg/s | $\mathbf{MT^{-1}}$ |
| $K_N$ | primary nucleation rate constant | 1/m$^3$s | $\mathbf{L^{-3}T^{-1}}$ |
| **k** | Boltzmann constant | $1.38 \times 10^{23}$ J/K | $\mathbf{ML^2T^{-2}\theta^{-1}}$ |
| $k_1$ | constant in equation 15.33 | 1/m$^3$s | $\mathbf{L^{-3}T^{-1}}$ |
| $k_2$ | constant in equation 15.34 | m/s | $\mathbf{LT^{-1}}$ |
| $k_3$ | constant in equation 15.35 | m$^{-(i+3)}$ s$^{(i-1)}$ | $\mathbf{L^{-(i+3)}T^{(i-1)}}$ |
| $k_4$ | constant in equation 15.38 | m$^{-(i+3)}$ s$^{(i-1)}$ | $\mathbf{L^{-(i+3)}T^{(i-1)}}$ |
| $k_d$ | diffusion mass transfer coefficient | kg/m$^2$s | $\mathbf{ML^{-2}T^{-1}}$ |
| $k_r$ | integration or reaction mass transfer coefficient | kg/m$^2$s | $\mathbf{ML^{-2}T^{-1}}$ |
| $l$ | exponent in equation 15.11 | — | — |
| $M$ | molecular weight of solute in solution | kg/kmol | $\mathbf{MN^{-1}}$ |
| $M'$ | molecular weight | kg/kmol | $\mathbf{MN^{-1}}$ |
| $M_g$ | molecular weight of inert gas | kg/kmol | $\mathbf{MN^{-1}}$ |
| $M_s$ | molecular weight of sublimed material | kg/kmol | $\mathbf{MN^{-1}}$ |
| $m$ | particle mass or mass deposited in time $t$ | kg | $\mathbf{M}$ |
| $m_s$ | mass of seeds | kg | $\mathbf{M}$ |
| $N$ | rotational speed of impeller | Hz | $\mathbf{T^{-1}}$ |
| $n$ | order of nucleation process or number of stages | — | — |
| $n'$ | population density of crystals | m$^{-4}$ | $\mathbf{L^{-4}}$ |
| $n_i$ | moles of ions/mole of electrolyte | — | — |
| $n^\circ$ | population density of nuclei | m$^{-4}$ | $\mathbf{L^{-4}}$ |
| $P$ | vapour pressure | N/m$^2$ | $\mathbf{ML^{-1}T^{-2}}$ |
| $P'$ | crystal production rate | kg/s | $\mathbf{MT^{-1}}$ |
| $P_g$ | partial pressure of inert gas | N/m$^2$ | $\mathbf{ML^{-1}T^{-2}}$ |
| $P_s$ | vapour pressure at surface | N/m$^2$ | $\mathbf{ML^{-1}T^{-2}}$ |
| $P'_s$ | partial pressure of vaporised material | N/m$^2$ | $\mathbf{ML^{-1}T^{-2}}$ |
| $P_t$ | total pressure | N/m$^2$ | $\mathbf{ML^{-1}T^{-2}}$ |
| $Q$ | heat load | W | $\mathbf{ML^2T^{-3}}$ |

| | | Units in SI System | Dimensions in M, N, L, T, $\theta$ |
|---|---|---|---|
| $q$ | heat of crystallisation | J/kg | $\mathbf{L}^2\mathbf{T}^{-2}$ |
| $\mathbf{R}$ | Universal Gas Constant | 8314 J/kmol K | $\mathbf{MN}^{-1}\mathbf{L}^2\mathbf{T}^{-2}\boldsymbol{\theta}^{-1}$ |
| $R$ | molecular mass of hydrate/molecular mass of anhydrous salt | — | — |
| $R_G$ | mass deposition rate | kg/m$^2$s | $\mathbf{ML}^{-2}\mathbf{T}^{-1}$ |
| $Re$ | Reynolds' number $(ud'\rho_v/\mu)$ | — | — |
| $r$ | radius of particle or equivalent sphere | m | $\mathbf{L}$ |
| $r'$ | particle size in equilibrium with bulk solution | m | $\mathbf{L}$ |
| $r_c$ | critical size of nucleus | m | $\mathbf{L}$ |
| $S$ | supersaturation ratio | — | — |
| $S'$ | mass sublimation rate | kg/s | $\mathbf{MT}^{-1}$ |
| $Sc$ | Schmidt Number $(\mu/\rho_v D)$ | — | — |
| $s$ | order of overall crystal growth | — | — |
| $T$ | temperature | K | $\boldsymbol{\theta}$ |
| $\Delta T_m$ | logarithmic mean temperature difference | deg K | $\boldsymbol{\theta}$ |
| $T_M$ | melting point | K | $\boldsymbol{\theta}$ |
| $T_s$ | surface temperature | K | $\boldsymbol{\theta}$ |
| $t$ | time | S | $\mathbf{T}$ |
| $t_b$ | batch time | s | $\mathbf{T}$ |
| $t_i$ | induction period | s | $\mathbf{T}$ |
| $t_r$ | residence time | s | $\mathbf{T}$ |
| $U$ | overall coefficient of heat transfer | W/m$^2$K | $\mathbf{MT}^{-3}\theta^{-1}$ |
| $u$ | gas velocity | m/s | $\mathbf{LT}^{-1}$ |
| $u'$ | mean linear growth velocity | m/s | $\mathbf{LT}^{-1}$ |
| $V'$ | volume of suspension of crystals | m$^3$ | $\mathbf{L}^3$ |
| $V_L$ | volume of liquor in vessel | m$^3$ | $\mathbf{L}^3$ |
| $V_W$ | volume of wash water | m$^3$ | $\mathbf{L}^3$ |
| $v$ | molar volume | m$^3$/kmol | $\mathbf{N}^{-1}\mathbf{L}^3$ |
| $v'$ | maximum theoretical vaporisation rate | kg/m$^2$s | $\mathbf{ML}^{-2}\mathbf{T}^{-1}$ |
| $v_g$ | molar volume of gas | m$^3$/kmol | $\mathbf{N}^{-1}\mathbf{L}^3$ |
| $v_1$ | molar volume of liquid | m$^3$/kmol | $\mathbf{N}^{-1}\mathbf{L}^3$ |
| $v_s$ | molar volume of solid | m$^3$/kmol | $\mathbf{N}^{-1}\mathbf{L}^3$ |
| $w_1$ | initial mass of solvent in liquor | kg | $\mathbf{M}$ |
| $w_2$ | final mass of solvent in liquor | kg | $\mathbf{M}$ |
| $x$ | solute concentration in terms of mole fraction | — | — |
| $x'$ | effective film thickness | m | $\mathbf{L}$ |
| $x_c$ | amount of component **A** in crystallised solid | kg | $\mathbf{L}$ |
| $y$ | yield of crystals | kg | $\mathbf{M}$ |
| $y_c$ | amount of component **B** in crystallised solid | kg | $\mathbf{M}$ |
| $z$ | kmol of gas produced by 1 kmol of electrolyte | — | — |
| $z_1, z_2$ | functions in equation 15.6 | 1/K | $\boldsymbol{\theta}^{-1}$ |
| $\alpha$ | shape factor in equation 15.16 | — | — |
| $\beta$ | shape factor in equation 15.16 | — | — |
| $\phi$ | relative supersaturation | — | — |
| $\gamma$ | ion activity coefficient | — | — |
| $\gamma\pm$ | mean activity coefficient | — | — |
| $\varphi$ | degree of supersaturation/equilibrium saturation | — | — |
| $\lambda$ | latent heat of vaporisation of solvent | J/kg | $\mathbf{L}^2\mathbf{T}^{-2}$ |
| $\lambda_f$ | latent heat of fusion | J/kg | $\mathbf{L}^2\mathbf{T}^{-2}$ |
| $\lambda_s$ | latent heat of sublimation | J/kg | $\mathbf{L}^2\mathbf{T}^{-2}$ |
| $\lambda_v$ | latent heat of vaporisation per unit mass | J/kg | $\mathbf{L}^2\mathbf{T}^{-2}$ |

| | | Units in SI System | Dimensions in **M, N, L, T, $\theta$** |
|---|---|---|---|
| $\lambda'_v$ | latent heat of vaporisation per mole | J/kmol | $\mathbf{MN^{-1}L^2T^{-2}}$ |
| $\mu$ | fluid viscosity | Ns/m$^2$ | $\mathbf{ML^{-1}T^{-1}}$ |
| $\rho$ | density of crystal | kg/m$^3$ | $\mathbf{ML^{-3}}$ |
| $\rho_g$ | density of inert gas | kg/m$^3$ | $\mathbf{ML^{-3}}$ |
| $\rho_s$ | density of solid or sublimed material | kg/m$^3$ | $\mathbf{ML^{-3}}$ |
| $\rho_v$ | density of vapour | kg/m$^3$ | $\mathbf{ML^{-3}}$ |
| $\sigma$ | interfacial tension of crystallisation surface | J/m$^2$ | $\mathbf{MT^{-2}}$ |

CHAPTER 16

# *Drying*

## 16.1. INTRODUCTION

The drying of materials is often the final operation in a manufacturing process, carried out immediately prior to packaging or dispatch. Drying refers to the final removal of water, or another solute, and the operation often follows evaporation, filtration, or crystallisation. In some cases, drying is an essential part of the manufacturing process, as for instance in paper making or in the seasoning of timber, although, in the majority of processing industries, drying is carried out for one or more of the following reasons:

(a) To reduce the cost of transport.
(b) To make a material more suitable for handling as, for example, with soap powders, dyestuffs and fertilisers.
(c) To provide definite properties, such as, for example, maintaining the free-flowing nature of salt.
(d) To remove moisture which may otherwise lead to corrosion. One example is the drying of gaseous fuels or benzene prior to chlorination.

With a crystalline product, it is essential that the crystals are not damaged during the drying process, and, in the case of pharmaceutical products, care must be taken to avoid contamination. Shrinkage, as with paper, cracking, as with wood, or loss of flavour, as with fruit, must also be prevented. With the exception of the partial drying of a material by squeezing in a press or the removal of water by adsorption, almost all drying processes involve the removal of water by vaporisation, which requires the addition of heat. In assessing the efficiency of a drying process, the effective utilisation of the heat supplied is the major consideration.

## 16.2. GENERAL PRINCIPLES

The moisture content of a material is usually expressed in terms of its water content as a percentage of the mass of the dry material, though moisture content is sometimes expressed on a wet basis, as in Example 16.3. If a material is exposed to air at a given temperature and humidity, the material will either lose or gain water until an equilibrium condition is established. This equilibrium moisture content varies widely with the moisture content and the temperature of the air, as shown in Figure 16.1. A non-porous insoluble solid, such as sand or china clay, has an equilibrium moisture content approaching zero

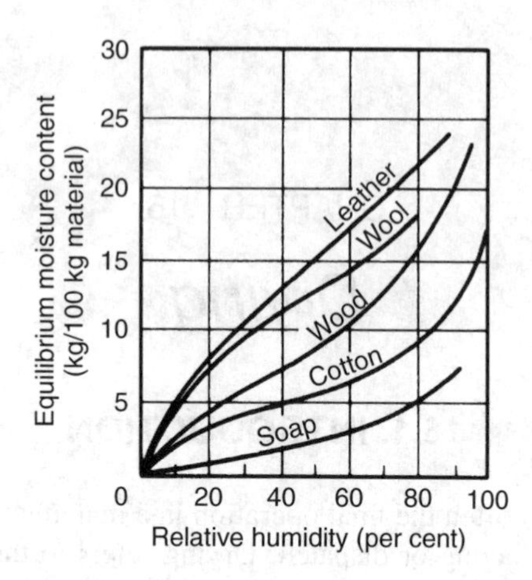


Figure 16.1. Equilibrium moisture content of a solid as a function of relative humidity at 293 K

for all humidities and temperatures, although many organic materials, such as wood, textiles, and leather, show wide variations of equilibrium moisture content. Moisture may be present in two forms:

*Bound moisture*. This is water retained so that it exerts a vapour pressure less than that of free water at the same temperature. Such water may be retained in small capillaries, adsorbed on surfaces, or as a solution in cell walls.

*Free moisture*. This is water which is in excess of the equilibrium moisture content.

The water removed by vaporisation is generally carried away by air or hot gases, and the ability of these gases to pick up the water is determined by their temperature and humidity. In designing dryers using air, the properties of the air–water system are essential, and these are detailed in Volume 1, Chapter 13, where the development of the humidity chart is described. For the *air–water system*, the following definitions are of importance:

*Humidity* $\mathscr{H}$, mass of water per unit mass of dry air.

Since:
$$\frac{\text{moles of water vapour}}{\text{moles of dry air}} = \frac{P_w}{(P - P_w)}$$

then:
$$\mathscr{H} = \frac{18 P_w}{29(P - P_w)}$$

where $P_w$ is the partial pressure of water vapour and $P$ is the total pressure.

*Humidity of saturated air* $\mathscr{H}_0$. This is the humidity of air when it is saturated with water vapour. The air then is in equilibrium with water at the given temperature and pressure.

*Percentage humidity*

$$= \frac{\text{Humidity of air}}{\text{Humidity of saturated air}} \times 100 = \frac{\mathscr{H}}{\mathscr{H}_0} \times 100$$

*Percentage relative humidity*, $\mathscr{R}$

$$= \frac{\text{Partial pressure of water vapour in air}}{\text{Vapour pressure of water at the same temperature}} \times 100$$

The distinction between *percentage humidity* and *percentage relative humidity* is of significance though, the difference in the values of the two quantities does not usually exceed 7 to 8 per cent. Reference may be made here to Volume 1, Section 13.2.1.

*Humid volume*, is the volume of unit mass of dry air and its associated vapour. Then, under ideal conditions, at atmospheric pressure:

$$\text{humid volume} = \frac{22.4}{29}\left(\frac{T}{273}\right) + \frac{22.4\mathscr{H}}{18}\left(\frac{T}{273}\right) \text{ m}^3/\text{kg}$$

where $T$ is in degrees K,

$$\text{or :} \quad \frac{359}{29}\left(\frac{T}{492}\right) + \frac{359\mathscr{H}}{18}\left(\frac{T}{492}\right) \text{ ft}^3/\text{lb}$$

where $T$ is in degrees Rankine.

*Saturated volume* is the volume of unit mass of dry air, together with the water vapour required to saturate it.

*Humid heat* is the heat required to raise unit mass of dry air and associated vapour through 1 degree K at constant pressure or $1.00 + 1.88\mathscr{H}$ kJ/kg K.

*Dew point* is the temperature at which condensation will first occur when air is cooled.

*Wet bulb temperature*. If a stream of air is passed rapidly over a water surface, vaporisation occurs, provided the temperature of the water is above the dew point of the air. The temperature of the water falls and heat flows from the air to the water. If the surface is sufficiently small for the condition of the air to change inappreciably and if the velocity is in excess of about 5 m/s, the water reaches the wet bulb temperature $\theta_w$ at equilibrium.

The rate of heat transfer from gas to liquid is given by:

$$Q = hA(\theta - \theta_w) \tag{16.1}$$

The mass rate of vaporisation is given by:

$$\begin{aligned} G_v &= \frac{h_D A M_w}{\mathbf{R}T}(P_{w0} - P_w) \\ &= \frac{h_D A M_A}{\mathbf{R}T}[(P - P_w)_{\text{mean}}(\mathscr{H}_w - \mathscr{H})] \\ &= h_D A \rho_A(\mathscr{H}_w - \mathscr{H}) \end{aligned} \tag{16.2}$$

The rate of heat transfer required to effect vaporisation at this rate is given by:

$$G_v = h_D A \rho_A(\mathscr{H}_w - \mathscr{H})\lambda \tag{16.3}$$

At equilibrium, the rates of heat transfer given by equations 16.1 and 16.3 must be equal, and hence:

$$\mathscr{H} - \mathscr{H}_w = -\frac{h}{h_D \rho_A \lambda}(\theta - \theta_w) \tag{16.4}$$

In this way, it is seen that the wet bulb temperature $\theta_w$ depends only on the temperature and humidity of the drying air.

In these equations:

- $h$ is the heat transfer coefficient,
- $h_D$ is the mass transfer coefficient,
- $A$ is the surface area,
- $\theta$ is the temperature of the air stream,
- $\theta_w$ is the wet bulb temperature,
- $P_{w0}$ is the vapour pressure of water at temperature $\theta_w$,
- $M_A$ is the molecular weight of air,
- $M_w$ is the molecular weight of water,
- $\mathbf{R}$ is the universal gas constant,
- $T$ is the absolute temperature,
- $\mathscr{H}$ is the humidity of the gas stream,
- $\mathscr{H}_w$ is the humidity of saturated air at temperature $\theta_w$,
- $\rho_A$ is the density of air at its mean partial pressure, and
- $\lambda$ is the latent heat of vaporisation of unit mass of water.

Equation 16.4 is identical with equation 13.8 in Volume 1, and reference may be made to that chapter for a more detailed discussion.

## 16.3. RATE OF DRYING

### 16.3.1. Drying periods

In drying, it is necessary to remove free moisture from the surface and also moisture from the interior of the material. If the change in moisture content for a material is determined as a function of time, a smooth curve is obtained from which the rate of drying at any given moisture content may be evaluated. The form of the drying rate curve varies with the structure and type of material, and two typical curves are shown in Figure 16.2. In curve 1, there are two well-defined zones: AB, where the rate of drying is constant and BC, where there is a steady fall in the rate of drying as the moisture content is reduced. The moisture content at the end of the constant rate period is represented by point B, and this is known as the *critical moisture content*. Curve 2 shows three stages, DE, EF and FC. The stage DE represents a constant rate period, and EF and FC are falling rate periods. In this case, the Section EF is a straight line, however, and only the portion FC is curved. Section EF is known as the first falling rate period and the final stage, shown as FC, as the second falling rate period. The drying of soap gives rise to a curve of type 1, and sand to a curve of type 2. A number of workers, including SHERWOOD[(1)] and NEWITT and co-workers[(2–7)], have contributed various theories on the rate of drying at these various stages.

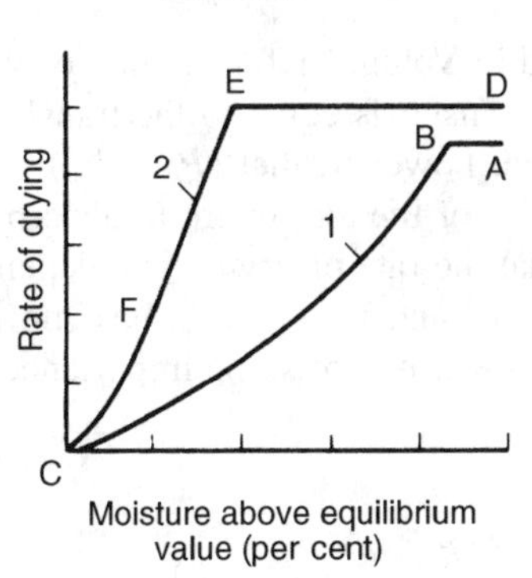


Figure 16.2. Rate of drying of a granular material

### *Constant rate period*

During the constant rate period, it is assumed that drying takes place from a saturated surface of the material by diffusion of the water vapour through a stationary air film into the air stream. GILLILAND[8] has shown that the rates of drying of a variety of materials in this stage are substantially the same as shown in Table 16.1.

Table 16.1. Evaporation rates for various materials under constant conditions[8]

| Material | Rate of evaporation $(kg/m^2\ h)$ | $(kg/m^2\ s)$ |
|---|---|---|
| Water | 2.7 | 0.00075 |
| Whiting pigment | 2.1 | 0.00058 |
| Brass filings | 2.4 | 0.00067 |
| Brass turnings | 2.4 | 0.00067 |
| Sand (fine) | 2.0–2.4 | 0.00055–0.00067 |
| Clays | 2.3–2.7 | 0.00064–0.00075 |

In order to calculate the rate of drying under these conditions, the relationships obtained in Volume 1 for diffusion of a vapour from a liquid surface into a gas may be used. The simplest equation of this type is:

$$W = k_G A(P_s - P_w) \tag{16.5}$$

where $k_G$ is the mass transfer coefficient.

Since the rate of transfer depends on the velocity $u$ of the air stream, raised to a power of about 0.8, then the mass rate of evaporation is:

$$W = k_G A(P_s - P_w)u^{0.8} \tag{16.6}$$

where: $A$ is the surface area,
$P_s$ is the vapour pressure of the water, and
$P_w$ is the partial pressure of water vapour in the air stream.

This type of equation, used in Volume 1 for the rate of vaporisation into an air stream, simply states that the rate of transfer is equal to the transfer coefficient multiplied by the driving force. It may be noted, however, that $(P_s - P_w)$ is not only a driving force, but it is also related to the capacity of the air stream to absorb moisture.

These equations suggest that the rate of drying is independent of the geometrical shape of the surface. Work by POWELL and GRIFFITHS[9] has shown, however, that the ratio of the length to the width of the surface is of some importance, and that the evaporation rate is given more accurately as:

(a) For values of $u = 1$–$3$ m/s:

$$W = 5.53 \times 10^{-9} L^{0.77} B (P_s - P_w)(1 + 61u^{0.85}) \text{ kg/s} \tag{16.7}$$

(b) For values of $u < 1$ m/s:

$$W = 3.72 \times 10^{-9} L^{0.73} B^{0.8} (P_s - P_w)(1 + 61u^{0.85}) \text{ kg/s} \tag{16.8}$$

where: $P_s$, the saturation pressure at the temperature of the surface (N/m$^2$),
$P_w$, the vapour pressure in the air stream (N/m$^2$), and
$L$ and $B$ are the length and width of the surface, respectively (m).

For most design purposes, it may be assumed that the rate of drying is proportional to the transfer coefficient multiplied by $(P_s - P_w)$. CHAKRAVORTY[10] has shown that, if the temperature of the surface is greater than that of the air stream, then $P_w$ may easily reach a value corresponding to saturation of the air. Under these conditions, the capacity of the air to take up moisture is zero, while the force causing evaporation is $(P_s - P_w)$. As a result, a mist will form and water may be redeposited on the surface. In all drying equipment, care must therefore be taken to ensure that the air or gas used does not become saturated with moisture at any stage.

The rate of drying in the constant rate period is given by:

$$W = \frac{dw}{dt} = \frac{hA\Delta T}{\lambda} = k_G A (P_s - P_w) \tag{16.9}$$

where: $W$ is the rate of loss of water,
$h$ is the heat transfer coefficient from air to the wet surface,
$\Delta T$ is the temperature difference between the air and the surface,
$\lambda$ is the latent heat of vaporisation per unit mass,
$k_G$ is the mass transfer coefficient for diffusion from the wet surface through the gas film,
$A$ is the area of interface for heat and mass transfer, and
$(P_s - P_w)$ is the difference between the vapour pressure of water at the surface and the partial pressure in the air.

It is more convenient to express the mass transfer coefficient in terms of a humidity difference, so that $k_G A(P_s - P_w) \simeq kA(\mathscr{H}_s - \mathscr{H})$. The rate of drying is thus determined by the values of $h$, $\Delta T$ and $A$, and is not influenced by the conditions inside the solid. $h$ depends on the air velocity and the direction of flow of the air, and it has been found that $h = CG'^{0.8}$ where $G'$ is the mass rate of flow of air in kg/s m$^2$. For air flowing parallel to plane surfaces, SHEPHERD *et al.*(11) have given the value of $C$ as 14.5 where the heat transfer coefficient is expressed in W/m$^2$ K.

If the gas temperature is high, then a considerable proportion of the heat will pass to the solid by radiation, and the heat transfer coefficient will increase. This may result in the temperature of the solid rising above the wet bulb temperature.

### *First falling-rate period*

The points B and E in Figure 16.2 represent conditions where the surface is no longer capable of supplying sufficient free moisture to saturate the air in contact with it. Under these conditions, the rate of drying depends very much on the mechanism by which the moisture from inside the material is transferred to the surface. In general, the curves in Figure 16.2 will apply, although for a type 1 solid, a simplified expression for the rate of drying in this period may be obtained.

### *Second falling-rate period*

At the conclusion of the first falling rate period it may be assumed that the surface is dry and that the plane of separation has moved into the solid. In this case, evaporation takes place from within the solid and the vapour reaches the surface by molecular diffusion through the material. The forces controlling the vapour diffusion determine the final rate of drying, and these are largely independent of the conditions outside the material.

## 16.3.2. Time for drying

If a material is dried by passing hot air over a surface which is initially wet, the rate of drying curve in its simplest form is represented by BCE, shown in Figure 16.3

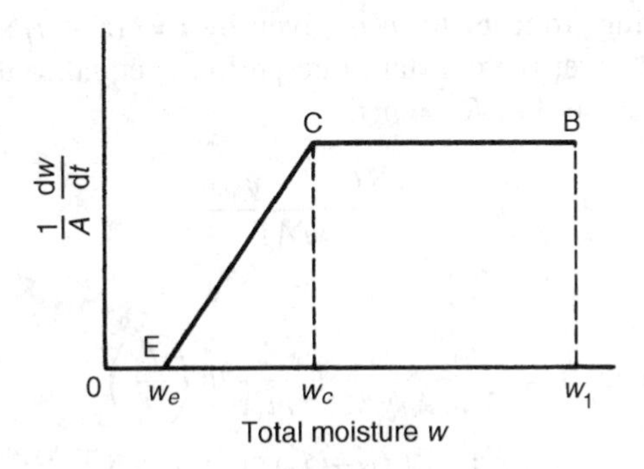


Figure 16.3. The use of a rate of drying curve in estimating the time for drying

where: $w$ is the total moisture,
$w_e$ is the equilibrium moisture content (point E),
$w - w_e$ is the free moisture content, and
$w_c$ is the critical moisture content (point C).

### Constant-rate period

During the period of drying from the initial moisture content $w_1$ to the critical moisture content $w_c$, the rate of drying is constant, and the time of drying $t_c$ is given by:

$$t_c = \frac{w_1 - w_c}{R_c A} \quad (16.10)$$

where: $R_c$ is the rate of drying per unit area in the constant rate period, and
$A$ is the area of exposed surface.

### Falling-rate period

During this period the rate of drying is, approximately, directly proportional to the free moisture content $(w - w_e)$, or:

$$-\left(\frac{1}{A}\right)\frac{\mathrm{d}w}{\mathrm{d}t} = m(w - w_e) = mf \quad \text{(say)} \quad (16.11)$$

Thus:

$$\frac{1}{mA}\int_{w_e}^{w}\frac{\mathrm{d}w}{(w - w_e)} = \int_0^{t_f}\mathrm{d}t$$

or:

$$\frac{1}{mA}\ln\left[\frac{w_c - w_e}{w - w_e}\right] = t_f$$

and:

$$t_f = \frac{1}{mA}\ln\left(\frac{f_c}{f}\right) \quad (16.12)$$

### Total time of drying

The total time $t$ of drying from $w_1$ to $w$ is given by $t = (t_c + t_f)$.

The rate of drying $R_c$ over the constant rate period is equal to the initial rate of drying in the falling rate period, so that $R_c = mf_c$.

Thus:

$$t_c = \frac{(w_1 - w_c)}{mAf_c} \quad (16.13)$$

and the total drying time,

$$t = \frac{(w_1 - w_c)}{mAf_c} + \frac{1}{mA}\ln\left(\frac{f_c}{f}\right)$$

$$= \frac{1}{mA}\left[\frac{(f_1 - f_c)}{f_c} + \ln\left(\frac{f_c}{f}\right)\right] \quad (16.14)$$

## Example 16.1

A wet solid is dried from 25 to 10 per cent moisture under constant drying conditions in 15 ks (4.17 h). If the critical and the equilibrium moisture contents are 15 and 5 per cent respectively, how long will it take to dry the solid from 30 to 8 per cent moisture under the same conditions?

## Solution

*For the first drying operation*:

$$w_1 = 0.25 \text{ kg/kg},\ w = 0.10 \text{ kg/kg},\ w_c = 0.15 \text{ kg/kg and } w_e = 0.05 \text{ kg/kg}$$

Thus:
$$f_1 = (w_1 - w_e) = (0.25 - 0.05) = 0.20 \text{ kg/kg}$$
$$f_c = (w_c - w_e) = (0.15 - 0.05) = 0.10 \text{ kg/kg}$$
$$f = (w - w_e) = (0.10 - 0.05) = 0.05 \text{ kg/kg}$$

From equation 16.14, the total drying time is:

$$t = (1/mA)[(f_1 - f_c)/f_c + \ln(f_c/f)]$$

or:
$$15 = (1/mA)[(0.20 - 0.10)/0.10 + \ln(0.10/0.05)]$$

and:
$$mA = 0.0667(1.0 + 0.693) = 0.113 \text{ kg/s}$$

*For the second drying operation*:

$$w_1 = 0.30 \text{ kg/kg},\ w = 0.08 \text{ kg/kg},\ w_c = 0.15 \text{ kg/kg and } w_e = 0.05 \text{ kg/kg}$$

Thus:
$$f_1 = (w_1 - w_e) = (0.30 - 0.05) = 0.25 \text{ kg/kg}$$
$$f_c = (w_c - w_e) = (0.15 - 0.05) = 0.10 \text{ kg/kg}$$
$$f = (w - w_e) = (0.08 - 0.05) = 0.03 \text{ kg/kg}$$

*The total drying time is then*:

$$t = (1/0.113)[(0.25 - 0.10)/0.10 + \ln(0.10/0.03)]$$
$$= 8.856(1.5 + 1.204)$$
$$= \underline{\underline{23.9 \text{ ks } (6.65 \text{ h})}}$$

## Example 16.2

Strips of material 10 mm thick are dried under constant drying conditions from 28 to 13 per cent moisture in 25 ks (7 h). If the equilibrium moisture content is 7 per cent, what is the time taken to dry 60 mm planks from 22 to 10 per cent moisture under the same conditions assuming no loss from the edges? All moistures are given on a wet basis.

The relation between $E$, the ratio of the average free moisture content at time $t$ to the initial free moisture content, and the parameter $J$ is given by:

| $E$ | 1 | 0.64 | 0.49 | 0.38 | 0.295 | 0.22 | 0.14 |
|---|---|---|---|---|---|---|---|
| $J$ | 0 | 0.1 | 0.2 | 0.3 | 0.5 | 0.6 | 0.7 |

It may be noted that $J = kt/l^2$, where $k$ is a constant, $t$ the time in ks and $2l$ the thickness of the sheet of material in millimetres.

## Solution

*For the 10 mm strips*

Initial free moisture content = (0.28 − 0.07) = 0.21 kg/kg.

Final free moisture content = (0.13 − 0.07) = 0.06 kg/kg.

Thus: when $t = 25$ ks, $E = (0.06/0.21) = 0.286$

and from Figure 16.4, a plot of the given data,

$$J = 0.52$$

Thus: $$0.52 = (k \times 25)/(10/2)^2$$

and: $$k = 0.52$$

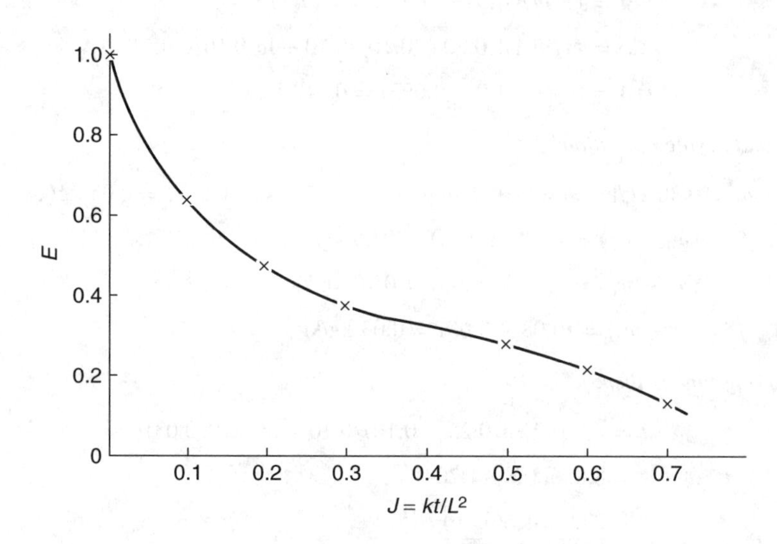


Figure 16.4. Drying data for Example 6.2

*For the 60 mm planks*

Initial free moisture content = (0.22 − 0.07) = 0.15 kg/kg.

Final free moisture content = (0.10 − 0.07) = 0.03 kg/kg.

$$E = (0.03/0.15) = 0.20$$

From Figure 16.4: $J = 0.63$

and hence: $t = Jl^2/k$

$$= 0.63(60/2)^2/0.52 = \underline{\underline{1090 \text{ ks}}} \text{ (12.6 days)}$$

## Example 16.3

A granular material containing 40 per cent moisture is fed to a countercurrent rotary dryer at a temperature of 295 K and is withdrawn at 305 K, containing 5 per cent moisture. The air supplied, which contains 0.006 kg water vapour/kg dry air, enters at 385 K and leaves at 310 K. The dryer handles 0.125 kg/s wet stock.

Assuming that radiation losses amount to 20 kJ/kg dry air used, determine the mass flowrate of dry air supplied to the dryer and the humidity of the exit air.

The latent heat of water vapour at 295 K = 2449 kJ/kg, specific heat capacity of dried material = 0.88 kJ/kg K, the specific heat capacity of dry air = 1.00 kJ/kg K, and the specific heat capacity of water vapour = 2.01 kJ/kg K.

## Solution

This example involves a heat balance over the system. 273 K will be chosen as the datum temperature, and it will be assumed that the flowrate of dry air = $G$ kg/s.

*Heat in:*

(a) *Air*

$G$ kg/s dry air enter with $0.006G$ kg/s water vapour and hence the heat content of this stream

$$= [(1.00G) + (0.006G \times 2.01)](385 - 273) = 113.35G \text{ kW}$$

(b) *Wet solid*

0.125 kg/s enter containing 0.40 kg water/kg wet solid, assuming the moisture is expressed on a wet basis.

Thus: mass flowrate of water $= (0.125 \times 0.40) = 0.050$ kg/s

and: mass flowrate of dry solid $= (0.125 - 0.050) = 0.075$ kg/s

Hence:

the heat content of this stream $= [(0.050 \times 4.18) + (0.075 \times 0.88)](295 - 273) = 6.05$ kW

*Heat out*:

(a) *Air*

Heat in exit air $= [(1.00\ G) + (0.006\ G \times 2.01)](310 - 273) = 37.45G$ kW.

Mass flowrate of dry solids = 0.075 kg/s containing 0.05 kg water/kg wet solids.

Hence:

water in the dried solids leaving $= (0.05 \times 0.075)/(1 + 0.05) = 0.0036$ kg/s

and:

the water evaporated into gas steam $= (0.050 - 0.0036) = 0.0464$ kg/s.

Assuming evaporation takes place at 295 K, then:

$$\text{heat in the water vapour} = 0.0464[2.01(310 - 295) + 2449 + 4.18(295 - 273)]$$

$$= 119.3 \text{ kW}$$

and:

$$\text{the total heat in this stream} = (119.30 + 37.45G) \text{ kW.}$$

(b) *Dried solids*

The dried solids contain 0.0036 kg/s water and hence heat content of this stream is:

$$= [(0.075 \times 0.88) + (0.0036 \times 4.18)/(305 - 273)] = 2.59 \text{ kW}$$

(c) *Losses*

These amount to 20 kJ/kg dry air or $20m$ kW.

*Heat balance*

$$(113.35\ G + 6.05) = (119.30 + 37.45\ G + 2.59 + 20\ G)$$

and:

$$G = \underline{\underline{2.07 \text{ kg/s}}}$$

$$\text{Water in the outlet air stream} = (0.006 \times 2.07) + 0.0464 = 0.0588 \text{ kg/s}$$

and:

$$\text{the humidity } \mathscr{H} = (0.0588/2.07) = \underline{\underline{0.0284 \text{ kg/kg dry air}}}$$

## 16.4. THE MECHANISM OF MOISTURE MOVEMENT DURING DRYING

### 16.4.1. Diffusion theory of drying

In the general form of the curve for the rate of drying of a solid shown in Figure 16.2, there are two and sometimes three distinct sections. During the constant-rate period, moisture vaporises into the air stream and the controlling factor is the transfer coefficient for diffusion across the gas film. It is important to understand how the moisture moves to the drying surface during the falling-rate period, and two models have been used to describe the physical nature of this process, the diffusion theory and the capillary theory. In the diffusion theory, the rate of movement of water to the air interface is governed by rate equations similar to those for heat transfer, whilst in the capillary theory the forces controlling the movement of water are capillary in origin, arising from the minute pore spaces between the individual particles.

#### *Falling rate period, diffusion control*

In the falling-rate period, the surface is no longer completely wetted and the rate of drying steadily falls. In the previous analysis, it has been assumed that the rate of drying per unit effective wetted area is a linear function of the water content, so that the rate of drying is given by:

$$\left(\frac{1}{A}\right)\frac{\mathrm{d}w}{\mathrm{d}t} = -m(w - w_e) \qquad \text{(equation 16.11)}$$

In many cases, however, the rate of drying is governed by the rate of internal movement of the moisture to the surface. It was initially assumed that this movement was a process of diffusion and would follow the same laws as heat transfer. This approach has been examined by a number of workers, and in particular by SHERWOOD[12] and NEWMAN[13].

Considering a slab with the edges coated to prevent evaporation, which is dried by evaporation from two opposite faces, the $Y$-direction being taken perpendicular to the drying face, the central plane being taken as $y = 0$, and the slab thickness $2l$, then on drying, the moisture movement by diffusion will be in the $Y$-direction, and hence from Volume 1, equation 10.66:

$$\frac{\partial C_w}{\partial t} = D_L \frac{\partial^2 C_w}{\partial y^2}$$

where $C_w$ is the concentration of water at any point and any time in the solid, and $D_L$ is the coefficient of diffusion for the liquid.

If $w$ is the liquid content of the solid, integrated over the whole depth, $w_1$ the initial content, and $w_e$ the equilibrium content, then:

$$\frac{(w - w_e)}{(w_1 - w_e)} = \frac{\text{Free liquid content at any time}}{\text{Initial free liquid content}}$$

SHERWOOD[12] and NEWMAN[13] have presented the following solution assuming an initially uniform water distribution and zero water-concentration at the surface once drying has started:

$$\frac{(w - w_e)}{(w_1 - w_e)} = \frac{8}{\pi^2}\left\{e^{-D_L t(\pi/2l)^2} + \frac{1}{9}e^{-9D_L t(\pi/2l)^2} + \frac{1}{25}e^{-25D_L t(\pi/2l)^2} + \cdots\right\} \tag{16.15}$$

This equation assumes an initially uniform distribution of moisture, and that the drying is from both surfaces. When drying is from one surface only, then $l$ is the total thickness. If the time of drying is long, then only the first term of the equation need be used and thus, differentiating equation 16.15 gives:

$$\frac{dw}{dt} = -\frac{2D_L}{l^2} e^{-\frac{D_L t\pi^2}{4l^2}}(w_1 - w_e) \tag{16.16}$$

In the drying of materials such as wood or clay, the moisture concentration at the end of the constant rate period is not uniform, and is more nearly parabolic. Sherwood has presented an analysis for this case, and has given experimental values for the drying of brick clay.

In this case, it is assumed that the rate of movement of water is proportional to a concentration gradient, and capillary and gravitational forces are neglected. Water may, however, flow from regions of low concentration to those of high concentration if the pore sizes are suitable, and for this and other reasons, CEAGLSKE and HOUGEN[14,15] have proposed a capillary theory which is now considered.

## 16.4.2. Capillary theory of drying

### *Principles of the theory*

The capillary theory of drying has been proposed in order to explain the movement of moisture in the bed during surface drying. The basic importance of the pore space between granular particles was first pointed out by SLICHTER[16] in connection with the movement of moisture in soils, and this work has been modified and considerably expanded by HAINES[17]. The principles are now outlined and applied to the problem of drying. Considering a systematic packing of uniform spherical particles, these may be arranged in six different regular ways, ranging from the most open to the closest packing. In the former, the spheres are arranged as if at the corners of a cube with each sphere touching six others. In the latter arrangement, each sphere rests in the hollow of three spheres in adjacent layers, and touches twelve other spheres. These configurations are shown in Figure 16.5. The densities of packing of the other four arrangements will lie between those illustrated.

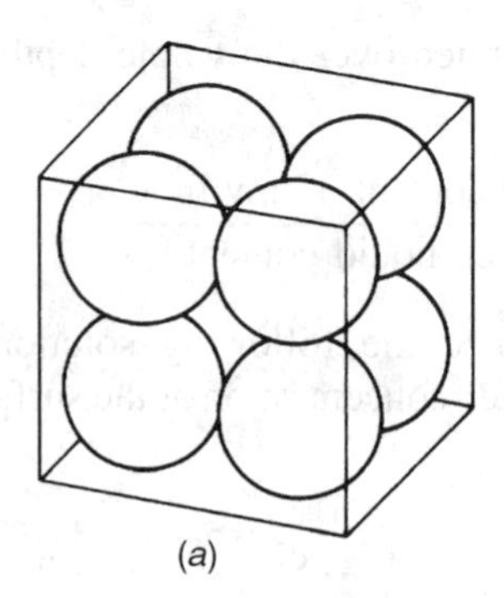


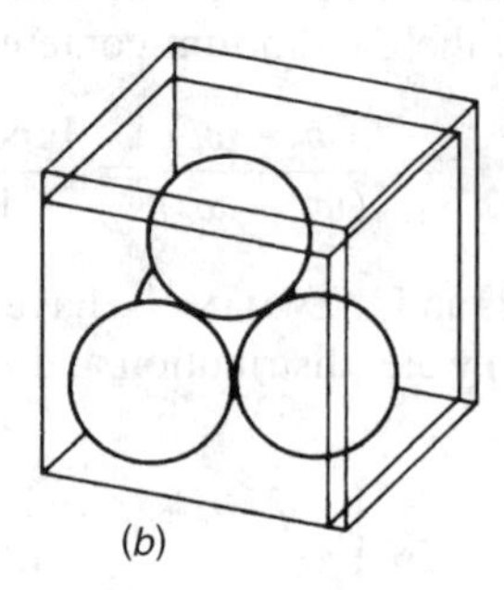


Figure 16.5. Packing of spherical particles. (*a*) Cubic arrangement, one sphere touching six others. (*b*) Rhombohedral packing, one sphere touching twelve others, with layers arranged in rhombic formation

In each case, a regular group of spheres surrounds a space which is called a pore, and the bed is made up of a series of these elemental groupings. The pores are connected together by passages of various sizes, the smallest portions of which are known as *waists*. The size of a pore is defined as the diameter of the largest sphere which can be fitted into it, and the size of a waist as the diameter of the inscribed circle. The sizes of the pores and waists will differ for each form of packing, as shown in Table 16.2.

Table 16.2. Properties of packing of spheres of radius $r$

| Packing arrangement | Pore space (per cent total volume) | Radius of pore | Radius of waist | Value of $x$ in equation 16.20 for: limiting suction potential of pores | entry suction potential of waists |
|---|---|---|---|---|---|
| Cubical | 47.64 | $0.700r$ | $0.414r$ | 2.86 | 4.82 |
| Rhombohedral | 25.95 | $0.288r$ | $0.155r$ | 6.90 | 12.90 |

The continuous variation in the diameter of each passage is the essential difference between a granular packing and a series of capillary tubes. If a clean capillary of uniform diameter $2r'$ is placed in a liquid, the level will rise in the capillary to a height $h_s$ given by:

$$h_s = \left(\frac{2\sigma}{r'\rho g}\right)\cos\alpha \tag{16.17}$$

where: $\rho$ is the density of the liquid,
$\sigma$ is the surface tension, and
$\alpha$ is the angle of contact.

A negative pressure, known as a *suction potential*, will exist in the liquid in the capillary. Immediately below the meniscus, the suction potential will be equivalent to the height of the liquid column $h_s$ and, if water is used, this will have the value:

$$h_s = \frac{2\sigma}{r'\rho g} \tag{16.18}$$

If equilibrium conditions exist, the suction potential $h_1$ at any other level in the liquid, a distance $z_1$ below the meniscus, will be given by:

$$h_s = h_1 + z_1 \tag{16.19}$$

Similarly, if a uniform capillary is filled to a height greater than $h_s$, as given by equation 16.17, and its lower end is immersed, the liquid column will recede to this height.

The non-uniform passages in a porous material will also display the same characteristics as a uniform capillary, with the important difference that the rise of water in the passages will be limited by the pore size, whilst the depletion of saturated passages will be controlled by the size of the waists. The height of rise is controlled by the pore size, since this approximates to the largest section of a varying capillary, whilst the depletion of water is controlled by the narrow waists which are capable of a higher suction potential than the pores.

The theoretical suction potential of a pore or waist containing water is given by:

$$h_t = \frac{x\sigma}{r\rho g} \tag{16.20}$$

where: $x$ is a factor depending on the type of packing, shown in Table 16.2, and
$r$ is the radius of the spheres.

For an idealised bed of uniform rhombohedrally packed spheres of radius $r$, for example, the waists are of radius $0.155r$, from Table 16.2, and the maximum theoretical suction potential of which such a waist is capable is:

$$\frac{2\sigma}{0.155r\rho g} = \frac{12.9\sigma}{r\rho g}$$

from which $x = 12.9$.

The maximum suction potential that can be developed by a waist is known as the *entry suction potential*. This is the controlling potential required to open a saturated pore protected by a meniscus in an adjoining waist and some values for $x$ are given in Table 16.2.

When a bed is composed of granular material with particles of mixed sizes, the suction potential cannot be calculated and it must be measured by methods such as those given by HAINES[17] and OLIVER and NEWITT[3].

### *Drying of a granular material according to the capillary theory*

If a bed of uniform spheres, initially saturated, is to be surface dried in a current of air of constant temperature, velocity and humidity, then the rate of drying is given by:

$$\frac{dw}{dt} = k_G A(P_{w0} - P_w) \tag{16.21}$$

where $P_{w0}$ is the saturation partial pressure of water vapour at the wet bulb temperature of the air, and $P_w$ is the partial pressure of the water vapour in the air stream. This rate of drying will remain constant so long as the inner surface of the "stationary" air film remains saturated.

As evaporation proceeds, the water surface recedes into the waists between the top layer of particles, and an increasing suction potential is developed in the liquid. When the menisci on the cubical waists, that is the largest, have receded to the narrowest section, the suction potential $h_s$ at the surface is equal to $4.82\sigma/r\rho g$, from Table 16.2. Further evaporation will result in $h_s$ increasing so that the menisci on the surface cubical waists will collapse, and the larger pores below will open. As $h_s$ steadily increases, the entry suction of progressively finer surface waists is reached, so that the menisci collapse into the adjacent pores which are thereby opened.

In considering the conditions below the surface, the suction potential $h_1$ a distance $z_1$ from the surface is given by:

$$h_s = h_1 + z_1 \qquad \text{(equation 16.19)}$$

The flow of water through waists surrounding an open pore is governed by the size of the waist as follows:

(a) If the size of the waist is such that its entry suction potential exceeds the suction potential at that level within the bed, it will remain full by the establishment of a meniscus therein, in equilibrium with the effective suction potential to which it is subjected. This waist will then protect adjoining full pores which cannot be opened until one of the waists to which it is connected collapses.

(b) If the size of the waist is such that its entry suction potential is less than the suction potential existing at that level, it will in turn collapse and open the adjoining pore. In addition, this successive collapse of pores and waists will progressively continue so long as the pores so opened expose waists having entry suction potentials of less than the suction potentials existing at that depth.

As drying proceeds, two processes take place simultaneously:

(a) The collapse of progressively finer surface waists, and the resulting opening of pores and waists connected to them, which they previously protected, and

(b) The collapse of further full waists within the bed adjoining opened pores, and the consequent opening of adjacent pores.

Even though the effective suction potential at a waist or pore within the bed may be in excess of its entry or limiting suction potential, this will not necessarily collapse or open. Such a waist can only collapse if it adjoins an opened pore, and the pore in question can only open upon the collapse of an adjoining waist.

*Effect of particle size.* Reducing the particle size in the bed will reduce the size of the pores and the waists, and will increase the entry suction potential of the waists. This increase means that the percentage variation in suction potentials with depth is reduced, and the moisture distribution is more uniform with small particles.

As the pore sizes are reduced, the frictional forces opposing the movement of water through these pores and waists may become significant, so that equation 16.19 is more accurately represented by:

$$h_s = h_1 + z_1 + h_f \qquad (16.22)$$

where $h_f$, the frictional head opposing the flow over a depth $z_1$ from the surface, will depend on the particle size. It has been found[2] that, with coarse particles when only low suction potentials are found, the gravity effect is important though $h_f$ is small, whilst with fine particles $h_f$ becomes large.

(a) For particles of 0.1–1 mm radius, the values of $h_1$ are independent of the rate of drying, and vary appreciably with depth. Frictional forces are, therefore, negligible whilst capillary and gravitational forces are in equilibrium throughout the bed and are the controlling forces. Under such circumstances the percentage moisture loss at the critical point at which the constant rate period ends is independent of the drying rate, and varies with the depth of bed.

(b) For particles of 0.001–0.01 mm radius, the values of $h_1$ vary only slightly with rate of drying and depth, indicating that both gravitational and frictional forces are negligible whilst capillary forces are controlling. The critical point here will be independent of drying rate and depth of bed.

(c) For particles of less than 0.001 mm (1 μm) radius, gravitational forces are negligible, whilst frictional forces are of increasing importance and capillary and frictional forces may then be controlling. In such circumstances, the percentage moisture loss at the critical point diminishes with increased rate of drying and depth of bed. With beds of very fine particles an additional factor comes into play. The very high suction potentials which are developed cause a sufficient reduction of the pressure for vaporisation of water to take place inside the bed. This internal vaporisation results in a breaking up of the continuous liquid phase and a consequent interruption in the free flow of liquid by capillary action. Hence, the rate of drying is still further reduced.

Some of the experimental data of NEWITT *et al.*[2] are illustrated in Figure 16.6.

## *Freeze drying*

Special considerations apply to the movement of moisture in freeze drying. Since the water is frozen, liquid flow under capillary action is impossible, and movement must be

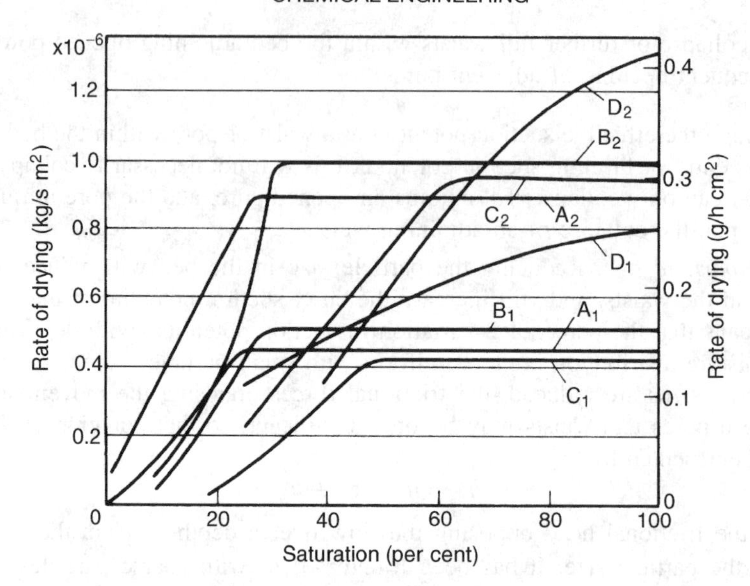


Figure 16.6. Rates of drying of various materials as a function of percentage saturation. A–60 μm glass spheres, bed 51 mm deep. B–23.5 μm silica flour, bed 51 mm deep. C–7.5 μm silica flour, bed 51 mm deep. D–2.5 μm silica flour, bed 65 mm deep. Subscripts: 1. Low drying rate 2. High drying rate

by vapour diffusion, analogous to the "second falling rate period" of the normal case. In addition, at very low pressures the mean free path of the water molecules may be comparable with the pore size of the material. In these circumstances the flow is said to be of the 'Knudsen' type, referred to in Volume 1, Section 10.1.

## 16.5. DRYING EQUIPMENT

### 16.5.1. Classification and selection of dryers

Because of the very wide range of dryer designs available, classification is a virtually impossible task. PARKER[18] takes into account, however, the means by which material is transferred through the dryer as a basis of his classification, with a view to presenting a guide to the selection of dryers. Probably the most thorough classification of dryer types has been made by KRÖLL[19] who has presented a decimalised system based on the following factors:

(a) Temperature and pressure in the dryer,
(b) The method of heating,
(c) The means by which moist material is transported through the dryer,
(d) Any mechanical aids aimed at improving drying,
(e) The method by which the air is circulated,
(f) The way in which the moist material is supported,
(g) The heating medium, and
(h) The nature of the wet feed and the way it is introduced into the dryer.

In selecting a dryer for a particular application, as SLOAN[20] has pointed out, two steps are of primary importance:

(a) A listing of the dryers which are capable of handling the material to be dried,
(b) Eliminating the more costly alternatives on the basis of annual costs, capital charges + operating costs. A summary of dryer types, together with cost data, has been presented by BACKHURST and HARKER[21] and the whole question of dryer selection is discussed further in Volume 6.

Once a group of possible dryers has been selected, the choice may be narrowed by deciding whether batch or continuous operation is to be employed and, in addition to restraints imposed by the nature of the material, whether heating by contact with a solid surface or directly by convection and radiation is preferred.

In general, continuous operation has the important advantage of ease of integration into the rest of the process coupled with a lower unit cost of drying. As the rate of throughput of material becomes smaller, however, the capital cost becomes the major component in the total running costs and the relative cheapness of batch plant becomes more attractive. This is illustrated in Figure 16.7 which is taken from the work of KEEY[22]. In general, throughputs of 5000 kg/day (0.06 kg/s) and over are best handled in batches whilst throughputs of 50,000 kg/day (0.06 kg/s) and over are better handled continuously. The ease of construction of a small batch dryer compared with the sophistication of the

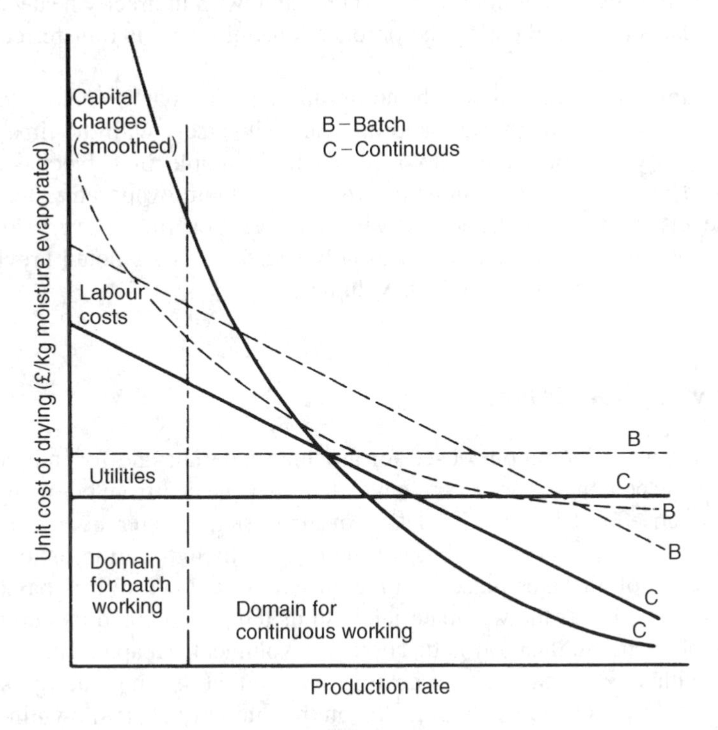


Figure 16.7. Variation of unit costs of drying with production rate[22]

continuous dryer should also be taken into account. In addition, a batch dryer is much more versatile and it can often be used for different materials. The humidity may be controlled during the drying operation, and this is especially important in cases where the humidity has to be maintained at different levels for varying periods of time.

*Direct heating*, in which the material is heated primarily by convection from hot gases has several advantages. Firstly, directly heated dryers are, in general, less costly, mainly because of the absence of tubes or jackets within which the heating medium must be contained. Secondly, it is possible to control the temperature of the gas within very fine limits, and indeed it is relatively simple to ensure that the material is not heated beyond a specified temperature. This is especially important with heat-sensitive materials. Against this, the overall thermal efficiency of directly heated dryers is generally low due to the loss of energy in the exhaust gas and, where an expensive solvent is evaporated from the solid, the operation is often difficult and costly. Losses also occur in the case of fluffy and powdery materials, and further problems are encountered where either the product or the solvent reacts with oxygen in the air.

A major cost in the operation of a dryer is in heating the air or gas. Frequently, the hot gases are produced by combustion of a fuel gas or atomised liquid, and considerable economy may be effected by using a combined heat and power system in which the hot gases are first passed through a turbine connected to an electrical generator.

Many of these disadvantages may be overcome by modifications to the design, although these increase the cost, and often an indirectly heated dryer may prove to be more economical. This is especially the case when thermal efficiency, solvent recovery or maximum cleanliness is of paramount importance and, with indirectly heated dryers, there is always the danger of overheating the product, since the heat is transferred through the material itself.

The maximum temperature at which the drying material may be held is controlled by the thermal sensitivity of the product and this varies inversely with the time of retention. Where lengthy drying times are employed, as for example in a batch shelf dryer, it is necessary to *operate under vacuum* in order to maintain evaporative temperatures at acceptable levels. In most continuous dryers, the retention time is very low, however, and operation at atmospheric pressure is usually satisfactory. As noted previously, dryer selection is considered in some detail in Volume 6.

### 16.5.2. Tray or shelf dryers

Tray or shelf dryers are commonly used for granular materials and for individual articles. The material is placed on a series of trays which may be heated from below by steam coils and drying is carried out by the circulation of air over the material. In some cases, the air is heated and then passed once through the oven, although, in the majority of dryers, some recirculation of air takes place, and the air is reheated before it is passed over each shelf. As air is passed over the wet material, both its temperature and its humidity change. This process of air humidification is discussed in Volume 1, Chapter 13.

If air of humidity $\mathscr{H}_1$ is passed over heating coils so that its temperature rises to $\theta_1$, this operation may be represented by the line AB on the humidity chart shown in Figure 16.8. This air then passes over the wet material and leaves at, say 90 per cent relative humidity,

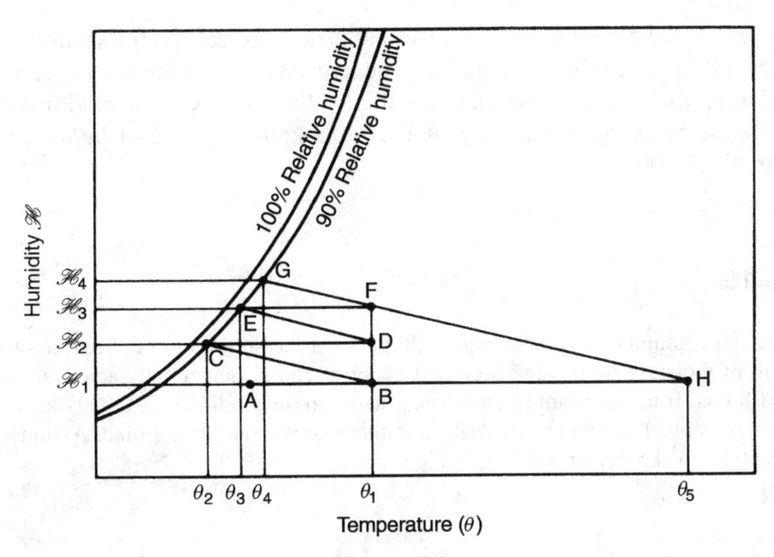


Figure 16.8. Drying with reheating of air

with its temperature falling to some value $\theta_2$. This change in the condition of the air is shown by the line BC, and the humidity has risen to $\mathscr{H}_2$. The wet-bulb temperature of the air will not change appreciably and therefore BC will coincide with an adiabatic cooling line. Each kg of air removes $(\mathscr{H}_2 - \mathscr{H}_1)$ kg of water, and the air required to remove a given amount of water from the material may easily be found. If the air at $\theta_2$ is now passed over a second series of heating coils and is heated to the original temperature $\theta_1$, the heating operation is shown by the line CD. This reheated air can then be passed over wet material on a second tray in the dryer, and pick up moisture until its relative humidity rises again to 90 per cent at a temperature $\theta_3$. This is at point E. In this way each kilogram of air has picked up water amounting to $(\mathscr{H}_3 - \mathscr{H}_1)$ kg of water. Reheating in this way may be effected a number of times, as shown in Figure 16.8, so that the moisture removed per kilogram of air can be considerably increased compared with that for a single pass. Thus, for three passes of air over the material, the total moisture removed is $(\mathscr{H}_4 - \mathscr{H}_1)$ kg/kg air.

If the air of humidity $\mathscr{H}_1$ had been heated initially to a temperature $\theta_5$, the same amount of moisture would have been removed by a single passage over the material, assuming that the air again leaves at a relative humidity of 90 per cent.

This reheating technique has two main advantages. Firstly, very much less air is required, because each kilogram of air picks up far more water than in a single stage system, and secondly, in order to pick up as much water in a single stage, it would be necessary to heat the air to a very much higher temperature. This reduction in the amount of air, needed simplifies the heating system, and reduces the tendency of the air to carry away any small particles.

A modern tray dryer consists of a well-insulated cabinet with integral fans and trays which are stacked on racks, or loaded on to trucks which are pushed into the dryer. Tray areas are 0.3–1 $m^2$ with a depth of material of 10–100 mm, depending on the particle size of the product. Air velocities of 1–10 m/s are used and, in order to conserve heat,

85–95 per cent of the air is recirculated. Even at these high values, the steam consumption may be 2.5–3.0 kg/kg moisture removed. The capacity of tray dryers depends on many factors including the nature of the material, the loading and external conditions, although for dyestuffs an evaporative capacity of 0.03–0.3 kg/m$^2$ ks (0.1–1 kg/m$^2$ h) has been quoted with air at 300–360 K[22].

## Example 16.4

A 100 kg batch of granular solids containing 30 per cent moisture is to be dried in a tray drier to 15.5 per cent of moisture by passing a current of air at 350 K tangentially across its surface at a velocity of 1.8 m/s. If the constant rate of drying under these conditions is 0.0007 kg/s m$^2$ and the critical moisture content is 15 per cent, calculate the approximate drying time. Assume the drying surface to be 0.03 m$^2$/kg dry mass.

## Solution

In 100 kg feed, mass of water = (100 × 30/100) = 30 kg

and: mass of dry solids = (100 − 30) = 70 kg

For $b$ kg water in the dried solids: $100b/(b + 70) = 15.5$

and the water in the product, $b = 12.8$ kg

Thus: initial moisture content, $w_1 = (30/70) = 0.429$ kg/kg dry solids

final moisture content, $w_2 = (12.8/70) = 0.183$ kg/kg dry solids

and water to be removed = (30 − 12.8) = 17.2 kg

The surface area available for drying = (0.03 × 70) = 2.1 m$^2$ and hence the rate of drying during the constant period = (0.0007 × 2.1) = 0.00147 kg/s.

As the final moisture content is above the critical value, all the drying is at this constant rate and the time of drying is:

$$t = (17.2/0.00147) = 11{,}700 \text{ s or } \underline{\underline{11.7 \text{ ks}}} \text{ (3.25 h)}$$

## 16.5.3. Tunnel Dryers

In tunnel dryers, a series of trays or trolleys is moved slowly through a long tunnel, which may or may not be heated, and drying takes place in a current of warm air. Tunnel dryers are used for drying paraffin wax, gelatine, soap, pottery ware, and wherever the throughput is so large that individual cabinet dryers would involve too much handling. Alternatively, material is placed on a belt conveyor passing through the tunnel, an arrangement which is well suited to vacuum operation. In typical tunnel arrangements, shown in Figure 16.9,

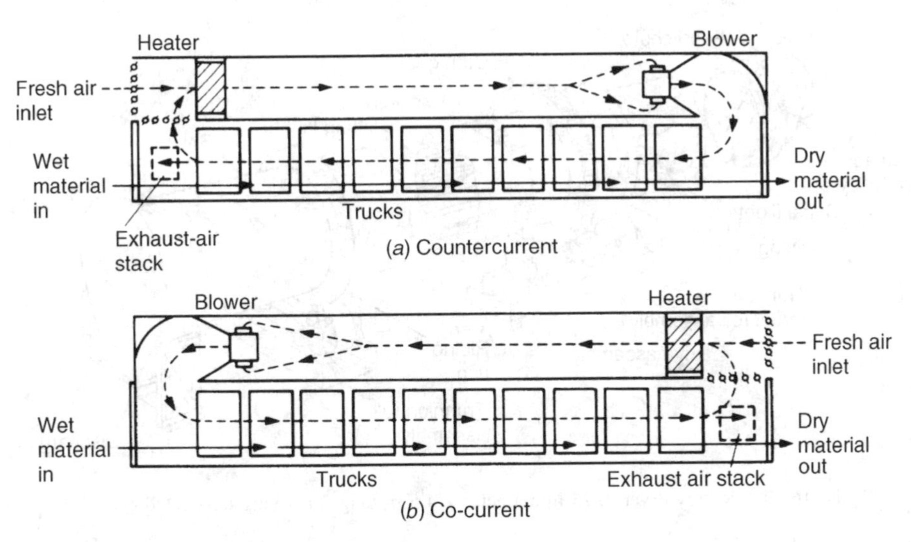


Figure 16.9. Arrangements for tunnel dryers

the construction is of block or sheet metal and the size varies over a wide range, with lengths sometimes exceeding 30 m.

## 16.5.4. Rotary Dryers

For the continuous drying of materials on a large scale, 0.3 kg/s (1 tonne/h) or greater, a rotary dryer, which consists of a relatively long cylindrical shell mounted on rollers and driven at a low speed, up to 0.4 Hz is suitable. The shell is supported at a small angle to the horizontal so that material fed in at the higher end will travel through the dryer under gravity, and hot gases or air used as the drying medium are fed in either at the upper end of the dryer to give co-current flow or at the discharge end of the machine to give countercurrent flow. One of two methods of heating is used:

(a) Direct heating, where the hot gases or air pass through the material in the dryer.
(b) Indirect heating, where the material is in an inner shell, heated externally by hot gases. Alternatively, steam may be fed to a series of tubes inside the shell of the dryer.

The shell of a rotary dryer is usually constructed by welding rolled plate, thick enough for the transmission of the torque required to cause rotation, and to support its own weight and the weight of material in the dryer. The shell is usually supported on large tyres which run on wide rollers, as shown in Figure 16.10, and although mild steel is the usual material of construction, alloy steels are used, and if necessary the shell may be coated with a plastics material to avoid contamination of the product.

With countercurrent operation, since the gases are often exhausted by a fan, there is a slight vacuum in the dryer, and dust-laden gases are in this way prevented from escaping.

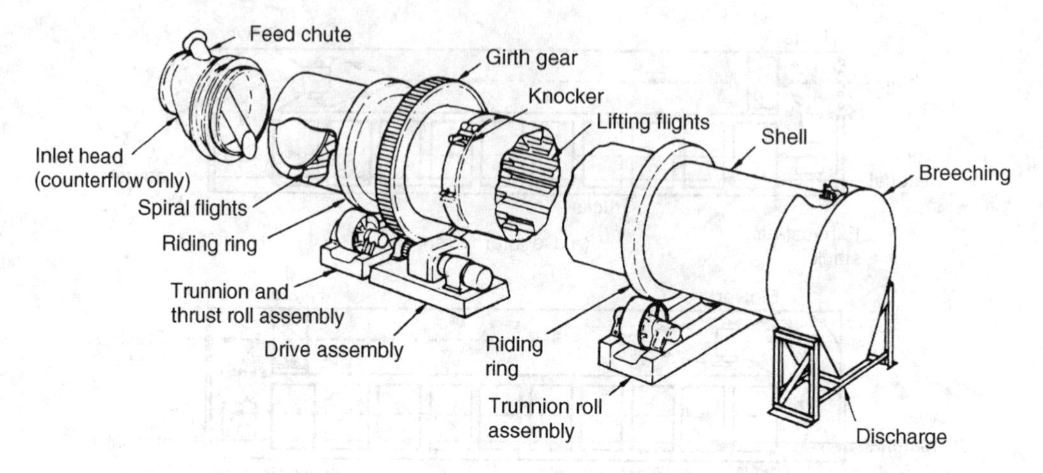


Figure 16.10. Rotary dryer, 0.75 m diameter × 4.5 m long for drying dessicated coconut[21]

This arrangement is suitable for sand, salt, ammonium nitrate and other inorganic salts, and is particularly convenient when the product is discharged at a high temperature. In this case, gas or oil firing is used and, where air is used as a drying medium, this may be filtered before heating, in order to minimise contamination of the product. As the gases leaving the dryer generally carry away very fine material, some form of cyclone or scrubber is usually fitted. Since the hot gases come into immediate contact with the dried material, the moisture content may be reduced to a minimum, though the charge may become excessively heated. Further, since the rate of heat transfer is a minimum at the feed end, a great deal of space is taken up with heating the material.

With co-current flow, the rate of passage of the material through the dryer tends to be greater since the gas is travelling in the same direction. Contact between the wet material and the inlet gases gives rise to rapid surface drying, and this is an advantage if the material tends to stick to the walls. This rapid surface drying is also helpful with materials containing water of crystallisation. The dried product leaves at a lower temperature than with countercurrent systems, and this may also be an advantage. The rapid lowering of the gas temperature as a result of immediate contact with the wet material also enables heat sensitive materials to be handled rather more satisfactorily.

Since the drying action arises mainly from direct contact with hot gases, some form of lifter is essential to distribute the material in the gas stream. This may take the form of flights, as shown in Figure 16.11, or of louvres. In the former case, the flights lift the material and then shower it across the gas stream, whilst in the latter the gas stream enters the shell along the louvres. In Figure 16.12 it may be seen that, in the rotary louvre dryer, the hot air enters through the louvres, and carries away the moisture at the end of the dryer. This is not strictly a co-current flow unit, but rather a through circulation unit, since the material continually meets fresh streams of air. The rotation of the shell, at about 0.05 Hz (3 rpm), maintains the material in agitation and conveys it through the dryer. Rotary dryers are 0.75–3.5 m in diameter and up to 9 m in length[23].

The thermal efficiency of rotary dryers is a function of the temperature levels, and ranges from 30 per cent in the handling of crystalline foodstuffs to 60–80 per cent in the

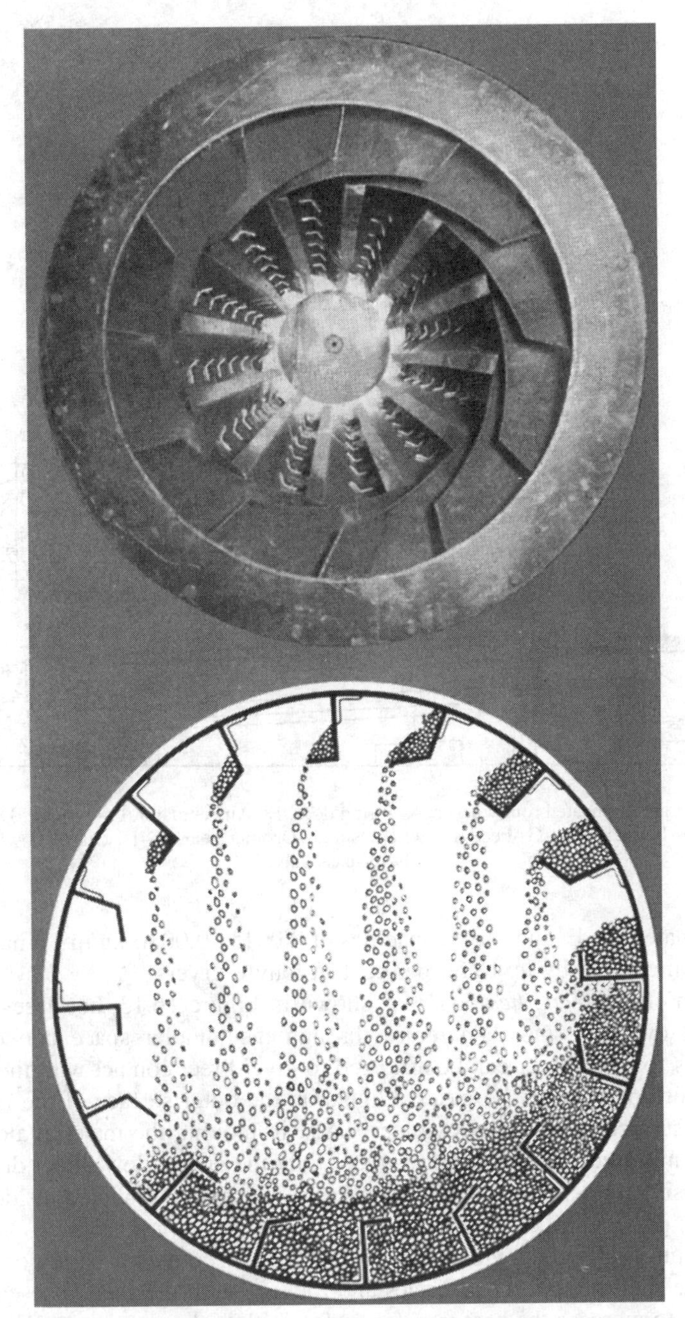

Figure 16.11. Helical and angled lifting flights in a rotary dryer

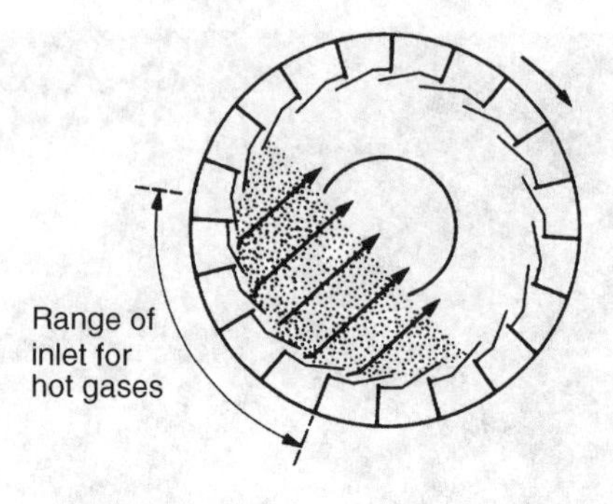


Figure 16.12. Air flow through a rotary louvre dryer

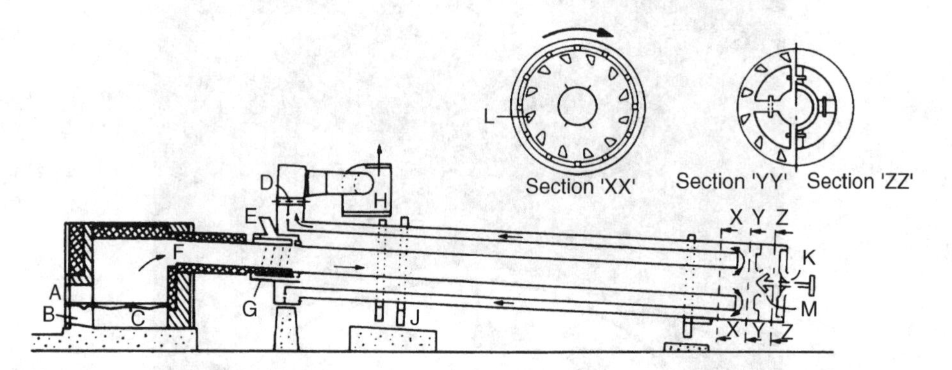


Figure 16.13. Indirectly heated rotary dryer. A–Firing door. B–Air regulator. C–Furnace. D–Control valves. E–Feed chute. F–Furnace flue. G–Feed screw. H–Fan. J–Driving gear. K–Discharge bowl. L–Duct lifters. M–By-pass valve

case of inert materials. Evaporative capacities of 0.0015–0.0080 $kg/m^3$ s may be achieved and these are increased by up to 50 per cent in louvre dryers.

In one form of *indirectly heated dryer*, shown in Figure 16.13, hot gases pass through the innermost cylinder, and then return through the annular space between the outer cylinders. This form of dryer can be arranged to give direct contact with the wet material during the return passage of the gases. Flights on the outer surface of the inner cylinder, and the inner surface of the outer cylinder, assist in moving the material along the dryer. This form of unit gives a better heat recovery than the single flow direct dryer, though it is more expensive. In a simpler arrangement, a single shell is mounted inside a brickwork chamber, through which the hot gases are introduced.

The steam-tube dryer, shown in Figure 16.14, incorporates a series of steam tubes, fitted along the shell in concentric circles and rotating with the shell. These tubes may be fitted with fins to increase the heat transfer surface although material may then stick to the tubes. The solids pass along the inclined shell, and leave through suitable ports at the other end. A small current of air is passed through the dryer to carry away the moisture, and the air leaves almost saturated. In this arrangement, the wet material comes in contact with very humid air, and surface drying is therefore minimised. This type of unit has a high thermal efficiency, and can be made from corrosion resisting materials without difficulty.

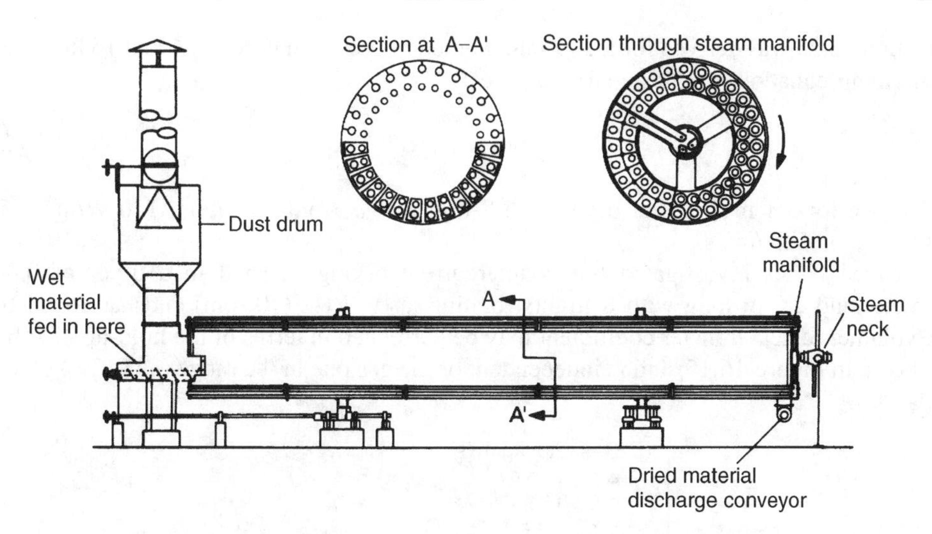


Figure 16.14. Steam-tube rotary dryer

## *Design considerations*

Many of the design problems associated with rotary dryers have been discussed by FRIEDMAN and MARSHALL[24], by PRUTTON *et al.*[25] and by MILLER *et al.*[26].

Heat from the air stream passes to the solid material during its fall through the air stream, and also from the hot walls of the shell, although the first mechanism is much the more important. The heat transfer equation may be written as:

$$Q = UaV\Delta T \tag{16.23}$$

where: $Q$ is the rate of heat transfer,
$U$ is the overall heat transfer coefficient,
$V$ is the volume of the dryer,
$a$ is the area of contact between the particles and the gas per unit volume of dryer, and
$\Delta T$ is the mean temperature difference between the gas and material.

The combined group $Ua$ has been shown[25,26] to be influenced by the feed rate of solids, the air rate and the properties of the material, and a useful approximation is given by:

$$Ua = \bar{\kappa}G'^{n}/D \tag{16.24}$$

where $\bar{\kappa}$ is a dimensional coefficient. Typical values for a 300 mm diameter dryer revolving at 0.08–0.58 Hz (5–35 rpm) show that $n = 0.67$ for specific gas rates in the range 0.37–1.86 kg/m$^2$, as given by SAEMAN[27]. The coefficient $\bar{\kappa}$ is a function of the number of flights and, using SI units, this is given approximately by:

$$\bar{\kappa} = 20(n_f - 1) \tag{16.25}$$

Equation 16.25 was derived for a 200 mm diameter dryer with between 6 and 16 flights[28]. Combining equations 16.24 and 16.25 gives:

$$Ua = 20(n_f - 1)G'^{0.67}/D \tag{16.26}$$

and hence for a 1 m diameter dryer with 8 flights, $Ua$ would be about 140 W/m$^3$ K for a gas rate of 1 kg/m$^2$ s.

SAEMAN[27] has investigated the countercurrent drying of sand in a dryer of 0.3 m diameter and 2.0 m long with 8 flights rotating at 0.17 Hz (10 rpm) and has found that the volumetric heat transfer coefficient may be correlated in terms of the hold-up of solids, as shown in Figure 16.15, and is independent of the gas rate in the range 0.25–20 kg/m$^2$ s.

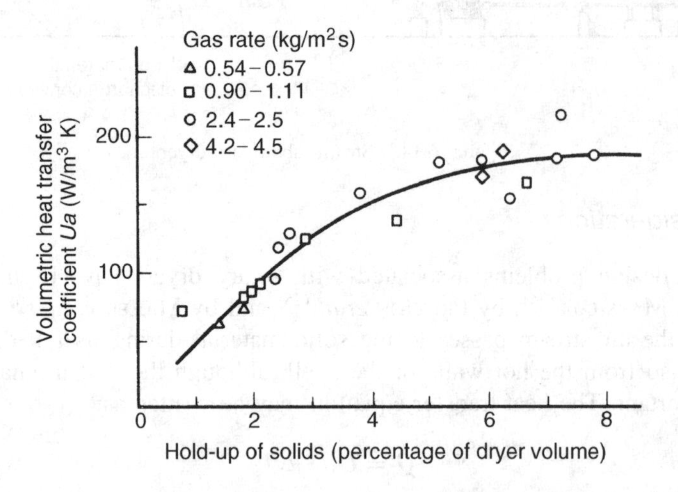


Figure 16.15. Correlation of volumetric heat-transfer coefficients with hold-up[27]

Other typical values of heat transfer coefficients achieved in drying operations are:

| *Application* | *Parameter* | *Value* | | *Reference* |
|---|---|---|---|---|
| Heated surface (fairly dry solid) | $U$ | 5–12 | W/m$^2$K | 28 |
| Unagitated dryers | $U$ | 5–29 | W/m$^2$K | 29 |
| Moderate agitation | $U$ | 29–85 | W/m$^2$K | 29 |
| High agitation | $U$ | 85–140 | W/m$^2$K | 29 |
| Cake-encrusted heating surface | $U$ | 5–29 | W/m$^2$K | 29 |
| Light powdery materials | $U\Delta T$ | 950 | W/m$^2$ | 24 |
| Coarse granular materials | $U\Delta T$ | 6300 | W/m$^2$ | 24 |

The hold-up in a rotary dryer varies with the feed rate, the number of flights, the shell diameter and the air rate. For zero air flow, FRIEDMAN and MARSHALL[24] give the hold-up as:

$$X = \frac{25.7F'}{SN^{0.9}D} \text{ per cent of dryer volume} \tag{16.27}$$

where: $D$ is the diameter of the drum (m),
$F'$ is the feed rate ($m^3/s\ m^2$),
$S$ is the slope of the dryer (m/m length),
$N$ is the rate of rotation (Hz), and
$X$ is the holdup, expressed as a percentage of the drum volume.

As the air flowrate is increased, $X$ changes and an empirical relation for the hold-up $X_a$ with air flow is:

$$X_a = X \pm KG' \tag{16.28}$$

although the values of $K$ are poorly defined. $X_a$ usually has a value of about 3 per cent, when working with a slope of about 0.1. In equation 16.28, the positive sign refers to countercurrent flow and the negative sign to co-current flow.

A more general approach has been made by SHARPLES *et al.*[30] who have solved differential moisture and heat-balance equations coupled with expressions for the forward transport of solids, allowing for solids being cascaded out of lifting baffles. Typical results, shown in Figure 16.16, were obtained using a 2.6 m diameter dryer, 16 m long with a 1° slope (60 mm/m). Commercial equipment is available with diameters up to 3 m and lengths up to 30 m, and hence the correlations outlined in this section must be used with caution beyond the range used in the experimental investigations.

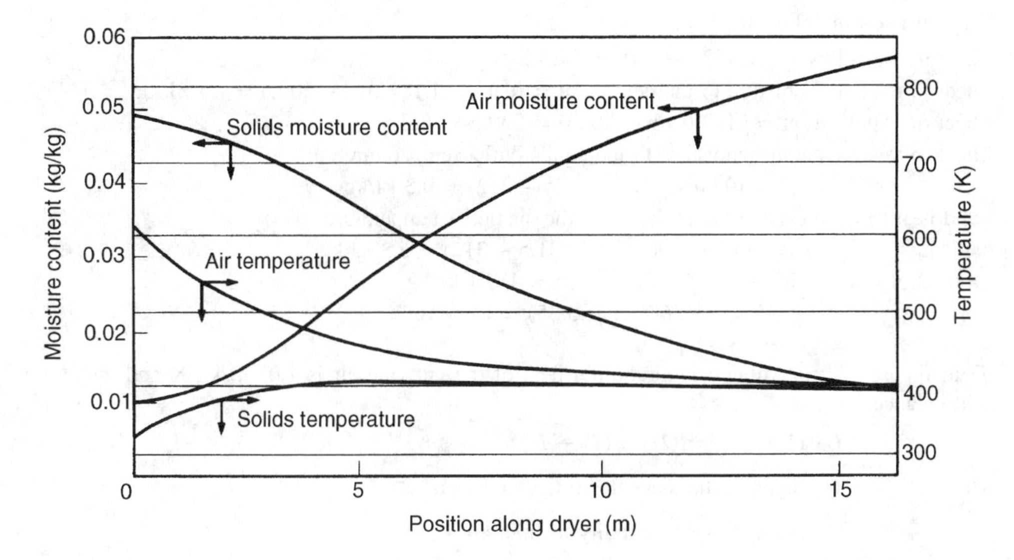


Figure 16.16. Conditions in a 2.6 m diameter × 16 m long rotary dryer with 1° slope. Co-current drying of 11.3 kg/s of fertiliser granules with 9.1 kg/s of air[30]

In general, the allowable mass velocity of the gas in a direct-contact rotary dryer depends on the dust content of the solids and is 0.55–7.0 $kg/m^2\ s$ with coarse particles. Inlet gas temperatures are typically 390–450 K for steam-heated air and 800–1100 K for flue gas, and the peripheral speed of the shell is 0.3–0.4 m/s.

## Example 16.5

A flow of 0.35 kg/s of a solid is to be dried from 15 per cent to 0.5 per cent moisture on a dry basis. The mean heat capacity of the solids is 2.2 kJ/kg deg K and it is proposed that a co-current adiabatic dryer should be used with the solids entering at 300 K and, because of the heat sensitive nature of the solids, leaving at 325 K. Hot air is available at 400 K with a humidity of 0.01 kg/kg dry air and the maximum allowable mass velocity of the air is 0.95 kg/m$^2$s. What diameter and length should be specified for the proposed dryer?

## Solution

With an inlet air temperature and humidity of 400 K and 0.01 kg/kg dry air respectively from Figure 13.4 in Volume 1, the inlet wet-bulb temperature = 312 K. If, not unreasonably, it is assumed that the number of transfer units is 1.5, then for adiabatic drying the outlet air temperature, $T_0$ is given by:

$$1.5 = \ln((400 - 312)/(T_0 - 312) \text{ and } T_0 = 331.5 \text{ K}.$$

The solids outlet temperature will be taken as the maximum allowable, 325 K.

From the steam tables in the Appendix, the latent heat of vaporisation of water at 312 K is 2410 kJ/kg. Again from steam tables, the specific heat capacity of water vapour = 1.88 kJ/kg K and that of the solids will be taken as 2.18 kJ/kg K.

For a mass flow of solids of 0.35 kg/s and inlet and outlet moisture contents of 0.15 and 0.005 kg/kg dry solids respectively, the mass of water evaporated = 0.35(0.15 − 0.005) = 0.0508 kg/s.

For unit mass of solids, the heat duty includes:

heat to the solids = 2.18(325 − 300) = 54.5 kJ/kg

heat to raise the moisture to the dew point = (0.15 × 4.187(312 − 300)) = 7.5 kJ/kg

heat of vaporisation = 2410(0.15 − 0.005) = 349.5 kJ/kg

heat to raise remaining moisture to the solids outlet temperature

= (0.005 × 4.187)(325 − 312) = 0.3 kJ/kg

and heat to raise evaporated moisture to the air outlet temperature

= (0.15 − 0.005)1.88(331.5 − 312) = 5.3 kJ/kg

a total of (54.5 + 7.5 + 349.5 + 0.3 + 5.3) = 417.1 kJ/kg

or: (417.1 × 0.35) = 146 kW

From Figure 13.4 in Volume 1, the humid heat of the entering air is 1.03 kJ/kg K and making a heat balance:

$$G_1\ (1 + \mathscr{H}_1) = Q/C_{p_1}(T_1 - T_2)$$

where: $G_1$ (kg/s) is the mass flowrate of inlet air,

$\mathscr{H}_1$ (kg/kg) is the humidity of inlet air,

$Q$ (kW) is the heat duty,

$C_{p_1}$ (kJ/kg K) is the humid heat of inlet air

and: $T_1$ and $T_2$ (K) are the inlet and outlet air temperatures respectively.

In this case:

$$G_1(1 + 0.01) = 146/(1.03(400 - 331.5)) = 2.07 \text{ kg/s}$$

and: mass flowrate of dry air, $G_a = (2.07/1.01) = 2.05$ kg/s

The humidity of the outlet air is then:

$$\mathscr{H}_2 = 0.01 + (0.0508/2.05) = 0.0347 \text{ kg/kg}$$

At a dry bulb temperature of 331.5 K, with a humidity of 0.0347 kg/kg, the wet-bulb temperature of the outlet air, from Figure 13.4 in Volume 1, is 312 K, the same as the inlet, which is the case for adiabatic drying.

The dryer diameter is then found from the allowable mass velocity of the air and the entering air flow and for a mass velocity of 0.95 kg/m$^2$s, the cross sectional area of the dryer is:

$$(2.07/0.95) = 2.18 \text{ m}^2$$

equivalent to a diameter of $[(4 \times 2.18)/\pi]^{0.5} = \underline{\underline{1.67 \text{ m}}}$

With a constant drying temperature of 312 K:

at the inlet: $\Delta T_1 = (400 - 312) = 88$ deg K

and at the outlet: $\Delta T_2 = (331.5 - 312) = 19.5$ deg K

and the logarithmic mean, $\Delta T_m = (88 - 19.5)/\ln(88/19.5) = 45.5$ deg K.

The length of the dryer, $L$ is then: $L = Q/(0.0625\pi D G'^{0.67} \Delta T_m)$

where $D$ (m) is the diameter

and $G'$ (kg/m$^2$s) is the air mass velocity.

In this case: $L = 146/[0.0625\pi \times 1.67(0.95)^{0.67} \times 45.5)] = \underline{\underline{10.1 \text{ m}}}$

This gives a length/diameter ratio of $(10.1/1.67) = 6$, which is a reasonable value for rotary dryers.

## 16.5.5. Drum dryers

If a solution or slurry is run on to a slowly rotating steam-heated drum, evaporation takes place and solids may be obtained in a dry form. This is the basic principle used in all drum dryers, some forms of which are illustrated in Figure 16.17. The feed to a single drum dryer may be of the dip, pan, or splash type. The dip-feed system, the earliest design, is still used where liquid can be picked up from a shallow pan. The agitator prevents settling of particles, and the spreader is sometimes used to produce a uniform coating on the drum. The knife, which is employed for removing the dried material, functions in a similar manner to the doctor blade on a rotary filter. If the material is dried to give a free-flowing powder, this comes away from the drum quite easily. The splash-feed type is used for materials, such as calcium arsenate, lead arsenate, and iron oxide, where a light fluffy product is desired. The revolving cylinder throws the liquor against the drum, and a uniform coating is formed with materials which do not stick to the hot surface of the drum.

Double drum dryers may be used in much the same way, and Figure 16.17 shows dip-feed and top-feed designs. Top-feed gives a larger capacity as a thicker coating is obtained. It is important to arrange for a uniform feed to a top-feed machine, and this may be effected by using a perforated pipe for solutions and a travelling trough for suspensions.

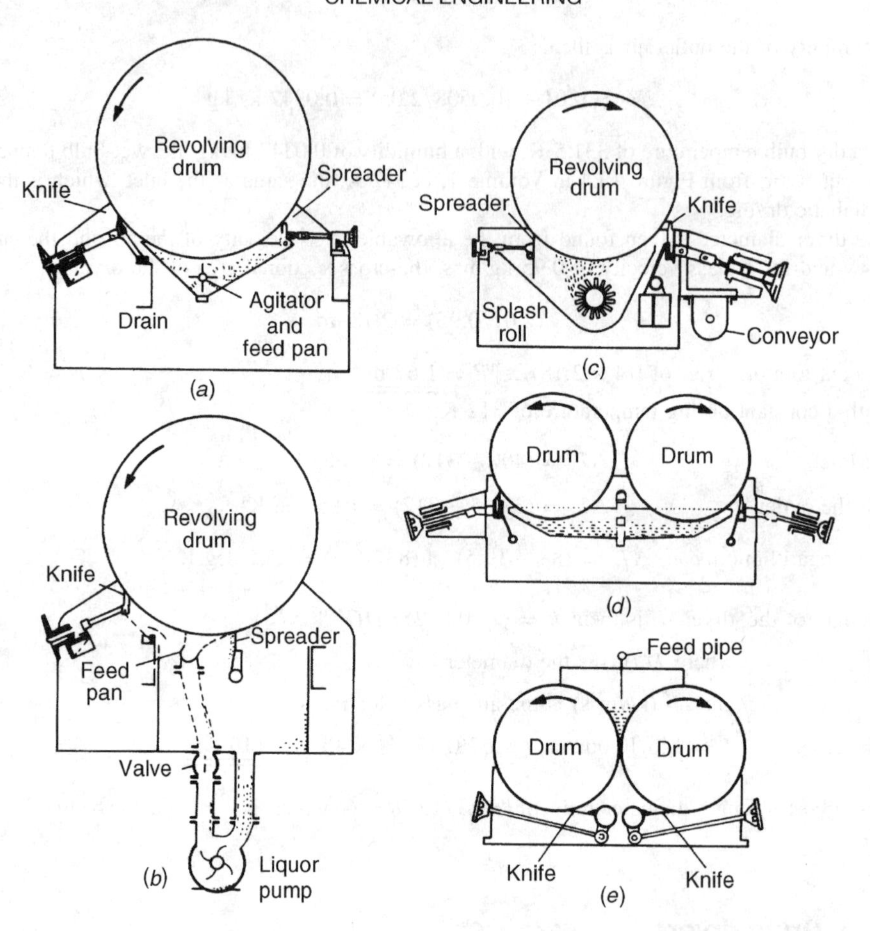


Figure 16.17. Methods of feeding drum dryers. (*a*) Single drum, dip-feed. (*b*) Single drum, pan-feed. (*c*) Single drum, splash-feed. (*d*) Double drum, dip-feed. (*e*) Double drum, top-feed

Drums are usually made from cast iron, although chromium-plated steel or alloy steel is often used where contamination of the product must be avoided, such as with pharmaceuticals or food products. Arrangements must be made for accurate adjustment of the separation of the drums, and the driving gears should be totally enclosed. A range of speeds is usually obtained by selecting the gears, rather than by fitting a variable speed drive. Removal of the steam condensate is important, and an internal syphon is often fitted to keep the drum free of condensate. In some cases, it is better for the drums to be rotated upwards at the point of closest proximity, and the knives are then fitted at the bottom. By this means, the dry material is kept away from the vapour evolved. Some indication of the sizes of this type of dryer is given in Table 16.3, and it may be noted that the surface of each drum is limited to about 35 $m^2$. This, coupled with developments in the design of spray dryers, renders the latter more economically attractive, especially where large throughputs are to be handled. For steam-heated drum dryers, normal evaporative capacities are proportional to the active drum area and are 0.003–0.02 $kg/m^2$ s, although higher rates are claimed for grooved drum dryers.

Table 16.3. Sizes of double drum dryers

| Drum dimensions Diameter (m) × Length (m) | Length (m) | Width (m) | Height (m) | Mass (kg) |
|---|---|---|---|---|
| 0.61 × 0.61 | 3.5 | 2.05 | 2.3 | 3,850 |
| 0.61 × 0.91 | 3.8 | 2.05 | 2.4 | 4,170 |
| 0.81 × 1.32 | 5.0 | 2.5 | 2.8 | 7,620 |
| 0.81 × 1.83 | 5.5 | 2.5 | 2.8 | 8,350 |
| 0.81 × 2.28 | 5.9 | 2.5 | 2.8 | 8,890 |
| 0.81 × 2.54 | 6.25 | 2.5 | 2.8 | 9,300 |
| 0.81 × 3.05 | 6.7 | 2.5 | 2.8 | 10,120 |
| 1.07 × 2.28 | 6.4 | 3.0 | 3.0 | 14,740 |
| 1.07 × 2.54 | 6.7 | 3.0 | 3.0 | 15,420 |
| 1.07 × 3.05 | 7.2 | 3.0 | 3.0 | 16,780 |
| 1.52 × 3.66 | 7.9 | 4.1 | 4.25 | 27,220 |

The solid is usually in contact with the hot metal for 6–15 s, short enough to prevent significant decomposition of heat sensitive materials, and heat transfer coefficients are 1–2 kW/m$^2$ K.

When the temperature of the drying material must be kept as low as possible, vacuum drying is used, and one form of vacuum dryer is shown in Figure 16.18. The dried material is collected in two screw conveyors and carried usually to two receivers so that one can be filled while the other is emptied.

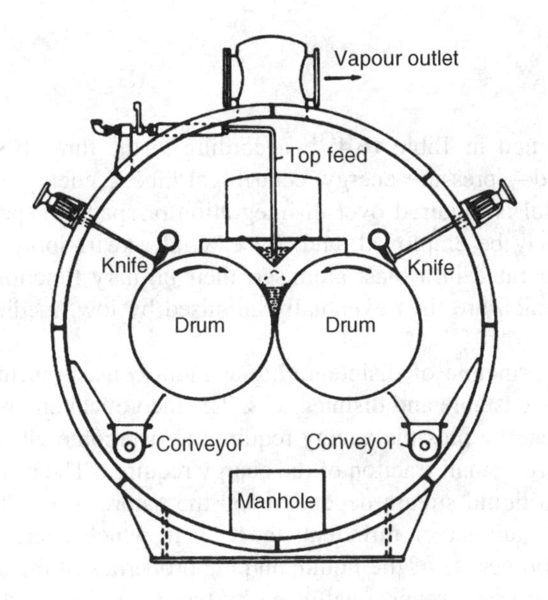


Figure 16.18. Vacuum drum dryer

## 16.5.6. Spray dryers

Water may be evaporated from a solution or a suspension of solid particles by spraying the mixture into a vessel through which a current of hot gases is passed. In this way, a large interfacial area is produced and consequently a high rate of evaporation is obtained. Drop temperatures remain below the wet bulb temperature of the drying gas until drying is almost complete, and the process thus affords a convenient means of drying substances which may deteriorate if their temperatures rise too high, such as milk, coffee, and plasma. Furthermore, because of the fine state of subdivision of the liquid, the dried material is obtained in a finely divided state.

In spray drying, it is necessary to atomise and distribute, under controlled conditions, a wide variety of liquids, the properties of which range from those of solutions, emulsions, and dispersions, to slurries and even gels. Most of the atomisers commonly employed are designed for simple liquids, that is mobile Newtonian liquids. When atomisers are employed for slurries, pastes, and liquids having anomalous properties, there is a great deterioration in performance and, in many cases, atomisers may be rapidly eroded and damaged so as to become useless. There is therefore much to be gained by considering various types and designs of atomiser so that a suitable selection can be made for the given duty.

The performance of a spray dryer or reaction system is critically dependent on the drop size produced by the atomiser and the manner in which the gaseous medium mixes with the drops. In this context an atomiser is defined as a device which causes liquid to be disintegrated into drops lying within a specified size range, and which controls their spatial distribution.

### *Atomisers*

Atomisers are classified in Table 16.4[31] according to the three basic forms of energy commonly employed — pressure energy, centrifugal kinetic energy, and gaseous energy. Where greater control is required over disintegration or spatial dispersion, combinations of atomiser types may be employed, and, for example, swirl-spray nozzles or spinning discs may be incorporated in a blast atomiser, their primary functions being to produce thin liquid sheets which are then eventually atomised by low, medium or high velocity gas streams.

The fundamental principle of disintegrating a liquid consists in increasing its surface area until it becomes unstable and disintegrates. The theoretical energy requirement is the increase in surface energy plus the energy required to overcome viscous forces, although in practice this is only a small fraction of the energy required. The process by which drops are produced from a liquid stream depends upon the nature of the flow in the atomiser, that is whether it is laminar or turbulent, the way in which energy is imparted to the liquid, the physical properties of the liquid, and the properties of the ambient atmosphere. The basic mechanism is, however, unaffected by these variables and consists essentially of the breaking down of unstable threads of liquid into rows of drops and conforms to the classical mechanism postulated by LORD RAYLEIGH[32]. This theory states that a free column of liquid is unstable if its length is greater than its circumference, and that, for a non-viscous liquid, the wavelength of that disturbance which will grow most rapidly in amplitude is 4.5 times the diameter. This corresponds to the formation of droplets of diameter approximately 1.89 times that of the jet $d_j$. WEBER[33] has shown that the

Table 16.4. Classification of atomisers[31]

| Pressure used in pressure atomisers | Centrifugal energy used in rotary atomisers | Gaseous energy used in twin-fluid or blast atomisers |
|---|---|---|
| Fan-spray nozzles<br>0.25–1.0 MN/m$^2$ | Spinning cups<br>6–30 m/s | External mixing<br>(fluid pressures independent) |
| Impact nozzles<br>Impinging jet nozzles<br>0.25–1.0 MN/m$^2$<br>Impact plate nozzles<br>Up to 3.0 MN/m$^2$<br>Deflector nozzles<br>7.0 MN/m$^2$ | Spinning discs<br>Flat discs<br>Saucer-shaped discs<br>Radial-vaned discs<br>Multiple discs<br>30–180 m/s | Internal mixing<br>(fluid pressures interdependent)<br><br>Low velocity<br>Gas velocity<br>30–120 m/s<br>Gas–liquid ratio<br>2–25 kg/kg |
| Swirl-spray nozzles<br>Hollow cone, full cone<br>0.4–70 MN/m$^2$ | | Medium velocity<br>Gas velocity<br>120–300 m/s<br>Gas–liquid ratio<br>0.2–1 kg/kg |
| Divergent pintle nozzles<br>0.25–7.0 MN/m$^2$<br>Fixed or vibrating pintle | | High velocity<br>Gas velocity<br>sonic or above<br>Gas–liquid ratio<br>0.2–1 kg/kg |

optimum wavelength for the disruption of jets of viscous liquid is:

$$\lambda_{\text{opt}} = \sqrt{(2\pi d_j)}\left[1 + \frac{3\mu}{\sqrt{(\rho\sigma d_j)}}\right]^{0.5} \tag{16.29}$$

A uniform thread will break down into a series of drops of uniform diameter, each separated by one or more satellite drops. Because of the heterogeneous character of the atomisation process, however, non-uniform threads are produced and this results in a wide range of drop sizes. An example of a part of a laminar sheet collapsing into a network of threads and drops is shown in Figure 16.19. Only when the formation and disintegration of threads are controlled can a homogeneous spray cloud be produced. One method by which this can be achieved is by using a rotary cup atomiser operating within a critical range of liquid flow rates and rotor speeds, although, as shown later, this range falls outside that normally employed in practice.

Although certain features are unique to particular atomiser types, many of the detailed mechanisms of disintegration are common to most forms of atomiser[34]. The most effective way of utilising energy imparted to a liquid is to arrange that the liquid mass has a large specific surface before it commences to break down into drops. Thus the primary function of an atomiser is to transpose bulk liquid into thin liquid sheets. Three modes of disintegration of such spray sheets have been established[35], namely *rim*, *perforated sheet*, and *wave*. Because of surface tension, the free edge of any sheet contracts into a thick rim, and *rim disintegration* occurs as it breaks up by instabilities analogous to those

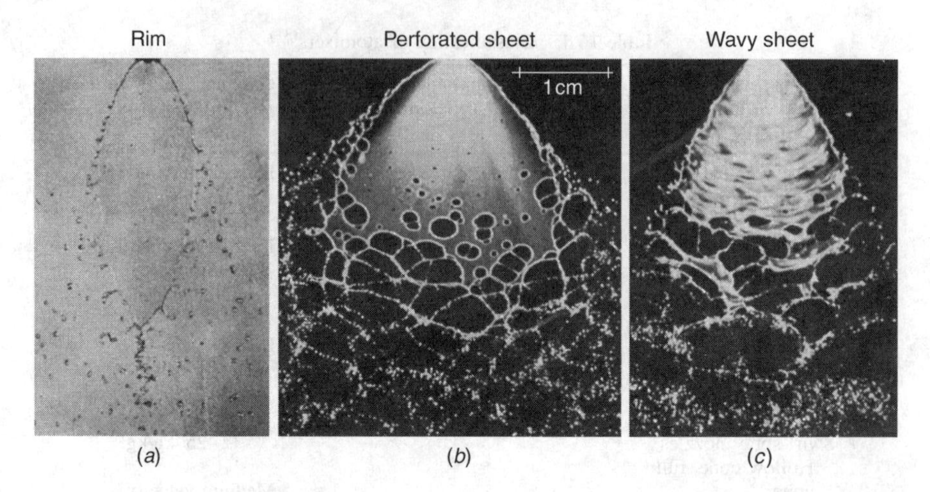


Figure 16.19. Modes of disintegration of liquid films

of free jets. Figure 16.19*a* illustrates a fan spray sheet and shows that, as the liquid in each edge moves along the curved boundary, the latter becomes disturbed and disintegrates. When this occurs, the resulting drops sustain the direction of flow of the edge at the point at which the drops are formed, and remain attached to the receding surface by thin threads which rapidly disintegrate into streams of drops.

In *perforated-sheet disintegration* shown in Figure 16.19*b*, small holes suddenly appear in the sheet as it advances into the atmosphere. They rapidly grow in size(36) until the thickening rims of adjacent holes coalesce to form threads of varying diameter. The threads finally break down into drops.

Disintegration can also occur through the superimposition of a wave motion on the sheet, as shown in Figure 16.19*c*. Sheets of liquid corresponding to half or full wavelengths of liquid are torn off and tend to draw up under the action of surface tension, although these may suffer disintegration by air action or liquid turbulence before a regular network of threads can be formed.

In a *pressure atomiser*, liquid is forced under pressure through an orifice, and the form of the resulting liquid sheet can be controlled by varying the direction of flow towards the orifice. By this method, flat and conical spray sheets can be produced. From the Bernoulli equation, given in Volume 1, Chapter 6, the mass rate of flow through a nozzle may be derived as:

$$G = C_D \rho A_N \sqrt{[2(-\Delta P)]\rho} \tag{16.30}$$

For a given nozzle and fluid, and an approximately constant coefficient of discharge, $C_D$, then:

$$G = \text{constant} \sqrt{(-\Delta P)} \tag{16.31}$$

The capacity of a nozzle is conveniently described by the *flow number*, **FN** a dimensional constant defined by:

$$\mathbf{FN} \equiv \frac{\text{Volumetric flow (gal/h)}}{\sqrt{\text{Pressure (lb/in}^2)}} = 2.08 \times 10^6 \frac{\text{Volume flow (m}^3\text{/s)}}{\sqrt{\text{Pressure (kN/m}^2)}} \tag{16.32}$$

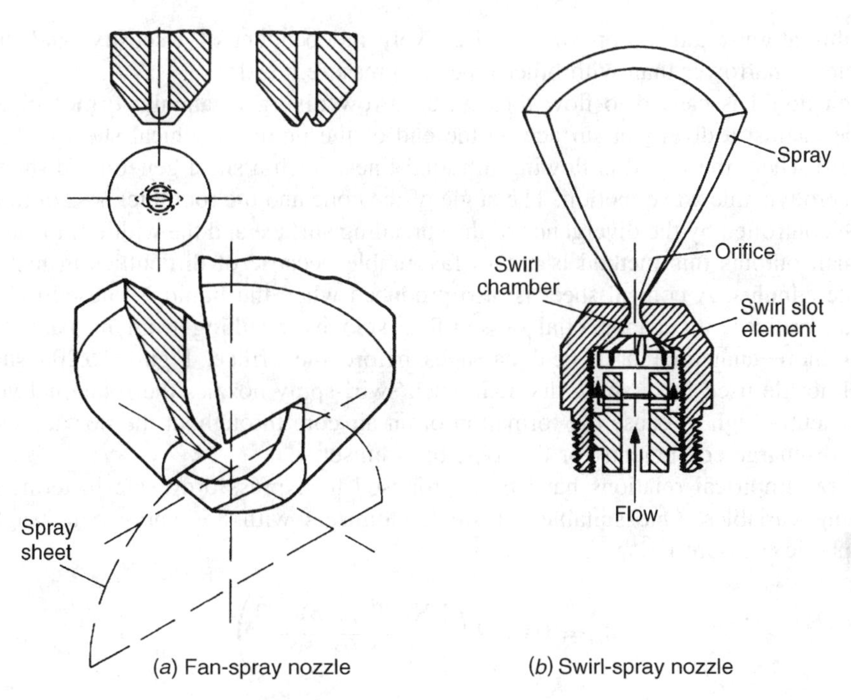


Figure 16.20. Pressure atomisers

In the fan-spray drop nozzle shown in Figure 16.20*a*, two streams of liquid are made to impinge behind an orifice by specially designed approach passages, and a sheet is formed in a plane perpendicular to the plane of the streams. The orifice runs full and, since the functional portion is sharp-edged, high discharge coefficients are obtained which are substantially constant over wide ranges of Reynolds number.

The influence of conditions on the droplet size where the spray sheet disintegrates through aerodynamic wave motion may be represented by the following expression proposed by DOMBROWSKI and MUNDAY[34] for ambient densities around normal atmospheric conditions:

$$d_m = \left(\frac{0.000156}{C_D}\right)\left[\frac{\mathbf{FN}\sigma\rho}{\sin\phi(-\Delta P)}\right]^{1/3}\rho_A^{-1/6} \tag{16.33}$$

where **FN** is given in equation 16.32, $\phi$ is the half-angle of the sheet, $-\Delta P$ is in kN/m$^2$ and all other quantities are expressed in SI units.

The principle of operation of the impinging jet nozzle resembles that of the fan spray nozzle with the exception that two or more independent jets are caused to impinge in the atmosphere. In impact atomisers, one jet is caused to strike against a solid surface, and for two jets impinging at 180°[34], using SI units:

$$d_m = 1.73\left(\frac{d_j^{0.75}}{u_l^{0.5}}\right)\left[\frac{\sigma}{\rho}\right]^{0.25} \tag{16.34}$$

With this atomiser, the drop size is effectively independent of viscosity, and the size spectrum is narrower than with other types of pressure nozzle.

When liquid is caused to flow through a narrow divergent annular orifice or around a pintle against a divergent surface on the end of the pintle, a conical sheet of liquid is produced where the liquid is flowing in radial lines. Such a sheet generally disintegrates by an aerodynamic wave motion. The angle of the cone and the root thickness of the sheet may be controlled by the divergence of the spreading surface and the width of the annulus. For small outputs this method is not so favourable because of difficulties in making an accurate annulus. A conical sheet is also produced when the liquid is caused to emerge from an orifice with a tangential or swirling velocity resulting from its path through one or more tangential or helical passages before the orifice. Figure 16.20*b* shows a typical nozzle used for a spray dryer. In such swirl-spray nozzles the rotational velocity is sufficiently high to cause the formation of an air core throughout the nozzle, resulting in low discharge coefficients for this type of atomiser.

Several empirical relations have been proposed to express drop size in terms of the operating variables. One suitable for small atomisers with 85° spray cone angles, at atmospheric pressure is[34]:

$$d_m = 0.0134\left(\frac{\mathbf{FN}^{0.209}(\mu/\rho)^{0.215}}{(-\Delta P)^{0.348}}\right) \tag{16.35}$$

where $-\Delta P$ is in kN/m$^2$, **FN** is given in equation 16.32, and all other quantities are in SI units.

Pressure nozzles are somewhat inflexible since large ranges of flowrate require excessive variations in differential pressure. For example, for an atomiser operating satisfactorily at 275 kN/m$^2$, a pressure differential of 17.25 MN/m$^2$ is required to increase the flowrate to ten times its initial value. These limitations, inherent in all pressure-type nozzles, have been overcome in swirl spray nozzles by the development of spill, duplex, multi-orifice, and variable port atomisers, in which ratios of maximum to minimum outputs in excess of 50 can be easily achieved[34].

In a *rotary atomiser*, liquid is fed on to a rotating surface and spread out by centrifugal force. Under normal operating conditions the liquid extends from the periphery in the form of a thin sheet which breaks down some distance away from the periphery, either freely by aerodynamic action or by the action of an additional gas blast. Since the accelerating force can be controlled independently, this type of atomiser is extremely versatile and it can handle successfully a wide range of feed rates with liquids having a wide range of properties. The rotating component may be a simple flat disc though slippage may then occur, and consequently it is more usual to use bowls, vaned discs, and slotted wheels, as shown in Figure 16.21. Diameters are 25–460 mm and small discs rotate at up to 1000 Hz while the larger discs rotate at up to 200 Hz with capacities of 1.5 kg/s. Where a coaxial gas blast is used to effect atomisation, lower speeds of the order of 50 Hz may be used. At very low flowrates, such as 30 mg/s, the liquid spreads out towards the cup lip where it forms a ring. As liquid continues to flow into the ring, its inertia increases, overcomes the restraining surface tension and viscous forces and is centrifuged off as discrete drops of uniform size, which initially remain attached to the rim by a fine attenuating thread. When the drop is finally detached, the thread breaks down into a chain of small satellite drops.

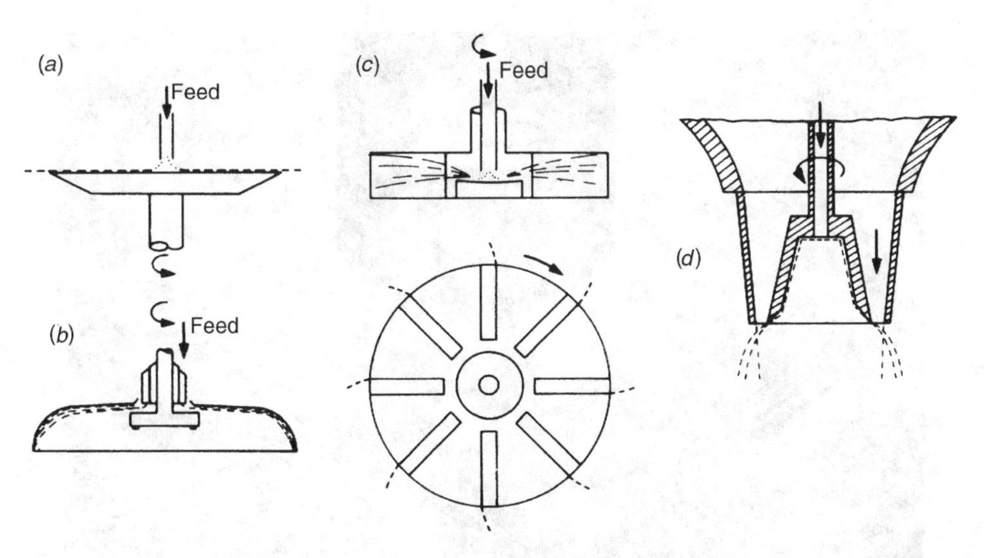


Figure 16.21. Characteristic rotary atomisers: (*a*) Sharp-edge flat disc; (*b*) Bowl; (*c*) Vaned disc; (*d*) Air-blast bowl atomiser

Since the satellite drops constitute only a small proportion of the total liquid flowrate, a cup operating under these conditions effectively produces a monodisperse spray. Under these conditions the drop size from sharp-edged discs has been given by WALTON and PREWETT[37] as:

$$d_m = \left(\frac{0.52}{N}\right)\sqrt{\left(\frac{\sigma}{D\rho}\right)} \qquad (16.36)$$

When the liquid flowrate is increased, the retaining threads grow in thickness and form long jets. As they extend into the atmosphere, these jets are stretched and finally break down into strings of drops. Under more practical ranges of flowrate, the jets are unable to remove all the liquid, the ring is forced away and a thin sheet extends around the cup lip and eventually breaks up into a polydisperse spray, as shown in Figure 16.22.

A far greater supply of energy for disintegration of the liquid jet can be provided by using a high-speed gas stream which impinges on a liquid jet or film. By this means a greater surface area is formed and drops of average size less than 20 μm can be produced. This is appreciably smaller than is possible by the methods previously described, although the energy requirements are much greater. The range of flowrates which can be used is wide, because the supply of liquid and the energy for atomisation can be controlled independently. Gas velocities ranging from 50 m/s to sonic velocity are common, and sometimes the gas is given a vortex motion.

Break-up of the jet occurs as follows. Ligaments of liquid are torn off, which collapse to form drops. These may be subsequently blown out into films, which in turn further collapse to give a fine spray. Generally, this spray has a small cone angle and is capable of penetrating far greater distances than the pressure nozzle. Small atomisers of this type have been used in spray-drying units of low capacities.

(*a*) low speed

(*b*) high speed

Figure 16.22. Spray sheet from a rotating cup

Where the gas is impacted on to a liquid jet, the mean droplet size is given approximately by[34]:

$$d_s = \frac{0.585}{u_r}\sqrt{\left(\frac{\sigma}{\rho}\right)} + 0.0017\left(\frac{\mu}{\sqrt{(\sigma\rho)}}\right)^{0.45}\left(\frac{1000}{j}\right)^{1.5} \tag{16.37}$$

where: $d_s$ is the surface mean diameter (m),
$j$ is the volumetric ratio of gas to liquid rates at the pressure of the surrounding atmosphere,
$u_r$ is the velocity of the gas relative to the liquid (m/s),
$\mu$ is the liquid viscosity (Ns/m$^2$),
$\rho$ is the liquid density (kg/m$^3$), and
$\sigma$ is the surface tension (N/m).

As FRASER *et al.* point out, more efficient atomisation is achieved if the liquid is spread out into a film before impact[36].

### *Drying of drops*

The amount of drying that a drop undergoes depends upon the rate of evaporation and the contact time, the latter depending upon the velocity of fall and the length of path through the dryer. The terminal velocity and the transfer rate depend upon the flow conditions around the drop. Because of the nature of the flow pattern, the latter also varies with angular position around the drop, although no practical design method has incorporated such detail and the drop is always treated as if it evaporates uniformly from all its surface.

There are two main periods of evaporation. When a drop is ejected from an atomiser its initial velocity relative to the surrounding gas is generally high and very high rates of transfer are achieved. The drop is rapidly decelerated to its terminal velocity, however, and the larger proportion of mass transfer takes place during the free-fall period. Little error is therefore incurred in basing the total evaporation time on this period.

An expression for the evaporation time for a pure liquid drop falling freely in air has been presented by MARSHALL[38,39]. For drop diameters less than 100 μm this may be simplified to give:

$$t = \frac{\rho\lambda(d_0^2 - d_t^2)}{8k_f \Delta T} \tag{16.38}$$

Sprays generally contain a wide range of drop sizes, and a stepwise procedure can be used[39] to determine the size spectrum as evaporation proceeds.

When single drops containing solids in suspension or solutions are suspended in hot gas streams, it is found that evaporation initially proceeds in accordance with equation 16.38 for pure liquid, although when solids deposition commences, a crust or solid film is rapidly formed which increases the resistance to transfer. Although this suggests the existence of a falling rate period similar to that found in tray-drying, the available published data indicate that it has little effect on the total drying time. As a result of crust formation, the dried particles may be in the form of hollow spheres.

### *Industrial spray dryers*

Spray dryers are used in a variety of applications where a fairly high grade product is to be made in granular form. In the drying chamber the gas and liquid streams are brought into contact, and the efficiency of mixing depends upon the flow patterns induced in the chamber. Rotating disc atomisers are most commonly used. Countercurrent dryers give the highest thermal efficiencies although product temperatures are higher in these units. This limits their use to materials which are not affected by overheating. Co-current dryers suffer from relatively low efficiencies, although they have the advantage of low product temperatures unless back-mixing occurs. In the case of materials which are extremely sensitive to heat, great care has to be taken in the design of the chamber to avoid overheating. Combustion gases are frequently used directly although, in some cases, such as the preparation of food products, indirectly heated air is used. Maximum temperatures are then normally limited to lower values than those with direct heating. Typical flow arrangements in spray drying are shown in Figure 16.23.

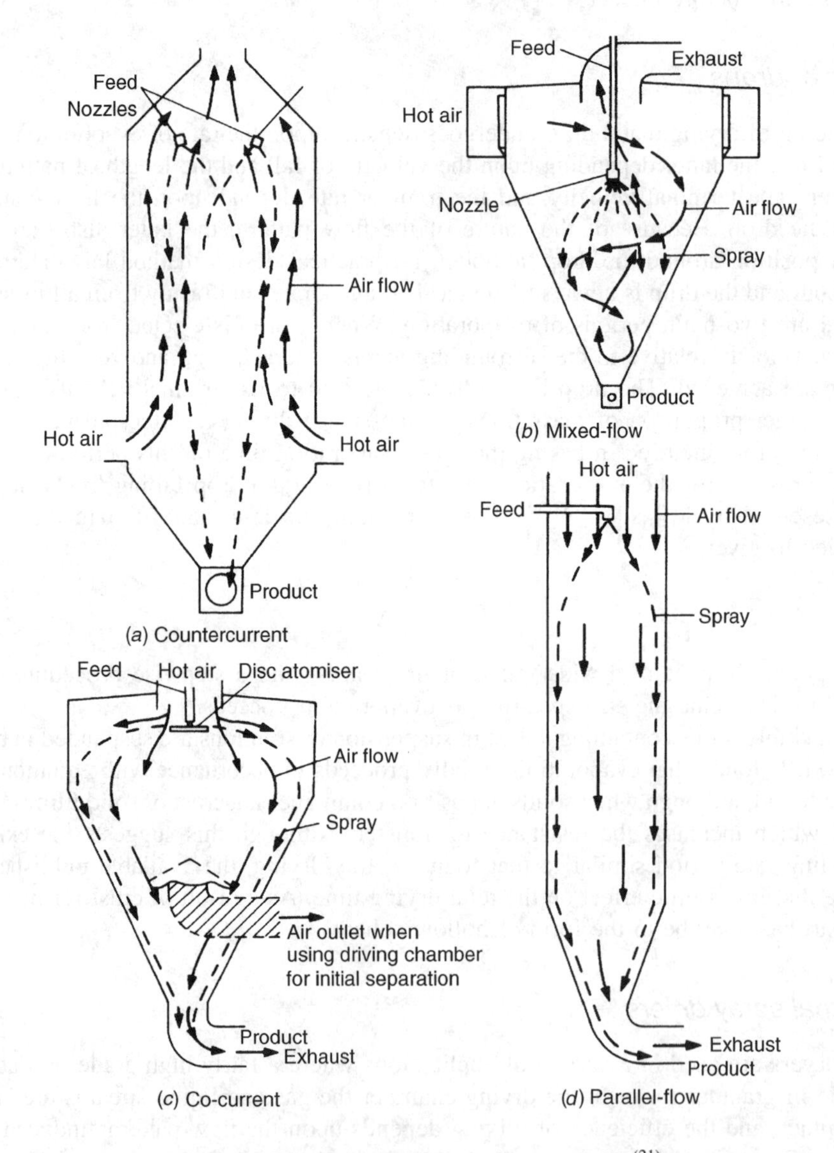


Figure 16.23. Flow arrangements in spray-drying[21]

The drying time and size of the particles are directly related to the droplet size, and therefore the initial formation of the spray is of great importance. The factors which govern the choice of atomisers for any specific drying application are principally dependent upon the characteristics of the liquid feed and upon the required drying characteristics of the drying chamber. A general guide is given in Table 16.5.

Table 16.5. Choice of atomiser

| | Atomiser | | | |
|---|---|---|---|---|
| | Pressure | Rotary | Twin-fluid | Spinning-cup plus gas blast |
| *Drying chamber* | | | | |
| Co-current | * | * | * | * |
| Countercurrent | * | | * | |
| *Feed* | | | | |
| Low viscosity solutions | * | | * | |
| High viscosity solutions | | | * | * |
| Slurries | | * | * | * |
| Pastes | | * | * | * |
| *Flexibility* | Flowrate $\alpha\sqrt{(-\Delta P)}$ | Flowrate independent of cup speed | Liquid flow independent of gas energy | Liquid flow independent of cup speed and gas energy |

Pressure nozzles are most suited to low viscosity liquids and, where possible, viscous liquids should be preheated to ensure the minimum viscosity at the nozzle. Because of their simplicity, pressure nozzles are also employed to atomise viscous liquids with a kinematic viscosity up to 0.01 $m^2/s$, depending upon the nozzle capacity. Under these conditions, injection pressures of up to 50 $MN/m^2$ may be required to produce the required particle size. With slurries, the resulting high liquid velocities may cause severe erosion of the orifice and thus necessitate frequent replacement.

Spinning discs are very suitable for slurries and pastes, while high viscosity liquids tend to produce a stringy product. Care must also be taken in design to minimise incrustation around the lip and subsequent out-of-balance as drying takes place.

The simple gas atomiser is inherently fairly flexible although it has not yet found widespread application. This is a result of its tendency to produce a dusty product containing a large proportion of very small particles.

Often, little difficulty is experienced in removing the majority of the dried product, though in most cases the smaller particles that may be carried over in the exit gases must be reclaimed. Cyclones are the simplest form of separator though bag filters or even electrostatic precipitators may be required. With heat-sensitive materials, and in cases where sterility is of prime importance, more elaborate methods are required. For example, cooling streams of air may be used to aid the extraction of product while maintaining the required low temperature. Mechanical aids are often incorporated to prevent particles adhering to the chamber walls, and, in one design, the cooling air also operates a revolving device which sweeps the walls.

In some cases all the product is conveyed from the dryer by the exhaust gases and collected outside the drying chamber. This method is liable to cause breakage of the particles though it is particularly suited for heat-sensitive materials which may deteriorate if left in contact with hot surfaces inside the dryer.

Spray drying has generally been regarded as a relatively expensive process, especially when indirect heating is used. The data given in Table 16.6 taken from GROSE and DUFFIELD[40] using 1990 costs, illustrate the cost penalties associated with indirect heating or with low inlet temperatures in direct heating.

Table 16.6. Operating costs for a spray dryer evaporating 0.28 kg/s (1 tonne/h) of water[40]

Basis: Air outlet temperature, 353 K
Cost of steam, £10/Mg
Cost of fuel oil, £150/Mg
Cost of power, £0.054/kWh (£15.0 × $10^{-6}$/kJ)

| | Air inlet temperature (K) | Steam flow (kg/s) | Oil flow (kg/s) | Power consumption (kW) | Cost (£/100 h) | | | Cost (£/Mg) |
|---|---|---|---|---|---|---|---|---|
| | | | | | Fuel | Power | Total | Total |
| Air heated: | | | | | | | | |
| Indirectly | 453 | 0.708 | — | 71 | 2550 | 384 | 2934 | 8150 |
| Direct combustion | 523 | — | 0.033 | 55 | 1800 | 296 | 2096 | 5820 |
| Direct combustion | 603 | — | 0.031 | 47 | 1650 | 254 | 1904 | 5300 |

QUINN[41] has drawn attention to the advantages with larger modern units using higher air inlet temperatures, 675 K for organic products and 925 K for inorganic products.

In spray dryers, using either a nozzle or rotating disc as the atomiser[41], volumetric evaporative capacities are 0.0003–0.0014 kg/m$^3$ s for cross-and co-current flows, with drying temperatures of 420–470 K. For handling large volumes of solutions, spray dryers are unsurpassed, and it is only at feed rates below 0.1 kg/s, that a drum dryer becomes more economic. Indeed the economy of spray drying improves with capacity until, at evaporative capacities of greater than 0.6 kg/s, the unit running cost is largely independent of scale.

In the jet spray dryer, cold feed is introduced[42] into preheated primary air which is blown through a nozzle at velocities up to 400 m/s. Very fine droplets are obtained with residence times of around 0.01 s, and an air temperature of 620 K. This equipment has been used for evaporating milk without adverse effect on flavour and, although operating costs are likely to be high, the system is well suited to the handling of heat-sensitive materials.

## 16.5.7. Pneumatic dryers

In pneumatic dryers, the material to be dried is kept in a state of fine division, so that the surface per unit volume is high and high rates of heat transfer are obtained. Solids are introduced into the dryer by some form of mechanical feeder, such as a rotating star wheel, or by an extrusion machine arranged with a high-speed guillotine to give short lengths of material, such as 5–10 mm. Hot gases from a furnace, or more frequently from an oil burner, are passed into the bottom of the dryer, and these pick up the particles and carry them up the column. The stream of particles leaves the dryer through a cyclone separator and the hot gases pass out of the system. In some instances, final collection of the fine particles is by way of a series of bag filters. The time of contact of particles with the gases is small, typically 5 s, even with a lengthy duct — and the particle temperature does not approach the temperature of the hot gas stream. In some cases the material is recycled, especially where bound moisture is involved. Evaporative capacities, which are high, are greatly affected by the solids–air ratio. Typical thermal and power requirements are given by QUINN[43] as 4.5 MJ/kg moisture evaporated and 0.2 MJ/kg, respectively.

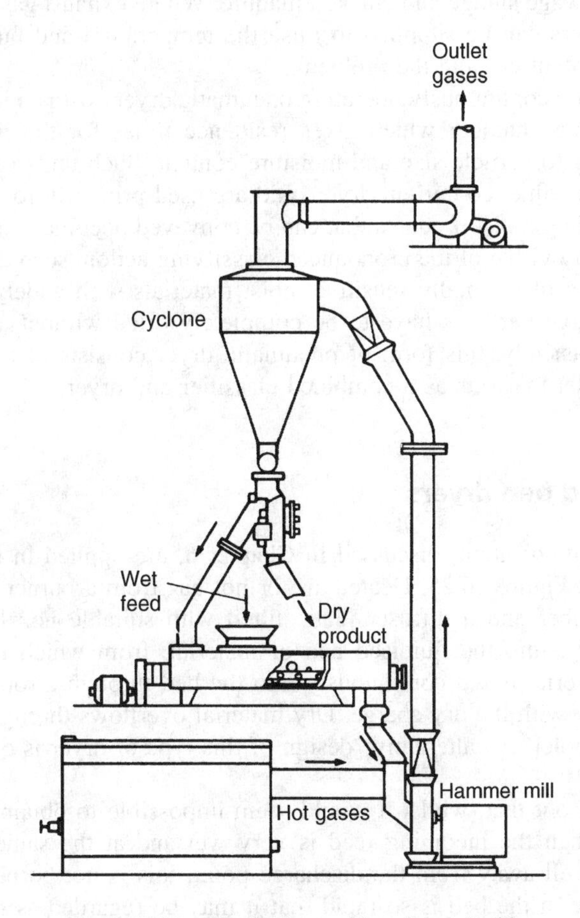


Figure 16.24. Air-lift dryer with an integral mill

A typical installation with the associated equipment is shown in Figure 16.24. Wet feed is delivered on to a bed of previously dried material in a double-paddle mixer to produce a friable mixture. This then passes to a cage mill where it comes into contact with hot gases from the furnace. Surface and some inherent moisture is immediately flash evaporated. The cage mill breaks up any agglomerates to ensure uniform drying of each individual particle. Gases and product are drawn up the drying column, where inherent moisture continues to be evaporated, before passing into the cyclone collector. Separated solids discharge through a rotary air lock into a dry divider which is set to recycle an adequate percentage of solids for conditioning the new wet feed in the double paddle mixer. Gases from the cyclone are vented to atmosphere through a suitable dust collector or wet scrubber. The system operates under suction and dust is therefore reduced to a minimum. A direct oil or gas fired furnace is generally employed and the heat input is controlled according to the vent stack gas temperature. Indirect heating may be used where contamination of the product is undesirable.

Materials handled include food products, chalk, coal, organic chemicals, clays, spent coffee grounds, sewage sludge and chicken manure. Where exhaust gases have unpleasant odours, after-burners can be supplied to raise the temperature and burn off the organic and particulate content causing the problem.

*Convex dryers* are continuously operating pneumatic dryers with an inherent classifying action in the drying chamber which gives residence times for the individual particles differing according to particle size and moisture content. Such units offer the processing advantages of short-time, co-current dryers and are used primarily for drying reasonably to highly free-flowing moist products that can be conveyed pneumatically and do not tend to stick together. By virtue of the pronounced classifying action, such dryers are also well suited to the drying of thermally sensitive moist materials with widely differing particle sizes where the large particles have to be completely dried without any overheating of the small ones. Basically, this form of pneumatic dryer consists of a truncated cyclone with a bottom outlet that acts as a combined classifier and dryer.

## 16.5.8. Fluidised bed dryers

The principles of fluidisation, discussed in Chapter 6, are applied in this type of dryer, shown typically in Figure 16.25. Heated air, or hot gas from a burner, is passed by way of a plenum chamber and a diffuser plate, fitted with suitable nozzles to prevent any back-flow of solids, into the fluidised bed of material, from which it passes to a dust separator. Wet material is fed continuously into the bed through a rotary valve, and this mixes immediately with the dry charge. Dry material overflows through a downcomer to an integral after-cooler. An alternative design of this type of dryer is one in which a thin bed is used.

QUINN[43] points out that, whilst it would seem impossible to obtain very low product moisture levels when the incoming feed is very wet and at the same time ensure that the feed point is well away from the discharge point, this is not borne out by operating experience. Mixing in the bed is so rapid that it may be regarded as homogeneous, and baffles or physical separation between feed and discharge points are largely ineffective. The very high mass-transfer rates achieved make it possible to maintain the whole bed in a dry condition. Some rectangular fluidised-bed dryers have separately fluidised compartments through which the solids move in sequence. The residence time is very similar for all particles and the units, known as *plug-flow* dryers, often employ cold air to effect product cooling in the last stage.

Many large-scale uses include the drying of fertilisers, plastics materials, foundry sand, and inorganic salts, and AGARWAL, DAVIS, and KING[44] describe a plant consisting of two units, each drying 10.5 kg/s of fine coal. Small fluidised-bed dryers also find use in, for example, the drying of tablet granulations in the pharmaceutical industry[45].

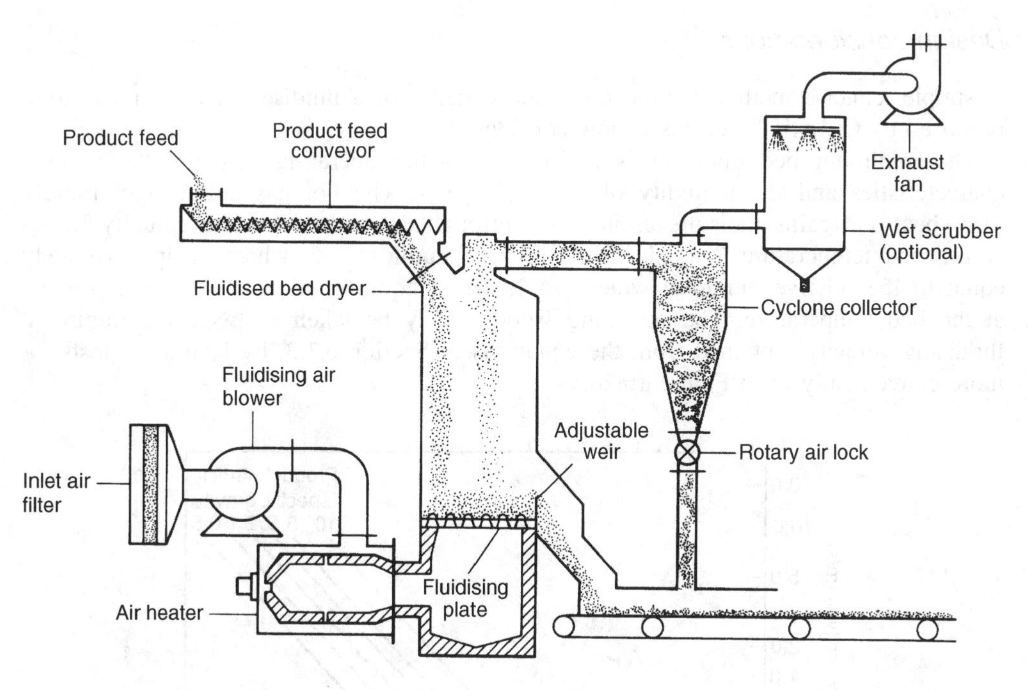


Figure 16.25. Flow diagram for a typical continuous fluidised-bed dryer

Dryers with grid areas of up to 14 $m^2$ have been built[46] and evaporative capacities vary from 0.02 kg/s $m^2$ grid area for the low-temperature drying of food grains to 0.3 kg/s $m^2$ for the drying of pulverised coal by direct contact with flue gases[44]. Specific air rates are usually 0.5–2.0 kg/s $m^2$ grid area and the total energy demand is 2.5–7.5 MJ/kg moisture evaporated. The exit gas is nearly always saturated with vapour for all allowable fluidisation velocities.

The choice between spray, pneumatic and fluidised dryers depends very much on the properties of the particles and some guidance in this respect is given in Table 16.7.

Table 16.7. Selection of spray, pneumatic and fluidised-bed dryers

| Particle property | Spray dryer | Pneumatic dryer | Fluidised-bed dryer |
|---|---|---|---|
| Initial moisture greater than 80 per cent | Yes | No | No |
| Too dry to pump | No | Yes | Yes |
| Wet enough to pump but moisture less than 80 per cent | Yes | Yes | No |
| Solids in dissolved state | Yes | No | No |
| Partially dry but sticky particles | Yes | No | No |
| Fragile particles | Yes | No | Possible |
| Very small particles | Yes | Yes | No |
| Residence time (s) | 3–10 | 1–10, often less than 1 | Widely variable, greater than 10 |
| Heat sensitive material | Yes | Yes | No |
| Relative drying speed | Third | First | Second |

### *Design considerations*

A simple, concise method for the preliminary sizing of a fluidised bed dryer has been proposed by CLARK[47] and this is now considered.

The minimum bed diameter is a function of the operating velocity, the particle characteristics and the humidity of the drying gas. The hot gas at the inlet rapidly loses heat and gains moisture as it passes through the bed which it eventually leaves at the bed temperature $T_b$ and with a relative humidity $\mathscr{R}$, which is approximately equal to the relative humidity which would be in equilibrium with the dried product at the bed temperature. The operating velocity may be taken as twice the minimum fluidising velocity, obtained from the equations in Section 6.1.3, by laboratory tests, or more conveniently from Figure 16.26.

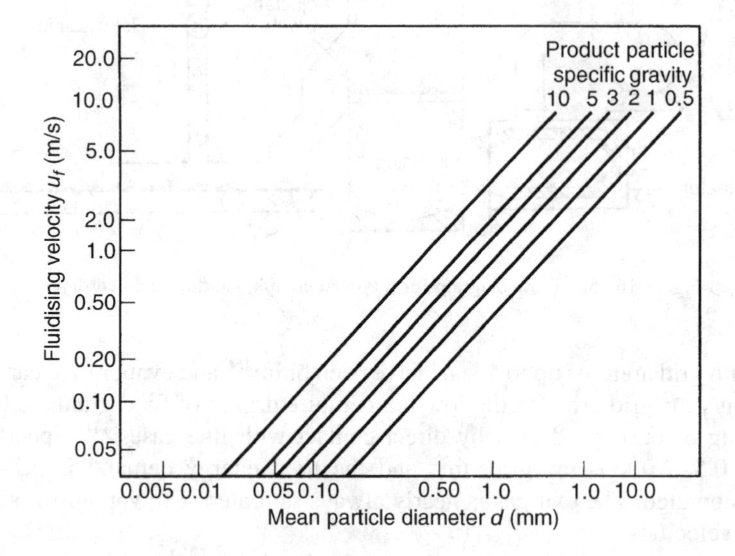


Figure 16.26. Superficial operating velocity in fluidised bed dryers[47]

For drying media other than air at approximately atmospheric pressure, the velocity obtained from Figure 16.26 should be multiplied by $0.00975\ \rho^{-0.29}\mu^{-0.43}$ where $\rho$ and $\mu$ are the density (kg/m$^3$) and the viscosity (Ns/m$^2$) of the fluidising gas.

From a mass balance across the bed:

$$(\mathscr{R}100)(P_w/P) = [W + (G/(1+\mathscr{H}).\mathscr{H}]\big/[W + (G/(1+\mathscr{H})(0.625+\mathscr{H})] \quad (16.39)$$

where: $\mathscr{R}$ = exit gas relative humidity (per cent),
$P_w$ = vapour pressure of water at the exit gas temperature (N/m$^2$),
$P$ = total static pressure of gases leaving the bed (N/m$^2$),
$W$ = evaporative capacity (kg/s),
$G$ = inlet flow of air (kg/s), and
$\mathscr{H}$ = humidity of inlet air (kg/kg dry air)

The term $G/(1+\mathscr{H})$ is in effect the flow of dry air, and 0.625 is approximately the ratio of molecular masses in the case of water and air.

Values of $P_w$ may be obtained from Figure 16.27 and, for indirect heating, $\mathscr{H}$ is the humidity of the ambient air. For direct heating, $\mathscr{H}$ may be obtained from Figure 16.28 which assumes that the air at 293 K, with a humidity of 0.01 kg/kg, is heated by the combustion of methane.

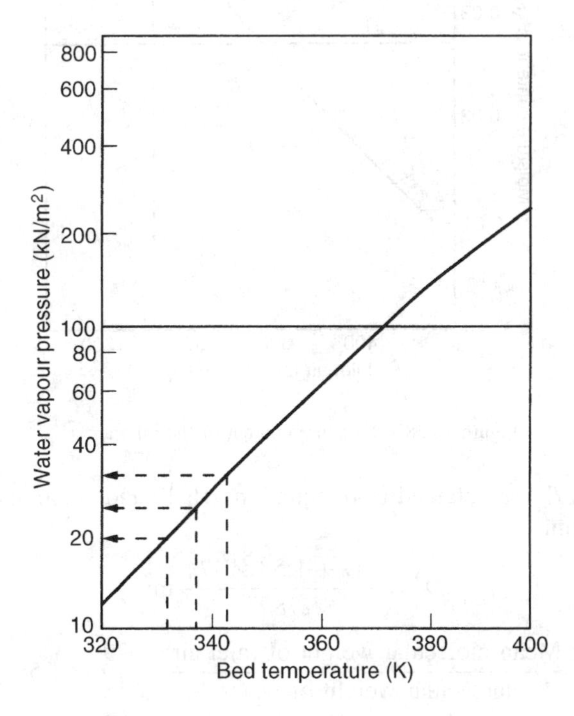


Figure 16.27. Water vapour pressure at the bed exit[47]

A heat balance (in W) across the bed gives:

$$C_m G(T_m - T_b) = \lambda_b W + C_f F(T_b - T_f) \tag{16.40}$$

where:
$C_m$ = mean thermal capacity of the gas between $T_m$ and $T_b$ (J/kg deg K) (1000 J/kg deg K for air),
$C_f$ = $[(X_f c_x + c_s)/(X_f + 1)]$ average thermal capacity of the wet solid between $T_f$ and $T_b$ (J/kg deg K),
$X_f$ = moisture content of wet feed (kg/kg dry solid),
$C_x$ = heat capacity of liquid being evaporated (J/kg deg K),
$C_s$ = heat capacity of dry solid (J/kg deg K),
$\lambda_b$ = mean latent heat of liquid at $T_b$ (J/kg),
$T_m, T_b, T_f$ = temperatures of inlet gas, bed and wet feed respectively (K), and
$F$ = rate of wet solid feed (kg/s)

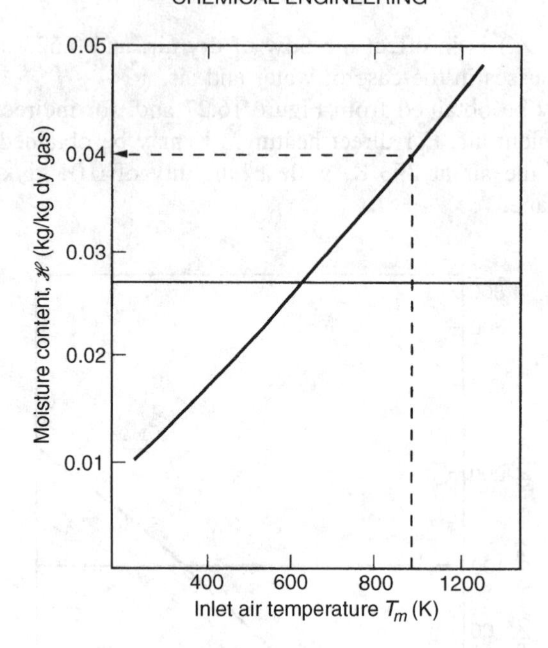


Figure 16.28. Moisture content of the inlet air[47]

Values of $G$ and $T_b$ are obtained from equations 16.39 and 16.40 and the bed diameter is then obtained from:

$$D^2 = \frac{(G + 1.58\ W)T_b}{278u_f}\ \text{m}^2 \tag{16.41}$$

noting that : $\dfrac{\text{Mean molecular weight of inlet air}}{\text{molecular weight of water}} = \dfrac{29}{18} \approx 1.58$

and:

$$\frac{\pi}{4}\left\{\frac{\text{(molecular weight or air)} \times \text{(datum temperature)}}{\text{kilomolecular volume}}\right\} = \frac{\pi}{4}\left(\frac{29 \times 273}{22.4}\right)$$

$$= 278\ \text{kg deg K/m}^3$$

## Example 16.6

A granular material of density 5000 kg/m$^3$ is to be dried in a fluidised bed dryer using directly heated air at 811 K. The particle size is 0.5 mm and 1.26 kg/s of water is to be removed from 12.6 kg/s of solid feed at 293 K. What diameter of bed should be specified?

$\lambda_b = 2325$ kJ/kg, ambient air is at 293 K, $\mathscr{H} = 0.01$ kg/kg and $C_f = 1.67$ kJ/kg deg K

## Solution

From Figure 16.28: $\mathscr{H} = 0.036$ kg/kg at 811 K

and the right-hand side of equation 16.39 is:

$$= (1.26 + 0.0347\ G)/(1.26 + 0.638\ G)$$

Taking $\mathscr{R}$ as 90 per cent and $P$ as 101.3 kN/m$^2$, then, for assumed values of $T_b$ of 321, 333 and 344 K:

$$P_w = 13, 20 \text{ and } 32 \text{ kN/m}^2, \text{ respectively}$$

and: $$G = 27.8, 12.9 \text{ and } 6.02 \text{ kg/s, respectively.}$$

Using equation 16.40, for $T_b = 321$, 333 and 344 K,

$$G = 7.16, 7.8 \text{ and } 7.54 \text{ kg/s respectively.}$$

Plotting $G$ against $T_b$ for each equation on the same axis, then:

$$G = 8.3 \text{ kg/s when } T_b = 340 \text{ K.}$$

From Figure 16.27, $u_f = 0.61$ m/s. Hence, in equation 16.41:

$$D^2 = 340(8.3 + (1.58 \times 1.26))/(278 \times 0.61)$$

and: $$\underline{\underline{D = 4.60 \text{ m}}}$$

This is a very large diameter bed and it may be worthwhile increasing the fluidising velocity provided any increased elutriation is acceptable. If $u_f$ is increased to 1.52 m, the diameter then becomes 2.88 m with a subsequent reduction in capital costs.

One development in the field of fluidised bed drying is what is known as *sub-fluidised conditioning*. A fluidised-bed dryer normally works with a maximum residence time of some 1200 s. If this is increased, then the spread of product residence times increases excessively due to the fact that axial and longitudinal mixing occur in the bed during the fluidisation process and, the more vigorous the fluidisation, the greater the spread of residence times. If fluidisation continues for too long, breakage and other product damage is likely to occur. A solution to this problem, developed by Ventilex, is sub-fluidised conditioning where forward movement of the product is created by a shaking mechanism in which the product is kept at the threshold of fluidisation. As a result, there is little or no axial or longitudinal mixing and the spread of residence time is kept to a minimum. In essence, this process provides a situation which approaches that of ideal plug flow, and many products can be conditioned by this combination of fluidised and sub-fluidised techniques. In addition, even less air is required for drying and thus small auxiliary equipment is required and energy requirements are less. Nominal bed thicknesses are 150–250 mm and long residence times, of up to 7 ks, are possible with a minimal time spread. In the drying of drugs, nuts, meats and rice, no degradation takes place and, in this respect, the process is comparable with traditional installations such as conveyor dryers, with the added benefits of fluidisation. The system is also applicable to the drying of minerals such as china clay and sand, and also to animal wastes and sludges. A typical installation together with a flow diagram is shown in Figure 16.29.

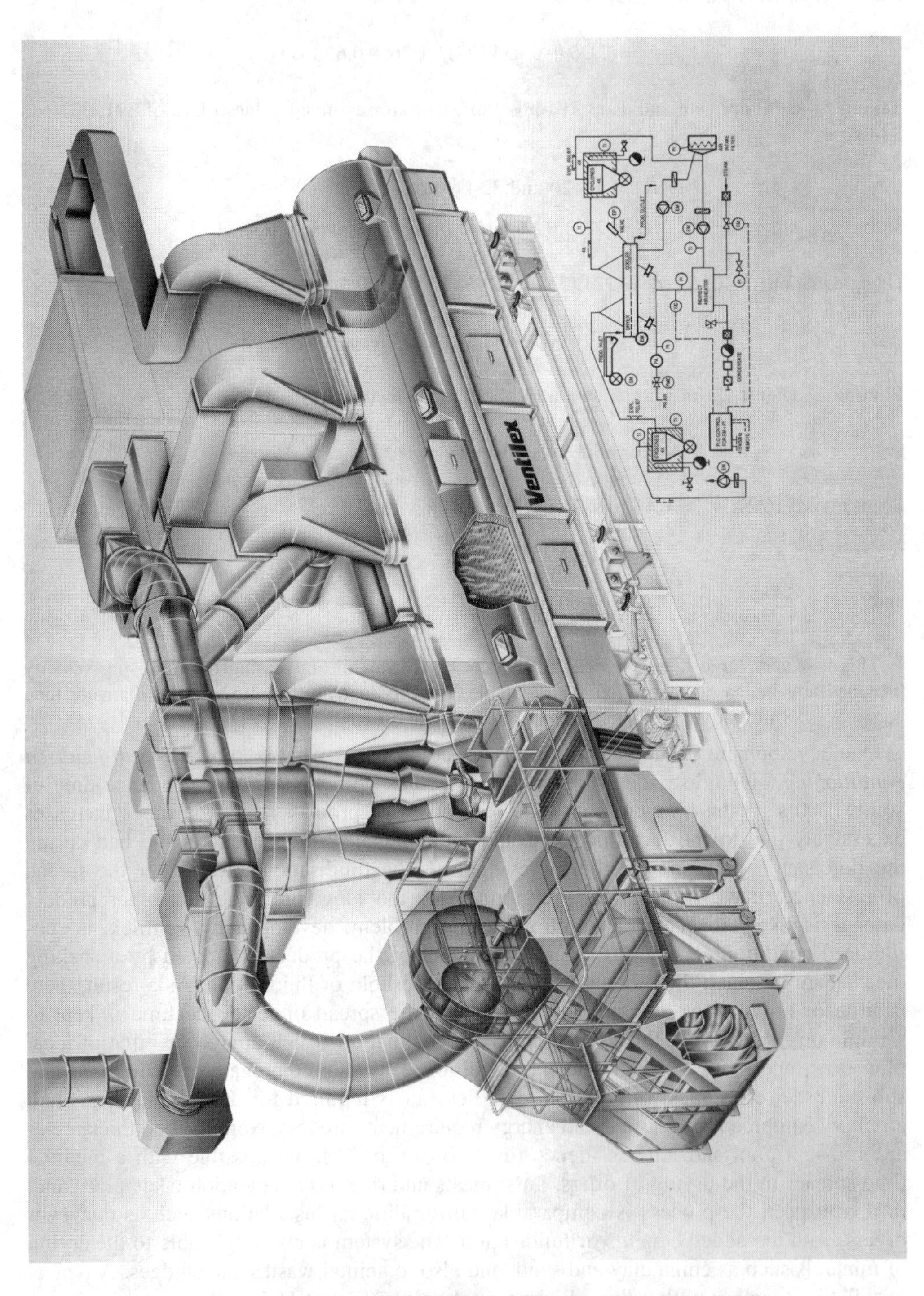


Figure 16.29. Unit for sub-fluidised drying

### 16.5.9. Turbo-shelf dryers

The handling of sticky materials can present difficulties, and one type of dryer which is useful for this type of material is the turbo-dryer. As shown in Figure 16.30, wet solid is fed in a thin layer to the top member of a series of annular shelves each made of a number of segmental plates with slots between them. These shelves rotate and, by means of suitably placed arms, the material is pushed through a slot on to a shelf below. After repeated movements, the solid leaves at the bottom of the dryer. The shelves are heated by a row of steam pipes, and in the centre there are three or more fans which suck the hot air over the material and remove it at the top.

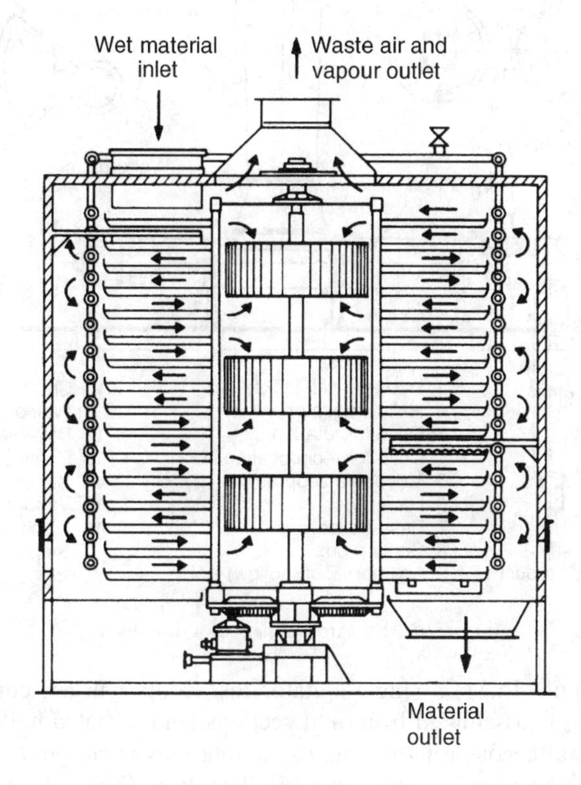


Figure 16.30. Turbo-shelf dryer

The accelerated drying induced by the raking of the material gives evaporative capacities of 0.0002–0.0014 kg/s $m^2$ shelf area which are comparable with those obtained by through circulation on perforated belts. Shelf areas are 0.7–200 $m^2$ in a single unit and the dryer may easily be converted to closed-circuit operation, either to prevent emission of fumes or in order to recover valuable solvents. Typical air velocities are 0.6–2.5 m/s, and the lower trays are often used to cool the dry solids. A turbo-dryer combines cross-circulation drying, as in a tray dryer, with drying by showering the particles through the hot air as they tumble from one tray to another.

## 16.5.10. Disc dryers

A disc dryer provides a further way in which pasty and sticky materials may be handled, and it can also cope satisfactorily with materials which tend to form a hard crust or pass through a rheologically-difficult phase during the drying operation.

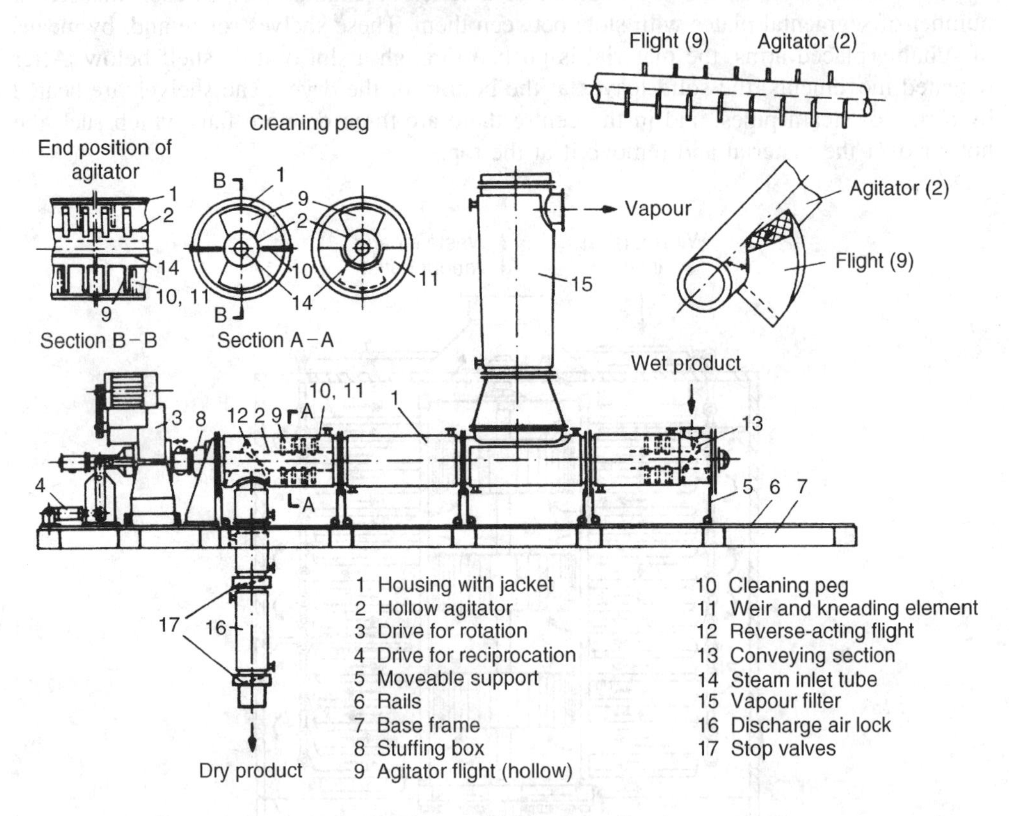


Figure 16.31. Arrangement of a disc dryer

As shown in Figure 16.31, a single-agitator disc contact dryer, consists of a heated cylindrical housing (1) assembled from unit sections, and a heated hollow agitating rotor (2) which has a simultaneous rotating and oscillating movement produced by means of a rotating drive (3) and a reciprocating drive (4). The drive (3 and 4) and stuffing box (8) are located at the dry product end, and the stuffing box is protected by a reverse acting flight (12). The wet product is introduced, drawn in by the screw flight (13) and continuously conveyed through the dryer. The vapour passes through the vapour filter (15) to the condenser. This vapour filter system has a back-scavenging action, and is specially designed for removing dust from vapour in vacuum drying plants. The hollow agitator (2) is fitted, over its whole length, with heated flights, which are arranged equidistantly in pairs. Between every two axially neighbouring agitator flights, there are, projecting inwards from the housing fixed wiping pegs (10) or annular weir/kneading elements (11) which extend inward to the agitator core.

The self-cleaning of the heating surfaces is achieved by the combined rotating-reciprocating movement of the agitator. The stationary elements (10 and 11) clean the faces of the agitator flights, at the end of each oscillating movement, during one rotation of the agitator. By the forward and backward movement, the edges of the agitator flights (9) clean the inner surface of the housing and the fixed elements projecting inwards to the core of the agitator clean the agitator (2). In all, about 95 per cent of the heating surface is cleaned.

The rotating and reciprocating motions need not be synchronised, because the individual agitator flights (9) oscillate only between two adjacent fixed elements (10 and 11). This is so that the speed of rotation, the frequency of reciprocation and the forward and backward speed of reciprocation can be adjusted independently of one another over a wide range of settings. The housing sections (1) which are supported by frames (5) on rails, can be drawn forward when cleaning is necessary. The transport of pasty products through the dryer may be achieved by the differential forward and backward oscillatory motion, combined with the action of the bevelled edges of the agitator flights.

*Contivac dryers*, for example, have heating surfaces of 4–60 $m^2$ and volumes of 0.1–3 $m^3$. They may be operated under vacuum or up to 400 $kN/m^2$ with heating fluids at 330 K–670 K. The evaporative capacity is 0.03–0.55 kg/s and the agitator speed ranges from 0.1 Hz (6 rpm) for rheologically difficult materials, to 1 Hz (60 rpm) for easier applications.

Operating on a similar principle is the *Buss paddle* dryer which effects batch drying of liquids, pasty and sandy materials. The paddle dryer, which is the main element of the drying unit. It consists of a horizontal, cylindrical housing which contains a paddle agitator in the form of a hollow shaft carrying agitator arms. The jacket, the hollow shaft and the agitator arms are steam-heated. The paddle agitator is driven by an electric motor and gear unit. A built-on torque bracket with a microswitch protects the paddle agitator against overloading. The product to be dried is passed through the charging nozzle into the paddle dryer, where it is distributed uniformly by the rotating paddle agitator. The drying proceeds under vacuum while intensive intermixing by the agitator causes continual renewal of the product particles in contact with the heated surfaces. This guarantees efficient heat transfer and uniform product quality. Vapours are purged of dust in passing through a vapour filter and are then fed to a condenser. Non-condensable gases are drawn off by the vacuum system. The vapour filter is equipped with a removable filter insert and, to prevent excessive pressure drop across the fabric from thick, and possibly moist dust build-up on the filter sacks, these are provided with a reverse jet arrangement, that is the individual filter sacks are cleaned in turn automatically during operation by short though powerful countercurrent blasts of steam blown through the filter sack. This serves to blow and shake off the dust layer and keep the filter sack dry. After the drying process is complete, the dried material may be cooled by applying cooling water to the dryer jacket and paddle agitator. The dryer is emptied by the arms of the paddle agitator, which are designed to move the material in the vessel towards the outlet when rotated backwards. The discharge outlet is specially constructed to prevent the formation of a plug of material. At 373 K, an evaporation rate of over 4 g/s $m^2$ can be achieved for a steam consumption of 1.5 kg/kg of evaporated water.

### 16.5.11. Centrifuge dryers

Where a product of a high quality and purity is required, this can, in many cases, be achieved only by washing the solids during a centrifugation process in order to flush out the suspension mother liquor or to dissolve salt crystals. The washed crystals are then usually dried in a separate physical unit operation involving different equipment and problems can be encountered due to the exposure to the atmosphere of the wet cake as it is transferred between the centrifuge and the dryer. These problems may be largely eliminated by using a combined centrifuge–dryer which it is claimed, has the following advantages:

(a) It is hermetically sealed from the environment and is therefore easy to inert.
(b) There is no human contact with the product at any stage of the operation and no possibility of foreign materials being introduced.
(c) The handling of the product is simple and gentle.
(d) Drying times can be reduced by using a jet-pulsed bed drying technique.
(e) The unit can be emptied completely with no residual cake in the basket or on the wall of the drum.

All these advantages are inherent in the FIMA, TZT centrifuge–dryer which is operated as follows.

The feed suspension is introduced into the centrifuge through a hollow drive shaft, and the liquid from the suspension is separated from the crystals and discharged from a sealed filter basket by passing it though a metal filter which retains the solids inside a sealed chamber. Any contaminants are then washed away from the filling pipe, the centrifuge basket and the separated solids by introducing wash liquid through the hollow drive shaft. This operation may be repeated as required. One advantage of the TZT system is that, with no internal pipes or structures that might accumulate unwashed solids, all the separated solids make contact with the wash water. The ring of separated solids is then removed from the wall of the centrifuge by introducing powerful gas blasts beneath the filter medium and the wet en-masse solids accumulate at the bottom of the filter drum. The drying process is achieved by rotating the drum at slow speeds and injecting heated drying gas as process into the closed centrifuge chamber. The moisture-laden gas leaves passing through the filter medium so that the solids are retained within the basket. Such a process is extremely gentle to the solids and dries at low product temperatures even when using higher gas temperatures as a result of the cooling effect of the surface evaporation of the moisture. Dry powder is discharged from the unit by opening the sealing centrifuge housing and fluidising the solids which then enter an integrated powder conveying system. In this system the gas used for drying may be recycled in a closed gas loop, or it may be discharged to atmosphere in an open system.

Units of this type have a capacity of 20–400 kg in terms of filling mass, filter areas of 0.37–2.4 $m^2$ and drum diameters of 400–1300 mm and lengths of 300–600 mm.

Although this is, in essence a batch operation, close reproducibility between batches overcomes many of the problems associated with batch identification which can be a problem with more conventional drying equipment.

## 16.6. SPECIALISED DRYING METHODS

### 16.6.1. Solvent drying

Two processes are used here:

*Superheated-solvent drying* in which a material containing non-aqueous moisture is dried by contact with superheated vapours of its own associated liquid, and,

*Solvent dehydration* in which water-wet substances are exposed to an atmosphere of a saturated organic solvent vapour.

The first of these has advantages where a material containing a flammable liquid such as butanol is involved. Drying is effected with a gas with an air–moisture ratio of 90 kg/kg moisture in order to ensure that the composition is well below the lower explosive limit. The heat requirement is as great as those for superheated solvent drying at the same gas outlet temperature of 400 K[48]. Superheated solvent drying in a fluidised bed has been used for the drying of polypropylene pellets to eliminate the need for water-washing and for fractionating. A flowsheet of the installation is shown[49] in Figure 16.32.

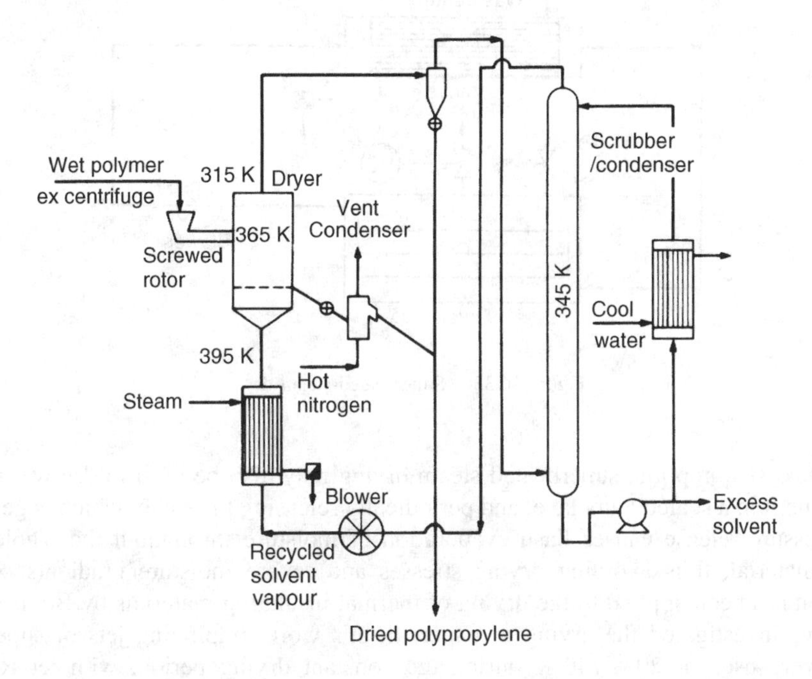


Figure 16.32. Flowsheet for drying polypropylene by fluidising with a hot stream of solvent vapour[49]

The most important applications of solvent dehydration lie within the field of kiln drying for seasoning timber where substantial reductions in drying times have been achieved[50].

### 16.6.2. Superheated steam drying

The replacement of air by superheated steam to take up evaporating moisture is attractive in that it provides a high temperature heat source which also gives rise to a much higher driving force for mass transfer since it does not become saturated at relatively low moisture contents as is the case with air. In the drying of foodstuffs, a further advantage is the fact that the steam is completely clean and there is much less oxidation damage. In the seasoning of timber, for example, drying times can be reduced quite significantly. Although the principles involved have been understood for some considerable time, as BASEL and GRAY[49] points out, applications have been limited to due to corrosion problems and the lack of suitable equipment. A flowsheet of a batch dryer using superheated steam is shown in Figure 16.33. The dryer is initially filled with air circulated by a blower, together with evaporated moisture. Any excess moisture is vented to atmosphere so that the air is gradually replaced by steam. For an evaporation rate of 10 kg/m$^3$ volume of the dryer chamber, the composition of the gas phase would reach about 90 per cent steam in about 600 s.

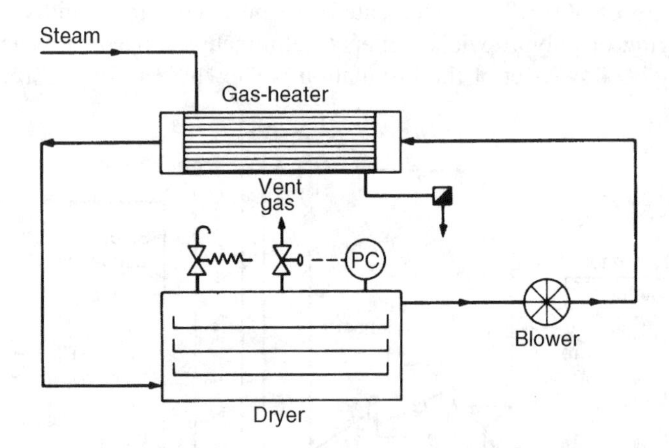


Figure 16.33. Superheated-steam dryer

As LUIKOV[51] reports, superheated steam drying may also be used to dry wet material by heating it in a sealed autoclave, and periodically releasing the steam which is generated. This pressure release causes flash evaporation of moisture throughout the whole extent of the material, thus avoiding drying stresses and severe moisture gradients. Such an operation has been applied to the drying of thermal insulating materials by BOND *et al.*[52] who have investigated the drying of paper. In this work, impinging jets of superheated steam were used at 293–740 K during the constant drying period, with jet Reynolds numbers, 100–12000. Above 450 K, steam drying was found to be much faster than air drying for the same mass velocity of gas. The specific blower power was found to be much lower than for air drying at temperatures of industrial importance. It was concluded that steam-impingement drying can lead to significant reductions in both capital investment and energy costs.

In tests on the drying of sand, WENZEL and WHITE[53] found that the use of steam rather than air did not alter the general characteristics of the drying process, and that the drying rate during the constant rate period was determined by the heat transfer rate. In these tests, the heat transferred by radiation from the steam and surrounding surfaces was 7.5–31 per cent of the total heat transferred, and coefficients of convective heat transfer were 13–100 W/m$^2$K. It was concluded that higher drying rates and greater thermal efficiencies are possible when drying with superheated steam as opposed to air, and that the choice of steam drying must depend on a balance between the savings in operating costs and the higher capital investment attributable to the higher temperatures and pressures.

SCHWARTZE and BRÖCKER[54], who has made a theoretical study of the evaporation of water into mixtures of superheated steam and air, has calculated the inversion temperature above which the evaporation rate into pure superheated steam is higher than that into dry air under otherwise similar conditions. The data obtained are given in Figure 16.34 which shows quite clearly the enhanced evaporation rates at gas temperatures above about 475 K. This inversion temperature is given by the point of intersection of the curves for evaporation rate with dry air and superheated steam. The Nusselt and Sherwood equations for heat and mass transfer coefficients for the relevant geometrical configuration, given in Volume 1, were used to calculate the evaporation rates, and these were found to be in excellent agreement with experimental data in the literature. TAYLOR and KRISHNA[55] points out that this approach may be used for a wide range of applications, and VIDAURRE and MARTINEZ[56] show that the model may be extended to include specialised applications, such as the evaporation of multicomponent liquids.

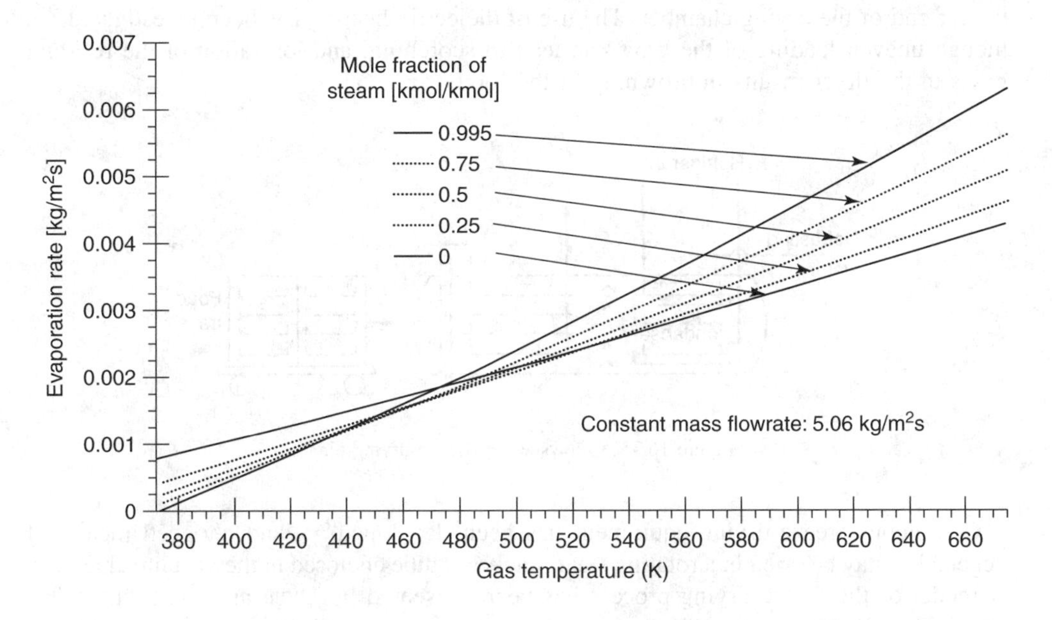


Figure 16.34. Inversion temperature with superheated steam drying[54]

### 16.6.3. Freeze drying

In this process, the material is first frozen and then dried by sublimation in a very high vacuum, 10–40 N/m$^2$, at a temperature of 240–260 K. During the sublimation of the ice, a dry surface layer is left, though this is not free to move because it has been frozen, and a honeycomb structure is formed. With normal vacuum drying of biological solutions containing dissolved salts, a high local salt concentration is formed at the skin, although with freeze drying this does not occur because of the freezing of the solid. Thus, freeze drying is useful not only as a method of working at low temperatures, but also as a method of avoiding surface hardening.

As described by CHAMBERS[57], heat has to be supplied to the material to provide the heat of sublimation. The rate of supply of heat should be such that the highest possible water vapour pressure is obtained at the ice surface without danger of melting the material. During this stage, well over 95 per cent of the water present should be removed, and in order to complete the drying, the material should be allowed to rise in temperature, to say ambient temperature. The great attraction of this technique is that it does not harm heat-sensitive materials, and it is suitable for the drying of penicillin and other biological materials. Initially, high costs restricted the application of the process though economic advances have reduced these considerably and foodstuffs are now freeze-dried on a large scale. MAGUIRE[58] has estimated a total cost of £0.06/kg of water evaporated, in a plant handling 0.5–0.75 kg/s (2–3 tonne/h) of meat.

A typical layout of a freeze-drying installation is shown in Figure 16.35. Heat is supplied either by conduction, or by radiation from hot platens which interleave with trays containing the product, and the sublimed moisture condenses on to a refrigerated coil at the far end of the drying chamber. The use of dielectric-heating has been investigated[59], though uneven loading of the trays can lead to scorching, and ionisation of the residual gases in the dryer results in browning of the food.

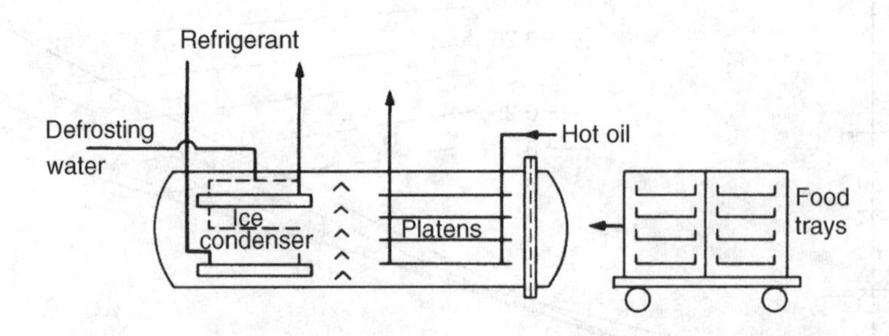


Figure 16.35. Flowsheet of freeze-drying plant

Continuous freeze-drying equipment has been developed[60], and chopped meat and vegetables may be dried in a rotating steam-jacketed tube enclosed in the vacuum chamber. A model of the freeze-drying process has been presented by DYER and SUNDERLAND[61], and further details are given in Section 15.8 on freeze crystallisation.

### 16.6.4. Flash drying

Conventional drying consists of mixing and heating the solids to achieve even drying, and simultaneously transporting the vapours away from the surface of the solids so as to maintain a high rate of mass transfer. Although these two stages can be achieved simultaneously by contacting the solids with a hot gas stream, heat-sensitive materials, such as foods and pharmaceuticals, may suffer thermal degradation, and other solids may lose some of their water of hydration when subjected to high temperatures. This problem may often be overcome by using a flash drying technique to ensure that the solids are in contact with hot gas in a highly turbulent environment for only a very short time, perhaps only for a few seconds.

A typical flash-drying process consists of a modified pneumatic conveyor in which the wet solids are introduced into a pipe through which they are transported in a high velocity hot gas stream. In such a process, sticky sludges must first be mixed with dry solids so that the resultant mixture breaks up into smaller particles in the gas stream. Such a process can be used for a wide range of applications[62]. In the drying of the gypsum by-product of flue gas desulphurisation, for example, operating at a minimum solids temperature allows the gypsum to retain its water of hydration, and in the conversion of sewerage sludge into dry fertiliser flash drying prevents the fertilizer from oxidising. Because waste flue gases can often be used as the drying medium, little or no external heat energy is required and, when the dried sludge is burned as a fuel, the heat generated provides much of the energy required for drying.

Even when back-mixed with dry solids, some materials are too sticky or paste-like to feed to a flash dryer, though these can often be handled in a spin dryer consisting of a high shear agitator inside a specially designed drying chamber. In this way, the wet paste may be subjected simultaneously to high shear to break up the solid mass, and to gas at a high velocity to dry the smaller particles. As the particles become small enough and dry enough, they are carried out of the chamber with the spent air and, as with conventional flash drying, the contact time is but a few seconds.

Neither flash nor spin-drying produces highly uniform particles and, even though the bulk of the particles remain at or near the wet bulb temperature, small particles are often over-dried and overheated. Particles of uniform size can be formed by spray drying, however, although it is then necessary that the feed should be in the form of a pumpable liquid which can be atomised so that the particles are evenly exposed to the hot air.

### 16.6.5. Partial-recycle dryers

The majority of drying operations depend on direct heating using a high flowrate of hot air and/or combustion products which is passed once through or over the wet material and then exhausted to the atmosphere. A variation of this arrangement is the closed-loop system where the entire air or gas stream is confined and recycled after condensing out the vapour. Such a system is justified only where the vaporised liquid or gas has to be recovered for economic or environmental reasons. In addition to the once-through and the closed-loop systems, there is a third class of dryers which incorporates the partial-recycle mode of operation. In this system, a substantial proportion, typically 40–60 per cent, of the outlet air is returned to the dryer in order to minimise the heating requirements and the amount of effluent treatment required. Conveyor, flash-pneumatic conveying-, fluidised-bed, rotary, spray and tray types of direct dryers can all be designed for closed-loop or partial-recycle operation and Cook[63] has claimed that recycling part of the exhaust stream has the following advantages:

(a) smaller collectors are needed for removing dried product from the exhaust gases,
(b) less heat is lost in the exhausts which leave, typically, at 340–420 K,
(c) since the total volume of exhaust gases is reduced, the cost of preventing undesirable gases or particles from entering the atmosphere is reduced,
(d) when using direct burners to heat the air, the level of oxygen content is minimised, thus enabling sensitive products to be handled and reducing potential fire and/or explosion hazards.

In conveyor and tray types of dryer, air is often recirculated inside the drying vessel in an attempt to save energy or to maintain a relatively high moisture content in the drying air. In other direct dryers such as flash, fluidised-bed, rotary and spray units, any recycle of exhaust air must be returned to the dryer using external ducting, the cost of which is offset by the net savings from the lower volumes of exhaust streams which have to be handled.

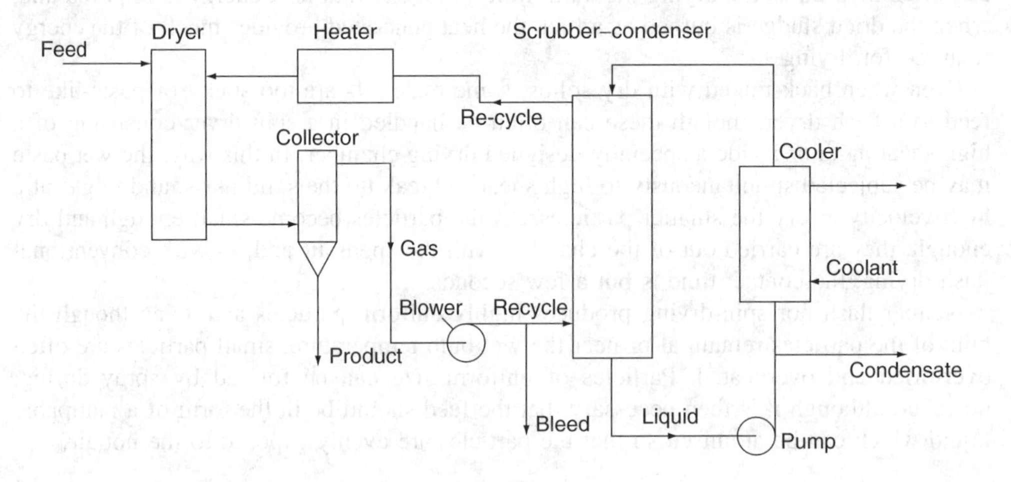


Figure 16.36. Partial-recycle drying systems

As shown in Figure 16.36, in a partial-recycle drying process, the total airflow leaving the dryer is first passed to a particulate or dust collector in order to remove entrained product or fines, and is then split into two streams — a bleed stream which is vented to atmosphere and a recycle stream which is returned to the system. The bleed stream is large enough to carry all of the moisture that has entered the system, and the recycle portion is returned to the inlet of either the heater or the drying vessel. In the condensing operation, shown in Figure 16.36, the exhaust stream from the dust collector is passed through a condenser, usually a wet scrubber, to remove most of the entrained moisture before it is returned to the inlet of the heater or the dryer. The bleed stream may also be passed through the condenser for gas-cleaning prior to discharge to atmosphere, thus removing moisture and thereby reducing the load on the condenser. Although condensers minimise the amount of bleed to the atmosphere and the moisture content of the drying air, they also cool the recycle stream and this increases costs due to the need to reheat the stream. The total volume of the bleed must be balanced by the heater output and any leakage into the system, noting that moisture can enter the dryer from three sources — the feed, combustion air and the supply air — and can leave in the bleed in the product and as condensate, if any. It may be noted that, in Figure 16.36, the scrubber–condenser also serves as a collector for particulates. If an oxygen content of less than 21 per cent by volume or a temperature above 270–640K is required, then a direct-fired heater may be used. If non-aqueous solvents are to be evaporated in the dryer, the effluent may require incineration or other treatment prior to discharge to the atmosphere, and dual blowers are often needed to maintain the required pressures throughout the system.

Cook[63] has presented data on the performance of a partial-recycle units based on a direct-fired burner using natural gas. These are compared with the once-through operation in Table 16.8.

Table 16.8. Comparison of Once-Through with Partial-Recycle Operation

| Example | Recycle mode | Burner outlet temperature (K) | Recycle (per cent) | Bleed (per cent) | Moisture (kg/kg air) | Heat Load (MW) | Air flow ($m^3/s$) |
|---|---|---|---|---|---|---|---|
| 1 | condensing | 1090 | 46.0 | 54.0 | 0.206 | 2.74 | 8.14 |
| 2 | condensing | 810 | 19.7 | 80.3 | 0.173 | 2.83 | 8.22 |
| 3 | condensing | 1370 | 58.9 | 41.1 | 0.222 | 2.68 | 8.11 |
| 4 | bleed only | 1160 | 46.0 | 54.0 | 0.325 | 2.52 | 7.92 |
| 5 | bleed only | 858 | 25.4 | 74.6 | 0.206 | 2.68 | 8.14 |
| 6 | once-through | 700 | 0.0 | 100.0 | 0.148 | 2.90 | 8.28 |

(Dryer inlet = 700 K, dryer outlet = 365 K)

In these systems, the total collection efficiencies of the dry product are 85 per cent for the drying vessel, 90 per cent for the cyclone collector and 98 per cent for the scrubber-condenser. The net efficiency of the system may be as high as 99.97 per cent if the scrubber effluent is considered as product. All the runs are based on 1.25 kg/s product and 0.75 kg/s evaporation at an elevation of 300 m above sea level. The total air flow is measured at the outlet before the stream is split into the recycle and bleed portions and, for such flows, the design of suitable fans is outlined by Jorgensen[64]. The calculations outlined here may be confirmed by the use of psychometric charts, and this procedure has been considered in some detail by Cook and Dumount[65].

In recent years, air-recycling in drying systems has become more widely adopted as a means of reducing total dryer discharges to the atmosphere. It would seem that this trend is likely to continue because of environmental concerns and the ever-increasing costs of treating emissions.

## 16.7. THE DRYING OF GASES

The drying of gases is carried out on a very large scale. The most important applications of the process are:

(a) in order to reduce the tendency of the gas to cause corrosion,
(b) preparation of dry gas for use in a chemical reaction, and
(c) reduction of the humidity of air in air-conditioning plants.

The problems involved in the drying or dehumidification of gases are referred to in Volume 1, Chapter 13, and the most important methods available are now summarised.

*Cooling*. A gas stream may be dehumidified by bringing it into contact with a cold liquid or a cold solid surface. If the temperature of the surface is lower than the dew point of the gas, condensation will take place, and the temperature of the surface will tend to rise by virtue of the liberation of latent heat: It is therefore necessary to remove heat constantly from the surface. Because a far larger interfacial surface can be produced with a liquid, it is usual to spray a liquid into the gas and then to cool it again before it is recycled. In many cases, countercurrent flow of the gas and liquid is obtained by introducing the liquid at the top of a column and allowing the gas to pass upwards.

*Compression*. The humidity of a gas may be reduced by compressing it, cooling it down to near its original temperature, and then draining off the water which has condensed. During compression, the partial pressure of the water vapour increases and condensation occurs as soon as the saturation value is exceeded.

*Liquid absorbents*. If the partial pressure of the water in the gas is greater than the equilibrium partial pressure at the surface of a liquid, condensation will take place as a result of contact between the gas and liquid. Thus, water vapour is frequently removed from a gas by bringing it into contact with concentrated sulphuric acid, phosphoric acid, or glycerol. Concentrated solutions of salts, such as calcium chloride, are also effective. The process may be carried out either in a packed column or in a spray chamber. Regeneration of the liquid is an essential part of the process, and this is usually effected by evaporation.

*Solid adsorbents and absorbents*. The use of silica gel or solid calcium chloride to remove water vapour from gases is a common operation in the laboratory. Moderately large units can be made, although the volume of packed space required is generally large because of the comparatively small transfer surface per unit volume. If the particle size is too small, the pressure drop through the material becomes excessive. The solid desiccants are regenerated by heating.

Gas is frequently dried by using a calcium chloride liquor containing about 0.56 kg calcium chloride/kg solution. The extent of recirculation of the liquor from the base of the tower is governed by heating effects, since the condensation of the water vapour gives rise to considerable heating. It is necessary to install heat exchangers to cool the liquor leaving the base of the tower.

In the contact plant for the manufacture of sulphuric acid, sulphuric acid is itself used for drying the air for the oxidation of the sulphur. When drying hydrocarbons such as benzene, it is sometimes convenient to pass the material through a bed of solid caustic soda, although, if the quantity is appreciable, this method is expensive.

The great advantage of materials such as silica gel and activated alumina is that they enable the gas to be almost completely dried. Thus, with silica gel, air may be dried down to a dew point of 203 K. Small silica gel containers are frequently used to prevent moisture condensation in the low pressure lines of pneumatic control installations.

## 16.8. FURTHER READING

BACKHURST, J. R. and HARKER, J. H.: *Process Plant Design* (Heinmann, London, 1973).

BACKHURST, J. R., HARKER, J. H., and PORTER, J. E.: *Problems in Heat and Mass Transfer* (Edward Arnold, London, 1974).

BAILEY, A. G.: *Electrostatic Spraying of Liquids* (John Wiley, New York, 1988).
BARCLAY, S. F.: *A Study of Drying* (Inst. Energy, London, 1957).
BRIDGMAN, M. J.: *Drying—Principles and Practice in the Process Industries* (Caxton, Christchurch, 1966).
BROWN, W. H.: *An Introduction to the Seasoning of Timber* (Pergamon Press, Oxford, 1965).
BRUIN, S. (ed): *Preconcentration and Drying of Food Materials* (Elsevier, London, 1988).
BUTCHER, C.: *Chem. Engr. London* No 450 (July 1988) 16. Fluid bed dryers.
CLARKE, R. J. and MACRAE, R. (eds): *Coffee*, Vol. 2: *Technology* (Elsevier, London, 1987).
KEEY, R. B.: *Drying Principles and Practice* (Pergamon Press, Oxford, 1972).
KEEY, R. B.: *Introduction to Industrial Drying Operations* (Pergamon Press, Oxford, 1978).
KEEY, R. B.: *Drying of Loose and Particulate Materials* (Hemisphere Publishing Corporation, London and Washington DC, 1991).
LUIKOV, A. V.: *Heat and Mass Transfer in Capillary-porous Bodies* (Pergamon Press, Oxford, 1966).
MASTERS, K.: *Spray Drying Handbook* (George Godwin, London, 1985).
MCCABE, W. L., SMITH J. C. and HARRIOTT, P.: *Unit Operations of Chemical Engineering*. 4th edn (McGraw-Hill, New York, 1984).
MUJUMDER, A. S. (ed): *Handbook of Industrial Drying*. (Marcel Dekker, London, 1987).
NONHEBEL, G.: *Gas Purification Processes* (Newnes, London, 1964).
NONHEBEL, G. and MOSS, A. A. H.: *Drying of Solids in the Chemical Industry* (Butterworth, London, 1971).
THIJSSEN, H. A. C. and RULKENS, W. H.: *Recent Developments in Freeze Drying* (Int. Inst. Refrigeration, Lausanne, 1969).
VANĚCĚK, J., MARKVART, M. and DRBOHLAV, R.: *Fluidized Bed Drying* (Leonard Hill, London, 1966).
WALAS, S. M.: *Chemical Process Equipment: Selection and Design* (Butterworth, London, 1989).
WILLIAMS-GARDNER, A.: *Industrial Drying* (George Godwin, London, 1971).

## 16.9. REFERENCES

1. SHERWOOD, T. K.: *Trans. Am. Inst. Chem. Eng.* **32** (1936) 150. The air drying of solids.
2. PEARSE, J. F., OLIVER, T. R., and NEWITT, D. M.: *Trans. Inst. Chem. Eng.* **27** (1949) 1. The mechanism of the drying of solids. Part I. The forces giving rise to movement of water in granular beds, during drying.
3. OLIVER, T. R. and NEWITT, D. M.: *Trans. Inst. Chem. Eng.* **27** (1949) 9. The mechanism of drying of solids. Part II. The measurement of suction potentials and moisture distribution in drying granular solids.
4. NEWITT, D. M. and COLEMAN, M.: *Trans. Inst. Chem. Eng.* **30** (1952) 28. The mechanism of drying of solids. Part III. The drying characteristics of China clay.
5. CORBEN, R. W. and NEWITT, D. M.: *Trans. Inst. Chem. Eng.* **33** (1955) 52. The mechanism of drying of solids. Part VI. The drying characteristics of porous granular material.
6. KING, A. R. and NEWITT, D. M.: *Trans. Inst. Chem. Eng.* **33** (1955) 64. The mechanism of drying of solids. Part VII. Drying with heat transfer by conduction.
7. CORBEN, R. W.: *University of London*, Ph.D. Thesis (1955). The mechanism of drying of solids. Part IV. A study of the effect of granulation on the rate of drying of a porous solid.
8. GILLILAND, E. R.: *Ind. Eng. Chem.* **30** (1938) 506. Fundamentals of drying and air conditioning.
9. POWELL, R. W. and GRIFFITHS, E.: *Trans. Inst. Chem. Eng.* **13** (1935) 175. The evaporation of water from plane and cylindrical surfaces.
10. CHAKRAVORTY, K. R.: *J. Imp. Coll. Chem. Eng. Soc.* **3** (1947) 46. Evaporation from free surfaces.
11. SHEPHERD, C. B., HADLOCK, C., and BREWER, R. C.: *Ind. Eng. Chem.* **30** (1938) 388. Drying materials in trays. Evaporation of surface moisture.
12. SHERWOOD, T. K.: *Ind. Eng. Chem.* **21** (1929) 12, 976. The drying of solids I, II.
13. NEWMAN, A. B.: *Trans. Am. Inst. Chem. Eng.* **27** (1931) 203. The drying of porous solids: diffusion and surface emission equations.
14. CEAGLSKE, N. H. and HOUGEN, O. A.: *Trans. Am. Inst. Chem. Eng.* **33** (1937) 283. The drying of granular solids.
15. CEAGLSKE, N. H. and HOUGEN, O. A.: *Ind. Eng. Chem.* **29** (1937) 805. Drying granular solids.
16. SLICHTER, C. S.: *U.S. Geol. Survey, 19th Annual Report* (1897-8) Part 2, 301. Theoretical investigation of the motion of ground waters.
17. HAINES, W. B.: *J. Agric. Science* **17** (1927) 264. Studies in the physical properties of soils. IV. A further contribution to the theory of capillary phenomena in soil.
18. PARKER, N. H.: *Chem. Eng. Albany* **70** No. 13 (24 June 1963) 115. Aids to dryer selection.
19. KRÖLL, K.: *Trockner, einteilen, ordnen, benennen, benummern, Schilde Schriftenreihe* **6** (Schilde Bad-Hersfeld, 1965).

20. SLOAN, C. E.: *Chem Eng. Albany* **74** No. 13 (19 June 1967) 169. Drying systems and equipment.
21. BACKHURST, J. R. and HARKER, J. H.: *Process Plant Design* (Heinemann, London 1973).
22. KEEY, R. B.: *Drying Principles and Practice* (Pergamon Press, Oxford, 1972).
23. ERISMAN, J. L.: *Ind. Eng. Chem.* **30** (1938) 996. Roto-louvre dryer.
24. FRIEDMAN, S. J. and MARSHALL, W. R.: *Chem. Eng. Prog.* **45** (1949) 482, 573. Studies in rotary drying.
25. PRUTTON, C. F., MILLER, C. O., and SCHUETTE, W. H.: *Trans. Am. Inst. Chem. Eng.* **38** (1942) 123, 251. Factors influencing the performance of rotary dryers.
26. MILLER, C. O., SMITH, B. A., and SCHUETTE, W. H.: *Trans. Am. Inst. Chem. Eng.* **38** (1942) 841. Factors influencing the performance of rotary dryers.
27. SAEMAN, W. C.: *Chem. Eng. Prog.* **58**, No. 6 (June 1962) 49–56. Air-solids interaction in rotary dryers and coolers.
28. FISHER, J. J.: *Ind. Eng. Chem.* **55** No. 2 (1963) 18. Low temperature drying in vacuum tumblers.
29. LAPPLE, W. C., CLARK, W. E. and DYBDAL, E. C.: *Chem. Eng. Albany* **62** No. 11 (1955) 177. Drying. Design and costs.
30. SHARPLES, K., GLIKEN, P. G. and WARNE, R.: *Trans. Inst. Chem. Eng.* **42** (1964) T275. Complete simulation of rotary dryers.
31. FRASER, R. P., EISENKLAM, P., and DOMBROWSKI, N.: *Brit. Chem. Eng.* **2** (1957) 414–17, 496–501, 536–43, and 610–13. Liquid atomisation in chemical engineering.
32. RAYLEIGH, Lord: *Proc. Lond. Math. Soc.* **10** (1878-9) 4–13. On the instability of jets.
33. WEBER, C.: *Z. angew. Math. Mech.* **11** (1931) 136–54. Zum Zerfall eines Flüssigkeitsstrahles.
34. DOMBROWSKI, N. and MUNDAY, G.: in *Biochemical and Biological Engineering Science*. Vol. 2. Ch. 16 Spray drying (Academic Press, 1967).
35. DOMBROWSKI, N. and FRASER, R. P.: *Phil. Trans.* **247A** (1954) 101. A photographic investigation into the disintegration of liquid sheets.
36. FRASER, R. P., EISENKLAM, P., DOMBROWSKI, N., and HASSON, D.: *A.I.Ch.E.Jl.* **8** (1962) 672–80. Drop formation from rapidly moving liquid sheets.
37. WALTON, W. H. and PREWETT, W. C.: *Proc. Phys. Soc.* **628** (1949) 341. The production of sprays and mists of uniform drop size by means of spinning disc type sprayers.
38. MARSHALL, W. R.: *Chem. Eng. Prog.* Monogr. Ser. No. 2, **50** (1954). Atomization and spray drying.
39. MARSHALL, W. R.: *Trans. Am. Soc. Mech. Eng.* **77** (1955) 1377. Heat and mass transfer in spray drying.
40. GROSE, J. W. and DUFFIELD, G. H.: *Chem. and Ind.* (1954) 1464. Chemical engineering methods in the food industry.
41. QUINN, J. J.: *Ind. Eng. Chem.* **57**, No. 1 (Jan. 1965) 35–37. The economics of spray drying.
42. BRADFORD, P. and BRIGGS, S. W.: *Chem. Eng. Prog.* **59**, No. 3 (1963) 76. Equipment for the food industry — 3. Jet spray drying.
43. QUINN, M. F.: *Ind. Eng. Chem.* **55**, No. 7 (July, 1963) 18–24. Fluidized bed dryers.
44. AGARWAL, J. C., DAVIS, W. L., and KING, D. T.: *Chem. Eng. Prog.* **58**, No. 11 (Nov. 1962) 85–90. Fluidized-bed coal dryer.
45. SCOTT, M. W., LIEBERMAN, H. A., RANKELL, A. S., CHOW, F. S., and JOHNSTON, G. W.: *J. Pharm. Sci.* **52**, No. 3 (Mar 1963) 284–91. Drying as a unit operation in the pharmaceutical industry. I. Drying of tablet granulations in fluidized beds.
46. VANĚČEK, V., MARKVART, M. and DRBOHLAV, R.: *Fluid Bed Drying* (Leonard Hill, 1966).
47. CLARK, W. E.: *Chem. Eng. Albany* **74** No. 6 (13 March 1967) 177. Fluid-bed drying.
48. CHU JU CHIN, LANE, A. M. and CONKER, D.: *Ind. Eng. Chem.* **45** (1953) 1586. Evaporation of liquids into their superheated vapours.
49. BASEL, L. and GRAY, E.: *Chem. Eng. Prog.* **58** (June 1962) 67. Superheated solvent drying in a fluidized bed.
50. ELLWOOD, E. L., GOTTSTEIN, J. W., and KAUMAN, W. G.: A Laboratory Study of the Vapour Drying Process. CSIRO Forest Prod. Div., Paper 14 (1961) 111.
51. LUIKOV, A. V.: *Heat and Mass Transfer in Capillary-porous Bodies* (Pergamon Press, Oxford, 1966). Oxford, 1966).
52. BOND, J. F., MUJUMDAR, A. S., VAN HEININGEN, A. R. P. and DOUGLAS, W. J. M.: *Can J. Chem. Eng.* **72** (1994) 452–456. Drying paper by impinging jets of superheated steam.
53. WENZEL, L. and WHITE, R. R.: *Ind. Eng. Chem. Proc. Des. Dev.* **9** (1970) 207–214. Drying granular solids by superheated steam.
54. SCHWARTZE, J. P. and BRÖCKER, S.: *Int. J. Heat Mass Trans.* **43** (2000) 1791–1800. The evaporation of water into air of different humidities and the inversion temperature phenomenon.
55. TAYLOR, R. and KRISHNA, R.: *Multicomponent Mass Transfer.* (Wiley, New York, 1993)
56. VIDAURRE, M. and MARTINEZ, J.: *A.I.Ch.E.Jl* **43** (3) (1997) 681–692. Continuous drying of a solid wetted with ternary mixtures.
57. CHAMBERS, H. H.: *Trans. Inst. Chem. Eng.* **27** (1949) 19. Vacuum freeze drying.

58. MAGUIRE, J. F.: *Food Eng.* **34**, No. 8 (Aug. 1962) 54–5 and **34**, No. 9 (Sept. 1962) 48–52. Freeze drying moves ahead in U.S.
59. HARPER, J. C., CHICHESTER, C.O., and ROBERTS, T. E.: *Agric. Eng.* **43** (1962) 78, 90. Freeze-drying of foods.
60. MAISTER, H. G., HEGER, E. N., and BOGARD, W. M.: *Ind. Eng. Chem.* **50** (1958) 623. Continuous freeze-drying of *Serratia marcescens*.
61. DYER, D. F. and SUNDERLAND, J. E.: *Chem. Eng. Sci.* **23** (1968) 965. The role of convection in drying.
62. FRANCE, S.: *Chem. Eng. Albany* (December 1996) 83–84. Delicately drying solids - in a flash.
63. COOK, E. M.: *Chem. Eng. Albany* (April 1996) 82–89. Process calculations for partial-recycle dryers.
64. JORGENSEN, R. (ed.): *Fan Engineering* (8th edn.) (Buffalo Forge Co., Buffalo, New York, 1983).
65. COOK, E. M. and DUMOUNT, H. D.: *Process Drying Practice* (McGraw-Hill, New York, 1991).

## 16.10. NOMENCLATURE

| | | Units in SI System | Dimensions in **M, N, L, T, θ** |
|---|---|---|---|
| $A$ | Area for heat transfer or evaporation | $m^2$ | $\mathbf{L}^2$ |
| $A_N$ | Area of nozzle or jet normal to direction of flow | $m^2$ | $\mathbf{L}^2$ |
| $a$ | Surface area per unit volume | $m^2/m^3$ | $\mathbf{L}^{-1}$ |
| $B$ | Width of surface | m | $\mathbf{L}$ |
| $C$ | Coefficient | $kg^{0.2}/m^{0.4}s^{0.2}$ | $\mathbf{M}^{0.2}\ \mathbf{L}^{-0.4}\ \mathbf{T}^{-0.2}$ |
| $C_D$ | Coefficient of discharge | — | — |
| $C_w$ | Water concentration of a point in solid | $kg/m^3$ | $\mathbf{ML}^{-3}$ |
| $C_f$ | Heat capacity of wet material | J/kg K | $\mathbf{L}^2\ \mathbf{T}^{-2}\ \theta^{-1}$ |
| $C_m$ | Mean heat capacity of gas | J/kg K | $\mathbf{L}^2\ \mathbf{T}^{-2}\ \theta^{-1}$ |
| $C_s$ | Heat capacity of dry solid | J/kg K | $\mathbf{L}^2\ \mathbf{T}^{-2}\ \theta^{-1}$ |
| $C_x$ | Heat capacity of liquid evaporated | J/kg K | $\mathbf{L}^2\ \mathbf{T}^{-2}\ \theta^{-1}$ |
| $D$ | Diameter of drum or disc | m | $\mathbf{L}$ |
| $D_L$ | Diffusion coefficient (liquid phase) | $m^2/s$ | $\mathbf{L}^2\ \mathbf{T}^{-1}$ |
| $d$ | Particle diameter | m | $\mathbf{L}$ |
| $d_j$ | Nozzle or jet diameter | m | $\mathbf{L}$ |
| $d_m$ | Main drop diameter | m | $\mathbf{L}$ |
| $d_s$ | Surface-mean diameter of drop | m | $\mathbf{L}$ |
| $d_t$ | Diameter of evaporating drop at time $t$ | m | $\mathbf{L}$ |
| $d_o$ | Initial diameter of drop | m | $\mathbf{L}$ |
| $F$ | Wet solid feed rate | kg/s | $\mathbf{MT}^{-1}$ |
| $F'$ | Volumetric rate of feed per unit cross-section | $m^3/sm^2$ | $\mathbf{LT}^{-1}$ |
| $f$ | Free moisture content | kg | $\mathbf{M}$ |
| $f_c$ | Free moisture content at critical condition | kg | $\mathbf{M}$ |
| $f_1$ | Initial free moisture content | kg | $\mathbf{M}$ |
| **FN** | Flow number of nozzle (equation 16.32) | — | — |
| $G$ | Mass flow in jet or nozzle *or* of gas | kg/s | $\mathbf{MT}^{-1}$ |
| $G_v$ | Mass rate of evaporation | kg/s | $\mathbf{MT}^{-1}$ |
| $G'$ | Mass rate of flow of air per unit cross-section | $kg/m^2\ s$ | $\mathbf{ML}^{-2}\ \mathbf{T}^{-1}$ |
| $g$ | Acceleration due to gravity | $m/s^2$ | $\mathbf{LT}^{-2}$ |
| $\mathscr{H}$ | Humidity | kg/kg | — |
| $\mathscr{H}_s$ | Humidity of saturated air at surface temperature | kg/kg | — |
| $\mathscr{H}_w$ | Humidity of saturated air at temperature $\theta_w$ | kg/kg | — |
| $\mathscr{H}_0$ | Humidity of saturated air | kg/kg | — |
| $h$ | Heat transfer coefficient | $W/m^2K$ | $\mathbf{MT}^{-3}\theta^{-1}$ |
| $h_D$ | Mass transfer coefficient | m/s | $\mathbf{LT}^{-1}$ |
| $h_f$ | Friction head over a distance $z_1$ from surface | m | $\mathbf{L}$ |
| $h_s$ | Suction potential immediately below meniscus | m | $\mathbf{L}$ |
| $h_t$ | Theoretical suction potential of pore or waist | m | $\mathbf{L}$ |
| $h_1$ | Suction potential at distance $z_1$ below meniscus | m | $\mathbf{L}$ |
| $j$ | Volumetric gas/liquid ratio | — | — |
| $K$ | Coefficient | $m^2s/kg$ | $\mathbf{M}^{-1}\ \mathbf{L}^2\ \mathbf{T}$ |

| | | Units in SI System | Dimensions in **M, N, L, T, $\theta$** |
|---|---|---|---|
| $K'$ | Coefficient | $s^{1.8}/m^{1.5}$ | $\mathbf{L^{-1.8} T^{1.8}}$ |
| $K''$ | Transfer coefficient ($h_D \rho_A$) | $kg/m^2 s$ | $\mathbf{ML^{-2} T^{-1}}$ |
| $k_G$ | Mass transfer coefficient | s/m | $\mathbf{L^{-1} T}$ |
| $k_f$ | Thermal conductivity of gas film at interface | W/mK | $\mathbf{MLT^{-3}\theta^{-1}}$ |
| $L$ | Length of surface | m | **L** |
| $l$ | Half thickness of slab | m | **L** |
| $M_A$ | Molecular weight of air | kg/kmol | $\mathbf{MN^{-1}}$ |
| $M_w$ | Molecular weight of water | kg/kmol | $\mathbf{MN^{-1}}$ |
| $m$ | Ratio of rate of drying per unit area to moisture content | $m^{-2}s^{-1}$ | $\mathbf{L^{-2} T^{-1}}$ |
| $N$ | Revolutions per unit time | Hz | $\mathbf{T^{-1}}$ |
| $n$ | Index | — | — |
| $n_f$ | Number of flights | — | — |
| $P$ | Total pressure | $N/m^2$ | $\mathbf{ML^{-1} T^{-2}}$ |
| $P_s$ | Vapour pressure of water at surface of material | $N/m^2$ | $\mathbf{ML^{-1} T^{-2}}$ |
| $P_w$ | Partial pressure of water vapour | $N/m^2$ | $\mathbf{ML^{-1} T^{-2}}$ |
| $P_{wo}$ | Partial pressure at surface of material at wet bulb temperature | $N/m^2$ | $\mathbf{ML^{-1} T^{-2}}$ |
| $-\Delta P$ | Pressure drop across nozzle | $N/m^2$ | $\mathbf{ML^{-1} T^{-2}}$ |
| $Q$ | Rate of heat transfer | W | $\mathbf{ML^2 T^{-3}}$ |
| $R_c$ | Rate of drying per unit area for constant rate period | $kg/m^2 s$ | $\mathbf{ML^{-2} T^{-1}}$ |
| $\mathbf{R}$ | Universal gas constant | 8314 J/kmol K | $\mathbf{MN^{-1} L^2 T^{-2}\theta^{-1}}$ |
| $\mathscr{R}$ | Exit gas relative humidity (per cent) | — | — |
| $r$ | Radius of sphere | m | **L** |
| $r'$ | Radius of capillary | m | **L** |
| $S$ | Slope of drum | — | — |
| $T$ | Absolute temperature | K | $\boldsymbol{\theta}$ |
| $T_b$ | Bed temperature | K | $\boldsymbol{\theta}$ |
| $T_f$ | Temperature of wet solids feed | K | $\boldsymbol{\theta}$ |
| $T_m$ | Mean temperature of inlet gas | K | $\boldsymbol{\theta}$ |
| $\Delta T$ | Temperature difference | K | $\boldsymbol{\theta}$ |
| $t$ | Time | s | **T** |
| $t_c$ | Time of constant rate period of drying | s | **T** |
| $t_f$ | Time of drying in falling rate period | s | **T** |
| $U$ | Overall heat transfer coefficient | $W/m^2 K$ | $\mathbf{MT^{-3}\theta^{-1}}$ |
| $u$ | Gas velocity | m/s | $\mathbf{LT^{-1}}$ |
| $u_f$ | Fluidising velocity | m/s | $\mathbf{LT^{-1}}$ |
| $u_l$ | Liquid velocity in jet or spray | m/s | $\mathbf{LT^{-1}}$ |
| $u_r$ | Velocity of gas relative to liquid | m/s | $\mathbf{LT^{-1}}$ |
| $V$ | Volume | $m^3$ | $\mathbf{L^3}$ |
| $W$ | Mass rate of evaporation | kg/s | $\mathbf{MT^{-1}}$ |
| $w$ | Total moisture | kg | **M** |
| $w_c$ | Critical moisture content | kg | **M** |
| $w_e$ | Equilibrium moisture content | kg | **M** |
| $w_1$ | Initial moisture content | kg | **M** |
| $X$ | Hold-up of drum | — | — |
| $X_a$ | Hold-up of drum with air flow | — | — |
| $X_f$ | Moisture content of wet feed | kg/kg | — |
| $x$ | Factor depending on type of packing | — | — |
| $y$ | Distance in direction of diffusion | m | **L** |
| $z_1$ | Distance below meniscus | m | **L** |
| $\alpha$ | Angle of contact | — | — |
| $\bar{\kappa}$ | Coefficient | $kg^{1-n}\ m^{2n}\ s^{n-3}/K$ | $\mathbf{M^{1-n} L^{2n} T^{n-3}\theta^{-1}}$ |
| $\theta$ | Gas temperature | K | $\boldsymbol{\theta}$ |
| $\theta_s$ | Surface temperature | K | $\boldsymbol{\theta}$ |
| $\theta_w$ | Wet bulb temperature | K | $\boldsymbol{\theta}$ |
| $\lambda$ | Latent heat vaporisation per unit mass | J/kg | $\mathbf{L^2 T^{-2}}$ |

| | | Units in SI System | Dimensions in **M, N, L, T, θ** |
|---|---|---|---|
| $\lambda_b$ | Mean latent heat vaporisation per unit mass of $T_b$ | kJ/kg | $\mathbf{L^2\ T^{-2}}$ |
| $\lambda_{opt}$ | Optimum wavelength for jet disruption | m | **L** |
| $\mu$ | Viscosity | $Ns/m^2$ | $\mathbf{ML^{-1}\ T^{-1}}$ |
| $\rho$ | Density | $kg/m^3$ | $\mathbf{ML^{-3}}$ |
| $\rho_A$ | Density of air at its mean partial pressure | $kg/m^3$ | $\mathbf{ML^{-3}}$ |
| $\sigma$ | Surface tension | $J/m^2$ | $\mathbf{MT^{-2}}$ |
| $\phi$ | Half-angle of spray cone or sheet | — | — |

CHAPTER 17

# Adsorption

## 17.1. INTRODUCTION

Although adsorption has been used as a physical-chemical process for many years, it is only over the last four decades that the process has developed to a stage where it is now a major industrial separation technique. In adsorption, molecules distribute themselves between two phases, one of which is a solid whilst the other may be a liquid or a gas. The only exception is in adsorption on to foams, a topic which is not considered in this chapter.

Unlike *absorption*, in which solute molecules diffuse from the bulk of a gas phase to the bulk of a liquid phase, in *adsorption*, molecules diffuse from the bulk of the fluid to the surface of the solid adsorbent forming a distinct adsorbed phase.

Typically, gas adsorbers are used for removing trace components from gas mixtures. The commonest example is the drying of gases in order to prevent corrosion, condensation or an unwanted side reaction. For items as diverse as electronic instruments and biscuits, sachets of adsorbent may be included in the packaging in order to keep the relative humidity low. In processes using volatile solvents, it is necessary to guard against the incidental loss of solvent carried away with the ventilating air and recovery may be effected by passing the air through a packed bed of adsorbent.

Adsorption may be equally effective in removing trace components from a liquid phase and may be used either to recover the component or simply to remove a noxious substance from an industrial effluent.

Any potential application of adsorption has to be considered along with alternatives such as distillation, absorption and liquid extraction. Each separation process exploits a difference between a property of the components to be separated. In distillation, it is volatility. In absorption, it is solubility. In extraction, it is a distribution coefficient. Separation by adsorption depends on one component being more readily adsorbed than another. The selection of a suitable process may also depend on the ease with which the separated components can be recovered. Separating *n*- and *iso*-paraffins by distillation requires a large number of stages because of the low relative volatility of the components. It may be economic, however, to use a selective adsorbent which separates on the basis of slight differences in mean molecular diameters, where for example, *n*- and *iso*-pentane have diameters of 0.489 and 0.558 nm respectively. When an adsorbent with pore size of 0.5 nm is exposed to a mixture of the gases, the smaller molecules diffuse to the adsorbent surface and are retained whilst the larger molecules are excluded. In another stage of the process, the retained molecules are desorbed by reducing the total pressure or increasing the temperature.

Most commercial processes for producing nitrogen and oxygen from air use the cryogenic distillation of liquefied air. There is also interest in adsorptive methods, particularly for moderate production rates. Some adsorbents take up nitrogen preferentially and can be used to generate an oxygen-rich gas containing 95 mole per cent of oxygen. Regeneration of the adsorbent yields a nitrogen-rich gas. It is also possible to separate the gases using an adsorbent with 0.3 nm pores. These allow oxygen molecules of 0.28 nm in size to diffuse rapidly on to the adsorption surface, whereas nitrogen molecules of 0.30 nm will diffuse more slowly. The resulting stream is a nitrogen-rich gas of up to 99 per cent purity. An oxygen stream of somewhat lower purity is obtained from the desorption stage.

Other applications of commercial adsorbents are given in Table 17.1, taken from the work of CRITTENDEN[1]. Some typical solvents which are readily recovered by adsorptive techniques are listed in Table 17.2, taken from information supplied by manufacturers.

All such processes suffer one disadvantage in that the capacity of the adsorbent for the adsorbate in question is limited. The adsorbent has to be removed at intervals from the process and regenerated, that is, restored to its original condition. For this reason, the adsorption unit was considered in early industrial applications to be more difficult to integrate with a continuous process than, say, a distillation column. Furthermore, it was difficult to manufacture adsorbents which had identical adsorptive properties from batch to batch. The design of a commercial adsorber and its operation had to be sufficiently flexible to cope with such variations.

These factors, together with the rather slow thermal regeneration that was common in early applications, resulted in the adsorber being an unpopular option with plant designers. Since a greater variety of adsorbents has become available, each tailor-made for a specific application, the situation has changed, particularly as faster alternatives to thermal regeneration are often possible.

Adsorption occurs when molecules diffusing in the fluid phase are held for a period of time by forces emanating from an adjacent surface. The surface represents a gross discontinuity in the structure of the solid, and atoms at the surface have a residue of molecular forces which are not satisfied by surrounding atoms such as those in the body of the structure. These residual or van der Waals forces are common to all surfaces and the only reason why certain solids are designated "adsorbents" is that they can be manufactured in a highly porous form, giving rise to a large internal surface. In comparison the external surface makes only a modest contribution to the total, even when the solid is finely divided. Iron oxide particles with a radius of 5 $\mu$m and a density of 5000 kg/m$^3$ have an external surface of 12,000 m$^2$/kg. A typical value for the total surface of commercial adsorbents is 400,000 m$^2$/kg.

The adsorption which results from the influence of van der Waals forces is essentially physical in nature. Because the forces are not strong, the adsorption may be easily reversed. In some systems, additional forces bind absorbed molecules to the solid surface. These are chemical in nature involving the exchange or sharing of electrons, or possibly molecules forming atoms or radicals. In such cases the term *chemisorption* is used to describe the phenomenon. This is less easily reversed than physical adsorption, and regeneration may be a problem. Chemisorption is restricted to just one layer of molecules on the surface, although it may be followed by additional layers of physically adsorbed molecules.

When molecules move from a bulk fluid to an adsorbed phase, they lose degrees of freedom and the free energy is reduced. Adsorption is always accompanied by the

Table 17.1. Typical applications of commercial adsorbents[1]

| Type | Typical applications |
|---|---|
| Silica gel | Drying of gases, refrigerants, organic solvents, transformer oils; desiccant in packings and double glazing; dew point control of natural gas. |
| Activated alumina | Drying of gases, organic solvents, transformer oils; removal of HCl from hydrogen; removal of fluorine and boron-fluorine compounds in alkylation processes. |
| Carbons | Nitrogen from air; hydrogen from syn-gas and hydrogenation processes; ethene from methane and hydrogen; vinyl chloride monomer (VCM) from air; removal of odours from gases; recovery of solvent vapours; removal of $SO_x$ and $NO_x$; purification of helium; clean-up of nuclear off-gases; decolourising of syrups, sugars and molasses; water purification, including removal of phenol, halogenated compounds, pesticides, caprolactam, chlorine. |
| Zeolites | Oxygen from air; drying of gases; removing water from azeotropes; sweetening sour gases and liquids; purification of hydrogen; separation of ammonia and hydrogen; recovery of carbon dioxide; separation of oxygen and argon; removal of acetylene, propane and butane from air; separation of xylenes and ethyl benzene; separation of normal from branched paraffins; separation of olefins and aromatics from paraffins; recovery of carbon monoxide from methane and hydrogen; purification of nuclear off-gases; separation of cresols; drying of refrigerants and organic liquids; separation of solvent systems; purification of silanes; pollution control, including removal of Hg, $NO_x$ and $SO_x$ from gases; recovery of fructose from corn syrup. |
| Polymers and resins | Water purification, including removal of phenol, chlorophenols, ketones, alcohols, aromatics, aniline, indene, polynuclear aromatics, nitro- and chlor-aromatics, PCB, pesticides, antibiotics, detergents, emulsifiers, wetting agents, kraftmill effluents, dyestuffs; recovery and purification of steroids, amino acids and polypeptides; separation of fatty acids from water and toluene; separation of aromatics from aliphatics; separation of hydroquinone from monomers; recovery of proteins and enzymes; removal of colours from syrups; removal of organics from hydrogen peroxide. |
| Clays (acid-treated and pillared) | Treatment of edible oils; removal of organic pigments; refining of mineral oils; removal of polychlorobiphenyl (PCB). |

Table 17.2. Properties of some solvents recoverable by adsorptive techniques

| | Molecular weight (kg/kmol) | Specific heat capacity at 293 K (kJ/kg K) | Density at 293 K (kg/m$^3$) | Latent heat evaporation (kJ/kg) | Boiling point (K) | Explosive limits in air at 293 K (per cent by volume) | | Solubility in water at 293 K (kg/m$^3$) |
|---|---|---|---|---|---|---|---|---|
| | | | | | | low | high | |
| acetone | 58.08 | 2.211 | 791.1 | 524.6 | 329.2 | 2.15 | 13.0 | ∞ |
| allyl alcohol | 58.08 | 2.784 | 853.5 | 687 | 369.9 | 2.5 | 18 | ∞ |
| *n*-amyl acetate | 130.18 | 1.926 | 876 | 293 | 421.0 | 3.6 | 16.7 | 1.8 |
| *iso*-amyl acetate | 130.18 | 1.9209 | 876 | 289 | 415.1 | 3.6 | — | 2.5 |
| *n*-amyl alcohol | 88.15 | 2.981 | 817 | 504.9 | 410.9 | 1.2 | — | 68 |
| *iso*-amyl alcohol | 88.15 | 2.872 | 812 | 441.3 | 404.3 | 1.2 | — | 32 |
| amyl chloride | 106.6 | — | 883 | — | 381.3 | — | — | Insol. |
| amylene | 70.13 | 1.181 | 656 | 314.05 | 309.39 | 1.6 | — | Insol. |
| benzene | 78.11 | 1.720 | 880.9 | 394.8 | 353.1 | 1.4 | 4.7 | 0.8 |
| *n*-butyl acetate | 116.16 | 1.922 | 884 | 309.4 | 399.5 | 1.7 | 15 | 10 |
| *iso*-butyl acetate | 116.16 | 1.921 | 872 | 308.82 | 390.2 | 2.4 | 10.5 | 6.7 |
| *n*-butyl alcohol | 74.12 | 2.885 | 809.7 | 600.0 | 390.75 | 1.45 | 11.25 | 78 |
| *iso*-butyl alcohol | 74.12 | 2.784 | 805.7 | 578.6 | 381.8 | 1.68 | — | 85 |
| carbon disulphide | 76.13 | 1.005 | 1267 | 351.7 | 319.25 | 1.0 | 50 | 2 |
| carbon tetrachloride | 153.84 | 0.846 | 1580 | 194.7 | 349.75 | Non-flammable | | 0.084 |
| cellosolve | 90.12 | 2.324 | 931.1 | — | 408.1 | — | — | ∞ |
| cellosolve methyl | 76.09 | 2.236 | 966.3 | 565 | 397.5 | — | — | ∞ |
| cellosolve acetate | 132.09 | — | 974.8 | — | 426.0 | — | — | 230 |
| chloroform | 119.39 | 0.942 | 1480 | 247 | 334.26 | Non-flammable | | 8 |
| cyclohexane | 84.16 | 2.081 | 778.4 | 360 | 353.75 | 1.35 | 8.35 | Insol. |
| cyclohexanol | 100.16 | 1.746 | 960 | 452 | 433.65 | — | — | 60 |
| cyclohexanone | 98.14 | 1.813 | 947.8 | — | 429.7 | 3.2 | 9.0 | 50 |
| cymene | 134.21 | 1.666 | 861.2 | 283.07 | 449.7 | — | — | Insol. |
| *n*-decane | 142.28 | 2.177 | 730.1 | 252.0 | 446.3 | 0.7 | — | Insol. |
| dichloroethylene | 96.95 | 1.235 | 1291 | 305.68 | 333.0 | 9.7 | 12.8 | Insol. |
| ether (diethyl) | 74.12 | 2.252 | 713.5 | 360.4 | 307.6 | 1.85 | 36.5 | 69 |
| ether (di-*n*-butyl) | 130.22 | — | 769.4 | 288.1 | 415.4 | — | — | 3 |
| ethyl acetate | 88.10 | 2.001 | 902.0 | 366.89 | 350.15 | 2.25 | 11.0 | 79.4 |
| ethyl alcohol | 56.07 | 2.462 | 789.3 | 855.4 | 351.32 | 3.3 | 19.0 | ∞ |
| ethyl bromide | 108.98 | 0.812 | 1450 | 250.87 | 311.4 | 6.7 | 11.2 | 9.1 |
| ethyl carbonate | 118.13 | 1.926 | 975.2 | 306 | 399.0 | — | — | V.sl.sol. |
| ethyl formate | 74.08 | 2.135 | 923.6 | 406 | 327.3 | 2.7 | 16.5 | 100 |
| ethyl nitrite | 75.07 | — | 900 | — | 290.0 | 3.0 | — | Insol. |
| ethylene dichloride | 96.97 | 1.298 | 1255.0 | 323.6 | 356.7 | 6.2 | 15.6 | 8.7 |
| ethylene oxide | 44.05 | — | 882 | 580.12 | 283.5 | 3.0 | 80 | ∞ |
| furfural | 96.08 | 1.537 | 1161 | 450.12 | 434.8 | 2.1 | — | 83 |
| *n*-heptane | 100.2 | 2.123 | 683.8 | 318 | 371.4 | 1.0 | 6.0 | 0.052 |
| *n*-hexane | 86.17 | 2.223 | 659.4 | 343 | 341.7 | 1.25 | 6.9 | 0.14 |
| methyl acetate | 74.08 | 2.093 | 927.2 | 437.1 | 330.8 | 4.1 | 13.9 | 240 |
| methyl alcohol | 32.04 | 2.500 | 792 | 1100.3 | 337.7 | 6.72 | 36.5 | 67.2 |
| methyl cyclohexanol | 114.18 | — | 925 | — | 438.0 | — | — | 11 |
| methyl cyclohexanone | 112.2 | 1.842 | 924 | — | 438.0 | — | — | 30 |
| methyl cyclohexane | 98.18 | 1.855 | 768 | 323 | 373.9 | 1.2 | — | Insol. |
| methyl ethyl ketone | 72.10 | 2.085 | 805.1 | 444 | 352.57 | 1.81 | 11.5 | 265 |
| methylene chloride | 84.93 | 1.089 | 1336 | 329.67 | 313.7 | Non-flammable | | 20 |
| monochlorobenzene | 112.56 | 1.256 | 1107.4 | 324.9 | 404.8 | — | — | 0.49 |
| naphthalene | 128.16 | 1.683 | 1152 | 316.1 | 491.0 | 0.8 | — | 0.04 |
| nonane | 128.25 | 2.106 | 718 | 274.2 | 422.5 | 0.74 | 2.9 | Insol. |
| octane | 114.23 | 2.114 | 702 | 296.8 | 398.6 | 0.95 | 3.2 | 0.015 |
| paraldehyde | 132.16 | 1.825 | 904 | 104.75 | 397.0 | 1.3 | — | 120 |
| *n*-pentane | 72.15 | 2.261 | 626 | 352 | 309.15 | 1.3 | 8.0 | 0.36 |

(*continued overleaf*)

Table 17.2. (*continued*)

| | Molecular weight (kg/kmol) | Specific heat capacity at 293 K (kJ/kg K) | Density at 293 K ($kg/m^3$) | Latent heat evaporation (kJ/kg) | Boiling point (K) | Explosive limits in air at 293 K (per cent by volume) | | Solubility in water at 293 K ($kg/m^3$) |
|---|---|---|---|---|---|---|---|---|
| | | | | | | low | high | |
| *iso*-pentane | 72.09 | 2.144 | 619 | 371.0 | 301.0 | 1.3 | 7.5 | Insol. |
| pentachloroethane | 202.33 | 0.900 | 1678 | 182.5 | 434.95 | Non-flammable | | 0.5 |
| perchloroethylene | 165.85 | 0.879 | 1662.6 | 209.8 | 393.8 | Non-flammable | | 0.4 |
| *n*-propyl acetate | 102.3 | 1.968 | 888.4 | 336.07 | 374.6 | 2.0 | 8.0 | 18.9 |
| *iso*-propyl acetate | 102.3 | 2.181 | 880 | 332.4 | 361.8 | 2.0 | 8.0 | 29 |
| *n*-propyl alcohol | 60.09 | 2.453 | 803.6 | 682 | 370.19 | 2.15 | 13.5 | ∞ |
| *iso*-propyl alcohol | 60.09 | 2.357 | 786.3 | 667.4 | 355.4 | 2.02 | — | ∞ |
| propylene dichloride | 112.99 | 1.398 | 1159.3 | 302.3 | 369.8 | 3.4 | 14.5 | 2.7 |
| pyridine | 79.10 | 1.637 | 978 | 449.59 | 388.3 | 1.8 | 12.4 | ∞ |
| tetrachloroethane | 167.86 | 1.130 | 1593 | 231.5 | 419.3 | Non-flammable | | 3.2 |
| tetrahydrofuran | 72.10 | 1.964 | 888 | 410.7 | 339.0 | 1.84 | 11.8 | ∞ |
| toluene | 92.13 | 1.641 | 871 | 360 | 383.0 | 1.3 | 7.0 | 0.47 |
| trichloroethylene | 131.4 | 0.934 | 1465.5 | 239.9 | 359.7 | Non-flammable | | 1.0 |
| xylene | 106.16 | 1.721 | 897 | 347.1 | 415.7 | 1.0 | 6.0 | Insol. |
| water | 18 | 4.183 | 998 | 2260.9 | 373.0 | Non-flammable | | ∞ |

liberation of heat. For physical adsorption, the amount of heat is similar in magnitude to the heat of condensation. For chemisorption it is greater and of an order of magnitude normally associated with a chemical reaction. If the heat of adsorption cannot be dispersed by cooling, the capacity of the adsorbent will be reduced as its temperature increases.

It is often convenient to think of adsorption as occurring in three stages as the adsorbate concentration increases. Firstly, a single layer of molecules builds up over the surface of the solid. This monolayer may be chemisorbed and associated with a change in free energy which is characteristic of the forces which hold it. As the fluid concentration is further increased, layers form by physical adsorption and the number of layers which form may be limited by the size of the pores. Finally, for adsorption from the gas phase, capillary condensation may occur in which capillaries become filled with condensed adsorbate, and its partial pressure reaches a critical value relative to the size of the pore.

Although the three stages are described as taking place in sequence, in practice, all three may be occurring simultaneously in different parts of the adsorbent since conditions are not uniform throughout. Generally, concentrations are higher at the outer surface of an adsorbent pellet than in the centre, at least until equilibrium conditions have been established. In addition, the pore structure will consist of a distribution of pore sizes and the spread of the distribution depends on the origin of the adsorbent and its conditions of manufacture.

## 17.2. THE NATURE OF ADSORBENTS

Adsorbents are available as irregular granules, extruded pellets and formed spheres. The size reflects the need to pack as much surface area as possible into a given volume of bed and at the same time minimise pressure drop for flow through the bed. Sizes of up to about 6 mm are common.

To be attractive commercially, an adsorbent should embody a number of features:

(a) it should have a large internal surface area.
(b) the area should be accessible through pores big enough to admit the molecules to be adsorbed. It is a bonus if the pores are also small enough to exclude molecules which it is desired not to adsorb.
(c) the adsorbent should be capable of being easily regenerated.
(d) the adsorbent should not age rapidly, that is lose its adsorptive capacity through continual recycling.
(e) the adbsorbent should be mechanically strong enough to withstand the bulk handling and vibration that are a feature of any industrial unit.

### 17.2.1. Molecular sieves

An increase in the use of adsorbers as a means of separation on a large scale is the result of the manufacturers' skill at developing and producing adsorbents which are tailored for specific tasks. First, by using naturally occurring zeolites and, later, synthesised members of that family of minerals, it has been possible to manufacture a range of adsorbents known collectively as *molecular sieves*. These have lattice structures composed of tetrahedra of silica and alumina arranged in various ways. The net effect is the formation of a cage-like structure with windows which admit only molecules less than a certain size as shown in Figure 17.1. By using different source materials and different conditions of manufacture, it is possible to produce a range of molecular sieves with access dimensions of 0.3 nm–1 nm. The dimensions are precise for a particular sieve because they derive from the crystal structure of that sieve. Some of the molecules admitted by different molecular sieves are given in Table 17.3 which is taken from the work of BARRER[2]. The crystallites of a sieve are about 10 μm in size and are aggregated for commercial use by mixing with a clay binder and extruding as pellets or rolling into spheres. The pelletising creates two other sets of pores, between crystallites and between pellets. Neither may add significantly to the adsorptive surface though each will influence rates of diffusion and pressure drop. It has been estimated by YANG[3] that there are about forty naturally occurring zeolites and that some one hundred and fifty have been synthesised.

The manufacture of molecular sieves has been reviewed in the literature, and particularly by BRECK[4], BARRER[5] and ROBERTS[6].

### 17.2.2. Activated carbon

In some of the earliest recorded examples of adsorption, activated carbon was used as the adsorbent. Naturally occurring carbonaceous materials such as coal, wood, coconut shells or bones are decomposed in an inert atmosphere at a temperature of about 800 K. Because the product will not be porous, it needs additional treatment or *activation* to generate a system of fine pores. The carbon may be produced in the activated state by treating the raw material with chemicals, such as zinc chloride or phosphoric acid, before carbonising. Alternatively, the carbon from the carbonising stage may be selectively

Table 17.3. Classification of some molecular sieves[2]

Molecular Size Increasing ⟶

| | | |
|---|---|---|
| Type 5 | He, Ne, A, CO<br>$H_2$, $O_2$, $N_2$,<br>$NH_3$, $H_2O$ | Size limit for Ca- and Ba-mordenites and levynite about here (~0.38 nm) |
| Type 4 | Kr, Xe<br>$CH_4$<br>$C_2H_6$<br>$CH_3OH$<br>$CH_3CN$<br>$CH_3NH_2$<br>$CH_3Cl$<br>$CH_3Br$<br>$CO_2$<br>$C_2H_2$<br>$CS_2$ | Size limit for Na-mordenite and Linde sieve 4 A about here (~0.4 nm) |
| Type 3 | $C_3H_8$<br>$n$-$C_4H_{10}$<br>$n$-$C_7H_{16}$<br>$n$-$C_{14}H_{30}$<br>etc.<br>$C_2H_5Cl$<br>$C_2H_5Br$<br>$C_2H_5OH$<br>$C_2H_5NH_2$<br>$CH_2Cl_2$<br>$CH_2Br_2$<br>$CHF_2Cl$<br>$CHF_3$<br>$(CH_3)_2NH$<br>$CH_3I$<br>$B_2H_6$<br><br>$CF_4$<br>$C_2F_6$<br>$CF_2Cl_2$<br>$CF_3Cl$<br>$CHFCl_2$ | Size limit for Ca-rich chabazite, Linde sieve 5 A, Ba-zeolite and gmelinite about here (~0.49 nm) |
| Type 2 | $SF_6$<br>$iso$-$C_4H_{10}$<br>$iso$-$C_5H_{12}$<br>$iso$-$C_8H_{18}$<br>etc.<br>$CHCl_3$<br>$CHBr_3$<br>$CHI_3$<br>$(CH_3)_2CHOH$<br>$(CH_3)_2CHCl$<br>$n$-$C_3F_8$<br>$n$-$C_4F_{10}$<br>$n$-$C_7F_{16}$<br>$B_5H_9$<br><br>$(CH_3)_3N$<br>$(C_2H_5)_3N$<br>$C(CH_3)_4$<br>$C(CH_3)_3Cl$<br>$C(CH_3)_3Br$<br>$C(CH_3)_3OH$<br>$CCl_4$<br>$CBr_4$<br>$C_2F_2Cl_4$<br><br>$C_6H_6$<br>$C_6H_5CH_3$<br>$C_6H_4(CH_3)_2$<br>Cyclohexane<br>Thiophen<br>Furan<br>Pyridine<br>Dioxane<br>$B_{10}H_{14}$<br><br>Naphthalene<br>Quinoline,<br>6-decyl-1, 2, 3, 4-tetrahydro-naphthalene,<br>2-butyl-1-hexyl indan<br>$C_6F_{11}CF_3$ | Size limit for Linde sieve 10X about here (~0.8 nm) |
| Type 1 | 1, 3, 5 triethyl benzene<br>1, 2, 3, 4, 5, 6, 7, 8, 13, 14, 15, 16-decahydro-chrysene | Size limit for Linde sieve 13X about here (~1.0 nm) |
| | $(n$-$C_4F_9)_3N$ | |

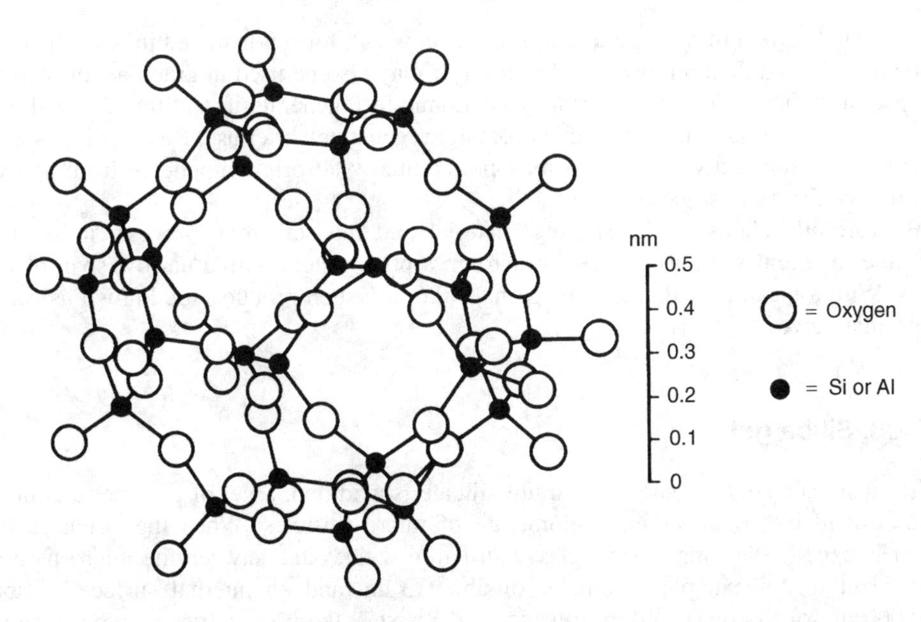


Figure 17.1. A cubo-octahedral unit composed of $SiO_4$ and $AlO_4$ tetrahedra[2]

oxidised at temperatures in excess of 1000 K in atmospheres of materials such as steam or carbon dioxide.

Activated carbon has a typical surface area of $10^6$ $m^2/kg$, mostly associated with a set of pores of about 2 nm in diameter. There is likely to be another set of pores of about 1000 nm in diameter, which do not contribute to the surface area. There may even be a third, intermediate set of pores which is particularly developed in carbons intended for use with liquids, as shown in Figure 17.2 which is taken from the work of SCHWEITZER[7].

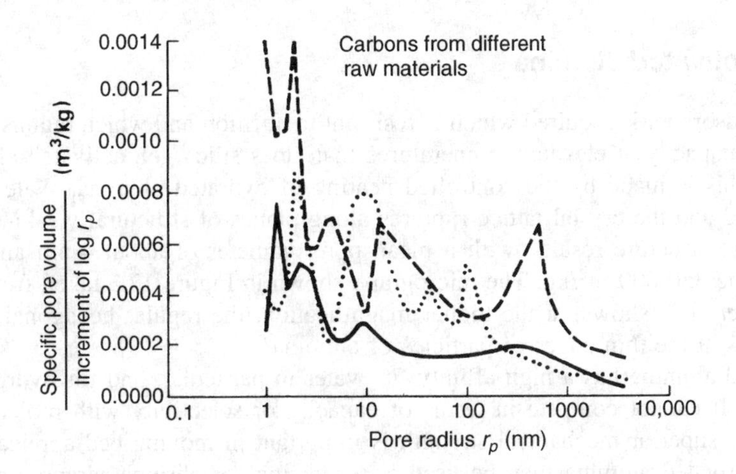


Figure 17.2. Typical pore volume distributions for three activated carbons used for liquid-phase processes[7]

Activated carbon may be used as a powder, in which form it is mixed in with the liquid to be treated, and then removed by filtration. It may also be used in granular form. When the use of carbon is low, it is normally economic to regenerate it, and this is usually the case with powder. Granular carbon is normally regenerated after use. Because it has a low affinity for water, activated carbon may preferentially adsorb components from aqueous solution or from moist gases.

By carefully choosing the starting material and the activating process, it has been possible in recent years to generate in carbon a pore system with a narrow span of pore sizes. With a mean pore diameter of perhaps 0.6 nm, such products are known as carbon molecular sieves.

### 17.2.3. Silica gel

When a silicate solution such as sodium silicate is acidified, a gel of polymeric colloidal silicilic acid is formed as an agglomerate of micro-particles. When the gel is heated, water is expelled leaving a hard, glassy structure with voids between the micro-particles equivalent to a mean pore diameter of about 3 nm and an internal surface of about 500,000 $m^2/kg$. As discussed by EVERETT and STONE[8] these properties may be varied by controlling the pH of the solution from which the gel is precipitated.

Silica gel is probably the adsorbent which is best known. Small sachets of gel are often included in packages of material that might deteriorate in a damp atmosphere. Sometimes a dye is added in the manufacturing process so that the gel changes colour as it becomes saturated.

Unlike the activated carbons, the surface of silica gel is hydrophilic and it is commonly used for drying gases and also finds applications where there is a requirement to remove unsaturated hydrocarbons. Silica gels are brittle and may be shattered by the rapid release of the heat of adsorption that accompanies contact with liquid water. For such applications, a tougher variety is available with a slightly lower surface area.

### 17.2.4. Activated alumina

When an adsorbent is required which is resistant to attrition and which retains more of its adsorptive capacity at elevated temperatures than does silica gel, activated alumina may be used. This is made by the controlled heating of hydrated alumina. Water molecules are expelled and the crystal lattice ruptures along planes of structural weakness. A well-defined pore structure results, with a mean pore diameter of about 4 nm and a surface area of some 350,000 $m^2/kg$. The micrograph shown in Figure 17.3 taken from the work of BOWEN *et al.*[9] shows, at the higher magnification, the regular hexagonal disposition of the pores at the thin edges of particles of alumina.

Activated alumina has a high affinity for water in particular, and for hydroxyl groups in general. It cannot compete in terms of capacity or selectivity with molecular sieves although its superior mechanical strength is important in moving-bed applications. As a powder, activated alumina may be used as a packing for chromatographic columns, as described in Chapter 19.

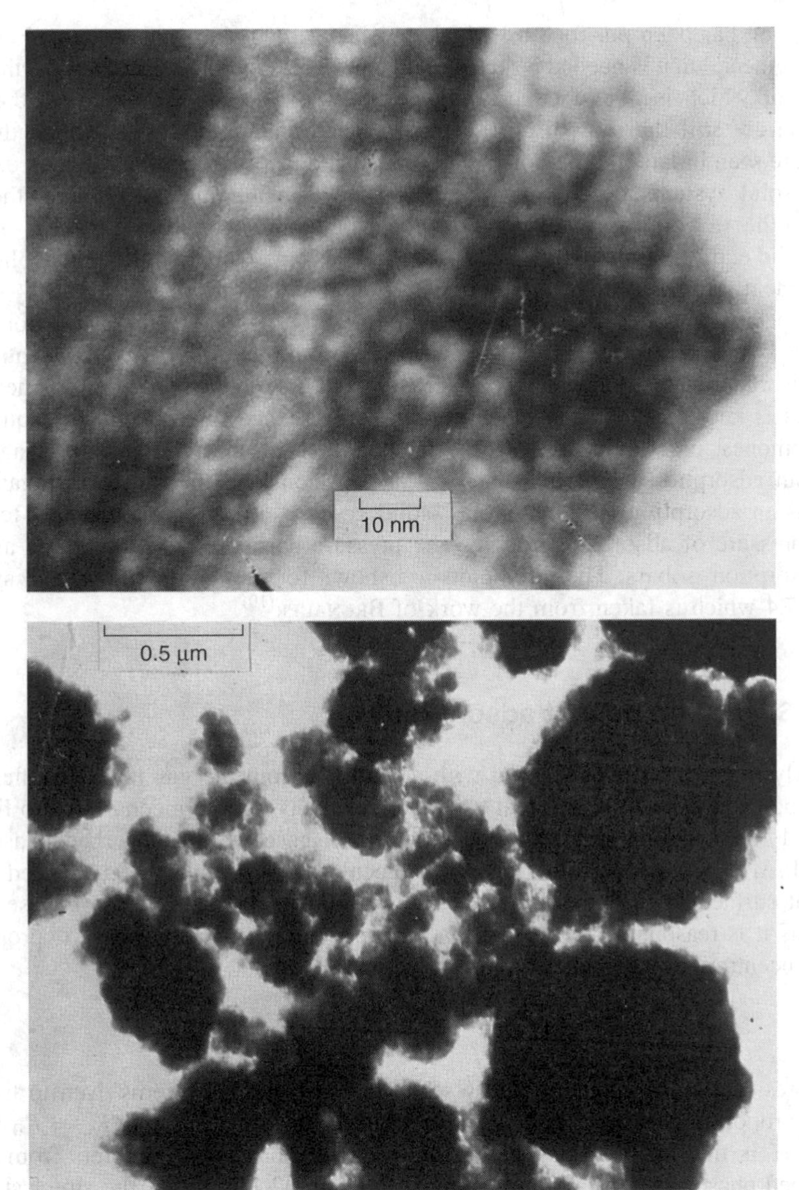


Figure 17.3. Electron micrographs of a commercial activated alumina at two magnifications[9]

## 17.3. ADSORPTION EQUILIBRIA

Much of the early work on the nature of adsorbents sought to explain the equilibrium capacity and the molecular forces involved. Adsorption equilibrium is a dynamic concept achieved when the rate at which molecules adsorb on to a surface is equal to the rate at which they desorb. The physical chemistry involved may be complex and no single theory of adsorption has been put forward which satisfactorily explains all systems. Fortunately for the engineer, what is needed is an accurate representation of equilibrium, the theoretical minutiae of which is not of concern. For this reason, some of the earliest theories of adsorption are still the most useful, even though the assumptions on which they were based were seen in later years to be not entirely valid. Most theories have been developed for gas–solid systems because the gaseous state is better understood than the liquid. Statistical theories are being developed which should apply equally well to gas–solid and liquid–solid equilibria, though these are not yet at a stage when they can be applied easily and confidently to the design of equipment.

The capacity of an adsorbent for a particular adsorbate involves the interaction of three properties — the concentration $C$ of the adsorbate in the fluid phase, the concentration $C_s$ of the adsorbate in the solid phase and the temperature $T$ of the system. If one of these properties is kept constant, the other two may be graphed to represent the equilibrium. The commonest practice is to keep the temperature constant and to plot $C$ against $C_s$ to give an adsorption isotherm. When $C_s$ is kept constant, the plot of $C$ against $T$ is known as an adsorption isostere. In gas–solid systems, it is often convenient to express $C$ as a pressure of adsorbate. Keeping the pressure constant and plotting $C_s$ against $T$ gives adsorption isobars. The three plots are shown for the ammonia-charcoal system in Figure 17.4 which is taken from the work of BRUNAUER[10].

## 17.3.1. Single component adsorption

Most early theories were concerned with adsorption from the gas phase. Sufficient was known about the behaviour of ideal gases for relatively simple mechanisms to be postulated, and for equations relating concentrations in gaseous and adsorbed phases to be proposed. At very low concentrations the molecules adsorbed are widely spaced over the adsorbent surface so that one molecule has no influence on another. For these limiting conditions it is reasonable to assume that the concentration in one phase is proportional to the concentration in the other, that is:

$$C_s = K_a C \tag{17.1}$$

This expression is analogous to Henry's Law for gas–liquid systems even to the extent that the proportionality constant obeys the van't Hoff equation and $K_a = K_0\,\mathrm{e}^{-\Delta H/\mathbf{R}T}$ where $\Delta H$ is the enthalpy change per mole of adsorbate as it transfers from gaseous to adsorbed phase. At constant temperature, equation 17.1 becomes the simplest form of adsorption isotherm. Unfortunately, few systems are so simple.

## 17.3.2. The Langmuir isotherm

At higher gas phase concentrations, the number of molecules absorbed soon increases to the point at which further adsorption is hindered by lack of space on the adsorbent surface. The rate of adsorption then becomes proportional to the empty surface available, as well as to the fluid concentration. At the same time as molecules are adsorbing, other molecules

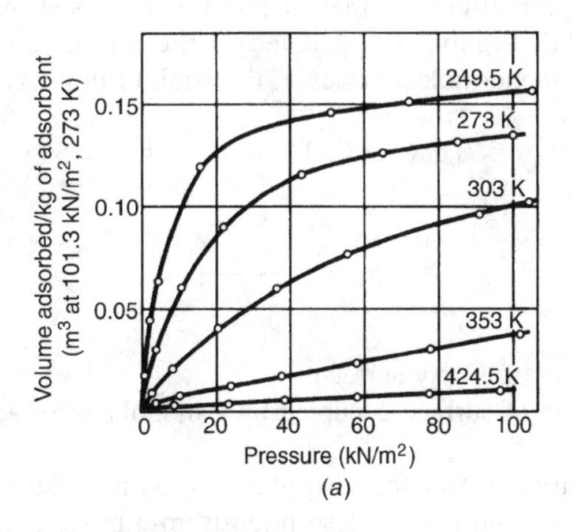


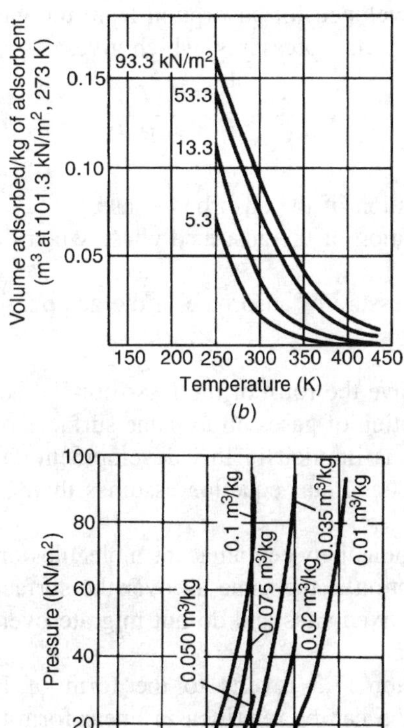


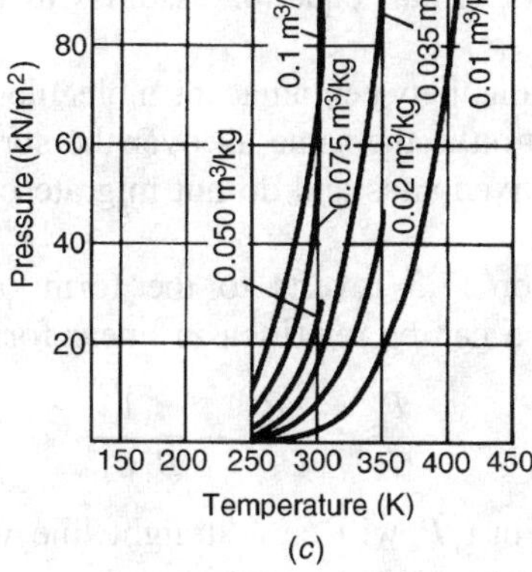


Figure 17.4. Equilibrium data for the adsorption of ammonia on charcoal[10]
(*a*) Adsorption isotherm (*b*) Adsorption isobar (*c*) Adsorption isostere

will be desorbing if they have sufficient activation energy. At a fixed temperature, the rate of desorption will be proportional to the surface area occupied by adsorbate. When the rates of adsorption and desorption are equal, a dynamic equilibrium exists. For adsorption which is confined to a mono-molecular layer, the equilibrium may be written as:

$$k_0 a_0 C = k_0(1 - a_1)C = k_1' a_1$$

or:

$$a_1 = \frac{B_0 C}{1 + B_0 C} \tag{17.2}$$

where: $a_0$ is the fraction of empty surface,
$a_1$ is the fraction of surface occupied by a monolayer of adsorbed molecules,
$B_0 = k_0/k_1'$,
$k_0$ is the velocity constant for adsorption on to empty surface, and
$k_1'$ is the velocity constant for desorption from a monolayer.

Equation 17.2 has been developed for adsorption from the gas phase. It is convenient to also express it in terms of partial pressures, which gives:

$$\frac{C_s}{C_{sm}} = \frac{B_1 P}{1 + B_1 P} \tag{17.3}$$

where: $C_s$ is the concentration of the adsorbed phase,
$C_{sm}$ is the concentration of the adsorbed phase when the monolayer is complete,
$B_1 = B_0/\mathbf{R}T$,and
$P$ is the partial pressure of adsorbate in the gas phase

.

Equations 17.2 and 17.3 have the form of the LANGMUIR[11] equation, developed in 1916, which describes the adsorption of gases on to plane surfaces of glass, mica and platinum. A number of assumptions is implicit in this development. As well as being limited to monolayer adsorption, the Langmuir equation assumes that:

(a) these are no interactions between adjacent molecules on the surface.
(b) the energy of adsorption is the same all over the surface.
(c) molecules adsorb at fixed sites and do not migrate over the surface.

When $B_1 P \ll 1$, equation 17.3 reverts to the form of Henry's Law, as given in equation 17.1. Equation 17.3 can be rewritten in linear form to give:

$$\frac{P}{C_s} = \frac{P}{C_{sm}} + \frac{1}{B_1 C_{sm}} \tag{17.4}$$

so that a plot of $P/C_s$ against $P$ will be a straight line when applied to a system that behaves in accordance with the Langmuir isotherm.

It has been shown experimentally, however, that many systems do not follow this isotherm and efforts continue to find an improved equation.

### 17.3.3. The BET isotherm

In 1938, BRUNAUER, EMMETT and TELLER[12] and EMMETT and DE WITT[13] developed what is now known as the BET theory. As in the case in Langmuir's isotherm, the theory is based on the concept of an adsorbed molecule which is not free to move over the surface, and which exerts no lateral forces on adjacent molecules of adsorbate. The BET theory does, however, allow different numbers of adsorbed layers to build up on different parts of the surface, although it assumes that the net amount of surface which is empty or which is associated with a monolayer, bilayer and so on is constant for any particular equilibrium condition. Monolayers are created by adsorption on to empty surface and by desorption from bilayers. Monolayers are lost both through desorption and through the adsorption of additional layers. The rate of adsorption is proportional to the frequency with which molecules strike the surface and the area of that surface. From the kinetic theory of gases, the frequency is proportional to the pressure of the molecules and hence:

The rate of adsorption on to empty surface $= k_0 a_0 P$, and
the rate of desorption from a monolayer $= k_1' a_1$

Desorption is an activated process. If $E_1$ is the excess energy required for one mole in the monolayer to overcome the surface forces, the proportion of molecules possessing such energy is $e^{-E_1/\mathbf{R}T}$. Hence the rate of desorption from a monolayer may be written as:

$$A_1' e^{-E_1/\mathbf{R}T} a_1$$

where $A_1'$ is the frequency factor for monolayer desorption.

The dynamic equilibrium of the monolayer is given by:

$$k_0 a_0 P + A_2' e^{-E_2/\mathbf{R}T} a_2 = k_1 a_1 P + A_1' e^{-E_1/\mathbf{R}T} a_1 \tag{17.5}$$

where $A_2'$ is the frequency factor for description from a bilayer, thus creating a monolayer.

Applying similar arguments to the empty surface, then:

$$k_0 a_0 P = A_1' e^{-E_1/\mathbf{R}T} a_1 \tag{17.6}$$

From equations 17.5 and 17.6:

$$k_1 a_1 P = A_2' e^{-E_2/\mathbf{R}T} a_2$$

$$a_1 = \frac{k_0}{A_1'} e^{E_1/\mathbf{R}T} a_0 P = \alpha_0 a_0 \tag{17.7}$$

and:

$$a_2 = \frac{k_1}{A_2'} e^{E_2/\mathbf{R}T} a_1 P = \beta a_1 \tag{17.8}$$

The BET theory assumes that the reasoning used for one or two layers of molecules may be extended to $n$ layers. It argues that energies of activation after the first layer are all equal to the latent heat of condensation, so that:

$$E_2 = E_3 = E_4 = \cdots = E_n = \lambda_M$$

Hence it may be assumed that $\beta$ is constant for layers after the first and:

$$a_i = \beta^{i-1} a_1 = B_2 \beta^i a_0$$

where $B_2 = \alpha_0/\beta$, and $a_i$ is the fraction of the surface area containing $i$ layers of adsorbate.

Since $a_0, a_1, \ldots$ are fractional areas, their summation over $n$ layers will be unity and:

$$1 = a_0 + \sum_{i=1}^{n} a_i$$

$$= a_0 + \sum_{i=1}^{n} B_2 \beta^i a_0 \tag{17.9}$$

The total volume of adsorbate associated with unit area of surface is given by:

$$v_s = v_s^1 \sum_{i=1}^{n} i a_i = v_s^1 \sum_{i=1}^{n} i B_2 \beta^i a_0 \tag{17.10}$$

where $v_s^1$ is the volume of adsorbate in a unit area of each layer.

Since $v_s^1$ does not change with $n$, a geometrically plane surface is implied. Strictly, equation 17.10 is not applicable to highly convex or concave surfaces. Equations 17.9 and 17.10 may be combined to give:

$$\frac{v_s}{v_s^1} = \frac{\displaystyle\sum_{i=1}^{n} i B_2 \beta^i a_0}{a_0 + \displaystyle\sum_{i=1}^{n} B_2 \beta^i a_0} \tag{17.11}$$

The numerator of equation 17.11 may be written as:

$$B_2 a_0 \beta \frac{d}{d\beta}\left(\sum_{i=1}^{n} \beta^i\right) = B_2 a_0 \beta \frac{d}{d\beta}\left\{\left(\frac{1-\beta^n}{1-\beta}\right)\beta\right\}$$

and the denominator as:

$$a_0\left[1 + B_2\beta\left(\frac{1-\beta^n}{1-\beta}\right)\right]$$

Substituting these values into equation 17.11 and rearranging gives:

$$\frac{v_s}{v_s^1} = \frac{B_2\beta}{1-\beta}\frac{[1-(n+1)\beta^n + n\beta^{n+1}]}{[1+(B_2-1)\beta - B_2\beta^{n+1}]} \tag{17.12}$$

On a flat unrestricted surface, there is no theoretical limit to the number of layers that can build up. When $n = \infty$, equation 17.12 becomes:

$$\frac{v_s}{v_s^1} = \frac{B_2\beta}{(1-\beta)(1-\beta+B_2\beta)} \tag{17.13}$$

When the pressure of the adsorbate in the gas phase is increased to the saturated vapour pressure, condensation occurs on the solid surface and $v_s/v_s^1$ approaches infinity. In equation 17.13, this condition corresponds to putting $\beta = 1$. It may be noted that putting $\beta = 1/(1 - B_2)$ is not helpful.

Hence from equation 17.8:

$$1 = \frac{k_1}{A_2'} e^{\lambda_M/\mathbf{R}T} \cdot P^0$$

where: $\lambda_M$ is the molar latent heat and
$P^0$ is the saturated vapour pressure.

Hence, from equation 17.8, $\beta = P/P^0$.

Equation 17.12 may be rewritten for unit mass of adsorbent instead of unit surface. This is known as the limited form of the BET equation which is:

$$\frac{V_s}{V_s^1} = B_2 \frac{P}{P^0} \frac{[1 - (n+1)(P/P^0)^n + n(P/P^0)^{n+1}]}{(1 - P/P^0)[1 + (B_2 - 1)(P/P^0) - B_2(P/P^0)^{n+1}]} \tag{17.14}$$

where $V_s^1$ is the volume of adsorbate contained in a monolayer spread over the surface area present in unit mass of adsorbent.

When $n = 1$, adsorption is confined to a monolayer and equation 17.14 reduces to the Langmuir equation.

When $n = \infty$, $(P/P^0)^n$ approaches zero and equation 17.13 may be rearranged in a convenient linear form to give:

$$\frac{P/P_0}{V(1 - P/P_0)} = \frac{1}{V^1 B_2} + \frac{B_2 - 1}{V^1 B_2}\left(\frac{P}{P^0}\right) \tag{17.15}$$

where $V$ and $V^1$ are the equivalent gas phase volumes of $V_s$ and $V_s^1$.

If a plot of the left-hand term against $P/P^0$ is linear, the experimental data may be said to fit the infinite form of the BET equation. From the slope and the intercept, $V^1$ and $B_2$ may be calculated.

The advantage of equation 17.14 is that it may be fitted to all known shapes of adsorption isotherm. In 1938, a classification of isotherms was proposed which consisted of the five shapes shown in Figure 17.5 which is taken from the work of BRUNAUER *et al.*[14]. Only gas–solid systems provide examples of all the shapes, and not all occur frequently. It is not possible to predict the shape of an isotherm for a given system, although it has been observed that some shapes are often associated with a particular adsorbent or adsorbate properties. Charcoal, with pores just a few molecules in diameter, almost always gives a Type I isotherm. A non-porous solid is likely to give a Type II isotherm. If the cohesive forces between adsorbate molecules are greater than the adhesive forces between adsorbate and adsorbent, a Type V isotherm is likely to be obtained for a porous adsorbent and a Type III isotherm for a non-porous one.

In some systems, three stages of adsorption may be discerned. In the activated alumina-air-water vapour system at normal temperature, the isotherm is found to be of Type IV. This consists of two regions which are concave to the gas concentration axis separated by a region which is convex. The concave region that occurs at low gas concentrations is usually associated with the formation of a single layer of adsorbate molecules over the

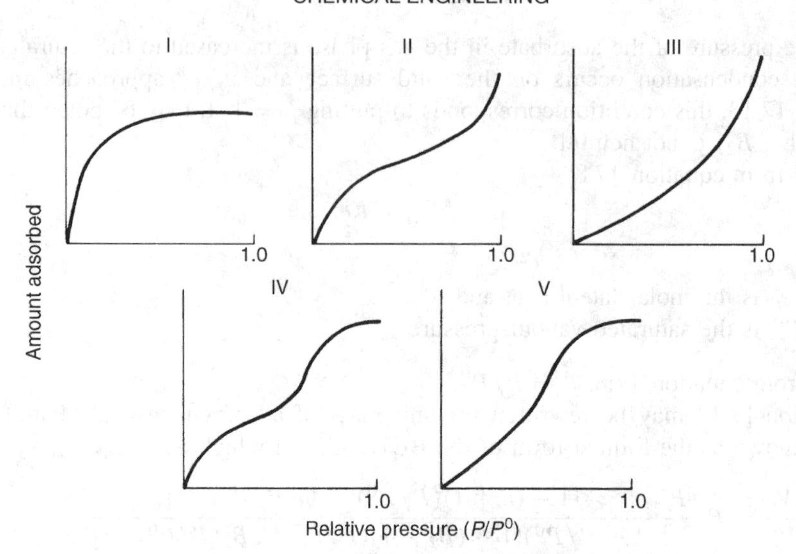


Figure 17.5. Classification of isotherms into five types of BRUNAUER, DEMING, DEMING and TELLER[14]

surface. The convex portion corresponds to the build-up of additional layers, whilst the other concave region is the result of condensation of adsorbate in the pores — so called *capillary condensation* as discussed earlier in this Section.

At low gas concentrations, whilst the monolayer is still incomplete, the absorbed molecules are relatively immobile. In the multilayer region, the adsorbed molecules behave more like a liquid film. The amount of capillary condensation that occurs depends on the pore sizes and their distribution, as well as on the concentration in the gas phase.

When $n = 1$, equation 17.14 represents a Type I isotherm.

When $n = \infty$, equation 17.14 represents a Type II, and the rarer Type III isotherm by choosing a suitable value for $B_2$. As $B_2$ is increased, the point of inflexion or "knee" of Type II becomes more prominent. This corresponds to an increasing tendency for the monolayer to become complete before a second layer starts. In the extreme case of an adsorbent whose surface is very uniform from an energy point of view, the adsorbate builds up in well-defined layers. This gives rise to a stepped isotherm, in which each step corresponds to another layer. When $B_2$ is less than 2, there is no point of inflexion and Type III isotherms are obtained. The condition $1 > B_2 > 0$ often corresponds to a tendency for molecules to adsorb in clusters rather than in complete layers.

The success of the BET equation in representing experimental data should not be regarded as a measure of the accuracy of the model on which it is based. Its capability of modelling the mobile multilayers of a Type IV isotherm is entirely fortuitous because, in the derivation of the equation, it is assumed that adsorbed molecules are immobile.

## Example 17.1

Spherical particles of 15 nm diameter and density 2290 kg/m$^3$ are pressed together to form a pellet. The following equilibrium data were obtained for the sorption of nitrogen at 77 K. Obtain estimates of the surface area of the pellet from the adsorption isotherm and compare the estimates with the geometric surface. The density of liquid nitrogen at 77 K is 808 kg/m$^3$.

| $P/P^0$ | 0.1 | 0.2 | 0.3 | 0.4 | 0.5 | 0.6 | 0.7 | 0.8 | 0.9 |
|---|---|---|---|---|---|---|---|---|---|
| m$^3$ liq $N_2$ $\times 10^6$/kg solid | 66.7 | 75.2 | 83.9 | 93.4 | 108.4 | 130.0 | 150.2 | 202.0 | 348.0 |

where $P$ is the pressure of the sorbate and $P^0$ is its vapour pressure at 77 K.

Use the following data:

$$\text{density of liquid nitrogen} = 808 \text{ kg/m}^3$$

$$\text{area occupied by one adsorbed molecule of nitrogen} = 0.162 \text{ nm}^2$$

$$\text{Avogadro Number} = 6.02 \times 10^{26} \text{ molecules/kmol}$$

## Solution

For 1 m$^3$ of pellet with a voidage $\varepsilon$, then:

$$\text{Number of particles} = (1-\varepsilon)/(\pi/6)(15 \times 10^{-9})^3$$

$$\text{Surface area per unit volume} = (1-\varepsilon)\pi(15 \times 10^{-9})^2/(\pi/6)(15 \times 10^{-9})^3$$

$$= 6(1-\varepsilon)/(15 \times 10^{-9}) \text{ m}^2/\text{m}^3$$

1 m$^3$ of pellet contains 2290 $(1-\varepsilon)$ kg solid and hence:

$$\text{specific surface} = 6(1-\varepsilon)/[(15 \times 10^{-9}(1-\varepsilon)2290]$$

$$= \underline{\underline{1.747 \times 10^5 \text{ m}^2/\text{kg}}}$$

(a) *Using the BET isotherm*

$$(P/P^0)/[V(1-P/P^0)] = 1/V'B + (B-1)(P/P^0)/V'B \quad \text{(equation 17.15)}$$

where $V$ and $V'$ are the liquid volumes of adsorbed nitrogen.

From the adsorption data given:

| $(P/P^0)$ | 0.1 | 0.2 | 0.3 | 0.4 | 0.5 | 0.6 |
|---|---|---|---|---|---|---|
| $V$ (m$^3$ liquid $N_2$/kg solid $\times 10^6$) | 66.7 | 75.2 | 83.9 | 93.4 | 108.4 | 130.0 |
| $((P/P^0)/V) \times 10^{-6}$ | 1500 | 2660 | 3576 | 4283 | 4613 | 4615 |
| $(P/P^0)/[V(1-P/P^0)]$ | 1666 | 3333 | 5109 | 7138 | 9226 | 11538 |

Plotting $(P/P^0)/[V(1-P/P^0)]$ against $(P/P^0)$, as shown in Figure 17.6, then:

$$\text{intercept, } 1/V'B = 300, \text{ and slope, } (B-1)/V'B = 13{,}902$$

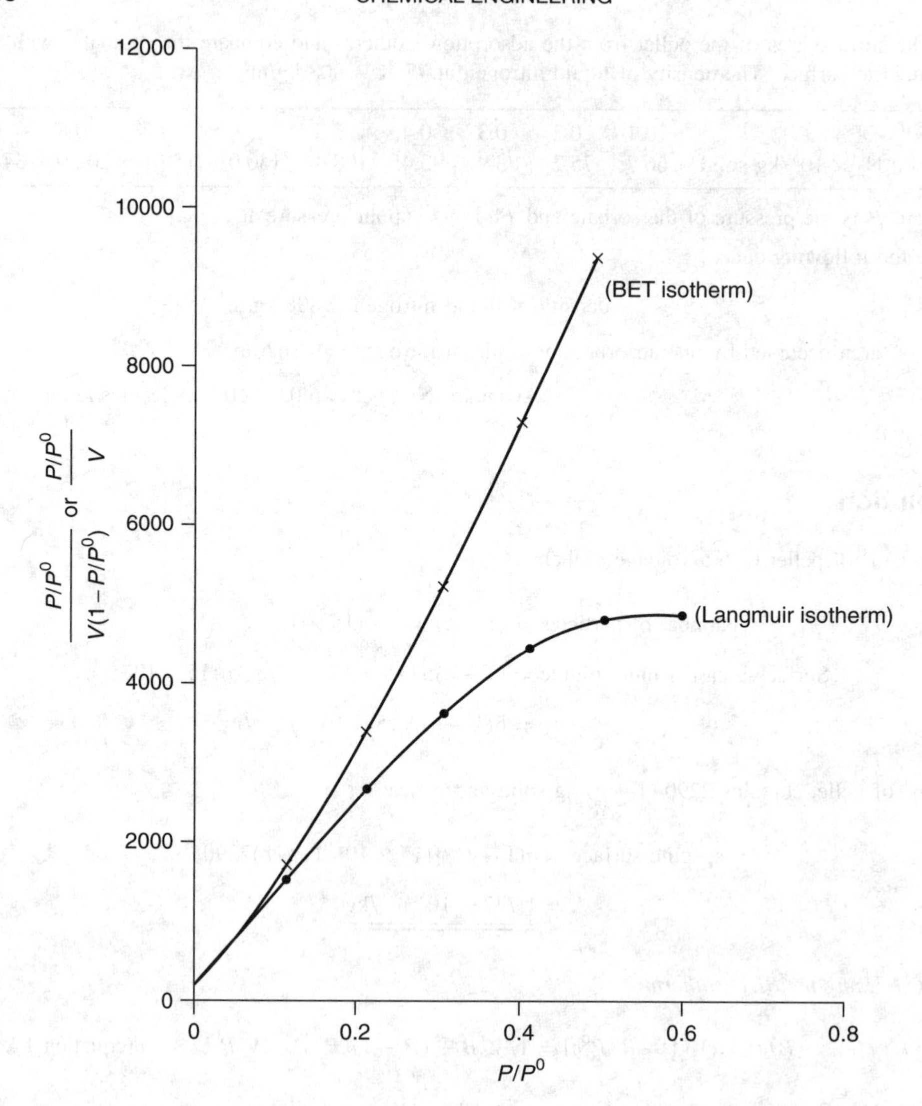


Figure 17.6. Adsorption isotherms for Example 17.1.

from which:

$$B = (13{,}902/300) + 1 = 47.34$$

and:

$$V' = 1/(300 \times 47.34) = 70.4 \times 10^{-6}\ \text{m}^3/\text{kg}.$$

$$\text{The total surface area} = [(70.4 \times 10^{-6} \times 808 \times 6.2 \times 10^{26} \times 0.162 \times 10^{-18})]/28$$

$$= \underline{\underline{2.040 \times 10^5\ \text{m}^2/\text{kg}}}.$$

(b) *Using the Langmuir form of the isotherm:*

Assuming this applies at low concentrations then, expressing pressure as the ratio $P/P^0$, and the amount adsorbed as a volume of liquid adsorbate, equation 17.4 becomes:

$$(P/P^0)/V = (P/P^0)/V' + 1/(B_2 V')$$

Thus, a plot of $(P/P^0)/V$ against $(P/P^0)$ will have a slope of $(1/V')$.

Thus, from Figure 17.6:

$$1/V' = 13{,}902$$

and:

$$V' = 71.9 \times 10^{-6}\,\text{m}^3/\text{kg}$$

which agrees with the value from the BET isotherm.

It may be noted that areas calculated from the isotherm are some 20 per cent greater than the geometric surface, probably due to the existence of some internal surface within the particles.

### 17.3.4. The Gibbs isotherm

An entirely different approach to equilibrium adsorption is to assume that adsorbed layers behave like liquid films, and that the adsorbed molecules are free to move over the surface. It is then possible to apply the equations of classical thermodynamics. The properties which determine the free energy of the film are pressure and temperature, the number of molecules contained and the area available to the film. The Gibbs free energy $G$ may be written as:

$$G = F(P, T, n_s, A_s) \tag{17.16}$$

Hence:

$$dG = \left(\frac{\partial G}{\partial P}\right)_{T,n_s,A_s}; dP + \left(\frac{\partial G}{\partial T}\right)_{P,n_s,A_s}; dT + \left(\frac{\partial G}{\partial n_s}\right)_{T,P,A_s}; dn_s + \left(\frac{\partial G}{\partial A_s}\right)_{T,P,n_s} dA_s \tag{17.17}$$

At constant temperature and pressure this becomes:

$$dG = \left(\frac{\partial G}{\partial n_s}\right) dn_s + \left(\frac{\partial G}{\partial A_s}\right) dA_s \tag{17.18}$$

$$= \mu_s\, dn_s - \Gamma\, dA_s \tag{17.19}$$

where: $\mu_s$ is the free energy per mole or chemical potential of the film, and
$\Gamma$ is defined as a two-dimensional or spreading pressure.

The total Gibbs free energy may be written as:

$$G = \mu_s n_s - \Gamma A_s \tag{17.20}$$

so that:

$$dG = \mu_s\, dn_s + n_s\, d\mu_s - \Gamma\, dA_s - A_s\, d\Gamma \tag{17.21}$$

A comparison of equations 17.19 and 17.21 shows that:

$$d\Gamma = \frac{n_s}{A_s}\, d\mu_s$$

If the gas phase is ideal and equilibrium exists between it and the sorbed phase then, by definition:

$$d\mu_s = d\mu_g = \mathbf{R}T\, d(\ln P) \tag{17.22}$$

where $\mu_g$ is the chemical potential of the gas.

Substituting for $d\mu_s$ gives:

$$d\Gamma = \frac{n_s}{A_s}\mathbf{R}T\, d(\ln P) \tag{17.23}$$

Equation 17.23 has the form of an adsorption isotherm since it relates the amount adsorbed to the corresponding pressure. This is known as the *Gibbs Adsorption Isotherm*. For it to be useful, an expression is required for $\Gamma$. Assuming an analogy between adsorbed and liquid films, HARKINS and JURA[15] have proposed that:

$$\Gamma = \alpha_1 - \beta_1 a_m \tag{17.24}$$

where $\alpha_1$, $\beta_1$, are constants, $a_m = A_s/\mathbf{N}n_s$, the area per molecule of adsorbate and $\mathbf{N}$ is the Avogadro number. Substituting for $d\Gamma$ in equation 17.23 and integrating, at constant $A_s$, from some condition $P_1$, $n_{s1}$ at which the adsorbed film becomes mobile, to an arbitrary coverage $n_s$ at pressure $P$, gives:

$$\ln\frac{P}{P_1} = \frac{1}{\mathbf{R}T}\frac{\beta_1 A_s^2}{2\mathbf{N}}\left(\frac{1}{n_{s1}^2} - \frac{1}{n_s^2}\right) \tag{17.25}$$

Equation 17.25 may be rewritten as:

$$\ln\frac{P}{P^0} = L' - \frac{M'}{V^2} \tag{17.26}$$

where $V$ is the volume occupied in the gas phase by $n_s$ moles of sorbate at a temperature $T$ and pressure $P$.

Thus:

$$L' = \ln\frac{P_1}{P^0} + \frac{1}{\mathbf{R}T}\frac{\beta_1 A_s^2}{2\mathbf{N}}\frac{1}{n_{s1}^2} \tag{17.27a}$$

and:

$$M' = \frac{1}{\mathbf{R}T}\frac{\beta_1 A_s^2}{2\mathbf{N}}\frac{1}{\rho_g^2} \tag{17.27b}$$

where $\rho_g$ is the molar density of the gas phase. Equation 17.26 is the *Harkins-Jura (H–J) equation*, which may be used to correlate adsorption data and to obtain an estimate of the surface area of an adsorbent.

The H–J model may be criticised because the comparison between an adsorbed film on a solid surface and a liquid film on a liquid surface does not stand detailed examination. Other equations of state have been used instead of equation 17.24. These are discussed in more detail by RUTHVEN[16].

Although making fundamentally different assumptions about the static or mobile nature of an adsorbed layer, the BET and H–J equations may often represent a set of experimental data equally well. As pointed out by GARG and RUTHVEN[17], their respective assumptions lead one to expect that the BET equation may be applied to low surface coverage when it is thought that adsorbed molecules do not move. The H–J equation may be applied to the middle range of relative pressures $(P/P^0)$ corresponding to the completion of a monolayer and the formation of additional mobile layers. Neither theory, it seems, is equipped to deal with the filling of micropores that occurs with the onset of capillary condensation at higher relative pressures. For this condition, a third theory of adsorption has been shown to be useful.

### 17.3.5. The potential theory

In this theory the adsorbed layers are considered to be contained in an *adsorption space* above the adsorbent surface. The space is composed of equipotential contours, the separation of the contours corresponding to a certain adsorbed volume, as shown in Figure 17.7. The theory was postulated in 1914 by POLANYI[18], who regarded the potential of a point in adsorption space as a measure of the work carried out by surface forces in bringing one mole of adsorbate to that point from infinity, or a point at such a distance from the surface that those forces exert no attraction. The work carried out depends on the phases involved. Polanyi considered three possibilities: (a) that the temperature of the system was well below the critical temperature of the adsorbate and the adsorbed phase could be regarded as liquid, (b) that the temperature was just below the critical temperature and the adsorbed phase was a mixture of vapour and liquid, (c) that the temperature was above the critical temperature and the adsorbed phase was a gas. Only the first possibility, the simplest and most common, is considered here.

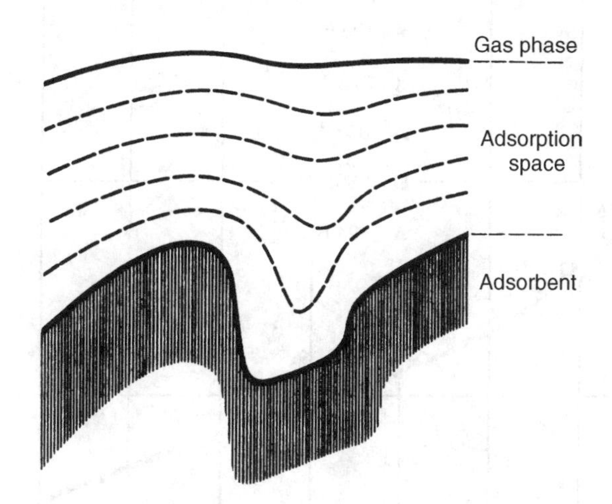


Figure 17.7. The concept of adsorption space used by Polanyi[18]

At temperatures below the critical point, the pressure at the adsorbent surface is the saturated vapour pressure of the liquid adsorbate, $P^0$. At the limit of adsorption space, the pressure exerted by the adsorbate molecules in the gas phase is $P$. For an ideal gas, the work done bringing a mole to the adsorbent surface is given by:

$$\varepsilon_p = \mathbf{R}T \ln P^0/P \qquad (17.28)$$

The potential theory postulates a unique relationship between the adsorption potential $\varepsilon_p$ and the volume of adsorbed phase contained between that equipotential surface and the solid. It is convenient to express the adsorbed volume as the corresponding volume in the gas phase.

Hence:
$$\varepsilon_p = \mathrm{f}(V) \tag{17.29}$$

where the function is assumed to be independent of temperature. Equations 17.28 and 17.29 may be combined to give an adsorption isotherm equation, although this is not explicit until the form of f($V$) has been specified. The form may be expressed graphically by plotting $\mathbf{R}T \ln P^0/P$ against $V$, giving what is called a characteristic curve as shown in Figure 17.8. The curve is valid for a particular adsorbent and adsorbate at any temperature and its validity may often be extended to other adsorbates on the same adsorbent. The adsorption potential must then be rewritten as:

$$\varepsilon_p = \beta_2 \mathrm{f}_1(V) \tag{17.30}$$

where $\beta_2$ is a coefficient of affinity. For many adsorbates, $\beta_2$ is proportional to the molar volume $V'_M$ of the liquid adsorbate at a temperature $T$. Values of $\beta_2$ for some hydrocarbons, relative to benzene, are given in Table 17.4 which is taken from the work

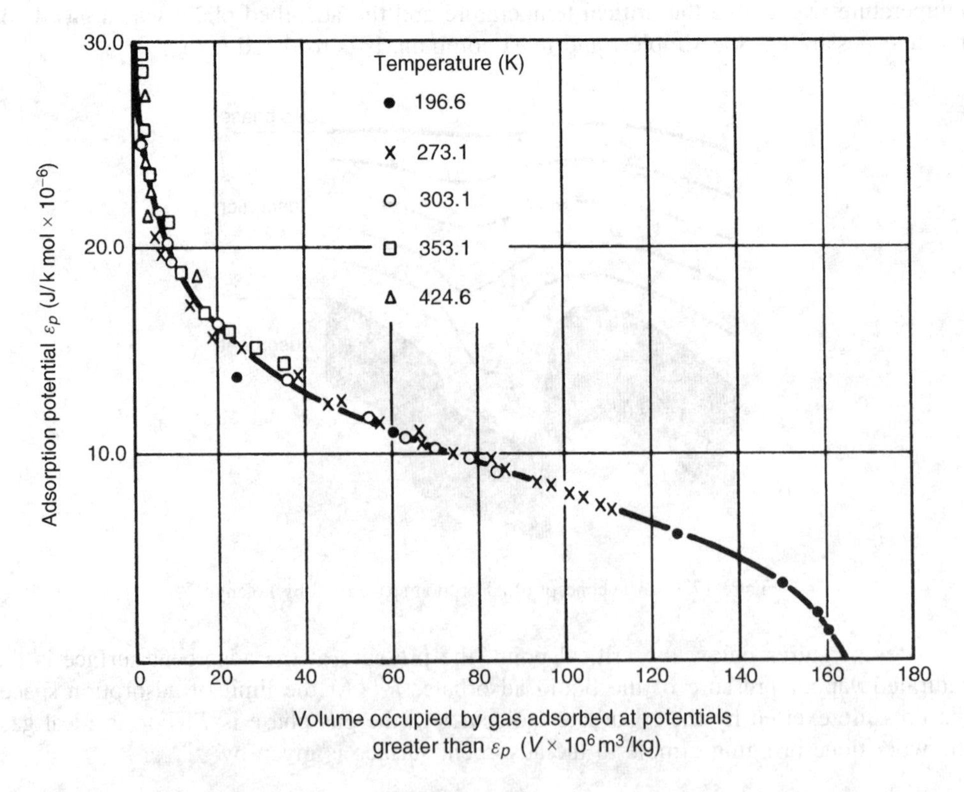


Figure 17.8. Characteristic curve for the adsorption of carbon dioxide on to charcoal[10]

Table 17.4. Values of coefficients of affinity relative to $\beta_2 = 1$ for benzene, and of $V'_M/V'_{M(\text{benzene})}$

| Vapour | $\beta_2$ expt | $V'_M/V'_{M(\text{benzene})}$ |
|---|---|---|
| $C_6H_6$ | 1 | 1 |
| $C_5H_{12}$ | 1.12 | 1.28 |
| $C_6H_{12}$ | 1.04 | 1.21 |
| $C_7H_{16}$ | 1.50 | 1.65 |
| $C_6H_5CH_3$ | 1.28 | 1.19 |
| $CH_3Cl$ | 0.56 | 0.59 |
| $CH_2Cl_2$ | 0.66 | 0.71 |
| $CHCl_3$ | 0.88 | 0.90 |
| $CCl_4$ | 1.07 | 1.09 |
| $C_2H_5Cl$ | 0.78 | 0.80 |
| $CH_3OH$ | 0.40 | 0.46 |
| $C_2H_5OH$ | 0.61 | 0.65 |
| HCOOH | 0.60 | 0.63 |
| $CH_3COOH$ | 0.97 | 0.96 |
| $(C_2H_5)_2O$ | 1.09 | 1.17 |
| $CH_3COCH_3$ | 0.88 | 0.82 |
| $CS_2$ | 0.70 | 0.68 |
| $CCl_3NO$ | 1.28 | 1.12 |
| $NH_3$ | 0.28 | 0.30 |

of YOUNG and CROWELL[19], with relative molar volumes listed for comparison. For the same volume $V$ of adsorbates 1 and 2:

$$\frac{\varepsilon_{p1}}{\varepsilon_{p2}} = \frac{\beta_{21}}{\beta_{22}} = \frac{V'_{M1}}{V'_{M2}} \tag{17.31}$$

so that the pressures exerted by a particular volume of adsorbate on the same adsorbent are related by:

$$\left(\frac{T}{V'_M}\ln\frac{P^0}{P}\right)_1 = \left(\frac{T}{V'_M}\ln\frac{P^0}{P}\right)_2 \tag{17.32}$$

Thus, from adsorption data for one gas, data for other gases on the same adsorbent may be found. Several other methods of plotting the characteristic curve have been proposed[3].

## 17.4. MULTICOMPONENT ADSORPTION

The three isotherms discussed, BET, (H–J based on Gibbs equation) and Polanyi's potential theory involve fundamentally different approaches to the problem. All have been developed for gas–solid systems and none is satisfactory in all cases. Many workers have attempted to improve these and have succeeded for particular systems. Adsorption from gas mixtures may often be represented by a modified form of the single adsorbate equation. The Langmuir equation, for example, has been applied to a mixture of $n''$ components[11].

The volume of the $i$th component $V_i$ which is adsorbed at a partial pressure $P_i$ is given by:

$$\frac{V_i}{V_i^1} = \frac{B_i P_i}{1 + \sum_{j=1}^{n''} B_j P_j} \tag{17.33}$$

where $V_i^1$ is the volume of $i$ contained in a monolayer.

Thermodynamic arguments indicate that $V^1$ should be the same for all species. A mean value $V^1$ is used, which for a binary mixture is given by:

$$\frac{1}{V^1} = \frac{y_1}{V_1^1} + \frac{y_2}{V_2^1} \tag{17.34}$$

where $y_1$, $y_2$ are mole fractions.

Equation 17.33 has a limited application. A fuller discussion of adsorption from gas mixtures can be found in the literature[3,16,19,20].

## 17.5. ADSORPTION FROM LIQUIDS

Adsorption from liquids is less well understood than adsorption from gases. In principle the equations derived for gases ought to be applicable to liquid systems, except when capillary condensation is occurring. In practice, some offer an empirical fit of the equilibrium data. One of the most popular adsorption isotherm equations used for liquids was proposed by FREUNDLICH[21] in 1926. Arising from a study of the adsorption of organic compounds from aqueous solutions on to charcoal, it was shown that the data could be correlated by an equation of the form:

$$C_s' = \alpha_2 (C^*)^{1/n} \tag{17.35}$$

where: $C_s'$ is mass of adsorbed solute per unit mass of charcoal, and
$C^*$ is the concentration of solute in solution in equilibrium with that on the solid.
$\alpha_2$ and $n$ are constants, the latter normally being greater than unity.

Although this was proposed originally as an empirical equation, the Freundlich isotherm was shown later to have some thermodynamic justification by GLUECKAUF and COATES[22]. It has also been modified for binary mixtures. In gas–solid systems, it is convenient to measure the amount adsorbed by noting the changes in pressures and volume of the gas phase, or the change in the mass of the solid. Neither approach is practicable in liquid–solid systems because the volume changes in the liquid phase are small, and because of the difficulty of distinguishing between the adsorbed phase and liquid clinging to the solid that is removed from solution. Instead, the amount adsorbed is calculated from the changes in concentration of a known volume of liquid after it has attained equilibrium with a known amount of adsorbent. Adsorption from solutions of non-electrolytes has been discussed by KIPLING[23].

## 17.6. STRUCTURE OF ADSORBENTS

The equilibrium capacity of an adsorbent for different molecules is one factor affecting its selectivity. Another is the structure of the system of pores which permeates the adsorbent.

This determines the size of molecules that can be admitted and the rate at which different molecules diffuse towards the surface. Molecular sieves, with their precise pore sizes, are uniquely capable of separating on the basis of molecular size. In addition, it is sometimes possible to exploit the different rates of diffusion of molecules to bring about their separation. A particularly important example referred to earlier, concerns the production of oxygen and nitrogen from air.

Although it is sometimes possible to view pores with an electron microscope and to obtain a measure of their diameter, it is difficult by this means to measure the distribution of sizes and impossible to measure the associated surface area. Adsorptive methods are used instead, employing some of the theories of adsorption explained previously.

### 17.6.1. Surface area

The simplest measure of surface area is that obtained by the so-called *point B method*. Many isotherms of Type II or IV show a straight section at intermediate relative pressures. The more pronounced the section, the more complete is the adsorbed monolayer before multilayer adsorption begins. As may be seen from Figure 17.9, the lower limit of the straight section is the point B which was identified by EMMETT[24] and YANG[3] as corresponding to a complete monolayer. This interpretation is accepted as a convenient empiricism.

The BET isotherm, represented by equation 17.14, may be rearranged as:

$$\frac{\phi_1}{V_s} = \frac{1}{V_s^1 B_2} + \frac{\theta_1}{V_s^1} \tag{17.36}$$

where:
$$\phi_1 = \frac{P/P^0[1 - (P/P^0)^n - n(P/P^0)^n(1 - P/P^0)]}{(1 - P/P^0)^2}$$

$$\theta_1 = \frac{P}{P^0}\frac{[1 - (P/P^0)^n]}{(1 - P/P^0)^n}$$

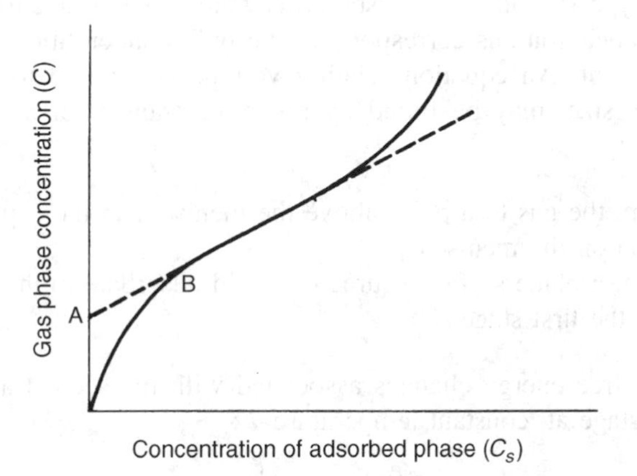


Figure 17.9. The "point B" method of estimating surface area

For a given value of $n$, a plot of $\phi_1/V_s$ against $\theta_1$ should give a straight line from the slope of which $V_s^1$ may be calculated. The simpler, infinite, form of the BET isotherm is given by equation 17.15. The appropriate plot gives a straight line, from the slope and intercept of which $V_1$ may be calculated. Equation 17.15 is most likely to apply at low relative pressures.

Equation 17.26, derived by HARKINS and JURA[15] may be plotted as ln $(P/P_0)$ against $1/V^2$ to give a straight line. The slope is proportional to $A_s^2$. The constant of proportionality may be found by using the same adsorbate on a solid of known surface area. Since the equation was derived for mobile layers and makes no provision for capillary condensation, it is most likely to fit data in the intermediate range of relative pressures.

### 17.6.2. Pore sizes

Having obtained a measure of surface area, a mean pore size may be calculated by simplifying the pore system into $n_p$ cylindrical pores per unit mass of adsorbent, of mean length $L_p$ and mean pore radius $r_p$.

Hence: Pore volume $V_p = n_p \pi r_p^2 L_p$

Surface area $A_p = 2n_p \pi r_p L_p$

and:

$$r_p = \frac{2V_p}{A_p} \tag{17.37}$$

This expression has been generalised by EVERETT and STONE[8] for any shape of capillary by including a shape factor $\gamma$ which takes a value that depends on the geometry of the capillary. This is unity for parallel sided fissures as well as for cylindrical pores.

Where there is a wide distribution of pore sizes and, possibly, quite separately developed pore systems, a mean size is not a sufficient measure. There are two methods of finding such distributions. In one a porosimeter is used, and in the other the hysteresis branch of an adsorption isotherm is utilised. Both require an understanding of the mechanism of capillary condensation.

The regions of Type IV and Type V isotherms which are concave to the gas-concentration axis at high concentrations correspond to the bulk condensation of adsorbate in the pores of the adsorbent. An equation relating volume condensed to partial pressure of adsorbate and pore size, may be found by assuming transfer of adsorbate to occur in three stages:

(i) transfer from the gas to a point above the meniscus in the capillary.
(ii) condensation on the meniscus.
(iii) transfer from a plane surface source of liquid adsorbate to the gas-phase in order to maintain the first stage.

At equilibrium the free energy changes associated with the second and third stages are zero. For the first stage at, constant temperature $T$:

$$\Delta G = \int_{P^0}^{P_m} V\,dP = n_s \mathbf{R}T \ln \frac{P_m}{P^0} \tag{17.38}$$

where: $n_s$ is the number of moles condensed,
$P_m$ is the vapour pressure over the meniscus, and
$P^0$ is the vapour pressure over the plane surface.

Another expression for $\Delta G$ may be obtained from consideration of the interfacial changes that occur as the capillaries fill. This takes the form:

$$\Delta G = \Delta A_p(\sigma_{sl} - \sigma_{sv}) = -\Delta A_p \sigma_{lv} \cos\phi \tag{17.39}$$

where: $\sigma$ represents surface tensions at the three interfaces of the solid ($s$), liquid ($l$) and vapour ($v$),
$\phi$ is the interfacial angle between liquid and solid, and
$\Delta A_p$ is the change in interfacial area.

Hence:

$$n_s \mathbf{R}T \ln \frac{P_m}{P^0} = -\Delta A_p \sigma_{lv} \cos\phi \tag{17.40}$$

If the transfer of $dn_s$ moles results in a change in interfacial area $dA_p$, then:

$$\frac{dn_s}{dA_p} = -\frac{\sigma_{lv}\cos\phi}{\mathbf{R}T \ln(P_m/P^0)} = \frac{1}{V_M}\frac{dV}{dA_p} \tag{17.41}$$

where: $dV$ is the volume occupied by $dn_s$ moles in the vapour phase, and
$V_M$ is the molar volume.

Thus:

$$\frac{dV}{dA_p} = \frac{V_M \sigma_{lv}\cos\phi}{\mathbf{R}T \ln(P^0/P_m)} \tag{17.42}$$

The left-hand side of equation 17.42 is a characteristic dimension of the pores in which condensation or evaporation occurs at an adsorbate pressure $P_m$. This depends on the geometry of the pores, as well as whether adsorption or desorption is occurring. Figure 17.10 illustrates the conditions in a cylindrical pore. For desorption occurring from the free surface of condensate:

$$dV = d(\pi r_p^2 L_p) = \pi r_p^2\, dL_p$$

$$dA_p = (2\pi r_p L_p) = 2\pi r_p\, dL_p$$

Hence:

$$\frac{dV}{dA_p} = \frac{r_p}{2} \tag{17.43}$$

and:

$$r_p = \frac{2V_M}{\mathbf{R}T}\frac{\sigma_{lv}\cos\phi}{\ln P^0/P} \tag{17.44}$$

Equation 17.44 is known as the *Kelvin equation*, and is considered to be valid for desorption.

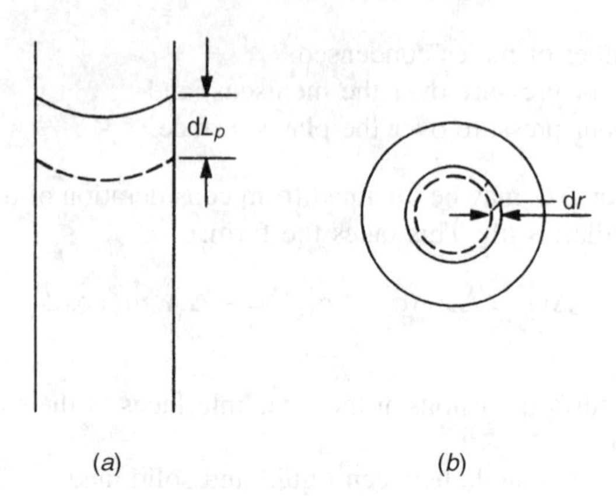


Figure 17.10. The capillary condensation equation applied to a cylindrical pore (*a*) for desorption (*b*) for adsorption

For adsorption on to existing layers of adsorbate on the cylindrical surface of the pores:

$$\mathrm{d}V = \mathrm{d}[\pi(r_p^2 - r^2)L_p] = -2\pi r L_p\,\mathrm{d}r$$

$$\mathrm{d}A_p = \mathrm{d}[2\pi r L_p] = 2\pi L_p\,\mathrm{d}r$$

Therefore:

$$r_p' = \frac{V_M}{\mathbf{R}T}\frac{\sigma_{lv}\cos\phi}{\ln P^0/P} \tag{17.45}$$

where $r_p'$ is the net radius of the pore.

Equation 17.45 is known as the *Cohan equation*[25]. Together with equation 17.44, this offers an explanation of the hysteresis effect which many isotherms exhibit at high relative concentrations of 0.4–0.95. Figure 17.11 shows an isotherm for nitrogen on activated-alumina. A difference between adsorption and desorption conditions over the middle-to-high range of relative pressures may be noted. When applied to this range, equation 17.44 or equation 17.45 may be used to estimate the distribution of pore sizes contained within the adsorbent. A method, such as that of CRANSTON and INKLEY[26] leads to the distribution shown in Figure 17.12.

Interpretation of the pore radii calculated in this way is complicated by the fact that adsorption is occurring on the surface of pores which are, as yet, too big for condensation to occur. Radii given by equations 17.44 or 17.45 are net values, and an allowance has to be made for any adsorbed layers present when condensation occurs.

The actual structure of an adsorbent may be very different from a model assuming independent cylindrical pores. It may be possible to allow for the difference by including a geometric factor in the simple equations, or it may be necessary to "view" the surface using, for example, a scanning electron microscope. Sometimes the results are a pleasing confirmation of the simple assumptions. Figure 17.3 shows a transmission electron micrograph of an activated alumina similar to that which gave rise to Figures 17.11 and 17.12. The thin edges of particles are sufficiently transparent to the electron beam to show up parallel sets of pores on a hexagonal array, and about 4 nm in diameter, very close to

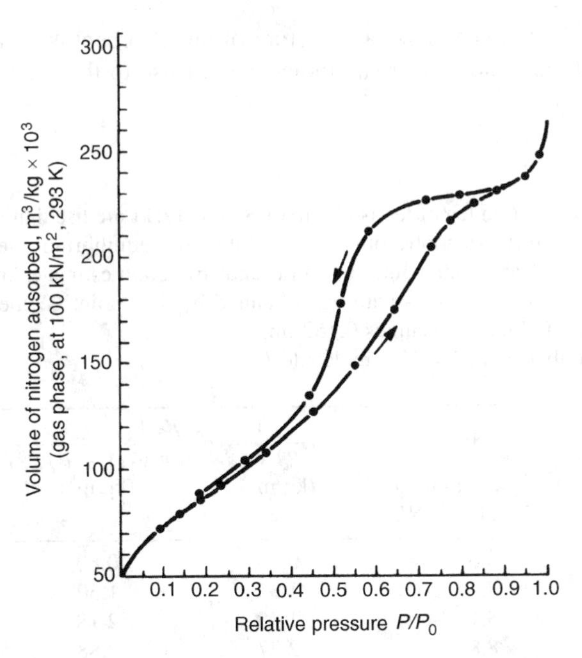


Figure 17.11. Nitrogen isotherm for activated alumina determined at the boiling point of liquid nitrogen

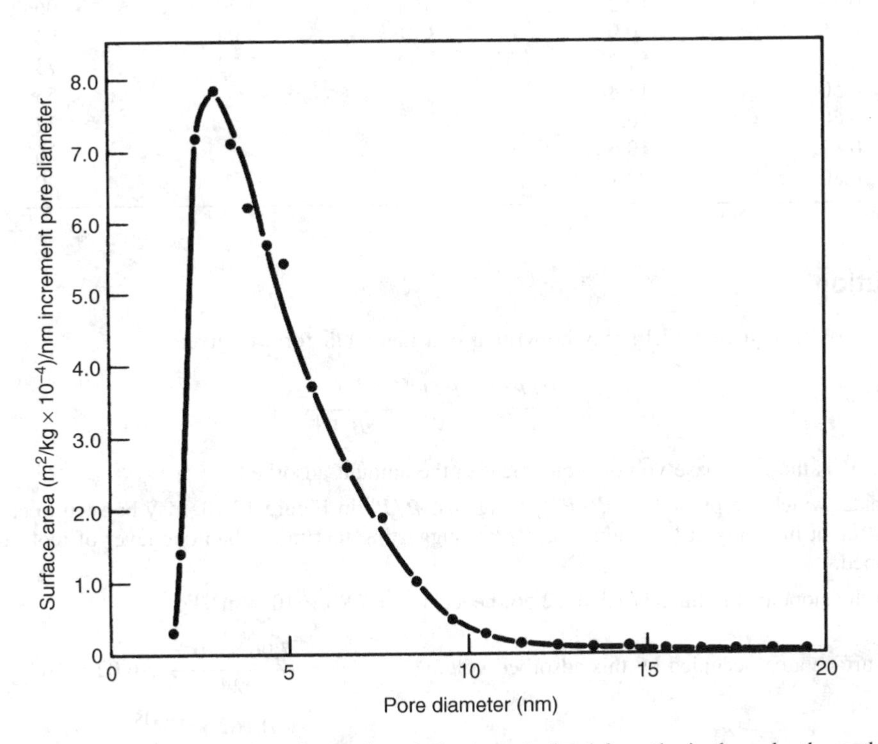


Figure 17.12. Pore size distribution of an activated alumina calculated from the isotherm by the method of Cranston and Inkley[26]

the peak in Figure 17.12. The physical structure of an adsorbent is an important factor in determining the rate at which adsorbate molecules diffuse to the adsorbent surface.

## Example 17.2

Data taken from the adsorption leg of the isotherm of Figure 17.11 are listed in the first two columns of the following table. Test the applicability of the following equilibrium theories: (a) Langmuir (b) infinite BET and (c) Harkins and Jura. From (a) and (b) obtain estimates of the surface area of the adsorbent and compare the values with that obtained by the "point B" method. One molecule of nitrogen adsorbed on alumina occupies 0.162 $nm^2$.

The Avogadro Number is $6.02 \times 10^{26}$ molecules/kmol.

| $P/P^0$ (—) | $V \times 10^2$ ($m^3 N_2$/kg alumina at 293 K and $10^5$ $N/m^2$) | $\frac{P}{P^0}\frac{1}{V}$ ($kg/m^3$) | $\frac{P}{P^0}\frac{1}{V}\frac{1}{(1-P/P^0)}$ ($kg/m^3$) | $\frac{1}{V^2}$ ($kg^2/m^6$) |
|---|---|---|---|---|
| 0.05 | 6.6 | 0.76 | 0.80 | 230 |
| 0.10 | 7.4 | 1.35 | 1.50 | 183 |
| 0.15 | 8.1 | 1.85 | 2.18 | 152 |
| 0.20 | 8.8 | 2.27 | 2.88 | 129 |
| 0.25 | 9.4 | 2.66 | 3.55 | 113 |
| 0.30 | 10.2 | 2.94 | 4.20 | 96 |
| 0.35 | 10.9 | 3.21 | 4.94 | 84 |
| 0.40 | 11.7 | 3.42 | 5.73 | 73 |
| 0.50 | 13.8 | | | 53 |
| 0.60 | 16.5 | | | 37 |
| 0.87 | 19.6 | | | 26 |
| 0.80 | 22.1 | | | 20 |

## Solution

(a) For $n = 1$, equation 17.14 may be written in a Langmuir form to give:

$$\frac{P/P^0}{V} = \frac{P/P^0}{V^1} + \frac{1}{B_2 V^1}$$

where $V$ is the gas phase volume equivalent to the amount adsorbed.

The data, which are plotted as $(P/P^0)/V$ against $P/P^0$ in Figure 17.13, may be seen to conform to a straight line only at low values of $P/P^0$, suggesting that more than one layer of molecules is adsorbed.

From the slope of the line, $1/V^1 = 12.56$, hence $V^1 = 7.96 \times 10^{-2}$ $m^3/kg$

The surface area occupied by this adsorbed volume $= \dfrac{7.96 \times 10^{-2}}{24} \times 6.02 \times 10^{26} \times 0.162 \times 10^{-18}$

$= \underline{\underline{323{,}000 \text{ m}^2/\text{kg}}}$

(b)
$$\frac{P/P^0}{V(1-P/P^0)} = \frac{1}{V^1 B_2} + \frac{B_2 - 1}{V^1 B_2}(P/P^0) \qquad \text{(equation 17.15)}$$

The left-hand side of equation 17.15 is plotted against $P/P^0$ in Figure 17.13. The data conform to a straight line over a wider range of relative pressure than was the case in (a).

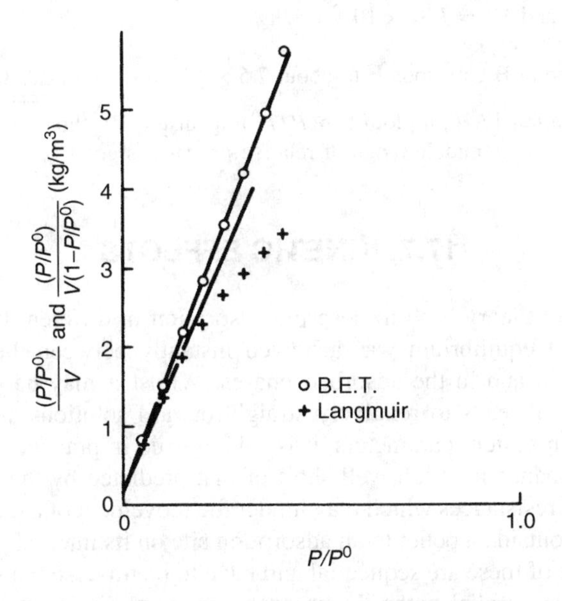


Figure 17.13. Langmuir and BET plots for Example 17.2

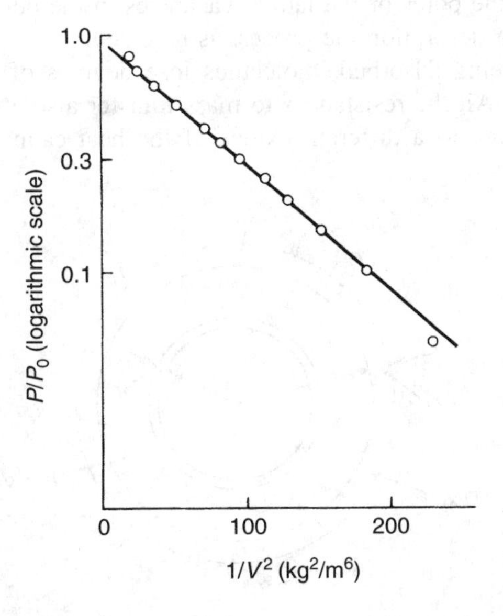


Figure 17.14. Harkins–Jura plot for Example 17.2

$$\text{Slope, } \left(\frac{B_2 - 1}{V^1 B_2}\right) = 14.05$$

$$\text{Intercept, } \left(\frac{1}{V^1 B_2}\right) = 0.2$$

Hence: $B_2 = 71.2$ and $V^1 = 7.02 \times 10^{-2}$ m$^3$/kg
Surface area $= 285{,}000$ m$^2$/kg
From Figure 17.11, point B corresponds to about $7.5 \times 10^{-2}$ m$^3$/kg or $\underline{\underline{305{,}000 \text{ m}^2/\text{kg}}}$

(c) According to equation 17.26, a plot of $\ln(P/P^0)$ against $1/V^2$ should be linear. Figure 17.14 shows that agreement in the middle range of relative pressure is good.

## 17.7. KINETIC EFFECTS

In much of the early theory used to describe adsorption in different kinds of equipment, it was assumed that equilibrium was achieved instantly between the concentrations of adsorbate in the fluid and in the adsorbed phases. Whilst it may be useful to make this assumption because it leads to relatively straightforward solutions and shows the interrelationship between system parameters, it is seldom true in practice. In large-scale plant particularly, performance may fall well short of that predicted by the equilibrium theory.

There are several resistances which may hinder the movement of a molecule of adsorbate from the bulk fluid outside a pellet to an adsorption site on its internal surface, as shown in Figure 17.15. Some of these are sequential and have to be traversed in series, whilst others derive from possible parallel paths. In broad terms, a molecule, under the influence of concentration gradients, diffuses from the turbulent bulk fluid through a laminar boundary layer around a solid pellet to its external surface. It then diffuses, by various possible mechanisms, through the pores or the lattice vacancies in the pellet until it is held by an adsorption site. During desorption the process is reversed.

In the process of being adsorbed, molecules lose degrees of freedom and a heat of adsorption is released. All the resistances to mass transfer also affect the transfer of heat out of the pellet, though to a different extent. If the heat cannot disperse fast enough,

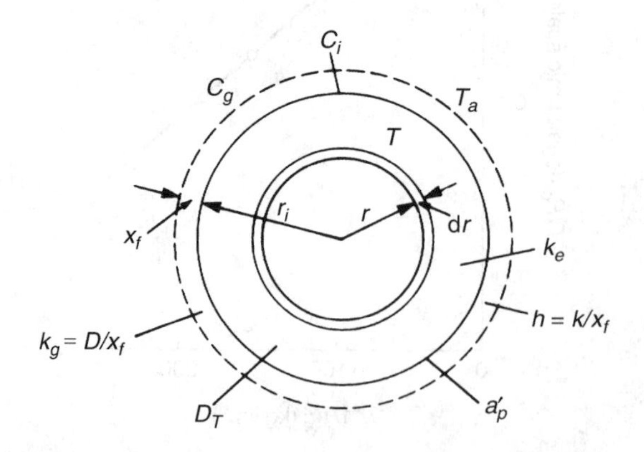


Figure 17.15. Adsorbate and energy transport in a spherical pellet

the temperature of the adsorbent increases and the effective capacity of the equipment is reduced.

### 17.7.1. Boundary film

There have been many studies of the effect of boundary films on mass and heat transfer to single pellets and in packed beds, including the work of RANZ and MARSHALL[27] and DWIVEDI and UPADHEY[28]. Other theories of mass and heat transfer are discussed in Volume 1, Chapter 10, although only the steady-state film-theory is considered here. It is assumed that the difference in concentration and temperature between the bulk fluid and the external surface of a pellet is confined to a narrow laminar boundary-layer in which the possibility of accumulation of adsorbate or of heat is neglected.

Adsorbate is transferred to an adsorbent at a rate which depends upon $a'_p$ the external surface area of the solid and upon $\Delta C$, the concentration difference across the boundary film.

Thus: Rate of mass transfer $= k_g a'_p \Delta C$

where $k_g$ is a mass transfer coefficient.
The rate at which heat is generated by mass transfer into the pellet is given approximately by:

$$\text{Rate of liberation of heat} = k_g a'_p \Delta C \Delta H$$

where $\Delta H$ is the heat of adsorption. When the temperature of the pellet has increased to the extent that heat is lost as fast as it is generated, then:

$$k_g a'_p \Delta C \Delta H = h a'_p \Delta T \tag{17.46}$$

where $h$ is a heat transfer coefficient
Since $k_g = D/x_f$ and $h = k/x_f$ then:

$$\Delta T = \frac{D}{k} \Delta H \Delta C \tag{17.47}$$

which may be written as:

$$\Delta T = \frac{1}{\rho c_{pg}} \frac{Pr}{Sc} \Delta H \Delta C \tag{17.48}$$

where: $D$ = fluid diffusivity,
$k$ = fluid thermal conductivity,
$x_f$ = thickness of boundary film,
$Pr$ = Prandtl Number $c_{pg}\mu/k$, and
$Sc$ = Schmidt Number $\mu/\rho D$

For gas mixtures, the ratio $Pr/Sc$ is about 0.2–5. $\Delta H/\rho c_{pg}$ is normally large. It follows that a small value of $\Delta C$ corresponds to a large value of $\Delta T$. Generally, the boundary film does not offer much resistance to mass transfer, except in relative terms during the unsteady state conditions that exist when a pellet is first exposed to a fluid. The boundary film may, however, offer a significant resistance to heat transfer.

A re-evaluation of published data for mass transfer carried out by WAKAO and FUNAZKRI[29] indicates that, for the range $3 < Re < 10^4$:

$$Sh = 2.0 + 1.1 Sc^{1/3} Re^{0.6} \tag{17.49}$$

where $Sh$ is the Sherwood number $(k_g d_p/D)$
$Re$ is the Reynolds number based on the superficial velocity through the bed, and pellet diameter. Equation 17.49 assumes that axial diffusion in the bed has been allowed for, which is not the case in most of the earlier work.

The corresponding equation for heat transfer, neglecting radiation, is:

$$Nu = 2.0 + 1.1 Pr^{1/3} Re^{0.6} \tag{17.50}$$

where $Nu$ is the Nusselt number $hd_p/k$.

### 17.7.2. Intra-pellet effects

After passing through the boundary layer, the molecules of adsorbate diffuse into the complex structure of the adsorbent pellet, which is composed of an intricate network of fine capillaries or interstitial vacancies in a solid lattice. The problem of diffusion through a porous solid has attracted a great deal of interest over the years and there is a fairly good understanding of the mechanisms involved, at least for gas phase diffusion. Here, diffusion within a single cylindrical pore is considered and, then, the pore is related to the pellet as a whole.

Molecules entering a pore move randomly between the pore-wall and the pore-space. Sometimes molecules are more closely associated with the wall and may be thought of as diffusing over the surface, subject to a surface concentration-difference and a surface diffusion-coefficient. At other times, molecules move in pore-space, where their behaviour depends on the mean free path of the gas and on the pore diameter. The mean free path $\lambda$, which is the average distance between collisions of gas molecules, may be predicted using the kinetic theory of gases, and given by GLASSTONE and LEWIS[30] as:

$$\lambda = \frac{1}{\sqrt{2} n' \pi \sigma^2} \tag{17.51}$$

where: $n'$ = number of molecules in unit volume of gas phase,
$\sigma$ = collision diameter.

At atmospheric pressure and 293 K:

$$n' = \frac{6 \times 10^{26}}{24.0} = 2.5 \times 10^{25} \text{ molecules/m}^3$$

The collision diameter of most molecules is about 0.2 nm.

Hence: $\lambda = 225$ nm.

In pores that are appreciably smaller than the mean free path, molecules tend to collide with the pore walls rather than with other molecules. Having collided with the wall, the molecules are momentarily retained and then released in a random direction. The coefficient, $D_k$, which controls this *Knudsen diffusion*, considered by SATTERFIELD[31], and in Volume 1, Chapter 3, may be derived from the kinetic theory to give:

$$D_k = \frac{2}{3}\left(\frac{8k'_B T}{\pi m_M}\right) r_p \tag{17.52}$$

or:

$$D_k = 1.0638\left(\frac{\mathbf{R}T}{M}\right) r_p \tag{17.53}$$

where: $r_p$ is the pore radius
$m_M$ is the mass of a molecule, and
$M$ is the molecular weight.

The flux due to Knudsen diffusion is given by:

$$N_{Ak} = -D_k \frac{\partial C_A}{\partial l} \tag{17.54}$$

where $l$ is the distance into the pore.

When the mean free path is small compared with pore diameter, the dominating experience of molecules is that of collision with other molecules in the gas phase. In that respect, the situation is much the same as that which exists in the bulk gas. The appropriate diffusion coefficient $D_m$ may be obtained from published experimental values, or calculated from a theoretical expression. For molecular diffusion in a binary gas, the Chapman and Enskog equation may be used, as discussed by BIRD, STEWART and LIGHTFOOT[32]. This takes the form:

$$D_m = \frac{1.88 \times 10^{-4} T^{3/2}}{P\sigma_{AB}^2 \Omega_{AB}}\left(\frac{1}{M_A} + \frac{1}{M_B}\right)^{1/2} \tag{17.55}$$

where: $D_m$ is the diffusion coefficient (m²/s),
$M_A$, $M_B$ are molecular weights of **A** and **B**,
$P$ is the pressure (N/m²),
$T$ is the temperature (K),
$\sigma_{AB}$ is a collision diameter (nm), and
$\Omega_{AB}$ is a dimensionless collision integral.

Bird Stewart and Lightfoot have published tables which give values of $\Omega$ and $\sigma$.

The contribution to the total flux that comes from bulk diffusion is given by:

$$N_{Am} = -D_m \frac{\partial C_A}{\partial l} \tag{17.56}$$

The total diffusion in any one adsorbent is complicated by the fact that there is no sudden transition from Knudsen to molecular diffusion at a particular pore size, and most adsorbents contain a range of pore sizes. Methods of obtaining a value for an average coefficient have been suggested. For example, EVANS *et al.*[33] have suggested that, in a binary system:

$$\frac{1}{D_{av}} = \frac{1}{D_k} + \frac{1}{D_m}\left[1 - \left(1 + \frac{N_B}{N_A}\right) y_A\right] \tag{17.57}$$

where $y_A$ is the mole fraction of one component. The final bracket allows for deviations from equimolecular counterdiffusion. Whilst both Knudsen and molecular diffusion may be regarded as alternative mechanisms in a particular pore, a third mechanism, namely surface diffusion, is a possible additional route by which molecules may enter a pellet. Surface diffusion is sometimes likened to movement in a liquid film, the surface diffusion coefficient $D_s$ increasing with surface coverage until the monolayer is complete and then remaining constant. Since it exists in parallel with other forms of diffusion, surface diffusion is additive in its effect and:

$$\alpha D_T = \alpha D_{av} + (1 - \alpha) J D_s \tag{17.58}$$

where: $\alpha$ is pore volume per unit volume of pellet[34], and
$J$ is a factor which has been shown to decrease with temperature.

Surface diffusion is only significant when the pores are small and $D_{av} = D_k$.
There is an energy of activation $E_s$ for surface diffusion which leads to a temperature dependence of an Arrhenius kind and:

$$D_s = D_{s0}\, e^{-E_s/\mathbf{R}T} \tag{17.59}$$

Equation 17.59 has been confirmed experimentally, suggesting that molecules move over a surface by "hopping" to adjacent adsorption sites. It may be assumed that this process involves a lower energy of activation than that required for complete desorption. The molecule will continue to "hop" until it finds a vacant adsorption site, thus explaining the increase of surface diffusion coefficient with coverage.

The Einstein equation for surface diffusion gives $D_s = \lambda_A^2/2\,\delta t$, where $\lambda_A$ is the average distance between vacant adsorption sites and $\delta t$ is the residence time on a site.

Experimental measurements of surface diffusion are usually calculated by subtracting from the measured total diffusion that predicted theoretically for Knudsen and molecular diffusion.

Using the total diffusion coefficient defined in equation 17.58, the flux into a pore may be written as:

$$N_A = -D_T \frac{\partial C_A}{\partial l} \tag{17.60}$$

The total diffusional flow into a pellet must allow for the fraction of pellet which is pore, and the tortuous path through the pore system.

In terms of an effective diffusivity $D_e$ and a mean concentration gradient across a porous medium of thickness $L$, the flux through the medium may be written as:

$$N_A = D_e(-\Delta C_A)/L \tag{17.61}$$

$N_A$, which refers to unit superficial area, may be expressed in terms of an interstitial flow $N_{Ai}$ and voidage $\varepsilon$ as:

$$N_A = \varepsilon N_{Ai} \tag{17.62}$$

A third flux $N_{Ac}$ may be defined which refers to the actual flow of molecules along the tortuous path of a pore of length $L_e$.

Thus:
$$N_{Ac} = D_T(-\Delta C_A/L_e) \tag{17.63}$$

The velocity of **A** along the direct axial path $= N_{Ai}/\bar{C}_A$
The velocity of **A** through a tortuous pore $= N_{Ac}/\bar{C}_A$
where $\bar{C}_A$ is the mean molar concentration of **A**.
The tortuous and direct paths are equivalent if the time taken for **A** to traverse each is the same.

Thus:
$$\frac{L}{N_{Ai}/\bar{C}_A} = \frac{L_e}{N_{Ac}/\bar{C}_A} \tag{17.64}$$

Hence:
$$N_{Ac} = N_{Ai}\frac{L_e}{L} = \frac{N_A}{\varepsilon}\frac{L_e}{L}$$

Substituting in equation 17.63:
$$N_A = D_T\varepsilon\frac{(-\Delta C_A)}{L}\left(\frac{L}{L_e}\right)^2 \tag{17.65}$$

From equations 17.61 and 17.65:
$$D_e = \frac{D_T\varepsilon}{(L_e/L)^2} = \frac{D_T\varepsilon}{\tau^2} \tag{17.66}$$

where $\tau$ is the tortuosity (actual pore length/superficial diffusion path) and $\tau^2$ is a tortuosity factor.

Values of $\tau^2$ of 2–6 are common. Occasionally, much higher values are quoted which are more likely to indicate shortcomings in the theory rather than highly tortuous paths. There is some confusion in the literature between $\tau^2$ and $\tau$, as discussed by EPSTEIN[35].

### 17.7.3. Adsorption

The final stage in getting a molecule from the bulk phase outside a pellet on to the interior surface is the adsorption step itself. Where adsorption is physical in nature, this step is unlikely to affect the overall rate. An equilibrium state is likely to exist between adsorbate molecules immediately above a surface and those on it. An adsorption isotherm such as equation 17.4 or 17.35 may be applied.

When chemisorption is involved, or when some additional surface chemical reaction occurs, the process is more complicated. The most common combinations of surface mechanisms have been expressed in the *Langmuir–Hinshelwood* relationships[36]. Since the adsorption process results in the net transfer of molecules from the gas to the adsorbed phase, it is accompanied by a bulk flow of fluid which keeps the total pressure constant. The effect is small and usually neglected. As adsorption proceeds, diffusing molecules may be denied access to parts of the internal surface because the pore system becomes blocked at critical points with condensate. Complex as the situation may be in theory, it may often be simplified by selecting what is termed the *controlling resistance*. This relates to the step whose resistance is overwhelmingly large in comparison with that for the other layers. The whole of the difference in adsorbate concentration, from its value in the bulk gas to its value at the adsorbent surface, occurs across the controlling resistance. It may be, however, that rate control is mixed, involving two or more resistances.

The relative importance of the boundary-film and intra-pellet diffusion in mass transfer is measured by the *Biot number*. On the assumption that there is no accumulation of adsorbate at the external surface of a pellet, then:

$$k_g(C_g - C_i) = D_e \left.\frac{\partial C}{\partial r}\right|_{r=r_i}$$

$$\frac{k_g r_i}{D_e} = \frac{1}{(1 - C_i/C_g)} \left.\frac{\partial(C/C_g)}{\partial(r/r_i)}\right|_{r=r_i} \tag{17.67}$$

where the left-hand dimensionless group $Bi$ is the *Biot number*. A high value of $Bi$ indicates that intra-pellet diffusion is controlling the rate of transport.

The discussion so far has concentrated on mass transfer. The transfer of the heat liberated on adsorption or consumed on desorption may also limit the rate process or the adsorbent capacity. Again the possible effects of the boundary-film and the intra-pellet thermal properties have to be considered. A Biot number for heat transfer is $hr_i/k_e$. In general, this is less than that for mass transfer because the boundary layer offers a greater resistance to heat transfer than it does to mass transfer, whilst the converse is true in the interior of the pellet.

## 17.8. ADSORPTION EQUIPMENT

The scale and complexity of an adsorption unit varies from a laboratory chromatographic column a few millimeters in diameter, as used for analysis, to a fluidised bed several metres in diameter, used for the recovery of solvent vapours, from a simple container in which an adsorbent and a liquid to be clarified are mixed, to a highly-automated moving-bed of solids in plug-flow.

All such units have one feature in common in that in all cases the adsorbent becomes saturated as the operation proceeds. For continuous operation, the spent adsorbent must be removed and replaced periodically and, since it is usually an expensive commodity, it must be regenerated, and restored as far as possible to its original condition.

In most systems, regeneration is carried out by heating the spent adsorbent in a suitable atmosphere. For some applications, regeneration at a reduced pressure without increasing the temperature is becoming increasingly common. The precise way in which adsorption and regeneration are achieved depends on the phases involved and the type of fluid–solid contacting employed. It is convenient to distinguish three types of contacting:

(a) Those in which the adsorbent and containing vessel are fixed whilst the inlet and outlet positions for process and regenerating streams are moved when the adsorbent becomes saturated. The fixed bed adsorber is an example of this arrangement. If continuous operation is required, the unit must consist of at least two beds, one of which is on-line whilst the other is being regenerated.

(b) Those in which the containing vessel is fixed, though the adsorbent moves with respect to it. Fresh adsorbent is fed in and spent adsorbent removed for regeneration at such a rate as to confine the adsorption within the vessel. This type of arrangement includes fluidised beds and moving beds with solids in plug flow.

(c) Those in which the adsorbent is fixed relative to the containing vessel which moves relative to fixed inlet and outlet positions for process and regenerating fluids. The rotary-bed adsorber is an example of such a unit.

## 17.8.1. Fixed or packed beds

When used as part of a commercial operation with gas or liquid mixtures, the single pellets discussed in the context of *rate processes* are consolidated in the form of packed beds. Usually the beds are stationary and the feed is switched to a second bed when the first becomes saturated. Whilst there are applications for moving-beds, as discussed later, only fixed-bed equipment will be considered, here as this is the most widely used type.

Figure 17.16$a$ depicts the way in which adsorbate is distributed along a bed, during an adsorption cycle. At the inlet end of the bed, the adsorbent has become saturated and is

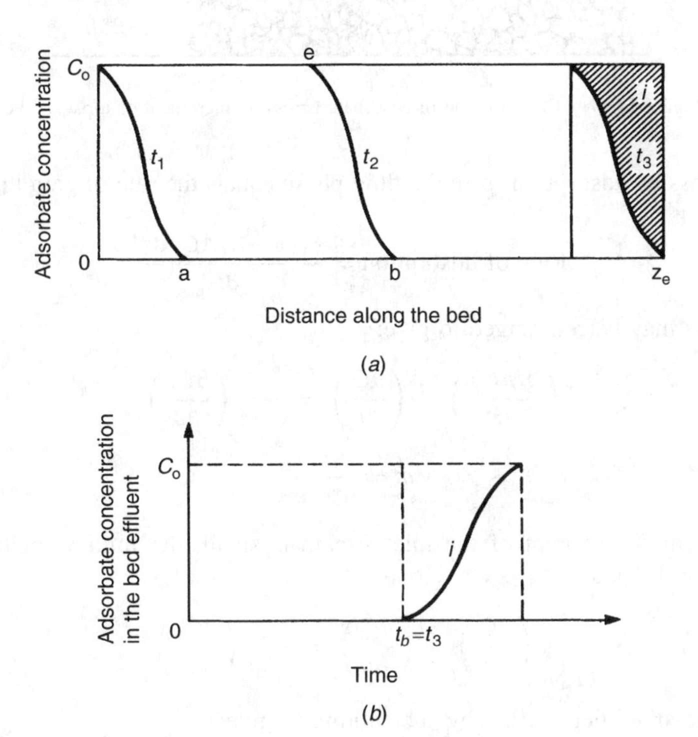


Figure 17.16. The distribution of adsorbate concentration in the fluid phase through a bed ($a$) Development and progression of an adsorption wave along a bed. ($b$) Breakthrough curve

in equilibrium with the adsorbate in the inlet fluid. At the exit end, the adsorbate content of the adsorbent is still at its initial value. In between, there is a reasonably well-defined mass-transfer zone in which the adsorbate concentration drops from the inlet to the exit value. This zone progresses through the bed as the run proceeds. At $t_1$ the zone is fully formed, $t_2$ is some intermediate time, and $t_3$ is the breakpoint time $t_b$ at which the zone begins to leave the column. For efficient operation the run must be stopped just before the breakpoint. If the run extends for too long, the breakpoint is exceeded and the effluent concentration rises sharply, as is shown in the breakthrough curve of Figure 17.16*b*.

A mass balance of adsorbate in the fluid flowing through an increment d$z$ of bed as shown in Figure 17.17 gives:

$$uA\varepsilon C - \left[uA\varepsilon C + \frac{\partial(uA\varepsilon C)}{\partial z}\mathrm{d}z\right] = \frac{\partial(\varepsilon AC\mathrm{d}z)}{\partial t} + \text{Loss} \tag{17.68}$$

INPUT − OUTPUT = ACCUMULATION + LOSS BY ADSORPTION

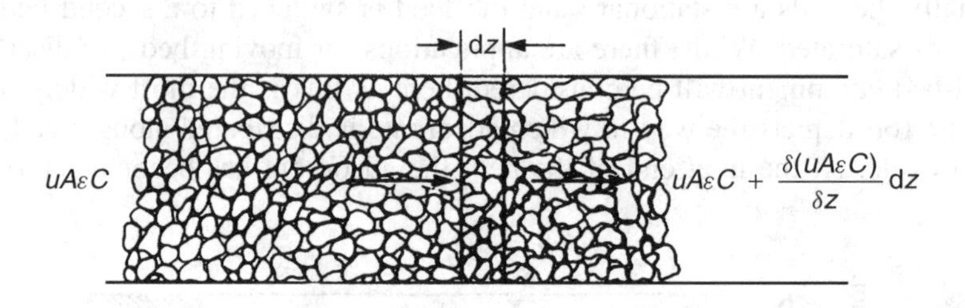


Figure 17.17. Conservation of adsorbate across an increment of a packed bed

The rate of loss by adsorption from the fluid phase equals the rate of gain in the adsorbed phase and:

$$\text{Rate of adsorption} = \frac{\partial[(1-\varepsilon)AC_s\,\mathrm{d}z]}{\partial t}$$

The equations may be rearranged to give:

$$\left(\frac{\partial(uC)}{\partial z}\right)_t + \left(\frac{\partial C}{\partial t}\right)_z = -\frac{1}{m}\left(\frac{\partial C_s}{\partial t}\right)_z \tag{17.69}$$

where:

$$m = \frac{\varepsilon}{1-\varepsilon}$$

When the adsorbate content of the inlet stream is small, the fluid velocity is virtually constant along the bed.

Therefore:

$$u\left(\frac{\partial C}{\partial z}\right)_t + \left(\frac{\partial C}{\partial t}\right)_z = -\frac{1}{m}\left(\frac{\partial C_s}{\partial t}\right)_z \tag{17.70}$$

which may be simplified further by substituting to give:

$$\chi = \frac{z}{mu} \quad \text{and} \quad t_1 = \left(t - \frac{z}{u}\right)$$

so that:

$$\frac{\partial C}{\partial \chi} = -\frac{\partial C_s}{\partial t_1} \tag{17.71}$$

Equation 17.70 includes explicitly the interpellet voidage. The intra-pellet voidage $\alpha$ contributed by the pores is subsumed in the term $C_s$, which is the mean adsorbate content on the pellet. $C_s$ varies along the bed although it is assumed to be constant at any radius at a particular distance from the inlet.

If $\alpha$ is included in the conservation equation, then this becomes:

$$u\frac{\partial C}{\partial z} + \frac{\partial C}{\partial t} = -\frac{1}{m}\frac{\partial}{\partial t}[(1-\alpha)C'_s + \alpha C'] \tag{17.72}$$

where: $C'$ is the mean adsorbate concentration in the fluid phase which is present in the pore volume of the pellet, and

$C'_s$ is the mean adsorbate concentration in the adsorbed phase in a pellet,

$C'$ is not equal to $C$, except in equilibrium operation, although it is likely to be in equilibrium with $C'_s$ through an adsorption isotherm $C' = f(C'_s)$. $C$ is normally expressed in moles per unit volume of fluid. In equation 17.70, consistent units for $C_s$ are moles per unit volume of pellet. In practice, $C_s$ is often quoted as mass of adsorbate per unit mass of adsorbate-free adsorbent ($C''_s$).

It then follows that:

$$C''_s = \frac{M}{\rho_p}C_s$$

where: $M$ is the molecular weight of adsorbate, and

$\rho_p$ is the pellet density.

The latter may be related to true solid density $\rho_s$ by:

$$\rho_p = (1-\alpha)\rho_s$$

and to the bed density, $\rho_B$, by:

$$\rho_B = (1-\varepsilon)\rho_p$$

Whilst it is convenient to refer to pellet mean concentrations when analysing the performance of agglomerates of pellets such as is found in a packed bed, in reality, both $C_s$ and $C$ decrease from the outside of the pellet to the centre, as shown in Figure 17.18.

Another important approximation implicit in equation 17.70 is that radial and longitudinal dispersions may be neglected. Radial concentration gradients are likely to be small. It has been shown that because of the greater bed voidage at the wall, and within about three pellet diameters of it, a peak in longitudinal velocity occurs near the wall and the breakthrough for wall-flow is earlier. For a bed/pellet diameter ratio of greater than 20, the effect is small. At low Reynolds numbers longitudinal dispersion may become important. This gives rise to axial mixing, which elongates the mass transfer zone and reduces separation efficiency. When it is necessary to take longitudinal dispersion into account, a form of Fick's Law is assumed to apply and the term $D_L(\partial^2 C/\partial z^2)$ is added to the left-hand side of equation 17.70. Values of $D_L$ may be calculated from published

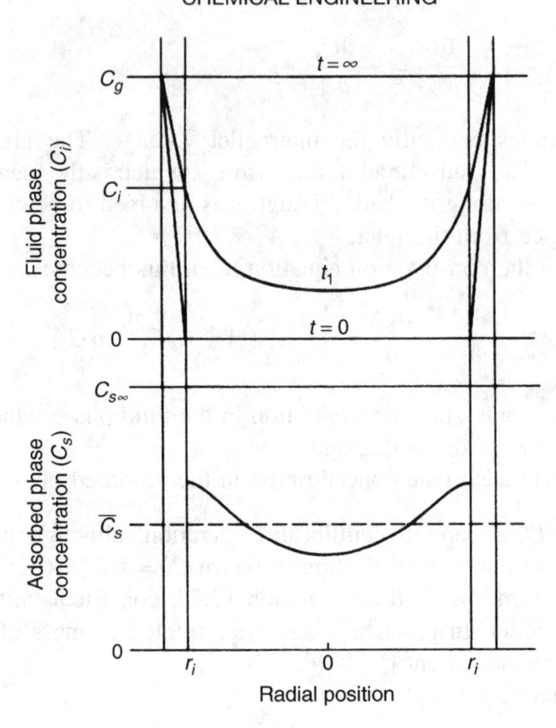


Figure 17.18. Distribution of adsorbate through a spherical pellet

correlations of Peclet Numbers. For gases, EDWARDS and RICHARDSON[37] have shown that:

$$\frac{1}{Pe} = \frac{0.73\varepsilon}{Re\,Sc} + \frac{1}{2\left(1 + \dfrac{9.5\varepsilon}{Re\,Sc}\right)} \tag{17.73}$$

where:

$$Pe = \frac{2ur_i}{D_L}$$

In liquids the effects of longitudinal dispersion are small, even at low Reynolds Numbers.

Many solutions are available for equation 17.70 and its refinements. Three cases are considered to illustrate the range of solutions. Firstly, it is assumed that the bed operates isothermally and that equilibrium is maintained between adsorbate concentrations in the fluid and on the solid. Secondly, the non-equilibrium isothermal case is considered and, finally, the non-equilibrium non-isothermal case.

## 17.8.2. Equilibrium, isothermal adsorption in a fixed bed, single adsorbate

At all positions in the bed, concentrations in the fluid and adsorbed phases are related by the adsorption isotherm. This implies that there is no resistance to the transfer of molecules of adsorbate from bulk fluid to adsorption site.

If the adsorption isotherm is written as $C_s = \mathrm{f}(C)$, equation 17.69 may be rewritten as:

$$u\frac{\partial C}{\partial z} + \frac{\partial C}{\partial t} = -\frac{1}{m}\frac{\partial C_s}{\partial C}\frac{\partial C}{\partial t}$$

$$= -\frac{1}{m}\mathrm{f}'(C)\frac{\partial C}{\partial t} \tag{17.74}$$

Also:

$$\left(\frac{\partial C}{\partial t}\right)_z = -\left(\frac{\partial C_s}{\partial z}\right)_t \cdot \left(\frac{\partial z}{\partial t}\right)_c$$

and:

$$\left\{u - \left[1 + \frac{1}{m}\mathrm{f}'(C)\right]\frac{\partial z}{\partial t}\right\}\frac{\partial C}{\partial z} = 0$$

Hence:

$$\left(\frac{\partial z}{\partial t}\right)_c = \frac{u}{1 + \frac{1}{m}\mathrm{f}'(C)} \tag{17.75}$$

Equation 17.75 is important as it illustrates, for the equilibrium case, a principle that applies also to the non-equilibrium cases more commonly encountered. The principle concerns the way in which the shape of the adsorption wave changes as it moves along the bed. If an isotherm is concave to the fluid concentration axis it is termed *favourable*, and points of high concentration in the adsorption wave move more rapidly than points of low concentration. Since it is physically impossible for points of high concentration to overtake points of low concentration, the effect is for the adsorption zone to become narrower as it moves along the bed. It is, therefore, termed *self-sharpening*.

An isotherm which is convex to the fluid concentration axis is termed *unfavourable*. This leads to an adsorption zone which gradually increases in length as it moves through the bed. For the case of a linear isotherm, the zone goes through the bed unchanged. Figure 17.19 illustrates the development of the zone for these three conditions.

Whilst the simple theory predicts that the adsorption zone associated with a favourable isotherm reduces to a step change in concentration, in practice finite resistance to mass transfer and the effect of longitudinal diffusion will result in a zone of finite and constant width being propagated. The property is important because it leads to simplified methods of sizing fixed beds. Figure 17.20 taken from the work of BOWEN and RIMMER[38] shows a typical isotherm for activated alumina and water. This has sections concave to the gas concentration axis at high and low concentrations and the middle section is convex. From the predictions of the equilibrium solutions, it might be expected that the portions of the adsorption wave corresponding to the extremes of the concentration range would sharpen as they moved through the column. The middle of the range should become longer. Experimental results are shown in Figure 17.21. These were obtained from a narrow, jacketed, laboratory column operating essentially isothermally, though not under equilibrium conditions. The figure does show the tendency for high and low concentrations to develop a constant pattern and for the middle range to spread as the wave progresses. In order to predict the breakpoint, only the leading zone has to be considered. Equation 17.75 may be integrated at constant $C$ to give:

$$\frac{ut}{1 + \frac{1}{m}\mathrm{f}'(C)} = z - z_0 \tag{17.76}$$

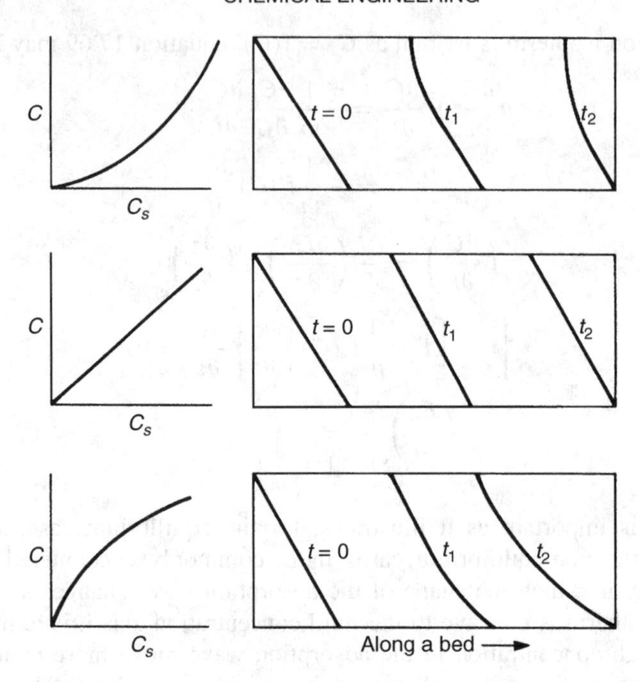


Figure 17.19. Effect of the shape of the isotherm on the development of an adsorption wave through a bed with the initial distribution of adsorbate shown at $t = 0$

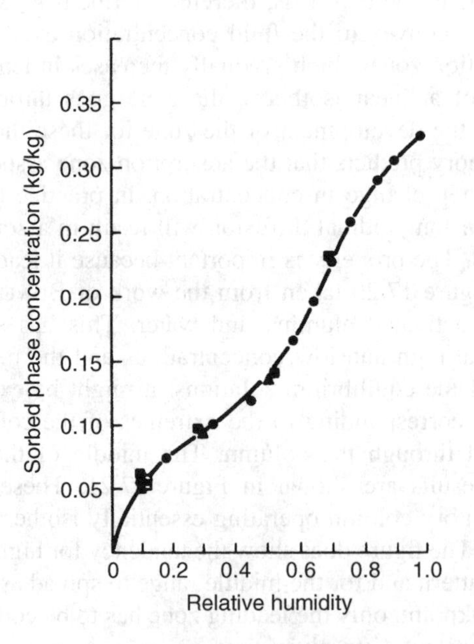


Figure 17.20. Adsorption isotherm for water vapour on activated alumina, plotted as a function of the relative humidity. Temperatures: ● 303 K, ■ 308 K, ▲ 315 K, □ 325 K, ○ 335 K

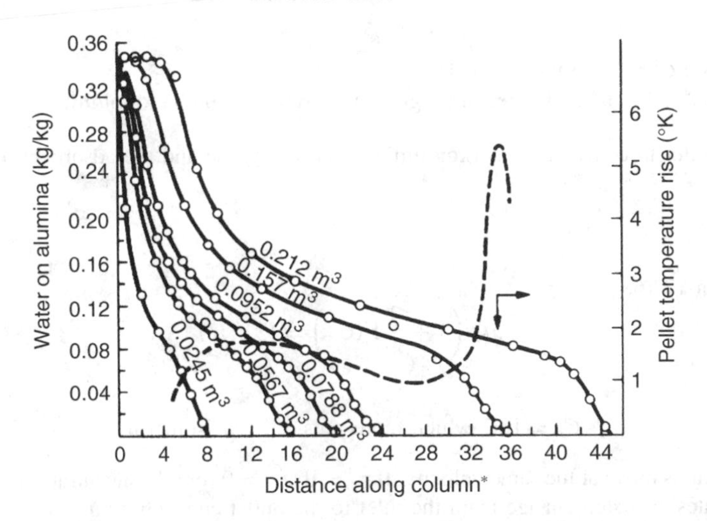


Figure 17.21. Distribution of adsorbate along a column operating quasi-isothermally with a Type IV isotherm. The dotted line is the temperature distribution at the breakpoint of a column of length 36 units. Air flow $8.8 \times 10^{-6}$ m$^3$/s. Temperature of jacket 303 K. Relative humidity of feed 95 per cent. The volumes to breakpoint of each bed are indicated on the curves.

*Each unit of distance is the length of bed containing 1 g of adsorbent[38]

where $z_0$ is the position of $C$ initially. For a bed initially free of adsorbate, $z_0$ is zero for all values of $C$.

The condition for which the bed is likely to operate at near equilibrium is when the feed rate is low. This is also the condition when longitudinal dispersion may be significant. Equilibrium solutions have been found by Lapidus and Amundson[39] and by Levenspiel and Bischoff[40] for this case. These take the form:

$$\frac{C}{C_0} = \frac{1}{2} \cdot \left\{1 + \operatorname{erf}\left[\left(\frac{uz}{4D_L}\right)^{1/2} \frac{(t - t_{min})}{(t\, t_{min})^{1/2}}\right]\right\} \tag{17.77}$$

where $t_{min}$ is the minimum time, under given flow conditions, to saturate a bed of unit cross-section and length $z$. $C_0$ is the constant concentration of adsorbate in the fluid entering the bed. The minimum time is given by:

$$t_{min} = \left(1 + \frac{1}{m}\frac{C_{s\infty}}{C_0}\right)\frac{z}{u} \tag{17.78}$$

where $C_{s\infty}$ is the concentration of adsorbed phase in equilibrium with $C_0$.

## Example 17.3

A solvent contaminated with 0.03 kmol/m$^3$ of a fatty acid is to be purified by passing it through a fixed bed of activated carbon which will adsorb the acid but not the solvent. If the operation is essentially isothermal and equilibrium is maintained between the liquid and the solid, calculate the length of a bed of 0.15 m diameter to give 3600 s (1 h) of operation when the fluid is fed at $1 \times 10^{-4}$ m$^3$/s. The bed is initially free of adsorbate, and the intergranular voidage is 0.4. Use an equilibrium, fixed-bed theory to obtain an answer for three types of isotherm:

(a) $C_s = 10\ C$
(b) $C_s = 3.0\ C^{0.3}$ — use the mean slope
(c) $C_s = 10^4\ C^2$ — take the breakthrough concentration as 0.003 kmol/m$^3$.

$C$ and $C_s$ refer to concentrations (kmol/m$^3$) in the gas phase and the adsorbent, respectively.

## Solution

From equation 17.76:

$$ut\Big/\left(1+\frac{1}{m}f'(C)\right) = (z - z_0)$$

*For case* (a):

$$C_s = 10C \text{ which represents a linear isotherm.}$$

All concentrations move at the same velocity. If $z_0 = 0$ at $t = 0$ for all concentrations, the adsorption wave propagates as a step change from the inlet to the outlet concentration,

$$f'(C_s) = 10$$

$$u = 1 \times 10^{-4}/[(\pi/4)(0.15^2\varepsilon)] \text{ m/s}$$

where $\varepsilon$ is the intergranular voidage $= 0.4$.

$$m = \varepsilon/(1-\varepsilon) = (0.4/0.6)$$

$$t = 3600\text{s}$$

Thus:

$$z = \left(\frac{4\times 10^{-4}}{\pi \times 0.15^2 \times 0.4}\right)\left(\frac{3600}{1+10(0.6/0.4)}\right)$$

$$= \underline{\underline{3.18 \text{ m}}}$$

It may be noted that, when the adsorption wave begins to emerge from the bed, the bed is saturated in equilibrium with the inlet concentration.

Hence:

$$uA\varepsilon tC_0 = zA[\varepsilon C_0 + (1-\varepsilon)C_s^*]$$

which is the same as that obtained by applying equation 17.76 to a linear isotherm.

*For case* (b):

$$C_s = 3C^{0.3} \text{ which represents a favourable isotherm.}$$

As $C$ increases, f($C$) decreases and points of higher concentrations are predicted to move a greater distance in a given time than lower concentrations. It is not possible for points of higher concentrations to overtake lower concentrations, and if $z_0 = 0$ for all concentrations, the adsorption wave will propagate as a step change similar to case $a$.

Hence:

$$z = ut\Big/\left[1+\frac{1}{m}\frac{C_s^*}{C_0}\right]$$

$$= \left(\frac{4\times 10^{-4}}{\pi \times 0.15^2 \times 0.4}\right)\left(\frac{3600}{1+(0.6/0.4)(3/C_0^{0.7})}\right)$$

$$= \underline{\underline{0.95 \text{ m}}}$$

*For case* (c):

$$C_s = 10^4 C^2 \quad \text{which represents an unfavourable isotherm.}$$

$$f'(C) = 2 \times 10^4\ C.$$

As $C$ increases, $f'(C)$ increases such that, in a given time, $z$ for lower concentrations is greater than for higher concentrations. Following the progress of the breakpoint concentration, $C = 0.003$ kmol/m$^3$, then:

$$f'(0.003) = 60$$

Hence:

$$z = \left(\frac{4 \times 10^{-4}}{\pi \times 0.15^2 \times 0.4}\right)\left(\frac{3600}{1 + (0.60/0.40)60}\right)$$

$$= \underline{\underline{0.55\ \text{m}}}$$

At breakpoint, the bed is far from saturated and:

$$\text{saturation} = \frac{100uA\varepsilon tC_0}{zAC_0\varepsilon\left[1 + \dfrac{1}{m}\cdot\dfrac{C_s^*}{C_0}\right]} = \frac{100ut}{z\left[1 + \dfrac{1}{m}\cdot\dfrac{C_s^*}{C_0}\right]}$$

$$= 100\left(\frac{4 \times 10^{-4} \times 3600}{\pi \times 0.15^2 \times 0.4}\right) \bigg/ [0.55(1 + (0.6/0.4)(9/0.03))]$$

$$= \underline{\underline{20.5\ \text{per cent}}}$$

### 17.8.3. Non-equilibrium adsorption — isothermal operation

Isothermal operation in a fixed bed may be achieved in a well-cooled laboratory column and also in large-scale equipment if the concentration of adsorbate is low and the release of heat of adsorption is not great. A third and rather specialised situation in which isothermal conditions may exist is that in which a component is adsorbed on to a surface already covered with a second component. If this second component is displaced by the first, its heat of desorption will "consume" the heat released when the adsorption of the first component occurs.

#### *Constant patterns analysis*

A constant wave-pattern develops when adsorption is governed by a favourable isotherm. In Figure 17.16*a*, a typical wave is assumed to move a distance d$z$ in a time d$t$. If the wave is already fully developed, it will retain its shape. A mass balance gives:

$$\varepsilon uAC_0\,dt = A[(1-\varepsilon)C_{s\infty} + \varepsilon C_0]\,dz$$

$$\frac{dz}{dt} = \frac{u}{1 + \dfrac{1}{m}\dfrac{C_{s\infty}}{C_0}} \tag{17.79}$$

where equation 17.79 is similar in form to the equilibrium equation 17.75, and identical to it if the isotherm is linear.

The mass balance may be carried out at any level of concentration within the zone to give:

$$\frac{dz}{dt} = \frac{u}{1 + \frac{1}{m}\frac{C_s}{C}} \tag{17.80}$$

For a constant pattern wave, all concentrations within the wave have the same velocity.

Therefore:

$$\frac{C_s}{C} = \frac{C_{s\infty}}{C_0} \tag{17.81}$$

This is the *constant-pattern simplification* that enables many solutions to be obtained from what might otherwise be complex rate equations. It represents a condition that is approached as the wave becomes fully developed and leads to what are termed *asymptotic solutions*.

Representing the mass balance in a fixed bed by:

$$\left(\frac{\partial C_s}{\partial \chi}\right)_{t_1} = -\left(\frac{\partial C_s}{\partial t_1}\right)_{\chi} \quad \text{(equation 17.71)}$$

and assuming a general rate expression:

$$\frac{\partial C_s}{\partial t_1} = \mathrm{G}(C, C_s) \tag{17.82}$$

where G denotes a function.

The constant pattern assumption gives:

$$\frac{\partial C_s}{\partial t_1} = \mathrm{G}_1(C_s) = \mathrm{G}_2(C) = -\frac{\partial C}{\partial \chi} \tag{17.83}$$

Equation 17.83 may be integrated to give, at constant $\chi$ to give:

or, at constant $t_1$ to give:

$$\left.\begin{aligned} \int \frac{dC_s}{\mathrm{G}_1(C_s)} &= \left[\int \frac{dC_s}{\mathrm{G}_1(C_s)}\right]_{t_1=0} + t_1 \\ \int \frac{dC_s}{\mathrm{G}_2(C)} &= \left[\int \frac{dC_s}{\mathrm{G}_2(C)}\right]_{x=0} - \chi \end{aligned}\right\} \tag{17.84}$$

If, for example, a rate expression may be written as:

$$\frac{\partial C_s}{\partial t_1} = k_g^0(C - C^*) \tag{17.85}$$

where $C^*$ is the fluid concentration in equilibrium with the mean solid concentration $C_s$, then, assuming a Langmuir equilibrium relationship similar to equation 17.3:

$$C_s = \frac{C_{sm} B_3 C^*}{1 + B_3 C^*} \tag{17.86}$$

Substituting in equation 17.85 for $C$ and $C^*$, from equations 17.81 and 17.86, respectively, gives the expression for $G_1(C_s)$ for the particular case.

### Rosen's solutions

Rate equations, such as equation 17.85, make no attempt to distinguish mechanisms of transfer within a pellet. All such mechanisms are taken into account within the rate constant $\boldsymbol{k}$. A more fundamental approach is to select the important factors and combine them to form a rate equation, with no regard to the mathematical complexity of the equation. In most cases this approach will lead to the necessity for numerical solutions although for some limiting conditions, useful analytical solutions are possible, particularly that presented by ROSEN[41].

A mass balance for diffusion of adsorbate into a spherical pellet may be written as:

$$\underbrace{\left[-4\pi r^2 D_e \frac{\partial C_r}{\partial r}\right]}_{\text{IN}} - \underbrace{\left[-4\pi r^2 D_e \frac{\partial C_r}{\partial r} + \frac{\partial\left(-4\pi r^2 D_e \dfrac{\partial C_r}{\partial r}\right)\mathrm{d}r}{\partial r}\right]}_{\text{OUT}}$$

$$= \underbrace{\left[4\pi r^2\,\mathrm{d}r \frac{\partial C_r}{\partial t} + 4\pi r^2\,\mathrm{d}r \frac{\partial C_{sr}}{\partial t}\right]}_{\text{ACCUMULATION}} \tag{17.87}$$

where $C_r$ and $C_{sr}$ are concentrations of adsorbate at a radius $r$.

If there is equilibrium between fluid in the pores and the adjacent surface, and if the equilibrium is linear then:

$$C_{sr} = K_a C_r \tag{17.88}$$

Hence the mass balance may be written as:

$$\frac{\partial C_r}{\partial t} = \frac{D_e}{1 + K_a}\left(\frac{\partial^2 C_r}{\partial r^2} + \frac{2}{r}\frac{\partial C_r}{\partial r}\right) \tag{17.89}$$

The total adsorbate in the pellet at any particular time may be written as:

$$4\pi \int (r^2 C_r + r^2 K_a C_r)\,\mathrm{d}r$$

so that the mean concentration $C_s$ is given by:

$$C_s = \frac{3}{r_i^3}(1 + K_a)\int r^2 C_r\,\mathrm{d}r \tag{17.90}$$

The rate at which adsorbate enters a pellet may be expressed in terms of the concentration driving force across the boundary film to give:

$$\frac{4}{3}\pi r_i^3 \frac{\partial C_s}{\partial t} = 4\pi r_i^2 k_g (C - C_i)$$

Thus:

$$\frac{\partial C_s}{\partial t} = \frac{3k_g}{r_i}(C - C_i) \tag{17.91}$$

where $C_i$ is in equilibrium with the solid concentration at $r = r_i$.

Solutions have been found by Rosen, using the fixed-bed equation 17.70 together with equations 17.90 and 17.91. The general solution takes the form:

$$\frac{C}{C_0} = \tfrac{1}{2} + \mathrm{F}(\lambda', \tau', \psi) \tag{17.92}$$

where the length parameter is:

$$\lambda' = \frac{3D_e K_a z}{m u r_i^2} \tag{17.93}$$

the time parameter is:

$$\tau' = \frac{2D_e}{r_i^2}\left(t - \frac{z}{u}\right) \tag{17.94}$$

and the resistance parameter is:

$$\psi = \frac{D_e K_a}{r_i k_g} \tag{17.95}$$

Solutions are available in tabulated and graphical form. Except for small values of $\lambda'$, the solution has the following convenient form:

$$\frac{C}{C_0} = \frac{1}{2}\left\{1 + \mathrm{erf}\left[\frac{\left(3\dfrac{\tau'}{2\lambda'} - 1\right)}{2\sqrt{[(1+5\psi)/5\lambda']}}\right]\right\} \tag{17.96}$$

RUTHVEN[16] gives a useful summary of other solutions.

## Example 17.4

A bed is packed with dry silica gel beads of mean diameter 1.72 mm to a density of 671 kg of gel/m$^3$ of bed. The density of a bead is 1266 kg/m$^3$ and the depth of packing is 0.305 m. Humid air containing 0.00267 kg of water/kg of dry air enters the bed at the rate of 0.129 kg of dry air/m$^2$ s. The temperature of the air is 300 K and the pressure is $1.024 \times 10^5$ N/m$^2$. The bed is assumed to operate isothermally. Use the method of Rosen to find the effluent concentration as a function of time. Equilibrium data for the silica gel are given by the curve in Figure 17.22. An appropriate value of $k_g$, the film mass transfer coefficient, is 0.0833 m/s.

## Solution

The length parameter $\lambda'$ is calculated from equation 17.83:

$$\lambda' = \frac{3D_e K_a z}{m u r_i^2}$$

$$K_a = \left(\frac{0.084}{0.00267}\right) \times \left(\frac{1266}{1.186}\right) = 3.36 \times 10^4$$

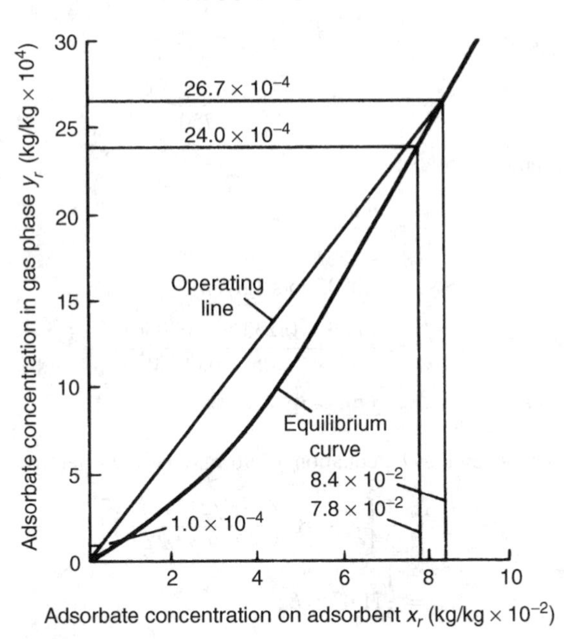


Figure 17.22. Adsorption isotherm for water vapour in air on silica gel

where $K_a$ is obtained from the mean slope of the isotherm between its origin and the point corresponding to the inlet concentration. This slope has been multiplied by the ratio of bead to gas densities to give $K_a$ in the same units as equation 17.88.

$$m = \frac{\varepsilon}{1-\varepsilon} = \frac{1266 - 671}{671} = 0.89$$

where $\varepsilon$ is the volume (and area) voidage between beads.

$$\varepsilon = \frac{1266 - 671}{1266} = 0.47$$

$$u = \frac{0.129}{0.47 \times 1.186} = 0.233 \text{ m/s}$$

where $u$ is the interstitial velocity of the air.

$$r_i = 0.86 \text{ mm}$$

$$z = 0.305 \text{ m}$$

$D_e$ is the diffusivity of sorbate referred to the sorbed phase (and this has a value of $10^{-10}$–$10^{-11}$ m$^2$/s). Substituting values gives $\lambda' = 18{,}500 - 1850$. The use of equation 17.96 is valid for large values of $\lambda'$. From equations 17.93 and 17.94:

$$\frac{\tau'}{\lambda'} = \frac{2\left(t - \dfrac{z}{u}\right)mu}{3K_a z} = \left(\frac{2 \times 0.89}{3 \times 3.36 \times 10^4}\right)\left(\frac{0.233t}{0.305} - 1\right)$$

where $t$ is expressed in hours.

Thus:

$$t = 20.55\frac{\tau'}{\lambda'} + \frac{1}{2750}$$

From equations 17.93 and 17.95:

$$\frac{\psi}{\lambda'} = \frac{mur_i}{3zk_g}$$

$$k_g = 0.0833 \text{ m/s (given)}$$

Thus:

$$\frac{\psi}{\lambda'} = \frac{(0.89 \times 0.233 \times 0.00086)}{(3 \times 0.305 \times 0.0833)}$$

$$= 2.36 \times 10^{-3}$$

For the relative values of $\lambda'$ and $\psi/\lambda'$, equation 17.96 may be rewritten as:

$$\frac{C}{C_0} = \frac{1}{2}\left[1 + \text{erf}\left(\frac{(3\tau'/2\lambda' - 1)}{2\sqrt{(\psi/\lambda')}}\right)\right]$$

$$= \frac{1}{2}[1 + \text{erf}E]$$

where:

$$E = \frac{(3\tau'/2\lambda' - 1)}{2\sqrt{(\psi/\lambda')}}$$

For selected values of $C/C_0$, the value of $E$ may be found from the tables of error functions given in the Appendix in Volume 1. From $E$ a ratio $\tau/\lambda'$ may be calculated and hence a corresponding time. These calculations are summarised as follows:

| $\frac{C}{C_0}$ | erf $E$ | $E$ | $\frac{\tau'}{\lambda'}$ | $t$ (h) |
|---|---|---|---|---|
| 0.024 | −0.952 | −1.40 | 0.576 | 11.8 |
| 0.045 | −0.910 | −1.20 | 0.589 | 12.1 |
| 0.079 | −0.842 | −1.00 | 0.602 | 12.4 |
| 0.24 | −0.520 | −0.50 | 0.635 | 13.0 |
| 0.50 | 0 | 0 | 0.667 | 13.7 |
| 0.715 | 0.430 | 0.40 | 0.693 | 14.2 |
| 0.92 | 0.840 | 1.00 | 0.732 | 15.0 |

## 17.8.4. Non-equilibrium adsorption — non-isothermal operation

When the effects of heats of adsorption cannot be ignored — the situation in most industrial adsorbers — equations representing heat transfer have to be solved simultaneously with those for mass transfer. All the resistances to mass transfer will also affect heat transfer although their relative importance will be different. Normally, the greatest resistance to mass transfer is found within the pellet and the smallest in the external boundary film. For heat transfer, the thermal conductivity of the pellet is normally greater than that of the boundary film so that temperatures through a pellet are fairly uniform. The temperature difference between bulk conditions outside a pellet and conditions within it occurs almost wholly across the boundary film.

The effects of non-isothermal adsorption in fixed beds have been examined by LEAVITT[42]. Non-isothermal adsorption of a single adsorbate from a carrier fluid leads to the complex wave front shown in Figure 17.23. As adsorption proceeds, the leading edge of the adsorption wave meets adsorbent that is essentially free of adsorbate. Adsorption occurs and the temperature of the adsorbent increases until a dynamic equilibrium is established between fluid and adsorbed phase at the prevailing temperature. These conditions persist through a first plateau zone until the rate of adsorption falls to a point where the temperature of the plateau zone cannot be sustained. As the temperature falls, more adsorption occurs until the second plateau zone is formed, in equilibrium with the incoming stream. The net result is that adiabatic, or near adiabatic, conditions can lead to the formation of two transfer zones separated by a zone in which conditions remain constant.

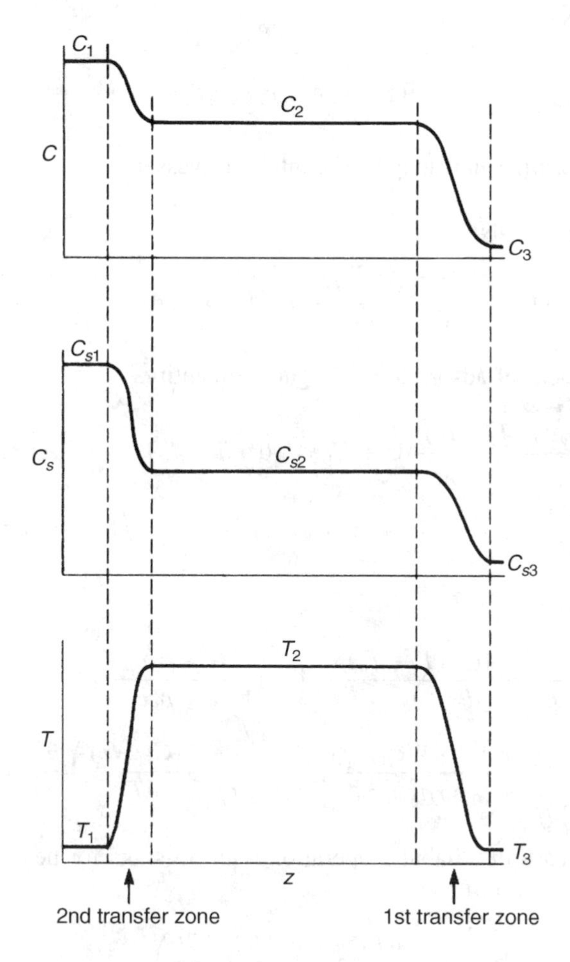


Figure 17.23. Idealised distributions of temperature and concentrations during adiabatic adsorption

The adiabatic profile may be further complicated by the shape of the isotherm. Under isothermal conditions, a favourable isotherm produces a single transfer zone, although an isotherm with favourable and unfavourable sections may generate a more complex profile, as shown in Figure 17.21.

A heat balance over an increment of bed d$z$ may be written as:

$$(\text{IN} - \text{OUT}) = -\frac{\partial}{\partial z}(u\varepsilon A\rho_g c_{pg}(T - T_0))\,dz - U_0\pi d_b\,dz(T - T_a) \tag{17.97}$$

where: $d_b$ is the diameter of the containing vessel,
$T_a$ is the temperature of the surroundings,
$T_0$ is the reference temperature, and
$U_0$ is the overall coefficient for heat transfer between the containing vessel and the surroundings

ACCUMULATION:

$$= \frac{\partial}{\partial t}(\varepsilon A\,dz\rho_g c_{pg}(T - T_0) + (1-\varepsilon)A\,dz\rho_p c_{ps}(T - T_0)) + \frac{\partial}{\partial t}(W\,dz c_{pw}(T - T_0)) \tag{17.98}$$

where $W$ is the mass per unit length of containing vessel.

LOSS BY ADSORPTION:

$$= (1-\varepsilon)A\,dz\frac{\partial(C_s\Delta H)}{\partial t} = (1-\varepsilon)A\,dz\frac{\partial(C_s\Delta H)}{\partial T}\frac{\partial T}{\partial t} \tag{17.99}$$

where $\Delta H$ is the heat of adsorption, a negative quantity.

Hence:
$$\frac{\partial(u\varepsilon A\rho_g c_{pg}(T - T_0))}{\partial z}dz + U_0\pi d_b\,dz(T - T_a) + \frac{\partial}{\partial t}(\varepsilon A\,dz\rho_g C_{pg}(T - T_0)$$
$$+(1-\varepsilon)A\,dz\rho_p c_{ps}(T - T_0) + W\,dz c_{pw}(T - T_0)) + (1-\varepsilon)A\,dz\frac{\partial(C_s\Delta H)}{\partial T}\frac{\partial T}{\partial t} = 0$$

For a cylindrical bed:

$$u\frac{\partial T}{\partial z} + \frac{4U_0}{\varepsilon d_b}\frac{(T - T_a)}{\rho_g c_{pg}} + \left(1 + \frac{1-\varepsilon}{\varepsilon}\frac{\rho_p c_{ps}}{\rho_g c_{pg}}\right.$$
$$\left. + \frac{4W c_{pw}}{\varepsilon\pi d_b^2\rho_g c_{pg}} + \frac{1-\varepsilon}{\varepsilon}\frac{1}{c_{pg}}\frac{\partial(C_s\Delta H)}{\partial T}\right)\frac{\partial T}{\partial t} = 0 \tag{17.100}$$

For the particular case of adiabatic operation with no sinks for heat in the wall, $U_0 = 0$ and $W = 0$.

Since:
$$\frac{\partial T}{\partial z} = -\left(\frac{\partial T}{\partial t}\right)\left(\frac{\partial t}{\partial z}\right)$$

Then:

$$\left.\frac{\partial z}{\partial t}\right|_T = u_T = \frac{u}{1 + \dfrac{1}{m}\cdot\left(\dfrac{\rho_p c_{ps}}{\rho_g c_{pg}} + \dfrac{1}{\rho_g c_{pg}} + \dfrac{\partial(C_s \Delta H)}{\partial T}\right)} \tag{17.101}$$

where $u_T$ is the velocity of a point of constant temperature. If the thermal wave is coherent, that is all points travel at the same velocity, then $u_T$ is the thermal wave velocity. This may be compared with the concentration wave velocity $u_c$, where:

$$u_c = \frac{u}{1 + \dfrac{1}{m}\dfrac{\partial C_s}{\partial C}} \qquad \text{(equation 17.75)}$$

If the velocities of the thermal and concentration waves are equal, then from equations 17.75 and 17.101:

$$\frac{\partial C_s}{\partial C} = \frac{\rho_p c_{ps}}{\rho_g c_{pg}} + \frac{1}{\rho_g c_{pg}} + \frac{\partial(C_s \Delta H)}{\partial T} \tag{17.102}$$

When equation 17.102 is applied to the finite difference between inlet concentration and plateau, and between plateau and exit concentration, as shown in Figure 17.23, it becomes:

$$\frac{C_{s1} - C_{s2}}{C_1 - C_2} = \frac{\rho_p c_{ps}}{\rho_g c_{pg}} + \frac{1}{\rho_g c_{pg}}\left(\frac{(C_s \Delta H)_1 - (C_s \Delta H)_2}{T_1 - T_2}\right) \tag{17.103}$$

$$\frac{C_{s2} - C_{s3}}{C_2 - C_3} = \frac{\rho_p c_{ps}}{\rho_g c_{pg}} + \frac{1}{\rho_g c_{pg}}\left(\frac{(C_s \Delta H)_2 - (C_s \Delta H)_3}{T_2 - T_3}\right) \tag{17.104}$$

In equations 17.103 and 17.104, $c_{ps}$ and $c_{pg}$ are the mean specific heats over the ranges of temperature and concentrations encountered. Properties with subscripts 1 and 3 are known from inlet and exit conditions respectively. If the plateau values represented by subscript 2 are in equilibrium, then the values $C_2$, $C_{s2}$ and $T_2$ may be found from the equations for any known form of the adsorption isotherm $C_{s2} = \mathrm{f}(C_2)$.

Equation 17.102 was derived on the assumption that concentration and thermal waves propagated at the same velocity. Amundson *et al.*[43] showed that it was possible for the temperatures generated in the bed to propagate as a pure thermal wave leading the concentration wave. A simplified criterion for this to occur can be obtained from equations 17.75 and 17.101. Since there is no adsorption term associated with a pure thermal wave, and if changes within the bed voids are small, then:

$$u_T = \frac{m\rho_g c_{pg}}{\rho_p c_{ps}} u \tag{17.105}$$

Also:

$$u_c = m\frac{\Delta C}{\Delta C_s} u \tag{17.106}$$

For a bed initially free of adsorbate, the thermal wave propagates more quickly than the concentration wave if:

$$\frac{\rho_g}{\rho_p}\frac{c_{pg}}{c_{ps}} > \frac{C_1}{C_{s1}} \tag{17.107}$$

Since $C_1/C_{s1}$ increases with temperature, $u_c$ (non-isothermal) $>$ $u_c$ (isothermal). It has been estimated that the rear wave moves at about two-thirds of the velocity of the front wave, so a more cautious criterion than equation 17.107 is given by:

$$\frac{\rho_g c_{pg}}{\rho_p c_{ps}} > \frac{3}{2}\frac{C_1}{C_{s1}} \tag{17.108}$$

where $C_{s1}$ is a function of $C_1$ and the maximum temperature $T_{2\max}$ of the plateau zone.

Equation 17.101 may be rearranged to give:

$$T_2 = T_1 + \frac{[(\Delta H)C_s]_2 - [(\Delta H)C_s]_1}{\rho_g c_{pg}\dfrac{\Delta C_s}{\Delta C} - \rho_p c_{ps}} \tag{17.109}$$

$T_2$ is a maximum when $C_{s2}$ is zero.

Hence:

$$T_{2\max} = T_1 + \frac{[(-\Delta H)C_s]_1}{\rho_g c_{pg}(C_s/C)_1 - \rho_p c_{ps}} \tag{17.110}$$

When equilibrium between the fluid and the solid cannot be assumed, it may still be possible to obtain analytical solutions for beds operating non-isothermally. In general, however, it will be necessary to look for numerical solutions. This problem has been summarised by RUTHVEN[16].

## 17.9. REGENERATION OF SPENT ADSORBENT

Most theory concerning the dynamics of an industrial adsorption unit is directed at sizing a bed for a given adsorption duty. In the design of a complete adsorption unit, however, it is most important to ensure that the spent adsorbent can be regenerated in a given time and that the total inventory of adsorbent is kept to a minimum. If spent adsorbent could be regenerated instantly, all units would consist of a single cylindrical-bed packed with a depth of pellets just sufficient to accommodate the adsorption zone. In practice, units vary in size and configuration because regeneration takes a finite time and what is an optimum arrangement for one application is not necessarily the optimum case for another.

In an ideal fixed-bed adsorber, the adsorption stage continues until the adsorbate wave is about to emerge from the bed and the effluent concentration begins to rise, as shown in Figure 17.16*b*. Conditions behind the adsorption zone are such that the adsorbent is more or less in equilibrium with the feed. The equilibrium condition has then to be changed for regeneration to occur. This is usually brought about by changing the temperature or the pressure so that the driving force, which had previously resulted in the movement of adsorbate from fluid to solid, is now reversed.

### 17.9.1. Thermal swing

The simplest and the most common way of regenerating an adsorbent in industrial applications is by heating. The vapour pressure exerted by the adsorbed phase increases with temperature, so that molecules desorb until a new equilibrium with the fluid phase is established. Figure 17.24 depicts adsorption isotherms for a lower temperature $T_1$ and a higher temperature $T_2$. For a fixed concentration $C$ in the fluid phase, the adsorbate concentration falls from $C_{s1}$ to $C_{s2}$ when the temperature is increased.

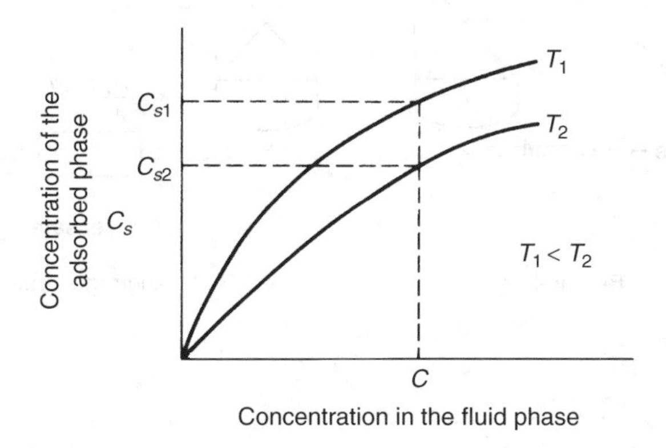


Figure 17.24. Thermal regeneration utilises the change in concentration that follows from a change in temperature

An adsorption unit using thermal swing regeneration usually consists of two packed beds, one on-line and one regenerating, as shown in Figure 17.25. Regenerating consists of heating, and purging to remove adsorbate. The arrangement is flexible and robust. The desorption temperature depends on the properties of the adsorbent and the adsorbates. Manufacturers normally recommend the best regenerating temperature for their particular adsorbent. Exceeding this temperature may accelerate the ageing processes which cause pores to coalesce and capacity to be reduced. Too low a temperature may result in incomplete regeneration so that the effluent concentration in subsequent adsorption stages will be higher than its design value. Hot spots may develop in the operation of a fixed bed, so particular care has to be taken to control temperature when handling flammable materials.

The relatively poor conductivity of a packed bed makes it difficult to get the heat of regeneration into the bed, either from a jacket or from coils embedded in the packing. This is more easily achieved by preheating the purge stream. Even in the best conditions, it takes time for the temperature of the bed to rise to the required level. Thermal regeneration is normally associated with long cycle times, measured in hours. Such cycles require large beds and, since the adsorption wave occupies only a small part of the bed on-line, the utilisation of the total adsorbent in the unit is low.

It is good operating practice to regenerate a bed in the reverse direction to that followed during adsorption. This ensures that the adsorbent at the end of the bed, which controls the quality of the treated stream, is that which is most thoroughly regenerated. CARTER(44) has quantified the effect and showed that regeneration is achieved in a shorter time in this way.

## 17.9.2. Plug-flow of solids

Better utilisation of adsorbent would be achieved if a unit could be designed in which adsorbent were removed for regeneration as soon as it became saturated, and even better

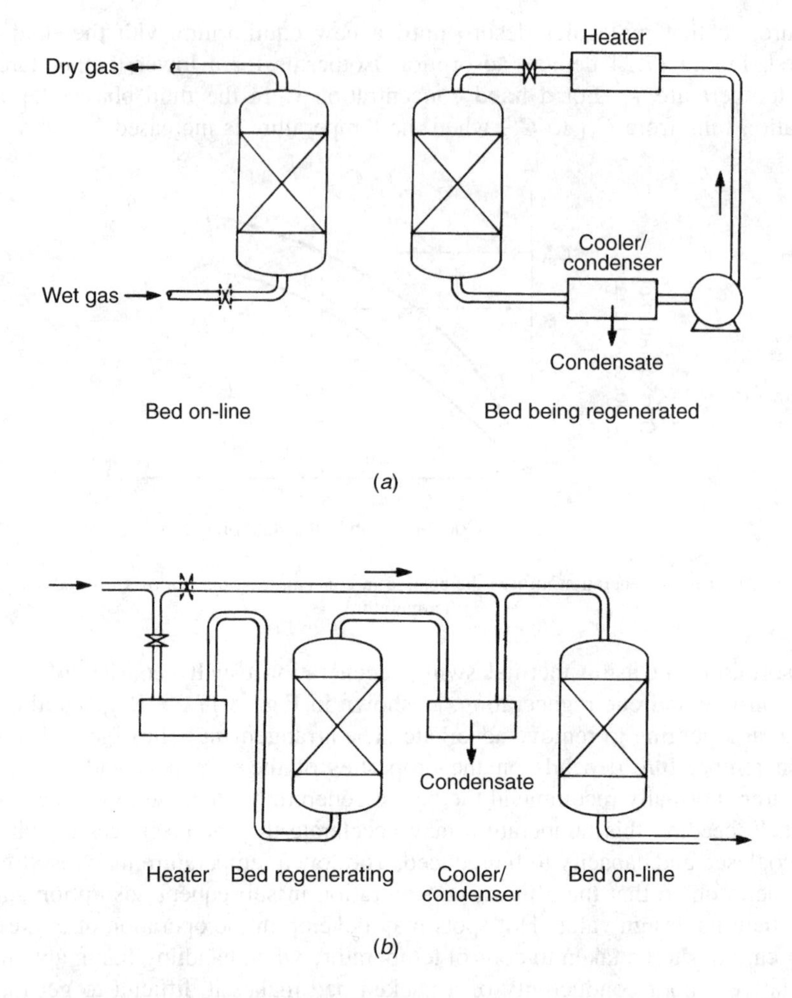


Figure 17.25. Typical arrangement of a two bed dryer. (*a*) Separate regeneration (*b*) Integrated regeneration

if the advancing adsorption wave were presented with only sufficient fresh adsorbent to contain the wave. Both characteristics are possible if the adsorbent moves countercurrent to the fluid at such a rate that the adsorption zone is stationary.

The earliest application of a moving bed in which solids moved with respect to the containing vessel was reported in the late 1940s. A typical application was the recovery of ethylene from gas composed mainly of hydrogen and methane, and with some propane and butane. The unit shown diagrammatically in Figure 17.26, taken from the work of BERG[45], is known as the *hypersorber*.

The mixture to be separated is fed to the centre of the column down which activated carbon moves slowly. Immediately above the feed, the rising gas is stripped of ethylene and heavier components leaving hydrogen, methane and any non-adsorbed gases to be discharged as a top-product. The adsorbent with its adsorbate continues down the column into an enriching section where it meets an upwards stream of recycled top-product. The

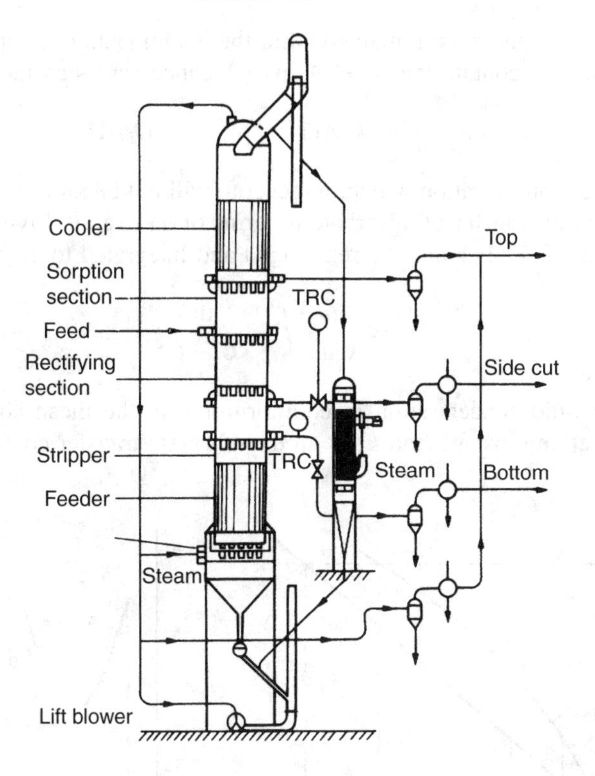


Figure 17.26. The hypersorber[45]

least-strongly adsorbed ethylene is desorbed and recovered in a side stream. The heavy components continue downwards on the carbon until these are also desorbed by steam, to be recovered as a bottom product. The carbon, now stripped of all the adsorbates, is lifted to the top of the column where it is cooled before the cycle starts again.

The rate at which the carbon is circulated may be controlled at entry to the lift-pipe. In normal operation, regenerating conditions do not maintain the carbon in its initial highly active state. Consequently, a small proportion of the regenerated carbon is steam-treated in a small column at a temperature of about 870 K. Some large hypersorbers, about 25 m high and 1.4 m in diameter, have been built for commercial operation. It seems that the units were beset from the beginning with problems of solids-handling. There was difficulty in maintaining an even flow of adsorbent and the problem of solids attrition and their subsequent loss as fines. Recently developed adsorbents, which are more selective and therefore more attractive as separating agents are, if anything, less resistant to attrition, and are unsuitable for moving-solid applications as a result.

In the 1960s there were attempts to use a moving bed of carbon to remove sulphur dioxide from flue gas on a pilot scale. As described by KATELL[46] and CARTELYOU[47], this *Reinluft* process was abandoned because of the problems caused by the carbon igniting in the presence of oxygen.

In order to design moving-bed equipment, the velocity of the adsorption zone relative to the solid has to be calculated. This gives the velocity at which the solids must move in plugflow in order that the zone remains within the equipment. The depth of packing, $z_a$, should be sufficient to contain the zone. A mass balance across an increment d$z$ gives:

$$u\varepsilon A\,\mathrm{d}C = k_g A(1-\varepsilon)a_p(C - C_i)\,\mathrm{d}z \tag{17.111}$$

$C_i$ is an interfacial concentration which, in general, will not be known. It is often possible to express the rate of transfer of adsorbate in terms of an overall driving force $(C - C^*)$. In this case, equation 17.111 may be rearranged and integrated to give:

$$z_a = \frac{mu}{k_g^0 a_p}\int_{C_B}^{C_E}\frac{\mathrm{d}C}{C - C^*} \tag{17.112}$$

where $C^*$ is the fluid concentration in equilibrium with the mean concentration of the adsorbed phase, at any instant, and $k_g^0$ is an overall mass transfer coefficient.

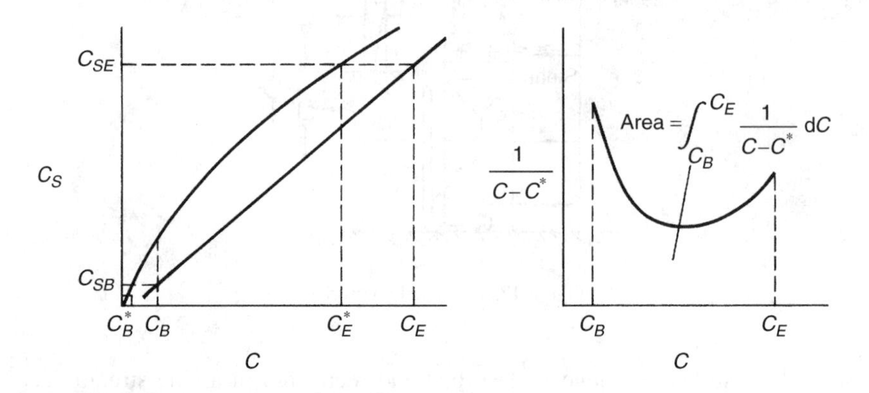


Figure 17.27. Graphical calculation of the number of transfer units

The integral in equation 17.112 may be evaluated numerically or graphically as shown in Figure 17.27. This is the *number of transfer units*, and the group outside the integral is the *height of a transfer unit*. The integral in equation 17.112 covers the whole concentration span of the adsorption process. If, instead, the limits are taken as from $C_B$ to an arbitrary concentration $C$, then the length $z'$ corresponding to $C$ is given by:

$$z' = \frac{mu}{k_g^0 a_p}\int_{C_B}^{C}\frac{\mathrm{d}C}{C - C^*} \tag{17.113}$$

and:

$$\frac{z'}{z_a} = \frac{\displaystyle\int_{C_B}^{C}\frac{\mathrm{d}C}{C - C^*}}{\displaystyle\int_{C_B}^{C_E}\frac{\mathrm{d}C}{C - C^*}} \tag{17.114}$$

In Figure 17.28, $C/C_E$ is plotted as a function of $z'/z_a$, and the unsaturated fraction $i$ of the adsorption zone may then be found. Since the zone in a moving bed is essentially the

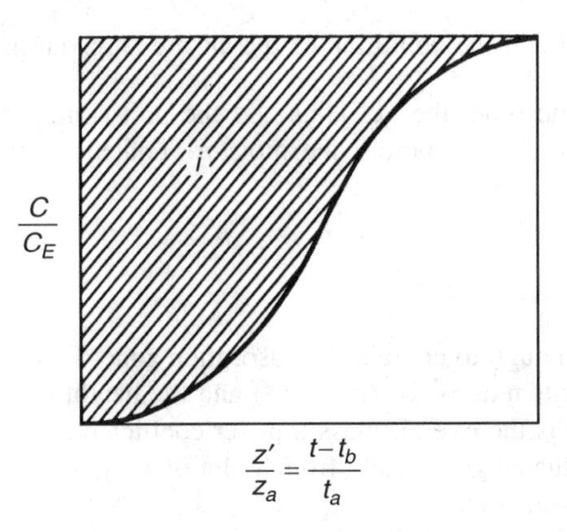


Figure 17.28. Dimensionless breakthrough curve showing fractional unsaturation of the adsorption zone

same as that which develops in a fixed bed, the time to break-point for the latter case, $t_b$, may be found from:

$$uA\varepsilon C_0 t_b = (z' - iz_a)C_{s\infty}\, A(1-\varepsilon) \tag{17.115}$$

The shape of the breakthrough wave, subsequent to the breakpoint, may then be determined from:

$$z' = \frac{(t-t_b)}{t_a} z_a \tag{17.116}$$

where $t_a$ is the time for the adsorption zone to move its own length, and $z'$ is measured from the inlet of the adsorption zone.

The method, which is illustrated in Example 17.5, applies only to isothermal beds in which the zone becomes fully-developed quite soon after the flow begins.

## Example 17.5

An adsorption unit is to be designed to dry air using silica gel. A moving-bed design is considered in which silica gel moves down a cylindrical column in plug flow while air flows up the column. Air enters the unit at the rate of 0.129 kg of dry air/m²s and with a humidity of 0.00267 kg water/kg dry air. It leaves essentially bone dry. There is equilibrium between air and gel at the entrance to and the exit from the adsorption zone. Experiments were carried out to find the relative resistances of the external gas film and pellet diffusion. Referred to a driving force expressed as mass ratios then:

(a) for the gas film, the coefficient $k_g a_z = 31.48 G'^{0.55}$ kg/m³s
 where $G'$ is the mass flowrate of dry air per unit cross-section of bed.
(b) for pellet diffusion, solid-film coefficent $k_p a_z = 0.964$ kg/m³s

where $a_z$ is the external area of adsorbent per unit volume of bed

(a) Using the transfer-unit concept, calculate the minimum length of packing which will reduce the moisture content of the air to 0.0001 kg water/kg dry air. At what rate should the gel travel through the bed?

The properties of the gel and the condition of the air are as given in Example 17.4.

(b) After operating for some time, the gel jams and the unit continues operating as a fixed bed. How long after jamming will it be before the moisture content of the effluent rises to half the inlet value?

## Solution

(a) The bed must be long enough to contain the adsorption zone. From equation 17.112 the number of transfer units may be written as: $\int dy_r/(y_r - y_r^*)$ and the height of a transfer unit in appropriate units is $G'/k_g^0 a_z$, where $k_g^0$ is the overall mass transfer coefficient.

The integral may be evaluated graphically from a plot of $1/(y_r - y_r^*)$ against $y_r$ over the concentration range of the adsorption zone.

A mass balance over a part of the bed gives an operating line:

$$y_r = \frac{G'_s}{G'}x_r + \left(y_{r\text{in}} - \frac{G'_s}{G'}x_{r\text{out}}\right)$$

This line, together with the equilibrium line, is similar to that shown in Figure 17.27. Corresponding values of $y$ and $y^*$ may be measured and the integral evaluated. The pinches between the operating and equilibrium lines which occur at each end of the zone prevent the end concentration from being used as the limits of the integration. If the lower limit of $y_r = 0.0001$ and the upper limit 0.0024, then from a graphical construction:

$$\int_{0.0001}^{0.0024} \frac{dy_r}{y_r - y_r^*} = 10.95 \text{ transfer units}$$

The height of a transfer unit is: $$\frac{G'}{k_g^0 a_z}$$ (from equation 17.112)

$k_g^0$ may be evaluated from the film coefficients, as discussed in Chapter 12.

When the zone is fully developed, each part will move at the same constant velocity. If $f'(C)$ is the mean slope of the isotherm over the range of concentrations of interest, then, in appropriate units:

$$[f'(C)]_{\text{mean}} = \frac{(8.4 \times 10^{-2} \times 1266)}{(2.67 \times 10^{-3} \times 1.186)}$$

$$m = \frac{\varepsilon}{1-\varepsilon} = \left(\frac{0.47}{0.53}\right) = 0.89$$

The inter-pellet air velocity = 0.233 m/s. The velocity $u_c$ with which the adsorption wave moves through the column may be obtained from equation 17.79.

Hence: $$u_c = 6.2 \times 10^{-6} \text{ m/s}$$

$$G'_s = u_c(1-\varepsilon)\rho_p$$

and: $$= 6.2 \times 10^{-6}(0.53)1266$$

$$= \underline{\underline{4.16 \times 10^{-3} \text{ kg/m}^2\text{s}}}$$

The rate at which clean adsorbent must be added and spent adsorbent removed in order to maintain steady state may also be found from an overall balance:

$$G_s'(0.084 - 0) = 0.129(0.00267)$$

$$G_s' = 4.11 \times 10^{-3} \text{ kg/m}^2\text{s}$$

(b) When the gel stops moving, the bed behaves as a fixed-bed already at its breakpoint. The concentration of water in the effluent begins to rise.
The time $t_a$ for the adsorption zone to move its own length $z_a$ is given by:

$$t_a = \frac{z_a}{u_c} = 8.3 \text{ h}$$

The time taken for a point at a distance $z'$ into the zone to emerge is given by:

$$t = \frac{z'}{z_a} t_a$$

where:

$$\frac{z'}{z_a} = \int_{y_r^B}^{y_r} \frac{dy_r}{y_r - y_r^*} \Bigg/ \int_{y_r^B}^{y_r^E} \frac{dy_r}{y_r - y_r^*}$$

The results for graphical integration are tabulated below.

$$\frac{1}{k_g^0 a_z} = \frac{1}{k_g a_z} + \frac{m'}{k_p a_z}$$

where $m'$ is the slope of the operating line $= \left(\frac{0.00267}{0.084}\right) = 0.0318$, as shown in Figure 17.22.

$$k_g a_z = 31.48 G'^{0.55} = 10.21 \text{ kg/m}^3 \text{ s}$$

Hence:

$$k_g^0 a_z = 7.64 \text{ kg/m}^3 \text{ s}$$

and:

$$\frac{G'}{k_g^0 a_z} = \left(\frac{0.129}{7.64}\right) = 0.0169 \text{ m}$$

The length of the adsorption zone $= (10.95 \times 0.0169) = 0.185$ m

Hence the minimum length of bed to contain the adsorption zone is 0.185 m. In practice, a somewhat greater length would be used to allow for variations in the length of the zone that might result from fluctuations in operating conditions. The data are summarised as follows:

| $y_r$ | $y_r^*$ | $\frac{1}{y_r - y_r^*}$ | $\int_{y_r^B}^{y_r} \frac{dy_r}{y_r - y_r^*}$ | $\frac{z'}{z_a}$ | $\frac{y_r}{y_{r0}}$ | $t$ (h) |
|---|---|---|---|---|---|---|
| 0.0001 | 0.00005 | 20,000 | 0 | 0 | 0.038 | 0 |
| 0.0002 | 0.00010 | 10,000 | 1.50 | 0.137 | 0.075 | 0.8 |
| 0.0006 | 0.00032 | 3570 | 4.00 | 0.362 | 0.225 | 3.1 |
| 0.0010 | 0.00062 | 2630 | 5.18 | 0.473 | 0.374 | 4.1 |
| 0.0014 | 0.00100 | 2500 | 6.13 | 0.560 | 0.525 | 4.5 |
| 0.0018 | 0.00133 | 3700 | 7.38 | 0.674 | 0.674 | 5.8 |
| 0.0022 | 0.00204 | 6250 | 9.33 | 0.852 | 0.825 | 7.1 |
| 0.0024 | 0.00230 | 10,000 | 10.95 | 1.000 | 0.899 | 7.7 |

By interpolation, $y_r/y_{r0} = 0.5$ when $t \approx 4.4$ h.

### 17.9.3. Rotary bed

Because of the difficulty of ensuring that the solid moves steadily and at a controlled rate with respect to the containing vessel, other equipment has been developed in which solid and vessel move together, relative to a fixed inlet for the feed and a fixed outlet for the product. Figure 17.29 shows the principle of operation of a rotary-bed adsorber used, for example, for solvent recovery from air on to activated-carbon. The activated-carbon is contained in a thick annular layer, divided into cells by radial partitions. Air can enter through most of the drum circumference and passes through the carbon layer to emerge free of solvent. The clean air leaves the equipment through a duct connected along the axis of rotation. As the drum rotates, the carbon enters a section in which it is exposed to steam. Steam flows from the inside to the outside of the annulus so that the inner layer of carbon, which determines the solvent content of effluent air, is regenerated as thoroughly as possible. Steam and solvent pass to condensers and the solvent recovered, either by decanting or by a process such as distillation. In the particular equipment shown, there is no separate provision for cooling the regenerated adsorbent; instead it is allowed to cool in contact with vapour-laden air and the adsorptive capacity may be lower as a result[16].

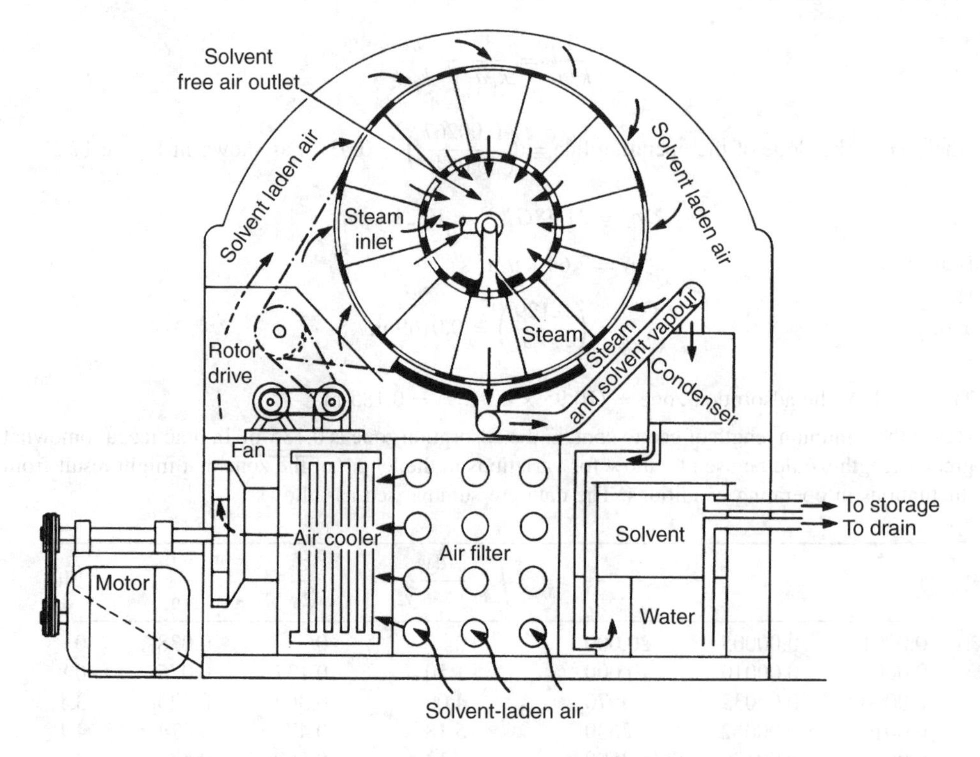


Figure 17.29. Rotary-bed adsorber

### 17.9.4. Moving access

In an interesting alternative to a moving-bed or a moving-container adsorber, a multiway valve effectively changes the position of the inlet and outlet valves relative to a fixed bed.

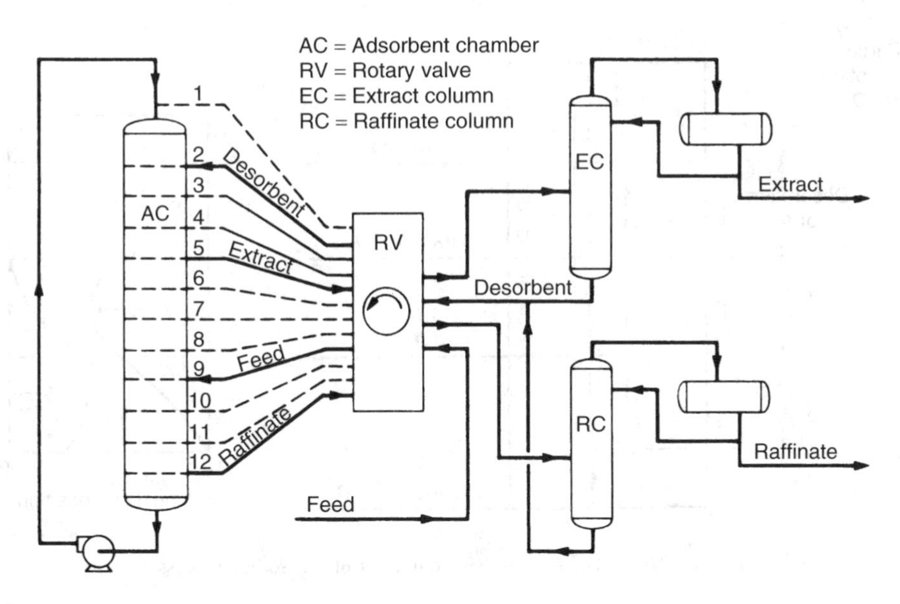


Figure 17.30. Sorbex: a number of small beds used with a rotary valve to simulate a moving bed

Such a system is shown diagrammatically in Figure 17.30 which shows a unit consisting of twelve small beds housed in one column. It may be assumed that it is fed with a mixture containing components **A** and **B**, the former being the more strongly absorbed of the two, with the desorption carried out using a third component **D**, the most strongly adsorbed component of all, and therefore capable of displacing **B** and **A** in that order. The valve rotates in a stepwise fashion, at regular intervals, with the positions of the inlets and the outlets for the process and regenerating streams moving to each numbered position in turn, thus simulating the behaviour of a moving bed as shown in Figure 17.31. As described by BROUGHTON[48] and JOHNSON and KABZA[49], the arrangement was developed by Universal Oil Products under the general name "Sorbex". The unit currently operates in the liquid phase, chiefly for separating *p*-xylene from $C_8$ aromatics, normal from branched and cyclo-paraffins, or olefins from mixture with paraffins. In principle, the unit may be used for gas-phase separations although, in either phase, success depends crucially on the proper working of the rotary valve.

### 17.9.5. Fluidised beds

Although the moving packed-bed has not yet achieved commercial success, another arrangement in which the solid moves with respect to the containing vessel, the fluidised bed, has fared better. Behaving essentially as a stirred tank reactor, the fluidised bed does not gives the ideal configuration for use as an adsorber. The solid is thoroughly mixed, so that the condition of the effluent is controlled by the mean adsorbate concentration on the solid, rather than by the initial concentration as in the case of a fixed bed. Nevertheless, there are other considerations which outweigh this disadvantage. Solid is easily added to and removed from fluidised bed. The pressure drop through a fluidised bed is effectively

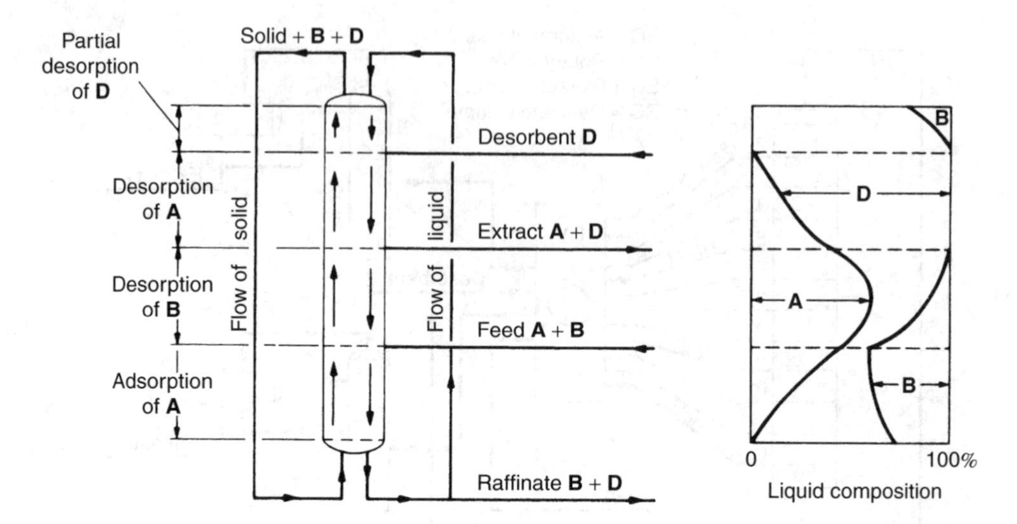


Figure 17.31. The moving-bed equivalent of the Sorbex process

constant over a wide range of fluid flowrates, and this makes it possible to treat materials at high flowrates in relatively compact equipment.

If the mean residence time in the fluidised bed is sufficiently long, it may be regarded as a single stage, from which streams of fluid and solid leave in equilibrium.

## 17.9.6. Compound beds

There is sometimes an advantage in using two kinds of adsorbent in an adsorption bed. Near the inlet would be an adsorbent with a high capacity at high concentrations, although it may have an unfavourable isotherm so that, on its own, the adsorption zone would then be unduly long, particularly if large pellets were used to minimise the pressure drop. If it is followed by a second bed of adsorbent with a highly favourable isotherm and a low mass transfer resistance, a short mass transfer zone will be sufficient to effect the required separation.

## 17.9.7. Pressure-swing regeneration

In thermal-swing regeneration, the bed may need a substantial time to reach the regeneration temperature. The high temperatures may also affect the product and accelerate the ageing processes in the adsorbent.

An alternative is to use pressure rather than temperature as the thermodynamic variable to be changed with adsorption taking place at high pressure and desorption at low pressure — hence the description *pressure–swing adsorption*. An arrangement utilising this principle was proposed by SKARSTROM[50,51]. Figure 17.32 shows a typical arrangement of a unit consisting of two fixed beds, one adsorbing and one regenerating. These functions are later reversed. A simple cycle consists of four steps. In step 1, high-pressure feed flows through bed A. Part of its effluent is expanded to the lower pressure, and then passed

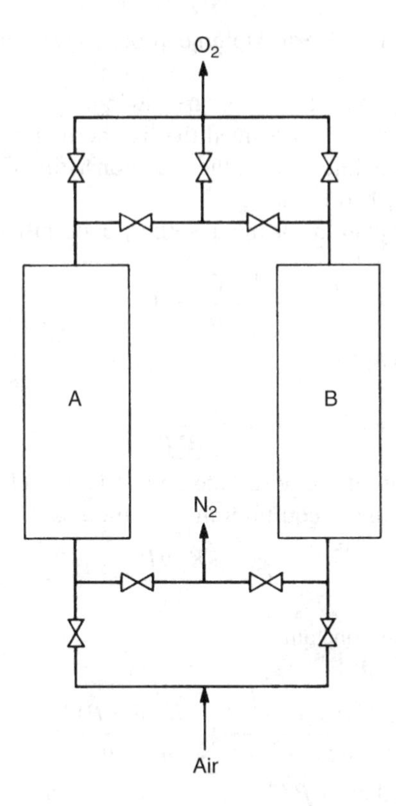


Figure 17.32. A two-bed unit using pressure swing regeneration for separating oxygen and nitrogen on a small scale

through bed B which it regenerates. In step 2, B is repressurised to the feed pressure using feed gas, while A is blown down to the purge pressure. Steps 3 and 4 follow the sequence of 1 and 2 except that the functions of beds A and B are reversed.

In large-scale equipment, more than two beds may be used so that pressure energy is better utilised. The regenerating effect of the purge stream depends on its volume rather than on its mass, so only a fraction of the high pressure effluent, say 20 per cent, is needed to achieve effective regeneration. Because changes in pressure can be brought about more rapidly than changes in temperature, pressure-swing regeneration can be used with shorter cycle-times than was possible with thermal-swing. This, in turn, allows smaller beds to be used and consequently a smaller inventory of adsorbent is needed in the system.

Pressure-swing regeneration is useful when the stream to be treated is needed at pressures above atmospheric, as for example in the case of instrument air. Pressure-swing units are compact and can readily be made portable. When the process stream is at atmospheric pressure or below, it may be possible to regenerate using a partial vacuum. When Skarstrom was patenting his "heatless" adsorber, GUERIN DE MONTGAREUIL and DOMINE[52] were patenting a system using vacuum regeneration. The original aim of both patents was to separate oxygen and nitrogen from air. With the range of adsorbents then available, neither process was particularly successful for that particular application. Skarstrom's equipment was, however, found to be suitable for the drying of gases, and, after some modifications, the Guerin-Domine process was applied successfully to the separation of air on a large scale.

A mathematical description of a pressure swing system has been presented by SHENDALMAN and MITCHELL[53] who assumed that isothermal equilibrium adsorption takes place and that the isotherm is linear, with the feed consisting of a single adsorbate at low concentration in a non-adsorbed carrier gas.

The mass conservation equation for the adsorber, over an increment d$z$ of bed may be written as:

$$\frac{\partial(uC)}{\partial z} + \frac{\partial C}{\partial t} + \frac{1}{m}\frac{\partial C_s}{\partial t} = 0 \qquad \text{(equation 17.69)}$$

where $m$ is the interpellet void ratio, $\varepsilon/(1-\varepsilon)$.

For an ideal gas:

$$C = \frac{yP}{\mathbf{R}T}$$

where $y$ is the mole fraction of the adsorbate and $P$ is the total pressure.
If the adsorbed phase is in linear equilibrium with the gas, then:

$$C_s = \frac{K_a yP}{\mathbf{R}T}$$

where $K_a$ is the equilibrium constant.
Substituting in equation 17.69:

$$\frac{\partial(uyP)}{\partial z} + \frac{\partial P}{\partial t} + \frac{K_a}{m}\frac{\partial(yP)}{\partial t} = 0 \qquad (17.118)$$

Neglecting the pressure gradient $\partial P/\partial z$:

$$\frac{\partial u}{\partial z} + u\frac{\partial \ln y}{\partial z} + \left(1 + \frac{K_a}{m}\right)\frac{\partial \ln P}{\partial t} + \left(1 + \frac{K_a}{m}\right)\frac{\partial \ln y}{\partial t} = 0 \qquad (17.119)$$

For a pure carrier gas, $y = 1$ and $K_a = 0$, and therefore:

$$\frac{\partial u}{\partial z} + \frac{\partial \ln P}{\partial t} = 0 \qquad (17.120)$$

Substituting in equation 17.119:

$$u\frac{\partial \ln y}{\partial z} + \frac{K_a}{m}\frac{\partial \ln P}{\partial t} + \left(1 + \frac{K_a}{m}\right)\frac{\partial \ln y}{\partial t} = 0 \qquad (17.121)$$

By the method of characteristics[54], equation 17.121 may be reduced to an ordinary differential equation to give:

$$\frac{\mathrm{d}z}{u} = \frac{\mathrm{d}t}{\left(1 + \dfrac{K_a}{m}\right)} = -\frac{\mathrm{d}\ln y}{\left(\dfrac{K_a}{m}\right)\left(\dfrac{\mathrm{d}\ln P}{\mathrm{d}t}\right)} \qquad (17.122)$$

from which:

$$\frac{\mathrm{d}z}{\mathrm{d}t} = \frac{u}{1 + \dfrac{1}{m}K_a} \qquad (17.123)$$

Equation 17.123 is a particular case of equation 17.75.

The left-hand side of equation 17.123 is the velocity of a point of fixed concentration on the adsorption wave. For a linear isotherm and if longitudinal diffusion is neglected, all points of concentration will move at the same velocity. Changing the pressure will affect $u$ and, to a lesser extent, $K_a$.

Pressure-swing regeneration is achieved by using a part of the high-pressure adsorber effluent for purging. The volume of purge must be such that the distance the adsorption wave moves at high pressure is completely reversed in the same time at low pressure. The requirement is normally satisfied by using a fraction of the high pressure effluent which is equal to the ratio of the low pressure to the high pressure.

The second characteristic that follows from equation 17.122 is:

$$\frac{\mathrm{d}\ln y}{\mathrm{d}\ln P} = \frac{-\dfrac{1}{m}K_a}{1 + \dfrac{K_a}{m}} \tag{17.124}$$

where $m$ is the inter-pellet void ratio $\varepsilon/(1-\varepsilon)$.

Integrating between the high and low pressure gives:

$$\frac{y_H}{y_L} = \left(\frac{P_L}{P_H}\right)^{(K_a/m)/[1+(K_a/m)]} \tag{17.125}$$

This relates the change in effluent concentration to the pressure ratio.

The change in the position of the characteristic that results from a pressure-swing is given by integrating equation 17.120 from the closed end of the bed where $u = 0$, $z = 0$.

Thus: $$u = -\left(\frac{\partial \ln P}{\partial t}\right) z \tag{17.126}$$

From equation 17.123: $$u = \left(1 + \frac{K_a}{m}\right)\frac{\mathrm{d}z}{\mathrm{d}t} = -\left(\frac{\mathrm{d}\ln P}{\mathrm{d}t}\right) z$$

which, on integration gives: $$\frac{z_H}{z_L} = \left(\frac{P_L}{P_H}\right)^{1/(1+K_a/m)} \tag{17.127}$$

where $z_H$ represents the distance moved by the adsorption front during the high pressure stage.

The net distance moved by the front, from the beginning of one position of low pressure to the next is given by:

$$\Delta z = z_L - z'_L = z_L - \left(\frac{P_H}{P_L}\right)^{1/(1+K_a/m)} z_H \tag{17.128}$$

If $\Delta z$ is negative, insufficient regeneration is occurring to sustain a condition of cyclic steady-state.

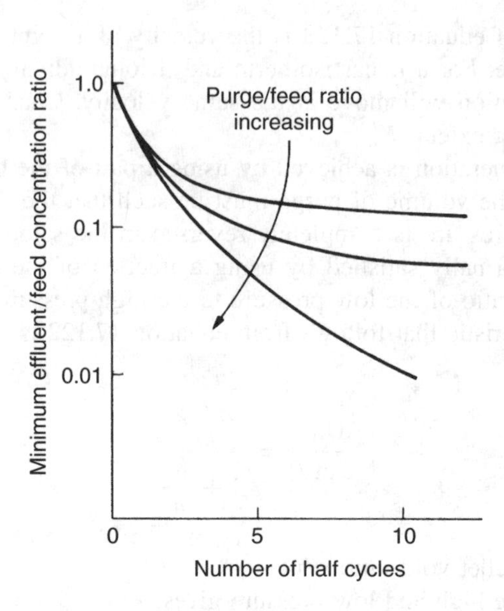


Figure 17.33. Effect of cycling and purge/feed volumetric ratio on minimum effluent/feed concentration ratio, using pressure-swing regeneration

Figure 17.33 shows how the purge–feed volumetric ratio and cycling affect effluent concentration. Ratios of 1.1–1.5 are normal.

## 17.9.8. Parametric pumping

When operated in a conventional mode, a fixed bed is fed with the stream to be processed until the breakpoint is reached. Thus, maximum use is made of the adsorptive capacity of the bed, without exceeding it. Regeneration is accomplished by changing a variable, such as temperature, pressure or concentration, and purging the bed in a countercurrent manner.

As described by WILHELM *et al.*(55), an alternative operating procedure has been developed in order to improve the separation obtained, where separation is defined as the ratio of concentrations in the upper and lower reservoirs, or in a reservoir and the feed. The technique has become known as *parametric pumping* because changing an operating parameter, such as temperature, may be considered as *pumping* the adsorbate into a reservoir at one end of a bed and, by difference, depleting the adsorbate in a reservoir at the other end.

### *Direct mode of operation*

Figure 17.34*a* shows a simple one-bed unit, operating in batch mode, and heated and cooled through a jacket. The arrangement is known as the *direct thermal mode* because heat is supplied to the whole length of bed at the same instant. Finite resistances to heat transfer will mean, however, that, in practice, the bed takes a finite time to reach the required temperature.

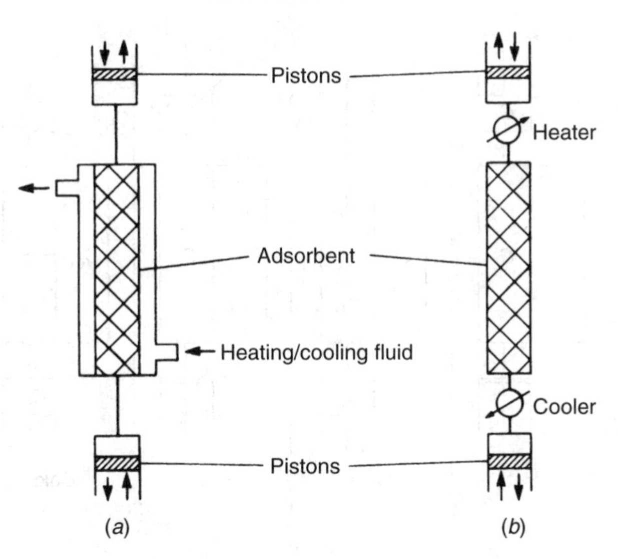


Figure 17.34. Parametric pumping, batch operation (*a*) Direct thermal mode (*b*) Recuperative thermal mode

To illustrate the principle, a number of assumptions will be made which are not realised in practice, although they enable the source of the separation to be identified. It is assumed that there is equilibrium at all times between fluid and the solid with which it is in contact. It is assumed that changes in temperature can be achieved instantaneously. It is further assumed that the adsorption front travels at such a velocity at the higher temperature that it traverses the length of the bed in the time allowed for upwards flow of the fluid. At the lower temperature, it is assumed that the adsorption wave travels half the distance in the same time with flow downwards. Each change may be regarded as taking place in two steps — first the temperature is changed and the new equilibrium established and then the pistons move to reverse the direction of flow. This simplified process is illustrated in Figure 17.35.

When the temperature of the bed is changed, the mass of adsorbate in an increment of bed $dz$ is conserved. Hence, for the first change from hot to cold:

$$A\,dz[\varepsilon C_F + (1-\varepsilon)C_{sF}] = A\,dz[\varepsilon C_c + (1-\varepsilon)C_{sc}] \tag{17.129}$$

For equilibrium operation and a linear isotherm, $C_{sc} = K_c C_c$ and $C_{sF} = K_H C_F$, and hence:

$$\frac{C_F}{C_c} = \frac{1 + \dfrac{1}{m}K_c}{1 + \dfrac{1}{m}K_H} \tag{17.130}$$

It may be seen from equation 17.123 that the right-hand side of equation 17.130 is the ratio of zone velocities, and:

$$\frac{C_F}{C_c} = \frac{u_H}{u_c} \tag{17.131}$$

It was assumed that $u_H/u_c = 2$, and hence $C_F = 2C_c$, $C_H = 2C_F$, $C_c = 2C_{cc}$ and so on.

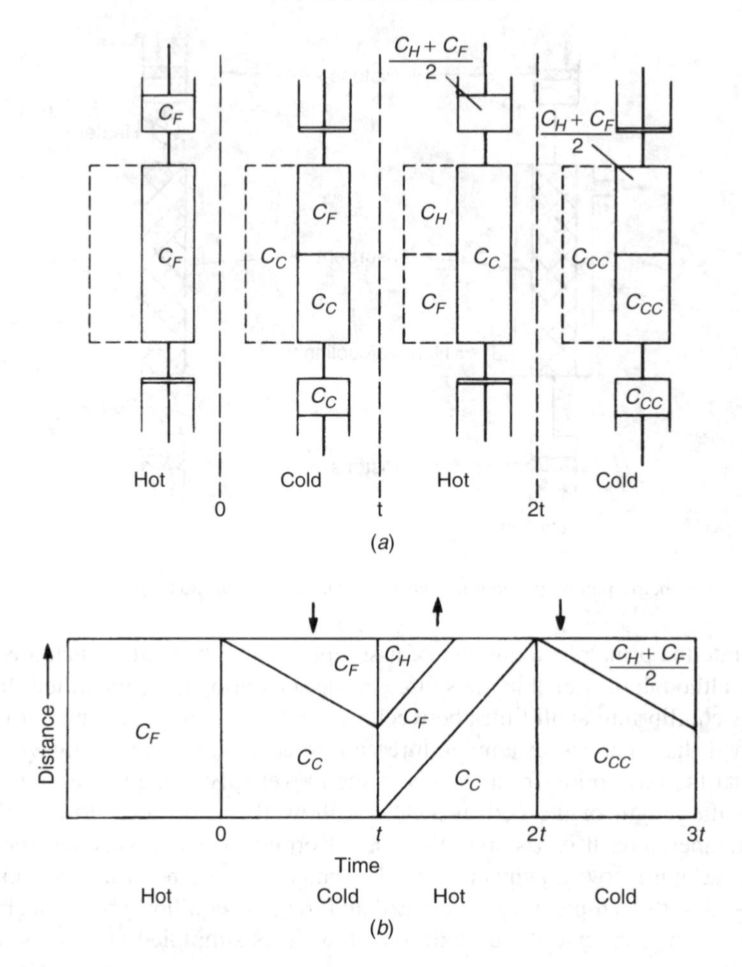


Figure 17.35. Ideal parapump, direct heating (*a*) Dotted section represents the condition immediately after a temperature change but before the pistons move (*b*) Position of the wave-front against time

In equation 17.129, $C_s$ is the mean adsorbate concentration over a pellet of adsorbent. It may be desirable to include the intra-pellet voidage $\alpha$ and intra-pellet concentrations $C'_F$, $C'_{sF}$ and so on to give:

$$\varepsilon C_F + (1-\varepsilon)\alpha C'_F + (1-\varepsilon)(1-\alpha)C'_{sF} = \varepsilon C_c + (1-\varepsilon)\alpha C'_c + (1-\varepsilon)(1-\alpha)C'_{sc}$$

At equilibrium:

$$C_F = C'_F, \quad C_c = C'_c,$$

$$C'_{sF} = K_H C_F, \quad C'_{sc} = K_c C_c$$

Hence equation 17.130 becomes:

$$\frac{C_F}{C_c} = \frac{1 + \dfrac{1}{m}[\alpha + (1-\alpha)K_c]}{1 + \dfrac{1}{m}[\alpha + (1-\alpha)K_H]} \tag{17.132}$$

It may be seen that even with only the two cycles shown, a significant difference has been achieved between the concentrations of the material in the two reservoirs $[0.5(C_H + C_F)/C_{cc} = 6]$. Separation is sometimes defined as the ratio of the upper reservoir concentration to that of the feed. In this case a value of 3/2 is obtained. Continuing the cycling process will increase the degree of separation without limit in this ideal case. In practice, thermal lags and diffusional processes make it impossible to sustain sharp differences of concentration, though separations giving a 100,000-fold change in concentrations have been achieved.

The process described so far has made maximum use of the adsorptive capacity of the bed in upwards flow. If the flow were continued beyond that point, the contents of the lower reservoir would pass unchanged into the top reservoir and the separation would be reduced. It may be more convenient, however, to use a shorter time for upwards flow. The effect of allowing the process to continue only until the concentration wave has reached two-thirds of the way along the bed is shown in Figure 17.36. As in Figure 17.35, $u_H = 2u_c$. After the second up-flow stage, the average concentration in the material that has left the top of the bed is given by:

$$\tfrac{1}{2}[(C_{HH} + C_H)/2 + C_F] = 2C_F \tag{17.133}$$

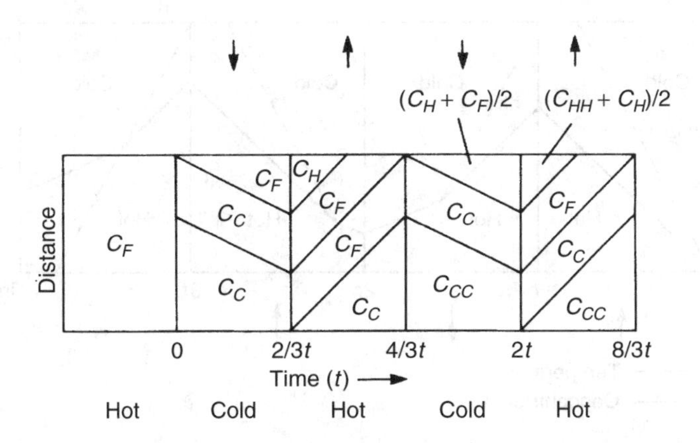


Figure 17.36. Effect of cycle-time on separation

The separation between upper and lower reservoirs in this case is 8. This compares with a value of 7 shown in Figure 17.35*b* after the same number of flow-reversals. The degree of separation depends on frequency of cycling as well as the total number of cycles.

### *Recuperative mode*

Figure 17.34*b* shows an alternative method of supplying heat in thermal parametric-pumping. Heat is supplied in upwards flow by passing feed from the lower reservoir through a heat exchanger. Cold operation in downward flow is achieved by cooling the feed from the top reservoir. Clearly, even when this method is idealised, the thermal wave takes a finite time to travel the length of the bed. The method is known as the *indirect* or *recuperative mode*, and is shown in Figure 17.34*b* as applied to a batch process.

The velocity with which a pure thermal wave travels through an insulated packed bed may be obtained from equation 17.100 by putting $U_0 = 0$ and $(\partial/\partial T)(C_s \Delta H) = 0$ to give:

$$\frac{u}{u_T} = \left[1 + \left(\frac{1}{m}\frac{\rho_p C_{ps}}{\rho_g C_{pg}}\right) + \left(\frac{4W C_{pw}}{\varepsilon \pi d_p^2 \rho_g C_{pg}}\right)\right] \tag{17.134}$$

It has been assumed that the gas and solid have the same temperature at any point, and that the fluid concentration is constant throughout a pellet at a value equal to that immediately outside the pellet. Within the limits of these assumptions, the thermal wave velocity $u_T$ is independent of temperature. As discussed in Section 17.8.4, the velocity of the thermal wave relative to that of the concentration wave can be positive, as it normally is in liquids, negative or zero.

Figure 17.37 shows a thermal wave plotted as a dotted line of distance against time. The velocity $u_c$ of the concentration wave depends on where it is in relation to the thermal wave, as can be seen by comparison with the full line in the Figure 17.37.

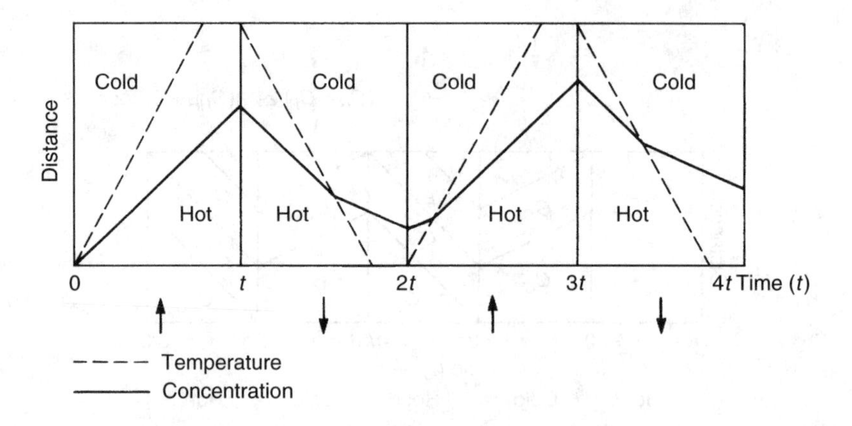


Figure 17.37. Wave propagation in recuperative mode

It may be shown that the ratio of the concentration in a hot zone to that in a cold zone for recuperative parametric pumping is given by:

$$\frac{C_H}{C_c} = \frac{1/u_c - 1/u_T}{1/u_H - 1/u_T} \tag{17.135}$$

For 'instant' heating and cooling, $u_T$ equals infinity and equation 17.135 becomes equivalent to equation 17.131 for the direct-heating mode.

The net movement upwards of a concentration wave is greater in the direct mode. Fewer cycles are needed to achieve a given separation. Nevertheless, the recuperative mode is probably the more convenient method to use on a commercial scale. Indeed, its equivalent is the only mode that can be used when other parameters, such as pH or pressure, are changed instead of temperature.

Many workers have demonstrated the effectiveness of parametric pumping in order to achieve separations in laboratory-scale equipment. It is mainly liquid systems that have been studied, using either temperature or pH as the control variable. Pressure parametric-pumping is described in a US patent and is discussed by YANG[3].

The principles of separation have been discussed using equilibrium theory. Finite resistances to heat and mass transfer will reduce the separation achieved.

## 17.9.9. Cycling-zone adsorption (CZA)

When parametric pumping was being developed, an alternative parameter-swinging technique was proposed by PIGFORD *et al.*[56] which is called *cycling-zone adsorption*. Instead of reversing the flow through a single bed as temperature is changed, a number of beds is used, connected in series, alternatively hot and cold. As with parametric pumping, heat may be supplied in the feed stream to each bed or through a jacket. The reversals of temperature remain, though each reversal of direction of the parametric pump corresponds to an additional stage of CZA. Figure 17.38 illustrates a two-bed unit. Given the same assumptions of ideality and, in particular, equilibrium and instant temperature changes, the sequence of events giving rise to concentration peaks and troughs in the effluent is shown in Figure 17.39. The effluent is switched between high and low concentration reservoirs.

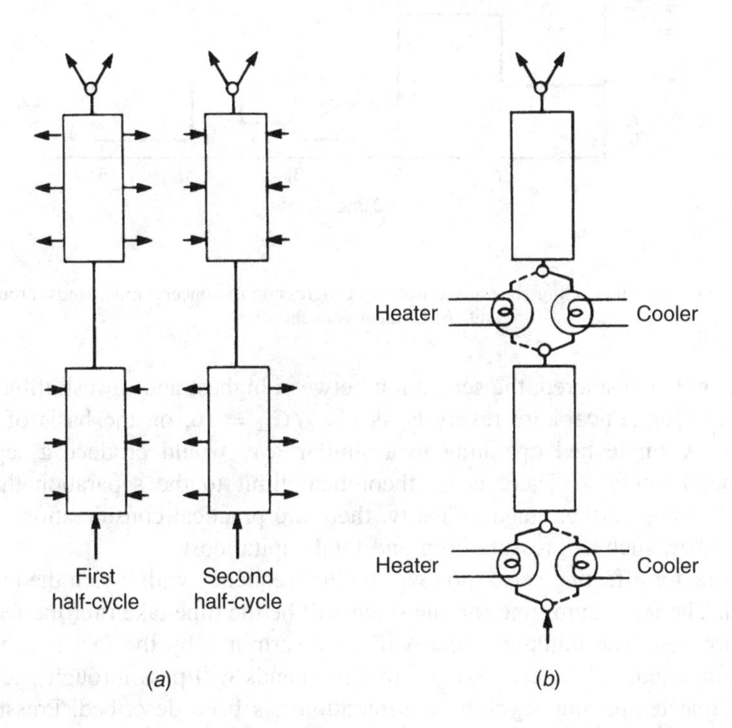


Figure 17.38. A two-bed cycling zone adsorption unit (*a*) Direct heating mode (*b*) Recuperative heating mode

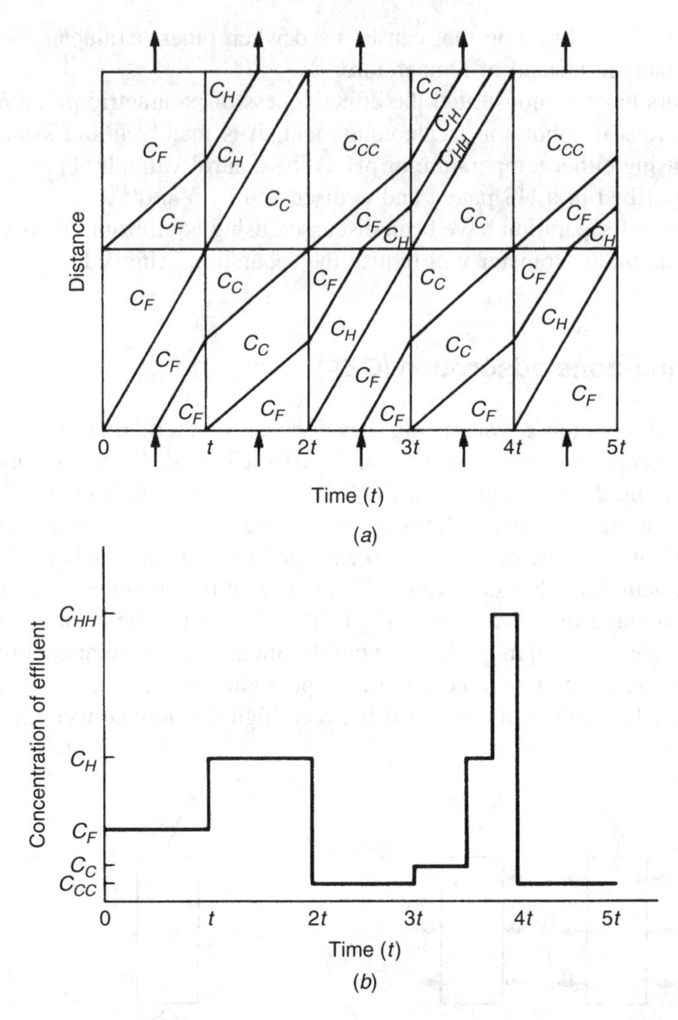


Figure 17.39. Direct mode cycling zone adsorption (*a*) Progression of concentration bands through a two-bed unit (*b*) Effluent concentration

In the example considered, the separation between highest and lowest effluent concentrations, after four temperature reversals, is $C_{HH}/C_{cc} = 16$, on the basis of the earlier assumptions. A single bed operating in a similar way would produce a separation of $C_H/C_c$ equal to only 4. There is no theoretical limit to the separation that may be achieved by adding further stages. Clearly, there are practical considerations which will limit the number, such as pressure drop and total capital cost.

The other factor affecting separation will be the frequency with which the temperatures are changed. The maximum time for one stage will be the time taken for the feed to break through a hot bed. The minimum time will be determined by the fact that, if there are too many temperature changes, the concentration bands will pass through unchanged.

The principle temperature-cycling of separation has been described. Pressure-cycling has been described by PLATT and LAVIE[57].

Further discussions of pressure swing adsorption, parametric pumping and cycling-zone adsorption have been presented by YANG[3], SCHWEITZER[7] and WANKAT[58].

## 17.10. FURTHER READING

Adsorption and Ion Exchange: *A.I.Ch.E. Symposium Series* No 259 **83** (1987) (see also numbers 14, 24, 69, 74, 80, 96, 117, 120, 134, 152, 165, 219, 233, 242).

*Adsorption and its Applications in Industry and Environmental Protection* Vol 1 Applications in Industry Vol II Applications in Environmental Protection (Elsevier, Amsterdam, 1999).

COONEY, D. O.: *Adsorption Design for Wastewater Treatment* (CRC Press, Lewis Publishers, Boca Raton, 1998).

CHEREMISINOFF, N. P. and CHEREMISINOFF, P. N.: *Carbon Adsorption for Pollution Control* (Prentice Hall, Englewood Cliff, New Jersey, 1993).

KARGER, J. and RUTHVEN, D. M.: *Diffusion in Zeolites and other Microporous Solids* (John Wiley & Sons, New York, 1992).

LE VAN, M. D., (ed.) *Fundamentals of Adsorption V* (Kluwer Academic Publishers, Norwell, 1996).

PERRY, R. H., GREEN, D. W., and MALONEY, J. O. (eds.): *Perry's Chemical Engineers' Handbook*. 7th edn (McGraw-Hill Book Company, New York, 1997).

RUTHVEN, D. M.: *Principles of Adsorption and Adsorption Processes* (Wiley, 1984).

RUTHVEN D. M., FAROOQ S. and KNAEBEL K. S.: *Pressure Swing Adsorption* (VCH Publishers, New York, 1994).

SLEJKO, F. L.: *Adsorption Technology* (Marcel Dekker, New York, 1985).

SUZUKI, M. (ed.): *Fundamentals of Adsorption IV* (Kodansha, Tokyo, 1993).

SUZUKI, M.: *Adsorption Engineering* (Elsevier, Amsterdam, 1990).

THOMAS, W. J. and CRITTENDEN, B. D.: *Adsorption, Technology and Design* (Butterworth-Heinemann, Oxford, 1998).

TIEN, Chi: *Adsorption Calculations and Modeling* (Butterworth, Boston, 1994).

VALENZUELA, D. Y. and MYERS A. I.: *Adsorption Equilibrium Data Handbook* (Prentice Hall, Englewood Cliffs, 1989).

WANKAT, P. C.: *Large Scale Adsorption and Chromatography (2* vols*)* (CRC Press Boca Raton, 1986).

YANG, R. T.: *Gas Separation by Adsorption Processes* (Butterworth, London 1987).

YANG, R. T.: *Gas Separation by Adsorption Processes* (Series on Chemical Engineering, Vol 1) (World Scientific Publishing Co, 1997).

## 17.11. REFERENCES

1. CRITTENDEN, B.: *The Chemical Engineer* No. 452 (Sept. 1988). 21. Selective adsorption.
2. BARRER, R. M.: *Brit. Chem. Eng.* **4** (1959) 267. New selective sorbents: porous crystals as molecular filters.
3. YANG, R. T.: *Gas Separation by Adsorption Processes* (Butterworth, London, 1987).
4. BRECK, D. W.: *Zeolite Molecular Sieves* (Wiley, New York, 1974).
5. BARRER, R. M.: *Zeolites and Clay Minerals* (Academic Press, London, 1978).
6. ROBERTS, C. W.: *Properties and Applications of Zeolites* (Chemical Society, London, 1979).
7. SCHWEITZER, P. A. (ed.): *Handbook of Separation Techniques for Chemical Engineers* 2nd edn. (McGraw-Hill, New York, 1988).
8. EVERETT, D. H. and STONE, F. S. (eds): *The Structure and Properties of Porous Materials* (Butterworth, Oxford, 1958).
9. BOWEN, J. H., BOWREY, R. and MALIN, A. S.: *J. Catalysis* **7** (March 1967) 457. A study of the surface area and structure of activated alumina by direct observation.
10. BRUNAUER, S.: *The Adsorption of Gases and Vapours* (Oxford U.P., Oxford, 1945).
11. LANGMUIR, I.: *J. Am. Chem. Soc.* **40** (1918) 1361. The adsorption of gases on plane surfaces of glass, mica and platinum.
12. BRUNAUER, S., EMMETT, P. H. and TELLER, E.: *J. Am. Chem. Soc.* **60** (1938) 309. Adsorption of gases in multimolecular layers (for errata see reference 13).
13. EMMETT, P. H. and DE WITT, T.: *Ind. Eng. Chem. (Anal.)* **13** (1941) 28. Determination of surface areas.
14. BRUNAUER, S., DEMING, L. S., DEMING, W. E. and TELLER, E.: *J. Am. Chem. Soc.* **62** (1940) 1723. On a theory of the van der Waals adsorption of gases.
15. HARKINS, W. D. and JURA, G.: *J. Chem. Phys.* **11** (1943) 431. An adsorption method for the determination of the area of a solid without the assumption of a molecular area and the area occupied by nitrogen molecules on the surface of solids. *J. Amer. Chem. Soc.* **66** (1944) 1366. Surface of solids. Part XIII.

16. RUTHVEN, D. M.: *Principles of Adsorption and Adsorption Processes* (Wiley, New York, 1984).
17. GARG, D. R. and RUTHVEN, D. M.: *A.I.Ch.E. Jl* **21** (1975) 200. Linear driving force approximation for diffusion controlled adsorption in molecular sieve columns.
18. POLANYI, M.: *Trans. Farad. Soc* **28** (1932) 316. Theory of adsorption of gases — introductory paper.
19. YOUNG, D. M. and CROWELL, A. D.: *Physical Adsorption of Gases* (Butterworth, 1962).
20. MERSMANN, A., MUNSTERMANN, U. and SCHADL, J.: *Germ. Chem. Eng.*, **7** (1984) 137. Separation of gas mixtures by adsorption.
21. FREUNDLICH, H.: *Colloid and Capillary Chemistry* (Methuen, London, 1926).
22. GLUECKAUF, E., and COATES, J. I.: *J. Chem. Soc.* (1947) 1315. Theory of chromatography.
23. KIPLING, J. J.: *Adsorption from Solutions of Non-Electrolytes* (Academic Press, 1965).
24. EMMETT, P. H.: *Advances in Colloid Science* **1** (1942) 1. The measurement of the surface areas of finely divided or porous solids by low temperature adsorption isotherms.
25. COHAN, L. H.: *J. Am. Chem. Soc.* **60** (1938) 433. Sorption hysteresis and vapour pressure of concave surfaces.
26. CRANSTON, R. W. and INKLEY, F. A.: *Advances in Catalysis* **9** (1957) 143. The determination of pore structures from nitrogen adsorption isotherms.
27. RANZ, W. E. and MARSHALL, W. R.: *Chem. Eng. Prog.* **48** (4) (1952) 173. Evaporation from drops. Part II.
28. DWIVEDI, P. N. and UPADHEY, S. N.: *Ind. Eng. Chem. Proc. Des. Dev.* **16** (1977) 157. Particle–fluid mass transfer in fixed and fluidized beds.
29. WAKAO, N. and FUNAZKRI, T.: *Chem. Eng. Sci.* **33** (1978) 1375. Effect of fluid dispersion coefficients on particle–fluid mass transfer coefficients in packed beds.
30. GLASSTONE, S. and LEWIS, D.: *Elements of Physical Chemistry* (Macmillan, 1970).
31. SATTERFIELD, C. N.: *Mass transfer in Heterogeneous Catalysis* (MIT Press, Boston, 1970).
32. BIRD, R. B., STEWART, W. E. and LIGHTFOOT, E. N.: *Transport Phenomena* (Wiley, New York, 1960).
33. EVANS, R. B., WATSON, G. M. and MASON, E. A.: *J. Chem. Phys.* **33** (1961) 2076. Gaseous diffusion in porous media at uniform pressure.
34. SCHNEIDER, P. and SMITH, J. M.: *A.I.Ch.E. Jl* **14** (1968) 762. Adsorption rate constants from chromatography.
35. EPSTEIN, N.: *Chem. Eng. Sci.* **44** (1989) 777. On tortuosity and the tortuosity factor in flow and diffusion through porous media.
36. HINSHELWOOD, C. N.: *The Kinetics of Chemical Change* (Clarendon Press, Oxford, 1940).
37. EDWARDS, M. F. and RICHARDSON, J. F.: *Chem. Eng. Sci.* **23** (1968) 109. Gas dispersion in packed beds.
38. BOWEN, J. H. and RIMMER, P. G.: *Trans. Inst. Chem. Eng.*, **50** (1972) 168. Design of fixed bed sorbers using a quadratic driving force equation.
39. LAPIDUS, L. and AMUNDSON, N. R.: *J. Phys. Chem.* **56** (1952) 373. Mathematics of adsorption in fixed beds — The rate determining steps in radial adsorption analysis; *ibid* **56** (1952) 984. The effect of longitudinal diffusion in ion exchange and chromatographic columns.
40. LEVENSPIEL, O. and BISCHOFF, K. B.: In *Advances in Chemical Engineering* Vol. 4 DREW, T. B., HOOPES, J. W. and VERMEULEN, T. (eds.) (Academic Press, 1963) 95. Patterns of flow in chemical process vessels.
41. ROSEN, J. B.: *J. Chem. Phys.* **20** (1965) 387. Kinetics of fixed bed systems for solid diffusion into spherical particles.
42. LEAVITT, F. W.: *Chem. Eng. Prog.* **58** (8) (1962) 54. Nonisothermal adsorption in large fixed beds.
43. AMUNDSON, N. R., ARIS, R. and SWANSON, R.: *Proc. Roy. Soc.* **286A** (1965) 129. On simple exchange waves in fixed beds.
44. CARTER, J. W.: *A.I.Ch.E. Jl.* **21** (1975) 380. On the regeneration of fixed adsorber beds.
45. BERG, C.: *Chem. Eng. Prog.* **47** (11) (1951) 585. Hypersorption design.
46. KATELL, S.: *Chem. Eng. Prog.* **62** (Oct. 1966) 67. Removing sulphur dioxide from flue gases.
47. CARTELYOU, C. G.: *Chem. Eng. Prog.* **65** (9) (1969) 69. Commercial processes for $SO_2$ removal.
48. BROUGHTON, D. B.: *Chem. Eng. Prog.* **64** (8) (1968) 60. Molex, history of a process.
49. JOHNSON, J. A. and KABZA, R. G.: *I. Chem. E. Annual Research Meeting, Swansea* (1990) Sorbex: Industrial scale adsorptive separation.
50. SKARSTROM, C. W.: *Annals N Y Acad. Sci.* **72** (1959) 751. Use of adsorption phenomena in automatic plant-type gas analysis.
51. SKARSTROM, C. W.: *US Patent* 2944627 (1960). Method and apparatus for fractionating a gaseous mixture by adsorption.
52. GUERIN DE MONTGAREUIL, P. and DOMINE, D.: *US Patent* 3155468 (1964). Process for separating a binary gaseous mixture by adsorption.
53. SHENDALMAN, L. H. and MITCHELL, J. E.: *Chem. Eng. Sci.* **27** (1972) 1449. A study of heatless adsorption in the model system $CO_2$ in He.
54. ACRIVOS, A.: *Ind. Eng. Chem.* **45** (1956) 703. Method of characteristics technique. Application to heat and mass transfer problems.

55. Wilhelm, R. H., Rice, A. W., Rolke, R. W. and Sweed, N. H.: *Ind. Eng. Chem. (Fund)* **7** (1968) 337. Parametric pumping.
56. Pigford, R. L., Baker, B. and Blum, D.: *Ind. Eng. Chem. (Fund)* **8** (1969) 144. An equilibrium theory of the parametric pump.
57. Platt, D. and Lavie, R.: *Chem. Eng. Sci*. **40** (1985) 733. Pressure cycle zone adsorption.
58. Wankat, P. C.: *Cyclic Separations—Parametric Pumping, Pressure Swing Adsorption and Cycling Zone Adsorption* (A.I.Ch.E. Modular Instruction Series, Module B6.11, 1986).

## 17.12. NOMENCLATURE

| | | Units in SI System | Dimensions in **M, N, L, T, $\theta$** |
|---|---|---|---|
| $A$ | Superficial cross-sectional area of bed | m$^2$ | $\mathbf{L}^2$ |
| $A_p$ | Interfacial area for condensate in a pore | m$^2$ | $\mathbf{L}^2$ |
| $A_s$ | Area of an adsorbed film | m$^2$ | $\mathbf{L}^2$ |
| $A'_1, A'_2$ | Arrhenius frequency factors for desorption | kmol/m$^2$s | $\mathbf{NL}^{-2}\ \mathbf{T}^{-1}$ |
| $a_m$ | Area occupied by one molecule in an adsorbed film | m$^2$ | $\mathbf{L}^2$ |
| $a_p$ | External area of adsorbent per unit volume of adsorbent | m$^{-1}$ | $\mathbf{L}^{-1}$ |
| $a'_p$ | External surface of a pellet | m$^2$ | $\mathbf{L}^2$ |
| $a_z$ | External area of adsorbent per unit volume of bed | m$^{-1}$ | $\mathbf{L}^{-1}$ |
| $a_0, a_1, a_2$ | Fraction of the adsorbent surface covered by no adsorbate, one, two layers | — | — |
| $B_0, B_1, B_i, B_j$ | Constants in Langmuir-type equations | various | |
| $B_2$ | $\alpha_0/\beta$ in equation 17.9 | — | — |
| $C$ | Concentration of adsorbate in the fluid | kmol/m$^3$ | $\mathbf{NL}^{-3}$ |
| $C_B$ | Concentration of adsorbate **B** in the fluid | kmol/m$^3$ | $\mathbf{NL}^{-3}$ |
| $C_c$ | Concentration of adsorbate after 1 cold stage | kmol/m$^3$ | $\mathbf{NL}^{-3}$ |
| $C'_c$ | Concentration of adsorbate within a pore after 1 cold stage | kmol/m$^3$ | $\mathbf{NL}^{-3}$ |
| $C_{cc}$ | Concentration of adsorbate after two cold stages | kmol/m$^3$ | $\mathbf{NL}^{-3}$ |
| $C_E$ | Concentration of adsorbate at the emergence of the adsorption zone | kmol/m$^3$ | $\mathbf{NL}^{-3}$ |
| $C_F$ | Concentration of adsorbate in feed conditions | kmol/m$^3$ | $\mathbf{NL}^{-3}$ |
| $C_i$ | Concentration of adsorbate at the exterior surface of a pellet | kmol/m$^3$ | $\mathbf{NL}^{-3}$ |
| $C_H, C_{HH}$ | Concentration of adsorbate after one, two hot stages | kmol/m$^3$ | $\mathbf{NL}^{-3}$ |
| $C_0$ | Concentration of adsorbate initially or at inlet | kmol/m$^3$ | $\mathbf{NL}^{-3}$ |
| $C'_p$ | Concentration of adsorbate in feed conditions within a pore | kmol/m$^3$ | $\mathbf{NL}^{-3}$ |
| $C_r, C_{sr}$ | Concentration of adsorbate at radius $r$ in the fluid, or in adsorbed phases | kmol/m$^3$ | $\mathbf{NL}^{-3}$ |
| $C_s$ | Concentration of adsorbed phase | kmol/m$^3$ | $\mathbf{NL}^{-3}$ |
| $C'_s$ | Concentration of adsorbed phase within a pore | kmol/m$^3$ | $\mathbf{NL}^{-3}$ |
| $C''_s$ | Mass concentration of adsorbed phase referred to adsorbate free adsorbent | kg/m$^3$ | $\mathbf{ML}^{-3}$ |
| $\bar{C}_s$ | Mean concentration of adsorbed phase | kmol/m$^3$ | $\mathbf{NL}^{-3}$ |
| $C'_{sc}$ | Concentration of adsorbed phase in a pore after one cold stage | kmol/m$^3$ | $\mathbf{NL}^{-3}$ |
| $C'_{SF}$ | Concentration of adsorbed phase in a pore in feed conditions | kmol/m$^3$ | $\mathbf{NL}^{-3}$ |
| $C_{sm}$ | Concentration of adsorbed phase in a monolayer | kmol/m$^3$ | $\mathbf{NL}^{-3}$ |
| $C_{s\infty}$ | Ultimate or maximum concentration of adsorbed phase | kmol/m$^3$ | $\mathbf{NL}^{-3}$ |

| | | Units in SI System | Dimensions in M, N, L, T, $\theta$ |
|---|---|---|---|
| $C^*$ | Concentration of adsorbate in equilibrium with $\bar{C}_s$ | kmol/m$^3$ | $\mathbf{NL^{-3}}$ |
| $c_{pg}$ | Specific heat capacity of the gas phase | J/kg K | $\mathbf{L^2T^{-2}}\theta^{-1}$ |
| $c_{ps}$ | Specific heat capacity of the adsorbent with adsorbate | J/kg K | $\mathbf{L^2T^{-2}}\theta^{-1}$ |
| $c_{pw}$ | Specific heat capacity of the wall | J/kg K | $\mathbf{L^2T^{-2}}\theta^{-1}$ |
| $D$ | Diffusivity | m$^2$/s | $\mathbf{L^2T^{-1}}$ |
| $D_{av}$ | Average diffusivity | m$^2$/s | $\mathbf{L^2T^{-1}}$ |
| $D_e$ | Effective diffusivity | m$^2$/s | $\mathbf{L^2T^{-1}}$ |
| $D_k$ | Knudsen diffusivity | m$^2$/s | $\mathbf{L^2T^{-1}}$ |
| $D_L$ | Longitudinal diffusivity | m$^2$/s | $\mathbf{L^2T^{-1}}$ |
| $D_M$ | Molecular diffusivity | m$^2$/s | $\mathbf{L^2T^{-1}}$ |
| $D_s$ | Surface diffusivity | m$^2$/s | $\mathbf{L^2T^{-1}}$ |
| $D_{s0}$ | Surface diffusivity in standard conditions | m$^2$/s | $\mathbf{L^2T^{-1}}$ |
| $D_T$ | Total diffusivity | m$^2$/s | $\mathbf{L^2T^{-1}}$ |
| $d_b$ | Bed diameter | m | $\mathbf{L}$ |
| $d_p$ | Pellet diameter | m | $\mathbf{L}$ |
| $E$ | Argument of an error function | — | — |
| $E_s$ | Energy of activation in surface diffusion | J/kmol | $\mathbf{MN^{-1}L^2T^{-2}}$ |
| $E_0, E_1 \dots E_n$ | Energy of activation of desorption from empty surface, monolayer etc. | J/kmol | $\mathbf{MN^{-1}L^2T^{-2}}$ |
| F | A function in equation 17.92 | — | — |
| $f'(C)$ | Slope of an adsorption isotherm | — | — |
| G | Function in equation 17.82 | — | — |
| $G$ | Gibbs free energy of an adsorbed film *or* | J | $\mathbf{ML^2T^{-2}}$ |
| | Mass flowrate of fluid | kg/s | $\mathbf{MT^{-1}}$ |
| $G', G'_s$ | Mass flowrate per unit cross sectional area of fluid, solid | kg/m$^2$s | $\mathbf{ML^{-2}T^{-1}}$ |
| $G_1, G_2$ | Functions in equation 17.83 | — | — |
| $H$ | Enthalpy per kmol | J/kmol | $\mathbf{MN^{-1}L^2T^{-2}}$ |
| $h$ | Film heat transfer coefficient | W/m$^2$ K | $\mathbf{MT^{-3}}\theta^{-1}$ |
| $i$ | Unsaturated fraction of an adsorption zone | — | — |
| $J$ | A factor in equation 17.58 | — | — |
| $K$ | An equilibrium constant based on activities | various | |
| $K_a, K_c, K_H$ | Adsorption equilibrium constants | — | — |
| $K_0$ | Adsorption equilibrium constant at a standard condition | — | — |
| $k$ | Thermal conductivity of fluid | W/mK | $\mathbf{MLT^{-3}}\theta^{-1}$ |
| $\boldsymbol{k}$ | Reaction rate constant (1st order) | s$^{-1}$ | $\mathbf{T^{-1}}$ |
| $k_B$ | Boltzmann constant | $1.3805 \times 10^{-23}$ J/K | $\mathbf{ML^2T^{-2}}\theta^{-1}$ |
| $k_e$ | Effective thermal conductivity of solid | W/mK | $\mathbf{MLT^{-3}}\theta^{-1}$ |
| $k_g$ | Film mass transfer coefficient | m/s | $\mathbf{LT^{-1}}$ |
| $k_g^o$ | Overall mass transfer coefficient | m/s | $\mathbf{LT^{-1}}$ |
| $k_p$ | Mass transfer coefficient based on 'solid film' | m/s | $\mathbf{LT^{-1}}$ |
| $\boldsymbol{k}_0, \boldsymbol{k}_1$ | Adsorption velocity constants for empty surface, monolayer | kmol/Ns | $\mathbf{M^{-1}NL^{-1}T}$ |
| $\boldsymbol{k}'_1, \boldsymbol{k}'_2$ | Desorption velocity constants for monolayer, bilayer | kmol/m$^2$s | $\mathbf{NL^{-2}T^{-1}}$ |
| $L$ | Length of porous medium | m | $\mathbf{L}$ |
| $L'$ | Constant in equation 17.26 | — | — |
| $L_e$ | Length of porous path | m | $\mathbf{L}$ |
| $L_p$ | Mean length of a pore | m | $\mathbf{L}$ |
| $l$ | Distance into a pore | m | $\mathbf{L}$ |
| $M$ | Molecular weight | kg/kmol | $\mathbf{MN^{-1}}$ |
| $M'$ | Constant in equation 17.26 | m$^{-6}$ | $\mathbf{L^{-6}}$ |
| $M_A, M_B$ | Molecular weights of **A**, **B** | kg/kmol | $\mathbf{MN^{-1}}$ |

| | | Units in SI System | Dimensions in **M, N, L, T, θ** |
|---|---|---|---|
| $m$ | Inter-pellet void ratio $\varepsilon/(1-\varepsilon)$ | — | — |
| $m'$ | Slope of an operating line | — | — |
| $m_M$ | Mass of one molecule | kg | **M** |
| **N** | Avogadro number ($6.023 \times 10^{26}$ molecules per kmol *or* $6.023 \times 10^{23}$ molecules per mol) | $kmol^{-1}$ | $\mathbf{N}^{-1}$ |
| $N_A, N_B$ | Flux of molecular species **A**, **B** | $kmol/m^2s$ | $\mathbf{NL^{-2}T^{-1}}$ |
| $N_{AC}$ | Interstitial flux of **A** | $kmol/m^2s$ | $\mathbf{NL^{-2}T^{-1}}$ |
| $N_{Ai}$ | Flux along a tortuous pore | $kmol/m^2s$ | $\mathbf{NL^{-2}T^{-1}}$ |
| $Nu$ | Nusselt number $(hd_p/k)$ | — | — |
| $n$ | Number of adsorbed layers | — | — |
| $n$ | Index in the Freundlich equation 17.35 | — | — |
| $n'$ | Number of molecules per unit volume of gas | $m^{-3}$ | $\mathbf{L}^{-3}$ |
| $n''$ | Number of components in equation 17.33 | — | — |
| $n_0$ | Initial number of mols of adsorbate | kmol | **N** |
| $n_p$ | Number of pores | — | — |
| $n_s$ | Number of mols of adsorption in a film or capillary | kmol | **N** |
| $n_{s1}$ | Number of mols of adsorbate at pressure $P_1$ | kmol | **N** |
| $P$ | Total pressure | $N/m^2$ | $\mathbf{ML^{-1}T^{-2}}$ |
| $\bar{P}$ | Mean pressure | $N/m^2$ | $\mathbf{ML^{-1}T^{-2}}$ |
| $P^0$ | Vapour pressure of the adsorbed phase | $N/m^2$ | $\mathbf{ML^{-1}T^{-2}}$ |
| $P_A$ | Partial pressure of **A** | $N/m^2$ | $\mathbf{ML^{-1}T^{-2}}$ |
| $Pe$ | Peclet number $(ud_p/D_L)$ | — | — |
| $P_H/P_L$ | High/low pressure ratio | — | — |
| $P_m$ | Vapour pressure over a meniscus | $N/m^2$ | $\mathbf{ML^{-1}T^{-2}}$ |
| $Pr$ | Prandtl number $(c_p\mu/k)$ | — | — |
| $P_1$ | Pressure at which the adsorbed film becomes mobile | $N/m^2$ | $\mathbf{ML^{-1}T^{-2}}$ |
| **R** | Gas constant | 8314 J/kmol K | $\mathbf{MN^{-1}L^2T^{-2}\theta^{-1}}$ |
| $r$ | Radius within a spherical pellet or pore | m | **L** |
| $r_i$ | Outer radius of a spherical pellet | m | **L** |
| $r_p$ | Radius of a pore | m | **L** |
| $r'_p$ | Net radius of a pore allowing for adsorbed layers | m | **L** |
| $Sc$ | Schmidt number $(\mu/\rho D)$ | — | — |
| $Sh$ | Sherwood number $(k_g d_p/D)$ | — | — |
| $T$ | Absolute temperature | K | $\boldsymbol{\theta}$ |
| $T_a$ | Ambient temperature | K | $\boldsymbol{\theta}$ |
| $T_0$ | Initial or reference temperature | K | $\boldsymbol{\theta}$ |
| $t$ | Time | s | **T** |
| $t_a$ | Time for an adsorption zone to move its own length | s | **T** |
| $t_b$ | Time to breakpoint | s | **T** |
| $t_1$ | Time parameter in equation 17.71 | s | **T** |
| $t_{min}$ | Minimum time to saturate unit cross section of a bed of length, $z$ | s | **T** |
| $U_0$ | Overall heat transfer coefficient | $W/m^2$ K | $\mathbf{MT^{-3}\theta^{-1}}$ |
| $u$ | Interpellet velocity of fluid | m/s | $\mathbf{LT^{-1}}$ |
| $u_c, u_H$ | Velocity of a point on an adsorption wave at lower/higher temperatures | m/s | $\mathbf{LT^{-1}}$ |
| $u_T$ | Velocity of a point on a thermal wave | m/s | $\mathbf{LT^{-1}}$ |
| $V$ | Volume of fluid *or* volume per unit mass of adsorbent | $m^3$ $(m^3/kg)$ | $\mathbf{L^3}$ $(\mathbf{M^{-1}L^3})$ |
| $V^1$ | Volume in the gas phase equivalent to an adsorbed monolayer | $m^3$ | $\mathbf{L^3}$ |
| $V_i$ | Volume of $i$th component | $m^3$ | $\mathbf{L^3}$ |
| $V_0$ | Initial volume of solution | $m^3$ | $\mathbf{L^3}$ |

| | | Units in SI System | Dimensions in M, N, L, T, $\theta$ |
|---|---|---|---|
| $V_p$ | Volume of a pore | $m^3$ | $\mathbf{L}^3$ |
| $V_s$ | Volume of the adsorbed phase | $m^3$ | $\mathbf{L}^3$ |
| $V_M$ | Molar volume of adsorbate | $m^3/kmol$ | $\mathbf{N}^{-1}\mathbf{L}^3$ |
| $V'_m$ | Molar volume of liquid adsorbate | $m^3/kmol$ | $\mathbf{N}^{-1}\mathbf{L}^3$ |
| $V_s^1$ | Volume of the adsorbed phase in a monolayer | $m^3$ | $\mathbf{L}^3$ |
| $v_s$ | Volume of adsorbate on unit area of surface | $m^3/m^2$ | $\mathbf{L}$ |
| $v_s^1$ | Volume of adsorbate on unit area of monolayer | $m^3/m^2$ | $\mathbf{L}$ |
| $W$ | Mass per unit length of containing vessel | kg/m | $\mathbf{ML}^{-1}$ |
| $x_f$ | Thickness of boundary film | m | $\mathbf{L}$ |
| $x_r$ | Mass ratio of adsorbed phase to adsorbent | — | — |
| $y$ | Mole fraction of adsorbate in the fluid phase | — | — |
| $y_H, y_L$ | Mole fractions in the effluent during high, low pressure stages of PSA | — | — |
| $y_r$ | Mass ratio of adsorbate to carrier fluid | — | — |
| $y_r^B, y_r^E$ | $y_r$ at breakpoint, after emergence of adsorption zone | — | — |
| $y_r^*$ | $y_r$ in equilibrium with mean concentration of adsorbed phase | — | — |
| $z$ | Distance along a bed | m | $\mathbf{L}$ |
| $z'$ | Position of concentration $C$ | m | $\mathbf{L}$ |
| $z_a$ | Length of an adsorption zone | m | $\mathbf{L}$ |
| $z_e$ | Length of a fixed bed | m | $\mathbf{L}$ |
| $z_H$ | Distance moved by an adsorption zone at high pressure | m | $\mathbf{L}$ |
| $z_L, z'_L$ | Distances moved by an adsorption zone in successive low pressure stages | m | $\mathbf{L}$ |
| $z_0$ | Initial position of a point on the adsorption wave | m | $\mathbf{L}$ |
| $\alpha$ | Intra-pellet void fraction | — | — |
| $\alpha_0, \alpha_1, \alpha_2$ | Constants in equations 17.7, 17.24, 17.35 | various | |
| $\beta, \beta_1, \beta_2$ | Constants in equations 17.8, 17.24, 17.30 | various | |
| $\Gamma$ | Two dimensional spreading pressure | N/m | $\mathbf{MT}^{-2}$ |
| $\Delta$ | Change in property | — | — |
| $\varepsilon$ | Interpellet void fraction | — | — |
| $\varepsilon_p$ | Adsorption potential | J/kmol | $\mathbf{MN}^{-1}\mathbf{L}^2\mathbf{T}^{-2}$ |
| $\theta_1$ | Constant in equation 17.36 | $m^3$ | $\mathbf{L}^3$ |
| $\Omega$ | Collision integral | — | — |
| $\lambda$ | Mean free path of a gas molecule | m | $\mathbf{L}$ |
| $\lambda_A$ | Average distance between adsorption sites | m | $\mathbf{L}$ |
| $\lambda_M$ | Molar latent heat | J/kmol | $\mathbf{MN}^{-1}\mathbf{L}^2\mathbf{T}^{-2}$ |
| $\lambda'$ | Length parameter in equation 17.93 | — | — |
| $\mu$ | Viscosity | $Ns/m^2$ | $\mathbf{ML}^{-1}\mathbf{T}^{-1}$ |
| $\mu_g$ | Chemical potential of a gas | J/kmol | $\mathbf{MN}^{-1}\mathbf{L}^2\mathbf{T}^{-2}$ |
| $\mu_s$ | Chemical potential of an adsorbed film | J/kmol | $\mathbf{MN}^{-1}\mathbf{L}^2\mathbf{T}^{-2}$ |
| $\rho_B$ | Mass or molar density of a packed bed | $kg/m^3$ ($kmol/m^3$) | $\mathbf{ML}^{-3}$ ($\mathbf{NL}^{-3}$) |
| $\rho_g$ | Mass (or molar) density of a gas | $kg/m^3$ ($kmol/m^3$) | $\mathbf{ML}^{-3}$ ($\mathbf{NL}^{-3}$) |
| $\rho_p$ | Mass (or molar density) of an adsorbent pellet | $kg/m^3$ ($kmol/m^3$) | $\mathbf{ML}^{-3}$ ($\mathbf{NL}^{-3}$) |
| $\rho_s$ | Mass (or molar) density of a solid | $kg/m^3$ ($kmol/m^3$) | $\mathbf{ML}^{-3}$ ($\mathbf{NL}^{-3}$) |
| $\sigma_{AB}$ | Collision diameter between **A** and **B** molecules | m | $\mathbf{L}$ |
| $\sigma_{sl}, \sigma_{sv}, \sigma_{lv}$ | Surface tensions at three interfaces in a solid–liquid–gas system | N/m | $\mathbf{MT}^{-2}$ |
| $\phi$ | Interfacial angle between liquid and solid | — | — |
| $\phi_1$ | Constant in equation 17.36 | $m^3$ | $\mathbf{L}^3$ |
| $\tau$ | Tortuosity, $L/L_e$ | — | — |
| $\tau'$ | Time parameter in equation 17.94 | — | — |
| $\varphi$ | A resistance parameter in equation 17.95 | — | — |
| $\chi$ | Time parameter in equations 17.71 and 17.83 | s | $\mathbf{T}$ |

CHAPTER 18

# Ion Exchange

## 18.1. INTRODUCTION

Ion exchange is a unit operation in its own right, often sharing theory with adsorption or chromatography, although it has its own special areas of application. The oldest and most enduring application of ion exchange is in water treatment, to soften or demineralise water before industrial use, to recover components from an aqueous effluent before it is discharged or recirculated, and this is discussed by ARDEN[1]. Ion exchange may also be used to separate ionic species in various liquids as discussed by HELFFRICH[2] and SCHWEITZER[3]. Ion exchangers can catalyse specific reactions or be suitable to use for chromatographic separations, although these last two applications are not discussed in this chapter. Applications of ion exchange membranes are considered in Chapter 8.

The modern history of ion exchange began in about 1850 when two English chemists, THOMPSON[4] and WAY[5], studied the exchange between ammonium ions in fertilisers and calcium ions in soil. The materials responsible for the exchange were shown later to be naturally occurring alumino-silicates[6]. History records very much earlier observations of the phenomenon and, for example, ARISTOTLE[7], in 330 BC, noted that sea-water loses some of its salt when allowed to percolate through some sands. Those who claim priority for MOSES[8] should note however that the process described may have been adsorption!

In the present context, the *exchange* is that of equivalent numbers of similarly charged ions, between an immobile phase, which may be a crystal lattice or a gel, and a liquid surrounding the immobile phase. If the exchanging ions are positively charged, the ion exchanger is termed *cationic*, and *anionic* if they are negatively charged. The rate at which ions diffuse between an exchanger and the liquid is determined, not only by the concentration differences in the two phases, but also by the necessity to maintain electroneutrality in both phases.

As well as occurring naturally, alumino-silicates are manufactured. Their structure is that of a framework of silicon, aluminium and oxygen atoms. If the framework contains water, then this may be driven off by heating, leaving a porous structure, access to which is controlled by "windows" of precise molecular dimensions. Larger molecules are excluded, hence the description "molecular sieve" as discussed in Chapter 17.

If the framework had originally contained not only water but also a salt solution, the drying process would leave positive and negative ions in the pores created by the loss of water. When immersed in a polar liquid, one or both ions may be free to move. It is often found that only one polarity of ion moves freely, the other being held firmly to the framework. An exchange of ions is then possible between the mobile ions in the exchanger and ions with like-charge in the surrounding liquid, as long as those ions are not too large and electro-neutrality is maintained.

## 18.2. ION EXCHANGE RESINS

A serious obstacle to using alumino-silicates as ion exchangers, is that they become unstable in the presence of mineral acids. It was not possible, therefore, to bring about exchanges involving hydrogen ions until acid-resisting exchangers had been developed. First, sulphonated coal and, later, synthetic phenol formaldehyde were shown to be capable of cation exchange. Nowadays, cross-linked polymers, known as resins, are used as the basic framework for most ion exchange processes, both cationic and anionic.

The base resin contains a styrene–divinylbenzene polymer, DVB. If styrene alone were used, the long chains it formed would disperse in organic solvents. The divinylbenzene provides cross-linking between the chains. When the cross-linked structure is immersed in an organic solvent, dispersion takes place only to the point at which the osmotic force of solvation is balanced by the restraining force of the stretched polymer structure.

When the styrene–DVB polymer is sulphonated, it becomes the cation exchanger which is polystyrene sulphonic acid, with exchangeable hydrogen ions. The framework of the resin has a fixed negative charge, so that no exchange can occur with the mobile negative ions outside the resin. Ions of the same polarity as the framework are termed *co-ions*. Those of opposite polarity have the potential to exchange and are called *counter-ions*. Resins for anion exchange may also be manufactured from polystyrene as a starting material, by treating with monochloroacetone and trimethylamine, for example. The structures of these particular resins are shown in Figure 18.1.

Both resins can be described as strongly ionic. Each is fully ionised so that all the counter-ions within the gel may be exchanged with similarly charged ions outside the gel, whatever the concentration of the latter.

## 18.3. RESIN CAPACITY

Various measures of the capacity of a resin for ion exchange are in common use. The *maximum capacity* measures the total number of exchangeable ions per unit mass of resin, commonly expressed in milliequivalents per gram (meq/g). The base unit of a polystyrene-sulphonic-acid polymer, as shown in Figure 18.1, has a molecular weight of approximately 184 kg/kmol. Each unit has one exchangeable hydrogen ion, so its maximum capacity is (1000/184) or 5.43 milliequivalents per gram.

The capacities of styrene-based anion exchangers are not so easily calculated because there may not be an anionic group on every benzene ring. Values of 2.5–4.0 meq/g are typical for strong anion resins.

It is the number of fixed ionic groups which determines the maximum exchange capacity of a resin although the extent to which that capacity may be exploited depends also on the chemical nature of those groups. Weak acid groups such as the carboxyl ion, $COO-$, ionise only at high pH. At low pH, that is a high concentration of hydrogen ions, they form undissociated COOH. Weak base groups such as $NH^+{}_3$ lose a proton when the pH is high, forming uncharged $NH_2$ ions. Consequently, for resins which are weakly ionic, the exploitable capacity depends on the pH of the liquid being treated. Figure 18.2 illustrates the expected dependence.

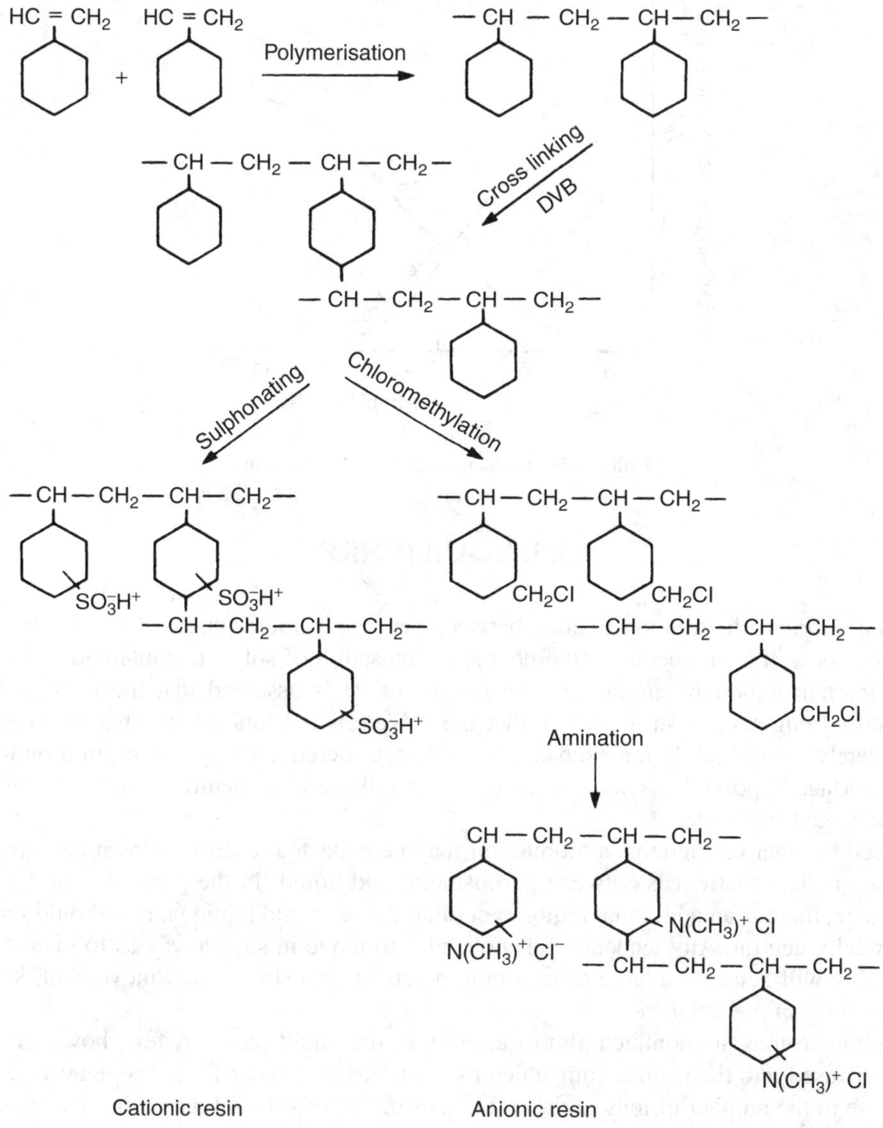

Figure 18.1. Formation of styrene-based cationic and anionic resins

When the resin is incompletely ionised, its *effective capacity* will be less than the maximum. If equilibrium between resin and liquid is not achieved, a *dynamic capacity* may be quoted which will depend on the contact time. When equipment is designed to contain the resin, it is convenient to use unit volume of water-swollen resin as the basis for expressing the capacity. For fixed-bed equipment, the *capacity at breakpoint* is sometimes quoted. This is the capacity per unit mass of bed, averaged over the whole bed, including the ion exchange zone, when the breakpoint is reached.

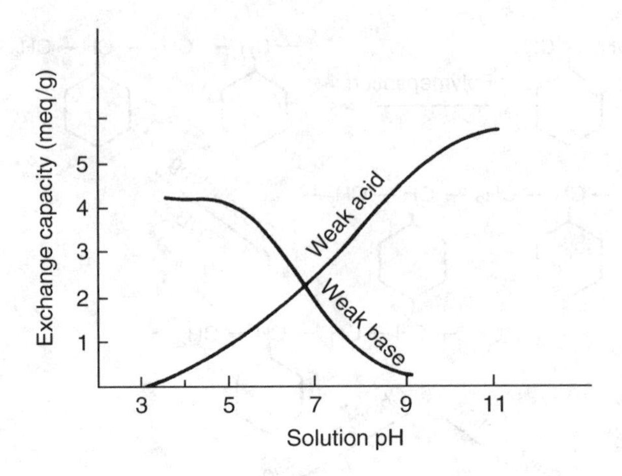


Figure 18.2. Exchange capacity of weak resins

## 18.4. EQUILIBRIUM

The equilibrium distribution of ions between resin and liquid phases depends on many factors. As well as temperature, the degree of ionisation of solvent, solute and resin may be important although, to simplify the discussion, it is assumed that the resin is fully ionised (strong acid or strong base), that the solvent is not ionised and that the solute is completely ionised. Only ion exchange itself is considered, although adsorption on to the resin surface is possible as well as diffusion into the resin of neutral groups of ions and of uncharged molecules.

Freed of other restrictions, a mobile ion may be expected to diffuse down any concentration gradient that exists between porous solid and liquid. In the particular case of ion exchange, there is an additional requirement that the resin and liquid phases should remain electrically neutral. Any tendency for molecules to move in such a way as to disturb this neutrality will generate a large electrostatic potential opposing further movement, known as the *Donnan potential.*

Mobile co-ions are confined almost entirely to the liquid phase. A few, however, may have diffused into the resin accompanied by neutralising counter-ions. The net effect is an increase in the number of ions in the resin, causing it to swell and increasing its exchange capacity slightly above that which arises from the fixed ionic groups alone. Swelling is a reproducible equilibrium characteristic of a resin, depending on its degree of cross-linking and its ion exchange capacity, as well as temperature and the solution composition. Polyvalent exchanging ions create more cross-linking within the resin and, therefore, produce less swelling than monovalent ions. $Al^{3+} < Ca^{2+} < Na^+$. Water swelling is generally the result of the hydration of ionic groups.

In considering ionic equilibria, it is convenient to write the exchange process in the form of a chemical equation. For example, a water-softening process designed to remove calcium ions from solution may be written as:

$$Ca^{++} + 2Cl^- + 2Na^+R_c{}^- = 2Na^+ + 2Cl^- + Ca^{++}(R_c{}^-)_2$$

where $R_c$ is a cationic exchange resin. When its capacity is depleted, the resin may be regenerated by immersing it in sodium chloride solution so that the reverse reaction takes place.

In general, the exchange of an ion **A** of valency $\upsilon_A$ in solution, for an ion **B** of valency $\upsilon_B$ on the cationic resin may be written as:

$$\upsilon_B C_A + \upsilon_A C_{SB}(R_c)_{\upsilon_B} = \upsilon_A C_B + \upsilon_B C_{SA}(R_c)_{\upsilon_A} \tag{18.1}$$

Including activity coefficients $\gamma$, the thermodynamic equilibrium constant $K$ becomes:

$$\begin{aligned} K &= \frac{(\gamma_B C_B)^{\upsilon_A}(\gamma_{SA} C_{SA})^{\upsilon_B}}{(\gamma_A C_A)^{\upsilon_B}(\gamma_{SB} C_{SB})^{\upsilon_A}} \\ &= \frac{(\gamma_B)^{\upsilon_A}(\gamma_{SA})^{\upsilon_B}}{(\gamma_A)^{\upsilon_B}(\gamma_{SB})^{\upsilon_A}} K_c \end{aligned} \tag{18.2}$$

where $K_c$ is the selectivity coefficient, a measure of preference for one ionic species. Defining ionic fractions as:

$$x_A = C_A/C_0 \quad \text{and} \quad y_A = C_{SA}/C_{S\infty}$$

then:

$$\begin{aligned} K_c &= \frac{C_B^{\upsilon_A} C_{SA}^{\upsilon_B}}{C_A^{\upsilon_B} C_{SB}^{\upsilon_A}} \\ &= \frac{(y_A/x_A)^{\upsilon_B}}{(y_B/x_B)^{\upsilon_A}} \left(\frac{C_0}{C_{S\infty}}\right)^{\upsilon_A - \upsilon_B} \end{aligned} \tag{18.3}$$

where: $C_0$ is the total ionic strength of the solution, and:
$C_{S\infty}$ is the exchangeable capacity of the resin.

In a dilute solution, the activity coefficients approach unity and $K_c$ approaches $K$. For the water softening, represented by the equation given previously, $\upsilon_A = 2$ and $\upsilon_B = 1$.

Hence:

$$\begin{aligned} K_c &= \frac{y_A/x_A}{(y_B/x_B)^2} \frac{C_0}{C_{S\infty}} \\ &= \frac{y_A/x_A}{[(1-y_A)/(1-x_A)]^2} \frac{C_0}{C_{S\infty}} \end{aligned} \tag{18.4}$$

Except when $\upsilon_A = \upsilon_B$, the selectivity coefficient depends on the total ionic concentrations of the resin and the liquid phases.

Another measure of the preference of an ion exchanger for one other ionic species is the *separation factor* $\alpha$. This is defined in a similar way to relative volatility in vapour–liquid binary systems, and is independent of the valencies of the ions.

Thus:

$$\alpha_B^A = \frac{y_A/x_A}{y_B/x_B} \tag{18.5}$$

When $K_c$ is greater than unity, the exchanger takes up ion **A** in preference to ion **B**. In general, the value of $K_c$ depends on the units chosen for the concentrations although it is more likely than $\alpha$ to remain constant when experimental conditions change. When **A** and **B** are monovalent ions, $K_c = \alpha$.

Equilibrium relationships may be plotted as $y$ against $x$ diagrams using equation 18.4. Figure 18.3 shows such a plot for $v_A = 2$ and $v_B = 1$. The group $K_c(C_{S\infty}/C_0)^{v_A - v_B}$ has values of 0.01–100.

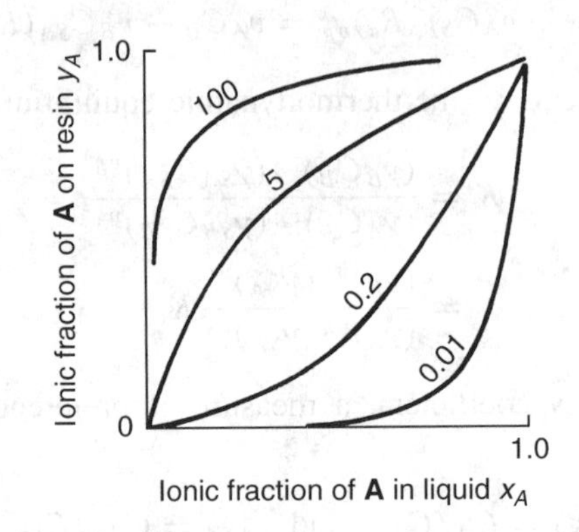


Figure 18.3. Equilibrium isotherm for the ion exchange $\mathbf{A} + 2\mathbf{B}(S) = \mathbf{A}(S) + 2\mathbf{B}$. The parameters are values of the group $K_c C_{S\infty}/C_0$

As is the case with adsorption isotherms, those curves in Figure 18.3 which are concave to the concentration axis for the mobile phase are termed *favourable* and lead to self-sharpening ion exchange waves.

Deciding which of several counter-ions will be preferably exchanged may be difficult without experimental work, although some general guidance may be given. The Donnan potential results in counter-ions with a high valency being exchanged preferentially. If there is a specific interaction between a counter-ion and a fixed ionic group, that ion will be preferred. Ions may be preferred because of their small size or shape.

Approximate selectivity coefficients for the exchange of various cations for lithium ions on a sulphonated polystyrene, a typically strong acid resin, are given in Table 18.1. The values are relative to Li = 1.0. The selectivity coefficient between two ions is the ratio of their selectivities relative to lithium. Hence, for a sodium–hydrogen exchange:

$$K_{Na^+H^+} = \left(\frac{2}{1.3}\right) \approx 1.5$$

Because these ions have the same valency, the selectivity coefficient is equal to the separation factor and a value greater than unity indicates that $Na^+$ adheres to the resin in preference to $H^+$. The same procedure may be used for exchange between di- and mono-valent ions although its validity is more questionable.

Thus:
$$K_{Ca^{++}Na^+} = \left(\frac{5.2}{2.0}\right) = 2.6$$

A similar table, shown in Table 18.2, is available for anion exchange although this is based on fewer data[3]. The problem is complicated by the fact that there are two types

Table 18.1. Selectivities on 8 per cent cross-linked strong acid resin for cations. Values are relative to lithium[3]

| | | | |
|---|---|---|---|
| $Li^{+}$ | 1.0 | $Zn^{2+}$ | 3.5 |
| $H^{+}$ | 1.3 | $Co^{2+}$ | 3.7 |
| $Na^{+}$ | 2.0 | $Cu^{2+}$ | 3.8 |
| $NH_4^{+}$ | 2.6 | $Cd^{2+}$ | 3.9 |
| $K^{+}$ | 2.9 | $Ba^{2+}$ | 4.0 |
| $Rb^{+}$ | 3.2 | $Mn^{2+}$ | 4.1 |
| $Cs^{+}$ | 3.3 | $Ni^{2+}$ | 3.9 |
| $Ag^{+}$ | 8.5 | $Ca^{2+}$ | 5.2 |
| $UO_2^{2+}$ | 2.5 | $Sr^{2+}$ | 6.5 |
| $Mg^{2+}$ | 3.3 | $Pb^{2+}$ | 9.9 |
| | | $Ba^{2+}$ | 11.5 |

Table 18.2. Selectivities on strong base resin[3]

| | | | |
|---|---|---|---|
| $I^{-}$ | 8 | $HCO_3^{-}$ | 0.4 |
| $NO_3^{-}$ | 4 | $CH_3COO^{-}$ | 0.2 |
| $Br^{-}$ | 3 | $F^{-}$ | 0.1 |
| $HSO_4^{-}$ | 1.6 | $OH^{-}$(II) | 0.06 |
| $NO_2^{-}$ | 1.3 | $SO_4^{2-}$ | 0.15 |
| $CN^{-}$ | 1.3 | $CO_3^{2-}$ | 0.03 |
| $Cl^{-}$ | 1.0 | $HPO_4^{2-}$ | 0.01 |
| $BrO_3^{-}$ | 1.0 | | |
| $OH^{-}$(I) | 0.65 | | |

of functional structure used in strong anion exchange resins and that anions in solution may exist in complex form. Nevertheless, the table provides some guidance when several systems are being compared.

### 18.4.1. Ion exclusion and retardation

As well as being used for ion exchange, resins may be used to separate ionic and non-ionic solutes in aqueous solution. A packed bed of resin is then filled with water and a sample of solution added. If water is then drained from the bed as more water is added to the top, the sample is eluted through the column. If the Donnan potential prevents the ionic components from entering the resin, there will be no effect on the non-ionic species. When a solution of HCl and $CH_3COOH$ is eluted through a bed of hydrogen and chloride resin, the HCl appears first in the effluent, followed by the $CH_3COOH$.

The process described is referred to as *ion-exclusion* as discussed by ASHER and SIMPSON[9]. The resins used are normal and the non-ionic molecules are assumed to be small enough to enter the pores. When large non-ionic molecules are involved, an alternative process called *ion-retardation* may be used, as discussed by HATCH *et al.*[10]. This requires a special resin of an amphoteric type known as a *snake cage polyelectrolyte*. The polyelectrolyte consists of a cross-linked polymer physically entrapping a tangle of linear polymers. For example, an anion exchange resin which is soaked in acrylic acid becomes entrapped when the acrylic acid is polymerised. The intricacy of the interweaving is such that counter-ions cannot be easily displaced by other counter-ions. On the other hand, ionic mobility within the resin maintains the electro-neutrality. The ionic molecule as a whole is absorbed by the resin in preference to the non-ionic molecule. When a solution of NaCl and sucrose is treated by the method of ion-retardation, the sucrose appears first in the effluent.

## 18.5. EXCHANGE KINETICS

It is insufficient to have data on the extent of the ion exchange at equilibrium only. The design of most equipment requires data on the amount of exchange between resin and liquid that will have occurred in a given contact time. The resistances to transfer commonly found in such a system are discussed in Chapter 17. It is necessary to consider the counter-diffusion of ions through a boundary film outside the resin and through the pores of the resin. The ion exchange process on the internal surface does not normally constitute a significant resistance.

It is theoretically possible that equilibrium between liquid and resin will be maintained at all points of contact. Liquid and solid concentrations are then related by the sorption isotherm. It is usual, however, that pellet or film diffusion will dominate or "control" the rate of exchange. It is also possible that control will be mixed, or will change as the ion exchange proceeds. In the latter case, the initial film-diffusion control will give way to pellet-diffusion control at a later stage.

### 18.5.1. Pellet diffusion

Ions moving through the body of an exchanger are subject to more constraints than are molecules moving through an uncharged porous solid. If the exchanging ions are equivalent and are of equal mobility, the complications are relatively minor and are associated with the tendency of the exchanger to swell and of some neutral groups of ions to diffuse. In the general case of ions with different valencies and mobilities, however, allowances have to be made for a diffusion potential arising from electrostatic differences, as well as the usual driving force due to concentration differences.

Exchange between counter-ions **A** in beds of resin and counter-ions **B** in a well-stirred solution may be represented by the Nernst-Planck equation as:

$$N_A = (N_A)_{\text{diff}} + (N_A)_{\text{elec}} = -D_A\left(\text{grad } C_A + \frac{v_A C_A \mathbf{F}}{\mathbf{R}T}\text{grad } \phi\right) \tag{18.6}$$

and similarly for **B**, where $\phi$ is the electrical potential and **F** is the Faraday constant. The requirements of maintaining electroneutrality and no net electric current may be expressed as:

$$\left.\begin{aligned} v_A N_A + v_B N_B &= 0 \\ v_A C_A + v_B C_B &= \text{constant} \end{aligned}\right\} \tag{18.7}$$

Thus:

$$v_A \text{grad } C_A + v_B \text{grad } C_B = 0$$

and equation 18.6 may be written as:

$$N_A = -\left[\frac{D_A D_B(v_A^2 C_A + v_B^2 C_B)}{v_A^2 C_A D_A + v_B^2 C_B D_B}\right]\text{grad } C_A \tag{18.8}$$

The term in the square bracket is an effective diffusion coefficient $D_{AB}$. In principle, this may be used together with a material balance to predict changes in concentration within a pellet. Algebraic solutions are more easily obtained when the effective diffusivity is constant. The conservation of counter-ions diffusing into a sphere may be expressed in terms of resin phase concentration $C_{Sr}$, which is a function of radius and time.

Thus:
$$\frac{\partial C_{Sr}}{\partial t} = \frac{1}{r^2}\frac{\partial}{\partial r}\left(r^2 D_R \frac{\partial C_{Sr}}{\partial r}\right) \tag{18.9}$$

where $D_R$ is the diffusivity referred to concentrations in the resin phase. If this is constant, the equation 18.9 can be rewritten as:

$$\frac{\partial C_{Sr}}{\partial t} = D_R\left(\frac{\partial^2 C_{Sr}}{\partial r^2} + \frac{2}{r}\frac{\partial C_S}{\partial r}\right) \tag{18.10}$$

This equation has been solved by EAGLE and SCOTT[11] for conditions of constant concentration outside the sphere and negligible resistance to mass transfer in the boundary film. The solution may be written in terms of a mean concentration through a sphere $C_s$, which is a function of time only, to give:

$$\frac{C_S - C_{S0}}{C_S^* - C_{S0}} = 1 - \frac{6}{\pi^2}\sum_{n=1}^{\infty}\frac{1}{n^2}\exp[-(D_R\pi^2 t)/n^2 r_i^2] \tag{18.11}$$

where $C_{S0}$ is the initial concentration on the resin and $C_S^*$ is the concentration on the resin in equilibrium with $C_0$, the constant concentration in the solution.

When $t$ is large, the summation may be restricted to one term and the equation becomes:

$$\frac{C_S - C_{S0}}{C_S^* - C_{S0}} = 1 - \frac{6}{\pi^2}\exp[-(D_R\pi^2 t)/r_i^2] \tag{18.12}$$

The corresponding rate equation may be found by taking the derivative with respect to time and rearranging to give:

$$\frac{dC_S}{dt} = \frac{\pi^2 D_R}{r_i^2}(C_S^* - C_S) \tag{18.13}$$

Equation 18.13 is a *linear driving force* equation. VERMEULEN[12] has suggested the following form that more accurately represents experimental data:

$$\frac{dC_S}{dt} = \frac{\kappa D_R}{r_i^2}\frac{(C_S^{*2} - C_S^2)}{2(C_S - C_{S0})} \tag{18.14}$$

This is known as the *quadratic driving force* equation. A plot of $\ln[1 - (C_S/C_S^*)^2]$ against $t$ gives a straight line and the diffusion factor $\kappa D_S/r_i^2$ may be obtained from the slope.

When ion exchange involves ions of different mobilities, the rate depends also on the relative positions of the ions. If the more mobile ion is diffusing out of the resin, the rate will be greater than if it is diffusing into the resin, when pellet-diffusion controls.

## Example 18.1

A single pellet of alumina is exposed to a flow of humid air at a constant temperature. The increase in mass of the pellet is followed automatically, yielding the following results:

| $t$(min) | 2 | 4 | 10 | 20 | 40 | 60 | 120 |
|---|---|---|---|---|---|---|---|
| $x_r$ (kg/kg) | 0.091 | 0.097 | 0.105 | 0.113 | 0.125 | 0.128 | 0.132 |

Assuming the effect of the external film is negligible, predict time $t$ against $x_r$ values for a pellet of twice the radius, where $x_r$ is the mass of adsorbed phase per unit mass of adsorbent.

## Solution

When pellet diffusion is dominant, assuming $C_{s0} = 0$ and the fluid concentration outside the pellet is constant, then:

$$\frac{C_S}{C_S^*} = 1 - \frac{6}{\pi^2}\exp[-(D_R\pi^2 t/r_i^2)] \qquad \text{(equation18.12)}$$

Hence a plot of ln $\left[1 - \frac{C_S}{C_S^*}\right]$ against $t$ should be linear.

For $r = r_i$, with $C_S^* = 0.132$ kg/kg then:

| $t$ (min) | $C_S$ (kg/kg) | $(C_S/C_S^*)$ | $1 - (C_S/C_S^*)$ |
|---|---|---|---|
| 2 | 0.091 | 0.69 | 0.31 |
| 4 | 0.097 | 0.73 | 0.27 |
| 10 | 0.105 | 0.80 | 0.20 |
| 20 | 0.113 | 0.86 | 0.14 |
| 40 | 0.125 | 0.95 | 0.05 |
| 60 | 0.128 | 0.97 | 0.03 |
| 120 | 0.132 | 1.0 | 0 |

These data are plotted in Figure 18.4, which confirms the linearity and from which:

$$\pi^2 D_R/r_i^2 = 0.043$$

For a pellet of twice the radius, that is $r = 2r_i$

$$\text{and the slope} = (-0.043/4) = -0.011$$

Thus, when the radius $= 2r_i$:

$$C_S/C_S^* = 1 - (6/\pi^2)\exp(0.011t) \qquad \text{(i)}$$

Alternatively, use may be made of the quadratic driving force equation:

$$\frac{dC_S}{dt} = \frac{\kappa D_R}{r_i^2}\frac{(C_S^{*2} - C_S^2)}{2(C_S - C_{S0})} \qquad \text{(equation 18.14)}$$

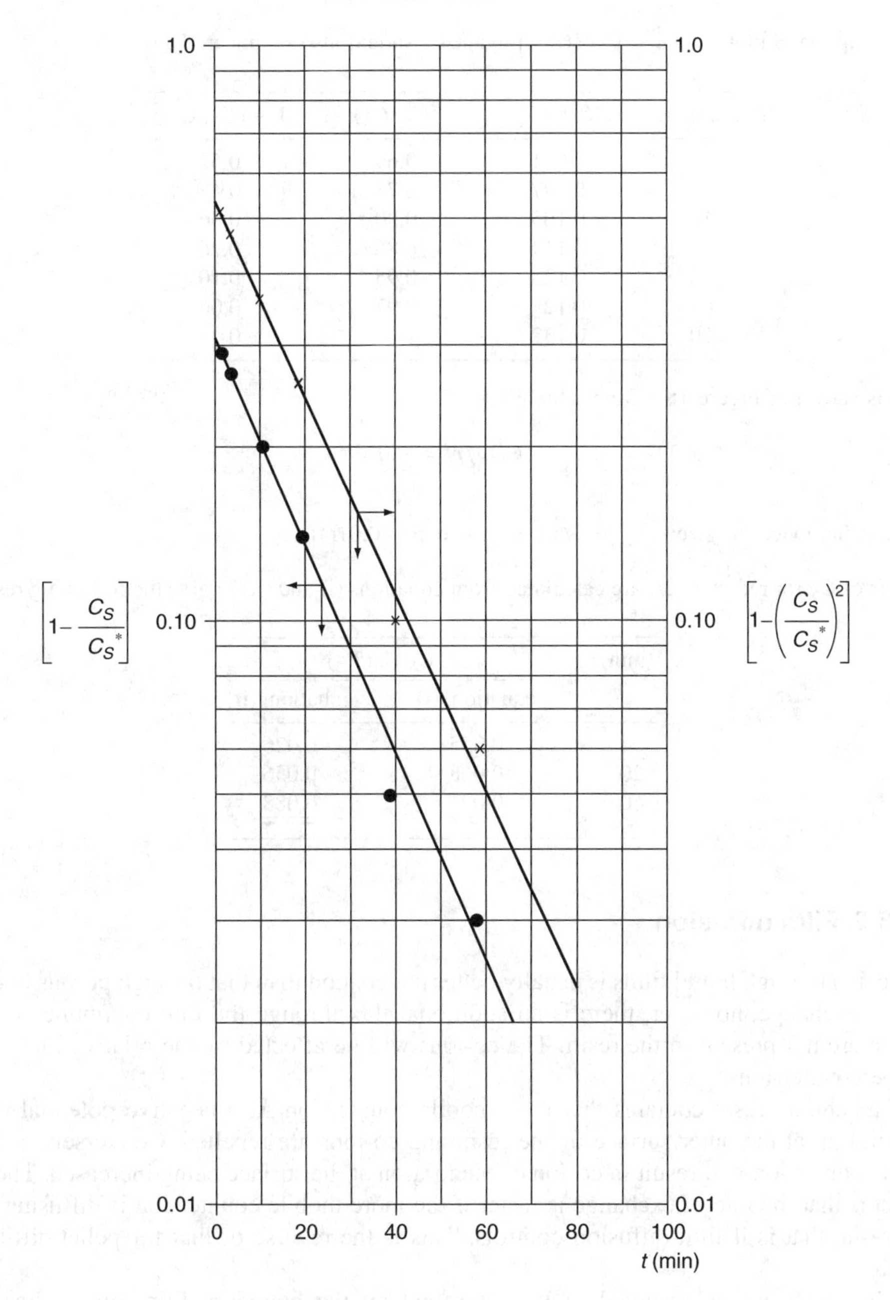


Figure 18.4. Data for Example 18.1

Integrating from the initial condition, $t = 0$ and $C_S = 0$, then:

$$C_S/C_S^* = \left[1 - \exp\left(-\frac{\kappa D_R t}{r_i^2}\right)\right]^{0.5}$$

indicating that a plot of $\ln[1 - (C_S/C_S^*)^2]$ against $t$ should also be linear. Thus:

| $t$(min) | $C_S$ (kg/kg) | $(C_S/C_S^*)$ | $1 - (C_S/C_S^*)^2$ |
|---|---|---|---|
| 2 | 0.091 | 0.69 | 0.52 |
| 4 | 0.097 | 0.73 | 0.47 |
| 10 | 0.105 | 0.80 | 0.36 |
| 20 | 0.113 | 0.86 | 0.26 |
| 40 | 0.125 | 0.95 | 0.10 |
| 60 | 0.128 | 0.97 | 0.06 |
| 120 | 0.132 | 1.0 | 0 |

This is shown in Figure 18.4 from which:

$$\kappa D_R/r_i^2 = 0.04$$

For a pellet twice the size:
$$C_S/C_S^* = [1 - \exp(-0.01t)]^{0.5} \tag{ii}$$

Values of $C_S$ for radius $= 2r_i$ are calculated from equations (i) and (ii) to give the following results:

| $t$(min) | $C_S$ (kg/kg) | |
|---|---|---|
| | equation (i) | equation (ii) |
| 4 | 0.055 | 0.026 |
| 20 | 0.068 | 0.056 |
| 60 | 0.091 | 0.088 |

## 18.5.2. Film diffusion

Diffusion through liquid films is usually better understood than that through porous bodies. In ion exchange, however, there is an additional flux through the film of mobile co-ions which are not present in the resin. The co-ions will be affected by the relative mobilities of the counter-ions.

If a cationic resin contains the more mobile counter-ion **A**, a negative potential tends to build up at the outer surface of the resin and co-ions are repelled. Conversely, a slow resin counter-ion will result in co-ion concentration at the surface being increased. The net effect is that the rate of exchange is faster if the more mobile counter-ion is diffusing into the resin, that is if film-diffusion controls. This is the reverse of that for pellet-diffusion control.

When diffusion is assumed to be controlled by the boundary film, by implication, all other resistances to diffusion are negligible. Therefore, concentrations are uniform through the solid and local equilibrium exists between fluid and solid. The whole of the concentration difference between bulk liquid and solid is confined to the film. The rate of transfer into a spherical pellet may then be expressed as:

$$4\pi r_i^2 k_l (C_{Ab} - C_A^*) \tag{18.15}$$

where $C_{Ab}$ is the concentration of the molecular species **A** in the liquid. It is assumed that the volume of liquid is large compared with the exchange capacity of the resin so that $C_{Ab}$ remains constant. $C_A^*$ is the concentration of **A** in the liquid at the outer surface of the pellet and it is assumed that this is in equilibrium with the mean concentration $C_{SA}$ on the pellet, an assumption which is strictly true only when transfer to the pellet is controlled by the external film-resistance.

The rate which may also be expressed as a rate of increase of that molecular species in the pellet is:

$$\frac{\mathrm{d}}{\mathrm{d}t}\left(\frac{4}{3}\pi r_i^3 C_{SA}\right) \tag{18.16}$$

Hence, from equations 18.15 and 18.16:

$$\frac{\mathrm{d}C_{SA}}{\mathrm{d}t} = \frac{3k_l}{r_i}(C_{Ab} - C_A^*) \tag{18.17}$$

and: $\quad C_{SA} = \mathrm{f}(C_A)$, the sorption isotherm.

If the equilibrium relationship between $C_A$ and $C_{SA}$ is linear so that $C_A^* = (1/b_A)C_{SA}$ and $C_{Ab} = (1/b_A)C_{S\infty}$, then equation 18.17 becomes:

$$\frac{\mathrm{d}C_{SA}}{\mathrm{d}t} = \frac{3k_l}{r_i b_A}(C_{S\infty} - C_{SA}) \tag{18.18}$$

For a solid initially free of **A**, equation 18.18 may be integrated to give:

$$\ln\left[\frac{C_{S\infty} - C_{SA}}{C_{S\infty}}\right] = \frac{-3k_l}{r_i b_A}t \tag{18.19}$$

The situation is more complicated when charged ions rather than uncharged molecules are transferring. In this case, a Nernst–Planck equation which includes terms for both counter-ions and mobile co-ions must be applied. The problem may be simplified by assuming that the counter-ions have equal mobility, when the relationship is:

$$\ln\left[\frac{C_{S\infty} - C_{SA}}{C_{S\infty}}\right] + \left(1 - \frac{1}{\alpha_B^A}\right)\left(\frac{C_{SA}}{C_{S\infty}}\right) = \frac{-3k_l}{r_i b_A}t \tag{18.20}$$

where $\alpha_B^A$ is the separation factor, which equals $b_A/b_B$. Equations 18.19 and 18.20 are identical when $\alpha_B^A = 1$. More complex systems are discussed by CRANK[13].

### 18.5.3. Ion exchange kinetics

In a theory of fixed bed performance for application to ion exchange columns, THOMAS[14] assumed that the rate was controlled by the ion-exchange step itself. A rate equation may be written as:

$$\frac{\mathrm{d}C_S}{\mathrm{d}t} = k\left[C(C_{S\infty} - C_S) - \frac{1}{K_i}C_S(C_0 - C)\right] \tag{18.21}$$

where: $k$ is the forward velocity constant of the exchange,
$C_{S\infty}$ is the total concentration of exchangeable ion in the resin,
$C, C_S$ are fluid and resin concentrations of counter-ion,
$C_0$ is the initial concentration in the fluid, and
$K_i$ is an equilibrium constant.

Although, in practice, ion exchange kinetics are unlikely to limit the rate, the solutions proposed by Thomas may be adapted to represent other controlling mechanisms, as discussed later.

### 18.5.4. Controlling diffusion

Whether film or pellet diffusion is rate-determining may be found experimentally. A pellet is immersed in an ionic solution and the change in concentration of the solution is measured with time. Before the exchange is complete, the pellet is taken out of the solution and held in air for a short period. If, after returning the pellet to the solution, the change in concentration continues smoothly from where it had stopped, then the rate of ion exchange is controlled by film-diffusion. If the resumed rate is higher than when the pellet was removed, the process is pellet-diffusion controlled. The buildup of concentration at the outer edges of the pellet which occurs when diffusion through the pellet is difficult, has been given time to disperse.

Various criteria have been developed to indicate whether film- or pellet-diffusion will be controlling. In one, proposed by Helffrich and Plesset[(15)], the times are compared for a pellet to become half-saturated under the hypothetical conditions of either film-diffusion control, $t_{f(1/2)}$, or pellet-diffusion control, $t_{p(1/2)}$.

From equation 18.20:

$$t_{f(1/2)} = (0.167 + 0.064\,\alpha_B^A)\frac{r_i b_A}{k_l \alpha_B^A} \tag{18.22}$$

From equation 18.11 and limiting the summation to the first order term in $t$, then:

$$t_{p(1/2)} = 0.03\frac{r_i^2}{D_R} \tag{18.23}$$

If $t_{f(1/2)} > t_{p(1/2)}$, then film-diffusion controls, and conversely. In an alternative approach proposed by Rimmer and Bowen[(16)], it was recommended that film-diffusion should be assumed to control until the rate predicted by equation 18.14 is less than that predicted by equation 18.17 when $C_A^* = 0$.

## 18.6. ION EXCHANGE EQUIPMENT

Equipment for ion exchange is selected on the basis of the method to be used for regenerating the spent resin. Regeneration has to be carried out with the minimum disruption of the process and at a minimum cost. At its simplest, equipment may consist of a vessel containing the liquid to be treated, possibly fitted with stirrer to ensure good mixing. Ion exchange beads, a few millimeters in diameter, are added. Counter-ions diffuse from the liquid to the resin against a counterflow of ions diffusing from resin to liquid. Rates are such as to keep both resin and solution electrically neutral.

### 18.6.1. Staged operations

In the simple batch process, conservation of counter-ions leaving the liquid may be written as:

$$V(C_0 - C) = R_v(C_S - C_{S0}) \tag{18.24}$$

where $V$ and $R_v$ refer to initial volumes of liquid and resin.

Hence:
$$C_S = \frac{-V}{R_v}C + \left(\frac{V}{R_v}C_0 + C_{S0}\right) \tag{18.25}$$

If the batch process behaves as an equilibrium stage, the phases in contact will achieve equilibrium.

If the equilibrium relationship is known, then:

$$C_S^* = \mathrm{f}(C) \tag{18.26}$$

and equations 18.25 and 18.26 may be solved. In Figure 18.5 it is assumed that $V$ and $R_v$ remain constant. It is sometimes convenient to use the fractional concentrations:

$$y = \frac{C_S}{C_{S\infty}} \quad \text{and} \quad x = \frac{C}{C_0} \tag{18.27}$$

where $C_{S\infty}$ is the maximum concentration of counter-ions on the resin.

Thus:
$$y = -D^*x + D^* + y_0 \tag{18.28}$$

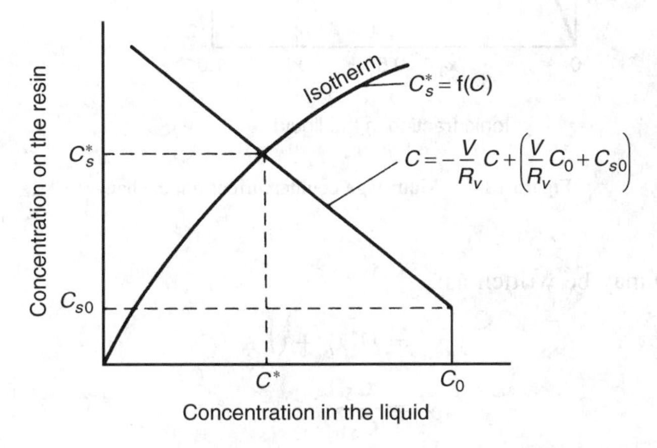


Figure 18.5. Graphical solution for a single batch stage

where a distribution coefficient $D^*$ is defined as:

$$D^* = VC_0/R_v C_{S\infty} \tag{18.29}$$

A plot of $y$ against $x$ gives a straight line of slope $-D^*$, passing through the point $(1, y_0)$. The intercept of the line with the equilibrium curve $y^* = f(x)$ gives the equilibrium condition that will be achieved in a single stage of mixing, starting from concentrations $(1, y_0)$, using volumes $V$ and $R_v$ of liquid and resin respectively.

If liquid and resin flow through a series of equilibrium stages at constant rates $\dot{V}$ and $\dot{R}_v$ respectively, as shown in Figure 18.6, a mass balance over the first $n$ stages gives:

$$\dot{V}C_0 + \dot{R}_v C_{S(n+1)} = \dot{R}_v C_{S1} + \dot{V}C_n \tag{18.30}$$

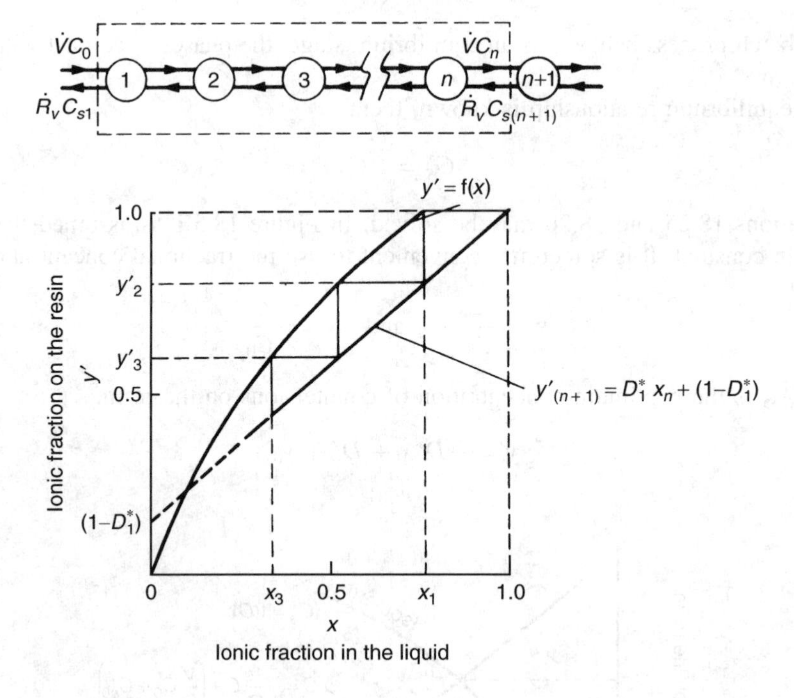


Figure 18.6. Multistage countercurrent ion-exchange

Equation 18.30 may be written as:

$$y'_{n+1} = D_1^* x_n + (1 - D_1^*) \tag{18.31}$$

where:

$$y' = \frac{C_S}{C_{S1}}$$

$$D_1^* = \dot{V}C_0/\dot{R}_v C_{S1}$$

If the ion exchange equilibrium isotherm is written in the form:

$$y'^* = f(x) \tag{18.32}$$

the total exchange obtained from a series of countercurrent equilibrium stages may be found by stepping off between the operating line, equation 18.31, and the equilibrium line, equation 18.32, as shown in Figure 18.6.

## 18.6.2. Fixed beds

Most ion exchange operations are carried out in fixed beds of resin contained in vertical cylindrical columns. The resin is supported on a grid fine enough to retain the pellets of resin, but sufficiently open so as not to hinder liquid flow. Sizes range from laboratory scale to industrial units 1–3 m in diameter and height. Liquid is fed to the top of the column through a distributor carefully designed and fitted to ensure an even flow over the whole cross-section of the bed. A mass transfer zone develops at the inlet to the bed in which the ion exchange takes place. As more feed is added, the zone travels through the bed and operation continues until unconverted material is about to emerge with the effluent. The bed has now reached the limit of its working capacity, its breakpoint, so the run is stopped and the resin regenerated.

Regenerating liquid is normally arranged to flow countercurrently to the feed direction thus ensuring that the end of the bed, which controls the effluent condition, is the most thoroughly regenerated. Fine particles from impurities entering with the feed and from attrition of resin pellets may accumulate in the bed. Such fines have to be removed from time to time by backwashing, so that the ion-exchange capacity is not reduced.

Backwashing may also be used for re-arranging the components of a mixed-resin bed. To demineralise water, for example, it is convenient to use a bed containing a random mixture of cationic and anionic resins. When either becomes exhausted, the bed is taken off-line and a back flow of untreated water is used to separate the resins into two layers according to their densities. Figure 18.7 shows such a process in which an anion layer, of density 1100 kg/m$^3$, rests on the cation layer, of density 1400 kg/m$^3$. The former is regenerated using a 5 per cent solution of caustic soda. After rinsing to remove residual caustic soda, 5 per cent hydrochloric acid is introduced above the cation layer for its regeneration. Finally, the resins are remixed using a flow of air and the bed is ready to be used again[1].

The design of fixed-bed ion exchangers shares a common theory with fixed-bed adsorbers, which are discussed in Chapter 17. In addition, THOMAS[14] has developed a theory of fixed-bed ion exchange based on equation 18.21. It assumed that diffusional resistances are negligible. Though this is now known to be unlikely, the general form of the solutions proposed by Thomas may be used for film- and pellet-diffusion control.

A material balance of counter-ions across an increment of bed may be written as:

$$u\frac{\partial C}{\partial z} + \frac{\partial C}{\partial t} = -\frac{1}{m}\frac{\partial C_S}{\partial t} \tag{18.33}$$

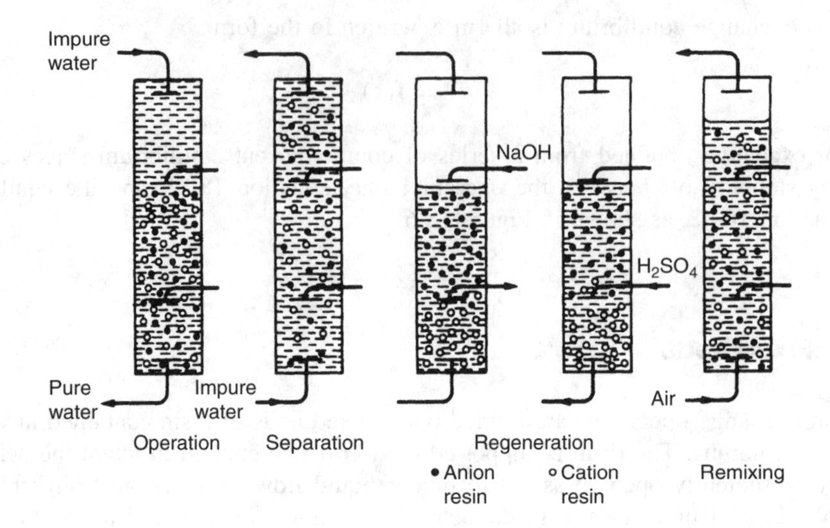


Figure 18.7. A mixed demineralising bed[1]

Defining distance and time variables, then:

$$\chi = \frac{kC_{S\infty}z}{mu}; \tau = kC_0\left(t - \frac{z}{u}\right) \tag{18.34}$$

Equations 18.33 and 18.21 may be written as:

$$\frac{\partial(C/C_0)}{\partial\chi} = -\frac{\partial(C_S/C_{S\infty})}{\partial\tau} \tag{18.35}$$

$$\frac{\partial(C_S/C_{S\infty})}{\partial\tau} = \frac{C}{C_0}\left(1 - \frac{C_S}{C_{S\infty}}\right) - \frac{1}{K_i}\frac{C_S}{C_{S\infty}}\left(1 - \frac{C}{C_0}\right) \tag{18.36}$$

The solutions given by Thomas are expressed as complex functions of $\chi$, $\tau$, and $K_i$. For design purposes, these are more conveniently presented in graphical form. VERMEULEN[12] and HESTER and VERMEULEN[17] have extended their use to include diffusion control.

When film-diffusion controls, the kinetics step, given by equation 18.36, is essentially at equilibrium. Hence:

$$\frac{C_i^*}{C_0}\left(1 - \frac{C_S}{C_{S\infty}}\right) = \frac{1}{K_i}\frac{C_S}{C_{S\infty}}\left(1 - \frac{C_i^*}{C_0}\right) \tag{18.37}$$

and:

$$C_i^* = \frac{C_S}{C_{S\infty}}\frac{r^*C_0}{[1 + (r^* - 1)C_S/C_{S\infty}]} \tag{18.38}$$

where $r^* = 1/K_i$, is an equilibrium parameter.

The rate-controlling step may be written as:

$$\frac{\partial C_S}{\partial t} = \frac{k_l a_z}{1 - \varepsilon}(C - C_i^*) \tag{18.39}$$

where $C_i^*$ is the fluid concentration at the exterior surface of the resin. This is assumed to be in equilibrium with the resin and uniform throughout the pellet.

Substituting in equation 18.37, then:

$$\frac{\partial(C_S/C_{S\infty})}{\partial\tau} = \frac{C}{C_S}\left(1 - \frac{C_S}{C_{S\infty}}\right) - r^*\left(1 - \frac{C}{C_S}\right)\frac{C_S}{C_{S\infty}} \tag{18.40}$$

where:

$$\tau = \frac{k_l a_z C_0}{(1-\varepsilon)[1 + (r^* - 1)\bar{C}_S/C_{S\infty}]C_{S\infty}}\left(t - \frac{z}{u}\right) \tag{18.41}$$

If a mean value $\bar{C}_S/C_{S\infty}$ is taken for $C_S/C_{S\infty}$ in the definition of $\tau$, equation 18.41 is a time parameter similar in form to equation 18.34.

Therefore the solutions found for the kinetics-controlling-condition may be used with the new time parameter for the case of film-diffusion control.

When *pellet-diffusion* controls the exchange rate, the rate is often expressed in terms of a hypothetical solid-film coefficient $k_p$ and a contrived driving force $(C_S^* - C_S)$ where $C_S^*$ is the concentration of the resin phase in equilibrium with $C$, and $C_S$ is the concentration of resin phase, averaged over the pellet.

Thus:

$$\frac{\partial C_S}{\partial t} = \frac{k_p a_z}{(1-\varepsilon)}(C_S^* - C_S) \tag{18.42}$$

The ion exchange step is at equilibrium, and hence:

$$C_S^* = \frac{C_{S\infty}}{(1 - r^*) + r^*(C_0/C)} \tag{18.43}$$

Substituting for $C_S^*$ in equation 18.42 gives:

$$\frac{\partial(C_S/C_{S\infty})}{\partial t} = \frac{k_p a_z}{[\bar{C}/C_0(1 - r^*) + r^*](1-\varepsilon)}\left[\frac{C}{C_0}\left(1 - \frac{C_S}{C_{S\infty}}\right) - r^*\frac{C_S}{C_{S\infty}}\left(1 - \frac{C}{C_0}\right)\right] \tag{18.44}$$

For a mean value of $\bar{C}/C_0$ outside the bracket, or several mean values for different concentration ranges, equation 18.44 has the same form as equation 18.36 with a new time parameter given by:

$$\tau = \frac{k_p a_z}{[(\bar{C}/C_0)(1 - r^*) + r^*](1-\varepsilon)}\left(t - \frac{Z}{u}\right) \tag{18.45}$$

The application to the design of fixed beds using graphed versions of the solution given by Thomas is discussed by HEISTER and VERMEULIN[17]. Other solutions for fixed beds, including those that apply to equilibrium operation, are discussed in Chapter 17.

### 18.6.3. Moving beds

In principle, all the moving bed devices discussed in Chapter 17 may be used for ion exchange. Ion exchange is largely a liquid-phase phenomenon in which the solids may be made to flow relatively easily when immersed in liquid. A method of moving resin discontinuously has been developed by Higgins, as shown in Figure 18.8, and described in KIRK-OTHMER[18] and by SETTER *et al.*[19]. For a short period in a cycle time, ranging from a few minutes to several hours, the resin is moved by pulses of liquid generated

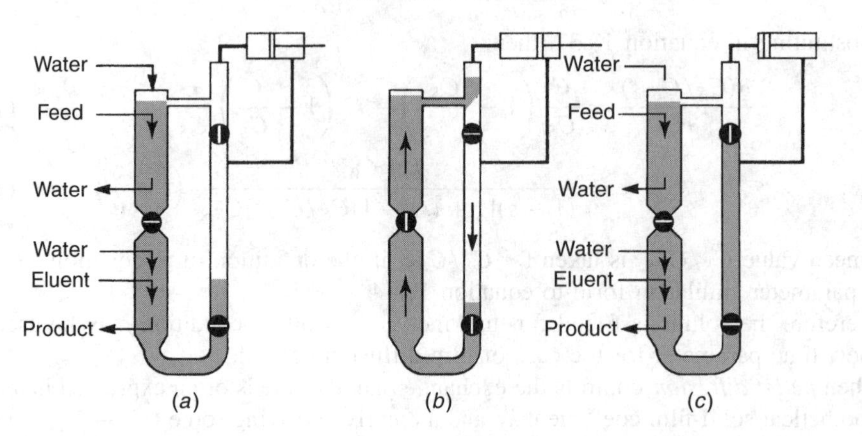


Figure 18.8. Principles of the Higgins contactor (*a*) Solution pumping (several minutes) (*b*) Resin movement (3–5 seconds) (*c*) Solution pumping (several minutes)

by a double-acting piston which simultaneously sucks liquid from the top of the column and delivers it to the bottom. The operational problems encountered with this equipment are mechanical, resulting from wear and tear on the valves and attrition of the resin. Replacement of the resin can be as high as 30 per cent each year when the equipment is used for water-softening. Other commercial applications include recovery of phosphoric acid from pickle liquor and the recovery of ammonium nitrate.

Fluidised beds are used commercially for ion exchange. These generally consist of a compartmented column with fluidisation in each compartment. The solid is moved periodically downwards from stage to stage and leaves at the bottom from which it passes to a separate column for regeneration. An arrangement of the Himsley-type is shown in Figure 18.9[18].

## Example 18.2

An acid solution containing 2 per cent by mass of $NaNO_3$ and an unknown concentration of $HNO_3$ is used to regenerate a strong acid resin. After sufficient acid had been passed over the resin for equilibrium to be attained, analysis showed that 10 per cent of resin sites were occupied by sodium ions. What was the concentration of $HNO_3$ in the solution, if its density were 1030 kg/m$^3$.

## Solution

$NaNO_3$ (Molecular weight = 85 kg/kmol)

$$\text{Concentration} = 2 \text{ per cent by mass}$$
$$= (20/85)(103/1000) = 0.242 \text{ kg/m}^3$$

$HNO_3$ (Molecular weight = 63 kg/kmol)

$$\text{Concentration} = p \text{ per cent}$$
$$\text{Concentration} = (10p/63)(1030/1000)$$
$$= 0.163p \text{ kg/m}^3$$

In the solution:
$$x_{Na^+} = 0.242/(0.242 + 0.163p)$$

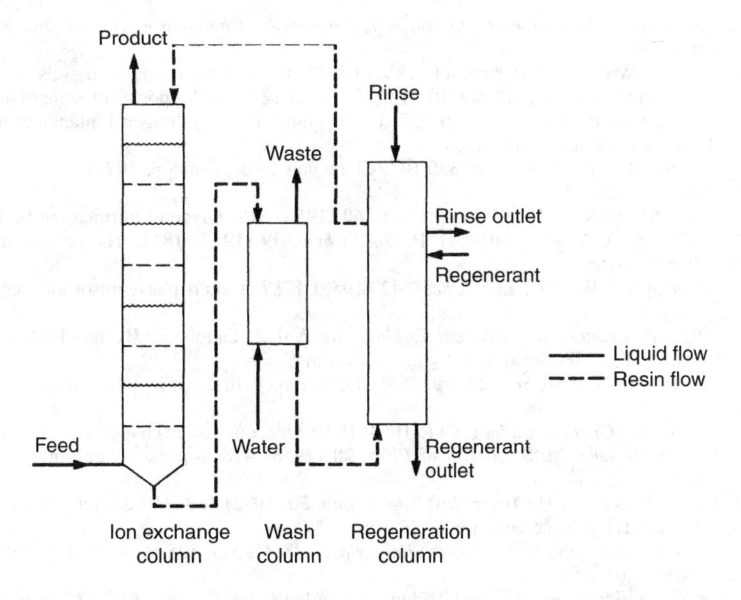


Figure 18.9. A staged fluidised-bed ion-exchange column of the Himsley type[18]

For univalent ion exchange, equation 18.3 becomes:

$$y_{Na^+}/(1 - y_{Na^+}) = K_{Na^+H^+}[x_{Na^+}/(1 - x_{Na^+})]$$

But: $y_{Na^+} = 0.1$

and from Table 18.1: $K_{Na^+H^+} = \left(\dfrac{2.0}{1.3}\right) = 1.5$

Thus: $(0.1/0.9) = 1.5[0.242/(0.242 + 0.163p)]/[0.163p/(0.242 + 0.163p)]$

and $p = \underline{\underline{20 \text{ per cent}}}$

## 18.7. FURTHER READING

DORFNER K. (ed.): *Ion Exchangers* (Walter De Gruyter, 1991).
KIRK-OTHMER: *Encyclopedia of Chemical Technology* Volume 13. *Ion Exchange* (Wiley, New York, 1981).
LIBIRTI, L. and MILLAR, J. R. (ed.) *Fundamentals and applications of Ion Exchange (NATO Asi Series' Series E Applied Sciences No. 98)* 1985.
MORINSKI, J. A. and MARCUS, V. (eds.): *Ion Exchange and Solvent Extraction (A series of Advances)*. (Marcel Dekker, New York, 2001).
SCHWEITZER, P. A. (ed.): *Handbook of Separation Techniques for Chemical Engineers*, 2nd edn (McGraw-Hill, New York, 1988).
SENGUPTA, A. K. (ed.): *Ion Exchange Technology: Advances in Environmental Pollution Control* Technomic Publishing Co, Lancaster, Pennsylvania, 1995.
SLATER M. J. (ed.): *Ion Exchange Advances* SCI conference IEX 92-Ion 1995.

## 18.8. REFERENCES

1. ARDEN, T. V.: *Water Purification by Ion Exchange* (Butterworth, London 1968).
2. HELFFRICH, F.: *Ion Exchange* (McGraw-Hill, New York 1962).

3. SCHWEITZER, P. A. (ed): *Handbook of Separation Techniques for Chemical Engineers*, 2nd edn. (McGraw-Hill, New York 1988).
4. THOMPSON, H. S.: *J. Roy. Agr. Soc. Eng.* **11** (1850) 68. On the absorbent power of soils.
5. WAY, J. T.: *J. Roy. Agr. Soc. Eng.* **11** (1850) 313, **13** (1852) 123. On the power of soils to absorb manure.
6. LEMBERG, J.: *Z. deut. geol. Ges.* **22** (1870) 355. Ueber einige Umwandlungen Finländischer Feldspat, **28** (1876) 519. Ueber Siliciumumwandlungen.
7. ARISTOTLE: *Works* vol. 7 p. 933b, about 330 BC (Clarendon Press, London, 1977).
8. MOSES: *Exodus* 15 vv 23–25.
9. ASHER, D. R. and SIMPSON, D. W.: *J. Phys. Chem.* **60** (1956) 518. Glycerol purification by ion exclusion.
10. HATCH, M. J., DILLON, J. A. and SMITH, H. B.: *Ind. Chem.* **49** (1957) 1812. The preparation and use of snake cage polyelectrolytes.
11. EAGLE, S. and SCOTT, J. W.: *Ind. Eng. Chem.* **42** (1950) 1287. Liquid phase adsorption equilibrium and kinetics.
12. VERMEULEN, T.: In *Advances in Chemical Engineering* Vol 2, DREW, T. B. and HOOPES, J. W. (eds.) (Academic Press, 1958) 148. Separation by adsorption methods.
13. CRANK, J.: *Discussions Faraday Soc.* **23** (1957) 99. Diffusion coefficients in solids, their measurement and significance.
14. THOMAS, H. C.: *J. Am. Chem. Soc.* **66** (1944) 1664. Heterogeneous ion exchange in a flowing system.
15. HELFFRICH, F. and PLESSET, M. S.: *J. Chem. Phys.* **28** (1958) 418. Ion exchange kinetics. A non-linear diffusion problem.
16. RIMMER, P. G. and BOWEN, J. H.: *Trans. Inst. Chem. Eng.* **50** (1972) 168. The design of fixed bed adsorbers using the quadratic driving force equation.
17. HEISTER, N. K. and VERMEULEN, T.: *Chem. Eng. Prog.* **48** (1952) 505. Saturated performance of ion exchange and adsorption columns.
18. KIRK-OTHMER: *Encyclopedia of Chemical Technology*, Vol. 13, *Ion Exchange* (Wiley, New York 1981).
19. SETTER, N. J., GOOGIN, J. M. and MARROW, G. B.: USAEC Report Y-1257 (9th July 1959). The recovery of uranium from reduction residues by semi-continuous ion exchange.

## 18.9. NOMENCLATURE

| | | Units in SI System | Dimensions in **M, N, L, T, $\theta$, A** |
|---|---|---|---|
| $a_z$ | External surface area of resin per unit volume of bed | $m^{-1}$ | $\mathbf{L}^{-1}$ |
| $b_A$ | Slope of a linear sorption isotherm in equation 18.18 | — | — |
| $C$ | Concentration of counter-ions in the liquid phase | kmol/m$^3$ | $\mathbf{NL}^{-3}$ |
| $\bar{C}$ | Mean concentration of counter-ions in the liquid phase | kmol/m$^3$ | $\mathbf{NL}^{-3}$ |
| $C_{Ab}$ | Concentration of counter-ions **A** in the bulk liquid | kmol/m$^3$ | $\mathbf{NL}^{-3}$ |
| $C_A^*$ | Concentration of counter-ions **A** in equilibrium with the mean concentration of counter-ions in the resin | kmol/m$^3$ | $\mathbf{NL}^{-3}$ |
| $C_i^*$ | Concentration of counter-ions in the liquid phase, in equilibrium at the external surface of the resin | kmol/m$^3$ | $\mathbf{NL}^{-3}$ |
| $C_n$ | Concentration of counter-ions in the liquid leaving the $n$th stage | kmol/m$^3$ | $\mathbf{NL}^{-3}$ |
| $C_0$ | Concentration of counter-ions in the liquid initially, or in the feed | kmol/m$^3$ | $\mathbf{NL}^{-3}$ |
| $C_S$ | Concentration of counter-ions in the resin phase | kmol/m$^3$ | $\mathbf{NL}^{-3}$ |
| $C_S^*$ | Concentration of counter-ions in the resin phase in equilibrium with the bulk liquid | kmol/m$^3$ | $\mathbf{NL}^{-3}$ |
| $\bar{C}_S$ | Mean concentration of counter-ions in the resin | kmol/m$^3$ | $\mathbf{NL}^{-3}$ |
| $C_{S1}, C_{S(n+1)}$ | Concentrations of counter-ions in resin streams leaving 1st, $(n+1)$th stage | kmol/m$^3$ | $\mathbf{NL}^{-3}$ |
| $C_{S0}$ | Concentration of counter-ions in the resin initially | kmol/m$^3$ | $\mathbf{NL}^{-3}$ |

| | | Units in SI System | Dimensions in **M, N, L, T, θ, A** |
|---|---|---|---|
| $C_{S\infty}$ | Ultimate concentration of counter-ions in the resin | kmol/m$^3$ | $\mathbf{NL^{-3}}$ |
| $D_A$ | Diffusivity of species **A** | m$^2$/s | $\mathbf{L^2\ T^{-1}}$ |
| $D_R$ | Diffusivity in the resin | m$^2$/s | $\mathbf{L^2\ T^{-1}}$ |
| $D^*$ | Distribution coefficient $VC_0/R_vC_{S\infty}$ | — | — |
| $D_1^*$ | Distribution coefficient $\dot{V}C_0/\dot{R}_vC_{S1}$ | — | — |
| **F** | Faraday constant | 9.6487× 10$^7$ C/kmol | $\mathbf{N^{-1}\ TA}$ |
| $f()$ | Various functions | — | — |
| $K$ | Ion exchange equilibrium constant | — | — |
| $K_c$ | Selectivity coefficient | — | — |
| $K_i$ | Equilibrium constant in equation 18.36 | — | — |
| $K_{Ca^+Na^+}$ | Selectivity between calcium ions and sodium ions in a cationic resin | — | — |
| $k$ | Velocity constant in equation 18.21 | m$^3$/kmol s | $\mathbf{N^{-1}\ L^3\ T^{-1}}$ |
| $k_l$ | Liquid-film mass transfer coefficient | m/s | $\mathbf{LT^{-1}}$ |
| $k_p$ | Hypothetical solid ‘film’ mass transfer coefficient | m/s | $\mathbf{LT^{-1}}$ |
| $m$ | $\varepsilon/(1-\varepsilon)$ | — | — |
| $N_A, N_B$ | Molar fluxes of **A**, **B** | kmol/m$^2$s | $\mathbf{NL^{-2}\ T^{-1}}$ |
| $n$ | An index (equation 18.11) *or* number of stages (Figure 18.5) | — | — |
| **R** | Gas constant | 8314 J/kmol K | $\mathbf{MN^{-1}\ L^2\ T^{-2}\theta^{-1}}$ |
| $R_c^-$ | Cationic resin | — | — |
| $R_v$ | Volume of resin | m$^3$ | $\mathbf{L^3}$ |
| $\dot{R}_v$ | Volumetric flowrate of resin | m$^3$/s | $\mathbf{L^3\ T^{-1}}$ |
| $r$ | Radius within a spherical pellet | m | **L** |
| $r_i$ | Outside radius of a spherical pellet | m | **L** |
| $r^*$ | Equilibrium parameter | — | — |
| $T$ | Temperature | K | $\boldsymbol{\theta}$ |
| $t$ | Time | s | **T** |
| $t_{f1/2}$ | Time for half saturation assuming film diffusion control | s | **T** |
| $t_{p1/2}$ | Time for half saturation assuming pellet diffusion control | s | **T** |
| $V$ | Volume of liquid | m$^3$ | $\mathbf{L^3}$ |
| $\dot{V}$ | Volume flowrate of liquid | m$^3$/s | $\mathbf{L^3\ T^{-1}}$ |
| $x$ | Ionic fraction in the liquid, $C/C_0$ | — | — |
| $y$ | Ionic fraction in the resin, $C_S/C_{S\infty}$ | — | — |
| $y'$ | Ionic ratio in the resin, $C_S/C_{S1}$ | — | — |
| $Z$ | Distance along a fixed bed | m | **L** |
| $\alpha_B^A$ | Separation factor of **A** relative to **B** | — | — |
| $\gamma, \gamma_s$ | Activity coefficient for liquid, resin | — | — |
| $\varepsilon$ | Interpellet voidage | — | — |
| $\nu$ | Valence | — | — |
| $\phi$ | Electric potential | V | $\mathbf{ML^2\ T^{-3}\ A^{-1}}$ |
| $\kappa$ | Constant in equation 18.14 | — | — |
| $\chi$ | Distance parameter in equation 18.34 | — | — |
| $\tau$ | Time parameter in equation 18.34 | — | — |

CHAPTER 19

# Chromatographic Separations

## 19.1. INTRODUCTION

Chromatographic methods of separation are distinguished by their high *selectivity*, that is their ability to separate components of closely similar physical and chemical properties. Many mixtures which are difficult to separate by other methods may be separated by chromatography. The range of materials which can be processed covers the entire spectrum of molecular weights, from hydrogen to proteins.

Chromatography plays several roles in the process industries. The design of a new plant begins with a proposed chemical route from starting materials to finished product, already tested at least on a small laboratory scale. In many cases, *analytical chromatography* will have been used in the laboratory to separate and identify the products in the mixtures produced by the proposed chemical process. In the biotechnology area, for example, the original starting material may be a natural product or a complex synthetic mixture which has been analysed by chromatography. Information on the choice of chromatographic stationary and mobile phases and operating conditions is therefore often already available and provides a basis for initial scale–up to an intermediate laboratory scale, *preparative chromatography*, and for the subsequent design of the commercial separation process, *production* or *large-scale chromatography*, as shown in Figure 19.1.

Chromatography also plays two other roles in process engineering. First, the detailed design of each unit operation in a process requires a knowledge of a variety of physical and chemical properties of the materials involved. Chromatographic techniques can provide rapid and accurate methods of measuring a great variety of thermodynamic, kinetic and other physico-chemical properties[1]. Secondly, chromatography is widely used for routine chemical analysis in quality control and for automated analysis of process streams in process control (*process chromatography*)[2,3]. This chapter, however, is concerned mainly with production chromatography as a unit operation whose use is increasing as the demand for high purity materials grows.

In chromatography the components of a mixture are separated as they pass through a column. The column contains a *stationary phase* which may be a packed bed of solid particles or a liquid with which the packing is impregnated. The mixture is carried through the column dissolved in a gas or liquid stream known as the *mobile phase, eluent* or *carrier*. Separation occurs because the differing distribution coefficients of the components of the mixture between the stationary and mobile phases result in differing velocities of travel.

Chromatographic methods are classified according to the nature of the mobile and stationary phases used. The terms *gas chromatography* (GC) and *liquid*

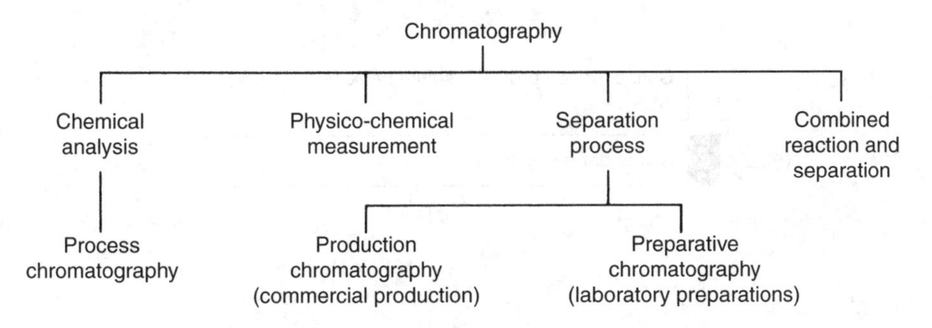


Figure 19.1. Uses of chromatography

*chromatography* (LC) refer to the nature of the mobile phase. The different types of stationary phase are described in Section 19.4.

Both GC and LC may be operated in one of several modes. The principal modes currently used for large-scale separations are elution, selective adsorption or desorption, and simulated countercurrent chromatography. In addition, reaction and separation can be combined in a single column with unique advantages. Elution is the most used and best developed form of the technique and is described first.

## 19.2. ELUTION CHROMATOGRAPHY

### 19.2.1. Principles

For convenience, the term *solute* is used to refer to a component of the feed mixture to be separated, regardless of the nature of the mobile and stationary phases: when the stationary phase is a solid surface, the "solute" might be better described as an adsorbate.

In elution chromatography, a discrete quantity (batch or "sample") of feed mixture is introduced into the column at the inlet. The mobile-phase flow causes the band of feed to migrate and split progressively into its component solute-bands or peaks, as shown in Figure 19.2. Emergence of the bands at the column outlet is monitored by a suitable detector, and the components are collected in sequence. The velocity at which each band travels through the column is normally less than that of the mobile phase and depends on the distribution coefficient of solute between the two phases. A solute 1 will travel faster than a solute 2, as shown in Figure 19.2 if it has a lower affinity for the stationary phase, or a higher affinity for the mobile phase, than solute 2. The principle is one of differential migration. This is the basis of the use of chromatography for (i) measurement of distribution coefficients from migration velocities and (ii) chemical analysis and separation of mixtures.

Continuous production is achieved in elution chromatography by repetitive batch injection or *cyclic batch elution*. The injection cycle is timed so that the emergence of the last-eluted component of one batch at the column exit is immediately followed by the emergence of the first-eluted component to be collected from the next batch as shown, for example, in Figure 19.2*b*. At any one time there is more than one batch moving through the column.

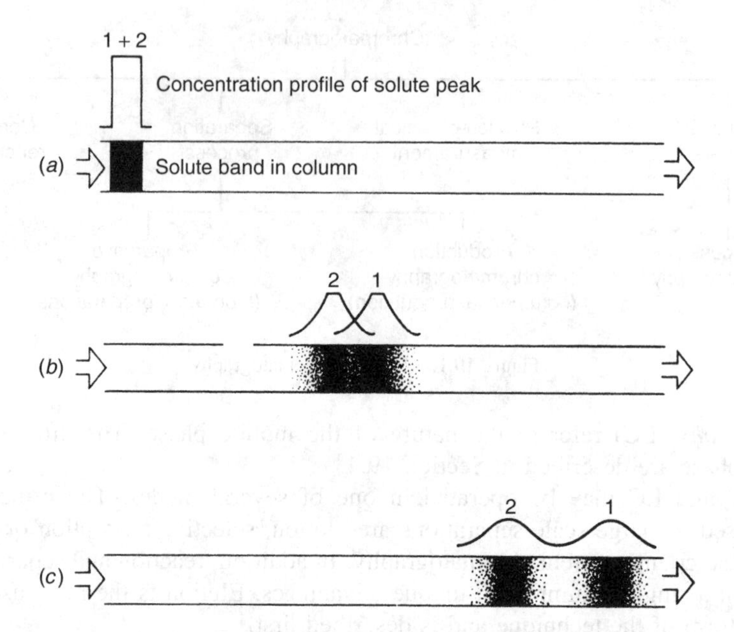


Figure 19.2. Separation of a mixture of two solute components 1 and 2 by elution chromatography. (*a*) Shows the solute band immediately after entry of the band into the column. (*b*) and (*c*) show the band separating into two component bands as it passes through the column

## 19.2.2. Retention theory

The theory of retention permits the calculation of the time the solute is *retained* in the column between injection and elution. The *retention time* $t_R$ is defined, as shown in Figure 19.3, in terms of a solute concentration–time plot, knwon as the *chromatogram* registered by a detector at the column outlet. A part of this time $t_M$ is required by the solute simply to pass through the mobile phase from inlet to outlet. The *adjusted retention time* $t'_R$ represents the extra retention due to repeated partitioning or distribution of the solute between mobile and stationary phases as the band migrates along the column. An analogy for the principle of retention by repeated partitioning has been well put by BAILEY and OLLIS[4]: "Imagine a number of coachmen who start together down a road lined with pubs. Obviously those with the greatest thirst will complete the journey last, while those with no taste for ale will progress rapidly!" This analogy provides a good basis for a simple physical argument which leads to an equation for the retention time of a solute.

The average molecule of a given solute moves repeatedly in and out of the stationary phase during its passage through the column. The molecule spends some of its time being swept along by the mobile phase and some of its time "sitting in a pub", that is in the stationary phase. If $R$ is the ratio of the total time spent by this average molecule in the mobile phase to the total time spent in the mobile and stationary phases, then:

$$R = \frac{t_M}{t_R} \tag{19.1}$$

$$= \frac{u_R}{u} \tag{19.2}$$

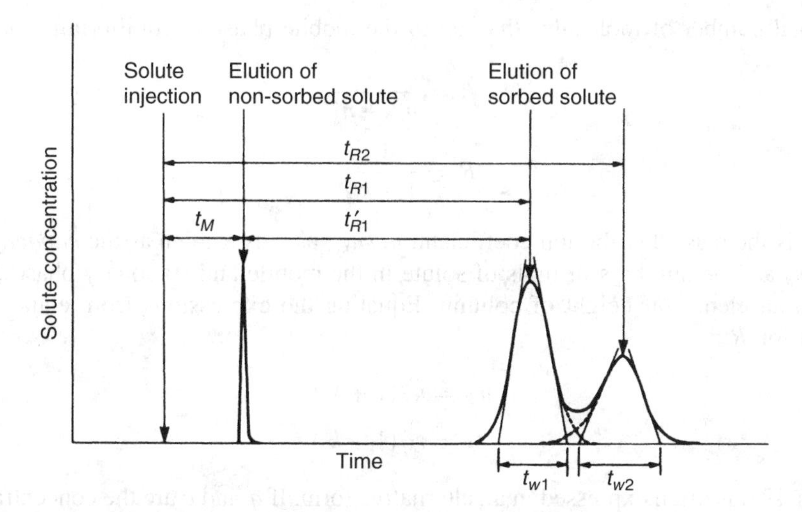


Figure 19.3. Chromatogram obtained by elution chromatography of a mixture of two solutes. The retention time $t_R$ is the time taken by a solute to pass through the column. $t_M$ is the mobile-phase holdup and is measured as the retention time of a non-sorbed solute. $t'_R$ is the adjusted retention time, the total time spent by the solute in the stationary phase; it is equal to $t_R - t_M$. $t_W$ is the width of a solute band at the baseline, i.e., the distance between the points of intersection of the baseline with tangents at the points of inflexion on the sides of the band

Here $u_R$ is the velocity at which the solute band moves along the column and $u$ is the velocity of the mobile phase; that is, $u =$ (superficial velocity)/$\varepsilon$, where superficial velocity is volumetric flow rate divided by cross-sectional area of column and $\varepsilon$ is the fractional volume of column occupied by mobile phase. Most column packings are porous, in which case $\varepsilon$ includes both interstitial and pore (intraparticle) voidage, as defined in the note to Table 19.1, and here $u$ is less than the interstitial velocity.

Table 19.1. Contributions to plate height

| Term in equation 19.10 | Mechanism |
|---|---|
| $A = 2\lambda d_p$ | Inequalities in patterns of flow in column packing. |
| $B/u = 2\gamma D_m/u$ | Axial diffusion in mobile phase. |
| $C_s u = \dfrac{2}{3}\dfrac{k' d_f^2}{(1+k')D_s}u$ | Lack of equilibrium between solute in the two phases due to slow mass transfer in stationary phase film. |
| $C_m u = \dfrac{\omega d_p^2}{D_m}u$ | Slow mass transfer of solute in mobile phase during passage of solute to and from interface with stationary phase (includes diffusion through stagnant mobile phase in intra-particle* pores). |

*Note. *Intra*-particle pores are pores *within* the particle. Inter-particle (interstitial) voidage refers to space *between* the particles, i.e. in the interstices of the packing.

The crux of the argument is that the average molecule is representative of a large number of identical molecules of the solute. Hence $R$, defined previously as the fractional time the average molecule spends in the mobile phase, may also be viewed as the fraction of the total number of molecules that are in the mobile phase at equilibrium; that is:

$$R = \frac{n_m}{n_m + n_s} \tag{19.3}$$

or:

$$R = \frac{1}{1 + k'} \tag{19.4}$$

where $k'$ is the mass distribution coefficient, $n_s/n_m$, usually known as the *capacity factor*. $n_m$ and $n_s$ are the numbers of mols of solute in the mobile and stationary phases, respectively, in an elemental height of column. Equating the expressions from equations 19.2 and 19.4 for $R$:

$$u_R = u/(1 + k') \tag{19.5}$$

or:

$$t_R = t_M(1 + k') \tag{19.6}$$

Equation 19.6 is often expressed in an alternative form. If $q$ and $c$ are the concentrations of solute in the stationary and mobile phases, respectively, and $K(= q/c)$ is the distribution coefficient of solute between the two phases, then:

$$\frac{n_s}{n_m} = \frac{1 - \varepsilon}{\varepsilon} K \tag{19.7}$$

and:

$$t_R = t_M\left(1 + \frac{1 - \varepsilon}{\varepsilon} K\right) \tag{19.8}$$

This equation allows the retention time (or solute band velocity) to be calculated from the equilibrium distribution coefficient $K$, and vice versa.

Equation 19.8 is the basic retention equation of elution chromatography. It is founded on the assumptions that the distribution isotherm, the plot of $q$ against $c$ at constant temperature, is linear and that equilibrium of the solute between phases is achieved instantaneously throughout the column.

More rigorous treatments of retention theory start from conservation of mass over a differential length of column[5,6]. They require the same assumptions as the simpler analysis presented here and lead to the same result, equation 19.8.

## 19.3. BAND BROADENING AND SEPARATION EFFICIENCY

The equations in the previous section relate only to the average molecule and so describe only the mean retention of a band. The mean retention is that of the peak of the band, as shown in Figure 19.3 if the band is symmetrical. In practice there is a spread of time about this mean due to several processes which tend to broaden the bands as they migrate through the column, as shown in the sequence (*a*)–(*c*) of Figure 19.2.

The band broadening may be characterised by a plate height, and its causes provide a basis for understanding why modern chromatography is such an efficient separation technique.

### 19.3.1. Plate height

In a hypothetical ideal column, a solute band would retain its initial profile unaltered as it migrated along the column. In a real, non-ideal, column an initially narrow band broadens by dispersion as it migrates. The band width is proportional to the square root of the distance travelled along the column.

The rate at which the band broadens depends on the *inefficiency* of the column. This is more precisely defined as the *height equivalent to a theoretical plate (HETP)*, discussed in detail in Chapter 11. MARTIN and SYNGE[7] introduced the concept when describing their invention of liquid chromatography in 1941. The HETP is defined as a unit of column length sufficient to bring the solute in the mobile phase issuing from it into equilibrium with that in the stationary phase throughout the unit. Plate models[6,8], using this concept, show that the HETP **H** of a column of length $L$ may be determined by injecting a very small sample of solute, measuring its retention $t_R$ and band width $t_w$ at the column outlet as shown in Figure 19.3, and using the relation:

$$\mathbf{H} = \frac{L}{N} = \frac{L}{16(t_R/t_w)^2} \tag{19.9}$$

The greater the ratio $t_R/t_w$, then the greater the number of theoretical plates $N$ in the column.

To maximise separation efficiency requires low $H$ and high $N$ values. In general terms this requires that the process of repeated partitioning and equilibration of the migrating solute is accomplished rapidly. The mobile and stationary phases must be mutually well-dispersed. This is achieved by packing the column with fine, porous particles providing a large surface area between the phases (0.5–4 $m^2/g$ in GC, 200–800 $m^2/g$ in LC). Liquid stationary phases are either coated as a very thin film (0.05–1 μm) on the surface of a porous *solid support* (GC) or chemically bonded to the support surface as a monomolecular layer (LC).

### 19.3.2. Band broadening processes and particle size of packing

There are several band-broadening processes operating in a chromatographic column. Each contributes a term to the plate height. Many equations for **H** have been proposed[9] since the original one of VAN DEEMTER, ZUIDERWEG and KLINKENBERG[10]. A widely used modern version of this equation is:

$$\mathbf{H} = A + \frac{B}{u} + C_s u + C_m u \tag{19.10}$$

The terms arise from the mechanisms of dispersion, or band-broadening, listed in Table 19.1.

In gas chromatography the $B$ and $C$ terms are usually larger than the $A$ term. A plot of plate height against velocity shows a minimum.

Diffusion rates in liquids (LC) are typically three to four orders of magnitude less than in gases (GC). The lower mobile-phase diffusivity $D_m$ affects two of the plate-height terms in liquid chromatography given in Table 19.1. First, the $B/u$ term is small. Secondly, the $C_m u$ term is large. The $C_s u$ term is small in many LC applications where the stationary phase is only a monolayer of "liquid" bonded to the surface of a solid support. Thus, overall, the $A$ and $C$ terms tend to be dominant in LC. $C_m$ would be larger than in GC if the low value of diffusivity $D_m$ were not compensated by a small particle diameter $d_p$. This is why much smaller particles (3–8 μm in analytical applications) are used in *high performance liquid chromatography* (HPLC) than in gas chromatography (120–300 μm). Smaller particles, however, entail much higher pressure drops, in the region of 40–400 bar (4–40 MN/m$^2$), across the column. Indeed, the abbreviation HPLC is often regarded as meaning *high pressure liquid chromatography*.

For reasons explained in Section 19.5.3, large-scale LC employs larger particles (10–70 μm) than analytical HPLC. This increases the $C_m u$ term by deepening the intraparticle pores containing stagnant mobile phase to be penetrated by the solute. (See note to Table 19.1). *Pellicular* packings, in which each particle contains only a superficially porous layer with an inner solid core, have been used, though the present trend is towards totally porous particles of 15–25 μm.

### 19.3.3. Resolution

The *resolution* between two solute components achieved by a column depends on the opposed effects of (a) the increasing separation of band centres and (b) the increasing band width as bands migrate along the column. The resolution $R_s$ is defined by:

$$R_s = \frac{t_{R2} - t_{R1}}{\frac{1}{2}(t_{w1} + t_{w2})} \tag{19.11}$$

where $t_{R1}$ and $t_{R2}$ are the retention times of the two components, and $t_{w1}$ and $t_{w2}$ are the widths of the bands measured as defined in Figure 19.3. For the two bands in Figure 19.3. $R_s$ is 1.26.

The resolution equation (19.11) may be expressed in a more useful form by introducing a *separation (selectivity) factor*:

$$\alpha = k'_2/k'_1 \tag{19.12}$$

Equations 19.6 and 19.12 may be manipulated to give:

$$t_{R2} - t_{R1} = t_M \left[ 2k' \left( \frac{\alpha + 1}{\alpha - 1} \right) \right] \tag{19.13}$$

where $k'$ is the mean capacity factor, $\frac{1}{2}(k'_1 + k'_2)$. This equation is used to substitute for the numerator in equation 19.11; equation 19.9 is similarly used to substitute for $t_{w1}$ and $t_{w2}$ is the denominator. The result is:

$$R_s = \left( \frac{(\alpha - 1)}{2(\alpha + 1)} \right) \left( \frac{k'}{1 + k'} \right) \sqrt{\mathbf{N}} \tag{19.14}$$

This is an important basic equation for elution chromatography. Although the equation needs modification to cope with the wide, concentrated, feed bands of preparative and production chromatography[11], it shows how separation is controlled by the parameters $\alpha$, $k'$ and $\mathbf{N}$. The great power of chromatography as a separation process lies in the high values that can readily be achieved for $\alpha$ and $\mathbf{N}$.

### 19.3.4. The separating power of chromatography

The selectivity or separation factor, $\alpha$, is a ratio of mass distribution coefficients given in equation 19.12, and so is a thermodynamic rather than a kinetic factor. The value of $\alpha$ depends mainly on the nature of the two solutes, on the stationary phase and, in liquid chromatography, the mobile phase. It is the analogue of the relative volatility $\alpha$ in distillation, considered in Chapter 11. If a liquid mixture has two components for which $\alpha$ is close to unity, that is $\alpha - 1 \ll 1$, they are difficult to separate by distillation because $\alpha$ cannot be controlled without introducing another constituent, as in azeotropic or extractive distillation, whose presence changes $\alpha$. Judicious choice of the stationary phase and in liquid chromatography the mobile phase too, can greatly enhance $(\alpha - 1)$ with a corresponding beneficial effect, as shown by equation 19.14, on resolution.

A second reason why very high resolution can be obtained in chromatography is that very large numbers of theoretical plates are readily achieved. If the column is well packed with particles having a narrow spread of sizes, the plate height is about twice the particle diameter(9,12). A typical large-scale GC or LC column will contain $10^3$–$10^4$ plates.

Equation 19.14 also shows that resolution increases with increasing capacity factor $k'$. Diminishing returns apply at high $k'$ and values of 1–5 are generally advocated.

## 19.4. TYPES OF CHROMATOGRAPHY

To separate compounds effectively requires an appropriate choice of the chromatographic mobile and stationary phases. Much chemical ingenuity has been exercised in devising a great many different types of chromatography, and only those relevant to production chromatography are described. The applications of GC and LC are distinguished in general terms, and then the principal types of GC, LC and their more recent stable-mate, supercritical fluid chromatography (SFC) are outlined.

### 19.4.1. Comparison of gas and liquid chromatography

A solute travelling along a GC column is in its vapour form while it is the mobile phase. Its retention is therefore inversely proportional to its vapour pressure. For optimum capacity factor $k'$ (Section 19.3.4) and production economics, solutes usually need to be subjected to chromatography at a temperature within 20–30 K of their boiling point[12,13]. Although extreme column temperatures have been used for chemical analysis, the range of stability of organic liquid phases restricts normal large-scale operation to solutes boiling at 270–520 K. Thus GC is best suited to separating materials of molecular weight below about 300. The column temperature must be chosen to match the materials being separated and close control is required for reproducible retention times.

In LC the mobile phase is a liquid and so has a density some three orders of magnitude greater than in GC. The denser mobile phase has a much greater solubilising capacity for compatible solutes, permitting higher molecular weight materials to be separated without excessive retention. Liquid chromatography is therefore used to separate liquid or solid mixtures of higher boiling point than those in GC, though there is some overlap in the ranges of molecular weight appropriate to each method. LC is the preferred method also for heat-sensitive materials because the advantage of lower retention allows operation at lower temperatures. Room temperature is often used and precise temperature control is not as important as in GC.

Unlike GC, LC requires no vaporiser to inject the materials to be separated into the mobile phase. Low energy consumption is often advanced as a virtue of LC., although the overall energy consumption in LC may be at least as high as in GC if the eluted components have to be separated from the mobile phase itself. This separation is usually achieved by evaporation or distillation, taking advantage of the normally large difference in volatility between the solute and mobile phase, which is frequently an organic solvent or water. The solvent must be of very high purity and completely free of non-volatile residues in order to avoid contaminating the product.

### 19.4.2. Gas chromatography (GC)

The principal types of chromatography involved in large-scale separations are summarised in Table 19.2, and some explanatory comments are pertinent.

Gas chromatography (GC) employs a gaseous mobile phase, known as the *carrier gas*. In *gas–liquid chromatography* (GLC) the stationary phase is a liquid held on the surface and in the pores of a nominally inert solid support. By far the most commonly used support is diatomaceous silica, in the form of pink crushed firebrick, white diatomite filter aids or proprietary variants. Typical surface areas of 0.5–4 $m^2/g$ give an equivalent film thickness of 0.05–1 μm for normal liquid/support loadings of 5–50 per cent by mass.

In *gas–solid chromatography* (GSC) the stationary phase is a solid adsorbent, such as silica or alumina. The associated virtues associated therewith, namely, cheapness and longevity, are insufficiently appreciated. The disadvantages, surface heterogeneity and irreproducibility, may be overcome by surface modification or coating with small amounts of liquid to reduce heterogeneity and improve reproducibility[14,15]. Porous polymers, for example polystyrene and divinyl benzene, are also available. Molecular sieves, discussed in Chapter 17, are used mainly to separate permanent gases.

### 19.4.3. Liquid chromatography (LC)

Many forms of liquid chromatography have been developed in order to separate different types of compound, as shown in Table 19.2.

*Bonded-phase chromatography* (BPC) is an improved form of the now virtually extinct liquid–liquid chromatography. The problem of mutual dissolution of the two liquid phases, and hence progressive loss of stationary phase during service, is solved by chemically bonding the stationary liquid as a monomolecular layer to the surface of a solid support. The support is usually silica gel, a porous, amorphous, rigid solid form of silica. Surface areas are 200–800 $m^2/g$ and average pore diameters are 5–25 nm. BPC exists in two forms. In *normal-phase* BPC the mobile phase is less polar than the stationary phase. The stationary phase is a polar-bonded organic which brings about retention of moderately to

Table 19.2. Main types of chromatography for large-scale separation

| | Name | Mobile phase | Stationary phase | Materials separated |
|---|---|---|---|---|
| Gas chromatography (GC) | Gas–Liquid (GLC) | Gas, e.g. $N_2$, $H_2$, He | Liquid-film coated on solid support | Gases and adequately volatile liquids (mol. wt. <300) of adequate thermal stability, e.g. hydrocarbons and their derivatives, solvents, inorganic gases, essential oils, some steroids and vitamins. |
| | Gas–Solid (GSC) | Gas, e.g. $N_2$, $H_2$, He | Solid adsorbent, e.g. alumina | |
| Liquid chromatography (LC) | Normal bonded phase (NP-BPC) | Nonpolar or slightly polar solvent, e.g. heptane/dichloromethane | Polar organic group, usually monolayer, e.g. $-C_2H_4CN$, chemically bonded to silica surface. | Almost every possible type of nonionic compound of mol. wt. 100–3000, and many synthetic or biopolymers of higher mol. wt. Some forms of RP-BPC suitable for ionic compounds. |
| | Reverse bonded phase (RP-BPC) | Polar solvents, e.g. water/methanol, acetonitrile. | Nonpolar (sometimes polar) organic moiety, usually monolayer, e.g. $n$-$C_8$, $n$-$C_{18}$, chemically bonded to silica surface. | |
| | Liquid–solid (LSC) (adsorption) | Polar or nonpolar solvent | Solid adsorbent, usually silica or alumina. | |
| | Ion-exchange (IEC) | Aqueous buffer or salt solution, sometimes with organic solvent modifier. | Organic polymer (e.g. poly-styrene-divinylbenzene) surface derivatised with ionic functional groups. (See also Section 19.6.2). | Organic or inorganic ions, e.g. rare earth elements, pharmaceuticals, amino acids, peptides, nucleic acids, proteins. Saccharides. |
| | Size exclusion (SEC) (gel permeation GPC) | Organic or aqueous | Porous gels of silica, synthetic polymers or biopolymers with exclusion limits from $10^2$ upto $10^8$ | Mainly synthetic and biopolymers of mol.wt >2000, but also smaller molecules. |
| | Affinity (AC) | Aqueous, usually buffered | Specific affine ligand bonded to support matrix. (See also Section 19.6.2). | Proteins (enzymes, antibodies, antigens, lectins), peptides, nucleic acids, oligonucleotides, viruses, cells. |
| | Hydrophobic interaction (HIC) | Aqueous, usually buffered | Apolar ligand (e.g. octylamino) bonded to support matrix. A form of AC: ligand complexes with apolar (hydrocarbon) sites on protein solute. | Usually proteins. |
| Supercritical fluid chromatography (SFC) | | Supercritical fluid, e.g. $CO_2$ | Liquid film coated on solid support; solid adsorbent; or bonded phase. | Materials of mol. wt. overlapping with both GC and LC. |
| Chiral chromatography (form of GC, LC or SFC) | | Gas, liquid or supercritical fluid | Chiral stationary phase (GC, LC and SFC); or nonchiral stat. phase with chiral eluent (in LC) | Racemates are separated into their enantionomers. |

strongly polar solutes. In *reverse-phase* BPC, the mobile phase is more polar than the stationary phase, which retains a great variety of solutes less polar than in the normal phase mode. Because of the cheapness and versatility of the often water-based mobile phase, the reverse-phase technique is now dominant amongst analytical and preparative LC methods. In larger scale separations, the superior and milder surface chemistry of reverse-phase BPC may need to be balanced against the even cheaper *liquid–solid (adsorption) chromatography* (LSC) packings.

The other LC techniques listed in Table 19.2 rely on mechanisms other than solution, and simple adsorption and need more explanation.

*Ion-exchange chromatography* (IEC) is carried out with packings that possess charge-bearing functional groups. The most common retention mechanism is simple ion-exchange of solute ions X and mobile phase ions Y with the charged groups R of the stationary phase:

$$X^- + R^+Y^- = Y^- + R^+X^- \quad \text{(anion exchange)}$$

$$X^+ + R^-Y^+ = Y^+ + R^-X^+ \quad \text{(cation exchange)}$$

Solute ions X that compete weakly with mobile phase eluent ions Y for ion exchanger sites R will be retained only slightly on the column. Solute ions that interact strongly with the ion-exchanger elute later in the chromatogram.

The ion-exchangers used in LC consist either of an organic polymer with ionic functional groups, or silica coated with an organic polymer with ionic functional groups. The types of functional groups used are the same as described in Chapter 18. Since IEC can be carried out with an aqueous mobile phase near physiological conditions, it is an important technique in the purification of sensitive biomolecules such as proteins.

*Size exclusion* or *gel permeation chromatography* (SEC, GPC) differs from all other methods in separating according to molecular size, rather than structural characteristics such as functional groups. The particulate packing contains pores of well-defined size. Solute molecules, too large to enter the pores, are totally excluded from the interior of the particles and elute first. Their molecular weights are said to exceed the *exclusion limit*. Smaller molecules permeate the particles to a varying extent, depending on molecular size, and are retained to varying extents in order of size. The technique is complementary to other methods which separate on the basis of molecular structural difference. SEC is often the first step in the resolution of a complex mixture by more than one LC method.

*Affinity chromatography* depends on the specific adsorption which results from molecular "recognition". The "lock-and-key" mechanism shown in Figure 19.4 involves a number of co-operating interactions within regions of high complementarity on two molecules. One of the two molecules is chemically bonded to the support *matrix* and forms the *ligand*. The other molecule is then the *ligate*, that is the solute subjected to chromatography. Because the adsorption complex results from a multisite mechanism, the interaction is very specific to the two molecular species concerned. *Bio-affinity chromatography* uses the variety of such complexes that occur in biological systems, for example, antibody–antigen, enzyme–inhibitor, enzyme–cofactor and hormone–receptor complexes, and specific base sequences for binding nucleic acids. In each case either component of the pair may be bound to the matrix in order to chromatograph the other. The principle of affinity chromatography is not restricted to purely biological *ligands*;

synthetic ligands, which are usually less *group-specific*, may be dyes, metal chelates or electron donors or acceptors[16]. All of these techniques have been almost exclusively applied to biological *ligates* (Table 19.2).

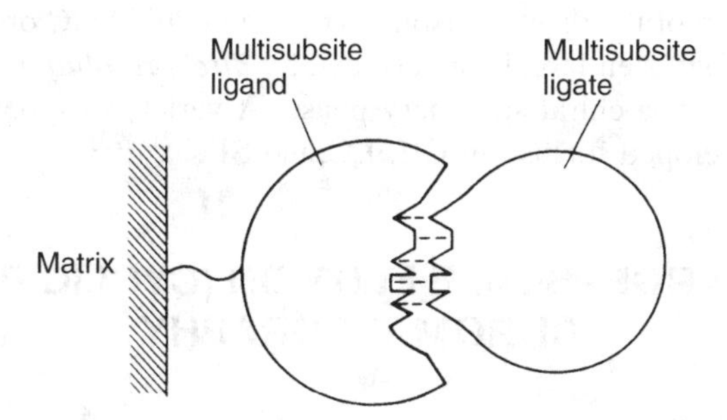


Figure 19.4. Principle of molecular recognition used in affinity chromatography[17]

## 19.4.4. Supercritical fluid chromatography (SFC)

Supercritical fluid chromatography (SFC) is intermediate between GC and LC. The mobile phase is a supercritical fluid, usually carbon dioxide at near ambient temperatures. As with LC, SFC is suitable for separating higher molecular weight materials than GC, and the low operating temperature is advantageous for heat-sensitive materials, as in LC. SFC separations, however, may be performed two or three times faster than LC because materials have higher diffusivities in a supercritical fluid than in a liquid, and so suffer less band-broadening from the effect of the $C_m u$ term in equation 19.10 and Table 19.1. A supercritical fluid may readily be separated from the required products merely by reducing the pressure, avoiding the costs of the mobile phase separation that are a main disadvantage of LC. Carbon dioxide is a particularly attractive mobile phase for large-scale separations because it is non-toxic, environmentally acceptable and cheap, and has a very convenient temperature and pressure at its critical point. Pure carbon dioxide is a non-polar mobile phase of low solvating power, however, and exclusive reliance on it has restricted SFC to niche applications involving a rather narrow range of low polarity compounds. This position now appears to be changing as a result, in part, of mixing polar organic solvents with carbon dioxide in the mobile phase. The applications of SFC may soon expand to match those of its more-established relative, supercritical fluid extraction, as discussed in Section 13.8.1[18–21].

## 19.4.5. Chiral chromatography

The optical activity of biologically-active chemicals is important to their activity and toxicology. Pure enantiomers, or optical isomers, of pharmaceuticals and agrochemicals can in many cases be made by enantiospecific synthesis. An alternative method is to use a less complicated synthesis followed by chromatographic resolution of the racemic mixture into its enantiomers.

Since all the physical properties of two given enantiomers are the same in the absence of a chiral, or optically active, medium, their chromatographic resolution needs a different approach from the relatively simple separation of geometrical isomers, stereoisomers or positional isomers. Two methods are used. The older technique of *indirect resolution*, requires conversion of the enantiomers to diastereoisomers using a suitable chiral reagent, followed by separation of the diastereoisomers on a non-chiral GC or LC stationary phase. This technique has now been largely superseded by *direct resolution*, using either a chiral mobile phase (in LC) or a chiral stationary phase. A variety of types of chiral stationary phase have been developed for use in GC, LC and SFC[21–23].

# 19.5. LARGE-SCALE ELUTION (CYCLIC BATCH) CHROMATOGRAPHY

## 19.5.1. Introduction

The three main modes of chromatographic operation are elution chromatography, selective adsorption/desorption, and simulated countercurrent chromatography. Of these, elution chromatography, used as a cyclic batch process, was the first to be developed for large-scale separations.

The distinction between preparative and production chromatography is sometimes drawn in terms of the scale of operation though it is really a matter of purpose. The term *preparative chromatography* is used to refer to the technical preparation of limited quantities of material in the laboratory, and *production chromatography* to continuous production for commercial purposes, where economics are important. The two are compared in Table 19.3.

Table 19.3. Comparison of production, preparative and analytical chromatography

| | | Production | Preparative | Analytical |
|---|---|---|---|---|
| Performance criteria and basis for design | | Product cost, that is throughput (cost/yr.), for specified purity | Throughput for specified purity, or simple diameter scale-up of known analytical separation | Resolution and speed of analysis |
| Processing rate | GC | 0.1–1.5 kg/h (100 mm dia.)<br>1–15 kg/h (300 mm dia.) | 0.1 g–1 kg batch | — |
| | LC | 0.02–0.3 kg/h (100 mm dia.)<br>0.2–3.0 kg/h (300 mm dia.) | 0.01 g–0.2 kg batch | — |
| Column diameter | GC | >80 mm | 5–120 mm | 0.1–5 mm |
| | LC | >30 mm | 3–40 mm | 0.3–5 mm |
| Column length | GC | 1–8 m | 0.6–8 m | >0.5 m |
| | LC | 0.2–2 m | 0.1–2 m | >0.03 m |
| Particle size | GC | 150–500 μm | 100–400 μm | 80–250 μm |
| | LC* | 20–70 μm | 8–60 μm | 3–8 μm |

* Except LC of proteins, etc. which are 30–300 μm.

Since elution is basically a batch technique, continuous production requires cyclic batch operation under automatic control. This is achieved, first, by using repetitive automatic injection of successive batches of the mixture to be separated and, secondly, by recycling the mobile phase, unless this is water or carbon dioxide.

## 19.5.2. Gas chromatography equipment

Figure 19.5 shows the general flow scheme of a production gas-chromatograph. The *carrier gas* (mobile phase) which is normally nitrogen, hydrogen or helium flows continuously. The feed mixture to be separated is vaporised and periodically injected into the gas stream. The duration of the injection period is usually 10–180 s. The feed band entering the column then has a rectangular concentration–time profile and is far larger in both duration and concentration than in analytical chromatography. As the separated fractions elute from the column, they are passed in turn to individual fraction collectors controlled by appropriate valve sequencing. Here they are condensed by cooling and any aerosol formed is trapped. The carrier gas is completely cleaned of traces of solute by passage through a packed bed of molecular sieve or activated carbon, recompressed to a pressure of, typically, 2–6 bar (200–600 $kN/m^2$) and recycled.

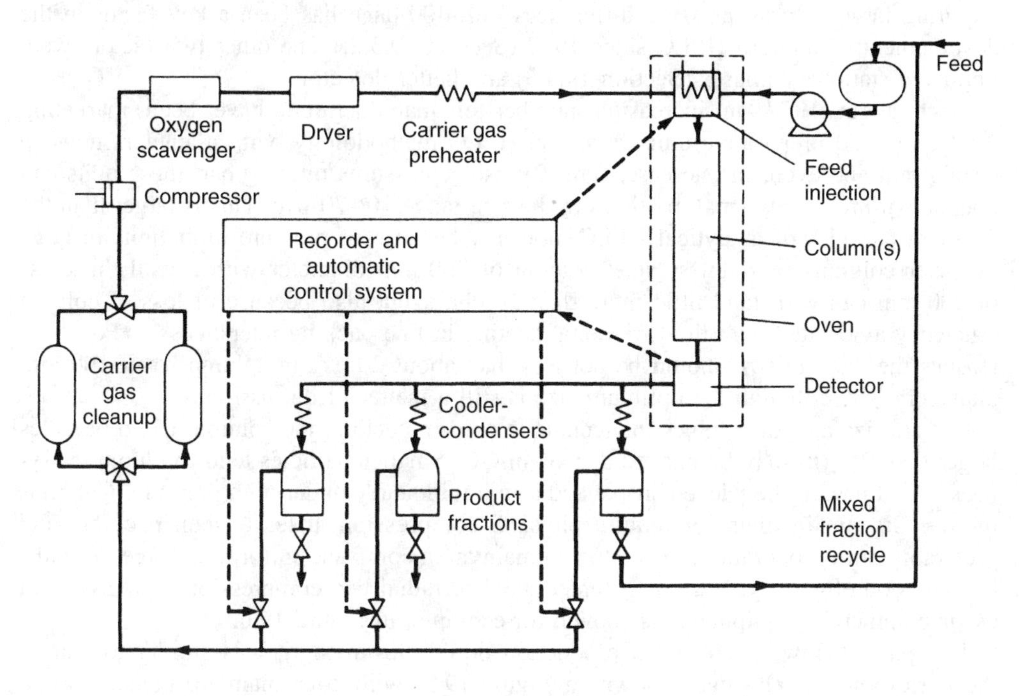


Figure 19.5. Production gas-chromatograph

The duration of the injection is a substantial fraction of the cycle time. An effective technique, therefore, is to run the feed vaporiser continuously and to direct the vapour/carrier mixture to one of, say, 2–5 parallel columns in sequence[24–26].

Production experience with large-scale gas-chromatography has been accumulated mainly since the mid-1970s, with columns up to about 400 mm diameter and throughputs up to about 100 tonne/yr, separating a variety of organic compounds[12,27–32]. These developments have been summarised elsewhere[33]. Several detailed descriptions have been given of equipment of various sizes[12,26,34–36]. There are many preparative chromatographs on the market and one company has offered production units with column diameters up to 600 mm[32,37]. For a 400 mm unit total production costs (capitalised plus direct operating costs) were projected in 1987 as £3.5/kg for a production rate of 10 kg/h[32]. Elf Aquitaine has devised an interesting process using a molecular sieve as stationary phase to separate 100,000 tonne/year of a light naphtha into an $n$-paraffin product that may be used as a petroleum feedstock and a high octane fraction, rich in *iso*-paraffins, for use in lead-free petrol[38].

## 19.5.3. Liquid chromatography equipment

Very large LC columns up to 1.2 m in diameter have been used on occasion since the 1940s. These were low pressure units with very large packing particles and were therefore relatively slow and inefficient. The use of smaller particles and higher velocities, both requiring large column pressure differences (40–400 bar), has been a key factor in the development of modern HPLC since 1967 (Section 19.3.2.) The other two factors were improved stationary phases (Section 19.4.3) and better detectors.

Since about 1977 an increasing number of manufacturers have been marketing production and preparative units based on HPLC methodology, with diameters now up to 600 mm and even, in one case, 2 m. By using pressures over 10 bar, these units can operate with relatively small particles in the size range 10–70 μm. This is larger than the 3–8 μm particles of analytical HPLC, not only because of pressure limitations in large diameter columns (a stainless steel column of 400 mm diameter with a wall thickness of 200 mm can withstand little more than 100 bar), but also because of loss of column efficiency associated with fluid frictional heating in fine packing interstices[39]. For these reasons the particle size should be not less than about 20 μm in columns over 100 mm diameter[39,40], although the optimum size is still a matter of discussion.

Particle size also determines the technique used in packing the column. Rigid particles larger than 25 μm may be packed dry, as in GC. Smaller particles tend to clump if dry-packed and need to be packed suspended as a liquid slurry under high pressure. Columns of over 30 mm in diameter require physical compression at least when packing, and preferably during operation, to create and maintain a compact, uniform bed free of voids, having good efficiency. Various patented axial or radial bed compression systems are in use in commercial equipment, as shown, for example, in Figure 19.6.

The general flow scheme of a production liquid chromatograph is similar to that of the corresponding GC unit, shown in Figure 19.5, with four main differences. First, thermostatting requirements for the column are less strict, and may sometimes even be dispensed with. Secondly, the feed is injected as a liquid, and not vaporised. Thirdly, if the product is to be separated from the mobile-phase solvent, distillation or evaporation and solvent recycle are incorporated in the loop[28,41,42]. Finally, the liquid streams are filtered to ensure column longevity, and de-aerated to prevent air bubbles forming.

Commercial LC units, available from a number of manufacturers, have been surveyed[39]. Some of the largest scale applications are summarised by WANKAT[43]

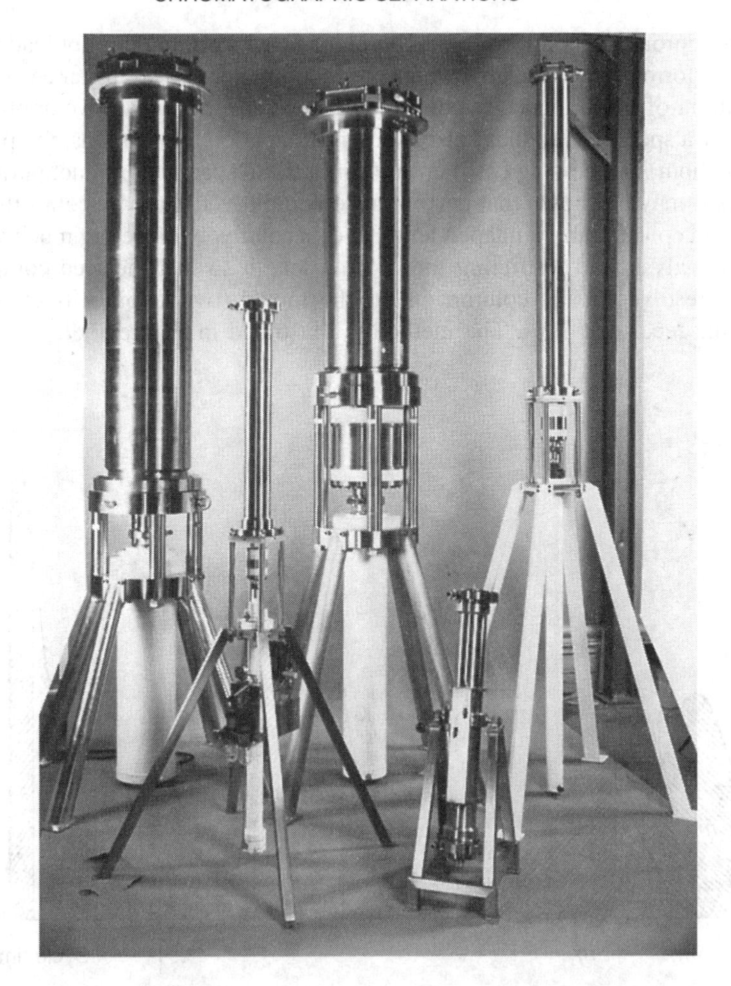

Figure 19.6. Stainless steel columns for liquid chromatography[37]. Two are 450 mm dia., one 150 mm dia. and two 110 mm dia. Pressure rating 70 bar. The devices below each column permit axial bed compression by piston

and many more applications exist for pharmaceuticals, biochemicals, fine chemicals and industrial chemicals. Production costs were estimated in 1987 at £20–170/kg for production rates of 0.06–2.3 kg/h in 150–450 mm diameter columns[44].

## 19.5.4. Process design and optimisation

The design of a production chromatograph is a complex exercise because of the many process variables involved. Some of the published work on optimisation proceeds from the unsatisfactory notion that a large-scale chromatograph is little more than a scaled-up analytical one into which large samples are injected. Thus the sample size is commonly chosen as the largest which does not excessively degrade resolution.

While this approach simplifies scale-up in preparative work, it does not lead to optimum large scale performance, especially when production economics are taken into account. The best criteria of performance for design purposes are, in preparative applications, the throughput for a specified product purity and, in production applications, the product cost, that is (throughput)/(processing cost per unit time), for a specified product purity[28,45–47]. On this basis it may be shown that performance is improved, first, by controlling the feed band-width and concentration independently and, secondly, by injecting much wider bands than used in analysis and restricting the column length so that the feed components are incompletely resolved at the column outlet; the mixed (overlapping) fractions are then recycled to the feedstock[47,48]. The method is illustrated in Figure 19.7.

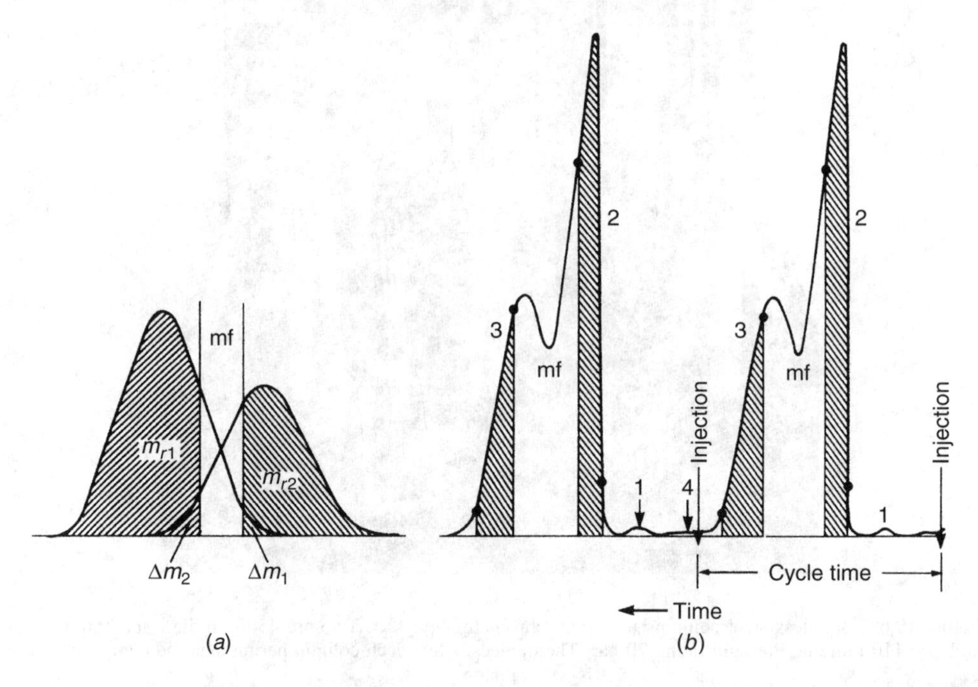


Figure 19.7. Cyclic batch elution chromatography: obtaining high product purity and high throughput by using incomplete resolution (overlapping bands) and recycling the mixed fraction (mf) to the feedstock (*a*) Control of band separation and cut points determines fractional impurities $\Delta m_2/m_{r1}$ and $\Delta m_1/m_{r2}$.[49] (*b*) Chromatogram for separation of pure cis- and trans- 1,3-pentadiene. Components: 1, isoprene; 2, trans- 1,3-pentadiene; 3, cis-1,3-pentadiene; 4, cyclo-pentadiene. Component 1 is eluted at almost the same time as component 4 of the previous injection[12]

Information on process design and optimisation is available in the literature already cited and in references for GC[49–50] and for LC[42,51–57]. The flow chart for design calculations shown in Figure 19.8 demonstrates the large number of variables involved and the relations between them. The problem of maintaining column efficiency as the column diameter increases, once a major difficulty in scale-up, is now generally regarded as solved by using a narrow particle-size range, good packing technique and efficient distributors[12,32,33,51].

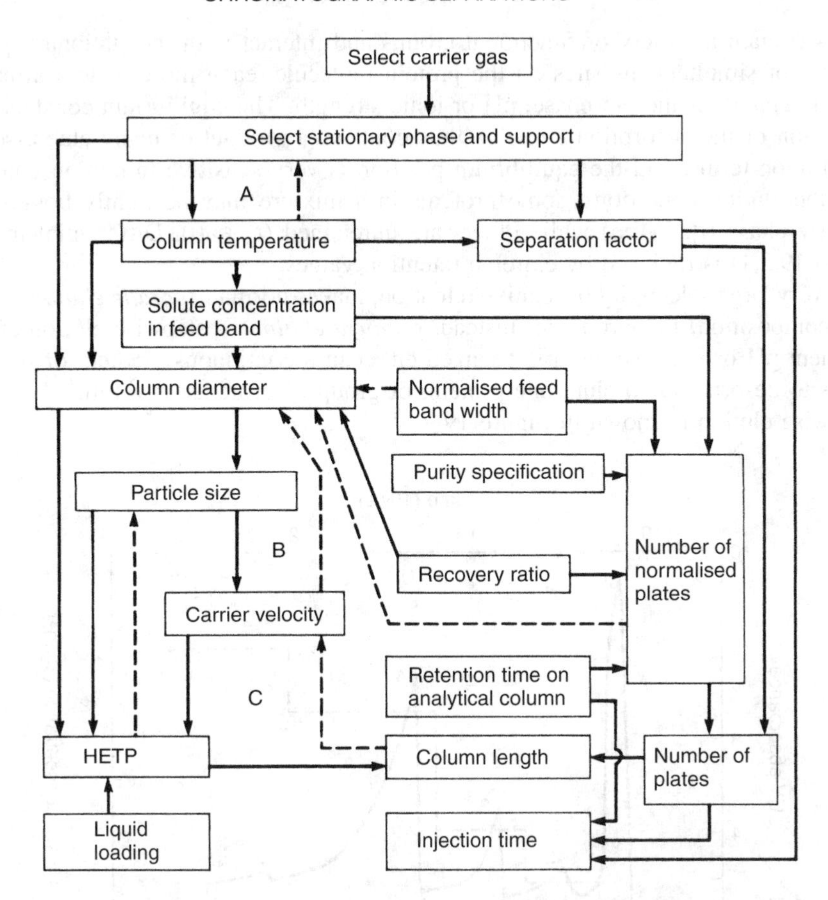


Figure 19.8. Flow chart for design of production and preparative gas chromatographs[13]. Feedback is shown by broken lines. A, B and C are principal calculation loops

## 19.6. SELECTIVE ADSORPTION OF PROTEINS

Chromatography is now an integral and indispensable part of the biotechnology industry. Most regulatory authorities expect at least one chromatographic step to be used in producing a protein-based pharmaceutical. Other techniques cannot provide the purity levels now required, which may sometimes reach 99.999 per cent. Moreover, other techniques, such as fractional precipitation, electrophoresis, isoelectric focusing and ultrafiltration, involve heat generation or strong shear forces. Chromatography usually permits milder processing conditions close to physiological, so avoiding denaturation, the unfolding of the protein molecule, and retaining biological activity.

### 19.6.1. Principles

Although elution chromatography is an efficient technique for resolving most mixtures, it is less appropriate for biological macromolecules such as proteins and nucleic acids.

Proteins contain a variety of functional groups and interact with the stationary phase at a number of simultaneous sites on the protein molecule, each more or less affected by change in *eluent*, or mobile-phase, pH or ionic strength. The equilibrium constant for the dissociation of the adsorption complex thus contains a product of many eluent-sensitive concentration terms, and the equilibrium position is very sensitive to elution conditions. Under the elution conditions, some proteins in a mixture may be tightly bound by the stationary phase ($t'_R \to \infty$) while others are unretained ($t'_R \approx 0$). Differential migration (Section 19.2.1) is replaced by extreme retention values.

This very large selectivity or relative retention, makes normal *isocratic elution* (constant eluent composition) inappropriate. Instead, *gradient elution* or stepwise elution are used. The eluent pH or ionic strength is changed either in a continuous gradient or in a series of steps to desorb and so elute one protein, or group of proteins, at a time. An example of stepwise elution is shown in Figure 19.9.

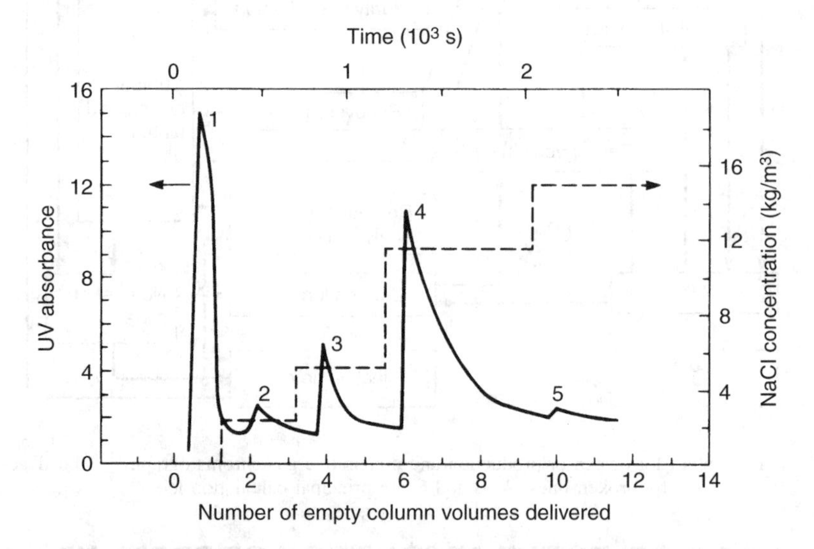


Figure 19.9. Chromatogram for stepwise elution of bovine serum albumin on a Vistec diethyl aminoethyl cellulose ion-exchanger, using stepwise increases in sodium chloride concentration in the mobile phase to achieve selective desorption. Proteins: 1, serum fraction not adsorbed by column (includes $\gamma$-gobulin); 2,3, transferrin, and so on; 4,5 albumin[57]

For large-scale separations, a further operational change is commonly adopted. Maximum production requires that the width of the protein band fed to the column be increased so that protein is adsorbed on the whole of the column packing before desorption[58]. The procedure is then essentially the same as the adsorption–desorption techniques described in Chapter 17, where both adsorption and desorption are pursued to breakthrough[59]. This procedure may be termed *selective adsorption* or *desorption* to distinguish it from elution, or differential migration. Although in principle a form of adsorption, the procedure is usually regarded as a form of chromatography because it uses the high-selectivity technology of the latter.

Affinity chromatography is conducted in the selective-adsorption mode, whereas ion-exchange chromatography is usually carried out in the selective-desorption mode. The reason is that affinity chromatography, being the more highly selective, normally relies on specificity in the adsorption step. Ion-exchange chromatography usually gives a multi-component adsorption which is resolved in the desorption step. Both procedures involve a cyclic batch process. Various aspects of the process have been analysed theoretically and optimised[59–63].

### 19.6.2. Practice

Large-scale separation of proteins may be carried out by *ion-exchange* (IEC), *affinity* (AC), *hydrophobic-interaction, size-exclusion* or *reverse-bonded-phase* (RP-BPC) chromatography, as given in Table 19.2. Whereas IEC and AC are best conducted in the selective-adsorption or desorption mode of operation, RP-BPC and SEC are conducted in the elution mode, often by gradient elution. RP-BPC has the disadvantages that the organic solvents often added to the aqueous eluent tend to denature proteins, and that the silica in the bonded-phase packing is chemically stable only at pH of 2–8.

Protein separations by IEC, AC and SEC differ from other separations in using soft polysaccharide-gel packings. Agarose, dextran and cellulose have the desired properties of hydrophilicity, to avoid denaturation, and large pore size, to permit access of the large protein molecules to the high interior surface area, but are soft and will not tolerate high pressure drops across the column. Particle sizes are therefore commonly quite large, and usually 30–300 μm. Improved, more rigid, supports are now widely available[64,65].

Chromatography is currently used in the downstream processing, or bioseparation, of biological macromolecules produced both from natural sources, such as albumin and various factors from blood plasma, and by biotechnology, as with monoclonal antibodies and recombinant DNA products like interferons, insulin, vaccines, growth hormones and tissue plasminogen activator. Downstream processing, which can account for 50–80 per cent of total production costs, involves integrated product separation and recovery trains[66,67]. Frequently both IEC, AC and SEC steps are included. Two examples are shown in Figure 19.10. In this context, one advantage of IEC and AC is that they can effect concentration from dilute solution at the same time as separation. This results from tight binding in the adsorption stage, coupled with freedom to control concentration in the desorption stage.

### 19.6.3. Expanded bed adsorption

A recent innovation in downstream processing is to use an expanded bed for adsorption. An expanded bed occurs over a narrow range of flow velocity, and is intermediate between a fixed bed and a fluidised bed. It resembles a fluidised bed in having a greater voidage than a fixed bed. This has the great advantage that crude, particulate-containing feedstocks will pass through the expanded bed without the suspended solids causing the blockages that would normally result with a packed bed. Proteins, for example, may be removed from fermentation broths and preparations of disrupted cells by adsorption in an expanded bed of a suitable adsorbent. After washing, the flow velocity is reduced so that the adsorbent settles into a fixed bed, and the proteins are recovered by changing the eluent, or mobile phase, composition. This technique can simplify downstream processing by replacing the

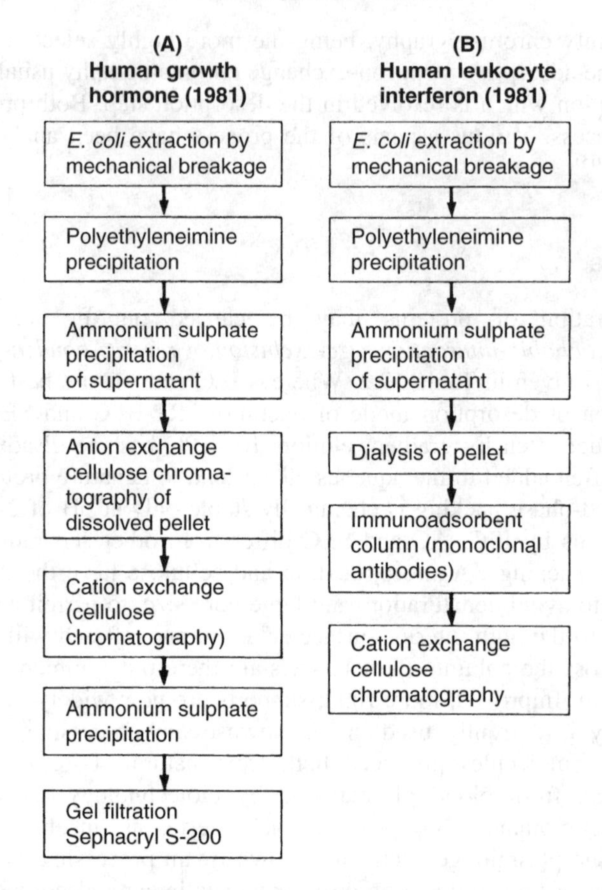


Figure 19.10. Summary of purification processes for two human proteins synthesised in recombinant *E. coli*[68]. The term gel filtration is sometimes used as an alternative to size-exclusion chromatography

early filtration, centrifugation and concentration steps with a single process. An expanded bed is preferred to a fully fluidised bed in this application because it suffers less bed-wide axial mixing, especially if suitable steps are taken to stabilise the bed in order to reduce axial mixing. A stable expanded bed resembles a fixed bed in providing many theoretical plates to the incoming protein which is thereby captured efficiently. The technique is variously called *expanded-bed adsorption* and *expanded-bed chromatography*[69].

## 19.7. SIMULATED COUNTERCURRENT TECHNIQUES

The preference for operating processes continuously wherever possible has led to the development of alternative modes of operation to cyclic-batch elution-chromatography and selective adsorption/desorption. There are several truly continuous modes, such as so-called *countercurrent chromatography* (CCC) using centrifugal fields[70], the continuous rotating annulus[71] and cross-flow devices[71]. These have recently seen extensive development at the preparative scale though not yet at larger scales. The production scale, however, is now well established in pseudo-continuous schemes which use a simulated moving-bed for countercurrent operation. In these, the solid or liquid "stationary phase" is in effect a moving bed flowing countercurrently to the liquid or gaseous "mobile phase". The feed mixture to be separated is fed continuously to a point in the middle of the column. The ratio of the flowrates of the two phases is chosen so that the feed splits into two fractions moving in opposite directions from the feed point. The basic principle of these moving-bed schemes has been described in Section 17.9, as shown in Figure 17.31.

The original moving-bed process was the *hypersorber*, described in Section 17.9.2. This is no longer used due to problems of solids-attrition and high rates of axial mixing[43,71]. These problems are nowadays avoided by simulating the movement of the bed using a fixed bed, and periodically moving the positions of entrance and exit of the process streams. The process is thus periodical rather than truly continuous.

The two main forms of simulated countercurrent/moving-bed processes are *Simulated Moving Bed Chromatography* (SMB) and the *Semi-Continuous Chromatographic Refiner* (SCCR). Both are now well established industrially, with at least 227 SMB trains of units world-wide by 2000. These are well suited to large-scale separation of simple binary mixtures where the purity requirements for the product are high, but not so demanding as to warrant use of elution chromatography or selective adsorption. Examples include the separation of xylenes, sugars and pharmaceutical enantiomers.

The SMB process was invented by Broughton in 1961 and developed by Universal Oil Products under the general name "Sorbex". Initially used for separating *n*-paraffins in bulk, it is now used for a variety of individual-isomer separations and class separations, and is currently attracting considerable interest for separating pharmaceutical enantiomers. The SMB process is described in Section 17.9.4 and in a growing literature[21,22,71–74].

The *Semi-Continuous Chromatographic Refiner* (SCCR), developed by Ganetsos and Barker[71], uses an array of twelve column sections (each, for example, 76 mm i.d. × 610 mm long) mounted in a circle. The column sections are connected in series (at top and bottom alternately) to form a closed loop (12 × 0.61 m = 7.32 m long). The loop is shown diagrammatically in Figure 19.11. Valved inlet and outlet ports are provided at top and bottom of each column section. The array is fixed, but countercurrent operation is simulated by periodically shifting the five process stream ports and the carrier fluid locks to the next adjacent column sections in one direction, the same as for fluid flow, round the array. This simulates bed movement in the opposite direction. The rate of port advancement is less than the velocity of the less strongly adsorbed component 1 through the bed, but greater than that of the more strongly adsorbed component 2. The components are therefore collected at ports at opposite "ends" of the loop.

Compared with elution chromatography, the advantage of both the SMB and SCCR forms of simulated countercurrent operation is that each product is taken off as soon as it is separated. The disadvantages are mechanical complexity and the fact that the number of pure products readily obtainable from one column is limited to two at most. Elution chromatography allows many components to be separated on one column. If no components are taken off before reaching the column exit, however, any components that are much more easily separated than the key components occupy space in

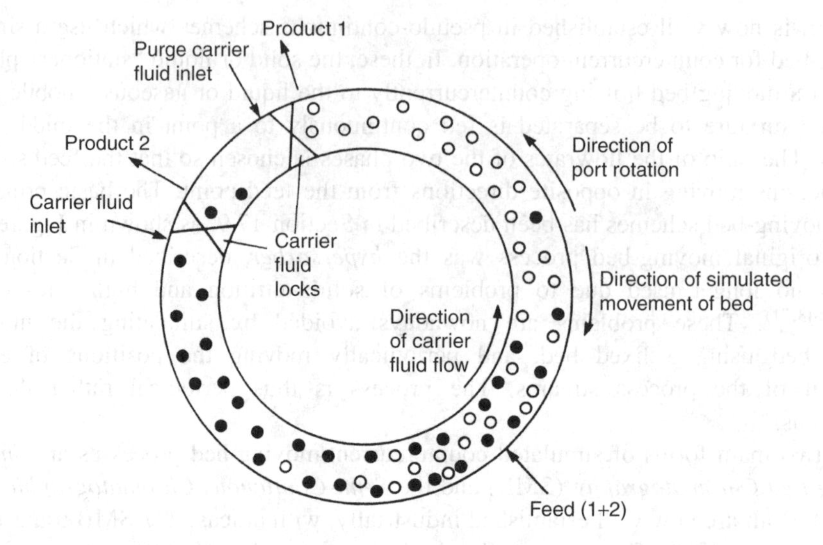


Figure 19.11. Diagrammatic representation of the principle of operation of a *Semi-Continuous Chromatographic Refiner*[75]. Open and closed circles show schematically the concentrations of the two components being separated. The distance from feed point to product 1 offtake and from feed point to carrier fluid inlet must be sufficient to allow the required purity levels to be achieved against the usual band-spreading process

the column, even after they have separated, and so raise the cycle time between injections and reduce throughput. This often-cited problem of elution chromatography may be overcome either by using column-switching methods[43] or by adopting window-diagram techniques. The latter permit the compositions of mixed stationary or mixed mobile phases to be chosen to optimise band spacing in the elution sequence[76]. Either approach allows the whole column to be engaged in separation at all times, as in simulated countercurrent operation.

## 19.8. COMBINED REACTION AND SEPARATION

When chemical reactions are carried out in chromatographic columns, reaction and separation occur simultaneously. The immediate advantages of combining these processes in a single column are that the reaction products are obtained in high purity, and both the capital and energy costs of the process can be lower than when the operations are conducted separately. A more subtle but important advantage is that, when the reaction is equilibrium-limited, much higher conversions than the equilibrium values can be achieved. If the reaction is of the type $\mathbf{A} \rightleftharpoons \mathbf{B} + \mathbf{C}$, on-column separation of **B** and **C** prevents the reverse reaction, so that **A** can be converted entirely into $\mathbf{B} + \mathbf{C}$.

Several types of reaction may be carried out in a chromatographic reactor. The reaction can be chemical or biochemical, taking place on the stationary phase, in the mobile phase, or both. The stationary phase must be chosen to have a good retention (affinity) for at least one component of the reaction system, and in some cases it has to act as a catalyst or catalyst support. Chromatographic reactors are particularly suited to enzyme-catalysed reactions such as the inversion of sucrose and biosynthesis of dextran, to various dissociation, isomer-interconversion and catalytic-cracking reactions, and to the selective production of intermediate species in reaction sequences, or where side reactions would otherwise interfere with production. Both elution-chromatography and simulated-moving-bed modes of operation are employed and have been extensively analysed[71,77,78].

## 19.9. COMPARISON WITH OTHER SEPARATION METHODS

In general, chromatography is a powerful but relatively expensive separation technique whose advantages and disadvantages need to be evaluated carefully in selecting a separation system. The expense arises chiefly from the packings and the need in many cases to recycle the mobile phase. GLC and some LC and SFC packings need occasional replacement. Published cost data are as yet limited.

The most obvious virtue of chromatographic methods is their great separating power. This results from the very small plate height and the enhanced relative volatility permitted by adding a third component, the stationary phase, to the mixture to be separated, as discussed in Section 19.3.4. Thus, gas chromatography employs columns typically an order of magnitude shorter than those required for the same separation by distillation[79]. On the other hand, GC throughputs are much less than in distillation for the same column diameter, unless the separation is difficult (low relative volatility), in which case the throughput in distillation is reduced by the need for a high reflux ratio. Overall, the balance of advantage probably lies with GC for difficult separations where the relative volatility, before enhancement with stationary phase, is less than about 1.2, while distillation is to be preferred when it is greater than 1.6. Chromatographic methods in general are advantageous for difficult separations and for separations where high product purities are specified.

The same advantages are exhibited by LC in comparison with techniques such as fractional crystallisation, liquid extraction, ultrafiltration and adsorption. It has already been pointed out (Section 19.6) that LC now plays a major part in bioseparations, where the technique needs to be integrated into the process train as part of a systems approach.

In choosing suitable methods to separate a given mixture of compounds, the general characteristics favouring a chromatographic method are:

(a) It can achieve difficult separations, as discussed in Section 19.3.4.
(b) It can meet high product purity specifications[46].
(c) It can separate heat-sensitive compounds (permitted by low residence times and absence of reflux, together with low temperatures in LC and SFC).
(d) It has relatively low energy consumption (cf. distillation).
(e) It can split an $n$-component mixture into $n$ pure components in one column instead of $(n - 1)$ columns (cf. distillation and other countercurrent processes).
(f) Product is often recovered in a non-toxic carrier or solvent (in GC, SFC, and LC with aqueous mobile phase) from which it is readily separated, if necessary.
(g) The technique is very versatile, and an appropriate type of chromatography may be chosen for most separations.
(h) It is well suited, though not restricted, to low volume, high value separations.

In this chapter the three main modes of large-scale chromatographic operation, and combined reaction and separation. Many useful but small-scale chromatographic methods have been omitted, as well as allied separation techniques which combine aspects of chromatographic principles or practice with aspects of adsorption, extraction, sedimentation or electrophoresis. Such is the pace of invention that novel processes related to chromatography are still being developed and described in the literature.

## 19.10. FURTHER READING

BRAITHWAITE, A. and SMITH, F. J.: *Chromatographic Methods*, 4th edn. (Chapman and Hall, 1986).
CONDER, J. R., in *New Developments in Gas Chromatography* (ed. Purnell, J. H.) pp. 137–186, Production-Scale Gas Chromatography (Wiley, 1973).
GANETSOS, G. and BARKER, P. E.: *Preparative And Production Scale Chromatography* (Marcel Dekker, 1993)
HAMADA, J. S.: *J. Chromatogr.* **760** (1997) 81. Large-scale high-performance liquid chromatography of enzymes for food applications. (Review)
SKEA, W. M. in *High Performance Liquid Chromatography* (eds. BROWN, P. R. and HARTWICK, R. A.), Chapter 12: Process High Performance Liquid Chromatography (Wiley, 1989).
SNYDER, L. R. and KIRKLAND, J. J.: *Introduction to Modern Liquid Chromatography*, 2nd edn. (Wiley, 1979).
SOFER, G. K. and NYSTRÖM, L. E.: *Process Chromatography* (Academic Press, 1989)
SUBRAMANIAN, G. (ed.): *Preparative and Process-Scale Liquid Chromatography* (Ellis Horwood, 1991)
SUBRAMANIAN, G. (ed.): *Process-Scale Liquid Chromatography* (VCH Verlagsgesellschaft, 1995)
YAMAMOTO, S., NAKANISHI, K. and MATSUNO, R.: *Ion-Exchange Chromatography of Proteins* (Marcel Dekker, 1988). Chapter 9: Large Scale Operation.

## 19.11. REFERENCES

1. CONDER, J. R. and YOUNG, C. L.: *Physicochemical Measurement by Gas Chromatography* (Wiley, 1979).
2. SYNOVEC, R. E., MOORE, L. K., RENN, C. N. and HANCOCK, D. O.: *Internat. Lab.* (Dec. 1989), 16. New directions in process liquid chromatography.
3. GUIOCHON, G. and GUILLEMIN, C. L.: *Quantitative Gas Chromatography for Laboratory Analysis and On-Line Process Control* (Elsevier, 1988).
4. BAILEY, J. E. and OLLIS, D. F.: *Biochemical Engineering Fundamentals*, 2nd edn. (McGraw-Hill, 1986).
5. LITTLEWOOD, A. B.: *Gas Chromatography*, 2nd edn. (Academic Press, 1970).
6. GLUECKAUF, E.: *Trans. Faraday Soc.* **51** (1955) 34. Theory of chromatography, Part 9: the theoretical plate concept in column separation.
7. MARTIN, A. J. P. and SYNGE, R. L. M.: *Biochem. J.* **35** (1941) 1358. A new form of chromatography employing two liquid phases.
8. PURNELL, J. H.: *Gas Chromatography* (Wiley, 1962).
9. JÖNSSON, J. Å., in *Chromatographic Theory and Basic Principles* (JÖNSSON, J. Å. ed.), Chapter 3, Dispersion and Peak Shapes in Chromatography (Marcel Dekker, 1987).
10. VAN DEEMTER, J. J., ZUIDERWEG, F. J. and KLINKENBERG, A.: *Chem. Eng. Sci.* **5** (1956) 271. Longitudinal diffusion and resistance to mass transfer as causes of nonideality in chromatography.
11. CONDER, J. R. and PURNELL, J. H.: *Chem. Eng. Sci.* **25** (1970) 353. Separation and throughput in production and preparative chromatography.
12. ROZ, B., BONMATI, R., HAGENBACH, G. and VALENTIN, P.: *J. Chromatog. Sci.* **14** (1976) 367. Practical operation of prep-scale gas chromatographic units.
13. CONDER, J. R.: *J. Chromatog.* **256** (1983) 381. Design procedure for preparative and production gas chromatography.
14. PHILLIPS, C. S. G. and SCOTT, C. G., in *Progress in Gas Chromatography* (PURNELL, J. H. ed.), pp. 121–152, Modified Solids for Gas–Solid Chromatography (Wiley, 1968).
15. AL-THAMIR, W. K., LAUB, R. J. and PURNELL, J. H.: *J. Chromatog.* **142** (1977) 3. Gas chromatographic separation of all $C_1$–$C_4$ hydrocarbons by multi-substrate gas–solid–liquid chromatography.
16. MOHR, P. and POMMERENING, K.: *Affinity Chromatography* (Marcel Dekker, 1985).
17. PORATH, J.: *J. Chrom.* **218** (1981) 241. Development of modern bioaffinity chromatography.

18. PERRUT, M.: *J. Chromatogr.* **A, 658** (1994) 293. Advances in supercritical fluid chromatographic processes. (Review).
19. JUSFORGUES, P.: In *Process-Scale Liquid Chromatography*, SUBRAMANIAN, G. (ed.) (VCH Verlagsgesellschaft, 1995) Chapter 7. Separation in large scale industrial supercritical fluid chromatography.
20. BEVAN, C. D., MELLISH, C. J.: In *Process-Scale Liquid Chromatography*, SUBRAMANIAN, G. (ed.) (VCH Verlagsgesellschaft, 1995) Chapter. 8. Scaling-up of supercritical fluid chromatography to large-scale applications.
21. SCHULTE, M. and STRUBE, J.: *J. Chromatogr*. **A, 906** (2001) 399. Preparative enantioseparation by simulated moving bed chromatography.
22. FRANCOTTE, E.: *J. Chromatogr*. **A, 906** (2001) 379. Enantioselective chromatography as a powerful alternative for the preparation of drug enantiomers.
23. WHITE, C. A.: In *Preparative and Process-Scale Liquid Chromatography*, SUBRAMANIAN, G. (ed.) (Ellis Horwood, 1991) Chapter 13. An Introduction to large-scale enantioseparation.
24. RENDELL, M.: *Process Engineering* (April 1975) 66. The real future for large-scale chromatography.
25. SAID, A. S.: *American Laboratory* (June 1983) 17; *ibid*: (August 1983) 38. Calculations of a continuous and preparative gas chromatograph.
26. SHINGARI, M. K., CONDER, J. R. and FRUITWALA, N. A.: *J. Chromatog.* **285** (1984) 409. Construction and operation of a pilot scale production gas chromatograph for separating heat-sensitive materials.
27. RYAN, J. M. and DIENES, G. L.: *Drug. and Cosmetic Industry* **99** (4) (1966) 60. Plant-scale gas chromatography.
28. TIMMINS, R. S., MIR, L. and RYAN, J. M.: *Chem. Engg. Albany*, **76** (19 May, 1969) 170. Large-scale chromatography: new separation tool.
29. VALENTIN, P., HAGENBACH, G., ROZ, B. and GUIOCHON, G.: In *Gas Chromatography* 1972, PERRY, S.G., ed. (Institute of Petroleum, 1973) p. 157, New Advances in the Operation of Large-Scale Gas Chromatographic Units.
30. BONMATI, R. and GUIOCHON, G.: *Perfumer and Flavourist* **3** (October, 1978) 17. Gas chromatography as an industrial process operation — application to essential oils.
31. SAKODYNSKII, K. I., VOLKOV, S. A., KOVANKO, Y. A., ZELVENSKII, V. U., REZNIKOV, V. I. and AVERIN, V. A., *J. Chromatog.* **204** (1981) 167. Design of and experience in operating technological preparative installations.
32. HILAIREAU, P. and COLIN, H.: *Chromatogr. Soc. Bull.* **26** (1987) 10. Gas chromatography: a real industrial separation process.
33. CONDER, J. R.: *Manuf. Chemist.* **55** (1984) 38. GC scales up to production.
34. CAREL, A. B., CLEMENT, R. E. and PERKINS, G.: *J. Chromatog. Sci.* **7** (1969) 218. Construction and operation of a gas chromatograph using sectional one foot diameter columns.
35. CAREL, A. B. and PERKINS, G.: *Analyt. Chim.* **34** (1966) 83. Gas chromatographic fractionation as a supplement and replacement for laboratory distillation.
36. GYIMESI, J. and SZEPESY, L.: *Chromatographia* **9** (1976) 195. Experiences with the development and operation of a large scale preparative gas chromatograph.
37. Literature of Prochrom SA (later Separex), B.P.9, 54250 Champigneulles, France.
38. BERNARD, J. R., GOURLIA, J-P. and GUTTIERREZ, J.: *Chem. Eng. Albany* **88** (18 May, 1981) 92. Separating paraffin isomers using chromatography.
39. VERZELE, M., DE CONINCK, M., VINDEVOGEL, J. and DEWAELE, C.: *J. Chromatog.* **450** (1988) 47. Column hardware in preparative liquid chromatography.
40. DONE, J. N.: *J. Chromatog.* **125** (1976) 43. Sample loading and efficiency in adsorption, partition and bonded-phase high-speed liquid chromatography.
41. PIRKLE, W. H.: cited in ref. (42).
42. HAYWOOD, P. A. and MUNRO, G., in *Developments in Chromatography*, Vol. 2 (KNAPMAN, C. E. H. ed.), Chapter 2, Preparative Scale Liquid Chromatography (Applied Science, 1980).
43. WANKAT, P. C.: In *Handbook of Separation Process Technology*, ROUSSEAU, R. W., ed. (Wiley, 1987) Chapter 14, Large-Scale Chromatography.
44. JONES, K.: *Symposium on Advances in Chromatography* (Chromatographic Society and Royal Society of Chemistry), Warrington, November 1986, reported in *Chrom. Soc. Bull.* **25** (1987) 23. Large-Scale HPLC.
45. HUPE, K. P.: *J. Chromatog. Sci.* **9** (1971) 11. Efficiency of production GLC.
46. CONDER, J. R.: In *New Developments in Gas Chromatography*, PURNELL, J. H., ed. (Wiley, 1973) 137–186, Production-Scale Gas Chromatography.
47. GAREIL, P. and ROSSET, R.: *Analysis* **10** (1982) 397. La chromatographie en phase liquide préparative par développment par élution.
48. CONDER, J. R. and SHINGARI, M. K.: *J. Chromatog. Sci.* **11** (1973) 525. Throughput and band overlap in production and preparative chromatography.

49. CHEH, C. H.: *J. Chromatogr.* **A, 658** (1994) 283. Advances in preparative gas chromatography for hydrogen isotope separation.
50. CONDER, J. R.: *Chromatographia* **8** (1975) 60. Performance optimisation for production gas chromatography.
51. MOSCARIELLO, J., PURDOM, G., COFFMAN, J., ROOT, T. W. and LIGHTFOOT, E. N.: *J. Chromatogr.* **A, 908** (2001) 131. Characterising the performance of industrial-scale columns.
52. KNOX, J. H. and PYPER, H. M.: *J. Chromatog.* **363** (1986) 1. Framework for maximising throughput in preparative liquid chromatography.
53. GRUSHKA, E., ed.: *Preparative-Scale Chromatography* (Dekker, 1989). Reprinted from *Sep. Sci. Technol.* **22** (Nos. 8–10), (1987) 1791–2110. (Several articles in this volume deal with optimisation).
54. COX, G. B. and SNYDER, L. R.: *LC-GC* **6** (1988) 894. Preparative and process-scale HPLC.
55. GOLSHAN-SHIRAZI, S. and GUIOCHON, G.: *J. Chromatog.* **A, 658** (1994) 149. Modelling of preparative liquid chromatography.
56. NICOUD, R. M. and PERRUT, M.: in COSTA, C. A. and CABRAL, J. S. (eds.) *Chromatographic and Membrane Processes in Biotechnology* (Kluwer Academic Publishers, 1991) p.381. Operating modes, scale-up and optimisation of chromatographic processes.
57. PORSCH, B.: *J. Chromatogr.* **A, 658** (1994) 179. Some specific problems in the practice of preparative high performance liquid chromatography.
58. LEAVER, G., CONDER, J. R. and HOWELL, J. A.: *Sep. Sci. Tech.* **22** (1987) 2037. A method development study of the production of albumin from animal blood by ion-exchange chromatography.
59. CHASE, H. A.: *J. Chromatog.* **297** (1984) 179. Prediction of the performance of preparative affinity chromatography.
60. ARNOLD, F. H., CHALMERS, J. J., SAUNDERS, M. S., CROUGHAN, M. S., BLANCH, H. W. and WILKE, C. R.: *ACS Sympos. Ser.* **271** (1985), Chap. 7. A rational approach to the scale-up of affinity chromatography.
61. HORSTMAN, B. J. and CHASE, H. A.: *Chem. Eng. Res. Des.* **67** (1989) 243. Modelling the affinity adsorption of immunoglobulin G to protein A immobilised to agarose matrices.
62. FERNANDEZ, M. A., LAUGHINGHOUSE, W. S., and CARTA, G.: *J. Chromatogr.* **A, 746** (1996) 184–198. Characterisation of protein adsorption by composite silicapolyacrylamide gel anion exchanges II: mass transfer in packed columns and predictability of breakthrough behaviour.
63. CONDER, J. R. and HAYEK, B. O.: *Biochem. Eng. J.* **6** (2000) 225. Adsorption and desorption kinetics of bovine serum albumin in ion-exchange and hydrophobic interaction chromatography on silica matrices.
64. LEONARD, M.: *J. Chromatogr.* B, **699** (1997) 3. New packing materials for protein chromatography.
65. ROPER, D. K. and LIGHTFOOT, E. N.: *J. Chromatogr.* A, **702** (1995) 3. Separation of biomolecules using adsorptive membranes.
66. STOWELL, J. D. BAILEY, P. J. and WINSTANLEY, D. J. (eds.): *Bioactive Microbial Products* 3: *Downstream Processing* (Academic Press, 1986).
67. VERRALL, M. S. and HUDSON, M. J. (eds.): *Separations for Biotechnology* (Ellis Horwood, 1987).
68. MCGREGOR, W. C.: *Ann. N. Y. Acad. Sci.* **413** (1983) 231. Large scale isolation and purification of proteins from recombinant *E. coli.*
69. ANSPACH, F. B., CURBELO, D., HARTMANN, R., GARKE, G. and DECKWER, W. D.: *J. Chromatogr.* A, **865** (1999) 129. Expanded-bed chromatography in primary protein purification.
70. MENET, J. M. and THIÉBAUT, D.: *Countercurrent Chromatography* (Marcel Dekker, 1999).
71. GANETSOS, G. and BARKER, P. E.: *Preparative and Production Scale Chromatography* (Marcel Dekker, 1993).
72. PYNNONEN, B.: *J. Chromatogr.* **A, 827** (1998) 143. Simulated moving-bed processing: escape from the high-cost box.
73. CHARTON, F. and NICOUD, R. M.: *J. Chromatogr.* **A, 702** (1995) 97. Complete design of a simulated moving bed.
74. AZEVEDO, D. C. S., PAIS, L. S. and RODRIGUES, A. E.: *J. Chromatogr.* **A, 865** (1999) 187. Enantiomers separation by simulated moving-bed chromatography.
75. BARKER, P. E., in *Developments in Chromatography Vol.* 1 (ed. KNAPMAN, C. E. H.), Chapter 2, Developments in Continuous Chromatographic Refining (Applied Science, 1978).
76. PURNELL, J. H.: *Phil. Trans. Roy. Soc. Lond.* **A305** (1982) 657. The current chromatographic scene.
77. COCA, J., ADRIO, G., JENG C. Y. and LANGER, S. H.: In *Preparative and Production-scale Chromatography*, GANETSOS, G. and BARKER, P. E. (eds.) (Marcel Dekker, 1993), Chapter 19. Gas and liquid chromatographic reactors.
78. ADACHI, S.: *J. Chromatogr.* **A, 658** (1994) 271. Simulated moving-bed chromatography for continuous separation of two components and its application to bioreactors.
79. CONDER, J. R. and FRUITWALA, N. A.: *Chem. Eng. Sci.* **36** (1981) 509. Comparison of plate numbers and column lengths in chromatography and distillation.

## 19.12. NOMENCLATURE

| | | Units in SI System | Dimensions in **M, N, L, T** |
|---|---|---|---|
| $A$ | Coefficient in equation 19.10 (Table 19.1) | m | **L** |
| $B$ | Coefficient of $1/u$ in equation 19.10 (Table 19.1) | $m^2/s$ | $\mathbf{L}^2\,\mathbf{T}^{-1}$ |
| $C_s, C_m$ | Coefficient of $u$ in equation 19.10 (Table 19.1) | s | **T** |
| $c$ | Concentration of solute in mobile phase | $kg/m^3$ | $\mathbf{ML}^{-3}$ |
| $D_m$ | Diffusivity of solute in mobile phase | $m^2/s$ | $\mathbf{L}^2\,\mathbf{T}^{-1}$ |
| $D_s$ | Diffusivity of solute in stationary phase | $m^2/s$ | $\mathbf{L}^2\,\mathbf{T}^{-1}$ |
| $d_f$ | Effective thickness of stationary liquid "film" | m | **L** |
| $d_p$ | Diameter of packing particles | m | **L** |
| **H** | Plate height | m | **L** |
| $K$ | Distribution coefficient of solute between the two phases | — | — |
| $k'$ | Capacity factor (mass distribution coefficient, $= n_s/n_m$); mean capacity factor for two solutes | — | — |
| $L$ | Column length | m | **L** |
| **N** | Number of theoretical plates in column | — | — |
| $n_m, n_s$ | Number of moles of solute in equilibrated mobile and stationary phases in elemental length of column | kmol | **N** |
| $q$ | Concentration of solute in/on stationary phase | $kg/m^3$ | $\mathbf{ML}^{-3}$ |
| $R$ | Retention ratio (equation 19.1) | — | — |
| $R_s$ | Resolution | — | — |
| $t_M$ | Mobile phase hold-up time in column | s | **T** |
| $t_R$ | Retention time of solute in column | s | **T** |
| $t'_R$ | Adjusted retention time, $= t_R - t_M$ | s | **T** |
| $t_w$ | Band width measured at base (Figure 19.3) | s | **T** |
| $u$ | Average mobile-phase velocity, = (superficial velocity)/$\varepsilon$ | m/s | $\mathbf{LT}^{-1}$ |
| $u_R$ | Velocity of migration of solute band | m/s | $\mathbf{LT}^{-1}$ |
| $\alpha$ | Separation factor (selectivity factor) (equation 19.12) | — | — |
| $\gamma$ | Obstruction (labyrinth) factor for diffusion through packed bed | — | — |
| $\varepsilon$ | Packing voidage (interstitial plus intraparticle) | — | — |
| $\lambda$ | "Eddy diffusion" constant in packed bed | — | — |
| $\omega$ | Packing geometry factor | — | — |

CHAPTER 20

# Product Design and Process Intensification

## 20.1. PRODUCT DESIGN

### 20.1.1. Introduction

Because, in its earlier years, chemical engineering was overshadowed by the requirements of the bulk chemical and petroleum industries, it was concerned with operations for the large volume manufacture of relatively low value materials of simple structures. Initially, chemical engineers made a major contribution in the development of separation processes — an area that was largely neglected and little understood by chemists. The study of the design and operation of chemical reactors came to the fore only in the early 1950's at a time when the importance of flow patterns and residence time distributions was only just being appreciated. Pioneering work in this field was carried out by DANCKWERTS[1,2] whose classic papers form the foundation for much of the later work. On reflection, it seems incredible that so little attention had been given to optimising the whole system, that is the reactor and the downstream processing. Many of the problems inherent in the separation of reactor products were attributable to the absence of any real attempt to design the reactor in such a way so as to maximise the yield of the desired product. In many ways, chemical engineers were the victims of their own success, in that they concentrated overmuch on the design of processes and paid little attention to the design of products for developing markets.

At this time, the turn of the millennium, interest has rapidly turned to meeting the needs and aspirations of an ever-demanding consumer industry that needs to supply products directly to the end-user. Bulk chemicals, as such, have always been predominantly intermediates forming the feedstock for the production of the final products to be used by the consumer industry, or by a proxy consumer in the case of health-care products. For bulk, or 'commodity' chemicals, price competition is severe and the tendency is for production to be located in those parts of the world where costs of labour, raw materials, energy and so on are low. Economy of scale has been an important factor with the result that production is tending to take place in a very few plants of high throughputs. The plants themselves are viable only if they operate with a very efficient utilisation of resources of all kinds and, at the same time, satisfy the requirements for safe operation and for being 'environmentally friendly'. Thus to work in this field is in no sense a 'soft option'. With bulk chemical operation, there is also a 'squeeze' which is becoming increasingly more severe. Customers are insisting on tighter specifications for final products while, at the same time, raw material quality is deteriorating as a result of the tendency to use up the best resources of raw materials first. The continuing challenge is therefore to make a better product from a lower quality starting material.

One feature of many commodity chemicals is that they are essentially intermediates used in the production of a wide range of consumer products. It was once suggested that the per capita rate of production of sulphuric acid was a measure of a country's prosperity, although the consumer demand for sulphuric acid itself is almost zero. Similarly styrene, most of which is converted into polystyrene, is hardly a saleable product in the retail market. The exception, on the other hand, is the market for materials that are finally utilised as fuels many of which are distributed to the final consumer following various degrees of 'polishing'. Even so, the greatest demand for fuels comes from the electrical supply industry which dominates the market, and not from individual consumers.

The shift towards synthetic speciality chemicals and products has been marked in recent years. There have been several driving forces at work. The first is the realisation, even by companies sitting on very large amounts of raw materials, that reserves are limited and that, in the long term anyway, it is in their interest to upgrade at least part of their reserves into higher value products. Thus, for instance, in the china clay industry, the vast bulk of whose product finds its application as a filler or surface-coating agent in the paper industry, kaolin is used to form lightweight structural materials on the one hand and support materials in chromatographic columns on the other. The importance of added-value is rapidly being appreciated. Many of the more sophisticated products are replacing materials from natural sources, and there is a large expansion in the sale of factory-made products both in the food and in the cosmetics industries. Ice-creams, now often of highly complex formations, are structured materials consisting of tiny air bubbles and fats dispersed in an essentially aqueous continuous phase. Selling air dispersed in water has always been a very attractive proposition! The structure, involving as it does, the size distribution, the concentration and the stability of air bubbles and the size and form of the ice crystals is of critical importance. These factors all contribute to give the desired rheology and structure, which are at least as important as the material composition. Tooth paste must have the correct rheology — it must not run out of the tube of its own volition, and yet it must be easy to extrude it on to the brush where it will remain in place until it is sheared when used on the teeth. In other words, it needs to be a shear-thinning material with a yield stress. These characteristics are discussed in Volume 1, Chapter 3. Face creams must above all have an appropriate texture, must be easy to apply, must stay in place and must 'feel right' — a the most difficult condition to define! Consumers are seldom concerned with the chemical make-up, except to satisfy themselves that it is not injurious, though they need to be satisfied that it is 'fit for purpose'. Providing the appropriate feel, whether it be for a foodstuff or cosmetic, depends on getting the correct microstructure, which in turn exerts a strong influence on the rheology of the product. Understanding the nature of the microstructure of the material is therefore of paramount importance.

In general, the required production rates of these more sophisticated materials are orders of magnitude less than those of the commodity materials referred to earlier. Some will be made in dedicated plant but many others may be manufactured in multi-product plant, which present new problems in the scheduling of efficient production. These include:

(a) the need to provide buffer to storage to cover demand when the equipment is being used for other products.

(b) the need for the plant to be designed so as to facilitate cleaning between runs,
(c) the batch times must be optimised — short runs mean that the downtime becomes unacceptably long, long runs mean that the buffer storage needs to be greater,
(d) sequencing of the runs on the various products can sometimes be arranged so that the first product in a sequence is the one which is the least tolerant to cross-run contamination and so on. This can mean that the intermediate cleaning operations within a sequence may need to be less through than those between sequences.

To quote from CUSSLER and MOGGRIDGE[3], 'Product design is the procedure by which customer needs are translated into commercial products'. This involves assessing the essential requirements in a material which will satisfy the customer, and then designing a material with the requisite physical and sometimes chemical and structural properties. There may well be a very large number of ways in which the needs can be met, possibly by using starting materials with widely different chemical compositions, and the final judgement will be based on a complex synthesis of considerations of competing attributes and costs of both raw materials and processing. Finally, the product must be 'safe' to use and must not have harmful environmental features.

In general, there is a good correlation between the selling price per unit mass of the material, $C$, and its production rate, $P$. DUNNILL[4] has produced a logarithmic plot of unit price against production rate for biochemical products, and this covered several orders of magnitude. At the high cost end are expensive pharmaceutical products and at the bottom cost limit is water (off the scale), with a wide range of products of intermediate complexity in between. The plot, which is reproduced in Figure 20.1, is seen to be represented by a straight line, of negative slope $-n$, which passes approximately through the majority of points giving an expression of the form:

$$C = \text{constant } P^{-n} \tag{20.1}$$

If $n$ were equal to unity, it would imply that the annual production or utilisation value of all the components considered was approximately constant, a somewhat striking situation!

Although Figure 20.1 is based on information relating only to the biochemical industry and the absolute level of prices and throughputs may be significantly out-of-date, the general trend is also applicable to other industrial areas as well — even to the production of motor vehicles!

## 20.1.2. Design of particulate products

EDWARDS and INSTONE[5], have reviewed the manufacture and use of the particulate products which, as they describe it, are made by the 'fast moving consumer goods' industry and used by consumers around the world. It is claimed that all the products of this sector have the following common features:

(a) The products are created from a range of raw materials to yield complex multiphase mixtures which include emulsions, suspensions, gels, agglomerates and so on.

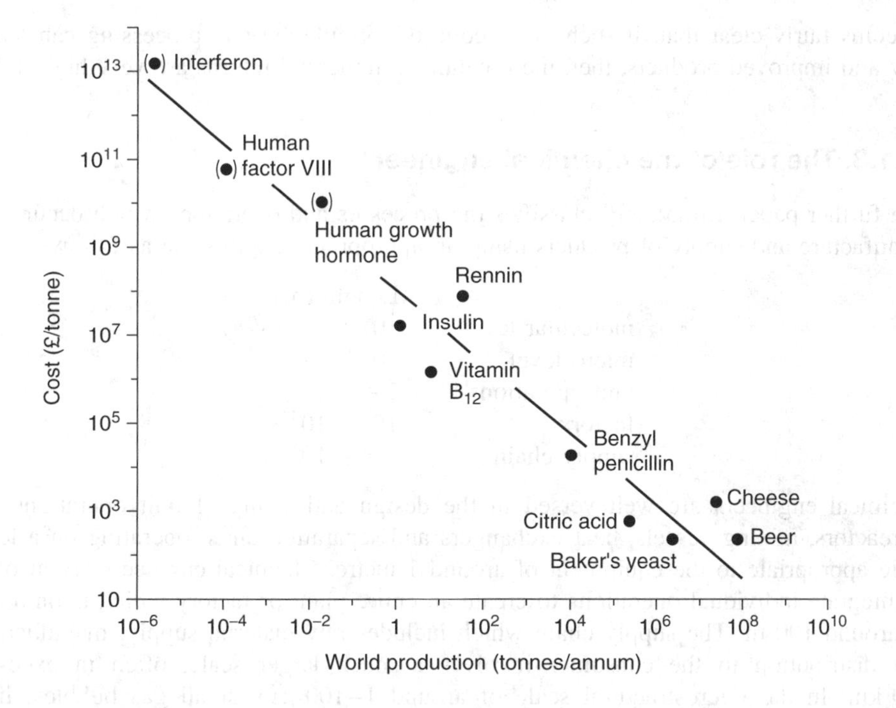


Figure 20.1. World production tonnages and prices per tonne of some products of biotechnology[(4)] (1983 data)

(b) During the processing of the product, a microstructure is created on the scale of 1–100 μm.
(c) This microstructure is usually required to be stable through the supply chain until the product is used by the consumer.
(d) It is the microstructure which determines the appearance of the product and its efficacy in use.
(e) The microstructure assembled during processing is destroyed during the use of the product.

Edwards and Instone discuss the production of particulate products by spray drying and by binder granulation since these two processes are widely used because of their capability to produce multi-component, designed granules aimed at meeting particular consumer needs. It is shown that the product microstructure resulting from such processing depends upon a complex dynamic interaction between the ingredients of the formulation and the processing conditions used during the manufacture. It is this microstructure which is generally destroyed when the product is used by the consumer and this is also a complex, dynamic interaction between the applications of conditions and the formulated microstructure. Edwards and Instone argue that the understanding of the control of microstructure in particular products is far from complete and that interdisciplinary research, ranging form measurement science, phase equilibrium and microstructure kinetics to process engineering is needed to advance knowledge in this demanding area.

It seems fairly clear that, if such knowledge of formulation and processing can lead to new and improved products, then the potential commercial returns are very high indeed.

### 20.1.3. The role of the chemical engineer

In a further paper, EDWARDS[6] classifies the processes and operations which occur in the manufacture and supply of products using an appropriate length scale as follows:

| | Length (m) |
|---|---|
| molecular level | $10^{-10} - 10^{-7}$ |
| micro level | $10^{-6} - 10^{-3}$ |
| unit operations | $1 - 10$ |
| factory | $10^2 - 10^3$ |
| supply chain | $10^3 - 10^6$ |

Chemical engineers are well versed in the design and sizing of unit operations such as reactors, mixing vessels, heat exchangers and separation units, operating on a length scale appropriate to the equipment of around 1 metre. Chemical engineers are also able to integrate individual operations to create an entire plant or factory which is on a scale of around 100 m. The supply chain which includes raw material supply, manufacturing and distribution to the consumer involves a much larger scale, often in excess of 100 km. In the microstructural scale of around 1–100 μm, small gas bubbles, liquid droplets and suspended fine particles in multiphase products are encountered together with microstructures created by surfactants, polymers, clays and so on. The molecular reactions that create this microstructure take place within an even finer level of scrutiny. The microstuctures of products can be very complex as, for example, when a product contains more than ten components and where processing can involve flows that create wide residence time distributions and varying stress and temperature levels. Edwards argues that chemical engineers can provide a key role in producing optimum microstructures provided they can link the physical and chemical sciences of microstructure formation with processing conditions. In this way, chemical engineers must, in addition to dealing with flow and heat transfer in complex equipment, be able to determine the associated product structure. In the supply chain, the total system from the raw material supply, through manufacturing to distribution must be optimised, ensuring that the desired microstructure is delivered intact to the consumer. Cost effective solutions within this supply chain require a systems engineering approach which, Edwards claims, chemical engineers are well placed to tackle if, in addition to the heartland of processing, the challenges of purchasing, material supply, packing activities and distribution can also be met.

### 20.1.4. Green chemistry

Increasing concern for the need to conserve and to use effectively world reserves of raw materials and, at the same time, to reduce the quantities of waste materials which are likely to have an adverse effect on the environment has led to pressure for the increased use of renewable resources and so-called 'green chemistry'. The principles of green chemistry have been enunciated by HAMLEY and POLIAKOFF[7] as follows:

(a) It is better to prevent waste than to treat or clean up waste after it is formed.
(b) Synthetic methods should be designed to maximise the incorporation of all materials used in the process into the final product.
(c) Wherever practicable, synthetic methodologies should be designed to use and generate substances that possess little or no toxicity to human health and the environment.
(d) Chemical products should be designed to preserve efficacy of function while reducing toxicity.
(e) The use of auxiliary substances, such as solvents and separation agents, should be made unnecessary wherever possible, and innocuous when used.
(f) Energy requirements should be recognised for their environmental and economic impacts and should be minimised. Synthetic methods should be conducted at as close as possible to ambient temperature and pressure.
(g) A raw material or feedstock should be renewable, rather than depleting, wherever this is technically and economically practicable.
(h) Unnecessary derivatisation (blocking group, protection/deprotection, temporary modification of physical/chemical processes) should be avoided wherever possible.
(i) Catalytic reagents (as selective as possible) are superior to stoichiometric reagents.
(j) Chemical products should be designed so that at the end of their function they do not persist in the environment but break down into innocuous degradation products.
(k) Analytical methodologies need to be developed further to allow for real-time in-process monitoring to minimise the potential for chemical accidents, including releases, explosions, and fires.

The current emphasis on the production of very high value products has led to a complete rethinking of the way in which processes are carried out. There are considerable gains to be achieved by carefully controlling the conditions in a chemical reactor in order to minimise the formation of unwanted products. In many cases, the product itself has an inhibitory effect on the progress of the reaction, and considerable gains in productivity can often be achieved in combining the reaction and separation stages into a single unit. The concept is not new in the sense that reactive distillation, in which, in effect, a chemical reaction takes place within a column with continuous separation of the products, has been practiced for many years. The technique is now being applied over a far wider range of conditions.

### 20.1.5. New processing techniques

Although the general principles of separation processes are applicable widely across the process industries, more specialised techniques are now being developed. Reference is made in Chapter 13 to the use of supercritical fluids, such as carbon dioxide, for the extraction of components from naturally produced materials in the food industry, and to the applications of aqueous two-phase systems of low interfacial tensions for the separation of the products from bioreactors, many of which will be degraded by the action of harsh organic solvents. In many cases, biochemical separations may involve separation processes of up to ten stages, possibly with each utilising a different technique. Very often, differences in both physical and chemical properties are utilised. Frequently the materials are stereo-isomers, differing only in the relative spatial orientations of the groups. With pharmaceutical products, near complete separation of the isomers is essential, one having the desired therapeutic properties and the other possibly being highly toxic, or even worse, as in the case of phthalidamide. Separations are then only possible by using 'molecular recognition' techniques and high resolution columns with the packing treated with a suitable ligand to which one of the isomers will selectively attach itself. This is sometimes referred to as a 'lock and key' situation.

For many of the new highly specialised products, both reactors and separation stages need to be designed for low holdups and rapid processing, and these conditions may frequently be achieved by carrying out the processes under conditions of 'accelerated gravity' by the deployment of centrifugal in place of the normal gravitational field. Under such conditions, retention times and holdups are low — the latter being of particular importance when multipurpose plant is used and where shorter cleaning times minimise the loss of material when changing from one product to the next. Furthermore, reduction of holdup may contribute to safer operation when the material has hazardous properties. The use of intensified fields for separation processes is not confined solely to gravitational fields. In many cases, differences in, say, electrical or magnetic properties may form the essential driving force for separation. Where significant differences in any single property are not sufficiently great, the use of combined fields, gravitational plus magnetic or electrical, for example, may provide the most satisfactory basis for separation.

In the following section, the use of intensified (gravitational) fields is given as an example of the way in which the operation of both reactors and separation units may be intensified and smaller, more efficient, units may be developed.

## 20.2. PROCESS INTENSIFICATION

### 20.2.1. Introduction

In many unit operations such as distillation, absorption and liquid–liquid extraction, fluids pass down columns solely under the force of gravity and this limits not only the flowrates that may be attained, but also rates of mass and also heat transfer. When a force other than gravity is utilised, such as a centrifugal force for example, then, in theory, there is no limit to the force which can be applied nor indeed to the increase in heat and mass transfer rates that may be achieved. A reduction in residence times is also possible and this leads to a decrease in the physical size of the plant itself. One fairly simple example of this reduction in plant size, or *process intensification* as it is termed, may be seen by comparing the relatively vast size of a settling tank with the very modest size of a centrifuge accomplishing the same task. The advantage of reducing plant size in this way has been prompted by the fairly recent requirement for processing units which are suitable for confined spaces such as, for example, on oil-drilling rigs. There are also important benefits to be gained, however, from reducing the size of land-based units and not least in minimising intrusions on the environment. Rotating devices probably provide the most important way of achieving this processing intensification and, where very thin films are produced on spinning discs, for example, these have the advantage that mass and heat transfer take place in this thin film. This means that the resistance to diffusion from the film to the bulk of the fluid is low. Similar considerations apply not only to physical operations but also to those systems involving a chemical reaction.

It is the aim of this section to offer but a brief introduction to this relatively new field of endeavour that will surely have a not inconsiderable role in shaping the whole future of both physical and chemical processing. It may be noted at the outset, however, that it is not the use of centrifugal forces which is a relatively new development, but rather their application in spinning discs and indeed, centrifugal devices are already widely used in processing, as indicated by the following examples:

(a) Centrifugal fluidised beds, as described in Section 6.3.6., are being used much more widely mainly due to the fact that the centrifugal field increases the minimum fluidising velocity and consequently increases the flowrates which can be handled. Much smoother fluidisation may also be achieved.
(b) In Chapter 13 on liquid–liquid extraction, various rotating devices are described which include the Podbielniak extractor, the Alfa-Laval centrifugal extractor and the Scheibel column.
(c) The whole of Chapter 9 is devoted to a discussion of the use of centrifugal forces for carrying out the processes of settling, thickening and filtration.

It is for this reason that the present chapter is in essence concerned with recent developments in which spinning discs are the dominant feature.

### 20.2.2. Principles and advantages of process intensification

Process intensification, pioneered by RAMSHAW[8] in the 1980's, may be defined as a strategy which aims to achieve process miniaturisation, reduction in capital cost, improved inherent safety and energy efficiency, and often improved product quality. In recent years, process intensification has been seen to provide processing flexibility, just-in-time manufacturing capabilities and opportunities for distributed manufacturing. In order to develop a fully intensified process plant, it is essential that all the unit operation systems, that is reactors, heat exchangers, distillation columns, separators and so on, should be intensified. Wherever possible, the aim should be to develop and use multi-functional modules for performing heat transfer, mass transfer, and separation duties. As discussed by STANKIEWICZ and MOULIJN[9], process intensification encompasses not only the development of novel, more compact equipment but also intensified methods of processing which may involve the use of ultrasonic and radiation energy sources.

Additional benefits of process intensification include improved intrinsic safety, simpler scale-up procedures, and increased energy efficiency. Adopting a process intensification approach can substantially improve the intrinsic safety of a process by significantly reducing the inventory of potentially hazardous chemicals in the processing unit. A further advantage of process intensification is that it allows the replacement of batch processing by small continuous reactors, which frequently give more efficient overall operation especially in the case of highly exothermic reactions where heat can be rapidly removed continuously as it is being released. The inherent safety aspect of process intensification, and its role in minimising hazards in the chemical and process industries, has been discussed in a recent article by HENDERSHOT[10], who has suggested that the design considerations to be taken into account in intensifying a process include:

(a) is the process based on batch or continuous technology?
(b) what is the rate-limiting step — heat transfer, mass transfer, mixing and so on?
(c) what are the appropriate intensification tools, modules and concepts?
(d) is it possible to eliminate solvents?
(e) is it possible to use supported catalysts?
(f) can pressure and temperature gradients be reduced?
(g) can the number of processing steps be reduced by using multifunctional modules?
(h) does it achieve the ultimate aim of enhancing transport rates by orders of magnitude?

Process intensification may also be seen as an ideal vehicle for performing chemical reactions based on what is known as 'green chemistry' as discussed in section 20.1.4. Intensification can provide appropriate reactor technologies for utilising heterogeneous catalysis, phase transfer catalysis, supercritical-fluid chemistry and ionic liquids. Process intensification allows chemical processes to be accelerated by using reactors such as spinning disc reactors, heat exchanger (HEX) reactors, oscillatory baffle reactors, microwave reactors, micro-reactors, cross-corrugated membrane reactors and catalytic plate reactors. For example, intensification may permit the use of higher reactant concentrations giving significantly beneficial effects with regard to the kinetics, selectivity and inventory. Often, due to limitations attributable to the heat and mass transfer resistances and inadequate mixing in the reactor, the effective concentrations of reactants are reduced resulting in slow rates, poor selectivity and the need for extensive downstream separation processing. Process intensification thus presents a range of exciting processing tools and opportunities in processing, which have seldom been explored in the past. It now offers the following characteristics:

(a) it gives every molecule approximately the same processing experience.
(b) it matches the mixing and transport rates to the reaction rate.
(c) significant enhancements are offered in heat and mass transfer rates.
(d) the reaction rate is limited only by the design and performance of the equipment.
(e) selectivity and yield are both improved.
(f) product quality and specification are both improved.
(g) a rapid grade change is possible because of the low hold-up and ease of cleaning the equipment.
(h) a rapid response to set-points is possible.
(i) for certain processes the laboratory-scale equipment may constitute the full-scale unit.

On the aesthetic side, it is likely that intensified process plants will be less intrusive on the environment, making them far less of an eyesore than the unsightly and massive constructions that are characteristic of present processing units. In some cases the plant may be mobile, thereby offering the opportunity for distributed manufacturing of chemicals close to the point of utilisation. This may reduce the quantities of hazardous products currently being transported by road and rail, thereby improving safety. The improved energy efficiency obtainable in intensified unit operations constitutes yet another highly attractive benefit in a world where there is overwhelming concern over the ever-growing demand on non-renewable energy resources, and also over the release of greenhouse gases such as carbon dioxide. In this respect, there is a great and urgent need for the development of new process technologies, which will utilise energy in an efficient manner, and process intensification may represent a positive step in this direction. Large enhancements in the rates of heat and mass transfer, two of the most fundamental and frequently encountered in process engineering, can be achieved in intensified units. Such improvements could permit processing times and the associated energy consumption to be dramatically reduced for a given operation. An additional advantage in the case of packed bed units is that the operating range is considerably increased by the use of a centrifugal force. This may be seen by an examination of Figure 4.21, where it is quite clear that an increase in '$g$' which appears in the ordinate will bring many operating points well into the region below the flooding curve.

Green *et al.*[11] have discussed and detailed a series of possible intensified processes that include nitration in a compact heat exchanger, a gas–liquid reaction using static mixers and hypochlorous acid production in a rotating packed bed. All of these clearly illustrate the benefits of process intensification that may be achieved in real processing situations. Keller and Bryan[12] have recently suggested that process intensification will dictate the future advancement of the chemicals and process industries. The process benefits that may be achieved based on green chemistry are shown in Figure 20.2.

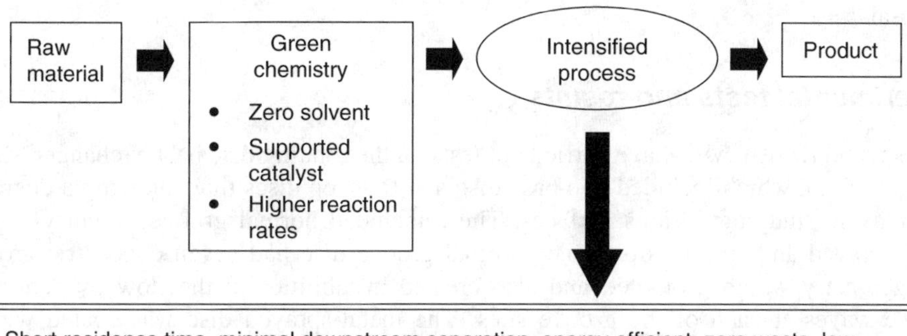


Figure 20.2. Process characteristics of an intensified plant using green chemistry

## 20.2.3. Heat transfer characteristics on rotating surfaces

### *Introduction*

Heat transfer has been identified by Reay[13] as an important area in which process intensification is expected to offer major benefits in terms of energy efficiency, pollution control and plant operating costs. So-called *passive* techniques including modifying the walls of a plant unit, for example, are routinely used to improve heat transfer coefficients in evaporation and condensation and to raise critical heat fluxes. The use of *active* methods, which offer high potential rewards in terms of efficiency and compactness, has been less well explored and less extensively applied. Of the several active methods in operation, the use of high gravity fields created by rotation is potentially the most rewarding since this offers the following advantages over other active techniques such as stirring, scraping or vibration:

(a) variable rotation speed offers a further degree of freedom in exchanger design and operation.
(b) the increased '*g*' coupled with built-in surface roughness factors, enhances film processes.
(c) because there is a self-cleaning action, rotating devices can handle liquids containing solids.
(d) reduced fluid residence times in the heating zone permit the processing of heat-sensitive fluids.

Heat transfer studies on smooth rotating surfaces have shown that, with thin films, heat transfer rates may be significantly enhanced, although BRAUNER and MARON[14] have shown that, where a fluid film flows over a surface, ripples may develop and these may be responsible for a marked improvement in the rates of both the heat and mass transfer. JACHUCK and RAMSHAW[15] have suggested that surface irregularities might enhance the heat transfer characteristics of a rotating surface still further and have investigated this proposal.

## *Experimental tests and results*

JACHUCK and RAMSHAW[15] have carried out tests on the rotating disc heat exchanger, shown in Figure 20.3, which included one base disc and four top discs thus allowing a degree of flexibility in studying various surfaces. These included normal groove, re-entry groove, metal-sprayed and smooth discs. The normal groove disc had seven concentric grooves of a geometry which promoted and also created instabilities in the flow by generating surface waves at each of the groove sites. The metal-sprayed disc was coated with an aluminium-bronze composite powder and the smooth surface disc was identical to that of the other top discs, except that the upper surface had no grooves or metal coating.

Thermocouples were connected to a data acquisition system by way of a slip-ring assembly incorporating a protective shroud, as shown in Figure 20.3, which was cooled by compressed air. The disc speed was recorded by an analogue tachometer and the bulk liquid temperature was measured by pressing the edge of a thin rubber strip against the surface of the disc. Due to its high velocity, the liquid was forced up the side of the rubber strip at the point of contact where there was a build-up of the liquid, large enough for its temperature to be measured by a thermocouple. This method was used since the film was too thin for a thermocouple to be inserted directly into the liquid, without it touching the disc surface and thereby giving an incorrect reading. It is estimated that this technique was accurate to within 0.1 deg K.

Knowing the heat flux, the disc surface temperature and the liquid film temperature, it was possible to calculate the average and local heat transfer coefficients for a variety

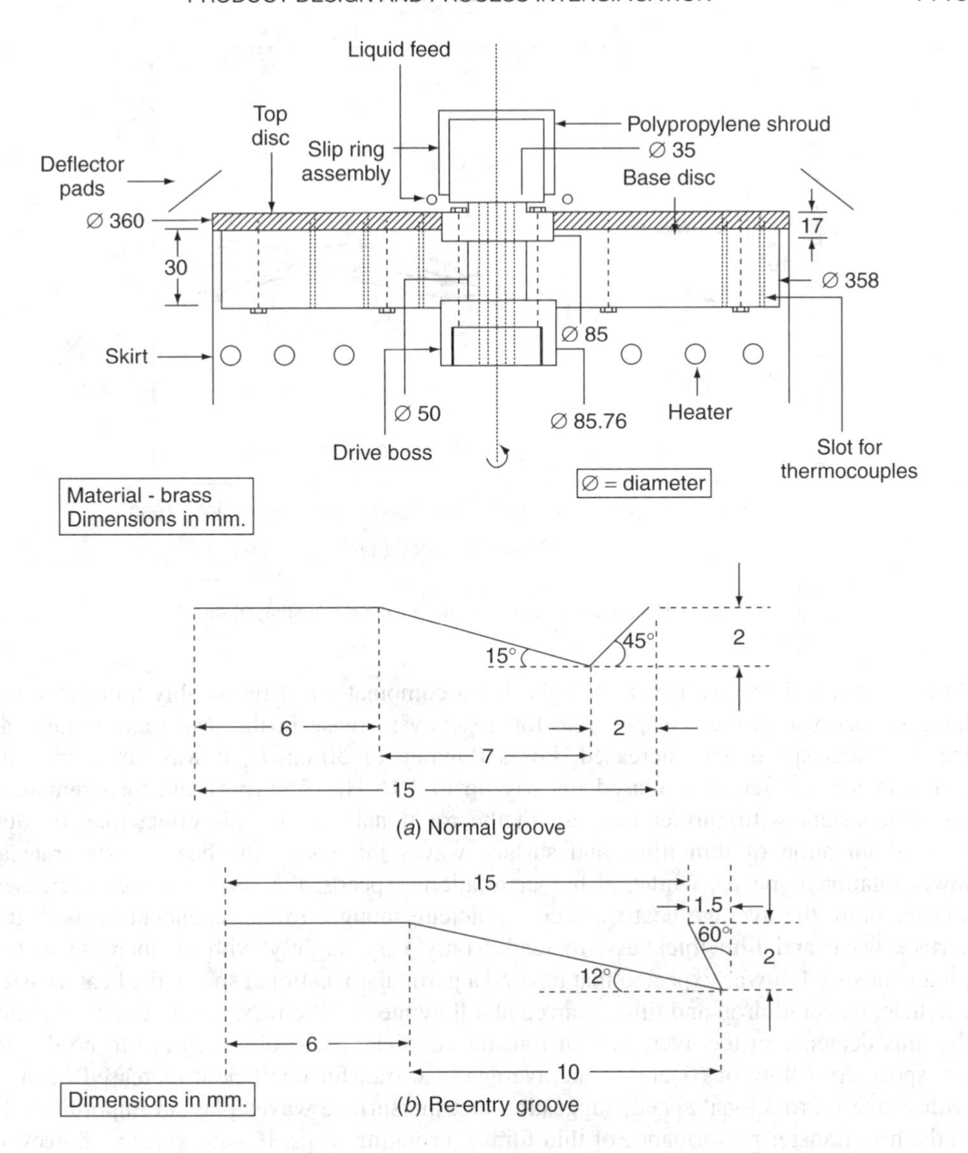


Figure 20.3. Rotating disc heat exchanger[15]

of flow conditions. For the *metal sprayed disc*, the results obtained are summarised in Figure 20.4. For a given rotational speed, increasing the feed flow rate from 30 to 67 $cm^3/s$ resulted in an increase in the mean heat transfer coefficient, as shown in Figure 20.4. Jachuck suggested that the heat transfer coefficient is dependent on both the liquid film thickness and the surface waves, or instabilities in the liquid film. It was considered that the best results would be achieved for conditions where thin films with large instabilities were formed to give high shear mixing. For a flow of 30 $cm^3/s$, the value of the coefficient decreased at rotational speeds in excess of 10.8 Hz (650 rpm) probably because extremely thin films flowed smoothly over the disc surface without generating any surface waves.

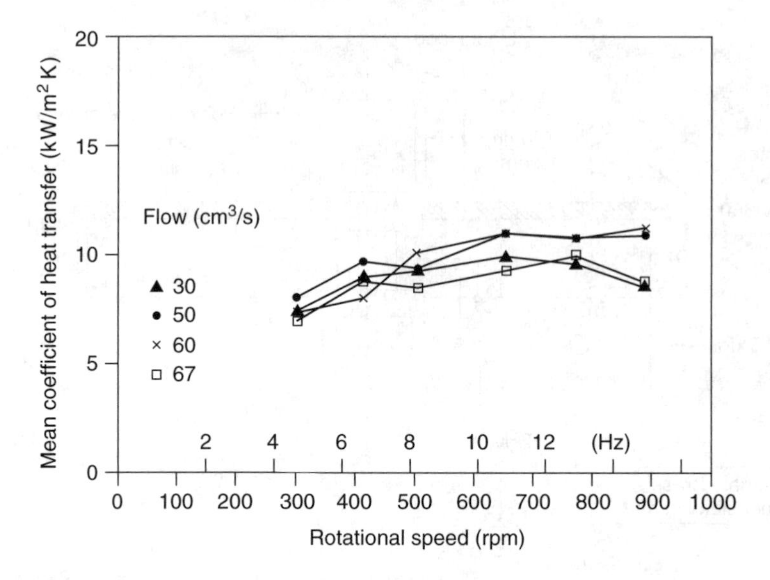


Figure 20.4. Heat transfer characteristics of the metal sprayed disc[15]

At low rotational speeds, it was thought that a combination of reasonably thin films and large surface waves were responsible for a steady increase in the heat transfer rates as the rotational speed was increased. For a flowrate of 50 $cm^3/s$, it was noted that the heat transfer coefficient increased linearly up to 10.8 Hz (650 rpm) and then remained almost constant with further increase in the rotational speed. This effect may be due to a combination of thin films and surface waves increasing the heat transfer rate at lower rotational speeds, whilst, at higher rotational speeds, the surface waves decreased and therefore the average heat transfer coefficient, thought to be dependent on both the surface wave and film thickness, increased only very slightly with an increase in the rotational speed. It was expected that beyond a particular rotational speed, the heat transfer coefficients would drop and this occurred at a flowrate of 30 $cm^3/s$. It is important to note that this decrease in the average heat transfer coefficient was observed not to be due to dry spots. At a flow of 67 $cm^3/s$, the average heat transfer coefficient increased linearly with increased rotational speed, suggesting that the surface waves play an important role in the heat transfer performance of thin films on rotating discs. It was expected, however, that for a flowrate of 67 $cm^3/s$, the average heat transfer coefficient would drop at very high rotational speeds, when the surface waves ceased to exist. Results for a flow of 70 $cm^3/s$ suggested that for a given rotational speed, the average heat transfer coefficient obtained was less than that for 67 $cm^3/s$. This suggests that there is a cut-off point in the feed flowrate, and therefore a compromise between the film thickness and the formation of surface waves should be made in order to achieve the best results.

Currently, there are no correlations available between the surface wave function, the film thickness and the average heat transfer coefficient that successfully describe the experimental results. The increase in the average heat transfer coefficient for an increasing rotational speed may be due to better shear mixing, resulting from thinner films and smaller and more concentrated surface waves. Similar phenomena have been observed by both

ELSAADI[16] and LIM[17] who explained the existence of the maximum value of the mass transfer coefficient by suggesting that, by increasing the flowrate, the film thickness would increase, thereby creating waves which would induce progressively more efficient mixing in the film. When the film thickness was increased beyond some optimum value, the waves would be unable to exert the levels of mixing required for the higher mass transfer rates.

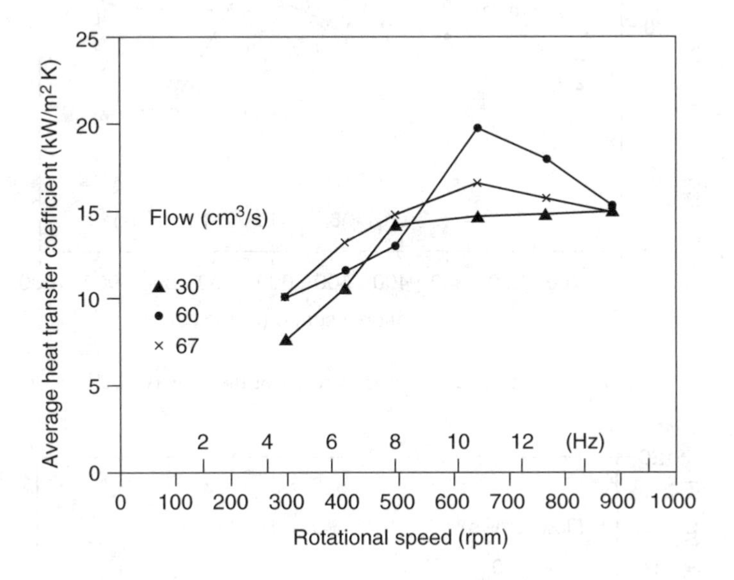


Figure 20.5. Heat transfer characteristics of the normal grooved disc[15]

The results from tests with the *normal grooved discs*, shown in Figure 20.5, indicate that the average heat transfer coefficient increased with an increase in the liquid flowrate, although only up to a certain point, above which the films were too thick and, therefore, the heat transfer performance decreased. The coefficients with the grooved disc were higher than those obtained with the metal-sprayed disc and this may be due to better mixing and the creation of surface waves by continual creation and breakdown of the boundary layer. The peak in the heat transfer profile may be due to the forward-mixing effect, although, ideally, a grooved disc should be operated under conditions where forward mixing does not take place. It may be that, for a viscous liquid melt, such as a polymer, higher rotational speeds could be used before the peak in the profile was experienced. Test results with the re-entry type of grooved disc are shown in Figure 20.6. It was observed that a considerable proportion of the liquid was being thrown off the disc, due to the re-entry effect of the disc. A proportion of the liquid that experienced a hydraulic jump, discussed in Volume 1, Chapter 3, at the grooves caused a forward-mixing effect, though most of the liquid was thrown off the disc. It may be concluded the re-entry groove design will be suitable for denser liquids, as it can provide effective mixing and create surface waves. Test results for the *smooth disc*, shown in Figure 20.7, again confirm that the heat transfer coefficient was dependent on both the surface waves and the film thickness of the liquid and that there was clearly an optimum flowrate above which there would be an

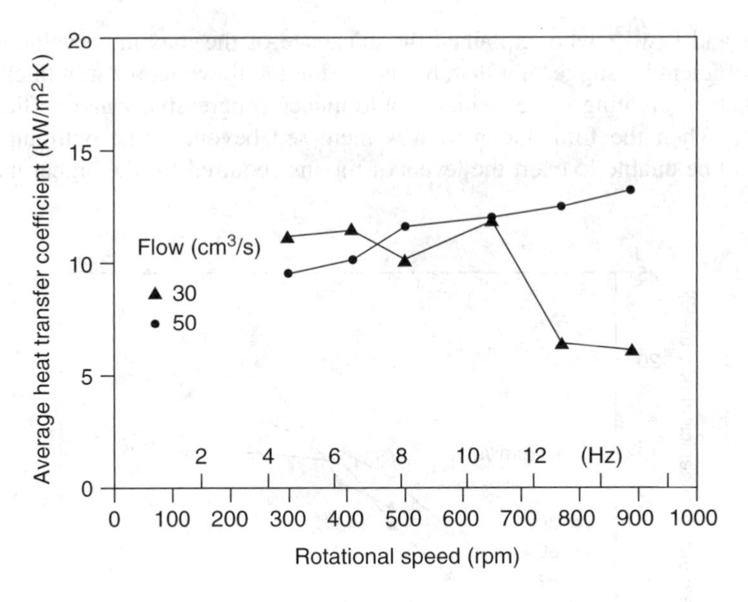


Figure 20.6. Heat transfer characteristics of the re-entry disc[15]

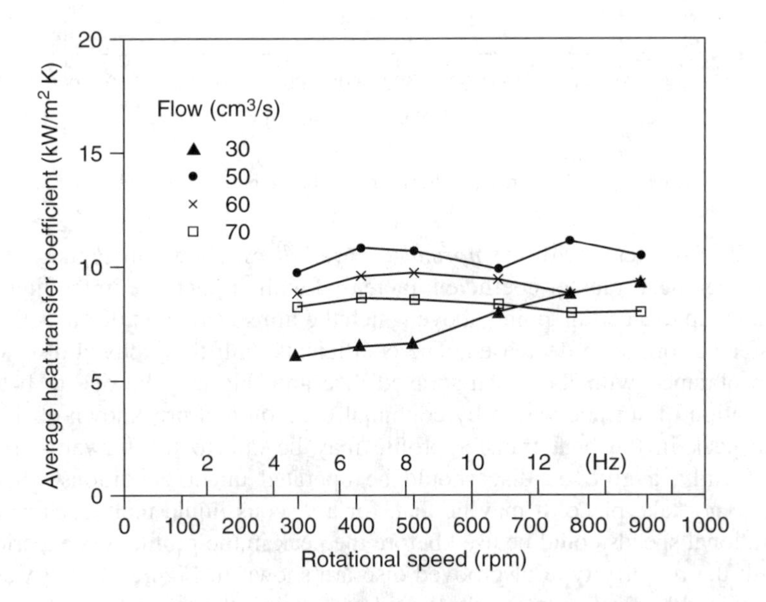


Figure 20.7. Heat transfer characteristics of the smooth disc[15]

adverse effect on the mean heat transfer coefficient. At higher rotational speeds, there was a sharp decrease in the performance of the metal-sprayed and the re-entry discs, although the normal grooved disc continued to perform well. Even though the performance of the metal-sprayed and the re-entry discs decreased at higher rotational speeds, they performed considerably better than the smooth disc.

A stroboscope operating at twice the rotational speed of the disc was used to obtain a clear view of the flow pattern on the spinning disc. Photographs were taken to study the flow behaviour. On increasing the speed of rotation, the films became thinner and, in addition, the number of surface waves increased. At lower rotational speeds, the surface waves were larger, whereas at higher speeds the wavelength of the waves was reduced and several smaller waves were generated. This effect created larger instabilities and therefore enhanced the heat transfer performance. For a given flowrate and speed of rotation, surface instabilities were a minimum for the smooth disc and a maximum for the grooved discs. It was difficult to differentiate visually between the performance of the normal and of the re-entry grooved disc although, at lower rotational speeds, the re-entry disc generated more waves than the normal grooved disc. This was probably the reason why the re-entry grooved disc performed better than the normal disc at very low rotational speeds. At higher speeds, the advantages of the re-entry disc were undermined, due to the forward-mixing effect. It was suggested that the presence of surface imperfections clearly promoted the formation of a large number of small waves, which created instabilities in the film and enhanced the heat transfer rate.

In summary, this investigation showed that:

(a) heat transfer coefficients were dependent on both the liquid-film thickness and the nature of the surface waves, or the instabilities, in the liquid film. Typically, the best results were achieved for conditions that resulted in thin films with large instabilities, which ensured high-shear mixing.
(b) mean heat transfer coefficients as high as 18 $kW/m^2K$ were achieved by using a grooved rotating surface.
(c) for a given flowrate, an increase in the rotational speed increased the average heat transfer coefficient, although at very high rotational speeds, the heat transfer coefficient decreased for the grooved and coated surfaces and showed very little change compared with that for a smooth disc.
(d) in general, a normal grooved disc performed better than the other tailored surfaces, although, at low rotational speeds the re-entry groove disc seemed to perform better. At higher speeds, the performance of a re-entry disc dropped because the liquid was thrown off the disc surface due to its extremely high velocity, as a result of which cold liquid from the centre of the disc mixed with that at the edge.
(e) Jachuck and Ramshaw suggested that, with viscous liquids, a re-entry disc would perform better than a normal grooved disc, even at higher rotational speeds.

## 20.2.4. Condensation in rotating devices

### *Introduction*

Certain sectors of industry are seeking to use lightweight and corrosion-resistant compact heat exchangers for condensation as well as for convection duties. This requirement is of particular interest in the aviation, automobile and domestic heating and ventilation industries. Also, recent interest in the concept of mobile chemical plants necessitates the use of lightweight compact heat exchangers. Such plants may well have an important role to play in the future of processing as they provide flexibility, improved inherent safety and tighter process control. One way of reducing the weight of a heat exchanger is to use polymeric materials in their construction. Although many polymers are poor thermal conductors and, with a wall thickness of 0.5–1 mm, the heat transfer performance can be adversely affected, a relatively new material, poly-ether-ether-ketone (PEEK) has many advantages in that it is a thermoplastic with a working temperature of up to 500 K, which is capable of being formed into a 100 μm thick film, can be easily corrugated and, depending on the geometry, can withstand differential pressures of up to 1 MN/m$^2$. In addition, PEEK has attractive chemical resistance properties, making it suitable for application in chemically aggressive areas. BURNS and JACHUCK[18] have explored the performance of this material, in the form of cross-corrugated films, for the condensation of water in the presence of a non-condensable gas.

## *Experimental tests and results*

JACHUCK[19] describes a compact heat exchanger formed from corrugated sheets of PEEK which provided a total surface area for heat transfer of approximately 0.125 m$^2$. The heat exchanger was mounted inside a thick Perspex box to minimise heat losses from the PEEK sheets. The gas stream was initially heated by use of heating tape wrapped around the piping and moisture was added to the gas by bubbling some or all of it through a bath of thermostatically controlled water. The pressure drop over the gas-side was measured using a water-filled U-tube connected to the entrance and exit chambers of the heat exchanger to give a resolution of 10 kN/m$^2$ and a range of up to 8000 kN/m$^2$.

Heat transfer characteristics of the system were expressed in terms of an overall heat transfer coefficient and it was found that condensation of water vapour within the system was the major component in the transfer of energy from the gas side. Thermal effectiveness, as defined in Volume 1, Chapter 9, was used to gauge the efficiency of the process. It was suggested that standard quantities, such as Nusselt number and number of transfer units were meaningless for this system due to the heat generation from the condensation process in the presence of a non-condensable gas. The tests carried out focused on the effects of gas and liquid flowrates on heat transfer performance for two different packing orientations. The humidity of the input gas was close to saturation for all the tests, and the outlet air stream was also saturated with water vapour. The packing orientations used differed only in the angle at which the flow met the corrugation. The primary purpose of the tests was to examine the influence of gas flowrate on the heat exchanger performance. In conventional systems, increased gas flowrates result in well-defined increases in heat transfer performance through the improved forced convective mixing occurring at high Reynolds numbers. In the case of condensation processes, however, much lower gas flowrates are encountered and hence there are lower Reynolds numbers and less turbulent mixing. The results are reported by BURNS and JACHUCK[18] and are shown in Figures 20.8 and 20.9. Overall heat transfer coefficients in the range of 70–370 W/m$^2$K were observed for the first packing configuration shown in Figure 20.8 and 60–300 W/m$^2$K for the second, shown in Figure 20.9, with little significant difference in performance for the different orientations. The greatest influence observed for both packing configurations was the liquid flowrate and, with the exception of the highest liquid flowrate in the first set of tests, little change was observed in the heat transfer coefficient as gas flowrate was increased. It was felt that, although changes in mixing

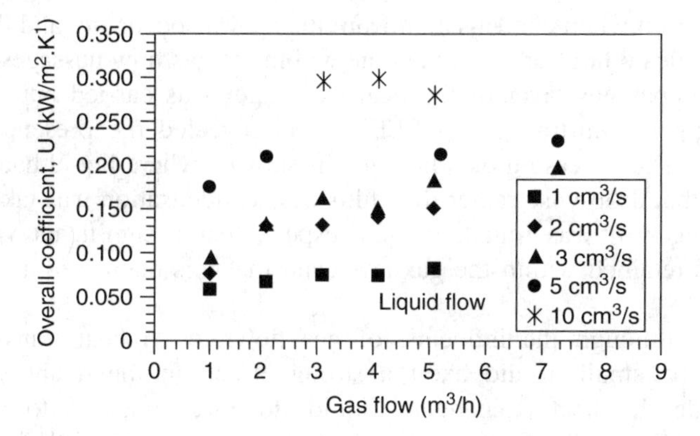


Figure 20.8. Overall heat transfer coefficients in the first configuration[18]

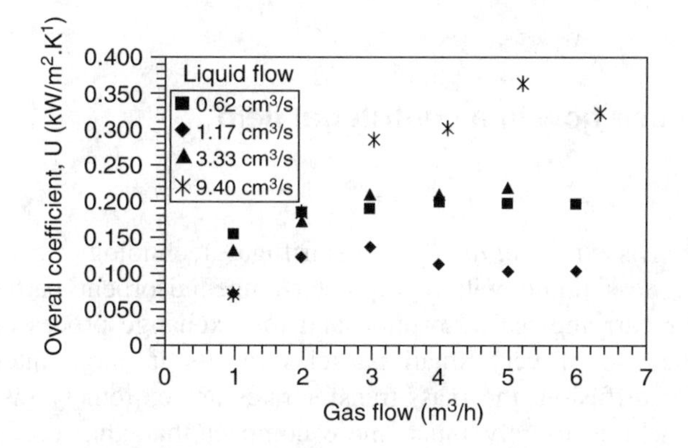


Figure 20.9. Overall heat transfer coefficients in the second configuration[18]

patterns on the liquid side should be negligible at the low Reynolds numbers prevailing (<33) for this flow, variations in the temperature distribution over the packing surface caused by changing the liquid flowrate might, however, have influenced the condensation rate and this would be the most likely explanation for the observed results. In summary, it may be concluded that heat transfer performance was not significantly influenced by corrugation orientation or gas flowrate, but was more strongly influenced by liquid flowrate over the range of conditions examined.

Thermal effectiveness values for the condensation process in both series of tests were in excess of 0.9 for most of these tests where the liquid/gas mass flow ratios were greater than 2. These high values were partly due to the small change in liquid temperature when the gas temperature was allowed to approach that of the liquid input, but more importantly due to the high heat transfer capability of the compact heat exchanger. For lower mass flow ratios, the effectiveness was only 0.7–0.9, due mainly to the greater warming of the cooling water. Values of the gas-side heat transfer coefficient which were calculated from overall heat transfer coefficients were 63–520 W/m$^2$ K. Pressure drops for the gas flow across the unit were measured for normal operating conditions and compared with measurements taken when using low humidity 'dry' air. This comparison showed a much higher pressure drop when condensation from the air was occurring and this suggests that there was a significant hold-up of condensate within the gas flow passages. Visual analysis of flow in the upper gas layer of the heat exchanger was carried out and photographs taken of the upper semi-transparent PEEK sheet revealed the presence of condensate within this layer. Dark areas on the photographs showed where liquid had collected and it was confirmed that drop-wise rather than film-wise condensation was taking place within the heat exchanger. It was found, in this respect, that a significant volume of liquid condensate was retained within the gas layer and that this decreased as the gas flowrate was increased.

In summary, although the influence of gas flowrate on heat transfer performance was shown to be small, it did exert a strong effect on the quantity of condensate remaining inside the heat exchanger. Liquid flowrate appeared to have a stronger influence on heat transfer performance and it was suggested that the temperature distribution over the PEEK film might have an additional effect on the condensation process.

## 20.2.5. Two-phase flow in a centrifugal field

### Introduction

As described by Bisschops *et al.*[20,21], centrifugal technology, using countercurrent contact of the process liquid with micrometer-range adsorbent particles constitutes a new technique for carrying out adsorption and ion-exchange processes in a centrifugal field. Because the use of very small particles results in large interfacial areas and short distances for diffusion, the mass-transfer rates are extremely fast, and centrifugal adsorption equipment is usually much more compact than that used for conventional countercurrent processes, which operate as either as fluidised-beds under gravity or as packed-beds. A more detailed description of the advantages of centrifugal adsorption technology over existing techniques has also been published by Bisschops *et al.*[22,23]. A critical aspect of this technology is that very small adsorbent particles are moving countercurrent to the liquid flow. Since centrifugal force increases with radius, the critical condition for countercurrent flow occurs at the solids feed position in the contact zone where the centrifugal force is lowest. In order to evaluate the hydrodynamic capacity of the rotor, the relation between the centrifugal force and the flowrates of both the liquid and solid phases must be known. If the liquid flow exceeds a certain maximum, solids will be rejected at the entrance. This phenomenon has been called 'flooding' by Elgin and Foust[24]. As the adsorbent particles move through the countercurrent flow region towards the rim of the rotor, they will experience an increase in the centrifugal force and, as Bisschops points out, this may affect the two-phase flow in either of the following two ways:

(a) The increase in terminal settling velocity with increasing centrifugal force leads to an increase in the slip velocity between particles and liquid. Since the flowrates of the two phases do not change over the contact zone, this will lead to an increase of the void fraction toward the periphery of the rotor.

(b) The increase in the centrifugal force acting upon the particles results in a denser fluidised bed, in which the particles experience more hindrance due to the inter-particle interactions. This leads to a decrease in slip velocity and an increase of particle concentration towards the periphery of the rotor.

The interfacial area in the contactor, which is directly related to the solids hold-up, strongly influences the mass transfer rate. To maximise the overall mass-transfer rate per unit volume of equipment, a high solids hold-up is necessary. On the other hand, the solids hold-up also influences the pressure drop over the contactor. The pressure drop has a hydrostatic and a dynamic component, both of which rise with increased solids hold-up. Since the adsorbent consists of extremely small particles, fluid friction between liquid and solids may lead to a relatively high dynamic pressure drop. The hydrostatic pressure drop is attributable to the density difference between the suspension in the contact zone and in the liquid.

DI FELICE[25] points out that liquid fluidised systems with large density differences between the particle and liquid phases may give rise to heterogeneous behaviour similar to that exhibited by gas-fluidised systems. According to Bisschops *et al.*, similar heterogeneous fluidisation phenomena can occur in centrifugal systems, where the action of the centrifugal force is similar to that of increasing the apparent density differences between particles and liquid, although this phenomenon does not seem to have been reported for centrifugal systems. Since the occurrence of heterogeneous flow may be detrimental to the desired countercurrent flow in centrifugal adsorption technology, experiments must he carefully checked for fluidisation regimes.

Bisschops *et al.* have investigated the hydrodynamic capacity and solids holdup in countercurrent two-phase flow in the centrifugal field, as well as the relation between the pressure drop and void fraction. Moreover, the analysis included a check as to whether the two-phase flow in the centrifugal field was homogeneous or heterogeneous.

### Experimental studies

Bisschops adopted the following strategy in this investigation:

(a) Verification of the homogeneous two-phase flow model under gravity with two solid phases — relatively large particles (>1 mm) in water with a small density difference and relatively small particles (100 μm) in water with a large density difference.
(b) Verification of the model in the centrifuge, using solid phases in water with different density differences and particle diameters.

Tests under gravity were carried out with for Maxazyme Gl Immob. biocatalyst particles in water, and small ballottini glass beads in tap water at ambient temperature. The expansion behaviour of the systems was investigated by fluidisation experiments in a Perspex column, shown in Figure 20.10, which is equipped with water manometers at several heights. Tests with centrifugal force were carried out with two devices: (a) the low-$g$ rotor and (b) the high-$g$ rotor.

The low-$g$ rotor, which is shown in Figure 20.11, consisted of a rotating disc, fitted with two straight horizontal columns. The rotor, fitted with a rotary seal, was contained

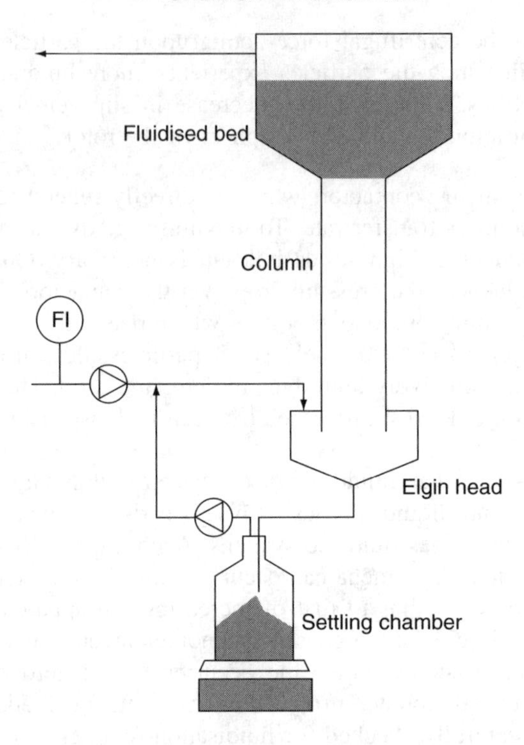


Figure 20.10. Experimental system for measuring flooding characteristics under gravity[20]

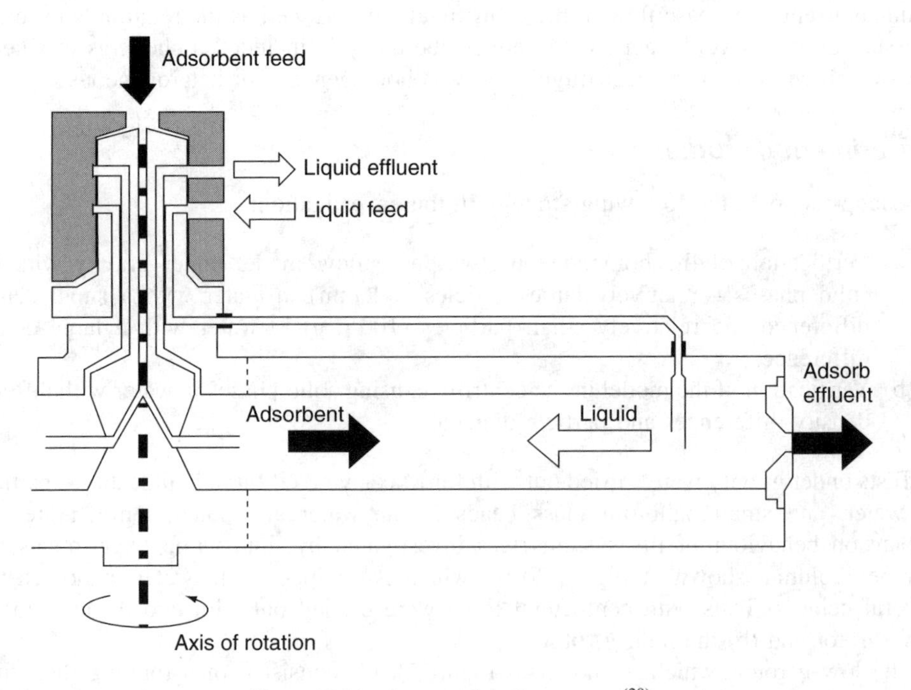


Figure 20.11. Layout of the low-$g$ rotor[20]

in a transparent Perspex drum to prevent leakage, and the liquid feed, liquid discharge, and solids feed were directed along the axis of rotation. The maximum speed of rotation was 4.2 Hz (250 rpm). As shown in Figure 20.12, the solid phase was fed to the rotor as a slurry from a fluidised bed and the solids left the rotor by way of the solids discharge, either in the form of a nozzle or as a backflow system from which they flowed to a separate settler.

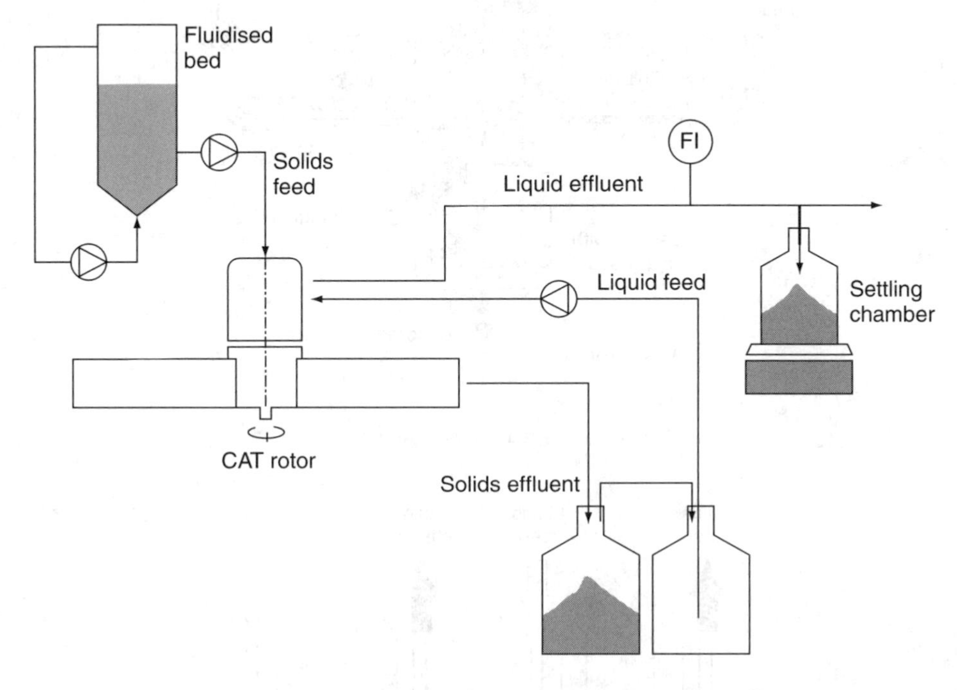


Figure 20.12. Equipment for experiments with low-$g$ rotor[20]

For tests under higher centrifugal accelerations, the high-$g$ rotor which was constructed, as shown in Figure 20.13, from stainless steel which allowed high speeds of rotation to be used, although the maximum was limited to 42 Hz (2500 rpm). Solids were fed to the rotor in a suspension containing approximately 30 per cent solids by volume, and two geometries for the solids discharge and liquid distributor head at the rim of the columns were tested as shown in Figure 20.14. The first distributor was designed such that the solids passed the liquid inlet and were collected and discharged at the periphery of the rotor. In order to obtain a stable solids discharge without clogging, the liquid fraction in the solids discharge flowrate had to exceed a certain minimum, usually about 10 per cent by volume. The second distributor was constructed with the solids discharge located closer to the axis of rotation than the liquid feed. This allowed a fluidised bed to accumulate in the contactor, from which a dense slurry could be extracted from the rotor. Initially, the rotor was operated with a very high solids feedrate, and the surplus solids were continuously rejected in the liquid effluent.

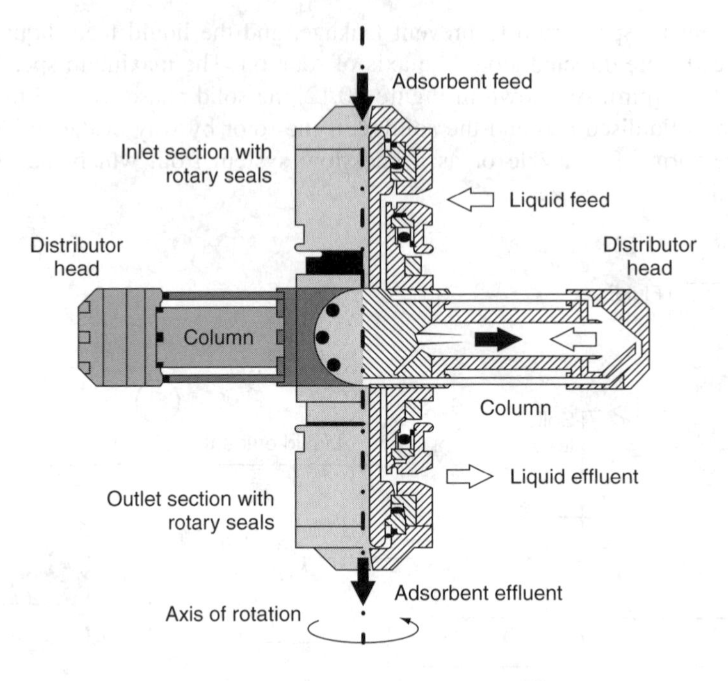


Figure 20.13. Layout of high-$g$ rotor[20]

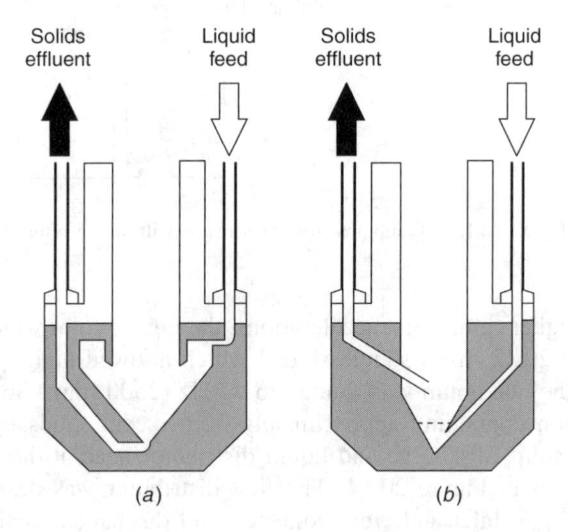


Figure 20.14. Layout of distributor heads applied in experiments with high-$g$ rotor[20]. (*a*) The Elgin-head distributor. (*b*) The fluidised-bed distributor

## *Results*

All the model systems investigated by Bisschops *et al.* fell within the completely homogeneous fluidisation regime modelled by Gibilaro and Hossain[26]. With the low-$g$ rotor and

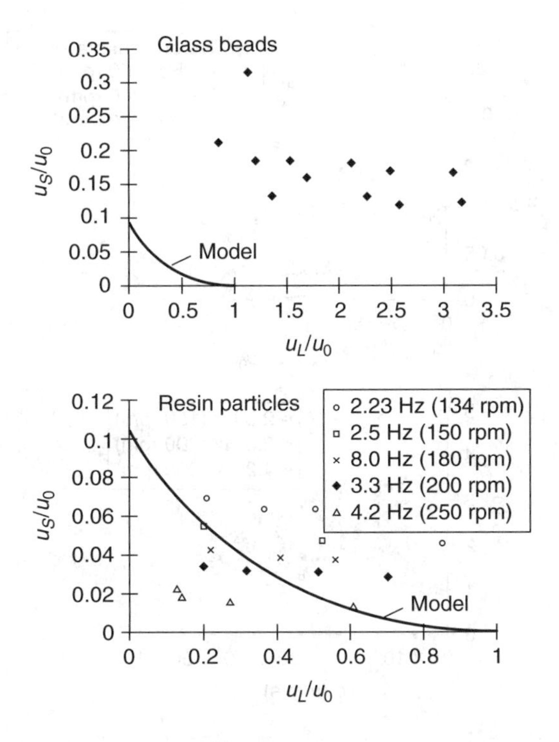


Figure 20.15. Experiment flooding data obtained in the low-$g$ rotor with nozzle discharge[20]

the nozzle discharge, the results for flooding conditions obtained from the tests are shown in Figure 20.15 in which $u_L$ is the superficial liquid velocity, $u_S$ is the superficial solids velocity, and $u_0$ is the terminal falling velocity of the particles in the gravitational field. For glass ballotini beads, much higher capacities were achieved than those predicted for a single particle at the solids inlet position. The capacity of the low-$g$ rotor was also tested with back-flow of the solids discharge and the results are compared with values predicted by the homogeneous flow model, as shown in Figure 20.16. Again the capacity of the rotor was higher than was expected. The capacity data, shown in Figure 20.17, show that the predicted maximum flow would be exceeded in practice.

Bisschops *et al.*[20] developed a model for describing countercurrent two-phase flow in a centrifugal field, based on homogeneous flow conditions. When the model was tested under normal gravity conditions, it was found that, for systems with relatively low-density differences between dispersed and continuous phase, it described the observations very accurately. For small glass Ballottini beads with a larger density, however, the model failed to describe the countercurrent flow pattern accurately. Large inhomogeneties were observed in this system, although the stability criterion of Gibilaro and Hossain was met. In addition, the model did not accurately describe the hydrodynamic capacity of the centrifugal rotor. When the rotor was equipped with a nozzle discharge, however, the capacity for ion-exchange resin of relatively low-density difference compared with water, was predicted to the correct order of magnitude, although the influence of the superficial liquid velocity was not taken into account, probably because the solids throughput depends

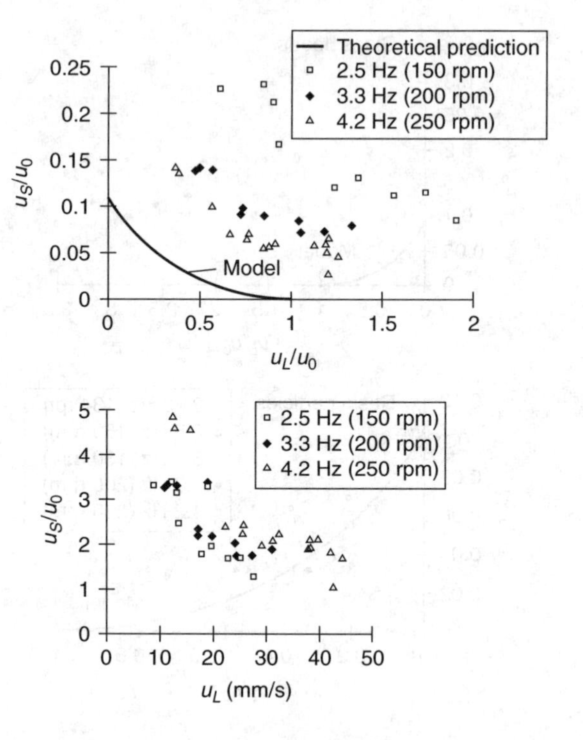


Figure 20.16. Experimental flooding data for the ion-exchange resin ($d$ = 190 μm, $\rho_s$ = 1, 250 kg/m$^3$) in the low-$g$ rotor with backflow of the solids discharge[20]

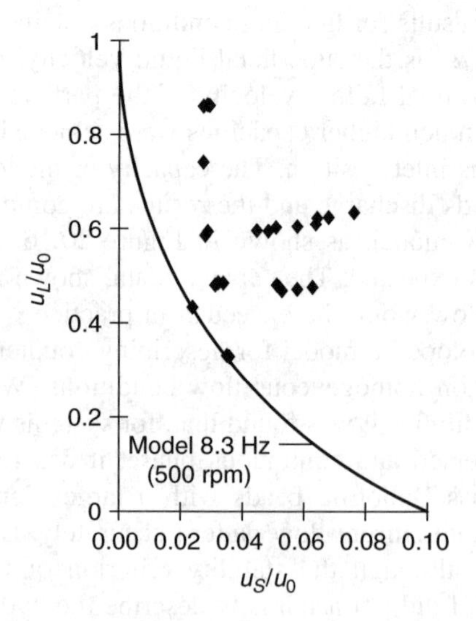


Figure 20.17. Experimental results obtained in the high-$g$ rotor in fluidized state under overload conditions at 500 rpm (8.3 Hz)[20]

entirely on the nozzle diameter and the pressure drop acting across the solids discharge port. The low-$g$ rotor also yielded much higher capacities than expected, possibly due to the formation of vortices, although deviations from the model became less as the centrifugal force was increased.

Bisschops concluded that the stability criterion for fluidised beds, as derived by Gibilaro and Hossian, cannot be applied to countercurrent systems and that it is uncertain which mode of transport dominates in countercurrent flow in a centrifugal field. It was found that vortices may not appear if the flow is unrestricted in a tangential direction, and for scale-up of the centrifugal adsorption technology, a contact zone that covers the entire periphery is preferred. Although this might lead to a decrease in the capacity of the rotor, it could reduce the degree of back-mixing, and thereby improve the overall separation performance of the process.

## 20.2.6. Spinning disc reactors (SDR)

### *Performance characteristics*

A further example of the benefits to be gained from process intensification is the spinning disc reactor, illustrated in Figure 20.18. The effect of centrifugal force is to produce highly sheared thin films, as shown in Figure 20.19, on the surfaces of rotating discs or cones. Extensive heat and mass transfer studies carried out by JACHUCK and RAMSHAW[27] using spinning disc devices have shown that convective heat transfer coefficients as high as 14 kW/m$^2$ K and values of overall mass transfer coefficients $-K_L$, for the liquid phase, as high as $30 \times 10^{-5}$ m/s and $K_G$, for the gas phase, as high as $12 \times 10^{-8}$ m/s and can be

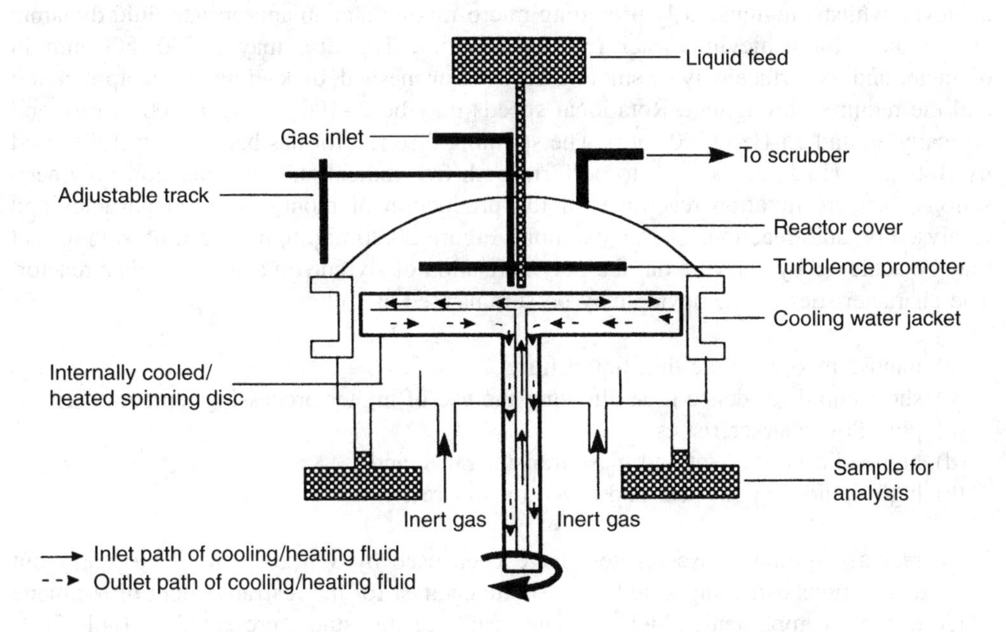


Figure 20.18. Schematic diagram of a spinning disc reactor[28]

Figure 20.19. View of sheared thin films on a spinning disc reactor[28]

achieved whilst simultaneously providing micro-mixing and an appropriate fluid dynamic environment for achieving faster reaction kinetics. The disc may be 60–500 mm in diameter and its surface may be smooth, grooved or meshed, depending on the application and the required throughput. Rotational speeds may be 2–100 Hz (120–6000 rpm), and typically around 25 Hz (1500 rpm). The spinning disc reactor has been successfully used by Boodhoo and Jachuck[28,29] to perform both free radical and condensation polymerisations, fast precipitation reactions for the production of mono-dispersed particles and catalysed organic reactions. As an example, Figure 20.20 highlights the time saving that may be achieved by carrying out the polymerisation of styrene on a spinning-disc reactor. The characteristics of the device may be summarised as follows:

(a) intense mixing in the thin liquid film,
(b) short liquid residence time allowing the use of higher processing temperatures,
(c) plug-flow characteristics,
(d) high solid-liquid heat and mass transfer rates, and
(e) high liquid–vapour heat and mass transfer rates.

More recently spinning disc reactors have been used by Wilson *et al.*[30] to carry out catalytic reactions using supported zinc triflate catalyst for the rearrangement of $\alpha$-pinene oxide to yield campholenic aldehyde. The results of this study, presented in Table 20.1, suggest that by using a supported catalyst on a spinning disc reactor it is possible to

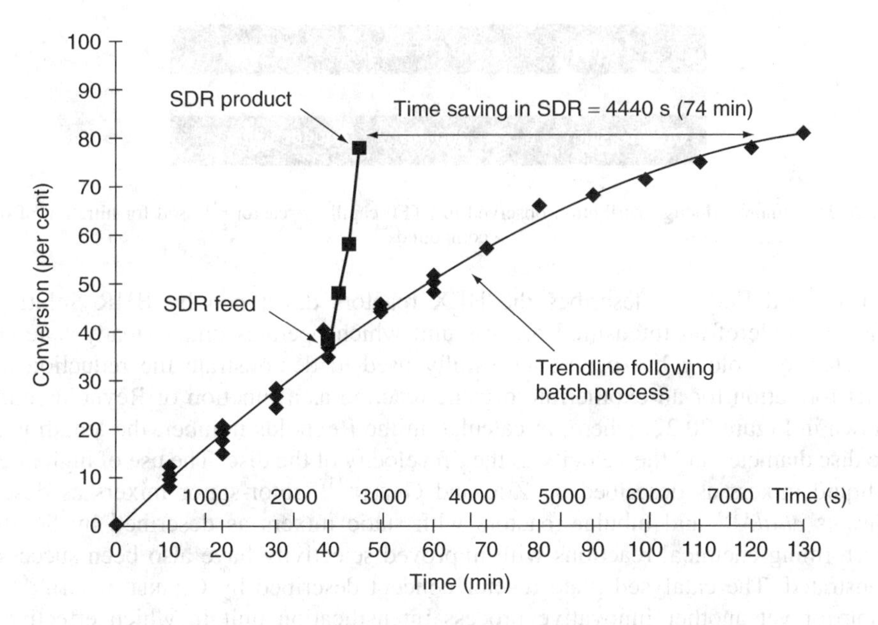


Figure 20.20. Process time saving for the polymerisation styrene in a spinning disc reactor with a disc of 1.08 m in diameter rotating at 6.7 Hz (400 rpm)[28]

Table 20.1. Performance of batch, catalysed spinning-disc reactors[30]

| Rearrangement of $\alpha$-pinene oxide to campholenic aldehyde | Batch Reactor | Catalysed SDR |
|---|---|---|
| Feed or total throughput ($cm^3$) | 100 | 100 |
| Conversion (per cent) | 50 | 95 |
| Yield (per cent) | 42 | 71 |
| Processing time (s) | 900 | 17 |

achieve faster reaction rates, improved yield and the elimination of the need for any downstream separation process for catalyst recovery.

## Micro-reactors

Improved methods of manufacturing at the micro-scale are opening up new avenues for the development of compact devices for performing a range of chemical processes from reactions to extraction and separation. Equipment achieving rapid heat and mass transfer within the sub-millimetre scale channels offers the possibility of a low inventory environment with a high degree of control over the chemical process. Micro-reactors based on this concept, as described by Burns and Ramshaw[31], can provide intrinsically safe environments for catalysed and non-catalysed fluid processing. One such example of this is the nitration of organic compounds in a PTFE capillary reactor where rapid heat transfer rates allow higher acid concentrations to be used at lower temperatures resulting in the formation of less organic oxidisation by-products.

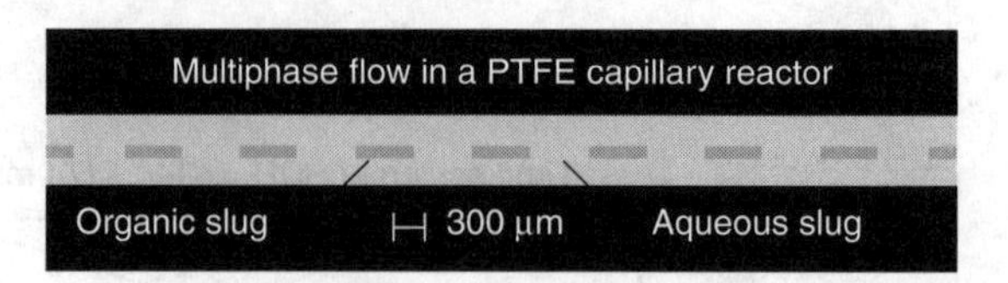


Figure 20.21. Improved slug distribution observed in PTFE capillary reactor[31] used for nitration of organic compounds

PHILLIPS and EDGE[32] describes the HEX reactors developed by BHR Solutions as another example of an intensified process unit which operates under steady-state conditions. This technology has been successfully used to demonstrate the reduction in by-product formation for an exothermic organic reaction as a function of Reynolds number, as shown in Figure 20.22, where, in calculating the Reynolds number, the length is taken as the disc diameter and the velocity as the tip velocity of the disc. The use of high intensity gas–liquid mixers as described by ZHU and GREEN[33], rotor-stator mixers as described by SPARKS *et al.*[34] and tubular reactors with static mixers as described by SCHUTZ[35] for performing chemical reactions with improved selectivity have also been successfully demonstrated. The catalysed plate reactor concept described by CHARLESWORTH[36] is an example of yet another innovative process intensification unit in which effective heat transfer is achieved by performing exothermic and endothermic reactions on opposite sides of a catalysed plate. A schematic diagram representing the coupling of methane steam-reforming and combustion of methane, is shown in Figure 20.23. This technology enables the size of equipment to be reduced by several orders of magnitude, and eliminates $NO_X$ emissions as the process is operated at lower temperatures.

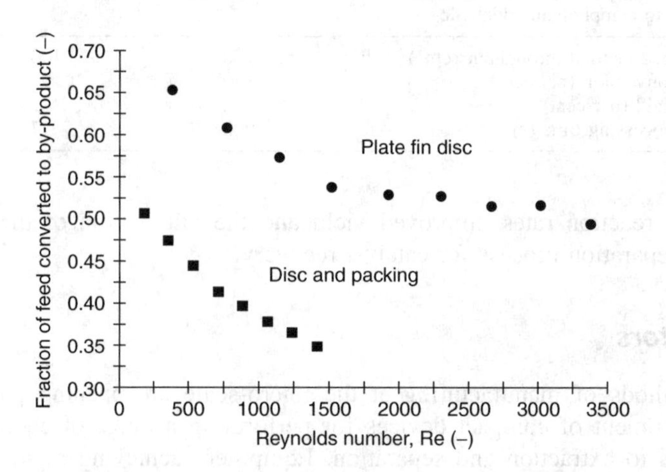


Figure 20.22. Use of HEX reactors for reducing by-product formation[32] (courtesy BHR Solutions)

## Cross-corrugated multi-functional membranes

HALL *et al.*[37] discuss the use of cross-corrugated membrane modules, illustrated in Figure 20.24, as offering the potential for developing multi-functional units for performing both reactions and product separation in one miniaturised module. The use of microporous

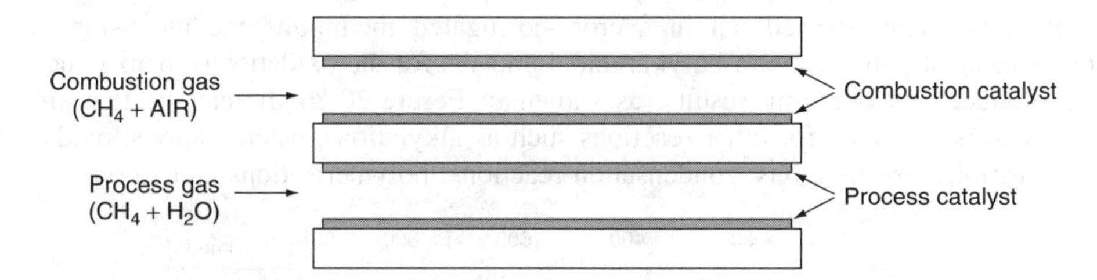


Figure 20.23. Exothermic and endothermic reactions using a catalysed plate reactor[33]

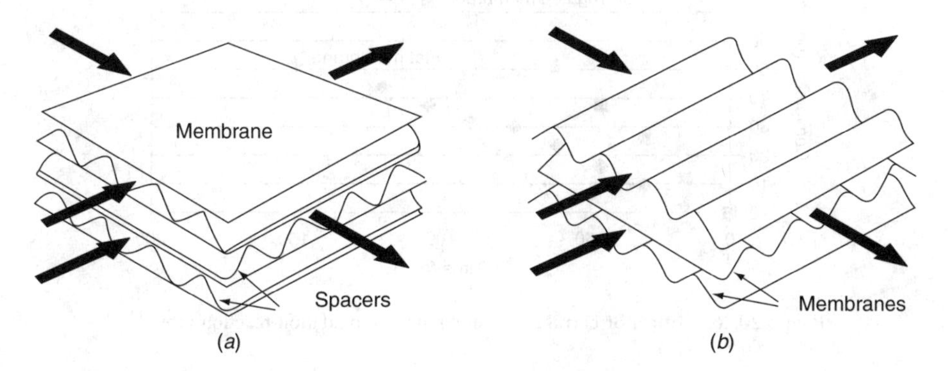


Figure 20.24. Membrane flow cell (*a*) Flat-sheet (*b*) Cross-corrugated form[37]

membranes in a reaction system is a relatively new method of solving various reaction–separation problems. Miscible fluids must be kept apart in order to control the overall reaction rate. Alternatively, the membrane may be designed to be permselective and to allow only a desired species to pass through it, such as the product of an organic synthesis, while holding back another, such as a by-product, and unreacted feed. This type of application may be successfully applied for organic reactions in the manufacture of pharmaceuticals, cosmetics, agricultural chemicals, dyes and flavouring ingredients by using the process of phase-transfer catalysis which is discussed by REUBEN and SJOBERG[38]. Figure 20.25 shows a representation of the early phase-transfer work of STARKS[39], in which 1-chloro-octane was reacted with aqueous cyanide to produce 1-cyano-octane. In a batch reactor, with no phase-transfer agent present, no reaction occurs, although with the addition of a small amount of quaternary ammonium salt, the reaction goes to completion in just a few hours.

In industry, many of these reactions are currently carried out in stirred tanks and the separation of the resulting mixture for product and catalyst recovery requires a significant

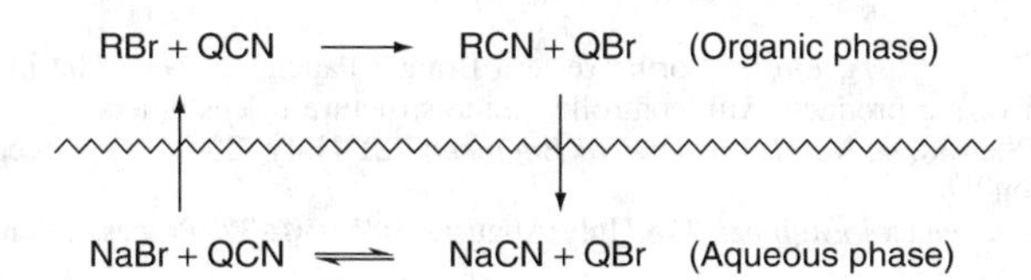


Figure 20.25. Phase-transfer catalysed cyanide displacement[39]

energy input. Tests carried out in a cross-corrugated membrane module using phase transfer catalysts, such as tetra-butyl ammonium salts for the oxidation of benzyl alcohol, have produced encouraging results, as shown in Figure 20.26. It may be that similar results can be obtained for other reactions such as alkylations, esterifications, oxidations and reductions, epoxidations, condensation reactions, polymerisations and so on.

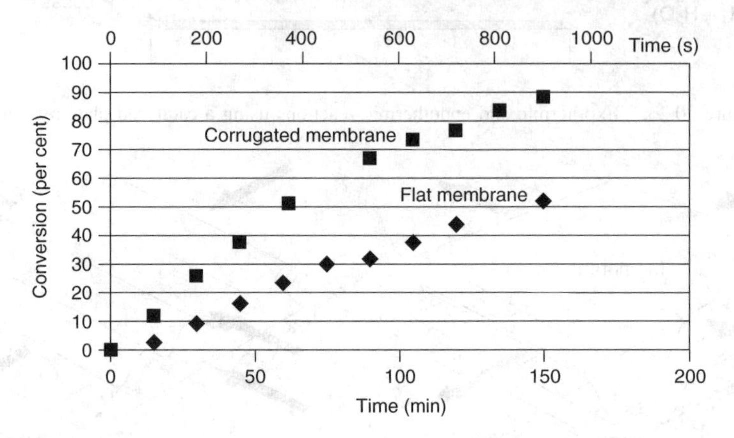


Figure 20.26. Effect of corrugated membranes on oxidation reaction rates[39]

# 20.3. FURTHER READING


BISSCHOPS, M. A. T., VAN DER WIELEN, L. A. M. and LUYBEN, K. Ch. A. M. in SEMEL, J. (ed): *Process Intensification in Practice, Applications and Opportunities* (BHR Group, London, 1999).

BISSCHOPS, M. A. T., VAN DER WIELEN, L. A. M. and LUYBEN, K. Ch. A. M. in GRIEG, J. A. (ed): *Ion Exchange Developments and Applications*, (IEX'96, SCI, London, 1996).

CUSSLER, E. L. and MOGGRIDGE, G. D.: *Chemical Product Design*. (Cambridge University Press, Cambridge, 2001).

FAVRE, E., MARCHAL-HEUSLER, L. and KIND, M.: *Chem. Eng. Res. Design* **80** (2002) 65 Chemical product engineering: Research and educational challenges.


# 20.4. REFERENCES


1. DANCKWERTS, P. V.: *Chem. Eng. Sci.* **2** (1953) 1. Continuous flow systems. Distribution of residence times. *Chem. Eng. Sci.* **7** (1958) 116. Measurement of molecular homogeneity in a mixture. *Chem. Eng. Sci.* **8** (1958) 93. The effect of incomplete mixing on homogeneous reactions. *A.I.Ch.E.Jl.* **1** (1955) 456. Gas absorption accompanied by chemical reaction.
2. DANCKWERTS, P. V.: *Gas–Liquid Reactions*. (McGraw-Hill, New York, 1970).
3. CUSSLER, E. L. and MOGGRIDGE, G. D.: *Chemical Product Design*. (Cambridge University Press, Cambridge, 2001).
4. DUNNILL, P.: *Biochem. Soc. Symposium* **48** (1983) 9–23. The future of biotechnology.
5. EDWARDS, M. F. and INSTONE, T.: *Particulate Products—Their Manufacture and Use.* (Unilever Research, Port Sunlight, 2000).
6. EDWARDS. M. F.: *Instn. Chem. Engrs.* North Western Branch Paper 8 (1998). The importance of chemical engineering in delivering products with controlled microstructure to customers.
7. HAMLEY, P. and POLIAKOFF, M.: *The Chemical Engineer* **721** (July, 2001) 27. Green chemistry: the next industrial revolution?
8. RAMSHAW, C.: *The Chemical Engineer* **416** (July/August, 1985) 30–32. Process intensification: a game for n-players.
9. STANKIEWICZ, A. I. and MOULIJN, J. A.: *Chem. Eng. Prog.* **96** (2) (2000) 8. Process intensification: transforming chemical engineering.

10. HENDERSHOT, D. C.: *Chem. Eng. Prog.* **96** (1) (2000) 35–40. Process minimisation: making plants safer.
11. GREEN, A., JOHNSON, B. and JOHN, A.: *Chem. Eng. (Albany)* **106** (13) (1999) 66–73. Process intensification magnifies profits.
12. KELLER, G. E. and BRYAN, P. F.: *Chem. Eng. Prog.* **96** (1) (2000) 41–50. Process engineering: moving in new directions.
13. REAY, D. A.: *Heat Recovery Systems & CHP* **11** (1) (1991) 1–4. Heat transfer enhancement - a review of techniques and their possible impact on energy efficiency in the U.K.
14. BRAUNER, N. and MARON, D. M.: *Chem. Eng. Sci.* **38** (5) (1983) 775–788. Modeling of wavy flow in inclined thin-films.
15. JACHUCK, R. J. J. and RAMSHAW, C.: *Heat Recovery Systems & CHP* **14** (5) (1994) 475. Process intensification: heat transfer characteristics of tailored rotating surfaces.
16. ELSAADI, M. S.: *Heat transfer to thin film liquids on closed rotating discs systems.* M.Phil. thesis, (University of Newcastle upon Tyne, U.K. 1992).
17. LIM, S. T.: *Hydrodynamics and mass transfer processes associated with the absorption of oxygen in liquid films flowing across a rotating disc,* Ph. D. thesis, (University of Newcastle upon Tyne, U.K., 1980).
18. BURNS, J. R. and JACHUCK, R. J. J.: *App. Thermal Eng.* **21** (2001) 495–510. Condensation studies using cross-corrugated polymer film compact heat exchanger.
19. JACHUCK, R. J. J.: *Opportunities presented by cross-corrugated polymer film compact heat exchangers* (Proceedings of the International Conference on Compact heat Exchangers and Enhancement Technology for the Process Industries, Banff, Canada, 1999) 243–250.
20. BISSCHOPS, M. A. T., VAN DER WIELEN, L. A. M. and LUYBEN, K. Ch. A. M.: *A.I.Ch.E.Jl* **47** (6) (2001) 1263–1277. Hydrodynamics of countercurrent two-phase flow in a centrifugal field.
21. BISSCHOPS, M. A. T., VAN DER WIELEN, L. A. M. and LUYBEN, K. Ch. A. M.: PCT Patent Application, PCT/NL 97/00121 (1997).
22. BISSCHOPS, M. A. T., VAN DER WIELEN, L. A. M. and LUYBEN, K. Ch. A. M. in GRIEG, J. A. (ed): *Ion Exchange Developments and Applications*, (IEX'96, SCI, London, 1996) 297. Operating conditions of centrifugal ion exchange.
23. BISSCHOPS, M. A. T., VAN DER WIELEN, L. A. M. and LUYBEN, K. Ch. A. M. in SEMEL, J. (ed): *Process Intensification in Practice, Applications and Opportunities* (BHR Group, London, 1999) 229. Centrifugal adsorption technology for the removal of volatile organic compounds from water.
24. ELGIN, J. C. and FOUST, H. C.: *Ind. Eng. Chem.* **42** (1950) 1127. Countercurrent flow of particles through moving continuous liquid; pressure drop and flooding in spray-type liquid towers.
25. DI FELICE, R.: *Chem. Eng. Sci*. **50** (1995) 1213. Review article 47: Hydrodynamics of liquid fluidization.
26. GIBILARO, L. G. and HOSSIAN, I.: *Can. J. Chem. Eng.* **64** (1986) 931. Aggregate behavior of liquid fluidized beds.
27. JACHUCK, R. J. J. and RAMSHAW, C.: *Process intensification: Spinning disc polymeriser* (IChemE Research Event, First European Conference, 1995).
28. BOODHOO, K. V. K. and JACHUCK, R. J. J.: *Appl. Therm. Eng.* **20** (2000) 1127. Process intensification: spinning disc reactor for styrene polymerisation.
29. BOODHOO, K. V. K. and JACHUCK, R. J. J.: *Green Chemistry* **4** (2000) 235–244. Process intensification: spinning disc reactor for condensation polymerisation.
30. WILSON, K., RENSON, A. and CLARK, J. H.: *Catalysis Letters* **61** (1–2) (1999) 51–55. Novel heterogeneous zinc triflate catalysts for the rearrangement of alpha-pinene oxide.
31. BURNS, J. R. and RAMSHAW, C.: *A microreactor for the nitration of benzene and toluene* (Proceedings of the 4th International Conference on Microreaction Technology, Atlanta GA, U.S.A., 2000) 133–140.
32. PHILLIPS, C. H. and EDGE, A. M.: *The reactor–heat exchanger* (6th. Heat Exchanger Action Group (HEXAG) meeting, Leatherhead, UK, October 1996).
33. ZHU, Z. M. and GREEN, A.: *Chem. Eng. Sci*. **47** (1992) 2847–2852. Use of high intensity gas–liquid mixers as reactors.
34. SPARKS, T. G., BROWN, D. E. and GREEN, A.: *Assessing rotor/stator mixers for rapid chemical reactions using overall power characteristics* (BHR conference series. Publication 18. Mechanical Engineering Publications Ltd. London, 1995).
35. SCHUTZ, J.: *Chem. Eng. Sci*. **43** (8) (1988) 1975. Agitated thin film reactors and tubular reactors with static mixers for rapid exothermic multiple reactions.
36. CHARLESWORTH, R. J.: *The steam reforming and combustion of methane on micro-thin catalysts for use in a catalytic plate reactor* (Ph. D. Thesis, University of Newcastle upon Tyne, 1996).
37. HALL, K., SCOTT, K. and JACHUCK, R. J. J.: *Mass transfer characteristics of cross-corrugated membranes.* (presented at ICoM '99, Toronto, June 1999).
38. REUBEN, B. and SJOBERG, K.: *Chem. Tech.* (May 1981) 315–320. Phase transfer catalysis in industry.
39. STARKS, C. M.: *J. Am. Chem. Soc.* **93** (1971) 195: Phase transfer catalysis I. Heterogeneous reactions involving anion transfer by quaternary phosphonium salts.

# Appendix

# A1. STEAM TABLES

Adapted from the
*Abridged Callendar Steam Tables*
by permission of Messrs Edward Arnold (Publishers) Ltd.

Table 1A. Properties of saturated steam (SI units)

| Absolute pressure (kN/m$^2$) | Temperature (°C) $\theta_s$ | Temperature (K) $T_s$ | Enthalpy per unit mass ($H_s$) (kJ/kg) water | latent | steam | Entropy per unit mass ($S_s$) (kJ/kg K) water | latent | steam | Specific volume ($v$) (m$^3$/kg) water | steam |
|---|---|---|---|---|---|---|---|---|---|---|
| *Datum:* Triple point of water | | | | | | | | | | |
| 0.611 | 0.01 | 273.16 | 0.0 | 2501.6 | 2501.6 | 0 | 9.1575 | 9.1575 | 0.0010002 | 206.16 |
| 1.0 | 6.98 | 280.13 | 29.3 | 2485.0 | 2514.4 | 0.1060 | 8.8706 | 8.9767 | 0.001000 | 129.21 |
| 2.0 | 17.51 | 290.66 | 73.5 | 2460.2 | 2533.6 | 0.2606 | 8.4640 | 8.7246 | 0.001001 | 67.01 |
| 3.0 | 24.10 | 297.25 | 101.0 | 2444.6 | 2545.6 | 0.3543 | 8.2242 | 8.5785 | 0.001003 | 45.67 |
| 4.0 | 28.98 | 302.13 | 121.4 | 2433.1 | 2554.5 | 0.4225 | 8.0530 | 8.4755 | 0.001004 | 34.80 |
| 5.0 | 32.90 | 306.05 | 137.8 | 2423.8 | 2561.6 | 0.4763 | 7.9197 | 8.3960 | 0.001005 | 28.19 |
| 6.0 | 36.18 | 309.33 | 151.5 | 2416.0 | 2567.5 | 0.5209 | 7.8103 | 8.3312 | 0.001006 | 23.74 |
| 7.0 | 39.03 | 312.18 | 163.4 | 2409.2 | 2572.6 | 0.5591 | 7.7176 | 8.2767 | 0.001007 | 20.53 |
| 8.0 | 41.54 | 314.69 | 173.9 | 2403.2 | 2577.1 | 0.5926 | 7.6370 | 8.2295 | 0.001008 | 18.10 |
| 9.0 | 43.79 | 316.94 | 183.3 | 2397.9 | 2581.1 | 0.6224 | 7.5657 | 8.1881 | 0.001009 | 16.20 |
| 10.0 | 45.83 | 318.98 | 191.8 | 2392.9 | 2584.8 | 0.6493 | 7.5018 | 8.1511 | 0.001010 | 14.67 |
| 12.0 | 49.45 | 322.60 | 206.9 | 2384.2 | 2591.2 | 0.6964 | 7.3908 | 8.0872 | 0.001012 | 12.36 |
| 14.0 | 52.58 | 325.73 | 220.0 | 2376.7 | 2596.7 | 0.7367 | 7.2966 | 8.0333 | 0.001013 | 10.69 |
| 16.0 | 55.34 | 328.49 | 231.6 | 2370.0 | 2601.6 | 0.7721 | 7.2148 | 7.9868 | 0.001015 | 9.43 |
| 18.0 | 57.83 | 330.98 | 242.0 | 2363.9 | 2605.9 | 0.8036 | 7.1423 | 7.9459 | 0.001016 | 8.45 |
| 20.0 | 60.09 | 333.24 | 251.5 | 2358.4 | 2609.9 | 0.8321 | 7.0773 | 7.9094 | 0.001017 | 7.65 |
| 25.0 | 64.99 | 338.14 | 272.0 | 2346.4 | 2618.3 | 0.8933 | 6.9390 | 7.8323 | 0.001020 | 6.20 |
| 30.0 | 69.13 | 342.28 | 289.3 | 2336.1 | 2625.4 | 0.9441 | 6.8254 | 7.7695 | 0.001022 | 5.23 |
| 35.0 | 72.71 | 345.86 | 304.3 | 2327.2 | 2631.5 | 0.9878 | 6.7288 | 7.7166 | 0.001025 | 4.53 |
| 40.0 | 75.89 | 349.04 | 317.7 | 2319.2 | 2636.9 | 1.0261 | 6.6448 | 7.6709 | 0.001027 | 3.99 |
| 45.0 | 78.74 | 351.89 | 329.6 | 2312.0 | 2641.7 | 1.0603 | 6.5703 | 7.6306 | 0.001028 | 3.58 |
| 50.0 | 81.35 | 354.50 | 340.6 | 2305.4 | 2646.0 | 1.0912 | 6.5035 | 7.5947 | 0.001030 | 3.24 |
| 60.0 | 85.95 | 359.10 | 359.9 | 2293.6 | 2653.6 | 1.1455 | 6.3872 | 7.5327 | 0.001033 | 2.73 |
| 70.0 | 89.96 | 363.11 | 376.8 | 2283.3 | 2660.1 | 1.1921 | 6.2883 | 7.4804 | 0.001036 | 2.37 |
| 80.0 | 93.51 | 366.66 | 391.7 | 2274.0 | 2665.8 | 1.2330 | 6.2022 | 7.4352 | 0.001039 | 2.09 |
| 90.0 | 96.71 | 369.86 | 405.2 | 2265.6 | 2670.9 | 1.2696 | 6.1258 | 7.3954 | 0.001041 | 1.87 |
| 100.0 | 99.63 | 372.78 | 417.5 | 2257.9 | 2675.4 | 1.3027 | 6.0571 | 7.3598 | 0.001043 | 1.69 |
| 101.325 | 100.00 | 373.15 | 419.1 | 2256.9 | 2676.0 | 1.3069 | 6.0485 | 7.3554 | 0.0010437 | 1.6730 |
| 105 | 101.00 | 374.15 | 423.3 | 2254.3 | 2677.6 | 1.3182 | 6.0252 | 7.3434 | 0.001045 | 1.618 |
| 110 | 102.32 | 375.47 | 428.8 | 2250.8 | 2679.6 | 1.3330 | 5.9947 | 7.3277 | 0.001046 | 1.549 |
| 115 | 103.59 | 376.74 | 434.2 | 2247.4 | 2681.6 | 1.3472 | 5.9655 | 7.3127 | 0.001047 | 1.486 |
| 120 | 104.81 | 377.96 | 439.4 | 2244.1 | 2683.4 | 1.3609 | 5.9375 | 7.2984 | 0.001048 | 1.428 |
| 125 | 105.99 | 379.14 | 444.4 | 2240.9 | 2685.2 | 1.3741 | 5.9106 | 7.2846 | 0.001049 | 1.375 |
| 130 | 107.13 | 380.28 | 449.2 | 2237.8 | 2687.0 | 1.3868 | 5.8847 | 7.2715 | 0.001050 | 1.325 |
| 135 | 108.24 | 381.39 | 453.9 | 2234.8 | 2688.7 | 1.3991 | 5.8597 | 7.2588 | 0.001050 | 1.279 |
| 140 | 109.32 | 382.47 | 458.4 | 2231.9 | 2690.3 | 1.4109 | 5.8356 | 7.2465 | 0.001051 | 1.236 |
| 145 | 110.36 | 383.51 | 462.8 | 2229.0 | 2691.8 | 1.4225 | 5.8123 | 7.2347 | 0.001052 | 1.196 |
| 150 | 111.37 | 384.52 | 467.1 | 2226.2 | 2693.4 | 1.4336 | 5.7897 | 7.2234 | 0.001053 | 1.159 |
| 155 | 112.36 | 385.51 | 471.3 | 2223.5 | 2694.8 | 1.4445 | 5.7679 | 7.2123 | 0.001054 | 1.124 |
| 160 | 113.32 | 386.47 | 475.4 | 2220.9 | 2696.2 | 1.4550 | 5.7467 | 7.2017 | 0.001055 | 1.091 |
| 165 | 114.26 | 387.41 | 479.4 | 2218.3 | 2697.6 | 1.4652 | 5.7261 | 7.1913 | 0.001056 | 1.060 |
| 170 | 115.17 | 388.32 | 483.2 | 2215.7 | 2699.0 | 1.4752 | 5.7061 | 7.1813 | 0.001056 | 1.031 |
| 175 | 116.06 | 389.21 | 487.0 | 2213.3 | 2700.3 | 1.4849 | 5.6867 | 7.1716 | 0.001057 | 1.003 |
| 180 | 116.93 | 390.08 | 490.7 | 2210.8 | 2701.5 | 1.4944 | 5.6677 | 7.1622 | 0.001058 | 0.977 |
| 185 | 117.79 | 390.94 | 494.3 | 2208.5 | 2702.8 | 1.5036 | 5.6493 | 7.1530 | 0.001059 | 0.952 |
| 190 | 118.62 | 391.77 | 497.9 | 2206.1 | 2704.0 | 1.5127 | 5.6313 | 7.1440 | 0.001059 | 0.929 |
| 195 | 119.43 | 392.58 | 501.3 | 2203.8 | 2705.1 | 1.5215 | 5.6138 | 7.1353 | 0.001060 | 0.907 |
| 200 | 120.23 | 393.38 | 504.7 | 2201.6 | 2706.3 | 1.5301 | 5.5967 | 7.1268 | 0.001061 | 0.885 |

(*continued overleaf*)

Table 1A. (*continued*)

| Absolute pressure ($kN/m^2$) | Temperature (°C) $\theta_s$ | (K) $T_s$ | Enthalpy per unit mass ($H_s$) (kJ/kg) water | latent | steam | Entropy per unit mass ($S_s$) (kJ/kg K) water | latent | steam | Specific volume ($v$) ($m^3$/kg) water | steam |
|---|---|---|---|---|---|---|---|---|---|---|
| 210 | 121.78 | 394.93 | 511.3 | 2197.2 | 2708.5 | 1.5468 | 5.5637 | 7.1105 | 0.001062 | 0.846 |
| 220 | 123.27 | 396.42 | 517.6 | 2193.0 | 2710.6 | 1.5628 | 5.5321 | 7.0949 | 0.001064 | 0.810 |
| 230 | 124.71 | 397.86 | 523.7 | 2188.9 | 2712.6 | 1.5781 | 5.5018 | 7.0800 | 0.001065 | 0.777 |
| 240 | 126.09 | 399.24 | 529.6 | 2184.9 | 2714.5 | 1.5929 | 5.4728 | 7.0657 | 0.001066 | 0.746 |
| 250 | 127.43 | 400.58 | 535.4 | 2181.0 | 2716.4 | 1.6072 | 5.4448 | 7.0520 | 0.001068 | 0.718 |
| 260 | 128.73 | 401.88 | 540.9 | 2177.3 | 2718.2 | 1.6209 | 5.4179 | 7.0389 | 0.001069 | 0.692 |
| 270 | 129.99 | 403.14 | 546.2 | 2173.6 | 2719.9 | 1.6342 | 5.3920 | 7.0262 | 0.001070 | 0.668 |
| 280 | 131.21 | 404.36 | 551.5 | 2170.1 | 2721.5 | 1.6471 | 5.3669 | 7.0140 | 0.001071 | 0.646 |
| 290 | 132.39 | 405.54 | 556.5 | 2166.6 | 2723.1 | 1.6596 | 5.3427 | 7.0022 | 0.001072 | 0.625 |
| 300 | 133.54 | 406.69 | 561.4 | 2163.2 | 2724.7 | 1.6717 | 5.3192 | 6.9909 | 0.001074 | 0.606 |
| 320 | 135.76 | 408.91 | 570.9 | 2156.7 | 2727.6 | 1.6948 | 5.2744 | 6.9692 | 0.001076 | 0.570 |
| 340 | 137.86 | 411.01 | 579.9 | 2150.4 | 2730.3 | 1.7168 | 5.2321 | 6.9489 | 0.001078 | 0.538 |
| 360 | 139.87 | 413.02 | 588.5 | 2144.4 | 2732.9 | 1.7376 | 5.1921 | 6.9297 | 0.001080 | 0.510 |
| 380 | 141.79 | 414.94 | 596.8 | 2138.6 | 2735.3 | 1.7575 | 5.1541 | 6.9115 | 0.001082 | 0.485 |
| 400 | 143.63 | 416.78 | 604.7 | 2132.9 | 2737.6 | 1.7764 | 5.1179 | 6.8943 | 0.001084 | 0.462 |
| 420 | 145.39 | 418.54 | 612.3 | 2127.5 | 2739.8 | 1.7946 | 5.0833 | 6.8779 | 0.001086 | 0.442 |
| 440 | 147.09 | 420.24 | 619.6 | 2122.3 | 2741.9 | 1.8120 | 5.0503 | 6.8622 | 0.001088 | 0.423 |
| 460 | 148.73 | 421.88 | 626.7 | 2117.2 | 2743.9 | 1.8287 | 5.0186 | 6.8473 | 0.001089 | 0.405 |
| 480 | 150.31 | 423.46 | 633.5 | 2112.2 | 2745.7 | 1.8448 | 4.9881 | 6.8329 | 0.001091 | 0.389 |
| 500 | 151.85 | 425.00 | 640.1 | 2107.4 | 2747.5 | 1.8604 | 4.9588 | 6.8192 | 0.001093 | 0.375 |
| 520 | 153.33 | 426.48 | 646.5 | 2102.7 | 2749.3 | 1.8754 | 4.9305 | 6.8059 | 0.001095 | 0.361 |
| 540 | 154.77 | 427.92 | 652.8 | 2098.1 | 2750.9 | 1.8899 | 4.9033 | 6.7932 | 0.001096 | 0.348 |
| 560 | 156.16 | 429.31 | 658.8 | 2093.7 | 2752.5 | 1.9040 | 4.8769 | 6.7809 | 0.001098 | 0.337 |
| 580 | 157.52 | 430.67 | 664.7 | 2089.3 | 2754.0 | 1.9176 | 4.8514 | 6.7690 | 0.001100 | 0.326 |
| 600 | 158.84 | 431.99 | 670.4 | 2085.0 | 2755.5 | 1.9308 | 4.8267 | 6.7575 | 0.001101 | 0.316 |
| 620 | 160.12 | 433.27 | 676.0 | 2080.8 | 2756.9 | 1.9437 | 4.8027 | 6.7464 | 0.001102 | 0.306 |
| 640 | 161.38 | 434.53 | 681.5 | 2076.7 | 2758.2 | 1.9562 | 4.7794 | 6.7356 | 0.001104 | 0.297 |
| 660 | 162.60 | 435.75 | 686.8 | 2072.7 | 2759.5 | 1.9684 | 4.7568 | 6.7252 | 0.001105 | 0.288 |
| 680 | 163.79 | 436.94 | 692.0 | 2068.8 | 2760.8 | 1.9803 | 4.7348 | 6.7150 | 0.001107 | 0.280 |
| 700 | 164.96 | 438.11 | 697.1 | 2064.9 | 2762.0 | 1.9918 | 4.7134 | 6.7052 | 0.001108 | 0.272 |
| 720 | 166.10 | 439.25 | 702.0 | 2061.1 | 2763.2 | 2.0031 | 4.6925 | 6.6956 | 0.001109 | 0.266 |
| 740 | 167.21 | 440.36 | 706.9 | 2057.4 | 2764.3 | 2.0141 | 4.6721 | 6.6862 | 0.001110 | 0.258 |
| 760 | 168.30 | 441.45 | 711.7 | 2053.7 | 2765.4 | 2.0249 | 4.6522 | 6.6771 | 0.001112 | 0.252 |
| 780 | 169.37 | 442.52 | 716.3 | 2050.1 | 2766.4 | 2.0354 | 4.6328 | 6.6683 | 0.001114 | 0.246 |
| 800 | 170.41 | 443.56 | 720.9 | 2046.5 | 2767.5 | 2.0457 | 4.6139 | 6.6596 | 0.001115 | 0.240 |
| 820 | 171.44 | 444.59 | 725.4 | 2043.0 | 2768.5 | 2.0558 | 4.5953 | 6.6511 | 0.001116 | 0.235 |
| 840 | 172.45 | 445.60 | 729.9 | 2039.6 | 2769.4 | 2.0657 | 4.5772 | 6.6429 | 0.001118 | 0.229 |
| 860 | 173.43 | 446.58 | 734.2 | 2036.2 | 2770.4 | 2.0753 | 4.5595 | 6.6348 | 0.001119 | 0.224 |
| 880 | 174.40 | 447.55 | 738.5 | 2032.8 | 2771.3 | 2.0848 | 4.5421 | 6.6269 | 0.001120 | 0.220 |
| 900 | 175.36 | 448.51 | 742.6 | 2029.5 | 2772.1 | 2.0941 | 4.5251 | 6.6192 | 0.001121 | 0.215 |
| 920 | 176.29 | 449.44 | 746.8 | 2026.2 | 2773.0 | 2.1033 | 4.5084 | 6.6116 | 0.001123 | 0.210 |
| 940 | 177.21 | 450.36 | 750.8 | 2023.0 | 2773.8 | 2.1122 | 4.4920 | 6.6042 | 0.001124 | 0.206 |
| 960 | 178.12 | 451.27 | 754.8 | 2019.8 | 2774.6 | 2.1210 | 4.4759 | 6.5969 | 0.001125 | 0.202 |
| 980 | 179.01 | 452.16 | 758.7 | 2016.7 | 2775.4 | 2.1297 | 4.4602 | 6.5898 | 0.001126 | 0.198 |
| 1000 | 179.88 | 453.03 | 762.6 | 2013.6 | 2776.2 | 2.1382 | 4.4447 | 6.5828 | 0.001127 | 0.194 |
| 1100 | 184.06 | 457.21 | 781.1 | 1998.6 | 2779.7 | 2.1786 | 4.3712 | 6.5498 | 0.001133 | 0.177 |
| 1200 | 187.96 | 461.11 | 798.4 | 1984.3 | 2782.7 | 2.2160 | 4.3034 | 6.5194 | 0.001139 | 0.163 |
| 1300 | 191.60 | 464.75 | 814.7 | 1970.7 | 2785.4 | 2.2509 | 4.2404 | 6.4913 | 0.001144 | 0.151 |
| 1400 | 195.04 | 468.19 | 830.1 | 1957.7 | 2787.8 | 2.2836 | 4.1815 | 6.4651 | 0.001149 | 0.141 |
| 1500 | 198.28 | 471.43 | 844.6 | 1945.3 | 2789.9 | 2.3144 | 4.1262 | 6.4406 | 0.001154 | 0.132 |
| 1600 | 201.37 | 474.52 | 858.5 | 1933.2 | 2791.7 | 2.3436 | 4.0740 | 6.4176 | 0.001159 | 0.124 |

Table 1A. (*continued*)

| Absolute pressure (kN/m²) | Temperature (°C) $\theta_s$ | Temperature (K) $T_s$ | Enthalpy per unit mass ($H_s$) (kJ/kg) water | latent | steam | Entropy per unit mass ($S_s$) (kJ/kg K) water | latent | steam | Specific volume ($v$) (m³/kg) water | steam |
|---|---|---|---|---|---|---|---|---|---|---|
| 1700 | 204.30 | 477.45 | 871.8 | 1921.6 | 2793.4 | 2.3712 | 4.0246 | 6.3958 | 0.001163 | 0.117 |
| 1800 | 207.11 | 480.26 | 884.5 | 1910.3 | 2794.8 | 2.3976 | 3.9776 | 6.3751 | 0.001168 | 0.110 |
| 1900 | 209.79 | 482.94 | 896.8 | 1899.3 | 2796.1 | 2.4227 | 3.9327 | 6.3555 | 0.001172 | 0.105 |
| 2000 | 212.37 | 485.52 | 908.6 | 1888.7 | 2797.2 | 2.4468 | 3.8899 | 6.3367 | 0.001177 | 0.0996 |
| 2200 | 217.24 | 490.39 | 930.9 | 1868.1 | 2799.1 | 2.4921 | 3.8094 | 6.3015 | 0.001185 | 0.0907 |
| 2400 | 221.78 | 494.93 | 951.9 | 1848.5 | 2800.4 | 2.5342 | 3.7348 | 6.2690 | 0.001193 | 0.0832 |
| 2600 | 226.03 | 499.18 | 971.7 | 1829.7 | 2801.4 | 2.5736 | 3.6652 | 6.2388 | 0.001201 | 0.0769 |
| 3000 | 233.84 | 506.99 | 1008.3 | 1794.0 | 2802.3 | 2.6455 | 3.5383 | 6.1838 | 0.001216 | 0.0666 |
| 3500 | 242.54 | 515.69 | 1049.7 | 1752.2 | 2802.0 | 2.7252 | 3.3976 | 6.1229 | 0.001235 | 0.0570 |
| 4000 | 250.33 | 523.48 | 1087.4 | 1712.9 | 2800.3 | 2.7965 | 3.2720 | 6.0685 | 0.001252 | 0.0498 |
| 4500 | 257.41 | 530.56 | 1122.1 | 1675.6 | 2797.7 | 2.8612 | 3.1579 | 6.0191 | 0.001269 | 0.0440 |
| 5000 | 263.92 | 537.07 | 1154.5 | 1639.7 | 2794.2 | 2.9207 | 3.0528 | 5.9735 | 0.001286 | 0.0394 |
| 6000 | 275.56 | 548.71 | 1213.7 | 1571.3 | 2785.0 | 3.0274 | 2.8633 | 5.8907 | 0.001319 | 0.0324 |
| 7000 | 285.80 | 558.95 | 1267.5 | 1506.0 | 2773.4 | 3.1220 | 2.6541 | 5.8161 | 0.001351 | 0.0274 |
| 8000 | 294.98 | 568.13 | 1317.2 | 1442.7 | 2759.9 | 3.2077 | 2.5393 | 5.7470 | 0.001384 | 0.0235 |
| 9000 | 303.31 | 576.46 | 1363.8 | 1380.8 | 2744.6 | 3.2867 | 2.3952 | 5.6820 | 0.001418 | 0.0205 |
| 10000 | 310.96 | 584.11 | 1408.1 | 1319.7 | 2727.7 | 3.3606 | 2.2592 | 5.6198 | 0.001453 | 0.0180 |
| 11000 | 318.04 | 591.19 | 1450.6 | 1258.8 | 2709.3 | 3.4304 | 2.1292 | 5.5596 | 0.001489 | 0.0160 |
| 12000 | 324.64 | 597.79 | 1491.7 | 1197.5 | 2698.2 | 3.4971 | 2.0032 | 5.5003 | 0.001527 | 0.0143 |
| 14000 | 336.63 | 609.78 | 1571.5 | 1070.9 | 2642.4 | 3.6241 | 1.7564 | 5.3804 | 0.0016105 | 0.01150 |
| 16000 | 347.32 | 620.47 | 1650.4 | 934.5 | 2584.9 | 3.7470 | 1.5063 | 5.2533 | 0.0017102 | 0.00931 |
| 18000 | 356.96 | 630.11 | 1734.8 | 779.0 | 2513.9 | 3.8766 | 1.2362 | 5.1127 | 0.0018399 | 0.007497 |
| 20000 | 365.71 | 638.86 | 1826.6 | 591.6 | 2418.2 | 4.0151 | 0.9259 | 4.9410 | 0.0020374 | 0.005875 |
| 22000 | 373.68 | 646.83 | 2010.3 | 186.3 | 2196.6 | 4.2934 | 0.2881 | 4.5814 | 0.0026675 | 0.003735 |
| 22120 | 374.15 | 647.30 | 2107.4 | 0 | 2107.4 | 4.4429 | 0 | 4.4429 | 0.0031700 | 0.003170 |

Table 1B. Properties of saturated steam (Centigrade and Fahrenheit units)

| Pressure | | Temperature | | Enthalpy per unit mass | | | | | | Entropy (Btu/lb°F) | | Specific volume (ft$^3$/lb) |
|---|---|---|---|---|---|---|---|---|---|---|---|---|
| | | | | Centigrade units (kcal/kg) | | | Fahrenheit units (Btu/lb) | | | | | |
| Absolute (lb/in.$^2$) | Vacuum (in. Hg) | (°C) | (°F) | Water | Latent | Steam | Water | Latent | Steam | Water | Steam | Steam |
| **0.5** | 28.99 | 26.42 | 79.6 | **26.45** | 582.50 | **608.95** | **47.6** | 1048.5 | **1096.1** | 0.0924 | 2.0367 | 643.0 |
| **0.6** | 28.79 | 29.57 | 85.3 | **29.58** | 580.76 | **610.34** | **53.2** | 1045.4 | **1098.6** | 0.1028 | 2.0214 | 540.6 |
| **0.7** | 28.58 | 32.28 | 90.1 | **32.28** | 579.27 | **611.55** | **58.1** | 1042.7 | **1100.8** | 0.1117 | 2.0082 | 466.6 |
| **0.8** | 28.38 | 34.67 | 94.4 | **34.66** | 577.95 | **612.61** | **62.4** | 1040.3 | **1102.7** | 0.1196 | 1.9970 | 411.7 |
| **0.9** | 28.17 | 36.80 | 98.2 | **36.80** | 576.74 | **613.54** | **66.2** | 1038.1 | **1104.3** | 0.1264 | 1.9871 | 368.7 |
| **1.0** | 27.97 | 38.74 | 101.7 | **38.74** | 575.60 | **614.34** | **69.7** | 1036.1 | **1105.8** | 0.1326 | 1.9783 | 334.0 |
| **1.1** | 27.76 | 40.52 | 104.9 | **40.52** | 574.57 | **615.09** | **72.9** | 1034.3 | **1107.2** | 0.1381 | 1.9702 | 305.2 |
| **1.2** | 27.56 | 42.17 | 107.9 | **42.17** | 573.63 | **615.80** | **75.9** | 1032.5 | **1108.4** | 0.1433 | 1.9630 | 281.1 |
| **1.3** | 27.35 | 43.70 | 110.7 | **43.70** | 572.75 | **616.45** | **78.7** | 1030.9 | **1109.6** | 0.1484 | 1.9563 | 260.5 |
| **1.4** | 27.15 | 45.14 | 113.3 | **45.12** | 571.94 | **617.06** | **81.3** | 1029.5 | **1110.8** | 0.1527 | 1.9501 | 243.0 |
| **1.5** | 26.95 | 46.49 | 115.7 | **46.45** | 571.16 | **617.61** | **83.7** | 1028.1 | **1111.8** | 0.1569 | 1.9442 | 228.0 |
| **1.6** | 26.74 | 47.77 | 118.0 | **47.73** | 570.41 | **618.14** | **86.0** | 1026.8 | **1112.8** | 0.1609 | 1.9387 | 214.3 |
| **1.7** | 26.54 | 48.98 | 120.2 | **48.94** | 569.71 | **618.65** | **88.2** | 1025.5 | **1113.7** | 0.1646 | 1.9336 | 202.5 |
| **1.8** | 26.33 | 50.13 | 122.2 | **50.08** | 569.06 | **619.14** | **90.2** | 1024.4 | **1114.6** | 0.1681 | 1.9288 | 191.8 |
| **1.9** | 26.13 | 51.22 | 124.2 | **51.16** | 568.47 | **619.63** | **92.1** | 1023.3 | **1115.4** | 0.1715 | 1.9243 | 182.3 |
| **2.0** | 25.92 | 52.27 | 126.1 | **52.22** | 567.89 | **620.11** | **94.0** | 1022.2 | **1116.2** | 0.1749 | 1.9200 | 173.7 |
| **3.0** | 23.88 | 60.83 | 141.5 | **60.78** | 562.89 | **623.67** | **109.4** | 1013.2 | **1122.6** | 0.2008 | 1.8869 | 118.7 |
| **4.0** | 21.84 | 67.23 | 153.0 | **67.20** | 559.29 | **626.49** | **121.0** | 1006.7 | **1127.7** | 0.2199 | 1.8632 | 90.63 |
| **5.0** | 19.80 | 72.38 | 162.3 | **72.36** | 556.24 | **628.60** | **130.2** | 1001.6 | **1131.8** | 0.2348 | 1.8449 | 73.52 |
| **6.0** | 17.76 | 76.72 | 170.1 | **76.71** | 553.62 | **630.33** | **138.1** | 996.6 | **1134.7** | 0.2473 | 1.8299 | 61.98 |
| **7.0** | 15.71 | 80.49 | 176.9 | **80.52** | 551.20 | **631.72** | **144.9** | 992.2 | **1137.1** | 0.2582 | 1.8176 | 53.64 |
| **8.0** | 13.67 | 83.84 | 182.9 | **83.89** | 549.16 | **633.05** | **151.0** | 988.5 | **1139.5** | 0.2676 | 1.8065 | 47.35 |
| **9.0** | 11.63 | 86.84 | 188.3 | **86.88** | 547.42 | **634.30** | **156.5** | 985.2 | **1141.7** | 0.2762 | 1.7968 | 42.40 |
| **10.0** | 9.59 | 89.58 | 193.2 | **89.61** | 545.82 | **635.43** | **161.3** | 982.5 | **1143.8** | 0.2836 | 1.7884 | 38.42 |
| **11.0** | 7.55 | 92.10 | 197.8 | **92.15** | 544.26 | **636.41** | **165.9** | 979.6 | **1145.5** | 0.2906 | 1.7807 | 35.14 |
| **12.0** | 5.50 | 94.44 | 202.0 | **94.50** | 542.75 | **637.25** | **170.1** | 976.9 | **1147.0** | 0.2970 | 1.7735 | 32.40 |
| **13.0** | 3.46 | 96.62 | 205.9 | **96.69** | 541.34 | **638.03** | **173.9** | 974.6 | **1148.5** | 0.3029 | 1.7672 | 30.05 |
| **14.0** | 1.42 | 98.65 | 209.6 | **98.73** | 540.06 | **638.79** | **177.7** | 972.2 | **1149.9** | 0.3086 | 1.7613 | 28.03 |
| **14.696** | Gauge (lb/in.$^2$) | 100.00 | 212.0 | **100.06** | 539.22 | **639.28** | **180.1** | 970.6 | **1150.7** | 0.3122 | 1.7574 | 26.80 |
| **15** | 0.3 | 100.57 | 213.0 | **100.65** | 538.9 | **639.5** | **181.2** | 970.0 | **1151.2** | 0.3137 | 1.7556 | 26.28 |
| **16** | 1.3 | 102.40 | 216.3 | **102.51** | 537.7 | **640.2** | **184.5** | 967.9 | **1152.4** | 0.3187 | 1.7505 | 24.74 |
| **17** | 2.3 | 104.13 | 219.5 | **104.27** | 536.5 | **640.8** | **187.6** | 965.9 | **1153.5** | 0.3231 | 1.7456 | 23.38 |
| **18** | 3.3 | 105.78 | 222.4 | **105.94** | 535.5 | **641.4** | **190.6** | 964.0 | **1154.6** | 0.3276 | 1.7411 | 22.17 |
| **19** | 4.3 | 107.36 | 225.2 | **107.53** | 534.5 | **642.0** | **193.5** | 962.2 | **1155.7** | 0.3319 | 1.7368 | 21.07 |
| **20** | 5.3 | 108.87 | 228.0 | **109.05** | 533.6 | **642.6** | **196.3** | 960.4 | **1156.7** | 0.3358 | 1.7327 | 20.09 |
| **21** | 6.3 | 110.32 | 230.6 | **110.53** | 532.6 | **643.1** | **198.9** | 958.8 | **1157.7** | 0.3396 | 1.7287 | 19.19 |
| **22** | 7.3 | 111.71 | 233.1 | **111.94** | 531.7 | **643.6** | **201.4** | 957.2 | **1158.6** | 0.3433 | 1.7250 | 18.38 |
| **23** | 8.3 | 113.05 | 235.5 | **113.30** | 530.8 | **644.1** | **203.9** | 955.6 | **1159.5** | 0.3468 | 1.7215 | 17.63 |
| **24** | 9.3 | 114.34 | 237.8 | **114.61** | 530.0 | **644.6** | **206.3** | 954.0 | **1160.3** | 0.3502 | 1.7181 | 16.94 |
| **25** | 10.3 | 115.59 | 240.1 | **115.87** | 529.2 | **645.1** | **208.6** | 952.5 | **1161.1** | 0.3534 | 1.7148 | 16.30 |
| **26** | 11.3 | 116.80 | 242.2 | **117.11** | 528.4 | **645.5** | **210.8** | 951.1 | **1161.9** | 0.3565 | 1.7118 | 15.72 |
| **27** | 12.3 | 117.97 | 244.4 | **118.31** | 527.6 | **645.9** | **212.9** | 949.7 | **1162.6** | 0.3595 | 1.7089 | 15.17 |
| **28** | 13.3 | 119.11 | 246.4 | **119.47** | 526.8 | **646.3** | **215.0** | 948.3 | **1163.3** | 0.3625 | 1.7060 | 14.67 |
| **29** | 14.3 | 120.21 | 248.4 | **120.58** | 526.1 | **646.7** | **217.0** | 947.0 | **1164.0** | 0.3654 | 1.7032 | 14.19 |
| **30** | 15.3 | 121.3 | 250.3 | **121.7** | 525.4 | **647.1** | **219.0** | 945.6 | **1164.6** | 0.3682 | 1.7004 | 13.73 |
| **32** | 17.3 | 123.3 | 254.0 | **123.8** | 524.1 | **647.9** | **222.7** | 943.1 | **1165.8** | 0.3735 | 1.6952 | 12.93 |
| **34** | 19.3 | 125.3 | 257.6 | **125.8** | 522.8 | **648.6** | **226.3** | 940.7 | **1167.0** | 0.3785 | 1.6905 | 12.21 |

Table 1B. (*continued*)

| Pressure | | Temperature | | Enthalpy per unit mass | | | | | | Entropy (Btu/lb°F) | | Specific volume ($ft^3$/lb) |
|---|---|---|---|---|---|---|---|---|---|---|---|---|
| | | | | Centigrade units (kcal/kg) | | | Fahrenheit units (Btu/lb) | | | | | |
| Absolute (lb/in.$^2$) | Vacuum (in. Hg) | (°C) | (°F) | Water | Latent | Steam | Water | Latent | Steam | Water | Steam | Steam |
| **36** | 21.3 | 127.2 | 260.9 | **127.7** | 521.5 | **649.2** | **229.7** | 938.5 | **1168.2** | 0.3833 | 1.6860 | 11.58 |
| **38** | 23.3 | 128.9 | 264.1 | **129.5** | 520.3 | **649.8** | **233.0** | 936.4 | **1169.4** | 0.3879 | 1.6817 | 11.02 |
| **40** | 25.3 | 130.7 | 267.2 | **131.2** | 519.2 | **650.4** | **236.1** | 934.4 | **1170.5** | 0.3923 | 1.6776 | 10.50 |
| **42** | 27.3 | 132.3 | 270.3 | **132.9** | 518.0 | **650.9** | **239.1** | 932.3 | **1171.4** | 0.3964 | 1.6737 | 10.30 |
| **44** | 29.3 | 133.9 | 273.1 | **134.5** | 516.9 | **651.4** | **242.0** | 930.3 | **1172.3** | 0.4003 | 1.6700 | 9.600 |
| **46** | 31.3 | 135.4 | 275.8 | **136.0** | 515.9 | **651.9** | **244.9** | 928.3 | **1173.2** | 0.4041 | 1.6664 | 9.209 |
| **48** | 33.3 | 136.9 | 278.5 | **137.5** | 514.8 | **652.3** | **247.6** | 926.4 | **1174.0** | 0.4077 | 1.6630 | 8.848 |
| **50** | 35.3 | 138.3 | 281.0 | **139.0** | 513.8 | **652.8** | **250.2** | 924.6 | **1174.8** | 0.4112 | 1.6597 | 8.516 |
| **52** | 37.3 | 139.7 | 283.5 | **140.4** | 512.8 | **653.2** | **252.7** | 922.9 | **1175.6** | 0.4146 | 1.6566 | 8.208 |
| **54** | 39.3 | 141.0 | 285.9 | **141.8** | 511.8 | **653.6** | **255.2** | 921.1 | **1176.3** | 0.4179 | 1.6536 | 7.922 |
| **56** | 41.3 | 142.3 | 288.3 | **143.1** | 510.9 | **654.0** | **257.6** | 919.4 | **1177.0** | 0.4211 | 1.6507 | 7.656 |
| **58** | 43.3 | 143.6 | 290.5 | **144.4** | 510.0 | **654.4** | **259.9** | 917.8 | **1177.7** | 0.4242 | 1.6478 | 7.407 |
| **60** | 45.3 | 144.9 | 292.7 | **145.6** | 509.2 | **654.8** | **262.2** | 916.2 | **1178.4** | 0.4272 | 1.6450 | 7.175 |
| **62** | 47.3 | 146.1 | 294.9 | **146.8** | 508.4 | **655.2** | **264.4** | 914.6 | **1179.0** | 0.4302 | 1.6423 | 6.957 |
| **64** | 49.3 | 147.3 | 296.9 | **148.0** | 507.6 | **655.6** | **266.5** | 913.1 | **1179.6** | 0.4331 | 1.6398 | 6.752 |
| **66** | 51.3 | 148.4 | 299.0 | **149.2** | 506.7 | **655.9** | **268.6** | 911.6 | **1180.2** | 0.4359 | 1.6374 | 6.560 |
| **68** | 53.3 | 149.5 | 301.0 | **150.3** | 505.9 | **656.2** | **270.7** | 910.1 | **1180.8** | 0.4386 | 1.6350 | 6.378 |
| **70** | 55.3 | 150.6 | 302.9 | **151.5** | 505.0 | **656.5** | **272.7** | 908.7 | **1181.4** | 0.4412 | 1.6327 | 6.206 |
| **72** | 57.3 | 151.6 | 304.8 | **152.6** | 504.2 | **656.8** | **274.6** | 907.4 | **1182.0** | 0.4437 | 1.6304 | 6.044 |
| **74** | 59.3 | 152.6 | 306.7 | **153.6** | 503.4 | **657.0** | **276.5** | 906.0 | **1182.5** | 0.4462 | 1.6282 | 5.890 |
| **76** | 61.3 | 153.6 | 308.5 | **154.7** | 502.6 | **657.3** | **278.4** | 904.6 | **1183.0** | 0.4486 | 1.6261 | 5.743 |
| **78** | 63.3 | 154.6 | 310.3 | **155.7** | 501.8 | **657.5** | **280.3** | 903.2 | **1183.5** | 0.4510 | 1.6240 | 5.604 |
| **80** | 65.3 | 155.6 | 312.0 | **156.7** | 501.1 | **657.8** | **282.1** | 901.9 | **1184.0** | 0.4533 | 1.6219 | 5.472 |
| **82** | 67.3 | 156.5 | 313.7 | **157.7** | 500.3 | **658.0** | **283.9** | 900.6 | **1184.5** | 0.4556 | 1.6199 | 5.346 |
| **84** | 69.3 | 157.5 | 315.4 | **158.6** | 499.6 | **658.2** | **285.6** | 899.4 | **1185.0** | 0.4579 | 1.6180 | 5.226 |
| **86** | 71.3 | 158.4 | 317.1 | **159.6** | 498.9 | **658.5** | **287.3** | 898.1 | **1185.4** | 0.4601 | 1.6161 | 5.110 |
| **88** | 73.3 | 159.4 | 318.7 | **160.5** | 498.3 | **658.8** | **289.0** | 896.8 | **1185.8** | 0.4622 | 1.6142 | 5.000 |
| **90** | 75.3 | 160.3 | 320.3 | **161.5** | 497.6 | **659.1** | **290.7** | 895.5 | **1186.2** | 0.4643 | 1.6124 | 4.896 |
| **92** | 77.3 | 161.2 | 321.9 | **162.4** | 496.9 | **659.3** | **292.3** | 894.3 | **1186.6** | 0.4664 | 1.6106 | 4.796 |
| **94** | 79.3 | 162.0 | 323.3 | **163.3** | 496.3 | **659.6** | **293.9** | 893.1 | **1187.0** | 0.4684 | 1.6088 | 4.699 |
| **96** | 81.3 | 162.8 | 324.8 | **164.1** | 495.7 | **659.8** | **295.5** | 891.9 | **1187.4** | 0.4704 | 1.6071 | 4.607 |
| **98** | 83.3 | 163.6 | 326.6 | **165.0** | 495.0 | **660.0** | **297.0** | 890.8 | **1187.8** | 0.4723 | 1.6054 | 4.519 |
| **100** | 85.3 | 164.4 | 327.8 | **165.8** | 494.3 | **660.1** | **298.5** | 889.7 | **1188.2** | 0.4742 | 1.6038 | 4.434 |
| **105** | 90.3 | 166.4 | 331.3 | **167.9** | 492.7 | **660.6** | **302.2** | 886.9 | **1189.1** | 0.4789 | 1.6000 | 4.230 |
| **110** | 95.3 | 168.2 | 334.8 | **169.8** | 491.2 | **661.0** | **305.7** | 884.2 | **1189.9** | 0.4833 | 1.5963 | 4.046 |
| **115** | 100.3 | 170.0 | 338.1 | **171.7** | 489.8 | **661.5** | **309.2** | 881.5 | **1190.7** | 0.4876 | 1.5927 | 3.880 |
| **120** | 105.3 | 171.8 | 341.3 | **173.6** | 488.3 | **661.9** | **312.5** | 878.9 | **1191.4** | 0.4918 | 1.5891 | 3.729 |
| **125** | 110.3 | 173.5 | 344.4 | **175.4** | 486.9 | **662.3** | **315.7** | 876.4 | **1192.1** | 0.4958 | 1.5856 | 3.587 |
| **130** | 115.3 | 175.2 | 347.3 | **177.1** | 485.6 | **662.7** | **318.8** | 874.0 | **1192.8** | 0.4997 | 1.5823 | 3.456 |
| **135** | 120.3 | 176.8 | 350.2 | **178.8** | 484.2 | **663.0** | **321.9** | 871.5 | **1193.4** | 0.5035 | 1.5792 | 3.335 |
| **140** | 125.3 | 178.3 | 353.0 | **180.5** | 482.9 | **663.4** | **324.9** | 869.1 | **1194.0** | 0.5071 | 1.5763 | 3.222 |
| **145** | 130.3 | 179.8 | 355.8 | **182.1** | 481.6 | **663.7** | **327.8** | 866.8 | **1194.6** | 0.5106 | 1.5733 | 3.116 |
| **150** | 135.3 | 181.3 | 358.4 | **183.7** | 480.3 | **664.0** | **330.6** | 864.5 | **1195.1** | 0.5140 | 1.5705 | 3.015 |

Table 1C. Enthalpy of superheated steam, $H$ (kJ/kg)

| Pressure | Saturation | | Temperature, $\theta$ (°C) | 100 | 200 | 300 | 400 | 500 | 600 | 700 | 800 |
|---|---|---|---|---|---|---|---|---|---|---|---|
| $P$ (kN/m$^2$) | $T_s$ (K) | $S_s$ (kJ/kg) | Temperature, $T$(K) | 373.15 | 473.15 | 573.15 | 673.15 | 773.15 | 873.15 | 973.15 | 1073.15 |
| 100 | 372.78 | 2675.4 | | 2676.0 | 2875.4 | 3074.6 | 3278.0 | 3488.0 | 3705.0 | 3928.0 | 4159.0 |
| 200 | 393.38 | 2706.3 | | | 2870.4 | 3072.0 | 3276.4 | 3487.0 | 3704.0 | 3927.0 | 4158.0 |
| 300 | 406.69 | 2724.7 | | | 2866.0 | 3069.7 | 3275.0 | 3486.0 | 3703.1 | 3927.0 | 4158.0 |
| 400 | 416.78 | 2737.6 | | | 2861.3 | 3067.0 | 3273.5 | 3485.0 | 3702.0 | 3926.0 | 4157.0 |
| 500 | 425.00 | 2747.5 | | | 2856.0 | 3064.8 | 3272.1 | 3484.0 | 3701.2 | 3926.0 | 4156.8 |
| 600 | 431.99 | 2755.5 | | | 2850.7 | 3062.0 | 3270.0 | 3483.0 | 3701.0 | 3925.0 | 4156.2 |
| 700 | 438.11 | 2762.0 | | | 2845.5 | 3059.5 | 3269.0 | 3482.6 | 3700.2 | 3924.0 | 4156.0 |
| 800 | 443.56 | 2767.5 | | | 2839.7 | 3057.0 | 3266.8 | 3480.4 | 3699.0 | 3923.8 | 4155.0 |
| 900 | 448.56 | 2772.1 | | | 2834.0 | 3055.0 | 3266.2 | 3479.5 | 3698.6 | 3923.0 | 4155.0 |
| 1000 | 453.03 | 2776.2 | | | 2828.7 | 3051.7 | 3264.3 | 3478.0 | 3697.5 | 3922.8 | 4154.0 |
| 2000 | 485.59 | 2797.2 | | | | 3024.8 | 3248.0 | 3467.0 | 3690.0 | 3916.0 | 4150.0 |
| 3000 | 506.98 | 2802.3 | | | | 2994.8 | 3231.7 | 3456.0 | 3681.6 | 3910.4 | 4145.0 |
| 4000 | 523.49 | 2800.3 | | | | 2962.0 | 3214.8 | 3445.0 | 3673.4 | 3904.0 | 4139.6 |
| 5000 | 537.09 | 2794.2 | | | | 2926.0 | 3196.9 | 3433.8 | 3665.4 | 3898.0 | 4135.5 |
| 6000 | 548.71 | 2785.0 | | | | 2886.0 | 3178.0 | 3421.7 | 3657.0 | 3891.8 | 4130.0 |
| 7000 | 558.95 | 2773.4 | | | | 2840.0 | 3159.1 | 3410.0 | 3648.8 | 3886.0 | 4124.8 |
| 8000 | 568.13 | 2759.9 | | | | 2785.0 | 3139.5 | 3398.0 | 3640.4 | 3880.8 | 4121.0 |
| 9000 | 576.46 | 2744.6 | | | | | 3119.0 | 3385.5 | 3632.0 | 3873.6 | 4116.0 |
| 10000 | 584.11 | 2727.7 | | | | | 3097.7 | 3373.6 | 3624.0 | 3867.2 | 4110.8 |
| 11000 | 591.19 | 2709.3 | | | | | 3075.6 | 3361.0 | 3615.5 | 3862.0 | 4106.0 |
| 12000 | 597.79 | 2698.2 | | | | | 3052.9 | 3349.0 | 3607.0 | 3855.3 | 4101.2 |

Table 1D. Entropy of superheated steam, $S$ (kJ/kg K)

| Pressure | Saturation | | Temperature, $\theta$ (°C) | 100 | 200 | 300 | 400 | 500 | 600 | 700 | 800 |
|---|---|---|---|---|---|---|---|---|---|---|---|
| $P$ (kN/m$^2$) | $T_s$ (K) | $H_s$ (kJ/kg) | Temperature, $T$(K) | 373.15 | 473.15 | 573.15 | 673.15 | 773.15 | 873.15 | 973.15 | 1073.15 |
| 100 | 372.78 | 7.3598 | | 7.362 | 7.834 | 8.216 | 8.544 | 8.834 | 9.100 | 9.344 | 9.565 |
| 200 | 393.38 | 7.1268 | | | 7.507 | 7.892 | 8.222 | 8.513 | 8.778 | 9.020 | 9.246 |
| 300 | 406.69 | 6.9909 | | | 7.312 | 7.702 | 8.033 | 8.325 | 8.591 | 8.833 | 9.057 |
| 400 | 416.78 | 6.8943 | | | 7.172 | 7.566 | 7.898 | 8.191 | 8.455 | 8.700 | 8.925 |
| 500 | 425.00 | 6.8192 | | | 7.060 | 7.460 | 7.794 | 8.087 | 8.352 | 8.596 | 8.820 |
| 600 | 431.99 | 6.7575 | | | 6.968 | 7.373 | 7.708 | 8.002 | 8.268 | 8.510 | 8.738 |
| 700 | 438.11 | 6.7052 | | | 6.888 | 7.298 | 7.635 | 7.930 | 8.195 | 8.438 | 8.665 |
| 800 | 443.56 | 6.6596 | | | 6.817 | 7.234 | 7.572 | 7.867 | 8.133 | 8.375 | 8.602 |
| 900 | 448.56 | 6.6192 | | | 6.753 | 7.176 | 7.515 | 7.812 | 8.077 | 8.321 | 8.550 |
| 1000 | 453.03 | 6.5828 | | | 6.695 | 7.124 | 7.465 | 7.762 | 8.028 | 8.272 | 8.502 |
| 2000 | 485.59 | 6.3367 | | | | 6.768 | 7.128 | 7.431 | 7.702 | 7.950 | 8.176 |
| 3000 | 506.98 | 6.1838 | | | | 6.541 | 6.922 | 7.233 | 7.508 | 7.756 | 7.985 |
| 4000 | 523.49 | 6.0685 | | | | 6.364 | 6.770 | 7.090 | 7.368 | 7.620 | 7.850 |
| 5000 | 537.07 | 5.9735 | | | | 6.211 | 6.647 | 6.977 | 7.258 | 7.510 | 7.744 |
| 6000 | 548.71 | 5.8907 | | | | 6.060 | 6.542 | 6.880 | 7.166 | 7.422 | 7.655 |
| 7000 | 558.95 | 5.8161 | | | | 5.933 | 6.450 | 6.798 | 7.088 | 7.345 | 7.581 |
| 8000 | 568.13 | 5.7470 | | | | 5.792 | 6.365 | 6.724 | 7.020 | 7.280 | 7.515 |
| 9000 | 576.46 | 5.6820 | | | | | 6.288 | 6.659 | 6.958 | 7.220 | 7.457 |
| 10000 | 584.11 | 5.6198 | | | | | 6.215 | 6.598 | 6.902 | 7.166 | 7.405 |
| 11000 | 591.19 | 5.5596 | | | | | 6.145 | 6.540 | 6.850 | 7.117 | 7.357 |
| 12000 | 597.79 | 5.5003 | | | | | 6.077 | 6.488 | 6.802 | 7.072 | 7.312 |

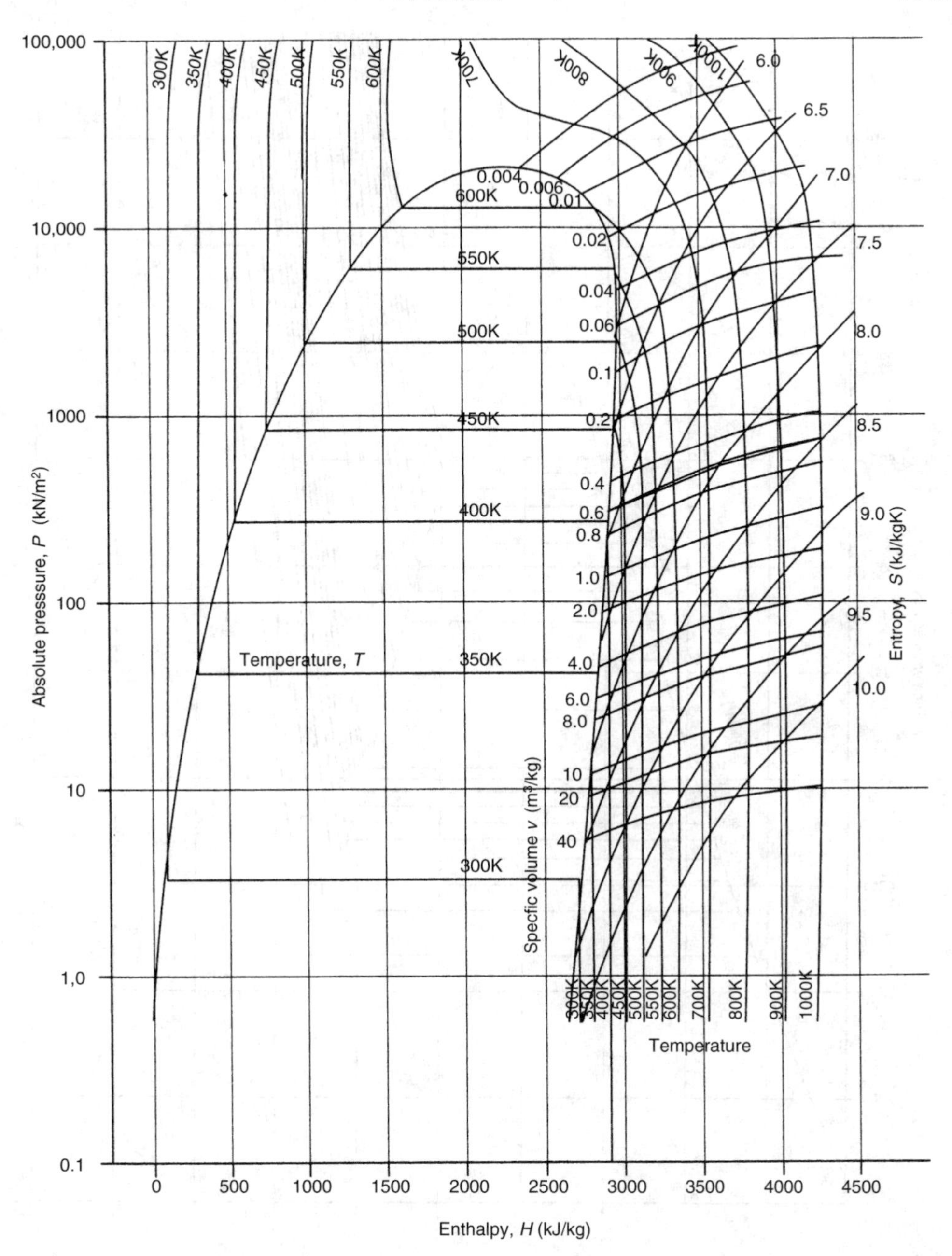


Figure 1A. Pressure–enthalpy diagram for water and steam

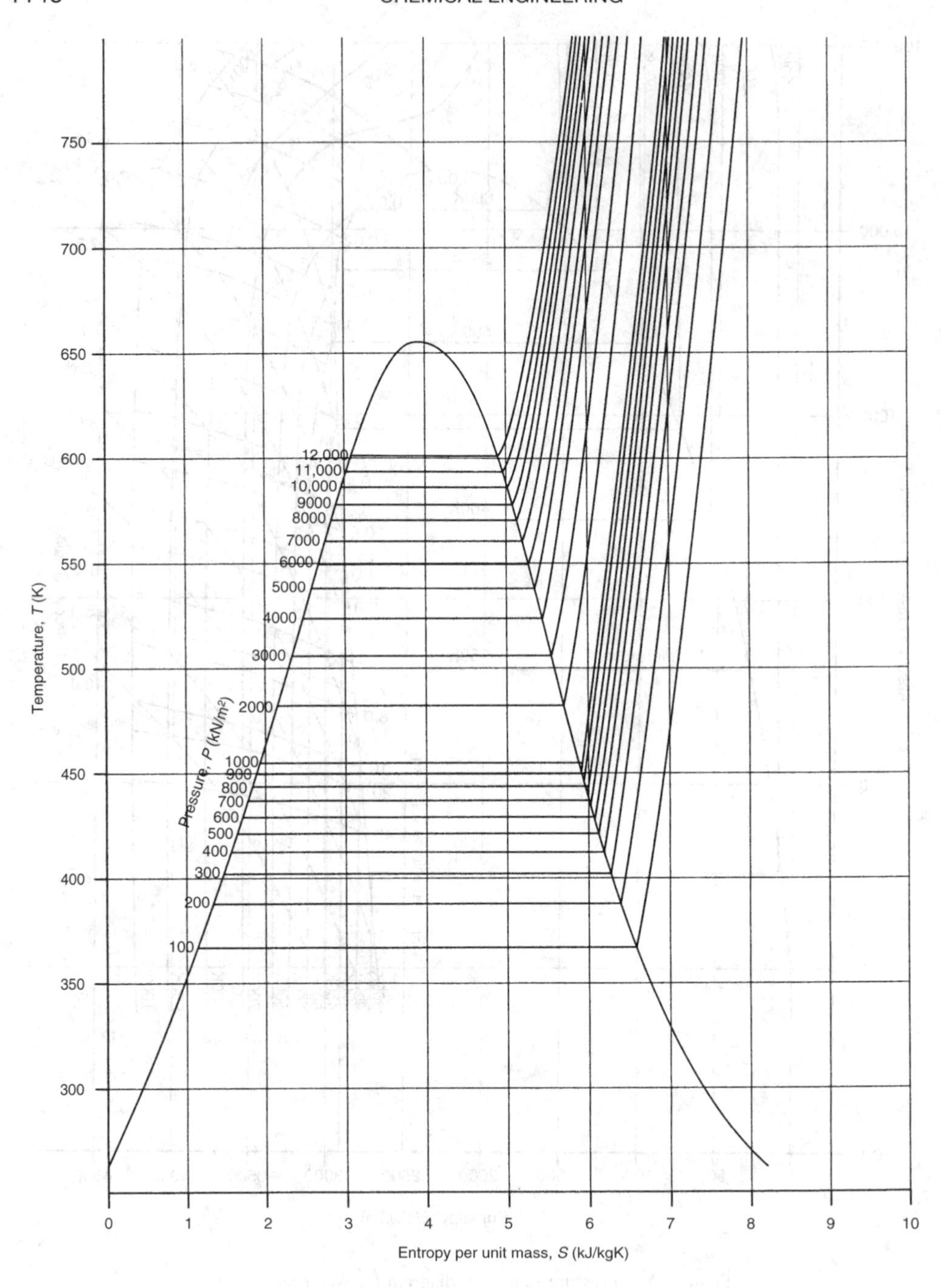


Figure 1B. Temperature–entropy diagram for water and steam

# Table A2. CONVERSION FACTORS FOR SOME COMMON SI UNITS

An asterisk (*) denotes an exact relationship.

|  |  | |  |
|---|---|---|---|
| Length | 1 in | : | 25.4 mm |
| | 1 ft | : | 0.3048 m |
| | 1 yd | : | 0.9144 m |
| | 1 mile | : | 1.6093 km |
| | *1 Å (Angstrom) | : | $10^{-10}$ m |
| Time | *1 min | : | 60 s |
| | *1 h | : | 3.6 ks |
| | *1 day | : | 86.4 ks |
| | 1 year | : | 31.5 Ms |
| Area | 1 $in^2$ | : | 645.16 $mm^2$ |
| | 1 $ft^2$ | : | 0.092903 $m^2$ |
| | 1 $yd^2$ | : | 0.83613 $m^2$ |
| | 1 $mile^2$ | : | 2.590 $km^2$ |
| | 1 acre | : | 4046.9 $m^2$ |
| Volume | 1 $in^3$ | : | 16.387 $cm^3$ |
| | 1 $ft^3$ | : | 0.02832 $m^3$ |
| | 1 $yd^3$ | : | 0.76453 $m^3$ |
| | 1 UK gal | : | 4546.1 $cm^3$ |
| | 1 US gal | : | 3785.4 $cm^3$ |
| Mass | 1 oz | : | 28.352 g |
| | *1 lb | : | 0.45359237 kg |
| | 1 cwt | : | 50.8023 kg |
| | 1 ton | : | 1016.06 kg |
| Force | 1 pdl | : | 0.13826 N |
| | 1 lbf | : | 4.4482 N |
| | *1 kgf | : | 9.80665 N |
| | 1 tonf | : | 9.9640 kN |
| | *1 dyn | : | $10^{-5}$ N |
| Temperature difference | *1 deg F (deg R) | : | $\frac{5}{9}$ deg C (deg K) |
| Energy (work, heat) | 1 ft lbf | : | 1.3558 J |
| | 1 ft pdl | : | 0.04214 J |
| | *1 cal (international table) | : | 4.1868 J |
| | *1 erg | : | $10^{-7}$ J |
| | 1 Btu | : | 1.05506 kJ |
| | 1 hp h | : | 2.6845 MJ |
| | *1 kWh | : | 3.6 MJ |
| | 1 therm | : | 105.51 MJ |
| | 1 thermie | : | 4.1855 MJ |
| Calorific value (volumetric) | 1 Btu/$ft^3$ | : | 37.259 kJ/$m^3$ |
| Velocity | 1 ft/s | : | 0.3048 m/s |
| | 1 mile/h | : | 0.44704 m/s |
| Volumetric flow | 1 $ft^3$/s | : | 0.028316 $m^3$/s |
| | 1 $ft^3$/h | : | 7.8658 $cm^3$/s |
| | 1 UK gal/h | : | 1.2628 $cm^3$/s |
| | 1 US gal/h | : | 1.0515 $cm^3$/s |
| Mass flow | 1 lb/h | : | 0.12600 g/s |
| | 1 ton/h | : | 0.28224 kg/s |
| Mass per unit area | 1 lb/$in^2$ | : | 703.07 kg/$m^2$ |

| | | | |
|---|---|---|---|
| | 1 lb/ft$^2$ | : | 4.8824 kg/m$^2$ |
| | 1 ton/sq mile | : | 392.30 kg/km$^2$ |
| Density | 1 lb/in$^3$ | : | 27.680 g/cm$^3$ |
| | 1 lb/ft$^3$ | : | 16.019 kg/m$^3$ |
| | 1 lb/UK gal | : | 99.776 kg/m$^3$ |
| | 1 lb/US gal | : | 119.83 kg/m$^3$ |
| Pressure | 1 lbf/in$^2$ | : | 6.8948 kN/m$^2$ |
| | 1 tonf/in$^2$ | : | 15.444 MN/m$^2$ |
| | 1 lbf/ft$^2$ | : | 47.880 N/m$^2$ |
| | *1 standard atmosphere | : | 101.325 kN/m$^2$ |
| | *1 atm (1 kgf/cm$^2$) | : | 98.0665 kN/m$^2$ |
| | *1 bar | : | $10^5$ N/m$^2$ |
| | 1 ft water | : | 2.9891 kN/m$^2$ |
| | 1 in. water | : | 249.09 N/m$^2$ |
| | 1 in. Hg | : | 3.3864 kN/m$^2$ |
| | 1 mm Hg (1 torr) | : | 133.32 N/m$^2$ |
| Power (heat flow) | 1 hp (British) | : | 745.70 W |
| | 1 hp (metric) | : | 735.50 W |
| | *1 erg/s | : | $10^{-7}$ W |
| | 1 ft lbf/s | : | 1.3558 W |
| | 1 Btu/h | : | 0.29307 W |
| | 1 ton of refrigeration | : | 3516.9 W |
| Moment of inertia | 1 lb ft$^2$ | : | 0.042140 kg m$^2$ |
| Momentum | 1 lb ft/s | : | 0.13826 kg m/s |
| Angular momentum | 1 lb ft$^2$/s | : | 0.042140 kg m$^2$/s |
| Viscosity, dynamic | *1 P (poise) | : | 0.1 N s/m$^2$ |
| | 1 lb/ft h | : | 0.41338 mN s/m$^2$ |
| | 1 lb/ft s | : | 1.4882 N s/m$^2$ |
| Viscosity, kinematic | *1 S (stokes) | : | $10^{-4}$ m$^2$/s |
| | 1 ft$^2$/h | : | 0.25806 cm$^2$/s |
| Surface energy (surface tension) | *1 erg/cm$^2$ | : | $10^{-3}$ J/m$^2$ |
| | *(1 dyn/cm) | : | ($10^{-3}$ N/m) |
| Mass flux density | 1 lb/h ft$^2$ | : | 1.3562 g/s m$^2$ |
| Heat flux density | 1 Btu/h ft$^2$ | : | 3.1546 W/m$^2$ |
| | *1 kcal/h m$^2$ | : | 1.163 W/m$^2$ |
| Heat transfer coefficient | 1 Btu/h ft$^2$ °F | : | 5.6783 W/m$^2$ K |
| Specific enthalpy (latent heat, etc.) | 1 Btu/lb | : | 2.326 kJ/kg |
| Specific heat capacity | 1 Btu/lb °F | : | 4.1868 kJ/kg K |
| Thermal conductivity | 1 Btu/h ft °F | : | 1.7307 W/mK |
| | *1 kcal/h m °C | : | 1.163 W/mK |

Taken from MULLIN, J. W.: *The Chemical Engineer* **211** (Sept. 1967), 176. SI units in chemical engineering.

# Problems

(Several of these questions have been taken from examination papers)

**1.1.** The size analysis of a powdered material in terms of is represented by a straight line from 0 per cent at 1 μm particle size to 100 per cent by mass at 101 μm particle size. Calculate the surface mean diameter of the particles constituting the system.

**1.2.** The equations giving the number distribution curve for a powdered material are $dn/dd = d$ for the size range 0–10 μm, and $dn/dd = 100,000/d^4$ for the size range 10–100 μm, where $d$ is in μm. Sketch the number, surface and mass distribution curves and calculate the surface mean diameter for the powder.

Explain briefly how the data for the construction of these curves may be obtained experimentally.

**1.3.** The fineness characteristic of a powder on a cumulative basis is represented by a straight line from the origin to 100 per cent undersize at particle size 50 μm. If the powder is initially dispersed uniformly in a column of liquid, calculate the proportion by mass which remains in suspension in the time from commencement of settling to that at which a 40 μm particle falls the total height of the column.

**1.4.** In a mixture of quartz of density 2650 kg/m$^3$ and galena of density 7500 kg/m$^3$, the sizes of the particles range from 0.0052 to 0.025 mm.

On separation in a hydraulic classifier under free settling conditions, three fractions are obtained, one consisting of quartz only, one a mixture of quartz and galena, and one of galena only. What are the ranges of sizes of particles of the two substances in the original mixture?

**1.5.** A mixture of quartz and galena of a size range from 0.015 mm to 0.065 mm is to be separated into two pure fractions using a hindered settling process. What is the minimum apparent density of the fluid that will give this separation? How will the viscosity of the bed affect the minimum required density? The density of galena is 7500 kg/m$^3$ and the density of quartz is 2650 kg/m$^3$.

**1.6.** The size distribution of a dust as measured by a microscope is as follows. Convert these data to obtain the distribution on a mass basis, and calculate the specific surface, assuming spherical particles of density 2650 kg/m$^3$.

| Size range (μm) | Number of particles in range (–) |
|---|---|
| 0–2 | 2000 |
| 2–4 | 600 |
| 4–8 | 140 |
| 8–12 | 40 |
| 12–16 | 15 |
| 16–20 | 5 |
| 20–24 | 2 |

**1.7.** The performance of a solids mixer was assessed by calculating the variance occurring in the mass fraction of a component amongst a selection of samples withdrawn from the mixture. The quality was tested at intervals of 30 s and the data obtained are:

| | | | | | |
|---|---|---|---|---|---|
| sample variance (–) | 0.025 | 0.006 | 0.015 | 0.018 | 0.019 |
| mixing time (s) | 30 | 60 | 90 | 120 | 150 |

If the component analysed represents 20 per cent of the mixture by mass and each of the samples removed contains approximately 100 particles, comment on the quality of the mixture produced and present the data in graphical form showing the variation of mixing index with time.

**1.8.** The size distribution by mass of the dust carried in a gas, together with the efficiency of collection over each size range is as follows:

| Size range (μm) | 0–5 | 5–10 | 10–20 | 20–40 | 40–80 | 80–160 |
|---|---|---|---|---|---|---|
| Mass (per cent) | 10 | 15 | 35 | 20 | 10 | 10 |
| Efficiency (per cent) | 20 | 40 | 80 | 90 | 95 | 100 |

Calculate the overall efficiency of the collector and the percentage by mass of the emitted dust that is smaller than 20 μm in diameter. If the dust burden is 18 $g/m^3$ at entry and the gas flow is 0.3 $m^3/s$, calculate the mass flow of dust emitted.

**1.9.** The collection efficiency of a cyclone is 45 per cent over the size range 0–5 μm, 80 per cent over the size range 5–10 μm, and 96 per cent for particles exceeding 10 μm. Calculate the efficiency of collection for a dust with a mass distribution of 50 per cent 0–5 μm, 30 per cent 5–10 μm and 20 per cent above 10 μm.

**1.10.** A sample of dust from the air in a factory is collected on a glass slide. If dust on the slide was deposited from one cubic centimetre of air, estimate the mass of dust in $g/m^3$ of air in the factory, given the number of particles in the various size ranges to be as follows:

| Size range (μm) | 0–1 | 1–2 | 2–4 | 4–6 | 6–10 | 10–14 |
|---|---|---|---|---|---|---|
| Number of particles (–) | 2000 | 1000 | 500 | 200 | 100 | 40 |

It may be assumed that the density of the dust is 2600 $kg/m^3$, and an appropriate allowance should be made for particle shape.

**1.11.** A cyclone separator 0.3 m in diameter and 1.2 m long, has a circular inlet 75 mm in diameter and an outlet of the same size. If the gas enters at a velocity of 1.5 m/s, at what particle size will the theoretical cut occur?

The viscosity of air is 0.018 mN $s/m^2$, the density of air is 1.3 $kg/m^3$ and the density of the particles is 2700 $kg/m^3$.

**2.1.** A material is crushed in a Blake jaw crusher such that the average size of particle is reduced from 50 mm to 10 mm, with the consumption of energy of 13.0 kW/(kg/s). What will be the consumption of energy needed to crush the same material of average size 75 mm to average size of 25 mm:

(a) assuming Rittinger's Law applies,
(b) assuming Kick's Law applies?

Which of these results would be regarded as being more reliable and why?

**2.2.** A crusher was used to crush a material with a compressive strength of 22.5 $MN/m^2$. The size of the feed was *minus* 50 mm, *plus* 40 mm and the power required was 13.0 kW/(kg/s). The screen analysis of the product was:

| Size of aperture (mm) | Amount of product (per cent) |
|---|---|
| through 6.0 | all |
| on 4.0 | 26 |
| on 2.0 | 18 |
| on 0.75 | 23 |
| on 0.50 | 8 |
| on 0.25 | 17 |
| on 0.125 | 3 |
| through 0.125 | 5 |

What power would be required to crush 1 kg/s of a material of compressive strength 45 $MN/m^2$ from a feed of *minus* 45 mm, *plus* 40 mm to a product of 0.50 mm average size?

**2.3.** A crusher reducing limestone of crushing strength 70 $MN/m^2$ from 6 mm diameter average size to 0.1 mm diameter average size, requires 9 kW. The same machine is used to crush dolomite at the same output from 6 mm diameter average size to a product consisting of 20 per cent with an average diameter of 0.25 mm, 60 per cent with an average diameter of 0.125 mm and a balance having an average diameter of 0.085 mm. Estimate the power required, assuming that the crushing strength of the dolomite is 100 $MN/m^2$ and that crushing follows Rittinger's Law.

**2.4.** If crushing rolls 1 m diameter are set so that the crushing surfaces are 12.5 mm apart and the angle of nip is 31°, what is the maximum size of particle which should be fed to the rolls?

If the actual capacity of the machine is 12 per cent of the theoretical, calculate the throughput in kg/s when running at 2.0 Hz if the working face of the rolls is 0.4 m long and the feed density is 2500 $kg/m^3$.

**2.5.** A crushing mill which reduces limestone from a mean particle size of 45 mm to the following product:

| Size (mm) | Amount of product (per cent) |
|---|---|
| 12.5 | 0.5 |
| 7.5 | 7.5 |
| 5.0 | 45.0 |
| 2.5 | 19.0 |
| 1.5 | 16.0 |
| 0.75 | 8.0 |
| 0.40 | 3.0 |
| 0.20 | 1.0 |

requires 21 kJ/kg of material crushed.

Calculate the power required to crush the same material at the same rate, from a feed having a mean size of 25 mm to a product with a mean size of 1 mm.

**2.6.** A ball-mill 1.2 m in diameter is run at 0.8 Hz and it is found that the mill is not working satisfactorily. Should any modification in the condition of operation be suggested?

**2.7.** 3 kW is supplied to a machine crushing material at the rate of 0.3 kg/s from 12.5 mm cubes to a product having the following sizes: 80 per cent 3.175 mm 10 per cent 2.5 mm and 10 per cent 2.25 mm.

What power should be supplied to this machine to crush 0.3 kg/s of the same material from 7.5 mm cube to 2.0 mm cube?

**3.1.** A finely ground mixture of galena and limestone in the proportion of 1 to 4 by mass, is subjected to elutriation by an upwardly flowing stream of water flowing at a velocity of 5 mm/s. Assuming that the size distribution for each material is the same, and is as shown in the following table, estimate the percentage of galena in the material carried away and in the material left behind. The viscosity of water is 1 mN $s/m^2$ and Stokes' equation may be used.

| Diameter (μm) | 20 | 30 | 40 | 50 | 60 | 70 | 80 | 100 |
|---|---|---|---|---|---|---|---|---|
| Undersize (per cent mass) | 15 | 28 | 48 | 54 | 64 | 72 | 78 | 88 |

The densities of galena and limestone are 7500 $kg/m^3$ and 2700 $kg/m^3$, respectively.

**3.2.** Calculate the terminal velocity of a steel ball, 2 mm diameter and of density 7870 $kg/m^3$ in an oil of density 900 $kg/m^3$ and viscosity 50 mN $s/m^2$.

**3.3.** What is the terminal settling velocity of a spherical steel particle of 0.40 mm diameter, in an oil of density 820 $kg/m^3$ and viscosity 10 mN $s/m^2$? The density of steel is 7870 $kg/m^3$.

**3.4.** What will be the terminal velocities of mica plates, 1 mm thick and ranging in area from 6 to 600 $mm^2$, settling in an oil of density 820 $kg/m^3$ and viscosity 10 mN $s/m^2$? The density of mica is 3000 $kg/m^3$.

**3.5.** A material of density 2500 $kg/m^3$ is fed to a size separation plant where the separating fluid is water which rises with a velocity of 1.2 m/s. The upward vertical component of the velocity of the particles is 6 m/s. How far will an approximately spherical particle, 6 mm diameter, rise relative to the walls of the plant before it comes to rest in the fluid?

**3.6.** A spherical glass particle is allowed to settle freely in water. If the particle starts initially from rest and if the value of the Reynolds number with respect to the particle is 0.1 when it has attained its terminal falling velocity, calculate:

(a) the distance travelled before the particle reaches 90 per cent of its terminal falling velocity,
(b) the time elapsed when the acceleration of the particle is one hundredth of its initial value.

**3.7.** In a hydraulic jig, a mixture of two solids is separated into its components by subjecting an aqueous slurry of the material to a pulsating motion, and allowing the particles to settle for a series of short time intervals such that their terminal falling velocities are not attained. Materials of densities 1800 and 2500 $kg/m^3$ whose particle size ranges from 0.3 mm to 3 mm diameter are to be separated. It may be assumed that the particles are approximately spherical and that Stokes' Law is applicable. Calculate the approximate maximum time interval for which the particles may be allowed to settle so that no particle of the less dense material falls a greater distance than any particle of the denser material. The viscosity of water is 1 mN $s/m^2$.

**3.8.** Two spheres of equal terminal falling velocities settle in water starting from rest starting at the same horizontal level. How far apart vertically will the particles be when they have both reached their terminal falling velocities? It may be assumed that Stokes' law is valid and this assumption should be checked.

The diameter of one sphere is 40 $\mu$m and its density is 1500 $kg/m^3$ and the density of the second sphere is 3000 $kg/m^3$. The density and viscosity of water are 1000 $kg/m^3$ and 1 mN $s/m^2$ respectively.

**3.9.** The size distribution of a powder is measured by sedimentation in a vessel having the sampling point 180 mm below the liquid surface. If the viscosity of the liquid is 1.2 mN $s/m^2$, and the densities of the powder and liquid are 2650 and 1000 $kg/m^3$ respectively, determine the time which must elapse before any sample will exclude particles larger than 20 $\mu$m.

If Stokes' law does not apply when the Reynolds number is greater than 0.2, what is the approximate maximum size of particle to which Stokes' Law may be applied under these conditions?

**3.10.** Calculate the distance a spherical particle of lead shot of diameter 0.1 mm settles in a glycerol/water mixture before it reaches 99 per cent of its terminal falling velocity.

The density of lead is 11 400 $kg/m^3$ and the density of liquid is 1000 $kg/m^3$. The viscosity of liquid is 10 mN $s/m^2$.

It may be assumed that the resistance force may be calculated from Stokes' Law and is equal to $3\pi\mu du$, where $u$ is the velocity of the particle relative to the liquid.

**3.11.** What is the mass of a sphere of material of density 7500 $kg/m^3$ which falls with a steady velocity of 0.6 m/s in a large deep tank of water?

**3.12.** Two ores, of densities 3700 and 9800 $kg/m^3$ are to be separated in water by a hydraulic classification method. If the particles are all of approximately the same shape and each is sufficiently large for the drag force to be proportional to the square of its velocity in the fluid, calculate the maximum ratio of sizes which can be separated if the particles attain their terminal falling velocities. Explain why a wider range of sizes can be separated if the time of settling is so small that the particles do not reach their terminal velocities.

An explicit expression should be obtained for the distance through which a particle will settle in a given time if it starts from rest and if the resistance force is proportional to the square of the velocity. The acceleration period should be taken into account.

**3.13.** Salt, of density 2350 kg/m$^3$, is charged to the top of a reactor containing a 3 m depth of aqueous liquid of density 1100 kg/m$^3$ and viscosity 2 mN s/m$^2$, and the crystals must dissolve completely before reaching the bottom. If the rate of dissolution of the crystals is given by:

$$-\frac{\mathrm{d}d}{\mathrm{d}t} = 3 \times 10^{-6} + 2 \times 10^{-4}u$$

where $d$ is the size of the crystal (m) at time $t$ (s) and $u$ is its velocity in the fluid (m/s), calculate the maximum size of crystal which should be charged. The inertia of the particles may be neglected and the resistance force may be taken as that given by Stokes' Law ($3\pi\mu du$) where $d$ is taken as the equivalent spherical diameter of the particle.

**3.14.** A balloon of mass 7 g is charged with hydrogen to a pressure of 104 kN/m$^2$. The balloon is released from ground level and, as it rises, hydrogen escapes in order to maintain a constant differential pressure of 2.7 kN/m$^2$, under which condition the diameter of the balloon is 0.3 m. If conditions are assumed to remain isothermal at 273 K as the balloon rises, what is the ultimate height reached and how long does it take to rise through the first 3000 m?

It may be assumed that the value of the Reynolds number with respect to the balloon exceeds 500 throughout and that the resistance coefficient is constant at 0.22. The inertia of the balloon may be neglected and at any moment, it may be assumed that it is rising at its equilibrium velocity.

**3.15.** A mixture of quartz and galena of densities 3700 and 9800 kg/m$^3$ respectively with a size range is 0.3 to 1 mm is to be separated by a sedimentation process. If Stokes' Law is applicable, what is the minimum density required for the liquid if the particles all settle at their terminal velocities?

A separating system using water as the liquid is considered in which the particles were to be allowed to settle for a series of short time intervals so that the smallest particle of galena settled a larger distance than the largest particle of quartz. What is the approximate maximum permissible settling period?

According to Stokes' Law, the resistance force $F$ acting on a particle of diameter $d$, settling at a velocity $u$ in a fluid of viscosity $\mu$ is given by:

$$F = 3\pi\mu du$$

The viscosity of water is 1 mN s/m$^2$.

**3.16.** A glass sphere, of diameter 6 mm and density 2600 kg/m$^3$, falls through a layer of oil of density 900 kg/m$^3$ into water. If the oil layer is sufficiently deep for the particle to have reached its free falling velocity in the oil, how far will it have penetrated into the water before its velocity is only 1 per cent above its free falling velocity in water? It may be assumed that the force on the particle is given by Newton's law and that the particle drag coefficient, $R'/\rho u^2 = 0.22$.

**3.17.** Two spherical particles, one of density 3000 kg/m$^3$ and diameter 20 μm, and the other of density 2000 kg/m$^3$ and diameter 30 μm, start settling from rest at the same horizontal level in a liquid of density 900 kg/m$^3$ and of viscosity 3 mN s/m$^2$. After what period of settling will the particles be again at the same horizontal level? It may be assumed that Stokes' Law is applicable, and the effect of mass acceleration of the liquid moved with each sphere may be ignored.

**3.18.** What will be the terminal velocity of a glass sphere 1 mm in diameter in water if the density of glass is 2500 kg/m$^3$?

**3.19.** What is the mass of a sphere of density 7500 kg/m$^3$ which has a terminal velocity of 0.7 m/s in a large tank of water?

**4.1.** In a contact sulphuric acid plant the secondary converter is a tray type converter, 2.3 m in diameter with the catalyst arranged in three layers, each 0.45 m thick. The catalyst is in the form of cylindrical pellets 9.5 mm in diameter and 9.5 mm long. The void fraction is 0.35. The gas enters the converter at 675 K and leaves at

720 K. Its inlet composition is:

$$SO_3\ 6.6,\quad SO_2\ 1.7,\quad O_2\ 10.0,\quad N_2\ 81.7 \text{ mole per cent}$$

and its exit composition is:

$$SO_3\ 8.2,\quad SO_2\ 0.2,\quad O_2\ 9.3,\quad N_2\ 82.3 \text{ mole per cent}$$

The gas flowrate is 0.68 kg/m$^2$s. Calculate the pressure drop through the converter. The viscosity of the gas is 0.032 mN s/m$^2$.

**4.2.** Two heat-sensitive organic liquids of average molecular weight of 155 kg/kmol are to be separated by vacuum distillation in a 100 mm diameter column packed with 6 mm stoneware Raschig rings. The number of theoretical plates required is 16 and it has been found that the HETP is 150 mm. If the product rate is 5 g/s at a reflux ratio of 8, calculate the pressure in the condenser so that the temperature in the still does not exceed 395 K, equivalent to a pressure of 8 kN/m$^2$. It may be assumed that $a = 800$ m$^2$/m$^3$, $\mu = 0.02$ mN s/m$^2$, $e = 0.72$ and that the temperature changes and the correction for liquid flow may be neglected.

**4.3.** A column 0.6 m diameter and 4 m high is, packed with 25 mm ceramic Raschig rings and used in a gas absorption process carried out at 101.3 kN/m$^2$ and 293 K. If the liquid and gas approximate to those of water and air respectively and their flowrates are 2.5 and 0.6 kg/m$^2$s, what is the pressure drop across the column? By how much may the liquid flow rate be increased before the column floods?

**4.4.** A packed column, 1.2 m in diameter and 9 m tall, is packed with 25 mm Raschig rings, and used for the vacuum distillation of a mixture of isomers of molecular weight 155 kg/kmol. The mean temperature is 373 K, the pressure at the top of the column is maintained at 0.13 kN/m$^2$ and the still pressure is 1.3–3.3 kN/m$^2$. Obtain an expression for the pressure drop on the assumption that this is not appreciably affected by the liquid flow and may be calculated using a modified form of Carman's equation. Show that, over the range of operating pressures used, the pressure drop is approximately directly proportional to the mass rate of flow of vapour, and calculate the pressure drop at a vapour rate of 0.125 kg/m$^2$. The specific surface of packing, $S = 190$ m$^2$/m$^3$, the mean voidage of bed, $e = 0.71$, the viscosity of vapour, $\mu = 0.018$ mN s/m$^2$ and the molecular volume = 22.4 m$^3$/kmol.

**5.1.** A slurry containing 5 kg of water/kg of solids is to be thickened to a sludge containing 1.5 kg of water/kg of solids in a continuous operation. Laboratory tests using five different concentrations of the slurry yielded the following results:

| concentration (kg water/kg solid) | 5.0 | 4.2 | 3.7 | 3.1 | 2.5 |
|---|---|---|---|---|---|
| rate of sedimentation (mm/s) | 0.17 | 0.10 | 0.08 | 0.06 | 0.042 |

Calculate the minimum area of a thickener to effect the separation of 0.6 kg/s of solids.

**5.2.** A slurry containing 5 kg of water/kg of solids is to be thickened to a sludge containing 1.5 kg of water/kg of solids in a continuous operation.

Laboratory tests using five different concentrations of the slurry yielded the following data:

| concentration (kg water/kg solid) | 5.0 | 4.2 | 3.7 | 3.1 | 2.5 |
|---|---|---|---|---|---|
| rate of sedimentation (mm/s) | 0.20 | 0.12 | 0.094 | 0.070 | 0.050 |

Calculate the minimum area of a thickener to effect the separation of 1.33 kg/s of solids.

**5.3.** When a suspension of uniform coarse particles settles under the action of gravity, the relation between the sedimentation velocity $u_c$ and the fractional volumetric concentration $C$ is given by:

$$\frac{u_c}{u_0} = (1 - C)^n,$$

where $n = 2.3$ and $u_0$ is the free falling velocity of the particles. Draw the curve of solids flux $\psi$ against concentration and determine the value of $C$ at which $\psi$ is a maximum and where the curve has a point of inflexion. What is implied about the settling characteristics of such a suspension from the Kynch theory? Comment on the validity of the Kynch theory for such a suspension.

**5.4.** For the sedimentation of a suspension of uniform fine particles in a liquid, the relation between observed sedimentation velocity $u_c$ and fractional volumetric concentration $C$ is given by:

$$\frac{u_c}{u_0} = (1 - C)^{4.8}$$

where $u_0$ is the free falling velocity of an individual particle.

Calculate the concentration at which the rate of deposition of particles per unit area is a maximum and determine this maximum flux for 0.1 mm spheres of glass of density 2600 kg/m$^3$ settling in water of density 1000 kg/m$^3$ and viscosity 1 mN s/m$^2$.

It may be assumed that the resistance force $F$ on an isolated sphere is given by Stokes' Law.

**5.5** A binary suspension consists of equal masses of spherical particles whose free falling velocities in the liquid are 1 mm/s and 2 mm/s respectively. The system is initially well mixed and the total volumetric concentration of solids is 20 percent. As sedimentation proceeds, a sharp interface forms between the clear liquid and suspension consisting only of small particles, and a second interface separates the suspension of fines from the mixed suspension. Choose a suitable model for the behaviour of the system and estimate the falling rates of the two interfaces. It may be assumed that the sedimentation velocity, $u_c$, in a concentrated suspension of voidage $e$ is related to the free falling velocity $u_0$ of the particles by:

$$(u_c/u_0) = e^{2.3}.$$

**6.1.** Oil, of density 900 kg/m$^3$ and viscosity 3 mN s/m$^2$, is passed vertically upwards through a bed of catalyst consisting of approximately spherical particles of diameter 0.1 mm and density 2600 kg/m$^3$. At approximately what mass rate of flow per unit area of bed will (a) fluidisation, and (b) transport of particles occur?

**6.2.** Calculate the minimum velocity at which spherical particles of density 1600 kg/m$^3$ and of diameter 1.5 mm will be fluidised by water in a tube of diameter 10 mm. Discuss the uncertainties in this calculation. The viscosity of water is 1 mN s/m$^2$ and Kozeny's constant is 5.

**6.3.** In a fluidised bed, *iso*-octane vapour is adsorbed from an air stream onto the surface of alumina microspheres. The mole fraction of iso-octane in the inlet gas is $1.442 \times 10^{-2}$ and the mole fraction in the outlet gas is found to vary with time as follows:

| Time from start (s) | Mole fraction in outlet gas ($\times 10^2$) |
|---|---|
| 250 | 0.223 |
| 500 | 0.601 |
| 750 | 0.857 |
| 1000 | 1.062 |
| 1250 | 1.207 |
| 1500 | 1.287 |
| 1750 | 1.338 |
| 2000 | 1.373 |

Show that the results may be interpreted on the assumptions that the solids are completely mixed, that the gas leaves in equilibrium with the solids and that the adsorption isotherm is linear over the range considered. If the flowrate of gas is $0.679 \times 10^{-6}$ kmol/s and the mass of solids in the bed is 4.66 g, calculate the slope of the adsorption isotherm. What evidence do the results provide concerning the flow pattern of the gas?

**6.4.** Cold particles of glass ballotini are fluidised with heated air in a bed in which a constant flow of particles is maintained in a horizontal direction. When steady conditions have been reached, the temperatures recorded by a bare thermocouple immersed in the bed are:

| Distance above bed support (mm) | Temperature (K) |
|---|---|
| 0 | 339.5 |
| 0.64 | 337.7 |
| 1.27 | 335.0 |
| 1.91 | 333.6 |
| 2.54 | 333.3 |
| 3.81 | 333.2 |

Calculate the coefficient for heat transfer between the gas and the particles, and the corresponding values of the particle Reynolds and Nusselt numbers. Comment on the results and on any assumptions made. The gas flowrate is 0.2 kg/m$^2$ s, the specific heat capacity of air is 0.88 kJ/kg K, the viscosity of air is 0.015 mN s/m$^2$, the particle diameter is 0.25 mm and the thermal conductivity of air 0.03 W/mK.

**6.5.** The relation between bed voidage $e$ and fluid velocity $u_c$ for particulate fluidisation of uniform particles which are small compared with the diameter of the containing vessel is given by:

$$\frac{u_c}{u_0} = e^n$$

where $u_0$ is the free falling velocity.

Discuss the variation of the index $n$ with flow conditions, indicating why this is independent of the Reynolds number $Re$ with respect to the particle at very low and very high values of $Re$. When are appreciable deviations from this relation observed with liquid fluidised systems?

For particles of glass ballotini with free falling velocities of 10 and 20 mm/s the index $n$ has a value of 2.39. If a mixture of equal volumes of the two particles is fluidised, what is the relation between the voidage and fluid velocity if it is assumed that complete segregation is obtained?

**6.6.** Obtain a relationship for the ratio of the terminal falling velocity of a particle to the minimum fluidising velocity for a bed of similar particles. It may be assumed that Stokes' Law and the Carman–Kozeny equation are applicable. What is the value of the ratio if the bed voidage at the minimum fluidising velocity is 0.4?

**6.7.** A bed consists of uniform spherical particles of diameter, 3 mm and density, 4200 kg/m$^3$. What will be the minimum fluidising velocity in a liquid of viscosity, 3 mN s/m$^2$ and density 1100 kg/m$^3$?

**6.8.** Ballotini particles, 0.25 mm in diameter, are fluidised by hot air flowing at the rate of 0.2 kg/m$^2$ cross-section of bed to give a bed of voidage 0.5 and a cross-flow of particles is maintained to remove the heat. Under steady state conditions, a small bare thermocouple immersed in the bed gives the following data:

| Distance above bed support (mm) | Temperature (°C) | Temperature (K) |
|---|---|---|
| 0 | 66.3 | 339.5 |
| 0.625 | 64.5 | 337.7 |
| 1.25 | 61.8 | 335.0 |
| 1.875 | 60.4 | 333.6 |
| 2.5 | 60.1 | 333.3 |
| 3.75 | 60.0 | 333.2 |

Assuming plug flow of the gas and complete mixing of the solids, calculate the coefficient for heat transfer between the particles and the gas. The specific heat capacity of air is 0.85 kJ/kg K.

A fluidised bed of total volume 0.1 m$^3$ containing the same particles is maintained at an approximately uniform temperature of 425 K by external heating, and a dilute aqueous solution at 375 K is fed to the bed at the rate of 0.1 kg/s so that the water is completely evaporated at atmospheric pressure. If the heat transfer coefficient is the same as that previously determined, what volumetric fraction of the bed is effectively carrying out the evaporation? The latent heat of vaporisation of water is 2.6 MJ/kg.

**6.9.** An electrically heated element of surface area 12 $cm^2$ is immersed so that it is in direct contact with a fluidised bed. The resistance of the element is measured as a function of the voltage applied to it giving the following data:

| Potential (V) | 1 | 2 | 3 | 4 | 5 | 6 |
|---|---|---|---|---|---|---|
| Resistance (ohms) | 15.47 | 15.63 | 15.91 | 16.32 | 16.83 | 17.48 |

The relation between resistance $R_w$ and temperature $T_w$ is:

$$\frac{R_w}{R_0} = 0.004T_w - 0.092$$

where $R_0$, the resistance of the wire at 273 K, is 14 ohms and $T_w$ is in K. Estimate the bed temperature and the value of the heat transfer coefficient between the surface and the bed.

**6.10.** (a) Explain why the sedimentation velocity of uniform coarse particles in a suspension decreases as the concentration is increased. Identify and, where possible, quantify the various factors involved.

(b) Discuss the similarities and differences in the hydrodynamics of a sedimenting suspension of uniform particles and of an evenly fluidised bed of the same particles in the liquid.

(c) A liquid fluidised bed consists of equal volumes of spherical particles 0.5 mm and 1.0 mm in diameter. The bed is fluidised and complete segregation of the two species occurs. When the liquid flow is stopped the particles settle to form a segregated two-layer bed. The liquid flow is then started again. When the velocity is such that the larger particles are at their incipient fluidisation point what, approximately, will be the voidage of the fluidised bed composed of the smaller particles?

It may be assumed that the drag force $F$ of the fluid on the particles under the free falling conditions is given by Stokes' law and that the relation between the fluidisation velocity $u_c$ and voidage, $e$, for particles of terminal velocity, $u_0$, is given by:

$$u_c/u_0 = e^{4.8}$$

For Stokes' law, the force $F$ on the particles is given by $F = 3\pi\mu d u_0$, where $d$ is the particle diameter and $\mu$ is the viscosity of the liquid.

**6.11.** The relation between the concentration of a suspension and its sedimentation velocity is of the same form as that between velocity and concentration in a fluidised bed. Explain this in terms of the hydrodynamics of the two systems.

A suspension of uniform spherical particles in a liquid is allowed to settle and, when the sedimentation velocity is measured as a function of concentration, the following results are obtained:

| Fractional volumetric concentration ($C$) | Sedimentation velocity ($u_c$ m/s) |
|---|---|
| 0.35 | 1.10 |
| 0.25 | 2.19 |
| 0.15 | 3.99 |
| 0.05 | 6.82 |

Estimate the terminal falling velocity $u_0$ of the particles at infinite dilution. On the assumption that Stokes' law is applicable, calculate the particle diameter $d$.

The particle density, $\rho_s = 2600$ kg/m$^3$, the liquid density, $\rho = 1000$ kg/m$^3$ and the liquid viscosity, $\mu =$ 0.1 Ns/m$^2$.

What will be the minimum fluidising velocity of the system? Stokes' law states that the force on a spherical particle $= 3\pi\mu d u_0$.

**6.12.** A mixture of two sizes of glass spheres of diameters 0.75 and 1.5 mm is fluidised by a liquid and complete segregation of the two species of particles occurs, with the smaller particles constituting the upper portion of the bed and the larger particles in the lower portion. When the voidage of the lower bed is 0.6, what will be the voidage of the upper bed?

The liquid velocity is increased until the smaller particles are completely transported from the bed. What is the minimum voidage of the lower bed at which this phenomenon will occur?

It may be assumed that the terminal falling velocities of both particles may be calculated from Stokes' law and that the relationship between the fluidisation velocity $u$ and the bed voidage $e$ is given by:

$$(u_c/u_0) = e^{4.6}$$

**6.13.** (a) Calculate the terminal falling velocities in water of glass particles of diameter 12 mm and density 2500 kg/m$^3$, and of metal particles of diameter 1.5 mm and density 7500 kg/m$^3$.

It may be assumed that the particles are spherical and that, in both cases, the friction factor, $R'/\rho u^2$ is constant at 0.22, where $R'$ is the force on the particle per unit of projected area of the particle, $\rho$ is the fluid density and $u$ the velocity of the particle relative to the fluid.

(b) Why is the sedimentation velocity lower when the particle concentration in the suspension is high? Compare the behaviour of the concentrated suspension of particles settling under gravity in a liquid with that of a fluidised bed of the same particles.

(c) At what water velocity will fluidised beds of the glass and metal particles have the same densities? The relation between the fluidisation velocity $u_c$ terminal velocity $u_0$ and bed voidage $e$ is given for both particles by:

$$(u_c/u_0) = e^{2.30}$$

**6.14.** Glass spheres are fluidised by water at a velocity equal to one half of their terminal falling velocities. Calculate:

(a) the density of the fluidised bed,
(b) the pressure gradient in the bed attributable to the presence of the particles.

The particles are 2 mm in diameter and have a density of 2500 kg/m$^3$. The density and viscosity of water are 1000 kg/m$^3$ and 1 mN s/m$^2$ respectively.

**7.1.** A slurry, containing 0.2 kg of solid/kg of water, is fed to a rotary drum filter, 0.6 m in diameter and 0.6 m long. The drum rotates at one revolution in 360 s and 20 per cent of the filtering surface is in contact with the slurry at any given instant. If filtrate is produced at the rate of 0.125 kg/s and the cake has a voidage of 0.5, what thickness of cake is formed when filtering at a pressure difference of 65 kN/m$^2$? The density of the solid is 3000 kg/m$^3$.

The rotary filter breaks down and the operation has to be carried out temporarily in a plate and frame press with frames 0.3 m square. The press takes 120 s to dismantle and 120 s to reassemble, and, in addition, 120 s is required to remove the cake from each frame. If filtration is to be carried out at the same overall rate as before, with an operating pressure difference of 275 kN/m$^2$, what is the minimum number of frames that must be used and what is the thickness of each? It may be assumed that the cakes are incompressible and the resistance of the filter media may be neglected.

**7.2.** A slurry containing 100 kg of whiting/m$^3$ of water, is filtered in a plate and frame press, which takes 900 s to dismantle, clean and re-assemble. If the filter cake is incompressible and has a voidage of 0.4, what is the optimum thickness of cake for a filtration pressure of 1000 kN/m$^2$? The density of the whiting is 3000 kg/m$^3$. If the cake is washed at 500 kN/m$^2$ and the total volume of wash water employed is 25 per cent of that of the filtrate, how is the optimum thickness of cake affected? The resistance of the filter medium may be neglected and the viscosity of water is 1 mN s/m$^2$. In an experiment, a pressure of 165 kN/m$^2$ produced a flow of water of 0.02 cm$^3$/s though a centimetre cube of filter cake.

**7.3.** A plate and frame press, gave a total of 8 m$^3$ of filtrate in 1800 s and 11.3 m$^3$ in 3600 s when filtration was stopped. Estimate the washing time if 3 m$^3$ of wash water is used. The resistance of the cloth may be neglected and a constant pressure is used throughout.

**7.4.** In the filtration of a sludge, the initial period is effected at a constant rate with the feed pump at full capacity, until the pressure differences reaches 400 kN/m$^2$. The pressure is then maintained at this value for a remainder of the filtration. The constant rate operation requires 900 s and one-third of the total filtrate is obtained during this period.

Neglecting the resistance of the filter medium, determine (a) the total filtration time and (b) the filtration cycle with the existing pump for a maximum daily capacity, if the time for removing the cake and reassembling the press is 1200 s. The cake is not washed.

**7.5.** A rotary filter, operating at 0.03 Hz, filters at the rate of 0.0075 m$^3$/s. Operating under the same vacuum and neglecting the resistance of the filter cloth, at what speed must the filter be operated to give a filtration rate of 0.0160 m$^3$/s?

**7.6.** A slurry is filtered in a plate and frame press containing 12 frames, each 0.3 m square and 25 mm thick. During the first 180 s, the filtration pressure is slowly raised to the final value of 400 kN/m$^2$ and, during this period, the rate of filtration is maintained constant. After the initial period, filtration is carried out at constant pressure and the cakes are completely formed in a further 900 s. The cakes are then washed with a pressure difference of 275 kN/m$^2$ for 600 s, using *thorough washing*. What is the volume of filtrate collected per cycle and how much wash water is used?

A sample of the slurry was tested, using a vacuum leaf filter of 0.05 m$^2$ filtering surface and a vacuum giving a pressure difference of 71.3 kN/m$^2$. The volume of filtrate collected in the first 300 s was 250 cm$^3$ and, after a further 300 s, an additional 150 cm$^3$ was collected. It may be assumed that cake is incompressible and the cloth resistance is the same in the leaf as in the filter press.

**7.7.** A sludge is filtered in a plate and frame press fitted with 25 mm frames. For the first 600 s the slurry pump runs at maximum capacity. During this period the pressure difference rises to 415 kN/m$^2$ and 25 per cent of the total filtrate is obtained. The filtration takes a further 3600 s to complete at constant pressure and 900 s is required for emptying and resetting the press.

It is found that, if the cloths are precoated with filter aid to a depth of 1.6 mm, the cloth resistance is reduced to 25 per cent of its former value. What will be the increase in the overall throughput of the press if the precoat can be applied in 180 s?

**7.8.** Filtration is carried out in a plate and frame filter press, with 20 frames 0.3 m square and 50 mm thick, and the rate of filtration is maintained constant for the first 300 s. During this period, the pressure is raised to 350 kN/m$^2$, and one-quarter of the total filtrate per cycle is obtained. At the end of the constant rate period, filtration is continued at a constant pressure of 350 kN/m$^2$ for a further 1800 s, after which the frames are full. The total volume of filtrate per cycle is 0.7 m$^3$ and dismantling and refitting of the press takes 500 s.

It is decided to use a rotary drum filter, 1.5 m long and 2.2 m in diameter, in place of the filter press. Assuming that the resistance of the cloth is the same in the two plants and that the filter cake is incompressible, calculate the speed of rotation of the drum which will result in the same overall rate of filtration as was obtained with the filter press. The filtration in the rotary filter is carried out at a constant pressure difference of 70 kN/m$^2$, and the filter operates with 25 per cent of the drum submerged in the slurry at any instant.

**7.9.** It is required to filter a slurry to produce 2.25 m$^3$ of filtrate per working day of 8 hours. The process is carried out in a plate and frame filter press with 0.45 m square frames and a working pressure of 450 kN/m$^2$. The pressure is built up slowly over a period of 300 s and, during this period, the rate of filtration is maintained constant.

When a sample of the slurry is filtered, using a pressure of 35 kN/m$^2$ on a single leaf filter of filtering area 0.05 m$^2$, 400 cm$^3$ of filtrate is collected in the first 300 s of filtration and a further 400 cm$^3$ is collected during the following 600 s. Assuming that the dismantling of the filter press, the removal of the cakes and the setting up again of the press takes an overall time of 300 s, plus an additional 180 s for each cake produced, what is the minimum number of frames that need be employed? The resistance of the filter cloth may be taken as the same in the laboratory tests as on the plant.

**7.10.** The relation between flow and head for a slurry pump may be represented approximately by a straight line, the maximum flow at zero head being 0.0015 $m^3/s$ and the maximum head at zero flow 760 m of liquid. Using this pump to feed a slurry to a pressure leaf filter:

(a) how long will it take to produce 1 $m^3$ of filtrate, and
(b) what will be the pressure across the filter after this time?

A sample of the slurry was filtered at a constant rate of 0.00015 $m^3/s$ through a leaf filter covered with a similar filter cloth but of one-tenth the area of the full scale unit and after 625 s the pressure drop across the filter was 360 m of liquid. After a further 480 s the pressure drop was 600 m of liquid.

**7.11.** A slurry containing 40 per cent by mass solid is to be filtered on a rotary drum filter 2 m diameter and 2 m long which normally operates with 40 per cent of its surface immersed in the slurry and under a pressure of 17 $kN/m^2$. A laboratory test on a sample of the slurry using a leaf filter of area 200 $cm^2$ and covered with a similar cloth to that on the drum, produced 300 $cm^3$ of filtrate in the first 60 s and 140 $cm^3$ in the next 60 s, when the leaf was under an absolute pressure of 17 $kN/m^2$. The bulk density of the dry cake was 1500 $kg/m^3$ and the density of the filtrate was 1000 $kg/m^3$. The minimum thickness of cake which could be readily removed from the cloth was 5 mm.

At what speed should the drum rotate for maximum throughput and what is this throughput in terms of the mass of the slurry fed to the unit per unit time?

**7.12.** A continuous rotary filter is required for an industrial process for the filtration of a suspension to produce 0.002 $m^3/s$ of filtrate. A sample was tested on a small laboratory filter of area 0.023 $m^2$ to which it was fed by means of a slurry pump to give filtrate at a constant rate of 12.5 $cm^3/s$. The pressure difference across the test filter increased from 14 $kN/m^2$ after 300 s filtration to 28 $kN/m^2$ after 900 s, at which time the cake thickness had reached 38 mm. What are suitable dimensions and operating conditions for the rotary filter, assuming that the resistance of the cloth used is one-half that on the test filter, and that the vacuum system is capable of maintaining a constant pressure difference of 70 $kN/m^2$ across the filter?

**7.13.** A rotary drum filter, 1.2 m diameter and 1.2 m long, handles 6.0 kg/s of slurry containing 10 per cent of solids when rotated at 0.005 Hz. By increasing the speed to 0.008 Hz it is found that it can then handle 7.2 kg/s. What will be the percentage change in the amount of wash water which may be applied to each kilogram of cake caused by the increased speed of rotation of the drum, and what is the theoretical maximum quantity of slurry which can be handled?

**7.14.** A rotary drum with a filter area of 3 $m^3$ operates with an internal pressure of 70 $kN/m^2$ below atmospheric and with 30 per cent of its surface submerged in the slurry. Calculate the rate of production of filtrate and the thickness of cake when it rotates at 0.0083 Hz, if the filter cake is incompressible and the filter cloth has a resistance equal to that of 1 mm of cake.

It is desired to increase the rate of filtration by raising the speed of rotation of the drum. If the thinnest cake that can be removed from the drum has a thickness of 5 mm, what is the maximum rate of filtration which can be achieved and what speed of rotation of the drum is required? The voidage of cake = 0.4, the specific resistance of cake = $2 \times 10^{12}$ $m^{-2}$, the density of solids = 2000 $kg/m^3$, the density of filtrate = 1000 $kg/m^3$, the viscosity of filtrate = $10^{-3}$ $Ns/m^2$ and the slurry concentration = 20 per cent by mass of solids.

**7.15.** A slurry containing 50 per cent by mass of solids of density 2600 $kg/m^3$ is to be filtered on a rotary drum filter, 2.25 m in diameter and 2.5 m long, which operates with 35 per cent of its surface immersed in the slurry and under a vacuum of 600 mm Hg. A laboratory test on a sample of the slurry, using a leaf filter with an area of 100 $cm^2$ and covered with a cloth similar to that used on the drum, produced 220 $cm^3$ of filtrate in the first minute and 120 $cm^3$ of filtrate in the next minute when the leaf was under a vacuum of 550 mm Hg. The bulk density of the wet cake was 1600 $kg/m^3$ and the density of the filtrate was 1000 $kg/m^3$.

On the assumption that the cake is incompressible and that 5 mm of cake is left behind on the drum, determine the theoretical maximum flowrate of filtrate obtainable. What drum speed will give a filtration rate of 80 per cent of the maximum?

**7.16.** A rotary filter which operates at a fixed vacuum gives a desired rate of filtration of a slurry when rotating at 0.033 Hz. By suitable treatment of the filter cloth with a filter aid, its effective resistance is halved and the required filtration rate is now achieved at a rotational speed of 0.0167 Hz (1 rpm). If, by further treatment, it is possible to reduce the effective cloth resistance to a quarter of the original value, what rotational speed is required? If the filter is now operated again at its original speed of 0.033 Hz (2 rpm), by what factor will the filtration rate be increased?

**8.1.** Obtain expressions for the optimum concentration for minimum process time in the diafiltration of a solution of protein content $S$ in an initial volume $V_0$,

(a) If the gel-polarisation model applies.
(b) If the osmotic pressure model applies.

It may be assumed that the extent of diafiltration is given by:

$$V_d = \frac{\text{Volume of liquid permeated}}{\text{Initial feed volume}} = \frac{V_p}{V_0}$$

**8.2.** In the ultrafiltration of a protein solution of concentration 0.01 kg/m$^3$, analysis of data on gel growth rate and wall concentration $C_w$ yields the second order relationship:

$$\frac{dl}{dt} = K_r C_w^2$$

where $l$ is gel thickness, and $K_r$ is a constant, $9.2 \times 10^{-6}$ m$^7$/kg$^2$s.

The water flux through the membrane may be described by:

$$J = \frac{|\Delta P|}{\mu_w R_m}$$

where $|\Delta P|$ is pressure difference, $R_m$ is membrane resistance and $\mu_w$ is the viscosity of water.

This equation may be modified for protein solutions to give:

$$J = \frac{|\Delta P|}{\mu_p\left(R_m + \dfrac{l}{P_g}\right)}$$

where $P_g$ is gel permeability, and $\mu_p$ is the viscosity of the permeate.

The gel permeability may be estimated from the Carman–Kozeny equation:

$$P_g = \left(\frac{d^2}{180}\right)\left(\frac{e^3}{(1-e)^2}\right)$$

where $d$ is particle diameter and $e$ is the porosity of the gel.

Calculate the gel thickness after 30 minutes operation.

| Data: | | | |
|---|---|---|---|
| Flux (mm/s) | 0.02 | 0.04 | 0.06 |
| $\lvert\Delta P\rvert$ (kN/m$^2$) | 20 | 40 | 60 |

| | | |
|---|---|---|
| Viscosity of water | = | 1.3 mNs/m$^2$ |
| Viscosity of permeate | = | 1.5 mNs/m$^2$ |
| Diameter of protein molecule | = | 20 nm |
| Operating pressure | = | 10 kN/m$^2$ |
| Porosity of gel | = | 0.5 |
| Mass transfer coefficient to gel $h_D$ | = | $1.26 \times 10^{-5}$ m/s |

**9.1.** If a centrifuge is 0.9 m diameter and rotates at 20 Hz, at what speed should a laboratory centrifuge of 150 mm diameter be run if it is to duplicate the performance of the large unit?

**9.2.** An aqueous suspension consisting of particles of density 2500 kg/m$^3$ in the size range 1–10 μm is introduced into a centrifuge with a basket 450 mm diameter rotating at 80 Hz. If the suspension forms a layer 75 mm thick in the basket, approximately how long will it take for the smallest particle to settle out?

**9.3.** A centrifuge basket 600 mm long and 100 mm internal diameter has a discharge weir 25 mm diameter. What is the maximum volumetric flow of liquid through the centrifuge such that when the basket is rotated at 200 Hz all particles of diameter greater than 1 μm are retained on the centrifuge wall? The retarding force on a particle moving liquid may be taken as $3\pi\mu du$, where $u$ is the particle velocity relative to the liquid $\mu$ is the liquid viscosity, and $d$ is the particle diameter. The density of the liquid is 1000 kg/m$^3$, the density of the solid is 2000 kg/m$^3$ and the viscosity of the liquid is 1.0 mN s/m$^2$. The inertia of the particle may be neglected.

**9.4.** When an aqueous slurry is filtered in a plate and frame press, fitted with two 50 mm thick frames each 150 mm square at a pressure difference of 350 kN/m$^2$, the frames are filled in 3600 s. The liquid in the slurry has the same density as water.

How long will it take to produce the same volume of filtrate as is obtained from a single cycle when using a centrifuge with a perforated basket 300 mm in diameter and 200 mm deep? The radius of the inner surface of the slurry is maintained constant at 75 mm and the speed of rotation is 65 Hz (3900 rpm).

It may be assumed that the filter cake is incompressible, that the resistance of the cloth is equivalent to 3 mm of cake in both cases, and that the liquid in the slurry has the same density as water.

**9.5.** A centrifuge with a phosphor bronze basket, 380 mm in diameter, is to be run at 67 Hz with a 75 mm layer of liquid of density 1200 kg/m$^3$ in the basket. What thickness of walls are required in the basket? The density of phosphor bronze is 8900 kg/m$^3$ and the maximum safe stress for phosphor bronze is 87.6 MN/m$^2$.

**10.1.** 0.4 kg/s of dry sea-shore sand, containing 1 per cent by mass of salt, is to be washed with 0.4 kg/s of fresh water running countercurrently to the sand through two classifiers in series. It may be assumed that perfect mixing of the sand and water occurs in each classifier and that the sand discharged from each classifier contains one part of water for every two of sand by mass. If the washed sand is dried in a kiln dryer, what percentage of salt will it retain? What wash rate would be required in a single classifier in order to wash the sand to the same extent?

**10.2.** Caustic soda is manufactured by the lime-soda process.

A solution of sodium carbonate in water containing 0.25 kg/s $Na_2CO_3$ is treated with the theoretical requirement of lime and, after the reaction is complete, the $CaCO_3$ sludge, containing by mass 1 part of $CaCO_3$ per 9 parts of water is fed continuously to three thickeners in series and is washed countercurrently. Calculate the necessary rate of feed of neutral water to the thickeners, so that the calcium carbonate, on drying, contains only 1 per cent of sodium hydroxide. The solid discharged from each thickener contains one part by mass of calcium carbonate to three of water. The concentrated wash liquid is mixed with the contents of the agitator before being fed to the first thickener.

**10.3.** How many stages are required for 98 per cent extraction of a material containing 18 per cent of extractable matter of density 2700 kg/m$^3$ and which requires 200 volumes of liquid per 100 volumes of solid for it to be capable of being pumped to the next stage? The strong solution is to have a concentration of 100 kg/m$^3$.

**10.4.** Soda ash is mixed with lime and the liquor from the second of three thickeners and passed to the first thickener where separation is effected. The quantity of this caustic solution leaving the first thickener is such as to yield 10 Mg (10 tonnes) of caustic soda per day of 24 hours. The solution contains 95 kg of caustic soda/1000 kg of water, whilst the sludge leaving each of the thickeners consists of one part of solids to one of liquid.

Determine:

(a) the mass of solids in the sludge,
(b) the mass of water admitted to the third thickener, and
(c) the percentages of caustic soda in the sludges leaving the respective thickeners.

**10.5.** Seeds, containing 20 per cent by mass of oil, are extracted in a countercurrent plant and 90 per cent of the oil is recovered in a solution containing 50 per cent by mass of oil. If the seeds are extracted with fresh solvent and 1 kg of solution is removed in the underflow in association with every 2 kg of insoluble matter, how many ideal stages are required?

**10.6.** It is desired to recover precipitated chalk from the causticising of soda ash. After decanting the liquor from the precipitators the sludge has the composition 5 per cent $CaCO_3$, 0.1 per cent NaOH and the balance water.

1000 Mg/day of this sludge is fed to two thickeners where it is washed with 200 Mg/day of neutral water. The pulp removed from the bottom of the thickeners contains 4 kg of water/kg of chalk. The pulp from the last thickener is taken to a rotary filter and concentrated to 50 per cent solids and the filtrate is returned to the system as wash water. Calculate the net percentage of $CaCO_3$ in the product after drying.

**10.7.** Barium carbonate is to be made by reacting sodium carbonate and barium sulphide. The quantities fed to the reaction agitators in 24 hours are 20 Mg of barium sulphide dissolved in 60 Mg of water, together with the theoretically necessary amount of sodium carbonate.

Three thickeners in series, are run on a countercurrent decantation system. Overflow from the second thickener goes to the agitators and overflow from the first thickener is to contain 10 per cent sodium sulphide. Sludge from all thickeners contains two parts water to one part barium carbonate by mass. How much sodium sulphide will remain in the dried barium carbonate precipitate?

**10.8.** In the production of caustic soda by the action of calcium hydroxide on sodium carbonate, 1 kg/s of sodium carbonate is treated with the theoretical quantity of lime. The sodium carbonate is made up as a 20 per cent solution. The material from the extractors is fed to a countercurrent washing system where it is treated with 2 kg/s of clean water. The washing thickeners are so arranged that the ratio of the volume of liquid discharged in the liquid offtake to that discharged with the solid is the same in all the thickeners and is equal to 4.0. How many thickeners must be arranged in series so that not more than 1 per cent of the sodium hydroxide discharged with the solid from the first thickener is wasted?

**10.9.** A plant produces 8640 tonne/day (100 kg/s) of titanium dioxide pigment which must be 99 per cent pure when dried. The pigment is produced by precipitation and the material, as prepared, is contaminated with 1 kg of salt solution containing 0.55 kg of salt/kg of pigment. The material is washed countercurrently with water in a number of thickeners arranged in series. How many thickeners will be required if water is added at the rate of 17,400 tonne/day (200 kg/s) and the solid discharged from each thickeners removes 0.5 kg of solvent/kg of pigment?

What will be the required number of thickeners if the amount of solution removed in association with the pigment varies with the concentration of the solution in the thickener as follows:

| kg solute/kg solution | 0 | 0.1 | 0.2 | 0.3 | 0.4 | 0.5 |
|---|---|---|---|---|---|---|
| kg solution/kg pigment | 0.30 | 0.32 | 0.34 | 0.36 | 0.38 | 0.40 |

The concentrated wash liquor is mixed with the material fed to the first thickener.

**10.10.** Prepared cottonseeds containing 35 per cent of extractable oil are fed to a continuous countercurrent extractor of the intermittent drainage type using hexane as the solvent. The extractor consists of ten sections and the section efficiency is 50 per cent. The entrainment, assumed constant, is 1 kg solution/kg solids. What will be the oil concentration in the outflowing solvent if the extractable oil content is to be reduced by 0.5 per cent by mass?

**10.11.** Seeds containing 25 per cent by mass of oil are extracted in a countercurrent plant and 90 per cent of the oil is to be recovered in a solution containing 50 per cent of oil. It has been found that the amount of solution removed in the underflow in association with every kilogram of insoluble matter is given by:

$$k = 0.7 + 0.5y_s + 3y_s^2$$

where $y_s$ is the concentration of the overflow solution in terms of mass fraction of solute. If the seeds are extracted with fresh solvent, how many ideal stages are required?

**10.12.** Halibut oil is extracted from granulated halibut livers in a countercurrent multibatch arrangement using ether as the solvent. The solids charge contains 0.35 kg oil/kg of exhausted livers and it is desired to obtain a 90 per cent oil recovery. How many theoretical stages are required if 50 kg of ether are used/100 kg of untreated solids. The entrainment data are:

| | | | | | | | | |
|---|---|---|---|---|---|---|---|---|
| Concentration of overflow (kg oil/kg solution) | 0 | 0.1 | 0.2 | 0.3 | 0.4 | 0.5 | 0.6 | 0.67 |
| Entrainment (kg solution/kg extracted livers) | 0.28 | 0.34 | 0.40 | 0.47 | 0.55 | 0.66 | 0.80 | 0.96 |

**11.1.** A liquid containing four components, **A**, **B**, **C** and **D**, with 0.3 mole fraction each of **A**, **B** and **C**, is to be continuously fractionated to give a top product of 0.9 mole fraction **A** and 0.1 mole fraction **B**. The bottoms are to contain not more than 0.5 mole fraction **A**. Estimate the minimum reflux ratio required for this separation, if the relative volatility of **A** to **B** is 2.0.

**11.2.** During the batch distillation of a binary mixture in a packed column the product contained 0.60 mole fraction of the more volatile component when the concentration in the still was 0.40 mole fraction. If the reflux ratio used was 20 : 1, and the vapour composition $y$ is related to the liquor composition $x$ by the equation $y = 1.035x$ over the range of concentration concerned, determine the number of ideal plates represented by the column. $x$ and $y$ are in mole fractions.

**11.3.** A mixture of water and ethyl alcohol containing 0.16 mole fraction alcohol is continuously distilled in a plate fractionating column to give a product containing 0.77 mole fraction alcohol and a waste of 0.02 mole fraction alcohol. It is proposed to withdraw 25 per cent of the alcohol in the entering steam as a side stream containing 0.50 mole fraction of alcohol.

Determine the number of theoretical plates required and the plate from which the side stream should be withdrawn if the feed is liquor at its boiling point and a reflux ratio of 2 is used.

**11.4.** In a mixture to be fed to a continuous distillation column, the mole fraction of phenol is 0.35, *o*-cresol is 0.15, *m*-cresol is 0.30 and xylenols is 0.20. A product is required with a mole fraction of phenol of 0.952, *o*-cresol 0.0474 and *m*-cresol 0.0006. If the volatility to *o*-cresol of phenol is 1.26 and of *m*-cresol is 0.70, estimate how many theoretical plates would be required at total reflux.

**11.5.** A continuous fractionating column, operating at atmospheric pressure, is to be designed to separate a mixture containing 15.67 per cent $CS_2$ and 84.33 per cent $CCl_4$ into an overhead product containing 91 per cent $CS_2$ and a waste of 97.3 per cent $CCl_4$, all by mass. A plate efficiency of 70 per cent and a reflux of 3.16 kmol/kmol of product may be assumed. Determine the number of plates required. The feed enters at 290 K with a specific heat capacity of 1.7 kJ/kg K and has a boiling point of 336 K. The latent heats of $CS_2$ and of $CCl_4$ are 25.9 MJ/kmol. The latent heat of $CS_2$ and $CCl_4$ is 25900 kJ/kmol.

| | | | | | | | | | | | |
|---|---|---|---|---|---|---|---|---|---|---|---|
| Mole per cent $CS_2$ in the vapour: | 0 | 8.23 | 15.55 | 26.6 | 33.2 | 49.5 | 63.4 | 74.7 | 82.9 | 87.8 | 93.2 |
| Mole per cent $CS_2$ in the liquor: | 0 | 2.96 | 6.15 | 11.06 | 14.35 | 25.85 | 39.0 | 53.18 | 66.30 | 75.75 | 86.04 |

**11.6.** A batch fractionation is carried out in a small column which has the separating power of 6 theoretical plates. The mixture consists of benzene and toluene containing 0.60 mole fraction of benzene. A distillate is required, of constant composition, of 0.98 mole fraction benzene, and the operation is discontinued when 83 per cent of the benzene charged has been removed as distillate. Estimate the reflux ratio needed at the start and finish of the distillation, if the relative volatility of benzene to toluene is 2.46.

**11.7.** A continuous fractionating column is required to separate a mixture containing 0.695 mole fraction *n*-heptane ($C_7H_{16}$) and 0.305 mole fraction *n*-octane ($C_8H_{18}$) into products of 99 mole per cent purity. The column is to operate at 101.3 $kN/m^2$ with a vapour velocity of 0.6 m/s. The feed is all liquid at its boiling-point, and this is supplied to the column at 1.25 kg/s. The boiling-point at the top of the column may be taken as 372 K, and the equilibrium data are:

| | | | | | | | | |
|---|---|---|---|---|---|---|---|---|
| mole fraction of heptane in vapour | 0.96 | 0.91 | 0.83 | 0.74 | 0.65 | 0.50 | 0.37 | 0.24 |
| mole fraction of heptane in liquid | 0.92 | 0.82 | 0.69 | 0.57 | 0.46 | 0.32 | 0.22 | 0.13 |

Determine the minimum reflux ratio required. What diameter column would be required if the reflux used were twice the minimum possible?

**11.8.** The vapour pressures of chlorobenzene and water are:

| Vapour pressure ($kN/m^2$) | 13.3 | 6.7 | 4.0 | 2.7 |
|---|---|---|---|---|
| Temperatures, (K) | | | | |
| Chlorobenzene | 343.6 | 326.9 | 315.9 | 307.7 |
| Water | 324.9 | 311.7 | 303.1 | 295.7 |

A still is operated at 18 $kN/m^2$ and steam is blown continuously into it. Estimate the temperature of the boiling liquid and the composition of the distillate if liquid water is present in the still.

**11.9.** The following values represent the equilibrium conditions in terms of mole fraction of benzene in benzene–toluene mixtures at their boiling-point:

| Liquid | 0.51 | 0.38 | 0.26 | 0.15 |
|---|---|---|---|---|
| Vapour | 0.72 | 0.60 | 0.45 | 0.30 |

If the liquid compositions on four adjacent plates in a column are 0.18, 0.28, 0.41 and 0.57 under conditions of total reflux, determine the plate efficiencies.

**11.10.** A continuous rectifying column handles a mixture consisting of 40 per cent of benzene by mass and 60 per cent of toluene at the rate of 4 kg/s, and separates it into a product containing 97 per cent of benzene and a liquid containing 98 per cent toluene. The feed is liquid at its boiling-point.

(a) Calculate the masses of distillate and waste liquor produced per unit time.
(b) If a reflux ratio of 3.5 is employed, how many plates are required in the rectifying part of the column?
(c) What is the actual number of plates if the plate-efficiency is 60 per cent?

| Mole fraction of benzene in liquid | 0.1 | 0.2 | 0.3 | 0.4 | 0.5 | 0.6 | 0.7 | 0.8 | 0.9 |
|---|---|---|---|---|---|---|---|---|---|
| Mole fraction of benzene in vapour | 0.22 | 0.38 | 0.51 | 0.63 | 0.7 | 0.78 | 0.85 | 0.91 | 0.96 |

**11.11.** A distillation column is fed with a mixture of benzene and toluene, in which the mole fraction of benzene is 0.35. The column is to yield a product in which the mole fraction of benzene is 0.95, when working with a reflux ratio of 3.2, and the waste from the column is not to exceed 0.05 mole fraction of benzene.

If the plate efficiency is 60 per cent, estimate the number of plates required and the position of the feed point. The relation between the mole fraction of benzene in liquid and in vapour is given by:

| Mole fraction of benzene in liquid | 0.1 | 0.2 | 0.3 | 0.4 | 0.5 | 0.6 | 0.7 | 0.8 | 0.9 |
|---|---|---|---|---|---|---|---|---|---|
| Mole fraction of benzene in vapour | 0.20 | 0.38 | 0.51 | 0.63 | 0.71 | 0.78 | 0.85 | 0.91 | 0.96 |

**11.12.** The relationship between the mole fraction of carbon disulphide in the liquid and in the vapour evolved from the mixture during the distillation of a carbon disulphide–carbon tetrachloride mixture is:

| $x$ | 0 | 0.20 | 0.40 | 0.60 | 0.80 | 1.00 |
|---|---|---|---|---|---|---|
| $y$ | 0 | 0.445 | 0.65 | 0.795 | 0.91 | 1.00 |

Determine graphically the theoretical number of plates required for the rectifying and stripping portions of the column, if the reflux ratio = 3, the slope of the fractionating line = 1.4, the purity of the product = 99 per cent and the percentage of carbon disulphide in the waste liquors = 1 per cent.

What is the minimum slope of the rectifying line in this case?

**11.13.** A fractionating column is required to distill a liquid containing 25 per cent benzene and 75 per cent toluene by mass, to give a product of 90 per cent benzene. A reflux ratio of 3.5 is to be used, and the feed will enter at its boiling point. If the plates used are 100 per cent efficient, calculate by the Lewis–Sorel method the composition of liquid on the third plate, and estimate the number of plates required using the Macabe-Thiele method.

**11.14.** A 50 mole per cent mixture of benzene and toluene is fractionated in a batch still which has the separating power of 8 theoretical plates. It is proposed to obtain a constant quality product with a mole per cent benzene of 95, and to continue the distillation until the still has a content of 10 mole per cent of benzene. What will be the range of reflux ratios used in the process? Show graphically the relation between the required reflux ratio and the amount of distillate removed.

**11.15.** The vapour composition on a plate of a distillation column is:

| | $C_1$ | $C_2$ | $i-C_3$ | $n-C_3$ | $i-C_4$ | $n-C_4$ |
|---|---|---|---|---|---|---|
| mole fraction | 0.025 | 0.205 | 0.210 | 0.465 | 0.045 | 0.050 |
| relative volatility | 36.5 | 7.4 | 3.0 | 2.7 | 1.3 | 1.0 |

What will be the composition of the liquid on the plate if it is in equilibrium with the vapour?

**11.16.** A liquor of 0.30 mole fraction of benzene and the rest toluene is fed to a continuous still to give a top product of 0.90 mole fraction benzene and a bottom product of 0.95 mole fraction toluene.

If the reflux ratio is 5.0, how many plates are required:

(a) if the feed is saturated vapour?
(b) if the feed is liquid at 283 K?

**11.17.** A mixture of alcohol and water containing 0.45 mole fraction of alcohol is to be continuously distilled in a column to give a top product of 0.825 mole fraction alcohol and a liquor at the bottom containing 0.05 mole fraction alcohol.

How many theoretical plates are required if the reflux ratio used is 3? Indicate on a diagram what is meant by the Murphree plate efficiency.

**11.18.** It is desired to separate 1 kg/s of an ammonia solution containing 30 per cent $NH_3$ by mass into 99.5 per cent liquid $NH_3$ and a residual weak solution containing 10 per cent $NH_3$. Assuming the feed to be at its boiling point, a column pressure of 1013 $kN/m^2$, a plate efficiency of 60 per cent and that 8 per cent excess over minimum reflux requirements is used, how many plates must be used in the column and how much heat is removed in the condenser and added in the boiler?

**11.19.** A mixture of 60 mole per cent benzene, 30 per cent of toluene and 10 per cent xylene is handled in a batch still. If the top product is to be 99 per cent benzene, determine:

(a) the liquid composition on each plate at total reflux,
(b) the composition on the 2nd and 4th plates for a reflux ratio $R = 1.5$,
(c) as for (b) but $R = 3$,
(d) as for (c) but $R = 5$,
(e) as for (d) but $R = 8$ and for the condition when the mole per cent benzene in the still is 10,
(f) as for (e) but with $R = 5$.

The relative volatility of benzene to toluene may be taken as 2.4, and of xylene to toluene as 0.43.

**11.20.** A continuous still is fed with a mixture of 0.5 mole fraction of the more volatile component, and gives a top product of 0.9 mole fraction of the more volatile component and a bottom product containing 0.10 mole fraction.

If the still operates with an $L_n/D$ ratio of 3.5 : 1, calculate by Sorel's method the composition of the liquid on the third theoretical plate from the top:

(a) for benzene–toluene, and
(b) for $n$-heptane–toluene.

**11.21.** A mixture of 40 mole per cent benzene with toluene is distilled in a column to give a product of 95 mole per cent benzene and a waste of 5 mole per cent benzene, using a reflux ratio of 4.

(a) Calculate by Sorel's method the composition on the second plate from the top.
(b) Using the McCabe and Thiele method, determine the number of plates required and the position of the feed if supplied to the column as liquid at the boiling point.
(c) Determine the minimum reflux ratio possible.
(d) Determine the minimum number of plates.
(e) If the feed is passed in at 288 K, determine the number of plates required using the same reflux ratio.

**11.22.** Determine the minimum reflux ratio using Fenske's equation and Colburn's rigorous method for the following three systems:

(a) 0.60 mole fraction $C_6$, 0.30 mole fraction $C_7$, and 0.10 mole fraction $C_8$ to give a product of 0.99 mole fraction $C_6$.

| | | | Mole fraction | Relative volatility $\alpha$ | $x_d$ |
|---|---|---|---|---|---|
| (b) | Components | **A** | 0.3 | 2 | 1.0 |
| | | **B** | 0.3 | 1 | — |
| | | **C** | 0.4 | 0.5 | — |
| (c) | Components | **A** | 0.25 | 2 | 1.0 |
| | | **B** | 0.25 | 1 | — |
| | | **C** | 0.25 | 0.5 | — |
| | | **D** | 0.25 | 0.25 | — |

**11.23.** A liquor consisting of phenol and cresols with some xylenols is fractionated to give a top product of 95.3 mole per cent phenol. The compositions of the top product and of the phenol in the bottoms are as follows:

| | Compositions (mole per cent) | | |
|---|---|---|---|
| | Feed | Top | Bottom |
| phenol | 35 | 95.3 | 5.24 |
| *o*-cresol | 15 | 4.55 | — |
| *m*-cresol | 30 | 0.15 | — |
| xylenols | 20 | — | — |
| | 100 | 100 | — |

If a reflux ratio of 10 is used,

(a) Complete the material balance over the still for a feed of 100 kmol.
(b) Calculate the composition on the second plate from the top.
(c) Calculate the composition on the second plate from the bottom.
(d) Calculate the minimum reflux ratio by Underwood's equation and by Colburn's approximation.

The heavy key is *m*-cresol and the light key is phenol.

**11.24.** A continuous fractionating column is to be designed to separate 2.5 kg/s of a mixture of 60 per cent toluene and 40 per cent benzene, so as to give an overhead of 97 per cent benzene and a bottom product containing 98 per cent toluene by mass. A reflux ratio of 3.5 kmol of reflux/kmol of product is to be used and the molar latent heat of benzene and toluene may be taken as 30 MJ/kmol.

Calculate:

(a) The mass of top and bottom products per unit time.
(b) The number of theoretical plates and position of feed if the feed is liquid at 295 K, of specific heat capacity 1.84 kJ/kg K.
(c) How much steam at 240 kN/m$^2$ is required in the still.
(d) What will be the required diameter of the column if it operates at atmospheric pressure and a vapour velocity of 1 m/s.

(e) The necessary diameter of the column if the vapour velocity is to be 0.75 m/s, based on free area of column.
(f) The minimum possible reflux ratio, and the minimum number of plates for a feed entering at its boiling-point.

**11.25.** For a system that obeys Raoult's law show that the relative volatility $\alpha_{AB}$ is $P_A^0/P_B^0$, where $P_A^0$ and $P_B^0$ are the vapour pressures of the components **A** and **B** at the given temperature. From vapour pressure curves of benzene, toluene, ethyl benzene and of *o*-, *m*- and *p*-xylenes, obtain a plot of the volatilities of each of the materials relative to *m*-xylene in the range 340–430 K.

**11.26.** A still has a liquor composition of *o*-xylene 10 per cent, *m*-xylene 65 per cent, *p*-xylene 17 per cent, benzene 4 per cent and ethyl benzene 4 per cent. How many plates at total reflux are required to give a product of 80 per cent *m*-xylene, and 14 per cent *p*-xylene? The data are given as mass per cent.

**11.27.** The vapour pressures of *n*-pentane and of *n*-hexane are:

| Pressure | | | | | | | | |
|---|---|---|---|---|---|---|---|---|
| ($kN/m^2$) | 1.3 | 2.6 | 5.3 | 8.0 | 13.3 | 26.6 | 53.2 | 101.3 |
| (mm Hg) | 10 | 20 | 40 | 60 | 100 | 200 | 400 | 760 |
| Temperature (K) | | | | | | | | |
| $C_5H_{12}$ | 223.1 | 233.0 | 244.0 | 257.0 | 260.6 | 275.1 | 291.7 | 309.3 |
| $C_6H_{14}$ | 248.2 | 259.1 | 270.9 | 278.6 | 289.0 | 304.8 | 322.8 | 341.9 |

The equilibrium data at atmospheric pressure are:

| $x =$ 0.1 | 0.2 | 0.3 | 0.4 | 0.5 | 0.6 | 0.7 | 0.8 | 0.9 |
|---|---|---|---|---|---|---|---|---|
| $y =$ 0.21 | 0.41 | 0.54 | 0.66 | 0.745 | 0.82 | 0.875 | 0.925 | 0.975 |

(a) Determine the relative volatility of pentane to hexane at 273, 293 and 313 K.
(b) A mixture containing 0.52 mole fraction pentane is to be distilled continuously to give a top product of 0.95 mole fraction pentane and a bottom of 0.1 mole fraction pentane. Determine the minimum number of plates that is the number of plates at total reflux by the graphical McCabe–Thiele method, and analytically by using the relative volatility method.
(c) Using the conditions in (b), determine the liquid composition on the second plate from the top by Lewis's method, if a reflux ratio of 2 is used.
(d) Using the conditions in (b), determine by the McCabe–Thiele method the total number of plates required, and the position of the feed.

It may be assumed that the feed is all liquid at its boiling-point.

**11.28.** The vapour pressures of *n*-pentane and *n*-hexane are given in Problem 11.27. Assuming that both Raoult's and Dalton's Laws are obeyed,

(a) Plot the equilibrium curve for a total pressure of 13.3 $kN/m^2$.
(b) Determine the relative volatility of pentane to hexane as a function of liquid composition for a total pressure of 13.3 $kN/m^2$.
(c) Estimate the error caused by assuming the relative volatility to be constant at its mean value.
(d) Would it be more advantageous to distil this mixture at a higher pressure?

**11.29.** It is desired to separate a binary mixture by simple distillation. If the feed mixture has a composition of 0.5 mole fraction, calculate the fraction it is necessary to vapourise in order to obtain:

(a) a product of composition 0.75 mole fraction, when using a continuous process, and
(b) a product whose composition is not less than 0.75 mole fraction at any instant, when using a batch process.

If the product of batch distillation is all collected in a single receiver, what is its mean composition?

It may be assumed that the equilibrium curve is given by:

$$y = 1.2x + 0.3$$

for liquid compositions in the range 0.3–0.8.

**11.30.** A liquor, consisting of phenol and cresols with some xylenol, is separated in a plate column. Given the following data complete the material balance:

| Component | Mole per cent | | |
|---|---|---|---|
| | Feed | Top | Bottom |
| $C_6H_5OH$ | 35 | 95.3 | 5.24 |
| $o$-$C_7H_7OH$ | 15 | 4.55 | — |
| $m$-$C_7H_7OH$ | 30 | 0.15 | — |
| $C_8H_9OH$ | 20 | — | — |
| Total | 100 | 100 | — |

Calculate:

(a) the composition on the second plate from the top,
(b) the composition on the second plate from the bottom.

A reflux ratio of 4 is used.

**11.31.** A mixture of 60, 30, and 10 mole per cent benzene, toluene, and xylene respectively is separated by a plate-column to give a top product containing at least 90 mole per cent benzene and a negligible amount of xylene, and a waste containing not more than 60 mole per cent toluene.

Using a reflux ratio of 4, and assuming that the feed is boiling liquid, determine the number of plates required in the column, and the approximate position of the feed plate.

The relative volatility of benzene to toluene is 2.4 and of xylene to toluene is 0.45, and it may be assumed that values are constant throughout the column.

**11.32.** It is desired to concentrate a mixture of ethyl alcohol and water from 40 mole per cent to 70 mole per cent alcohol. A continuous fractionating column, 1.2 m in diameter and having 10 plates is available. It is known that the optimum superficial vapour velocity in the column at atmosphere pressure is 1 m/s, giving an overall plate efficiency of 50 per cent.

Assuming that the mixture is fed to the column as a boiling liquid and using a reflux ratio of twice the minimum value possible, determine the feed plate and the rate at which the mixture can be separated.

Equilibria data:

| | | | | | | | | | |
|---|---|---|---|---|---|---|---|---|---|
| Mole fraction alcohol in liquid | 0.1 | 0.2 | 0.3 | 0.4 | 0.5 | 0.6 | 0.7 | 0.8 | 0.89 |
| Mole fraction alcohol in vapour | 0.43 | 0.526 | 0.577 | 0.615 | 0.655 | 0.70 | 0.754 | 0.82 | 0.89 |

**12.1.** Tests are made on the absorption of carbon dioxide from a carbon dioxide–air mixture in caustic soda solution of concentration 2.5 kmol/m$^3$, using a 250 mm diameter tower packed to a height of 3 m with 19 mm Raschig rings.

The results obtained at atmospheric pressure are:

Gas rate, $G' = 0.34$ kg/m$^2$s. Liquid rate, $L' = 3.94$ kg/m$^2$s.

The carbon dioxide in the inlet gas is 315 parts per million and in the exit gas 31 parts per million.

What is the value of the overall gas transfer coefficient $K_G a$?

**12.2.** An acetone–air mixture containing 0.015 mole fraction of acetone has the mole fraction reduced to 1 per cent of this value by countercurrent absorption with water in a packed tower. The gas flowrate $G'$ is 1 kg/m$^2$s of air and the water flowrate entering is 1.6 kg/m$^2$s. For this system, Henry's law holds and $y_e = 1.75x$, where

$y_e$ is the mole fraction of acetone in the vapour in equilibrium with a mole fraction $x$ in the liquid. How many overall transfer units are required?

**12.3.** An oil containing 2.55 mole per cent of a hydrocarbon is stripped by running the oil down a column up which live steam is passed, so that 4 kmole of steam are used 100 kmol of oil stripped. Determine the number of theoretical plates required to reduce the hydrocarbon content to 0.05 mole per cent, assuming that the oil is non-volatile. The vapour–liquid relation of the hydrocarbon in the oil is given by $y_e = 33x$, where $y_e$ is the mole fraction in the vapour and $x$ the mole fraction in the liquid. The temperature is maintained constant by internal heating, so that steam does not condense in the tower.

**12.4.** Gas, from a petroleum distillation column, has its concentration of $H_2S$ reduced from 0.03 kmol $H_2S$/kmol of inert hydrocarbon gas to 1 per cent of this value, by scrubbing with a triethanolamine-water solvent in a countercurrent tower, operating at 300 K and at atmospheric pressure.

$H_2S$ is soluble in such a solution and the equilibrium relation may be taken as $Y = 2X$, where $Y$ is kmol of $H_2S$ kmol inert gas and $X$ is kmol of $H_2S$/kmol of solvent.

The solvent enters the tower free of $H_2S$ and leaves containing 0.013 kmol of $H_2S$/kmol of solvent. If the flow of inert hydrocarbon gas is 0.015 kmol/m$^2$s of tower cross-section and the gas-phase resistance controls the process, calculate:

(a) the height of the absorber necessary, and
(b) the number of transfer units required.

The overall coefficient for absorption $K''_G a$ may be taken as 0.04 kmol/s m$^3$ of tower volume (unit driving force in $Y$).

**12.5.** It is known that the overall liquid transfer coefficient $K_L a$ for absorption of $SO_2$ in water in a column is 0.003 kmol/s m$^3$ (kmol/m$^3$). Obtain an expression for the overall liquid film coefficient $K_L a$ for absorption of $NH_3$ in water in the same apparatus using the same water and gas rates. The diffusivities of $SO_2$ and $NH_3$ in air at 273 K are 0.103 and 0.170 cm$^2$/s. $SO_2$ dissolves in water, so that Henry's constant $\mathscr{H}$ is equal to 50 (kN/m$^2$)/(kmol/m$^3$). All the data refer to 273 K.

**12.6.** A packed tower is used for absorbing sulphur dioxide from air by means of a 0.5 N caustic soda solution. At an air flow of 2 kg/m$^2$s, corresponding to a Reynolds number of 5160, the friction factor $R/\rho u^2$ is found to be 0.020.

Calculate the mass transfer coefficient in kg $SO_2$/s m$^2$ (kN/m$^2$) under these conditions if the tower is at atmospheric pressure. At the temperature of absorption the diffusion coefficient of $SO_2$ is $0.116 \times 10^{-4}$ m$^2$/s, the viscosity of the gas is 0.018 mN s/m$^2$ and the density of the gas stream is 1.154 kg/m$^3$.

**12.7.** In an absorption tower, ammonia is absorbed from air at atmospheric pressure by acetic acid. The flowrate of 2 kg/m$^2$s in a test corresponds to a Reynolds number of 5100 and hence a friction factor $R/\rho u^2$ of 0.020. At the temperature of absorption the viscosity of the gas stream is 0.018 mN s/m$^2$, the density is 1.154 kg/m$^3$ and the diffusion coefficient of ammonia in air is $1.96 \times 10^{-5}$ m$^2$/s.

Determine the mass transfer coefficient through the gas film in kg/m$^2$ s (kN/m$^2$).

**12.8.** Acetone is to be recovered from a 5 per cent acetone–air mixture by scrubbing with water in a packed tower using countercurrent flow. The liquid rate is 0.85 kg/m$^2$s and the gas rate is 0.5 kg/m$^2$s.

The overall absorption coefficient $K_G a$ may be taken as $1.5 \times 10^{-4}$ kmol/[m$^3$s(kN/m$^2$)] partial pressure difference] and the gas-film resistance controls the process.

What height of tower is required tower to remove 98 per cent of the acetone? The equilibrium data for the mixture are:

| | | | | |
|---|---|---|---|---|
| Mole fraction acetone in gas | 0.0099 | 0.0196 | 0.0361 | 0.0400 |
| Mole fraction acetone in liquid | 0.0076 | 0.0156 | 0.0306 | 0.0333 |

**12.9.** Ammonia is to be removed from 10 per cent ammonia–air mixture by countercurrent scrubbing with water in a packed tower at 293 K so that 99 per cent of the ammonia is removed when working at a total pressure of 101.3 kN/m$^2$. If the gas rate is 0.95 kg/m$^2$s of tower cross-section and the liquid rate is 0.65 kg/m$^2$ s, what is the necessary height of the tower if the absorption coefficient $K_G a = 0.001$ kmol/m$^3$s (kN/m$^2$) partial pressure difference. The equilibrium data are:

| Concentration (kmol $NH_3$/kmol water) | 0.021 | 0.031 | 0.042 | 0.053 | 0.079 | 0.106 | 0.150 |
|---|---|---|---|---|---|---|---|
| Partial pressure $NH_3$ (mm Hg) | 12.0 | 18.2 | 24.9 | 31.7 | 50.0 | 69.6 | 114.0 |
| (kN/m$^2$) | 1.6 | 2.4 | 3.3 | 4.2 | 6.7 | 9.3 | 15.2 |

**12.10.** Sulphur dioxide is recovered from a smelter gas containing 3.5 per cent by volume of $SO_2$, by scrubbing it with water in a countercurrent absorption tower. The gas is fed into the bottom of the tower, and in the exit gas from the top the $SO_2$ exerts a partial pressure of 1.14 kN/m$^2$. The water fed to the top of the tower is free from $SO_2$, and the exit liquor from the base contains 0.001145 kmol $SO_2$/kmol water. The process takes place at 293 K, at which the vapour pressure of water is 2.3 kN/m$^2$. The water rate is 0.43 kmol/s.

If the area of the tower is 1.85 m$^2$ and the overall coefficient of absorption for these conditions $K''_L a$ is 0.19 kmol $SO_2$/s m$^3$ (kmol of $SO_2$ per kmol $H_2O$), what is the height of the column required?

The equilibrium data for $SO_2$ and water at 293 K are:

| kmol $SO_2$/1000 kmol $H_2O$ | 0.056 | 0.14 | 0.28 | 0.42 | 0.56 | 0.84 | 1.405 |
|---|---|---|---|---|---|---|---|
| kmol $SO_2$/1000 kmol inert gas | 0.7 | 1.6 | 4.3 | 7.9 | 11.6 | 19.4 | 35.3 |

**12.11.** Ammonia is removed from a 10 per cent ammonia–air mixture by scrubbing with water in a packed tower, so that 99.9 per cent of the ammonia is removed. What is the required height of tower? The gas enters at 1.2 kg/m$^2$s, the water rate is 0.94 kg/m$^2$s and $K_G a$ is 0.0008 kmol/s m$^3$ (kN/m$^2$).

**12.12.** A soluble gas is absorbed from a dilute gas–air mixture by countercurrent scrubbing with a solvent in a packed tower. If the liquid fed to the top of the tower contains no solute, show that the number of transfer units required is given by:

$$\mathbf{N} = \frac{1}{\left[1 - \dfrac{mG_m}{L_m}\right]} \ln\left[\left(1 - \frac{mG_m}{L_m}\right)\frac{y_1}{y_2} + \frac{mG_m}{L_m}\right]$$

where $G_m$ and $L_m$ are the flowrates of the gas and liquid in kmol/s m$^2$ tower area, and $y_1$ and $y_2$ the mole fraction of the gas at the inlet and outlet of the column. The equilibrium relation between the gas and liquid is represented by a straight line with the equation $y_e = mx$, where $y_e$ is the mole fraction in the gas in equilibrium with mole fraction $x$ in the liquid.

In a given process, it is desired to recover 90 per cent of the solute by using 50 per cent more liquid than the minimum necessary. If the HTU of the proposed tower is 0.6 m, what height of packing will be required?

**12.13.** A paraffin hydrocarbon of molecular weight 114 kg/kmol at 373 K, is to be separated from a mixture with a non-volatile organic compound of molecular weight 135 kg/kmol by stripping with steam. The liquor contains 8 per cent of the paraffin by mass and this is to be reduced to 0.08 per cent using an upward flow of steam saturated at 373 K. If three times the minimum amount of steam is used, how many theoretical stages will be required? The vapour pressure of the paraffin at 373 K is 53 kN/m$^2$ and the process takes place at atmospheric pressure. It may be assumed that the system obeys Raoult's law.

**12.14.** Benzene is to be absorbed from coal gas by means of a wash oil. The inlet gas contains 3 per cent by volume of benzene, and the exit gas should not contain more than 0.02 per cent benzene by volume. The suggested oil circulation rate is 480 kg oil/100 m$^3$ of inlet gas measured at NTP. The wash-oil enters the tower solute-free. If the overall height of a transfer unit based on the gas phase is 1.4 m, determine the minimum height of the tower which is required to carry out the absorption. The equilibrium data are:

| | | | | | | |
|---|---|---|---|---|---|---|
| Benzene in oil (per cent by mass) | 0.05 | 0.01 | 0.50 | 1.0 | 2.0 | 3.0 |
| Equilibrium partial pressure of benzene in gas | | | | | | |
| ($kN/m^2$) | 0.013 | 0.033 | 0.20 | 0.53 | 1.33 | 3.33 |
| (mm Hg) | 0.1 | 0.25 | 1.5 | 4.0 | 10.0 | 25.0 |

**12.15.** Ammonia is to be recovered from a 5 per cent by volume ammonia–air mixture by scrubbing with water in a packed tower. The gas rate is 1.25 $m^3/s\ m^2$ measured at NTP and the liquid rate is 1.95 $kg/m^2s$. The temperature of the inlet gas is 298 K and of the inlet water 293 K. The mass transfer coefficient is $K_G a = 0.113$ $kmol/m^3$ s (mole ratio difference) and the total pressure is 101.3 $kN/m^2$. What is the height of the tower to remove 95 per cent of the ammonia. The equilibrium data and the heats of solutions are:

| | | | | | |
|---|---|---|---|---|---|
| Mole fraction in liquid | 0.005 | 0.01 | 0.015 | 0.02 | 0.03 |
| Integral heat of solution (kJ/kmol of solution) | 181 | 363 | 544 | 723 | 1084 |
| Equilibrium partial pressures: ($kN/m^2$) | | | | | |
| at 293 K | 0.4 | 0.77 | 1.16 | 1.55 | 2.33 |
| at 298 K | 0.48 | 0.97 | 1.43 | 1.92 | 2.93 |
| at 303 K | 0.61 | 1.28 | 1.83 | 2.47 | 3.86 |

Adiabatic conditions may be assumed and heat transfer between phases neglected.

**12.16.** A bubble-cap column with 30 plates is to be used to remove $n$-pentane from solvent oil by means of steam stripping. The inlet oil contains 6 kmol of $n$-pentane/100 kmol of pure oil and it is desired to reduce the solute content of 0.1 kmol/100 kmol of solvent. Assuming isothermal operation and an overall plate efficiency of 30 per cent, what is the specific steam consumption, that is kmol of steam required/kmol of solvent oil treated, and the ratio of the specific and minimum steam consumptions. How many plates would be required if this ratio is 2.0?

The equilibrium relation for the system may be taken as $Y_e = 3.0X$, where $Y_e$ and $X$ are expressed in mole ratios of pentane in the gas and liquid phases respectively.

**12.17.** A mixture of ammonia and air is scrubbed in a plate column with fresh water. If the ammonia concentration is reduced from 5 per cent to 0.01 per cent, and the water and air rates are 0.65 and 0.40 $kg/m^2s$, respectively, how many theoretical plates are required? The equilibrium relationship may be written as $Y = X$, where $X$ is the mole ratio in the liquid phase.

**13.1.** Tests are made on the extraction of acetic acid from a dilute aqueous solution by means of a ketone in a small spray tower of diameter 46 mm and effective height of 1090 mm with the aqueous phase run into the top of the tower. The ketone enters free from acid at the rate of 0.0014 $m^3/s\ m^2$, and leaves with an acid concentration of 0.38 $kmol/m^3$. The concentration in the aqueous phase falls from 1.19 to 0.82 $kmol/m^3$.

Calculate the overall extraction coefficient based on the concentrations in the ketone phase, and the height of the corresponding overall transfer unit.

The equilibrium conditions are expressed by:

(Concentration of acid in ketone phase) = 0.548 (Concentration of acid in aqueous phase).

**13.2.** A laboratory test is carried out into the extraction of acetic acid from dilute aqueous solution, by means of methyl iso-butyl ketone, using a spray tower of 47 mm diameter and 1080 mm high. The aqueous liquor is run into the top of the tower and the ketone enters at the bottom.

The ketone enters at the rate of 0.0022 $m^3/s\ m^2$ of tower cross-section. It contains no acetic acid, and leaves with a concentration of 0.21 $kmol/m^3$. The aqueous phase flows at the rate of 0.0013 $m^3/s\ m^2$ of tower cross-section, and enters containing 0.68 kmol $acid/m^3$.

Calculate the overall extraction coefficient based on the driving force in the ketone phase. What is the corresponding value of the overall HTU, based on the ketone phase?

Using units of $kmol/m^3$, the equilibrium relationship under these conditions may be taken as:

(Concentration of acid in the ketone phase) = 0.548 (concentration in the aqueous phase.)

**13.3.** Propionic acid is extracted with water from a dilute solution in benzene, by bubbling the benzene phase into the bottom of a tower to which water is fed at the top. The tower is 1.2 m high and 0.14 $m^2$ in area, the drop volume is 0.12 $cm^3$, and the velocity of rise is 12 cm/s. From laboratory tests, the value of $K_w$ for forming drops is $7.6 \times 10^{-5}$ kmol/s $m^2$ $(kmol/m^3)$ and for rising drops $K_w = 4.2 \times 10^{-5}$ kmol/s $m^2$ $(kmol/m^3)$.

What is the value of $K_w a$ for the tower in $kmol/sm^3$ $(kmol/m^3)$?

**13.4.** A 50 per cent solution of solute **C** in solvent **A** is extracted with a second solvent **B** in a countercurrent multiple contact extraction unit. The mass of **B** is 25 per cent that of the feed solution, and the equilibrium data are:

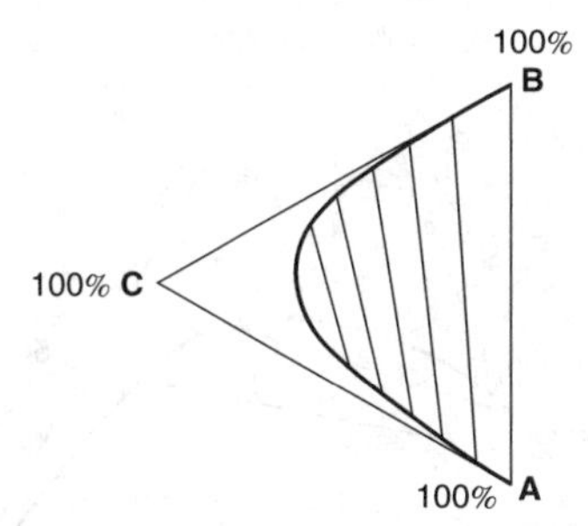


Determine the number of ideal stages required, and the mass and concentration of the first extract if the final raffinate contains 15 per cent of solute **C**.

**13.5.** A solution of 5 per cent acetaldehyde in toluene is to be extracted with water in a five stage co-current unit. If 25 kg water/100 kg feed is used, what is the mass of acetaldehyde extracted and the final concentration? The equilibrium relation is given by:

(kg acetaldehyde/kg water) = 2.20(kg acetaldehyde/kg toluene)

**13.6.** If a drop is formed in an immiscible liquid, show that the average surface available during formation of the drop is $12\pi r^2/5$, where $r$ is the radius of the drop, and that the average time of exposure of the interface is $3t_f/5$, where $t_f$ is the time of formation of the drop.

**13.7.** In the extraction of acetic acid from an aqueous solution in benzene in a packed column of height 1.4 m and cross-sectional area 0.0045 $m^2$, the concentrations measured at the inlet and outlet of the column are:

acid concentration in the inlet water phase, $C_{w2}$ = 0.69 $kmol/m^3$

acid concentration in the outlet water phase, $C_{w1}$ = 0.685 $kmol/m^3$

flowrate of benzene phase = $5.7 \times 10^{-6}$ $m^3/s$ = $1.27 \times 10^{-3}$ $m^3/m^2s$

inlet benzene phase concentration, $C_{B1}$ = 0.0040 $kmol/m^3$

outlet benzene phase concentration, $C_{B2}$ = 0.0115 $kmol/m^3$

The equilibrium relationship for this system is:

$$C_B^*/C_w^* = 0.247.$$

Determine the overall transfer coefficient and the height of the transfer unit.

**13.8** It is required to design a spray tower for the extraction of benzoic acid from solution in benzene.

Tests have been carried out on the rate of extraction of benzoic acid from a dilute solution in benzene to water, in which the benzene phase was bubbled into the base of a 25 mm diameter column and the water fed to the top of the column. The rate of mass transfer was measured during the formation of the bubbles in the water phase and during the rise of the bubbles up the column. For conditions where the drop volume was 0.12 $cm^3$ and the velocity of rise 12.5 cm/s, the value of $K_w$ for the period of drop formation was 0.000075 kmol/s $m^2$ (kmol/$m^3$), and for the period of rise 0.000046 kmol/s $m^2$ (kmol/s $m^3$).

If these conditions of drop formation and rise are reproduced in a spray tower of 1.8 m in height and 0.04 $m^2$ cross-sectional area, what is the transfer coefficient, $K_w a$, kmol/s $m^3$ (kmol/$m^3$), where $a$ represents the interfacial area in $m^2$/unit volume of the column? The benzene phase enters at the flowrate of 38 $cm^3$/s.

**13.9.** It is proposed to reduce the concentration of acetaldehyde in aqueous solution from 50 per cent to 5 per cent by mass, by extraction with solvent **S** at 293 K. If a countercurrent multiple contact process is adopted and 0.025 kg/s of the solution is treated with an equal quantity of solvent, determine the number of theoretical stages required and the mass and concentration of the extract from the first stage.

The equilibrium relationship for this system at 293 K is as follows:

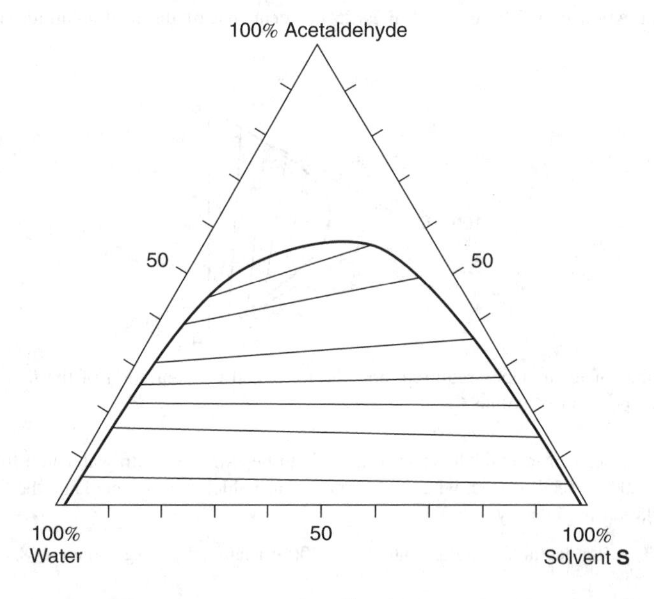


**13.10.** 160 $cm^3$/s of a solvent **S** is used to treat 400 $cm^3$/s of a 10 per cent by mass solution of **A** in **B**, in a three-stage countercurrent multiple contact liquid–liquid extraction plant. What is the composition of the final raffinate?

Using the same total amount of solvent, evenly distributed between the three stages, what would be the composition of the final raffinate, if the equipment were used in a simple multiple contact arrangement?

Equilibrium data:

| | | | |
|---|---|---|---|
| kg **A**/kg **B** | 0.05 | 0.10 | 0.15 |
| kg **A**/kg **S** | 0.069 | 0.159 | 0.258 |
| Densities (kg/$m^3$): | $\rho_A = 1200$, | $\rho_B = 1000$, | $\rho_C = 800$ |

**13.11.** In order to extract acetic acid from a dilute aqueous solution with isopropyl ether, the two immiscible phases are passed countercurrently through a packed column 3 m in length and 75 mm in diameter. It is found that if 0.5 kg/$m^2$ of the pure ether is used to extract 0.25 kg/$m^2$ s of 4.0 per cent acid by mass, then the ether phase leaves the column with a concentration of 1.0 per cent acid by mass. Calculate:

(a) the number of overall transfer units, based on the raffinate phase, and
(b) the overall extraction coefficient, based on the raffinate phase.

The equilibrium relationship is given by: (kg acid/kg isopropyl ether) = 0.3 (kg acid/kg water).

**13.12.** It is proposed to recover material **A** from an aqueous effluent by washing it with a solvent **S** and separating the resulting two phases. The light product phase will contain **A** and the solvent **S** and the heavy phase will contain **A** and water. Show that the most economical solvent rate, $W$ (kg/s) is given by:

$$W = [(F^2 a x_0)/mb)]^{0.5} - F/m$$

where the feedrate is $F$ kg/s water containing $x_0$ kg **A**/kg water, the value of **A** in the solvent product phase = £$a$/kg **A**, the cost of solvent **S** = £$b$/kg **S** and the equilibrium data are given by:

$$(\text{kg } \mathbf{A}/\text{kg } \mathbf{S})_{\text{product phase}} = m(\text{kg } \mathbf{A}/\text{kg water})_{\text{water phase}}$$

where $a$, $b$ and $m$ are constants.

**14.1.** A single-effect evaporator is used to concentrate 7 kg/s of a solution from 10 to 50 per cent of solids. Steam is available at 205 kN/m$^2$ and evaporation takes place at 13.5 kN/m$^2$. If the overall heat transfer coefficient is 3 kW/m$^2$ K, calculate the heating surface required and the amount of steam used if the feed to the evaporator is at 294 K and the condensate leaves the heating space at 352.7 K. The specific heat capacity of a 10 per cent solution is 3.76 kJ/kg K, the specific heat capacity of a 50 per cent solution is 3.14 kJ/kg K.

**14.2.** A solution containing 10 per cent of caustic soda has to be concentrated to a 35 per cent solution at the rate of 180,000 kg/day during a year of 300 working days. A suitable single-effect evaporator for this purpose, neglecting the condensing plant, costs £1600 and for a multiple-effect evaporator the cost may be taken as £1600 $N$, where $N$ is the number of effects.

Boiler steam may be purchased at £0.2/1000 kg and the vapour produced may be assumed to be 0.85 $N$ kg/kg of boiler steam. Assuming that interest on capital, depreciation, and other fixed charges amount to 45 per cent of the capital involved per annum, and that the cost of labour is constant and independent of the number of effects employed, determine the number of effects which, based on the data given, will give the maximum economy.

**14.3.** Saturated steam leaves an evaporator at atmospheric pressure and is compressed by means of saturated steam at 1135 kN/m$^2$ in a steam jet to a pressure of 135 kN/m$^2$. If 1 kg of the high pressure steam compresses 1.6 kg of the evaporated atmospheric steam, what is the efficiency of the compressor?

**14.4.** A single-effect evaporator operates at 13 kN/m$^2$. What will be the heating surface necessary to concentrate 1.25 kg/s of 10 per cent caustic soda to 41 per cent, assuming a value of the overall heat transfer coefficient $U$ of 1.25 kW/m$^2$ K, using steam at 390 K? The heating surface is 1.2 m below the liquid level.

The boiling-point rise of solution is 30 deg K, the feed temperature is 291 K, the specific heat capacity of the feed is 4.0 kJ/kg K, the specific heat capacity of the product is 3.26 kJ/kg K and the density of the boiling liquid is 1390 kg/m$^3$.

**14.5.** Distilled water is produced from sea-water by evaporation in a single-effect evaporator, working on the vapour compression system. The vapour produced is compressed by a mechanical compressor of 50 per cent efficiency, and then returned to the calandria of the evaporator. Extra steam, dry and saturated at 650 kN/m$^2$, is bled into the steam space through a throttling valve. The distilled water is withdrawn as condensate from the steam space. 50 per cent of the sea-water is evaporated in the plant. The energy supplied in addition to that necessary to compress the vapour may be assumed to appear as superheat in the vapour. Using the following data, calculate the quantity of extra steam required in kg/s.

The production rate of distillate is 0.125 kg/s, the pressure in the vapour space is 101.3 kN/m$^2$, the temperature difference from steam to liquor is 8 deg K, the boiling point rise of sea-water is 1.1 deg K and the specific heat capacity of sea-water is 4.18 kJ/kg K.

The sea water enters the evaporator at 344 K through an external heater.

**14.6.** It is claimed that a jet booster requires 0.06 kg/s of dry and saturated steam at 700 kN/m$^2$ to compress 0.125 kg/s of dry and saturated vapour from 3.5 kN/m$^2$ to 14.0 kN/m$^2$. Is this claim reasonable?

**14.7.** A forward-feed double-effect evaporator, having 10 $m^2$ of heating surface in each effect, is used to concentrate 0.4 kg/s of caustic soda solution from 10 per cent by mass. During a particular run, when the feed is at 328 K, the pressures in the two calandrias are 375 and 180 kN/$m^2$ respectively, while the condenser operates at 15 kN/$m^2$. For these conditions, calculate:

(a) the load on the condenser.
(b) the steam economy, and
(c) the overall heat transfer coefficient in each effect.

Would there have been any advantages in using backward feed in this case? Heat losses to the surroundings are negligible.

Physical properties of caustic soda solutions:

| Solids concentration (per cent by mass) | Boiling point rise (deg K) | Specific heat capacity (kJ/kg K) | Heat of dilution (kJ/kg) |
|---|---|---|---|
| 10 | 1.6 | 3.85 | 0 |
| 20 | 6.1 | 3.72 | 2.3 |
| 30 | 15.0 | 3.64 | 9.3 |
| 50 | 41.6 | 3.22 | 220 |

**14.8.** A 12 per cent glycerol–water mixture is produced as a secondary product in a continuous process plant and flows from the reactor at 4.5 MN/$m^2$ and at 525 K. Suggest, with preliminary calculations, a method of concentration to 75 per cent glycerol, in a plant which has no low-pressure steam available.

**14.9.** A forward feed double-effect vertical evaporator, with equal heating areas in each effect, is fed with 5 kg/s of a liquor of specific heat capacity of 4.18 kJ/kg K, and with no boiling point rise, so that 50 per cent of the feed liquor is evaporated. The overall heat transfer coefficient in the second effect is 75 per cent of that in the first effect. Steam is fed at 395 K and the boiling point in the second effect is 373 K. The feed is heated by an external heater to the boiling point in the first effect.

It is decided to bleed off 0.25 kg/s of vapour from the vapour line to the second effect for use in another process. If the feed is still heated to the boiling point of the first effect by external means, what will be the change in steam consumption of the evaporator unit? For the purpose of calculation, the latent heat of the vapours and of the steam may both be taken as 2230 kJ/kg.

**14.10.** A liquor containing 15 per cent solids is concentrated to 55 per cent solids in a double-effect evaporator, operating at a pressure in the second effect of 18 kN/$m^2$. No crystals are formed. The flowrate of feed is 2.5 kg/s at 375 K with a specific heat capacity of 3.75 kJ/kg K. The boiling-point rise of the concentrated liquor is 6 deg K and the steam fed to the first effect is at 240 kN/$m^2$. The overall heat transfer coefficients in the first and second effects are 1.8 and 0.63 kW/$m^2$ K, respectively. If the heat transfer area is to be the same in each effect, what areas should be specified?

**14.11.** Liquor containing 5 per cent solids is fed at 340 K to a four-effect evaporator. Forward feed is used to give a product containing 28.5 per cent solids. Do the following figures indicate normal operation? If not, why not?

| Effect | 1 | 2 | 3 | 4 |
|---|---|---|---|---|
| solids entering (per cent) | 5.0 | 6.6 | 9.1 | 13.1 |
| Temperature in steam chest (K) | 382 | 374 | 367 | 357.5 |
| Temperature of boiling solution (K) | 369.5 | 364.5 | 359.6 | 336.6 |

**14.12.** 1.25 kg/s of a solution is concentrated from 10 to 50 per cent solids in a triple-effect evaporator using steam at 393 K and a vacuum such that the boiling point in the last effect is 325 K. If the feed is initially at 297 K and backward feed is used, what is the steam consumption, the temperature distribution in the system and the heat transfer area in each effect, each effect being identical?

For the purpose of calculation, it may be assumed that the specific heat capacity is 4.18 kJ/kg K, that there is no boiling point rise, and that the latent heat of vaporisation is constant at 2330 kJ/kg over the temperature range in the system. The overall heat transfer coefficients may be taken as 2.5, 2.0 and 1.6 kW/m$^2$ K in the first, second and third effects, respectively.

**14.13.** A triple-effect evaporator concentrates a liquid with no appreciable elevation of boiling point. If the temperature of the steam to the first effect is 395 K, and vacuum is applied to the third effect so that the boiling point is 325 K, what are the approximate boiling points in the three effects? The overall heat transfer coefficients may be taken as 3.1, 2.3, 1.3 kW/m$^2$ K in the three effects, respectively.

**14.14.** A three-stage evaporator is fed with 1.25 kg/s of a liquor which is concentrated from 10 to 40 per cent solids by mass. The heat transfer coefficients may be taken as 3.1, 2.5 and 1.7 kW/m$^2$ K, respectively, in each effect. Calculate the steam flow at 170 kN/m$^2$ and the temperature distribution in the three effects, if:

(a) the feed is at 294 K, and
(b) the feed is at 355 K.

Forward feed is used in each case and the values of $U$ are the same for the two systems. The boiling point in the third effect is 325 K and the liquor has no boiling point rise.

**14.15.** An evaporator operating on the thermo-recompression principle employs a steam ejector to maintain atmospheric pressure over the boiling liquid. The ejector uses 0.14 kg/s of steam at 650 kN/m$^2$, and superheated by 100 deg K, and produces a pressure in the steam chest of 205 kN/m$^2$. A condenser removes surplus vapour from the atmospheric pressure line. What is the capacity and economy of the system, and how could the economy be improved?

The feed enters the evaporator at 293 K and the concentrated liquor is withdrawn at the rate of 0.025 kg/s. The concentrated liquor exhibits a boiling point rise of 10 deg K. Heat losses to the surroundings are negligible.

For the ejector, the nozzle efficiency is 0.95, the efficiency of momentum transfer is 0.80 and the efficiency of compression is 0.90.

**14.16.** A single-effect evaporator is used to concentrate 0.075 kg/s of a 10 per cent caustic soda liquor to 30 per cent. The unit employs forced circulation in which the liquor is pumped through the vertical tubes of the calandria which are 32 mm o.d. by 28 mm i.d., and 1.2 m long. Steam is supplied at 394 K, dry and saturated, and the boiling point rise of the 30 per cent solution is 15 deg K. If the overall heat transfer coefficient is 1.75 kW/m$^2$ K, how many tubes should be used, and what material of construction would be specified for the evaporator? The latent heat of vaporisation under these conditions is 2270 kJ/kg.

**14.17.** A steam-jet booster compresses 0.1 kg/s of dry and saturated vapour from 3.4 kN/m$^2$ to 13.4 kN/m$^2$. High pressure steam consumption is 0.05 kg/s at 700 kN/m$^2$. (a) What must be the condition of the high pressure steam for the booster discharge to be superheated through 20 deg K? (b) What is the overall efficiency of the booster if the compression efficiency is 100 per cent?

**14.18.** A triple-effect backward-feed evaporator concentrates 5 kg/s of liquor from 10 per cent to 50 per cent solids. Steam is available at 375 kN/m$^2$ and the condenser operates at 13.5 kN/m$^2$. What is the area required in each effect, assumed equal, and the economy of the unit?

The specific heat capacity is 4.18 kJ/kg K at all concentrations and there is no boiling-point rise. The overall heat transfer coefficients are 2.3, 2.0 and 1.7 kW/m$^2$ K respectively in the three effects, and the feed enters the third effect at 300 K.

**14.19.** A double-effect climbing film evaporator is connected so that the feed passes through two preheaters, one heated by vapour from the first effect and the other by vapour from the second effect. The condensate from the first effect is passed into the steam space of the second. The temperature of the feed is initially 289 K, 348 K after the first heater and 383 K and after the second heater. The vapour temperature in the first effect is 398 K and in the second 373 K. The flowrate of feed is 0.25 kg/s and the steam is dry and saturated at 413 K. What is the economy of the unit if the evaporation rate is 0.125 kg/s?

**14.20.** A triple-effect evaporator is fed with 5 kg/s of a liquor containing 15 per cent solids. The concentration in the last effect, which operates at 13.5 kN/m$^2$, is 60 per cent solids. If the overall heat transfer coefficients are 2.5, 2.0 and 1.1 kW/m$^2$ K, respectively, and the steam is fed at 388 K to the first effect, determine the temperature distribution and the area of heating surface required in each effect, assuming the calandrias are identical. What is the economy and what is the heat load on the condenser? The feed temperature is 294 K and the specific heat capacity of all liquors is 4.18 kJ/kg K

If the unit is run as a backward-feed system, in which the coefficients are 2.3, 2.0 and 1.6 kW/m$^2$ K, determine the new temperatures, the heat economy and the heating surface required under these conditions.

**14.21.** A double-effect forward-feed evaporator is required to give a product which contains 50.0 per cent by mass of solids. Each effect has 10 m$^2$ of heating surface and the heat transfer coefficients are 2.8 and 1.7 kW/m$^2$ K in the first and second effects respectively. Dry and saturated steam is available at 375 kN/m$^2$ and the condenser operates at 13.5 kN/m$^2$. The concentrated solution exhibits a boiling-point rise of 3 deg K. What is the maximum permissible feed rate if the feed contains 10 per cent solids and is at a temperature of 310 K? The latent heat of vapourisation is 2330 kJ/kg and the specific heat capacity is 4.18 kJ/kg K under all conditions.

**14.22.** For the concentration of fruit juice by evaporation, it is proposed to use a falling-film evaporator and to incorporate a heat-pump cycle with ammonia as the medium. The ammonia in vapour form enters the evaporator at 312 K and the water is evaporated from the juices at 287 K. The ammonia in the vapour–liquid mixture enters the condenser at 278 K and the vapour then passes to the compressor. It is estimated that the work in compressing the ammonia is 150 kJ/kg of ammonia and that 2.28 kg of ammonia is cycled/kilogram of water evaporated. The following proposals are made for driving the compressor:

(a) To use a diesel engine drive taking 0.4 kg of fuel/MJ. The calorific value of the fuel is 42 MJ/kg, and the cost £0.02/kg.
(b) To pass steam, costing £0.01/10 kg, through a turbine which operates at 70 per cent isentropic efficiency, between 700 and 101.3 kN/m$^2$.

Explain by means of a diagram how this plant will work, illustrating all necessary major items of equipment. Which method for driving the compressor is to be preferred?

**14.23.** A double-effect forward-feed evaporator is required to give a product consisting of 30 per cent crystals and a mother liquor containing 40 per cent by mass of dissolved solids. Heat transfer coefficients are 2.8 and 1.7 kW/m$^2$ K in the first and second effects respectively. Dry saturated steam is supplied at 375 kN/m$^2$ and the condenser operates at 13.5 kN/m$^2$.

(a) What area of heating surface is required in each effect assuming the effects are identical, if the feed rate is 0.6 kg/s of liquor, containing 20 per cent by mass of dissolved solids, and the feed temperature is 313 K?
(b) What is the pressure above the boiling liquid in the first effect?

The specific heat capacity may be taken as constant at 4.18 kJ/kg K, and the effects of boiling-point rise and of hydrostatic head may be neglected.

**14.24.** 1.9 kg/s of a liquid containing 10 per cent by mass of dissolved solids is fed at 338 K to a forward-feed double-effect evaporator. The product consists of 25 per cent by mass of solids and a mother liquor containing 25 per cent by mass of dissolved solids. The steam fed to the first effect is dry and saturated at 240 kN/m$^2$ and the pressure in the second effect is 20 kN/m$^2$. The specific heat capacity of the solid may be taken as 2.5 kJ/kg K both in solid form and in solution, and the heat of solution may be neglected. The mother liquor exhibits a boiling-point rise of 6 deg K. If the two effects are identical, what area is required if the heat transfer coefficients in the first and second effects are 1.7 and 1.1 kW/m$^2$K, respectively?

**14.25.** 2.5 kg/s of a solution at 288 K containing 10 per cent of dissolved solids is fed to a forward-feed double-effect evaporator, operating at 14 kN/m$^2$ in the last effect. If the product is to consist of a liquid containing 50 per cent by mass of dissolved solids and dry saturated steam is fed to the steam coils, what should be the pressure of the steam? The surface in each effect is 50 $m^2$ and the coefficients for heat transfer in the first and second effects are 2.8 and 1.7 kW/$m^2$ K, respectively. It may be assumed that the concentrated solution exhibits a boiling-point rise of 5 deg K, that the latent heat has a constant value of 2260 kJ/kg and that the specific heat capacity of the liquid stream is constant at 3.75 kJ/kg K.

**14.26.** A salt solution at 293 K is fed at the rate of 6.3 kg/s to a forward-feed triple-effect evaporator and is concentrated from 2 per cent to 10 per cent of solids. Saturated steam at 170 kN/$m^2$ is introduced into the calandria of the first effect and a pressure of 34 kN/$m^2$ is maintained in the last effect. If the heat transfer coefficients in the three effects are 1.7, 1.4 and 1.1 kW/$m^2$ K, respectively, and the specific heat capacity of the liquid is approximately 4 kJ/kg K, what area is required if each effect is identical? Condensate may be assumed to leave at the vapour temperature at each stage, and the effects of boiling point rise may be neglected. The latent heat of vaporisation may be taken as constant throughout.

**14.27.** A single-effect evaporator with a heating surface area of 10 $m^2$ is used to concentrate NaOH solution at 0.38 kg/s from 10 per cent to 33.33 per cent by mass. The feed enters at 338 K and its specific heat capacity is 3.2 kJ/kg K. The pressure in the vapour space is 13.5 kN/$m^2$ and 0.3 kg/s of steam is used from a supply at 375 K. Calculate:

(a) The apparent overall heat transfer coefficient.
(b) The coefficient corrected for boiling point rise of dissolved solids.
(c) The corrected coefficient if the depth of liquid is 1.5 m.

**14.28.** An evaporator, working at atmospheric pressure, is to concentrate a solution from 5 per cent to 20 per cent solids at the rate of 1.25 kg/s. The solution, which has a specific heat capacity of 4.18 kJ/kg K, is fed to the evaporator at 295 K and boils at 380 K. Dry saturated steam at 240 kN/$m^2$ is fed to the calandria, and the condensate leaves at the temperature of the condensing stream. If the heat transfer coefficient is 2.3 kW/$m^2$ K, what is the required area of heat transfer surface and how much steam is required? The latent heat of vaporisation of the solution may be taken as being equal to that of water.

**15.1.** A saturated solution containing 1500 kg of potassium chloride at 360 K is cooled in an open tank to 290 K. If the density of the solution is 1200 kg/$m^3$, the solubility of potassium chloride is 53.55 kg/100 kg water at 360 K and 34.5 at 290 K calculate:

(a) the capacity of the tank required, and
(b) the mass of crystals obtained, neglecting loss of water by evaporation.

**15.2.** Explain how fractional crystallisation may be applied to a mixture of sodium chloride and sodium nitrate, given the following data. At 293 K, the solubility of sodium chloride is 36 kg/100 kg water and of sodium nitrate 88 kg/100 kg water. Whilst at this temperature, a saturated solution comprising both salts will contain 25 kg sodium chloride and 59 kg sodium nitrate/100 kg of water. At 373 K these values, again per 100 kg of water, are 40 and 176, and 17 and 160 kg respectively.

**15.3.** 10 Mg (10 tonne) of a solution containing 0.3 kg $Na_2CO_3$/kg solution is cooled slowly to 293 K to form crystals of $Na_2CO_3.10H_2O$. What is the yield of crystals if the solubility of $Na_2CO_3$ at 293 K is 21.5 kg/100 kg water and during cooling 3 per cent of the original solution is lost by evaporation?

**15.4.** The heat required when 1 kmol of $MgSo_4.7H_2O$ is absorbed isothermally at 291 K in a large mass of water is 13.3 MJ. What is the heat of crystallisation per unit mass of the salt?

**15.5.** A solution of 500 kg of $Na_2SO_4$ in 2500 kg water is cooled from 333 K to 283 K in an agitated mild steel vessel of mass 750 kg. At 283 K, the solubility of the anhydrous salt is 8.9 kg/100 kg water and the stable crystalline phase is $Na_2SO_4.10H_2O$. At 291 K, the heat of solution is −78.5 MJ/kmol and the heat capacities of the solution and mild steel are 3.6 and 0.5 kJ/kg deg K respectively. If, during cooling, 2 per cent of the water initially present is lost by evaporation, estimate the heat which must be removed.

**15.6.** A batch of 1500 kg of saturated potassium chloride solution is cooled from 360 K to 290 K in an unagitated tank. If the solubilities of KCl are 53 and 34 kg/100 kg water at 360 K and 290 K respectively and water losses due to evaporation may be neglected, what is the yield of crystals?

**15.7.** Glauber's salt, $Na_2SO_4.10H_2O$, is to be produced in a Swenson–Walker crystalliser by cooling to 290 K a solution of anhydrous $Na_2SO_4$ which is saturated at 300 K. If cooling water enters and leaves the unit at 280 K and 290 K, respectively, and evaporation is negligible, how many sections of crystalliser, each 3 m long, will be required to process 0.25 kg/s of the product? The solubilities of anhydrous $Na_2SO_4$ in water are 40 and 14 kg/100 kg water at 300 K and 290 K respectively, the mean heat capacity of the liquor is 3.8 kJ/kg K and the heat of crystallisation is 230 kJ/kg. For the crystalliser, the available heat transfer area is 3 $m^2$/m length, the overall coefficient of heat transfer is 0.15 kW/$m^2$ K and the molecular weights are $Na_2SO_4{\cdot}10H_2O = 322$ kg/kmol and $Na_2SO_4 = 142$ kg/kmol.

**15.8.** What is the evaporation rate and yield of $CH_3COONa.3H_2O$ from a continuous evaporative-crystalliser operating at 1 kN/$m^2$ when it is fed with 1 kg/s of a 50 per cent by mass aqueous solution of sodium acetate at 350 K? The boiling-point elevation of the solution is 10 deg K and the heat of crystallisation is 150 kJ/kg. The mean heat capacity of the solution is 3.5 kJ/kg K and, at 1 kN/$m^2$, water boils at 280 K at which the latent heat of vaporisation is 2.482 MJ/kg. Over the range 270–305 K, the solubility of sodium acetate in water $s$ at $T$(K) is given approximately by:

$$s = 0.61T - 132.4 \text{ kg/100 kg water.}$$

Molecular weights: $CH_3COONa.3H_2O = 136$ kg/kmol, $CH_3COONa = 82$ kg/kmol.

**16.1.** A wet solid is dried from 25 per cent to 10 per cent moisture, under constant drying conditions in 15 ks (4.17h). If the equilibrium moisture content is 5 per cent and the critical moisture content is 15 per cent, how long will it take to dry to 8 per cent moisture under the same conditions?

**16.2.** Strips of material 10 mm thick are dried under constant drying conditions from 28 to 13 per cent moisture in 25 ks. If the equilibrium moisture content is 7 per cent, what is the time taken to dry 60 mm planks from 22 to 10 per cent moisture under the same conditions, assuming no loss from the edges? All moistures are expressed on the wet basis. The relation between $E$, the ratio of the average free moisture content at time $t$ to the initial free moisture content, and the parameter $J$ is given by:

| $E$ | 1 | 0.64 | 0.49 | 0.38 | 0.295 | 0.22 | 0.14 |
|---|---|---|---|---|---|---|---|
| $J$ | 0 | 0.1 | 0.2 | 0.3 | 0.5 | 0.6 | 0.7 |

It may be noted that $J = kt/l^2$, where $k$ is a constant, $t$ is the time in ks and $2l$ is the thickness of the sheet of material in mm.

**16.3.** A granular material containing 40 per cent moisture is fed to a countercurrent rotary dryer at 295 K and is withdrawn at 305 K containing 5 per cent moisture. The air supplied, which contains 0.006 kg water vapour/kg of dry air, enters at 385 K and leaves at 310 K. The dryer handles 0.125 kg/s wet stock.

Assuming that radiation losses amount to 20 kJ/kg of dry air used, determine the mass flowrate of dry air supplied to the dryer and the humidity of the outlet air.

The latent heat of water vapour at 295 K is 2449 kJ/kg, the specific heat capacity of dried material is 0.88 kJ/kg K, the specific heat capacity of dry air is 1.00 kJ/kg K and the specific heat capacity of water vapour is 2.01 kJ/kg K.

**16.4.** 1 Mg (1 tonne) of dry mass of a non-porous solid is dried under constant drying conditions with air at a velocity of 0.75 m/s. The area of surface drying is 55 $m^2$. If the initial rate of drying is 0.3 g/$m^2$s, how long will it take to dry the material from 0.15 to 0.025 kg water/kg dry solid? The critical moisture content of the material may be taken as 0.125 kg water/kg dry solid. If the air velocity were increased to 4.0 m/s, what would be the anticipated saving in time if the process is surface-evaporation controlled?

**16.5.** A 100 kg batch of granular solids containing 30 per cent of moisture is to be dried in a tray dryer to 15.5 per cent of moisture by passing a current of air at 350 K tangentially across its surface at the velocity of 1.8 m/s. If the constant rate of drying under these conditions is 0.7 g/s $m^2$ and the critical moisture content is 15 per cent, calculate the approximate drying time. It may be assumed that the area of the drying surface is 0.03 $m^2$/kg dry mass.

**16.6.** A flow of 0.35 kg/s of a solid is to be dried from 15 per cent to 0.5 per cent moisture based on a dry basis. The mean heat capacity of the solids is 2.2 kJ/kg deg K. It is proposed that a co-current adiabatic dryer should be used with the solids entering at 300 K and, because of the heat sensitive nature of the solids, leaving at 325 K. Hot air is available at 400 K with a humidity of 0.01 kg/kg dry air, and the maximum allowable mass velocity of the air is 0.95 kg/$m^2$s. What diameter and length should be specified for the proposed dryer?

**16.7.** 0.126 kg/s of a product containing 4 per cent water is produced in a dryer from a wet feed containing 42 per cent water on a wet basis. Ambient air at 294 K and 40 per cent relative humidity is heated to 366 K in a preheater before entering the dryer which it leaves at 60 per cent relative humidity. Assuming that the dryer operates adiabatically, what is the amount of air supplied to the preheater and the heat required in the preheater?

How will these values be affected if the air enters the dryer at 340 K and sufficient heat is supplied within the dryer so that the air leaves also at 340 K and again with a relative humidity of 60 per cent?

**16.8.** A wet solid is dried from 40 to 8 per cent moisture in 20 ks. If the critical and the equilibrium moisture contents are 15 and 4 per cent respectively, how long will it take to dry the solid to 5 per cent moisture under the same conditions? All moisture contents are on a dry basis.

**16.9.** A solid is to be dried from 1 kg water/kg dry solids to 0.01 kg water/kg dry solids in a tray dryer consisting of a single tier of 50 trays, each 0.02 m deep and 0.7 m square completely filled with wet material. The mean air temperature is 350 K and the relative humidity across the trays may be taken as constant at 10 per cent. The mean air velocity is 2.0 m/s and the convective coefficient of heat transfer is given by:

$$h_c = 14.3G'^{0.8} \quad \text{W/m}^2\text{ deg K}$$

where $G'$ is the mass velocity of the air in kg/$m^2$s. The critical and equilibrium moisture contents of the solid are 0.3 and 0 kg water/kg dry solids respectively and the density of the solid is 6000 kg/$m^3$. Assuming that the drying is by convection from the top surface of the trays only, what is the drying time?

**16.10.** Skeins of a synthetic fibre are dried from 46 per cent to 8.5 per cent moisture on a wet basis in a 10 m long tunnel dryer by a countercurrent flow of hot air. The air mass velocity, $G'$, is 1.36 kg/$m^2$s and the inlet conditions are 355 K and a humidity of 0.03 kg moisture/kg dry air. The air temperature is maintained at 355 K throughout the dryer by internal heating and, at the outlet, the humidity of the air is 0.08 kg moisture/kg dry air. The equilibrium moisture content is given by:

$$w_e = 0.25 \text{ (per cent relative humidity)}$$

and the drying rate by:

$$R = 1.34 \times 10^{-4} G'^{1.47}(w - w_c)(\mathscr{H}_w - \mathscr{H}) \quad \text{kg/s kg dry fibres}$$

where $\mathscr{H}$ is the humidity of the dry air and $\mathscr{H}_w$ the saturation humidity at the wet bulb temperature. Data relating $w$, $\mathscr{H}$ and $\mathscr{H}_w$ are as follows:

| $w$ (kg/kg dry fibre) | $\mathscr{H}$ (kg/kg dry air) | $\mathscr{H}_w$ (kg/kg dry air) | relative humidity (per cent) |
|---|---|---|---|
| 0.852 | 0.080 | 0.095 | 22.4 |
| 0.80 | 0.0767 | 0.092 | 21.5 |
| 0.60 | 0.0635 | 0.079 | 18.2 |
| 0.40 | 0.0503 | 0.068 | 14.6 |
| 0.20 | 0.0371 | 0.055 | 11.1 |
| 0.093 | 0.030 | 0.049 | 9.0 |

At what velocity should the skeins be passed through the dryer?

**17.1.** Spherical particles of 15 nm diameter and density of 2290 $kg/m^3$ are pressed together to form a pellet. The following equilibrium data were obtained for the sorption of nitrogen at 77 K. Obtain estimates of the surface area of the pellet from the adsorption isotherm and compare the estimates with the geometric surface. The density of liquid nitrogen at 77 K is 808 $kg/m^3$.

| $P/P^0$ | 0.1 | 0.2 | 0.3 | 0.4 | 0.5 | 0.6 | 0.7 | 0.8 | 0.9 |
|---|---|---|---|---|---|---|---|---|---|
| $m^3$ liquid $N_2 \times 10^6$/kg solid | 66.7 | 75.2 | 83.9 | 93.4 | 108.4 | 130.0 | 150.2 | 202.0 | 348.0 |

where $P$ is the pressure of the sorbate in the gas and $P^0$ is its vapour pressure at 77 K.

**17.2.** A 1 $m^3$ volume of a mixture of air and acetone vapour is at a temperature of 303 K and a total pressure of 100 $kN/m^2$. If the relative saturation of the air by acetone is 40 per cent, how much activated carbon must be added to the space in order to reduce the value to 5 per cent at 303 K?

If 1.6 kg carbon is added, what is relative saturation of the equilibrium mixture assuming the temperature to be unchanged?

The vapour pressure of acetone at 303 K is 37.9 $kN/m^2$ and the adsorption equilibrium data for acetone on carbon at 303 K are:

| Partial pressure acetone $\times 10^{-2}$ ($N/m^2$) | 0 | 5 | 10 | 30 | 50 | 90 |
|---|---|---|---|---|---|---|
| $x_r$ (kg acetone/kg carbon) | 0 | 0.14 | 0.19 | 0.27 | 0.31 | 0.35 |

**17.3.** A solvent, contaminated with 0.03 $kmol/m^3$ of a fatty acid, is to be purified by passing it through a fixed bed of activated carbon to adsorb the acid but not the solvent. If the operation is essentially isothermal and equilibrium is maintained between the liquid and the solid, calculate the length of a bed of 0.15 m diameter to give one hour's operation when the fluid is fed at $1 \times 10^{-4}$ $m^3/s$. The bed is free of adsorbate initially and the intergranular voidage is 0.4. Use an equilibrium, fixed-bed theory to obtain the length for three types of isotherm:

(a) $C_s = 10\ C$.
(b) $C_s = 3.0\ C^{0.3}$ (use the mean slope).
(c) $C_s = 10^4\ C^2$ (the breakthrough concentration is 0.003 $kmol/m^3$).

$C$ and $C_s$ refer to concentrations in $kmol/m^3$ in the gas phase and the absorbent, respectively.

**18.1** A solution is passed over a single pellet of resin and the temperature is maintained constant. The take-up of exchanged ion is followed automatically and the following results are obtained:

| $t$ (min) | 2 | 4 | 10 | 20 | 40 | 60 | 120 |
|---|---|---|---|---|---|---|---|
| $x_r$ (kg/kg) | 0.091 | 0.097 | 0.105 | 0.113 | 0.125 | 0.128 | 0.132 |

On the assumption that the resistance to mass transfer in the external film is negligible, predict the values of $x_r$, the mass of sorbed phase per unit mass of resin, as a function of time $t$, for a pellet of a resin twice the radius.

**19.1.** Describe the principle of separation involved in elution chromatography and derive the retention equation:

$$t_R = t_M \left[1 + \left(\frac{1-\varepsilon}{\varepsilon}\right) K\right]$$

where $t_R$ is the retention time of solute in the column, $t_M$ is the mobile phase hold-up time in the column. $\varepsilon$ is the packing voidage and $K$ is the distribution coefficient.

**19.2.** In chemical analysis, chromatography may permit the separation of more than a hundred components from a mixture in a single run. Explain why chromatography offers such a large separating power. In production chromatography, the complete separation of a mixture containing more than a few components is likely to involve two or three columns for optimum economic performance. Why is this?

**19.3.** By using the chromatogram in Figure 19.3, show that $k'_1 = 3.65$, $k'_2 = 4.83$, $\alpha = 1.32$, $R_s = 1.26$ and $N = 500$. Show also that, if $\varepsilon = 0.8$ and $L = 1.0$ m, then $K_1 = 14.6$, $K_2 = 19.3$ and $H = 2.0$ mm, where $R$ is the retention ratio, $R_s$ is the resolution, $d$ is the obstruction factor, $H$ is the plate height, $K_1$ and $K_2$ are the distribution coefficients, $k'_1$ and $k'_2$ are the capacity factors, $\varepsilon$ is the packing voidage, $L$ is the length of the column, and $N$ is the number of theoretical plates. Calculate the ratio of plate height to particle diameter to confirm that the column is inefficient, as might be anticipated from the wide bands in Figure 19.3. It may be assumed that the particle size is that of a typical GC column as given in Table 19.3.

**19.4.** Suggest one or more types of chromatography to separate each of the following mixtures:

(a) $\alpha$- and $\beta$-pinenes
(b) blood serum proteins
(c) hexane isomers
(d) purification of cefonicid, a synthetic $\beta$-lactam antibiotic.

# Index